Phylogeny of the Living World—B

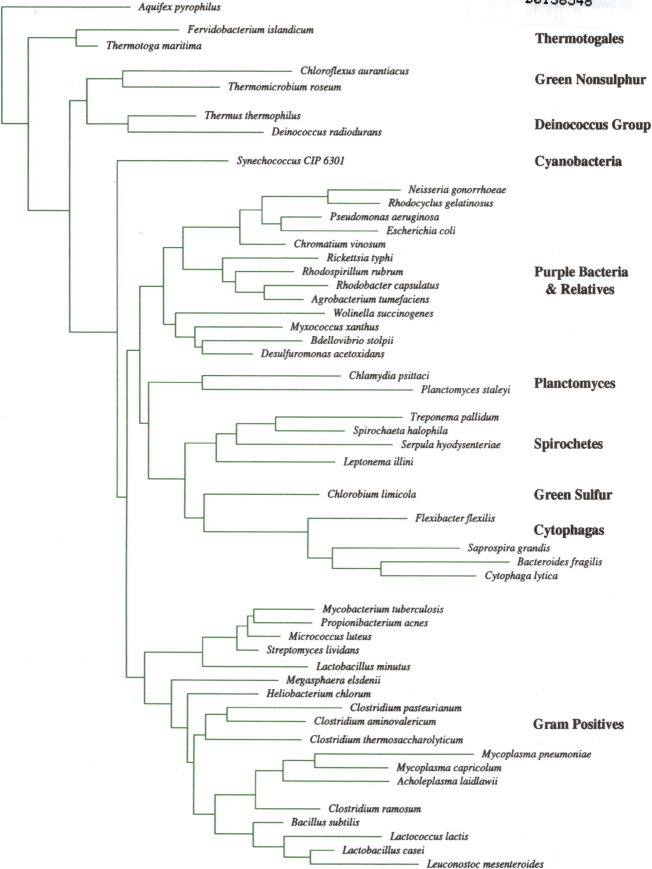

PHYLOGENETIC TREE OF BACTERIA. This tree is derived from 16S ribosomal RNA sequences. Eleven major groups of Bacteria can be defined as indicated. Compare with Figure 18.11. *Data of Carl R. Woese.*

Biology
of Microorganisms

Thomas D. Brock was trained initially in botany, mycology, and yeast physiology (Ph.D., Ohio State University, 1952). He then spent five years as a research microbiologist at the Upjohn Company, where he was engaged in the search for new antibiotics. In his academic career he has taught at Western Reserve University, Indiana University, and the University of Wisconsin–Madison where he is now the E.B. Fred Professor of Natural Sciences–Emeritus. His teaching has involved a wide variety of courses, including general microbiology, mycology, microbial genetics, medical microbiology, microbial ecology, and microbial diversity. His research has been equally broad in scope, but for the past 30 years has focused on the broad area of microbial ecology. He published his first book with Prentice Hall in 1961 (*Milestones in Microbiology*), and is now the author or coauthor of over 20 books as well as close to 300 research papers. Among his major research accomplishments is the discovery of hyperthermophilic bacteria, including *Thermus aquaticus*, the source of the key enzyme in the polymerase chain reaction (PCR). He lives in Madison, WI, with his wife, Kathie, and two children, Emily and Brian. Beside his textbook duties, he continues to write scholarly monographs on various aspects of the history of microbiology.

Michael T. Madigan received a bachelor's degree in biology and education from Wisconsin State University at Stevens Point in 1971 and M.S. and Ph.D. degrees in 1974 and 1976, respectively, from the University of Wisconsin, Madison, Department of Bacteriology. His graduate work involved study of the biology of hot spring photosynthetic bacteria under the direction of Thomas D. Brock. Following three years of postdoctoral training in the Department of Microbiology, Indiana University, where he worked on photosynthetic bacteria with Howard Gest, he moved to Southern Illinois University at Carbondale, where he is now Professor of Microbiology. He has been a coauthor of *Biology of Microorganisms* since the 4th edition (1984) and teaches courses in introductory microbiology and bacterial diversity. In 1988 he was selected as the outstanding teacher in the College of Science. His research has dealt almost exclusively with anoxygenic phototrophic bacteria, especially those representatives that inhabit extreme environments, and has focused on the nitrogen-fixing and major photosynthetic properties of these organisms. He has published nearly 70 research papers and has coedited a major treatise on photosynthetic bacteria. His nonscientific interests include reading, hiking, and caring for his dogs. He lives on a quiet lake about five miles from the SIU campus with his wife, Nancy, two golden retrievers, Andy (see photo) and Marcy, and King (horse).

John Martinko attended The Cleveland State University and majored in Biology with a Chemistry minor. As an undergraduate student he participated in a unique cooperative education program gaining research experience in several microbiology and immunology laboratories. He then worked for two years at Case Western Reserve University as a Laboratory Manager, continuing his cooperative education research on the structure and serology of *Streptococcus pyogenes* cell wall antigens and on the epidemiology of streptococcal infections. He next went to the State University of New York (SUNY) at Buffalo where he did research on antibody specificity for his M.A. and Ph.D. in Microbiology. As a postdoctoral fellow, he worked at Albert Einstein College of Medicine in New York on the structure of transplantation antigens. Since 1981, he has been in the Department of Microbiology at Southern Illinois University at Carbondale where he is currently an Associate Professor. His research interests include the structure, genetics, and evolution of transplantation antigens. Other research interests include the molecular evolution of blood group antigens and the effects of stress on the immune response. His major teaching duties include undergraduate and graduate courses in immunology and a team-taught general microbiology course, where he is responsible for the immunology, host-defense, and infectious disease portions. He lives in Carbondale with his wife, Judy, a grade school science teacher, and their three daughters, Martha, Helen, and Sarah. He is an avid cyclist and coaches his daughters' soccer teams.

Jack Parker received his bachelors degree in biology and also received his doctoral degree in a biology program (Ph.D., Purdue University, 1973). However, his research project dealt with bacterial physiology and he completed his Ph.D. research while in the microbiology department at the University of Michigan. Following this he spent four years studying bacterial genetics at York University in Toronto, Ontario. He has taught courses in bacterial genetics, general genetics, molecular biology, and molecular genetics, and has participated in courses in introductory microbiology, medical microbiology, and virology primarily at Southern Illinois at Carbondale, where he is now Professor and Chair of the Department of Microbiology. His research has been in the broad area of molecular genetics and gene expression and for the last 15 years has been focused most specifically on studies of how cells control the accuracy of protein synthesis. He is the author of approximately 50 research papers. His home is on the edge of the Shawnee National Forest in deep southern Illinois where he lives with his wife, Beth, and three children, Justine, D'Arcy and Grant.

Biology of Microorganisms

SEVENTH EDITION

Thomas D. Brock
University of Wisconsin—Madison

Michael T. Madigan
Southern Illinois University—Carbondale

John M. Martinko
Southern Illinois University—Carbondale

Jack Parker
Southern Illinois University—Carbondale

Prentice Hall, *Englewood Cliffs, New Jersey 07632*

Library of Congress Cataloging-in-Publication Data
Biology of microorganisms / Thomas D. Brock . . . [et al.]. -- 7th ed.
    p. cm.
    Rev. ed. of: Biology of microorganisms / Thomas D. Brock, Michael
T. Madigan. 6th ed. c1991.
    Includes bibliographical references and index.
    ISBN 0-13-042169-3
    1. Microbiology.   I. Brock, Thomas D. II. Brock, Thomas D.
Biology of microorganisms.
QR41.2.B77   1994
576--dc20   93-8549
                                                             CIP

Editions of the *Biology of Microorganisms*

First Edition, 1970, Thomas D. Brock
Second Edition, 1974, Thomas D. Brock
Third Edition, 1979, Thomas D. Brock
Fourth Edition, 1984, Thomas D. Brock, David W. Smith, and Michael T. Madigan
Fifth Edition, 1988, Thomas D. Brock and Michael T. Madigan
Sixth Edition, 1991, Thomas D. Brock and Michael T. Madigan
Seventh Edition, 1994, Thomas D. Brock, Michael T. Madigan,
John M. Martinko, and Jack Parker

Executive Editor: David Kendric Brake
Editorial/production supervision: Susan Fisher
Design Director: Florence Dara Silverman
Interior and cover design: Lisa A. Jones
Page layout: Lisa A. Jones and Meryl Poweski
Manufacturing buyer: Trudy Pisciotti
Supplements Editor: Mary Hornby
Editorial Assistant: Mary DeLuca
Art work by Hans & Cassady, Inc., Westerville, OH 43081
Cover art: *Crustose Lichens* by Dwight R. Kuhn, Photography
(Credits and acknowledgments appear on pp. 878 and 879,
which constitute a continuation of the copyright page.)

Printed in the United States of America
10   9   8   7   6   5   4   3   2

ISBN 0-13-042169-3

Prentice-Hall International (UK) Limited, *London*
Prentice-Hall of Australia Pty. Limited, *Sydney*
Prentice-Hall Canada Inc., *Toronto*
Prentice-Hall Hispanoamericana, S.A., *Mexico*
Prentice-Hall of India Private Limited, *New Delhi*
Prentice-Hall of Japan, Inc., *Tokyo*
Simon & Schuster Asia Pte. Ltd., *Singapore*
Editora Prentice-Hall do Brasil, Ltda., *Rio de Janeiro*

Thomas D. Brock dedicates this book to all of the students who, over the past two and one half decades, have used this book as their first introduction to the fascinating field of microbiology.

■

Michael T. Madigan dedicates this book to his wife, Nancy, whose love, encouragement, and support has guided him through four editions of *Biology of Microorganisms.*

■

John M. Martinko dedicates this book to his wife, Judy, whose patience, understanding, and vision stimulated and inspired him through the creative process.

■

Jack Parker wishes to dedicate this book to his wife, Elizabeth, *femina sine qua non.*

Overview

Biology of Microorganisms, SEVENTH EDITION

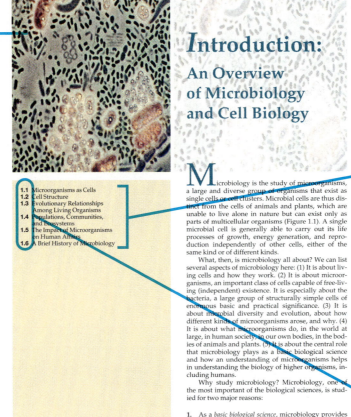

Colorful Visuals
full color photographs
throughout the book

1

Introduction:

An Overview
of Microbiology
and Cell Biology

1.1 Microorganisms as Cells
1.2 Cell Structure
1.3 Evolutionary Relationships
    Among Living Organisms
1.4 Populations, Communities,
    and Ecosystems
1.5 The Impact of Microorganisms
    on Human Affairs
1.6 A Brief History of Microbiology

Microbiology is the study of microorganisms, a large and diverse group of organisms that exist as single cells or cell clusters. Microbial cells are thus distinct from the cells of animals and plants, which are unable to live alone in nature but can exist only as parts of multicellular organisms (Figure 1.1). A single microbial cell is generally able to carry out its life processes of growth, energy generation, and reproduction independently of other cells, either of the same kind or of different kinds.

What, then, is microbiology all about? We can list several aspects of microbiology here: (1) It is about living cells and how they work. (2) It is about microorganisms, an important class of cells capable of free-living (independent) existence. It is especially about the bacteria, a large group of structurally simple cells of enormous basic and practical significance. (3) It is about microbial diversity and evolution, about how different kinds of microorganisms arose, and why. (4) It is about what microorganisms do, in the world at large, in human society, in our own bodies, in the bodies of animals and plants. (5) It is about the central role that microbiology plays as a basic biological science and how an understanding of microorganisms helps in understanding the biology of higher organisms, including humans.

Why study microbiology? Microbiology, one of the most important of the biological sciences, is studied for two major reasons:

1. As a *basic biological science*, microbiology provides some of the most accessible research tools for probing the nature of life processes. Our most so-

Chapter Outline
provides a sketch
of the chapter content

Section Numbers
used for convenient
reference

1

Miniglossary for Chapter 4

AEROBE a microorganism able to use O_2 in respiration

ALLOSTERIC an enzyme that contains two combining sites, the active site (where substrate binds) and the allosteric site (where an effector molecule binds)

ANABOLISM the sum total of all biosynthetic reactions in the cell

ANAEROBE a microorganism unable to use O_2 in respiration and which may even be harmed or killed by O_2

ASEPTIC TECHNIQUE methods for maintaining sterile culture media and other sterile objects free from contamination during manipulations

AUTOTROPH an organism capable of biosynthesizing all cell material from CO_2 as the sole carbon source

CATABOLISM biochemical reactions leading to the production of useable energy (usually ATP) by the cell

CHEMOLITHOTROPH an organism which uses inorganic chemicals as energy sources (electron donors)

CHEMOORGANOTROPH an organism which uses organic chemicals as energy sources (electron donor)

COENZYME a small nonprotein molecule which participates in a catalytic reaction as part of an enzyme

CULTURE MEDIA an aqueous solution of various nutrients suitable for the growth of microorganisms

ELECTRON ACCEPTOR a substance that can accept electrons from some other substance, thereby becoming reduced in the process

ELECTRON DONOR a substance that can donate electrons to some electron acceptor, thereby becoming oxidized in the process

ENZYME a protein that has the ability to speed up (catalyze) a specific chemical reaction

FACULTATIVE in reference to oxygen utilization, an organism capable of growing under either aerobic or anaerobic conditions

FERMENTATION anaerobic catabolism of an organic compound in which the compound serves as both an electron donor and an electron acceptor and in which ATP is produced by substrate-level phosphorylation

HETEROTROPH an organism requiring organic compounds as a carbon source

OXIDATIVE (ELECTRON TRANSPORT) PHOSPHORYLATION the nonphototrophic production of ATP at the expense of a proton motive force formed by electron transport

PHOTOTROPH an organism capable of using light as an energy source

PROTON MOTIVE FORCE the energy available from establishment of a proton gradient formed across a membrane

PURE CULTURE a microbial culture containing a single kind of microorganism

REDUCTION POTENTIAL the inherent tendency (measured in volts) of a compound to donate electrons

RESPIRATION the process in which a compound is oxidized with O_2 serving as the terminal electron acceptor, usually accompanied by ATP production by oxidative phosphorylation

STERILE absence of all living organisms and viruses

SUBSTRATE-LEVEL PHOSPHORYLATION production of ATP by the direct transfer of a high-energy phosphate molecule from a phosphorylated organic compound to ADP

Miniglossary
new feature that defines key terms; set off by color panel

est. For instance, one important waste product produced by yeast during catabolism is *ethanol*, the key constituent of alcoholic beverages (wine, beer, whiskey, etc.). We will have more to say about the formation of ethanol by yeast later in this chapter.

> Metabolism is the study of the chemical reactions that are carried out in cells. Metabolism involves two basic kinds of chemical transformations, building up (biosynthetic) processes, called anabolism, and breaking down processes, called catabolism, which usually result in energy release. During metabolism, cells take in nutrients, convert them into cell components and obtain energy from them, and excrete waste products into the environment. Two kinds of energy sources can be used by cells—light and chemicals.

4.2   Energy

Energy is defined as the ability to do ... chapter, we discuss how living organism... cal energy. **Chemical energy** is the ene... when organic or inorganic compounds ... The first law of thermodynamics tells u... can be converted from one form to and... neither be created nor destroyed. Beca... forms are interconvertible, energy is ...

niently expressed by a single energy unit. Although a variety of units exist, in biology the most commonly used energy units are the kilocalorie (kcal) and the kilojoule (kJ). A kilocalorie is defined as the quantity of heat energy necessary to raise the temperature of 1 kilogram of water by 1°C. One kilocalorie is equivalent to 4.184 kilojoules. Because the kJ is widely used in microbial energetics, we will use the convention kilojoule throughout this book.

Free energy

Chemical reactions are accompanied by *changes* in energy. The amount of energy involved in a chemical reaction is expressed in terms of the gain or loss of energy during the reaction. There are two expressions of the amount of energy released during a chemical reaction, abbreviated H and G. H, called *enthalpy*, expresses the total amount of energy released during a chemical reaction. However, some of the energy re...

In Brief
reinforces key points in a section; highlighted in color

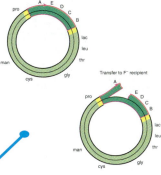

FIGURE 7.26 Breakage of the Hfr chromosome at the origin and transfer of DNA to the recipient. Replication occurs during transfer, as illustrated in Figure 7.23.

be selected against on agar plates so that only recombinants can grow (see Figure 7.24). The recipient is usually resistant to an antibiotic and is auxotrophic for one or more nutritional characters. The donor is antibiotic sensitive but prototrophic for the nutritional characters. With proper agar media, only the recombinants can grow and the large background of nonrecombinants is eliminated.

The oriented transfer of chromosomal genes from Hfr to F⁻ is most clearly shown by a procedure called **interrupted mating**. The mating pairs are rather weakly joined and can be separated by agitation in a mixer or blender. If mixtures of Hfr and F⁻ cells are agitated at various times after mixing and the genetic recombinants scored, it is found that the longer the time between pairing and agitation, the more genes of the Hfr will appear in the F⁻ recombinant. In addition, gene transfer always occurs in a specific order in a specific Hfr. As shown in Figure 7.28, genes present closer to the origin enter the F⁻ first and are always present in a higher percentage of the recombinants than genes that enter late. In addition to showing that gene transfer from donor to recipient is a sequential process, experiments of this kind provide a method of

Full-color Diagrams
color coded to help the student learn

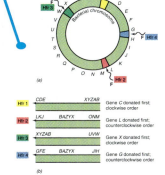

FIGURE 7.27 Manner of formation of different Hfr strains, which donate genes in different orders and from different origins. The bacterial chromosome is a circle (a) that can open at various insertion sequences, at which F plasmids become inserted. The gene orders are shown in part (b).

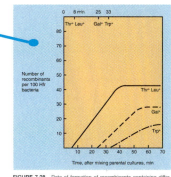

FIGURE 7.28 Rate of formation of recombinants containing different genes after mixing Hfr and F⁻ bacteria by the process known as interrupted mating. The location of the genes along the Hfr chromosome is shown in the small diagram. Note that the genes closest to the origin (0 minutes) are the first ones detected in the recombinants. The experiment is done by mixing Hfr and F⁻ cells under conditions in which essentially all Hfr cells find mates. The F⁻ recipient was streptomycin resistant but auxotrophic for the markers being scored. The Hfr donor was streptomycin sensitive. At various times, samples of the mixture are shaken violently to separate the mating pairs and plated on a selective medium in which only the recombinants can grow and form colonies.

Section Numbers
used for convenient
assignment and
cross reference

Bold-faced Key Terms
also defined in the
glossary

Excellent Micrographs
keyed to colorful line art

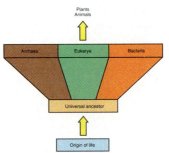

1.3 Evolutionary Relationships Among Living Organisms

Although cells can be differentiated structurally as being either prokaryotic or eukaryotic, cell structure does not imply *evolutionary* relationship. However, phylogenetic (evolutionary) relationships among microorganisms can now be determined, and Chapter 18 describes the methods used to discern such relationships. These methods rely on nucleic acid sequence comparisons, in particular the sequences of *ribosomal RNA*—structural RNA of the ribosome, the key cell structure involved in translation (see Section 1.1). Indeed, one of the great recent discoveries in biology is that changes in the sequence of bases in ribosomal RNA (ultimately dictated by mutations in the DNA which code for ribosomal RNA) can be used as a measure of evolutionary relationships among cells.

From studies of ribosomal RNA sequences, three evolutionarily distinct cellular lineages can be defined, two of which are prokaryotic in structure and one of which is eukaryotic. The groups have been given the names **Bacteria, Archaea,** and **Eukarya** (Figure 1.11).* However, despite the fact that both are structurally prokaryotic at the molecular level, Bacteria and Archaea are as evolutionarily distinct from one another as either group is from the Eukarya. All three groups are thought to have diverged from a common ancestral organism, the "universal ancestor," early in the history of life on earth (Figure 1.12). Because the cells of higher animals and plants are all eukaryotic, it follows that eukaryotic microorganisms were the ancestors of multicellular organisms, whereas Bacteria and Archaea represent evolutionary branches that never evolved past the microbial stage.

Classification

In addition to understanding and appreciating the phylogenetic origins of cellular organisms, it is impor-

tant for a variety of reasons to be able to identify and classify microorganisms. For example, rapid identification of a particular human disease-causing microorganism (pathogen) is usually essential to know how to treat an infected person. We discuss many methods for identifying microorganisms in Chapter 13, Clinical Microbiology. Several criteria have been used to characterize microorganisms, and at present, both phylogenetic standing and other cellular characteristics are used for classification purposes. After close study of the structure and function of a microorganism, including its genetics, metabolism, behavior, and other key properties, it is usually possible to recognize a group of characteristics unique to a given organism. Once an organism has been defined by such a set of characteristics, it can be given a name.

Microbiologists use the binomial system of nomenclature first developed by Linnaeus for plants and animals. The *genus* name is applied to a number of related organisms; each different type of organism within the genus has a *species* name. Genus and species names are always used together to describe a specific type of organism, whether it be a single cell or a group of such cells. (In writing, the genus and species names are either underlined or printed in italics.)

FIGURE 1.12 Cellular evolution as deduced from ribosomal RNA sequencing.

FIGURE 1.11 The major types of cellular life forms.

*To avoid the perception that all p[...] (which indeed they are not), the term [...] placed *eubacteria* and *archaebacteria*, re[...] the three taxonomic domains (the hi[...] Chapter 18 and Figure 1.12) of cellula[...] of organisms is the *Eukarya* (all of wh[...] ever, the word *bacteria* is so firmly e[...] crobiology that it is unlikely to ever [...] word *prokaryote*. This fact thus create[...] bacteria (prokaryotes) still Bacteria (a[...]

FIGURE 3.8
Representative cell shapes (morphology) in prokaryotes. Next to each drawing is a phase photomicrograph showing an example of that morphology. Organisms are coccus, *Thiocapsa roseopersicina* (diameter of a single cell = 1.5 μm); rod, *Desulfuromonas acetoxidans* (diameter = 1 μm); spirillum, *Rhodospirillum rubrum* (diameter = 1 μm); spirochete, *Spirochaeta stenostrepta* (diameter = 0.25 μm); budding/appendaged, *Rhodomicrobium vannielii* (diameter = 1.2 μm). Photos courtesy of Norbert Pfennig.

Coccus

Rod

Spirillum

Spirochete

Stalk   Hypha
Budding and appendaged bacteria

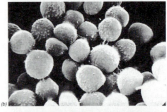

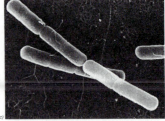

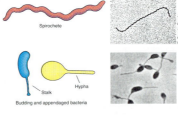

cells. One kind of organelle found in most eukaryotes is the **mitochondrion** (plural, **mitochondria**) (Figure 3.10; see also Figure 3.70). Mitochondria are the organelles within which the energy-generating functions of the cell occur. The energy produced in mitochondria is then used throughout the cell.

Algae are eukaryotic microorganisms that carry out the process of *photosynthesis*. In these organisms, as well as in green plants, an additional type of organelle is found: the **chloroplast**. The chloroplast is green and is the site where chlorophyll is localized and where the light-gathering functions involved in photosynthesis occur (see Figure 3.2a).

Size of microbial cells and the significance of smallness

Prokaryotes vary in size from cells as small as 0.1–0.2 μm in width to those more than 50 μm in diameter; a few very large prokaryotes, such as the surgeonfish symbiont *Euplopiscium fishelsoni*, are known to be

FIGURE 3.9 Spherical and rod-shaped bacteria which associate in characteristically different ways as viewed by scanning electron microscopy. (a) *Streptococcus*, a chain-forming organism which divides only in one plane. (b) *Staphylococcus*, a coccus which divides in more than one plane. Cells of both organisms are about 1 μm in diameter. (c) Chains of the rod-shaped bacterium *Bacillus*. Cells are about 0.8 μm in diameter.

Selenocysteine: The 21st Amino Acid

The genetic code has codons for 20 amino acids that are assembled into proteins during translation. However, many proteins contain other amino acids. In fact, there are well over 100 different amino acids found in at least a few proteins. Until recently, it was thought that these "extra" amino acids were made by modifying one of the standard amino acids *after* it was incorporated into protein, a process called *posttranslational modification*. However, it is now clear that one of these extra amino acids is put into protein by the translational machinery itself. This one exception is *selenocysteine*.

Selenocysteine has the same structure as cysteine, but it has a selenium atom rather than a sulfur atom. It was known for some time that a few proteins contained this unusual amino acid. For example, *E. coli* makes two different formate dehydrogenase enzymes and both contain a single selenocysteine residue. When the gene encoding one of these enzymes was sequenced, it was found that the codon that corresponded to the selenocysteine was a UGA. UGA is normally an efficient stop codon in *E. coli*, but it has now been demonstrated that UGA can be translated directly as selenocysteine in certain mRNA molecules, not only in *E. coli* but also in other prokaryotes and in eukaryotes, including humans. Therefore, selenocysteine is the 21st amino acid known to be encoded by the genetic code.

How can a codon sometimes be a stop codon and sometimes a sense codon *in the same chromosome*? The answer apparently lies in the *context* of the codon, the sequence of the bases surrounding the UGA codon and in their secondary structure. In certain contexts, the translational machinery interprets UGA as "selenocysteine." In all other contexts, UGA means "stop translation." Selenocysteine has its own tRNA (as do all the standard amino acids) and also has a special protein factor that brings only this tRNA to the ribosome.

Selenocysteine is even more readily oxidized than cysteine. Therefore, enzymes that contain this amino acid must be protected from oxygen. It has been proposed that UGA might once have been a normal sense codon, calling only for selenocysteine, but that the increase in oxygen in our environment following the evolution of photosynthesis (see Chapter 18) selected for proteins that contained cysteine (whose codons are UGU and UGC). This allowed the coding assignment of UGA to be altered, except in a few special cases.

Boxes highlighted in color present current issues in microbiology

Table 5.5   Variations in the genetic code*

Codon	Universal code	Other nuclear codes			Other mitochondrial codes		
		Mycoplasma	*Paramecium*	*Euplotes*	Yeast	Protozoans	Mammals
UGA	Stop	Tryptophan	Stop	Cysteine	Tryptophan	Tryptophan	Tryptophan
UAA/UAG	Stop	Stop	Glutamine	Stop	Stop	Stop	Stop
AUA	Isoleucine	Isoleucine	Isoleucine	Isoleucine	Methionine	Methionine	Methionine
CUA	Leucine	Leucine	Leucine	Leucine	Threonine	Leucine	Leucine
AGA/AGG	Arginine	Arginine	Arginine	Arginine	Arginine	Arginine	Stop

The universal genetic code is used in the chromosomes of most cells, chloroplasts, plant mitochondria, and their viruses and plasmids. A few organisms use slightly different codes in their chromosomes (in the nucleus). The examples of these other nuclear codes are from Mycoplasma (Bacteria) and two different ciliated protozoans (Eukarya). All nonplant mitochondria use variations of the universal code, whereas plant mitochondria use the universal code. The examples here are only a few of the different types known.

Tables highlighted in color summarize important information

Overlapping genes

Although the evidence is strong that the nucleotide sequence specifying one product is separate and distinct from the sequence specifying another product, studies on the small bacterial virus φX174 have shown that this virus has insufficient gen... nonoverlapping genes to code fo... necessary for its reproduction, an... omy is introduced by using the sa... the coding of more than one pro... made possible by reading of the ... quence in two different reading f... different sites. There are now kno... teresting patterns of overlapping g... occur in viruses (see Chapter 6).

The complete genetic code, usually expressed in terms of mRNA rather than DNA, is known. A single amino

---

Study Questions

1. Calculate the surface/volume ratio of a spherical cell 10 μm in diameter, a cell 1 μm in diameter, and a rod-shaped cell 0.5 μm in diameter by 2 μm long. What are the consequences of these differences in surface/volume ratio for cell function?

2. From what you know about the nature of the bacterial cell wall, explain why a rod-shaped bacterial cell becomes a *spherical* structure when its wall is removed under conditions such that cell lysis cannot occur.

3. Describe in a single sentence the manner by which a unit membrane is formed from phospholipid molecules. How would the membrane formation process differ if the phospholipid molecules were dissolved in a nonaqueous solvent rather than water? How does a *membrane vesicle* differ from a unit membrane? How is such a vesicle *similar* to a unit membrane?

4. Explain in a single sentence why ionized molecules do not readily pass through the membrane barrier of a cell. How *do* such molecules get through the cytoplasmic membrane?

5. Why is the bacterial cell wall rigid layer called "peptidoglycan"? What are the chemical reasons for the rigidity that is conferred on the cell wall by the peptidoglycan structure?

6. Since a single peptidoglycan molecule is very thin, explain in chemical terms how the very *thick* peptidoglycan-containing cell wall of Gram-positive Bacteria is formed.

7. Both lysozyme and penicillin bring about bacterial cell lysis, but by different mechanisms. Describe the mechanism by which each of these agents causes cell lysis. Be sure to include in your explanation an exact description of why *lysis* occurs.

8. What is the bacterial periplasm? What types of Bacteria have a periplasm and of what significance is the periplasmic space?

9. List several functions for the outer wall layer in Gram-negative Bacteria.

10. What is osmosis? Why does it occur? Of what significance is osmosis for the stability of the bacterial cell?

11. Write a clear explanation (two or three sentences) for why sucrose is able to stabilize bacterial cells from lysis by lysozyme.

12. In a few sentences, write an explanation for how a motile bacterium is able to "sense" the direction of an attractant and move toward it.

13. In a few sentences, indicate how the bacterial endospore differs from the vegetative cell in structure, chemical composition, and ability to resist extreme environmental conditions.

14. The discovery of the bacterial endospore was of great practical importance. Why?

15. Describe one chemical difference between the eukaryotic and prokaryotic membrane. Can you offer an explanation for why this chemical difference might exist?

16. Describe a major chemical difference between membranes of Bacteria and Archaea.

17. Water molecules penetrate cytoplasmic membranes fairly readily but hydrogen ions (protons) do not, even though a proton is smaller than a water molecule. Why?

18. List three kinds of *membrane proteins* and give a short explanation for the function of each.

19. Compare and contrast the following processes: *passive diffusion, facilitated diffusion, group translocation*, and *active transport*. For each of these processes, include a discussion of specificity, energy requirement, and transport against a concentration gradient.

20. Describe the structure and function of a bacterial flagellum.

21. How does a *cilium* differ from a flagellum? In what way(s) are they similar?

22. What types of cytoplasmic inclusions are formed by prokaryotes? How does an inclusion of PHB differ from a magnetosome in composition and metabolic role?

23. List several properties by which Bacteria, mitochondria, and chloroplasts are similar. List several ways in which they differ.

24. List two ways in which the DNA of the eukaryote differs from the DNA of the prokaryote. List two ways in which the DNA in these two kinds of organisms is similar.

25. Set up a table following the format of Table 3.5 with the second and third columns blank, and then fill in the blanks. As you do so, think back to the figures in this chapter that illustrate the properties being considered.

26. How does the eukaryotic nucleus differ from the prokaryotic nucleoid? In what ways are these two structures similar?

Study Questions challenge the student's mastery of the chapter concepts

Supplementary Readings

Alberts, B., D. Bray, J. Lewis, M. Raff, K. Roberts, and **J. D. Watson.** 1989. *The Molecular Biology of the Cell*, 2nd edition. Garland Publishing, New York. A substantial textbook with extensive coverage of cell structure, with emphasis on the eukaryotic cell.

Cole, J. A., C. Dow, and **S. Mohan** (eds.) 1992. *Prokaryotic Structure and Function: A New Perspective*. Soc. Gen Microbiol. Symp. 47. Cambridge University Press, New York. Chapters, written by experts, on the structure, function, and regulation of synthesis of bacterial components.

Cooper, S. 1991. *Bacterial Growth and Division*. Academic Press, San Diego, CA. A consideration of replication of the bacterial chromosome in relation to cell division processes. Also covers major processes of cell division in eukaryotic cells.

Neidhardt, F. C., J. L. Ingraham, and **M. Schaechter.** 1990. *Physiology of the Bacterial Cell—A Molecular Approach*. Sinauer Associates, Inc., Sunderland, MA. A very readable textbook of bacterial physiology with an emphasis on molecular aspects of cell structure/function, growth, and genetics.

Riley, M. (ed.) 1990. *The Bacterial Chromosome*. American Society for Microbiology, Washington, D.C. A detailed treatment of the structure and processing of bacterial DNA.

Schlegel, H.G. 1993. *General Microbiology*, 7th edition. Cambridge University Press, New York. An excellent overview of cell structure and many other aspects of microbiology for the beginning student.

Watson, J. D., N. H. Hopkins, J. W. Roberts, J. A. Steitz, and **A. M. Weiner.** 1987. *Molecular Biology of the Gene*, 4th edition. Benjamin/Cummings, Menlo Park. This excellent textbook has a useful overview of the principles of cell biology.

Supplementary Readings offer further reading for in-depth pursuit of a topic

Brief Contents

Contents

Chapter 7   Microbial Genetics   237

Chapter 8   Genetic Engineering and Biotechnology   287

Chapter 9   Growth and Its Control   321

Chapter 10   Industrial Microbiology   361

Chapter 11   Host–Parasite Relationships   397

Chapter 12   Immunology and Immunity   435

Preface

The authors are proud to present the seventh edition of *Biology of Microorganisms* to students and instructors of microbiology. The seventh edition is a landmark for this book, as nearly a quarter of a century has elapsed since the first edition of *Biology of Microorganisms* was written by the senior author and published by Prentice Hall in 1970. Obviously, much has changed in the field of microbiology in 25 years. But through the years the goal of each edition of this book has remained the same: to cover the field of microbiology as it is currently understood while at the same time emphasizing basic microbiological principles. The seventh edition of *Biology of Microorganisms* carries on this tradition with the most complete and current treatment of microbiology available today.

For the seventh edition, two new coauthors have been added to cover the important areas of genetics/molecular biology (Jack Parker) and medical microbiology/immunology (John Martinko). Microbiology has been advancing so rapidly that it has become virtually impossible for only two authors to write a textbook that is complete, accurate, current, and at a level that falls in the tradition of *Biology of Microorganisms*. The new coauthors are active teachers and researchers (see "About the Authors"), and along with Michael Madigan, are on the faculty of the Department of Microbiology, Southern Illinois University. Thus, this book was written by experienced teachers who regularly bring the excitement of research into the classroom. In addition, having all but one of the authors in the same place has allowed us to work together closely in the preparation of this book, yielding what we feel is a balanced and eminently readable treatment of microbiology from cover to cover.

As with previous editions, the seventh edition of *Biology of Microorganisms* has undergone extensive changes. The genetics revolution of the past 20 years has spawned new and powerful research tools that have impacted on virtually every area of microbiology and offer new solutions to applied problems in medicine, agriculture, and the environment. All these areas are covered in this new edition, using our popular approach of treating the basic science along with the applications of the science. However, users of previous editions of this book can rest assured that the seventh edition of *Biology of Microorganisms* retains its emphasis on the microorganisms themselves. Although important research tools, microorganisms are significant in their own right and thus their basic biology and ecological activities remain well covered in this book.

Because the organization and coverage of the sixth edition of *Biology of Microorganisms* has worked well in practice, we have not added any new chapters to the seventh edition. However, every chapter has seen revision, and in many chapters there have been extensive revisions. Over 480 pieces of full color art are either new or revised for this edition, to keep pace with the rapid developments in microbiology. As in previous editions, we have taken every effort to produce a book that maintains the authority, clarity, and breadth of coverage that instructors of courses in the biomedical sciences have come to expect.

A variety of pedagogical aids are built into this book. As usual, relevant and challenging **Study Questions** can be found at the end of each chapter. Many questions are new to this edition and are written in the tradition of the ones that have proven so successful in previous editions: The questions deemphasize rote memorization and instead make the student go back into the chapter, review, interpret, and analyze the material, and then synthesize information to arrive at an answer. As usual, **Supplementary Readings** are listed for students and instructors who wish to probe the most current seminal reference sources for a more detailed treatment of the material than is possible in an introductory textbook.

The **In Brief** segments, so popular when introduced in the sixth edition, are now an integral part of this book, and have been expanded.

Totally new in the seventh edition is a concept we call the **Miniglossary**. Terminology is important for understanding any science, and microbiology is no exception. We believe that if key terms are clearly understood, reading comprehension will automatically improve. Thus we have two glossaries in *Biology of Microorganisms*. The main glossary in the back of the book remains the storehouse for all major terms used in the book. However, we have pulled out the truly seminal terms from the main glossary and arranged them, chapter by chapter, in the form of "miniglossaries." Each miniglossary contains about 20 succinctly defined terms and can be found in the big yellow box on the pages immediately following the chapter opening. Such ready access to the definitions of key terms should quickly build a student's vocabulary and strengthen the learning process. We thus believe that regular use of the miniglossaries will make students masters of terminology, rather than the other way around. Finally, as has been the case since the first edition of *Biology of Microorganisms* rolled off the presses nearly 25 years ago, the seventh edition of this book contains a comprehensive index.

Highlights of the 7th edition include:

Chapter 1—Introduction to Microbiology and Cell Biology maintains the thrust of the previous edition, introducing the student to the basic biology of the cell and to the importance of evolutionary theory to microbiology. New material will be found on the evolution of microbial cells and microbial diversity, the applications of microbiology to our everyday lives, and historical developments in microbiology during the twentieth century.

Chapter 2—Cell Chemistry maintains its important early position in the book as the chemical primer every student needs to master. The emphasis remains on understanding the fundamental chemistry of cellular macromolecules.

Chapter 3—Cell Biology once again emphasizes the structure of the prokaryotic cell, with a completely rewritten introductory section that begins with a consideration of microscopes, the major tools for dissecting cell structure.

Cell 4—Metabolism, Biosynthesis, and Nutrition maintains the organization of the previous edition but is updated in several important areas, such as the mechanism of membrane-mediated ATP synthesis and alternative modes of energy generation.

Chapter 5—Macromolecules and Molecular Genetics has been heavily revised with several pieces of new art to portray the many new discoveries in this major area of microbiology. As in previous editions, the essentials of molecular genetics are clearly presented along with exciting and informative boxes. New material on alternative genetic codes and global regulatory networks rounds out this up-to-the-minute and very exciting chapter.

Chapter 6—Viruses takes the approach of the sixth edition but has been streamlined somewhat to better focus on the essentials of virology. This chapter ends with a new section on viroids and prions—virus-like particles that are not viruses—comparing and contrasting these interesting genetic elements with viruses and emphasizing their unique replication features and pathogenic nature.

Chapter 7—Microbial Genetics builds on the material in Chapters 5 and 6 to give a modern picture of the major genetic phenomena observed in prokaryotes. Coverage of both classical bacterial genetics and the molecular phenomena behind it is supported by several new pieces of art and boxes.

Chapter 8—Genetic Engineering and Biotechnology has a new title from that of the sixth edition, reflecting the heavy revision this fast moving area has required. Material on major products of the biotechnology industry from Chapter 10 of the sixth edition has been updated and moved here to better unite the practical results of biotechnology with the basic science that underlies it. Transgenic plants and animals, human blood proteins, gene therapy in human medicine—all of the major uses of recombinant DNA technology—are discussed in a way that is certain to generate excitement.

Chapter 9—Growth and Its Control emphasizes the fundamental principles of exponential growth and the effect of the environment on growth, as well as the practical consequences of microbial growth, using food spoilage as an example. New material on the mathematical treatment of microbial growth, water

potential and compatible solutes, and microbial spoilage of fresh foods, have been added.

Chapter 10—Industrial Microbiology has a new title and has been condensed by removal of the biotechnology material; the chapter now focuses more on traditional industrial fermentations. Antibiotic production remains the driving force of this chapter, reflecting the importance of the multibillion dollar pharmaceutical industry to clinical medicine.

Chapter 11—Host-Parasite Relationships maintains the long tradition of this book in covering microbial interactions with humans. The focus remains on the struggle between the host and the parasite and the major weapons both possess to maintain health or induce disease.

Chapter 12—Immunology and Immunity has been extensively revised in line with the fact that this area of microbiology, along with molecular genetics, is today one of the most rapidly moving areas in microbiology and perhaps in all of biology. This chapter has seen considerable reorganization and rewriting and contains many pieces of new art, supported by new boxes like "Over-the-counter Immunodiagnostic Kits" (for example, the human pregnancy test). New material on immune tolerance and vaccination strategies helps tie the basic science of immunology to the applications of immunology so widespread in clinical medicine.

Chapter 13—Clinical and Diagnostic Microbiology continues to keep students and instructors up to date on the constantly evolving field of clinical and diagnostic microbiology. Several highly innovative clinical diagnostic methods have emerged from advances in genetics and immunology and are described in detail in this chapter. Because these methods are rapidly changing the way hospital microbiology is done, this chapter will surely be important for courses with a health professions emphasis.

Chapter 14—Epidemiology and Public Health Microbiology although short, sets the stage for consideration of specific diseases in Chapter 15 by outlining the principles of disease transmission and the concepts of public health.

Chapter 15—Major Microbial Diseases maintains broad coverage of microbial diseases using an ecological approach of grouping individual diseases by their mode of transmission. Coverage of AIDS has been greatly updated and expanded, but there is also new material on pathogenic fungi, toxic shock, rabies, Lyme disease, and malaria, all supported by excellent color photos and disease statistics that reflect the worldwide prevalence and trends of these diseases.

Chapter 16—Metabolic Diversity begins the strong theme of microbial diversity that runs through the remainder of the book. Greatly updated material on photosynthesis, fermentation, and several other key mechanisms of energy generation, highlight this chapter.

Chapter 17—Microbial Ecology emphasizes the importance of understanding the scope and dimensions of microbial habitats and the activities of microorganisms in their habitats. New material on molecular methods for microbial ecology, the microbial biogeochemistry of mercury, reductive dechlorination, and bacterial insecticides, are just some of the more heavily updated sections in this chapter.

Chapter 18—Molecular Systematics and Microbial Evolution has been greatly revised from the sixth edition. The current revolution in microbial phylogenetics is thoroughly covered in this chapter, setting the stage for detailed treatment of the organisms themselves in Chapters 19–21. Look for new material on the metabolism of primitive organisms and a greatly updated phylogenetics discussion.

Chapters 19–21—These "organisms" chapters remain an integral part of *Biology of Microorganisms*. This material has served well in the past as supplemental material for introductory microbiology courses and also as the foundation of "organisms courses" in bacterial/microbial diversity. Every section of these three chapters has seen updating and Chapter 20 on the Archaea has been extensively revised. These chapters also serve as a reference source for advanced courses in microbiology and have been well received ever since they were introduced.

Several Supplements accompany this textbook. These include a Student Study Guide, which highlights key topics and contains a wealth of additional objective and subjective review questions. For the instructor, a valuable **Instructor's Manual** is available which describes various ways to structure an introductory microbiology course using *Biology of Microorganisms* as text, and gives answers to all the Study Questions. In addition, a set of 200 color transparencies or color slides (twice as many as was available with the sixth edition) will assist those adopting this book to organize and present class lectures.

The **Contemporary View** program sponsored jointly by Prentice Hall and *The New York Times,* and initiated with the sixth edition of *Biology of Microorganisms,* complements the new edition as well. Through this program, core subject matter from the text is supplemented by articles describing microbiology in "real life" situations pared from the pages of *The New York Times.* Comtemporary View helps strengthen the connection between the classroom and the real world, as students see how important an understanding of microbiology is to activities in their everyday lives.

This book is not the product of just the four authors. Many scientists gave valuable reviews of an earlier draft and others made special efforts to provide color photographs taken directly from their laboratory research. We are extremely grateful for their efforts. These include:

Laurie Achenbach, Southern Illinois University
Peter T. Borgia, Southern Illinois University
Dawn Cardascia, Earth Science Support Office, NASA
John H. Caster, Southern Illinois University
David Clark, Southern Illinois University
Morris Cooper, Southern Illinois University
Frank Dazzo, Michigan State University
Stephen Edmondson, Southern Illinois University
George Feher, University of California, San Diego
Martin Flajnik, University of Miami
Allan Konopka, Purdue University
Dino Moras, Universite Louis Pasteur
Ed Moticka, Southern Illinois University
Norman Pace, Indiana University
Elizabeth Parker
Norbert Pfennig, University of Konstanz, Germany
D. J. Read, University of Sheffield, England
Gordon Sauer, University of Kansas
Jolynn F. Smith, Southern Illinois University
Daniel E. Snyder, Animal Health Division, Pfizer Laboratories, Inc.
Donald P. Stahley, University of Iowa
Thomas A. Steitz, Yale University
Mark L. Tamplin, University of Florida
Howard Temin, University of Wisconsin
David M. Ward, Montana State University
Carl R. Woese, University of Illinois
Stephen H. Zinder, Cornell University

In addition, the Photo Credits at the end of the book lists the many more who have provided photographs or other materials.

The authors are grateful to all the people at Prentice Hall who have contributed in significant ways to this edition, including in particular David K. Brake (editorial) and Susan Fisher (production). We also wish to acknowledge the expert contributions of Kathy Walker (Renton, WA) to copyediting, Barbara Littlewood (Madison, WI), who composed the index, and Toni Gower, Southern Illinois University, for her expert word processing skills.

T.D.B.
M.T.M.
J.M.M.
J.P.

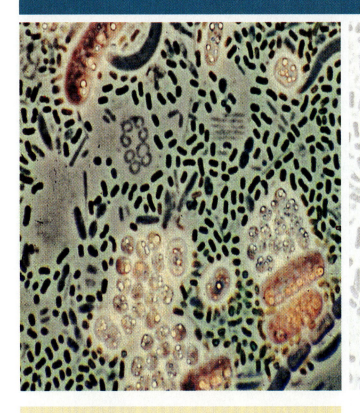

Introduction:

An Overview of Microbiology and Cell Biology

Microbiology is the study of microorganisms, a large and diverse group of organisms that exist as single cells or cell clusters. Microbial cells are thus distinct from the cells of animals and plants, which are unable to live alone in nature but can exist only as parts of multicellular organisms (Figure 1.1). A single microbial cell is generally able to carry out its life processes of growth, energy generation, and reproduction independently of other cells, either of the same kind or of different kinds.

What, then, is microbiology all about? We can list several aspects of microbiology here: (1) It is about living cells and how they work. (2) It is about microorganisms, an important class of cells capable of free-living (independent) existence. It is especially about the bacteria, a large group of structurally simple cells of enormous basic and practical significance. (3) It is about microbial diversity and evolution, about how different kinds of microorganisms arose, and why. (4) It is about what microorganisms do, in the world at large, in human society, in our own bodies, in the bodies of animals and plants. (5) It is about the central role that microbiology plays as a basic biological science and how an understanding of microorganisms helps in understanding the biology of higher organisms, including humans.

Why study microbiology? Microbiology, one of the most important of the biological sciences, is studied for two major reasons:

1. As a *basic biological science*, microbiology provides some of the most accessible research tools for probing the nature of life processes. Our most so-

(b)

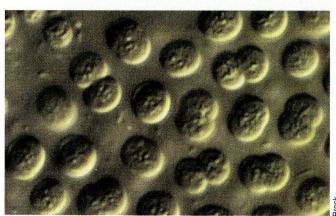

(c)

FIGURE 1.1 Living organisms are composed of cells. (a) Plants and (b) animals are composed of many cells; they are called *multicellular*. A single plant or animal cell cannot have an independent existence; each of its cells is dependent on the other. (c) Microorganisms are free-living cells. A single microbial cell can have an independent existence.

phisticated understanding of the chemical and physical principles behind living processes has arisen from studies using microorganisms.

2. As an *applied biological science*, microbiology deals with many important practical problems in medicine, agriculture, and industry. Some of the most important diseases of humans, animals, and plants are caused by microorganisms. Microorganisms play major roles in soil fertility and animal production. Many large-scale industrial processes are microbially based, which has led to development of a whole new discipline, *biotechnology*.

In this book, both the basic and applied aspects of microbiology are covered in an integrated fashion. We discuss the experimental basis of microbiology, the general principles of cell structure and function, the classification and diversity of microorganisms, biochemical processes in cells, and the genetic basis of microbial growth and evolution. From an applied viewpoint, we discuss disease processes in humans that are caused by microorganisms, the nature of the immune response, the roles of microorganisms in food and agriculture, and industrial and biotechnological processes employing microorganisms.

The material in this textbook serves as a foundation for advanced work in microbiology. It also serves as a basis for further studies in cell biology, biochemistry, molecular biology, and genetics. Although a student may begin to study microbiology primarily because of its applied problems, the basic concepts learned will serve as a foundation for advanced study in many areas of contemporary biology. A firm grasp of microbiological principles will also serve as a basis for understanding biological processes in higher organisms, including humans.

> **Microbiology is the study of microorganisms, those organisms that exist in nature as single cells. Microbiology is a broad biological discipline with strong roots in both basic and applied science.**

Miniglossary for Chapter 1

ARCHAEA   a group of phylogenetically related prokaryotes distinct from Bacteria

BACTERIA   a group of phylogenetically related prokaryotes distinct from Archaea

CELL   the fundamental unit of living matter

CYTOPLASM   the fluid portion of a cell, bounded by the cell membrane but excluding the nucleus (if present)

DNA   deoxyribonucleic acid, the hereditary material of cells and some viruses

ECOLOGY   the study of organisms in their natural environments

ENZYMES   protein catalysts that function to speed up chemical reactions

EUKARYA   all eukaryotic organisms

EUKARYOTE   a cell possessing a membrane bound nucleus and usually other organelles

GENETICS   heredity and variation of living organisms

METABOLISM   all biochemical reactions in a cell

PROKARYOTE   a cell lacking a true nucleus

RNA   ribonucleic acid, involved in protein synthesis as messenger RNA, transfer RNA, and ribosomal RNA

SPONTANEOUS GENERATION   the hypothesis that living organisms can originate from nonliving matter

STERILE   absence of all living organisms and viruses

TRANSCRIPTION   the synthesis of RNA using a DNA template

TRANSLATION   the synthesis of proteins using the genetic information in RNA as a template

1.1   Microorganisms as Cells

The **cell** is the fundamental unit of all living matter. A single cell is an entity, isolated from other cells by a cell membrane (and perhaps a cell wall) and containing within it a variety of chemical materials and subcellular structures (Figure 1.2). The **cell membrane** is the *barrier* which separates the inside of the cell from the outside. Inside the cell membrane are the various structures and chemicals which make it possible for the cell to function. Key structures are the **nucleus** or **nucleoid**, where the *information* needed to make more cells is stored, and the **cytoplasm**, where the *machinery* for cell growth and function is present.

All cells contain certain types of complex chemical components: **proteins, nucleic acids, lipids**, and **polysaccharides**. Because these chemical components are common throughout the living world, it is thought that all cells have descended from a single common ancestor, the *universal ancestor*. Through billions of years of evolution, the tremendous diversity of cell types that exist today has arisen.

Although each kind of cell has a definite structure and size, a cell is a dynamic unit, constantly undergoing change and replacing its parts. Even when it is not growing, a cell is continually taking materials from its environment and working them into its own fabric. At the same time, it continuously discards waste products into its environment. A cell is thus an *open system*, forever changing, yet generally remaining the same.

The hallmarks of a cell

A living cell is a complex chemical system. What are the characteristics that set living cells apart from nonliving chemical systems? We list five major characteristics here (Figure 1.3).

1. **Self-feeding or nutrition**. Cells take up chemicals from the environment, transform these chemicals from one form to another, release energy, and eliminate waste products.

2. **Self-replication or growth**. Cells are capable of directing their own synthesis. A cell grows and divides, forming two cells, each nearly identical to the original cell.

3. **Differentiation**. Many cells can undergo changes in form or function, a process called *differentiation*. When a cell differentiates, certain substances or structures that were not formed previously are now formed, or substances or structures that had been formed previously are no longer formed. Cell differentiation is often a part of a cellular life cycle, in which cells form specialized structures involved in sexual reproduction, dispersal, or survival when confronted with unfavorable conditions.

4. **Chemical signaling**. Cells can often *interact* or *communicate* with other cells, generally by means of **chemical signals**. Multicellular organisms, such as plants and animals, are composed of many different cell types, arranged to form tissues and organs, that have arisen as a result of differentiation from single cells. In multicellular organisms, complex interactions between these different cell types lead to the behavior and function of these cells. One of the striking things about the cells of multicellular organisms is that they are incapable of independent existence in nature, but only exist as part of a whole plant or animal. This interdependence of the cells of higher organisms is one of the hallmarks of multicellular life. In the microbial world chemical communication occurs, although it is less highly developed.

5. **Evolution**. Unlike inanimate structures, unicellular and multicellular organisms evolve. By this is meant that hereditary changes (which occur at low but regular rates in all cells) can influence the overall fitness of the cell or higher organism in a positive or negative way. The result of evolution is selection for those organisms best suited to life in a particular environment.

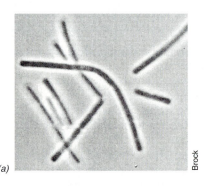

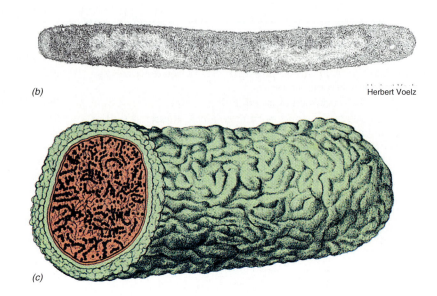

FIGURE 1.2 Cells. (a) Photomicrograph of bacterial cells, as seen in the light microscope. (b) Cross section through a bacterial cell, as viewed with an electron microscope. The two lighter areas represent the nucleoid. (c) An artist's interpretation in three dimensions of a bacterial cell, showing the characteristic wrinkled outer structure.

The improbable cell

A fundamental law of physics is that the universe is moving constantly toward a condition of greater disorder, with molecules and atoms becoming arranged in a random manner. If the universe is random, life then seems like a miracle, since the living cell is anything but random, being a highly ordered, exceedingly improbable (nonrandom) structure. More careful analysis of cells, however, convinces one that there is no divergence between biology and physics. How can this be? First, we note that the idea of randomness implies that the physical system is in *equilibrium* with its surroundings. A living cell, on the other hand, is definitely *not* in equilibrium with its surroundings. We say that the living cell is a *nonequilibrium* system. How is it possible for a cell to maintain this nonequilibrium condition?

As we have seen, a living cell is actually an *open system*, a system in which energy is taken in from the surroundings and used to maintain cell structure. Viewed in this way, a living cell can be considered as a *chemical system that works*. A cell carries out energy transformations, and some of this energy is used to maintain the structure of the cell itself. What happens if the cell runs out of energy? Its structure deteriorates and it eventually dies.

Thus, for a cell to function as a cell, its structure must be maintained. Indeed, *the basis of cell function is cell structure*. Biological information, the information needed to produce a new cell, is structural. The importance of structure as a foundation of life is further emphasized when we recall that cells are *self-replicating systems*. Cell reproduces cell. Therefore, making a living cell is a matter of making the right structure. This principle, which is the basis of biology, was enunciated many years ago by the famous German cellular pathologist Rudolph Virchow: *"Omnis cellula e cellula."* "Every cell from a cell." If every cell comes from a preexisting cell, where did the first cell come from? In some way, the first cell must have come from a noncell, something before the cell, a procellular structure. We discuss the origin of life and the evolution of cells in some detail in Chapter 18.

The nonrandom nature of a living cell is shown most dramatically by an analysis of its chemical composition and a comparison of that chemical composition with the chemical composition of the earth. The average chemical composition of a living cell is quite different from the average chemical composition of the earth. The cell therefore is not a random assortment of chemical elements found on earth, but instead a *selective* chemical system primarily composed of C, H, O, N, S, and P, the major elements of life (Figure 1.4). This further emphasizes the special or nonrandom nature of a living cell.

However, because a cell is a nonrandom structure it does not necessarily follow that life is a miracle. Life is indeed an improbable phenomenon. Once the first cell arose, however, subsequent events appear highly probable, a result of the chemical reactions that cells are inherently able to carry out.

Astronomers tell us that there are probably vast numbers of planets in the universe with earth-like conditions on which life might arise or might have arisen. If life is an almost inevitable event when the proper physical and chemical conditions are available, then we can anticipate that there are other planets in the universe with living organisms similar to those found on earth. The precise assemblage of organisms would be different, of course, because so much of evolution depends upon accidents of history, and the time stage and evolution on these other planets would likely be earlier or later than those on earth. But if we were able to sample some of these other planets, it is likely we would not find any surprises; as we will see, major chemical principles govern the structure and function of a highly diverse array of cell types here on earth.

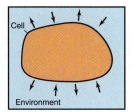

1. Self-feeding (nutrition)
Uptake of chemicals from the environment and elimination of wastes into the environment.

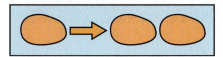

2. Self-replication (growth)
Chemicals from the environment are turned into new cells under the direction of preexisting cells.

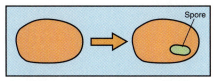

3. Differentiation
Formation of a new cell structure such as a spore, usually as part of a cellular *life cycle*

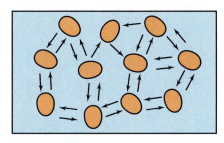

4. Chemical signaling
Cells *communicate* or *interact* primarily by means of chemicals which are released or taken up.

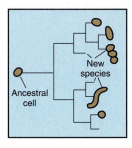

5. Evolution
Cells evolve to display new biological properties. Phylogenetic trees show the evolutionary relationships between cells.

FIGURE 1.3   The hallmarks of cellular life.

There are five principle characteristics of living cells: self-feeding (nutrition), self-replication (growth), differentiation (change in form or function), chemical signaling (communication with other cells by means of chemical substances), and evolution, the introduction of hereditary changes as a result of natural selection.

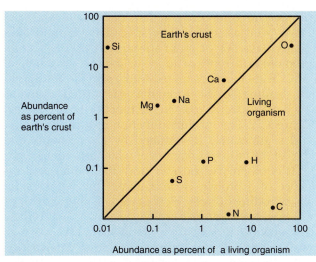

FIGURE 1.4   Chemical differences between a living organism and the earth. Note that the key elements C, H, O, N, P, and S are much more abundant in living organisms than in nonliving matter. Thus, living organisms *concentrate* these elements from the environment.

Cells as machines and coding devices

Cells can be studied as *machines* that carry out chemical transformations. In this view, the cell is a *chemical machine* that converts energy from one form to another, breaking molecules down into smaller units, building up larger molecules from smaller ones and carrying out many other kinds of chemical transformations (a molecule is defined as two or more atoms chemically bonded to one another). The term **metabolism** is used to refer to the collective series of chemical processes which occur in living organisms, both biosynthetic and degradative. When we speak of metabolic reactions, we mean *chemical* reactions occurring in living organisms.

Cells can also be studied as *coding devices*, analogous to computers, possessors of information which is either passed on to offspring or is processed into another form. The term **genetics** is used to describe the heredity and variation of living organisms and to describe the mechanisms by which these are brought about. Depending upon our interest, and upon the problem at hand, we may study cells either as chemical machines or as coding devices. In the following we discuss these two attributes briefly.

If a cell is a chemical machine, what are the components which make it function? The components of the cell's chemical machine are **enzymes**, protein molecules capable of catalyzing specific chemical reactions. Whether or not a cell can carry out a particular chemical reaction will depend on the presence in the cell of an enzyme that catalyzes the reaction. Specificity of enzymes is generally very high, so that even very closely related chemical reactions in the cell are catalyzed by separate enzymes. The specificity of an enzyme is determined primarily by its structure. As proteins, enzymes consist of long chains of amino acids (20 different amino acids) which are connected in specific and highly precise ways. It is the amino acid *sequence* of an enzyme that determines its structure as well as its catalytic specificity. Proteins frequently have 300 or more amino acid residues, composed of the 20 types of amino acids. The long chain of amino acids becomes folded into a specific configuration, leading to the formation of various domains which play specific roles in the function of the protein (Figure 1.5). The machine function of a cell (metabolism) is ultimately determined by the amounts and types of the various enzymes of which it is composed.

It is when we consider how the amino acid sequence of a protein is determined that we consider cells as coding devices. How is the cell able to arrange amino acids into precise sequences, each constituting a separate kind of protein? To understand this, we must consider the cell as a device for storing information and converting it into the appropriate form. In this context, the cell can be viewed as a repository of protein sequence information stored in a coded manner.

This code, called the *genetic code*, is stored in a sequence of bases in the hereditary molecule, **deoxyribonucleic acid, DNA** (Figure 1.6). DNA is present in the cell as two long molecules that are intertwined to form a helix, the famous DNA double helix. Each half of the helix is chemically *complementary* to the other, the two molecules being held together by highly specific pairing of the bases of which the molecules are constructed. The genetic code is contained in the *sequence* of bases in the molecule. Each amino acid in a protein is encoded by a sequence of *three* bases, and the genetic code is thus said to be a *triplet* code.

DNA undergoes two major cellular transformations. First, all DNA molecules in a cell undergo *replication* before cell division, allowing for each new cell to receive a complete set of genetic instructions. Replication involves copying the information from the

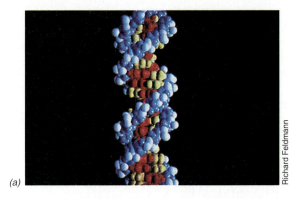

(a)

Richard Feldmann

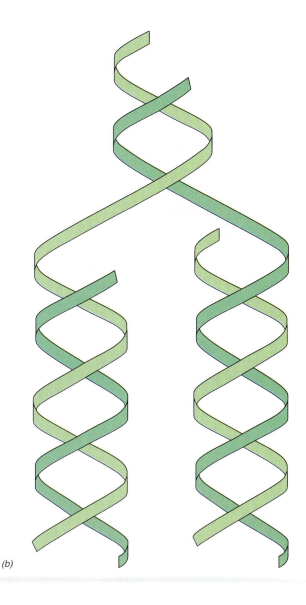

(b)

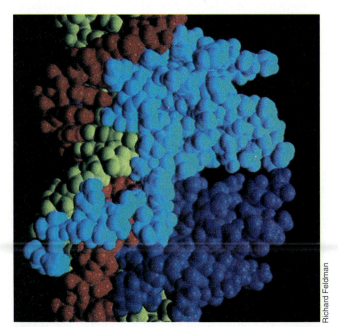

FIGURE 1.5   The structure of a protein as shown in a computer-generated model. Different domains (regions) of the protein are shown in different colors.

Richard Feldman

FIGURE 1.6   DNA: deoxyribonucleic acid. (a) A computer model of DNA. The blue colored atoms represent the sugar/phosphate backbone of the double helix. The bases which hold the two chains together are shown in red and yellow. Only a small part of the DNA double helix is shown. (b) The DNA replication process. Each strand of the DNA double helix is copied into a complementary strand.

DNA molecule into new molecules. As each strand of the DNA double helix opens, a new complement is made by the complementary base pairing just mentioned (Figure 1.6b). Second, the genetic code written in the DNA must be *translated*: the DNA base sequence will dictate the specific amino acid sequence of the protein. This translation process is carried out by a special and highly complex translation machinery in the cell. The translation apparatus is at the very core of cell function.

The translation system

In human language, it is primarily a convenience if a text is translated from one language to another. In the language of the cell, however, translation is essential if the cell is to function. This is so because the genetic information in the DNA molecule is only a repository; it is incapable of being interpreted directly. A protein is only made using the translation system as an intermediate. The translation apparatus is thus the central and most basic attribute of the cell.

Because of the central role of the translation system, it is difficult to change it. In fact, the translation system, although differing in certain details, is essentially the same in all kinds of organisms. It is very likely that in the origin of life the translation system was one of the first things to arise and, once formed, was retained in essentially the same form throughout the long history of life on earth. Interestingly, the translation apparatus cannot use DNA directly, but the information in DNA must first be transcribed into RNA. This may also be a reflection of the origin of life on earth.

Although we will describe the steps in protein synthesis in some detail in Chapter 5, at the moment we note that the production of proteins requires two processes (Figure 1.7):

1. **Transcription**, the formation of **messenger ribonucleic acid (RNA)** molecules, which contain a complementary copy of the genetic information stored in DNA.
2. **Translation**, a process occurring on **ribosomes**, structures composed of another form of RNA and several proteins, on which the translation process actually occurs.

In the translation process, other RNA molecules called **transfer RNA** act as the critical adaptor molecules of the translation process, recognizing both the amino acid and the genetic code in messenger RNA. The messenger RNA combines with ribosomes, forming a complex, and the amino acid-carrying transfer RNA molecules move up to the messenger RNA one at a time and the amino acid being carried is transferred onto the growing protein chain. For each protein, there are three parts of the genetic code, the **start site**, which is the triplet coding unit that indicates where the translation process should begin, the **gene**,

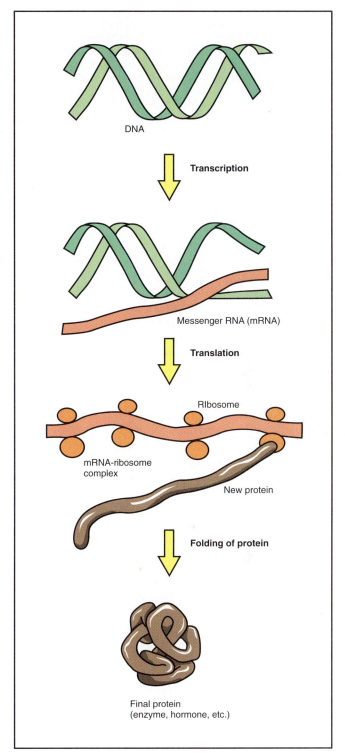

FIGURE 1.7 The two processes required for protein synthesis, transcription and translation.

a sequence of bases in DNA that, when transcribed into RNA and read three bases at a time, specifies the amino acid sequences, and the **stop site**, the triplet coding unit that indicates where the translation process should stop.

> Cells are chemical machines that convert energy from one form to another. The chemical reactions of cells are carried out by specialized proteins called enzymes. Cells are also coding devices, translating information in the genes (DNA base sequences) into the structure of proteins (amino acid sequences). The translation process is mediated by messenger RNA, ribosomes, and transfer RNA.

Growth of the cell

The connection between the two attributes of a cell, its machine function and its coding function, is expressed through the process of *cell growth*. A living cell grows in size, then divides and forms two cells. In the orderly growth process that results in two cells being formed from a single cell, all the constituents of the cell are doubled in amount. Growth in size requires the functioning of the chemical machinery of the cell, such as energy transformation and other metabolic reactions. But each of the two cells must contain *all* the information necessary for the formation of more cells, so that during cell growth and division, there must be a duplication of the genetic information. Thus, it is not just that the DNA content doubles when one cell turns into two, but that the *precise* sequence of bases in the DNA must be copied (see Figure 1.6). The fidelity of this copying function is very high, so that the two progeny of a single cell are identical in DNA base sequence to the parent. We say that the two progeny are *genetically identical* to the parent (top part of Figure 1.8).

However, mistakes in copying DNA do occur occasionally, so that one of the two offspring cells may not always be identical to the parent (bottom part of Figure 1.8). The mistakes that are made in copying DNA are called **mutations**. A mutation is a change in the sequence of bases in the DNA molecule which is passed on to one or more offspring. In some cases, a mutation has no detectable effect, but in most cases, the mutation results in the formation of a malfunctioning protein (or no protein at all), so that the cell with the mutation is defective, possibly either dying or deteriorating. Such cells eventually disappear from the population and are no longer seen. Thus, mutations are generally harmful. However, in rare cases the mutation may result in the formation of an improved protein, for example an enzyme better able to function than that of the parent cell. The cell containing this altered enzyme thus has a *selective advantage* and may eventually replace, as a result of further cell divisions, the parent type. The survival value of a mutation usually depends on the environment in which the organism is living. In some environments, a mutation may be advantageous, and in other cases it may be harmful.

Genes and evolution

The process of natural variation through mutation and the consequent effects on the cell which we have just discussed is called **natural selection**, a phenomenon that is at the basis of the process of **evolution**. Although Charles Darwin first proposed the theory of evolution from observations of higher organisms, Darwinian evolution is demonstrated most dramatically and effectively at the microbial level. Not only has microbial evolution provided some of the most critical support for the theory of evolution, it has led to some very important practical consequences. For instance, disease-causing microorganisms have arisen which are resistant to certain antibiotics, and therefore the diseases caused by these microorganisms cannot be treated with these antibiotics. We see this as a dramatic and very practical result of microbial evolution. Many other important consequences of natural selection will be discussed later in this book. We note here most importantly that the vast diversity of microorganisms, as well as the diversity of higher organisms, is due to the process of evolution through mutation and natural selection.

Genetic potential of the cell

As we have noted, the DNA coding sequence which specifies the amino acid sequence of a specific protein molecule is called a *gene*. How many proteins are there in a cell? A simple bacterial cell, the organism *Escherichia coli*, has about 1900 different *kinds* of proteins

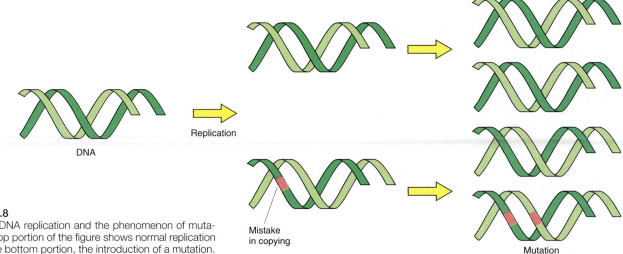

FIGURE 1.8
Fidelity of DNA replication and the phenomenon of mutation. The top portion of the figure shows normal replication events; the bottom portion, the introduction of a mutation.

DNA

Replication

Mistake in copying

Mutation

and contains a total of about 2.4 million protein molecules. Some proteins in the *E. coli* cell are present in very large numbers, even greater than 100,000 molecules per cell, whereas other proteins are present in the cell only in a very small number of molecules. Thus, the cell has some means of controlling the *expression* of its genes so that not all genes are expressed to the same extent or at the same time.

How many genes does a cell have? If we determine the amount of DNA per cell, we find that our *E. coli* cell has about 4.6 million base pairs of DNA (4700 kilobases). If a single gene is about 1.1 kilobases in length, that means that the cell has about 4200 genes, if all the DNA coded for protein. However, we know that some of the DNA does not code for proteins and we thus estimate that the bacterial cell has around 3000 or fewer total *coding* genes, of which only about 1000–2000 appear to be expressed at any particular time.

> Cells are unusual mathematicians since they multiply by dividing. Cell division requires that all of the constituents of the cell, including the genes, are duplicated. A single bacterial cell can have more than 4000 genes. Occasional mistakes are made in the duplication of genes, resulting in mutations. Evolution occurs because mutants that are better adapted to the environment than the parent are favored and are selected. The vast diversity of life on earth has resulted from this process of natural selection.

1.2 Cell Structure

What is the structure of a cell? All cells have a barrier separating inside from outside which is called the **cytoplasmic membrane** (Figure 1.9). It is through the cell membrane that all food materials and other substances of vital importance to the cell pass in, and it is through this membrane that waste materials and other cell products pass out. If the membrane is damaged, the interior contents of the cell leak out, and the cell usually dies. As we shall see, some drugs and other chemical agents damage the cytoplasmic membrane and, in this way, bring about the destruction of cells.

The cytoplasmic membrane is a very thin, highly flexible layer and is structurally weak. By itself, it usually cannot hold the cell together, and an additional stronger layer, called the **cell wall**, is usually necessary (Figure 1.9). The wall is a relatively rigid layer which is present outside the membrane and protects the membrane and strengthens the cell. Plant cells and most microorganisms have such rigid cell walls. Animal cells, however, do not have walls; these cells have developed other means of support and protection.

Within a cell, and bounded by the cytoplasmic membrane, is a complicated mixture of substances and structures called the **cytoplasm.** These materials and structures, bathed in water, carry out the functions of the cell. The major components of the cytoplasm, other than water, include macromolecules

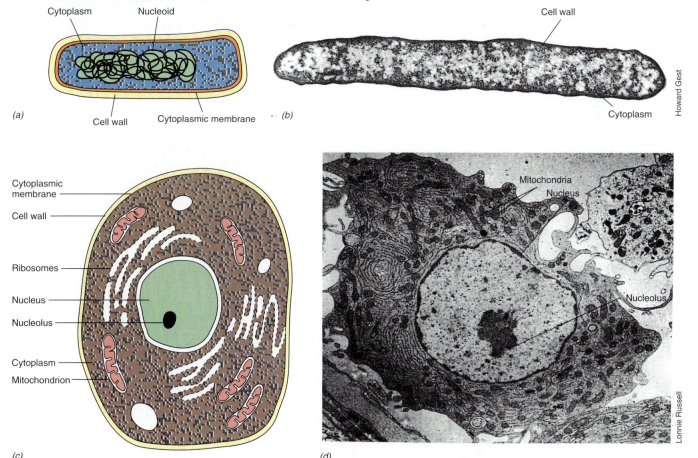

FIGURE 1.9 Internal structure of microbial cells. (a) Diagram of a prokaryote. (b) Electron micrograph of a prokaryote. The cell is about 1 μm in diameter. (c) Diagram of a eukaryote. (d) Electron micrograph of a eukaryote (animal cell). The cell is about 25 μm in diameter.

(proteins, nucleic acids, polysaccharides, and lipids), ribosomes (a key part of the cell translation machinery), small organic molecules (mainly precursors of macromolecules), and various inorganic ions.

Prokaryotic and eukaryotic cells

Upon careful study of the structures of cells, two basic types have been recognized which differ greatly in structure. These are called **prokaryote** and **eukaryote** (Figure 1.9a and b). These two types of cells are structurally very different. The major structural difference between prokaryotes and eukaryotes, other than size, is their different nuclear structures. The eukaryotic nucleus is bounded by a nuclear membrane, contains several DNA molecules, and undergoes division by the well-known process of **mitosis**. By contrast, the prokaryotic nuclear region, called the *nucleoid*, is not surrounded by a membrane and consists of a single DNA molecule whose division is nonmitotic. **Bacteria** and **Archaea** (see Section 1.3) are the only prokaryotes. There are several groups of eukaryotic microorganisms, including **algae**, **fungi**, and **protozoa**. In addition, all higher life forms (plants and animals) are constructed of eukaryotic cells.

Microorganisms in general are very small. A prokaryote of typical size is about 1 to 5 μm long, and thus completely invisible to the naked eye. To illustrate how small a bacterium is, consider that 500 bacteria 1 μm in length could be placed end to end across the period at the end of this sentence.

Viruses are not cells. Viruses lack many of the attributes of cells, of which the most important is that they are not dynamic open systems. A single virus particle is a static structure, quite stable and unable to change or replace its parts. Only when it is associated with a cell does a virus become able to replicate and acquire some of the attributes of a living system. Thus, unlike cells, viruses have no metabolism of their own. Although viruses have genetic information (either DNA or RNA), they lack the translation apparatus and use the cell machinery for protein synthesis. Viruses are known to infect various organisms including microorganisms. Some size comparisons of viruses and cells are shown in Figure 1.10. Although many viruses cause disease in the organisms they infect, virus infection does not always lead to disease; many viruses cause inapparent infections.

All cells have a critical barrier, the cytoplasmic membrane, that separates the cytoplasm from its surroundings. The most important structure inside the cell is the nucleus or nucleoid, the site of the genetic information, DNA. Cells can be divided into two large groups depending on the organization of their DNA and several other properties. In eukaryotes, the nucleus contains several DNA molecules and is bounded by a nuclear membrane; nuclear division occurs by the process of mitosis. There is no nuclear membrane in prokaryotes, the DNA being present inside the cell in a single molecule whose replication is nonmitotic.

Typical animal cell

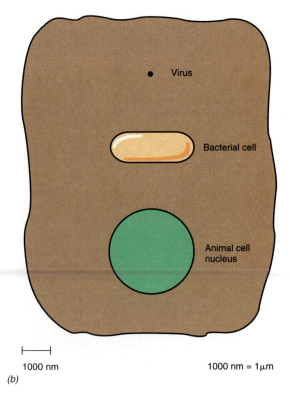

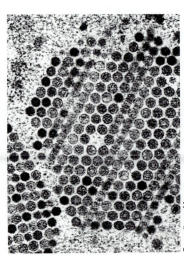

FIGURE 1.10
Virus size and structure. (a) Particles of adenovirus, a virus which causes respiratory infections in humans. A single virus particle is about 100 nm in diameter. (b) The size of a virus in comparison to a bacterial and animal cell.

(a)

(b)

Councilman Morgan

1000 nm

1000 nm = 1μm

1.3   Evolutionary Relationships Among Living Organisms

Although cells can be differentiated structurally as being either prokaryotic or eukaryotic, cell structure does not imply *evolutionary* relationship. However, phylogenetic (evolutionary) relationships among microorganisms can now be determined, and Chapter 18 describes the methods used to discern such relationships. These methods rely on nucleic acid sequence comparisons, in particular the sequences of *ribosomal RNA*—structural RNA of the ribosome, the key cell structure involved in translation (see Section 1.1). Indeed, one of the great recent discoveries in biology is that changes in the sequence of bases in ribosomal RNA (ultimately dictated by mutations in the DNA which code for ribosomal RNA) can be used as a measure of evolutionary relationships among cells.

From studies of ribosomal RNA sequences, three evolutionarily distinct cellular lineages can be defined, two of which are prokaryotic in structure and one of which is eukaryotic. The groups have been given the names **Bacteria, Archaea,** and **Eukarya** (Figure 1.11).* However, despite the fact that both are structurally prokaryotic at the molecular level, Bacteria and Archaea are as evolutionarily distinct from one another as either group is from the Eukarya. All three groups are thought to have diverged from a common ancestral organism, the "universal ancestor," early in the history of life on earth (Figure 1.12). Because the cells of higher animals and plants are all eukaryotic, it follows that eukaryotic microorganisms were the ancestors of multicellular organisms, whereas Bacteria and Archaea represent evolutionary branches that never evolved past the microbial stage.

Classification

In addition to understanding and appreciating the phylogenetic origins of cellular organisms, it is impor-

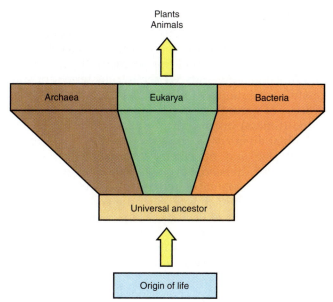

FIGURE 1.12   Cellular evolution as deduced from ribosomal RNA sequencing.

tant for a variety of reasons to be able to identify and classify microorganisms. For example, rapid identification of a particular human disease-causing microorganism (pathogen) is usually essential to know how to treat an infected person. We discuss many methods for identifying microorganisms in Chapter 13, Clinical Microbiology. Several criteria have been used to characterize microorganisms, and at present, both phylogenetic standing and other cellular characteristics are used for classification purposes. After close study of the structure and function of a microorganism, including its genetics, metabolism, behavior, and other key properties, it is usually possible to recognize a group of characteristics unique to a given organism. Once an organism has been defined by such a set of characteristics, it can be given a name.

Microbiologists use the binomial system of nomenclature first developed by Linnaeus for plants and animals. The *genus* name is applied to a number of related organisms; each different type of organism within the genus has a *species* name. Genus and species names are always used together to describe a specific type of organism, whether it be a single cell or a group of such cells. (In writing, the genus and species names are either underlined or printed in italics.)

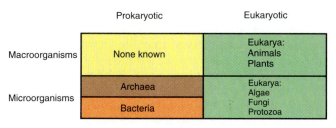

FIGURE 1.11   The major types of cellular life forms.

*To avoid the perception that all prokaryotes are closely related (which indeed they are not), the terms *Bacteria* and *Archaea* have replaced *eubacteria* and *archaebacteria,* respectively, to describe two of the three taxonomic domains (the highest of biological taxons; see Chapter 18 and Figure 1.12) of cellular organisms. The third group of organisms is the *Eukarya* (all of which are eukaryotic cells). However, the word *bacteria* is so firmly entrenched in the science of microbiology that it is unlikely to ever lose its synonymity with the word *prokaryote.* This fact thus creates a semantic dilemma. Are all bacteria (prokaryotes) still Bacteria (a phylogenetic unit)? No. In this

book, the word *Bacteria* (with a capital "B") will be reserved for describing organisms in a *phylogenetic sense,* that is, as members of the domain Bacteria. In contrast, the word *bacteria* (lowercase "b") will appear many times in this book and is used to refer to prokaryotes in general, without reference to phylogeny. Because detailed discussion of the Archaea is reserved for Chapter 20, most references to bacteria in chapters preceding Chapter 20 will be to Bacteria. However, no phylogenetic implication should be attached to the words *bacteria* or *bacterium* (lowercase "b") no matter where they are used in this book.

Microbial diversity

An understanding of microbial diversity requires an appreciation of the evolutionary roots of cells. Because evolution has shaped all life on earth, the structural and functional diversity we see in cells represent successful evolutionary events which, through the process of natural selection, have conferred survival value (fitness) on the microorganisms extant today. Microbial diversity can be seen in terms of variations in cell size and morphology, metabolic strategies, motility, cell division, developmental biology, adaptation to environmental extremes, and many other structural and functional aspects of the cell. We describe the major groups of Bacteria, Archaea, and Eukarya in Chapters 19, 20, and 21, respectively, and consider there many of the aspects that make microbial life so amazingly diverse. Here we give only a quick overview of things to come.

Twelve major evolutionary branches occur within the **Bacteria,** including all disease-causing (pathogenic) prokaryotes and most of the bacteria commonly found in soil, water, animal digestive tracts, and many other environments. Some of these organisms can use light as an energy source in a process called *photosynthesis*, others rely on organic chemicals as energy sources, and some can even use inorganic chemicals as fuel to drive cellular processes. Most of the prokaryotes encountered in everyday life are phylogenetically Bacteria, and some have evolved special structures such as spores for aiding survival. Oxic environments (those containing O_2) as well as various anoxic habitats are inhabited by various species of Bacteria.

With those prokaryotes called **Archaea**, in contrast, we see a much different picture. Most Archaea are anaerobes, cells incapable of living in air. Many also thrive under unusual growth conditions, inhabiting what humans would consider extreme environments: hot springs (to temperatures *above* the boiling point of water), extremely salty bodies of water, and highly acidic or alkaline soils and water. Indeed, species of Archaea currently define the limits of biological tolerance to physiochemical extremes. Certain Archaea also show unusual biochemical features, chief among them being the methanogens, which are prokaryotes that produce methane (natural gas) as an integral part of their energy metabolism.

Among those **Eukarya** that are microorganisms are the algae, fungi, and protozoa. *Algae* contain chlorophyll, a green pigment that serves as a light-gathering molecule, making it possible for them to carry out photosynthesis. Algae are common in aquatic habitats and can also be found in soil. *Fungi*—the molds, yeasts, and mushrooms—lack chlorophyll and obtain their energy from organic compounds in soil and water. Fungi are thought to play a major role in the breakdown of dead organic matter in these and other environments. *Protozoa* are colorless and motile Eukarya that obtain food by ingesting other organisms or organic particles. Protozoa lack the cell walls of algae and fungi and in this respect resemble animal cells. Many protozoa are free-living microorganisms, but several cause disease in humans and other animals.

> Living organisms can be divided into prokaryotes and eukaryotes based on cell structure, but from an evolutionary viewpoint, living organisms form three major groups: Bacteria, Archaea, and Eukarya. Although Bacteria and Archaea are prokaryotes, from an evolutionary point of view the Bacteria and Archaea are no more closely related to each other than they are to the Eukarya. Structural and functional diversity among microorganisms is substantial.

1.4   Populations, Communities, and Ecosystems

Up to now, we have been discussing cells as if they lived in isolation in their environments. Nothing could be further from the truth. Cells live in nature in association with other cells, in assemblages which we call **populations**. Such populations are composed of groups of related cells, generally derived by successive cell divisions from a single parent cell. The location in the environment where a population lives is called the **habitat** of that population. **Ecology** is the discipline which deals with the study of living organisms in their natural environments.

In nature, populations of cells rarely live alone. Rather, populations of cells live in association with other populations of cells, in assemblages which are called **communities** (Figure 1.13). Frequently, populations in communities interact, either in beneficial ways or in harmful ways. If two populations interact in a beneficial way, these populations will then maintain themselves better when together than when separate. In such cases we speak of the cooperative nature of the populations.

In other cases, two populations living in the same habitat may interact in a way which is harmful to one of the populations. If such harmful interaction occurs, the population which is harmed will be reduced in number, or even replaced. If the effect is severe enough, the population may be completely eliminated.

The living organisms in a habitat also interact with the physical and chemical environment of that habitat. Habitats differ markedly in their physical and chemical characteristics, and a habitat which is favorable for the growth of one organism may be harmful for another organism. Thus, the community which we see in any given habitat will be determined to a great extent by the physical and chemical characteristics of that environment.

In addition, the organisms of the habitat modify the physical and chemical properties of the environment. Organisms carrying out metabolic processes remove chemical consituents from the environment and use these constituents as energy or nutrient sources.

(a)

Brock

(b)

D. E. Caldwell

FIGURE 1.13 Examples of microbial communities. (a) A dense community of algae and cyanobacteria in surface waters of a nutrient-rich lake. (b) Photomicrograph of a bacterial community that developed in the depths of a small lake, showing cells of various sizes.

At the same time, organisms excrete waste products of their metabolism into the environment. Therefore, as time progresses the environment is gradually changed through life processes. We speak collectively of the living organisms together with the physical and chemical consituents of the environment as an **ecosystem**. We can also think of the earth as a whole as an ecosystem; this global ecosystem is sometimes called the **biosphere** (Figure 1.14). The biosphere interacts with two major regions of the earth, the **atmosphere**, and the **lithosphere**.

Since single microbial cells are too small to be seen with the unaided eye, our knowledge of microorganisms in nature begins with studies using the microscope. Examination of natural materials such as soil, mud, water, spoiling food, human bodily excretions, and other living and dead material under the microscope reveals that such materials teem with vast numbers of very tiny cells. Although such tiny cells seem harmless, we can state that microorganisms are *small but powerful*. Even though a single cell can never, by itself, cause a harmful effect that we can perceive, this single cell is often capable of multiplying rapidly and producing a massive number of progeny which may be capable of harm. One of the most important and impressive attributes of microorganisms is *rapid growth* (multiplication), so that single cells do not remain single very long, but can quickly develop into *populations* that can significantly change the chemical properties of a habitat. Thus, although microorgan-

NASA

FIGURE 1.14 The global biosphere. A view of Earth taken from a satellite orbiting the moon.

isms may seem to occupy inconsequential niches in nature, they are extremely important components of ecosystems. Microorganisms also have very serious effects on higher plants and animals. In later chapters, after we have learned some of the details of microbial structure and function, we will discuss again the ways in which microorganisms affect animals, plants, and the whole global ecosystem. For now, we consider an overview of the activities of microorganisms, especially as they impact on humans.

> In general, microorganisms exist in nature in populations that interact with other populations in microbial communities. The activities of microbial communities can greatly affect the chemical and physical properties of their habitats.

1.5   The Impact of Microorganisms on Human Affairs

One goal of the microbiologist is to understand how microorganisms work, and through this understanding to devise ways that benefits may be increased and damages curtailed. Microbiologists have been eminently successful in achieving these goals, and microbiology has played a major role in the advancement of human health and welfare.

Microorganisms as disease agents

One measure of the microbiologist's success is shown by the statistics in Figure 1.15, which compare the present causes of death in the United States to those at the beginning of this century. At the beginning of the century, the major causes of death were infectious diseases; currently, such diseases are of only minor importance. Control of infectious disease has come as a result of our comprehensive understanding of disease processes. As we will see later in this chapter, microbiology had its beginnings in these studies of disease.

However, although we now live in a world where pathogenic microorganisms are mostly under control, for the individual dying slowly of acquired immune deficiency syndrome (AIDS), or the cancer patient whose immune system has been devastated as a result of treatment with an anti-cancer drug, microorganisms can still be the major threat to survival. Although such tragic situations barely appear in our health statistics, they are of no less concern. Further, microbial diseases still constitute the major causes of death in many of the developing countries of the world. Although eradication of smallpox from the world has been a stunning triumph for medical science, millions still die yearly from such pervasive illnesses as malaria, tuberculosis, cholera, African sleeping sickness, and severe diarrheal diseases.

Thus, microorganisms are still serious threats to human existence. But on the other hand we must emphasize that most microorganisms are *not* harmful to humans. In fact, most microorganisms cause no harm at all and instead are actually *beneficial*, carrying out processes that are of immense value to human society. Even in the health-care industry, microorganisms play beneficial roles. For instance, the pharmaceutical industry is a multi-billion dollar industry built in part on the large-scale production of antibiotics by microorganisms. We will have much to say about antibiotics in this book, but note here that current worldwide antibiotic production is over *100,000 tons* per year! A number of other major pharmaceutical products are also derived, at least in part, from the activities of microorganisms.

Microorganisms and agriculture

Our whole system of *agriculture* depends in many important ways on microbial activities. A number of major crops are members of a plant group called the **legumes**, which live in close association with special bacteria which form structures called *nodules* on their roots. In these root nodules, atmospheric nitrogen (N_2) is converted into fixed nitrogen compounds that the plants can use for growth. In this way, the activities of the root nodule bacteria reduce the need for costly plant fertilizer. Millions of acres and billions of dollars are annually committed to the cultivation of these leguminous crop plants. Also of major agricultural importance are the microorganisms that are essential for the digestive process in ruminant animals such as cattle and sheep. These important farm animals have a special digestive organ called the **rumen** in which microorganisms carry out the digestive process. Many billions of dollars of meat and milk production are linked to these rumen microorganisms. Without these microorganisms, cattle and sheep production would be virtually impossible. Microorganisms also play key roles in the cycling of important nutrients in plant nutrition, particularly carbon, nitrogen, and sulfur. Microbial activities in soil and water convert these elements into forms that are readily accessible to plants. In addition to benefits to agriculture, microorganisms also cause harmful effects. Animal and plant diseases due to microorganisms have major economic impact.

Microorganisms and the food industry

Once food crops are produced, they must be delivered in wholesome form to consumers. Microorganisms play important roles in the *food industry*. We note first that because of food spoilage vast amounts of money are wasted every year. The canning, frozen-food, and dried-food industries exist to prepare foods in such ways that they will not spoil. Food preservation is a $30 billion industry today.

However, not all microorganisms have harmful effects on foods. Dairy products manufactured, at least in part, via microbial activity include cheese, yogurt, and buttermilk, all products of major economic value. Sauerkraut, pickles, and some sausages also owe their existence to microbial activity. Baked goods are made using yeast. Even more pervasive in our society are the alcoholic beverages, also based on the activities of yeast, which build a $70 billion per year industry.

All of these applications of microorganisms in food and agriculture are of ancient origin, but microbiology has not rested on the past. Consider, for in-

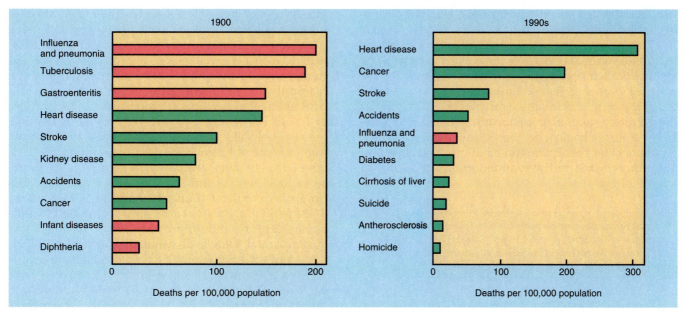

FIGURE 1.15   Death rates for the 10 leading causes of death in the United States: 1900 and 1990s. Infectious diseases were the leading causes of death in 1900, whereas today they are much less important. Microbial diseases are shown in red, nonmicrobial diseases in green. Data from the National Center for Health Statistics.

stance, the microorganism's contribution to a carbonated soft drink. The major sugar in many soft drinks is *fructose*, produced from corn starch via microbial activity (see Section 10.10). Over 17 billion pounds of corn sweeteners are produced each year! In diet soft drinks, the artificial sweetener *aspartame* is a combination of two amino acids, both produced microbiologically. Finally, the *citric acid* added to many soft drinks to give them tang and bite is produced in a large-scale industrial process using a fungus.

Microorganisms and energy

Our complex industrial society is energy-driven, and here also microorganisms play major roles. Much natural gas (methane) is a product of bacterial action, arising from the activities of methanogenic bacteria. It is harvested worldwide in vast amounts as a primary fuel. A few other mineral and energy products are also the result of microbial activity, but of even greater interest is the relationship of microorganisms to the petroleum industry. Crude oil is subject to vigorous microbial attack, and drilling, recovery, and storage of crude oil all have to be done under conditions that minimize microbial damage. Petroleum today is a $100 billion industry.

Because human activity will result in the complete consumption of available fossil fuels during the next century, we must seek new ways to supply the energy needs of society. In the future, microorganisms may provide major alternative energy sources. Photosynthetic microorganisms can harvest light energy for the production of **biomass**, energy stored in living organisms. Microbial biomass and existing waste materials such as domestic refuse, surplus grain, and animal wastes, can be converted to "biofuels," such as meth-

ane and ethanol, by other microorganisms. Other microbial products may be used as "chemical feedstocks," the chemicals from which synthetic materials are manufactured, and which are now also commonly derived from petroleum. Human activity is also decreasing the supplies of other substances, such as metals, and microorganisms are being used increasingly to recover metals from low grade ores.

Microorganisms and the future

The above recital of the benefits of microorganisms is only the beginning. One of the most exciting new areas of microbiology is **biotechnology**. In the broad sense, biotechnology entails the use of microorganisms in large-scale industrial processes, but by "biotechnology" today we usually mean the application of genetic procedures, generally to create novel microorganisms capable of synthesizing specific products of high commercial value. Biotechnology is highly dependent on **genetic engineering**, the discipline that concerns the artificial manipulation of genes and their products.

Genes from human sources, for instance, can be broken into pieces, modified, and added to or subtracted from, using microorganisms and their enzymes as precise and sophisticated molecular tools. It is even possible to make completely artificial genes using genetic engineering techniques. Once the desired gene has been selected or created, it can be inserted into a microorganism where it can be made to reproduce and make the desired gene product. For instance, human insulin, a hormone found in abnormally low amounts in people with the disease diabetes, has now been produced microbiologically with the human insulin gene engineered into a microorganism. We discuss genetic engineering and biotechnology in detail in Chapter 8.

The overwhelming influence of microorganisms in human society is clear. We have many reasons to be aware of microorganisms and their activities. As the eminent French scientist Louis Pasteur, one of the founders of microbiology, expressed it: "The role of the infinitely small in nature is infinitely large." Therefore, before we begin a detailed study of microbiology, let us consider briefly the contributions which Pasteur and others made to the foundation of the science of microbiology.

> **Microorganisms can be both beneficial and harmful to humans. Although we tend to emphasize the harmful microorganisms (infectious disease agents), many more microorganisms are beneficial than harmful.**

1.6   A Brief History of Microbiology

Although the existence of creatures too small to be seen with the eye had long been suspected, their discovery was linked to the invention of the microscope. Robert Hooke described the fruiting structures of molds in 1664 (Figure 1.16), but the first person to see

microorganisms in any detail was the Dutch amateur microscope builder Antoni van Leeuwenhoek, who used simple microscopes of his own construction (Figure 1.17a). Leeuwenhoek's microscopes were extremely crude by today's standards, but by careful manipulation and focusing he was able to see organisms as small as bacteria. He reported his observations in a series of letters to the Royal Society of London, which published them in English translation. Drawings of some of Leeuwenhoek's "wee animalcules," as he referred to them, are shown in Figure 1.17b. His observations were confirmed by other workers, but progress in understanding the nature and importance of these tiny organisms came slowly. Only in the nineteenth century did improved microscopes become available and widely distributed. During its history, the science of microbiology has traditionally taken the greatest steps forward when better microscopes have been developed, for these have enabled scientists to penetrate ever deeper into the mysteries of the cell.

Microbiology as a science did not develop until the latter part of the nineteenth century. This long delay occurred because, in addition to microscopy, certain basic techniques for the study of microorganisms needed to be devised. In the nineteenth century, investigation of two perplexing questions led to the devel-

(b)

(a)   (c)

FIGURE 1.16   (a) The microscope used by Robert Hooke. (b,c) Two drawings by Robert Hooke that represent one of the first microscopic descriptions of microorganisms. (b) A blue mold growing on the surface of leather; the round structures contain spores of the mold. (c) A mold growing on the surface of an aging and deteriorating rose leaf.

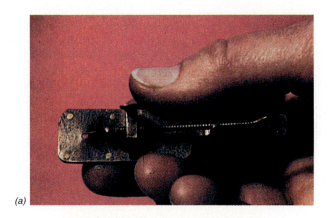

(a)

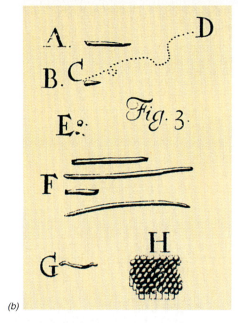

(b)

FIGURE 1.17 (a) Photograph of a replica of Leeuwenhoek's microscope. (b) Leeuwenhoek's drawings of bacteria, published in 1684. Even from these crude drawings we can recognize several morphological types of common bacteria. Those lettered A, C, F, and G are rod-shaped; E, spherical or coccus-shaped; H, cocci packets.

opment of these techniques and laid the foundation of microbiological science: (1) Does spontaneous generation occur? (2) What is the nature of contagious disease? By the end of the nineteenth century both questions were answered, and the science of microbiology was firmly established as a distinct and growing field.

Spontaneous generation

The basic idea of spontaneous generation can easily be understood. If food is allowed to stand for some time, it putrefies. When the putrefied material is examined microscopically, it is found to be teeming with bacteria. Where do these bacteria come from, since they are not seen in fresh food? Some people said they developed from seeds or germs that had entered the food from the air, whereas others said that they arose spontaneously, from nonliving materials.

Spontaneous generation would mean that life could arise from something nonliving, and many people could not imagine something so complex as a liv-

ing cell arising spontaneously from nonliving materials. The most powerful opponent of spontaneous generation was the French chemist Louis Pasteur, whose work on this problem was the most exacting and convincing. Pasteur first showed that structures were present in air that resembled closely the microorganisms seen in putrefying materials. He did this by passing air through guncotton filters, the fibers of which stopped solid particles. After the guncotton was dissolved in a mixture of alcohol and ether, the particles that it had trapped fell to the bottom of the liquid and were examined on a microscope slide. Pasteur found that in ordinary air there exists constantly a variety of solid structures ranging in size from 0.01 mm to more than 1.0 mm. Many of these structures resemble the spores of common molds, the cysts of protozoa, and various other microbial cells. As many as 20 to 30 of them were found in 15 liters of ordinary air, and they could not be distinguished from the organisms found in much larger numbers in putrefying materials. Pasteur concluded that the organisms found in putrefying materials originated from the organized bodies present in the air. He postulated that these bodies are constantly being deposited on all objects. If this conclusion was correct, it would mean that, if food were treated to destroy all the living organisms contaminating it, then it should not putrefy.

Pasteur used heat to eliminate contaminants, since it had already been established that heat effectively killed living organisms. In fact, many workers had shown that if a nutrient solution was sealed in a glass flask and heated to boiling, it never putrefied. The proponents of spontaneous generation criticized such experiments by declaring that fresh air was necessary for spontaneous generation and that the air itself inside the sealed flask was affected in some way by heating so that it would no longer support spontaneous generation. Pasteur skirted this objection simply and brilliantly by constructing a swan-necked flask, now called the "Pasteur flask" (Figure 1.18). In such a flask putrefying materials could be heated to boiling; after the flask is cooled, air could reenter but the bends in the neck prevented particulate matter, bacteria, or other microorganisms from getting in the main body of the flask. Material sterilized in such a flask did not putrefy, and no microorganisms ever appeared as long as the neck of the flask remained intact. If the neck was broken, however, putrefaction occurred and the liquid soon teemed with microorganisms. This simple experiment served to effectively settle the controversy surrounding the theory of spontaneous generation.

Killing all the bacteria or other microorganisms in or on objects is a process we now call **sterilization**, and the procedures that Pasteur and others used were eventually refined and carried over into microbiological research. Disproving the theory of spontaneous generation thus led to the development of effective sterilization procedures, without which microbiology as a science could not have developed.

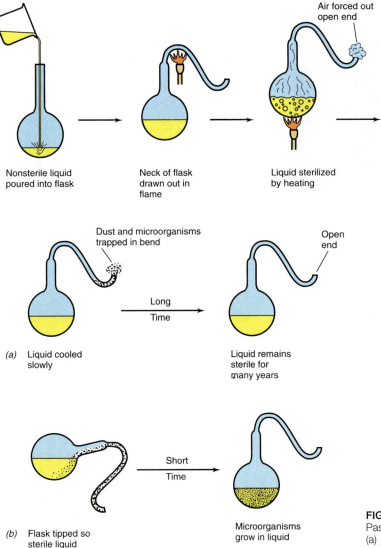

Air forced out
open end

Nonsterile liquid
poured into flask

Neck of flask
drawn out in
flame

Liquid sterilized
by heating

Dust and microorganisms
trapped in bend

Open
end

Long
Time

(a)   Liquid cooled
slowly

Liquid remains
sterile for
many years

Short
Time

(b)   Flask tipped so
sterile liquid
contacts micro-
organism-laden dust

Microorganisms
grow in liquid

FIGURE 1.18
Pasteur's experiment with the swan-necked flask.
(a) If the flask remains upright, no microbial growth
occurs. (b) If microorganisms trapped in the neck
reach the sterile liquid, they grow rapidly.

It was later shown that flasks and other vessels could be protected from contamination by cotton stoppers, which permit the exchange of air but not particulate matter. The principles of **aseptic technique**, the handling of sterile objects so as to maintain their sterility, are the first procedures learned by the novice microbiologist. Food science also owes a debt to Pasteur, as his principles are applied in the canning and preservation of many foods.

Although Pasteur was successful in sterilizing materials with simple boiling, some workers found that boiling was insufficient. We now know that the failure resulted from the presence in these materials of bacteria which formed unusually heat-resistant structures called *endospores*. Initial work on the endospore was carried out by two men: John Tyndall in England, and Ferdinand Cohn in Germany. Both scientists observed that some preparations, such as the fruit juice solutions used by Pasteur, were relatively easy to sterilize, requiring only 5 minutes of boiling, whereas others were not sterilized by much longer periods of boiling,

sometimes even hours. Notably difficult to sterilize were hay infusions. In addition, once hay had been brought into the laboratory, even the sugar solutions could no longer be reliably sterilized by hours of boiling. Cohn performed detailed microscopic observations and discovered endospores inside cells of old cultures of species of *Bacillus*. Cohn and another German scientist, Robert Koch, applied this observation to the study of disease, as discussed below. Bacterial endospores are the most heat-resistant living structures known and most sterilization methods are designed to kill these spores.

The germ theory of disease

Proof that microorganisms could cause disease provided the greatest impetus for the development of the science of microbiology. Indeed, even in the sixteenth century it was thought that something could be transmitted from a diseased person to a well person to induce in the latter the disease of the former. Many diseases seemed to spread through populations and were called *contagious*; the unknown thing that did the spreading was called the *contagion*. After the discovery of microorganisms, it was more or less widely held that these organisms were responsible for contagious diseases, but proof was lacking. Discoveries by Ignaz Semmelweis and Joseph Lister provided some evidence for the importance of microorganisms in causing human diseases, but it was not until the work of Robert Koch, a physician, that the *germ theory of disease* was placed on a firm footing.

In his early work, published in 1876, Koch studied *anthrax*, a disease of cattle, which sometimes also occurs in humans. Anthrax is caused by a spore-forming bacterium now called *Bacillus anthracis*, and the blood of an animal infected with anthrax teems with cells of this large bacterium. Koch established by careful microscopy that the bacteria were always present in the blood of an animal that was succumbing to the disease. However, the mere association of the bacterium with the disease did not prove that it actually *caused* the disease; it might instead be a *result* of the disease. Therefore, Koch demonstrated that it was possible to take a small amount of blood from a diseased animal and inject it into another animal, which subsequently became diseased and died. He could then take blood from this second animal, inject it into another, and again obtain the characteristic disease symptoms. By repeating this process as often as 20 times, successively transferring small amounts of blood containing bacteria from one animal to another, he proved that the bacteria did indeed cause anthrax: the twentieth animal died just as rapidly as the first; and in each case Koch could demonstrate by microscopy that the blood of the dying animal contained large numbers of the spore-forming bacterium.

However, Koch carried this experiment further. He found that the bacteria could also be cultivated in nutrient fluids outside the animal body, and even after many transfers in culture the bacteria could still cause the disease when reinoculated into an animal. Bacteria from a diseased animal and bacteria in culture both induced the same disease symptoms upon injection. On the basis of these and other experiments Koch formulated the following criteria, now called **Koch's postulates**, for proving that a specific type of microorganism causes a specific disease:

1. The organism should be constantly present in animals suffering from the disease and should not be present in healthy individuals.

2. The organism must be cultivated in pure culture away from the animal body.

3. Such a culture, when inoculated into susceptible animals, should initiate the characteristic disease symptoms.

4. The organism should be reisolated from these experimental animals and cultured again in the laboratory, after which it should still be the same as the original organism.

Koch's postulates not only supplied a means of demonstrating that specific organisms cause specific diseases but also provided a tremendous spur for the development of the science of microbiology by stressing the importance of laboratory culture.

Pure cultures

As we have noted, in order to successfully study the activities of a microorganism, such as a microorganism which causes a disease, one must be sure that it alone is present in culture. That is, the culture must be *pure*. With objects as small as microorganisms, ascertaining purity is not easy, for even a very tiny sample of blood or animal fluid may contain several kinds of organisms that may all grow together in culture. Koch realized the importance of pure cultures. He developed several ingenious methods of obtaining them, of which the most useful is that involving the isolation of single *colonies* (Figure 1.19). Koch observed that when a solid nutrient surface, such as a potato slice, was exposed to air and then incubated, bacterial colonies developed, each having a characteristic shape and color. He inferred that each colony had arisen from a single bacterial cell that fell on the surface, found suitable nutrients, and began to multiply. Because the solid surface prevented the bacteria from moving around, all of the offspring of the initial cell remained together, and when a large enough number of organisms was present, the mass of cells became visible to the naked eye. He assumed that colonies with different shapes and colors were derived from different kinds of microorganisms. When the cells of a single colony were spread out on a fresh surface, many colonies developed, each with the same shape and color as the original.

Koch realized that this discovery provided a simple way of obtaining pure cultures: he found that if mixed cultures were spread on solid nutrient surfaces, the individual cells were so far apart that the colonies they produced did not mingle. Many organisms could not grow on potato slices, so he devised semisolid media, in which gelatin was added to a nutrient fluid such as blood serum in order to solidify it. When the gelatin-containing fluid was warmed, it liquefied and could be poured out on glass plates; upon cooling, the solidified medium could be inoculated. Later *agar* (a

FIGURE 1.19 Bacterial colonies are visible on the surface of a culture medium solidified with agar. Note the distinct and characteristic colors of different colonies.

material derived from seaweed) was found to be a better solidifying agent than gelatin, and this substance is widely used today.

In the 20 years following the formulation of Koch's postulates, the causal agents of a wide variety of contagious diseases were isolated. These discoveries led to the development of successful treatments for the prevention and cure of many infectious diseases and contributed to the development of modern medical practice. The impact of Koch's work has been felt throughout the world.

It is important to realize that Koch's postulates have relevance beyond identifying organisms which cause specific diseases. The essential general conclusion to be gathered is that *specific organisms have specific effects*. This principle that different organisms have unique biological activities was important in establishing microbiology as an independent biological science. By the beginning of the twentieth century, the disciplines of bacteriology and microbiology were on a firm footing.

Developments of microbiology in the twentieth century

In the twentieth century, the field of microbiology has developed rapidly in two separate directions—applied and basic. On the applied side, the practical advances made by Koch and the German school led to extensive developments in *medical microbiology* and *immunology* in the early part of the century, with the discovery of many new bacterial pathogens and the working out of the principles by which these pathogens infect the body and are in turn resisted by the body's defenses. Other early practical advances were in the field of *agricultural microbiology*, which led to an understanding of microbial processes in the soil that are beneficial or harmful to plant growth. Later in the twentieth century, such studies in soil microbiology have led to the discovery of important uses of microorganisms, such as in the formation of *antibiotics* and *industrial chemicals*. This has led, especially after World War II, to the field of *industrial microbiology*.

Finally, soil microbiology has also provided an important foundation for studies of microbial processes in water bodies such as lakes, rivers, and oceans, studies classified under the field of *aquatic microbiology*. One branch of aquatic microbiology deals with the development of processes for providing safe water for human society. The handling of human wastes, especially domestic sewage, has required the development of large-scale engineering processes for sewage treatment, most of which are microbial. Thus, the field of *sanitary microbiology* has developed, which is of importance not only to biologists but also to engineers whose responsibility is the design of these large-scale

processes. To provide safe drinking water, procedures for eliminating harmful bacteria from water supplies have been developed, which are classified under the field of *drinking water microbiology*. By the late twentieth century, all of these subdisciplines of applied microbiology mentioned here have coalesced into the field called *microbial ecology*, which will be discussed in Chapter 17.

In addition to the applied aspects of microbiology which have provided such important advances for human society, extensive developments have occurred in our understanding of the basic principles of microbial function. Although these developments in basic microbiology have been of interest in themselves, they have provided important support for the development of applied microbiology. Applied microbiology uncovers problems that must be answered by basic research, whereas basic microbiology provides information that can be used to advance applied microbiology. Thus, basic and applied microbiology develop together since new understanding of basic microbiology makes possible the perfection of applied microbiology.

In the early part of the twentieth century, the most important developments in basic microbiology involved the discovery of new kinds of bacteria and their proper classification (*bacterial taxonomy*). Bacterial classification required a study of the nutrients that bacteria consume and the products that they make, studies comprising a part of the field of *bacterial physiology*. One part of physiology that became of major importance as the twentieth century progressed involved the study of the physical and chemical structure of bacteria, studies included in the field of *bacterial cytology* (the word *cytology* refers to the study of the cell). Another major development from physiology was the study of bacterial enzymes and the chemical reactions which they carry out, a field called *bacterial biochemistry*.

Another very important area of basic research has involved the study of heredity and the variation that bacteria undergo during their growth and development, studies that fall under the discipline of *bacterial genetics*. Although some ideas of bacterial variation were known early in the twentieth century, it was not until the discovery of sex in bacteria in about 1950 that bacterial genetics really became a major field of study. Bacterial genetics, biochemistry, and physiology developed mainly during the 1950s, leading by the early 1960s to an advanced understanding of DNA, RNA, and protein synthesis. The field of *molecular biology* arose to a great extent from these bacterial studies.

Another important development in the twentieth century involved the study of viruses. Although disease-causing viruses were first discovered at the end of the nineteenth century, it was not until the middle of the twentieth century that the true nature of viruses was determined. Much of this work involved the study of viruses that infect bacteria, called *bacteriophage*. An important development was the realization that virus infection was analogous to genetic transfer, and the relationships between viruses and other genetic elements was worked out primarily from research on bacteriophage.

By the 1970s, our knowledge of the basic processes of bacterial physiology, biochemistry, and genetics had advanced to such a great extent that it was now possible to manipulate the genetic material of cells experimentally, using bacteria as tools. It also became possible to introduce genetic material (DNA) from foreign sources into bacteria and control its replication and characteristics. This led to the development of the field of *biotechnology*. Although biotechnology originally arose from basic studies, its use to promote human welfare required the application of the principles of physiology and industrial microbiology, a good example of how basic and applied research advance together. Also at about this same time, *molecular sequencing* developed as a tool to discern phylogenetic relationships among prokaryotes, which led to revolutionary new concepts in the field of biological classification and to the first true understanding of the evolutionary history of microorganisms.

The overwhelming influence of microorganisms in human society is thus clear. We have many reasons to be aware of microorganisms and their activities. As Pasteur said, "The role of the infinitely small is infinitely large."

We have now briefly outlined the practical significance of microbiology. But this book begins with a careful discussion of the basic properties and activities of microoganisms. Only after a solid understanding of basic microbiology has been established is it possible to discuss the applications intelligently. Thus, the material in this book moves from the basic to the applied, and from the simple to the more complex.

Louis Pasteur's work on spontaneous generation led to the development of methods for the control of the growth of microorganisms. Pasteur showed that microorganisms could be destroyed by heat, resulting in the sterilization of growth-supporting materials. Robert Koch developed a set of postulates, which provided an experimental approach to the study of infectious microorganisms. Koch is the founder of the "germ theory of disease," and developed the first methods for the growth of pure cultures of microorganisms. In the twentieth century, basic and applied aspects of microbiology have worked hand in hand to yield a number of important practical advances and the revolution in molecular biology.

Study Questions

1. List five key properties associated with the living state. Can you think of inanimate systems that might exhibit any of these properties? Explain.

2. Cells can be thought of either as machines or as coding devices. Explain how these two attributes of a cell differ. Can you imagine a cell existing that lacks one of these two attributes?

3. In unicellular forms, such as bacteria and yeasts, the cell and the organism are one and the same. Most multicellular organisms, such as human beings, consist of an indefinite but very large number of cells. From a single yeast cell, by division, billions of cells can arise, and we can get a yeast cake. From a fertilized human egg, billions of cells can arise and we can get a baby. It is obvious that the billions of cells making up a yeast cake differ in a fairly fundamental way from the billions of cells making up a baby. Describe briefly this fundamental difference and discuss its significance.

4. Compare and contrast the prokaryotic and the eukaryotic cell. List three properties that are common to both cell types. List three properties that are different in these two cell types.

5. List three ways in which viruses differ from cells.

6. Translation is a key cellular process. What is meant by the word *translation* and how does the process occur?

7. Pasteur's experiments on spontaneous generation were of enormous importance for the advance of microbiology. They had impact on the following areas: (a) methodology of microbiology; (b) ideas on the origin of life; (c) preservation of food. Explain briefly how the impact of his experiments was felt in each of these areas.

8. Why was the development of sterilization methods of crucial importance for the science of microbiology?

9. If spontaneous generation was an everyday occurrence, would sterilization be possible?

10. Explain the principle behind the use of the Pasteur flask in studies on spontaneous generation.

11. If you were an opponent of Pasteur, how might you attempt to refute his experiments on spontaneous generation?

12. What is a pure culture, and why was a knowledge of how to obtain one important for the development of microbiology?

13. Explain why the invention of solid media was of great importance in the development of microbiology.

14. What evidence can you present that supports the idea that microorganisms are far more beneficial than harmful to humans?

Supplementary Readings

Brock, T. D. 1988. *Robert Koch: A Life in Medicine and Bacteriology*. Science Tech Publishers, Madison, WI. The standard biography of the discoverer of the germ theory of disease.

Brock, T. D. (ed.) 1989. *Microbes and Infectious Diseases*. Scientific American Books, New York. Historical and modern ideas about the role of microorganisms as disease-causing agents.

Brock, T. D. 1975. *Milestones in Microbiology*. American Society for Microbiology, Washington, D. C. The key papers of Pasteur, Koch, and others are translated, edited, and annotated for the beginning student.

Bulloch, W. 1935. *The History of Bacteriology*. Oxford University Press, London. The standard history of bacteriology; emphasis on medical aspects.

Dobell, C., (ed. and trans.) *Antoni van Leeuwenhoek and his "little animals"*. Constable and Co., London (1960, Dover Publications, New York). An introduction to Leeuwenhoek's life, his work, and his times.

Dubos, R. 1988. *Pasteur and Modern Science*. Science Tech Publishers, Madison, WI. A brief, well-illustrated biography of Louis Pasteur.

Lederberg, J. (ed.) 1992. *Encyclopedia of Microbiology*. Academic Press, San Diego, CA. A four-volume, 2100-page treatise on all aspects of microbiology.

Postgate, J. 1992. *Microbes and Man*, third edition. Cambridge University Press, Cambridge, England. A short treatment of microorganisms and how they affect the lives of humans.

Postlethwait, J. H., and **J. L. Hopson.** 1992. *The Nature of Life*. McGraw-Hill, Inc., New York. An elementary treatment of biological principles.

Raven, P. H. and **G. B. Johnson**. 1992. *Biology*, third edition. Times Mirror/Mosby College Publishing, St. Louis, MO. An excellent modern treatment of biological principles.

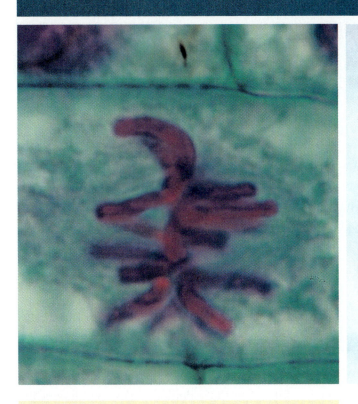

Cell Chemistry

To understand microbiology today one must have some understanding of the chemical processes taking place within cells. As pointed out in Chapter 1, living cells are governed by the same chemical and physical principles that dictate the properties of nonliving matter. However, as self-replicating entities, cells contain a variety of molecules not usually found in inanimate structures. The chemical nature of these molecules is the subject of this chapter. We begin by introducing chemical principles of atoms, molecules, and atomic bonding, and proceed through a discussion of the biochemistry of macromolecules, the building blocks of the cell. The student may find it useful to refer to material in this chapter from time to time while mastering the principles of cellular metabolism (Chapter 4) and genetics, virology, and molecular biology (Chapters 5 through 8).

2.1 Atoms

Atoms are the basic units of matter and can be defined as the smallest particles of an element that can enter into chemical combinations. The atoms of different elements have characteristic chemical and physical properties that are determined by the number of constituent particles of which they are composed. Atoms consist of particles called **protons, electrons**, and **neutrons**. Protons are *positively* charged and electrons are *negatively* charged; neutrons, as their name implies, are *uncharged* particles. About twenty different elements are found in cells. However only six elements are the most abundant in living systems: hydrogen, carbon, nitrogen, oxygen, phosphorus, and sulfur. The atomic structure of these six important elements is shown in Table 2.1.

Table 2.1   The atomic structure of the major biologically relevant elements

Element	Atomic number	Atomic weight	Protons	Neutrons	Electrons	Electrons in outer shell	Simplified structure*
Hydrogen	1	1	1	0	1	1	H •
Carbon	6	12	6	6	6	4	• C •
Nitrogen	7	14	7	7	7	5	• N •
Oxygen	8	16	8	8	8	6	: O :
Phosphorus	15	31	15	16	15	5	• P •
Sulfur	16	32	16	16	16	6	: S :

Only the outer shell electrons are depicted in the simplified structure.

The **nucleus** of an atom contains protons and neutrons and constitutes the majority of the atomic mass. Each proton carries an electronic charge of +1, and the number of protons in the atoms of an element correspond to the **atomic number** of that element. The **atomic weight** of an element is the total mass of protons and neutrons in the atoms of that element. Hence the atomic number of carbon is 6 and its atomic weight is 12 (Table 2.1). Although the number of protons in the atoms of a given element is constant, the number of *neutrons* may vary, giving rise to **isotopes** of a given element. Thus sulfur atoms always contain 16 protons, but exist in nature in three isotopic forms. Sulfur-32 (^{32}S), the most abundant form, contains 16 protons and 16 neutrons, sulfur-34 (^{34}S), a less abundant sulfur isotope, contains 16 protons and 18 neutrons, and sulfur-35 (^{35}S) a radioactive isotope or **radioisotope**, contains 16 protons and 19 neutrons. Unlike ^{32}S and ^{34}S, the atomic nucleus of ^{35}S is highly unstable and spontaneously disintegrates, releasing an electron (beta particle). Despite possessing different atomic weights, all of these forms of sulfur are chemically alike because they have the same number of protons and the same number of electrons in their outer shell (Table 2.1).

Electrons are negatively charged particles of negligible mass and are present in atoms in numbers equal to protons. Electrons orbit the nucleus in "shells," zones of electron density at increasing distance from the nucleus of the atom. The chemical properties of elements are for the most part determined by the number of electrons in the outermost shell of their atoms (see Table 2.1). The outermost shell of most elements of atomic number 8 or greater contains up to 8 electrons. The potential of an element for combination with other elements to form molecules is to a major

degree governed by whether or not the outer electron shell is full or partially depleted. For the biologically relevant elements, the most stable electron configuration in the outer shell of atoms is eight, except for the first shell which holds only two electrons. Table 2.1 shows that the outer shells of atoms of the major bioelements, hydrogen, carbon, nitrogen, oxygen, phosphorus, and sulfur, are all incomplete; these elements thus readily combine with other atoms to yield molecules in which all atoms achieve stable outer electron shells.

> The uniqueness of each type of chemical element is determined by the number and distribution of protons and neutrons in the nucleus and of electrons in the outer shells of its atoms.

2.2   Molecules and Chemical Bonding

When atoms combine with one another by chemically bonding, the resulting combination is referred to as a **molecule**. Several types of chemical bonds are known, including ionic, covalent, and hydrogen bonds. **Ionic bonds** are bonds in which electrons are *not* equally shared between two atoms, resulting in charged molecules. For example, in NaCl, the atom of chlorine is much more electronegative (electron attracting) than that of sodium, and formally removes the electron from the outer shell of sodium to form, in effect, Na^+Cl^-. The electrical attraction between the two atoms holds the molecule together. Ionic bonds are biologically important in nucleic acid/protein interactions and a few other molecular interactions. However, ionic bonds are not significant in the bonding of the important bioelements listed in Table 2.1 and thus

Miniglossary for Chapter 2

ATOMIC WEIGHT   the total mass of protons and neutrons in an atom

COVALENT BOND   a chemical bond in which electrons are shared equally between two atoms

DENATURATION   destruction of the folding properties of a protein leading (usually) to loss of biological activity

GLYCOSIDIC BOND   a type of covalent bond that links sugar units together in a polysaccharide

HYDROGEN BOND   a weak chemical bond between a hydrogen atom and a second, more electronegative element, usually an oxygen or nitrogen atom

ISOTOPE   one form of an element in which the number of neutrons varies from other forms of the same element

LIPID   glycerol bonded to fatty acids and other groups such as phosphate by an ester or ether linkage

MACROMOLECULE   polymer of covalently linked monomeric units

MOLECULE   two or more atoms chemically bonded to one another

NONPOLAR   possessing hydrophobic (water-repelling) characteristics and not easily dissolved in water

NUCLEOTIDE   a monomer of a nucleic acid containing a nitrogen base (adenine, guanine, cytosine, thymine, or uracil), a molecule of phosphate, and a sugar, either ribose (in RNA) or deoxyribose (in DNA)

PEPTIDE BOND   a type of covalent bond joining amino acids in a polypeptide

PHOSPHODIESTER BOND   a type of covalent bond linking

nucleotides together in a polynucleotide

POLAR   possessing hydrophilic characteristics and generally water soluble

POLYNUCLEOTIDE   a polymer of nucleotides bonded to one another by phosphodiester bonds

POLYPEPTIDE   a polymer of amino acids bonded to one another by peptide bonds

POLYSACCHARIDE   a polymer of sugar units bonded to one another by glycosidic bonds

PRIMARY STRUCTURE   in an informational macromolecule, such as a polypeptide, the precise sequence of monomeric units

PROTEIN   a polypeptide or group of polypeptides that form a molecule of specific biological function

QUATERNARY STRUCTURE   in proteins, the number and arrangement of individual polypeptides in the final protein molecule

RADIOISOTOPE   an isotope of an element that undergoes spontaneous decay with the release of radioactive particles

SECONDARY STRUCTURE   the initial pattern of folding of a polypeptide or a polynucleotide, usually dictated by opportunities for hydrogen bonding

STEREOISOMER   different forms of the same molecule that are mirror images of one another

TERTIARY STRUCTURE   the final folded structure of a polypeptide that has previously attained secondary structure

---

they will not be further discussed here. Biologically relevant molecules are held together by *covalent* and *hydrogen* bonds.

Covalent bonds

Bonds in which electrons are more or less shared equally between two atoms are referred to as **covalent bonds**. Consider the element hydrogen. A hydrogen atom contains a single electron and readily combines with itself to form a molecule of hydrogen gas, H_2 (Figure 2.1a). The H_2 molecule consists of two hydrogen atoms, each containing two shared electrons, the maximum allowed in the initial electron shell surrounding any nucleus. The atoms in the H_2 molecule are held together by covalent bonds. By comparison, carbon has an outer electron shell containing four electrons and can accept four additional electrons in this shell. Hence, a carbon atom readily combines with four hydrogen atoms, leading to the formation of methane (Figure 2.1b) a simple *organic* (carbon-containing) compound. By sharing electrons in this fashion, the carbon atom and the hydrogen atoms achieve stable outer electron shells: eight electrons in the outer shell for carbon, and two each for the hydrogens. Likewise, when water forms from the elements hydrogen and oxygen (Figure 2.1c) the outer electron shells in both elements reach the stable configuration.

If more than one electron pair is shared between two atoms, double or even triple bonds are formed (Figure 2.2). Double bonds occur most frequently as

$C=O$ or $C=C$ bonds and are found in a variety of important biological molecules. For triple bonds, three electron pairs are shared between two atoms. Triple bonds are rare in biological molecules, but microorganisms occasionally metabolize triply bonded compounds, the best example being N_2 (Figure 2.2), which is reduced in nitrogen fixation (see Section 16.26).

Hydrogen bonds

A second type of bond that is extremely important in biological systems is the **hydrogen bond**. Hydrogen bonds form between hydrogen atoms and relatively electronegative elements like oxygen and nitrogen. Hydrogen bonds are weak bonds. However, when many hydrogen bonds are formed within and between macromolecules, the overall stability of the molecule is greatly increased.

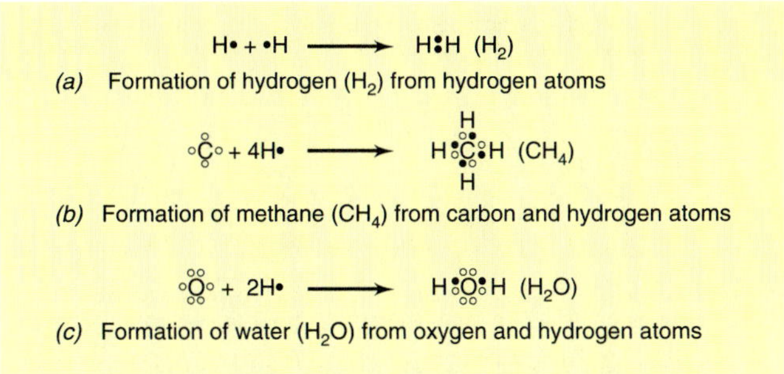

(a) Formation of hydrogen (H_2) from hydrogen atoms

(b) Formation of methane (CH_4) from carbon and hydrogen atoms

(c) Formation of water (H_2O) from oxygen and hydrogen atoms

FIGURE 2.1   Formation of typical single covalently bonded molecules.

Water molecules readily undergo hydrogen bonding (Figure 2.3). Since an oxygen atom is relatively electronegative and a hydrogen atom is not, the covalent bond between oxygen and hydrogen is one in which the shared electrons in the outer shells orbit nearer the oxygen nucleus than the hydrogen nucleus. This creates a *slight* electrical charge separation, oxygen slightly negative, hydrogen slightly positive. The latter attracts the oxygen of a second water molecule near the positive charge, in effect creating a positively charged "bridge" between the two electronegative oxygens of adjacent water molecules. The most common hydrogen bonds in macromolecules are those which involve O—H$^+$—O$^-$, O$^-$—H$^+$—$^-$N, and N$^-$—H$^+$—$^-$N interactions in proteins and nucleic acids (Figure 2.3). We will see later that hydrogen bonds play major roles in the biological properties of macromolecules, especially in the higher order structure of proteins and nucleic acids.

Other atomic interactions

Molecules are involved in other types of atomic interactions not referred to as formal chemical bonds. **Van der Waals forces** are nonspecific *attractive* forces which occur when the distance between two atoms is reduced to the range of 3–4Å. Van der Waals forces occur because of momentary charge asymmetries around atoms due to electron movement. If atoms become closer than 3–4Å, however, overlap of the atom's electron shells causes repulsive forces to occur. Van der Waals forces can play a significant role in the bind-

ing of substrates to enzymes (see Section 4.3) and in protein/nucleic acid interactions.

Hydrophobic interactions can also be important in the properties of biologically relevant molecules. Hydrophobic interactions occur because nonpolar (water repelling) molecules tend to cluster together in an aqueous environment. Because of this, nonpolar portions of a macromolecule will tend to associate, as will polar portions of a macromolecule (but for the opposite reason). Hydrophobic interactions are a major consideration in the folding of macromolecules such as proteins, and play an important role in the binding of substrates to enzymes.

Bond strengths and the importance of carbon

The relative strength of chemical bonds can be measured by the energy required to dissociate them. Table 2.2 lists the bond energies of a number of biologically

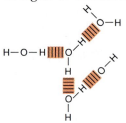

1. Hydrogen bonding between water molecules

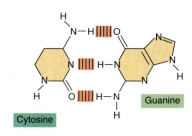

2. Hydrogen bonds between amino acids in protein chain

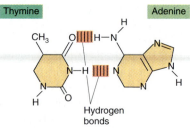

3. Hydrogen bonds between bases in DNA

FIGURE 2.3 Hydrogen bonding of water and some biologically relevant molecules.

Formation of ethylene, a double-bonded carbon compound

Formation of acetylene, a triple-bonded carbon compound

Carbon dioxide   Nitrogen   Phosphate

Some other simple compounds with double or triple bonds

Peptide bond   Cytosine (nitrogen   Phenylalanine (amino
of proteins   base of DNA and RNA)   acid in proteins)

More complex compounds with double bonds

FIGURE 2.2   Covalent bonding of some biologically important molecules containing double or triple bonds.

Table 2.2 Energies of some covalent bonds and noncovalent interactions*	
Covalent bonds	**Bond energy**
Single bonds	
H—H	436
C—H	411
C—O	369
C—N	294
C—S	260
Double bonds	
C=C	616
C=O	704
O=O	402
Triple bonds	
C≡C	805
N≡N	955
Noncovalent interactions	**Energy**
Hydrogen bonds	4.2–8.4
Hydrophobic interactions	4.2–8.4
van der Waal's attractions	4.2–8.4

The bond energies are given as the amount of heat, in kilojoules per mole, needed to break the bonds.

drogen bonds will form and dissociate in the cell spontaneously and rapidly, whereas covalent bonds will only be formed and broken by specific chemical reactions brought about by enzymes (see Section 4.3). Noncovalent interactions such as van der Waals forces and hydrophobic bonds are also very weak. However, like hydrogen bonds, these interactions become significant when large numbers of them develop either within or between various molecules.

All cellular structures contain an abundance of carbon in chemical combination with other elements. Carbon is a unique element in that it is able to combine not only with many other elements but also with itself, thus forming larger chemical structures of considerable diversity and complexity. An enormous number of different carbon compounds have been identified in various biological systems. For example, many biological compounds contain a carbon atom bonded to an oxygen atom (Table 2.3). Differences in the carbon-oxygen bond (single or double) and the nature of the atoms surrounding the carbon-oxygen bond dictates the chemical properties of the molecule. Hence, carboxylic acids are chemically distinct from alcohols and each plays specific roles in cellular biochemistry (Table 2.3).

relevant chemical bonds. Note that double and triple bonds are much stronger than single bonds and that hydrogen bonds are by comparison very weak. Hy-

> **Hydrogen bonds and other types of weak bonds play important roles in the way biomolecules function. Carbon is at the core of cell structure.**

Table 2.3 Carbon–oxygen compounds of biochemical importance		
Chemical species	**Structure**	**Biological importance**
Carboxylic acid	$\underset{\displaystyle -C-OH}{\overset{\displaystyle O}{\parallel}}$	Organic, amino, and fatty acids
Aldehyde	$\underset{\displaystyle -C-H}{\overset{\displaystyle O}{\parallel}}$	Functional group of reducing sugars such as glucose
Alcohol	$-\overset{\displaystyle H}{\underset{\displaystyle H}{C}}-OH$	Lipids, carbohydrates
Keto	$\underset{\displaystyle -C-}{\overset{\displaystyle O}{\parallel}}$	Pyruvate, citric acid cycle intermediates
Ester	$-\overset{H}{\underset{H}{C}}-O-\overset{O}{\overset{\parallel}{C}}-$	Lipids of Bacteria and Eukarya, amino acid attachment to tRNAs
Ether	$-\overset{H}{\underset{H}{C}}-O-\overset{H}{\underset{H}{C}}-$	Lipids of Archaea, sphingolipids

2.3   Water as a Biological Solvent

The chemistry of life occurs in water. Microbial cells are 70–90% water by weight, and all chemical reactions which occur in the cytoplasm of a cell take place in this aqueous environment. Water is an ideal biological solvent. Although pure water is electrically neutral, having an equal number of electrons and protons (see Figure 2.1), water molecules contain two elements, oxygen and hydrogen, of quite different electronegativity, and bonding between these elements leads to charge asymmetry (Figure 2.3a) and the property of *polarity*. The polar properties of water make it an excellent solvent because many biologically important molecules are themselves polar and thus readily dissolve in water. As we will see in Chapter 3, dissolved substances are continually passed into and out of the cell through transport activities of the cytoplasmic membrane.

The polar properties of water allow for ready hydrogen bonding, which we have seen is important in the overall structure of macromolecules like proteins and nucleic acids (Figure 2.3). Water forms three-dimensional networks with other molecules including macromolecules and, by so doing, spatially positions these molecules for potential interactions. But in addition to hydrogen bonding, the polar nature of water makes it highly *cohesive*, meaning that water molecules tend to have a high affinity for one another and will form chemically ordered arrangements in which hydrogen bonds are constantly forming, breaking, and reforming. The cohesive nature of water is responsible for many of its biologically important properties, such as high surface tension and high specific heat. Also, the fact that water expands upon freezing to yield a less dense solid form has profound effects on life in temperate or polar aquatic environments.

The high polarity of water is also beneficial to the cell because it tends to force *nonpolar* substances to aggregate and remain together. We will see a practical example of this in the next chapter when we study membrane structure and function. Membranes contain nonpolar substances such as lipids, which aggregate and function as a barrier to the flow of polar molecules in and out of the cell.

Life originated in water, and indeed anywhere on earth where liquid water exists, microorganisms are likely to be found. The unique chemical properties of water make it an ideal biological solvent, and therefore it is not surprising that water is the solvent for all cellular chemistry. Indeed, water is intimately involved in several biochemical reactions that take place in the cell. We now consider the major substances dissolved in the water of a cell.

Life could not exist without liquid water; the polar properties of water make it an excellent solvent for biological systems.

2.4   Small Molecules: Monomers

The main chemical components of cells are structures called **macromolecules**. Macromolecules are built up of individual "building blocks" which are connected in specific ways. A single building block is called a **monomer**, and the macromolecule is called a **polymer**. There are only a few types of polymers important in cell biochemistry, and each is made of a characteristic set of monomers.

Monomers are organic compounds containing up to about 30 carbon atoms and are grouped in classes according to their chemical properties. There are four classes of *monomers* to be considered here: *sugars*, the monomeric constituents of polysaccharides; *fatty acids*, the monomeric units of lipids; *nucleotides*, the basic units of the nucleic acids (DNA and RNA); and *amino acids*, the monomeric constituents of the proteins. The four classes of macromolecules can be subdivided into "informational" and "noninformational" types. Nucleic acids and proteins are considered informational macromolecules because the *sequence* of monomeric units within them is highly specific and carries biological information and the means to process this information. Lipids and polysaccharides, on the other hand, are not informational, because the sequence of monomers in these polymers is frequently highly repetitive and the sequence itself is generally of no functional importance. However, it should be appreciated that both the nature *and* the sequence of the monomeric units of any macromolecule is important in distinguishing it chemically from related macromolecules.

Macromolecules are polymers of monomers. Important cellular macromolecules are nucleic acids, proteins, polysaccharides, and lipids.

2.5   From Carbohydrates to Polysaccharides

Carbohydrates (sugars) are organic compounds containing carbon, hydrogen, and oxygen in the ratio of 1:2:1. The structural formula for glucose, the most abundant of all sugars, is $C_6H_{12}O_6$ (Figure 2.4). The most biologically relevant carbohydrates are those containing 4, 5, 6, and 7 carbon atoms (designated as C_4, C_5, C_6, and C_7). C_5 sugars (pentoses) are of special significance because of their roles as the structural backbones of nucleic acids. Likewise, C_6 sugars (hexoses) are the monomeric constituents of cell wall polymers and energy reserves. Figure 2.4 shows the structural formulas of a few common sugars. Derivatives of simple carbohydrates can be formed by replacing one or more of the hydroxyl groups with other chemical species. For example, the important bacterial cell wall polymer **peptidoglycan** (see Section 3.5) contains the

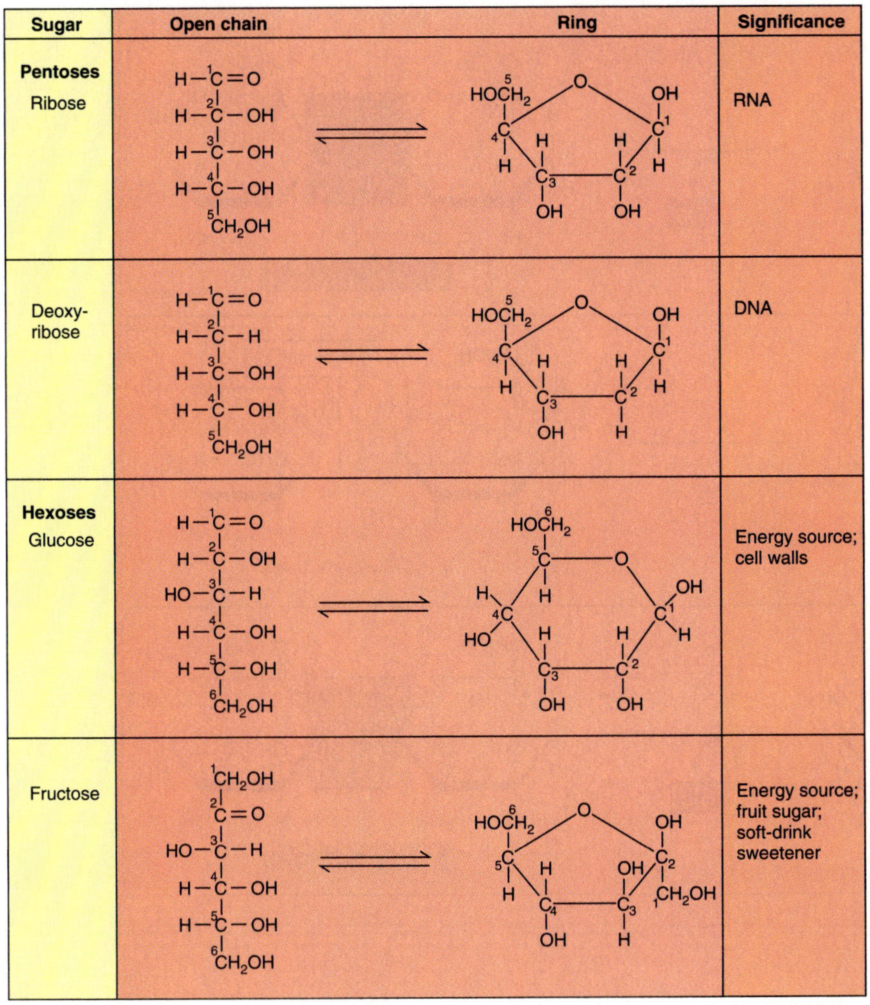

Sugar	Open chain	Ring	Significance
Pentoses Ribose	$H-\overset{1}{C}=O$ / $H-\overset{2}{C}-OH$ / $H-\overset{3}{C}-OH$ / $H-\overset{4}{C}-OH$ / $\overset{5}{C}H_2OH$	(ring structure)	RNA
Deoxy-ribose	$H-\overset{1}{C}=O$ / $H-\overset{2}{C}-H$ / $H-\overset{3}{C}-OH$ / $H-\overset{4}{C}-OH$ / $\overset{5}{C}H_2OH$	(ring structure)	DNA
Hexoses Glucose	$H-\overset{1}{C}=O$ / $H-\overset{2}{C}-OH$ / $HO-\overset{3}{C}-H$ / $H-\overset{4}{C}-OH$ / $H-\overset{5}{C}-OH$ / $\overset{6}{C}H_2OH$	(ring structure)	Energy source; cell walls
Fructose	$\overset{1}{C}H_2OH$ / $\overset{2}{C}=O$ / $HO-\overset{3}{C}-H$ / $H-\overset{4}{C}-OH$ / $H-\overset{5}{C}-OH$ / $\overset{6}{C}H_2OH$	(ring structure)	Energy source; fruit sugar; soft-drink sweetener

FIGURE 2.4
Structural formulas of a few common sugars. The formulas can be represented in two alternate ways, open chain and ring. The open chain is easier to visualize but the ring form is the commonly used structure.

glucose derivatives *N*-acetylglucosamine and *N*-acetylmuramic acid (Figure 2.5). Besides sugar derivatives, sugars having the same *structural* formula can still differ in their *stereoisomeric* properties (see Section 2.9). Hence, a large number of different sugars are potentially available to the cell for the construction of polysaccharides.

Polysaccharides are high-molecular-weight carbohydrates containing many (sometimes hundreds or even thousands) monomeric units connected to one another by a type of covalent bond referred to as a **glycosidic bond** (Figure 2.6). If two sugar units (monosaccharides) are joined by glycosidic linkage, the resulting molecule is called a **disaccharide**. The addition of one more monosaccharide would yield a **trisaccharide**, several more an **oligosaccharide**, and an extremely long chain of monosaccharides in glycosidic linkage is called a **polysaccharide**.

The glycosidic bond can exist in two different orientations, referred to as alpha (α) and beta (β) (Figure 2.6*a*). Polysaccharides with a repeating structure composed of glucose units linked between carbons 1 and 4 in the alpha orientation (for example, glycogen and starch, Figure 2.6*b*) function as important carbon and energy reserves in bacteria, plants, and animals. Alter-

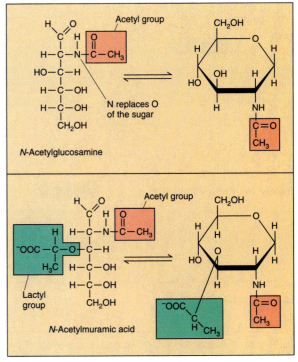

FIGURE 2.5   Sugar derivatives found in the cell walls of most Bacteria. Note that the parent structure is *glucose* in both cases.

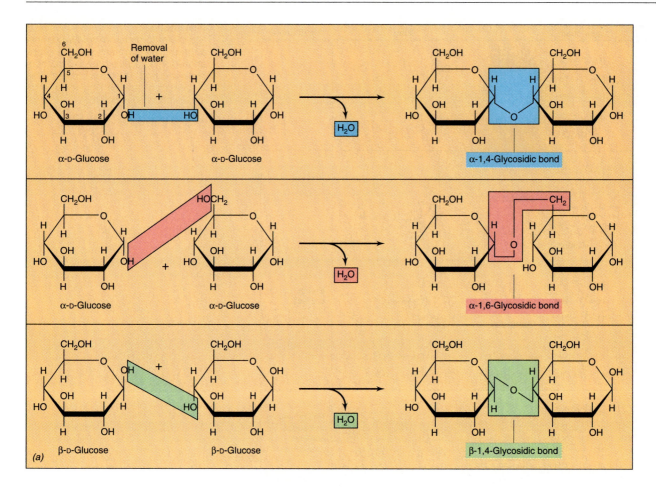

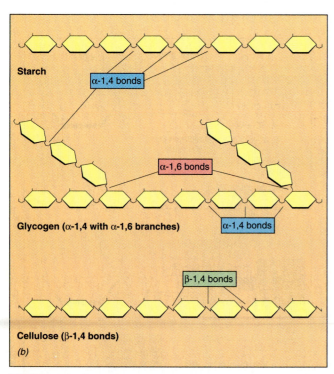

FIGURE 2.6 Polysaccharides. (a) Formation of polysaccharides from monosaccharides. (b) General structures of some common polysaccharides.

natively, glucose units joined by beta 1–4 linkages are present in cellulose (Figure 2.6*b*), a plant and algal cell wall component functionally unrelated to glycogen or starch. Thus, even though starch and cellulose are both composed solely of glucose units, their functional properties are entirely different as a result of the different configurations, α and β, of their glycosidic bonds. Polysaccharides can also combine with other classes of macromolecules, such as protein or lipid, to form **glycoproteins** or **glycolipids**. These compounds play important roles in cell membranes as cell surface receptor molecules. The compounds reside on the external surfaces of the membrane where they are in contact with the environment. Glycolipids also constitute a major portion of the cell wall of Gram-negative Bacteria, and as such impart a number of unique surface properties on these organisms (see Section 3.5).

Sugars (carbohydrates), combine into long polymers called polysaccharides. Cell walls of most prokaryotes and of plants are composed of polysaccharides. Many energy storage compounds, including starch and glycogen, are also polysaccharides. Polysaccharides can also contain other molecules such as protein or lipid forming complex polysaccharides.

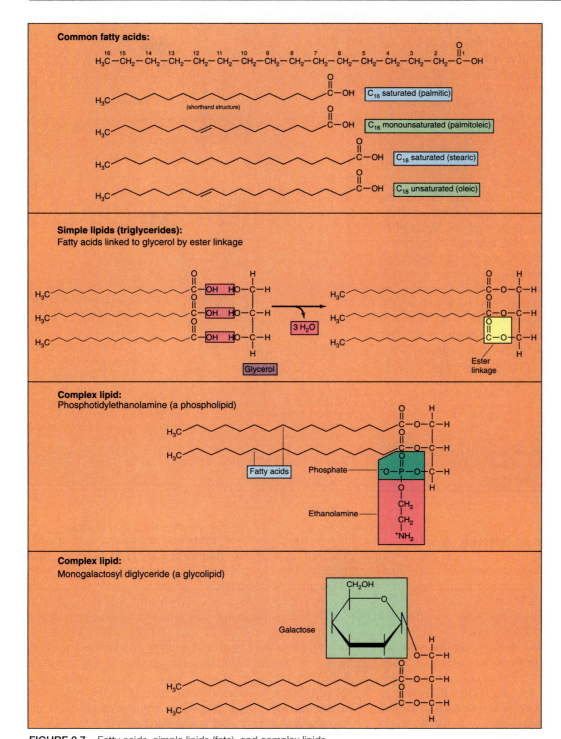

FIGURE 2.7   Fatty acids, simple lipids (fats), and complex lipids.

2.6   From Fatty Acids to Lipids

Fatty acids are the main constituents of **lipids**. Fatty acids have interesting chemical properties because they contain both highly hydrophobic (water repelling) and highly hydrophilic (water soluble) regions. Palmitate,* for example (Figure 2.7), is a 16 carbon fatty acid composed of a chain of 15 saturated (fully hydrogenated) carbon atoms and a single car-

boxylic acid group. Other common fatty acids are stearic (C_{18} saturated) and oleic (C_{18} monounsaturated) acids (Figure 2.7).

*Fatty acids can exist in both protonated (RCOOH) and unprotonated (RCOO$^-$) forms, depending on pH. Because at pH 7 fatty acids are generally unprotonated, this is indicated by adding the suffix -*ate* to the root term for the fatty acid. Thus, palmitic acid is $C_{15}H_{31}COOH$, while palmitate is $C_{15}H_{31}COO^-$.

Simple lipids (fats) consist of fatty acids bonded to the C_3 alcohol *glycerol* (Figure 2.7). Simple lipids are frequently referred to as **triglycerides** because three fatty acids are linked to the glycerol molecule.

Complex lipids are simple lipids which contain additional elements such as phosphate, nitrogen, or sulfur, or small hydrophilic carbon compounds such as sugars, ethanolamine, serine, or choline (Figure 2.7). **Phospholipids** are a very important class of complex lipids, as they play a major structural role in the cytoplasmic membrane (see Section 3.3).

The chemical properties of lipids make them ideal structural components of membranes. Because they are *amphipathic*, that is, show properties of hydrophobicity and hydrophilicity, lipids aggregate in membranes with the hydrophilic portions toward the external or internal (cytoplasmic) environment, while maintaining their hydrophobic portions away from the aqueous milieu (see Section 3.3). Such structures are ideal permeability barriers because of the inability of water-soluble substances to flow through the hydrophobic fatty acid portion of the lipids. Indeed, the major function of the cytoplasmic membrane is to serve as the barrier to the diffusion of substances into or out of the cell.

> **Lipids contain both hydrophobic and hydrophilic units; their chemical properties make them ideal structural components for cell membranes.**

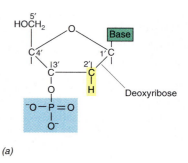

(a)

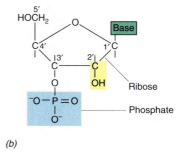

(b)

FIGURE 2.8   Nucleotides. (a) Of DNA. (b) Of RNA. Note the numbering system employed and the chemical differences on carbon atom 2' between deoxyribose and ribose. Both deoxyribonucleotides (DNA) and ribonucleotides (RNA) contain a 3' phosphate.

2.7   From Nucleotides to Nucleic Acids

The nucleic acids, **DNA** (deoxyribonucleic acid), and **RNA** (ribonucleic acid), are polymers of monomers called **nucleotides**. DNA and RNA are thus both **polynucleotides**. DNA carries the genetic blueprint for the cell and RNA acts as an intermediary molecule to convert the blueprint into defined amino acid sequences in proteins. Despite these important cellular functions, nucleic acids are composed of only a relatively few simple building blocks. Each nucleotide is composed of three separate units: a five carbon sugar, either ribose (in RNA) or deoxyribose (in DNA), a nitrogen base, and a molecule of phosphate, PO_4^{3-}. Figure 2.8 shows schematic drawings of single nucleotides of DNA and RNA.

Nucleotides

The nitrogen bases of nucleic acids belong to either of two chemical classes. *Purine* bases, **adenine** and **guanine**, contain two fused carbon–nitrogen rings while the *pyrimidine* bases **thymine, cytosine,** and **uracil,** contain a single six-membered carbon–nitrogen ring (Figure 2.9). Guanine, adenine, and cytosine are found in both DNA and RNA; thymine is present (with minor exceptions) only in DNA, uracil is present only in RNA. In a nucleotide, a base is attached to a pentose sugar by glycosidic linkage between carbon atom number 1 of the sugar and a nitrogen atom of the base, either nitrogen atom labeled as atom 1 (pyrimidine base) or nitrogen atom 9 (purine base). Without a phosphate, a base bonded to its sugar is referred to as a **nucleoside**. Nucleotides are thus nucleosides containing one or more phosphates (Figure 2.10).

Nucleotides play other roles in the cell besides their major role as the constituents of nucleic acids. Nucleotides, especially adenosine triphosphate (ATP, Figure 2.10), can act as carriers of chemical energy and can release sufficient energy during the hydrolytic cleavage of a phosphate bond to drive energy requiring reactions in the cell (see Section 4.6). Other nucleotides or nucleotide derivatives function in oxidation-reduction reactions in the cell (see Section 4.5) as

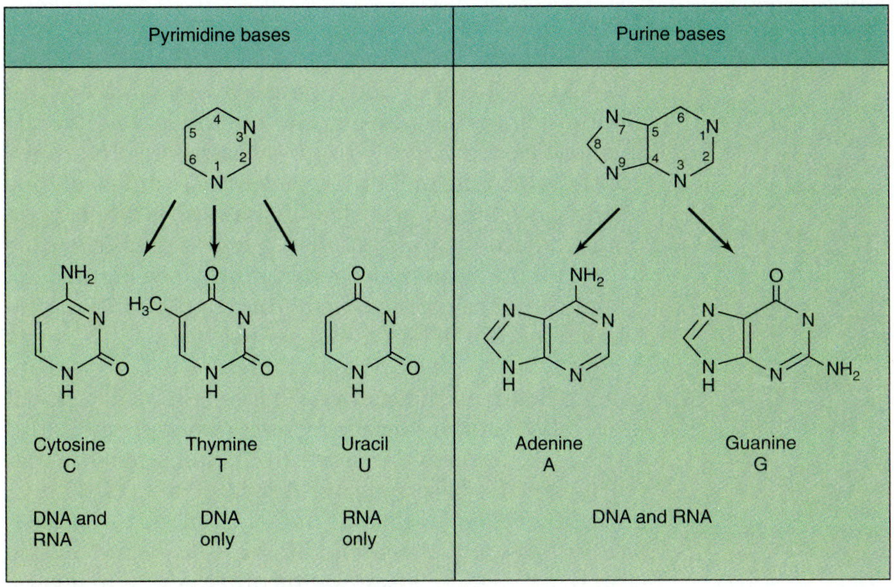

Pyrimidine bases	Purine bases

Cytosine
C

Thymine
T

Uracil
U

Adenine
A

Guanine
G

DNA and
RNA

DNA
only

RNA
only

DNA and RNA

FIGURE 2.9   Structure of bases of DNA and RNA. The letters C, T, U, A, and G, are often used to refer to the individual bases.

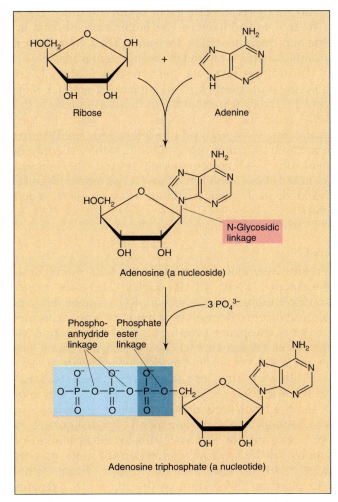

Ribose

Adenine

N-Glycosidic
linkage

Adenosine (a nucleoside)

3 PO$_4$$^{3-}$

Phospho-
anhydride
linkage

Phosphate
ester
linkage

Adenosine triphosphate (a nucleotide)

FIGURE 2.10   Components of the important nucleoside triphosphate, adenosine triphosphate (ATP).

carriers of sugars in the biosynthesis of polysaccharides (see Section 4.17), and as regulatory molecules, inhibiting or stimulating the activities of certain enzymes or metabolic events. However, here we are discussing the role of a nucleotide is as a building block of nucleic acid, the major *informational* junction of nucleotides.

Nucleic acids

Nucleic acids are long polymers in which nucleotides are covalently bonded to one another in a defined sequence, forming structures called **polynucleotides**. The backbone of the nucleic acid is a polymer in which sugar and phosphate moieties alternate (Figure 2.11). Hence, when we refer to a specific *sequence* of nucleotides in a nucleic acid, we are really referring to the variable portions of the nucleotide—the bases; the sugar-phosphate backbone is always the same.

The precise sequence of nucleotides in a DNA or RNA molecule is referred to as the molecule's *primary structure.* The sequence of bases in a DNA or RNA molecule is *informational*, representing the genetic information necessary to reproduce an identical copy of the organism. We will see later that the replication of DNA and the production of RNA are highly complex processes (see Chapter 5) and that a virtually error-free mechanism is necessary to insure the faithful transfer of genetic traits from one generation to another.

In chemical terms nucleic acids are composed of nucleotides covalently attached to one another via phosphate from carbon 3 (referred to as the 3′ [3

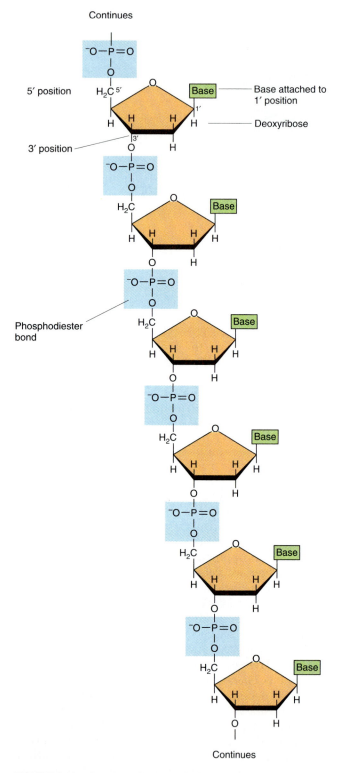

FIGURE 2.11 Structure of part of a DNA chain.

prime] carbon) of one sugar to carbon 5 (5') of the adjacent sugar** (Figure 2.11). The phosphate linkage is chemically a **phosphodiester**, since a single phosphate is connected by ester linkage to two separate sugars.

**Because the ring structure in the base is numbered also, the prime numbering system is used to refer to positions of carbon atoms on the sugars.

DNA

Each cellular chromosome contains two strands of DNA, each strand containing several million nucleotides linked by phosphodiester bonds. The strands themselves associate with one another by hydrogen bonds which form between the nucleotides of one strand and the nucleotides of the other. When positioned adjacent to one another, purine and pyrimidine bases can undergo hydrogen bonding (see Figure 2.3). Chemically, the most stable hydrogen bonding configuration occurs when guanine (G) forms hydrogen bonds with cytosine (C), and adenine (A) forms hydrogen bonds with thymine (T) (see Figure 2.3). It will be recalled that hydrogen bonds, although individually weak, collectively serve to stabilize macromolecules. Specific base pairing, A with T and G with C, means that the two strands of DNA will be *complementary* in base sequence; wherever a G is found in one strand, a C will be found in the other, and wherever a T is present in one strand its complementary strand will have an A. The molar amounts of guanine and cytosine are therefore *identical* in double-stranded DNA from any source; likewise the molar amounts of adenine and thymine will be the same. The two opposing strands are thus arranged in complementary fashion.

In addition, however, the two strands of the DNA molecule are arranged in an *antiparallel* manner (Figure 2.12). This means that the individual strands of the DNA molecule are arranged in a "head to toe" arrangement—one strand (the left in Figure 2.12) runs in a 5'→3' direction top to bottom, while its complement runs 5'→3' bottom to top (Figure 2.12). It should be noted that although DNA is generally double-stranded, some viruses contain only single-stranded DNA.

RNA

With the exception of certain viruses that contain double-stranded RNA, all ribonucleic acids are *single-stranded* molecules. However, RNA molecules can fold back upon themselves in regions where complementary base pairing can occur to form a variety of highly folded structures. The pattern of folding observed in RNA is referred to as its *secondary structure*.

RNA plays three crucial roles in the cell. **Messenger RNA** (mRNA) contains the genetic information of DNA in a single-stranded molecule *complementary* in base sequence to a portion of the base sequence of DNA. The mRNA sequence directs incorporation of amino acids into a growing polypeptide in the process called *translation* (see Sections 5.6–5.8). These events take place on the surface of the ribosome. **Transfer RNA** (tRNA) molecules are the "adaptor" molecules in protein synthesis. Transfer RNAs are so named because of their dual specificities, one for a specific three-base sequence on mRNA (the codon) and the other for a specific amino acid. The tRNA molecule effectively adapts the genetic information from the language of nucleotides to the language of amino acids,

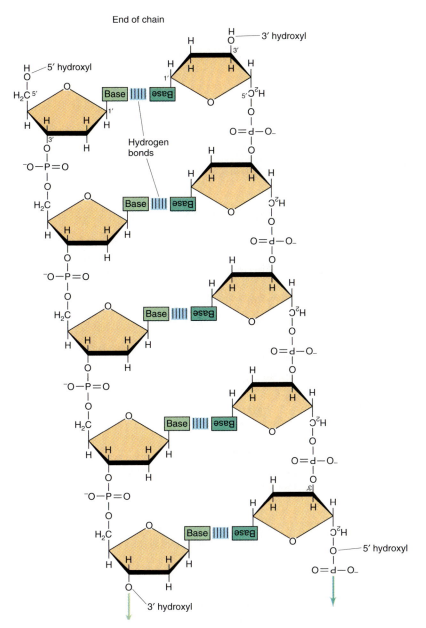

FIGURE 2.12
DNA structure. Complementary and antiparallel nature of DNA. Note that one chain ends in a 5' hydroxyl group whereas the other ends in a 3' hydroxyl.

the building blocks of proteins. **Ribosomal RNA (rRNA)** molecules (several distinct types are known), are important structural and catalytic components of the ribosome, the protein-synthesizing system of the cell. These various RNA molecules are discussed in detail in Chapter 5.

> The informational content of a nucleic acid is determined by the sequence of bases along the polynucleotide chain. Both RNA and DNA are informational macromolecules. RNA can often fold into various configurations to obtain secondary structure.

2.8   From Amino Acids to Proteins

Amino acids are the monomeric units of proteins. Most amino acids consist only of carbon, hydrogen, oxygen, and nitrogen, but two of the twenty common amino acids found in cells also contain sulfur atoms.

All amino acids contain two important functional groups, a *carboxylic acid* group (—COOH) and an *amino* group (—NH$_2$). These groups are functionally important because covalent bonds between the carbon of the carboxyl group of one amino acid and the nitrogen of the amino group of a second amino acid (with elimination of a molecule of water), forms the **peptide bond**, a type of covalent bond characteristic of proteins (see Figure 2.14).

All amino acids conform to the general structure shown in Figure 2.13. Amino acids differ in the nature of the side group (abbreviated "R" in Figure 2.13) attached to the alpha carbon. The alpha carbon is the carbon atom *immediately adjacent* to the carbon atom of the carboxylic acid group. The side chains on the alpha carbon vary considerably, from as simple as a hydrogen atom in the amino acid glycine, to aromatic ringed structures in amino acids such as phenylalanine (Figure 2.13). The chemical properties of an amino acid are to a major degree governed by the na-

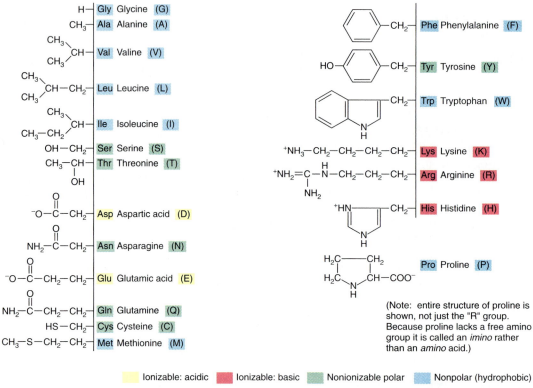

FIGURE 2.13 Structure of the 20 common amino acids. The three-letter codes for the amino acids are to the left of the names, the one-letter codes are in parentheses to the right of the names.

ture of the side chain, and thus amino acids which show similar chemical properties are grouped into amino acid "families," as shown in Figure 2.13. For example, the side chain may itself contain a carboxylic acid group, such as in aspartic acid or glutamic acid, rendering the amino acid acidic. Alternatively, several amino acids contain nonpolar hydrophobic side chains and are grouped together as nonpolar amino acids. The amino acid cysteine contains a sulfhydryl group (—SH) which is frequently important in connecting one chain of amino acids to another by *disulfide linkage* (R—S—S—R).

The large number of chemically distinct amino acids available makes it possible for cells to produce an enormous number of unique proteins with widely different biochemical properties. For example, proteins that are in direct contact with highly hydrophobic regions of the cell, such as proteins embedded in the lipid-rich cytoplasmic membrane, generally contain higher overall proportions of hydrophobic amino acids (or contain regions extremely rich in hydrophobic amino acids) than proteins which function in the aqueous environment of the cytoplasm.

Structure of proteins: primary and secondary structure

Proteins play key roles in cell function. Two kinds of proteins are recognized in cells: *catalytic proteins* (enzymes) and *structural proteins*. Enzymes serve as catalysts for the wide variety of chemical reactions which occur in cells (see Chapter 4). Structural proteins are those which become integral parts of the structures of cells, in membranes, walls, and cytoplasmic components. In essence, a cell is what it is because of the kinds of proteins which it contains. Therefore, an understanding of protein structure is essential for an understanding of cell function.

Proteins are polymers of various lengths containing defined sequences of amino acids covalently bonded by peptide linkages. The combination of two amino acids head to tail (carboxyl group to amino group) along with the elimination of water generates the peptide bond (Figure 2.14). Two amino acids connected together constitute a *dipeptide*, three amino acids a *tripeptide*, and so on. Many amino acids covalently linked via peptide bonds constitute a **polypeptide**. Proteins consist of one or more polypeptides. The

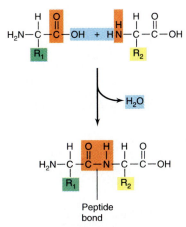

FIGURE 2.14   Peptide bond formation.

mary structure will allow only certain types of higher order structure to occur. The juxtaposition of α-carbon side groups dictated by the primary structure forces the polypeptide to twist and fold in a specific way. This twisting and coiling of the polypeptide molecule leads to the formation of the **secondary structure** of the protein. Hydrogen bonds, the weak noncovalent linkages discussed earlier (see Section 2.2), play important roles in the type of secondary structure that a protein will attain.

A typical secondary structure for many polypeptides is called the *α-helix* (Figure 2.15). Imagine a linear polypeptide wound around a cylinder. Under these conditions oxygen and nitrogen atoms from different amino acids would become positioned close enough together in the twisted structure to allow for hydrogen bonding to occur. This opportunity for H-bonding (and the inherent stability associated with it) helps direct many polypeptides to take on an α-helix secondary structure (Figure 2.15*a*).

Many polypeptides conform to a different type of secondary structure referred to as the β-sheet. In the β-sheet, the chain of amino acids in the polypeptide folds back and forth upon itself instead of forming a helix; this type of folding exposes hydrogen atoms which can undergo extensive hydrogen bonding (Figure 2.15*b*). Some polypeptides contain both regions of α-helix and regions of β-sheet secondary structure, the

number of amino acids varies from one protein to another. Proteins with as few as 15 and as many as 10,000 amino acids are known. Since proteins may vary in their composition, sequence, and number of amino acids, it is easy to see that enormous variation in protein types is possible.

All proteins are folded and show complex arrangements of structure. The linear array of amino acids is referred to as the **primary structure** of the polypeptide. In many ways the primary structure of a polypeptide is most important, because a given pri-

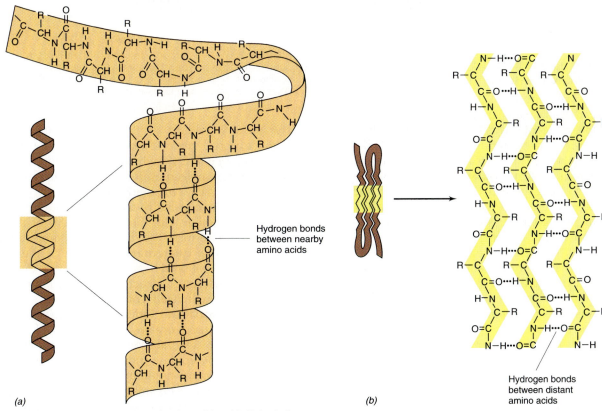

(a)   *(b)*

FIGURE 2.15   *Secondary structure of polypeptides. (a) Alpha-helix secondary structure. Note that hydrogen bonding does not involve the R groups but instead occurs between the atoms involved in the peptide bonds. (b) Beta-sheet secondary structure.*

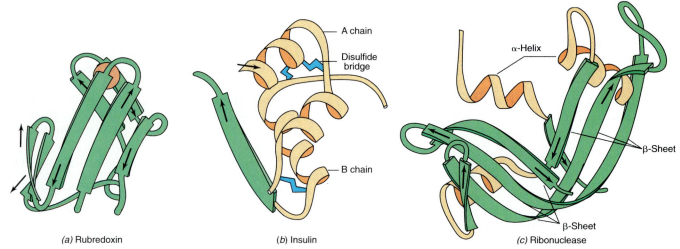

(a) Rubredoxin   (b) Insulin   (c) Ribonuclease

FIGURE 2.16 Tertiary structure of polypeptides showing where regions of alpha-helix or beta-sheet secondary structure might be located. (a) Rubredoxin. (b) Insulin, a protein containing two polypeptide chains. Note how disulfide linkages (—S—S—) may help in dictating folding patterns. (c) Ribonuclease, a large protein with several regions of alpha helix and beta sheet.

type of folding being determined by the available hydrogen-bonding opportunities (which is ultimately dictated by the primary structure—the amino acid sequence—of the polypeptide). Since β-sheet secondary structure generally yields a rather rigid structure while α-helical secondary structures are usually more flexible, the secondary structure of a given polypeptide will to some degree dictate a functional role for the protein in the cell. Many polypeptides fold into two or more segments, each displaying α-helix or β-sheet secondary structure. These segments, referred to as *domains*, represent regions of the polypeptide that have specific functions in the final protein molecule.

Structure of proteins: tertiary and quaternary structure

Once a polypeptide has achieved a given secondary structure it folds back upon itself to form an even more stable molecule. This folding leads to the formation of the **tertiary structure** of the protein. Like secondary structure, tertiary structure of a protein is ultimately determined by primary structure, but tertiary structure is also governed to some extent by the secondary structure of the molecule. As a result of the formation of secondary structure, the side chain of each amino acid in the polypeptide is positioned in a specific way. If additional hydrogen bonds, covalent bonds, or other atomic interactions are able to form, the polypeptide will fold to accommodate them and attain a unique three-dimensional shape (Figure 2.16).

Frequently a polypeptide will fold in such a way that adjacent sulfhydryl (—SH) groups of cysteine residues are exposed (Figure 2.16b). These free —SH groups can join covalently to form a disulfide (—S—S—) bridge between the two amino acids. If the two cysteine residues are located in different polypeptide chains of a protein, the disulfide bond physically links the two molecules (Figure 2.16b). By contrast, a single polypeptide can undergo folding itself if two cysteine residues form a disulfide linkage within the molecule.

FIGURE 2.17 Quaternary structure of hemoglobin, a protein containing four polypeptide subunits. There are two kinds of polypeptide in hemoglobin, α chains (top) and β chains bottom. Separate colors are used to distinguish the four chains.

The tertiary folding of the polypeptide ultimately forms exposed regions or grooves in the molecule (Figures 2.16 and 2.17) which may be of importance in binding other molecules (for example, the binding of a substrate to an enzyme, see Section 4.3).

If a protein consists of more than one polypeptide, and many proteins do, the arrangement of polypeptide subunits to form the final protein molecule is referred to as the **quaternary structure** of the protein (Figure 2.17). It should be remembered that in proteins showing quaternary structure, each subunit of the final protein will itself contain primary, secondary, and tertiary structure. Some proteins displaying quaternary structure contain many identical subunits; others contain several nonidentical subunits, while still others may contain more than one identical subunit and a second nonidentical subunit in the final protein molecule. The subunits of multisubunit proteins are held together either by noncovalent interactions (hydrogen bonding, van der Waals forces, or hydrophobic interactions) or by covalent linkages, generally intersubunit disulfide bonds.

Denaturation of proteins

When proteins are exposed to extremes of heat or pH, or to certain chemicals which affect their folding properties, they are said to undergo **denaturation** (Figure 2.18). In general, the biological properties of a protein are lost when it is denatured. When proteins are denatured, peptide bonds are generally unaffected, and the sequence of amino acids (primary structure) in the polypeptide remains unchanged. However, denaturation causes the polypeptide chain to unfold, destroying higher order structure of the molecule, in particular hydrogen bonds. The denatured polypeptide retains its primary structure because it is held together by covalent peptide bonds. Depending upon the severity of the denaturing conditions, refolding of the polypeptide may occur after removing the denaturant (Figure 2.18). However, the fact that denaturation is generally associated with the loss of biological activity of the protein, clearly shows that biological activity is not inherent in the primary structure of proteins, but instead is a result of the unique folding of the molecule as ultimately directed by primary structure. Folding of a polypeptide therefore accomplishes two things: (1) the polypeptide obtains a unique shape which is compatible with a *specific* biological function, and (2) the folding process converts the molecule into its most chemically stable form.

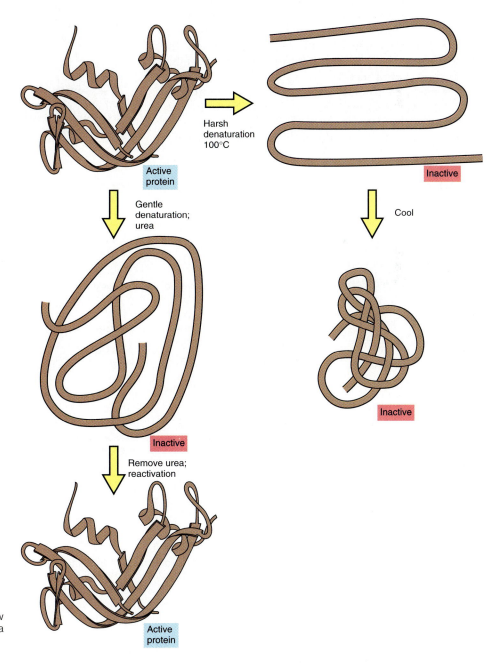

Active protein

Harsh denaturation 100°C

Inactive

Gentle denaturation; urea

Cool

Inactive

Inactive

Remove urea; reactivation

Active protein

FIGURE 2.18
Denaturation of a protein. Note how harsh denaturation generally yields a permanently destroyed molecule.

The nature of a cell depends on the number and kinds of its various protein constituents and an understanding of protein structure is essential for an understanding of cell function. The building blocks of proteins are amino acids, which combine to form polypeptides. The primary structure of a protein is determined by its amino acid sequence, but it is the folding of the polypeptide that determines how the protein will function in the cell.

2.9 Stereoisomerism

It was mentioned in reference to carbohydrates in Section 2.5 that two molecules may have the same molecular formula but exist in different structural forms. These related, but not identical molecules are referred to as **isomers** (Figure 2.19). Isomers are important in biology, especially in the chemistry of sugars. For example, *Escherichia coli* grows well on glucose as a carbon and energy source, but will not grow on the closely related sugar allose (Figure 2.19*a*), presumably because it cannot incorporate or metabolize this hexose isomer. Many isomers of common sugars are found as constituents of the cell walls of Bacteria (Section 3.5) and Archaea (Section 20.2). Sugar isomers are also known which contain both the same molecular

and structural formulas except that one is the "mirror image" of the other, just as the left hand is the mirror image of the right. In carbohydrate chemistry two identical sugars which are mirror images of one another are called **stereoisomers** or **enantiaomers** and have been given the designation D and L (Figure 2.19*b*). D-sugars predominate in biological systems.

Stereoisomerism is also important in protein chemistry. As with sugars, amino acids can exist as D or L stereoisomers. However, in the case of protein, life has evolved to use the L form rather than the D (Figure 2.19*c* and see box). D-amino acids are found occasionally in nature, most commonly in the cell wall polymer, peptidoglycan (see Section 3.5), and in certain peptide antibiotics (see Section 9.17). Prokaryotes are equipped to handle the conversion of D-amino acids to the L form by way of enzymes that specifically catalyze this transformation. Cells contain enzymes called **racemases** whose function is to convert the unusual form (L-sugar or D-amino acid) into the readily metabolizable form, the D-sugar or L-amino acid.

The property of stereoisomerism has profound biological implications. Amino acids in cells are of the L stereoisomer whereas sugars are generally of the D stereoisomer.

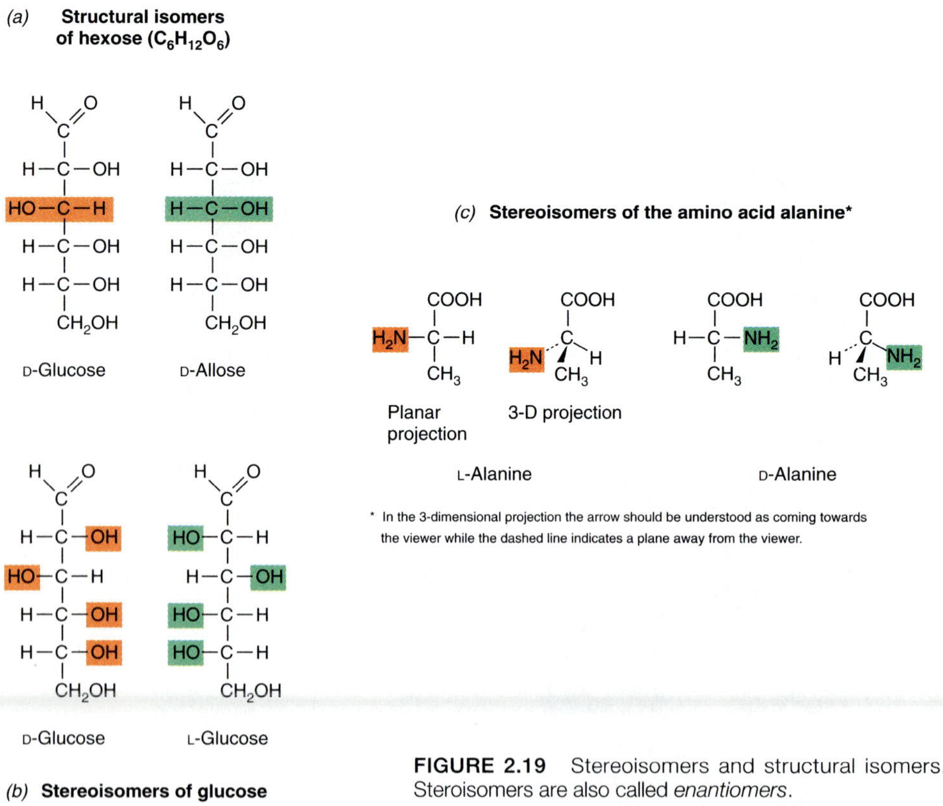

(a) **Structural isomers of hexose ($C_6H_{12}O_6$)**

(c) **Stereoisomers of the amino acid alanine***

Planar projection   3-D projection

L-Alanine   D-Alanine

* In the 3-dimensional projection the arrow should be understood as coming towards the viewer while the dashed line indicates a plane away from the viewer.

(b) **Stereoisomers of glucose**

FIGURE 2.19 Stereoisomers and structural isomers. Stereoisomers are also called *enantiomers*.

Chemical Isomers and Living Organisms

Louis Pasteur was trained as a chemist and his first research work was on the chemistry of stereoisomers. He was one of the first people to recognize the significance of stereoisomers in living organisms, and his interest in this problem led him into biology, a field that would occupy him for the rest of his life. For Pasteur, the fact that living organisms discriminated between stereoisomers was of profound philosophical significance. He saw life processes as inherently asymmetric, whereas nonliving chemical processes were not asymmetric. Pasteur's vision of the asymmetry of life has been well confirmed over the more than 100 years since he did his work. In fact, when Pasteur began to study spontaneous generation (see Chapter 1), he used his knowledge of the asymmetry of living organisms to bolster his confidence that spontaneous generation could not exist.

Why is it that the amino acids of proteins in living organisms are of only the L configuration, whereas when the same amino acids are made chemically they consist of mixtures of D and L forms? We know that this is because chemical reactions in living organisms are carried out by asymmetric catalysts, the **enzymes**. Asymmetry begets asymmetry, and since enzymes are asymmetric, they introduce asymmetry into the molecules they produce. (It should be noted that D-amino acids are found in a few biological molecules, most notably bacterial cell walls, but are absent from cellular proteins.)

The above statement begs the question, of course, of how the *first* asymmetry arose in the living world. Was it by chance or by design? Is there somewhere in the universe a planet with living organisms whose proteins are comprised of D-amino acids? A planet, perhaps, where everything got started along the opposite asymmetry to that of life on earth, and in which life is now trapped forever in this opposite asymmetry? Possibly. We will return to the whole question of evolution and the origin of life in Chapter 18.

Isomers such as the L-amino acids will gradually, with time, racemize spontaneously, resulting in a mixture of D and L forms. Organic material derived from dead organisms, occurring as part of fossils, gradually becomes a racemic mixture. In a million years or so, an equal amount of D- and L-amino acids will be left. The amino acids of fossils consist of variable proportions of L- and D-amino acids, the L/D ratio depending upon how long the organism has been dead. And, from knowledge of the rate at which amino acids spontaneously racemize, the L/D ratio of a fossil can actually be used as a way of obtaining its approximate age.

Study Questions

1. What is the relationship of protons to neutrons and electrons in atoms? Which of these particles carries an electrical charge?

2. Compare and contrast covalent and noncovalent bonds. Give three examples of noncovalent interactions.

3. What is a molecule? How many atoms are in a molecule of hydrogen gas? In a molecule of glucose?

4. Why are the weak noncovalent bonds of special importance for the molecules of living organisms?

5. Compare and contrast *monomer* and *polymer*. Give three examples of biologically important polymers and list the key constituents of each.

6. Polymers are sometimes called macromolecules. Why? Contrast *informational* and *noninformational* macromolecules. Give two examples of each kind.

7. RNA and DNA are similar types of macromolecules but they show distinct differences. List three ways in which RNA differs chemically or physically from DNA.

8. *Complementarity* is an important feature of the two polynucleotides of the DNA double helix. What is meant by complementarity and how is it expressed chemically in DNA? Are the two DNA molecules held together by covalent or noncovalent bonds?

9. Give at least three reasons why water is an ideal solvent in biological systems.

10. Write a general structure for an *amino acid* and explain how the R groups of the amino acid differ. What is the importance of R groups to final protein structure? Show how two amino acids can be connected in the formation of a *peptide bond*. Does the peptide bond involve covalent or noncovalent interactions?

11. Define each of the following as it relates to protein structure: primary, secondary, tertiary, quaternary.

12. Describe briefly the effects of heat on protein structure. Heat sensitivity of proteins is one reason why most living organisms do not live at high temperatures, yet a few prokaryotes not only live but thrive at temperatures up to boiling. Can you offer an explanation, based on your knowledge of protein structure? (You should be able to refine your answer after covering the material in Chapter 9.)

Supplementary Readings

Alberts, B., D. Bray, J. Lewis, M. Raff, K. Roberts, and **J. D. Watson**. 1989. *Molecular Biology of the Cell*. Second edition. Garland Publishing, Inc., New York. An excellent textbook of cell biology with emphasis on the molecular aspects of cells. Chapters 1–3 give a detailed account of cell chemistry and macromolecular structure. Highly recommended for advanced reading.

Mathews, C. K., and **K. E. van Holde.** 1990. *Biochemistry*. Benjamin-Cummings Publishing Co., Redwood City, CA. A detailed treatment of biochemistry.

Stryer, L. 1988. *Biochemistry*. Third edition. W. H. Freeman and Company, San Francisco. One of the best general biochemistry texts with excellent art work making the material come alive. Highly recommended.

Watson, J. D., N. H. Hopkins, J. W. Roberts, J. Argetsinger Steitz, and **A. M. Weiner**. 1987. *Molecular Biology of the Gene*. Fourth edition. Benjamin/Cummings Publishing Co., Redwood City, CA. Chapters 4 and 5 consider aspects of cell chemistry, in particular nucleic acid structure and functions.

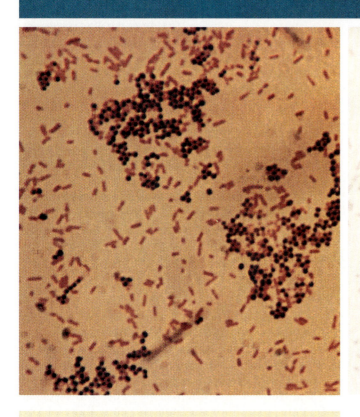

3

Cell Biology

In this chapter we present current understanding of structure and function relationships in microbial cells. We emphasize the biology of the prokaryotic cell, in particular, cells of Bacteria, but we compare and contrast prokaryotes with eukaryotes and discuss in detail a few eukaryotic cell structures.

Cells, like houses, are built by connecting simple building blocks in various ways to create more complex structures. We discussed the chemical nature of cellular building blocks in Chapter 2 and emphasized how these simple structures could be polymerized to form macromolecules. Cells are basically assemblages of macromolecules. Despite great diversity in the chemical composition of the macromolecules found in different cells, from a structural perspective, all cells solve a number of biological problems in common ways. For example, virtually all cells employ a lipid bilayer as the structural foundation of the cytoplasmic membrane. Although the bilayer may vary in its exact chemical composition from species to species, all cytoplasmic membranes have the same basic structure and all serve as permeability barriers. Also, ribosomes, the structures on which proteins are synthesized in all organisms, differ somewhat in chemical detail from species to species but not in cellular function. Thus, *unity* exists in many structure and function relationships in a wide variety of cells.

Because cells are microscopic, we begin this chapter with a discussion of microscopes and microscopy. Historically it was the microscope that first revealed the secrets of cell structure, and even today it remains a powerful tool in studies of cell biology.

3.1 Microscopes and Microscopy

Microscopic examination of microorganisms makes use of either the **light microscope** or the **electron microscope**. For most routine use, the light microscope is used, whereas for special research purposes, especially in studies on internal cell structure, the electron microscope is used in addition to the light microscope. All microscopes employ the principle that specific lenses will magnify the image of a cell such that details of its structure are most apparent. In addition to magnification, however, is *resolution*, the ability to distinquish two adjacent points as separate. Although magnification can be increased virtually without limit, resolution cannot; resolution is dictated by the physical properties of light. It is thus resolution and not magnification that ultimately defines the limits of what we are able to see with a microscope. We begin our discussion with the light microscope, for which the limits of resolution are about 0.2 μm (200 nm), and then proceed to describe the electron microscope, for which resolution is improved over that of the light microscope by about 1000-fold.

The compound light microscope

The **light microscope** has been of crucial importance for the development of microbiology as a science and remains a basic tool of routine microbiological research. Several types of light microscopes are commonly used in microbiology: *bright-field, phase-contrast,* and *fluorescence.* The **bright-field microscope** is most commonly used in elementary biology and microbiology courses and consists of two series of lenses (objective lens and ocular lens) that function together to resolve the image (Figure 3.1) With this microscope, specimens are visualized because of the differences in contrast that exist between them and the surrounding medium. Contrast differences arise because cells absorb or scatter light in varying degrees. Many bacterial cells are difficult to see well with the bright-field microscope because of their lack of contrast with the surrounding medium. Pigmented organisms may be an exception, however, because the color of the organism adds contrast, thus improving visualization of the cells (Figure 3.2).

The **phase-contrast** microscope was developed to improve contrast differences between cells and the surrounding medium to make it possible to see cells without staining (Figure 3.3). It is based on the principle that cells differ in refractive index from their surrounding medium and hence bend some of the light rays that pass through them. This difference can be used to create an image of much greater contrast than can be obtained in the bright-field microscope alone. The phase-contrast microscope is widely employed in research and teaching applications because it can be used to observe wet mount (living) preparations as well as stained preparations.

The **fluorescence microscope** is used to visualize specimens that fluoresce, that is, emit light of one color when light of another color shines upon them. Fluorescence occurs either because of the presence within cells of naturally fluorescent substances (for ex-

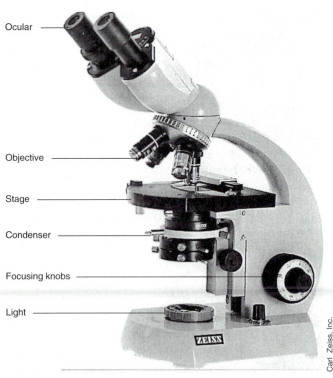

FIGURE 3.1 A modern compound light microscope. Various key parts of the microscope are labeled.

Ocular

Objective

Stage

Condenser

Focusing knobs

Light

Carl Zeiss, Inc.

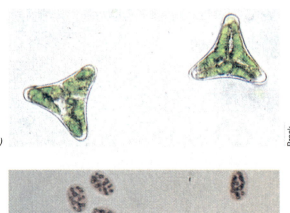

(a)

Brock

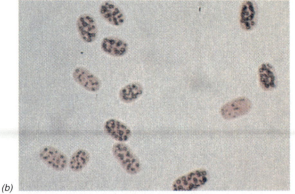

(b)

Norbert Pfennig

FIGURE 3.2 Photomicrographs of pigmented microorganisms by bright-field microscopy. (a) A green alga. (b) A purple phototrophic bacterium. The algal cells are about 15 μm in diameter, and the bacterial cells are about 5 μm in diameter.

Miniglossary for Chapter 3

ACTIVE TRANSPORT an energy-dependent transport process in which a substance is transported across the cytoplasmic membrane at the expense of ATP or a membrane potential

CHEMOTAXIS movement of an organism toward (*positive*) or away from (*negative*) a chemical gradient

CHLOROPLAST the chlorophyll-containing photosynthetic organelle of eukaryotic photosynthetic organisms

CHROMOSOME a DNA molecule, usually circular in prokaryotes and linear in eukaryotes, carrying genes essential to cellular function

CYTOPLASMIC MEMBRANE the permeability barrier of the cell, separating the cytoplasm from the environment

ENDOSPORE a highly heat resistant, thick-walled, differentiated cell produced by certain Gram-positive bacteria

EUKARYOTE a cell containing a membrane-bound nucleus and usually other organelles

FLAGELLUM a long, thin cellular appendage capable of rotation in prokaryotic cells and responsible for swimming motility

GAS VESICLES gas-filled cytoplasmic structures bounded by protein and conferring buoyancy upon cells

GRAM-NEGATIVE a prokaryotic cell whose cell wall contains relatively little peptidoglycan but contains an outer membrane composed of lipopolysaccharide (LPS), lipoprotein, and other complex macromolecules

GRAM-POSITIVE a prokaryotic cell whose cell wall consists chiefly of peptidoglycan and lacks the outer membrane of Gram-negative cells

GROUP TRANSLOCATION an energy-dependent transport process in which the substance transported is chemically modified during the transport process

LIPOPOLYSACCHARIDE (LPS) lipid in combination with polysaccharide and protein forming the major portion of the cell wall in Gram-negative bacteria

MAGNETOSOMES particles of magnetite (Fe_3O_4) organized into nonunit membrane-bound structures in the cytoplasm of magnetotactic bacteria

MITOCHONDRION (MITOCHONDRIA) an organelle found in most eukaryotic cells in which respiration and energy generation occurs

NUCLEOID an aggregated state of the circular chromosome of prokaryotic cells

ORGANELLE a unit membrane-bound structure found in the cytoplasm of eukaryotic cells

PEPTIDOGLYCAN a polysaccharide composed of alternating repeats of acetylglucosamine and acetylmuramic acid with the latter in adjacent layers cross-linked by short peptides

PERIPLASM a gel-like region between the outer surface of the cytoplasmic membrane and the inner surface of the lipopolysaccharide layer of Gram-negative cells

POLY-β-HYDROXYBUTYRATE (PHB) a common storage material of prokaryotic cells consisting of a polymer of β-hydroxybutyrate or other β-alkanoic acids

PROKARYOTE a cell that lacks a membrane-bound nucleus and that usually has a single circular DNA molecule as its chromosome

PROTOPLAST an osmotically protected cell whose cell wall has been removed

RIBOSOME small particles composed of RNAs and proteins that function in protein synthesis

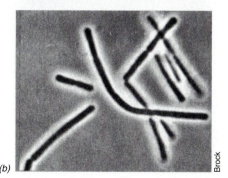

(a)

Brock

(b)

Brock

FIGURE 3.3 Photomicrographs of the same field of bacterial cells by (a) bright-field microscopy and (b) phase-contrast microscopy.

ample, chlorophyll, see Figure 3.71*a*) or because the cells have been treated with a fluorescent dye. Fluorescence microscopy is widely used in clinical diagnostic microbiology and also in microbial ecology (see Figures 13.8, 13.9, and 17.3)

Staining

Dyes can be used to stain cells and increase their contrast so that they can be more easily seen in the bright-field microscope. Dyes are organic compounds, and each class of dye has an affinity for specific cellular materials. Many commonly used dyes in microbiology are positively charged (cationic) and combine strongly with negatively charged cellular constituents such as nucleic acids and acidic polysaccharides. Examples of cationic dyes include *methylene blue*, *crystal violet*, and *safranin*. Since cell surfaces are generally negatively charged, these dyes combine with structures on the surfaces of cells and hence are excellent general stains.

The simplest staining procedures are done with dried preparations (Figure 3.4). A slide containing a dried suspension of microorganisms is flooded for a minute or two with a dilute solution of a dye, rinsed several times in water, and blotted dry. It is usual to observe dried stained preparations of bacteria with a high power (oil-immersion) lens (Figure 3.4).

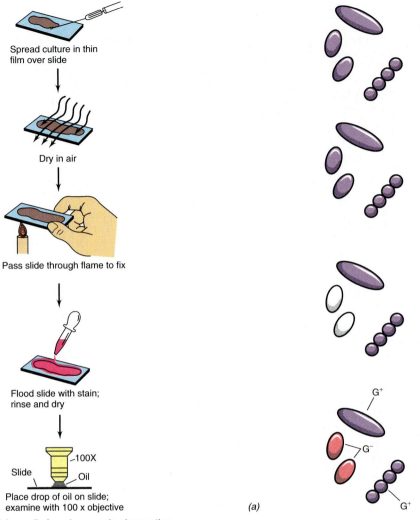

Spread culture in thin film over slide

Dry in air

Pass slide through flame to fix

Flood slide with stain; rinse and dry

Slide — 100X — Oil

Place drop of oil on slide; examine with 100 x objective

FIGURE 3.4 Staining cells for microscopic observation.

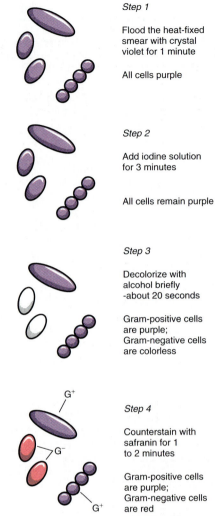

Step 1

Flood the heat-fixed smear with crystal violet for 1 minute

All cells purple

Step 2

Add iodine solution for 3 minutes

All cells remain purple

Step 3

Decolorize with alcohol briefly -about 20 seconds

Gram-positive cells are purple; Gram-negative cells are colorless

Step 4

Counterstain with safranin for 1 to 2 minutes

Gram-positive cells are purple; Gram-negative cells are red

G^+ G^- G^+

(a)

Differential stains are so named because they are used in procedures that do not stain all kinds of cells equally. An important differential staining procedure widely used in bacteriology is the *Gram stain* (Figure 3.5*a*). On the basis of their reaction to the Gram stain, bacteria can be divided into two major groups: **Gram-positive** and **Gram-negative**. After Gram staining, Gram-positive bacteria appear purple and Gram-negative bacteria appear red (Figure 3.5*b*). This difference in reaction to the Gram stain arises because of differences in the cell wall structure of Gram-positive and Gram-negative cells (as discussed later in this chapter). The Gram stain is one of the most useful staining procedures in the bacteriological laboratory; it is almost essential in identifying an unknown bacterium to know first whether it is Gram-positive or Gram-negative.

The electron microscope

Electron microscopes are widely used for studying the detailed structure of cells. To study the internal structure of cells, a **transmission electron microscope (TEM)** is essential. In the TEM, electrons are used instead of light rays and electromagnets function as

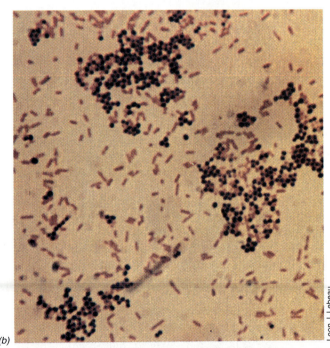

(b)

Leon J. Lebeau

FIGURE 3.5 The Gram stain. (a) Steps in the Gram stain procedure. (b) Photomicrograph of Bacteria which are Gram-positive (blue-purple) and Gram-negative (pink-red). The species are *Staphylococcus aureus* and *Escherichia coli*, respectively.

lenses, the whole system operating in a high vacuum (Figure 3.6). The resolving power of the electron microscope is much greater than that of the light microscope, and thus the electron microscope enables one to see many structures of even molecular size, such as proteins and nucleic acids (see Figure 3.65). However, electron beams do not penetrate very well, and if one is interested in seeing internal cell structure, even a single cell is too thick to be viewed directly. Consequently, special techniques of *thin sectioning* are needed to prepare specimens for the electron microscope. A single bacterial cell, for instance, is cut into many very thin slices that are then examined individually with the electron microscope (see Figure 3.7a). To obtain sufficient contrast, the preparations are treated with special electron-microscope stains such as osmic acid, permanganate, uranium, lanthanum, or lead. Because these substances are composed of atoms of high atomic weight, they scatter electrons well and thus improve contrast (see Figure 3.7a).

If only the *external* features of an organism need be observed, thin sections are not necessary, and intact cells can be examined directly with the **scanning electron microscope (SEM)** (Figures 3.6 and 3.7b). With this tool, the specimen is coated with a thin film of metal such as gold. An electron beam from the SEM is then directed down on the specimen and scans back and forth across it. Electrons scattered by the metal are collected, and they activate a viewing screen to produce an image (Figures 3.7b and 3.9). In the SEM, even fairly large specimens can be observed and the depth of field is extremely good. A wide range of magnifications can be obtained with the SEM, from as low as $15\times$ up to about $100,000\times$, but only the *surface* of an object can be visualized. All electron microscopes are fitted with cameras to allow a photograph, called an *electron micrograph*, to be taken.

> Microscopes are essential for microbiological studies. Various types of microscopes exist, including those that employ visible light (bright-field, phase-contrast, and fluorescence microscopes) or electron beams (transmission and scanning electron microscopes) as the illuminating source. In light microscopy, staining greatly increases the contrast of cells and the stains applied may be either simple or differential ones. Special procedures are necessary to prepare cells for viewing with the electron microscope.

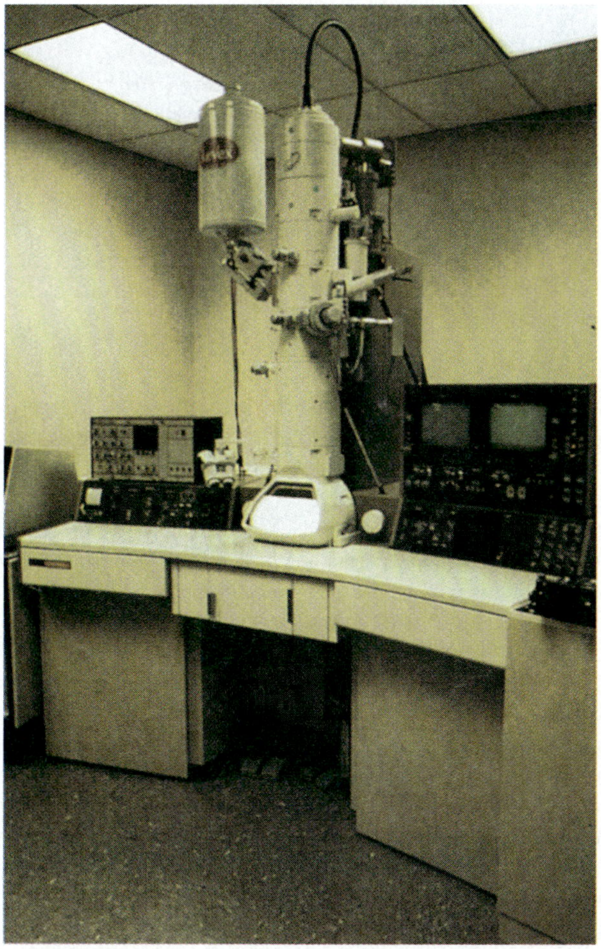

3.2 Overview of Cell Structure: Prokaryotes and Eukaryotes

Careful study with light and electron microscopes has revealed the detailed structure of cells, and we discuss here the major structures of each cell type.

The prokaryotic cell

A typical prokaryotic cell of either Bacteria or Archaea generally has the following major structures: cell wall, cytoplasmic membrane, ribosomes, inclusions, and nucleoid (Figure 3.7a). What are these structures and what are their functions?

The **cytoplasmic membrane** is the critical permeability barrier, separating the inside from the outside of the cell. The **cell wall** is a rigid structure outside the cytoplasmic membrane which provides support and protection from osmotic lysis. **Ribosomes** are small particles composed of protein and ribonucleic acid (RNA), which are visible in the transmission electron microscope. A single prokaryotic cell may have as many as 10,000 ribosomes. Ribosomes are part of the translation apparatus, and synthesis of cell proteins takes place upon these structures. (We discuss ribosomes, proteins, and RNA in detail in Chapter 5.) Prokaryotes occasionally contain **inclusions** consisting of storage material made up of compounds of carbon, nitrogen, sulfur, or phosphorus. Such inclusions can be formed when these nutrients are in excess in the environment and serve the cells as repositories of these nutrients when limitations occur.

The nuclear region of the prokaryotic cell differs quite significantly from that of the eukaryotic cell. Prokaryotic cells do not possess a true nucleus, the

FIGURE 3.6 A modern electron microscope. This instrument encompasses both transmission and scanning electron microscope functions.

JEOL Company

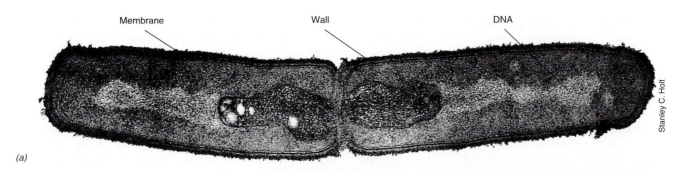

Membrane   Wall   DNA

(a)

Stanley C. Holt

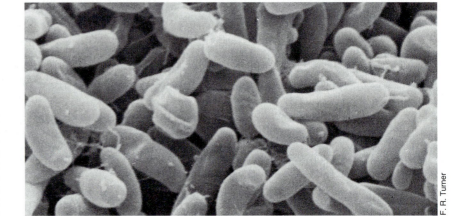

FIGURE 3.7
Electron micrographs of bacterial cells taken with the (a) transmission and (b) scanning electron microscopes. (a) Thin section of a typical Gram-positive bacterium, *Bacillus subtilis*. The cell has just divided and two membrane-containing structures are attached to the cross wall. Note the light region in the middle which is DNA. The cell is about 0.8 μm in diameter. (b) Cells of the phototrophic bacterium *Rhodospirillum sodomense*. A single cell is about 0.75 μm wide.

(b)

F. R. Turner

function of the nucleus being carried out by a single molecule of deoxyribonucleic acid (DNA). DNA is present in a more or less free state within the prokaryotic cell but is often seen in electron micrographs (Figure 3.7a) in an aggregated form referred to as the **nucleoid**. In analogy to the eukaryote, the DNA molecule of the prokaryote is called a **chromosome**.

Many, but not all, bacteria are able to move. Movement of a prokaryotic cell is usually by means of a structure called a **flagellum** (pural, **flagella**). Each flagellum consists of a single coiled tube of protein. The *rotation* of flagella propels the cell through liquids. Bacterial flagella are too small to be seen with the light microcope without special staining techniques, but are readily visible with the electron microscope (see Figures 3.38 and 3.39).

Morphology of prokaryotes

Several distinct shapes of bacteria can be recognized and have been given different names. Schematic examples of some of these bacterial shapes along with phase photomicrographs are shown in Figure 3.8. A bacterium that is spherical or ovoid in morphology is called a **coccus** (plural, **cocci**). A bacterium with a cylindrical shape is called a **rod**. Some rods are curved, frequently forming spiral-shaped patterns and are then called **spirilla**.

In many prokaryotes, the cells remain together in groups or clusters after division, and the arrangements in these groups are often characteristic of different organisms. For instance, cocci or rods may occur in long chains (Figure 3.9a). Some cocci form thin sheets of cells, while others occur in three-dimensional cubes

or irregular grapelike clusters (Figure 3.9b). Several groups of bacteria are immediately recognizable by their unusual shapes. Examples include **spirochetes**, which are coiled bacteria, and **appendaged bacteria**, which possess extensions of their cells as long tubes or stalks (Figure 3.8). It should be noted that the morphologies of prokaryotic cells shown in Figures 3.8 and 3.9 are *representative* ones; many variations of each of these basic morphological types have been found in newly isolated organisms.

The eukaryotic cell

Eukaryotic cells are larger and more complex in structure than prokaryotic cells, and a key difference is that eukaryotes contain *true nuclei*. The **nucleus** is a special membrane-enclosed structure within which DNA is located (Figure 3.10). The DNA in the nucleus is organized into **chromosomes**, structures that remain essentially invisible except at the time of cell division. Before cell division occurs, the chromosomes are duplicated and then condense, become thicker, and then undergo division as the nucleus divides. The process of nuclear division in eukaryotes is called **mitosis** (see Figure 3.68), which is a complex but highly organized process. Two identical daughter cells result from the division of one parent cell; each daughter cell receives a nucleus with an identical set of chromosomes.

Eukaryotic cells also contain distinct structures called **organelles**, within which important cellular functions occur (Figure 3.10). Organelles are absent from prokaryotes, although major physiological processes that take place in organelles, such as respiration or photosynthesis, may still occur in prokaryotic

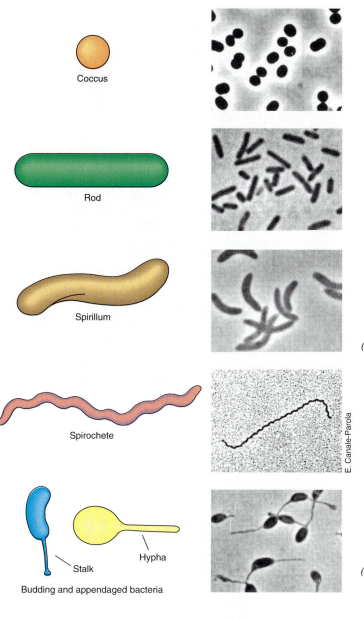

Coccus

Rod

Spirillum

Spirochete

Stalk   Hypha

Budding and appendaged bacteria

E. Canale-Parola

FIGURE 3.8
Representative cell shapes (morphology) in prokaryotes. Next to each drawing is a phase photomicrograph showing an example of that morphology. Organisms are coccus, *Thiocapsa roseopersicina* (diameter of a single cell = 1.5 μm); rod, *Desulfuromonas acetoxidans* (diameter = 1 μm); spirillum, *Rhodospirillum rubrum* (diameter = 1 μm); spirochete, *Spirochaeta stenostrepta* (diameter = 0.25 μm); budding/appendaged, *Rhodomicrobium vannielii* (diameter = 1.2 μm). Photos courtesy of Norbert Pfennig.

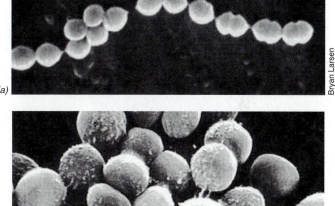

(a)   Bryan Larsen

(b)   A. Umeda

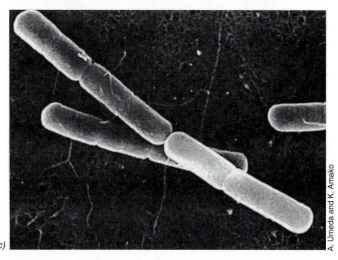

(c)   A. Umeda and K. Amako

FIGURE 3.9 Spherical and rod-shaped bacteria which associate in characteristically different ways as viewed by scanning electron microscopy. (a) *Streptococcus*, a chain-forming organism which divides only in one plane. (b) *Staphylococcus*, a coccus which divides in more than one plane. Cells of both organisms are about 1 μm in diameter. (c) Chains of the rod-shaped bacterium *Bacillus*. Cells are about 0.8 μm in diameter.

cells. One kind of organelle found in most eukaryotes is the **mitochondrion** (plural, **mitochondria**) (Figure 3.10; see also Figure 3.70). Mitochondria are the organelles within which the energy-generating functions of the cell occur. The energy produced in mitochondria is then used throughout the cell.

Algae are eukaryotic microorganisms that carry out the process of *photosynthesis*. In these organisms, as well as in green plants, an additional type of organelle is found: the **chloroplast**. The chloroplast is green and is the site where chlorophyll is localized and where the light-gathering functions involved in photosynthesis occur (see Figure 3.2a).

Size of microbial cells and the significance of smallness

Prokaryotes vary in size from cells as small as 0.1–0.2 μm in width to those more than 50 μm in diameter; a few very large prokaryotes, such as the surgeonfish symbiont *Euplopiscium fishelsoni*, are known to be

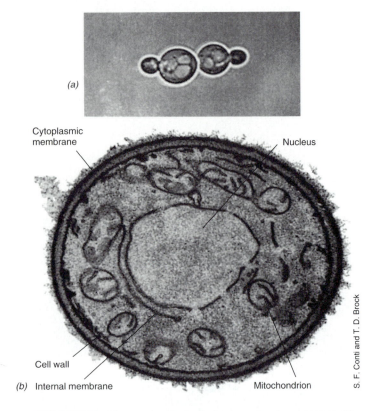

(a)

Cytoplasmic
membrane

Nucleus

Cell wall

(b) Internal membrane

Mitochondrion

S. F. Conti and T. D. Brock

FIGURE 3.10   The yeast cell, a typical eukaryotic microorganism. (a) Photomicrograph of a yeast cell in the process of budding. (b) Electron micrograph of a thin section through a yeast cell, showing various structures. A single cell is about 8 μm in diameter.

The small size of prokaryotes also affects their ecology and evolution. We discuss in Chapter 17 how high cell numbers of rapidly metabolizing prokaryotes can catalyze major physiochemical changes in an ecosystem. Such changes may have dramatic effects on other organisms, both microorganisms and macroorganisms, that constitute the ecosystem. Furthermore, the evolutionary process, driven by mutations and natural selection in all living systems, can proceed more quickly in prokaryotes than in eukaryotes. This is because the large cell populations typical of prokaryotes coupled with their haploid genetic makeup (a *single* copy of each gene is present, unlike most eukaryotic cells which have *two* copies of each gene) provide ideal conditions for rapid evolutionary change. Later chapters will show several examples of the rapidity with which evolution can occur in bacteria, and it should be recalled there that the small size and genetic properties of the prokaryotic cell is what sets the stage for these events.

We now consider several cell structures in detail. Our objectives are to describe the chemical building blocks of these cell structures and to discover how they are connected in a way that leads to a defined cellular function. We begin with the cell membrane, a structure that is critical to life processes in the cell.

50–100 μm in diameter and over 0.5 mm in length (Figure 3.11). However, the dimensions of an average rod-shaped prokaryote, the bacterium *Escherichia coli*, for example, are about 1 × 3 μm (Figure 3.11). Typical eukaryotic cells may be 2 to over 200 μm in diameter. Thus, prokaryotes are very small cells compared to eukaryotes, and the small size of prokaryotes actually dictates a number of their biological properties. For example, the rate at which nutrients and waste products pass into and out of a cell, a factor that can greatly affect cellular metabolic rates and growth rates, is in general *inversely* proportional to cell size. This is because transport rates are to some degree a function of the amount of *membrane surface* available, and relative to cell volume, small cells have more surface available than do large cells. This point can be seen most readily in the case of a sphere, in which the *volume* is a function of the cube of the radius ($V = \frac{4}{3}\pi r^3$), while the *surface area* is a function of the square of the radius ($SA = 4\pi r^2$). The surface to volume ratio of a sphere can thus be expressed as $3/r$ (Figure 3.12). A cell with a smaller r value therefore has a *higher* ratio of surface area to volume than a larger cell and thus can have more efficient exchange with its surroundings than a large cell. This advantage of the small cell frequently translates into more rapid growth rates and larger populations of prokaryotic cells than eukaryotic cells in typical microbial habitats.

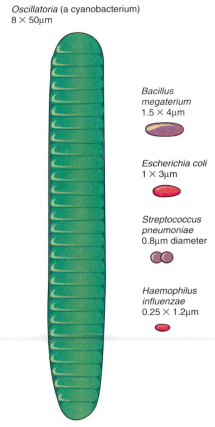

Oscillatoria (a cyanobacterium)
8 × 50μm

Bacillus megaterium
1.5 × 4μm

Escherichia coli
1 × 3μm

Streptococcus pneumoniae
0.8μm diameter

Haemophilus influenzae
0.25 × 1.2μm

FIGURE 3.11   Comparisons of sizes of a variety of prokaryotic microorganisms.

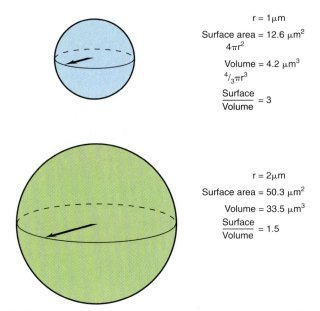

FIGURE 3.12   As a cell increases in size, its surface area to volume ratio decreases.

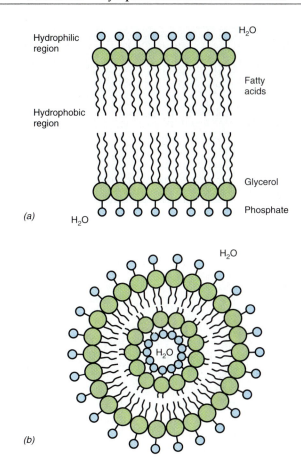

FIGURE 3.13   (a) Fundamental structure of phospholipid bilayer. (b) Phospholipid membrane vesicle—a stable arrangement of lipids in water.

> Although prokaryotes and eukaryotes are distinguished by nuclear structure, other important differences exist between these two cell types. Prokaryotes are smaller in size than eukaryotes, and eukaryotes contain a membrane-bound nucleus and organelles within which many important functions are carried out. The most prominent eukaryotic organelles are the mitochondrion, involved in energy generation, and the chloroplast, involved in photosynthesis. The small size of prokaryotic cells affects their physiology, growth rate, ecology, and evolution.

3.3   Cytoplasmic Membrane: Structure

The **cytoplasmic membrane** is a thin structure that completely surrounds the cell. Only 8 nm thick, this vital structure is the critical barrier separating the inside of the cell (the cytoplasm) from its environment. If the membrane is broken, the integrity of the cell is destroyed, the internal contents leak into the environment, and the cell dies. The cytoplasmic membrane is also a *highly selective barrier*, enabling a cell to concentrate specific metabolites and excrete waste materials.

Chemical composition of membranes

The general structure of most biological membranes is a **phospholipid bilayer** (Figure 3.13). As discussed in Section 2.6, phospholipids contain both highly hydrophobic (fatty acid) and relatively hydrophilic (glycerol) moieties, and can exist in many different chemical forms as a result of variation in the nature of the fatty acids or phosphate-containing groups attached to the glycerol backbone. As phospholipids aggregate in an aqueous solution, they tend to form bilayer structures spontaneously—the fatty acids point inward toward each other in a hydrophobic environment, while the hydrophilic portions remain exposed to the aqueous external environment (Figure 3.13).

In experimental studies of membrane function, **membrane vesicles** have been used extensively as models of the cytoplasmic membrane. Membrane vesicles are miniature versions of the cytoplasmic membrane, except that they do not surround all the cytoplasmic components of the living cell, but instead trap only a small amount of cytoplasm when they form from a larger membrane. Membrane vesicles form spontaneously when a bacterial cell is broken under carefully controlled conditions. The membrane fragments generated during cell breakage aggregate in spherical structures because of the hydrophobic/hydrophilic bonding of phospholipids (Figure 3.13b). Therefore the bilayer character of membranes probably represents the most stable arrangement of lipid molecules in an aqueous environment.

Thin sections of the cytoplasmic membrane can be visualized with the electron microscope; a representative example is seen in Figure 3.14a. To prepare the membrane for electron microscopy, cells must first be treated with osmic acid or some other electron-dense material that combines with hydrophilic components of the membrane (Figure 3.14b). By careful high-resolution electron microscopy, the cytoplasmic membrane appears as two light-colored lines separated by

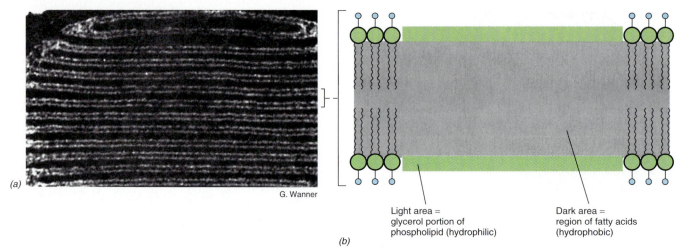

G. Wanner

(a)

Light area =
glycerol portion of
phospholipid (hydrophilic)

Dark area =
region of fatty acids
(hydrophobic)

(b)

FIGURE 3.14 The cytoplasmic membrane. (a) Electron micrograph of photosynthetic membrane stacks that derive from the cytoplasmic membrane in the phototrophic bacterium *Ectothiorhodospira halochloris*. Note the distinct lipid bilayers. Each bilayer is about 8 nm thick. (b) Enlarged schematic view of a single membrane bilayer shown in (a).

a darker area (Figure 3.14*a*). This unit membrane consists of a phospholipid (see Figure 2.7) bilayer and proteins embedded within it (Figure 3.15). The major proteins of the cell membrane generally have very hydrophobic external surfaces in the regions of the protein that make intimate association with the highly nonpolar fatty acid chains. It is thought that virtually all integral membrane proteins actually span the bilayer and thus have surfaces exposed on both the inside and the outside of the cell (Figure 3.15). Such proteins are referred to as *transmembrane* proteins. The

overall structure of the cytoplasmic membrane is stabilized by hydrogen bonds and hydrophobic interactions. In addition, cations such as Mg^{2+} and Ca^{2+} also help stabilize the membrane by combining ionically with negative charges of the phospholipids.

In some prokaryotes, complex molecules called *hopanoids* are present in the cytoplasmic membrane. Because of their structural similarity to the sterols of eukaryotic membranes (Figure 3.16), it is assumed that hopanoids function in a similar way to sterols to strengthen the prokaryotic cytoplasmic membrane.

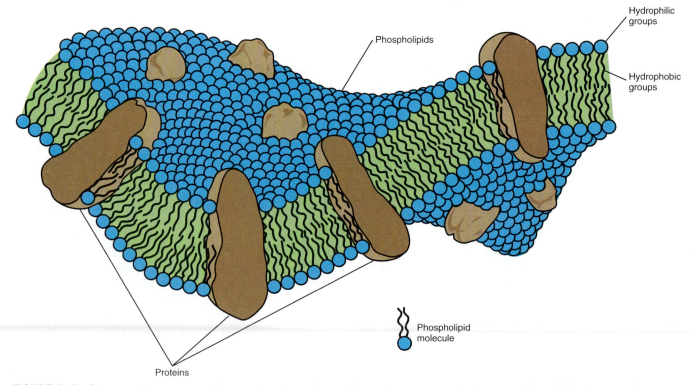

Phospholipids

Hydrophilic groups

Hydrophobic groups

Phospholipid molecule

Proteins

FIGURE 3.15 Diagram of the structure of the cytoplasmic membrane. The matrix is composed of phospholipids, with the hydrophobic groups directed inward and the hydrophilic groups toward the outside, where they associate with water. Embedded in the matrix are proteins that have considerable hydrophobic character in the region that traverses the fatty acid bilayer. Hydrophilic proteins and other charged substances, such as metal ions, may be attached to the hydrophilic surfaces. Although there are some chemical differences, the overall structure of the cytoplasmic membrane shown is similar in both prokaryotes and eukaryotes.

Eukaryotic cytoplasmic membranes

One major difference in chemical composition of membranes between eukaryotic and prokaryotic cells is that the eukaryotes have **sterols** in their membranes (Figure 3.16). Sterols are absent from the membranes of virtually all prokaryotes. Depending on the cell type, sterols can make up from 5 to 25 percent of the total lipids of eukaryotic membranes. Sterols are rigid, planar molecules, whereas fatty acids are flexible. The association of sterols with the membrane serves to stabilize its structure and make it less flexible. It has been shown that when artificial lipid bilayers are supplemented with sterols, they are much less leaky than when they are composed of pure phospholipid. Membrane rigidity may be necessary in eukaryotes because they are much larger than prokaryotes and thus must endure greater physical stresses on the membrane, necessitating a more rigid membrane structure in order to keep the cell stable and functional. One group of antibiotics, the polyenes (filipin, nystatin, and candicidin, for example) react with sterols and destabilize the membrane. It is of interest that these antibiotics are active against eukaryotes but generally do not affect prokaryotes, probably because the latter lack sterols in their membranes. However, some bacteria of the mycoplasma group (Section 19.27), which lack a cell wall, require sterol for growth. The sterol becomes incorporated into the cytoplasmic membrane and probably stabilizes the membrane structure. Mycoplasmas are inhibited by polyene antibiotics, in contrast to the insensitivity to these antibiotics displayed by all other prokaryotes.

Archaeal membranes

The lipids of Archaea are chemically unique. Unlike the lipids of Bacteria and Eucarya in which *ester* linkages bond the fatty acids to the glycerol molecule (Figure 3.17a; see also Section 2.6), archaeal lipids have *ether* linkages between glycerol and the hydrophobic side chains (Figure 3.17b). In addition, instead of containing fatty acids, archaeal lipids have side chains composed of repeating units of the branched-chain hydrocarbon, *isoprene* (Figure 3.17c). Because of these unique features, the structure of archaeal membranes differ somewhat from membranes composed of glycerol and fatty acids. We consider the details of the novel structure and properties of archaeal membranes in Chapter 20.

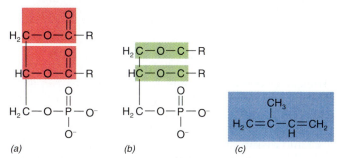

(a)   *(b)*   *(c)*

FIGURE 3.17 Chemical linkages in lipids. (a) The ester linkage, as found in the lipids of Bacteria and Eukarya. (b) The ether linkage of lipids from Archaea. (c) Isoprene, the parent structure of the hydrophobic side chains of archaeal lipids.

Other features of the cytoplasmic membrane

The outer surface of the cytoplasmic membrane faces the environment and in certain bacteria makes contact with a variety of proteins that serve to bind substrates or process large molecules for transport into the cell (periplasmic proteins; see discussion in Section 3.5). The inner side of the cytoplasmic membrane faces the cytoplasm and interacts with proteins involved in energy-yielding reactions and other important cellular functions. Some proteins, such as those in the periplasm (see Section 3.5) and some cytoplasmic proteins, may associate quite firmly with the surface of the membrane and actually function as if they were membrane bound proteins. Although not themselves integral membrane proteins, such proteins usually interact directly with integral membrane proteins in various cellular processes. Some of these *membrane-associated* proteins have been shown to be lipoproteins and contain a lipid tail on the amino terminus of the protein which serves to anchor the protein into the membrane.

Although appearing rigid when viewed diagrammatically (Figure 3.15), the cytoplasmic membrane is actually quite fluid; phospholipid and protein molecules have significant freedom to move about the membrane surface. Measurements of membrane viscosities indicate that membranes have a viscosity approximating that of a light grade oil, and thus the

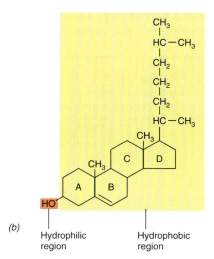

(a)

(b)

Hydrophilic region   Hydrophobic region

FIGURE 3.16 (a) The general structure of a sterol. All sterols contain the same four rings, labeled A, B, C, and D. (b) The structure of cholesterol.

movement of components within the membrane is not surprising. In summary, the cytoplasmic membrane can be thought of as a *fluid mosaic* in which globular proteins oriented in a specific manner span a highly mobile yet ordered phospholipid bilayer. This arrangement confers a number of important functional properties on membranes; we discuss these properties now.

> The cytoplasmic membrane, a highly selective barrier, is constructed principally of lipid, within which certain proteins are embedded. The hydrophilic (water-loving) portions of membrane lipids associate with the cytoplasm or are exposed to the environment and the hydrophobic (water-fearing) portions of the lipids aggregate together. This association of hydrophobic groups results in the formation of a thin but highly specific lipid bilayer.

3.4 Cytoplasmic Membrane: Function

The cytoplasmic membrane is not just the barrier separating inside from outside. It has a major role in cell function. It is through the cytoplasmic membrane that all nutrients must pass, and it is also through the cytoplasmic membrane that all waste products leave the cell. The term **permeability** is used to refer to the property of membranes that permits movement of molecules back and forth across the cytoplasmic membrane. However, most molecules that move through the membrane do not move passively; the cytoplasmic membrane is *selectively permeable*. Further, the cell itself plays an active role in the movement of molecules across the membrane; molecules are **transported** through the membrane. In the present section we discuss the permeability and transport properties of membranes.

Osmosis

The interior of the cell (the *cytoplasm*) consists of an aqueous solution of salts, sugars, amino acids, vitamins, coenzymes, and a wide variety of other soluble materials. In contrast, most bacterial habitats have solute concentrations considerably lower than that within the cell, which is approximately 10 mM (millimolar) (Figure 3.18*a*). Water naturally passes from regions of low solute concentration to regions of high solute concentrations in a process called **osmosis**. Thus, there is a constant tendency throughout the life of the cell for water to enter, and the cell would swell and burst were it not for the strength of the cell wall, to be described in the next section. If the solute concentration is *higher* in the medium than in the cell, water flows out, and the cell becomes dehydrated, a process called **plasmolysis** (Figure 3.18*b*).

The cytoplasmic membrane as a permeability barrier

The hydrophobic nature of the cytoplasmic membrane allows it to function as a tight barrier, so that the passive movement of polar solutes does not readily occur.

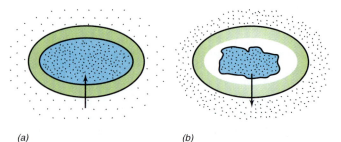

(a)   *(b)*

FIGURE 3.18 Osmosis and plasmolysis. (a) Normal situation with a bacterial cell containing a higher solute concentration than the surrounding medium. The tendency for water to enter the cell is counteracted by the cell wall. (b) Plasmolysis. The cell has become dehydrated by loss of water due to the high external solute concentration. Arrows indicate the direction of water flow.

Some small nonpolar and fat-soluble substances, such as fatty acids, alcohols, and benzene, may enter the cell readily by becoming dissolved in the lipid phase of the membrane. However, charged molecules, such as organic acids, amino acids, and inorganic salts, which are hydrophilic, do not readily pass the membrane barrier, but instead must be specifically transported. Even a substance as small as a hydrogen ion, H^+, does not readily breach the cytoplasmic membrane passively because it is always hydrated, occurring in solution as the charged hydronium ion (H_3O^+) (see Section 4.11 for a discussion of hydrogen ion transport). One molecule which does freely penetrate the membrane is water itself, which is sufficiently small and uncharged to pass between phospholipid molecules. The relative permeability of a few biologically important substances is shown in Table 3.1. As can be seen, most substances will not passively enter the cell, and thus transport processes are critical to cellular function.

Transport across biological membranes

Although intact membranes do not allow the free diffusion of various polar molecules such as sugars, amino acids, ions, and the like, these substances do pass through the membrane by other means and can be concentrated to over 1000 times that of the external environment through the action of **membrane trans-**

Table 3.1	Relative permeability rates of various biologically important molecules or ions through artificial lipid bilayers	
Substance	**Rate of permeability***
Water	100
Glycerol	0.1
Tryptophan	0.001
Glucose	0.001
Chloride ion (Cl⁻)	0.000001
Potassium ion (K⁺)	0.0000001
Sodium ion (Na⁺)	0.00000001

*Relative scale—permeability with respect to permeability of water, given as 100.

port proteins. Three classes of transport proteins have been identified. **Uniporters** are proteins that transport a substance from one side of the membrane to the other (Figure 3.19). The other two classes of transport proteins move the substance of interest across the membrane along with a second substance required for transport of the first. They are thus *cotransport* proteins. **Symporters** are membrane proteins that carry both substances across the membrane in the same direction (Figure 3.19). **Antiporters** transport one substance across the membrane in one direction while transporting the second substance in the opposite direction (Figure 3.19).

The necessity for carrier-mediated transport mechanisms in microorganisms can readily be appreciated. If diffusion were the only type of transport mechanism available, cells would not be able to acquire the proper concentrations of solutes. In diffusion processes, both the rate of uptake and the intracellular level are proportional to the external concentration. Carrier-mediated transport mechanisms overcome this problem by enabling the cell to accumulate solutes *against* a concentration gradient. As shown in Figure 3.20, carrier-mediated transport shows a saturation effect: if the concentration of substrate in the medium is high enough to saturate the carrier, which is frequently the case even at quite low substrate concentrations, the rate of uptake (and often the internal level as well) becomes maximal.

One characteristic of carrier-mediated transport processes is the *highly specific* nature of the transport event. The binding and carrying of a substance across the membrane resembles an enzyme reaction (see Section 4.3). Certain carrier proteins react with only a single kind of molecule, but many show affinities for a chemical *class* of molecules. For instance, there are carriers which transport certain, usually related, amino

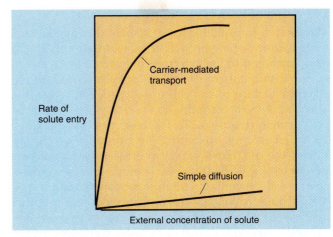

FIGURE 3.20 Relationship between uptake rate and external concentration in passive uptake and active transport. Note that in the carrier-mediated process the uptake rate shows saturation at relatively low external concentrations.

acids but not others. This economy in uptake reduces the need for separate transport proteins for every single amino acid.

Compounds called **ionophores** destroy the selective permeability properties of membranes. Ionophores are small hydrophobic molecules that dissolve in lipid bilayers and allow the passive diffusion of ionized substances into or out of the cell. Certain antibiotics are effective ionophores. *Valinomycin*, for example, acts as a potassium channel, transporting potassium down the electrochemical gradient until the concentration is equal on both sides of the membrane. *Tyrocidin*, another ionophore (Figure 3.21), is a small hydrophobic peptide that can form a transmembrane channel and allow massive leakage of various monovalent cations. Since the formation of concentration gradients are essential in cell function, ionophores are obviously lethal agents to cells.

> **Most chemicals do not move through the cytoplasmic membrane passively; rather they are transported. The cytoplasmic membrane is selectively permeable and contains specific carrier proteins that participate in the transport process.**

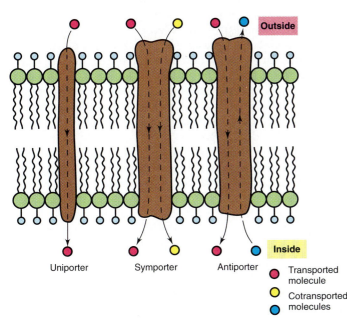

FIGURE 3.19 Schematic drawing showing the operation of various types of transport proteins.

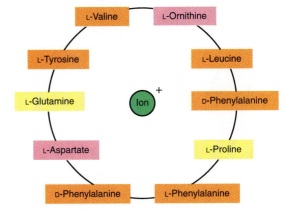

FIGURE 3.21 Structure of the peptide ionophore (and antibiotic) tyrocidin. Note the predominance of nonpolar and aromatic amino acids. The ion is carried in the center of the molecule.

The action of membrane transport proteins

Membrane transport proteins span the membrane, with portions of the protein being exposed to both the cytoplasm and to the external environment. Arranged in this fashion, it is possible for solutes bound on the external surface of the cell to be carried through the membrane by a conformational change in the transport protein (Figure 3.22). The conformational change serves as a "gate," releasing the substance on the inside of the cell. However, even if the substance is carried in by a specific transport protein, the law of mass action will prevent any significant concentration gradient from forming and the concentration of the solute inside the cell will be about the same as that outside the cell. This type of *carrier-mediated* diffusion, which does not involve expenditure of energy, is known as **facilitated diffusion**.

Most transport processes are linked to the expenditure of energy and result in a much higher concentration of the transported molecule inside than outside the cell. If a solute is transported by an energy-dependent process, then energy can be used to pump the solute *against* the concentration gradient. Energy can be derived from either high energy phosphate compounds, such as ATP (see Section 4.6), or by the dissi-

pation of a gradient of protons or Na^+ ions across the membrane. The ion gradients are themselves established during energy-releasing reactions in the cell (see Chapters 4 and 16) and can be used as a source of potential energy to drive the uptake of solutes against the concentration gradient (see below).

Two major mechanisms of energy-linked transport are known. **Group translocation** is the process whereby a substance is transported while simultaneously being chemically modified, generally by phosphorylation. Alternatively, in the process of **active transport** the substance can accumulate to high concentration in the cytoplasm in chemically unaltered form. Active transport requires energy and is linked to the dissipation of ion gradients mentioned previously or to ATP.

Group translocation

These are transport processes in which the substance is *chemically altered* in the course of passage across the membrane. Since the product that appears inside the cell is chemically different from the external substrate, no actual concentration gradient is produced across the membrane. The best studied cases of group translocation involve transport of the sugars glucose, mannose, fructose, *N*-acetylglucosamine, and β-glucosides, which are *phosphorylated* during transport by the **phosphotransferase system**.

The phosphotransferase system in the bacterium *Escherichia coli* is composed of 24 proteins, at least 4 of which are necessary to transport a given sugar. The proteins in the phosphotransferase system are themselves alternately phosphorylated in a cascading fashion until an integral membrane transport protein called Enzyme II_c receives the phosphate group and phosphorylates the sugar in the actual transport process (Figure 3.23). The high energy phosphate bond for the phosphotransferase system comes from a key metabolic intermediate called *phosphoenol pyruvate*. A small protein called HPr, the enzyme that phosphorylates it (Enzyme I), and Enzyme II_a are cytoplasmic proteins, whereas Enzyme II_b and II_c are membrane proteins (Figure 3.23). HPr and Enzyme I are nonspecific components of the phosphotransferase system and participate in all phosphotransferase reactions, whereas specific Enzymes II exist for the uptake of each individual sugar.

Besides actually transporting various sugars and sugar derivatives across the membrane, the phosphotransferase system also plays important roles in *regulating* transport of sugars. For example, in *E. coli*, if glucose and mannose (another sugar transported by group translocation) are simultaneously present, the phospho-carrier protein preferentially phosphorylates glucose. This spares the cell the task of transporting both sugars when one is adequate as a carbon or energy source. Other substances transported by group translocation processes include purines, pyrimidines, and fatty acids. However, many substances, including several sugars, are not taken up by the phosphotrans-

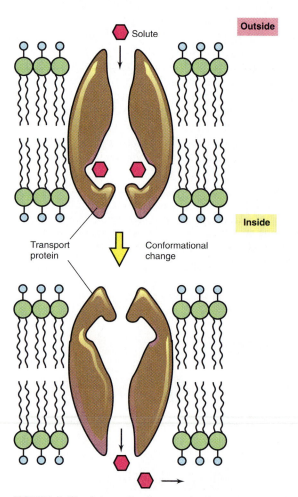

FIGURE 3.22 Schematic diagram showing how a conformational change in a transmembrane protein could serve to drive the transport event.

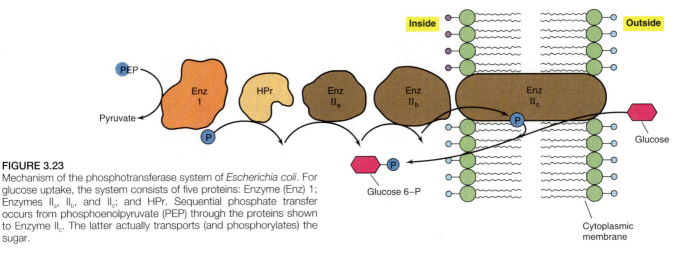

FIGURE 3.23
Mechanism of the phosphotransferase system of *Escherichia coli*. For glucose uptake, the system consists of five proteins: Enzyme (Enz) 1; Enzymes II$_a$, II$_b$, and II$_c$; and HPr. Sequential phosphate transfer occurs from phosphoenolpyruvate (PEP) through the proteins shown to Enzyme II$_c$. The latter actually transports (and phosphorylates) the sugar.

ferase system but instead are accumulated by the process of active transport.

Evidence supporting the mechanism of the phosphotransferase system has come from studies on membrane vesicles (Figure 3.13*b*). Vesicles will carry out transport reactions and have the virtue that substances can be placed inside the vesicles at the time of closure, so that movement in either direction can be studied. If a sugar is placed inside the vesicles along with the enzymes of the phosphotransferase system, it is not phosphorylated. However, if a sugar is applied externally, it becomes phosphorylated when it is transported into the vesicles. This experiment shows that phosphorylation occurs by the phosphotransferase system *during* the actual transport event.

Active transport

This is an energy-dependent pump in which the substance being transported combines with a membrane-bound carrier which then releases the *chemically unchanged* substance inside the cell. Since the substance is not altered during the transport process, if it is not consumed in cell reactions, its concentration inside may reach many times the external concentration. Substances transported by active transport include some sugars, most amino acids and organic acids, and a number of inorganic ions such as sulfate, phosphate, and potassium. Glucose is taken up by active transport processes in some bacteria and by the phosphotransferase system in others (Table 3.2).

As in any other pump, active transport requires that work be performed. In bacteria, the energy for driving the pump comes from ATP in the case of some transporters, or more commonly from the separation of hydrogen ions (protons) across the membrane, called the *proton motive force* (see Section 4.11). Energy released from the breakdown of organic or inorganic compounds, or from the energy of light, is used to establish a separation of protons across the membrane, with the proton concentration highest outside the cell and lowest inside (see Section 4.11). This results in an energized membrane as depicted in Figure 3.24. It is the electrochemical potential residing in the proton motive force that drives the uptake of nutrients by active transport (Figure 3.24). Each membrane carrier involved in active transport has specific sites for both its

Table 3.2	Mechanism of glucose uptake by various bacteria	
Phosphotransferase system	**Active transport**	
Escherichia coli	*Pseudomonas aeruginosa*	
Bacillus subtilis	*Azotobacter vinelandii*	
Clostridium pasteurianum	*Micrococcus luteus*	
Staphylococcus aureus	*Mycobacterium smegmatis*	

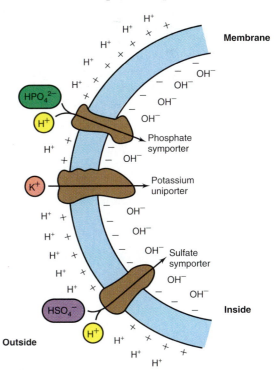

FIGURE 3.24   Use of ion separation in the proton motive force, in this case the separation of cytrons from hydroxyl ions across the membrane, to transport inorganic ions by specific transport proteins. Note that there is both a separation of protons and of electrical charge.

substrate (for example, glucose or potassium) and a proton (or protons). As the substrate is taken up, protons move across the membrane and the proton motive force is diminished (Figure 3.24).

The proton motive force serves as the energy link between membrane transporters and the metabolic machinery, making it possible for the carriers to "pump" nutrients inward. Cations, such as K⁺, may be actively transported into the cytoplasm by uniporters in response to the proton motive force, since the interior of the cell is negative when the membrane is energized (Figure 3.24). Uptake of anions probably occurs together with protons by symporters, so that it is effectively the undissociated acid that enters the cell (Figure 3.24). Transport of uncharged molecules such as sugars or amino acids can also be linked to the electrical charge differences: the symporter transports both the substrate and one or more protons. Proton pumps linked to transport are key constituents of all prokaryotic membranes and are also present in the inner membranes of mitochondria and chloroplasts.

Those substances taken up by active transport but not linked to dissipation of a proton gradient use the energy of ATP to drive the transport reaction. For example, in *E. coli*, lactose is actively transported at the expense of a proton motive force, while the related disaccharide maltose is actively transported at the expense of ATP.

Ion gradients in eukaryotes

In the cytoplasmic membrane of animal cells, a second type of ion pump is largely responsible for the energy required to transport nutrients by active transport. As we will see in Chapter 4, adenosine triphosphate (ATP) is a key energy carrier in cell function. An enzyme, the Na⁺/K⁺ ATPase, pumps Na⁺ out of the cell while pumping K⁺ in. A Na⁺/K⁺ gradient is thus established that, analogous to the proton separation of bacteria, represents work done on the system (Figure 3.25). ATP produced from energy-yielding reactions

in the cell powers the Na⁺/K⁺ pump and then the transport of solutes, glucose for example, is driven by the energy released during passage of Na⁺/K⁺ across the membrane. The glucose transport protein in this case could be a symporter which binds glucose and Na⁺ and carries them into the cell. The Na⁺/K⁺ pump also functions in the uptake of amino acids in animal cells.

Plant and algal cytoplasmic membranes also contain Na⁺/K⁺ pumps, however solute transport in plant cells and in algae is primarily driven by a proton pump.

> **Three kinds of specific transport processes have been recognized: facilitated diffusion, group translocation, and active transport. In facilitated diffusion, there is no accumulation against a concentration gradient, the carrier serving simply as a specific "gate," allowing only certain substances to pass. In group translocation, the compound transported is chemically modified, generally by phosphorylation. Active transport is an energy-dependent process in which the transported substance is chemically unchanged, but is pulled across the membrane by action of a proton motive force or other ion gradient.**

3.5 The Cell Wall of Bacteria: Structure and Function

Because of the concentration of dissolved solutes inside the bacterial cell, a considerable turgor pressure develops, estimated at 2 atmospheres in a bacterium like *Escherichia coli*; this is roughly the same as the pressure in an automobile tire. To withstand these pressures, bacteria contain **cell walls**, which also function to give shape and rigidity to the cell. The prokaryotic cell wall is difficult to visualize well with the light microscope, but can be readily seen in thin sections of cells with the electron microscope.

Bacteria can be divided into two major groups, called **Gram-positive** and **Gram-negative**. The original distinction between Gram-positive and Gram-negative was based on a special staining procedure, the *Gram stain* (see Section 3.1), but differences in cell wall structure are at the base of these differences in the Gram staining reaction. Gram-positive and Gram-negative cells differ markedly in the appearance of their cell walls, as is shown in Figure 3.26. The Gram-negative cell wall is a multilayered structure and quite complex, while the Gram-positive cell wall consists of primarily a single type of molecule and is often much thicker. Close examination of Figure 3.26 shows that there is also a significant textural difference between the surfaces of Gram-positive and Gram-negative Bacteria, as revealed by the scanning electron microscope.

Although various Archaea and even certain Eukarya stain Gram-positive or Gram-negative, the cell wall of Bacteria is chemically unique and of immense practical importance. For these reasons we focus in this section on the cell walls of Bacteria.

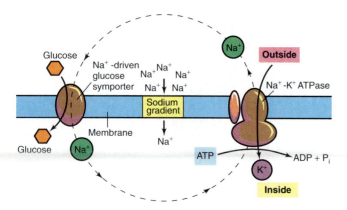

FIGURE 3.25 Transport of metabolites linked to the Na⁺/K⁺ pump in the membranes of animal cells. The ATPase pumps Na⁺ out of the cell while K⁺ is pumped in, and the transport reactions are driven by the energy released from the reaction ATP → ADP+Pᵢ.

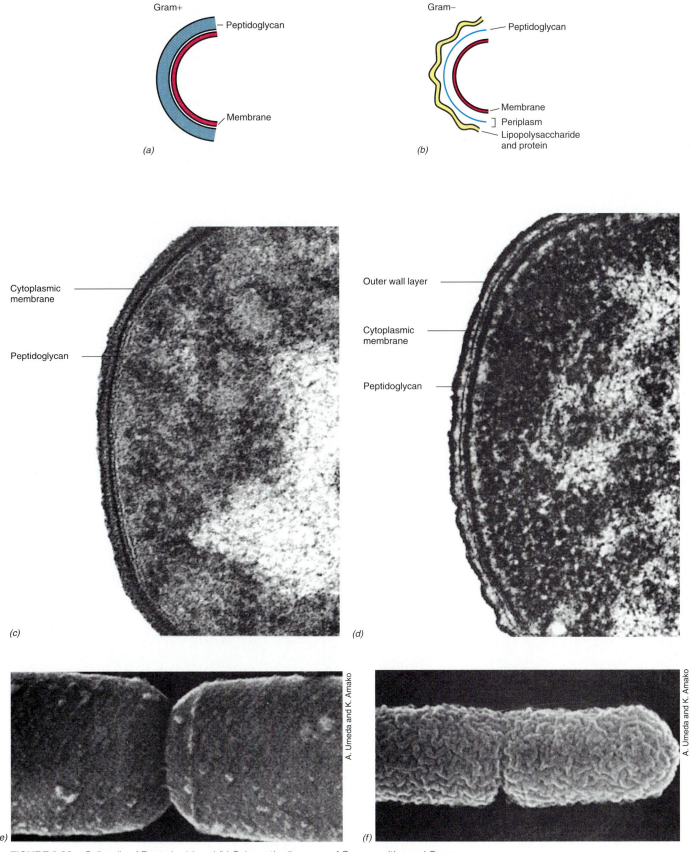

FIGURE 3.26   Cell walls of Bacteria. (a) and (b) Schematic diagrams of Gram-positive and Gram-negative cell walls. (c) Electron micrograph showing the cell wall of a Gram-positive bacterium, *Arthrobacter crystallopoietes*. (d) Gram-negative bacterium, *Leucothrix mucor*. (e) and (f) Scanning electron micrographs of Gram-positive (*Bacillus subtilis*) and Gram-negative (*Escherichia coli*) Bacteria. Note the surface texture in the cells shown in (e) and (f). A single cell of *B. subtilis* or *E. coli* is about 1 μm in diameter.

The rigid layer: peptidoglycan

In the cell walls of Bacteria there is one layer which is primarily responsible for the strength of the wall. In most Bacteria, additional layers are present outside this rigid layer. The rigid layer of both Gram-negative and Gram-positive Bacteria is very similar in chemical composition. Called **peptidoglycan** (or **murein**), this layer is a thin sheet composed of two sugar derivatives, N-acetylglucosamine and N-acetylmuramic acid, and a small group of amino acids, consisting of L-alanine, D-alanine, D-glutamic acid, and either lysine or diaminopimelic acid (DAP) (Figure 3.27). These constituents are connected to form a repeating structure, the *glycan tetrapeptide* (Figure 3.28).

The basic structure of peptidoglycan is in reality a thin sheet in which the glycan chains formed by the sugars are connected by *peptide cross-links* formed by the amino acids. The glycosidic bonds connecting the sugars in the glycan chains are very strong, but these chains alone cannot provide rigidity in all directions.

The full strength of the peptidoglycan structure is only realized when these chains are cross-linked by amino acids. This cross-linking occurs to characteristically different extents in different Bacteria, with greater rigidity coming from more complete cross-linking. In Gram-negative Bacteria, cross-linkage usually occurs by direct peptide linkage of the amino group of diaminopimelic acid to the carboxyl group of the terminal D-alanine (Figure 3.29a). In Gram-positive Bacteria, cross-linkage is usually by a peptide interbridge, the kinds and numbers of cross-linking amino acids varying from organism to organism. In *Staphylococcus aureus*, the best studied Gram-positive organism, each interbridge peptide consists of five molecules of the amino acid glycine connected by peptide bonds (Figure 3.29b). The overall structure of a peptidoglycan molecule is shown in Figure 3.29c. The shape of a cell is thought to be determined by the lengths of the peptidoglycan chains and by the manner and extent of cross-linking of the chains.

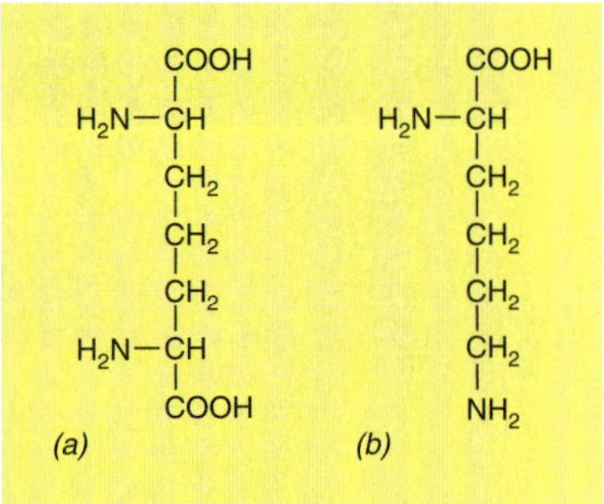

Diversity in peptidoglycan

Peptidoglycan is present only in Bacteria; the sugar N-acetylmuramic acid and the amino acid diaminopimelic acid (DAP) are never found in the cell walls of Archaea or Eukarya. However, not all Bacteria have DAP in their peptidoglycan. This amino acid is present in all Gram-negative Bacteria and in some Gram-positive species, but most Gram-positive cocci have lysine instead of DAP, and a few other Gram-positive Bacteria have other amino acids. Another unusual feature of the bacterial cell wall is the presence of two amino acids that have the D configuration,

FIGURE 3.27   (a) Diaminopimelic acid. (b) Lysine.

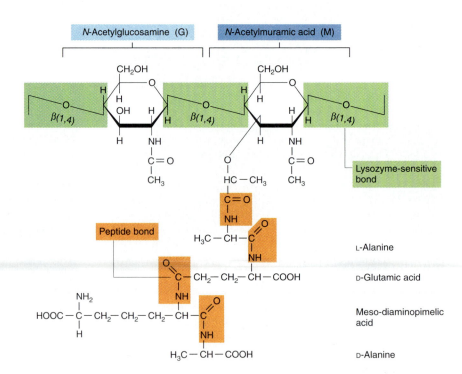

FIGURE 3.28
Structure of one of the repeating units of the peptidoglycan cell-wall structure, the glycan tetrapeptide. The structure given is that found in *Escherichia coli* and most other Gram-negative Bacteria. In some Bacteria, other amino acids are found.

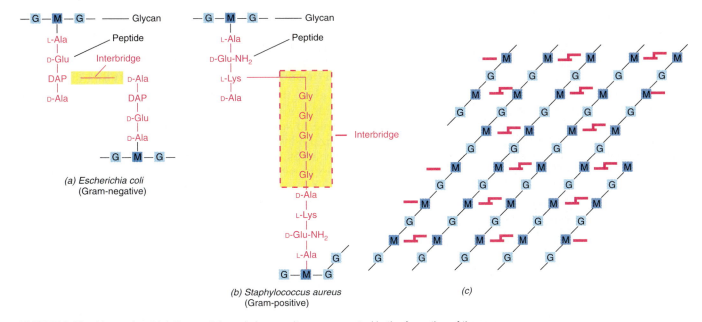

FIGURE 3.29   Manner in which the peptide and glycan units are connected in the formation of the peptidoglycan sheet. (a) Direct interbridge in Gram-negative Bacteria. (b) Glycine interbridge in *Staphylococcus aureus* (Gram-positive). (c) Overall structure of peptidoglycan. The diagram depicts several ribbons of peptidoglycan cross-linked to one another. To visualize an entire single layer of peptidoglycan, imagine these cross-linked ribbons extending around a cylinder or sphere representing the cell. G, *N*-acetylglucosamine; M, *N*-acetylmuramic acid; bold lines in (c) indicate peptide cross-links.

D-alanine and D-glutamic acid. As we have seen in Chapter 2, in proteins amino acids are always of the L configuration.

Several generalizations regarding peptidoglycan structure can be made. The glycan portion is uniform, with only the sugars *N*-acetylglucosamine and *N*-acetylmuramic acid being present, and these sugars are always connected in β-1,4 linkage. The tetrapeptide of the repeating unit shows major variation only in one amino acid, the lysine-diaminopimelic acid alternation. However, the D-glutamic acid at position 2 can be hydroxylated in some organisms, while substitutions occur in amino acids at positions 1 and 3 in a few others.

Almost 100 different peptidoglycan types are known and the greatest variation among them occurs in the interbridge. Any of the amino acids present in the tetrapeptide can also occur in the interbridge, but in addition, a number of other amino acids can be found in the interbridge, such as glycine, threonine, serine, and aspartic acid. However, certain amino acids are never found in the interbridge: branched-chain amino acids, aromatic amino acids, sulfur-containing amino acids, and histidine, arginine, and proline.

In Gram-positive Bacteria, as much as 90 percent of the wall consists of peptidoglycan, although another kind of constituent, teichoic acid (discussed below), is usually present in small amounts. In most Gram-positive Bacteria, about 20 layers of peptidoglycan are present. In Gram-negative Bacteria, only about 10 percent of the wall is peptidoglycan, the remainder being present in an outer wall layer outside the peptidoglycan layer, as discussed below. In *E. coli* and several other Gram-negative Bacteria, only a single layer of peptidoglycan exists.

> The cell walls of Bacteria but not Archaea contain a unique polysaccharide called peptidoglycan. This material consists of alternating repeats of the glucose derivatives *N*-acetylglucosamine and *N*-acetylmuramic acid, with muramic acid residues in adjacent rows of peptidoglycan cross-linked by short peptides. The cross-links in peptidoglycan form a two-dimensional web that can readily bear stress in any direction.

Outer-membrane (lipopolysaccharide) layer

Gram-negative Bacteria contain an outer wall layer made of **lipopolysaccharide**. This layer is effectively a second lipid bilayer, but is not constructed solely of phospholipid, as is the cytoplasmic membrane, but also contains polysaccharide and protein. The lipid and polysaccharide are intimately linked in the outer layer to form specific lipopolysaccharide structures. Because of the presence of lipopolysaccharide, the outer layer is frequently called the **lipopolysaccharide** or **LPS layer**.

Although complex, the chemical structures of some LPS layers are now understood. As seen in Figure 3.30, the polysaccharide consists of two portions, the core polysaccharide and the *O*-polysaccharide. In *Salmonella*, where it has been best studied, the **core polysaccharide** consists of ketodeoxyoctonate, seven-carbon sugars (heptoses), glucose, galactose, and *N*-acetylglucosamine. Connected to the core is the *O*-

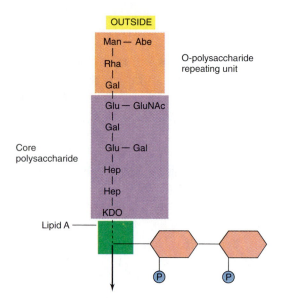

FIGURE 3.30 Structure of the lipopolysaccharide of *Salmonella*. KDO, ketodeoxyoctonate; Hep, heptose; Glu, glucose; Gal, galactose; GluNac, *N*-acetylglucosamine; Rha, rhamnose; Man, mannose; Abe, abequose. Attached to the lipid A is a disaccharide of GluNac residues. The entire structure comprises the endotoxin complex (see Section 11.10).

polysaccharide, which usually contains galactose, glucose, rhamnose, and mannose (all six-carbon sugars) as well as one or more unusual dideoxy sugars such as abequose, colitose, paratose, or tyvelose. These sugars are connected in four- or five-membered sequences, which often are branched. When the sugar sequences are repeated, the long *O*-polysaccharide is formed.

The relationship of the *O*-polysaccharide to the rest of the LPS layer is shown in Figure 3.31. The lipid portion of the lipopolysaccharide, referred to as **lipid A**, is not a glycerol lipid, but instead the fatty acids are connected by ester amine linkage to a disaccharide composed of *N*-acetylglucosamine phosphate (Figure 3.30). Fatty acids commonly found in lipid A include caproic, lauric, myristic, palmitic, and stearic acids. In the outer membrane, the LPS associates with various proteins to form the *outer* half of the unit membrane structure. A **lipoprotein** complex is found on the *inner* side of the outer membrane of a number of Gram-negative Bacteria (Figure 3.31). This lipoprotein is a small (~7200 molecular weight) protein that serves as an anchor between the outer membrane and peptidoglycan. In the *outer* leaf of the outer membrane, LPS replaces phospholipids; the latter are predominantly found in the inner leaf (Figure 3.31).

One important biological property of the outer membrane layer of many Gram-negative Bacteria is that it is frequently *toxic* to animals. Gram-negative Bacteria that are pathogenic for humans and other mammals include members of the genera *Salmonella*, *Shigella*, and *Escherichia*. The toxic property of the outer membrane layer of these bacteria is responsible for some of the symptoms of infection which these bacteria bring about. The toxic properties are associated with part of the lipopolysaccharide (LPS) layer of these organisms. The term *endotoxin* is frequently used to refer to this toxic component of LPS, as we will discuss in Section 11.10.

Permeability in the Gram-negative cell

Unlike the cytoplasmic membrane, the outer membrane of Gram-negative Bacteria is relatively permeable to small molecules, even though it is basically a lipid bilayer. How, then, can molecules move so readily across the outer membrane? Proteins called **porins** are present in the outer membrane of Gram-negative Bacteria, and these proteins serve as membrane channels for the entrance and exit of hydrophilic low-molecular-weight substances (Figure 3.31). Several porins have now been identified, and both specific and nonspecific classes are known. *Nonspecific porins* form water-filled channels through which small substances of any type can pass. By contrast, some porins are highly specific because they contain a specific binding site for one or more substances. The largest porins allow entry of substances up to 5000 molecular weight.

Structural studies have shown that most porins are proteins containing three identical subunits. Porins are transmembrane proteins (Figure 3.31) and associate to form small membrane holes about 1 nm in diameter. Apparently a mechanism exists for opening and closing the pores, because resistance to certain antibiotics is related to porin structure. Presumably, such porins can be closed to prevent antibiotic uptake.

The outer membrane is thus relatively permeable to small hydrophilic molecules. However, it is not permeable to protons as it must prevent protons, ejected across the cytoplasmic membrane during formation of the proton motive force, from drifting away from the cell. In addition, the outer membrane is *not* permeable to enzymes or other large molecules. In fact, one of the main functions of the outer layer may be to keep certain enzymes, which are present outside the cytoplasmic membrane, from diffusing away from the cell. These enzymes are present in a region called the **periplasm** (see Figures 3.31 and 3.32). In *E. coli* this space between the outer surface of the cytoplasmic membrane and the inner surface of the LPS-containing outer membrane, occupies a distance of some 12–15 nm and is gel-like in consistency, presumably because of the abundance of periplasmic proteins found there (Figure 3.32). The periplasm of Gram-negative Bacteria generally contains three types of proteins: *hydrolytic enzymes*, which function in the initial degradation of food molecules, *binding proteins*, that begin the process of transporting substrates, and *chemoreceptors*, which are proteins involved in the chemotaxis response (see Section 3.8). Periplasmic binding proteins function to bind a substance and bring it to the membrane-bound carrier; these processes are probably not linked to the proton gradient but may instead use ATP as an energy source. Binding proteins seem to be absent in Gram-positive Bacteria, which also lack a lipopolysaccharide layer and a defined periplasmic space.

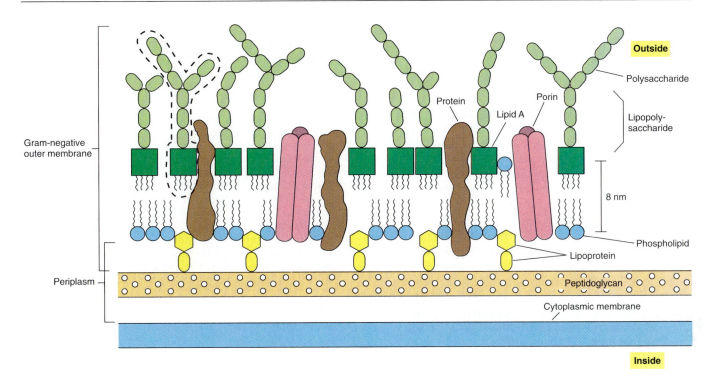

FIGURE 3.31   Arrangement of lipopolysaccharide, lipid A, phospholipid, porins, and lipoprotein in the Gram-negative bacterium *Salmonella*. Dashed lines circle the endotoxin complex (see Section 11.10).

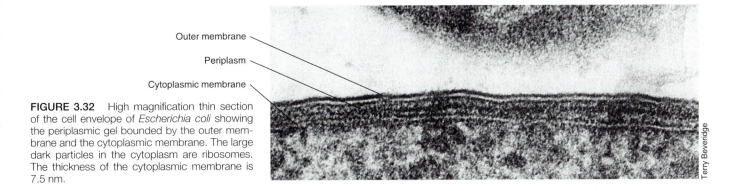

FIGURE 3.32   High magnification thin section of the cell envelope of *Escherichia coli* showing the periplasmic gel bounded by the outer membrane and the cytoplasmic membrane. The large dark particles in the cytoplasm are ribosomes. The thickness of the cytoplasmic membrane is 7.5 nm.

Teichoic acids

Although Gram-positive Bacteria do not have a lipopolysaccharide outer layer, they generally do have acidic polysaccharides attached to their cell wall called **teichoic acids** (from the Greek word *teichos*, meaning "wall"). The term *teichoic acids* includes all wall, membrane, or capsular polymers containing glycerophosphate or ribitol phosphate residues. These polyols are connected by phosphate esters and usually have other sugars and D-alanine attached (Figure 3.33). Because they are negatively charged, teichoic acids are partially responsible for the negative charge of the cell surface as a whole and may function to effect passage of ions through the cell wall. Certain glycerol-containing acids are bound to membrane lipids of Gram-positive Bacteria; because these teichoic acids are intimately associated with lipid, they have been called *lipoteichoic acids*.

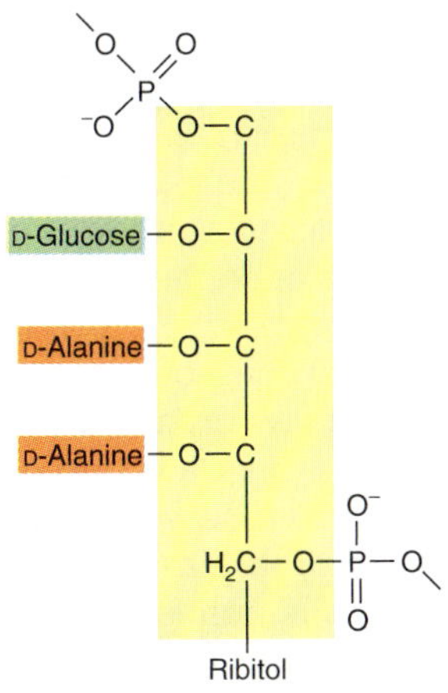

FIGURE 3.33   Structure of the ribitol teichoic acid of *Bacillus subtilis*. The teichoic acid is a polymer of the repeating ribitol units shown here.

Relationship of cell-wall structure to the Gram stain

Are the structural differences between the cell walls of Gram-positive and Gram-negative Bacteria responsible in any way for the Gram-stain reaction? In the Gram stain (see Section 3.1), an insoluble crystal violet–iodine complex is formed inside the cell and this complex is extracted by alcohol from Gram-*negative* but not from Gram-*positive* Bacteria. Gram-positive Bacteria, which have very thick cell walls consisting of several layers of peptidoglycan, become dehydrated by the alcohol. This causes the pores in the walls to close, preventing the insoluble crystal violet-iodine complex from escaping. In Gram-negative Bacteria, the solvent readily penetrates the outer layer, and the thin peptidoglycan layer also does not prevent solvent passage. However, the Gram reaction is not related directly to cell-wall chemistry since yeasts, which have a thick cell wall but of an entirely different chemical composition, also stain Gram-positive. Thus it is not the chemical constituents but the *physical structure* of the wall that confers Gram-positivity.

Osmosis, lysis, and protoplast formation

In addition to conferring shape, the cell wall is essential in maintaining the structural integrity of the cell. This can be dramatically illustrated by treating a suspension of Bacteria with the enzyme **lysozyme**. Lysozyme is found in tears, saliva, other body fluids, and egg white. It hydrolyzes peptidoglycan by breaking the glycosidic bonds between *N*-acetylglucosamine and *N*-acetylmuramic acid, thereby weakening the wall. Water then enters, the cell swells, and eventually bursts, a process called **lysis**.

If the proper concentration of a solute that does not penetrate the cell, such as sucrose, is added to the medium, the solute concentration outside the cell balances that inside. Under these conditions, lysozyme still digests peptidoglycan, but lysis does not occur, and an intact **protoplast** is formed. If such sucrose-stabilized protoplasts are placed in water, lysis occurs immediately.

Osmotically stabilized protoplasts are always spherical in liquid medium, even if derived from rod-shaped organisms, further emphasizing that the shape of the intact cell is conferred by the cell wall. However, the wall is not an inelastic structure, and is able to stretch and contract to some extent. Thus the cell wall might be viewed as somewhat like an automobile tire.

Although most prokaryotes cannot survive without their cell walls, a few organisms are able to do so; these are the mycoplasmas, a group of organisms that cause certain infectious diseases. Mycoplasmas are essentially free-living protoplasts, and are able to survive without cell walls either because they have unusually tough membranes or because they live in osmotically protected habitats, such as the animal body. Certain mycoplasmas have sterols in their cell membranes, which lends strength and rigidity to this structure. Mycoplasmas will be discussed in Section 19.27.

> Gram-negative Bacteria contain, in addition to peptidoglycan, an outer membrane consisting of lipopolysaccharide, protein, and lipoprotein. This second lipid bilayer is more permeable than the cytoplasmic membrane because of pores formed by proteins called porins. The region between the inner and outer membranes in Gram-negative Bacteria is referred to as the periplasm and contains a number of proteins involved in important cellular functions. Gram-positive Bacteria lack an outer membrane but may contain additional cell wall components such as teichoic acids. Cells in which the cell wall has been removed are called protoplasts.

3.6   Cell-Wall Synthesis and Cell Division

When a cell enlarges during the division process, new cell-wall synthesis must take place, and this new wall material must be added in some way to the preexisting wall without loss of structural integrity. This process can occur in different ways, as shown in Figure 3.34*a*. Small openings in the macromolecular

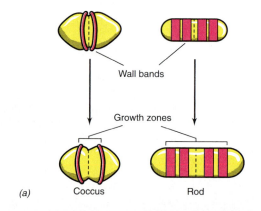

(a)   Coccus   Rod

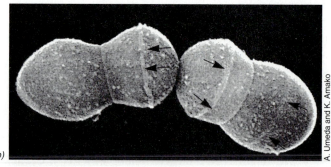

(b)

FIGURE 3.34   (a) Localization of new cell-wall synthesis during cell division. In cocci, new cell-wall synthesis is localized at only one point, whereas in rod-shaped bacteria it occurs at several locations along the cell wall. (b) Scanning electron micrograph of *Streptococcus hemolyticus* showing division bridges (arrows). A single cell is about 1 μm in diameter.

structure of the wall are created by enzymes called **autolysins**, similar to lysozyme, that are produced within the cell. New wall material is then added across the openings. Obviously, new wall synthesis occurs at several points along the surface of the growing cell (Figure 3.34). The junction between new and old peptidoglycan forms a ridge on the cell surface of Gram-positive Bacteria (Figure 3.34b) analogous to a scar. Thus, it is essential that new peptidoglycan somehow be spliced onto preexisting peptidoglycan *before* severing bonds within the latter to ensure that cell turgor pressure does not burst the cell wall at a splice point. If this does not occur, a process of spontaneous lysis called **autolysis** can occur.

Biosynthesis of peptidoglycan

The peptidoglycan layer can be thought of as a stress-bearing fabric, much like a sheet of rubber. Synthesis of new peptidoglycan during cell growth involves controlled cutting by autolysins of bonds connecting small areas of preexisting peptidoglycan, with the simultaneous insertion of new pieces of peptidoglycan, much as a patch would be inserted into a piece of woven fabric. As this process continues during cell division, cell volume increases until a cross wall (septum) forms and the cell divides into two cells.

The biosynthesis of peptidoglycan is of great interest and practical importance. Most information is known about the synthesis of the cell wall of *Staphylococcus aureus*, and the discussion here will be restricted to this organism.

Two carrier molecules participate in peptidoglycan synthesis, uridine diphosphate and a lipid carrier. The lipid carrier, called **bactoprenol**, is a C_{55} isoprenoid alcohol which is connected via phosphodiester linkage to N-acetylmuramic acid to which a pentapeptide is attached (Figure 3.35). The second amino sugar of peptidoglycan, N-acetylglucosamine, is then added followed by addition of the pentaglycine bridge.

The assembly of polymers such as peptidoglycan *outside* the cytoplasmic membrane presents special problems of transport and control. Bactoprenol is involved in transport of the peptidoglycan building blocks across the membrane, where the peptidoglycan

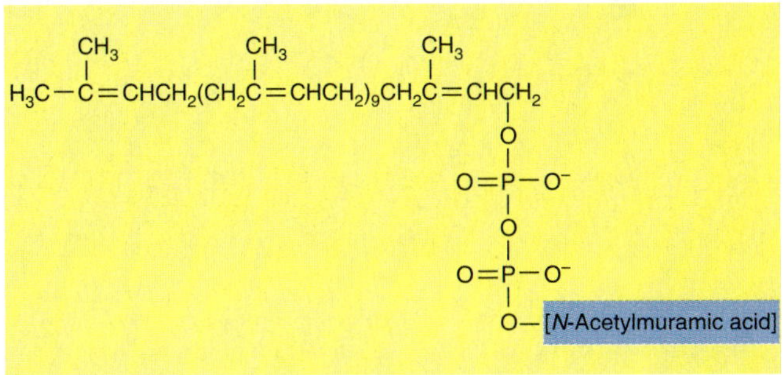

is then inserted into a growing point of the cell wall (Figure 3.36). The function of bactoprenol is to render sugar intermediates sufficiently hydrophobic so that they will pass through the hydrophobic cytoplasmic membrane. The lipid carrier inserts the disaccharide pentapeptide complex into the glycan backbone and then moves back inside the cell to pick up another peptidoglycan precursor unit (Figure 3.36).

The final step in cell-wall synthesis is the formation of the peptide cross-links between adjacent glycan chains. The formation of the peptide cross-links involves an unusual type of peptide bond formation, called **transpeptidation**, which is also noteworthy because it is the reaction inhibited by the antibiotic *penicillin*. This cross-linking reaction involves peptide formation with one of several different amino acids, depending on the organism involved. In Gram-negative Bacteria, such as *Escherichia coli*, the cross-linking is between diaminopimelic acid (DAP) on one peptide and D-alanine on an adjacent peptide (Figure 3.36b). Initially, there are *two* D-alanine groups at the end of the peptidoglycan precursor, but one D-alanine group is split off during the transpeptidation reaction. The peptide bond between the two molecules of D-alanine serves to *activate* the subterminal D-alanine, thereby favoring its reaction with the DAP. This reaction occurs outside the cytoplasmic membrane, where energy is not available, and the transpeptidation reaction replaces the requirement for energy input. In *Staphylococcus aureus* the transpeptidation reaction occurs via the pentaglycine bridge.

Inhibition of transpeptidation by penicillin thus leads to the formation of a weakened peptidoglycan. Further damage to the cell, resulting in lysis and death, occurs as autolysins continue to act, but because new peptidoglycan cross-links cannot occur, the cell wall becomes progressively weaker and osmotic lysis occurs. As with lysozyme, lysis by penicillin can be prevented by adding an osmotic stabilizing agent such as sucrose. Under such conditions, continued growth in the presence of penicillin leads to the formation of protoplasts. Penicillin-induced lysis only occurs with *growing* cells. In nongrowing cells, action of autolysins does not occur, so that breakdown of the cell-wall peptidoglycan is prevented. It is fascinating to consider that one of the key developments in human medicine, the discovery of penicillin, is linked to a specific biochemical reaction involved in cell-wall synthesis, transpeptidation.

FIGURE 3.35 Bactoprenol (undecaprenolphosphate), the lipid carrier of the cell-wall peptidoglycan building blocks.

New cell wall is synthesized during bacterial growth by inserting new glycan units into preexisting wall material. A long-chain alcohol called bactoprenol facilitates transport of new glycan units through the cytoplasmic membrane to become part of the growing cell wall. The action of enzymes such as lysozyme and antibiotics such as penicillin, results in a weakening of the peptidoglycan layer and the ultimate death of the cell.

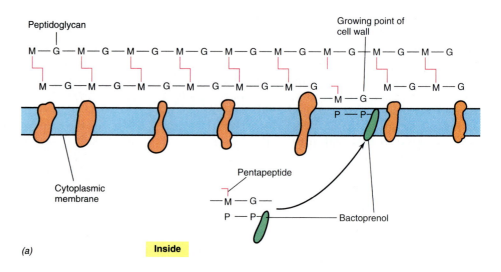

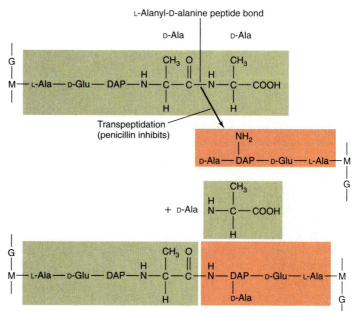

FIGURE 3.36
Peptidoglycan synthesis. (a) Transport of peptidoglycan precursors across the cytoplasmic membrane to the growing point of the cell wall. (b) The transpeptidation reaction that leads to the final cross-linking of two peptidoglycan chains. Penicillin inhibits this reaction.

3.7   Flagella and Motility

Many prokaryotes are motile, and this ability to move independently is usually due to a special structure, the **flagellum** (plural, **flagella**; Figure 3.37). Certain bacterial cells can move along solid surfaces by *gliding* (see Section 19.13), and certain aquatic microorganisms can regulate their position in the water column by gas-filled structures called gas vesicles (see Sections 3.10 and 19.2). However, the majority of motile prokaryotes move by means of flagella. Motility allows the cell to reach different regions of its microenvironment. In the struggle for survival, movement to a new location may mean the difference between survival and death of the cell. But, as in any physical process, cell movement is closely tied to an energy expenditure. We begin now with a detailed consideration of flagellar motility in prokaryotes.

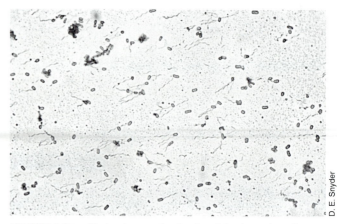

FIGURE 3.37   Photomicrograph of cells of the bacterium *Proteus mirabilis* stained to show flagella.

Bacterial flagella

Bacterial flagella are long, thin appendages free at one end and attached to the cell at the other end. They are so thin (about 20 nm) that a single flagellum can never be seen directly with the light microscope, but only after staining with special flagella stains which increase their diameter (Figure 3.37; see also Figure 3.39). Flagella are also readily seen with the electron microscope (Figure 3.38).

Flagella are arranged differently on different bacteria. In **polar flagellation** the flagella are attached at one or both ends of the cell. Occasionally a tuft (group) of flagella may arise at one end of the cell, an arrangement called "lophotrichous" (*lopho-* means "tuft"; *trichous* means "hair") (Figure 3.39). Tufts of flagella of this type can often be seen in the living state by dark-field microscopy, a type of microscopy in which light reaches the specimen from the sides only. The only light reaching the lens is that scattered by the specimen, which thus appears light on a dark background (Figure 3.40). In **peritrichous flagellation** the flagella are inserted at many places around the cell surface (*peri* means "around"). The type of flagellation is often used as a characteristic in the classification of bacteria (see Figure 3.39 and Section 18.10).

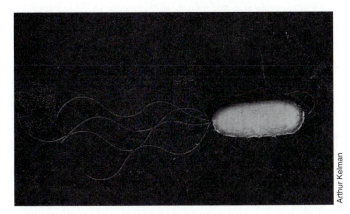

FIGURE 3.38 Electron micrograph of a bacterial cell, showing flagella. The cell is approximately 3 μm long.

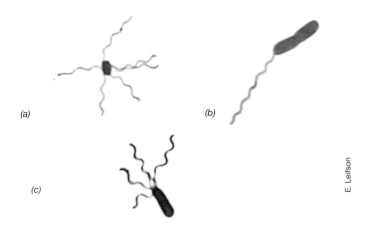

(a)

(b)

(c)

FIGURE 3.39 Light photomicrographs of prokaryotes containing different flagellar arrangements. Cells are stained with the Leifson flagella stain. (a) Peritrichous. (b) Polar. (c) Lophotrichous.

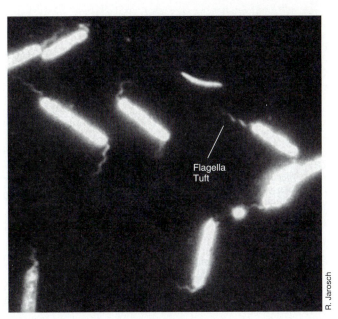

Flagella Tuft

R. Jarosch

FIGURE 3.40 Dark-field photomicrograph of a group of large rod-shaped bacteria with flagellar tufts at each pole. A single cell is about 2 μm wide.

Flagellar structure

Flagella are not straight but helically shaped; when flattened, they show a constant length between two adjacent curves, called the *wavelength*, and this wavelength is constant for each species (Figure 3.38). Bacterial flagella are composed of protein subunits; the protein is called **flagellin**. The shape and wavelength of the flagellum are determined by the structure of the flagellin protein, and a change in the structure of the flagellin can lead to a change in the morphology of the flagellum.

The basal region of the flagellum is different in structure from the rest of the flagellum (Figure 3.41). There is a wider region at the base of the flagellum called the *hook*. Attached to the hook is the **basal body**, a motor-like structure involved in the connection of the flagellar apparatus to the cell envelope. The hook and basal body are composed of proteins different from those of the flagellum itself. The basal body consists of a small central rod which passes through a system of rings. In Gram-negative Bacteria, the outer pair of rings is associated with the lipopolysaccharide and peptidoglycan layers of the cell wall, and the inner pair of rings is located within or just above the cytoplasmic membrane (Figure 3.41). In Gram-positive Bacteria, which lack the outer lipopolysaccharide layer, only the inner pair of rings is present.

Several genes are required for flagellar synthesis and subsequent mobility. In *Escherichia coli* and *Salmonella typhimurium*, where studies have been most extensive, over 40 genes, called the *fla* genes, necessary for flagella synthesis have been identified. These *fla* genes serve several functions, including encoding structural proteins of the flagellar apparatus, export of flagellar components through the membrane to the outside of the cell, and regulation of the many bio-

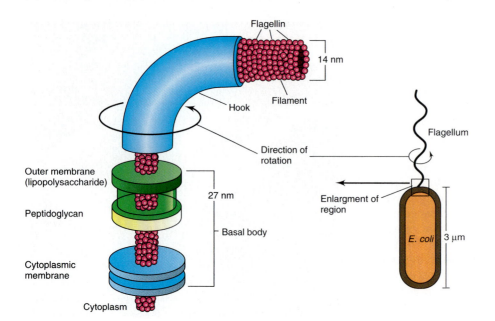

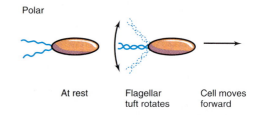

FIGURE 3.41
An interpretive drawing of the manner of attachment of the flagellum in a Gram-negative bacterium. Although cells of *Escherichia coli* are peritrichously flagellated, for simplicity, only a single flagellum is shown.

chemical events surrounding synthesis of new flagella. Control of flagella synthesis is tightly regulated in the cell both by metabolic factors and by signals emerging from the cell division cycle.

Flagellar movement

How is motion imparted to the flagellum? Each individual flagellum is actually a rigid structure, which does not flex but moves by *rotation*, in the manner of a propeller. Visual evidence for this conclusion can be obtained by observing the behavior of motile cells that are tethered by their flagella to microscope slides. Such cells rotate around the point of attachment at rates of revolution consistent with those inferred for flagellar movement in free-swimming cells.

It seems likely that the rotary motion of the flagellum is imparted from the basal body, which acts like a motor. It is likely that the two inner rings located at the membrane rotate in relation to each other, so that (to extend the motor analogy) one of the rings could be considered to be the rotor, the other the stator. The energy required for rotation of the flagellum comes from the proton motive force (see Sections 3.4 and 4.11). Dissipation of the proton gradient creates a force which rotates the flagellum counterclockwise and propels the cell through the liquid. Flagella do not rotate at a constant speed, and the maximum rate of flagellar rotation is about 200 revolutions per second (12,000 rpm).

The motions of polar and lophotrichous organisms are different from those of peritrichous organisms. Peritrichously flagellated organisms generally move and rotate in a straight line in a slow, stately fashion. Polar organisms, on the other hand, move more rapidly, spinning around and dashing from place to place. The different behavior of flagella on polar and peritrichous organisms is illustrated in Figure 3.42.

The average velocities of bacteria vary from about 20 μm to 80 μm/s. A speed of 50 μm/s is equivalent to

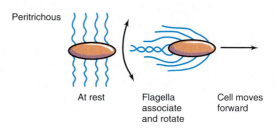

FIGURE 3.42 Manner of flagellar movement in polarly and peritrichously flagellated organisms.

0.0001 mile/h, which seems slow. However, it is more reasonable to compare velocities in terms of number of cell lengths moved per second. The fastest animal, the cheetah, is about 4 ft long and moves at a maximum rate of 70 miles/h, or about 25 lengths per second. The rates of bacterial movement, often around 10 or more lengths per second, are about as significant as those of higher organisms.

Flagellar growth

The individual flagellum grows not from the base, as does an animal hair, but from the tip. Flagellin molecules formed in the cell pass up through the hollow core of the flagellum and add on at the terminal end. The synthesis of a flagellum from its flagellin protein molecules occurs by a process called *self-assembly*: all of the information for the final structure of the flagellum resides in the protein subunits themselves. Growth of the flagellum occurs more or less continuously until a maximum length is reached; however, if a portion of the tip is broken off, it is regenerated.

When a cell divides, the two daughter cells must acquire a full complement of flagella. In polarly flagellated organisms, the process of cell division probably occurs as shown in Figure 3.43, the new flagellum forming at the location where cell division has just occurred. In a monopolarly flagellated cell, the two poles of the cell probably differ in some way so that the flagellum is formed at one pole and not at the other. In peritrichously flagellated organisms, at cell division preexisting flagella are distributed equally between the two daughter cells, and new flagella are synthesized and fill in the gaps.

Motility in eukaryotes

Many eukaryotic cells are motile, and two types of organelles of motility are recognized: flagella and cilia (Figure 3.44). **Flagella,** as just discussed, are long, filamentous structures but in eukaryotes they move in a whiplike manner instead of rotating like the flagella of prokaryotes. Eukaryotic flagella are much larger than those of prokaryotes and are composed of protein structures called **microtubules**. Movement is imparted to the flagellum by the coordinated sliding of the several microtubules present within each flagellum. Energy for microtubule sliding is supplied by ATP.

Cilia are similar to eukaryotic flagella in fine structure but differ in being shorter and more numerous (Figure 3.44). Cilia function much as do oars in a rowboat; these rigid structures beat in synchrony to impart rapid movement to the ciliated cell. In microorganisms, cilia are found primarily in one group of protozoa called the *ciliates*; we discuss the biology of these organisms in Chapter 21.

> Motility in cells is most commonly associated with the presence of special organelles called flagella or cilia. In prokaryotes, the flagellum is a submicroscopic structure composed of a single type of protein called flagellin. The bacterial flagellum is a rigid helix that is inserted at the base of the cell and rotates, at the expense of a proton gradient, to cause cellular movement. In eukaryotes, flagella move by a whiplike motion imparted by sliding microtubules.

3.8   Chemotaxis in Bacteria

Chemotaxis is the movement of an organism toward or away from a chemical. Positive chemotaxis refers to movement *toward* a chemical and is usually exhibited when the chemical is of some benefit to the cell (for example, a nutrient). Negative chemotaxis is movement *away* from a chemical, usually one that is harmful. Chemicals that induce positive chemotaxis are called **attractants**, and chemicals that induce negative chemotaxis are called **repellents**. Chemotaxis is a behavioral phenomenon and suggests some kind of primitive nervous response. Although both prokaryotes and eukaryotes exhibit chemotactic responses, the phenomenon has been studied most extensively in motile bacteria, in particular *Escherichia coli* and *Salmonella typhimurium* because of the availability of sophisticated genetic methods.

Experimental measurements of chemotaxis

Bacterial chemotaxis can be most easily demonstrated by immersing a small glass capillary containing an attractant into a suspension of motile bacteria which does not contain the attractant. From the tip of the capillary, a chemical gradient is set up into the surrounding medium, with the concentration of chemical gradually decreasing with distance from the tip. If the capillary contains an attractant, the bacteria will move toward the capillary, forming a swarm around the

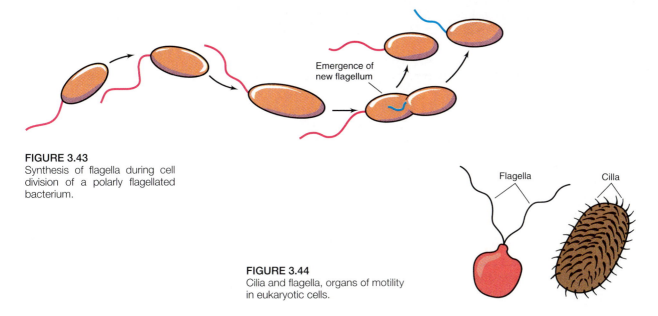

FIGURE 3.43
Synthesis of flagella during cell division of a polarly flagellated bacterium.

Emergence of new flagellum

FIGURE 3.44
Cilia and flagella, organs of motility in eukaryotic cells.

Flagella

Cilla

open tip (Figure 3.45); subsequently, many of the motile bacteria will move into the capillary. Of course, some bacteria will move into the capillary even if it contains a solution of the same composition as the medium, because of random movements; but if an attractant is within, the concentration of bacteria within the capillary can be many times higher than the external concentration. On the other hand, if the capillary contains a repellent, the concentration of bacteria within the capillary will be considerably less than the concentration outside. The problem is to determine how these tiny cells are able to "sense" the chemical gradient and move toward or away from it.

To explain chemotaxis, we must first consider the behavior of a single bacterial cell when it is moving in a chemical gradient. When viewed under the microscope, the movement of single cells is extremely hard to follow with the eye, because it is so rapid. To follow single cells a special microscope, called a **tracking microscope**, has been constructed. It automatically moves a small chamber containing the bacteria so as to keep a particular cell fixed in space, and a record is generated in three dimensions of the position of the chamber with time. An example of the kind of data obtained with this technique is shown in Figure 3.46. When movements of a large number of cells are analyzed with the tracking microscope, bacterial behavior can be divided into two actions, called **runs** and **tumbles**. When the organism runs, it swims steadily in a gently curved path. Then it tumbles, that is, stops and jiggles in place. Then it runs off again in a new direction. A tumble is a random event, but occurs on the

average of about once per second and lasts about a tenth of a second. Following the tumble, the direction for the next run is almost random. Thus, by means of runs and tumbles, the organism moves randomly but does not go anywhere. If a chemical gradient of an attractant is present, the random movements become biased. As the organism experiences higher concentrations of the attractant, tumbles are less frequent, and the runs are longer (Figure 3.46*b*). The net result of this situation is that the organism moves up the concentration gradient by increasing the length of runs that take it in a direction favoring a higher concentration of the attractant.

Because prokaryotes are so small, they do not respond to the *spatial* gradient of an attractant or a repellent, but instead respond to the *temporal* gradient that develops as they move through the medium. In other words, if the bacterium is, by chance, moving toward the capillary, it experiences with time a progressively higher concentration of the substance, and this temporal increase in chemical concentration brings into play the mechanism to decrease tumbles, thus increasing runs.

In peritrichous organisms, such as *E. coli* and *S. typhimurium*, forward movement occurs when the flagellar bundle comes together. For forward movement, flagella within the bundle rotate counterclockwise, and the cell is propelled in a direction away from the flagella (Figure 3.42). This behavior results in a run. On the other hand, when the flagella rotate clockwise, the flagellar bundle falls apart and each flagellum pulls, resulting in cessation of forward motion and ini-

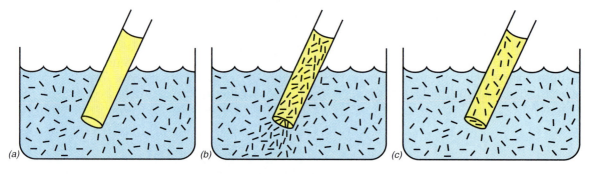

(a)   (b)   (c)

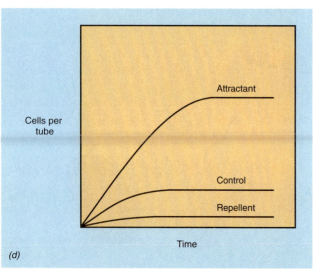

(d)

FIGURE 3.45 Capillary technique for studying chemotaxis in bacteria. (a) Insertion of capillary into a bacterial suspension. (b) Accumulation of bacteria in a capillary containing an attractant. (c) Control capillary contains a salt solution which is neither an attractant nor a repellent. Cell concentration inside the capillary becomes the same as that outside. (d) Time course showing cell numbers in capillaries containing various chemicals.

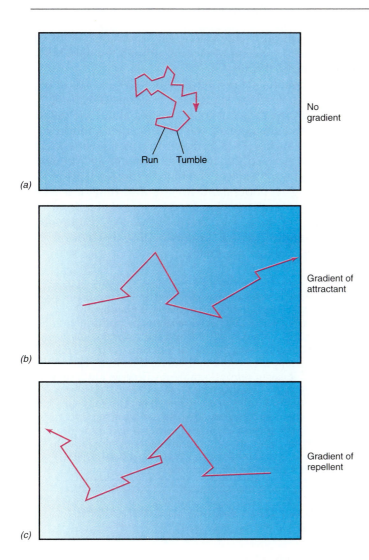

(a)

(b)

(c)

No gradient

Run   Tumble

Gradient of attractant

Gradient of repellent

FIGURE 3.46
Diagrammatic representation of *Escherichia coli* movement, as analyzed with the tracking microscope. These drawings are two-dimensional projections of the three-dimensional movement. (a) Random movement of a cell in a uniform chemical field. Each run is followed by a tumble, and the tumbles occur fairly frequently. (b) Directed movement toward a chemical attractant. The runs still go off in random directions, but when the run is up the chemical gradient, the tumbles occur less frequently, and the runs are longer. When runs occur away from the chemical, tumbles occur more frequently. The net result is movement toward the chemical. (c) Directed movement away from a chemical repellent.

tiation of a tumble. Bacterial flagella are rigid helices, and the direction of twist of the helix is left-handed, which results in the flagella coming together in a bundle when rotated counterclockwise and flying apart when rotated clockwise. Thus the chemical sensing mechanism controls the direction of flagellar rotation.

Mechanism of chemotaxis

How do bacteria use temporal changes in chemical concentrations to control flagellar rotation? Bacteria contain proteins that can sense the presence of attractants and repellants. These proteins allow the cell to sense whether, over time, the concentration of the substance increases or decreases as the cell moves. The cell thus responds to the *change* in concentration rather than the *absolute* concentration of the chemical stimulus. The sensory proteins are called *methyl-accepting chemotaxis proteins* (**MCPs**), also called **transducers**. In *E. coli*, four different MCPs have been identified, and each is a transmembrane protein (Figure 3.47). Each MCP can sense a variety of compounds. For example,

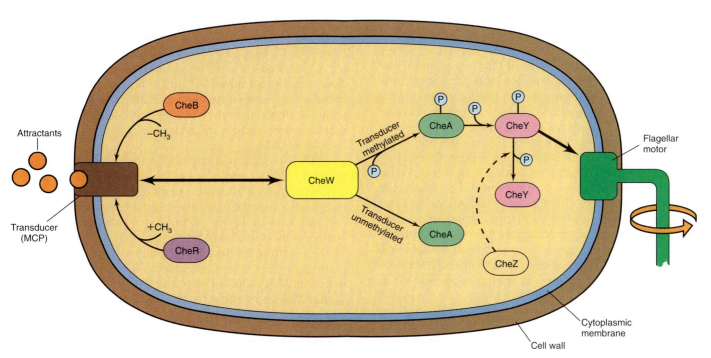

FIGURE 3.47   Interactions of transducers, chemotaxis (Che) proteins, and the flagellar motor in bacterial chemotaxis. Methylation of a methyl-accepting chemotaxis protein (MCP) causes CheW to phosphorylate CheA which then phosphorylates CheY. The latter interacts directly with the flagellar motor switch. CheZ affects the phosphorylation state of CheY, and CheB serves to slowly demethylate MCPs.

the *Tar* transducer of *E. coli* can sense the attractants aspartate and maltose as well as repellants such as the heavy metals cobalt and nickel.

MCPs bind attractants or repellents directly, or in some cases indirectly, through interactions with periplasmic binding proteins that have previously bound the stimulus molecule. Binding of an attractant or repellant sets in play a series of interactions with cytoplasmic proteins that eventually affect flagellar motion. MCPs are so named because they can be alternately methylated or demethylated in response to attractant and repellant concentrations, and the methylation state of an MCP ultimately governs which direction the flagellum will rotate. If rotation of the flagellum is *counterclockwise*, the cell continues to move in a run. If the flagellum rotates *clockwise*, however, the cell will tumble.

How does the methylation state of a transducer affect motility in the cell? Transducers are constantly methylated at a slow rate by a protein in the cell called CheR. Methylation occurs using the methyl donor *S*-adenosyl methionine. Methylation affects the secondary structure of the MCP, and at a high level of methylation, the MCP folds in such a way that prolonged runs are signaled. If the concentration of an attractant does not increase, a methyl-removing protein, CheB, gradually removes methyl groups from an MCP. This action tends to favor tumbling of the cell. However, between the level of the transducer and the flagellar motor switch, several other proteins are involved. These proteins are ones that can be alternately phosphorylated and dephosphorylated, and the phosphorylation state of these proteins affects their function.

The current model for flagellar control (Figure 3.47) shows that a cytoplasmic protein referred to as CheW interacts with transducers. If the transducer is demethylated, CheW activates a second protein called CheA, causing it to become phosphorylated. CheA then directs phosphorylation of CheY, a protein that interacts directly with the motor switching apparatus of the flagella (the motor switch itself consist of proteins encoded by *fla* genes). Thus, CheY is the central protein in the system because it serves as the *response regulator* for chemotaxis, governing the direction of rotation of the flagellum. When CheY is phosphorylated, the flagellar motor switches from a counterclockwise to a clockwise rotation, causing the cell to tumble. If unphosphorylated, CheY's flagellar motor rotates counterclockwise and the cell undergoes a run. Another protein, CheZ, dephosphorylates CheY, returning it to a form promoting runs instead of tumbles. What controls activity of CheZ is unknown. The major events governing chemotactic responses are shown in Figure 3.47.

Repellants presumably stimulate tumbles until a cell senses a decreasing concentration gradient of the repellant, after which runs are favored. Increasing concentrations of repellants and decreasing concentrations of attractants tend to stimulate transducer demethylation, while decreasing gradients of repellants and increasing attractants stimulate methylation and continued runs.

A chemotactic response is thus the end result of a cascading series of regulatory reactions (Figure 3.47). Although much is left to be learned about the details of bacterial chemotaxis, the use of genetics to identify key proteins involved in the process has greatly accelerated the pace of research. Interest in bacterial chemotaxis continues to be high because scientists are beginning to understand for the first time at a truly molecular level the biochemistry of a behavioral response. Results of studies on chemotaxis might even have significance for understanding behavioral responses in higher organisms.

> Motile bacteria often exhibit attraction or repulsion when placed in chemical gradients, a phenomenon called chemotaxis. Bacterial motion can be divided into two phases, called the "run" and the "tumble." Bacteria do not "decide" which direction to run, but move randomly, stopping occasionally to tumble. Also, bacteria do not "sense" chemical gradients directly, but if they are moving in the direction of an attractant, runs are longer than tumbles, the end result being a movement up the chemical gradient. Several proteins regulate the chemotactic response, including transmembrane receptor proteins capable of being methylated and demethylated and cytoplasmic proteins that can be alternately phosphorylated and dephosphorylated.

3.9 Bacterial Cell Surface Structures and Cell Inclusions

Fimbriae and pili

Fimbriae and pili are structurally similar to flagella but are not involved in motility. **Fimbriae** are considerably shorter than flagella and are more numerous (Figure 3.48) but, like flagella, consist of protein. Not all organisms have fimbriae, and the ability to produce them is an inherited trait. The functions of fimbriae are not known for certain in all cases, but there is some evidence that they enable organisms to stick to inert surfaces, or to form pellicles or scums on the surfaces of liquids.

Pili are similar structurally to fimbriae but are generally longer and only one or a few pili are present on the surface. Pili can be visualized under the electron microscope because they serve as specific receptors for certain types of virus particles, and when coated with virus can be easily seen (Figure 3.49). There is strong evidence that pili are involved in the mating process in some bacteria, as will be discussed in Section 7.10. Pili are also involved in attachment to human tissues by some pathogenic bacteria.

Capsules and slime layers

Many prokaryotic organisms secrete on their surfaces slimy or gummy materials (Figure 3.50). A variety of structures consist of polysaccharide and a few consist of protein. The terms **capsule** and **slime layer** are sometimes used to describe polysaccharide layers but the more general term **glycocalyx** is also applied. Glycocalyx is defined as the polysaccharide-containing material lying outside the cell. The glycocalyx varies in different organisms, but usually contains glycoproteins and a large number of different polysaccharides, including polyalcohols and amino sugars. The glycocalyx may be thick or thin, rigid or flexible, depending on its chemical nature in a specific organism. The rigid layers are organized in a tight matrix which excludes particles, such as India ink; this form is referred to as a

capsule. If the glycocalyx is more easily deformed, it will not exclude particles and is more difficult to see; this arrangement is referred to as a slime layer.

Glycocalyx layers serve several functions in bacteria. Outer polysaccharide layers play an important role in the *attachment* of certain pathogenic microorganisms to their hosts. As we will see in Section 11.6, pathogenic microorganisms that enter the animal body by specific routes usually do so because of binding reactions which occur between outer cell surface components (such as the glycocalyx) and specific host tissues. The glycocalyx plays other roles as well. There

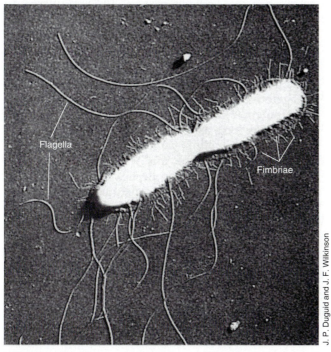

FIGURE 3.48   Electron micrograph of a dividing cell of *Salmonella typhi*, showing flagella and fimbriae. A single cell is about 0.9 μm in diameter.

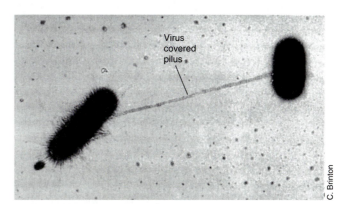

FIGURE 3.49   The presence of pili on an *Escherichia coli* cell is revealed by the use of viruses that specifically adhere to the pilus. The cell is about 0.8 μm in diameter.

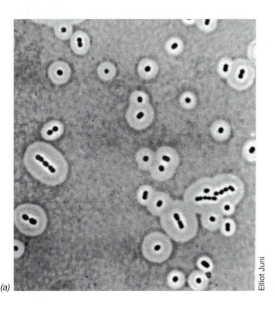

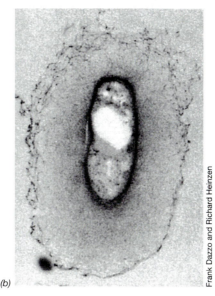

FIGURE 3.50
Bacterial capsules. (a) Demonstration of the presence of a capsule in *Acinetobacter sp.* by negative staining with India ink observed by phase-contrast microscopy. The India ink does not penetrate the capsule so that it is revealed in outline as a light structure on a dark background. (b) Electron micrograph of a thin section of a *Rhizobium trifolii* cell stained with ruthenium red to reveal the capsule. The diameter of the cell proper (not including the capsule) is about 0.7 μm.

is some evidence that encapsulated bacteria are more difficult for phagocytic cells of the immune system (see Section 11.13) to recognize and subsequently destroy. In addition, since outer polysaccharide layers probably bind a significant amount of water, there is reason to believe a glycocalyx layer plays some role in resistance to desiccation.

Inclusions and storage products

Granules or other inclusions are often seen within cells. Their nature differs in different organisms, but they almost always function in the storage of energy or serve as a reservoir of structural building blocks. Inclusions can often be seen directly with the light microscope using special staining procedures, but their contrast can usually be increased by using dyes. Inclusions often show up very well with the electron microscope (Figure 3.51). Most cellular inclusions are bounded by a thin nonunit membrane consisting of lipid which separates the inclusion from the cytoplasm proper.

In prokaryotic organisms, one of the most common inclusion bodies consists of **poly-β-hydroxybutyric acid (PHB)**, a lipid-like compound that is formed from β-hydroxybutyric acid units (Figure 3.51). The monomers of this acid are connected by ester linkages,

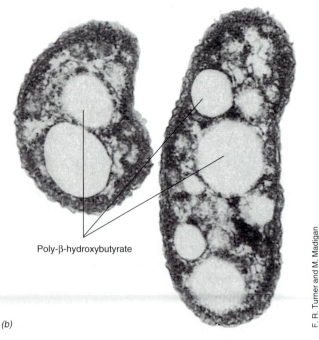

(a)

Poly-β-hydroxybutyrate

(b)

F. R. Turner and M. Madigan

FIGURE 3.51 Poly-β-hydroxybutyrate (PHB). (a) Chemical structure of PHB, a common poly-β-hydroxyalkanoate. A monomeric unit is shaded. Other alkanoate polymers are made by substituting longer chain hydrocarbons for the —CH₃ group on the β carbon. (b) Electron micrograph of a thin section of cells of the phototrophic bacterium *Rhodospirillum sodomense* containing granules of PHB.

forming the long PHB polymer, and these polymers aggregate into granules. The length of the monomer in the polymer can vary considerably, from as short as C-4 to as long as C-18 in certain organisms. The collective term *poly-β-hydroxyalkanoate* (PHA) has been coined to describe this whole class of storage polymers. The physical properties of polyesters containing different monomeric constituents vary considerably and many PHAs have plastic-like consistencies. Because of this, there is considerable interest in the commercial exploitation of PHAs as biodegradable plastic substitutes (see Section 17.20). A wide variety of prokaryotes, including representatives of both the Bacteria and the Archaea, produce PHAs. Eukarya do not naturally produce PHAs.

With the electron microscope the positions of these granules can often be seen as light areas that do not scatter electrons, surrounded by the nonunit membrane (Figure 3.51b). The granules have an affinity for fat-soluble dyes such as Sudan black and can be identified tentatively with the light microscope by staining with this compound. Poly-β-hydroxybutyric acid can be positively identified by extraction and chemical analysis. The PHB granules are a storage depot for carbon and energy.

Another storage product is **glycogen**, which is a starchlike polymer of glucose subunits (we discussed the chemistry of glycogen in Section 2.5). Glycogen granules are usually smaller than PHB granules, and can only be seen with the electron microscope, but the presence of glycogen in a cell can be detected in the light microscope because the cell appears a red-brown color when treated with dilute iodine, due to a glycogen–iodine reaction. Like PHB, glycogen is a storage depot for carbon and energy.

Many microorganisms accumulate large reserves of inorganic phosphate in the form of granules of **polyphosphate**. These granules are stained by many basic dyes; one of these dyes, toluidine blue, becomes reddish violet in color when combined with polyphosphate. This phenomenon is called *metachromasy* (color change), and granules that stain in this manner are often called **metachromatic granules**.

A variety of prokaryotes are capable of oxidizing reduced sulfur compounds such as hydrogen sulfide, thiosulfate and the like. These oxidations are linked to either reactions of energy metabolism (Sections 16.8 and 16.10) or biosynthesis (Section 16.7), but in both instances **elemental sulfur** frequently accumulates inside the cell in large readily visible granules (see Sections 16.10 and 19.1). The granules of elemental sulfur remain as long as a source of reduced sulfur is still present. However, as the reduced sulfur source becomes limiting, the sulfur in the granules is oxidized, usually to sulfate, and the granules slowly disappear as this reaction proceeds.

Magnetosomes are intracellular crystal particles of the iron oxide magnetite, Fe_3O_4 (Figure 3.52). Magnetosomes impart a permanent magnetic dipole on a cell, allowing it to respond to a magnetic field. Bacte-

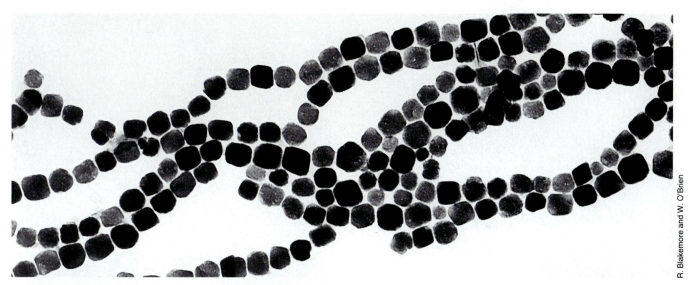

R. Blakemore and W. O'Brien

FIGURE 3.52   Magnetosomes. Magnetic particles of Fe_3O_4 isolated from the magnetotactic bacterium *Aquaspirillum magnetotacticum*. Each particle is about 50 nm in length. See also Figure 19.47.

ria that produce magnetosomes exhibit *magnetotaxis*, the process of orienting and migrating along geomagnetic field lines (see Section 19.11 for more on magnetotactic bacteria). Magnetosomes are surrounded by a membrane containing phospholipids, proteins, and glycoproteins. Magnetosome membrane proteins probably play a role in precipitating Fe^{3+} (brought into the cell in soluble form by chelating agents), as Fe_3O_4 in the developing magnetosome. The morphology of magnetosomes appears to be species specific, varying in shape from square to rectangular to spike-shaped in certain bacteria.

Magnetosomes have been described in a variety of different primarily aquatic bacteria, and have even been found in some algae (eukaryotes). Algal magnetosomes presumably function to make the algal cells magnetotactic as in bacteria. Measurements of the magnetic moment of magnetotactic algal cells indicates that the magnetic force within these cells is much greater than that of magnetotactic bacteria, which is consistent with the larger size of the algal cells.

3.10 Gas Vesicles

A number of prokaryotic organisms that live a floating existence in lakes and the sea produce **gas vesicles**, which confer buoyancy upon the cells. Gas vesicles are a means of motility, allowing cells to float up and down in a water column in response to environmental factors. The most dramatic instances of flotation due to gas vesicles are seen in cyanobacteria that form massive accumulations (blooms) in lakes (Figure 3.53). Gas-vesiculate cells rise to the surface of the lake and are blown by winds into dense masses. Gas vesicles are also present in certain purple and green phototrophic bacteria (see Section 19.1) and in some nonphototrophic bacteria that live in lakes and ponds. Some Archaea also contain gas vesicles.

Brock

FIGURE 3.53   Flotation of cyanobacteria from a bloom on a nutrient-rich lake, caused by the presence of gas vesicles.

Gas vesicles are spindle-shaped structures made of protein, hollow but rigid, that are of variable lengths and diameter. Gas vesicles in different organisms vary in length from about 300 to 700 nm and in width from 60 to 110 nm, but the vesicles of any given organism are more or less of constant size. They are present in the cytoplasm and may number from a few to hundreds per cell. The gas vesicle membrane is an exception to the rule that membranes are composed of lipid bilayers. The gas vesicle membrane is composed only of protein, and consists of repeating protein subunits that are aligned to form a rigid ribbed structure cross-linked by a second type of protein. The gas vesicle membrane is about 2 nm thick and is impermeable to water and solutes, but permeable to gases, so that it exists as a gas-filled structure surrounded by the constituents of the cytoplasm (Figure 3.54).

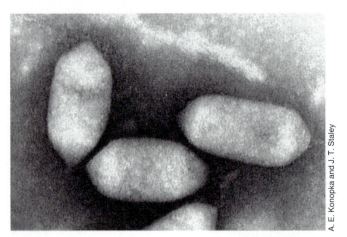

A. E. Konopka and J. T. Staley

FIGURE 3.54 Electron micrographs of gas vesicles purified from the bacterium *Microcyclus aquaticus* and examined in negatively stained preparations. A single gas vesicle is about 100 nm in diameter.

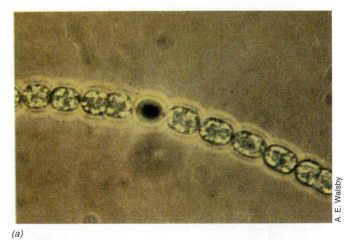

(a)

A. E. Walsby

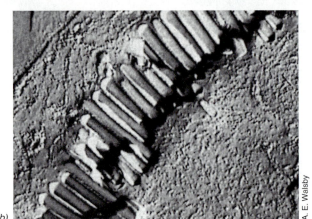

(b)

A. E. Walsby

FIGURE 3.55 Gas vesicles of the cyanobacterium *Anabaena flos-aquae*. (a) The cell in the center (a heterocyst) lacks gas vesicles. In the other cells, the vesicles group together as phase-bright objects that scatter light. (b) Freeze-fracture preparations of cells, showing gas vesicles.

The rigidity of the gas vesicle membrane is essential for the structure to resist the pressures exerted on it from without; it is probably for this reason that it is composed of a protein able to form a rigid membrane rather than of lipid, which would form a fluid and a highly mobile membrane. However, even the gas vesicle membrane cannot resist high hydrostatic pressure, and can be collapsed, leading to a loss of buoyancy. Once collapsed, gas vesicles cannot be reinflated. The presence of gas vesicles can be determined by either bright-field or phase-contrast microscopy (Figure 3.55), but their identity is never certain unless they disappear when the cells are subjected to high hydrostatic pressure.

Gas vesicle structure

Gas vesicles contain only two different types of protein. The major gas vesicle protein, called GVPa, is a 7.5-kilodalton (kD) highly hydrophobic protein. GVPa makes up 97 percent of the total protein of the gas vesicle. The second protein, called GVPc, is a 22-kD protein and is present in much smaller amounts; the function of GVPc protein is to strengthen the shell of the gas vesicle. Gas vesicles are constructed of several copies of the GVPa protein aligned as parallel ribs forming a water-tight surface. GVPa protein folds as a β-sheet and thus gives considerable rigidity to the overall vesicle structure. Studies have shown that most of the hydrophobic amino acids of GVPa protein face the gas surface side of the vesicles, whereas more hydrophilic amino acids face the cytoplasm. The ribs of GVPa protein are strengthened by GVPc protein, which acts as a cross-linker, binding several GVPa ribs together. One can draw an analogy between the ribs of gas vesicles and the reinforcement of a garden hose; the nylon cross-stitching wrapped around the hose greatly strengthens it against breakage from the high pressure of water.

The composition of the gas inside the gas vesicle is the same as the gas in which the organism is suspended, and gas is present in the vesicle at about 1 atm pressure. Because the gas vesicle attains a density about 5 to 20 percent of that of the cell proper, when gas vesicles are inflated they decrease the density of the cell, thereby increasing its buoyancy. Aquatic phototrophic organisms in particular benefit from this motility strategy because it allows them to adjust their position rapidly in a water column to regions where the light intensity for photosynthesis is optimal.

Studies of genes coding for the gas vesicle proteins GVPa and GVPc from taxonomically diverse gas-vesiculate bacteria have shown a remarkable degree of DNA sequence homology, suggesting great evolutionary conservation in these unusual structures. Although the final shape of the gas vesicle can vary in different organisms from long and thin to short and fat (compare Figures 3.54 and 3.55b), these differences are not due to major differences in the primary structure of gas vesicle proteins in each case, but instead are a function of how the proteins are arranged to form the intact vesicle. The observation has been made that long and narrow gas vesicles are typically found in organisms that reside at great depths, while short fat vesicles predominate in organisms that inhabit upper layers of the water column. Although less buoyant, long narrow vesicles are structurally stronger than short fat ones, consistent with the fact that gas-vesiculate organisms living at great depths, where water pressure is greater, would require vesicles of greater strength.

3.11   Endospores

Certain Bacteria produce special structures called **endospores** within their cells (Figure 3.56). Endospores are very resistant to heat and cannot be destroyed easily, even by harsh chemicals. Endospore-forming bacteria are found most commonly in the soil, and virtually any sample of soil will have some endospores present.

The discovery of bacterial endospores was of immense importance to microbiology because the knowledge of such remarkably heat-resistant forms was essential for the development of adequate methods of sterilization, not only of culture media but also of foods and other perishable products (see Section 1.6). Although many organisms other than bacteria form spores, the bacterial endospore is unique in its degree of heat resistance. Endospores are also resistant to other harmful agents such as drying, radiation, acids, and chemical disinfectants. The life cycle of a spore-forming organism is illustrated in Figure 3.57.

Endospores (so called because the spore is formed *within* the cell) are readily seen under the light microscope as strongly refractile bodies (see Figure 3.56). Spores are very impermeable to dyes, so that occasionally they are seen as unstained regions within cells that have been stained with basic dyes such as methylene blue. To stain spores specifically, special spore-staining procedures must be used. The structure of the spore as seen with the electron microscope is vastly different from that of the vegetative cell, as shown in Figure 3.58. The structure of the spore is much more complex than that of the vegetative cell in that it has many layers. The outermost layer is the **exosporium**, a thin, delicate covering. Within this is the **spore coat**, which is composed of a layer or layers of protein. Below the spore coat is the **cortex**, which consists of loosely cross-linked peptidoglycan, and inside the cortex is the **core**, which contains the usual cell wall (core wall), cytoplasmic membrane, nucleoid and so on.

Thus the spore differs structurally from the vegetative cell primarily in the kinds of structures found outside the core wall.

One chemical substance that is characteristic of endospores but is not present in vegetative cells is **dipicolinic acid** (Figure 3.59). This substance has been found in all endospores examined, and it is located in the core. Spores are also high in calcium ions, most of which are in combination with dipicolinic acid. The calcium–dipicolinic acid complex of the core represents up to 10 percent of the dry weight of the endospore. The calcium plus dipicolinic acid complex,

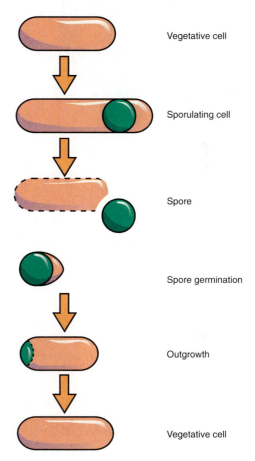

Vegetative cell

Sporulating cell

Spore

Spore germination

Outgrowth

Vegetative cell

FIGURE 3.57   Life cycle of an endospore-forming bacterium.

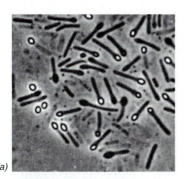

(a)

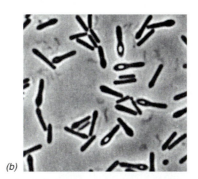

(b)

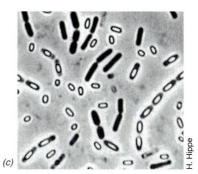

(c)

H. Hippe

FIGURE 3.56   The bacterial endospore. Phase contrast photomicrographs illustrating several types of endospore morphologies and intracellular locations. (a) Terminal. (b) Subterminal. (c) Central.

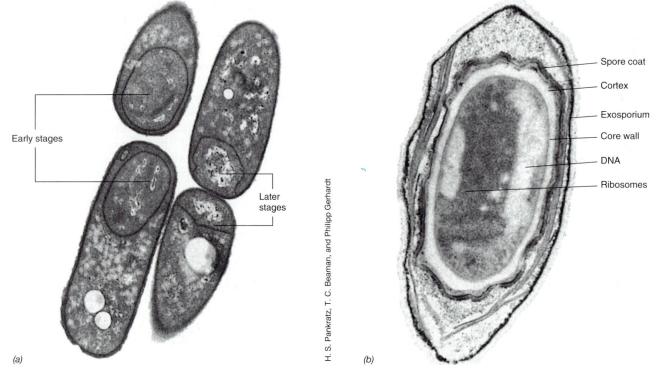

Early stages

Later stages

(a)

Spore coat
Cortex
Exosporium
Core wall
DNA
Ribosomes

(b)

H. S. Pankratz, T. C. Beaman, and Philipp Gerhardt

H. S. Pankratz, T. C. Beaman, and Philipp Gerhardt

FIGURE 3.58 Electron microscopy of the bacterial spore. (a) Formation of spores within mother cells (sporangia) of *Bacillus megaterium*. A single cell is about 1.5 μm in diameter. (b) Mature free spore.

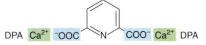

DPA Ca²⁺ ⁻OOC — COO⁻ Ca²⁺ DPA

FIGURE 3.59 Dipicolinic acid (DPA). Ca^{2+} ions associate with the carboxyl groups to form a complex.

along with the fact that water is pressed out of the developing spore, play major roles in conferring the unusual heat resistance to bacterial endospores.

An endospore is able to remain dormant for many years, but it can convert back into a vegetative cell. This process involves three steps: activation, germination, and outgrowth (Figure 3.60). *Activation* is accomplished by heating freshly formed endospores for several minutes at a sublethal but elevated temperature or by storing spore suspensions for weeks or months at 4°C or room temperature. Activated spores are conditioned to germinate when placed in the presence of specific nutrients. *Germination*, usually a rapid process

(minutes), involves loss of refractility of the spore, increased ability to be stained by dyes, and loss of resistance to heat and chemicals. Loss from the spore of calcium dipicolinate and cortex components occurs during this stage. The next stage, *outgrowth*, involves visible swelling due to water uptake and synthesis of new RNA, proteins, and DNA. The cell emerges from the broken spore coat and eventually begins to divide.

Endospore formation

During endospore formation, a vegetative cell is converted to a nongrowing, heat-resistant structure—the endospore. The differences between the endospore and the vegetative cell are profound (Table 3.3), and sporulation involves a very complex series of events. Bacterial sporulation does not occur when cells are dividing exponentially but only when growth ceases owing to the exhaustion of an essential nutrient. Thus, cells of *Bacillus*, a typical endospore-forming bac-

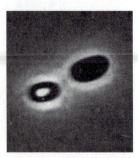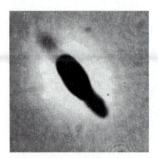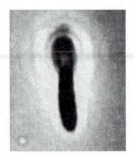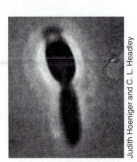

Judith Hoeniger and C. L. Headley

Figure 3.60 Conversion of the endospore into a vegetative cell; photomicrographs showing the sequence of events.

Table 3.3   Differences between endospores and vegetative cells

Characteristic	Vegetative cell	Endospore
Structure	Typical Gram-positive cell	Thick spore cortex Spore coat Exosporium
Microscopic appearance	Nonrefractile	Refractile
Calcium	Low	High
Dipicolinic acid	Absent	Present
Enzymatic activity	High	Low
Metabolism (O_2 uptake)	High	Low or absent
Macromolecular synthesis	Present	Absent
mRNA	Present	Low or absent
Heat resistance	Low	High
Radiation resistance	Low	High
Resistance to chemicals and acids	Low	High
Stainability by dyes	Stainable	Stainable only with special methods
Action of lysozyme	Sensitive	Resistant

terium, cease vegetative growth and begin sporulation when a key nutrient such as the carbon or nitrogen source becomes limiting.

Many genetically directed changes in the cell underlie the conversion from vegetative growth to sporulation, and the whole process can be viewed as a type of *cellular differentiation*. The structural changes occurring in sporulating cells of *Bacillus* are shown in Figure 3.61, and the process can be divided into seven stages. Genetic studies of mutants of *Bacillus*, each blocked at one of the various stages of sporulation shown in Figure 3.61, have shown that as many as 200 genes are involved in sporulation-associated events. Sporulation requires that the synthesis of some proteins involved in vegetative cell functions cease and that specific spore proteins be made. This is accomplished by activation of spore-specific genes (called *spo* genes) in response to an environmental trigger to sporulate. Spore proteins encoded by the *spo* genes cause the differentiation process to occur (Figure 3.61), resulting in formation of the mature endospore.

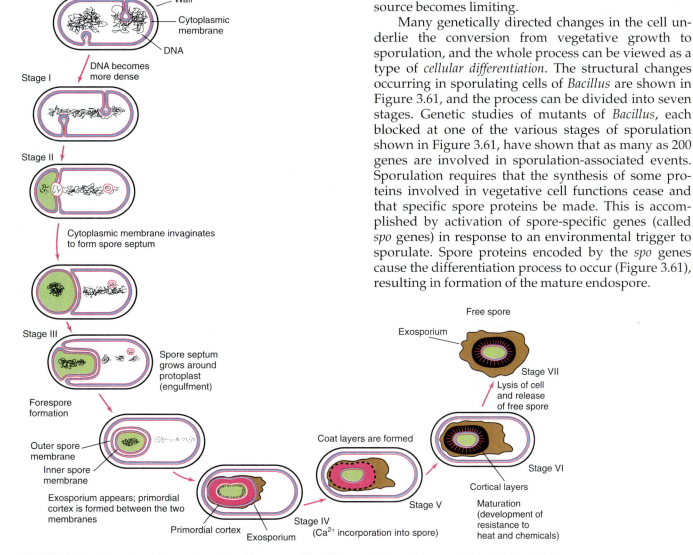

FIGURE 3.61   Stages in endospore formation. The stages listed (0 through VII) are those most clearly distinguishable microscopically and are used in studies on the kinetics of the sporulation process.

One of the most important bacterial structures is the endospore, a very heat-resistant resting cell. Endospores, which are formed only by certain kinds of Bacteria, are also very resistant to chemicals, radiation, and drying. The endospore can remain dormant many years, but if placed in the right conditions, can be triggered to convert back into a vegetative cell, a process called spore germination. Endospore formation occurs in response to nutrient exhaustion. It is initiated by the activation of spore-specific genes that encode spore-specific proteins that actually carry out the differentiation process from vegetative cell to free mature endospore.

3.12   Arrangement of DNA in Prokaryotes

We learned in Chapter 2 that DNA is a double-stranded molecule formed of complementary antiparallel strands of polynucleotides. Although in the eukaryote DNA is present in chromosomes, in the prokaryote, DNA is usually present primarily as a naked DNA molecule, arranged as a covalently closed *circular* molecule which is extensively folded and twisted to fit into the cell.

The total amount of DNA in the chromosome of a bacterium such as *Escherichia coli* is about 4700 kilobase pairs. Not surprisingly, this is considerably less than that of eukaryotic cells, but it is greater than that of viruses or organelles (Figure 3.62). Bacterial DNA is not surrounded by a membrane typical of the eukaryotic nucleus, although it does tend to aggregate as a distinct structure within the cell and is visible when observed with the electron microscope (Figure 3.63).

The bulk if not all of the DNA in the prokaryotic cell is present in a single molecule called the **bacterial chromosome**. The term **nucleoid** is used to describe aggregated DNA in the prokaryotic cell (Figure 3.63), and under special staining conditions the nucleoid can actually be observed in cells examined with just the light microscope (Figure 3.64). Ribosomes are absent in the region of the cell where the bacterial nucleoid is

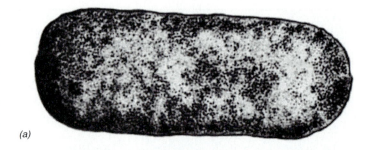

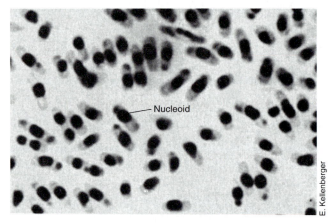

(a)

(b)

E. Kellenberger and J. A. Hobot

FIGURE 3.63   The bacterial nucleoid. (a) Transmission electron micrograph of a thin section of *Escherichia coli*. (b) Same as (a), but with the nucleoid outlined.

Nucleoid

E. Kellenberger

FIGURE 3.64   Light photomicrograph of cells of *Escherichia coli* treated in such a way as to make the nucleoid visible.

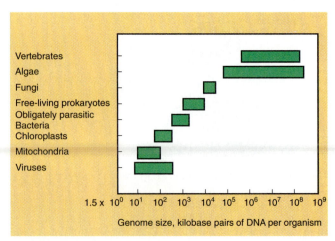

| Vertebrates |
| Algae |
| Fungi |
| Free-living prokaryotes |
| Obligately parasitic Bacteria |
| Chloroplasts |
| Mitochondria |
| Viruses |

1.5 × 10^0 10^1 10^2 10^3 10^4 10^5 10^6 10^7 10^8 10^9

Genome size, kilobase pairs of DNA per organism

FIGURE 3.62   Range of genome sizes in various groups of organisms.

situated, probably because nucleoid DNA exists in a gel-like form which would tend to exclude particulate matter. In addition to the nucleoid, one or more small circular DNA molecules, called *plasmids*, may be present in prokaryotic cells.

Supercoiling and copy number

If gently lysed, DNA can be released from prokaryotic cells (Figure 3.65) and the extensive folding and twisting necessary to store the DNA in the cell becomes readily apparent. The amount of twisting and folding can be appreciated when it is considered that the 4.7 *million* base pairs in the genome of *Escherichia coli*, if opened and linearized, would be about 1 *millimeter* in length, yet the *E. coli* cell is only about 2–3 *micrometers* long! To package this much DNA into the cell requires that the DNA be supercoiled (Figure 3.66). Supercoiled DNA takes on a considerably more compact

B. Arnold-Schulz-Gahmen

FIGURE 3.65
Electron micrograph of an isolated nucleoid from a cell of *Escherichia coli*. Cells were gently lysed to allow the highly compacted nucleoid to emerge intact.

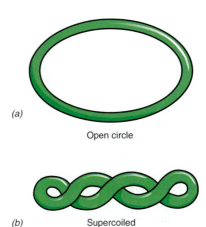

(a)   Open circle

(b)   Supercoiled

FIGURE 3.66   The bacterial chromosome. (a) Open circular form. (b) Supercoiled form. Note that in either case the DNA is present in a covalently closed form, typical of most prokaryotes.

shape than its freely circularized counterpart. However, supercoiling introduces some interesting biochemical problems for proteins that act on DNA, and we will see in Chapter 5 that the supercoiled form of DNA in the bacterial cell can undergo winding and unwinding as transcription and DNA replication processes proceed in the growing cell.

Actively growing prokaryotic cells usually contain multiple copies or partially completed copies (generally two to four) of the bacterial chromosome, and only when cell growth has ceased does the chromosome number approach one per cell. The reason for this has to do with the fact that rapidly growing cells can actually divide faster than the DNA replication machinery can make new copies of the bacterial chromosome. Thus, to ensure that a complete copy of the bacterial chromosome is ready for each daughter cell at cell division, new rounds of DNA synthesis are initiated before the old round is completed. This leads to multiple copies or partial copies in rapidly growing cells.

Gene transfer in prokaryotes

The prokaryotic cell is normally *haploid* in genetic complement, meaning that a single copy of each gene is present. Nevertheless, extensive genetic exchange processes exist in prokaryotic cells, but the mechanisms are quite distinct from the processes in eukaryotes. First, the process is quite fragmentary, almost

never involving whole chromosome complements of the two cells. Second, the DNA is transferred in only one direction, from a donor to a recipient. Third, the mechanisms by which DNA transfer occurs are specialized. Three distinct types of mechanisms for DNA transfer have been recognized: (1) *conjugation*, in which DNA transfer occurs as a result of cell-to-cell contact (the process that most closely resembles sex in eukaryotes); (2) *transduction*, in which DNA transfer is mediated by viruses; and (3) *transformation*, in which free DNA is involved. In transformation, the donor cell generally lyses, releasing DNA into the medium, and some of this free (naked) DNA is taken up by recipient cells. All three mechanisms of gene transfer have been shown to occur in certain Archaea as well as Bacteria. We discuss the details of these various DNA transfer processes in Chapter 7.

> The DNA of the prokaryotic cell exists in a very long, single, circular molecule, called the bacterial chromosome, which is present in the cell in a highly aggregated state called the nucleoid. The nucleoid is not surrounded by a membrane and is present free in the cytoplasm. It is in a highly supercoiled form. Several mechanisms of genetic exchange are known in prokaryotes including conjugation, transformation, and transduction.

3.13   The Eukaryotic Nucleus

One of the distinguishing characteristics of a eukaryote is that its DNA is organized in chromosomes contained in a membrane-enclosed structure, the nucleus. In many eukaryotic cells the nucleus is a large organelle many micrometers in diameter, easily visible with the light microscope, even without staining. In smaller eukaryotes, however, special staining procedures often are required to see the nucleus.

The key genetic processes of DNA replication and RNA synthesis (transcription) occur in the nucleus whereas the process of protein synthesis (translation) occurs in the cytoplasm. We will learn in Chapter 5 of the distinctly different arrangement of DNA in genes in eukaryotic and prokaryotic cells. Eukaryotic cells contain much more DNA than prokaryotes (see Figure 3.62), and much of this DNA serves no known coding function. In eukaryotes, genes are commonly split into **exons** (coding regions) and **introns** (noncoding regions). Following the transfer of the genetic blueprint into the intermediary carrier, RNA, in the process of transcription (see Section 5.6), eukaryotic RNAs are "processed" in the nucleus by removing any RNA that does not actually code for a protein product. Thus, the nucleus serves as both a storehouse and processing factory of genetic information. In prokaryotes, which lack a membrane-bounded nucleus, the processes of transcription and translation are tightly coupled (see Section 5.10), and the extensive RNA processing typical of eukaryotes does not occur.

Nuclear structure

The nuclear membrane consists of a pair of parallel unit membranes separated by a space of variable thickness. The inner membrane is usually a simple sac, but the outer membrane is in many places continuous with the cytoplasmic membrane. The dual membrane arrangement does, however, facilitate functional specificity because the inner and outer membranes specialize in interactions with the nucleoplasm and cytoplasm, respectively. The nuclear membrane contains many pores (Figure 3.67), which are formed from holes in both unit membranes at places where the inner and outer membranes are joined. The pores are about 9 nm wide and permit the facile passage in and out of the nucleus of macromolecules up to about 60,000 molecular weight. However, certain proteins larger than 60,000 molecular weight, such as DNA and RNA polymerases (which are up to 200,000 in molecular weight) are also able to pass into the nucleus from their sites of synthesis in the cytoplasm. Whether these pass through nuclear pores or are threaded through the lipid bilayer by specific transport proteins is not known. A typical animal cell nucleus contains 3000 to 4000 nuclear pores.

A structure often seen within the nucleus is the **nucleolus**, an area rich in ribonucleic acid (RNA) and which is the site of ribosomal RNA synthesis. The

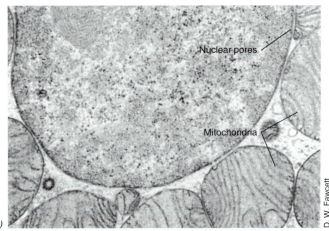

FIGURE 3.67   The nucleus and nuclear pores. (a) Electron micrograph of a yeast cell by the freeze-etch technique, showing a surface view of the nucleus. The cell is about 8 μm wide. (b) Thin section of mouse adipose tissue showing a portion of the nucleus and several mitochondria. The nucleus is about 2 μm wide. Note the pores in the nuclear membrane in both (a) and (b).

small and large subunits of the eukaryotic ribosome are synthesized in the nucleolus and are exported to the cytoplasm where the complete (80S) ribosome is assembled and functions in protein synthesis.

Chromosomes and DNA

As we discussed in the previous section, the DNA of prokaryotes is contained primarily in a single molecule of free DNA, whereas in eukaryotes, DNA is present in more complex structures, the *chromosomes*. **Chromosome** means "colored body," for chromosomes were first seen as structures colored by certain stains. Many chromosome stains involve dyes that react strongly with basic (that is, cationic) proteins called **histones**, which in eukaryotes often are attached to the DNA. Chromosomes also usually contain small amounts of RNA.

During cell division, the nucleus divides following a doubling of the chromosome number, a process called **mitosis** (see Figure 3.68), yielding two cells, each with a full complement of chromosomes. Histones and small protein tubes called microtubules play important roles in the mitotic process. Histones are spaced along the DNA double helix at regular intervals, the DNA itself being wound around each histone molecule. The packing forms a discrete structure called a **nucleosome** (Figure 3.69). Nucleosomes aggregate and form a fibrous material called **chromatin**. Chromatin itself can be compacted by folding and looping to eventually form the intact chromosome. Because DNA is negatively charged (due to the large number of phosphate groups present), there is a strong tendency for various parts of the molecule to repel each other. Histones neutralize some of these negative charges, permitting contraction of the chromosomes. Microtubules function to form the *spindle apparatus*, which is the actual structure that moves chromosomes to the two poles of the dividing cell. (Figure 3.68).

In each chromosome, the DNA is a single *linear* molecule to which histones (and other proteins) are attached. In yeast (and probably many other microorganisms), the length of the DNA in a single chromosome is actually shorter than a linearized prokaryotic chromosome. For instance, the total amount of DNA per yeast cell is only three times that of *Escherichia coli*, but yeast has 17 chromosomes, so that the average yeast DNA molecule is much shorter than the *E. coli* chromosome. In higher organisms, however, the

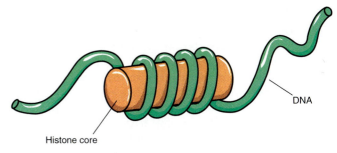

FIGURE 3.69   Packing of DNA around histones in the formation of a nucleosome, typical of the arrangement of DNA in eukaryotic cells.

length of the DNA molecule in a single chromosome is many times longer than the prokaryotic chromosome if it were opened and linearized.

The DNA content per nucleus varies from species to species, in much the same way as does nuclear size. In addition, the chromosome number also varies greatly, from just a few to many hundreds. Another variable feature is genome size, which is the actual amount of DNA per cell. The genome size of various eukaryotes was compared with that of viruses and prokaryotes in Figure 3.62.

Sexual reproduction in eukaryotic microorganisms involves the coming together and fusing of two cells called **gametes**, which are analogous to the sperm and egg of multicellular eukaryotes. The gametes form a single cell called a **zygote**, and the nucleus of the zygote usually results from fusion of the nuclei of the two gametes. The zygote nucleus thus has twice the chromosome complement of the gametes. The chromosome number of the gametes is called the **haploid** number, and the zygote then has twice that many, the **diploid** number.

> The DNA of the eukaryotic cell is present in a number of separate linear molecules, each associated with histone proteins (forming the nucleosome) and formed into structures called chromosomes. The chromosomes are present in a membrane-bounded structure called the nucleus, which is separated from the cytoplasm during most of the cell life cycle. Each organism has a specific number of chromosomes, with the haploid number existing in sperm or eggs and the diploid number in the zygote.

3.14   Eukaryotic Organelles

Eukaryotic cells have a number of important functions localized in discrete bodies called **organelles**. The two most important organelles are *mitochondria*, in which energy metabolism is carried out, and *chloroplasts*, in which the process of photosynthesis is carried out in plants and algae. We discuss these two organelles in more detail here.

(a)

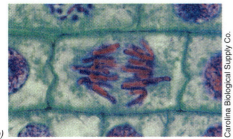

(b)

Carolina Biological Supply Co.

FIGURE 3.68   Mitosis, as seen in the light microscope. These are onion root tip cells which have been stained to reveal nucleic acid and chromosomes. (a) *Metaphase*. Chromosomes are paired in the center of the cell. (b) *Anaphase*. Chromosomes are separating.

Mitochondria

In eukaryotic cells the processes of respiration and oxidative phosphorylation (a mechanism of ATP formation, see Section 4.10) are localized in membrane-enclosed structures, the **mitochondria** (singular, **mitochondrion**). Mitochondria are of prokaryotic size, and can be rod-shaped or nearly spherical (Figure 3.70; see also Figure 3.10). A typical animal cell such as a liver cell can contain 1000 mitochondria, but the number per cell depends somewhat on the cell type and size; a yeast cell may have as few as two mitochondria per cell. The mitochondrial membrane, which lacks sterols, is much less rigid than a cell's cytoplasmic membrane. Mitochondria therefore show a considerable plasticity which makes their shape as seen in electron micrographs highly variable (Figure 3.70).

The mitochondrial membrane is constructed in a manner similar to other membranes; a bilayer of phospholipid with embedded proteins. However, unlike the cytoplasmic membrane, the mitochondrial membrane is rather permeable. In the outer membrane, protein channels are present that allow passage of any molecule of molecular weight less than approximately 10,000. It is for this reason that ATP, produced within the mitochondrion, can move to the cytoplasm where it is used in energy-requiring reactions. In addition to the outer membrane, mitochondria possess a system of folded inner membranes called *cristae*. These inner membranes, formed by invagination of the outer membrane, are the site of enzymes involved in respiration and ATP production and of specific transport proteins that regulate the passage of metabolites into and out of the *matrix* of the mitochondrion (Figure 3.70). The matrix contains a number of enzymes involved in the oxidation of organic compounds, in particular, enzymes of the tricarboxylic acid cycle (see Section 4.12). The mitochondrion can thus be viewed as the energy storehouse of the cell.

Chloroplasts

Chloroplasts are chlorophyll-containing organelles found in all eukaryotic organisms able to carry out photosynthesis. Chloroplasts of many algae are quite

FIGURE 3.70
Structure of the mitochondrion. (a) A diagram showing the overall structure of the mitochondrion. (b) and (c) Transmission electron micrographs of mitochondria from rat tissue showing the variability in morphology of typical mitochondria.

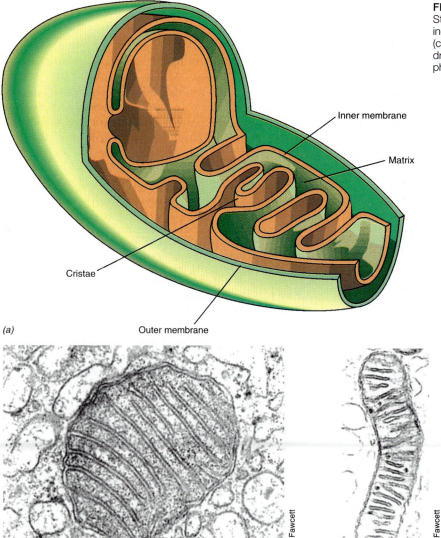

Inner membrane

Matrix

Cristae

(a)

Outer membrane

(b)   D. W. Fawcett

(c)   D. W. Fawcett

large and hence are readily visible with the light microscope (Figure 3.71). The size, shape, and number of chloroplasts vary markedly but, unlike mitochondria, they are generally much larger than bacteria.

Like mitochondria, chloroplasts have a very permeable outer membrane, a much less permeable inner membrane, and an intermembrane space. The inner membrane surrounds the lumen of the chloroplast, called the *stroma*, but is not folded into cristae like the inner membrane of the mitochondrion. Instead chlorophyll, photosynthetic-specific proteins, the photosynthetic electron transport chain, and all other components needed for photosynthesis are located in a series of flattened membrane discs called **thylakoids** (Figure 3.72). The thylakoid membrane is highly impermeable to ions and other metabolites, as befits a membrane whose function is to pump out protons to establish the proton gradient necessary for ATP synthesis (see Section 4.11). In green algae and green plants, thylakoids are usually associated in stacks of discrete structural units called *grana* (see Figure 16.5).

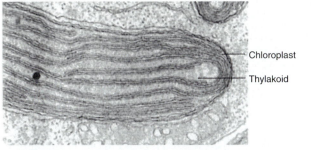

FIGURE 3.72 Electron micrograph showing a chloroplast of the alga *Ochromonas danica*. Note the thylakoids.

The chloroplast stroma contains large amounts of the enzyme ribulose bisphosphate carboxylase, called RubisCO for short. This enzyme is the key enzyme of the Calvin cycle, the series of reactions by which most photosynthetic organisms convert CO_2 into organic form (see Section 16.7). RubisCO makes up over 50 percent of the total chloroplast protein and produces phosphoglyceric acid, a key compound in the biosynthesis of glucose (see Sections 4.17 and 16.7). The permeability of the outer chloroplast membrane allows glucose and ATP produced during photosynthesis to diffuse into the cytoplasm where they can be used to build new cell material.

Two key organelles of eukaryotes are the mitochondrion, involved in energy generation, and the chloroplast, involved in photosynthesis. Both organelles have their own internal membrane systems and contain a number of key cellular enzymes.

3.15 Relationships of Chloroplasts and Mitochondria to Bacteria

On the basis of their relative autonomy and morphological resemblance to bacteria, it was suggested long ago that mitochondria and chloroplasts were descendents of ancient prokaryotic organisms. This theory of *endosymbiosis* (*endo* means "within") says that eukaryotes arose from the engulfment of a prokaryotic cell by a larger cell. Several pieces of evidence strongly suggest that this scenario is correct:

1. Mitochondria and chloroplasts contain DNA. Although most of their functions are encoded by nuclear DNA, a few organellar components are coded from within the organellar genome, most notably ribosomal RNAs, transfer RNAs, and certain proteins of the respiratory chain. Mitochondrial and chloroplast DNA exists in *covalently closed circular form*, as it does in prokaryotes, and is generally present in more than one copy (see Section 3.12). Mitochondrial DNA can be visualized in cells by special staining methods (Figure 3.73).

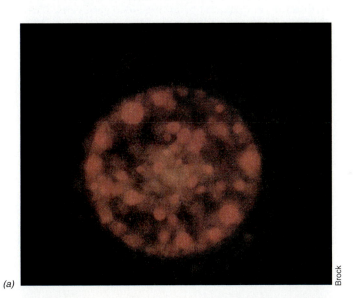

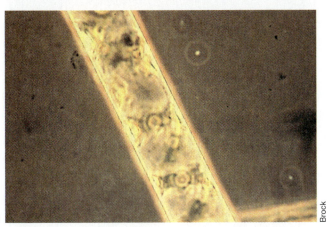

FIGURE 3.71 Photomicrographs of algal cells showing the presence of chloroplasts. (a) Fluorescence photomicrograph of the diatom *Stephanodiscus*. The chlorophyll in the chloroplasts absorbs light and fluoresces red. (b) Phase contrast photomicrograph of *Spirogyra* showing the characteristic spiral chloroplasts.

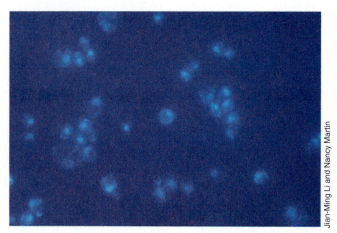

Jian-Ming Li and Nancy Martin

FIGURE 3.73   Cells of the yeast *Saccharomyces cerevisiae* stained to show mitochondrial DNA. Each mitochondrion has two to four circular chromosomes which stain blue with the fluorescent dye used. See also Figure 3.10.

2. Mitochondria and chloroplasts contain their own ribosomes. Ribosomes, the cells' protein synthesis "factories" (see Section 5.8), exist in either a large form (80S) typical of the cytoplasm of eukaryotic cells or a smaller form (70S) unique to prokaryotes. Mitochondrial and chloroplast ribosomes are 70S in size, the same as that of prokaryotes.

3. Many of the antibiotics that kill or inhibit Bacteria by specifically interfering with 70S ribosome function, for example, streptomycin, also inhibit protein synthesis in mitochondria and chloroplasts.

4. Phylogenetic studies using comparative ribosomal RNA sequencing (see Section 18.4) have shown convincingly that the chloroplast and mitochondrion are related to Bacteria. These studies have clearly shown that the modern eukaryotic cell arose from an association of two organisms. The mitochondrion and chloroplast are descendents of

different groups of Bacteria because their ribosomal RNA sequences closely match those of certain species of Bacteria. The same techniques show that the cytoplasmic component of eukaryotes evolved totally independently.

Presumably organelles evolved (following endosymbiotic events) by the progressive loss of more and more of their genetic independence, eventually becoming functionally specialized and dependent on their cytoplasmic host cell. The result is the chloroplasts and mitochondria we see today. Although the endosymbiotic hypothesis can probably never be rigorously proven, substantial molecular evidence remains in organelles today to identify them as having once been Bacteria.

> Because the genetic systems, ribosome size, and antibiotic sensitivity of organelles resemble those of prokaryotes more than eukaryotes, it is thought that organelles have evolved from Bacteria that established residence within the cytoplasm of primitive eukaryotic cells. This theory is known as endosymbiosis.

3.16   Comparisons of the Prokaryotic and Eukaryotic Cell

At this stage it might be useful to draw some comparisons between the prokaryotic and eukaryotic cell. It should be clear by now that there are profound differences in the internal structure of these two cell types. One important distinction is that eukaryotes have many types of cellular functions segregated into membrane-containing structures (organelles). We discussed mitochondria and chloroplasts earlier and Table 3.4 lists a number of other membranous structures.

Table 3.4   Membrane-containing structures in eukaryotes

Structure	Characteristics	Function
Mitochondria	Prokaryotic in size, complex internal membrane arrays	Energy generation: respiration
Chloroplasts	Green, chlorophyll-containing, many shapes, often quite large	Photosynthesis
Endoplasmic reticulum	Not a distinct organelle, extensive array of internal membranes; sites of ribosomes	Protein synthesis
Golgi bodies	Membrane aggregates of distinct structure	Secretion of enzymes and other macromolecules
Vacuoles	Round, membrane-enclosed bodies of low density	Food digestion: food vacuoles; waste product excretion: contractile vacuoles
Lysosomes	Submicroscopic membrane-enclosed particles	Contain and release digestive enzymes
Peroxisomes	Submicroscopic membrane-enclosed particles	Photorespiration in plants
Glyoxysomes	Submicroscopic membrane-enclosed particles	Enzymes of glyoxylate cycle
Nucleus	Large, generally centrally located	Contains genetic material

Table 3.5 groups these differences into several categories of which the most important are nuclear structure and function, cytoplasmic structure and organization, and forms of motility.

Although Table 3.5 emphasizes the great *structural* differences between prokaryotes and eukaryotes, we should return to the theme that began this chapter. All cells contain proteins, nucleic acids, polysaccharides, and lipids, and all use many of the same kinds of metabolic machinery. Chemical differences in the building blocks and variations in the assembly of macromolecules to form cells lead to the considerable structural and functional diversity we see in living organisms today.

We now turn to the important aspects of energy metabolism and nutrition. How does the cell create structurally complex molecules that are eventually assembled to make new cells? We consider these problems and how cells solve them in the next chapter.

Table 3.5   Comparison of the prokaryotic and eukaryotic cell

Properties	Prokaryotes	Eukaryotes
Groups	Bacteria, Archaea	Eukarya: Algae, fungi, protozoa, plants, animals
Nuclear structure and function:		
Nuclear membrane	Absent	Present
Nucleolus	Absent	Present
DNA	Single molecule, not complexed with histones (other DNA in plasmids)	Present in several chromosomes, usually complexed with histones
Division	No mitosis	Mitosis; mitotic apparatus with microtubular spindle
Sexual reproduction	Fragmentary process, unidirectional; no meiosis; usually only portions of genetic complement reassorted	Regular process; meiosis; reassortment of whole chromosome complement
Introns in genes	Rare	Common
Cytoplasmic structure and organization:		
Cytoplasmic membrane	Usually lacks sterols	Sterols usually present
Internal membranes	Relatively simple; limited to specific groups	Complex; endoplasmic reticulum; Golgi apparatus
Ribosomes	70S in size	80S, except for ribosomes of mitochondria and chloroplasts, which are 70S
Membranous organelles	Absent	Several
Respiratory system	Part of cytoplasmic membrane; mitochondria absent	In mitochondria
Photosynthetic pigments	In internal membranes or chlorosomes; chloroplasts absent	In chloroplasts
Cell walls	Present (in most), composed of peptidoglycan, other polysaccharides, protein, glycoprotein	Present in plants, algae, fungi; absent in animals, most protozoa; usually polysaccharide
Endospores	Present (in some), very heat resistant	Absent
Gas vesicles	Present (in some)	Absent
Forms of motility:		
Flagellar movement	Flagella of submicroscopic size; each flagellum composed of one fiber of molecular dimensions; flagella rotate	Flagella or cilia; microscopic size; composed of microtubules; do not rotate
Nonflagellar movement	Gliding motility; gas vesicle-mediated	Cytoplasmic streaming and amoeboid movement; gliding motility
Microtubules	Absent	Widespread: present in flagella, cilia, basal bodies, mitotic spindle apparatus, centrioles
Size	Generally small, usually <2 μm in diameter	Usually larger, 2 to >100 μm in diameter

Study Questions

1. Calculate the surface/volume ratio of a spherical cell 10 μm in diameter, a cell 1 μm in diameter, and a rod-shaped cell 0.5 *μm* in diameter by 2 *μm* long. What are the consequences of these differences in surface/volume ratio for cell function?

2. From what you know about the nature of the bacterial cell wall, explain why a rod-shaped bacterial cell becomes a *spherical* structure when its wall is removed under conditions such that cell lysis cannot occur.

3. Describe in a single sentence the manner by which a unit membrane is formed from phospholipid molecules. How would the membrane formation process differ if the phospholipid molecules were dissolved in a nonaqueous solvent rather than water? How does a *membrane vesicle* differ from a unit membrane? How is such a vesicle *similar* to a unit membrane?

4. Explain in a single sentence why ionized molecules do not readily pass through the membrane barrier of a cell. How *do* such molecules get through the cytoplasmic membrane?

5. Why is the bacterial cell wall rigid layer called "peptidoglycan"? What are the chemical reasons for the rigidity that is conferred on the cell wall by the peptidoglycan structure?

6. Since a single peptidoglycan molecule is very thin, explain in chemical terms how the very *thick* peptidoglycan-containing cell wall of Gram-positive Bacteria is formed.

7. Both lysozyme and penicillin bring about bacterial cell lysis, but by different mechanisms. Describe the mechanism by which each of these agents causes cell lysis. Be sure to include in your explanation an exact description of why *lysis* occurs.

8. What is the bacterial periplasm? What types of Bacteria have a periplasm and of what significance is the periplasmic space?

9. List several functions for the outer wall layer in Gram-negative Bacteria.

10. What is osmosis? Why does it occur? Of what significance is osmosis for the stability of the bacterial cell?

11. Write a clear explanation (two or three sentences) for why sucrose is able to stabilize bacterial cells from lysis by lysozyme.

12. In a few sentences, write an explanation for how a motile bacterium is able to "sense" the direction of an attractant and move toward it.

13. In a few sentences, indicate how the bacterial endospore differs from the vegetative cell in structure, chemical composition, and ability to resist extreme environmental conditions.

14. The discovery of the bacterial endospore was of great practical importance. Why?

15. Describe one chemical difference between the eukaryotic and prokaryotic membrane. Can you offer an explanation for why this chemical difference might exist?

16. Describe a major chemical difference between membranes of Bacteria and Archaea.

17. Water molecules penetrate cytoplasmic membranes fairly readily but hydrogen ions (protons) do not, even though a proton is smaller than a water molecule. Why?

18. List three kinds of *membrane proteins* and give a short explanation for the function of each.

19. Compare and contrast the following processes: *passive diffusion, facilitated diffusion, group translocation*, and *active transport*. For each of these processes, include a discussion of specificity, energy requirement, and transport against a concentration gradient.

20. Describe the structure and function of a bacterial flagellum.

21. How does a *cilium* differ from a flagellum? In what way(s) are they similar?

22. What types of cytoplasmic inclusions are formed by prokaryotes? How does an inclusion of PHB differ from a magnetosome in composition and metabolic role?

23. List several properties by which Bacteria, mitochondria, and chloroplasts are similar. List several ways in which they differ.

24. List two ways in which the DNA of the eukaryote differs from the DNA of the prokaryote. List two ways in which the DNA in these two kinds of organisms is similar.

25. Set up a table following the format of Table 3.5 with the second and third columns blank, and then fill in the blanks. As you do so, think back to the figures in this chapter that illustrate the properties being considered.

26. How does the eukaryotic nucleus differ from the prokaryotic nucleoid? In what ways are these two structures similar?

Supplementary Readings

Alberts, B., D. Bray, J. Lewis, M. Raff, K. Roberts, and J. D. Watson. 1989. *The Molecular Biology of the Cell*, 2nd edition. Garland Publishing, New York. A substantial textbook with extensive coverage of cell structure, with emphasis on the eukaryotic cell.

Cole, J. A., C. Dow, and S. Mohan (eds.) 1992. *Prokaryotic Stucture and Function: A New Perspective*. Soc. Gen Microbiol. Symp. 47. Cambridge University Press, New York. Chapters, written by experts, on the structure, function, and regulation of synthesis of bacterial cell components.

Cooper, S. 1991. *Bacterial Growth and Division*. Academic Press, San Diego, CA. A consideration of replication of the bacterial chromosome in relation to cell division processes. Also covers major processes of cell division in eukaryotic cells.

Neidhardt, F. C., J. L. Ingraham, and M. Schaechter. 1990. *Physiology of the Bacterial Cell—A Molecular Approach*. Sin-

auer Associates, Inc., Sunderland, MA. A very readable textbook of bacterial physiology with an emphasis on molecular aspects of cell structure/function, growth, and genetics.

Riley, M. (ed.). 1990. *The Bacterial Chromosome*. American Society for Microbiology, Washington, D.C. A detailed treatment of the structure and processing of bacterial DNA.

Schlegel, H.G. 1993. *General Microbiology*, 7th edition. Cambridge University Press, New York. An excellent overview of cell structure and many other aspects of microbiology for the beginning student.

Watson, J. D., N. H. Hopkins, J. W. Roberts, J. A. Steitz, and A. M. Weiner. 1987. *Molecular Biology of the Gene*, 4th edition. Benjamin/Cummings, Menlo Park. This excellent textbook has a useful overview of the principles of cell biology.

Metabolism, Biosynthesis, and Nutrition

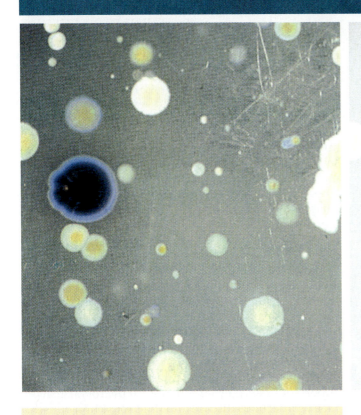

A key feature of a living organism is its ability to organize molecules and chemical reactions into specific structures and systematic sequences. The ultimate expression of this organization is the ability of a living organism to replicate itself. The term **metabolism** is used to refer to all the chemical processes taking place within a cell. The word *metabolism* is derived from the Greek word *metabole*, which means "change," and we can think of a cell as continually changing as it carries out its life processes. Although the cell appears under the microscope to be a fixed and stable structure, it is actually a dynamic entity, continually undergoing change, as a result of all the chemical reactions which are constantly taking place.

Microbial cells are built of chemical substances of a wide variety of types, and when a cell grows, all of these chemical constituents increase in amount. The basic chemical elements of a cell come from outside the cell, from the environment, but these chemical elements are transformed by the cell into the characteristic constituents of which the cell is composed.

The chemicals from the environment of which a cell is built are called **nutrients**. Nutrients are taken up into the cell and are changed into cell constituents. This process by which a cell is built up from the simple nutrients obtained from its environment is called **anabolism**. Because anabolism results in the biochemical synthesis of new material, it is often called **biosynthesis**.

Biosynthesis is an energy-requiring process, and each cell must thus have a means of obtaining energy. The energy source is obtained from the environment, and two kinds of energy sources are used: light and chemicals. Although a number of organisms obtain their energy from light, most microorganisms obtain energy from the oxidation of chemical compounds. Chemicals used as energy sources are broken down into simpler constituents, and as this breakdown occurs, energy is released. The process by which chemicals are broken down and energy released is called **catabolism**. Cells also need energy for other cell functions, such as cell movement (motility) and transport of nutrients.

We thus see that there are two basic kinds of chemical transformation processes occurring in cells, the building-up processes, called *anabolism*, and the breaking-down processes, called *catabolism*. Metabolism is thus the collective result of anabolic and catabolic reactions.

4.1   Overview of Metabolism

A simplified view of cell metabolism is shown in Figure 4.1, which depicts how catabolic reactions supply energy needed for cell functions, and how anabolic (biosynthetic) reactions bring about the synthesis of cell components from nutrients. Note that in anabolism, nutrients from the environment are converted into *cell components*, whereas in catabolism, energy sources from the environment are converted into *waste products*. Catabolic reactions result in the *release* of energy whereas anabolic reactions result in the *consumption* of energy. In this chapter, we will consider some of the anabolic and catabolic processes used by microorganisms.

It is conventional to place microorganisms into metabolic classes, depending on the sources of energy which they use. All of the terms used to describe these classes employ the combining form *troph*, derived from a Greek word meaning "to feed." Thus, organisms which use *light* as an energy source are called **phototrophs** (*photo* is from the Greek for "light"). Organisms that use *chemicals* as energy sources are called **chemotrophs.** Most of the organisms we deal with in microbiology use *organic* compounds as energy sources and thus are called **chemoorganotrophs.** The material in the present chapter will deal with the metabolism of organic energy sources, and we reserve for Chapter 16 a discussion of the utilization of light and inorganic chemicals as energy sources.

A knowledge of cell metabolism is useful in understanding the biochemistry of microbial growth. Energy is needed for macromolecular synthesis and for the variety of chemical reactions needed for cell growth. Also, a knowledge of metabolism aids in developing useful laboratory procedures for culturing microorganisms, and in developing suitable procedures for preventing the growth of unwanted microorganisms. We will discuss some of these procedures at the end of this chapter. Because many of the important practical consequences of microbial growth, such as infectious disease and the production of useful products by microorganisms, are linked to microbial metabolism, a knowledge of cellular metabolism is of great use in applied and medical microbiology. Even the formation of metabolic waste products is of inter-

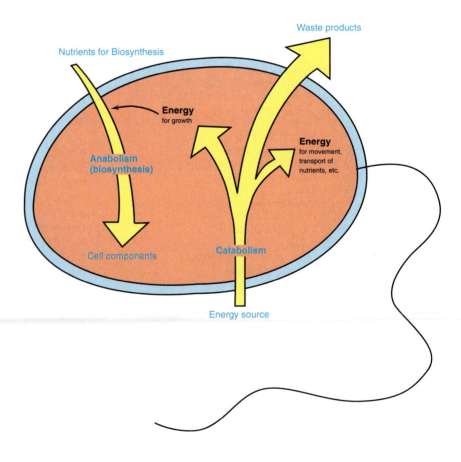

FIGURE 4.1
A simplified view of the major
features of cell metabolism.

Miniglossary for Chapter 4

AEROBE   a microorganism able to use O_2 in respiration

ALLOSTERIC   an enzyme that contains two combining sites, the active site (where substrate binds) and the allosteric site (where an effector molecule binds)

ANABOLISM   the sum total of all biosynthetic reactions in the cell

ANAEROBE   a microorganism unable to use O_2 in respiration and which may even be harmed or killed by O_2

ASEPTIC TECHNIQUE   methods for maintaining sterile culture media and other sterile objects free from contamination during manipulations

AUTOTROPH   an organism capable of biosynthesizing all cell material from CO_2 as the sole carbon source

CATABOLISM   biochemical reactions leading to the production of useable energy (usually ATP) by the cell

CHEMOLITHOTROPH   an organism which uses inorganic chemicals as energy sources (electron donors)

CHEMOORGANOTROPH   an organism which uses organic chemicals as energy sources (electron donor)

COENZYME   a small nonprotein molecule which participates in a catalytic reaction as part of an enzyme

CULTURE MEDIA   an aqueous solution of various nutrients suitable for the growth of microorganisms

ELECTRON ACCEPTOR   a substance that can accept electrons from some other substance, thereby becoming reduced in the process

ELECTRON DONOR   a substance that can donate electrons to some electron acceptor, thereby becoming oxidized in the process

ENZYME   a protein that has the ability to speed up (catalyze) a specific chemical reaction

FACULTATIVE   in reference to oxygen utilization, an organism capable of growing under either aerobic or anaerobic conditions

FERMENTATION   anaerobic catabolism of an organic compound in which the compound serves as both an electron donor and an electron acceptor and in which ATP is produced by substrate-level phosphorylation

HETEROTROPH   an organism requiring organic compounds as a carbon source

OXIDATIVE (ELECTRON TRANSPORT) PHOSPHORYLATION the nonphototrophic production of ATP at the expense of a proton motive force formed by electron transport

PHOTOTROPH   an organism capable of using light as an energy source

PROTON MOTIVE FORCE   the energy available from establishment of a proton gradient formed across a membrane

PURE CULTURE   a microbial culture containing a single kind of microorganism

REDUCTION POTENTIAL   the inherent tendency (measured in volts) of a compound to donate electrons

RESPIRATION   the process in which a compound is oxidized with O_2 serving as the terminal electron acceptor, usually accompanied by ATP production by oxidative phosphorylation

STERILE   absence of all living organisms and viruses

SUBSTRATE-LEVEL PHOSPHORYLATION   production of ATP by the direct transfer of a high-energy phosphate molecule from a phosphorylated organic compound to ADP

est. For instance, one important waste product produced by yeast during catabolism is *ethanol*, the key constituent of alcoholic beverages (wine, beer, whiskey, etc.). We will have more to say about the formation of ethanol by yeast later in this chapter.

> Metabolism is the study of the chemical reactions that are carried out in cells. Metabolism involves two basic kinds of chemical transformations, building up (biosynthetic) processes, called anabolism, and breaking down processes, called catabolism, which usually result in energy release. During metabolism, cells take in nutrients, convert them into cell components and obtain energy from them, and excrete waste products into the environment. Two kinds of energy sources can be used by cells—light and chemicals.

4.2   Energy

Energy is defined as the ability to do work. In this chapter, we discuss how living organisms use chemical energy. **Chemical energy** is the energy released when organic or inorganic compounds are oxidized. The first law of thermodynamics tells us that energy can be converted from one form to another, but can neither be created nor destroyed. Because its many forms are interconvertible, energy is most conve-

niently expressed by a single energy unit. Although a variety of units exist, in biology the most commonly used energy units are the kilocalorie (kcal) and the kilojoule (kJ). A kilocalorie is defined as the quantity of heat energy necessary to raise the temperature of 1 kilogram of water by 1°C. One kilocalorie is equivalent to 4.184 kilojoules. Because the kJ is widely used in microbial energetics, we will use the convention kilojoule throughout this book.

Free energy

Chemical reactions are accompanied by *changes* in energy. The amount of energy involved in a chemical reaction is expressed in terms of the gain or loss of energy during the reaction. There are two expressions of the amount of energy released during a chemical reaction, abbreviated H and G. H, called *enthalpy*, expresses the total amount of energy released during a chemical reaction. However, some of the energy released is not available to do useful work, but instead is lost as heat energy. G, called **free energy**, is used to express the energy released *that is available to do useful work*. The change in free energy during a reaction is expressed as $\Delta G^{0\prime}$, where the symbol Δ should be read to mean "change in." Reactant and product concentrations and pH (when H^+ is a reactant or product) will affect the observed free energy changes. The superscripts ^{0} and $\prime$ mean that a given free-energy value was obtained under "standard" conditions: pH 7, 25°C, all reactants and products initially at 1 molar concentration. If in the reaction

$$A + B \rightarrow C + D$$

the $\Delta G^{0\prime}$ is *negative*, then free energy is released and the reaction as written will occur spontaneously; such reactions are called **exergonic**. If, on the other hand, $\Delta G^{0\prime}$ is *positive*, the reaction will not occur spontaneously, but instead the reverse reaction (to the left) will occur spontaneously; such reactions are called **endergonic**.

In an exergonic reaction, the reaction proceeds until the concentration of products builds up, and then the reverse reaction, the conversion of products back to reactants, increases. An equilibrium is eventually reached in which the forward and reverse reactions are exactly balanced. This balanced condition does not mean that reactant and product occur in equal concentrations. The concentration of products and reactants at equilibrium is related to the free energy of the reaction. If the reaction proceeds with a *large* negative $\Delta G^{0\prime}$, then the equilibrium is far toward the products and very little of the reactants will remain. In contrast, if the reaction proceeds with a *small* negative $\Delta G^{0\prime}$, then at equilibrium there are nearly equal amounts of products and reactants. By determining the concentrations of products and reactants at equilibrium, it is possible to calculate the free-energy yield of any reaction (see Appendix 1 for further details).

In addition to speaking of the free energy *yield* of reactions, it is also necessary to talk about the free energy *of* individual substances. This is the *free energy of formation*, the energy yielded or energy required for the *formation* of a given molecule from the elements. By convention, the free energy of formation (G^{0}_{f}) of the elements (for instance, C, H_2, N_2) is zero. If the formation of a *compound* from the elements proceeds exergonically, then the free energy of formation of the compound is negative (energy is released), whereas if the reaction is endergonic (energy is required) then the free energy of formation of the compound is positive. A few examples of free energies of formation are given in Table 4.1. For most compounds G^{0}_{f} is *negative*,

reflecting the fact that compounds tend to form spontaneously from the elements. Again, the relative probabilities of different reactions (formations in this case) can be derived from comparison of the respective energies of formation. Thus we see that glucose (G^{0}_{f}, -917.22 kJ/mole) is more likely to form from carbon, hydrogen, and oxygen than is methane (G^{0}_{f}, -50.75 kJ/mole) to form from carbon and hydrogen. The positive G^{0}_{f} for nitrous oxide ($+104.18$ kJ/mole) tells us that this molecule will not form spontaneously but rather will decompose to nitrogen and oxygen. The free energies of formation of a variety of compounds of microbiological interest are given in Appendix 1.

Using free energies of formation, it is possible to calculate the *change* in free energy occurring in a given reaction. For a simple reaction, such as $A + B \rightarrow C + D$, $\Delta G^{0\prime}$ is calculated by subtracting the *sum* of the free energies of formation of the reactants (in this case A and B) from the products (C and D). Thus:

$$\Delta G^{0\prime} \text{ of } A + B \rightarrow C + D$$
$$= G^{0}_{f}[C + D] - G^{0}_{f}[A + B]$$

The saying "products minus reactants" summarizes the necessary steps for calculating changes in free energy during chemical reactions. However, it is necessary to balance the reaction chemically before free energy calculations can be made. Appendix 1 details the steps involved in calculating free energies for any hypothetical reaction.

> The chemical reactions of the cell are accompanied by changes in energy. The free energy of a reaction can be expressed quantitatively in terms of the kilojoules of energy used up or given off. A chemical reaction can occur with the release of free energy, in which case it is called exergonic, or with the consumption of free energy, in which case it is called endergonic.

4.3   Activation Energy, Catalysis, and Enzymes

Free-energy calculations tell us only what conditions will prevail when the reaction or system is at equilibrium; they do not tell us how long it will take for equilibrium to be reached. The formation of water from gaseous oxygen and hydrogen is a good example. The energetics of this reaction is quite favorable (free energy of formation of -237.17 kJ/mole). However, if we were to simply mix O_2 and H_2 together, no measurable formation of water would occur within our lifetime. The explanation is that the rearrangement of oxygen and hydrogen atoms to form water requires that the chemical bonds of the reactants be broken first. The breaking of bonds requires energy and this energy is referred to as **activation energy**. Activation energy is the amount of energy (in kJ) required to bring all molecules in a chemical reaction to the reactive state. For a reaction that proceeds with a net release of free energy (that is, an exergonic reaction), the situation is as diagrammed in Figure 4.2.

Table 4.1 Free energies of formation of a few compounds of biological interest	
Compound	**Free energy of formation**[a]
H_2O	-237.17
CO_2	-394.4
H_2	0
NH_4^{+}	-79.37
N_2O	$+104.18$
Glucose	-917.22
CH_4	-50.75

[a]*The free-energy values (G^{0}_{f}) are in kilojoules/mole.*

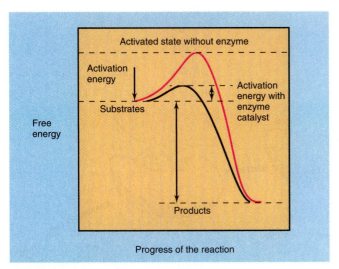

FIGURE 4.2   The concept of activation energy. Chemical reactions may not proceed spontaneously even though energy would be released, because the reactants must first be activated. Once activation has occurred, the reaction then proceeds spontaneously. Catalysts such as enzymes lower the required activation energy.

The idea of activation energy leads us to the concept of catalysis. A **catalyst** is a substance which serves to *lower* the activation energy of a reaction. A catalyst serves to *increase* the rate of reaction even though it itself is not changed. It is important to note that catalysts do not affect the energetics of a reaction; catalysts only affect the *speed* at which reactions proceed.

Most reactions in living organisms would not occur at appreciable rates without catalysis. The catalysts of biological reactions are proteins called **enzymes**. Enzymes are highly specific in the reactions which they catalyze. That is, each enzyme catalyzes only a *single type* of chemical reaction, or in the case of certain enzymes, a class of closely related reactions. This specificity is related to the precise three-dimensional structure of the enzyme molecule. In an enzyme-catalyzed reaction, the enzyme temporarily combines with the reactant, which is termed a **substrate** (S) of the enzyme, forming an **enzyme-substrate complex**. Then, as the reaction proceeds, the **product** (P) is released and the enzyme (E) is returned to its original state:

$$E + S \rightleftharpoons E\text{—}S \rightleftharpoons E + P$$

The enzyme is generally much larger than the substrate(s) and the combination of enzyme and substrate(s) usually depends on weak bonds, such as hydrogen bonds, van der Waal's forces, and hydrophobic interactions (see Section 2.8) to join the enzyme to the substrate. The small portion of the enzyme to which substrates bind is referred to as the **active site** of the enzyme.

Enzyme catalysis

The catalytic power of enzymes is impressive. Enzymes typically increase the rate of chemical reactions some 10^8 to 10^{20} times the rate that would occur spontaneously. To catalyze a specific reaction, an enzyme

must do two things: (1) bind the correct substrate, and (2) position the substrate relative to the catalytically active groups at the enzyme's active site. Binding of substrate to enzyme produces the enzyme-substrate complex (Figure 4.3). This serves to align reactive groups and places strain on specific bonds in the substrate(s). The result of enzyme-substrate complex formation is a reduction in the activation energy required to make the reaction proceed (Figure 4.2) with the conversion of substrate(s) to product(s). These steps are summarized diagrammatically in Figure 4.3.

Note that the reaction depicted in Figure 4.2 is exergonic, since the free energy of formation of the substrate is *greater* than that of the product. Enzymes can also catalyze endergonic reactions, converting energy-poor substrates into energy-rich products. In this case, not only must an activation energy barrier be overcome, but sufficient free energy must be put *into* the system to raise the energy level of the substrates to that of the products. Although theoretically all enzymes are reversible in their action, in practice, enzymes catalyzing highly exergonic or highly endergonic reactions are essentially unidirectional. If a particularly exergonic reaction needs to be reversed during cellular metabolism, a distinctly different enzyme is frequently involved in the reaction.

Structure of enzymes

As we have discussed, enzymes are proteins, polymers of subunits called amino acids (see Section 2.8). Each enzyme has a specific three-dimensional shape. The linear array of amino acids (primary structure) will fold and twist into a specific configuration to achieve secondary and tertiary structure. The precise conformation of an enzyme may be seen more easily in a computer-generated space-filling model (Figure 4.4). In this example, the large cleft is the site where the substrate binds (the active site).

Being proteins, enzymes are also subject to the effects of physical and chemical variables, most notably temperature and pH. For example, enzymes from organisms capable of growth at high temperatures (thermophiles and hyperthermophiles, see Section 9.9) are frequently quite heat stable, while those from nonthermophiles are rapidly inactivated by heat. Such differences arise from variations in amino acid *sequence* in different enzymes. Recall that differences in amino acid sequence eventually translate into differences in *folding* of the protein. A specifically folded protein will thus assume specific binding and physical properties.

Many enzymes contain small nonprotein molecules which participate in the catalytic function. These small enzyme-associated molecules are divided into two categories on the basis of the nature of the association with the enzyme, *prosthetic groups* and *coenzymes*. **Prosthetic groups** are bound very tightly to their enzymes, usually permanently. If an enzyme contains a prosthetic group, the protein part alone is called an *apoenzyme* and the complete enzyme, formed when the prosthetic group is attached, is called a *holoenzyme*. The heme group present in cytochromes is an example

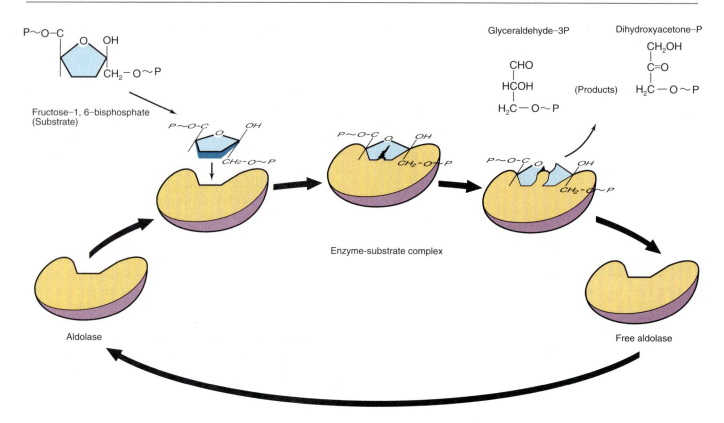

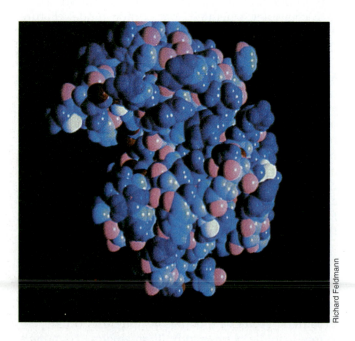

FIGURE 4.3   The catalytic cycle of an enzyme as depicted for the enzyme fructose bisphosphate aldolase. This enzyme catalyzes the following reaction: fructose-1,6-bisphosphate ⇌ glyceraldehyde-3-phosphate + dihydroxyacetone phosphate. Following binding of fructose 1,6-bisphosphate in the formation of the enzyme–substrate complex, the shape of the enzyme is altered, placing strain on certain bonds of the substrate which break and yield the two products.

FIGURE 4.4   Computer-generated space-filling model of the enzyme lysozyme. The substrate-binding site (active site) is in the large cleft on the left side of the model. See Section 3.5 for a discussion of the mode of action of lysozyme.

of a prosthetic group; cytochromes will be described in detail later in this chapter. **Coenzymes** are bound rather loosely to enzymes and a single coenzyme molecule may associate with a number of different enzymes at different times during growth. Coenzymes serve as intermediate carriers of small molecules from one enzyme to another. Most coenzymes are derivatives of vitamins.

Enzymes are named either for the substrate they bind or for the chemical reaction which they catalyze, by addition of the combining form *-ase*. Thus cellul*ase* is an enzyme that attacks cellulose, glucose oxid*ase* is an enzyme that catalyzes the oxidation of glucose, and ribonucle*ase* is an enzyme that decomposes ribonucleic acid.

> The reactants in a chemical reaction must first be activated before the reaction can take place and this requires the input of activation energy. The amount of activation energy required can be decreased by the use of a catalyst and the catalysts of living cells are called enzymes. Enzymes are proteins which are highly specific in the reactions which they catalyze and this specificity resides in the folding pattern of the polypeptide(s) in the protein.

4.4   Oxidation–Reduction

The utilization of chemical energy in living organisms involves **oxidation–reduction** (sometimes called **redox**) **reactions**. Chemically, an oxidation is defined as the *removal* of an electron or electrons from a substance. A reduction is defined as the *addition* of an electron (or electrons) to a substance. In biochemistry, oxidations and reductions frequently involve the transfer of not just electrons, but whole hydrogen atoms. A hydrogen atom (H) consists of an electron plus a proton. When the electron is removed, the hydrogen atom becomes a *proton* (or hydrogen ion, H^+). We will on occasion need to distinguish between oxidation-reduction reactions which involve electrons only or hydrogen atoms only, but the energetic concepts developed here will use the transfer of hydrogen atoms as primary examples.

Many oxidation–reduction reactions do *not* involve molecular oxygen (O_2). Instead, it is the transfer of hydrogen atoms (or electrons) that is important. For example, hydrogen gas, H_2, can release electrons and hydrogen ions (protons) and become oxidized:

$$H_2 \rightarrow 2e^- + 2H^+$$

However, electrons cannot exist alone in solution; they must be part of atoms or molecules. The equation as drawn thus gives us chemical information but does not itself represent a real reaction. The above reaction is only a *half reaction*, a term which implies the need for a second reaction. This is because for any oxidation to occur, a subsequent reduction must also occur. For example, the oxidation of H_2 could be coupled to the reduction of many different substances including O_2 in a second reaction:

$$\tfrac{1}{2}O_2 + 2e^- + 2H^+ \rightarrow H_2O$$

This half reaction, which is a reduction, when coupled to the oxidation of H_2, yields the following overall balanced reaction:

$$H_2 + \tfrac{1}{2}O_2 \rightarrow H_2O$$

In reactions of this type, we will refer to the substance *oxidized*, in this case H_2, as the **electron donor**, and the substance *reduced*, in this case O_2, as the **electron acceptor** (Figure 4.5). The key to understanding biological oxidations and reductions is to keep straight the proper half reactions—there must always be one reaction which involves an electron donor and another reaction involving an electron acceptor.

Reduction potentials

Substances vary in their tendencies to give up electrons and become oxidized or to accept electrons and become reduced. This tendency is expressed as the **reduction potential (E_0)** of the substance. This potential is measured electrically in reference to a standard substance, H_2. By relating all potentials to a standard, it is possible to express potentials for various half reactions on a single scale, making possible easy compar-

$$H_2 \rightarrow 2e^- + \boxed{2H^+}$$

Electron donating half-reaction

$$\tfrac{1}{2}O_2 + 2e^- \rightarrow \boxed{O^{2-}}$$

Electron accepting half-reaction

$$\boxed{2H^+} + \boxed{O^{2-}} \rightarrow H_2O$$

Formation of water

$$H_2 + \tfrac{1}{2}O_2 \rightarrow H_2O$$

Net reaction

H_2 is the reductant (electron donor)
It becomes oxidized

O_2 is the oxidant (electron acceptor)
It becomes reduced

FIGURE 4.5   Example of a coupled oxidation–reduction reaction.

isons between various reactions. By convention, reduction potentials are expressed for half reactions written with the oxidant on the left, that is, as *reductions*. Thus, oxidant + $e^- \rightarrow$ reduced product. If protons are involved in the reaction, as is often the case, then the reduction potential will to some extent be influenced by the hydrogen ion concentration (pH). By convention in biology, reduction potentials are given for neutrality (pH 7), since the cytoplasm of the cell is neutral or nearly so. Using these conventions, at pH 7 the reduction potential (E_0') of

$$\tfrac{1}{2}O_2 + 2H^+ + 2e^- \rightarrow H_2O$$

is +0.816 V, and that of

$$2H^+ + 2e^- \rightarrow H_2$$

is −0.421 V.

Oxidation–reduction pairs and coupled reactions

Most molecules can serve as either electron donors or electron acceptors under different circumstances, depending upon what other substances they react with. The same atom on each side of the arrow in the half reactions can be thought of as representing an oxidation-reduction (O–R) pair, such as $2H^+/H_2$ or $\tfrac{1}{2}O_2/H_2O$. When writing an O–R pair, the *oxidized* form will always be placed on the left.

In constructing coupled oxidation–reduction reactions from their constituent half reactions, it is simplest to remember that the reduced substance of an O–R pair whose reduction potential is more negative *donates* electrons to the oxidized substance of an O–R pair whose potential is more positive. Thus, in the redox pair $2H^+/H_2$ which has a potential of −0.42 V, H_2 has a great tendency to *donate* electrons. On the other hand, in the redox pair $\tfrac{1}{2}O_2/H_2O$, which has a potential of +0.82 V, H_2O has a very slight tendency to donate electrons, but O_2 has a great tendency to *accept* electrons. It follows then that in the coupled reaction of H_2 and O_2, H_2 will serve as the electron *donor*, and become oxidized, and O_2 will serve as the electron *acceptor* and become reduced (Figure 4.5). Even though by chemical convention both half reactions are written as reductions, in an O–R reaction one of the two half reactions must be written as an oxidation and therefore will proceed in the reverse direction. Thus note that in the reaction shown in Figure 4.5, the oxidation of H_2 to $2H^+ + 2e^-$ is reversed from the formal half reaction, written as a reduction.

The electron tower

A convenient way of viewing electron transfer in biological systems is to imagine a vertical tower (Figure 4.6). The tower represents the range of reduction potentials for O–R pairs, from the most negative at the top to the most positive at the bottom. The reduced substance in the pair at the top of the tower has the *greatest* amount of potential energy (roughly, the energy that it took to lift the substance to the top), and

the reduced substance at the bottom of the tower has the *least* amount of potential energy. On the other hand, the oxidized substance in the O–R pair at the top of the tower has the *least* tendency to accept electrons, whereas the oxidized substance in the pair at the bottom of the tower has the *greatest* tendency to accept electrons.

As electrons from the electron donor at the top of the tower fall, they can be "caught" by acceptors at various levels. The difference in electrical potential between two substances is expressed as $\Delta E_0'$. The farther the electrons drop before they are caught, the greater the amount of energy released; that is, $\Delta E_0'$ *is proportional to* $\Delta G^{0'}$. O_2, at the bottom of the tower, is the final acceptor (or put in other terms, is the most powerful oxidizing agent). In the middle of the tower, the O–R pairs can act as either electron donors or acceptors. For instance, the $2H^+/H_2$ couple has a reduction potential of -0.42 volts. The fumarate/succinate couple has a potential of $+0.02$ volts. Hence, the oxidation of hydrogen can be coupled to the reduction of fumarate:

$$H_2 + \text{fumarate} \rightarrow \text{succinate}$$

On the other hand, the oxidation of succinate to fumarate can be coupled to the reduction of NO_3^- or $\frac{1}{2}O_2$:

$$\text{Succinate} + NO_3^- \rightarrow \text{fumarate} + NO_2^-$$
$$\text{Succinate} + \frac{1}{2}O_2 \rightarrow \text{fumarate} + H_2O$$

Hence, under conditions in which oxygen is absent (called *anaerobic* or *anoxic*) in the presence of H_2, fumarate can act as an electron acceptor (producing succinate), and under other conditions (e.g., anaerobic in the presence of NO_3^-, or aerobic) succinate can act as an electron donor (producing fumarate). Indeed, all of the transformations involving fumarate and succinate described here are carried out by various microorganisms under certain nutritional and environmental conditions.

In catabolism the electron donor is often referred to as an **energy source**. It is necessary to remember, however, that it is the *coupled* oxidation–reduction reaction which actually releases energy. As discussed in the context of the electron tower, the amount of energy released in an O–R reaction depends on the nature of *both* the electron donor and the electron acceptor: the greater the difference between reduction potentials of the two half reactions, the more energy there will be released upon their coupling (see Appendix 1).

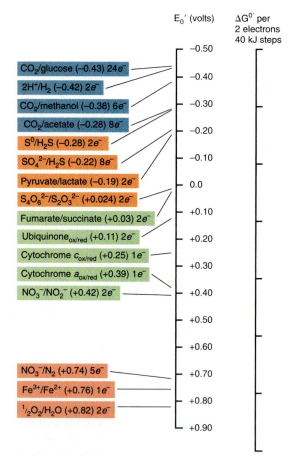

E_0' (volts) $\Delta G^{0'}$ per 2 electrons 40 kJ steps

- CO_2/glucose (−0.43) $24e^-$
- $2H^+/H_2$ (−0.42) $2e^-$
- CO_2/methanol (−0.38) $6e^-$
- CO_2/acetate (−0.28) $8e^-$
- S^0/H_2S (−0.28) $2e^-$
- SO_4^{2-}/H_2S (−0.22) $8e^-$
- Pyruvate/lactate (−0.19) $2e^-$
- $S_4O_6^{2-}/S_2O_3^{2-}$ (+0.024) $2e^-$
- Fumarate/succinate (+0.03) $2e^-$
- Ubiquinone$_{ox/red}$ (+0.11) $2e^-$
- Cytochrome $c_{ox/red}$ (+0.25) $1e^-$
- Cytochrome $a_{ox/red}$ (+0.39) $1e^-$
- NO_3^-/NO_2^- (+0.42) $2e^-$
- NO_3^-/N_2 (+0.74) $5e^-$
- Fe^{3+}/Fe^{2+} (+0.76) $1e^-$
- $\frac{1}{2}O_2/H_2O$ (+0.82) $2e^-$

Scale: −0.50, −0.40, −0.30, −0.20, −0.10, 0.0, +0.10, +0.20, +0.30, +0.40, +0.50, +0.60, +0.70, +0.80, +0.90

FIGURE 4.6   The electron tower. Redox pairs are arranged from the strongest reductants (negative reduction potentials) at the top to the strongest oxidants (positive reduction potentials) at the bottom. As electrons are donated from the top of the tower, they can be "caught" by acceptors at various levels. The farther the electrons fall before they are caught, the greater the difference in reduction potential between electron donor and electron acceptor, and the more energy is released. On the right, the energy released is given in 40 kJ/mol steps, assuming two electron transfers in each case. Some of the half-reactions indicated involve the transfer of several electrons, for example, the CO_2/glucose couple. This reaction is included as a biologically important example of an overall process. The actual reduction of CO_2 to glucose involves several smaller redox reactions.

> Oxidation–reduction reactions, which are involved in the energy-yielding reactions of cells, involve the transfer of electrons from one reactant to another. The energy source, which is the electron donor, gives up one or more electrons, which are transferred to an electron acceptor. In this process, the electron donor is oxidized and the electron acceptor is reduced. One of the most common electron acceptors of living organisms is molecular oxygen. The tendency of a compound to accept or release electrons is expressed quantitatively by its reduction potential.

4.5 Electron Carriers

In the cell, the transfer of electrons in an oxidation-reduction reaction from donor to acceptor involves one or more intermediates, referred to as **carriers**. When such carriers are used, we refer to the initial donor as the **primary electron donor** and to the final acceptor as the **terminal electron acceptor**. The net energy change of the complete reaction sequence is determined by the *difference* in reduction potentials between the primary donor and the terminal acceptor. The transfer of electrons through the intermediates involves a series of oxidation-reduction reactions, but the energy change from these individual steps must add up to the value obtained by considering only the starting and ending compounds.

The intermediate electron carriers may be divided into two general classes: those freely diffusible, and those firmly attached to enzymes in the cytoplasmic membrane. The fixed carriers function in membrane-associated electron transport reactions and are discussed in Section 4.10. Freely diffusible carriers include the coenzymes NAD$^+$ (nicotinamide-adenine dinucleotide) and NADP$^+$ (NAD-phosphate) (Figure 4.7). NAD$^+$ and NADP$^+$ are *hydrogen atom* carriers and always transfer two hydrogen atoms to the next carrier in the chain. Such hydrogen atom transfer is referred to as a dehydrogenation.*

The reduction potential of the NAD$^+$/NADH (or NADP$^+$/NADPH) pair is −0.32 V, which places them fairly high on the electron tower; that is, NADH (or NADPH) is a good electron *donor*. Although the NAD$^+$ and NADP$^+$ couples have the same reduction potentials, they generally serve in different capacities in the cell. NAD$^+$ is directly involved in energy-generating (catabolic) reactions, whereas NADP$^+$ is involved primarily in biosynthetic (anabolic) reactions.

Coenzymes increase the diversity of possible O–R reactions by making it possible for chemically dissimilar molecules to be coupled as initial electron donor and ultimate electron acceptor, the coenzyme acting as intermediary. As we have discussed, most biological reactions are catalyzed by specific enzymes which can only react with a limited range of substrates. Oxidation-reduction reactions may be considered to proceed in three stages: removal of electrons from the primary donor, transfer of electrons through one or a series of electron carriers, and addition of electrons to the terminal acceptor. Each step in the reaction is catalyzed by a different enzyme, each of which binds to its substrate and to its specific coenzyme. Figure 4.8 is a schematic diagram showing the functioning of the coenzyme NAD$^+$. Note that after a coenzyme has performed its chemical function in one reaction it can diffuse through the cytoplasm until it collides with another enzyme that requires the coenzyme in that form. Following conversion of the coenzyme back to its original form, the whole process can be repeated again.

---

*Strictly speaking NAD$^+$ or NADP$^+$ carry two electrons and one proton, the second H$^+$ being released to solution. Therefore, NAD$^+$ + 2e$^-$ + 2H$^+$ actually yields NADH + H$^+$. However, for simplicity, we will write "NADH + H$^+$" as NADH.

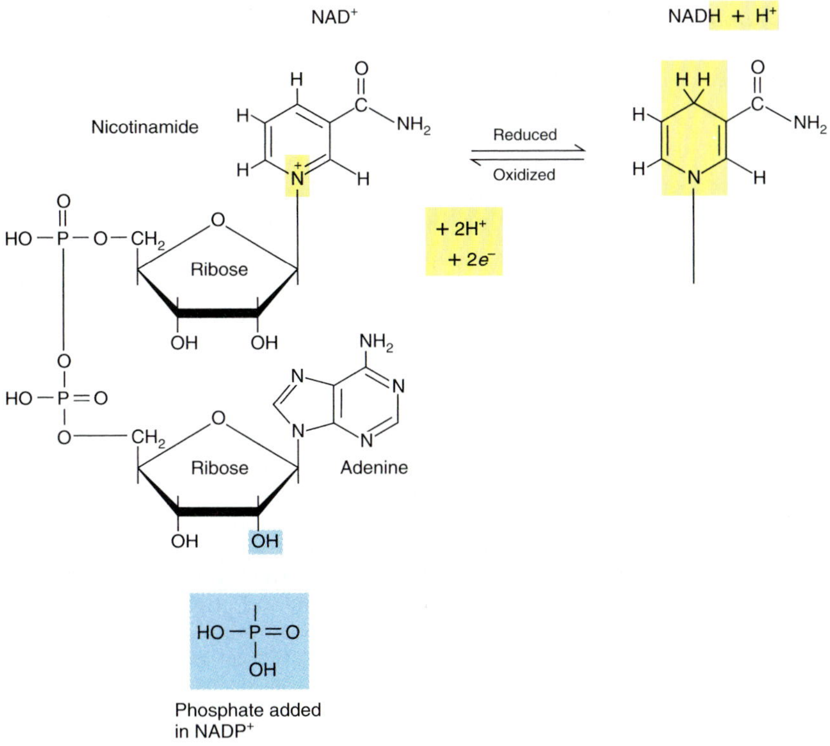

> The transfer of electrons from donor to acceptor in a cell involves the participation of one or more electron carriers. Some electron carriers are fixed in membranous structures whereas others are freely diffusible, transferring electrons from one place to another in the cell. Two of the most common electron carriers are NAD$^+$ and NADP$^+$.

FIGURE 4.7
Structure of the oxidation–reduction coenzyme nicotinamide adenine dinucleotide (NAD$^+$). In NADP$^+$, a phosphate group is present, as indicated. Both NAD$^+$ and NADP$^+$ undergo oxidation–reduction as shown.

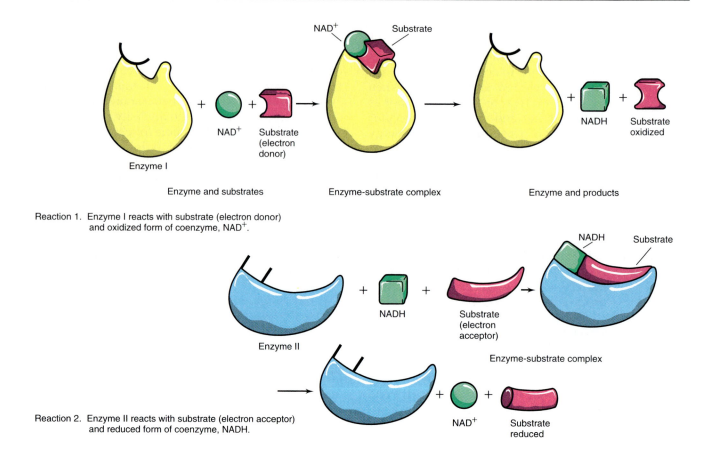

Reaction 1. Enzyme I reacts with substrate (electron donor)
and oxidized form of coenzyme, NAD⁺.

Reaction 2. Enzyme II reacts with substrate (electron acceptor)
and reduced form of coenzyme, NADH.

FIGURE 4.8   Schematic example of an oxidation–reduction reaction involving the coenzyme $NAD^+/NADH$.

4.6   High Energy Phosphate Compounds and Adenosine Triphosphate (ATP)

Energy released as a result of oxidation-reduction reactions must be conserved for cell functions. In living organisms, chemical energy released in O–R reactions is most commonly transferred to a variety of phosphate compounds in the form of **high-energy phosphate bonds**, and these compounds then serve as intermediaries in the conversion of energy into useful work.

In phosphorylated compounds, phosphate groups are attached via oxygen atoms in *ester* or *anhydride* linkage, as illustrated in Figure 4.9. However, not all phosphate ester linkages are high-energy bonds. As a means of expressing the energy of phosphate bonds, the free energy released when water is added and the bond is hydrolyzed can be given. As seen in Figure 4.9, the $\Delta G^{0'}$ of hydrolysis of the phosphate bond in glucose-6-phosphate is only -13.8 kJ/mole, whereas the $\Delta G^{0'}$ of hydrolysis of the phosphate bond in phosphoenolpyruvate is -51.6 kJ/mole, almost four times that of glucose-6-phosphate. Thus phosphoenolpyruvate is considered a *high-energy compound* while glucose-6-phosphate is not. (Free energies of hydrolysis for a number of other high-energy phosphate compounds are given in Appendix 1.)

Adenosine triphosphate (ATP)

The most important high energy phosphate compound in living organisms is adenosine triphosphate (ATP). ATP consists of the ribonucleoside adenosine, to which three phosphate molecules are bonded in series (Figure 4.9). ATP serves as the prime energy carrier in living organisms, being generated during exergonic reactions and being used to drive endergonic reactions. From the structure of ATP (Figure 4.9) it can be seen that two of the phosphate bonds of ATP have high free energies of hydrolysis.

It should be emphasized that although we express the energy of the high-energy phosphate bonds in terms of the free energy of hydrolysis, in actuality it is undesirable for these bonds to hydrolyze in cells in the absence of a second reaction which can utilize the energy released, since the free energy of hydrolysis will then be lost to the cell as heat. The free energy of high energy phosphate bonds is generally used to drive biosynthetic reactions and other aspects of cell function through carefully regulated processes in which the energy released from ATP hydrolysis is coupled to energy-requiring reactions.

> The energy released in oxidation-reduction reactions is not wasted but is conserved by the formation of certain biochemical compounds that contain high-energy phosphate bonds. One of the most common high-energy phosphate compounds is ATP, which serves as a prime energy carrier in the cell.

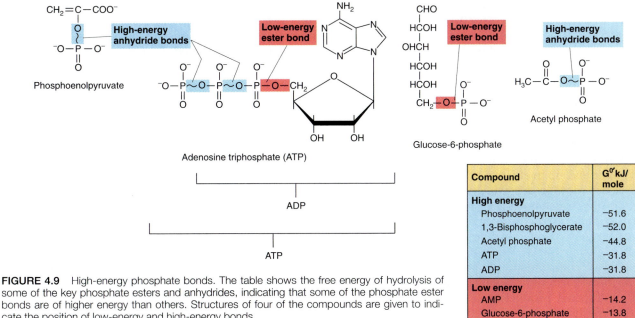

FIGURE 4.9 High-energy phosphate bonds. The table shows the free energy of hydrolysis of some of the key phosphate esters and anhydrides, indicating that some of the phosphate ester bonds are of higher energy than others. Structures of four of the compounds are given to indicate the position of low-energy and high-energy bonds.

Compound	$G^{0'}$kJ/mole
High energy	
Phosphoenolpyruvate	−51.6
1,3-Bisphosphoglycerate	−52.0
Acetyl phosphate	−44.8
ATP	−31.8
ADP	−31.8
Low energy	
AMP	−14.2
Glucose-6-phosphate	−13.8

4.7   Energy Release in Biological Systems

Organisms use a wide variety of energy sources in the synthesis of the high-energy phosphate bonds in ATP. Both chemical energy and light energy are used to synthesize ATP. Among chemical substances, both organic and inorganic compounds may serve as electron donors. Similarly, a number of different electron acceptors can be used in coupled redox reactions, including both organic and inorganic compounds. In the following pages we will be concerned with the mechanisms by which ATP is synthesized as a result of oxidation-reduction reactions involving *organic* compounds. Metabolism of organic compounds is the source of energy for all animals and for the vast majority of microorganisms and we begin our study of energy yielding reactions here. In Chapter 16 we will consider mechanisms of energy generation from light and inorganic chemicals.

The series of reactions involving the oxidation of a compound is called a **biochemical pathway**. The pathways for the oxidation of organic compounds and conservation of energy in ATP can be divided into two major groups: (1) **fermentation**, in which the O–R process occurs in the *absence* of any added terminal electron acceptors; and (2) **respiration**, in which molecular oxygen or some other oxidant serves as the terminal electron acceptor.

4.8   Fermentation

In the absence of externally supplied electron acceptors, many organisms perform *internally balanced* oxidation-reduction reactions of organic compounds with the release of energy, a process called **fermentation**. There are many different types of fermentations (see Chapter 16), but under fermentative conditions only *partial* oxidation of the carbon atoms of the organic compound occurs and therefore only a small amount of the potential energy available is released. The oxidation in a fermentation is coupled to the subsequent reduction of an organic compound generated from catabolism of the initial fermentable substrate; thus, no externally supplied electron acceptor is required (Figure 4.10).

ATP is produced in fermentations by a process called **substrate-level phosphorylation**. In substrate-level phosphorylation, ATP is synthesized during specific enzymatic steps in the catabolism of the organic compound. This is in contrast to **oxidative** (or **electron transport**) **phosphorylation** (discussed later), where ATP is produced via membrane-mediated events not connected directly to the metabolism of specific substrates.

An example of fermentation is the catabolism of glucose by a lactic acid bacterium:

$$\text{glucose} \rightarrow 2 \text{ lactic acid}$$
$$C_6H_{12}O_6 \rightarrow 2C_3H_6O_3$$

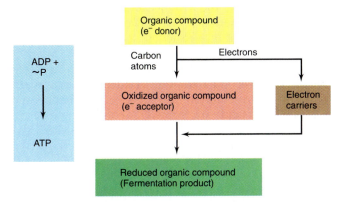

FIGURE 4.10 Carbon and electron flow in fermentation. Note that there is no externally supplied electron acceptor.

Note that this is a balanced reaction and that the product, lactic acid, has the same proportion of hydrogen and oxygen atoms as glucose. Likewise, a similar situation exists in the catabolism of glucose by yeast in the absence of oxygen:

$$\text{glucose} \rightarrow 2 \text{ ethanol} + 2 \text{ carbon dioxide}$$
$$C_6H_{12}O_6 \rightarrow 2C_2H_6O + 2CO_2$$

Note that in this reaction some of the carbon atoms end up in CO_2, a more *oxidized* form than the carbon atoms in the starting molecule, glucose, while other carbon atoms end up in ethanol, which is more *reduced* (that is, it has more hydrogens and electrons per carbon atom) than glucose. In each fermentation, oxidation–reduction is internally balanced.

The energy released in the fermentation of glucose to ethanol or lactic acid (-235.08 and -118.4 kJ/mole, respectively) is conserved by substrate-level phosphorylations in the form of high-energy phosphate bonds in ATP, with a net production of *two* such bonds in each case. We discuss now the details of the fermentation of glucose and the manner in which some of the energy released is conserved in high-energy phosphate bonds.

> Some compounds can be used as energy sources even without the presence of an externally added electron acceptor in a process called fermentation. A sugar such as glucose is a good example of a fermentable chemical. A fermentation is an internally balanced oxidation–reduction reaction in which some atoms of the energy source become more reduced while others become more oxidized, and energy is produced by substrate-level phosphorylation. Only a small amount of energy is released during glucose fermentation, most of the energy remaining in the reduced fermentation product.

Glucose fermentation: oxidation and ATP production

The biochemical pathway for the breakdown of glucose is actually quite simple. It can be divided into three major stages (Figure 4.11). Stage I is a series of preparatory rearrangements, reactions that do not involve oxidation–reduction and do not release energy, but which lead to the production of two molecules of the key intermediate, *glyceraldehyde-3-phosphate*. In Stage II, oxidation–reduction occurs, high-energy phosphate bond energy is produced in the form of ATP, and two molecules of pyruvate are formed. In Stage III, a second oxidation–reduction reaction occurs, and fermentation products (e.g., ethanol and CO_2, or lactic acid) are formed. A common biochemical pathway from glucose to pyruvate, is called **glycolysis**, and it is also called the Embden–Meyerhof (EM) pathway after its discoverers.

In Stage I, glucose is phosphorylated by ATP yielding glucose-6-phosphate; the latter can also be formed when glucose is transported by the phosphotransferase system (see Section 3.4). Phosphorylation

reactions of this sort often occur preliminary to oxidation. The initial phosphorylation of glucose *activates* the molecule for the subsequent reactions. Glucose-6-phosphate is converted into its isomer, fructose-6-phosphate, and a second phosphorylation leads to the production of *fructose-1,6-bisphosphate,* which is a key intermediate product. The enzyme *aldolase* catalyzes the splitting of fructose-1,6-bisphosphate into two three-carbon molecules, glyceraldehyde-3-phosphate and its isomer, dihydroxyacetone phosphate.* Note that thus far, there have been no oxidation-reduction reactions, and that all the reactions, including the consumption of ATP, proceed without electron transfers.

The oxidation reaction of glycolysis occurs in the conversion of glyceraldehyde-3-phosphate to 1,3-bisphosphoglyceric acid. In this reaction (which occurs twice, once for each molecule of glyceraldehyde-3-phosphate), an enzyme involving the coenzyme NAD^+ accepts two hydrogen atoms and NAD^+ is converted into NADH. Simultaneously, each glyceraldehyde-3-P molecule is phosphorylated by addition of a molecule of inorganic phosphate. This energetically favorable reaction in which inorganic phosphate has been converted to organic form sets the stage for the next process, the step in which ATP is actually formed. High-energy bond formation is possible because each of the phosphates on a molecule of 1,3-bisphosphoglyceric acid represents a high-energy phosphate bond (see Figure 4.9). The synthesis of ATP occurs when each molecule of 1,3-bisphosphoglyceric acid is converted to 3-phosphoglyceric acid, and later on in the pathway, when each molecule of phosphoenolpyruvate is converted to pyruvate.

In glycolysis, *two* ATP molecules are consumed in the two phosphorylations of glucose, and *four* ATP molecules are synthesized (two from each 1,3-bisphosphoglyceric acid converted to pyruvate). Thus the *net gain* to the organism is two molecules of ATP per molecule of glucose fermented.

Glucose fermentation: reductive steps

During the formation of two molecules of 1,3-bisphosphoglyceric acid, two molecules of NAD^+ were reduced to NADH (see Figure 4.11). However, a cell contains only a small amount of NAD^+, and if all of it were converted to NADH, the oxidation of glucose would stop; the continued oxidation of glyceraldehyde-3-phosphate can proceed only if there is a molecule of NAD^+ present to accept released electrons. This "roadblock" is overcome in fermentation by the oxidation of NADH back to NAD^+ through reactions involving the reduction of pyruvate to any of a variety of **fermentation products**. In the case of yeast, pyruvate is reduced to ethanol with the release of CO_2. In lactic acid bacteria (or in muscle tissue rendered anaerobic by vigorous exercise) pyruvate is reduced to lactic acid (see lower

---

*There is an enzyme that catalyzes the interconversion of dihydroxyacetone phosphate and glyceraldehyde-3-phosphate. For simplicity, we consider here only glyceraldehyde-3-phosphate since it is the compound which is further metabolized.

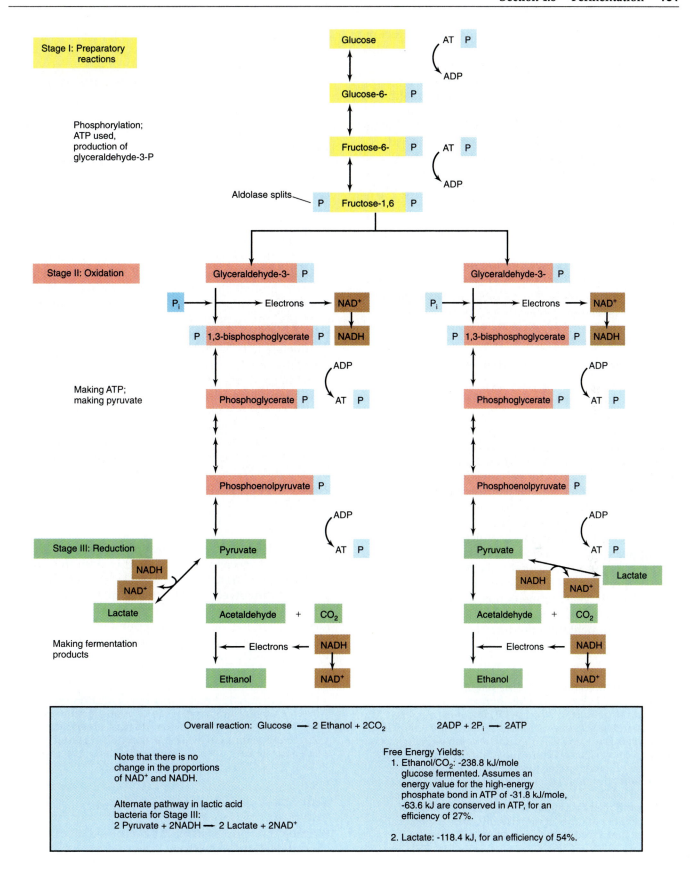

FIGURE 4.11   Embden–Meyerhof pathway, the sequence of enzymatic reactions in the conversion of glucose to pyruvate and then to fermentation products. In the example shown (from yeast), ethanol and CO_2 are the fermentation products. In many organisms, lactate is a common fermentation product. The steps from glucose to pyruvate are sometimes collectively called glycolysis.

part of Figure 4.11). Many routes of pyruvate reduction in various fermentative prokaryotes are known (discussed in Chapter 16), but the net result is the same; NADH must be returned to the oxidized form, NAD^+, for the energy-yielding reactions of fermentation to continue. As a diffusible coenzyme, NADH can move away from the enzyme which oxidizes glyceraldehyde-3-phosphate, attach to an enzyme which reduces pyruvate to lactic acid (or to some other product), and diffuse away once again following conversion to NAD^+ to repeat the cycle all over again (see Figure 4.8 for details of this mechanism).

In any energy-yielding process, oxidation must balance reduction, and there must be an electron acceptor for each electron removed. In this case, the *reduction* of NAD^+ at one enzymatic step is coupled with its *oxidation* at another. The final product(s) must also be in oxidation–reduction balance with the starting substrate, glucose. Hence, the products discussed here, ethanol plus CO_2 or lactic acid, are in electrical balance with the starting glucose.

Glucose fermentation: net and practical results

The ultimate result of glycolysis is the consumption of glucose, the net synthesis of two ATP, and the production of fermentation products. For the organism the crucial product is ATP, which is used in a wide variety of energy-requiring reactions, and fermentation products are merely waste products. However, the latter substances would hardly be considered waste products by the distiller, the brewer, or the cheesemaker. The anaerobic fermentation of glucose by yeast is a means of producing ethanol, the desired product in alcoholic beverages, and the production of lactic acid from glucose by lactic acid bacteria is the initial step in the production of fermented milk products including cheeses. For the baker, on the other hand, the production of CO_2 by yeast fermentation is the essential step in the leavening of bread. We discuss industrial production of fermentation products for commercial benefit in Chapter 10.

> ATP is produced during glucose fermentation by a process called glycolysis. This involves a series of biochemical reactions in which ATP functions as energy carrier and NAD^+ as electron carrier. The end result of glycolysis is the release of a small amount of energy that is used for various cell functions, and the loss of a large amount of energy in the form of fermentation products. Common fermentation products of glycolysis include ethanol (produced by yeast and certain bacteria), lactic acid (produced by animal muscle and many bacteria), and a variety of other acids, alcohols, and gaseous substances (produced by a variety of other bacteria).

4.9 Respiration

We have discussed above the catabolism of glucose as it occurs in the *absence* of external electron acceptors. A relatively small amount of energy is released in this process (and few ATP molecules synthesized). This

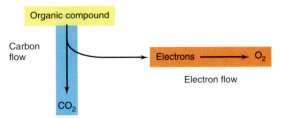

FIGURE 4.12 Carbon and electron flow in aerobic respiration. Note that an externally supplied electron acceptor (O_2) is present.

small energy release may be understood in terms of the formal principles of oxidation–reduction reactions. Fermentation processes yield little energy for two reasons: (1) the carbon atoms in the starting compound are only partially oxidized, and (2) the difference in reduction potentials between the primary electron donor and terminal electron acceptor is small. However, if O_2 or some other *external* terminal acceptor is present, all the substrate molecules can be oxidized completely to CO_2, and a far *higher* yield of ATP is theoretically possible. The process by which a compound is oxidized using O_2 as external electron acceptor is called **aerobic respiration** (Figure 4.12).

The greater energy release during respiration occurs because respiring cells surmount the two limitations just listed for fermentation: (1) the carbon atoms in the starting compound can be completely oxidized to CO_2; and (2) the terminal electron acceptor has a relatively positive reduction potential, leading to a large net difference in potentials between primary donor and terminal acceptor and therefore the synthesis of much ATP.

Our discussion of respiration will deal with the biochemical mechanisms involved in both the carbon and electron transformations: (1) the biochemical pathways involved in the transformation of organic carbon to CO_2 and (2) the way electrons are transferred from the organic compound to the terminal electron acceptor, driving ATP synthesis (Figure 4.12). We begin with a discussion of electron flow.

> Aerobic respiration is an oxidation–reduction process in which molecular oxygen serves as an electron acceptor. Because of oxygen's high reduction potential, the maximum amount of available energy is released from an energy source when oxygen is the electron acceptor.

4.10 Electron Transport Systems

Electron transport systems are composed of *membrane-associated* electron carriers. These systems have two basic functions: (1) to accept electrons from an electron donor and transfer them to an electron acceptor and (2) to conserve some of the energy released during electron transfer for synthesis of ATP.

Several types of oxidation–reduction enzymes are involved in electron transport: (1) NADH dehydrogenases, which transfer hydrogen atoms from NADH;

(2) riboflavin-containing electron carriers, generally called flavoproteins (such as FAD); (3) iron–sulfur proteins, and (4) cytochromes, which are proteins containing an iron–porphyrin ring called *heme*. In addition, one class of nonprotein electron carriers is known, the lipid soluble quinones, sometimes called coenzymes Q. These electron transport components diffuse freely through the membrane, generally transferring electrons from iron–sulfur proteins to cytochromes. We discuss each of these classes of electron transport components below.

NADH dehydrogenases are proteins bound to the inside surface of the cell membrane. They accept electrons from NADH, generated in various cellular reactions, and pass two hydrogen atoms to flavoproteins.

Flavoproteins are proteins containing a derivative of riboflavin (Figure 4.13); the flavin portion, which is bound to a protein, is the prosthetic group, which is alternately reduced as it accepts hydrogen atoms, and oxidized when electrons are passed on. Note that flavoproteins *accept* hydrogen atoms and *donate* electrons; we will consider what happens to the two protons later. Two flavins are known, flavin mononucleotide (FMN) and flavin-adenine dinucleotide (FAD), in which FMN is linked to ribose and adenine through a second phosphate. Riboflavin, also called vitamin B_2, is a required organic growth factor for some organisms.

The **cytochromes** are proteins with iron-containing porphyrin ring prosthetic groups attached to them (Figure 4.14). They undergo oxidation and reduction through loss or gain of a *single electron* by the iron atom at the center of the cytochrome:

$$\text{Cytochrome–Fe}^{2+} \rightleftharpoons \text{Cytochrome–Fe}^{3+} + e^-$$

Several classes of cytochromes are known, differing in their reduction potentials. One cytochrome can transfer electrons to another that has a more positive reduction potential and can itself accept electrons from a cytochrome with a less positive reduction potential. The different cytochromes are designated by letters, such as cytochrome *a*, cytochrome *b*, cytochrome *c*. The cytochromes of one organism may differ slightly from those of another, so that there are designations such as cytochrome a_1, cytochrome a_2, and so on. Occasionally, cytochromes form tight complexes with other cytochromes or with iron–sulfur proteins. An example

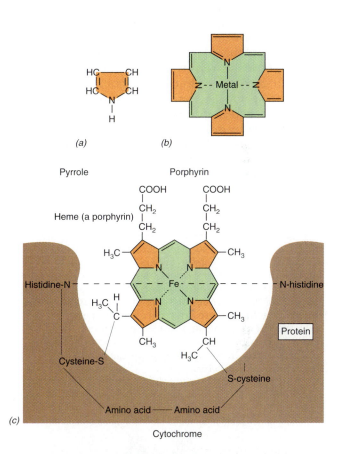

FIGURE 4.14 Cytochrome and its structure. (a) Structure of the pyrrole ring. (b) Four pyrrole rings are condensed, leading to the formation of the porphyrin ring. Various metals can be incorporated into the porphyrin ring system. (c) In the cytochromes, the porphyrin ring is covalently linked via disulfide bridges to cysteine molecules in the protein. Note the presence of iron in the center of the ring. (d) Computer-generated model of cytochrome *c*. The protein completely surrounds the porphyrin ring (light color) in the center.

FIGURE 4.13 Flavin mononucleotide (FMN) (riboflavin phosphate). Note that $2H^+$ are taken up when the flavin becomes reduced, and $2H^+$ are given off when the flavin becomes oxidized.

of such a complex is the cytochrome bc_1 complex, which contains two different b-type cytochromes and one c-type cytochrome. It plays an important role in energy metabolism (see next section).

In addition to the cytochromes, where iron is bound to heme, several **nonheme iron–sulfur proteins** are associated with the electron transport chain in various places. Several arrangements of sulfur and iron have been found in different nonheme iron–sulfur proteins, but the Fe_2S_2 and Fe_4S_4 clusters are the most common (Figure 4.15). The iron atoms are bonded to free sulfur and to the protein via sulfur atoms from cysteine residues (Figure 4.15). *Ferredoxin*, a common iron–sulfur protein in biological systems, has an Fe_2S_2 configuration. The reduction potentials of iron–sulfur proteins vary over a wide range depending on the number of iron atoms and sulfur atoms present and how the iron centers are attached to protein. Thus different iron–sulfur proteins can serve at different points in the electron transport process. Like cytochromes, iron–sulfur proteins carry *electrons only*, not hydrogen atoms.

The **quinones** (Figure 4.16) are lipid-soluble substances involved in electron transport. Some quinones found in bacteria are related to vitamin K, a growth factor for higher animals. Like flavoproteins, quinones serve as *hydrogen atom* acceptors and *electron* donors.

> **Electron transport systems consist of a series of membrane-associated electron carriers that function in an integrated fashion to carry electrons from the energy source to an external electron acceptor such as oxygen. Except for the quinones, the carriers are bound to membrane proteins and are arranged in the electron transport system in the order of their reduction potentials, from most negative to most positive. Key components of the electron transport system are flavoproteins, quinones, and cytochromes.**

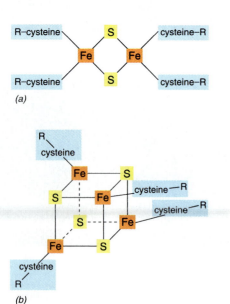

FIGURE 4.15 Arrangement of the iron–sulfur centers of nonheme iron–sulfur proteins. (a) Fe_2S_2 center. (b) Fe_4S_4 center. The cysteine linkages are from the protein portion of the molecule.

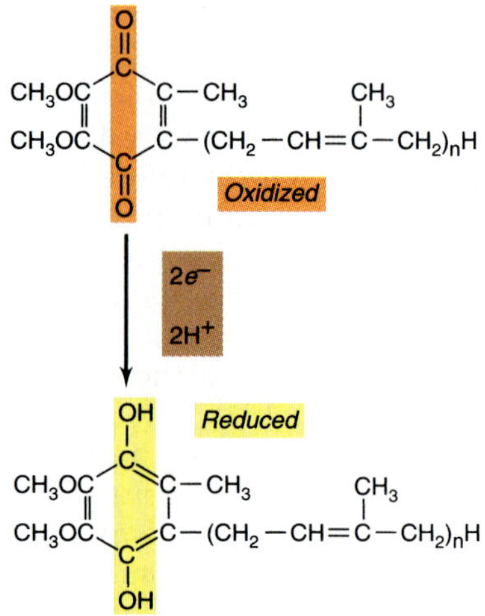

FIGURE 4.16 Structure of oxidized and reduced forms of coenzyme Q, a quinone. The five-carbon unit in the side chain (an isoprenoid) occurs in a number of multiples. In bacteria, the most common number is $n = 6$; in higher organisms, $n = 10$. Note that $2H^+$ are taken up when the quinone becomes reduced, and $2H^+$ are given off when the quinone becomes oxidized. An intermediate form, the semiquinone (one H more reduced than oxidized quinone) is formed during the reduction of a quinone (see Figure 4.18).

4.11   Energy Conservation in the Electron Transport System

The overall process of electron transport in the electron transport chain is shown in Figure 4.17. During electron transport, ATP is produced by the process of oxidative phosphorylation. The production of ATP is linked directly to the establishment of a *proton gradient* across the membrane, electron transport reactions serving to establish the gradient. We now consider the details of this process.

The proton motive force: chemiosmosis

To understand the manner by which electron transport is linked to ATP synthesis, we must first discuss the manner in which the electron transport system is oriented in the cell membrane. The overall structure of the membrane was outlined in Section 3.3 (see Figure 3.15). It was shown there that proteins are embedded in the lipid bilayer and that the orientation of proteins in the membrane is such that most have access to both the outside and the inside of the cell (transmembrane proteins). The electron transport carriers discussed above are oriented in the membrane in such a way that a *separation* of protons from electrons occurs during the transport process. Hydrogen atoms, removed from hydrogen atom carriers such as NADH, are separated into electrons and protons, the electrons returning to the cytoplasmic side of the membrane by specific carriers, and the protons being extruded outside the cell (Figure 4.18). At the end of the electron transport chain, the electrons are passed to the final electron acceptor (in the case of aerobic respiration, this is O_2) and reduce it.

When O_2 is reduced to H_2O it requires H^+ from the cytoplasm to complete the reaction, and these protons originate from the dissociation of water into H^+ and OH^-; $H_2O \rightarrow H^+ + OH^-$. The use of H^+ in reduction of

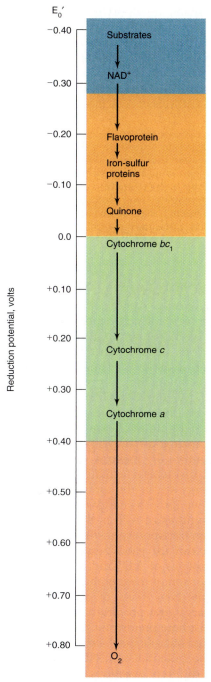

FIGURE 4.17   One example of an electron transport system, leading to the transfer of electrons from substrate to O_2. This particular sequence is typical of the electron transport chain of the mitochondrion and some Bacteria (for example, *Paracoccus denitrificans*). The chain in *Escherichia coli* lacks cytochromes *c* and *a*, and instead electrons go directly from cytochrome *b* to cytochrome *o* or *d* which act as terminal oxidases. By breaking up the complete oxidation into a series of discrete steps, energy conservation is possible through proton gradient formation, and ATP synthesis can occur.

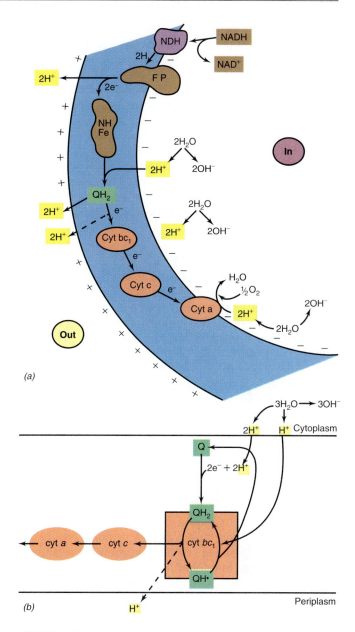

FIGURE 4.18   Generation of a proton motive force. (a) Orientation of electron carriers in the bacterial membrane, showing the manner in which electron transport can lead to charge separation and generation of a proton gradient. The + and − charges originate from the protons (H^+) or hydroxyl ions (OH^-) accumulating on opposite sides of the membrane. NDH, NADH dehydrogenase; FP, flavoprotein; NHFe, nonheme iron–sulfur protein; Q, coenzyme Q; cyt, cytochromes. IN and OUT refer to the cytoplasm and the environment, respectively. (b) The Q cycle. Electrons shuttle between QH_2 and the cytochrome bc_1 complex. For each two molecules of QH_2 oxidized to two QH· (semiquinone), two electrons enter the later stages of the electron transport chain. The bc_1 complex can disproportionate two QH· to QH_2 plus Q, with the uptake of a proton from the cytoplasm. This mechanism serves to increase the number of protons pumped at the Q-bc_1 site.

O_2 to H_2O and the extrusion of H^+ cause a net formation of OH^- on the *inside* of the membrane. Despite their small size, neither H^+ nor OH^- freely pass through the membrane so that equilibrium cannot be spontaneously restored. Thus, although electron transport to O_2 can be thought of as producing water, what is actually produced are the *elements* of water, H^+ and OH^-, which accumulate on opposite sides of the

membrane. The net result is the generation of a *pH gradient* and an electrical potential across the membrane, with the *inside* of the cytoplasm electrically negative and alkaline, and the *outside* of the membrane electrically positive and acidic. This pH gradient and electrical potential cause the membrane to be energized (something like a battery), and this electrical energy can be used by the cell.

In the same way that the energized state of a battery is expressed as its electromotive force (in volts), the energized state of a membrane can be expressed as a **proton motive force** (also in volts). The energized state of the membrane induced as a result of electron transport processes can be used directly to do useful work such as ion transport (see Section 3.4) or flagellar rotation (Figure 3.41), or it can be used to drive the formation of high-energy phosphate bonds in ATP, as will be described below. The idea of a proton gradient driving ATP synthesis was first proposed as the *chemiosmotic theory* in 1961 by Peter Mitchell of England; Mitchell later received the Nobel Prize for this important contribution.

Generation of the proton motive force

The key steps involved in proton gradient formation across the membrane involve the activities of the flavin enzymes, quinones, and the cytochrome bc_1 complex (Figure 4.18). The cytochrome bc_1 complex contains several proteins and is present in the electron transport chain of most organisms able to respire. It also plays a role in photosynthetic electron flow (see Chapter 16). The major function of the cytochrome bc_1 complex is to transfer electrons from quinones to cytochrome c linked to the translocation of protons across the membrane. Carriers such as the cytochrome bc_1 complex are oriented vectorially (directionally) in the membrane in such a way that protons are discharged to the environment when electrons are transferred to an acceptor, resulting in the accumulation of OH^- in the cytoplasm (Figure 4.18).

The series of oxidation–reduction reactions occurring during electron transport may be analyzed by examining each pair of carriers sequentially (Figure 4.18*a*). Following the donation of two hydrogen atoms from NADH to FAD, two H^+ are extruded when FADH donates two electrons (only) to an iron–sulfur protein. Two protons are taken up from the cytoplasm when the nonheme iron protein reduces a molecule of quinone called *coenzyme Q*. Coenzyme Q passes electrons one at a time to the cytochrome bc_1 complex. Reduced coenzyme Q (QH_2) donates one electron to the bc_1 complex, extruding a proton and converting QH_2 to $QH\cdot$, the semiquinone form of coenzyme Q. The semiquinone can be reduced to QH_2 by one of the b-type cytochromes in the bc_1 complex, along with the uptake of a proton. For every two $QH\cdot$ molecules that enter the complex, one is reduced to QH_2 while the other is oxidized to Q. This "Q cycle" (Figure 4.18*b*), acts to increase the number of protons extruded across the membrane at the Q-bc_1 site. Electrons travel from the bc_1 complex to cytochrome c and cytochrome a, the latter of which serves in conjunction with the terminal oxidase (Figure 4.18*b*). Finally, the reduction of $^1/_2 O_2$ to H_2O consumes two H^+, but these of course are not extruded. All of the required protons except those donated by FADH are generated from the dissociation of water yielding OH^- on the inside surface of the membrane (Figure 4.18).

The electron transport scheme shown in Figure 4.18 is just one of many different carrier sequences observed in different organisms. However, the important feature of all of them is the generation of a *proton gradient*, acidic outside and alkaline inside. The gradient results in a proton motive force which actually drives the synthesis of ATP (see next section).

> When electrons are transported through an electron transport system that is embedded in a membrane, protons (hydrogen ions) are transported through the membrane, leading to the formation of a proton gradient and proton motive force. This proton motive force energizes the membrane in a manner analogous to the way an electromotive force energizes a battery.

The proton motive force and ATP formation

How is the proton motive force used to synthesize ATP? An important component of this process is a membrane-bound enzyme called *ATPase*, which contains two parts, a multi-subunit headpiece present on the inside of the membrane, and a proton-conducting tailpiece that spans the membrane (Figure 4.19). This enzyme catalyzes a reversible reaction between ATP and ADP + P_i (inorganic phosphate) as shown in Figure 4.19*b*. Operating in one direction, this enzyme catalyzes the formation of ATP by allowing the controlled reentry of protons across the energized membrane. Just as the formation of the proton gradient was energy *driven*, the controlled dissipation of the proton motive force is energy *releasing*, some of which is used to synthesize ATP in the process called oxidative (electron transport) phosphorylation.

ATPase can also catalyze the reverse reaction, that is, the hydrolysis of ATP and the extrusion of $3H^+$ to the outer portion of the membrane (Figure 4.19*b*). This results in the conversion of phosphate bond energy into the energy of a proton motive force. Thus, high-energy phosphate bonds and the proton motive force can be looked at as *different forms of cell energy*. Because a membrane potential is used by the cell to drive a number of different reactions, transport and motility being chief among them, ATPases are present even in organisms which do not carry out oxidative phosphorylation, such as lactic acid bacteria.

Supporting evidence for operation of the electron transport chain and development of a proton motive force comes from studies of various chemicals that affect these processes. Two classes of chemicals are known: **inhibitors** and **uncouplers**. Inhibitors block electron flow and thus ATP synthesis. Examples include carbon monoxide (CO), which prevents the reduction of O_2 to H_2O, and cyanide (CN^-) or azide (N_3^-), which bind tightly to cytochromes and block electron transport. In contrast, uncouplers inhibit ATP synthesis *without* affecting electron transport. All of these agents, such as dinitrophenol and dicumarol, are lipid-soluble substances that destroy proton gradients, thereby pro-

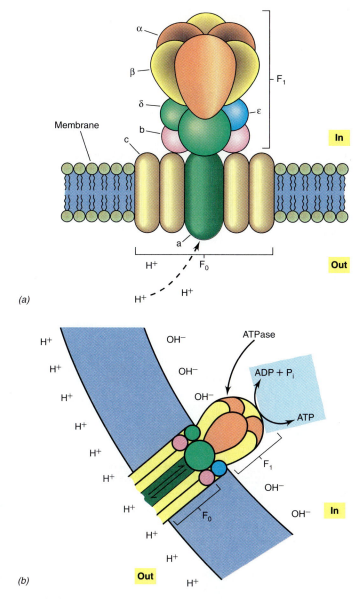

(a)

(b)

FIGURE 4.19 Structure and function of the membrane-bound ATPase, which acts as a proton channel between the cytoplasm and the cell exterior (environment). (a) Structure. F1 consists of five polypeptides, α (3 copies), β (3 copies), γ, δ, and ε. This is the catalytic protein, responsible for the interconversion of ATP and ADP + P$_i$. F$_0$ is integrated in the membrane and consists of three polypeptides, a, b (2 copies), and c (4 copies). It is responsible for channeling protons across the membrane. (b) ATPase action. As protons flow in, the dissipation of the proton gradient drives ATP synthesis from ADP + P$_i$. The reverse reaction, ATP → ADP + P$_i$, drives the extrusion of protons to the cell exterior.

moting the leakage of protons across the membrane. The latter results in dissipation of the proton motive force and hence inhibition of ATP synthesis.

Contrast of respiration between prokaryotes and eukaryotes

So far we have been discussing electron transport and the proton motive force in abstract terms. However, in the cell these processes occur in membrane structures and there are important differences between prokaryotes and eukaryotes in where these processes occur.

In prokaryotes, the electron transport components are embedded in the cytoplasmic membrane, and the development of a proton motive force occurs across this membrane. Thus, in prokaryotes, protons are extruded outside the cell into the environment (in Gram-negative prokaryotes, protons are extruded to the periplasm), resulting in a mild acidification of the external milieu.

In eukaryotes, the processes of electron transport and ATP synthesis occur in the membranes of the mitochondrion. As we saw in Section 3.14, the mitochondrion is a membrane-bounded structure which also has extensive internal membranes and it is across these internal membranes that the proton motive force develops. The ATPase structure (Figure 4.19) is a part of this internal membrane and ATP synthesis occurs within the mitochondrial matrix. The ATP synthesized then diffuses into the cytoplasm through the permeable external membrane, where it is used in various biosynthetic reactions.

> The cell is able to utilize its proton motive force to generate ATP because of the presence in the membrane of ATPases, which catalyze a reversible reaction between ATP, ADP, and inorganic phosphate (ATP $\rightleftharpoons$ ADP + P$_i$). The proton motive force is used to drive this ATPase in the direction of ATP synthesis.

4.12 Carbon Flow: The Tricarboxylic Acid Cycle

We now consider the metabolic aspects of carbon flow in respiration. The early steps in the respiration of glucose involve the same biochemical steps as those of glycolysis (see Figure 4.11). As we noted, a key intermediate in glycolysis is pyruvate. Whereas in fermentation, pyruvate is converted to fermentation products, in respiration, pyruvate is oxidized fully to CO_2. One important pathway by which pyruvate is completely oxidized to CO_2 is called the **tricarboxylic acid cycle** (TCA cycle) as outlined in Figure 4.20.

Pyruvate is first decarboxylated, leading to the production of one molecule of NADH and an acetyl moiety coupled to coenzyme A (acetyl–CoA, Figure 4.21). The acetyl group of acetyl–CoA combines with the four-carbon compound oxalacetate, leading to the formation of citric acid, a six-carbon organic acid, the energy of the high-energy acetyl–CoA bond being used to drive this synthesis (Figure 4.21). Dehydration, decarboxylation, and oxidation reactions follow, and two additional CO_2 molecules are released. Ultimately, oxalacetate is regenerated and can serve again as an acetyl acceptor, thus completing the cycle.

For each pyruvate molecule oxidized through the cycle, three CO_2 molecules are released (Figure 4.20), one during the formation of acetyl–CoA, one by the decarboxylation of isocitrate, and one by the decarboxylation of α-ketoglutarate. As in fermentation, the electrons released during the oxidation of intermedi-

FIGURE 4.20
The tricarboxylic acid cycle. The 3-carbon compound, pyruvate, is oxidized to CO_2, with the electrons used to make NADH and FADH. The reoxidation of NADH and FADH in the electron transport chain leads to the generation of a proton motive force and thus to the synthesis of ATP. The cycle actually begins when the 2-carbon compound acetyl–CoA condenses with the 4-carbon compound oxalacetate, to form the 6-carbon compound citrate. Through a series of oxidations and transformations, this 6-carbon compound is ultimately converted back to the 4-carbon compound oxalacetate, which then passes through another cycle with addition of the next molecule of acetyl–CoA. The overall balance sheet is shown at the bottom. The reducing equivalents formed (NADH and FADH) are shown on the left and right sides of the cycle.

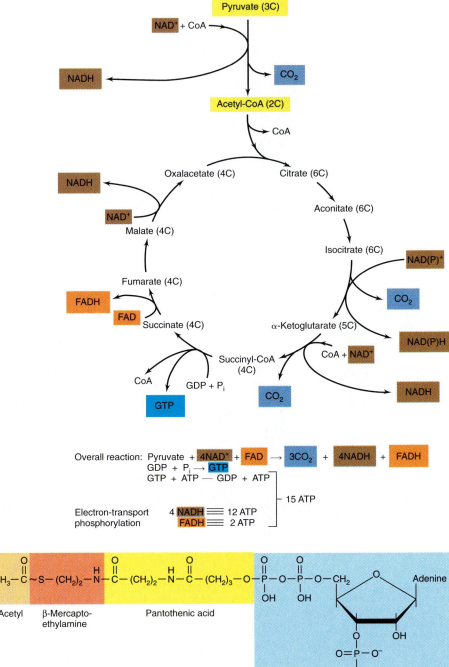

Overall reaction: Pyruvate + 4NAD⁺ + FAD ⟶ 3CO₂ + 4NADH + FADH
GDP + P_i ⟶ GTP
GTP + ATP — GDP + ATP
⎫ 15 ATP

Electron-transport phosphorylation 4 NADH ≡ 12 ATP
FADH ≡ 2 ATP

FIGURE 4.21
Structure of acetyl–CoA. The coenzyme A portion consists of the β-Mercaptoethylamine/Pantothenic acid/ADP complex.

Acetyl   β-Mercapto-ethylamine   Pantothenic acid   Adenine

3′-Phospho ADP

ates in the TCA cycle are usually transferred initially to enzymes containing the coenzyme NAD⁺. However, respiration differs from fermentation in the manner in which NADH is oxidized. In respiration, the electrons from NADH, instead of being transferred to an intermediate such as pyruvate, are transferred to oxygen or other terminal electron acceptors through the action of the *electron transport system* just described. Thus, unlike fermentation, the presence of an electron acceptor in respiration allows for the complete oxidation of glucose to CO_2, with a much greater yield of energy.

Respiration involves the complete oxidation of an organic energy source to carbon dioxide and water, with much greater energy release than during fermentation. Pyruvate is a key intermediate in the metabolism of carbon compounds and its oxidation after conversion to acetyl–CoA involves the action of the tricarboxylic acid cycle. Electrons removed during the oxidation of pyruvate are transferred to NAD⁺ and through the electron transport system to oxygen. This process is linked to the production of ATP.

4.13   The Balance Sheet of Aerobic Respiration

The net result of reactions of the tricarboxylic acid cycle is the complete oxidation of pyruvic acid to three molecules of CO_2 with the production of four molecules of NADH and one molecule of FADH. As we have seen, the NADH and FADH molecules can be reoxidized through the electron transport system, yielding up to three ATP molecules per molecule of NADH and two ATP per molecule of FADH. In addition, the oxidation of α-ketoglutarate to succinate involves a substrate-level phosphorylation, producing guanosine triphosphate (GTP), which is later converted to ATP. Thus a total of 15 ATP molecules can be synthesized for each turn of the cycle. Since the oxidation of glucose yields two molecules of pyruvic acid, a total of 30 molecules of ATP can be synthesized in the citric acid cycle. Also, when oxygen is available, the two NADH molecules produced during glycolysis can be reoxidized by the electron transport system, yielding six more molecules of ATP. Finally, two molecules of ATP are produced by substrate-level phosphorylation during the conversion of glucose to pyruvic acid. Thus, aerobes can form up to 38 ATP molecules from one glucose molecule in contrast to the 2 molecules of ATP produced fermentatively.

If we assume that the high-energy phosphate bond of ATP has an energy of about 31.8 kJ/mole, then 1208 kJ of energy can be converted to high-energy phosphate bonds in ATP by the complete oxidation of glucose to CO_2 and H_2O. Since free energy calculations show that the total amount of energy available from the complete oxidation of glucose by oxygen is 2822 kJ/mole, aerobic respiration is about 43 percent efficient, the rest of the energy being lost as heat.

If an organism were 100 percent efficient, all of the energy yield of a biochemical reaction would be conserved in the form of high-energy bonds or a membrane potential. However, organisms are not 100% efficient and a part of the energy yield is not conserved but is lost as heat. Thus we must distinguish between energy *release*—the total energy yield of a reaction—and energy *conservation*—that energy available to the organism to do work. Interestingly, although the *yield* of ATP from fermentation is very low, the *efficiency* of fermentations can be reasonably high. The lactic fermentation [glucose $\rightarrow$ 2 lactate], for example, releases 118 kJ/mole and drives the synthesis of two ATPs, for an efficiency of over 50%. Thus, fermentations release relatively *low amounts* of energy, but the conversion process is thermodynamically efficient.

In addition to its function as an energy-yielding mechanism, the tricarboxylic acid cycle also provides key intermediates for biosynthesis. Oxalacetate and α-ketoglutarate lead to several amino acids (see Section 4.18), succinyl–CoA is the starting point for porphyrin biosynthesis, and acetyl–CoA provides the material for fatty acid synthesis.

> Although only two high-energy phosphate bonds are conserved as ATP during fermentation of one glucose molecule, as many as 38 ATPs can be formed during aerobic respiration. Some energy is lost as heat, so that the amount of energy conserved in respiration is about 40% or less of the total amount of energy in the energy source.

The Lowly Yeast Cell

The aerobic and anaerobic processes of energy generation may seem dull and prosaic, but they are at the basis of one of the most exciting discoveries of the human race, alcoholic fermentation. Although many prokaryotes form alcohol, it is a simple eukaryote which is most commonly exploited, the yeast *Saccharomyces cerevisiae*. Found in various sugar-rich environments such as fruit juices and nectar, yeasts have the ability to carry out the two opposing modes of metabolism discussed in this chapter, fermentation and respiration. When oxygen is present, yeasts grow efficiently on the sugar substrate, making yeast cells and CO_2. However, when O_2 is absent, yeasts switch to an anaerobic metabolism, resulting in a reduced cell yield but significant amounts of alcohol. Every home wine maker or brewer is an amateur microbiologist, even without realizing it (see box, Chapter 10). When grapes are squeezed to make juice, small numbers of yeast cells present on the grapes in the vineyard are transferred to the must. During the first several days of the wine-making process, these yeast cells grow by respiration, but consume O_2, making the juice anoxic. As soon as the oxygen is depleted, fermentation can begin and the process of alcohol formation takes over. This switch from aerobic to anaerobic metabolism is crucial, and special care must be taken to make sure air is kept out of the fermenting vessel. Lots of things can go wrong in wine making. The wrong yeast might get started, insufficient sugar in the juice will result in too low an alcohol concentration, and spoilage can occur by growth of bad yeasts or bacteria. Wine is only one of many products made with yeast. Others include beer, whisky, vodka, and gin. Even alcohol for motor fuel is made with yeast in parts of the world where sugar is plentiful and petroleum is in short supply (such as Brazil). Yeast also serves as the leavening agent in bread, although here it is not the alcohol that is important but CO_2, the other product of the alcohol fermentation. We discuss yeast in some detail in Chapter 10.

4.14   Turnover of ATP and the Role of Energy-Storage Compounds

It can be calculated that for the synthesis of 1 gram of cell material (wet weight), about 20 millimoles (mmoles) of ATP would need to be consumed. Since the intracellular concentration of ATP is only about 2 millimolar, ATP obviously has only a *catalytic* role during growth; it is continually being broken down and resynthesized. It has been calculated that during the time that a cell doubles, its ATP pool must turn over about 10,000 times.

Related to this short life of ATP is the fact that if ATP is not immediately used for energy, growth, and biosynthesis, it is hydrolyzed by reactions not yielding energy. For long-term storage of energy, most organisms produce insoluble organic polymers that can later be oxidized for the production of ATP. The glucose polymers starch and glycogen are produced by many microorganisms, both prokaryotic and eukaryotic, and poly-β-hydroxybutyrate (PHB) and other polyhydroxyalkanoates (see Section 17.20) are produced by many prokaryotes. These polymers often are deposited within the cells in large granules that can be seen with the light or electron microscope (see Section 3.9). In the absence of an external energy source, the cell may then oxidize this energy-storage material and thus be able to maintain itself, even under starvation conditions.

Polymer formation yields a twofold advantage for the cell. Not only are carbon and potential energy stored in a stable form, but insoluble polymers have little effect on the internal osmotic pressure of cells. If the same number of units were present as monomers in the cell, the high solute concentration would increase cellular osmotic pressure, resulting in the inflow of water and possible swelling and lysis. A certain amount of energy is lost when a polymer is formed from monomers, but this disadvantage is more than offset by the benefits to the cell.

4.15   Alternate Modes of Energy Generation

This chapter deals only with energy generation by either fermentation or respiration of organic compounds, but microorganisms have many other possibilities for obtaining energy. These alternate modes of energy metabolism form the theme of Chapter 16, but will be summarized here to provide an overview of catabolic processes.

One alternate mode of energy generation is a variation on respiration in which electron acceptors *other than* oxygen are used. Because of the analogy to respiration, these processes are classified under the heading of **anaerobic respiration**. Electron acceptors used in anaerobic respiration include nitrate (NO_3^-), sulfate (SO_4^{2-}), carbonate (CO_3^{2-}), and even certain organic compounds. Because of their positions on the electron tower (see Figure 4.6), less energy is released when

these electron acceptors are used instead of oxygen. However, the utilization of these alternate electron acceptors permits nonfermentative microorganisms to develop in environments where oxygen is absent. The contrasts between aerobic and anaerobic respiration are presented in Figure 4.22.

A second mode of energy generation involves the use of *inorganic* rather than organic chemicals. Organisms able to use inorganic chemicals as energy sources are a type of chemotroph called **chemolithotrophs** (literally, *rock eating*). Examples of inorganic energy sources include hydrogen sulfide (H_2S), hydrogen gas (H_2), and ammonia (NH_3). Chemolithotrophic metabo-

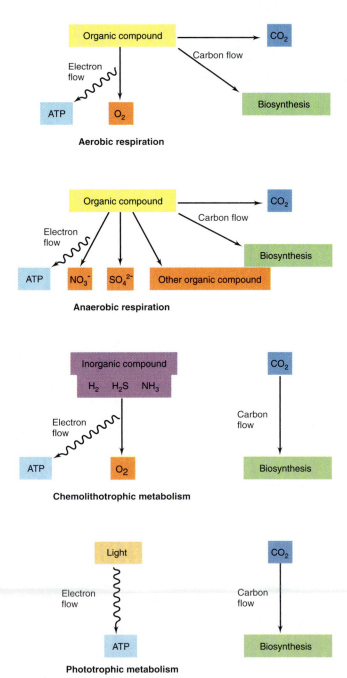

FIGURE 4.22   Electron and carbon flow in aerobic respiration, anaerobic respiration, chemolithotrophic metabolism, and phototrophic metabolism.

lism most commonly involves aerobic respiratory processes such as those described in this chapter, but using an inorganic energy source rather than an organic one (Figure 4.22). Chemolithotrophs have electron transport components like chemoorganotrophs and form a proton motive force. However, one important distinction between chemolithotrophs and chemoorganotrophs is in their sources of *carbon* for biosynthesis. Chemoorganotrophs can generally use compounds such as glucose as carbon sources as well as energy sources, but chemolithotrophs cannot, of course, use their inorganic energy compounds as sources of carbon. Most chemolithotrophs utilize carbon dioxide as carbon source and are, hence, **autotrophs**. Organisms that use organic compounds as carbon sources are called **heterotrophs.**

A large number of microorganisms, as well as higher plants, are *photosynthetic*, using light as an energy source. We call such organisms **phototrophs** (literally, *light eating*). The mechanisms by which light is used as an energy source are unique and complex, but the underlying result is the generation of a proton-motive force which can be used in the synthesis of ATP. Most phototrophs use energy conserved in ATP for the assimilation of carbon dioxide as the carbon source for biosynthesis. Such phototrophs are also, hence, autotrophs. However, as we will see in Chapter 16, photosynthesis in microorganisms has some special features and complications. There are, for instance, two types of photosynthesis in microorganisms, one form similar to that of higher plants, and a unique type of photosynthesis found only in certain prokaryotes.

One overall conclusion that can be stated at this point is that in terms of energy metabolism, microorganisms show an amazing diversity. However, with the exception of fermentations, this diversity is centered around various means of generating a proton gradient and the resulting proton motive force. In the present chapter, we have presented the simplified picture that is obtained if one only examines the energy metabolism of common chemoorganotrophs. The fascinating metabolic diversity in the prokaryotic world will be considered in detail in Chapters 16 through 20.

> Carbon storage polymers are common in prokaryotes and can serve as a source of carbon and energy during periods of nutrient deprivation. Electron acceptors other than oxygen can serve as terminal electron acceptors for energy generation. Since oxygen is not present, this is called anaerobic respiration. Chemolithotrophs use inorganic compounds as sources of energy, while phototrophs use light as energy source.

4.16   Biosynthetic Pathways: Anabolism

We have just discussed **catabolism**, the biochemical processes by which microorganisms obtain energy from organic compounds. We now turn to the other side of the coin, **anabolism**, the biochemical processes

by which microorganisms build up the vast array of chemical substances of which they are composed (Figure 4.1). The chemical composition of the cell is outlined in Table 4.2 (see Section 4.21). Note that the dominant constituents (other than water) are macromolecules. Macromolecules, as we know (see Sections 2.5 through 2.8), are made by the polymerization of monomers. Polysaccharides are polymers built up of sugar monomers, proteins are built up from amino acids, and nucleic acids are polymers of nucleotides. We now summarize the key biosynthetic reactions which form these small molecule building blocks. Note that in this chapter we are discussing only the biosynthesis of *small molecules* (monomers). The biosynthesis of *informational macromolecules* is discussed in the next chapter.

Energy for anabolism is provided by ATP or the proton motive force. Energy conserved during catabolic reactions in ATP and the proton gradient is consumed during biosynthesis (Figure 4.23). However, although energy is required in many biosynthetic reactions, the focus of biosynthesis is not on energy but on carbon, and on the intermediates occurring during the buildup of cell constituents from simple starting materials. In Sections 4.8 and 4.12 we described the intermediates in the oxidation of glucose in both glycolysis and the tricarboxylic acid cycle; in those sections our main concern was not with the ultimate fate of these intermediates but with how their transformations led to the formation of ATP. Now we are concerned with how these intermediates are used in biosynthesis. Some of the same intermediates that are formed during catabolism are used during anabolism. These key intermediates, summarized in Figure 4.24, are actually very few in number.

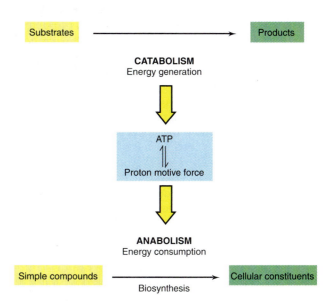

FIGURE 4.23   *Scheme of anabolism and catabolism showing the key role of ATP and the proton motive force in integrating the processes.*

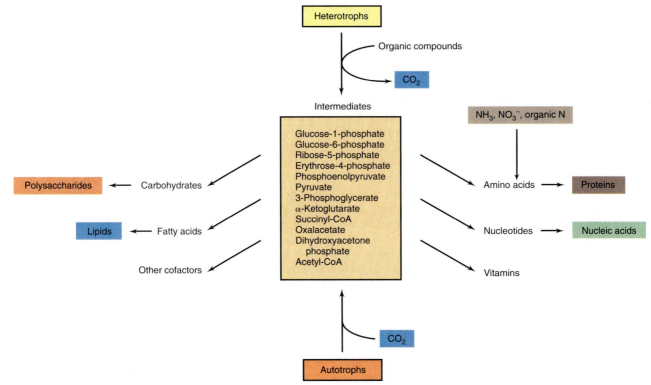

FIGURE 4.24 The key central metabolic intermediates produced in catabolism and used in anabolism.

> Anabolism, also called biosynthesis, is the process by which the cell builds up the vast array of chemical substances that are needed for growth. The ATP produced during catabolism is consumed during anabolism.

4.17 Sugar Metabolism

A small but significant amount of a cell is composed of polysaccharides (see Table 4.2 for the chemical composition of a cell). Polysaccharides are key constituents of the cell walls of many organisms, and in Bacteria, the peptidoglycan cell wall has a polysaccharide backbone. In addition, cells often store energy in the form of the polysaccharides **glycogen** and **starch**. The sugars in polysaccharides are primarily six carbon sugars called *hexoses*. The most common hexose in polysaccharides is **glucose**, which, as we have seen, is also an excellent energy source for many microorganisms. In addition to hexoses, two five-carbon sugars, called *pentoses*, are key constituents of nucleic acids, **ribose** in RNA and **deoxyribose** in DNA. It is thus convenient to separate our discussion into two parts, dealing separately with hexoses and pentoses.

Hexoses

Six-carbon sugars needed for biosynthesis can be obtained either from the environment or can be synthesized within the cell from nonsugar starting materials. A summary of the main pathways is given in Figure 4.25. It can be seen that two key intermediates of hexose metabolism are **glucose-6-phosphate** and **uridine diphosphoglucose (UDP-glucose)**. We already have discussed glucose-6-phosphate in some detail when

discussing the process of glycolysis (see Figure 4.11). Glucose-6-phosphate serves as a key intermediate in the oxidation and fermentation of glucose as an energy source, in which case it is converted either to carbon dioxide or fermentation products. Glucose-6-phosphate can also feed into the pathways for polysaccharide synthesis, in which case it is converted to UDP-glucose (UDPG), a nucleoside diphosphate sugar. (The structure of UDPG is uracil-ribose-P-P-glucose. Note that there is an additional role for uracil; it is also one of the bases in RNA.) UDPG is an activated form of glucose which is synthesized from uridine triphosphate (UTP) and glucose-1-phosphate, as outlined in Figure 4.25. UDPG serves as the starting material not only for the synthesis of glucose-containing polysaccharides, but for the synthesis of other nucleoside diphosphate sugars needed in biosynthesis. Thus, while glucose-6-phosphate is the central intermediate in glucose *catabolism*, UDPG is the central intermediate in glucose *anabolism*.

When a microorganism has a hexose available as an energy and carbon source, this hexose can also generally serve as the precursor of the hexoses needed for biosynthesis. However, many microorganisms are able to grow using nonhexose carbon sources. In such situations, the hexose needed for biosynthesis of cell walls and other hexose-containing polymers must be synthesized within the cell by a process called **gluconeogenesis**. Gluconeogenesis is the creation of new glucose molecules from noncarbohydrate precursors. The starting material for gluconeogenesis is phosphoenolpyruvate (PEP), one of the key intermediates in glycolysis. As we saw in Section 4.7 (see Figure 4.11), PEP is formed in the lower part of the glycolytic path-

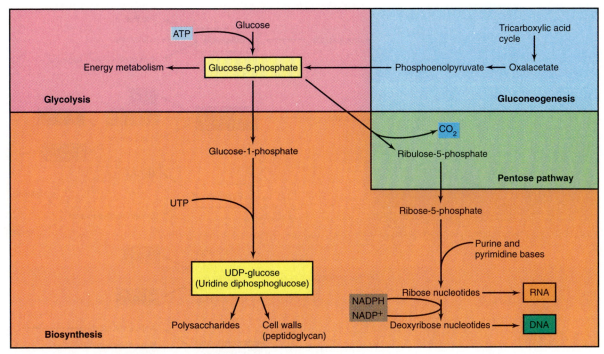

FIGURE 4.25   Hexose and pentose biosynthesis: summary of the main routes.

way, and by *reversal* of this pathway (but using a different enzyme than that used in glycolysis), glucose-6-phosphate can be formed. Where does the PEP needed for gluconeogenesis come from? There are a number of ways in which PEP can be formed, but a major one is by decarboxylation of oxalacetate, which itself is a key intermediate of the tricarboxylic acid cycle (see Figure 4.20).

Pentoses

In most instances, the five-carbon sugars are formed by the removal of one carbon atom from a hexose (Figure 4.25). Several different pathways are known. One of the common pathways is the oxidative decarboxylation of glucose-6-phosphate, yielding CO_2 and the five-carbon intermediate ribulose-5-phosphate. Ribulose-5-phosphate is converted to two other pentose sugars, one of which is ribose-5-phosphate, the pentose needed for nucleotide synthesis. Once a ribose nucleotide is formed, it can feed directly into RNA synthesis. For DNA synthesis, deoxyribose is needed and this is synthesized by enzymatic reduction (using NADPH) of the 2' oxygen on the ribose of a ribose nucleotide. These various reactions are summarized in Figure 4.25.

Although sugars are important energy sources for most cells, they are also needed in biosynthesis of cell walls, nucleic acids, etc. If sugars are not available as a nutrient from the environment, the cell must synthesize them from other sources. The biosynthetic reactions of sugar synthesis sometimes involve different enzymes and coenzymes than those involved in sugar breakdown. Key sugars in the cell contain six-carbons (hexoses) or five-carbons (pentoses).

4.18   Amino Acid Biosynthesis

As we have noted, there are 20 amino acids common in proteins. Those organisms that cannot obtain some or all amino acids preformed from the environment must synthesize them from other sources. The structures of the amino acids were discussed in Section 2.8 and illustrated in Figure 2.13. There are two aspects of amino acid biosynthesis: the synthesis of the **carbon skeleton** of each amino acid, and the manner by which the amino group is incorporated.

Amino acids can be grouped into *families* based on the precursor used for the synthesis of the carbon skeleton and these families are summarized in Figure 4.26. Note that the precursor molecules are ones which we have encountered before in glycolysis and the TCA cycle, and were summarized in Figure 4.24.

The attachment of the amino group to the carbon skeleton can either be done at the very end, after the carbon skeleton is completely synthesized, or at some intermediate step. For instance, **glutamate**, the first amino acid of the glutamate family, is formed by direct addition of the amino group to the carbon skeleton. Other amino acids generally get their amino groups from glutamate or some other amino donor by a type of reaction called a *transamination* (see below).

What is the ultimate source of the amino group for the amino acids? Many microorganisms utilize **ammonia** as the source of nitrogen, and there are several enzymes which are able to catalyze the addition of ammonia to a carbon skeleton. One of the most widespread enzymes involved in ammonia incorporation is **glutamate dehydrogenase**, an enzyme which uses NAD(P)H as a source of reducing power for the synthesis of glutamate from α-ketoglutaric acid and ammonia. The role of glutamate dehydrogenase in amino

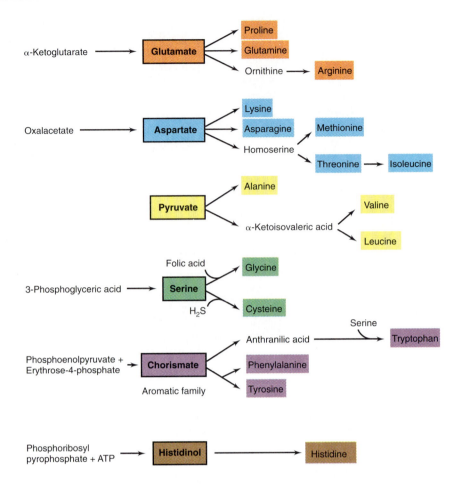

FIGURE 4.26
Amino acid families. The amino acids on the right are derived from the starting materials on the left. The compounds in boxes are the parent molecules of each amino acid family.

acid biosynthesis is illustrated in Figure 4.27a. Once ammonia has been incorporated into the amino group of glutamate, the amino group can be transferred to other carbon skeletons by enzymes called **transaminases**. In the transamination process, the amino group of glutamate is exchanged for the keto group of an α-keto acid, leading to the formation of a new amino acid (from the α-keto acid) and regenerating α-ketoglutaric acid from glutamate (Figure 4.27b). By successive functioning of glutamate dehydrogenase and transaminase, ammonia can be assimilated into a variety of amino acids (Figure 4.27c). Not all organisms use the pathway illustrated in Figure 4.27. In some cases, other amino acid dehydrogenases replace glutamate dehydrogenase, and in still other cases a completely different enzyme, *glutamine synthetase*, is involved in ammonia assimilation. These alternate pathways are discussed in Chapter 16, and the complete pathways of biosynthesis can be found in biochemistry textbooks.

> The 20 amino acids present in proteins are synthesized in a series of complex biosynthetic reactions which start with a few key intermediates. Amino acids can be grouped into families based on the precursor used for synthesis of the carbon skeleton.

4.19   Purine and Pyrimidine Biosynthesis

Purines and pyrimidines are components of nucleic acids, as well as of many vitamins and coenzymes (for example, ATP, NAD^+). The terminology of purines and pyrimidines was discussed in Section 2.7, where the structures of the key compounds were presented. The biosynthesis of purines and pyrimidines is surprisingly complex, and the details are beyond the scope of the present chapter and can be found in biochemistry textbooks. Here, we provide simply an overview.

Purine synthesis

The purine ring is built up almost atom by atom using carbons and nitrogens derived from amino acids, CO_2, and formyl groups (themselves derived from the amino acid serine), as outlined in Figure 4.28a. The starting material for purine biosynthesis is a ribose-phosphate compound to which the various atoms are added. Thus, in the case of purine nucleosides, the sugar residue is present from the beginning.

The two carbon atoms in the purine ring derived from formyl groups are added in reactions involving a coenzyme derivative of the vitamin *folic acid*. This coenzyme is involved in a number of reactions involving addition of one-carbon units, and is of additional interest

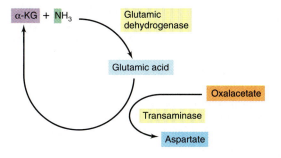

(a) *Assimilation* of ammonia into glutamic acid, catalyzed by the enzyme *glutamic dehydrogenase*

(b) Transfer of the amino group of glutamic acid via *transamination* leads to the synthesis of other amino acids, in this case aspartic acid.

(c) Cycle of ammonia assimilation via combined action of glutamic dehydrogenase and transaminase

FIGURE 4.27   *Pathway of ammonia incorporation into amino acids.*

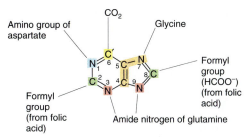

(a)   The basic precursors of the purine skeleton

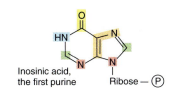

(b)

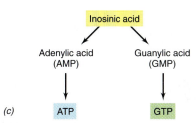

(c)

FIGURE 4.28   Purine biosynthesis.

before the sugar is added to form a nucleoside. After the addition of ribose and phosphate to orotic acid, the other pyrimidines are successively synthesized (Figure 4.29*b*). Complete details of pyrimidine biosynthesis are beyond this text. The biosynthesis of the deoxyribose sugar was discussed briefly in Section 4.17.

> Biosynthesis of the key components of the nucleic acids—the purines and the pyrimidines—is a complex process involving several enzymes and several carbon skeleton donors.

4.20   Regulation of Enzyme Activity

In the previous pages we have discussed a few of the key enzymatic reactions occurring during anabolism and catabolism. There are, of course, hundreds of different enzymatic reactions occurring simultaneously during a single cycle of cell growth. One thing that should be very clear is that not all these enzymatic reactions occur in the cell to the same extent. Some compounds are needed in large amounts and the reactions involved in their synthesis must therefore occur in large amounts. However, other compounds are needed only in small amounts and the reactions involved in their synthesis must occur only in small amounts. The need for regulation of biochemical reactions is clear. How does regulation occur?

Enzyme amount or enzyme activity?
There are two major modes of regulation, one which controls the amount (or even the complete presence or

because a number of microorganisms cannot synthesize their own folic acid but must have it provided from the environment. Additionally, certain antimicrobial drugs called *sulfonamides* are specific inhibitors of folic acid synthesis. Sulfonamide inhibition of growth can be almost completely overcome by providing the end products of pathways in which folic acid serves as a coenzyme, among which is a purine (see Section 9.17 for a detailed discussion of sulfonamide action).

The first purine ring formed is that of *inosinic acid*, which serves as an intermediate for the formation of the two key purine derivatives, adenylic acid and guanylic acid, as shown in Figure 4.28*c*.

Pyrimidine biosynthesis
The origin of the atoms of the pyrimidine ring is shown in Figure 4.29*a*. In contrast to the purine ring, the complete pyrimidine ring (orotic acid) is built up

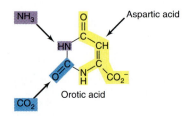

(a) Basic precursors of the pyrimidine skeleton

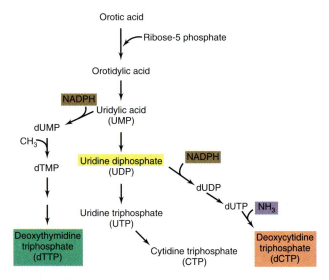

(b) Biosynthesis of the pyrimidine nucleotides

FIGURE 4.29   Pyrimidine biosynthesis. Formation of deoxythymidine and deoxycytidine occur via reduction with NADPH, as shown by the enzyme ribonucleotide reductase. The methyl group of thymidine is donated by methyltetrahydrofolate. The amino group of cytidine is donated by glutamine.

absence) of an enzyme, and the other which controls the *activity* of preexisting enzyme. Regulation of enzyme amount, by the phenomena called **induction** and **repression**, occurs at the *gene* level in prokaryotes and will be discussed after we have discussed molecular biology and protein synthesis in Chapter 5. We discuss here briefly the processes involved in regulating the *activity* of preformed enzymes.

Product inhibition

A simple mechanism by which an enzymatic reaction may be regulated is by a process called *product inhibition*. As we have seen, an enzyme combines with its *substrate*, the reaction occurs, and the *product* is released. Because enzymatic reactions are generally *equilibrium reactions*, as product builds up the reaction can occur in the *reverse* direction, from product to substrate. Product inhibition occurs primarily if the product of the enzymatic reaction is not used in subsequent reactions. But if the product is itself the substrate for another enzyme, then it will be continually removed and product inhibition will not occur. Note that in product inhibition it is the enzyme which *formed* the product which is also *inhibited* by the product.

Feedback inhibition

A major mechanism for the control of enzymatic activity involves the phenomenon of **feedback inhibition**.

Feedback inhibition is seen primarily in the regulation of entire biosynthetic pathways, such as the pathway involved in the synthesis of an amino acid or purine. As we have seen, such pathways involve many enzymatic steps, and the final product, the amino acid or nucleotide, is many steps removed from the starting substrate. Yet, this final product is able to feed back to the first step in the pathway and regulate its own biosynthesis. How?

In feedback inhibition the amino acid or other end product of the biosynthetic pathway inhibits the activity of the *first* enzyme in this pathway. Thus, as the end product builds up in the cell, its further synthesis is inhibited. If the end product is used up, however, synthesis can resume (Figure 4.30).

How is it possible for the end product to inhibit the activity of an enzyme that acts on a substrate quite unrelated to it? This occurs because of a property of the inhibited enzyme known as **allostery**. An allosteric enzyme has two important combining sites, the *active* site, where the substrate binds, and the *allosteric* site, where the inhibitor (sometimes called an "effector") binds reversibly. When an inhibitor binds, generally noncovalently, at the allosteric site, the conformation of the enzyme molecule changes so that the substrate no longer binds efficiently at the active site (Figure 4.31). When the concentration of the inhibitor falls, equilibrium favors the dissociation of the inhibitor from the allosteric site, returning the active site to its catalytic shape. Allosteric enzymes are very common in both anabolic and catabolic pathways and are especially important in branched pathways. For example, the amino acids proline and arginine are both synthesized from glutamic acid. Figure 4.32 shows that these two amino acids can control the first enzyme unique to their own synthesis without affecting the other, so that a surplus of proline, for example, will not cause the organism to be starved for arginine.

In addition, some biosynthetic pathways are regulated by the use of **isozymes** (short for isofunctional enzymes: *iso* means same or constant). These enzymes catalyze the same reaction, but are subject to different regulatory control. An example is the synthesis of the aromatic amino acids as summarized in Figure 4.26

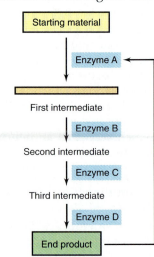

FIGURE 4.30
Feedback inhibition. The activity of the first enzyme of the pathway is inhibited by the end product, thus controlling production of end product.

FIGURE 4.31
Mechanism of enzyme inhibition by an allosteric effector. When the effector combines with the allosteric site, the conformation of the enzyme is altered so that the substrate can no longer bind.

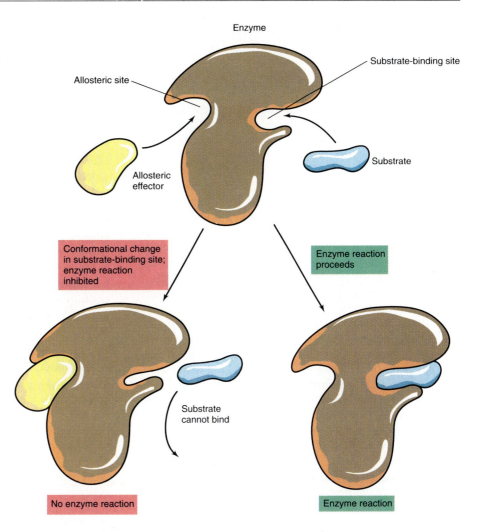

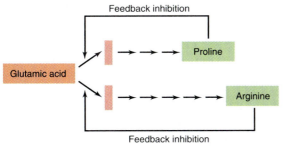

FIGURE 4.32   Feedback inhibition (dashed arrows) in a branched biosynthetic pathway. A key intermediate in the pathway is shown in pink.

(see also Figure 10.3). Three different isozymes catalyze the first reaction in this pathway and each enzyme is regulated independently by each of the three different end product amino acids. Unlike the earlier examples of feedback inhibition where inhibitors completely stopped an enzyme activity, in this case the total amount of the initial enzyme activity is diminished in a stepwise fashion and falls to zero only when all three products are present in excess.

Covalent modification

Several examples are known in bacteria in which an enzyme is regulated by being *covalently modified*, usually by addition or deletion of some small organic

molecule. As in the case of allosteric proteins, the covalent binding of the modifying group changes the conformation of the protein and can dramatically affect its catalytic activity. Removal of the modifying group returns the enzyme to an active state. Many examples of covalent modification regulatory systems are known but the best characterized are enzymes whose activity is affected by attachment of the nucleotides AMP or ADP, by attachment of inorganic phosphate, or by methylation.

These examples are representative of the major regulatory patterns observed; there are other, rather elegant examples of regulation now known. The phenomenon of enzyme regulation may have evolved because efficient control of the rate of enzyme *activity* enables an organism to quickly adapt to changing environments.

> The vast array of metabolic reactions do not all occur in the cell at the same rate. Metabolic reactions are regulated through control of the activities of the enzymes that catalyze these reactions. An important type of regulation of enzyme activity is feedback inhibition, in which the final product of a biosynthetic pathway feeds back and inhibits the first enzyme unique to that pathway. Covalent modification is a regulatory mechanism to temporarily inactivate a specific enzyme.

4.21 Microbial Nutrition

Chemical composition of a cell

During our discussion of cellular structure in Chapter 3, some aspects of the chemical makeup of various cell constituents were presented. An overview of the chemical structures themselves was presented in Chapter 2. It may be useful at this point to summarize what is known about the overall chemical composition of a cell (Table 4.2). Cells contain large numbers of small molecules, inorganic ions, and organic substances, and large amounts of macromolecules, the most important of which are the proteins and nucleic acids. The cell may obtain most of the small molecules it needs from the environment in preformed condition, whereas macromolecules are synthesized in the cell. However, many small molecules are also synthesized within the cell, from substances obtained from the environment.

Although there are many naturally occurring elements, virtually the whole mass of a cell consists of only four types of atoms: carbon, oxygen, hydrogen, and nitrogen (see Figure 1.4). In addition, a number of other atoms are quantitatively less important, but functionally very important. These include phosphorus, calcium, magnesium, sulfur, iron, zinc, manganese, copper, molybdenum, and cobalt, which are present in microbial cells, but in far lesser amounts than C, H, O, and N. With the exception of molybdenum and tungsten, the elements found in living organisms are all of atomic number 30 or less.

Water accounts for the bulk of the weight of a cell (about 90 percent) and macromolecules together make up most of the remaining dry weight, proteins being

the most abundant class of macromolecules (Table 4.2). Inorganic ions and monomers constitute the balance of the cell dry weight. Chemically, therefore, living organisms have evolved around a relatively few types of chemical elements. How are these elements obtained by cells? We consider this problem now.

Nutrition

Substances in the environment used by organisms for catabolism and anabolism are called **nutrients**. Nutrients can be divided into two classes: (1) *macronutrients*, which are required in large amounts; and (2) *micronutrients*, which are required in small amounts. Some nutrients are the building blocks from which the cell makes macromolecules and other important structures, whereas other nutrients serve only for energy generation without being incorporated directly into cellular material; sometimes a nutrient can play both roles. We begin with a discussion of the major macronutrients, carbon and nitrogen.

Major macronutrients

Most prokaryotes require an organic compound of some sort as their source of **carbon**. Nutritional studies have shown that many bacteria can assimilate a variety of different organic carbon compounds and use them to make new cell material. Amino acids, fatty acids, organic acids, sugars, organic nitrogen bases, aromatic compounds, and countless other organic compounds have been shown to be used by one bacterium or another.

After carbon, the next most abundant element in the cell is **nitrogen**. A typical bacterial cell is 12 to 15

Table 4.2 Chemical composition of a prokaryotic cell[a]

Molecule	Percent of dry weight[b]	Molecules per cell	Different kinds
Water	—	—	1
Total macromolecules	96	24,609,802	~2500
Protein	55	2,350,000	~1850
Polysaccharide	5	4,300	2[c]
Lipid	9.1	22,000,000	4[d]
DNA	3.1	2.1	1
RNA	20.5	255,500	~660
Total monomers	3.5		~350
Amino acids and precursors	0.5		~100
Sugars and precursors	2		~50
Nucleotides and precursors	0.5		~200
Inorganic ions	1		18
Totals	100%		

[a]Data taken from Neidhardt, F. C. et al. (eds.) 1987. Escherichia coli and Salmonella typhimurium—Cellular and Molecular Biology. American Society for Microbiology, Washington, D.C.
[b]Dry weight of an actively growing cell of E. coli $\cong 2.8 \times 10^{-13}$ g.
[c]Assuming peptidoglycan and glycogen to be the major polysaccharides present.
[d]There are several classes of phospholipids, each of which exists in many kinds because of variability in fatty acid composition between species and because of different growth conditions.

percent nitrogen (by dry weight) and nitrogen is a major constituent of proteins and nucleic acids. Nitrogen is also present in the complex polysaccharide *peptidoglycan*, the rigid layer of the cell wall of most bacteria (see Section 3.5). Nitrogen can be found in nature in both organic and inorganic forms. As a major constituent of amino acids and nitrogenous bases, nitrogen is available to organisms as organic nitrogen from the breakdown and mineralization of dead organisms. However, the bulk of available nitrogen in nature is in *inorganic* form, either as ammonia (NH_3) or nitrate (NO_3^-) (see Section 17.14). Most bacteria are capable of using ammonia as the sole nitrogen source and many can also use nitrate. The atmosphere contains a vast store of nitrogen as nitrogen gas, N_2. However, nitrogen gas only serves as a nitrogen source for certain bacteria, the nitrogen-fixing bacteria (see Sections 16.26, 17.14, and 17.24).

Other macronutrients

Phosphorus occurs in nature in the form of organic and inorganic phosphates and is used by the cell primarily in nucleic acids and phospholipids. Most microorganisms utilize inorganic phosphate (PO_4^{3-}) for growth. Organic phosphates occur very often in nature, however, and they can be utilized following the action of cell enzymes called *phosphatases*. The latter hydrolyze the organic phosphate ester, releasing free inorganic phosphate.

Sulfur is absolutely required by organisms because of its structural role in the amino acids cysteine and methionine (see Section 2.8) and because it is present in a number of vitamins (thiamine, biotin, and lipoic acid). Sulfur undergoes a number of chemical transformations in nature carried out exclusively by microorganisms (see Section 17.15), and is available to organisms in a variety of forms. Most cell sulfur originates from inorganic sources, either sulfate (SO_4^{2-}) or sulfide (HS^-).

Potassium is required by all organisms. A variety of enzymes, including some of those involved in protein synthesis, are specifically activated by potassium. **Magnesium** functions to stabilize ribosomes, cell membranes, and nucleic acids, and is also required for the activity of many enzymes, especially those involving phosphate transfer. Thus, relatively large amounts of magnesium are required for growth. **Calcium**, which is not essential for the growth of many microorganisms, can help in stabilizing the bacterial cell wall and plays a key role in the heat stability of endospores.

Sodium is required by some but not all organisms, and its need may reflect the habitat of the organism. For example, seawater has a high sodium content, and marine microorganisms generally require sodium for growth, whereas closely related freshwater forms are usually able to grow in the absence of sodium.

Iron is required in fairly large amounts, although not at the level of other macronutrients. Iron is found

Table 4.3	Common forms of the major elements needed for biosynthesis of cell components
Element	**Usual form in the environment**
C	Carbon dioxide (CO_2)
	Organic compounds
H	Water (H_2O)
	Organic compounds
O	Water (H_2O)
	Oxygen gas (O_2)
N	Ammonia (NH_3)
	Nitrate (NO_3^-)
	Organic compounds (e.g., amino acids)
P	Phosphate (PO_4^{3-})
S	Hydrogen sulfide (H_2S)
	Sulfate (SO_4^{2-})
	Organic compounds (e.g., cysteine)
K	K^+
Mg	Mg^{2+}
Ca	Ca^{2+}
Na	Na^+
Fe	Fe^{3+}
	Organic iron complexes

in a number of proteins in the cell, especially those involved in respiration (see below).

Common sources of macronutrients are summarized in Table 4.3.

Micronutrients (trace elements)

Although required in just tiny amounts, micronutrients are nevertheless just as critical to the overall nutrition of a microorganism as are macronutrients.

Cobalt is needed only for the formation of vitamin B_{12}, and if this vitamin is added to the medium, cobalt is no longer needed. **Zinc** plays a structural role in many enzymes including carbonic anhydrase, alcohol dehydrogenase, RNA and DNA polymerases, and other DNA binding proteins. **Molybdenum** is present in certain enzymes called *molybdoflavoproteins*, which are involved in assimilatory nitrate reduction. It is also present in nitrogenase, the enzyme involved in N_2 reduction (N_2 fixation, see Section 16.26). **Copper** plays a role in certain enzymes involved in respiration. Like iron, the copper ion is a site for reaction with O_2. **Manganese** is an activator of many enzymes and is also found in certain superoxide dismutases, enzymes of critical importance for detoxifying toxic forms of oxygen (see Section 9.12). Manganese also plays an important role in green plant photosynthesis. **Nickel** is present in enzymes called *hydrogenases* that function to either take up or evolve H_2. **Tungsten** and **selenium** are required by those prokaryotes capable of metabolizing formate and are present as part of the enzyme formate dehydrogenase.

The hundreds of chemical compounds present inside a living cell are formed from a relatively small number of starting materials that are available as nutrients from the environment. Each compound is formed from one or a few basic chemical elements that are obtained from the various nutrients. Elements required in fairly large amounts, called macronutrients, include carbon, hydrogen, oxygen, nitrogen, phosphorus, sulfur, potassium, magnesium, sodium, and iron. Micronutrients, or trace elements, include copper, cobalt, molybdenum, manganese, nickel, and zinc.

4.22   Laboratory Culture of Microorganisms

We now bring together our discussion of microbial nutrition by considering a practical application of nutritional principles: the culture of microorganisms in the laboratory.

Although we can get an idea of what microorganisms look like from a microscopic study, we can only get an idea of what microorganisms *do* by studying their activities in pure cultures. A **pure culture** is a culture consisting of only one type of microorganism. In order to obtain a pure culture, we must be able to grow the organism in the laboratory. This requires that we provide the organism with the proper nutrients and environmental conditions so that it can grow. It is also essential that we be able to keep other organisms from entering the pure culture. Such unwanted organisms, called **contaminants**, are ubiquitous, and microbiological technique revolves around the avoidance of contaminants. Once we have isolated a pure culture, we can then proceed to a study of the characteristics of the organism and a determination of its capabilities.

What are the conditions required for microbial growth? Microorganisms are cultured in water to which appropriate nutrients have been added. The aqueous solution containing such necessary nutrients is called a **culture medium**. The nutrients present in the culture medium provide the microbial cell with those ingredients required for the cell to produce more cells like itself. Besides an energy source, which could be an organic or inorganic chemical, or light, a culture medium must have a source of carbon, nitrogen, and the other macro- and micronutrients discussed in the previous section. We now briefly consider some of the important aspects of culturing microorganisms.

Preparing culture media

A knowledge of microbial nutrition allows the microbiologist to culture microorganisms in the laboratory. In a general sense, the nutrient requirements of all organisms are the same; that is, all organisms require the macro- and micronutrients discussed earlier in this chapter. But more specifically, certain organisms require certain nutrients in specific forms. Thus, the proportion and type of each nutrient in a culture medium

can vary dramatically for growth of different microorganisms. Literally thousands of different recipes for culture media have been published, and most experienced microbiologists pay careful attention to the nutritional aspects of the organism or organisms they are growing. Table 4.4 lists the common chemical forms by which many of the nutrients are supplied in culture media.

Two classes of culture media are used in microbiology: **chemically defined (synthetic)** or **undefined (complex)**. Chemically defined media are prepared by adding precise amounts of pure inorganic or organic chemicals to distilled water. Therefore, the *exact* chemical composition of a defined medium is known. In many cases, however, knowledge of the exact composition is not critical. In these instances undefined media may suffice, or may even be advantageous. Complex media employ crude digests of substances such as casein (milk protein), beef, soybeans, yeast cells, or a number of other highly nutritious, yet chemically undefined, substances. Such digests are available commercially in powdered form and can be weighed out rapidly and added to culture media. However, an important disadvantage of using complex media is the loss of control over the precise nutrient specifications of the medium.

Culture media can be prepared for use either in a liquid state, or in a gel (semisolid) state. A liquid culture medium is converted to the semisolid state by addition of a gelling agent. **Agar** is the most commonly used gelling agent. Agar is manufactured from certain

Table 4.4	Nutrient requirements of microorganisms and common means to satisfy them in culture
Nutrient	**Chemical form supplied in culture media**
Carbon	*Organic*—defined media: glucose, acetate, pyruvate, malate, hundreds of other compounds; complex media: yeast extract, beef extract, peptone, many other complex digests; *Inorganic*—CO_2, HCO_3^-
Nitrogen	*Organic*—Amino acids, nitrogenous bases; *Inorganic*—NH_4Cl, $(NH_4)_2SO_4$, KNO_3, N_2
Phosphorus	KH_2PO_4, Na_2HPO_4
Sulfur	Na_2SO_4, H_2S
Potassium	KCl, K_2HPO_4
Magnesium	$MgCl_2$, $MgSO_4$
Sodium	$NaCl$
Iron	$FeCl_3$, $Fe(NH_4)(SO_4)_2$, iron chelates
Micronutrients	$CoCl_2$, $ZnCl_2$, Na_2MoO_4, $CuCl_2$, $MnSO_4$, $NiCl_2$, Na_2SeO_4, Na_2WO_4, Na_2VO_4

seaweeds and is not a nutrient for most microorganisms. Culture media containing agar are dispensed in flat, covered dishes called **Petri plates,** where microbial cells can grow and form visible masses called *colonies* (Figure 4.33).

Provision of trace elements

The micronutrients (trace elements) are special cases in the preparation of culture media. As we have seen, most of these minerals are required in extremely small amounts. In chemically defined media, trace elements are usually added in small amounts from a stock solution containing a mixture of the metals discussed in Section 4.21. In complex media, trace elements generally do not need to be added because they are already present in the digests employed.

Iron is a special case, as it is sometimes considered a trace mineral but is required in larger amounts than other trace minerals. As we have seen, iron plays a major role in cellular respiration, being a key component of the cytochromes and iron–sulfur proteins involved in electron transport (see Section 4.10). Because most inorganic iron salts are highly insoluble, providing an adequate supply of iron to a culture medium presents some difficulties. One way of supplying iron is to complex it with an organic chemical called a **chelating agent**. Two common chelating agents used to provide iron to culture media are ethylene diamine tetraacetic acid (EDTA) and nitrilotriacetic acid (NTA).

Many organisms produce specific iron chelators called **ironophores**, which solubilize iron salts and transport iron into the cell. One major group of ironophores are derivatives of **hydroxamic acid**, which bind iron very strongly, as shown in Figure 4.34. Once the iron–hydroxamate complex has passed into the cell, the iron is released and the hydroxamate can pass out of the cell and be utilized again for iron transport. In some bacteria, the iron-binding compounds are not hydroxamates but phenolic acids. Enteric bacteria such as *Escherichia coli* and *Salmonella* spp. produce complex phenolic derivatives called **enterobactins**. These compounds are catechol derivatives with a high binding affinity for iron. The iron-binding enterobactin of *E. coli* is shown in Figure 4.35. As we will see in Chapter 11, availability of iron has important consequences in the ability of many harmful (pathogenic) bacteria to grow in the body.

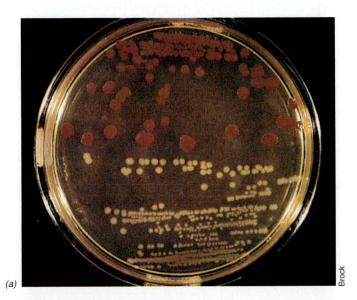
(a)

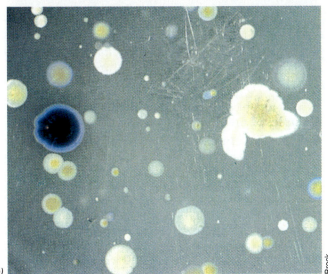
(b)

FIGURE 4.33   An agar-containing medium in a Petri plate. Bacterial colonies are visible on the surface of the agar. Note the distinct and characteristic colors. (a) Red and white colonies. (b) Purple, yellow, and white colonies.

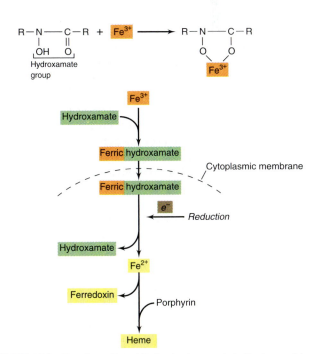

FIGURE 4.34   Function of iron-binding hydroxamate in the iron nutrition of an organism living in an environment low in iron.

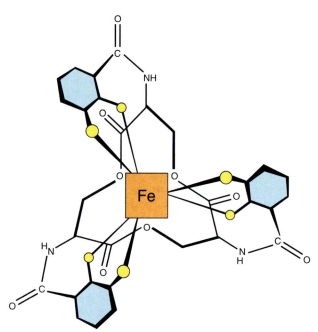

FIGURE 4.35   Ferric enterobactin, an iron chelator of *Escherichia coli*. The oxygen atoms of each catechol molecule are highlighted in yellow.

In order to understand the functions of microorganisms, they must be cultured in the laboratory. Culture media are prepared which provide the essential nutrients in forms that the organism can utilize. In synthetic culture media, all of the elements are provided from known chemical compounds. In complex media, undefined materials such as plant or animal extracts provide the necessary ingredients.

Growth factors

Growth factors are specific organic compounds that are required in very small amounts and that cannot be synthesized by some cells. Substances frequently serving as growth factors are vitamins, amino acids, purines, and pyrimidines. Although most microorganisms are able to synthesize all of these compounds, certain others require them preformed from the environment. Hence, they must be added to the culture medium. If a complex digest such as yeast extract or peptone is used in a culture medium, most or all of the potential growth factors will already be supplied, but in synthetic media, provision of proper growth factors is often of great importance.

Vitamins are the most commonly needed growth factors. Vitamins are defined as organic compounds required in small amounts for growth and function that do not serve as either energy sources or building blocks of macromolecules. Most vitamins function as parts of coenzymes (see, for instance, Figures 4.7, 4.13, and 4.21) and these are summarized in Table 4.5. Many microorganisms are able to synthesize all of the components of their coenzymes, but some are unable to do so and must be provided with certain parts of these coenzymes in the form of vitamins. The lactic acid bacteria, which include the genera *Streptococcus*

Table 4.5   Vitamins and their functions

Vitamin	Function
p-Aminobenzoic acid	Precursor of folic acid
Folic acid	One-carbon metabolism; methyl group transfer
Biotin	Fatty acid biosynthesis; β-decarboxylations; CO_2 fixation
Cobalamin (B_{12})	Reduction of and transfer of single carbon fragments; synthesis of deoxyribose
Lipoic acid	Transfer of acyl groups in decarboxylation of pyruvate and α-ketoglutarate
Nicotinic acid (niacin)	Precursor of NAD^+; electron transfer in oxidation-reduction reactions
Pantothenic acid	Precursor of coenzyme A; activation of acetyl and other acyl derivatives
Riboflavin	Precursor of FMN, FAD in flavoproteins involved in electron transport
Thiamine (B_1)	α-Decarboxylations; transketolase
Vitamins B_6 (pyridoxal-pyridoxamine group)	Amino acid and keto acid transformations
Vitamin K group; quinones	Electron transport; synthesis of sphingolipids
Hydroxamates	Iron-binding compounds; solubilization of iron and transport into cell
Coenzyme M (Co-M)	Required by certain methanogens; plays a role in methanogenesis

and *Lactobacillus* (see Section 19.24), are renowned for their complex vitamin requirements, which are even greater than those of humans. The vitamins most commonly required by microorganisms are thiamine (vitamin B_1), biotin, pyridoxine (vitamin B_6), and cobalamin (vitamin B_{12}).

Examples of microbial culture media

We can now summarize the material of this section by presenting the detailed chemical compositions of two culture media, one defined, one complex (Table 4.6). The defined culture medium is one that will support the growth of *E. coli* and a number of other enteric bacteria. The biosynthetic capabilities of *E. coli* are quite impressive, as shown by the fact that only a *single* organic compound, glucose, need be added to the medium.

The other culture medium presented in Table 4.6 is one that satisfies the requirements of a bacterium that has complex nutritional requirements, the lactic acid bacterium *Leuconostoc mesenteroides*. Which organism has more biosynthetic capacity, *E. coli* or *L. mesen-*

Table 4.6   Examples of synthetic culture media for microorganisms with simple and complex nutritional requirements

Culture medium for *Escherichia coli*		Culture medium for *Leuconostoc mesenteroides*	
K_2HPO_4	7.0 g	K_2HPO_4	0.6 g
KH_2PO_4	2.0 g	KH_2PO_4	0.6 g
$(NH_4)_2SO_4$	1.0 g	NH_4Cl	3 g
$MgSO_4$	0.1 g	$MgSO_4$	0.1 g
$CaCl_2$	0.02 g	Glucose	25 g
Glucose	4–10 g	Sodium acetate	20 g
Trace elements (Fe, Co, Mn, Zn, Cu, Ni, Mo)	2–10 µg each		
Distilled water	1000 ml	Amino acids (alanine, arginine, asparagine, aspartate, cysteine, glutamate, glutamine, glycine, histidine, isoleucine, leucine, lysine, methionine, phenylalanine, proline, serine, threonine, tryptophan, tyrosine, valine)	100–200 µg of each
		Purines and Pyrimidines (adenine, guanine, uracil, xanthine)	10 mg of each
		Vitamins (biotin, folate, nicotinic acid, pyridoxal, pyridoxamine, pyridoxine, riboflavin, thiamine, pantothenate, para-aminobenzoic acid)	0.01–1 mg of each
		Trace elements	2–10 µg each
		Distilled water	1000 ml

teroides? Obviously, *E. coli*, since its ability to grow on a simple defined culture medium means that it has the ability to synthesize *all* of its organic cellular constituents from a single carbon compound. By contrast, *L. mesenteroides* has multiple growth factor requirements, indicative of limited biosynthetic capacity.

> Many microorganisms will only grow if specific organic compounds, called growth factors, are added to the culture medium. Common growth factors include various vitamins and amino acids.

Environmental factors

In addition to appropriate nutrients, it is essential that environmental factors be adjusted appropriately for each organism to be cultured. One of the most important environmental factors that must be adjusted is **temperature**. Each microorganism has a defined temperature range over which it is capable of growing and if a temperature too low or too high for this organism is used, satisfactory growth will not be obtained. The **optimum temperature** is the temperature at which the organism grows the fastest. The optimum growth temperature of different bacteria can vary dramatically (see Section 9.9), but as a general rule, it can be stated that the most appropriate temperature for culture of a microorganism is near the temperature of the habitat in which that microorganism is growing.

A proper acidity or alkalinity must be provided for each organism. Acidity or alkalinity of a solution is expressed by its **pH** on a scale in which neutrality is pH 7. Those pH values that are less than 7 are acidic and those greater than 7 are alkaline. Usually the final step in preparing culture media is to adjust the pH to the proper value for growth of the organism(s) of interest.

Microorganisms vary in their need for, or tolerance of, **oxygen**. Microorganisms can be divided into several groups, depending on the effect of oxygen. Microorganisms able to use oxygen are called **aerobes**. Some organisms are unable to use oxygen and may actually be harmed by it. Organisms unable to use oxygen are called **anaerobes** and those harmed by it are called **obligate anaerobes**. For the culture of anaerobes, the main problem is to exclude oxygen. Because oxygen is ubiquitous in the air, it is not a trivial task to culture microorganisms under anaerobic conditions. Bottles or tubes filled completely to the top with culture medium and provided with tightly fitting stoppers will provide anaerobic conditions for organisms that are not too sensitive to small amounts of oxygen. It is also possible to add a chemical to the culture medium which reacts with oxygen and therefore removes it from the culture medium. Such a substance is called a *reducing agent*.

A number of organisms, called **facultative aerobes**, can grow either in the presence or absence of oxygen. The bacterium *E. coli* is a good example of a facultative aerobe. *E. coli* can grow anaerobically by fermentation, producing various acids, alcohol, and CO_2, or it can grow aerobically by respiration, producing only CO_2.

In addition to appropriate nutrients, culture of microorganisms require appropriate environmental conditions. Suitable temperature, pH, and oxygen conditions must be arranged. Organisms that require oxygen for growth are called **aerobes**. Some organisms, called **anaerobes**, are able to grow in the absence of oxygen. Facultative aerobes are able to grow under either aerobic or anaerobic conditions.

4.23 Sterilization, Aseptic Technique, and Pure Cultures

We have considered what must be put into a medium and how to prepare it. Before proceeding to the actual use of media to grow cultures of microorganisms, we must consider how to exclude unwanted or contaminant organisms. Microorganisms are everywhere. Because of their small sizes, they are easily dispersed in the air and on surfaces (such as the human skin). Therefore, we must **sterilize** the culture medium soon after its preparation to eliminate microorganisms already contaminating it. It is equally important to take precautions during the subsequent handling of a sterile culture medium, to exclude from the medium all but the organisms desired. Thus, other materials which come into contact with culture media must also be sterilized. It is also necessary to sterilize contaminated materials in order to control the spread of unwanted bacteria.

Sterilization and aseptic technique

The most common way of sterilizing culture media is by heat. However, since heat may also cause harmful changes in many of the ingredients of culture media, it is desirable to keep the heating time as short as possible. The best procedure is to heat under pressure, since at pressures above atmospheric the temperature can be increased above 100°C without boiling, thus decreasing the time necessary to make the medium sterile. Devices for heating under pressure are called **autoclaves** (see Figure 9.26). The conditions used to sterilize in an autoclave are generally 15 minutes at 15 psi pressure (121°C).

The technique used in the prevention of contamination during manipulations of cultures and sterile culture media is called **aseptic technique**. Its mastery is required for success in the microbiology laboratory, and it is one of the first methods learned by the novice microbiologist. Airborne contaminants are the most common problem, since air always contains dust particles that generally have a community of microorganisms. When containers are opened, they must be handled in such a way that contaminant-laden air does not enter. This is best done by keeping the containers at an angle so that most of the opening is not exposed to the air. Manipulations should preferably be carried out in a dust-free room in which air currents are absent. Aseptic transfer of a culture from one tube of medium to another is usually accomplished with an inoculating loop or needle which has been sterilized

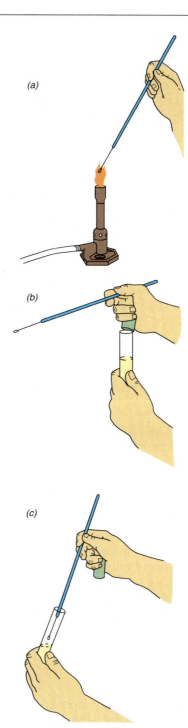

FIGURE 4.36 Aseptic transfer. (a) Loop is heated until red-hot. (b) Tube is uncapped and loop is cooled in air briefly. (c) Sample is removed and tube is recapped. Sample is transferred to a sterile tube. Loop is reheated before being taken out of service.

by incineration in a flame (Figure 4.36). Cultures in which growth has taken place can also be transferred to the surface of agar plates (Figure 4.37), where colonies develop from the growth and division of single cells. Picking and restreaking from an isolated colony is a major method of obtaining pure cultures.

Maintaining pure cultures

Once a pure culture is obtained, it must be kept pure. One of the most frequent ways in which erroneous results and conclusions are obtained in microbiology is

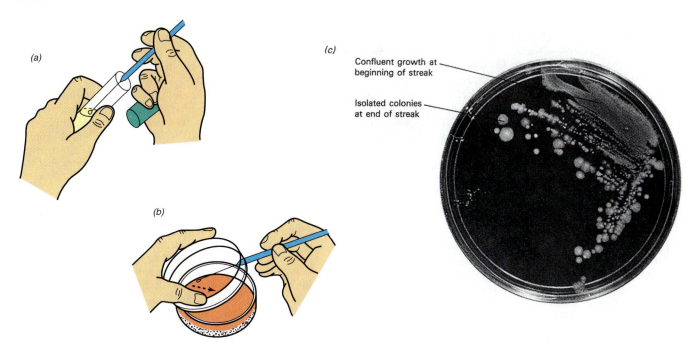

FIGURE 4.37 Method of making a streak plate to obtain pure cultures. (a) Loopful of inoculum is removed from tube. (b) Streak is made over a sterile agar plate, spreading out the organisms. (c) Appearance of the streaked plate after incubation. Note the presence of isolated colonies. It is from such well-isolated colonies that pure cultures can usually be obtained.

by the use of contaminated cultures. Cultures of organisms of interest that are maintained in the laboratory for study and reference are called *stock cultures*. The stock culture must be maintained so that it is free from contamination, retains viability, and remains genetically homogeneous. Cultures which are infrequently used can be purchased from a commercial culture collection (see Section 10.1). For long-term storage of cultures, they may be frozen in *liquid nitrogen*, at which low temperature viability is effectively maintained. Many cultures can also be dried by *freeze-drying* (*lyophilization*) and preserved almost indefinitely in the dried state.

Now that we have developed a basic understanding of metabolism and have explored methods for culturing microorganisms, let us turn our attention to genes and how they work, because metabolic reactions of all types and actual growth of the organism are ultimately directed by genetic events. We now consider how this happens directly at the DNA level.

> Successful cultivation of microorganisms can only be done if aseptic technique is practiced. This requires the preparation of sterile culture media and the manipulation of cultures under such conditions that contamination with foreign organisms does not occur. Pure cultures can be maintained in a nongrowing state for long periods by ultracold freezing or freeze drying.

Study Questions

1. In the following list of substances, indicate which ones could serve as energy sources: Fe^{2+}, O_2, CO_2, NH_4^+, SO_4^{2-}, NO_2^-, NO_3^-, H_2S, glucose, methane, Fe^{3+}. In this same list, indicate which compounds could serve as electron acceptors. As both donor and acceptor.

2. The following is a series of coupled electron donors and electron acceptors. Using the data given in Figure 4.6, order this series from most energy-yielding to least energy-yielding. H_2/Fe^{3+}, H_2S/O_2, methanol/NO_3^- (producing NO_2^-), H_2/O_2, Fe^{2+}/O_2, NO_2^-/Fe^{3+}, H_2S/NO_3^-.

3. What is an electron carrier? Give three examples of electron carriers and indicate their oxidized and reduced forms.

4. Which of the following reactions is exergonic and which endergonic: glucose + ATP yielding glucose-6-phosphate + ADP; acetyl phosphate yielding acetate + phosphate; ethanol + carbon dioxide yielding glucose; H_2 + $\frac{1}{2}O_2$ yielding H_2O. H_2O yielding H_2 + $\frac{1}{2}O_2$; Fe^{2+} + SO_4^{2-} yielding Fe^{3+} + H_2S.

5. Give an example of an electron donor and external electron acceptor which could function in each of the following processes: fermentation, respiration, anaerobic respiration.

6. An experimenter has isolated a mutant yeast which is blocked in glycolysis at the step between acetaldehyde and ethanol. This organism is no longer able to grow anaerobically on glucose but is still able to grow when O_2 is present. Give a possible biochemical explanation for this observation.

7. Iron plays an important role within the cell in energy generating processes. Give three examples in which iron plays a role as an electron carrier. How is iron provided as a nutrient in culture media?

8. In the following list indicate which substance is a coenzyme and which is a prosthetic group: nicotinamide adenine dinucleotide (NAD^+), adenosine diphosphate (ADP), heme, riboflavin, iron (in ferredoxin).

9. Explain how the cytoplasmic membrane is critical for the generation of a proton motive force.

10. The synthesis of ATP from ADP and inorganic phosphate is an endergonic reaction. How then is the membrane-bound ATPase able to bring about ATP synthesis?

11. The chemicals dinitrophenol and cyanide both affect the energy-generation process, but in quite different ways. Compare and contrast the modes of action of these two chemicals.

12. Much more energy is available from glucose respiration than from glucose fermentation. However, the law of conservation of energy states that energy is neither created nor destroyed. Can you give an explanation for where all the energy in the glucose molecules that was not released in the fermenting organism might have gone to?

13. Work through the energy balance sheets for fermentation and respiration and account for all sites of ATP synthesis. Organisms can obtain about 15 times more ATP when growing aerobically on glucose than anaerobically. Write one sentence which accounts for this difference.

14. Cells frequently produce energy-storage polymers, either organic or inorganic. Since ATP is the primary energy currency of the cell, why is some of the energy from ATP converted into storage polymers? Why are polymers formed rather than equivalent monomers?

15. Explain why CO_2 could not serve as an energy source. Explain why it *could* serve as an electron acceptor. When CO_2 serves as an electron acceptor, to what substance is it often converted?

16. In aerobic respiration, O_2 serves as the ultimate electron acceptor. To what substance is O_2 always converted as it accepts electrons?

17. Knowing the function of the cytochrome system, could you imagine an organism that could live if it completely lacked a cytochrome system? How?

18. Give in simplified form the equation for the oxidation-reduction reaction which takes place in the cytochrome molecule. Can you suggest a role for the portions of the cytochrome molecule which are not involved in oxidation-reduction?

19. Indicate the electron donor (energy source) and the electron acceptor for each of the following cases: oxidation of glucose with air, reduction of nitrate in the presence of glucose, oxidation of glucose in the absence of air, oxidation of H_2S in air.

20. A substance such as glucose can serve in an aerobic organism both as an energy source and as a building block of cell substance. What are the names of the two types of metabolic processes involved in these two disparate functions? What is the fate of the carbon atoms of those glucose molecules that are used in energy generation? List three groups of carbon compounds derived from glucose that are building blocks of cell substance.

21. Pyruvate is an important intermediate in both catabolic and anabolic reactions. Describe one catabolic reaction in which pyruvate serves as a key intermediate. Describe an anabolic reaction in which pyruvate is converted to an amino acid.

22. Examine the data in Table 4.2. Why is it accurate to say that a bacterial cell consists primarily of macromolecules? What class of macromolecules constitutes the largest weight in the cell? What class of macromolecules is present in the cell in the largest number? From what you know about energy, where would you say that the largest amount of energy in the cell is stored?

23. There are a number of bacteria which are able to obtain energy and building blocks by attacking other bacteria. From what you know about bacterial cell structure, energy generation, and biosynthesis, describe in biochemical terms the processes by which a whole bacterial cell could be consumed by another bacterium for nutrition, biosynthesis, and energy generation. In this regard, the information given in Table 4.2 should be of value.

24. Can you suggest why it is necessary that polysaccharides such as starch and cellulose must be digested *outside* the cell whereas disaccharides such as lactose and sucrose can be digested *inside* the cell?

25. There are bacteria which *synthesize* glycogen and there are also bacteria which *utilize* glycogen. However the biochemical pathways for glycogen synthesis and utilization are distinct. Describe these two distinct types of pathways. Can you see any advantage in terms of cell function if separate pathways are used for these two processes?

26. The role of NAD^+ in energy generation has been emphasized. $NADP^+$ also plays important roles as a coenzyme in cell function. List several of these roles and give a rationale for why $NADP^+$ might be used instead of NAD^+.

27. The cell wall peptidoglycan of *E. coli* contains two important hexose sugars. When *E. coli* is growing on glucose, some of the glucose molecules can be converted directly into these cell wall sugars. However, when *E. coli* is growing on alanine, these sugars must be synthesized *de novo*. Give an overall scheme (exact biochemical compounds are not needed) by which alanine can be converted into hexose. What is this process called?

28. Figure 4.24 indicates that there are only a few intermediate compounds which serve as the starting points for anabolism. For each of the amino acid families, list the intermediates from Figure 4.24 which are involved in the biosynthetic reactions.

29. Purine and pyrimidine metabolism can be approached in the same way that amino acid metabolism was just approached in the above question. List the comparable intermediates in purine and pyrimidine metabolism.

30. Define *contaminant, aseptic,* and *culture medium*.

31. Define *chemoorganotroph, chemolithotroph, phototroph, autotroph,* and *heterotroph*.

32. Define *sterilization*. Describe a common means of sterilizing culture media.

33. What are ironophores? How do ironophores function and why are they necessary?

Supplementary Readings

Atlas, R. M. 1993. *Handbook of Microbiological Media.* CRC Press, Boca Raton, FL. A comprehensive listing of media for the growth of bacteria and some tips on preparing culture media.

Dawes, E. A. 1986. *Microbial Energetics.* Blackie and Sons, Ltd., Glasgow. An excellent short monograph dealing specifically with the energetics of bacteria and how energetic principles relate to growth.

Dawes, I. W. and **I. W. Sutherland**. 1992. *Microbial Physiology,* 2nd edition. Blackwell Scientific Publications, London. A short but detailed account of bacterial physiology.

Gottschalk, G. 1986. *Bacterial Metabolism*, 2nd edition. Springer-Verlag, New York. A college level text dealing exclusively with bacterial topics. Contains an in depth treatment of anaerobic metabolism.

Harold, F. M. 1986. *The Vital Force: A Study of Bioenergetics.* W. H. Freeman and Co., New York. An excellent treatment of the principles of bioenergetics, especially in terms of membrane-mediated events. Highly recommended.

Neidhardt, F. C., J. L. Ingraham, and **M. Schaechter**. 1990. *Physiology of the Bacterial Cell—A Molecular Approach.* Sinauer Associates, Inc., Sunderland, MA. A very readable account of bacterial physiology.

Schlegel, H.G. and **B. Bowien**. 1989. *Autotrophic Bacteria.* Science Tech Publishers, Madison, Wisconsin. The definitive book on phototrophic and chemolithotrophic metabolism.

Stryer, L. 1988. *Biochemistry*, 3rd edition. W. H. Freeman and Co., San Francisco. Excellent general coverage of basic energetics, enzymes, and other aspects of biochemistry.

Zubay, G. 1993. *Biochemistry*. Third edition. Wm. C. Brown, Dubuque, IA. A standard textbook of biochemistry with a lot of metabolic details.

5

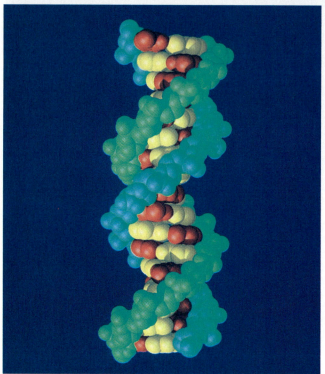

Macromolecules and Molecular Genetics

We now begin a study of the genetics of microorganisms that will extend over the next four chapters. **Genetics** is the discipline that deals with the mechanisms by which traits are passed from one organism to another. As we noted in Chapter 1, two hallmarks of life are *energy transformation* and *information flow*. In the previous chapter we dealt with the problem of energy transformation: *metabolism*. In the present chapter and in those that follow, we deal with the problem of information flow: *genetics*.

The study of molecular genetics is central to an understanding of the variability of organisms and the evolution of species. Since biological information flow is the basis of cellular function, genetics is also a major research tool in attempts to understand the molecular mechanisms by which cells function. Genetics and biochemistry work together in the continuing quest to discern the ultimate basis of life.

Genetics also provides us with approaches to the construction of new organisms of potential use in human affairs. As such, genetics has provided us with some of the most important advances in agriculture, medicine, and industry. An understanding of genetic mechanisms makes it possible for researchers to manipulate species and construct new organisms. It also makes it possible for scientists and physicians to develop means for controlling the important infectious diseases of humankind. We will have much to say about the application of genetics to human affairs in subsequent chapters.

What is a gene and what is its function?

A **gene** can be simply defined as an entity that specifies the structure of a single polypeptide chain. We

Miniglossary for Chapter 5

ACTIVATOR PROTEIN   a regulatory protein which binds to specific sites on DNA and stimulates transcription; involved in positive control

ANTICODON   a sequence of three bases in a tRNA molecule that base pairs with a codon during protein synthesis

ANTIPARALLEL   in reference to double-stranded DNA, one strand runs 5′→ 3′ and the other 3′ → 5′

CHROMOSOME   a genetic element, usually circular in prokaryotes and linear in eukaryotes, carrying genes essential to cellular function

CODON   a sequence of three bases in mRNA that codes for an amino acid

EXON   the coding DNA sequences in a split gene (contrast with introns)

GENE   a segment of DNA specifying a protein (via mRNA), a tRNA, or a rRNA

GENOME   the total complement of genes contained in a cell or virus

HYBRIDIZATION   formation of a duplex nucleic acid molecule with strands derived from different sources by complementary base pairing

INTRON   the intervening noncoding DNA sequences in a split gene (contrast with exon)

MESSENGER RNA (mRNA)   an RNA molecule which contains the genetic information necessary to encode a particular protein

OPERON   a cluster of genes whose expression is controlled by a single operator

PRIMARY TRANSCRIPT   an unprocessed RNA molecule which is the direct product of transcription

PRIMER   a molecule (usually a polynucleotide), to which DNA polymerase can attach the first nucleotide during DNA replication

PROMOTER   a site on DNA to which RNA polymerase can bind and begin transcription

REPLICATION   synthesis of DNA using DNA as a template

REPRESSOR PROTEIN   a regulatory protein which binds to specific sites on DNA and blocks transcription; involved in negative control

RESTRICTION ENZYME   an enzyme which recognizes and makes double-stranded breaks at specific DNA sequences

RIBOSOMAL RNA (rRNA)   types of RNA found in the ribosome; some participate actively in the process of protein synthesis

RIBOZYME   an RNA molecule which can catalyze chemical reactions

RNA PROCESSING   the conversion of a precursor RNA into its mature form

SEMICONSERVATIVE REPLICATION   DNA synthesis yielding new double helices, each consisting of one parental and one progeny strand

TRANSCRIPTION   the synthesis of RNA using a DNA template

TRANSFER RNA (tRNA)   an adaptor molecule used in translation which has specificity for both a particular amino acid and for one or more codons

TRANSLATION   the synthesis of protein using the genetic information in messenger RNA as a template

---

discussed the chemistry of proteins in Chapter 2 and noted that proteins consist of polypeptides and a polypeptide is composed of a series of amino acids connected in peptide linkage. There are 20 different amino acids present in proteins, and a single polypeptide chain will usually have several hundred amino acid residues (see Section 2.8). The gene is the element of information which specifies the *sequence* of amino acids of the protein. The main purpose of this chapter is to explain and expand on this definition.

Genetic phenomena involve three types of macromolecules: *deoxyribonucleic acid (DNA)*, the genetic material of the cell; *ribonucleic acid (RNA)*, the intermediary or messenger; and *protein*, the functional entities of the living cell. During growth, all three types of macromolecules are synthesized. DNA is *replicated*, leading to the formation of exact copies. The information in DNA is also *transcribed* into complementary sequences of nucleotide bases in RNA; this RNA, containing the *information* for the amino acid sequence of the protein, is called *messenger RNA (mRNA)*. Messenger RNA is then *translated*, using the specific protein-synthesizing machinery of the ribosomes, and the translation product is a **polypeptide** (Figure 5.1). Because the steps from DNA to RNA to protein involve the transfer of information, these macromolecules are often called **in-**

formational macromolecules to distinguish them from macromolecules such as polysaccharides and lipids which are large but are not informational.

In what form is the information stored in the informational macromolecules? In DNA and RNA the information is encoded in the *base sequence* of the

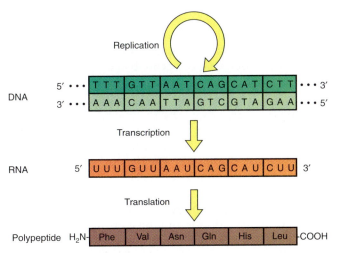

FIGURE 5.1   Synthesis of the three types of informational macromolecules. Note that in any particular region, only one of the two strands of the DNA double helix is transcribed.

purine and pyrimidine bases of the polynucleotide chain. When we discuss the information content of a nucleic acid we thus speak of the *coding* properties of this material. The amino acid sequence of the polypeptide is *coded* by the sequence of purine and pyrimidine bases within the nucleic acid, with *three* bases coding for a single amino acid. We will discuss this coding function in detail in this chapter.

Prokaryotic and eukaryotic genetics

The basic pattern of information flow, replication, transcription, and translation is the same in prokaryotic and eukaryotic cells. However, prokaryotes have relatively simple genetic systems. Their chromosomes are single DNA molecules (see Section 3.12). Mechanisms for the transfer of genes from one prokaryotic cell to another are also simple and easy to study. Eukaryotes, even the simplest eukaryotic microorganisms, have more DNA (and more genes) and are much more complex structurally. Their chromosomes are present in larger numbers (see Section 3.13), and they have fairly complex mechanisms of sexual reproduction for bringing about genetic recombination between organisms. In this chapter we will be making contrasts between prokaryotes and eukaryotes. We will first develop prokaryotic genetics and then highlight those features that are different in eukaryotes.

> The key informational macromolecules in prokaryotic and eukaryotic cells are DNA, RNA, and protein. Molecular genetics deals with the mechanisms by which the information in DNA is replicated, and by which it can be transcribed into RNA and translated into specific proteins.

5.1 Overview of Molecular Genetics

What molecular changes do informational macromolecules undergo during cell growth and division? A cell is an integrated system containing a large number of specific macromolecules. When a cell divides and forms two cells, all of these macromolecules are duplicated. The fidelity of duplication is very high, although occasional errors do occur. The molecular processes underlying cell growth can be divided into a number of stages which are described briefly here.

1. *Replication.* The DNA molecule is a **double helix** of two long chains (see Section 2.7). During replication, DNA, containing the master genetic blueprint, duplicates. The products of DNA replication are two molecules, each a double helix, the two strands thus becoming four strands.

2. *Transcription.* DNA does not function directly in protein synthesis, but through an RNA intermediate. The transfer of the information to RNA is called **transcription**, and the RNA molecule carrying the information is called **messenger RNA (mRNA)**. In prokaryotes messenger RNA molecules frequently contain the instructions for making more than one protein. In most cases, at any particular location on the chromosome, only one strand of the DNA is transcribed, and the information of this strand is then contained in the mRNA. Some regions of DNA do not encode proteins but rather contain information for other types of RNA, such as **transfer RNA (tRNA)** and **ribosomal RNA (rRNA)**. Therefore, we must expand our definition of a gene to include a region of DNA that encodes one of these types of RNA. As we shall see, these other types of RNA molecules also have important functions in the cell.

3. *Translation.* The specific sequence of amino acids in each protein is directed by a specific sequence of bases in the mRNA (which was transcribed from the DNA). This information in the nucleic acids is present as a **genetic code**. It takes *three* bases to code for a single amino acid, and each triplet of bases is called a **codon**. There is a direct correspondence between the base sequence of a gene and the amino acid sequence of a polypeptide (Figure 5.1). The genetic code is actually translated into protein by means of the protein-synthesizing system. This system consists of **ribosomes** (which are themselves made up of proteins and rRNA), transfer RNA, and a number of enzymes. The ribosomes are the structures to which messenger RNA attaches. Transfer RNA is the key link between codon and amino acid. There is one or more separate tRNA molecules corresponding to each amino acid, and the tRNA has a triplet of three bases, the **anticodon**, which is *complementary* to the codon of the messenger RNA. An enzyme brings about the attachment of the correct amino acid to the correct tRNA.

4. *Regulation.* Not all proteins are synthesized at equal rates. Complex systems of regulation exist which control the rates of synthesis of proteins. Some proteins, called *inducible proteins*, are synthesized only when small molecules, called **inducers**, are *present*. Usually the inducer is a substrate or chemically resembles the substrate of the enzyme, and induction thus ensures that the enzyme is only formed when it is needed. Another class of enzymes, called *repressible enzymes*, is synthesized only in the *absence* of specific small molecules, generally biosynthetic products. Enzyme repression and enzyme induction have as their basis the same underlying mechanism, which is the regulation of mRNA synthesis.

5. *Mutation.* A change in a single base in a gene changes the codon and can lead to a change in the amino acid in the peptide. Changes could also occur in the other DNA sequences, such as those involved in regulation. Any such changes in the DNA sequence, which are frequently detrimental, are called **mutations**. Mutations can arise from errors in DNA replication or as a result of damage to DNA.

Microorganisms carrying mutations (*mutants*) are essential in genetic research. They make possible genetic crosses between related organisms, which are essential for locating (**mapping**) genes to specific regions of DNA. In addition, study of mutant strains is often very important in determining the function of the normal (wild type) gene.

Gene structure in prokaryotes and eukaryotes

We emphasized in Chapter 3 the basic differences in organization of DNA in prokaryotes and eukaryotes. In summary, the typical bacterial genome consists of a single, covalently closed *circular* molecule of DNA distributed in the cytoplasm of the cell, and the eukaryotic genome consists of several *linear* pieces of DNA which are present in individual chromosomes in the cell nucleus. Each eukaryotic chromosome consists of

a single DNA molecule bound to proteins called **histones** (see Section 3.13). In contrast to prokaryotes, transcription and translation are spatially separated in eukaryotes. Transcription occurs in the nucleus and the RNA molecules move to the cytoplasm for translation. The genes of eukaryotes are frequently split into two or more coding regions, with noncoding regions separating the coding regions. The coding sequences are called **exons**, and the intervening noncoding regions, **introns**. Both intron and exon regions are transcribed into RNA, and the functional mRNA is subsequently formed by enzymatic removal of noncoding regions. A summary contrasting genetic phenomena in prokaryotes and eukaryotes is given in Figure 5.2.

Although the chromosome structure and the genetic processes of eukaryotes differ markedly from those of prokaryotes, those of eukaryotic organelles such as mitochondria and chloroplasts much more

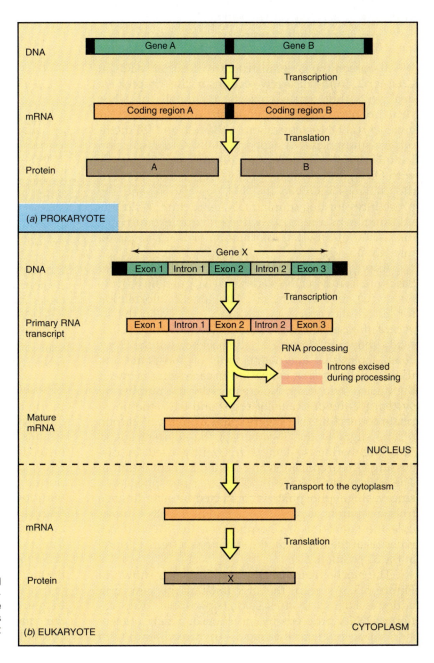

FIGURE 5.2
Contrast of information transfer in prokaryotes and eukaryotes. (a) Prokaryote. A single mRNA often contains more than one coding region (such mRNAs are called *polycistronic*). (b) Eukaryote. Noncoding regions (*introns*) are removed from the primary RNA transcript before translation.

closely resemble those of prokaryotes. From an evolutionary viewpoint, genetic and macromolecular studies provide strong support for the hypothesis that mitochondria and chloroplasts of eukaryotes arose from prokaryotic cells by a process of endosymbiosis (see Sections 3.14, 5.8, 7.15, and 18.3).

The three key processes of macromolecular synthesis are replication, the copying of genetic DNA; transcription, the transfer of the base sequence of DNA into its complement in messenger RNA; and translation, the synthesis of specific proteins using messenger RNA as template. Although the basic processes are the same in both prokaryotes and eukaryotes, the organization of the genetic information is more complex in eukaryotes since many genes have distinct coding regions (exons) and noncoding regions (introns).

5.2   DNA Structure and DNA Binding Proteins

We discussed the general structure of nucleic acids in Chapter 2. In the next few sections of this chapter we deal with three main subjects: (1) details of DNA structure necessary for an understanding of molecular genetics, (2) DNA replication in the cell, and (3) methods for studying DNA experimentally. Our discussion includes information on the interactions of proteins with DNA, the chemistry of DNA replication, some of the enzymes involved in DNA replication, and the procedures for determining DNA structure. With this information as a basis, we will then be able to turn to a discussion of how the information of DNA is copied into RNA.

As we have noted, only four different nucleic acid bases are found in DNA: adenine (A), guanine (G), cytosine (C), and thymine (T). The genetic information for all cellular processes is stored in DNA in the *sequence* of bases along the polynucleotide chain. As already shown in Figure 2.11, the backbone of the DNA chain consists of alternating units of phosphate and the sugar *deoxyribose*; connected to each sugar is one of the nucleic acid *bases*. Note especially the numbering system for the positions of sugar and base; the phosphate connecting two sugars spans from the 3'-carbon of one sugar to the 5'-carbon of the adjacent sugar. This numbering system is frequently used in discussing DNA replication and should be kept in mind. The phosphate linkage in DNA is a phospho*diester*, since a single phosphate is connected by ester linkage to two separate sugars. At one end of the DNA molecule the sugar has a phosphate on the 5'-hydroxyl, whereas at the other end the sugar has a free hydroxyl at the 3'-position.

The biochemistry of DNA replication is shown in Figure 5.3. As seen, the precursor of the new unit added is a deoxyribonucleoside *tri*phosphate. Replication of DNA proceeds by insertion of a new nucleoside triphosphate at the free 3'- (hydroxyl) end, with

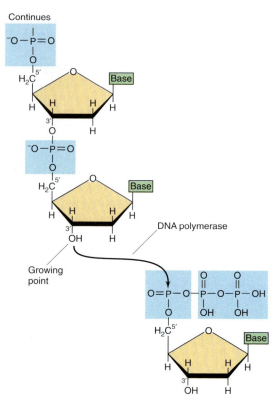

FIGURE 5.3 Structure of the DNA chain and mechanism of growth by addition from a deoxyribonucleotide triphosphate at the 3'-end of the chain. Growth always proceeds from the 5'-phosphate to the 3'-hydroxyl end. The enzyme DNA polymerase catalyzes the addition reaction. The four deoxyribonucleotides that serve as precursors are deoxythymidine triphosphate (dTTP), deoxyadenosine triphosphate (dATP), deoxyguanosine triphosphate (dGTP), and deoxycytidine triphosphate (dCTP). The two terminal phosphates of the triphosphate are split off as pyrophosphate (PP$_i$). Thus, two high-energy phosphate bonds are consumed upon the addition of a single nucleotide.

the subsequent loss of two phosphates (generating a deoxyribonucleoside *mono*phosphate); thus DNA synthesis *always* proceeds toward the 3'-end of the molecule (5' → 3'). As we will see, this requirement that DNA synthesis always proceeds 5' → 3' has important consequences in the replication of double-stranded DNA for both cells and viruses.

DNA as a double helix

In the chromosome, DNA does not exist as a single-stranded polynucleotide, but as two polynucleotide strands which are not identical in base sequence but instead are **complementary**. The complementarity of DNA arises because of the specific pairing of the purine and pyrimidine bases: adenine always pairs with thymine and guanine always pairs with cytosine (Figure 5.4). This **double-stranded** molecule is arranged in a helix, the **double helix** (Figure 5.5). In this double helix, DNA has two distinct grooves, the *major groove* and the *minor groove*. There are many important proteins which interact specifically with DNA (as we shall see later in this chapter and in Chapter 6). In general, these proteins interact predominantly with the major groove, where there is a considerable amount of space. Because of the regularity of the dou-

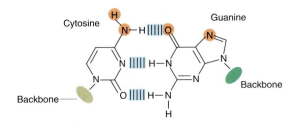

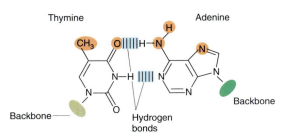

FIGURE 5.4 Specific pairing between adenine (A) and thymine (T) and between guanine (G) and cytosine (C) via hydrogen bonds. These two base pairs are the base pairs typically found in double-stranded DNA. Atoms found in the major groove of the double helix and which interact with proteins are highlighted in red. The deoxyribose phosphate backbones of the two strands of DNA are also indicated.

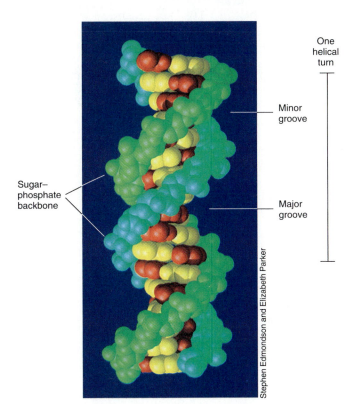

FIGURE 5.5 A computer model of a short segment of DNA showing the overall arrangement of the double helix. One of the sugar–phosphate backbones is shown in blue and the other in green. The pyrimidine bases are shown in red and the purines in yellow. Note the locations of the major and minor grooves. The model was produced using software from the Computer Graphics Laboratory, University of California at San Francisco.

ble helix, some atoms of the bases are always exposed in the major groove (and some in the minor groove). Atoms in the major groove that are known to be important in interactions with proteins are shown in Figure 5.4.

The size of a DNA molecule can be expressed in terms of its *molecular weight*, but since a single nucleotide has a molecular weight of around 330, and since DNA molecules are many nucleotides long, the molecular weight mounts up rapidly. (The nucleic acid in even small viruses, for instance, may have molecular weights in the millions; the DNA in cells in the billions.) A more convenient way of expressing the sizes of DNA molecules is in terms of the *number of thousands* of nucleotide bases per molecule. Thus, a DNA molecule with 1000 bases would contain 1 *kilobase* (1 *kb*) of DNA. If the DNA were a double helix, then one would speak of *kilobase pairs*. Thus, a double helix 5000 bases in length would have a length of 5 kilobase pairs. The bacterium *Escherichia coli* has about 4700 kilobase pairs of DNA in its chromosome.

Supercoiled DNA

Large DNA molecules, representing hundreds to millions of base pairs, are often represented in figures very simply as rods or, as in the case of the bacterial chromosome, uniform circles. This conventional form of illustration may give the impression that DNA is a rather rigid molecule. However, a rigid molecule could not be packed into a cell. For instance, the DNA of *Escherichia coli* is 1000 times longer than the *E. coli* cell itself! Such packaging problems also occur in viral DNA and the DNA of eukaryotic cells. How is it possible to pack so much DNA into such a little space? The solution: *supercoiling*. Supercoiling is a state in which the double-stranded DNA molecules are further twisted. Figure 5.6 shows a diagram of how this could happen in a circular DNA duplex. Supercoiling puts the DNA molecule under torsion. (Take a rubber band and twist it about itself. This twisting generates a tightly coiled structure which is under considerable torsion. This torsion is only held, however, if the circular structure is maintained. Cut the twisted rubber band and see what happens!) DNA can be supercoiled in either a *positive* or *negative* direction. **Negative su-**

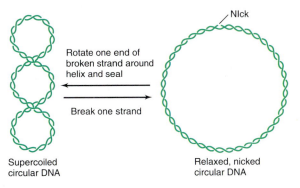

FIGURE 5.6 Supercoiled circular DNA and relaxed, nicked circular DNA interconversions. A nick is a break in a phosphodiester bond of one strand.

percoiling occurs when the DNA is twisted about its axis in the *opposite* direction from that of the right-handed double helix. It is in this form that supercoiled DNA is predominantly found in nature.

How is supercoiling brought about? In eukaryotic chromosomes, the formation of the nucleosome (see Section 3.13) introduces negative supercoils. In Bacteria, there is a special enzyme called *DNA gyrase,* **topoisomerase II**, which introduces negative supercoils. The process can be thought to occur in several stages. First, the circular DNA molecule is twisted, then a break occurs where the two chains come together, then the broken double helix is resealed on the opposite side of the intact strand (Figure 5.7a). Note the derivation of the name *topoisomerase. Topology* is the branch of mathematics which deals with the properties of geometrical figures that are unaltered when the figures are twisted or contorted. We are dealing here with the topology of DNA, and a topoisomerase is an enzyme which affects this topology. Of some interest is the fact that two antibiotics which act on Bacteria, *nalidixic acid* and *novobiocin*, inhibit the action of DNA gyrase.

There is another enzyme that is able to *remove* supercoiling in DNA. This enzyme, called *topoisomerase I*, introduces a single strand break in the DNA and causes the passage of one single strand of the double helix through the other, as illustrated in Figure 5.7b. Through the action of these topoisomerases, the DNA molecule can be alternately coiled and relaxed. Because coiling is necessary for packing the DNA into the confines of a cell and relaxing is necessary so that DNA can be replicated, these two complementary processes clearly play an important role in the behavior of DNA in the cell. In addition, however, supercoiling is known to affect gene expression. Certain genes are more actively transcribed when DNA is supercoiled, whereas transcription of other genes is inhibited by excessive supercoiling.

As was shown in Figure 5.6, because of the torsion in the supercoiled DNA molecule, a break in the backbone (a **nick**) of either strand allows the DNA to return to the relaxed state. Linear DNA, as in eukaryotic chromosomes, is prevented from returning to the relaxed state by the proteins bound to it. To prevent the entire bacterial chromosome from becoming relaxed

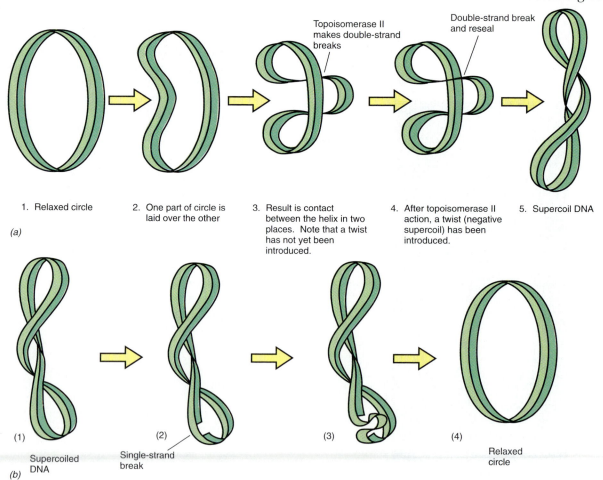

1. Relaxed circle   2. One part of circle is laid over the other   3. Result is contact between the helix in two places. Note that a twist has not yet been introduced.   4. After topoisomerase II action, a twist (negative supercoil) has been introduced.   5. Supercoil DNA

(a)

Topoisomerase II makes double-strand breaks

Double-strand break and reseal

(1)   (2)   (3)   (4)

Supercoiled DNA   Single-strand break   Relaxed circle

(b)

FIGURE 5.7   Introduction of supercoiling in a circular DNA. (a) By action of topoisomerase II, which makes double-strand breaks. (1) Relaxed circle. (2) One part of circle is laid over the other. (3) Result is contact between the double helix at two places. Note that as yet a twist has not been introduced. (4) After topoisomerase II action, a twist (negative supercoil) has been introduced in the DNA. (b) Topoisomerase I passes one single strand through another, resulting in the removal of supercoils in DNA. (1) Supercoiled DNA. (2) Single strand break. (3) Other strand is passed through the break, and the broken strand is resealed. (4) DNA untwists into a relaxed configuration.

every time a nick is made, the chromosome contains approximately 50 *supercoiled domains*. A nick in the DNA in one of these domains will not relax the DNA in the others. It is unclear what holds the DNA in these domains, but it is likely to involve proteins.

> DNA is generally arranged as a double-stranded molecule which assumes a helical configuration. Proteins interact specifically within the major and minor grooves of this helix. The very long DNA molecule is able to be packaged into the cell because it is supercoiled. Supercoiling is an important feature of nearly all chromosomes, whether they be circular, as in prokaryotes, or linear, as in eukaryotic cells. In Bacteria this supercoiling is brought about by enzymes called topoisomerases.

Other important features of DNA structure

As we have noted, complementary base pairing results in the association of the two single strands of DNA in the double helix. For the complementary base pairs to form, the strands must be **antiparallel** (one must be going 5' to 3' and the other 3' to 5'; see Section 2.7). The sequence of DNA in most regions of the chromosomes of microorganisms is primarily determined by the nature of the genetic code and the amino acid sequences of the proteins encoded by the regions. However, there are frequently base sequences in DNA that are present not because of their coding properties but because they influence the secondary structure of DNA, or the way in which DNA interacts with proteins.

Long DNA molecules are quite flexible, but stretches of DNA less than 100 bp (base pairs) are much more rigid. Some short segments of DNA can be bent by proteins which interact with them. However, certain sequences themselves result in bends in the DNA. The sequences of this **bent DNA** often involve several runs of five or six adenines (in the same strand) each separated by four or five bases; an example of such a sequence is shown in Figure 5.8.

Short repeated sequences are often found in DNA molecules. Many proteins have been found which interact with regions of DNA containing repeated sequences (see later in this section) but which are repeated in inverse orientation. This type of repeat is called an **inverted repeat**. Inverted repeats give the DNA sequence a twofold symmetry. As shown in Figure 5.9, nearby inverted repeats could theoretically lead to the formation of **stem-loop** (cruciform) structures in DNA. (Note that the stem of a stem-loop is a short double helix with normal base pairing and antiparallel strands.)

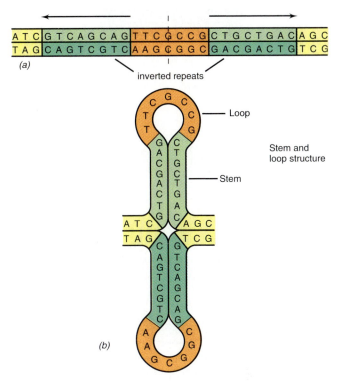

FIGURE 5.9   Inverted repeats and the formation of a stem-loop structure. (a) Nearby inverted repeats in DNA. The arrows indicate the symmetry around the imaginary axis (dashed line). (b) Formation of stem-loop structures (cruciform structures) by pairing of complementary bases on the *same* strand.

Secondary structure formed by base pairing within a single strand of nucleic acid is very important in certain types of RNA. In ribosomal RNA, for example, stems and loops of various shapes give the molecule its final higher order structure (see Section 18.4).

The *ends* of linear DNA molecules can also have interesting sequences which lead to changes in structure. Some DNA molecules have single-stranded regions at each end which are complementary. This leads to the possibility that the two ends can find each other and associate by complementary base pairing, as illustrated in Figure 5.10 for the formation of a circle. DNA which has single-strand complementary sequences at the ends is said to have "sticky ends." Some linear DNA molecules have **hairpin** structures at each end. A hairpin is like a stem-loop but with almost no loop. Hairpins could be formed from a single-stranded region at the end of a molecule which contained an inverted repeat, as illustrated in Figure 5.11.

Because of the enormous length of the DNA in a cell, it can almost never be handled experimentally as a complete unit. The mere manipulation of DNA in the test tube leads to its fragmentation into molecules of smaller size. The need to study the size and shape of DNA is evident. Some of the tools for studying DNA are presented in the Nucleic Acids box in this chapter. As described, a useful technique for studying the sizes of DNA molecules is electrophoresis. As noted, the nucleic acid molecules migrate through the pores of the gel at rates depending upon their molecular weight or molecular shape. Small molecules, or

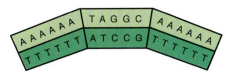

FIGURE 5.8   Double-stranded DNA with runs of five or six A's can form a bent structure.

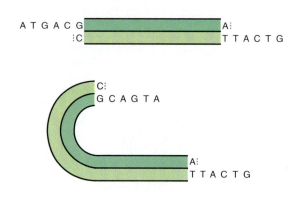

FIGURE 5.10 Linear DNA with complementary single-stranded ends ("sticky ends") can cyclize by base pairing of the complementary ends.

FIGURE 5.11 A hairpin structure at one end of a linear DNA molecule. If the linear DNA had been completely double-stranded, the sequences shown in green would have been inverted repeats.

compact molecules, migrate more rapidly than large or loose molecules. In one figure (see Nucleic Acids box), a number of DNA fragments have been separated out in the gel.

Interactions of chemicals with DNA

A number of organic chemicals interact specifically with DNA, altering its structure and affecting its biological properties. Many chemicals associate with DNA by becoming inserted between adjacent base pairs along the chain, a phenomenon called *intercalation* (Figure 5.12). Chemicals which intercalate are generally planar molecules which can fit between adjacent bases without causing disruption of hydrogen bonding. However, the base pairs must become unstacked vertically to allow for intercalation, so that the sugar–phosphate backbone is distorted and the regular helical structure is altered.

Examples of intercalating chemicals are the acridine dyes such as *acriflavine*, *acridine orange* (Figure 5.12), and *ethidium bromide* (see Nucleic Acids box). Such compounds serve as useful tools in studying the structure and function of DNA and in visualizing DNA experimentally. Some of the planar molecules which intercalate into DNA are cancer-inducing agents, *carcinogens*, or mutation-inducing agents, *mutagens*. We will discuss carcinogens and mutagens later (see Sections 7.3 and 7.4).

Some antibiotics combine strongly and specifically with DNA. We will also discuss later the details of how DNA-binding antibiotics act, but note here one important group of antibiotics, the *actinomycins*. These antibiotics not only intercalate into the double helix, but have peptide side chains which attach to the major groove (Figure 5.12), effectively inhibiting both DNA replication and transcription.

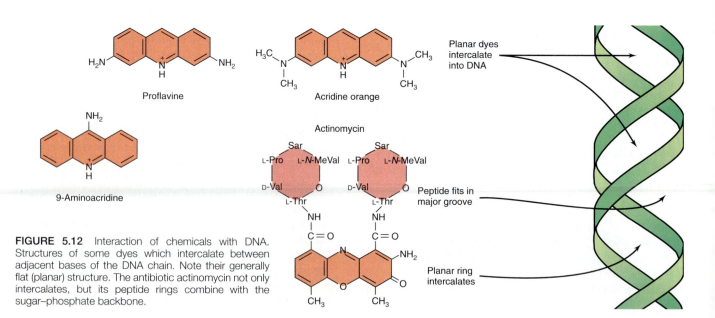

FIGURE 5.12 Interaction of chemicals with DNA. Structures of some dyes which intercalate between adjacent bases of the DNA chain. Note their generally flat (planar) structure. The antibiotic actinomycin not only intercalates, but its peptide rings combine with the sugar–phosphate backbone.

Working with DNA and RNA: The Tools

Our knowledge of molecular biology and genetics has depended on the development of adequate research tools. Advances in knowledge of how nucleic acids work have generally been tied to the development of new methods. We discuss here some of these methods.

1. Extraction and purification of DNA   The first requirement is a sample of DNA free of other cellular chemicals. The steps in the purification of DNA are shown here. The aqueous solution in the final step is treated with RNAse to remove RNA. Proteins are then removed by use of denaturing solvents (usually phenol). By repeating the purification steps a number of times, a solution can be obtained that is virtually free of any components other than DNA.

Note that the solution of DNA obtained never consists of native DNA molecules of the length found in the cell. The purification process causes the DNA to break down into fragments of various (random) lengths. If the DNA has been handled gently during purification, the lengths of the fragments will be about one-hundredth of the length of the whole chromosome.

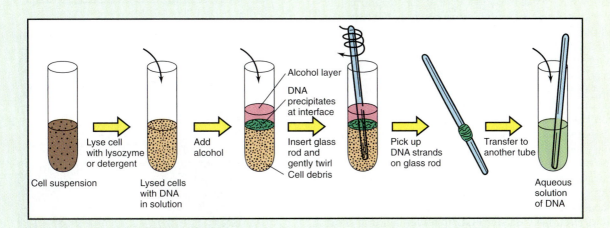

Cell suspension → Lyse cell with lysozyme or detergent → Lysed cells with DNA in solution → Add alcohol → Alcohol layer / DNA precipitates at interface / Insert glass rod and gently twirl / Cell debris → Pick up DNA strands on glass rod → Transfer to another tube → Aqueous solution of DNA

2. Detecting the presence of DNA   There are several methods for detecting the presence of DNA in a solution. One of the most widely used is by its absorption of ultraviolet radiation. DNA strongly absorbs ultraviolet radiation at a wavelength of 260 nm. The absorption is due to the purine and pyrimidine bases. As seen in the figure, double-stranded DNA absorbs less strongly than single-stranded DNA. This is because the interaction between the bases on the opposite strands of the double-stranded DNA (hydrogen bonding) reduces the ultraviolet absorbance.

3. Density-gradient centrifugation of DNA   DNA molecules vary in density, depending on their exact chemical composition. DNA molecules with higher content of guanine plus cytosine (GC) are denser than molecules with low GC. The density of DNA can be determined by centrifugation at very high speed in a *gradient* of cesium chloride (CsCl). The DNA solution is layered on top of a solution of CsCl and centrifuged at high speed for several days, until equilibrium is reached. The CsCl forms a density gradient from top to bottom of the tube and DNA molecules form bands at appropriate densities. At equilibrium, the DNA molecules become positioned in the gradient at positions corresponding to their densities. Following the addition of ethidium bromide to make the DNA fluoresce (see below), observation of the centrifuge tube with ultraviolet radiation after the centrifugation reveals bands of DNA. This method is called the *buoyant density method* and permits both determination of density and separation of molecules of differing density.

4. Gel electrophoresis   One of the most widespread methods of studying nucleic acids is gel electrophoresis. Introduction of electrophoresis methods has revolutionized research on molecular genetics. *Electrophoresis* is the procedure by which charged molecules are allowed to migrate in an electrical field, the rate of migration being determined by the size of the

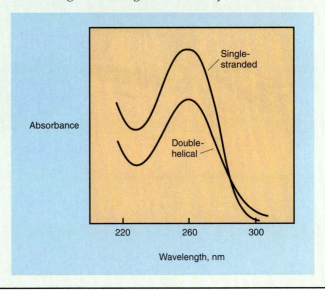

Absorbance / Single-stranded / Double-helical / 220   260   300 / Wavelength, nm

molecules and their electrical charge. In gel electrophoresis, the nucleic acid is suspended in a gel, usually made of polyacrylamide or agarose. The gel is a complex network of fibrils and the *pore size* of the gel can be controlled by the way in which the gel is prepared. The nucleic acid molecules migrate through the pores of the gel at rates dependant on their molecular weight and molecular shape. Small molecules, or compact molecules, migrate more rapidly than large or loose molecules. After a defined period of time of migration (usually a few hours), the locations of the DNA molecules in the gel are assessed by making the DNA molecules fluorescent and observing the gel with ultraviolet radiation.

Shown here is a photograph of an electrophoresis apparatus. The horizontal frame, made of lucite plastic, holds the gel. The ends of the gel are immersed in buffer which makes an electrical connection to the power supply (shown in foreground). The gel is observed after electrophoresis by use of ultraviolet radiation. In each lane, a mixture of DNA fragments had been applied. A computerized scanner can be used to locate the positions of the DNA bands.

Separation of DNA fragments by these relatively simple means has proven extremely useful. However, large molecules (greater than 40,000 base pairs) are not separated from one another. New electrophoretic techniques have been developed that allow separation of large fragments. One of these methods, called *pulse field gel electrophoresis* or **PFGE**, involves sending short pulses of electricity to an array of electrodes surrounding the agarose gel. PFGE and related techniques are very valuable for analyzing DNA molecules the size of those found in a small eukaryotic chromosome.

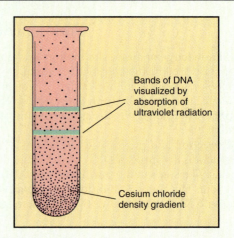

Bands of DNA visualized by absorption of ultraviolet radiation

Cesium chloride density gradient

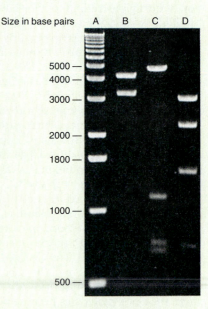

Applying a sample to a gel

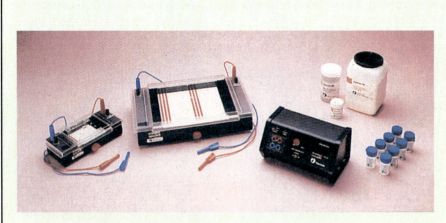

Apparatus for gel electrophoresis of nucleic acids

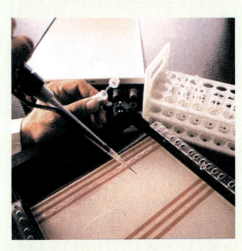

Size in base pairs   A   B   C   D

5000
4000
3000

2000

1800

1000

500

A typical gel

5. Detecting DNA by fluorescence   When nucleic acids are treated with dyes which are fluorescent and which are able to combine firmly with the nucleic acid chain, the nucleic acid is rendered fluorescent. The dye *ethidium bromide* is widely used to render DNA fluorescent because it combines tightly within the DNA molecule. Ethidium bromide interacts with double-stranded DNA. If the DNA is now observed with an ultraviolet source, it will fluoresce.

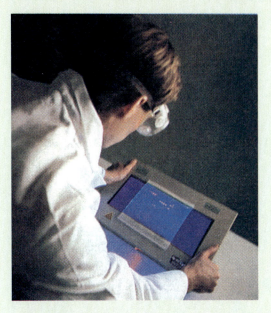

Using fluorescence to locate nucleic acid bands on a gel

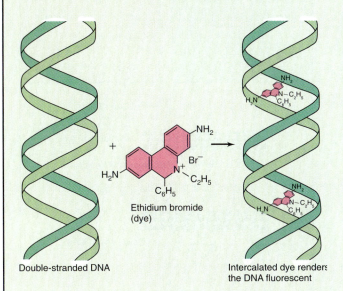

Ethidium bromide (dye)

Double-stranded DNA

Intercalated dye renders the DNA fluorescent

6. Making nucleic acids radioactive   Radioactivity is widely used in nucleic acid research because radioactivity can be detected in extremely tiny amounts. Radioactive nucleic acids can either be detected directly with Geiger or scintillation counters, or indirectly via their effect on photographic film (autoradiography). Autoradiography of radioactive nucleic acids is one of the most widely used techniques in molecular genetics because it can be applied to the detection of nucleic acid molecules during gel electrophoresis.

a. A nucleic acid can be made radioactive by incorporation of radioactive phosphate during nucleic acid synthesis. If radioactive phosphate is added to a culture while nucleic acid synthesis is taking place, the newly synthesized nucleic acid becomes radioactively labeled.

$$^{32}PO_4 \rightarrow {}^{32}P\text{-labeled nucleotides} \rightarrow$$
$$^{32}P\text{-labeled nucleic acid}$$

b. End-labeling of DNA that contains a free hydroxyl group at the 5′ position can be done, using radioactive ATP labeled in the third phosphate. The enzyme polynucleotide kinase specifically attaches the third phosphate of ATP to the free hydroxyl group at the 5′ end of the molecule. End-labeling is an extremely useful technique as it permits labeling of preformed molecules. By tracing the radioactivity

through subsequent chemical steps, the end of the molecule can be followed.

$$^{32}P\text{-P-P-adenosine} + \text{HO-deoxyribose-DNA} \rightarrow$$
$$^{32}P\text{-O-deoxyribose-DNA} + \text{ADP}$$

7. Effect of temperature on nucleic acids   As we have noted, double-stranded nucleic acid molecules are held together by large numbers of weak (hydrogen) bonds. These bonds break when the nucleic acid is heated, but the covalent bonds holding the polynucleotide chains together are unaffected. As shown in part 2 above, double-stranded molecules show lower ultraviolet absorbance than single-stranded molecules. Therefore, if the ultraviolet absorbance of a nucleic acid solution is measured while it is being heated, the increase in absorbance when the double-stranded molecules are converted into single-stranded molecules will show the temperature at which strand separation occurs. Strand separation brought about by heat is generally called *melting*. The stronger the double strands are held together, the higher will be the temperature of melting. Because guanine-cytosine base pairs are stronger than adenine-thymine base pairs (three hydrogen bonds for GC pairs, only two for AT pairs), the higher the GC content of a nucleic acid, the higher the melting temperature. The taxonomic significance of determining the G + C content of an organism's DNA is discussed in Chapter 18. The figure shows the change in absorbance at 260 nm when a solution containing double-stranded DNA is gradually heated. The mid-point of the transition, called T_m, is a function of the GC content of the DNA. If the heated DNA is allowed to cool slowly, the double-stranded native DNA may reform.

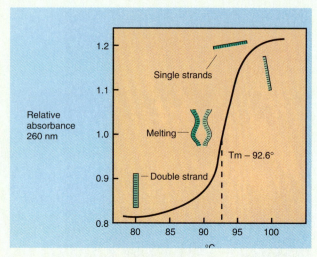

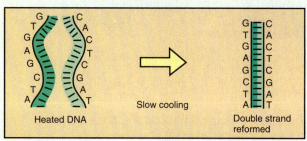

Heated DNA → Slow cooling → Double strand reformed

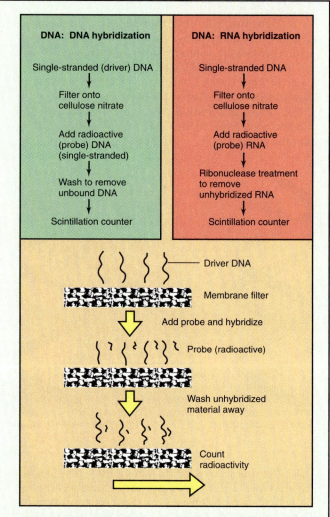

8. Nucleic acid hybridization *Hybridization* is the artificial construction of a double-stranded nucleic acid by complementary base pairing of two single-stranded nucleic acids. If a DNA solution that has been heated (see above) is allowed to cool slowly, many of the complementary strands will reassociate and the original double-stranded complex reforms, a process called *reannealing*. The reannealing only occurs if the base sequences of the two strands are complementary. Thus, nucleic acid hybridization permits the formation of artificial double-stranded hybrids of either DNA, RNA, or DNA-RNA. Nucleic acid hybridization provides a powerful tool for studying the genetic relatedness between nucleic acids. It also permits the detection of pieces of nucleic acid that are complementary to a single-stranded molecule of known sequence. Such a single-stranded molecule of known sequence is called a **probe**. For instance, a radioactive nucleic acid probe can be used to *locate*, in an unknown mixture, a nucleic acid sequence complementary to the probe. Detection of nucleic acid hybridization is usually done with membrane filters constructed of nitrocellulose. Single-stranded DNA is first bound to the filter, and then the probe is added. Probe that does not base-pair to the DNA on the filter is then washed off. If necessary, hybridization conditions can be manipulated to favor the formation of DNA:DNA or DNA:RNA hybrids. Nucleic acid hybridization is a powerful tool in genetics.

Hybridization can also be done after gel electrophoresis. The nucleic acid molecules are transferred by blotting from the gel to a sheet of membrane filter

material, and the probe is then added to the filter. The procedure when DNA is in the gel and RNA or DNA is the probe is often called a *Southern blot procedure*, named for the scientist E.M. Southern, who first developed it. When RNA is in the gel and DNA or RNA is the probe, the procedure is called a *Northern blot*. A Western blot (sometimes called an immunoblot) involves protein-antibody binding rather than nucleic acids; see Section 13.9. The figure also shows the use of a nucleic acid probe to search for complementary sequences in a mixture. The DNA fragments have been spread out by gel electrophoresis and then transferred to the membrane filter. The RNA probe, which is radioactively labeled, is allowed to reanneal to the DNA on the filter and its position is determined by autoradiography.

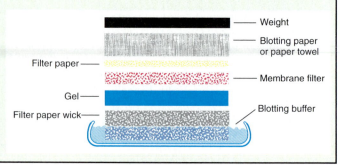

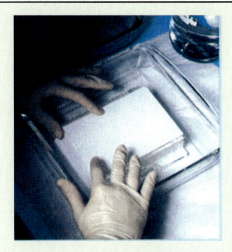

Laying the membrane filter on the gel

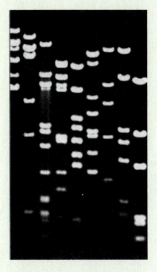

Agarose gel electrophoresis of DNA molecules. Purified molecules of a virus DNA were treated with restriction enzymes and then subject to electrophoresis.

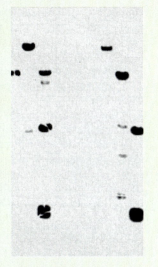

Southern blotting of the DNA gel shown to the left. After blotting, hybridization with radioactively labeled mRNA was carried out. The positions of the bands have been detected by X-ray autoradiography. Note that only some of the DNA fragments have sequences complementary to the labeled mRNAs.

9. Determining the sequence of DNA   Although the base sequences of both DNA and RNA can be determined, it turns out for chemical reasons that it is easier to sequence DNA. Even automated machines are now available for determining the sequences of DNA molecules. Appropriate treatments are used to generate DNA fragments that end at the four bases and that are radioactive. Then the fragments are subjected to electrophoresis so that molecules with one nucleotide difference in length are separated on the gel. This electrophoresis procedure involves *four* separate lanes, one for fragments ending at each of the four bases of the DNA, adenine, guanine, cytosine, and thymine. The positions of these fragments are located by autoradiography and from a knowledge of which base is represented by each lane, the sequence of the DNA can be read off.

Two different procedures have been developed to accomplish the above, called the *Maxam–Gilbert* and the *Sanger dideoxy* procedures. In the Maxam–Gilbert procedure chemicals are used which break the DNA preferentially at each of the four nucleotide bases, under conditions in which only one break per chain is made. (Thus, there are four separate test tubes prepared.)

In the Sanger dideoxy procedure the sequence is actually determined by making a *copy* of the single-stranded DNA, using the enzyme *DNA polymerase*. This enzyme uses deoxyribonucleoside triphosphates as substrates and adds them to a *primer*. In the incubation mixtures (four separate test tubes) are small amounts of each of the dideoxy analogs of the deoxyribonucleoside triphosphates. Because the dideoxy sugar lacks the 3′ hydroxyl, continued lengthening of the chain cannot occur. The dideoxy analog thus acts as a *specific chain termination reagent*. Fragments of variable length are obtained, depending on the incubation conditions. The nucleic acid fragments formed are radioactive from using either a radioactive primer or a radioactive deoxynucleoside triphosphate in the reactions. Electrophoresis of these fragments is then carried out and the positions of the radioactive bands determined by autoradiography.

Normal deoxynucleotide

Dideoxy analog   Missing OH — H

DNA chain

Direction of chain growth

No free 3′-OH, replication will stop at this point

By aligning the four dideoxynucleotide lanes and noting the vertical position of each fragment relative to its neighbor, the sequence of the DNA copy can be read directly from the gel.

Another approach based on the Sanger principle is to use fluorescent labels instead of radioactivity, one fluorescent color for each of the four bases. Then the electrophoresis can be done in one lane instead of four, with the fragments allowed to run off the bottom of the gel, where their fluorescence color is measured with a special laser fluorimeter. This procedure makes it possible to *automate* the sequencing process.

A major advantage of the Sanger method is that it can be used to sequence RNA as well as DNA. To sequence RNA, a single-stranded DNA copy is made (using the RNA as the template) by the enzyme reverse transcriptase. By making the single-stranded DNA in the presence of dideoxynucleotides, various sized DNA fragments are generated suitable for Sanger-type sequencing. From the sequence of the DNA, the RNA sequence is deduced by base-pairing rules. The Sanger method has been instrumental in rapidly sequencing ribosomal RNAs for use in studies on microbial evolution (see Chapter 18).

For determining the DNA sequence of a long molecule, such as a whole gene, it is necessary to proceed in stages. First, the DNA is broken into small overlapping fragments and the sequence of each fragment determined. Using the overlaps as a guide, the sequence of the whole molecule can be deduced.

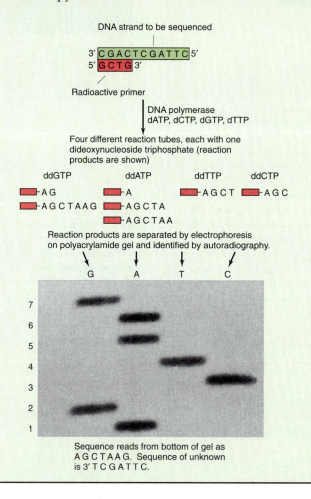

DNA strand to be sequenced

3′ CGACTCGATTC 5′
5′ GCTG 3′

Radioactive primer

DNA polymerase
dATP, dCTP, dGTP, dTTP

Four different reaction tubes, each with one dideoxynucleoside triphosphate (reaction products are shown)

ddGTP	ddATP	ddTTP	ddCTP
-AG	-A	-AGCT	-AGC
-AGCTAAG	-AGCTA		
	-AGCTAA		

Reaction products are separated by electrophoresis on polyacrylamide gel and identified by autoradiography.

G   A   T   C

Sequence reads from bottom of gel as
A G C T A A G. Sequence of unknown
is 3′ T C G A T T C.

Hybridization of nucleic acids

Hybridization is the artificial construction of a double-stranded nucleic acid by complementary base pairing of two single-stranded nucleic acids. The procedure for constructing nucleic acid hybrids is shown in the Nucleic Acids box. Both DNA:DNA and DNA:RNA hybrids can be made. There must be a high degree of complementarity between two single-stranded nucleic acid molecules if they are to form a stable hybrid. In the most common use of hybridization, one of the molecules is used as a radioactive *probe* to detect a specific nucleic acid sequence and formation of hybrids is detected by observing the formation of double-stranded molecules containing radioactivity.

One of the most common uses of hybridization is to detect DNA sequences that are complementary to mRNA molecules. Detection of DNA:RNA hybridization is usually done with membrane filters made of nitrocellulose or nylon. The DNA is first denatured (made single stranded) and immobilized on the filter. The membrane is then treated to prevent any nonspecific nucleic acid binding. Following this, a radioactive RNA probe is added. This RNA will be able to remain attached to the membrane only if it can base pair with

Ser – Gly – Arg – Gly – Lys – Gly – Gly – Lys – Gly – Leu –
Gly – Lys – Gly – Gly – Ala – Lys – Arg – His – Arg – Lys –
Val – Leu – Arg – Asp – Asn – Ile – Gln – Gly – Ile – Thr –
Lys – Pro – Ala – Ile – Arg – Arg – Leu – Ala – Arg – Arg –
Gly – Gly – Val – Lys – Arg – Ile – Ser – Gly – Leu – Ile –
Tyr – Glu – Glu – Thr – Arg – Gly – Val – Leu – Lys – Val –
Phe – Leu – Glu – Asn – Val – Ile – Arg – Asp – Ala – Val –
Thr – Tyr – Thr – Glu – His – Ala – Lys – Arg – Lys – Thr –
Val – Thr – Ala – Met – Asp – Val – Val – Tyr – Ala – Leu –
Lys – Arg – Gln – Gly – Arg – Thr – Leu – Tyr – Gly – Phe –
Gly – Gly

FIGURE 5.13   Structure of a histone protein. The positively charged amino acids are marked.

the immobilized DNA. After appropriate incubation, the unhybridized RNA is washed out and the radioactivity still bound to the filter is measured (see Nucleic Acids box).

Interaction of proteins with nucleic acids

Of great importance is the interaction of proteins with nucleic acids. Protein–nucleic acid interactions are central to replication, transcription, and translation, and to the regulation of these processes. Two general kinds of protein–nucleic acid interactions are noted: nonspecific and specific, depending upon whether the protein will attach *anywhere* along the nucleic acid, or whether the interaction is sequence specific. As an example of proteins that do *not* interact in a sequence-specific fashion, we mention the **histones**, proteins which are extremely important in the structure of the eukaryotic chromosome (see Section 3.14), although less significant in prokaryotes. Histones are relatively small proteins that have a high proportion of positively charged amino acids (arginine, lysine, histidine) (Figure 5.13). DNA, as we have noted, is a polynucleotide and has a high proportion of negatively charged phosphate groups making it a negatively charged molecule. These phosphate groups are on the outside of the DNA double helix. Histones, because of their positive charge, combine strongly and relatively

nonspecifically with the negatively charged DNA. In the eukaryotic cell there is generally enough histone so that all of the phosphate groups of the DNA are covered. Association of histones with DNA leads to the formation of nucleosomes, the unit particles of the eukaryotic chromosome (see Section 3.13).

There are also a number of proteins that interact with DNA in a *sequence-specific* manner. These interactions occur by association of the amino acid side chains of the proteins with the bases as well as with the phosphate and sugar molecules of the DNA. The major groove, because of its size, is an important site of protein binding. Figure 5.4 shows several of the atoms of the base pairs that are found in the major groove and known to interact with proteins. In order to achieve *specificity* in such interactions, the protein must interact simultaneously with more than one nucleic acid base, frequently several. We have already described a structure in DNA called an *inverted repeat* (see Figure 5.9). Such inverted repeats are frequently the locations at which protein molecules combine specifically with DNA (Figure 5.14). Note that this interaction does not involve the formation of cruciform structures in the DNA. Proteins which interact specifically with DNA are frequently *dimers*, composed of two identical polypeptide chains. On each polypeptide chain is a region, called a *domain*, which will interact specifically with a region of DNA in the major groove. A consideration of this type of interaction provides an explanation for the fact that such proteins interact with inverted repeats: in this way, *each* of the polypeptides of the protein dimer combines with each of the DNA strands (Figure 5.15). Note, however, that the protein *does not directly read base pairs on* the DNA.

5′   TGTGTGGAATTGTGAGCGGATAACAATTTCACACA   3′
3′   ACACACCTTAACACTCGCCTATTGTTAAAGTGTGT   5′

FIGURE 5.14   Nucleotide sequence of the operator gene of the lactose operon. Nearby inverted repeats, which are sites in which the *lac* repressor makes contact with the DNA, are shown in shaded boxes.

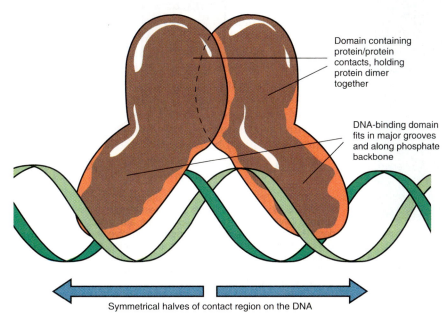

Domain containing protein/protein contacts, holding protein dimer together

DNA-binding domain fits in major grooves and along phosphate backbone

Symmetrical halves of contact region on the DNA

FIGURE 5.15
A protein dimer combines specifically with *two sites* on the DNA.

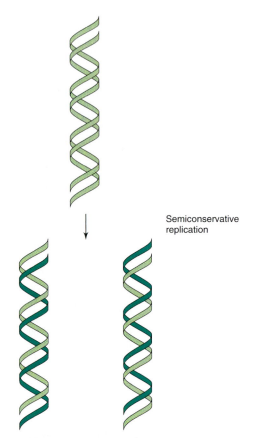

Semiconservative replication

FIGURE 5.19 DNA replication is a semiconservative process. In both prokaryotes and eukaryotes the process is always semiconservative. Note that the new double strands each contain one new and one old strand.

The chemistry of DNA, the nature of its precursors, and the activities of the enzymes involved in replication places some important restrictions on the manner by which this new strand is synthesized. The precursor of each new nucleotide in the chain is a nucleoside *tri*phosphate, of which the two terminal phosphates are removed and the internal phosphate is attached covalently to deoxyribose of the growing chain (see Figure 5.3). The addition of the nucleotide to the growing chain requires the presence of a free hydroxyl group, and such a free hydroxyl group is only available at the 3'-end of the molecule. This chemical restriction leads to an important law which is at the basis of many facets of DNA replication: *DNA replication always proceeds from the end with the 5'-phosphate to the 3'-hydroxyl end, the 5'-phosphate of the incoming nucleotide being attached to the 3'-hydroxyl of the previously added nucleotide.*

The enzymes which catalyze the addition of the nucleotides are called **DNA polymerases**. All DNA polymerases synthesize new DNA in the 5' to 3' direction. *However, no known DNA polymerase can begin a*

new chain. All of these enzymes can only add a nucleotide onto a preexisting 3'-OH group. Therefore, for a *new* chain to be started, there must be a **primer**, a site at which the DNA polymerase can attach the first nucleotide. In most cases this primer is a short stretch of *RNA.*

When the double helix opens up at the beginning of replication, an RNA polymerizing enzyme acts first, resulting in the formation of this RNA primer. A specific RNA-polymerizing enzyme, called *primase*, participates in primer synthesis by laying down a short stretch of RNA. At the growing end of this RNA primer is a 3'-OH group to which DNA polymerase can add the first deoxyribonucleotide. Once priming has begun, continued extension of the molecule occurs as *DNA* rather than RNA. Thus, the newly synthesized molecule has a structure such as that shown in Figure 5.20. The primer must eventually be removed, as we shall see.

To understand the complete replication of a double-stranded DNA molecule, it is easiest to choose an actual example and see how it is replicated. Most of the information on the mechanism of DNA replication has been obtained from the bacterium *Escherichia coli*, and the following discussion will deal primarily with this organism.

> **Both strands of the DNA helix serve as templates for the synthesis of two new strands. The two progeny double helices each contain one parental strand and one new strand. The new strands are elongated by always adding on to the 3'-end. DNA polymerases cannot start new strands. Therefore, new strands must start with a primer, which is usually RNA.**

Initiation of DNA synthesis

As is the case for most prokaryotes, the chromosome of *E. coli* is a circular DNA molecule. Also like most Bacteria, there is a single location on this chromosome where DNA synthesis is initiated, the so-called **origin of replication**. The origin of replication consists of a specific sequence of about 300 bases which is recognized by specific initiation proteins. At the origin of replication, the DNA double helix opens up and the initiation of DNA replication occurs on the two single strands. As replication proceeds, the site of replication, called the **replication fork**, moves down the DNA.

Replication is frequently bidirectional from the origin of replication, as shown in Figure 5.21a, and therefore there are *two* replication forks moving in opposite directions. In circular DNA, bidirectional replication leads to the formation of characteristic structures, visible under the electron microscope, called

FIGURE 5.20
Structure of the RNA-DNA combination which results at the initiation of DNA synthesis.

DNA

Old   3'
New   PPP-5'   3'-OH
      RNA primer   DNA   5'

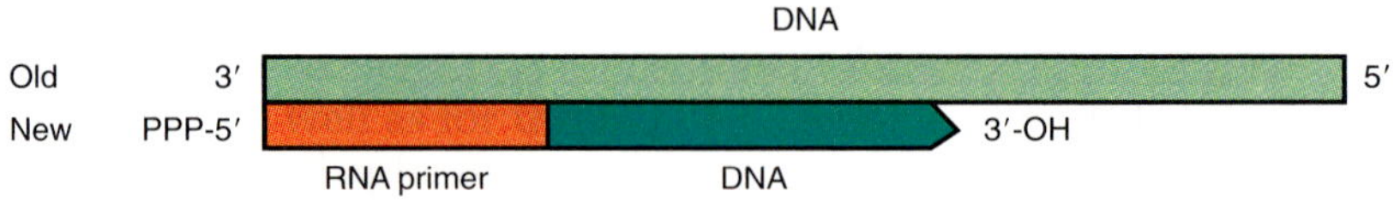

Ser – Gly – [Arg] – Gly – [Lys] – Gly – Gly – [Lys] – Gly – Leu –
Gly – [Lys] – Gly – Gly – Ala – [Lys] – [Arg] – [His] – [Arg] – [Lys] –
Val – Leu – [Arg] – Asp – Asn – Ile – Gln – Gly – Ile – Thr –
[Lys] – Pro – Ala – Ile – [Arg] – [Arg] – Leu – Ala – [Arg] – [Arg] –
Gly – Gly – Val – [Lys] – [Arg] – Ile – Ser – Gly – Leu – Ile –
Tyr – Glu – Glu – Thr – [Arg] – Gly – Val – Leu – [Lys] – Val –
Phe – Leu – Glu – Asn – Val – Ile – [Arg] – Asp – Ala – Val –
Thr – Tyr – Thr – Glu – [His] – Ala – [Lys] – [Arg] – [Lys] – Thr –
Val – Thr – Ala – Met – Asp – Val – Val – Tyr – Ala – Leu –
[Lys] – [Arg] – Gln – Gly – [Arg] – Thr – Leu – Tyr – Gly – Phe –
Gly – Gly

FIGURE 5.13   Structure of a histone protein. The positively charged amino acids are marked.

Interaction of proteins with nucleic acids

the immobilized DNA. After appropriate incubation, the unhybridized RNA is washed out and the radioactivity still bound to the filter is measured (see Nucleic Acids box).

Of great importance is the interaction of proteins with nucleic acids. Protein–nucleic acid interactions are central to replication, transcription, and translation, and to the regulation of these processes. Two general kinds of protein–nucleic acid interactions are noted: nonspecific and specific, depending upon whether the protein will attach *anywhere* along the nucleic acid, or whether the interaction is sequence specific. As an example of proteins that do *not* interact in a sequence-specific fashion, we mention the **histones**, proteins which are extremely important in the structure of the eukaryotic chromosome (see Section 3.14), although less significant in prokaryotes. Histones are relatively small proteins that have a high proportion of positively charged amino acids (arginine, lysine, histidine) (Figure 5.13). DNA, as we have noted, is a polynucleotide and has a high proportion of negatively charged phosphate groups making it a negatively charged molecule. These phosphate groups are on the outside of the DNA double helix. Histones, because of their positive charge, combine strongly and relatively

nonspecifically with the negatively charged DNA. In the eukaryotic cell there is generally enough histone so that all of the phosphate groups of the DNA are covered. Association of histones with DNA leads to the formation of nucleosomes, the unit particles of the eukaryotic chromosome (see Section 3.13).

There are also a number of proteins that interact with DNA in a *sequence-specific* manner. These interactions occur by association of the amino acid side chains of the proteins with the bases as well as with the phosphate and sugar molecules of the DNA. The major groove, because of its size, is an important site of protein binding. Figure 5.4 shows several of the atoms of the base pairs that are found in the major groove and known to interact with proteins. In order to achieve *specificity* in such interactions, the protein must interact simultaneously with more than one nucleic acid base, frequently several. We have already described a structure in DNA called an *inverted repeat* (see Figure 5.9). Such inverted repeats are frequently the locations at which protein molecules combine specifically with DNA (Figure 5.14). Note that this interaction does not involve the formation of cruciform structures in the DNA. Proteins which interact specifically with DNA are frequently *dimers*, composed of two identical polypeptide chains. On each polypeptide chain is a region, called a *domain*, which will interact specifically with a region of DNA in the major groove. A consideration of this type of interaction provides an explanation for the fact that such proteins interact with inverted repeats: in this way, *each* of the polypeptides of the protein dimer combines with each of the DNA strands (Figure 5.15). Note, however, that the protein *does not directly read base pairs on* the DNA.

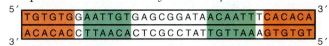

FIGURE 5.14   Nucleotide sequence of the operator gene of the lactose operon. Nearby inverted repeats, which are sites in which the *lac* repressor makes contact with the DNA, are shown in shaded boxes.

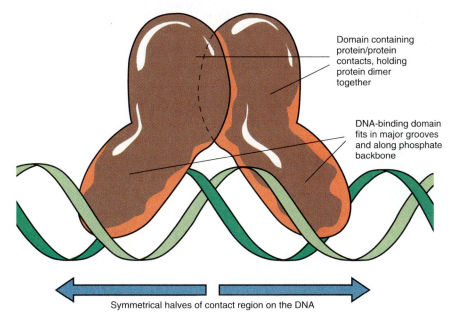

Domain containing protein/protein contacts, holding protein dimer together

DNA-binding domain fits in major grooves and along phosphate backbone

Symmetrical halves of contact region on the DNA

FIGURE 5.15
A protein dimer combines specifically with *two sites* on the DNA.

Rather, it recognizes *contact points* that are associated with specific base sequences.

Studies of several DNA binding proteins from both prokaryotes and eukaryotes have revealed a few types of common protein substructures which are apparently critical for proper binding of many of these proteins to DNA. One of these is termed the *helix-turn-helix motif* (Figure 5.16). The "helix-turn-helix" consists of a stretch of amino acids which form an α-helix secondary structure (the so-called recognition helix), which is joined to a short stretch of three amino acids, the first of which is usually a glycine that functions to "turn" the protein (Figure 5.16*a*). The other end of the "turn" is connected to a second helix which stabilizes the first by interacting hydrophobically with it. Recognition of specific DNA sequences occurs by a combination of noncovalent interactions including hydrogen bonds and van der Waals contacts (see Section 2.2) between the protein and base pairs on the DNA. Many different DNA binding proteins from Bacteria show the helix-turn-helix structure, including many repres-

sor proteins such as the bacteriophage lambda repressor (Figure 5.16*b*) and the *lac* and *trp* repressors of *Escherichia coli* (see Section 5.10).

Two other protein substructures are also commonly found in DNA binding proteins. One of these, the *zinc finger*, is frequently found in eukaryotic regulatory proteins that bind to DNA. The zinc finger is a substructure of protein which, as its name implies, binds a zinc ion (Figure 5.17*a*). It seems most likely that part of the "finger" of amino acids that is created forms an α-helix and this interacts with the DNA in the major groove. There are typically at least two such fingers on the protein involved in binding. The other protein secondary structure commonly found in DNA binding proteins is the *leucine zipper*. This substructure is formed by the side chains on leucine residues spaced every seven amino acids, and it somewhat resembles a zipper. Unlike the helix-turn-helix and the zinc finger, the leucine zipper does not seem to inter-

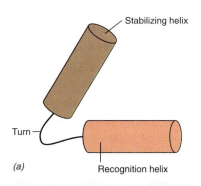

(a)

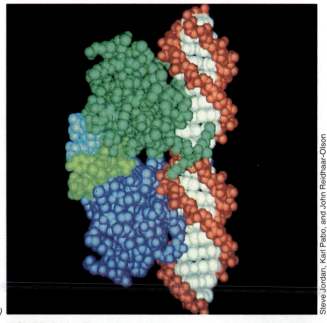

(b)

FIGURE 5.16   The helix-turn-helix structure of some DNA binding proteins. (a) A simple model of the helix-turn-helix elements. (b) A computer model of the bacteriophage lambda repressor, a typical helix-turn-helix protein, bound to its operator gene. One subunit of the dimeric repressor is shown in dark blue and the other in dark green. The light colors represent regions on the dimers involved in subunit interactions.

Steve Jordan, Karl Pabo, and John Reidhaar-Olson

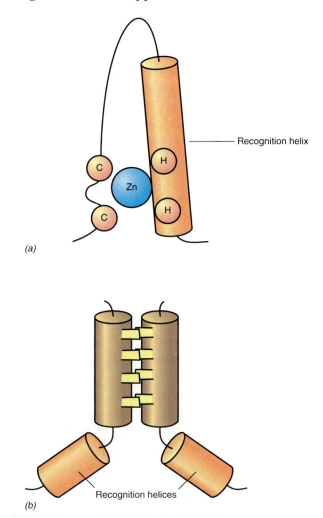

(a)

(b)

FIGURE 5.17   Simple models of protein substructures found in eukaryotic DNA-binding proteins. α-Helices are represented by cylinders. Recognition helices are the domains involved in DNA binding. (a) The zinc finger structure. The amino acids holding the Zn²⁺ ion always include at least two cysteine residues (C) with the other residues being histidine (H). (b) The leucine zipper structure. The leucine residues (shown in yellow) are always spaced exactly every seven amino acids. The interaction of the leucine side chains helps hold the two helices together.

act with DNA itself, but serves to hold two other α-helices in the correct position (Figure 5.17b).

Once a protein combines at a specific site on the DNA, a number of outcomes can occur. In some cases, all the protein does is *block* some other process, such as transcription (see Section 5.6). In other cases, the protein is an enzyme which carries out some specific action on the DNA, such as RNA polymerase, which makes RNA using DNA as the template. However, one group of such proteins are the *restriction enzymes*, enzymes which specifically *cut* DNA at sites near where they combine. We discuss these interesting and useful enzymes in the next section.

> **Two complementary, antiparallel strands of DNA form a regular double-helical structure. The strands of a double-helical DNA molecule can be separated experimentally by chemicals or heat. Two complementary single strands can hybridize to form a stable double-stranded molecule. RNA can also hybridize with single-stranded DNA. Hybridization can be used to measure the degree of sequence homology of two single-stranded nucleic acid molecules and is an important tool in modern molecular genetics. Recognition occurs between specific sequences in nucleic acids and certain proteins (DNA-binding proteins) because of specific protein–nucleic acid interactions, that affect the structure and function of nucleic acids.**

5.3   Restriction Enzymes and Their Action on DNA

Organisms are occasionally faced with the problem of coping with foreign DNA, generally derived from viruses, that may derange cellular metabolism or initiate processes leading to cell death. Although a number of mechanisms for coping with foreign DNA exist, one of the most dramatic is enzymatic destruction. The enzymes involved in the destruction of foreign DNA are remarkably specific in their action, an essential property if destruction of cellular DNA is to be avoided. One class of highly specific enzymes are called **restriction endonucleases**. Restriction endonucleases combine with DNA only at sites with specific sequences of bases. Most restriction enzymes not only recognize specific sequences of bases but also make double-stranded breaks *within* these sequences. Many of these sequences exhibit twofold symmetry around a given point. Thus one restriction endonuclease from *Escherichia coli*, called *Eco*RI, has the following recognition sequence:

$$5' \ldots \text{G} \overset{\downarrow}{-} \text{A} - \text{A} - \text{T} - \text{T} - \text{C} - \ldots 3'$$

$$3' \ldots \text{C} - \text{T} - \text{T} - \text{A} - \text{A} \underset{\uparrow}{-} \text{G} - \ldots 5'$$

The cleavage sites are indicated by arrows, and the axis of symmetry by a dashed line. Note that the two strands have the same sequence if one is read

from the left and the other from the right (or, in terms of polynucleotide strands, if both are read 5' to 3' or both read 3' to 5'). Such a structure is called a **palindrome**. (A palindrome is a sequence of characters which reads the same when read from either right or left. For instance, *Sex at noon taxes* or *Able was I ere I saw Elba*. The term *palindrome* is derived from the Greek meaning "to run back again.")

Many restriction enzymes are composed of two identical polypeptide subunits, each of which recognizes and cuts the sequence on a single strand. Since the sequences recognized by restriction enzymes are relatively short, and frequently palindromic, such enzymes will always make *double*-stranded breaks, and such double-stranded breaks are not subject to correction by repair enzymes. This ensures that an invading nucleic acid will be destroyed.

Almost all known restriction enzymes are from prokaryotes. The recognition sequences and cutting sites for a few restriction enzymes are given in Table 5.1. Note that the recognition sites for the enzymes listed are 4, 5, 6, and 8 base pairs. In a "random" DNA molecule, one would expect any 4 base-pair sequence to occur approximately once every 256 base pairs based on the probability of $1/4 \times 1/4 \times 1/4 \times 1/4$ (assuming each base pair is equally probable in the DNA). Therefore, such an enzyme would cut a large DNA molecule into many specific fragments. A specific 6 base-pair sequence should appear every 4096 base pairs in random DNA, and an 8 base-pair sequence should appear only once about every 1 million base pairs (a megabase pair). The *E. coli* chromosome, which is about 4.7 megabase pairs, is cut 21 times by the enzyme *Not*I, which recognizes an 8 base pair sequence indicating that in this chromosome, the *Not*I recognition sequence is used slightly more often than one might have predicted.

A large number of restriction enzymes are now known and more are being sought. The reason for this is not just that they are very interesting enzymes. They are also of great importance in DNA research. Note that the enzyme *Eco*RI can cut *any* double-stranded DNA that has its recognition sequence, and it will cut only at that sequence. This enables scientists to cut large DNA molecules into smaller fragments. Such fragments with defined termini, created as a result of the action of specific restriction enzymes, are amenable to determination of nucleotide sequences, thus permitting the working out of the complete sequence of DNA molecules (see Nucleic Acids box).

Another use of certain restriction enzymes is that they permit the conversion of DNA molecules into fragments which can be joined by DNA ligase (see Section 5.4). This enables laboratory researchers to clone DNA, as will be discussed in Chapter 8.

Modification: protection from restriction

An integral part of the cell's restriction mechanism is the chemical **modification** of the specific sequences on its *own* DNA so that these sequences are not attacked

Table 5.1	Recognition sequences of a few restriction endonucleases	
Organism	**Enzyme designation**	**Recognition sequence***
Bacillus subtilis	*Bsu*RI	GG↓C̆C
Brevibacterium albidum	*Bal*I	TGG↓C̆CA
Escherichia coli	*Eco*RI	G↓AÅTTC
Escherichia coli	*Eco*RII	↓C̆CAGG (Not a palindromic sequence)
Haemophilus haemolyticus	*Hha*I	GC̆G↓C
Haemophilus influenzae	*Hind*II	GTPy↓PuAC̆
Haemophilus influenzae	*Hind*III	A↓AGCTT
Nocardia otitidis–caviarum	*Not*I	GC↓GGC̆CGC
Thermus aquaticus	*Taq*I	T↓CGÅ

*Arrows indicate the sites of enzymatic attack. Asterisks indicate the site of methylation (modification). G = guanine; C = cytosine; A = adenine; T = thymine; Pu = any purine; Py = any pyrimidine. Only the 5' → 3' sequence is shown.

by the cell's own restriction enzymes. Such modification generally involves *methylation* of specific bases within the recognition sequence so that the restriction nuclease can no longer act. Thus, for each restriction enzyme there must also be a modification enzyme, the two enzymes being closely associated. For example, the sequence recognized by the *Eco*RII restriction enzyme (also see Table 5.1) is:

$$C–C–A–G–G$$
$$G–G–T–C–C$$

and modification of this sequence results in methylation of two cytosines:

$$\overset{m}{C}–C–A–G–G$$
$$G–G–T–C–\underset{m}{C}$$

Note that a given nucleotide sequence can be a substrate for *either* a restriction enzyme *or* a modification enzyme but not both. This is because modification makes the sequence unreactive with the restriction enzyme, and action of the restriction enzyme destroys the recognition site of the modification enzyme.

Restriction enzymes are such important tools in modern molecular genetic research that they have become widely available commercially. A number of companies purify and market restriction enzymes with a variety of specificities. If the DNA sequence of a particular region of a molecule is known, a research worker can generally obtain a restriction enzyme that will cut in this region. On the other hand, if it is known that a particular piece of DNA has been cut by

a particular restriction enzyme, then, by reference to the base sequence which that restriction enzyme cuts, it is possible to deduce the base sequence around that site. This provides a powerful tool for studying DNA molecules.

Restriction enzyme analysis of DNA

As noted, a DNA molecule can be cut at a specific location by a given restriction enzyme. Because the base sequences recognized by most restriction enzymes are four to six nucleotides long, there will generally be only a limited number of such sequences in a piece of DNA. After cleaving the DNA (Figure 5.18a), the fragments can be separated by agarose gel electrophoresis, as shown in Figure 5.18b. The distance migrated by any band of DNA in such a gel can be determined by calibrating the electrophoresis system with DNA molecules of known size. By judicious use of several restriction enzymes of different specificities, and by use of overlapping fragments, it is possible to construct a **restriction enzyme map** in which the positions cut by each of the several restriction enzymes can be designated (Figure 5.18c).

Several procedures are now available for determining the base sequences of DNA molecules. In fact, automated machines are available for sequencing DNA. Details are presented in the Nucleic Acids box. By successively determining the sequences of small overlapping fragments of DNA, it is possible to determine the sequences of very large pieces of DNA. The sequences are now known for thousands of genes, as well as for the complete genome of many viruses (see Chapter 6).

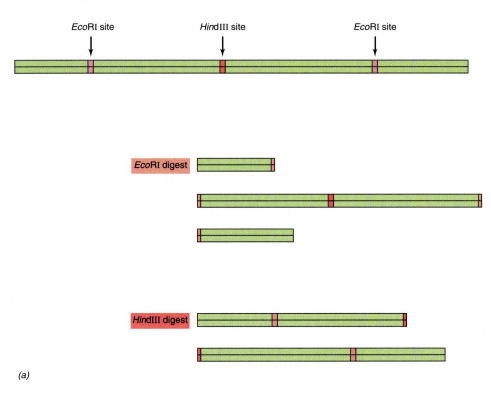

(a)

FIGURE 5.18
Restriction enzyme analysis of DNA. (a) A DNA molecule with 2 *Eco*RI and 1 *Hind*III restriction sites. (b) Results of electrophoresis of digests with each of the enzymes separately, and a "double digest" with both enzymes. The lane with the standards allows one to determine the size of each fragment (and of the entire molecule, which in this case is 480 base pairs). Note that comparison of the results of the single and double digests indicates that the 300 base pair fragment generated by *Eco*RI digestion must have the *Hind*III site within it. (c) By comparing the fragments generated, it is possible to deduce this map. If you ignore the original figure given in part (a) and make a map just using the data in part (b), you can also get another map that is identical to this but rotated 180°. Both maps are correct unless you have more information that defines which end of the molecule is left and which is right.

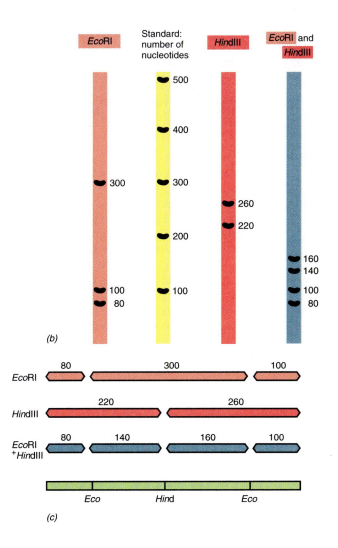

(b)

(c)

Restriction enzymes are cellular enzymes which recognize specific short base sequences in DNA and make two single-stranded breaks at locations within the recognition sites. A restriction enzyme does not affect the cell that produces it because its own DNA is methylated at the recognition site by a modification enzyme specific for that site. Thus, restriction and modification enzymes constitute a specific system for protecting the cell's own DNA but destroying foreign DNA. Restriction enzymes are important research tools in modern molecular genetics.

5.4   DNA Replication

The problem of DNA replication can be simply put: the nucleotide base sequence residing in each long molecule of the DNA double helix must be precisely duplicated to form a copy of the original molecule. The cell has solved this seemingly complex problem in an elegant fashion: by means of *complementary base pairing*. As we have discussed (see Figure 5.4), adenine pairs specifically with thymine and guanine pairs with cytosine. If the DNA double helix opens up, a new strand can be synthesized as the complement of each of the parental strands. As shown in Figure 5.19, replication is **semiconservative**, the two resulting double helices consisting of one progeny and one parental strand.

Templates and Primers

The DNA molecule that is copied to form a complement is called a *template*. A template is a preformed pattern which is copied, but the *new* DNA molecule is not covalently connected to the *old* DNA molecule.

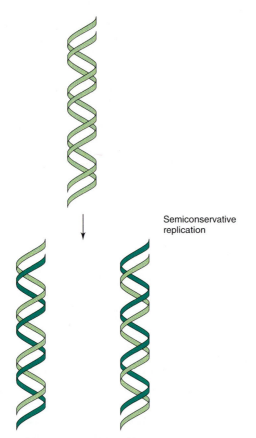

Semiconservative replication

FIGURE 5.19 DNA replication is a semiconservative process. In both prokaryotes and eukaryotes the process is always semiconservative. Note that the new double strands each contain one new and one old strand.

The chemistry of DNA, the nature of its precursors, and the activities of the enzymes involved in replication places some important restrictions on the manner by which this new strand is synthesized. The precursor of each new nucleotide in the chain is a nucleoside *tri*phosphate, of which the two terminal phosphates are removed and the internal phosphate is attached covalently to deoxyribose of the growing chain (see Figure 5.3). The addition of the nucleotide to the growing chain requires the presence of a free hydroxyl group, and such a free hydroxyl group is only available at the 3'-end of the molecule. This chemical restriction leads to an important law which is at the basis of many facets of DNA replication: *DNA replication always proceeds from the end with the 5'-phosphate to the 3'-hydroxyl end, the 5'-phosphate of the incoming nucleotide being attached to the 3'-hydroxyl of the previously added nucleotide.*

The enzymes which catalyze the addition of the nucleotides are called **DNA polymerases**. All DNA polymerases synthesize new DNA in the 5' to 3' direction. *However, no known DNA polymerase can begin a*

new chain. All of these enzymes can only add a nucleotide onto a preexisting 3'-OH group. Therefore, for a *new* chain to be started, there must be a **primer**, a site at which the DNA polymerase can attach the first nucleotide. In most cases this primer is a short stretch of *RNA*.

When the double helix opens up at the beginning of replication, an RNA polymerizing enzyme acts first, resulting in the formation of this RNA primer. A specific RNA-polymerizing enzyme, called *primase*, participates in primer synthesis by laying down a short stretch of RNA. At the growing end of this RNA primer is a 3'-OH group to which DNA polymerase can add the first deoxyribonucleotide. Once priming has begun, continued extension of the molecule occurs as *DNA* rather than RNA. Thus, the newly synthesized molecule has a structure such as that shown in Figure 5.20. The primer must eventually be removed, as we shall see.

To understand the complete replication of a double-stranded DNA molecule, it is easiest to choose an actual example and see how it is replicated. Most of the information on the mechanism of DNA replication has been obtained from the bacterium *Escherichia coli*, and the following discussion will deal primarily with this organism.

> Both strands of the DNA helix serve as templates for the synthesis of two new strands. The two progeny double helices each contain one parental strand and one new strand. The new strands are elongated by always adding on to the 3'-end. DNA polymerases cannot start new strands. Therefore, new strands must start with a primer, which is usually RNA.

Initiation of DNA synthesis

As is the case for most prokaryotes, the chromosome of *E. coli* is a circular DNA molecule. Also like most Bacteria, there is a single location on this chromosome where DNA synthesis is initiated, the so-called **origin of replication**. The origin of replication consists of a specific sequence of about 300 bases which is recognized by specific initiation proteins. At the origin of replication, the DNA double helix opens up and the initiation of DNA replication occurs on the two single strands. As replication proceeds, the site of replication, called the **replication fork**, moves down the DNA.

Replication is frequently bidirectional from the origin of replication, as shown in Figure 5.21*a*, and therefore there are *two* replication forks moving in opposite directions. In circular DNA, bidirectional replication leads to the formation of characteristic structures, visible under the electron microscope, called

FIGURE 5.20
Structure of the RNA-DNA combination which results at the initiation of DNA synthesis.

Old   3'   DNA   5'

New   PPP-5'   3'-OH

RNA primer   DNA

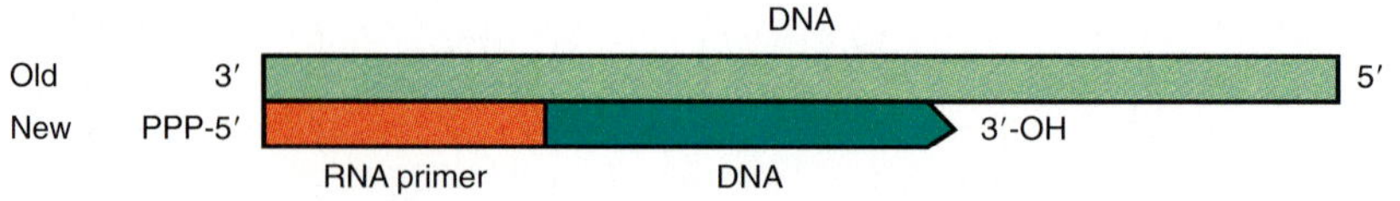

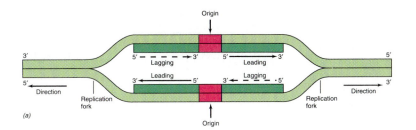

FIGURE 5.21
(a) Direction of replication of DNA is frequently bidirectional from the origin. (b) In circular DNA, replication leads to the formation of structures resembling the Greek letter theta (θ).

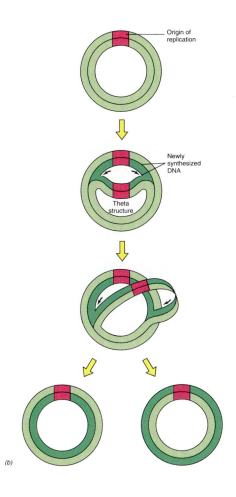

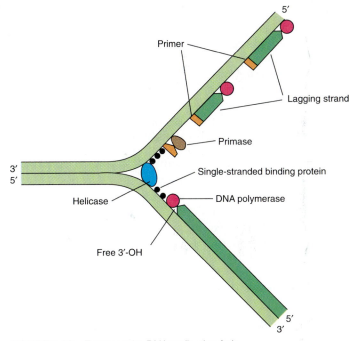

FIGURE 5.22   Events at the DNA replication fork.

theta structures (Figure 5.21*b*). Most large DNA molecules, whether from prokaryotes or eukaryotes, have bidirectional replication from fixed origins. A single eukaryotic chromosome has many origins. This is not simply because the DNA is longer because, as we have seen, this is not always the case (Section 3.13). It may reflect the fact that DNA polymerases from eukaryotes do not replicate as fast as the prokaryotic enzymes. DNA replication is carefully regulated, and the site where this regulation takes place is the origin.

Leading and lagging strands

There are three different DNA polymerases in *E. coli*, called DNA polymerase I, II, and III. It is DNA polymerase III that is the primary enzyme of replication at the replicating forks. However, several other enzymes are also involved. The details of events at the replica-

tion fork are illustrated in Figure 5.22. At the replication fork, the DNA double helix unwinds and a small single-stranded region is formed by the action of specific proteins called *helicases*. Helicases are ATP-dependent enzymes that hydrolyze ATP as they move down the helix in advance of the replicating fork. The single-stranded region generated is complexed with a special protein, the *single-strand binding protein*, which stabilizes the single-stranded DNA, preventing the formation of intrastrand hydrogen bonds.

Figure 5.22 reveals an important difference between replication of the two strands, which arises from the fact that DNA replication always proceeds from 5'-phosphate to 3'-hydroxyl (always adding a *new* nucleotide to the 3'-OH of the growing chain). On the strand growing from the 5'-phosphate → 3'-hydroxyl, called the **leading strand**, DNA synthesis can occur *continuously*, because there is always a free 3'-OH at the replication fork to which a new nucleotide can be added. But on the opposite strand, called the **lagging strand**, DNA synthesis must occur *discontinuously* (because there is no 3'-OH at the replication fork to which a new nucleotide can attach). Where is the 3'-OH on this strand? At the *opposite* end, *away* from the growing point. Therefore, on the lagging strand, a small (11 bases) RNA primer must be synthesized by primase to provide free 3'-OH groups. After synthesizing the primer, primase is replaced by the enzyme DNA polymerase III. Then deoxyribonucleotides are added until DNA polymerase III reaches the previously synthesized DNA.

At this point, DNA polymerase III stops and is released from the DNA. The next enzyme that is involved, *DNA polymerase I*, has more than one activity. It can obviously synthesize DNA. However, at the same time it is adding on nucleotides to the 3'-OH, it has an *exonuclease* activity that removes the RNA primer from in front of it (Figure 5.23). When the primer has been removed and replaced with DNA, DNA polymerase I is then released. The last phosphodiester bond is made by an enzyme called *DNA ligase*. (This enzyme can seal any nicks made in DNAs which have a 5'-phosphate and 3'-OH, and along with DNA polymerase I is also involved in DNA repair.)

While DNA synthesis is continuing at the replication fork, changes in the coiling of the DNA will be occurring, modified by unwinding enzymes and topoisomerases (see Section 5.2). Unwinding is obviously an essential feature of DNA replication, and because supercoiled DNA is under strain, it unwinds more easily than DNA that is not supercoiled. Thus, by regulating the degree of supercoiling, topoisomerases regulate the process of replication (and also transcription, see later).

Fidelity of DNA replication: proofreading

Errors in DNA replication introduce mutations. Mutation rates in living organisms are remarkably low; between 10^{-8} and 10^{-11} errors per base pair inserted. Part of the reason for this accuracy is because DNA polymerase actually gets *two* chances to incorporate the correct base at a given site. The first chance occurs when complementary bases are inserted by base pairing rules, A with T and G with C, using the template strand as the pattern. The second chance occurs because of a second enzymatic activity, referred to as **proofreading**, associated with DNA polymerase III (Figure 5.24). In addition to inserting nucleotides in the replicating strand, DNA polymerase also contains a 3' → 5' *exonuclease* activity which can remove a misinserted nucleotide and replace it with the correct nucleotide. Proofreading activity is summoned if an incorrect base is inserted, since misinsertion creates unstable base pairing. The proofreading activity makes it less likely that the misinserted nucleotide will remain bound to the active site on DNA polymerase. This allows the exonuclease activity time to remove

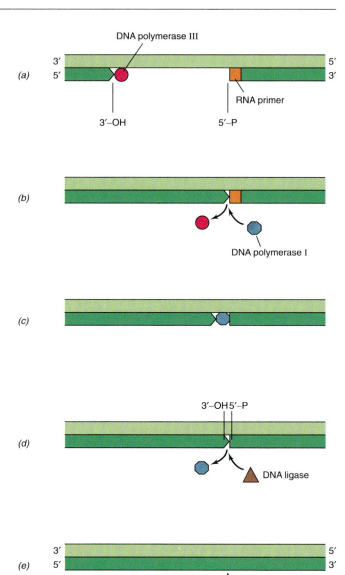

FIGURE 5.23 Sealing two fragments on the lagging strand. (a) DNA polymerase III is synthesizing DNA in the 5' to 3' direction toward the RNA primer of a previously synthesized fragment on the lagging strand. (b) Upon reaching the fragment, DNA polymerase I replaces III. (c) DNA polymerase I continues synthesizing DNA while removing the RNA primer from the previous fragment. (d) DNA ligase replaces DNA polymerase III after the primer has been removed. (e) DNA ligase seals the two fragments together.

the misinserted base and gives the polymerase activity a second chance to insert the correct base (Figure 5.24). (Note that the proofreading exonuclease activity is the *opposite* of the 5' → 3' exonuclease activity of DNA polymerase I that is used to remove the primer from "in front" of the polymerase.)

Exonuclease proofreading occurs in prokaryotes, eukaryotes, and viral DNA replication systems. In addition to exonucleolytic proofreading capabilities, prokaryotes and eukaryotes contain *endonucleolytic* proteins capable of removing a misinserted nucleotide long after DNA polymerase has passed the point of the error (see Sections 7.1 and 7.2). The combination of endonucleolytic and exonucleolytic (proofreading) ac-

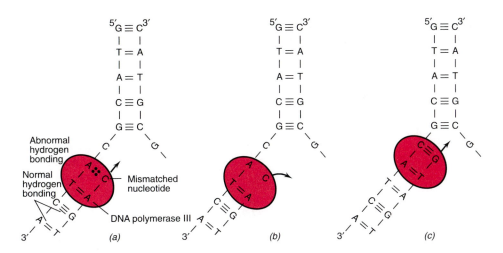

FIGURE 5.24
Proofreading by the 3′ → 5′ exonuclease activity of DNA polymerase III. (a) A mismatch in base pairing at the terminal base pair causes the polymerase to pause briefly. This is a signal for the proofreading activity (b) to excise the mismatched nucleotide, after which the correct base is incorporated (c) by polymerase activity.

tivities ensures nearly error-free replication of the extremely long DNA sequences which make up genomic DNA.

We have mentioned a number of enzymes and other proteins which combine or act on DNA. A summary of some of these enzymes is given in Table 5.2.

> DNA synthesis begins at a unique location called the origin of replication. The double helix is unwound by helicase and is stabilized by single-stranded binding protein. DNA polymerases all make DNA by adding onto a 3′-hydroxyl group. To start a new chain, a short RNA primer must first be made by primase. Extension of the DNA occurs continuously on the leading strand, but discontinuously on the lagging strand. Most errors in base-pairing are corrected by proofreading functions associated with the action of DNA polymerase.

5.5   Genetic Elements

A genetic element is a particle or structure containing genetic material. A number of different kinds of genetic elements have been recognized. Although the main genetic element is the *chromosome*, other genetic elements are found and play important roles in gene function in both prokaryotes and eukaryotes (Table 5.3). Two key properties of genetic elements are (1) their ability to self-replicate, and (2) their genetic coding properties.

Although single DNA molecules can be thought of as genetic elements, most genetic elements have a more complex structure than pure DNA. This structure is often associated with the condensation of DNA into compact bodies, a process necessary to fit the extremely long DNA molecules into the confines of a

Table 5.2   Enzymes affecting DNA

Enzyme	Action	Function in the cell
Restriction endonuclease	Cuts DNA at specific base sequences	Destroys foreign DNA
DNA ligase	Links DNA molecules	Completes replication process
DNA polymerase I	Attaches nucleotides to the growing DNA molecule, removes RNA primers	Fills gaps in DNA, primarily for DNA repair, and removes primers
DNA polymerase III	Attaches nucleotides to the growing DNA molecule; proofreads each inserted nucleotide	Replication of DNA
DNA gyrase (topoisomerase II)	Increases the twisting pattern of DNA, promoting supercoiling	Maintains compact structure of DNA
DNA helicase	Binds to DNA near replicating fork	Promotes DNA strand separation
DNAse	Degrades DNA to nucleotides	General destruction of DNA
DNA methylase	Places methyl groups on DNA bases, thus inhibiting restriction endonuclease action	Modifies cellular DNA so that it is not affected by its own restriction endonuclease
Primase	Makes short RNA chains using a DNA template	Needed to make primer to be used by DNA polymerase

Table 5.3 Kinds of genetic elements	
Element	**Description**
Prokaryote	
Chromosome	Extremely long, usually circular double-stranded DNA molecule
Plasmid	Relatively short, usually circular double-stranded DNA molecule
Virus	Single- or double-stranded DNA or RNA molecule
Transposable element	Double-stranded DNA molecule always found within another DNA molecule
Eukaryote	
Chromosome	Extremely long linear double-stranded DNA molecule
Plasmid	Relatively short circular or linear double-stranded DNA molecule
Mitochondrion/ Chloroplast	Intermediate length DNA molecules, usually circular
Virus	Single- or double-stranded DNA or RNA molecules
Transposable element	Double-stranded DNA molecule always found within another DNA molecule

reasonable space. However, when replication or transcription occurs, the condensed molecule must, at least temporarily, be unfolded.

The chromosome

The single prokaryotic chromosome contains most if not all of the genetic information of the bacterial cell. The chromosomes of many different kinds of Bacteria have been examined, and until recently, all were found to be circular like that of *E. coli.* However, it has now been shown that the Bacterium *Borrelia burgdorferi,* the causative agent of Lyme disease (see Section 15.10), has a linear chromosome (the ends of which may be in hairpin repeats like those shown in Figure 5.11).

In eukaryotes, a chromosome is a linear DNA molecule containing genes, which has a special DNA sequence called a *telomere* at each end and a *centromere* somewhere between the telomeres. Centromeres are important for partitioning the chromosomes during cell division. As we shall see, telomeres play an important role in the replication of these molecules. In eukaryotic cells, the **genome,** the total complement of genes contained in a cell or virus, is contained on more than one chromosome. The number of chromosomes is constant within a species but varies widely between species. The overall structure of a eukaryotic chromosome can be visualized as a series of compact, highly folded units called **nucleosomes,** each containing about 200 DNA base pairs, separated by linkers of less extensively complexed DNA (see Section 3.13 and Figure 3.69). Within each nucleosome, the DNA molecule

is wound around a cluster of histone molecules organized in a precise and repeatable pattern. One function of the nucleosome structure is to permit packing of the long DNA molecules within the cell. For instance, if all of the DNA of the chromosomes of a human cell were stretched end to end, the length would be close to 2 meters, yet this DNA is packed into 46 chromosomes in a total length of only 200 μm.

In prokaryotes, there is no membrane separating the chromosome from the cytoplasm (see Section 3.2), and there is the possibility of close association between the production of messenger RNA from the DNA template and the protein-synthesizing machinery (ribosomes, transfer RNA, and so on). In eukaryotes, chromosomes are located inside the nucleus, and only at the time of division, when the nuclear membrane breaks down, are the chromosomes free in the cell. Because of this partitioning of chromosomes within the nucleus, transcription and translation are spatially separated (see Figure 5.2). Transcription of DNA occurs within the nucleus and the messenger RNA molecules are transported out of the nucleus to the cytoplasm, where translation occurs. In prokaryotes, transcription and translation can occur simultaneously, and it is even possible for translation to be initiated at one end of a messenger RNA molecule before transcription is complete at the other end (see, for example, Figure 5.47). On the other hand, in eukaryotes, the initial RNA molecule, called the **primary transcript**, is extensively modified before it is finally translated, as we will discuss in Section 5.6.

The chromosomal organization in eukaryotes involves two features not generally found in prokaryotes:

1. *Split genes.* Many eukaryotic genes are discontinuous, with noncoding DNA sequences inserted between the sequences that actually code for a single polypeptide (see Figure 5.2). These noncoding, intervening sequences are called **introns**, and the coding sequences are called **exons**. The number of introns per gene is variable, and ranges from none to over 50. During transcription, both introns and exons are copied, and the intron sequences are subsequently cut out and removed when the messenger RNA is processed into its final form in the cytoplasm.

2. *Repetitive sequences.* Eukaryotes generally contain much more DNA per genome than is needed to code for all of the proteins required for cell function. Eukaryotic DNA can be divided into several classes. **Single copy DNA** contains the coding sequences for the main proteins of the cell (and the associated intron DNA). **Moderately repetitive DNA**, found in a few to relatively large numbers of copies, codes for some major macromolecules of the cell: histones, immunoglobulins (involved in immune mechanisms, as discussed in Chapter 12), ribosomal RNA, and transfer RNA. **Highly repetitive (satellite) DNA** is found in a

very large number of copies. In humans, about 20 to 30 percent of the DNA is found in repetitive sequences, and almost all eukaryotic DNAs studied have some repeated sequences. The function of most highly repetitive DNA is unknown.

Nonchromosomal genetic elements

A number of genetic elements which are *not* part of the cell's chromosome have been recognized. Some nonchromosomal genetic elements which we will discuss briefly here include viruses, plasmids, mitochondria, and chloroplasts.

Viruses are genetic elements, either DNA or RNA, which control their own replication and transfer from cell to cell. (The viral genome is also referred to as a chromosome but it is distinct from the cellular chromosomes.) Viruses are of special interest because they are often (but not always) responsible for disease states. We discuss viruses in Chapter 6 and virus diseases in Chapter 15.

Plasmids are small double-stranded DNA molecules that exist and replicate separately from the chromosome (see Section 7.8). Although most plasmids are circular, some are linear. Plasmids differ from viruses in two ways: (1) they do not cause cellular damage (generally they are beneficial), and (2) viruses have extracellular forms, whereas plasmids do not. Although plasmids have been recognized in only a few eukaryotes, they have been found in most prokaryotic species. Some plasmids are excellent genetic vectors, and find wide use in gene manipulation and genetic engineering, as outlined in Chapter 8.

Many prokaryotes seem to contain one or more plasmids in addition to their chromosome. Some plasmids contain genes whose protein products can confer important properties on the host cell, such as resistance to antibiotics. Many plasmids are rather small (a few kilobase pairs), but some are quite large (several megabases). None are as large as the chromosome, however. From this information, you might think that in prokaryotes a chromosome is simply defined as the largest genetic element in the cell. Although the definition of what constitutes a chromosome in Bacteria is still controversial, it is coming to mean a genetic element that contains genes whose products are involved in essential metabolic steps *under all growth conditions.* Such genes are sometimes referred to as "housekeeping genes." For instance, a gene encoding DNA gyrase is always required by a cell, whereas a gene that enables a bacterium to be resistant to an antibiotic is only required under certain conditions (the presence of the antibiotic). It has not yet been conclusively demonstrated that any prokaryote has more than one chromosome by this definition, but it is possible that some do.

Mitochondria and **chloroplasts** are nonchromosomal genetic elements found in eukaryotes. As we discussed in Section 3.15, the mitochondrion is the site of respiratory enzymes, and plays a major role in energy generation in most eukaryotes. The chloroplast is a green, chlorophyll-containing structure which is the site of photosynthetic ATP formation. From a genetic viewpoint, mitochondria and chloroplasts can be viewed as independently replicating genetic elements. However, these organelles are much more complex than plasmids and viruses, since they contain not only DNA, but a complete machinery for protein synthesis, including 70S ribosomes, transfer RNA, and all of the other components necessary for translation and formation of functional proteins. One intriguing feature of mitochondria and chloroplasts is that, despite the fact that they contain many genes and a complete translation system, their existence is not independent from the chromosomes, since most proteins in them are coded not by organelle DNA but by chromosomal DNA. We discuss the genetics of mitochondria in Section 7.15.

Transposable elements are pieces of DNA having the ability to move from one site on a chromosome to another. Transposable elements are found in prokaryotes and eukaryotes and play important roles in genetic variation. There are three types of transposable elements: insertion sequences, transposons, and some special viruses. *Insertion sequences* are the simplest type and carry no genetic information other than that required for them to move into new locations. *Transposons* are larger and contain other genes. We discuss both of these types in more detail in Chapter 7. In Chapter 6 we discuss a virus, Mu, that is also a transposable element. The unique feature of transposable elements is that *they all replicate as part of some other molecule of DNA.* In spite of this, some of these elements are clearly "self-replicating" in the sense that they control their own replication and as such fit our definition of a genetic element. Others might simply be considered "jumping genes."

Replicating linear genetic elements

We used a circular DNA molecule for discussing the steps in DNA replication. Circular DNA molecules are common; most bacterial chromosomes are circular, as are most plasmids and some viruses. Almost all of the steps in replication are identical whether the chromosome is linear or circular. However, there is one problem with replication of linear genetic elements that circular ones do not have, and the problem is at the extreme 5'-end of each strand. To understand the problem, refer back to Figure 5.20. Imagine that the left end of the DNA in this diagram is actually one end of a linear chromosome. Even if the RNA primer were very short and there was a special enzyme to remove it, no DNA polymerase could replace it with DNA since all DNA polymerases require a primer. Therefore, if nothing was done, the DNA molecule would become shorter each time it was replicated. Obviously, genetic elements that are linear have solved this problem!

In fact, there are many solutions to this problem. Some viruses having linear chromosomes actually circularize themselves by their "sticky ends," as shown in Figure 5.10. Some other viruses have direct repeats at each end of their chromosomes. A recombination process (joining together of different DNA molecules) uses the repeats to join several partially replicated DNA molecules together into a very large molecule from which perfect copies are cut by endonucleases (see Sections 6.11 and 6.12). Several types of viruses and many linear plasmids solve the problem of replicating linear DNA by not using an *RNA* primer but rather a *protein* primer. Although all DNA polymerases must add each nucleotide to a free -OH group, some DNA polymerases can add the first base onto an -OH group found on specific proteins that bind to the ends of these linear chromosomes (Figure 5.25). These proteins are encoded by the plasmid or virus, and they function to recognize the ends of the chromosomes. These protein primers are not removed, so these particular types of plasmids and viruses have proteins covalently attached to the 5'-ends of their DNA.

None of these methods of replicating linear DNA are used to complete the ends of eukaryotic chromosomes (telomeres). Telomeres of eukaryotic chromosomes contain repetitive DNA: a short sequence (often 6 base pairs) tandemly repeated from 20 to several hundred times (Figure 5.26a). The sequences from different eukaryotes are closely related, and one strand always has several guanines. This guanine-rich sequence can be added on to the 3'-end of a DNA molecule by an interesting enzyme called **telomerase** (see

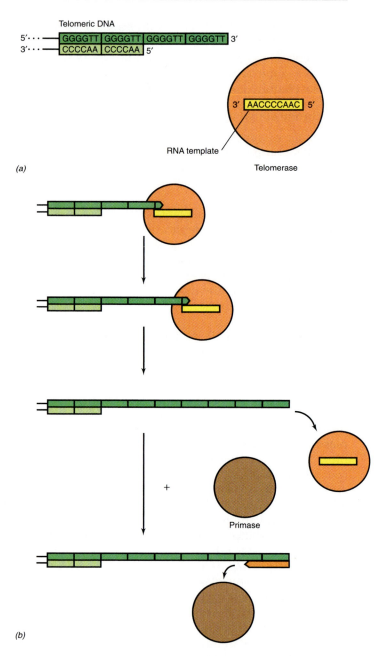

(a)

(b)

FIGURE 5.26 Model for the action of the telomerase at one end of a eukaryotic chromosome. (a) A diagram of the sequence of the end of the DNA in a telomere, with four of the guanine-rich repeats, and the enzyme telomerase which contains a short RNA template. (b) Steps in the elongation of the guanine-rich strand catalyzed by telomerase. After telomerase finishes, the lagging strand can be primed with an RNA primer by primase. The next step (not shown) would be completion of the lagging strand by DNA polymerase and ligase.

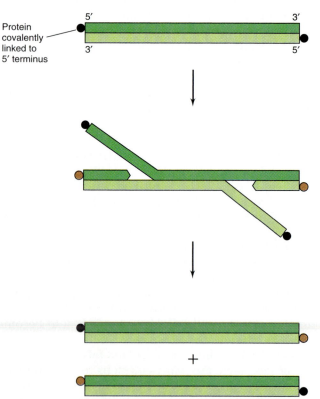

FIGURE 5.25 Replication of linear DNA using protein primers. The new strands of DNA are primed by proteins that stay covalently attached to the 5' ends.

Figure 5.26). Telomerases add onto the 3'-ends of linear DNA. *They do not need a DNA template because they contain a small RNA template as a cofactor.* These enzymes can work repetitively to make a long extension. Once this extension is long enough, the other strand can be primed with an RNA primer in the normal fashion. The telomeres do not need to be a precise number of repeats long, just long enough to ensure that no genetic information becomes lost during DNA replication.

In addition to the chromosomes, a number of other genetic elements exist in cells. Plasmids are DNA molecules that exist separately from the nuclear structure of the cell. Mitochondria and chloroplasts are eukaryotic organelles that are independently replicating but are still dependent on the nucleus for some of their functions. Viruses are genetic elements, either DNA or RNA, which control their own replication. Transposable elements exist as a part of other genetic elements.

5.6   RNA Structure and Function

Ribonucleic acid (RNA) plays a number of important roles in the expression of genetic information in the cell. Three major types of RNA have been recognized: **messenger RNA (mRNA), transfer RNA (tRNA), and ribosomal RNA (rRNA)**. There are three key differences between the chemistry of RNA and that of DNA: (1) RNA has the sugar *ribose* instead of *deoxyribose*; (2) RNA has the base *uracil* instead of the base *thymine*; and (3) except in certain viruses, RNA is not double stranded. A change from *deoxyribose* to *ribose* affects some of the chemical properties of a nucleic acid, and enzymes which affect DNA in general have no effect on RNA, and vice versa. The change from *thymine* to *uracil* does not affect base pairing, as the two nucleotide bases pair with adenine equally well.

It should be emphasized that RNA acts at two levels, genetic and functional. At the *genetic* level, RNA can carry the genetic information from DNA (mRNA; or in the case of RNA viruses, play a direct genetic function). At the *functional* level, RNA acts as a macromolecule in its own right, serving a structural role in ribosomes (rRNA) or an amino acid transfer role in protein synthesis (tRNA). Some RNA even has catalytic (enzymatic) activity. We discuss the various roles of RNA in this section.

Transcription

The transcription of the genetic information from DNA to RNA is carried out through the action of the enzyme **RNA polymerase**, which catalyzes the formation of phosphodiester bonds between ribonucleotides. RNA polymerase requires the presence of DNA, which acts as a template. The precursors of RNA are the ribonucleoside triphosphates, ATP, GTP, UTP, and CTP. The chemistry of RNA synthesis is much like the chemistry of DNA synthesis (Figure 5.3). During elongation of an RNA chain, the nucleotides are added to the 3'-OH of the ribose of the preceding nucleotide, which are polymerized with the release of the two high-energy phosphate bonds. Thus, in RNA synthesis (as in DNA synthesis), the overall direction of chain growth is from the 5'-end to the 3'-end. Unlike DNA polymerase, however, *RNA polymerase can start chains* (the initial nucleotide in an RNA chain then retains all three phosphates). The first base in the RNA is almost always a purine, either adenine or guanine.

In most cases, the DNA template for RNA polymerase is a double-stranded DNA molecule, but only *one* of the two strands is transcribed for any given gene. The enzyme RNA polymerase differs markedly between Bacteria, Archaea, and Eukarya. The following discussion deals only with RNA polymerase from Bacteria; which has the simplest structure (and for which the most is known). See Section 18.9 for a comparison of RNA polymerases in different organisms.

All RNA polymerases studied from Bacteria are complex enzymes with closely related subunit structures. The enzyme from *Escherichia coli* has four different types of protein subunits designated β, β', α, and σ (sigma), with α appearing in two copies. The subunits interact to form the active enzyme, but the sigma factor is not as tightly bound as the others, and easily dissociates, leading to the formation of what is called the *core enzyme* ($\alpha_2\beta\beta'$). The core enzyme alone can catalyze the formation of RNA, and the role of sigma is in the *recognition* of the appropriate site on the DNA for the initiation of RNA synthesis. The process of RNA synthesis involving RNA polymerase and sigma is illustrated in Figure 5.27.

RNA polymerase is a large protein, and forms contacts with the DNA over many bases simultaneously. We have discussed the general problem of protein–nucleic acid interactions in Section 5.2 and noted that the interactions are often sequence specific. How does the RNA polymerase recognize the proper region to *start* the transcription process? The binding of the RNA polymerase to the DNA occurs at particular sites called **promoters**. Note that only *one* strand of the DNA double helix is transcribed at a time. Which strand is transcribed is determined by the orientation of the promoter sequence. RNA polymerase travels away from the promoter region, synthesizing RNA as it moves.

Once the RNA polymerase has bound, the process of transcription can proceed. In this process, the DNA double helix at the promoter is *opened up* by the RNA polymerase (Figure 5.27). As the polymerase moves, it causes the DNA to unwind in short segments, transcription of these segments occurs, and the DNA double helix closes up again. As a result of this transient unwinding, the bases of the template strand are *exposed* and then can be copied into the RNA complement. Thus, the promoter *points* the RNA polymerase in one or the other direction. If a region of DNA has two nearby promoters on opposite strands, then transcription from one of the promoters will occur in one direction (on one of the strands) and transcription from the other will occur in the opposite direction (on the other strand).

Once a small portion of RNA has been formed, the sigma factor dissociates; most of the elongation is therefore carried out by the core enzyme alone (Figure 5.27). Thus, sigma is involved only in the formation of the initial RNA polymerase–DNA complex. As the newly synthesized RNA dissociates from the DNA, the opened DNA closes into the original double helix.

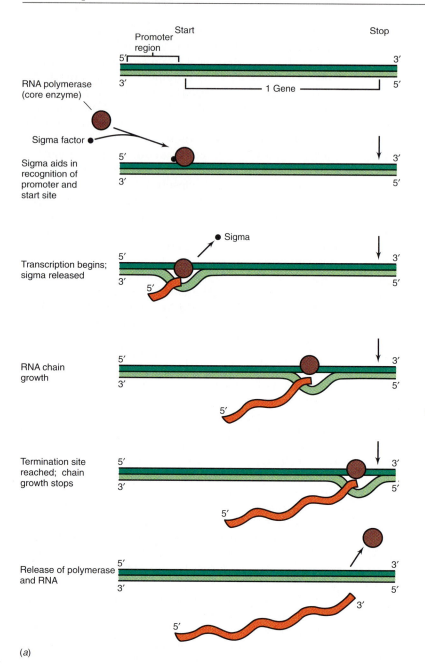

(a)

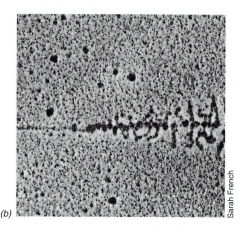

(b)

FIGURE 5.27

Transcription. (a) Steps in messenger RNA synthesis. The start and stop sites are specific nucleotide sequences on the DNA. RNA polymerase moves down the DNA chain, causing temporary opening of the double helix and transcription of one of the DNA strands. When a termination site is reached chain growth stops and the mRNA and polymerase are released. (b) Electron micrograph of transcription occurring along a gene on the *E. coli* chromosome. The region of active transcription represents about 2 kilobase pairs of DNA. Transcription proceeds from left to right.

As important as initiation of transcription is *termination* of transcription. **Termination** of RNA synthesis occurs at specific base sequences on the DNA. A common termination sequence on the DNA is one containing an inverted repeat with a central nonrepeating segment (see Section 5.2 and Figure 5.9 for an explanation of inverted repeats). When such a DNA sequence is transcribed, the RNA can form a stem-loop structure by intrastrand base pairing (Figure 5.28). When such stem-loop structures in the RNA are followed by runs of uridines, they are effective **transcription terminators**. Other termination sites are regions where a GC-rich sequence is followed by an AT-rich sequence. Such kinds of structures lead to termination without addition of any extra factors.

Another type of transcription terminator has been recognized which involves a distinct protein called *rho*. Rho does not bind to RNA polymerase or to DNA, but binds tightly to RNA and moves down the chain toward the RNA polymerase–DNA complex. Once RNA polymerase has paused at a termination site, rho can then cause the RNA and polymerase to leave the DNA, thus terminating transcription. Whether rho-dependent or rho-independent, transcription termination is ultimately determined by *specific nucleotide sequences on the DNA*.

Promoters

As we have noted, the promoter plays a key role in the initiation of RNA synthesis. Promoters are specific DNA sequences at which RNA polymerase enzymes attach. The sequences of a large number of promoters from a variety of organisms have been determined. Figure 5.29 shows the sequence of a few promoters

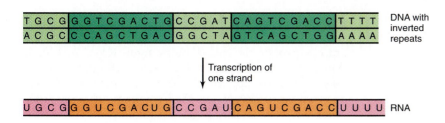

DNA with inverted repeats

Transcription of one strand

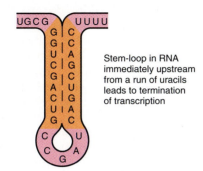

RNA

Stem-loop in RNA immediately upstream from a run of uracils leads to termination of transcription

FIGURE 5.28
Inverted repeats in transcribed DNA leads to formation of a stem-loop structure in the RNA, which can result in termination of transcription.

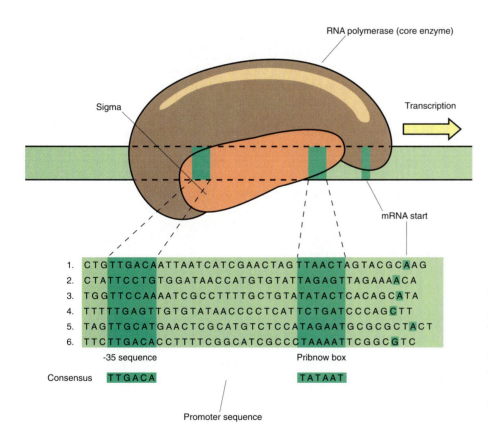

RNA polymerase (core enzyme)

Sigma

Transcription

mRNA start

1.	CTGTTGACAATTAATCATCGAACTAGTTAACTAGTACGCAAG
2.	CTATTCCTGTGGATAACCATGTGTATTAGAGTTAGAAAACA
3.	TGGTTCCAAAATCGCCTTTTGCTGTATATACTCACAGCATA
4.	TTTTTGAGTTGTGTATAACCCCTCATTCTGATCCCAGCTT
5.	TAGTTGCATGAACTCGCATGTCTCCATAGAATGCGCGCTACT
6.	TTCTTGACACCTTTTCGGCATCGCCCTAAAATTCGGCGTC

-35 sequence   Pribnow box

Consensus   TTGACA   TATAAT

Promoter sequence

FIGURE 5.29
The interaction of RNA polymerase with the promoter. Shown below the diagram are 6 different promoter sequences identified in *E. coli*. The contacts of the RNA polymerase with the −35 sequence and the Pribnow box are shown. Transcription begins at a unique base just downstream from the Pribnow box. Below the actual sequences at the −35 and Pribnow box regions are consensus sequences derived from comparing many promoters.

from *E. coli*. It is the sigma factor, as part of the RNA polymerase, which recognizes these promoters.

A single organism can have several different sigma factors, and these can recognize different promoter sequences. All the sequences in Figure 5.29 are recognized by the same sigma factor, the major sigma factor in *E. coli*. If you examine the sequences, you will see that they are not identical. However, two sequences within the promoter region *are highly conserved* between promoters, and it is these that are recognized by sigma. Both sequences are upstream of the

start of transcription. One is a region 10 bases before the start of transcription, the −10 region (called the *Pribnow box*). Notice that, although each promoter is slightly different, many bases are the same. When comparing the −10 regions of all the promoters recognized by this sigma to determine which base occurs most often at each position, one arrives at the *consensus sequence* of TATAAT. In our example, each promoter has from three to five matches for these bases. The second region of conserved sequence is about 35 bases from the start of transcription. The consensus sequence in the −35 region is TTGACA. Once again, most of the sequences are not *exactly* the same as the consensus sequence.

Other sigma factors in other organisms are sometimes much more specific; very little leeway is allowed in the critical bases that are recognized. In *E. coli,* promoters that are most like consensus are usually more effective in binding RNA polymerase. The more effective promoters are called *strong promoters* and are of considerable value in genetic engineering, as will be discussed in Chapter 8.

> The three major types of RNA are messenger RNA (mRNA), transfer RNA (tRNA), and ribosomal RNA (rRNA). The transcription of RNA from the DNA involves the enzyme RNA polymerase, which recognizes a specific start site on the DNA called the promoter. Only regions of DNA that are preceded by a promoter can be transcribed into RNA.

Specific inhibitors of RNA polymerase action

A number of antibiotics and synthetic chemicals have been shown to specifically inhibit RNA synthesis. For example, a group of antibiotics called the *rifamycins* inhibits by attacking the beta subunit of the RNA polymerase enzyme. Rifamycin has marked specificity for prokaryotes, but also inhibits RNA synthesis in chloroplasts and mitochondria of some eukaryotes. Rifamycin has been an especially useful tool in studying nucleic acid synthesis in virus-infected cells. A group of antibiotics called the *streptovaricins* is related to the rifamycins in structure and function. *Streptolydigin* is an antibiotic that also inhibits transcription by binding to the beta subunit of RNA polymerase, but it binds to a different site than rifamycin. Another chemical, *amanitin,* inhibits RNA synthesis in eukaryotes without affecting prokaryotes. Eukaryotes have three different types of RNA polymerase; one for synthesizing most rRNA, one for synthesizing mRNA, and one for tRNA. Amanitin specifically inhibits RNA polymerase II, the one that synthesizes mRNA.

Actinomycin (see Figure 5.12) inhibits RNA synthesis by combining with DNA and blocking elongation. Actinomycin binds most strongly to DNA at guanine-cytosine base pairs, fitting into the major groove on the double strand where RNA is synthesized.

Messenger RNA and operons

The RNA carrying the information that is translated into a protein is called **messenger RNA (mRNA)**. Most mRNA, in both prokaryotes and eukaryotes, is unstable and is degraded by cellular nucleases. This is in contrast to rRNA and tRNA, which are sometimes referred to as stable RNA. In prokaryotes, a single mRNA molecule often codes for more than one protein (see Figure 5.2). In prokaryotic genetic elements, genes coding for related enzymes are often clustered together. In these situations the RNA polymerase proceeds down the chain and transcribes the whole series of genes into a single long mRNA molecule. An mRNA coding for such a group of genes is called a **polygenic mRNA** or a **polycistronic mRNA** (see Section 7.5). Subsequently, when this polycistronic mRNA participates in protein synthesis (see Section 5.8), several polypeptides coded by a single mRNA can be synthesized at one time.

We will discuss regulation of mRNA synthesis later in this chapter, but introduce here the concept of the operon. An **operon** is a complete unit of gene expression, generally involving genes coding for several polypeptides on a polycistronic mRNA or genes coding for ribosomal RNA. In some cases, the transcription of the mRNA for an operon is under the control of a specific region of the DNA, the **operator**, which is adjacent to the coding region of the first gene in the operon. The operator is nearby and may overlap with the promoter (Figure 5.30). How does the operator participate in the regulation of the transcription process? Within the cell are specific proteins, known as **repressor proteins**, which are able to bind to specific operators. If the repressor protein is attached to the operator, then transcription of that operon cannot occur. We discuss the details of repressor action in Section 5.10.

> One of the most important kinds of RNA is messenger RNA, which carries the genetic information for the synthesis of specific proteins. Transcription of several genes into a single mRNA molecule may occur, so that the mRNA may contain the information for more than one polypeptide. Such genes that are transcribed together from a single promoter constitute an operon. Often the transcription of such related genes is regulated by blocking the synthesis of the whole mRNA.

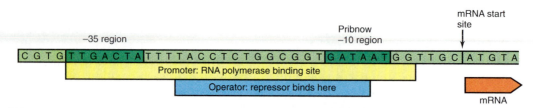

FIGURE 5 30   DNA base sequence of an operator/promoter region. Note how the operator region overlaps the location where RNA polymerase binds. (The sequence shown is one of the operators of bacteriophage lambda.)

RNA Processing

Transcription can produce several types of RNA: messenger RNA, transfer RNA, and ribosomal RNA. In prokaryotes, a transcript of a protein-encoding gene (a mRNA) is used directly to make protein, but transcripts of many other genes generally need to be trimmed to reach final form. In eukaryotes, all transcripts (including mRNAs) are trimmed before being used. The conversion of a *precursor* RNA into a *mature* RNA is called **RNA processing**. For instance, in prokaryotes, tRNAs and rRNAs are made initially as long precursor molecules which are then cut at several places to make the final mature RNAs. In eukaryotes, and less commonly in prokaryotes, mRNA also undergoes extensive processing. As discussed in Section 5.5, the genes of eukaryotes are often split, with noncoding intervening sequences, *introns*, separating the coding regions called *exons*. The primary RNA transcript

must be extensively processed to remove the noncoding regions before the translation process can be initiated (Figure 5.31). Although, as shown in Figure 5.31a, prokaryote mRNA is generally translated directly, a few introns have been discovered in prokaryotes and in bacteriophage.

A complex containing several different ribonucleoproteins (each contains both a small RNA and several proteins) called a **spliceosome**, is responsible for removal of introns from pre-messenger RNAs to form the processed message in eukaryotic cells. The spliceosome is a highly complex structure capable of removing introns and joining adjacent exons to form a mature mRNA. The spliceosome is unique to the eukaryotic nucleus. Some introns (particularly those found in tRNA genes and in genes of the mitochondria or chloroplasts) are removed by a different process just involving proteins. Several introns, in-

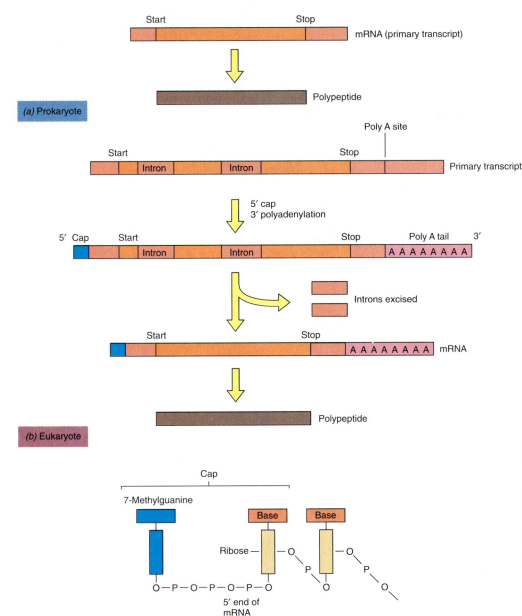

(a) Prokaryote

(b) Eukaryote

(c)

FIGURE 5.31
Contrast of mRNA in prokaryotes and eukaryotes. (a) Prokaryote: messenger RNA is synthesized in final form. (b) In eukaryotes, the mature mRNA is often derived from a primary transcript by a variety of processing steps. These include adding a modified guanine cap at the 5'-phosphate end and adding a poly A tail after a specific cut at the 3'-hydroxyl end. Any noncoding intervening sequences (introns) are also excised by splicing. (c) Structure of the cap.

cluding all of those that are found in Bacteria and bacteriophage, are *self-splicing* (ribozymes; see next subsection).

Another major feature of eukaryotic mRNA is that after transcription but before transfer to the cytoplasm, the mRNA molecules have added to them a long stretch of adenine nucleotides at the 3'-end, the *poly A tail*, and a methylated guanine nucleotide at the 5'-phosphate end, called the *cap* (Figure 5.31c). Also, a number of the bases in mRNA are methylated after transcription and the 2'-hydroxyl group of the ribose is occasionally methylated.

We thus see that RNA synthesis is a complex and dynamic process that involves considerably more than the simple transcription of a DNA template.

Ribozymes

We emphasized in Chapter 4 the role of proteins as biochemical catalysts. Many important cellular processes involve ribonucleoproteins, complexes including both RNA and protein. The role of RNA in such complexes was *assumed* to be structural (a place for the proteins to bind) or involved in base pairing with other nucleic acids. As we have seen, there is a short RNA molecule in the enzyme *telomerase* that functions as a template. However, it has now been shown that certain types of RNA can act as *enzymes* as well. Catalytic RNAs, referred to as **ribozymes**, are involved in a number of important cellular reactions. RNA enzymes work like protein enzymes in that an "active site" exists that binds the substrate and catalyzes formation of a product. Ribozymes have been discovered in both prokaryotes and eukaryotes, and in organelles, and others have now been synthesized in laboratories. Studies show that some very short RNAs, as few as 19 bases, can catalyze biochemical reactions.

Most ribozymes are **self-splicing introns**. They are *RNA-splicing enzymes* that remove themselves from an RNA molecule while joining adjacent exons together. In one well-studied case of splicing in a ribosomal RNA in *Tetrahymena* (a protozoan), a 413-nucleotide *intron* acts as a ribozyme and splices itself out of a longer precursor rRNA, joining two adjacent exons to form the final rRNA (Figure 5.32). The intron ribozyme acts as a sequence-specific endoribonuclease, and once removed from the precursor RNA, circularizes with the further removal of a short oligonucleotide fragment (Figure 5.32). This particular type of self-splicing intron is widespread in nature and is the only type known in Bacteria and bacteriophage. Absolute proof that these ribozymal transformations occur in the absence of specific protein has come from experiments in which the gene for the entire precursor rRNA from *Tetrahymena* has been transferred to *E. coli* where the segment could be transcribed. This transcribed segment carries out the splicing reaction in the complete absence of *Tetrahymena* proteins.

Self-splicing introns differ from most protein enzymes in that they normally can only act once. However, there is another ribozyme, RNAse P, that can act

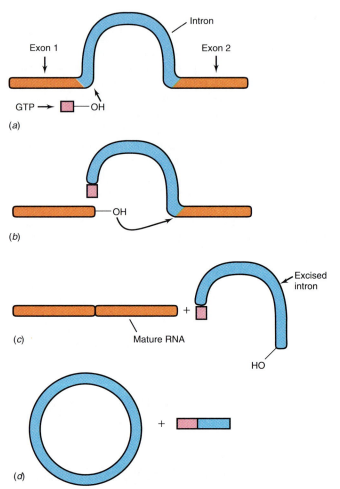

FIGURE 5.32 Self-splicing ribozymal intron of the protozoan *Tetrahymena*. (a) A ribosomal RNA precursor contains a 413 nucleotide intron. (b) Following the addition of GTP (or the nucleoside guanosine) the intron splices itself out and joins the two exons. (c) The intron is spliced out. (d) The intron circularizes with the loss of a 15-nucleotide fragment.

repeatedly on many different substrate molecules because it does not digest itself in the reaction. RNAse P is a ribonucleoprotein, but the small RNA (377 nucleotides in *E. coli*) is the catalytic component, not the protein. As is the case for proteins with enzymatic activity, all ribozymes must be folded into the proper structure for activity. In some cases, this structure might be supplied by the secondary structure of the RNA itself. In others, like RNAse P, specific proteins may help keep the RNA in the active conformation. RNAse P functions in the cell to modify primary transcripts coding for transfer RNAs (see next section).

The discovery of ribozymes has caused a reevaluation of other cellular processes that involve RNA. For instance, it is now clear that ribosomal RNA plays an active role in protein synthesis (see Section 5.8). But clearly most enzymes are protein. Why do ribozymes exist? It has been proposed that ribozymes are the vestigial remains of a simpler form of life, "RNA life," which may have predated the era of proteins as the cell's major catalysts. We discuss this concept in more detail in Chapter 18.

> RNA molecules are often modified after transcription by removal of noncoding regions, an operation called RNA processing. In some cases, the enzyme involved in processing is not a protein but a distinct RNA molecule itself, called a ribozyme.

5.7   Transfer RNA

The translation of the RNA message into the amino acid sequence of protein is brought about through the action of transfer RNA (tRNA) (Figure 5.33). Transfer RNA is an adaptor molecule having two specificities, one for a codon on mRNA, the other for an amino acid. The transfer RNA and its specific amino acid are brought together by means of specific enzymes that ensure that a particular tRNA receives its correct amino acid. These enzymes, called **amino acid activating enzymes** or **aminoacyl-tRNA synthetases**, have the important function of recognizing *both* the amino acid *and* the specific tRNA for that amino acid.

The detailed structure of tRNA is now well understood. There are about 60 different types of tRNA in bacterial cells and 100 to 110 in mammalian cells. Transfer RNA molecules are short, single-stranded molecules, with lengths (among different tRNAs) of 73 to 93 nucleotides. When compared, it has been found that certain bases and secondary structures are constant for all tRNAs, and there are other parts that are variable. Transfer RNA molecules also contain some purine and pyrimidine bases differing slightly from the normal bases found in RNA in that they are chemically modified, often methylated. Some of these unusual bases are pseudouridine, inosine, dihydrouridine, ribothymidine, methyl guanosine, dimethyl guanosine, and methyl inosine. These modifications are added to the bases after transcription. Base modifi-

cations as well as other types of processing (see earlier) are necessary to make a functional tRNA from the transcript of a tRNA-encoding gene. Although the molecular structure of tRNA is single-stranded, there are extensive double-stranded regions within the molecule as a result of internal base pairing when the molecule folds back on itself.

The structure of tRNA is generally drawn in cloverleaf fashion, as in Figure 5.33a. Some regions of secondary structures are given names having to do with either the bases most often found there (the TψC loop and the D loop) or with specific functions (anticodon loop and acceptor end). The three-dimensional structure of a tRNA is more clearly shown in Figure 5.33b. Note that bases that appear widely separated in the cloverleaf model are actually close together when viewed in three dimensions. This means that some of the bases in the "loops" are actually paired.

One of the variable parts of the tRNA molecule contains the **anticodon**, the site recognizing the codon on the mRNA. The anticodon is found in the *anticodon loop*, shown in Figure 5.33. There are just *three* nucleotides in the anticodon loop that are specifically involved in the recognition process and which base pair with the codon (see Section 5.9). Other portions of the tRNA interact with the ribosome (both rRNA and protein), other protein factors, and with the activating enzyme. At the 3'-end of the chain, where a sequence of nucleotides projects from the rest of the molecule, the sequence of these three nucleotides is always the same, cytosine-cytosine-adenine (CCA), and it is to the ribose sugar of the terminal A that the amino acid is covalently attached, via an ester linkage. From this acceptor portion of the tRNA, the amino acid is transferred to the growing polypeptide chain on the ribosome by a mechanism that will be described in the next section.

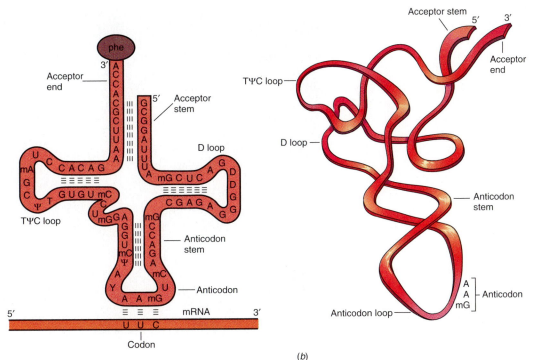

FIGURE 5.33
Structure of a transfer RNA, yeast phenylalanine tRNA. (a) The conventional cloverleaf structure. The amino acid is attached to the ribose of the terminal A at the acceptor end. Abbreviations: A, adenine; C, cytosine; U, uracil; G, guanine; ψ, pseudouracil; D, dihydrouracil; m, methyl; Y, a modified purine. (b) In actuality the molecule folds so that the D loop and TψC loops are close together and associate by hydrophobic interactions.

(a)

(b)

Recognition, activation, and charging

Recognition of the correct tRNA by an aminoacyl-tRNA synthetase involves specific contacts between key regions of the nucleic acid and particular amino acids of its respective synthetase (Figure 5.34). As might be expected because of the unique sequence in this region, the *anticodon* of the tRNA is important in recognition by the synthetase. However, other contact sites between the tRNA and the synthetase are also important. Studies of tRNA binding to aminoacyl tRNA synthetases in which specific bases in the tRNA have been changed by genetic mutation have shown that only a small number of key nucleotides in a tRNA besides the anticodon region are involved in recognition; the key recognition nucleotides are often part of the acceptor stem of the tRNA molecule (see Figure 5.33). In a few cases, recognition of a tRNA by its cognate synthetase is totally independent of the anticodon region. It should be emphasized at this point that the fidelity of this recognition process is crucial,

for if the wrong amino acid is attached to the tRNA, it may be inserted in the improper place in the polypeptide, leading to the synthesis of a faulty protein.

The specific chemical reaction between amino acid and tRNA catalyzed by the aminoacyl-tRNA synthetase first involves *activation* of the amino acid by reaction with ATP:

$$\text{Amino acid} + \text{ATP} \rightleftharpoons \text{aminoacyl-AMP} + \text{P--P}$$

The aminoacyl-AMP intermediate formed normally remains bound to the enzyme until collision with the appropriate tRNA molecule, and the activated amino acid is then transferred to the tRNA to form a *charged* tRNA:

$$\text{Aminoacyl-AMP} + \text{tRNA} \rightleftharpoons$$
$$\text{aminoacyl-tRNA} + \text{AMP}$$

The pyrophosphate (P--P) formed in the first reaction is split by a pyrophosphatase, forming two molecules of inorganic phosphate. Since ATP is used and AMP is

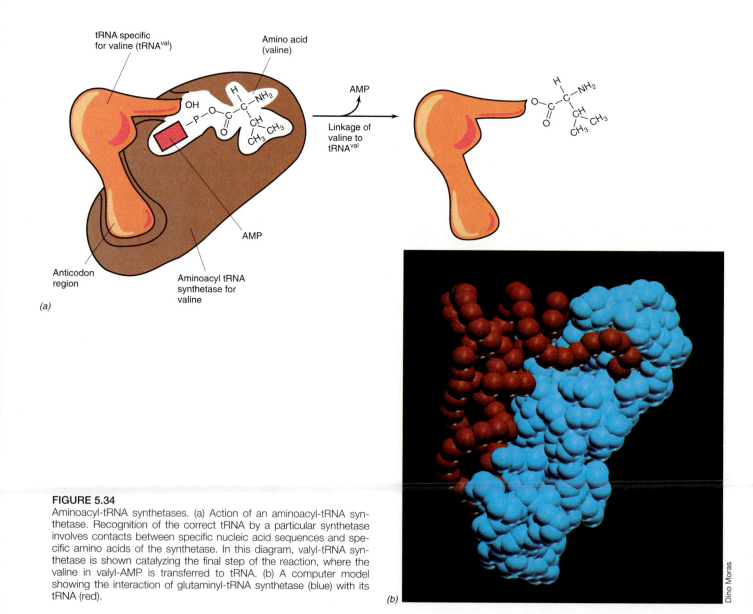

FIGURE 5.34

Aminoacyl-tRNA synthetases. (a) Action of an aminoacyl-tRNA synthetase. Recognition of the correct tRNA by a particular synthetase involves contacts between specific nucleic acid sequences and specific amino acids of the synthetase. In this diagram, valyl-tRNA synthetase is shown catalyzing the final step of the reaction, where the valine in valyl-AMP is transferred to tRNA. (b) A computer model showing the interaction of glutaminyl-tRNA synthetase (blue) with its tRNA (red).

Dino Moras

formed, a total of *two* high-energy phosphate bonds are required for the activation of an amino acid and charging a tRNA. Once activation and charging have occurred, the aminoacyl-tRNA leaves the synthetase and is brought to the ribosome by a protein factor. The mechanism of protein synthesis is discussed in the next section.

> One or more transfer RNAs exist for each amino acid found in protein and these tRNAs have the unique role of adapting the amino acid to the anticodon. Enzymes called aminoacyl-tRNA synthetases catalyze this critical function of protein synthesis. Once the correct amino acid is attached to its tRNA, further specificity resides only in the codon–anticodon interaction.

5.8   Translation: The Process of Protein Synthesis

It is the amino acid *sequence* that determines the structure (and ultimately the function) of the final active protein. The key objective of protein synthesis is thus the placement of the proper amino acid at the proper place in the polypeptide chain. This is the role of the protein-synthesizing machinery of the cell.

Steps in protein synthesis

Ribosomes are the site of protein synthesis. Each ribosome is constructed of two subunits. In prokaryotes, the ribosome subunits are of 30S (Svedberg units) and 50S, yielding intact 70S ribosomes.* Each subunit is itself a ribonucleoprotein complex made up of specific ribosomal RNAs and ribosomal proteins. The 30S sub-

*The numbers 30S, 50S, and 70S refer to the sedimentation coefficients of ribosome subunits or intact ribosomes when subjected to centrifugal force in an ultracentrifuge.

unit contains 16S rRNA and about 21 proteins, while the 50S subunit contains 5S and 23S rRNA and about 34 proteins (Figure 5.35). In *E. coli*, there are 53 different ribosomal proteins, most present at one copy per ribosome. The actual synthesis of a protein involves a complex cycle in which the various ribosomal components play specific roles.

Although a continuous process, protein synthesis can be thought of as occurring in a number of discrete steps: **initiation, elongation, termination-release**, and **polypeptide folding**. The first two steps are outlined in Figure 5.36. In addition to mRNA, tRNA, and ribosomes, the process involves a number of proteins designated initiation, elongation, and termination factors; guanosine triphosphate (GTP) provides energy for the process.

Initiation of protein synthesis

Initiation always begins with a free 30S ribosome subunit, and an **initiation complex** forms consisting of a 30S ribosome subunit, mRNA, formylmethionine tRNA, and initiation factors. Guanosine triphosphate (GTP) is required for this step. To this initiation complex a 50S ribosome subunit is added to make the active 70S ribosome. At the end of the translational process, the released ribosome separates again into 30S and 50S subunits. Just preceding the initiation codon on the mRNA is a sequence of from three to nine nucleotides (the so-called **Shine–Dalgarno sequence**) which is involved in the binding of the mRNA to the ribosome. This ribosome binding site at the 5'-end of the mRNA is complementary to the 3'-end of the 16S RNA of the ribosome, and it is thought that base pairing ensures effective formation of the ribosome–mRNA complex.

The presence of the Shine–Dalgarno site on the mRNA and its specific interaction with 16S rRNA allow prokaryotic ribosomes to use polycistronic

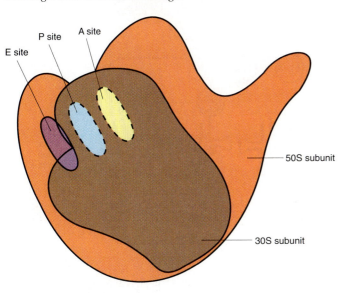

Ribosome structure*		
Property	**Prokaryote**	**Eukaryote**
Overall size	70S	80S
Small subunit	30S	40S
Number of proteins	~21	~30
RNA size (number of bases)	16S (1500)	18S (2300)
Large subunit	50S	60S
Number of proteins	~34	~50
RNA size (number of bases)	23S (2900) 5S (120)	28S (4200) 5.8S (160) 5S(120)

*Ribosomes of mitochondria and chloroplasts of eukaryotes are similar to prokaryotic ribosomes.

FIGURE 5.35   Structure of the ribosome, showing the position of the A or acceptor site (right), the P or peptide site, and the E or exit site. Details of the ribosome subunits are shown in the table. The size of the ribosome particle or ribosomal RNA is expressed in Svedberg units (S), a measure of the rate at which the component sediments in the ultracentrifuge.

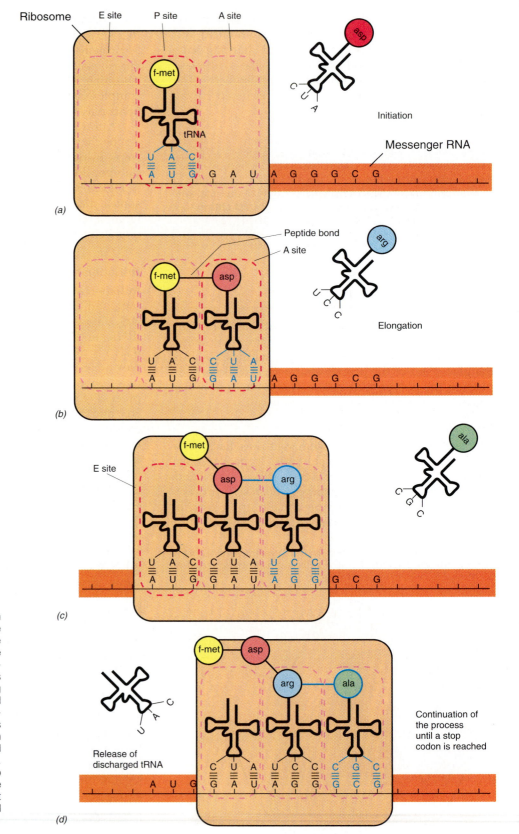

FIGURE 5.36
Translation of the in-formation from messenger RNA (mRNA) into the amino acid sequence of protein. The whole process occurs on the surface of the ribosome. (a) Interaction between codon and anticodon brings into position the correct tRNA carrying the amino acid. (b) A peptide bond forms between amino acids on adjacent tRNA molecules. (c) and (d) As the ribosome moves to the next codon a new tRNA attaches and the old tRNA, now free of amino acid, leaves. The result is a growing chain of amino acids. Although part *d* of this figure shows all three sites occupied at once, it is likely that only two sites will be occupied at the same time.

mRNA because the ribosome can find each initiation site within a message. Eukaryotic ribosomes typically recognize an mRNA by its 5'-cap and initiate only at the first possible initiation codon. Therefore, they normally cannot translate polycistronic mRNA.

Initiation always begins with a special initiator aminoacyl-tRNA binding to the **start codon**, AUG. In Bacteria this is **formylmethionine** tRNA. Subsequently, the formyl group at the N-terminal end of the polypeptide is removed; the terminal amino acid of the completed protein is hence methionine. In Eukarya and Archaea, initiation begins with methionine instead of formylmethionine. (Although all proteins are *initiated* with a methionine, this amino acid is often removed by a specific protease after translation.)

Elongation and termination

The mRNA is threaded through the ribosome primarily bound to the 30S subunit. The ribosome contains other sites where the tRNAs interact. Two of these sites are located primarily on the 50S subunit, and they are termed the P-site and the A-site (see Figure 5.36). The A-site, the **acceptor** site, is the site where the new AA–tRNA first attaches. The P-site, the **peptide** site, is the site where the growing peptide is held by a tRNA. During peptide bond formation, the peptide moves to the tRNA at the A-site as a new peptide bond is formed. Several soluble (nonribosomal) elongation factors are required for **elongation**, as well as additional molecules of GTP (to simplify Figure 5.36, the elongation factors are omitted and only a portion of the ribosome is shown). The tRNA that holds the peptide must now be removed (translocated) from the A-site to the P-site, thus opening up the A-site for another AA–tRNA. Translocation requires a specific elongation factor and one molecule of GTP per each tRNA translocated. At each translocation step the message is advanced three nucleotides, exposing a new codon in the ribosome A-site. It had been thought that translocation caused the empty tRNA to be released from the ribosome. However, it now appears that translocation pushes this empty tRNA to a third site, called the E-site. It is from this **exit** site that the tRNA is actually released from the ribosome.

When several ribosomes are simultaneously translating a single message, the complex is called a **polysome** (Figure 5.37). Polysomes increase the speed and efficiency of mRNA translation, and because each ribosome acts independently of the others, each ribosome in a polysome complex can make a complete polypeptide (Figure 5.37). Note in Figure 5.37 how ribosomes closest to the 5′-end (the beginning) of the mRNA molecule have short polypeptides attached to them because only a few codons have been read, while ribosomes closest to the 3′-end of the message have nearly finished polypeptides. Note also how polypeptides begin to fold almost immediately as they are formed on each ribosome (Figure 5.37).

The **termination** of protein synthesis occurs when a codon is reached which does not specify an AA–tRNA. There are three codons of this type and they are called **stop** or **nonsense codons** (see Section 5.9); they serve as the stopping points for protein synthesis. No tRNA binds to a stop codon, but instead proteins called *release factors* read the chain-terminating signal and serve to cleave the attached polypeptide from the terminal tRNA. Following this, the ribosome dissociates, and the subunits are then free to form new initiation complexes.

Role of ribosomal RNA in protein synthesis

Ribosomes are composed of a series of proteins and RNAs (see Figure 5.35). Although the RNA plays a *structural* role, serving as a support for ribosomal proteins, it also plays a *functional* role. Ribosomal RNA is intimately involved in all stages of protein synthesis, from initiation to termination. Messenger RNAs bind to the small subunit because of base pairing between the ribosome binding sequence (the Shine–Dalgarno sequence) upstream of the start codon on the mRNA and a complementary sequence in 16S rRNA.

There is strong evidence that mRNA and rRNA interactions also occur during elongation, perhaps being involved in maintaining the correct reading frame. The 23S rRNA seems to play a role in translocation, and the elongation factors are known to interact with 23S rRNA. The 16S rRNA is also involved in termination, possibly interacting with the mRNA or through interactions with the release factors.

Strong evidence also exists for a role for rRNA in ribosome subunit association, as well as for tRNA positioning in the decoding (A- or P-, see Figure 5.36) sites on the ribosome and even in catalyzing peptide bond formation. Because 16S rRNA and 23S rRNA reside in different subunits of the ribosome, it is thought that complementary base sequences of these two RNA species or RNA and protein interactions are the "glue" that holds the two subunits together. Charged tRNA's that enter the ribosome recognize the correct codon by codon:anticodon base pairing, but they are also physically attached to the ribosome by interactions of the anticodon stem-loop of the tRNA with specific locations within 16S rRNA. The acceptor end of the tRNA interacts with the 23S rRNA. Finally, the peptidyl transferase reaction (the actual formation of peptide

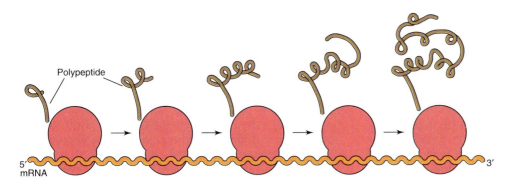

Polypeptide

5′
mRNA   3′

FIGURE 5.37
Translation by several ribosomes on a single messenger RNA (polysome). Note how the ribosomes nearest the 5′-end of the message are at an earlier stage in the translation process.

bonds which occurs on the 50S subunit of the ribosome) is associated with 23S rRNA. It is possible that the reaction itself is catalyzed by ribozymal activity (see Section 5.6), although specific ribosomal proteins may also be involved.

Ribosomal RNA thus plays a major role in translation. The role of the many proteins present in the ribosome, although less clear, may be as facilitators of RNA function by stabilizing or positioning the key functional sequences in the various ribosomal RNAs. Ribosomal function is clearly dependent upon the major RNA species present.

Secretory proteins

Many proteins are used *outside* the cell and must somehow get from the site of synthesis on ribosomes through the cytoplasmic membrane. In prokaryotes, periplasmic enzymes and extracellular enzymes are secretory proteins, and in eukaryotes various digestive enzymes and enzymes of the lysosome are in this category.

How is it possible for a cell to selectively transfer some proteins across a membrane, while leaving most proteins in place in the cytoplasm? This is explained by the **signal hypothesis**, which states that secretory proteins are synthesized with an extra N-terminal peptide sequence, some 15 to 20 amino acids in length, which is called the **signal sequence**. In this signal sequence, hydrophobic amino acids predominate and permit the enzyme to be threaded through the hydrophobic lipid membrane. In many cases, the ribosomes that synthesize secretory proteins are bound directly to the cytoplasmic membrane, so that the protein is formed and passes through the membrane simultaneously. Once the protein has been secreted, the signal sequence is removed by a peptidase enzyme, an example of the process of **posttranslational modification**.

The study of protein secretion has important practical implications for genetic engineering (see Chapter 8). If bacteria are genetically engineered to serve as agents for the production of foreign proteins, it is desirable to manipulate the signal sequence in order to arrange for the desired protein to be excreted so that it can be readily isolated and purified.

Effect of antibiotics on protein synthesis

A large number of antibiotics inhibit protein synthesis by interacting with the ribosome. These interactions are quite specific, and many have been shown to involve rRNA. Several of these antibiotics are medically useful, and several are also effective research tools because they are specific for different steps in protein synthesis. For instance, *streptomycin* inhibits initiation, whereas *puromycin, chloramphenicol, cycloheximide,* and *tetracycline* inhibit elongation. Even when two antibiotics inhibit the same overall step in protein synthesis, the mechanisms of the inhibition can be quite different. Puromycin binds to the A-site on the ribosome, and the growing peptide is transferred to it instead of to an AA-tRNA. The puromycin–peptide is then re-

leased from the ribosome, halting elongation. Chloramphenicol inhibits elongation by blocking the formation of the peptide bond.

Many antibiotics specifically inhibit ribosomes of organisms from only one or two of the phylogenetic domains. Of the antibiotics just listed, chloramphenicol and streptomycin are specific for the ribosomes of Bacteria and cycloheximide for ribosomes of Eukarya. Since mitochondria and chloroplasts have ribosomes of the prokaryotic type, it is of interest that antibiotics inhibiting protein synthesis in Bacteria also generally inhibit protein synthesis in mitochondria and chloroplasts. This is one piece of evidence supporting the hypothesis that mitochondria and chloroplasts were originally derived by intracellular infection of a eukaryotic cell by prokaryotes (see Sections 3.14 and 18.3). The fact that an antibiotic inhibits eukaryotic ribosomes as well as those from Bacteria does not necessarily mean it cannot be used in medicine. Although tetracycline also inhibits eukaryotic ribosomes, it apparently does not do so at the concentrations that result from administering it during antibacterial therapy.

One interesting inhibitor of protein synthesis is diphtheria toxin, an agent involved in the pathogenesis of the disease diphtheria. As discussed in Section 11.8, diphtheria toxin inactivates an elongation factor and hence is a powerful inhibitor of protein synthesis in eukaryotes and Archaea. The overall pattern of antibiotic sensitivity of the Archaea is unlike either the Bacteria or the Eukarya (see Chapter 18).

> The ribosome plays a key role in the translation process, bringing together mRNA and amino acid-charged tRNAs. There are two sites on the ribosome, the acceptor site, where the charged tRNA first combines, and the peptide site, where the growing polypeptide chain is held. There is also a third site where the tRNA exits the ribosome. During each step of amino acid addition, the message advances three nucleotides (one codon) and the tRNA moves from the acceptor to the peptide site. Termination of protein synthesis occurs when a unique codon, called a nonsense codon, which does not code for any amino acid, is reached. A major target of antibiotic action is the ribosome.

5.9 The Genetic Code

As noted, a triplet of three bases encodes a specific amino acid. It is conventional to present the genetic code in mRNA rather than in the DNA because it is with mRNA that the translation process occurs. The 64 possible codons of mRNA are presented in Table 5.4. Note that in addition to the codons specifying the various amino acids, there are also special codons to start (AUG) and to stop (UAA, UAG, UGA) translation.

Perhaps the most interesting feature of the genetic code is that a single amino acid may be encoded by several different but related base triplets. This means that in most cases there is no one-to-one correspon-

Table 5.4   The genetic code as expressed by triplet base sequences of mRNA*

Codon	Amino acid	Codon	Amino acid	Codon	Amino acid	Codon	Amino acid
UUU	Phenylalanine	CUU	Leucine	GUU	Valine	AUU	Isoleucine
UUC	Phenylalanine	CUC	Leucine	GUC	Valine	AUC	Isoleucine
UUG	Leucine	CUG	Leucine	GUG	Valine	AUG (start)†	Methionine
UUA	Leucine	CUA	Leucine	GUA	Valine	AUA	Isoleucine
UCU	Serine	CCU	Proline	GCU	Alanine	ACU	Threonine
UCC	Serine	CCC	Proline	GCC	Alanine	ACC	Threonine
UCG	Serine	CCG	Proline	GCG	Alanine	ACG	Threonine
UCA	Serine	CCA	Proline	GCA	Alanine	ACA	Threonine
UGU	Cysteine	CGU	Arginine	GGU	Glycine	AGU	Serine
UGC	Cysteine	CGC	Arginine	GGC	Glycine	AGC	Serine
UGG	Tryptophan	CGG	Arginine	GGG	Glycine	AGG	Arginine
UGA	None (stop signal)	CGA	Arginine	GGA	Glycine	AGA	Arginine
UAU	Tyrosine	CAU	Histidine	GAU	Aspartic	AAU	Asparagine
UAC	Tyrosine	CAC	Histidine	GAC	Aspartic	AAC	Asparagine
UAG	None (stop signal)	CAG	Glutamine	GAG	Glutamic	AAG	Lysine
UAA	None (stop signal)	CAA	Glutamine	GAA	Glutamic	AAA	Lysine

*The codons in DNA are complementary to those given here. Thus U here is complementary to the A in DNA, C is complementary to G, G to C, and A to T. The nucleotide on the left is at the 5'-end of the triplet.
†AUG codes for N-formylmethionine at the beginning of mRNAs of Bacteria.

dence between the amino acid and the codon—knowing the amino acid at a given location does not mean that the codon at that location is automatically known.* The property of a code in which there is no one-to-one correspondence between word and code is called **degeneracy** (a term derived from the technical field of cryptography).

What is the significance of degeneracy? Degeneracy means that either (1) a single tRNA molecule can pair with more than one codon or (2) there is more than one tRNA for each amino acid. Actually, both situations are true. For some amino acids there are more than one tRNA molecule (there are, for instance, six different tRNA molecules in *E. coli* which carry the amino acid leucine). However, there is not a tRNA corresponding to every possible anticodon. In some cases, tRNA molecules form *standard* base pairs at only the first two positions of the codon, tolerating unusual base pairs at the third position. This apparent mismatch phenomenon, called **wobble**, is illustrated in Figure 5.38. The pairing between G and U is allowed at the wobble position.

Start and stop codons

As seen in Table 5.4, a few triplets do not correspond to any amino acid. These triplets (UAA, UAG, UGA) are the **nonsense** or **stop codons** and they signal the termination of translation of the gene coding for a specific protein (see Section 5.8).

What is the mechanism by which the proper *starting point* for translation is found? The message is read by reading the **start codon, AUG,** which at the beginning of the message, codes for the amino acid *N*-formylmethionine (or methionine in Eukarya and Archaea). The importance of having a well-defined starting point is readily understood if we consider that with a triplet code it is absolutely essential that translation begin at the correct location, since if it does not, the whole **reading frame** would be shifted and an entirely different protein (or no protein at all) would be formed (see later).

As we discussed in Section 5.8, ribosomes from Bacteria recognize a specific AUG as a start codon with the aid of an upstream sequence called the Shine–Dalgarno sequence. This extra help at initiation explains why a few messages from Bacteria actually use other codons, such as GUG, for a start codon. However, even these unusual start codons specify *N*-formylmethionine.

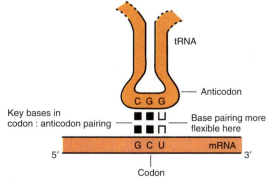

FIGURE 5.38   The wobble concept: base pairing is more flexible for the third base of the codon.

*The reverse is true, however. Knowing the DNA codon, one can specify the amino acid in the protein (assuming the proper reading frame is known). This permits the determination of amino acid sequences from DNA base sequences.

Open reading frames

We have shown that a protein is coded by a specific segment of DNA, but that for the protein to be synthesized, the DNA segment must first be transcribed into mRNA. The transcription of DNA into mRNA requires the presence of a promoter just upstream of the transcription start site. How does an experimenter know that the transcribed mRNA codes for a protein? One approach is to examine the base sequence of the mRNA to see whether there is a *start codon*, such as AUG, present near the beginning of the sequence. If such a start codon is followed by a long base sequence before a *stop codon* is present, then it is likely that this mRNA codes for a protein. Such a base sequence is called an **open reading frame (ORF)**, because the start codon sets the proper reading frame for translation and a stop codon in the same frame will end translation. A computer can be programmed to scan long base sequences in DNA data bases to look for open reading frames. The search for ORFs is very useful in genetic engineering (see Chapter 8) when one has isolated and sequenced an unknown piece of DNA and is not certain whether or not it codes for protein. The computer search for ORFs permits the researcher to locate genes that were previously unsuspected.

In a sense, the term *open reading frame* is equivalent to the term *gene*. If a piece of DNA lacks a detectable open reading frame, it is unlikely that this piece of DNA codes for protein. In eukaryotic cells especially, a large amount of noncoding DNA is present.

Other genetic codes

When the genetic code was cracked during the 1960s, all the prokaryotes and eukaryotes examined were found to use the same code. When the mRNA for mammalian hemoglobin was given to the *E. coli* protein synthesizing machinery (such as ribosomes and tRNA), mammalian hemoglobin was synthesized. Therefore, the genetic code appeared to be a **universal code** in that the exact same code was used by all living systems. Once techniques became available to sequence DNA easily, genes from many organisms began to be sequenced. However, comparison of these DNA sequences to the amino acid sequence of the proteins they encode has led to a few unexpected surprises. One of these, the discovery of introns in eukaryotic genes, was very surprising but did not change our ideas about the genetic code. However, it has also been discovered that some organelles and some cells use genetic codes that are slight variations of the "universal" genetic code (see Selenocysteine box).

The original findings of these alternative codes were in the genomes of mitochondria. So far as is known, only the mitochondria of plants use the universal code without change. The other organelles in plants, the chloroplasts, also use the standard code. The mitochondria of all other eukaryotes use codes with one or a few slight differences. A few of these variations are shown in Table 5.5. Note that there is *not* simply a mitochondrial code, although there are a few common themes, such as the general use of UGA

as a tryptophan codon. It is also clear that these alternate codes are very closely related to the universal code and are almost certainly derived from it evolutionarily. Several examples of cells are now known whose chromosomes also use slightly different codes, and a few examples of these are also given in Table 5.5. Note that all of these alternative chromosomal genetic codes have different assignments for what are normally stop codons. These organisms simply have fewer nonsense codons because one or two are now read as sense.

If every codon has an assignment, you might imagine that it is very difficult to change the genetic code in an organism. For instance, the change of AUA from an isoleucine codon to a methionine codon means that every protein that once had an isoleucine encoded by AUA now has a methionine at this position. Such a protein may not function normally. This may not be a severe problem if *codon usage* is not random. After the genetic code had been worked out by biochemists and before any genes had actually been sequenced, it was assumed that the degenerate codons for an amino acid would be used at an equal frequency. This is another assumption that DNA sequencing has shown to be incorrect! Codon usage is highly biased, and this bias changes from organism to organism. In *E. coli,* for instance, only about 1 out of 20 isoleucine residues is encoded by an AUA, the other 19 being encoded by AUU and AUC. It is thought that one of the steps that can lead to codon reassignment is that the codon becomes rarely used in a genome. This is easier to achieve in mitochondria because they have very small genomes (see Section 7.15).

Mistranslation

Another problem in the translation of the genetic code is that errors sometimes occur. This means that a codon may be "read" improperly and the wrong amino acid inserted. Amino acids whose codons differ by only a single base, for example, phenylalanine (UUU) and leucine (UUA), are most likely to be mistranslated because occasionally only two of the three bases are involved in codon-anticodon recognition due to "wobble" (see Figure 5.38). In rare instances leucine may be added to the growing polypeptide instead of phenylalanine even when the codon is UUU. In the normal cell these rare errors occur in only a small number of all the protein molecules and hence have no detrimental effect. For example, experimental measurements of mistranslation have shown that only one in 10^3–10^4 codons are misread. However, certain antibiotics which act on ribosomes, such as streptomycin and neomycin, increase translation errors to such an extent that many protein molecules in the cell are abnormal and the cell can no longer function properly.

Other types of translational errors can also occur, such as a ribosome shifting into the wrong reading frame or erroneously reading a stop codon as a sense codon. Amazingly, certain genetic elements seem to have evolved to take advantge of such translational "errors" to make essential proteins (see Section 6.22).

Selenocysteine: The 21st Amino Acid

The genetic code has codons for 20 amino acids that are assembled into proteins during translation. However, many proteins contain other amino acids. In fact, there are well over 100 different amino acids found in at least a few proteins. Until recently, it was thought that these "extra" amino acids were made by modifying one of the standard amino acids *after* it was incorporated into protein, a process called *posttranslational modification*. However, it is now clear that one of these extra amino acids is put into protein by the translational machinery itself. This one exception is *selenocysteine*.

Selenocysteine has the same structure as cysteine, but it has a selenium atom rather than a sulfur atom. It was known for some time that a few proteins contained this unusual amino acid. For example, *E. coli* makes two different formate dehydrogenase enzymes and both contain a single selenocysteine residue. When the gene encoding one of these enzymes was sequenced, it was found that the codon that corresponded to the selenocysteine was a UGA. UGA is normally an efficient stop codon in *E. coli*, but it has now been demonstrated that UGA can be translated directly as selenocysteine in certain mRNA molecules, not only in *E. coli* but also in other prokaryotes and in

eukaryotes, including humans. Therefore, selenocysteine is the 21st amino acid known to be encoded by the genetic code.

How can a codon sometimes be a stop codon and sometimes a sense codon *in the same chromosome?* The answer apparently lies in the *context* of the codon, the sequence of the bases surrounding the UGA codon and in their secondary structure. In certain contexts, the translational machinery interprets UGA as "selenocysteine." In all other contexts, UGA means "stop translation." Selenocysteine has its own tRNA (as do all the standard amino acids) and also has a special protein factor that brings only this tRNA to the ribosome.

Selenocysteine is even more readily oxidized than cysteine. Therefore, enzymes that contain this amino acid must be protected from oxygen. It has been proposed that UGA might once have been a normal sense codon, calling only for selenocysteine, but that the increase in oxygen in our environment following the evolution of photosynthesis (see Chapter 18) selected for proteins that contained cysteine (whose codons are UGU and UGC). This allowed the coding assignment of UGA to be altered, except in a few special cases.

Table 5.5   Variations in the genetic code*

Codon	Universal code	Other nuclear codes			Other mitochondrial codes		
		Mycoplasma	*Paramecium*	*Euplotes*	Yeast	Protozoans	Mammals
UGA	Stop	Tryptophan	Stop	Cysteine	Tryptophan	Tryptophan	Tryptophan
UAA/UAG	Stop	Stop	Glutamine	Stop	Stop	Stop	Stop
AUA	Isoleucine	Isoleucine	Isoleucine	Isoleucine	Methionine	Methionine	Methionine
CUA	Leucine	Leucine	Leucine	Leucine	Threonine	Leucine	Leucine
AGA/AGG	Arginine	Arginine	Arginine	Arginine	Arginine	Arginine	Stop

*The universal genetic code is used in the chromosomes of most cells, chloroplasts, plant mitochondria, and their viruses and plasmids. A few organisms use slightly different codes in their chromosomes (in the nucleus). The examples of these other nuclear codes are from Mycoplasma (Bacteria) and two different ciliated protozoans (Eukarya). All nonplant mitochondria use variations of the universal code, whereas plant mitochondria use the universal code. The examples here are only a few of the different types known.

Overlapping genes

Although the evidence is strong that the nucleotide sequence specifying one product is separate and distinct from the sequence specifying another product, studies on the small bacterial virus φX174 have shown that this virus has insufficient genetic information in nonoverlapping genes to code for all of the proteins necessary for its reproduction, and that genetic economy is introduced by using the same piece of DNA for the coding of more than one product. It is a process made possible by reading of the same nucleotide sequence in two different reading frames, beginning at different sites. There are now known a number of interesting patterns of overlapping genes, most of which occur in viruses (see Chapter 6).

The complete genetic code, usually expressed in terms of mRNA rather than DNA, is known. A single amino acid may be encoded by several different but related codons, and for each codon there is a specific tRNA. In addition to the nonsense or stop codons, there is also a specific start codon, that signals the location where the translation process should begin. The start codon always codes for a specific amino acid, *N*-formylmethionine (in Bacteria), or methionine (in Archaea and Eukarya). If a start codon did not exist, translation might not begin in the proper reading frame and faulty proteins would be made.

5.10   Regulation of Protein Synthesis: Induction and Repression

Not all enzymes are synthesized by a cell in the same amounts; some enzymes are present in far greater copy number than others. Clearly, then, the cell is able to *regulate* enzyme synthesis. Several different mechanisms for controlling enzyme synthesis are known in bacteria and all of them are greatly influenced by the *environment* in which the organism is growing, in particular by the presence or absence of specific small molecules. These molecules can interact with specific proteins to control transcription, or more rarely, translation. We begin our discussion of enzyme regulation by describing repression and induction, simple forms of regulation that govern gene expression in a variety of operons in bacteria.

Enzyme repression

Often the enzymes catalyzing the synthesis of a specific product are not synthesized if this product is present in the medium. For example, the enzymes involved in the formation of the amino acid arginine are synthesized only when arginine is *not* present in the culture medium; external arginine *represses* the synthesis of these enzymes. As can be seen in Figure 5.39, if arginine is added to a culture growing exponentially in a medium devoid of arginine, growth continues at the previous rate, but the formation of the enzymes involved in arginine synthesis stops. Note that this is a *specific* effect, as the syntheses of all other enzymes in the cell continue at the same rates as previously.

Enzyme repression is a very widespread phenomenon in bacteria—it occurs as a means of controlling the synthesis of a wide variety of enzymes involved in the biosynthesis of amino acids, purines, and pyrimidines. In almost all cases it is the final product of a particular biosynthetic pathway that represses the enzymes of this pathway. In these cases repression is quite specific, and the process usually has no effect on the synthesis of enzymes other than those involved in the specific biosynthetic pathway. The value to the organism of enzyme repression is obvious, since it effectively ensures that the organism does not waste energy synthesizing unneeded enzymes.

Enzyme induction

A phenomenon complementary to repression is enzyme induction, the synthesis of an enzyme only when its substrate is present. Figure 5.40 shows this process in the case of the enzyme β-galactosidase, which is involved in utilization of the sugar lactose. If lactose is absent from the medium the enzyme is not synthesized, but synthesis begins almost immediately after lactose is added. Enzymes involved in the catabolism of carbon and energy sources are often inducible. Again, one can see the value to the organism of such a mechanism, as it provides a means whereby the organism does not synthesize an enzyme until it is needed.

The substance that initiates enzyme induction is called an **inducer**, and a substance that represses enzyme production is called a **corepressor**; these substances, which are always small molecules, are often

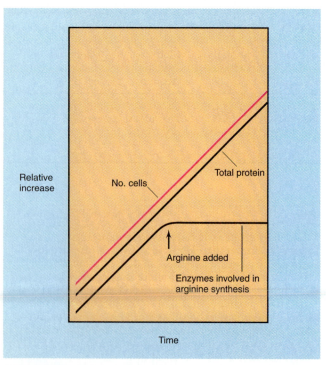

FIGURE 5.39   Repression of enzymes involved in arginine synthesis by addition of arginine to the medium. Note that the rate of total protein synthesis remains unchanged.

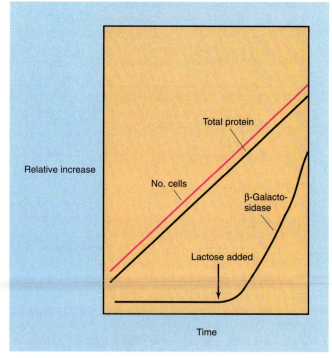

FIGURE 5.40   Induction of the enzyme β-galactosidase upon the addition of lactose to the medium. Note that the rate of total protein synthesis remains unchanged.

Inducers and Repressors

The phenomenon of enzyme induction has had a long history in microbial physiology, and detailed studies of this phenomenon have played a major role in understanding macromolecular synthesis and its regulation. The concepts that have been developed from studies on enzyme induction in bacteria have also been widely applied to eukaryotic cells and to an understanding of cancer. Emile Duclaux, an associate of Pasteur, first reported in 1899 that a fungus, *Aspergillus niger*, only produced the enzyme invertase (which hydrolyzes sucrose) when growing on a sucrose medium. Similar observations were made by other workers in the early 1900s. H. Karström of Finland first studied these phenomena systematically in the 1930s. He found that certain enzymes were always present, irrespective of the culture medium, whereas other enzymes were only formed when their substrates were present. Karström termed the enzymes formed all the time *constitutive*, and those formed only when substrate was present *adaptive*. In the 1940s, the French microbiologist Jacques Monod began a systematic study of bacterial growth and discovered the phenomenon he termed *diauxic growth*. From his study of diauxy, Monod turned to a more detailed study of the process of enzyme adaptation. For many years, the interpretation of enzyme adaptation had been complicated by the fact that the substrate of the enzyme seemed to be involved in the process. However, with the discovery of nonmetabolizable substances, such as thiomethylgalactoside (TMG) and isopropylthiogalactoside (IPTG), which brought about enzyme formation even though they were not substrates of the enzyme, it became clear that enzyme adaptation and enzyme function were two different things. Because the word *adaptation* has certain confusing connotations (among other things, adaptation does not distinguish between the selection of mutant strains from a culture and the synthesis of a new enzyme in a preexisting genotype), the term has been abandoned, and the word *induction* is used instead. Among the most critical experiments carried out by Monod and his colleagues was that which showed that enzyme induction resulted in the *synthesis of a new protein* and not the activation of some preexisting protein. This suggested that the enzyme

inducer was somehow causing differential gene action.

While the phenomenon of enzyme induction had been known for a long time, enzyme repression, the specific inhibition of enzyme synthesis, was not discovered until 1953, being reported simultaneously by the laboratories of Monod, Edward A. Adelberg and H. Edwin Umbarger in Berkeley, and Donald D. Woods at Oxford, England. It soon became clear that enzyme induction and repression, although having opposite effects, were manifestations of a similar mechanism. Biochemical studies on enzyme synthesis provided considerable insight into the phenomena of induction and repression, but by themselves they probably never would have led to a final understanding of the picture. The introduction of the techniques of bacterial genetics was essential, and some of the aspects of this will be discussed in Chapter 7. Among the major contributors to this genetic approach was Francois Jacob, a student and colleague of Monod at the Pasteur Institute. Jacob isolated a large number of mutants of *Escherichia coli* in the lactose pathway and analyzed these genetically. The results led to the conclusion that induction and repression were under the control of specific proteins, called *repressors*, which were encoded by regulatory genes. These regulatory genes were associated with, but were distinct from, the structural genes coding for the specific enzyme proteins. Finally, Walter Gilbert and Benno Müller-Hill at Harvard University developed a cell-free system of assaying for repressor protein, and carried out a biochemical isolation and purification process for the repressor. The proof that the repressor was indeed a protein not only confirmed the theory of Jacob and Monod, but provided the approaches necessary for studying the manner in which proteins are able to specifically interact with defined sequences on DNA molecules. Biochemical and genetic studies on RNA polymerase then clarified the manner in which the information in DNA is transcribed into RNA. The most significant result of this important fundamental research is that it showed that regulation of enzyme synthesis in bacteria occurs at the level of transcription rather than at the level of translation.

---

collectively called **effectors**. Not all inducers and corepressors are substrates or end products of the enzymes involved. For example, *analogs* of these substances may induce or repress even though they are not substrates of the enzyme. Thiomethylgalactoside (TMG), for instance, is an inducer of β-galactosidase even though it cannot be hydrolyzed by the enzyme. In nature, however, inducers and corepressors are probably normal cell metabolites.

Mechanism of induction and repression

Enzyme repression or induction acts at the level of transcription; enzyme synthesis is controlled by initi-

ating or terminating mRNA production for a particular enzyme or group of enzymes. How can inducers and corepressors affect transcription in such a specific manner? They do this indirectly by combining with specific regulatory proteins which then in turn affect mRNA synthesis. In the case of a repressible enzyme, the corepressor (for example, arginine) combines with a specific **repressor protein**, the arginine repressor, that is present in the cell (Figure 5.41). The repressor protein is an allosteric protein (see Section 4.20), its conformation being altered when the corepressor combines with it. This altered repressor protein can then combine with a specific region of the DNA at the

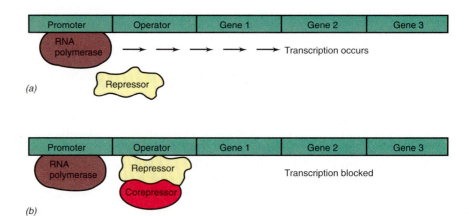

FIGURE 5.41
The process of enzyme repression. (a) Transcription of the operon occurs because the repressor is unable to bind to the operator. (b) After a corepressor (small molecule) binds to the repressor, the repressor now binds to the operator and blocks transcription; mRNA and the proteins it codes for are not made.

initial end of the gene, the **operator region**, adjacent to which synthesis of mRNA is initiated (see Section 5.6). If this occurs, the synthesis of mRNA is blocked, and the protein or proteins specified by this mRNA cannot be synthesized. If the mRNA is polycistronic, *all* of the proteins coded for by this mRNA will be repressed. A series of genes all regulated by one operator is called an **operon**.

Enzyme induction can also be controlled by a **repressor**. In this case, the situation described above is reversed. The specific repressor protein is active in the *absence* of the inducer, completely blocking the synthesis of mRNA, but when the inducer is added it combines with the repressor protein and inactivates it. Inhibition of mRNA synthesis being overcome, the enzyme or enzymes can be made (Figure 5.42). All systems involving repressors have the same underlying mechanism, the *inhibition* of the synthesis of mRNA by the action of specific repressor proteins that are themselves under the control of specific small-molecule inducers and repressors.

It should be emphasized that not all enzymes of the cell are controlled by simple induction or repression and that the synthesis of some enzymes is not strongly controlled at all. Enzymes whose level of synthesis is about the same under all growth conditions are called *constitutive enzymes*. Constitutive enzymes

are generally key cellular enzymes required for growth under all nutritional conditions and are thus synthesized continuously in the growing cell.

Contrasts between repression and feedback inhibition

We have presented in this section some of the details by which the *synthesis* of enzymes is regulated. It should be clear that if a needed enzyme is not synthesized, then the process which this enzyme catalyzes cannot occur in the cell. Thus regulation of enzyme *synthesis* provides an important mechanism for regulating cell metabolism. However, regulation of enzyme synthesis is a relatively slow mechanism of control, since a period of time is needed after synthesis of an enzyme is started before that enzyme is present in the cell in sufficient amounts to affect metabolism.

We described in Section 4.20 other mechanisms of regulation of enzyme action, such as *feedback inhibition*, which affect not enzyme *synthesis* but enzyme *activity*. In feedback inhibition, one of the metabolites of a pathway inhibits a critical step (frequently an early step) in that pathway. In contrast to induction and repression, feedback inhibition provides an *immediate* mechanism of metabolic control, since it acts on *preexisting* enzyme. Thus, feedback inhibition can be viewed as a mechanism for finely regulating cell me-

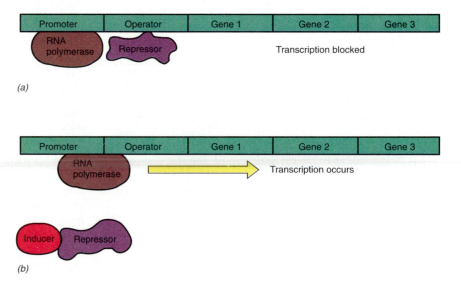

FIGURE 5.42
The process of enzyme induction using a repressor. (a) A repressor protein binds to the operator region and blocks the action of RNA polymerase. (b) Inducer molecule binds to the repressor and inactivates it. Transcription by RNA polymerase occurs and a mRNA for that operon is formed.

tabolism, whereas induction and repression are coarser mechanisms. Working together, these mechanisms result in an efficient regulation of cell metabolism, so that energy is not wasted by carrying out unnecessary reactions.

> The amount of an enzyme in the cell can be controlled by increasing (induction) or decreasing (repression) the amount of mRNA that encodes the enzyme. This transcriptional regulation involves regulatory proteins that bind to DNA and to small molecules called effectors. For one type of transcriptional regulation, the regulatory protein is called a repressor and it functions by inhibiting mRNA synthesis. While repression reduces the amount of enzyme being made, feedback inhibition reduces the activity of existing enzyme.

5.11   Regulation of Protein Synthesis: Positive Control

Repression constitutes a kind of regulation called **negative control**. The controlling element—the repressor protein—brings about the *repression* of mRNA synthesis. Even though the repressor has a negative role, a system using a repressor can control enzyme induction, as we saw with β-galactosidase. However, another type of control has also been recognized which is called **positive control**. In positive control, a regulator protein *promotes* the binding of RNA polymerase, thus acting to *increase* mRNA synthesis. We will now consider a system that involves positive regulation, the regulation of maltose catabolism in *Escherichia coli*.

The maltose regulon

The enzymes for the utilization of the sugar maltose in *E. coli* are synthesized only after the addition of maltose to the medium. The pattern of induction of these enzymes follows that shown for β-galactosidase in Figure 5.40, but in this case it is maltose, not lactose, that is the inducer. The synthesis of the enzymes for maltose utilization is controlled at the level of transcription, but by an **activator protein**, not by a repressor. The *maltose activator protein* cannot bind to the DNA unless the protein first binds maltose, the effector. When the activator protein does bind to DNA, it allows RNA polymerase to begin transcription (Figure 5.43). Activators, like repressors, recognize specific sequences on the DNA. The sequence that serves as the binding site of the activator is not called an operator but an *activator binding site*. Nonetheless, the genes controlled by this activator binding site *are* called an operon.

In negative control, the repressor binds to the operator and blocks transcription. How does an activator protein work? Positively controlled promoters have nucleotide sequences that are not close matches to the consensus sequence (see Figure 5.29). Even with the correct sigma factor, the RNA polymerase has difficulty recognizing these promoters. The activator protein, when bound to DNA, helps the RNA polymerase

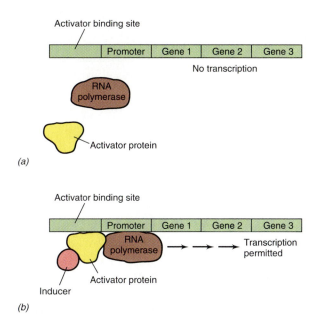

(a)

(b)

FIGURE 5.43   Positive control of enzyme induction. (a) In the absence of an inducer, neither the activator protein nor the RNA polymerase can bind to the DNA. (b) Inducer molecule binds to the activator protein, which in turn binds to the activator binding site. This allows RNA polymerase to bind to the promoter and begin transcription.

either recognize the promoter or helps it begin transcription. The activator protein may cause a change in the structure of the DNA, perhaps by bending it (Figure 5.44), allowing the RNA polymerase to make the correct contacts with the DNA. The activator protein may also interact directly with the RNA polymerase. This could either happen when the activator binding site is close to the promoter (Figure 5.45a) or when it is several hundred base pairs away from the promoter (Figure 5.45b).

The genes needed for maltose utilization are spread out in several operons, each of which has an

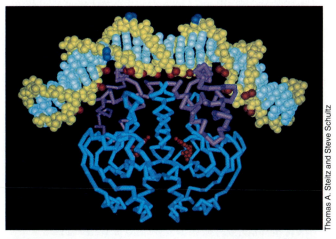

FIGURE 5.44   A computer model of the interaction of a positive regulatory protein with DNA. This figure shows the cyclic AMP binding protein, a regulatory protein involved in the control of several operons (see Section 5.13). The α-carbon backbone of this protein is shown in blue and purple. The protein is shown binding to a DNA double helix, which is shown in yellow and light blue. Note that binding of this protein to DNA has caused the DNA to be bent by almost 90°.

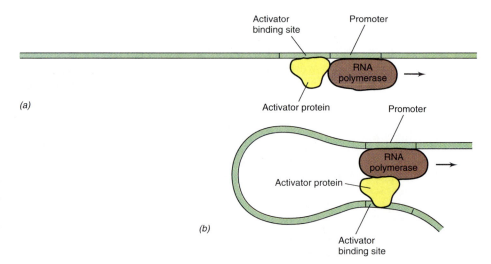

FIGURE 5.45
Some activator proteins interact with RNA polymerase. (a) The activator binding site is near the promoter. (b) The activator binding site is several hundred base pairs from the promoter. In this case, the DNA must be looped to allow the activator and the RNA polymerase to contact.

activator binding site to which the maltose activator protein can bind. Therefore, the maltose activator protein actually controls more than one operon. When more than one operon is under the primary control of the same regulatory protein, these operons are collectively known as a **regulon**. Therefore, the enzymes for maltose utilization are encoded by the *maltose regulon*. Regulons are also known for operons under negative control. The arginine biosynthetic enzymes (mentioned in Section 5.10) are encoded by the *arginine regulon* whose operons are all under the control of the arginine repressor protein.

Many genes in *E. coli* have promoters under positive control, and many have promoters under negative control. However, there are other types of regulation known. In addition, many genes (perhaps most genes) either have a promoter with multiple types of control or they have more than one promoter, each with its own control system! We next discuss a type of regulation found in prokaryotes where regulation of transcription involves a brief period of translation (attenuation). Then we turn our attention to global control networks and describe how cells can regulate many genes in response to particular environmental conditions.

> Both negative and positive regulators of transcription are known. Negative regulators, called repressors, are protein molecules that bind at specific regions of the DNA called operators, and block mRNA synthesis. The activity of the repressor can be modified by the presence of substances called effectors. In the case of enzyme repression, the effector promotes the binding of the repressor to the operator, thus blocking mRNA synthesis. For negative control of enzyme induction, the effector prevents the repressor from binding to the DNA, thus allowing mRNA synthesis. Positive regulators of transcription are called activator proteins which bind to activator binding sites and stimulate transcription by RNA polymerase. Activator protein activity is also modified by effectors. For positive control of enzyme induction, the effector promotes the binding of the activator protein and thus stimulates mRNA synthesis. A series of genes under the control of a single operator is called an operon. When two or more operons are under control of the same regulatory protein, they are collectively called a regulon.

5.12   Attenuation

Another element of control, called **attenuation**, has been recognized in some operons controlling the biosynthesis of amino acids. The best studied case is that involving biosynthesis of the amino acid *tryptophan*. The tryptophan operon contains structural genes for five proteins of the tryptophan biosynthetic pathway, plus the promoter and regulatory sequences at the beginning of the operon (Figure 5.46). Like many operons, the tryptophan operon has more than one type of regulation. One type is repression, and one of the regulatory sequences is an operator to which the tryptophan repressor can bind. In addition to promoter and operator regions, there is a sequence called the **leader sequence**, within which is a region called the **attenuator**, which codes for a polypeptide that contains tandem tryptophan codons near its terminus (Figure 5.46). If tryptophan is plentiful in the cell, the leader peptide will be synthesized. On the other hand,

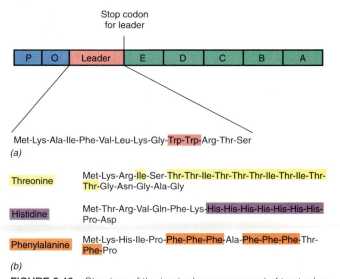

FIGURE 5.46   Structure of the tryptophan operon and of tryptophan and other leader peptides. (a) Arrangement of the tryptophan operon. Note that the leader is a short peptide containing two tryptophan residues near its terminus. (b) Amino acid sequence of leader peptides synthesized in some other amino acid biosynthetic operons. Because isoleucine is made from threonine, it is an important constituent of the threonine leader peptide.

if tryptophan is in short supply, the tryptophan-rich leader peptide will *not* be synthesized. The striking fact is that synthesis of the leader peptide results in *termination* of transcription of the tryptophan structural genes, whereas if synthesis of the leader peptide is blocked by tryptophan deficiency, transcription of the tryptophan structural genes can occur.

How does *translation* of the leader peptide regulate *transcription* of the tryptophan genes downstream? This can be explained by considering that these two processes in prokaryotic cells are occurring

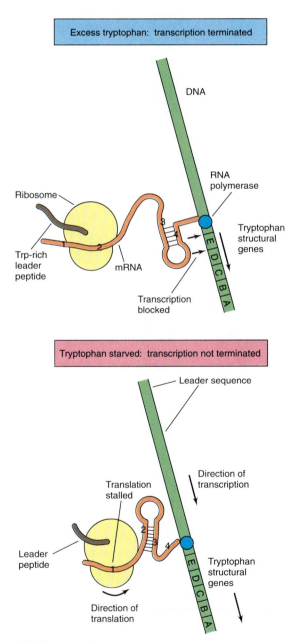

FIGURE 5.47   Control of transcription of tryptophan operon structural genes by attenuation in *Escherichia coli*. The leader peptide is coded by regions 1 and 2 of the mRNA. Two regions of the growing mRNA chain are able to form double-stranded loops, 2:3 and 3:4. Under conditions of excess tryptophan, the ribosome translates the complete leader peptide, so that region 2 cannot pair with region 3. Regions 3 and 4 then pair to form a loop which blocks RNA polymerase. If translation is stalled due to tryptophan starvation, loop formation via 2:3 pairing occurs, loop 3:4 does not form, and transcription proceeds past the leader sequence.

virtually simultaneously (Figure 5.47). Thus, while *transcription* of downstream DNA sequences is still proceeding, *translation* of sequences already transcribed has begun. Apparently, as the mRNA is released from the DNA, the ribosome binds to it and translation begins. Attenuation occurs (RNA polymerase stops transcription) because a portion of the newly formed mRNA folds into a double-stranded loop which signals cessation of RNA polymerase action (see Figure 5.28). The stem and loop structures formed by mRNA are brought about because two stretches of nucleotide bases near each other are complementary and can thus base pair. If tryptophan is plentiful, the ribosome will translate the leader sequence until it comes to the stop codon. The remainder of the leader RNA can then assume a stem-loop, a *transcription pause site*, which is followed by a uracil-rich sequence that actually causes termination. However, if tryptophan is in short supply, the ribosome pauses at a tryptophan codon before the stop codon; this allows an alternative stem-loop to form (sites 2 and 3 in Figure 5.47). This stem-loop is *not* a termination signal and it effectively prevents the terminator (sites 3 and 4 in Figure 5.47) from forming. RNA polymerase then moves past the nonfolded termination site and begins transcription of the tryptophan structural genes. Thus we see that in attenuation there is a highly integrated system in which transcription and translation interact, with the rate of transcription being influenced by the rate of translation.

Thus, in the tryptophan biosynthetic pathway, two distinct mechanisms for the regulation of transcription exist, repression and attenuation. Repression is a mechanism that has large effects on the rate of enzyme synthesis, whereas attenuation brings about a finer control. Working together, these two mechanisms precisely regulate the synthesis of tryptophan biosynthetic enzymes, and hence the biosynthesis of tryptophan. Attenuation has also been shown to occur in the biosynthetic pathways for histidine, threonine-isoleucine, phenylalanine, and several other amino acids and essential metabolites as well. As shown in Figure 5.46b, the leader peptide for each amino acid biosynthetic operon is rich in that particular amino acid. The *his* operon is dramatic in this regard because its leader contains *seven* histidines in a row near the end of the peptide (Figure 5.46b).

> **Attenuation is a type of regulation that depends on the fact that in prokaryotes transcription and translation are linked processes. If translation of the leader peptide occurs normally (no amino acid limitation), then the newly formed mRNA folds into a double-stranded structure that causes RNA polymerase to stop before it transcribes the operon. However, if translation is slowed because of the attenuation process (by a specific amino acid limitation), the transcription terminator does not form and the entire operon is transcribed.**

5.13 Global Regulation

Often an organism needs to regulate many different genes simultaneously in response to a change in its environment. For instance, when the bacterium *Escherichia coli* is starved for phosphate, over 80 different genes are transcribed in response, bringing about the synthesis of new proteins. These proteins play roles in adapting the bacterium to a phosphate-deficient environment. There are several such sets of genes in *E. coli* whose products are required to respond to particular conditions (Table 5.6). Because these control mechanisms operate on a wide cellular basis, they are referred to as *global control systems* and may include one or more regulons. Some scientists use the term *stimulon* to refer to a group of genes that becomes active in response to an environmental signal. Other scientists use the term *modulon* to emphasize that global control often modulates (adjusts) other controls on the same genes.

In addition to allowing an organism to respond to a signal by activating a network of genes, global regulation can be used to prevent some genes from responding unnecessarily. For instance, Sections 5.10 and 5.11 covered how the enzymes for lactose or maltose utilization can be induced by adding either lactose or maltose to the growth medium. However, it would be wasteful to induce these enzymes if the cells were already growing on a carbon source that they could use more efficiently. In fact, one of the global regulatory networks, **catabolite repression**, prevents this problem.

Catabolite repression

In catabolite repression the syntheses of a variety of unrelated enzymes, primarily catabolic, are inhibited when cells are grown in a medium that contains an energy source such as glucose. Catabolite repression has been called the **glucose effect** because glucose was the first substance shown to initiate it, although in some organisms glucose does not cause this form of enzyme repression. Catabolite repression occurs when the organism is offered a catabolizable energy source in the presence of a more readily catabolizable energy source, such as glucose.

One consequence of catabolite repression is that it can lead to so-called **diauxic growth** if the two energy sources are present in the medium at the same time and if the enzyme needed for utilization of one of the energy sources is subject to catabolite repression. In diauxic growth, the organism grows first on one energy source, and there is then a temporary cessation before growth is resumed on the other energy source. This phenomenon is illustrated in Figure 5.48 for growth on a mixture of glucose and lactose. The enzyme β-galactosidase, which is responsible for utilization of lactose, is inducible, but its synthesis is also subject to catabolite repression. Thus, as long as glucose is present in the medium, β-galactosidase is not synthesized; the organism grows only on the glucose and leaves the lactose untouched. When the glucose is exhausted, catabolite repression is abolished. After a

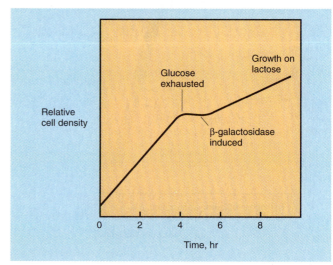

FIGURE 5.48 Diauxic growth on a mixture of glucose and lactose. Glucose represses the synthesis of β-galactosidase. After glucose is exhausted, a lag occurs until β-galactosidase is synthesized, and then growth can resume on lactose.

Table 5.6   A few of the global control systems known in *Escherichia coli**

System	Signal	Primary activity of regulatory protein	Number of genes regulated
Aerobic respiration	Presence of O_2	Repressor	20
Anaerobic respiration	Lack of O_2	Activator	20+
Catabolite repression	Cyclic AMP concentration	Activator	300+
Heat shock	Temperature	Alternative sigma	17
Nitrogen utilization	NH_3 limitation	Activator/alternative sigma	12+
Oxidative stress	Oxidizing agent	Activator	12+
SOS response	Damaged DNA	Repressor	17

*For many of the global control systems, regulation is complex. A single regulatory protein can play more than one role. For instance, the regulatory protein for anaerobic respiration, FNR, is an activator protein for many promoters but a repressor for others. Regulation can also be indirect or require more than one regulatory protein. Note that the activator for the nitrogen utilization system activates promoters recognized by an alternative sigma. Many genes are regulated by more than one global control system.

lag, β-galactosidase is synthesized and growth on lactose can occur. Notice that Figure 5.48 shows that the cells grow more rapidly on glucose. Thus, catabolite repression ensures that the cells use the *best* carbon source first.

How does catabolite repression work? Catabolite repression involves control of transcription by an activator protein (see Section 5.11). In the case of catabolite-repressible enzymes, binding of RNA polymerase only occurs if another protein, called **catabolite activator protein (CAP)**, has bound first. An allosteric protein, CAP only binds if it has first bound a small molecule called *cyclic adenosine monophosphate* or **cyclic AMP** (see Figure 5.44). Cyclic AMP (Figure 5.49) has been shown to be a key element in a variety of control systems, not only in bacteria but in higher organisms. Cyclic AMP is synthesized from ATP by an enzyme called *adenylate cyclase*, and glucose inhibits the synthesis of cyclic AMP and stimulates its transport out of the cell. When glucose is present, the cyclic AMP level in the cell is low, and binding of RNA polymerase to the promoter does not occur. Thus, catabolite repression is really a result of a deficiency of cyclic AMP, and can be overcome by adding this compound to the medium.

Although this may sound like a simple positive regulatory system (as in Figure 5.43), each of the operons that CAP controls is *also* under control of a specific regulatory protein. Therefore, catabolite repression modulates several unrelated regulatory systems and thus is an example of global control. As long as glucose is present, catabolite repression prevents expression of all other catabolic operons under this global controlling element. The complete regulatory sequence of the *lactose operon* is shown in Figure 5.50. For transcription to occur, two requirements must be met: (1) the level of cyclic AMP must be high enough so that the CAP protein binds to the CAP binding site, and (2) there must be an inducer such as lactose present so that the lactose repressor does not block transcription by binding to the operator.

Cyclic AMP has a number of regulatory roles in eukaryotes that do not involve catabolite repression and is also an extracellular signal for the aggregation process in certain cellular slime molds (see Section 21.3).

Other global control networks

Genes belonging to global control systems do not all use a simple combination of repressors or activators to achieve regulation. Several involve *alternative sigma factors*, including some shown in Table 5.6, and in these cases, regulation is brought about by changing either the amount or the activity of these factors. Most genes in *E. coli* require the sigma factor referred to as σ^{70} (the superscript 70 indicates the size of this protein, 70 kD) for transcription and have promoters like those shown in Figure 5.29. The genes that are induced by an increase in temperature (heat shock) have promoters with a quite different sequence, and RNA polymerase requires a different sigma factor (σ^{32}) to recognize them. It is the amount of this alternative sigma factor in the cell that regulates the *heat shock response*, and the amount of σ^{32} itself is not controlled by transcription but by the stability of the factor.

Most proteins are very stable; once made they continue to perform their functions and are passed along at cell division. However, some proteins are unstable. They are recognized by enzymes in the cell called **proteases** and are rapidly degraded. In *E. coli*, σ^{32} is normally degraded within a minute or two after

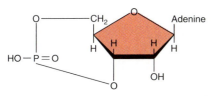

FIGURE 5.49   Cyclic adenosine monophosphate (cyclic AMP).

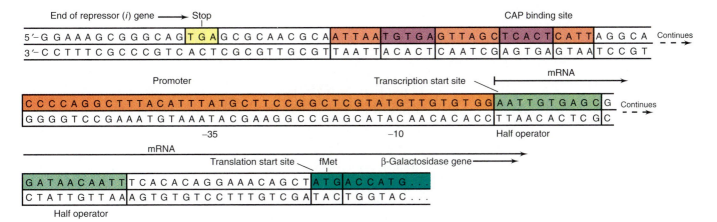

FIGURE 5.50   The genetic elements involved in regulation of the lactose operon. The first gene in this operon encodes the enzyme β-galactosidase, which breaks down lactose. The operon contains two other genes that are also involved with lactose metabolism. Notice that the two halves of the operator (where the repressor would bind) are almost perfect inverted repeats. There are also inverted repeats in the CAP binding site (shown in darker boxes).

Antisense Nucleic Acid

Regulation of the synthesis of proteins often involves transcriptional control—whether or not the mRNA for a protein is made. Less often, genes are controlled at the level of translation—whether or not the mRNA will be read. Most control networks, whether they are transcriptional or translational, utilize regulatory proteins. However, it is now clear that in some cases it is a regulatory *RNA*, not a regulatory *protein*, that is involved. One type of regulatory RNA, called **antisense RNA**, is known to be used in the regulation of several different bacterial genes. Antisense RNA acts by forming base pairs with a complementary, or sense, strand of RNA. If the sense RNA is mRNA, the resulting double-stranded structure can prevent translation. Antisense RNA could be synthesized from the same gene as the sense RNA by having a second promoter oriented in the opposite direction of the first or by having a second gene with the promoter at the other end. Antisense RNA does not have to be used only to regulate the synthesis of a protein. In some plasmids, antisense RNA controls the initiation of DNA synthesis.

Antisense nucleic acids can be specifically designed and synthesized by scientists in the laboratory and delivered directly to cells. These short (15 to 25 nucleotides) synthetic chains (oligonucleotides) are usually made of DNA rather than RNA. Their sequence can be made to allow them to bind to a specific mRNA and prevent translation (or allow the molecule to be recognized by nucleases). Antisense nucleic acid can also bind to the DNA in the nucleus and prevent transcription. The latter is possible because some DNA can form a *triple* helix! The "extra" strand (the oligonucleotide) forms specific interactions with those parts of the bases that are in the major groove of a normal double helix to give **triplex DNA**. (Not all DNA sequences can form triple helices, at least not without the aid of special enzymes.)

Synthetic antisense oligonucleotides can be designed to be extremely specific, whether they bind to a message or to the regulatory region of a gene. The specificity arises because a sequence of only 20 bases should occur no more often than once in 10^{12} bases of "random" DNA. Therefore, it is unlikely that the antisense RNA would bind to anything other than its known target in any cell. This specificity might allow antisense nucleic acids to become an important new type of antibiotic, and this possibility is being pursued by a number of pharmaceutical companies. Antisense nucleic acids could be designed to be used against specific viruses or to regulate particular genes in either disease-causing (pathogenic) organisms or human tumor cells. The possible utility of these molecules is just one example of how understanding the structure of a gene may have very practical applications (see Chapter 8).

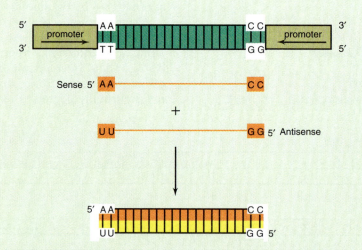

A gene with promoters at either end. If the RNA made by the promoter on the left is the sense RNA, then the RNA made by the promoter on the right is the antisense RNA. If both RNA molecules are made, they will form a duplex. Only a relatively short region of overlap is necessary for strong base pairing. Usually the antisense RNA is shorter than the sense RNA, and therefore the second promoter is actually within the gene.

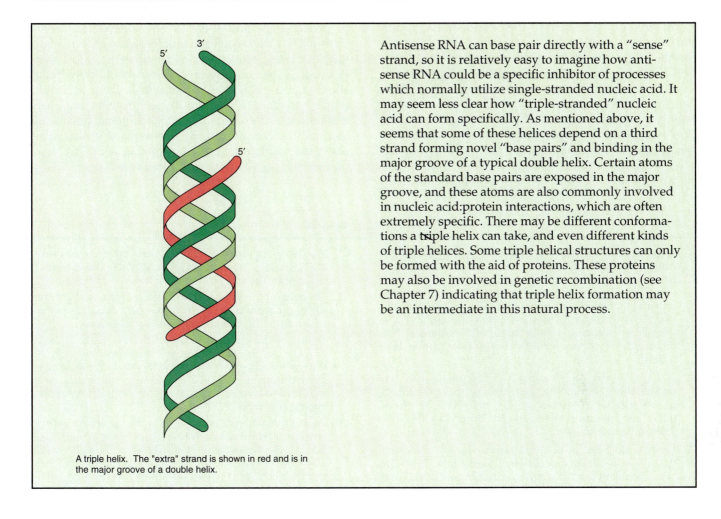

Antisense RNA can base pair directly with a "sense" strand, so it is relatively easy to imagine how anti-sense RNA could be a specific inhibitor of processes which normally utilize single-stranded nucleic acid. It may seem less clear how "triple-stranded" nucleic acid can form specifically. As mentioned above, it seems that some of these helices depend on a third strand forming novel "base pairs" and binding in the major groove of a typical double helix. Certain atoms of the standard base pairs are exposed in the major groove, and these atoms are also commonly involved in nucleic acid:protein interactions, which are often extremely specific. There may be different conformations a triple helix can take, and even different kinds of triple helices. Some triple helical structures can only be formed with the aid of proteins. These proteins may also be involved in genetic recombination (see Chapter 7) indicating that triple helix formation may be an intermediate in this natural process.

A triple helix. The "extra" strand is shown in red and is in the major groove of a double helix.

it is synthesized. When cells experience a heat shock, this degradation process is inhibited. This means there will be more σ^{32}, and therefore it can direct more RNA polymerase to more heat shock promoters.

Global regulatory systems must have a common way to transmit a signal from the environment to the gene(s). We have seen how this is accomplished with catabolite repression, but in the case of the heat shock response, how does a bacterium know what the temperature is? This mechanism seems to involve one of the heat shock proteins, a protein called DnaK. DnaK is essential for the normal growth of *E. coli* at any temperature, but the amount that is synthesized is increased by heat shock (recall that genes under the control of a global control system are also usually regulated in other ways). DnaK somehow helps other proteins fold properly and is also involved in the degradation of σ^{32}. Possibly when the temperature is increased, the activity of DnaK is directed more toward folding proteins and less is available for the degradation of σ^{32}. Certainly an increase in temperature could influence the formation of the correct secondary and tertiary structures of proteins or even cause them to denature slightly (see Section 2.8). How-

ever, since the amount of DnaK increases as part of the heat shock response, it eventually builds up and σ^{32} is degraded again, bringing the cell back to its normal state. The protein DnaK is a **chaperonin**, one of a group of proteins called *molecular chaperones*. Some of these proteins not only aid other proteins in folding correctly but are also involved in the assembly of multi-subunit structures.

Many genes belong to more than one global control system and therefore can have several overlapping regulatory systems. Comprehending how these global systems are interconnected and interact will eventually allow us to understand how a cell regulates its overall metabolism.

5.14   Contrasts in Gene Expression between Prokaryotes and Eukaryotes

Because of the lack of compartmentation in prokaryotes, the processes of transcription and translation are *coupled*. Also, the messenger RNA of prokaryotes is frequently polycistronic, with more than one protein being translated from the same message.

In eukaryotes, on the other hand, transcription and translation take place in separate parts of the cell and the integration of these processes seen in prokaryotes is lacking. How about induction and repression in eukaryotes? Although many eukaryotes do exhibit repression, there is no good evidence for the kind of negative control so commonly found in prokaryotes. However, positive control mechanisms are common in eukaryotes. If operons exist in eukaryotes, they involve the control of only single enzymes, rather than the multi-enzyme control systems so commonly seen in prokaryotes, and there is no evidence for polycistronic RNA molecules in eukaryotes except in a few viruses. These are translated inefficiently. However, eukaryotes occasionally and viruses commonly make single large proteins (polyproteins) from a single monocistronic mRNA and these proteins are then cleaved into several smaller active proteins. Posttranslational cleavage is rare in prokaryotes. Eukaryotes also have regulation involving splicing of mRNA, regulation that does not exist in prokaryotes.

Study Questions

1. A gene can be defined rather simply in both functional (informational) and chemical terms. Write a one-sentence definition for each of these qualities. What problem do you have with your chemical definition of a gene when you consider the phenomenon of split genes? Do you have a problem with the functional definition of a gene when you consider the coding of tRNA and rRNA molecules? If so, how would you alter your simple definition?

2. DNA molecules which are A-T rich separate into two strands more easily when the temperature is raised than do DNA molecules which are G-C rich. Write an explanation for this observation based on the properties of A-T and G-C base pairing.

3. From a biochemical point of view, why is it essential that DNA replication occurs in the direction $5' \rightarrow 3'$? What problem does this pose for replication of the two separate strands of DNA?

4. Describe the mechanism by which the two enzymes DNA polymerase and DNA ligase function together to effect DNA synthesis.

5. Circular DNA molecules, such as those of most bacterial chromosomes, circumvent one problem encountered with replication of linear DNA. What is this problem? Also, having a circular chromosome results in a new problem: the two daughter chromosomes are interlocked after replication is completed. Is there any type of enzyme discussed in this chapter that might help separate these molecules so they could be partitioned?

6. What are restriction enzymes? What is the prime function of a restriction enzyme in the cell which produces it (that is, why do cells have restriction enzymes)? How is it that the restriction enzyme in a cell does not cause degradation of that cell's DNA?

7. Nucleic acid hybridization is at the basis of modern genetic techniques. Write a short explanation for each of the following statements:
   a. The strength of the DNA hybrid is greater if two long than if two short DNA molecules are involved.
   b. Even if the base sequences are not *exactly* complementary, hybridization can still occur between two relatively long DNA molecules, but this is less likely to occur if the molecules are short.
   c. Suppose you had available a pure mRNA and were given a pure double-stranded DNA containing the gene which coded for this mRNA. How could you isolate the DNA strand against which the mRNA had been made? (Assume you have some way of immobilizing the mRNA.)

8. We have indicated three ways in which eukaryotic mRNA differs from prokaryotic mRNA. Write a short description of each of these three ways.

9. Describe one way in which mitochondria and chloroplasts are similar to bacterial plasmids and one way in which they differ, from a genetic point of view.

10. What would be the consequence (in terms of both mRNA and protein synthesis) if the *promoter* region for the gene coding for the enzyme β-galactosidase was deleted from the DNA? If the *base sequence* of the promoter were changed so that the binding of RNA polymerase was weaker?

11. From your understanding of how the antibiotic *actinomycin* acts, draw a graph to indicate what would happen to β-galactosidase synthesis if *actinomycin* were added just before the inducer lactose. If the *actinomycin* were added a short while after the inducer was added?

12. What would be the consequence (in terms of protein synthesis) if one base of the anticodon of a specific tRNA was changed to another base? Why would it matter which base was changed? Why would it matter what it was changed to? Why might it matter which amino acid was charged to the normal tRNA?

13. Explain why a *nonsense* codon functions as a stopping point for protein synthesis. Why doesn't a nonsense codon also function as a stopping point for mRNA synthesis?

14. Many bacterial mRNA molecules are polycistronic, each mRNA coding for more than one protein. Imagine an mRNA which codes for two proteins with an intervening noncoding region between the two coding regions. From your understanding of how the translation process works, explain why the end result would be two separate proteins rather than one mixed (hybrid) protein.

15. The start and stop sites for mRNA synthesis (on the DNA) are different than the start and stop sites for protein synthesis (on the mRNA). Explain.

16. Imagine you had isolated a new antibiotic which inhibited the translocation process in protein synthesis. Explain what would happen, in terms of mRNA synthesis, polypeptide synthesis, and cell growth, when such an antibiotic was added to a growing culture.

17. If you looked at the sequence of a short stretch of bases from the middle of a particular mRNA, you should be able to identify three different reading frames (and determine the sequence of the protein encoded by each). However, if this mRNA was in a cell, each ribosome that translates it would use only one of those reading frames. How do ribosomes determine which reading frame is correct?

18. What is diauxic growth? Why is there a lag in growth after glucose is exhausted before growth on lactose commences (what is happening during the lag)?

19. Overlapping genes make it possible for a small piece of DNA to code for more than one protein. What would be the result if a small deletion occurred in the middle of a DNA molecule that was involved in an overlapping gene?

20. What would be the result (in terms of protein synthesis) if RNA polymerase initiated *transcription* 1 base upstream of its normal starting point? Why? What would be the result (in terms of protein synthesis) if *translation* began 1 base downstream of its normal starting point? Why?

21. Two methods for determining the sequence of a DNA molecule are the Maxam-Gilbert and the Sanger methods. In what ways do these two methods differ? In what ways are they similar?

22. Suppose you were going to determine the DNA sequence of a gene of 2000 nucleotides. This is too long for direct determination by the Maxam–Gilbert or Sanger method. Describe a procedure you could use that would make it possible to determine the sequence of this gene in stages. In completing your answer, write a diagram that lists in order the various steps you would use.

23. What is a topoisomerase and what is its function? Contrast the actions of topoisomerase I and II.

24. Describe in chemical terms the manner by which a protein such as RNA polymerase or *lac* repressor combines in a *sequence-specific* manner with the DNA. Explain how symmetry in the DNA relates to the symmetry of the protein molecule.

25. Digestion of a short piece of *linear* DNA 1500 base pairs long with the restriction enzyme *Taq*I resulted in fragments of 200, 500, and 800 base pairs. When the same DNA was digested with the restriction enzyme *Hind*III, fragments of 300 and 1200 base pairs were obtained. Simultaneous digestion with both enzymes gave fragments of 100, 200, 500, and 700 base pairs. From these data, prepare a restriction map of the 1500 base pair piece of DNA. (Of what help was it to know that this piece of DNA was linear?)

26. A classic experiment to show that DNA replication was semiconservative was that of Mathew Meselson and Franklin Stahl. These workers used the heavy isotope of nitrogen, N^{15}, to label DNA during the replication process. DNA labeled with N^{15} can be separated from regular DNA by ultracentrifugation. In the Meselson-Stahl experiment, cells whose DNA was fully labeled with N^{15} were transferred to a medium containing regular (light) N^{14}. As replication proceeded, DNA containing light nitrogen was obtained and could be separated in the ultracentrifuge. Three kinds of molecules can be anticipated: both strands heavy (initial parent), half heavy–half light, and both strands light. Describe the anticipated result of this experiment after the first round of replication. Label each of the two strands obtained as to whether it would be heavy or light. Describe the anticipated results after a further round of replication.

27. A structure commonly seen in circular DNA during replication is called a *theta structure*. Draw a diagram of the replication process and show how a theta structure could arise.

28. Describe the general structure of a helix-turn-helix DNA binding protein. How do such proteins recognize specific nucleic acid sequences?

29. Why are errors in DNA replication so rare? What additional enzyme activity (other than polymerization), is associated with DNA polymerase III and how does this serve to reduce errors?

30. What are ribozymes and what types of biochemical reactions are they generally associated with?

31. What are aminoacyl-tRNA synthetases and what types of reactions do they carry out? Approximately how many different types of these enzymes are present in the cell? How does a synthetase recognize its correct substrates?

32. Besides its role as a structural component of the ribosome, what other role(s) are played by ribosomal RNA in protein synthesis?

33. The maltose regulon is inducible and is regulated by an activator protein. The *lac* operon is inducible but is regulated by a repressor protein. Explain how induction can be brought about by either positive control (activator protein) or negative control (repressor protein).

34. Distinguish between an operon and a regulon.

35. Many genes are under multiple control systems. In the *lac* operon, there is lactose-specific regulation and regulation by a global control system. Describe how each of the controls on the lactose operon actually functions and discuss why you think both systems are necessary.

36. Describe how regulation by attenuation works. What is actually being "attenuated"? Why hasn't this type of control been found in eukaryotes?

Supplementary Readings

Alberts, B., D. Bray, J. Lewis, M. Raff, K. Roberts, and **J. D. Watson**. 1989. *Molecular Biology of the Cell*, 2nd edition. This book has become a standard reference book for cellular molecular biology. Although the emphasis on eukaryotes is strong, prokaryotes are not slighted.

Darnell, J., H. Lodish, and **D. Baltimore**. 1990. *Molecular Cell Biology*, 2nd edition. Scientific American Books, New York. One of the "big" cell biology textbooks, with strong emphasis on macromolecular processes in eukaryotes.

Freifelder, D., and **G. M. Malacinski**. 1993. *Essentials of Molecular Biology*, 2nd edition. Jones and Bartlett, Boston. An excellent introductory textbook, emphasizing the fundamental principles of gene structure and expression.

Hill, W. E., A. Dahlberg, R. A. Garrett, P. B. Moore, D. Schlessinger, and **R. Warner**. 1990. *The Ribosome: Structure, Function, and Evolution*. American Society for Microbiology, Washington, D.C. A compendium of current research on the biology of the ribosome.

Kornberg, A., and **T. A. Baker**. 1992. *DNA Replication,* 2nd edition. W. H. Freeman and Co., New York. An advanced textbook covering the biochemistry, molecular biology, and genetics of DNA replication in prokaryotes and eukaryotes.

Lewin, B. 1990. *Genes IV*. Cell Press, Cambridge, MA. A very readable account of information flow in biological systems. Covers both prokaryotes and eukaryotes.

Singer, M., and **P. Berg.** 1991. *Genes and Genomes.* University Science Books, Mill Valley, CA. An advanced textbook of molecular biology focusing on gene structure and expression in both prokaryotes and eukaryotes.

Stryer, L. 1988. *Biochemistry*, 3rd edition. W. H. Freeman & Co., New York. A biochemistry text with a nice treatment of elementary molecular biology, including regulation.

Watson, J. D., N. H. Hopkins, J. W. Roberts, J. A. Steitz, and **A. M. Weiner**. 1987. *Molecular Biology of the Gene,* 4th edition. Benjamin-Cummings Publishing Co., Menlo Park, CA. This book has been the standard textbook on macromolecules and DNA for over 20 years.

Zubay, G. 1993. *Biochemistry*, 3rd edition. Wm. C. Brown, Dubuque, IA. A biochemistry text which includes considerable material on molecular biology.

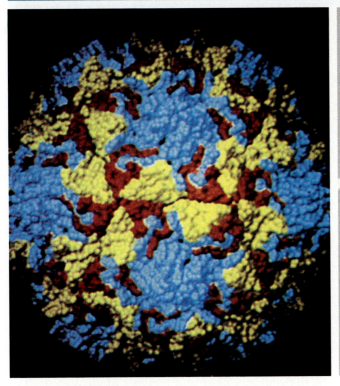

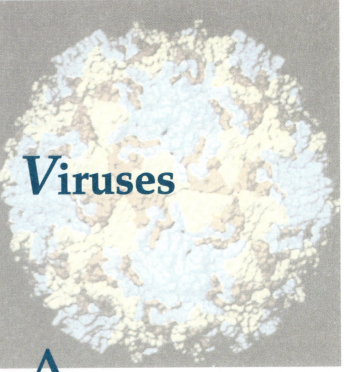

6

Viruses

A **virus** is a noncellular genetic element that enlists a cell for its own replication and is characterized by also having an extracellular state. In the extracellular state, a virus is a submicroscopic particle containing nucleic acid surrounded by protein and occasionally containing other macromolecular components. In this extracellular state, the **virus particle**, also called the **virion**, is metabolically inert and does not carry out respiratory or biosynthetic functions. The virion is the structure by which the virus genome is carried from the cell in which the virion has been produced to another cell where the viral nucleic acid can be introduced. Once in the new cell, the **intracellular state** is initiated. In the intracellular state, **virus reproduction** occurs: the virus genome is produced and the components that make up the virus coat are synthesized. When a virus genome is introduced into a cell and reproduces, the process is called **infection**. The cell that a virus can infect and in which it can replicate is called a **host**. Viral genomes are very limited in size, and they encode primarily those functions that they cannot adapt from their hosts. Therefore, during reproduction inside a cell, there is a heavy dependence on host-cell structural and metabolic components. The virus redirects preexisting host machinery and metabolic functions necessary for virus replication.

Viruses are not the only type of genetic element that take advantage of the metabolic machinery encoded by the cell's own chromosomes (see Section 5.5). Like these other elements, viruses can confer important new properties on their host cell. These properties will be inherited when the host cell divides, if each new cell

Miniglossary for Chapter 6

BACTERIOPHAGE   a virus that infects prokaryotic cells

LYSOGEN   a bacterium containing a prophage

MINUS (NEGATIVE)-STRAND NUCLEIC ACID   an RNA or DNA strand which has the opposite sense of (is complementary to) the mRNA of a virus

ONCOGENE   a gene whose expression causes formation of a tumor

PLAQUE   a zone of lysis or cell inhibition caused by virus infection of a lawn of sensitive cells

PLUS-STRAND NUCLEIC ACID   an RNA or DNA strand which has the same sense as the mRNA of a virus

PRION   an infectious agent whose extracellular form may contain no nucleic acid

PROVIRUS (PROPHAGE)   the genome of a temperate virus when it is replicating with, and usually integrated into, the host chromosome

RETROVIRUS   a virus whose RNA genome has a DNA intermediate as part of its replication cycle

REVERSE TRANSCRIPTION   the process of copying information found in RNA into DNA

TEMPERATE VIRUS   a virus whose genome is able to replicate along with that of its host and not cause cell death in a state called lysogeny

TRANSFORMATION   a process by which a normal cell becomes a cancer cell (but see alternative usage in Chapter 7)

VIRION   the complete virus particle; the nucleic acid surrounded by a protein coat and in some cases other material

VIRULENT VIRUS   a virus which lyses or kills the host cell after infection; a nontemperate virus

VIRUS   a genetic element containing either RNA or DNA which replicates in cells but is characterized by having an extracellular state

also inherits the viral genome. These changes are often not harmful and may even be beneficial. However, viruses, unlike genetic elements such as plasmids (see Sections 5.5 and 7.8), have an extracellular form that enables them to transmit themselves from one host to another. This extracellular form has enabled some viruses to replicate themselves in a host in a way that is destructive to the host cell. This destructive replication accounts for the fact that some viruses are agents of disease. In many cases, whether a virus causes disease or hereditary change depends upon the host cell and on the environmental conditions.

Unlike cells, all of which have double-stranded DNA as their genetic material, viruses can have either DNA or RNA as their genetic material and it can be either single-stranded or double-stranded. Viruses are sometimes divided into two types based on whether they have DNA or RNA as their genetic material, and *all* viruses contain one or the other in the virion. However, there is a third group of viruses that use *both* DNA and RNA as their genetic material but at different stages of their reproductive cycle (Figure 6.1). These latter include the retroviruses, which contain an RNA genome in the virion but which replicate through a DNA intermediate, and the human hepatitis B virus, which contains DNA in the virion but has an RNA intermediate in replication. These classes can be further subdivided into whether the nucleic acid in the virion is single- or double-stranded (Figure 6.1).

Viruses can also be classified on the basis of the hosts they infect. Thus, we have animal viruses, plant viruses, and bacterial viruses. Bacterial viruses, sometimes called *bacteriophages* (or *phage* for short, from the Greek *phagein* meaning *to eat*), have been studied primarily as convenient model systems for research on the molecular biology and genetics of virus reproduction. Many of the basic concepts of virology were first worked out with bacterial viruses and subsequently applied to viruses of higher organisms. Because of their frequent medical importance, *animal viruses* have been extensively studied. The two groups of animal viruses most studied are those infecting insects and those infecting warm-blooded animals. *Plant viruses* are often important in agriculture but have been less studied than animal viruses. In this chapter, we discuss the structure, replication, and genetics of viruses infecting bacteria and warm-blooded animals.

To summarize,

1. In the virion, the **virus genome** consists of either RNA or DNA. The genome is surrounded by a *coat* of *protein* (and occasionally other material). The virus genome together with the coat is called the **virus particle** or **virion**.

2. Viruses lack independent metabolism. They multiply only inside living cells, using the host cell metabolic machinery. Some virus particles do contain enzymes, however, that are under the genetic

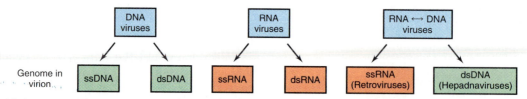

FIGURE 6.1   Viral genomes. The genomes of viruses can be composed of either DNA or RNA, and some use both as their genomic material at different stages in their life cycle. However, only one type of nucleic acid is found in the virion of any particular type of virus. This can be single-stranded (ss), double-stranded (ds), or in the case of the hepadnaviruses, partially double-stranded.

control of the virus genome. Such enzymes are only produced during the infection cycle.

3. When a virus multiplies, the genome becomes released from the coat. This occurs during the infection process. Once released, the viral genome undergoes replication and directs the synthesis of new coat proteins. New virus particles are then reconstructed from the virus genome and virus coat.

This chapter is divided into three parts. The first part deals with basic concepts of virus structure and function. The second part deals with the nature and manner of multiplication of the bacterial viruses (bacteriophages). In this second part we introduce the basic molecular biology of virus multiplication. The third part deals with important groups of animal viruses, with emphasis on molecular aspects of animal virus multiplication.

6.1 Nature of the Virion

Virions vary widely in size and shape. Viruses are smaller than cells, ranging in size from 0.02 μm to 0.3 μm. A common unit of measure for viruses is the *nanometer* (abbreviated nm), which is 1000 times smaller than 1 μm and one million times smaller than 1 mm. Smallpox virus, one of the largest viruses, is about 200 nm in diameter; polio virus, one of the smallest, is only 28 nm in diameter.

As we have stated, some viruses contain RNA, others DNA. Viral *genomes* are also smaller than those of cells. Most bacterial genomes are between 1000 and 9000 kilobase pairs of DNA, with the smallest known being about 590 kilobase pairs. (Interestingly, the bacteria with the smallest genomes are, like viruses, parasites that replicate in other cells; see Sections 19.22 and 19.23.) However, one of the largest known viral genomes, that of vaccinia, is only 190 kilobase pairs. The sizes of the genomes of a few representative types of viruses are given in Table 6.1.

The Name "Virus"

The word *virus* originally referred to any poisonous emanation, such as the venom of a snake, and later came to be used more specifically for the causative agent of any infectious disease. Pasteur often referred to bacteria that caused infectious diseases as viruses. By the end of the nineteenth century, a large number of bacteria had been isolated and shown to be causal agents of specific infectious diseases, but there were some diseases for which a bacterial cause had not been shown. One of these was foot-and-mouth disease, a serious skin disease of animals. In 1898, Friedrich Loeffler and Paul Frosch presented the first evidence that the cause of foot-and-mouth disease was an agent so small that it could pass through filters that could hold back all known bacteria. That the agent was not an ordinary toxin could be shown by the fact that it was active at very low dilution and could be transmitted in filtered material from animal to animal. Loeffler and Frosch concluded "that the activity of the filtrate is not due to the presence in it of a soluble substance, but due to the presence of a causal agent capable of reproducing. This agent must then be obviously so small that the pores of a filter which will hold back the smallest bacterium will still allow it to pass. . . . If it is confirmed by further studies . . . that the action of the filtrate . . . is actually due to the presence of such a minute living being, this brings up the thought that the causal agents of a large number of other infectious diseases . . . which up to now have been sought in vain, may also belong to this smallest group of organisms."

A year later, the Dutch microbiologist Martinus Beijerinck published his work on tobacco mosaic disease, a crippling leaf disease of tobaccos and tomatoes. In 1892, D. Ivanowsky of Russia had first shown that the causal agent of tobacco mosaic disease was filterable, but Beijerinck went much further and provided strong evidence that although the causal agent was filterable, it had many of the properties of a living organism. He called the agent a *Contagium vivum fluidum*, a living germ that is soluble. He postulated that the agent must be incorporated into the living protoplasm of the cell in order to reproduce, and that its reproduction must be brought about with the reproduction of the cell. This postulate comes very close to our current understanding of how viruses reproduce. Beijerinck also noted that there were other plant diseases for which causal agents had not been isolated, and these might also be caused by filterable agents. Soon a number of other filterable agents were shown to be the causes of both plant and animal diseases. Such agents came to be called **filterable viruses**, but as further work on these agents was carried out, the word "filterable" was gradually dropped. Today, the original meaning of "virus" has been forgotten, and the word is now used to refer to the kinds of agents discussed in this chapter. Bacterial viruses were first discovered by the British scientist F. W. Twort in 1915, and independently by the French scientist F. d'Herelle in 1917, who called them *bacteriophages* (from the combining form *phago* meaning "to eat"). Although bacteriophages are viruses, the name "phage" is still widely used to refer to this particular class of filterable infectious agents.

Table 6.1   Some types of viral genomes*

| Virus | Host | Viral genome | | | |
		Type of nucleic acid in virion	Structure	Number of molecules	Size
H-1 parvovirus	Animals	Single-stranded DNA	Linear	1	5,176 bases
Simian virus 40 (SV40)	Animals	Double-stranded DNA	Circular	1	5,224 base pairs
Poliovirus	Animals	Single-stranded RNA	Linear	1	7,433 bases
Cauliflower mosaic virus	Plants	Double-stranded DNA	Circular	1	8,025 base pairs
Cowpea mosaic virus	Plants	Single-stranded RNA	Linear	2 different	9,370 bases (total)
Reovirus type 3	Animals	Double-stranded RNA	Linear	10 different	23,549 base pairs (total)
Bacteriophage λ	Bacteria	Double-stranded DNA	Linear	1	48,514 base pairs
Herpes simplex virus type 1	Animals	Double-stranded DNA	Linear	1	152,260 base pairs

The sizes of the viral genomes chosen for this table are known accurately because they have been sequenced. However, this accuracy can be misleading since only a particular strain or isolate of a virus was sequenced. Therefore, the sequence and exact number of bases for other isolates may be slightly different. No attempt has been made to choose the largest and smallest viruses known, but rather to give a fairly representative sampling of the sizes and structures of the genomes of viruses containing both single-stranded and double-stranded RNA and DNA.

The structures of virions (virus particles) are quite diverse. Viruses vary widely in size, shape, and chemical composition. The nucleic acid of the virion is always located within the particle, surrounded by a protein coat called the *capsid*. The terms *coat, shell,* and *capsid* are often used interchangeably to refer to this outer layer. The protein coat is always formed of a number of individual protein molecules, called *protein subunits* (sometimes called *capsomeres*), which are arranged in a precise and highly repetitive pattern around the nucleic acid (Figure 6.2). A few viruses have only a single kind of protein subunit, but most viruses have several chemically distinct kinds of protein subunits which are themselves associated in specific ways to form larger assemblies called *morphological units*. It is the morphological unit which is seen with the electron microscope. Genetic economy dictates that the variety of virus proteins be kept small, since many virus genomes do not have sufficient genetic information to code for a large number of different kinds of proteins.

The information for proper aggregation of the protein subunits into the morphological units is contained within the structure of the subunits themselves, and the overall process of assembly is thus called **self-assembly**. For many viruses, this self-assembly process is assisted by *molecular chaperones*, proteins that assist in folding and assembly but that themselves are not a part of the final structure. A single virion generally has a large number of morphological units.

The complete complex of nucleic acid and protein, packaged in the virus particle, is called the virus **nucleocapsid**. Although the virus structure just described is frequently the total structure of a virus particle, a number of viruses have more complex

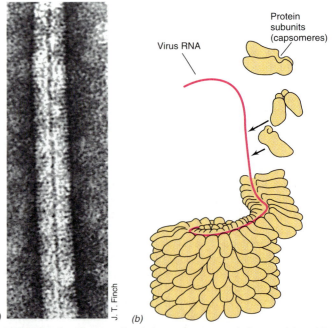

(a)   (b)

J. T. Finch

FIGURE 6.2   An example of the arrangement of virus nucleic acid and protein coat in a simple virus, tobacco mosaic virus. (a) Electron micrograph at high resolution of a portion of the virus particle. (b) Assembly of the tobacco mosaic virion. The RNA assumes a helical configuration surrounded by the protein capsomeres. The center of the particle is hollow.

structures. These viruses are *enveloped* viruses, in which the nucleocapsid is enclosed in a membrane (Figure 6.3). (Viruses without membranes are sometimes called *naked* viruses.) *Virus membranes* are generally lipid bilayer membranes (see Section 3.3), but associated with these membranes are often virus-specific proteins. Inside the virion are often one or more

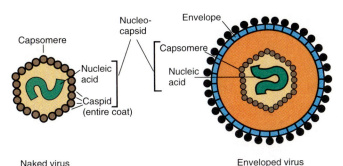

Capsomere

Nucleo-
capsid

Envelope

Capsomere

Nucleic
acid

Nucleic
acid

Caspid
(entire coat)

Naked virus

Enveloped virus

FIGURE 6.3 Comparison of naked and enveloped virus, two basic types of virus particles.

virus-specific enzymes. Such enzymes usually play a role during the infection and replication process, as we will discuss later in this chapter.

Virus symmetry

The nucleocapsids of viruses are constructed in highly symmetrical ways. Symmetry refers to the way in which the protein morphological units are arranged in the virus shell. When a symmetrical structure is rotated around an axis, the same form is seen again after a certain number of degrees of rotation. Two kinds of symmetry are recognized in viruses which correspond to the two primary shapes, rod and spherical. Rod-shaped viruses have helical symmetry and spherical viruses have icosahedral symmetry.

A typical virus with **helical symmetry** is the tobacco mosaic virus (TMV) that was illustrated in Figure 6.2. This is an RNA virus in which the 2130 identical protein subunits (each 158 amino acids in length) are arranged in a helix. In TMV, the helix has 16 1/2 subunits per turn and the overall dimensions of the virion are 18 × 300 nm. The lengths of helical viruses are determined by the length of the nucleic acid, but the width of the helical virus particle is determined by the size and packing of the protein subunits.

An **icosahedron** is a symmetrical structure roughly spherical in shape which has 20 faces. Icosahedral symmetry is the most efficient arrangement for subunits in a closed shell because it uses the smallest number of units to build a shell. The simplest arrangement of morphological units is three per face, for a total of 60 units per virus particle. The three units at each face can be either identical or different. Most viruses have more nucleic acid than can be packed into a shell made of just 60 morphological units. The next possible structure which permits close packing contains 180 units and many viruses have shells with this configuration. Other known configurations involve 240 units and 420 units.

To help understand icosahedral symmetry, a model can be made following the instructions given in Figure 6.4. When discussing symmetry, one speaks of *axes of rotation*. A flat triangle shape, for instance, has one three-fold axis of symmetry, since there are three

possible rotations that will lead to the exact configuration seen originally. Three dimensional objects such as viruses can have more than one axis of symmetry. An icosahedron, for instance, has three different axes of symmetry, two-fold, three-fold, and five-fold (see Figure 6.4). When a rod is placed through the two-fold axis of symmetry (one of the edges) in the model, the model can be turned once around this axis (1/2 way or 180°) to obtain the same configuration again. When the rod is placed through one of the three-fold axes of symmetry (one of the faces), the model can be turned three times, and if the rod is placed through one of the five-fold axes of symmetry (one of the vertices) the model can be turned five times.

In all cases, the characteristic structure of the virus is determined by the structure of the protein subunits of which it is constructed. Self-assembly leads to the final virus particle. An electron micrograph of a typical icosahedral virus is shown in Figure 6.5a.

Enveloped viruses

Many viruses have complex membranous structures surrounding the nucleocapsid (Figure 6.5b). Enveloped viruses are common in the animal world (for example, influenza virus), but some enveloped bacterial viruses are also known. The virus envelope consists of a lipid bilayer with proteins, usually glycoproteins, embedded in it. Although the glycoproteins of the virus membrane are encoded by the virus, the lipids are derived from the membranes of the host cell; proteins of the host cell membrane are somehow excluded. The symmetry of enveloped viruses is expressed not in terms of the virion as a whole but in terms of the nucleocapsid present inside the virus membrane.

What is the function of the membrane in a virus particle? We will discuss this in detail later but note that because of its location in the virion, the membrane is the structural component of the virus particle that interacts first with the cell. The specificity of virus infection, and some aspects of virus penetration, are controlled in part by characteristics of virus membranes.

Complex viruses

Some virions are even more complex, being composed of several separate parts, with separate shapes and symmetries. The most complicated viruses in terms of structure are some of the bacterial viruses, which possess not only icosahedral heads but helical tails (Figure 6.5c). In some bacterial viruses, such as the T4 virus of *Escherichia coli*, the tail itself is a complex structure. For instance, T4 has almost 20 separate proteins in the tail, and the T4 head has several more proteins. In such complex viruses, assembly is also complex. For instance, in T4 the complete tail is formed as a subassembly, and then the tail is added to the DNA-containing head. Finally, tail fibers formed from another protein are added to make the mature, infectious

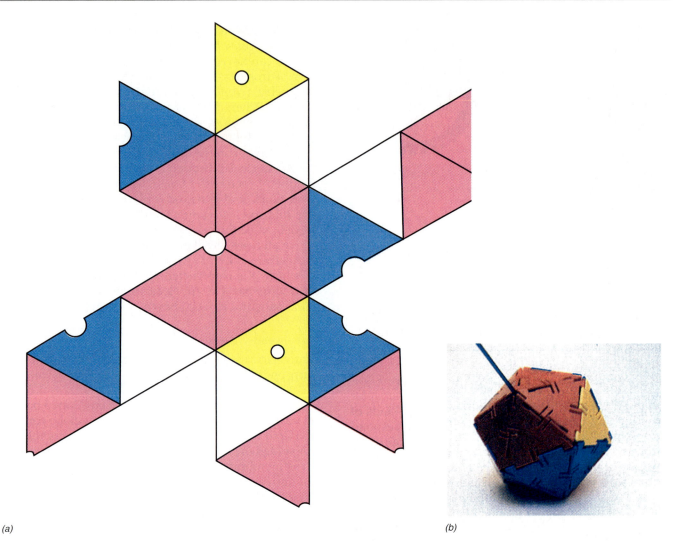

(a)   *(b)*

FIGURE 6.4   Demonstration of icosahedral symmetry. (a) A pattern that can be used to make a model of an icosahedral virus. Make a photocopy of this figure, cut it out, and fold at every line. Then tape the adjoining faces together to make a three-dimensional structure. (A copy machine that enlarges will provide a copy that can be more easily folded.) When the structure is folded, the axes of symmetry will be evident. Cut the paper at these axes, as marked, before folding. A metal rod or wire can then be inserted completely through the center of the model to study each axis. (b) Photograph of a model showing a rod through the axis of fivefold symmetry.

virus particle (see the discussion of T4 assembly in Section 6.11).

The virus genome

We have stated that the virus genome consists of either DNA or RNA. Viruses differ in size, amount, and character of their nucleic acid (see Table 6.1). Both single-stranded and double-stranded nucleic acid is found in viruses, and the amount of nucleic acid per virion may vary greatly from one virus type to another. In general, in enveloped viruses the nucleic acid constitutes only a small part of the mass of the virus particle (1–2 percent), whereas in nonenveloped viruses the percent of the particle which is nucleic acid is much larger, often 25–50 percent.

Interestingly, the nucleic acid in some viruses is not present in a single molecule, the genome being segmented into several or many molecules. For in-stance, *retroviruses*—causal agents of some cancers and AIDS, among other diseases—have two identical RNA molecules, influenza virus has eight RNA molecules, and some other animal viruses have even more RNA molecules than this. The manner in which these various pieces of nucleic acid are replicated in the cell and then assembled into mature virions is of considerable interest—how do all these nucleic acid pieces end up together in one particle? We will consider this question in Section 6.14.

Enzymes in viruses

We have stated that virions do not carry out metabolic processes. Outside of a host cell, a virion is metabolically inert. However, some virions do contain enzymes which play roles in the infection process. For instance, many viruses contain their own nucleic acid polymerases which transcribe the viral nucleic acid

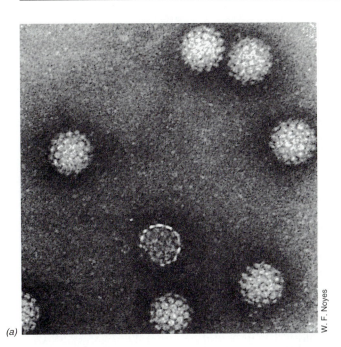

(a)

W. F. Noyes

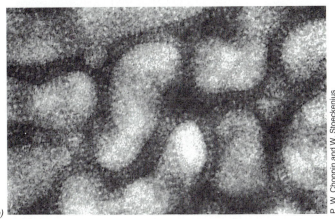

(b)

P. W. Choppin and W. Stoeckenius

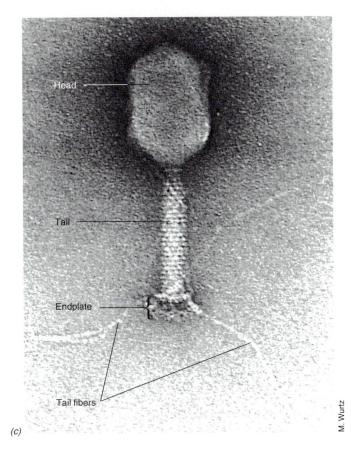

(c)

M. Wurtz

FIGURE 6.5 Electron micrographs of various virus particles. (a) Human wart virus, a virus with icosahedral symmetry. The individual particles are about 55 nm in diameter. (b) Influenza virus, an enveloped virus. The individual particles are about 80 nm in diameter. (c) Bacterial virus (bacteriophage) T4 of *Escherichia coli*. Note the complex structure. The tail components are involved in attachment of the virion to the host and infection of the nucleic acid. The head is about 60 nm in diameter.

into messenger RNA once the infection process has begun. The retroviruses are RNA viruses which replicate inside the cell as DNA intermediates. These viruses possess an enzyme, an RNA-dependent DNA polymerase called *reverse transcriptase*, which transcribes the information in the incoming RNA into a DNA intermediate. A number of viruses contain enzymes which aid in entering cells or in release of the virus from the host cells in the final stages of the infection process. One group of such enzymes, called *neuraminadases*, breaks down glycosidic bonds of glycoproteins and glycolipids of the connective tissue of animal cells, thus aiding in the liberation of the virus. Virions infecting some bacteria possess an enzyme, *lysozyme* (see Section 3.5), which makes a small hole in the bacterial cell wall that allows the viral nucleic acid to get in. This same enzyme is produced in large amounts in the later stages of infection, causing lysis of the host cell and release of the virions. We will discuss some of these enzymes in more detail later.

In the virion of a naked virus, only nucleic acid (DNA or RNA) and protein are present, with the nucleic acid on the inside; the whole unit is called the nucleocapsid. Enveloped viruses have one or more lipoprotein layers surrounding the nucleocapsid. The nucleocapsid is arranged in a symmetrical fashion, with a precise number and arrangement of protein subunits surrounding the virus nucleic acid. Although viruses are metabolically inert, in some viruses, one or more enzymes are present within the virus particle. Such enzymes play a role in the initial stages of the infection process.

6.2   The Virus Host

Because viruses only replicate inside living cells, research on viruses requires use of appropriate hosts. For the study of bacterial viruses, pure cultures are used either in liquid or on semi-solid (agar) media. Because bacteria are so easy to culture, it is quite easy to study bacterial viruses and this is why such detailed knowledge of bacterial virus reproduction is available.

With animal viruses, the initial host may be a whole animal which is susceptible to the virus, but for research purposes it is desirable to have a more manageable host. Many animal viruses can be cultivated in *tissue* or *cell cultures*, and the use of such cultures has enormously facilitated research on animal viruses.

Cell cultures

A cell culture is obtained by promoting growth of cells taken from an organ of the experimental animal. Cell cultures are generally obtained by aseptically removing pieces of the tissue in question, dissociating the cells by treatment with an enzyme which breaks apart the intercellular cement, and spreading the resulting suspension out on the bottom of a flat surface, such as

a bottle or Petri dish. The cells generally produce glycoprotein-like materials that permit them to adhere to glass surfaces. The thin layer of cells adhering to the glass or plastic dish, called a *monolayer*, is then overlayed with a suitable culture medium and the culture incubated. The culture media used for cell cultures are generally quite complex, employing a number of amino acids and vitamins, salts, glucose, and a bicarbonate buffer system. To obtain best growth, addition of a small amount of blood serum is usually necessary, and several antibiotics are generally added to prevent bacterial contamination.

Some cell cultures prepared in this way will grow indefinitely, and can be established as *permanent cell lines*. Such cell cultures are most convenient for virus research because cell material is continuously available for research purposes. In other cases, indefinite growth does not occur but the culture may remain alive for a number of days. Such cultures, called *primary cell cultures*, may still be useful for virus research, although of course new cultures will have to be prepared from fresh sources from time to time.

In some cases, cell culture monolayers cannot be obtained, but whole organs, or pieces of organs, can be cultured. Such **organ cultures** may still be useful in virus research, since they permit growth of viruses under more or less controlled laboratory conditions.

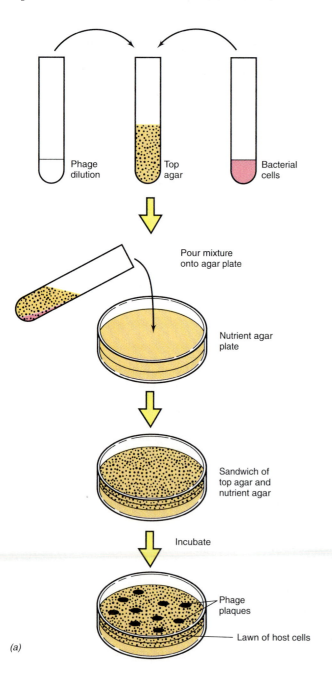

(a)

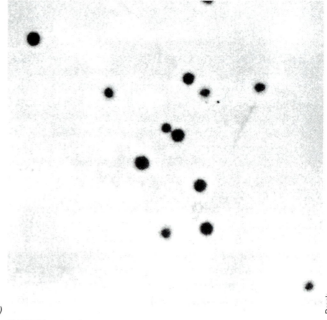

(b)

FIGURE 6.6 Quantification of bacterial virus by plaque assay using the agar overlay technique. A dilution of a suspension containing the virus material is mixed in a small amount of melted agar with the sensitive host bacteria, and the mixture poured on the surface of a nutrient agar plate. The host bacteria, which have been spread uniformly throughout the top agar layer, begin to grow, and after overnight incubation will form a *lawn* of confluent growth. Each virus particle that attaches to a cell and reproduces may cause cell lysis, and the virus particles released can spread to adjacent cells in the agar, infect them, be reproduced, and again lead to lysis and release. The size of the plaque formed depends on the virus, the host, and conditions of culture. (b) Photograph of plaques. The plaques shown are about 1–2 mm in diameter.

6.3   Quantification of Viruses

In order to obtain any significant understanding of the nature of viruses and virus replication, it is necessary to be able to *quantify* the number of virus particles. Virus particles are almost always too small to be seen under the light microscope. Although virus particles can be observed under the electron microscope, the use of this instrument is cumbersome for routine study. In general, viruses are quantified by measuring their effects on the host cells which they infect. It is common to speak of a *virus infectious unit*, which is the smallest unit that causes a detectable effect when placed with a susceptible host. By determining the number of infectious units per volume of fluid, a measure of virus quantity can be obtained. We discuss here several approaches to assessment of the virus infectious unit.

Plaque assay

When a virus particle initiates an infection upon a layer or lawn of host cells which is growing spread out on a flat surface, a zone of *lysis* or *growth inhibition* may occur which results in a clear area in the lawn of growing host cells. This clearing is called a **plaque** and it is assumed that each plaque has originated from replication events that began with one virion.

Plaques are essentially "windows" in the lawn of confluent cell growth. With bacterial viruses, plaques may be obtained when virus particles are mixed into a thin layer of host bacteria which is spread out as an agar overlay on the surface of an agar medium (Figure 6.6*a*). During incubation of the culture, the bacteria grow and form a turbid layer which is visible to the naked eye. However, wherever a successful virus in-fection has been initiated, lysis of the cells occurs, resulting in the formation of a clear zone, called a *plaque* (Figure 6.6*b*).

The plaque procedure also permits the isolation of pure virus strains, since if a plaque has arisen from one virus particle, all the virions in this plaque are probably genetically identical. Some of the particles from this plaque can be picked and inoculated into a fresh bacterial culture to establish a pure virus line. The development of the plaque assay technique was as important for the advance of virology as was Koch's development of solid media (Section 1.5) for bacteriology.

Plaques may be obtained for animal viruses by using animal cell-culture systems as hosts. A monolayer of cultured animal cells is prepared on a plate or flat bottle and the virus suspension overlayed. Plaques are revealed by zones of destruction of the animal cells (Figure 6.7).

In some cases, the virus may not actually destroy the cells, but cause changes in morphology or growth rate which can be recognized. For instance, tumor viruses may not destroy cells but cause the cells to grow faster than uninfected cells, a phenomenon called *transformation*. As we have noted, the general arrangement of cells in a tissue culture is a monolayer. This is because growth generally ceases when the cells, as a result of growth, come in contact with each other (a phenomenon known as *contact inhibition*). Transformed cells have altered growth requirements and continue to grow, piling up to form a small *focus of growth* (called a *focus of infection* when the transformation has been brought about by virus infection) (Figure 6.8). By counting foci of infection, a quantitative measure of virus may be obtained. We discuss cancer viruses in more detail later in this chapter.

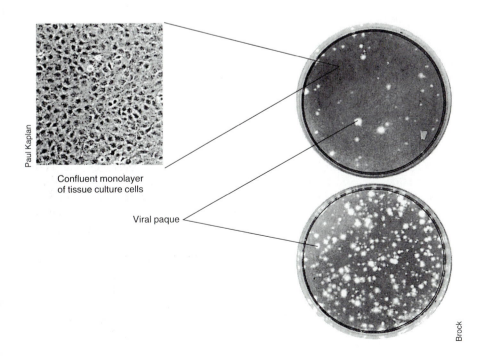

Confluent monolayer of tissue culture cells

Viral paque

Paul Kaplan

Brock

FIGURE 6.7
Cell cultures in monolayers within Petri plates. Note the presence of plaques where virus-induced cell lysis has occurred. Also shown is a photomicrograph of a cell culture.

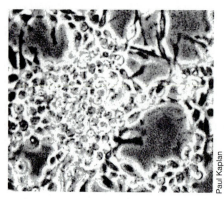

Paul Kaplan

FIGURE 6.8 Microscopic appearance of a cell culture in which some of the cells have been transformed. The normal cells are elongated because they spread out on the glass culture dish. The transformed cells have lost contact inhibition and have piled up to make a small clump.

Efficiency of plating

One important concept in quantitative virology involves the idea of *efficiency of plating*. Counts made by plaque assay are always lower than counts made with the electron microscope. The efficiency with which virions infect host cells is rarely 100 percent and may often be considerably less. This does not mean that the virions that have not caused infection are inactive, although this is sometimes the case. It may merely mean that under the conditions used, successful infection with these particles has not occurred. Although with bacterial viruses, efficiency of plating is often higher than 50 percent, with many animal viruses it may be very low, 0.1 or 1 percent. Why virus particles vary in infectivity is not well understood. It is possible that the conditions used for quantification are not optimal. Because it is technically difficult to count virus particles with the electron microscope, it is difficult to assess the actual efficiency of plating, but the concept is important in both research and medical practice. Because the efficiency of plating is rarely close to 100 percent, when the plaque method is used to quantify virus, it is accurate to express the concentration (called the *titer*) of the virus suspension not as the absolute number of virion units but as the number of *plaque-forming units*.

Animal infectivity methods

Some viruses do not cause recognizable effects in cell cultures but cause death in the whole animal. In such cases, quantification can only be done by some sort of titration in infected animals. The general procedure is to carry out a serial dilution of the unknown sample, generally at ten-fold dilutions, and samples of each dilution are injected into numbers of sensitive animals. After a suitable incubation period, the fraction of dead and live animals at each dilution is tabulated and an *end point dilution* is calculated. This is the dilution at which, for example, *half* of the injected animals die. Although such serial dilution methods are much more

cumbersome and much less accurate than cell culture methods, they may be essential for the study of certain types of viruses.

> Although it requires only a single virion to initiate an infectious cycle, not all virus particles are equally infectious. One of the most accurate ways of measuring virus infectivity is by the plaque assay. Plaques are clear zones that develop on layers or lawns of host cells, each plaque due to infection by a single virus particle. The virus plaque is analogous to the bacterial colony.

6.4   General Features of Virus Reproduction

The basic problem of virus replication can be simply put: the virus must induce a living host cell to synthesize all of the essential components needed to make more virus particles. These components must then be assembled into the proper structure, and the new virions must escape from the cell and infect other cells. The various phases of this replication process in a bacteriophage can be categorized in seven steps (Figure 6.9):

1. **Attachment** (adsorption) of the virion to a susceptible host cell.
2. **Penetration** (injection) of the virion or its nucleic acid into the cell.
3. **Early steps in replication** of the virus nucleic acid, in which the host cell biosynthetic machinery is altered as a prelude to virus nucleic acid synthesis. Virus-specific enzymes may be made.
4. **Replication** of the virus nucleic acid.
5. **Synthesis of protein subunits** of the virus coat.
6. **Assembly** of protein subunits (and membrane components in enveloped viruses) and **packaging** of nucleic acid into new virus particles.
7. **Release** of mature virions from the cell (lysis).

These stages in virus replication are recognized when virus particles infect cells in culture and are illustrated in Figure 6.10, which exhibits what is called a **one-step growth curve.** In the first few minutes after infection, a period called the *eclipse* occurs, in which the virus nucleic acid has become separated from its protein coat so that the virion no longer exists as an infectious entity. Although virus nucleic acid may be infectious, the infectivity of virus nucleic acid is many times lower than that of whole virions because the machinery for bringing the virus genome into the cell is lacking. Also, outside the virion the nucleic acid is no longer protected from deleterious activities of the environment as it was when it was inside the protein coat.

The eclipse is the period during which the stages of virus multiplication occur. This is called the *latent period*, because no infectious virions are evident. Fi-

Large-Scale Purification of Viruses

Modern virus research often requires the preparation of large amounts of pure virus particles for chemical studies or for preparation of antibodies. We describe here briefly the steps involved in the purification of poliovirus, one of the most widely studied viruses affecting humans.

Before laboratory work is begun with a human virus, it is essential to immunize all laboratory workers against the virus. Vaccines for polio are widely available and most workers will probably already be immune, but reimmunization is an essential precaution.

The steps in virus purification can be briefly listed: (1) Growth of the virus in large-scale equipment; (2) removal of the virus-enriched fluid; (3) concentration of the virus material by precipitation; (4) final purification of the virus by density-gradient centrifugation procedures. We now present some details of each of these steps. The infectivity of the virus needs to be checked during the steps of purification.

1. Growth of the virus in large-scale equipment In order to obtain sufficient virus for chemical studies, large volumes of virus must be prepared. Distinct procedures are available for the cultivation of each kind of virus. Poliovirus is cultivated in a human or primate cell culture system, either on monolayers of cells growing in large bottles or in cell suspension in bottles that are gently agitated. Under favorable virus growth conditions, between 10 and 1000 infectious units of virus are produced per host cell, leading to titers of 10^7 to 10^9 infectious units per milliliter.

2. Removal of virus-enriched culture fluid Virus growth and release is generally accompanied by cell degeneration, so that at the time virus replication is complete, most of the cells in the culture have become converted into cell debris. To obtain the complete release of all virus particles, the cells are further disrupted by several freeze/thaw cycles. The virus-rich fluid is then pipetted from the culture bottles and transferred to centrifuge tubes. A low-speed centrifugation removes large particles of cell debris.

3. Concentration of the virus particles by precipitation Because viruses are proteinaceous, they can be precipitated by methods that will precipitate proteins. Poliovirus can be precipitated with the use of high concentrations of ammonium chloride (0.4 g per ml of culture fluid). The procedure is carried out in the cold (4°C), the ammonium chloride being added to the culture fluid slowly with stirring. The solution will become cloudy with precipitated virus proteins. The precipitate is sedimented by low-speed centrifugation (2000 × gravity for 1 to 2 hours). The precipitate containing the virus particles is then resuspended in a small volume of phosphate-buffered saline. This procedure concentrates the virus about 10-fold and over 99 percent of the virus particles are recovered in the precipitate.

4. Final purification by density-gradient cetrifugation Poliovirus can be purified either by centrifugation through a density gradient of sucrose or by cesium chloride density gradient centrifugation (described in the box in Chapter 5). With the latter procedure, the virus particles are banded in the centrifuge tube in a manner similar to that illustrated for DNA in the box in Chapter 5. The centrifugation must be carried out at very high speed, 120,000× gravity, and for many hours, preferably overnight. The liquid in the centrifuge tube is removed through the bottom of the tube as small drops, each of which is assayed for virus titer. Under appropriate centrifugation conditions, all of the virus particles will be contained in one or two fractions, and should be extremely pure.

5. Crystallization of polio virus If large amounts of highly purified virus are allowed to concentrate under appropriate conditions, crystals of virus can be obtained. Such crystals provide favorable material for chemical analysis, and can be subjected to structural analysis by X ray diffraction. The computer-generated model of poliovirus (illustrated in this chapter) was obtained using highly purified poliovirus crystals.

---

nally, maturation begins as the newly synthesized nucleic acid molecules become packaged inside protein coats. During the *maturation* phase, the titer of active virions inside the cell rises dramatically. At the end of maturation, *release* of mature virions occurs, either as a result of cell lysis or because of some budding or excretion process. The number of virions released, called the *burst size*, will vary with the particular virus and the particular host cell, and can range from a few to a few thousand. The timing of this overall virus replication cycle varies from 20–30 minutes in many bacterial viruses to 8–40 hours in most animal viruses. We now consider each of the steps of the virus multiplication cycle in more detail.

> The virus life cycle can be divided into seven stages: attachment (adsorption), penetration (injection), early protein synthesis, nucleic acid replication, synthesis of virus protein subunits, assembly of mature virions, and virus release. The various stages of the cycle can be measured quantitatively by means of a one-step growth curve.

6.5   Steps in Virus Multiplication

Before discussing the stages of virus multiplication, we must return briefly to a consideration of the virus genome. Although many viruses have a double-

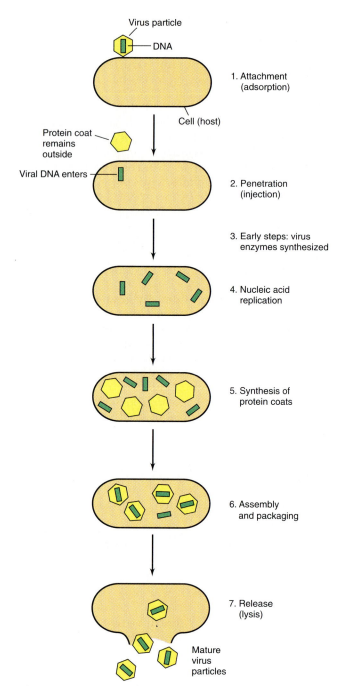

FIGURE 6.9 The replication cycle of a bacterial virus. The general stages of virus replication are indicated.

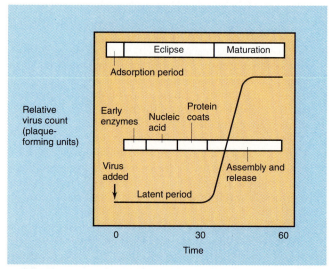

FIGURE 6.10 The one-step growth curve of virus replication. This graph displays the results of a single round of viral multiplication in a population of cells. Following adsorption, the infectivity of the virus particles disappears, a phenomenon called *eclipse*. This is due to the uncoating of the virus particles. During the *latent period*, replication of viral nucleic acid and protein occurs. The *maturation period* follows, when virus nucleic acid and protein are assembled into mature virus particles. At this time, if the cells are broken up, active virus can be detected. Finally, *release* occurs, either with or without cell lysis. The timing of the one-step growth cycle varies with the virus and host. With many bacterial viruses, the whole cycle may be complete in 30–60 minutes, whereas with animal viruses 12–24 hours are usually required for a complete cycle.

stranded DNA genome, a great many do not, and these other types of genomes include not only single-stranded DNA but both single-stranded and double-stranded RNA. Furthermore, we have mentioned that some viruses have one type of nucleic acid in the virion but use another as a replicative intermediate. All these "unusual" genomes present problems in understanding virus multiplication because they involve information transfers, such as RNA to RNA or RNA to DNA, that host enzymes do not perform.

As we have noted, the outcome of a virus infection is the synthesis of viral nucleic acid and viral protein coats. In effect, the virus takes over the biosynthetic machinery of the host and uses it for its own synthesis. A few enzymes needed for virus replication may be present in the virion and may be introduced into the cell during the infection process, but the host supplies everything else: energy-generating system, ribosomes, amino-acid activating enzymes, transfer RNA (with a few exceptions), and all soluble factors. The virus genome codes for all new proteins. Such proteins would include the coat protein subunits (of which there are generally more than one kind) plus any new virus-specific enzymes.

Attachment

There is a high specificity in the interaction between virus and host. The most common basis for host specificity involves the attachment process. The virus particle itself has one or more proteins on the outside which interact with specific cell surface components called *receptors*. The receptors on the cell surface are normal surface components of the host, such as proteins, polysaccharides, or lipoprotein-polysaccharide complexes, to which the virion attaches. In the absence of the receptor site, the virus cannot adsorb, and hence cannot infect. If the receptor site is altered, the host may become resistant to virus infection. However, mutants of the virus can also arise which are able to adsorb to resistant hosts.

In general, virus receptors carry out normal functions in the cell. For example, in Bacteria some phage receptors are pili or flagella, others are cell-envelope components, and others are transport binding proteins. The receptor for influenza virus is a glycoprotein found on red blood cells and on cells of the mucous membrane of susceptible animals, whereas the receptor site of poliovirus is a cell surface lipoprotein. However, many animal and plant viruses do not have specific attachment sites at all and the virus enters passively as a result of phagocytosis or some other endocytotic process.

Penetration

The means by which the virus penetrates into the cell depends on the nature of the host cell, especially on its surface structures. Cells with cell walls, such as bacteria, are infected in a different manner from animal cells, which lack a cell wall. The most complicated penetration mechanisms have been found in viruses that infect bacteria. The bacteriophage T4, which infects *Escherichia coli*, can be used as an example.

The structure of the bacterial virus T4 was shown in Figure 6.5c. The virion has a **head**, within which the viral DNA is folded, and a long, fairly complex **tail**, at the end of which is a series of tail fibers. During the attachment process, the virus particles first attach to cells by means of the tail fibers (Figure 6.11). These tail fibers then contract, and the core of the tail makes contact with the cell envelope of the bacterium. The action of a lysozyme-like enzyme results in the formation of a small hole. The tail sheath contracts and the DNA of the virus passes into the cell through a hole in the tip of the tail, the majority of the coat protein remaining outside. The DNA of T4 has a total length of about 50 μm, whereas the dimensions of the head of the T4 virion are 0.095 μm by 0.065 μm. This means that the DNA must be highly folded and packed very tightly within the head.

With animal cells, the *whole virion* penetrates the cell, being carried inside by endocytosis (phagocytosis or pinocytosis), an active cellular process. We describe some of these processes in detail later in this chapter.

> The attachment of a virion to a host cell is a highly specific process, involving the interaction of receptors on the host with which proteins on the surface of the virus particle interact. Only after attachment has occurred can the virus or its nucleic acid penetrate the host cell.

Virus restriction and modification by the host

We have already seen that one form of host resistance to virus arises when there is no receptor site on the cell surface to which the virus can attach. Another and more specific kind of host resistance occurs in prokaryotes and involves destruction of the viral nucleic acid after it has been injected. This destruction is brought about by host enzymes that cleave the viral DNA at one or several places, thus preventing its replication. This phenomenon is called *restriction*, and is part of a general host mechanism to prevent the invasion of foreign nucleic acid. We discussed *restriction enzymes* and their action in some detail in Section 5.3 and noted that their cellular role was in defense against foreign DNA. Restriction enzymes are highly specific, attacking only certain sequences (generally four or six base pairs). The host protects its own DNA from the action of restriction enzymes by *modifying* its DNA at the sites where the restriction enzymes will act. Modification of host DNA is brought about by methylation of purine or pyrimidine bases.

FIGURE 6.11
Attachment of T4 bacteriophage particle to the cell wall of *E. coli* and injection of DNA: (a) Unattached particle. (b) Attachment to the wall by the long tail fibers. (c) Contact of cell wall by the tail pins. (d) Contraction of the tail sheath and injection of the DNA.

Some viruses can overcome host restriction mechanisms by modifications of their nucleic acids so that they are no longer subject to enzymatic attack. Two kinds of chemical modifications of viral DNA have been recognized, glucosylation and methylation. For instance the T-even bacteriophages (T2, T4, and T6) have their DNA glucosylated to varying degrees, and the glucosylation prevents or greatly reduces endonuclease attack. Many other viral nucleic acids have been found to be modified by methylation but glucosylation has been found only in the T-even bacteriophages. It should be emphasized that modification of viral nucleic acid occurs after replication has occurred and the modified bases are not copied directly. Other viruses, such as the bacteriophages T3 and T7, avoid restriction by encoding proteins that inhibit the host restriction systems. Some hosts have multiple restriction and methylation systems that help in preventing infection by viruses that can circumvent only one of them.

The ability to modify nucleic acid is not found in all host strains that support the growth of a given virus. Thus, when bacteriophage lambda is grown on *E. coli* strain C it is not modified (*E. coli* strain C lacks modification and restriction enzymes), and therefore nucleic acid of virus grown on strain C is destroyed when it enters *E. coli* strain K-12, which does have the restriction enzyme. However, strain K-12 also has the modification enzyme, and, if lambda is grown on K-12, its nucleic acid is modified and it will infect both strains K-12 and C equally well.

However, not all restriction systems recognize unmodified DNA. Host restriction systems are also known that restrict only *modified* DNA! Clearly the

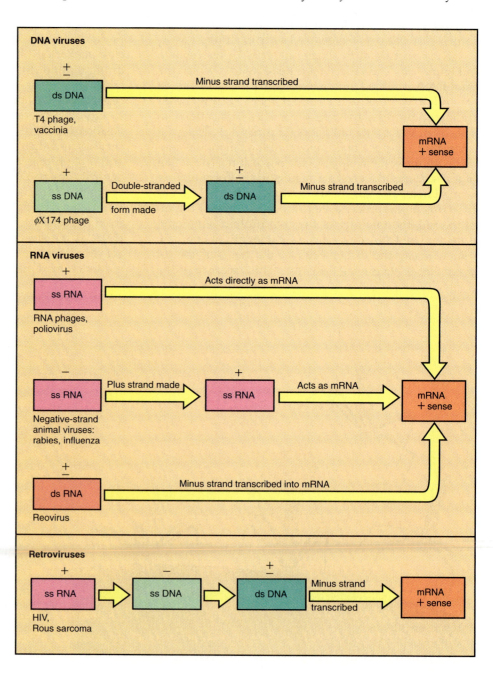

FIGURE 6.12

Formation of mRNA after infection of cells by viruses of different types. The chemical sense of the mRNA is considered as plus (+). The senses of the various virus nucleic acids are indicated as + if the same as mRNA, as − if opposite, or as ± if double-stranded. Examples are indicated next to the virus nucleic acid.

host containing this enzyme does *not* contain the modification enzyme. However, this host is protected from infection by a virus that was modified during reproduction in its previous host strain.

Hosts also contain other DNA methylases, and these can methylate either its DNA or that of infecting viruses. Some of these methylases may be involved in DNA repair or in gene regulation, but others may offer protection to *host* DNA. This could be during DNA transfer to other cells during genetic recombination (see Chapter 7) or because some viruses themselves encode restriction systems.

As we discussed in Chapter 5, a knowledge of modification and restriction systems is of considerable practical utility in studying DNA chemistry. We discuss the use of restriction enzymes in genetic engineering in Chapter 8. Restriction–modification systems are confined almost exclusively to prokaryotes. At least one-fourth of all bacteria have restriction–modification systems. As we will see in Chapters 11 and 12, higher organisms have other methods of resisting infection.

> The virus nucleic acid is foreign to the host and the restriction–modification system of the host, which recognizes and destroys foreign DNA, is one means of defense against virus infection.

Virus messenger RNA

In order for the new virus-specific proteins to be made from the virus genome, it is necessary for new virus-specific RNA molecules to be made. Exactly how the virus brings about new mRNA synthesis depends upon the type of virus, upon whether its genetic material is RNA or DNA, and whether it is single-stranded or double-stranded. The essential features of mRNA synthesis were discussed in Chapter 5, and it was shown that mRNA represents a complementary copy made by RNA polymerase of one of the two strands of the DNA double helix. Which copy is read into mRNA depends upon the location of the appropriate promoter, since the promoter points the direction that the RNA polymerase will follow. In cells uninfected with virus, all mRNA is made on the DNA template, but when viruses are present the situation is obviously different.

A wide variety of modes of viral mRNA synthesis are outlined in Figure 6.12. Because the nucleic acid sequence of the mRNA can be translated directly into protein it is by convention considered to be of the *plus* (+) configuration (Figure 6.13). The sequence of the viral genome nucleic acid is then indicated by a *plus* if it is the same as the mRNA and a *minus* if it is of opposite sense. As seen in Figure 6.12, if the virus has double-stranded DNA (ds DNA), then mRNA synthesis can proceed directly as in uninfected cells. However, if the virus has a single-stranded DNA (ss DNA), then it is first converted to ds DNA and the latter serves as

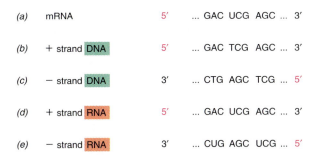

FIGURE 6.13 Comparison of sequences of nucleic acids with different configurations. The strands of nucleic acid in viral genomes are called *plus* (+) if they are in the same configuration as mRNA and *minus* (−) if they are in the opposite configuration of mRNA. (a) The direction and sequence of an mRNA. (b) The sequence of a + strand of a viral DNA genome that could encode that mRNA (using a − strand DNA intermediate). (c) A − strand of a viral DNA genome that could encode the mRNA. (d) and (e) The + and − strands of RNA genomes that could encode the mRNA. Note that a single-stranded + strand RNA genome has the same sequence and orientation as the mRNA. The 5′-end of each nucleic acid strand is highlighted in red.

the template for mRNA synthesis by the RNA polymerase of the cell.

For RNA viruses, a virus-specific RNA-dependent RNA polymerase is needed, since the cell RNA polymerase is DNA dependent (requires a DNA template) and will generally not copy RNA. Several different types of viruses contain RNA as their genome, and each requires a different strategy for producing mRNA. The simplest case is the positive-strand RNA viruses in which the single incoming viral RNA strand is the *plus* strand and hence serves directly as mRNA. In addition to the other required proteins, this mRNA encodes the virus-specific RNA polymerase. This polymerase first makes complementary *minus* strands and then uses these as templates to make more plus strands. For the negative-strand RNA viruses (whose virion contains only the minus strand) or the double-stranded RNA viruses, the situation is more complicated. In neither case can the incoming RNA serve as mRNA, and therefore mRNA must be synthesized first. However, as mentioned earlier, cells do not typically have an RNA polymerase capable of this. To circumvent this problem, these viruses contain some of this enzyme in their virions, and it is injected into the cell along with the genomic RNA. Therefore, in these cases, the complementary plus strand is synthesized by this RNA-dependent RNA polymerase and used as message.

Retroviruses (causal agents of certain kinds of cancers and AIDS) are RNA viruses that replicate through a DNA intermediate. The process of copying the information found in RNA into DNA is called **reverse transcription**, and thus these viruses require an enzyme called **reverse transcriptase**. (Telomerase is a type of reverse transcriptase: see Section 5.5.) In spite of the fact that the incoming RNA of retroviruses is the plus strand, it is not used as message, and therefore these viruses must carry reverse transcriptase in their virions. After infection, the virion ss RNA is copied to a double-stranded DNA (through an ss DNA interme-diate) and the ds DNA then serves as the template for mRNA synthesis (thus: ss RNA → ss DNA → ds DNA). Retrovirus replication is of unusual complexity and is discussed in Section 6.22.

Viral proteins

Once viral mRNA is made, viral proteins (for exam-ple, enzymes and capsomeres) can be synthesized. The proteins synthesized as a result of virus infection can be grouped into three broad categories:

1. proteins (usually enzymes) synthesized soon after infection, called the **early proteins**, which are nec-essary for the replication of virus nucleic acid;

2. proteins synthesized later, called the **late proteins**, which include the proteins of the virus coat;

3. *lytic proteins* that open the host cell and release free virus particles.

Generally, both the time of appearance and the amount of these three groups of virus proteins are reg-ulated. The early proteins are enzymes which, because they act catalytically, are synthesized in smaller amounts and the late proteins, often structural, are made in much larger amounts.

Virus infection obviously upsets the regulatory mechanisms of the host, since there is a marked over-production of viral nucleic acid and protein in the in-fected cell. In some cases, virus infection causes a com-plete shutdown of host macromolecular synthesis, while in other cases host synthesis proceeds concur-rently with virus synthesis. In either case, the regula-tion of virus synthesis is under the control of the virus rather than the host. There are several elements of this control which are similar to the host regulatory mech-anisms discussed in Chapter 5, but there are also some uniquely viral regulatory mechanisms. We discuss various regulatory mechanisms when we consider the individual viruses later in this chapter.

> Before replication of viral nucleic acid can occur, new virus proteins are often needed. These are encoded by messenger RNA molecules made from the virus genome. In the case of some RNA viruses, the viral RNA itself acts as mRNA. In other cases, the virus genome serves as a template for the formation of viral mRNA and certain essential enzymes are contained in the virion. Retroviruses show an unusually complex multiplication cycle since the RNA of the virus particle is copied into DNA for replication within the cell, and then back to RNA when mature virus particles are formed.

6.6   Overview of Bacterial Viruses

Various kinds of bacterial viruses are illustrated in Figure 6.14. Most of the bacterial viruses which have been studied in any detail infect Bacteria of the enteric

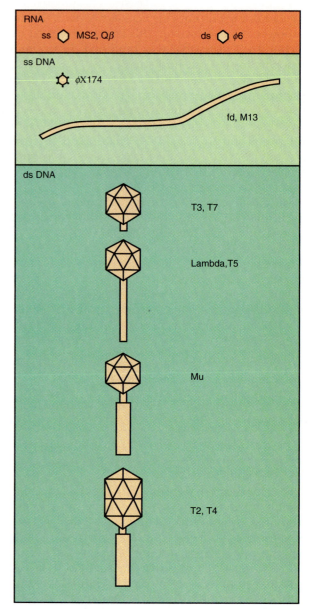

FIGURE 6.14   Schematic representations of the main types of bac-terial viruses. Those discussed in detail are fd, M13, φX174, MS2, T4, lambda, T7, and Mu. Sizes are to approximate scale.

group, such as *Escherichia coli* and *Salmonella ty-phimurium*. However, viruses are known that infect a variety of prokaryotes, both Bacteria and Archaea. A few bacterial viruses have lipid envelopes but most do not. However, many bacterial viruses are structurally complex, with head and complex tail structures. As we illustrated in Figure 6.11, the tail is involved in the injection of the nucleic acid into the cell.

We now discuss some of the bacterial viruses for which molecular details of the multiplication process are known. Although these bacterial viruses were first studied as *model systems* for understanding general features of virus multiplication, many of them now serve as convenient tools for *genetic engineering* (dis-cussed in Chapter 8). Thus, the information on bacter-ial viruses is not only valuable as background for the discussion of animal viruses, but is essential for the material presented in the next two chapters on microbial genetics and genetic engineering.

It should be clear that a great diversity of viruses exists. It should therefore not be surprising that there is also a great diversity in the manner by which virus multiplication occurs. In the present chapter, we are only able to present some of the major types of virus replication patterns and must skip some of the interesting exceptional cases.

6.7   RNA Bacteriophages

A number of bacterial viruses have RNA genomes. The best-known bacterial RNA viruses have single-stranded RNA. Interestingly, the bacterial RNA viruses known in the enteric bacteria group infect only bacterial cells which behave as gene donors (males) in genetic recombination (the interesting concept of male and female bacteria is discussed in Chapter 7). This restriction to male bacterial cells arises because these viruses infect bacteria by attaching to *male-specific pili* (Figure 6.15). Since such pili are absent on female cells, these RNA viruses are unable to attach to the females, and hence do not initiate infection in females.

The bacterial RNA viruses are all quite small, about 26 nm in size, and they are all icosahedral, with 180 copies of coat protein per virus particle. The complete nucleotide sequences of several RNA phage genomes are known. In the RNA phage MS2, which infects *Escherichia coli*, the viral RNA is 3569 nucleotides long. The RNA strand in the virion has the plus (+) sense, acting directly as mRNA upon entry into the cell (see Figure 6.12).

The genetic map of MS2 is shown in Figure 6.16*a* and the flow of events of MS2 multiplication is shown in Figure 6.16*b*. The infecting RNA goes to the host ribosome, where it is translated into four proteins. These four proteins are the **maturation protein** (A-protein, present in the mature virus particle as a single copy), **coat protein, lysis protein** (involved in the lysis

process which results in release of mature virus particles), and **RNA replicase**, the enzyme which brings about the replication of the viral RNA. Interestingly, the RNA replicase is a composite protein, composed partly of a virus-encoded polypeptide and partly of host polypeptides. The host proteins involved in the formation of active viral replicase are *ribosomal protein S1* (one of the subunits of the 30S ribosome), and specific elongation factors, involved in the translation process. Thus, the virus appears to employ host proteins that normally have entirely distinct functions and use them to make an active viral replicase.

As noted, the viral RNA is of the *plus* (+) sense and can thus be translated immediately. After RNA replicase is synthesized, it in turn can synthesize RNA of *minus* (−) sense using the infecting RNA as template (see Figure 6.12). After *minus* RNA has been synthesized, more *plus* RNA is made using this *minus* RNA as template. The newly made *plus* RNA strands now serve as messengers for continued virus protein synthesis. The gene for the maturation protein is at the 5′ end of the RNA. Translation of the gene coding for the maturation protein (needed in only one copy per virus particle), occurs only from the nascent form of the *plus*-strand RNA as the replication process occurs. In this way, the amount of maturation protein needed is limited. As the virus RNA is made, it folds into a complex form with extensive secondary structure. Of the four AUG start sites, the most accessible to the translation process is that for the coat protein. As coat protein molecules increase in number in the cell, they combine with the RNA around the AUG start site for the replicase protein, effectively turning off synthesis of replicase. Thus, the major virus protein synthesized is coat protein, which is needed in 180 copies per RNA molecule (MS2 is a virus of icosahedral symmetry) (see Section 6.1).

Another interesting feature of bacteriophage MS2 is that the fourth virus protein, the *lysis* protein, is coded by a gene which *overlaps* with both the coat protein gene and the replicase gene (see genetic map in

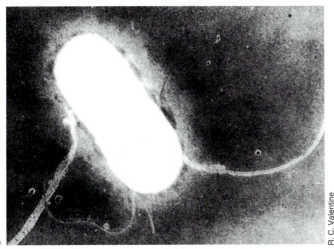

(a)

R. C. Valentine

(b)

FIGURE 6.15 Electron micrographs of a male bacterial cell of *Escherichia coli* infected with a small RNA phage. (a) Note that the phage particles have attached to the pilus (male-specific pilus). (b) Close up of a pilus with virus particles.

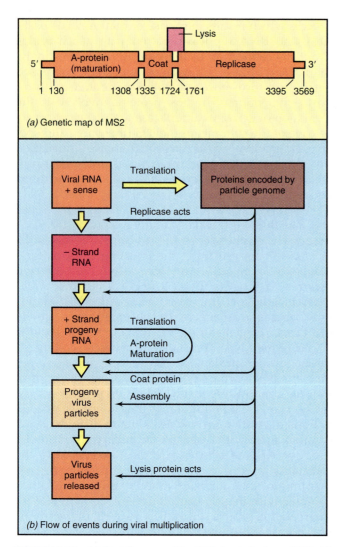

FIGURE 6.16   (a) Genetic map of the RNA bacteriophage MS2. (b) Flow of events during multiplication. The numbers in (a) refer to the nucleotide positions on the RNA.

A variety of RNA viruses which infect bacteria are known. These viruses are all very small and only infect bacteria that are gene donors (males). The RNA of these small bacterial viruses acts directly as mRNA and encodes only a few virus proteins.

6.8   Single-Stranded Icosahedral DNA Bacteriophages

A number of small bacterial viruses have genomes consisting of single-stranded DNA in circular configuration. These viruses are very small, about 25 nm in diameter, and the principle building block of the protein coat is a single protein present in 60 copies (the minimum number of protein subunits possible in an icosahedral virus), to which are attached at the vertices of the icosahedron several other proteins which make up spike-like structures (see Figure 6.14). In contrast to the RNA viruses, much of the enzymatic machinery for the replication of DNA already exists in the cell. These small DNA viruses possess only a limited amount of genetic information in their genomes, and the host cell DNA replication machinery is used in the replication of virus DNA.

Phage φX174 and the phenomenon of overlapping genes

The most extensively studied virus of this group is the phage designated φX174, which infects *Escherichia coli*. φX174 is of special interest because it was the first genetic element shown to have **overlapping genes**. As we have noted in Chapter 5, the genomes of cells are organized in contiguous fashion, with the gene coding for each protein separate from that for all other genes. In very small viruses such as φX174 there is insufficient DNA to code for all virus-specific proteins. φX174 has solved this problem by the use of overlapping genes. Thus, parts of certain virus nucleotide sequences are read more than once in different reading frames (Figure 6.17). It should be noted that although the use of overlapping genes makes possible more efficient use of genetic information, it seriously complicates the evolution process, since a mutation in a region of gene overlap may affect two genes simultaneously.

As seen in the genetic map of φX174, the sequences of genes D and E overlap each other, gene E being contained completely *within* gene D. In addition, the termination codon of gene D overlaps the initiation codon of gene J by one nucleotide. The reading frame of gene E is therefore in a different phase (starting point) from that of gene D. Obviously, any mutation in gene E will also lead to an alteration in the sequence of gene D, but whether a given mutation affects one or both proteins will depend on the exact nature of the alteration (because the genetic code is degenerate; see Section 5.9). Several other instances of gene overlap occur in the φX174 genome (Figure 6.17). Additionally, a small gene A-protein, called A*-protein, is formed by *reinitiation of translation* (not tran-

Figure 6.16a). The start of this lysis gene is not directly accessible to ribosomes because of the secondary structure found in the RNA. When the ribosome terminates synthesis of the coat protein gene, the secondary structure in this region is disrupted, and sometimes this disruption allows a ribosome to begin reading the lysis gene. By restricting the efficiency of translation in this way, premature lysis of the cell is probably avoided. Only after sufficient copies of coat protein are available for the assembly of mature virus particles does lysis commence. (In another RNA phage, Qβ, the maturation protein itself also functions as a lysis protein, and a separate lysis gene as such is not present.)

Ultimately, phage assembly occurs and release of virions from the cell occurs as a result of cell lysis. The features of replication of these simple RNA viruses are themselves fairly simple. The viral RNA itself functions as an mRNA and regulation occurs primarily by way of controlling access of ribosomes to the appropriate start sites on the viral RNA.

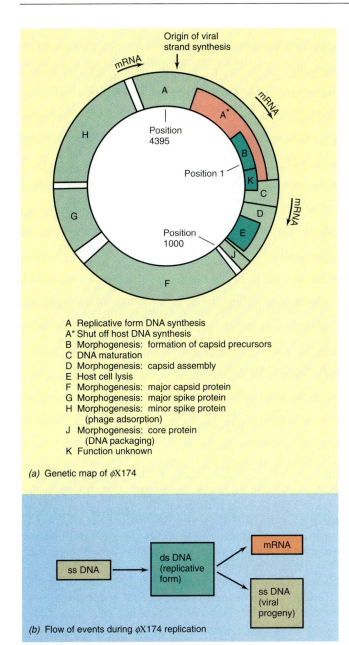

A Replicative form DNA synthesis
A* Shut off host DNA synthesis
B Morphogenesis: formation of capsid precursors
C DNA maturation
D Morphogenesis: capsid assembly
E Host cell lysis
F Morphogenesis: major capsid protein
G Morphogenesis: major spike protein
H Morphogenesis: minor spike protein
     (phage adsorption)
J Morphogenesis: core protein
     (DNA packaging)
K Function unknown

(a) Genetic map of φX174

(b) Flow of events during φX174 replication

FIGURE 6.17 Bacteriophage φX174, a single-stranded DNA phage. (a) Genetic map. Note the regions of gene overlap (A/B, K/B, K/C, K/A, A/C, and D/E). Intergenic regions are not colored. Protein A* is formed using only a part of the coding sequence of gene A by reinitiation of translation (see text). (b) Flow of events in φX174 multiplication.

scription) within the mRNA of gene A, with A*-protein being read and terminated from the same mRNA reading frame as A-protein but starting at a different codon.

DNA replication by the rolling circle mechanism

The DNA of φX174 consists of a circular single-stranded molecule of 5386 nucleotide residues. The DNA of φX174 was the first DNA to be completely sequenced, a remarkable achievement when it was accomplished by Frederick Sanger and colleagues in 1977. Now, DNA sequencing is a routine procedure (see Nucleic Acids box in Chapter 5). The replication

process of such a circular single-stranded DNA molecule is of considerable general interest, since cellular DNA replicates always in the double-stranded configuration (see Section 5.4). The DNA strand in single-stranded DNA phages is of the plus (+) sense. Upon infection, this viral plus strand becomes separated from the protein coat; entrance into the cell is accompanied by the conversion of this single-stranded DNA into a double-stranded form called the **replicative form (RF)** DNA (Figure 6.17b). Cell-coded proteins involved in the conversion of viral DNA into RF consist of the enzymes *primase, DNA polymerase, ligase,* and *gyrase.* No virus-encoded proteins are involved in the conversion of single-stranded DNA to RF. The RF is a closed, double-stranded, circular DNA which has extensive supercoiling.

As we discussed in Section 5.4, DNA replication differs between the leading strand and the lagging strand of the DNA double helix. In cells, replication of the lagging strand involves the formation of short *RNA primers* by action of an enzyme called *primase.* Such RNA primers are made at intervals on the lagging strand and are then removed and replaced with DNA by DNA polymerase (see Figure 5.23). In φX174, however, replication begins with a single-stranded closed circle, a rather atypical situation. To replicate this DNA, primase brings about the synthesis of a short RNA primer, beginning at one or more specific initiation sites on the DNA. DNA is then synthesized by DNA polymerase III, and the primer is removed and replaced by DNA polymerase I, exactly as in the case of a lagging strand. This results in the formation of the complete double-stranded RF.

Once the RF is formed, DNA replication occurs by conventional semiconservative replication, involving theta-form intermediates (see Figure 5.21) resulting in the formation of new RF molecules. However, the formation of single-stranded viral genomes involves a different type of replication mechanism called **rolling circle replication** (Figure 6.18). The rolling circle arises because one strand is nicked and the 3'-end of this nick is used to prime synthesis of a new strand. Continued rotation of the circle leads to the synthesis of a linear single-stranded structure. Note that synthesis is asymmetric because only one of the strands is serving as template. In φX174, synthesis begins when the protein encoded by gene A, called *gene A-protein,* cleaves the plus strand of the RF. When the growing viral strand reaches unit length (5386 residues for φX174), gene A-protein cleaves and then ligates the two ends of the newly synthesized single strand to give a circular single-stranded DNA.

Many other viruses and some plasmids also use rolling circle replication. We will see that this type of replication can also be used to synthesize double-stranded DNA.

Transcription and translation for φX174

Viral mRNA synthesis is directed by the RF. Synthesis of mRNA begins at several major promoters on the

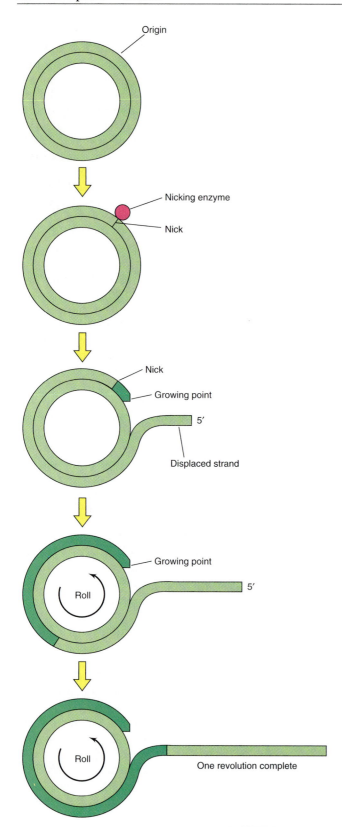

Origin

Nicking enzyme

Nick

Nick

Growing point

5′

Displaced strand

Growing point

5′

Roll

Roll

One revolution complete

FIGURE 6.18 Rolling circle replication. Replication is initiated at the origin by nicking one strand of the DNA. In the case of φX174, this strand is the *plus* strand, and the nick is made by the gene A-protein. After one new progeny strand has been synthesized (one revolution of the circle), the gene A-protein will cleave the new strand and ligate its two ends together (not shown).

RF, and terminates at a number of sites (see map, Figure 6.17). The polycistronic mRNA molecules are then translated into the various phage proteins. As we have noted, A-protein and A*-protein are both made from the *same gene*, A*-protein arising as a result of translation of a secondary initiation site internal to the A mRNA. Further, as we have noted, several proteins are made from mRNA transcripts formed from different reading frames from the same DNA sequences (overlapping genes). One can truly be impressed by the efficiency with which such a small genome as that of φX174 can have multiple uses.

Ultimately, assembly of mature virus particles occurs. Release of virions from the cell occurs as a result of cell lysis, which involves the participation of gene E protein.

> The bacterial DNA virus φX174 is so small that it can only code for all its essential functions if its small single-stranded DNA genome is translated in different reading frames. This virus provided the first example of overlapping genes. The production of progeny viral DNA involves a rolling circle mechanism.

6.9 Single-Stranded Filamentous DNA Bacteriophages

Quite distinct from φX174 are the filamentous DNA phages, which have helical rather than icosahedral symmetry. The most studied member of this group is phage M13, which infects *Escherichia coli*, but related phages include f1 and fd. As with the small RNA viruses, these filamentous DNA phages only infect male cells, entering after attachment to the male-specific pilus. Interestingly, even though these phages are linear (filamentous) they possess *circular* single-stranded DNA. The DNA is not self-complementary, however, so that the two adjacent halves of the molecule which run up and down the virus particle form loops at the ends but exhibit very little if any base pairing. Phage M13 has found extensive use as a cloning vector and DNA sequencing vehicle in genetic engineering (see Section 8.4). The virion of M13 is only 6 nm in diameter but is 860 nm long. These filamentous DNA phages have the additional unique property of being released from the cell *without* killing the host cell. Thus, a cell infected with phage M13 or fd can continue to grow, all the while releasing virus particles. Virus infection causes a slowing of cell growth, but otherwise a cell is able to coexist with its virus. Plaques are thus seen only as areas of reduced cell growth in the bacterial lawn.

Many aspects of DNA replication in filamentous phages are similar to that of φX174. The unique property of release without cell killing occurs by a budding process in which the virus particle is always released from the cell with the end containing the A protein first (Figure 6.19). Interestingly, the orientation of the virus particle across the cytoplasmic membrane is the

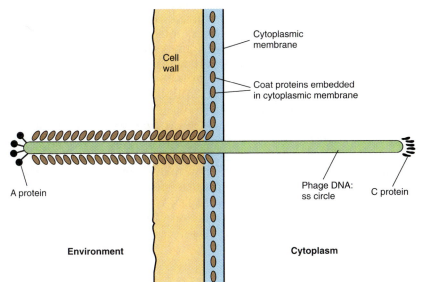

Cytoplasmic
membrane

Cell
wall

Coat proteins embedded
in cytoplasmic membrane

A protein

Phage DNA:
ss circle

C protein

Environment

Cytoplasm

FIGURE 6.19
Illustration of the manner by which the virion of a filamentous single-stranded phage (such as M13 or fd)
leaves an infected cell without lysis. The A protein
passes first through the membrane at a site on the
membrane where coat protein molecules have first
become imbedded. The intracellular circular DNA is
coated with dimers of another phage protein which is
displaced by coat protein as the DNA passes through
the intact cytoplasmic membrane.

same for its entry and exit from the cell. There is no accumulation of intracellular virus particles; the assembly of mature virions occurs on the inner surface of the
cytoplasmic membrane and virus assembly is coupled
with the budding process.

Several features of these phages make them useful
as cloning and DNA sequencing vehicles. First, they
have single-stranded DNA, which means that sequencing can be carried out by the Sanger dideoxynucleotide method (see Nucleic Acids box, Chapter 5).
Second, as long as infected cells are kept in the growing state they can be maintained indefinitely with
cloned DNA, so that a continuous source of the cloned
DNA is available. Third, there is an intergenic space
which does not code for protein and can be replaced
by variable amounts of foreign DNA.

6.10   Double-Stranded DNA
Bacteriophages: T7

Many bacterial viruses have genomes containing double-stranded DNA. Such viruses were the first bacterial viruses discovered, and have been the most extensively studied. With such a range of double-stranded
DNA viruses, a wide variety of replication systems are
present. In this and the next sections, we discuss the
best studied and most representative of the group, T4
and T7. The simpler, T7, will be discussed first.

Bacteriophage T7 and its close relative T3 are relatively small DNA viruses that infect *Escherichia coli*.
(Some strains of *Shigella* and *Pasteurella* are also hosts
for phage T7.) The virus particle has an icosahedral
head and a very small tail (see Figure 6.14). The virus
particle is fairly complex, with five different proteins
in the head and three to six different proteins in the
tail. One tail protein, the tail fiber protein, is the means
by which the virus particle attaches to the bacterial cell
surface.

Genome and genetic map
The nucleic acid of the T7 genome is a linear double-
stranded molecule of 39,936 base pairs. The complete
genome has been sequenced, and the sequence information has permitted discernment of gene structure
and features of gene regulation. About 92 percent of
the DNA of T7 codes for proteins. At least 25 separate
genes have been characterized, but not all genes are
separately coded on the DNA. Gene overlap occurs for
several genes through translation in different reading
frames and through internal translational reinitiation
with one or more genes in the same reading frame.
Further genetic economy is achieved by internal frame
shifts within certain genes to yield longer proteins.

The genetic map of T7 is shown in Figure 6.20. The
order of the genes influences the regulation of virus
multiplication. When the virion attaches to the bacterial cell, the DNA is injected in a linear fashion, with
the genes at the "left end" of the genetic map entering
the cell first. Several genes at the left end of the DNA
are transcribed immediately by a cell RNA polymerase, using three closely spaced promoters, generating a set of overlapping polycistronic mRNA molecules. These mRNA molecules are then cleaved by a
specific cell RNase at five sites, thus generating
smaller mRNA molecules which code for one to four
proteins each. One of these proteins inhibits the host
restriction system. Note that this protein is synthesized before the entire T7 genome enters the cell. Another one of these proteins is a viral RNA polymerase.
Two other early mRNA molecules code for proteins
which stop the action of host RNA polymerase, thus
turning off the transcription of the early genes as well
as the transcription of host genes. Thus, a host RNA
polymerase is used just to copy the first few genes and
to make the mRNA that codes for the phage-specific
RNA polymerase; this phage-specific RNA polymerase is then involved in the major transcription
processes of the phage. This T7 RNA polymerase uses

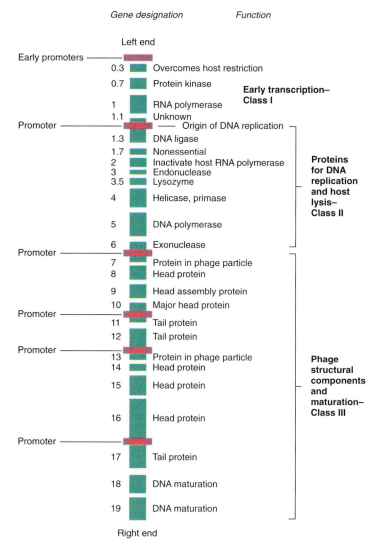

Gene designation   Function

Left end

Early promoters ——————

0.3	Overcomes host restriction
0.7	Protein kinase
1	RNA polymerase
1.1	Unknown

Early transcription–Class I

Promoter —————— Origin of DNA replication

1.3	DNA ligase
1.7	Nonessential
2	Inactivate host RNA polymerase
3	Endonuclease
3.5	Lysozyme
4	Helicase, primase
5	DNA polymerase
6	Exonuclease

Proteins for DNA replication and host lysis–Class II

Promoter ——————

7	Protein in phage particle
8	Head protein
9	Head assembly protein
10	Major head protein

Promoter ——————

| 11 | Tail protein |
| 12 | Tail protein |

Promoter ——————

13	Protein in phage particle
14	Head protein
15	Head protein
16	Head protein

Promoter ——————

17	Tail protein
18	DNA maturation
19	DNA maturation

Phage structural components and maturation–Class III

Right end

FIGURE 6.20 Genetic map of phage T7, showing gene numbers, approximate sizes, and functions of the gene products. Transcription can be divided into three segments (see text). The genes are designated as numbers.

proteins that are involved in T7 DNA synthesis, the formation of virus coat proteins, and assembly. Three classes of T7 proteins are formed: class I, made 4–8 minutes after infection, which use the cell RNA polymerase; class II, made 6–15 minutes after infection, which are made from T7 RNA polymerase and are involved in DNA metabolism; class III, made from 6 minutes to lysis, which are transcribed by T7 RNA polymerase and which code for phage assembly and coat protein (Figure 6.20). This sort of sequential pattern, commonly seen in many large double-stranded DNA phages, results in an efficient channeling of host resources, first toward DNA metabolism and replication, then on to formation of mature virions and release of virus by cell lysis.

DNA replication

DNA replication in T7 begins at an origin of replication (shown in Figure 6.20) at which DNA synthesis is initiated, and DNA synthesis proceeds *bidirectionally* from this origin (Figure 6.21). In both directions, an RNA primer (not shown in the figure) is involved, but the enzyme involved in the synthesis of this primer is different for primer synthesis in the leftward and rightward direction. In the *rightward direction*, the RNA primer is synthesized by T7 RNA polymerase, whereas in the *leftward direction*, a virus-specific enzyme, T7 primase (gene 4 protein) is used. Both primers are then elongated by T7 DNA polymerase. Replicating molecules of T7 DNA can be recognized under the electron microscope by their characteristic structures. Because the origin of replication is near the left end, Y-shaped molecules are frequently seen, and earlier in replication, bubble-shaped molecules appear (Figure 6.21).

A structural feature of the T7 DNA which is important in DNA replication is that there is a *direct terminal repeat* of 160 base pairs at the ends of the molecule. In order to replicate DNA near the 5'-terminus, RNA primer molecules have to be removed before replication is complete. There is thus an unreplicated portion of the T7 DNA at the 5'-terminus of each strand (see lower part of Figure 6.21a). As discussed in Section 5.5, genetic elements with linear DNA genomes have a variety of strategies to solve this problem in DNA replication. The strategy employed by T7 involves the repeated sequences at its ends. The opposite single 3'-strands on two separate DNA molecules, being complementary, can pair with these 5'-strands, forming a DNA molecule twice as long as the original T7 DNA (Figure 6.21b). The unreplicated portions of this end-to-end bimolecular structure are then completed through the action of DNA polymerase and DNA ligase, resulting in a *linear bimolecule*, called a *concatamer*. Continued replication can lead to concatamers of considerable length, but ultimately a cutting enzyme slices each concatamer at a specific site, resulting in the formation of virus-sized linear molecules with terminal repeats (Figure 6.21c).

only phage-specific promoters that are distributed along the left-center and center portions of the genome (see Figure 6.20). It is thus seen that regulation of T7 has both negative and positive control: *negative*, by means of the formation of proteins that stop host RNA polymerase and thus shut off transcription of the early T7 genes that are recognized by this enzyme, and *positive*, by means of the formation of the new RNA polymerase which recognizes the rest of the T7 promoters. We also note that T7 is an example of a virus which strongly affects host transcription and translation processes, by producing proteins which turn off transcription of host genes. The virus also has genes coding for enzymes which degrade host cell DNA, and nucleotides from such degraded DNA end up in virus progeny. Obviously, such a virus has profound pathological effects on its host cell.

As seen in the genetic map, the genes after gene 1.1, transcribed by the T7 RNA polymerase, code for

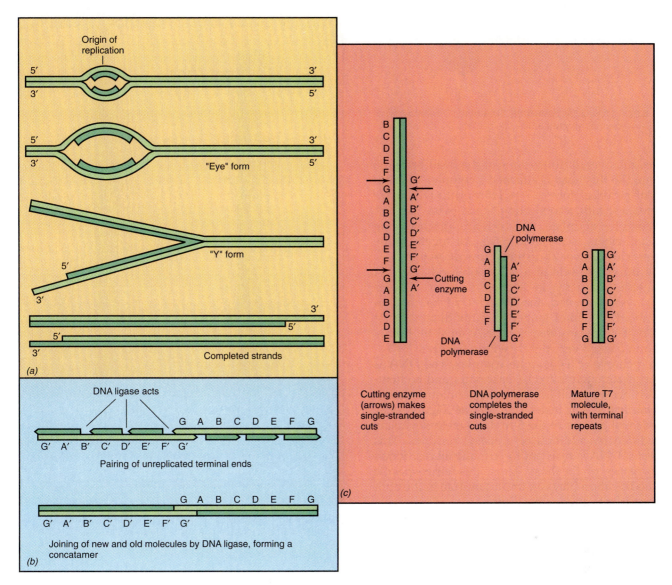

FIGURE 6.21   DNA replication in bacteriophage T7, a linear, double-stranded DNA. (a) Bidi-rectional replication of linear T7 DNA giving rise to intermediate "eye" and "Y" forms. (b) Forma-tion of concatamers by joining DNA molecules at the unreplicated terminal ends. The designation of the gene is arbitrary. (c) Production of mature viral DNA molecules from long T7 concatamers by action of cutting enzyme, an endonuclease. *Left*: the enzyme makes single-stranded cuts of specific sequences (arrows); *center*: DNA polymerase completes the single-stranded ends; *right*: the mature T7 molecule, with terminal repeats.

We thus see that T7 has a much more complex replication scheme than that seen for the other bacter-ial viruses discussed earlier.

T7 is one of the best studied double-stranded DNA viruses. The complete nucleic acid sequence of its genome is known, and most of the proteins encoded for by this nucleic acid have been identified. Replication of T7 follows a sequential pattern, with early genes being transcribed using a host RNA polymerase, followed by synthesis of a viral RNA polymerase which transcribes the later genes. The DNA is replicated bidirectionally from a single origin of replication.

6.11   Large Double-Stranded DNA Bacteriophages

One of the most extensively studied groups of DNA viruses is the group called the *T-even phages*, which in-clude the phages T2, T4, and T6. These phages are among the largest and most complicated in terms of both structure and manner of replication. In the pre-sent section, we will discuss primarily bacteriophage T4, the representative of this group for which the most information is available.

The virion of phage T4 is structurally complex (see Figure 6.5c). It consists of an icosahedral head which is elongated by the addition of one or two extra bands of

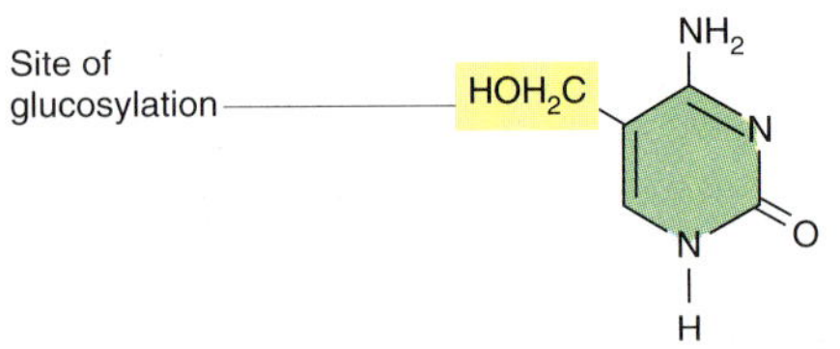

Site of glucosylation

FIGURE 6.22 The unique base in the DNA of the T-even bacteriophages, 5-hydroxymethylcytosine. The site of glucosylation is shown.

proteins, the overall dimensions of the head being 85 × 110 nm. To this head is attached a complex tail consisting of a helical tube (25 × 110 nm) to which are connected a sheath, a connecting "neck" with "collar," and a complex endplate, to which are attached long jointed tail fibers (see Figure 6.5c). All together, the virus particle has over 25 distinct types of proteins.

The infection cycle of a susceptible cell with T4 was described briefly in Section 6.4 (see Figures 6.9–6.11). As we noted, the DNA of T4 has a total length about 650 times longer than the dimension of the head. This means that the DNA must be highly folded and packed very tightly within the head.

The genome of T4 is quite complex. The DNA is large, approximately 1.7×10^5 base pairs, and is chemically distinct from cell DNA, having a unique base, *5-hydroxymethylcytosine* instead of *cytosine* (Figure 6.22). The hydroxyl groups of the 5-hydroxymethylcytosine are modified by addition of glucosyl residues. This *glucosylated* DNA is resistant to virtually all restriction endonucleases of the host. Thus, this virus-specific DNA modification plays an important role in the ability of the virus to attack a host cell.

Genetic map and DNA replication in T4

The genetic map of T4 is generally represented as a circle, even though the DNA itself is linear. This "genetic circularity" arises because the DNA of the phage exhibits a phenomenon called *circular permutation*. This arises because in different T4 phage particles, the sequence of bases at each end differs (although for a given molecule the same base sequence occurs at both

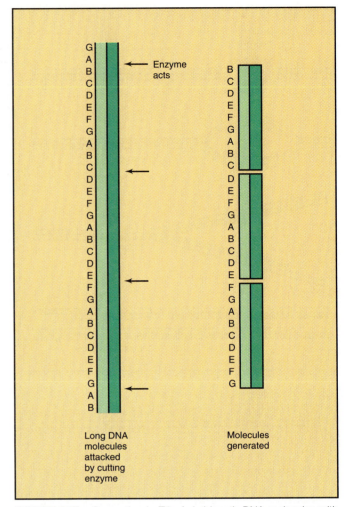

Long DNA molecules attacked by cutting enzyme

Molecules generated

Enzyme acts

FIGURE 6.23 Generation in T4 of viral length DNA molecules with permuted sequences by a cutting enzyme, which cuts off constant lengths of DNA irrespective of the sequence. Left, arrows, sites of enzyme attack. Right, molecules generated.

The Phage Group

Historically, the T-even phages provided the study material for early research of the "Phage Group," a group of research workers from various universities and research institutions who spent their summers working together at the Cold Spring Harbor Laboratory on Long Island. The key members of the Phage Group, Max Delbrück, Salvador Luria, and A. D. Hershey, subsequently shared the Nobel Prize for their pioneering work. Among important concepts first uncovered from research on the T-even phages: only the nucleic acid of the virus entered the cell during infection (a discovery which provided key support for the hypothesis that DNA is the genetic material); the existence of genetic recombination in viruses; the phenomenon of restriction and modification (which led to the discovery of the restriction enzymes so important for genetic engineering); the presence in viruses of unique virus-encoded gene functions; the distinction between early and late viral functions; the phenomenon of phenotypic mixing. The first ideas of how viruses cause killing of host cells were also developed from research on the T-even phages. Selecting the T-even phages as a model system and concentrating work on them was greatly responsible for the remarkable success of the Phage Group.

ends since, like T7, T4 also has terminal repeats). This structure, a consequence of the way the T4 DNA replicates, results in an appearance of genetic circularity even though the DNA itself is linear.

The process of DNA replication in T4 is similar to that in T7, but in T4, the cutting enzyme which forms virus-sized fragments does not recognize specific locations on the long molecule, but rather cuts off head-full packages of DNA irrespective of the sequence (Figure 6.23). Thus, each virus DNA molecule not only contains repetitious ends, but the nucleotide sequences at the ends of different molecules are different. Each molecule contains slightly more than one complete copy of the entire genome (Figure 6.23). As shown, the cutting process results in the formation of DNA molecules with permuted sequences at the ends.

Transcription, translation, and regulation in phage T4

In bacteriophage T4, the details of regulation of replication are more complex than those of T7 and involve primarily *positive* control. T4 is a much larger phage than T7 and has many more genes and phage functions. And, as previously mentioned, the DNA of T4 contains the unusual base, 5-hydroxymethylcytosine (Figure 6.22) and some of the OH groups of this base are glucosylated. Thus enzymes for the synthesis of this unusual base and for its glucosylation must be formed after phage infection, as well as formation of an enzyme that breaks down the normal DNA precursor deoxycytidine triphosphate. In addition, T4 codes for a number of enzymes that have functions similar to those host enzymes in DNA replication, but are formed in larger amounts, thus permitting faster synthesis of T4-specific DNA. In all, T4 codes for over 20 new proteins that are synthesized early after infection. It also codes for the synthesis of several new tRNAs, whose function is presumably to read T4 mRNA more efficiently.

Overall, the T4 genes can be divided into two groups, one encoding early proteins and the other, late proteins (Figure 6.24). The **early proteins** are the enzymes involved in DNA replication and transcription. The early genes are often subdivided into *immediate early*, which are transcribed immediately upon injection of the DNA, and *delayed early*, in which transcription begins 1 to 2 minutes after infection. The **late proteins** are the head and tail proteins and the enzymes involved in liberating the mature phage particles from the cell.

In T4, there is no evidence for a new phage-specific RNA polymerase, as in T7. The control of T4 mRNA synthesis involves the production of proteins that modify the specificity of the host RNA polymerase so that it recognizes different phage promoters. The early promoter, present at the beginning of the T4 genome, is read directly by the host RNA polymerase, and involves the function of host *sigma* factor. Host RNA polymerase moves down the chain until it reaches a stop signal. One of the early proteins blocks host sigma factor action. The early protein combines with the RNA polymerase core enzyme, and when this protein builds up, initiation of early phage genes is stopped. The RNA polymerase cores are now available to combine with new phage-specific sigma factors which control the transcription of the delayed early and late genes.

Assembly and lysis

In the case of phage T4, the entire lytic cycle takes about 25 minutes (Figure 6.24). Assembly of heads and tails occurs independently, DNA is packaged into the assembled head, and the tail and tail fibers are added later. Exit of the virus from the cell occurs as a result of cell lysis (Figure 6.24). The phage codes for a lytic enzyme, the *T4 lysozyme*, which attacks the peptidoglycan of the host cell. The burst size of the virus (the average number of phage particles per cell) de-

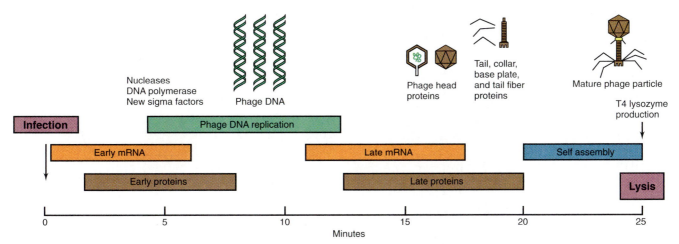

FIGURE 6.24   Time course of events in phage T4 infection. Following injection of DNA, early and middle mRNA is produced which codes for nucleases, DNA polymerase, new phage-specific sigma factors, and various other proteins involved in DNA replication. Late mRNA codes for structural proteins of the phage virion and for T4 lysozyme, needed to lyse the cell and release new phage particles.

pends upon how rapidly lysis occurs. If lysis occurs early, then a smaller burst size occurs, whereas slower lysis leads to a higher burst size. The wild-type phage exhibits the phenomenon of *lysis inhibition*, and therefore has a large burst size, but *rapid lysis mutants*, in which lysis occurs early, show smaller burst sizes.

6.12   Temperate Bacterial Viruses: Lysogeny and Lambda

Most of the bacterial viruses described above are called **virulent** viruses, since they usually kill (lyse) the cells they infect. However, many other bacterial viruses, although also able to kill cells, frequently have more subtle effects. Such viruses are called **temperate**. These viruses can enter into a state called **lysogeny**, where most phage genes are not expressed, and the phage genome is replicated in synchrony with the host chromosome. Thus, the phage genome is duplicated along with the host material at the time of cell division, being passed from one generation of bacteria to the next. Under certain conditions these bacteria, called **lysogens**, can spontaneously produce virions of the temperate virus, which can be detected by their ability to infect a closely related strain of bacteria. Lysogeny is probably of ecological importance because most bacteria isolated from nature are lysogenic for one or more bacteriophages.

If the host simply makes a copy of the viral DNA, lysis does not occur because none of the required viral proteins are produced; but if complete virion particles are produced, then the host cell lyses. In a culture of lysogens at any one time, only a small fraction of the cells, 0.1 to 0.0001 percent, produce virus and lyse, while the majority of the cells neither produce virus nor lyse. Although only rarely do cells of a lysogenic

FIGURE 6.25
Consequences of infection by a temperate bacteriophage. The alternatives upon infection are integration of the virus DNA into the host DNA (lysogenization) or replication and release of mature virus (lysis). The lysogenic cell can also be induced to produce mature virus and lyse.

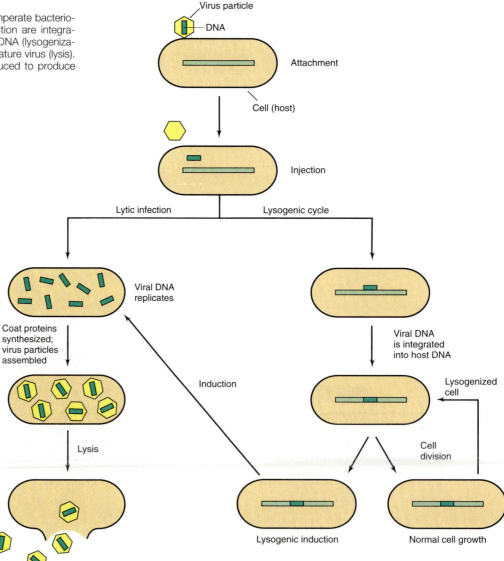

strain actually produce virus, every cell has the potential for virus production. Lysogeny can thus be considered a genetic trait of a bacterial strain.

An overall view of the life cycle of a temperate bacteriophage is shown in Figure 6.25. The temperate virus does not exist in its mature, infectious state inside the cell, but rather in a latent form, called the **provirus** or **prophage** state. In the example shown in Figure 6.25, the prophage is integrated into the bacterial chromosome. In considering virulent viruses, we learned that the DNA of the virulent virus contains information for the synthesis of a number of enzymes and other proteins essential to virus reproduction. The prophage of the temperate virus carries similar information, but in the lysogen this information remains dormant because the expression of the virus genes is blocked through the action of a specific repressor coded for by the virus. However, in some cases, the repressor can be inactivated, and then virus reproduction occurs, the cell lyses, and virus particles are released.

A lysogenic culture can be treated so that most or all of the cells produce virus and lyse. Such treatment, called **lysogenic induction** (center, Figure 6.25), usually involves the use of agents such as ultraviolet radiation, nitrogen mustards, or X-rays, known to damage DNA and activate the SOS system (see Section 7.3). However, not all prophages are inducible; in some temperate viruses, prophage expression occurs only by natural events.

Although a lysogenic bacterium may be susceptible to infection by other viruses, it cannot be infected by virus particles of the type for which it is lysogenic. This **immunity**, which is characteristic of lysogenized cells, is conferred by the intracellular repression mechanism under the control of virus genes.

It is sometimes possible to eliminate the temperate virus (to "cure" the host) by heavy irradiation or treatment with nitrogen mustards. Among the few survivors may be some cells that have been cured. Presumably the treatment causes the prophage to excise from the host chromosome and be lost during subsequent cell growth. Such a cured strain is no longer immune to the virus and can serve as a suitable host for study of virus replication.

Consequences of temperate virus infection

What happens when a temperate virus infects a nonlysogenic organism? The virus may inject its DNA and initiate a reproductive cycle similar to that described for virulent viruses (left side, Figure 6.25, see also Figure 6.9), with the infected cell lysing and releasing more virus particles. Alternatively, when the virus injects its DNA, **lysogenization** may occur instead: the viral DNA becomes a prophage and the host bacterium is converted into a lysogenic bacterium (right side, Figure 6.25). In lysogenization the infected cell thus becomes genetically changed. Sensitive cells can undergo either lysis or lysogenization; which of

these occurs is often determined by the action of a complex repression system, as will be described below. We thus see that the temperate virus can have a dual existence. Under one set of conditions, it is an independent entity able to control its own replication, but when its DNA is integrated into the host genetic material, replication is then under the control of the host.

Regulation of lambda reproduction

One of the best studied temperate phages is lambda, which infects *E. coli*, and our knowledge of the molecular mechanisms involved in lysogenization and lytic processes in this phage is very advanced. Morphologically, lambda particles look like those of many other bacteriophages (Figure 6.26). The virus particle has an icosahedral head 64 nm in diameter, and a tail 150 nm long which has helical symmetry. Attached to the tail is a single 23 nm long fiber. In addition to the major proteins of the coat, there are a number of minor coat proteins.

The nucleic acid of lambda consists of a linear double-stranded DNA molecule, but at the 5′-terminus of each of the single strands is a single-stranded tail 12 nucleotides long. These single-stranded ends are complementary (the ends of the DNA are said to be *cohesive*). Thus, when the two ends of the DNA are free in the host cell, they associate and the genome forms a double-stranded circle. In the circular form the DNA contains 48,502 base pairs, and its complete sequence is known.

Lysis or lysogenization?

If lysogeny occurs, then the phage genes are maintained stably in the lysogenic state until a *switch* occurs and they are converted with high efficiency into a second state in which *lytic growth* occurs. But what controls whether a particular infection will be lytic or lysogenic?

The lambda genome has two sets of genes, one controlling lytic growth, the other lysogenic growth. Upon infection, genes promoting both lytic growth

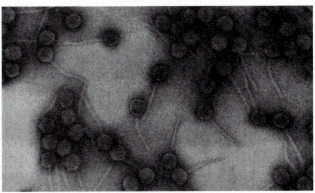

FIGURE 6.26   Electron micrograph by negative staining of bacteriophage lambda particles. The head of each particle is about 65 nm in diameter.

FIGURE 6.27
Genetic and molecular map of lambda. The genes are designated by letters; *att*, attachment site for phage to host chromosome. Genes of special interest: *cI*, repressor protein; O_R, operator right; O_L, operator left; *cro*, gene for second repressor; N, positive regulator counteracting rho-dependent termination. J through U are genes that code for tail proteins. Genes Z through A code for head proteins. The regulatory region of lambda (shown in yellow) is positioned at the top of this circular map. It is also known as the immunity region and contains the *cI* gene. The site created when the cohesive ends of the lambda genome join is called *cos* (shown in blue). Early transcription in lambda is primarily leftward (counterclockwise) from promoter left (near O_L on map) and rightward (clockwise) from promoter right (near O_R on map). The main leftward transcript is labeled L1, and the main early rightward transcript is labeled R1. The late rightward transcript, which encodes head and tail proteins and proteins for lytic function, is labeled R2 and begins at a promoter near the Q gene. The transcript labeled L2 is the positively regulated transcript that encodes the repressor protein.

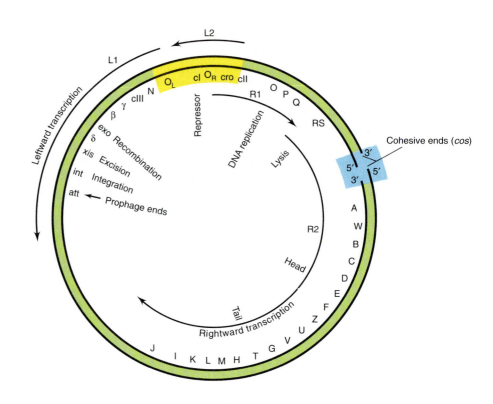

and lysogenic integration are expressed. Which pathway succeeds is determined by the competing action of these early gene products and by the influence of host factors. To understand how this *genetic switch* controlling lysogeny and lysis works, we need to present the genetic map of lambda (Figure 6.27). The genetic map, although actually linear, can be oriented as a circle because of the cyclization via the cohesive ends mentioned above. The lambda map consists of several operons, each of which controls a set of related functions. Upon injection, transcription of the phage genes occurs. The product of the *cI* gene is the lambda repressor. If repressor accumulates in the cell before lytic functions are expressed, lytic reproduction is blocked. The repressor protein blocks the transcription of all later lambda genes, thus preventing expression of the genes involved in the lytic cycle.

Two of the lambda proteins synthesized early in infection are involved in positive regulation of the *cI* gene. The reason that lysis is not always blocked by repressor is that the positive regulatory proteins made by lambda are themselves regulated by the host and by another lambda protein called Cro (see following). If the positive regulatory proteins do not accumulate, repressor is not made and the lytic cycle is completed. The lambda protein called Cro, is coded by a gene called *cro*. The gene *cro* is located almost adjacent to the gene *cI* which codes for lambda repressor (Figure 6.28).

The key to the genetic switch lies in the close proximity of the regulator genes for repressor and *cro* pro-

tein. These two genes are transcribed in opposite directions, beginning at different start points. In the region separating these two genes are two kinds of sites, promoters and operators, to which each of the proteins of the switch can bind. When lambda repressor is bound to its operator, it covers the *cro* promoter, whereas when Cro is bound, it covers one of the *cI* promoters. As we have noted (see Section 5.6), the direction in which transcription occurs on a DNA double-strand (and hence which of the two strands is read) depends upon the promoter. A promoter essentially points the RNA polymerase in the proper direction. In the case of lambda, the *cI* promoter points RNA polymerase "leftward," whereas *cro* promoter points the polymerase "rightward."

In a lysogenic cell, only one phage gene is expressed continuously, the gene which codes for the lambda repressor protein. This repressor protein binds to two operators on the lambda DNA and thereby turns off the transcription of *all the other genes*

FIGURE 6.28 Two back-to-back promoters in the region of *cI* and *cro* control the genetic switch in lambda. When the *cI* protein is present, it activates its own synthesis and *blocks* transcription of *cro*. When *cI* is inactivated, transcription of *cro* can occur, resulting in the lytic cycle. The *cI* (repressor) protein combines with the operator, O_R. See Figure 6.27 for the location of this region on the lambda genome.

of the phage genome. This is the *negative* control function of the lambda repressor. In addition, the lambda repressor turns on its *own* synthesis. This is the *positive* control function of the lambda repressor. Thus, by promoting its own synthesis, the lambda repressor ensures that no other genes except the gene coding for itself is made. In a lysogenic cell, there will usually be only one copy of the lambda genome, but about 100 active molecules of repressor protein. Therefore, there is almost always excess repressor to bind to lambda DNA and prevent the transcription of the genes necessary for lambda growth and reproduction.

Lytic growth of lambda after induction

How, then, does lambda virus multiplication occur? In a lysogenic cell, multiplication of lambda occurs only after the repressor is inactivated. As we have noted, agents which induce lysogenic cells to produce phage are agents which damage DNA, such as ultraviolet irradiation, X-rays, or DNA-damaging chemicals such as the nitrogen mustards. Upon DNA damage, a host defense mechanism called the SOS response (see Section 7.3) is brought into play (Figure 6.29). An array of 10–20 bacterial genes is turned on, some of which help the bacterium survive radiation. However, one result of DNA damage is that a bacterial protein called RecA (normally involved in genetic recombination) is turned into a special kind of protease which participates in the destruction of the lambda repressor. With lambda repressor destroyed, the inhibition of expression of lambda lytic genes is abolished. We should note that the protease activity of RecA, brought about by DNA damage, normally plays an important role in the cell's response to DNA damaging agents, by participating in the breakdown of a host protein, LexA, which represses a set of host genes involved in DNA repair (Section 7.3). Induction of bacteriophage lambda is thus an indirect consequence of the SOS response.

Once the lambda repressor has been inactivated, the positive and negative control exerted by this repressor are abolished, and new transcriptional events can be initiated. These inevitably lead to lysis because even if the lambda repressor is made it is inactivated.

The lambda system provides one of the best studied examples of a genetic switch, in which one or the other of two competing genetic functions occurs. Which of the two genetic functions prevails will depend initially on chance events, but once one of the two functions has become established, it prevents the action of the other. Only under unusual circumstances, such as when induction occurs, would the dominant genetic function be superseded.

Integration

Integration of lambda DNA into the host chromosome occurs at a unique site on the *E. coli* genome. Integration occurs by insertion of the virus DNA into the host genome (thus effectively lengthening the host genome by the length of the virus DNA). As illustrated in Figure 6.30, upon injection, the cohesive ends of the linear lambda molecule find each other and form a circle, and it is this circular DNA which becomes integrated into the host genome (the site created when these ends join is called *cos*). To establish lysogeny, genes *cI* and *int* (Figure 6.27) must be expressed. As we have noted, the *cI* gene product is a protein which represses early transcription and thus shuts off transcription of all later genes. The integration process requires the product of the *int* gene, which is a site-specific topoisomerase catalyzing recombination of the phage and bacterial attachment sites (labeled *att* in Figures 6.27 and 6.30).

During cell growth, the lambda repression system prevents the expression of the integrated lambda genes except for the gene *cI*, which codes for the lambda repressor. During host DNA replication, the integrated lambda DNA is replicated along with the rest of the host genome, and transmitted to progeny cells. When release from repression occurs (see above), the lambda productive cycle occurs. In order to be excised from the chromosome, *excisionase* (the product of the *xis* gene) and the *int* gene product are required.

Replication

Replication of lambda DNA occurs in two distinct fashions during different parts of the phage production cycle. Initially, after infection (or after liberation of lambda DNA from the host *chromosome*) the lambda

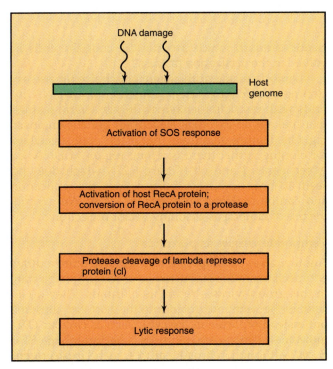

FIGURE 6.29 Activation of the host SOS response leads to *lysis* of a lysogenic cell.

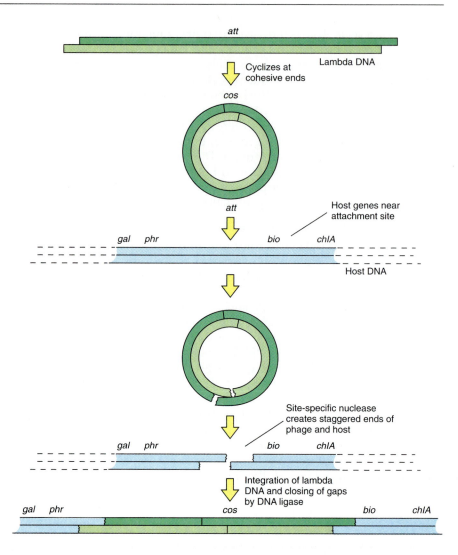

FIGURE 6.30
Integration of lambda DNA into the host. See the genetic map, Figure 6.27, for details of the gene order. Integration always occurs at a specific site on the host DNA, involving a specific attachment site (*att*) on the phage. Some of the host genes near the attachment site are given. A site-specific enzyme (integrase) is involved, and specific pairing of the complementary ends results in integration of phage DNA.

DNA replicates in a circular fashion, but subsequently linear concatamers are formed, which replicate in a different way. Replication is initiated at a site close to gene *O* (Figure 6.27) and from there proceeds in opposite directions (bidirectional symmetrical replication), terminating when the two replication forks meet. In the second stage, generation of long linear concatamers occurs, and replication occurs in an asymmetric way by **rolling circle replication** (see Figure 6.18). In this mechanism, replication proceeds in one direction only and can result in very long chains of replicated DNA. Unlike the replication of ϕX174 DNA (see Figure 6.18), however, rolling circle replication of lambda involves synthesis of both strands (Figure 6.31). This mechanism is efficient in permitting extensive, rapid, relatively uncontrolled DNA replication; thus it is of value in the later stages of the phage replication cycle when large amounts of DNA are needed to form mature virions. The long concatamers formed are then cut into virus-sized lengths by a DNA cutting enzyme. In the case of lambda, the cutting enzyme makes staggered breaks at spcific sites on the two strands, twelve nu-

cleotides apart, which provide the cohesive ends involved in the cyclization process.

Lambda is one of the agents of choice for use as a cloning vector for artificial construction of DNA hybrids with restriction enzymes. It has several features that make it an excellent system for genetic engineering. One feature of lambda that makes it of special use for cloning is that there is a long region of DNA, between genes *J* and *att* (Figure 6.27), which does not seem to have any essential functions for replication, and can be replaced with foreign DNA. We describe the use of lambda as a cloning vector in Section 8.3.

Temperate viruses as plasmids

Another class of temperate viruses have quite a different mechanism for maintenance of the prophage state. In this group, viruses resemble plasmids. They do *not* actually become integrated into the host chromosome, but instead replicate in the cytoplasm as circular DNA molecules. Among such viruses is bacteriophage P1 of *E. coli*. Although in broad features, such viruses resemble the temperate viruses such as lambda just dis-

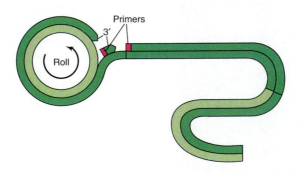

FIGURE 6.31
A late stage in the rolling circle replication of lambda. Both strands of DNA are being copied at the replicating fork, and two copies of the genome have already been synthesized. Note that this synthesis is *asymmetric* since one of the parental strands continues to serve as a template and the other is used only once.

cussed, at the level of virus replication they are, of course, quite different. Interestingly, although the plasmid prophage is not physically connected to the host DNA, phage DNA replication is closely coordinated with cell division, since only one copy of the prophage is present per host chromosome. The phage repressor is somehow involved in this regulation process.

> Temperate viruses do not always cause death of the cells they infect. The infected cell sometimes survives because the virus genome becomes a prophage (and replicates with the host chromosome), and the lytic genes of the prophage are kept under the control of a virus-encoded repressor. Sometimes spontaneously, or more commonly as a result of the action of DNA damaging agents such as radiation, a host DNA repair system capable of destroying the virus repressor protein can be activated, releasing prophage from host control and resulting in virus multiplication and lysis of the host cell. Host cells carrying temperate viruses are called lysogenic because they may lyse and release virus particles.

6.13   A Transposable Phage: Bacteriophage Mu

One of the more interesting bacteriophages is that called **Mu**, which has the unusual property of replicating as a movable genetic element (transposable element; see Section 5.5 and also Chapter 7). This phage is called Mu because it is a *mutator* phage, inducing mutations in a host genome into which it becomes integrated. This mutagenic property of Mu arises be-

cause the genome of the virus can become inserted into the middle of host genes, causing these genes to become inactive (and hence the host which has become infected with Mu behaves as a mutant). Mu is a useful phage because it can be used to generate a wide variety of bacterial mutants very easily. Also, as we will discuss in Chapter 8, Mu can be used in genetic engineering.

A transposable element is a piece of DNA which has the ability to move from one site to another as a discrete element. They are found in both prokaryotes and eukaryotes, and play important roles in genetic variation (see Section 7.11 for a detailed discussion of transposition). There are three types of transposable elements: *insertion elements, transposons,* and viruses like Mu (see Section 5.5). Mu is a very large transposable element, carrying a number of Mu genes involved in Mu multiplication.

Structure and genetic map of Mu

Structurally, bacteriophage Mu is a large double-stranded DNA virus, with an icosahedral head, a helical tail, and six tail fibers (Figure 6.32). The genetic map of Mu is shown in Figure 6.33*a*. It can be seen that the bulk of the genetic information is involved in the synthesis of the head and tail proteins, but that important genes at each end are involved in replication and immunity. The DNA molecule found within the virion is approximately 39 kilobase pairs long, but only 37.2 kilobase pairs is the actual Mu genome. This is because both ends of this DNA molecule contain host DNA. At the left end of the Mu DNA are 50 to 150 base pairs of host DNA and at the right end are 1 to 2 kilobase pairs of host DNA. These host DNA sequences are not unique and represent DNA adjacent to the location where Mu had become inserted into the genome of its previous host. When a Mu phage parti-

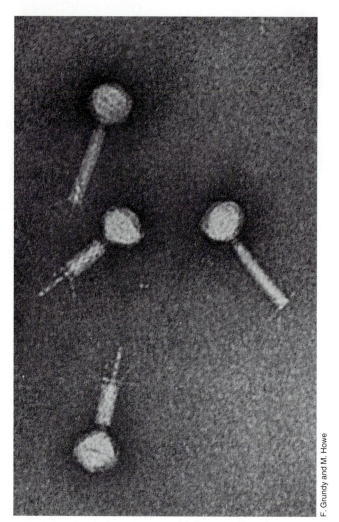

F. Grundy and M. Howe

FIGURE 6.32 Electron micrograph of virions of bacteriophage Mu, the mutator phage.

phage. Since adsorption to the host cell is controlled by the specificity of the tail fibers, the host range of Mu is determined by which orientation of this invertible segment is present in the phage. If the G segment is in the orientation designated G⁺, then the phage particle will infect *Escherichia coli* strain K-12. If the G segment is in the G⁻ orientation, then the phage particle will infect *E. coli* strain C or several other species of enteric bacteria. The two tail fiber proteins are encoded on opposite strands within this small G segment. Left of the G segment is a promoter that directs transcription into the G segment. In the orientation G⁺, the promoter directing transcription of S and U is active, whereas in the orientation G⁻, a different promoter directs transcription of genes S' and U' on the opposite strand.

Replication of Mu

Upon infection of a host cell by Mu, the DNA is injected, and is protected from host restriction by a modification system in which about 15 percent of the adenine residues are acetoamidated. In contrast with lambda, integration of Mu DNA into the host genome is essential for both lytic and lysogenic growth. Integration requires the activity of the A gene product, which is a transposase enzyme. At the site where the Mu DNA becomes integrated, a 5 base pair duplication of the host DNA arises at the target site. As shown in Figure 6.33b, this host DNA duplication arises because staggered cuts are made in the host DNA at the point Mu becomes inserted, and the resulting single-stranded segments are converted into the double-stranded form as part of the integration process.

Lytic growth of Mu can occur either upon initial infection, if the *c* gene repressor is not formed, or by induction of a lysogen. In either case, replication of Mu DNA involves repeated transposition of Mu to multiple sites on the host genome. Initially, transcription of only the early genes of Mu occurs, but after gene C protein, a positive activator of late RNA synthesis, is expressed, the synthesis of the Mu head and tail proteins occurs. Eventually, expression of the lytic function occurs and mature phage particles are released.

Mutations and modified Mu phages

Because Mu integrates at a wide variety of host sites, it can be used to induce mutations at many locations. Also, Mu can be used to carry into the cell genes that have been derived from other host cells, a form of in vivo genetic engineering. In addition, modified Mu phage have been made artificially in which some of the lytic functions of Mu have been deleted. These phages, called Mini-Mu, are deleted for significant portions of Mu but have the ends of the phage in normal orientation. Mini-Mu phages are usually defective, unable to form plaques, and their presence must be ascertained by the presence of other genes which they carry. One set of Mini-Mu phages containing the β-galactosidase gene of the host (called Mu*d-lac*, *d* for

cle is formed, a length of DNA containing the Mu genome just large enough to fill the phage head is cut out of the host, beginning at the left end. The DNA is rolled in until the head is full but the place at the right end where the DNA is cut varies from one phage particle to another. For that reason, as shown on the genetic map, there is a variable sequence of host DNA at the right-hand end of the phage (right of the *attR* site) which represents the *host* DNA that has become packaged into the phage head. Each virion arising from a single infected cell will have a different amount of host DNA, and the host DNA base sequence in each virion from the same cell will be different. In some cases, completely empty Mu heads become filled with purely host DNA. Such particles can transfer host genes from one cell to another, a process called *transduction* (see Section 7.7).

As shown in the genetic map (Figure 6.33), a specific segment of the Mu genome called G (distinct from the G gene) is invertible, being present in either the orientation designated SU, or in the inverted orientation U'S'. The orientation of this segment determines the kind of tail fibers that are made for the

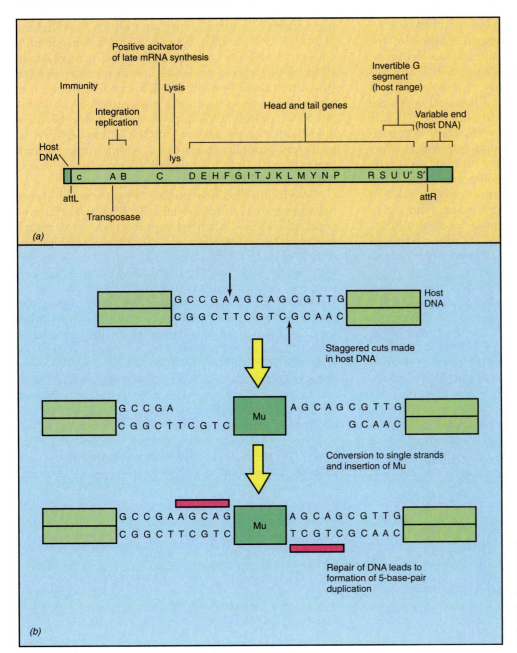

(a)

(b)

FIGURE 6.33
Bacteriophage Mu. (a) Genetic map of Mu. See text for details. (Confusingly, there are two G's, the G gene and the invertible G segment. These are different G's.) (b) Integration of Mu into the host DNA, showing the generation of a five-base-pair duplication of host DNA.

defective) can be detected in the integrated state if the lac gene is oriented properly in relation to a host promoter. Under these conditions, the host cell will form the enzyme β-galactosidase, which can be detected in colonies by a special color indicator (see Figure 7.1c). β-galactosidase-positive colonies from a β-galactosidase-negative host are thus an indication that Mu*d-lac* infection has occurred. We thus see that phage Mu provides a useful tool for geneticists, as well as being an interesting bacteriophage in its own right.

6.14   Overview of Animal Viruses

We have discussed in a general way the nature of animal viruses in the first part of this chapter. Now we discuss in some detail the structure and molecular biology of a number of important animal viruses.

Viruses will be discussed which illustrate different ways of replicating, and both RNA and DNA viruses will be covered. One important group of animal viruses, those called the *retroviruses*, have both an RNA and a DNA phase of replication. Retroviruses are especially interesting not only because of their unusual mode of replication, but because retroviruses cause such important diseases as certain *cancers* and *acquired immune deficiency syndrome* (AIDS).

Before beginning our discussion of the manner of replication of animal viruses, we should remind ourselves of the important differences which exist between eukaryotic and prokaryotic cells. Since virus replication makes use of the biosynthetic machinery of the host, these differences in cellular organization and function imply differences in the way the viruses themselves replicate.

Molecular processes in prokaryotes and eukaryotes

Prokaryotes do not show compartmentation of the biosynthetic processes. The genome of a bacterium relates directly to the cytoplasm of the cell. Transcription into mRNA can lead directly to translation, and the processes of transcription and translation are not carried out in separate compartments (see Figure 5.2a). Animal cells, being eukaryotic, show compartmentation of the transcription and translation processes. Transcription of the genome into mRNA occurs in the nucleus, whereas translation occurs in the cytoplasm (see Figure 5.2b). Furthermore, as discussed in Section 5.6, the transcripts from eukaryotic genes must be processed before they can be used as mRNA. This processing usually involves adding a **poly (A) tail** to the 3'-end and a methylated guanosine triphosphate, called the **cap**, to the 5'-end. The cap is required for binding of the mRNA to the ribosome, and the poly (A) tail may be involved in subsequent RNA processing and transfer of the mature mRNA from the nucleus to the cytoplasm. All of the protein-synthesizing machinery of the eukaryotic cell, the ribosomes, tRNA molecules, and accessory components, is in the cytoplasm, and the mature mRNA associates with the protein-synthesizing apparatus once it leaves the nucleus.

Also, as we have discussed in Sections 5.1 and 5.6, the genes of eukaryotes are often *split*, with noncoding regions called *introns* separating coding regions (*exons*). Transcription of both the coding and noncod-

ing regions of a gene occurs, and an RNA, called the *primary RNA transcript*, is formed and is subsequently converted into the mature mRNA by a mechanism called *RNA splicing*, in which the introns are excised (the cap and tail remain after RNA processing is complete). After processing, the mature mRNA is translated into protein. One important distinction between eukaryotic and prokaryotic mRNA is that prokaryotic mRNA is generally *polycistronic*, coding for more than one polypeptide in a single mRNA molecule, whereas eukaryotic mRNA is monocistronic.

We might also note another important difference between animal and bacterial cells. Bacterial cells have rigid cell walls containing peptidoglycan and associated substances (see Section 3.5). Animal cells, on the other hand, lack cell walls. This difference is important for the way by which the virus genome enters and exits the cell. In prokaryotes, the protein coat of the virus remains on the outside of the cell and only the nucleic acid enters. In animal viruses, on the other hand, uptake of the virus often occurs by endocytosis (pinocytosis or phagocytosis), processes which are characteristic of animal cells, so that the whole virus particle enters the cell (Figure 6.34). The separation of animal virus genomes from their protein coats then occurs inside the cell. In addition we will see that in eukaryotic cells there are two different sites in which viral replication can occur; the nucleus and the cytoplasm. We will see examples of both types as we consider the major animal viruses.

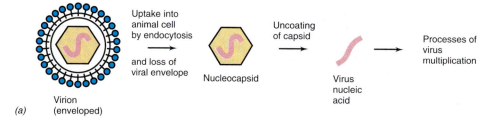

(a) Virion (enveloped) → Uptake into animal cell by endocytosis and loss of viral envelope → Nucleocapsid → Uncoating of capsid → Virus nucleic acid → Processes of virus multiplication

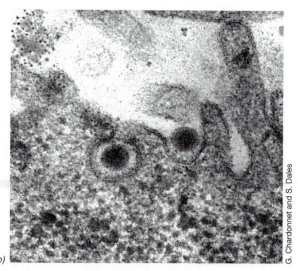

(b)

G. Chardonnet and S. Dales

FIGURE 6.34
Uptake of an enveloped virion by an animal cell. (a) The process by which the viral nucleocapsid is separated from its envelope. (b) Electron micrograph of adenovirus virions entering a cell. Each particle is about 70 nm in diameter.

Classification of animal viruses

Various types of animal viruses are illustrated in Figure 6.35. Note that the major criteria used in classifying animal viruses are the type of nucleic acid, the presence or absence of an envelope, and, for certain families, the manner of replication. Most of the animal viruses which have been studied in any detail are those which have been amenable to cultivation in cell cultures. Animal viruses are known with either single-stranded or double-stranded DNA or RNA. Some animal viruses are enveloped, others are naked. Size varies greatly, from those large enough to be just visible in the light microscope, to those so tiny that they are hard to see even in the electron microscope. In the following sections, we will discuss characteristics and

manner of multiplication of some of the most important and best studied animal viruses.

Consequences of virus infection in animal cells

Viruses can have varied effects on cells. **Lytic infection** results in the destruction of the host cell (Figure 6.36). However, there are several other possible effects following viral infection of animal cells. In the case of enveloped viruses, release of virions, which occurs by a kind of budding process, may be slow and the host cell may not be lysed. The cell may remain alive and continue to produce virus over a long period of time. Such infections are referred to as **persistent infections** (Figure 6.36). Viruses may also cause **latent infection** of a host. In a latent infection, there is a delay between

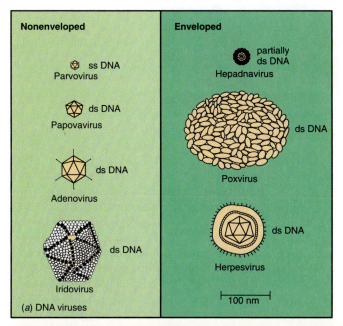

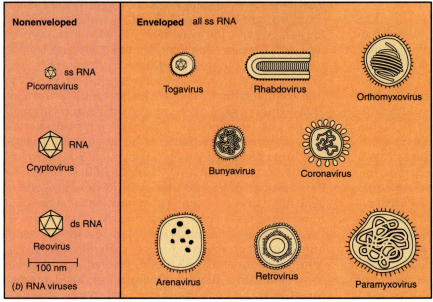

FIGURE 6.35
The shapes and relative sizes of vertebrate viruses of the major taxonomic groups. The hepadnavirus genome has one complete DNA strand and part of its complement. Bar = 100 nm.

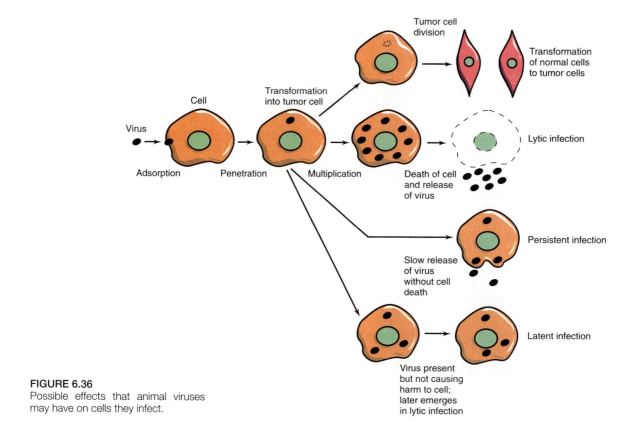

FIGURE 6.36
Possible effects that animal viruses
may have on cells they infect.

infection by the virus and the appearance of symptoms. Fever blisters (cold sores), caused by the herpes simplex virus (see Section 6.19), result from a latent viral infection; the symptoms reappear sporadically as the virus emerges from latency. The latent stage in viral infection of an animal cell is generally not due to the integration of the viral genome into the genome of the animal cell, as is the case with latent infections by temperate bacteriophages.

Viruses and cancer

A number of animal viruses have the potential to change a cell from a normal one to a cancer or tumor cell (Figure 6.36 and Table 6.2). **Cancer** is a cellular phenomenon of uncontrolled growth. Most cells in a mature animal, although alive, do not divide extensively, apparently because of the presence of growth-inhibiting factors that prevent them from initiating cell division. As noted in Section 6.3, infection by certain types of animal viruses leads to a process called **transformation**, during which growth becomes uncontrolled. One of the key differences between normal cells and cancer cells is that the latter have different requirements for growth factors. Rapidly growing cells pile up into accumulations that are visible in culture as **foci of infection**. Because cancerous cells in the animal body have fewer growth requirements, they grow profusely, leading to the formation of large masses of cells, called *tumors*. The term **neoplasm** is often used in the medical literature to describe malignant tumors.

Not all tumors are seriously harmful. The body is able to wall off some tumors so that they do not spread; such noninvasive tumors are said to be **benign**. Other tumors, called **malignant**, invade the body and destroy normal body tissues and organs. In advanced stages of cancer, malignant tumors may develop the ability to spread to other parts of the body and initiate new tumors, a process called **metastasis**.

How does a normal cell become cancerous? The growth and division of normal cells is regulated by at least two types of genes. The first type, called *proto-oncogenes*, promote growth, but they are controlled by the second type, the growth-restraining *tumor-suppressor genes*. Changes in either or both types can lead to uncontrolled cell growth and therefore to cancer. The process can be broken down into several stages. In the first step, *initiation*, genetic changes in the cell occur. This step may be induced by certain chemicals, called *carcinogens* (see Section 7.3), or by physical stimuli, such as ultraviolet radiation or X-rays. Certain viruses also bring about the genetic change that results in initiation of tumor formation. This initiation event can be the activation of a proto-oncogene into an **oncogene** (a gene that causes a tumor), or the inactivation of a tumor-suppressor gene. Once initiation has occurred, the potentially cancerous cell may remain dormant, but under certain conditions, generally involving some environmental alteration, the cell may become converted into a tumor cell, a process called **promotion**. Once a cell has been promoted to the cancerous condition, continued cell division can result in the formation of a tumor.

Table 6.2   Some human cancers that may be caused by viruses

Cancer	Virus	Family	Genome in virion
Adult T-cell leukemia	Human T-cell leukemia virus (type I)	Retrovirus	RNA
Burkitt's lymphoma	Epstein-Barr virus	Herpes	DNA
Nasopharyngeal carcinoma	Epstein-Barr virus	Herpes	DNA
Hepatocellular carcinoma (liver cancer)	Hepatitis B virus	Hepadna	DNA
Skin and cervical cancers	Papilloma virus	Papova	DNA

Although the ability of viruses to cause tumors in animals has been proved for many years, the relationship of viruses to cancer in humans has, in most cases, been uncertain. It is difficult to prove the viral origin of a human cancer because of the difficulties of carrying out the necessary experimentation. However, it is now well established that certain specific kinds of human tumors do have a viral origin. A summary of some of the human cancers with definite viral origins is given in Table 6.2. In addition, some viral infections can lead indirectly to an increased risk of cancer, apparently by weakening the immune system's ability to detect and destroy transformed cells. This might be why infection with the retrovirus that causes AIDS increases the risk for developing certain cancers.

> Animal viruses can be either naked or enveloped, and contain either single-stranded or double-stranded DNA or RNA. Multiplication of animal viruses differs in significant ways from the multiplication of bacterial viruses, because of differences in the way macromolecular synthesis occurs in eukaryotes as opposed to prokaryotes. Not all infections of animal host cells result in cell lysis or death. In some cases, latent infection occurs, the virus remaining infectious but dormant inside the host, and appearing spontaneously at a later time. Some animal viruses cause transformation of host cells to the cancerous state.

6.15   Positive-Strand RNA Animal Viruses

An important group of animal viruses is the *picornavirus family*, which contains such important viruses infecting humans as the *polioviruses*, the *cold viruses*, and the *hepatitis A virus*. The first animal virus discovered, foot-and-mouth disease virus, is also a picornavirus. These viruses are called *picornaviruses* because they are very small viruses (30 nm in diameter, *pico* means small) and contain single-stranded RNA. The virus particle has a simple icosahedral structure with 60 morphological units per virion, each unit consisting of four distinct proteins (Figure 6.37a).

In poliovirus, the RNA is a linear molecule of about 7500 bases in length (see Table 6.1). At the 5'-terminus of the viral RNA is a protein, called the *VPg protein*, which is attached covalently to the RNA. At the 3'-terminus of the RNA is a poly (A) tail. RNA molecules which lack the poly (A) tail are not infectious, but the VPg protein is not essential for infectivity.

An overview of the manner of multiplication of poliovirus is illustrated in Figure 6.37b. The RNA of the virus acts directly as a messenger RNA. It is therefore of plus sense, as we outlined in Figure 6.12. The virus RNA is monocistronic but codes for all of the proteins of the virus in a single *polyprotein* that is later cleaved into the individual proteins. The coat proteins are coded for by sequences at the 5'-end of the molecule and the proteins necessary for replication are coded for at the 3'-end. The whole replication process occurs in the cytoplasm.

At initial infection, the virus particle attaches to a specific receptor on the surface of a sensitive cell and enters the cell. Once inside the cell, the virus particle is uncoated, and the free RNA associates with ribosomes to form a polysome. The viral RNA is then translated from a single initiation point several hundred bases from the 5'-end into a large protein precursor, the polyprotein mentioned above, which has a molecular weight of about 240,000. This giant protein is then cleaved into about 20 smaller proteins (including cleavage intermediates), among which are the four *structural proteins* of the virus particle, the RNA-linked *protein VPg*, an *RNA polymerase* which is responsible for synthesis of minus-strand RNA, and at least one *virus-encoded protease* which carries out the cleavage process. This cleavage process, called *posttranslational cleavage*, occurs in a wide variety of animal viruses as well as in normal cell metabolism in animal cells.

Replication of poliovirus

Replication of viral RNA begins within a short time after infection and is catalyzed by the RNA-dependent RNA polymerase (replicase) made in the process described in the previous paragraph. This replicase transcribes the viral RNA, of plus complementarity, into an RNA molecule of minus complementarity. This minus strand then serves as a template for repeated transcription of progeny plus strands. Some of the progeny plus strands may again be transcribed into minus strands, and as many as 1000 minus strands may subsequently be present in the cell. From these minus strands, as many as a million plus strands may ultimately be formed. Both the plus and minus strands

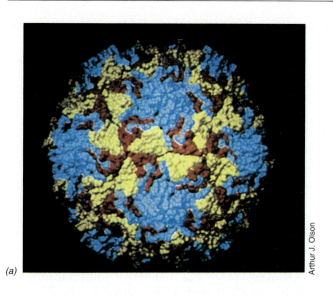

(a)

Arthur J. Olson

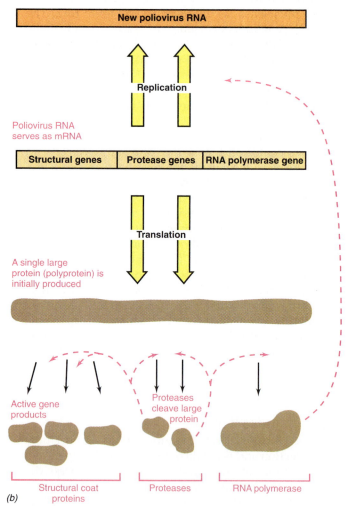

(b)

FIGURE 6.37 (a) The poliovirus particle. This is a computer model based on electron diffraction analysis of virus crystals. The various structural proteins are shown in distinct colors. (b) The reproduction of poliovirus. The single-stranded RNA of the virus is translated directly as a messenger RNA, with the production of one large protein molecule. This protein is cleaved, leading to the production of the active viral proteins, including the structural coat proteins and the RNA polymerase which brings about the replication of the poliovirus RNA. The assembly of intact poliovirus from coat protein molecules and RNA then follows.

become covalently linked to the tiny protein VPg (only 22 amino acids long), and it is thought that VPg serves as a primer for transcription or is added after synthesis. Viral RNA molecules which serve as mRNA lack VPg and are not assembled into mature virus particles.

Once virus multiplication begins, host RNA and protein syntheses are inhibited. Host protein synthesis is inhibited as a result of cleavage of a host protein, the cap-binding protein complex required for translation of capped mRNAs, by a protease. Virus coat proteins are made by cleavage of precursor molecules which are much longer than the protein subunits of the virus particle.

At one time, polio was a major infectious disease of humans, but the development of an effective vaccine (see Chapter 12) has brought the disease completely under control.

> In small RNA viruses such as polio, the viral RNA acts directly as a single messenger RNA, causing the production of a long polyprotein which is broken down by enzymes into the numerous small proteins necessary for nucleic acid multiplication and virus assembly.

6.16 Negative-Strand RNA Viruses

In a number of RNA viruses of animals, the RNA does not serve directly as a messenger, but is transcribed into a complement which functions as the mRNA. As discussed in Section 6.5, it is conventional to express the complementarity of the mRNA as *plus*, so that if the viral genomic RNA is of opposite complementarity, it is called *minus*. This group of viruses is then called *minus* strand or *negative-strand* RNA viruses. We discuss here two important negative-strand viruses; rhabdoviruses, including rabies virus, and orthomyxoviruses, including influenza virus.

Rhabdoviruses

The most important human pathogen which is a negative-stranded RNA virus is the rabies virus, which causes the important disease rabies in animals and humans (see Section 15.8). Rabies virus is called a *rhabdovirus*, from *rhabdo* meaning *rod*, which refers to the shape of the virus particle. Another rhabdovirus which has been extensively studied is vesicular stomatitis virus (VSV, Figure 6.38), a virus which causes the disease *vesicular stomatitis* in cattle, pigs, horses, and sometimes in humans. Many rhabdoviruses, such as potato yellow dwarf virus, infect both insects and plants and can cause important agricultural problems.

The rhabdoviruses are enveloped viruses, with an extensive and rather complex lipid envelope surrounding the nucleocapsid (Figure 6.38). In animal rhabdoviruses the virus particle is bullet-shaped, about 70 nm in diameter and 175 nm long. The nucleocapsid is helically symmetrical and makes up only a small part of the virus particle weight (about 2–3 per-

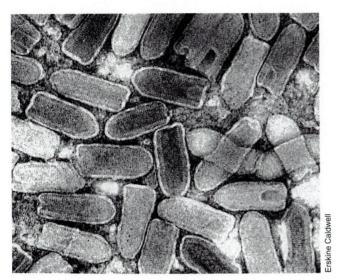

Erskine Caldwell

FIGURE 6.38   Electron micrograph of a rhabdovirus (vesicular stomatitis virus). A particle is about 65 nm in diameter.

cent of the virion is RNA, in contrast to the 30 percent or more RNA content in nonenveloped RNA viruses). Not only does the virus contain an extensive lipid envelope, but it contains several enzymes which are essential for the infection process: *RNA-dependent RNA polymerase, RNA methylase*, and some capping enzymes. As discussed in Section 6.5, the presence of RNA polymerase is essential because the genome of these negative-strand viruses cannot act as messengers directly, but must first be transcribed into the plus complement, and host enzymes which transcribe RNA into RNA do not exist.

The RNA of the rhabdoviruses is transcribed in-

side the cytoplasm of the cell into two distinct kinds of RNA (Figure 6.39). The first type of RNA synthesis results in a series of messenger RNAs which are made from the various genes of the virus (VSV has 5). The second is a plus strand RNA which is a *copy* of the complete viral genome (the VSV genome is 11,162 nucleotides long). These long plus strand RNAs then serve as templates for the synthesis of the *negative-strand* RNA molecules of the new crop of virions (yet to come). Each mRNA is monocistronic, coding for a single protein. A mechanism exists to ensure that transcription stops at the end of each virus gene and that a series of adenylic acid residues (a poly (A) tail) is put on the end of the RNA. Once the mRNA for the virus RNA polymerase is made in this primary transcription process, synthesis of the virus RNA polymerase can begin, leading to the formation of many *plus*-strand RNA molecules, both messengers and genomic (viral) RNA templates.

Translation of viral mRNAs leads to the synthesis of viral coat proteins, and copying of full-length *plus*-strand RNA leads to the formation of full-length *negative*-strand molecules. *Assembly* of an enveloped virus is considerably more complex than assembly of a simple virus particle. Two kinds of coat proteins are formed, *nucleocapsid proteins* and *envelope proteins*. The nucleocapsid is formed first, by association of the nucleocapsid protein molecules around the viral RNA. These nucleocapsid protein molecules are synthesized on ribosome complexes in the cytoplasm.

The *envelope proteins* of the virus are synthesized in quite a different way than the nucleocapsid proteins. First, we should note that the envelope proteins of the virus, like other proteins that associate with membranes, possess hydrophobic amino acid leader

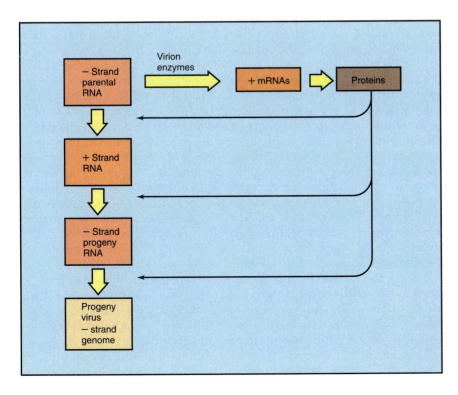

FIGURE 6.39
Flow of events during multiplication of a negative-stranded RNA virus.

sequences 15- to 20-amino acids long at their amino terminal ends, which are cleaved off as soon as they have penetrated the lipid bilayer. Hydrophobic leader sequences ensure that viral envelope proteins will associate with membranes. The envelope proteins are synthesized on ribosome complexes that are themselves associated with membranes. As these proteins are synthesized, sugar residues are added, leading to the formation of *glycoproteins*. Such glycoproteins, characteristic of membrane-associated proteins, are transported to the cytoplasmic membrane, where they replace host membrane proteins. Nucleocapsids then migrate to the areas on the cytoplasmic membrane where these virus-specific glycoproteins exist, recognizing the virus glycoproteins with great specificity. The nucleocapsids then become aligned with the glycoproteins and bud through them, becoming coated by the glycoproteins in the process. The final result is an enveloped virus particle which has a nucleocapsid center and a surrounding membrane whose lipid is derived from the host cell but whose membrane proteins are encoded by the virus. The budding process itself does not cause detectable damage to the cell, which may continue to release virus particles in this way for a considerable period of time. (Host damage does occur, but is brought about by other unknown factors.)

Influenza and other orthomyxoviruses

Another group of negative-strand viruses of great importance is the group called the *orthomyxoviruses*, which contains the important human virus *influenza*. The term *myxo* refers to the fact that these viruses interact with the *mucus* or *slime* of cell surfaces. In the case of influenza virus, this mucus is at the mucous membrane of the respiratory tract, as these viruses are transmitted primarily by the respiratory route (see Section 15.4). The term *ortho* has been added to the influenza virus group to distinguish this group from another group of negative-strand viruses, the *paramyxovirus group*. The paramyxoviruses, which include such important human viruses as those causing mumps and measles, are actually similar in molecular biology to rhabdoviruses. The orthomyxoviruses have been extensively studied over many years, beginning with early work during and after the 1918 pandemic of influenza which caused the deaths of millions of people world-wide (see Section 15.4).

The orthomyxoviruses are enveloped viruses in which the viral RNA is present in the virus in a number of separate pieces. The genome of the orthomyxoviruses is thus said to be a **segmented genome**. In the case of influenza A virus, the genome is segmented into *eight* linear single-stranded molecules ranging in size from 890 to 2341 nucleotides. The influenza virus nucleocapsid is of helical symmetry, about 6–9 nm in diameter and about 60 nm long. This nucleocapsid is embedded in an envelope which has a number of viral-specific proteins as well as lipid derived from the host (Figure 6.40).

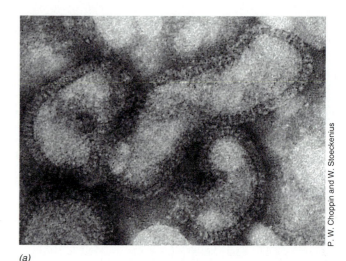

(a)

P. W. Choppin and W. Stoeckenius

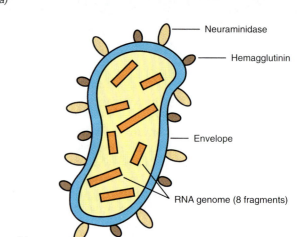

- Neuraminidase
- Hemagglutinin
- Envelope
- RNA genome (8 fragments)

(b)

FIGURE 6.40 *Influenza virion structure. (a) Electron micrograph. (b) Diagram, showing some of the components.*

Because of the way influenza virus buds as it leaves the cell, the virus has no defined shape and is said to be *polymorphic* (Figure 6.40a). There are spikes on the outside of the envelope which interact with the host cell surface. One spike is called a *hemagglutinin* because it causes agglutination of red blood cells. (Agglutination is a process by which cells are caused to clump when they are mixed with an antibody or other protein or polysaccharide molecule which combines specifically with a substance on the cell surface, as described in Sections 12.12 and 13.8.) If the cells undergoing agglutination are red blood cells, then the process is called *hemagglutination*. (*Hema* is the combining form referring to *blood*.) The red blood cell is not the type of host cell which the virus normally infects, but contains on its surface the same type of membrane component, chemically characterized as *sialic acid*, which the mucous membrane cells of the respiratory tract contain. Thus, the red blood cell is merely a convenient cell type for measurement of agglutination activity. An important feature of the influenza virus hemagglutinin is that antibody directed against this hemagglutinin will *prevent* the virus from infecting a cell. Thus, antibody directed against the hemagglutinin *neutralizes* the virus, and this is the mechanism by which immunity to influenza is brought about during the immunization process (see Sections 12.15 and 15.4).

A second type of spike on the virus surface is an enzyme called *neuraminidase* (see Figure 6.40). Neuraminidase breaks down the sialic acid component of the cytoplasmic membrane, which is a derivative of neuraminic acid. Neuraminidase appears to function primarily in the virus assembly process, destroying host membrane sialic acid that would otherwise block assembly or become incorporated into the mature virus particle.

In addition to the neuraminidase, the virus possesses two other enzymes, *RNA-dependent RNA polymerase*, which is involved in the conversion of negative to positive strand (as already discussed for the rhabdoviruses), and an *RNA endonuclease*, which cuts a primer from capped mRNA precursors.

The virus particle enters via the process of *endocytosis*. Once inside the cytoplasm, the nucleocapsid becomes separated from the envelope and migrates to the nucleus. *Replication* of the viral nucleic acid then occurs in the nucleus. *Uncoating* results in the activation of the virus RNA polymerase. The mRNA molecules are then *transcribed* in the nucleus from the virus RNA, using oligonucleotide primers which have been cut from the 5'-ends of newly synthesized capped cellular mRNAs. Thus, the viral mRNAs have 5'-caps. The poly (A) tails of the viral mRNAs are added and the virus mRNA molecules move to the cytoplasm.

Although influenza virus RNA replicates in the nucleus, influenza virus *proteins* are synthesized in the cytoplasm. There are *ten* virus proteins encoded by the *eight* segments of the virus genome (Table 6.3). The six mRNAs transcribed from segments 1 through 6 each encode a single protein, while the other two segments (numbers 7 and 8 in Table 6.3) each encode two proteins. This is not done by using true polycistronic mRNA as in prokaryotes because eukaryotic ribosomes typically only recognize the AUG codon closest to the 5'-end of the mRNA as a start codon (see Section 5.8). Therefore, they can only make one protein from a given RNA. The original full-length mRNA transcribed from segments 7 and 8 are each translated to give one protein. In each case, an additional protein is translated from these messages after they have been processed by the host's RNA splicing machinery. Like overlapping genes, this is another example of how RNA viruses make maximum use of their small genome size.

Some of these proteins are involved in virus RNA replication and others are structural proteins of the virion. The overall strategy of virus RNA synthesis resembles that of the rhabdoviruses, with primary transcription resulting in the formation of *plus*-stranded templates for the formation of progeny *minus*-stranded molecules. Details of assembly are still uncertain. One possibility is that the nucleocapsid proteins are transported from the cytoplasm to the cell nucleus, where *assembly* of the nucleocapsids occurs. The assembled nucleocapsids then migrate to the *cytoplasmic membrane* where the hemagglutinin and neuraminidase are present. The formation of the complete enveloped virus particle occurs by a budding out process, as was described for the rhabdoviruses. The neuraminidase plays a role in permitting the liberation of the virus particle from the cell, and virus liberation does not occur if neuraminidase activity is inhibited. It is thought that the neuraminidase removes neuraminic acid from oligosaccharide units of the

Table 6.3   The eight influenza virus RNA segments and the proteins they encode

Segment	Number of nucleotides	Protein coded for	Size of protein (mol wt)	Approx. percentage of virus particle protein	Function of protein
1	2341	P1	96,000	1	Initiation of transcription
2	2300	P2	87,000	1	Cap-binding protein
3	2233	P3	85,000	1	Elongation of transcription
4	1765	HA (HA1)	36,000	25	Hemagglutinin spike. HA is
		HA (HA2)	27,000		converted into HA1 and HA2.
5	1565	NP	56,000	30	Structural protein of helical nucleocapsid
6	1413	NA	50,000	4	Neuraminidase
7	1027	M1	27,000	38	Matrix protein
		M2	11,000		Nonstructural protein of unknown function
8	890	NS1	26,000	trace	Nonstructural protein of unknown function
		NS2	12,000	trace	Nonstructural protein of unknown function

hemagglutinin protein, thus eliminating the affinity which the virus has for the host cell surface.

The segmented genome of the influenza virus has some important practical consequences. Influenza virus, and other viruses of this family, exhibit a phenomenon called **antigenic shift**, in which pieces of the RNA genome from two genetically distinct strains that have infected the same cell become associated. This results in a change in the surface antigens (coat proteins) of the virus, making the virus resistant to antibody that has been formed as a result of an immunization process (see Section 15.4). This antigenic shift makes it possible for the newly formed virus to infect hosts that the parent could not have infected. Antigenic shift is thought to bring about major epidemics of influenza.

> In a number of RNA viruses, called negative-strand viruses, the virus RNA does not act directly as messenger, but is copied into mRNA by an RNA-dependent RNA polymerase present in the viral particle. An important negative-stranded virus is influenza virus.

6.17   Double-Stranded RNA Viruses: Reoviruses

The reoviruses, an important family of animal viruses, have a genome consisting of *double-stranded RNA*. The name *reovirus* is an acronym, derived from the terms *r*espiratory, *e*nteric, and *o*rphan. The term *orphan* was applied because the first viruses of this group to be isolated from humans were not associated with any specific disease syndrome. However, viruses of the reovirus group are known to infect a variety of mammals as well as insects and plants. The mammalian reoviruses first isolated are now classified in the virus genus *Reovirus*. To date, no specific human syndrome has been associated with this genus, although these viruses are commonly isolated from persons with an array of inconsequential respiratory and/or gastrointestinal illnesses (as well as being isolated from apparently healthy people). A more important genus of reoviruses is the genus *Rotavirus*, a member of which has been implicated in an important diarrheal disease of infants. *Rotavirus* is probably the most common cause of diarrhea in infants from 6 to 24 months of age. Rotaviruses are also known to cause diarrhea in young animals. A reovirus of the genus *Orbivirus* causes Colorado tick fever in humans. Other orbiviruses cause important veterinary diseases such as equine encephalosis and blue-tongue disease in sheep.

As noted, the RNA of the reoviruses is double stranded. This is the only group of animal viruses with double-stranded RNA, all other RNA virus groups having single-stranded RNA (Figure 6.35). The reovirus particle consists of a nonenveloped nucleocapsid 60–80 nm in diameter, with a *double* shell of

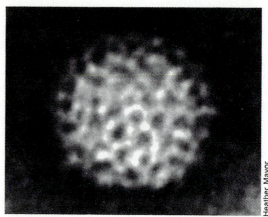

FIGURE 6.41   A reovirus particle. The diameter of the particle is about 70 nm.

icosahedral symmetry (Figure 6.41). Three proteins make up the outer capsid and five proteins the inner capsid. Present within the virus particle are several enzymes: *RNA-dependent RNA polymerase, nucleotide phosphohydrolase*, and enzymes that participate in the capping of messenger RNA, the *RNA methyltransferase* and *guanyl transferase*.

However, the unique thing about the reoviruses is their *genome*, which consists of 10–12 molecules of double-stranded RNA. Generally, each molecule of RNA in the genome codes for a single protein, although in a few cases the protein formed is cleaved to give the final product. However, one of the mRNAs produced actually encodes two proteins, and the RNA does not have to be modified to do so. Apparently a ribosome sometimes "misses" the start codon for the first gene in this message and travels on to the start codon of the second gene. As is true for many generalizations in the field of biology (such as "eukaryotic ribosomes initiate at the first AUG codon in mRNA"), continued investigation often uncovers exceptions. These exceptions are usually first found in viruses and often seem related to their small genome size. Replication of reovirus RNA occurs exclusively in the *cytoplasm* of the host. The double-stranded RNA is inactive as mRNA and the first step in reovirus replication is *transcription*, using the minus strand as a template to make mRNA.

Replication of the reovirus seems to occur within an intracellular equivalent of the viral core, called the *subviral particle*, which remains intact in the cell. Each of the 10 capped single-stranded plus RNAs is assembled into this double-stranded RNA synthesizing body. The capped single-stranded plus RNAs act as templates for the synthesis of the progeny minus genomic RNAs, yielding progeny double-stranded viral RNAs. The progeny double-stranded RNAs are further encapsidated and when enough viral capsid proteins are present, mature virions are assembled.

In the initial infection process, the virion binds to a cellular protein. Once attachment has occurred, the virus enters the cell and is transported into lysosomes. Within the lysosome the outer shell of the virus particle is modified by removal of two proteins and cleavage of another by lysosomal enzymes. This uncoating process activates the viral RNA-dependent RNA polymerase and hence initiates the viral replication process.

6.18 Replication of DNA Viruses of Animals

Most animal viruses with DNA genomes contain double-stranded DNA (one group, the parvoviruses, contains single-stranded DNA). Among these DNA viruses of animals, the four major families are the papovaviruses, the herpesviruses, the pox viruses, and the adenoviruses. Of these, all replicate in the nucleus except for the pox viruses, which have the unique character (for DNA viruses) of replicating in the cytoplasm. In the present section, we discuss the replication of each of these families briefly.

Papovaviruses: SV40

Some viruses of the papovavirus group have the interesting property of inducing cancer in animals. One of the DNA viruses of animals, the virus SV40, was first isolated from monkeys, and it was thus called *Simian Virus 40*. SV40 was one of the first genetic elements to be studied by genetic engineering techniques, and has been extensively used as a *vector* for moving genes into cells. When a virus of the papovavirus group infects a host cell, one of two modes of replication can occur, depending upon the type of host cell. In some types of host cells, known as *permissive* cells, virus infection results in the formation of new virions and the lysis of the host cell. In other types of host cells, known as *nonpermissive*, efficient multiplication does not occur, but the virus DNA becomes integrated into some of the host cells, thereby creating new, genetically altered cells. Such cells frequently show loss of growth inhibition and are called *transformed* cells.

The SV40 virion is a simple nonenveloped particle 45 nm in diameter with an icosahedral head containing 72 protein subunits. The capsid has one major protein and two minor proteins, and there are no enzymes in the virus particle. Four host-derived *histone proteins* are found complexed with the viral DNA. We have discussed histone proteins during our discussion of chromosome structure (see Section 3.13) and have noted that histones play a role in neutralizing the negative charge originating from the phosphates of chromosomal DNA and aid in the packing of the DNA into more compact configurations.

The genome of SV40 consists of one molecule of double-stranded DNA of 5243 base pairs. The DNA is circular (Figure 6.42) and exists in a supercoiled configuration within the virion. The complete base sequence of SV40 has been determined and the genetic map is known in some detail (Figure 6.43). The genome of SV40 has a single *Eco*RI site, which is useful in cloning DNA.

The nucleic acid is synthesized in the nucleus but the proteins are synthesized in the cytoplasm. Final assembly of the virus particle occurs in the nucleus. The replication phase of these viruses can be divided into two distinct phases, *early* and *late*. During the *early phase*, about 50 percent of the viral DNA, in the region called the early genes, is transcribed. A single RNA molecule, the primary transcript, is made by cellular RNA polymerase, but it is processed into *two species of mRNA*, a large one and a small one. The DNA of SV40 has *introns* that are excised out of the primary RNA transcript. In the cytoplasm, viral mRNA is translated with the formation of two proteins. One of these proteins, the T-antigen, binds to the site on the parental DNA which is the *origin of replication*.

The viral DNA of SV40 is too small to code for its own DNA polymerases; host DNA polymerases are used. Replication occurs in a bidirectional fashion (so-called *theta* replication, see Figure 5.21) from a single replication origin. The process involves the same events that have already been described for host cell

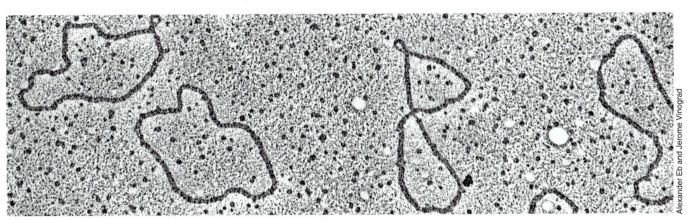

FIGURE 6.42 Electron micrograph of circular DNA from a tumor virus. The contour length of each circle is about 1.5 μm.

Alexander Eb and Jerome Vinograd

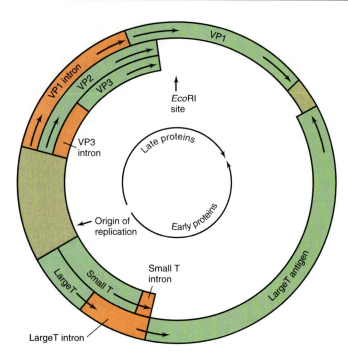

FIGURE 6.43 Genetic map of SV40. VP1, VP2, and VP3 are the genes coding for the three proteins that make up the coat of SV40.

DNA replication (see Section 5.4): RNA primer synthesis, formation of discontinuous DNA fragments on the lagging strand, gap filling, ligase action, and supercoiling the DNA through the action of gyrase. Because of the circular nature of SV40 DNA, the formation of concatamers does not occur.

Late SV40 mRNA molecules are synthesized using the strand complementary to that used for early mRNA synthesis (see Figure 6.43). Transcription begins at several positions in a small region near the origin of replication, leading to the synthesis of very large primary RNA transcripts. These RNAs are then processed by splicing and polyadenylation to give multiple forms of mRNA corresponding to the three coat proteins. These genes overlap; part of the nucleotide sequence contains information for all three proteins. These late mRNA molecules are transported to the cytoplasm and translated into the viral coat proteins. These proteins are then transported back into the nucleus, where *virion assembly* takes place. The mechanism by which virions are released from the cell is not known.

In *nonpermissive hosts,* transformation to *tumor cells* can occur. In such hosts, the early phase of viral replication occurs but the late phase does not. Once the viral DNA has entered the nucleus and transcription of the early genes has taken place, there is a great stimulation in all of the host cell's biosynthetic activities involved in cellular DNA replication and mitosis. However, no replication of viral DNA occurs. The characteristic of nonpermissive cells is that viral replication cannot occur in them but infection can occur. It is generally found that if a cell type can support viral multiplication, then the cell will not be transformed; if

the virus cannot multiply and the cell can be infected, then the cell is often transformed. In the transformed cell, the viral DNA becomes stably integrated into the DNA of the host cell (Figure 6.44). In the integrated state, the viral DNA can only replicate as a *cellular gene.* Integration can occur at many sites in the cellular and viral genome. In this integrated form, two viral proteins are made which are essential for the maintenance of a stably integrated viral DNA, but no viral structural proteins are synthesized. Some transformed cells can become converted into cells capable of producing virus, a process which probably involves excision of the viral genome from the host genome. It is also possible to detect host cells which have reverted to the noncancerous state, not by having lost the viral DNA but by changing the host regulatory processes in such a way that the cell no longer replicates like a transformed cell.

A study of the manner of replication of SV40 virus has given some important insights into the manner by which viruses bring about the cancerous state in host cells. However, we note that DNA viruses of other groups can also cause the cancerous transformation. Also, the important family of viruses, the *retroviruses,* are also cancer viruses but have a completely different mode of replication (see Section 6.22).

> A number of viruses causing cancer and other conditions in animal cells are double-stranded DNA viruses. In a permissive host, the virus may cause death and lysis, but in a nonpermissive host, it may cause transformation of the infected host cell to a tumor cell.

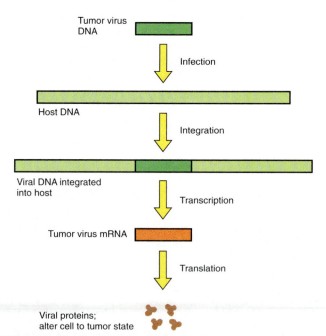

FIGURE 6.44 General scheme of molecular events involved in cell transformation by a DNA tumor virus such as SV40. All or portions of viral DNA are incorporated into host cell DNA. The viral genes that encode transforming information are transcribed and processed to viral mRNA molecules that are transported to the cytoplasm. Here they are translated to form transforming proteins or T antigens that code for functions that can convert host cells into cancer cells.

6.19   Herpesviruses

The herpesviruses are a large group of double-stranded DNA viruses which cause a wide variety of diseases in humans and animals, including fever blisters (cold sores), venereal herpes, chickenpox, shingles, and infectious mononucleosis; a number of these diseases are discussed in Chapter 15. Some herpesviruses also cause cancer. One of the interesting features of some herpesviruses is their ability to remain *latent* in the body for long periods of time, becoming active only under conditions of stress. Both *herpes simplex*, the virus that causes *fever blisters*, and *varicella-zoster virus*, the cause of *chicken pox* and *shingles*, are able to remain latent in the neurons of the sensory ganglia, from which they are able to emerge to cause infections of the skin.

An important group of herpesviruses are tumorigenic, causing clinical forms of cancer. One herpesvirus which is tumorigenic is the *Epstein-Barr virus*, which causes *Burkitt's lymphoma*, a common tumor among children in Central Africa and New Guinea. Burkitt's lymphoma was among the first human cancers to have been linked to virus infection (see Table 6.2).

The **herpesvirus particle** is structurally complex, consisting of four distinct morphologic units. In herpes simplex type 1, an enveloped virus of about 150 nm in diameter (Figure 6.45a), the center of the virus is an electron dense *core* which consists of *double-stranded* DNA. Surrounding this core is the nucleocapsid, of icosahedral symmetry, which consists of 162 capsomeres, each of which is composed of a number of distinct proteins. Outside the nucleocapsid is an electron dense amorphous material which is called the *tegument*, a fibrous structure which is unique to the herpesviruses. Surrounding the tegument is an *envelope* whose outer surface contains many small *spikes*. A large number of separate proteins are present within the virion, but not all of them have been characterized.

The *genome* of herpes simplex type I virus consists of one large *linear* double-stranded DNA molecule of 152,260 base pairs with a molecular weight of about 100×10^6 (about 30 times that of SV40). As a further indication of the complexity of herpes simplex, the DNA sequence of this virus indicates that it codes for at least 70 separate proteins.

Infection occurs by attachment of virus particles to specific cell receptors, and, following fusion of the cytoplasmic membrane with the virus envelope, the nucleocapsid is released into the cell. The nucleocapsids are transported to the nucleus, where viral DNA is uncoated. Components of the virus particle inhibit macromolecular synthesis by the host.

There are *three classes of messenger RNAs*: immediate early (also called alpha, α), which codes for five regulatory proteins; delayed early (also called beta, β), which codes for DNA replication proteins, including thymidine kinase and DNA polymerase; and late (also called gamma, γ), which codes for the proteins of the virus particle (Figure 6.45b). During the *immediate early stage*, about one-third of the viral genome is tran-

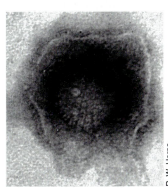

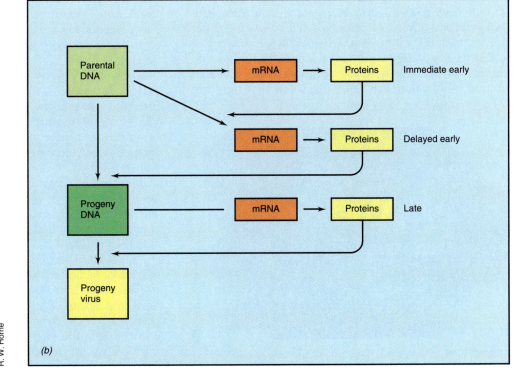

FIGURE 6.45   Herpesvirus. (a) Electron micrograph of a herpesvirus particle. The diameter of the particle is about 150 nm. (b) Flow of events in multiplication of herpes simplex virus.

scribed by a host cell RNA polymerase. Early mRNA codes for certain positive-acting regulatory proteins which appear to stimulate the synthesis of the delayed early proteins. The second stage, *delayed early*, only occurs after the early proteins have been made. During this stage, about 40 percent of the viral genome is transcribed. Among the 10 proteins characterized from the delayed early stage are a *DNA polymerase*, enzymes involved in synthesis of *deoxyribonucleotides*, and a *DNA-binding protein*. These enzymes are all involved in the process of viral DNA replication.

Herpes viral DNA synthesis itself takes place in the nucleus. After infection, the herpes genome apparently circularizes (remarkably like bacteriophage lambda; see Section 6.12) and replicates by a rolling circle mechanism (see Figure 6.31). However, three origins seem to be involved. Long concatamers are formed which become processed to viral length DNA during the assembly process itself (in a manner similar to that described for DNA bacteriophages; see Sections 6.11 and 6.12).

Viral nucleocapsids are assembled in the nucleus and acquisition of the virus envelope occurs via a budding process through the *inner membrane of the nucleus*. Mature virions are subsequently released through the endoplasmic reticulum to the outside of the cell. Thus, the assembly of this enveloped DNA virus differs markedly from that of the enveloped RNA viruses, which were assembled on the *cytoplasmic* membrane instead of the nuclear membrane.

6.20   Pox Viruses

These are the most complex and largest animal viruses known (Figure 6.46) and have some characteristics that approach those of primitive cells. However, the pox viruses, like all viruses, are not able to metabolize and thus depend upon the host for the complete machinery of protein synthesis. These viruses are also unique in that they are DNA viruses which replicate

in the *cytoplasm*. Thus, a host cell infected with a pox virus exhibits *DNA synthesis outside of the nucleus*, something that otherwise only occurs in intracellular organelles such as mitochondria.

Pox viruses have been important medically as well as historically. *Smallpox* was the first virus to be studied in any detail, and was the first virus for which a vaccine was developed (described by Edward Jenner in 1798). By diligent application of this vaccine on a worldwide basis, the disease smallpox has been *eradicated*, the first infectious disease to have been eliminated in this fashion. Other pox viruses of importance are *cowpox* and *rabbit myxomatosis virus*, an important infectious agent of rabbits and one which was intentionally used in an attempt to control the Australian rabbit population (see Section 14.6). Some pox viruses also cause tumors, but these tumors are generally benign.

The pox viruses are very large, so large that they can actually be seen under the light microscope. Most research has been done with *vaccinia virus*, the source of smallpox vaccine. The vaccinia particle is a brick-shaped structure about $400 \times 240 \times 200$ nm in size. The virus particle is covered on its outer surface with tubules or filaments arranged in a membrane-like pattern, although the virus does not have a lipid membrane since the outer envelope consists of protein. Within the virion there are two lateral bodies of unknown composition and a core, the *nucleocapsid*, which contains DNA bounded by a layer of protein subunits.

The pox virus genome consists of double-stranded DNA. The vaccinia virus genome has about 185 kilobase pairs and contains between 150 and 200 genes. The pox DNA is interesting because the two strands of the double helix are crosslinked at the ends, due to the formation of phosphodiester bonds between adjacent strands (as in the hairpin structure shown in Figure 5.11).

Vaccinia virions are taken up into cells via a phagocytic process, from which the cores are liberated into the cytoplasm. Interestingly, *uncoating* of the virus genome requires the action of a new protein that is synthesized after infection. This protein is encoded by viral DNA and the gene specifying this protein is transcribed by an *RNA polymerase* that is present *within* the virus particle. In addition to this uncoating gene, a number of other viral genes are transcribed. The primary transcripts are turned into mRNAs by capping and polyadenylation while they are still inside the virus core.

Once the vaccinia DNA is fully uncoated, the formation of *inclusion bodies* within the cytoplasm begins. Within these inclusion bodies, transcription, replication, and encapsidation into progeny virus particles occurs. Each infecting virion initiates its own inclusion body, so that the number of inclusions depends on the multiplicity of infection. Progeny DNA molecules form a pool from which individual molecules are in-

FIGURE 6.46 Electron micrograph of a negatively stained vaccinia virus virion. The virion is approximately 400 nm (0.4 μm) long.

D. Dales and F. Fenner

corporated into virions. Mature virions accumulate in the cytoplasm. There seems to be no specific release mechanism and most virions are released only when the infected cell disintegrates.

Pox viruses and recombinant vaccines

Vaccinia virus has been used as a host for genetically altered proteins of other viruses, permitting the construction of genetically engineered vaccines. As we will see in Chapter 12, a vaccine is a substance capable of eliciting an immune response in an animal and serves to protect the animal from future infection with the same agent. Vaccinia virus causes no serious health effects in humans but is highly immunogenic. Molecular cloning methods have been used to express key viral proteins of influenza virus, rabies virus, herpes simplex type 1 virus, and hepatitis B virus in vaccinia virus virions, and then the latter used as a vaccine (see Section 8.13). Thus far very promising results have been achieved in the battle against diseases caused by the viruses listed above, and research is progressing on using vaccinia virus as a vehicle for cloned viral proteins from HIV, the human immunodeficiency virus. Such an approach, if successful, could result in a safe and effective AIDS vaccine.

A similar vaccine delivery system using adenovirus (see next section) as vehicle has been developed because, like vaccinia virus, adenoviruses are of little health consequence to humans.

> Pox viruses are DNA viruses with an unusual property: they replicate in the cytoplasm of host cells. Smallpox, and its relative vaccinia (cowpox), were the first viruses to be studied in any detail. Vaccinia virus causes only a weak infection in humans, and has been used for hundreds of years as a vaccine for smallpox. Vaccinia virus is now being used in genetic engineering experiments to introduce foreign genes into animal or human cells. This opens up the possibility for correcting inborn genetic deficiencies (gene therapy) and for the production of completely safe vaccines.

6.21   Adenoviruses

The adenoviruses are a major family of icosahedral DNA-containing viruses which have unique molecular biological properties. The term *adeno* is derived from the Latin for *gland* and refers to the fact that these viruses were first isolated from the tonsils and adenoid glands of humans. Adenoviruses cause generally mild respiratory infections in humans, and a number of such viruses are isolated from apparently healthy individuals.

The *genomes* of the adenoviruses consist of linear double-stranded DNA of about 36 kilobase pairs. Attached in covalent linkage to the 5'-terminus of the DNA is a protein component which is essential for infectivity of the DNA. The DNA has inverted terminal repeats of 100–1800 base pairs (this varies with the virus strain). The DNA of the adenoviruses is six to seven times the size of the DNA of SV40.

Replication of the virus DNA occurs in the *nucleus* (Figure 6.47). After the virus particle has been transported to the nucleus, the core is released and converted into a viral DNA-histone complex. *Early transcription* is carried out by an RNA polymerase of the host and a number of primary transcripts are made. The transcripts are spliced, capped, and polyadenylated, giving several different mRNAs.

The early proteins are involved in regulation of DNA replication; the later proteins are the virus coat proteins. *Viral DNA replication* uses a virus-encoded protein as a primer and another virus-encoded protein as DNA polymerase. For the replication of a *linear double-stranded* DNA molecule such as that of adenovirus, initiation of replication could begin at either end or at both ends simultaneously (see Figure 5.25). In the case of the adenoviruses, replication begins at *either* end, the two strands being replicated asynchronously. The products of a round of replication are double-stranded and single-stranded molecules. The latter then cyclizes by means of the inverted terminal repeats, and a new

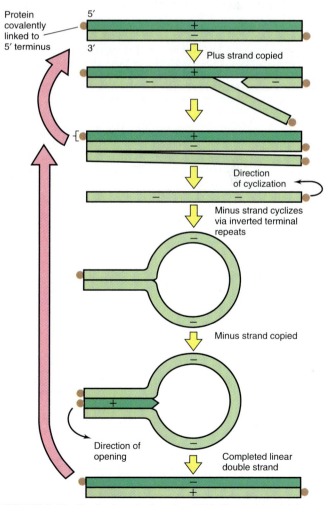

FIGURE 6.47   Replication of adenovirus DNA. See text for details.

complementary strand is synthesized beginning from the 5'-end, the products being another double-stranded molecule (Figure 6.47). This mechanism of replication is interesting because it does not involve the formation of discontinuous fragments of DNA on the lagging strand, as occurs in conventional DNA replication.

6.22   Retroviruses

We now discuss one of the most interesting and complex families of animal viruses, the **retroviruses**. The term *retro* means *backwards*, and is derived from the fact that these viruses appear to have a backwards mode of nucleic acid replication. The retroviruses are RNA viruses but they *replicate by means of a DNA intermediate*. The retroviruses are of interest for a number of reasons. First, they were the first viruses shown to cause *cancer*, and have been studied most extensively for their carcinogenic characteristics. Second, one retrovirus, the one causing *acquired immune deficiency syndrome (AIDS)*, has only been known since the early 1980s but has already become a major public health problem. Third, the retrovirus genome can become specifically integrated into the host genome, by way of

the DNA intermediate, and this integration process is being studied as a means of introducing *foreign* genes into a host, a process called *gene therapy*. Finally, the retroviruses have an enzyme, called *reverse transcriptase*, which copies RNA sequences into DNA, and this enzyme has become a major tool in *genetic engineering*.

As we will see, the retroviruses have some properties like those of RNA viruses and some like those of DNA viruses. They resemble to a considerable extent movable genetic elements and are sometimes considered to be *escaped cellular transposable elements*. In this respect, the retroviruses resemble bacterial viruses such as *Mu* (see Section 6.13). We should note that the use of reverse transcription is not restricted to the retroviruses, because hepatitis B virus (a human virus) and cauliflower mosaic virus (a plant virus) also use reverse transcription in their replication processes. But in contrast to the retroviruses, these latter viruses encapsidate the DNA genome rather than the RNA genome as retroviruses do. Some transposable elements of eukaryotes, called *retrotransposons*, also encode and use reverse transcriptase as part of their replication cycle. (These retrotransposons also make virus-like particles, but the particles remain inside the cell; retrotransposons can be considered viruses that

FIGURE 6.48
Retrovirus structure and function. (a) Structure of a retrovirus. (b) Structure of a retrovirus genome, showing the manner by which the two identical viral RNA molecules are held together. (c) Detailed genetic map, showing the location of the long terminal repeats (LTR). See text for details.

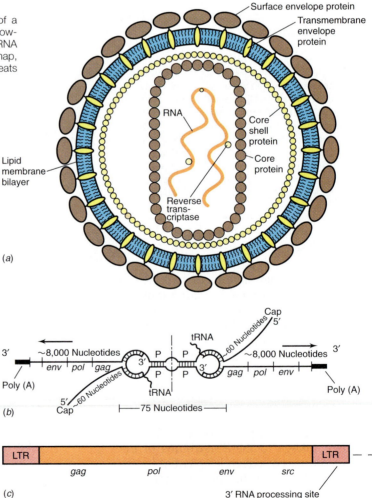

became trapped.) In addition, reverse transcriptases capable of producing small multicopy DNA (msDNA) with an RNA template have been discovered in myxobacteria and *E. coli*. The reverse transcriptase in bacteria is encoded as part of a short genetic element called a *retron*. Although many copies of the msDNA are made (which contain their RNA template covalently attached), their function is unknown.

The retroviruses are enveloped viruses of uncertain symmetry (Figure 6.48*a*). There are a number of proteins in the virus coat, including at least two envelope proteins and typically seven internal proteins, four of which are structural and three enzymatic. The enzymatic activities found in the virus particle are *reverse transcriptase, DNA endonuclease (integrase)*, and a *protease*.

Features of retroviral genomes and replication

The *genome* of the retrovirus is unique. It consists of *two* identical single-stranded RNA molecules, of *plus* complementarity, which are hydrogen-bonded together by means of base pairing with specific cellular tRNA molecules (Figure 6.48*b*). Each of the two RNAs is 8.5–9.5 kb in length. The 5' terminus of the RNA is capped and the 3' terminus is polyadenylated, so that the RNA is capable of acting directly as mRNA but is not used as such. A *genetic map* of the most studied retrovirus, that of Rous sarcoma, is shown in Figure 6.48*c*. Key regions of this map are those called *gag*, coding for internal structural proteins; *pol*, coding for reverse transcriptase and integrase; *env*, coding for envelope proteins; and *src*, a cancer transformation gene carried by certain retroviruses. The long terminal repeat (LTR) shown on the map plays an essential role in the replication process.

The overall process of replication of a retrovirus can be summarized in the following steps (Figure 6.49):

1. **Entrance** into the cell.

2. **Reverse transcription** of *one* of the two RNA genomes into a single-stranded DNA which is subsequently converted to a linear double-stranded DNA by reverse transcriptase.

3. **Integration** of the DNA copy into the host genome.

4. **Transcription** of the viral DNA, leading to the formation of viral mRNAs and progeny viral RNA.

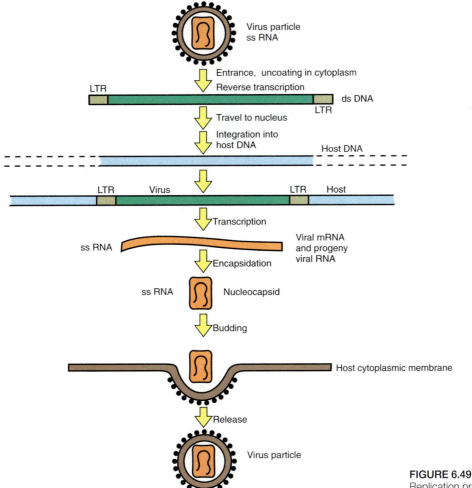

FIGURE 6.49
Replication process of a retrovirus.

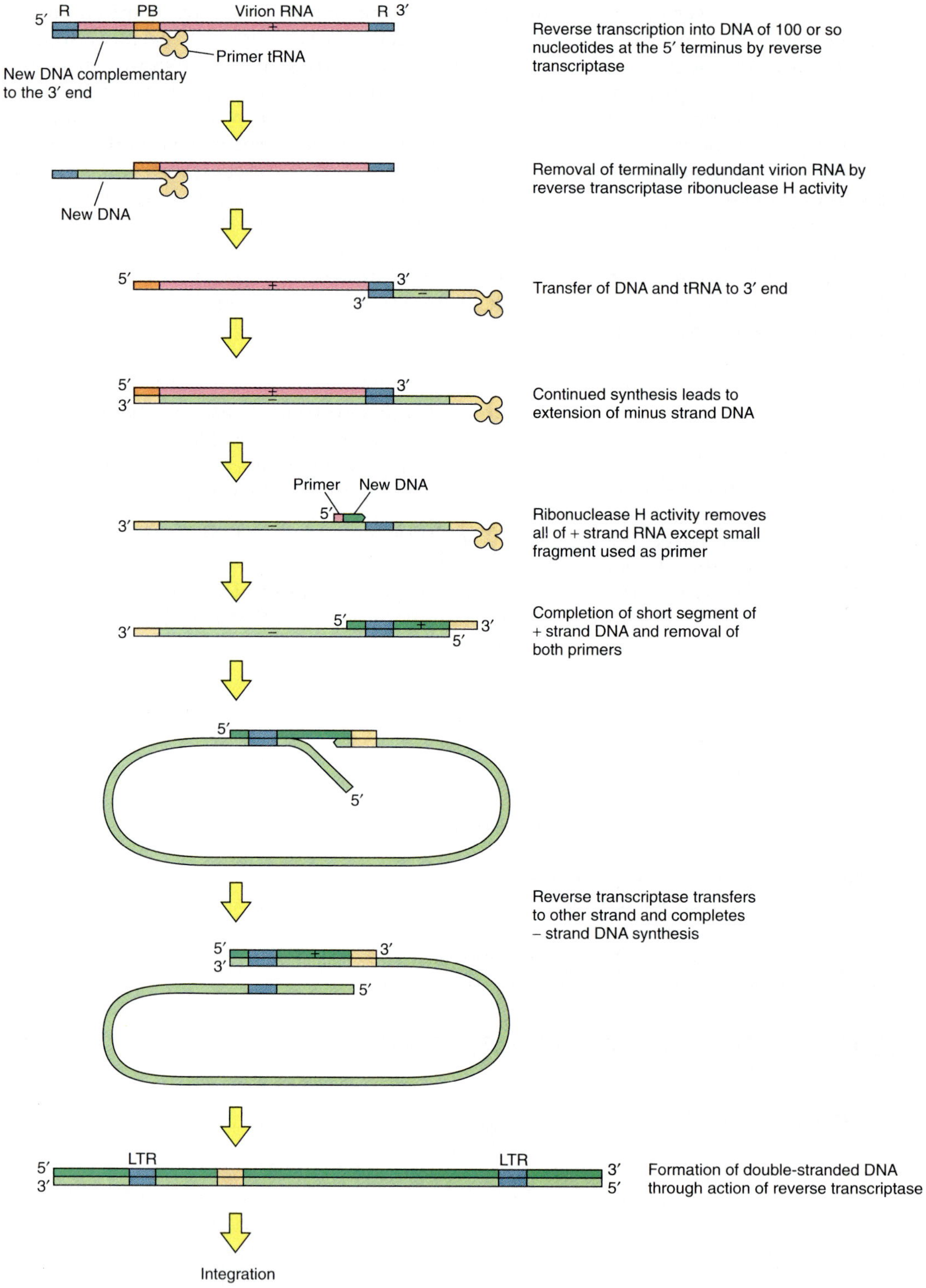
5′
R
PB
Virion RNA
R 3′
+
Primer tRNA
New DNA complementary
to the 3′ end
Reverse transcription into DNA of 100 or so nucleotides at the 5′ terminus by reverse transcriptase
New DNA
Removal of terminally redundant virion RNA by reverse transcriptase ribonuclease H activity
5′
3′
3′
Transfer of DNA and tRNA to 3′ end
Continued synthesis leads to extension of minus strand DNA
Primer
New DNA
Ribonuclease H activity removes all of + strand RNA except small fragment used as primer
Completion of short segment of + strand DNA and removal of both primers
Reverse transcriptase transfers to other strand and completes – strand DNA synthesis
LTR
LTR
Formation of double-stranded DNA through action of reverse transcriptase
Integration

FIGURE 6.50 Overall steps in the formation of double-stranded DNA from retrovirus single-stranded RNA. The sequences labeled R on the RNA are direct repeats found at either end. The sequence labeled PB is where the primer (tRNA) binds. Note that the process of DNA synthesis has yielded longer direct repeats on the DNA than were originally on the RNA. These are called long terminal repeats, or LTR.

5. **Encapsidation** of the viral RNA into nucleocapsids in the cytoplasm.
6. **Budding** of enveloped virions at the cytoplasmic membrane and release from the cell.

We now discuss some aspects of the retrovirus multiplication process in detail. As we have noted, the first step after the entry of the RNA genome into the cell is reverse transcription: conversion of RNA into a DNA copy, using the enzyme reverse transcriptase present in the virion. The DNA formed is a linear double-stranded molecule and is synthesized in the cytoplasm. An outline of the reverse transcription of viral RNA into DNA is given in Figure 6.50.

The enzyme reverse transcriptase is essentially a DNA polymerase, but it actually shows *three* enzymatic activities: (1) synthesis of DNA with an RNA template (reverse transcription), (2) synthesis of DNA with a DNA template, and (3) ribonuclease H activity (an activity that degrades the RNA strand of an RNA:DNA hybrid). Like all DNA polymerases, reverse transcriptase needs a primer for DNA synthesis. The primer for retrovirus reverse transcription is a specific *cellular transfer RNA (tRNA)*. The type of tRNA used as primer depends on the virus and is brought into the virion from the last host cell. In the case of Rous sarcoma virus, the tRNA used is the *tryptophan* tRNA. Using the tRNA primer, the 100 or so nucleotides at the 5′-terminus of the RNA are reverse transcribed into DNA. Once transcription reaches the 5′-end of the RNA, the transcription process stops. In order to copy the remaining RNA, which is the bulk of the RNA of the virus, a different mechanism comes into play. First, terminally redundant RNA sequences at the 5′-end of the molecule are removed by the action of another enzymatic activity of reverse transcriptase, *ribonuclease H*. This leads to the formation of a small single-stranded DNA which is complementary to the RNA segment at the *other end* of the viral RNA. The small single-stranded piece of DNA then hybridizes with the other end of the viral RNA molecule, where copying of the viral RNA sequences continues. As summarized in Figure 6.50, continued action of reverse transcriptase and ribonuclease H leads to the formation of a double-stranded DNA molecule which has long terminal repeats (LTR) at each end (Figure 6.48c). These LTRs contain strong promoters of transcription and are involved in the integration process. The *integration* of the viral DNA into the host genome is analogous to the integration of Mu (see Section 6.13) or a bacterial transposon into a bacterial genome. Integration can occur anywhere in the cellular DNA and once integrated, the element, now called a *provirus*, is a stable genetic element.

The integrated proviral DNA is transcribed by a cellular RNA polymerase into transcripts that are capped and polyadenylated. These RNA transcripts may either be encapsidated into virus particles or they may be processed and translated into virus proteins. Some virus proteins are made initially as a large primary *gag* protein which is split by proteolytic action into the capsid proteins. Occasionally, read-through past the *gag* region (involving either inserting an amino acid in response to a stop codon or a shift in reading frame by the ribosome) leads to the translation of *pol*, the reverse transcriptase gene. Other proteins are synthesized from spliced transcripts.

When the virus proteins have accumulated in sufficient amounts, *assembly* of the nucleocapsid can occur. This encapsidation process leads to the formation of nucleocapsids which move to the cytoplasmic membrane for final assembly into the enveloped virus particles.

Not all retroviruses cause cancer, but tumorigenic retroviruses are quite common. Some tumorigenic retroviruses known cause *sarcomas or acute leukemia*; they possess high oncogenic potential. Infection with one of these viruses can cause cellular transformation, leading to the formation of a tumor. Why are these viruses tumorigenic? It appears that they possess a transforming gene, or *oncogene*, which codes for a protein that brings about cellular transformation (see Section 6.14). This gene, known in Rous sarcoma virus as the *src* gene (*src* for *sarcoma*, see Figure 6.48c), encodes a phosphoprotein which possesses protein kinase activity. Protein kinases bring about the phosphorylation of proteins, and protein phosphorylation is one mechanism for regulating the activity of proteins in eukaryotes. Depending on the protein, phosphorylation can either increase or decrease the activity of a protein. Transforming genes analogous to *src* have also been detected in human cancer cells. Interestingly, similar genes have also been detected in *normal* (that is, noncancerous) cells. These cell sequences are the proto-oncogenes (see Section 6.14) and have been found not only in mammalian cells but also in the cells of insects and yeast, suggesting that these sequences are of fundamental importance in the regulation of cell growth. Retroviruses are able to incorporate such normal sequences, which become altered and are abnormally expressed. Retroviruses are thus the agents by which such genes are transferred from cell to cell.

Genetic engineering with retroviruses occurs naturally by the incorporation of oncogene or proto-oncogene sequences. It is also possible to use modified retroviruses to incorporate foreign genes into cells. This is possible because substitution of such foreign genes for essential virus genes can lead to the production of virus particles carrying the foreign gene. Such particles are capable of being integrated into the host genome, but are incapable of replicating or causing cancer.

As we have noted, one of the newly discovered retroviruses is HIV, the virus causing AIDS. Although retroviruses had been known for a long time, it is only in the past decade or so that retroviruses have been found to infect humans. HIV, the AIDS virus, infects a specific cell type in the human, a kind of T lymphocyte that is vital for proper functioning of the immune system. We discuss the medical and immunological aspects of AIDS in Sections 14.4 and 15.7.

Because viruses are not cells, but depend on cells for their replication, viral diseases pose serious medical problems; it is frequently difficult to prevent antiviral drugs from doing some damage to host cells. Despite this, certain chemotherapeutic strategies have been devised for use against viral pathogens, including retroviruses. We discuss these in some detail along with the chemotherapy of other microbial diseases, in Section 11.17.

> **Retroviruses, which are enveloped viruses, have very complex life cycles since they are RNA viruses that replicate by means of a DNA intermediate. Important retroviruses cause cancer and acquired immune deficiency syndrome (AIDS). The retrovirus particle contains a special enzyme, reverse transcriptase, which copies the information from its RNA genome into DNA. The DNA becomes integrated into the host chromosome, in the manner of a temperate virus or transposon, where it replicates for many generations. Under certain conditions, the retrovirus DNA escapes from the host chromosome, becomes converted back into RNA, and forms mature virions which leave the host by budding off the cell membrane. Because genetic engineering occurs naturally with retroviruses, modified (harmless) retroviruses are also considered as possible agents for gene therapy.**

6.23   Viroids and Prions

In this chapter, we have discussed some representative viral groups. Recall that our definition of a virus was a genetic element that subverts the normal cellular process for its own replication and that has an extracellular form. There are a few known entities whose properties are at variance with this definition and which most scientists would not actually consider to be viruses. Two of the most important of these are *viroids* and *prions.*

Viroids are small, circular single-stranded RNA molecules that are the smallest known pathogens (ranging from Coconut cadang-cadang viroid which is 246 nucleotides in size to Citrus exocortis viroid which is 375 nucleotides). Viroids cause a number of very important crop diseases. The extracellular form of the viroid is naked RNA—there is no capsid of any kind. Even more interestingly, *the RNA molecule contains no protein-encoding genes,* and therefore the viroid is totally dependent on host function for its replication. Although the viroid RNA is a single-stranded circle, there is such considerable secondary structure possible that it resembles a short double-stranded molecule with closed ends (Figure 6.51). The viroid molecule seems to be replicated in the host cell nucleus, and its structure, which somewhat mimics DNA, apparently allows it to be replicated by the host RNA polymerase.

Section 6.22 mentioned that in certain aspects, retroviruses resemble "escaped" cellular transposable elements. Viroids are sometimes considered escaped introns and, like self-splicing introns (see Section 5.6), appear to be remnants of an RNA world.

FIGURE 6.51
Structure of viroids, showing how single-stranded circular RNA can form a seemingly double-stranded structure by intrastrand base pairing.

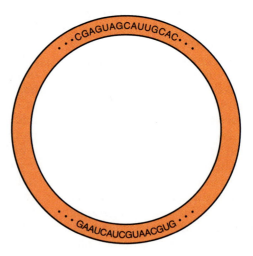

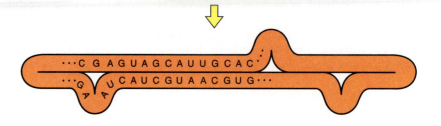

Prions represent the other extreme from viroids. They do have a distinct extracellular form, but the extracellular form may be *entirely protein.* It apparently does not contain any nucleic acid, or if it does, the molecule is not long enough to encode the single kind of protein of which the prion is composed. However, the prion protein particle is infectious, and various prions are known to cause a variety of diseases in animals, such as scrapie in sheep, bovine spongiform encephalopathy in cattle, and kuru and Creutzfeldt–Jakob disease in humans.

In addition to serious disease, prion infection results in production of more copies of the prion protein. Unless prions violate the central pattern of genetic information flow discussed in Chapter 5, this protein *must* be encoded by nucleic acid. Indeed, it has been discovered that the host cell contains a gene on one of its chromosomes that encodes a protein very similar to the prion protein. The host protein is normally produced and is found mostly in neurons. Apparently, the incoming prion modifies this host protein either during or after synthesis. This modification may involve a small fragment of nucleic acid either already in the cell or carried into the cell by the prion (it

is possible that a very small and undetectable nucleic acid is in the prion particle). Therefore, prions do not simply subvert host enzymes but somehow cause a normal host gene to produce more copies of the pathogenic protein itself.

Viroids and prions do more than stretch our definition of a virus. They also demonstrate both the unexpected ways that genetic elements can replicate and the unexpected ways they can subvert the host cell. They are, of course, also of interest to us because they cause disease.

> Viroids are circular single-stranded RNA molecules that encode no proteins and are completely dependent on host-encoded enzymes. They are the smallest known pathogens. Unlike viruses, their extracellular form is the same as their intracellular form and they have no protein coat. Prions have an extracellular form that does contain protein, but it does not contain the nucleic acid that encodes that protein. The gene that encodes the prion protein is found in the host cell and the prion somehow modifies this protein product. Although they are very simple elements, like viruses, both prions and viroids are infective and are reproduced inside cells.

Study Questions

1. Define the term *host* as it relates to viruses.
2. Define *virus*. What are the minimal features needed to fit your definition?
3. Under some conditions, it is possible to obtain nucleic acid-free protein coats (*capsids*) of certain viruses. These capsids look very similar under the electron microscope to complete virions. What does this fact tell you about the role of the virus nucleic acid in the virus assembly process? Would you expect such virions to be infectious? Why?
4. Write a paragraph describing the events which occur on an agar plate containing a bacterial lawn which result when a single bacteriophage particle causes the formation of a *bacteriophage plaque.*
5. Suppose you had been the first person to discover bacterial viruses. You observe clear plaques on a bacterial lawn and hypothesize that these plaques arise because of the action of a virus (remember that viruses affecting plants and animals had already been characterized). Describe the experiments you would carry out: (1) to show that the clear zones (plaques) were really due to an infectious agent; (2) that the agent was subcellular. On one of your plates, two different types of plaque appear. Describe the experiments you would carry out to show that each plaque type is due to a separate virus type.
6. *Chemotherapeutic agents* are lacking for most virus diseases. From what you know about the stages of virus multiplication, write (in general terms) how you might try to assay (detect) a chemotherapeutic agent which acts at each of these stages?
7. Several stages in *virus multiplication* involve specific host biochemical machinery. Why is it less likely that successful (that is, nontoxic) chemotherapeutic agents could be found that would affect these stages?

8. How might an agent which affects the *virus release process* still be an effective chemotherapeutic agent, even though cells which have passed through the *assembly* stages are effectively dead?
9. What causes the viral plaques that appear on a bacterial lawn to stop growing larger?
10. Describe how a *restriction endonuclease* might play a role in resistance to bacteriophage infection. Why could a restriction endonuclease play such a role whereas a generalized DNase could not?
11. Put together a diagram describing mRNA synthesis and nucleic acid replication for each of the following virus types: RNA tumor virus, reovirus, poliovirus, T4 phage, ϕX174 phage. For each diagram, be sure to indicate the complementarity of both the virus nucleic acid and its mRNA.
12. Although the RNA in the retrovirus virion is the plus strand and has both caps and tails, it is not used as mRNA. How could its translation be prevented?
13. Examine the graph which illustrates a one-step growth curve (Figure 6.10). Several stages are listed. Answer the following questions in reference to the stages indicated on the graph:
    a. In which stage does attachment of virions to cells occur?
    b. In which stage are early enzymes formed?
    c. In which stage does virus nucleic acid replication predominantly occur?
    d. In which stage does cell lysis occur?
    e. In which stage would the self-assembly process predominantly occur?
14. Suppose you wanted to determine whether a bacterial culture (strain) you were studying was a *lysogen*. You

have no information about the genetics of this strain, but you have access to a large number of other strains of the same bacterial species. How would you proceed? Suppose you had no other strains of the same species. Is there any way you might get an idea of lysogenicity?

15. One characteristic of *temperate bacteriophages* is that they cause turbid rather than clear plaques on bacterial lawns. Can you think why this might be? (Remember the process by which a bacterial plaque develops.)

16. *Virulent mutants* of temperate bacterial viruses are known. How might you detect such virulent mutants based on observations of their plaque morphology? How might such mutants differ genetically from the wild type? (Examine the genetic map of lambda.)

17. Suppose a mutant of lambda temperate bacteriophage was isolated which lacked the *att* site. How might its phenotype be expressed? Suppose a host mutant were isolated (that is, a bacterial mutant) which lacked the site on its DNA where the lambda DNA attached. What result would you predict when an infection of this host with *lambda* virus was brought about?

18. Describe the role which the *lambda repressor protein* plays in the lambda infection process.

19. From what you know about cell function and virus function, would you say viruses are living or dead? Why?

20. How does the existence of the prion alter the validity of the definition of a virus which you gave in question 2?

21. A number of small viruses have developed unusual features that make the same genetic material serve more than one function. Several such approaches have been discussed in this chapter. List three.

22. Describe the process by which a single-stranded filamentous DNA bacteriophage is released from the cell. Why is it that even though this process does not result in lysis, plaques can be detected in conventional bacteriophage plaque assays?

23. Compare and contrast DNA replication in ϕX174, the M13-fd group, T7, Mu, and adenovirus.

24. The T-even bacteriophages have developed a different *modification-restriction system* than that found in other systems. What is this system and how does it work?

25. *Replication of DNA in bacteriophage Mu* differs significantly from replication of DNA in the other double-stranded DNA phages described. Discuss. Why is it stated that Mu has some similarities to a transposon?

26. Compare and contrast virus multiplication in *positive-strand* and *negative-strand* RNA viruses of animals.

27. What is unique about reovirus genomes and what special problems does this introduce for nucleic acid replication?

Supplementary Readings

Fenner, F., and **A. Gibbs** (eds.). 1988. *Portraits of Viruses—A History of Virology.* Karger, Basel. A series of chapters written by experts, each dealing with the historical development and primary concepts of a major group of viruses.

Fields, B. N. and **D. M. Knipe** (eds.). 1991. *Fundamental Virology*, 2nd edition. Raven Press, New York. A large and detailed treatment of animal and human viruses. Each chapter written by an expert.

Fraenkel-Conrat, H., **P. C. Kimball**, and **J. A. Levy**. 1988. *Virology*, 2nd edition. Prentice-Hall, Englewood Cliffs, N.J. A short textbook of virology.

Ptashne, M. 1992. *A Genetic Switch*, 2nd edition. Blackwell Scientific Publications, Palo Alto, CA. An excellent short book which explains how lambda is regulated and also describes eukaryotic gene regulation.

Semancik, J. S. (ed.). 1987. *Viroids and viroid-like pathogens.* CRC Press, Boca Raton, FL. An overview of the molecular biology and pathogenicity of sub-viral agents including a short chapter on prions.

Watson, J. D., **N. H. Hopkins, J. W. Roberts, J. A. Steitz**, and **A. M. Weiner**. 1987. *Molecular Biology of the Gene*, 4th edition. Benjamin-Cummings Publishing Company, Menlo Park, CA. Has extensive coverage of how viruses replicate.

Microbial Genetics

Now that we have introduced the main features of molecular genetics of cells and viruses, we can turn to a discussion of specific aspects of microbial genetics. In this chapter we discuss mutation and explain how genetic material is transferred from one organism to another. Gene transfer can occur in a number of different ways, and if it is accompanied by genetic recombination, it can lead to the formation of new organisms.

Mutation is the inherited change in the base sequence of the nucleic acid comprising the genome of an organism. **Genetic recombination** is the process by which genetic elements contained in two separate genomes are brought together in one unit. Through this mechanism, new combinations of genes can arise even in the absence of mutation. Since the genetic elements brought together may enable the organism to carry out some new function, genetic recombination can result in adaptation to changing environments. While mutation usually brings about only a very small amount of genetic change in a cell, genetic recombination usually involves much larger changes. Entire genes, sets of genes, or even whole chromosomes, are transferred between organisms. As a result, offspring are not exactly like either parent; they are *hybrids* and contain a combination of the traits exhibited by each parent. In eukaryotes, genetic recombination is a regular process which occurs as a result of sexual reproduction (see Section 3.13). Prokaryotes do not carry out an exactly analogous process, but they do have important ways of undergoing genetic exchange. There are distinct differences in genetic exchange and recombination between eukaryotes and prokaryotes.

Genetic recombination is an important research tool in dissecting the genetic structure of an organism. It is also of major importance in the construction of new organisms for practical applications, a major activity in the field of genetic engineering. We present the basic principles of microbial genetics in this chapter and then show in the next chapter how these principles apply to research in genetic engineering.

Importance

Microbial genetics is important for a number of reasons:

1. Gene function is at the basis of cell function, and basic research in microbial genetics is necessary to understand how microorganisms function.

2. Microorganisms provide relatively simple systems for studying genetic phenomena, and are thus useful tools in attempts to decipher the mechanisms underlying the genetics of all organisms.

3. Microorganisms are used for the isolation and duplication of specific genes from other organisms, a technique called **molecular cloning**. In molecular cloning, genes are manipulated and placed in a microorganism where they can be induced to increase in number.

4. Microorganisms produce many substances of value in industry, such as antibiotics, and genetic manipulations can be used to increase yields and improve manufacturing processes. Also, genes of higher organisms that specify the production of particular substances, such as human insulin, can be transferred by molecular cloning into microorganisms, and the latter used for the production of these useful substances. The use of genetically modified microorganisms in large-scale industrial processes is an important part of the field of **biotechnology** (see Chapter 8).

5. Many diseases are caused by microorganisms, and genetic traits underlie these harmful activities. By understanding the genetics of disease-causing microorganisms we can more readily control them and prevent their growth in the body. Viruses, although in many cases agents of disease, can be thought of as genetic elements, and understanding the genetics of viruses helps us to control virus disease.

6. Some of the types of genetic transfer that occur in prokaryotes, particularly conjugation (see Section 7.10), also play important roles in the spread of genes that confer properties such as resistance to antibiotics. Understanding such processes can help us to determine how genes can be transferred from one organism to another, even from one species to another.

To detect genetic exchange between two organisms, it is necessary to employ *genetic markers* whose transfer can be detected. Genetically altered strains are used for this purpose, the alteration(s) being due to one or more mutations in the DNA of the organism. We begin our chapter on microbial genetics with a consideration of the molecular mechanism of mutation and the properties of mutant microorganisms as a prelude to our discussion of genetic exchange.

7.1   Mutations and Mutants

As previously mentioned, a *mutation* is a heritable change in the base sequence of the DNA of an organism. A strain carrying such a change is called a *mutant*. A mutant will by definition differ from its parental strain in **genotype**, the precise sequence of nucleotides in the DNA of a genome. But in addition, the visible properties of the mutant, its **phenotype**, may also be altered relative to the parental strain. It is common to refer to a strain isolated from nature as a *wild-type* strain. Mutant derivatives can be obtained from either wild-type strains or from a strain derived from the wild type, for example, another mutant.

Depending on the mutation, a mutant may or may not show an altered phenotype from its parent. By convention in microbial genetics, the *genotype* of an organism is designated by three lower case letters followed by a capital letter (all in italics) indicating the particular gene involved. For example, the *hisC* gene of *Escherichia coli* codes for a protein which could be called the HisC protein. In this case this protein (an enzyme in the biosynthetic pathway of histidine) is usually referred to by the name histidinol-phosphate aminotransferase, which describes its enzymatic activity. However, some proteins, such as the RecA protein (see Sections 6.12, 7.3, and 7.5), do not have other names because enzyme functions can be difficult to describe in a few words. Mutations in the *hisC* gene would be designated as *hisC1*, *hisC2*, and so on, the numbers referring to the order of isolation of the mutant strains.

The *phenotype* of an organism is designated by a capital letter followed by two small letters, with either a plus or minus superscript to indicate the presence or absence of that property. For example, a His+ strain of *E. coli* is capable of making its own histidine while a His− strain is not.

Isolation of mutants

We can distinguish between two kinds of mutations, selectable and nonselectable. An example of a nonselectable mutation is that of loss of color in a pigmented organism (Figure 7.1a). Such colonies usually have neither an advantage nor a disadvantage over the pigmented parent colonies when grown on agar plates (there may be a selective advantage for pigmented organisms in nature, however). This means that the only way we can detect such mutations is to examine large numbers of colonies and look for the "different" ones.

Nonselectable mutants must be found by **screening** a large population of organisms, and the mutant phenotype may not be as easy to recognize as the dif-

Miniglossary for Chapter 7

AUXOTROPH   an organism which has developed a nutritional requirement through mutation

CONJUGATION   transfer of genes from one prokaryotic cell to another by a mechanism involving cell-to-cell contact and a plasmid

DIPLOID   a eukaryotic cell or organism which contains two sets of chromosomes

ELECTROPORATION   the use of an electrical pulse to induce cells to take up free DNA

GAMETES   in eukaryotic organisms, the haploid germ cells which result from meiosis

GENETIC MAP   the arrangement of genes on a chromosome

GENOTYPE   the precise genetic make-up of an organism

HAPLOID   a cell or organism which has only one set of chromosomes

MUTAGENS   agents which cause mutation

MUTANT   an organism whose genome carries a mutation

MUTATION   an inheritable change in the base sequence of the DNA of an organism

PHENOTYPE   the observable characteristics of an organism

PLASMID   an extrachromosomal genetic element which has no extracellular form

POINT MUTATION   a mutation which involves one or only a very few base pairs

RECOMBINATION   the process by which parts or all of DNA molecules from two separate sources are exchanged or brought together into a single unit

SELECTION   placing organisms under conditions where the growth of those with a particular genotype will be favored

TRANSDUCTION   transfer of host genes from one cell to another by a virus

TRANSFORMATION   transfer of bacterial genes involving free DNA (but see alternative usage in Chapter 6)

TRANSPOSABLE ELEMENT   a genetic element which has the ability to move (transpose) from one site on a chromosome to another

TRANSPOSON   a type of transposable element which carries genes in addition to those involved in transposition

ference between pigmented and nonpigmented colonies. A *selectable* mutation, on the other hand, confers upon the mutant an advantage under certain environmental conditions, so that the progeny of the mutant cell are able to outgrow and replace the par-

ent. An example of a selectable mutation is drug resistance: an antibiotic-resistant mutant can grow in the presence of antibiotic concentrations that inhibit or kill the parent (Figure 7.1*b*). It is relatively easy to detect and isolate selectable mutants by choosing the appro-

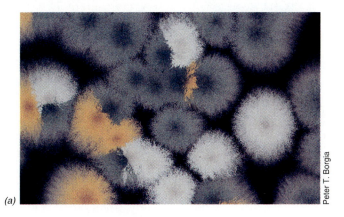

(a)

Peter T. Borgia

FIGURE 7.1

Observation of several kinds of mutants. (a) Pigmented mutants and nonpigmented mutants of the fungus *Aspergillus nidulans*. The wild-type has a green pigment. The white or colorless mutants make no pigment, whereas the yellow mutants cannot convert the pigment they do make to the normal color. (b) Development of antibiotic-resistant mutants within the inhibition zone of an antibiotic assay disk. (c) Colonies of *Escherichia coli* have been mutated by a derivative of the bacteriophage Mu (see Section 6.13) on agar plates that allow detection of cells producing β-galactosidase. The Mu derivative carries the gene for this enzyme, but the enzyme is not produced unless Mu inserts into the *E. coli* chromosome in the proper orientation next to a promoter. Blue colonies are producing β-galactosidase, while sectored colonies contain some cells where Mu is not inserted in the correct orientation.

Mutants

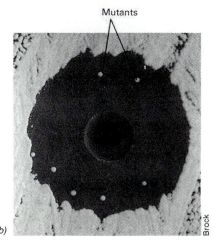

(b)

Brock

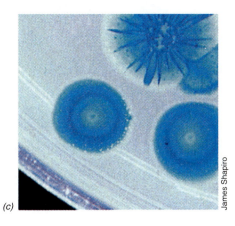

(c)

James Shapiro

priate environmental conditions. Therefore, **selection** is an extremely powerful genetic tool, allowing the isolation of a single mutant from a population containing millions of parental organisms.

Virtually any characteristic of a microorganism can be changed through mutation. Nutritional mutants can be detected by the technique of **replica plating** (Figure 7.2a). Using sterile velveteen cloth, an imprint of colonies from a master plate is made onto an agar plate lacking the nutrient. The colonies of the parental type will grow normally, whereas those of the mutant will not. Thus, the inability of a colony to grow on the replica plate (Figure 7.2b) will be a signal that it is a mutant. The colony on the master plate corresponding to the vacant spot on the replica plate (Fig-

ure 7.2b) can then be picked, purified, and characterized. A nutritional mutant that has a requirement for a growth factor is often called an **auxotroph**, and the wild-type parent from which the auxotroph was derived is called a **prototroph**. Although of great utility, replica plating is a screening process, and it can be laborious to isolate mutants by screening.

An ingenious method which is widely used to isolate mutants that require amino acids or other growth factors is the **penicillin-selection method**. Ordinarily, mutants that require growth factors are at a disadvantage in competition with the parent cells, so that there is no direct way of isolating them. However, penicillin kills only *growing* cells, so that if penicillin is added to a population growing in a medium lacking the growth

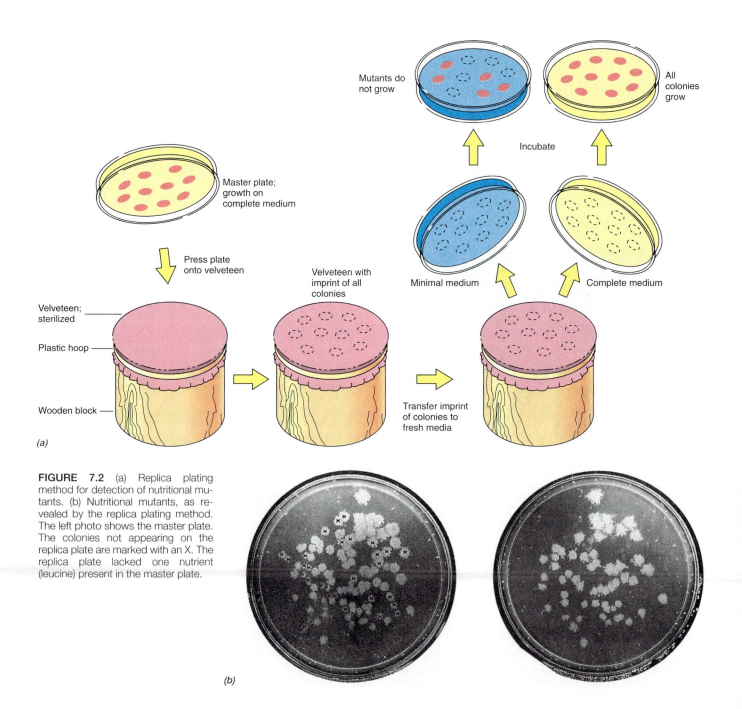

FIGURE 7.2 (a) Replica plating method for detection of nutritional mutants. (b) Nutritional mutants, as revealed by the replica plating method. The left photo shows the master plate. The colonies not appearing on the replica plate are marked with an X. The replica plate lacked one nutrient (leucine) present in the master plate.

Table 7.1   Kinds of mutants

Description	Nature of change	Detection of mutant
Nonmotile	Loss of flagella; nonfunctional flagella	Compact colonies instead of flat, spreading colonies
Noncapsulated	Loss or modification of surface capsule	Small, rough colonies instead of larger, smooth colonies
Rough colony	Loss or change in lipopolysaccharide outer layer	Granular, irregular colonies instead of smooth, glistening colonies
Nutritional	Loss of enzyme in biosynthetic pathway	Inability to grow on medium lacking the nutrient
Sugar fermentation	Loss of enzyme in degradative pathway	Do not produce color change on agar containing sugar and a pH indicator
Drug resistant	Impermeability to drug or drug target is altered or drug is detoxified	Growth on medium containing a growth-inhibitory concentration of the drug
Virus resistant	Loss of virus receptor	Growth in presence of large amounts of virus
Temperature sensitive	Alteration of any essential protein so that it is more heat sensitive	Inability to grow at a temperature normally supporting growth (e.g., 40°C) but still growing at a lower temperature (e.g., 30°C)
Pigmentless	Loss of enzyme in biosynthetic pathway leading to loss of one or more pigments	Detect visually by mutant being different color or colorless
Cold sensitive	Alteration in an essential protein so that it is inactivated at low temperature	Inability to grow at a low temperature (e.g., 20°C) that normally supports growth

factor required by the desired mutant, the parent cells will be killed, whereas the nongrowing mutant cells will be unaffected. Thus, after preliminary incubation in the absence of growth factor in penicillin-containing medium, the population is washed free of penicillin and transferred to plates containing the growth factor. Among the colonies that grow up (including some wild-type cells that have escaped penicillin killing) should be some growth factor mutants. Penicillin selection is a kind of *negative selection;* the selection is not for the mutant but against the parental type.

Some of the most common kinds of mutants and the means by which they are detected are listed in Table 7.1.

> Mutation, a heritable change in the DNA, can lead to a change in phenotype. Selectable mutations are those which give the mutant a growth advantage under certain environmental conditions and are especially useful for genetics research.

7.2   Molecular Basis of Mutation

As previously mentioned, mutations arise because of changes in the *base sequence* in the DNA. In most cases, mutations that occur in the base sequence of the DNA lead to changes in the organism; these changes are mostly harmful, although beneficial changes do occur occasionally.

Mutation can be either spontaneous or induced. **Spontaneous mutations** can occur as a result of the action of natural radiation (cosmic rays, etc.) which alter the structure of bases in the DNA. Spontaneous mutations can also occur during replication, as a result of errors in the pairing of bases, leading to changes in

the replicated DNA. In fact, such errors occur at a frequency of about 10^{-7} to 10^{-11} per base pair during a single round of replication (and a typical gene has about 1000 base pairs). Thus, in a normal, fully grown culture of organisms having approximately 10^8 cells/ml, there will probably be a number of different mutants in each milliliter of culture.

Mutations involving one (or a very few) base pairs are sometimes referred to as **point mutations**. Point mutations can result in *base pair substitutions* in the DNA or the insertion or deletion of a base pair (called *microinsertions* and *microdeletions*). As is the case with all mutations, the phenotypic change that comes about because of a point mutation depends on exactly where the mutation took place in the gene, what the nucleotide change was, and what product the gene normally encodes.

Base pair substitutions

If a point mutation occurs within a gene that encodes a protein, any change in the phenotype of the cell is almost certainly the result of a change in the amino acid *sequence* of the protein being produced. Figure 7.3 shows a number of base pair substitutions that could occur in a short region of DNA within a gene that encodes a protein. The error in the DNA is transcribed into mRNA, and this erroneous mRNA in turn is used as a template and translated into protein. (Because only one strand of the DNA is used as template for the mRNA, an A:T base pair does not have the same meaning as a T:A base pair.) The triplet code that directs the insertion of an amino acid via a transfer RNA will thus be incorrect. What are the consequences of base substitutions?

In interpreting the results of mutation, we must first recall that the genetic code is degenerate (see Sec-

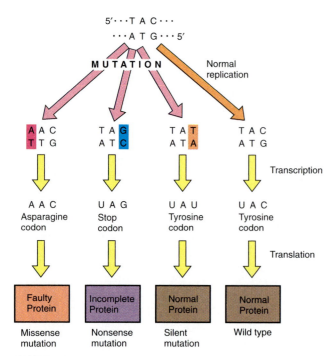

FIGURE 7.3   Possible effects of base pair substitution in a gene coding for a protein: three different protein products from changes in the DNA for a single codon.

premature termination of translation, leading to an incomplete protein which would almost certainly not be functional (Figure 7.3). Mutations of this type are called **nonsense mutations** because the change is to a nonsense codon (see Section 5.9).

Thus, not all mutations that cause amino acid substitution necessarily lead to nonfunctional proteins. The outcome depends on where in the polypeptide chain the substitution has occurred, and on how it affects the folding and the catalytic activity of the protein. A missense mutation can lead to an enzyme which is temperature sensitive, and this type of mutation is termed a **temperature-sensitive mutation**. For instance, temperature-sensitive mutants of bacteria are known that function normally at 30°C but cannot grow at 40°C, although the wild-type grows fine at both temperatures. Such mutations are also referred to as **conditionally lethal**, since the bacteria cannot grow under one condition but can under another. Temperature-sensitive phenotypes often result because the mutant protein can maintain its correct conformation at the low temperature but becomes partially unfolded (denatured) at the high temperature.

Frameshift mutations

Since the genetic code is read from one end in consecutive blocks of three bases, any deletion or insertion of a base pair results in a **reading frame shift**, and the translation of the gene is completely upset (Figure 7.4). Partial restoration of gene function can often be accomplished by insertion of another base pair near the one deleted (one kind of suppressor mutation, see below). After correction, depending on the exact amino acids coded by the still faulty region and the region of the protein involved, the protein formed may have some biological activity or even be completely normal.

It is important to remember that microinsertions or microdeletions are only frameshift mutations if they occur in the part of a protein-encoding gene that includes the reading frame. A single base pair insertion in the promoter of a gene could lead to a dramatic change in the ability of the gene to function, but it would not be a frameshift mutation. (Similarly, base pair substitutions that are not within the reading frame are not missense or nonsense mutations.)

Back mutations or reversions

Point mutations are reversible. A *revertant* is operationally defined as a strain in which the wild-type

tion 5.9). Because of degeneracy, not all mutations in protein-encoding genes result in changes in protein. This is illustrated in Figure 7.3, which shows several possible results when the DNA that encodes a single *tyrosine codon* undergoes mutation. As seen, a change in the RNA from UAC to UAU would have no apparent effect, since UAU is an additional tyrosine codon. Mutations that give rise to such changes are called **silent mutations**. Note that silent mutations almost always occur in the *third base* of the codon (arginine and leucine can also have silent mutations in the first position). As seen in Table 5.4, changes in the first or second base of the triplet much more often lead to significant changes in the protein. For instance, a single base change from UAC to AAC (Figure 7.3) would result in a change in the protein from tyrosine to asparagine. This is referred to as a **missense mutation**, because the chemical "sense" (sequence of amino acids) in the ensuing polypeptide has changed. If the change occurred at a critical point in the polypeptide chain, the protein could be inactive, or of reduced activity. Another possible result of a base pair substitution would be the formation of a *stop codon*, which would result in

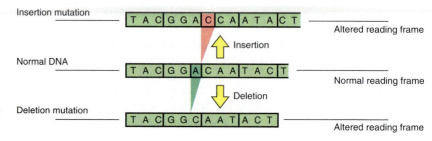

FIGURE 7.4
Shifts in reading frame caused by insertion or deletion mutations.

phenotype that was lost in the mutant is restored. Revertants can be of two types. In *same-site revertants*, the mutation that restores activity occurs at the same site at which the original mutation occurred. (If the back mutation is not only at the same site but also leads to the wild-type sequence, it is called a *true revertant*.) In *second-site revertants*, the mutation occurs at some different site in the DNA.

Second-site mutations may cause restoration of wild-type phenotype because of several types of **suppressor mutations** that restore the original phenotype. Suppressor mutations are new mutations that compensate for the effect of the original mutation. Several types of suppressor mutations are known: (1) a mutation somewhere else in the same gene can restore enzyme function, such as in a reading-frame shift mutation; (2) a mutation in another gene may restore the wild-type phenotype; and (3) the mutation may result in the production of another enzyme that can replace the mutant one by introducing a metabolic pathway different from that used by the mutant enzyme. In this last type no production of the original enzyme occurs although it does in the other types.

Mutations involving many base pairs

Deletions are mutations in which a region of the DNA has been eliminated. As we have discussed, microdeletions, the removal of one or a very few base pairs (Figure 7.4), are often frameshift mutations and can inactivate a gene. However, deletions can also involve the loss of hundreds or thousands of base pairs. Deletion of a large segment of the DNA results in complete loss of function of any gene that may be involved. Some deletions are so large that they involve several genes (if any of the genes are essential, the mutation will be lethal). Such deletions cannot be restored through further mutations, but only through genetic recombination. Indeed, one way in which large deletions are distinguished from point mutations is that the latter are reversible through further mutations, whereas the former are not.

Insertions occur when new bases are added to the DNA. Insertions, like deletions, can involve only a single base or many bases. Microinsertions result from replication errors as do deletions, but larger insertions arise as a result of mistakes that occur during genetic recombination. Many insertion mutations are due to the insertion of specific identifiable DNA sequences 700 to 1400 base pairs in length called **insertion sequences**, a type of transposable element (see Section 5.5). The behavior of such insertion sequences is discussed in detail in Section 7.12. Insertions inactivate the gene in which they occur, and even large ones can revert by further mutation.

Other types of large-scale mutations exist that also seem to involve rearrangements brought about by mistakes in recombination. These include **translocations**, in which a large section of chromosomal DNA is moved to a new location (and in eukaryotes sometimes to a different chromosome), and **inversions**, in which the orientation of a particular segment of DNA is reversed with respect to the surrounding DNA.

Rates of mutation

There are wide variations in the rates at which various kinds of mutations occur. Some types of mutations occur so very rarely that they are almost impossible to detect whereas others occur so frequently that they often present difficulties for the experimenter who is trying to maintain a genetically stable stock culture.

Spontaneous mutations in a gene occur at frequencies around 10^{-6} per generation. This means that there is one chance in 1,000,000 that a mutation will arise at some location in a given gene during one cell cycle. Transposition events may occur more frequently, around 10^{-4}. On the other hand, the occurrence of a nonsense mutation is less frequent, 10^{-6} to 10^{-8}, because only a few codons can mutate to nonsense codons. Missense mutations also occur at about the same frequency as nonsense mutations. Unless the mutant is selectable, the experimental detection of events of such rarity is obviously difficult and much of the skill of the microbial geneticist is applied to increasing the efficiency of mutation detection. As we will see in the next section, it is possible to significantly increase the rate of mutation by the use of mutagenic treatments.

> **Mutations, which can be either spontaneous or induced, arise because of changes in the base sequence of the DNA. A point mutation, which is due to a change in a single base pair, can lead to a single amino acid change in a protein, or to no change at all, depending on the particular codon involved. In a nonsense mutation, the codon becomes a stop codon and an incomplete protein is made. Deletions and insertions result in more dramatic changes in the DNA, including frameshift mutations, and often result in complete loss of phenotype.**

7.3   Mutagens

It is now well established that a wide variety of chemical and physical agents can induce mutations. We discuss some of the major categories and their actions here.

Chemical mutagens

An overview of some of the major chemical mutagens and their modes of action are given in Table 7.2. Several classes of chemical mutagens exist. A variety of chemical mutagens are **base analogs**, resembling DNA purine and pyrimidine bases in structure, yet showing faulty pairing properties (Figure 7.5). When one of these base analogs is incorporated into DNA, replication may occur normally most of the time, but occasional copying errors occur, resulting in the incor-

Table 7.2   Chemical and physical mutagens and their modes of action

Agent	Action	Result
Base analogs:		
5-Bromouracil	Incorporated like T; occasional faulty pairing with G	A–T pair → G–C pair Occasionally G–C → A–T
2-Aminopurine	Incorporated like A; faulty pairing with C	A–T → G–C Occasionally G–C → A–T
Chemicals reacting with DNA:		
Nitrous acid (HNO$_2$)	Deaminates A,C	A–T → G–C and G–C → A–T
Hydroxylamine (NH$_2$OH)	Reacts with C	G–C → A–T
Alkylating agents:		
Monofunctional (e.g., ethyl methane sulfonate)	Put methyl on G; faulty pairing with T	G–C → A–T
Bifunctional (e.g., nitrogen mustards, mitomycin, nitrosoguanidine)	Cross-link DNA strands; faulty region excised by DNase	Both point mutations and deletions
Intercalative dyes (e.g., acridines, ethidium bromide)	Insert between two base pairs	Microinsertions and microdeletions
Radiation:		
Ultraviolet	Pyrimidine dimer formation	Repair may lead to error or deletion
Ionizing radiation (e.g., X-rays)	Free-radical attack on DNA, breaking chain	Repair may lead to error or deletion

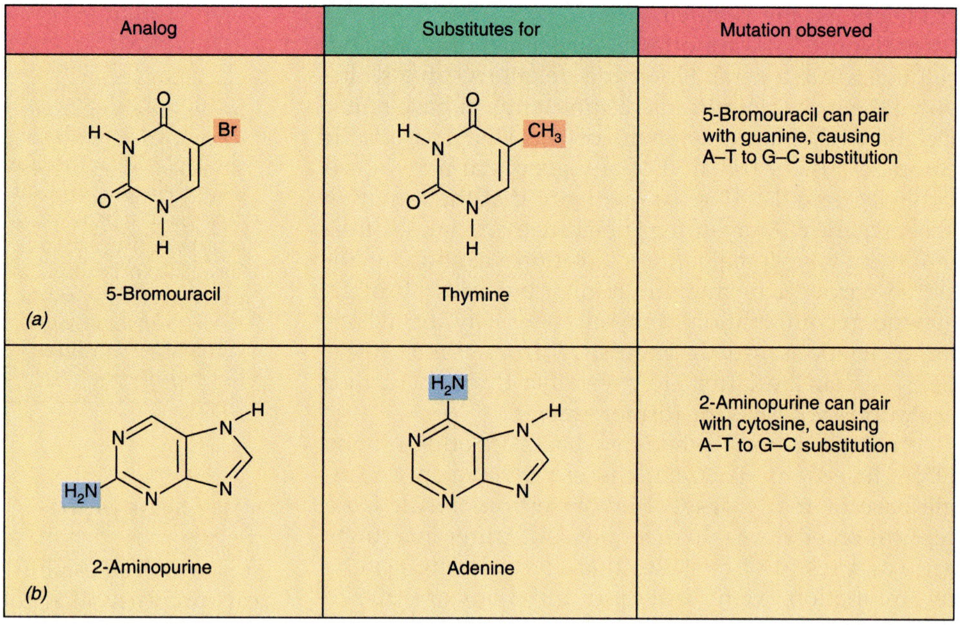

FIGURE 7.5
Structure of two common nucleotide base analogs used to induce mutations, and the normal nucleic acid bases they substitute for.

poration of the wrong base into the copied strand. During subsequent segregation of this strand, the mutation is revealed.

A variety of chemicals react directly on DNA, causing chemical changes in one or another base, which results in faulty pairing or other changes (Table 7.2). Alkylating agents such as nitrosoguanidine, for example, are powerful mutagens, and generally induce mutations at higher frequency than base analogs.

Such chemicals differ in their action from the base analogs in that the chemicals reacting on DNA are able to introduce direct changes even in nonreplicating DNA, whereas the base analogs only act after incorporation during replication. One interesting group of chemicals, the acridines, are planar molecules which act as *intercalating agents*. These mutagens become inserted between two DNA base pairs, thereby pushing them apart. During replication, an extra base

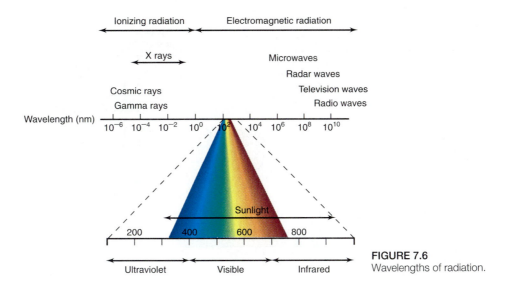

FIGURE 7.6
Wavelengths of radiation.

can then be inserted in acridine-treated DNA, lengthening the DNA by one base and thus shifting the reading frame.

Radiation

Several forms of radiation are highly mutagenic. We can divide mutagenic radiation into two main categories, ionizing and nonionizing (electromagnetic) (Figure 7.6). Although both kinds of radiation are used in microbial genetics, *nonionizing* radiations find the widest use and will be discussed first.

The purine and pyrimidine bases of the nucleic acids absorb ultraviolet (UV) radiation strongly, and the absorption maximum for DNA and RNA is at 260 nm (see Nucleic Acids box, Chapter 5). Proteins also absorb UV, but have a peak at 280 nm, due to the absorption of the aromatic amino acids (tryptophan, phenylalanine, tyrosine). It is now well established that killing of cells by UV radiation is due primarily to its action on DNA so that UV radiation at 260 nm is most effective as a lethal agent. Although several effects are known, one well-established effect is the induction in DNA of **pyrimidine dimers**, a state in which two adjacent pyrimidine bases become covalently joined, so that during replication of the DNA the probability of DNA polymerase inserting an incorrect nucleotide at this position is greatly increased.

The type of UV radiation source most frequently used in mutation work is the germicidal lamp, which emits large amounts of UV radiation in the 260 nm region. A dose of UV radiation is used which brings about 90–95 percent killing of the cell population, and mutants are then looked for in the survivors. If much higher doses of radiation are used, the number of viable cells will be too low, whereas if lower doses are used, insufficient damage to the DNA will have been induced. UV radiation is a very useful tool in isolating mutants of microbial cultures.

Ionizing radiation

Ionizing radiation is a more powerful form of radiation, and includes short wavelength rays such as X-rays, cosmic rays, and gamma rays (Figure 7.6). These radiations cause water and other substances to ionize, and mutagenic effects are brought about indirectly through this ionization. Among the potent chemical species formed by ionizing radiation are chemical free radicals, of which the most important is the hydroxyl radical, $OH\cdot$. Free radicals react with and inactivate macromolecules in the cell, of which the most important is DNA. DNA is probably no more sensitive to ionizing radiation than other macromolecules, but since each DNA molecule contains only one copy of most genes, inactivation can have a permanent effect. At low doses of ionizing radiation, only a few hits on DNA occur, but at higher doses multiple hits occur, leading to the death of the cell. In contrast to UV radiation, ionizing radiation penetrates readily through glass and other materials. Because of this, ionizing radiation is used frequently to induce mutations in animals and plants (where its penetrating power makes it possible to reach the germ cells of these organisms readily), but because ionizing radiation is more dangerous to use and is less readily available, it finds less use with microorganisms (where penetration with UV is not a problem).

Mutations that arise from DNA repair

Recall that a mutation is an *inheritable* change in the DNA. Therefore, if an error in DNA synthesis can be corrected before the cell divides, there will be no mutation. Furthermore, some DNA damage clearly cannot be replicated and therefore cannot itself be a mutation. For instance, if a DNA molecule contains pyrimidine dimers, it will not be replicated to give two DNA molecules each containing pyrimidine dimers. If such DNA damage cannot be repaired, cells

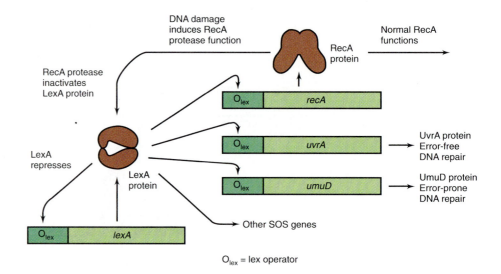

FIGURE 7.7
Mechanism of the SOS response. DNA damage results in conversion of RecA protein into a protease which cleaves LexA protein. LexA protein normally represses the activities of the *recA* gene and the DNA repair genes *uvrA* and *umuD*. With LexA inactivated, these genes become active.

often die. Most cells have a variety of different DNA repair processes to correct mistakes or repair damage. Many of these DNA repair systems do not make mistakes. However, some processes seem to be *error prone*, and it is the repair process itself that introduces the mutation.

Many kinds of mutations arise as a result of faulty repair of damage induced in DNA by some of the various agents just discussed. A complex cellular mechanism, called the **SOS regulatory system**, is activated as a result of DNA damage, initiating a number of DNA repair processes. However, in the SOS system, some DNA repair occurs in the absence of template instruction, which results in the creation of many errors, hence many mutations.

In the SOS regulatory system, DNA damage serves as a distress signal to the cell, resulting in the coordinate derepression (induction) of a number of cellular functions involved in DNA repair. The SOS system is normally repressed by a protein called the LexA protein, but LexA is inactivated by RecA, a protease that is activated as a result of DNA damage (Figure 7.7). Since one of the DNA repair mechanisms of the SOS system is inherently error prone, many mutations arise. Thus, through the SOS regulatory system, DNA damage by various agents such as chemicals and radiation leads to mutagenesis.

The SOS system senses the presence in the cell of DNA damage and the repair mechanisms are activated. But once the DNA damage has been repaired, the SOS system is switched off, and further mutagenesis ceases. In addition to its effect on cellular mutagenesis, the SOS regulatory system plays a central role in the regulation of temperate virus replication, as was discussed in Section 6.12.

It should be emphasized that not all DNA repair occurs in the absence of template instruction. Cells generally have many DNA repair systems which require template instruction and lead to proper DNA repair. These systems apparently work most of the time, but are not sufficient to repair the large amounts of damage done by some of the agents mentioned above.

Biological mutagens

Mutations can be introduced without the use of harsh chemical or physical agents through the process of *transposon mutagenesis*. We discuss the details of transposon mutagenesis later in this chapter (Section 7.11) and note here only that if insertion of a transposable element occurs *within* a gene, it will generally result in loss of gene function. For example, bacteriophage Mu, discussed previously (see Section 6.13), can serve as a mutagen by disrupting the coding sequence of a gene into which it inserts (see Figure 7.1c). Because transposable elements can enter the chromosome at various locations, transposons are widely used by microbial geneticists as mutagenic agents.

Site-directed mutagenesis

So far, the mutations that we have been discussing have been randomly directed at the genome of the microbial cell. Recombinant DNA technology and the use of synthetic DNA make it possible to induce *specific* mutations in *specific* genes. The procedures for carrying out mutagenesis of specific sites in the genome are called *site-directed mutagenesis* and will be discussed in detail in Chapter 8. Here we briefly indicate the overall principle of site-directed mutagenesis.

If the DNA containing a specific gene has been isolated and its sequence determined, it is then possible to construct a modified form of this gene in which a specific base or series of bases has been changed. Such a modified DNA can now be inserted into a recipient cell and *mutants* selected. Such mutants will then differ by the desired change, and will represent specific site-directed mutations. Site-directed mutagenesis has many uses in microbial genetics and molecular biology and has been especially useful for structure/function studies of enzymes and other proteins (see Section 8.11).

> Mutagens are chemical or physical agents which affect DNA, usually by altering the base sequence. Mutagens are useful in the laboratory because they make it easier to isolate mutants. Although most mutagens act randomly, certain methods are available for inducing changes in specific genes or at specific sites in genes. Alterations in DNA are not mutations unless they can be inherited. Some DNA damage can lead to cell death if not repaired.

7.4   Mutagenesis and Carcinogenesis: The Ames Test

A practical use of mutant bacterial strains has been developed to identify potentially hazardous chemicals in the environment. Because the sensitivity by which selectable mutants can be detected in large populations of bacteria is very high, bacteria can be used as screening agents for the potential mutagenicity of chemicals. This is relevant because it has been found that many mutagenic chemicals are also carcinogenic, capable of causing cancer in animals or humans.

The variety of chemicals, both natural and artificial, which the human population comes into contact with through industrial exposure is enormous. There is considerable need for simple tests to ascertain the safety of such compounds. There is good evidence that a large proportion of human cancers have environmental causes, most likely through the agency of various chemicals, making the detection of chemical carcinogens urgent. It does not necessarily follow that because a compound is mutagenic, it will also be carcinogenic. The correlation, however, is quite high, and the knowledge that a compound is mutagenic in a bacterial system serves as a warning of possible danger. Similarly, the fact that a compound is not mutagenic in a bacterial system does not mean that it is not carcinogenic, since the bacterial system cannot detect all compounds active in higher animals. The development of bacterial tests for carcinogenic screening has been carried out primarily by a group at the University of California in Berkeley, under the direction of Bruce Ames, and the mutagenicity test for carcinogens is sometimes called the **Ames test** (Figure 7.8).

The standard way to test chemicals for mutagenesis has been to measure the rate of *back* mutation (reversion) in strains of bacteria that are auxotrophic for some nutrient. It is important, of course, that the original mutation be a point mutation, so that reversion can occur. When cells of such an auxotrophic strain are spread on a medium lacking the required nutrient (for example, an amino acid or vitamin), no growth will occur, and even very large populations of cells

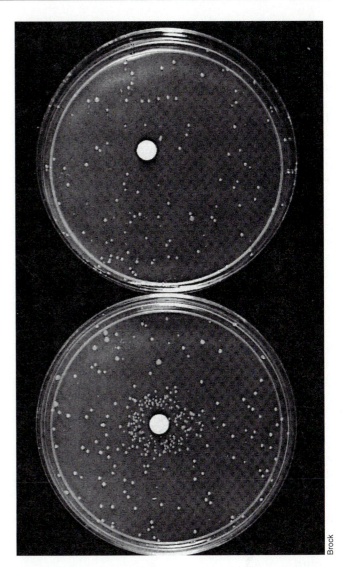

FIGURE 7.8   The Ames test is used to evaluate the mutagenicity of a chemical. Both plates were inoculated with a culture of a histidine-requiring mutant of *Salmonella typhimurium*. The medium does not contain histidine, so that only cells that revert back to wild type can grow. Spontaneous revertants appear on both plates, but the chemical on the filter paper disc in the test plate (bottom) has caused an increase in the mutation rate, as shown by the large number of colonies surrounding the disk. Revertants are not seen very close to the disk because the concentration of the mutagen is so high there that it is lethal.

can be spread on the plate without formation of visible colonies. However, if back mutants are present, those cells will be able to form colonies. Thus if 10^8 cells are spread on the surface of a single plate, even as few as 10 to 20 back mutants (revertants) can be detected by the 10 to 20 colonies they will form. Histidine auxotrophs of *Salmonella typhimurium* (Figure 7.8) and tryptophan auxotrophs of *Escherichia coli* have been the major tools for the Ames test, but a test has also been designed in which the induction of a phage lambda lysogen is used as an assay of DNA damage.

Although the simple testing of chemicals for mutagenesis in bacteria has been carried out for a long time, two elements have been introduced in the Ames test to make it a much more powerful test. The first of these is the use of strains of bacteria lacking DNA repair enzymes, so that any damage that might be induced in DNA is not corrected. The second important element in the Ames test is the use of liver enzyme preparations to convert the chemicals into their active mutagenic (and carcinogenic) forms. It has been well established that many potent carcinogens are not directly carcinogenic or mutagenic, but undergo chemical changes in the human body, which convert them into active substances. These changes take place primarily in the liver, where enzymes (mixed-function oxygenases) normally involved in detoxification cause formation of epoxides or other activated forms of the compounds, which are then highly reactive with DNA.

In the Ames test, a preparation of enzymes from rat liver is first used to activate the compound. Next the activated complex is taken up on a filter-paper disk, which is placed in the center of a plate on which the proper bacterial strain has been overlayed. After overnight incubation, the mutagenicity of the compound can be detected by looking for a halo of back mutations in the area around the paper disk (Figure 7.8). It is always necessary, of course, to carry out this test with several different concentrations of the compound and with appropriate positive and negative controls, because compounds vary in their mutagenic activity and are lethal at higher levels. A wide variety of chemicals has been subjected to the Ames test, and it has become one of the most useful prescreens to determine the potential carcinogenicity of a compound.

7.5 Genetic Recombination

Genetic recombination is the process by which genetic elements from two separate sources are brought together in a single unit. At the molecular level, recombination can be thought of as the movement of genetic information (nucleic acid sequences) from one molecule of nucleic acid to another. We focus here on **general** or **homologous recombination**, which results in genetic exchange between *homologous* DNA sequences from two different sources. Homologous DNA sequences have the same or nearly the same sequence; therefore, base-pairing can occur over an extended length of the two DNA molecules.

Molecular events in general recombination

At the molecular level, recombination has been studied mostly in prokaryotes and viruses. In Bacteria, general recombination involves the participation of a specific protein called the RecA protein, which is specified by the *recA* gene. When the RecA protein binds to single-stranded DNA, the complex forms a helical structure that facilitates recombination. Bacteria which are mutant in *recA* show markedly decreased levels of general recombination.

As noted, general recombination involves the *pairing* of DNA molecules over long stretches. An overall molecular mechanism of general recombination is shown in Figure 7.9. The process begins with a *nick* in one of the DNA molecules. This nicked strand must be displaced from the other strand by proteins having helicase activity (see Section 5.4). Single-stranded binding protein (see Section 5.4) then binds to the resulting single-stranded segment. Next, the RecA protein binds to the single-stranded fragment and positions it in such a way that annealing occurs with a complementary sequence in the adjacent duplex, simultaneously displacing the resident strand (Figure 7.9). This process is often referred to as *strand invasion*. Following pairing, *exchange* of homologous DNA molecules can occur, leading to the formation of recombinant DNA structures. This process also involves DNA polymerase and ligase.

Note that this mechanism for the formation of recombinant DNA structures is a completely natural mechanism which occurs extensively within the cell. Whether or not it leads to the formation of new genotypes will depend upon whether the two molecules undergoing recombination differ genetically in regions outside the region of recombination. Within limits, general recombination can be thought of as occurring at random sites throughout the genome. Thus, the probability of recombination occurring between two genes is proportional to their distance. This fact is useful for genetic mapping. By recombinational analysis it is possible to map the position of genes on chromosomes, since the *farther* two genes are apart, the more likely they will show recombination.

Until the advent of molecular techniques such as restriction enzyme mapping (see Section 5.3) and DNA sequencing (see Nucleic Acids box in Chapter 5), recombinational analysis was the only method available for ordering genes on chromosomes and determining the distance between them. It is interesting to note that the order and relative distances as measured by mapping using recombination have been largely confirmed using modern molecular techniques.

For new genotypes to arise as a result of general recombination, it is essential that the two homologous sequences be genetically distinct. This would be the case in a diploid eukaryotic cell (see Section 7.13), which has two sets of chromosomes, one from each parent. The two distinct molecules are brought together as the result of sexual reproduction, a process that occurs as part of the regular life cycles of most eukaryotic organisms (see Section 7.13). In prokaryotes, genetically distinct DNA molecules are brought

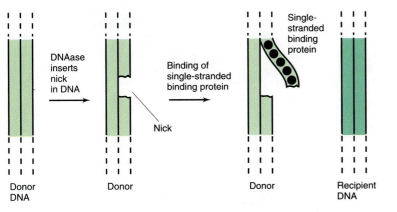

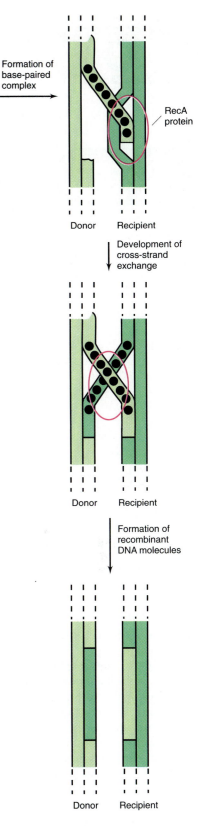

FIGURE 7.9   A simplified version of one molecular mechanism of genetic recombination. Homologous DNA molecules pair and exchange DNA segments. The mechanism involves breakage and reunion of paired segments. Two of the proteins involved, a single-stranded binding protein and the RecA protein, are shown. The diagram is not to scale: pairing can occur over hundreds or thousands of bases.

together in different ways, but the process of genetic recombination is no less important. Recombination can also be critical in the life cycle of some viruses. Chapter 6 showed that certain bacteriophages, such as T7 and T4 (see Sections 6.10 and 6.11), require homologous recombination as a step during DNA replication.

In prokaryotes genetic recombination is observed because of the transfer to a recipient cell of a fragment of genetically different DNA derived from a donor cell; next, the integration of this DNA fragment or its copy into the genome of the recipient cell must occur. We shall define briefly here the three means by which the DNA fragment is introduced into the recipient: (1) **transformation** is a process by which free DNA is inserted directly into a competent recipient cell; (2) **transduction** involves the transfer of bacterial DNA

from one bacterium to another within a temperate or defective virion; and (3) **conjugation (mating)** involves DNA transfer via actual cell-to-cell contact between the recipient and the donor cell. These processes are contrasted in Figure 7.10 and will be discussed in detail later in this chapter.

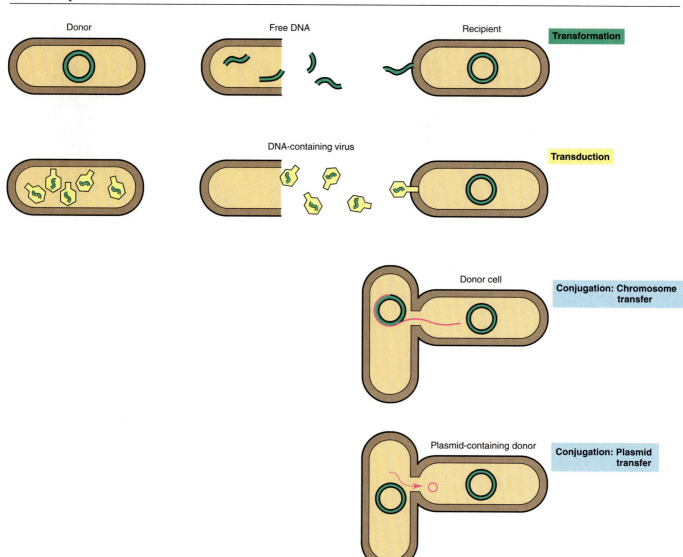

Donor Free DNA Recipient **Transformation**

DNA-containing virus **Transduction**

Donor cell **Conjugation: Chromosome transfer**

Plasmid-containing donor **Conjugation: Plasmid transfer**

FIGURE 7.10
Processes by which DNA is transferred from donor to recipient bacterial cell. Just the initial steps in transfer are shown. For details of how the DNA is integrated into the recipient, see text.

Detection of recombination

In order to detect physical exchange of DNA segments, the cells resulting from recombination must be phenotypically different from the parents. In crosses involving microorganisms, one must usually use as recipients strains that lack some selectable characteristic that the recombinants will possess. For instance, the recipient may not be able to grow on a particular medium, and genetic recombinants are selected which can. Various kinds of selectable and nonselectable markers (such as drug resistance, nutritional requirements, and so on) were discussed in Section 7.1. The exceedingly great sensitivity of the selection process is shown by the fact that 10^8 or more bacterial cells can be spread on a single plate and, if proper selective conditions are used, no parental colonies will appear, whereas even a few recombinants can form colonies (Figure 7.11). The only requirement is that the *reverse* mutation rate for the selected characteristic

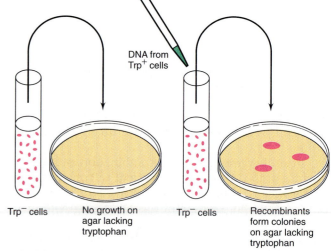

DNA from Trp⁺ cells

Trp⁻ cells | No growth on agar lacking tryptophan | Trp⁻ cells | Recombinants form colonies on agar lacking tryptophan

FIGURE 7.11
How a selective medium can be used to detect rare genetic recombinants among a large population of nonrecombinants. On the selective medium only the rare recombinants form colonies. Procedures such as this, which offer high resolution for genetic analyses, can ordinarily be used only with microorganisms.

must be low, since revertants will also form colonies. This problem can often be overcome by using double mutants, since it will be very unlikely that two back mutations will occur in the same cell. Much of the skill of the bacterial geneticist is exhibited in the choice of proper mutants and selective media for efficient detection of genetic recombination. Because selection is so powerful and because crosses can be made using billions of individual cells, recombinational analysis is a very important tool to the microbial geneticist.

Complementation

When two mutant strains are genetically crossed (mated), homologous recombination can yield a wild-type recombinant unless both mutations include changes in exactly the same base pairs. Therefore, if two different Trp⁻ *E. coli* (strains that require the amino acid tryptophan) are crossed and Trp⁺ recombinants are obtained, it is clear that the mutations in the two strains did not include the same base pairs. However, this experiment cannot detect whether the mutations were in the same gene. This can be determined by a type of experiment called a **complementation test**. The two mutations could be on the same or on separate DNA molecules. If the two mutations are each on separate DNA molecules, they would be said to be in **trans** configuration. On the other hand, if the two mutations were on the *same* DNA molecule, then they would be said to be in **cis** configuration.

Two mutations complement one another (that is, give a wild-type phenotype) when they are present in a diploid only if they are present in different genes. This is shown diagrammatically in Figure 7.12. Mutations can complement because the genes encode proteins. If each homologous DNA molecule contributes a different required gene, then the cell will have all the enzymes that it requires to synthesize tryptophan. Notice that complementation *does not* involve recombination. To do the test, the mutations must be in *trans*. (If one molecule has both mutations, the other is wild type and should be sufficient itself to confer the wild-type phenotype. If it does not, then one of the mutations must exert its phenotype even in the presence of the wild-type gene. Such a mutation would be said to be *dominant*.)

This type of complementation test, called a *cis–trans test*, is used to define whether two mutations are in the same genetic (functional) unit. The genetic unit defined by the cis–trans test is sometimes called a **cistron** (a term essentially equivalent to a gene). As noted, two mutations in the *same* cistron *cannot* complement each other, so that if complementation is found to exist, this implies that the two mutations lie in *different* cistrons (that is, different genes). The term *cistron* is now rarely used except when describing whether a mRNA has the genetic information from one gene (monocistronic mRNA) or from more than one gene (polycistronic mRNA) (see Section 5.6).

Genetic recombination arises when two genetically distinct elements are combined together in the same element. Recombination is an important evolutionary process and cells have specific mechanisms for ensuring that recombination takes place. In eukaryotes, genetic recombination occurs as a consequence of the sexual cycle. Mechanisms of recombination also occur in prokaryotes but involve DNA transfer during the processes of transformation, transduction, and conjugation.

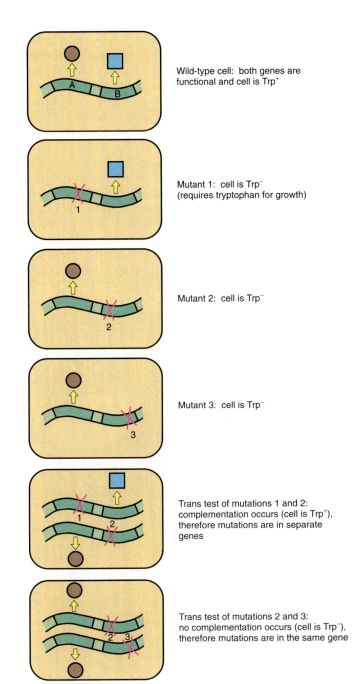

FIGURE 7.12 Complementation analysis. Mutations 1, 2, and 3 each lead to the same phenotype, a requirement for tryptophan. Complementation analysis indicates that mutations 2 and 3 are in one gene and that mutation 1 is in another. The protein products of both genes (A and B) must be required to synthesize tryptophan.

7.6 Genetic Transformation

As we have noted, genetic transformation is a process by which free DNA is incorporated into a recipient cell and brings about genetic change. The discovery of genetic transformation in bacteria was one of the outstanding events in biology, as it led to experiments proving without a doubt that DNA was the genetic material (see The Origins of Bacterial Genetics box). This discovery became the keystone of molecular biology and modern genetics.

A number of bacteria have been found to be transformable, including both Gram-negative and Gram-positive species. However, even within transformable genera, only certain strains or species are transformable. Since the DNA of prokaryotes is present in the cell as a long single molecule, when the cell is gently lysed, the DNA pours out (Figure 7.13). Because of its extreme length (1700 μm in *Bacillus subtilis*), the DNA molecule breaks easily; even after gentle extraction it fragments into 100 or more pieces (*B. subtilis* DNA of 5×10^6 base pairs is converted into fragments of about 15 kilobase pairs). Since the DNA which corresponds to an average gene is about 1000 nucleotides, each of the fragments of purified DNA will have about 15 genes. Any cell will usually incorporate only one or a few DNA fragments so that only a small proportion of the genes of one cell can be transferred to another by a single transformation event.

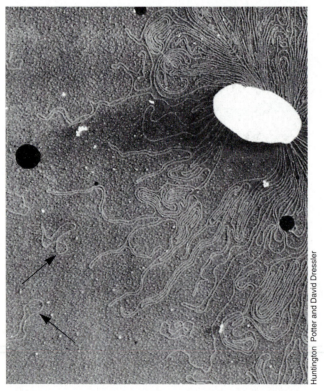

FIGURE 7.13 The bacterial chromosome and bacterial plasmids, as shown in the electron microscope. The plasmids (arrows) are the circular structures, much smaller than the main chromosomal DNA. The cell (large white structure) was broken gently so that the DNA would remain intact.

Huntington Potter and David Dressler

Competence

A cell that is able to take up a molecule of DNA and be transformed is said to be **competent**. Only certain strains are competent; the ability seems to be an inherited property of the organism. Competence in most naturally transformable bacteria is regulated, and special proteins play a role in the uptake and processing of DNA. These competence-specific proteins may include a membrane-associated DNA-binding protein, a cell wall autolysin, and various nucleases. In both *Bacillus subtilis* and *Streptococcus pneumoniae*, induction of competence is dependent on the medium and growth stage of the culture. In *Bacillus*, about 20 percent of the cells become competent and stay that way for several hours. However, in *Streptococcus*, 100 percent of the cells can become competent, but only for a few minutes during the growth cycle.

Uptake of DNA

Bacteria differ in the form in which DNA is taken up. In *Haemophilus*, which is Gram-negative, for example, only double-stranded DNA is taken up into the cell despite the fact that only single-stranded segments actually become incorporated into the genome by recombination. In Gram-positive Bacteria such as *Streptococcus* and *Bacillus*, by contrast, only a single DNA strand is taken up, while the complementary strand is simultaneously degraded. However, in all cases, double-stranded DNA binds more effectively to the cells.

During the transformation process, competent bacteria first bind DNA reversibly; soon, however, the binding becomes irreversible. Competent cells bind much more DNA than do noncompetent cells—as much as 1000 times more. As we noted earlier, the sizes of the transforming fragments are much smaller than that of the whole genome and this DNA is further degraded during the uptake process. In *Streptococcus pneumoniae* each cell can bind only about 10 molecules of double-stranded DNA of 15–20 kilobase pairs each. However, as they are taken up these are converted to single-stranded pieces of about 8 kilobases. The DNA fragments in the mixture compete with each other for uptake, and if excess DNA that does not contain the genetic marker is added, a decrease in the number of transformants occurs. In preparations of transforming DNA, only about one out of 100 to 200 DNA fragments contains the marker being studied. Thus at high concentrations of DNA, the competition between DNA molecules results in saturation of the system so that even under the best conditions it is impossible to transform all of the cells in a population for a given genetic marker. The maximum frequency of transformation that has so far been obtained is about 20 percent of the population; actually the values usually obtained are between 0.1 and 1.0 percent. The minimum concentration of DNA yielding detectable transformants is about 0.00001 μg/ml (1×10^{-5} μg/ml), which is so low that it is undetectable chemically.

Origins of Bacterial Genetics*

Although genetic recombination in eukaryotes had been known for a long time, the discovery of genetic recombination in bacteria by transformation, transduction, and conjugation has been a relatively recent event. Of the three recombination processes, the discovery of transformation was the most significant as it provided the first direct evidence that DNA is the genetic material. The first evidence of bacterial transformation was obtained by the British scientist Fred Griffith in the late 1920s. Griffith was working with *Streptococcus pneumoniae* (pneumococcus), a bacterium which owes its ability to invade the body in part due to the presence of a polysaccharide capsule. Mutants can be isolated which lack this capsule and are thus unable to cause infection; such mutants are called R strains, because their colonies appear rough on agar, in contrast to the smooth appearance of capsulated strains. A mouse infected with only a few cells of an S (smooth) strain will succumb in a day or two to pneumococcus infection, whereas even large numbers of R cells will not cause death when injected. Griffith showed that if heat-killed S cells were injected along with living R cells, a fatal infection ensued, and the bacteria isolated from the dead mouse were S types. A number of different polysaccharide capsules were known in different pneumococcus S strains, and it was possible to do this experiment with heat-killed S cells from a type different from that from which the R strain was derived. Since the isolated living S cells always had the capsule type of the heat-killed S cells, the R cells had been transformed into a new type, and the process had all the properties of a genetic event. The molecular explanation for the transformation of pneumococcus types was provided by Oswald T. Avery and his associates at Rockefeller Institute in New York, by a series of studies carried out during the 1930s, culminating in the now classic paper by Avery, McCarty, and MacLeod in 1944. Avery and his co-workers showed that under certain conditions the transformation process could be carried out in the test tube rather than the mouse, and that a cell-free extract of heat-killed cells could induce transformation. By a long series of painstaking biochemical experiments, the active fraction of cell-free extracts was purified and was shown to consist of DNA. The transforming activity of purified DNA preparations was very high, and only very small amounts of material were necessary. Subsequently, Rolin Hotchkiss, Harriet Ephrussi, and others at Rockefeller showed that transformation could occur in pneumococcus not only for capsular characteristics, but for other genetic characteristics of the organism, such as antibiotic resistance and sugar fermentation. In the 1950s, transformation was also shown to occur in *Haemophilus, Neisseria, Bacillus*, and a variety of other organisms. In 1953, James Watson and Francis Crick announced their model for the structure of DNA, providing a theoretical framework for how DNA could serve as the genetic material.

Thus, two types of studies, the bacteriological and biochemical ones of Avery, and physical-chemical ones of Watson and Crick, solidified the concept of DNA as the genetic material. In the subsequent years, this work has led to the whole field of molecular genetics.

Although bacterial transformation resulted from an essentially accidental discovery, bacterial conjugation was initially shown to occur by Joshua Lederberg and E.L. Tatum in 1946, through experiments carefully designed to determine if a sexual process might occur in bacteria. Because it appeared that the process, if present, would be quite rare (no microscopic evidence for bacterial mating had ever been seen, although such evidence can easily be obtained in eukaryotes), Lederberg developed a method which involved the use of nutritional mutants of *Escherichia coli*. Fortunately, he isolated these mutants in strain K-12, one of the few wild-type strains now known to contain the F plasmid. The principle was to mix two strains, one requiring biotin and methionine, the other requiring threonine and leucine, and plate the mixture on a minimal medium lacking all four growth factors. Neither parental type could grow on this medium, but any recombinants could, and when about 10^8 cells were plated, a small but significant number of colonies was obtained. Strains with two separate nutritional requirements were employed since it would be unlikely that spontaneous back mutation of both genes would occur in a single cell. Thus the only explanation for the phenomenon was some sort of genetic recombination. To show that the process required cell-to-cell contact, and hence could not be a type of transformation, it was shown that when culture filtrates or extracts were separated by a sintered glass disk, permeable to macromolecules but not to cells, recombination did not occur. Although initially conjugation appeared to be a very rare event, by the early 1950s a strain of *E. coli* had been isolated by the Italian scientist L. L. Cavalli-Sforza, while he was working in Lederberg's laboratory, that showed a high frequency of recombination. The British physician William Hayes, who independently isolated an Hfr strain, then showed that genetic transfer during mating between Hfr and F⁻ was a one-way event, with the Hfr serving as donor. The interrupted mating experiment and the demonstration of the circular genetic map of *E. coli* were then carried out by Elie Wollman and Francois Jacob, working with Jacques Monod at the Pasteur Institute in Paris. The distinction between Hfr and F⁺ was made by Lederberg, who also showed that F⁺ behaved in an infectious manner. Lederberg coined the term *plasmid* in the 1950s, to describe such apparently extrachromosomal genetic elements, although the term did not find wide usage until the 1970s, when infectious drug resistance became a major medical problem.

Bacterial transduction was discovered by the American scientist Norton Zinder when he was working at the University of Wisconsin as a graduate student with Lederberg on genetic recombination in *Salmonella typhimurium*. The original motivation for this work was to show that conjugation occurred in an organism other than *E. coli*, and the techniques involved isolation of mutants and quantification of recombination by observing colony growth on minimal medium. However, although evidence of recombination was obtained, it could be shown that cell-to-cell contact was not required. Although this suggested a type of transformation, the process was not affected by DNase, and the gene transfer agent behaved like a bacteriophage. The gene transfer agent could be purified by the same procedures used to purify virus particles, and transduction occurred only with recipient cells that had receptor sites for the virus in question. Further, transducing activity could be eliminated by treatment of a lysate with substances able to adsorb

the virus, such as sensitive cells or antibodies. Thus, in all cases, transducing activity and virus activity behaved in similar ways. Zinder and Lederberg coined the word "transduction" to refer to any genetic recombinational process that was only fragmentary and did not involve cell-to-cell contact, intending in this way to encompass processes involving either free DNA (transformation) or phage, but subsequently the word *transduction* has been applied only to virus-mediated genetic transfer.

*Key references to this section: Dubos, R. 1976. *The Professor, the Institute, and DNA*. Rockefeller University Press, New York. An account of Oswald T. Avery's research career, culminating in the discovery that DNA was the genetic material. Also provides brief historical background on other aspects of bacterial genetics. McCarty, M. 1985. *The Transforming Principle: Discovering that Genes are Made of DNA*. W. W. Norton, New York. A personal account of one of the key members of Avery's team.

Interestingly, in *Haemophilus influenzae* there is a requirement that the DNA fragment have a particular 11 base pair sequence for irreversible binding and uptake to occur. This sequence is found at an unexpectedly high frequency in the *Haemophilus* chromosome. Evidence such as this, and the fact that at least certain bacteria become competent in their natural environment, suggest that transformation is not just a laboratory artifact but may play an important role in gene transfer in nature.

Integration of transforming DNA

Transforming DNA is bound at the cell surface by a DNA-binding protein, after which either the entire double-stranded fragment is taken up or a nuclease degrades one strand while the other is taken up (Figure 7.14). After uptake, the DNA associates with a competence-specific protein which remains attached to the DNA, presumably preventing it from nuclease attack, until it reaches the chromosome where RecA protein takes over. The DNA is then integrated into the genome of the recipient by recombinational processes (Figure 7.14; see also Figure 7.9). During replication of this hybrid DNA, one parental and one recombinant DNA molecule is formed. Upon segregation at cell division, the latter will be present in the transformed cell, which is now genetically altered as compared to the parental type. The above discussion pertains only to small pieces of *linear* DNA. The transformation of plasmid DNA generally occurs in the absence of recombination between the plasmid and bacterial chromosome.

Transfection

Bacteria can be transformed with DNA extracted from a *bacterial virus* rather than from another bacterium, a process known as **transfection**. If the DNA is from a

lytic bacteriophage, transfection can be measured by the standard phage plaque assay (see Section 6.3). Transfection has become a useful tool for studying the mechanism of transformation and recombination because the small size of phage genomes allows for the isolation of a nearly homogeneous population of DNA molecules. By contrast, in conventional transformation, the transforming DNA is generally a random assortment of chromosomal DNA of various lengths and this tends to complicate experiments designed to study the mechanism of transformation.

Artificially induced competence

High efficiency natural transformation is only found in a few bacteria; *Bacillus*, *Streptococcus*, *Haemophilus*, and *Neisseria*, for example, are easily transformed. Many bacteria are transformed only poorly or not at all. Determination of how to induce competence in such bacteria may involve considerable empirical study, with variation in culture medium, temperature, and other factors. The nature of the cell surface must be of importance in determining whether a cell can take up DNA, and the presence or absence of intracellular nucleases are also important. To transfer DNA into cells for genetic engineering (see Chapter 8), it was necessary to find a way to make *Escherichia coli*, a Gram-negative organism, competent. It has been found that if *E. coli* is treated with high concentrations of calcium ions and then stored in the cold, it becomes transformable at low efficiency. With proper procedures it is possible to select *E. coli* transformants for chromosomal genes, but at low frequency. *E. coli* treated in this manner takes up double-stranded DNA, and therefore transformation by plasmid DNA is more efficient since no recombination is required. Why the calcium treatment works is not known, but this procedure also works with some other Gram-neg-

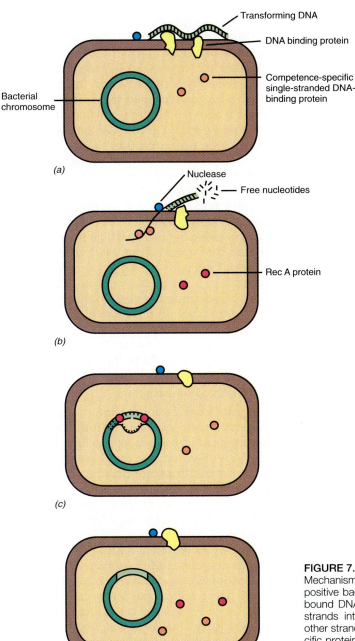

Transforming DNA

DNA binding protein

Competence-specific single-stranded DNA-binding protein

Bacterial chromosome

(a)

Nuclease

Free nucleotides

Rec A protein

(b)

(c)

FIGURE 7.14
Mechanism of DNA transfer by transformation in a Gram-positive bacterium. (a) Binding of free DNA by a membrane-bound DNA-binding protein. (b) Passage of one of the two strands into the cell while nuclease activity degrades the other strand. (c) The single strand in the cell is bound by specific proteins and recombination with homologous regions of the bacterial chromosome mediated by RecA protein occurs. (d) Transformed cell.

(d)

ative Bacteria. However, such methods of artificially induced competence are rapidly being supplanted by a new method termed *electroporation*.

DNA transfer by electroporation

Small pores are produced in the membranes of cells that are exposed to pulsed electrical fields. If DNA molecules are present outside the cells during the electrical pulse, they can then enter the cells through these pores. This process is called **electroporation**. Electroporation requires a sophisticated power supply since the pulses must be carefully controlled and last for only milliseconds. This technique has now been used to transport DNA into a large number of different species of bacteria, both Gram-negative and Gram-

positive. Additionally, electroporation allows an experimenter to transfer a plasmid directly from one cell to another if both are present during electroporation. Therefore, electroporation allows small molecules of DNA to come out of cells as well as to go in! This type of "transformation" eliminates the steps required to isolate the plasmid from the first strain before introducing it into the second.

Transformation (transfection) of eukaryotic cells

Eukaryotic microorganisms and animal and plant cells can take up DNA in a process which resembles bacterial transformation. Because the word *transformation* in mammalian cells is used to describe the conversion of cells to the malignant (tumorous, cancerous) state (see

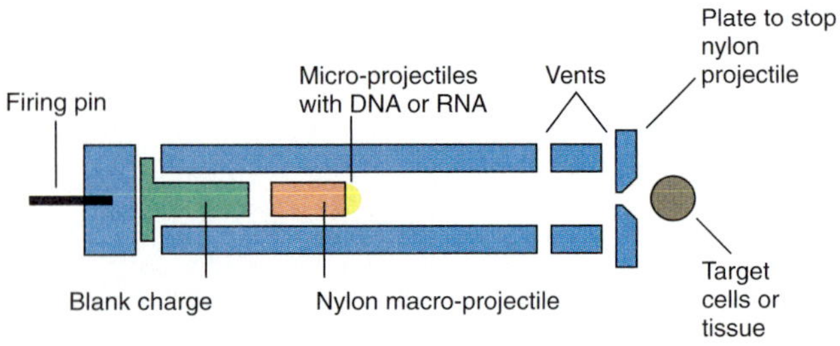

FIGURE 7.15 Nucleic acid gun for transfection of eukaryotic cells. The inner workings of the gun show how nucleic acids attached to metal pellets are projected at target cells.

Section 6.14), the introduction of DNA into mammalian cells has been called *transfection* (a term with another meaning in bacterial systems; see above).

Transfection of cultured animal cells was originally accomplished by precipitating DNA in such a way that the cells would take it up by phagocytosis (see Section 11.13) since they do not have cell walls. In yeast, where the introduction of cloned DNA is popular in genetic engineering, transfection at low efficiencies can be mediated by treating cells with enzymes which partially destroy the cell wall, generating spheroplasts, and then adding DNA in the presence of Ca^{2+} and polyethylene glycol (which serves to permeabilize the membrane).

Other methods of artificially induced competence are also used in eukaryotes. As in the case of bacteria, electroporation is becoming widely applied to all types of eukaryotic cells and can be used whether or not the cell wall is removed.

In addition to electroporation, a high velocity microprojectile "gun" has been developed for incorporating DNA into cells. The original **particle gun** operates somewhat like a conventional shotgun. A small steel cylinder containing a gun powder charge is used to fire nucleic acid–coated particles at the target cells (Figure 7.15). The particles bombard the cell, piercing cell walls and membranes without actually killing the cells. The nucleic acid entering the cells can then recombine with host DNA. The particle gun has been used successfully to transfect yeast, algae, a variety of plant cells, and even mitochondria and chloroplasts. The particle gun is very useful because, unlike electroporation, it can be used on intact tissue such as plant seeds.

Certain bacteria exhibit competence, a state in which cells are able to take up free DNA released by other bacteria. This process is called transformation. Competence depends upon the presence of a specific membrane-associated DNA uptake system. Transformation of chromosomal DNA involves uptake of small fragments of DNA; thus only a relatively small amount of DNA can be acquired in a single event. Certain laboratory procedures have been developed that make it possible to introduce DNA into completely unrelated organisms, even eukaryotes. Electroporation involves modification of the cytoplasmic membrane by treatment with an electrical field to facilitate DNA uptake, and the particle gun actually shoots DNA into the cell.

7.7 Transduction

In transduction, DNA is transferred from cell to cell through the agency of viruses. Genetic transfer of host genes by viruses can occur in two ways. In the first, called **generalized transduction**, host genes derived from virtually any portion of the host genome become a part of the DNA of the mature virus particle in place of, or in addition to, the virus genome. The second, called **specialized transduction**, occurs only in some temperate viruses; a specific group of host genes is integrated directly into the virus genome—usually replacing some of the virus genes—and is transferred to the recipient during lysogenization. The transducing virus particle in both specialized and generalized transduction is *defective* as a virus because bacterial genes have replaced some necessary viral genes.

Transduction has been found to occur in a variety of bacteria. Not all phages will transduce, and not all bacteria are transducible; but the phenomenon is sufficiently widespread for us to assume that it plays an important role in genetic transfer in nature.

Generalized transduction

In generalized transduction, virtually any genetic marker can be transferred from donor to recipient. Generalized transduction was first discovered and extensively studied in the bacterium *Salmonella typhimurium* with phage P22, but is known to occur in *Escherichia coli* and many other bacteria. An example of how *transducing particles* may be formed is given in Figure 7.16. When the population of sensitive bacteria is infected with a phage, the events of the phage lytic cycle may be initiated. During a lytic infection, the enzymes responsible for packaging viral DNA into the bacteriophage sometimes accidentally package host DNA. The resulting particle is called a *transducing particle*. Upon lysis of the cell, these particles are released along with normal virions, so that the lysate contains a mixture of normal and transducing virions. Because transducing particles cannot initiate a normal viral infection (they contain no viral DNA), they are said to be *defective*. When this lysate is used to infect a population of recipient cells, most of the cells become infected with normal virus. However, a small proportion of the population receives transducing particles, whose DNA can now undergo genetic recombination with the host DNA. Since only a small proportion of the particles in the lysate are of the defective transducing type and since each of those contains only a small fragment of donor DNA, the probability of a transducing particle containing a particular gene is quite low, and usually only about 1 cell in 10^6 to 10^8 is transduced for a given marker.

Phages that form transducing particles can be either temperate or virulent, the main requirements being that they have a DNA packaging mechanism that permits accidental recognition of host DNA and that packaging occurs before the host genome is com-

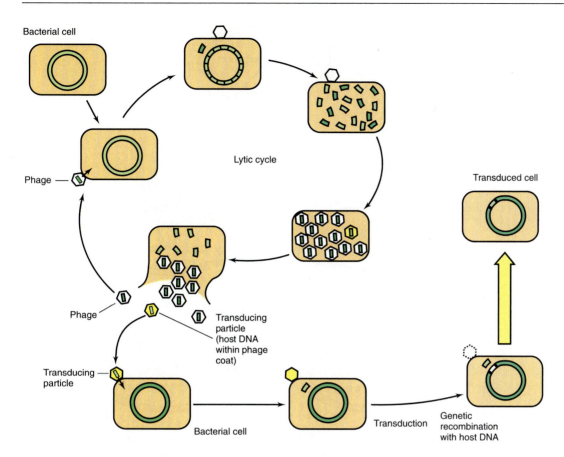

FIGURE 7.16 Generalized transduction: one possible mechanism by which virus (phage) particles containing host DNA can be formed.

pletely degraded. The detection of transduction is most certain when the multiplicity of phage to host is low, so that a host cell is infected with only a single phage particle; with multiple infection, the cell may be killed by the normal particles.

Specialized transduction

Generalized transduction is a rare genetic event, but in the specialized transduction to be discussed now, a very efficient transfer by phage of a specific set of host genes can be arranged. The example we shall use, which was the first to be discovered and is the best understood today, involves transduction of the galactose genes by the temperate phage lambda of *E. coli*.

As we discussed in Section 6.12, when a cell is lysogenized by lambda, the phage genome becomes integrated into the host DNA at a specific site. The region in which lambda integrates is immediately adjacent to the cluster of host genes that control the enzymes involved in galactose utilization (Figure 6.30), and the DNA of lambda is inserted into the host DNA at that site. From then on, viral DNA replication is under host control. Upon induction, (for example, by ultraviolet radiation) the viral DNA separates from the host DNA by a process that is the reverse of integration (Figure 7.17). Ordinarily when the lysogenic cell is induced, the lambda DNA is excised as a unit. Under rare conditions, however, the phage genome is excised incorrectly. Some of the adjacent bacterial

genes (the galactose cluster) are excised along with phage DNA. At the same time, some phage genes are left behind. One type of altered phage particle, called **lambda dgal** or *λdgal* (*dgal* means "defective galactose"), is defective because of the phage genes lost and does not make mature phage. However, if another phage called a **helper** is used together with *λdgal* in a mixed infection, then the defective phage can be replicated and can transduce the galactose genes. The role of the helper phage is to provide those functions missing in the defective particles. Thus the culture lysate obtained contains a few *λdgal* particles mixed in with a large number of normal lambda virions.

If a galactose-negative bacterial culture is infected at high multiplicity with such a lysate, and *gal*⁺ transductants are selected, many are double lysogens, carrying both lambda and *λdgal*. When such a double lysogen is induced, a lysate is produced containing about equal numbers of lambda and *λdgal*. Such a lysate can transduce at high efficiency, although only for a restricted group of *gal* genes.

If the phage is to be viable, there is a maximum limit to the amount of phage DNA that can be replaced with host DNA, since sufficient phage DNA must be retained in order to provide the information for production of the phage protein coat and for other phage proteins needed for lysis and lysogenization. However, if a helper phage is used together with the defective phage in a mixed infection, then even less in-

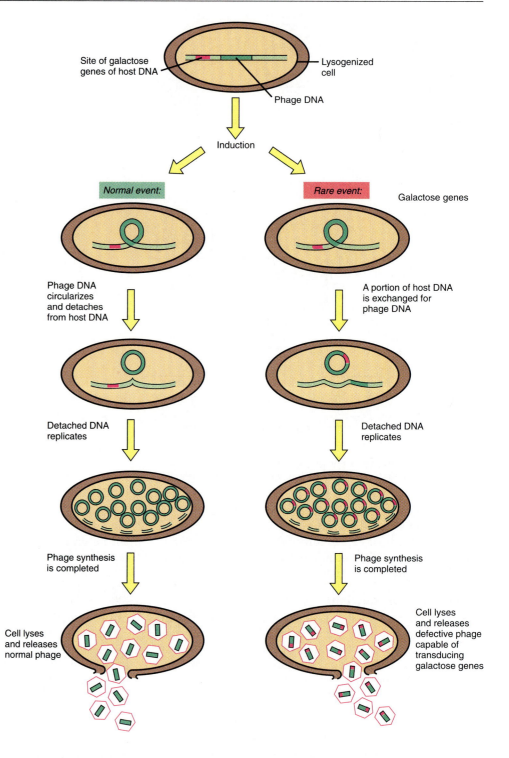

FIGURE 7.17
Normal lytic events and the production of particles transducing the galactose genes in an *E. coli* cell lysogenic for lambda virus.

formation is needed in the defective phage for transduction. Only the *att* (attachment) region, the *cos* site (cohesive ends, for packaging), and the replication origin of the lambda genome are needed for production of a transducing particle, provided a helper is used (see the genetic map of lambda in Section 6.12).

One important distinction between specialized and generalized transduction is in how the transducing lysate can be formed. In specialized transduction this *must* occur by induction of a lysogenic cell whereas in generalized transduction it can occur either in this way or by infection of a nonlysogenic cell

by the temperate phage, with subsequent phage replication and cell lysis.

Although we have discussed specialized transduction only in the lambda-*gal* system, phage lambda and its relative φ80 have been widely used to form specialized transducing phages covering many specific regions of the *E. coli* genome.

Phage conversion

This is a phenomenon analogous in some ways to specialized transduction. When a normal temperate phage (that is, a nondefective one) lysogenizes a cell and its DNA is converted into the prophage state, the lysogenic cell is immune to further infection by the same type of phage. This acquisition of immunity can be considered a change in phenotype. In certain cases other phenotypic alterations can be detected in the lysogenized cell, which seem to be unrelated to the phage immunity system. Such a change, which is brought about through lysogenization by a normal temperate phage, is called **phage conversion**.

Two cases of conversion have been especially well studied. One involves a change in structure of a polysaccharide on the cell surface of *Salmonella anatum* upon lysogenization with phage ϵ^{15}. The second involves the conversion of nontoxin-producing strains of *Corynebacterium diphtheriae* to toxin-producing (pathogenic) strains, upon lysogenization with phage β (we discuss the disease diphtheria in Section 15.2). In these situations the information for production of these new materials is apparently an integral part of the phage genome and hence is automatically and exclusively transferred upon infection by the phage and lysogenization.

Lysogeny probably carries a strong selective value for the host cell, since it confers resistance to infection by viruses of the same type. Phage conversion seems also to be of considerable evolutionary significance, since it results in efficient genetic alteration of host cells. Many bacteria isolated from nature are lysogenic. It seems reasonable to conclude, therefore, that lysogeny is the normal state of affairs and may often be essential for survival of the host in nature.

> As genetic elements, viruses can transfer not only their own genomes but also fragments of cellular genomes. In generalized transduction, defective virus particles randomly incorporate fragments of the cell DNA; virtually any gene of the donor can be transferred, but the efficiency is low. In specialized transduction, the DNA of a temperate virus excises incorrectly and brings adjacent host gene(s) along with it; only genes close to the integration point of the virus are transduced but the efficiency may be high.

7.8 Plasmids

Before we discuss the third method of genetic transfer, **conjugation**, we must discuss another kind of genetic element called the **plasmid** (see Section 5.5). Plasmids are circular genetic elements that reproduce autonomously and have an extrachromosomal existence (Figure 7.13). An outline of various aspects of plasmid biology is given in Figure 7.18. Naturally occurring plasmids vary in size from approximately 1 to over 1000 kilobase pairs. Although most plasmids are circular double-stranded DNA, some are linear double-stranded DNA.

Many plasmids can be transmitted from cell to cell by means of the conjugation process, which is dis-

cussed below. Some plasmids also have the ability to become integrated into the chromosome, and under such conditions their replication comes under control of the chromosome.

Plasmids can carry a variety of different genes, for example, genes that control the production of toxins or provide the host with resistance to antibiotics, heavy metals, and other inhibitory compounds. Some plasmids carry genes for the catabolism of unusual substrates, such as aromatic compounds and pesticides. Many plasmids also carry genes that control the process of conjugation, such as genes which alter the cell surface to permit cell-to-cell contact and genes which bring about the transfer of DNA from one cell to another. Plasmids which govern their own transfer by cell-to-cell contact are called **conjugative**, but not all plasmids are conjugative. Transmissability by conjugation is controlled by a set of genes within the plasmid called the *tra region*. The presence of a *tra* region in a plasmid can have another important consequence if the plasmid becomes integrated into the chromosome. In that case, the plasmid can *mobilize* the transfer of chromosomal DNA from one cell to another. Strains of bacteria that transfer large amounts of chromosomal DNA during conjugation are called *Hfr* (high frequency of recombination). Although considerably rarer than bacteria transferring only plasmid DNA, Hfr bacteria are of considerable interest and importance, since the study of Hfr bacteria permits an analysis of the genetic organization of the whole bacterial chromosome.

Physical evidence of plasmids

The typical plasmid is a circular double-stranded DNA molecule, less than one-twentieth the size of the chromosome. When isolated, this circular double helix is twisted about itself in a form called a **supercoil** (see Figure 5.6). A single break (called a "nick") in one of the two strands causes the supercoil to convert to an open circular form, and if breaks occur in both strands at the same place, a linear duplex structure will be formed. Most of the plasmid DNA isolated from cells is in the supercoiled configuration, which would be the most compact form within the cell.

Isolation of plasmid DNA can generally be readily accomplished by making use of certain physical properties of supercoiled DNA molecules, which can be separated from other types of DNA using an ultracentrifuge. Plasmid DNA molecules of different sizes can also be readily separated by electrophoresis on agarose gels (see Nucleic Acids box, Chapter 5). This technique provides a good opportunity to analyze the action of endonucleases and other enzymes on plasmid DNA.

Plasmid DNA can be observed under the electron microscope, and one type of evidence for the existence and molecular size of a plasmid is observation of its structure and measurement of its length in electron micrographs (see Figure 7.13).

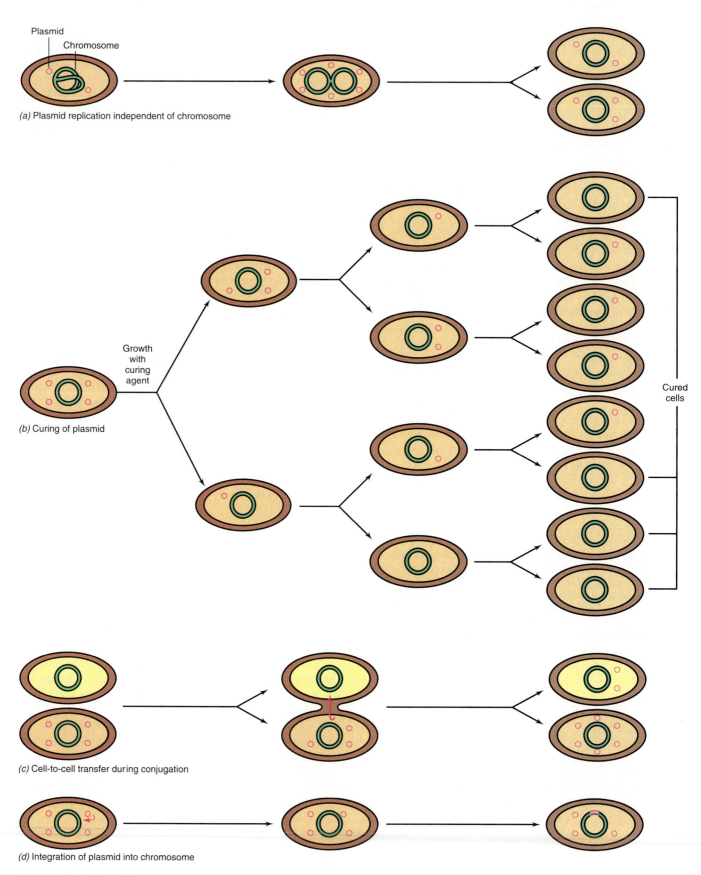

(a) Plasmid replication independent of chromosome

Growth
with
curing
agent

(b) Curing of plasmid

Cured
cells

(c) Cell-to-cell transfer during conjugation

(d) Integration of plasmid into chromosome

FIGURE 7.18 Plasmid biology.

Curing of plasmids

One of the common features of plasmids is that they can be eliminated from host cells by various treatments. This process, termed **curing**, apparently results from inhibition of plasmid replication without parallel inhibition of chromosome replication, and as a result of cell division the plasmid is diluted out (Figure 7.18). Curing may occur spontaneously, but it is greatly increased by use of acridine dyes which become inserted into DNA, as well as by other treatments that affect DNA, such as use of ethidium bromide, ultraviolet or ionizing radiation, and heavy metals. Electroporation may also be used to cure a cell of plasmids.

Structure and replication of plasmids

One plasmid about which much is known is the **F plasmid** of *Escherichia coli*. The F plasmid is conjugative and can be transferred from a donor cell to a recipient cell by itself or along with donor chromosomal DNA. The F plasmid is a circular molecule of 100 kilobase pairs and contains various genes for replication and transfer processes (Figure 7.19). Because the *tra* region is large, about 30 kilobase pairs of the F plasmid, conjugative plasmids are generally much larger than nonconjugative plasmids. In addition, the F plasmid contains a number of specific nucleic acid sequences which allow F to recombine with the bacterial chromosome to form Hfr strains (Figure 7.19). Plasmids having the ability to integrate into host chromosomes are called *episomes*.

Most plasmids in Gram-negative Bacteria replicate in a manner similar to that already described for the chromosome (see Section 5.4). This involves initiation of replication at an origin and bidirectional replication around the circle, giving a *theta* intermediate. Because of the small size of plasmid DNA relative to the chromosome, the whole replication process occurs very quickly, perhaps in one-tenth or less of the total time of the cell division cycle. The enzymes involved in plasmid replication are normal cell enzymes, so that the genetic elements within the plasmid itself, which control its replication, may be concerned primarily with the control of the timing of the initiation process and with the apportionment of the replicated plasmids between daughter cells.

Most plasmids of Gram-positive Bacteria replicate by a rolling circle mechanism similar to that used by the phage ϕX174 (see Section 6.8 and Figure 6.18). This mechanism gives rise to a single-stranded intermediate, and thus these plasmids are sometimes referred to as single-stranded DNA plasmids. Most of the linear plasmids now known replicate using a mechanism involving a protein bound to the 5'-end of each strand which is used in priming DNA synthesis (see Figures 5.25 and 6.47). Although plasmids are independently replicating genetic elements, the number of plasmid molecules per cell is determined to some extent by the cell and by environmental conditions. Some plasmids, such as F, are present in the cell in only one to three copies, whereas others may be present in as many as 100 copies.

> Plasmids, which vary widely in size, can be detected by their size and mobility in gel electrophoresis and can also be observed under the electron microscope. Many plasmids are conjugative, able to transfer themselves from cell to cell as a result of cell-to-cell contact. In some cases, plasmids can also mobilize chromosomal DNA and cause its transfer from one cell to another.

7.9   Types of Plasmids and Their Biological Significance

A very large number of plasmids are known; approximately 300 different naturally occurring ones have been isolated from strains of *Escherichia coli* alone. It should be clear from our discussion of plasmids that these genetic elements are of great importance as tools for understanding a wide variety of genetic phenomena in prokaryotes. Plasmids also help explain a number of biological and ecological phenomena. Although plasmids do not carry genes that are essential to the host under all conditions (see Section 5.5), the presence of plasmids in a cell can have a profound influence on the cell's phenotype. In some cases, plasmids encode for properties that we think of as fundamental to the bacterium in question, for example, the ability of *Rhizobium* to interact with plants (see Section 17.24). Because plasmids can be large and may carry many

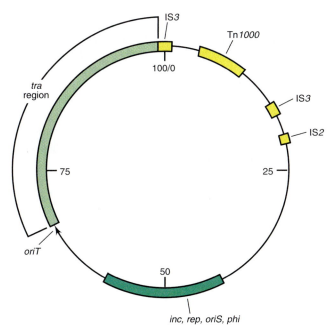

FIGURE 7.19   Genetic map of the F (fertility) plasmid of *Escherichia coli*. The numbers on the interior show the size of the plasmid in kilobase pairs. The location of key F genes are shown: *tra*, transfer functions; *oriT*, origin of transfer; *oriS*, origin of replication; *inc*, incompatability group; *rep*, replication functions; *phi*, phage inhibition. The regions shown in yellow on F are transposable elements where integration into identical elements on the bacterial chromosome can occur and lead to the formation of different Hfr strains.

different genes, it is not always a simple matter to classify a plasmid into a simple phenotypic category. As we shall see, a single plasmid may confer many different phenotypes on its host cell.

Conjugative plasmids and pili

We have already mentioned that conjugative plasmids are those with the capacity to transfer themselves from one cell to another by cell-to-cell contact. The genes responsible for this are carried by the plasmids in the *tra* region. Many of the genes in this region have to do with the synthesis of pili. At least two kinds of pili, called F and I pili, are known to be involved in cell-to-cell transfer of plasmids. Two classes of RNA phages are known to infect cells which carry transmissable plasmids. These phages can be used to demonstrate the presence on the cell of either F or I pili (see Figures 6.15 and 7.22). The two kinds of pili can also be distinguished immunologically. F pili are involved in the transfer of F plasmid and some antibiotic resistance plasmids. I pili are involved in the transfer of other antibiotic resistance plasmids, colicin-determining plasmids, and others.

Resistance plasmids

Among the most widespread and well-studied groups of plasmids are the *resistance plasmids* (*R plasmids*), which confer resistance to antibiotics and various other inhibitors of growth. R plasmids were first discovered in Japan in strains of enteric bacteria which had gained resistance to a number of antibiotics (multiple resistance), and have since been found in other parts of the world. The emergence of bacteria resistant to several antibiotics is of considerable medical significance, and was correlated with the increasing use of antibiotics for the treatment of infectious diseases. Soon after these resistant strains were isolated it was shown that they could transfer resistance to sensitive strains via cell-to-cell contact. This is probably one of the reasons for the rapid rise of multiply resistant strains, since it would be unlikely for resistance to a number of antibiotics to develop simultaneously by mutation and selection. The infectious nature of the R plasmids permits rapid spread of the characteristic through populations (see Figure 7.18c).

A variety of antibiotic-resistance genes can be carried by an R plasmid. In general, these genes encode proteins that either inactivate the antibiotic or affect its uptake into the cell (see following). Plasmid R100, for example, is an 89.3 kilobase pair plasmid (Figure 7.20) that carries resistance genes for sulfonamides, streptomycin/spectinomycin, fusidic acid, chloramphenicol, and tetracycline. R100 also carries several genes conferring resistance to mercury (see Section 17.18). R100 can transfer itself between enteric bacteria of the genera *Escherichia*, *Klebsiella*, *Proteus*, *Salmonella*, and *Shigella*, but will not transfer to the nonenteric bacterium *Pseudomonas*. R plasmids with genes for resistance to kanamycin, penicillin, tetracycline, and neo-

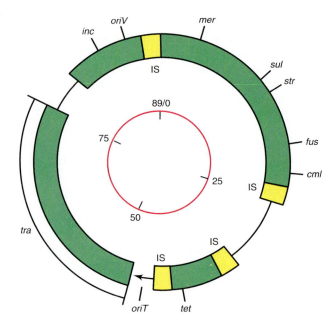

FIGURE 7.20 Genetic map of the resistance plasmid R100. The inner circle shows the size of the plasmid in kilobase pairs. The outer circle shows the location of major antibiotic resistance genes and other key functions. *inc*, incompatibility genes; *oriV*, origin of replication site; *oriT*, origin of conjugative transfer; *mer*, mercuric ion resistance; *sul*, sulfonamide resistance; *tet*, tetracycline resistance; *tra*, transfer functions. The locations of insertion sequences (IS) are also shown.

mycin are also known. Many drug-resistant elements on R plasmids, such as those on R100, are transposable elements and can be used in transposon mutagenesis (see Section 7.12).

Genes for characteristics not related to antibiotic resistance are also carried by R plasmids. Paramount among these is the *tra* region necessary for conjugative transfer, but R plasmids also carry genes permitting their own replication and genes controlling production of proteins that prevent the introduction of other related plasmids (Figure 7.20). Thus the presence of one R plasmid inhibits the introduction of another of the same type, a phenomenon known as *incompatibility* (see further discussion below).

Because R plasmids are able to undergo genetic recombination, genes from two R plasmids can be integrated into one. Plasmid recombination is one means by which multiply drug-resistant organisms might have first arisen. The genes in an R plasmid are carried in a definite order (see Figure 7.20), as are chromosomal genes, and this order can be mapped by genetic recombination. Many R plasmids and F plasmids have related *tra* regions, and recombination can occur between F and R plasmids.

Biochemical mechanism of resistance mediated by R plasmids

In the laboratory antibiotic-resistant cells are often isolated from cultures that were predominantly antibiotic sensitive. The resistance of these isolates is usually due to mutations in *chromosomal* genes. On the other hand, the majority of drug-resistant bacteria isolated from patients contain the drug-resistance genes on R plasmids. The biochemical mechanism of R plasmid resistance is different from that of chromosomal resistance. In most cases, antibiotic resistance mediated by chromosomal genes arises because of a modification of the *target* of antibiotic action (e.g., a ribosome).

By contrast, R plasmid resistance is in most cases due to the presence in the R plasmid of genes coding for new enzymes which *inactivate* the drug (Figure 7.21) or genes which code for enzymes that either prevent uptake of the drug or actively pump it out. For instance, a number of antibiotics are known which have similar chemical structures, containing aminoglycoside units. Among the aminoglycoside antibiotics are streptomycin, neomycin, kanamycin, and spectinomycin. Strains carrying R plasmids conferring resistance to these antibiotics contain enzymes that chemically modify these aminoglycoside antibiotics, either by phosphorylation, acetylation, or adenylylation, the modified drug then lacking antibiotic activity (Figure 7.21). In the case of the penicillins, R plasmid resistance is due to the formation of penicillinase (β-lactamase), which splits the β-lactam ring, thus destroying the molecule. Chloramphenicol resistance mediated by an R plasmid arises because of the presence of an enzyme that acetylates the antibiotic. Thus, the fact that R plasmids can confer multiple antibiotic resistance does not imply that the mode of action of the R plasmid genes is similar. The presence of multiple antibiotic resistance is due to the fact that a single R plasmid contains a variety of genes coding for different antibiotic inactivating enzymes.

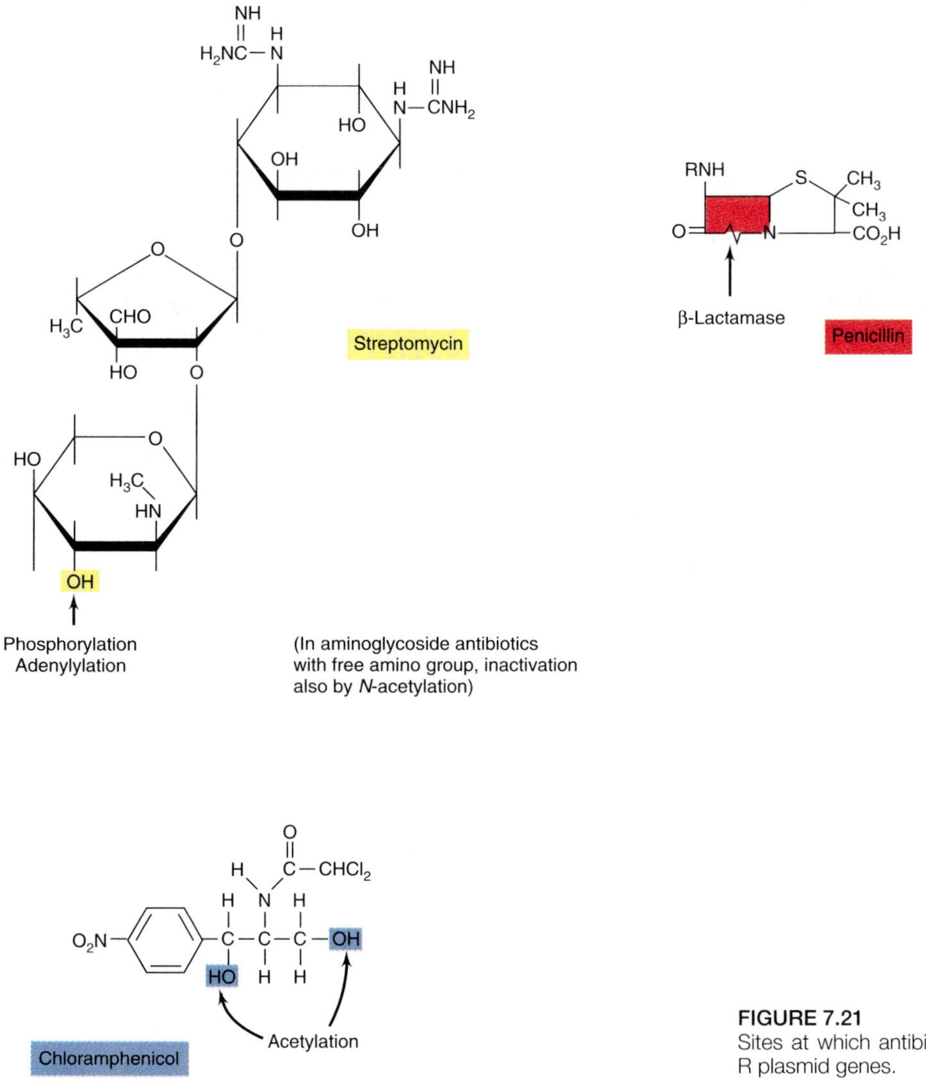

FIGURE 7.21
Sites at which antibiotics are attacked by enzymes coded for by R plasmid genes.

> Plasmids often carry antibiotic resistance genes and a single plasmid may confer resistance to multiple antibiotics. Conjugative plasmids that confer multiple antibiotic resistance are called resistance plasmids (R plasmids), and the evolution of R plasmids paralleled the introduction of antibiotics into medicine.

Toxins and other virulence characteristics

We will discuss in Chapter 11 the physiological and genetic characteristics of microorganisms that enable them to colonize hosts and set up infections, which can lead to harm. In the present context, we merely note the two major characteristics involved in virulence: (1) the ability of microorganisms to attach to and colonize specific sites in the host; and (2) the formation of substances (toxins, enzymes, and other molecules), which cause damage to the host. It has now been well established that in several pathogenic bacteria each of these virulence characteristics is carried on plasmids. For example, enteropathogenic strains of *E. coli* are characterized by an ability to colonize the small intestine and to produce a toxin that causes symptoms of diarrhea. Colonization requires the presence of a cell surface protein called the colonization factor antigen (CFA), encoded by a plasmid, which confers on the cells the ability to attach to epithelial cells of the intestine. At least two toxins in enteropathogenic *E. coli* are known to be coded for by a plasmid: the *hemolysin,* which lyses red blood cells, and the *enterotoxin,* which induces extensive secretion of water and salts into the bowel. It is the enterotoxin which is responsible for the induction of diarrhea, as will be discussed in Chapter 11. Since other virulence factors of *E. coli* are carried by chromosomal genes, it is not clear why these toxin genes are carried by plasmids, but the existence of such plasmids raises the question of how widespread plasmid-related virulence is in microorganisms.

In *Staphylococcus aureus*, some virulence-conferring properties are known to be plasmid-linked; *S. aureus* is noteworthy for the variety of enzymes and other extracellular proteins it produces that are involved in its virulence, and the production of *coagulase, hemolysin, fibrinolysin,* and *enterotoxin* is thought to be plasmid-linked. In addition, the yellow pigment in *S. aureus*, which is probably involved in its ability to resist the destructive action of singlet oxygen in the phagocyte (see Sections 11.13 and 11.14), is plasmid-linked. *S. aureus* is also a notorious hospital-borne (nosocomial) pathogen, and multiple antibiotic resistance, encoded by plasmids, is common in this species (see Section 14.7). Virulence factors from a variety of bacteria are known to be encoded on plasmids, and others are known to be encoded by other types of *mobile genetic elements:* transposons and bacteriophages.

Bacteriocins

Many bacteria produce agents that inhibit or kill closely related species or even different strains of the same species; these agents are called **bacteriocins** to distinguish them from the antibiotics, which have a wider spectrum of activity. Bacteriocins are composed of protein and in some cases are coded for by plasmids. Bacteriocins are named in accordance with the species of organism that produces them. Thus, in *E. coli* we have *colicins,* coded by Col plasmids, *Bacillus subtilis* produces *subtilisin,* and so on. Since the colicins are the best studied bacteriocins, we limit our discussion here to them.

The Col plasmids of *E. coli* code for various colicins (Table 7.3). Colicins released from a producing cell bind to specific receptors on the surface of susceptible cells. The receptors for colicins are generally entities whose normal function is to transport some substance, frequently a growth factor or micronutrient, through the outer membrane (the lipopolysaccharide layer) of the cell. Colicins kill cells by disrupting some critical cell function. Many colicins, such as B, E1, I, K, and V (Table 7.3), form channels in the cell membrane which allow potassium ions and protons to leak out, leading to a loss of the cell's energy-forming ability. However, colicin E2 is a DNA endonuclease that can cleave cellular DNA, and colicin E3 is a nuclease that cuts at a specific site in 16S rRNA and inactivates ribosomes (Table 7.3). Col plasmids can be either conjuga-

Table 7.3   Properties of some colicins and the plasmids that encode them

Colicin	Mode of action	Normal function of colicin receptor	Plasmid	Size of plasmid (kilobase pairs)
Colicins encoded by conjugative plasmids				
B	Causes membrane leakage	Enterochelin transport	ColB-K77	106
I	Causes membrane leakage	Iron transport	ColIb-P9	94
V	Causes membrane leakage	Iron transport	ColV-B188	80
Colicins encoded by nonconjugative plasmids				
E1	Causes membrane leakage	Vitamin B_{12} transport	ColE1-16	9
E2	Cleaves DNA	Vitamin B_{12} transport	ColE2-P9	8
E3	Cleaves 16S rRNA	Vitamin B_{12} transport	ColE3-CA38	8
K	Causes membrane leakage	Nucleoside transport	ColK-K235	9

tive or nonconjugative, the latter generally being smaller in size (Table 7.3). The transmission of most conjugative Col plasmids involves either the F or I pilus.

Incompatibility of plasmids

A phenomenon of considerable importance both in research on plasmids and in the evolution and ecology of plasmids is compatibility. When a plasmid is inserted into a cell which already carries another plasmid, a common observation is that the second plasmid may not be maintained and is lost during subsequent cell replication. The two plasmids are said to be **incompatible**. Plasmid incompatibility is controlled by genes in the plasmid, which are involved in regulating DNA replication. A number of incompatibility (Inc) groups have been recognized, the plasmids of one incompatibility group excluding each other but being able to coexist with plasmids from other groups. Plasmids of one incompatibility group are related to one another. Therefore, although a bacterial cell may contain different kinds of plasmids, they are not closely related because they must be compatible.

Biological and taxonomic significance of plasmids

A summary of plasmids which have been identified and organisms for which evidence of plasmids exist is given in Table 7.4. It is likely that virtually all prokaryotic groups possess plasmids. Techniques are now available for detecting the presence in an organism of plasmid-like DNA, and these techniques have been applied to a wide variety of bacteria. In some cases, virtually every strain tested has been shown to have plasmid-like DNA. Almost certainly, such DNA molecules control some genetic functions for the cell, although it is frequently difficult to relate the presence

of the physical structure with a genetic function or functions. The only certain way of showing this is to transfer the plasmid to another strain, where the DNA can be distinguished by its physical characteristics, and to show that at the same time some genetic characteristic of the donor has also been transferred. This can usually be accomplished easily with conjugative plasmids, which are known in both Gram-negative and Gram-positive Bacteria.

However, as we have noted, many plasmids are not conjugative, and the demonstration of the existence of plasmids is more difficult. Evidence for plasmids comes from curing experiments, and from the transfer of plasmids via transduction. On the other hand, conjugative transfer of plasmids is extremely common in Gram-negative Bacteria, not only within the enteric group, which has been so widely studied, but in other groups, such as the pseudomonads. In fact, some pseudomonad plasmids are transferrable to a wide variety of other Gram-negative Bacteria. Some pseudomonad plasmids have been shown to transfer the genetic information for biochemical pathways for the degradation of unusual organic compounds, such as camphor, octane, and naphthalene. These substances are highly insoluble, and the plasmid-coded enzymes convert them into more soluble substrates that are on the main line of biochemical catabolism, such as acetate, pyruvate, and isobutyrate, the catabolism of which is encoded by chromosomal genes.

Some conjugative plasmids can transfer genetic information between distantly related organisms. Conjugative plasmids have been shown to transfer between Gram-negative and Gram-positive Bacteria, between Bacteria and plant cells, and between Bacteria and fungi. Even if the plasmid cannot replicate in the new host, the transfer of the DNA itself could have

---

Table 7.4   Types of plasmids*

Type	Organisms
Conjugative plasmids	F plasmid, *Escherichia coli*; pfdm, K, *Pseudomonas*; P, *Vibrio cholerae*; SCP, *Streptomyces*
R plasmids:	
Wide variety of antibiotics	Enteric bacteria, *Staphylococcus*
Resistance to mercury, cadmium, nickel, cobalt, zinc, arsenic	
Bacteriocin and antibiotic production	Enteric bacteria; *Clostridium; Streptomyces*
Physiological functions:	
Lactose, sucrose, urea utilization, nitrogen fixation	Enteric bacteria
Degradation of octane, camphor, naphthalene, salicylate	*Pseudomonas*
Pigment production	*Erwinia, Staphylococcus*
Nodulation and symbiotic nitrogen fixation	*Rhizobium*
Virulence plasmids:	
Enterotoxin, K antigen, endotoxin	*Escherichia coli*
Tumorigenic plasmid	*Agrobacterium tumefaciens*
Adherence to teeth (dextran)	*Streptococcus mutans*
Coagulase, hemolysin, fibrinolysin, enterotoxin	*Staphylococcus aureus*

*Plasmids have now been found in most bacterial genera.

important evolutionary consequences, as well as being involved in pathogenic processes if it can recombine into the genome of the new host.

Origin of resistance plasmids

Although specific evidence for the origin of multiple drug-resistance R plasmids is not available, a number of lines of circumstantial evidence suggest that plasmids with R plasmid type character existed before the antibiotic era and that the widespread clinical use of antibiotics provided selective conditions for the spread of R plasmids with one or more antibiotic-resistance genes. Indeed, a strain of *E. coli* which was freeze-dried in 1946 was found to contain a plasmid with genes conferring resistance to tetracycline and streptomycin, even though neither of these antibiotics were used clinically until several years later. Also, strains carrying R plasmid genes for resistance to a number of semisynthetic penicillins were shown to exist before the semisynthetic penicillins had been synthesized. Of perhaps even more ecological significance, R plasmids conferring antibiotic resistance have been detected in some nonpathogenic Gram-negative soil Bacteria. In the soil, such resistance may confer selective advantage, since major antibiotic-producing organisms (*Streptomyces, Penicillium*) are also normal soil organisms. Thus, it seems reasonable to conclude that R plasmids are not a recent phenomenon, but existed to a small extent in the natural bacterial population before the antibiotic era, and that widespread use of antibiotics provided selective conditions for the rapid spread of these R plasmids. The evolution of R plasmids is thus completely understandable, but the implication for the continued success of human medicine in combatting rampant infection is frightening.

Engineered plasmids

The techniques of genetic engineering, discussed in the next chapter, have made possible the construction in the laboratory of a limitless number of new, artificial plasmids. Incorporation into artificial plasmids of genes from a wide variety of sources has made possible the transfer of genetic material across virtually any species barrier. It is even possible to synthesize completely new genes and introduce them into plasmids. Such artificial plasmids are useful tools in understanding plasmid structure and function, as well as for the more practical aims of genetic engineering that are discussed in the next chapter. In order to create an artificial plasmid, the main requirement is that the plasmid genes involved in replication and maintenance of the plasmid in the cell be connected to the new material of interest. Artificial plasmids are introduced into appropriate hosts via transformation and are selected by means of antibiotic resistance characters. The ability to create artificial plasmids has greatly expanded the possibilities for plasmid research.

We now turn our attention to the details of conjugation and show how certain plasmids can act to mobilize the bacterial chromosome, allowing transfer of chromosomal genes from donor to recipient.

> The genetic information that plasmids carry is generally not essential for cell function but may confer selective growth advantage under certain conditions. Examples include toxins, bacteriocins, enzymes for degradation of unusual organic compounds, and special metabolic pathways. Plasmids are of great utility in genetic engineering because they can be manipulated so readily.

7.10 Conjugation and Chromosome Mobilization

Bacterial conjugation (mating) is a process of genetic transfer that involves cell-to-cell contact. The genetic material transferred may be a plasmid, or it may be a portion of the chromosome mobilized by a plasmid. Mechanisms of conjugative transfer may differ depending on the plasmid involved, but most plasmids

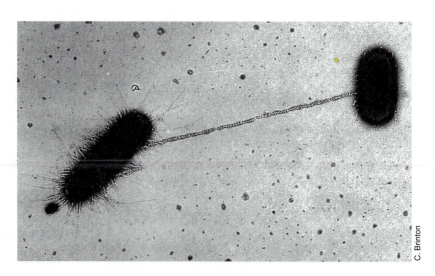

FIGURE 7.22
Direct contact between two conjugating bacteria is first made via a pilus. The cells are then drawn together for the actual transfer of DNA.

C. Brinton

in Gram-negative Bacteria seem to use a mechanism similar to that used by the F plasmid. In conjugation, one cell, the *donor,* transmits genetic information to another cell, the *recipient.* In the conjugation process, specific pairing between donor and recipient cells must occur. The donor cell, by virtue of its possession of a conjugative plasmid, possesses a surface structure, the **sex pilus** (Figure 7.22), which is involved in pair formation. Conjugative plasmids possess the genetic information to code for sex pili and for some proteins needed for DNA transfer. The sex pili make possible specific contact between donor and recipient cells and then retract, pulling the two cells together so that a conjugation bridge can form, through or on which DNA passes from one cell to another. Although the details are not completely understood, it is thought that there is actual membrane fusion between donor and recipient cells, and that this in some way triggers DNA transfer. Although the recipient cells lack sex pili, they have receptors on their surface that are recognized by the sex pili.

Mechanism of DNA transfer during conjugation

DNA synthesis is necessary for DNA transfer to occur, and the evidence suggests that one of the DNA strands is derived from the donor cell and the other is newly synthesized in the recipient during the transfer process. A mechanism of DNA synthesis in certain bacteriophages, called **rolling circle replication**, was presented in Figures 6.18 and 6.31. This model best explains DNA transfer during conjugation, and a possible mechanism for this process is outlined in Figure 7.23. The whole series of events is probably triggered by cell-to-cell contact, at which time the plasmid DNA circle is nicked and one parental strand is transferred. As this transfer occurs, DNA synthesis by the rolling circle mechanism replaces the transferred strand in the donor. A complementary DNA strand is also made in the recipient. The model accounts for the fact that if the DNA of the donor is labeled, some labeled DNA is transferred to the recipient, but only a *single* labeled strand is transferred. Therefore, at the end of the process, both donor and recipient possess completely formed plasmids.

The high efficiency of the plasmid DNA transfer process is shown by the fact that under appropriate conditions, virtually every recipient cell that pairs acquires a plasmid. If the plasmid genes can be expressed in the recipient, the recipient itself becomes a donor and can transfer the plasmid to other recipients. In this fashion, conjugative plasmids can spread rapidly between populations, behaving in the manner of infectious agents. The infectious nature of this phenomenon is of major ecological significance, since a few plasmid-positive cells introduced into an appropriate population of recipients can, if they contain genes which confer a selective advantage, convert the whole recipient population into a plasmid-bearing population in a short period of time. The widespread

occurrence of infectious drug resistance in clinical medicine has led to some serious problems in the chemotherapy of infectious disease (see Section 11.16).

The F plasmid of *Escherichia coli* (see Section 7.8) is not only conjugative, but has the special property of being able to mobilize the chromosome so that it can be transferred during cell-to-cell contact. When the F plasmid is integrated into the chromosome, large blocks of chromosomal genes can be transmitted, and genetic recombination between donor and recipient is then very extensive. As mentioned earlier, bacterial strains that possess a chromosome-integrated F plasmid and do show such extensive genetic recombination are called Hfr (for high frequency of recombination). When the F plasmid is not integrated into the chromosome, it behaves as a conjugative plasmid. Cells possessing an unintegrated F plasmid are called F^+, and strains which can act as recipients for F^+ or Hfr are called F^-. F^- cells lack the F plasmid; in general, cells that contain a plasmid are very poor recipients for the same or closely related plasmids.

We thus see that the presence of the F plasmid results in three distinct alterations in the properties of a cell: (1) ability to synthesize the F pilus, (2) mobilization of DNA for transfer to another cell, and (3) alteration of surface receptors so that the cell is no longer able to behave as a recipient in conjugation.

Selection for recombinants formed as the result of mating of an Hfr with an F^- strain is accomplished by plating the mating mixture on culture media that allow growth of only the recombinant cells with the desired genotype. For instance, in the experiment shown in Figure 7.24, an Hfr donor which is sensitive to streptomycin (Str^s) and contains wild-type genes coding for enzymes needed for synthesis of the amino acids threonine and leucine (Thr^+ and Leu^+) and for utilization of the energy source lactose (Lac^+) is mated with a recipient cell which is mutant for these genes but is resistant to streptomycin (Str^r). The selective medium is a minimal medium containing streptomycin so that only recipient cells can grow. The composition of each selective medium is varied depending on which genotypic characteristics are desired in the recombinant, as shown in the figure. The frequency of the process is measured by counting the colonies which grow on the selective medium.

Formation and behavior of Hfr strains

As noted above, an F plasmid can become integrated into the chromosome and mobilize it for conjugation. There are several specific sites in the chromosome at which F plasmids can be integrated, and these sites, called IS (for "insertion sequence"), represent regions of homology between chromosome and F plasmid DNA (see Section 7.11 for a discussion of insertion sequences). As seen in Figure 7.25, integration of an F plasmid involves insertion in the chromosome at the specific site. In the particular Hfr shown, the integration site is between the chromosomal genes *pro* and

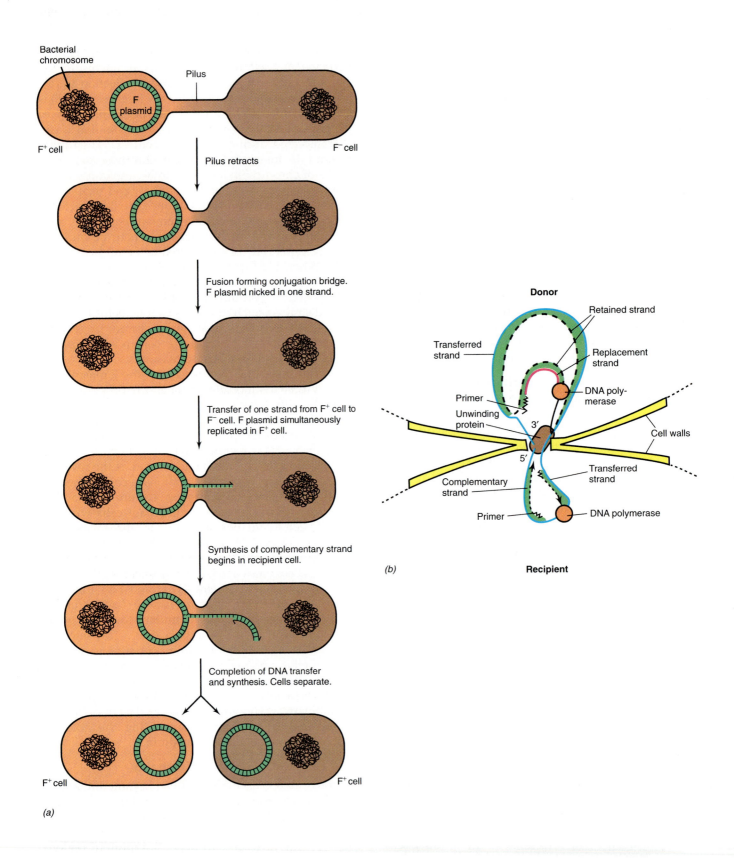

FIGURE 7.23 Transfer of plasmid DNA by conjugation. (a) In this example, the F plasmid of an F⁺ cell is being transferred to an F⁻ recipient cell. Note the mechanism of rolling circle replication. Compare with Figure 6.3. (b) Details of the replication and transfer process.

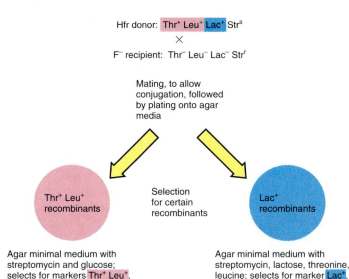

FIGURE 7.24 Laboratory procedure for the detection of genetic conjugation. Symbols: Thr, threonine; Leu, leucine; Lac, lactose; Str, streptomycin. Note that each medium selects for specific classes of recombinants.

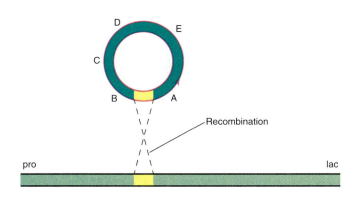

FIGURE 7.25 Integration of an F plasmid into the chromosome with the formation of an Hfr. The insertion of the F plasmid occurs at a variety of specific sites where IS elements are located, the one here being between chromosomal genes *pro* and *lac*. The letters on the F plasmid represent arbitrary genes. The arrow indicates the origin of transfer, with the arrow as the leading end. The site in the F plasmid at which pairing with the chromosome occurs is between A and B. Thus, in this Hfr *pro* would be the first chromosomal gene to be transferred and *lac* would be among the last.

lac. The site on the F plasmid (*oriT, origin of transfer*) at which transfer will initiate during conjugation is indicated by an arrow. At the time of specific cell pairing, the chromosome is opened (nicked) at the origin, and the host genes are inserted into the recipient beginning with the gene downstream from the origin (Figure 7.26). Essentially, the mechanism of transfer is the

same as that for just the F plasmid, but here the plasmid is part of the chromosome.

Hfr strains thus arise as a result of the integration of the F plasmid into the chromosome. Since a number of distinct insertion sites are present, a number of distinct Hfr strains are possible. A given Hfr strain always donates genes in the same order, beginning with the same position, but Hfr strains of independent origin transfer genes in different sequences. During cell division, the DNA of the Hfr replicates normally, but at the time of pairing with an F⁻ cell, a DNA strand from the Hfr is transferred to the F⁻ cell, and replication occurs by the rolling circle process. After transfer, the Hfr strain still remains Hfr, since it has retained a copy of the transferred genetic material.

Usually, because of breakage of the DNA strand during transfer, only a *part* of the donor chromosome is transferred. Since only a part of the chromosome is transferred, it cannot replicate in the recipient cell. Therefore, donor genes normally cannot be detected unless recombination between the incoming fragment and the recipient chromosome takes place.

Although Hfr strains transmit chromosomal genes at high frequency, they usually do not convert F⁻ cells to F⁺ or Hfr, because the entire F plasmid is only rarely transferred. On the other hand, F⁺ cells efficiently convert F⁻ to F⁺ because of the infectious nature of the F plasmid.

At some insertion sites, the F plasmid is integrated with the origin in one direction, whereas at other sites the origin is in the opposite direction. The direction in which the F factor is inserted determines which of the chromosomal genes will be inserted first into the recipient. The manner in which a variety of Hfr strains can arise is illustrated in Figure 7.27. By use of various Hfr strains, it has been possible in *E. coli* to determine the arrangement and orientation of a large number of chromosomal genes, as will be described in Section 7.12.

Transfer of chromosomal genes to the F plasmid

Occasionally integrated F plasmids may be excised from the chromosome, and the possibility exists for the incorporation at that time of *chromosomal* genes into the liberated F plasmid. Such F plasmids containing chromosomal genes are called *F′ (F prime) plasmids*. These F′ plasmids differ from normal F plasmids in that they contain identifiable chromosomal genes, and they transfer these genes at high frequency to recipients. F′-mediated transfer resembles specialized transduction in that only a restricted group of chromosomal genes can be transferred. It is often with F′ plasmids that complementation tests (see Section 7.5) are done in *E. coli*.

Oriented transfer of Hfr and the phenomenon of interrupted mating

The detection of bacterial conjugation is usually done by use of parental strains having properties that can

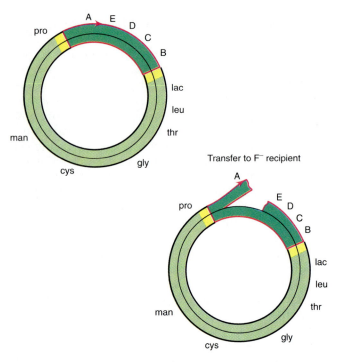

FIGURE 7.26 Breakage of the Hfr chromosome at the origin and transfer of DNA to the recipient. Replication occurs during transfer, as illustrated in Figure 7.23.

be selected against on agar plates so that only recombinants can grow (see Figure 7.24). The recipient is usually resistant to an antibiotic and is auxotrophic for one or more nutritional characters. The donor is antibiotic sensitive but prototrophic for the nutritional characters. With proper agar media, only the recombinants can grow and the large background of nonrecombinants is eliminated.

The oriented transfer of chromosomal genes from Hfr to F⁻ is most clearly shown by a procedure called **interrupted mating**. The mating pairs are rather weakly joined and can be separated by agitation in a mixer or blender. If mixtures of Hfr and F⁻ cells are agitated at various times after mixing and the genetic recombinants scored, it is found that the longer the time between pairing and agitation, the more genes of the Hfr will appear in the F⁻ recombinant. In addition, gene transfer always occurs in a specific order in a specific Hfr. As shown in Figure 7.28, genes present closer to the origin enter the F⁻ first and are always present in a higher percentage of the recombinants than genes that enter late. In addition to showing that gene transfer from donor to recipient is a sequential process, experiments of this kind provide a method of

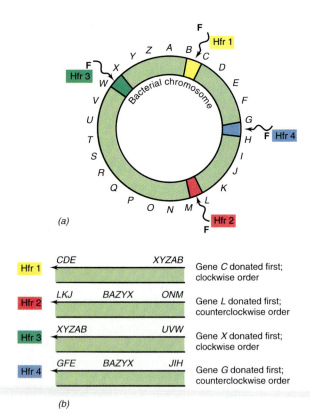

FIGURE 7.27 Manner of formation of different Hfr strains, which donate genes in different orders and from different origins. The bacterial chromosome is a circle (a) that can open at various insertion sequences, at which F plasmids become inserted. The gene orders are shown in part (b).

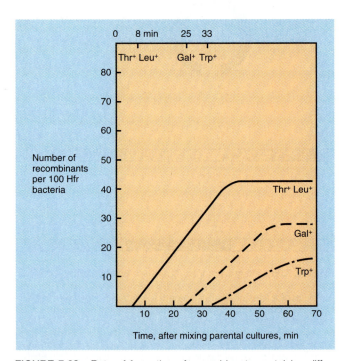

FIGURE 7.28 Rate of formation of recombinants containing different genes after mixing Hfr and F⁻ bacteria by the process known as interrupted mating. The location of the genes along the Hfr chromosome is shown in the small diagram. Note that the genes closest to the origin (0 minutes) are the first ones detected in the recombinants. The experiment is done by mixing Hfr and F⁻ cells under conditions in which essentially all Hfr cells find mates. The F⁻ recipient was streptomycin resistant but auxotrophic for the markers being scored. The Hfr donor was streptomycin sensitive. At various times, samples of the mixture are shaken violently to separate the mating pairs and plated on a selective medium in which only the recombinants can grow and form colonies.

determining the order of the genes on the bacterial DNA (genetic mapping). The arrangement of gene loci on the chromosome is called a **genetic map** (see Section 7.12).

Genetic recombination between Hfr genes and F⁻ genes in the F⁻ cell requires the presence of enzymes in the recipient cell. This has been shown by the isolation of mutants of F⁻ strains, which are unable to form recombinants when mated with Hfr. These mutants are Rec⁻ (recombination-minus) and are deficient in the RecA protein because of a mutation in the *recA* gene (see Figure 7.9).

It is important to remember that recombination is not the same as DNA transfer. Both F plasmids and F' plasmids are transferred normally to a Rec⁻ cell. Indeed, experiments have shown that although chromosomal DNA from an Hfr is also transferred normally to a Rec⁻ recipient, recombination does not take place after transfer.

> Conjugative plasmids, like the F plasmid, carry genes that control the conjugation process. The ability to conjugate is associated in the donor with the presence of a special structure, the sex pilus, which participates in the initial stage of cell-to-cell contact. Two kinds of donors have been recognized, F⁺, in which the F plasmid exists as a typical plasmid and transfers only itself, and Hfr, in which the integrated F plasmid mobilizes the transfer of chromosomal DNA. After mating cells have come in contact, DNA synthesis is initiated and a single strand of DNA is transferred from donor to recipient. DNA synthesis in both donor and recipient maintains the double-stranded nature of the DNA and ensures that no genetic information is lost from the donor. Transfer occurs sequentially beginning at a single point on the donor chromosome, but transfer of the whole chromosome occurs only rarely, since the mating pairs usually break apart before transfer is complete. By interrupting the transfer process experimentally, it is possible to map the order of the gene transfer process and hence also map the order of the genes on the chromosome.

7.11   Transposons and Insertion Sequences

We will see in Section 7.12 that the order of genes on a bacterial chromosome can be determined by the methods of gene transfer we have discussed, just as genes in eukaryotic chromosomes can be mapped by mating experiments. However, the exact arrangement of the genes along a chromosome is not necessarily permanently fixed; some genes are capable of moving around under certain conditions. The process by which a gene moves from one place to another in the genome is called **transposition** and is an important process in evolution and in genetic analysis.

We should emphasize that transposition typically is a *rare* event, occurring at frequencies of 10^{-5} to 10^{-7} per generation. Thus, the genes of living organisms are relatively stable.

In addition, not all genes are capable of transposition. Rather, transposition of genes is linked to the presence of special genetic elements called *transposable elements*.

Transposition was originally discovered in corn (maize) and then later in Bacteria owing to the extremely sensitive types of genetic analysis available in these organisms. It has been shown by using DNA hybridization and sequencing techniques (see the Nucleic Acids box in Chapter 5) that DNA sequences with the properties of transposable elements are widespread in nature. However, in organisms with poorly characterized genetic systems, it can be difficult to detect transposition (since it is a rare event) and actually prove that a sequence is a transposable element.

Transposable elements

As discussed in Section 5.5, there are three types of transposable elements in Bacteria: *insertion sequences, transposons,* and some special viruses (such as Mu; see Section 6.13). A brief summary of different types of transposable elements found in both prokaryotes and eukaryotes is given in Table 7.5.

In prokaryotes, insertion sequences are the simplest type and carry no genetic information other than that required for them to move to new locations. Insertion sequences are short segments of DNA, about 1000 nucleotides long, that can become integrated at specific sites on the genome. Insertion sequences, abbreviated IS, are found in both chromosomal and plasmid DNA, as well as in certain bacteriophages. Several distinct IS elements have been characterized, and each is designated by a number which identifies its type: IS*1*, IS*2*, IS*3*, etc. IS elements are scattered about the chromosome, and strains vary in the number and frequency of these elements. For instance, one strain of *Escherichia coli* has five copies of IS*2* and five copies of IS*3*. The F plasmid also carries these insertion sequences (see Figure 7.19), and it is homologous recom-

Table 7.5 Transposable elements	
Prokaryote	**Eukaryote**
Insertion sequence: IS	Yeast: *sigma*
Transposon: Tn	Yeast: Ty
	Fruit fly: copia, P
	Maize: Ac
Virus: Mu	Retrovirus: Rous sarcoma, Human immunodeficiency virus (AIDS)

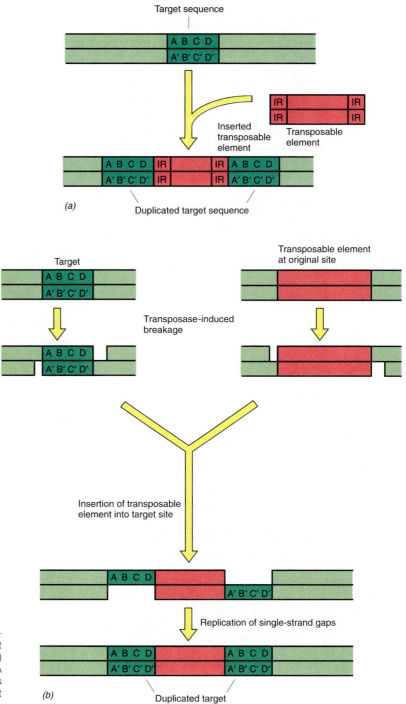

FIGURE 7.29
The transposition process. (a) Insertion of a transposable element generates a duplication of the target sequence. Note the presence of inverted repeats (IR) at the ends of the transposable element. (b) A schematic diagram indicating how target sequences might be duplicated. For simplicity, the IR's are not marked.

bination between identical sequences on the plasmid and the chromosome (*not* transposition) that allows the F plasmid to integrate into the bacterial chromosome and mobilize it (see Section 7.10).

Transposons are larger than insertion sequences and carry other genes, some of them conferring important properties on the organism carrying them. These often include drug-resistance markers and other easily selectable genes. In addition, in certain Gram-positive Bacteria, there are *conjugative transposons*. These transposons have genes allowing them not only to move from one location on a bacterial genome to

another but also to transfer themselves from one bacterium to another.

Some transposons are actually composite structures containing a gene or group of genes lying between two identical insertion sequences. The existence of such *composite transposons* indicates that new transposons probably continue to arise in cells that have insertion sequences.

Both IS elements and transposons have short *inverted terminal repeats* at the ends of their DNA (remember that these "ends" are continuous with whatever DNA the element is inserted into). These repeats range in length from about 40 base pairs in simple IS elements to greater than 1000 base pairs in composite transposons; each IS has a specific number of base pairs in its terminal repeats. Such inverted terminal repeats are involved in the transposition process (see below).

When a transposable element becomes inserted into another DNA (the target DNA), a short sequence in the target DNA at the site of integration is duplicated. This target DNA sequence was not present in the transposon but the transposable element has brought about a duplication of this DNA by the insertion process (Figure 7.29a). The duplication of the target sequence apparently arises because single-stranded breaks are generated by the transposase (Figure 7.29b). The transposon is then attached to the single-stranded ends that have been generated, and repair of the single-strand portions results in the duplication.

Certain transposable elements prefer certain sequences as target sites, but others, including the bacteriophage Mu, can insert themselves almost randomly (for a representation of Mu insertion, see Figure 6.33).

> Although the chromosome itself is a highly stable structure, certain genetic elements are able to jump from one location to another on the genome, a process called transposition. Transposition involves the participation of special DNA base sequences called transposable elements which control their own movement. Insertion sequences are transposable elements that contain only genes connected with transposition, while transposons contain other genes. Antibiotic-resistance genes present in transposons can be used as markers to follow the movement of transposons from place to place.

The mechanism of transposition

As mentioned previously, the inverted repeats found at the ends of transposable elements are essential for transposition. The other essential component is an enzyme called *transposase*, which recognizes these repeats. This enzyme is usually encoded by the transposable element, although some very simple IS elements use an enzyme encoded by another genetic element. The transposase apparently recognizes, cuts, and eventually ligates the DNA during transposition (see Figure 7.29).

Two mechanisms of transposition are known, called *conservative* and *replicative*. In conservative transposition, such as can occur in the transposon Tn5, the transposable element is excised from one location in the chromosome and becomes reinserted at a second location. The copy number of a conservative transposon therefore remains at one. By contrast, replicative transposons, such as bacteriophage Mu (see Section 6.13), become duplicated, a new copy becomes inserted at another location. Thus, after the transposition event is completed, *one* copy of the transposing element *remains* at the original site and *an-*

other copy is found at the new site. During this whole transposition process, the source transposon *remains* at its original site; at no time does the source transposon become free in the cell.

Although many of the molecular details of transposition are uncertain, and different transposable elements appear to have different mechanisms, one model for replicative transposition is illustrated in Figure 7.30. As seen, single-strand cuts are made at the ends of the transposon (at the sites of the inverted repeats), and staggered single-strand cuts are made at the target site. The transposon now becomes joined to

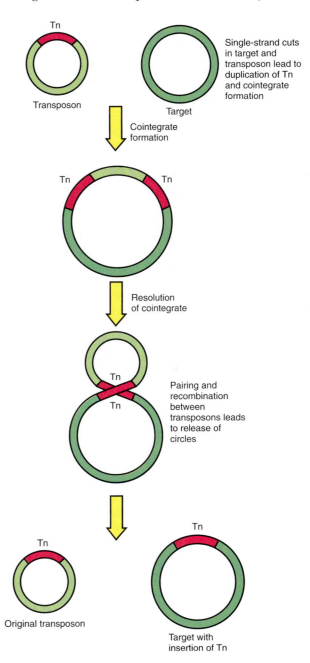

FIGURE 7.30   Replicative transposition. After the formation of single-strand cuts, a cointegrate structure arises by association of the two molecules. After recombination, resolution of the cointegrate structure leads to the release of the original transposon and duplication of the transposon in the target molecule. See Figure 7.29b for an explanation of how the duplication process might occur.

the target site via the single-stranded ends, leading to the formation of a composite structure called a *cointegrate*. Replication repair then fills in the single-strand gaps in the target site. This process results in the formation of *direct repeats* in the target site at the ends of the transposon (in addition to the inverted repeats of the transposon). The final event is the *resolution* of the cointegrate structure, leading to the release of the original transposon and the presence of a new copy of the transposon at the target site. Now that the transposon is present at the new target site, it can also serve as another source of transposition.

It should be emphasized that transposition is essentially a *recombination event*, but one which does not occur between homologous sequences or use the regular genetic recombination system of the cell. It involves the special protein *transposase* rather than the RecA protein which is involved in general recombination. Because this recombination involves a *specific base sequence*, it is called *site-specific recombination* (in contrast to homologous recombination discussed earlier in this chapter).

Transposon mutagenesis

If the insertion site for a transposable element is *within* a gene, insertion of the transposon will result in loss of linear continuity of the gene, leading to mutation (Figure 7.31). Transposons thus provide a facile means of creating mutants throughout the chromosome. The most convenient element for **transposon mutagenesis** is one containing an antibiotic resistance gene. Clones containing the transposon can then be selected by the isolation of antibiotic resistant colonies. If the antibiotic-resistant clones are selected on rich medium on which all auxotrophs can grow, they can be subsequently screened on minimal medium supplemented with various growth factors, to determine if a growth factor is required.

Transposons are also useful for incorporating an auxotrophic gene marker into a wild-type organism. Normally, auxotrophic recombinants cannot be isolated by positive selection, but if the auxotrophic marker to be introduced contains a transposable element with an antibiotic resistance marker, then one can select for antibiotic-resistant clones, a positive selection procedure, and automatically obtain clones which have incorporated the auxotrophic marker.

Two transposons widely used for mutagenesis are Tn*5*, which confers neomycin and kanamycin resist-

ance, and Tn*10,* which contains a marker for tetracycline resistance.

> When a transposon moves, it becomes inserted at the new site. Some transposons can insert almost randomly and their target sites will usually be within a gene. Such an insertion results in the inactivation of the gene. Because the presence of the transposon itself can be followed by its antibiotic resistance properties, selection of antibiotic resistant cells after transposition can be used to isolate a wide variety of mutants. Transposon mutagenesis thus provides a useful tool for creating mutants throughout the chromosome.

Invertible DNA and the phenomenon of phase variation

An interesting phenomenon of genetic change that has been recognized in some bacteria involves the **inversion** of a segment of DNA from one orientation to the other. When the segment is oriented in one direction, a particular gene is expressed, whereas when it is oriented in the opposite direction, a different gene is expressed. This "flip-flop" mechanism provides an interesting example of regulation of gene activity. The process by which gene inversion occurs is another example of *site-specific recombination* (see above).

The best studied case of gene inversion is that called *phase variation*, which has been well studied in bacteria of the genus *Salmonella*. These enteric bacteria are motile by means of peritrichous flagella. As we have noted (see Section 3.7), bacterial flagella are composed of a single type of protein. As a result of phase variation, the flagellar protein can be of one or two separate types. Each *Salmonella* cell has two genes, H1 and H2, coding for the two different flagellar proteins, but only one of the two genes is expressed at any one time. Thus, an individual bacterial cell will make either H1-type flagella or H2-type flagella.

The invertible element involved in expression of the flagellar proteins is a 970 base pair segment in the H2 gene (Figure 7.32). When the invertible segment is in one orientation, the H2 gene is transcribed, but in addition, another gene is transcribed which codes for a protein which represses transcription of gene H1. Thus, when H2 is expressed, H1 is turned off. On the other hand, when the invertible segment is in the opposite orientation, the genes for H2 and the H1 repressor are no longer expressed, so that H1 can now be transcribed and expressed.

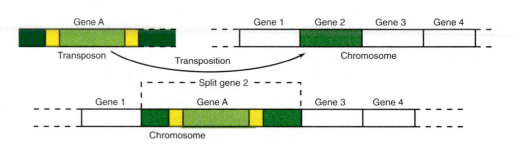

FIGURE 7.31
Transposon mutagenesis. The transposon moves into the middle of gene 2. Gene 2 is now split by the transposon and is inactivated. Gene A of the transposon will be expressed in both locations.

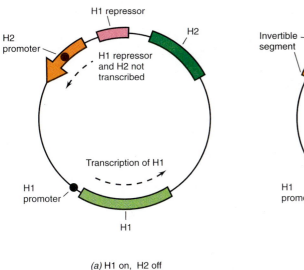

(a) H1 on, H2 off

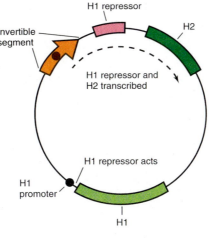

(b) H2 on, H1 off

FIGURE 7.32
Site-specific inversion, the mechanism by which phase variation in *Salmonella* flagella is brought about. The dotted lines show the location and direction of transcription. (a) When the invertible segment is in one orientation, the H2 promoter points away from the H1 repressor and H2 gene; H1 is expressed. (b) In the opposite orientation, H1 repressor is made, turning off H1. At the same time, H2 is expressed. The result is a "flip-flop" between two alternate states.

Invertible segments are also known in other genetic systems. We described an invertible segment which is involved in host range of bacteriophage Mu in Section 6.13. Regulation by rearrangement also occurs in eukaryotes: we discuss the regulation of mating type in yeast in Section 7.14 and we discuss the complex genetic rearrangements involved in the production of antibodies (defense against infection) in Section 12.8.

7.12   Overview of the Bacterial Genetic Map

The three mechanisms of genetic exchange described in this chapter, transformation, transduction, and conjugation, can be used to map the locations of various genes (actually mutations in genes) on the chromosome. Genes are generally mapped to a particular region of the chromosome using conjugation. Different Hfr strains are used which initiate DNA transfer from different parts of the chromosome, depending on where the F plasmid inserts. By using Hfr strains with origins in different sites, it is possible to map the whole bacterial gene complement. A circular reference map for *Escherichia coli* strain K-12 is shown in Figure 7.33. The map distances are given in minutes of transfer, with 100 minutes for the whole chromosome and with "zero time" arbitrarily set as that at which the first genetic transfer (the *threonine* operon) can be detected using the original Hfr strain.

Conjugation experiments do not permit ordering genes that are closely linked to each other. Generalized transduction is used for more fine structure mapping of the *E. coli* chromosome. Bacteriophage P1 can carry fragments of DNA equivalent to about 2 minutes on the map and has proved very useful for mapping genes. Specialized transducing phages, usually obtained by recombinant DNA techniques (Chapter 8), are also commonly used to study gene organization.

By using a combination of conjugation and trans-

duction, over 1400 genes were positioned on the circular chromosome (only a few of which are shown in Figure 7.33). Except for genes grouped in operons, there does not seem to be any pattern to the location of specific genes on the *E. coli* genetic map. While some sets of related genes, for example, genes involved in biosynthesis of the amino acid tryptophan (*trp* genes) are tightly clustered, other sets, for example, the genes involved in arginine biosynthesis (*arg* genes), are scattered around the chromosome (Figure 7.33).

Gene clusters and operons

Mapping of the genes that control the enzymes of a single biochemical pathway has shown that these genes are often clustered or closely linked in prokaryotes. The gene clusters for biochemical pathways on the *E. coli* chromosome are shown in Figure 7.33; letters are used to indicate the genes for the specific enzymes of the pathway (for instance, the galactose cluster at minute 17, with genes *K*, *T*, and *E*). It has also been found that all of the enzymes of a single gene cluster are often affected simultaneously by induction or repression. These related observations have been of considerable importance in the development of the *operon concept*, explaining not only how enzyme induction or repression can be brought about, but also how the synthesis of a series of related enzymes can be simultaneously controlled (Section 5.6).

The transcription of some operons proceeds in one direction along the chromosome, whereas with other operons transcription is in the opposite direction. The direction of mRNA transcription of the operons listed in Figure 7.33 is shown by arrows above the gene clusters. Since transcription always occurs in a 5' to 3' direction (see Section 5.6), this implies that transcription off of either strand of DNA can occur and thus that operons exist on each of the strands of the DNA double helix. In *E. coli* it appears that approximately equal numbers of operons exist on both strands.

Although many enzymes controlling specific bio-

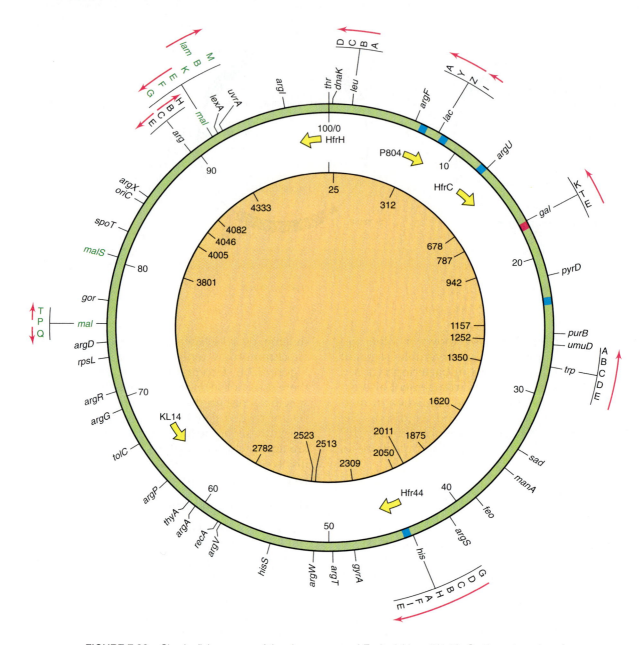

FIGURE 7.33 Circular linkage map of the chromosome of *Escherichia coli* K-12. On the outer edge of the map, the locations of a few of the mapped genes are indicated. A few operons are also shown, along with the direction in which they are transcribed. Along the inner edge of the map, the numbers from 0 to 100 refer to map position in minutes. The origin of DNA replication is marked *oriC* (84.3 minutes), and replication proceeds bidirectionally from this point. The inner circle shows the locations, in kilobase pairs, of the sequences recognized by the restriction enzyme *Not*I. Note that 0 minutes and 0 kilobase pairs are both, by convention, at the *thr* locus. The origins and directions of transfer of a few Hfr strains are also shown (arrows). The positions where five copies of the transposable element IS3 have been located in a particular strain are shown in blue. This element is also found in two copies on the F plasmid and is involved in Hfr formation. The position of the site where the bacteriophage lambda prophage integrates is shown in red. If the prophage was present, it would add an extra 48.5 kilobase pairs (slightly over 1 minute) to the map. The genes of the maltose regulon (see Section 5.11), which includes several operons, are shown in green. Although most genes in this regulon have an abbreviation beginning with *mal*, note that one of the genes is *lamB*. This gene encodes a membrane protein involved in maltose uptake by the cell, but the protein is also the receptor for bacteriophage lambda. The gene *rpsL* (73 minutes) encodes a ribosomal protein. The gene was once called *str* because mutations in this gene lead to streptomycin resistance.

Escherichia coli, the Best Known Prokaryote

By far, the best known prokaryotic organism is the intestinal bacterium *Escherichia coli*. Indeed, there are those who say that we know more about the biology of *E. coli* than about any other living organism, even including the human *Homo sapiens*. The structure and function of *E. coli* is often considered the archetype of *all* living organisms. Why is *E. coli* so well known? There is nothing especially unusual about this organism. It is a run-of-the-mill prokaryote and a common, if minor, inhabitant of the human intestine.

E. coli is so well known primarily because it is easy to work with in the laboratory. Even those who have difficulty with sterile technique and other bacteriological procedures can generally work with *E. coli* without difficulty. *E. coli* grows rapidly, has simple nutritional requirements, and does a fair bit of interesting biochemistry and physiology. But it is something else which has made it so useful in modern biology: "sex." *E. coli* was the first bacterium in which the process of conjugation was discovered, being first recognized in 1946 by Joshua Lederberg and Edward L. Tatum. At the time of its discovery, conjugation seemed very similar to sexual reproduction in eukaryotes. Although we now know that these are quite different, conjugation is an extremely important genetic tool. With conjugation, the possibility of doing real genetics was available. Genetic crosses could be carried out and genetic properties analyzed. Another valuable property of *E. coli* is its ability to support the growth of a whole range of bacterial viruses, which made it possible to study in detail the nature of viruses and virus multiplication. Thus, with its favorable laboratory properties, its suitability for studies of virology, and

the ability to do genetic crosses, biochemists and molecular biologists were able to probe deep into the nature of life, their work leading ultimately to the sophisticated understanding we now have of molecular biology and virology (see Chapters 5–8).

Escherichia coli was first isolated in 1885 by the German bacteriologist, Theodor Escherich, as a normal inhabitant of the intestinal tract. Escherich named the organism *Bacterium coli*, the name reflecting the rod shape of the cell (*Bacterium* means rod-shaped) and its intestinal habitat (*coli* for colon). The genus name *Bacterium* subsequently was changed to *Escherichia* in honor of its discoverer. Although Escherich's strains, and most other strains of *E. coli*, are harmless, some *E. coli* strains are pathogenic, causing diarrhea and urinary tract infections. As we will see in Chapter 11, the pathogenic *E. coli* strains differ in significant ways from the harmless intestinal strains.

It is curious that although *E. coli* is the organism of choice for genetic engineering and biotechnological research (see Chapter 8), it has never been a major organism of use in industrial microbiology. The large-scale cultivation of living organisms for industrial purposes (see Chapter 10) has involved a quite different group of microorganisms. For complex and not especially revealing reasons, *E. coli* is not nearly as suitable for large-scale cultivation as yeast, *Streptomyces*, and *Bacillus*. Indeed, there were many who worried about the possible use of *E. coli* in industry, because of its potential pathogenicity. However, industrialists have "tamed" *E. coli* for industrial purposes, although many other organisms are also used.

---

chemical pathways are linked and their synthesis is regulated by operons, this is not always the case. For some pathways, individual enzymes are encoded by genes at different locations on the chromosome, with the same operator present at each site. Such a collection of genes is termed a *regulon* (see Section 5.11), and coordinate regulation may occur because of a shared regulatory mechanism.

In addition to minutes, the map in Figure 7.33 also contains size units in kilobase pairs and shows the location of some restriction enzyme recognition sites. Maps using restriction enzymes (see Section 5.3), molecular cloning of genes (see Chapter 8), and DNA sequencing are now becoming widely used tools of the geneticist. These techniques have allowed the "map" of *E. coli* to be nearly completed. Over 1900 genes have actually been identified and located, and the chromosome has been over 60 percent sequenced.

Our knowledge of *E. coli* is now very extensive. This knowledge grew from the fact that *E. coli* was the first bacterium studied that had a relatively simple

system for mapping genes (see *Escherichia coli*, The Best Known Prokaryote box).

This does not mean that genetic studies of *E. coli* are almost over. To understand what a gene does, it is almost always necessary to isolate mutants, map their mutations, and determine their effects on the organism. Knowledge of the sequence of a genome is a powerful tool, but it does not in itself explain how an organism functions.

Organization of chromosomes from Bacteria

As discussed in Chapter 5, most Bacteria have a single, circular double-stranded DNA molecule as their only chromosome. The *E. coli* chromosome of 4700 kilobase pairs seems slightly above average in size, but less than 100 different species have been examined. The chromosome of *Mycoplasma pneumoniae* is small at 790 kilobase pairs, while the bacterium *Myxococcus xanthus* has an unusually large chromosome of 9500 kilobase pairs. The latter chromosome is two-thirds the size of the entire genome of the yeast, which is a eukaryote with 17 different chromosomes (see Section 7.14).

The chromosomes of several Bacteria are being analyzed by genetic techniques similar to those used for *E. coli,* and comparative studies are being done. Bacteria closely related to *E. coli* such as *Salmonella* have very similar arrangements of genes, while distantly related bacteria such as *Bacillus* do not. Even the nature and arrangement of genes found in the same operon from two different Bacteria can differ. Although there may be no general rules governing gene location, there do seem to be some constraints. Many Bacteria have a similar organization of genes near the origin of DNA synthesis, and similarities also exist at the terminus of DNA synthesis. Also, in at least some Bacteria, including *E. coli,* it appears that for certain genes or regions it is advantageous for the direction of transcription to be the same as that for the movement of the DNA replication fork. In these cases, an inversion mutation can be very deleterious to the organism.

The pattern of gene location along the chromosome may give clues to the evolution of the chromosome, and it seems clear that in many Bacteria, transposition and other large-scale rearrangements have been important. Good evidence from a number of Bacteria also shows that some groups of genes have been acquired from extrachromosomal genetic elements such as plasmids and bacteriophage.

> **Conjugation and transduction are important methods used to map genes on the chromosomes of Bacteria. These techniques, coupled with restriction enzyme analysis, cloning, and DNA sequencing, have allowed the placing of 1900 genes on the *E. coli* chromosome, which is over 60 percent sequenced. Although there are few rules governing gene location, the genes encoding the enzymes for many biochemical pathways are often found tightly linked in operons in Bacteria.**

7.13   Genetics in Eukaryotic Microorganisms

We now turn from a discussion of genetic mechanisms in Bacteria to a consideration of analogous processes in eukaryotes, using yeast as a model system. Genetic recombination in eukaryotes differs in many ways from that in prokaryotes. The complex nuclear organization of eukaryotes and the existence in each nucleus of a number of chromosomes lead to more regular mechanisms of gene assortment and segregation. Unlike prokaryotes, whose genomes are single DNA molecules, eukaryotic genomes are segmented into a number of chromosomes. While prokaryotic chromosomes are usually circular, eukaryotic chromosomes contain linear DNA molecules.

Eukaryotic cells can be of two types, depending on the chromosome number: in the *haploid* phase the number of chromosomes per cell is called *n*, and in the *diploid* phase, *2n* (Figure 7.34). Thus, in the yeast *Saccharomyces cerevisiae* a haploid cell contains 17 chromo-

somes and the diploid 34. In humans, haploid and diploid cells contain 23 and 46 chromosomes, respectively. Multicellular plants and animals are usually diploid, with the haploid phase present only in the germ cells (sperm and eggs, also called *gametes*), whose life spans are transitory. Eukaryotic microorganisms can be either haploid or diploid. Vegetative cells of *S. cerevisiae* can be either haploid or diploid (Figure 7.34). In most eukaryotic microorganisms, the diploid phase is transitory.

In diploid cells, two copies of each gene are present, one on each of the two homologous chromosomes. The term *allele* is used to refer to the two alternate forms of the *same gene* present on the two homologous chromosomes. If the allele on one chromosome has a mutation preventing the normal product from being expressed, the allele on the other chromosome can continue to be expressed, so that the effect of the mutation may not be evident. Thus, the expression of one allele may be masked by the other. The gene that is expressed is said to be *dominant* to the other allele, which is said to be *recessive*. Diploidy presents difficulties in genetic research since isolation of mutants is much easier in a haploid cell because only one form of the gene is present, and the effect of a mutation can be directly determined.

Meiosis

Mitosis is the process after DNA replication in which chromosomes condense, divide, and are sorted into two identical sets, one for each daughter cell. By contrast, **meiosis** is the process by which the change from the diploid to the haploid state is brought about (see Section 3.13). Meiosis involves two divisions. During the first meiotic division, the two sets of homologous chromosomes are segregated into two separate cells, so that the number of chromosomes is reduced from *2n* to *n*, yielding haploid cells that are the precursors of the germ cells. The second meiotic division is similar to a mitotic division but involves *n* chromosomes in each of two daughter cells. The products of meiosis are four haploid gametes. By definition, eggs are formed by females and sperm are formed by males, but in many microorganisms there is no clear sexual distinction, but only different mating types.

If the diploid homologs are genetically identical for a particular allele, the cell is said to be *homozygous*. Under such conditions, the haploid cells formed by meiosis will also be identical for that gene. However, if the two homologs differ, the cell is said to be *heterozygous*. Under such conditions, the four haploid gametes will not be identical for that gene.

Once gametes are formed, **mating** of gametes of different type can occur, leading to the formation of diploids. The first diploid cell formed as a result of mating of haploid gametes is called a *zygote*. The diploid cell formed can then become the forerunner of a new population of genetically identical diploid organisms or, as is the case with most eukaryotic mi-

Asexual reproduction

Haploid

Germination

Interaction of opposite mating types

Ascospores

Cell fusion

Ascus

Nuclear fusion

Asexual reproduction

Meiosis

Diploid

FIGURE 7.34
Life cycle of a typical yeast, *Saccharomyces cerevisiae*. A haploid cell of *S. cerevisiae* contains 17 chromosomes.

croorganisms, meiosis can generate haploid cells which can divide vegetatively.

7.14   Yeast Genetics

More is known about the molecular genetics of yeast than about almost any other eukaryote. This is because yeast is easily grown in the laboratory and is an extremely favorable organism for studies of genetic phenomena. The yeast that has been most studied is *Saccharomyces cerevisiae*, the common baker and brewer's yeast. This yeast is not only a useful model system for studies on eukaryotic genetics, but is an important organism of commerce, and therefore studies of its genetics can be expected to have practical significance (see Section 10.14). An important facet of yeast genetics, discussed in Section 7.15, is the study of mitochondrial inheritance.

A yeast grows as a single cell, and each haploid yeast cell is capable of serving as a gamete. Because yeast is unicellular and can be grown as a haploid, isolation of mutants is straightforward, and a large variety of mutant types are known. By mating mutants, genetic analyses of yeast can be carried out.

The life cycle of a typical yeast was shown in Figure 7.34. Many yeasts have two separate *mating types*, which can be considered analogous to male and female. However, the two mating types of a yeast are alike in structure and can only be differentiated by allowing them to mate. Upon mating of opposite types, a diploid cell is formed. In many yeasts, this diploid cell is capable of growing vegetatively, leading to the formation of a population of genetically identical, albeit diploid, cells. Under certain conditions, diploid cells of such a population can undergo meiosis and form haploid gametes. Two distinct types of gametes are formed, of opposite mating type. From a single diploid cell, a structure containing four such gametes is formed, two of each mating type. The cell in which the gametes are formed is called an *ascus*, and the cells within the ascus are called *ascospores*. Yeast ascospores generally possess some resistance to environmental conditions.

One important advantage of yeast is that genetic analysis is fairly straightforward. After mating and ascospore formation has occurred, the experimenter can dissect the four ascospores from the ascus, use each ascospore as the forerunner of a separate culture, and analyze the cultures so obtained for phenotypic characters that were in the parent. In this way, it is possible to *map* genes in yeast. Another advantage of yeast is that it can be transformed using exogenous DNA, and plasmids are available for use in genetic engineering.

Haploid cells of *Saccharomyces cerevisiae* have 17 chromosomes, and extensive genetic maps have been prepared for these chromosomes. One of the yeast chromosomes, number 17, is apparently very small and may not even be present in all strains; thus, there is some debate as to whether it is actually a chromosome. The other 16 chromosomes range in size from 245 to 2200 kilobase pairs, for a total genome size of 14 *mega*base (10^6) pairs. This is only about three times that of *Escherichia coli*. A few of the yeast chromosomes, each of which is smaller than the single *E. coli* chromosome, have been completely sequenced.

Mating type genetics of yeast

As we have noted, yeast has two mating types which are indistinguishable except through their behavior in mating. The two mating types of *Saccharomyces cerevisiae* are designated α and a. Cells of type α mate only with cells of type a and whether a cell is α or a is itself determined genetically. However, although a yeast cell line generally remains either α or a, haploid yeast cells are periodically able to *switch* their mating type from one to the other. (One consequence of this switching is that a pure culture of a single mating type can ultimately form diploids. Note that it is not appropriate to call a yeast cell *bisexual*. Under a given set of conditions, it is either one or the other mating type, never both.)

The switch in mating type from α to a and back to α has at its basis the behavior of a mobile genetic element, somewhat reminiscent of transposition or phase variation in *Salmonella* (see Section 7.11 for discussion of both topics). The phenomenon in yeast is illustrated in Figure 7.35. There is a single active genetic locus, called the MAT (for *mating type*) locus, at which either gene α or gene a can be inserted. At this active locus, the MAT promoter controls the transcription of whichever mating type gene is present. Thus, if gene α is at that locus, then the cell is mating type α, whereas if gene a is at that locus, the cell is mating type a. Somewhere else on the yeast genome are copies of both genes, α and a, which are not expressed. These *silent copies* serve as the source of the gene that is inserted when the switch occurs. When the switch occurs, the appropriate gene, α or a, is copied from its

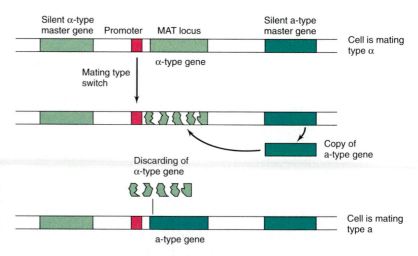

FIGURE 7.35
The cassette mechanism that is involved in the switch in yeast from mating type α to a and back again. Whichever "cassette" is inserted at the active locus (reading head) determines the mating type. The process shown is reversible, so that type a can revert to type α.

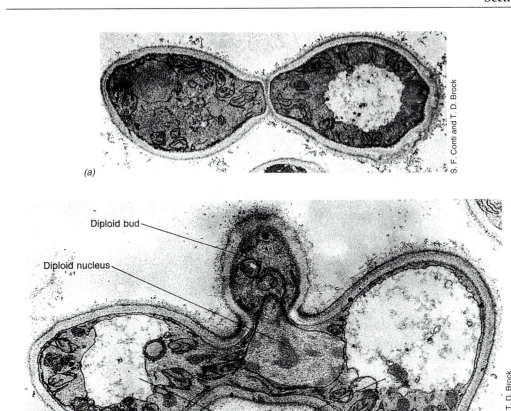

(a)

Diploid bud

Diploid nucleus

Former haploid cells

(b)

S. F. Conti and T. D. Brock

FIGURE 7.36
Electron micrographs of the mating process in a yeast, *Hansenula wingei*. (a) Two cells have fused at the point of contact and have sent out protuberances toward each other. (b) Late stage of mating. The nuclei of the two cells have fused and the diploid bud has formed at a right angle to the conjugation tube. This bud will eventually separate and become the forerunner of a diploid cell line.

silent site and then inserted into the MAT location, *replacing* the gene which is already present. Thus, the old gene is excised out of the locus and discarded and the new gene is inserted. This mechanism has been called the *cassette mechanism*, because each gene can be considered analogous to a tape cassette that is inserted into the "reading head," the place on the chromosome where active transcription takes place.

What do genes α and a do? At least one function has been identified. Yeast cells that are undergoing mating excrete *peptide hormones* called α factor and *a* factor. These hormones bind to cells of opposite mating type and bring about changes in the cell surfaces of these cells so that cells of opposite mating type will associate and fuse. It seems that α cells have receptors on their surfaces only for *a* factor, whereas a cells have receptors only for α factor. Once two cells of opposite mating type have associated, a complex series of events is initiated which leads to fusion of these two cells as well as their nuclei, resulting in the formation of a diploid zygote (Figure 7.36).

Yeast plasmids

We have defined plasmids as DNA molecules which replicate independently from chromosomal DNA (see Section 5.5 and Section 7.8). The DNA of plasmids is generally circular. Plasmids also exist in eukaryotes. One eukaryotic plasmid that has been well studied is the so-called 2 μm circle in yeast. The 2 μm plasmid is a 6300 base pair circular DNA molecule that replicates to high copy number (~30) in cells of *S. cerevisiae*. Although no genes have been mapped to the 2 μm circle, it has all the characteristics of a plasmid.

Most yeasts contain the 2 μm circle DNA. It is an autonomously replicating ring which is found in multiple copies within the nucleus. The 2 μm circle is packaged within the nucleus into nucleosomes, using histone protein derived from the nucleus. The 2 μm circle has its own origin of replication, but uses replication proteins derived, at least in part, from the nucleus. Segregation of the plasmid from mother to daughter cell during mitosis is random, but if the number of 2 μm circles drops to a low value, an increased rate of replication can bring the copy number back up to 30 to 50 per cell. There is no evidence that this plasmid ever becomes integrated into the nuclear chromosomal DNA.

Although the 2 μm plasmid is a useful model for studying DNA replication in yeast, its greatest value appears to be as a vector for cloning foreign genes into yeast. By use of appropriate treatment (see Section 7.6), it is possible to transform yeast cells using this plasmid, and hence to incorporate genes of interest.

The value of vectors such as this for genetic engineering is discussed in Chapter 8.

We discuss mitochondrial inheritance in yeast in the next section.

7.15 Mitochondrial Genetics

Although most genetic characters are contributed by chromosomal genes, mitochondria and chloroplasts contain separate genetic systems that are required for their own replication. In addition, mitochondria and chloroplasts have their own translation systems which are needed to synthesize the few proteins which these organelles make independently of the nuclear system.

As a eukaryotic cell grows, the number of mitochondria and chloroplasts increases by a process of organellar division—an existing organelle divides and forms two new organelles. Mitochondria and chloroplasts are never made *de novo* (that is, from scratch); they always arise from preexisting mitochondria and chloroplasts. Thus, if a cell lacks an organelle, its offspring will lack this same type of organelle. Because mitochondria and chloroplasts only arise from preexisting organelles, the genetic characteristics of the organelle in a new cell are determined by the genetic characteristics of the organelle of its parent.

Mitochondrial inheritance in yeast

One of the most interesting genetic analyses using yeast is that involving *mitochondrial inheritance*. It has been possible to develop mutant yeast strains in which mutations have occurred in the mitochondrion rather than in the nucleus, and by genetic analysis, the inheritance of the mitochondria themselves can be studied. Because inheritance of genetic characteristics via mitochondria occurs outside the nucleus, and outside the process of mitosis and meiosis, it is a form of **cytoplasmic inheritance** (sometimes called *non-Mendelian inheritance*, to indicate that it does not follow Mendel's laws.)

An example of the manner in which mitochondrial characteristics are inherited in yeast is shown in Figure 7.37. As seen, when two yeast cells of opposite mating type which also differ in mitochondrial characteristics are mated, the outcome depends upon which of the two types of mitochondria replicates most rapidly during subsequent cell divisions. In one class of mitochondrial mutants, called *petite*, large deletions in the mitochondrial DNA have led to abolition of all mitochondrial protein synthesis. Such mitochondria are nonfunctional, and yeast cells containing such mitochondria are unable to carry out respiration, although they are still able to grow anaerobically by fermentation processes. The term *petite* comes from the fact that these mutants produce small colonies on agar both aerobically and anaerobically; wild-type cells form larger colonies aerobically. If a petite yeast cell is mated with a wild-type yeast cell, the diploid zygote will contain both types of mitochondria, but since the

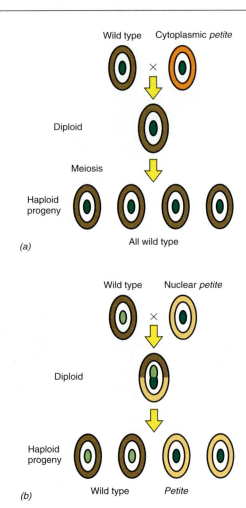

FIGURE 7.37 Mitochondrial inheritance. Outcome of crosses between petite and wild-type yeast: (a) cytoplasmic petite and (b) nuclear petite.

mutant mitochondria do not replicate as effectively as the wild-type mitochondria, they lose out in the competition during subsequent cell divisions. Ultimately, the cell lines derived from each of the four ascospores will all be wild type (Figure 7.37a). On the other hand, there are also petite mutants of yeast in which the mutation has occurred in a nuclear gene (since most of the proteins of the mitochondria are coded for by nuclear genes rather than mitochondrial genes). Crosses involving nuclear petites show conventional Mendelian inheritance (Figure 7.37b).

Like most mitochondria, that of yeast encodes relatively few of the required mitochondrial proteins. Interestingly, most of the yeast mitochondrial DNA does not code for proteins; its function is unknown. Proteins coded for by yeast mitochondrial DNA include cytochrome *b*, cytochrome *c* oxidase, ATPase, and one ribosome protein. In addition, a number of tRNA molecules are encoded by yeast mitochondrial DNA, as are the ribosomal RNA molecules. There is some reason to believe that during evolution, some genes have moved from mitochondrial DNA to nuclear DNA, and vice versa (see Section 7.10).

Organellar genomes

The genomes of most mitochondria and chloroplasts exist as circular DNA molecules. Organellar genome sizes vary dramatically among various eukaryotes. Chloroplast genomes fall within a narrow range from 120 to 200 kilobase pairs, but mitochondrial genomes vary much more, from as low as 16 to as high as 2500 kilobase pairs. Oddly, the mitochondria of higher animals have smaller genomes than do the mitochondria of many eukaryotic microorganisms. The human mitochondrial genome has been completely sequenced and contains 16,569 base pairs, whereas that of yeast is about five times larger. Plant mitochondrial genomes are often large and complex.

The genetic map of the human mitochondrion is shown in Figure 7.38. By convention, the outer strand is called the *heavy* strand and its complement is called the *light* strand. Both strands are completely transcribed. However, as shown, one strand codes for most of the proteins made in the mitochondrion, as well as for a number of tRNAs, and the two major species of ribosomal RNA. The other strand codes mainly for other tRNAs (Figure 7.38). Genes are tightly packed in the human mitochondrial genome, with genes for tRNAs acting as spacers between sequences coding for protein or rRNA (Figure 7.38). By contrast, the yeast mitochondrial genome (see above) has about the same number of genes as mammalian mitochondria, but because of its larger size, the yeast mitochondrial genome has lots of spacer (noncoding) DNA, and several of its genes contain intervening sequences (introns). However, like the human mitochondrial genome, most of the genes coding for proteins in the yeast mitochondria reside on just one of the two strands.

Mitochondrial genomes have some surprising molecular features. Whereas prokaryotic or eukaryotic nuclear genomes code for 32 or more different tRNAs, only 22 tRNAs are coded for in most mitochondria. The economy of the mitochondrion in this regard is governed by the fact that only a *single* tRNA is used to code for any amino acid specified by four different

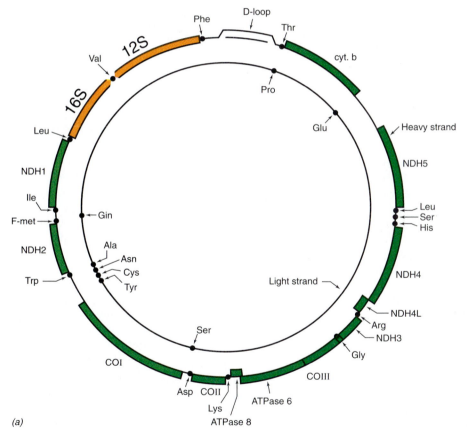

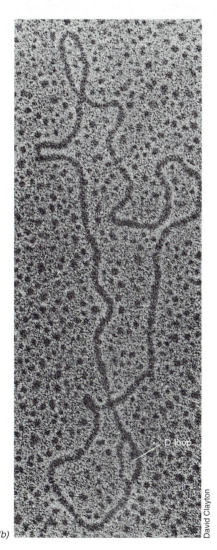

FIGURE 7.38 (a) Circular genetic map of the human mitochondrion. The complete genome contains 16,569 base pairs. The protein-encoding and ribosomal RNA genes are shown in different colors. A dark circle indicates the position of a tRNA gene. The D loop is the site of the origin of replication of the heavy strand. The location of various genes is shown, cyt *b*, cytochrome *b*; NDH 1-5, components of the NADH dehydrogenase complex; COI-III, subunits of the cytochrome oxidase complex; ATPase 6,8, polypeptides of the mitochondrial ATPase complex. The three letter code for amino acids (see Section 2.8) is used to designate a corresponding tRNA. Note that only serine and leucine, each of which has six codons, require more than one tRNA. (b) Electron micrograph of human mitochondrial DNA. Note the circular nature of the DNA. The D-loop is shown by an arrow.

codons; this is in contrast to nonmitochondrial systems where the wobble rules (see Section 5.9) demand at least two tRNAs for amino acids coded for by four different codons.

Protein synthesis in organelles

Chloroplasts and plant mitochondria use the same genetic code as that used in most prokaryotes and in the cytoplasm of most eukaryotes. This code has been referred to as the *universal code.* However, as discussed in Section 5.9, the mitochondria of other organisms use codes that differ from this universal code and from each other (see Table 5.5). One common feature of these altered codes is the use of UGA as a tryptophan codon rather than as a termination codon. This lack of universality in the code was first discovered in mitochondria, but (as discussed in Section 5.9) nonuniversal codes are now also known to be used by some cellular genomes.

Proteins coded for by the mitochondrial genome include components of the mitochondrial respiratory chain such as cytochromes, dehydrogenases, and ATPases. However, most mitochondrial proteins are not coded by mitochondrial genes but are transported into the organelle from their sites of synthesis in the cytoplasm. We noted in Section 3.14 that organelles possess rather permeable outer membranes and that transport of macromolecules and other substances that would not normally traverse the cytoplasmic membrane can occur. Of the 300 to 400 proteins known to reside in mitochondria, only 13 of these are coded for by mitochondrial genes in the human and only 8 in yeast.

Organellar ribosomes resemble prokaryotic ribosomes in size, although there is some variance, from as small as 50S to as high as 80S (prokaryotic ribosomes are 70S; see Section 5.8). However, an important feature of organellar ribosomes is their sensitivity to antibiotics that affect protein synthesis in Bacteria. Thus, protein synthesis in mitochondria and chloroplasts is inhibited by antibiotics such as chloramphenicol and tetracycline. This phenomenon suggests a close relationship between organelles and Bacteria and is one of several lines of evidence which supports the hypothesis that eukaryotic organelles are derived from *endosymbiotic Bacteria,* as discussed in more detail in Sections 3.15 and 18.3.

Because there are many subtle differences between different mitochondrial genetic codes and because some are identical to the universal code, it is most likely that these changes in the genetic code occurred during the genomic simplification that took place after the endosymbiotic events in evolution (see Section 18.3).

Table 7.6	Some key features of the genetic and translation systems of eukaryotic organelles
System	**Features**
Genome	Generally small, circular, double-stranded DNA (linear genomes present in the mitochondria of a few species) Histones absent DNA mostly coding and tightly packed in mammalian mitochondria; mostly noncoding with lots of spacer DNA in yeast mitochondria
Genetic code	Some differences from the universal code (see Table 5.5)
Translation apparatus	Ribosomes 70S-like; strong relationship in 16S rRNA sequences between organelles and Bacteria Sensitive to antibiotics that affect ribosomes of Bacteria Polypeptides start with formylmethionine (like in Bacteria)

Table 7.6 summarizes some of the key features of the genetic and protein synthesis systems of organelles. We should emphasize that unlike mitochondria and chloroplasts, many of the other structures found in eukaryotes do not arise by division of preexisting structures and do not have their own genetic systems. Thus, structures such as lysosomes, Golgi apparatus, flagella, cilia, etc., all arise *de novo* and are completely under nuclear control. Only mitochondria and chloroplasts seem to have their own genetic systems, albeit only to a partial extent.

Although mitochondria and chloroplasts are independent genetic elements, they contain only part of the genetic information for their own function and replication, the rest coming from the chromosome. The genomes of these organelles are usually circular DNA molecules, thus resembling bacterial chromosomes or plasmids, rather than eukaryotic chromosomes. The genetic code of mitochondria from cells other than plants differs from that of the cells of most eukaryotes and prokaryotes, whereas the translation system resembles that of Bacteria. The molecular evidence supports the hypothesis that mitochondria and chloroplasts evolved in ancient times from endosymbiotic Bacteria.

Study Questions

1. Write a one sentence definition of the term *genotype*. Do the same for the term *phenotype*. Does the phenotype of an organism automatically change when a change in genotype occurs? Why or why not? Can phenotype change without a change in genotype? In both cases, give some examples to support your answer.

2. What are silent mutations and why do they occur? From your knowledge of the genetic code, why do you think most silent mutations affect the *third* position of the codon?

3. Although a large number of mutagenic chemicals are known, no chemical is known to induce mutations in a single gene (gene-specific mutagenesis). From what you know about mutagens, explain why it is unlikely that a gene-specific chemical mutagen could be found.

4. Explain how it is possible for a frameshift mutation early in a gene to be corrected by another frameshift mutation farther along the gene.

5. What is site-specific mutagenesis? How can this procedure target specific genes for mutagenesis?

6. A constitutive mutant is a strain which continuously makes a protein which in the wild-type is inducible. Describe two ways in which a change in a DNA molecule could lead to the development of a constitutive mutant. How could these two types of constitutive mutants be distinguished genetically?

7. Explain why an *Escherichia coli* strain that is His⁻ is an auxotroph and one that is Lac⁻ is not. (Hint: think about what *E. coli* does with histidine and lactose.)

8. Frameshift mutations never occur in genes which encode tRNAs. Explain why this statement is true.

9. One type of *suppressor mutation* involves a change in tRNA. Draw a diagram with coding sequences and amino acid sequences which indicates how this occurs.

10. Describe the principle behind the Ames test. How is the test run in practice? From your knowledge of how mutants are isolated, why is the back mutation procedure used in the Ames test preferable to a forward mutation procedure?

11. What is the net result of genetic recombination? Why is it more likely that the farther two genes are apart on a chromosome, the more likely they will show recombination?

12. Why is it difficult in a single experiment using transformation to transfer a large number of genes to a cell?

13. From what you know about cell wall structure of Bacteria (Chapter 3), explain the problem a DNA molecule would encounter if the transformation process were to occur.

14. Suppose you were given the task of developing a genetic transduction system for an industrial organism of interest. You have a large collection of bacterial strains but no phages. Describe the steps you would use to develop such a transduction system. (Hint: you must not only obtain transducing phage but also a collection of bacterial mutants to be able to follow genetic markers.)

15. The physical presence of a plasmid can be determined by use of electrophoresis on gels. How could you use such a procedure to show that a curing agent is able to eliminate the plasmid from the bacterial cell?

16. Two bacterial strains are available, one resistant to streptomycin, the other resistant to penicillin. Suppose the resistance to one of these two antibiotics was chromosomal, the other associated with a conjugative plasmid. Describe experiments that would determine which antibiotic resistance was plasmid encoded.

17. How could you determine whether a plasmid conferring antibiotic resistance had become integrated into the chromosome?

18. Describe an experiment that could be used to show that a conjugative plasmid was associated with the presence of a particular type of pilus.

19. Explain how it is possible to use the *interrupted mating procedure* to determine the relative order of genes on a bacterial chromosome.

20. Compare and contrast the processes of transduction and transfection as they apply to *Bacteria*.

21. Design an experiment that would help decide whether a genetic transfer process in a particular bacterium is transformation, transduction, or conjugation. Assume that the following tools are available: appropriate mutants and selective media; DNase (an enzyme that destroys naked DNA); two kinds of filters, one capable of retaining bacteria and bacterial viruses but not free DNA, the other capable of retaining bacteria only; and a glass chamber in which the filters can be inserted to separate the chambers into two compartments. Give the experimental set-up and the anticipated results for each of the three gene transfer processes.

22. Explain why the insertion of a transposon leads to mutation.

23. The most useful transposons for isolating a variety of bacterial mutants are transposons which contain antibiotic-resistance genes. Why are such transposons of such great use for this purpose?

24. Compare and contrast the genetic mechanisms in prokaryotes and eukaryotes. Is there an analogy to the mitochondrial genetic element in prokaryotes?

25. Compare and contrast the "flip-flop" process for *Salmonella* phase variation and the "cassette" mechanism for yeast mating type switching.

26. Some retroviruses (Chapter 6) seem to be capable of acting as transducing viruses. Use your knowledge of the life cycle of these viruses to explain whether you think they would more likely participate in generalized or specialized transduction.

27. Describe the key genes on the F plasmid of *E. coli*. Give at least two reasons why F⁻ cells are unable to act as donors in conjugation.

28. Make a list of the different phenotypic traits linked to plasmids. Which of these have medical significance? Which have potential environmental significance?

29. Describe how certain functions in the yeast cell can be transmitted in non-Mendelian fashion.

Supplementary Readings

Alberts, B., D. Bray, J. Lewis, M. Raff, K. Roberts, and **J. D. Watson**. 1989. *Molecular Biology of the Cell*, second edition. Garland Publishing Co., New York. Good comparative coverage of prokaryotic and eukaryotic genetics, with extensive treatment of genetic mechanisms in yeast.

Beckwith, J. and **T. J. Silhavy**. 1992. *The Power of Bacterial Genetics: A Literature-based Course*. Cold Spring Harbor Laboratory Press, Cold Spring Harbor, N.Y. An excellent advanced textbook on prokaryotic genetics containing reprints of the original papers in several key areas as well as commentary and questions.

Berg, D. E. and **M. M. Howe**. (eds.) 1989. *Mobile DNA*. American Society for Microbiology, Washington, D.C. A very complete treatment of transposition and other means of moving DNA around.

Berg, P. and **M. Singer**. 1993. *Dealing with Genes: The Language of Heredity*. University Science Books, Mill Valley, CA. An excellent introductory text on modern genetics, stressing molecular biology.

Brock, T. D. 1990. *The Emergence of Bacterial Genetics*. Cold Spring Harbor Laboratory Press, Cold Spring Harbor, NY. A detailed history of the early days of bacterial genetics.

Bryan, L. E. (ed.) 1989. *Microbial Resistance to Drugs*. Springer-Verlag, Berlin. An excellent series of chapters on antibiotic action and resistance. Highly recommended for its thorough treatment of the genetics of antibiotic resistance.

Dale, J. W. 1989. *Molecular Genetics of Bacteria*. John Wiley & Sons, New York. A brief textbook of bacterial genetics covering most of the topics discussed in this chapter and in Chapter 5.

Levy, S. B. and **R. V. Miller** (eds.) 1989. *Gene Transfer in the Environment*. McGraw-Hill Publishing Company, New York. Reviews the major mechanisms of gene transfer in bacteria and gives examples of genetic processes in prokaryotes occurring in soil, water, plants, and animals.

Miller, J. H. 1992. *A Short Course in Bacterial Genetics: A Laboratory Manual and Handbook for Escherichia coli and Related Bacteria*. Cold Spring Harbor Laboratory Press, Cold Spring Harbor, N.Y. This book is an excellent source for information on the entire range of genetic techniques used in current research in bacterial genetics.

Neidhardt, F. C., J. L. Ingraham, K. B. Low, B. Magasanik, M. Schaechter, and **H. E. Umbarger** (eds.) 1987. *Escherichia coli and Salmonella typhimurium: Cellular and Molecular Biology*. American Society for Microbiology, Washington, D.C. This two-volume set contains excellent information on mutations, mapping, and genetic techniques in these important organisms.

Sebald, M. (ed.) *Genetics and Molecular Biology of Anaerobic Bacteria*. 1992. Springer-Verlag, New York. A comprehensive resource book on the genetics and molecular biology of anaerobic prokaryotes, both Bacteria and Archaea.

Streips, U. N. and **R. E. Yasbin** (eds.) 1991. *Modern Microbial Genetics*. Wiley-Liss, New York. An advanced textbook describing genetic analysis of *Escherichia coli* and other Bacteria, as well as the genetic systems of some lower eukaryotes.

8

Genetic Engineering and Biotechnology

Che concepts of molecular genetics, described in the three preceding chapters, have made possible the development of sophisticated procedures for the isolation, manipulation, and expression of genetic material, a field called **genetic engineering**. Genetic engineering has applications in both basic and applied research. In basic research, genetic engineering techniques are used to study the mechanisms of gene replication and expression in prokaryotes, eukaryotes, and their viruses. Some of the most important basic discoveries of molecular genetics were made using genetic engineering techniques. For applied research, genetic engineering permits the development of microbial cultures capable of producing valuable products, such as human insulin, human growth hormone, interferon, vaccines, and industrial enzymes. Genetic engineering for commercial application, sometimes called **biotechnology**, seems to have limitless potential.

Underlying genetic engineering is the isolation and purification of specific genes, a process called **gene cloning**. Having large amounts of pure DNA allows characterization and manipulation of genes and their products. With cloned genes, we can determine the *nucleotide sequence* of a gene, from which we can derive, through the genetic code, the *amino acid sequence* of its protein product. The cloned DNA itself can be used as a *probe* to determine the structure of more complex DNA molecules like the human genome. By taking genes from organisms that are difficult or dangerous to work with and moving them into well-characterized and safe microorganisms, valuable biological substances can be produced cheaply and in quantities that were unthinkable until the advent of genetic engineering. By changing the sequence of a cloned gene in a predetermined way, genetic engineers can literally design new, possibly useful biological products that are simply unavailable from natural sources.

In this chapter we discuss the tools and processes involved in creating and purifying desired genes by techniques using *in vitro* recombination, or **recombinant DNA**. We then describe how genetic engineering is used to produce large quantities of desired gene products, and how the products themselves can be altered by **site-directed mutagenesis**. We present examples of practical results derived from genetic engineering, and conclude this chapter with a review of the principles that underlie biotechnology.

8.1　Gene Cloning Systems

Gene cloning is at the base of most genetic engineering procedures. The purpose of gene cloning is to isolate large quantities of specific genes in pure form. While it might be theoretically possible to physically isolate pure DNA fragments with single genes from a restriction enzyme digest of chromosomal DNA, a little reflection will demonstrate the impracticality of such an approach. Consider that even for a genetically simple organism like *Escherichia coli*, a specific gene represents 1–2 kilobases out of a genome of about 4,700 kb. An average gene in *Escherichia coli* is thus less than *0.05 percent* of the total DNA in the cell. In humans the problem is even worse, since the average genes are not much bigger than in *Escherichia coli*, but the genome is 1000 times larger! In contrast, the DNA of bacteriophage *lambda* is only 50 kb, and the DNA of some plasmids is less than 5 kb. In these genetic elements, a single average gene constitutes 2–40 percent of the DNA.

Thus, the basic strategy of gene cloning is to move the desired gene from a large, complex genome to a small, simple one. Fortunately, our knowledge of DNA chemistry and enzymology allows us to break and join DNA molecules *in vitro*. This process is known as *in vitro* **recombination**. Restriction enzymes, DNA ligase (see Sections 5.3 and 5.4), and synthetic DNA (see Section 8.8) are important tools used for *in vitro* recombination.

Gene cloning can be divided into several steps:

1. Isolation and fragmentation of the source DNA. This can be total genomic DNA from an organism of interest, DNA synthesized from an RNA template by reverse transcriptase (see Section 6.22), or even DNA synthesized from nucleotides *in vitro*. If genomic DNA is the source, it is generally cut with restriction enzymes to give a mixture of fragments.

2. Joining the DNA fragments to a **cloning vector** with DNA ligase. The small, independently replicating genetic elements used to replicate genes are known as **cloning vectors**. Cloning vectors are generally designed to allow recombination of foreign DNA at a restriction site that cuts the vector in a way that does not affect its replication. If the source DNA and the vector are cut with the same restriction enzyme, then joining can be mediated by annealing of the single-stranded regions called "sticky ends" (see Section 5.3 and Figure 5.10). Blunt ends generated by different restriction enzymes can also be joined, and different sticky ends or blunt ends can be joined by the use of synthetic DNA **linkers** or **adapters**. The properties of cloning vectors are discussed in Sections 8.2–8.4.

3. Incorporation into a **host**. The recombinant DNA molecule made in a test tube can be introduced into a host organism by DNA transformation or by infection with phage particles made by *in vitro* packaging. Incorporation of the DNA into the host usually yields a mixture of clones. Some cells contain the desired cloned gene while other cells contain other clones that were generated by joining the source DNA to the vector. Such a mixture is known as a **DNA library** or a **gene library**, since many different clones can be purified from the mixture, each containing different cloned DNA segments from the source organism.

4. Detection and purification of the desired clone. Often one of the most difficult tasks is finding the right clone in a mixture that may contain thousands of others. Techniques for finding the right clone will be discussed in Section 8.6.

5. Production of large numbers of cells or bacteriophage containing the desired clone for isolation and study of the cloned DNA.

> The isolation of large quantities of a specific gene by gene cloning is usually done in bacteria using a plasmid or phage as the cloning vector. Restriction enzymes and DNA ligase are used in an *in vitro* recombination procedure to produce the hybrid DNA molecule. Once introduced into a suitable bacterial host by transformation, the target DNA can be produced in large amounts under the control of the cloning vector.

8.2　Plasmids as Cloning Vectors

We have discussed plasmids in some detail in Sections 7.8–7.10. Plasmids have very useful properties as cloning vectors. These properties include: (1) small size, which makes the DNA easy to isolate and manipulate; (2) circular DNA, which makes the DNA more stable during chemical isolation; (3) independent origin of replication, so that plasmid replication in the cell proceeds independent from direct chromosomal control; (4) multiple copy number, so that they can be present in the cell in several or numerous copies, making amplification of the DNA possible; (5) selectable markers such as antibiotic resistance genes are pres-

Miniglossary for Chapter 8

BIOTECHNOLOGY   the use of living organisms to carry out defined chemical processes for industrial application

CLONING VECTOR   genetic elements into which genes can be recombined and replicated

DNA FINGERPRINTING   use of the techniques of genetic engineering to determine the origin of DNA in a sample of tissue

DNA LIBRARY   a collection of cloned DNA fragments which in total contain cloned genes representing the entire genome of an organism; also called a gene library

EXPRESSION VECTOR   a cloning vector which contains the necessary regulatory sequences to allow transcription and translation of cloned genes

GENE CLONING   isolation and incorporation of a gene into a vector where it can be replicated

GENE DISRUPTION   use of both *in vitro* and *in vivo* recombination to substitute an easily selected mutant gene for a wild-type gene

GENE THERAPY   replacement or augmentation of a dysfunctional gene for medical purposes

NUCLEIC ACID PROBE   a strand of nucleic acid which can be labeled and used to hybridize to a complementary molecule from a mixture of other nucleic acids

POLYMERASE CHAIN REACTION (PCR)   a method used to amplify a specific DNA sequence *in vitro* by repeated cycles of synthesis using specific primers and DNA polymerase

SHUTTLE VECTOR   a cloning vector which can replicate in two or more dissimilar hosts

SITE-DIRECTED MUTAGENESIS   a technique whereby a gene with a specific mutation can be constructed *in vitro*

T-DNA   the segment of the *Agrobacterium* Ti plasmid which is transferred to plant cells

Ti PLASMID   a plasmid in *Agrobacterium* species capable of transferring genes from bacteria to plants

TRANSGENIC ORGANISMS   plants or animals which stably pass on cloned DNA that has been inserted into them

---

ent, making detection and selection of plasmid-containing clones easier.

Although in the natural environment conjugative plasmids are generally transferred by cell-to-cell contact, plasmid cloning vectors generally have been modified to prevent their transfer conjugatively, for safety reasons. However, transfer in the laboratory can be brought about by transformation procedures. Depending on the host-plasmid system, replication of the plasmid may be under tight cellular control, in which case only a few copies are made, or under relaxed cellular control, in which case a large number of copies are made. Achievement of high copy number is often important in gene cloning, and by proper selection of the host-plasmid system and manipulation of cellular macromolecule synthesis, plasmid copy numbers of several thousand per cell can be obtained.

An example of a suitable cloning plasmid is pBR322, which replicates in *Escherichia coli* (Figure 8.1). Plasmid pBR322 has a number of characteristics that make it suitable as a cloning vehicle:

1. It is relatively small, only 4361 base pairs.

2. It is stably maintained in its host (*Escherichia coli*) in relatively high copy number, 20–30 copies/cell.

3. It can be amplified to a very high number (1000–3000 copies per cell, about 40 percent of the genome!) by inhibition of protein synthesis by the addition of chloramphenicol.

4. It is easy to isolate in the supercoiled form by a variety of simple techniques (see the various methods described in the Nucleic Acids box, Chapter 5).

5. A reasonable amount of foreign DNA can be inserted, although inserts of more than 10 kilobases lead to plasmid instability.

6. The complete base sequence of this plasmid is

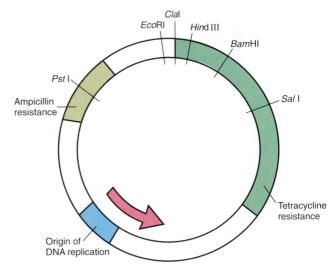

FIGURE 8.1   The structure of plasmid pBR322, a typical cloning vector, showing the essential features. The arrow shows the direction of DNA replication from the origin.

known, making it possible to locate sites where restriction enzymes can act.

7. There are *single* cleavage sites for various restriction enzymes such as *Pst*I, *Sal*I, *Eco*RI, *Hind*III, *Bam*HI, and several others. The *Eco*RI site is between the antibiotic resistance genes of the plasmid. It is important that only a single recognition site for at least one restriction enzyme is available, so that treatment with that enzyme will open the plasmid but not cut it into pieces.

8. It has two antibiotic resistance markers, ampicillin and tetracycline, which permit ready selection of hosts containing the plasmid.

9. It can be placed into cells easily by transformation.

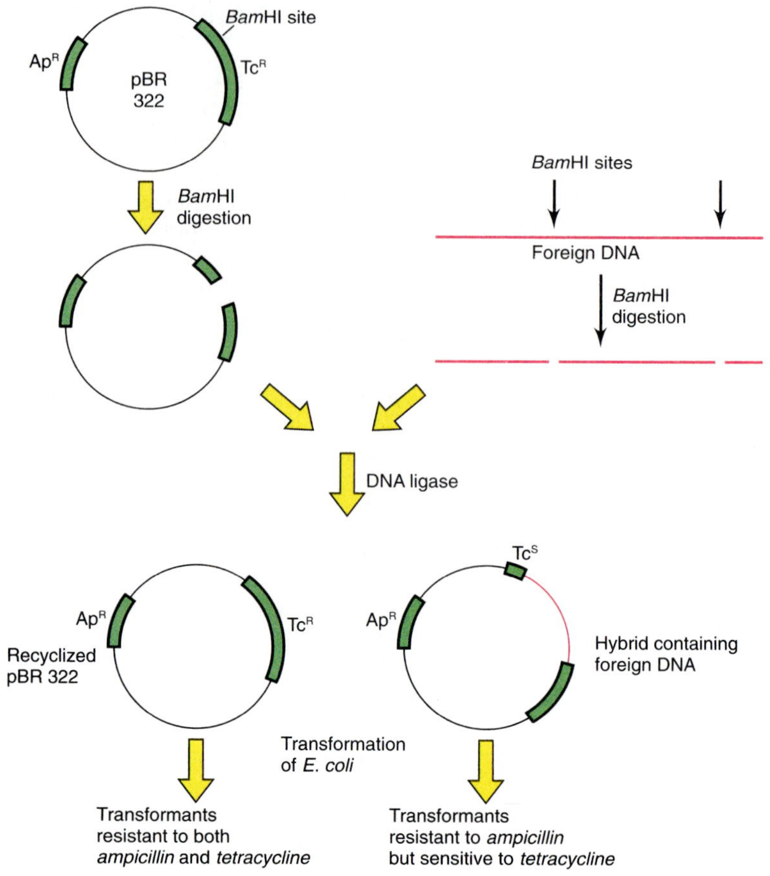

FIGURE 8.2
The use of plasmid pBR322 as a cloning vector, showing how insertion of foreign DNA causes inactivation of tetracycline resistance, permitting easy isolation of transformants containing the cloned DNA fragment.

The use of plasmid pBR322 in gene cloning is shown in Figure 8.2. As seen, the *Bam*HI site is within the gene for tetracycline resistance, and the *Pst*I site is within the gene for ampicillin resistance. If a piece of foreign DNA is inserted into one of these sites, the antibiotic resistance conferred by the gene containing this site is lost, a phenomenon called **insertional inactivation**. Insertional inactivation is used to detect the presence of foreign DNA within the plasmid. Thus, if pBR322 is digested with *Bam*HI, linked with foreign DNA, and then transformed bacterial clones are isolated, those clones which are both ampicillin resistant and tetracycline resistant *lack* the foreign DNA (the plasmid incorporated into these cells represents vector DNA that had recyclized without picking up foreign DNA). However, those cells still *resistant* to ampicillin but *sensitive* to tetracycline *contain* the plasmid with inserted foreign DNA. Since ampicillin resistance and tetracycline resistance can be determined independently on agar plates, isolation of bacteria containing the desired clones and elimination of cells not containing the plasmid can readily be accomplished.

The first plasmid cloning vectors used were naturally occurring plasmids. The plasmid pBR322 represents a later generation of cloning vectors, themselves constructed *in vitro*. The tetracycline resistance gene of pBR322 came from one plasmid, the replication origin from another, and the ampicillin resistance gene came from the transposon Tn3 (see Section 7.11). There are now newer generations of plasmid vectors that have

been engineered to have even more useful features and are even simpler to use. These new features almost always include a **polylinker** or **multiple cloning site,** a short segment of DNA with many restriction sites unique to the vector. This polylinker is usually contained in the coding region of a gene where insertional inactivation is very easy to monitor. Such features are also found in bacteriophage vectors, and a specific example is discussed in Section 8.4.

Cloning in plasmids such as pBR322 is a versatile and fairly general procedure of wide use in genetic engineering, particularly when the fragment to be cloned is fairly small. Plasmids are the best cloning vectors if the *expression* of the cloned gene is desired, as will be discussed in Section 8.7.

> Plasmids are useful cloning vectors because they are easy to isolate and purify and are able to multiply to high copy numbers in bacterial cells. Antibiotic resistance genes of the plasmid are used to identify bacterial cells containing the plasmid.

8.3 Bacteriophages as Cloning Vectors

We have discussed bacteriophage lambda in Section 6.12 and the fact that it could be used as a specialized transducing phage in Section 7.7. During specialized transduction, lambda acts as a vector but the recombination occurs in the cell, not in a test tube. Lambda can also be used as a cloning vector for *in vitro* recombination. It is a particularly useful cloning vector because its molecular genetics are well known, it can hold larger amounts of DNA than most plasmids, and DNA can be efficiently packaged into phage particles *in vitro*. These can be used to infect suitable host cells, and infection is much more efficient than transformation (transfection). Lambda has a complex genetic map (see Figure 6.27) and a large number of genes. However, the central third of the lambda genome, between genes *J* and *N*, is not essential and can be substituted with foreign DNA.

Modified lambda phages

Wild-type lambda is not suitable as a cloning vector because it has too many restriction enzyme sites. To avoid this difficulty, modified lambda phages have been constructed which can be used in cloning. In one set of modified lambda phages, the so-called Charon phages, unwanted restriction enzyme sites have been removed by point mutation, deletion, or substitution. In those variants that have only a single restriction site, a foreign piece of DNA can be *inserted*, whereas in variants with two sites, the foreign DNA can *replace* the lambda DNA. The latter variants, called **replacement vectors**, are especially useful for cloning large DNA fragments.

Figure 8.3 shows some of the essential features of a wild-type lambda and two of the Charon vectors. Whereas wild-type lambda contains five *Eco*RI sites, Charon 4A contains three and Charon 16 only one. Charon 4A is used as replacement vector; the two small interior fragments are cut out and discarded during cloning. With Charon 16, the DNA to be cloned is inserted at the single *Eco*RI site. Both Charon 4A and 16 contain deletions (not shown in the figure) which

not only remove some sites found in the wild-type lambda but make the genome smaller. This allows the cloning of larger DNA fragments.

Both vectors also contain substitution mutations, which are shown in Figure 8.3. One of the substitutions is the gene for β-galactosidase. When the vectors replicate on a lactose-negative (Lac$^-$) strain of *Escherichia coli*, β-galactosidase is synthesized from the phage gene and the presence of lactose-positive (Lac$^+$) plaques can be detected by use of a color indicator agar (see Section 8.4). If a foreign gene is inserted *into* the β-galactosidase gene, the Lac$^+$ character is lost. Such Lac$^-$ plaques can be readily detected as colorless plaques among a background of colored plaques.

Steps in cloning with lambda

Cloning with lambda replacement vectors involves the following steps (Figure 8.4):

1. Isolation of the vector DNA from phage particles and digestion with the appropriate restriction enzyme.

2. Connection of the two lambda fragments to fragments of foreign DNA using DNA ligase. Conditions are chosen so that molecules are formed of a length suitable for packaging into phage particles.

3. Packaging of the DNA by adding cell extracts containing the head and tail proteins and allowing the formation of viable phage particles.

4. Infection of *E. coli* and isolation of phage clones by picking plaques on a host strain.

5. Checking recombinant phage for the presence of the desired foreign DNA sequence by nucleic acid hybridization procedures or observation of genetic properties.

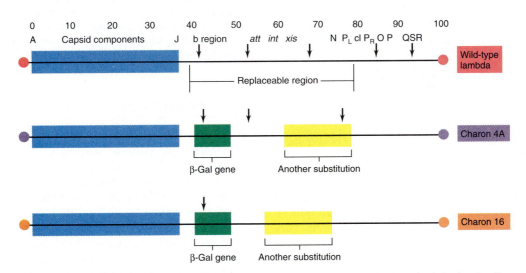

FIGURE 8.3   Gene cloning with lambda. Abbreviated genetic map of bacteriophage lambda showing the cohesive ends as circles (see also Figure 6.28). Charon 4A and 16 are both derivatives of lambda, which have various substitutions and deletions in the nonessential region. One of the substitutions in each case is a gene (β-Gal) which codes for the enzyme β-galactosidase, which permits detection of clones containing this phage. Whereas the wild-type lambda genome is 48.5 kilobase pairs, that for Charon 4A is 45.4 and that for Charon 16 is 41.7 kilobase pairs. The arrows shown above the maps of each phage indicate the sites recognized by the restriction enzyme *Eco*RI.

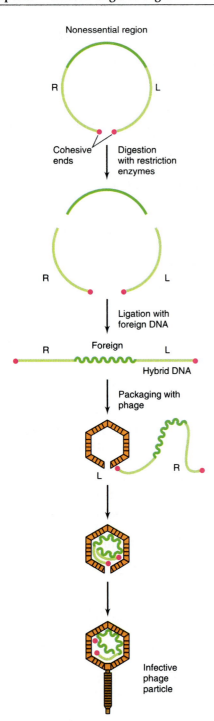

Nonessential region

Cohesive ends

Digestion with restriction enzymes

Ligation with foreign DNA

Foreign

Hybrid DNA

Packaging with phage

Infective phage particle

FIGURE 8.4 The use of bacteriophage lambda as a cloning vector. (See text for details.)

Selection of recombinants is less of a problem with lambda replacement vectors (such as Charon 4A) than with plasmids because (1) the efficiency of transfer of recombinant DNA into the cell by lambda is very high, and (2) lambda fragments which have not received new DNA are too small to be incorporated into phage particles.

Although lambda is a useful cloning vector, there are limits on how much DNA can be inserted. Viability of phage particles is low if the DNA is longer than 105 percent of normal lambda DNA, and some

lambda genes cannot be discarded and still maintain the vector's ability to replicate. Therefore, really large DNA fragments (greater than 20 kilobases) cannot be efficiently cloned.

Cosmids

A related vector that employs specific lambda genes is called a **cosmid**. Cosmids are plasmid vectors containing foreign DNA plus only the *cos* (cohesive end) site from the lambda genome. These *cos* sites are required for packaging DNA into lambda virions. Cosmids are constructed from plasmids containing cloned DNA by ligating the lambda *cos* region to the plasmid DNA. The modified plasmid can then be packaged into lambda virions *in vitro* as described above, and the phage particles used to transduce *E. coli*. Cosmid construction avoids the necessity of having to transform *E. coli*, which at best is an inefficient process (see Section 7.6).

One major advantage of cosmids is that they can be used to clone large fragments of DNA. Therefore, fewer clones are needed to get representation of the whole genetic element. This has been especially useful in the cloning of genes from eukaryotic chromosomes, where large amounts of DNA are involved. Another advantage of cosmids is that the DNA can be stored in phage particles instead of in plasmids. Phage particles are much more stable than plasmids, so that the recombinant DNA can be kept for long periods of time (gene banking).

> Bacteriophages such as lambda are useful cloning vectors because the phage particle can be used as a means for introducing hybrid DNA into the bacterial cell. Because lambda forms a large number of copies in the infected cell, it is an efficient cloning vector, and a larger amount of foreign DNA can be cloned with lambda than with many plasmids. Plasmid vectors containing the lambda *cos* sites are called cosmids, and these can carry a large fragment of foreign DNA.

8.4 Other Vectors

A number of other vectors have been developed which are useful for various purposes in recombinant DNA technology. It is understandable that in the early phases of the development of recombinant DNA techniques, the vectors used were those that had been in most widespread use in other types of genetic research. However, it does not necessarily follow that a genetic system that is useful in basic research automatically provides the best system for the development of a cloning vector. In addition, vectors are sometimes required that have specific applications beyond those needed for simply cloning a DNA fragment. In this section, we discuss briefly several other kinds of vectors and describe some useful modifications.

Yeast artificial chromosomes and the human genome project

As mentioned in Sections 8.2 and 8.3, plasmid vectors usually contain less DNA than lambda vectors, and lambda vectors less than cosmid vectors. When making *gene libraries* of bacteria or simple eukaryotes, lambda or cosmid vectors can be used and the entire library will contain at most a few thousand different clones. The size of a gene library (the number of clones) required to contain the complete genome of an organism depends on both the size of the genome and the size insert that can be placed in the vector. If the average size of the insert is 20 kilobases, then it will take many more clones to make a library of a typical mammalian genome (3×10^9 base pairs) than of a bacterial genome (4×10^6 base pairs). Currently, there is a concerted effort in the international genetics community to map, clone, and sequence the entire human haploid genome—the *Human Genome Project*. This is clearly a much larger undertaking than sequencing the *Escherichia coli* chromosome (see Section 7.12). It is useful in such a project to have a cloning vector that can hold very large segments of DNA so that the size of the initial gene library can be limited. Such vectors have been developed and are called **yeast artificial chromosomes,** or **YACs** for short.

These vectors have been designed to replicate in yeast like normal chromosomes, but they have sites where DNA can be inserted. To function like normal eukaryotic chromosomes, YACs must have an *origin of DNA replication, telomeres* at the ends of the chromosome (see Section 5.5), and a *centromere* (the section of the chromosome required for segregation during mitosis). They must also contain a cloning site and a gene that can be used for selection after transformation into the host. Figure 8.5 shows a diagram of a YAC vector into which foreign DNA has been cloned. YAC vectors are themselves only about 10 kilobase pairs, but they can have 200 to 800 kilobase pairs of cloned DNA inserted. After identifying a particular gene or region in the cloned DNA on a YAC, this gene can be *subcloned* into a plasmid or bacteriophage vector for more detailed analysis.

Vectors for DNA Sequencing: Bacteriophage M13

We have discussed the nature and life cycle of bacteriophage M13 in Section 6.9. This filamentous phage contains single-stranded DNA, and replicates without killing its host. Mature particles of M13 are released from host cells by a budding process, and it is possible to obtain infected cultures which can provide continuous sources of phage DNA. An important feature of M13 is its single-stranded DNA. In the Sanger procedure for DNA sequencing (see Nucleic Acids box, Chapter 5), single-stranded DNA is needed, and DNA cloned into M13 thus provides a ready source of this single-stranded DNA. Also, single-stranded DNA is very useful as a probe for detecting other nucleic acid sequences in transfer procedures such as the Southern blot, and M13 permits ready production of such single-stranded DNA probes.

However, in order to use M13 for cloning, a double-stranded form must be available, since restriction enzymes only work on double-stranded DNA. Double-stranded M13 DNA can be obtained from infected cells, since M13 replicates in the host as a double-stranded *replicative form* (see Section 6.9).

Most of the genome of wild-type M13 contains genetic information essential for virus replication. However, there is a small region called the intergenic sequence, which may be used as a cloning site. Variable lengths of foreign DNA, up to about 5 kilobase pairs, can be cloned without affecting phage viability—as the genome gets larger, the virion gets longer. M13mp18 is a derivative of M13 in which the intergenic region has been modified to facilitate cloning. A map of this vector is shown in Figure 8.6a.

One modification is the insertion of a functional fragment of *lacZ*, the *E. coli* gene that encodes the enzyme β-galactosidase. Therefore, cells infected with M13mp18 can easily be detected by their color on indicator plates (see Figure 8.6b). This *lacZ* gene has itself been modified to contain a 54 base pair DNA fragment called a *polylinker*. The polylinker contains several restriction sites unique in M13 and can therefore be used for cloning. The polylinker is inserted into the beginning of the coding portion of the *lacZ* gene. This small insertion is in-frame, and the extra 18 amino acids do not affect the activity of the enzyme encoded by the gene. However, insertion of additional DNA into the polylinker during cloning inactivates the gene. Phages that contain additional DNA inserts give rise to colorless plaques, and it is therefore very simple to identify clones (Figure 8.6b and c). A similar construct is used in lambda cloning vectors to allow identification of cells containing cloned DNA.

How then are M13 vectors used in cloning? The

FIGURE 8.5 Diagram of a yeast artificial chromosome (YAC) containing foreign DNA. The foreign DNA was cloned into the YAC vector at a *Not*I restriction site. The telomeres at the end of the YAC are labeled TEL and the centromere CEN. The origin of replication is labeled ARS (for Autonomous Replication Sequence). For this vector, the gene used for selection is called URA3. The host into which the clone is transformed has a mutation in that gene so that it normally requires uracil for growth (Ura⁻). Host cells containing this YAC become Ura⁺. The diagram is not drawn to scale; the inserted DNA would normally be 200–800 kilobase pairs long and the vector about 10 kilobase pairs.

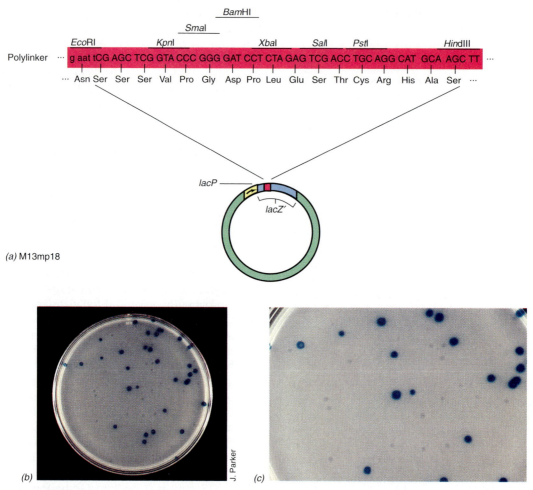

FIGURE 8.6 (a) A partial map of M13mp18, a derivative of M13 constructed for use as a cloning vector. The vector contains the *lac* promoter and a gene, *lacZ'*, which encodes a functional part of β-galactosidase. At the beginning of this gene is a polylinker that contains several restriction sites but maintains the proper reading frame. The amino acids encoded by the polylinker are shown. Most DNA fragments cloned into the polylinker will disrupt the *lacZ'* gene and abolish β-galactosidase activity. (b) A plate with plaques formed by M13mp18 and by clones made using this vector on a lawn of sensitive bacteria plated on a medium containing the chemical 5-bromo-4-chloro-3-indolyl-β-D-galactopyranoside, called X-gal. When β-galactosidase hydrolyzes X-gal, it releases a relatively insoluble blue dye (this same compound was used to make the plate shown in Figure 7.1c). Many plaques on this plate are blue, indicating the presence of vector without cloned DNA. However, many of the plaques are colorless, indicating that foreign DNA has been inserted into the vector and the *lacZ'* gene has been disrupted. (c) An enlargement of a portion of this plate.

replicative double-stranded DNA is isolated from the infected host and treated with a restriction enzyme. The foreign DNA, also double stranded, is treated with the same restriction enzyme. Upon ligation, double-stranded M13 molecules are obtained which contain the foreign DNA. When these molecules are introduced into the cell by transformation, they replicate and in time produce mature bacteriophage particles containing single-stranded DNA molecules. Only one strand of DNA is packaged into mature phage. Which of the two foreign strands the mature phage will contain will depend upon the *orientation* in which the strand was inserted. Since foreign DNA can be inserted (in separate phage molecules) in either orientation, *both* strands of the foreign DNA can be cloned.

The single-stranded M13 DNA containing the foreign DNA can then be used in DNA sequencing. Since the base sequence where the foreign DNA is inserted is known (based on the specificity of the restriction enzyme used), it is possible to use a synthetic oligonucleotide complementary to this region as a primer, and hence determine the sequence of the whole DNA downstream from this point. In this way, M13 derivatives have proved extremely useful for sequencing foreign DNA, even rather long molecules. (The sequence of the whole bacteriophage lambda DNA, 48,514 bases in length, was determined rather quickly in this way.) A modified system for producing single-stranded DNA has been constructed by placing an M13 origin of replication into a plasmid vector. These plasmids can be used for genetic manipulation in the conventional way and single-stranded DNA can be produced at will by infecting cells containing such a plasmid with an M13 helper phage.

Vectors have also been constructed that are hybrids between a filamentous phage, like M13, and a plasmid. Such vectors are called **phagemids**. They contain both phage and plasmid origins of replication. Normally replication is from the plasmid origin, but when a cell containing a phagemid is infected with a wild-type phage, the phage origin on the vector is used to synthesize single-stranded DNA from the vector (and whatever cloned gene it might carry). This single-stranded DNA is packaged into virions and can easily be isolated and used for sequencing. Usually, phagemids can stably carry a larger fragment of cloned DNA then a typical M13-derived vector.

Other specialized vectors

One important class of vectors, called **expression vectors,** are used when it is desired to obtain synthesis of the protein coded for by the foreign gene cloned into the vector. We discuss expression vectors in Section 8.7. Related to expression vectors are **secretion vectors,** in which the protein product is not only expressed but secreted (excreted) from the cell.

For a number of reasons it is useful to be able to move DNA between completely unrelated organisms. This would be the case, for instance, when a gene of interest is not present in wild-type *E. coli* but is present in another species of bacteria. To move DNA between unrelated organisms, a **shuttle vector** is used. A shuttle vector is one that can replicate in two different organisms. Like most specialized vectors, shuttle vectors have themselves been constructed using recombinant DNA techniques. Shuttle vectors have been developed which will replicate in both *E. coli* and *Bacillus subtilis*, *E. coli* and yeast, and *E. coli* and mammalian cells, as well as in many other pairs of organisms.

8.5   Hosts for Cloning Vectors

The ideal characteristics of a host for cloned genes are: rapid growth, capable of growth in cheap culture medium, not harmful or pathogenic, transformable by DNA, and stable in culture. The most useful hosts for cloning are microorganisms which grow well and for which a lot of genetic information is available, such as the Bacteria *Escherichia coli* and *Bacillus subtilis* and the yeast *Saccharomyces cerevisiae*.

Prokaryotic hosts

Although most molecular cloning has been done in *E. coli*, there are perceived disadvantages in using this host. *E. coli* presents dangers for large-scale production of products derived from cloned DNA, since it is found in the human intestinal tract and is potentially pathogenic. Also, even nonpathogenic strains produce endotoxins which could contaminate products, an especially bad situation with pharmaceutical injectibles. Finally, *E. coli* retains extracellular proteins in the periplasmic space, making isolation and purification potentially difficult. However, modified *E. coli* strains have been developed in which most of these problems have been eliminated. Because of the extensive knowledge of its genetics and biochemistry, *E. coli* remains the organism of choice for most cloning studies.

The Gram-positive organism *Bacillus subtilis* can also be used as a host. *B. subtilis* is not potentially pathogenic, does not produce endotoxin, and excretes proteins into the medium. Although the technology for cloning in *B. subtilis* is not nearly as well developed as for *E. coli*, plasmids and phages suitable for cloning have been developed and transformation is a well-developed procedure in *B. subtilis*. Disadvantages of using *B. subtilis* as a cloning host exist, however. Plasmid instability is a real problem, and it is hard to maintain plasmid replication over many culture transfers. Also, foreign DNA is not well maintained in *B. subtilis* cells, so that the cloned DNA is often unexpectedly lost. Adapting a bacterium for use as a host for cloning experiments is not always simple.

Often organisms used as hosts for cloning must have specific genotypes to be of use. For instance, if the vector carries the gene for β-galactosidase, then the host must have a mutation in this gene. Since M13 only infects bacteria with F pili (see Section 7.9), then hosts used with M13 derived vectors contain the F plasmid. These types of considerations, and others such as the ability to select for transformants, must be taken into account whether the host is prokaryotic or eukaryotic.

Eukaryotic hosts

Cloning in *eukaryotic microorganisms* has some important uses, especially for understanding details of gene regulation in eukaryotic systems. The yeast *Saccharomyces cerevisiae* is the best known genetically (see Section 7.14), and is being extensively studied as a cloning host. Plasmid vectors, as well as YACs, have been developed for yeast, and transformation using genetically engineered DNA can be accomplished. The ability to clone appropriate genetic material in yeast will advance our understanding of the complex transcription and translation systems of eukaryotes and should provide a better foundation for basic research.

For many purposes, gene cloning in *mammalian cells* would be desirable. Mammalian cell culture systems can be handled in some ways like microbial cultures, and find wide use in research on human genetics, cancer, infectious disease, and physiology. In addition, DNA can be introduced into mammalian cells by transfection or electroporation (see Section 7.6). The DNA virus SV40 (see Section 6.18), a virus causing tumors in primates, has been developed as a cloning vector into human tissue culture lines. SV40 virus has double-stranded *circular* DNA and the entire nucleotide sequence is known. Derivatives of SV40 which do not induce cancer have been developed which permit cloning of mammalian genes, and expression of these genes has been obtained. SV40 or similar mammalian cloning vectors should prove very useful in understanding the events involved in gene expression in these complex organisms.

The *retroviruses* (see Section 6.22) can also be used

to introduce genes into mammalian cells, since these viruses replicate through a DNA form which becomes integrated into the host chromosome. In addition, *vaccinia virus*, a large double-stranded DNA virus, has been used for cloning a variety of viral genes for vaccine purposes (see Section 8.13).

One important advantage of eukaryotic cells as hosts for cloning vectors is that they already possess the complex RNA and posttranslational processing systems that are involved in the production of gene products in higher organisms, so that these systems do not have to be engineered into the vector as they need to be if production of the desired product is to be carried out in a prokaryote (posttranslational processing, in particular, can create some gene cloning problems, see Section 8.10).

A disadvantage of mammalian cells as hosts is that they are expensive and difficult to produce under large-scale conditions, and expression levels of cloned genes is often low. Insect cell lines are simpler to grow, and vectors have been developed from an insect DNA virus, the baculovirus. Very high levels of expression can be achieved using baculovirus vectors.

8.6   Finding the Right Clone

A crucial step in recombinant DNA technology is finding the right clone among the mixture of clones that have been created by the recombinant DNA procedure. The foreign DNA used in the cloning procedure will typically contain a large number of genes, only one or a few of which may be the genes of interest. Sections 8.3 and 8.4 discussed how one can select for hosts containing a plasmid vector by selecting for a vector marker, such as antibiotic resistance, so that only these cells form colonies. For host cells containing a viral vector, one simply looks for plaques. We also discussed how these colonies or plaques can be screened for vectors that contain foreign DNA inserts by looking for the inactivation of a vector gene, often that for another antibiotic resistance in the case of a plasmid (see Figure 8.2), or for β-galactosidase in the case of a virus (see Figure 8.6). However, one is then left with the biggest challenge: selecting the clone that has the gene of interest. Procedures must be available for examining colonies of bacteria or plaques of infected cells growing on agar plates and detecting those few which contain small amounts of the protein or gene of interest. It is the purpose of the present section to discuss possible approaches to finding the right clone. We consider first the situation in which the gene is *expressed* (that is, the protein is synthesized) in the cloning host. Then we discuss the situation, rather common, in which the gene is not expressed and we must look for the DNA itself.

If the foreign gene is expressed in the cloning host

If the foreign gene is expressed (that is, the protein product is synthesized) in the cloning host, then procedures can be used which look for the presence of this protein in recombinant colonies. The cloning host must *not* itself produce the protein being studied. If we are looking for clones which express the gene, then we are looking for the rare colonies in which this protein is present. If the protein is one that the cloning host normally produces, then the host used must be defective, that is, mutant, for the gene of interest. Then, when the foreign gene is incorporated, the expression of this foreign gene can be detected by complementation (see Section 7.5). Obviously, if the host already expressed a protein with the same activity, there would be a large background of this activity against which the protein produced via the foreign gene could not be detected. If the protein is a eukaryotic protein that is not normally produced in host bacteria, then the host may be naturally defective. A striking example is the cloning of luciferase genes from various types of bioluminescent beetles into *Escherichia coli*; in the dark, the clones glow in various colors depending on the type of luciferase present (see later, Figure 8.19).

Antibody as a method of detecting the protein

If the protein does not have a readily detectable function, then a different approach is needed. It involves the use of an antibody as a reagent which is specific for the protein of interest. We will discuss antibodies and immunology in Chapter 12. For our present purposes, we note that an antibody is a serum protein produced by a mammalian system which combines in a highly specific way with another protein, the *antigen*. In the present case, the protein of interest is the antigen, and this protein is used to produce an antibody in an experimental animal. Since the antibody combines specifically with the antigen, if the antigen is present in one or more colonies on the plate, then the locations of these colonies can be determined by observing the binding of the antibody. Because only a small amount of the protein (antigen) will be present in the colonies, only a small amount of antibody will be bound, so that a highly sensitive procedure for detecting bound antibody must be available. In practice, a system involving radioactive labelling is used, and the presence of bound antibody is detected by means of autoradiography using X-ray film. Other extremely sensitive techniques for measuring antigen:antibody interactions are discussed in Section 12.9.

The whole procedure is outlined in Figure 8.7a. As seen, the replica plating procedure (see Figure 7.2) is used to make a duplicate of the master plate, but the duplication is done onto a membrane filter and all of the manipulations are done with this filter. After the duplicate colonies have grown up, they are partially lysed to release the protein (antigen) of interest. The antibody is then added and the antibody–antigen reaction allowed to proceed. Unbound antibody is then washed off and a radioactive agent is then added which is specific for the antibody. A piece of X-ray film is placed over the filter and allowed to become exposed. If a radioactive colony is present, a spot on the

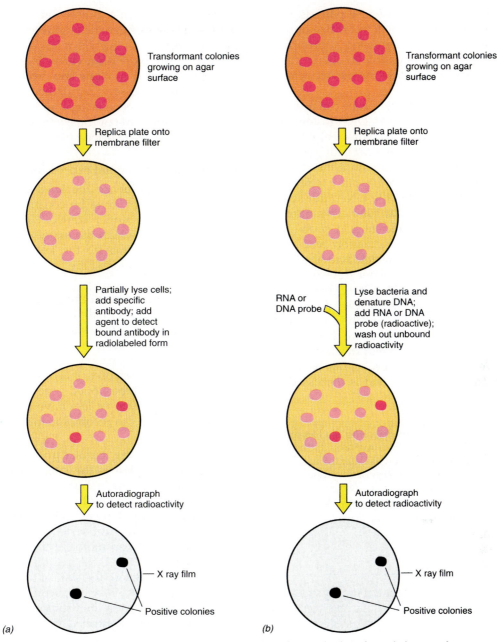

FIGURE 8.7   Finding the right clone. (a) Method for detecting production of protein by use of specific antibody. (b) Method for detection of recombinant clones by colony hybridization with a radioactive nucleic acid probe.

X-ray film will be observed after the X-ray film is developed. The location of this spot on the film corresponds to a location on the master plate where a colony is present which produces the protein. This colony can now be picked from the master plate and cultured.

One limitation of this procedure is that an antibody must be available which is *specific* for the protein in question. As we will see in Chapter 12, production of antibody can be readily done by injecting the protein (antigen) into an animal, but the protein injected must be pure, otherwise more than one antibody will be formed. Thus, one must have previously purified the protein. (The use of monoclonal antibodies simplifies this requirement a little; see Section 12.11.)

Nucleic acid probes: searching for the gene itself

Suppose the gene is not expressed in the cloning host or that no assay or antibody is available for the gene product. How does one detect its presence in colonies? The most general way is to use a **nucleic acid probe** which contains a key part of the base sequence of the gene of interest. As we have discussed (see Section 5.2), nucleic acid hybridization can be used as a specific means of detecting polynucleotides with specific sequences. Either DNA or RNA can be used as probe. The general procedure is to label the nucleic acid probe, usually with radioactive phosphate, and allow single-stranded probe to hybridize with single-stranded nucleic acid derived from the cloned DNA.

Because of specific complementary base pairing, two single-stranded polynucleotides will hybridize only if they are fairly complementary. By using appropriate hybridization conditions, it is possible to obtain binding of the radioactive probe only to the nucleic acid of interest.

The way in which a nucleic acid probe can be used to detect the presence of recombinant DNA in colonies is shown in Figure 8.7b. The procedure, **colony hybridization**, again makes use of replica plating to produce a duplicate of the master plate on a membrane filter. The cells on the filter are lysed in place to release their nucleic acid and to convert the DNA into a single-stranded form and fix it to the filter. This filter is then treated with radioactive nucleic acid probe (either RNA or DNA) to allow hybridization, and after removal of unbound radioactive nucleic acid, the filter is subjected to autoradiography. After development, the X-ray film is examined for spots. These correspond to locations on the membrane where the radioactive probe hybridized the DNA from a particular colony. Colonies corresponding to these spots are then picked and studied further. A modification of this procedure, avoiding the use of a radioactive probe, has been developed for clinical microbiology and is discussed in Section 13.11.

> **Special procedures are needed for detecting the foreign gene in the cloning host. If the gene is expressed, the presence of the foreign protein itself as detected by either its activity or by reaction with specific antibodies, is evidence that the gene is present. However, if the gene is not expressed, then its presence can be detected by use of a nucleic acid probe.**

8.7 Expression Vectors

For practical applications it is essential that systems be available in which the cloned genes can be *expressed*. Organisms have complex regulatory systems, and many genes are not expressed all of the time. One of the major goals of genetic engineering is the development of vectors in which *high levels* of gene expression can occur. An **expression vector** is a vector which not only can be used to clone the desired gene but also contains the necessary regulatory sequences so that expression of the gene is kept under control of the genetic engineer. Some of the elements involved in gene expression are summarized in Figure 8.8.

Requirements of a good expression system

Many factors influence the level of expression of a gene, and a vector must be constructed in which all these factors are under control. In addition, a host must be used in which the expression vector is most effective. We summarize the key requirements of a good expression system here.

1. **Number of copies of the gene per cell**. In general, more product is made if many copies of the gene are present. Vectors such as small plasmids (for example, pBR322) are valuable because they can replicate to a high copy number. Sometimes, often for research purposes, it is desirable to have only a single copy of the cloned gene in the cell. For these cases, *integrating vectors* have been developed so that the gene can recombine into the host chromosome.

2. **Strength of the transcriptional promoter**. The promoter region is the site at which binding of RNA polymerase first occurs (see Section 5.6), and native promoters in different genes vary considerably in RNA polymerase binding strength. For engineering a practical system, it is important to include a *strong* promoter in the expression vector. For bacteria, the DNA region around 10 and 35 nucleotides before the start of transcription (called the −10 and −35 *regions*, see Figure 5.29) is especially important in the promoter. Many *Escherichia coli* genes are controlled by relatively weak pro-

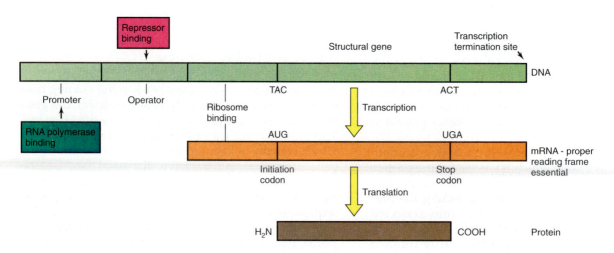

FIGURE 8.8 Factors affecting the expression of cloned genes in bacteria. Sequences and signals that must be appropriate for high levels of gene expression are indicated.

moters, and promoters from eukaryotes and some other prokaryotes function poorly or not at all in *E. coli*. Strong *E. coli* promoters that have been used in the construction of expression vectors include *lac*uv 5 (which normally controls β-galactosidase), *trp* (which normally controls tryptophan synthetase), *tac* (a synthetic hybrid of the −35 region of the *trp* promoter and the −10 region of the *lac* promoter), lambda P_L (which normally regulates lambda virus production), and *ompF* (which regulates production of an *o*uter *mem*brane *p*rotein).

A novel regulatory system has been created using the bacteriophage T7 promoter and RNA polymerase. When T7 infects *E. coli* it codes for its own RNA polymerase which recognizes only T7 promoters, thus effectively shutting down host transcription (see Section 6.10). In expression vectors it is possible to place expression of cloned genes under control of a T7 promoter. However, if this is done it is necessary to engineer into the plasmid the gene for T7 RNA polymerase as well. The latter is placed under control of an easily regulated promoter such as that of lambda or *lac*. Expression of the cloned gene(s) occurs shortly after T7 RNA polymerase transcription has been switched on. Because it recognizes only T7 promoters, T7 RNA polymerase transcribes only the cloned genes; all other host genes remain untranscribed.

3. **Presence of the bacterial ribosome binding site**. The transcribed mRNA must bind firmly to the ribosome if translation is to begin, and an early part of the transcript contains the ribosome-binding site (Shine-Dalgarno sequence, Figure 8.8 and Section 5.8). Bacterial ribosome binding sites are not found in eukaryotic genes, and it is thus essential that the bacterial region be present in the cloned gene if high levels of gene expression are to be obtained. Part of the requirement for proper ribosome binding is the necessity for a proper distance between the ribosome binding site and the translation initiation codon. If these sites are too close or too far apart, the gene will be translated at low efficiency.

4. **Proper reading frame**. In some cases, the ribosome binding site and even the initiation codon for the gene to be cloned are part of the expression vector. Because of the way the source DNA is fused into such a vector, three possible reading frames (see Section 7.2) could be obtained, only one of which is satisfactory. One approach that can be used if the correct frame is not known is the use of three vectors, each having the restriction site into which new DNA will be inserted positioned such that the insert will be in a different reading frame. The gene fragment is inserted into all three vectors and the one which gives proper expression is selected by testing.

5. **Codon usage**. There is more than one codon for most of the 20 amino acids (see Table 5.4), and some codons are used more frequently than others. Codon usage is partly a function of the concentration of the appropriate tRNA in the cell. A codon frequently used in a mammalian cell may be used less frequently in the organism in which the gene is being cloned. Insertion of the appropriate codon would be difficult because it would have to be changed in all locations in the gene. However, this can be done if necessary by using synthetic DNA and site-directed mutagenesis (see Sections 7.3 and 8.11) to create a gene more amenable to the codon usage patterns of the host.

6. **Fate of the protein** after it is produced. Some proteins are susceptible to degradation by intracellular proteases and may be destroyed before they can be isolated. Excreted proteins must have the signal sequence attached (see Section 5.8) if they are to move through the cytoplasmic membrane. Some eukaryotic proteins are toxic to the prokaryotic host, and the host for the cloning vector may be killed before a sufficient amount of the product is synthesized. Further engineering of either the host or the vector may be necessary to eliminate these problems.

The skill of the genetic engineer is thus essential in the construction of an appropriate vector which can be (1) efficiently incorporated into the proper host, (2) replicated to high copy number, (3) efficiently transcribed, and (4) efficiently translated. Many mammalian proteins are completely unexpressed when their genes are first cloned in *E. coli*, but expression can sometimes be achieved with appropriate manipulation of the vector. The best example is the production on a commercial scale of human insulin in *E. coli*, as described in Section 8.13.

Role of regulatory switches in expression vectors

For maximum production of a protein from a cloned gene, it is usually undesirable to design a vector that permits the gene to be transcribed and translated at all times. There are several reasons for this. As mentioned earlier, some proteins that are of commercial interest are toxic to the bacterial hosts. In addition, some expression systems, such as that involving the T7 promoter, are so powerful that normal host genes cannot be expressed. In either of these cases, it is very desirable that the synthesis of the protein be under the direct control of the experimenter. The ideal situation is to be able to grow the culture containing the expression vector until a large population of cells is obtained, each containing a large copy number of the vector, and then turn on expression in all copies simultaneously by manipulation of a regulatory switch.

We have discussed regulatory controls of gene expression in Sections 5.10 through 5.13. Recall the major importance of the repressor/operator system in regulating gene transcription. A strong repressor can completely block the synthesis of the proteins under its control by binding to the operator region. Repressor function can be turned off at the chosen time by adding an inducer, allowing the transcription of the genes controlled by the operator.

For the repressor-operator system to work as a regulatory switch for the production of a foreign protein, it is desirable to retain in the expression vector a fragment of the structural gene and the operator controlled by the repressor, to which the cloned gene is fused. This permits proper arrangement of the sequence of genetic elements: promoter-operator-ribosome binding site-structural gene, so that efficient transcription and translation can occur. The presence of a fragment of the normal protein can help render the foreign protein stable and capable of being excreted.

The construction of plasmid expression vectors containing the regulatory components of the *lac* operon provides one means of providing a suitable regulatory switch. As we have discussed in Section 5.10, the *lac* operon is switched on by inducers such as lactose or related β-galactosides. Phasing of cell growth and protein synthesis can thus be achieved by allowing growth to proceed in the absence of inducer until a suitable cell density is achieved, and then adding inducer to bring about synthesis of the desired proteins. Plasmids have been constructed containing the *lac* promoter, ribosome binding site, and operator. When the desired gene is inserted into such a system, expression can then be achieved by adding *lac* inducer.

Another regulatory system that has been used to construct expression vectors is the *tryptophan operon*. Although repressed by tryptophan, the *trp* operon can be induced by adding a tryptophan analog (such as β-indolacrylic acid) which brings about an apparent tryptophan deficiency.

We also mentioned that the bacteriophage lambda promoter P_L is also sometimes used in expression systems. This promoter is kept turned off by having the lambda repressor protein in the cell (see Section 6.12). Typically the lambda repressor is encoded by a mutant gene (carried by the vector or by a prophage in the host) and is temperature sensitive. By raising the temperature of the culture to the proper value (usually 8 to 10°C higher than the growth temperature), the lambda repressor is inactivated and transcription from P_L begins.

Using such expression systems, one can produce very high levels of foreign proteins in *Escherichia coli*. In many cases the desired protein exceeds 100,000 molecules per cell and makes up over 20 percent of the protein molecules in a cell.

8.8 Synthetic DNA

Techniques are available for the synthesis of short fragments of DNA of specified base sequence. **Synthetic DNA** is widely used in molecular genetics, especially in genetic engineering, but also in basic research. The procedures for synthesis of DNA can be completely automated so that an oligonucleotide of 30–35 bases can be easily made in a few hours, and oligonucleotides of well over 100 bases in length can be made if necessary. For the synthesis of longer polynucleotides, the oligonucleotide fragments can be joined enzymatically, using DNA ligase.

DNA is synthesized in a *solid-phase procedure*, in which the first nucleotide in the chain is fastened to an insoluble porous support (such as a silica gel with particles around 50 μm in size). The overall procedure, the chemical details of which need not concern us here, is shown in Figure 8.9. Several chemical steps are needed for the addition of each nucleotide. After each step is completed, the reaction mixtures are flushed out of the solid support, and the series of reactions repeated for the addition of the next nucleotide. Once the desired length is achieved, the oligonucleotide is removed from the solid phase support and purified to eliminate byproducts and contaminents.

Synthetic DNA molecules are widely used as

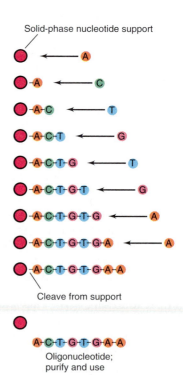

FIGURE 8.9 Solid-phase procedure for synthesis of a DNA fragment of defined sequence. Chemical synthesis proceeds by adding one nucleotide at a time to the growing chain.

Not all cloned genes are expressed at high efficiency in foreign hosts. Expression vectors are special cloning vectors that contain various regulatory elements necessary for obtaining high levels of gene expression. Regulatory switches are also useful in expression vectors since they can be used by the investigator to turn on expression at the most favorable time in the growth cycle.

probes in genetic engineering, to detect, via nucleic acid hybridization, specific DNA sequences (see Figure 8.7b). Synthetic DNA is also used extensively in basic research on the molecular genetics of viruses and cells. We will describe later (see Section 8.11) how synthetic DNA is used in a procedure called **site-directed mutagenesis** to create mutations at specific locations on the genome. Finally, synthetic DNA is used extensively as a source of DNA primers for the polymerase chain reaction (see next section).

8.9 Amplifying DNA: The Polymerase Chain Reaction

Conventional gene cloning methods can be considered *in vivo* DNA amplifying tools. However, the development of synthetic DNA has spawned a new method for the rapid amplification of DNA *in vitro*, the **polymerase chain reaction**. The polymerase chain reaction, or **PCR**, can multiply DNA molecules by up to one billion-fold in the test tube, yielding large amounts of specific genes for cloning, sequencing, or mutagenesis purposes. PCR makes use of the enzyme *DNA polymerase*, which copies DNA molecules (see Section 5.4).

The PCR technique requires that the nucleotide sequence of a portion of the desired gene be known. This is necessary because short oligonucleotide *primers* complementary to sequences in the gene or genes of interest must be available for PCR to work. The steps in PCR amplification of DNA are as follows. (1) Two oligonucleotide primers which flank the target DNA (Figure 8.10b) are made on an oligonucleotide synthesizer and added in great excess to heat-denatured target DNA (Figure 8.10a). (2) As the mixture cools, the excess of primers relative to the target DNA ensures that most target strands will anneal to a primer and not to each other (Figure 8.10b). (3) DNA polymerase is then added and this enzyme extends the primers using the target strands as template (Figure 8.10c). (4) After an appropriate incubation period, the mixture is heated again to separate the strands. The mixture is then cooled to allow the primers to hybridize with complementary regions of newly synthesized DNA and the whole process is repeated (Figure 8.10e).

Thus, each PCR "cycle" involves the following: (1) heat denaturation of double-stranded target DNA, (2) cooling to allow hybridization of specific primers to target DNA, and (3) primer extension by the action of DNA polymerase (Figure 8.10). Note in Figure 8.10 how the extension products of one primer can serve as a template for the other primer in the next cycle. The beauty of the PCR technique lies in the fact that each cycle literally *doubles* the content of the original target DNA. In practice, 20–30 cycles are usually run, yielding a 10^6–10^9-fold increase in the target sequence (Figure 8.10f).

The original PCR technique employed *Escherichia coli* DNA polymerase, but because of the high temperatures needed to denature the double-stranded copies

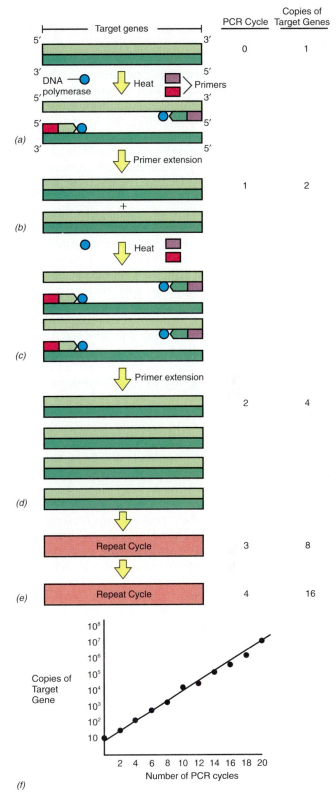

FIGURE 8.10 The polymerase chain reaction for amplifying specific DNA sequences. (a) Target DNA is heated to separate the strands, and a large excess of two oligonucleotide primers, one complementary to the target strand and one to the complementary strand, is added along with DNA polymerase. (b) Following primer annealing, primer extension yields a copy of the original double-stranded DNA. (c) Further heating, primer annealing, and primer extension yields a second double-stranded DNA. (d) The second double-stranded DNA. (e) Two additional PCR cycles yield 8 and 16 copies, respectively, of the original DNA sequence. (f) Effect of running 20 PCR cycles on a DNA preparation originally containing 10 copies of a target gene. Note that the graph is semilogarithmic.

of DNA being made, the polymerase itself also was denatured and had to be replenished every cycle. This tended to limit the number of cycles that could be run and was very expensive. This problem was solved by employing a thermostable DNA polymerase isolated from the thermophilic Bacterium *Thermus aquaticus*. DNA polymerase from *T. aquaticus*, known as *Taq polymerase*, is stable to 95°C and thus is unaffected by the denaturation step employed in the PCR reaction. The use of *Taq* DNA polymerase also increased the *specificity* of the PCR reaction because the DNA is copied at 72°C rather than 37°C. At high temperatures, nonspecific hybridization of primers to nontarget DNA rarely occurs, thus making the product of *Taq* PCR more homogenous than what was obtained using the *E. coli* enzyme.

One problem with the *Taq* polymerase is that it has no proofreading function (see Section 5.4) and consequently makes more mistakes than the *E. coli* enzyme. DNA polymerases from hyperthermophilic marine Archaeans have been isolated that do have proofreading activity, and these are now being used for PCR (see Section 9.9).

Because a number of highly repetitive steps are involved in the PCR technique, machines have been developed that can be programmed to run through heating and cooling cycles automatically. Since each cycle requires only about five minutes, the automated procedure allows for large amplifications in only a few hours (by contrast, such amplification by *in vivo* cloning methods would take several days). To supply the demand for thermostable DNA polymerase in the growing PCR and DNA sequencing markets, the genes for these enzymes have been cloned into *E. coli* and produced in large quantities; the cost of doing PCR is now just a fraction of what it was when the technique was first introduced.

The polymerase chain reaction has found many practical uses. PCR is extremely valuable for cloning DNA because the gene or genes of interest, identified by hybridization to a primer, can be amplified prior to cloning. As a tool for DNA sequencing, PCR yields large amounts of specific DNA sequences, which greatly improves the resolution of sequencing proto-

cols. PCR can also be used to produce large amounts of mutated DNA. If made sufficiently long, primers containing one or a few base mismatchs will still anneal to the target gene; the introduced mutation can then be amplified by PCR (this is a form of site-directed mutagenesis, see Section 8.11). The mutated DNA can be used to transform cells, generating mutant derivatives.

Because the primers used do not have to be perfectly complementary, PCR is routinely used in comparative or evolutionary studies to isolate genes from a variety of sources, where the DNA has already been cloned (and sequenced) from one organism. In these cases the primers are made to regions of the gene which are thought to be conserved throughout a wide variety of organisms. Because of the sensitivity of PCR, it has been used to amplify and clone DNA from sources such as mummified human remains and even samples of extinct plants and animals.

PCR can also be used to amplify very small quantities of DNA present in a sample. Using appropriate primers one can find and identify a single bacterial cell in a sample even if large numbers of other species are present. PCR has also been used to help develop *DNA fingerprinting*, a powerful technique which can permit identification of individuals, or relationships between individuals, from small samples of their DNA (see DNA Fingerprinting box).

The polymerase chain reaction, a unique procedure for amplifying DNA *in vitro*, makes use of a heat stable DNA polymerase from thermophilic prokaryotes. Heat is used to denature the DNA into two single-stranded molecules, each of which is copied by the polymerase. Beginning with a small oligonucleotide that serves as a primer for the target DNA to be amplified, the polymerase copies the complete DNA to which the primer associates. After a single copy cycle, the newly formed double strands are separated by heat again, and a new round of copying permitted. At each thermal cycle, the amount of target DNA doubles. Under appropriate conditions, a billion-fold increase in the target DNA can be obtained.

DNA Fingerprinting

The techniques used in molecular genetics and genetic engineering are not only useful in determining the evolutionary relationships among organisms (see Chapter 18) but also in determining relationships among individuals and in establishing whether a tissue sample, even a very small one, came from a particular individual. These latter techniques are called *DNA fingerprinting*. DNA fingerprinting is made possible both by the technology that allows precise detection and amplification of very small amounts of DNA

and by the fact that higher organisms contain repetitive DNA sequences that can exist in different numbers and patterns in the genome.

As mentioned in Section 5.5, the genomes of higher eukaryotes contain a very large amount of repetitive DNA. Some of these repeats exist in families of related sequences scattered around the genome, and members of these families have been cloned and sequenced. In order to be useful for identification purposes, a DNA sequence must have a reasonable chance of differing among different groups in a population of organisms. One family of these repetitive sequences was found to vary not simply as to sequence but also as to how many repeats of an individual sequence occurred at a single site on a particular chromosome. These sequences are called *variable number of tandem repeats,* or *VNTR,* and several have been identified.

The use of VNTRs in DNA fingerprinting is illustrated in the accompanying figure. It shows two different alleles (alternative states of the same gene) of a eukaryotic chromosome which differ only in how many copies of the repeated sequence are present. Since the VNTR DNA has been sequenced, it is known what restriction enzyme sites are *not* found in a particular VNTR. Digestion with such an enzyme then releases the complete VNTR intact. When the DNA from two chromosomes with a different number of repeats at this particular locus is digested, the restriction fragments containing this DNA will differ in size. (Such a difference is called a *restriction fragment length polymorphism,* or *RFLP.*) This DNA can be separated by gel electrophoresis, and the VNTR-containing fragments detected by Southern blotting (see the Nucleic Acids box, Chapter 5) using a probe made from the cloned VNTR sequence. Having only a single difference in one band is not a very precise way of identifying an individual. Since higher eukaryotes are diploid, there will be two copies of the VNTRs which may or may not be the same. Also, most VNTR sequences are families and several different loci can be detected with the same probe. In addition, it is possible to probe the di-

gested DNA simultaneously for several different VNTR markers. Using these methods, it has been estimated that the probability of identifying a particular individual by comparing two different DNA samples is very high. There is some controversy about exactly how high the probability is with any given protocol, but it is acknowledged that DNA fingerprinting is a very powerful technique.

Notice that PCR does not need to be used in DNA fingerprinting. However, the use of PCR is essential when the amount of DNA in the sample is very small—such as that found in the cells on the root of a single hair. The use of PCR in DNA fingerprinting is also shown in the figure. To use PCR, it is necessary to know the sequences surrounding a particular site that contains a VNTR so that primers can be synthesized. However, since PCR only amplifies the DNA between the primers, one does not have to cut the DNA with restriction enzymes before running it on a gel. With enough cycles of amplification, it is also sometimes possible to detect the PCR-generated bands by simply staining the gels rather than using a hybridization probe.

Research labs also use DNA fingerprinting to screen tissue culture cells to determine if the cells are the correct line or from the correct animal. Although it might seem unlikely, tissue culture cells can become contaminated with other cell lines, and there are many fewer tests that can distinguish different species of mammalian cells than there are those that can differentiate species of bacteria.

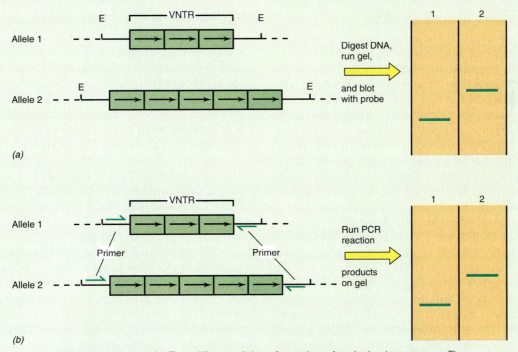

DNA fingerprinting. (a) Two different alleles of a region of a single chromosome. The alleles differ only in the number of repeats in the VNTR. DNA from cells containing these chromosomes can be cut with the restriction enzyme *Eco*RI (which does not cut within the VNTR) and the fragments separated on an agarose gel. The fragments containing the VNTR are then identified after Southern blotting by hybridization with a probe specific to the VNTR. (For simplicity, the figure shows only the result from individuals whose two chromosomes each have the same allele at this site.) (b) The same alleles, but this time with primers that could be used to amplify the VNTR segments by PCR. The products of the PCR reaction can be loaded directly onto the gel without restriction digestion.

8.10 Cloning and Expression of Mammalian Genes in Bacteria

We have now laid out the principles behind the development of systems for obtaining the production of foreign genes in an organism such as *Escherichia coli*. How are these principles put together in the engineering of a desired expression system? In the human or some other complex organism, the sought-for gene will be buried within a large mass of DNA containing unwanted genes, as well as repetitive DNA sequences that have no coding function at all (see Section 5.5). Also, most genes from higher eukaryotes are split, with noncoding introns interspersed among the coding exons (see Section 5.1). How can a gene be selected from this complex mixture? Although a number of approaches are available, this is the least well developed stage of genetic engineering. Not only skill, but also intuition and good luck play big roles in a successful outcome. A summary of approaches and procedures is given in Figure 8.11.

We start first with a consideration of the desired product and where it is produced in the human body. A human being is a highly differentiated organism, and many genes are expressed only in certain organs or tissues. The hormone *insulin*, for instance, is produced only in the pancreas. Of course, the insulin gene is found in all tissues and organs, but it is only *expressed* in this one organ.

Reaching the gene via messenger RNA

One approach is to get to the gene through its mRNA. A major advantage of using mRNA is that the noncoding information present in the DNA (introns) has been removed (see Section 5.6). The isolated mRNA is used to make complementary DNA (cDNA) by means of reverse transcription (see Section 6.22). It is likely that a tissue expressing the gene will contain large amounts of the desired mRNA, although except in rare cases this will certainly not be the only mRNA produced. In a fortunate situation, where a single mRNA dominates a tissue type, extraction of mRNA from that tissue provides a useful starting point for gene cloning.

In a typical mammalian cell, about 80 to 85 percent of the RNA is ribosomal, 10 to 15 percent is transfer RNA and other low-molecular-weight RNAs, and 1 to 5 percent is messenger RNA. Although low in abundance, the mRNA in a eukaryote is identifiable because of the poly A tails found at the 3′-end (see Section 5.6). In maturing red blood cells, for instance, where virtually the only protein made is the globin portion of hemoglobin, from 50 to 90 percent of the poly A-containing cytoplasmic RNA consists of globin mRNA. By passing a poly A-rich RNA extract over a chromatographic column containing poly T fragments (linked to a cellulose support), most of the mRNA of the cell can be separated from the other cellular RNA by the specific pairing of A and T bases. Elution of the RNA from the column then gives a preparation greatly enriched in mRNA.

Once the RNA message has been isolated, it is necessary to convert the information into DNA. This is accomplished by use of the enzyme *reverse transcriptase*, which we have discussed in Section 6.22. This remarkable enzyme, an essential component of retrovirus replication, copies information from RNA into DNA (Figure 8.12). As we noted, this enzyme requires a primer in order for it to begin working (in retrovirus infection the primer is a tRNA). In the present procedure, an oligo-dT primer is used which is complementary to the poly A tail of the isolated mRNA. The oligo-dT primer is hybridized with the mRNA and then reverse transcriptase is allowed to act (Figure 8.12). As seen, the newly synthesized DNA copy has a hairpin loop at its end, arising because when the enzyme completes copying the mRNA it starts to copy the newly synthesized DNA. This hairpin loop, which is probably an artifact of the test tube reaction, provides a convenient primer for the synthesis of the second DNA strand. The resultant double-stranded DNA, with the hairpin loop intact, is then cleaved by a single-strand-specific nuclease to produce the desired double-stranded DNA, one strand of which is complementary to the mRNA. This double-stranded DNA (the gene of interest) can then be inserted into a plasmid or other vector for cloning. The detection of specific clones makes use of the procedures discussed in Section 8.6.

Reaching the gene via the protein

If for some reason, mRNA for the gene of interest cannot be obtained, other approaches are possible. Two such approaches, reverse translation and polyribosome precipitation, will be discussed here.

The most widely used method of cloning low-abundance mRNA is to make a **synthetic DNA** which is complementary to part of the mRNA, and then use this DNA as a *probe* in a Northern blot procedure (see Nucleic Acids box, Chapter 5) to pull out by hybridization the mRNA of interest. This requires that a partial or complete sequence of the protein be known. Then, from a consideration of the genetic code, the nucleotide sequence of a section of the DNA is deduced, and this piece of DNA is synthesized.

The procedure by which the nucleotide sequence is deduced from the protein sequence is called **reverse translation**. (Note that *reverse translation* is not a cellular process but a mental exercise of the genetic engineer.) The procedure of reverse translation is illustrated in Figure 8.13. From the genetic code, the nucleotide sequence of a section of the DNA is deduced, and this piece of DNA is synthesized. Unfortunately, degeneracy of the genetic code (see Section 5.9) somewhat complicates the problem. Most amino acids are coded for by more than one codon, and codon usage varies from organism to organism. The best section of DNA to synthesize is one which corresponds to

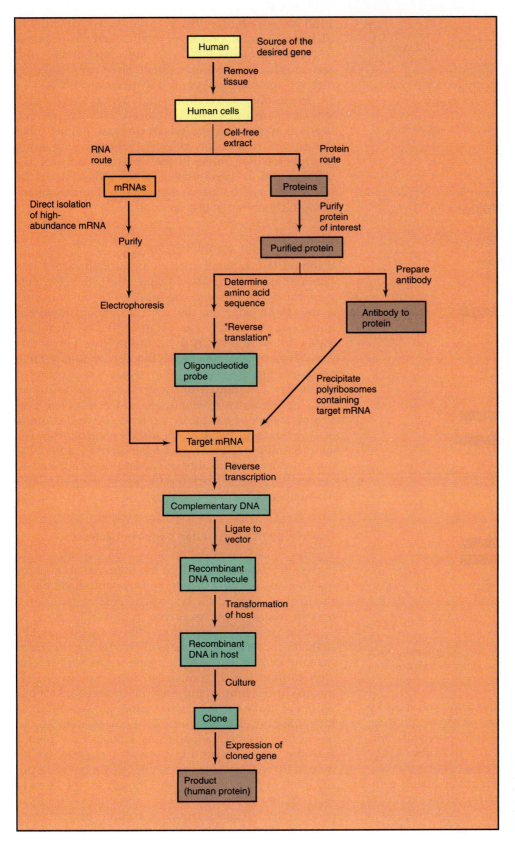

FIGURE 8.11
Several routes to the isolation and expression of mammalian genes in prokaryotes. The colors indicate whether a step involves DNA (green), RNA (orange), or protein (brown).

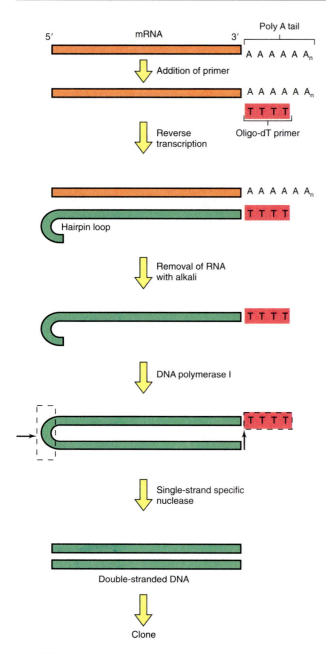

5′ mRNA 3′ Poly A tail

A A A A A A_n

⬇ Addition of primer

A A A A A A_n

T T T T

Oligo-dT primer

⬇ Reverse transcription

A A A A A A_n

T T T T

Hairpin loop

⬇ Removal of RNA with alkali

T T T T

⬇ DNA polymerase I

T T T T

⬇ Single-strand specific nuclease

Double-stranded DNA

⬇ Clone

FIGURE 8.12 Steps in the synthesis of complementary DNA (cDNA) from an isolated mRNA, using the retroviral enzyme reverse transcriptase.

a part of the protein rich in amino acids specified by only a single codon (methionine, AUG; tryptophan, UGG) or by two codons (for example, phenylalanine; UUU, UUC; tyrosine; UAU, UAC; histidine; CAU, CAC) since this will increase the chances that the synthesized DNA will be complementary or nearly complementary to the mRNA of interest. If the complete amino acid sequence of the protein is not known, then the sequence used is generally one at the amino terminus of the protein, since it is at the amino terminus that sequencing of the protein begins.

Polyribosome precipitation involves the separation from the tissue of *ribosome complexes* that are in the process of synthesizing the desired protein, using an antibody specific for this protein. As we describe in Chapter 12, antibodies are proteins made in response to foreign proteins which are able to combine with and specifically precipitate them. How is an antibody directed against a *protein* used to detect an *mRNA*?

In the ribosome complex, polypeptides of the protein of interest are still complexed with the protein-synthesizing machinery (which contains, among other things, the sought-for mRNA). When the antibody precipitates the protein in the ribosome complex, the mRNA also is precipitated. After isolation of the mRNA, it can be used to prepare complementary DNA as described in this section.

Synthesis of the complete gene

If the protein is small enough, or is of sufficient economic interest to justify a major effort, the complete gene can be synthesized chemically; obviously this requires knowledge of the complete amino acid sequence of the protein (see Figure 8.9). Chemical synthesis not only permits the acquisition of genes that cannot be obtained otherwise, but also permits synthesis of *modified genes* which may make new proteins of utility. The techniques for the synthesis of DNA molecules are now well developed, and it is possible to synthesize genes coding for proteins 100 to 200 amino acid residues in length (300 to 600 nucleotides). The synthetic approach was used for the production of the human hormone insulin in bacteria, as discussed in Section 8.13. We will discuss the use of synthetic DNA in mutagenesis in Section 8.11.

Expression of mammalian genes in bacteria

For the cloned mammalian gene to be expressed in a bacterium, it is essential that the mammalian gene be inserted adjacent to a strong *promoter* and that a bacterial *ribosome-binding site* be present. Also, it is essential that the *reading frame* be correct.

One method of providing a ribosome-binding site in the proper reading frame is to arrange for the mammalian DNA sequence to be expressed as part of a **fusion protein** which contains a short prokaryotic sequence at the amino end and the desired eukaryotic sequence at the carboxyl end. Although less desirable for some purposes, fusion proteins are often more stable in bacteria than unmodified eukaryotic proteins, and the bacterial peptide portion can often be removed by chemical (or enzymatic) treatment after purification of the fused protein. One advantage of making a fusion protein is that the bacterial portion can contain the bacterial sequence coding for the *signal peptide* that enables transport of the protein across the cytoplasmic membrane (see Section 5.8), making possible the development of a bacterial system which not only synthesizes the mammalian protein but actually excretes it.

To obtain expression of an *unfused* protein, it is es-

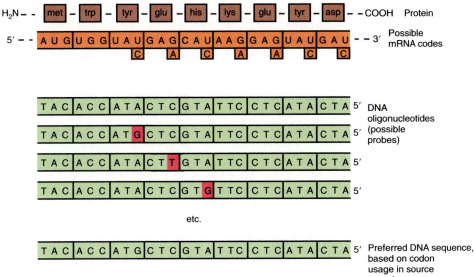

FIGURE 8.13
Reverse translation: deducing the best sequence of an oligonucleotide probe from the amino acid sequence of the protein. Because of degeneracy, many probes are possible. If codon usage by the same organism is known, then a preferred sequence can be selected. It is not essential that complete accuracy be achieved since a small amount of mismatch can be tolerated.

sential that the eukaryotic gene be inserted into the vector so that its initiation codon is just downstream from the ribosome-binding site. The ribosome-binding site is provided by the expression vector. In many cases the initiation codon is also from the vector. This is because in many mammalian proteins, the mature and active form lacks some of the amino acids present at the initiation site because they are cleaved after synthesis by a posttranslational modification. Thus, to obtain synthesis of the active form of the protein directly, it is necessary to fuse only the sequence coding for the final protein to the initiation codon. If high efficiency of translation is to be obtained, an intervening region of noncoding DNA of the proper length must be placed between the ribosome-binding site and the initiation codon. Several plasmids have been constructed which contain built-in promoters and suitable restriction enzyme sites, so that the proper tailoring of coding sequence to initiation site and ribosome-binding site can be achieved. Human proteins that have been expressed at high yield under the control of bacterial regulatory systems include human growth hormone, insulin, virus antigens, interferon, and somatostatin (see Section 8.13).

> Numerous mammalian proteins have medical value and can be produced in bacteria once the genes have been cloned into appropriate expression vectors. To detect and isolate the gene from the large amount of DNA in the mammalian cell, it is necessary to use special procedures. One approach is to isolate the mRNA from the mammalian cell and make a complementary DNA by use of the enzyme reverse transcriptase. Another approach is to synthesize a nucleic acid probe for the gene in question and use this probe to detect the cloned sequence in bacterial colonies. Under some conditions, the complete mammalian gene can be synthesized chemically and this synthetic gene cloned in bacteria.

8.11 *In Vitro* and Site-Directed Mutagenesis

Recombinant DNA technology has opened up a whole new field of mutagenesis. Whereas conventional mutagens (see Section 7.3) act at random, by use of synthetic DNA and recombinant DNA techniques it is possible to introduce mutations at *precisely determined sites* on genes. Proteins made from strains carrying such mutations can be expected to have different properties than the wild-type proteins, properties that may be predicted from a knowledge of protein structure.

Site-directed mutagenesis

This approach is relatively simple and very powerful. The basic procedure is to synthesize a short oligodeoxyribonucleotide containing the desired base change and to allow this to pair with a single-stranded DNA containing the gene of interest. Pairing will be complete except for the short region of mismatch. Then the short single-stranded fragment of the synthetic oligonucleotide is extended using DNA polymerase, thus copying the rest of the gene. The double-stranded molecule obtained is inserted into a cloning host by transformation and mutants selected by a procedure already described (see Section 8.6). The mutant obtained is then used in the production of the modified (mutant) protein.

The whole procedure of site-directed mutagenesis is illustrated in Figure 8.14. Several modifications of this technique have been developed to increase the ratio of mutants recovered. As seen, one must begin with the gene of interest cloned into a single-stranded DNA. A widely used vector for site-directed mutagenesis is bacteriophage M13, which, as we have seen (Sections 6.10 and 8.4), has some properties that are useful in recombinant DNA technology. The target

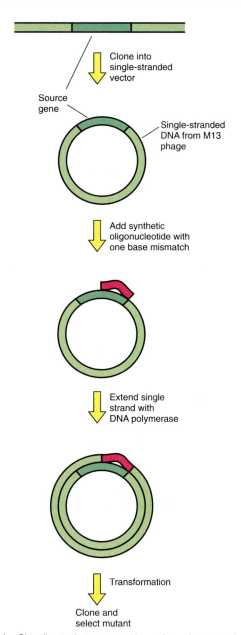

FIGURE 8.14 Site-directed mutagenesis, using short synthetic oligodeoxyribonucleotide fragments.

DNA is cloned into M13, from which single-stranded DNA can be purified with ease. Since cells infected with this phage remain alive, a ready source of DNA is available.

Total synthesis of mutant genes

Synthetic DNA technology can be used to synthesize a complete gene with a genetic change inserted at the desired location. Although a complete gene of 1000 or so nucleotides would be difficult to synthesize in one piece, it is possible to synthesize smaller portions of the gene and then to link these together to produce the final gene. By addition of appropriate sites at the ends of the gene, this gene can then be linked into a vector and cloned. The possibilities of this approach seem virtually limitless, although because of the expense involved, it is essential that one have a carefully consid-

ered rationale for the particular sequence. Typically, however, it is more advantageous to synthesize only a small part of a gene and use this to replace the same part of the wild-type gene, a process called *cassette mutagenesis*.

Cassette mutagenesis and gene disruption

Because of the large number of restriction enzymes commercially available and therefore the large number of different DNA sequences that can be cut, it is usually possible to find several different restriction sites in the gene of interest. If sites for the appropriate enzyme are not found in the gene, or in the precise location required, they can also be added by site-directed mutagenesis (see Figure 8.14). If restriction sites are close together, the intervening DNA fragment can be cut out and replaced by a synthetic DNA fragment in which one or more of the bases have been changed. These synthetic fragments are called *cassettes* (or cartridges), and so the process is known as **cassette mutagenesis**.

Insertion mutations can also be generated by simply inserting a cassette at a single site. When using cassettes to replace sections of genes, the cassettes are typically the same size as wild-type DNA fragments. However, the cassette used for making insertion mutations can be almost any size and can even be an entire gene. In fact, cassettes that encode proteins which confer a particular antibiotic resistance on the host are commonly used. This type of cassette mutagenesis is used in a process called **gene disruption**. The process of gene disruption is illustrated in Figure 8.15. In this case, a fragment carrying a gene conferring kanamycin resistance, the *kan* cassette, is inserted at a restriction site in a cloned gene. The vector carrying this mutant gene is then linearized by being cut with a different restriction enzyme, and the linear DNA is transformed into the host with kanamycin resistance selected. The linearized plasmid cannot replicate, so resistant cells likely arise by homologous recombination (see Section 7.5) between the mutated gene on the plasmid and the wild-type gene on the chromosome.

The cells have not only gained kanamycin resistance, they have also lost the function of the gene in which the *kan* cassette was inserted. Therefore, these mutations are also called "knockout" mutations. This process is similar to searching for insertion mutations made by transposons (see Section 7.11), but in this case the geneticist chooses exactly which gene will receive the mutation. Note that gene disruption in haploid organisms will only yield viable cells if the

> Synthetic DNA molecules of desired sequence can be made *in vitro* and used to construct a mutated gene directly, or to change specific base pairs within a gene via site-directed mutagenesis. Genes can also be disrupted by inserting DNA fragments, called cassettes, into them. The inserted cassette eliminates the function of the wild-type gene while conferring a new, and usually selectable, phenotype on the cell.

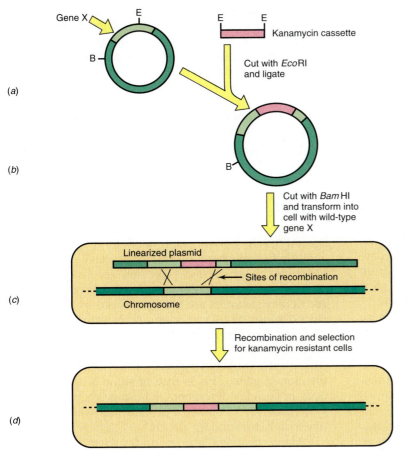

FIGURE 8.15
Gene disruption using cassette mutagenesis. (a) A plasmid containing a cloned wild-type copy of gene X is cut with the restriction enzyme *Eco*RI and mixed with a DNA fragment (the kanamycin cassette) which contains a gene capable of conferring kanamycin resistance upon a cell and which has been obtained using the same restriction enzyme. The cut plasmid and the cassette are ligated. (b) The product of the ligation is a plasmid which now contains the kanamycin cassette as an insertion mutation within gene X. This new plasmid is now cut with a further restriction enzyme, *Bam*HI, and transformed into a cell containing a wild-type gene X. (c) The transformed cell contains the linearized plasmid with a disrupted gene X and its own chromosome with a wild-type copy of the gene. In some cells, homologous recombination will occur between the wild type and mutant forms of gene X. (d) Cells that can grow in the presence of kanamycin must have the kanamycin cassette recombined into their chromosome since the linearized plasmid cannot replicate. These cells now have only a single disrupted copy of gene X. This disruption will typically abolish all gene X function.

disrupted gene is unessential. Gene disruption experiments are now used as one mechanism to find out whether a gene is essential. Methods of obtaining gene disruption have been developed for higher organisms, including mice.

8.12   Practical Applications of Genetic Engineering

We note here some of the main practical applications of recombinant DNA technology and then describe some of the applications in more detail in Sections 8.13, 8.14, and 8.15.

Genetic engineering has many commercial and practical applications. A few main areas of interest for commercial development are as follows:

1. **Microbial fermentations.** A number of important products are made industrially using microorganisms, of which the antibiotics are the most significant (see Section 10.6). Genetic engineering procedures can be used to manipulate the antibiotic-producing organism in order to obtain increased yields or produce modified antibiotics.

2. **Virus vaccines.** A vaccine is a material which can induce immunity to an infectious agent (see Section 12.15). Frequently, killed virus preparations are used as vaccines and there is always a potential danger to the patient if the virus has not been completely inactivated. Since the active ingredient in the killed virus vaccine is the protein coat, it would be desirable to produce the protein coat separately from the rest of the virus particle. By genetic engineering, viral coat protein genes can be cloned and expressed in bacteria or in nonpathogenic viruses, making possible the development of safe, convenient vaccines (see Section 8.13).

3. **Mammalian proteins.** A number of mammalian proteins are of great medical and commercial interest. Some of these are discussed in Section 8.13. In the case of human proteins, commercial production by direct isolation from tissues or fluids is complicated and expensive, or even impossible. By cloning the gene for a human protein in bacteria, its commercial production is possible.

4. **Transgenic plants and animals.** In addition to the production of valuable *products* by microbial means, genetic engineering promises the advent of genetically altered whole plants and animals. Some of these developments are discussed in Sections 8.14 and 8.15. Such organisms, referred to as *transgenic*, hold great promise for boosting agricultural productivity, altering the nutritional quality of meats and vegetables, and producing certain proteins not readily produced by genetically engineered microorganisms. By introducing cloned DNA into the fertilized eggs of animals, or directly into plant cells grown in tissue culture, it is now possible to grow genetically altered higher organisms.

5. **Environmental biotechnology.** Because of the enormous metabolic diversity of Bacteria (to be discussed in Chapter 16), a large gene pool exists in bacteria from natural habitats. In some cases these genes code for proteins that degrade environmental pollutants. Genes for the biodegradation of many toxic wastes and wastewater pollutants have been shown to exist in natural isolates of bacteria (see Section 17.20). Genetic engineering is beginning to tap these resources for the purpose of environmental cleanup. In many cases the gene donors are bacterial strains isolated from contaminated waste sites. Some examples include genes for the biodegradation of chlorinated pesticides, like 2,4,5-trichlorophenoxyacetic acid (2,4,5-T), chlorobenzenes and related chlorophenolics, napthalene, toluene, anilines, and various hydrocarbons. The desired genes are isolated from species of *Pseudomonas*, *Alcaligenes*, and a few other Bacteria, and then cloned into plasmids. Some plasmids have been constructed containing genes for the biodegradation of several different toxic chemicals.

6. **Gene regulation and gene therapy.** The first use of genetic engineering in the biotechnology industry was primarily for producing useful gene products more easily, or for creating transgenic organisms. As we discuss in the following sections, this approach is yielding great benefits. However, much current research in this area involves the creation of new products ("designer drugs") and controlling the expression of specific genes. Much research is currently being directed toward design of antisense RNA (see the Antisense Nucleic Acid box, Chapter 5) and ribozymes (see Section 5.6), which regulate the expression of specific genes. (Much research is also directed toward delivering these "drugs" to, and getting them inside, the correct cells and tissues.) An additional area of research showing great promise is the use of genetic engineering to treat genetic disease.

Despite the exciting promise of genetic engineering in biotechnology, getting a product to market is an enormous undertaking. Other than the obvious problems of correctly cloning and expressing the gene of interest in a microorganism, generally a bacterium or a yeast, and purifying the desired product, related matters such as clinical trials and governmental approval must be considered. Any microbially synthesized product intended for human use must pass extensive clinical trials. For example, human insulin produced microbially by recombinant DNA technology (see Section 8.13) had to pass strict clinical trials with human volunteers despite the fact that microbially produced insulin could be shown to be identical to the protein made in humans. If all goes well in clinical trials, federal approval is usually obtained, in the United States by action of the Food and Drug Administration (FDA), but this can be a time-consuming process. At present,

hundreds of products are under development or in various stages of clinical trials and nearing final approval, and the next decade should see the release of many of these compounds for human benefit. We now consider a few of these major products.

8.13 Production of Mammalian Products and Vaccines by Genetically Engineered Microorganisms

One of the first practical applications of genetic engineering was the use of easily grown bacteria to produce proteins whose genes were from organisms that are more difficult or expensive to grow. Although the special DNA polymerases used in the polymerase chain reaction were originally isolated from thermophilic bacteria, they are now produced in *Escherichia coli* from cloned genes (see Section 8.9). Most restriction enzymes are also produced in *E. coli* from cloned genes. Similarly, many proteins used industrially are now produced from cloned genes, and in some cases, the protein itself has been altered by using site-specific mutagenesis (see Section 8.11) to change the cloned gene. This is true of some of the enzymes used in laundry detergents, whose genes were altered systematically to obtain a protein that was less sensitive to denaturation by bleach.

Many proteins and peptides from mammalian cells have high pharmaceutical value. However, these proteins are usually present in very small amounts in normal tissue, and it is therefore extremely costly to purify them. Another of the first efforts of the biotechnology industry was to use genetic engineering to produce these proteins in microorganisms. A few examples of these proteins and their uses are given in Table 8.1. This section discusses how this was accomplished in the case of the hormone *insulin*. We also discuss how genetic engineering has been used to facilitate the production of virus vaccines.

Hormones

Many hormones are peptides or small proteins, and several hormones are listed in Table 8.1. These molecules are extremely important in controlling mammalian metabolism and have important therapeutic uses. One of the most dramatic examples of the value of genetic engineering is the production of **human insulin**. Genetically engineered human insulin was the first "biotech drug" to hit the market. Insulin is a protein produced in the pancreas that is vital for the regulation of carbohydrate metabolism in the body. Diabetes, a disease characterized by insulin deficiency, afflicts millions of people. The standard treatment for diabetes is periodic injections or oral administration of insulin, and because insulins of most mammals are similar in structure, it is possible to treat human diabetes by use of insulin isolated commercially from beef or pork pancreas. However, nonhuman insulin is not as effective as human insulin, and the isolation

Table 8.1 Major biotechnological products made by recombinant DNA techniques*

Product	Function
Blood Proteins	
Tissue plasminogen activator	Dissolves clots
Factors VII, VIII, IX	Promote clotting
Erythropoietin	Red cell growth
Urokinase	Blood clotting
Hormones	
Insulin	Treatment of diabetes
Human growth hormone	Treat dwarfism
Epidermal growth factor	Wound healing
Parathyroid hormone	Calcium regulation
β-Endorphin	Pain relief
Bone growth factor	Osteoporosis
Atriopeptin	Diuretic, antihypertensive
Immune modulators	
Interferons	Antiviral and potential anticancer agents
Interleukin-2	T-cell stimulator
Tumor necrosis factor	Antitumor agent
Colony stimulating factor	Treatment of infections and cancer
Lysozyme	Anti-inflammatory
Vaccines	
Hepatitis B	Prevention of serum hepatitis
Cytomegalovirus	Prevention of infection
Measles	Prevention of measles
Cholera	Prevention of cholera
AIDS	Prevention of AIDS
Rabies	Prevention of rabies

Although research is currently progressing in all of these areas, not all of the listed products are yet on the market.

process is expensive and complex. Cloning of a human insulin "gene" in bacteria has hence been carried out.

Producing hormones such as insulin in genetically engineered microorganisms is not simply a matter of cloning a gene (or cDNA, see Section 8.10) in an expression vector. This is because many of these hormones are only small fragments of the polypeptide encoded by the gene. Insulin in its active form consists of two polypeptides (A and B) connected by disulfide bridges (Figure 8.16a). These two polypeptides are coded by separate parts of a single insulin gene. The insulin gene codes for *preproinsulin*, a longer polypeptide containing a signal sequence (involved in excretion of the protein), the A and B polypeptides of the active insulin molecule, and a connecting polypeptide that is absent from mature insulin. *Proinsulin* is formed from preproinsulin and the conversion of proinsulin to insulin involves the enzymatic cleavage

of the connecting polypeptide from the A and B chains.

Two approaches have been used to obtain production of human insulin in bacteria: (1) production of proinsulin and conversion to insulin by chemical cleavage, and (2) production in two separate bacterial cultures of the A and B chains, and joining of the two chains chemically to produce insulin. Because the insulin protein is fairly small, it was more convenient with either approach to synthesize the proper DNA sequence chemically rather than to attempt to isolate the insulin gene from human tissue. There are 63 bases coding for the A chain and 90 bases coding for the B chain (Figure 8.16b). In proinsulin there are an additional 105 bases for the peptide which connects the A and B chains. When the polynucleotides were synthesized, suitable restriction enzyme sites were placed at each end so that the polynucleotides could be ligated to plasmid pBR322. To obtain effective expression, the synthesized genes were inserted next to a suitable *E. coli* promoter, either *lac* or *trp*, but in a manner such that the insulin fragment was synthesized as part of the β-galactosidase or tryptophan synthetase protein as a *fusion protein*. An important advantage of making the fusion protein is that the fusion product is much more stable in *E. coli* than insulin itself. The *trp* fusion in particular results in the formation of an insoluble protein that precipitates inside *E. coli*, thus preventing *E. coli* proteases from breaking it down. Finally, a nucleotide triplet coding for methionine was placed at the region joining the *trp* or *lac* gene to the insulin gene. The reason for this is that the chemical reagent *cyanogen bromide* specifically cleaves polypeptide chains at methionine residues, permitting recovery of the insulin product once the fused protein has been isolated from the bacteria. Insulin itself does not contain methionine and hence is unaffected by cyanogen bromide treatment.

If the proinsulin route is used, then the proinsulin isolated from the bacteria via cyanogen bromide treatment is converted to insulin by disulfide bond formation, followed by enzymatic removal of the connecting peptide of proinsulin. Proinsulin naturally folds so that the cysteine residues are opposite each other (Figure 8.16a), and chemical treatment then causes the formation of disulfide cross links. Once this has been accomplished, the connecting peptide can be removed by treatment with the proteases trypsin and carboxypeptidase B, which have no effect on insulin itself.

If insulin is produced by way of the separate A and B peptides, then each of the fusion proteins is isolated from a separate bacterial culture and the chains then released by cyanogen bromide cleavage. The cleaved chains are then connected by use of chemical treatment that results in disulfide bond formation. Under appropriate conditions, a yield of at least 60 percent of the theoretical yield can be obtained.

The final product, biosynthetic human insulin, is identical in all respects to insulin purified from the

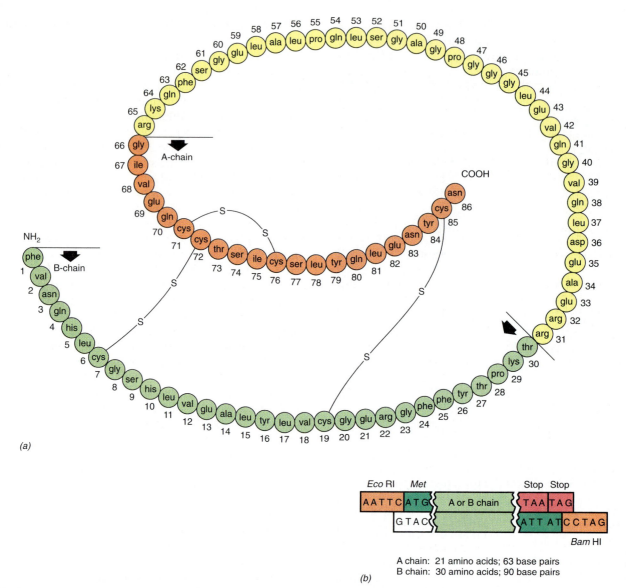

(a)

Eco RI   Met   A or B chain   Stop Stop

```
AATTCATG    A or B chain    TAATAG
    GTAC                    ATTATCCTAG
```
                                    Bam HI

A chain: 21 amino acids; 63 base pairs
B chain: 30 amino acids; 90 base pairs

(b)

FIGURE 8.16 Genetic engineering for the production of human insulin in bacteria. (a) Structure of human proinsulin. The peptide shown in yellow must be removed from between the A-chain and B-chain in order to make insulin. (b) Chemical synthesis of the insulin gene and suitable linkers, permitting cloning and expression. The synthesized fragments were linked via restriction sites *Eco*RI and *Bam*HI to plasmid pBR322. The methionine coding sequence was inserted to permit chemical cleavage of the A and B chains from the fused protein made in the bacteria, since the reagent cyanogen bromide specifically cleaves at methionine residues and insulin does not contain methionine. Two stop signals were incorporated at the downstream (carboxyl) end of the coding sequence.

human pancreas and is being marketed commercially.

The production of human insulin led the way in the biotechnological production of hormones. Microbially produced human insulin is cheaper to make and just as effective as porcine or bovine insulins, the major sources of insulin for diabetics before the advent of biotechnology. Other important hormones produced by recombinant microorganisms include **human growth hormone** for treating dwarfism, **epidermal growth factor** for stimulating wound healing, **bone growth factors** for treating osteoporosis, and **animal growth hormone** for stimulating growth of livestock animals in order to reduce feed costs and get the animals to market sooner.

Virus vaccines

Vaccines are suspensions of killed or modified pathogenic microorganisms or specific fractions isolated from the microorganisms that when injected into an animal produce immunity to a particular disease. As we discussed in Chapter 6, a virus particle consists of a nucleic acid core surrounded by a coat containing one or more specific proteins. For vaccine purposes, it is the protein coat that is of interest because it generally elicits the greatest antibody production (see Section 12.15), and thus the genes for the virus coat protein are cloned. The importance of the recombinant vaccines, of course, is the fact that they can replace the killed or inactivated viral suspensions.

Genetic engineering has proved successful in the development of some *subunit vaccines*. These vaccines contain only a specific *subunit* of a protein from the pathogenic organism (usually a coat protein), and recombinant DNA techniques are used to produce these subunits in microorganisms. The highly immunogenic coat proteins are purified and used in high dosage to elicit a rapid and high level of immunity with no possibility of transmitting infection. The steps for viral gene cloning are those outlined in the previous sections: fragmentation of viral DNA by restriction enzymes; cloning viral coat protein genes into a suitable vector; providing for proper promoters, reading frame, and ribosome-binding sites; and reinsertion and expression of the viral genes in a microorganism. Unfortunately, when *E. coli* is used as the cloning host, the vaccines are often poorly immunogenic and fail to protect animals from subsequent infection with the virus. The problem involves the fact that many key viral coat proteins are posttranslationally modified, generally by the addition of sugar residues (glycosylation), when the virus replicates in the host. However, the recombinant proteins produced by *E. coli* or other Bacteria are unglycosylated, and apparently glycosylation is necessary for the proteins to be immunologically active. Therefore, a eukaryotic host is used.

The first recombinant subunit vaccine approved for use in humans was made using yeast. The gene encoding a surface protein from Hepatitis B virus was cloned into a high copy yeast expression vector. The protein was produced and formed aggregates very similar to those found in patients infected with the virus. These aggregates were purified and used to vaccinate people against Hepatitis B virus. Subunit vaccines to a large variety of viruses (Table 8.2) and pathogenic organisms are also being developed using genetic engineering. In addition to using yeast, insect cells and even cultured mammalian cells are being used as hosts. To obtain the correct pattern of glycosylation or other modifications of the protein, it is often important to use a host that is closely related to humans.

Many laboratories are working on subunit vaccines to the virus that causes AIDS (human immunodeficiency virus, or HIV). Because of the serious consequences associated with AIDS, a safe, effective AIDS vaccine would find a huge worldwide market and thus the stakes in this area are high. The HIV virus has

a number of different coat or envelope proteins, and genes for these have been cloned, or constructed, and have been expressed using a variety of vectors in a variety of different hosts. One of the vectors used for cloning into mammalian cells is the *vaccinia virus*.

Cloning in vaccinia virus is done using an *E. coli* plasmid containing a fragment of the vaccinia virus thymidine kinase gene (Figure 8.17a). An appropriate foreign DNA is inserted into this plasmid, and the recombinant plasmid is mixed with wild-type vaccinia virus DNA (Figure 8.17b). If homologous recombination occurs between plasmid DNA and vaccinia genomic DNA (Figure 8.17c), recombinant virions can be obtained which contain only part of the thymidine kinase gene. This results in an *inactive* thymidine kinase in recombinant vaccinia virions, and because thymidine kinase is inhibited by the compound 5-bromodeoxyuridine, recombinant vaccinia virions can be selected by allowing viral replication to occur in the presence of this inhibitor (Figure 8.17d). Although such recombinant viruses no longer express thymidine kinase, they can still infect human cells and they do express the foreign genes that have been cloned into them. Indeed, some recombinant vaccinia viruses can carry genes from four different viruses!

Vaccinia virus itself is generally not pathogenic for humans (vaccinia virus was originally used as a vaccine against the related virus smallpox). However, vaccinia virus is not completely benign (it causes severe complications in some people), and therefore more research must be done before such vaccines can be used in humans.

Genetically engineered vaccines are likely to become increasingly common because (1) they are safer than normal attenuated or killed virus vaccines, (2) they are more reproducible because their genetic makeup can be carefully monitored, and (3) they can be administered in high doses without fear of side effects.

Other proteins and other products

Table 8.1 lists a number of mammalian proteins other than hormones and vaccines. A number of these are proteins involved in blood clotting and other blood processes. Chief among these are tissue plasminogen activator and clotting factors VII, VIII, and IX. **Tissue plasminogen activator (TPA)** is a protein found in the blood that acts to scavenge and dissolve old blood clots in the final stages of the healing process. The clinical usefulness of TPA is primarily with heart patients or anyone suffering from poor circulation because of excessive clotting tendencies. TPA can be administered following cardiac bypass, transplant, or other open heart surgeries to prevent the development of pulmonary embolisms which are often life threatening. Heart disease is a leading cause of death in many developed countries, so microbially produced TPA promises to be in high demand.

In contrast to TPA, blood clotting factors VII, VIII, and IX are critically important for the *formation* of

---

Table 8.2   Viral proteins that have been expressed in recombinant vaccines
Hepatitis B virus surface antigen
Rabies virus glycoprotein G
Herpes simplex virus glycoprotein D
Epstein–Barr virus glycoprotein
Influenza virus hemagglutinin
Vesicular stomatitis virus nucleoprotein
Human respiratory syncytial virus glycoprotein G
Human immunodeficiency virus (HIV) envelope protein

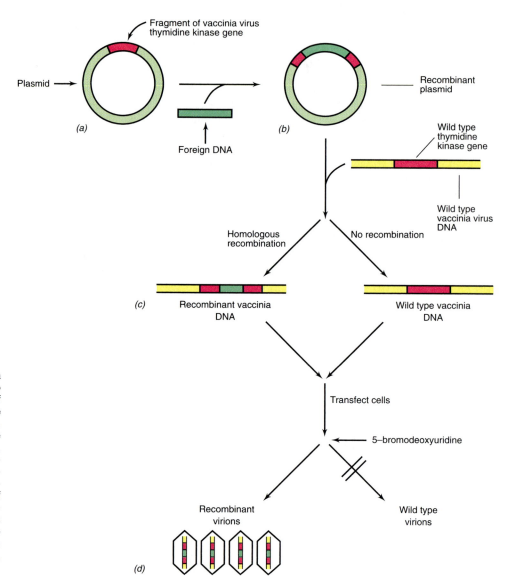

FIGURE 8.17
Production of recombinant vaccinia virus. (a) Foreign DNA is cloned into a plasmid that contains a small piece of the vaccinia virus thymidine kinase gene. (b) Recombinant plasmid formed. (c) The latter is mixed with wild-type vaccinia DNA. If recombination occurs, recombinant vaccinia DNA can be produced. (d) The mixture is used to transfect host cells and viral replication is allowed to occur in the presence of 5-bromodeoxyuridine, a thymidine kinase inhibitor. Only recombinant virions develop under these conditions. If the recombinant vaccinia virions contain genes for other viral coat proteins, these may be expressed.

blood clots. Hemophiliacs suffer from a deficiency of one or more clotting factors and can be readily treated with the microbially produced product. Recombinant clotting factors take on added significance when one considers that hemophiliacs have in the past been treated with concentrated clotting factor extracts from pooled human blood, some of which was contaminated with the AIDS virus: this has put hemophiliacs at high risk for contracting AIDS. A final blood protein of high biotechnological interest is **erythropoietin**, a protein that stimulates red blood cell production. In clinical trials erythropoietin has shown considerable promise for treating anemia.

A number of proteins have roles as anticancer agents or immune modulators. **Interferons** are a series of proteins made by animal cells in response to viral infection (see Section 11.17) or immune activation in the case of one type of interferon. *Alpha interferon* in particular may be useful as an anticancer agent. Treatment of some tumor cells with alpha interferon causes the tumor cells to express specific tumor-associated antigens. This phenomenon is of great benefit in can-

cer therapy because following treatment with alpha interferon, monoclonal antibodies directed against tumor antigens can be used as vehicles for delivering toxic drugs specifically to tumor cells. **Interleukin-2** is a protein that stimulates proliferation and differentiation of T lymphocytes. In combination with two additional proteins, **tumor necrosis factor** and **granulocyte macrophage colony stimulating factor**, interleukin-2 has shown some potential for treating certain forms of cancer through stimulation of the body's own immune system to attack tumor cells.

Even monoclonal antibodies are now produced in microorganisms by genetic engineering (we discuss monoclonal antibodies in some detail in Section 12.11).

The first generation of products made by genetic engineering were primarily the protein products of cloned genes, and there is still a great deal to be done with this approach. Added applications come from being able to use site-directed mutagenesis on the cloned gene so that new products with new attributes can be generated. It must also be remembered that molecules such as antibiotics are synthesized in cells in biochemical pathways using a series of enzymes (proteins). These enzymes can be modified so that new antibiotics can be developed.

> The first human protein produced commercially using engineered bacteria was human insulin, but numerous other hormones and other human proteins are now being produced. Many proteins found in the human that were formerly extremely expensive to produce because they were found in human tissues in only small amounts can now be produced in very large amounts from the cloned gene in a suitable expression system. In addition to useful pharmaceuticals such as anticancer agents and immune modulators, even vaccines can be produced using genetic engineering.

8.14 Genetic Engineering in Plant Agriculture

Classical genetic improvement of plants has generally been a slow and difficult task, but recombinant DNA technology promises revolutionary changes. It is possible to use plant tissue culture techniques to select clones of plant cells that have been genetically altered, and then, with proper treatments, induce these cell cultures to make whole plants which can be propagated vegetatively or by seeds. Recombinant DNA technology enters into this approach because one can transform plant cells with free DNA by either electroporation or particle gun methods (see Section 7.6) or insert foreign DNA into plant cells directly via the bacterium *Agrobacterium tumefaciens*. This organism causes the plant disease *crown gall* by transferring specific genes to the plant (see Section 17.23), and plant genetic engineers have used this natural transformation system as a vehicle for the introduction of foreign DNA into plants.

Vectors for cloning in plants

The Gram-negative plant pathogen *A. tumefaciens* contains a large plasmid called the *Ti plasmid* that is responsible for its virulence. The plasmid contains genes which mobilize DNA for transfer to the plant (for details of the disease process and genetic events, see Section 17.23). The segment of the Ti plasmid DNA that is actually transferred to the plant is called *T-DNA*. The sequences at the ends of the T-DNA are essential for transfer, and the DNA to be transferred must be between these ends. One type of vector that has been constructed and is used for the actual transfer of genes to plants is called a *binary vector*. The word *binary* means consisting of two parts, and a binary vector must always be used in conjunction with another plasmid for the cloned gene to be transferred to the plant. The vector contains the two ends of the T-DNA on either side of the site used for cloning and an antibiotic resistance marker that can be used in plants. It also contains an origin of replication so that it can replicate in both *A. tumefaciens* and in *Escherichia coli* (the latter serves as host for cloning work), and another antibi-

otic resistance marker that is expressed in bacteria (Figure 8.18). The DNA to be cloned is inserted into the vector that is then transformed into *E. coli*. It is then transferred to *A. tumefaciens* (usually by conjugation).

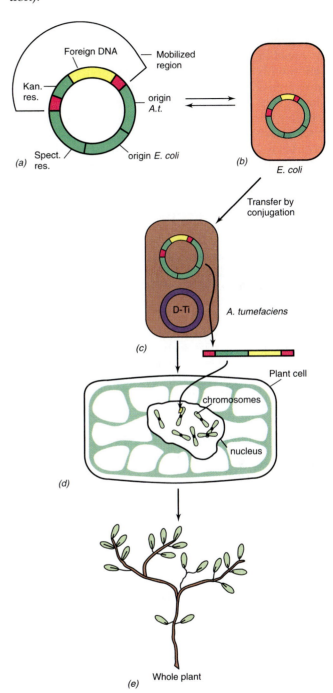

FIGURE 8.18 Production of transgenic plants using *Agrobacterium tumefaciens*. (a) Generalized plant transfection vector containing ends of T-DNA (in red), foreign DNA (in yellow), origin of replication elements for both *E. coli* and *A. tumefaciens*, and spectinomycin and kanamycin resistance markers. The kanamycin resistance marker can be selected for in plants. (b) The vector can be put into *E. coli* for cloning purposes and then transferred to *A. tumefaciens* by conjugation. (c) The resident Ti plasmid used for transferring the vector to the plant (D-Ti) is itself genetically engineered to remove key pathogenesis genes. (d) However, D-Ti can mobilize the T-DNA region of the vector for transfer to plant cells grown in tissue culture. From the recombinant cell, whole plants can be regenerated.

Final transfer to the plant depends on the *A. tumefaciens* containing a Ti plasmid since the vector does not contain the genes required for T-DNA transfer. The cloned DNA and the kanamycin resistance marker of the vector can be mobilized by the Ti plasmid and transferred into a plant cell (Figure 8.18; note here that the Ti plasmid used for this purpose has been genetically altered in such a way as to prevent the onset of disease in the plant). Following recombination with host chromosomes, the foreign DNA can be expressed to confer new properties on the plant. Many genes are not expressed efficiently in plants unless they are cloned into an expression vector that contains a plant promoter. Some of the promoters that have been used in constructing plant expression vectors include some normally found in T-DNA and a promoter from cauliflower mosaic virus, a plant DNA virus (see Table 6.1).

Using *A. tumefaciens*, a number of transgenic plants have been produced. Most successes have come with herbaceous plants (dicots) such as tomato, potato, tobacco, soybean, alfalfa, cotton, and lettuce, but *A. tumefaciens* has also been used to produce transgenic woody dicots such as poplar, walnut, and apple. Transgenic crop plants from the grass family (monocots) are difficult to generate using *A. tumefaciens*, but other methods of introducing DNA, such as electroporation or the microprojectile gun (see Section 7.6), seem to work well here. In addition, some plant vectors have been developed by modifying plant viruses, but these have not yet proved as useful as the vectors derived from the Ti plasmid.

Applications in plant biotechnology

Major areas targeted for genetic improvement in plants include herbicide, insect, and microbial disease resistance. Herbicide resistance can be obtained by genetically engineering the crop plant to no longer respond to the toxic chemical. Many herbicides act by inhibiting a key plant enzyme or protein necessary for growth. For example, the herbicide *glyphosate* kills plants by inhibiting the activity of an enzyme necessary for making aromatic amino acids. Such an herbicide kills both weeds and crop plants of course, and thus must be used as a "preemergence herbicide," that is, before the crop plants emerge from the ground. However, glyphosate-resistant strains of the Gramnegative bacterium *Salmonella* (glyphosate inhibits the bacterial enzyme as well) have been obtained using standard genetic techniques and the glyphosate resistance genes have been transferred to plants, using the Ti plasmid as a vehicle. When sprayed with glyphosate, plants containing the bacterial gene grow as well as unsprayed control plants.

Novel means of insect resistance have been genetically introduced into plants. One of the most promising has been the toxic protein genes of *Bacillus thuringiensis*. This organism produces a crystalline protein (see Section 17.22), called *Bt-toxin,* that is toxic to moth and butterfly larvae, and certain strains of *B. thuringiensis* produce additional proteins toxic to beetle and fly larvae and mosquitoes. Biotechnologists are using several different approaches to enhance the use of Bt-toxin for pest control in plants.

One approach is to develop a single Bt-toxin that is effective against many different insects. This can be done because the protein has separate domains for its specificity and its toxic function. The toxic domain is highly conserved in all of the various Bt-toxins. Genetic engineers are attempting to make a gene that will encode a Bt-toxin carrying one toxic domain and several different specificity domains. Such a toxin could be applied directly to a number of different plants. Another approach is to transfer the gene to bacteria that normally live (harmlessly) in plant tissue and have the toxin produced by these bacteria in these tissues. An even more effective approach may be to transfer the gene directly into the plant genome. This has already been done with agricultural plants such as cotton, potatoes, and tomatoes, and these transgenic plants have shown increased resistance to the caterpillar pests that feed on the control plants.

However, there have been reports of insects that have gained resistance to Bt-toxin. Resistance to insecticides and herbicides is a common problem in agriculture, and the fact that a product has been produced by genetic engineering does not give it any magical properties. This emphasizes that many approaches must be used for pest control, and the Bt-toxin is only one of many that is being developed by biotechnologists.

Genetic engineering has also been used to protect plants from virus infection. For example, it has been discovered that transgenic plants that express the coat protein gene of a virus become resistant to infection by that virus. Although the mechanism of resistance is unknown, the presence of viral coat protein in plant cells apparently interferes with the uncoating of viral particles containing that coat protein and this interrupts the virus replication cycle.

Not all genetic engineering is directed toward making plants disease resistant. Genetic engineering can be used in a variety of ways for developing mutant strains of plants with desired characteristics such as delayed spoilage. In addition, transgenic plants can be genetically engineered to produce commercial or pharmaceutical products, as has been done with microorganisms (see Section 8.13) and animals (see Section 8.15). Crop plants such as tobacco and tomatoes have been engineered to produce a number of different products, such as the human protein interferon and the bacterial protein α-amylase, which is used in the food processing industry to degrade starch (see Section 10.11). Transgenic plants can be used to produce animal antibodies in quantity (such plant-made antibodies are sometimes called "plantibodies") and even granules of a polyester used to manufacture plastics.

Plant hosts can be useful for producing these types of products because plants typically modify proteins correctly and because crop plants can be efficiently grown and harvested. The future of plant biotechnology is exciting. Besides the topics discussed above, improvements in plant drought and salinity resistance are on the horizon. In addition, cloned gene technology has given plant biologists powerful tools for studying gene expression in plants.

8.15   Genetic Engineering in Animal and Human Genetics

This section covers only a few highlights of some of the huge number of uses of genetic engineering in animal and human genetics. Some of these applications have to do with producing products, but more have to do with understanding gene function in mammals and in curing or treating genetic disease.

Transgenic animals

Using recombinant DNA technology and microinjection techniques to deliver cloned genes to fertilized eggs, several foreign genes have been expressed in both laboratory research animals and in species important in commercial animal industries. Transgenic animals have become increasingly important in basic biomedical research for studying gene regulation and developmental biology. However, many applied aspects of transgenic animals are also of interest.

One approach is to improve the productivity or disease resistance of the animal, as in the case of agricultural plants. However, transgenic animals are also being used to produce proteins of pharmaceutical value—a process some scientists have called "pharming." Transgenic animals may be useful for producing human proteins that require posttranslational modifications for activity, such as certain blood clotting enzymes; many proteins of this type are not produced in an active form by microorganisms. Also, some proteins have been engineered to be secreted into the animals' milk, which can be readily collected and processed. These proteins include α-1-antitrypsin (used to treat lung disease) produced in sheep and tissue plasminogen activator (used to dissolve blood clots) produced in goats. Transgenic pigs have been developed that produce large quantities of human hemoglobin. This hemoglobin can easily be isolated and processed for use as a blood substitute. The production of transgenic animals for research and commercial purposes seems likely to continue as an important area in biotechnology. Microinjection techniques or SV40 virus or retroviruses have now been used as vehicles to deliver foreign DNA to embryos, and transgenic chickens, cows, fish, pigs, rabbits, and sheep have now been produced.

Human genetics

Conventional genetics, involving genetic crosses or mutagenesis, cannot be done with humans. Therefore, in spite of the obvious interest, our understanding of human genetics has lagged considerably behind our understanding of the genetics of many other organisms. The advent of genetic engineering has quite simply revolutionized studies of the human genome. A detailed discussion of human genetics is beyond the scope of this book, but a few general remarks on the utility of recombinant DNA technology can be made. We have already mentioned the use of PCR for DNA fingerprinting (see the DNA Fingerprinting box), and considerable efforts are being made to clone and sequence the entire human genome (see Section 8.4).

However, there are some applications of genetic engineering of the human genome which are directed at treating human disease. A vast number of genetic diseases are known, but except in rare cases, little was known until recently about their molecular bases. By use of recombinant DNA technology, coupled with conventional genetic studies (following family inheritance, etc.), it is possible to localize particular defects to particular chromosomes and to particular locations on chromosomes. With the use of recombinant DNA technology, it is possible to clone the region containing the genetic defect and then to make comparisons between the base sequence in the normal gene and in genetically altered chromosomes. From such studies, even in the absence of knowledge of the enzyme defect, it has been possible to obtain information about the genetic change. Many genes, including those for Huntington's disease, cystic fibrosis, and Duchennes' muscular dystrophy, have been localized using these techniques.

Genetic engineering is being used to provide treatments for some of these diseases using *gene therapy*. In gene therapy, a nonfunctional or dysfunctional gene is augmented or replaced by a functional gene. Not all gene therapy is being designed toward treating genetic disease; a considerable effort has been directed toward protocols for treating cancer. Major obstacles to this approach exist in trying to target the correct cells for gene therapy and in successfully transfecting cell lines that will perpetuate the genetic alteration.

The first genetic disease for which an approved gene therapy technique was used is a severe combined immune deficiency caused by the absence of adenosine deaminase (ADA, a enzyme involved in purine metabolism) in bone marrow cells. The gene therapy involves using a retrovirus as a vector to insert a good copy of the ADA gene into T lymphocytes (cells that are part of the immune system; see Section 12.2) removed from the patient and then placing these "corrected" cells back in the body (the retrovirus also carries a marker gene, resistance to neomycin, so that cells carrying the inserted retrovirus can be selected and identified). Since T lymphocytes have a limited life span, it is necessary to repeat the therapy every month or two. Attempts are being made to insert the gene into the stem cells of the bone marrow (which continue to divide) and effect a true cure for the disease.

Several other gene therapy treatments, some using other virus vectors, are currently being tested. It is important to note that in the protocols being tested, the defective gene is not replaced. Rather, the retrovirus (and the good copy of the gene) simply integrates somewhere in the human genome of these cells. Actual gene replacements in germ line cells (cells that give rise to gametes) can be accomplished with some mammals, although the techniques of isolating individual animals with these changes cannot readily be applied to humans. However, attempts to change the germ cells of an individual would raise many ethical and societal questions beyond simple questions of experimental protocols.

> Genetic engineering is providing vast opportunities for improving processes of economic value and in treating disease. Medicine, agriculture, and industry are all finding important uses of cloning and gene expression technology.

8.16 Genetic Engineering as a Microbial Research Tool

Up until now we have mostly presented the principles of genetic engineering in the light of practical goals and processes. However, genetic engineering technology finds many uses in basic research on microorganisms. Gene cloning and the engineering of new microbial strains provides some of the best ways in which to understand basic microbial processes such as structure-function relationships, cell growth, enzyme regulation, and microbial ecology. A novel microbial strain can be created which differs from the wild type in a defined manner, and the underlying process can then be studied. In this way, one can observe the importance of a particular gene product for a basic microbial process under precisely controlled conditions. Gene cloning, restriction mapping, and sequencing also allow geneticists to quickly map and study the genomes of newly identified organisms.

We have discussed how cloned genes can be subjected to site-specific mutagenesis and how gene disruptions or knockout mutations can be introduced into a gene. These mutated genes can then be introduced into the microbial chromosome so that the mutant organism can be studied. Genetic engineering also allows the researcher to "tag" a gene so that it is easier to study. For instance, if the gene product is difficult to assay or has no known function, it is often very difficult to know under what conditions the cell makes this protein. One can also use this technique to tag the organism itself so that it can be traced in the environment.

In these cases, a gene or its promoter can be fused to a *reporter gene*. Reporter genes are simple to assay, and the gene encoding β-galactosidase (whose activity can be detected on indicator plates and also assayed easily by biochemical means) is commonly used as a reporter. However, there are other choices. For instance, *Escherichia coli* has been engineered in such a way that it produces the luciferase enzyme of the bacterium *Photobacterium* or that of several different luminescent beetles (Figure 8.19). Because of the presence

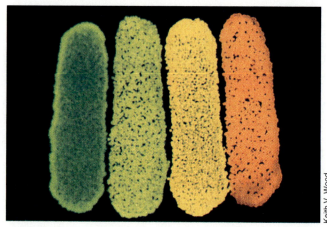

FIGURE 8.19 Bioluminescence from colony streaks of four strains of *E. coli*, each containing cloned luciferase genes from a different species of click beetle. Each luciferase emits light of a different color.

Keith V. Wood

of this enzyme, the engineered *E. coli* becomes luminescent, and its colonies therefore glow in the dark. One can then detect colonies of the engineered *E. coli* on agar plates by their luminescence among a large background of other colonies. This glow-in-the-dark system can be used in a wide variety of genetically engineered microorganisms.

Another approach to tracking organisms through the environment has been to introduce the β-galactosidase gene and use indicator agar plates to search for colonies producing this enzyme. An example of this is the introduction of the gene encoding β-galactosidase into the bacterium *Pseudomonas fluorescens*, a soil organism normally unable to catabolize lactose, to serve as a marker for tracking genetically engineered strains of this organism released into the environment.

8.17 Summary of Principles at the Basis of Genetic Engineering

We have presented the fundamentals of genetic engineering and have shown how the approaches used have been derived from an understanding of basic concepts of molecular genetics. We now summarize the principles of genetic engineering by relating current knowledge back to the basic information presented in Chapters 5 through 7.

The following developments were essential for the development of genetic engineering and their interrelationships are diagrammed in Figure 8.20:

1. **DNA chemistry**: development of procedures for isolation, sequencing, and synthesis of DNA.
2. **DNA enzymology**: discovery of restriction endonucleases, DNA ligases, and DNA polymerases.
3. **DNA replication**: understanding how DNA replication occurs, and the importance of DNA vectors capable of independent replication.

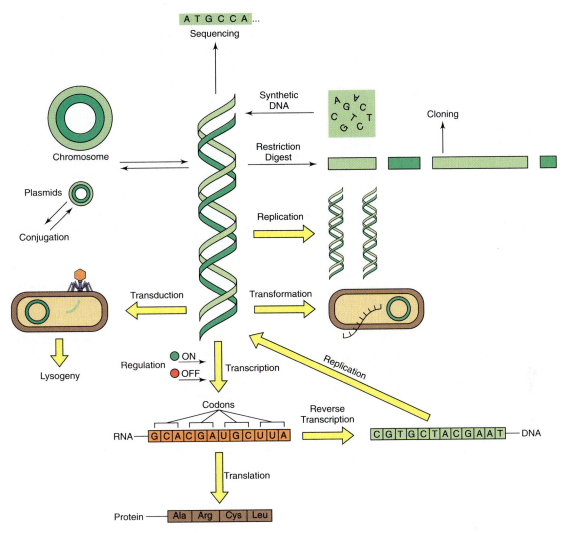

FIGURE 8.20 Summary of the fundamental processes underlying genetic engineering.

4. **Plasmids**: discovery of plasmids, and determination of the mechanisms by which plasmids replicate.

5. **Temperate bacteriophage**: understanding how replication and/or integration is controlled in the DNA of temperate bacteriophages.

6. **Transformation**: discovery of methods for getting free DNA into cells.

7. **RNA chemistry and enzymology**: understanding how to work with messenger RNA, how eukaryotic mRNA is constructed, and the importance of RNA processing in the formation of mature eukaryotic mRNA. Development of methods to synthesize RNA.

8. **Reverse transcription**: the discovery of the enzyme *reverse transcriptase* in retroviruses and its development as a means for transcribing information from mRNA back into DNA.

9. **Regulation**: understanding the factors involved in the regulation of transcription, including the discovery of promoter sites and operon control.

10. **Translation**: understanding the steps involved in translation, the importance of ribosome-binding sites on the mRNA, the role of the initiation codon, and the importance of a proper reading frame.

11. **Protein chemistry**: development of methods for isolation, purification, assay, and sequencing of proteins.

12. **Protein excretion and posttranslational modification**: understanding how proteins are built with signal sequences that are removed during or after excretion. Discovery of other kinds of posttranslational modification of proteins, such as the removal of polypeptides at the initiation end of the protein.

13. **The genetic code**: the discovery of the genetic code and the determination that it was the same in almost all organisms. The understanding of the importance of proper reading frame and that certain codons were less frequently used in some organisms than in others.

Study Questions

1. What are the characteristics of plasmids that make them especially useful for gene cloning? Why aren't all plasmids equally useful for cloning?

2. Most of the lambda derivatives that are used as cloning vectors are virulent. Based on your knowledge of lambda genetics (see Section 6.12) and the structure of the lambda Charon vectors, discuss why these viruses are no longer temperate. Why might it be useful to also have a temperate bacteriophage as a cloning vector?

3. What are the essential features of a cloning vector? Expression vector? Why aren't all cloning vectors also expression vectors?

4. If *insertional inactivation* is used to detect the presence of an introduced plasmid in a bacterial cell, why is it desirable to have *two* antibiotic-resistance markers in the plasmid?

5. If a plasmid is used as a cloning vector, why is it essential that only a few sites be present that are recognized by the restriction enzyme used for the cloning process? What would be the result if the plasmid contained a large number of recognition sites for this restriction enzyme?

6. Suppose you were given the task of constructing a plasmid suitable for gene cloning in an organism of industrial interest. List the characteristics which such a plasmid should have. List the steps you would use to develop such a plasmid.

7. When using the plasmid vector pUC18, cloning DNA fragments made using the restriction enzyme *Bam*HI inactivate the gene encoding β-galactosidase. When using pBR322, the same fragments inactivate the gene conferring tetracycline resistance. Both vectors contain a gene for ampicillin resistance which is used to select for bacterial transformants after making the recombinant molecules. Explain why it is much more efficient to use pUC18 rather than pBR322 as a cloning vector. (Hint: The increased efficiency has to do with finding the cells that contain vectors with cloned DNA.)

8. Suppose you were working with an organism that was intractable to genetic transformation with free DNA. How might you still be able to use this organism as a cloning host?

9. Suppose you have just determined the DNA base sequence for an especially strong promoter in *E. coli* and you are interested in incorporating this sequence into an expression vector. Describe the steps you would use. What precautions would be necessary to be sure that this promoter actually worked as expected in its new location?

10. Explain why the use of a regulatory switch is desirable for the large-scale production of a protein.

11. Suppose you are interested in using Southern blot analysis to detect the presence of nitrogenase genes (*nif*) in a newly isolated organism. You have cloned the *nif* genes from a known nitrogen-fixing organism and can obtain large amounts of *nif* DNA in radioactive form. How would you use Southern blot analysis to detect *nif* genes in your new organism? Could you conclude that if *nif* were present in your new organism, that this organism was capable of nitrogen fixation? Why?

12. Compare and contrast reverse transcription and reverse translation. Which of these processes would you not expect to occur in a mammalian organism?

13. You have just discovered a fantastic protein in human blood which is an effective cure for cancer, but it is only present in extremely small amounts. Describe the steps you would use to obtain production of this protein in *E. coli*.

14. How has bacteriophage T7 been used for expressing foreign genes in *E. coli* and what desirable features does this regulatory system possess?

15. Describe the basic principles of gene amplification using the polymerase chain reaction (PCR). How have thermophilic bacteria simplified the use of PCR? How many PCR cycles would need to be run to get 1000 copies of a specific target sequence (starting from one molecule of double-stranded DNA)?

16. What is a subunit vaccine and why are subunit vaccines considered a safer way of conferring immunity to viral pathogens than attenuated virus vaccines?

17. What is the Ti plasmid and how has it been of use in genetic engineering?

18. How do transgenic plants and animals differ from plants and animals modified by conventional breeding techniques?

19. What advantages might there be in using a transgenic plant rather than a transgenic animal to produce a protein?

Supplementary Readings

Drlica, K. 1992. *Understanding DNA and Gene Cloning: A Guide for the Curious,* 2nd edition. John Wiley & Sons, New York. A simple, introductory guide to genetic engineering.

Innis, M. A., D. H. Gelfand, J. J. Sninsky, and **T. J. White**. 1990. *PCR Protocols: A Guide to Methods and Applications.* Academic Press, New York. A laboratory manual of methods and practical applications of the polymerase chain reaction.

Primrose, S. B. 1991. *Molecular Biotechnology,* 2nd edition. Blackwell Scientific Publications, Oxford, England. A concise and well-illustrated overview of genetic engineering and biotechnology.

Sambrook, J., E. F. Fritsch, and **T. Maniatis**. 1989. *Molecular Cloning: A Laboratory Manual,* 2nd edition. Cold Spring Harbor Laboratory, Cold Spring Harbor, N.Y. A detailed manual in three volumes describing many procedures for cloning and obtaining expression of cloned genes. Brief discussions of the principles behind each procedure.

Watson, J. D., M. Gilman, J. Witkowski, and **M. Zoller**. 1992. *Recombinant DNA,* 2nd edition. Scientific American Books, distributed by W. H. Freeman and Co., New York. An excellent introductory text devoted to the principles and applications of genetic engineering.

9

Growth and Its Control

We have thus far discussed the molecular biology and biochemistry of the microorganism, and now we place these concepts in the broader context of how a cell grows. In microbiology, **growth** is defined as *an increase in the number of cells*. Growth is an essential component of microbial function, as any given cell has only a finite life span in nature, and the species is maintained only as a result of continued growth of the population. In most practical situations involving microorganisms, we are also concerned with growth. Control of microbial action requires a knowledge of microbial growth.

In the present chapter, we present first some general ideas about the biological basis of microbial growth, then discuss the growth rate of microbial populations, then consider some of the factors influencing growth, and finally consider some of the practical aspects. Most of our discussion concerns the bacterial cell, since most information is available about bacteria and most practical aspects deal with control of bacterial growth.

9.1 Cell Growth

The bacterial cell is essentially a synthetic machine which is able to duplicate itself. The synthetic processes of bacterial cell growth involve as many as 2000 chemical reactions of a wide variety of types. Some of these reactions involve energy transformations. Other reactions involve biosynthesis of small molecules, the building blocks of macromolecules, as well as the various cofactors and coenzymes needed for enzymatic reactions. These various metabolic reactions were discussed in Chapter 4.

Miniglossary for Chapter 9

ACIDOPHILE an organism that grows best at low pH

AEROBE an organism that can use O_2 in respiration

ALKALIPHILE an organism that grows best at high pH

ANAEROBE an organism that cannot use O_2 in respiration and whose growth may be inhibited by O_2

ANTIBIOTIC chemical substances produced by microorganisms that kill or inhibit the growth of other microorganisms

ANTIMICROBIAL AGENT a chemical that kills or inhibits the growth of microorganisms

ANTISEPTIC antimicrobial agents that are sufficiently nontoxic to be applied on living tissues

CHEMOSTAT a device that allows for the continuous culture of microorganisms in which both growth rate and cell number can be controlled independently

CHEMOTHERAPEUTIC AGENT an antimicrobial agent that can be used internally

COMPATIBLE SOLUTE a molecule accumulated in the cytoplasm for adjustment of water activity but which does not inhibit biochemical processes

DISINFECTANT antimicrobial agents used only on inanimate objects

EXPONENTIAL GROWTH growth of a microorganism where the cell number doubles within a fixed time period

FERMENTED FOOD a food product preserved by acids or alcohol produced during microbial growth

GENERATION TIME the time required for a population of microbial cells to double

GROWTH an increase in cell number

HALOPHILE a microorganism that requires NaCl for growth

HYPERTHERMOPHILE a microorganism that has a growth temperature optimum of 80°C or greater

LAG PHASE a period preceding the exponential growth phase when cells may be metabolizing but are not yet growing

LYSIS loss of cellular integrity with release of cytoplasmic constituents

MESOPHILE an organism that grows best at temperatures between 20 and 45°C

PASTEURIZATION heat treatment of a liquid to reduce the total number of organisms present

PSYCHROPHILE an organism with a growth temperature optimum of 15°C or lower and a maximum growth temperature below 20°C

PSYCHROTOLERANT an organism capable of growth at low temperatures but whose growth temperature optimum is above 20°C

STATIONARY PHASE the period immediately following exponential growth when the growth rate of the population falls to zero

THERMOPHILE an organism whose growth temperature optimum lies between 45 and 80°C

VIABLE capable of reproducing

XEROPHILE an organism able to live, or that lives best, in very dry environments

---

However, the main reactions of cell synthesis are *polymerization reactions*, the processes by which polymers (macromolecules) are made from monomers. The major reactions of macromolecular synthesis have been discussed in Chapter 5: DNA synthesis, RNA synthesis, and protein synthesis. Once polymers are made, the stage is set for the final events of cell growth: assembly of macromolecules and formation of cellular structures such as the cell wall, cytoplasmic membrane, flagella, ribosomes, inclusion bodies, enzyme complexes, and so forth.

In most microorganisms, growth of an individual cell continues until the cell divides into two new cells, a process called *binary fission* (*binary* to express the fact that *two* cells have arisen from one cell). In a growing culture of a rod-shaped bacterium such as *Escherichia coli*, for example, cells are observed to elongate to approximately twice the length of an average cell and then form a partition which eventually separates the cell into two daughter cells (Figure 9.1). This partition is referred to as a *septum* and is the result of the inward growth of the cytoplasmic membrane and cell wall from opposing directions until the two daughter cells are pinched off (Figure 9.1). During the growth cycle all cellular constituents increase in number such that each daughter cell receives a complete chromosome and sufficient copies of all other macromolecules, monomers, and inorganic ions to exist as an independent cell. Partitioning of the replicated DNA molecule between the two daughter cells depends on the DNA

remaining attached to membranes during division, with septum formation leading to separation of chromosome copies, one to each daughter cell (Figure 9.1).

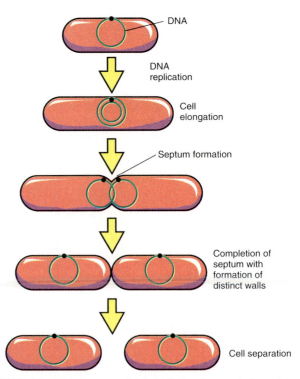

FIGURE 9.1 The process of binary fission in a rod-shaped bacterium.

The time required for a complete growth cycle in bacteria is highly variable and is dependent upon a number of factors, both nutritional and genetic. Under the best nutritional conditions the bacterium *E. coli* can complete the cycle in about 20 minutes; a few bacteria can grow even faster than this but many grow much slower. The control of cell division is a complex process and appears intimately tied to chromosomal replication events.

> **Most studies of microbial growth deal with an increase in the number of cells rather than in the size of individual cells. Growth of most microorganisms occurs by binary fission. Cell division and chromosome replication are usually coordinately regulated.**

9.2   Population Growth

As we have mentioned, *growth* is defined as an increase in the *number* of microbial cells in a population, which can also be measured as an increase in microbial *mass*. **Growth rate** is the change in cell number or mass *per unit time*. During this cell-division cycle, all the structural components of the cell double. The interval for the formation of two cells from one is called a **generation**, and the time required for this to occur is called the **generation time**. The generation time is thus the time required for the cell number to double. Because of this, the generation time is also sometimes called the *doubling time*. Note that during a single generation, both the cell number and cell mass have doubled. Generation times vary widely among organisms. Many bacteria have generation times of 1–3 hours, but a few very rapidly growing organisms are known that divide in as little as 10 minutes and others have generation times of several hours or even days.

A growth experiment beginning with a single cell having a doubling time of 30 minutes is presented in Figure 9.2. This pattern of population increase, where the number of cells *doubles* during each unit time period, is referred to as **exponential growth**. When the cell number from such an experiment is graphed on arithmetic coordinates as a function of elapsed time, one obtains a curve with a constantly increasing slope (Figure 9.2*b*). Examination of curved lines is not convenient, thus population growth data are usually converted to a logarithmic value by taking the logarithm of each data point. The $\log_{10}$ values are presented in Figure 9.2*b* in a graph in which cell number is plotted logarithmically and time is plotted arithmetically (a *semilogarithmic* graph), resulting in a straight line. This straight line function is an immediate indicator that the cells are growing exponentially. This semilogarithmic graph is also convenient and simple to use for estimating generation times from a set of results. The doubling time may be read directly from the graph (Figure 9.3).

One of the characteristics of exponential growth is that the *rate* of increase in cell number is slow initially

Time (hours)	Number of cells
0	1
0.5	2
1	4
1.5	8
2	16
2.5	32
3	64
3.5	128
4	256
4.5	512
5	1024
5.5	2048
6	4096
.	.
.	.
.	.
10	1,048,576

(a)

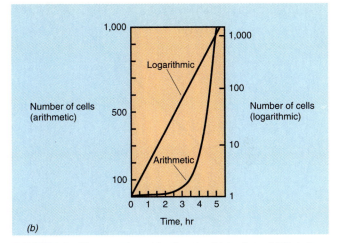

(b)

FIGURE 9.2   The rate of growth of a microbial culture. (a) Data for a population that doubles every 30 minutes. (b) Data plotted on an arithmetic (left ordinant) and a logarithmic (right ordinant) scale.

but increases at an ever faster rate. This results, in the later stages, in an explosive increase in cell numbers. For example, in the experiment of Figure 9.2, the *rate* of cell production in the first 30 minutes of growth is one cell per 30 minutes. However, between 4 hours and 4.5 hours of growth, the rate of cell production is considerably faster at 256 cells per 30 minutes. A practical implication of exponential growth is that when a nonsterile product such as milk is allowed to stand under conditions such that microbial growth can occur, a few hours during the early stages of exponential growth are not detrimental, whereas standing *for the same length of time* during the later stages is disastrous (see Section 9.18).

Calculating generation times

For many purposes in microbiology, it is useful to know the generation time of a microbial population during exponential growth, and such information can be obtained from a mathematical analysis of cell numbers data such as those in Figure 9.2. The increase in cell numbers that occurs in an exponentially growing bacterial culture is a simple geometric progression of the number 2. As two cells double (to become four cells), we can express this as $2^1 \rightarrow 2^2$. As four cells become eight, we express this as $2^2 \rightarrow 2^3$, and so on. Be-

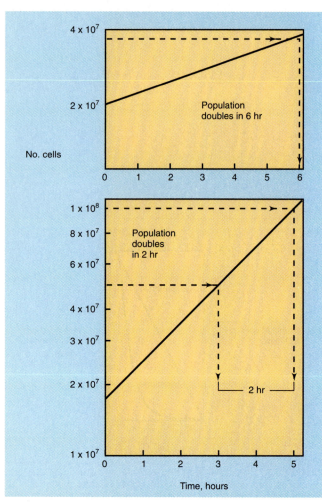

FIGURE 9.3 Method of estimating the generation times of exponentially growing populations with generation times of 2 and 6 hours, respectively. All numbers are expressed in scientific notation, that is, 10,000,000 is 1×10^7, 60,000,000 is 6×10^7, and so on.

With n now expressed in terms of readily measurable quantities, N and N_o, generation times can be calculated. As an example of how to perform a calculation, we use actual data from the lower graph in Figure 9.3. The generation time of 2 hours, which in this case was determined directly from the graph, can also be derived from the facts that $N = 10^8$, $N_o = 5 \times 10^7$, and $t = 2$. Thus,

$$n = \frac{\log (10^8) - \log (5 \times 10^7)}{0.301} = \frac{8 - 7.69}{0.301} = 1$$

Thus, the generation time $(t/n) = 2/1 = 2$.

Armed with knowledge of n and t, one can calculate g for different microorganisms growing exponentially under different culture conditions. Generation times in microbiological research are often useful as indications of the physiological state of a cell population and are frequently used to test the negative or positive effect of some treatment on the bacterial culture. Moreover, from knowledge of the generation time of a bacterium and the number of cells originally present (see Section 9.3 for how cell numbers are determined), it is possible to obtain a population of cells of known number by simply incubating the culture for a specific period of time. Thus, cells in the "mid-logarithmic" phase of growth (often desirable for study of enzymes and other components isolated from the cells) can easily be obtained by knowledge of the culture generation time under a given set of conditions.

We now turn our attention to how determinations of cell numbers N and N_o in the previous equations can be made. As you might expect, these data are crucial to any analyses of exponential growth, and a variety of methods, both direct and indirect, are available to derive these cell numbers.

> Microbial populations show a characteristic type of growth pattern called exponential growth, which is best seen by plotting the number of cells at various time periods on a semilogarithmic graph. One of the characteristics of exponential growth is that it can result in an explosive increase in cell number, which may have important practical consequences. From knowledge of initial and final cell numbers and the time of exponential growth, the generation time of the cell population can be calculated directly.

9.3   Measurement of Growth

Population growth is measured by following changes in number of cells or weight of cell mass. There are several methods for counting cell numbers or estimating cell mass, suited to different organisms or different problems.

Total cell count

The number of cells in a population can be measured by counting under the microscope, a method called the **direct microscopic count**. Two kinds of direct microscopic counts are done, either on samples dried on

cause of this geometric progression, there is a direct relationship between the number of cells present in a culture initially and the number present after a period of exponential growth:

$$N = N_o 2^n$$

where N = the final cell number, N_o = the initial cell number, and n = the *number of generations* that have occurred in the period of exponential growth. The generation time, g, of the cell population is calculated as t/n, where t is simply the hours or minutes of exponential growth. Thus, from a knowledge of the initial and final cell numbers in an exponentially growing cell population, it is possible to calculate n, and from n and knowledge of t, the generation time, g.

To express the equation $N = N_o 2^n$ in terms of n, the following transformations are necessary:

$$N = N_o 2^n$$
$$\log N = \log N_o + n \log 2$$
$$\log N - \log N_o = n \log 2$$
$$n = \frac{\log (N) - \log (N_o)}{\log 2} = \frac{\log (N) - \log (N_o)}{0.301}$$

slides or on samples in liquid. With liquid samples, special *counting chambers* must be used. In such a counting chamber, a grid is marked on the surface of the glass slide, with squares of known small area (Figure 9.4). Over each square on the grid is a volume of known size, very small but precisely measured. The number of cells per unit area of grid can be counted under the microscope, giving a measure of the number of cells per small chamber volume. Converting this value to the number of cells per milliliter of suspension is easily done by multiplying by a conversion factor based on the volume of the chamber sample (Figure 9.4).

Direct microscopic counting is a quick way of estimating microbial cell number. However, it has certain limitations: (1) Dead cells are not distinguished from living cells. (2) Small cells are difficult to see under the

microscope, and some cells are probably missed. (3) Precision is difficult to achieve. (4) A phase-contrast microscope is required when the sample is not stained. (5) The method is not suitable for cell suspensions of low density. With bacteria, if a cell suspension has less than 10^6 cells per milliliter, few if any bacteria will be seen in the microscope field.

Viable count

In the method just described, both living and dead cells are counted. In many cases we are interested in counting only live cells, and for this purpose *viable* cell counting methods have been developed. A viable cell is defined as one that is able to divide and form offspring, and the usual way to perform a viable count is to determine the number of cells in the sample capable of forming *colonies* on a suitable agar medium. For this reason, the viable count is often called the **plate count**, or **colony count**. The assumption made in this type of counting procedure is that *each viable cell will yield one colony*.

There are two ways of performing a plate count: the spread plate method and the pour plate method (Figure 9.5). With the **spread plate method**, a volume of an appropriately diluted culture usually no greater than 0.1 ml is spread over the surface of an agar plate, using a sterile glass spreader. The plate is then incubated until the colonies appear, and the number of colonies is counted. It is important that the surface of the plate be dry so that the liquid that is spread soaks in. Volumes greater than 0.1 ml are rarely used since the excess liquid will not soak in and may cause the colonies to coalesce as they form, making them difficult to count. In the **pour plate method** (Figure 9.5), a known volume (usually 0.1–1.0 ml) of culture is pipetted into a sterile petri plate; melted agar medium is then added and mixed well by gently swirling the plate on the table top. Because the sample is mixed with the molten agar medium, a larger volume can be used than with the spread plate; however, with the pour plate method the organism to be counted must be able to briefly withstand the temperature of melted agar, 45°C.

With both the spread plate and pour plate methods, it is important that the number of colonies developing on the plates not be too large, since on crowded plates some cells may not form colonies and the count will thus be erroneous. It is also essential that the number of colonies not be too small, or the statistical significance of the calculated count will be low. The usual practice, which is most valid statistically, is to count colonies only on those plates that have between 30 and 300 colonies.

To obtain the appropriate colony number, the sample to be counted must almost always be diluted. Since one rarely knows the approximate viable count ahead of time, it is usually necessary to make more than one dilution. Several ten-fold dilutions of the sample are commonly used (Figure 9.6). To make a ten-fold (10^{-1}) dilution, one can mix 0.5 ml of sample

Ridges which support cover slip

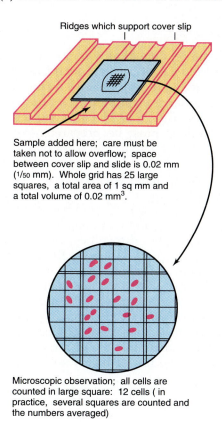

Sample added here; care must be taken not to allow overflow; space between cover slip and slide is 0.02 mm ($\frac{1}{50}$ mm). Whole grid has 25 large squares, a total area of 1 sq mm and a total volume of 0.02 mm³.

Microscopic observation; all cells are counted in large square: 12 cells (in practice, several squares are counted and the numbers averaged)

To calculate number per milliliter of sample:
12 cells x 25 squares x 50 x 10³
= 1.5 x 10⁷

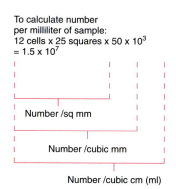

Number /sq mm

Number /cubic mm

Number /cubic cm (ml)

FIGURE 9.4   Direct microscopic counting procedure using the Petroff–Hausser counting chamber.

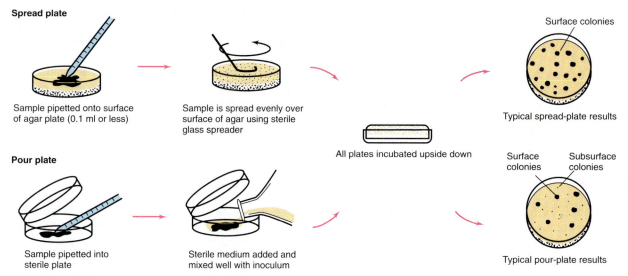

Spread plate

Sample pipetted onto surface of agar plate (0.1 ml or less)

Sample is spread evenly over surface of agar using sterile glass spreader

All plates incubated upside down

Surface colonies

Typical spread-plate results

Pour plate

Sample pipetted into sterile plate

Sterile medium added and mixed well with inoculum

Surface colonies   Subsurface colonies

Typical pour-plate results

FIGURE 9.5   Two methods of performing a viable count (plate count).

with 4.5 ml of diluent, or 1.0 ml with 9.0 ml diluent. If a hundred-fold (10^{-2}) dilution is needed, 0.05 ml can be mixed with 4.95 ml diluent, or 0.1 ml with 9.9 ml diluent. Alternatively, a 10^{-2} dilution could be made by doing two successive ten-fold dilutions. In most cases, such *serial dilutions* are needed to reach the final dilution desired. Thus, if a 10^{-6} ($1/10^6$) dilution is needed, this can be achieved by making three successive 10^{-2} ($1/10^2$) dilutions or six successive 10^{-1} dilutions (Figure 9.6).

The number of colonies obtained in a viable count will depend not only on the inoculum size but also on the suitability of the culture medium and the incubation conditions used; also, it will depend on the length of incubation. The cells deposited on the plate will not all develop into colonies at the same rate, and if a short incubation time is used, less than the maximum number of colonies will be obtained. Furthermore, the size of colonies often varies. If some tiny colonies develop, they may be missed during the counting. It is usual to determine incubation conditions (medium,

temperature, time) that will give the maximum number of colonies of a given organism and then to use these conditions throughout. Viable counts can be subject to large error, and if accurate counts are desired, great care must be taken and replicate plates of key dilutions must be prepared. Note that two or more cells in a clump will form only a single colony, so that a viable count may be erroneously low. To more clearly state the result, viable counts are often expressed as the number of *colony-forming units* obtained rather than as the number of *viable cells* (since a colony-forming unit may contain one or more cells).

Despite the difficulties involved with viable counting, the procedure gives the best information on the number of viable cells, so it is widely used. In food, dairy, medical, and aquatic microbiology, viable counts are used routinely. The method has the virtue of high sensitivity: samples containing very few cells can be counted, thus permitting sensitive detection of microbial contamination of products or materials. Moreover, the use of highly selective culture media

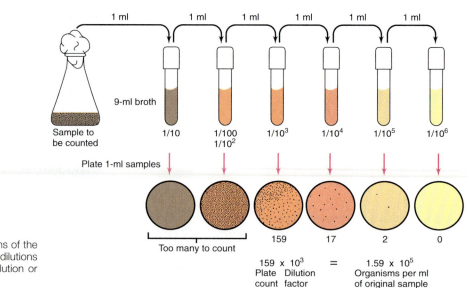

FIGURE 9.6
Procedure for viable count using serial dilutions of the sample. The sterile liquid used for making dilutions can simply be water, but a balanced salt solution or growth medium may yield a higher recovery.

1 ml   1 ml   1 ml   1 ml   1 ml   1 ml

9-ml broth

Sample to be counted

1/10   1/100   1/10^3 1/10^4 1/10^5 1/10^6
        1/10^2

Plate 1-ml samples

Too many to count   159   17   2   0

$$159 \times 10^3 = 1.59 \times 10^5$$

Plate Dilution   Organisms per ml
count factor   of original sample

(see Section 13.2) in viable counting procedures allows for the counting of only particular cell types in a mixed population of microorganisms.

Cell mass

For many studies it is desirable to estimate the *mass* of cells in a culture rather than the number. Net mass can be measured by centrifuging the cells and weighing the pellet of cells obtained. Dry mass is measured by drying the centrifuged cell mass before weighing, usually by placing it overnight in an oven at 100–105°C. Dry mass of bacterial cells is usually about 10 to 20 percent of the wet mass.

A simpler and very useful method for obtaining a relative estimate of cell mass is by use of *turbidity* measurements. A cell suspension looks cloudy (turbid) to the eye because each cell scatters light. The more cell material present, the more the suspension scatters light and the more turbid it will be. Turbidity can be measured with an electrically operated device called a *colorimeter*, or *spectrophotometer*. With such a device, the turbidity is expressed in units of *absorbance*. For unicellular organisms, absorbance is proportional (within certain limits) to cell number as well as to cell mass and thus turbidity readings can be used as a substitute for counting. To perform cell counts in this way, a standard curve must first be prepared for each organism studied, relating a direct measurement of cell number (microscopic or viable count) to cell mass or absorbance. Turbidity is a much less sensitive way of measuring cell density than is viable counting but has the virtues that it is quick, easy, and does not disturb or destroy the sample. Such measurements are used widely to follow the rate of growth of cultures, since the same sample can be checked repeatedly.

> Growth is measured by the change in number of cells with time. Cell counts done microscopically measure the total number of cells in a population, while viable cell counts measure only the living population. Viable counts are done by placing dilutions of the population on agar plates and counting the number of colonies that occur after incubation. Measurements of cell mass such as turbidity are indirect but very useful measures of cell growth.

9.4   The Growth Cycle of Populations

The data presented in Figure 9.2 reflect only part of the growth cycle of a microbial population. A typical *growth curve* for a population of cells is illustrated in Figure 9.7. This growth curve can be divided into several distinct phases, called the **lag phase, exponential phase, stationary phase**, and **death phase**.

Lag phase

When a microbial population is inoculated into a fresh medium, growth usually does not begin immediately, but only after a period of time called the *lag phase*, which may be brief or extended, depending on conditions. If an exponentially growing culture is inoculated into the same medium under the same conditions of growth, a lag is not seen, and exponential growth continues at the same rate. However, if the inoculum is taken from an old (stationary phase) culture and inoculated into the same medium, a lag usually occurs even if all of the cells in the inoculum are viable. This is because the cells are usually depleted of various essential coenzymes or other cell constituents, and time is required for resynthesis. A lag also ensues when the inoculum consists of cells that have been

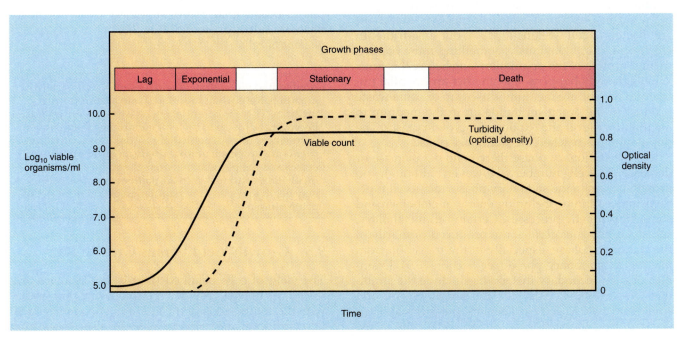

FIGURE 9.7   Typical growth curve for a bacterial population. The optical density is a measure of the turbidity of the bacterial culture and is obtained by placing the cell suspension in a spectrophotometer.

damaged (but not killed) by treatment with heat, radiation, or toxic chemicals, due to the time required for the cells to repair the damage.

A lag is also observed when a population is transferred from a rich culture medium to a poorer one. This happens because for growth to occur in a particular culture medium the cells must have a complete complement of enzymes for the synthesis of the essential metabolites not present in that medium. On transfer to a new medium, time is required for synthesis of the new enzymes.

Exponential phase

The *exponential phase* of growth has already been discussed. As noted, it is a consequence of the fact that each cell divides to form two cells, each of which also divides to form two more cells, and so on. Most unicellular microorganisms grow exponentially, but rates of exponential growth vary greatly. For instance, the organism causing typhoid fever, *Salmonella typhi*, grows very rapidly in culture, with a generation time of 20–30 minutes, whereas the tubercle bacterium, *Mycobacterium tuberculosis*, grows slowly, with only one or two doublings per day. The rate of exponential growth is influenced by environmental conditions (temperature, composition of the culture medium) as well as by genetic characteristics of the organism itself. In general, prokaryotes grow faster than eukaryotic microorganisms, and small eukaryotes grow faster than large ones.

Stationary phase

In a closed system, exponential growth cannot occur indefinitely. One can calculate that a single bacterium with a generation time of 20 minutes would, if it continued to grow exponentially for 48 hours, produce a population that weighed about 4000 times the weight of the earth! This is particularly impressive since a single bacterial cell weighs only about one-trillionth (10^{-12}) of a gram. Obviously, something must happen to limit growth of the population long before this time. What generally happens is that either an essential nutrient of the culture medium is used up or some waste product of the organism builds up in the medium to an inhibitory level and exponential growth ceases. The population has reached the **stationary phase**.

In the stationary phase there is no net increase or decrease in cell number. However, although no growth occurs in the stationary phase, many cell functions may continue, including energy metabolism and some biosynthetic processes. Certain cell metabolites, called *secondary metabolites*, are produced primarily in the stationary phase, especially in the transition from late log to stationary phase (see Section 10.3). Examples of such secondary metabolites include the antibiotics (for example, penicillin and streptomycin) and some enzymes. In some organisms, growth may even occur during the stationary phase; some cells in the population grow while others die, the two processes balancing out so that no net increase or decrease in cell

number occurs (this is called *cryptic growth*). In endospore-forming bacteria, the endospore is produced after the culture has entered the stationary phase.

Studies with *Escherichia coli* have identified several genes that are necessary for survival of cells that have entered stationary phase. Some of these genes, called *sur* (for *survival*) genes, have as yet unknown functions, but mutations in *sur* genes lead to rapid cell death as cells enter stationary phase. Other stationary phase-specific genes include ones for specific RNA polymerase sigma factors, which direct transcription of genes only expressed in the stationary phase, and genes that code for proteins that somehow protect the starving cell from oxidative damage. Because in nature it is likely that many bacterial cells are in a nongrowing or very slow growing state, it is not surprising that several genes have evolved to deal with conditions in which the cells in a population have reached stationary phase or for other reasons are growing at extremely low rates.

Death phase

If incubation continues after a population reaches the stationary phase, the cells may remain alive and continue to metabolize, but they may also die. If the latter occurs, the population is said to be in the *death phase*. During this death phase the total count (as measured by a direct microscopic count) may remain constant but the viable count slowly decreases. In some cases death is accompanied by cell **lysis**, leading to a decrease in direct microscopic count and turbidity with a concurrent drop in viable count. A discussion of the effects of germicidal chemicals on viability will be found later in this chapter.

It should be emphasized that the phases of the bacterial growth curve are reflections of the events in a population, not of individual cells. The terms lag phase, exponential phase, stationary phase, or death phase do not apply to individual cells but only to *populations* of cells.

A microbial population generally shows a characteristic growth pattern when inoculated into a fresh culture medium. There is an initial lag phase while the cells are becoming adjusted to the new conditions, then growth commences in an exponential fashion. As essential nutrients are depleted, or toxic products build up, growth ceases and the population enters the stationary phase. If incubation continues into the stationary phase, cells may begin to die and the population is said to be in the death phase.

9.5 Growth Yield Versus Growth Rate

The yield of cells obtained (that is, the biomass produced) in a given growth experiment can vary greatly with the composition of the medium and environmental conditions and, of course, yield also varies from one organism to another. As pointed out in Section 4.13, fermentative and aerobic organisms obtain vastly

Table 9.1   Molar growth yields for anaerobic growth of Bacteria and yeast using glucose as electron donor

Organism	$Y_{substrate}$ (grams dry weight per mole substrate used)	ATP yield (moles ATP per mole of substrate)*	Y_{ATP} (grams dry weight per mole ATP)
Zymomonas mobilis	9	1	9
Streptococcus faecalis	20	2	10
Lactococcus lactis	19.5	2	9.8
Lactobacillus plantarum	18.8	2	9.4
Saccharomyces cerevisiae	18.8	2	9.4
Klebsiella pneumoniae	29	3	9.6
Escherichia coli	26	3	8.6

ATP yield is based on a knowledge of the biochemical pathway by which glucose is fermented in the different organisms.

different amounts of ATP from the catabolism of the same energy source. These major differences in energy yield are directly reflected in *cell yield*, because the yield of cells obtained is directly proportional to the amount of ATP produced,

The parameter Y_{ATP} is used to describe the cell yield obtained per ATP generated (Table 9.1). To calculate a Y_{ATP} value, one must first measure the grams of cell mass produced (dry weight) from the catabolism of a known amount of energy source. Experimental determinations of Y_{ATP} are done most easily with fermentative organisms because the amount of ATP produced per molecule of substrate fermented can be accurately calculated from knowledge of the catabolic pathway used. For example, the fermentation of glucose to lactic acid via glycolysis leads to the formulation of two ATP per glucose (Section 4.8). Table 9.1 presents yield data and Y_{ATP} values from some fermentative organisms. Note that the Y_{ATP} values in Table 9.1 are quite similar, between 9 and 10, even though the ATP yield per mole of substrate varies from 1 to 3. This suggests that the energy costs of assembling macromolecules (which are the major energy costs for a growing microbial cell, see Chapter 5) are much the same for all microorganisms.

Effect of nutrient concentration on growth rate and growth yield

Nutrient concentration can affect both the growth *rate* and growth *yield* of a microorganism. At very low concentrations of a given nutrient, the rate of growth is reduced, whereas at moderate and higher levels of nutrient, growth rates may not be affected while cell yield continues to increase (Figure 9.8).

The effect of nutrient concentration on cell yield is easy to understand; if the nutrient is converted into cell material, the more nutrient present, the greater the cell yield (Figure 9.8). The reason for the reduction in growth rate at very low nutrient concentrations is less certain, but is probably because the nutrient cannot be transported into the cell at sufficiently rapid rates to satisfy all the metabolic demands for the nutrient; this slows the rate of metabolic reactions and thus growth rate. Evidence for this includes the fact that the shape of the curve relating growth rate to nutrient concentration resembles a saturation process of the kind also seen for active transport (see Figure 3.20).

We now consider a practical use of our knowledge of growth rates and cell yields in the operation of a type of culture apparatus called the *chemostat*.

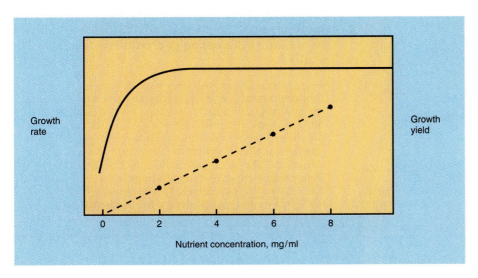

FIGURE 9.8
Relationship between nutrient concentration, growth rate (solid line), and growth yield (dashed line). At low nutrient concentrations both growth rate and growth yield are affected.

Cell growth requires the expenditure of energy, and the amount of growth that will occur depends on the amount of ATP that can be produced from the energy source.

9.6   Continuous Culture

Our discussion of population growth thus far has been confined to closed or *batch cultures*, growth occurring in a fixed volume of a culture medium that is continually being altered by the actions of the growing organisms until it is no longer suitable for growth. In the early stages of exponential growth in batch cultures, conditions may remain relatively constant, but in later stages when cell numbers become quite large, drastic changes usually occur. For many studies, it is desirable to keep cultures in constant environments for long periods, and this is done by employing *continuous cultures*. A continuous culture is essentially a flow system of constant volume to which medium is added continuously and from which continuous removal of any overflow can occur. Once such a system is in equilibrium, cell number and nutrient status remain *constant*, and the system is said to be in **steady state**.

The most common type of continuous culture device used is called a **chemostat** (Figure 9.9), which permits control of both the population density and the growth rate of the culture. Two elements are used in

the control of a chemostat—the *dilution rate* and the *concentration of a limiting nutrient*, such as a carbon or nitrogen source. As discussed, at low concentrations of an essential nutrient, growth rate is proportional to the nutrient concentration. However, at these low nutrient concentrations, the nutrient is quickly used up by growth. In batch cultures, growth ceases at the time the nutrient is used up, but in the chemostat, continuous addition of fresh medium containing the limiting nutrient permits continued growth. However, because the limiting nutrient is quickly assimilated, its concentration in the chemostat vessel itself is always virtually zero (see Figure 9.10).

Effects of varying dilution rate and concentration of the growth-limiting nutrient are given in Figure 9.10. As seen, there are rather wide limits over which dilution rate will control growth rate, although at both very low and very high dilution rates the steady state breaks down. At high dilution rates, the organism cannot grow fast enough to keep up with its dilution, and the culture is washed out of the chemostat. At the other extreme, at very low dilution rates, a large fraction of the cells may die from starvation, since the limiting nutrient is not being added fast enough to permit maintenance of cell metabolism. There is probably a minimum amount of energy necessary to maintain cell structure and integrity, called **maintenance energy**, and nutrients used for maintenance energy are not available for biosynthesis and cell growth. Thus at very low dilution rates steady-state conditions will not be maintained, and the population will slowly wash out.

The *cell density* (cells/ml) in the chemostat is controlled by the level of the limiting nutrient. If the concentration of this nutrient in the incoming medium is raised, with dilution rate remaining constant, cell density will increase although growth rate will remain the same and the steady-state concentration of the nutrient in the culture vessel will still be virtually zero. Thus, by adjusting dilution rate and nutrient level, the experimenter can obtain at will a variety of population densities growing at a variety of growth rates. The actual shape of the curve for bacterial concentration given in Figure 9.10 will depend on the organism, the environmental conditions, and the limiting nutrient used. Further discussion of chemostats occurs in Appendix 2.

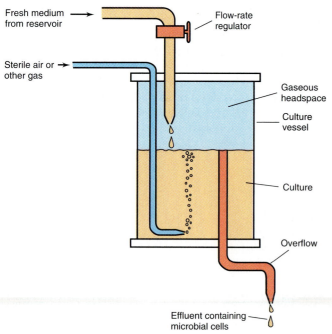

FIGURE 9.9 Schematic for a continuous culture device (chemostat). In such a device, the population density is controlled by the concentration of limiting nutrient in the reservoir, and the growth rate is controlled by the flow rate. Both parameters can be set by the experimenter.

Fresh medium from reservoir

Flow-rate regulator

Sterile air or other gas

Gaseous headspace

Culture vessel

Culture

Overflow

Effluent containing microbial cells

Continuous culture devices (chemostats) are a means of maintaining cell populations in exponential growth for long periods. The principle of a continuous culture device is to slowly replenish the culture vessel with fresh medium while removing used medium and cells at the same rate. The rate at which this dilution occurs governs the growth rate of the cell population in the culture vessel. The population size in a chemostat, however, is not governed by the dilution rate but instead by the concentration of the growth-limiting nutrient entering the vessel.

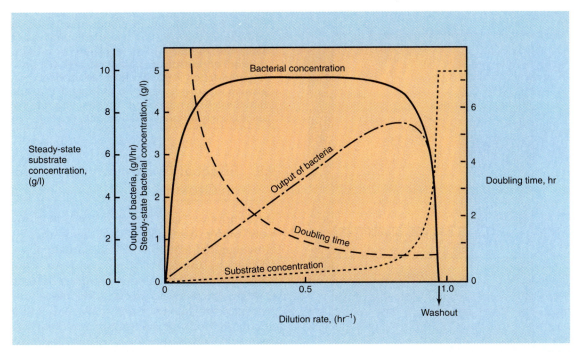

FIGURE 9.10   Steady-state relationships in the chemostat. The dilution rate is determined from the flow rate and the volume of the culture vessel. Thus, with a vessel of 1000 ml and a flow rate through the vessel of 500 ml/hr, the dilution rate would be 0.5 hr^{-1}. Note that at high dilution rates, growth cannot balance dilution, the population washes out, and the substrate concentration rises to the concentration in the medium reservoir (since there are no bacteria to use the inflowing substrate). However, throughout most of the range of dilution rates shown, the population density remains constant and the substrate concentration remains at a very low value. Note that although the population density remains constant, the growth rate (doubling time) varies over a wide range. Thus, the experimenter can obtain populations with widely varying growth rates, without affecting population density. See Appendix 2 for further details of chemostat operation.

9.7   Growth and Macromolecular Synthesis

Balanced and unbalanced growth

During exponential growth, all biochemical constituents are being synthesized at the same relative rates (Figure 9.11a), a condition called *balanced growth.* Figure 9.11b shows what happens to the growth rates and the rates of synthesis of RNA, DNA, and protein if the culture is transferred to a different medium, in which growth can occur at a faster rate than previously. (These are called "step-up" conditions.) Immediately upon transfer to the richer medium the *rate* of RNA synthesis increases, and somewhat later the rates of DNA and protein synthesis also increase. The rate of cell division also steps up after a longer lapse of time, and eventually the rates of synthesis of all components are in balance again. In the initial period after the transfer to the new medium we have conditions of *unbalanced* growth since not all cell constituents are being synthesized at the same rate. If the reverse experiment (a "step down") is performed, the rate of RNA synthesis decreases immediately, while the rate of DNA and protein synthesis and the rate of cell division continue at the previous (more rapid) rate, later decreasing. These results suggest that, since the rate of

RNA synthesis is the first to change during step-up or step-down conditions, it is the key factor controlling growth rate.

Although there are several kinds of RNA in the cell (see below), the largest fraction consists of RNA in ribosomes. During the step-up conditions, there is an increased rate of synthesis of ribosomes, leading to a greater number of ribosomes per cell. Since protein synthesis occurs on ribosomes, it is only understandable that the rate of protein synthesis should start to change after the rate of ribosome synthesis increases. The efficiency with which a ribosome acts in protein synthesis appears to remain relatively constant at different growth rates. Thus, protein synthesis and consequently growth itself is controlled at the most basic level by the *number* of ribosomes present per cell.

DNA and RNA synthesis and the cell cycle

As noted in Section 5.4, DNA synthesis is initiated at a single point on the chromosome, and proceeds bidirectionally around the circle. It takes about 40 minutes for a single round of DNA replication to occur in *Escherichia coli.* In cells growing with a generation time of 60 minutes, the 40-minute DNA replication phase (referred to as the S phase) is followed by a 20-minute period free of DNA synthesis (referred to as the D phase). The surprising thing is that the duration of the

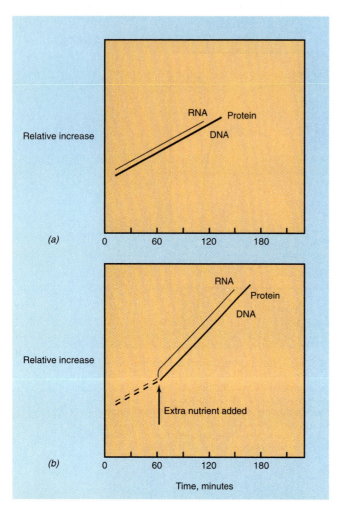

(a)
(b)

Time, minutes

FIGURE 9.11   Changes in RNA, DNA, and protein content of a population during growth: (a) balanced growth, (b) step-up conditions. The arrow indicates the time at which extra nutrient was added to speed up growth. Note that, during the step-up, the rate of RNA synthesis increases first, followed by increases in rates of protein and DNA syntheses.

S and D phases are invariant for cells growing with generation times between 20 and 60 minutes. When generation times are less than 60 minutes, several replicating forks of DNA may be present in the cell at one time, the exact number depending on how fast the organism is growing. When the DNA copies are partitioned into the daughter cells at cell division, they are usually already partly replicated before the new round of replication begins.

About 80 percent of the total RNA in a growing cell is ribosomal RNA, and most of the remainder is transfer RNA. Despite its importance, messenger RNA is only a small fraction of the total (~4%), and in prokaryotes, most mRNA is unstable, being rapidly synthesized, used for several rounds of protein synthesis, and then broken down (the half life of most bacterial mRNA molecules is less than three minutes). Thus in studies of RNA synthesis during growth, it is primarily ribosomal RNA that is examined. As opposed to DNA, ribosomal RNA synthesis is essentially continuous throughout the growth cycle and, in contrast to mRNA, rRNA is very stable.

During balanced growth, all macromolecular constituents of the microbial population increase at the same rate, but the growth rate itself is controlled by the number of ribosomes present per cell. If a slowly growing population is given a boost of nutrients so that its growth rate increases, the first thing that happens is that the cells synthesize more ribosomes. DNA synthesis is closely aligned with the cell cycle.

9.8   Effect of Environmental Factors on Growth

Up to now we have described growth of microorganisms under essentially ideal laboratory conditions. However, the activities of microorganisms are greatly affected by the chemical and physical conditions of their environments. Understanding environmental influences helps us to explain the distribution of microorganisms in nature and makes it possible for us to devise methods for controlling microbial activities and destroying undesirable organisms. Not all organisms respond equally to a given environmental factor. In fact, an environmental condition may be harmful to one organism, and actually beneficial to another. Organisms can tolerate some adverse conditions under which they cannot grow, and hence we must distinguish between the effects of environmental conditions on the *viability* of an organism and effects on growth, differentiation, and reproduction.

9.9   Temperature

Temperature is one of the most important environmental factors influencing the growth and survival of organisms. It can affect living organisms in either of two opposing ways. As temperature rises, chemical and enzymatic reactions in the cell proceed at more rapid rates and growth becomes faster. However, above a certain temperature, proteins, nucleic acids, and other cellular components may be irreversibly denatured. Thus, as the temperature is increased within a given range, growth and metabolic function increase up to a point where inactivation reactions set in. Above this point, cell functions fall sharply to zero. Thus, we find that for every organism there is a **minimum temperature** below which growth no longer occurs, an **optimum temperature** at which growth is most rapid, and a **maximum temperature** above which growth is not possible (Figure 9.12). The optimum temperature is always nearer the *maximum* than the minimum. These three temperatures, often called the **cardinal temperatures**, are generally characteristic for each type of organism, but are not completely fixed, as they can be modified by other factors of the environment.

The maximum growth temperature of a given organism most likely reflects the inactivation discussed above. However, the factors controlling an organism's minimum growth temperature are not as clear. As

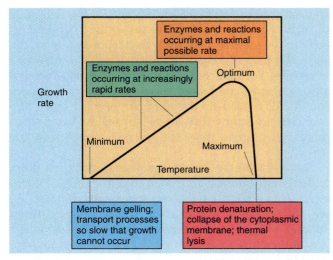

FIGURE 9.12 Effect of temperature on growth rate and the molecular consequences.

mentioned earlier (Section 3.4), the cytoplasmic membrane must be in a fluid state for proper functioning. Perhaps the minimum temperature of an organism results from "freezing" of the cytoplasmic membrane so that it no longer functions properly in nutrient transport or proton gradient formation. This explanation is supported by experiments in which the minimum temperature for an organism can be altered to some extent by causing it to incorporate different fatty acids into its phospholipids, thereby changing the fluidity. It is also observed that the cardinal temperatures of different microorganisms differ widely; some organisms have temperature optima as low as 5–10°C and some over 100°C. The temperature range throughout which growth occurs is even wider than this, from below freezing to greater than boiling (the archaean *Pyrodictium brockii* has a temperature maximum of 110°C and a related organism will grow up to 113°C!). However, no single organism will grow over this whole temperature range and the usual range for a given organism is about 30–40 degrees, although some have a much broader temperature range than others.

Although there is a continuum of organisms, from those with very low temperature optima to those with high temperature optima, it is possible to broadly distinguish *four groups* of organisms: **psychrophiles**, with low temperature optima, **mesophiles**, with mid-range temperature optima, **thermophiles**, with high temperature optima, and **hyperthermophiles**, with very high temperature optima (Figure 9.13). These temperature distinctions are made for convenience and the precise numbers should not be taken as absolutes. Mesophiles are found in warm-blooded animals and in terrestrial and aquatic environments in temperate and tropical latitudes. Psychrophiles and thermophiles are found in unusually cold or unusually hot environments, respectively. Hyperthermophiles are found in hot springs, geysers, or deep-sea vents (see below).

Cold environments

Much of the earth's surface experiences fairly low temperatures. The oceans, which make up over half of the earth's surface, have an average temperature of 5°C, and the depths of the open oceans have constant temperatures of around 1–2°C. Vast land areas of the Arctic and Antarctic are permanently frozen, or are unfrozen only for a few weeks in summer. These cold environments are rarely sterile, and some microorganisms can be found alive and growing at any low temperature at which liquid water still exists. Even in many frozen materials there are usually microscopic pockets of liquid water present where microorganisms can grow. It is important to distinguish between environments that are cold *throughout* the year and those that are cold *only* in winter. The latter, characteristic of continental temperate climates, may have summer temperatures as high as 40°C, and winter temperatures far below 0°C. Such highly variable environments are much less favorable for cold-adapted organisms than are the constantly cold environments found in polar regions, at high altitudes, and in the depths of the oceans.

As noted earlier, organisms that are able to grow

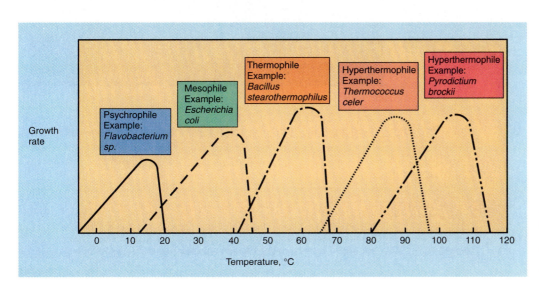

FIGURE 9.13
Relation of temperature to growth rates of a typical psychrophile, a typical mesophile, a typical thermophile, and two different hyperthermophiles.

at low temperatures are called **psychrophiles**. A psychrophile can be defined as an organism with an optimal temperature for growth of 15°C or lower, a maximum growth temperature below 20°C, and a minimal temperature for growth at 0°C or lower. Organisms which grow at 0°C but have optima of 20–40°C are called *psychrotolerant*.

Psychrophiles are found in environments that are constantly cold, and they may be rapidly killed even by brief warming to room temperature. For this reason, their laboratory study is unusually difficult, as great care must be taken to ensure that they never warm up during sampling, transport to the laboratory, plating, pipetting, or other manipulations. All media and equipment must be precooled before use, and the work must be carried out in a cold room or in a refrigerator chest. Because of these technical problems, psychrophiles have not been well studied. However, enough work has been done for us to know that psychrophiles exist and that they are an interesting and diverse group, including Bacteria, fungi, and algae from a variety of genera and species.

Some of the best studied psychrophiles have been the eukaryotic algae that grow in dense masses within and under the ice in polar regions. Psychrophilic algae are also often seen on the surfaces of snowfields and glaciers in such large numbers that they impart a distinctive red or green coloration to the surface (Figure 9.14a). The most common snow alga is the eukaryote *Chlamydomonas nivalis*; its brilliant red spores are responsible for the red color (Figure 9.14b). The alga probably grows within the snow as a green-pigmented vegetative cell, and then sporulates; as the snow dissipates by melting, erosion, and vaporization, the spores become concentrated on the surface. Snow algae are most commonly seen on melting permanent snowfields in mid- to late summer, and are especially common in sunny dry areas, probably because in more rainy areas they are washed away from the snowfields.

Psychrotolerant microorganisms are much more widely distributed than psychrophiles and can be isolated from soils and water in temperate climates. As we noted, they grow best at a temperature between 20° and 40°C. Since temperate environments do warm in summer it is understandable that they cannot support the heat-sensitive psychrophiles, the warming essentially providing a selective force favoring psychrotolerant species and excluding psychrophilic forms. It should be emphasized that although psychrotolerant microorganisms do grow at 0°C, they do not grow very well, and one must often wait several weeks before visible growth is seen in culture media. Various genera of Bacteria, fungi, algae, and protozoa have members that are psychrotolerant.

Meat, milk and other dairy products, cider, vegetables, and fruits, when stored in refrigerated areas, provide excellent habitats for the growth of psychrotolerant microorganisms. Growth of Bacteria and fungi in foods at low temperatures can lead to changes

(a)

Katherine M. Brock

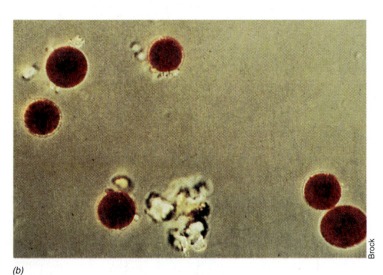

(b)

Brock

FIGURE 9.14 Snow algae. (a) Snowbank in the Sierra Nevada, California, with red coloration caused by the presence of snow algae. Pink snow such as this is common on summer snow banks at high altitudes throughout the world. (b) Photomicrograph of red-pigmented spores of the snow alga *Chlamydomonas nivalis*.

in the quality of the food and eventually to spoilage. The lower the temperature, the less rapidly does spoilage occur, but only when food is frozen solid is microbial growth impossible. The frozen food industry owes its great development to the much greater keeping qualities of frozen foods over merely refrigerated foods.

It is not completely understood how some organisms can grow well at low temperatures that prevent the growth of other organisms. At least one factor is that psychrophiles, as opposed to other organisms, have enzymes that are able to catalyze reactions more efficiently at low temperatures. Perhaps as a consequence of this, the enzymes of psychrophiles are very sensitive to higher temperatures, being rapidly inactivated at temperatures of 30–40°C. Another feature of psychrophiles is that their active-transport processes function well at low temperatures, thus making it possible for the organisms to effectively concentrate essential nutrients. In the same sense as we discussed minimum growth temperatures earlier in this section, it is thought that this efficient low-temperature transport mechanism is due to peculiarities of the cytoplasmic membrane. Psychrophiles grown at low temperatures adjust their lipid compositions to contain a higher content of *unsaturated* fatty acids in their cytoplasmic membrane, and a consequence of this is that the membrane remains semifluid at low temperatures. Membranes with predominantly saturated fatty acids would be expected to become waxy and nonfunctional at low temperatures and would become semifluid and functional only at higher temperatures.

Freezing

Despite the ability of some organisms to grow at low temperatures, there is a lower limit below which reproduction is impossible. Pure water freezes at 0°C and seawater at −2.5°C, but freezing is not continuous and microscopic pockets of water continue to exist at much lower temperatures. Although cellular growth ceases at about −30°C, enzymatic reactions can occur, albeit slowly, at even lower temperatures, and the lower limit for biochemical reactions in aqueous systems is around −140°C. Although freezing prevents microbial growth it does not always cause microbial death. In fact, freezing is one of the best ways of preserving many microbial cultures for later study and is used quite extensively in research laboratories.

The medium in which the cells are suspended considerably affects sensitivity to freezing. Water-miscible liquids such as glycerol and dimethylsulfoxide, when added at about 10 percent (final concentration) to the suspending medium, penetrate the cells and protect by reducing the severity of dehydration effects and preventing ice crystal formation. Macromolecules such as serum albumin, dextran, and polyvinylpyrrolidone, added at 10^{-5} to 10^{-3} M, do not penetrate the cells and probably protect by combining with the cell surface and preventing freezing damage

to the cell membrane. The practical significance of this is that sensitive cells can often be preserved in the frozen state by selecting the proper suspending medium.

High temperature environments

Organisms whose growth temperature optimum is above 45°C are called **thermophiles** and those whose optimum is above 80°C are called **hyperthermophiles**. Temperatures as high as these are found in nature only in certain restricted areas. For example, soils subject to full sunlight are often heated to temperatures above 50°C at midday, and some soils may become warmed even to 70°C, although a few centimeters under the surface the temperature is much lower. Fermenting materials such as compost piles and silage usually reach temperatures of 60–65°C. However, the most extensive and extreme high temperature environments are found in nature in association with volcanic phenomena.

Many hot springs have temperatures around boiling, and steam vents (fumaroles) may reach 150–500°C. Geothermal vents in the bottom of the ocean have temperatures of 350°C or greater (see Section 17.10 for a discussion of the deep sea vents). Hot springs occur throughout the world but are especially concentrated in the western United States, New Zealand, Iceland, Japan, the Mediterranean region, Indonesia, Central America, and central Africa. The area with the largest single concentration of hot springs in the world is Yellowstone National Park, Wyoming. Although some springs vary in temperature, others are very constant, not varying more than 1–2°C over many years.

Many of these springs are at the boiling point for the altitude (92–93°C at Yellowstone, 99–100°C at locations where the springs are close to sea level). As the water overflows the edges of the spring and flows away from the source, it gradually cools, setting up a *thermal gradient*. Along this gradient, microorganisms grow (Figure 9.15), with different species growing in the different temperature ranges. By studying the species distribution along such thermal gradients and by examining hot springs and other thermal habitats at different temperatures around the world, it is possible to determine the upper temperature limits for each kind of microorganism (Table 9.2). From this information we conclude that (1) prokaryotic organisms in general are able to grow at temperatures higher than those at which eukaryotes can grow; (2) the most thermophilic of all prokaryotes are certain Archaea; (3) nonphototrophic organisms are able to grow at higher temperatures than can phototrophic forms; and (4) structurally less complex organisms can grow at higher temperatures than can more complex organisms. However, it should be emphasized that not all organisms from a group are able to grow near the upper limits for that group. Usually only a relatively few species or genera are able to function successfully near the upper temperature limit.

Brock

FIGURE 9.15
Cyanobacterial growth in hot springs in Yellowstone National Park. (a) Aerial photograph of a very large boiling spring, Grand Prismatic Spring. The orange color in the outflow channel is due to the rich carotenoid pigments of bacteria and cyanobacteria. (b) Characteristic V-shaped pattern formed by cyanobacteria at the upper temperature for photosynthetic life, 70–74°C. The pattern develops because the water cools more rapidly at the edges than in the center of the channel. The spring flows from the back of the picture toward the foreground. *(a)*

Brock

(b)

Molecular adaptations to thermophily

How can organisms survive and grow at these high temperatures? First, their enzymes and other proteins are much more stable to heat than are those of mesophiles, and these macromolecules actually function *optimally* at high temperature. How is heat stability achieved? Studies of thermophilic enzymes have shown that they differ very little in amino acid sequence from the same enzyme in a mesophile. It ap-

pears that a critical amino acid substitution in one or more locations in the enzyme allows it to fold in a different way and thereby withstand the denaturing effects of heat. In addition, the protein-synthesizing machinery (that is, ribosomes and other constituents) of thermophiles, as well as such structures as the cytoplasmic membrane, are likewise heat stable. We mentioned earlier that psychrophiles have membrane lipids rich in unsaturated fatty acids, thus making the membranes fluid and functional at low temperatures. Conversely, thermophiles have membrane lipids rich in saturated fatty acids, thus allowing the membranes to remain stable and functional at high temperatures. Saturated fatty acids form much stronger hydrophobic bonds than do unsaturated fatty acids, which accounts for the membrane stability. Hyperthermophiles, virtually all of which are Archaea, do not contain fatty acids in the lipids of their membranes but instead have hydrocarbons of various lengths composed of repeating units of the five-carbon compound phytane (see Figure 20.1) bonded by ether linkage to glycerolphosphate. We discuss the details of the unique membrane architecture of hyperthermophilic Archaea in Chapter 20 (see Section 20.1).

Why is it that some groups of microorganisms have representatives that will grow at high temperatures, while others do not? The absence of eukaryotes at temperatures above 60°C may be due to the inherent thermolability of the organellar membranes of these organisms, especially those present in the nucleus, mitochondria, and chloroplast. The mitochondrial membrane seems especially sensitive to heat, and mitochondria literally disappear when cells are heated to a few degrees above their maximum temperature. The organellar membranes of eukaryotes must be fairly fluid to permit the passage of high molecular weight components (see Section 3.14). Messenger

Table 9.2	Presently known upper temperature limits for growth of living organisms	
Group		**Upper temperature limits (°C)**
Animals		
Fish and other aquatic vertebrates		38
Insects		45–50
Ostracods (crustaceans)		49–50
Plants		
Vascular plants		45
Mosses		50
Eukaryotic Microorganisms		
Protozoa		56
Algae		55–60
Fungi		60–62
Prokaryotes		
Bacteria		
Cyanobacteria		70–74
Anoxygenic phototrophic bacteria		70–73
Chemoorganotrophic bacteria		90
Archaea		
Hyperthermophilic methanogens		110
Sulfur-dependent hyperthermophiles		113

RNA and ribosomal RNA, for instance, are made in the nucleus and must move through the nuclear membrane to the cytoplasm. It is conceivable that eukaryotes cannot synthesize a functional nuclear membrane that is still stable at high temperatures.

The absence of phototrophic prokaryotes at temperatures above 70°–73°C may be related to the thermolability of the photosynthetic membrane system, a structure absent from chemoorganotrophs and chemolithotrophs.

Ecology of thermophiles and hyperthermophiles

Prokaryotes can grow over the complete range of temperatures in which life is possible, but no one organism can grow over this whole range. As mentioned above, each organism is limited to a restricted range of perhaps 30°C, and it can grow well only within a still narrower range. In most boiling hot springs (Figure 9.16), a variety of hyperthermophiles are usually present. The growth of such organisms can be studied by immersing microscope slides into the spring and retrieving them after a few days. Microscopic examination of the slides generally reveals small or large colonies of prokaryotes (Figure 9.16*b*) that have developed from single bacterial cells which attached to and grew on the glass surface.

(a)

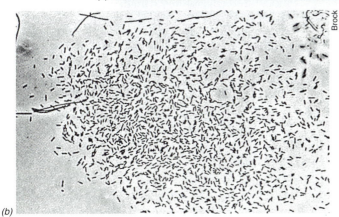
(b)

FIGURE 9.16 Bacterial growth in boiling water. (a) A typical small boiling spring in Yellowstone National Park. This spring is superheated, having a temperature 1–2°C above the boiling point. The mineral deposits around the spring consist mainly of silica. (b) Photomicrograph of a bacterial microcolony that developed on a microscope slide immersed in a boiling spring such as that shown above.

Table 9.3	Major groups of thermophilic and hyperthermophilic prokaryotes and their temperature ranges	
Genus	**Temperature range (°C)**	
Bacteria		
Phototrophic Bacteria		
Cyanobacteria	55–70 (One strain, 74)	
Purple bacteria	45–60	
Green bacteria	40–73	
Gram-positive Bacteria		
Bacillus	50–70	
Clostridium	50–75	
Lactic acid bacteria	50–65	
Actinomycetes	55–75	
Other Bacteria		
Thiobacillus	50–60	
Spirochete	54	
Desulfotomaculum	37–55	
Gram-negative aerobes	50–75	
Gram-negative anaerobes	50–75	
Thermotoga	55–90	
Thermus	60–80	
Archaea		
Methanogens	45–110	
Sulfur-dependent hyperthermophiles	60–113	
Thermoplasma	37–60	

Ecological studies of organisms living in boiling springs have shown that growth rates are fairly rapid, and doubling times of as short as one hour have been found. By use of radioactively labeled substrates it has been possible to show that prokaryotes living at these high temperatures have optimum growth temperatures near those of their environments. Both aerobic and anaerobic thermophiles and hyperthermophiles have been found and many morphological and physiological types exist (Table 9.3). Some hyperthermophiles show growth temperature optima greater than 100°C and thus must be grown in the laboratory in pressurized vessels to prevent boiling.

Thermophilic prokaryotes with growth temperature optima below 80°C have been widely found in artificial thermal environments. The hot water heater, domestic or industrial, usually has a temperature of 55–80°C and is a favorable habitat for the growth of thermophilic prokaryotes. Organisms resembling *Thermus aquaticus*, a common hot spring organism (Bacteria), have been isolated from the hot water of many installations. Electric power plants, hot industrial process water, and other artificial thermal sources probably also provide sites where thermophiles can grow.

Thermophiles and biotechnology

Thermophilic microorganisms are of interest for more than just basic biological reasons. Thermophiles offer some major advantages for biotechnological processes, many of which run more rapidly and efficiently at high temperatures. Increasing temperatures in-

crease the diffusion rate and solubilities of non-gaseous compounds and tend to discourage microbial contamination. Microbial processes carried out at high temperatures also eliminate or greatly reduce cooling costs, which may be a significant part of the total cost of running a large scale microbial process at lower temperatures.

Enzymes from thermophiles are capable of catalyzing biochemical reactions at high temperatures and are generally more stable than enzymes from mesophiles, thus prolonging the shelf life of enzyme preparations. A practical example of a thermophilic enzyme of great applied importance is the DNA polymerase isolated from the thermophile, *Thermus aquaticus*. *Taq polymerase*, as this enzyme is known, has been used to automate the repetitive steps involved in the *polymerase chain reaction (PCR)* technique, a method of amplifying specific DNA sequences (see Section 8.9). Because *Taq* polymerase does not denature at the high temperatures needed to melt DNA in the PCR method, it is possible to perform several repetitive melting and polymerization steps without having to add fresh DNA polymerase, as would be the case if enzymes from mesophilic microorganisms were used. This is very cost effective and has allowed the method to become totally automated. DNA polymerase isolated from the hyperthermophilic archaean *Pyrococcus furiosus* (growth temperature optimum, 100°C; see Section 20.6), called *Pfu* polymerase (or "vent polymerase"), is also widely used and is even more thermally stable than *Taq* polymerase. *Pfu* polymerase is also less prone to errors (that is, it has a higher fidelity) than *Taq* polymerase, making it a particularly good enzyme when accuracy is crucial. Other uses of thermophilic enzymes and thermal stable microbial processes are known, and many more will undoubtedly unfold as the biotechnology field expands.

> Temperature has a large effect on growth rate. Each organism shows a characteristic temperature optimum, a maximum, and a minimum. Thermophiles and hyperthermophiles grow at high temperatures (in areas like hot springs, geysers, and industrial heated wastes), psychrophiles and psychrotolerants at low temperatures (in snow fields, glaciers, polar regions and refrigerated foods), and mesophiles at moderate temperatures (such as the mammalian body). The mechanisms by which organisms adapt to their preferred temperatures are of great theoretical as well as practical interest.

9.10   Acidity and Alkalinity (pH)

Acidity or alkalinity of a solution is expressed by its **pH** on a scale in which neutrality is pH 7 (Figure 9.17). Those pH values that are less than 7 are *acidic* and those greater than 7 are *alkaline* (or *basic*). It is important to remember that pH is a *logarithmic function*; a change of one pH unit represents a *ten-fold* change in hydrogen ion concentration. Thus vinegar (pH near 2) and household ammonia (near pH 11) differ in hydrogen ion concentration by one billion fold.

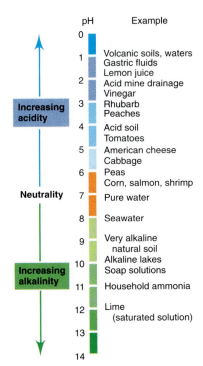

pH   Example

0

1   Volcanic soils, waters
    Gastric fluids
    Lemon juice
2   Acid mine drainage
    Vinegar
3   Rhubarb
    Peaches
4   Acid soil
    Tomatoes
5   American cheese
    Cabbage
6   Peas
    Corn, salmon, shrimp
7   Pure water
8   Seawater
9   Very alkaline
      natural soil
    Alkaline lakes
10 Soap solutions
11 Household ammonia
12 Lime
      (saturated solution)
13
14

Increasing acidity

Neutrality

Increasing alkalinity

FIGURE 9.17   The pH scale.

The pH of a culture medium is adjusted by adding an alkaline compound, such as sodium hydroxide, if the pH is too acid, or an acidic compound, such as hydrochloric acid, if the pH is too alkaline. Because microorganisms usually cause changes in the pH of their environments as they grow, it is often desirable to add a pH *buffer* to the culture medium, which acts to keep the pH relatively constant. Such pH buffers work only over a narrow pH range; hence, different buffers must be selected for different pH regions. For near neutral pH ranges (pH 6 to 7.5), phosphate makes an excellent buffer. Indicator dyes are often added directly to culture media and can then indicate not only the initial pH of the medium but also changes in pH that result from the growth or activity of microorganisms.

pH and microbial growth

Each organism has a pH range within which growth is possible, and usually has a well-defined pH optimum. Most natural environments have pH values between 5 and 9, and organisms with optima in this range are most common. Only a few species can grow at pH values of less than 2 or greater than 10. Organisms that live at low pH are called **acidophiles**. Fungi as a group tend to be more acid-tolerant than bacteria. Many fungi grow optimally at pH 5 or below, and a few grow quite well at pH values as low as 2, although their interior pH is close to neutrality. However, several bacteria are also acidophilic. In fact, some of these bacteria are *obligate* acidophiles, unable to grow at all at neutral pH. Obligately acidophilic bacteria include several species of *Thiobacillus* and several genera of Archaea, including *Sulfolobus* and *Thermoplasma*. *Thiobacillus* and *Sulfolobus* exhibit an interesting property related to their acidophilic nature: they oxidize sulfide minerals and produce sulfuric acid. We discuss

the role of these organisms in mining processes in Section 17.17. It is strange to consider that for obligate acidophiles, a neutral pH is actually toxic! Probably the most critical factor for obligate acidophily is the cytoplasmic membrane. When the pH is raised to neutrality, the cytoplasmic membrane of obligate acidophilic bacteria actually dissolves, and the cells lyse, suggesting that high concentrations of hydrogen ions are required for membrane stability.

A few organisms are *alkaliphilic* because they have high pH optima, sometimes as high as pH 10–11. Alkaliphilic microorganisms are usually found in highly basic habitats such as soda lakes and high carbonate soils. Most alkaliphilic prokaryotes studied have been aerobic nonmarine bacteria, and many are *Bacillus* species. Some extremely alkaliphilic bacteria are also halophilic (salt-loving), and most of these are Archaea (see Section 20.4).

9.11   Water Availability

As we saw in Chapter 2, water is the solvent of life. All organisms require water for life, and water availability is one of the most important factors affecting the growth of microorganisms in natural environments. Water availability does not depend only on the water content of the environment because various solid substances and surfaces are able to absorb water molecules more or less tightly and hence render them unavailable. Also, solutes such as salts and sugars that are dissolved in water have an affinity for water, and the water associated with such solutes becomes unavailable to organisms.

Water availability is generally expressed in physical terms such as **water activity** or **water potential**. In food microbiology where water availability is frequently of major concern (see Section 9.18), the concept of water activity is used. Water activity, abbreviated a_w, is expressed as a ratio of the vapor pressure of the air in equilibrium with a substance or solution divided by the vapor pressure at the same temperature of pure water. Thus values of a_w vary between 0 and 1. Some representative values are given in Table 9.4. Water activities in agricultural soils generally range between 0.90 and 1.00.

Water diffuses from a region of high water concentration (low solute concentration) to a region of lower water concentration (higher solute concentration). Thus, if pure water and a salt solution are separated by a semipermeable membrane, water will diffuse *from* the pure water *into* the salt solution. The process by which water diffuses across a semipermeable membrane is called *osmosis* (see Section 3.3). In most cases, the cytoplasm of a cell has a higher solute concentration than the environment, so that water will tend to diffuse into the cell. However, if a cell is present in an environment of low water activity, there will be a tendency for water to flow out of the cell. Thus, when a cell is immersed in a solution of low water activity, such as a salt or sugar solution, it *loses*

Table 9.4   Water activity of several substances

Water activity, a_w	Material	Some organisms growing at stated water activity
1.000	Pure water	*Caulobacter, Spirillum*
0.995	Human blood	*Streptococcus, Escherichia*
0.980	Seawater	*Pseudomonas, Vibrio*
0.950	Bread	Most Gram-positive rods
0.900	Maple syrup, ham	Gram-positive cocci
0.850	Salami	*Saccharomyces rouxii* (yeast)
0.800	Fruit cake, jams	*Saccharomyces bailii, Penicillium* (fungus)
0.750	Salt lake, salt fish	*Halobacterium, Halococcus*
0.700	Cereals, candy, dried fruit	*Xeromyces bisporus* and other xerophilic fungi

water to the environment and plasmolyses. The salt or sugar solution can, in effect, be considered analogous to a *dry* environment.

From an ecological viewpoint, osmotic effects are of interest mainly in habitats with high concentrations of salts. Seawater contains about 3 percent sodium chloride plus small amounts of many other minerals and elements. Microorganisms found in the sea usually have a specific requirement for the sodium ion, in addition to living optimally at the water activity of seawater. Such organisms are often called **halophiles**. Marine microorganisms are generally inhibited at both higher and lower concentrations of salt (Figure 9.18). Marine microorganisms require sodium ions for the stability of the cytoplasmic membrane, and in addition many of their enzymes require sodium ions for activity.

Most organisms are unable to cope with environments of very low water activity and either die or remain dormant. A few specialized organisms are able to live under conditions of low water activity, and these organisms are important in applied microbiology. Organisms capable of living in very salty environments are called **extreme halophiles**. Organisms able to live in environments high in sugar are called **osmophiles**, and organisms able to live in very dry environments are called **xerophiles**. Examples of these various organisms are given in Table 9.4.

Surviving conditions of reduced water potential

How do these organisms grow under conditions of low water activity? When an organism grows in a medium with a low water activity, it must perform work to extract water from the medium. However, organisms are not able to transport water molecules themselves; water movement is strictly a physical process. Thus, a cell can only obtain water by increasing its internal solute concentration. An increase in internal solute concentration can be done either by pumping ions into the cell from the environment, or by synthesizing or concentrating an organic solute. Organisms are known which have each of these mechanisms. Several examples are given below.

The solute which is used inside the cell for adjustment of cytoplasmic water activity must be noninhibitory to biochemical processes within the cell; such compounds are called **compatible solutes**. Several different compatible solutes are known in microorganisms (Table 9.5). These substances are all highly water-soluble sugars or sugar alcohols, other alcohols, amino acids or their derivatives, or in the case of extremely halophilic Archaea, potassium ions (Table 9.5). Compatible solutes are either synthesized by the microorganisms directly, or in some cases (such as glycine betaine or K+) accumulated from the environment. The concentration of compatible solutes in the cell is dictated by the level of external solutes, and in each organism the maximal amount of compatible solute(s) made or accumulated is a genetically directed function, resulting in different organisms tolerating different ranges of water potential (Tables 9.4 and 9.5). Thus, nonhalotolerant, halotolerant, halophilic, and extremely halophilic microorganisms are to some extent defined by their ability to produce or accumulate compatible solutes.

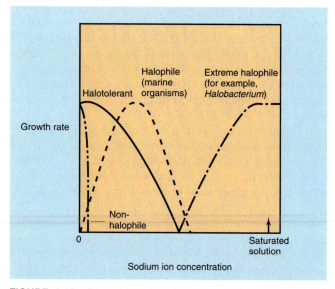

FIGURE 9.18 Effect of sodium ion concentration on growth of microorganisms of different salt tolerances. The optimum NaCl concentration for marine microorganisms is about 3 percent; for extreme halophiles, it is between 12 and 30 percent.

Table 9.5   Compatible solutes of microorganisms

Organism	Major solute(s) accumulated	Minimum a_w for growth
Bacteria, nonphototrophic	Glycine betaine, proline (mainly Gram-positive), glutamate (mainly Gram-negative)	0.97–0.90
Freshwater cyanobacteria	Sucrose	0.98
Marine cyanobacteria	α-Glucosylglycerol	0.92
Marine algae	Mannitol, various glycosides, proline, dimethylsulfonium propionate	0.92
Salt lake cyanobacteria	Glycine betaine	0.90–0.75
Halophilic anoxygenic phototrophic Bacteria (*Ectothiorhodospira* and *Rhodospirillum* species)	Glycine betaine, ectoine, trehalose	0.90–0.75
Halophilic Archaea	K^+	0.75
Dunaliella (halophilic green alga)	Glycerol and sorbitol	0.75
Xerophilic yeasts	Glycerol	0.83–0.62
Xerophilic filamentous fungi	Glycerol	0.72–0.61

Gram-positive cocci of the genus *Staphylococcus* are notoriously salt-tolerant bacteria (in fact, a common isolation procedure for them is to use media containing 7.5 percent NaCl), and these organisms use the amino acid proline as a compatible solute. Betaine is a derivative of the amino acid glycine in which the protons on the amino group are substituted with three methyl groups; this leaves a permanent positive charge on the N atom, which increases its solubility. Glycine betaine is widely distributed as a compatible solute, especially among halophilic Bacteria and cyanobacteria (Table 9.5). Some extremely halophilic members of the phototrophic bacterial genus *Ectothiorhodospira*, most of which inhabit salt lakes and soda lakes (see Section 20.4), produce a novel compatible solute called *ectoine*, which is a derivative of the cyclic amino acid proline. A variety of glycosides are produced by marine algae, but with rare exception they accumulate in only low amounts because the cells are not very halophilic. Xerophilic yeasts and the halophilic green alga *Dunnaliella salina*, which lives in extremely salty lakes such as Great Salt Lake, produce mainly glycerol as compatible solute. Other examples of compatible solutes are listed in Table 9.5.

9.12   Oxygen

As discussed in Chapter 4, microorganisms vary in their need for, or tolerance of, oxygen. Microorganisms can be divided into several groups depending on the effect of oxygen, as outlined in Table 9.6. Organisms that lack a respiratory system cannot use oxygen as a terminal electron acceptor. Such organisms are called **anaerobes**, but there are two kinds of anaerobes; **aerotolerant anaerobes**, which can tolerate oxygen and grow in its presence even though they cannot use it, and **obligate (or strict) anaerobes**, which are killed by oxygen. Although the reason obligate anaerobes are killed by oxygen is not clear, one possibility is that obligate anaerobes are unable to detoxify some of

Table 9.6   Terms used to describe O_2 relationships of microorganisms

Group	O_2 effect
Aerobes	
Obligate	Required
Facultative	Not required, but growth better with O_2
Microaerophilic	Required, but at levels lower than atmospheric
Anaerobes	
Aerotolerant	Not required, and growth no better when O_2 present
Obligate (strict) anaerobes	Harmful or lethal

the products of oxygen metabolism. When oxygen is reduced, several toxic products, hydrogen peroxide (H_2O_2), superoxide (O_2^-), and hydroxyl radical ($OH\cdot$) are formed. Aerobes have enzymes that decompose these products, whereas anaerobes seem to lack all or some of these enzymes.

Microbial culture and the effects of oxygen

For the growth of many aerobes, it is necessary to provide extensive aeration. This is because O_2 is only poorly soluble in water, and the O_2 used up by the organisms during growth is not replaced fast enough by diffusion from the air. Forced aeration of cultures is therefore frequently desirable and can be achieved either by vigorously shaking the flask or tube on a shaker or by bubbling sterilized air into the medium through a fine glass tube or porous glass disc. Usually aerobes grow much better with forced aeration than when O_2 is provided by simple diffusion.

For anaerobic culture, the problem is to *exclude*, not provide, oxygen. And because oxygen is ubiquitous in the air, special methods are needed to culture anaerobic microorganisms. Obligate anaerobes vary in their sensitivity to oxygen, and a number of procedures are available for reducing the O$_2$ content of cultures—some simple and suitable mainly for less sensitive organisms, others more complex but necessary for growth of strict anaerobes.

Bottles or tubes filled completely to the top with culture medium and provided with tightly fitting stoppers will provide anaerobic conditions for organisms not too sensitive to small amounts of oxygen. It is also possible to add a chemical which reacts with oxygen and excludes it from the culture medium. Such a substance is called a *reducing agent* because it reduces oxygen to water. A good example is *thioglycolate*, which is added to a medium commonly used to test whether an organism is aerobic, facultative, or anaerobic (Figure 9.19). After thioglycolate reacts with oxygen throughout the tube, oxygen can only penetrate near the top of the tube where the medium contacts air. Obligate aerobes grow only at the top of such tubes. Facultative organisms grow throughout the tube, but best near the top. Anaerobes grow only near the bottom of the tube, where oxygen cannot penetrate. A redox indicator dye, such as *resazurin*, is usually added to the medium because the dye will change color in the presence of oxygen and thereby indicate the degree of penetration of oxygen into the medium.

To remove all traces of O$_2$ for the culture of very fastidious anaerobes, it is possible to place an O$_2$-consuming gas in a jar holding the tubes or plates. One of the simplest devices for this is the *anaerobic jar*, a heavy-walled jar with a gas-tight seal, within which tubes, plates, or other containers to be incubated are placed. The air in the jar is replaced with hydrogen gas (H$_2$), and in the presence of a chemical catalyst the traces of O$_2$ left in the vessel or culture medium are consumed, thus leading to anaerobic conditions. (See Section 13.1 for a discussion of the use of the anaerobic jar.)

For strict anaerobes, such as the methanogens, it is necessary not only to carefully remove all traces of O$_2$ but also to carry out all manipulations of cultures in an anaerobic atmosphere, as these organisms are frequently killed by even brief exposure to O$_2$. In these cases, a culture medium is first boiled to render it oxygen free, and then a reducing agent such as H$_2$S is added and the mixture is sealed under an oxygen-free gas. All manipulations are carried out under a tiny jet of oxygen-free hydrogen or nitrogen gas which is directed into the culture vessel when it is open, thus driving out any O$_2$ that might enter. For extensive research on anaerobes, special boxes fitted with gloves, called *anaerobic glove boxes*, permit work with open cultures in completely anaerobic atmospheres (Figure 9.20).

Toxic forms of oxygen

To understand oxygen toxicity, it is necessary to consider the chemistry of oxygen. Molecular oxygen, O$_2$, is unique among diatomic elements in that two of its outer orbital electrons are unpaired. It is partly because it has two unpaired electrons that oxygen has such a high reduction potential and is such a powerful oxidizing agent. The unactivated ground state of molecular oxygen, which is the most common, is called **triplet oxygen**.

Singlet oxygen is a higher energy form of molecular oxygen. Singlet oxygen is extremely reactive and is one of the main forms of oxygen that is toxic to living organisms. Singlet oxygen is produced chemically in a variety of ways and is an atmospheric pollutant, being one of the components present in smog. Singlet oxygen may also be produced biochemically, either spontaneously or via specific enzyme systems. The most common biochemical means of singlet-oxygen production involves a reaction of triplet oxygen with visible light. Another important means of generating singlet oxygen is through the lactoperoxidase and myeloperoxidase enzyme systems. These systems are present in milk, saliva, and body cells (phagocytes) that eat and digest invading microorganisms (see Section 11.13). The high reactivity of singlet oxygen means that if it is present in a biological system, a wide variety of uncontrolled and undesirable oxidation reactions can occur, leading to oxidative destruction of vital cell components. Organisms that frequently encounter singlet oxygen, such as airborne bacteria or phototrophic microorganisms, contain pigments called *carotenoids*, which act to quench the effects of singlet oxygen, converting the molecule into a nontoxic form.

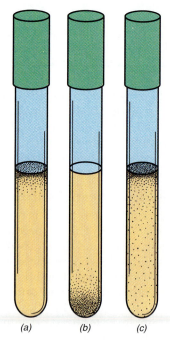

(a)   (b)   (c)

FIGURE 9.19 Aerobic, anaerobic, and facultative growth, as revealed by the position of microbial colonies within tubes of a culture medium. A small amount of agar has been added to keep the liquid from becoming disturbed. (a) Oxygen only penetrates a short distance into the tube, so that obligate aerobes grow only at the surface. (b) Anaerobes, being sensitive to oxygen, only grow away from the surface. (c) Facultative aerobes are able to grow either in the presence or absence of oxygen and thus grow throughout the tube.

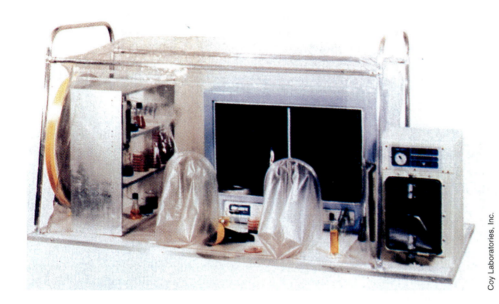

Coy Laboratories, Inc.

FIGURE 9.20
Anaerobic glove bag for manipulating and incubating cultures under anaerobic conditions. The airlock on the right, which can be evacuated and filled with oxygen-free gas, serves as a port for adding and removing materials to and from the glove bag.

Other highly toxic forms of oxygen are also produced in cells. For example, the reduction of O_2 to $2H_2O$, which occurs during respiration (see Section 4.12), requires addition of four electrons (Figure 9.21). This reduction usually occurs by single electron steps, and the first product formed in the reduction of O_2 is **superoxide** anion, O_2^-, a potentially toxic form of oxygen. Superoxide is probably formed transiently in small amounts during normal respiratory processes, and it is also produced in the presence of light following one-electron transfers to oxygen. Flavins, flavoproteins, quinones, thiols, and iron–sulfur proteins (see Section 4.10) all carry out one-electron reductions of oxygen to superoxide. Superoxide is highly reactive and can cause oxidative destruction of lipids and other biochemical components. It has the longest life of the various oxygen intermediates and may even pass from one cell to another. As we will see below, superoxide is important in explaining the oxygen sensitivity of obligate anaerobes.

The next product in the stepwise reduction of oxygen is **peroxide**, O_2^{2-}. Peroxide anion is most familiar in the form of hydrogen peroxide, H_2O_2, which is sufficiently stable that it can be used as an item of chemical commerce. Peroxide is commonly formed biochemically during respiratory processes by a two-electron reduction of O_2, generally mediated by flavoproteins.

Hydrogen peroxide is produced in small amounts by almost all organisms growing aerobically.

Hydroxyl radical, $OH\cdot$, is the most reactive of the various oxygen intermediates. Although it may have a half-life in the cell of only 10^{-9} seconds, the hydroxyl radical is the most potent oxidizing agent known and is capable of attacking any of the organic substances present in cells. It most frequently exerts damage by destroying enzymes and membranes, nicking DNA, and ultimately leading to cell lysis. Hydroxyl radical is formed as a result of the action of ionizing radiation, and is probably one of the main agents in the killing of cells by X rays and gamma rays. Hydroxyl radical can also be produced from H_2O_2 as shown in Figure 9.21.

Enzymes that act on oxygen derivatives

With such an array of toxic oxygen derivatives, it is perhaps not surprising that organisms have developed enzymes that destroy certain oxygen products (Figure 9.22). The most common enzyme in this category is **catalase**, which acts on hydrogen peroxide, the activity of which is illustrated in Figure 9.23. Another enzyme that acts on hydrogen peroxide is **peroxidase**

$O_2 + e^- \longrightarrow O_2^-$ Superoxide

$O_2^- + e^- + 2H^+ \longrightarrow H_2O_2$ Hydrogen peroxide

$H_2O_2 + e^- + H^+ \longrightarrow H_2O + OH\cdot$ Hydroxyl radical

$OH\cdot + e^- + H^+ \longrightarrow H_2O$ Water

Overall: $O_2 + 4e^- + 4H^+ \longrightarrow 2H_2O$

FIGURE 9.21   Four-electron reduction of O_2 to water by stepwise addition of electrons. All of the intermediates formed are reactive and toxic to cells.

(a) Catalase:
$H_2O_2 + H_2O_2 \longrightarrow 2H_2O + O_2$

(b) Peroxidase:
$H_2O_2 + NADH + H^+ \longrightarrow 2H_2O + NAD^+$

(c) Superoxide dismutase:
$O_2^- + O_2^- + 2H^+ \longrightarrow H_2O_2 + O_2$

FIGURE 9.22   Enzymes acting on toxic oxygen species. (a) Catalases and (b) peroxidases are generally porphyrin-containing proteins, although some flavoproteins may act in this manner. (c) Superoxide dismutases are metal-containing proteins, the metal being either copper, zinc, manganese, or iron.

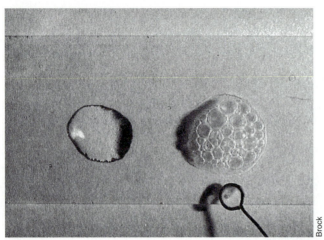

FIGURE 9.23 Method for testing a microbial culture for the presence of catalase. A heavy loopful of cells from an agar culture was mixed on a slide with a drop of 30 percent hydrogen peroxide. The immediate appearance of bubbles is indicative of the presence of catalase. The bubbles are O_2 produced by the reaction

$$H_2O_2 + H_2O_2 \rightarrow 2H_2O + O_2$$

(Figure 9.22*b*), which requires the presence of a reductant, usually NADH. Superoxide is destroyed by the enzyme **superoxide dismutase** (Figure 9.22*c*), which combines two molecules of superoxide to form one molecule of hydrogen peroxide and one molecule of oxygen. Superoxide dismutase and catalase working together can thus bring about the conversion of superoxide back to oxygen. No enzymatic system exists to deal with hydroxyl radicals, probably because of their transient nature in the cell. However, removal of H_2O_2 from cells probably protects the cells in part by preventing the formation of hydroxyl radicals.

Anaerobic microorganisms

Anoxic environments include muds and other sediments of lakes, rivers, and oceans; bogs and marshes; waterlogged soils; canned foods; intestinal tracts of animals; the oral cavity of animals, especially around the teeth; certain sewage treatment systems; deep underground areas such as oil pockets; and some underground waters (see Chapter 17 for description of some of these habitats). In most of these habitats, anoxic conditions and the accompanying low reduction potentials are due to the activities of organisms, mainly bacteria, that consume oxygen during respiration and produce highly reducing substances, such as H_2 and H_2S. If no replacement oxygen is available, the habitat becomes anaerobic.

So far as is known, obligate anaerobiosis occurs in three groups of microorganisms: a wide variety of prokaryotes, a few fungi, and a few protozoa. One of the best known groups of obligately anaerobic Bacteria belong to the genus *Clostridium*, a group of Gram-positive spore-forming rods. Clostridia are widespread in soil, lake sediments, and intestinal tracts, and are often responsible for spoilage of canned foods (see Sections 9.18 and 19.26). Other obligately anaerobic bacteria are found among the methanogens, the sulfate-reducing and homoacetogenic bacteria (see Chapters 19 and 20), and many of the bacteria which inhabit the animal gut (see Section 11.4). Among obligate anaerobes, however, the sensitivity to oxygen varies greatly; some organisms are able to tolerate traces of oxygen whereas others are not.

Oxygen is a peculiar environmental factor because it can be both essential and toxic. Aerobes require oxygen to live but anaerobes do not, and may even be killed by O_2. Several of the intermediates formed when oxygen is reduced to water are extremely toxic, including the hydroxyl radical (OH·), hydrogen peroxide (H_2O_2), and superoxide (O_2^-), but enzymes that will neutralize most of these toxic products also exist. The culture of anaerobes requires that special precautions be taken to exclude oxygen from their environment.

9.13 Control of Microbial Growth: Heat Sterilization

Thus far we have been discussing environmental factors that influence growth, with emphasis on *promoting* the growth process. However, the principles that have been presented can be put to practical use in the *control* of microbial growth. Control can be effected either by *killing* organisms or by *inhibiting* their growth. The complete killing of all organisms is called **sterilization** and is brought about by use of heat, radiation, or chemicals. Sterilization means the complete destruction of all microbial cells present; once a product is sterilized, it will remain sterile indefinitely if it is properly sealed (see the illustration of Pasteur's swan-necked flask in Figure 1.18).

Heat sterilization

One of the most important and widely used agents for sterilization is heat. Let us consider the effects of temperature on viability. As temperature rises past the maximum temperature for growth, lethal effects become apparent. As shown in Figure 9.24, death from

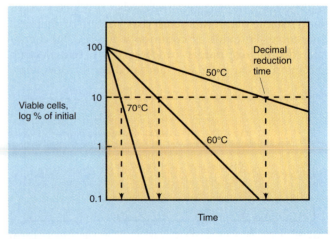

FIGURE 9.24 Effect of temperature on viability of a mesophilic bacterium.

heating is an exponential or first-order function and occurs more rapidly as temperature is raised. The first-order relationship shown in Figure 9.24 means that the rate of death is proportional at any instant only to the concentration of organisms at that instant and that the time taken for a definite fraction (for example, 90 percent) of the cells to be killed is independent of the initial concentration. These facts have important practical consequences. If we wish to *sterilize* a population, it will take longer at lower temperatures than at higher temperatures. It is thus necessary to select the time and temperature that will sterilize under stated conditions. The nature of the heat is also important, as moist heat has better penetrating power than dry heat.

The most useful way of characterizing heat inactivation is to determine the *time* at which a defined fraction of the population is killed. For a variety of reasons the time required for a ten-fold reduction in the population density at a given temperature, called the **decimal reduction time** or *D*, is the most useful parameter. Over the range of temperatures usually used in food sterilization, the relationship between *D* and temperature is essentially exponential, so that when the logarithm of *D* is plotted against temperature, a straight line is obtained (Figure 9.25). The slope of the line provides a quantitative measure of the sensitivity of the organism to heat under the conditions used, and the graph can be used in calculating process times for sterilization, such as in canning operations.

Although accurate, determination of decimal reduction times is a fairly lengthy procedure, since it requires making a number of viable count measurements. A less satisfactory but easier way of characterizing the heat sensitivity of an organism is to determine the **thermal death time**, the *time* at which all cells in a suspension are killed at a given temperature. This is done simply by heating samples of this suspension for different times, mixing the heated suspensions with culture medium, and incubating. When all cells are killed, no growth will be evident in the incubated samples. Of course, the thermal death time determined in this way will depend on the size of the population tested, since a longer time will be required to kill all cells in a large population than in a small one. If the number of cells is standardized, then it is possible to compare the heat sensitivities of different organisms by comparing their thermal death times. When the logarithm of the thermal death time is graphed versus temperature, a straight line similar to that shown in Figure 9.24 is obtained.

Spores and heat sterilization

Vegetative cells and bacterial endospores vary considerably in heat resistance. For instance, in the autoclave, a temperature of 121°C is normally reached. Under these conditions, endospores may require 4 to 5 minutes for a decimal reduction, whereas vegetative cells may require only 0.1 to 0.5 minute at 65°C for a decimal reduction. Thus, the practice of heat sterilization revolves around procedures for killing endospores.

The nature of the medium in which heating takes place also influences the killing of both vegetative cells and spores. Microbial death is more rapid at acidic pH values, and for this reason acid foods such as tomatoes, fruits, and pickles are much easier to sterilize than more neutral pH foods such as corn and beans. High concentrations of sugars, proteins, and fats decrease heat penetrability and usually increase the resistance of organisms to heat, while high salt concentrations may either increase or decrease heat resistance, depending on the organism. Dry cells (and spores) are more heat resistant than moist ones; for this reason, heat sterilization of dry objects always requires much higher temperatures and longer times than does sterilization of moist objects.

Bacterial endospores (see Section 3.11) are the most heat-resistant structures known: they are able to survive conditions of heating that would rapidly kill vegetative cells of the same species. It appears that a major factor in heat resistance is the amount and state of *water* within the endospore. During endospore formation (see Section 3.11), the protoplasm is reduced to a minimum volume as a result of the accumulation of Ca^{2+} and synthesis of dipicolinic acid, which lead to formation of a gel-like structure. At this stage, the thick cortex forms around the protoplast and contraction of the cortex results in a shrunken protoplast low in water. The water content of the protoplast determines the degree of heat resistance of the spore. If endospores have high water content, they have low heat resistance, and the heat resistance of spores can be varied by altering the water content of the spores. Water moves freely in and out of spores, so that it is not the impermeability of the spore coat that affects

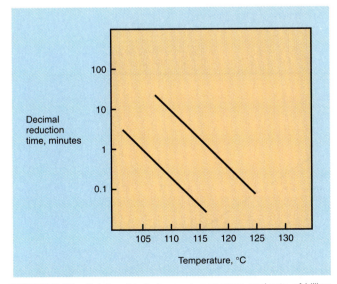

FIGURE 9.25   Relationship between temperature and rate of killing as indicated by the decimal reduction time for two different microorganisms. Data for such a graph are obtained from curves such as those given in Figure 9.24. The upper line in the figure represents data for a very heat resistant organism.

water content, but the physical state of the spore protoplast, that is, the degree of gel-like structure.

The autoclave

We now consider the **autoclave**, a device of immense practical value to microbiology, where the use of moist heat ensures killing of all microorganisms, including heat-resistant endospores.

Sterilization by heat involves treatment that results in the complete destruction of all organisms, and since bacterial endospores are ubiquitous, sterilization procedures are designed to eliminate them. This requires heating at temperatures above boiling and the use of steam under pressure (Figure 9.26a). The usual procedure is to heat at 1.1 kg/cm² (15 lb/in²) steam pressure, which yields a temperature of 121°C. Heating is usually done in an autoclave, a sealed device allowing for the entrance of steam under pressure (Figure 9.26a), although if only small batches of material need be sterilized, a home pressure cooker is quite satisfactory and much less expensive. At 121°C, the time

of autoclaving to achieve sterilization is generally considered to be 10–15 minutes (Figure 9.26b). If bulky objects are being sterilized, heat transfer to the interior will be slow, and the heating time must be sufficiently long so that the material is at 121°C for 10–15 minutes. Longer than normal times are also required when large volumes of liquids are being autoclaved because large volumes take longer to reach sterilization temperatures. Note that it is not the *pressure* of the autoclave that kills the microorganisms but the *high temperature* that can be achieved when steam is placed under pressure.

Pasteurization and sterilization

Pasteurization is a process using mild heat to reduce the microbial populations in milk and other foods that are exceptionally heat sensitive. It is named for Louis Pasteur, who first used heat for controlling the spoilage of wine (see The Origin of Pasteurization box). Pasteurization is not synonymous with sterilization since not all organisms are killed. Originally, pasteurization of milk was used to kill pathogenic bacteria, especially the organisms causing tuberculosis, brucellosis, Q fever, and typhoid fever, but it was discovered that the keeping qualities of milk were also improved following pasteurization. Today, since milk rarely comes from cows infected with the pathogens mentioned above, pasteurization is used primarily because it improves keeping qualities of milk, milk products, and various other beverages and food products (see box).

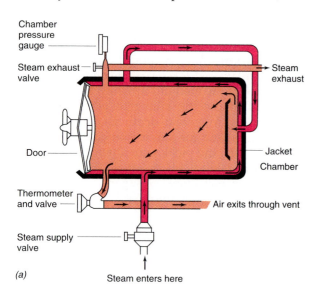

(a)

FIGURE 9.26
Use of the autoclave for sterilization. (a) The flow of steam through an autoclave. (b) A typical autoclave cycle. Shown is the sterilization of a fairly bulky object. Note that the temperature of the object rises more slowly than the temperature of the autoclave.

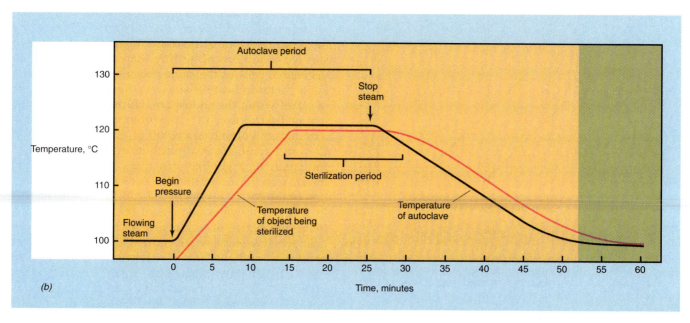

(b)

The Origin of Pasteurization

Pasteur's name is forever linked in the public mind with the process of pasteurization. The background of the development of the pasteurization process has been nicely discussed by René Dubos in his book on Pasteur's life.* We quote from this book here:

The demonstration that microbes do not generate spontaneously encouraged the development of techniques to destroy them and to prevent or minimize subsequent contamination. Immediately these advances brought about profound technological changes in the preparation and preservation of food products and subsequently in other industrial processes as well . . . .

It was soon discovered that the introduction of microorganisms in biological products can be minimized by an intelligent and rigorous control of the technological operations, but cannot be prevented entirely. The problem therefore was to inhibit the further development of these organisms after they had been introduced into the product. To this end, Pasteur first tried to add a variety of antiseptics, but the results were mediocre and, after much hesitation, he considered the possibility of using heat as a sterilizing agent.

Pasteur's first studies of heat as a preserving agent were carried out with wine. Pasteur had grown up in one of the best wine districts in France, and, as a connoisseur of the beverage, was much disturbed at the thought that heating might alter its flavor

and bouquet. He therefore proceeded with very great caution and eventually convinced himself that heating at 55°C would not alter appreciably the bouquet of the wine . . . . These considerations led to the process of partial sterilization, which soon became known the world over under the name of 'pasteurization,' and which was found applicable to wine, beer, cider, vinegar, milk, and countless other perishable beverages, foods, and organic products.

It was characteristic of Pasteur that he did not remain satisfied with formulating the theoretical basis of heat sterilization, but took an active interest in designing industrial equipment adapted to the heating of fluids in large volumes and at low cost. His treatises on vinegar, wine, and beer are illustrated with drawings and photographs of this type of equipment, and describe in detail the operations involved in the process. The word 'pasteurization' is, indeed, a symbol of his scientific life; it recalls the part he played in establishing the theoretical basis of the germ theory, and the phenomenal effort that he devoted to making it useful to his fellow humans. It reminds us also of his well-known statement: "There are no such things as pure and applied science—there are only science, and the application of science."

*René Dubos, *Pasteur and Modern Science*. 1988. Science-Tech Publishers, Madison, WI.

Pasteurization of milk is usually achieved by passing the milk continuously through a heat exchanger where its temperature is raised quickly to 71°C and held there for 15 seconds, and then is quickly cooled, a process called **flash pasteurization**. Occasionally pasteurization is done by heating milk in bulk at 63–66°C for 30 minutes and then quickly cooling. However, flash pasteurization is more satisfactory in that it alters the flavor less, kills heat-resistant organisms better, and can be carried out on a continuous-flow basis, thus making it adaptable to large dairy operations. Modern dairies almost always use flash

pasteurization methods, and ever shorter exposure and higher temperature flash procedures are becoming common in the dairy industry.

9.14   Filter Sterilization

Although heat is the most common and effective way of sterilizing liquids, it cannot be used for the sterilization of heat-sensitive fluids. An especially valuable technique for sterilizing such materials is filtration. The filter is a device with pores too small for the passage of microorganisms but large enough to allow the passage of the liquid. The range of particles involved in sterilization is rather large. Some of the largest microbial cells are greater than 10 μm in diameter, but at the lower end of the size scale certain bacteria are less than 0.3 μm in diameter. Filters are also frequently used in the field of virology, and here we are dealing with even smaller particles—as small as 10 nm. (Historically, filtration was used to demonstrate the existence of virus-sized infectious particles; see The Name *Virus* box in Chapter 6.)

There are three major types of filters, which are illustrated in Figure 9.27. One of the oldest types of filters used is the **depth filter**. A depth filter is a fibrous

Sterilization is the complete killing of all organisms. In many practical situations, an appropriate sterilization process must be used. The most widely used method for sterilization involves the application of heat. The temperature for heat sterilization is selected to eliminate the most heat-resistant organisms in the material, usually bacterial endospores. For routine sterilization, an autoclave is used; this permits application of heat under pressure at temperatures above the boiling point of water. Pasteurization does not sterilize but rids the material of pathogenic microorganisms and reduces the load of microorganisms in a material to prolong its storage life.

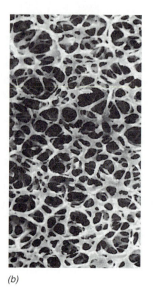
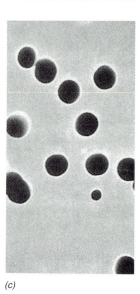

FIGURE 9.27
Comparison of the structure of (a) a depth filter, (b) a conventional membrane filter, and (c) a Nuclepore filter.

(a)   (b)   (c)

sheet or mat made from paper, asbestos, or glass fibers, which is constructed of a random array of overlapping fibers (Figure 9.27a). The depth filter functions primarily because particles get trapped in the tortuous paths created throughout the depth of the structure. Although it is possible to manufacture depth filters that will remove particles as small as viruses and bacteria, such filters have considerable limitations. In practice, depth filters are often used as *prefilters* to remove larger particles from a solution so that clogging does not occur in the final filter sterilization process. They are also used in the filter sterilization of air in industrial processes (see Section 10.5).

The most common type of filter for sterilization in the field of microbiology is the **membrane filter** (Figure 9.27b). The membrane filter is a tough disc, generally composed of cellulose acetate or cellulose nitrate, which is manufactured in such a way that it contains a large number of tiny holes. The membrane filter differs from the depth filter in that the membrane filter functions more like a sieve, trapping many of the particles on the surface of the filter. Membrane filters are open structures with about 80–85 percent of the filter occupied by space. This openness provides for a relatively high fluid flow rate. However, the conventional membrane filter does not function completely as a sieve, since many particles penetrate into the filter matrix before being removed.

The third type of filter in common use is the **nucleation track (Nuclepore) filter**. These filters are created by treating very thin polycarbonate films (10 μm in thickness) with nuclear radiation and then etching the film with a chemical. The radiation causes localized damage in the film and the etching chemical then acts to enlarge these damaged locations into holes. The sizes of the holes can be precisely controlled by the strength of the etching solution and the time allowed for etching to proceed. A typical Nuclepore filter has very uniform holes that are arranged almost vertically through the thin film (Figure 9.27c). Nuclepore filters function as true sieves, removing all parti-

cles greater than the size of the holes. However, because of their lower porosity, the flow rate through Nuclepore filters is lower than through membrane filters, and clogging occurs more rapidly. Nuclepore filters are very commonly used in scanning electron microscopy of microorganisms, since an organism of interest can be easily removed from a liquid by filtration, and the particles are all held in a uniform plane on the top of the filter where they can be readily studied in the microscope (Figure 9.28).

The use of a membrane filter for the sterilization of a liquid is illustrated in Figure 9.29. The filter apparatus is generally sterilized separately from the filter, and the apparatus assembled aseptically at the time of filtration. The arrangement shown in Figure 9.29 is suitable for small volumes of liquid. For large-volume sterile filtration, the membrane filter material is arranged in a cartridge and placed in a stainless steel housing. Large volume filtration of heat-sensitive flu-

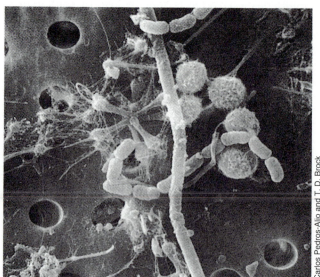

Carlos Pedros-Alio and T. D. Brock

FIGURE 9.28 Scanning electron micrograph of aquatic bacteria and algae trapped on a Nuclepore filter. The pore size is 5 μm.

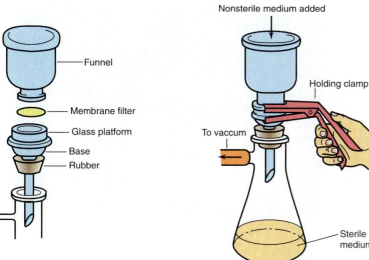

Nonsterile medium added

Funnel

Membrane filter

Glass platform

Base

Rubber

Holding clamp

To vaccum

Sterile medium

FIGURE 9.29
Use of a membrane filter assembly for sterilization of a liquid.

ids is very commonly done in the pharmaceutical industry.

Most membrane filters can be autoclaved to achieve sterility and then handled aseptically when assembling the filtering apparatus (Figure 9.29). However, small, reusable plastic cartridges are also available that can be autoclaved after the insertion of a membrane filter. Filtration is then done by using a syringe to force the liquid through the sterile filtration apparatus into a sterile collection vessel.

9.15 Chemical Control of Microbial Growth

Besides heat, the growth of microorganisms can also be controlled with chemical agents. An **antimicrobial agent** is a chemical that kills or inhibits the growth of microorganisms. Such a substance may be either a synthetic chemical or a natural product. Agents that kill organisms are often called *cidal agents*, with a prefix indicating the kind of organism killed. Thus, we have **bactericidal, fungicidal**, and **algicidal** agents. A bactericidal agent, or bactericide, kills bacteria. It may or may not kill other kinds of microorganisms. Agents that do not kill but only inhibit growth are called *static agents*, and we can speak of **bacteriostatic, fungistatic**, and **algistatic** agents.

The distinction between a static and cidal agent is often arbitrary since an agent that is cidal at high concentrations may only be static at lower concentrations. To be effective, a static agent must be continuously present, since if it is removed or its activity neutralized, the organisms present may initiate growth if conditions are favorable.

Antimicrobial agents can vary in their **selective toxicity**. Some act in a rather nonselective manner and have similar effects on all types of cells. Others are far more selective and are more toxic to microorganisms than to animal tissues. Antimicrobial agents with selective toxicity are especially useful as *chemotherapeutic*

agents in treating infectious diseases, as they can be used to kill disease-causing microorganisms without harming the host. They will be described later in this chapter.

Effect of antimicrobial agents on growth

Antimicrobial agents affect growth in a variety of ways, and a study of the action of these agents in relation to the growth curve is of considerable aid in understanding their modes of action. Three distinct kinds of effects can be observed when an antimicrobial agent is added to an exponentially growing bacterial culture: *bacteriostatic, bactericidal*, and *bacteriolytic*. As described in the preceding section, a **bacteriostatic** effect is observed when growth is inhibited, but no killing occurs (Figure 9.30a). Bacteriostatic agents are frequently inhibitors of protein synthesis and act by binding to ribosomes. The binding, however, is not tight, and when the concentration of the agent is lowered, the agent becomes free from the ribosome and growth is resumed. The mode of action of protein synthesis inhibitors is discussed in Section 5.8. **Bactericidal** agents kill cells, but lysis or cell rupture does not occur (Figure 9.30b). Bactericidal agents are a class of chemical agents that generally bind tightly to their cellular targets and are not removed by dilution. **Bacteriolytic** agents induce killing by cell lysis, which is observed as a decrease in cell numbers or in turbidity after the agent is added (Figure 9.30c). Bacteriolytic agents include antibiotics that inhibit cell wall synthesis, such as penicillin (see Sections 3.5 and 9.17), as well as agents that damage the cytoplasmic membrane.

Measuring antimicrobial activity

Antimicrobial activity is measured by determining the smallest amount of agent needed to inhibit the growth of a test organism, a value called the **minimum inhibitory concentration (MIC)**. In one method of determining the MIC, a series of culture tubes is prepared, each tube containing medium with a different concen-

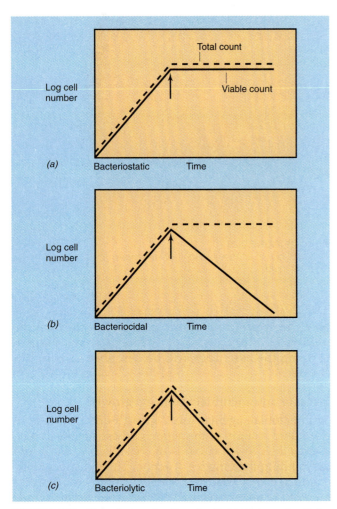

FIGURE 9.30 Three types of action of antimicrobial agents. At the time indicated by the arrow, a growth-inhibitory concentration was added to the exponentially growing culture. Note the relationships between viable and total cell counts.

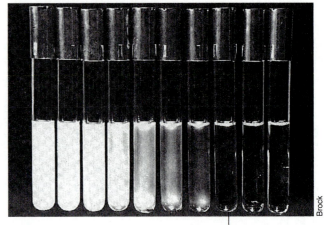

Minimum inhibitory concentration

FIGURE 9.31 Antibiotic assay by tube dilution, permitting detection of the *minimum inhibitory concentration, MIC*. A series of increasing concentrations of antibiotic is prepared in the culture medium. Each tube is inoculated and incubation allowed to proceed. Growth (turbidity) occurs in those tubes with antibiotic concentrations below the MIC.

The test plate may be inoculated by pouring an overlay of agar containing the test organism, or by swabbing the medium surface with a broth culture of the test organism. Known amounts of the antimicrobial agent are added to filter paper discs which are then placed on the surface of the agar. During incubation, the agent diffuses from the filter paper into the agar; the further it gets from the filter paper, the smaller the concentration of the agent. At some distance from the disc the MIC is reached. Past this point growth occurs, but closer to the disc growth is absent. A **zone of inhibition** is thus created, and its size can be measured with a ruler; the diameter will be proportional to the amount of antimicrobial agent added to the disc and to the overall effectiveness of the agent.

> Chemicals called antimicrobial agents are often used to control microbial growth. Chemicals that kill organisms are called cidal agents; those that inhibit their growth are called static agents. The value of a chemical agent is assessed by determining the minimum concentration necessary to kill or inhibit growth and by determining whether it exhibits selective toxicity.

9.16 Disinfectants and Antiseptics

Disinfectants and **antiseptics** are antimicrobial agents used in quite different situations. *Disinfectants* are chemicals that kill microorganisms and are used on inanimate objects. *Antiseptics*, on the other hand, are chemical agents that kill or inhibit growth of microorganisms and that are sufficiently nontoxic to be applied to living tissues (Table 9.7). The quantitative aspects of the killing of microorganisms with chemical agents were discussed in the previous section. Chemical antimicrobial agents, which are frequently referred to as **germicides**, have wide use in situations where it is impractical to use heat for sterilization. Hospitals

tration of the agent, and all tubes of the series are inoculated. After incubation, the tubes in which growth does *not* occur (indicated by absence of visible turbidity) are noted, and the MIC is thus determined (Figure 9.31). This simple and effective procedure is often called the *tube dilution technique*. The MIC is not an absolute constant for a given agent, since it is affected by the nature of the test organism used, the inoculum size, the composition of the culture medium, the incubation time, and the conditions of incubation, such as temperature, pH, and aeration. If all conditions are rigorously standardized, it is possible to compare different antimicrobials and determine which is most effective against a given organism or to assess the activity of a single agent against a variety of organisms. Note that the tube dilution method does not distinguish between a cidal and a static agent, since the agent is present in the culture medium throughout the entire incubation period.

Another commonly used procedure for studying antimicrobial action is the **agar diffusion method** (Figure 9.32). A Petri plate containing an agar medium evenly inoculated with the test organism is prepared.

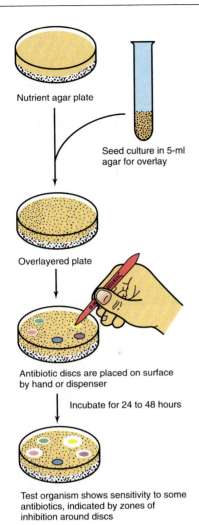

Nutrient agar plate

Seed culture in 5-ml agar for overlay

Overlayered plate

Antibiotic discs are placed on surface by hand or dispenser

Incubate for 24 to 48 hours

Test organism shows sensitivity to some antibiotics, indicated by zones of inhibition around discs

FIGURE 9.32 Agar diffusion method for assaying antibiotic activity.

frequently find it necessary to sterilize heat-sensitive materials, such as surgical instruments, thermometers, lensed instruments, polyethylene tubing and catheters, and inhalation and anaesthesia equipment. For these applications, *cold sterilization* is commonly used. Cold sterilization is performed in enclosed devices that resemble autoclaves but employ chemical agents such as ethylene oxide, formaldehyde, or hydrogen peroxide. In the food industry, floors, walls, and surfaces of equipment must often be treated with germicides to reduce the load of microorganisms. Drinking water is commonly disinfected to reduce or eliminate any potentially harmful organisms, and treated wastewater is generally disinfected before it is discharged into the environment.

Although the testing of germicides in laboratory situations is relatively straightforward, in practical cases the determination of efficacy is often very difficult. This is because many germicides are neutralized by organic materials, so that germicidal concentrations are not maintained for sufficient time. Further, bacteria and other microorganisms are often encased in particles, and the penetration of a chemical agent to the viable cells may be slow or absent. Also, bacterial spores are much more resistant to any germicides than are vegetative cells. Thus germicide effectiveness must ultimately be determined under the intended conditions of use. It should be emphasized that germicidal treatments do not necessarily sterilize. **Sterility** is defined as the complete absence of living organisms and their viruses, and sterilization with chemicals often requires long contact periods under special conditions. In most cases, the use of germicides ensures only that

Table 9.7	Antiseptics and disinfectants	
Agent	**Use in health-related fields**	**Mode of action**
Antiseptics		
Organic mercurials	Skin	Combines with SH groups of proteins
Silver nitrate	Eyes of newborn to prevent blindness due to infection by *Neisseria gonorrhoeae*	Protein precipitant
Iodine solution	Skin	Iodinates tyrosine residues of proteins; oxidizing agent
Alcohol (70% ethanol in water)	Skin	Lipid solvent and protein denaturant
Bis-phenols (hexachlorophene)	Soaps, lotions, body deodorants	Disrupts cell membrane
Cationic detergents (quaternary ammonium compounds)	Soaps, lotions	Interact with phospholipids of membrane
Hydrogen peroxide (3% solution)	Skin	Oxidizing agent
Disinfectants		
Mercuric dichloride	Tables, bench tops, floors	Combines with SH groups
Copper sulfate	Algicide in swimming pools, water supplies	Protein precipitant
Iodine solution	Medical instruments	Iodinates tyrosine residues
Chlorine gas	Purification of water supplies	Oxidizing agent
Chlorine compounds	Dairy, food industry equipment	Oxidizing agent
Phenolic compounds	Surfaces	Protein denaturant
Cationic detergents (quaternary ammonium compounds)	Medical instruments; food, dairy equipment	Interact with phospholipids
Ethylene oxide (gas)	Temperature-sensitive laboratory materials such as plastics	Alkylating agent
Ozone	Purifying drinking water	Strong oxidizing agent

Table 9.8   Industries that use chemicals to control microbial growth

Industry	Chemicals	How used
Paper	Organic mercurials, phenolics	To prevent microbial growth during manufacture
Leather	Heavy metals, phenolics	Antimicrobial agents are present in the final product
Plastic	Cationic detergents	To prevent growth of bacteria on aqueous dispersions of plastics
Textile	Heavy metals, phenolics	To prevent microbial deterioration of fabrics exposed in the environment such as awnings, tents
Wood	Phenolics	To prevent deterioration of wooden structures
Metal working	Cationic detergents	To prevent growth of bacteria in aqueous cutting emulsions
Petroleum	Mercurics, phenolics, cationic detergents	To prevent growth of bacteria during recovery and storage of petroleum and petroleum products
Air conditioning	Chlorine, phenolics	Prevent growth of bacteria (e.g. *Legionella*) in cooling towers
Electrical power	Chlorine	Prevent growth of bacteria in condensors and cooling towers
Nuclear	Chlorine	Prevent growth of radiation-resistant bacteria in nuclear reactors

the microbial load is reduced significantly, although perhaps with the hope that pathogenic organisms are completely eliminated. However, bacterial endospores as well as vegetative cells such as those of *Mycobacterium tuberculosis*, the causal agent of tuberculosis, are very resistant to the action of many germicides, so that the complete elimination of pathogens by germicidal treatment may not always occur. A summary of the most widely used germicides and their modes of action, is given in Table 9.7.

In addition to their wide use in the medical field where concern is great about the potential hazards of infectious wastes, antiseptic and disinfectant chemicals find many important uses in industry. In industry, the concern is generally to prevent microbial deterioration of materials and the use of antimicrobial agents is therefore quite extensive. This frequently leads to toxic waste problems when large amounts of antimicrobial agents are released into the environment. Table 9.8 summarizes some of the industries in which chemicals are used to control microbial growth.

9.17   Chemotherapeutic Agents

The discussion above has dealt primarily with chemical agents that are used to inhibit microbial growth *outside* the human body. Most of the chemicals mentioned were too toxic to be used in the body, although antiseptics can be used on the skin. For control of in-

fectious disease, agents which can be used internally are essential. Such agents are called **chemotherapeutic agents**, and they have played major roles in modern medicine (see Microbiology and "Magic Bullets" box). We discuss the uses of chemotherapeutic agents for treatment of infectious disease in Section 11.12. Here we present simply some general concepts about the agents themselves. The key requirement of a successful chemotherapeutic agent is *selective toxicity*, the ability to inhibit bacteria or other microorganisms without affecting the body (see box).

Sulfa drugs and other growth factor analogs

In Section 4.21 we discussed growth factors and defined them as specific chemical substances required in the medium because the organism cannot synthesize them. Substances exist that are related to growth factors and that act to block the utilization of the growth factor. These "growth factor analogs" usually are structurally similar to the growth factors in question but are sufficiently different so that they cannot duplicate the work of the growth factor in the cell. The first of these to be discovered were the *sulfa drugs*, the first modern chemotherapeutic agents to specifically inhibit the growth of bacteria (see box); they have since proved highly successful in the treatment of certain diseases. The simplest sulfa drug is **sulfanilamide** (Figure 9.33*a*). Sulfanilamide acts as an analog of *p*-aminobenzoic acid (Figure 9.33*b*), which is itself a

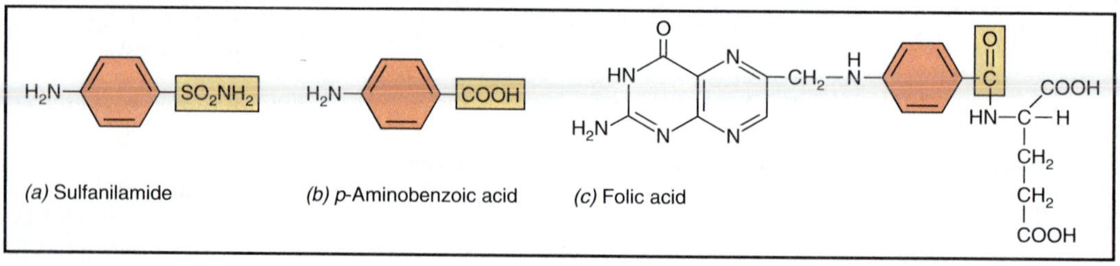

FIGURE 9.33   (a) The simplest sulfa drug, sulfanilamide. (b) It is an analog of *p*-aminobenzoic acid, which itself is part of (c) the growth factor folic acid.

Microbiology and "Magic Bullets"

The development of chemotherapeutic agents probably has had a greater impact on clinical medicine than any other discovery. Although a variety of natural chemical agents had been used earlier, the real advances in work with chemotherapeutic agents began with the German scientist Paul Ehrlich. In the early 1900s, Ehrlich developed the concept of selective toxicity. He began his work by studying the staining of microorganisms with dyes and observed that some dyes stained microorganisms but not animal tissue. He assumed that if a dye did not stain a tissue, the dye molecules were unable to combine with any of the cell constituents. He then reasoned that if such a dye had toxic properties, it should not affect the animal cells because it could not combine with them, but it should attack the microbial cells. In an infected animal, chemicals of this sort should behave like "magic bullets," striking the pathogen but missing the host. It was not necessary that the chemical be a dye; it only needed to be selective in its binding properties. Ehrlich proceeded to test large numbers of chemicals for selectivity and discovered the first chemotherapeutic agents, of which Salvarsan, a drug for the cure of syphilis, was the most famous.

However, no chemical agents were discovered that were able to affect the vast majority of infectious agents until the 1930s, when Gerhard Domagk discovered the sulfa drugs. The discovery of the sulfas came about as a direct offshoot of the approach used by Ehrlich—the large-scale screening of chemicals for activity in infectious diseases in experimental animals. Domagk, at the Bayer Chemical Company in Germany, tested a large variety of synthetic organic chemicals, mainly dyes, for their ability to cure streptococcal infections in mice. The first active compound was Prontosil, which had the intriguing property that it was active in mice but had no activity against streptococci grown in the test tube. It was then discovered that in the animal body, Prontosil broke down to sulfanilamide, which was the actual active agent. With this discovery, it was possible to embark on a program of synthesis based on the sulfanilamide structure, which yielded a large number of active drugs having various medical uses even today. D. D. Woods in England then showed that p-aminobenzoic acid specifically counteracted the inhibitory action of sulfanilamide, and he also showed that the streptococci required p-aminobenzoic acid for growth. This led to the concept of the growth factor analog, which enabled chemists to pursue the synthesis of a wide variety of chemotherapeutic agents.

Despite the successes of the sulfa drugs, most infectious diseases were still not under chemical control. It took the discovery of the first antibiotic, penicillin, by Alexander Fleming, a Scottish physician engaged in research at St. Mary's Hospital in London, to point investigators in the right direction. Fleming's first paper on penicillin, published in 1929, begins as follows:

> While working with staphylococcus variants a number of culture plates were set aside on the laboratory bench and examined from time to time. In the examination these plates were necessarily exposed to the air and they became contaminated with various micro-organisms. It was noticed that around a large colony of contaminating mould the staphylococcus colonies became transparent and were obviously undergoing lysis. Subcultures of this mould were made and experiments conducted with a view of ascertaining something of the properties of the bacteriolytic substance which had evidently been formed in the mould culture and which had diffused into the surrounding medium.

Fleming characterized the product, and since it was produced by a fungus of the genus *Penicillium,* gave it the name *penicillin.* His work, however, did not include a process for large-scale production nor did it show that penicillin was effective in the treatment of infectious disease. This was later done by a group of British scientists at Oxford University, headed by Howard Florey, who began their work in 1939, motivated in part by the impending World War II and the knowledge that infectious disease was at that time the leading cause of death among soldiers on the battlefield. Florey and his colleagues developed methods for the analysis and testing of penicillin and for its production in large quantities. They then proceeded to test penicillin against bacterial infections in humans. Penicillin was dramatically effective in controlling staphylococcal and pneumococcal infections and was also more effective for streptococcal infections than the sulfa drugs. With the effectiveness of penicillin demonstrated and the war in Europe becoming more intense, Florey brought cultures of the penicillin-producing fungus to the United States in 1941. He persuaded the U.S. government to create a large-scale research program, which led to a joint effort of the pharmaceutical industry, the U.S. Department of Agriculture at its laboratory in Peoria, Illinois, and several universities. By the end of World War II, penicillin was available in large amounts, for civilian as well as military use. As soon as the war was over, pharmaceutical companies entered into commercial production of penicillin on a competitive basis and began to look for other antibiotics. Success was quick and dramatic, and the impact on medicine has been close to phenomenal. Infant and child mortality have been greatly reduced, and many diseases that at one time had high fatality rates are now no more than medical curiosities.

part of the vitamin folic acid (Figure 9.33c). In organisms that synthesize their own folic acid, sulfanilamide acts by blocking the synthesis of folic acid. Sulfanilamide is active against Bacteria but not against higher animals because Bacteria synthesize their own folic acid, whereas higher animals must obtain folic acid preformed in their diet.

The concept that a chemical substance can act as a competitive inhibitor of an essential growth factor has had far-reaching effects on chemotherapeutic research, and today analogs are known for various vitamins, amino acids, purines, pyrimidines, and other compounds. A few examples are given in Figure 9.34. In these examples, the analog has been formed by addition of fluorine or bromine. Fluorine is a relatively small atom and does not alter the overall shape of the molecule, but it changes the chemical properties sufficiently so that the compound does not act normally in cell metabolism. Fluorouracil resembles the nucleic-acid base uracil; bromouracil resembles another base, thymine (chemicals that resemble bases, such as bromouracil, are also used as mutagens; see Section 7.3).

Analogs of purines and pyrimidines also act as inhibitors of virus replication. We discuss in Section 11.17 a few antiviral compounds which function in this way. It is of interest that the original concept of competitive inhibition was developed from studies on the mode of action of sulfa drugs (see box).

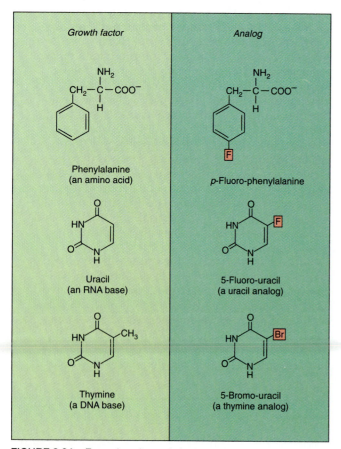

FIGURE 9.34 Examples of growth factor analogs.

Antibiotics

Antibiotics are chemical substances produced by certain microorganisms which are active against other microorganisms. Antibiotics constitute a special class of chemotherapeutic agents, distinguished most importantly by the fact that they are natural products (products of microbial activity) rather than synthetic chemicals (products of human activity). Antibiotics constitute one of the most important classes of substances produced by large-scale microbial processes. The industrial aspects of antibiotics, how antibiotics are discovered, and some details of specific antibiotics, are discussed in Chapter 10. The uses of antibiotics in treatment of disease are discussed in Chapter 11. Here we present a broad overview of the antibiotics and describe some of the methods of testing antibiotic activity.

A very large number of antibiotics have been discovered, but probably less than 1 percent of them have been of great practical value in medicine. Those few which have been useful have had a dramatic impact on the treatment of many infectious diseases. Further, some antibiotics can be made more effective by chemical modification; these are said to be *semisynthetic*.

The sensitivity of microorganisms to antibiotics varies (Figure 9.35). Gram-positive Bacteria are usually more sensitive to antibiotics than are Gram-negative Bacteria, although some antibiotics act only on Gram-negative Bacteria. An antibiotic that acts upon both Gram-positive and Gram-negative Bacteria is called a **broad-spectrum antibiotic**. In general, a broad-spectrum antibiotic will find wider medical usage than a *narrow-spectrum antibiotic*, which acts on only a single group of organisms. A narrow-spectrum antibiotic may, however, be quite valuable for the control of microorganisms that fail to respond to other antibiotics. Some antibiotics have an extremely limited spectrum of action, being effective for only one or a few bacterial species.

Antibiotics and other chemotherapeutic agents can be grouped based on chemical structure or on mode of action (Figure 9.36). In Bacteria, the important targets of chemotherapeutic action are the cell wall, the cytoplasmic membrane, and the biosynthetic processes of protein and nucleic acid synthesis. Some chemotherapeutic agents, such as the sulfa drugs, work because they mimic important growth factors needed in cell metabolism (see box). Examples of each major type are given in Figure 9.36.

Antibiotic resistance

It is obvious from the discussion above that not all antibiotics act against all microorganisms. Some microorganisms are resistant to some antibiotics. Antibiotic resistance can be an inherent property of a microorganism, or it can be acquired. There are several reasons why microorganisms may have inherent resistance to an antibiotic. (1) The organism may lack the structure which an antibiotic inhibits. For instance,

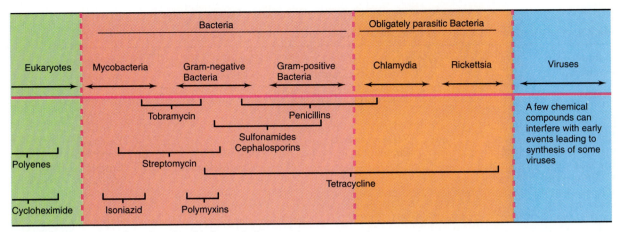

FIGURE 9.35 Ranges of action of antibiotics and other agents.

some bacteria, such as mycoplasmas, lack a typical bacterial cell wall and are resistant to penicillins. (2) The organism may be impermeable to the antibiotic. (3) The organism may be able to alter the antibiotic to an inactive form. (4) The organism may modify the *target* of the antibiotic. (5) By genetic change, alteration may occur in a metabolic pathway in which an antimicrobial agent blocks. (6) The organism may be able to pump out any antibiotic entering the cell (efflux).

We give some specific examples of bacterial resis-

tance to antibiotics in Table 9.9. Also, as discussed in Section 7.9, antibiotic resistance can be genetically coded for by the microorganism at either the chromosomal or the plasmid level, specific types of resistance typically having a genetic basis in one location or the other (Table 9.9). Because of the development of antibiotic resistance, testing of bacteria isolated from clinical material for antibiotic sensitivity must be carried out. Details of the sensitivity testing of clinical isolates is described in Sections 11.16 and 13.3.

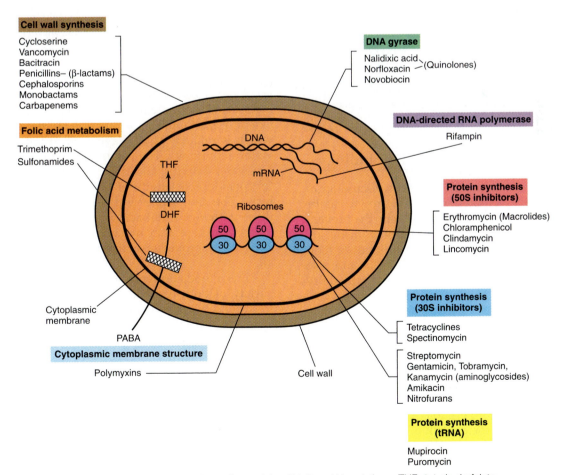

FIGURE 9.36 Mode of action of major antibacterial antibiotics. Abbreviations: THF, tetrahydrofolate; DHF, dihydrofolate; mRNA, messenger RNA; tRNA, transfer RNA.

Table 9.9   Mechanisms of bacterial resistance to antibiotics

Resistance mechanism	Antibiotic example	Genetic basis of resistance	Mechanism present in
Reduced permeability	Penicillins	Chromosomal	*Pseudomonas aeruginosa* Enteric Bacteria
Efflux	Tetracyclines	Plasmid	Enteric Bacteria
Inactivation of antibiotic (for example, penicillinase; modifying enzymes methylases, acetylases, and phosphorylases; and others)	Penicillins	Plasmid and chromosomal	*Staphylococcus aureus* Enteric Bacteria *Neisseria gonorrhoeae*
	Chloramphenicol	Plasmid and chromosomal	*Staphylococcus aureus* Enteric Bacteria
Alteration of target (for example, RNA polymerase, rifamycin; ribosome, erythromycin, and streptomycin; DNA gyrase, quinolones)	Aminoglycosides Erythromycin	Plasmid Chromosomal	*Staphylococcus aureus* *Staphylococcus aureus* Enteric Bacteria
	Rifamycin Streptomycin Norfloxacin		Enteric Bacteria Enteric Bacteria *Staphylococcus aureus*
Development of resistant biochemical pathway	Sulfonamides	Chromosomal	Enteric Bacteria *Staphylococcus aureus*

Many chemicals are too toxic to be used on or in the human body but chemotherapeutic agents are chemicals that can be used in this way. Chemotherapeutic agents have the property of selective toxicity, affecting microbial cells without significantly affecting cells of the body. Antibiotics constitute a class of chemotherapeutic agents that are produced by microorganisms themselves.

9.18   Some Aspects of Food Microbiology

We complete this chapter with a short discussion of how some of the principles of microbial growth can be applied to an important practical problem, the prevention of microbial growth in foods. Microorganisms are responsible for some of the most serious kinds of food poisonings and toxicities (see Section 15.13) and also cause spoilage of a wide variety of food and dairy products. Because of this, a knowledge of microbial growth in foods is of great value.

Food spoilage

Food spoilage is probably the most serious economic problem in the food processing industry. Foods are subject to attack by microorganisms in a variety of ways, and such attack is generally harmful to the quality of the food. Foods are organic and hence provide adequate nutrients for the growth of a wide variety of chemoorganotrophic bacteria. What determines whether a food will support the growth of a microorganism? The physical and chemical characteristics of the food and how it is stored determine its degree of sus-

ceptibility to microbial attack. Foods can be classified into three major categories: (1) *highly perishable foods* such as meats, fish, poultry, eggs, milk, and most fruits and vegetables; (2) *semiperishable foods* such as potatoes, some apples, and nuts; and (3) *stable* or *nonperishable foods* such as sugar, flour, rice, and dry beans. How do these three categories differ? To a great extent, perishability is a function of *moisture content*, which is related, as we have seen in Section 9.11, to water activity, a_w. The stable foods are the ones with *low* water activity, and these can generally be stored for considerable lengths of time without any undue effects. The perishable and semiperishable foods are those with *high* water activity. In order for these latter foods to be kept, they must be stored under conditions such that microbial growth is slow or does not occur at all.

Fresh foods undergo spoilage by the activities of a variety of both bacteria and fungi, but each type of fresh food is typically attacked only by particular microorganisms (Table 9.10). This is because the chemical properties of foods vary widely, and different foods select from the indigenous microflora spoilage organisms that are best able to utilize the nutrients available. For example, enteric bacteria are rarely implicated in fruit or vegetable spoilage but can spoil meat products because their habitat is the gut of warm-blooded animals and they can easily contaminate the meat when the animal is slaughtered. Likewise, lactic acid bacteria, the most common microorganisms in dairy products, are the major spoilage organisms of milk and milk products. *Pseudomonas*

Table 9.10   Microbial spoilage of fresh food*

Food product	Type of microorganism	Common spoilage organisms
Fruits and vegetables	Bacteria	*Erwinia, Pseudomonas, Corynebacteria* (mainly vegetable pathogens; will rarely spoil fruit)
	Fungi	*Aspergillis, Botrytis, Geotrichium, Rhizopus, Penicillium, Cladosporium, Alternaria, Phytophora,* various yeasts
Fresh meat, poultry, and seafood	Bacteria	*Acinetobacter, Aeromonas, Pseudomonas, Micrococcus, Achromobacter, Flavobacterium, Proteus, Salmonella*
	Fungi	*Cladosporium, Mucor, Rhizopus, Penicillium, Geotrichium, Sporotrichium, Candida, Torula, Rhodotorula*
Milk	Bacteria	*Streptococcus, Leuconostoc, Lactococcus, Lactobacillus, Pseudomonas, Proteus*
High sugar foods	Bacteria	*Clostridium, Bacillus, Flavobacterium*
	Fungi	*Saccharomyces, Torula, Penicillium*

Several other organisms not listed here have also been isolated, but the organisms listed are the most commonly observed spoilage agents of the fresh foods indicated.

species inhabit both soil and the animal body and are thus widely involved in the spoilage of various fresh foods (Table 9.10).

Food spoilage is defined as any change in the visual appearance, smell, or taste of a food product that makes it unacceptable to the consumer. From a health standpoint, spoiled food is not necessarily food that is unsafe to eat. However, unpalatable food will not be purchased by the average consumer and thus preventing food spoilage is a major goal of the food industry.

In discussing microbial growth in foods, it is useful to recall the standard bacterial growth curve, as was illustrated in Figure 9.7. The lag phase may be of variable duration in a food, depending upon the contaminating organism and its previous growth history. The rate of growth during the exponential phase will depend upon the temperature, the available nutrients, and other conditions of growth. The time required for the population density to reach a significant level in a given food product will depend on both the initial inoculum and how rapidly growth occurs during the ex-

ponential phase. However, it is only when the microbial population density reaches a substantial level that harmful effects are usually observed. Indeed, throughout much of the exponential phase of growth, population densities may be low enough that no perceptible effect can be observed, but because of the nature of exponential growth (see Figure 9.2), it is only the *last* doubling or two that leads to problems. Thus, for much of the period of microbial growth in a food, the observer is unaware of impending problems.

Food preservation

Besides moisture, one of the most crucial factors affecting microbial growth in food is **temperature** (see Section 9.9). In general, the *lower* the temperature of storage, the *less* rapid the spoilage rate, although, as we have seen, psychrophilic microorganisms are able to grow well at refrigerator temperatures. Therefore, storage of perishable food products for long periods of time is only possible at temperatures below freezing. Freezing greatly alters the physical structure of many

When the Pasteurizer Fails

Within a two-week period in the spring of 1985, about 18,000 persons in northern Illinois and surrounding states experienced severe intestinal disease due to infection with a bacterium called *Salmonella typhimurium*. The outbreak was traced to milk provided by a single dairy plant operated by a large grocery store chain. The bacterium was easily cultivated from patients and from the milk, using a culture medium selective for *Salmonella*. All of the cultures obtained were of a single strain which was highly resistant to the antibiotics ampicillin and tetracycline. The outbreak was the largest in the United States associated with a single *Salmonella* strain and illustrates how a

widely consumed product, once contaminated, can spread microbial disease. *Salmonella* is found in dairy cattle and in raw milk, but is killed by the pasteurization process. Careful engineering study at the suspected dairy plant showed that a defective valve had allowed unpasteurized milk to contaminate pasteurized milk. The economic impact of this outbreak on the company was enormous: millions of dollars were forfeited in lost revenue, legal fees were huge, and the milk processing plant closed its doors permanently. This story illustrates two points: the pasteurization process in a large dairy is complex, and antibiotic-resistant Bacteria are pervasive in modern society.

food products and therefore cannot be universally used, but it is widely and successfully used for the preservation of meats and many vegetables and fruits. Freezers providing a temperature of −20°C are most commonly used for storing frozen products. At −20°C, storage for weeks or months is possible, but even at this temperature some microbial growth may occur, usually in pockets of liquid water trapped within the frozen mass. Also, nonmicrobial chemical changes in the food may still occur. For very long-term storage, lower temperatures than −20°C are necessary, such as −80°C (dry ice temperature), but maintenance at such low temperatures is expensive and consequently is not used for routine food storage.

Another major factor affecting microbial growth in food is **pH** or **acidity**. Foods vary widely in pH, although most are neutral or acidic. As we have seen (in Section 9.10), microorganisms differ in their ability to grow under acidic conditions. It is an important practical matter that most food spoilage bacteria do not grow at pH values below 5. It is for this reason that acid is often used in food preservation, in the process called *pickling*. Foods commonly pickled include cucumbers (sweet, sour, and dill pickles), cabbage (sauerkraut), and some meats and fruits. The food can be made acid either by addition of vinegar or by allowing acidity to develop directly in the food through microbial action, in which case the product is called a **fermented food**. The microorganisms involved in food fermentations are acid-tolerant bacteria, the lactic acid bacteria, the acetic acid bacteria, and the propionic acid bacteria. But even these bacteria are not able to continue developing once the pH drops below about pH 4, so that the food fermentation is self-limiting. Vinegar, frequently added to food to lower the pH, is essentially dilute acetic acid. Vinegar itself is a product of the action of acetic acid bacteria; its industrial production is discussed in Section 10.12.

Since microorganisms do not grow at low water activities, microbial growth can be controlled by lowering the water activity of the product by drying or by adding salt. Sun drying is the least expensive way of

drying foods and is appropriate if the climate is right. Some foods can be successfully dried with heat, although generally with significant changes in flavor or quality. The least damaging way of drying foods is freeze-drying (lyophilization), but this is quite expensive and can only be justified if the food has a high economic value. Milk, meats, fish, vegetables, fruits, and eggs are all preserved by drying.

A number of foods are preserved by addition of salt or sugar to lower water activity. Foods preserved by addition of sugar are mainly fruits (jams, jellies, and preserves). Salted products are primarily meats and fish. Sausage and ham are preserved by salt, although these products vary in water activity depending on how much salt is added and how much the meat has been dried. Several famous sausages such as *landjaeger* can be stored indefinitely in the absence of refrigeration, but are fairly salty to eat.

Canning

Canning is a process in which a food is sealed and heated so as to kill all living organisms, or at least to assure that there will be no growth of residual organisms in the can. Canning is hence a type of heat sterilization, and the principles already presented in Section 9.13 apply. When the can is properly sealed and heated, the food should remain stable and unspoiled indefinitely, even when stored in the absence of refrigeration. Home-canned foods are usually prepared in glass containers whereas commercial products are most often in tin-coated steel cans. In any canning process, the *seal* on the can or jar is the most critical part. The heating process itself is done by submersion in water, generally under pressure.

The temperature–time relationships for canning depend upon the type of food, especially its pH, the size of the container, and the consistency or bulkiness of the food. Because heat must penetrate completely to the center of the food within the can, heating times must be longer for large cans or very viscous foods. Acid foods can often be canned effectively by heating just to boiling, 100°C, whereas nonacid foods must be

FIGURE 9.37
Changes in cans as a result of microbial spoilage. (a) Normal can; note that the top of the can is indented due to negative pressure (vacuum) inside. (b) Slight swell resulting from minimal gas production. Note that the lid is slightly raised. (c) Severe swell due to extensive gas production. Note the great deformation of the can. (d) The can shown in (c) was dropped and the gas pressure resulted in a violent explosion. Note that the lid has been torn apart.

(a)   (b)   (c)   (d)

heated to autoclave temperatures. It is not desirable to heat foods much longer than necessary, since prolonged heating affects the nutritional and eating qualities of the food. It should also be noted that the term *sterilize* is not completely accurate when used in relation to canning. The numbers of organisms are reduced greatly during the heating process, and in most cases the product is probably sterile, but in fact if the initial load of organisms in the food is high, then not every cell may be killed. Heating times long enough to guarantee absolute sterility of every can would alter the food so greatly that it would likely be unpalatable. Thus, a fine balance is generally kept between reduction in microbial numbers, eradication of all pathogens, and palatability.

The environment inside a can is anaerobic and microbial growth in a canned food frequently is because of the activity of fermentative organisms which produce extensive amounts of gas. This results in a change in the physical appearance of the can, which is often visible by observation of the outside of the can (Figure 9.37). If microbial growth is extensive enough, the gas pressure build-up can actually cause the can to explode. Because many of the anaerobic bacteria which grow in canned foods are powerful toxin producers (see Section 11.8 for a discussion of bacterial exotoxins), food from a visibly altered can should never be eaten.

Chemical food preservation

Although chemicals should never be used in place of careful food sanitation, there are a number of chemical antimicrobial agents that are used commercially to control microbial growth in foods. While a large number of antimicrobial chemicals exist that are probably suitable food preservatives, only a relatively few are classified by the U.S. Food and Drug Administration as "Generally Recognized as Safe" and find wide application in the food industry. These are summarized in Table 9.11. Many of these chemicals, like sodium propionate, have been used for many years with no evidence of human toxicity. Others, like nitrites, ethylene or propylene oxides, or antibiotics, are more controversial food supplements because of evidence that these compounds are detrimental to human health.

Nitrites can react at acid pH with secondary amines in the body to form *nitrosamines*, a class of po-

Table 9.11   Chemical food preservatives

Chemical	Foods
Sodium or calcium propionate	Bread
Sodium benzoate	Carbonated beverages, fruit juices, pickles, margarine, preserves
Sorbic acid	Citrus products, cheese, pickles, salads
Sulfur dioxide, sulfites, bisulfites	Dried fruits and vegetables; wine
Formaldehyde (from food-smoking process)	Meat, fish
Ethylene and propylene oxides	Spices, dried fruits, nuts
Sodium nitrite	Smoked ham, bacon

tentially carcinogenic chemical compounds. Ethylene and propylene oxides are alkylating agents and therefore mutagenic (see Section 7.3), and there is some suggestion that these compounds are also carcinogenic. Finally, antibiotic residues in meats and poultry have been questioned because these antibiotics might select for antibiotic resistant strains of bacteria in the human gut.

Because of lengthy and costly testing programs for any new chemical proposed as a food preservative today, it is unlikely that new compounds will be added to the list in Table 9.11 in the near future. An alternative to chemical preservatives, preservation by ionizing radiation, is beginning to make headway in the food industry, and commercial irradiators are now available to treat various food products.

> Food spoilage is one of the most important microbial processes, and much of food microbiology deals with the development of methods for keeping microorganisms from growing in food during its processing and storage. Foods vary considerably in their sensitivity to microbial growth, acid foods being much more resistant to spoilage than those of neutral or alkaline pH. Microbial growth in foods can be controlled by use of heat or chemical agents, but only a few chemical agents are approved for direct addition to foods.

Study Questions

1. Define growth and describe one direct and one indirect method by which growth can be measured. Make sure that the methods you choose agree with your definition.

2. Describe the growth pattern of a population of bacterial cells from the time this population is first inoculated into fresh medium. How may the growth pattern differ when it is measured by total count or by viable count?

3. Describe briefly the process by which a single cell develops into a visible colony on an agar plate. With this explanation as a background, describe the principle behind the viable count method.

4. Starting with four bacterial cells in a rich nutrient medium, with a one-hour lag phase and a 20-minute generation time, how many cells would there be after one hour? After two hours? After two hours if one of the initial four cells was dead?

5. How is the population density controlled in a chemostat? In a chemostat, if the dilution rate is increased, will

the growth rate increase, decrease, or stay the same? Will the population density increase, decrease, or stay the same? If the concentration of the limiting substrate in the medium reservoir is increased, will the growth rate increase, decrease, or stay the same? Will the population density increase, decrease, or stay the same?

6. Examine the graph describing the relationship between growth rate and temperature. Give an explanation, in biochemical terms, of why the optimum temperature for an organism is closer to its maximum than its minimum.

7. Would you expect to find a psychrophilic microorganism alive in a hot spring? Why? It is frequently possible to isolate thermophilic microorganisms from cold water environments. Give an explanation of how this can be.

8. Even though hyperthermophilic bacteria are resistant to killing by the temperatures used in the food canning process, they never present a problem in food spoilage. Why?

9. Give an explanation of why a culture of *Streptococcus* might not exhibit a simple first-order thermal death curve. (Hint: think about the morphology of this bacterium as discussed in Chapter 3.)

10. Even though heat sterilization is quite effective with many foods, give an explanation of why it is still important to begin with a food that has been prepared under relatively clean conditions so that its microbial load is low.

11. What is *thermal death time*? How would the presence of bacterial endospores affect the thermal death time?

12. Write an explanation in molecular terms for how an obligate halophile is able to make water molecules flow *into* the cell.

13. Contrast an aerotolerant and an obligate anaerobe in terms of sensitivity and ability to grow in the presence of oxygen (O_2). How does an aerotolerant anaerobe differ from a microaerophile?

14. In terms of oxygen relations (obligate anaerobe, obligate aerobe, facultative aerobe, microaerophile) list the kinds of organisms that you would expect to find in each of the following environments of the human body: skin, heart muscle, teeth, stomach, lungs, large intestines.

15. Compare and contrast the enzymes *catalase* and *superoxide dismutase* from the following points of view: substrates, oxygen products, organisms containing them, role in oxygen tolerance of the cell.

16. How would you distinguish bacteriostatic from bactericidal actions of an antibiotic? Bactericidal from bacteriolytic?

17. Define *antibiotic*. How do antibiotics differ from other antimicrobial agents?

18. Briefly describe two different ways to determine the *minimum inhibitory concentration* of an antibiotic. Explain why this is not necessarily the same concentration needed to inhibit microorganisms within the body.

19. What is the principle behind preservation of the following food products: corn, yogurt, pickles, and raisins.

Supplementary Readings

American Waterworks Association. 1991. *Standard Methods for the Examination of Water and Wastewater*, 18th edition. American Waterworks Association, Washington, D.C. A useful reference providing methods commonly used in measuring the numbers of microorganisms.

Banwart, G. J. 1989. *Basic Food Microbiology*, 2nd edition. Van Nostrand Rheinhold, New York. A standard textbook of food microbiology.

Block, S. S., (ed.) 1991. *Disinfection, Sterilization, and Preservation*, 4th edition. Lea and Febiger, Philadelphia. A standard reference on chemical disinfection, with separate chapters on each of the different groups of agents.

Brock, T. D., (ed.) 1986. *Thermophiles: General, Molecular, and Applied Microbiology*. John Wiley & Sons, New York. The best treatment available of thermophilic bacteria, with emphasis on hyperthermophiles.

Denyer, S. P., and **W. B. Hugo,** (eds.) 1991. *Chemical Biocides*. Blackwell Scientific Publications, London. Chapters written by experts on the effects of antibiotics and other chemicals on bacterial growth and metabolism.

Kristjansson, J. K., (ed.) 1992. *Thermophilic Bacteria*. CRC Press, Boca Raton, FL. An overview of different types of thermophilic and hyperthermophilic prokaryotes.

Lederberg, J., (ed.) 1992. *Encyclopedia of Microbiology*. Academic Press, San Diego. Many sections of this treatise on microbiology deal with topics related to bacterial growth.

Lennette, E. H., A. Balows, W. J. Hausler, Jr., and **H. J. Shadomy.** 1991. *Manual of Clinical Microbiology*, 5th edition. American Society for Microbiology, Washington, D.C. A good reference for information on measurement of antibiotic activity and sensitivity.

Monod, J. 1949. The growth of bacterial cultures. *Annu. Rev. Microbiol.* 3:371–394. The classic review of the bacterial growth curve.

Neidhardt, F. C., J. L. Ingraham, and **M. Schaechter.** 1990. *Physiology of the Bacterial Cell*. Sinauer Associates, Sunderland, MA. A textbook of bacterial physiology that deals with all aspects of bacterial growth.

Vanderzant, C., and **D. F. Splittstoesser,** (eds.) 1992. *Compendium of Methods for the Microbiological Examination of Foods*, 3rd edition. American Public Health Association, Washington, D.C. A treatise on the microorganisms involved in food spoilage and how to detect them in food products.

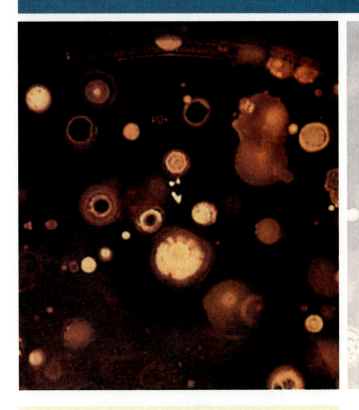

Industrial Microbiology

Industrial microbiology is the discipline that uses microorganisms, usually grown on a large scale, to produce valuable commercial products. Industrial microbiology originally began with alcoholic fermentation processes, such as those for making beer and wine. Subsequently, microbial processes were developed for the production of pharmaceutical agents (such as antibiotics), food additives (such as amino acids), enzymes, and chemicals such as butanol and citric acid. All of these industrial microbiological processes are enhancements of metabolic reactions that microorganisms were already capable of carrying out, with the goal in most cases of simply overproducing the product of interest. In research on such processes, the industrial microbiologist's main task is to modify the organism, usually by traditional genetic methods, or to modify the process so that the *highest yield* of the desired product is obtained.

But now, in addition to traditional industrial microbiology, a new era has unfolded—that of *microbial biotechnology*. In biotechnology, methods for gene manipulation have given rise to new microbial products, most of which are not naturally produced by microorganisms. Microbial production arises from the fact that the microorganisms have been genetically engineered to make the desired product. In biotechnology, the microbiologist employs genetic engineering methods to actually *create new organisms* containing new complements of genes that enable them to make novel products, usually of high commercial value.

In the present chapter, we focus on traditional industrial microbiological processes, emphasizing the unique problems and requirements of large-scale industrial cultivation of microorganisms. We discuss a number of large-scale microbial processes and concentrate on the key industrial products of antibiotics, foods, food supplements, and alcohol, with emphasis on those aspects of these processes that are of special interest to the industrial microbiologist. The biotechnological aspects of industrial microbiology were covered in Chapter 8, where the discussion focused on the gene manipulations necessary to produce "biotech drugs" and related products and on the nature of some of the products themselves.

10.1   Industrial Microorganisms

Not all microorganisms find industrial use. Whereas microorganisms isolated from nature exhibit cell growth as their main physiological property, industrial microorganisms are organisms which have been selected carefully so that they manufacture one or more specific *products*. Even if the industrial microorganism is one which has been isolated by traditional techniques, it becomes a highly "modified" organism before it enters large-scale industries. To a great extent, industrial microorganisms are metabolic specialists, capable of producing specifically and to high yield particular metabolites. In order to achieve this high metabolic specialization, industrial strains are genetically altered by mutation or recombination. Minor metabolic pathways are usually repressed or eliminated. Metabolic imbalance frequently is present. Industrial microorganisms may show many altered cellular and biochemical properties. Although industrial strains may grow satisfactorily under the highly specialized conditions of the industrial fermentor, they may show poor growth properties in competitive environments in nature.

Where do industrial strains come from?

The ultimate source of all strains of industrial microorganisms is, of course, the natural environment. But through the years, as large-scale microbial processes have become perfected, a number of industrial strains have been deposited in *culture collections*. When a new industrial process is patented, the applicant for the patent is required to deposit a strain capable of carrying out the process in a recognized culture collection. There are a number of culture collections which serve as the repositories of microbial cultures (Table 10.1). Although these culture collections can serve as ready sources of cultures, it should be understood that most industrial companies will be reluctant to deposit their *best* cultures in culture collections.

Strain improvement

As we have noted, the initial source of an industrial microorganism is the natural environment, but the original isolate is greatly modified in the laboratory. As a result of this modification, progressive improvement in the yield of a product can be anticipated. The most dramatic example of such progressive improvement is that of penicillin, the antibiotic produced by the fungus *Penicillium chrysogenum*. When penicillin was first produced on a large scale (see Microbiology and "Magic Bullets" box in Chapter 9), yields of 1–10 µg/ml were obtained. Over the years, as a result of strain improvement coupled with changes in the medium and growth conditions, the yield of penicillin has been increased to about 50,000 µg/ml! It is of interest that all of this 50,000-fold increase in yield was obtained by mutation and selection; no genetic engineering manipulations were involved. The introduction of new genetic techniques has led to further, albeit much more modest, yield increases.

What are the requisites of an industrial microorganism?

A microorganism suitable for industrial use must, of course, produce the substance of interest, but there is much more than that. The organism must be available in pure culture, must be genetically stable, and must

---

Table 10.1   Culture collections that supply cultures of industrial microorganisms*

Abbreviation	Name	Location
ATCC	American Type Culture Collection	Rockville, MD, United States
CBS	Centraalbureau voor Schimmelculturen	Baarn, The Netherlands
CCM	Czechoslovak Collection of Microorganisms	J.E. Purkyne University, Brno, Czech Republic
CDDA	Canadian Department of Agriculture	Ottawa, Canada
CIP	Collection of the Institut Pasteur	Paris, France
CMI	Commonwealth Mycological Institute	Kew, United Kingdom
DSM	Deutsche Sammlung von Mikroorganismen	Braunschweig, Germany
FAT	Faculty of Agriculture, Tokyo University	Tokyo, Japan
IAM	Institute of Applied Microbiology	University of Tokyo, Japan
NCIB	National Collection of Industrial Bacteria	Aberdeen, Scotland
NCTC	National Collection of Type Cultures	London, United Kingdom
NRRL	Northern Regional Research Laboratory	Peoria, IL, United States

*Listed here are just a few of the general culture collections. Many universities and research laboratories maintain collections of specific microbial groups.

Miniglossary for Chapter 10

AMINOGLYCOSIDES a group of antibiotics including streptomycin, containing amino sugars linked by glycosidic bonds

β-LACTAMS a group of antibiotics including penicillin, containing the four-membered heterocyclic β-lactam ring

BIOCONVERSION the use of microorganisms to carry out a chemical reaction that is more costly or not feasible nonbiologically

BREWING the manufacture of alcoholic beverages such as beer from the fermentation of malted grains

COMMODITY CHEMICAL chemicals such as ethanol that have low monetary value and are thus sold primarily in bulk

DISTILLED BEVERAGE a beverage containing alcohol concentrated by distillation

FERMENTATION in an industrial context, any large scale microbial process whether carried out aerobically or anaerobically

FERMENTOR the tank in which an industrial fermentation is carried out

IMMOBILIZED ENZYME an enzyme attached to a solid support over which substrate is passed and converted into product

PRIMARY METABOLITE a metabolite excreted during the growth phase

SCALE-UP conversion of an industrial process from a small laboratory setup to a large commercial fermentation

SECONDARY METABOLITE a metabolite excreted at the end of the primary growth phase and into the stationary phase

SEMISYNTHETIC PENICILLIN a penicillin produced using components derived from both microbial fermentation and from chemical syntheses

SINGLE-CELL PROTEIN protein derived from microbial cells for use as food or a food supplement

TETRACYCLINES a class of antibiotics containing the four-membered naphthacene ring

WINE a product of the alcoholic fermentation of fruit juices, usually grape juice, by yeast

grow in large-scale culture. It must also be possible to maintain cultures of the organism for a long period of time in the laboratory and in the industrial plant. The culture should preferably produce spores or some other reproductive cell form so that the organism can be easily inoculated into large fermentors.

An important characteristic is that the industrial organism should grow rapidly and produce the desired product in a relatively short period of time. The organism must also be able to grow in a relatively inexpensive liquid culture medium obtainable in bulk quantities. Many industrial microbiological processes use waste carbon from other industries as major or supplemental ingredients for large-scale culture media. These include *corn steep liquor* (a product of the corn wet milling industry that is rich in nitrogen and growth factors), *whey* (a waste liquid of the dairy industry containing lactose and minerals), and other industrial waste materials having high organic carbon contents. In addition, an industrial microorganism should not be harmful to humans or economically important animals or plants. Because of the large population size in the industrial fermentor, and the virtual impossibility of avoiding contamination of the environment outside the fermentor, a pathogen would present potentially disastrous problems.

Another important requisite of an industrial microorganism is that it should be possible to remove the microbial cells from the culture medium relatively easily. In the laboratory, cells are removed from culture media primarily by centrifugation, but centrifugation may be difficult or expensive on a large scale. The most favorable industrial organisms are those of large cell size, since larger cells settle rapidly from a culture or can be easily filtered out with relatively inexpensive filter materials. Fungi, yeasts, and filamentous bacteria are preferred. Unicellular bacteria, because of their small size, are difficult to separate from a culture fluid.

Finally, an industrial microorganism should be amenable to genetic manipulation. In industrial microbiology, increased yields have been obtained genetically primarily by means of mutation and selection. It is also desirable for the industrial organism to be capable of genetic recombination, either by a sexual or by some sort of parasexual process. Genetic recombination permits the incorporation in a single genome of genetic traits from more than one organism. However, many industrial strains have been greatly improved by mutation and selection without any use of genetic recombination.

> Only certain microorganisms are suitable to be used for industrial purposes, since an industrial microorganism must grow readily in large-scale equipment. Although an industrial microorganism must first of all manufacture the product of interest, it must also grow rapidly and on relatively cheap culture media. Industrial microorganisms must also be readily manipulated genetically, so that strain improvement can be accomplished. Lastly, an industrial microorganism must not be pathogenic or produce any toxic products.

10.2   Kinds of Industrial Products

Microbial products of industrial interest can be of several major types:

1. **Microbial cells** themselves may be the desired product. This is the case for *yeast* cultivated for food and baking, and for *mushrooms* cultivated for food. Processes have also been developed for rais-

ing *bacteria* and *algae* as food sources, but these processes have not yet become commercially significant. The term *single-cell protein* is sometimes used to refer to microbial cells as industrial products, because the protein content of the microorganisms is high and is often the characteristic of most commercial interest. Another group of industrial products in which the microbial cells themselves are of interest are those that are used for *inoculation*. For instance, bacteria of the genera *Rhizobium* and *Bradyrhizobium* are inoculated onto leguminous seeds in order to encourage the formation of nitrogen-fixing root nodules (see Section 17.24), and cultures of *lactic acid bacteria* are sold as inoculants for fermented dairy and sausage products.

2. **Enzymes** produced by microorganisms may be the desired product. A number of important enzymes used commercially are produced in large scale by microbial processes, including starch-digesting enzymes (amylases), protein-digesting enzymes (proteases, rennin), and fat-digesting enzymes (lipases). An important industrial enzyme is *glucose isomerase*, used in large amounts industrially for the production from glucose of fructose sweetener. Another important microbial enzyme is *penicillin acylase*, used industrially in the manufacture of semi-synthetic penicillins (see Section 10.7).

3. **Pharmaceutical products** constitute some of the most important industrial products. Pharmacologically active agents such as antibiotics, steroids, and alkaloids are generally in the category called *secondary metabolites*. Secondary metabolites are compounds that are not produced during the primary growth phase but are produced when the culture is entering stationary phase. Secondary metabolites are some of the most important and interesting industrial products, and an understanding of the nature of secondary metabolism is important in developing new processes (see Section 10.3). It is also in the pharmaceutical industry that some of the most important applications of genetic engineering have been used, although most of these advancements have not involved antibiotics, but specific proteins of medical interest (see Section 8.13 for details).

4. **Specialty chemicals and food additives** include the *amino acids*, a number of which are produced economically in large amounts via microbial processes. The artificial sweetener **aspartame** is a dipeptide of *aspartic acid* and *phenylalanine* and both component amino acids are made by microbial fermentation. The other amino acids most commonly produced microbially are: *glutamic acid*, *lysine*, and *tryptophan*. Several vitamins are also produced microbially, including riboflavin, vitamin B_{12}, and ascorbic acid (vitamin C).

5. **Commodity chemicals** are defined as those chemicals that have low monetary value and hence are sold primarily in bulk. Commodity chemicals differ from specialty chemicals, discussed above, in that the latter sell at significantly higher prices. Commodity chemicals are used primarily as starting materials for chemical synthesis of more complex molecules. The commodity chemicals that are made microbiologically are ethanol, acetic acid, lactic acid, and glycerol. Of these, *ethanol* is the most important. Ethanol is produced commercially by the action of yeast. Yeast is able to ferment sugars to a mixture of ethanol and carbon dioxide to a high yield, generally over 90 percent (see Figure 4.11). Fermentation ethanol has been used both as a chemical feedstock for other chemical syntheses and as a source of motor fuel (gasohol).

Other industrial products originating from microorganisms are generally produced on a smaller scale than the products discussed here and are frequently produced by genetically engineered microorganisms. These "biotech" products were discussed in detail in Chapter 8.

> The industrial product may be the microbial cells themselves, or an enzyme produced by the cells, but most commonly the industrial product is a microbial metabolite. The most important groups of industrial products are those used in the pharmaceutical industry, such as the antibiotics, and specialty and commodity chemicals.

10.3   Growth and Product Formation in Industrial Processes

In Section 9.4, we discussed the microbial growth process and described the various stages: lag, log, and stationary phase. In the present section, we discuss the microbial growth process as it occurs in an *industrial* process. We are concerned here primarily with those processes in which a microbial metabolite is the desired product. There are two basic types of microbial metabolite, those called *primary* and those called *secondary*. A *primary metabolite* is one that is formed during the primary growth phase of the microorganism whereas a *secondary metabolite* is one that is formed near the end of the growth phase, frequently at or near the stationary phase of growth. The contrast between a primary metabolite and a secondary metabolite is illustrated in Figure 10.1.

Primary microbial metabolites

A typical microbial process in which the product is formed during the primary growth phase is the *alcohol (ethanol) fermentation*.* Ethanol is a product of anaero-

---

*In industrial microbiology, the term *fermentation* refers to *any* large-scale microbial process, whether or not it is biochemically a fermentation. In fact, most industrial fermentations are aerobic. The *tank* in which the industrial fermentation is carried out is called a fermen*tor*; the microorganism involved is the fermen*ter*.

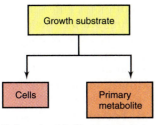

Cells and metabolite are produced more or less simultaneously

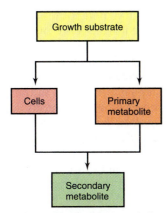

After cells and primary metabolite are produced, the cells convert the primary metabolite to a secondary metabolite

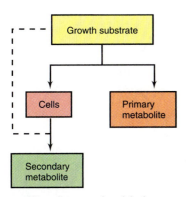

After cells are produced, further growth substrate is converted into a secondary metabolite

FIGURE 10.1   Contrast between primary and secondary metabolites.

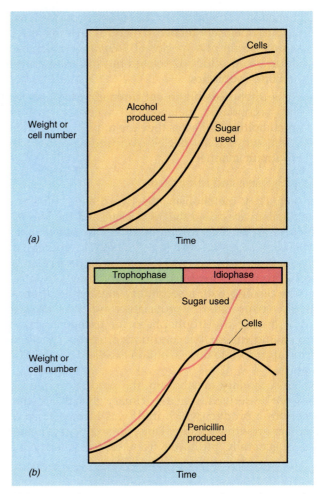

FIGURE 10.2   Contrasts between primary and secondary metabolism. (a) Primary metabolism: the formation of alcohol by yeast. (b) Secondary metabolism: the formation of penicillin by the fungus *Penicillium chrysogenum*, showing the separation of the growth phase (trophophase) and the production phase (idiophase). Note that in (b), most of the product is produced *after* growth has entered the stationary phase.

bic metabolism of yeast and certain bacteria (see Section 4.8), and is formed as part of energy metabolism. Since growth can only occur if energy production can occur, ethanol formation occurs in parallel with growth. A typical alcohol fermentation, showing the formation of microbial cells, ethanol, and sugar utilization, is illustrated in Figure 10.2*a*.

Secondary microbial metabolism

A more interesting, albeit more complex, type of microbial industrial process is one in which the desired product is not produced during the primary growth phase, but at or near the onset of the *stationary* phase. Metabolites produced during the stationary phase are called **secondary metabolites** and are some of the most common and important metabolites of industrial interest.

The best known and most extensively studied secondary metabolites are the antibiotics, and the kinetics of the penicillin process are shown in Figure 10.2*b*.

Whereas primary metabolism is generally similar in all cells, secondary metabolism shows distinct differences from one organism to another. The following characteristics of secondary metabolites have been recognized:

1. Each secondary metabolite is only formed by a relatively few organisms.

2. Secondary metabolites are seemingly not essential for growth and reproduction.

3. The formation of secondary metabolites is extremely dependent upon growth conditions, especially on the composition of the medium. Repression of secondary metabolite formation frequently occurs.

4. Secondary metabolites are often produced as a group of closely related structures. For instance, a single strain of a species of *Streptomyces* has been found to produce 32 related but different anthracycline antibiotics.

5. It is often possible to get dramatic *overproduction* of secondary metabolites, whereas primary metabolites, linked as they are to primary metabolism, can usually not be overproduced in such a dramatic manner.

Trophophase and idiophase

In secondary metabolism, the two distinct phases of metabolism are called the *trophophase* and the *idiophase* (Figure 10.2b). The trophophase is the *growth* phase (*tropho-* is a prefix meaning "growth"), whereas the metabolite production phase is the idiophase (*idio-* is a prefix from the Greek *idios* meaning "one's own"). If we are dealing with a secondary metabolite, then we should ensure that appropriate conditions are provided during the trophophase for excellent growth and should ensure that conditions are properly altered at the appropriate time to obtain excellent product formation.

In secondary metabolism, the product in question may not be derived from the primary growth substrate, but from a product which itself was formed from the primary growth substrate. Thus, the secondary metabolite is generally produced from several intermediate products that accumulate, either in the culture medium or in the cells, during primary metabolism.

One characteristic of secondary metabolites is that the enzymes involved in the production of the secondary metabolite are regulated separately from the enzymes of primary metabolism. In some cases, specific *inducers of secondary metabolite* production have been identified. For instance, a specific inducer has been identified for *streptomycin* production, a compound called *A-factor* (see Section 10.8).

Relationship between primary and secondary metabolism

Most secondary metabolites are complex organic molecules that require a large number of specific enzymatic reactions for synthesis. For instance, it is known that at least 72 separate enzymatic steps are involved in the synthesis of the antibiotic *tetracycline* and over 25 steps are involved in the synthesis of *erythromycin*, none of which are reactions occurring during primary metabolism. The metabolic pathways of these secondary metabolites do arise out of primary metabolism, however, since the starting materials for secondary metabolism come from the major biosynthetic pathways. This is summarized in Figure 10.3 which shows the interrelationship of the main primary metabolic pathway for aromatic amino acid synthesis with the secondary metabolic pathways for a variety of antibiotics.

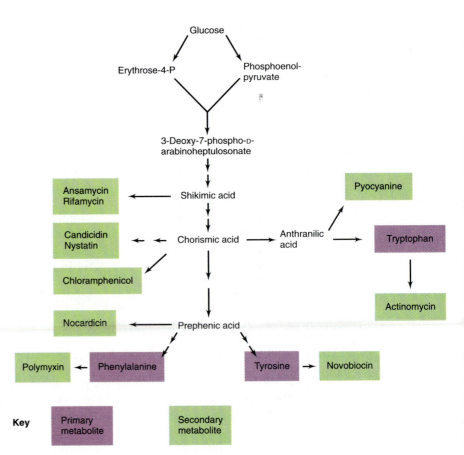

FIGURE 10.3
Relationship of the primary metabolic pathway for the synthesis of the aromatic amino acids (see Section 4.18) and formation of a variety of secondary metabolites containing aromatic rings. Note that this is a composite scheme of processes occurring in a variety of microorganisms: no one organism produces all these secondary metabolites.

Microbial metabolites can be divided into two major categories, primary and secondary. A primary metabolite is produced during the major growth phase of the organism and generally increases in amount along with the cells themselves. A secondary metabolite is produced at or near the onset of the stationary phase. Most economically important microbial products are secondary metabolites, and an understanding of how the synthesis of these unusual products is regulated is critical for developing effective large-scale processes.

10.4 Characteristics of Large-Scale Fermentations

The vessel in which the industrial process is carried out is called a *fermentor*. Fermentors can vary in size from the small 5–10 liter laboratory scale (Figure 10.4*a*) to the enormous 500,000 liter industrial scale. The size of the fermentor used depends on the process and how it is operated. Processes operated in batch mode require larger fermentors than processes operated

continuously or semi-continuously. A summary of fermentor sizes for some common microbial fermentations is given in Table 10.2.

Industrial fermentors can be divided into two major classes, those for *anaerobic* processes and those for *aerobic* processes. Anaerobic fermentors require little special equipment except for removal of heat generated during the fermentation, whereas aerobic fermentors require much more elaborate equipment to ensure that mixing and adequate aeration are achieved. Since most industrial processes are aerobic, the present discussion will be confined to aerobic fermentors.

Construction of an aerobic fermentor

Large-scale industrial fermentors are almost always constructed of stainless steel. Such a fermentor is essentially a large cylinder, closed at the top and bottom, into which various pipes and valves have been fitted (Figure 10.4*b*). Because sterilization of the culture medium and removal of heat are vital for successful operation, the fermentor generally has an external *cooling jacket* through which steam or cooling water

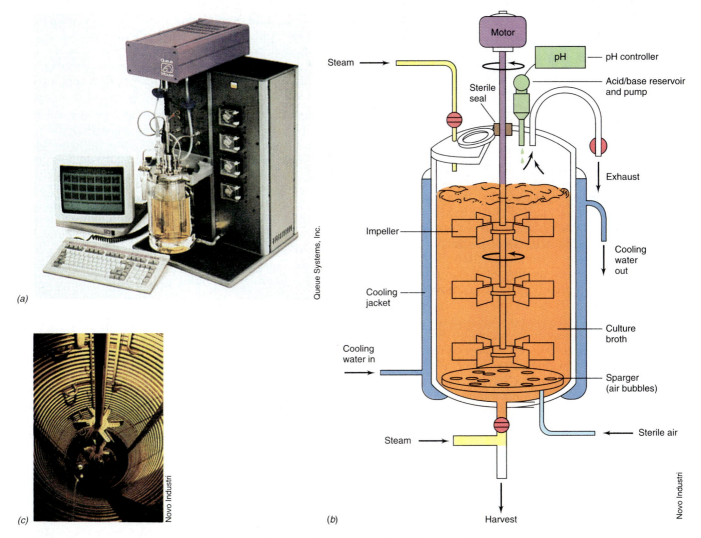

FIGURE 10.4 An industrial fermentor. (a) Photograph of a small research model. (b) Diagram of a fermentor, illustrating construction and facilities for aeration and process control. (c) Photograph of the inside of a large fermentor, showing the impeller and internal heating/cooling coils.

Table 10.2 Fermentor sizes for various processes	
Size of fermentor (liters)	**Product**
1–20,000	Diagnostic enzymes, substances for molecular biology
40–80,000	Some enzymes, antibiotics
100–150,000	Penicillin, aminoglycoside antibiotics, proteases, amylases, steroid transformations, amino acids
200,000–500,000	Amino acids (glutamic acid)

can be run. For very large fermentors, insufficient heat transfer occurs through the jacket so that *internal coils* must be provided, through which either steam or cooling water can be piped.

An important part of the fermentor is the aeration system. With large-scale equipment, transfer of oxygen from the gas to the liquid is a very difficult process and elaborate precautions must be taken to ensure proper aeration. Oxygen is poorly soluble in water, and in a fermentor with a high microbial population density, there is tremendous oxygen demand by the culture. Two separate installations are used to ensure adequate aeration: an aeration device, called a *sparger*, and a stirring device called an *impeller* (Figure 10.4*b*). The sparger is a device, often a series of holes in a metal ring or a nozzle, through which air can be passed into the fermentor under high pressure. The air enters the fermentor as a series of tiny bubbles, from which the oxygen passes by diffusion into the liquid. A good sparging system should ensure that the bubbles are of very small size, so that diffusion of oxygen from the bubble into the liquid can occur readily.

In small fermentors use of a sparger alone may be sufficient to ensure adequate aeration, but in industrial-sized fermentors, *stirring* of the fermentor with an impeller is essential (Figure 10.4*c*). Stirring accomplishes two things: it mixes the gas bubbles through the liquid; and it mixes the organism through the liquid, thus ensuring uniform access of the microbial cells to the nutrients. One of the most common stirrers is a flat blade or flat disk, fastened to a center shaft, which is rotated at high rate by a shaft attached to a motor. In order to ensure most effective mixing by the impeller, *baffles* are generally installed vertically along the inside diameter of the fermentor. As it is stirred by the impeller, the liquid passes the baffles and is broken up into smaller patches. Fluid dynamics in fermentors is extremely complex but is important for the industrial microbiologist to understand, because effective design and operation of a fermentor depend on adequate mixing.

The shaft which drives the impeller is attached to the motor by way of a drive shaft which must penetrate into the fermentor from outside. Because of the need to maintain sterility, it is vital that the seal which connects the drive shaft to the motor be arranged so that contaminants cannot pass through (see Figure 10.4*b*). A typical large-scale fermentor installation is shown in Figure 10.5.

Process control and monitoring

Any microbial process must be monitored to ensure that it is proceeding properly, but it is especially important that industrial fermentors be monitored carefully, because there is such a major expense involved. In most cases, it is not only necessary to measure growth and product formation, but to *control* the process by altering environmental parameters as the process proceeds. Environmental factors that are frequently monitored include oxygen concentration, pH, cell mass, and product concentration. It is also essential, in many cases, to control the fermentor for foaming, and to adjust the temperature (either cooling or heating, depending upon the process).

Computers now play important roles in fermentor process control (Figure 10.5*b*). They are used in two major ways: (1) in *acquisition of data* that reflect the changes taking place during the fermentation process, and (2) in *control* of various environmental factors that must be adjusted or altered during the fermentation.

During the growth and product formation process in a large-scale fermentation, it is essential, if the fermentation is to be operated properly, that data be obtained as the process is actually taking place. Data acquisition occurring in this way is spoken of as **on-line acquisition** and is of special value when such data can be processed by a computer. An even more sophisticated use of computers is in **on-line control** of the fermentation process. For instance, it may be desirable to change one of the environmental parameters as the fermentation progresses, or to feed a nutrient at a rate which exactly balances growth. The computer can be used to process the data used to measure growth, and then to *decide* (from instructions provided) when and how much nutrient to add. In this way, nutrient is added when needed, and not before, thus avoiding potential diversion of nutrient from the desired product into unwanted products.

Finally, computers can be used to *model* fermentation processes. In this use, a mathematical model of the fermentation process must be developed, a model which incorporates equations describing microbial growth and product formation. The value of using a computer to model the fermentation process is that one can test the effect of various parameters on growth and product yield quickly and interactively,

Large-scale industrial fermentations present several engineering problems. In aerobic processes, ensuring adequate oxygen availability in a large industrial tank is a difficult problem, requiring installation for stirring and aeration. The microbial process must be continuously monitored, usually with computers, in order to ensure satisfactory yields of the desired product.

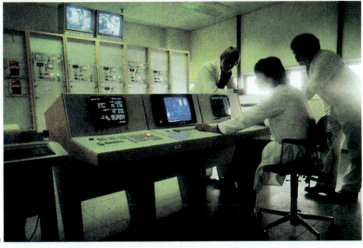

FIGURE 10.5
(a) A large industrial fermentation plant. Only the tops of the fermentors are visible. (b) Computer control room for a large fermentation plant.

and then make modifications in the parameters to see how they affect the process. In this way, many variations in the fermentation can be studied inexpensively at the computer terminal, rather than expensively in the pilot plant or industrial plant.

10.5   Scale-Up of the Fermentation Process

One of the most important and complicated aspects of industrial microbiology is the transfer of a process from small-scale laboratory equipment to large-scale commercial equipment, a procedure called **scale-up**. An understanding of the problems of scale-up are extremely important, because rarely does a microbial process behave the same way in large-scale fermentors as in small-scale laboratory equipment (Figure 10.6).

Why does a microbial process differ between large-scale and small-scale equipment? Mixing and aeration are much easier to accomplish in the small laboratory flask than in the large industrial fermentor. As the *size* of the equipment is increased, the *surface–volume* ratio changes (the principle of surface to volume ratio was explained in Figure 3.12); the large fermentor has much more volume for a given surface area. Since gas transfer and mixing depend

more on surface exposed than on fermentor volume, it is obviously more difficult to mix the big tank than the small flask. Oxygen transfer especially is much more difficult to obtain in a large fermentor, again due to the different surface–volume ratio. Because most industrial fermentations are aerobic, effective oxygen transfer is essential. With the rich culture media used in industrial processes, high biomass is obtained, leading to high oxygen demand. If aeration is reduced, even for a short period, the culture may experience partial anaerobiosis, with serious consequences in terms of product yield.

Scale-up of an industrial process is the task of the *biochemical engineer*, who is familiar with gas transfer, fluid dynamics, mixing, and thermodynamics. The role of the industrial microbiologist in the scale-up process is to work closely with the biochemical engineer to ensure that all parameters needed for a successful fermentation are understood, and that microbial strains appropriate for a large-scale fermentation are available. Very often it is found that a strain or culture medium which works well in a laboratory flask does not work well in the big fermentor. For instance, a strain may produce a polymer which increases markedly the viscosity of the culture, thus indirectly affecting aeration efficiency. The industrial microbiol-

Elmer L. Gaden, Jr.

Elmer L. Gaden, Jr.

(a)

(b)

FIGURE 10.6 (a) A bank of small research fermentors, used in process development. (b) A large bank of outdoor industrial scale fermentors (240 m³) used in commercial production of amino acids in Japan.

ogist would then be required to develop a strain which did not cause this increase in viscosity.

In transferring an industrial process from the laboratory to the commercial fermentor, several stages can be envisioned. (1) Experiments in the *laboratory flask* which are generally the first indication that a process of commercial interest is possible. (2) The *laboratory fermentor*, a small-scale fermentor, generally of glass and generally of 5–10 liter size, in which the first efforts at scale-up are made (Figures 10.5a and 10.6a). In the laboratory fermentor, it is possible to test variations in medium, temperature, pH, and so on inexpensively, since little cost is involved for either equipment or culture medium. (3) The *pilot plant stage*, usually carried out in equipment of 300–3000 liter size. Here, the conditions more closely approach the commercial scale, however cost is not yet a major factor. In the pilot plant fermentor, careful instrumentation and computer control is desirable, so that the conditions most similar to those in the laboratory fermentor can be obtained. (4) The *commercial fermentor* itself, generally of 10,000–500,000 liters (Figure 10.6b).

It is generally found in scale-up studies on aerobic fermentations that the oxygen rate in the fermentor is best kept *constant* as the size of the fermentor is increased. Thus, if an oxygen transfer rate of 200 millimoles O_2/liter/hr is required to obtain optimal yield in a small fermentor, then stirring and aeration in the large fermentor should be adjusted to ensure this same rate. To do this, it will probably require much more rapid stirring as well as a higher pressure of the inlet air. Because stirring is a mechanical process that

can be monitored in terms of *power*, one approach to scale-up is to maintain constant power to the fermentor when going from small- to large-scale equipment.

We now begin a consideration of the industrial production of microbial products, beginning with the antibiotics. Antibiotic production is a huge industry worldwide and one in which many important principles of large-scale microbial cultures were first developed.

> Because of changes in surface–volume relationships when moving from laboratory to large-scale equipment, the scale-up of an industrial process requires considerable research. For aerobic processes, the most important parameter to control during scale-up is the oxygen transfer rate.

10.6   Antibiotics: Isolation and Characterization

Of the microbial products manufactured commercially, probably the most important are the antibiotics. As we have discussed in Section 9.17, antibiotics are chemical substances that are produced by microorganisms which kill or inhibit the growth of other microorganisms. The development of antibiotics as agents for treatment of infectious disease has probably had more impact on the practice of medicine than any other single development.

Antibiotics are products of secondary metabolism. Although their yields are relatively low in most industrial fermentations, because of their high therapeutic activity and consequently high economic value, they

Table 10.3   Some antibiotics produced commercially

Antibiotic	Producing microorganism	Type of microorganism
Bacitracin	*Bacillus licheniformis*	Endospore-forming bacterium
Cephalosporin	*Cephalosporium* sp.	Fungus
Chloramphenicol	Chemical synthesis (formerly produced microbially by *Streptomyces venezuelae*)	
Cycloheximide	*Streptomyces griseus*	Actinomycete
Cycloserine	*S. orchidaceus*	Actinomycete
Erythromycin	*S. erythreus*	Actinomycete
Griseofulvin	*Penicillium griseofulvin*	Fungus
Kanamycin	*S. kanamyceticus*	Actinomycete
Lincomycin	*S. lincolnensis*	Actinomycete
Neomycin	*S. fradiae*	Actinomycete
Nystatin	*S. noursei*	Actinomycete
Penicillin	*Penicillium chrysogenum*	Fungus
Polymyxin B	*Bacillus polymyxa*	Endospore-forming bacterium
Streptomycin	*S. griseus*	Actinomycete
Tetracycline	*S. rimosus*	Actinomycete

can be produced commercially by microbial fermentation. Many antibiotics can be synthesized chemically, but because of the chemical complexity of the antibiotics and the great expense attendant on chemical synthesis, rarely is it possible for chemical synthesis to compete with microbial fermentation.

Commercially useful antibiotics are produced primarily by filamentous fungi and by Bacteria of the actinomycete group. A listing of the most important antibiotics produced by large-scale industrial fermentation is given in Table 10.3. Frequently, a number of chemically related antibiotics exist, so that *families* of antibiotics are known. Antibiotics can hence be classified according to their chemical structure, as outlined in Table 10.4. Most antibiotics employed medically are used to treat bacterial diseases, although a few antibiotics are known which are effective against fungal diseases. The economic significance of the antibiotics is shown by the fact that over 100,000 *tons* of antibiotics are produced per year, with gross sales of nearly $5 billion.

Search for new antibiotics

Over 8000 antibiotic substances are known, and several hundred antibiotics are discovered yearly. Are there more antibiotics waiting to be discovered? Almost certainly yes, because most of the microorganisms which have been examined for ability to produce antibiotics are members of a few genera, such as *Streptomyces*, *Penicillium*, and *Bacillus*. Many antibiotic researchers believe that a vast number of new antibiotics will be discovered if other groups of microorganisms are examined. It seems also likely that genetic engineering techniques will permit the artificial construction of new antibiotics, once the details of the gene structure of antibiotic-producing microorganisms have been discerned.

However, the main way in which new antibiotics have been discovered has been by *screening*. In the screening approach, a large number of isolates of possible antibiotic-producing microorganisms are obtained from nature in pure culture, and these isolates are then tested for antibiotic production by seeing whether they produce any diffusible materials that are inhibitory to the growth of test bacteria. The test bacteria used are selected from a variety of bacterial types, but are chosen to be representative of or related to bacterial pathogens. The classical procedure for testing new microbial isolates for antibiotic production is the cross-streak method, first used by Fleming in his pioneering studies on penicillin (see box in Chapter 9 and Figure 10.7). Those isolates which show evidence of antibiotic production are then studied further to determine if the antibiotics they produce are new. In most current screening programs, most of the isolates obtained produce *known* antibiotics, so that the microbiologist must quickly identify producers of known antibiotics and discard them. Once an organism producing a new antibiotic is discovered, the antibiotic is produced in large amounts, purified, and tested for toxicity and therapeutic activity in infected animals. Most new antibiotics *fail* these animal tests, but a few pass them successfully. Ultimately, a very few of these new antibiotics prove useful medically and are produced commercially.

Steps toward commercial production

An antibiotic that is to be produced commercially must first be produced successfully in large-scale industrial fermentors. We have discussed in general the problems of scale-up earlier in this chapter. One of the most important tasks is the development of efficient purification methods. Because of the relatively small amounts of antibiotic present in the fermentation liquid, elaborate methods for the extraction and purification of the antibiotic are necessary (Figure 10.8). If the antibiotic is soluble in an organic solvent which is immiscible in water, it may be relatively simple to purify cillins is penicillin G, which serves as the source of the 6-APA nucleus. Pencillin G is split either chemically or enzymatically (using penicillin acylase) and the 6-APA obtained is then coupled chemically to another side chain. As seen in Figure 10.10, the semisynthetic penicillins have significantly improved medical properties.

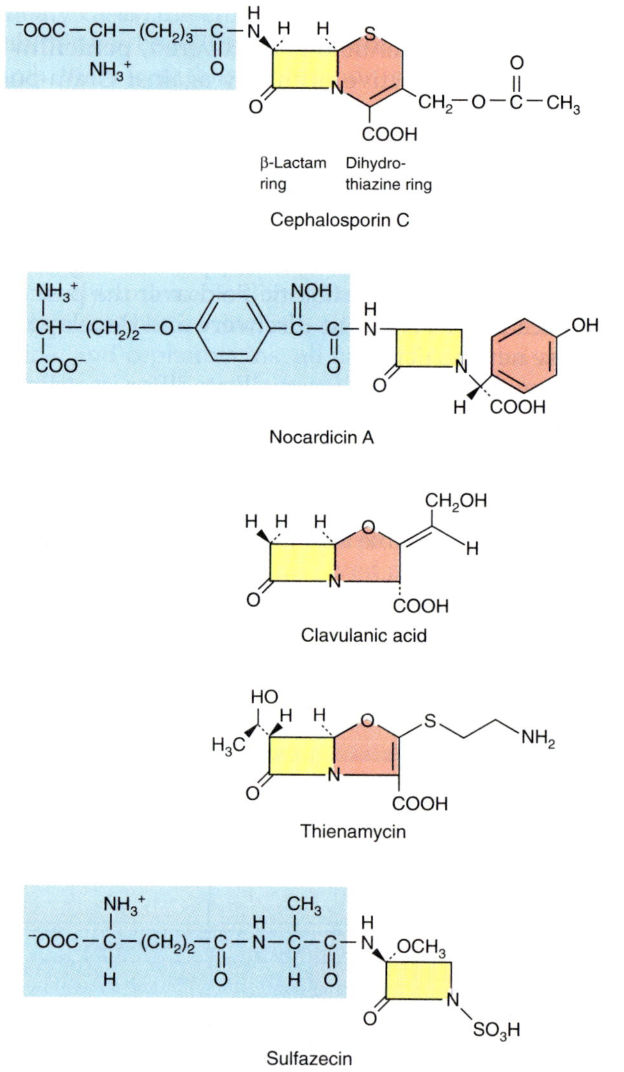

Production methods for β-lactam antibiotics

Penicillin G is produced using a submerged fermentation process in 40,000–200,000 liter fermentors. Pencillin production is a highly aerobic process and efficient aeration is necessary. Pencillin is a typical secondary metabolite, as illustrated in Figures 10.2b and 10.11. During the growth phase, very little penicillin is produced but once the carbon source has been exhausted, the penicillin production phase begins. By feeding with various culture medium components, the production phase can be extended for several days.

Certain culture media permit high penicillin yields whereas other culture media are quite effective for growth but lead to little product formation. A major ingredient of most penicillin production media is **corn steep liquor**. This component serves as the nitrogen source, as well as a source of other growth factors. The carbon source is generally *lactose*, which is not used as rapidly as glucose and does not cause catabolite repression. Glucose can be used as a carbon

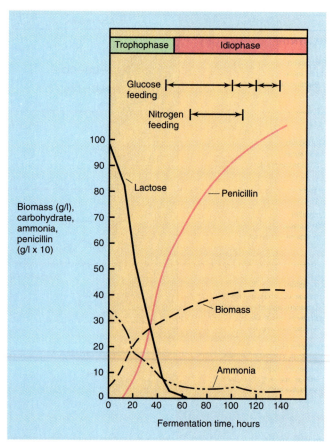

FIGURE 10.11 Kinetics of the penicillin fermentation with *Penicillium chrysogenum*.

FIGURE 10.12 Newer β-lactam antibiotics. The parent β-lactam ring of some of these antibiotics is shown in Figure 10.9.

source, but when glucose is used, it must be fed *slowly*, to avoid catabolite repression. The acyl side chain of penicillin G is the phenylacyl moiety, and yields of penicillin G can be markedly increased if phenylacetic acid or phenoxyacetic acid is fed continuously as a precursor. If a pencillin with another acyl side chain is desired, then its appropriate precursor should be fed.

Penicillin is excreted into the medium and less than 1 percent remains with the fungal filaments. After the fungal mycelium is removed by filtration, the pH of the medium is lowered and the antibiotic extracted from the filtered broth with amyl or butyl acetate. After concentration into the solvent, the antibiotic is back-extracted into an alkaline aqueous medium, concentrated further, and crystallized. Highly purified penicillin can be readily obtained in this way. When penicillin first became available commercially, its price was extremely high, but it is frequently said today that the package that the penicillin is in costs more than the ingredients within!

Cephalosporins are β-lactam antibiotics containing a dihydrothiazine instead of a thiazolidine ring sys-

Table 10.3   Some antibiotics produced commercially

Antibiotic	Producing microorganism	Type of microorganism
Bacitracin	*Bacillus licheniformis*	Endospore-forming bacterium
Cephalosporin	*Cephalosporium* sp.	Fungus
Chloramphenicol	Chemical synthesis (formerly produced microbially by *Streptomyces venezuelae*)	
Cycloheximide	*Streptomyces griseus*	Actinomycete
Cycloserine	*S. orchidaceus*	Actinomycete
Erythromycin	*S. erythreus*	Actinomycete
Griseofulvin	*Penicillium griseofulvin*	Fungus
Kanamycin	*S. kanamyceticus*	Actinomycete
Lincomycin	*S. lincolnensis*	Actinomycete
Neomycin	*S. fradiae*	Actinomycete
Nystatin	*S. noursei*	Actinomycete
Penicillin	*Penicillium chrysogenum*	Fungus
Polymyxin B	*Bacillus polymyxa*	Endospore-forming bacterium
Streptomycin	*S. griseus*	Actinomycete
Tetracycline	*S. rimosus*	Actinomycete

can be produced commercially by microbial fermentation. Many antibiotics can be synthesized chemically, but because of the chemical complexity of the antibiotics and the great expense attendant on chemical synthesis, rarely is it possible for chemical synthesis to compete with microbial fermentation.

Commercially useful antibiotics are produced primarily by filamentous fungi and by Bacteria of the actinomycete group. A listing of the most important antibiotics produced by large-scale industrial fermentation is given in Table 10.3. Frequently, a number of chemically related antibiotics exist, so that *families* of antibiotics are known. Antibiotics can hence be classified according to their chemical structure, as outlined in Table 10.4. Most antibiotics employed medically are used to treat bacterial diseases, although a few antibiotics are known which are effective against fungal diseases. The economic significance of the antibiotics is shown by the fact that over 100,000 *tons* of antibiotics are produced per year, with gross sales of nearly $5 billion.

Search for new antibiotics

Over 8000 antibiotic substances are known, and several hundred antibiotics are discovered yearly. Are there more antibiotics waiting to be discovered? Almost certainly yes, because most of the microorganisms which have been examined for ability to produce antibiotics are members of a few genera, such as *Streptomyces*, *Penicillium*, and *Bacillus*. Many antibiotic researchers believe that a vast number of new antibiotics will be discovered if other groups of microorganisms are examined. It seems also likely that genetic engineering techniques will permit the artificial construction of new antibiotics, once the details of the gene structure of antibiotic-producing microorganisms have been discerned.

However, the main way in which new antibiotics have been discovered has been by *screening*. In the screening approach, a large number of isolates of possible antibiotic-producing microorganisms are obtained from nature in pure culture, and these isolates are then tested for antibiotic production by seeing whether they produce any diffusible materials that are inhibitory to the growth of test bacteria. The test bacteria used are selected from a variety of bacterial types, but are chosen to be representative of or related to bacterial pathogens. The classical procedure for testing new microbial isolates for antibiotic production is the cross-streak method, first used by Fleming in his pioneering studies on penicillin (see box in Chapter 9 and Figure 10.7). Those isolates which show evidence of antibiotic production are then studied further to determine if the antibiotics they produce are new. In most current screening programs, most of the isolates obtained produce *known* antibiotics, so that the microbiologist must quickly identify producers of known antibiotics and discard them. Once an organism producing a new antibiotic is discovered, the antibiotic is produced in large amounts, purified, and tested for toxicity and therapeutic activity in infected animals. Most new antibiotics *fail* these animal tests, but a few pass them successfully. Ultimately, a very few of these new antibiotics prove useful medically and are produced commercially.

Steps toward commercial production

An antibiotic that is to be produced commercially must first be produced successfully in large-scale industrial fermentors. We have discussed in general the problems of scale-up earlier in this chapter. One of the most important tasks is the development of efficient purification methods. Because of the relatively small amounts of antibiotic present in the fermentation liquid, elaborate methods for the extraction and purification of the antibiotic are necessary (Figure 10.8). If the antibiotic is soluble in an organic solvent which is immiscible in water, it may be relatively simple to purify

Table 10.4   Classification of antibiotics according to their chemical structure. An example of each is given in parentheses

1. **Carbohydrate-containing antibiotics**
   - Pure sugars   (Nojirimycin)
   - Aminoglycosides   (Streptomycin)
   - Orthosomycins   (Everninomicin)
   - N-Glycosides   (Streptothricin)
   - C-Glycosides   (Vancomycin)
   - Glycolipids   (Moenomycin)
2. **Macrocyclic lactones**
   - Macrolide antibiotics   (Erythromycin)
   - Polyene antibiotics   (Candicidin)
   - Ansamycins   (Rifamycin)
   - Macrotetrolides   (Tetranactin)
3. **Quinones and related antibiotics**
   - Tetracyclines   (Tetracycline)
   - Anthracyclines   (Adriamycin)
   - Naphthoquinones   (Actinorhodin)
   - Benzoquinones   (Mitomycin)
4. **Amino acid and peptide antibiotics**
   - Amino acid derivatives   (Cycloserine)
   - β-Lactam antibiotics   (Penicillin)
   - Peptide antibiotics   (Bacitracin)
   - Chromopeptides   (Actinomycins)
   - Depsipeptides   (Valinomycin)
   - Chelate-forming peptides   (Bleomycins)
5. **Heterocyclic antibiotics containing nitrogen**
   - Nucleoside antibiotics   (Polyoxins)
6. **Heterocyclic antibiotics containing oxygen**
   - Polyether antibiotics   (Monensin)
7. **Alicyclic derivatives**
   - Cycloalkane derivatives   (Cycloheximide)
   - Steroid antibiotics   (Fusidic acid)
8. **Aromatic antibiotics**
   - Benzene derivatives   (Chloramphenicol)
   - Condensed aromatic antibiotics   (Griseofulvin)
   - Aromatic ether   (Novobiocin)
9. **Aliphatic antibiotics**
   - Compounds containing phosphorous   (Fosfomycins)
10. **Quinolone antibiotics**
   - 4-Quinolone   (Nalidixic acid)
   - Fluro-4-quinolones   (Norfloxacin, Ciprofloxacin)

the antibiotic by extracting the antibiotic into a small volume of the solvent and thus concentrating the antibiotic. If the antibiotic is not solvent-soluble, then it must be removed from the fermentation liquid by adsorption, ion exchange, or chemical precipitation. In all cases, the goal is to obtain a crystalline product of high purity, although some antibiotics do not crystallize readily and are difficult to purify. A related problem is that cultures often produce other end products, including other antibiotics, and it is essential to end up with a product consisting of only a single antibiotic. The purification chemist may be required to develop methods for eliminating undesirable byproducts, but in some cases it may be necessary for the microbiologist to find strains that do not produce such undesirable chemicals.

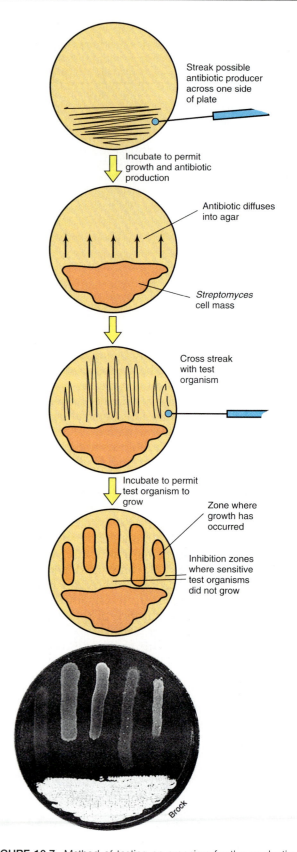

FIGURE 10.7 Method of testing an organism for the production of antibiotics. The producer (*Streptomyces*) was streaked across one-third of the plate and the plate incubated. After good growth was obtained, the test bacteria were streaked perpendicular to the *Streptomyces*, and the plate was further incubated. The failure of several organisms to grow near the *Streptomyces* indicates that the *Streptomyces* produced an antibiotic active against these bacteria. Test organisms left to right: *Escherichia coli*, *Bacillus subtilis*, *Staphylococcus aureus*, *Klebsiella pneumoniae*, *Mycobacterium smegmatis*.

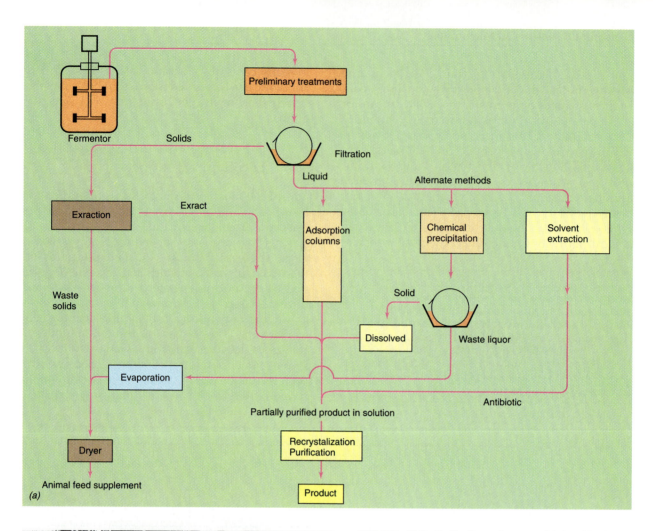

FIGURE 10.8 Purification of an antibiotic. (a) Overall process of extraction and purification. (b) Installation for the solvent extraction of an antibiotic from fermentation broth.

Rarely do antibiotic-producing strains just isolated from nature produce the desired antibiotic at sufficiently high concentration so that commercial production can begin immediately. One of the main tasks of the industrial microbiologist is thus to isolate new *high-yielding strains*. The industrial microbiologist has made very significant contributions to the antibiotic industry by developing high-yielding processes. As we noted earlier, the yield of penicillin was increased over 50,000-fold by strain selection and appropriate medium development. Strain selection involves mutagenesis of the initial culture, plating of mutant types, and testing of these mutants for antibiotic production. In most cases, mutants produce *less* antibiotic than the parent, so that only rarely would a higher-yielding strain be obtained.

In recent years, the development of genetic engineering techniques has greatly improved the proce-

dures for seeking high-yielding strains. The technique of *gene amplification* makes it possible to place additional copies of genes of interest into a cell by means of a vector such as a plasmid. Alterations in regulatory processes also may permit increased yields. However, one difficulty with using genetic procedures for increasing antibiotic yield is that the biosynthetic pathways for the synthesis of most antibiotics involve large numbers of steps with many genes, and it is not clear which genes should be altered to increase yields. Thus, it is critical that the rate-limiting step in a given biochemical pathway be identified by basic research.

> The industrial production of antibiotics begins with screening microbial habitats, generally soils, for antibiotic producers. Once new producers are identified, purification and chemical analyses of the antimicrobial agent are made. If the new antibiotic is biologically active *in vivo*, the industrial microbiologist may seek high-yielding strains or may genetically modify the wild-type isolate to increase yields to levels acceptable for commercial development.

In many cases, the efficiency of synthesis of a specific antibiotic has benefitted more from basic metabolic research than from blind mutation and selection programs.

10.7   β-Lactam Antibiotics: Penicillin and Cephalosporins

One of the most important groups of antibiotics, both historically and medically, is the β-lactam group. The β-lactam antibiotics include the penicillins, cephalosporins, and cephamycins, all medically useful antibiotics. These antibiotics are called β-lactams because they contain the β-lactam ring system (Figures 10.9 and 10.10), a complex heterocyclic ring system. As we have discussed in Section 3.5, β-lactam antibiotics act by inhibiting peptidoglycan synthesis in the cell walls of Bacteria. The target of these antibiotics is the transpeptidation reaction involved in the cross-linking step of peptidoglycan biosynthesis. Because this reaction is unique to Bacteria, the β-lactam antibiotics have high specificity and relatively low toxicity.

Basic structures	Antibiotics	Most important producing species
Penam 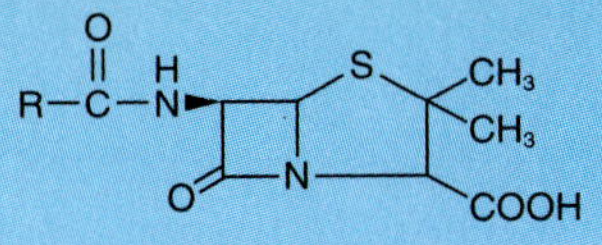	Penicillins	*Penicillium chrysogenum* *Aspergillus nidulans* *Cephalosporium acremonium* *Streptomyces clavuligerus*
Ceph-3-em 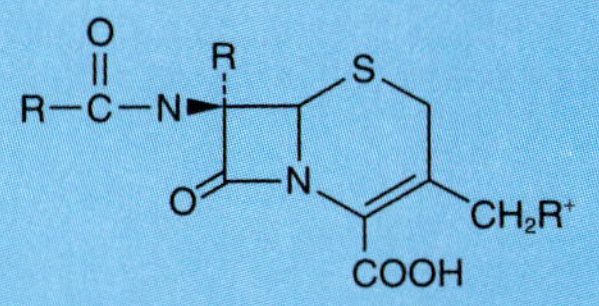	Cephalosporins 7-Methoxycephalosporins	*Cephalosporium acremonium* *Nocardia lactamdurans* *Streptomyces clavuligerus*
Clavam 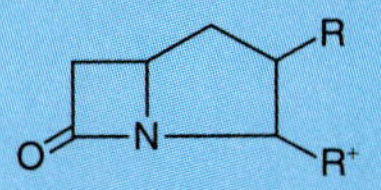	Clavulanic acids	*Streptomyces clavuligerus*
Carbapenem 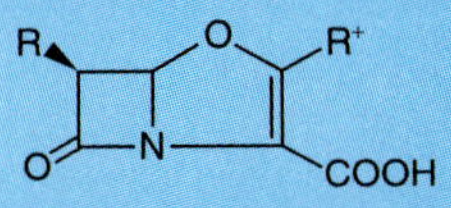	Thienamycins Olivanic acids Epithienamycins	*Streptomyces cattleya* *Streptomyces olivaceus* *Streptomyces flavogriseus*
Monolactam 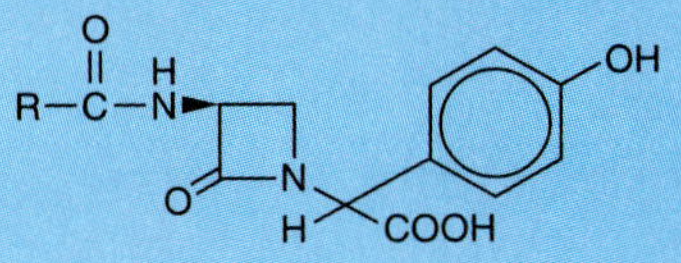	Nocardicins	*Nocardia uniformis* subsp. *tsuyamanesis*
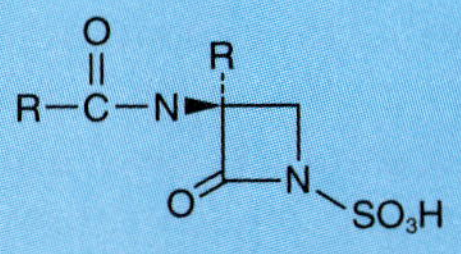	Monobactams	*Gluconobacter* sp. *Chromobacterium violaceum* *Agrobacterium radiobacter* *Pseudomonas acidophila* *Pseudomonas mesoacidophila* *Flexibacter* sp. *Acetobacter* sp.

FIGURE 10.9 The basic structures of the naturally occurring β-lactam antibiotics. The positions where chemical substitutions can occur are indicated by R.

Types of penicillin

The first β-lactam antibiotic discovered, **penicillin G** (Figure 10.10), is active primarily against Gram-positive Bacteria. Its action is restricted to Gram-positive Bacteria primarily because Gram-negative Bacteria are impermeable to the antibiotic. As a result of extensive research, a vast number of new penicillins have been discovered, some of which are quite effective against Gram-negative Bacteria. One of the most significant developments in the antibiotic field over the past several decades has been the discovery and development of these new penicillins.

How do the various penicillins differ in structure and how are new penicillins developed? As noted, the basic structure of the penicillins is *6-aminopenicillanic acid* (6-APA), which consists of a thiazolidine ring with a condensed β-lactam ring (Figure 10.10). The 6-APA carries a variable acyl moiety (side chain) in posi-

tion 6. If the penicillin fermentation is carried out without addition of side-chain precursors, the **natural penicillins** are produced (Figure 10.10, left side). Of the natural penicillins listed in Figure 10.10, only benzylpenicillin (penicillin G) is therapeutically useful. The fermentation can be better controlled by adding to the broth a *side-chain precursor*, so that only *one* desired penicillin is produced. Over 100 such **biosynthetic penicillins** have been produced in this way. In commercial processes, however, only penicillin G, penicillin V, and very limited amounts of penicillin O are produced (Figure 10.10, lower left).

In order to produce the most useful penicillins, those with activity against Gram-negative Bacteria, a combined fermentation and chemical approach is used which leads to the production of **semisynthetic penicillins** (Figure 10.10, right side). The starting material for the production of such semisynthetic peni-

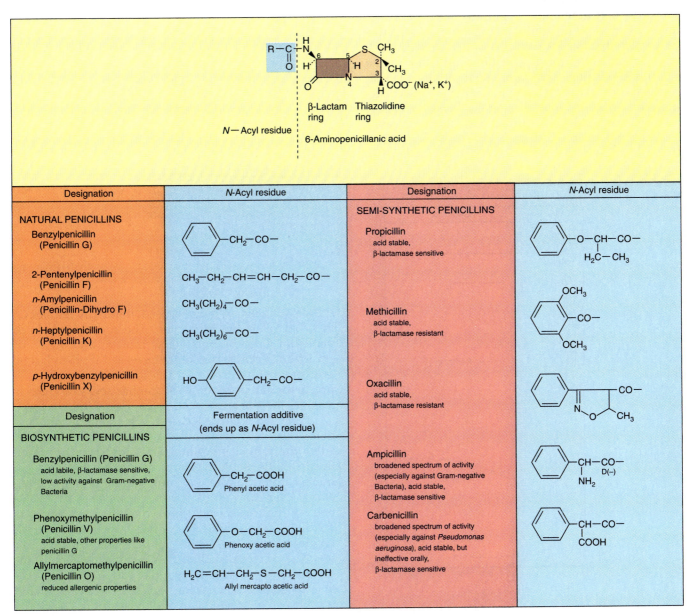

FIGURE 10.10 Structure of some of the natural penicillins, the most important biosynthetic penicillins, and several semisynthetic penicillins.

cillins is penicillin G, which serves as the source of the 6-APA nucleus. Pencillin G is split either chemically or enzymatically (using penicillin acylase) and the 6-APA obtained is then coupled chemically to another side chain. As seen in Figure 10.10, the semisynthetic penicillins have significantly improved medical properties.

Production methods for β-lactam antibiotics

Penicillin G is produced using a submerged fermentation process in 40,000–200,000 liter fermentors. Pencillin production is a highly aerobic process and efficient aeration is necessary. Pencillin is a typical secondary metabolite, as illustrated in Figures 10.2b and 10.11. During the growth phase, very little penicillin is produced but once the carbon source has been exhausted, the penicillin production phase begins. By feeding with various culture medium components, the production phase can be extended for several days.

Certain culture media permit high penicillin yields whereas other culture media are quite effective for growth but lead to little product formation. A major ingredient of most penicillin production media is **corn steep liquor**. This component serves as the nitrogen source, as well as a source of other growth factors. The carbon source is generally *lactose*, which is not used as rapidly as glucose and does not cause catabolite repression. Glucose can be used as a carbon

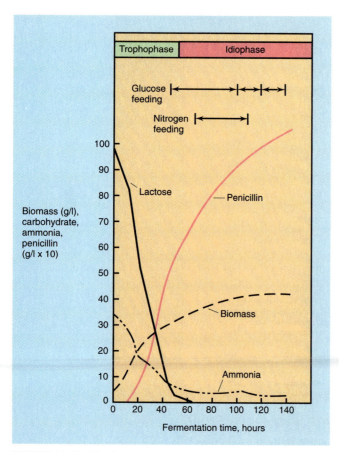

FIGURE 10.11 Kinetics of the penicillin fermentation with *Penicillium chrysogenum*.

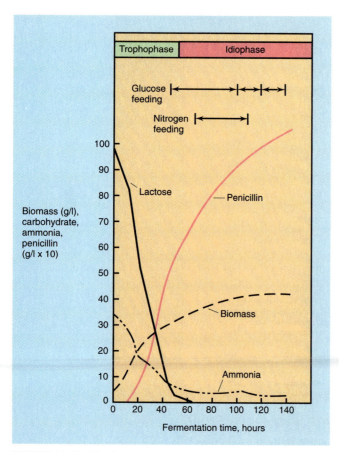

FIGURE 10.12 Newer β-lactam antibiotics. The parent β-lactam ring of some of these antibiotics is shown in Figure 10.9.

source, but when glucose is used, it must be fed *slowly*, to avoid catabolite repression. The acyl side chain of penicillin G is the phenylacyl moiety, and yields of penicillin G can be markedly increased if phenylacetic acid or phenoxyacetic acid is fed continuously as a precursor. If a penicillin with another acyl side chain is desired, then its appropriate precursor should be fed.

Penicillin is excreted into the medium and less than 1 percent remains with the fungal filaments. After the fungal mycelium is removed by filtration, the pH of the medium is lowered and the antibiotic extracted from the filtered broth with amyl or butyl acetate. After concentration into the solvent, the antibiotic is back-extracted into an alkaline aqueous medium, concentrated further, and crystallized. Highly purified penicillin can be readily obtained in this way. When penicillin first became available commercially, its price was extremely high, but it is frequently said today that the package that the penicillin is in costs more than the ingredients within!

Cephalosporins are β-lactam antibiotics containing a dihydrothiazine instead of a thiazolidine ring sys-

tem (Figure 10.12). Cephalosporins were first discovered as products of the fungus *Cephalosporium acremonium*, but a number of other fungi also produce antibiotics with this ring system. In addition, a number of semi-synthetic cephalosporins are produced. Cephalosporins are valued not only because of their low toxicity but also because they are broad-spectrum antibiotics (see Section 9.17).

Intensive screening for new β-lactam antibiotics has led to the development of compounds whose structures are different from both the penicillins and the cephalosporins. Included in this category are nocardicin, clavulanic acid, and thienamycin (Figure 10.12). Clavulanic acid is of particular interest because, although it is not very effective as an antibiotic by itself, it inhibits the activity of β-lactamases. These enzymes are produced by certain bacteria that destroy β-lactam antibiotics rendering them ineffective in treating a particular disease. Thus, when used in combination with β-lactamase-sensitive penicillins and cephalosporins, clavulanic acid causes a distinct increase in the activity of these antibiotics.

10.8   Antibiotics from Prokaryotes

Many antibiotics active against prokaryotes are actually produced by prokaryotes as well. These include the aminoglycosides, the macrolides, and the tetracyclines, among many others. Many of these antibiotics have major clinical applications, and their production is therefore a significant process in the pharmaceutical industry.

Aminoglycoside antibiotics

Aminoglycosides are antibiotics that contain amino sugars bonded by glycosidic linkage (see Section 2.5) to other amino sugars. A number of clinically useful antibiotics are aminoglycosides, including *streptomycin* (see Section 7.9) and its relatives, *kanamycin, gentamicin*, and *neomycin*. The aminoglycoside antibiotics are used clinically primarily against Gram-negative Bacteria. Streptomycin has also been used extensively in the treatment of tuberculosis. Historically, the discovery of the value of streptomycin for tuberculosis was a major medical advance, as this was the first antibiotic discovered capable of controlling this dreaded infectious disease. However, none of the aminoglycoside antibiotics find as wide use today as they formerly did. Streptomycin has been supplanted by several synthetic chemicals for tuberculosis, due to the fact that streptomycin causes several serious side effects and because bacterial resistance readily develops. The use of aminoglycosides for Gram-negative infections has been less significant since the development of the semisynthetic penicillins (see above) and the tetracyclines (see below). The aminoglycoside antibiotics are now considered reserve antibiotics, used primarily when other antibiotics fail. We discussed the mode of action of streptomycin in Section 9.17.

One of the interesting features of the aminoglycoside antibiotics is the *regulation of their biosynthesis*. As seen in Figure 10.13, the three moieties of streptomycin are synthesized by separate pathways, and the subunits brought together at the end. The final intermediate in the pathway, streptomycin-P, is biologically inactive but is activated by the removal of a phosphate molecule (Figure 10.13). Streptomycin is synthesized as a typical secondary metabolite. At least several of the enzymes involved are synthesized only at the end of the trophophase.

One aspect of the regulation involves the production of an inducer called *Factor A* (Figure 10.14). Factor A is not chemically related to streptomycin, but is involved somehow in carbohydrate metabolism. Key enzymes in streptomycin biosynthesis are not synthesized until Factor A concentration builds up, thus explaining how this factor might play a role in secondary metabolism. During the growth phase, Factor A that is synthesized is excreted and gradually builds up in the medium. Only when the concentration reaches a critical level does Factor A begin to act and induce the synthesis of the key streptomycin biosynthesis enzymes. The importance of Factor A for streptomycin biosynthesis is shown by the fact that Factor A-negative mutants lose their capacity for streptomycin biosynthesis, but if synthetic Factor A is added to such mutants, streptomycin biosynthesis is restored. However, Factor A is not itself a precursor of streptomycin, as shown by the fact that the addition of a tiny amount of pure Factor A, 1 μg, can lead to the production of 1 g of streptomycin. It should be noted, however, that streptomycin biosynthesis is not regulated *solely* by Factor A, since other regulatory processes also affect streptomycin biosynthesis.

Macrolide antibiotics

Macrolide antibiotics contain large lactone rings connected to sugar moieties (Figure 10.15). Variation in both the macrolide ring and the sugar moieties are known, so that a large variety of macrolide antibiotics exist. The best known macrolide antibiotic is *erythromycin*, but other macrolides include *oleandomycin*, *spiramycin*, and *tylosin*. We discussed the mode of action of erythromycin in Section 9.17. Erythromycin is commonly used clinically in place of penicillin in those patients allergic to penicillin or other β-lactam antibiotics. Erythromycin has been particularly valuable in treating cases of legionellosis (see Section 15.2), because of the exquisite sensitivity of the causative agent, the bacterium *Legionella pneumophila*, to this antibiotic.

The complexity of macrolide antibiotics is evident from their structure. Over 25 unique enzymatic steps are known to be involved in erythromycin biosynthesis. Regulation of biosynthesis occurs in a number of ways. Glucose and phosphate inhibition are known to occur, and erythronolide B, one of the intermediates in erythromycin biosynthesis, inhibits its own production. End product inhibition of certain key enzymes by erythromycin itself is also known to occur.

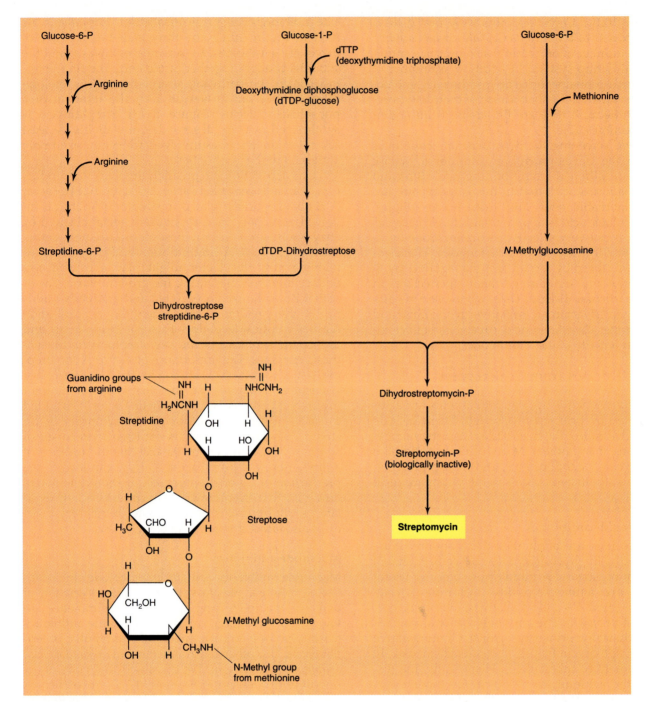

FIGURE 10.13 Streptomycin biosynthesis in *Streptomyces griseus*.

Tetracyclines

The tetracyclines are an important group of antibiotics which find widespread medical use. They were some of the first so-called *broad-spectrum* antibiotics, inhibiting almost all Gram-positive and Gram-negative Bacteria. The basic structure of the tetracyclines consists of a naphthacene ring system (Figure 10.16). To this ring is added any of several constituents. *Chlortetracycline*, for instance, has a chlorine atom whereas *oxytetracycline* has an additional hydroxyl (OH) group and no chlorine (Figure 10.16). All three of these antibiotics are produced microbiologically, but there are also semisynthetic tetracyclines on the market, in which other constituents have been inserted chemically into the naphthacene ring system. The mode of action of the tetracyclines was discussed in Section 9.17.

The tetracyclines and the β-lactam antibiotics are the two most important groups of antibiotics in the medical field. The tetracyclines also find use in veterinary medicine, and in some countries are also used as nutritional supplements for poultry and swine. At one time, chlortetracycline was used to preserve fish, being added to the ice with which fish were refrigerated when they were caught at sea, but such nonmedical uses of medically important antibiotics are now discouraged because of the potential danger of development of antibiotic resistance (see Section 11.16).

The biosynthesis of a tetracycline involves a large number of enzymatic steps. In the case of chlortetracycline, as many as 72 intermediate products may be involved, most of which are only known in a very general way. Studies on the genetics of *Streptomyces aureofaciens*, the producer of chlortetracycline, has shown that over 300 genes are involved! With such a large number of genes, regulation of antibiotic biosynthesis is obviously quite complex. Rational approaches to the development of high-yielding strains are probably also far in the future.

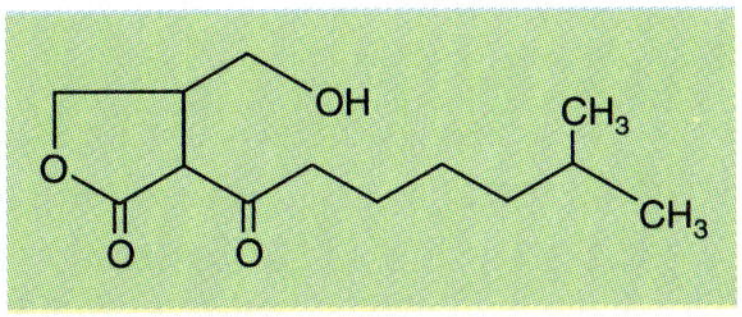

Repression of chlortetracycline synthesis by both glucose and phosphate is known to occur. Phosphate repression is especially significant, so that the culture medium used in commercial production must be run with restricted phosphate concentrations. A production scheme for chlortetracycline is shown in Figure 10.17.

FIGURE 10.14 Structure of Factor A from *Streptomyces griseus*.

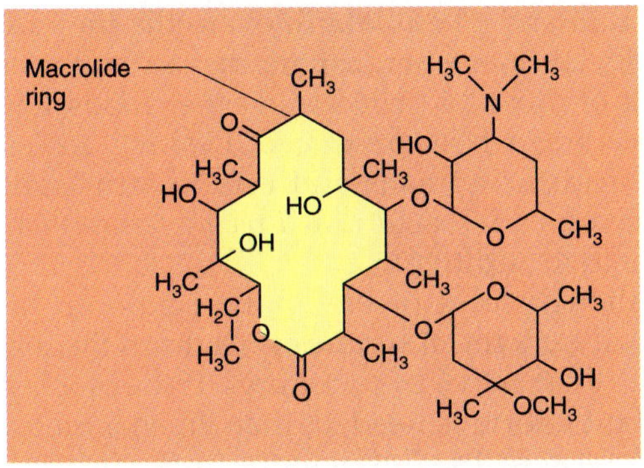

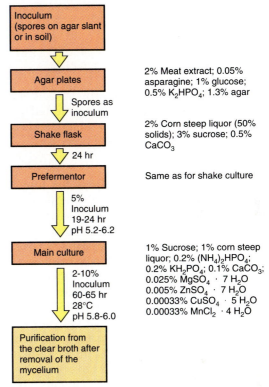

FIGURE 10.15 Structure of erythromycin, a typical macrolide antibiotic.

FIGURE 10.17 Production scheme for chlortetracycline with *Streptomyces aureofaciens*.

Inoculum (spores on agar slant or in soil)

Agar plates — 2% Meat extract; 0.05% asparagine; 1% glucose; 0.5% K$_2$HPO$_4$; 1.3% agar

Spores as inoculum

Shake flask — 2% Corn steep liquor (50% solids); 3% sucrose; 0.5% CaCO$_3$

24 hr

Prefermentor — Same as for shake culture

5% Inoculum 19-24 hr pH 5.2-6.2

Main culture — 1% Sucrose; 1% corn steep liquor; 0.2% (NH$_4$)$_2$HPO$_4$; 0.2% KH$_2$PO$_4$; 0.1% CaCO$_3$; 0.025% MgSO$_4$ · 7 H$_2$O 0.005% ZnSO$_4$ · 7 H$_2$O 0.00033% CuSO$_4$ · 5 H$_2$O 0.00033% MnCl$_2$ · 4 H$_2$O

2-10% Inoculum 60-65 hr 28°C pH 5.8-6.0

Purification from the clear broth after removal of the mycelium

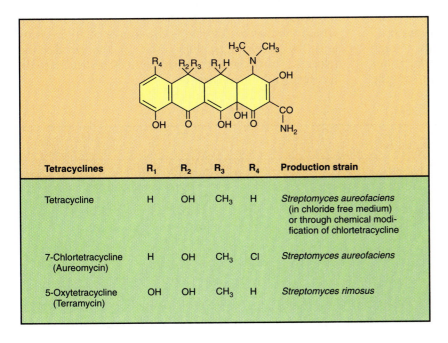

Tetracyclines	R$_1$	R$_2$	R$_3$	R$_4$	Production strain
Tetracycline	H	OH	CH$_3$	H	*Streptomyces aureofaciens* (in chloride free medium) or through chemical modification of chlortetracycline
7-Chlortetracycline (Aureomycin)	H	OH	CH$_3$	Cl	*Streptomyces aureofaciens*
5-Oxytetracycline (Terramycin)	OH	OH	CH$_3$	H	*Streptomyces rimosus*

FIGURE 10.16
Structure of tetracycline and important derivatives.

> The most important antibiotics are those containing the β-lactam ring, which include the penicillins and cephalosporins. Combined research by microbiologists and chemists has led to the production of semisynthetic β-lactams, in which the β-lactam ring is produced microbially and an appropriate side chain is added chemically. The aminoglycosides, macrolides, and tetracycline antibiotics are structurally complex molecules produced by prokaryotes and active against other prokaryotes. Erythromycin, a macrolide antibiotic, and the various tetracyclines are used widely in clinical medicine.

10.9   Vitamins and Amino Acids

Vitamins and amino acids are growth factors that are often used pharmaceutically or are added to foods. Several important vitamins and amino acids are produced commercially by microbial processes.

Vitamins

Vitamins are used as supplements for human food and animal feeds. Production of vitamins is second only to antibiotics in terms of total sales of pharmaceuticals—nearly $1 billion per year. Most vitamins are made commercially by chemical synthesis. However, a few are too complicated to be synthesized inexpensively but fortunately they can be made by microbial fermentation. Vitamin B_{12} and riboflavin are the most important of this class of vitamins.

Vitamin B_{12} is synthesized in nature exclusively by microorganisms. The requirements of animals for this vitamin are satisfied by food intake or by absorption of the vitamin produced in the gut of the animal by intestinal microorganisms. Humans, however, must obtain vitamin B_{12} from food or as a vitamin supplement, since even if it is synthesized by microorganisms in the large intestine, it does not pass from the large intestine into the blood stream. Plants do not produce or use vitamin B_{12}. Microbial strains are used that have been specifically selected for their high yields of the vitamin. Members of the bacterial genus *Propionibacterium* give yields of the vitamin ranging

from 19 to 23 mg/liter in a two-stage process, while another bacterium, *Pseudomonas denitrificans*, produces 60 mg/liter in a one-stage process which uses sugar-beet molasses as the carbon source. Vitamin B_{12} contains cobalt as an essential part of its structure, and yields of the vitamin are greatly increased by addition of cobalt to the culture medium.

Riboflavin is synthesized by many microorganisms, including bacteria, yeasts, and fungi. The fungus *Ashbya gossypii* produces a huge amount of this vitamin (up to 7 g/liter) and is therefore used for most of the microbial production processes. In spite of this good yield, there is great economic competition between this microbiological process and chemical synthesis.

Amino acids

Amino acids have extensive uses in the food industry, as feed additives, in medicine, and as starting materials in the chemical industry (Table 10.5). The most important commercial amino acid is **glutamic acid**, which is used as a flavor enhancer (monosodium glutamate, MSG). Two other important amino acids, **aspartic acid** and **phenylalanine**, are the ingredients of the artificial sweetener, **aspartame**, an important constituent of diet soft drinks and other foods which are sold as sugar-free products. **Lysine**, an essential amino acid for humans and certain farm animals, is produced by the bacterium *Brevibacterium flavum* for use as a food additive.

Although most of the amino acids can be made chemically, chemical synthesis results in the formation of optically inactive D, L-mixtures. If the biochemically important L-form is needed, then an enzymatic or microbiological method of manufacturing is needed (see Figure 10.6b). Microbiological production of amino acids can be either by *direct fermentation*, in which the microorganism produces the amino acid in a standard fermentation process, or by *enzymatic synthesis*, in which the microorganism is the source of an enzyme, and the enzyme is then used in the production process. The present discussion will be restricted to direct fermentation processes.

Table 10.5   Amino acids used in the food industry

Amino acid	Foods	Purpose
Glutamate (monosodium glutamate, MSG)	Various foods	Flavor enhancer; meat tenderizer
Aspartate and alanine	Fruit juices	"Round off" taste
Glycine	Sweetened foods	Improve flavor
Cysteine	Bread	Improves quality
	Fruit juices	Antioxidant
Tryptophan + histidine	Various foods, dried milk	Antioxidant, prevents rancidity
Aspartame (made from phenylalanine + aspartic acid)	Soft drinks, etc.	Low-calorie sweetener
Lysine	Bread (Japan)	Nutritive additive
Methionine	Soy products	Nutritive additive

Regulation of amino acid biosynthesis

We discussed amino acid biosynthesis and regulation by feedback inhibition and repression in Chapters 4 and 5. Because the amino acids are used by microorganisms as building blocks of proteins, strict regulation of amino acid production occurs. Overproduction of an amino acid in a normal growth process would be wasteful of energy and nutrients. The role of the industrial microbiologist is to find ways to bypass the microbial regulatory mechanisms.

As we discussed (see Section 4.18), the biosynthesis of each amino acid involves a number of steps, generally beginning with a central component of cell metabolism such as a citric acid cycle intermediate or a glycolysis intermediate. It has been determined from studies on the biosynthesis of the various amino acids that the *first* enzymatic step unique to a particular biosynthetic pathway is usually subject to feedback inhibition by the end product of this pathway (allostery, see Section 4.20 and Figures 4.30 and 4.31). Thus, as the concentration of an amino acid increases, the activity of the enzyme leading to the synthesis of this amino acid is inhibited, resulting in the reduction in synthesis.

One way to achieve overproduction of a desired amino acid is to obtain a *mutant* in which the first unique enzyme in that amino acid pathway is no longer subject to feedback inhibition. This can often be done by isolating a mutant that is *resistant* to a growth inhibitory analog of that amino acid. Such an approach works because the amino acid analog is recognized by the allosteric site of the biosynthetic enzyme, so that the amino acid analog inhibits growth of the organism. A mutant able to grow in the presence of the analog is often one in which the first biosynthetic enzyme is resistant to feedback inhibition. Thus, if one wishes to obtain a mutant capable of overproducing tryptophan, a mutant resistant to a tryptophan analog such as 5-methyl tryptophan could be isolated. In some cases, it may be desirable to isolate mutants resistant to a variety of amino acid analogs.

In addition to feedback-resistant mutants, mutants no longer sensitive to repression can also be isolated. As we have noted (see Section 5.10), repression results from the interaction of the amino acid (or other small molecule metabolite) with a repressor protein, activating the repressor and causing it to inhibit transcription of the mRNA for that gene. A mutant which either lacks the repressor altogether or makes a repressor unable to bind its corepressor, will make the mRNA all the time and hence should *overproduce* the enzyme involved in the amino acid biosynthesis.

A third factor that is very important, if a commercial production process is to be obtained, is to obtain *excretion* of the amino acid into the culture medium. In general, organisms do not excrete essential metabolites such as amino acids. By arranging for excretion, unusually high concentrations, which might cause feedback inhibition or repression even in resistant mutants, will not occur inside the cells. An interesting procedure for obtaining high excretion rate of an amino acid is that used for the production of *glutamic acid*. Production and excretion of excess glutamic acid is dependent upon cell permeability. The organism producing glutamic acid, *Corynebacterium glutamicum*, requires the vitamin *biotin*, an essential cofactor in fatty acid biosynthesis. Deficiency in biotin leads to membrane damage (as a result of poor phospholipid production), and under these conditions, intracellular glutamic acid is excreted. The medium used for commercial glutamic acid production thus contains sufficient biotin to obtain good growth, after which biotin deficiency ensues and glutamic acid is excreted.

> Microbial products used in foods and as food supplements include vitamins and amino acids. Vitamins produced microbially include vitamin B_{12} and riboflavin. The most important amino acids produced commercially are glutamic acid, used as a flavor enhancer, and aspartic acid and phenylalanine, used in the manufacture of the artificial sweetener aspartame. High yields of amino acids are obtained by modifying the regulatory signals for the particular amino acid biosynthetic pathway, so that overproduction of the amino acid occurs. Feedback inhibition-resistant mutants are commonly employed in vitamin/amino acid production.

10.10   Microbial Bioconversion

One of the most far-reaching discoveries in industrial microbiology is the understanding that microorganisms can be used to carry out specific chemical reactions that are beyond the capabilities of organic chemistry. The use of microorganisms for this purpose is called **bioconversion** and involves the growth of the organism in large fermentors, followed by the addition at an appropriate time of the chemical to be converted. Following a further incubation period during which the chemical is acted upon by the organism, the fermentation broth is extracted, and the desired product purified. Although in principle bioconversion may be used for a wide variety of processes, its major practical use has been in the production of certain steroid hormones (Figure 10.18).

We discussed the role of sterols in eukaryotic membranes in Section 3.3. Steroids, which are derivatives of sterols, are important hormones in animals which regulate various metabolic processes. Some steroids are also used as drugs in human medicine. One group, the *adrenal cortical steroids*, reduces inflammation and hence are effective in controlling the symptoms of arthritis and allergy. Another group, the estrogens and androgenic steroids, are involved in human fertility, and some of these can be used in the control of fertility. Steroids can be obtained by complete chemical synthesis, but this is a complicated and expensive process. Certain key steps in chemical synthesis can be carried out more efficiently by microorganisms, and commercial production of steroids usually has at least one microbial step.

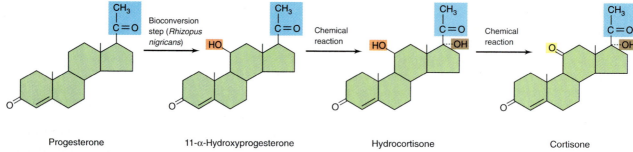

FIGURE 10.18 Cortisone production using a microorganism. The first reaction is a typical microbial bioconversion, the formation of 11-α-hydroxyprogesterone from progesterone. This highly specific oxidation, carried out by the fungus *Rhizopus nigricans*, bypasses a difficult chemical synthesis. All the other steps, from progesterone to the steroid hormone cortisone, are performed chemically.

10.11 Enzyme Production by Microorganisms

Each organism produces a large variety of enzymes, most of which are produced only in small amounts and are involved in cellular processes. However, certain enzymes are produced in much larger amounts by some organisms, and instead of being held within the cell, they are excreted into the medium. Extracellular enzymes are usually capable of digesting insoluble nutrient materials such as cellulose, protein, and starch, the products of digestion then being transported into the cell, where they are used as nutrients for growth. Some of these extracellular enzymes are used in the food, dairy, pharmaceutical, and textile industries and are produced in large amounts by microbial synthesis (Table 10.6). They are especially useful because of their specificity and efficiency when catalyzing reactions of interest at moderate temperature and pH. Similar reactions achieved by chemical means

would generally require extreme conditions of temperature or pH and be less efficient and less specific.

Enzymes are produced commercially from both fungi and bacteria. The production process is usually aerobic, and culture media similar to those used in antibiotic fermentations are employed. The enzyme itself is generally formed in only small amounts during the active growth phase but accumulates in large amounts during the stationary phase of growth. As we have seen (Section 5.10), induced enzymes are produced only when an appropriate inducer is present in the medium. The potential for the production of useful enzymes has improved markedly in recent years because of the increased ease with which genes can be manipulated.

The microbial enzymes produced in the largest amounts on an industrial basis are the bacterial proteases, used as additives in laundry detergents. By 1969, 80 percent of all laundry detergents contained enzymes, chiefly proteases, but also amylases, lipases, reductases, and other enzymes. Many of these en-

Table 10.6 Microbial enzymes and their application			
Enzyme	**Source**	**Application**	**Industry**
Amylase (starch digesting)	Fungi	Bread	Baking
	Bacteria	Starch coatings	Paper
	Fungi	Syrup and glucose manufacture	Food
	Bacteria	Cold-swelling laundry starch	Starch
	Fungi	Digestive aid	Pharmaceutical
	Bacteria	Removal of coatings (desizing)	Textile
	Bacteria	Removal of stains; detergents	Laundry
Protease (protein-digesting)	Fungi	Bread	Baking
	Bacteria	Spot removal	Dry cleaning
	Bacteria	Meat tenderizing	Meat
	Bacteria	Wound cleansing	Medicine
	Bacteria	Desizing	Textile
	Bacteria	Household detergent	Laundry
Invertase (sucrose-digesting)	Yeast	Soft-center candies	Candy
Glucose oxidase	Fungi	Glucose removal, oxygen removal	Food
		Test paper for diabetes	Pharmaceutical
Glucose isomerase	Bacteria	High-fructose corn syrup	Soft drink
Pectinase	Fungi	Pressing, clarification	Wine, fruit juice
Rennin	Fungi	Coagulation of milk	Cheese
Cellulase	Bacteria	Fabric softening/brightening; detergent	Laundry

zymes are isolated from alkaliphilic bacteria, mainly species of *Bacillus*. These enzymes, which have pH optima between 9 and 10, remain active at the alkaline pH of laundry detergent solutions.

Other important enzymes manufactured commercially are amylases and glucoamylases, which are used in the production of glucose from starch. The glucose so produced can then be acted upon by glucose isomerase to produce fructose (which is sweeter than either glucose or sucrose) resulting in the final production of a high-fructose sweetener from corn, wheat, or potato starch. The use of this process in the food industry has been increasing, especially in the production of soft drinks.

Three reactions, each catalyzed by a separate microbial enzyme, operate in sequence in the conversion of corn starch into the product called **high-fructose corn syrup**.

1. The enzyme *α*-**amylase** brings about the initial attack on the starch polysaccharide, shortening the chain, and reducing the viscosity of the polymer. This is called the *thinning reaction*.
2. The enzyme **glucoamylase** produces glucose monomers from the shortened polysaccharides, a process called *saccharification*.
3. The enzyme **glucose isomerase** brings about the final conversion of glucose to fructose, a process called *isomerization*.

All three enzymes are produced industrially by microbial fermentation. The end product of this series of reactions is a syrup containing about equal amounts of glucose and fructose which can be added directly to soft drinks and other food products.

The demand for high-fructose corn syrup is likely to increase. To date, the enzyme processes in use have been developed without the use of genetic engineering, but it is likely that recombinant-DNA technology will not only permit increased production of the present enzymes, but also will permit the development of completely new enzyme processes. The soft drink may continue to appear in its familiar container, but its contents will represent a refined product of the biotechnologist's art.

Another enzyme of commercial significance is **microbial rennin**. It has been used in place of calf's rennin for cheese production since 1965. It is much simpler and less expensive to produce than calf's rennin, and seems to be equally effective in cheese production. Recent research has also revealed that various species of hyperthermophilic Archaea produce enzymes of commercial interest. Besides the *Taq* and *Pfu* DNA polymerases mentioned in Sections 8.9 and 9.8, extremely thermally stable proteases, amylases, cellulases, and xylanases have been identified from hyperthermophiles. It is possible that these enzymes could find commercial application in high temperature versions of the industrial fermentations listed in Table 10.6.

Immobilized enzymes

For use in industrial processes, it is frequently desirable to convert soluble enzymes into some sort of immobilized state. Immobilization not only makes it easier to carry out the enzymatic reaction under large-scale conditions, it generally stabilizes the enzyme to denaturation. There are three basic approaches to enzyme immobilization (Figure 10.19):

1. **Cross-linkage (polymerization)** of enzyme molecules. Linkage of enzyme molecules with each other is usually done by chemical reaction with a bifunctional cross-linking agent such as glutaraldehyde. Cross-linking of enzymes involves the chemical reaction of amino groups of the enzyme protein with glutaraldehyde (Figure 10.19*b*).

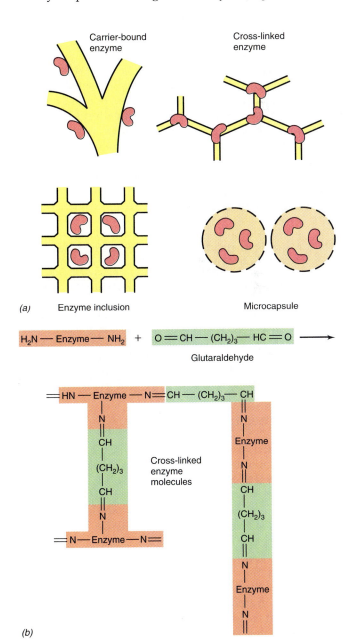

FIGURE 10.19 Immobilized enzymes. (a) Procedures for the immobilization of enzymes. (b) Procedure for cross-linking with glutaraldehyde.

If the reaction is carried out properly, the enzyme molecules can be linked in such a way that most enzymatic activity is maintained.

2. **Bonding** of the enzyme to a carrier. The bonding can be through adsorption, ionic bonding, or covalent bonding. Carriers used include modified celluloses, activated carbon, clay minerals, aluminum oxide, and glass beads.

3. **Enzyme inclusion**, which involves incorporation of the enzyme into a *semi-permeable membrane*. Enzymes can be enclosed inside microcapsules, gels, semi-permeable polymer membranes, or fibrous polymers such as cellulose acetate.

Each of these procedures has advantages and disadvantages and the procedure used would depend on the enzyme and on the particular industrial application.

Immobilized cells

In some cases it is not necessary to use purified enzyme. Rather, enzyme-rich *cells* can themselves be immobilized and the industrial process operated continuously. An example is the immobilization of glucose isomerase-containing cells of *Bacillus coagulans* by cross-linking with glutaraldehyde. The glucose syrup is passed through columns containing the immobilized cells and the resulting fructose syrup used. Much smaller installations are generally used for such continuous-flow processes.

> Microorganisms are ideal for the large-scale production of enzymes. Many enzymes are used in the laundry industry to remove stains from clothing. The production of high-fructose corn syrup, an important sweetening ingredient in soft drinks and food products, involves the participation of three microbial enzymes, of which glucose isomerase, which converts glucose into the sweeter sugar fructose, is the most important. When an enzyme is used in a large-scale process, it is usually desirable to immobilize it by chemically bonding it to an inert substrate.

10.12 Vinegar

Vinegar is the product resulting from the conversion of ethyl alcohol to acetic acid by **acetic acid bacteria**, members of the genera *Acetobacter* and *Gluconobacter*. Vinegar can be produced from any alcoholic substance, although the usual starting materials are wine or alcoholic apple juice (cider). Vinegar can also be produced from a mixture of pure alcohol in water, in which case it is called *distilled vinegar*, the term *distilled* referring to the alcohol from which the product is made rather than the vinegar itself. Vinegar is used as a flavoring ingredient in salads and other foods, and because of its acidity, it is also used in pickling. Meats and vegetables properly pickled in vinegar can be stored unrefrigerated for years.

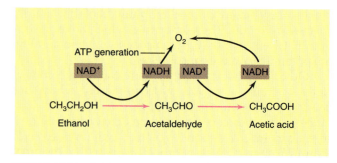

FIGURE 10.20   Oxidation of ethanol to acetic acid.

The aerobic *acetic acid bacteria* are an interesting group of bacteria (see Section 19.17); however, do not confuse these aerobic acetic acid bacteria with the anaerobic *homoacetogenic* bacteria (Section 19.9). The aerobic acetic acid bacteria differ from most other aerobes in that they do not oxidize their energy sources completely to CO_2 and water (Figure 10.20). Thus, when provided with ethyl alcohol as electron donor, they oxidize it only to acetic acid, which accumulates in the medium. Acetic acid bacteria are quite acid tolerant and are not killed by the acidity that they produce. There is a high oxygen demand during growth, and the main problem in the production of vinegar is to ensure sufficient aeration of the medium.

There are three different processes for the production of vinegar. The **open-vat** or **Orleans method** was the original process and is still used in France where it was developed. Wine is placed in shallow vats with considerable exposure to the air, and the acetic acid bacteria develop as a slimy layer on the top of the liquid. This process is not very efficient, since the only place that the bacteria come in contact with both the air and substrate is at the surface. The second process is the **trickle method** in which the contact between the bacteria, air, and substrate is increased by trickling the alcoholic liquid over beechwood twigs or wood shavings that are packed loosely in a vat or column while a stream of air enters at the bottom and passes upward. The bacteria grow upon the surface of the wood shavings and thus are maximally exposed both to air and liquid. The vat is called a vinegar generator (Figure 10.21), and the whole process is operated in a continuous fashion. The life of the wood shavings in a vinegar generator is long, from 5 to 30 years, depending on the kind of alcoholic liquid used in the process.

The third vinegar process is the **bubble method**. This is basically a submerged fermentation process, such as already described for antibiotic production. Efficient aeration is even more important with vinegar than with antibiotics, and special highly efficient aeration systems have been devised. The process is operated in a continuous fashion: alcoholic liquid is added at a rate just sufficient to balance removal of vinegar. The efficiency of the process is high, and 90–98 percent of the alcohol is converted to acid. One disadvantage

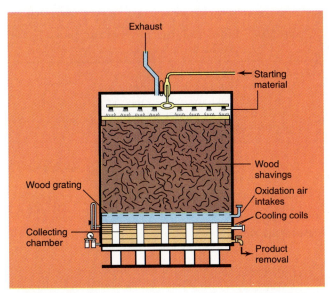

FIGURE 10.21 Diagram of one kind of vinegar generator. The alcoholic juice is allowed to trickle through the wood shavings and air is passed up through the shavings from the bottom. Acetic acid bacteria develop on the wood shavings and convert alcohol to acetic acid. The acetic acid solution accumulates in the collecting chamber and is removed periodically. The process can be run semicontinuously.

of the bubble method is that the product must undergo more filtering to remove the bacteria, whereas in the open-vat and trickle methods the product is virtually free of bacteria since the cells are bound in the slimy layer in the former and adhere to the wood chips in the latter.

Although acetic acid can be easily made chemically from alcohol, the microbial product, vinegar, is a distinctive material, the flavor being due in part to other substances present in the starting material. For this reason, the fermentation process has not been supplanted by a chemical process.

> The active ingredient in vinegar is acetic acid, which is produced by an acetic acid bacterium oxidizing an alcohol-containing fruit juice. Adequate aeration is the most important consideration in assuring a successful vinegar process.

10.13   Citric Acid and Other Organic Compounds

Many organic chemicals are produced by microorganisms in sufficient yields so that they can be manufactured commercially by fermentation. *Citric acid*, used widely in foods and beverages, *itaconic acid*, used in the manufacture of acrylic resins, and *gluconic acid*, used in the form of calcium gluconate to treat calcium deficiencies in humans and industrially as a washing and softening agent, are produced by fungi. *Sorbose*, which is produced when *Acetobacter* oxidizes sorbitol, is used in the manufacture of *ascorbic acid*, vitamin C. (In fact, this sorbitol-sorbose reaction is the only biological step in the otherwise entirely nonbiological

chemical synthesis of ascorbic acid.) *Gibberellin*, a plant growth hormone used to stimulate growth of plants, is produced by a fungus. *Dihydroxyacetone*, produced by allowing *Acetobacter* to oxidize glycerol, is used as a suntanning agent. *Dextran*, a gum used as a blood-plasma extender and as a biochemical reagent, and *lactic acid*, used in the food industry to acidify foods and beverages, are produced by lactic acid bacteria. *Acetone* and *butanol* can be produced in fermentations by *Clostridium acetobutylicum* but are now produced mainly from petroleum by strictly chemical synthesis.

Citric acid

Of the foregoing, citric acid is perhaps the most interesting product to consider here since it was one of the earliest successful aerobic fermentation products. Citric acid was formerly made commercially in Italy and Sicily by chemical purification from citrus fruits, and for many years, Italy held a world monopoly on citric acid, which resulted in relatively high prices. This monopoly was broken when the microbiological process using the fungus *Aspergillus niger* was developed, and the price of citric acid fell drastically. Today, virtually all citric acid is produced by fermentation. Citric acid is a *primary* metabolic product formed in the tricarboxylic acid cycle (Figure 4.20), but with certain organisms excretion of large amounts of citric acid can be obtained. The process is carried out in large aerated fermentors, using a molasses-ammonium salt medium. One of the key requirements for high citric acid yields is that the medium must be *low in iron* since the citric acid is produced by the fungus as a chelator specifically to scavenge iron from an iron-poor environment; therefore, most of the iron is removed from the medium before it is used.

The media used for citric acid production have been highly perfected over the many years that the commercial process has been underway. A variety of starting materials can be used as carbohydrate sources: starch from potatoes, starch hydrolysates, glucose syrup from saccharified starch, sucrose of different levels of purity, sugar cane syrup with two-thirds of the sucrose converted into invert sugar (a mixture of glucose and levulose), sugar cane molasses, and sugar beet molasses. If starch is used, amylases formed by the producing fungus or added to the fermentation broth hydrolyze the starch to sugars.

In the trophophase of the citric acid fermentation, part of the added sugar is used for the production of mycelium and part is converted through respiration into CO_2. In the idiophase, the rest of the sugar is converted into the organic acids, and during this phase there is a minimal loss through respiration (Figure 10.22).

Citric acid is produced by both surface and submerged processes. Surface processes can be further subdivided according to the state of the culture medium used: solid or liquid. Surface processes em-

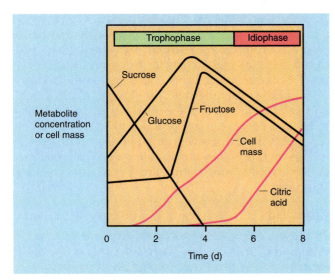

FIGURE 10.22 Kinetics of the citric acid fermentation.

Table 10.7 Industrial uses of yeast and yeast products

Production of yeast cells
Baker's yeast, for bread making
Dried food yeast, for food supplements
Dried feed yeast, for animal feeds
Yeast products
Yeast extract, for culture media
B vitamins, vitamin D
Enzymes for food industry; invertase, galactosidase
Biochemicals for research; ATP, NAD+, RNA
Fermentation products from yeast
Ethanol, for industrial alcohol
Glycerol
Beverage alcohol
Beer
Wine
Distilled beverages
Whiskey
Brandy
Vodka
Rum

ploying solid substrates may use either wheat bran or pulp from sweet potato starch production as a culture medium. The solid surface process takes 5–8 days after which the entire solution is extracted with hot water to isolate the citric acid.

Surface processes using liquid nutrient solutions are the oldest production methods and account for 20 percent of the world's supply of citric acid. However, 80 percent of the world's supply of citric acid is produced by submerged processes. Three factors are especially important for production in submerged processes: (a) quality of the material used to construct the fermentor, (b) mycelium structure, and (c) oxygen supply. Fermentors for citric acid production are generally constructed of stainless steel. This is because at the low pH values generated by citric acid production, the heavy metals leached from normal steel fermentor walls can inhibit the formation of citric acid.

Historically, the development of a submerged process for citric acid was of great importance because it was the first *aerobic* industrial fermentation. The technology for manufacturing aerobic fermentors was perfected with the citric acid process. This technology was then applied to penicillin and the other important antibiotic fermentations. Thus, we owe some of our current success with large-scale production of antibiotics to the pioneering work that was done on the citric acid fermentation.

> A number of organic chemicals are produced commercially by use of microorganisms, of which the most important economically is citric acid, produced by certain fungi.

10.14 Yeasts in Industry

Yeasts are the most important and the most extensively used microorganisms in industry (Table 10.7). They are cultured for the cells themselves, for cell

components, and for the end products that they produce during the alcoholic fermentation. Yeast cells are used in the manufacture of bread, and also as sources of food, vitamins, and other growth factors. Large-scale fermentation by yeast is responsible for the production of alcohol for industrial purposes but yeast is better known for its role in the manufacture of alcoholic beverages: beer, wine, and liquors. Production of yeast cells and production of alcohol by yeast are two quite different processes industrially, in that the first process requires the presence of oxygen for maximum production of cell material and hence is an *aerobic* process, whereas the alcoholic fermentation is *anaerobic* and takes place only in the absence of oxygen. However, the same or similar species of yeasts are used in virtually all industrial processes. The yeast *Saccharomyces cerevisiae* was derived from wild yeast used in ancient times for the manufacture of wine and beer. The yeasts currently used are descendants of the early *S. cerevisiae*. Since they have been cultivated in laboratories for such a long time, there has been ample opportunity for selection of strains according to particular desirable properties. It is possible to genetically alter yeasts in the laboratory, using genetic exchange methods to produce new strains that contain desirable qualities from two separate parent strains (see Section 7.14). By the techniques of genetic engineering, it is now also possible to improve strains by direct intervention.

Baker's yeast

The baker uses yeast as a leavening agent in the rising of the dough prior to baking. A secondary contribution of yeast to bread is its flavor. In the leavening process, the yeast is mixed with the moist dough in the presence of a small amount of sugar. The yeast converts the sugar to alcohol and CO_2, and the gaseous CO_2 expands, causing the dough to rise. When the bread is baked, the heat drives off the CO_2 (and incidentally, the alcohol) and holes are left within the bread mass, thus giving bread its characteristic light texture. That yeast contributes more to bread than CO_2 is shown by the fact that dough raised with baking powder, a chemical source of CO_2, produces a quite different product than dough raised by yeast. Only the latter bears the name *bread*.

In early times, the bread maker obtained yeast from a nearby brewery, since yeast is a by-product of the brewing of beer. Today, however, baker's yeast is specifically produced for bread making. The yeast is cultured in large aerated fermentors in a medium containing molasses as a major ingredient. Molasses, a by-product of sugar refining from beets or cane, still contains large amounts of sugar that serve as the source of carbon and energy. Molasses also contains minerals, vitamins, and amino acids used by the yeast. To make a complete medium for yeast growth, phosphoric acid (a phosphorus source) and ammonium sulfate (a source of nitrogen and sulfur) are added.

Fermentation vessels for baker's yeast production range from 40,000 to 200,000 liters. Beginning with the pure stock culture, several intermediate stages are needed to build up the inoculum to a size sufficient to inoculate the final stage (Figure 10.23). Fermentors and accessory equipment are made of stainless steel and are sterilized by high-pressure steam. The actual operation of the fermentor requires special control to obtain the maximum amount of yeast. It is undesirable to add all the molasses to the tank at once, since this results in a sugar excess, and the yeast converts some of this surplus sugar to alcohol rather than turning it into yeast cells. Therefore, only a small amount of the molasses is added initially, and then as the yeast grows and consumes this sugar, more is added.

At the end of the growth period, the yeast cells are recovered from the broth by centrifugation. The cells are usually washed by dilution with water and recentrifuged until they are light in color. Baker's yeast is marketed in two ways, either as compressed cakes or as a dry powder. *Compressed yeast* cakes are made by mixing the centrifuged yeast with emulsifying agents, starch, and other additives that give the yeast a suitable consistency and reasonable shelf life and the product is then formed into cubes or blocks of various sizes for domestic or commercial use. A yeast cake will contain about 70 percent moisture and about 2×10^{10} cells per gram. Compressed yeast must be stored in the refrigerator so that its activity is maintained. Yeast marketed in the dry state for baking is usually called *active dry yeast*. The washed yeast is mixed with additives and dried under vacuum at 25–45°C for a six-hour period, until its moisture is reduced to about 8 percent. It is then packed in airtight containers, such as fiber drums, cartons, or multiwall

> Yeasts are produced for use in the baking industry, where they are the key agent in the raising of bread. Baker's yeast is produced in very large scale aerated tanks, using molasses as the principal energy and carbon source.

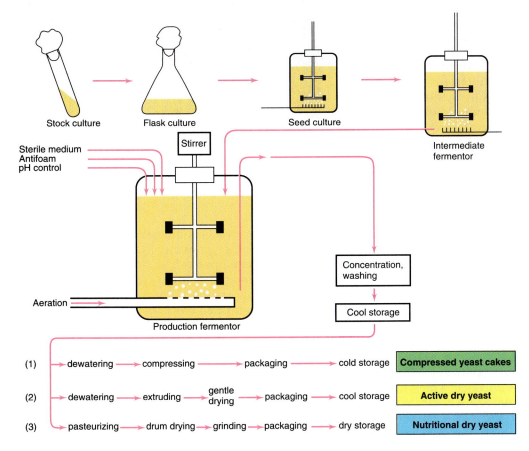

FIGURE 10.23
Stages in commercial yeast production.

bags, sometimes under a nitrogen atmosphere to promote long shelf life. Active dry yeast does not exhibit as great a leavening action as compressed fresh yeast but has a much longer shelf life.

10.15 Alcohol and Alcoholic Beverages

The use of yeast for the production of alcoholic beverages is an ancient process. Most fruit juices undergo a natural fermentation caused by the wild yeasts that are present on the fruit. From these natural fermentations, yeasts have been selected for more controlled production, and today, alcoholic-beverage production is a large industry worldwide. The most important alcoholic beverages are *wine*, produced by the fermentation of fruit juice; *beer*, produced by the fermentation of malted grains; and *distilled beverages*, produced by concentrating alcohol from a fermentation by distillation. The biochemistry of alcohol fermentation by yeast was discussed in Section 4.8.

Wine

Wine is a product of the alcoholic fermentation by yeast of fruit juices or other materials that are high in sugar. Most wine is made from grapes, and unless otherwise specified, the word *wine* refers to the product resulting from the fermentation of grape juice. Wine manufacture occurs in parts of the world where grapes can be most economically grown. The greatest wine-producing countries, in order of decreasing volume of production, are Italy, France, Spain, Algeria, Argentina, Portugal, and the United States. Wine manufacture originated in Egypt and Mesopotamia well before 2000 B.C. and spread from there throughout the Mediterranean region, which is still the largest wine-producing area in the world. Other parts of the world where wine is extensively produced often have a climate similar to that of the Mediterranean, for example, California (Figure 10.24), Chile, South Africa, and Australia. There are a great number of different wines, and their quality and character vary considerably. *Dry wines* are wines in which the sugars of the juice are practically all fermented, whereas in *sweet wines*, some of the sugar is left or additional sugar is added after the fermentation. A *fortified wine* is one to which brandy or some other alcoholic spirit is added after the fermentation; sherry and port are the best-known fortified wines. A *sparkling wine* is one in which considerable carbon dioxide is present, arising from a final fermentation by the yeast directly in the bottle.

The yeasts involved in wine fermentation are of two types: the so-called wild yeasts, which are present on the grapes as they are taken from the field and are transferred to the juice, and the cultivated wine yeast, *Saccharomyces ellipsoideus*, which is added to the juice to begin the fermentation. One important distinction between wild yeasts and the cultivated wine yeast is their alcohol tolerance. Most wild yeasts can tolerate only about 4 percent alcohol and when the alcohol concentration reaches this point, the fermentation

The Christian Brothers Winery

FIGURE 10.24 Commercial wine making. (a) Equipment for transporting grapes into the winery for crushing. (b) Large tanks where the main wine fermentation takes place. (c) Barrels where the aging process takes place.

stops. The wine yeast can tolerate up to 14 percent alcohol before it stops growing. In unfortified wine, the final alcoholic content reached is determined partly by the alcohol tolerance of the yeast and partly by the amount of sugar present in the juice. The alcohol content of most unfortified wines ranges from 8 to 14 percent. Fortified wines such as sherry have an alcohol content as high as 20 percent, but this is achieved by adding distilled spirits such as brandy. In addition to the lower alcohol content produced, wild yeasts do not produce some of the flavor components considered desirable in the final product, and hence the presence and growth of wild yeasts during fermentation is unwanted.

Wine production

The production of wine begins in the early fall with the harvesting of grapes. The grapes are crushed by machine, and the juice, called *must*, is squeezed out. Depending on the grapes used and on how the must is prepared, either white or red wine may be produced (Figure 10.25). A white wine is made either from white grapes or from the juice of red grapes from which the skins, containing the red coloring matter, have been removed. In the making of red wine, the *pomace* (skins, seeds and pieces of stem) is left in during the fermentation. In addition to the color difference, red wine has a stronger flavor than white because of the presence of larger amounts of chemicals called *tannins*, which are extracted into the juice from the grape skins during the fermentation.

It is the practice in many wineries to kill the wild yeasts present in the must by adding sulfur dioxide (listed on the bottle as "sulfites") at a level of about 100 ppm. The cultivated wine yeast is resistant to this concentration of sulfur dioxide and is added as a starter culture from a pure culture grown on sterilized or pasteurized grape juice. During the initial stages, air is present in the liquid and rapid aerobic growth of the yeast occurs; then, as the air is used up, anaerobic conditions develop and alcohol production begins. The fermentation may be carried out in vats of various sizes, from 50-gallon casks to 55,000-gallon tanks,

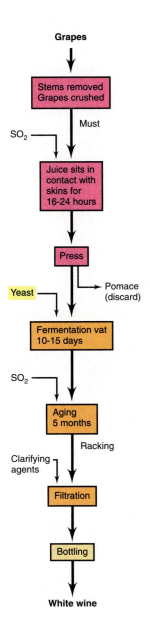

(a)

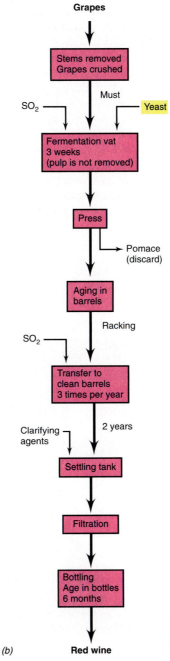

(b)

FIGURE 10.25
Production of wine. (a) White wine. (b) Red wine.

made of oak, cement, stone, or glass-lined metal (see Figure 10.24*b*). Temperature control during the fermentation is important, since heat produced during metabolism potentially raises the temperature above the point where yeast can function. Temperatures must be kept below 29°C, and the finest wines are produced at lower temperatures, from 21 to 24°C. Temperature control is achieved by using jacketed tanks through which cold water is circulated. The fermentor must be constructed so that the large amount of carbon dioxide produced during the fermentation can escape but air cannot enter; this is often accomplished by fitting the tank with a special one-way valve.

With a red wine, after three to five days of fermentation, sufficient tannin and color have been extracted from the pomace, and the wine is drawn off for further fermentation in a new tank, usually for another week or two. The next step is called *racking*; the wine is separated from the sediment (called *lees*), which contains the yeast and organic precipitate, and is then stored at lower temperature for aging, flavor development, and further clarification. The final clarification may be hastened by addition of materials called *fining agents*, such as casein, tannin, or bentonite clay, or the wine may be filtered through diatomaceous earth, asbestos, or membrane filters. The wine is then bottled and either stored for further aging or sold. Red wine is usually aged for several years or more after bottling (see Figure 10.24*c*), but white wine is usually sold without much aging. During the aging process, complex chemical changes occur, resulting in improvement in flavor and odor, or *bouquet*.

Brewing

The manufacture of alcoholic beverages made from malted grains is called *brewing*. Typical malt beverages include beer, ale, porter, and stout. *Malt* is prepared from germinated barley seeds, and it contains natural enzymes that digest the starch of grains and convert it into sugar. Since brewing yeasts are unable to digest starch, the malting process is essential for the preparation of a fermentable material from cereal grains. Malted beverages are made in many parts of the world but are most common in areas with cooler climates where cereal grains grow well and where wine grapes grow poorly.

The fermentable liquid from which beer and ale are made is prepared by a process called *mashing*. The grain of the mash may consist only of malt, or other grains such as corn, rice, or wheat may be added. The mixture of ingredients in the mash is cooked and allowed to steep in a large mash tub at warm temperatures. There are a number of different methods of mashing, involving heating at different temperatures for various lengths of time; the particular combination of temperature and time used will considerably influence the character of the final product. During the heating period, enzymes from the malt cause digestion of the starches and liberate sugars and dextrins, which are fermented by the yeast. Proteins and amino acids are also liberated into the liquid, as are other nutrient ingredients necessary for the growth of yeast.

After cooking, the aqueous extract, which is called *wort*, is separated by filtration from the husks and other grain residues of the mash. *Hops*, a herb that is derived from the female flowers of the hops plant, is added to the wort at this stage. Hops is a flavoring ingredient, but it also has antimicrobial properties, which probably help to prevent contamination in the subsequent fermentation. The wort is then boiled for several hours (Figure 10.26), during which time desired ingredients are extracted from the hops, proteins present in the wort that are undesirable from the point of view of beer stability are coagulated and removed,

(a)

(b)

FIGURE 10.26 Brewing beer. (a) The copper brew kettle is being filled with wort. (b) The aging process is carried out in these large tanks.

Home Brew

The skills of the brewer can be applied by anyone who is willing to learn sterile technique and the principles of microbiology. The amateur brewer can make many kinds of beer, from English bitters and India pale ale to German bock and Russian Imperial stout. The necessary equipment and supplies can be purchased from a local beer and winemakers shop (the Home Wine and Beer Trade Association, 604 N. Miller Road, Valrico, FL 33594, can supply the address of a nearby shop).

The brewing process can be divided into three basic stages: making the wort, carrying out the fermentation, bottling and aging. The character of the brew depends upon many factors: the proportion of malt, sugar, hops, and grain; the kind of yeast; the temperature and duration of the fermentation; and how the aging process is carried out. The instructions provided here are for a simple and relatively fool-proof beer (so-called single-stage fermentation).

The fermentor itself consists of a 20 liter (5 gallon) glass jar or carboy which can be fitted with a tightly-fitting closure. In order to have a good quality beer, it is essential that *everything* be sterilized that comes into contact with the wort. This includes the fermentor, tubing, stirring spoon, and bottles. The best procedure is to use a sterilizing rinse consisting of 50–60 ml of liquid bleach in 20 liters of water. Soak the items for 15 minutes, then rinse lightly with hot water or air dry.

1. **Making the wort**. In commercial brewing, the wort is made by producing fermentable sugars and yeast nutrients from malt, sugar, and hops. The process is complex and relatively difficult to carry out satisfactorily. Many home brewers do make their own wort from malt, but a reasonably satisfactory beer can be made with hop-flavored malt extract purchased ready-made. Malt extracts come in a variety of flavors and colors, and the kind of beer will depend upon the type of malt extract used. A simple recipe for making the wort uses 5–6 pounds (2.25–2.75 kg) of hop-flavored malt extract and 20 liters of water. The malt extract and 6 liters of water are brought to a boil for 15 minutes in an enamel or stainless steel container (aluminum heating kettles must be avoided because of the inhibiting action of metals leached from aluminum containers). The hot wort is then poured into 14–15 liters of clean, cold water which has already been added to the fermentor. After the temperature has dropped below 30°C the yeast can be added to initiate the fermentation.

2. **Carrying out the fermentation**. The process by which yeast is added to the wort is called *pitching*. Brewer's yeast can be purchased as active dry yeast from the home brew supplier. Different yeasts are available for producing different kinds of beer. If the brewing is carried out in the summer time when the temperature is higher, a yeast suitable for a high-temperature fermentation should be used. Add two packs of fresh beer yeast to the cooled wort and cover the fermentor with a rubber stopper into which a plastic hose has been inserted. The hose is directed into a bucket containing water. During the initial 2–3 days of the fermentation, large amounts of CO_2 will be given off which will exit through the hose. The water trap is to prevent wild yeasts or bacteria from the air from getting back into the fermentor. After about 3 days, the activity will diminish as the fermentable sugars are used up. At this time, the rubber stopper and hose are replaced with an inexpensive fermentation lock. The fermentation lock, which can be purchased at the home brew store, prevents contamination while permiting the small amount of gas still being produced to escape. Allow the beer to ferment for 7–10 days at 10–15°C or higher. The fermentation should begin within 24 hours after pitching the yeast. If it does not, the yeast used may not have been active, or the temperature too high when pitching was carried out.

3. **Bottling and aging**. The fermentation should be allowed to proceed for the full 7–10 days, even if the vigorous fermentation action has ceased earlier. Most of the yeast should have settled to the bottom of the fermentor. Carefully siphon the beer off the yeast layer, allowing it to run into glass beer bottles. The bottles themselves should have been sanitized first. Take care that the yeast at the bottom of the fermentor does not get stirred up, and leave the yeast-rich liquid at the bottom. The bottles used should accept standard crown caps, and new, clean caps should be used. Before capping, add 3/4 teaspoon of corn sugar syrup to each bottle. Be certain not to add more than 3/4 teaspoon of syrup, because if excess sugar is added, the build up of carbon dioxide in the bottles may cause them to burst. Once the bottles are capped, turn each one upside down once to mix the sugar syrup, then allow the beer to age upright at room temperature for at least 7–10 days. If another large container is available, a better way of adding the sugar is to siphon the beer into this second container, add the proper amount of sugar for the whole brew, dissolve, and then siphon into the bottles. After this aging period, the beer may be stored at cooler temperature.

All homemade beer has a natural yeast sediment in the bottom of the bottles. When drinking the beer, leave this residue for the beer fairy. The beer will improve if it is allowed to age for several weeks. Aging tends to make beer smoother.

Prosit!

and the wort is sterilized. Heating is accomplished either by passing steam through a jacketed kettle or by direct heating of the kettle from below by fire. Then the wort is filtered again, cooled, and then transferred to the fermentation vessel.

Brewery yeast strains are of two major types: the top-fermenting and the bottom-fermenting yeasts. The main distinction between the two is that **top-fermenting yeasts** remain uniformly distributed in the fermenting wort and are carried to the top by the CO_2 gas generated during the fermentation, whereas **bottom yeasts** settle to the bottom. Top yeasts are used in the brewing of ales, and bottom yeasts are used to make the lager beers. The bottom yeasts are usually given the species designation *Saccharomyces carlsbergensis*, and the top yeasts are called *S. cerevisiae*. Fermentation by top yeasts usually occurs at higher temperatures (14–23°C) than does that by bottom yeasts (6–12°C) and is accomplished in a shorter period of time (5 to 7 days for top fermentation versus 8 to 14 days for bottom fermentation). After completion of lager beer fermentation by bottom yeast (Figure 10.26b), the beer is pumped off into large tanks where it is stored at a cold temperature (about −1°C) for several weeks (in German, *lager* means "to store"). Lager beer is the most widely manufactured type of beer and is made by large breweries in the United States, Germany, Scandinavia, the Netherlands, and the Czech Republic. Top-fermented ale is almost exclusively a product of England and certain former British colonies. After its fermentation, the clarified ale is stored at a higher temperature (4–8°C), which assists in the development of the characteristic ale flavor.

Distilled alcoholic beverages

Distilled alcoholic beverages are made by heating a fermented liquid at a high temperature which volatilizes most of the alcohol. The alcohol is then condensed and collected, a process called *distilling*. A product much higher in alcohol content can be obtained than is possible by direct fermentation. Virtually any alcoholic liquid can be distilled, and each yields a characteristic distilled beverage. The distillation of malt brews yields *whiskey*, distilled wine yields *brandy*, distillation of fermented molasses yields *rum*, and the distillation of fermented grain or potatoes yields vodka.

The distillate contains not only alcohol but also other volatile products arising either from the yeast fermentation or from the mash itself. Some of these other products are desirable flavor ingredients, whereas others are undesirable substances called *fusel oils*. To eliminate the latter, the distilled product is almost always aged, usually in wood barrels. During the aging process, fusel oils are removed, and desirable new flavor ingredients develop. The fresh distillate is usually colorless, whereas the aged product is brown or yellow. The character of the final product is partly determined by the manner and length of aging, and the whole process of manufacturing distilled alcoholic beverages is highly complex. To a great extent, the process is carried out by traditional methods that have been found to yield a particular product, rather than by scientifically proven methods. Whiskey was originally almost exclusively an Anglo-Saxon (or Gaelic) product. A number of distinct whiskeys exist, usually associated with a country or region. Each of these has a characteristic flavor, owing to the local practices of fermenting, distilling, and aging. Even the word has local spellings, "whisky" being the Scottish, English, and Canadian spelling, and "whiskey" the Irish and United States spelling.

> Alcoholic beverages are produced by yeast from sugar under anaerobic conditions. Wine is produced from grape juice and beer from malted grain. Although yeast plays the critical role in the production of alcoholic beverages, much of the flavor develops during aging processes that occur after the yeast has been removed.

10.16   Food from Microorganisms

Microorganisms can be grown to produce food for humans, and we will discuss here the production of food, feed yeast, and mushrooms. In recent years, there has been considerable interest in the expanded production of microorganisms as food, especially in parts of the world where conventional sources of food are in short supply. Perhaps the most important potential use of microorganisms is not as a complete diet for humans but as a *protein supplement*. It is usually protein that is in shortest supply in food, and it is in the production of protein that microorganisms are perhaps the most successful. In many cases, microbial cells contain greater than 50 percent protein, and in at least some species, this is complete protein, that is, it contains sufficient amounts of all of the amino acids essential to humans. The protein produced by microorganisms as food has been called *single-cell protein* to distinguish it from the protein produced by multicellular animals and plants.

The only organism presently used as a source of single-cell protein is yeast, as already mentioned, but algae, bacteria, and fungi have also been considered. The following are desirable properties that an organism should possess to be most useful as a source of single-cell protein: (1) rapid growth, (2) simple and inexpensive medium, (3) efficient use of energy source, (4) simple culture system, (5) simple processing and separation of cells, (6) nonpathogenic, (7) harmless when eaten, (8) good flavor, (9) high digestibility, and (10) high nutrient content. Unfortunately, no organism currently known meets all these criteria.

Food and feed yeast

Yeast produced as food for humans or as feed for animals can be manufactured in much the same way as described for baker's yeast, or it may be a by-product of brewing or distilling. The yeast is heat-killed and usually dried. To be acceptable as food, dried yeast must be of proper flavor, color, and nutritional composition and must be free of contamination. The nutritional value of yeast to humans or animals is lower than that of meat or milk. However, yeast is one of the richest sources of vitamins of the B vitamin group. Although yeast is high in protein, this protein is deficient in several amino acids essential for humans, most especially the sulfur-containing amino acids. Animals fed on a diet in which all their protein comes from yeast do not grow as well as animals fed on milk or meat protein. Because of these deficiencies, yeast is used primarily as a food *supplement* rather than a food *source* for humans, being added to wheat or corn flour to increase the nutritional value of these foods.

Mushrooms

Several kinds of *fungi* are sources of human food, of which the most important are the mushrooms. Mushrooms are a group of filamentous fungi that form large complicated structures called **fruiting bodies** (Figure 10.27). The fruiting body is commonly called the *mushroom*. The fruiting body is formed through the association of a large number of individual hyphae to form a mycelium (Figure 10.27).

During most of its existence, the mushroom fungus lives as a simple mycelium, growing within soil, leaf litter, or decaying logs. However, when environmental conditions are favorable, the fruiting body develops, beginning first as a small button-shaped structure underground and then expanding into the full-grown fruiting body that we see above ground (Figure 10.27). The nutrients for growth come from organic matter in the soil and are taken up by the hyphal filaments which, like the roots of a plant, feed the growing fruiting body. Sexual spores, called **basidiospores**, are formed, borne on the underside of the fruiting body on flat plates called **gills** (Figure 10.27). The spore is the agent of dispersal of mushrooms and is carried away by the wind. If it alights in a favorable place, the spore will germinate and initiate the growth of new hyphae, mycelium, and fruiting body.

The mushroom commercially available in most parts of the world is *Agaricus bisporus*, and it is generally cultivated in mushroom farms. The organism is grown in special beds, usually in buildings where temperature and humidity are carefully controlled (Figure 10.28*a*). Since light is not necessary, mushrooms may even be grown in basements of homes, or in caves. Beds are prepared by mixing soil with a material very rich in organic matter, such as horse ma-

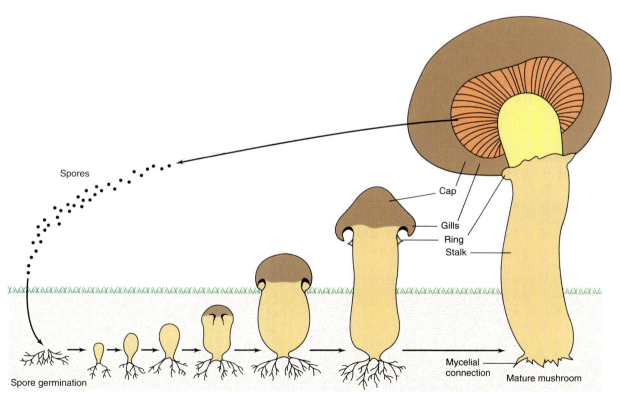

FIGURE 10.27 Mushroom life cycle, showing how the fruiting body develops from underground hyphae.

American Mushroom Institute

American Mushroom Institute

Bob Harris

Bob Harris

FIGURE 10.28 Commercial mushroom production. (a) An installation for *Agaricus bisporus*, the common commercial mushroom of the western world. (b) Close up of a mushroom flush. (c) Shiitake, *Lentinus edulus*, the most common commercial mushroom of the orient, but finding increasing production in the west. A large Japanese installation where the mushroom is cultivated on hardwood logs. (d) Close-up of the shiitake mushroom.

nure, and these beds are then inoculated with mushroom *spawn*. The spawn is actually a pure culture of the mushroom fungus that has been grown in large bottles on an organic-rich medium. In the bed, the mycelium grows and spreads through the substrate, and after several weeks it is ready for the next step, the induction of mushroom formation. This is accomplished by adding to the surface of the bed a layer of soil called *casing soil*. The appearance of mushrooms on the surface of the bed is called a *flush* (Figure 10.28b), and when flushing occurs, the mushrooms must be collected immediately while still fresh. After collection they are packaged and kept cool until brought to market. Several flushes will take place on a single bed, and after the last flush the bed must be cleaned out and the process begun again.

Another widely cultured mushroom is **shiitake**, *Lentinus edulus*. The most widely cultivated mushroom in the Orient, shiitake is now finding expanding interest in North America. Shiitake is a cellulose-digesting fungus that grows well on hardwood trees and is cultivated on small logs (Figure 10.28c). The logs are soaked in water to hydrate them, then inocu-

lated by inserting plugs of spawn into small holes drilled in the logs. The fungus grows through the log and after about a year forms a flush of fruiting bodies (see Figure 10.28d). Shiitake has the advantage that it can be cultivated on waste or scrap wood. Some people find it to be much tastier than *Agaricus bisporus*.

Although mushrooms make flavorful food, their digestibility and nutritional value are not very high. They are low in protein and deficient in certain essential amino acids; they are also not exceptionally rich in vitamins. The mushrooms and the filamentous fungi are definitely inferior to yeast as food sources, although they serve as valuable flavoring ingredients.

Many microorganisms are high in protein and could be used as sources of protein food for humans or animals. However, the economics of producing single-cell protein are unsatisfactory when compared with protein production from meat or grain. Yeast is an exception, and is used extensively as a food supplement. The most important food produced microbially is the mushroom, which is produced not for its protein but for its flavor.

Study Questions

1. In what ways do industrial microorganisms differ from conventional microorganisms? In what ways are they similar?

2. Describe some of the techniques that can be used to improve strains of industrial microorganisms.

3. List three major types of industrial products that can be obtained with microorganisms and give two examples of each type.

4. Give an example of a *commodity chemical* produced by a microorganism and describe briefly the process by which this chemical is manufactured.

5. Compare and contrast *primary* and *secondary metabolites* and give an example of each. List several molecular explanations for why some metabolites are secondary rather than primary.

6. Define *trophophase* and *idiophase*.

7. How does an industrial fermentor differ from a laboratory culture vessel? How does a fermentor differ from a fermenter?

8. Discuss the problems of *scale up* from the viewpoints of *aeration, sterilization,* and *process control*. Why is sterility so much more important in an industrial fermentor than in a laboratory fermentor?

9. Describe the vinegar process and give two ways by which aeration is achieved.

10. List five examples of *antibiotics* that are important industrially. For each of these antibiotics, list the producing organism, the general chemical structure, and the mode of action.

11. List the various stages that occur during the industrial production of an antibiotic.

12. Why are the β-lactam antibiotics so important medically? Compare and contrast *natural, biosynthetic,* and *semisynthetic* β-lactam antibiotics.

13. Describe briefly the unique aspect of the regulation of *streptomycin biosynthesis*.

14. List the unique characteristics of the *tetracycline* antibiotics.

15. Explain why the yield of *vitamin B_{12}* can be markedly improved by addition of cobalt to the fermentation medium.

16. What unusual characteristics must an organism have if it is to overproduce and excrete an amino acid such as *glutamic acid*?

17. Define *microbial bioconversion* and give an example. Explain why the chemical reactions involved in microbial bioconversions are preferably carried out microbially rather than chemically.

18. List four different kinds of *enzymes* that are produced commercially. For each enzyme, list the organism that is used in commerical production, the action of the enzyme, and how the enzyme is used in commerce.

19. Describe the stages involved in the production of *high-fructose syrup* and explain the role of an enzyme in each step. How is high fructose syrup used in the food industry?

20. Why is it desirable to *immobilize* enzymes? Give examples of three different immobilization procedures and describe how each is carried out.

21. In this chapter we have discussed at least six *foods or food-related products* that are manufactured at least in part with microorganisms. List as many of these six as you can and for each explain precisely the microbial role.

22. Why are yeasts of such great industrial importance?

23. In this chapter we have discussed at least four *beverages* that are manufactured at least in part with microorganisms. List as many of these as you can and for each explain precisely the microbial role.

24. In what way is the manufacture of *beer* similar to the manufacture of *wine*? In what ways do these two processes differ? How does the production of distilled alcoholic beverages differ from that of beer and wine?

Supplementary Readings

Baumberg, S., I. Hunter, and **M. Rhodes** (eds.). 1989. *Microbial Products: New Approaches.* Cambridge University Press, Cambridge. Coverage of antibiotic, amino acid, and recombinant protein production, and discussion of the impact of molecular biology on the future of biotechnology.

Crueger, W. and **A. Crueger.** 1990. *Biotechnology. A Textbook of Industrial Microbiology,* 2nd edition. English edition edited by Thomas D. Brock. Sinauer Associates, Sunderland, MA. A concise textbook of industrial microbiology which emphasizes the economically important microbial processes. Has excellent chapters on the various large-scale processes and a good chapter on fermentor design and scale-up.

Demain, A. L. and **N. A. Solomon** (eds.). 1985. *Biology of Industrial Microorganisms.* Benjamin-Cummings Publishing Co., Menlo Park, CA. Each chapter in this book deals with a different group or genus of industrial microorganisms, with emphasis on the physiological properties of industrial significance. Both fungi and bacteria are discussed.

Demain, A. L. and **N. A. Solomon** (eds.). 1986. *Manual of Industrial Microbiology and Biotechnology.* American Society for Microbiology, Washington, D.C. One of the respected manuals of the American Society for Microbiology, this detailed work represents a collection of chapters written by more than 30 authors. The emphasis is on principles that apply to a wide variety of processes. An advanced work of considerable value to the specialist in industrial microbiology.

Hershberger, C. L., S. W. Queener, and **G. Hegeman** (eds.). 1989. *Genetics and Molecular Biology of Industrial Microorganisms.* American Society for Microbiology, Washington, D.C. Detailed reviews on the genetics of biosynthesis of most major antibiotics and on the molecular biology of industrially useful Bacteria.

Hugo, W. B. and **A. D. Russell**. 1992. *Pharmaceutical Microbiology*, 5th edition. Blackwell Scientific Publications. Oxford, England. Detailed treatment of antibiotic production, and microbial contamination problems in large scale production processes.

Leatham, G. F. and **M. E. Himmel** (eds.). 1991. *Enzymes in Biomass Conversion*. American Chemical Society, Washington, D.C. A detailed treatment of the major enzymes in industrial use and their production.

Page, M. I. (ed.). 1992. *The Chemistry of β-Lactams*. Chapman & Hall, New York. An authoritative review of all aspects of β-lactam antibiotics including mode of action, structural features, and microbial resistance.

Rose, A. H. and **J. S. Harrison**, (eds.). 1989. *Metabolism and Physiology of Yeasts*. Academic Press, New York. An advanced treatise on yeast physiology.

Scragg, A. (ed.). 1988. *Biotechnology for Engineers*. Halsted Press: a division of John Wiley & Sons, New York. An elementary and very readable account of industrial processes with a fair bit of microbial physiology to accompany the applied material.

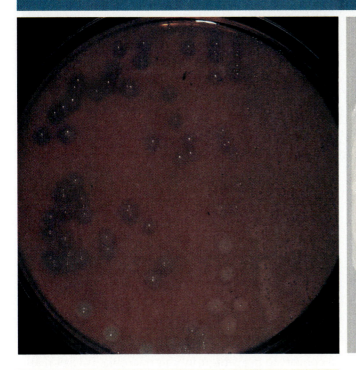

Host–Parasite Relationships

This chapter begins a major section of the book that deals with the roles of microorganisms in infectious disease. Infectious disease is the most important applied aspect of microbiology and is the one about which the most is known. In this chapter we discuss the general principles regarding how microorganisms grow in the animal body. In the next chapter, we discuss immunology, which concerns the specific processes by which the animal body counteracts infectious disease. Then in Chapter 13 we discuss the use of techniques for the diagnosis of infectious disease and for the selection of proper chemotherapeutic procedures. This is followed in Chapter 14 by a general discussion of how infectious disease spreads through populations and how a knowledge of the spread of disease can aid in disease control. Finally, Chapter 15 presents discussion of the major microbial diseases.

A word about the organisms we will be dealing with. Infectious diseases are caused by viruses, bacteria, fungi, and protozoa. We have already discussed viruses and how they interact with their hosts in Chapter 6. This chapter outlines our knowledge of how microorganisms affect the body, with emphasis on the Bacteria, the most important microbial pathogens of humans and animals. Although viruses will not be discussed in this chapter, they will be discussed in the following chapters along with the bacteria, fungi, and protozoa.

The animal body is in continual contact with microorganisms. Literally billions of bacterial cells are present in and on the human body and most play beneficial, indeed sometimes even essential, roles in the overall health of the person. These organisms are collectively referred to as the "normal" flora, and represent species which have developed an intimate relationship with certain tissues of the animal body.

A **parasite** is an organism that lives on or in a second organism, called the **host**. In some cases, the parasite has little or no harmful effect on the host and its presence may be inapparent. In some cases, however, the parasite brings about damage or harm to the host: such organisms are called **pathogens**. The relationship between host and parasite is dynamic, since each modifies the activities and functions of the other. The outcome of the host–parasite relationship depends on the **pathogenicity** of the parasite, that is, on its ability to inflict damage, and on the resistance or susceptibility of the host. The term **virulence** is a quantitative term used to indicate the *degree* of pathogenicity of the parasite and is usually expressed as the dose or cell number that will elicit a pathological response within a given time period. Neither the virulence of the parasite nor the resistance of the host are constant factors, however, each varying under the influence of external factors or as a result of the host–parasite relationship itself.

Infection refers to the growth of microorganisms in the host. *Infection is not synonymous with disease* because infection does not always lead to injury of the host, even if the pathogen is potentially virulent. In a diseased state, the host is harmed in some way, whereas infection refers to any situation in which a microorganism is established and growing in a host, whether or not the host is harmed.

The ability to cause infectious disease is one of the most dramatic properties of microorganisms. The understanding of the physiological and biochemical basis of infectious disease has led to therapeutic and preventive measures that have had far-reaching influence on medicine and human affairs. We begin this chapter by considering the normal flora of the healthy human adult. By understanding the microbial ecology of the human body we will be in a better position to appreciate the competitive forces which govern the success or failure of a potential pathogen in initiating disease. We end this chapter by discussing general mechanisms that the host uses to limit parasitism and prevent permanent tissue damage and then go on to discuss artificial mechanisms to limit the effects of the infections.

> Organisms that grow in or on other organisms are called parasites. Infection is the process by which a parasite becomes established and grows in its host. If the parasite causes harm, it is called a pathogen. Most pathogens are eventually rejected by a variety of host defense mechanisms.

11.1   Microbial Interactions with Higher Organisms

Animal bodies provide favorable environments for the growth of many microorganisms. Animals are rich in organic nutrients and growth factors required by chemoorganotrophs, they provide relatively constant conditions of pH and osmotic pressure, and warm-

FIGURE 11.1 Potential bacterial interactions on mucous membranes. (a) Loose association. (b) Adhesion. (c) Invasion into submucosal cells.

blooded animals have highly constant temperatures. However, an animal body should not be considered as one uniform microbial environment throughout. Each region or organ differs chemically and physically from other regions and thus provides a selective environment where certain kinds of microorganisms are favored over others. The skin, respiratory tract, gastrointestinal tract, and so on, each provide a wide variety of chemical and physical conditions in which different microorganisms can grow selectively. Further, animals possess a variety of defense mechanisms that act in concert to prevent or inhibit microbial invasion and growth. The microorganisms that ultimately colonize the host successfully are those that have developed ways of circumventing these defense mechanisms.

Infections frequently begin at sites in the animal body called *mucous membranes*. Mucous membranes are found throughout the body including the mouth, pharynx, esophagus, and the urinary, respiratory, and gastrointestinal tracts. Mucous membranes consist of single or multiple layers of *epithelial cells*, tightly packed cells which exist in direct contact with the external environment. Mucous membranes are frequently coated with a protective layer of mucus, primarily glycoproteins, which serves to protect epithelial cells. When bacteria contact host tissues at mucous membranes, they may associate either loosely or firmly. If they associate loosely with the mucosal surface they are usually swept away by physical processes, but they may also attach specifically to the epithelial surface as a result of specific cell-cell recognition between pathogen and host. From there, actual tissue infection may follow. When this occurs, the mucosal barrier is breached, allowing the pathogen to invade deeper tissues (Figure 11.1).

Microorganisms are almost always found in those regions of the body exposed to the outside world, such as the skin, oral cavity, respiratory tract, intestinal tract, and genitourinary tract. They are not normally found in the organs, blood, and lymph systems of the body; if microorganisms are found in significant quantities in any of these latter areas, it is usually indicative of a disease state.

> Animal bodies provide favorable environments for the growth of many microorganisms. In the initial infection, the parasite often becomes associated first with mucous membranes in various locations of the body.

Miniglossary for Chapter 11

ADHERENCE a property of bacteria which allows them to stick to surfaces

COLONIZATION multiplication of a pathogen after it has gained access to host tissues

DENTAL PLAQUE bacterial cells encased in a matrix of extracellular polymers and salivary products, found on the teeth

ENDOTOXIN the lipopolysaccharide portion of the cell wall of certain Gram-negative Bacteria which acts as a toxin when solubilized

ENTEROTOXIN a protein released by an organism as it grows (see exotoxin) that has its effect on the small intestine

EXOTOXIN a protein released by an organism as it grows which has toxic effects on the host

GLYCOCALYX bacterial polysaccharides important for attachment of bacterial cells to host structures and to one another

HOST an organism which harbors a parasite

IMMUNITY the ability of an organism to resist infection

IMMUNODEFICIENCY having a dysfunctional or completely nonfunctional immune system

INFLAMMATION host response to injury or infection, characterized by redness, swelling, heat, and pain

LEUKOCYTES nucleated cells found in the blood (white blood cells)

LOWER RESPIRATORY TRACT trachea, bronchi, and lungs

LYMPH a fluid similar to blood, but which lacks red blood cells and which travels through a separate circulatory system (the lymphatic system) containing lymph nodes that filter out particulate materials such as bacterial cells

LYMPHOCYTES a subset of nucleated cells found in the blood that are involved in the immune response

MACROPHAGES a type of large leukocyte which has phagocytic properties

NATURAL KILLER (NK) CELL a specialized lymphocyte which recognizes and destroys foreign cells or infected host cells in a nonspecific manner

PATHOGEN a parasite which does harm to a host

PATHOGENICITY the ability of a parasite to inflict damage on the host

PLASMA the liquid noncellular portion of blood

POLYMORPHONUCLEAR LEUKOCYTE (PMN) a type of leukocyte exhibiting phagocytic properties, a granular cytoplasm, and a multilobed nucleus

SERUM the liquid portion of the blood with clotting proteins and cells removed

UPPER RESPIRATORY TRACT the nasopharynx, oral cavity, and throat

11.2   Normal Flora of the Skin

An average human adult has about two square meters of skin surface that can vary greatly in chemical composition and moisture content. Figure 11.2 shows the anatomy of the skin and suggests some regions in which bacteria may live. The skin surface (epidermis) itself is not a favorable place for microbial growth, as it is subject to periodic drying. Only in certain areas of the body, such as the scalp, face, ear, underarm regions, genitourinary and anal regions, and palms and interdigital spaces of the toes, are moisture conditions on the skin surface sufficiently high to support resident microbial populations; in these regions characteristic surface microbial floras do exist.

Most skin microorganisms are associated directly or indirectly with the sweat glands, of which there are several kinds. The **eccrine glands** are not associated with hair follicles and are rather unevenly distributed over the body, with denser concentrations on the palms, finger pads, and soles of the feet. They are the main glands responsible for the perspiration associated with body cooling. Eccrine glands seem to be relatively devoid of microorganisms, perhaps because of the extensive flow of fluid, since when the flow of an eccrine gland is blocked, bacterial invasion and multiplication do occur. The **apocrine glands** are more restricted in their distribution, being confined mainly to the underarm and genital regions, the nipples, and the umbilicus. They are inactive in childhood and become fully functional only at puberty. Bacterial populations on the surface of the skin in these warm, humid places

are relatively high, in contrast to the situation on the smooth surface skin. *Underarm odor* develops as a result of bacterial activity on the secretions of the apocrines; aseptically collected apocrine secretion is odorless but develops odor upon inoculation with certain bacteria isolated from the skin. Each hair follicle is associated with a **sebaceous gland** which secretes a lubricant fluid. Hair follicles provide an attractive habitat for microorganisms; a variety of aerobic and anaerobic bacteria, yeasts, and filamentous fungi inhabit these regions, mostly within the area just below the surface of the skin. The secretions of the skin

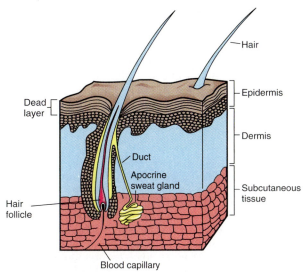

FIGURE 11.2 Anatomy of the human skin. Microorganisms are associated primarily with the sweat ducts and the hair follicles.

glands are rich in microbial nutrients. Urea, amino acids, salts, lactic acid, and lipids are present in considerable amounts. The pH of human secretions is almost always acidic, the usual range being between pH 4 and 6.

The microorganisms of the normal flora of the skin can be characterized as either **transients** or **residents**. The skin as an external organ is continually being inoculated with transients, virtually all of which are unable to multiply and usually die. Residents are organisms that are able to multiply, not merely survive, on the skin. The normal flora of the skin consists primarily of Gram-positive Bacteria restricted to a few groups. These include several species of *Staphylococcus* and a variety of both aerobic and anaerobic corynebacteria. Of the latter, *Propionibacterium acnes* is ordinarily a harmless resident but can incite or contribute to the condition known as *acne*. Gram-negative Bacteria are almost always minor constituents of the normal flora, even though such intestinal organisms as *Escherichia coli* are being continually inoculated onto the surface of the skin by fecal contamination. *Acinetobacter* is an exception and is one of the few Gram-negative Bacteria commonly found on skin. It is thought that the lack of colonization of Gram-negative Bacteria on the skin is due to their inability to compete with Gram-positive organisms that are better adapted to the dry conditions of the skin; if the latter are eliminated by antibiotic treatment, the Gram-negative Bacteria can flourish. Yeasts are uncommon on the skin surface, but the lipophilic yeast *Pityrosporum ovalis* is occasionally found on the scalp.

Although the resident microflora remains more or less constant, various factors can affect the nature and extent of the normal flora: (1) The weather may cause an increase in temperature and humidity, which increases the density of the skin microflora. (2) Age has an effect and young children have a more varied microflora and carry more Gram-negative Bacteria and potential pathogens than adults. (3) Hospitalized patients have greater numbers of pathogens and antibiotic-resistant organisms than normal people. (4) Personal hygienic habits influence the resident microflora and unclean individuals usually have higher population densities. Organisms that subsequently die when introduced onto the skin generally succumb from either the skin's low moisture content or low pH (due to organic acid content). Those organisms that survive and grow are able to resist these adverse chemical conditions.

> The skin is not a favorable habitat for most microorganisms but a few specialized types are able to live as residents. The most unfavorable environmental factors of the skin are its dryness and acidity. Microorganisms grow best in those parts of the body where moisture is highest, such as under the arms or between the toes. Gram-positive Bacteria are much more common on the surface of the skin than are Gram-negative Bacteria.

11.3   Normal Flora of the Oral Cavity

The oral cavity is one of the more complex and heterogeneous microbial habitats in the body. This cavity includes the teeth and tongue and the central space that they fill. The teeth can be viewed as a surface upon which saliva and materials derived from the food adsorb, rather than as a direct source of microbial nutrients. Although saliva is the most pervasive source of microbial nutrients in the oral cavity, it is not an especially good microbial culture medium. Saliva contains about 0.5 percent dissolved solids, about half of which are inorganic (mostly chloride, bicarbonate, phosphate, sodium, calcium, potassium, and trace elements); the predominant organic constituents of saliva are proteins, such as salivary enzymes, mucoproteins, and some serum proteins. Small amounts of carbohydrates, urea, ammonia, amino acids, and vitamins are also present. A number of antibacterial substances have been identified in saliva, of which the most important are the enzymes *lysozyme* and *lactoperoxidase*. Lysozyme is an enzyme which cleaves glycosidic linkages in peptidoglycan, the strengthening agent in the bacterial cell wall, leading to weakening of the wall and cell lysis (see Section 3.5). Lactoperoxidase, an enzyme present in both milk and saliva, kills bacteria by a reaction involving chloride ions and H_2O_2, in which singlet oxygen is probably generated (see Sections 9.12 and 11.13). The pH of saliva is controlled primarily by a bicarbonate buffering system ($H_2CO_3 \rightleftharpoons H^+ + HCO_3^-$) and varies between 5.7 and 7.0, with a mean pH near 6.7. The composition of saliva varies from individual to individual, and even within the same individual variations due to physiological and emotional factors are seen. Despite the activity of antibacterial substances, the presence of food particles and epithelial debris makes the oral cavity a very favorable microbial habitat.

The teeth and dental plaque

The tooth consists of a mineral matrix of calcium phosphate crystals (enamel), within which the living tissue of the tooth (dentin and pulp) is present (Figure 11.3). The teeth are of considerable importance in determining the nature of the microbial flora. Bacteria found in the mouth during the first year of life (when teeth are absent) are predominantly aerotolerant anaerobes such as streptococci and lactobacilli, but a variety of other bacteria, including some aerobes, can occur in small numbers. When the teeth appear, there is a pronounced shift in the balance of the microflora toward anaerobes, and a variety of bacteria specifically adapted for growth on surfaces and in crevices of the teeth develop. A film called **dental plaque** forms on the surface of the teeth. It consists mainly of bacterial cells in a matrix composed of extracellular polymers and salivary products. Dental plaque can be readily observed by staining with dyes such as basic fuchsin or erythrosin; substances that stain the plaque are called *disclosing agents*. If effective tooth brushing is practiced,

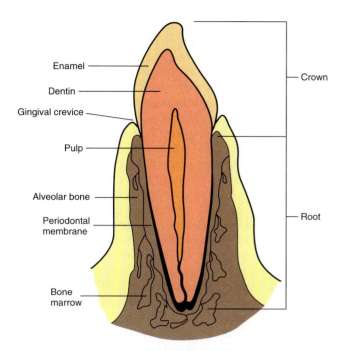

FIGURE 11.3   Section through a tooth showing the surrounding tissues which anchor the tooth in the gum.

dental plaque is present only in crevices which are protected from the brush, but plaque rapidly accumulates if brushing is stopped (Figure 11.4).

Dental plaque consists mainly of filamentous bacteria closely packed and extending out perpendicular to the surface of the tooth, embedded in an amorphous matrix. These filamentous organisms are usually classified as *Fusobacterium* species. They are obligate anaerobes on initial isolation but after subculturing become microaerophilic; they ferment carbohydrates to lactic acid. Associated with these predominant filamentous organisms are streptococci, spirochetes, diphtheroids, Gram-negative cocci, and others. The anaerobic nature of the flora may seem surprising,

considering that the mouth has good accessibility to oxygen. It is likely that anaerobiosis develops through the action of facultative bacteria growing aerobically upon organic materials on the tooth, since the dense matrix of the plaque greatly decreases the diffusion of oxygen onto the tooth surface. The microbial populations of the dental plaque are thus seen to exist in a microenvironment partly of their own making, and are probably able to maintain themselves in the face of wide variations in the macroenvironment of the oral cavity.

The bacterial colonization of smooth tooth surfaces occurs as a result of firm attachment of single bacterial cells, followed by growth in the form of microcolonies. Beginning with a freshly cleaned tooth surface, the first event is the formation of a thin organic film several micrometers thick, as a result of the attachment of acidic glycoproteins from the saliva. This film then provides a firmer attachment site for the colonization and growth of bacterial microcolonies (Figure 11.5). The colonization of this glycoprotein film is highly specific, and only a few species of *Strep-*

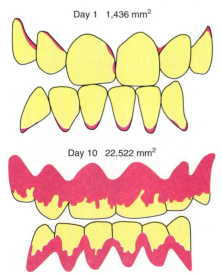

FIGURE 11.4   Distribution of dental plaque, as revealed by use of a disclosing agent, on brushed (top) and unbrushed (bottom) teeth. The numbers give the total area of dental plaque.

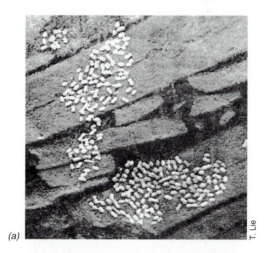

(a)

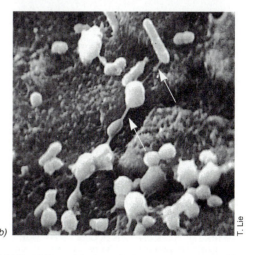

(b)

FIGURE 11.5   (a) Bacterial microcolonies growing on a model tooth surface inserted into the mouth for 6 hours. (b) Higher magnification of the preparation in part (a). Note the diverse morphology of the organisms present and the slime material (arrows) holding the organisms together.

tococcus (primarily *S. sanguis*, *S. sobrinus*, and *S. mitis*) are involved. As a result of extensive growth of these organisms, a thick bacterial zone is formed (Figure 11.6*a* and *b*). Subsequently, the plaque is colonized by other organisms, and in heavy plaque filamentous organisms such as the strictly anaerobic *Actinomyces* species may predominate.

In the accretion of dental plaque there is good evidence that various oral bacteria *coaggregate* to form the plaque matrix. Coaggregation is defined as the recognition and binding of complementary surface molecules on bacterial cells, resulting in firm *adherence* of the organisms to each other. Interestingly, coaggregation is not limited to interactions between cells of the same species. If the surface receptors of two organisms are complementary, coaggregation of different species or even different genera can occur in the formation of dental plaque.

Dental caries

As dental plaque accumulates and acid products are formed, tooth decay (dental caries) is the usual result. The role of the oral flora in tooth decay has now been well established through studies on germ-free animals, although the precise mechanisms of the process are still under study. The smooth surfaces of the teeth that are exposed to frequent cleaning by the tongue, cheek, saliva, or toothbrush or to the abrasive action of food mastication are relatively resistant to dental caries. The tooth surfaces in crevices, where food particles can be retained, are the sites where tooth decay predominates. The shape of the teeth is an important factor in the degree to which such crevices develop. For instance, dogs are highly resistant to tooth decay because the shape of their teeth does not favor retention of food. Diets high in sugars are especially cariogenic because lactic acid bacteria ferment the sugars to lactic acid, which causes decalcification of the enamel (see Figure 11.3) of the tooth. Once breakdown of the hard tissue has begun, proteolysis of the matrix of the tooth enamel occurs, through the action of proteolytic enzymes released by bacteria. Microorganisms penetrate further into the decomposing matrix, but the later stages of the process may be exceedingly slow and are often highly complex. The structure of the calcified tissue also plays an important role in the extent of dental caries. Incorporation of fluoride into the calcium phosphate crystal matrix makes the matrix more resistant to decalcification by acid. Thus, fluorides are used in drinking water and dentifrices to aid in controlling tooth decay.

Two organisms that have been implicated in dental caries are *Streptococcus sobrinus* and *Streptococcus mutans*, both lactic acid-producing bacteria. *S. sobrinus* is able to colonize smooth tooth surfaces because of its specific affinity for salivary glycoproteins (Figure 11.6), and this organism is probably the primary organism involved in decay of smooth surfaces. *S. mutans* is found predominantly in crevices and small fissures, and its ability to attach seems to be related to its

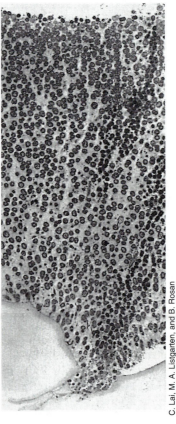

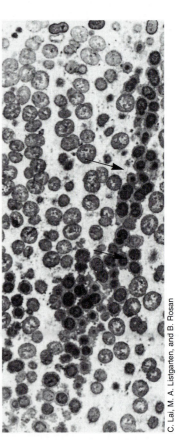

FIGURE 11.6
Electron micrographs of thin sections of dental plaque. Bottom is the base of the plaque; top is the portion exposed to the oral cavity. (a) Low-power electron micrograph. Organisms are predominantly streptococci. The species *Streptococcus sobrinus* has been labeled by an antibody-microchemical technique, and these cells appear darker than the rest. They are seen as two distinct chains (arrows). The total thickness of the plaque layer shown is about 50 μm. (b) Higher power electron micrograph showing the region with *S. sobrinus* cells (dark; arrow). Note the extensive glycocalyx (see Section 11.7) surrounding the *S. sobrinus* cells.

(a)

(b)

C. Lai, M. A. Listgarten, and B. Rosan

C. Lai, M. A. Listgarten, and B. Rosan

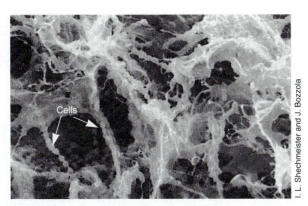

FIGURE 11.7 Scanning electron micrograph of the cariogenic bacterium *Streptococcus mutans*. The sticky dextran material can be seen as masses of filamentous particles. Individual cells are about 1 μm in diameter.

I. L. Shechmeister and J. Bozzola

ability to produce a dextran polysaccharide, which is strongly adhesive (Figure 11.7). *S. mutans* only produces dextran when sucrose is present, by means of the enzyme *dextransucrase*:

$$n \text{ sucrose} \rightarrow (\text{glucose})_n + n \text{ fructose}$$
$$\textit{Dextran}$$

Sucrose is common table sugar and is very prevalent in the diet of most Western Europeans and North Americans. Its ability to act as a substrate for dextransucrase is one reason that sucrose is highly cariogenic.

Susceptibility to tooth decay varies greatly among individuals and is affected by inherent traits in the individual as well as by diet and other extraneous factors. Studies of the distribution of oral streptococci have shown a direct correlation between the presence of *S. mutans*, and to a lesser degree *S. sobrinus*, in humans and the extent of dental caries. In the United States and Western Europe, for example, 80–90 percent of all people have their teeth colonized by *S. mutans* and dental caries is a nearly universal phenomenon. By contrast, dental caries do not occur in Tanzanian children, presumably because of dietary factors, and *S. mutans* is absent from the plaque of these individuals.

Although tooth decay is an infectious disease, we tend to place it in a different category from other infectious diseases. However, microorganisms in the mouth can also cause infections that are more typical disease states, such as periodontal disease, gingivitis, infections of the tooth pulp (abscesses), and so on.

A number of specialized bacteria grow profusely on tooth surfaces, forming a thick layer of cells which is called dental plaque. Cells of *Streptococcus* species are most common in dental plaque. Plaque-forming microorganisms produce extracellular substances such as polysaccharides which help them adhere to the tooth surface and to each other. Although plaque need not be harmful, dental caries (tooth decay) is a common condition caused by plaque-forming microorganisms.

11.4   Normal Flora of the Gastrointestinal Tract

The general anatomy of the gastrointestinal tract is shown in Figure 11.8. The human gastrointestinal tract, the site of food digestion, consists of the stomach, small intestine, and large intestine. The pH of stomach fluids is low, about pH 2. The stomach can thus be viewed as a microbiological barrier against entry of foreign bacteria into the intestinal tract. Although the bacterial count of the stomach contents is generally low, the walls of the stomach are often heavily colonized with bacteria. These are primarily acid-tolerant lactobacilli and streptococci and can be seen in large numbers in histological sections or by scanning electron microscopy of the stomach epithelium (Figure 11.9). These bacteria appear very early after birth and are well established by the first week. In humans, certain abnormal conditions such as cancer of the stomach produce higher pH values; under such conditions a characteristic microflora consisting of yeasts and bacteria (genera *Sarcina* and *Lactobacillus*) may develop.

The **small intestine** is separated into two parts, the *duodenum* and the *ileum*. The former, adjacent to the stomach, is fairly acidic and resembles the stomach in its microbial flora, although it may lack heavy populations on the epithelium. From the duodenum to the ileum the pH gradually becomes more alkaline and bacterial numbers increase. In the lower ileum, bacteria are found in the intestinal cavity (the lumen), mixed with digestive material. Cell numbers of 10^5–10^7 per gram are common.

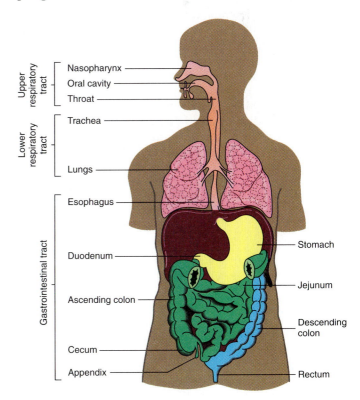

FIGURE 11.8 The gastrointestinal and respiratory tracts.

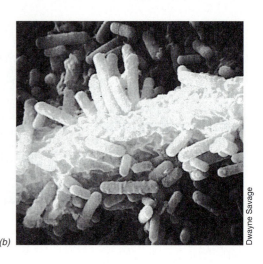

FIGURE 11.9
Light microscopy and scanning electron microscopy of the stomach microflora of a mouse. (a) Section through a Gram-stained preparation of the stomach wall of a 14-day-old mouse, showing extensive development of lactic acid bacteria (*Lactobacillus* sp.) in association with the epithelial layer. (b) Bacterial community on the surface of the keratinized epithelium of the stomach of an adult mouse as observed by scanning electron microscopy. Individual cells are 3–4 μm in length.

(a) *(b)*

Dwayne Savage

Dwayne Savage

In the **large intestine**, bacteria are present in enormous numbers, so much so that this region can be viewed as a specialized fermentation vessel. Many bacteria live within the lumen itself, probably using as nutrients some products of the digestion of food. Facultative aerobes, such as *Escherichia*, are present but in smaller numbers than other bacteria; total counts of facultative aerobes are generally less than 10^7 per gram of intestinal contents. The activities of facultative aerobes consume any oxygen present, making the environment of the large intestine strictly anaerobic and favorable for the profuse growth of obligate anaerobes. Many of these anaerobes are long, thin, Gram-negative rods, with tapering ends (called *fusiform*) and can be seen with the scanning electron microscope attached end-on to small indentations in the intestinal wall (Figure 11.10). Other obligate anaerobes include species of *Clostridium* and *Bacteroides*. The total number of obligate anaerobes is enormous. Counts of 10^{10}–10^{11} cells per gram of intestinal contents are not uncommon, with various species of *Bacteroides* accounting for the majority of intestinal obligate anaerobes. In addition, *Enterococcus faecalis* is almost always present in significant numbers.

The intestinal flora of the newborn becomes established early. In breast-fed human infants the flora is often fairly simple, consisting largely of *Bifidobac-*terium sp. In bottle-fed infants the flora is usually more complex. The flora is conditioned partly by the fact that the infant's main source of food is milk, which is high in the sugar lactose. The reason that the flora of breast-fed infants differs from that of bottle-fed ones is not completely understood, but it is known that human milk contains a disaccharide amino sugar that is required as a growth factor by *Bifidobacterium*. As the infant ages and its diet changes, the composition of the intestinal flora also changes, ultimately approaching that of the adult.

The normal flora of the gastrointestinal tract varies among species. For example, in guinea pigs lactobacilli make up 80 percent of the intestinal flora, whereas the same organisms are only minor components of human gastrointestinal flora. The gut flora in humans can also vary qualitatively depending on the diet. Persons who consume a considerable amount of meat show higher numbers of *Bacteroides* and lower numbers of coliforms and lactic acid bacteria than those on a vegetable diet. An overview of microorganisms of the gastrointestinal tract is given in Figure 11.11.

The intestinal flora has a profound influence on the animal, carrying out a wide variety of metabolic reactions (Table 11.1). Not all microorganisms carry out these reactions, and changes in the intestinal flora due to diet or disease may thus affect the ani-

FIGURE 11.10
Scanning electron micrographs of the microbial community on the surface of the columnar epithelium in the mouse ileum. (a) An overview at low magnification. Note the long, filamentous bacteria lying on the surface. (b) Higher magnification, showing several filaments attached at a single depression. Note that the attachment is at the end of the filaments only. Individual cells are 10–15 μm in length.

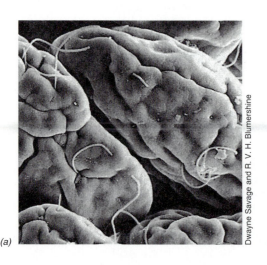
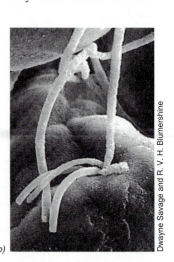

(a) *(b)*

Dwayne Savage and R. V. H. Blumershine

Dwayne Savage and R. V. H. Blumershine

Table 11.1	Biochemical and metabolic contributions of the intestinal microflora
Vitamin synthesis	Product: thiamine, riboflavin, pyridoxine, B_{12}, K
Gas production	Product: CO_2, CH_4, H_2
Odor production	Product: H_2S, NH_3, amines, indole, skatole, butyric acid
Organic acid production	Product: acetic, propionic, butyric acids
Glycosidase reactions	Enzyme: β-glucuronidase, β-galactosidase, β-glucosidase, α-glucosidase, α-galactosidase
Steroid metabolism	Process: esterification, dehydroxylation, oxidation, reduction, inversion

mal. Of special note in Table 11.1 are the roles of the intestinal flora in modifying the bile acids. Bile acids are steroids produced in the liver and excreted into the intestine via the gall bladder. Their role is to promote emulsification of fats in the diet so that the fats can be effectively digested. Intestinal microorganisms cause a variety of transformations of these bile acids so that the materials excreted in the feces are quite different from the original bile acids. Other products of microbial fermentation are the odor-producing substances listed in Table 11.1. Composition of the intestinal microflora as well as diet influences the amount of gas and the amount of odoriferous materials present.

During the passage of food through the gastrointestinal tract, water is withdrawn from the digested material, and it gradually becomes more concentrated and is converted into feces. Bacteria make up about one-third the weight of fecal matter. Organisms living in the lumen of the large intestine are continuously being displaced downward by the flow of material, and if bacterial numbers are to be maintained, those

bacteria that are lost must be replaced by new growth. Thus, the large intestine resembles a chemostat (see Section 9.6). The time needed for passage of material through the complete gastrointestinal tract is about 24 hours in humans; the growth rate of bacteria in the lumen is one to two doublings per day.

When an antibiotic is given orally it may inhibit the growth of the normal flora as well as pathogens; continued movement of the intestinal contents then leads to loss of the preexisting bacteria and the virtual sterilization of the intestinal tract. In the absence of the normal flora the environmental conditions of the large intestine change, and opportunistic microorganisms such as antibiotic-resistant *Staphylococcus*, *Proteus*, or the yeast *Candida albicans* which usually do not grow in the intestinal tract because they cannot compete with the normal flora, may become established. Occasionally, establishment of these opportunistic pathogens can lead to a harmful alteration in digestive function or even to disease. After antibiotic therapy stops, the normal flora eventually becomes reestablished, but often only after a considerable period.

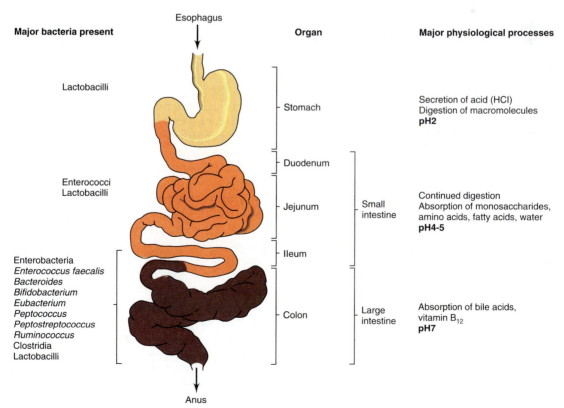

FIGURE 11.11   A detailed view of the human gastrointestinal tract showing functions and the distribution of microorganisms.

Intestinal gas

The gas produced within the intestines, called *flatus*, is the result of the action of fermentative and methanogenic microorganisms. Some foods which are poorly absorbed in the stomach and intestines can be metabolized by fermentative bacteria resulting in the production of hydrogen (H_2) and carbon dioxide (CO_2). In many individuals, methanogenic microorganisms then convert some of the H_2 and CO_2 to methane gas (CH_4). Methane is produced by a number of different prokaryotes called *methanogens* which are all members of the Archaea (see Section 20.5). Although all humans have some methanogens in their intestinal tract, about one-third of the population has an active methane-producing microbial flora in the intestines. Curiously, there is no pattern, such as inheritance, age, or diet, which can be associated with the presence in the intestinal tract of methanogens.

Normal adult humans expel a few hundred milliliters of gas per day via the rectum and in their breath. More than half of this gas is nitrogen (N_2) which enters the body in swallowed air and passes unchanged into the intestines, but other constituents are microbially produced (Table 11.1). Diet can have a dramatic effect on the amount and type of gas produced. If large quantities of beans or certain other high-polysaccharide foods are consumed, total gas production increases about tenfold. It is thought that particular polysaccharides that are not digested by humans pass into the intestines, where fermentative bacteria convert them to H_2 and CO_2; these can then serve as substrates for methanogenesis.

> Although the gastrointestinal tract is a major barrier to the penetration of microorganisms into the body, it also provides a number of favorable habitats for microbial growth. Because the stomach is acidic, it is a barrier to infection; few organisms are able to grow there. The intestinal tract is neutral to alkaline in pH and is a major site of microbial growth. In the large intestine, bacteria are present in enormous numbers, and they are eliminated from the body with the feces. Although most bacteria in the large intestine are harmless, a few harmful types may be present, but their numbers are kept low by the large numbers of nonpathogenic normal flora.

11.5 Normal Flora of Other Body Regions

Virtually all mucous membranes have a characteristic group of microorganisms that inhabit them. These organisms are part of the normal local environment and are characteristic of healthy tissue. In many cases, potentially pathogenic microorganisms cannot colonize mucous membranes because of the presence of the normal resident microorganisms. In this section, we discuss two such mucosal environments and their resident microorganisms.

Respiratory tract

The anatomy of the respiratory tract was shown in Figure 11.8 (see also Figure 15.2). In the **upper respiratory tract** (nasopharynx, oral cavity, and throat) microorganisms live primarily in areas bathed with the secretions of the mucous membranes. Bacteria enter the upper respiratory tract from the air during breathing, but most of these are trapped in the nasal passages and expelled again with the nasal secretions. The resident organisms most commonly found are staphylococci, streptococci, diphtheroid bacilli, and Gram-negative cocci. Potentially harmful bacteria, such as *Staphylococcus aureus*, *Streptococcus pneumoniae*, *Streptococcus pyogenes*, and *Corynebacterium diphtheriae* are often part of the normal flora of the nasopharynx of healthy individuals.

The **lower respiratory tract** (trachea, bronchi, and lungs) is essentially sterile, in spite of the large numbers of organisms potentially able to reach this region during breathing. Dust particles, which are fairly large, are filtered out in the upper respiratory tract. As the air passes into the lower respiratory tract, its rate of movement decreases markedly, and organisms settle onto the walls of the passages. These walls are lined with ciliated epithelium, and the cilia, beating upward, push bacteria and other particulate matter toward the upper respiratory tract where they are then expelled in the saliva and nasal secretions. Only droplet nuclei smaller than about 10 μm in diameter are able to reach the lungs.

Genitourinary tract

The main anatomical features of the male and female genitourinary tracts are shown in Figure 11.12*a*. In both male and female, the bladder itself is usually sterile, but the epithelial cells lining the urethra are colonized by facultatively aerobic Gram-negative rods and cocci. These organisms, including *Escherichia coli* and *Proteus mirabilis* and others, can occasionally become *opportunistic pathogens*. These organisms are normally present in the body or in the local environment, but they are not pathogenic under normal circumstances. Changes in the body, such as local pH changes or decreases in immune system functions (see Chapter 12), allow the organisms to multiply and become pathogenic. Such organisms frequently cause urinary tract infections, especially in women.

The vagina of the adult female generally is weakly acidic and contains significant amounts of the polysaccharide glycogen. *Lactobacillus acidophilus* ferments glycogen and produces acid. It is usually present in the vagina and may be responsible for the acidity (Figure 11.12*b*). Other organisms—yeasts (*Torulopsis* and *Candida* species), streptococci, and *E. coli*—may also be present. Before puberty, the female vagina is alkaline and does not produce glycogen, *L. acidophilus* is absent, and the flora consists predominantly of staphylococci, streptococci, diphtheroids, and *E. coli*. After menopause, glycogen disappears, the pH rises, and the flora again resembles that found before puberty.

11.6 Entry of the Pathogen into the Host

A pathogen must gain access to host tissues and multiply before damage can be done. In most cases this requires that the organism penetrate the skin, mucous membranes, or intestinal epithelium, surfaces that normally act as microbial barriers (see Figure 11.1).

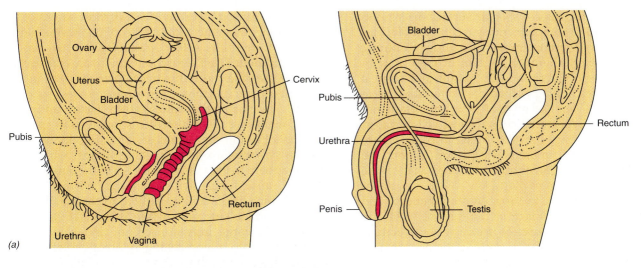

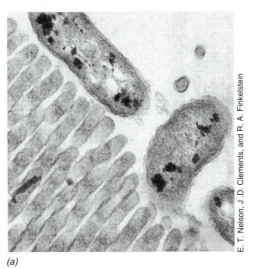

John Durham

FIGURE 11.12
(a) The genitourinary tracts of the human female and male, showing regions (color) where microorganisms often grow. (b) Gram stain of *Lactobacillus acidophilus*, the predominant organism in the vagina of women.

Passage through the skin into subcutaneous layers almost always occurs through wounds; in rare instances pathogens penetrate through the unbroken skin.

Specific adherence

Most microbial infections begin on the mucous membranes of the respiratory, alimentary, or genitourinary tracts. There is considerable evidence that bacteria or viruses able to initiate infection are able to adhere specifically to epithelial cells (Figure 11.13). The evidence for specificity is of several types. First, there is *tissue specificity*. An infecting microorganism does not adhere to all epithelial cells equally, but shows selectivity by adhering to cells in the particular region of the body where it normally gains entrance. For example, *Neisseria gonorrhoeae*, the causative agent of the sexually transmitted disease gonorrhea, adheres much more strongly to urogenital epithelia than to other tissues. Second, there is *host specificity*. A bacterial strain that normally infects humans will adhere more strongly to the appropriate human epithelial cells than to similar cells in another animal (for example, the rat), whereas a strain which specifically colonizes the rat will adhere more firmly to rat cells than to human cells.

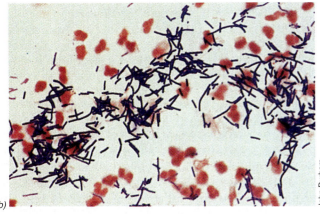

E. T. Nelson, J. D. Clements, and R. A. Finkelstein

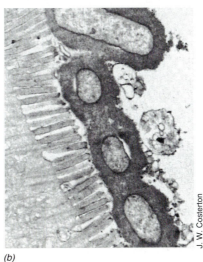

J. W. Costerton

FIGURE 11.13
Adherence of pathogens to animal tissues. (a) Transmission electron micrograph of a thin section of *Vibrio cholerae* adhering to the brush border of rabbit villi. Note the absence of the outer layer (glycocalyx). (b) Enteropathogenic *Escherichia coli* in fatal model infection in the newborn calf. The bacterial cells are attached to the brush border of calf villi via an extensive glycocalyx.

(a) (b)

Many bacteria possess specific surface macromolecules that bind to complementary receptor molecules on the surfaces of certain animal cells, thus promoting specific and firm adherence. Certain of these macromolecules are polysaccharide in nature and form a sticky meshwork of fibers called the bacterial **glycocalyx** (Figure 11.13*b*). The glycocalyx is important not only in attaching bacterial cells to host cell surfaces, but also in adherence between bacterial cells (Figure 11.13*b*). In addition, fimbriae (see Section 3.9) may be important in the attachment process. For instance, the fimbriae of *N. gonorrhoeae* play a key role in the attachment of this organism to urogenital epithelium, and fimbriated strains of *Escherichia coli* (Figure 11.14) are much more frequent causes of urinary tract infections than strains lacking fimbriae. Among the best characterized fimbriae are the so-called *type I fimbriae* of enteric bacteria (*Escherichia, Klebsiella, Salmonella, Shigella*). Type I fimbriae are 0.2–1 μm in length and 7 nm wide and are uniformly distributed on the surface of cells (Figure 11.14). Type I fimbriae function in attachment by binding mannose residues of specific host cell glycoproteins to initiate the attachment event.

Evidence of the specific interaction between mucosal epithelium and the pathogen comes from studies on diarrhea caused by *Escherichia coli*. Most strains of *E. coli* are nonpathogenic and are part of the normal flora of the *large* intestine. A few strains (only a handful of the 160 different *E. coli* serotypes known) are *enteropathogenic*, possessing the ability to colonize the *small* intestine and initiate diarrhea. Such strains possess specific surface structures, the **colonization factor antigens (CFA)**, which are fimbrial proteins involved in specific attachment to intestinal mucosa. Thus two kinds of *E. coli* can be recognized: pathogenic strains, which are able to adhere to the mucosal surface of the small intestine and cause disease symptoms, and "normal" *E. coli* strains which are unable to adhere to the small intestine or produce enterotoxin (see Section 11.9), which grow in the large intestine (cecum and colon), and which enter into a long-lasting symbiotic relationship with the mammalian host. A summary of major factors in microbial adherence is given in Table 11.2.

Penetration

In some diseases, the pathogen can remain localized at the mucosal surface and initiate damage by liberating toxins, as described later. Examples include whooping cough, caused by *Bordetella pertussis*, diphtheria, caused by *Corynebacterium diphtheriae*, and cholera, caused by *Vibrio cholerae*. However, in most cases the pathogen penetrates the epithelium and either grows in the submucosa, or spreads to other parts of the body where growth is initiated. Access to the interior of the body may occur in those areas where the lymphatic system (see Section 11.13) is near the surface, such as the nasopharyngeal region, the tonsils, and the lymphoid follicles of the intestine. Many times small breaks or lesions in the mucous membrane in one of these regions will permit an initial entry. Motility and chemotaxis may be of some value to the invader, although many pathogens are nonmotile. Among factors inhibiting establishment of the pathogen is the normal microbial flora itself, with which the pathogen must compete for nutrients and living space. If the normal flora is altered or eliminated by antimicrobial therapy, successful colonization by a pathogen may be easier to accomplish.

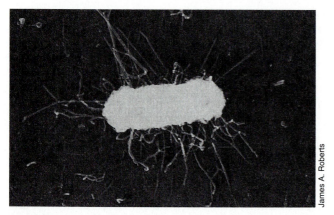

James A. Roberts

FIGURE 11.14 Shadow-cast electron micrograph of the bacterium *Escherichia coli* showing type P fimbriae and two flagella. Type P fimbriae resemble type I fimbriae but are somewhat longer. The cell shown is about 0.5 μm wide.

Table 11.2	Major adherence factors used to facilitate attachment of microbial pathogens to host tissues*	
Factor	**Example**
Sticky outer capsule (glycocalyx)	Enterotoxic *Escherichia coli* (ETEC)—glycocalyx promotes adherence to the brush border of intestinal villi
	Dental caries—binding to tooth surface by *Streptococcus mutans*
Adherence proteins	M-protein on surface of *Streptococcus pyogenes* binds receptor on respiratory mucosa
Lipoteichoic acid	Along with M-protein of *Streptococcus pyogenes*—facilitates binding to respiratory mucosal receptor
Fimbriae (pili)	Gonorrhea—pili on *Neisseria gonorrhoeae* facilitate binding to urogenital epithelium
	Salmonellosis—Type I fimbriae facilitate binding to small intestinal epithelium
	Enterotoxic *Escherichia coli* (ETEC)—colonization factor antigens (CFA), which are fimbrial, facilitate binding to small intestinal epithelium

For the most part, receptor sites on host tissues are glycoproteins or complex lipids such as gangliosides or globosides.

If infection is to occur, the invading microorganisms must first gain access to host tissues and multiply. There is a marked specificity for the site at which a particular parasite will invade, most parasites initiating infection by attaching to mucous membranes in specific organs or tissues. Attachment of a microorganism generally requires the presence of specific surface macromolecules such as fimbriae and complementary receptor molecules on host tissues. Penetration of the pathogen may be important for disease initiation.

11.7   Colonization and Growth

If a pathogen gains access to tissues, it must multiply, a process called *colonization*. Colonization requires that the pathogen bind to specific tissue surface receptors and overcome any chemical defenses produced by the tissues. The initial inoculum is rarely sufficient to cause damage; a pathogen must *grow* within host tissues in order to produce a disease. If the pathogen is to grow it must find appropriate nutrients and environmental conditions in the host. Temperature, pH, and reduction potential are environmental factors that affect pathogen growth, but of most importance is the availability of microbial nutrients in host tissues. Although it might seem that a vertebrate animal would be a nutritional paradise for microorganisms, not all nutrients are in plentiful supply. Soluble nutrients such as sugars, amino acids, and organic acids may often be in short supply and organisms able to utilize high molecular weight components such as glycogen may be favored. Not all vitamins and other growth factors are necessarily in adequate supply in all tissues at all times. *Brucella abortus*, for example, can grow slowly in most tissues of infected cattle, but grows very rapidly in the placenta, where it causes abortion. This specificity is due to the elevated concentration of erythritol found in the placenta, a nutrient which greatly stimulates growth of *B. abortus*.

Trace elements may also be in short supply and can influence establishment of the pathogen. In the latter category, the most evidence exists for the influence of **iron** on microbial growth. Specific proteins called *transferrin* and *lactoferrin*, present in animals, bind iron tightly and transfer it through the body. Such is the affinity of these proteins for iron that microbial iron deficiency may be common; indeed, administration of a soluble iron salt to an infected animal may greatly increase the virulence of some pathogens. As we noted in Section 4.22, many bacteria produce iron-chelating compounds (ironophores) which help them to obtain iron from the environment. Some iron chelators isolated from pathogenic bacteria are sufficiently strong iron-binding compounds that they can actually remove iron from animal iron-binding proteins. For example, a siderophore called *aerobactin*, produced by certain strains of *E. coli* and coded for by the Col V plasmid (see Section 7.9), is able to readily remove iron bound to transferrin.

Localization in the body

After initial entry, the organism often remains localized and multiplies, producing a small focus of infection such as the boil, carbuncle, or pimple that commonly arises from *Staphylococcus* infections of the skin. Alternatively, the organisms may pass through the lymphatic vessels and be deposited in lymph nodes. If an organism reaches the blood it will be distributed to distant parts of the body, usually concentrating in the liver or spleen. Spread of the pathogen through the blood and lymph systems can result in a generalized (systemic) infection of the body, with the organism growing in a variety of tissues. If extensive bacterial growth in tissues occurs, some of the organisms are usually shed into the bloodstream in large numbers, a condition called **bacteremia**. Generalized infections of this type almost always start as a localized infection at a specific organ site, but fortunately are quite rare.

Enzymes involved in invasion

Streptococci, staphylococci, pneumococci, and certain clostridia produce **hyaluronidase**, an enzyme that promotes spreading of organisms in tissues by breaking down hyaluronic acid, a polysaccharide that functions in the body as a tissue cement. Production of this enzyme may therefore enable these organisms to spread from an initial focus. Streptococci and staphylococci also produce a vast array of proteases, nucleases, and lipases which serve to depolymerize host proteins, nucleic acids, and fats, respectively. Clostridia that cause gas gangrene produce **collagenase**, which breaks down the collagen network supporting the tissues; the resulting dissolution of tissue is a factor in enabling these organisms to spread through the body. Fibrin clots are often formed by the host in a region of microbial invasion and serve to wall off the organism and prevent its spread through the body. Some organisms are able to produce fibrinolytic enzymes to dissolve these clots and make further invasion possible. One such fibrinolytic substance, produced by streptococci, is known as **streptokinase**. On the other hand, some organisms produce enzymes that actually promote fibrin clotting, which causes localization of the organism rather than its spread. The best studied fibrin-clotting enzyme is **coagulase**, produced by pathogenic staphylococci, which causes the fibrin material to be deposited on the cocci and may offer them protection from attack by host cells. The fibrin matrix produced as a result of coagulase activity probably accounts for localization of many staphylococcal infections in boils and pimples (see Figure 15.6).

Various pathogens produce proteins that are able to act on the animal cytoplasmic membrane, causing cell lysis and hence cell death. The action of these toxins is most easily detected with red blood cells (erythrocytes), hence they are often called **hemolysins**; in probably all cases, however, they also work on cells other than erythrocytes. The production of such toxins is most readily demonstrated in the laboratory by streaking the organism on a *blood agar plate*. During

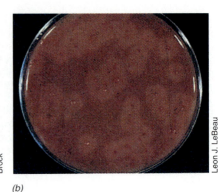

FIGURE 11.15 (a) Zones of hemolysis around colonies of *Streptococcus pyogenes* growing on a blood agar plate. (b) Action of phospholipase around colonies of *Clostridium perfringens*, growing on an agar medium containing egg yolk.

(a)   *(b)*

growth of the colonies, some of the hemolysin is released and lyses the surrounding red blood cells, typically clearing a zone of hemolysis (Figure 11.15*a*). Some hemolysins have been shown to be enzymes that attack the phospholipid of the host cytoplasmic membrane. Because the phospholipid lecithin (phosphatidylcholine) is often used as a substrate, these enzymes are called **lecithinases** or **phospholipases** (Figure 11.15*b*). Since the cytoplasmic membranes of all organisms, both prokaryotes and eukaryotes, contain phospholipids, hemolysins that are phospholipases sometimes destroy bacterial as well as animal cytoplasmic membranes. Some hemolysins are not phospholipases, however. Streptolysin O, a hemolysin produced by streptococci, affects the sterols of the host cytoplasmic membrane, and its action is neutralized by addition of cholesterol or other sterols. **Leukocidins** are lytic agents capable of lysing white blood cells and hence serve to decrease host resistance (see Section 11.13).

> The initial inoculum of a pathogenic microorganism usually contains too few organisms to be harmful to the host until the microorganisms proliferate. During growth, many pathogens produce enzymes that destroy host tissue components. This process provides increased access to sources of nutrients for pathogen growth. Microbial enzymes such as hyaluronidase, streptokinase, and collagenase attack structural components of host tissues and promote invasion of the pathogen from the site of initial infection.

11.8   Exotoxins

The ways in which pathogens bring about damage to the host are diverse. Only rarely are symptoms of a disease due simply to the presence of large numbers of microorganisms. Although a large mass of cells can block vessels or heart valves or clog the air passages of the lungs, in most cases pathogens produce *toxins* that are responsible for all or much of the host damage.

Toxins released extracellularly as the organism grows are called **exotoxins**. These toxins may travel from a focus of infection to distant parts of the body and cause damage in regions far removed from the site of microbial growth. Table 11.3 provides a summary of the properties and actions of some of the best known exotoxins.

Diphtheria toxin

The toxin produced by *Corynebacterium diphtheriae*, the causal agent of diphtheria, was the first exotoxin to be discovered. It is a protein of molecular weight 62,000 and differs markedly in its action on different animal species; rats and mice are relatively resistant, whereas humans, rabbits, guinea pigs, and birds are susceptible. Diphtheria toxin is very potent in its action, with only a single molecule being required to cause the death of a single cell. The toxin binds irreversibly to the cell, and within a few hours the cell loses its ability to synthesize protein because the toxin interferes with the protein synthesis process by blocking transfer of an amino acid from a tRNA to the growing peptide chain. The toxin specifically inactivates one of the elongation factors (elongation factor 2) involved in growth of the polypeptide chain (see Figure 11.16) by catalyzing the attachment of the adenosine diphosphate ribose moiety of NAD^+ to the elongation protein (Figure 11.16). The elongation protein is ADP-ribosylated at a single amino acid residue, a modified histidine molecule called *diphthamide*; following ADP-ribosylation the activity of elongation factor 2 drops dramatically and protein synthesis stops.

Diphtheria toxin is formed only by strains of *C. diphtheriae* that are lysogenized by a bacteriophage called phage β, and the toxin production is encoded by genes present in the phage genome. Mutants of phage β that produce altered toxins have been isolated. Nontoxigenic and hence nonpathogenic strains of *C. diphtheriae* can be converted to pathogenic strains by infection with the β phage (the process of phage conversion, see Section 7.7).

The gene coding for diphtheria toxin has been cloned and sequenced. The toxin as excreted by *C. diphtheriae* cells is a single polypeptide containing 535 amino acids. Following binding to the host cell, the polypeptide is cleaved by a protease into two fragments. Fragment A (193 amino acids) enters the cell and acts to disrupt protein synthesis, while the remaining piece, Fragment B (342 amino acids), is discarded. Before cleavage, Fragment B is believed to

Table 11.3   Exotoxins produced by certain bacteria pathogenic for humans

Organism	Disease	Toxin	Action
Clostridium botulinum	Botulism	Neurotoxin	Flaccid paralysis
C. tetani	Tetanus	Neurotoxin	Spastic paralysis
C. perfringens	Gas gangrene, food poisoning	α-Toxin	Hemolysis (lecithinase)
		β-Toxin	Hemolysis
		γ-Toxin	Hemolysis
		δ-Toxin	Hemolysis
		θ-Toxin	Hemolysis (cardiotoxin)
		κ-Toxin	Collagenase
		λ-Toxin	Protease
		Enterotoxin	Alters permeability of intestinal epithelium
Corynebacterium diphtheriae	Diphtheria	Diphtheria toxin	Inhibits protein synthesis in eukaryotes and in Archaea (see Figure 11.16)
Staphylococcus aureus	Pyogenic (pus-forming) infections (boils, etc.), respiratory infections, food poisoning, toxic shock syndrome, scalded skin syndrome	α-Toxin	Hemolysis
		Toxic shock syndrome toxin	Systemic shock
		Exfoliating toxin	Peeling of skin, shock
		Leukocidin	Destroys leukocytes
		β-Toxin`	Hemolysis
		γ-Toxin	Kills cells
		δ-Toxin	Hemolysis, leukolysis
		Enterotoxin	Induces vomiting, diarrhea
Streptococcus pyogenes	Pyogenic infections, tonsillitis, scarlet fever	Streptolysin O	Hemolysis
		Streptolysin S	Hemolysis
		Erythrogenic toxin	Causes scarlet fever rash
Vibrio cholerae	Cholera	Enterotoxin	Induces fluid loss from intestinal cells (see Figure 11.19)
Escherichia coli (enteropathogenic strains only)	Gastroenteritis	Enterotoxin	Induces fluid loss from intestinal cells (see Figure 11.18)
Bacillus cereus	Food poisoning	Enterotoxin	Induces fluid loss from intestinal cells
Shigella dysenteriae	Bacterial dysentery	Neurotoxin	Paralysis, hemorrhage
Yersinia pestis	Plague	Plague toxin	Kills cells
Bordetella pertussis	Whooping cough	Whooping cough toxin	Kills cells
Pseudomonas aeruginosa	Various *P. aeruginosa* infections	Exotoxin A	Kills cells

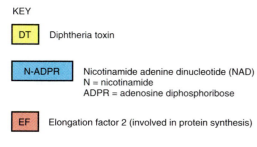

KEY

DT   Diphtheria toxin

N-ADPR   Nicotinamide adenine dinucleotide (NAD)
N = nicotinamide
ADPR = adenosine diphosphoribose

EF   Elongation factor 2 (involved in protein synthesis)

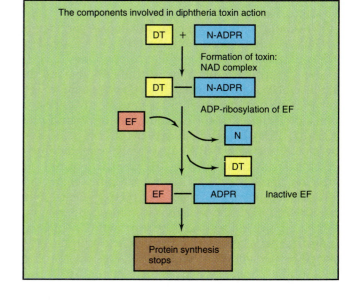

FIGURE 11.16 Catalysis by diphtheria toxin of attachment of the adenyldiphosphoribosyl (ADPR) portion of NAD⁺ to elongation factor 2, leading to inhibition of protein synthesis.

promote specific binding of the toxin to the host cell, and, following cleavage, it assists in the entry of Fragment A into the host cytoplasm.

A nongenetic factor of significance for toxin production is the concentration of *iron* present in the environment in which the bacteria are growing. In media containing sufficient iron for optimal growth no toxin is produced, whereas when the iron concentration is reduced to growth-limiting levels, toxin production occurs. Some evidence exists that the role of iron is to bind to a regulatory protein in *C. diphtheriae* (that is, act as a negative control element, see Section 5.10), and that the iron-binding protein combines with a control region of the DNA of β phage and prevents expression of the diphtheria toxin gene. When iron is absent, the regulatory protein does not act, and toxin synthesis can occur. Mutants in the regulatory region are known which synthesize toxin even in the presence of high iron concentrations. The disease diphtheria is discussed in Section 15.2.

Exotoxin A of *Pseudomonas aeruginosa* has been shown to have an action quite similar to diphtheria toxin, also transferring the ADP-ribosyl portion of NAD^+ to elongation factor 2. However, the two toxins have different receptors, as shown by the fact that mice or rats are naturally resistant to diphtheria toxin but remain sensitive to *P. aeruginosa* exotoxin A.

Tetanus and botulinum toxins

These toxins are produced by two species of obligately anaerobic bacteria, *Clostridium tetani* and *C. botulinum*, which are normal soil organisms that occasionally become involved in disease situations in animals. *Clostridium tetani* grows in the body in deep wound punctures that become anaerobic, and although it does not invade the body from the initial site of infection, the toxin it produces can spread and cause a severe neurological disturbance, which can result in death. *Clostridium botulinum* rarely grows directly in the body, but it does grow and produce toxin in improperly preserved foods. Ingestion of the food without proper heat treatment results in neurological disease and death.

Tetanus toxin is a protein of molecular weight 150,000, containing two polypeptides. Upon entry into the central nervous system this toxin becomes fixed to nerve synapses, binding specifically to one of the lipids. Tetanus toxin blocks the activity of a nerve factor that allows relaxation of the muscles, resulting in constant firing of motor neurons. To understand how it induces its characteristic spastic paralysis, it is essential to discuss briefly how normal nerve function in muscle contraction operates. In normal muscle contraction, nerve impulses from the brain travel through the spinal cord and initiate a sequence of events at the motor end plate (the structure forming a junction between a muscle fiber and its motor nerve), which results in muscle contraction. Further, each muscle is affected by two neurons, one that activates, and another that inhibits the muscle. When one set of muscles con-

tracts, the opposing set of muscles becomes stretched. A stretch-sensitive receptor in the opposing muscles would cause neurons enervating these muscles to fire and oppose the stretching, but the firing is inhibited by impulses from an inhibitory nerve. Thus in normal muscle action when one set of muscles contracts the opposing set of muscles relaxes. Tetanus toxin binds specifically to these inhibitory motorneurons and blocks the inhibitory action resulting in contraction of both sets of muscles. This leads to a spastic paralysis, as the two types of muscles attempt to oppose each other (Figure 11.17). If the muscles of the mouth are involved, the prolonged spasm restricts the mouth's movement, resulting in the condition known as **lockjaw**. If the respiratory muscles are involved, death may be due to asphyxiation.

Botulinum toxin is actually a series of seven related toxins that are the most poisonous substances known. One milligram of pure botulinum toxin is enough to kill more than 1 million guinea pigs. Of the seven distinct botulinum toxins described, the syntheses of at least two of these are encoded by genes residing on lysogenic bacteriophages specific for *C. botulinum*. The major toxin is a protein of about 150,000 molecular weight which readily forms complexes with nontoxic botulinum proteins to give an active form of the toxin of almost 1 million molecular weight. Toxicity occurs due to binding of the toxin to presynaptic terminal membranes at the nerve-muscle junction, blocking the release of acetylcholine. Since transmission of the nerve impulse to the muscle is by means of acetylcholine action, muscle contraction is inhibited, causing a flaccid paralysis. The fatality rate from botulism poisoning can approach 100 percent, but can be significantly reduced by quick administration of antibody against the toxin (antitoxin, see Section 12.15) and by use of an artificial respirator to prevent breathing failure. Death in cases of botulism is usually due to respiratory or cardiac failure. The mode of action of tetanus and botulinum toxins is contrasted in Figure 11.17.

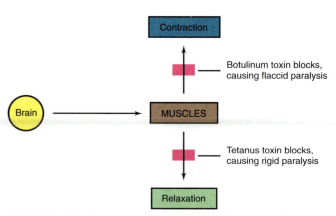

FIGURE 11.17 Summary of the mode of action of botulinum and tetanus toxins.

Production of specific toxins by the pathogen often results in damage to the host. Some of the most potent poisons known are exotoxins produced by pathogenic bacteria. Each exotoxin acts on specific cell types or against specific macromolecular components of the host. Diphtheria toxin specifically inhibits protein synthesis and tetanus and botulinum toxins affect nerve transmission in motor nerves.

11.9   Enterotoxins

Enterotoxins are exotoxins that act on the *small* intestine, generally causing massive secretion of fluid into the intestinal lumen, leading to the symptoms of diarrhea. The action of enterotoxins has been most commonly studied using animal models, of which the most precise is the ligated ileal loop. For most work, rabbits have been used, but pigs have also been studied. The procedure is done under anaesthesia, an incision being made and segments of the ileum being tied off with sutures. Experimental materials such as cultures, culture filtrates, or purified toxins, are injected into one or more of the segments (called loops), and control segments are either uninjected or are injected with sterile saline. After 18 to 24 hours, the animal is sacrificed and the ileal loops removed. The segments with positive enterotoxin action are swollen (Figure 11.18), but for more precise evaluation, the fluid from each segment is removed and its volume determined. Because of the expense and time involved in the ileal loop assay, a number of tissue culture assays have been developed for routine assays of enterotoxin. However, it is always essential to check these simpler models with the ileal loop model, since the latter duplicates the clinical action of the enterotoxin. Certain enterotoxins, such as the heat-stable enterotoxin of enterotoxic strains of *Escherichia coli* (see Section 15.13), can be assayed by measuring accumulated fluid in the stomachs of mice.

Enterotoxins are produced by a variety of bacteria, including the food-poisoning organisms *Staphylococcus aureus*, *Clostridium perfringens*, and *Bacillus cereus*, and the intestinal pathogens *Vibrio cholerae*, *Escherichia coli*, and *Salmonella enteritidis*. The *E. coli* enterotoxin is plasmid encoded. It is likely that this plasmid also codes for synthesis of the specific surface antigens that are essential for attachment of enteropathogenic *E. coli* to intestinal epithelial cells (see Section 11.6).

Cholera toxin

The enterotoxin produced by *V. cholerae*, the causal agent of cholera, is the best understood. Cholera toxin is a protein consisting of three polypeptides, the A_1, A_2, and B polypeptide chains of 82,000 total molecular weight. Chains A_1 and A_2 are connected together covalently by a disulfide bridge to make a dimer called subunit A, and this is loosely associated with a variable number of B chains. The B subunit contains the binding site by which the cholera toxin combines specifically with the ganglioside GM1 (a complex glycolipid) in the epithelial cytoplasmic membrane (Figure 11.19*a*), but the B subunit itself does not cause alteration in membrane permeability. Rather, the toxic action of cholera toxin is found in the A_1 chain, which activates the cellular enzyme *adenyl cyclase*, causing the conversion of ATP to cyclic AMP.

As we have discussed in Section 5.13, cyclic AMP is a specific mediator of a variety of regulatory systems in cells. In mammals, cyclic AMP is involved in the action of a variety of hormones, as well as in synaptic transmission in the nervous system, and in inflammatory and immune reactions of tissues, including allergies. Although the A_1 subunit of cholera toxin is responsible for activation of adenyl cyclase, A_1 must first be activated by an enzymatic activity of the cell, which requires NAD^+ and ATP. In the action of cholera enterotoxin, the increased cyclic AMP levels bring about the active secretion of chloride and bicarbonate ions from the mucosal cells into the intestinal lumen. This change in ionic balance leads to the secretion into the lumen of large amounts of water (Figure 11.19*b*). In the acute phase of cholera, the rate of secretion of water into the small intestine is greater than reabsorption of water by the large intestine, so that massive fluid loss occurs. Cholera victims generally die from extreme dehydration, and the best treatment for the disease is the oral administration of electrolyte solutions containing solutes (Figure 11.19*c*) to replace the lost fluid and ions.

It is of some interest that at the molecular level cholera enterotoxin has a mode of action (formation of cyclic AMP) identical to some normal mammalian hormones, and it has been suggested that cholera toxin may represent an ancestral hormone. Since cholera enterotoxin activates adenyl cyclase in a vari-

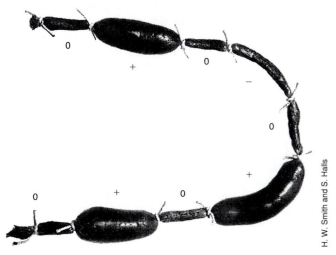

FIGURE 11.18 Action of *Escherichia coli* enterotoxin in the ligated ileum of the pig. The various segments were isolated by sutures, and inoculations made. Inoculated with enterotoxin-producing culture: + segments. Inoculated with culture not producing enterotoxin: − segments. Uninoculated segments: 0 segments. Enterotoxin action can be seen visually, but for precise quantification, the amount of fluid accumulating in each segment can be determined by removal with a syringe.

H. W. Smith and S. Halls

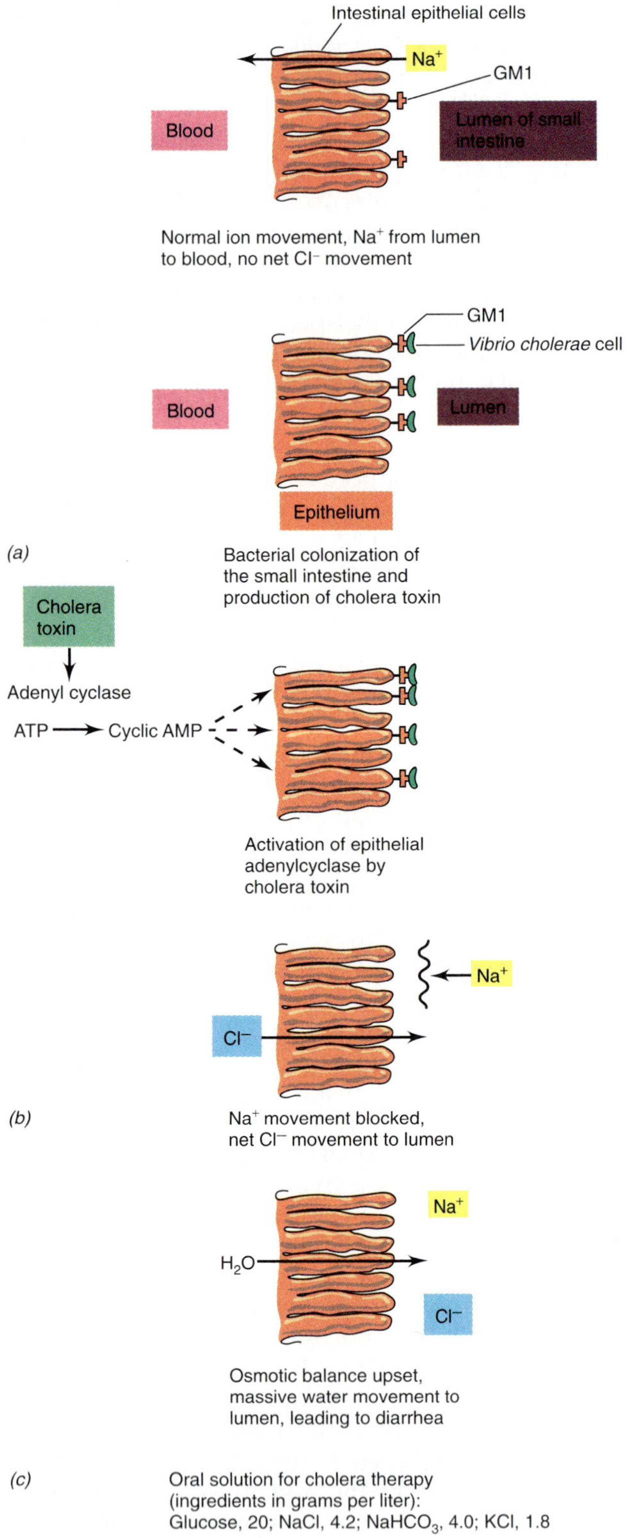

Normal ion movement, Na$^+$ from lumen to blood, no net Cl$^-$ movement

(a)

Bacterial colonization of the small intestine and production of cholera toxin

Activation of epithelial adenylcyclase by cholera toxin

(b)

Na$^+$ movement blocked, net Cl$^-$ movement to lumen

Osmotic balance upset, massive water movement to lumen, leading to diarrhea

(c) Oral solution for cholera therapy (ingredients in grams per liter): Glucose, 20; NaCl, 4.2; NaHCO$_3$, 4.0; KCl, 1.8

FIGURE 11.19 Action of cholera enterotoxin.

ety of cells and tissues, pathological manifestations of cholera toxin thus seem to be related more to the specific site at which it binds, the epithelial cells of the small intestine, than to its activation of adenyl cyclase. Indeed, it has been shown that purified B subunits devoid of adenyl cyclase activity can actually prevent the action of cholera enterotoxin, if they are administered

first, because they bind to the specific cholera receptors on the mucosal cells and block the binding of the complete toxin.

Genetic studies of cholera toxin have shown that the cholera enterotoxin is encoded by two genes, *ctxA* and *ctxB*. Cloning of these genes in *E. coli* has shown that expression of *ctxA* and *ctxB* is controlled by a positive regulatory element, a protein coded for by the *toxR* gene. The *toxR* gene product is a transmembrane protein that controls not only cholera toxin production but several other important virulence factors, such as outer membrane proteins and pili required for successful colonization of *V. cholerae* in the small intestine. Additional regulatory elements are probably also involved in cholera toxin production, as a number of mutations in the *V. cholerae* genome have been identified that result in either elevated or decreased toxin production. In addition, studies with human volunteers have shown that ingestion of cells of *ctxAB*-deletion mutants of *V. cholerae* still cause mild cholera symptoms, suggesting that substances other then just cholera toxin may also be involved in the pathogenesis of cholera.

There is good evidence that the enterotoxins produced by enteropathogenic *Escherichia* and *Salmonella* have similar modes of action to cholera toxin, and it is of interest that antibody against cholera enterotoxin also inactivates these other enterotoxins. Molecular cloning and sequencing of the cholera toxin genes *ctxAB* further support this relationship: cholera toxin genes show greater than 75 percent sequence homology with the genes encoding the heat-labile enterotoxin produced by enteropathogenic *E. coli*. Also, the active component of *Escherichia* enterotoxin is activated by a cellular enzyme system requiring ATP and NAD$^+$. As discussed in Section 7.9, *Escherichia* enterotoxin is controlled by a conjugative plasmid, but the enterotoxin gene of *V. cholerae* is chromosomal, although transmissible by conjugation. However, the enterotoxins produced by the food-poisoning bacteria (*S. aureus, C. perfringens, B. cereus*) may be quite different in their modes of action (for example, see Section 12.14 for a discussion of *S. aureus* toxic mechanisms), since their action is at least partly systemic and is not limited to alterations in intestinal permeability alone.

> **Enterotoxins are exotoxins that act specifically on the small intestine, causing changes in intestinal permeability that lead to diarrhea. Many food-poisoning microorganisms produce enterotoxins. The massive diarrhea observed in cholera is due to the action of the cholera toxin.**

11.10 Endotoxins

Gram-negative Bacteria produce lipopolysaccharides as part of the outer layer of their cell walls (Figures 3.30 and 3.31), which under many conditions are toxic.

Table 11.4   Basic properties of exotoxins and endotoxins

Property	Exotoxins	Endotoxins
Chemical properties	Proteins, excreted by certain Gram-positive or Gram-negative Bacteria; generally heat labile	Lipopolysaccharide/lipoprotein complexes (see Figures 3.30 and 3.31); released upon cell lysis as part of the outer membrane of Gram-negative Bacteria; extremely heat stable
Mode of action or symptomology	Specific; either cytotoxin, enterotoxin, or neurotoxin with defined specific action on cells or tissues	General; fever, diarrhea, vomiting
Toxicity	Highly toxic, often fatal	Weakly toxic, rarely fatal
Immunogenicity	Highly immunogenic; stimulate the production of neutralizing antibody (antitoxin)	Relatively poor immunogen; immune response not sufficient to neutralize toxin
Toxoid potential	Treatment of toxin with formaldehyde will destroy toxicity, but treated toxin (toxoid) remains immunogenic	None
Fever potential	Do not produce fever in host	Pyrogenic, often produce fever in host

These are called **endotoxins** because they are generally cell bound and released in large amounts only when cells lyse. In most cases, *endotoxin* can be equated with lipopolysaccharide toxin. Endotoxins have been studied primarily in the genera *Escherichia, Shigella,* and especially *Salmonella.* The major differences between exotoxins and endotoxins are listed in Table 11.4.

Endotoxin structure and function

When injected into an animal, endotoxins cause a variety of physiological effects. Fever is an almost universal symptom, because endotoxin stimulates host cells to release proteins called *endogenous pyrogens,* which affect the temperature-controlling center of the brain. In addition, however, the animal may develop diarrhea, experience a rapid decrease in lymphocyte, leukocyte, and platelet numbers, and enter into a generalized inflammatory state. Large doses of endotoxin can cause death, primarily through hemorrhagic shock and tissue necrosis. In general, the toxicity of endotoxins is much *lower* than that of exotoxins. For instance, in the mouse the amount of endotoxin required to kill 50% of a population of test animals (the so called LD_{50}) is 200 to 400 μg per mouse, whereas the LD_{50} for botulinum toxin is about 25 picograms (pg) per mouse, about 10 million times less! (A picogram is 10^{-12} g or 10^{-6} μg.)

The overall structure of lipopolysaccharide was diagrammed in Figure 3.31. Lipopolysaccharide consists of lipid A, a core polysaccharide, which in *Salmonella* is the same for many species, consisting of ketodeoxyoctonate, seven-carbon sugars (heptoses), glucose, galactose, and N-acetylglucosamine, and the *O-polysaccharide,* a highly variable molecule which usually contains galactose, glucose, rhamnose, and mannose, and generally contains one or more unusual dideoxy sugars such as abequose, colitose, paratose, or tyvelose (Figure 3.30). The sugars of the O-polysaccharide are connected in four- to five-sugar sequences (often branched) which then repeat to form the complete molecule (Figures 3.30, 3.31). Lipid A is not a normal glycerol lipid, but instead the fatty acids are connected by ester linkage to N-acetylglucosamine. Fatty acids frequently found in the lipid include β-hydroxymyristic, lauric, myristic, and palmitic acids.

Purification of lipopolysaccharide fractions has shown that it is the *lipid A complex* that is responsible for toxicity and that the polysaccharide acts mainly to render the lipid water soluble. However, animal studies have shown that the entire endotoxin complex, which contains both polysaccharide and lipid (Figures 3.30 and 3.31), is required to obtain a response.

Limulus assay for endotoxin

Because endotoxins are fever inducers, pharmaceuticals such as antibiotics and glucose solutions to be administered intravenously must be endotoxin-free. An endotoxin assay of very high sensitivity has been developed using suspensions of amoebocytes from the horseshoe crab, *Limulus polyphemus.* Although the mechanism of this assay is not known, endotoxin specifically causes the degranulation and lysis of the amoebocytes (Figure 11.20), and the degree of reaction can be measured by a spectrophotometer, an instrument that compares turbidities of various samples. A measurable reaction can be obtained with as little as 10–20 pg/ml of lipopolysaccharide. Apparently the active component of the *Limulus* extract reacts with the lipid component of lipopolysaccharide. The *Limulus* assay has been used to detect the presence of minute quantities of endotoxin in serum, cerebrospinal fluid, drinking water, and fluids used for injection.

The limulus test is so exquisitely sensitive that when it is used, considerable care must be taken to avoid contamination of the equipment, solutions, and reagents, since Gram-negative Bacteria abound in the laboratory and clinical environment, for example, as contaminants in the distilled water. In clinical work,

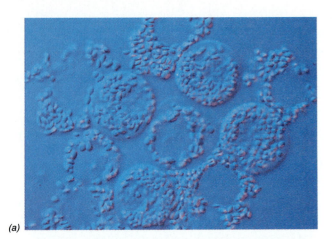

(a)

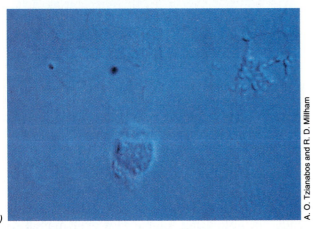

(b)

A. O. Tzianabos and R. D. Millham

FIGURE 11.20 Photomicrographs of *Limulus* amoebocytes. (a) Normal amoebocytes. (b) Amoebocytes following exposure to bacterial lipopolysaccharide. Treatment with lipopolysaccharide causes degranulation of the cells and this response can be used as an assay for lipopolysaccharide content.

detection of endotoxin by the *Limulus* assay in serum or cerebrospinal fluid can be taken as evidence of Gram-negative infection of these body fluids.

> **Many Gram-negative Bacteria produce toxic components that are part of the outer wall layer (lipopolysaccharide). Endotoxins are released from cells primarily when cell lysis occurs. Fever is a symptom of endotoxin action.**

11.11 Virulence and Attenuation

Virulence is a quantitative term that refers to the relative ability of a parasite to cause disease, that is, its *degree* of pathogenicity. Virulence is determined by the *invasiveness* of the organism and by its *toxigenicity*. Both are quantitative properties and may vary over a wide range from very high to very low. An organism that is only weakly invasive may still be virulent if it is highly toxigenic. A good example of this is the organism *Clostridium tetani*. The cells of this organism rarely leave the wound where they were first deposited; yet they are able to bring about death of the host because they produce the potent tetanus exotoxin, which can move to distant parts of the body and initiate paralysis. On the other hand, a weakly toxigenic organism

may still be able to produce disease if it is highly invasive. *Streptococcus pneumoniae* is not known to produce any potent toxin, but is able to cause extensive damage and even death because it is highly invasive, being able to grow in lung tissues in enormous numbers and initiate responses in the host that lead to disturbance of lung function. These two organisms exemplify the extremes of invasiveness and toxigenicity; most pathogens fall somewhere between these two extremes.

Several **virulence factors** have been identified in pathogenesis. In *Salmonella*, for example, a genus in which genetic studies can be readily done, a variety of virulence factors are known. Toxin production contributes to the virulence of *Salmonella* sp., and at least three toxins are produced: enterotoxin, endotoxin, and *cytotoxin*. Enterotoxins and endotoxins were discussed in Sections 11.9 and 11.10, respectively. Cytotoxin acts by inhibiting host cell protein synthesis, and because it is associated with the cell surface, it may also be involved in *adherence*, a key virulence factor for any pathogen, which allows *Salmonella* to bind to epithelial cells. Other factors involved in adherence are the cell surface polysaccharide O antigen and the flagellar Vi antigen (Figures 3.30 and 3.31). Fimbriae may also enhance adherence. *Invasion factors* include the O and Vi antigens. These invasion factors are important because they prevent killing by *phagocytes*, a group of white blood cells that normally ingest and kill bacteria (see Section 11.13). *Salmonella* is thought to establish infections through *intracellular parasitism*, the practice of residing in host cells, eventually growing and destroying those cells, and spreading to other cells. A plasmid-borne virulence factor is responsible for persistence and spread in most species of *Salmonella*.

To summarize, it appears that in *Salmonella*, and probably in most other pathogens as well, a host of virulence factors act in concert to initiate a successful infection. Although the genetics of many of these factors remain unknown, experiments in which transposon-generated mutants of *S. typhimurium* were screened for virulence showed that a substantial amount, about 4 percent, of the genome codes for functions necessary for virulence in a mouse model system. Figure 11.21 summarizes the known virulence factors in *Salmonella*.

The virulence of a pathogen can be estimated from experimental studies of the LD_{50}. Highly virulent pathogens frequently show little difference in the number of cells required to kill 100 percent of the population as compared to the number required to kill 50 percent of the population. This is illustrated in Figure 11.22 for experimental *Streptococcus* and *Salmonella* infections in mice. Only a few cells of *Streptococcus pneumoniae* are required to establish a fatal infection in mice. Less than 100 cells per mouse are necessary to kill every member of a test population once the virulence of a particular strain has been established (Figure 11.22). In fact, the LD_{50} for this organism is hard to ascertain because so few organisms are needed to produce a lethal infection. By contrast, the LD_{50} for *Salmo-*

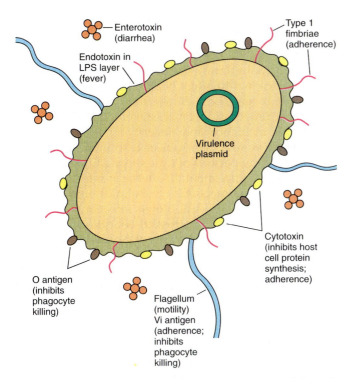

FIGURE 11.21   Summary of virulence factors important in *Salmonella* pathogenesis. See text for discussion.

nella typhimurium, also a mouse pathogen but a much less virulent one, is much higher than for *S. pneumoniae*, and the number of cells required to kill 100 percent of the population is much higher than the LD_{50}.

When pathogens are kept in laboratory culture and not passed through animals for long periods, their virulence is often decreased or even completely lost. Such organisms are said to be **attenuated.** Attenuation probably occurs because nonvirulent mutants grow faster and, through successive transfers to fresh media, such mutants are selectively favored. Attenuation

often occurs more readily when culture conditions are not optimal for the species. If an attenuated culture is reinoculated into an animal, virulent organisms are sometimes reisolated, but in many cases loss of virulence is permanent. Attenuated strains find frequent use in the production of vaccines, especially viral vaccines (see Section 12.15). Measles and mumps vaccines, for example, are composed of attenuated viruses, as is the rabies vaccine given to domesticated animals.

> Virulence is a quantitative measure of the degree of pathogenicity. Virulence is determined by the invasiveness and toxigenicity of the pathogen. Factors influencing invasiveness include ability to adhere to tissues, ability to move through the host, and ability to grow on the nutrients present at the site of infection. Tissue destruction caused by toxins leads to the formation of nutrients that promote further microbial growth.

11.12   Nonspecific Host Defenses

Pathogens are generally not part of the normal flora, and the host is able to prevent the growth of pathogens without affecting the normal flora. Many of the mechanisms responsible for the suppression of pathogens are not clearly understood. However, animals have a number of innate "resistance factors" which serve to suppress the growth of pathogenic microorganisms. These resistance factors can be divided into two categories: *specific* host defenses, which are directed against individual species or strains of pathogens, and *nonspecific* host defenses, directed against a variety of pathogens. In this chapter we consider the major *nonspecific* host defenses that have been identified as important in preserving the healthy state of the host. In the following chapter we consider *specific* host defenses—the immune response.

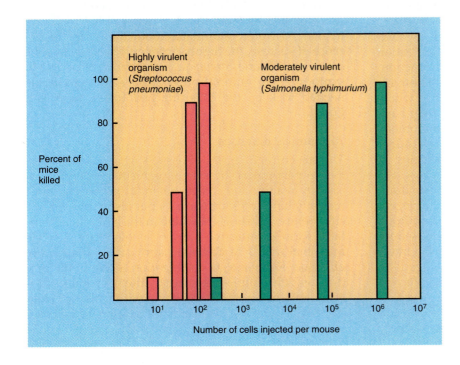

FIGURE 11.22

Comparison of the number of cells of *Streptococcus pneumoniae* or *Salmonella typhimurium* required to kill various fractions of a mouse population.

Natural host resistance

The ability of a particular pathogen to cause disease in different animal species is highly variable. In *rabies*, for instance, death usually occurs in all species of mammals once symptoms of the disease develop. Nevertheless, certain animal species are much more susceptible to rabies than others. Racoons and skunks, for example, are extremely susceptible to rabies infection as compared to opossums, which are rarely linked to cases of rabies in wild animals (see data of Table 14.4). *Anthrax*, which can infect a variety of animals but in particular herbivores, causes disease symptoms varying from mild pustules in humans to a fatal blood poisoning in cattle. Birds, on the other hand, are naturally resistant to anthrax. Also, diseases of warm-blooded animals are rarely transmitted to cold-blooded species and vice-versa. Why should this be so?

Resistance to certain diseases and susceptibility to others is an innate property of a given species and is probably governed by a number of different complex and interdependent factors. Differences in physiology and nutrition from species to species as well as anatomical differences are important here, as is variation in the structure of tissue surface receptors as discussed below. The net result, however, is the interesting and sometimes perplexing observation that different animal species, occasionally even closely related species, may show completely different susceptibilities to the same disease agent.

Age, stress, and diet

Age is an important factor in susceptibility to infectious disease. Infectious diseases are frequently more common in the very young and in the aged. In the infant, for example, development of an intestinal microflora occurs quite quickly, but the normal flora of a young infant is not the same as that of the adult. Before the development of an adult flora, and especially in the days immediately following birth, pathogens have a greater opportunity to become established and produce disease. Thus, diarrhea caused by enteropathogenic strains of *Escherichia coli* (see Section 15.13) or *Pseudomonas aeruginosa* is frequently encountered in infants under the age of one year. These organisms can be transmitted from the mother where they may be causing no ill effects because they have established a stable residency as part of the mother's flora. The undeveloped state of the infant's microflora provides poor competition for pathogenic species. In the elderly, certain diseases may develop because of a malfunctioning immune system or because of anatomical changes associated with old age. For example, enlargement of the prostate gland, a common condition in elderly males, is often associated with increased incidence of urinary tract infections. Many of the infections characteristic of young children or the aged are thus uncommon in healthy adults. The latter have a well-developed, characteristic normal flora (see Sections 11.4 and 11.5), and have already developed immunity to a number of infectious agents. The stability of the gut flora in healthy adults is therefore a major barrier to colonization of pathogens.

Stress can predispose a normally healthy adult to disease. Although not well understood, the effects of stress are known to affect health. In studies with rats and mice it has been shown that fatigue, exertion, poor diet, dehydration, or drastic climatic changes increase the incidence and severity of infectious diseases. For example, rats subjected to intense physical activity for long periods of time show a higher mortality rate from experimental *Salmonella* infections than do well-rested animals. Also, subclinical undiagnosed infections can become full-blown infections when the animal is placed under stressful conditions. It is hypothesized that the interaction of hormones with the immune system may play a role in stress-mediated disease. It is well known that hormonal balances change dramatically when an animal is placed under stressful conditions. The hormone *cortisone*, for example, is produced at much higher levels in times of stress than during calm periods, and this hormone is an effective anti-inflammatory agent. Suppression of inflammation removes one of the defenses an animal has against disease (see Section 11.15).

Diet plays a role in host resistance. The correlation between famine and infectious disease has been known for centuries. Protein shortages may alter the composition of the normal flora, thus allowing opportunistic pathogens a better chance to multiply. At the other extreme, overeating may be harmful as well. Studies of clostridial diseases of sheep, in particular bloats caused by excessive gas accumulation, indicate that constant overeating can affect the composition of the normal flora, leading to massive development of bacterial species normally present in low numbers.

Not eating a particular substance needed by a pathogen may serve to prevent disease. The best example here is the effect sucrose has on the development of dental caries. As explained in Section 11.3, absence of sucrose from the diet (along with good oral hygiene) virtually eliminates tooth decay. In the absence of sucrose, the highly cariogenic bacteria *Streptococcus mutans* and *Streptococcus sobrinus* are unable to synthesize the gummy outer surface polysaccharide needed to keep the cells attached to the teeth and gums.

Anatomical defenses

The structural integrity of tissue surfaces poses a barrier to penetration by microorganisms. When tissues remain healthy and intact, potential pathogens must not only bind to tissue surfaces but also grow at these sites before traveling elsewhere in the body. Intact surfaces form an effective barrier to colonization, while microbial access to damaged surfaces is more easily obtained. Resistance to colonization is due to the production of harmful substances by the host and to various mechanical actions which serve to disrupt colonization. A summary of the major anatomical defenses is shown in Figure 11.23.

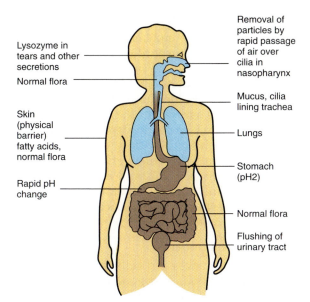

Lysozyme in tears and other secretions

Normal flora

Skin (physical barrier) fatty acids, normal flora

Rapid pH change

Removal of particles by rapid passage of air over cilia in nasopharynx

Mucus, cilia lining trachea

Lungs

Stomach (pH2)

Normal flora

Flushing of urinary tract

FIGURE 11.23 Summary of anatomical barriers to infection.

The **skin** is an effective barrier to the penetration of microorganisms. *Sebaceous glands* in the skin (see Section 11.2) secrete fatty acids and lactic acid which lower skin pH and inhibit colonization of pathogenic bacteria. Microorganisms inhaled through the *nose* or *mouth* are removed by the action of ciliated epithelial cells in the mucous surfaces of the nasopharynx and tracheal regions. Cilia beat and push bacterial cells upward until they are caught in oral secretions and are either expectorated or are swallowed and killed in the stomach. Potential pathogens entering the host via the oral route must first survive the *acidity* of the stomach (which is around pH 2) and then successfully compete with the increasingly abundant resident microflora present in the small intestine (which is around pH 5) and finally into the large intestine (pH 6–7). The latter organ contains bacterial numbers around 10^{10} per gram of intestinal contents in the healthy adult (see Section 11.4).

In a healthy adult, the kidney and the surface of the eye are constantly bathed with secretions containing *lysozyme* which markedly reduce microbial populations. Other tissues such as the spleen, thymus, and brain secrete basic proteins which have antibacterial activity. Extracellular tissues such as blood plasma also contain bactericidal substances. For example, serum proteins called β-lysins bind and destroy microbial cells. β-lysins are basic proteins which act by disrupting the bacterial cytoplasmic membrane leading to leakage of cytoplasmic constituents and death of the cell.

The chemical and physical barriers to microbial colonization present in the normal host combine to prevent routine colonization of host tissues by pathogenic microorganisms. However effective these defenses may be, certain pathogens are able to overcome them, especially in a weakened host. Damage to physical barriers and deleterious changes in other nonspecific defenses can quickly lead to growth of the pathogen and initiation of disease.

Tissue specificity

Unless introduced into the circulatory system where spread to other tissues is readily possible, most pathogens must first establish themselves at the site of infection. If the site is not compatible with their nutritional and environmental needs the organisms will not multiply. Thus, if cells of *Clostridium tetani* were ingested they would not bring about the disease tetanus since they are killed by the acidity of the stomach. If, on the other hand, *C. tetani* cells were introduced into a deep flesh wound, the organism would grow in the anaerobic zones created by localized tissue destruction and produce the potent toxin characteristic of tetanus (see Section 11.8). By contrast, enteric bacteria such as *Salmonella* and *Shigella* do not cause skin infections but successfully colonize the intestinal tract. Table 11.5 summarizes a number of examples of tissue specificity.

The compromised host

We see that there is a complex array of properties that are brought into play in the body at the onset of microbial infection. In addition to the nonspecific factors just discussed, there is the complex and highly specific immune system which will be discussed in the next chapter. Due to various circumstances, one or more of these resistance mechanisms may be lost, thus increasing the probability of infection. The term *compromised host* is used to refer to hosts in which one or more re-

Table 11.5 Tissue specificity as a factor in bacterial disease		
Disease	**Tissue infected**	**Organism**
Diphtheria	Throat epithelium	*Corynebacterium diphtheriae*
Gonorrhea	Urogenital epithelium	*Neisseria gonorrhoeae*
Cholera	Small intestine epithelium	*Vibrio cholerae*
Pyelonephritis	Kidney medulla	*Proteus* sp.
Dental caries	Oral epithelium	*Streptococcus mutans, Streptococcus sobrinus*
Spontaneous abortion (cattle)	Placenta	*Brucella abortus*
Acquired immune deficiency syndrome (AIDS)	T helper lymphocytes	Human immunodeficiency virus (HIV)
Malaria	Blood (erythrocytes)	*Plasmodium falciparum*

sistance mechanisms is malfunctioning and in which the probability of infection is therefore increased.

Hospital patients are often compromised hosts. Many hospital procedures such as catheterization, hypodermic injection, spinal puncture, and biopsy carry with them the risk of introducing pathogens into the patient. Surgical procedures are a major hazard, since not only are highly susceptible parts of the body exposed to sources of contamination but the stress of surgery often diminishes the resistance of the patient to infection. Finally, in organ transplant procedures, drugs are used which suppress the immune system (to prevent rejection of the transplant, as discussed in Sections 12.5 and 12.6), and these immunosuppressive drugs will greatly increase susceptibility to infection. Thus, many hospital patients with noninfectious primary ailments (cancer, heart disease) die of microbial infection because they are compromised hosts.

But compromised hosts exist even outside the hospital. Smoking, excess consumption of alcohol, intravenous drug usage, lack of sleep and poor nutrition are all conditions that can, in certain situations, produce a compromised host. Infection itself may lead to the compromise of the host. For instance, the virus causing acquired immune deficiency syndrome (AIDS) destroys one type of cell involved in the immune response (T helper cells, see Section 12.2). AIDS patients are hence unable to mount effective resistance to infection; death is generally due to some infectious agent.

Finally, there are certain genetic conditions that may compromise the host. Several genetic diseases are known which eliminate important parts of the immune system. Individuals with such conditions frequently die at an early age, not from the genetic condition itself, but from microbial infection.

Medical practice, including antimicrobial therapy, has greatly reduced the risk of infection in compromised hosts, but has not eliminated it.

> The animal host has many lines of defense against infection. Barriers to microbial penetration include the skin, which is a physical barrier, and the stomach, which is a chemical barrier because of its low pH. Nonspecific factors such as age, diet, and stress all influence infection, and medical treatments may make the host more susceptible. The term "compromised host" refers to a host in which one or more of these nonspecific resistance mechanisms is malfunctioning.

11.13   Interaction of Pathogens with Phagocytic Cells

The host is anything but passive to attack by invading microorganisms. Besides the defenses discussed in Section 11.12, other mechanisms of host defense against infection exist, which generally can be classified as **cellular** and **humoral**. *Humoral* defenses involve agents that are soluble (or at least noncellular). Such agents include antibodies, a large class of specific

protein molecules which will be discussed in the next chapter, and several enzymes and chemical substances that act as nonspecific antimicrobial agents. The best known enzyme involved in defense is *lysozyme*, which is found in tears, nasal secretions, saliva, mucus, and tissue fluids. As discussed in Section 3.5, lysozyme hydrolyzes the peptidoglycan layer of Bacteria, causing lysis and death. *Cellular* defense mechanisms involve the activity of *phagocytes* (literally, "cells that eat"), which are able to ingest and destroy invading microorganisms. Phagocytes are found both in tissues and in body fluids, such as blood and lymph. In this section we shall concentrate our discussion on cellular defenses.

Blood and its components

Many of the substances and cells involved in defense are found in the blood. Additionally, changes in blood constituents and properties are sensitive reflections of disease states, and because blood is a readily available material for clinical analysis, many analytical procedures involve sampling of blood. Blood consists of cellular and noncellular components (see Table 11.6). The most numerous cells in the blood are *red blood cells (erythrocytes)*, which are nonnucleated cells that function to carry oxygen from the lungs to the tissues. The **white blood cells**, or **leukocytes**, include a variety of phagocytic cells such as **monocytes**, as well as cells called **lymphocytes** which are involved in antibody production and cell-mediated immunity. Red blood cells outnumber leukocytes by roughly a factor of one thousand. **Platelets** are small cell-like constituents that lack a nucleus and play an important role in preventing leakage of blood from a damaged blood vessel. Platelets clump together to form a temporary plug in a damaged vessel until a permanent clot forms through the action of various clotting agents, some of which are released from the platelets themselves.

When the cells and platelets are removed from blood, the remaining fluid is called **plasma**. An important component of plasma is fibrinogen, a clotting agent, which undergoes a complex set of reactions during the formation of the fibrin clot. Clotting can be prevented by addition of an anticoagulant, such as

Table 11.6 Major formed elements and numbers in normal human blood	
Cell type	**Cells per ml**
Erythrocytes	$4.2–6.2 \times 10^9$
Leukocytes	$4.5–11 \times 10^6$
Lymphocytes	$1.0–4.8 \times 10^6$
Monocytes	up to 8.0×10^5
Platelets	$1.5–4.0 \times 10^8$

Source: Henry, J. B. 1991. Clinical Diagnosis and Management by Laboratory Methods, 18th edition. W. B. Saunders Company, Philadelphia.

potassium oxalate, potassium citrate, heparin, or sodium polyanetholsulfonate. Plasma is stable only when such an anticoagulant is added. When plasma or blood is allowed to clot (no anticoagulant added), the fluid components left behind, called **serum**, consist of all of the proteins and other dissolved materials of the plasma except fibrin. Since serum contains antibodies, it is widely used in immunological investigations (see Chapter 12).

Blood is pumped by the heart through a network of arteries and capillaries to various parts of the body and is returned through the veins (Figure 11.24). The circulatory system carries not only nutrients (including O_2) but also the components of the blood which are involved in host resistance to infection. At the same time, the circulatory system facilitates the spread of pathogens to various parts of the body.

Lymph is a fluid similar to blood, but lacks red blood cells. There is a separate circulatory system for lymph, called the **lymphatic system**, within which lymph flows (Figure 11.24). Fluids in tissues drain into lymphatic capillaries, then into **lymph nodes** found at various locations throughout the system, which filter out microorganisms and other particulate materials. Specialized white blood cells found in abundance in the lymphatic system, called **macrophages**, actually carry out the filtering action, as will be described below. In addition to filtering foreign particles, lymph nodes may be sites of infection, since organisms that are collected there by the filtering mechanisms may then proliferate. Lymphatic fluid eventually flows into the circulatory system via the thoracic duct.

Lymphocytes, another special type of cell found within the lymphatic system, are involved in the im-

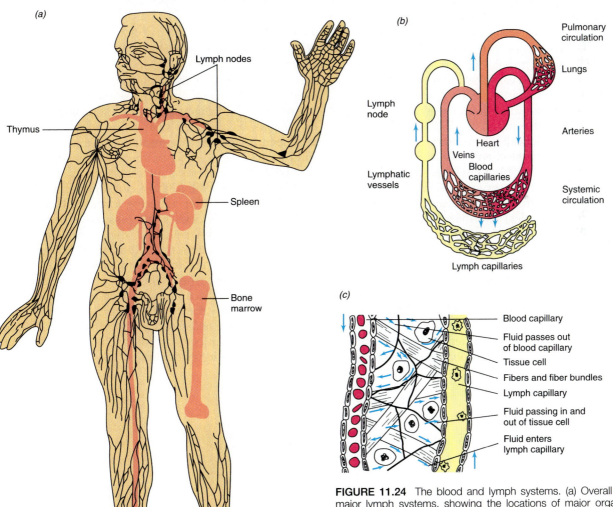

FIGURE 11.24 The blood and lymph systems. (a) Overall view of the major lymph systems, showing the locations of major organs. (b) Diagrammatic relationship between the lymph and blood systems. Blood flows from the veins to the heart, then to the lungs where it becomes oxygenated, then through the arteries to the tissues. (c) Connection between the blood and lymph systems is shown microscopically. Both blood and lymph capillaries are closed vessels but are permeable to water and salts. Changes in capillary permeability, as a result of infection or other source of inflammation, can greatly increase the movement of fluid into the tissue spaces.

mune response. Their critical role in specific immunity will be discussed in Chapter 12.

However, one type of lymphocyte does not exhibit specific immunity. This lymphocyte, called a *natural killer* or *NK* cell, is capable of recognizing nonhost cells, such as bacteria and fungi, or altered host cells, such as tumor cells and virus-infected cells. After contact with the target cell, the NK cell secretes lytic proteins called *perforins* that penetrate the membrane of the target cell, resulting in lysis and death.

Phagocytes

Some of the leukocytes found in whole blood are phagocytes, and phagocytes are also found in various tissues and fluids of the body. Phagocytes are usually actively motile by amoeboid action. Attracted to microorganisms by chemotactic phenomena, the phagocytes engulf the microbial cells and kill and digest them. One group of phagocytes, the **polymorphonuclear leukocytes** (sometimes abbreviated PMN), are small, actively motile cells containing many membranous granules called *lysosomes* (Figure 11.25a). These granules contain several bactericidal substances and enzymes, such as hydrogen peroxide, lysozyme, proteases, phosphatases, nucleases, and lipases. PMN leukocytes are short-lived cells that are found predominantly in the bloodstream and bone marrow and appear in large numbers during the acute phase of an infection. They can move rapidly, up to 40 μm/min, and are attracted chemotactically to bacteria and cellular components by immune mechanisms (see Section 12.6). Because they appear in the blood in large numbers during acute infection, they can serve as indicators of infection. Formation of granulocytes (the group of phagocytes containing granules) is severely retarded by ionizing radiation, leading to severe bacteremia from bacteria of the normal intestinal flora, such as *Escherichia coli*, in animals exposed to high doses of radiation.

The other group of phagocytes, the **macrophages**, are phagocytic cells that can readily be distinguished from PMN leukocytes by their nuclear morphology (Figure 11.25b). Macrophages are fixed at tissue sites in the circulatory or lymphatic system. **Monocytes** are circulating macrophages. Macrophages and mono-

cytes contain few lysosomes and hence do not appear granular when stained. Macrophages and monocytes are long-lived cells that play roles in both the acute and chronic phases of infection and in antibody formation (see Section 12.8).

The total system of phagocytic cells and organs is often called the **reticuloendothelial system** (RE system); it consists of the circulating PMN leukocytes and monocytes, and phagocytes fixed to connective tissue or the endothelial layer of blood capillaries.

> The host has a number of systems that act in prevention of microbial growth. An important humoral (non-cellular) factor is lysozyme, an enzyme which hydrolyzes the cell wall peptidoglycan of Bacteria. Cellular factors in resistance include the various phagocytes which ingest invading microorganisms and kill them. Two important classes of phagocytes are polymorphonuclear leukocytes, which are short-lived cells involved in the acute phase of an infection, and macrophages, which are long-lived cells that play a role both in chronic infection and in antibody formation.

Phagocytosis

As noted, phagocytes are attracted chemotactically to invading microorganisms. Some aspects of the chemotactic process involve the action of the complement system and lymphocytes, and will be discussed in Chapter 12. As a result of chemotactic attraction, large numbers of phagocytes are often seen around foci of infection, and such phagocytic accumulations are often the first indication of the presence of infection. Phagocytes work best when they can trap a microbial cell upon a surface, such as a vessel wall, a fibrin clot, or even particulate macromolecules. After adhering to the cell, an invagination of the phagocyte's cytoplasmic membrane envelopes the foreign cell, and the entire complex is pinched off and enters the cytoplasm in a phagocytic vacuole called a *phagosome*. The microbial cell is then released in a region of the phagocyte-containing granules, and the latter are disrupted and release enzymes, which digest and destroy the invader (Figure 11.26).

During the process of phagocytosis, the metabolism of the granulocyte converts from aerobic pathways to anaerobic fermentation. Glycolysis results in

Polymorphonuclear
leukocyte (PMN)

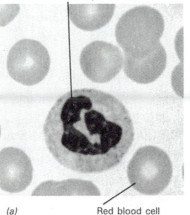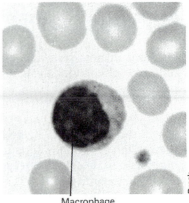

FIGURE 11.25
Two major phagocytic cell types. (a) Polymorphonuclear (PMN) leukocyte. (b) Monocyte. The red blood cells are about 8 μm in diameter.

(a)   Red blood cell   *(b)*   Macrophage (monocyte)

Brock

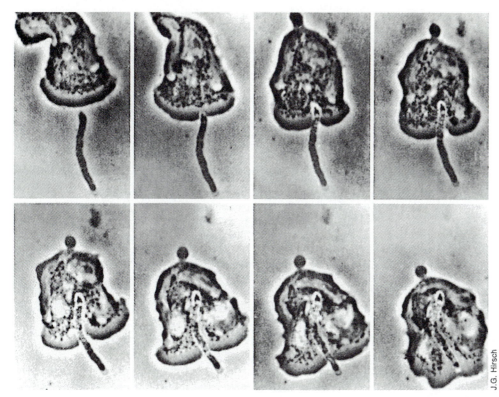

J.G. Hirsch

FIGURE 11.26
Phagocytosis: engulfment and digestion of a chain of *Bacillus megaterium* cells by a human phagocyte, observed by phase-contrast microscopy. The bacterial chain is about 18–20 μm long.

the formation of lactic acid and a consequent drop in pH; this lowered pH is partly responsible for the death of the microbial cell, and it also provides an environment in which the hydrolytic enzymes, all of which have acid pH optima, can act more effectively. The initial act of phagocytosis conditions a cell so that it is more efficient in subsequent phagocytic action—a cell that has recently phagocytized can take up bacteria about 10 times better than a cell that has not.

Oxygen-dependent phagocytic killing

As we discussed in Section 9.12, various biochemical reactions can lead to the formation of toxic forms of oxygen: hydrogen peroxide (H_2O_2), superoxide anion (O_2^-), hydroxyl radicals ($OH\cdot$), and singlet oxygen ($^1O_2$). Phagocytic cells make use of toxic forms of oxygen in killing ingested bacterial cells. Superoxide, formed by the reduction of O_2 by NADPH oxidase, reacts at the acid pH of the phagocyte to yield singlet oxygen and hydrogen peroxide (H_2O_2). The phagocytic enzyme myeloperoxidase forms hypochlorous acid (HOCl) from chloride ions and H_2O_2, and the HOCl reacts with a second molecule of H_2O_2 to yield additional singlet oxygen. The combined action of these oxygen-dependent phagocyte enzymes can form sufficient levels of toxic oxygen species to actually kill ingested bacterial cells by oxidizing key cellular constituents. These reactions occur within the phagosome; this insures that constituents of the phagocytic cell itself are not damaged by the toxic oxygen produced. The action of phagocytic cells in oxygen-mediated killing is summarized in Figure 11.27.

As we discussed in Section 9.12, carotenoids remove singlet oxygen, and bacteria containing carotenoids are much more resistant to killing within

phagocytes than those that do not contain carotenoids. It is of interest that *Staphylococcus aureus*, a pathogen which commonly causes infections where large numbers of phagocytes consequently develop, contains yellow carotenoids (hence the name *aureus*), and this bacterium is quite resistant to phagocytic killing.

Other intracellular pathogens use *cell wall-associated glycolipids* to scavenge oxygen radicals. For example, in the disease leprosy, caused by *Mycobacterium leprae*, cells of *M. leprae* readily persist and grow within phagocytic cells. *M. leprae* cells produce large amounts of a phenolic glycolipid in their cell walls and it has been shown that this material is highly effective in removing hydroxyl radicals and superoxide anions, the most damaging of the toxic oxygen species produced by phagocytic cells. Similar glycolipids have been identified in the protozoal parasite *Leishmania donovani*, the causitive agent of leishmaniasis. Related phenolic compounds may be present in the tubercle bacillus, *Mycobacterium tuberculosis*, also a notorious intracellular pathogen.

Leukocidins

Some pathogens produce proteins called **leukocidins** which destroy phagocytes. In such cases the pathogen is not killed when ingested but instead kills the phagocyte and is released alive. Pathogenic streptococci and staphylococci are the major leukocidin producers. Destroyed phagocytes make up much of the material of pus, and organisms that produce leukocidins are therefore usually *pyogenic* (pus-forming) and bring about characteristic abscesses. One leukocidin produced by certain strains of *S. aureus* binds to the cytoplasmic membrane of phagocytic cells and increases permeability. This protein also stimulates membrane

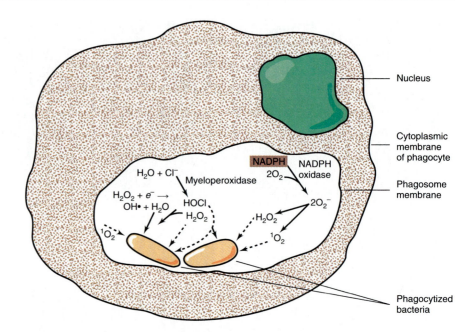

FIGURE 11.27
Action of phagocyte enzymes in generating toxic oxygen species. These include hydrogen peroxide (H_2O_2), the hydroxide radical ($OH\cdot$), hypochlorous acid (HOCl), and singlet oxygen ($^1O_2$).

fusions between the cytoplasmic membrane and the membranes which surround lytic granules; such fusions effectively transport the contents of the granules outside the phagocytic cell and away from the invading pathogen.

Capsules and phagocytosis

An important mechanism for defense against phagocytosis is the bacterial capsule (see Section 3.9). Capsulated bacteria are frequently highly resistant to phagocytosis, apparently because the capsule somehow prevents the adherence of the phagocyte to the bacterial cell. The clearest case of the importance of a capsule in permitting invasion is that of *Streptococcus pneumoniae*. If only a few cells of a capsulated strain of this species are injected into a mouse, an infection is initiated that leads to death in a few days (see Figure 11.22). On the other hand, noncapsulated mutants of capsulated strains are completely avirulent, and injection of even large numbers of bacteria usually causes no disease. If a noncapsulated strain is transformed genetically with DNA from a capsulated strain (see Origin of Bacterial Genetics box in Chapter 7), both capsulation and virulence are restored at the same time. Furthermore, enzymatic removal of the capsule renders the organism noninvasive. Surface components other than capsules can also inhibit phagocytosis. For instance, pathogenic streptococci produce on both the cell surface and fimbriae a specific protein called the *M protein*, which apparently alters the surface properties of the cell in such a way that the phagocyte cannot act.

> Some pathogens produce specific capsules or other cell-surface components that prevent phagocytosis. Other pathogens produce leukocidins, substances that are able to kill a phagocyte even after the pathogen has been eaten. Invasiveness is often associated with the production of such substances.

11.14 Growth of the Pathogen Intracellularly

One group of organisms, the **intracellular parasites**, are readily phagocytized but are not killed, nor do they kill the phagocyte. Instead, they can remain alive for long periods of time and can even reproduce within the phagocyte. A number of major diseases are caused by intracellular pathogens (Table 11.7). Most of these pathogens have active scavenging systems for removing toxic oxygen species (see Section 11.13). In most situations the pathogen enters the host cell in the phagocytic vacuole (phagosome), and the pathogen grows within this structure until the phagosome bursts. Intracellular parasites are a diverse group. Some, such as *Mycobacterium tuberculosis*, *Salmonella typhi* (cause of typhoid fever), and *Brucella* species are facultative parasites, and can live either intracellularly or extracellularly. In acute infections they multiply in the extracellular body fluids, but in chronic conditions they may live only intracellularly. Intracellular seclusion of certain pathogens may be one of the means by which people who are chronic disease "carriers" remain infectious despite being clinically free of symptoms. When growing intracellularly the organism is protected from immune mechanisms of the host and is less susceptible to drug therapy. Some important intracellular parasites are unable to grow outside of living cells and are called *obligate intracellular parasites*. Included in this category are the viruses, the chlamydias, the rickettsias, and some protozoa, such as the one that causes malaria (see Sections 15.9–11, 19.22, and 19.23).

The intracellular environment differs markedly from the extracellular environment in physical and chemical characteristics (Table 11.8), being much richer in organic nutrients and cofactors; the advantages to a parasite of living intracellularly are thus partly nutritional. In addition to the normal microbial

Table 11.7   Major diseases caused by pathogens capable of intracellular growth

Disease	Organism
Rocky Mountain spotted fever	*Rickettsia rickettsii*
Typhus	*Rickettsia prowazekii*
Chlamydia venereal syndromes	*Chlamydia trachomatis*
Trachoma	*Chlamydia trachomatis*
Brucellosis	*Brucella abortus*
Tuberculosis	*Mycobacterium tuberculosis*
Leprosy	*Mycobacterium leprae*
Listeriosis	*Listeria monocytogenes*
Leishmaniasis	*Leishmania donovani*
Malaria	*Plasmodium falciparum*
Tularemia	*Francisella tularensis*
Typhoid fever	*Salmonella typhi*

Table 11.8   Major chemical differences between the intracellular and extracellular mammalian environment

Character	Intracellular	Extracellular (Plasma)
Ca^{2+}	Very low	$0.0025\ M$
Mg^{2+}	$0.02\ M$	$0.0015\ M$
Na^+	$0.01\ M$	$0.15\ M$
K^+	$0.15\ M$	$0.005\ M$
Cl^-	$0.003\ M$	$0.1\ M$
HPO_4^{2-}	$0.05\ M$	$0.002\ M$
HCO_3^-	$0.01\ M$	$0.03\ M$
Overall ionic strength	High	About 75% of intracellular ionic strength
Organic: nutrients cofactors, acids, and high-energy compounds	High levels	Relatively low levels

nutrients, the intracellular environment also provides ATP, which is completely absent extracellularly; some of the obligate intracellular parasites are unable to generate their own ATP from organic compounds and require preformed ATP provided by the host (see Section 19.23).

Some intracellular pathogens (for example, *Listeria monocytogenes* and *Rickettsia rickettsii*) grow within the cell nucleus as well as the cytoplasm. The intracellular growth of *Rickettsia rickettsii* has been studied in some detail. The organism is able to penetrate the cytoplasmic membrane readily and initiate growth (Figure 11.28a and b), but progeny are not restricted to the intracellular environment, and readily pass out of the cytoplasmic membrane without causing damage to the host cell. Thus, despite sustained growth of the intracellular parasite, massive accumulation within the

host cell does not always occur. The cells released from the cytoplasm readily infect adjacent cells, thus resulting in a rapidly spreading infection. However, if growth occurs in the nucleus, the progeny do not leave, so that sustained growth results in an extensive bacterial colonization of the nucleus (Figure 11.28c). Thus the rickettsial traffic is bidirectional across the cytoplasmic membrane and dominantly monodirectional across the nuclear membrane. One significant aspect of these observations is that release of rickettsial cells from the host can occur without apparent damage to the host cell.

Some microorganisms are able to survive and grow inside phagocytes, and are thus able to set up long-term intracellular infections. In such protected sites, these pathogens can cause chronic infections that are very difficult to treat. Tuberculosis, typhoid fever, and brucellosis are three examples of such chronic diseases.

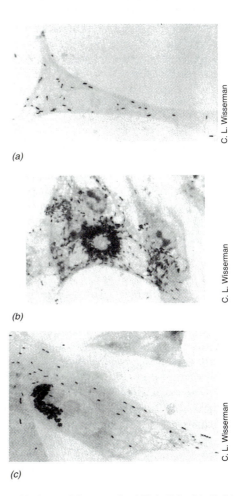

(a)

(b)

(c)

C. L. Wisserman

FIGURE 11.28 Intracellular growth of *Rickettsia rickettsii* in chicken embryo cells. The preparations were stained and examined by the light microscope. (a) Early stage of growth, 28 hours after infection. Note the generally dispersed cytoplasmic distribution. (b) Extensive intracytoplasmic colonies after 120 hours incubation. The dark mass represents extensive development of rickettsiae in a ring or doughnut radially arranged around a cytoplasmic vacuole. (c) Intranuclear growth, after 45 hours of incubation. Note the compact intranuclear mass and the dispersed cytoplasmic distribution. Not all cells showed intranuclear growth.

11.15 Inflammation and Fever

The tissues of the animal react to infection and to mechanical injury by an **inflammatory response**, the characteristic symptoms of which are redness, swelling, heat, and pain. The initial effect of the foreign stimulus is to cause local dilation of blood vessels and an increase in capillary permeability. This results in an increase flow of blood and passage of fluid out of the circulatory system into the tissues, causing swelling (edema). Phagocytes also pass through the capillary walls to the inflamed area. Initially granulocytes appear, followed later by monocytes. Within the inflamed area, a fibrin clot is usually formed, which in many instances will localize the invading microorganism. Pathogens that produce fibrinolytic enzymes may be able to escape and continue to invade the body.

Inflammation is one of the most important and ubiquitous aspects of host defense against invading microorganisms and is present in a small way virtually continuously. However, inflammation is also an important aspect of microbial pathogenesis since the inflammatory response elicited by an invading microorganism can result in considerable host damage.

Fever

The healthy human body maintains a surprisingly constant temperature. Over an average 24 hour period, body temperature varies only over the narrow range of 1–1.5°C. However, individuals vary in their "normal" temperatures, and although 37°C is assumed, normal temperature in some individuals may be as low as 36°C or as high as 38°C. Also, body temperature varies with amount of physical activity, and can be as much as 2°C below normal in sleep and as much as 4°C above normal during strenuous exercise.

Fever is defined as an abnormal increase in body temperature. Although fever can be caused by noninfectious disease, most fevers are caused by infection. At least one reason why fever occurs during many infectious diseases is that certain products of pathogenic organisms are pyrogenic (fever-inducing). The most well-studied pyrogenic agents are the endotoxins of Gram-negative Bacteria, which are often exceedingly potent in eliciting fever (see Section 11.10). However, many organisms which do not produce endotoxins are able to cause fever upon infection. Then, proteins known as *endogenous pyrogens* are released by phagocytic and other white blood cells during the phagocytosis process. Slight temperature increases benefit the host by accelerating phagocytic and antibody responses, while strong fevers of 40°C (104°F) or greater may benefit the pathogen if host tissues are further damaged.

Three kinds of characteristic fever patterns have been recognized in infectious disease. (1) *Continuous fever* is that condition in which the body temperature remains elevated over a whole 24-hour period and the total range of variation in temperature is less than 1°C. Continuous fever is seen in *typhoid fever* and *typhus*

fever. (2) A *remittent fever* is one in which the body temperature is abnormal over the whole of a 24-hour period, and the daily range shows swings greater than 1°C. This occurs in some *pyogenic infections* and *tuberculosis.* (3) An *intermittent fever* is one in which the temperature is normal for part of the day, and then rises above normal. Most infectious diseases elicit some intermittent fever, and the condition is a diagnostic characteristic of malaria, a protozoan infection. *Relapsing fever,* caused by various *Borrelia* species (see Section 19.12) is an intermittent fever in which the temperature remains normal for a long period of time, followed by a new attack of fever. This is characteristic of an extensive but incomplete recovery from an infectious disease, the fever arising when the disease flairs up again.

> Presence of foreign materials in the body, including microorganisms, leads to the development of inflammatory responses. Inflammation results from local dilation of blood capillaries, leading to accumulation of fluid and phagocytes at the site of attack. Another characteristic of infection is fever, which generally results from the formation of pyrogenic proteins by the pathogen.

11.16 Chemotherapy

Chemical agents able to cure infectious diseases are generally called **chemotherapeutic agents**. The hallmark of a chemotherapeutic agent is *selective toxicity,* that is, toxicity to the pathogen but not to the host. We have discussed antimicrobial action in some detail in Chapters 5 and 9, and industrial aspects in Chapter 10. Here we discuss the general principles involved in the use of antimicrobial agents in the therapy of infectious diseases caused by microorganisms other than viruses. In the next section we discuss the special problems inherent in viral chemotherapy. The action of antimicrobial agents in the test tube (*in vitro*) may be quite different from the action of such agents in the animal (*in vivo*). An agent highly effective *in vitro* may be completely ineffective *in vivo*, for the animal body is not a neutral environment for chemical agents, and many agents are modified so that they are no longer biologically active.

Drug distribution and metabolism in the body

When a drug is administered it becomes distributed through various compartments of the body. After initial absorption into the blood, as much as 90 percent of the drug may be bound to plasma proteins. This binding is reversible, and the bound drug is not inactivated, but can be looked upon as a reservoir to be released when the concentration of the free drug is lowered. In the tissues, the drug may be metabolized, and the metabolites are generally less active than the administered drug. From the viewpoint of the body, the drug is a foreign agent, and metabolic systems function to metabolize and detoxify many drugs. Detoxification enzymes are generally less well developed in infants, so that drugs may be much more toxic (per unit body weight) in infants than in older individuals.

Excretion of the drug and its metabolites generally occurs rapidly. Two main routes of excretion exist: renal (kidney) excretion to the urine, and hepatic (liver) excretion to the bile, which is excreted with the feces.

The characteristic pattern of drug absorption and excretion is illustrated by data for penicillin in Figure 11.29. As seen, after injection there is a rapid increase in concentration of antibiotic in the blood, followed by a gradual fall in concentration as the drug is excreted. The antibiotic concentration in the blood reaches a peak within 15 to 30 minutes, and quickly thereafter large amounts of antibiotic appear in the urine. Within 5 hours about 60 percent of the injected dose has been eliminated in the urine. Such data indicate that if bacteriostatic or bactericidal concentrations of an antibiotic are to be maintained in the body, *periodic doses* must be given.

Drug toxicity

Virtually all drugs have some toxicity (that is, cause some harm) to the host, and a knowledge of host toxicity is vital for the intelligent use of chemotherapeutic agents. A variety of toxic reactions can be observed within the host, including intestinal disturbances, kidney damage, and deafness. Two broad classes of toxicity are recognized: acute and chronic. Acute toxicity is expressed by pathological manifestations observed within a few hours after administration of a single dose. Generally, acute toxicity occurs as a result of drug overdosage.

Chronic toxicity is expressed by gradual changes, which take place during continuous administration of a drug over a long period of time. A variety of toxic manifestations require considerable time to develop, and if the course of treatment with the agent must be extended, chronic toxicity must be taken into consideration. For example, extended treatment with the antibiotic streptomycin in humans is thought to cause inner ear problems which can result in deafness. However, for treatment of infectious disease, very long courses of treatment are the exception rather than the rule, although for tuberculosis and certain other chronic infectious diseases, therapy may continue over many months or years. Also, some antimicrobial agents are used continuously as prophylactic agents (to prevent future attacks of the disease) in certain high-risk patients, and under these conditions, chronic toxicity becomes an important consideration. However, some manifestations of chronic toxicity may be exhibited in even as short a time as 1–2 weeks.

Although antibiotics are usually selective in their action, their toxicity for animals and humans varies, and a knowledge of antibiotic toxicity is essential for the wise use of antibiotics in medicine. Antibiotics that act against specific components of prokaryotes (for example, peptidoglycan, whose synthesis is specifically inhibited by penicillin), are usually not very toxic for eukaryotic host cells. On the other hand, it is more difficult to selectively injure protozoan or fungal cells in the human body because both the infecting microbial cells and host cells are eukaryotic. However, even antibiotics which are nontoxic to humans may elicit allergic responses in some people. For example, 5–10 percent of the human population is allergic to penicillin and must not be treated with this antibiotic. Thus, a wide variety of factors determine the usefulness of an antibiotic.

Testing of antimicrobial agents *in vivo*

Once the acute and chronic toxicity of a new antimicrobial agent has been determined, *in vivo* tests in experimental animals are initiated. Animal models can be developed for many infectious diseases, and such models are then used to evaluate chemotherapeutic agents. The mouse is the preferred experimental animal because it is a mammal and because it is relatively inexpensive and easy to maintain and manipulate in large numbers. Groups of mice can be infected with a pathogen and the effect of drug administration observed. The *in vivo* testing of antibiotics in experimen-

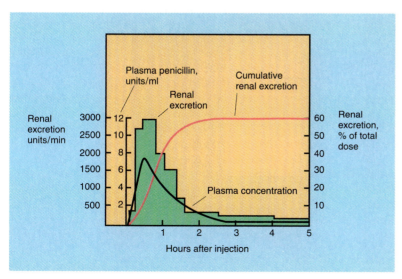

FIGURE 11.29
Penicillin absorption and excretion in a normal adult human. One intramuscular injection of 300,000 units of aqueous penicillin G was given at zero time. Samples of blood and urine were assayed periodically for antibiotic concentration. The results are average values found in humans with good renal function. The peak plasma level is reached within 15–30 minutes, and about 60 percent of the dose was excreted in the urine within 5 hours.

tal animals is an important procedure, and much can be learned about drug action from careful observation of such studies. The effect of the drug can be measured either by simply counting the number of infected animals that do not die after various periods of time, or by actually measuring the growth of the pathogen in the animal, and determining if the drug inhibits growth. It should be emphasized, however, that animal models do not closely mimic most human infections, since animal models make use of highly virulent strains, and large doses of pathogens are often used, so that the course of the infection is quick and severe. However, these animal models provide precise, reproducible systems for quantitatively evaluating the action of the antibiotic.

Antibiotic resistance

Pathogens may become resistant to an antibiotic by mutation or by acquiring a drug-resistant plasmid (see Section 9.17 and Table 9.9). In the presence of the drug, the altered pathogen has a selective advantage and may replace the parent type. Resistance can develop to virtually all chemotherapeutic agents and is known to occur *in vivo* as well as *in vitro* (Figure 11.30). The resistant strain may be just as virulent as the parent, may not be controllable by other chemotherapeutic agents, and may be passed on to other individuals.

It is now well established that the inappropriate, extensive use of antibiotics is leading to the rapid development of antibiotic resistance in disease-causing microorganisms. Parallel to the history of discovery and clinical use of the many known antibiotics has been the emergence of bacteria which resist their action. This is, in fact, a major reason why we continually seek new antibiotics and attempt to modify exist-

ing ones through chemical alterations (see Section 10.8). There are numerous examples of the association between the use of antibiotics and the development of resistance. The example in Figure 11.31a shows a correlation between the number of tons of antibiotics used and the percentage of bacteria resistant to each antibiotic which were isolated from patients with diarrheal disease. In general, the more an antibiotic was used, the more bacteria were resistant to it.

There are many examples of diseases in which the drug recommended for treatment has changed due to the increased resistance of the microorganism causing the disease. A classic example is the development of resistance to penicillin in *Neisseria gonorrhoeae*, the bacterium which causes gonorrhea (Figure 11.31b). The explanation for these observations is probably an ecological one. To grow in the presence of an antibiotic an organism must develop resistance to it. Thus, resistant microorganisms are *selected* by the presence of the antibiotic in the environment.

Surveys have shown that antibiotics are used far more often than is necessary. As a result, almost all pathogenic microorganisms have developed resistance to at least some antibiotics since widespread use of antimicrobial chemotherapy began in the 1950s (Figure 11.32). Penicillin and sulfa drugs, the first widely used chemotherapeutic agents, are not as widely used today because many pathogens have acquired at least some resistance to them. Even the few organisms that are still uniformly sensitive to penicillin, such as *Streptococcus pyogenes* (the bacterium that causes strep throat, scarlet fever, and rheumatic fever; see Section 15.2), now need significantly more penicillin for successful treatment than a decade ago. Other indiscriminant, nonessential uses of antibiotics

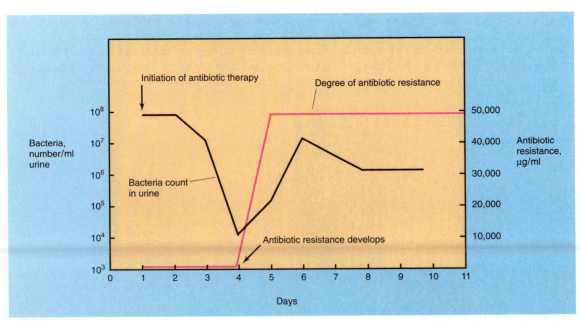

FIGURE 11.30 Development of an antibiotic-resistant mutant in a patient undergoing antibiotic therapy. The patient, suffering from chronic pyelonephritis caused by a Gram-negative Bacterium, was treated with streptomycin by intramuscular injection. Antibiotic resistance is expressed as the minimum inhibitory concentration of the antibiotic in micrograms per milliliter.

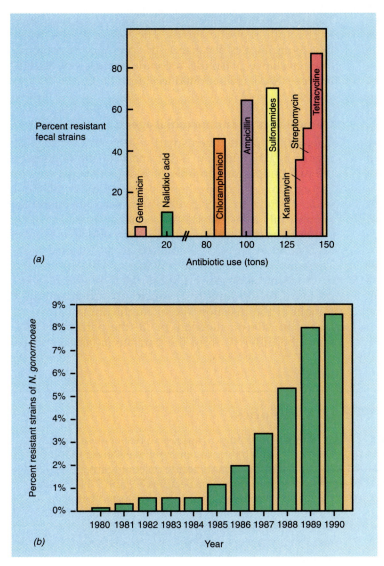

FIGURE 11.31
The emergence of antibiotic-resistant bacteria. (a) Relationship between antibiotic use and the percentage of bacteria isolated from diarrheal patients resistant to the antibiotic. Note that those antibiotics which have been used in the largest amounts, as indicated by the amount of antibiotic produced commercially, are those for which antibiotic-resistant strains are most frequent. (b) Percentage of reported cases of gonorrhea caused by antibiotic-resistant strains. The actual number of reported antibiotic-resistant cases in 1985 was 9000. This number rose to 59,000 in 1990. Greater than 95 percent of the reported antibiotic-resistant cases are due to penicillinase-producing strains of *Neisseria gonorrhoeae*. (Source: Centers for Disease Control, Atlanta, GA.)

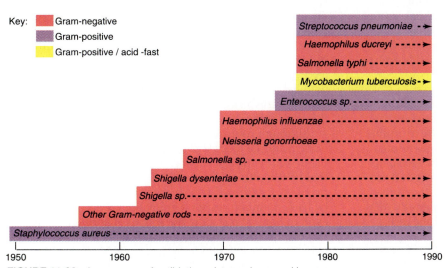

FIGURE 11.32   Appearance of antibiotic resistance in several human pathogens since the beginning of antibiotic therapy. (Adapted from M. L. Cohen, 1992, *Science* 257, p. 1050.)

may have worsened this situation. For example, antibiotics are used extensively in agricultural practices both as growth-promoting substances in animal feeds and as prophylactics (to prevent the occurrence of disease rather than to treat an existing one). Several recent food poisoning outbreaks have been blamed on the use of antibiotics in animal feeds (see Nonmedical Uses of Antibiotics box). By overloading various environments with antibiotics, rapid development of drug resistance may result. Resistance can be minimized if drugs are used only for serious diseases and are given in sufficiently high doses so that the microbial population level is reduced before mutants have a chance to appear. Resistance can also be minimized by combining two unrelated chemotherapeutic agents, since it is likely that a mutant resistant to one will still be sensitive to the other. However, with the increasing prevalence of resistance plasmids in pathogenic bacteria (see Section 7.9), multiple antibiotic therapy is proving less attractive as a clinically useful strategem. Some organisms, such as the streptococci, do not seem to develop drug resistance readily *in vivo*, whereas the staphylococci are notorious for developing resistance to chemotherapeutic agents. The rise of strains resistant to antibiotics because of the presence of conjugative resistance plasmids, a common means of acquiring resistance, is illustrated in Figure 11.33.

Clinical uses of antimicrobial agents

Chemotherapeutic agents vary widely in the range of organisms they attack. A summary of the actions of the most important antimicrobial agents was given in Figure 9.36. Although there are presently available a wide variety of relatively nontoxic antimicrobial agents, only a restricted number actually are in use in chemotherapy. Many agents that are effective against organisms *in vitro* have no effect in an infected host. There are several possible explanations for this: (1) the drug might be destroyed, inactivated, bound to body proteins, or too rapidly excreted; (2) the drug might remain at the injection site or might not penetrate as far as the site of infection; or (3) the parasite *in vivo* might be different from the parasite *in vitro*, perhaps

---

Nonmedical Uses of Antibiotics

A major nonmedical use of antibiotics in the United States is their addition to animal feed. The addition of low levels of antibiotics to animal feeds stimulates animal growth, shortening the period required to get the animal to market. For example, addition of 25 mg of penicillin per pound of chicken feed saves 2 billion pounds (900 million kg) of feed a year, due to more rapid weight gains and feeding efficiency. The antibiotics probably act by inhibiting organisms responsible for low-grade infections and by reducing intestinal epithelial inflammation. Studies with germ-free animals have borne this out. The growth of germ-free animals is not accelerated by antibiotic-supplemented feed. In addition, the intestinal wall of the normal animal is much thicker than that of the germ-free animal, probably because of low level inflammation caused by the bacterial flora. The lessening of inflammation in the gut wall of animals fed low levels of antibiotics probably promotes nutrient uptake and could account for the more efficient utilization of feed observed.

The problem with low level application of antibiotics to animal feeds is that an antibiotic resistant microflora is selected by the constant exposure to antibiotics. The use of antibiotics in animal feed therefore serves to expand the gene pool of antibiotic resistance in nature. Since some of the members of the gut flora of animals also inhabit the human gut, the transmission of resistant flora from animals to humans is thus a real possibility. Indeed, studies of antibiotic resistance in human gut flora have shown that many strains of human enteric bacteria are multiply resistant. This is especially true of those who work in animal husbandry.

Molecular biological studies of resistant strains of *Salmonella* isolated from poultry have shown that the resistance determinants reside on conjugative plasmids or transposons (see Sections 7.9 and 7.11). Resistance determinants are rapidly transferred between different species and even between different genera. Resistant organisms can then be transmitted to humans in contaminated meat or by contact with live animals.

Unfortunately, long-term studies of animals previously fed antibiotics and then put on antibiotic-free rations have shown that antibiotic resistant bacteria are not quickly lost from the gut. It is postulated that resistance genes have become part of stable plasmids in the gut flora, and, in the absence of counterselective forces, these resistance determinants have been maintained and will probably remain a part of the gut flora of animals for some time, even if supplementation of feeds with antibiotics was to stop immediately. Nonmedical use of antibiotics has therefore reinforced, albeit painfully, a simple lesson in microbial ecology: the environment selects the best adapted species.

Although continued use of clinically useful antibiotics in animal feeds will undoubtedly serve to widen the dissemination of resistance genes, it is not clear that halting this practice will effectively solve the problem. Continued veterinary use of antibiotics may by itself maintain resistant animal microflora. However, in hope of reducing the spread of antibiotic resistance in Europe, most European countries have banned the use of antibiotics in animal feeds. In the United States large amounts of antibiotics continue to be used in the cattle, poultry, and swine industries.

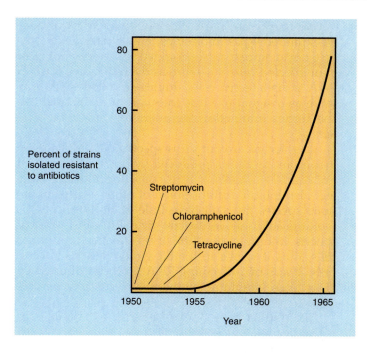

FIGURE 11.33 Rise in antibiotic-resistant *Shigella* in Japan following introduction of antibiotic therapy. Arrows indicate the years when the three antibiotics were introduced. These antibiotic resistances are commonly plasmid borne.

showing different physiological properties or possibly growing intracellularly, where it is protected from the action of the drug.

In medical practice the decision on whether or not to use a chemotherapeutic agent should be made after consideration of the following factors. (1) Is the organism sensitive to the agent as shown by antibiotic-sensitivity tests? (2) Are the symptoms due to the organism itself or to a toxin? If to the latter, then attack on the organism alone will not suffice to effect a cure. (3) Is it possible for the drug to reach the site of infection? Many skin infections are hard to treat because the drug will not penetrate to the infected site; again, many drugs will not enter the spinal fluid and therefore cannot be used in meningitis. Also, many agents cannot enter host cells and are thus ineffective against intracellular pathogens. (4) Is the patient allergic to the drug? (5) Will the chemotherapeutic agent cause adverse side effects or interfere with the host's defense mechanisms?

The decision to treat an infection with a certain antibiotic is usually reached after consideration of all the questions posed above. In general, the physician will use the antibiotic that is most effective at the lowest effective concentration. To guide such decisions a wealth of data exists on the antibiotic susceptibility of routinely encountered pathogens. However, because of the development of resistance, it cannot be assumed that any particular strain causing an infection in humans will be sensitive to a particular antimicrobial agent. Thus, it is essential that sensitivity be determined for each isolated pathogen. Procedures involved in determination of sensitivity to antimicrobial agents are discussed in Chapter 13 (Clinical Microbiology).

> Antibiotics and other chemotherapeutic agents can be used in the treatment of microbial infection, although not all such agents are suitable for therapy. It is essential that the toxicity of the chemical agent be sufficiently low and that the agent is absorbed and distributed well throughout the body. An important side effect of the use of chemotherapy is the development of resistance by the pathogen to the chemotherapeutic agent.

11.17   Viral Chemotherapy

Because viruses are not cells, the inhibition of viruses poses special problems to chemotherapy. However, from the detailed knowledge of the molecular events involved in viral replication (see Chapter 6), some clues have emerged as to how virus infections can be controlled. In this section we discuss two approaches to virus therapy that have been studied in some detail—chemical inhibition and interferon action.

Chemical inhibition of animal viruses

Since a virus depends on its host cell for most aspects of virus replication, it is difficult to inhibit virus multiplication without at the same time affecting the host cell itself. Because of this, the spectacular medical successes with antibacterial agents have not been followed by similar success in the search for specific antiviral agents. A few antiviral compounds are successful in controlling virus infections in laboratory situations (Table 11.9) and certain of these have been used in restricted clinical cases; but no substance has yet been found with more than limited practical use.

One interesting inhibitor listed in Table 11.9 is rifamycin, which is an inhibitor of RNA polymerase in Bacteria but not in eukaryotes or Archaea (see Section 5.6). The RNA polymerase of vaccinia and other poxviruses is also inhibited by rifamycin, since this antibiotic specifically inhibits the replication of these viruses, although it has no effect on a wide range of other viruses affecting animal cells.

Another interesting chemical is *azidothymidine* (*AZT*), an inhibitor of retroviruses such as the virus which causes acquired immune deficiency syndrome (AIDS). Azidothymidine is chemically related to thymidine but is a dideoxy derivative, lacking the 3′ hydroxyl (thus analogous to the dideoxynucleotides used in the Sanger DNA sequencing technique—see Nucleic Acids box, Chapter 5). AZT inhibits multiplication of retroviruses by blocking the synthesis of the DNA intermediate (reverse transcription) and is used to inhibit multiplication of the human immunodeficiency virus (HIV), the causative agent of AIDS (see Section 15.7).

Interferon

Interferons are antiviral substances produced by many animal cells in response to infection by certain viruses. They are low-molecular-weight proteins that prevent viral multiplication. They were first discovered in the course of studies on virus interference, a phenomenon whereby infection with one virus interferes with subsequent infection with another virus, hence the name *interferon*. It has been found that interferons are formed in response not only to live virus but also to virus inactivated by radiation or to viral nucleic acid. Interferon is produced in larger amounts by cells infected with viruses of low virulence, whereas little is produced against highly virulent viruses. Apparently the virulent viruses inhibit cell protein synthesis before any interferon can be pro-

duced. Interferon is also induced by a variety of double-stranded RNA molecules, either natural or synthetic, and since double-stranded RNA does not exist in uninfected cells but exists as the replicative form in RNA virus-infected cells, double-stranded RNA may serve as a signal of virus infection in the animal cell and brings into action the interferon-producing system.

Interferons are not virus specific but *host* specific; that is, an interferon produced by one type of animal (for example, chicken) in response to influenza virus will also inhibit multiplication of other viruses in the same species but will have little or no effect on the multiplication of influenza virus in other animal species. Interferon has little or no effect on uninfected cells; thus it seems to inhibit viral synthesis specifically. It acts by preventing RNA synthesis directed by virus, thus inhibiting synthesis of virus-specific proteins.

Interferons have been of interest as possible antiviral agents, and possibly also as anticancer agents. Their use as therapeutic agents was long hindered by the difficulty and expense of producing large quantities, but genetic engineering techniques (see Section 8.13) have now made possible the production of interferon on a commercial scale, and many clinical trials are under way.

> Viruses make use of the host metabolic machinery for multiplication and because of this it is much more difficult to use chemotherapy for inhibition of virus growth, although a few chemical agents are known which inhibit virus-specific functions. Animal cells themselves produce antiviral protein substances called interferons which affect certain stages of the virus multiplication process. Cloned gene technology has now made interferons available in amounts needed to test their clinical effectiveness.

Table 11.9   Stages of virus replication at which chemical inhibition of virus action is known to occur

Stage of replication	Chemical	Virus
Free virus	Kethoxal	Influenza virus
Adsorption	None known	—
Entry of nucleic acid (uncoating)	Amantadine	Influenza virus
	Carbobenzoxypeptides	Measles
	3-Methylisoxazole compounds	Rhinoviruses (cold viruses)
Nucleic acid replication	Benzimidazole, guanidine	Poliovirus
	5-Fluorodeoxyuridine (FUDR)	Herpesvirus
	5-Iododeoxyuridine (IUDR)	Herpesvirus
	Acyclovir	Herpesvirus, *Varicella zoster*
	Rifamycin	Vaccinia virus
	Azidothymidine (AZT)	Retrovirus (Human immunodeficiency virus)
	Dideoxyinosine (ddI)	Retrovirus (Human immunodeficiency virus)
Maturation (or late protein synthesis)	Isatin-thiosemicarbazone	Smallpox virus
Release	None known	—

Study Questions

1. Distinguish between a parasite and a pathogen. Between infection and disease.

2. Which parts of the human body are normally heavily colonized with microorganisms? What parts are normally devoid of microorganisms?

3. Distinguish between the resident and transient microflora of a body habitat. How could you distinguish between resident and transient microorganisms experimentally?

4. Why is a hair follicle a likely site of microbial entry into the body through the skin? What other ways can you imagine that bacteria might enter the body through the skin?

5. Write a brief discussion of the steps involved in the development of dental plaque.

6. Obligately anaerobic bacteria are very common in the large intestine, yet they are only able to grow there if facultatively aerobic bacteria are also present. Explain.

7. Certain antibiotics, even antibiotics whose mode of action is bacteriostatic rather than bactericidal, sterilize the intestinal tract. Explain how a bacteriostatic antibiotic could bring about this result.

8. Design an experiment to show the likely route of infection of a urinary tract pathogen.

9. What microbial structures have been implicated as important adherence factors for pathogenic microorganisms?

10. Describe the ways in which enteropathogenic strains of *E. coli* differ from normal strains of *E. coli*; include a discussion of structural and ecological variables.

11. What do the substances hyaluronidase, collagenase, streptokinase, and coagulase have in common? What is the mode of action of each substance in promoting disease?

12. Define and contrast: exotoxin, enterotoxin, endotoxin. Give two examples of each and the name of the organism producing each.

13. Although mutants incapable of producing exotoxins are relatively easy to isolate, those incapable of producing endotoxin are much more difficult to isolate. From what you know of the structure and function of these two classes of toxins, explain why this difference might be.

14. For each of the following exotoxins, describe (a) the organism producing it; (b) its mode of action on the host; (c) its role in pathogenicity; (d) how its effects can be counteracted: 1) diphtheria toxin; 2) tetanus toxin; 3) botulinum toxin; 4) cholera toxin.

15. Describe the manner by which fibrin formed by the host in response to a bacterial infection can reduce the severity of that infection. How do bacteria counteract host resistance due to fibrin clotting?

16. Compare and contrast the cell composition and protein content of whole blood, serum, and plasma. How are plasma and serum prepared from whole blood?

17. Compare and contrast monocytes, granulocytes, and lymphocytes. What are the principal roles of each in the body?

18. Although most bacteria ingested into phagocytic cells are killed, some types actually replicate. Explain. Give three examples of bacteria capable of growth in phagocytes.

19. A number of new antibiotics which show promise as clinically useful agents based on *in vitro* tests turn out to be worthless in treating the same bacterium *in vivo*. List at least three reasons why this might be.

20. Explain why the following statement is erroneous: Since the discovery of antibiotics it is no longer necessary to immunize persons for any bacterial disease because antibiotics can control any outbreaks of the disease.

21. What significance for our understanding of infectious diseases do you see in the fact that large numbers of bacteria and viruses may occur in a healthy human being?

22. When considering the inhibition of virus penetration by chemical agents, *naked* and *enveloped* viruses could be expected to behave differently. Why?

23. How can you explain the rising incidence of drug-resistant *Neisseria gonorrhoeae* isolates? Do you expect to see similar trends with other pathogens? Why or why not?

Supplementary Readings

Ayoub, E. M., G. H. Cassell, W. C. Branche, Jr., and **T. J. Henry** (eds.) 1990. *Microbial Determinants of Virulence and Host Response.* American Society for Microbiology, Washington, D.C. Coverage of the molecular biology of factors involved in bacterial and fungal pathogenicity.

Balows, A., W. J. Hausler, K. L. Herrmann, H. D. Isenberg, and **H. J. Shadomy** (eds.) 1991. *Manual of Clinical Microbiology,* 5th edition. American Society for Microbiology, Washington, D.C. A comprehensive collection of information concerning isolation, culture, and diagnosis of pathogenic microorganisms. An excellent reference source.

Bryan, L. E. (ed.) 1989. *Microbial Resistance to Drugs.* Springer-Verlag, Berlin. An excellent source of information on the mechanisms of antimicrobial agent resistance and methods for testing clinical isolates for resistance.

Davis, B. D., R. Dulbecco, H. N. Eisen, and **H. S. Ginsberg.** 1990. *Microbiology,* 4th edition. Harper & Row, Hagerstown, MD. A brief treatment of host–parasite relationships can be found in this standard medical school textbook. A good reference source for information on diseases caused by specific groups of organisms.

Friedman, H., T. W. Klein, M. Nakano, and **A. Nowotny** (eds.) 1990. *Endotoxin.* Plenum Press, New York. An advanced treatise on endotoxin structure and biological activity.

Grubb, R. T., Midtredt, and **E. Norin** (eds.) 1989. *The Regulatory and Protective Role of the Normal Flora.* Stockton Press, New York. Review articles on the beneficial aspects of the normal flora of animals.

Henry, J. B. 1991. *Clinical Diagnosis and Management by Laboratory Methods*, 18th edition. W. B. Saunders Company, Philadelphia. A standard technical medical reference source.

Iglewski, B, and **V. L. Clark** (eds.) 1990. *Molecular Basis of Bacterial Pathogenesis.* Academic Press, New York. A multi-authored volume on mechanisms of bacterial pathogenesis, especially in Gram-negative Bacteria.

Kreier, J. P., and **R. F. Mortensen.** 1990. *Infection, Resistance, and Immunity.* Harper & Row, New York. An elementary textbook of host–parasite relationships and immune mechanisms.

Roth, J. A. (ed.) 1988. *Virulence Mechanisms of Bacterial Pathogens.* American Society for Microbiology, Washington, D.C. A collection of chapters, written by experts, on mechanisms of bacterial adherence, colonization and invasion, mechanisms of pathogen resistance to host defenses, and host strategies to overcome bacterial virulence.

Schaechter, M., G. Medoff, and **D. Schlessinger** (eds.) 1989. *Mechanisms of Microbial Disease.* Williams and Wilkins, Baltimore, MD. This medical school textbook has a good section on host–parasite relationships.

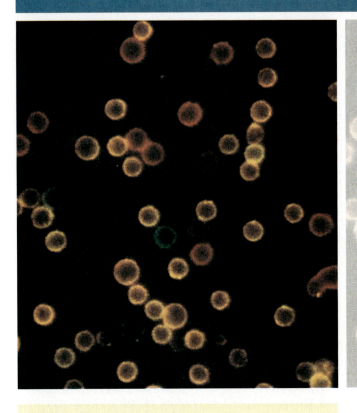

12

Immunology and Immunity

Vertebrate animals possess a highly sophisticated mechanism for developing specific resistance to a variety of **pathogens** (microorganisms that cause disease), including many bacteria, viruses, and fungi. This mechanism, known as the **immune response**, is elicited by many foreign molecules, collectively known as **immunogens**. These foreign molecules are usually macromolecular components of pathogens, such as cell surface proteins. When these foreign molecules are recognized by the immune system, they are known as **antigens**. As a result of antigen introduction, individual immune cells are stimulated to produce and secrete antigen-reactive proteins known as **antibodies** or **immunoglobulins**, a process called **humoral immunity**. In addition to this humoral response, antigen-specific **T cells** are also activated. These cells also interact with antigens in several processes collectively known as **cellular immunity**.

The immune response has three major characteristics: *specificity*, *memory*, and *tolerance*. First, the **specificity** of the antigen–antibody or antigen–T cell interaction is unlike any other host-resistance mechanism. As seen in Chapter 11, phagocytosis, inflammation, and other nonspecific host-resistance mechanisms develop immediately against virtually any invading microorganism, even those the host has never before encountered. In the immune response, however, each new microorganism must interact with the immune system before a response occurs. In most cases, no immune response can be detected for several days after initial antigen contact. However, when the response occurs, it is directed solely to that particular microorganism. Second, once the immune system produces a specific type of antibody or activated T cell, *challenge*, or restimulation, by the same microorganism results in rapid production of large amounts of the same antibody or large numbers of T cells. This capacity for **memory** allows the host to resist subsequent reinfection by pathogens and to be protected by **vaccination**, the practice of inoculating the host with inactive or weakened pathogens to stimulate immunity. Finally, **tolerance**, a state of specific immune unresponsiveness, occurs because macromolecules in the host are also potential antigens. These host molecules would be damaged if specific antibodies and activated T cells were produced that recognized them. Through tolerance, the host immune response learns to distinguish between foreign (nonself) macromolecules and host (self) macromolecules and to interact appropriately with them.

Figure 12.1 presents an overview of the immune response and highlights the activities of humoral and cellular immunity. We first discuss the cells and molecules of the immune system, the molecular basis for antigen recognition, and then explain how cells interact to provide specific disease resistance, using both humoral and cellular mechanisms.

Several features of the immune response are listed here and are discussed in more detail in the chapter:

1. Many, but not all, foreign macromolecules elicit an immune response.
2. The immune response is very specific: antibodies or activated T cells made in response to one antigen generally do not react with other antigens.
3. In virtually every case, an immune response to a foreign macromolecule occurs only if the animal is challenged with the same foreign substance.
4. Not all immune reactions are beneficial to the host: some, such as those involved in hypersensitivity and autoimmunity, can be harmful to the host.
5. Immune reactions are generated in response to foreign macromolecules, but not ordinarily against macromolecules of the animal's own tissues; thus, the host immune system is able to distinguish between its own (*self*) antigens and foreign (*nonself*) antigens.
6. Microorganisms and viruses that invade the host contain large numbers of different macromolecules that can act as antigens. Purposeful immunization with these antigens usually results in specific immune responses that can aid in the prevention and control of specific diseases.
7. The exquisite specificity of antigen–antibody reactions makes them useful for many research and diagnostic procedures.

> The immune response is a specific reaction by the body to the presence of foreign material, generally macromolecules. The substance which induces the immune response is called an antigen or immunogen. As a result of antigen stimulation, proteins called antibodies or immunoglobulins are produced, and specific cells called activated T cells are formed. Microorganisms and viruses capable of invading the body contain numerous antigens and the immune response participates in the prevention and control of infectious disease.

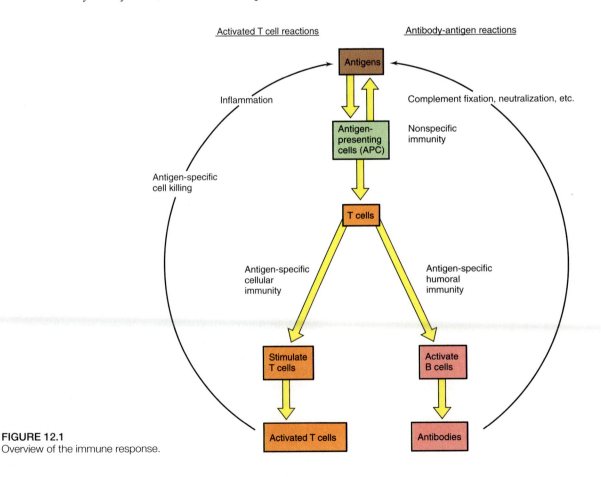

FIGURE 12.1
Overview of the immune response.

Miniglossary for Chapter 12

ANTIBODY   a soluble protein, produced by B cells, which interacts with antigen; also called immunoglobulin

ANTIGEN   a molecule capable of interacting with specific components of the immune system

ANTIGENIC DETERMINANT   that portion of an antigen that is reactive with a specific antibody or T cell receptor; also called an epitope

ANTIGEN-PRESENTING CELL (APC)   any cell that functions to present antigen to a T cell

AUTOANTIBODY   an antibody that reacts to self antigens

B CELL   a lymphocyte which produces immunoglobulin

CELL-MEDIATED IMMUNITY   immunity resulting from direct interaction with antigen-specific T cells

CLASS I MHC PROTEINS   antigen-presenting molecules found on all nucleated vertebrate cells

CLASS II MHC PROTEINS   antigen-presenting molecules found primarily on macrophages and B cells

CLONAL SELECTION   a theory that each B or T cell, when stimulated by antigen, produces copies of itself

COMPLEMENT   a series of proteins which react in a sequential manner with antibody–antigen complexes to amplify or potentiate their activity

CYTOKINE   a soluble immune response modulator produced by cells other than lymphocytes, usually phagocytic cells

DOMAIN   a region of a protein having a distinct function

ELISA   *enzyme linked immunosorbent assay*

HAPTEN   a low molecular weight substance that combines with specific antibodies, but which is incapable of eliciting an immune response by itself

HUMORAL IMMUNITY   immunity resulting from direct interaction with antibodies

HYBRIDOMA   the artificially fused product of two unrelated cells which exhibits properties of both cells; used to produce monoclonal antibodies

HYPERSENSITIVITY   activity of the immune response leading to damage to host tissues, referred to as allergies

IMMUNOGEN   a molecule capable of eliciting an immune response

IMMUNOGLOBULIN (Ig)   a soluble protein produced by B cells, which interacts with antigens; also called antibody

IMMUNOLOGIC MEMORY   ability to rapidly produce large quantities of specific immune cells or antibodies after subsequent exposure to a previously encountered antigen

LYMPHOKINE   a soluble immune response modulator produced by lymphocytes

MAJOR HISTOCOMPATIBILITY COMPLEX (MHC)   a genetic complex responsible for encoding several cell surface proteins important in antigen presentation

MONOCLONAL ANTIBODY   an antibody which is the product of a single B cell clone

POLYCLONAL ANTISERUM   a serum containing antibodies derived from many B cell clones, as occurs in a normal immune response

PRIMARY ANTIBODY RESPONSE   antibodies made upon first exposure to antigen; mostly of the class IgM

RIA   *radioimmunoassay*

SECONDARY ANTIBODY RESPONSE   antibody made upon second (subsequent) exposure to antigen; mostly of the class IgG

SEROLOGY   the study of antigen–antibody reactions *in vitro*

SPECIFICITY   the ability of the immune response to interact with individual antigens

T CELL   a lymphocyte responsible for antigen-specific cellular interactions

T CELL RECEPTOR (TCR)   antigen-specific receptor protein on the surface of T cells

TOLERANCE   inability to make an immune response to specific antigens

VACCINATION   inoculation of a host with inactive or weakened pathogens to stimulate immunity

12.1   Immunogens and Antigens

Immunogens are substances that, when administered to an animal in the appropriate manner, induce an immune response. The immune response may involve either antibody production, the activation of specific immunologically competent cells (called *activated T cells*), or both. **Antigens** are substances that react with either antibodies or antigen-specific receptors on activated T cells, known as **T cell receptors (TCR),** and most antigens are also immunogens. However, some substances are recognized by immune systems while not being true immunogens. For example, **haptens** are low-molecular-weight substances that combine with specific antibody molecules but do not by themselves induce antibody formation. Haptens include such molecules as sugars, amino acids, and small polymers.

An enormous variety of macromolecules can act as immunogens under appropriate conditions. These include virtually all proteins and lipoproteins, many polysaccharides, some nucleic acids, and certain of the teichoic acids. One important requirement is that the molecules must be of fairly *high* molecular weight, usually greater than 10,000. However, the antibody or TCR is directed not against the antigenic macromolecule as a whole, but only against distinct portions of the molecule that are called its **antigenic determinants** or **epitopes** (Figure 12.2). Chemically, antigenic determinants include sugars, amino acid side chains, organic acids and bases, hydrocarbons, and aromatic groups. Antibodies are formed most readily to deter-

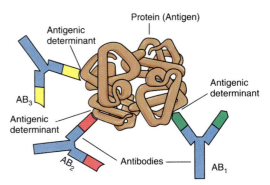

FIGURE 12.2 Antigens and antigenic determinants. Antigens may contain several different antigenic determinants, each capable of reacting with a specific antibody.

minants that project from the foreign molecule or to terminal residues of a polymer chain. In proteins, for example, the majority of antibodies are made to surface determinants, because the surface contains a continuum of antigenic sites. A region of as few as four or five amino acids can define an antigenic determinant on a protein. Also, the surface of a protein can and frequently will have many overlapping antigenic determinants. A cell or virus is a mosaic of proteins, polysaccharides, and other macromolecules, each of which is a potential antigen. Each antigen of the cell is also a mosaic of side chains and residues, each of which is a potential antigenic determinant. The immunological response to an invading microorganism is thus a complex phenomenon.

In general, the specificity of antibodies is comparable to that of enzymes, which are also able to distinguish between closely related substances. For instance, antibodies can distinguish between the sugars glucose and galactose, which differ only in the position of the hydroxyl group on carbon 4. However, specificity is not absolute, and an antibody may react at least to some extent with epitopes related to the one that induced its formation. The antigen which induced the antibody is called the **homologous antigen,** and others, if any, that react with the antibody are called **heterologous antigens**.

> An immunogen is a substance that induces an immune response; if the immunogen reacts with either antibody or activated T cell it is called an antigen. Although the reaction between antigen and antibody or TCR is highly specific, it is not directed at the antigen as a whole, but just against one or more restricted portions of the antigen called its antigenic determinants. Although the specificity of the immune response is comparable to the specificity of enzyme and substrate, antibodies or TCRs can sometimes react with heterologous antigens.

12.2   Cells of the Immune System

The immune system involves a number of organs and cell types that interact in various ways to elicit the final immune response. Organs of the immune system are located throughout the body (see Figure 11.24). A key cell involved in immune responses is a type of white blood cell called a **lymphocyte**; any tissue that contains lymphocytes and participates in the immune response is referred to as a *lymphoid tissue*.

Lymphocytes arise from undifferentiated stem cells in the bone marrow, the soft tissue in the hollow shafts of long bones. Stem cells differentiate into functionally distinct cell types during a maturation process that takes place in association with specific tissues (Figure 12.3). Although at least a dozen different ma-

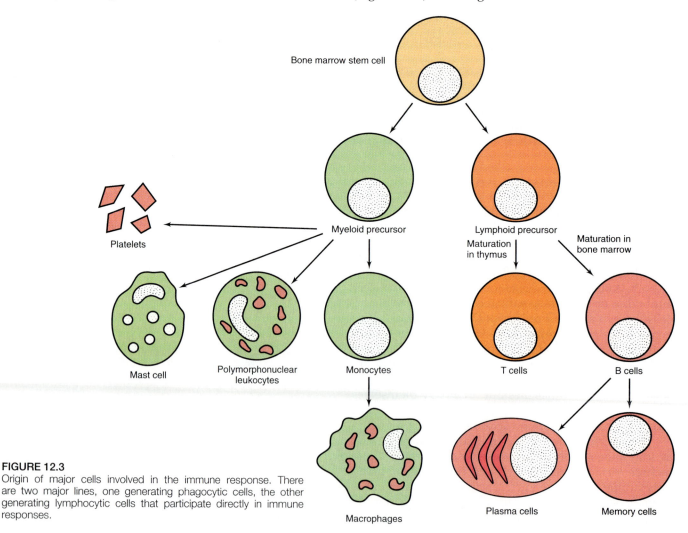

FIGURE 12.3
Origin of major cells involved in the immune response. There are two major lines, one generating phagocytic cells, the other generating lymphocytic cells that participate directly in immune responses.

ture cell types develop from the stem cells, we limit our discussion here to lymphocytes and macrophages because it is these cells that are primarily involved in specific immune responses.

Lymphocytes

Lymphocytes are dispersed throughout the body in the blood circulation and lymphatic circulation systems (see Figure 11.24) and are one of the most prevalent mammalian cell types (the average adult human has about 10^{12} lymphocytes). Two types of lymphocytes, *B lymphocytes* (or *B cells*) and *T lymphocytes* (or *T cells*), are involved in immune responses. Both B and T cells are derived from stem cells in the bone marrow. The differentiation of stem cells into mature lymphocytes is determined by the organ in which they become established (Figure 12.3). B cells mature in the bone marrow (hence, the designation "B") and T cells mature in the thymus (thus, the designation "T").

Because of their role in the initial development and maturation of B and T cells, the bone marrow and thymus are called *primary lymphoid organs*. After maturation, B and T cells are dispersed throughout the body via the blood and lymphatic circulatory systems. Mature T and B cells come to reside in the lymph

nodes or spleen (see Figure 11.24), which are collectively known as *secondary lymphoid organs*. These organs are positioned in the blood and lymph circulatory systems and act as filters. The macrophages in the spleen or lymph nodes trap antigen as it passes through, and the B and T cells in these organs produce an immune response.

B cells are responsible for antigen interaction and antibody production. B lymphocytes are distinguished from T lymphocytes by the presence of antibody molecules on their surface. This surface antibody is a copy of the single type of antibody that the B cell will produce later in its development.

The situation with T cells, however, is more complex. All T cells have antigen-specific TCRs on their surface and thus interact specifically with antigen. However, several functionally distinct subsets of T cells have been identified. Two major subpopulations are distinguished from each other by the presence of either CD4 or CD8, which are T lymphocyte-specific cell surface proteins (Figure 12.4). The CD4 subpopulation is further subdivided into two functional subsets. The first of these is the *T helper* or T_H subset. T_H cells are known to stimulate B lymphocytes to produce large amounts of immunoglobulin; in most cases,

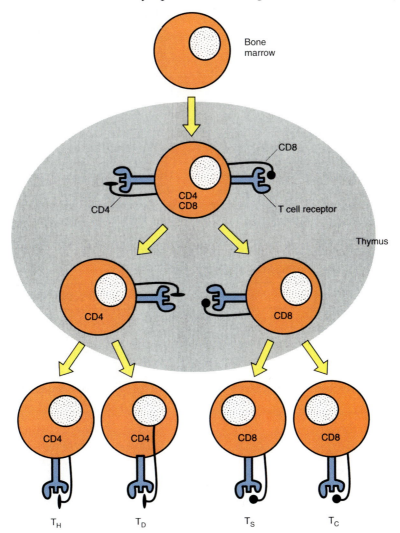

FIGURE 12.4   T cell subsets. T cell subsets develop in the thymus from bone marrow precursor cells.

little if any antibody is made by B cells without T_H interaction (see Section 12.8). The *T delayed-type hypersensitivity*, or T_D cell subset, is also CD4. T_D cells participate in T cell-mediated reactions but do not interact with B cells; T_D cells are responsible for recruiting and activating nonspecific effector cells, such as macrophages (see later). The second major T cell subpopulation, the CD8 cells, also contain at least two functional subsets. The first of these is the *T cytotoxic cell*, or T_C, which interacts with and directly destroys cells having antigen on their surface. The other CD8 subpopulation is the *T suppressor cell*, or T_S cell, which, as its name implies, serves to regulate the immune response by suppressing the action of immune cells such as B cells. Table 12.1 compares B and T lymphocytes with respect to development, surface antigens, and functions.

Macrophages

Macrophages are large phagocytic cells capable of ingesting antigens and destroying them as well as cooperating with lymphocytes in the production of specific antibody (see Section 11.13 and Figures 11.25 and 11.26). Phagocytic cells are of two types: **monocytes** and **polymorphonuclear granulocytes**. Monocytes can differentiate to become macrophages (Figure 12.3). The term *macrophage* is generally used to describe phagocytes that are *fixed* to tissue surfaces, while the term *monocyte* is used to describe freely circulating phagocytes. Macrophages are abundant in lymphoid tissue and in the spleen, while monocytes are abundant in the blood and lymph.

If an antigen penetrates an epithelial surface, it will eventually come in contact with phagocytic cells. As discussed in Section 11.13, phagocytic cells such as macrophages can engulf particles as large as bacterial cells and kill them by releasing lytic substances (proteases, nucleases, lipases, and lysozyme) from cytoplasmic vesicles within the cell. Lysis of a bacterial cell releases a variety of distinct bacterial antigens within the macrophage, and these, as well as antigens in-

gested directly, can be processed by the macrophage and used to initiate the early steps in antibody synthesis. Macrophages serve as **antigen-presenting cells (APC)** because they embed foreign antigens on their cell surfaces where they can come into contact with specific T cells and B cells; antigen recognition by T cells and B cells is the first step in antibody production (see Section 12.8).

Macrophages and monocytes are *nonspecific* cells. Unlike T cells and B cells, phagocytic cells cannot distinguish among antigens; any foreign substance is ingested, whether antigenic or not. However, since many macromolecules are antigens and foreign cells contain numerous antigens, many of the particles phagocytized by macrophages or monocytes will be antigenic. The action of macrophages in antigen processing and presentation is a key step in the overall process of antibody production because the vast majority of antigens can only stimulate lymphocytes through intermediary nonspecific APCs such as macrophages.

Precursors of cells that are involved in the immune response arise in the bone marrow from pleuripotent stem cells. The cells differentiate and develop to form functionally distinct cell types. Macrophages are cells that engulf foreign particles and macromolecules and digest them. They are nonspecific in their action, and their role is to *prepare* the antigen for specific immune responses. Lymphocytes are involved specifically in the immune response, differentiating into T cells and into the B cells that will become immunoglobulin-producing cells.

12.3 Immunoglobulins (Antibodies)

The next three sections of this chapter discuss antigen-specific receptors involved in the immune system. We begin with immunoglobulins because we understand their structure and function in detail. They serve as the

---

Table 12.1 Comparison of B and T lymphocytes	
T cells	**B cells**
Origin: bone marrow	Origin: bone marrow
Maturation: thymus	Maturation: bone marrow
Long-lived: months to years	Long lived: months to years, or short-lived: days to weeks
Mobile	Relatively immobile (stationary)
T cell receptor (TCR) on surface	Complement receptors
	Immunoglobulins on surface
Restricted antigenic specificity	Restricted antigenic specificity
Proliferate upon antigenic stimulation	Proliferate upon antigenic stimulation into plasma cells and memory cells
Produce lymphokines	Synthesize immunoglobulin (antibody)
Show delayed hypersensitivity (T_D cells)	
Help in immunoglobulin production by B cells (T_H cells)	
Perform as killer T lymphocytes in cell-mediated immunity (T_C cells)	
Control immune response (T_S cells)	

model for an antigen-specific receptor molecule. *Immunoglobulins (antibodies)* are protein molecules that are able to combine with antigenic determinants. They are found predominantly in the serum fraction of the blood, although they may also be found in other body fluids, as well as in milk. Serum is the fluid portion of the blood that is left when the blood cells and the materials responsible for clotting (fibrin, platelets, and various cofactors, see Section 11.13) are removed. Serum containing antigen-specific antibody is often called **antiserum**. Immunoglobulins (abbreviated Ig) can be separated into five major classes on the basis of their physical, chemical, and immunological properties: **IgG, IgA, IgM, IgD,** and **IgE** (Table 12.2). Immunoglobulin class IgG has been further resolved into four immunologically distinct subclasses called IgG_1, IgG_2, IgG_3, and IgG_4. The basis for this separation is discussed below. Antibody molecules specific for a given antigenic determinant can be found in each of the several classes, even in a single immunized individual. Upon initial immunization, the first immunoglobulin to appear is IgM, a pentameric immunoglobulin with a molecular weight of about 970,000; IgG appears later. In most individuals about 80 percent of the immunoglobulins are IgG proteins, and these have therefore been studied most extensively.

Immunoglobulin structure

Immunoglobulin G is the most common circulating antibody and thus we will discuss its structure in some detail. Immunoglobulin G has a molecular weight of about 146,000 and is composed of four polypeptide chains (Figure 12.5). Both intrachain and interchain disulfide (S—S) bridges are present (Figure 12.5). The two light (short) chains are identical in amino acid sequence, as are the two heavy (longer) chains. The molecule as a whole is thus symmetrical (Figure 12.5). Each light chain consists of about 212 amino acids and each heavy chain consists of about 450 amino acids.

When an IgG molecule is treated with the proteolytic enzyme *papain*, it breaks into several fragments. The two fragments containing the complete light chain plus the amino terminal half of the heavy chain are the portions that combine with antigen and are called *Fab* fragments (*fragment of antigen binding*). The fragment containing the carboxy terminal half of both heavy chains, called *Fc* (fragment *crystallizable*), does *not* combine with antigen (Figure 12.5c). Therefore, each antibody molecule of the IgG class contains *two antigen combining sites* (and is thus *bivalent*). This bivalency is of considerable importance in understanding the man-

Table 12.2 Properties of human immunoglobulins

Class desig- nation	Molecular weight	Proportion of total antibody (percent)	Concentration in serum (mg/ml)	Antigen binding sites	Properties	Distribution
IgG	146,000	80	13	2	Major circulating antibody; four subclasses exist: IgG_1, IgG_2, IgG_3, IgG_4; binds complement weakly; antitoxin	Extracellular fluid; blood and lymph; crosses placenta
IgM	970,000 (pentamer)	6	1.5	10	First antibody to appear after immunization; binds complement strongly	Blood and lymph; B lymphocyte surfaces (as monomer)
IgA	160,000 385,000 (secretory form)	13	3 0.05	2 4	Major secretory antibody	Secretions (saliva, colostrum, serum), cellular and blood fluids; exists as a monomer in serum and as a dimer in secretions
IgD	184,000	1	0.03	2	Minor circulating antibody; heat labile; high carbohydrate content	Blood and lymph; lymphocyte surfaces
IgE	188,000	0.002	0.00005	2	Involved in allergic reactions: contains mast cell binding fragment	Blood and lymph only

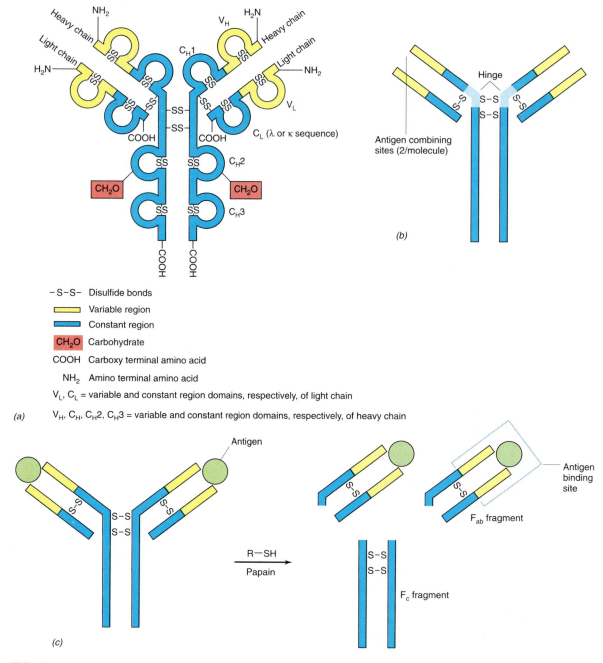

- -S-S- Disulfide bonds
- Variable region
- Constant region
- CH₂O Carbohydrate
- COOH Carboxy terminal amino acid
- NH₂ Amino terminal amino acid
- V_L, C_L = variable and constant region domains, respectively, of light chain
- V_H, C_H, C_H2, C_H3 = variable and constant region domains, respectively, of heavy chain

(a)

(b)

(c)

FIGURE 12.5 Structure of immunoglobulin G (IgG). (a) Structure showing disulfide linkages within and between chains. (b) Alternative structural diagram which deletes the intrachain disulfide bonds and carbohydrates to simplify the diagram. (c) Effect of papain treatment on immunoglobulin structure. R—SH represents one of several organic thiols which will react with selected S—S bonds of the immunoglobulin molecule.

ner in which some antigen–antibody reactions occur (see Section 12.12). The antigen binding site is found in a small region of the amino terminal portion of both the heavy and the light chains (Figure 12.5). Immunoglobulins also contain small amounts of complex carbohydrates consisting mainly of hexose and hexosamine, which are attached to portions of the heavy chain (Figure 12.5); the carbohydrate is not involved in the antigen binding site.

Although the view of the IgG molecule shown in Figure 12.5 is adequate for conveying the general structure of this molecule, since immunoglobulins are proteins they are twisted and folded in their final con-

formation and assume a complex three dimensional structure. This fact is illustrated in Figure 12.6 where a computer generated model and an electron micrograph of a single IgG molecule are shown. Note that although the basic Y-shaped structure is apparent in both views of the IgG molecule shown in Figure 12.6, the twisting and folding characteristic of the secondary and tertiary structure of proteins is readily apparent in Figure 12.6a. We will see later how the higher order structure of the polypeptide chains that combine to form an intact immunoglobulin expose a unique binding site on each antibody molecule and how this binding site is ultimately responsible for the specificity of antigen–antibody reactions.

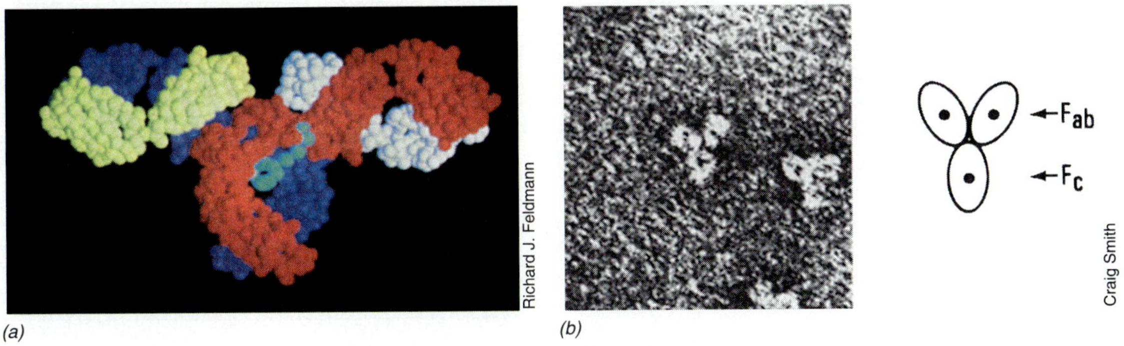

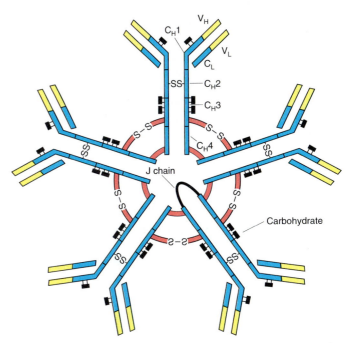

FIGURE 12.6 Three-dimensional views of immunoglobulin IgG. (a) Computer-generated model of IgG, the heavy chains are shown in red and dark blue, the light chains in green and light blue. The light blue near the F_c portion of the two heavy chains represents attached carbohydrate. (b) Electron micrograph of a single molecule of IgG.

Light chains of IgG

Each IgG light chain contains two amino acid regions, the *variable region* and the *constant region.* The sequence of amino acids in a major portion of the light chains of immunoglobulins of the class IgG are frequently identical, even in IgGs directed against completely different antigenic determinants. This is because the amino acid sequence in the carboxy terminal half of the light chain constitutes one of two specific sequences, referred to as the *lambda* (λ) sequence or the *kappa* (κ) sequence. One IgG molecule will have either two λ chains or two κ chains but never one of each (Figure 12.5a). By contrast, light chain *variable* regions always differ in amino acid sequence from one IgG molecule to the next unless both molecules are produced by the same cell or clone of cells.

Heavy chains of IgG

Each IgG heavy chain contains four amino acid regions, one variable and three constant regions (referred to as variable and constant "domains," respectively). Analogous to the situation that exists in the light chain, all immunoglobulins of the IgG class have a portion of their heavy chain (the carboxy terminal region) in which the amino acid sequence is identical (C_H1, C_H2, and C_H3, see Figure 12.5) from one IgG molecule to another. In addition, each heavy chain has a region in the amino terminal end (antigen binding site V_H, Figure 12.5) where considerable amino acid sequence variation occurs from one IgG to the next. The great specificity of a given antibody molecule for a particular antigen thus lies in the unique three-dimensional structure of the antigen binding site dictated by the amino acid sequence in the variable regions of the heavy and light chains (see Figure 12.6).

Other classes of immunoglobulins

How do immunoglobulins of the other classes differ from IgG? The heavy-chain *constant* region of a given immunoglobulin molecule defines its class and can have one of five amino acid sequences: gamma, alpha, mu, delta, or epsilon. These sequences constitute the carboxy terminal three-fourths of the heavy chains of immunoglobulins of the class IgG, IgA, IgM, IgD, or IgE, respectively (Figure 12.5a). Each antibody of the class IgM, for example, will contain a stretch of amino acids in its heavy chain which constitutes the mu sequence. If two immunoglobulins of *different classes* react with the same antigenic determinant, then the variable regions of their heavy and light chains would be identical, but their class-determining sequences, specific to their *heavy* chains, would be different. It is not unusual in a typical immune response to observe the production of antibodies of two different classes to the same antigenic determinant.

The structure of **immunoglobulin M** (IgM) is shown in Figure 12.7. It is usually found as an aggregate of five immunoglobulin molecules attached as shown in Figure 12.7 by short peptides called *J chains* and accounts for 5–10 percent of the total serum immunoglobulins. Each heavy chain of IgM contains an extra constant region domain (C_H4), and IgM in gen-

FIGURE 12.7 Structure of IgM, a large immunoglobulin with five molecules (a pentamer). Note that each heavy chain has four rather than three constant regions and that the five molecules are themselves held together by disulfide bonds. Also note that 10 antigen binding sites are available.

eral is very carbohydrate rich. IgM is the first class of immunoglobulin made in a typical immune response to a bacterial infection, but immunoglobulins of this class are generally of low affinity. The latter problem is compensated to some degree, however, by the high *valency* of the pentameric IgM molecule; 10 binding sites are available for interaction with antigen (Table 12.2 and Figure 12.7). The term *avidity* is used to describe the *strength* of multivalent antigen-binding molecules; thus, IgM is said to be of *low* affinity but *high* avidity.

Immunoglobulin A (IgA) is of interest because it is present in body secretions. IgA is the dominant antibody in all fluids bathing organs and systems in contact with the outside world: saliva, tears, breast milk and colostrum, gastrointestinal secretions, and mucus secretions of the respiratory and genitourinary tracts. IgA is also present in serum, but the IgA of secretions has an altered molecular structure, consisting of a dimeric immunoglobulin attached to a protein high in carbohydrate, called the *secretory piece*, and a J chain peptide (Figure 12.8). These proteins help hold the dimeric immunoglobulin molecule together and possibly aid in the passage of IgA into secretions.

Immunoglobulin E (IgE) is found in serum in extremely small amounts (in an average human about 1 of every 50,000 serum immunoglobulin molecules is IgE). Despite its low concentration, it is important since immediate-type hypersensitivities (allergies; see Section 12.14) are mediated by IgE. The molecular weight of an IgE molecule is significantly higher than that of other immunoglobulins (Table 12.2) because, like IgM, it contains an additional constant region. This region is thought to function in binding IgE to mast cell surfaces (see Section 12.14), an important prerequisite for certain allergic reactions.

Immunoglobulin D (IgD) is also present in low concentrations and its function in the overall immune response is unclear. Experiments have shown that IgD is abundant on the surfaces of antibody-producing cells (B lymphocytes, see next section) and IgD may play a role along with monomeric IgM in binding antigen as a signal to the lymphocyte to begin antibody production.

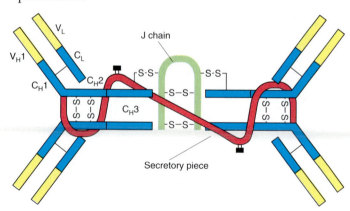

FIGURE 12.8 Structure of human secretory immunoglobulin A (IgA). Although most IgA is monomeric, the secretory form of IgA is a dimeric molecule.

> Immunoglobulins (antibodies) are protein molecules that are found in serum and certain other body fluids. The most common immunoglobulin, called IgG, contains certain polypeptide regions, called constant regions, that are the same from one IgG to another, and other regions, called variable, that are different. The antigen-binding site of IgG resides in the variable region. Since IgG has two antigen binding sites, it is called bivalent and each IgG molecule can combine with two antigen molecules. Other classes of immunoglobulins include IgM, IgA, IgE, and IgD.

12.4   T Cell Receptors

As we have seen, T cells play a variety of complex roles in the overall immune response. Although T cells do not *produce* antibody, they do *recognize* antigen, and this recognition process is due to specific antigen receptor molecules located on T cell surfaces called **T cell receptors (TCR)**. T cell receptors have immunoglobulin-like antigen specificity, but unlike most immunoglobulins, they are integrated into the T cell membrane. Both CD4 and CD8 lymphocytes have TCRs on their cell surface. Thus, it appears that all T cells are equipped to recognize specific antigens by virtue of their TCRs. How is this accomplished at the molecular level?

Structure of the T cell antigen receptor

T cells are highly specific. For example, a different T_C cell must be available to distinguish each virus or virus strain which infects the body (see Section 15.4). Thus, TCRs must be highly specific; their specificity must at least equal that of antibodies. Although TCRs are not antibody molecules, they resemble antibody molecules in many ways. Indeed, TCRs and antibodies have much in common, and in evolutionary terms are clearly related molecules (see Section 12.9).

The TCR consists of two disulfide-linked peptides, called alpha (α) and beta (β) on 85 percent of mature T cells. The remaining T cells have analogous polypeptide chains designated gamma (γ) and delta (δ). Both types of TCR contain regions of highly variable amino acid sequences. These *variable* regions, or domains (which are the amino terminal portion of each of the two polypeptide chains), combine to form the actual antigen-binding site, much as the combination of the heavy and light chain variable regions forms the antigen-binding site in immunoglobulins. Each TCR polypeptide also contains one constant domain, in which the amino acid sequence is invariant from chain to chain within a corresponding type. The variable and constant domains are roughly equivalent in size to the immunoglobulin domains (see Section 12.3). Thus, all α chain constant region sequences from any TCR are the same. A comparison of a TCR with an analogous antibody molecule is shown in Figure 12.9.

T cell receptors are integral cell membrane proteins. Studies of polypeptide folding have shown that the secondary structure of the TCR *constant* region

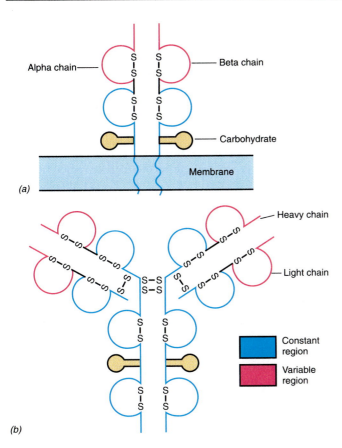

FIGURE 12.9 Structural comparison of (a) the T cell receptor with (b) an immunoglobulin. Note the presence of variable regions.

consists of relatively rigid β sheets (see Section 2.8), while the *variable* region (antigen-binding site) includes α helices, with a much greater capacity to flex and fold (the transmembrane "anchors" are also α helices). The α helix form of secondary structure is ideal for generating highly specific binding sites. The constant and variable regions of antibody molecules also show β sheet and α helical folding, respectively.

> T cells recognize antigen because of the presence on their surfaces of specific T cell receptors that have immunoglobulin-like specificity. TCRs are integral cell membrane proteins embedded in the cytoplasmic membrane in such a way that the variable portion is exposed to the exterior, where it can combine with the antigen.

12.5   Histocompatibility Antigens

Antibodies recognize antigens in solution. However, although they have much in common structurally with antibodies, TCRs can only recognize an antigen that is bound to a set of *self* proteins found on the surface of normal cells. These proteins are encoded by a genetic region, present in all vertebrates, called the *major histocompatibility complex (MHC)*. MHC proteins are produced by a number of genes in this complex and are collectively called *human leukocyte antigens*, or *HLAs*. MHC molecules were first discovered as the

major target molecules for transplantation rejection; if tissues from one animal, a donor, are immunologically rejected when transplanted to another animal, a recipient, then their MHC antigens are different. We now know that they act as antigen-presenting molecules and interact specifically with both the antigen and the TCR. Thus, MHC proteins are a *third* set of antigen-binding molecules and play an integral role in the immune response. We focus first on the structure and genetics of the MHC proteins in humans. Then we investigate the function of the MHC molecules.

Structure of major human histocompatibility complex proteins

The MHC genes encode two distinct types of proteins known as *class I* and *class II*. The entire MHC has several genes that encode class I proteins and additional genes that encode class II proteins. These class I and class II molecules are cell surface proteins and are intimately involved in immune recognition events. Class I MHC proteins are found on the surfaces of *all* nucleated cells. Class II MHC proteins are found only on the surface of B lymphocytes, macrophages, and other APCs. The reasons for this differential distribution will become apparent when we discuss the function of these molecules.

Class I MHC proteins consist of two polypeptides (Figure 12.10a), one of which is encoded in the MHC gene region. The other class I polypeptide, called *β-2 microglobulin* ($\beta_2 m$), is encoded by a non-MHC gene. The MHC-encoded polypeptide is a glycoprotein

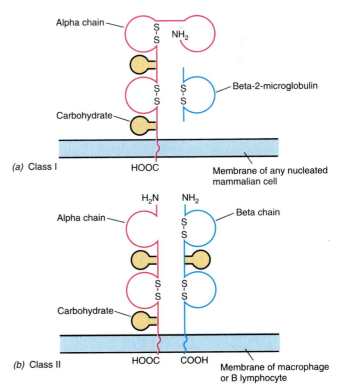

FIGURE 12.10 Structure of major histocompatibility (MHC) proteins. (a) Class I type. (b) Class II type. Note that class I molecules are present on the surfaces of all nucleated cells and that class II molecules are only present on certain specialized APC cells.

firmly anchored in the cell membrane; β_2m is a small polypeptide noncovalently linked to class I (Figure 12.10*a*). Class II MHC proteins consist of two noncovalently linked glycosylated polypeptides, called α and β. Like class I molecules, these polypeptides are embedded in the cytoplasmic membrane and project outward from the cell surface (Figure 12.10*b*).

Detailed studies of the crystal structure of class I MHC molecules have revealed a distinctive shape that suggests how these proteins interact with antigens and the cells of the immune system. The α chain of a class I molecule forms three separate extracellular domains. Between two of these domains, a large groove exists, and it is within this groove that the MHC molecule binds foreign antigens (Figure 12.11). As we will see, the binding of a foreign antigen in the groove of an MHC protein and the interaction of this foreign antigen–MHC complex with a TCR triggers the immune response.

MHC proteins are not structurally identical *within* a given species. Different individuals within a species often show subtle differences in the amino acid sequence of their MHC molecules. In humans, for example, there may be 100 different MHC proteins for each gene position, but each individual has only two proteins expressed; one protein is of paternal origin and one is of maternal origin. These limited sequence variations, referred to as *polymorphisms*, occur in both classes of histocompatibility antigens. Major structural changes in MHC proteins always exist *among* different animal species. These differences, both within and among species, are the major reason why tissues transplanted from one individual to another are recognized as nonself and are usually rejected.

Functions of MHC Proteins

MHC proteins serve as molecular reference points that permit the body to distinguish self from nonself. In a normal animal, T cells, through their TCRs, are constantly interacting with proteins or other potential antigens. It is critical that T cells be able to discriminate self from nonself antigens. The T cell, through its TCR, binds to MHC molecules and can then recognize foreign antigens embedded in the MHC structure; a T cell cannot recognize a foreign antigen unless it is presented in the context of an MHC protein.

How does this happen? In general, when a foreign antigen is taken up by a host cell, the cell "processes" or degrades it. This processed antigen then becomes embedded, or bound, to the MHC protein, and the two are passed through the cytoplasmic membrane and become attached to the surface of the cell. Two distinct antigen-processing schemes are known, one for class I antigen presentation and one for class II antigen presentation (Figure 12.12). In the class I scheme (Figure 12.12*a*), antigens that are manufactured by host degradation reactions are bound by class I proteins in the endoplasmic reticulum. This method of antigen contact is very important in virus infections, where the host cell is manufacturing viral proteins. Degraded viral peptides, which are nonself, then complex with class I proteins and move to the cell surface, where they are recognized by peptide-specific T cells through the antigen- and MHC-specific TCR, with the aid of the CD8 molecule. The T cells, in turn, activate immune response effectors and destroy the virus-containing cell.

In effect, the MHC molecules act as a *platform* on which the foreign antigen is bound. For example, viral infection of an animal cell leads to the embedding of viral antigens in class I molecules on the infected cell's surface (Figure 12.12*a*). T_C cells are constantly surveying the entire cell population looking for cells that express foreign antigens. Normal, healthy cells all express class I proteins on their surface, but the class I molecules contain self peptides, which are not recognized by the T cells. However, the T cells *do* recognize the virus-infected cell because it exhibits the nonself viral antigen bound by the self class I MHC molecule (Figure 12.12*a*). Thus, the TCR on the surface of the T cells interacts with both antigen-specific (nonself) and MHC molecule-specific (self) sites.

A second antigen presentation scheme involves the class II molecule (Figure 12.12*b*). In this case, class II molecules, complete with a self peptide called *Ii*, or invariant chain, line the cell vacuoles (endosomes) that degrade antigens phagocytized by APCs. When phagocytized foreign antigens enter the endosome,

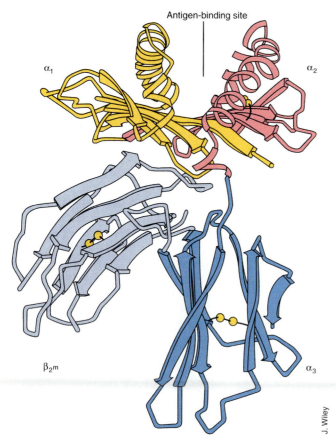

Antigen-binding site

α_1 α_2

β_2m α_3

J. Wiley

FIGURE 12.11 Three-dimensional structure of a MHC class I molecule. Note the antigen-binding site which exists as a groove between the α_1 and α_2 alpha helices of the protein. α_3 is the portion of the molecule that is nearest to the cytoplasmic membrane. β_2m, β-2 microglobulin.

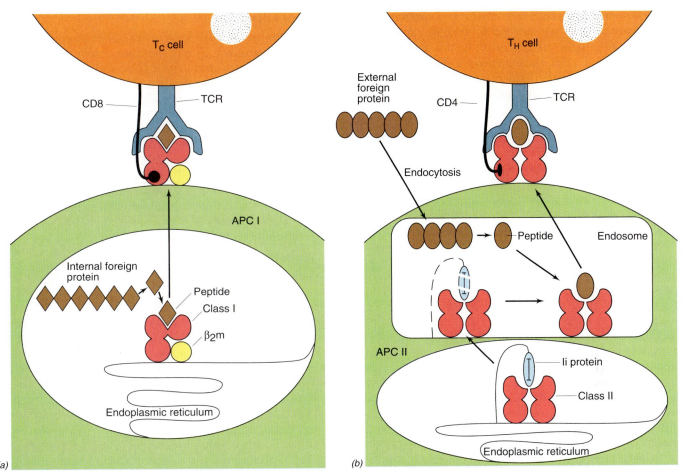

FIGURE 12.12   Antigen presentation by class I and class II MHC proteins. (a) In the class I antigen presentation pathway, the class I proteins are made and assembled in the endoplasmic reticulum, along with other proteins, including viral and tumor antigens. Some of these proteins are digested in the endoplasmic reticulum. Digested peptides then bind to class I and are expressed on the cell surface, where they interact with TCRs on the surface of T_C cells. The CD8 coreceptor on the T_C cell also engages the class I MHC, resulting in a stronger complex. The T_C cells then release lymphokines and cytotoxins, proteins that kill the target cell. Any nucleated cell can act as an APC for class I. (b) In the class II antigen presentation pathway, class II proteins are produced in the endoplasmic reticulum and are assembled with a blocking protein, Ii (invariant chain), which prevents class II from complexing with other peptides made in the endoplasmic reticulum. Class II then goes to the endosome where the Ii and foreign proteins, imported from outside the cell (by endocytosis), are digested. The class II protein then binds to the processed foreign peptides and the complex is expressed on the cell surface, where it interacts with TCRs and the CD4 coreceptor on T_H cells. The T_H cells then release lymphokines that act on B cells to cause antibody production. The APC II are phagocytes and B cells.

they are digested by proteolytic enzymes, along with the Ii. The foreign peptides are then bound by the class II molecule, and the whole complex is eventually expressed on the external cytoplasmic membrane, where it is presented to the T_H cells. The T_H cells, through the TCR and the CD4 coreceptor, then recognize the class II MHC–foreign peptide complex on the surface of the APCs. The T_H cell is activated by contact with foreign antigen and secretes molecules that stimulate antibody production by specific B cell clones.

> Proteins encoded in the major histocompatibility complex (MHC) serve as specific markers or molecular reference points that permit the T cells to distinguish self from nonself. These proteins are present on the surfaces of APCs and serve as a platform to which the foreign antigen is bound. Under these conditions, a cell containing the foreign antigen, presented to T cells in the context of an MHC protein, is recognized as nonself and becomes the target of the immune response.

12.6   Cell-Mediated Immunity

The term *cell-mediated immunity* was originally used to describe immune phenomena that did not involve the action of immunoglobulins. However, cellular immunity now must be considered in a broader context, because even the production of immunoglobulins involves the activities of many different cell types. The term *cellular immunity* is therefore now used to describe any immune response in which cells of the immune system are directly involved, but in which antibody production or activity is of minor importance.

Cellular immunity differs from antibody-mediated immunity in that the immune response cannot be transferred from animal to animal by antibodies or serum containing antibodies, but only by lymphocytes removed from the blood. The lymphocytes that function in transfer of cellular immunity are activated T cells. Most of the lymphocytes circulating in the blood are T cells; B cells are not generally circulating but instead are localized in lymphoid tissues.

The overall series of events that occur when an animal is exposed to an antigen is detailed in Figure 12.13. We now turn our attention to the activities of two subsets of T cells, the cytotoxic and delayed type hypersensitivity T cells, and natural killer cells, and discuss the role each plays in cellular immunity.

Cytotoxic T lymphocytes

In addition to macrophages, a subset of T lymphocytes, called *cytotoxic T cells* (T_C cells), also are involved in the destruction of foreign cells. As previously mentioned in connection with the major histocompatibility complex (MHC, see Section 12.5), T_C cells recognize foreign antigens embedded in MHC class I molecules. Any cell carrying a foreign antigen, such as those introduced by incompatible tissue grafts, or cells harboring viruses, can be lysed by T_C cells (recognition of viral antigens by T_C cells was discussed in Section 12.5 in connection with the T cell receptor). If tissue grafts are made between two different species, or even between two strains of the same species with insufficiently similar histocompatibility antigens, T_C cells respond by killing the foreign tissue, resulting in rejection of the graft. In humans, tissue cross matching to ensure that the major histocompatibility antigens are identical (or nearly so) in donor and recipient is now a routine clinical procedure and is required for successful skin grafting or organ transplants.

Contact between a T_C cell and the target cell is required for lysis, although the exact mechanism of lysis is not known. Presumably, lysis is the result of toxic substances released from the T_C cell. Interestingly, in contrast to the nonspecific nature of activated macrophage killing, T_C cells are only effective at lysing the specific target cells containing the foreign antigen; killing of other foreign cells occurs at much lower frequencies.

Natural killer cells

Killer cells are an additional class of lymphocytes distinct from cytotoxic T cells that play a role in destroying foreign cells. *Natural killer (NK)* cells are neither T cells nor B cells. Nevertheless, NK cells resemble T_C cells in their ability to kill foreign cells, but differ from T_C cells in that they are able to kill in the *absence* of stimulation by specific antigen. NK cells are capable of destroying malignant and virus-infected cells *in vitro* without previous exposure or contact with the foreign antigen. However, like T_C cells, NK cells must first bind to the cell containing a foreign antigen before releasing a cytotoxic factor that actually kills the cells. The receptors involved in lysis of targets by NK cells are completely unknown.

Lymphokines and cytokines

T cells, macrophages, other leukocytes, and NK cells can also act to eliminate invading microorganisms and other foreign material by releasing molecules called **lymphokines** if produced by lymphocytes or **cytokines** if produced by other cell types. These modulators of the immune response can have several functions. For example, *interleukin-1, interleukin-2, interleukin-3, interleukin-4,* and interferons are involved in the regulation of other immune cells. Interleukin-2, produced by antigen-activated T lymphocytes, primarily activates T cells, while interleukin-4, another T cell lymphokine, causes B cell proliferation and enhanced antibody synthesis. Other lymphokines, such as *chemotactic factor, macrophage activation factor,* and *migration inhibition factor,* act to modulate the function of phagocytic cells. Chemotactic factor attracts macrophages to the site of antigen concentration, macrophage activation factor stimulates macrophages to become more effective killers, and migration inhibition factor prevents the migration of macrophages away from the antigen. *Colony-stimulat-*

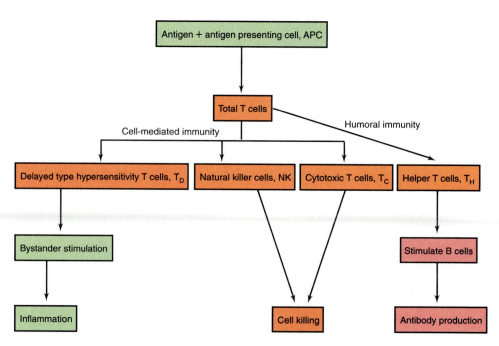

FIGURE 12.13
Overview of cell-mediated immunity and its relationship to humoral immunity.

ing factor causes phagocytic white blood cells of all types to differentiate and divide. Other cytokines serve to induce inflammatory responses or serve as proteins toxic to tumor cells (Figure 12.14).

Macrophage activation

Macrophages play a central role in both antibody-mediated and cell-mediated immunity. As illustrated in Figure 12.12, macrophages bind, process and present antigen to T cells. As phagocytic cells, however, macrophages also take up and kill certain foreign cells by themselves, and this ability can be stimulated by T cells. A key property of activated macrophages is that they can kill intracellular bacteria that would normally multiply. As we have noted (Section 11.14), some bacteria are able to survive and multiply within macrophages, whereas most bacteria taken into macrophages are killed and digested. Bacteria multiplying within macrophages include *Mycobacterium tuberculosis, M. leprae* (causal agents of tuberculosis and leprosy, respectively), *Listeria monocytogenes* (causal agent of listeriosis), and various *Brucella* species (causal agents of undulant fever and infectious abortion). Animals given a moderate dose of *M. tuberculosis* are able to overcome the infection and become immune, because of the development of a T cell-mediated immune response. Surprisingly, such immunized animals also phagocytize and kill unrelated organisms such as *Listeria*, and it can be shown that macrophages in the immunized animal have been activated so that they more readily kill the secondary invader.

It is of some interest that macrophages not only kill foreign pathogens, but are also involved in the destruction of foreign mammalian cells. This shows up in the development of transplantation immunity and is a major problem in the transplantation of tissues from one person to another. Tumor cells contain some specific antigens not seen on normal cells, and tumors function like self-inflicted transplants. There is considerable evidence that tumor cells are normally recognized as foreign and are destroyed by macrophages, NK cells, and T_C cells.

> Cellular immunity involves the formation of activated T cells that contain antigen-combining TCRs on their surfaces. A variety of activated T cells exist, each with a specialized function in the immune response. Activated T cells also serve as killer cells, destroying cancer cells and the foreign cells of transplantation. Activated T cells also produce lymphokines, which are specialized proteins that participate in immune responses. The nonantigen-specific NK lymphocytes are also important for destroying malignant and virus-infected cells.

12.7 Clonal Selection and Immune Tolerance

The previous sections dealt with specific antigen receptors involved in the immune response. We now turn to the roles these receptors play in the immune response, especially as they relate to signals controlling the activities of B and T cells.

Clonal Selection

The **clonal selection theory** states that each antigen-reactive B cell or T cell has only a single type, or specificity, of antigen-specific receptor on its surface. When stimulated by specific antigen, each cell is capable of dividing, making a copy of itself. The antigen-driven B and T cells continue to divide, while cells that are not antigen stimulated do not divide. The key feature of

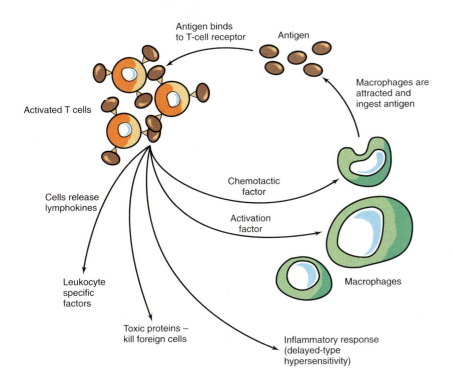

FIGURE 12.14
Lymphokine activity in cellular immunity.

clonal selection is that each cell divides and makes an exact copy of itself. Thus, one antigen-reactive cell produces multiple copies, or *clones*, of itself after antigen contact. Since each precursor cell produces its own unique antigen receptor, the resulting clones produce many copies of the same receptor (Figure 12.15).

Because of the infinite variety of antigens available, a large number of antigen-reactive cells must be available in the body, and each cell is capable of expanding into an antigen-reactive clone. However, antigen-reactive cells must avoid interactions and subsequent immune reactions with *self* antigens in the host. How does this occur? The clonal selection theory again provides an answer. Ultimately, the immune response has the ability to discriminate between foreign invaders (nonself) and host (self) tissue by deleting or inactivating self-reactive clones. This capacity for antigen-specific immune *un*responsiveness to self antigens is known as **tolerance** and occurs in both T and B cells. In Section 12.14 (Immune Diseases), we discuss the failure of tolerance and the subsequent development of autoimmunity.

Immune tolerance

The development of T cell tolerance occurs in the thymus, a walnut-sized organ that lies on top of the heart. The thymus has a central role in the maturation and development of T cells and is therefore a primary lymphoid organ. In the first T cell maturation stage, called *positive selection*, lymphocytes that will become T cells leave the bone marrow (Figure 12.16) and enter the thymus from the lymphatic ducts. After entry, some of these immature T cells interact, using their TCRs to bind to self MHC molecules on the thymus. The T cells that do not bind MHC molecules are then programmed to die, while the remaining T cells, which

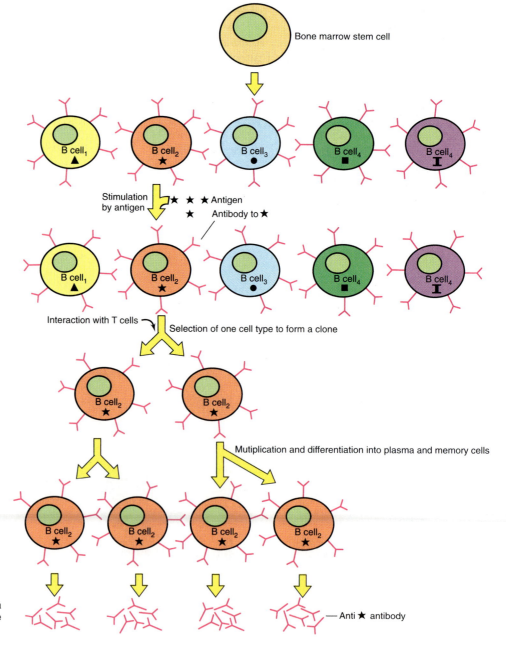

FIGURE 12.15
Clonal selection. The expansion of a particular antibody-producing cell line after stimulation with antigen.

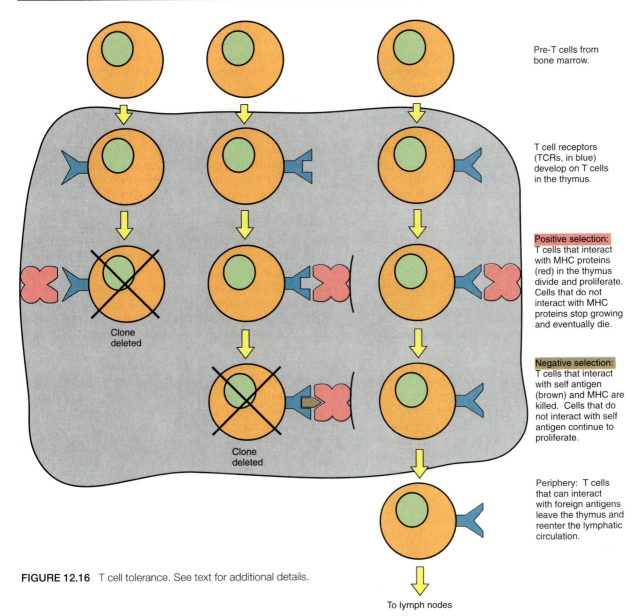

Pre-T cells from bone marrow.

T cell receptors (TCRs, in blue) develop on T cells in the thymus.

Positive selection: T cells that interact with MHC proteins (red) in the thymus divide and proliferate. Cells that do not interact with MHC proteins stop growing and eventually die.

Negative selection: T cells that interact with self antigen (brown) and MHC are killed. Cells that do not interact with self antigen continue to proliferate.

Periphery: T cells that can interact with foreign antigens leave the thymus and reenter the lymphatic circulation.

Clone deleted

Clone deleted

To lymph nodes

FIGURE 12.16   T cell tolerance. See text for additional details.

bind thymic MHC proteins, continue to mature. Thus, positive selection is a primary sorting procedure that retains T cells that can recognize self MHC proteins and deletes those T cells that cannot recognize self MHC proteins.

The second stage of T cell development is termed *negative selection*. After proliferation, positively selected T cells continue to react with MHC molecules, which (as we saw in Section 12.5) are complexed with antigen. The antigens in the thymus can be of self origin. T cells that react with self antigens are potentially dangerous because they can lead to destruction of self tissue. Therefore, these *autoreactive* T cells must be eliminated. Most autoreactive T cells bind tightly to thymus tissues and cannot leave; they remain bound to the thymus and eventually die. T cells that interact with nonself antigens (plus MHC), however, do not bind to the thymus as tightly, presumably because no nonself antigens are available to make a tight bond. These T cells leave the thymus and migrate to the spleen and lymph nodes, where they can contact for-

eign antigens and interact with other APCs and B lymphocytes (Figure 12.16).

This mechanism for inducing tolerance is called *clonal deletion*; precursors of T cell clones that are not useful, or even harmful, are deleted during development. There is also evidence for another tolerizing process. In some cases, self-reactive T cell clones somehow avoid clonal deletion, but become unresponsive to self antigen through interactions in the secondary lymphoid organs. This mechanism is known as *clonal anergy* or *clonal paralysis*. However, these clones remain in circulation and, in certain individuals, can be reactivated later in life. These reactivated self-reactive clones are often associated with autoimmune diseases (see Section 12.14).

The acquisition of immune tolerance in B cells is also necessary because antibodies produced by self-reactive B cells (**autoantibodies**) may damage host tissue (see Section 12.14). The mechanisms for achieving B cell tolerance parallel those already seen in T cell tolerance.

B cells also undergo clonal deletion; many self-reactive B cells are eliminated during development. In addition, clonal anergy also plays a role; cells that are self-reactive do not develop when exposed to antigen. Several explanations are possible for induction of B cell clonal anergy. It can result when high concentrations of self antigens are always present. These antigens constantly occupy all the B cell surface receptors, making the B cell unable to react with antigen in a normal manner. B cell anergy can also result if antigen-reactive T cells are not available (because they were eliminated during T cell maturation) to help B cells in the production of antibodies.

> The sole function of the thymus is to produce T cell clones that react only with nonself antigens. Most self-reactive T cells are clonally deleted. Other T cells as well as B cells are anergized in the secondary lymphoid organs, but occasionally became reactivated.

12.8   Mechanism of Immunoglobulin Formation

How is it possible for animals to produce such a large variety of specific proteins, the immunoglobulins, in response to invasion by foreign macromolecules, viruses, and cells? The problem of antibody formation can be divided into two separate questions: (1) How do cells of the immune system *interact* to produce antibodies when stimulated by an antigen? (2) How is antibody *diversity* generated?

The production of immunoglobulins in response to stimulation by antigen is a complex process involving T cells, B cells, and antigen-presenting cells (APCs such as macrophages), and involves the intimate interaction of the various cell surface molecules discussed previously. As we have described, APCs act nonspecifically to ingest foreign antigens. Once the antigen is processed, the APC presents it to T cell: B cell pairs. Following recognition, the T cell then releases substances that stimulate B cells to begin antibody synthesis. Although the initial interaction between antigen and the APC may be nonspecific, all further steps in the process of antibody production are *highly* specific; receptor molecules on the surfaces of both T cells (T cell receptor, see Section 12.4) and B cells (antibody molecules, see Section 12.3) help ensure that antibodies made in response to an antigen will indeed be specific for that antigen.

The genetic control of antibody production is also a highly complex process. It has been estimated that an animal is capable of making over one billion structurally distinct antibody molecules. Although at first this seems like an enormous demand on an organism's genetic coding potential, we will see that only a relatively small number of genes is required to code for this immense antibody diversity because of a phenomenon known as **gene rearrangement**. During development of lymphocytic cells in the bone marrow, gene rearrangements and deletions occur in B cells to eventually yield two complete transcriptional units, one of which codes for the synthesis of a specific heavy chain and one for a light chain of the antibody molecule. The number of possible gene rearrangements, even of a relatively small number of genes, is sufficient to account for the diversity of immunoglobulin molecules (see Section 12.9). Similar rearrangements occur during T cell development to yield the diversity of T cell receptors observed. We now detail the steps in antibody production, beginning with the injection of an antigen and ending with the production of a specific antibody that will react with that antigen.

Exposure to antigen

In considering the mechanism of antibody formation, we must first explore what happens to the antigen in the whole animal body. Antigens are carried to all parts of the body by the blood and lymph systems, which were illustrated in Figure 11.24. The main sites of antigen localization in the body are the lymph nodes, the spleen, and the liver. It has been well established that antibodies are formed in both the spleen and lymph nodes; the liver is not involved. If the antigen is injected intravenously, the spleen is the site of greatest antibody formation, whereas subcutaneous, intradermal, and intraperitoneal injections lead to antibody formation in lymph nodes. Fragments of lymph node or spleen from immunized animals can continue to produce antibody when placed in tissue culture or when injected into other, nonimmunized, animals.

Following the first introduction of an antigen, there is a lapse of time (latent period) before any antibody appears in the blood, followed by a gradual increase in *titer* (that is, concentration) and then a slow fall. This reaction to a single injection is called the **primary antibody response** (Figure 12.17). When a second exposure to antigen is made some days or weeks later, the titer rises rapidly to a maximum 10 to 100 times above the level achieved following the first exposure. This large rise in antibody titer is referred to as the **secondary antibody response** (see Figure 12.17). With time the titer slowly drops again, but later injections can bring it back up. The secondary response is the basis for the vaccination procedure known as a "booster shot," (for example, the yearly rabies shot given to domestic animals) to maintain high levels of circulating antibody specific for a certain antigen.

Antibody production

How do B cells, T cells, and APCs cooperate to produce immunoglobulins? First, antigen must be presented by APCs. While APCs for antibody production may be macrophages, the B cell itself is probably the most efficient APC. B cell APCs, through their cell surface immunoglobulin antigen receptors, are capable of binding antigen and internalizing it by a process known as *endocytosis*. The antigen is then processed (digested) in endosomes (see Figure 12.12b), and the resulting peptides are put back on the surface, bound to

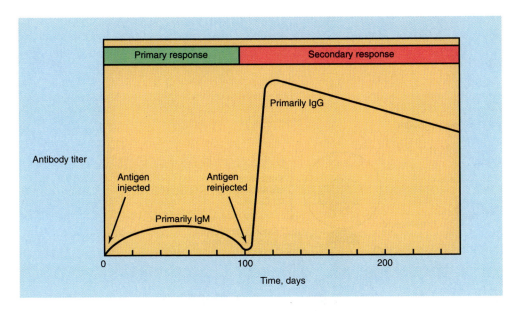

FIGURE 12.17
Primary and secondary antibody responses.

a class II MHC antigen (Figure 12.18). Macrophage APCs work essentially the same way except that macrophages have no antigen-specific receptor. However, macrophages are sticky and efficiently adhere to and phacocytize foreign antigens. The antigens are then processed and presented by class II antigens on the surface, as is the case for B cell APCs. Next, the APC presents the antigen-MHC complex to the T cell via the TCR. The class II MHC molecules on the APCs interact specifically with the CD4 molecule on T_H cells (Figure 12.12b), focusing the antigen presentation to the T_H cells. The T_H cell then triggers the B cell through this antigen-mediated contact, resulting in the release of interleukins, to divide and form many copies of itself, resulting in a B cell clone. This clone then produces only one specificity of antibody molecule.

Further differentiation of the activated B cell clone then occurs, resulting in the formation of large antibody-secreting cells called **plasma cells** and special B cells called **memory cells** (Figure 12.18). Plasma cells are relatively short-lived (less than one week) but excrete large amounts of antibody during this period. Memory cells, in contrast, are very long-lived cells, and upon reexposure to the initial stimulating antigen, they quickly transform into plasma cells and begin secreting antibody. This accounts for *immunological memory*, also known as the secondary response, and results in the rapid and more abundant antibody production observed after repeated antigenic stimulation (see Figure 12.17).

Certain antigens can stimulate low-level antibody production in the *absence* of previous T cell interactions (these are the so-called T-independent antigens). Most T-independent antigens are large polymeric molecules with repeating antigenic determinants (for example, polysaccharides). The immunoglobulins produced to T-independent antigens are usually of the IgM class and are of low affinity. In addition, B cells which respond to T-independent antigens do not have immunologic memory.

The discussion above describes the general principles of antibody production. Each antigen (strictly speaking, each antigenic determinant) will catalyze, through the action of APCs and T_H helper cells, the growth of a *different* B cell line, the B cell type in each case being genetically capable of producing antibodies that react specifically with that antigen. In this fashion the normal animal can respond to perhaps as many as one billion distinct antigens by developing a specific B cell clone in response to stimulation by antigen.

We turn now to the genetic orchestration of immune events and investigate the molecular diversity of antibodies and T cell receptors at the DNA level.

> Specific antibody production is a complex process initiated by antigen contact, through an APC, with an antigen-specific T_H cell. The T_H cell, in turn, signals an antigen-specific B cell to produce antibody. The antigen acts as a selective agent, causing proliferation of only those clones of cells that react with that particular antigen. Many of the clonal progeny remain active for years as memory cells and can be immediately stimulated after reexposure to antigen. Thus, a very focused, efficient, and specific response to the initiating antigen is generated and maintained.

12.9   Genetics and Evolution of Immunoglobulins and T Cell Receptors

Gene organization

If one B lymphocyte produces immunoglobulins of a single specificity, it should theoretically require only one gene to code for the two identical light chains and one gene for the two identical heavy chains. However, this is not the case. A single light or heavy chain is actually encoded by several genes that undergo a complex series of rearrangements as B cells mature. How does this series of events occur?

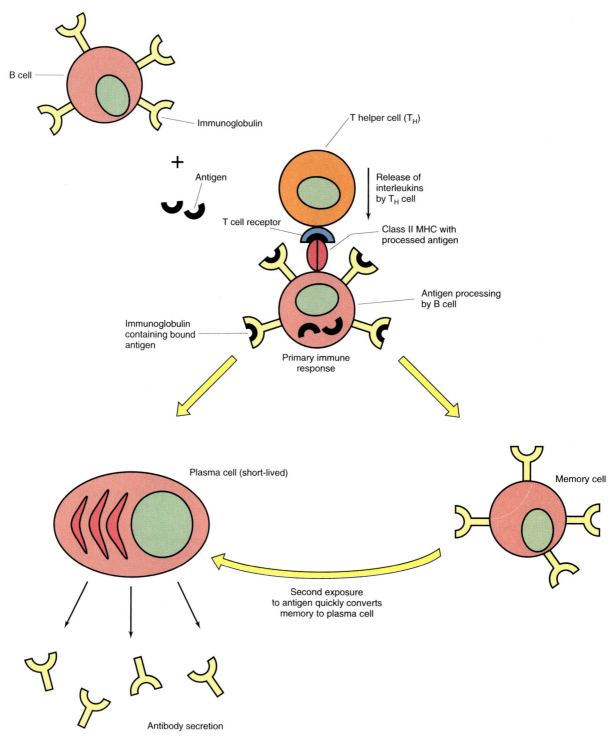

FIGURE 12.18 T cell –B cell interaction and antibody production. The B cell acts as an APC and concentrates antigen with specific immunoglobulin receptors. After processing, antigen is presented to the T_H cell by a class II MHC molecule. The T_H cell, in turn, signals the same B cell to proliferate and form plasma cells (antibody producers) or memory cells.

To understand the genetics of immunoglobulin synthesis, it is necessary to examine the structure of immunoglobulins in more detail. As illustrated in Figure 12.5, variation in amino acid sequence exists in the *variable region* of different immunoglobulins. Further, amino acid variability is especially apparent in several so-called **hypervariable regions** (Figure 12.19). It is at these hypervariable sites that combination with antigen actually occurs. Each variable region in the light and heavy chains has three hypervariable regions. A portion of the third hypervariable region on the heavy chain is coded for by a distinct gene called the D (for "diversity") gene, with the first two hypervariable sites being coded for by the variable region gene itself (Figure 12.20). In addition, at the site of joining between the diversity region and the constant region genes, there is a stretch of nucleotides, about 40 bases in length, called the J (joining) region that is coded for by a distinct gene (*J gene*). Thus, the third hypervariable region is encoded by a portion of the V gene and

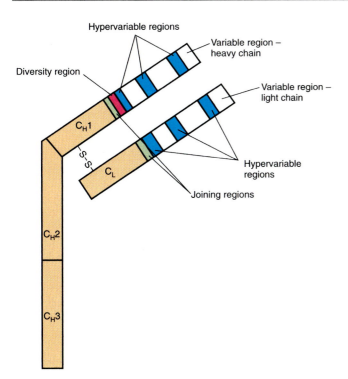

FIGURE 12.19
Detailed structure of variable regions of immunoglobulin light and heavy chains. Only one-half of a typical immunoglobulin molecule is shown. C_H, C_L, constant regions of heavy and light chains, respectively.

the entire D and J genes. Finally, the class-defining constant region of the immunoglobulin molecule is encoded by its own gene, the C (constant) gene. Light chains are encoded by their own variable region genes, joining region genes, and constant region genes, but do not contain diversity regions as in the heavy chains (Figure 12.20).

Gene rearrangements

The origin of cells of the immune system was presented in Figure 12.3. All the genetic diversity required to make antibodies exists in each lymphocyte as it is formed in the stem cells of the bone marrow. Each *immature* B cell contains about 150 light-chain variable region genes and five distinct joining sequences, while 100–200 variable region genes, 4 joining sequences, and about 50 diversity region genes exist for the heavy chains (Figure 12.20). In addition, 5 heavy-chain, constant-region genes and 2 light-chain, constant-region genes are present. The V, J, D, and C genes are not located adjacent to one another, but are separated by noncoding sequences (introns) typical of gene arrangements in eukaryotes (see Section 5.1). During maturation of B lymphocytes, genetic recombination in each B cell occurs, resulting in the construction of an *active heavy chain gene* and an *active light chain gene* that are transcribed and translated to make the heavy and light chains of the immunoglobulin molecule specific for that B cell (Figure 12.20). *Randomly selected* V, J, and D segments are fused by enzymes that delete all intervening DNA. The active gene (containing an intervening sequence between the VDJ gene segment and the constant region gene segment) is transcribed, and the resulting primary RNA transcript is spliced to yield the final mRNA (Figure 12.20).

The final light and heavy gene complement of a given B cell is a matter of *chance rearrangement,* and all possible gene combinations are equally probable. The number of possible gene combinations is such that as many as one *billion* different antibody molecules can be made. By the time each B cell has matured, however, the gene rearrangements will yield a B cell capable of making immunoglobulins with only a *single* specificity. Put another way, each B cell makes only immunoglobulins with identical variable regions.

Additional antibody diversity occurs as the result of mutations arising during formation of the active heavy chain gene. The DNA joining mechanism appears to be rather imprecise and frequently varies the site of VDJ fusions by a few nucleotides. This, of course, is sufficient to change an amino acid or two, and this apparently sloppy mechanism leads to even greater antibody diversity.

It is truly impressive that such an enormous number of molecules can be encoded by a relatively small amount of DNA. The discovery of gene rearrangements in lymphocytes has had a marked impact on our understanding of eukaryotic genetics. It had always been assumed that every cell of a multicellular organism was genetically identical (barring any somatic cell mutation, of course) and that the differences between cell types were a function of differential gene expression. This clearly is not the case with B lymphocytes, because several million (if not billion) genetically distinct lymphocytes, a product of *somatic cell* recombination, exist in vertebrates. It is not known whether lymphocytes constitute the only eukaryotic cell line in which this type of gene rearrangement occurs, but this simple, elegant mechanism demonstrates the power of gene shuffling for generating molecular diversity.

The selection of certain B cells as forerunners of an immunoglobulin-producing clone is, to a major de-

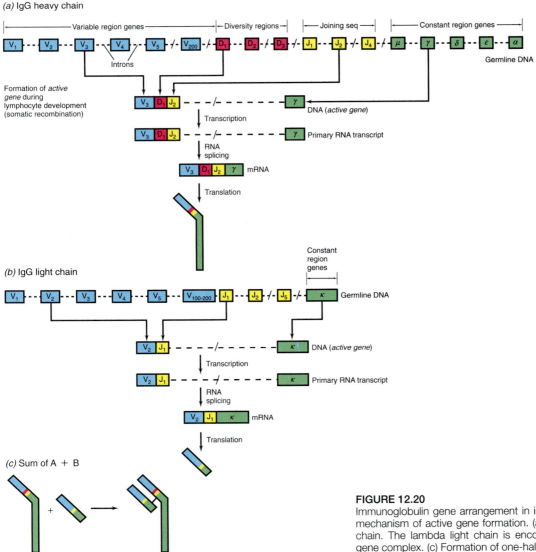

(a) IgG heavy chain

(b) IgG light chain

(c) Sum of A + B

FIGURE 12.20
Immunoglobulin gene arrangement in immature lymphocytes and the mechanism of active gene formation. (a) Heavy chain. (b) Kappa light chain. The lambda light chain is encoded in a completely separate gene complex. (c) Formation of one-half an antibody molecule.

gree, a function of the antigenic history of the animal. From the diverse pool of mature B lymphocytes discussed above, specific B cells are stimulated by antigen to expand and form clones of B cells, all of identical genetic makeup (Figure 12.15). The clonal selection theory predicts that many B cell types will remain forever "silent" within the animal's body due to lack of exposure to the correct antigen. Presumably, however, this large pool of antibody diversity remains available for future selection if immunogens are encountered.

Genetics of T cell receptors

We discussed the detailed structure of the T cell receptor in Section 12.4. Recall that TCRs recognize antigen fragments embedded in MHC molecules on the cell surface. Therefore, TCRs must be able to recognize both conserved regions of the MHC molecule and a specific antigenic determinant. How is this accounted for in genetic terms?

Like immunoglobulins, TCRs contain regions of "constant" and regions of "variable" amino acid sequences. Therefore, it is not surprising that, as for immunoglobulins, a number of genes are involved in cod-

ing for the variable regions of TCRs. In the mouse, the α-chain has about 100 variable region genes and 50 joining segments, while the β chain has 25 variable genes, 2 diversity genes, and 12 joining segments. In addition, several small additions of from 1 to 6 nucleotides can be inserted between variable, diversity, and joining gene segments in TCRs such that the number of possible sequence combinations is enormous, probably on the order of 10^{15}, which is even greater than the immunoglobulin combinations thought possible!

From the structural and genetic similarities between immunoglobulins and TCRs, it is easy to imagine that they are evolutionarily related molecules, and we consider this possibility now.

Evolution of molecules of the immune system

From an evolutionary standpoint, the three categories of immunologically specific molecules, MHC proteins, TCRs, and immunoglobulins, share a number of common structural features, and undoubtedly represent increasingly complex stages in the evolution of the immune system. Molecular cloning and sequencing of genes coding for immunoglobulins and TCRs strongly

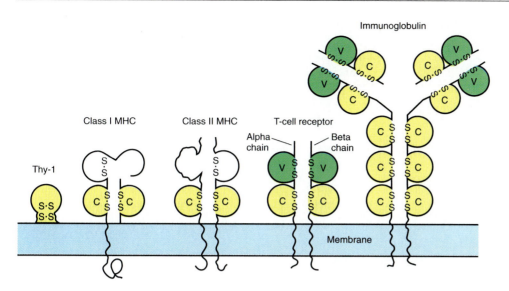

Immunoglobulin

Class I MHC

Class II MHC

T-cell receptor

Alpha chain

Beta chain

Thy-1

Membrane

FIGURE 12.21
Comparative structure of several molecules of the immune system. Note those regions of homologous or nearly homologous amino acid sequences that strongly suggest evolutionary relatedness. C = constant regions; V = variable regions.

suggests that a *family* of genes has evolved from a primordial gene or set of genes that originally encoded very simple, and presumably much less specific, immune molecules. This can be visualized by comparing the structures of the molecules discussed thus far in this chapter (Figure 12.21).

Because the molecules of the immune system are obviously related in an evolutionary sense, the term gene "superfamily" has been proposed to describe the genes which code for these molecules. This *immunoglobulin gene superfamily* presumably arose from evolutionary selection on genes coding for primitive immune molecules in ancient organisms, and this genetic progression is apparent today when the immune system of various animals is compared. For example, a comparison of immune-like reactions in vertebrates and invertebrates suggests that only portions of the gene superfamily exist in invertebrates. Invertebrates are capable of eliciting simple immune responses, such as secreting antibacterial substances and carrying out phagocytosis. Sea urchins, for example, secrete lymphokines and proteins with antibacterial activity similar to antibodies and are also capable of phagocytosis. However, invertebrates do not show T cell- or B cell-mediated immune responses. Although some higher invertebrates can actually reject grafted foreign tissues, suggesting that MHC-like molecules may be present on their cell surfaces, it is only in the vertebrates that we see the highly developed MHC-dependent immune functions, and the complex genetic system required to code for the many molecules involved.

B and T cells capable of recognizing a specific antigen arise spontaneously (by chance) as undifferentiated stem cells in the lymphoid tissue, the result of gene rearrangement. Only the variable portion of the immunoglobulin or T cell receptor is specific for the antigen. Actual contact with antigen occurs in the so-called hypervariable region of immunoglobulins or TCRs, which are highly specific in their binding properties.

12.10   The Complement System

In addition to the antigen–antibody and cell-mediated immune reactions described above, other types of reactions exist that are important in immunity. Before these reactions are discussed, however, it will be necessary to introduce another group of serum substances, called **complement**. Complement acts in concert with specific antibody to bring about several kinds of antigen–antibody reactions that would not otherwise occur. These reactions are described in Table 12.3.

Complement is not an antibody but is a series of enzymes found in blood serum that attack bacterial cells or other foreign cells, causing lysis or leakage of cellular constituents as a result of damage to the cell membrane. These enzymes, found even in unimmunized individuals, are normally inactive, but become active when an antibody-antigen reaction occurs. In fact, a major function of antibody is to recognize invading cells and activate the complement system for attack. There is considerable economy in an arrangement such as this, since a wide variety of antibodies, each specific for a single antigen, can call into action the complement enzymatic machinery; thus the body does not need separate enzymes to attack each kind of invading agent.

Some reactions in which complement participates include (1) bacterial lysis, especially in Gram-negative Bacteria, when specific antibody combines with antigen on bacterial cells in the presence of complement; (2) microbial killing, even in the absence of lysis; and (3) phagocytosis, which may not occur during infection if the invading microorganism possesses a capsule or other surface structure that prevents the phagocyte from acting. When specific antibody combines with the cell in the presence of complement, the cell is changed in such a way that phagocytosis can occur. (This process in which antibody plus complement renders a cell susceptible to phagocytosis is called **opsonization**.)

Table 12.3   Types of antigen–antibody reactions

Location of antigen	Accessory factors required	Reaction observed
Soluble	None	Precipitation
On cell or inert particle	None	Agglutination
Flagellum	None	Immobilization or agglutination
On bacterial cell	Complement	Lysis
On bacterial cell	Complement	Killing
On erythrocyte	Complement	Hemolysis
Toxin	None	Neutralization
Virus	None	Neutralization
On bacterial cell	Phagocyte, complement	Phagocytosis (opsonization)

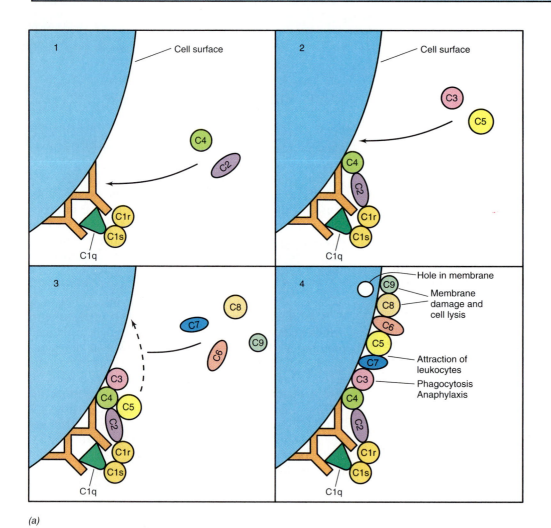

(a)

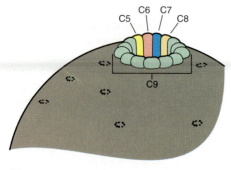

(b)

FIGURE 12.22 The complement system: (a) The sequence, orientation, and activity of the various components as they interact to lyse a cell. Panel 1, binding of the antibody recognition unit C1q, and other C1 proteins; panel 2, the C4-C2 complex; panel 3, the C4-C2-C3-C5 complex, after activation the C5 unit travels to an adjacent membrane site; panel 4, binding of C6, C7, C8, and C9, resulting in membrane damage. (b) Three-dimensional view of the hole formed by complement components C5–C9.

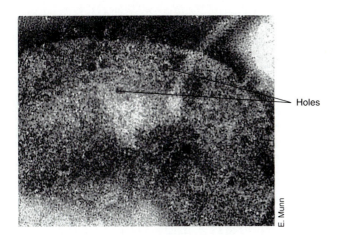

Holes

E. Munn

FIGURE 12.23
Electron micrograph of a negatively stained preparation of *Salmonella paratyphi*, showing holes created in the cell envelope as a result of reaction between cell-envelope antigens, specific antibody, and complement.

Activation of the complement system

Complement is a system of 11 proteins, designated C1, C2, C3, and so on. Activation of complement occurs only with antibodies of the IgG and IgM classes. When such antibodies combine with their respective antigens, especially on cell surfaces, the antibodies assume new conformations in their constant region domains, which allow them to fix (attach to) the ever-present complement proteins (Figure 12.22). The complement system acts in a cascade fashion, activation of one component resulting in the activation of the next. In summary, the steps are (1) binding of antibody to antigen (initiation); (2) recognition of antigen-antibody complex by C1; (3) C4-C2 binding to an adjacent membrane site; (4) activation of C3; (5) formation of the C5-C6-C7 complex, causing attraction of leukocytes; and (6) formation of the C8-C9 complex causing cell lysis.

As seen in Figure 12.22, complement reactions at the C3 level result in chemotactic attraction of phagocytes to invading agents, and to phagocytosis following opsonization. Reaction at C5 also leads to leukocyte attraction. The terminal series of reactions from C5 through C9 results in cell lysis and death. Lysis itself results from destruction of the cytoplasmic membrane, leading to the formation of holes through which cytoplasm can leak (Figure 12.23).

A variety of complement reactions

The complement system is involved in other immunological reactions besides those leading to the destruction of invading organisms. The reactions of importance here include certain aspects of the inflammatory response and the production of proteins called **anaphylatoxins** which produce symptoms resembling those of allergic responses (see Section 12.14). Anaphylatoxins also cause release of histamines following binding to tissue mast cells. Histamine induces inflammation and increases the permeability of capillaries (Section 11.15), enabling leukocytes and fluid to escape into the tissue cells. One serious type of allergic reaction, **anaphylactic shock** (see Section 12.14), may involve components of the complement system, although most such allergic responses involve the activity of immunoglobulins of the IgE class .

Complement is necessary in the bactericidal and lytic actions of antibodies against many Gram-negative Bacteria. (Interestingly, Gram-positive Bacteria are not killed by specific antibody, in either the presence or absence of complement, although Gram-positive Bacteria are opsonized.) Death of Gram-negative Bacteria probably involves antibodies against antigens on the surface of the cell; complement perhaps brings about an actual change in the cell surface, possibly by an enzyme-like reaction, after antibodies have prepared the way. No cytocidal or lytic effect is seen when cells and complement are mixed alone, whereas if antibody has bound to cells first, death or lysis occurs rapidly after complement is added.

As noted, complement is necessary for opsonization, the promotion of phagocytosis by antibodies against bacterial capsules. This does not require the whole complement pathway, but only the proteins through C3. Opsonization occurs because the binding of C3 causes the cells to adhere to phagocytes. Phagocytes have C3 receptors on their surfaces and it is presumably these receptor molecules which are involved in adherence of the C3-coated bacterial cell to the phagocyte.

12.11 Polyclonal and Monoclonal Antibody Production

In the whole animal, a typical immune response results in the production of a broad spectrum of immunoglobulin molecules of various classes, affinities, and specificities for the determinants present on the antigen. The immunoglobulins directed toward a particular determinant will represent only a portion of the total antibody pool. Each immunoglobulin would be produced by a single clone that had proliferated from antigen stimulation and an antiserum containing such a mixture of antibodies is thus referred to as a **polyclonal antiserum**. However, it is possible to produce antibodies of only a *single* specificity, derived from a single B cell clone. A variety of such monospecific antibodies, called **monoclonal antibodies**, have been generated for use in research and clinical medicine.

Over-the-Counter Immunodiagnostic Kits

A number of manufacturers now market immunodiagnostic kits for use by the general public. For example, virtually every pharmacy now carries several brands of pregnancy tests and many also stock kits to determine the onset of ovulation in women.

These kits are all based on principles involving detection of a hormone excreted in the urine which is bound by a monoclonal antibody. All pregnancy tests detect the presence of human chorionic gonadotropin (HCG). When the ovum is fertilized, it produces HCG, which functions to maintain the corpus luteum and sustain pregnancy. As pregnancy continues, the new embryo produces and releases increasing amounts of HCG. The hormone, by this time being produced in massive amounts, is released into the bloodstream, removed by the kidney, and secreted into the urine. Normal menstruation, which occurs about 10 to 14 days after ovulation if the ovum is not fertilized, does not take place, and at this time, the level of HCG present in the urine of the pregnant woman is high enough to be detected easily by these tests. Thus, detectable HCG in the urine means that the individual is pregnant. The tests are advertised to be sensitive enough to detect pregnancy on the first day of a missed period, *not* on the first day of a pregnancy.

The key to marketing these kits is their simplicity. Since most people using the tests are not trained in standard laboratory procedures, the test methods must be straightforward and the results easy to interpret. Typically, the patient must first provide a urine sample. In the easiest test, the sample urine is simply applied to a test strip. In others, the first step involves pouring the sample through a filter or immersing a "dipstick" in the urine, followed by exposing the filter or dipstick to a second solution. The results are interpreted as a simple color change: a positive test turns the strip, dipstick, or filter from white to pink (see figures) within 30 minutes or less. Each test varies in methodology, but all share certain principles. At some point, the HCG in the urine is bound to an HCG-specific monoclonal antibody that is immobilized on a solid support—the dipstick, test strip, or filter. Next, a second HCG-specific monoclonal antibody, provided in a solution for the dipstick or filter test and already soaked onto the test strip for the one-step test, reacts with the immobilized antibody–HCG complex. This second antibody is chemically linked to colloidal gold particles. If this second antibody reacts with the HCG, it becomes immobilized, and the whole complex appears colored.

These kinds of tests, all modifications of sol particle immunoassay (SPIA) tests, are also extensively used in doctor's offices to help diagnose diseases. For example, a number of available tests are specific for cell surface antigens on *Streptococcus pyogenes*, which causes strep throat. Formerly, the only way to positively identify a streptococcal infection was to culture and identify the organism, a process that involved skilled personnel and a laboratory. Because of the requirement for incubation of the culture, results were not available for at least 16 hours. Today, results are available in minutes, using material from a throat swab as the antigen source. Several test kits are also available to detect serum antibodies that react with self proteins. These kits are useful for helping diagnose autoimmune diseases such as rheumatoid arthritis.

These tests have several advantages. They are simple to perform, require no technical expertise, are relatively inexpensive, and are reasonably accurate. They also have a long shelf life and no special storage conditions. However, they also have disadvantages. They are qualitative tests, which means that they can only be used when the outcome is absolute. For example, one is either pregnant or not pregnant. However, quantitative measures of antigen levels might be desirable in some cases, such as in the tests that identify autoimmune antibodies. In addition, many manufacturers do not provide adequate positive and negative controls to ensure that the test works properly. Thus, results should always be interpreted with caution, and negative or positive results may be in error due to improper test procedures or faulty test reagents.

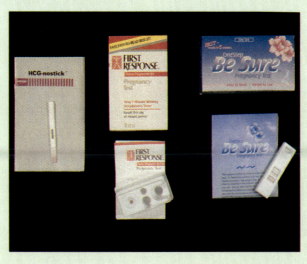

Table 12.4 Characteristics of monoclonal and polyclonal antibody production	
Polyclonal	**Monoclonal**
Contains antibodies recognizing many determinants on an antigen	Contains an antibody recognizing only a single determinant
Various classes of antibodies are present (IgG, IgM, etc.)	Single class of antibody produced
To make a specific system, highly purified antigen is necessary	Can make a specific system using an impure antigen
Reproducibility and standardization difficult	Highly reproducible

Table 12.4 compares the properties of antibodies prepared against an antigen in the usual way—by preparing a polyclonal antiserum—with monoclonal antibodies.

Each B cell clone produces a monoclonal antibody—it is genetically programmed to do so (see Section 12.9). But we consider here how to isolate and grow a single B cell clone for experimental production of monoclonal antibodies. Normal lymphocytes are not easy to grow and maintain in cell culture, hence this approach has not been found useful. However, B lymphocytes can be fused with myeloma (tumor) cells to form hybrid cell lines that will grow in culture and retain the ability to produce antibodies. This technique of B cell/myeloma cell fusion is known as the **hybridoma technique** and is summarized in Figure 12.24.

How are hybridomas capable of producing monoclonal antibodies made? First, a mouse is immunized with the antigen of interest and a period of weeks allowed for specific B cell clones to proliferate and begin producing antibody by the normal sequence of events. Then, spleen tissue, rich in B lymphocytes, is removed from the mouse and the B cells are fused with myeloma cells (Figure 12.25). Although true fusions represent only a small fraction of the total cell population remaining in the mixture, addition of the compounds hypoxanthine, aminopterin, and thymidine to the medium (the so-called HAT medium) strongly selects for *fused* hybrids. This occurs because unfused myeloma cells are unable to use the metabolites hypoxanthine and thymidine to bypass a metabolic block caused by aminopterin, a cell poison. By contrast, fused hybrid cells are able to use hypoxanthine and thymidine and thus grow normally in HAT medium. Unfused B cells die off in a week or two because they are unable to grow in culture for more than a few cell divisions. Following fusion, the most important job is to identify the clones of interest; that is, if individual fused cells are placed in the wells of microtiter plates and allowed to grow and produce antibody (Figure 12.25), which ones are producing the antibodies of interest?

A variety of assay techniques (see Sections 12.12 and 12.13) can be used to identify clones producing monoclonal antibodies. In a typical experiment, several distinct clones are isolated, each making a monoclonal antibody to a different determinant on the antigen. Once the clones of interest are identified they can be grown in cell culture, or they can be injected into mice and be perpetuated as a mouse myeloma tumor. As a mouse tumor line, hybridomas secrete large

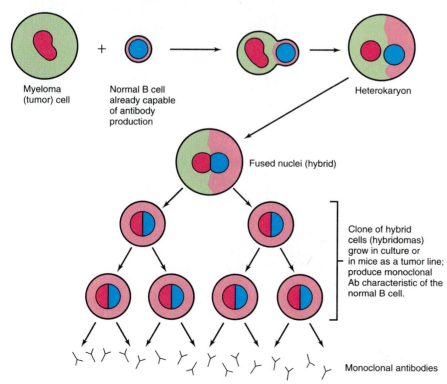

Myeloma (tumor) cell

Normal B cell already capable of antibody production

Heterokaryon

Fused nuclei (hybrid)

Clone of hybrid cells (hybridomas) grow in culture or in mice as a tumor line; produce monoclonal Ab characteristic of the normal B cell.

Monoclonal antibodies

FIGURE 12.24
Basic theory behind monoclonal antibody formation.

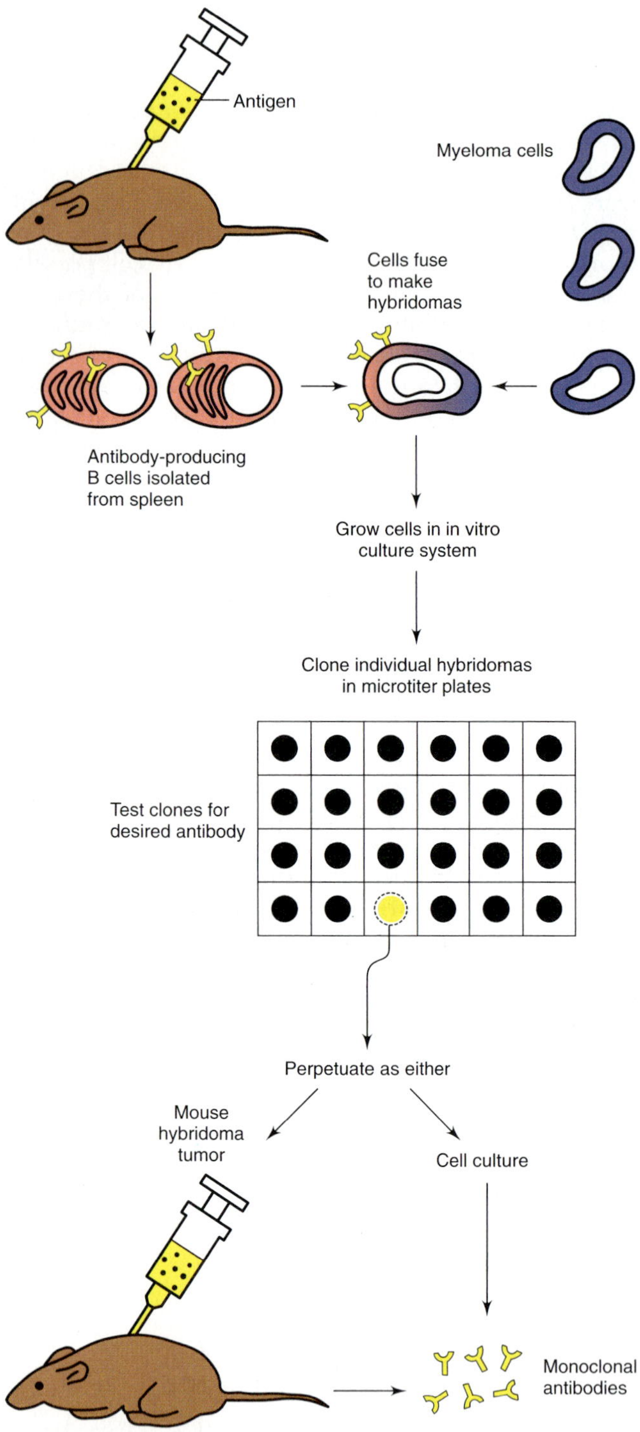

Antigen

Myeloma cells

Cells fuse to make hybridomas

Antibody-producing B cells isolated from spleen

Grow cells in in vitro culture system

Clone individual hybridomas in microtiter plates

Test clones for desired antibody

Perpetuate as either

Mouse hybridoma tumor

Cell culture

Monoclonal antibodies

FIGURE 12.25 Production of monoclonal antibodies.

amounts of monoclonal antibody (in mice over 10 milligrams of pure monoclonal antibody can be obtained per milliliter of mouse peritoneal fluid). Specific hybridomas of interest can also be stored frozen and later thawed and injected back into mice or grown in tissue culture when more of a specific monoclonal antibody is needed.

Monoclonal antibodies are extremely useful reagents for research and medical science. For research purposes, monoclonal antibodies directed against specific cell markers, for example, the CD4 and CD8 lymphocyte markers (see Section 12.2), can be used to *separate* mixtures of these cells from one an-

other. By attaching a fluorescent dye to the monoclonal antibody, those cells to which the monoclonal antibody binds will now fluoresce (Figure 12.26). Fluorescing cells can be separated from their nonfluorescing counterparts with a fluorescence activated cell sorter (FACS), an instrument employing a laser beam to place an electrical charge on fluorescing molecules (in this case fluorescing cells). Following laser exposure, the application of an electrical field to the cell mixture will orient fluorescing and nonfluorescing cells to opposite ends of the electrical field where they can be collected in separate receptacles. FACS machines have given the immunologist the ability to separate complex mixtures of immune cells into highly enriched populations of one type or another for use in immunological research.

Other research applications of monoclonal antibodies include studies of the active site of enzymes or the combining site of antibodies themselves by observing the effect on an enzyme's activity or an antibody's binding ability when treated with monoclonal antibodies to specific determinants on the molecule. Monoclonal antibodies show great promise in the immunological typing of bacteria and in the identification of cells containing foreign surface antigens (for

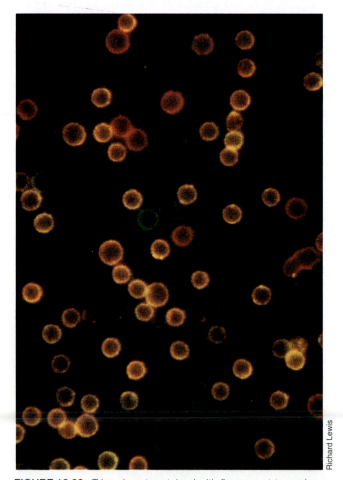

Richard Lewis

FIGURE 12.26 T lymphocytes stained with fluorescent-tagged monoclonal antibodies to specific surface markers. Yellow-green cells are cytotoxic (CD8) T cells, red cells are T helper (CD4) cells. The different-colored cells can be separated from one another by a fluorescence-activated cell sorter to yield homogeneous populations of different cell types.

example, a virus infected cell). Monoclonal antibodies have also been used in genetic engineering for measuring levels of gene products not detectable by other methods.

Of considerable importance are the applications of monoclonal antibodies to clinical diagnostics and medical therapeutics. We discuss the use of monoclonal antibodies in clinical diagnostics in Section 13.6. In therapeutics, perhaps the boldest application of monoclonal antibodies may be in the detection and treatment of human malignancies. Malignant cells contain a variety of surface antigens not expressed on the surfaces of normal cells. These include differentiation antigens expressed during fetal development but not in the normal adult, and unique glycolipid cell surface antigens (the latter were originally detected by monoclonal antibodies prepared against random tumor cell surface antigens). Because monoclonal antibodies prepared against cancer-related antigenic determinants should be able to distinguish between normal and malignant cells, monoclonal antibodies may serve as vehicles for toxins or radioisotopes to malignant cells. Such specificity would greatly improve cancer chemotherapy, because damage to normal cells by the toxic agents necessary to kill cancerous cells is one of the major problems with conventional cancer chemotherapy.

Monoclonal antibodies have also been prepared that can distinguish human transplantation antigens, thus improving the specificity of the tissue matching process critical for successful transplantations. Monoclonal antibodies also show great promise for increasing the specificity of a variety of conventional clinical tests including blood typing, rheumatoid factor determination, and even pregnancy determination; the latter test employs monoclonal antibodies prepared against specific hormones associated with the pregnant state (see Over-the-Counter box). Other monoclonals are being developed for *in vivo* diagnostics. For example, radioactively tagged monoclonal antibodies have been used experimentally to detect exposed myosin in heart muscle; myosin is a major protein in heart muscle cells that normally only becomes detectable following damage to the heart muscle. A myosin-specific monoclonal antibody would therefore be useful for diagnosing the extent of heart damage following a heart attack.

As with any clinical reagent, the greater the specificity the more useful it is in clinical diagnostics. Monoclonal antibodies now offer the physician and clinical microbiologist the ultimate in immunological specificity. (See Urine Testing for Drug Abuse box for the use of monoclonal antibodies for detecting illicit drugs.)

The hybridoma technique permits the selection and perpetuation of specific B cell clones by fusion of a B cell to a tumor cell. Propagation of these clones results in production of large quantities of monospecific antibody, which are reactive against a single antigenic determinant. Monoclonal antibodies are useful for diagnostic tests and immune therapy.

12.12   Antigen–Antibody Reactions

Antigen–antibody reactions are most easily studied *in vitro* using preparations of antigens and antisera. The study of antigen–antibody reactions *in vitro* is called **serology** and is especially important in clinical diagnostic microbiology. A variety of different kinds of serological reactions can be observed (Table 12.3), depending on the natures of the antigen and antibody and on the conditions chosen for reaction. Serology has many ramifications, only a few of which will be discussed here.

Regardless of the final visible or otherwise measurable **result**, the principal of all antigen–antibody reactions lies in the *specific* combination of determinants on the antigen with the *variable* region of the antibody molecule (Figure 12.27). The combining site of an antibody molecule measures about 2 by 3 nm and is capable of binding to a small number of amino acids, about 10–15, of a protein antigen. Recall that the recognition of antigen by antibody is ultimately governed by the higher order structure of the polypeptide chains in the antibody molecule. The folding itself is dictated by the primary structure—the amino acid sequence—which in turn is controlled by the genetic constitution of the B cell. The end result of this complex process is a specific antibody molecule which combines with a specific antigenic determinant as shown in Figure 12.27. This combination serves to distort the antigen sufficiently to either reduce or even totally eliminate its biological activity.

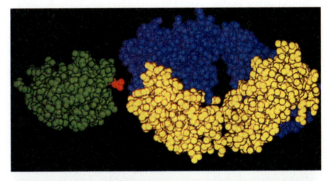

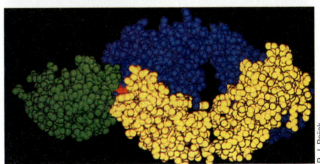

FIGURE 12.27 Structure of the combining site of an antigen and immunoglobulin. The antigen (lysozyme) is in green. The variable region of the immunoglobulin heavy chain is shown in blue, the light chain in yellow. The amino acid shown in red is a glutamine residue of lysozyme. The glutamine residue fits into a pocket on the immunoglobulin molecule, but the basis of antigen–antibody recognition involves contacts made between several other amino acids on both the immunoglobulin and antigen as well.

R. J. Poljak

Urine Testing for Drug Abuse

Many government agencies and private employers carry out drug testing programs in an effort to control drug abuse in the workplace. If a person uses a drug, metabolites of the drug are excreted into the urine. Testing urine for the presence of such metabolites will thus permit detection of drug use. Such drug testing is an example of how science can be applied to legal matters in the discipline called forensic science. Drugs that are commonly tested include cannabinoids (found in marijuana), cocaine, phencyclidine, amphetamines, propoxyphene, benzodiazepine, opiates, steroids, and barbiturates.

Because of the minute quantities of drugs or drug metabolites that occur in the urine, extremely sensitive detection methods are needed, but these methods must also be extremely specific. Immunological procedures are among the most sensitive and specific methods known for testing urine. Two types of immunoassays are usually employed for drug testing, radioimmunoassay (RIA) and enzyme-linked immunosorbent assay (ELISA). For a drug immunoassay, an antibody must first be prepared against the drug or drug metabolite to be assayed. The antibody thus specifically recognizes an antigenic determinant of the drug. The assay is based on the principle of *competition* between a labeled antigen present in the drug-testing system and an unlabeled antigen (the drug or drug metabolite in the urine) for binding sites on the specific antibody. In both the RIA and ELISA tests, the *greater* the amount of drug in the urine, the *less* labelled drug will be bound.

In RIA drug testing, known amounts of radiolabeled drug are added to a urine sample together with known amounts of an antibody specific for the drug being tested. Following standard RIA procedures, the presence of the drug is measured by determining the radioactivity of the antibody bound to a solid substrate. As the concentration of drug in the urine goes up, the radioactivity bound goes down. Concentrations of drug as low as 1–5 μg/ml can be detected and the time to complete an assay is from 1 to 5 hours. Although the RIA test is extremely sensitive, it has the disadvantage that a radioactive material must be used, and the reagents and instrumentation are costly. However, the RIA test is very suitable for large-scale screening programs, since automatic pipetting and counting procedures can be used.

In the ELISA drug testing procedure, the antigen (drug) is covalently linked to an enzyme, commonly glucose-6-phosphate dehydrogenase, whose activity can be detected by a simple and sensitive colorimetric assay. The assay is set up such that if the enzyme, via its linked drug ligand, becomes associated with the antibody, it loses its enzymatic activity, but if it remains free it can react with its substrate. Urine is mixed with a reagent consisting of antibody to the drug, the glucose-6-phosphate dehydrogenase–drug derivative, and glucose-6-phosphate. If drug is present in the urine, it competes with the enzyme/drug derivative for specific antibody and more enzyme is then free to react with its substrate. Thus, the more

drug present in the subject's urine, the more intense the color. The ELISA procedure has the advantage that the analysis time is short, and the color change can be detected simply. Also, potentially hazardous radioactive materials are not required. However, the ELISA procedure is sometimes less sensitive than the RIA procedure. The ELISA procedure is especially valuable where a small number of samples are to be analyzed at the site where the urine is collected (for instance, away from a laboratory). In some cases, the reagents have been incorporated into paper strips, providing means for urine testing that can be used even by those lacking laboratory experience (see photo). For large-scale testing, robotic devices have been manufactured capable of processing up to 18,000 urine samples per hour with each sample identified by a bar code identification sticker.

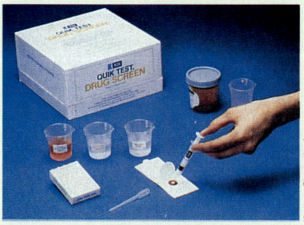

Keystone Diagnostics, Inc.

Either polyclonal or monoclonal antibodies can be used for urine testing. Monoclonal antibodies have the advantage of better specificity, but may not necessarily have as high an affinity for the drug as a selected polyclonal antibody. The selection of an antibody will depend on the specificity and sensitivity needed, the cost of the product, and the ease of use.

One problem with the use of any of these immunological methods for urine testing is cross reaction to other chemicals that might be innocently present in the urine. For instance, the pain killer ibuprofen will cross react with marijuana metabolites in some tests, and the antihistamine drug diphenhydramine can cross react in the ELISA test for methadone. Because of such false-positive reactions, it is recommended that positive tests be confirmed by independent tests, such as the use of gas chromatography or high-performance liquid chromatography. Highly sensitive combined mass spectrometer/gas chromatographic methods are also available and have proven especially valuable for identifying exotic drugs or drugs that elicit a poor immune response.

Immunological tests for drugs in the urine have provided a new and potentially widespread technique for controlling the use of illegal drugs in the workplace. Urine drug testing is another example of the power and utility of immunology in modern society.

Some typical antigen–antibody reactions are illustrated in Figure 12.28.

Neutralization

Neutralization of microbial toxins by specific antibody can occur when toxin molecules and antibody molecules directed against the toxin combine in such a way that the active portion of the toxin that is responsible for cell damage is blocked (Figure 12.28a). Neutralization reactions of this type are known for a wide variety of exotoxins, including most of those listed in Table 11.3. An antiserum containing an antibody that neutralizes a toxin is referred to as an **antitoxin**. Reactions of a similar nature also occur between viruses and their specific antibodies. For instance, antibodies directed against the protein coats of viruses may pre-

vent the adsorption of the viruses to host cells. This is one means by which a viral vaccine can act—if the virus is unable to enter host cells it cannot initiate infection.

Precipitation

Since antibody molecules generally have two combining sites (that is, they are bivalent; see Figure 12.5), it is possible for each site to combine with a separate antigen molecule, and if the antigen also has more than one combining site, a **precipitate** may develop consisting of aggregates of antibody and antigen molecules (Figure 12.28b). Because they are easily observed *in vitro*, precipitation reactions are very useful serological tests, especially in the quantitative measurement of antibody concentrations. Precipitation occurs maxi-

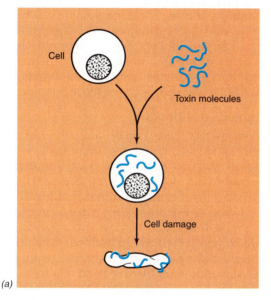

(a)

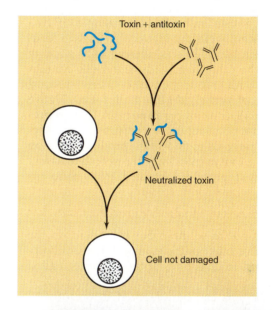

(b)

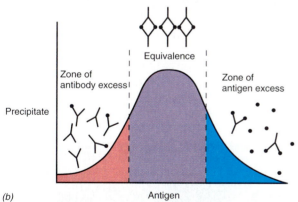

(d)

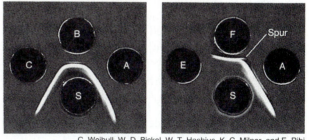

(c)

C. Weibull, W. D. Bickel, W. T. Hashius, K. C. Milner, and E. Ribi

Blood Type	Percent of population (United States)	Serum	
		Anti A	**Anti B**
Type O	47	No aggl.	No aggl.
Type A	42	Aggl.	No aggl.
Type B	8	No aggl.	Aggl.
Type AB	3	Aggl.	Aggl.

FIGURE 12.28 Some antigen–antibody reactions. (a) Neutralization of toxin by antitoxin. (b) Precipitation reaction between soluble antigen and antibody. The graph shows the extent of precipitation as a function of antigen and antibody concentration. (c) Precipitation in an agar gel, a process known as immunodiffusion. Wells in agar labeled S contained antibodies to cells of *Proteus mirabilis*. Wells labeled A, B, and C contained extracts of *P. mirabilis* cells. Left, a line of identity is observed. Right, antigen E does not react and antigen F shows a line of partial identity. (d) Agglutination. Typing of human blood in the ABO system. Left, typical agglutination pattern. Right, table of expected results with different blood types.

mally only when there are optimal proportions of the two reacting substances since, when either reactant occurs in excess, the formation of large antigen-antibody aggregates is not possible (Figure 12.28b).

Precipitation reactions carried out in agar gels, referred to as *immunodiffusion*, have proved to be of considerable utility in the study of the specificity of antigen-antibody reactions. Both antigen and antibody diffuse out from separate wells cut in the agar gel and precipitation bands form in the region where antibody and antigen meet in optimal proportions (Figure 12.28c). The shapes of the precipitation bands and the distance from the wells are characteristic for the reacting substances, and it is possible to determine whether two antigens reacting against antibodies in an antiserum are identical or not by observing the bands formed when the two antigens are placed in adjacent wells near the antiserum well. For example, if two antigens in adjacent wells are identical, they will form a fused precipitin band. This is referred to as a *line of identity*. If, on the other hand, adjacent wells contain one antigen in common but one well contains a second antigen, a *line of partial identity* will form (Figure 12.28c). The small protruding precipitin line (representing a reaction between the antiserum and the second antigen) is referred to as a *spur*. Immunodiffusion is often used as a tool in biochemical research to assess the relatedness of proteins obtained from different sources.

Agglutination

If an antigen is not in solution but is present on the surface of a cell or other particle, an antigen–antibody reaction can lead to clumping of the particles, called **agglutination**.

Soluble antigens can be adsorbed to or coupled chemically to cells or other particulate structures such as latex beads and they can then be detected by agglutination reactions, the cell or particle serving only as an inert carrier. This greatly increases the ability to detect the presence of antibodies against soluble antigens, since agglutination is much more sensitive than precipitation.

Agglutination of red blood cells is referred to as *hemagglutination*. Red blood cells contain a variety of antigens, and people differ in the antigens present on their red blood cells. The classification of blood antigens is called *blood typing* and is a good example of an agglutination reaction (Figure 12.28d). The principle of blood typing is that antibodies, specific for particular erythrocyte surface antigens, cause red blood cells to visibly clump. The human red cell antigens called A, B, and D are most commonly tested for. Humans also contain antibodies to certain blood group antigens. Type A humans contain antibodies to group B antigens, while Type B individuals have antibodies to group A antigens. Type AB individuals contain neither A nor B antibodies while Type O's have both. Antibodies against A and B antigens are called *natural antibodies*, since they appear to be produced even in the absence of apparent antigenic stimulation.

The reason for blood typing before blood transfusion is to prevent the red blood cell agglutination that would occur if blood containing a particular antigen were transfused into a person containing an antibody against this antigen. If agglutination occurred, the clumps of cells could lodge in blood vessels or arteries and block the flow of blood, causing serious illness or death; if the antibodies caused hemolysis through the action of complement, anemia could result.

The presence or absence of the A and B antigens or antibodies is the basis for the ABO blood grouping assay (Figure 12.28d). A small drop of blood is placed on a microscope slide containing antiserum prepared against either the A or B antigens of human erythrocytes (Figure 12.28d). These commercially available high titer antisera are obtained from human donors who have been immunized to these antigens by natural or artificial means. The agglutination reaction observed in the ABO blood grouping test is rapid and easy to interpret (Figure 12.28d), hence this simple assay is used routinely to group human blood. At least 15 different blood group systems have been described in humans, the ABO system being one of the most important.

Fluorescent antibodies

Antibody molecules can be made fluorescent by covalently attaching them to fluorescent organic compounds such as rhodamine B or fluorescein isothiocyanate. This does not alter the specificity of the antibody significantly, but makes it possible to detect the antibody bound to cell or tissue surface antigens by use of the fluorescence microscope (Figure 12.29). Cells to which fluorescent antibodies have bound emit a bright fluorescent color, usually green, red, or yellow, depending on the dye used. Fluorescent antibodies have been of considerable utility in diagnostic microbiology since they permit the study of immunological reactions on single cells. The fluorescent-antibody technique is very useful in microbial ecology as one of the few methods for identifying microbial cells directly in the natural environments in which they live (see Section 17.4).

Two distinct fluorescent antibody procedures, the **direct** and the **indirect** staining methods, are used

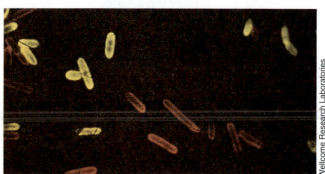

FIGURE 12.29 Fluorescent antibody reactions. Cells of *Clostridium septicum* were treated with antibody conjugated with fluorescein isothiocyanate, which fluoresces green. Cells of *Clostridium chauvei* were stained with antibody conjugated with rhodamine B, which fluoresces red.

Wellcome Research Laboratories

(Figure 12.30). In the direct method, the antibody against the organism itself is fluorescent. In the indirect method, the presence of a nonfluorescent antibody on the surface of the cell is detected by the use of a fluorescent antibody directed against the nonfluorescent antibody (Figure 12.30). This is done by immunizing one animal species, say a goat, with antibodies from a second species, say a rabbit, and then conjugating the fluorescent dye to the goat antibodies. The fluorescent goat anti-rabbit antibodies can then be used to detect the presence of rabbit immunoglobulin previously bound to cells. One of the advantages of the indirect staining method is that it eliminates the need to make a fluorescent antibody for each antigen of interest.

Metal-conjugated antibodies: immunoelectron microscopy

Antibodies to which heavy metals have been chemically conjugated can be used to locate antigens in cells by electron microscopy. This is possible because heavy metals scatter the electron beam of the electron microscope. This technique, called *immunoelectron microscopy*, is used primarily in research where there is a need to determine where a specific antigen (usually a protein) is localized in a particular region of the cell (Figure 12.31). Cells, following chemical fixation and other preparations necessary for observation by the electron microscope, are treated with antibodies containing a heavy metal, usually gold or platinum. These electron-dense metals will scatter light and thus the presence of bound antibody can be detected by black spots in the preparation (Figure 12.31).

In immunoelectron microscopy, although the cell is dead and chemically fixed, its antigens retain sufficient specificity and antibodies still react with little

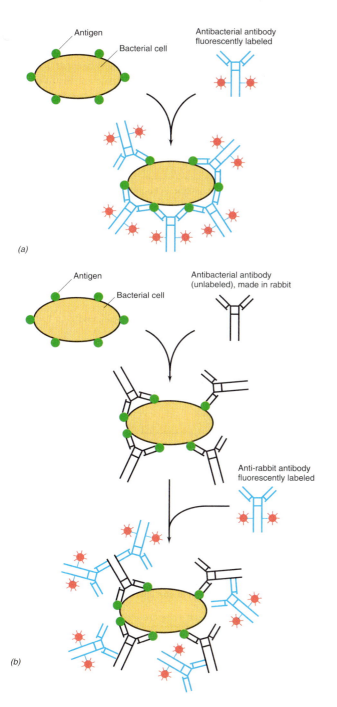

FIGURE 12.30 Methods of using fluorescent antibodies to detect bacterial surface antigens. (a) Direct staining method. (b) Indirect staining method.

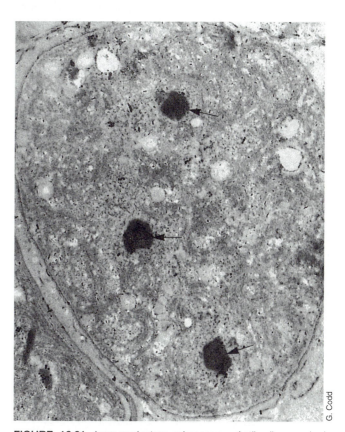

FIGURE 12.31 Immunoelectron microscopy. Antibodies made in rabbits to the enzyme ribulose-1,5-bisphosphate carboxylase from the cyanobacterium *Chlorogloeopsis fritschii* were added to thin sections of *C. fritschii* and the preparation treated with goat anti-rabbit IgG conjugated to 20 nm colloidal gold particles. The concentration of the particles around large inclusions called carboxysomes (arrows) indicate that these are sites of large amounts of the enzyme.

nonspecific cross-reaction. Immunoelectron microscopy has been used extensively to pinpoint the location of enzymes in cells, especially those suspected to be associated with the cytoplasmic membrane or some other internal structure (Figure 12.31).

> A number of different antigen–antibody reactions have been detected. The specific type of reaction depends on the location and condition of the antigen and the presence of accessory factors. In neutralization reactions, the antibody combines with a toxin or virus coat and prevents its action. In precipitation, a soluble antigen aggregates with a bivalent immunoglobulin to form a precipitate. In agglutination, the bivalent immunoglobulin interacts with an antigen present on the surface of a cell or particle, causing the formation of large cellular or particulate clumps. Fluorescent antibodies are immunoglobins to which fluorescent dyes have been artificially coupled; their interaction with antigen can be detected using a fluorescence microscope. Heavy metals can be attached to antibodies and the complexes visualized in the electron microscope.

12.13   ELISA and Radioimmunoassay

The specificity of antibodies are such that the limiting factor in most of the immunological reactions discussed thus far is not *specificity*, but *sensitivity*. Because of their exquisite sensitivity, radioimmunoassay (RIA) and enzyme-linked immunosorbent assay (ELISA) are two widely used immunological techniques. These methods employ radioisotopes and enzymes, respectively, to detect antibody molecules. Because radioactivity and the products of certain enzymatic reactions can be measured in very small amounts, the attachment of radioactive or enzyme ligands to antibody molecules serves to decrease the amount of antigen–antibody complex required in order to detect that a reaction has occurred. This increased sensitivity has been extremely helpful in clinical diagnostics and research, and has opened the door to the development of a variety of new immunological tests, previously impossible because the methods available were not sufficiently sensitive (see boxes on Urine Testing for Drug Abuse and Over-the-Counter Immunodiagnostic Kits).

ELISA

The covalent attachment of enzymes to antibody molecules creates an immunological tool possessing both high specificity and high sensitivity. The technique, called **ELISA** (for *e*nzyme-*l*inked *i*mmuno*s*orbent *a*ssay), makes use of antibodies to which enzymes have been covalently bound such that the enzyme's catalytic properties and the antibody's specificity are unaltered. Typical linked enzymes include peroxidase, alkaline phosphatase, and β-galactosidase, all of which catalyze reactions whose products are colored and can be measured in very low amounts.

Two basic ELISA methodologies have been developed, one for detecting antigen (*direct* ELISA) and the other for antibodies (*indirect* ELISA). For detecting antigens such as virus particles from a blood or fecal

sample, the direct ELISA method is used. In this procedure the antigen is "trapped" between two layers of antibodies (Figure 12.32). The specimen is added to the wells of a microtiter plate (see Microtiter box) previously coated with antibodies specific for the antigen to be detected. If the antigen (virus particle) is present in the sample it will be trapped by the antigen binding sites on the antibodies. After washing unbound material away, a second antibody containing a conjugated enzyme is added. The second antibody is also specific for the antigen so it will bind to any remaining exposed determinants. Following a wash, the enzyme activity of the bound material in each microtiter well is determined by adding the substrate of the enzyme. The color formed is proportional to the amount of antigen originally present (Figure 12.32).

To detect *antibodies* in human serum, an indirect ELISA is used (Figure 12.33). The microtiter wells in this case are coated with antigen and a serum sample

1. Antibody to antigen is bound to microtiter well

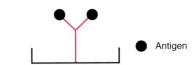

2. Sample added as source of antigen; wash

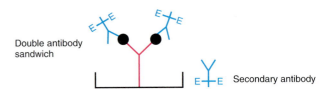

3. Antibody labeled with enzyme (E) (secondary antibody) added. Enzyme-labeled antibody binds to antigen; wash

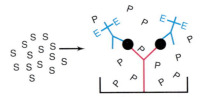

4. Substrate (S) for enzyme is added. Colored product of enzyme formed (P); color is proportional to amount of antigen

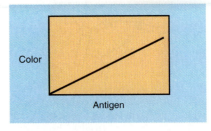

FIGURE 12.32 Enzyme-linked immunosorbent assay (ELISA). Direct ELISA for detecting antigen.

is added. If antibodies to the antigen are present in the serum, they will bind to the antigen in the wells. Following a wash, a second antibody is added as the "detecting" antibody. The second antibody is a rabbit or goat anti-human IgG antibody containing a conjugated enzyme. Following the addition of enzyme substrate, a color is formed, and the amount of circulating human antibody to the specific antigen is quantitated from the intensity of the color reaction (Figure 12.33).

Since ELISAs require very little in the way of expensive equipment, generate no radioactive waste, and are highly sensitive, they are widely used in clinical laboratories. Direct ELISAs have been developed for detecting hormones, drugs, viruses, and a variety of other substances or disease agents in human blood and other tissues. Indirect ELISAs have been developed to detect serum antibodies to *Treponema pallidum* (the causative agent of syphilis), *Vibrio cholerae* (cholera), *Salmonella* species, German measles virus, and HIV (the virus which causes acquired immune deficiency syndrome, AIDS). New ELISAs are being developed and marketed constantly. The specificity, rapidity, and inexpensive nature of ELISAs have insured their prominence in clinical and veterinary medicine. Several additional clinical examples of ELISA technology are given in Section 13.7.

Radioimmunoassay

Radioimmunoassay (*RIA*) employs radioisotopes instead of enzymes as antibody conjugates. The isotope

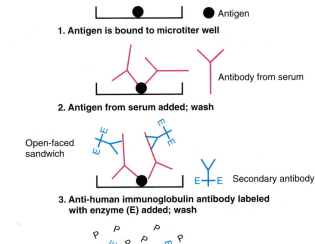

1. **Antigen is bound to microtiter well**

Antibody from serum

2. **Antigen from serum added; wash**

Open-faced sandwich

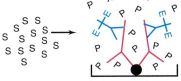

Secondary antibody

3. **Anti-human immunoglobulin antibody labeled with enzyme (E) added; wash**

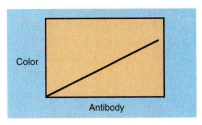

4. **Add enzyme substrate (S). Color formed (P) is proportional to amount of antibody in serum**

Color

Antibody

FIGURE 12.33   Indirect ELISA for detecting specific antibodies.

The Microtiter Plate and Immunology

A number of immunoassays have been developed that require that either the antigen or the antibody be bound to a solid support. A variety of solid phase carriers such as latex beads or plastic tubes have been used, but the plastic, disposable *microtiter plate* has proved the most successful solid support in modern immunoassays. Techniques such as the ELISA test and radioimmunoassay were designed to be used with microtiter plates.

Microtiter plates are small plastic trays containing a number of wells. The plates are made of polystyrene

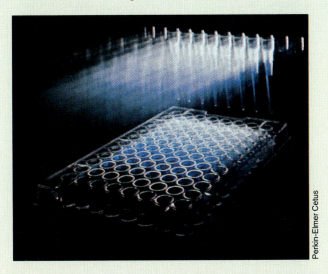

Perkin-Elmer Cetus

or polypropylene, both of which are transparent plastics. Proteins, either antigens or antibodies, adsorb to the plastic surface of the microtiter plate as a result of interactions between hydrophobic regions of the protein and the nonpolar plastic surface. Once bound, the proteins are not easily removed by washing or other manipulations and thus bound proteins can be treated with various reagents and washed repeatedly without being removed from the plate surface. The standard microtiter plate contains 96 wells, which allows for several replicates of each sample to be run. For quantitation, several dilutions are prepared. Although only small amounts of protein can be adsorbed to each well, the sensitivity of ELISAs or radioimmunoassays are such that only small amounts of antigen or antibody are needed anyway.

Automatic machines are available for adding reagents and preparing dilutions. Depending upon whether the microtiter plate assay is an ELISA or a radioimmunoassay, special machines are available to read the adsorbance of a colored product (ELISA) or amount of radioactivity (radioimmunoassay) in each of the wells of the plate. Such automation allows these immunoassays to be performed both rapidly and routinely at relatively low cost. The microtiter plate has therefore helped to revolutionize research and diagnostic immunoassays. ELISAs in particular have become part of the everyday activities of the clinical laboratory (see Section 13.7).

Iodine-125 is the most commonly used detecting system, as proteins can be readily iodinated without disrupting their specificity. RIA is used clinically to measure rare serum proteins such as human growth hormone, glucagon, vasopressin, testosterone, and insulin present in humans in extremely small amounts (Figure 12.34) and also in some urine tests for drug abuse (see Urine Testing box). In most cases a *direct* RIA is employed. The direct assay is a two-step procedure. First, radioactive antigen-specific antibodies are added to a series of microtiter wells containing known concentrations of pure antigen (such as a hormone) which is first bound to the wells. The radioactivity in each of these standard wells is then measured. Next, the antigen sample from a patient is allowed to bind to another well and radioactive antibodies are added and measured, as before. The amount of radioactivity bound by the patient sample is then compared to a standard plot generated from the binding data obtained using the pure antigen, and the concentration of antigen in the patient serum is interpolated from the standard plot (Figure 12.34). RIA has the advantage of rapidity and exquisite sensitivity, but concern over the disposal of radioactive wastes limits the use of RIA to applications where ELISA technology is not sufficiently sensitive.

An immunoglobulin can be linked to a specific enzyme and the reaction of the enzyme can be used to detect the presence of the immunoglobulin. This technique, called enzyme-linked immunosorbent assay (ELISA), finds wide use in research and diagnostic medicine. Another procedure for increasing the sensitivity of detection of an immune reaction is radioimmunoassay, a procedure in which the immunoglobulin is made radioactive. Radioimmunoassay is used primarily in research but permits the assay of tiny amounts of rare serum proteins such as hormones.

12.14   Immune Diseases

Immune diseases occur when the immune system reacts in a way that results in host damage. In this section, we first deal with **hypersensitivity**, the overreaction of the cell-mediated immune response to cause *delayed-type hypersensitivity* or the antibody-mediated immune response to cause *immediate hypersensitivity*. We next focus on the increasingly important **autoimmune diseases**, which result from a breakdown of tolerance (see Section 12.7) and inappropriate reactions to self antigens. Finally, we look at a group of toxins called **superantigens**, which are produced by bacteria and result in massive immune cell stimulation and host damage.

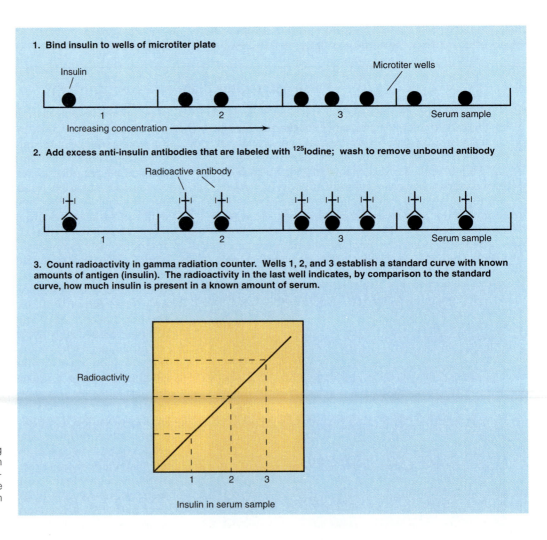

FIGURE 12.34
Radioimmunoassay (RIA). Using RIA to detect insulin levels in human serum. Following establishment of a standard curve, the insulin concentration in a serum sample can be estimated.

Delayed-type hypersensitivity

This form of hypersensitivity is cell-mediated and involves the activities of a special subset of T cells called T delayed-type hypersensitivity (T_D) cells, with symptoms appearing several hours to a few days after exposure to the eliciting agent. Delayed-type hypersensitivity reactions occur against invasion by certain microorganisms and as a result of skin contact with certain sensitizing chemicals. The latter phenomenon, known as **contact dermatitis**, is responsible for the majority of common allergic skin disorders in humans, including those to poison ivy, cosmetics, and certain drugs or chemicals. Shortly after exposure to the agent the skin feels itchy at the site of contact, and within several hours reddening and swelling appear, indicative of a general inflammatory response. Macrophages and white blood cells are attracted to the site of contact by lymphokines secreted by T_D cells, and localized tissue destruction occurs due to the lytic and phagocytic activities of these cells.

A good example of a delayed-type hypersensitivity is the development of immunity to the causal agent of tuberculosis, *Mycobacterium tuberculosis* (Figure 12.35). This cellular immune response was first discovered by Robert Koch during his classic work on tuberculosis and has been widely studied. Antigens derived from the bacterium, when injected subcutaneously into an animal previously immunized with the same antigen, elicit a characteristic skin reaction which develops only after a period of 24 to 48 hours. (In contrast, skin reactions to antibody-mediated responses, as seen in conventional allergic reactions discussed below, develop almost immediately after antigen injection.) In the region of the injected antigen, T_D cells become stimulated by the antigen and release lymphokines, which attract large numbers of macrophages. The macrophages are responsible for the ingestion and digestion of the invading antigen. The characteristic skin reaction seen at the site of injection is a result of an inflammatory response arising as a result of the release of lymphokines by activated T cells.

This skin response serves as the basis for the **tuberculin test** for determining prior exposure to *M. tuberculosis* (Figure 12.35).

A number of microbial infections elicit delayed-type hypersensitivity reactions. In addition to tuberculosis, these include leprosy, brucellosis, psittacosis (all caused by Bacteria), mumps (caused by a virus), and coccidioidomycosis, histoplasmosis, and blastomycosis (caused by fungi). In all of these cellular immune reactions, characteristic skin reactions are elicited upon injection of antigens derived from the pathogens, and these skin reactions can be used in diagnosis of prior exposure to the pathogen.

Immediate-type hypersensitivity

Another form of hypersensitivity is known in which an immune response occurs more quickly than in the delayed-type response. This form of hypersensitivity is referred to as **immediate-type hypersensitivity** and is mediated by *antibodies* instead of activated T cells. Immediate-type hypersensitivities commonly involve **allergies** and, depending on the individual and the antigen, may cause mild or extremely severe, even life-threatening reactions. Antigens that cause these hypersensitivities are known as *allergens*.

From 10 to 20 percent of the human population suffer from immediate-type hypersensitivities, involving allergic (*anaphylactic*) reactions to specific allergens such as pollens, animal dander, and a variety of other agents (Table 12.5). In a typical anaphylactic reaction an allergen will elicit (upon first exposure) the production of immunoglobulins of the class IgE. Instead of circulating like immunoglobulins of the class IgG or IgM, IgE molecules tend to become attached via their cell binding fragment to the surfaces of mast cells and basophils (Figure 12.36). **Mast cells** are nonmotile connective tissue cells found adjacent to capillaries throughout the body, and **basophils** are motile white blood cells (leukocytes) that make up about 1 percent of the total leukocyte population. Upon subsequent exposure to antigen, the IgE molecules attached to these cells bind the antigen. This triggers the release of several allergic mediators from mast cells and basophils. An antigen must bridge at least two IgE molecules on the cell to initiate release of these active substances.

The primary chemical mediators released from mast cells and basophils are **histamine** and **serotonin** (both are modified amino acids), but most other medi-

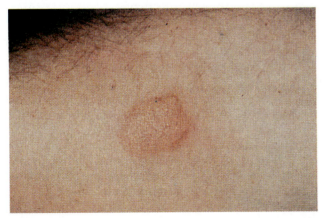

FIGURE 12.35 Cell-mediated immunity. A positive tuberculin test, typical for delayed hypersensitivity, and the result of the action of T_D effector cells.

Table 12.5   Common immediate-type hypersensitivity allergens
Pollen and fungal spores (hay fever)
Insect venoms
Penicillin and other drugs
Certain foods
Animal dander
Mites in house dust

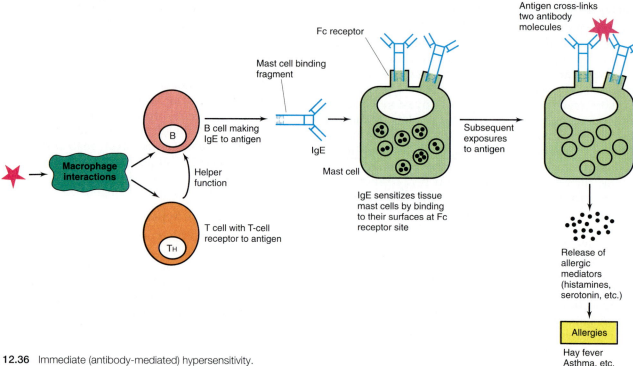

FIGURE 12.36 Immediate (antibody-mediated) hypersensitivity.

ators characterized are small peptides. The release of histamine and serotonin causes dilation of blood vessels and contraction of smooth muscle which initiate the typical symptoms of the immediate-type allergic response. These symptoms include, among others, difficulty in breathing, flushed skin, copious mucus production, sneezing, and itchy, watery eyes. In general, the symptoms are relatively short-lived, but once initially sensitized by an allergen, an individual can respond repeatedly upon subsequent exposure to the antigen. Depending on the individual, the magnitude of the anaphylactic reactions may vary from mild symptoms (or none), to such severe symptoms that the individual goes into **anaphylactic shock**. In humans, the latter is characterized by severe respiratory distress, capillary dilation (causing a sharp drop in blood pressure), and flushing and itching. If severe cases of anaphylactic shock are not treated immediately with large doses of adrenalin to counter smooth muscle contraction and promote breathing, death can occur.

Autoimmune diseases

We saw in Section 12.7 how T and B cells that were destined to react with self antigens are eliminated or anergized during the process of lymphocyte maturation. However, in some individuals, the anergized T and B lymphocytes become reactivated. This resurgence of self-reactive clones leads to immunological disorders referred to as *autoimmune diseases*. Most are slow, progressive disorders in which the function of some specific organ or set of tissues becomes increasingly poorer with time (Table 12.6). Depending on the specific disorder, autoimmunity may involve autoantibodies or a cellular immune response to self constituents. Certain autoimmune diseases are highly

organ specific. For example, in *Hashimoto's disease*, autoantibodies are made against thyroglobulin, the major iodine-containing protein in the thyroid. In *juvenile diabetes*, autoantibodies against the insulin-producing cells, the islets of Langerhans, are observed. Such antibodies may cause the disease by destroying the cells, or they may be the result of antigens released from islet cells damaged by other mechanisms. *Systemic lupus erythematosis (SLE)* involves a large-scale production of autoantibodies against many self constituents, including DNA. Organ-specific autoimmune diseases are sometimes more easily controlled clinically because the product of organ function, such as thyroxin in hypothyroidism or insulin in diabetes, can often be supplied in pure form from another source. More generalized syndromes such as SLE can only be controlled by immunosuppressive therapy, but this approach is not without risk because of the increased chance of opportunistic infections.

Evidence is accumulating that heredity has an important influence on the incidence, type, and severity of autoimmune diseases. An inherited tendency to develop certain autoimmune diseases is known to exist; many autoimmune diseases have a strong correlation with the presence or absence of certain MHC antigens (see Section 12.5). Studies of model autoimmune diseases in mice support such a genetic link, but the conditions necessary for developing autoimmunity are also known to be dependent on other factors, including hormone levels and the presence of infectious agents such as pathogenic bacteria and viruses.

Superantigens

In Sections 11.8 and 11.9, we discussed the effects of bacterial exotoxins and enterotoxins on the host. As we

Table 12.6   Some autoimmune diseases of humans

Disease	Organ or area affected	Mechanism
Juvenile diabetes	Pancreas	Autoantibodies against surface and cytoplasmic antigens of islets of Langerhans
Myasthenia gravis	Skeletal muscle	Autoantibodies against acetylcholine receptors on skeletal muscle
Goodpasture's syndrome	Kidney	Autoantibodies against basement membrane of kidney glomeruli
Rheumatoid arthritis	Cartilage	Autoantibodies against self IgG antibodies, leading to cartilage breakdown
Hashimoto's disease (hypothyroidism)	Thyroid	Autoantibodies to thyroid surface antigens
Male infertility (some cases)	Sperm cells	Autoantibodies agglutinate host sperm cells
Pernicious anemia	Intrinsic factor	Autoantibodies prevent absorption of vitamin B_{12}
Systemic lupus erythematosis	DNA, cardiolipin, nucleoprotein, blood clotting factors	Massive autoantibody response to various cellular constituents
Addison's disease	Adrenal glands	Autoantibodies to adrenal cell antigens
Allergic encephalomyelitis	Brain	Cell-mediated response against brain tissue
Multiple sclerosis	Brain	Cell-mediated and autoantibody response against central nervous system

noted, not all exotoxins or even enterotoxins work by the method described for cholera toxin (see Figure 11.19). In fact, a completely different mechanism of action is used by a family of exotoxins known as **superantigens**. These toxins are produced by the staphylococci and streptococci. They include *toxic shock syndrome toxin*, *staphylococcal enterotoxin*, and *exfoliating toxin* (see Table 11.3), and all have a toxic mechanism dependent on the immune system. These antigens all bind to the β chain of the T cell receptor (see Section 12.4) at a site that is outside the normal antigen-binding site. In some cases, this binding may stimulate greater than 10 percent of the normal T cells in the affected individual. Such massive stimulation results in simultaneous participation of all of these T cells in an overwhelming cell-mediated response characterized by systemic inflammatory effects, sometimes resulting in generalized shock (such as toxic shock syndrome). This way of stimulating the host immune system to instigate host damage may be a common pathogenic strategy for the staphylococci and streptococci.

12.15   Immunization against Infectious Diseases

Although many aspects of immunological reactions do not concern immunity to infectious disease, the major role of antibodies in the body is in protecting the animal from the consequences of infection. The importance of antibodies in disease resistance is shown dramatically in individuals with the genetic disorder **agammaglobulinemia**, in whom antibodies are not produced because their B cells are defective. Such individuals are unusually sensitive to infectious diseases, especially those involving bacterial infections,

and in the days before antibiotic therapy, few of them survived infancy. The general lack of an antibody response is also observed in those suffering from acquired immune deficiency syndrome (AIDS). However, in this case the problem is not due to defective B cells. Instead, AIDS patients suffer from a virtually total cessation of CD4 (primarily T_H) cell activities (see Section 15.7). Although T cells themselves do not make antibodies, the crucial importance of T cells in the production of antibodies is clearly evident in AIDS patients: the inability to mount an antibody response leads to the eventual death of AIDS patients from infectious diseases (see Section 15.7).

The induction of specific immunity to infectious diseases provided one of the first real triumphs of the scientific method in medicine, and was one of the outstanding contributions of microbiology to the treatment and prevention of infectious diseases. An animal or human may be brought into a state of immunity to a disease in either of two distinct ways. (1) The individual may be given injections of an antigen that is known to induce formation of antibodies, a type of immunity known as **active immunity** since the individual in question produced the antibodies itself. (2) Alternatively, the individual may receive injections of an antiserum that was derived from another individual, who had previously formed antibodies against the antigen in question. The second type is called **passive immunity** since the individual receiving the antibodies played no active part in the antibody-producing process. These two forms of immunity are contrasted in Table 12.7.

An important distinction between active and passive immunity is that in *active* immunity the immunized individual is fundamentally changed, since it is

Table 12.7 Comparison of passive and active immunity	
Active immunity	**Passive immunity**
Exposure to antigen; immunity achieved by injecting antigen	No exposure to antigen; immunity achieved by injecting antibodies to antigen
Antibodies made by individual achieving immunity	Antibodies made in a secondary host
Immune system activated to antigen; immunological memory in effect	No immunological memory
Antibody titer can remain high through subsequent boosters	Antibody titer progressively decays
Immune state develops over a period of weeks	Immune state develops immediately

able to continue to make the antibody in question, and it will exhibit a secondary or booster response if it later receives another injection of the antigen in question (Figure 12.17). Active immunity often may remain throughout life. A *passively* immunized individual will never have more antibodies than it received in the initial injection, and these antibodies will gradually disappear from the body; moreover, a later inoculation with the antigen will not elicit a booster response. Active immunity is usually used as a *prophylactic* measure, to protect a person against future attack by a pathogen. Passive immunity is usually *therapeutic*, designed to cure a person who is presently suffering from the disease. For example, tetanus toxoid (see the following section) actively immunizes an individual against future encounters with *Clostridium tetani* exotoxin, while tetanus antiserum (antitoxin; see below) is administered to passively immunize an individual suspected of coming in contact with *C. tetani* exotoxin via growth of the organism in a penetrating wound.

Vaccination

The material used in inducing active immunity, the antigen or mixture of antigens, is known as a **vaccine**. However, to induce active immunity to toxin-caused diseases, it is clearly not desirable to inject the toxin itself; to overcome this problem, many exotoxins can be modified chemically so that they retain their antigenicity but are no longer toxic. Such a modified exotoxin is called a **toxoid**. One of the common ways of converting toxin to toxoid is by treating it with formaldehyde, which blocks some of the free amino groups of the toxin. Toxoids are usually not such efficient antigens as the original exotoxin but they can be given safely and in high doses. When immunization against whole microorganisms is necessary, such as for endotoxin-producing organisms, the microorganism in question may first be killed by agents such as formaldehyde, phenol, or heat, and the dead cells then injected. Endotoxin-caused diseases for which vaccines are made routinely are whooping cough and typhoid. Formaldehyde treatment is also used to inactive viruses in preparing some vaccines, such as the Salk polio vaccine.

Immunization with live cells or virus is usually

more effective than with dead or inactivated material. Often it is possible to isolate a mutant strain of a pathogen which has lost its virulence but which still retains the immunizing antigens; strains of this type are called **attenuated strains** (see Section 11.11).

A summary of vaccines available for use in humans is given in Table 12.8.

Vaccination practices

Infants possess antibodies derived from their mothers and hence are relatively immune to infectious disease during the first six months of life. However, it is desirable to immunize infants for key infectious diseases as soon as possible, so that their own *active* immunity can replace the *passive* immunity received from the mother. However, infants have a rather poorly developed ability to form antibodies, so that immunization is not begun until a few months after birth. As discussed in Section 12.8, a single injection of antigen does not lead to a high antibody titer; it is desirable therefore to use a series of injections, so that a *high titer* of antibody is developed. A typical vaccination schedule for children from two months of age to the mid-teens is given in Table 12.9.

The importance of immunization procedures in controlling infectious diseases is well established. Upon introduction of a specific immunization procedure into a population, the incidence of the disease often drops markedly (Figure 12.37). The degree of immunity obtained by vaccination varies greatly, depending on the individual and on the quality and quantity of the vaccine. However, rarely is life-long immunity achieved by means of a single injection, or even a series of injections, and the population of antibody-producing cells induced by immunization will gradually disappear from the body. One way in which antigenic stimulation occurs even in the absence of immunization is by nonsymptomatic or minor infections. A natural infection will result in a rapid booster response, leading to both a further increase in activated antibody-producing cells and to production of antibody, which will attack the invading pathogen. It is not known how long immunity will last in the complete absence of antigenic stimulation, the immune period varying with different antigens. However, active

Table 12.8 Available vaccines for infectious diseases in humans	
Disease	**Type of vaccine used**
Bacterial diseases	
Diphtheria	Toxoid
Tetanus	Toxoid
Pertussis	Killed bacteria (*Bordetella pertussis*)
Typhoid fever	Killed bacteria (*Salmonella typhi*)
Paratyphoid fever	Killed bacteria (*Salmonella paratyphi*)
Cholera	Killed cells or cell extract (*Vibrio cholerae*)
Plague	Killed cells or cell extract (*Yersinia pestis*)
Tuberculosis	Attenuated strain of *Mycobacterium tuberculosis* (BCG)
Meningitis	Purified polysaccharide from *Neisseria meningitidis*
Bacterial pneumonia	Purified polysaccharide from *Streptococcus pneumoniae*
Typhus fever	Killed bacteria (*Rickettsia prowazekii*)
Viral diseases	
Yellow fever	Attenuated virus
Measles	Attenuated virus
Mumps	Attenuated virus
Rubella	Attenuated virus
Polio	Attenuated virus (Sabin) or inactivated virus (Salk)
Influenza	Inactivated virus
Rabies	Inactivated virus (human) or attenuated virus (dogs and other animals)
Hepatitus B	Recombinant DNA vaccine

immunity to certain pathogens can last many years, and in some cases, a lifetime.

Immunization procedures are not only beneficial to the individual, but are effective public health procedures, since disease spreads poorly through a population in which a large proportion of the individuals are immune (see Section 14.6).

Passive immunity

The material used in inducing passive immunity—the serum containing antibodies—is known as a *serum*, an *antiserum*, or an *antitoxin* (the last applies to a serum containing antibodies directed against a toxin). Antisera are obtained either from large-sized immunized animals, such as the horse, or from humans who have high antibody **titers** (quantities). These individuals are said to be **hyperimmune**. The antiserum or antitoxin is standardized to contain a known antibody titer; a sufficient number of units of antiserum must be inoculated to neutralize any antigen that might be present in the body. Sometimes the immunoglobulin fraction of pooled human serum is used as a source of antibodies. This contains a wide variety of antibodies that normal people have formed through the years by artificial or natural exposure to various antigens. Pooled sera are used when hyperimmune antisera are not available, but in recent years the routine use of pooled sera is declining because of the threat of contamination (for example, with the HIV virus, the cause of AIDS).

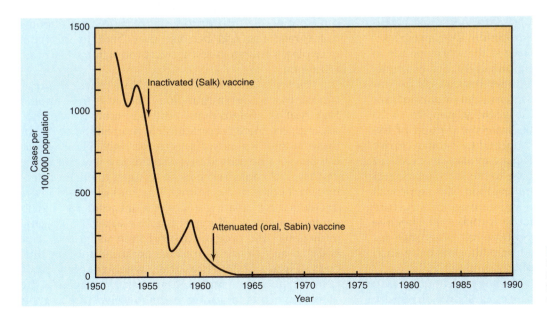

FIGURE 12.37

Cases of polio in the United States 1950–1990, showing the consequences of the introduction of polio vaccine.

Table 12.9	Some recommended immunization procedures for infants and children*	
Age	**Immunization**	**Comments****
During the first year of life	Diphtheria, pertussis, tetanus (DPT)	Given at 2, 4, and 6 months
	Oral polio (OP)	Given at 2, 4, and sometimes 6 months
During the second year	Mumps, measles, rubella (MMR)	Given at 15 months
	DPT booster	Given at 18 months
	OP booster	Given at 18 months
At 4–6 years	DPT booster, OP booster	May be given at 3 years on entering nursery school
At 14–16 years	Adult-type tetanus and diphtheria booster	Pertussis omitted

*Data from "General recommendations on immunization," Centers for Disease Control.
**These are recommended practices in some areas. Actual practice may vary in your location.

The most common use of passive immunization is in the prevention of infectious hepatitis (due to hepatitis A virus, see Section 15.13). Pooled human immunoglobulin, often referred to as *gamma globulin*, contains fairly high titers of antibody against hepatitis A virus, due to the fact that infection with hepatitis A virus is widespread in the population. Travelers to areas where the incidence of infectious hepatitis is high, such as North and tropical Africa, the Middle East, Asia, and parts of South America, may be given prophylactic doses of pooled human gamma globulin. A single dose of 0.02 ml/kg body weight should protect for up to two months, but for more prolonged exposures doses at repeated intervals should be given. Pooled human gamma globulin may also be of value in the therapy of infectious hepatitis, if given early in the incubation period.

Alternate vaccine strategies

Most vaccines currently in use are produced from whole organisms or toxoids, as described earlier in this section. However, there are several other methods for producing antigens suitable for vaccination purposes. Vaccines made by some of these methods are replacing more traditional vaccines.

The simplest alternate approach to vaccine development is the use of *synthetic peptides*. To make a vaccine, a peptide can be synthesized that corresponds to a known epitope on an infectious agent. For example, the structure of the protein antigen responsible for immunity to the foot-and-mouth virus, an important animal pathogen, is known. A synthetic peptide of 20 amino acids constituting the antigenic portion of the protein has been made and attached to suitable carrier molecules. This synthetic vaccine evokes an excellent neutralizing antibody response to the foot-and-mouth virus. However, as a general method, this approach has one major problem: the entire antigenic structure of the protein must be known to make an effective vaccine. Although this condition has been met for foot-and-mouth virus, very few pathogens have such a well-defined antigenic profile.

Sophisticated molecular biology techniques can also be used to make vaccines. As shown in Figure 8.17, genes that encode antigens from virtually any virus can be cloned into the vaccinia virus genome and expressed. The new *genetically engineered* vaccinia virus can then be used directly to induce immunity to the product of the cloned gene. Such a vaccine for the hepatitis B surface antigen is now in use. This method depends on the availability of the cloned gene that encodes the antigen and also on the ability of the vaccinia virus to express the cloned gene as an antigenic protein. The use of recombinant DNA methods to develop vaccines was discussed in Section 8.13.

Finally, a novel experimental approach to vaccine production has resulted from our knowledge of antibody structure. The foreign antigen binding site of an antibody molecule can itself act as an antigen to elicit a new antibody response, even in an animal that generated the first antibody. The variable region of an antibody molecule is referred to as its *idiotype*. Because they are structurally unique (recall there are about 10^9 different variable regions) idiotypes can serve as *antigens* in the same animal that produced them, yielding *anti-idiotypic* antibodies (Figure 12.38). If the variable region of an anti-idiotypic antibody is *complementary* to the idiotype, this region should structurally resemble the original antigen itself (Figure 12.38). Anti-idiotypic antibodies appear to be a normal part of the immune response and because they bind to idiotype, may play a role in curbing overproduction of specific idiotypes.

Idiotypes and anti-idiotypes have practical implications in vaccine immunology. Because the variable region of an anti-idiotypic antibody mimics the original antigen, there is considerable interest in the use of these antibodies as *antigen-free vaccines*. How would this work? The anti-idiotypic antibody will not cause disease, but could serve as an antigen to elicit protective antibodies in the animal. Vaccines of this type are developed by isolating a monoclonal antibody (which contains a single idiotype) reactive against a specific pathogenic microorganism, and then using this antibody as an *antigen* to produce an anti-idiotypic antibody (Figure 12.38). The latter, when properly purified, can then be employed as a vaccine against the pathogen. Thus far, antigen-free vaccines are only in

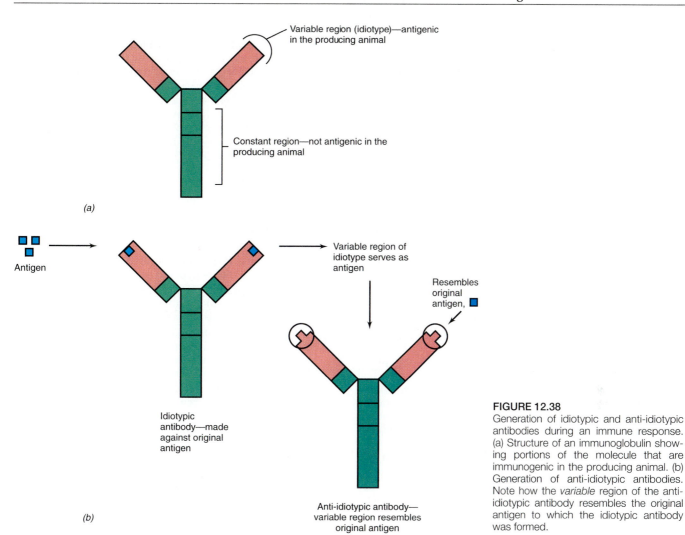

FIGURE 12.38
Generation of idiotypic and anti-idiotypic antibodies during an immune response. (a) Structure of an immunoglobulin showing portions of the molecule that are immunogenic in the producing animal. (b) Generation of anti-idiotypic antibodies. Note how the *variable* region of the anti-idiotypic antibody resembles the original antigen to which the idiotypic antibody was formed.

Labels in figure:
- Variable region (idiotype)—antigenic in the producing animal
- Constant region—not antigenic in the producing animal
- (a)
- Antigen
- Variable region of idiotype serves as antigen
- Resembles original antigen,
- Idiotypic antibody—made against original antigen
- Anti-idiotypic antibody— variable region resembles original antigen
- (b)

the experimental stage, but promising results have been obtained in animal studies. If antigen-free vaccines can be developed for a number of human diseases, genetic engineering might also enter the picture because cloned genes coding for anti-idiotypic antibodies (obtained by standard genetic engineering methods, see Chapter 8) would allow for antigen-free vaccines to be produced in microorganisms.

Antigen-free vaccines based on anti-idiotypic antibodies represent a novel and safe means of disease prophylaxis. This therapeutic alternative has emerged from basic knowledge of the structure of antibody molecules and the nature of the immune response.

> The immune response plays a major role in resistance to infectious disease. Two kinds of immunity can be characterized, active immunity, in which the body is induced to produce its own immune response, and passive immunity, in which the immune response is produced in another individual and transferred to the body in serum or cells. Vaccination is the procedure by which active immunity is brought about. Vaccines include inactivated toxins (toxoids), inactivated viruses, killed pathogenic microorganisms, and weakened (attenuated) live viruses or other microorganisms. Methods for producing vaccines based on artificially prepared epitopes, molecular cloning of antigenic fragments, and the use of antibody images to mimic antigens are being used to make vaccines safer and more efficient.

Study Questions

1. What are the major differences between a host-resistance mechanism such as phagocytosis and humoral and cell-mediated immunity?

2. What substances provoke antibody responses? What substances do not? What general properties are necessary for substances to provoke antibody responses?

3. Define antigen. Immunogen. Immunoglobulin. T cell receptor. Can you define antibody without references to antigen and vice versa?

4. Compare and contrast T and B lymphocytes with macrophages in terms of function, location, and immunological specificity. What is the origin of all of these cell types, and how do they end up taking on such different functions?

5. Draw the structure of an IgG molecule, labeling the heavy and light chains. Which portion of the molecule binds antigen? Which portion defines the class of the immunoglobulin? Which portions of the molecule have identical amino acid sequences? Be specific and be sure to consider all possible sequence identities and non-identities within the molecule.

6. What is the major immunological role of each class of immunoglobulin?

7. Describe the T-cell receptor (TCR) and compare it to IgG.

8. Compare and contrast MHC class I and class II antigens as to bodily location, composition, and overall dimensions of the molecules.

9. Describe how MHC class I antigens are thought to participate in the overall immune response.

10. Describe the clonal selection theory.

11. What is the basis of immune recognition that allows the lymphocytes to distinguish between self and nonself?

12. Why do animals normally not react immunologically against their own antigens, whereas the same antigens injected into a different animal species elicit a strong immunological response?

13. What is clonal anergy? How does it affect antibody production?

14. Discuss the major steps in the production of antibody starting with the introduction of antigen into the animal and resulting in a humoral immune response. What is the importance of MHC antigens in antibody production?

15. Compare and contrast an undifferentiated B cell, a memory cell, and a plasma cell. Which cell type(s) actually secretes significant amounts of antibody?

16. Write a one-sentence definition that clearly describes the nature and function of complement.

17. Why is it possible in an emergency to use human type O blood in transfusions to any blood type (type O is the universal donor), while people with type AB blood can receive blood from any blood type (type AB is the universal recipient)?

18. What is the major advantage of an ELISA or radioimmunoassay over a precipitation reaction? Which assay(s) requires more antigen and antibody?

19. Describe the differences between a direct and an indirect ELISA procedure. What are the advantages of the indirect test?

20. What major advantages does the ELISA test have over radioimmunoassay?

21. Compare and contrast the various types of T cells and other cells involved in cell-mediated immunity.

22. What are lymphokines and what do they do?

23. What are the major differences between delayed-type and immediate-type hypersensitivities?

24. List three autoimmune diseases. In each case, what is the immunological manifestation of the disease and how can the disease be treated?

25. Compare and contrast active and passive immunity from the following points of view: manner of induction (substance used), function in infectious disease resistance, long-term retention, and value in vaccination.

26. Why are infants not vaccinated immediately after birth? Discuss at least three reasons.

27. List five common vaccination procedures and indicate the nature of the antigen, how it is prepared, and the way in which antibody against this antigen confers immunity.

28. List the strategies for vaccine production. What are the advantages and disadvantages of each? Which methods have been successfully used to make currently available vaccines?

Supplementary Readings

Bittle, J. L., and **F. L. Murphy** (eds.) 1989. *Vaccine Biotechnology.* Academic Press, New York. New methods of producing vaccines and novel vaccination technology and principles.

Davis, B. D., R. Dulbecco, H. N. Eisen, and **H. S. Ginsberg.** 1990. *Microbiology,* 4th edition. Harper and Row, Hagerstown, MD. A detailed treatment of immunology can be found in this standard medical school textbook.

Honjo, T., F. W. Alt, and **T. H. Rabbitts** (eds.) 1989. *Immunoglobulin genes.* Academic Press, New York. An advanced level treatment of immunoglobulin gene structure and rearrangements.

Kimball, J. W. 1990. *Introduction to Immunology,* 3rd edition. Macmillan Publishing Co., New York. An excellent immunology text suitable for undergraduates, covering all major topics in immunology.

Mayer, R. J. and **J. H. Walker.** 1987. *Immunochemical Methods in Cell and Molecular Biology.* Academic Press, New York. An excellent source of modern immunological methods, describing procedures in detail.

Roitt, I. M. 1991. *Essential Immunology,* 7th edition. Blackwell Scientific, Oxford. Probably the best short treatment of immunology, with emphasis on the general processes of immune functions rather than molecular details.

Roitt, I. M., J. Brostoff, and **D. K. Male.** 1993. *Immunology,* 3rd edition. C. V. Mosby Co., St. Louis. A well-illustrated textbook of immunology for advanced undergraduates.

Sell, S. 1987. *Basic Immunology.* Elsevier, New York. A basic textbook of immunology.

Tizard, I. R. 1992. *Immunology: An Introduction,* 3rd edition. Saunders College Publishing, New York. A good introductory immunology textbook.

13

Clinical and Diagnostic Microbiology

The most important activity of the microbiologist in medicine is to isolate and identify the causal agents of infectious disease. This major area of microbiology is called **clinical** or **diagnostic microbiology**. There is increasing awareness of the importance of precise identification of the pathogen for proper treatment of infectious disease, and new sophisticated methods are being continually developed. Clinical laboratories are generally able to isolate, identify, and determine the antibiotic sensitivity of most routinely encountered pathogenic bacteria within 48 hours of sampling. However, recent advances in rapid diagnostic methods have made it possible to identify some pathogens in less than one day, and in only a few hours in certain cases. Diagnostic methods based on immunology and nucleic acid probes have reduced this time even more, and in many cases a pathogen can be positively identified without culturing the organism at all. Molecular methods have also greatly improved the diagnosis of viral and protozoal infections, diseases that are typically difficult to pinpoint because of the difficulty of culturing the agent. The clinical microbiologist works with and advises the physician in matters relating to the diagnosis and treatment of infectious diseases.

13.1 Isolation of Pathogens from Clinical Specimens

The physician, on the basis of careful examination of the patient, may suspect that an infectious disease is present. Samples of infected tissues or fluids are then collected for microbiological, immunological, and

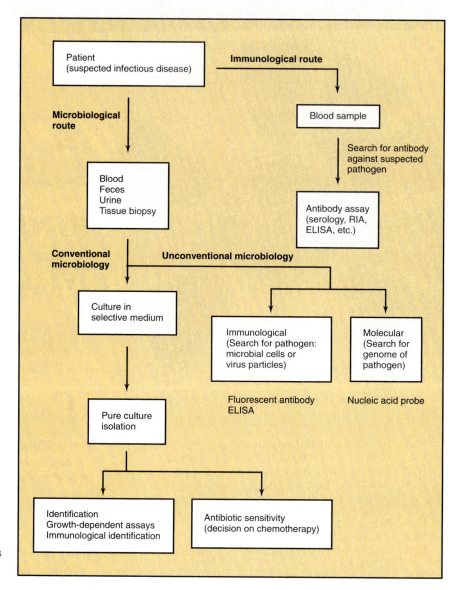

FIGURE 13.1
Clinical and diagnostic methods
used for infectious disease.

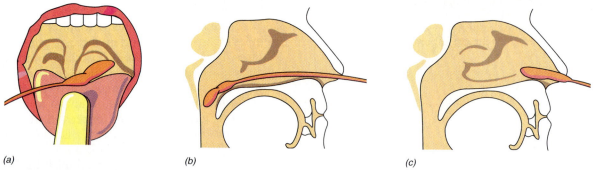

(a)   (b)   (c)

FIGURE 13.2 Methods for obtaining specimens from the upper respiratory tract.
(a) Throat swab. (b) Nasopharyngeal swab passed through the nose. (c) Swabbing the
inside of the nose.

molecular biological analyses (Figure 13.1). Depending upon the kind of infection, materials collected may include blood, urine, feces, sputum, cerebrospinal fluid, or pus. A sterile swab may be passed across a suspected infected area (Figure 13.2). The swab is then streaked over the surface of an agar plate or placed directly in a liquid culture medium. Small pieces of living tissue may be aseptically removed (biopsy). Table 13.1 summarizes current recommendations. If clinically relevant organisms are to be isolated and a correct diagnosis made, care must be taken in obtaining samples of clinical specimens. The physician must ensure that the specimen is removed from the *actual site of the infection*. Recovery or detection of pathogens

Miniglossary for Chapter 13

ABSCESS a localized infection characterized by production of pus

AGGLUTINATION reaction between antibody and particle-bound antigen, resulting in visible clumping of the particles

BACTEREMIA the presence of bacteria in the blood

COMPLEMENT FIXATION the consumption of complement by an antibody–antigen reaction

FLUORESCENT ANTIBODY covalent modification of an antibody molecule with a fluorescent dye; the dye makes the antibody visible under fluorescent light

GONOCOCCUS *Neisseria gonorrhoeae*, the Gram-negative diplococcus which causes gonorrhea

IMMUNOBLOT (WESTERN BLOT) electrophoresis of proteins followed by transfer to a membrane and detection by addition of specific antibodies

NUCLEIC ACID PROBE in clinical microbiology, short oligonucleotides of unique sequence used as hybridization probes for identifying pathogens (see Chapter 8 for general usage)

PRECIPITATION reaction between antibody and a soluble antigen resulting in a visible, insoluble complex

SEPTICEMIA blood infection

TITER in an immunological context, the quantity of antibody present in a solution

may not be possible if insufficient inoculum is available. The sample must also be taken under aseptic conditions so that contamination is avoided. Care must also be taken to ensure that metabolic requirements for certain organisms, such as anoxic conditions, are maintained. Once taken, the sample is analyzed as soon as possible. If it cannot be analyzed immediately, it is usually refrigerated to slow down deterioration. In the rest of this section, we describe some of the most common microbiological procedures used in the clinical laboratory.

Blood cultures

Bacteremia means the presence of bacteria in the blood (see Section 11.13). Because of the extreme danger associated with bacteremia, the rapid and accurate identification of the causative agent(s) is one of the most important duties of the clinical microbiologist. Bacteria are cleared from the bloodstream rapidly. Therefore bacteremia is uncommon in healthy individuals, and presence of bacteria in the blood is gener-

ally indicative of systemic infection. The most common pathogens found in blood include *Pseudomonas aeruginosa*, enteric bacteria, especially *Escherichia coli* and *Klebsiella pneumoniae*, and the Gram-positive cocci *Staphylococcus aureus* and *Streptococcus pyogenes*. The classic type of blood infection is **septicemia**, resulting from a virulent organism entering the blood from a focus of infection, multiplying, and traveling to various body tissues to initiate new infections. Septicemia is indicated by the presence of severe systemic symptoms, usually with fever and chills, followed by prostration. In many disease situations, culture of the blood provides the only immediate way of isolating and identifying the causal agent, and diagnosis may thus depend on careful and proper blood culture.

The standard blood culture procedure is to remove 10 ml of blood aseptically from a vein and inject it into a blood culture bottle containing an anticoagulant and an all-purpose culture medium. Two cultures are set up with one bottle being incubated aerobically and one anaerobically. Media used are all relatively

Table 13.1   Recommended agar media for primary isolation purposes in a clinical microbiology laboratory[a]

Specimen	Blood agar	Enteric agar	CA	TM	Anaerobic
Fluids: chest, abdomen, pericardium	+	+	+	−	+
Feces: rectal swabs	+	+	+	−	−
Surgical tissue biopsies: lung, lymph nodes	+	+	−	−	+
Throat: swabs, sputum, tonsil, nasopharynx	+	+	+	−	−
Genitourinary swabs: urethra, vagina, cervix	+	+	+	+	−
Urine	+	+	−	−	−
Blood	+	−	−	−	+
Swabs: wounds, abscesses, exudates	+	+	+	−	+

[a]Data from Balows, A., W. J. Hausler, K. L. Herrman, H. D. Isenberg, and H. J. Shadomy. 1991. *Manual of Clinical Microbiology*, 5th edition, ASM, Washington, D.C.
[b]Descriptions: Blood agar, 5 percent whole sheep blood added to trypticase soy agar; Enteric agar, either Eosin-Methylene Blue (EMB) agar, or MacConkey agar; CA, chocolate (heated blood) agar; TM, Thayer-Martin agar; Anaerobic, thioglycolate-containing blood agar or supplemented thioglycolate agar incubated anaerobically.

rich, containing protein digests and other complex ingredients. Blood culture bottles are incubated at 35°C and examined daily for up to seven days. Most clinically significant bacteria are recovered within this period. Some blood isolation systems employ a chemical that lyses red and white blood cells, releasing potential intracellular pathogens which might otherwise be overlooked.

Because a certain amount of skin contamination is unavoidable during initial drawing of the blood, even under optimal conditions a contamination rate of 2–3 percent can be expected. Because of this, identification of positive blood cultures is essential, and contamination may be indicated if certain organisms commonly found on the skin are isolated, such as *Staphylococcus epidermidis*, coryneform bacteria, or propionibacteria, although even these organisms can occasionally cause infection of the wall of the heart (subacute bacterial endocarditis). Thus considerable microbiological and clinical experience is necessary in interpreting blood cultures.

Urine cultures

Urinary tract infections are very common, and because the causal agents are often identical or similar to bacteria of the normal flora (for example, *E. coli*), considerable care must be taken in the bacteriological analysis of urine. Since urine supports extensive bacterial growth under many conditions, fairly high cell numbers are often found in urinary infection. In most cases, the infection occurs as a result of an organism ascending the urethra from the outside. Occasionally, even the bladder may become infected. Urinary tract infections are the most common form of *nosocomial* (hospital-acquired) infection (see Section 14.7).

Significant urinary infection generally results in bacterial counts of 10^5 or more organisms per milliliter of a clean-voided midstream specimen, whereas in the absence of infection, contamination of the urine from the external genitalia (almost unavoidable to some extent) results in less than 10^3 organisms per milliliter. The most common organisms are members of the enteric bacteria, with *E. coli* accounting for about 90 percent of the cases. Other organisms include *Klebsiella*, *Enterobacter*, *Proteus*, *Pseudomonas*, and *Enterococcus faecalis*. *Neisseria gonorrhoeae*, the causal agent of gonorrhea, does not grow in the urine itself, but in the urethral epithelium, and must be diagnosed by different methods (see below).

Because of the relatively high number of organisms found in a significant urinary infection, direct microscopic examination of the urine is of considerable value and is recommended as part of the basic procedure. A small drop of urine is allowed to dry on a microscope slide and a Gram stain performed. If more than 10^5 organisms per milliliter are present, there are usually one or more bacteria per microscope field

(100 × objective). Another common and rapid means of screening urine samples for suspected bacterial pathogens is with a dipstick containing reagents to detect nitrite. If significant numbers (>10^5/ml) of enteric bacteria are present in the urine sample, nitrate in the urine will be reduced to nitrite; nitrite can then be detected chemically at very low concentrations.

To culture potential urinary tract pathogens, two media are used; blood agar as a nonselective general medium, and a medium selective for enteric bacteria, such as MacConkey or EMB (*e*osin–*m*ethylene–*b*lue) agar (see Section 13.2). These specialized media permit the initial differentiation of lactose fermenters from nonfermenters, and the growth of Gram-positive organisms such as *Staphylococcus* (a common skin contaminant) is inhibited. Organisms isolated can be identified, and antibiotic susceptibility tests performed. Experienced clinical microbiologists may make a tentative identification of an isolate by observing the color and morphology of colonies of the suspected pathogen grown on various selective media as described in Table 13.2. Such an identification must be followed up with more detailed analyses, of course, but clinical microbiologists will use this information in conjunction with more detailed test results in order to make a positive identification. The immunological/nucleic acid probes described in Sections 13.4 through 13.11 allow for identification of the pathogen on molecular rather than physiological or morphological grounds.

If a urinary tract infection is suspected but no organism(s) can be cultured, a highly sensitive urine screening test is available that detects very low numbers of bacterial cells in urine by measuring the amount of ATP present; ATP concentration is proportional to cell biomass. Using firefly luciferin–luciferase, the concentration of ATP in a urine sample can be accurately determined in less than one hour, and the number of cells per ml calculated. The test is very useful for diagnosing low-grade urinary tract infections where microscopy or attempts to culture an organism may not reveal the presence of bacteria.

Fecal cultures

Proper collection and preservation of feces is important in the isolation of intestinal pathogens. During storage, the pH of feces drops and thus an extended delay between sampling and sample processing must be avoided. This is especially critical for the isolation of *Shigella* and *Salmonella* species, both of which are rather sensitive to acid pH. Samples, collected from feces freshly voided into a sterile plastic cup, are placed in a vial containing phosphate buffer for transport to the lab. If a patient has a bloody or pus-containing stool, this material is always sampled; such discharges generally contain a large number of the organisms of interest. In the case of suspected food- or water-borne infections, fecal samples should be inocu-

Table 13.2   **Colony characteristics of frequently isolated Gram-negative rods cultured on various clinically useful agars[a]**

Organism	Agar media[b]			
	EMB	MC	SS	BS
Escherichia coli	Dark center with greenish metallic sheen	Red or pink	Red to pink	Mostly inhibited
Enterobacter	Similar to E. coli, but colonies are larger	Red or pink	White or beige	Mucoid colonies with silver sheen
Klebsiella	Large mucoid, brownish	Pink	Red to pink	Mostly inhibited
Proteus	Translucent, colorless	Transparent, colorless	Black center, clear periphery	Green
Pseudomonas	Translucent, colorless to gold	Transparent, colorless	Mostly inhibited	No growth
Salmonella	Translucent, colorless to gold	Translucent, colorless	Opaque	Black to dark green
Shigella	Translucent, colorless to gold	Transparent, colorless	Opaque	Brown or inhibited

[a]Adapted from Balows, A., W. J. Hausler, K. L. Herrman, H.D. Isenberg, and H. J. Shadomy. 1991. Manual of Clinical Microbiology, 5th ed. American Society for Microbiology, Washington, D.C.
[b]BS, Bismuth Sulfite agar, EMB, Eosin-Methylene Blue agar; MC, MacConkey agar; SS, Salmonella-Shigella agar.

lated into a variety of selective media (see Section 13.2) for the isolation of specific bacteria or characterization of intestinal parasites. Positive identifications would be made by the techniques described in Sections 13.5–13.11.

Wounds and abscesses

Infections associated with traumatic injuries such as animal or human bites, burns, cuts, or the penetration of foreign objects, must be carefully sampled in order to recover the relevant pathogen. This is because wound infections and abscesses are frequently contaminated with members of the normal flora. Swab samples of such lesions are frequently misleading. The best sampling method is to aspirate purulent (pus-containing) lesions with a sterile syringe and needle following disinfection of the skin surface with 70 percent ethyl or isopropyl alcohol. Internal purulent discharges are usually sampled by biopsy or in tissues removed in surgery.

A variety of pathogens can be associated with wound infections, and because some of these are anaerobes, samples should be transported from the collection site under anaerobic conditions. A common pathogen associated with purulent discharges is S. aureus, but enteric bacteria, P. aeruginosa, and the anaerobes Bacteroides and Clostridium species are also commonly encountered. The major isolation media are blood agar, selective media for enteric bacteria

(see Table 13.1), and blood agar containing additional supplements and reducing agents for obligate anaerobes.

Genital specimens and the laboratory diagnosis of gonorrhea

In males, a purulent urethral discharge is the classic symptom of the sexually-transmitted disease, gonorrhea. If no discharge is present, a suitable sample can be obtained using a sterile narrow-diameter cotton swab which is inserted into the anterior urethra, left in place a few seconds to absorb any exudate, and then removed for culture of N. gonorrhoeae, the causative agent of gonorrhea. Alternatively, a sample of the first early morning urine of an infected individual usually contains viable cells of N. gonorrhoeae. In females suspected of having gonorrhea or other genital infections, samples are generally obtained by swab from the cervix and the urethra.

Gonorrhea is one of the most common infectious diseases in adults, and clinical microbiological procedures are central to its diagnosis. N. gonorrhoeae (referred to clinically as the gonococcus) colonizes mucosal surfaces of the urethra, uterine cervix, anal canal, throat, and conjunctiva. The organism is quite sensitive to drying, and because of this it is transmitted almost exclusively by direct person-to-person contact, usually by sexual intercourse. The major goal of public health measures to control gonorrhea involves identi-

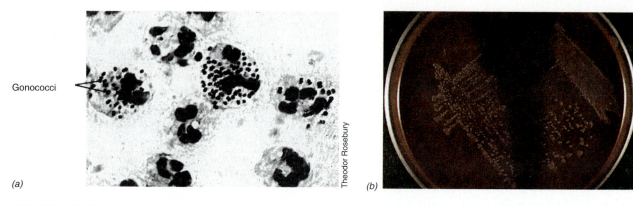

Gonococci

Theodor Rosebury

Leon J. LeBeau

(a)

(b)

FIGURE 13.3 (a) Photomicrograph of *Neisseria gonorrhoeae* within human polymorphonuclear leukocytes from a cervical smear. Note how many cells are in pairs as diplococci (arrows). (b) Colonies of *N. gonorrhoeae* growing on Thayer-Martin agar. The plate has been stained in the middle with a reagent that turns colonies blue if cells contain cytochrome *c* (the oxidase test).

fication of asymptomatic carriers, and this requires microbiological analysis.

Because the gonococcus is a Gram-negative coccus, and similar organisms are not very common in the normal flora of the urogenital tract, direct microscopy of Gram-stained material is of value. For example, observation of Gram-negative diplococci in a urethral discharge or in a vaginal or cervical smear is considered presumptive evidence for gonorrhea. During visualization under the oil-immersion lens, attention should be paid to polymorphonuclear leukocytes (granulocytes), since, in acute gonorrhea, such cells will frequently contain groups of Gram-negative, kidney-shaped diplococci (Figure 13.3*a*), and there will be few if any other types of organisms present.

Cultural procedures have a higher degree of sensitivity than microscopical analyses. Most media for the culture of *N. gonorrhoeae* contain heated blood or hemoglobin (referred to as *chocolate agar* because of its deep brown appearance), the heating causing the formation of a precipitated material, which is quite effective in absorbing toxic products present in the medium. A second primary isolation medium, called *Thayer-Martin agar*, also is used for isolation of *N. gonorrhoeae* (Figure 13.3*b*). This medium incorporates the antibiotics vancomycin, nystatin, and colistin, to which most clinical isolates of *N. gonorrhoeae* are naturally resistant.

After streaking, the plates must be incubated in a humid environment in an atmosphere containing 3 to 7 percent CO_2 (CO_2 is required for growth of gonococci). The plates are examined after 24 and 48 hours, and portions of colonies should be immediately tested by the oxidase test, since all *Neisseria* are oxidase-positive (see Section 13.2). Oxidase-positive Gram-negative diplococci growing on chocolate agar can be presumed to be gonococci if the inoculum was derived from genitourinary sources, but definite identification requires determination of carbohydrate utilization patterns or immunological or nucleic acid probe tests (see Sections 13.4 and 13.11).

A rapid test employing chromogenic substrates has been developed for differentiating *N. gonorrhoeae*

from other species of *Neisseria*. Starting from colonies on plates, various species can be differentiated by the color reaction obtained following incubation of the test medium with cells. The test is designed to assay the presence of specific enzymes present in one species of *Neisseria* but absent in the others. The enzymes act on substrates that yield colored products. A simple, sensitive, and highly specific nucleic acid probe test (see Section 13.11) has also been developed for identifying *N. gonorrhoeae*.

Culture of anaerobes

Obligately anaerobic bacteria are common causes of infection and will be completely missed in clinical diagnosis unless special precautions are taken for their isolation and culture. We have discussed anaerobes in general in Section 9.13, and we noted that many anaerobes are extremely susceptible to oxygen. Because of this, specimen collection, handling, and processing require special attention if it is thought that an obligate anaerobe is involved. There are several habitats in the body (for example, the oral cavity and the intestinal tract, see Sections 11.3 and 11.4) that are generally anoxic, and in which obligately anaerobic bacteria can be found as part of the normal flora. However, other parts of the body can become anoxic as a result of tissue injury or trauma, which results in reduction of blood supply to the injured site, and such anaerobic sites can then become available for colonization by obligate anaerobes. In general, pathogenic anaerobic bacteria are part of the normal flora and are only opportunistic pathogens, although two important pathogenic anaerobes, *Clostridium tetani* (causal agent of tetanus) and *C. perfringens* (causal agent of gas gangrene and one type of food poisoning), both endospore-forming Bacteria, are predominantly soil organisms.

With anaerobic culture, the microbiologist is presented not only with the usual problems of obtaining and maintaining an uncontaminated specimen, but also with ensuring that the specimen not come in contact with air. Samples, collected by suction or biopsy, must be immediately placed in a tube containing oxygen-free gas, preferably containing a small amount of a dilute salts solution with a reducing agent such as thioglycolate and the redox indicator, resazurin. This dye is colorless when reduced, and becomes pink when oxidized, thus quickly indicating any oxygen contamination of the specimen. If a proper anaerobic transport tube is not available, the syringe itself can be used to transport the specimen, the needle being inserted into a sterile rubber stopper so that no air is drawn into the syringe.

For anaerobic incubation, agar plates are placed in a sealed jar which is made anoxic by either replacing the atmosphere in the jar with an oxygen-free gas mixture (a mixture of N_2 and CO_2 is frequently employed), or by adding some compound to the enclosed vessel which removes O_2 from the atmosphere (Figure 13.4). Alternative means of isolating anaerobes include the use of culture media containing reducing agents or the use of anaerobic glove boxes. The latter are large gas impermeable bags filled with an oxygen-free gas such as nitrogen or hydrogen and which are fitted with an airlock for inserting and removing cultures (see Figure 9.20). The advantage of an anaerobic glove box is that manipulations can be done as one would normally perform them on a laboratory bench. However, because of their expense, anaerobic glove boxes are not used extensively in clinical laboratories, but are in widespread use in research laboratories which specialize in anaerobic microorganisms (see Figure 9.20).

In general, media for anaerobes do not differ greatly from those used for aerobes, except that they are generally richer in organic constituents, and contain reducing agents (usually cysteine or thioglycolate) and a redox indicator such as resazurin. Once positive cultures have been obtained, it is essential

that they be characterized and identified, to be certain that the isolate is not a member of the normal flora.

> Microbial culture provides a sensitive means for detecting the presence of a potential pathogen, but specific methods must be used for each organism and for each part of the body to be examined. Anaerobes must obviously be cultivated in media from which all oxygen is excluded. Depending on the case, blood, urine, feces, or tissue may be sampled and cultured. In the case of the important sexually transmitted disease gonorrhea, bacterial culture provides the most definitive means of diagnosis.

13.2   Growth-Dependent Identification Methods

If the inoculation of a primary medium results in bacterial growth, the clinical microbiologist goes to work immediately trying to identify the organism or organisms present. Identification of a clinical isolate can frequently be made using a variety of growth-dependent assays. We discuss some of these methods here.

Growth on selective and differential media

Based on growth in primary isolation media, it is frequently possible to restrict the unknown pathogen to a relatively few choices. Clinical microbiologists have at their disposable dozens of diagnostically useful culture media in which unknowns can be inoculated. In many large hospitals and clinics, media are purchased from commercial sources, which ensures quality control and reliable testing in different clinical settings (Figure 13.5). Many of these media are available in miniaturized kits containing a number of different media in separate wells, all of which can be inoculated at one time (Figure 13.5d and e).

The battery of media employed are either selective, differential, or both. A *selective medium* is one to which compounds have been added to selectively inhibit the growth of certain microorganisms but not others. A *differential medium* is one to which some sort of indicator, usually a dye, has been added, which allows the clinician to differentiate between various chemical reactions carried out during growth. Eosin-methylene blue (EMB) agar, for example, is a widely used selective *and* differential medium. EMB agar is used for the isolation of Gram-negative enteric Bacteria. The methylene blue is present to inhibit Gram-positive Bacteria; although the mechanism is unclear, small amounts of this dye effectively inhibit the growth of most Gram-positive Bacteria. Eosin is a dye which responds to changes in pH, going from colorless to black under acidic conditions. EMB contains lactose and sucrose, but not glucose, as energy sources, and colonies of lactose-fermenting (generally enteric) bacteria, such as *Escherichia coli*, *Klebsiella*, and *Enterobacter*, acidify the medium and have black centers with a greenish sheen. Colonies of lactose nonfer-

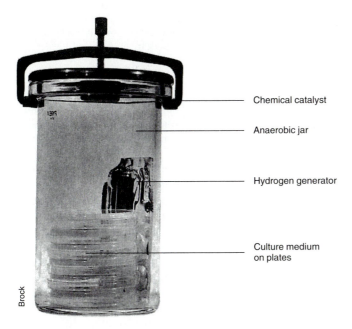

Chemical catalyst

Anaerobic jar

Hydrogen generator

Culture medium on plates

Brock

FIGURE 13.4 Sealed jar for incubating cultures under anoxic conditions.

(a)

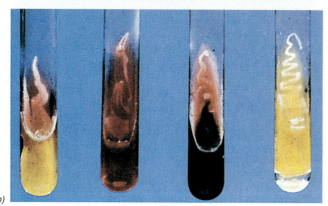

(b)

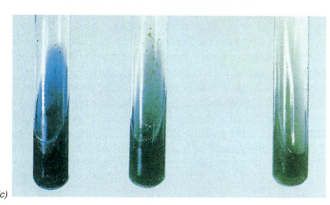

(c)

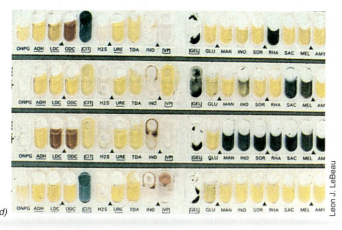

(d)

(e)

FIGURE 13.5 Growth-dependent diagnostic methods used for the identification of clinical isolates by color changes in various diagnostic media. (a) Use of a differential medium to assess sugar fermentation. Acid production is indicated by color change of the pH-indicating dye added to the liquid medium. If gas production occurs, a bubble appears in the inserted vial in each tube. From left to right: acid, acid and gas, negative, uninoculated. (b) A conventional diagnostic test for enteric bacteria in a medium called *triple sugar iron agar*. The medium is inoculated both on the surface of the slant and by stabbing into the butt. The medium contains a small amount of glucose and a large amount of lactose and sucrose. Organisms able to ferment only the glucose will cause acid formation only in the butt, whereas lactose or sucrose-fermenting organisms will also cause acid formation in the top. Gas formation is indicated by the breaking up of the agar in the butt. Hydrogen sulfide formation (from either protein degradation or from reduction of thiosulfate in the medium) is indicated by a blackening due to reaction of the H_2S with ferrous iron in the medium. From left to right: fermentation of glucose only; no reaction; hydrogen sulfide formation; fermentation of glucose and another sugar. (c) Measurement of citrate utilization by *Salmonella* on Simmons citrate agar. The change in pH causes a change in the color of the indicating dye. From left to right: positive, negative, uninoculated. (d) Media kits used for the rapid identification of clinical isolates. The principle is the same as in *a* but the whole arrangement has been miniaturized so that a number of tests can be run at the same time. Four separate strips, each with a separate culture, are shown. (e) Another arrangement of a miniaturized test kit. This one permits diagnosis of sugar utilization in nonfermentative organisms.

menters, such as *Salmonella*, *Shigella*, or *Pseudomonas*, are translucent or pink.

In the battery of tests performed to help identify an organism, many different biochemical reactions can be measured. The most important tests are summarized in Table 13.3. These tests measure the presence or absence of *enzymes* involved in catabolism of the substrate or substrates added to the differential medium. Fermentation of sugars is measured by incorporating pH dyes which change color upon acidification (Figure 13.5a). Production of hydrogen gas and/or carbon dioxide during sugar fermentation is assayed by observing gas production, either in gas collection vials or in agar (Figure 13.5a,b). Hydrogen sulfide production is indicated following growth in a medium containing ferric iron. If sulfide is produced, ferric iron complexes with H_2S to form a black precipitate of iron sulfide (Figure 13.5b). Utilization of citric acid, a six carbon acid containing three carboxylic acid groups, is accompanied by a pH rise, and a specific dye incorporated into this test medium changes color as conditions become alkaline (Figure 13.5c). Hundreds of differential tests have been developed for clinical use, but only about 20 are used routinely (Figure 13.5d).

Table 13.3 Important clinical diagnostic tests for bacteria

Test	Principle	Procedure	Most common use
Carbohydrate fermentation	Acid and/or gas during fermentative growth with sugars or sugar alcohols	Broth medium with carbohydrate and phenol red as pH indicator; inverted tube for gas	Enteric bacteria differentiation (also several other genera or species separations with some individual sugars, see Figure 13.5)
Catalase	Enzyme decomposes hydrogen peroxide, H_2O_2	Add drop of H_2O_2 to dense culture and look for bubbles (O_2) (Figure 9.23)	*Bacillus* (+) from *Clostridium* (−); *Streptococcus* (−) from *Micrococcus/Staphylococcus* (+)
Citrate utilization	Utilization of citrate as sole carbon source, results in alkalinization of medium	Citrate medium with bromthymol blue as pH indicator, look for intense blue color (alkaline pH)	*Klebsiella-Enterobacter* (+) from *Escherichia* (−), *Edwardsiella* (−) from *Salmonella* (+) (see Figure 13.5)
Coagulase	Enzyme causes clotting of blood plasma	Mix dense liquid suspension of bacteria with plasma, incubate, and look for clot	*Staphylococcus aureus* (+) from *S. epidermidis* (−)
Decarboxylases (lysine, ornithine, arginine)	Decarboxylation of amino acid releases CO_2 and amine	Medium enriched with amino acid. Bromcresol purple pH indicator. Alkaline pH if enzyme action, indicator becomes purple	Aid in determining bacterial group among the enteric bacteria
β-Galactosidase (ONPG) test	Orthonitrophenyl-β-galactoside (ONPG) is an artificial substrate for the enzyme. When hydrolyzed, nitrophenol (yellow) is formed	Incubate heavy suspension of lysed culture with ONPG, look for yellow color	*Citrobacter* and *Arizona* (+) from *Salmonella* (−). Identifying some *Shigella* and *Pseudomonas* species
Gelatin liquefaction	Many proteases hydrolyze gelatin and destroy the gel	Incubate in broth with 12% gelatin. Cool to check for gel formation. If gelatin hydrolyzed, tube remains liquid upon cooling	To aid in identification of *Serratia, Pseudomonas, Flavobacterium, Clostridium*
Hydrogen sulfide (H_2S) production	H_2S produced by breakdown of sulfur amino acids or reduction of thiosulfate	H_2S detected in iron-rich medium from formation of black ferrous sulfide (many variants: Kliger's iron agar, triple sugar iron agar, also detect carbohydrate fermentation)	In enteric bacteria, to aid in identifying *Salmonella, Arizona, Edwardsiella*, and *Proteus* (see Figure 13.5)
Indole test	Tryptophan from proteins converted to indole	Detect indole in culture medium with dimethyl-aminobenzaldehyde (red color)	To distinguish *Escherichia* (+) from *Klebsiella-Enterobacter* (−); *Edwardsiella* (+) from *Salmonella* (−)
Methyl red test	Mixed-acid fermenters produce sufficient acid to lower pH below 4.3	Glucose-broth medium. Add methyl red indicator to a sample after incubation	To differentiate *Escherichia* (+, culture red) from *Enterobacter* and *Klebsiella* (usually −, culture yellow)
Nitrate reduction	Nitrate as alternate electron acceptor, reduced to NO_2^- or N_2	Broth with nitrate. After incubation, detect nitrite with α-naphthylamine-sulfanilic acid (red color). If negative, confirm that NO_3^- still present by adding zinc dust to reduce NO_3^- to NO_2^-. If no color after zinc, then $NO_3^- \rightarrow N_2$	To aid in identification of enteric bacteria (usually +)

Table 13.3 *(continued)*

Test	Principle	Procedure	Most common use
Oxidase test	Cytochrome *c* oxidizes artificial electron acceptor: tetramethyl (or dimethyl)-*p*-phenylenediamine	Broth or agar. Oxidase-positive colonies on agar can be detected by flooding plate with reagent and looking for blue or brown colonies	To separate *Neisseria* and *Moraxella* (+) from *Acinetobacter* (−). To separate enteric bacteria (all −) from pseudomonads (+). To aid in identification of *Aeromonas* (+)
Oxidation–fermentation (O/F) test	Some organisms produce acid only when growing aerobically	Acid production in top part of sugar-containing culture tube; soft agar used to restrict mixing during incubation	To differentiate *Micrococcus* (aerobic acid production only) from *Staphylococcus* (acid anaerobically). To characterize *Pseudomonas* (aerobic acid production) from enteric bacteria (acid anaerobically)
Phenylalanine deaminase test	Deamination produces phenylpyruvic acid, which is detected in a colorimetric test	Medium enriched in phenylalanine. After growth, add ferric chloride reagent, look for green color	To characterize the genus *Proteus* and the *Providencia* group
Starch hydrolysis	Iodine-iodide gives blue color with starch	Grow organism on plate containing starch. Flood plate with Gram's iodine, look for clear zones around colonies	To identify typical starch hydrolyzers such as *Bacillus* spp.
Urease test	Urea (H_2N—CO—NH_2) split to $2NH_3 + CO_2$	Medium with 2% urea and phenol red indicator. Ammonia release raises pH, intense pink-red color	To distinguish *Klebsiella* (+) from *Escherichia* (−). To distinguish *Proteus* (+) from *Providencia* (−)
Voges–Proskauer test	Acetoin produced from sugar fermentation	Chemical test for acetoin using α-napthol	To separate *Klebsiella* and *Enterobacter* (+) from *Escherichia* (−). To characterize members of genus *Bacillus*

The typical reaction patterns for large numbers of strains of various pathogens have been published, and in the modern clinical microbiology laboratory, all of this information is stored in a computer. The results of the differential tests on an unknown pathogen are entered and the computer will make the best match by comparing the characteristics of the unknown with the species in the databank. For many organisms, as few as three or four key tests are all that are required to make an unambiguous identification. In cases of a dubious match, however, more sophisticated identification procedures may be called for, especially if the course of chemotherapy is a critical one.

Making a diagnosis

Many companies market their own versions of growth-dependent rapid identification systems (Figure 13.5*d* and *e*). Such systems are frequently designed for use in identifying enteric bacteria, because enterics are frequently implicated in routine urinary tract and intestinal infections (see Section 13.1).

Other growth-dependent rapid identification kits are available for other bacterial groups or even for sin-gle bacterial species. For example, commercial kits containing a battery of tests have been developed for specifically identifying the following bacteria: *Staphylococcus aureus*, *Streptococcus pyogenes*, *Neisseria gonorrhoeae*, *Haemophilus influenzae*, and *Mycobacterium tuberculosis*. Other kits are available for identification of the pathogenic yeasts *Candida albicans* and *Cryptococcus neoformans*.

The decision to use a specific diagnostic test is usually made by the clinical microbiologist following consultation with the physician. The clinical microbiologist will take into consideration the nature of the clinical specimen, basic characteristics (especially the Gram stain) of pure cultures obtained, and previous experience with similar cases. For example, an enteric

> Traditional methods for identifying pathogens depend on measuring metabolic changes induced as a result of growth. Numerous differential media are available in which color changes occur as a result of microbial growth. These growth-dependent methods provide rapid and reasonably accurate means of diagnosing many infectious diseases.

identification kit would be useless in identifying a Gram-positive coccus isolated from an abscess. Instead, an *S. aureus* or *S. pyogenes* kit would be used to make a positive identification.

13.3   Testing Cultures for Antibiotic Sensitivity

In medical practice, microbial cultures are isolated from diseased patients to confirm diagnosis and to aid in decisions on therapy. Determination of the sensitivity of microbial isolates to antimicrobial agents is one of the most important tasks of the clinical microbiologist.

We discussed the principles for the measurement of antimicrobial activity in Section 9.15. The sensitivity of a culture can be most easily determined by an agar diffusion method. Federal regulations of the Food and Drug Administration (FDA) now control the procedures used for sensitivity testing in the United States, and similar regulations exist in other countries. The recommended procedure is called the *Kirby–Bauer method*, after the workers who developed it (Figure

13.6). A plate of suitable culture medium is inoculated by spreading a sample of culture evenly across the agar surface. Filter paper discs containing known concentrations of different antimicrobial agents are then placed on the plate. The concentration of each agent on the disc is specified so that zone diameters of appropriate size will develop to indicate sensitivity or resistance. After incubation, the presence of inhibition zones around the discs of the different agents is noted. Table 13.4 presents typical zone sizes for several antibiotics. Zones observed on the plate are measured and compared to standard data, to determine if the isolate is truly sensitive to a given antibiotic.

Another procedure for antibiotic sensitivity testing involves an *antibiotic dilution assay*, either in culture tubes or in the wells of a microtiter plate (Figure 13.6e). (We discussed the use of microtiter plates in a box in Chapter 12.) A series of two-fold dilutions of each antibiotic are made in the wells and then all wells are inoculated with the same test organism. After incubation, the inhibition of growth by the various antibiotics can be observed by measuring turbidity. Sensitivity can be expressed as the *highest dilution* which completely inhibits growth.

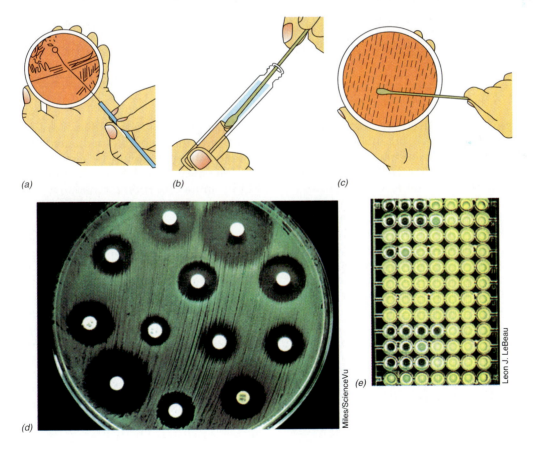

(a)   *(b)*   *(c)*

Miles/ScienceVu

Leon J. LeBeau

(d)   *(e)*

FIGURE 13.6   The Kirby–Bauer procedure for determining the sensitivity of an organism to antibiotics. (a) A colony is picked from an agar plate. It is inoculated into a tube of liquid culture medium and allowed to grow to a specified density. (b) A swab is dipped in the liquid culture. (c) The swab is streaked evenly over a plate of sterile agar medium. (d) Discs containing known amounts of different antibiotics are placed on the plate. After incubation, inhibition zones are observed. The susceptibility of the organism is determined by reference to a chart of zone sizes (see Table 13.4). (e) Antibiotic sensitivity determined by the dilution method on a microtiter plate. The organism is *Pseudomonas aeruginosa*. Each row has a different antibiotic. The use of the microtiter plate enables automation of these tests.

Table 13.4 Zone sizes for some antimicrobial disk susceptibility tests

Antibiotic	Amount on disc	Inhibition zone diameter (mm)[a]		
		Resistant	Intermediate	Sensitive
Ampicillin[b]	10 µg	11 or less	12–13	14 or more
Ampicillin[c]	10 µg	28 or less	—	29 or more
Cephoxitin	30 µg	14 or less	15–17	18 or more
Cephalothin	30 µg	14 or less	15–17	18 or more
Chloramphenicol	30 µg	12 or less	13–17	18 or more
Clindamycin	2 µg	14 or less	15–16	17 or more
Erythromycin	15 µg	13 or less	14–17	18 or more
Gentamicin	10 µg	12 or less	13–14	15 or more
Kanamycin	30 µg	13 or less	14–17	18 or more
Methicillin[c]	5 µg	9 or less	10–13	14 or more
Neomycin	30 µg	12 or less	13–16	17 or more
Nitrofurantoin	300 µg	14 or less	15–16	17 or more
Penicillin G[d]	10 Units	28 or less	—	29 or more
Penicillin G[e]	10 Units	11 or less	12–21	22 or more
Polymyxin B	300 Units	8 or less	9–11	12 or more
Streptomycin	10 µg	11 or less	12–14	15 or more
Tetracycline	30 µg	14 or less	15–18	19 or more
Trimethoprim-sulfamethoxazole	1.25/23.75 µg	10 or less	11–15	16 or more
Tobramycin	10 µg	12 or less	13–14	15 or more

[a]See Figure 13.6d for an illustration of a typical test.
[b]For Gram-negative organisms and enterococci.
[c]For staphylococci and highly penicillin-sensitive organisms.
[d]For staphylococci.
[e]For organisms other than staphylococci. Includes some organisms, such as enterococci and some Gram-negative rods, that may cause systemic infections treatable by high doses of Penicillin G.

Because of the widespread occurrence of antibiotic resistance (see Section 11.16), an antibiotic sensitivity test is essential for an appropriate isolate from each patient. Data such as those in Table 13.4 are useful to the physician in choosing the best antibiotic for a specific bacterial infection. Fortunately, many potentially serious pathogens are highly susceptible to a number of different antibiotics and this allows the physician considerable latitude in the course of treatment. Most *Pseudomonas aeruginosa* infections, on the other hand, are very difficult to treat with the majority of common antibiotics, and the physician's choice is limited to a rather restricted group of drugs.

> Antibiotics are widely effective for the treatment of infectious diseases but it is essential to determine which antibiotics a pathogen is susceptible to. The Kirby-Bauer method is a rapid and relatively precise method for determining antibiotic sensitivity by observing growth-inhibition zones on agar plates. The size of the inhibition zone around a filter paper disk containing the antibiotic indicates the sensitivity of the culture.

13.4 Antibody Titers and the Diagnosis of Infectious Disease

In the diagnosis of an infectious disease, isolation of the pathogen is not always possible, and one alternative is to measure antibody *titer* to a suspected pathogen. The principle here is that if an individual is infected with the suspected pathogen, the antibody titer to that pathogen should be elevated. Antibody titer can be measured by agglutination, precipitation, ELISA, immobilization methods, or complement fixation methods, depending upon the situation. The general procedure is to set up a series of dilutions of serum (usually twofold dilutions: 1:2, 1:4, 1:8, 1:16, 1:32, and so on) and to determine the *highest* dilution at which the antigen-antibody reaction occurs.

A *single* measure of antibody titer does not indicate active infection. Many antibodies remain at high titer for long times after infection; to establish that an acute illness is due to a particular agent, it is essential to show a *rise* in antibody titer in successive samples of serum from the same patient. Frequently, the antibody titer will be low during the acute stage of the infection and rise during convalescence (Figure 13.7). Such a rise in antibody titer is the best indication that the illness is due to the suspected agent. Measurement of such a rise in antibody titer is also useful in diagnosis of infectious diseases of a rather chronic nature, such as typhoid fever and brucellosis. In some cases, however, the mere presence of antibody may be sufficient to indicate infection. This is the case for a pathogen which is quite rare in a population, so that the presence of antibody is sufficient to indicate that the individual has experienced an infection. A relevant example here is AIDS. As discussed in Section

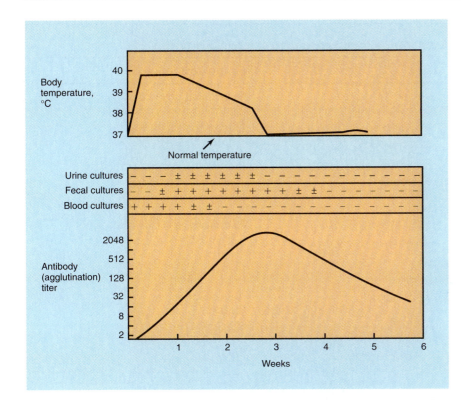

FIGURE 13.7
The course of infection in a typical untreated typhoid fever patient. Measurement of body temperature provides a measure of the course of clinical symptoms. The antibody titer was measured by determining the highest dilution (two-fold series) causing agglutination of a test strain of *Salmonella typhi*. Presence of viable bacteria in blood, feces, and urine was determined from periodic cultures. Note that the pathogen clears from the blood as the antibody titer rises, and clearance from feces and urine requires longer time. Body temperature gradually drops to normal as the antibody titer rises. The data given do not represent a single patient, but are a composite of the picture seen in large numbers of patients.

13.7, an extremely sensitive and highly reliable ELISA test is now available for detecting antibodies to human immunodeficiency virus (HIV), the AIDS virus. Upon infection with HIV, an antibody response is mounted, but, unlike most diseases where antibody titers *increase* in the later stages of the disease, in AIDS the loss of T-helper cell function (see Section 15.7) actually causes HIV-specific antibody titers to *decrease* as the disease progresses. Nevertheless, the exquisite sensitivity of ELISA allows detection of even very low antibody titers, and thus this immunoassay can be used to routinely screen blood samples for evidence of HIV infection.

Another point to emphasize is that not all infections result in formation of systemic antibody. If a pathogen is extremely localized in its action, there may be little induction of an immunological response and no rise in antibody titer, even if the pathogen is proliferating profusely at its site of infection. A good example is the disease gonorrhea. Infection with *Neisseria gonorrhoeae*, the causative agent of gonorrhea, does not elicit a systemic immune response and thus reinfection of a cured individual is not uncommon (see Sections 13.1 and 15.6 for further discussion of gonorrhea). It should also be pointed out that presence of antibody in the serum may have been due to vaccination. In fact, measurement of rise in antibody titer

---

> **Immunology can aid in diagnosis of infectious disease. During the active phase of an infection, there is usually a rise in antibody titer against the pathogen, and by measuring this increase during the course of the disease, a diagnosis can be accomplished.**

during vaccination is one of the best ways of indicating that the vaccine being used is effective.

Some of the most common clinical immunological procedures are outlined in Table 13.5.

13.5 Fluorescent Antibodies

As we have seen in Chapter 12, antibodies are highly specific in their interactions with antigens. This specificity can be used in many ways in the diagnosis of infectious disease and a wide variety of immunological methods are employed by the clinical microbiologist. Some immunological methods can be applied directly to a clinical specimen, thus avoiding the need for a culture of the organism.

We described the theory of fluorescent antibodies in Section 12.12. Recall that fluorescent dyes can be attached to antibody molecules. By injecting whole bacterial cells or cell fractions (for example, cell walls, surface proteins, flagella, pili) of known pathogens into experimental animals, banks of reference antisera can be prepared which can be made fluorescent for diagnostic use. Commercially available antisera prepared against several different antigens from known pathogenic microorganisms are readily available for clinical use. In the fluorescent antibody test, a smear of the suspected pathogen is allowed to react with a specific fluorescent antibody and observed under a fluorescent microscope. If the pathogen contains surface antigens against which the fluorescent antiserum was prepared (i.e., the suspected pathogen is identical to or immunologically closely related to the cells used to generate the antibodies), the cells will fluoresce (Fig-

Table 13.5   Some clinical immunological procedures for identification of infectious agents

Pathogen or disease	Antigen	Serological procedure*
Streptococcus (group A)	Streptolysin O (exotoxin)	Neutralization of hemolysis
	DNase (extracellular protein)	Neutralization of enzyme
Neisseria meningitidis	Capsular polysaccharide	Passive hemagglutination (*N. meningitidis* polysaccharide adsorbed to red cells)
	N. meningitidis cells	Indirect fluorescent antibody
Salmonella	O or H antigen	Agglutination (Widal test)
		ELISA
Vibrio cholerae	O antigen	Agglutination
		Bactericidal test (in presence of complement)
		ELISA
Borellia burgdorferi (Lyme disease)	Flagellin	ELISA
	Surface proteins	Immunoblot
Brucella	Cell wall antigen	Agglutination
		ELISA
Corynebacterium diphtheriae	Toxin	Skin test (Schick test)
Mycobacterium tuberculosis	Tuberculin (partially purified bacterial proteins, PPD)	Skin test (Tuberculin test)
		ELISA
Syphilis (*Treponema pallidum*)	Cardiolipin-lecithin-cholesterol	Flocculation [Venereal Disease Research Laboratory (VDRL) test]
	T. pallidum antigens	ELISA
	T. pallidum cells	Indirect fluorescent antibody (FTA)
Rickettsial diseases (Q fever, typhus, Rocky Mountain Spotted fever)	Killed rickettsial cells	Complement-based assay or cell agglutination tests
		ELISA
Influenza virus	Influenza virus suspensions	Complement-based assay
	Nasopharynx cells containing influenza virus	Immunofluorescence
AIDS	Human immunodeficiency virus (HIV)	ELISA
		Immunoblot
Pneumocystis carinii	*P. carinii* cells	Immunofluorescence

Except for the skin tests and the immunofluorescent tests, the serum of the patient is assayed for antibody against the specific antigen by the methods shown.

ure 13.8). Organisms immunologically unrelated to the control organism generally do not react or react only weakly.

Fluorescent antibodies can also be applied directly to infected host tissues, permitting diagnosis long before primary isolation techniques yield a suspected organism. For example, in diagnosing legionellosis (see Section 15.2) a positive diagnosis can be made by staining biopsied lung tissue with fluorescent antibodies prepared against cell walls of *Legionella pneumophila*, the causative agent of legionellosis. Likewise, a fluorescent antibody against the capsule of *Bacillus anthracis* can be used in the microscopic diagnosis for anthrax. Fluorescent antibody reactions can also be used in diagnosis of viral infection (Figure 13.8*b*), and in a variety of noninfectious diseases. For example, in identifying cell types expressing a particular antigen, such as malignant cells, fluorescent antibodies may be very valuable in following the course of the disease (Figure 13.9). Fluorescent antibodies therefore provide useful, rapid, and highly specific clinical tests that can

often be of great benefit in guiding the physician to the proper diagnosis.

Immunodiagnoses using fluorescent antibody techniques are not without their pitfalls, however. Nonspecific staining can be a problem because of surface antigens that may *cross-react* between various bacterial species, some of which may be members of the normal flora. This is particularly a problem among enteric bacteria, where antigens derived from lipopolysaccharides are frequently sufficiently similar among species to cause binding or partial binding of the fluorescent probe. The clinical microbiologist must therefore be careful to perform all the necessary controls using nonspecific (preimmune) sera, and confirm all positive immunofluorescent findings by other immunological or microbiological tests.

Fluorescent antibodies can be used for quick and relatively accurate identification of pathogens even in the absence of culture, and can also be used to trace the presence of a pathogen in body tissues.

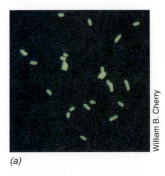
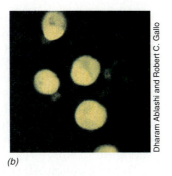

(a)   *(b)*

William B. Cherry

Dharam Ablashi and Robert C. Gallo

FIGURE 13.8 Examples of the use of fluorescent antibodies in clinical microbiology. (a) Immunofluorescent stained cells of *Legionella pneumophila*, the cause of legionellosis. (b) Detection of virus-infected cells by immunofluorescence. Human B lymphotrophic virus (HBLV) infected spleen cells were incubated with serum which contained antibodies to HBLV from a patient with a lymphoproliferative disorder. Cells were then treated with fluorescein isothiocyanate-conjugated antihuman IgG antibodies. HBLV-infected cells fluoresce bright yellow. Cells in the background did not react with patient's serum.

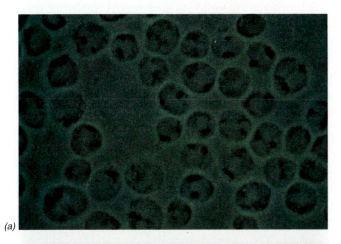

(a)

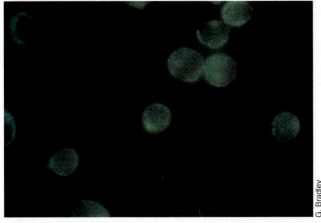

(b)

G. Bradley

FIGURE 13.9 Use of fluorescent antibodies in noninfectious disease diagnostics. (a) Human leukemic cells, some of which are sensitive to a toxic anticancer drug and some which are not, appear indistinguishable. (b) When the cells in (a) were treated with a fluorescent monoclonal antibody that binds specifically to a protein found only on the surface of drug-resistant cells, the latter fluoresce while drug-sensitive cells do not.

13.6   Monoclonal Antibodies

Monoclonal antibody technology has greatly reduced the problems of cross-reactivity in immunodiagnoses. We discussed monoclonal antibodies in Section 12.11. As noted, a monoclonal antibody is generally highly specific for a *single* antigenic determinant and hence is very useful in immunodiagnostics. Fluorescein-labeled monoclonal antibodies made against a variety of pathogens are now available. For example, fluorescent antibodies against *Chlamydia trachomatis* membrane proteins can be used to detect this organism in host tissues (*C. trachomatis* causes a variety of sexually transmitted diseases as well as trachoma, a serious eye disease). These monoclonal antibodies react with *C. trachomatis* but are so specific they fail to react with even the closely related species, *C. psittaci. C. trachomatis* is an obligate intracellular parasite and not easily cultured for this reason. Use of the fluorescent anti-*C. trachomatis* monoclonal antibody on cervical scrapings or urethral or vaginal exudates makes positive identification of chlamydial infections almost routine.

Other monoclonal antibodies have been developed against outer membrane proteins of *Neisseria gonorrhoeae*. These probes are not only monospecific, reacting only with *N. gonorrhoeae*, but are also capable of differentiating *strains* of *N. gonorrhoeae*. The use of fluorescent monoclonal antibodies therefore eliminates much of the cross-reactivity problem observed when polyclonal sera are used.

The power of monoclonal antibodies in microbiological diagnostics should be emphasized. With presently available technology, it is possible to generate monoclonal antibodies which react with only a certain bacterial species, or even with only a certain *strain* of a species, as mentioned above. In addition, viral antigens can be detected with the appropriate monoclonals. For example, both fluorescent and enzyme-conjugated monoclonal antibodies (subsequently assayed by ELISA, see next section) have been developed for the diagnosis of herpes infections and the typing of herpes virus obtained from clinical specimens. Hence, monoclonal antibodies are useful for broad screening purposes as well as highly detailed analyses. Monoclonal antibodies have also been widely used in noninfectious disease diagnoses as well (see Figure 13.9).

> **Monoclonal antibodies provide great specificity and reproducibility. They have revolutionized the use of antibodies in clinical diagnostics.**

13.7   Clinically Useful ELISA Tests

The enzyme-linked immunosorbent assay (ELISA), described in Section 12.13, has been frequently applied to problems in clinical diagnostics. Both direct and indirect ELISAs have been developed. Recall that *direct* ELISAs employ enzyme-conjugated antibodies, frequently monoclonal antibodies, to detect *antigen* in a patient specimen, while *indirect* ELISAs employ enzyme-conjugated antibodies to recognize *antibodies* from a patient's serum which have previously bound a specific antigen. In either case, the high specificity, extreme sensitivity, and low cost of ELISA tests make them excellent tools for clinical diagnostics.

The AIDS ELISA

Since the discovery that the causative agent of AIDS, the human immunodeficiency virus (HIV) (see Section 15.7), could be transmitted by bodily fluids including blood, a screening test was needed to test blood samples to ensure that HIV was not being inadvertently transmitted during blood transfusions or through the transfer of blood products. An ELISA test is used for

the routine screening of blood for signs of exposure to HIV (and hence possible AIDS).

The AIDS ELISA is an *indirect* ELISA designed to measure *antibodies* to the HIV virus present in serum. Even though AIDS is a disease that severely cripples the body's immune system, initial infection with HIV leads to the production of antibodies to several HIV antigens, in particular those of the HIV envelope. These antibodies can be detected by the AIDS ELISA test (Figure 13.10).

To carry out an AIDS ELISA, microtiter plates are first coated with a disrupted preparation of HIV particles; about 200 ng of disrupted HIV is all that is required. Following a brief incubation period to ensure binding of the antigens to the surface of the microtiter wells, a diluted serum sample is added and the mix-

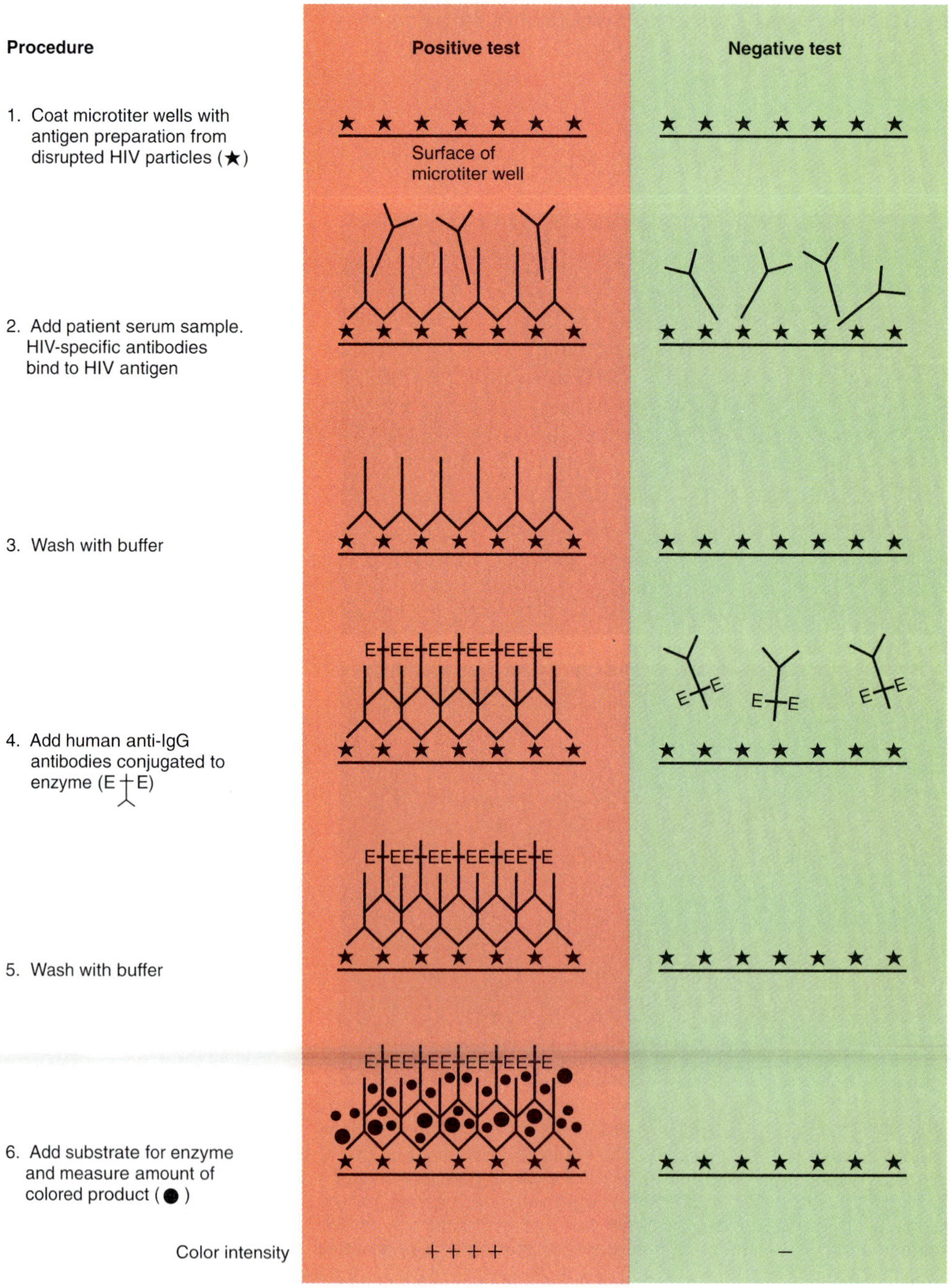

FIGURE 13.10
Indirect ELISA test for detecting antibodies to human immunodeficiency virus (HIV), the causal agent of acquired immune deficiency syndrome (AIDS).

Procedure

1. Coat microtiter wells with antigen preparation from disrupted HIV particles (★)

2. Add patient serum sample. HIV-specific antibodies bind to HIV antigen

3. Wash with buffer

4. Add human anti-IgG antibodies conjugated to enzyme (E╪E)

5. Wash with buffer

6. Add substrate for enzyme and measure amount of colored product (●)

Color intensity

ture incubated to allow any potential HIV-specific antibodies to bind to HIV antigens. To detect the presence of antigen-antibody complexes, a second antibody is then added. This second antibody is an enzyme-conjugated antihuman IgG preparation. Following a brief incubation period with the second antibody and a washing step to remove any unbound second antibody, the enzyme activity is assayed (the antihuman IgG antibodies will bind to any HIV-specific IgG antibodies previously bound to the HIV antigen preparation). A color is obtained in the enzyme assay in proportion to the amount of antihuman IgG antibody bound (Figure 13.10). The binding of the second antibody is an indication that antibodies from the patient's serum recognized the HIV antigens, and thus that the patient has antibodies to HIV (and hence, has been exposed to HIV). Control sera (known to be HIV-negative) are run in parallel with any samples to measure the extent of background absorbance in the assay.

The AIDS ELISA is a rapid, highly sensitive and specific method for detecting exposure to HIV, and a number of companies have developed their own versions of the test. Since ELISAs in general are highly adaptable to mass screening and automation and require very little in the way of equipment, the AIDS ELISA is used as a standard screening method for blood. However, this test method can give erroneous results under certain circumstances.

One drawback to using the AIDS ELISA is that the test occasionally gives false positive results. Because a number of factors can contribute to these results, none of which are related to exposure to HIV, all positive AIDS ELISA tests are confirmed by another independent test, usually the Western blot (immunoblot) test (see Section 13.9). A positive AIDS Western blot test after a positive AIDS ELISA is considered proof of HIV infection.

A final drawback to the AIDS ELISA is the possibility of obtaining false negative results. As we saw in Section 12.8, it takes the immune system some time to develop an effective antibody response with a detectable antibody titer. In the case of HIV infection, this lag time is estimated to be six weeks to a year. Therefore, individuals who have been recently infected with HIV may not yet be producing detectable amounts of antibody when they are tested. Another reason for a false negative result in the AIDS ELISA is the total destruction of the immune system seen in advanced cases of AIDS; if no immune cells are left in the body, no antibodies can be made and the ELISA test is not useful. However, at this stage of disease, a clinical diagnosis is possible and the ELISA test is only useful as a confirmatory indicator.

Other ELISA tests of clinical importance

Besides the ELISA test for AIDS, literally hundreds of clinically useful ELISAs have been developed. Some of these are *direct* ELISAs for detecting antigens. In these tests, antibody to the antigen of interest is bound to the well of a microtiter plate (see later, Figure 13.14). These antibodies are used to trap antigen from the clinical specimen. A second antibody containing a conjugated enzyme serves as the detection system. Direct ELISAs to detect bacterial toxins such as cholera toxin, enteropathogenic *Escherichia coli* toxin, and *Staphylococcus aureus* enterotoxin have also been developed. Viruses currently detected using ELISA techniques include rotavirus, HIV, hepatitis viruses, rubella virus, bunyavirus, measles and mumps viruses, and parainfluenza virus.

Besides the application to AIDS antibody detection, *indirect* ELISAs have been developed for detecting antibodies to a variety of clinically important bacteria. Although not meant to be a complete list, ELISAs for detecting serum antibodies to *Salmonella* (gastrointestinal problems), *Yersinia* (plague), *Brucella* (brucellosis), a variety of rickettsias (Rocky Mountain Spotted Fever, typhus, Q-fever), *Vibrio cholerae* (cholera), *Mycobacterium tuberculosis* (tuberculosis), *Mycobacterium leprae* (leprosy), *Legionella pneumophila* (legionellosis), *Borellia burgdorferi* (Lyme disease), and *Treponema pallidum* (syphilis) have been developed. ELISAs have also been developed for detecting antibodies to *Candida* (yeast), and antibodies to a variety of parasites, including those causing amebiasis, Chagas' disease, schistosomiasis, toxoplasmosis, and malaria.

The speed, low cost, lack of radioactive waste, and long shelf life make ELISA tests particularly attractive for many laboratories. But it is the extreme *sensitivity* of ELISAs that really make them important immunodiagnostic tools. New ELISA tests are marketed each year and many of them are rapidly replacing older methods.

> **ELISA tests are extremely sensitive and can be automated, thus permitting their widespread use as routine diagnostic procedures. The ELISA test for the virus causing AIDS has been used to screen blood in blood banks and thus avoid the use of contaminated blood for transfusions.**

13.8 Agglutination Tests Used in the Clinical Laboratory

Agglutination is due to the binding of a particulate antigen by antibody. Agglutination reactions were discussed in Section 12.12, the well known ABO blood grouping reaction serving as a prime example. However, many other agglutination reactions have been adapted for clinical use for the detection of antigens or antibodies associated with certain disease. Although not as sensitive a test as most ELISAs, agglutinations remain useful in clinical diagnostics as inexpensive and highly specific and rapid immunoassays.

Latex bead agglutination

The agglutination of antigen- or antibody-coated latex beads by complementary antibody or antigen from a patient is a typical method of rapid diagnosis. Small

(0.8 μm) latex beads coated with a specific antigen or antibody are mixed with patient serum on a microscope slide and incubated for a short period. If the antibody complementary to the molecule bound to the bead surface is present in the patient's serum, the milky-white latex suspension will be visibly agglutinated. Latex agglutination is also used to detect bacterial surface antigens by mixing a small amount of a bacterial colony with antibody-coated latex beads. For example, a commercially available suspension of latex beads containing antibodies to Protein A and Clumping Factor, two molecules found exclusively on the surface of *Staphylococcus aureus*, is virtually 100 percent accurate in identifying clinical isolates of *S. aureus*. Unlike traditional tests for *S. aureus*, many of which are growth-dependent assays, identification of *S. aureus* by the latex bead assay takes only 30 seconds. Other latex bead agglutination assays have been developed to identify pathogenic streptococci, *Neisseria gonorrhoeae*, *Haemophilus influenzae*, and the yeasts *Cryptococcus neoformans* and *Candida albicans*.

A very widely used latex agglutination assay is that for detecting specific serum antibodies for *rheumatoid factor*, an antibody directed against the body's own immunoglobulins, which is associated with the autoimmune disease *rheumatoid arthritis* (see Section 12.14). Latex beads coated with human immunoglobulin are mixed with whole blood and agglutination scored versus positive and negative control sera run in parallel. Latex bead assays are simple and specific. In addition, the inexpensive nature of the assays make them suitable for large scale screening purposes as well; the widespread use of the rheumatoid test is a good example of this. Because they require no expensive equipment, they should be especially useful in developing countries.

Some agglutination assays use a suspension of activated charcoal as the carrier. For example, a rapid diagnostic test for detection of the virus *Herpes simplex*, frequently associated with oral fever blisters or genital sores (see Section 15.6), employs anti-*H. simplex* virus antibodies adsorbed to small particles of activated charcoal. Cotton swabs used to sample suspected herpes lesions are placed in a buffer solution and samples of this are used to test for charcoal agglutination. A positive test is indicated by visible clumping of the charcoal into large black aggregates. Because of the specificity of the antiserum used (and here, naturally, monoclonal antibodies are ideal), complicating cross-reactions with related pathogens are not a problem. Like latex beads, charcoal agglutination tests can be rapid and cost effective diagnostic tools.

> Latex bead agglutination is a sensitive adaptation of a traditional agglutination test and finds use for certain diagnoses. Agglutination assays are rapid, inexpensive and available for identifying a variety of common pathogens.

13.9   Immunoblot Procedures

Antibodies can be used in clinical diagnostics to identify *proteins* associated with specific pathogens. The procedure employs two important biochemical techniques discussed previously: (1) the separation of proteins on polyacrylamide gels, and (2) the transfer (blotting) of proteins from gels to nitrocellulose paper (see Nucleic Acids box, Chapter 5). Since the latter technique, originally devised to transfer DNA from gels to nitrocellulose paper was termed a "Southern" blot, protein blotting and the subsequent identification of the proteins by specific antibodies is sometimes called the "Western" blot technique.

The immunoblot is a very sensitive method for detecting specific proteins in complex mixtures. In the first step of an immunoblot, a protein mixture is subjected to electrophoresis on a polyacrylamide gel. This separates the proteins into several distinct bands, each of which represents a single protein type of specific molecular weight (Figure 13.11). The proteins are then transferred to nitrocellulose paper by an electrophoretic transfer process that forces the proteins out of the gel and onto the paper. At this point, antibodies raised against a protein or group of proteins from a pathogen are added to the nitrocellulose blot. Following a short incubation period to allow the antibodies to bind, a radioactive marker which binds antigen–antibody complexes is added. The most common radioactive marker used is *Staphylococcus* protein A iodinated with radioactive iodine, ^{125}I. Protein A has a strong affinity for antigen–antibody complexes and binds firmly to them. Once the radioactive marker has bound, its vertical position on the blot can be detected by exposing the nitrocellulose blot to X-ray film; the gamma rays emitted by the ^{125}I expose the film only in the region where the radioactive antibody has bound to antigen-antibody complexes (Figure 13.11).

For clinical applications where the use of radioactivity is not advisable, immunoblots employ ELISA technology (see Section 13.7) for detection of bound antigen-antibody complexes. Following treatment of the blotted proteins with specific antibody, the paper is washed and then treated with a second antibody which binds to the first. For example, if antibodies from a human were used in the first step, then the second antibody could be a rabbit anti-human antibody. Covalently attached to this second antibody is an enzyme. The original antigen-antibody complexes are visualized when the enzyme is assayed because the product of the enzyme reaction leaves a colored product on the nitrocellulose filter at any spot where rabbit antibodies bound to the human antibodies. By comparing the location of the color on the nitrocellulose paper with the position of spots run from controls, a protein associated with a given pathogen can be positively identified; presence of the specific protein is evidence of exposure to the pathogen.

1. Denature proteins by boiling in detergent

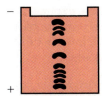

2. Subject mixture to electrophoresis; proteins separate by molecular weight

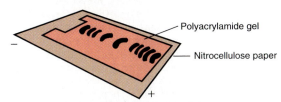

— Polyacrylamide gel

— Nitrocellulose paper

3. Blot the separated proteins from the gel to nitrocellulose paper

— Antibodies (Y) bound to protein

4. Treat nitrocellulose paper containing blotted proteins with antibodies; each antibody recognizes and binds to a specific protein

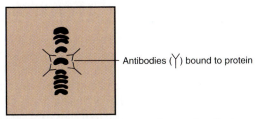

125I   125I

E E
E E

5. Add marker to bind to antigen: antibody complexes, either (left) radioactive *Staphylococcus* protein A–125I, or (right) antibody containing conjugated enzyme

X-ray film

Nitrocellulose with enzyme-produced colored spot

(a)

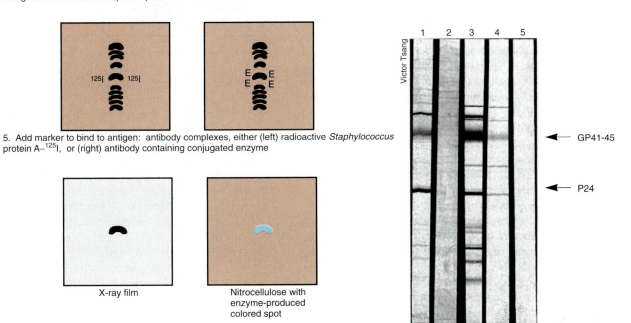

(b)

Victor Tsang

1 2 3 4 5

◄— GP41-45

◄— P24

FIGURE 13.11 The Western blot (immunoblot) and its use in the diagnosis of HIV infection. (a) Protocol for an immunoblot. (b) Developed HIV immunoblot. The proteins P24 and GP41-45 are coat proteins of the virus and are diagnostic for HIV. Lane 1, positive control serum (from known AIDS patients); lane 2, negative control serum (from healthy volunteer); lane 3, strong positive from patient sample; lane 4, weak positive from patient sample; lane 5, reagent blank to check for background binding.

The AIDS immunoblot

Immunoblots have had a significant clinical impact on the diagnosis and confirmation of cases of AIDS. Since an immunoblot is more laborious and costly than the ELISA test, AIDS ELISAs have been widely used for screening purposes. However, AIDS ELISAs occasionally yield false positive results. Thus, an immunoblot is used to confirm the ELISA results.

Like the AIDS ELISA, the AIDS immunoblot is designed to detect the presence of *antibodies* to HIV in a serum sample. To perform an AIDS immunoblot, a purified preparation of HIV is treated with the detergent sodium dodecyl sulfate (SDS), which solubilizes HIV proteins and also renders the virus inactive. HIV proteins are then resolved by polyacrylamide gel electrophoresis. The HIV proteins are then blotted from the gel onto sheets of nitrocellulose paper (Figure 13.11*b*). At least seven major HIV proteins are resolved by electrophoresis and two of them, designated P24 and GP41-45, are used as specific diagnostic proteins in the AIDS immunoblot. Protein P24 is the HIV core protein and proteins GP41-45 are HIV coat proteins (the structure of HIV is discussed in Section 6.22).

Following blotting of the proteins, the nitrocellulose strips are incubated with a serum sample previously identified as HIV positive by the AIDS ELISA. If the sample is truly HIV positive, antibodies against HIV proteins will be present and will bind to the HIV proteins separated on the nitrocellulose paper (Figure 13.11*b*). To detect whether antibodies from the serum sample have bound to HIV antigens, a *detecting antibody*, anti-human IgG conjugated to the enzyme peroxidase, is added to the strips. If detecting antibody binds, the activity of the conjugated enzyme will form a brown precipitate on the strip at the site of antibody binding (Figure 13.11*b*). A confirmation of HIV-positiveness is made if the positions of the bands obtained in strips using a serum sample correspond to the position of bands formed from a serum sample of a known HIV-infected patient; negative control sera must show no reaction (Figure 13.11*b*).

Although the intensity of the bands obtained in the AIDS immunoblot varies somewhat from sample to sample (see Figure 13.11*b*), the interpretation of an AIDS immunoblot is generally unequivocal, and thus the test is valuable in distinguishing false positives from real positives initially identified by ELISA. To make the AIDS immunoblot clinically accessible, nitrocellulose strips containing inactivated HIV antigens (previously separated by electrophoresis) are available commercially. These strips can be incubated directly with a serum sample and subsequently treated with the detecting antibody. This alleviates the need for clinical microbiology laboratories to grow their own HIV as a source of antigen for the AIDS immunoblot.

The immunoblot technique is considered the definitive test for serodiagnosis of HIV infection and is virtually always used to confirm positive AIDS ELISA tests (see Section 13.7). This technique is also used to determine the specificity of screening tests for the Lyme disease antibody (see Table 13.5). However, because of the expense, technical requirements, and time involved, immunoblot tests are not likely to supplant the rapid, low cost ELISA methods for general screening purposes.

> Immunoblot procedures permit the detection of specific antigens or antibodies in mixtures, such as in blood serum. Electrophoresis is used to separate out the mixture of antigens (or antibodies) and an immunological reagent is then used to detect the entity of interest. Blotting simply transfers the proteins from the electrophoresis gel in an undisturbed orientation to a carrier where the immunological reaction can be carried out. Immunoblotting is widely used in AIDS diagnostics and in research.

13.10   Plasmid Fingerprinting

With recent developments in nucleic acid technology, it is now possible to use DNA in the clinical laboratory as a diagnostic tool for identifying suspected pathogenic bacteria. We describe in Section 13.11 diagnostic methods based on nucleic acid *sequences* and hybridization analyses, but discuss here a phenotypic method based on plasmid compositions.

Plasmid fingerprinting involves the characterization of plasmids obtained from cells of a suspected pathogen. Most bacteria contain one or more plasmids and enteric bacteria typically contain several. Each type of plasmid is a covalently closed circle of DNA and has a characteristic molecular weight. Plasmids of different sizes can be separated by electrophoresis on agarose gels (see Nucleic Acids box, Chapter 5). To do a plasmid fingerprint, cells are lysed and the plasmid DNA separated from chromosomal DNA by a specific extraction procedure. The DNA is subject to electrophoresis on the agarose gel and then the gel is stained with ethidium bromide to visualize the DNA (Figure 13.12). The rate of migration of the plasmid DNA in the gel is *inversely* proportional to its molecular weight; that is, small plasmids migrate farther than large plasmids. The positions and number of plasmid bands in the gel (the plasmid "fingerprint") are characteristic of a given species, and the greater the number of plasmids present, the easier species differentiation becomes.

The plasmid fingerprint of a given organism is diagnostically useful; major differences in plasmid size and/or number between the isolated pathogen and reference strains are generally indicative of little taxonomic affiliation. Identical plasmid fingerprints of reference and sample organisms are highly suggestive of relatedness, but cannot be interpreted as definitive. Problems can be encountered if the particular strain of pathogen tested has picked up plasmids from other species or has lost plasmids characteristic of the reference strain. Plasmid fingerprinting is therefore used as an adjunct method, complementing other methods of

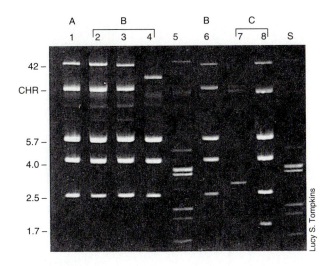

FIGURE 13.12 Plasmid fingerprints of *Staphylococcus epidermidis* isolates recovered from three patients suffering from prosthetic valve endocarditis. Patient A had the same strain of *S. epidermidis* as Patient B, but Patient B had, in addition, one unique strain (lane 4). Patient C showed two unique strains. The numbers on the left of the gel refer to the molecular size of the major plasmids in kilobase pairs. The CHR band is fragmented chromosomal DNA. Lanes 5 and S are molecular weight standards.

identification, especially if the latter produce ambiguous results. Of importance to the clinical laboratory, however, is the fact that the *sensitivity* of plasmid fingerprinting is very high. For a typical analysis, all that is needed is about 1 ml of a liquid culture or several colonies from a plate.

13.11 Nucleic Acid Probes in Clinical Diagnostics

The emergence of molecular biology has given rise to new molecular tools that are rapidly being adapted to the field of diagnostic microbiology. *DNA diagnostics*, as this area has come to be known, is revolutionizing the whole approach to identifying and monitoring infectious diseases, genetic and malignant diseases, and other medical conditions such as coronary artery disease and diabetes. The new approach is one in which *genotypic* rather than *phenotypic* factors are of importance. The power of DNA diagnostics is a consequence of two facts: (1) nucleic acids can be rapidly and sensitively measured, and (2) the *sequence* of nucleotides in a given DNA molecule is so specific that hybridization analyses can be used for reliable clinical diagnoses.

Automation is helping to make DNA analysis virtually routine in the clinical setting. Automated DNA extractors, polymerase chain reaction machines (see Section 8.9), DNA sequencers, and pulsed field gel electrophoresis equipment (for separating large DNA segments such as whole chromosomes; see Nucleic Acids box, Chapter 5) are all available for use in the diagnosis of diseases. However, it is the **nucleic acid probe** that is the major molecular biological tool in clinical laboratories, so we limit our discussion here to this method.

Nucleic acid probes

One of the most powerful analytical tools available to clinical microbiologists is *nucleic acid hybridization*. Instead of detecting a whole organism or its products (for example, antigens), hybridization detects the presence or absence of *specific DNA sequences* associated with a specific organism. To identify a microorganism through DNA analysis, the clinical microbiologist must have available a *nucleic acid probe* to that microorganism, a *single strand* of DNA containing sequences unique to the organism. If a microorganism in a clinical specimen contains DNA sequences complementary to the probe, the two sequences can hybridize (following appropriate sample preparation to yield single-stranded DNA from the microorganism) forming a *double-stranded* molecule (Figure 13.13). To detect that a reaction has occurred, the probe is labeled with a *reporter molecule*, either a radioisotope, an enzyme, or a fluorescent compound that can be measured in small amounts following hybridization (Figure 13.13). Depending on the reporter used (radioisotopes are the most sensitive), as little as 0.25 μg of DNA per sample can be detected.

Nucleic acid probes offer many advantages over clinical immunological assays. Nucleic acids are much more stable than proteins to high temperatures, high pH, organic solvents, and other chemicals. This means that a clinical sample can be treated in a relatively harsh manner so as to destroy most interfering material, leaving behind the nucleic acid. In addition, nucleic acid probes are more defined entities than antibodies; the composition of a probe can be accurately checked by sequence analysis and new molecules of the probe can be produced in DNA synthesizers whenever necessary.

Nucleic acid probes are also very sensitive. With current technology it is possible to detect less than 1 microgram of nucleic acid per sample. This translates into about 10^6 bacterial cells or virus particles. Although probes used in this fashion are not as sensitive as direct culture (where as few as 1–10 cells per sample can be detected), probe methods are of obvious benefit in situations where culture of the organism is difficult or even impossible. Probe sensitivity can be increased by more than 1 million times if the target DNA is first amplified using the polymerase chain reaction (PCR, see Section 8.9). With PCR, only a single cell is theoretically necessary for detection by a probe. We now consider how nucleic acid probes work in actual practice.

Probes in the clinical laboratory

In most clinical probe assays, colonies from plates or pieces of infected tissue are treated with strong alkali to lyse the cells and partially denature the DNA, forming single-stranded molecules (Figure 13.13a). This mixture is then affixed to a filter or left in solution (for dipstick assays, see later) and the labeled probe added. Hybridization is allowed to occur at a temperature at which considerable sequence homology be-

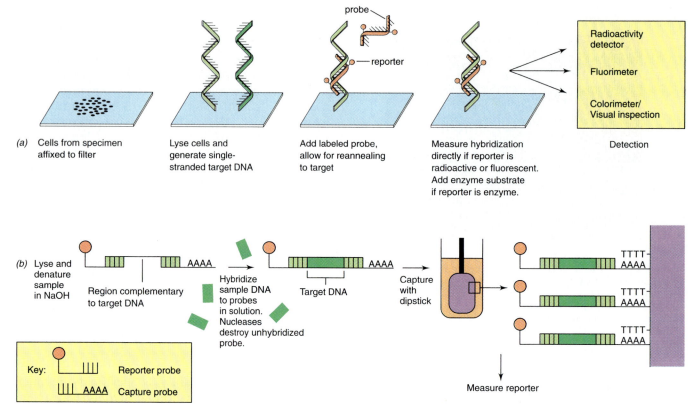

FIGURE 13.13 Nucleic acid probe methodology in clinical diagnostics. (a) Membrane filter assay. The detecting system (reporter) can be either a radioisotope, fluorescent dye, or an enzyme. (b) Dipstick assays. In the dipstick assay a dual reporter/capture probe is used. The capture probe contains a poly-dA tail that will hybridize to a poly-dT oligonucleotide affixed to the dipstick.

tween target DNA and probe DNA is necessary to form a stable duplex (the actual temperature used in a given probe assay is governed by the length and GC content of the probe and target DNA). Following a wash to remove any unhybridized probe DNA, the extent of hybridization that occurred is measured using the reporter molecule attached to the probe. Depending on how the probe was labeled, this involves measurement of radioactivity, enzyme activity, or fluorescence from an attached dye.

Nucleic acid probes have been marketed for the identification of several major microbial pathogens (Table 13.6). However, in addition to their clinical usefulness, probes are finding widespread application in food industries and in food regulatory agencies. Because culture of the pathogens is not required, probe detection systems can be used to routinely monitor foods for their content of important pathogens such as *Salmonella* and *Staphylococcus*. In probe assays of food, an enrichment period is usually employed to allow very low numbers of cells in the food to multiply to a sufficient number to be detectable by the probe.

Probes designed for use in the food industry employ probe dipsticks to remove hybridized DNA from solution. Two component probes are used here, one serving as a *reporter probe* and the other as a *capture probe* (Figure 13.13b). Following hybridization of the

reporter/capture probe to target DNA, the dipstick, which contains a sequence complementary to the capture probe (usually poly dT to capture poly dA on the probe, see Figure 13.13b) is inserted into the hybridization solution and it traps hybridized DNA for removal and measurement (Figure 13.13b).

Probes to detect certain cancer viruses are also being developed. For example, a probe is now available to detect DNA sequences unique to human papilloma viruses. These viruses sometimes cause skin and cervical cancer in humans (see Section 6.14) and a specific group of papilloma viruses causes genital warts (see Section 15.6). In women, an increased incidence of genital papilloma virus infection is associated with an increased risk of cervical cancer. The DNA probe developed for papilloma viruses can be used to search by hybridization for papilloma virus sequences in tissues removed during a cervical exam. Early detection and treatment of papilloma virus infections decreases the risk of cervical cancer.

The development of new nucleic acid probes is a major activity in pharmaceutical, biotechnology, and clinical diagnostic companies. Although developing probes for diseases for which no probe yet exists is a top priority, a major goal of probe research and development is to make existing probes even more specific and to continue to simplify the procedures necessary

Table 13.6   Nucleic acid probes for clinically significant microorganisms

Pathogen	Disease
Bacteria	
Legionella pneumophila	Pneumonia
Neisseria gonorrhoeae	Gonorrhea
Mycoplasma pneumoniae	Pneumonia
Mycoplasma hominis	Urinary tract infections; pelvic inflammatory disease
Mycobacterium tuberculosis	Tuberculosis
Mycobacterium avium	Tuberculosis
Chlamydia trachomatis	Venereal syndromes; trachoma
Listeria monocytogenes	Listeriosis
Salmonella spp.	Gastrointestinal disorders
Shigella spp.	Gastrointestinal disorders
Escherichia coli (enteropathogenic strains)	Gastrointestinal disorders
Yersinia enterocolitica	Gastroenteritis
Staphylococcus aureus	Purulent discharges (boils, blisters, pus-forming skin infections)
Streptococcus pyogenes	Scarlet, rheumatic fever; "strep" throat
Viruses	
Hepatitis viruses A, B, C, delta	Hepatitis
Human immunodeficiency virus (HIV)	AIDS
Cytomegalovirus	Congenital viral infections; mononucleosis
Epstein–Barr virus	Burkitts lymphoma and other lymphoproliferative diseases
Herpes virus (Types I and II)	Cold sores; genital herpes
Human papilloma virus	Genital warts; cervical cancer
Protozoa	
Leishmania donovanii	Leishmaniasis
Plasmodium spp.	Malaria
Pneumocystis carinii	Pneumonia
Trypanosoma spp.	Trypanosomiasis

for implementing probe-based assays in the clinical setting. In addition, some companies are marketing nucleic acid probes for identification of key veterinary pathogens. A major impediment to more extensive use of nucleic acid probes is the high cost per test. However, as automated methods and new tests become available, this problem will be overcome. Another problem is the lack of specific probes for many pathogens. However, recent advances in our understanding of bacterial phylogenetics based on 16S rRNA sequencing (see Chapter 18) should allow new and more specific nucleic acid probes to be constructed.

Nucleic acid hybridization provides a powerful tool for disease diagnosis. A probe specific for the virus or microbial pathogen must be available. In addition to their use in clinical diagnosis, probes are also finding wide application in food and environmental monitoring.

13.12   Diagnostic Virology

The growth-dependent procedures that we have discussed so far in this chapter are suitable only for microbial infections in which it is possible to cultivate a suspected pathogen in a sterile culture medium. What about viruses? We discussed in Chapter 6 the wide variety of viruses which infect animals (including humans) and we discuss some of the important viral diseases of humans in Chapter 15. Here we present laboratory procedures that can be used in the diagnosis of viral diseases. Briefly, these methods include direct isolation, immunological analyses, nucleic acid probes, and electron microscopy.

First, we should note that laboratory cultivation of viruses from clinical materials is more difficult, time-consuming, and specialized than the cultivation of other microbial pathogens. This is because viruses only grow in living cells. We discussed in Section 6.2 the use of cell cultures for the growth of viruses, and such cultures are most commonly used in diagnostic virology. A common cell line is a human diploid fibroblast culture called WI-38, which grows rapidly and reproducibly in cell culture medium. Another cell line sometimes used is HeLa, a cell culture derived initially from a human cancer. This cell line has been maintained *in vitro* for so many successive transfers that it has greatly changed its character (it is no longer diploid, for instance). In addition to these two cell lines, which can be maintained in the laboratory indefinitely, cultures are also made from Rhesus monkey kidneys. Monkey kidney cell lines are called *primary*, because they are not maintained by successive transfer in the laboratory. Primary monkey kidney cells support the rapid growth of a number of viruses infecting humans and are therefore of value in initial isolation of unknown viruses, but are less easy for routine use in the clinical laboratory (a steady supply of Rhesus monkeys is needed, for instance). It should be emphasized that diagnostic virology is much more technical than diagnostic microbiology and is therefore only carried out in specialized laboratories (such as those associated with large medical centers).

Although the best diagnostic technique for most viral infections is actual isolation of the virus, this is not practical in most clinical settings. Instead, several immunological tests and nucleic acid probes for viruses have been developed. Most of the immunological tests are either direct ELISAs (see Sections 12.13 and 13.7) that detect viral particles (Figure 13.14), or fluorescent antibody methods in which antibodies made against viral antigens are used to detect cells containing viruses (see Section 13.5 and Figure 13.8). Most ELISAs in viral diagnostics are *direct* assays, designed to detect the virus itself from a suitably treated specimen (Figure 13.14). Viruses currently detectable by direct ELISA include rotavirus, hepatitis viruses, rubella virus, bunyavirus, measles virus, mumps virus, and parainfluenza virus. Agglutination tests are also available in which anti-viral antibodies conju-

gated to latex beads or activated charcoal (see Section 13.8) are used to test for agglutination of viral antigens released from lysed tissue samples. Nucleic acid probes for detection and identification of clinically significant viruses are also a major diagnostic tool in clinical virology (see Section 13.11).

In addition to the tools described above, diagnostic virology can be done by electron microscopy in major clinical hospitals having the appropriate equipment. Because many viruses have distinctive morphologies (see Chapter 6), their presence in clinical specimens can often be detected by observing the sample under the electron microscope (Figure 13.15). In most specimens the virus particles must first be concentrated and separated from human tissues, and a variety of techniques, generally employing centrifugation and filtration, are used to obtain a sample enriched in virus particles. Although not as reliable as immunological or nucleic acid probe methods, the observation of virus particles of a specific morphology in

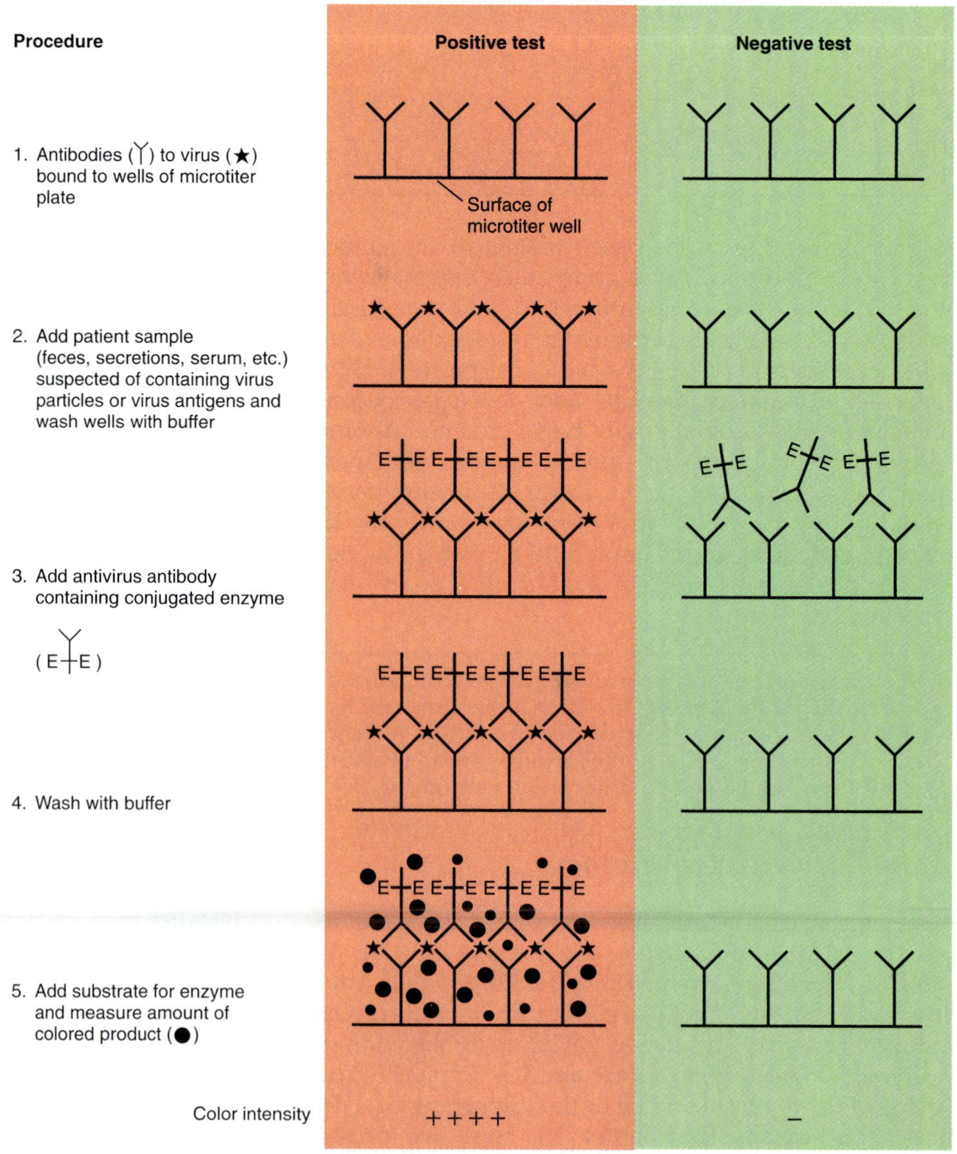

Procedure	Positive test	Negative test

1. Antibodies (Y) to virus (★) bound to wells of microtiter plate — Surface of microtiter well

2. Add patient sample (feces, secretions, serum, etc.) suspected of containing virus particles or virus antigens and wash wells with buffer

3. Add antivirus antibody containing conjugated enzyme

   (E┼E)

4. Wash with buffer

5. Add substrate for enzyme and measure amount of colored product (●)

Color intensity   ++++   —

FIGURE 13.14
Detection of viruses by a direct ELISA test.

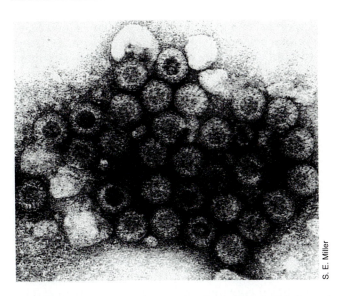

S. E. Miller

FIGURE 13.15 Electron microscopic observation of clinical specimens to detect viruses. Human rotavirus from an infant fecal sample. The distinct spherical nature of the virus, coupled with the source, are highly diagnostic criteria. Each rotavirus particle is approximately 75 nm in diameter.

a particular type of human tissue can often serve as presumptive evidence of a certain disease. Treating the sample with antibodies prepared against particular viruses can be used to increase the sensitivity and specificity of this method; such antibodies cause the viral particles to agglutinate, and this makes them easier to distinguish from cellular debris under the electron microscope. Viruses can also be visualized following treatment of specimens with antiviral antibodies containing conjugated heavy metals (see Section 12.12). The electron microscope can thus serve as a provisional means of identifying viruses directly from patient specimens, and offers the major advan-

tage of speed. Using negative staining techniques (Figure 13.15), results from electron microscopic analysis can be available 20 minutes after collection of the specimen.

A summary of a few laboratory procedures used in diagnostic virology is given in Table 13.7. Most of these procedures are used only under special circumstances. For routine virus infections, diagnosis will be made by assessing symptoms or by immunological or other indirect means. For example, testing for HIV infection involves the ELISA test (see Section 13.7) and the immunoblot test (see Section 13.9). Both these methods detect the presence of *antibodies* to HIV, not HIV itself; there is currently no *routine* method used to detect the presence of HIV directly.

> Viruses present special diagnostic problems because they are difficult to cultivate. Cell cultures can be used to detect the presence of some viruses, but a suitable cell culture may not always be available for a virus of interest. Because they can be used without culture, nucleic acid probes and immunological reagents are finding use in the clinical laboratory for viruses.

13.13   Safety in the Clinical Laboratory

By their very nature, clinical laboratories are areas in which potentially dangerous biological specimens must be handled on a routine basis. Hence, a defined protocol for handling clinical samples must be established to avoid laboratory accidents.

Studies of laboratory-associated infections have indicated that most such infections do not result from known exposures or accidents, but instead from routine handling of patient specimens. The two most

Table 13.7 Some laboratory procedures used in diagnostic virology*			
Condition	**Possible viral cause**	**Samples to obtain**	**Inoculation procedure**
Upper respiratory infection	Rhinovirus Coronavirus Adenovirus	Nasopharyngeal or tracheal fluid (aspirate)	Human fibroblast culture
Pneumonia	Influenza	Nasopharyngeal fluid or swab	Human fibroblast cultures or embryonated eggs
Measles	Measles virus	Nasopharyngeal fluid or swab	Monkey kidney cells
Vesicular rash	Herpes simplex	Vesicular fluid by aspiration	Human fibroblast culture
Diarrhea	Rotavirus (infants) Norwalk agent (adults)	Feces or rectal swab	Look for characteristic virus particles with the electron microscope (see Figure 13.15)
Nonbacterial meningitis	Enterovirus Mumps Herpes simplex	Spinal fluid	Human fibroblast or monkey kidney cultures

*Immunological methods and nucleic acid probe methods are also widely used in the diagnosis of viral infections (see Sections 13.4–13.11).

common causes of laboratory accidents are ignorance and carelessness. Infectious aerosols, generated during processing of the specimen, are the most likely cause of laboratory infections. In attempts to minimize the exposure of clinicians to infectious agents and to thereby reduce the number of nonaccident-associated laboratory infections, well-run clinical laboratories stress the following safety rules:

1. Laboratories handling hazardous materials restrict access to laboratory and support personnel. These individuals should have knowledge of the biological risks involved in the clinical laboratory and act accordingly.

2. Effective procedures for decontaminating infectious materials or wastes, including specimens, syringes and needles, inoculated media, bacterial cultures, tissue cultures, experimental animals, glassware, instruments, and surfaces must be in place and be practiced without compromise. A 10 percent chlorine bleach solution is recommended for disinfecting spilled infectious material.

3. Personnel working with hazardous infectious agents or vaccines (for example, rabies, polio, or diphtheria-pertussis-tetanus vaccines) must be properly vaccinated against the agent.

4. All clinical specimens should be considered potentially infectious and handled in the appropriate manner. This is especially important for preventing laboratory-acquired hepatitis, because of the relative frequency with which hepatitis viruses are present in clinical specimens.

5. All pipetting must be done with automatic pipetting devices (not by mouth), and devices such as syringes, needles, and clinical centrifuges must always be used with proper biological containment equipment.

6. Animals should only be handled by trained laboratory personnel and anesthetics or tranquilizers should be used to avoid injury to both personnel and animals.

7. Laboratory personnel should always wear safety clothing including laboratory coats or uniforms, and should practice good personal hygiene in respect to hand washing. Eating and drinking are never permitted in the clinical laboratory.

8. Because of the special risks associated with AIDS, all clinical specimens should be treated as if they contain HIV (which they might). Latex or vinyl gloves should be worn whenever handling specimens of *any* kind. Masks must be worn any time there is a possibility of generating an aerosol during specimen preparation. Needles must not be reopened, bent, or broken; they should be placed in a labeled container for this purpose which can be sealed and autoclaved before disposal.

The safety rules outlined above should be the norm for all clinical laboratories. Specialized clinical laboratories may have additional rules to ensure a safe work environment. For example, if laboratory personnel handle extremely hazardous airborne pathogens (such as the causative agent of tuberculosis, *Mycobacterium tuberculosis*) on a routine basis, the laboratory would be fitted with special features, such as negatively pressurized rooms and air filters, to prevent the accidental release of the pathogen from the laboratory. In the final analysis, however, it is the attitude of the personnel that make the laboratory a safe or an unsafe place to work. Any clinical laboratory is a potentially hazardous place for untrained personnel or those unwilling to take the necessary steps to prevent laboratory-acquired infection.

Study Questions

1. Describe the standard procedure for obtaining and culturing a blood sample for evidence of bacteremia. What would you conclude if the organism found in a blood culture was a *Staphylococcus* species?

2. Why is the *number* of bacterial cells in urine rather than simply the *presence* of cells in urine of significance to the microbiological analysis of urine? What organism is responsible for most urinary tract infections?

3. Describe the microscopic and cultural evidence required to make a diagnosis of gonorrhea. What is the advantage of Thayer-Martin agar over chocolate agar for the isolation of *Neisseria gonorrhoeae*?

4. Describe the major procedures used for culturing anaerobic microorganisms in liquid medium or on Petri plates. Why is it important to process all clinical samples quickly, especially for isolation of anaerobes?

5. What is a selective medium? How does a selective medium differ from a differential medium? Is EMB agar selective or differential? How is it used in the clinical microbiology laboratory?

6. Compare and contrast the changes undergone by pH dyes in the tests for carbohydrate fermentation and in the test for citrate utilization. Could the same dye be used in both tests? Why or why not?

7. What are the advantages of rapid identification systems as clinical diagnostic procedures? Discuss at least three advantages.

8. Why is it not a common medical practice to treat an infectious disease with antibiotics before isolating the suspected pathogen? What steps are usually taken before antibiotic treatment begins?

9. Design a fluorescent antibody assay for confirming an initial diagnosis of "strep throat" (*Streptococcus pyogenes* is the causative agent of strep throat). Discuss all aspects of the assay including preparation of antisera, necessary controls, and clinical interpretation.

10. What are the advantages of using fluorescent monoclonal antibodies instead of polyclonal antibodies in clinical diagnostics?

11. Design an ELISA test for detecting Hepatitis A virus in

feces. Would such a test require anti-human IgG antibodies? Why or why not?

12. What is the function of latex beads in the agglutination tests developed for several pathogens? What advantages do latex agglutination assays have over ELISA? What disadvantages?

13. What advantages does the AIDS Western blot have over the AIDS ELISA for the diagnosis of the HIV infection? Why is the AIDS Western blot only used for confirmation of HIV infection and not for screening?

14. What is a plasmid fingerprint and how is it obtained? What does a plasmid fingerprint say about sequence homology between plasmids?

15. What information is essential for design of a pathogen-specific nucleotide probe? Where can one obtain such information? Is this information available for all pathogens?

16. What are the major advantages of using DNA probes in diagnostic microbiology? Discuss at least four aspects of probe technology that benefit clinical medicine.

17. For a typical bacterial infection, describe how antibodies to the pathogen can be used in the diagnosis of the infectious disease. In what instances might such methods be preferable to conventional diagnostic techniques?

18. Explain why a *rise* in antibody titer is a better indication of active infectious disease than the mere *presence* of antibody in the serum. Exactly how is a rise in antibody titer measured in a patient?

19. For nucleic acid probe work, what is a "reporter" molecule and why is it necessary?

20. How can nucleic acid probe technology be made as *sensitive* a technique as direct culture methods, assuming that the latter is successful when only a few cells are present in a sample?

21. Design an ELISA for detection of a virus found in high numbers in a body fluid such as saliva.

22. What is a primary cell line? Why do some animal viruses grow in primary cell lines but not in cell lines such as HeLa cells?

23. How are most laboratory-associated infections contracted? What actions can be taken to prevent them?

Supplementary Readings

Balows, A., W. J. Hausler, K. L. Herrman, H. D. Isenberg, and H. J. Shadomy. 1991. *Manual of Clinical Microbiology*, 5th edition. American Society for Microbiology, Washington, D. C. The standard reference work of clinical microbiology.

Baron, E. J., and **S. M. Finegold**. 1994. *Bailey and Scott's Diagnostic Microbiology*, 9th edition. The C. V. Mosby Company, St. Louis, MO. A good source of clinical microbiological methods and diagnostic and identification procedures.

Baron, S. 1991. *Medical Microbiology*, 3rd edition. Churchill Livingstone, NY. A standard clinical microbiology text.

Brooks, G. F., J. S. Butel, L. N. Ornston, E. Jawetz, J. L. Melnick, and **E. A. Adelberg**. 1991. *Jawetz, Melnick and Adelberg's Medical Microbiology*, 19th edition. Appleton and Lange, Norwalk, CT. A brief, up-to-date treatment of infectious diseases and clinical information.

Conte, J. E., Jr., and **S. L. Barriere**. 1992. *Manual of Antibiotics and Infectious Diseases,* 7th edition. Lea and Febiger, Philadelphia. An excellent source of antibiotic treatment guidelines and other clinically relevant information about infectious diseases.

Davis, B. D., R. Dulbecco, H. N. Eisen, and **H. S. Ginsburg**. 1990. *Microbiology*, 4th edition. J. B. Lipincott Company, Philadelphia. A classic reference and medical school text.

Haukenes, G., L. R. Haahaim, and **J. R. Pattison** (eds.) 1990. *A Practical Guide to Clinical Virology*. John Wiley and Sons, New York. A short paperback "handbook" of clinical identification and treatment of viral pathogens.

Joklik, W. K., H. P. Willet, D. B. Amos, and **C. M. Wilfert**. 1992. *Zinsser Microbiology*, 20th edition. Appleton and Lange, East Norwalk, CT. A comprehensive overview of medical microbiology, including immunological and scientific principles.

Koneman, E. W., S. D. Allen, V. R. Dowell, Jr., W. M. Janda, H. M. Sommers, and **W. C. Winn, Jr.** 1992. *Color Atlas and Textbook of Diagnostic Microbiology*. 4th edition. A well-illustrated primary reference on clinical microbiology.

Luciw, P. A., and **K. S. Steimer** (eds.) 1989. *HIV Detection by Genetic Engineering Methods*. A detailed description of the major molecular methods for detecting HIV.

Macario, A. J. L., and **E. Conway de Macario** (eds.) 1990. *Gene Probes for Bacteria*. Academic Press, Orlando, FL. A comprehensive account of the clinical applications of nucleic acid probes for pathogenic bacteria.

Rayburn, S. R. 1990. *The Foundations for Laboratory Safety: A Guide for the Biomedical Laboratory*. Springer-Verlag, New York. An up-to-date and well written textbook on laboratory safety. Excellent treatment of the hazards of biological agents.

Stine, G. J. 1993. *Acquired Immune Deficiency Syndrome*. Prentice Hall, Englewood Cliffs, NJ. An excellent monograph discussing various aspects of AIDS biology, pathology, and societal issues.

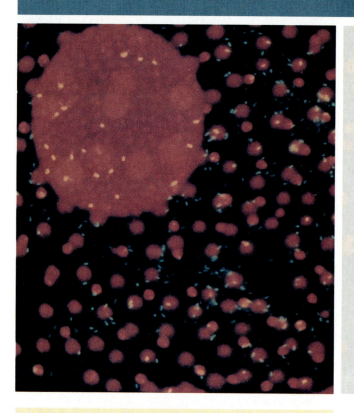

14

*E*pidemiology and Public Health Microbiology

In Chapters 11 and 12 we considered the general principles of how microorganisms cause infectious disease and how the host responds to microbial onslaught. In Chapter 13, we discussed the methods and procedures used in the diagnosis of microbial diseases. In this chapter, we consider how a pathogen spreads from an infected individual to others in a population. Thus, we are dealing here with *public health*. In the next chapter we consider the diseases themselves. An understanding of the principles put forward in this chapter is vital in attempts to control the spread of infectious disease.

One measure of the success of microbiology in the control of infectious disease was shown by the data presented in Figure 1.15, which compared the present causes of death in the United States with those at the beginning of the century. Many microbial diseases are no longer the threat to public health they once were in developed countries. However, in developing countries, which include about 75 percent of the world's population, infectious diseases are still a major problem. Worldwide, infectious diseases still account for over 30 percent of the total of 50 million annual deaths. Table 14.1 shows the most prevalent causes of death. Note that, for many of these diseases (for example, tuberculosis, measles, and whooping cough), effective vaccines are manufactured (see Table 12.8) but are often not available or are not used outside of developed countries. Finally, there is growing evidence that, because of the ease of travel between countries, many infectious diseases will remain an important public health problem in all countries. The current AIDS epidemic, which has apparently spread world-

Table 14.1   Infectious diseases: the leading human killers

Cause of death	Estimated yearly deaths*	Infectious agents
Acute respiratory infections	6,900,000	Bacteria, viruses, protozoa, fungi
Diarrheal diseases	4,200,000	Bacteria, viruses
Tuberculosis	3,300,000	Bacteria
Malaria	1,000,000–2,000,000	Protozoa
Hepatitis	1,000,000–2,000,000	Viruses
Measles	220,000	Virus
Meningitis, bacterial	200,000	Bacteria
Schistosomiasis	200,000	Parasitic worm
Pertussis (whooping cough)	100,000	Bacterium
Amoebiasis	40,000–100,000	Protozoa
Hookworm	50,000–60,000	Parasitic worm
Rabies	35,000	Virus
Yellow fever	30,000	Virus
African trypanosomiasis (sleeping sickness)	20,000 or more	Protozoan

*Data represent estimates for yearly worldwide total deaths from each disease during the last decade. There are approximately 50 million deaths per year, worldwide, from all causes.
Source: World Health Organization, 1992.

wide in 20 years or less, is only one example of the devastating consequences of an infectious disease in a global theater. There is also growing concern regarding the emergence and spread of antibiotic-resistant tuberculosis and gonorrhea in developed countries. Thus, eradication or even effective control of infectious diseases must involve scientific, economic, political, and educational solutions, and ultimately, global cooperation.

14.1   The Science of Epidemiology

The most visible aspect of microbial disease is the actual diseased individual. However, individuals do not live alone, and when we consider infectious diseases in populations, some new factors arise. The study of the occurrence, distribution, and control of infectious disease in populations is part of the field of **epidemiology**.

To continue existing in nature the pathogen must be able to grow and reproduce. For this reason, an important aspect of the epidemiology of any disease is a consideration of the natural history of the pathogen. In most cases the pathogen cannot grow outside the host, and if the host dies, the pathogen will also die. Pathogens that kill the host before they are transmitted to a new host would thus become extinct. This raises the question of why pathogens occasionally kill their hosts. Actually, a well-adapted parasite lives in harmony with its host, taking only what it needs for existence, and causing only a minimum of harm. Serious host damage most often occurs when new races of pathogens arise for which the host has not developed resistance, or when the resistance of the host changes because of the factors discussed in Chapter 11. Pathogens are selective forces in the evolution of the

host, just as hosts are selective forces in the evolution of pathogens. When equilibrium between host and pathogen exists, both are able to live more or less harmoniously together.

The epidemiologist is concerned with methods for tracing the spread of pathogens. By use of these methods, the epidemiologist can assess the microbial safety of various agents and trace the path of a disease to identify its origin and mode of transmission. The epidemiologist relies heavily on data obtained from clinical studies, disease reporting, insurance questionnaires, and interviews with patients in tracing the pattern of a disease in nature. The science of epidemiology has been referred to as "medical ecology," because the study of a disease in populations is really a study of a disease in its natural environment. This is in contrast to the clinical or laboratory study of disease, where the focus is on treating the individual patient. However, knowledge of both the clinical aspects and ecological aspects of a given disease are important if public health measures to control diseases are to be effective.

> Epidemiology deals with the distribution and spread of disease through populations. For infectious disease, the epidemiologist tries, by understanding the spread of the pathogen through the population, to develop methods for the control of infectious disease. Public health depends on the methods and procedures of the epidemiologist.

14.2   Epidemiological Terminology

A number of terms having specific meanings are used by the epidemiologist to describe patterns of disease. The **prevalence** of a disease in a population is defined

Miniglossary for Chapter 14

ACUTE   short-term infection usually characterized by dramatic onset and rapid recovery

CARRIER   subclinically infected individuals who may spread a disease

CHRONIC   long-term infection

COMMON-SOURCE EPIDEMIC   an epidemic resulting from infection of a large number of people from a single contaminated source

ENDEMIC   disease constantly present, usually in low numbers

EPIDEMIC   the occurrence of a disease in unusually high numbers in a localized region

EPIDEMIOLOGY   the study of the occurrence, distribution, and control of infectious diseases

FOMITES   inanimate objects which, when contaminated with a viable pathogen, can transfer the pathogen to a host

HERD IMMUNITY   resistance of a group to a pathogen due to immunity of a large portion of the group

INCIDENCE   the number of cases of disease in a population

MORBIDITY   incidence of illness in a population

MORTALITY   incidence of death in a population

NOSOCOMIAL INFECTION   hospital-acquired infection

OUTBREAK   the occurrence of a large number of cases of a disease in a short period of time

PANDEMIC   a worldwide epidemic

PERSON-TO-PERSON EPIDEMIC   an epidemic resulting from person-to-person contact, characterized by a gradual rise and fall in numbers of cases

PREVALENCE   the proportion or percentage of individuals in the population having a disease

PUBLIC HEALTH   the health of the population as a whole

QUARANTINE   the practice of restricting the movement of individuals with highly contagious serious infections to prevent spread of the disease

RESERVOIR   sites in which viable infectious agents remain and from which infection of individuals may occur

VECTOR   a living agent which transfers a pathogen (note alternative usage in Chapter 8)

VEHICLE   nonliving source of pathogens that infect large numbers of individuals; common vehicles are food and water

ZOONOSIS   a disease which occurs primarily in animals but can be transmitted to humans

---

as the proportion (or percentage) of diseased individuals in a population at any one time. The **incidence** of a disease is the *number* of diseased individuals in a population at risk. A disease is said to be **epidemic** when it occurs in an unusually high number of individuals in a community at the same time; a **pandemic** is a widely distributed epidemic (Figure 14.1). By contrast, an **endemic** disease is one which is constantly present, usually at low incidence, in a population. In an endemic disease, the pathogen may not be highly virulent, or the majority of the individuals may be immune, so that the incidence of disease is low. However, as long as an endemic situation lasts, there will remain a few individuals who may serve as reservoirs of infection.

Sporadic cases of a disease occur when individual cases are recorded in geographically remote areas, implying that the incidents are not related. A disease **outbreak**, on the other hand, occurs when a number of cases are observed, usually in a relatively short period of time, in an area previously experiencing only sporadic cases of the disease. Finally, *subclinical infection* is used to describe diseased individuals who show no or only mild symptoms. Subclinically infected individuals are frequently identified as **carriers** of a particular disease, because even though they themselves show few or perhaps no symptoms, they may still be actively carrying and shedding the pathogenic agent (see Section 14.3).

Mortality and morbidity

In practice, the incidence and prevalence of disease is determined by obtaining statistics of illness and death. From these data a picture of the public health in a population can be obtained. The population under consideration could range in size from the total global population of humans down to the population of a localized region of a country or district. Public health varies from region to region, as well as with time; thus a picture of public health at a given moment provides only an instantaneous picture of the situation. By continuing to examine health statistics over many years, it is possible to assess the value of various public health protective measures influencing the incidence of disease.

Mortality expresses the incidence of *death* in the population. Infectious diseases were the major causes of death in 1900 in developed countries (see Figure 1.15), whereas currently they are of much less significance, and noninfectious diseases such as heart disease and cancer are of greater importance. However, the current situation could rapidly change if a breakdown in public health measures were to occur and, in fact, does not mirror the worldwide situation. In developing countries, infectious diseases are still the major killers (Table 14.1).

Morbidity refers to the incidence of *disease* in populations and includes both fatal and nonfatal diseases. Clearly, morbidity statistics will define the health of the population more precisely than will mortality statistics, since many diseases that affect health in important ways have only a low mortality. The major causes of illness are quite different from the major causes of death. Major illnesses are acute respiratory diseases (the common cold, for instance) or acute digestive system conditions, which are generally due to infectious causes.

In terms of clinical symptoms, the course of a typical disease can be divided into stages:

1. *Infection*, when the organism begins to grow in the host.

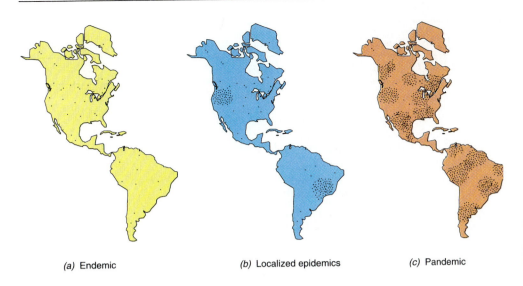

(a) Endemic   *(b)* Localized epidemics   *(c)* Pandemic

FIGURE 14.1
Types of disease incidence. Each dot represents several cases of a particular disease.

2. *Incubation period*, the time between infection and the appearance of disease symptoms. Some diseases have short incubation periods; others, longer ones. The incubation period for a given disease is determined by a variety of factors including inoculum size, virulence of pathogen, resistance of host, and distance of site of entrance from focus of infection.

3. *Prodromal period*, a short period sometimes following incubation in which the first symptoms such as headache and feeling of illness appear.

4. *Acute period*, when the disease is at its height, with overt symptoms such as fever and chills.

5. *Decline period*, during which disease symptoms are subsiding, the temperature falls, usually following a period of intense sweating, and a feeling of well-being develops. The decline may be rapid (within 1 day), in which case it is said to occur by *crisis*, or it may be slower, extending over several days, in which case it is said to be by *lysis*.

6. *Convalescent period*, during which the patient regains strength and returns to normal.

During the later stages of the infection cycle, the immune mechanisms of the host become increasingly important, and in most cases complete recovery from the disease requires and results in active immunity.

> **When a disease is constantly present in a population in relatively low numbers, it is said to be endemic. In an epidemic, an unusually high incidence of the disease occurs. The effect of the disease in a population can be manifested in death (mortality), but there are commonly many more infected individuals than those dying of the disease. Clinical disease symptoms can be divided into several phases.**

14.3   Disease Reservoirs

Reservoirs are sites in which viable infectious agents remain alive and from which infection of individuals may occur. Reservoirs may be either animate or inani-

mate. Table 14.2 lists some common human diseases and their reservoirs. Some pathogens are primarily saprophytic (living on dead matter) and only incidentally infect and cause disease. A good example is *Clostridium tetani* (the causal agent of tetanus) whose normal habitat is the soil. Infection by this organism is not essential for its continued existence and can be considered only an accidental event; if there were no susceptible hosts, *C. tetani* would still survive in nature.

Of most interest, from the viewpoint of epidemiology, are pathogens whose only reservoirs are living organisms. In these cases, the reservoir is an essential component of the cycle by which the infectious agent maintains itself in nature. Some infections occur only in humans, so that maintenance of the cycle involves person-to-person transmission. This type of pathogen cycle is the most common for most viral and bacterial respiratory diseases, including sexually transmitted diseases, staphylococcal and streptococcal infections, diphtheria, typhoid fever, and mumps.

Zoonosis

A large number of infectious diseases that occur in humans also occur in animals. A disease that occurs primarily in animals but is occasionally transmitted to humans is called a **zoonosis**. Because public health measures for animal populations are much less highly developed than for humans, the infection rate for these diseases will be much higher in animals, and animal-to-animal transmission will be the rule. However, occasionally transmission will be from animal to human. It would be less likely, in such diseases, for transmission to also occur from person to person. Thus the maintenance of the pathogen in nature depends on animal-to-animal transfer. It should be obvious that control of a zoonosis in the human population in no way eliminates it as a public health problem. Indeed, more effective human control can generally be achieved through elimination of the disease in the animal reservoir. Marked success has been achieved in the control of two of the diseases that occur primarily in domestic animals, bovine tuberculosis and brucel-

Table 14.2 Major reservoirs of microbial infection for a number of human diseases	
Disease	**Major reservoir of infection**
Anthrax, brucellosis	Cattle, swine, goats, horses, sheep
Salmonellosis	Domestic and wild animals, especially poultry, whole eggs and egg products; water polluted with sewage
Botulism	Soil, contaminated food
Giardiasis	Beaver, muskrat, marmot; water polluted with animal feces
Malaria	*Anopheles* mosquito
Plague	Wild rodents
Psittacosis	Parakeets, parrots, pigeons, other birds
Rabies	Wild and domestic carnivores, especially skunks and raccoons
Respiratory diseases (viral, bacterial)	Infected humans, human carriers
Rocky Mountain Spotted Fever	Ticks, rabbits, mice
Syphilis, gonorrhea, AIDS, other sexually transmitted diseases	Infected humans
Tetanus	Soil, intestine
Tuberculosis	Humans, dairy cattle
Tularemia	Wild animals, especially rabbits
Typhoid fever	Human carriers, infected humans

losis. Pasteurization of milk was of considerable importance in the prevention of the spread of bovine tuberculosis to humans, since milk was the main vehicle of transmission.

Certain infectious diseases have more complex cycles, involving an obligate transfer from animal to human to animal. These are due to organisms with complex life cycles and are either metazoans (for example, tapeworms) or protozoa (for example, malaria, see Section 15.11). In such cases, control of the disease in the population can be either through control in humans or in the alternate animal host.

Carriers

A carrier is an infected individual not showing obvious signs of clinical disease. Carriers are potential sources of infection for others, and are thus of considerable significance in understanding the spread of disease. Carriers may be individuals in the incubation period of the disease, in which case the carrier state precedes the development of actual symptoms. Carriers of this sort are prime sources of infectious agents for respiratory infections, since they are not yet aware of their infection and so will not be taking any precautions against infecting others. Such persons can be considered **acute carriers**, because the carrier state will only last for a short while. Also significant from the public health standpoint are chronic carriers, who may remain infected for long periods of time. **Chronic carriers** may be either individuals who had a clinical disease and recovered, or they may have a subclinical infection that has remained inapparent. These individuals may be perfectly healthy,

The Tragic Case of Typhoid Mary

The classic example of a chronic carrier was the woman known as "Typhoid Mary," a cook in New York City and Long Island in the early part of this century. Typhoid Mary (her real name was Mary Mallon) was employed in a number of households and institutions, and as a cook she was in a central position to infect large numbers of people. Eventually she was tracked down by Dr. George Soper after an extensive epidemiological investigation of a number of typhoid outbreaks revealed that she was the likely source of contamination. When her feces was examined bacteriologically, it was found that she had very high numbers of the typhoid bacterium, *Salmonella typhi*. She remained a carrier for many years, probably because her gallbladder was infected, and organisms were continuously being excreted from there into her intestine. Public health authorities offered to remove her gall-

bladder but she refused the operation, and to prevent her from continuing to serve as a source of infection she was imprisoned. After almost three years in prison, she was released on the pledge that she would not cook or handle food for others and that she was to report to the health department every three months. She promptly disappeared, changed her name, and cooked in hotels, restaurants, and sanitariums, leaving behind a wake of typhoid fever. After five years she was captured as a result of the investigation of an epidemic at a New York hospital. She was again arrested and imprisoned and remained in custody on North Brother Island in the East River of New York City for 23 years. She died in 1938, 32 years after epidemiologists had first discovered she was a chronic typhoid carrier.

but they harbor and spread viable pathogens (see box on Tragic Case of Typhoid Mary).

Carriers can be identified by routine surveys of populations, using cultural, radiological (chest X ray), or immunological techniques. In general, carriers are only sought among groups of individuals who may be sources of infection for the public at large, such as food handlers in restaurants, grocery stores, or food processing plants. Two diseases in which carriers have been of most significance are typhoid fever and tuber-

culosis, and routine surveys of food handlers for inapparent cases of these diseases are sometimes made. (Foodborne diseases are considered in Section 15.13.)

14.4   Epidemiology of AIDS: An Example of How Epidemiological Research Is Done

Cases of acquired immune deficiency syndrome (AIDS), the viral-mediated infectious disease that severely cripples the body's immune system (see Section 15.7), were first reported in the United States in 1981. As seen in Figure 14.2, the number of new AIDS cases diagnosed nearly *each year* in the United States has risen significantly. *At least 60,000 new cases of AIDS will be diagnosed annually* in the foreseeable future. Initial case-control studies suggested an unusually high AIDS prevalence among homosexual men and intravenous drug abusers. This in turn strongly implicated a trans-

> To understand how epidemics develop, it is important to know the reservoirs of the pathogen. Some pathogens can exist even in the absence of a human host, with reservoirs in soil, water, or animals. Certain pathogens are confined only to the human host and are maintained only by transmission from one person to another. An understanding of person-to-person transmission and the phenomenon of disease carriers can aid in developing methods for controlling infectious disease.

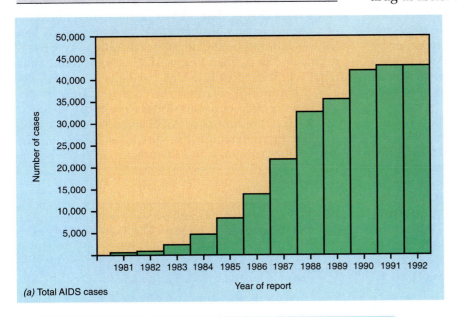

(a) Total AIDS cases

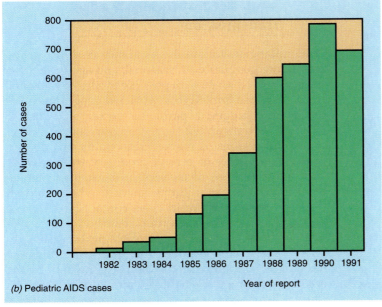

(b) Pediatric AIDS cases

FIGURE 14.2

Number of (a) new cases of acquired immune deficiency syndrome (AIDS) in the United States, and (b) pediatric AIDS cases. Data are from the Centers for Disease Control, Atlanta, GA, through 1992 and include approximately 250,000 total cases. The World Health Organization estimates that there have been more than 1 million AIDS cases worldwide, and up to 10 million people are currently infected with HIV.

missible agent, presumably transferred during sexual activity or by contaminated needles. The finding that individuals requiring blood or blood products were also at high risk strengthened the case for a transmissible agent (Figure 14.3).

Soon after the discovery of human immunodeficiency virus (HIV), the causal agent of AIDS, laboratory tests were developed to detect antibodies to the virus in serum (see Sections 13.7 and 13.9). This made possible more extensive surveys of the incidence of HIV infection in different populations, and also served as a screening method to ensure that new cases of AIDS were not transmitted by blood transfusions. Such tests revealed a fourth group of individuals at high risk for AIDS—the children of mothers who were themselves at high risk for AIDS.

Statistics such as those in Figure 14.3 are typical of an agent transmissible by sexual activity or by blood, and the association of AIDS cases with *specific* groups was actually an important epidemiological finding. The identification of these high risk groups implied that AIDS was *not* transmitted from person to person by casual contact, such as the respiratory route, or by contaminated food or water. Instead, epidemiological findings pointed clearly to bodily fluids, primarily blood and semen, as the major vehicles for transmission of HIV.

In the United States AIDS has affected mainly homosexual men (Figure 14.3), but intravenous drug users represent an increasing proportion of AIDS cases. Subsequent epidemiological surveys have shown that homosexual men with multiple sexual partners are more likely to contract AIDS than monogamous homosexual males. In addition, the fastest growing category for new AIDS cases is the heterosexual contact group. This undoubtedly reflects the increased probability of contacting an infected individual when engaging in any sexual activity with multiple partners.

The incidence of AIDS in hemophiliacs has been reduced greatly in recent years (Figure 14.3). The latter is due not only to screening of the blood supply but also because many blood clotting factors needed by hemophiliacs can withstand a heat treatment sufficient to inactivate HIV. Pediatric AIDS cases are still a

major concern. HIV can be transmitted to the fetus by infected mothers and probably also in mother's milk. All infants born to HIV-infected mothers have maternally derived antibodies to HIV in their blood, but a clear diagnosis of AIDS in infants must wait a year or more after birth, because some infants showing HIV antibodies at birth do not go on to develop HIV infection.

Epidemiological studies of AIDS in Africa, where the disease is thought to have originated, have clearly shown that transmission of AIDS is not linked to particular sexual practices, such as homosexuality, but instead to person-to-person transfer of HIV-infected fluids. In Africa heterosexual transmission of AIDS predominates, presumably due to transfer of the virus in semen. Unfortunately, reliable global statistics on AIDS are not available because of differences in reporting in various countries. Current estimates are around 1 million AIDS cases worldwide with up to 10 million infected people who do not (yet) show clinical symptoms; that is, they are carriers. The identification of high risk groups through epidemiological study has led to the development of public education campaigns to inform the public of how AIDS is transmitted and what activities constitute high risk behavior (see AIDS box, Chapter 15). Because no cure for AIDS is yet available, public education offers the most effective approach to the control of AIDS and serves as a major weapon for preventing the spread of infection. We discuss the pathology and therapy of AIDS in Section 15.7.

> AIDS is one of the best studied and newest epidemics of infectious disease in humans. The knowledge of how AIDS spreads in the human population has helped greatly in slowing the spread of this very serious infectious disease.

14.5 Modes of Infectious Disease Transmission

Epidemiologists follow the incidence of a disease by correlating geographical, seasonal, and age-group distribution of a disease with possible modes of transmission. A disease limited to a restricted geographical location may suggest a particular vector; malaria, for example, is transmitted by a mosquito found mainly in tropical regions. A marked seasonality to a disease is often indicative of certain modes of transmission, such as in the case of measles or chickenpox, where the number of cases jumps sharply when children enter school and come in close contact (see Figures 14.4 and 15.21). The age-group distribution of a disease can be an important epidemiological statistic, frequently suggesting or eliminating particular routes of transmission. Sometimes the data accumulated by the epidemiologist are for noninfectious diseases such as cancer, heart, and lung disease because the general

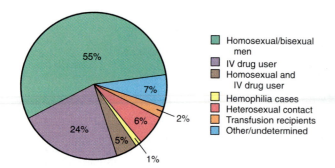

FIGURE 14.3 Distribution of AIDS cases by risk group in the United States, 1991. Data are from the Centers for Disease Control, Atlanta.

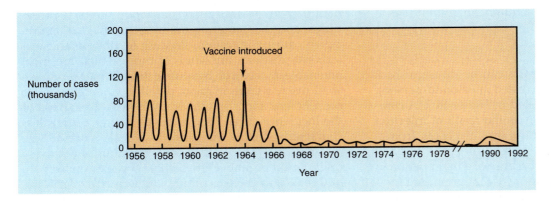

FIGURE 14.4
Cycles of disease, as illustrated by measles. Reported cases of measles by 4-week periods in the United States. The sharp decline in incidence after introduction of immunization is also seen.

study of disease statistics will sometimes point to a likely cause or contributing factor.

Different pathogens have different modes of transmission, which are usually related to the habitats of the organisms in the body. For instance, respiratory pathogens are generally airborne, whereas intestinal pathogens are spread by food or water. If the pathogen is to survive in nature, it must undergo transmission from one host to another. Thus pathogens generally have evolved features or mechanisms that permit or ensure transmittal. Transmission involves three stages: (1) escape from the host, (2) travel, and (3) entry into a new host. We give here a brief overview of transmission mechanisms and several of these will be discussed in detail for certain diseases in the next chapter.

Direct host-to-host transmission

Host-to-host transmission occurs whenever an infected host transmits the disease to a susceptible host by any of a number of means. These include transmission by the respiratory route and by direct contact. Transmission by infectious droplets is the most frequent means by which upper respiratory infections such as the common cold or influenza are propagated. However, some pathogens are so sensitive to environmental influences that they are unable to survive for significant periods of time away from the host. Such pathogens are transmitted from host-to-host by direct contact. The best examples of pathogens transmitted in this way are those responsible for sexually transmitted diseases, such as *Treponema pallidum* (syphilis) and *Neisseria gonorrhoeae* (gonorrhea). These agents are extremely sensitive to drying and do not survive away from the body, even for a few moments. Intimate person-to-person contact, such as kissing or sexual intercourse, provides a direct means for the transmission of such pathogens. However, it should be obvious that such intimate transfer can only occur if the viable pathogen is present on the transmitting person at the body site that comes in direct contact with that of the recipient. Thus the pathogens causing sexually transmitted diseases live in genitalia, the mouth, or the anus, since these are the sites involved in intimate person-to-person contact.

Direct contact is also involved in the transmittal of

skin pathogens, such as staphylococci (boils and pimples) or fungi (ringworm). Because these pathogens are relatively resistant to environmental influences such as drying, intimate person-to-person contact is not the only means of transmission, as it is with the sexually transmitted diseases. It could also be considered that many respiratory pathogens are transmitted by direct means, since they are spread by droplets emitted as a result of sneezing or coughing, and many of these droplets do not remain airborne for long. Transmission, therefore, almost always requires close, although not necessarily intimate, person-to-person contact.

Indirect host-to-host transmission

Indirect transmission can occur by either living or inanimate means. Living agents transmitting pathogens are called **vectors**; they are generally arthropods (for example, insects, mites, or fleas) or vertebrates (for example, dogs, rodents). Arthropod vectors may not actually be hosts for the disease, but merely carriers of the agent from one host to another. Large numbers of arthropods obtain nourishment by biting, and if the pathogen is present in the blood, the arthropod vector will ingest the pathogen and may transmit it when biting another individual. In some cases, the pathogen actually replicates in the arthropod, which is then considered an alternate host, and such replication leads to a buildup of the inoculum, increasing the probability that a subsequent bite will lead to infection.

Inanimate agents such as bedding, toys, books, or surgical instruments, which come in contact with people only casually, can also transmit disease. These inanimate objects are collectively referred to as **fomites**. Food and water are referred to as disease **vehicles**. Fomites can also be disease vehicles, but major epidemics originating from a single source are usually traced to food or water, since these are actively con-

---

A pathogen can be transmitted directly from one host to another, or indirectly by means of another living agent called a vector. Insects and other arthropods are important vectors for human disease. Pathogens can also be transmitted by inanimate objects (fomites) and disease vehicles such as food and water.

sumed in large amounts by a number of individuals in a population.

The nature of epidemics

Two major types of epidemics can be distinguished: common source and propagated (person-to-person). These two types are contrasted in Figure 14.5. A **common-source epidemic** arises as the result of infection (or intoxication) of a large number of people from a contaminated source, such as food or water. Usually such contamination occurs because of a malfunction in some aspect of the distribution system providing food or water to the population. Foodborne and water-borne diseases are primarily *intestinal* diseases, the pathogen leaving the body in fecal material, contami-

nating food or water via improper sanitary procedures, and then entering the intestinal tract of the recipient during ingestion. Because foodborne and waterborne diseases are some of those which are most amenable to control by public health measures, we shall discuss them in some detail in Chapter 15 (also see Cholera and Typhoid Mary boxes). The shape of the incidence curve for a common-source outbreak is characterized by a sharp rise to a peak, since a large number of individuals succumb within a relatively brief period of time (Figure 14.5). The common-source outbreak also declines rapidly, although the decline is less abrupt than the rise. Cases continue to be reported for a period of time approximately equal to the duration of one incubation period of the disease.

Snow on Cholera

The importance of drinking water as a vehicle for the spread of cholera was first shown in 1855 by British physician John Snow, who at that time was without any knowledge of the bacterial causation of the disease. Snow's study is one of the great classics of epidemiology and serves as a model for how a careful study can lead to clear and meaningful conclusions.

In London, the water supplies to different parts of the city were from different sources and were transmitted in different ways. In a large area south of the Thames River, across the river from Westminster Abbey and Parliament Building, the water was supplied to houses by two competing private water companies, the Southwark and Vauxhall Company, and the Lambeth Company. It was the water of the former company that was the major vehicle for the transmission of cholera. When Snow began to suspect the water supply of the Southwark and Vauxhall Company, he made a careful survey of the residence of every death in this district and determined which company supplied the water to that residence. In some parts of the area served by these two companies, each had a monopoly, but in a fairly large area the two companies competed directly, each having run independent water pipes along the various streets. Houses had the option of connecting with either supply, and the distribution of houses between the two companies was random. The clear-cut results of Snow's survey were completely convincing, even to those skeptical about the importance of polluted water in the transmission of cholera: in the first seven weeks of the epidemic, there were 315 deaths per 10,000 houses supplied by the Southwark and Vauxhall Company, and only 37 per 10,000 houses supplied by the Lambeth Company. In the rest of London, there were 59 deaths per 10,000 houses, showing that those supplied by the Lambeth Company had fewer deaths than the general population. In the districts where each company had exclusive rights, it could of course be argued that it was not the water, but some other factor (soil, air, general layout of houses and so on), which might have

been responsible for the differences in disease incidence, but in the districts where the two companies competed, all of these other factors were the same, yet the incidence was high for those supplied with Southwark and Vauxhall water and low for those supplied with Lambeth water. Snow attempted to relate these differences in disease incidence to the sources of the waters used by the two companies. Since he suspected that the excrement and evacuations from cholera patients were highly infectious, he considered that sewage contamination of the water supply might exist. In those days, sewage treatment did not exist and raw sewage was dumped directly into the Thames River. The Southwark and Vauxhall Company obtained its water supply from the Thames right in the heart of London, where much opportunity for sewage contamination could occur, while the Lambeth Company obtained its water from a point on the river considerably above the city, and hence was relatively free of pollution. It is almost certain that it was this difference in source which accounted for the difference in disease incidence. In Snow's words:

As there is no difference whatever, either in the houses or the people receiving the supply of the two Water Companies, or in any of the physical conditions with which they are surrounded, it is obvious that no experiment could have been devised which would more thoroughly test the effect of water supply on the progress of cholera than this. . . . The experiment, too, was on the grandest scale. No fewer than three hundred thousand people of both sexes, of every age and occupation, and of every rank and station, from gentlefolk down to the very poor, were divided into two groups without their choice, and, in most cases, without their knowledge; one group being supplied with water containing the sewage of London, and, amongst it, whatever might have come from cholera patients, the other group having water quite free from such impurity.

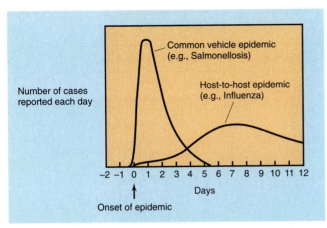

FIGURE 14.5 Origins of epidemics. The shape of the epidemic curve helps to distinguish the likely origin. In a common source epidemic, such as from contaminated food or water, the curve is characterized by a sharp rise to a peak, with a rapid decline, which is less abrupt than the rise. Cases continue to be reported for a period approximately equal to the duration of one incubation period of the disease. In a host-to-host epidemic, the curve is characterized by a relatively slow, progressive rise, and the cases will continue to be reported over a period equivalent to several incubation periods of the disease.

In a **propagated** or **person-to-person epidemic**, the curve of disease incidence shows a relatively slow, progressive rise (Figure 14.5), and a gradual decline. Cases continue to be reported over a period of time equivalent to several incubation periods of the disease. The epidemic may have been initiated by the introduction of a single infected individual into a susceptible population, and this individual has infected one or a few people in the population. Then the pathogen replicated in these individuals, reached a communicable stage, and was transferred to others, where it replicated and again became communicable. Table 14.3 summarizes the major epidemic diseases observed today.

> The rate at which a disease develops can be used to determine the epidemic type. When the source of the infection is a common source such as food or water, there is a sharp rise in disease incidence, followed by an equally sharp decline. For a person-to-person epidemic, the incidence rises more slowly, and also declines more slowly. By making a graph of disease incidence versus time, the type of epidemic can be determined.

14.6   Nature of the Host Community

The colonization of a susceptible, unimmunized host by a parasite may first lead to an explosive infection and an epidemic. As the host population develops resistance, however, the spread of the parasite is checked, and eventually a *balance* is reached in which host and parasite are in equilibrium. A subsequent genetic change in the parasite could lead to the formation of a more virulent form, which would then initi-

ate another explosive epidemic until the host again responded, and another balance was reached. In effect, the host and parasite are affecting each other's evolution; that is, the host and parasite are *coevolving*.

Introduction of a parasite into a previously unexposed population

An excellent experimental example of coevolution of host and parasite occurred when a virus was intentionally introduced for purposes of population control in the wild rabbits of Australia.

The wild rabbit was introduced into Australia from Europe in 1859 and quickly spread until it was overrunning large parts of the continent. Myxoma virus was discovered in South American rabbits, which are of a different species. In South America the virus and its hosts are apparently in equilibrium, as the virus causes only minor symptoms. However, this same virus was found to be extremely virulent in the European rabbit and almost always caused a fatal infection. The virus is spread from rabbit to rabbit by mosquitoes and other biting insects, and hence is capable of rapid spread in areas where appropriate insect vectors are present.

The virus was introduced into Australian rabbits in 1950, with the aim of controlling the rabbit population. Within several months, the virus was well established in the population and spread over an area in Australia as large as all of Western Europe. The disease showed a marked seasonal pattern, rising to a peak in the summer when the mosquito vectors were present and declining in the winter. Research on the epidemiology of myxoma virus was initiated as a model of a virus-induced epidemic by Australian workers. Isolations of the virus from wild rabbits were carried out, and the virus strains isolated were characterized for virulence with laboratory rabbits. At the same time, baby rabbits were removed from their dens before infection could occur and reared in the laboratory. Then these rabbits were challenged with standard virulent strains of myxoma virus to determine their susceptibility. The results of this large scale model study are shown in Figure 14.6.

During the early years of the epidemic over 99 percent of the infected rabbits were dying. However, within six years both the virus and the rabbit population had changed. Over this interval rabbit mortality had dropped to about 84 percent, and the virus isolated was of decreased virulence. In addition, changes were noted in the resistance of the rabbit. In parts of Australia where the virus had been first introduced, the remaining rabbit population had been subjected to selective pressure by the virus for several years. As seen in Figure 14.6, within years the resistance of the rabbits had increased dramatically. This resistance was due to innate changes in the rabbit population and not to immunological responses, for the rabbits tested had been removed from their mothers at birth and had never been in contact with the virus. Their re-

Table 14.3   Some major epidemic diseases and their control

Disease	Infective organism	Infection sources	Control measures
Common-vehicle epidemics[a]			
Typhoid fever	*Salmonella typhi*	Contaminated food and water	Decontamination of public water sources; pasteurization of milk; vaccination
Paratyphoid	*Salmonella paratyphi*	Contaminated food and water	Decontamination of public water sources, food vendors and handlers; vaccination
Bacillary dysentery	*Shigella dysenteriae*	Contaminated food and water	Detection and control of carriers; inspection of food handlers; decontamination of water supplies
Brucellosis	*Brucella melitensis*	Milk or meat from infected animals	Pasteurization of milk; control of infection in animals
Cholera	*Vibrio cholerae*	Contaminated water	Decontamination of public water sources; vaccination
Host-to-host epidemics *Respiratory*			
Diphtheria	*Corynebacterium diphtheriae*	Human cases and carriers; infected food and fomites	Immunization with diphtheria toxoid; quarantine of infected individuals
Meningococcal meningitis	*Neisseria meningitidis*	Human cases and carriers	Treatment of all exposed with sulfadiazine (if isolated strain is susceptible)
Pneumococcal pneumonia	*Streptococcus pneumoniae*	Human carriers	Treatment and rest for all upper respiratory infections; isolation of cases for period of communicability
Tuberculosis	*Mycobacterium tuberculosis*	Sputum from human cases; contaminated milk	Early and extensive treatment of all cases with isoniazid; pasteurization of milk
Whooping cough	*Bordetella pertussis*	Human cases	Immunization with *B. pertussis* vaccine; case isolation
Measles	Measles virus	Human cases	Measles vaccine
German measles	Rubella virus	Human cases	Rubella vaccine; avoidance of contact between infected individual and pregnant women
Influenza	Flu virus	Human cases	Vaccine (recommended in certain cases only)
Sexually transmitted			
Gonorrhea	*Neisseria gonorrhoeae*	Urethral and vaginal secretions	Chemotherapy of carriers and potential contacts; case tracing and treatment[b]
Syphilis	*Treponema pallidum*	Infected exudate or blood	Case identification by serological tests; treatment of seropositive individuals
Chlamydia	*Chlamydia trachomatis*	Urethral, vaginal, and anal secretions	Testing for *C. trachomatis* during routine pelvic examinations; treatment of infected individuals and all potential contacts
Trichomoniasis	*Trichomonas vaginalis*	Urethral, vaginal, prostate secretions	Chemotherapy of infected individuals and contacts
AIDS	HIV (human immunodeficiency virus)	Infected exudate or blood	Treatment with antiviral replication inhibitors (not curative)

[a]*Some common-vehicle diseases can also be spread from host to host.*
[b]*Sexually transmitted diseases can also be controlled by effective use of condoms.*

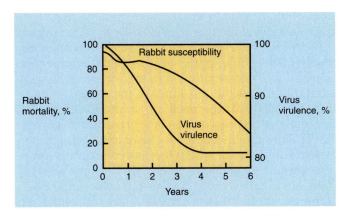

FIGURE 14.6 Changes in virulence of myxoma virus and in susceptibility of the Australian rabbit during the years after the virus was first introduced into Australia in 1950. Virus virulence is given as the average mortality in a standard laboratory breed of rabbits, for virus recovered from the field during each year. Rabbit susceptibility was determined by removing young rabbits from their dens and challenging with a virus strain of moderately high virulence, which killed 90–95 percent of normal laboratory rabbits.

sistance was due to some genetic change in the animal, which made it less susceptible to myxoma virus.

As a result of introduction of myxoma virus, the Australian rabbit population was greatly controlled, but the genetic changes in virus and host served to prevent a complete eradication of the rabbit from Australia. A steady state rabbit population of about 20 percent of that present before the introduction of myxoma virus was observed (Figure 14.6). The virus was thus a major factor in population control, but was unable to totally eliminate rabbits because of coevolutionary events in both host and parasite. From the present point of view, the significance of the Australian experiment is that it reveals how quickly an equilibrium is reached between host and parasite. The manner in which the malaria parasite has affected biochemical evolution in humans is another case in point and will be discussed in Section 15.11.

Herd immunity

An analysis of the immune state of a group is of great importance in understanding the role of immunity in the development of epidemics. **Herd immunity** is a concept used to explain the resistance of a group to invasion and spread of an infectious agent due to immunity of a high proportion of the members of the group. If the proportion of immune individuals is sufficiently high then the whole population will be protected. The fraction of resistant individuals necessary to prevent an epidemic is higher for a highly virulent agent or one with a long period of infectivity, and lower for a mildly virulent agent or one with a brief period of infectivity.

The proportion of the population that must be immune to prevent infection of the rest of the population can be estimated from data on poliovirus immunization in the United States. From epidemiological studies of the incidence of polio in large populations, it ap-

pears that if a population is 70 percent immunized, polio will be essentially absent in the population. Clearly, these immunized individuals are protecting the rest of the population. For a highly infectious disease such as influenza, the proportion of immune individuals necessary to confer herd immunity is higher, about 90 to 95 percent. A value of about 70 percent has also been estimated for diphtheria, but further study of several small diphtheria outbreaks has shown that in densely settled areas a much higher proportion may have to be immunized to prevent development of an epidemic. Apparently in dense populations, person-to-person transmission can occur even if the agent is not highly infectious. In the case of diphtheria an additional complication arises because immunized persons can still harbor the pathogen (inapparent infection) and thus act as chronic carriers, serving as potential sources of infection.

Cycles of disease

The concepts of propagated epidemics and herd immunity can also explain why certain diseases occur in *cycles*. A good example of a cyclical disease is measles, which, in the days before the measles vaccine, occurred in a high proportion of school children. Since the measles virus is transmitted by the respiratory route, its infectivity is high in crowded situations such as schools. Upon entry into school at age five, most children would be susceptible, so that upon the introduction of measles virus into the school, an explosive propagated epidemic would result. Virtually every individual would become infected and develop immunity, and as the immune population built up, the epidemic died down. Measles showed an annual cycle (Figure 14.4) probably because a new group of nonimmune children arrived each year; the phasing of the epidemic would be related to the time of the year at which school began after the summer vacation. The consequences of vaccination for the development of measles epidemics are also illustrated in Figure 14.4.

> Immune mechanisms play an important role in epidemics. If a large fraction of a population is immune, then it is more difficult for the disease to spread, a phenomenon called herd immunity. A disease cycle arises when a large nonimmune population is present; the disease incidence rises as infection occurs but then decreases as infected individuals become immune.

14.7 Hospital-Acquired (Nosocomial) Infections

The hospital may not only be a place where sick people get well, it may also be a place where sick people get sicker. The fact is that cross-infection from patient to patient or from hospital personnel to patients presents a constant hazard. Hospital infections are often called *nosocomial infections* (*nosocomium* is the Latin word for hospital), and occur in about 5 percent of all

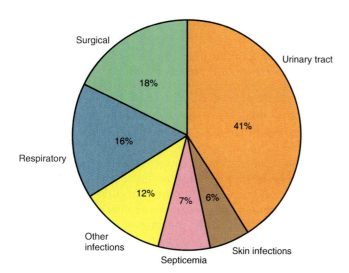

FIGURE 14.7 Major types of nosocomial infection.

patients admitted. Hospital infections are partly due to the prevalence of diseased patients, but are largely due to the presence of pathogenic microorganisms which are selected by the hospital environment.

Hospitals are especially hazardous for the following reasons. (1) Many patients have weakened resistance to infectious disease because of their illness (compromised hosts; see Section 11.12). (2) Hospitals must of necessity treat patients suffering from infectious disease, and these patients may be reservoirs of highly virulent pathogens. (3) The crowding of patients in rooms and wards increases the chance of cross-infection. (4) There is much movement of hospital personnel from patient to patient, increasing the probability of transfer of pathogens. (5) Many hospital procedures, such as catheterization, hypodermic injection, spinal puncture, and removal of tissue samples (biopsy) or fluids, carry with them the risk of introducing pathogens to the patient. (6) In maternity wards of hospitals, newborn infants are unusually

susceptible to certain kinds of infection because they lack well-developed immune systems. (7) Surgical procedures are a major hazard, since not only are highly susceptible parts of the body exposed to sources of contamination but the stress of surgery often diminishes the resistance of the patient to infection. (8) Many drugs used for immunosuppression (for instance, in organ transplant procedures) increase susceptibility to infection. (9) Use of antibiotics to control infection carries with it the risk of selecting antibiotic-resistant organisms, which then may not be easily controlled if they cause further infection. Figure 14.7 summarizes the most common types of hospital-acquired infections.

Hospital pathogens

Although a variety of pathogens are encountered in the hospital environment, a relatively limited number cause the majority of hospital infections. *Escherichia coli*, presumably introduced from the normal flora, is the most common cause of urinary tract infections in hospitals but other enteric bacteria and *Pseudomonas aeruginosa* (see below) are often implicated as well (Figure 14.8). *Enterococcus faecalis* is also a common urinary tract pathogen; the yeast *Candida albicans* is also encountered (Figure 14.8).

One of the most important and widespread hospital pathogens is *Staphylococcus aureus*. It is most commonly associated with skin, surgical, and lower respiratory tract infections, and is a particular problem in infections acquired by newborns in the hospital (Figure 14.8b). Certain strains of unusual virulence have been so widely associated with hospital infections that they are sometimes designated "hospital staphylococci." The habitat of these staphylococci is the upper respiratory tract, usually the nasal passages, and they often become established as "normal flora" in hospital personnel. In such healthy personnel the organism may cause no disease, but these symptomless carriers

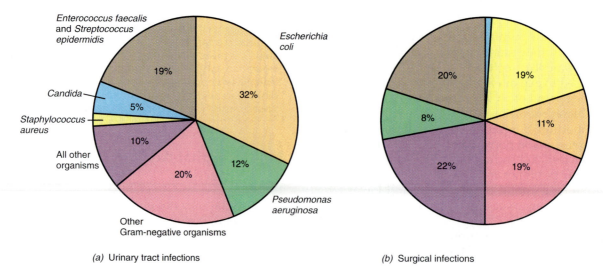

(a) Urinary tract infections

(b) Surgical infections

FIGURE 14.8 Most commonly isolated nosocomial pathogens. (a) In urinary tract infections. (b) In surgical infections. Color coding in part (b) is the same as in part (a).

may be a source of infection for susceptible patients. Since staphylococci are resistant to drying, they survive for long periods on dust particles and other inanimate objects and can subsequently infect patients. Because of the potential seriousness of infection with hospital staphylococci, careful application of principles of hospital sanitation is necessary.

Pseudomonas aeruginosa is important in causing infections of the lower respiratory and urinary tracts (Figure 14.8). It is also an important cause of infections in burn patients who have lost their primary barrier to skin infection. *P. aeruginosa* exhibits one of the most significant features complicating the treatment of nosocomial infections, *drug resistance*. Isolates of this bacterium from patients with hospital infections are commonly resistant to many antibiotics. A somewhat lower degree of resistance has been noted among *S. aureus* isolates (with occasional highly resistant strains not uncommon), whereas *E. coli* isolates generally remain sensitive to antibiotics. Antibiotic-resistant pathogenic bacteria in hospitals generally contain plasmids coding for *multiple drug resistance*, as described in Section 7.9.

> Because hospital patients often have impaired immune systems or are unusually sensitive to infectious disease, it is of great importance to control the spread of disease through the hospital. Certain pathogens that are ordinarily not important in normal populations may cause serious disease in hospitals.

14.8   Public Health Measures for the Control of Epidemics

An understanding of the epidemiology of an infectious disease generally makes it possible to develop methods for the control of the disease. **Public health** refers to the health of the population as a whole, and to the activities of public health authorities in the control of epidemics. It should be emphasized that the incidence of many infectious diseases has dropped dramatically, especially in developed countries, over the past 100 years not because of any specific control methods, but because of general increases in the well-being of the population. Better nutrition, less crowded living quarters, and lighter work loads have probably done as much as public health measures to control diseases such as tuberculosis. However, diseases such as typhoid fever, diphtheria, brucellosis, and poliomyelitis owe their low incidence to active and specific public health measures. Finally, one infectious disease, gonorrhea, is much more prevalent today than it was 50 years ago. This increased incidence can be linked directly to changes in sexual behavior of the population.

Overall public health depends on application of a series of control measures to prevent the spread of infectious disease and we discuss these here.

Controls directed against the reservoir

If the disease occurs primarily in *domestic animals*, then infection of humans can be prevented if the disease is eliminated from the infected animal population. Immunization procedures or slaughter of infected animals may be used to eliminate the disease in animals. These procedures have been quite effective in eliminating brucellosis and bovine tuberculosis from humans. Not incidentally, the health of the domestic animal population is also enhanced, with likely economic benefits to the farmer.

When the reservoir is a *wild animal* (for example, plague, see Section 15.12), then eradication is much more difficult. **Rabies** is a disease that occurs both in wild and domestic animals, but is transmitted to domestic animals primarily by wild animals. Thus control of rabies can be achieved by immunization of domestic animals, although this will never lead to complete eradication of the disease. Indeed, statistics show that the majority of rabies cases are in wild rather than domestic animals, at least in the United States (Table 14.4).

If the reservoir is an *insect* (such as a mosquito in the case of malaria) effective control of the disease can be maintained by the use of chemical insecticides or other lethal agents to eliminate the reservoir. However, such use must be balanced with environmental concerns about the use of toxic or carcinogenic chemicals—in some cases the elimination of one public health problem only creates another.

When *humans* are the reservoir (for example, gonorrhea), then control and eradication are much more difficult, especially if there are asymptomatic carriers.

Controls directed against transmission of the pathogen

If the organism is transmitted via food or water, then public health procedures can be instituted to prevent either contamination of these vehicles or to destroy

Table 14.4	Reported yearly cases of rabies in animals, United States*		
Domestic animals	**Cases**	**Wild animals**	**Cases**
Cattle	171	Skunks	1791
Cats	192	Bats	638
Dogs	128	Foxes	183
Other domestic	59	Racoons	1463
		Other wild	99
Total:	550	Total:	4174

Total domestic and wild: 4724
Percent domestic: 11.6%
Percent wild: 88.4%

From Centers for Disease Control, Atlanta, GA. Data are typical of recent years.

the pathogen in the vehicle. Water purification methods (see Section 15.14) have been responsible for dramatic reductions in the incidence of typhoid fever, and the pasteurization of milk has helped in the control of bovine tuberculosis in humans. Food protection laws have been devised that greatly decrease the probability of transmission of a number of enteric pathogens to humans. Transmission of respiratory pathogens is much more difficult to prevent. Attempts at chemical disinfection of air have been unsuccessful. In Japan, many individuals wear face masks when they have upper respiratory infections to prevent transmission to others, but such methods, although effective, are voluntary, and would be difficult to institute as public health measures.

Vaccination

Immunization of the host has been the prime means by which smallpox, diphtheria, tetanus, pertussis (whooping cough), and poliomyelitis have been controlled. As we have discussed in Section 14.6, 100 percent immunization is not necessary in order to control the disease in a population, although the percentage needed to ensure control varies with the virulence of the pathogen and with the condition of the population (for example, crowding).

The application of public health measures can lead to a confrontation between the rights of the private individual and the welfare of the population, and some resistance exists to the universal application of public health measures, such as immunization, which directly affect the individual. Over the past several decades, the proportion of children vaccinated for diphtheria, tetanus, pertussis, and polio has been slowly decreasing, apparently because the public has become less fearful of contracting these diseases due to their very low incidence in the population. However, since none of these diseases has been eradicated from the United States (indeed, the reservoir of tetanus is the soil, so that it will never be eradicated), and with a decrease in proportion of individuals immunized, the protection afforded by herd immunity (see Section 14.6) may be overcome, and diseases such as diphtheria, pertussis, and polio could reappear in epidemic form.

Also serious is the fact that many *adults* are inadequately immunized, either because they received low-titer vaccines when they were children or because their immunity has gradually disappeared with age. Surveys in the United States have indicated that up to 80 percent of adults may lack solid immunity to important childhood diseases. When these so-called childhood diseases occur in adults, they can have severe effects. If a woman contracts rubella (a viral disease; see Section 15.4) during pregnancy, the unborn child can be seriously impaired. Measles and polio are much more serious diseases in adults than in children.

All adults are advised to review their immunization status, checking their medical records (if available) to ascertain dates of vaccinations. *Tetanus* vaccinations should be given routinely at least every 10 years. Surveys of adult populations have shown that more than 10 percent of adults under the age of 40 and over 50 percent of those over 60 are not protected. *Measles* immunity also needs to be reviewed. People born before 1957 probably had measles as children and are immune. Those born after 1956 may have been vaccinated but the effectiveness of early vaccines was variable and solid immunity may not be present, especially if the vaccine was given before one year of age. Revaccination for polio is not recommended for adults unless they are traveling to countries in Africa and Asia where polio is still prevalent.

Vaccination practices and procedures have been discussed in Chapter 12 and those for particular infections will be discussed in Chapter 15.

Quarantine

Quarantine involves the limitation of the freedom of movement of individuals with active infections to prevent spread of disease to other members of the population. The time limit of quarantine is the longest period of communicability of the given disease. Quarantine must be done in such a manner that effective contact of the infected individual with those not exposed is prevented. Quarantine is not as severe a measure as strict isolation, which is used for unusually infectious diseases in hospital situations.

By international agreement six diseases are considered quarantinable: smallpox, cholera, plague, yellow fever, typhoid fever, and relapsing fever. Although smallpox has been eliminated from the world, quarantine for the other five diseases is still mandated. Each of these is considered a highly serious disease and particularly communicable. Thus it is essential to quarantine an infected individual for the period of communicability.

Public health has played a major role in reducing the incidence of many important diseases. Foodborne and waterborne diseases have been virtually eliminated from many countries by strict government regulation. Vaccination not only protects the individual but helps to control the spread of the disease through the population (herd immunity). For extremely serious diseases, isolation of infected individuals (quarantine) may be necessary.

14.9   Global Health Considerations

The United States can be considered typical of those countries where public-health protection is highly developed. Other countries with similar characteristics include Japan, Australia, New Zealand, Israel, and the European countries. In quite another category as far as infectious disease is concerned are the developing countries, a category that includes most of the coun-

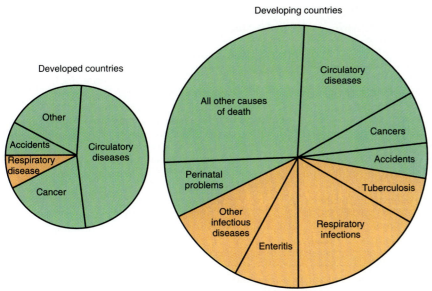

Developing countries

Developed countries

Other

Accidents

Respiratory disease

Cancer

Circulatory diseases

All other causes of death

Circulatory diseases

Cancers

Accidents

Tuberculosis

Perinatal problems

Other infectious diseases

Enteritis

Respiratory infections

FIGURE 14.9
Leading causes of death in developed and developing countries. Infectious diseases are shown in orange and noninfectious diseases in green. The sizes of the circles are proportional to the relative number of deaths which occurred. The data are typical of recent years.

tries in Africa, Central and South America, and Asia. In these countries, infectious diseases are still major causes of death (see Table 14.1).

Infectious disease in developing countries

There is a sharp contrast in the degree of importance of infectious diseases as causes of death in developing versus developed nations. In developing regions of the world, infectious diseases account for between 30 and 50 percent of deaths whereas infectious diseases account for about 4 to 8 percent of deaths in developed regions (see Figure 14.9). Diseases which were leading causes of death in the United States nearly a century ago, such as tuberculosis and gastroenteritis, are still leading causes of death in developing countries today (Figure 14.9). Furthermore, the majority of deaths due to infectious disease in developing regions occur among infants and children. Thus, the average age of individuals dying due to infectious diseases in developing versus developed countries is also dramatically different.

The distinct differences in the health status of people in different regions of the world is due in part to general nutritional deficiency in individuals in developing countries and to a lower overall standard of living. As discussed in Section 11.12, factors such as physical stress and diet play important roles in the ability of the host to ward off infection. Thus, it is not surprising that in developing countries death from infection is more likely. In addition, the generally lower levels of public health protection and lack of economic resources to implement widespread vaccination and food and water surveillance programs in developing countries make infection more likely in the first place. Statistics on disease in developed countries show that control of many diseases is possible. However, statistics on the worldwide incidence of disease show that infectious disease remains an important problem about which we must all be concerned.

Travel to endemic areas

The high incidence of disease in many parts of the world is also a concern for people traveling to such areas. It is possible to be immunized against many of the diseases which are endemic in foreign countries. Some typical recommendations for immunization for those traveling abroad are shown in Table 14.5. Many foreign countries currently require immunization certificates for yellow fever, but many other immunizations are only recommended for people who are expected to be at high risk. There is also risk in many parts of the world of exposure to diseases for which there is no effective vaccine available. These include amebiasis, dengue fever, encephalitis, giardiasis, malaria, and typhus. Travelers are advised to take reasonable precautions such as avoiding insect or animal bites, drinking only water which has been properly treated and eating food properly stored and prepared, and undergoing chemotherapeutic programs after exposure is suspected.

The epidemiologist and world public health

We have seen in this chapter that the epidemiologist is in some ways a medical sleuth, tracking down the source(s) and modes of transmission of infectious agents, sometimes even on a worldwide basis. Through the efforts of epidemiologists and government officials, most developed countries in the world today enjoy a high level of overall public health. Unfortunately, despite the efforts of international organizations such as the World Health Organization (WHO) and other health officials, serious infectious diseases are still a way of life in developing countries.

However, new diseases such as AIDS and legionellosis, tuberculosis, and well-entrenched diseases such as influenza and gonorrhea, keep health scientists busy in the developed countries as well. Infectious diseases are therefore still a global problem,

Table 14.5   Immunizations required or recommended for travel to developing countries*

Disease	Destination	Recommendation
Cholera	Many central African nations, India, Pakistan, S. Korea, Albania, Malta	*Vaccination recommended* if entering from or continuing to endemic areas
Yellow fever	Tropical and subtropical countries, worldwide	*Vaccination often required* for entry; or if entering from or continuing to endemic areas
Plague	Mostly rural mountainous or upland areas of Africa, Asia and South America	*Vaccination recommended* if direct contact with rodents is anticipated
Infectious hepatitis (A)	Specific tropical areas and many developing countries	*Passive immunization recommended* for long stays
Serum hepatitis (B)	Africa, Indochina, eastern and southern Europe, countries in the former Soviet Union, Central and South America	*Vaccination recommended*
Typhoid fever	Many African, Asian, Central and South American countries	*Vaccination recommended*

*From Health Information for International Travelers, 1992, U.S. Department of Health and Human Services. Vaccinations are not required, but are recommended, for diphtheria, pertussis, tetanus, polio, measles, mumps, and rubella. Most U.S. citizens are already immunized through normal immunization practices.

and many diseases will probably be impossible to totally eradicate, although their worldwide impact may be kept quite low. In the following chapter we discuss a number of common infectious diseases and emphasize the microbiology and ecology of the disease process in each case.

> **Although infectious disease has been partially controlled in developed countries, it is still the major cause of death in developing countries. Travelers to such areas may need to be vaccinated for some of these widespread infectious diseases.**

Study Questions

1. How would a physician and an epidemiologist likely differ in their perspective on infectious diseases?

2. How could a disease have relatively low incidence but at the same time be highly prevalent?

3. Distinguish between *morbidity* and *mortality*. How would you get statistics on morbidity? Mortality?

4. What is an *epidemic*? How is it distinguished from an *endemic* disease? *Pandemic*?

5. What is the disease status of someone who is identified as a disease *carrier*?

6. What are the possibilities for control of an infectious disease in which the primary reservoir is a domestic animal? A wild animal such as a rat? A wild animal such as a racoon?

7. In the United States, even a small number of cases of the disease polio would be considered to be an epidemic whereas if even a large number of cases of chickenpox occurred it would not likely be considered an epidemic. Explain.

8. What are the consequences for public health control of infectious disease for those diseases in which most cases occur as inapparent infections? How would you determine the incidence of a disease which caused many inapparent infections (refer back to Table 13.5 for a clue).

9. What lines of evidence led epidemiologists to conclude that the disease *AIDS* was not transmitted by casual contact between humans?

10. Compare and contrast the epidemiological picture of a disease like measles with a disease like botulism.

11. What is a disease *vector*? What is a disease *vehicle*? What is a *fomite*? Briefly discuss three different diseases in which you correctly use these terms.

12. Based on the consideration of four kinds of individuals: infected individuals, immune individuals, uninfected individuals, and carriers; describe how a disease cycle might occur for a respiratory pathogen.

13. Although *myxoma* virus is highly fatal for rabbits, explain why the introduction of this virus into the Australian rabbit population did not kill off all the rabbits.

14. If most of the population is immune for a particular infectious disease, the unimmunized remainder of the population rarely becomes ill even though they are not immunized. Explain.

15. How does the concept of *herd immunity* depend to some extent on the virulence of the pathogen?

16. *Rabies* is almost nonexistent in dogs (see Table 14.4), yet it is still necessary to ensure that all dogs are vaccinated. Explain.

17. Why does a disease like chickenpox show a periodic cycle whereas other diseases, such as sexually transmitted diseases, do not?

18. What are *nosocomial* infections? Why are they so common? How has the use of antibiotics contributed to the incidence, severity, and type of nosocomial infections?

19. Why is *Staphylococcus aureus* a common nosocomial pathogen? *Pseudomonas aeruginosa?*

20. What are the major public health measures taken to prevent disease epidemics? Which measures are most effective? Which methods do you think would be the most expensive and why?

21. Why would it be necessary to quarantine an individual suffering from cholera, assuming that a case was discovered in your community?

22. What are the major reasons why public health measures are more effective and the overall level of public health higher in developed countries than in developing countries?

Supplementary Readings

Alexander, J. J., H. L. Gabelnick, and **J. M. Spieler** (eds.) 1990. *Heterosexual Transmission of AIDS*. John Wiley and Sons, New York. Covers the epidemiology of AIDS transmission by means other than homosexual and intravenous drug activities.

Brachman, P. S. 1991. *Bacterial Infections of Humans: Epidemiology and Control*. Plenum Medical Books, New York. The epidemiology of bacterial diseases.

Brock, T. D. (ed.) 1989. *Microorganisms: From Smallpox to Lyme Disease*. W. H. Freeman, New York. A collection of fascinating articles on infectious disease from *Scientific American* magazine.

Centers for Disease Control. *Morbidity and Mortality Weekly Report*. Atlanta, GA. Available from Massachusetts Medical Society, Waltham, MA. Issued weekly. This publication, available in most large libraries, gives an instantaneous view of epidemiological problems in the United States. Lists incidence of all major infectious diseases by geographic region and gives updates on particular disease problems. An *Annual Summary* is published yearly providing an overview of the incidence of past and present infectious diseases. A good source of graphs, tables, and statistical analyses of disease trends.

Dunsmore, D. J. 1986. *Safety Measures for Use in Outbreaks of Communicable Disease*. World Health Organization, Geneva, Switzerland. A compendium of procedures used to protect public health during disease epidemics.

Fenner, F. 1983. Biological control as exemplified by smallpox eradication and myxomatosis. *Proceedings of the Royal Society of London B* 218:259–285. An excellent historical account of the control of the Australian rabbit population and the global eradication of smallpox.

Hardy, A. 1991. *Incidence and Impact of Selected Infectious Diseases in Childhood*. U.S. Dept. of Health and Human Services, Hyattsville, MD. Epidemiology and statistics of childhood diseases in the United States.

Kurstak, E. 1990. *Virus Variability, Epidemiology and Control*. Plenum Medical Books, New York. An advanced treatise on the epidemiology of several viral diseases.

Snow, J. 1936. *Snow on Cholera*. A reprint of two papers by John Snow, M.D. The Commonwealth Fund, New York. A reprinting of Snow's 1855 study of cholera, the first epidemiological investigation, and a fascinating detective story. Highly recommended.

Stine, G. J. 1993. *Acquired Immune Deficiency Syndrome*. Prentice Hall, Englewood Cliffs, NJ. An excellent monograph describing the current status of the AIDS epidemic.

Streiner, D. L., G. R. Norman, and **H. Munroe-Blum.** 1989. *Epidemiology*. B. C. Decker, Toronto. A short book covering the basic principles of epidemiology.

Walker, D. H. (ed.) 1993. *Global Infectious Diseases; Prevention, Control, and Eradication*. Springer-Verlag, New York. Review of the epidemiology, ecology, pathogenesis, and immunity of several infectious diseases of worldwide significance.

Wilson, M. E. 1991. *A World Guide to Infections: Disease, Distribution, Diagnosis*. Oxford University Press, New York. Epidemiological considerations of immigration and travel.

World Health Organization. 1992. *Global Estimates for Health Situation Assessment and Projections 1992*. WHO, Geneva, Switzerland. An excellent source of epidemiological data from worldwide sources, including information concerning the most prevalent infectious diseases and predictions for future disease problems.

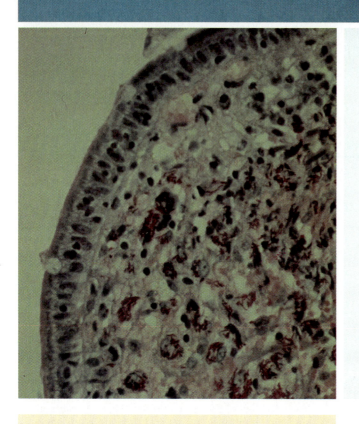

Major Microbial Diseases

Within the microbial world only a relatively few species are pathogens. The majority of microorganisms carry out essential activities in nature and many are closely associated with plants or animals in stable, beneficial relations (see Chapter 17). However, as we have seen, pathogenic organisms can have profound effects on animals and humans. In this chapter we consider the major human diseases and will group together those microbial diseases that affect a specific human organ system. We will continually make reference to *modes of transmission* in our discussion of diseases, because the pathology of a disease is best considered in light of its ecology. For example, streptococcal sore throat and influenza are diseases whose etiological agents are completely different, one a bacterium and the other a virus. Yet, both diseases are transmitted from person to person primarily by an airborne route. Hence, if we group diseases by their modes of transmission, what at first looks like a long list of unrelated diseases quickly shrinks to a short list of ecologically related diseases. We will also refer back to material in Chapters 11 to 14 from time to time, because the biology of specific virulence factors and the nature of the immune responses are considered in detail in these chapters.

15.1 Airborne Transmission of Pathogens

Air is not a suitable medium for the growth of microorganisms; organisms found in air are derived from soil, water, plants, animals, people, or other sources. In outdoor air, soil organisms predominate. Microbial numbers indoors are considerably higher than those outdoors, and the organisms are mostly those commonly found in the human respiratory tract.

Miniglossary for Chapter 15

ANTIGENIC DRIFT   minor changes in antigens due to gene mutation in influenza virus

ANTIGENIC SHIFT   major changes in antigens due to gene reassortment in influenza virus

CD4 CELLS   T helper (T_H) cells, which are targets for HIV infection

CHLORINATION   a highly effective disinfectant procedure for drinking water using chlorine gas or other chlorine-containing compounds

COLIFORM   a large group of Gram-negative, facultative Bacteria

CONGENITAL SYPHILIS   syphilis contracted by an infant from its mother during birth

FOOD INFECTION   infection resulting from ingestion of contaminated food

FOOD POISONING   disease resulting from ingestion of a bacterial exotoxin

LYME DISEASE   a tick-transmitted disease caused by the spirochete *Borrelia burgdorferi*

MYCOSES   infections caused by fungi

RHEUMATIC FEVER   an inflammatory autoimmune disease triggered by an immune response to infection by *Streptococcus pyogenes*

RICKETTSIAS   obligate intracellular parasites which cause a variety of diseases including typhus and Rocky Mountain spotted fever

SCARLET FEVER   characteristic reddish rash resulting from an exotoxin produced by *Streptococcus pyogenes*

SEXUALLY TRANSMITTED DISEASE (STD)   a disease whose usual means of transmission is by sexual contact

SICKLE-CELL ANEMIA   a genetic trait which confers resistance to malaria, but which causes a reduction in numbers of red blood cells

TOXIC SHOCK SYNDROME   acute shock resulting from a host response to an exotoxin produced by *Staphylococcus aureus*

TUBERCULIN TEST   a skin test for previous infection with *Mycobacterium tuberculosis*

Windblown dust carries with it significant microbial populations, which can travel long distances. Most of these organisms survive only poorly in air, so that effective transmittal to a suitable habitat (another human) occurs only over short distances. However, certain human pathogens (*Staphylococcus, Streptococcus*) are able to survive under dry conditions fairly well and so may remain alive in dust for long periods of time. Gram-positive Bacteria are in general more resistant to drying than Gram-negative Bacteria, and it may be for this reason that Gram-positive Bacteria are more likely to be involved in air dispersal. Spore-forming bacteria are also resistant to drying, but are generally not present in humans in the spore form. Other airborne microorganisms found to be derived from soil are also Gram-positive (for example, *Micrococcus*). The reason Gram-positive Bacteria are more resistant to drying than Gram-negative ones is probably because the thicker and more rigid Gram-positive cell wall confers some drought resistance.

An enormous number of droplets of moisture are expelled during sneezing (Figure 15.1), and a considerable number are expelled during coughing or even merely talking. Each infectious droplet has a size of about 10 µm and contains one or two bacteria. The speed of the droplet movement is about 100 m/sec (over 200 miles/hr) in a sneeze and about 16 to 48 m/sec during coughing or loud talking. The number of bacteria in a single sneeze varies from 10,000 to 100,000. Because of the small size of the droplets, the moisture evaporates quickly in the air, leaving behind a nucleus of organic matter and mucus, to which bacterial cells are attached.

Respiratory infection

The average human breathes several million cubic feet of air in a lifetime, much of it containing microorganism-laden dust, which is a potential source of inoculum for upper-respiratory infections caused by streptococci and staphylococci resistant to drying. The speed at which air moves through the respiratory tract varies, and in the lower respiratory tract the rate is quite slow. As the air slows down, particles in it stop moving and settle. The larger particles settle first and the smaller ones later, and as seen in Figure 15.2, in the tiny bronchioles of the lung only particles below 3 µm will be present. As also shown in Figure 15.2, different organisms reach different levels in the tract, thus accounting for the differences in the kinds of infections that occur in the upper and lower respiratory tracts.

> **Although microorganisms do not grow in air, some respiratory pathogens are sufficiently resistant to drying that they can survive and be transmitted through air. This is especially true of Gram-positive Bacteria.**

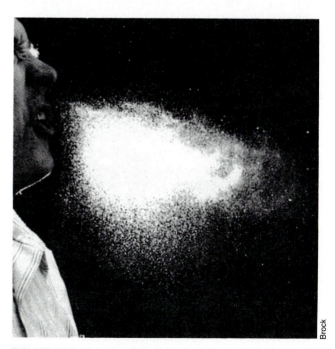

Brock

FIGURE 15.1   High speed photograph of an unstifled sneeze.

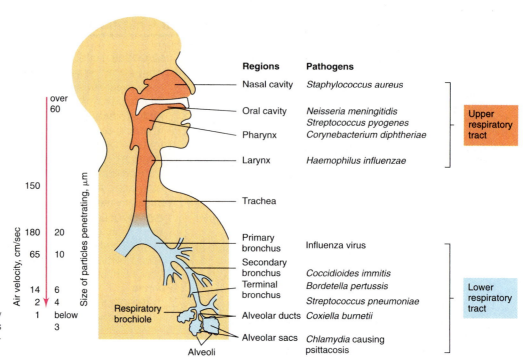

FIGURE 15.2
Characteristics of the respiratory system of humans and locations at which various organisms generally initiate infections.

Regions	Pathogens	
Nasal cavity	*Staphylococcus aureus*	Upper respiratory tract
Oral cavity	*Neisseria meningitidis*	
Pharynx	*Streptococcus pyogenes*	
	Corynebacterium diphtheriae	
Larynx	*Haemophilus influenzae*	
Trachea		
Primary bronchus	Influenza virus	Lower respiratory tract
Secondary bronchus	*Coccidioides immitis*	
Terminal bronchus	*Bordetella pertussis*	
	Streptococcus pneumoniae	
Alveolar ducts	*Coxiella burnetii*	
Alveolar sacs	*Chlamydia* causing psittacosis	

Air velocity, cm/sec — over 60, 150, 180 (20), 65 (10), 14 (6), 2 (4), 1 (below 3)

Size of particles penetrating, μm

15.2 Respiratory Infections: Bacterial

A variety of bacterial pathogens affect the respiratory tract and, as discussed above, because their mode of transmission is air, they are predominantly Gram-positive Bacteria. Because secondary problems associated with an initial bacterial respiratory infection can often be quite serious, it is important to swiftly diagnose and treat these infections to prevent damage to host tissues. Fortunately, most respiratory bacterial pathogens respond readily to antibiotic therapy and many can also be controlled by immunization programs. Nevertheless, bacterial respiratory infections are still rather common and we begin here with a consideration of one of the most common bacterial respiratory pathogens, *Streptococcus*.

Streptococcal diseases

Streptococci are Gram-positive cocci which typically grow in elongated chains (see Section 19.25). Several species of streptococci are potent human pathogens. Of particular importance are *S. pyogenes* and *S. pneumoniae*.

Streptococcus pyogenes and *S. pneumoniae* are frequently isolated from the nose of healthy adults. Although numbers of *S. pyogenes* and *S. pneumoniae* are usually low here, if the host's defenses are weakened, or a new, highly virulent strain is introduced, acute streptococcal bacterial infections are possible. *S. pyogenes* is the leading cause of streptococcal pharyngitis, so-called strep throat (the pharynx is the tube which connects the oral cavity to the larynx and the esophagus, see Figure 15.2). Most isolates from clinical cases of strep throat produce a toxin which lyses red blood cells, a condition called *β-hemolysis* (see Figure 11.15). Streptococcal pharyngitis is characterized by intense inflammation of the mucous membranes of the throat,

a mild fever, and a general feeling of malaise. *S. pyogenes* can also cause related infections of the inner ear (otitis media), the tonsils (tonsillitis), the mammary glands (mastitis), and infections of the superficial layers of the skin, a condition referred to as impetigo (however, most cases of impetigo are caused by *Staphylococcus aureus*) (Figure 15.3).

About half of the clinical cases of severe sore throat turn out to be due to *S. pyogenes*, the remainder being of viral origin. An accurate diagnosis is important since if the sore throat is due to a virus, the appli-

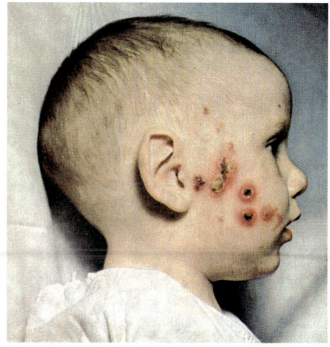

FIGURE 15.3 Typical lesions of impetigo, commonly caused by *Streptococcus pyogenes* or *Staphylococcus aureus*.

Franklin H. Top

cation of antibacterial antibiotics would be useless, whereas if the sore throat is due to *S. pyogenes*, immediate antibiotic therapy is indicated. This is because occasional cases of streptococcal sore throat can lead to more serious streptococcal syndromes such as scarlet fever, rheumatic fever, and acute glomerulonephritis.

Certain strains of *S. pyogenes* carry a lysogenic bacteriophage which codes for production of a potent exotoxin, responsible for most of the symptoms of *scarlet fever*. The erythrogenic toxin causes a pink-red rash to develop (Figure 15.4) and also acts to damage small blood vessels and initiate fever. The condition is acute and easily treated with antibiotics.

Untreated or insufficiently treated cases of *S. pyogenes* infection may lead to severe *delayed sequelae*, or follow-up diseases. **Rheumatic fever**, one of these delayed sequelae, is caused by strains of *S. pyogenes* which contain cell surface antigens sufficiently similar to certain human cell-surface antigens that when an immune response to the invading pathogen is made, the antibodies produced also damage host tissues, in particular those of the heart, joints, and kidneys. In essence, rheumatic fever is a type of autoimmune disease—antibodies reacting with self constituents (see Section 12.14). Rheumatic fever patients must keep a close watch in later life for symptoms of heart or joint damage.

Another potential delayed sequela of *S. pyogenes* infection is *acute glomerulonephritis*, a painful disease of the kidney. This is an immune complex disease (see Section 12.14) resulting from the formation of streptococcal antigen–antibody complexes in the bloodstream during the recovery phase of a streptococcal infection. These immune complexes lodge in the *glomeruli*, or fil-

tration membranes of the kidney, causing inflammation of the kidney (*nephritis*) accompanied by severe kidney pain. Within several days, these complexes are usually dissolved and the patient quickly returns to normal. However, as for rheumatic fever, timely, adequate antibiotic treatment of the initial *S. pyogenes* infection prevents the disease from occurring.

The other major pathogenic streptococcal species, *S. pneumoniae*, causes lung infections which often develop as secondary infections to other respiratory disorders. A characteristic of *S. pneumoniae* is that cells are typically present in pairs (or short chains) and are surrounded by a large capsule (Figure 15.5). The capsule enables the cells to resist phagocytosis, and capsulated strains of *S. pneumoniae* are hence very invasive. Cells invade alveolar tissues (lower respiratory tract) of the lung and elicit a strong host inflammatory response. Reduced lung function can result from accumulation of phagocytic cells and fluid, and the *S. pneumoniae* cells can spread from the focus of infection as a bacteremia, resulting in bone and inner ear infections and occasionally endocarditis. Pneumococcal pneumonia is a serious infection, untreated cases having a mortality rate of about 30 percent. At this time, most strains of *S. pneumoniae* and *S. pyogenes* respond dramatically to penicillin therapy. However, although penicillin remains the clinical drug of choice, an increasing incidence of penicillinase-producing streptococci has been reported in recent years, and erythromycin is the next best therapeutic agent.

The potential for serious host damage following a streptococcal sore throat has encouraged the development of clinical methods for rapid identification of *S. pyogenes*. At least two immunological techniques have been developed thus far that can detect surface proteins unique to *S. pyogenes* by latex bead agglutination or by fluorescent antibody staining (see Sections 13.5 and 13.8) directly from a patient throat swab. In addition, a nucleic acid probe test is available to detect DNA sequences unique to *S. pyogenes* in host tissues. Such procedures allow the confident initiation of im-

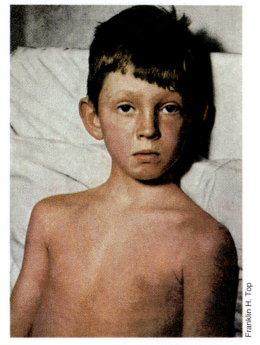

FIGURE 15.4   The typical rash of scarlet fever, due to the action of the erythrogenic toxin produced by *Streptococcus pyogenes*.

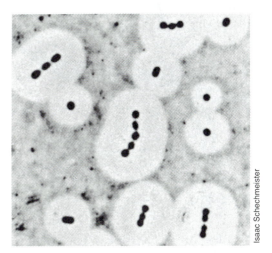

FIGURE 15.5   India ink negatively stained preparation of cells of *Streptococcus pneumoniae*. Note the extensive capsule surrounding the cells.

mediate antibiotic therapy in order to avoid complications such as rheumatic fever. However, the definitive diagnostic confirmation for infection by pathogenic streptococci is a positive culture from the throat (*S. pyogenes*) or the sputum (*S. pneumoniae*).

> Two major respiratory diseases caused by streptococci are streptococcal sore throat and pneumonia. Under certain conditions, streptococcal sore throat can develop into more serious conditions such as scarlet fever and rheumatic fever.

Staphylococcus

The genus *Staphylococcus* contains common pathogens of humans and animals, including some that are occasionally serious pathogens. Staphylococci are Gram-positive cocci which divide in several planes to form irregular clumps (see Figure 19.80). Staphylococci are relatively resistant to drying and hence can be readily dispersed in dust particles through the air. In humans, two species are important: *S. epidermidis*, a nonpigmented, nonpathogenic form usually found on the skin or mucous membranes; and *S. aureus*, a yellow-pigmented form associated with pathological conditions, including boils, pimples, and impetigo (Figures 15.3 and 15.6), but also pneumonia, osteomyelitis, carditis, meningitis, and arthritis.

Those strains of *S. aureus* most frequently causing human disease produce a number of extracellular enzymes or toxins (see Sections 11.8 and 11.9). At least four different *hemolysins* have been recognized, a single strain often being capable of producing more than one hemolysin, and the production of these is responsible for the hemolysis seen around colonies on blood agar plates. *S. aureus* is also capable of producing an *enterotoxin*, commonly associated with foodborne illness (see Sections 11.9 and 15.13).

Another substance produced by *S. aureus* is *coagulase*, an enzyme-like factor that causes fibrin to coagulate and form a clot (see Section 11.8). The production of coagulase is generally associated with pathogenicity. It seems likely that blood clotting induced by coagulase results in the accumulation of fibrin around the bacterial cells and thus renders them resistant to phagocytosis (Figure 15.6). In addition, the formation of such fibrin clots results in the walling off of the infected area, making it difficult for host defense agents to come into contact with the bacteria (Figure 15.6). Most *S. aureus* strains also produce *leukocidin*, which causes the destruction of leukocytes, allowing the *S. aureus* cells to escape phagocytosis unharmed. Production of leukocidin in skin lesions such as boils and pimples results in much cell destruction and is one of the factors responsible for pus formation (Figure 15.6). Other extracellular factors produced by some strains of *S. aureus* include proteolytic enzymes, hyaluronidase, fibrinolysin, lipase, ribonuclease, and deoxyribonuclease.

The most common habitat of *S. aureus* is the upper respiratory tract, especially the nose and throat and the surface of the skin, and many healthy people are carriers of this organism. Most infants become infected during the first week of life and usually acquire the strain associated with the mother or with another close human contact. In most cases, these strains do not cause pathological conditions, and serious staphylococcal infections occur only when the resistance of the host is low because of hormonal changes, debilitating illness, wounds to the skin's surface, or treatment with steroids or other anti-inflammatory drugs. Hospital epidemics have occurred in recent years (see Sec-

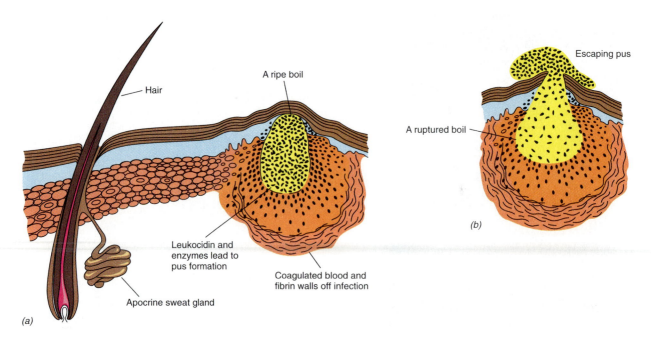

FIGURE 15.6 The structure of a boil. (a) Staphylococci initiate a localized infection of the skin and become walled off by coagulated blood and fibrin. (b) The rupture of the boil releases pus and bacteria.

tion 14.7), which have usually involved antibiotic resistant strains. Extensive use of antibiotics has resulted in the natural selection of resistant strains of *S. aureus*. Hospital epidemics with these antibiotic-resistant staphylococci may occur when patients whose resistance to infection is lowered (owing to other diseases, surgical procedures, or drug therapy) receive cells from hospital personnel who are normal carriers of antibiotic-resistant staphylococci. Control of such hospital epidemics requires careful attention to the maintenance of asepsis.

Certain strains of *S. aureus* have been implicated as the agents responsible for **toxic shock syndrome (TSS)**, a severe result of staphylococcal infection characterized by high fever, rash, vomiting, and diarrhea. Toxic shock has occurred in menstruating women and has been associated with use of tampons. In addition, several cases of toxic shock have been reported in both men and women from staphylococcal infections following surgery. In females, blood and mucus in the vagina become colonized by hemolytic *S. aureus* from the skin, and the presence of a tampon concentrates this material, creating ideal microbial growth conditions.

The symptoms of TSS result indirectly from an exotoxin called *toxic shock toxin*. This toxin is a superantigen (see Section 12.14). The exotoxin is released by the growing staphylococci and binds to the antigen-presenting proteins encoded by the major histocompatibility complex (MHC) and to the T cell receptors (TCR). Unlike a conventional antigen, however, the binding is outside the normal antigen-binding site. Binding of MHC and TCR molecules by the toxin in this nonantigen-specific manner results in the activation of large numbers (perhaps greater than 10 percent of the total population) of T cells at one time. These activated cells then release lymphokines, which in turn activate even more T cells! These hyperstimulated cells then release chemical mediators that produce a systemic inflammatory reaction, which is recognized as TSS. Since 1981, the incidence of TSS has declined sharply, primarily due to changes in absorbent materials used in tampons, proper use of super absorbent tampons, and public awareness of TSS symptoms.

Staphylococcal enterotoxin A, which causes by far the most prevalent form of food poisoning in the United States, is also a superantigen. Presumably, after ingestion of toxin-contaminated food, the toxin stimulates T cells localized along the intestine, resulting in a massive inappropriate T cell response, release of mediators, and increased permeability of the intestine. The final outcome is the severe but short-lived diarrhea and vomiting associated with this type of food poisoning (see Section 15.13).

Finally, appropriate antibiotic therapy for *S. aureus* infections is a problem. Although many "wild-type" infections are easily treatable with penicillin, nosocomial (hospital-acquired) infections due to *S. aureus* often possess drug-resistance plasmids (see Section 14.7). Therefore, disease-producing isolates of *S.*

aureus should be individually checked for antibiotic sensitivity.

> **Although staphylococci are often harmless inhabitants of the upper respiratory tract, several serious diseases can result from infection, including some caused by staphylococcal toxins that act as superantigens.**

Diphtheria

Corynebacterium diphtheriae, the causative agent of diphtheria, is a Gram-positive irregular rod (see also Section 19.29 for detailed description of coryneform bacteria). *C. diphtheriae* enters the body via the respiratory route with cells lodging in the throat and tonsils. Although limited information is available concerning the mechanism of adherence of *C. diphtheriae* to these tissues, the organism produces a neuraminidase capable of splitting *N*-acetylneuraminic acid (a component of glycoproteins found on animal cell surfaces), and this may enhance the invasion process. The inflammatory response of throat tissues to *C. diphtheriae* infection results in formation of a characteristic lesion called a *pseudomembrane* (Figure 15.7), which consists

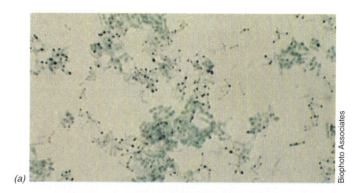
(a)
Biophoto Associates

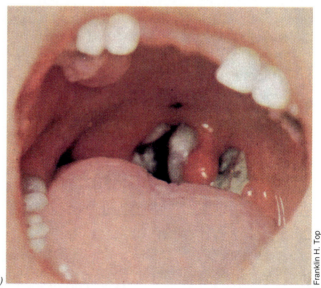
(b)
Franklin H. Top

FIGURE 15.7 Diphtheria. (a) Cells of *Corynebacterium diphtheriae* stained to show metachromatic (polyphosphate) granules. (b) Pseudomembrane in an active case of diphtheria caused by the bacterium *Corynebacterium diphtheriae*.

Discoverers of the Main Bacterial Pathogens

The history of the discovery of the microbial role in infectious disease has been described briefly in Chapter 1. Once the concept of specific microbial disease agents was clarified, and the procedures for culture of microorganisms were developed, it was a relatively simple procedure to isolate a large number of micro-bial pathogens. The two decades after the enunciation of Koch's postulates were indeed fruitful for medical microbiology. The rapid development of this field is indicated by the table below, which lists the main bacterial pathogens isolated during the "golden age of bacteriology."

Year	Disease	Organism	Discoverer
1877	Anthrax	*Bacillus anthracis*	Koch, R.
1878	Suppuration	*Staphylococcus*	Koch, R.
1879	Gonorrhea	*Neisseria gonorrhoeae*	Neisser, A. L. S.
1880	Typhoid fever	*Salmonella typhi*	Eberth, C. J.
1881	Suppuration	*Streptococcus*	Ogston, A.
1882	Tuberculosis	*Mycobacterium tuberculosis*	Koch, R.
1883	Cholera	*Vibrio cholerae*	Koch, R.
1883	Diphtheria	*Corynebacterium diphtheriae*	Klebs, T. A. E.
1884	Tetanus	*Clostridium tetani*	Nicolaier, A.
1885	Diarrhea	*Escherichia coli*	Escherich, T.
1886	Pneumonia	*Streptococcus pneumoniae*	Fraenkel, A.
1887	Meningitis	*Neisseria meningitidis*	Weichselbaum, A.
1888	Food poisoning	*Salmonella enteritidis*	Gaertner, A. A. H.
1892	Gas gangrene	*Clostridium perfringens*	Welch, W. H.
1894	Plague	*Yersinia pestis*	Kitasato, S., Yersin, A. J. E. (independently)
1896	Botulism	*Clostridium botulinum*	van Ermengem, E. M. P.
1898	Dysentery	*Shigella dysenteriae*	Shiga, K.
1900	Paratyphoid	*Salmonella paratyphi*	Schottmüller, H.
1903	Syphilis	*Treponema pallidum*	Schaudinn, F. R., and Hoffman, E.
1906	Whooping cough	*Bordetella pertussis*	Bordet, J., and Gengou, O.

of damaged host cells and cells of *C. diphtheriae*. As described in Section 11.8, certain strains of *C. diphtheriae* are lysogenized by bacteriophage β, and these strains produce a powerful exotoxin, the *diphtheria toxin*. Diphtheria toxin inhibits eukaryotic protein synthesis and thus kills cells (see Sections 11.8 and 18.9).

The pseudomembrane that forms in diphtheria may block the passage of air, and death from diphtheria is usually due to a combination of the effects of partial suffocation and tissue destruction by exotoxin. Although at one time a major cause of childhood mortality, diphtheria is now rarely encountered because an effective vaccine is available. This vaccine is made by treating the diphtheria exotoxin with formalin to yield an immunogenic, yet nontoxic, toxoid (see Section 12.15 for a general discussion of toxoids).

Diphtheria toxoid is part of the so-called DTP (diphtheria, tetanus, pertussis) vaccine, administered several times in the first year of life (see Section 12.15). To determine whether an individual is immune to diphtheria, the *Schick Test* has been used. This test involves injecting a diluted diphtheria toxin suspension under the skin, usually on the forearm, and observing for formation of a reddening inflammatory response due to neutralization of the toxin by circulating antibodies. A patient diagnosed as having diphtheria by culture of *C. diphtheriae* from a pseudomembrane in the throat is usually treated simultaneously with antibiotics and diphtheria antitoxin (an antitoxin contains neutralizing antibodies formed in another animal, see Section 12.12 for a discussion of antitoxins). Penicillin, erythromycin, or gentamicin are generally effective in diphtheria therapy. Studies of recent outbreaks of diphtheria have shown that early administration of both antibiotics and antitoxin is necessary for effective control of the disease.

Diphtheria is caused by the Gram-positive Bacterium *Corynebacterium diphtheriae*. A standard early childhood vaccine (DTP) is very effective in preventing this very serious respiratory disease.

Legionellosis (Legionnaires' disease)

This disease, caused by the organism *Legionella pneumophila*, derives its name from the fact that it was first recognized as a disease entity from an outbreak of pneumonia occurring during a convention of the American Legion in the summer of 1976. *Legionella* is a thin Gram-negative rod (Figure 15.8) with complex nutritional requirements, including an unusually high iron requirement, and is immunologically distinct from any known pathogen associated with respiratory infections. *Legionella* can be detected by immunofluo-

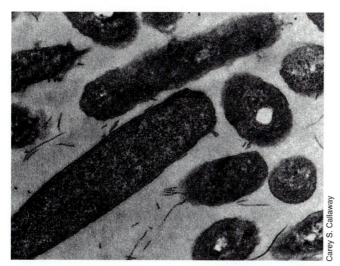

FIGURE 15.8 Transmission electron micrograph of *Legionella pneumophila*. Cells are approximately 0.6 μm in diameter.

rescence techniques (see Section 13.5 and Figure 13.8) and can be isolated from many terrestrial and aquatic habitats as well as from patients suffering from legionellosis. Epidemiological studies have shown that *Legionella* is a common inhabitant of cooling towers of air conditioning units and that infectious *Legionella* aerosols from such sources can spread the organisms to humans. Curiously, although apparently spread via an airborne route, no evidence of direct person-to-person transmission of *Legionella* has been obtained. Consistent with these findings is the fact that cases of legionellosis tend to peak in mid to late summer months when air conditioners are most extensively used. This is in contrast to an airborne disease such as chickenpox, which is spread from person to person and peaks in the winter months (see Figure 15.21) when people are more frequently indoors and in close contact. More recently, a trend to late fall/early winter peaks of Legionnaires' disease has been observed, and if this continues, it may suggest alternative vehicles for transmission of the disease. Overall, the incidence of legionellosis has been steadily increasing (Figure 15.9).

Legionella infections may be totally asymptomatic and occasionally result in only mild symptoms such as

headache and fever. The majority of cases of *Legionella* pneumonia are in elderly individuals whose resistance has been previously compromised. In addition, certain serotypes of *Legionella* (eight are known) are more strongly associated with the pneumonic form of the illness than others. Prior to the onset of pneumonia, intestinal disorders are common, followed by high fever, chills, and muscle aches. These symptoms precede the dry cough and chest and abdominal pains typical of legionellosis. Death, if it occurs, is usually due to respiratory failure. Clinical detection of *L. pneumophila* is now straightforward, because of fluorescent antibody and other highly specific immunological probes (see Figure 13.8a). *Legionella pneumophila* is sensitive to the antibiotics rifampicin and erythromycin, and intravenous administration of erythromycin is the treatment of choice in most cases.

> Legionellosis is not currently a major public health problem, but its incidence is on the rise. It remains to be seen whether this rise simply reflects increased familiarity and precision in diagnosing the disease or whether *Legionella* is becoming a more prevalent respiratory pathogen in artificial microenvironments.

Whooping cough

Whooping cough (pertussis) is an acute, highly infectious respiratory disease generally observed in children under one year of age. Whooping cough is caused by a small, Gram-negative, strictly aerobic coccobacillus, *Bordetella pertussis*. The organism attaches to cells of the upper respiratory tract by producing a specific adherence factor called *filamentous hemagglutinin antigen* which recognizes a complementary molecule on the surface of host cells. Once attached, *B. pertussis* grows and produces pertussis exotoxin that induces synthesis of cyclic adenosine monophosphate (cyclic AMP, see Figure 5.49), which is at least partially responsible for events that lead to host tissue damage. *B. pertussis* also produces an endotoxin, which also may induce some of the symptoms of whooping cough. Clinically, whooping cough is characterized by a recurrent, violent cough that usually lasts up to six weeks. The spasmodic coughing gives the disease its name, for a whooping sound results from the patient inhaling in deep breaths to obtain sufficient air.

A vaccine consisting of killed whole cells of *B. pertussis* is part of the routinely administered DTP vaccine (see Section 12.15) and thus fewer than 5000 cases of whooping cough are reported each year in the United States. Because of undesirable side effects of pertussis vaccine, including local swelling and redness, fever, and occasional more serious problems such as encephalitis and convulsions, a "second generation" pertussis vaccine, containing purified cell fractions of *B. pertussis* rather than whole cells, is now licensed for use in the United States and is already in wide use in Japan.

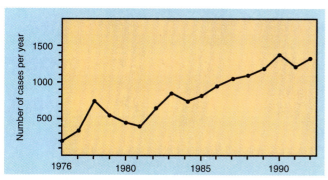

FIGURE 15.9 Annual incidence of legionellosis in the United States. Note the overall rise in the total number of annual cases.

Diagnosis of whooping cough can be made by fluorescent antibody staining of throat smears or by actually culturing the organism. For culture of *B. pertussis*, the "cough-plate" method is used. The patient is asked to cough directly into a blood-glycerol-potato extract agar plate (although not selective, this medium supports good recovery of *B. pertussis*). Alternatively, throat and nose swabs (if streaked immediately after sampling) can be used. β-hemolytic colonies containing small Gram-negative coccobacilli are tested for *B. pertussis* by a latex bead agglutination test, or are stained with an anti-*B. pertussis* fluorescent antibody for positive identification (see Sections 13.5 and 13.8). Cultures of *B. pertussis* are killed by ampicillin, tetracycline, and erythromycin, though antibiotics alone do not seem to be sufficient to kill the pathogen *in vivo*. Since a patient with whooping cough remains infectious for up to two weeks following commencement of antibiotic therapy, it is generally considered that the immune response is as important, if not more so, than antibiotics, in the elimination of *B. pertussis* from the body.

> While the overall incidence of whooping cough remains low in the population, there has been a disturbing increase in the number of annual cases in the last decade. From 1975 to 1982, there was an average of less than 2000 cases per year but that number has risen to 5000 cases per year at present. Up to 60 percent of preschool children in some cities are inadequately immunized, creating a major potential public health threat.

15.3   *Mycobacterium* and Tuberculosis

The famous German microbiologist Robert Koch isolated and described the causative agent of tuberculosis, *Mycobacterium tuberculosis*, in 1882 (see Discoverers box). At one time, tuberculosis was the single most important infectious disease of humans and accounted for one-seventh of all deaths. At present in the United States, over 25,000 new cases of tuberculosis are diagnosed each year and about 2000 deaths occur. Worldwide, tuberculosis still accounts for over three million deaths per year (see Table 14.1). In recent years, there has been a slight but alarming annual increase in the number of tuberculosis cases in the United States, due at least in part to the elevated incidence of tuberculosis in AIDS patients. After influenza and pneumonia, tuberculosis is the leading cause of death by infectious disease in the United States.

The microbiology of *M. tuberculosis* is discussed in Section 19.31. The interaction of the human host and *M. tuberculosis* is extremely complex, being determined in part by the virulence of the strain but probably more importantly by the specific and nonspecific resistance of the host. Cell-mediated immunity plays an important role in the development of disease symptoms. It is convenient to distinguish between two kinds of human tuberculosis infection: *primary* and *postprimary* (or reinfection). Primary infection is the first infection that an individual receives and often results from inhalation of droplets containing viable bacteria derived from an individual with an active pulmonary infection. Dust particles that have become contaminated from sputum of tubercular individuals are another source of primary infection. The bacteria settle in the lungs and grow. A delayed-type hypersensitivity reaction (see Sections 12.6 and 12.14) results in the formation of aggregates of activated macrophages, called *tubercles*, characteristic of tuberculosis. However, the bacteria are often able to survive and grow to some extent within the macrophages. In a few individuals with low resistance, the bacteria are not effectively controlled, and an acute pulmonary infection occurs, which leads to the extensive destruction of lung tissue, the spread of the bacteria to other parts of the body, and death.

In most cases of tuberculosis, however, acute infection does not occur, and the infection remains localized and is usually inapparent; later it subsides. But this initial infection hypersensitizes the individual to the bacteria or their products and consequently alters the response of the individual to subsequent infections. A diagnostic test, called the **tuberculin test**, can be used to measure this hypersensitivity. If *tuberculin*, a protein fraction extracted from *M. tuberculosis*, is injected intradermally into a hypersensitive individual, it elicits a localized immune reaction at the site of injection characterized by *induration* (hardening) and *edema* (swelling) at the site 1–3 days after injection (see Figure 12.35). An individual exhibiting this reaction is said to be *tuberculin positive*, and many healthy adults give positive reactions as a result of previous inapparent infections. A positive tuberculin test does not indicate active disease but only that the individual has been exposed to the organism at some time.

It is in tuberculin-positive individuals that the postprimary type of tuberculosis infection can occur. When renewed pulmonary infections occur in tuberculin-positive individuals, they are usually chronic infections that involve destruction of lung tissue, followed by partial healing and a slow spread of the lesions within the lungs. Spots of destroyed tissue may be revealed by X-ray examination (Figure 15.10),

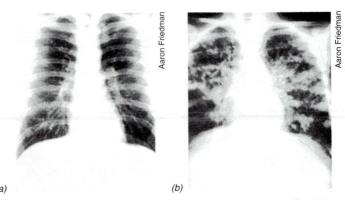

FIGURE 15.10   X-ray photographs. (a) Normal chest X ray. The faint white lines are arteries and other blood vessels. The heart is visible as a white bulge in the lower right quadrant. (b) An advanced case of pulmonary tuberculosis; white patches indicate areas of disease.

but viable bacteria are found in the sputum only in individuals with extensive tissue destruction. In many cases, symptoms in tuberculin-positive individuals are a result of reactivation and growth of bacteria that have remained alive and dormant in the lungs for long periods of time. Malnutrition, overcrowding, stress, and hormonal imbalance often are factors predisposing an individual to reinfection.

Chemotherapy of tuberculosis has been a major factor in control of the disease. The initial success in chemotherapy occurred with the introduction of streptomycin, but the real revolution in tuberculosis treatment came with the discovery of isonicotinic acid hydrazide (isoniazid or INH) (Figure 15.11), a nicotinamide derivative virtually specific for mycobacteria. This agent is not only effective and free from toxicity, but inexpensive and readily absorbed when given orally. Although the mode of action of INH is not completely understood, it is known that it affects in some way the synthesis of mycolic acid by *Mycobacterium* (mycolic acid is a complex lipid which complexes to the peptidoglycan of the mycobacterial cell wall, see Section 19.31). INH may act by mimicking the activity of a structurally related molecule, nicotinamide (Figure 15.11), becoming incorporated in place of nicotinamide, thus inactivating enzymes requiring this compound for activity. Treatment of mycobacteria with very small amounts of INH (as little as 5 picomoles per 10^9 cells) results in complete inhibition of mycolic acid synthesis, and continued incubation results in a complete loss of outer membrane areas of the cell, a loss of cellular integrity, and death. Following treatment with INH, mycobacteria lose their acid-alcohol fastness, in keeping with the role of mycolic acid in this staining property (see Section 19.31). However, mycobacterial resistance to INH and other drugs is increasing at an alarming rate, especially in AIDS patients (see Section 15.7).

Other pathogenic mycobacteria

Another species of the genus *Mycobacterium*, *M. bovis*, is pathogenic for both humans and other animals. A common pathogen of dairy cattle, *M. bovis* enters humans via the intestinal tract typically from the ingestion of raw milk. After a localized intestinal infection, the organism eventually spreads to the respiratory tract and initiates the classic symptoms of tuberculosis. The question as to whether *M. bovis* is really a different organism than *M. tuberculosis* is unclear because of the nearly 100 percent homology observed in hybridization of the two organisms' DNA. Pasteuriza-

tion of milk and elimination of diseased cattle have essentially eradicated bovine-to-human transmission of tuberculosis.

Mycobacterium leprae is the causative agent of the ancient and dreaded disease, *leprosy* (Hansen's disease). Unfortunately, *M. leprae* has never been grown on artificial media. It can be grown in mice but the typical human symptoms of leprosy are not observed. The only experimental animal which has been successfully used is the armadillo. The symptoms of leprosy are the characteristic folded, bulb-like lesions on the body, especially on the face and extremities (Figure 15.12) due to growth of *M. leprae* cells in the skin. In severe cases the disfiguring lesions lead to destruction of peripheral nerves and loss of motor function. Leprosy can be treated with the drug dapsone (disulfone 4,4'-sulfonylbistenzamine), although, as for tuberculosis, extended drug therapy is required to effect a cure.

Despite ancient myths, leprosy is not a highly contagious disease. Little is known of the pathogenicity of *M. leprae* or even of its mode of transmission, though it would seem likely that it is transmitted by direct contact. The bacterium grows within macrophages, causing an intracellular infection which can result in an enormous population of bacteria within the skin. In many areas of the world the incidence of leprosy is very low, although in ancient times it was apparently much more common, perhaps due to crowding and poor sanitation. In the United States, fewer than 300 cases of leprosy are diagnosed per year. In other areas, such as tropical areas, the incidence is much higher; leprosy remains a medical problem for some 15 million people worldwide.

> Although tuberculosis is no longer as prevalent in developed countries as it was 50–100 years ago, it is still one of the most common causes of death in developing countries and appears to be increasing in the United States. Many of the symptoms of tuberculosis arise because of delayed hypersensitivity reactions to the causal agent, *Mycobacterium tuberculosis*. A similar situation exists for *M. leprae*, the causative agent of leprosy.

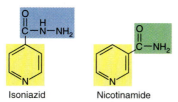

FIGURE 15.11 Structure of Isoniazid (isonicotinic acid hydrazide), an effective chemotherapeutic agent for tuberculosis. Note the structural relationship to nicotinamide.

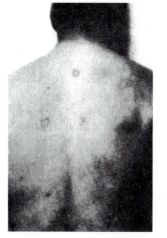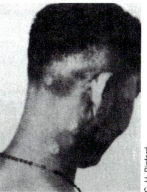

FIGURE 15.12 Leprosy lesions on the skin due to infection with *Mycobacterium leprae*.

15.4   Respiratory Infections: Viral

As was pointed out in Chapter 11, viruses, by their very nature, are less easily controlled by chemotherapeutic means than bacteria or other living microorganisms. Since the growth of viruses is intimately tied to host cell functions, it is difficult to specifically attack viruses by medical therapy without causing at least some harm to host cells as well. Not surprisingly, therefore, we see that the most prevalent infectious diseases today are of viral etiology (Figure 15.13). On the other hand, many viral diseases are acute, self-limiting infections that are rarely fatal in the normal healthy adult. In addition, serious viral diseases such as smallpox and rabies have been effectively controlled by immunization. We begin here by describing two widely experienced viral infections, the common cold and influenza, and proceed to discuss measles, mumps, and chickenpox; these viral diseases are all transmitted in infectious droplets by an airborne route.

The common cold

The common cold is one of the most prevalent diseases of children and adults. The symptoms include rhinitis (inflammation of the nasal region, especially the mucous membranes), nasal obstruction, watery nasal discharges, and a general feeling of malaise. Rhinoviruses (single-stranded RNA viruses of the picornavirus group, see Figure 15.14a and Section 6.15), are the dominant etiologic agents of the common cold. Over 100 different serotypes of rhinoviruses are known and hence long-term immunity to the common cold via vaccination or previous exposure is not to be expected. Another group of single-stranded RNA viruses, the coronaviruses, (Figure 15.14b), are responsible for about 15 percent of all colds in adults. A variety of other viruses including adenoviruses, coxsackie viruses, respiratory syncytial virus, and orthomyxoviruses, are responsible for about 10 percent of common colds.

Although airborne transmission of the common cold virus is suspected of being a major means of spreading the infection, transmission experiments with human volunteers suggest that direct contact and/or fomite transmission (see Section 14.5) is also

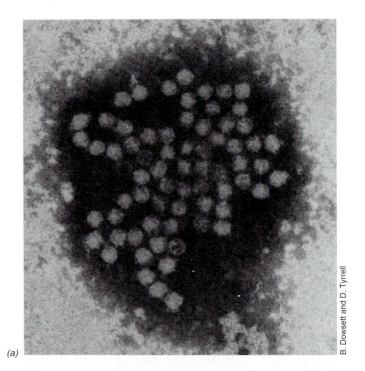

(a)

(b)

B. Dowsett and D. Tyrrell

Heather Davies and D. Tyrrell

FIGURE 15.14 Electron micrographs of some common cold viruses. (a) Human rhinovirus. (b) Human coronavirus. Each rhinovirus virion is about 30 nm in diameter. Each coronavirus virion is about 60 nm in diameter.

an important means of transmission. Some progress is being made in developing drugs that will prevent rhinovirus multiplication. Although a series of drugs effective against other viruses have proven relatively ineffective against rhinoviruses, a pyrazidine derivative (Figure 15.15) has proven promising in preventing colds in human volunteers. This drug, administered as a nasal spray, binds tightly to rhinovirus particles and inhibits uncoating of the virus and thus replication. Interferon treatment has been associated with negative side effects that make use of this drug currently unacceptable. Symptoms of the common cold, especially nasal discharges, can be moderated by a variety of antihistamine and decongestant drugs.

Influenza

Influenza is caused by an RNA virus of the orthomyxovirus group (see Section 6.16). Influenza virus is an enveloped virus, the RNA genome being surrounded by an envelope made up of protein, a lipid bilayer, and

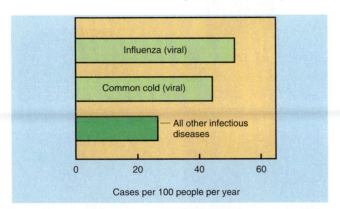

Influenza (viral)

Common cold (viral)

All other infectious diseases

0   20   40   60

Cases per 100 people per year

FIGURE 15.13 Viruses are the leading causes of acute illness in the United States. The data are typical of recent years.

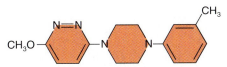

FIGURE 15.15 The structure of 3-methoxy-6-[4-(3-methylphenyl)-1-piperazinyl] pyrazidine, an antirhinovirus compound. This drug prevents uncoating of the rhinovirus virion (see Figure 15.14a) and has been tested as a treatment to prevent the onset of symptoms of the common cold.

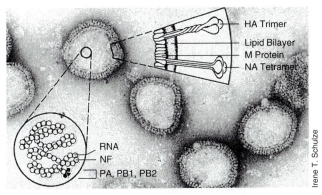

FIGURE 15.16 Electron micrograph of the influenza virus, showing the location of the major viral coat proteins and the nucleic acid. Each virion is about 100 nm in diameter. HA, hemagglutinin, three copies make up the HA coat spike; NA, neuraminidase, four copies make up the NA coat spike; M, coat protein; NP, nucleoprotein; PA, PB1, PB2, other internal proteins some of which may have enzymatic functions.

external glycoproteins (Figure 15.16; see also Figure 6.40 and Table 6.3). Human influenza virus exists in nature only in humans. It is transmitted from person to person through the air, primarily in droplets expelled during coughing and sneezing. The virus infects the mucous membranes of the upper respiratory tract and occasionally invades the lungs. Symptoms include a low grade fever for 3 to 7 days, chills, fatigue, headache, and general aching (see Cold or Flu box). Recovery is usually spontaneous and rapid. Most of the serious consequences of influenza infection do not occur because of the viral infection, but because bacterial invaders may be able to set up secondary infections in persons whose resistance has been lowered. Especially in infants and elderly people, influenza is often followed by bacterial pneumonia; death, if it occurs, is usually due to the bacterial infection.

Influenza often occurs in pandemics. Early pandemics, of which the one in 1918 is the most famous, occurred before knowledge was sufficiently advanced to make careful analysis possible, but the 1957 pandemic of the so-called Asiatic flu provided an opportunity for careful study of how a worldwide epidemic develops (Figure 15.17). The epidemic probably arose when a mutant virus strain of marked virulence and differing from all previous strains in antigenicity, appeared in the population. Since immunity to this strain was not present, the virus was able to advance rapidly throughout the world. It first appeared in the interior of China in late February 1957 and by early April had been brought to Hong Kong by refugees. It spread from Hong Kong along air and naval routes and was apparently transferred to San Diego, California, by naval ships. An outbreak occurred in Newport, Rhode Island, on a naval vessel in May. Other outbreaks occurred in various parts of the United States. Peak incidence occurred in the last two weeks of October, during which time 22 million new cases developed. From this period on there was a progressive decline.

The genetic material of influenza virus, single-stranded RNA, is arranged in a highly unusual manner. As discussed in Section 6.16, the influenza virus genome is *segmented*, with genes found on each of eight distinct fragments of its single-stranded RNA (see Figure 6.40 and Table 6.3). Such an arrangement facilitates the reassortment of genes with genes from a different strain of influenza virus, since more than one strain of influenza virus can infect a cell at one time. This allows recombinant viruses to arise at frequent

Is it a Cold or Is it the Flu?

The symptoms of a common cold and symptoms of "the flu" (influenza) often seem similar, but the two diseases are distinct and caused by quite different viruses. A typical common cold caused by a rhinovirus is associated with nasal discharges, cough, chills, and perhaps a sore throat. Influenza, caused by an orthomyxovirus, is generally associated with a different set of symptoms. Although either condition may make one feel miserable for a period, colds are usually of shorter duration and the symptoms are milder. The following can serve as a guideline for determining whether you have "caught a cold" or "caught the flu":

Symptoms	Common cold	Influenza
Fever	Rare	Common (39–40°C); sudden onset
Headache	Rare	Common
General malaise	Slight	Common; often quite severe; can last several weeks
Nasal discharge	Common and abundant	Less common; usually not abundant
Sore throat	Common	Much less common
Vomiting and/or diarrhea	Rare	Common

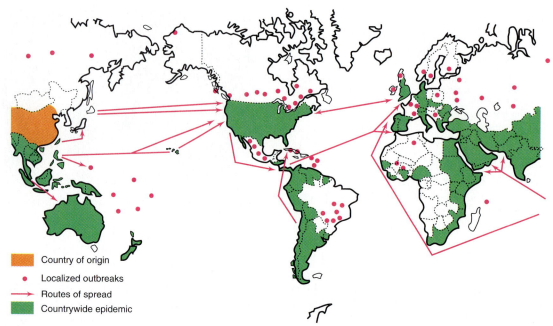

FIGURE 15.17   Route of spread of a major influenza epidemic, the Asian flu pandemic of 1957.

intervals and is probably responsible for the generally unsuccessful attempts to completely control influenza by vaccination. This reassortment of genes in different strains of influenza virus usually manifests itself in the phenomenon called *antigenic shift*. Antigenic shift refers to modifications in the protein coat of the virions, especially to two proteins important in the attachment and eventual release of virus from host cells, hemagglutinin and neuraminidase, respectively (see Figures 6.40 and 15.16). Immunity to influenza in humans is largely dependent on the production of secretory antibody (IgA, see Section 12.3), especially to antigenic determinants of the hemagglutinin and neuraminidase proteins.

Once a strain of influenza virus has passed through the population, a majority of the people will be immune to that strain and it will be impossible for a strain of similar antigenic type to cause an epidemic for about three years. During this time, the hemagglutinin and neuraminidase antigens exhibit frequent minor antigenic variation because of genetic mutations that result in the change of one or more amino acids. This phenomenon is known as *antigenic drift* and is responsible for the recurrence of minor epidemics of influenza in a two- to three-year cycle. There is some suggestion that in 1918 (the year of the unusually serious *pandemic*, or worldwide epidemic), the strain may have originated from a related virus which infects swine (swine flu), and the 1957 strain may have arisen from a similar animal reservoir (perhaps a wild animal) somewhere in Asia. Although a vaccine can be prepared to any strain, the large number of strains and the phenomena of antigenic shift and antigenic drift make it difficult to prevent influenza epidemics.

Control of influenza epidemics can be carried out by vaccination, but this is complicated by the fact that so many different virus strains may cause the disease. Obviously, the vaccine must be derived from the strain causing the epidemic. Vaccines prepared from several different virus strains can be mixed, producing what is called a *polyvalent* vaccine. If a new strain has arisen, vaccine for it will of course not be immediately available, but by careful worldwide surveillance, it is possible to obtain cultures of the major strain(s) of influenza virus before the disease has reached epidemic proportions and thus to produce vaccine ahead of an epidemic. Because this is a costly endeavor, the vaccination strategy used in many countries is to recommend influenza vaccination only for those most likely to succumb to severe or fatal secondary illnesses, such as the aged and those suffering from chronic debilitating diseases, and to health care workers. Duration of immunity for influenza is not long, usually only for a few years (and only to the strain(s) used in the vaccine), so that revaccination is necessary when a new epidemic occurs.

Influenza may also be controlled by the use of the chemical *amantadine*. This drug has been used as a chemoprophylactic agent to prevent the spread of influenza to those at high risk. It may also help to shorten the course of infection somewhat. The alleviation of symptoms of influenza by the use of aspirin is not recommended, as there is evidence of a link between aspirin treatment of influenza and Reye's Syndrome (a rare but occasionally fatal affliction involving the central nervous system) in children.

Measles

Measles (rubeola) virus causes an acute highly infectious childhood disease characterized by nasal discharges, redness of the eyes, cough, and fever. The measles virus is a paramyxovirus (see Section 6.16), and enters the body in the nose and throat by airborne transmission and quickly leads to systemic viremia. As the disease progresses, the fever and cough intensify and a rash appears (Figure 15.18); in most cases measles lasts a total of 7–10 days. Circulating antibodies to measles virus are measurable about 5 days after initiation of infection, and the activities of both serum antibodies and cytotoxic T lymphocytes (see Sections 12.3 and 12.6) combine to eliminate the virus from the system. As a consequence of measles, a variety of complications may occur, including inner ear infection, pneumonia, and, in rare cases, measles encephalomyelitis. The latter can cause neurological disorders and a form of epilepsy. Measles encephalomyelitis is a serious complication of measles infection because it has a mortality rate of nearly 20 percent.

Although once a common childhood illness, measles generally occurs nowadays in rather isolated outbreaks because of widespread vaccination programs begun in the mid 1960s (Figure 15.19*a*). In the United States most school systems require proof of measles vaccination before allowing children to enroll. The potential complications and highly infectious nature of measles mandates that this disease be tightly controlled. Active immunization is done with the MMR (measles, mumps, and rubella) vaccine (see Tables 12.8 and 12.9). A childhood case of measles generally confers life-long immunity to reinfection.

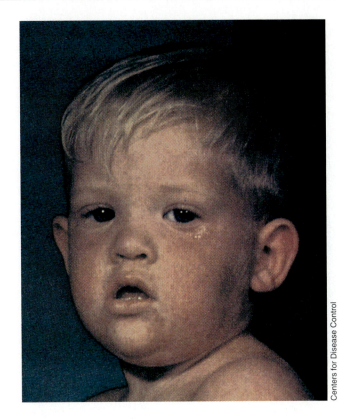

FIGURE 15.18 Typical rash associated with measles in children.

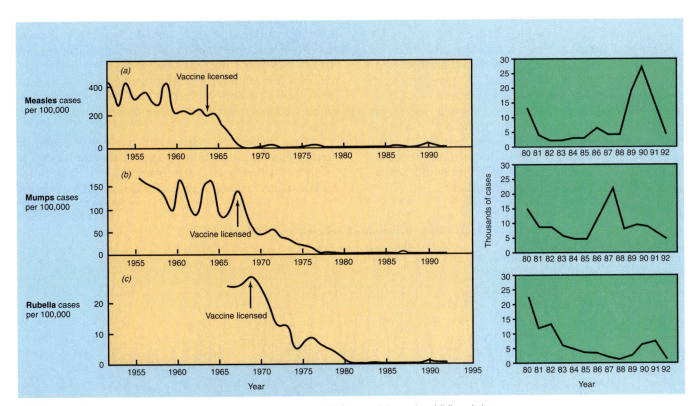

FIGURE 15.19 The effect of vaccines on the incidence in the United States of the major childhood virus diseases now controlled by the MMR (measles, mumps, rubella) vaccine. (a) Measles. (b) Mumps. (c) Rubella. The insets show a more detailed picture of these diseases in the past decade.

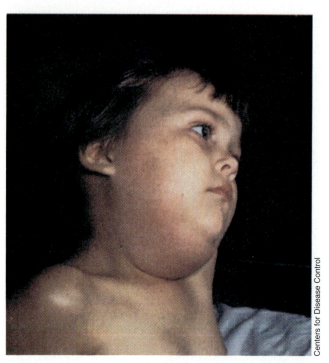

FIGURE 15.20 Typical glandular swelling associated with mumps.

Mumps

Mumps is caused by a different paramyxovirus than that causing measles, but shares with the disease measles a highly infectious character. Mumps is spread by airborne droplets, and the disease is characterized by inflammation of the salivary glands leading to swelling of the jaws and neck (Figure 15.20). The virus spreads through the bloodstream and may infect other organs including the brain, testes, and pancreas. The host immune response produces antibodies to mumps virus surface proteins and this generally leads to a quick recovery. An attenuated mumps vaccine is highly effective in preventing the disease (see Figure 15.19*b*). Hence, like measles, the incidence of mumps has also been greatly reduced in the last two decades, with mumps epidemics usually restricted to those individuals who for one reason or another did not receive the MMR vaccine during childhood.

Rubella

Rubella (*German measles*) is caused by a positive-strand RNA virus of the togavirus group (see Section 6.15). The symptoms of the disease resemble those of measles, but are generally milder. German measles is less contagious than true measles and thus a good proportion of the population has never been infected. However, during the first 3 months of pregnancy rubella virus can infect the fetus by placental transmission and cause a host of serious fetal abnormalities. Rubella can cause stillbirth, or deafness, heart and eye defects, and brain damage in live births. Thus, it is important that pregnant women not be exposed to, vaccinated against, or contract German measles during this period. For this reason, routine childhood vaccination against German measles should be practiced. An attenuated virus vaccine is administered with attenuated measles and mumps viruses in the MMR vaccine mentioned above (see Figure 15.19*c* and Table 12.9).

Chickenpox and shingles

Chickenpox (varicella) is a common childhood disease caused by a herpes virus (see Section 6.19). Chickenpox is highly contagious and is transmitted by infectious droplets, especially when susceptible individuals are in close contact. The incidence of chickenpox shows a disease cycle typical of a respiratory infection (Figure 15.21). In school children, for example, close confinement during the winter months leads to the spread of chickenpox by airborne secretions from infected classmates and through contact with contaminated fomites. The virus enters the respiratory tract, multiplies, and is quickly disseminated via the bloodstream resulting in a systemic papular rash that quickly heals, rarely leaving disfiguring marks (Figure 15.22). For unknown reasons, early chickenpox vaccines have been poorly immunogenic, but a successful and highly protective vaccine has been marketed in Japan and a similar vaccine is available in the United States.

The chickenpox virus can remain dormant in nerve cells for years with no apparent symptoms. The virus occasionally migrates from this reservoir to the skin surface causing a painful skin infection referred

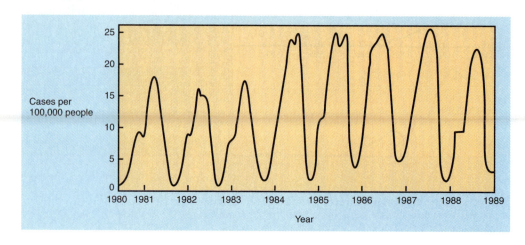

FIGURE 15.21
Reported cases of chickenpox by month in the United States, 1981–1989. Note the marked early spring seasonality of chickenpox typical of a disease transmitted by an airborne route.

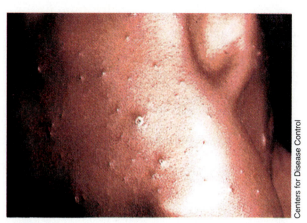

Centers for Disease Control

FIGURE 15.22 Mild papular rash associated with the disease chickenpox.

to as **shingles** (zoster). Shingles most commonly strikes immune suppressed individuals or the elderly. Studies with human volunteers suggest that T cells are important in destroying the virus. The prophylactic use of human hyperimmune globulin prepared against the virus is useful for preventing the onset of symptoms of shingles. Such therapy is only advised for patients where secondary infections occasionally associated with shingles, such as pneumonia or encephalitis, may be life threatening.

> There are numerous respiratory infections caused by viruses. Although the most prevalent is the common cold, the most serious is influenza, which often occurs in pandemics. These diseases all share a common mode of transmission, the airborne route.

15.5 Why Are Respiratory Infections so Common?

Although we have described a number of different respiratory infections, some bacterial and some viral, what do they have in common? Respiratory infections are generally transmitted by *airborne droplets*. The infectious agents occur in finely dispersed form in air, dust, or on the surfaces of fomites, and the diseases typically show marked seasonality. In addition, because they are easily transmitted by normal human activities, respiratory infections are the most common infectious diseases observed today (at least in developed countries).

Respiratory infections are frequently difficult to control from a public health standpoint. There are many reasons for this. First, many respiratory infections are caused by viruses; antibiotics, and indeed most other therapeutic agents, are completely ineffective in controlling viral diseases. Second, most respiratory diseases are acute, showing a rapid onset of symptoms and a rapid recovery. Although this may seem a desirable outcome, in many ways it isn't. Patients suffering from the common cold, for example, are frequently communicable before symptoms sug-

gest they are actually ill; chickenpox can spread in explosive fashion among school children for the same reason. Third, effective vaccines are not available for a number of respiratory diseases. This can be due to unusual molecular biological properties of the pathogen. For example, the antigenic drift and genomic rearrangements of the influenza virus generate new strains (from an immunological standpoint) every few years. Also, unusually large numbers of genetic variants of a particular pathogen (for example, the more than 100 serotypes of rhinovirus, the agent responsible for most common colds) makes vaccine production very difficult. In addition, vaccines against certain respiratory pathogens, for example against *Streptococcus pyogenes*, might actually harm the host due to cross-reacting antigens on bacterial and host cell surfaces (see Section 12.14). Fourth, many respiratory infections are transmitted from healthy carriers to susceptible hosts. This is very common in the case of streptococcal infections, because about 20 percent of the adult population carry, apparently without ill effect, strains of *Streptococcus pyogenes* in their nose and throat.

Respiratory diseases therefore represent a difficult challenge for public health officials. Since many respiratory diseases cause discomfort but are not life threatening, the tendency for many individuals suffering a respiratory illness is to simply "wait it out," while carrying on normal everyday activities. Unfortunately, such individuals are often unaware that they are actively infecting others. Because of this and the other problems discussed above, no practical means of control exist for many respiratory infections. It indeed appears that pathogens transmitted by the respiratory route have evolved a highly successful strategy for maintaining themselves in nature.

15.6 Sexually Transmitted Diseases

Various sexually transmitted diseases (STDs) caused by bacteria, viruses, and protozoans are known. Table 15.1 summarizes the major sexually transmitted diseases seen today. STDs are of major medical importance worldwide.

Control of STDs presents an unusually difficult public health problem for several reasons. First, nearly one-third of all cases involve teenagers. Sexual activity in this age group is on the rise, and sexually active young people today are more likely to have more than one sex partner, further complicating the disease picture. Second, many STDs initially cause no symptoms, or the symptoms that develop may be confused with the symptoms of other diseases not of a sexually transmitted nature. Third, the social stigma attached to STDs tends to reduce the motivation for some individuals to seek prompt medical care; this is especially true of young people.

On the other hand, there are many reasons to seek medical treatment of STDs. First of all, virtually all sexually transmitted diseases are curable (AIDS being

Table 15.1   Summary of some sexually transmitted diseases and treatment guidelines

Disease	Causative organisms[a]	Recommended treatment[b]
Gonorrhea	*Neisseria gonorrhoeae* (B)	Ceftriaxone plus doxycycline
Syphilis	*Treponema pallidum* (B)	Penicillin
Chlamydia trachomatis infections	*Chlamydia trachomatis* (B)	Doxycycline
Nongonococcal urethritis	*C. trachomatis* (B) or *Ureaplasma urealyticum* (B) or *Trichomonas vaginalis* (P)	Doxycycline or tetracycline
Lymphogranuloma venereum	*C. trachomatis* (B)	Doxycycline
Chancroid	*Haemophilus ducreyi* (B)	Erythromycin
Genital herpes	Herpes simplex type 2 (V)	No known cure; symptoms can be controlled with topical application of acyclovir (see Figure 15.29).
Genital warts	Papilloma virus (certain strains)	No known cure; warts can be removed surgically, chemically, or by cryotherapy.
Trichomoniasis	*T. vaginalis* (P)	Metronidazole
Acquired immune deficiency syndrome (AIDS)	Human immunodeficiency virus (HIV)	No known cure; nucleotide base analogs clinically useful in some treatments (see Table 15.2).
Pelvic inflammatory disease	*N. gonorrhoeae* (B) or *C. trachomatis* (B)	Cefoxitin plus doxycycline
Vulvovaginal candidiasis	*Candida albicans* (F)	Miconazole nitrate

[a]*B = bacterium, V = virus, P = protozoan, F = fungus.*
[b]*U.S. Department of Health and Human Services, Public Health Service Recommendation. From sexually transmitted diseases treatment guidelines, Morbidity and Mortality Weekly Report, vol. 38, no. S-8, Sept. 1989.*

a notable exception). As shown in Table 15.1, various antibiotic and chemotherapeutic agents are highly effective in treating most STDs. Second, the long-term health consequences of many STDs can be quite serious. For example, certain STDs can cause secondary health problems such as pelvic inflammatory disease (a major cause of infertility), cervical cancer, and heart and nerve damage. Pregnant women infected with a sexually transmitted pathogen can pass the agent on to the fetus, resulting in birth defects or even still-births. Third, diagnosis and treatment of one STD often reveals the presence of a second, inapparent infection that could still be transmitted after treatment for the first (inapparent chlamydial infections in individuals treated for gonorrhea, for example).

Despite the fact that most sexually transmitted diseases can be controlled, the incidence of many of

these diseases is still quite high; sexually transmitted diseases are obviously a social as well as a medical problem (Figure 15.23). We begin our discussion here with the disease gonorrhea, because despite the value of antibiotics in treating this disease, the prevalence of inapparent infections and the use of birth control pills have made gonorrhea the most widespread of the reportable sexually transmitted diseases.

Gonorrhea

Gonorrhea is one of the most widespread human diseases, and in spite of the availability of excellent treatment it is still a common disease (Figure 15.23). The disease symptoms of gonorrhea are quite different in the male and female. In the female the symptoms are usually a mild vaginitis that is difficult to distinguish from vaginal infections caused by other organisms,

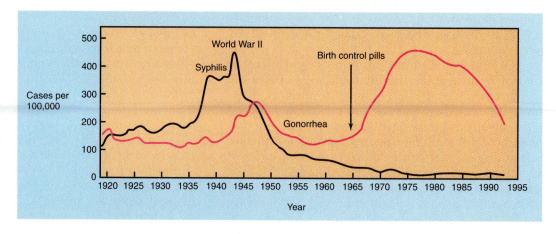

FIGURE 15.23 Sexually transmitted diseases: reported cases of gonorrhea and syphilis per 100,000 population, United States. After 1940, military cases are excluded.

and the infection may easily go unnoticed; in the male, however, the organism causes a painful infection of the urethral canal (refer to Figure 11.12 for a description of the genitourinary tract). The causative agent of gonorrhea, *Neisseria gonorrhoeae*, is killed quite rapidly by drying, sunlight, and ultraviolet light; this extreme sensitivity probably explains in part the sexually transmitted nature of the disease, the organism being transmitted from person to person only by intimate direct contact. In addition to gonorrhea, the organism also causes eye infections in the newborn and adult. Infants may become infected in the eyes during birth from diseased mothers. Prophylactic treatment of the eyes of all newborns with silver nitrate or an ointment containing penicillin is generally mandatory and has helped to control infection in infants. Complications arising from gonorrhea include pelvic inflammatory disease and damage to heart valves and joint tissues.

We discussed the clinical microbiology of *N. gonorrhoeae* in Section 13.1, and the general bacteriology of the genus *Neisseria* is discussed in Section 19.21. The pathogen enters the body by way of the mucous membranes of the genitourinary tract (Figure 15.24*a*), being transmitted during sexual intercourse. Treatment of the infection with penicillin has been successful in the past, with a single injection usually resulting in elimination of the organism and complete cure. For many years, all isolates of *N. gonorrhoeae* were found to be sensitive to penicillin or ampicillin (a semisynthetic penicillin), and sensitivity testing was not necessary. However, strains of *N. gonorrhoeae* resistant to penicillin are now widespread, and this resistance is due to a plasmid-encoded penicillinase. The incidence of penicillinase-producing *N. gonorrhoeae* has increased dramatically since the discovery of these strains in the mid-1970s (Figure 15.24*b*). In the United States, over eight percent of all clinical isolates are penicillinase-producing. Fortunately, however, the majority of penicillinase-producing strains respond to alternative antibiotic therapy, with a single dose of ceftriaxone, followed by doxycycline for seven days being the current recommended antibacterial regimen. Doxycycline, a tetracycline derivative, is given because it is also an antichlamydial agent (Table 15.1) and nearly 50 percent of gonorrhea patients are also infected with the hard-to-diagnose *Chlamydia trachomatis* organism.

Despite the ease with which gonorrhea can be cured, the incidence of gonococcus infection remains relatively high. The reasons for this are threefold. (1) Acquired immunity does not exist, hence repeated reinfection is possible (whether this is due to lack of local immunity or to the fact that at least 16 distinct serotypes of *N. gonorrhoeae* have been isolated is not known). (2) The widespread use of oral contraceptives. The latter cause a mimicking of the pregnant state, which results, among other things, in a lack of glycogen production in the vagina and a raising of the vaginal pH. Lactic acid bacteria, normally found in the adult vagina (see Section 11.5) fail to develop under such circumstances, and this allows *N. gonorrhoeae* transmitted from an infected partner to colonize more easily than in an acidic vagina containing lactobacilli. (3) Symptoms in the female are frequently such that the disease may go totally unrecognized, and a promiscuous infected female can serve as a reservoir for the infection of many males. The disease can be controlled if the sexual contacts of infected persons are

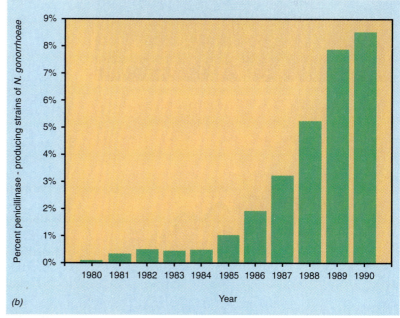

FIGURE 15.24 The causative agent of gonorrhea, *Neisseria gonorrhoeae*, and the incidence of penicillinase production in this organism. (a) Scanning electron micrograph of the microvilli of human fallopian tube mucosa showing how cells of *N. gonorrhoeae* attach to the surfaces of epithelial cells. Note the distinct diplococcus morphology. Cells of *N. gonorrhoreae* are about 0.8 μm in diameter. (b) Reported penicillinase-producing *Neisseria gonorrhoeae* (PPNG) in the United States in different years.

quickly identified and treated, but it is often difficult to obtain the necessary information and even more difficult to arrange treatment.

Syphilis

The sexually transmitted disease syphilis is potentially much more serious than gonorrhea, but because of differences in pathobiology, the incidence of syphilis in the United States is much lower (Figure 15.23). Syphilis is caused by a spirochete, *Treponema pallidum*, an organism that has been very difficulty to cultivate (Figure 15.25). We mentioned the clinical immunology and diagnostic methods for syphilis in Table 13.5, and the biology of the spirochetes is discussed in Section 19.12.

The disease syphilis exhibits variable symptoms. The organism does not pass through unbroken skin, and initial infection most probably takes place through tiny breaks in the epidermal layer. In the male, initial infection is usually on the penis; in the female it is most often in the vagina, cervix, or perineal region. In about 10 percent of cases, infection is extragenital, usually in the oral region. During pregnancy,

(a)

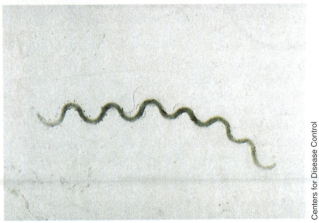

(b)

Theodor Rosebury

Centers for Disease Control

FIGURE 15.25 The spirochete of syphilis, *Treponema pallidum*. (a) Dark field microscopy of an exudate. *T. pallidum* cells measure 0.15 μm in width by 10–15 μm in length. (b) Shadow-cast electron micrograph of a cell of *T. pallidum*. Note the axial fibrils, typical of spirochetes.

the organism can be transmitted from an infected woman to the fetus; the disease acquired in this way by an infant is called **congenital syphilis**. *T. pallidum* multiplies at the initial site of entry and a characteristic *primary* lesion known as a **chancre** (Figure 15.26) is formed within two weeks to two months. Dark field microscopy of the exudate from syphilitic chancres often reveals the actively motile spirochetes (Figure 15.25a). In most cases the chancre heals spontaneously and the organisms disappear from the site. Some cells, however, spread from the initial site to various parts of the body, such as the mucous membranes, the eyes, joints, bones, or central nervous system, and extensive multiplication occurs. A hypersensitive reaction to the treponeme takes place, which is revealed by the development of a generalized skin rash; this rash is the key symptom of the *secondary* stage of the disease. At this stage the patient's condition may be highly infectious, but eventually the organism disappears from secondary lesions and infectiousness ceases.

The subsequent course of the disease in the absence of treatment is highly variable. About one-fourth of infected individuals undergo a spontaneous cure and another one-fourth do not exhibit any further symptoms, although the infection may persist. In about half of the patients the disease enters the *tertiary* stage, with symptoms ranging from relatively mild infections of the skin and bone to serious or fatal infections of the cardiovascular system or central nervous system. Involvement of the nervous system is the most serious phase of the illness, since generalized paralysis or other severe neurological damage may result. In the tertiary stage very few organisms are present, and most of the symptoms probably result from delayed hypersensitivity reactions (see Sections 12.6 and 12.14) to the spirochetes.

Penicillin is highly effective in syphilis therapy and the early stages of the disease can usually be controlled by a series of injections over a period of one to two weeks. In the secondary and tertiary stages, treatment must extend for longer periods of time. However, even though syphilis is easily diagnosed and effectively treated, the number of individuals in the United States with infectious primary and secondary syphilis has been *rising* at a steady rate during the last decade. The disease is now concentrated in sexually promiscuous heterosexuals, individuals of low socioeconomic status, and illegal drug users.

Chlamydial infections

Although not officially a *reportable* disease as are gonorrhea and syphilis, a host of sexually transmitted syndromes can be ascribed to the obligate intracellular parasite *Chlamydia trachomatis* (Figure 15.27), with the total incidence of such diseases probably greatly outnumbering that of gonorrhea or syphilis. *C. trachomatis* also causes a serious disease of the eye called *trachoma* (see Section 19.23), but the strains responsible for venereal infections consist of a group of *C. trachomatis* serotypes distinct from those causing trachoma.

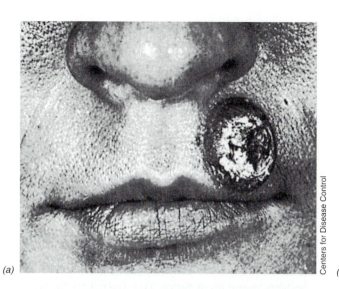

Centers for Disease Control

S. Olansky and L. W. Shaffer

FIGURE 15.26
Primary syphilis lesions. (a) Chancre on lip. (b) Several chancres on penis.

Morris D. Cooper

FIGURE 15.27 Cells of *Chlamydia trachomatis* attached to human fallopian tube tissues. (a) Cells attached to microvilli of fallopian tube. (b) Damaged fallopian tube containing a cell of *C. trachomatis* (arrow) in the lesion.

Chlamydial infections may also be transmitted to the newborn from contamination in the birth canal, causing newborn conjunctivitis and respiratory distress.

Chlamydial nongonococcal urethritis (NGU) is one of the most frequently observed sexually transmitted diseases today. *C. trachomatis* causes urethritis in males and urethritis, cervicitis, and pelvic inflammatory disease in females. In both the male and female inapparent chlamydial infections are common, and this undoubtedly accounts for the prevalence of chlamydial infections in sexually active adults. In a small percentage of cases, NGU can lead to serious complications, including testicular swelling and prostate inflammation in men and pelvic inflammatory diseases and fallopian tube damage in women. In the latter instance, cells of *C. trachomatis* attach to microvilli of fallopian tube cells, enter, multiply, and eventually lyse the cells (Figure 15.27b). Untreated, this condition can cause infertility.

Chlamydial NGU is relatively difficult to diagnose by traditional isolation and identification methods. To expedite diagnoses, a variety of immunological tests have been developed for identifying *C. trachomatis* from a vaginal/pelvic swab or from discharges. These clinical tests include fluorescent monoclonal antibodies and various ELISA tests for detecting specific *C. trachomatis* antigens. If a chlamydial infection is suspected, treatment is begun with doxycycline or tetracycline; penicillin is ineffective against most strains of *C. trachomatis* because the organisms lack peptidoglycan, the target of penicillin (see Section 19.23).

Chlamydial NGU is frequently observed as a secondary event following gonorrhea infection. If both *N. gonorrhoeae* and *C. trachomatis* are transmitted to a new host in a single event, treatment of gonorrhea with ceftriaxone is usually successful, but does not eliminate the chlamydia. Although cured of gonorrhea, such patients are still communicable for chlamydia and eventually experience an apparent recurrence of gonorrhea which is instead a case of NGU. It is therefore advisable for patients undergoing gonorrhea therapy to be tested for chlamydial infections as well, because a latent chlamydial infection will require additional chemotherapy, usually with doxycycline, a tetracycline derivative.

Lymphogranuloma venereum is a sexually transmitted disease also caused by a specific strain of *C. trachomatis*. The disease, which occurs most frequently in males, consists of a swelling of the lymph nodes in and about the groin. From the infected lymph nodes chlamydial cells may travel to the rectum and cause a painful inflammation of rectal tissues called *proctitis*. Because of the potential for regional lymph node damage and the complications of proctitis, lymphogranuloma venereum is considered to be one of the most serious sexually transmitted chlamydial syndromes.

Herpes

We discussed the molecular biology of herpesviruses in Section 6.19. Herpes simplex viruses are responsible for fever blisters and cold sores, but can also cause genital infections. There are two main types of herpesviruses and many serotypes of each main type. **Herpesvirus type 1** (HV1) is generally associated with cold sores and fever blisters in and around the mouth and lips (Figure 15.28*a*). The incubation period of HV1 infections is short (three to five days) and the lesions heal without treatment in two to three weeks. Relapses of HV1 infections are relatively common and it is thought that the virus is spread primarily via contact with infectious lesions. Latent herpes infections are apparently quite common, with the virus persisting in low numbers in nerve tissue. Recurrent acute herpes infections are due to a periodic triggering of virus activity by unknown causes.

Herpesvirus type 2 (HV2) infections are associated primarily with the anogenital region, where the virus causes painful blisters on the penis of males or the cervix, vulva, or vagina of females (Figure 15.28*b*). HV2 infections are transmitted by direct sexual contact, and the disease is most easily transmitted during the active blister stage rather than during periods of inapparent (presumably latent) infection.

Genital herpes infections are presently incurable, although a limited number of drugs have been successful in controlling the infectious blister stages. The guanine analog **acyclovir** (Figure 15.29) in ointment form is particularly effective in limiting the shed of active virus from blisters and promoting the healing of blisterous lesions. Acyclovir acts by specifically interfering with herpesvirus DNA polymerase, hence inhibiting viral DNA replication.

The overall health implications of genital herpes infections are not yet understood. Oral herpes is quite common and apparently has no harmful effects on the host beyond the oral blisters. However, epidemiological studies have shown a significant correlation between *genital* herpes infections and cervical cancer in females. In addition, herpesvirus type 2 can be transmitted to the newborn at birth by contact with herpetic lesions in the birth canal. The disease in the newborn varies from latent infections with no apparent damage to systemic disease which can result in brain damage or even death. To avoid infection in the newborn, delivery by caesarean section is advised for pregnant women showing genital herpes lesions.

Trichomoniasis

Nongonococcal urethritis may also be caused by infections with the protozoan *Trichomonas vaginalis* (Figure 15.28*c*). Although many protozoa produce resting cells called *cysts*, *T. vaginalis* does not produce cysts and thus transmission must be from person to person, generally by sexual intercourse. However, cells of *T. vaginalis* can survive a few hours outside the host, provided they do not dry out. Thus transmission of *T. vaginalis* by contaminated toilet seats, sauna benches, and paper towels occasionally occurs. *T. vaginalis* infects the vagina in women, the prostate and seminal vesicles of men, and the urethra of both males and females.

Many cases of trichomoniasis are totally asymptomatic. In fact, asymptomatic cases are the rule in males. In women trichomoniasis is characterized by a vaginal discharge, vaginitis, and painful urination.

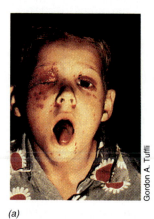
Gordon A. Tuffi

(a)

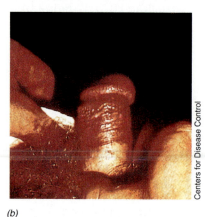
Centers for Disease Control

(b)

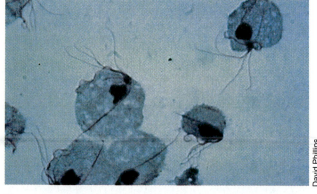
David Phillips

(c)

FIGURE 15.28 Nonbacterial sexually transmitted pathogens: Herpes virus and *Trichomonas*. (a) A severe case of herpes blisters on the face due to infection with herpesvirus type I. (b) Genital herpes due to infection on the penis with herpesvirus type II. (c) Cells of the flagellated protozoan, *Trichomonas vaginalis*.

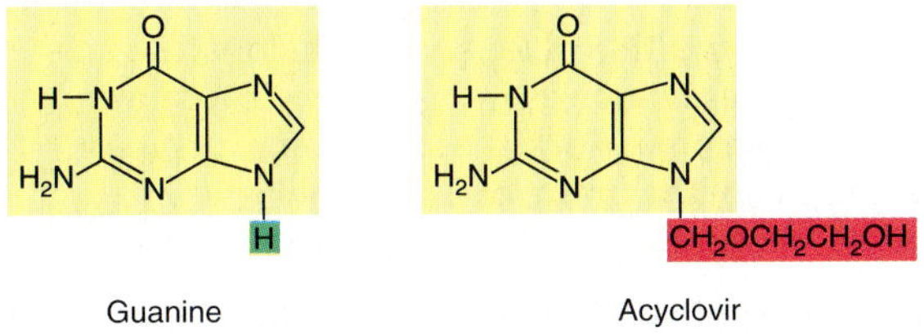

FIGURE 15.29 Structure of guanine and the guanine analog acyclovir. Acyclovir has been used therapeutically to control genital herpes (HV2) blisters. See text for mode of action.

The organism is commonly found in females, surveys indicating from 25 to 50 percent of sexually active women being infected; only about 5 percent of men are infected. Trichomoniasis is diagnosed by preparing and microscopically examining a wet mount of fluid discharged from the patient for motile protozoa. The antiprotozoal drug *metronidazole* is particularly effective in treating trichomoniasis.

It is important in the control of trichomoniasis that the male partner of an infected female be examined for *T. vaginalis* and treated if necessary, since promiscuous asymptomatic males can serve as reservoirs, transmitting the infection to several females. Transmission of *T. vaginalis* between either infected partner can be prevented by the use of condoms.

Ecology of sexually transmitted diseases

Thus far we have seen that a variety of microorganisms can cause sexually transmitted diseases. What properties do these organisms share in common that limit their distribution to the human genitourinary tract and their mode of transmission to sexual activity? For one, unlike respiratory infections where large numbers of infectious particles may be expelled by an individual, sexually transmitted pathogens are generally *not* shed in large numbers other than during sexual activity. Consequently, transmission will be limited to physical contact, generally during sexual intercourse. In addition, many sexually transmitted pathogens are very sensitive to drying. Their habitat, the human genitourinary tract, is generally a moist environment. Thus, these organisms colonize moist niches, and have apparently lost the ability to survive outside the animal host.

We now consider the most serious of all sexually transmitted diseases, AIDS. We will see that this disease has other modes of transmission as well (see also Section 14.4), but unlike the diseases we have discussed thus far, AIDS is of unusual health concern because the disease is invariably fatal.

> Pathogens causing sexually transmitted diseases (STDs) are generally rather sensitive to environmental conditions and do not survive outside the human body. Viruses, bacteria, fungi, and protozoa are all causes of STDs. Despite the fact that the most common STDs are readily treated by antibiotics, they are still frequent, partly because of the reluctance of many infected individuals to obtain treatment.

15.7   Acquired Immune Deficiency Syndrome (AIDS)

Because of its fatal consequences, the disease AIDS has received considerable attention since it was first recognized as a clinical syndrome in 1981. We discussed the epidemiology of AIDS in Section 14.4 and the laboratory diagnosis of AIDS in Sections 13.7 and 13.9. Over 300,000 cases of AIDS have been reported in the United States, and the outlook is for this number to rise dramatically unless an effective treatment becomes available soon (see Figure 14.2).

AIDS was first suspected of being a disease of the immune system when a startling increase in the number of so-called opportunistic infections was observed (see box). *Opportunistic infections* are defined as infections rarely observed in humans with normal immune responses (see Section 11.12 for a discussion of the *compromised host*). Opportunistic pathogens in effect take advantage of the "opportunity" afforded by reduced host resistance to initiate an infection that would otherwise be checked by the host's immune system. The occurrence of opportunistic infections in otherwise healthy individuals was one of the first epidemiological clues that a new immune disorder was appearing in the population. Opportunistic infections are important in the overall clinical picture of AIDS because they are the leading cause of death in AIDS patients.

The most common AIDS-associated opportunistic infections include pneumonia caused by the protozoan *Pneumocystis carinii* (Figure 15.30*a*) and other protozoal infections such as cryptosporidiosis, caused by *Cryptosporidium* species (Figure 15.30*b*), and toxoplasmosis, caused by *Toxoplasma gondii* (Figure 15.30*c*), systemic yeast infections due to *Cryptococcus neoformans* (Figure 15.30*d*), *Candida albicans* (Figure 15.30*e*), and *Histoplasma capsulatum* (Figure 15.30*f*), viral infections due to herpes simplex (Figure 6.45*a*) or cytomegalovirus, tuberculosis and other mycobacterial infections (Figure 15.30*g*), and enteric helminthic infections due to *Strongyloides stercoralis* (Figure 15.30*h*). However, *Pneumocystis* pneumonia is by far the most common opportunistic disease encountered, being observed at some time in nearly two-thirds of all cases of clinical AIDS.

Besides the opportunistic infections associated with many AIDS cases, a rare form of cancer called *Kaposi's sarcoma* is also observed in many AIDS patients. Kaposi's sarcoma is a cancer of the cells lining blood vessel walls and is diagnosed by the characteristic purplish patches it leaves on the surface of the skin (Figure 15.31). The incidence of Kaposi's sarcoma is much higher in homosexual men suffering from AIDS than in the other high risk groups identified (see Figure 14.3), but the reasons for this are not known.

Human immunodeficiency virus

The disease AIDS is caused by human immunodeficiency virus (HIV). HIV is a retrovirus (see Section 6.22) containing 9,749 nucleotides in each of its two

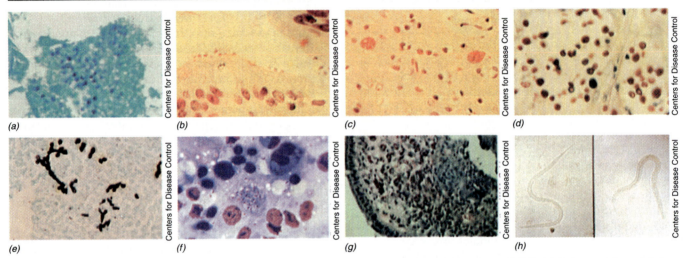

FIGURE 15.30 Opportunistic pathogens associated with cases of acquired immune deficiency syndrome (AIDS). (a) *Pneumocystis carinii*, from patient with pulmonary pneumocystosis. (b) *Cryptosporidium* sp., from biopsy of small intestine. (c) *Toxoplasma gondii*, from brain tissue of patient with toxoplasmosis. (d) *Cryptococcus neoformans*, from liver tissue of patient with cryptococcosis. (e) *Candida albicans*, from heart tissue of patient with systemic *Candida* infection. (f) *Histoplasma capsulatum*, from liver tissue of patient with histoplasmosis. (g) Mycobacterial infection of small bowel, acid-fast stain. (h) *Strongyloides stercoralis*, filariform larvae.

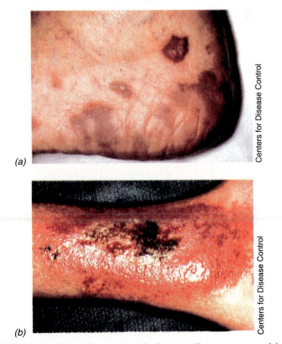

FIGURE 15.31 Kaposi's sarcoma lesions as they appear on (a) the heel and lateral foot, and (b) the distal leg and ankle.

identical single-stranded RNA genomes. Using the enzyme *reverse transcriptase*, which is present in the intact virion, HIV forms a complementary single-stranded DNA molecule using RNA as a template, and converts the cDNA formed into double-stranded DNA which can enter the host cell genome. A major host cell of HIV is the CD4 class of T lymphocyte (T-helper cells, see Section 12.2), and HIV in some way prevents normal division processes in these cells. Thus, immune functions can be greatly compromised in those infected with HIV and this leads to development of opportunistic infections. We now consider the AIDS disease process itself and see how HIV effectively dismantles the immune system.

HIV: T lymphocyte interactions

HIV has the specific capacity to infect the CD4 class of T lymphocytes; infected CD4 lymphocytes produce high numbers of HIV particles about 90–120 nm in diameter (Figure 15.32). Recall that CD4 lymphocytes include the T-helper (T$_H$) and T-delayed type hypersensitivity (T$_D$) subsets (Section 12.2). CD4 cells respond to foreign antigens bound in association with class II

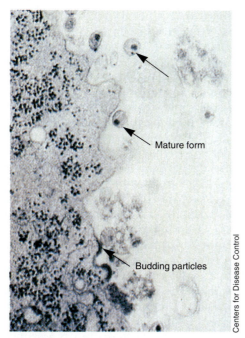

FIGURE 15.32 Transmission electron micrograph of a thin section of a lymphocyte releasing HIV. Cells were from a hemophiliac patient who developed AIDS. HIV particles are 90–120 nm in diameter.

major histocompatibility (MHC) antigens on the surface of macrophages and B cells (see Section 12.5). Once activated by a properly presented antigen, normal T_H cells secrete lymphokines, helper factors, and other immune modulators which, together with contact-mediated signals, stimulate particular clones of B-lymphocytes to multiply and secrete antibody specific for the given antigen (see Section 12.8).

The specificity of HIV for CD4 T cells is due to the fact that the CD4 molecule acts as a *cell surface receptor* for HIV. CD4 normally plays a role as a recognition molecule on cell surfaces by binding to the class II MHC molecule on the antigen-presenting cell (Figure 12.12b). However, sufficient complementarity exists between CD4 and coat proteins of HIV for CD4 to act as a vehicle for entrance of HIV into the T lymphocyte. Binding of HIV to host cells takes place when CD4 interacts with a major envelope protein of HIV, called *gp120* (Figure 15.33). This interaction causes the membrane of HIV and the host cytoplasmic membrane to fuse with release of the nucleocapsid (nucleic acid and reverse transcriptase) of HIV into the cell (Figure 15.33).

Besides this one class of T lymphocytes, several other cell types in the human body contain the CD4 molecule on their surface and can thus also be infected with HIV. Monocytes and macrophages express the CD4 molecule and can be infected, apparently without ill effect, and serve as a reservoir of HIV to spread the infection to T lymphocytes. A small percentage of B lymphocytes can also be infected. Certain human brain cells and intestinal cells as well as a variety of malignant cell cultures derived from human brain or bowel tissues can also be infected with HIV, although many of these cells express the CD4 molecule on their surfaces in only trace amounts. However, it is infection of the CD4 class of T lymphocytes by HIV that eventually causes the destruction of the immune system associated with AIDS.

In cases of clinical AIDS, lymphocytes bearing the CD4 molecule (mainly T_H cells) are greatly reduced in number. Exactly how HIV causes this reduction is still not known. However, unlike lytic animal viruses, HIV usually does not immediately kill and lyse its host cell. Following reverse transcription to produce DNA from the RNA genome, the cDNA formed can integrate into host chromosomal DNA and exists as a provirus. If this occurs, the cell may show no outward sign of infection and HIV DNA can remain in a latent state for long periods. Eventually, however, productive virus synthesis occurs and new HIV particles are produced

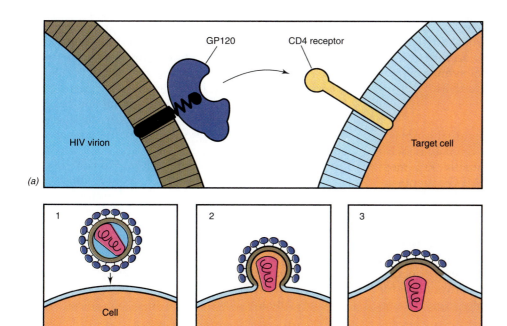

FIGURE 15.33
Process of infection with human immunodeficiency virus. (a) Uptake of HIV by CD4 cells by interaction of HIV protein gp120 and the CD4 molecule on host cells. (b) Fusion of the HIV envelope with the host cell envelope and entry of the nucleocapsid.

and released from the cell. For unknown reasons, T cells producing HIV no longer divide and eventually die.

Accelerated destruction of CD4 cells occurs following the processing of HIV antigens by infected T cells. Such cells embed molecules of gp120 from HIV particles on their cell surfaces. The embedded gp120 protein on the infected cells then sticks to uninfected T cells by binding to the CD4 molecule. Eventually, numerous cells of each type fuse to produce multinucleate giant cells called *syncytia*. One HIV-infected T cell may eventually bind and fuse with up to 50 uninfected T cells. Shortly after syncytia formation occurs, the resulting cells lose immune function and die.

The end result of HIV infection is that CD4 cells progressively decline in number. This has serious health consequences. In a normal human, CD4 cells constitute about 70 percent of the total T cell pool; in AIDS patients, the number of CD4 cells steadily decreases, and by the time opportunistic infections set in, CD4 cells may be totally absent (Figure 15.34). As CD4 cells decline in number, there is a concomitant loss in the immune modulators they produce, such as interleukin-2 and other T cell and B cell proliferative factors. This leads to a gradual reduction in all types of lymphocytes, effectively shutting down the immune system in those suffering from clinical AIDS. This loss of both humoral and cellular immune function is readily apparent in the opportunistic infections observed. Systemic infections by fungi and mycobacteria (Figure 15.30) point to a loss in T_D cells (T_D cells are CD4 cells involved in cellular immunity and are discussed in Section 12.6). Other opportunistic infections, such as the various viral infections associated with AIDS, indicate the loss of humoral immunity; decline in antibody production is due to the loss of T_H cells necessary to stimulate antibody production by B cells.

In the final analysis it is the severe crippling of the immune system that places AIDS patients in life-threatening situations. Without a reservoir of healthy T lymphocytes, the body is unable to fight off even very minor infections. The progressive loss in T cell function in an average AIDS case, from initial infection with HIV to the complete loss of an immune response and the corresponding clinical symptoms observed at each stage, is shown in Figure 15.34.

Diagnosis and treatment of AIDS

The prognosis of an HIV-infected individual is not encouraging. Opportunistic pathogens or malignancies (see Figures 15.30 and 15.31) eventually kill most AIDS patients. Long-term studies of AIDS patients indicate that the average person infected with HIV progresses through several stages of decreasing immune function, with CD4 cells dropping from a normal average of 600–700 per cubic millimeter of blood to near zero over a period of five to seven years (Figure 15.34). Although the *rate* of decline in immune function varies from one HIV-infected individual to another for reasons that are not yet clear, virtually every person

infected with HIV will eventually contract AIDS, usually within ten years of initial infection, and will eventually die from AIDS.

No *cure* for HIV infection is known, although research is intense in the areas of vaccine production and chemotherapy. Several drugs have been identified as helpful in delaying symptoms of AIDS and in some cases in prolonging the life of those infected with HIV (Table 15.2). The most promising drugs discovered thus far are chemicals which inhibit the activity of *reverse transcriptase*, the enzyme which converts the genetic information (which resides in single-stranded RNA) of retroviruses, such as HIV, into a complementary DNA copy (see Section 6.22). For example, the nucleotide base analog **azidothymidine** (AZT) (Figure 15.35) is an effective inhibitor of HIV replication because it closely resembles thymidine but lacks the correct attachment point for the next nucleotide in the chain and thus serves as a *DNA chain terminator*. The dideoxynucleotide analogs (Table 15.2) also function as chain terminators (recall that dideoxynucleotides were used to interrupt nucleic acid synthesis in DNA sequencing as well, see Nucleic Acids box in Chapter 5). Other AIDS drugs have uncertain modes of action but have simply been observed to improve the clinical picture of AIDS (Table 15.2). A major problem observed with many AIDS drugs is that the compounds are frequently toxic when used on a long-term basis. In addition, strains of HIV that are resistant to AZT have already been identified: clearly, long-term use of any drug will eventually result in decreased pathogen susceptibility.

AIDS vaccines

The genetic variability of HIV has thus far hampered the development of an AIDS vaccine. One vaccine strategy is to make antibodies to the envelope protein, gp120, and use these antibodies to block CD4-gp120 interactions (see Figure 15.33a) and thus block infection. However, this approach has not proven successful thus far because of the propensity for the gene encoding gp120 to mutate, forming immunological variants of the protein that are not recognized by antibodies made to a different antigenic form. The most impressive results from clinical immunization trials have emerged from *subunit vaccines* (see Section 8.13), where genes for several HIV envelope proteins have been engineered into vaccinia virus or adenovirus particles. Using these harmless viruses as expression vectors and vehicles for delivery of HIV antigens, several subunit vaccines have been shown to elicit a potent humoral and cellular immune response to HIV.

Other potential vaccines include *killed intact HIV* (for use only in those already infected with the virus) and *anti-idiotypic antibodies*. In the latter connection, antibodies made against the receptor for HIV (CD4), should, when used as an *antigen*, raise a population of anti-idiotypic antibodies whose antigen-binding site would structurally resemble CD4 (we discussed anti-idiotypic antibodies in Section 12.15). Experimentally, this has already been shown to occur. In sufficient

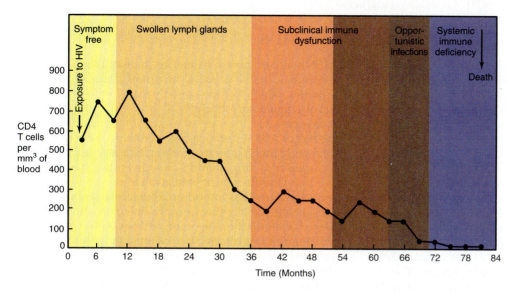

FIGURE 15.34
Decline in CD4 T lymphocytes in blood of a young male AIDS patient showing the typical progressive loss in CD4 cells and immune function.

Table 15.2 Some chemotherapeutic strategies for AIDS

Drug	Mechanism of action and comments
Azidothymidine (AZT or Zidovudine)	Reverse transcriptase inhibitor (chain terminator); available as prescription drug; increases survival time and reduces incidence of opportunistic infection in AIDS patients; toxic to bone marrow cells
Dideoxycytidine (ddC)/ dideoxyinosine (ddI or didanosine)	Reverse transcriptase inhibitors (chain terminators); alternatives to AZT
Phosphonoformate	Reverse transcriptase inhibitor; also active against some viral opportunistic pathogens
Dextran sulfate	Inhibits binding of HIV to CD4 receptor; little or no toxicity
Alpha interferon	Reduces HIV budding from infected cells; strong antitumor agent against Kaposi's sarcoma
Ampligen	Interferon inducer; essentially nontoxic
Soluble CD4	Inhibits binding of HIV to T cells; genetically engineered form of CD4, the HIV receptor
Phosphorothioate oligodeoxynucleotides complementary to HIV genome sequences	Serves as "antisense" strand of DNA (see Antisense Nucleic Acid box, Chapter 5). Binds HIV RNA and because of the unique chemistry of antisense DNA, prevents viral protein synthesis

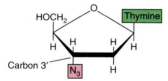

FIGURE 15.35 Structure of azidothymidine (AZT), a drug used for AIDS therapy. Because AZT lacks a hydroxyl group on the 3' carbon, replication ceases when it is incorporated into the growing DNA chain. AZT is also known as Zidovudine (ZDV).

quantity, anti-idiotypic antibodies could thus serve to scavenge HIV particles from the body by gp120-CD4 type interactions. A related approach is to use the CD4 receptor itself as an AIDS vaccine. Referred to as "soluble CD4" (and a product of recombinant DNA technology), this protein could interact with HIV and block its spread in the body (Table 15.2).

Although progress is being made, no effective AIDS vaccine is currently available. Additionally, it should be remembered that an AIDS vaccine, even if successful, would most probably not be useful in *treating* most patients that already have the disease. As we have seen, many of these individuals lack significant immune function and thus would not respond to a vaccine. Those recently infected with HIV would have a better chance of benefiting from an AIDS vaccine. Thus at present, despite considerable advances in our molecular understanding of HIV and in our clinical understanding of the AIDS disease process, public education about AIDS and avoidance of high risk behavior are still the major tools to combat AIDS today (see Sexual Activity and AIDS box).

Clinical detection of HIV infection

Exposure to HIV can be sensitively diagnosed by immunological means. Both radiommunoassay (RIA, see Section 12.13), and ELISA tests (see Section 12.13 and 13.7) have been developed for screening blood samples to detect anti-HIV antibodies. The ELISA test has proven particularly valuable for large-scale screening

Sexual Activity and AIDS

Sexual promiscuity has always been associated with sexually transmitted diseases, but the rise of the AIDS epidemic, discussed in this chapter and elsewhere in this book, has focused new attention on the dangers of multiple sex partners and on the high risk associated with certain sex practices. AIDS, caused by the human immunodeficiency virus (HIV), is only one type of sexually transmitted disease. Others include gonorrhea, syphilis, herpes simplex, nonspecific urethritis (caused by *Chlamydia*), protozoal vaginitis (caused by *Trichomonas vaginalis*), fungal vaginitis (caused by *Candida albicans*), and venereal warts (caused by the human papilloma virus). Some of these sexually transmitted diseases have been associated with human society for all of recorded history. The unique aspect of AIDS is that it is new and that it is almost uniformly fatal. There are neither drugs to cure AIDS nor vaccines to prevent it, and it is unlikely that either drugs or vaccines will be available in the near future. We do not at this time know how extensive the AIDS epidemic is, since the long latent period means that many people now infected (and perhaps infectious) have not yet exhibited symptoms. Some public health officials believe that the AIDS cases so far seen are only a small part of the AIDS cases that will eventually appear, and that over the next 10 years there will be a massive increase in the incidence of the disease, as latent cases develop into full-blown AIDS.

Because AIDS is linked to certain sex practices, prevention means avoidance of these sex practices. The United States Surgeon General has issued a report which makes specific recommendations that individuals can follow if they wish to reduce the likelihood of AIDS infection. Among the recommendations are

1. Avoid mouth contact with penis, vagina, or rectum.
2. Avoid all sexual activities which could cause cuts or tears in the linings of the rectum, vagina, or penis.
3. Avoid sexual activities with individuals from high-risk groups. These include prostitutes (both male and female), homosexual or bisexual individuals, and intravenous drug users.
4. If a person has had sex with a member of one of the high-risk groups, a blood test should be done to determine if infection with HIV has occurred. If the test is positive, then it is essential that sexual partners of an HIV-positive individual be protected by use of a condom during sexual intercourse.

It is important to emphasize that AIDS is *not* just a disease of male homosexuals. In certain cultures, AIDS is as common in women as in men. The disease seems to be linked primarily to promiscuous sexual activities, which include not only male homosexuality but also female prostitution.

Is it possible, then, to have sex without incurring the risk of AIDS? Certain sex practices do seem to be inherently much safer than others. Safe sex practices include dry kissing (mouths closed), mutual masturbation (in the absence of breaks in the skin), and intercourse protected by a condom. Dangerous sex practices include wet kissing (mouths open), masturbation where breaks in the skin occur, oral sex (either male or female), and unprotected sexual intercourse (either vaginal or anal). The U.S. Surgeon General has strongly recommended that if the health status of the partner is unknown that a condom be used for all sex practices in which exchange of bodily fluids occurs.

The AIDS epidemic has focussed anew on the condom, a device that has been used for many years. Condoms have always played two roles in sexual activity: disease protection, and prevention of pregnancy. Although it should be firmly stated that the best way of avoiding AIDS is avoidance of dangerous sex practices, if sexual intercourse is to be carried out with an individual whose infectiousness is unknown, then a latex condom should be used. The U.S. Surgeon General strongly recommends the use of condoms for all extramarital sexual activity. In certain countries, advertising campaigns to promote the use of condoms are already widespread.

It is very important to emphasize that moralistic statements alone (prescriptions for monogamy, abstinence, avoidance of sexual activity outside of matrimony), will *not* control the AIDS epidemic. Epidemiological studies on all previously known sexually transmitted diseases have shown that fear of disease is not, by itself, sufficient to prevent sexual activities that put an individual at risk for a sexually transmitted disease. The sex drive in some individuals is so strong that it will suppress the fear of disease, even a disease like AIDS. Every individual must therefore take the responsibility for protecting himself or herself from this widespread and extremely dangerous infectious disease.

For more information on prevention of AIDS, see the *Surgeon General's Report on Acquired Immune Deficiency Syndrome*, U.S. Department of Health and Human Services. For more information on protection against AIDS, the Public Health Service has established a toll-free telephone number, called the PHS AIDS Hotline. The number to call is 800-342-2437.

---

of donated blood to prevent transfusion-associated HIV. Statistics have shown that about 0.25 percent (2–3 per thousand) of all blood donated by volunteer donors in the United States tests HIV-positive in the ELISA assay. A positive AIDS ELISA test must be confirmed by a second procedure called immunoblotting (Western blotting), a technique which combines the analytical tools of protein purification and immunology (see Section 13.9). Although not foolproof because both methods can miss HIV carriers who have become infected but who have not yet produced antibodies to HIV, these immunological tools have served to greatly improve the safety of the blood supply. Statistics suggest that the current risk of contracting AIDS through contaminated blood or blood products is very low. Sexual promiscuity and group intravenous drug use are the major routes to HIV infection today (see Figure 14.3).

> AIDS is quickly becoming one of the most prevalent infectious diseases in the human population. Concern about HIV infection is particularly great because virtually all infected individuals eventually progress to AIDS, which has a 100 percent fatality rate. Death is not due to direct effects of HIV, but since HIV destroys a central part of the immune system, opportunistic pathogens usually kill the host. There is still no effective vaccine for HIV. Likewise, the few drugs that inhibit HIV multiplication are also toxic to the host. The best and only prevention for the spread of HIV infection is through total avoidance of risky behavior such as intravenous drug use (needle sharing) and unsafe sexual practices.

15.8   Animal-Transmitted Diseases: Rabies

Animals can contract a number of infectious diseases, some of which can be passed to humans. Through the use of effective vaccination practices and good veterinary care, domestic animal populations are generally maintained in good overall health. But when a new pathogen, or a new strain of a pathogen currently under control, is introduced into the domestic animal population, infections can spread quickly, leading to substantial economic losses. This situation is different in the wild animal population. Wild animals cannot be routinely vaccinated, and do not receive veterinary care. Thus animal diseases (zoonoses) cycle through wild animal populations on a periodic basis.

Most zoonoses are of little serious consequence to human health. Rabies is an exception. Rabies is one of the best examples of a disease which occurs primarily in animals, but under certain conditions occurs in humans. The major reservoir of rabies in the United States is in wild animals, primarily carnivores. However, a significant number of cases are still seen in domestic animals (Figure 15.36). Surprisingly, rabies is still a major disease in humans as well: worldwide, approximately 35,000 people die every year from this disease, primarily in developing countries (see Table 14.1). Rabies is caused by a single-stranded RNA virus of the rhabdovirus family (negative-strand viruses, see Section 6.16 and Figure 6.38), which attacks the central nervous system of most warm-blooded animals, almost invariably leading to death if not treated. The virus enters the body through a bite wound from a rabid animal. The virus multiplies at the site of inoculation and then travels to the central nervous system. The incubation period for the onset of symptoms is highly variable, depending on the size, location, and depth of the wound, and the actual number of viral particles transmitted in the bite. In dogs, the incubation period averages 10–14 days. In humans, up to nine months may elapse before the onset of rabies symptoms. The virus proliferates in the brain (especially in the thalamus and hypothalamus), leading to fever, excitation, dilation of the pupils, excessive salivation, and anxiety. A fear of swallowing (hydrophobia) develops from uncontrollable spasms of the throat muscles; death eventually results from respiratory paralysis.

A rabies vaccine was developed by Louis Pasteur in his classic work on immunology, and today such highly effective rabies vaccines for both animal and human hosts are available that an average of less than one case of human rabies is reported in the United States each year. Because of the long incubation period, a rabies vaccine can even be used in therapy since treatment of an infected human before the onset

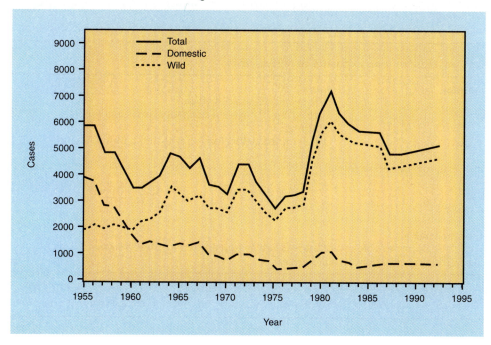

FIGURE 15.36
Number of rabies cases in wild and domestic animals in the United States. Most cases of rabies occur in wild animal populations.

of symptoms is almost always successful in preventing rabies. An animal which makes an unprovoked bite of a human is generally held 10 days for observation and if no signs of rabies are apparent, treatment of the human with rabies vaccine or rabies immune serum (antirabies virus antibodies obtained from a hyperimmune individual) is generally not undertaken. However, if the animal is rabid, the patient will be passively immunized with rabies immune globulin (injected at both the site of the bite and intramuscularly) and also will be vaccinated with inactivated rabies virus, preferably virus grown in human cell lines.

A "second generation" rabies vaccine is now under development. This vaccine is prepared by recombinant DNA techniques involving the cloning and expression in *Escherichia coli* of highly immunogenic rabies virus surface glycoproteins. Subunit vaccines (see Section 8.13) are also in development. In these, rabies virus genes coding for coat proteins are expressed in vaccinia virus. Such vaccines would be a completely safe means of controlling rabies in both animals and humans. A summary of guidelines for treating possible human exposure to rabies is shown in Table 15.3.

Rabies has been effectively controlled in domestic animal populations by vaccination, but it is still a problem in the United States because of the apparently large wild animal reservoir of rabies virus (see Table 14.4). All dogs and cats should be vaccinated beginning at 3 months of age and booster inoculations given yearly. Because of the highly fatal nature of the disease and the effective treatment available, any unprovoked animal bite must be taken seriously. A wild animal suspected of being rabid should be captured, sacrificed, and examined immediately for evidence of rabies in order to expedite treatment of those exposed. Characteristic virus inclusion bodies in the cytoplasm of nerve cells, called *Negri bodies*, are taken as confirmation of rabies. Fluorescent antibody preparations (see Sections 12.12 and 13.5) which recognize rabies-infected brain tissue are also very useful for confirming a diagnosis of rabies.

> Although rabies occurs primarily in wild animals, it can be transmitted to domestic animals and from them to humans. Vaccination of dogs and cats is a central part of the control of the spread of rabies to humans. In developing countries, rabies is still an important human disease.

15.9   Insect- and Tick-Transmitted Diseases: Rickettsias

The rickettsias are small bacteria that have a strictly intracellular existence in vertebrates, usually in mammals, and are also associated at some point in their natural cycle with blood-sucking arthropods such as fleas, lice, or ticks. We discuss the biology of rickettsias in Section 19.22. Rickettsias cause a variety of diseases in humans and animals, of which the most important are typhus fever, Rocky Mountain spotted fever, scrub typhus (tsutsu-gamushi disease), and Q fever. Rickettsias take their name from Howard Ricketts, a scientist at the University of Chicago who first provided evidence for their existence, but who unfortunately died from infection with the rickettsia that causes typhus fever, *Rickettsia prowazekii*. Rickettsias have not been cultured in nonliving media but can be cultured in laboratory animals, lice, mammalian tissue cultures, and the yolk sac of chick embryos. In animals, growth takes place primarily in phagocytic cells (see Section 11.14).

Typhus fever (epidemic typhus)

Typhus fever is caused by *R. prowazekii*. Epidemic typhus is transmitted from host to host by the common body or head louse. Typhus can be a serious disease.

Table 15.3   Guidelines for treating possible human exposure to rabies virus

I. Unprovoked bite by a domestic animal

Animal suspected of rabies	*Animal not suspected of rabies*
1. Sacrifice animal and test for rabies	1. Hold for 10 days—if no symptoms, do not treat human
2. Begin treatment of human immediately*	2. If symptoms develop—treat human immediately*

II. Bite by wild carnivore (e.g., skunk, bat, fox, raccoon, coyote)
Regard animal as rabid.
1. Sacrifice animal and test for rabies.
2. Begin treatment of human immediately.*

III. Bite by wild rodent, squirrel, livestock, rabbit
Consult local or state public health officials about possible recent cases of rabies transmitted by these animals (these animals rarely transmit rabies). If no reports, do not treat human.

*All bites should be thoroughly cleansed with soap and water. Treatment is generally a combination of rabies immune globulin and human diploid cell rabies vaccine (five injections intramuscularly).

During World War I an epidemic of typhus spread throughout eastern Europe and claimed almost three million lives. Typhus has frequently been a problem in military troops during wartime. Cells of *R. prowazekii* are introduced through the skin when the puncture caused by the louse bite becomes contaminated with louse feces, the major source of rickettsial cells. During an incubation period of from one to three weeks, the organism multiples inside cells lining the small blood vessels, following which symptoms of typhus (fever, headache, and general body weakness) begin to appear. Five to nine days later a characteristic *rash* is observed which begins in the armpits and generally spreads throughout the body *except* for the face, palms of the hands, and soles of the feet. If untreated, complications from typhus may develop involving damage to the central nervous system, lungs, kidneys, and heart. Several drugs are effective against rickettsias, tetracycline and chloramphenicol being the most commonly used to control *R. prowazekii*.

Murine typhus, caused by *R. typhi*, is much less common than epidemic typhus. The disease occurs primarily in rodents and is transmitted to humans by feces from the rat flea. Symptoms are similar to those of epidemic typhus, including the formation of a rash. Unlike epidemic typhus, however, murine typhus is rarely fatal, although kidney problems are observed in untreated cases. Prevention of murine typhus depends on effective rat control and thus the disease is more common in areas of poor sanitation.

Rocky Mountain spotted fever

Rocky Mountain spotted fever was first recognized in the western United States around the turn of the century, but is actually more common today in the southeastern United States. Rocky Mountain spotted fever is caused by *Rickettsia rickettsii* and is transmitted to humans by various species of ticks, most commonly the dog and wood ticks (Figure 15.37). Humans acquire the pathogen from tick fecal matter which gets injected into the body during a bite, or by rubbing infectious material into the skin by scratching. Cells of *R. rickettsii*, unlike other rickettsias, grow within the nucleus of the host cell as well as in host cell cytoplasm (Figure 15.37a,b). Following an incubation period of 3–12 days, an abrupt onset of symptoms occurs, including fever and a severe headache. Three to five days later, a rash is observed that is present on the palms of the hands and soles of the feet (Figure 15.37c) (the location of rickettsial rashes are of some diagnostic value). Gastrointestinal problems such as diarrhea or vomiting are usually observed as well, and the clinical symptoms of Rocky Mountain spotted fever may exist for over two weeks if the disease is left untreated. Tetracycline or chloramphenicol generally promotes a prompt recovery from Rocky Mountain spotted fever if administered early in the course of the infection.

Q fever

Q fever is a pneumonia-like infection caused by an obligate intracellular parasite, *Coxiella burnetii*, related to the rickettsias (see Section 19.22). Although not transmitted to humans directly by an insect bite, the agent of Q fever is transmitted to animals by insect bites, and various arthropod species serve as a reservoir of infection (the "Q" in Q fever stands for *query*, because

(a)
Willy Burgdorfer

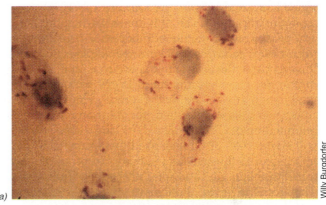
(b)
Willy Burgdorfer

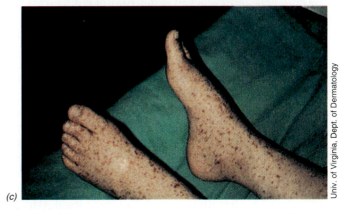
(c)
Univ. of Virginia, Dept. of Dermatology

FIGURE 15.37 *Rickettsia rickettsii*, the causative agent of Rocky Mountain spotted fever. (a) Cells of *R. rickettsii*, growing in the cytoplasm and nucleus of tick hemocytes. Individual cells are about 0.4 μm in diameter. (b) Cells of *R. rickettsii* in a granular hemocyte of an infected wood tick, *Dermacentor andersoni*. Transmission electron micrograph, bar = 1 μm. (c) Rash of the disease on the feet.

when cases of the disease were first observed, no pathogenic agent could be clearly implicated). Domestic animals generally have inapparent infections, but may shed large quantities of *C. burnetii* in their urine, feces, milk, and other secretions.

Cattle ticks are frequently implicated in transmitting the infection to dairy herds, where the disease Q fever can be a considerable occupational hazard to farmers and others involved in animal husbandry. Outbreaks of Q fever in humans have usually been traced to improperly pasteurized milk. With the recognition of Q fever as a potential milk pathogen, dairy processing facilities have modified their pasteurization protocol to ensure that pasteurized milk is *Coxiella*-free. Because it is a somewhat thermotolerant organism, the "holding method" of milk pasteurization, originally set by federal law at 60°C for 20 minutes, was raised to 62.8°C for 30 minutes to ensure elimination of *C. burnetii*.

Symptoms of Q fever are quite variable. They include an influenza-like illness, prolonged fever, headache and chills, chest pains, and pneumonia. Although Q fever is rarely fatal, complications arising from a *C. burnetii* infection may be severe. In particular, endocarditis (inflammation of the lining of heart and heart valves) is associated with many cases of Q fever. Q fever endocarditis frequently occurs many months or even years after the primary infection, the organisms remaining dormant in liver cells during the interim. Damage to heart valves as a result of Q fever endocarditis weakens the heart and may necessitate heart valve replacement surgery in later life.

An obligate intracellular parasite, *C. burnetii* is difficult to isolate and culture. However, a diagnosis of Q fever can readily be made by immunological tests designed to measure host antibodies to the organism (an indirect ELISA test has made the diagnosis of Q fever an almost routine procedure). *C. burnetii* infections respond quite dramatically to the antibiotic tetracycline, and therapy is usually begun quickly in any suspected human case of Q fever in order to prevent heart damage by latent *C. burnetii* infections. Q fever is one of the infectious diseases that has been studied as a possible agent for biological warfare.

Diagnosis and control of rickettsial diseases

In the past, rickettsial infections have been difficult to diagnose because the characteristic rash associated with many rickettsial diseases may be mistaken for measles, scarlet fever, or adverse drug reactions. Clinical confirmation of rickettsial diseases has now been greatly aided by the introduction of specific immunological reagents. These include polyclonal or monoclonal antibodies that can detect rickettsial surface antigens by immunofluorescence, latex bead agglutination assays, and ELISA analyses (see Sections 13.5, 13.7, and 13.8 for descriptions of the principles of these assays). Control of most rickettsial diseases requires control of the vectors: lice, fleas, and ticks. For humans traveling in wooded or grassy areas, the use of insect

repellants on the exposed extremities usually prevents tick attachment. Firmly attached ticks should be removed gently with forceps, care being taken to remove all of the mouth parts. Solvents, such as gasoline or ethanol applied to a tick from a saturated swab will usually expedite removal of the tick. Although a vaccine is available for the prevention of typhus, the few cases reported do not warrant its general administration. No vaccines are currently available for the prevention of Rocky Mountain spotted fever or Q fever. However an experimental Q fever vaccine, using a mixed *C. burnetii* antigen preparation, is now undergoing clinical trials in human volunteers and thus far looks promising.

> Rickettsias are obligate intracellular parasitic Bacteria which are often transmitted by insect or tick bite. Most rickettsial infections are readily controlled by antibiotic therapy.

15.10   Tick-Transmitted Diseases: Lyme Disease

Lyme disease is a tick-borne disease which affects humans and other animals. Lyme disease was named for Lyme, Connecticut, where cases were first recognized, and has rapidly become the most prevalent tick-borne disease in the United States. Lyme disease is caused by a spirochete, *Borellia burgdorferi* (Figure 15.38), which is spread primarily by the deer tick, *Ixodes dammini*, but can also be spread by the common dog (wood) tick and other types of ticks (Figure 15.39). The ticks that carry *B. burgdorferi* cells feed on the blood of birds, domesticated animals, various wild animals, and occasionally humans; deer and the white-footed field mouse are prime hosts of the tick. Complicating the disease picture in humans is the fact that the

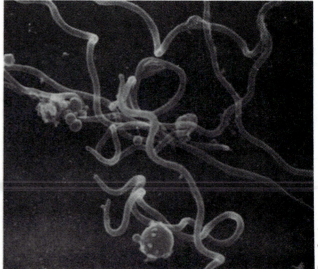

FIGURE 15.38 Electron micrograph of the Lyme spirochete, *Borrelia burgdorferi*. The diameter of a single cell is approximately 0.4 μm.

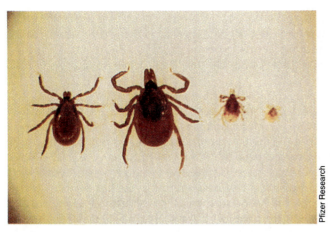

FIGURE 15.39 Deer ticks (*Ixodes dammini*), the major vectors of Lyme disease. Left to right, male and female adult ticks, nymph, and larval forms. The length of an adult female is about 3 mm.

deer tick is much smaller than many other types of ticks and thus is easy to overlook (Figure 15.39). Unlike the case with other tick-borne diseases, a very high percentage (up to 50 percent in certain regions of the northeast) of all deer ticks have been found to carry *B. burgdorferi* cells. Thus, extended contact with a vector gives a high probability of disease transmission.

Although most cases of Lyme disease have been reported from the northeastern and upper midwestern United States, cases have been reported in nearly every state, and Lyme disease is rapidly spreading west and south. For example, in California, Lyme disease is becoming more prevalent; different species of ticks and mammals (in this case, the dusky-footed woodrat) have been implicated as nonhuman reservoirs. Figure 15.40 shows the recent spread of Lyme

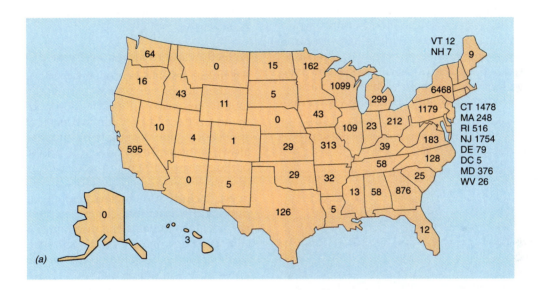

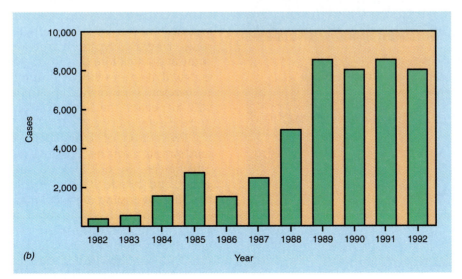

FIGURE 15.40 Incidence of Lyme disease in the United States. (a) Geographic incidence for 1989 and 1990 combined. Lyme disease is spreading south and west from its original focus in the northeast. (b) Total number by year of reported cases of Lyme disease. Note that the rate of increase has slowed, and the number of new cases is relatively constant in recent years. Data are from the Centers for Disease Control, Atlanta, GA.

disease into virtually all of the continental United States and the alarming and rapid rise in the total number of cases.

Transmission and pathogenesis of Lyme disease

Cells of *B. burgdorferi* are transmitted to humans while the tick is obtaining a blood meal (Figure 15.41*a*). A systemic infection develops leading to the main symptoms of Lyme disease, which include an acute headache, backache, chills, and fatigue. In about 75 percent of all cases a large rash is observed at the site of the tick bite (Figure 15.41*b*). If a correct diagnosis is made at this point, Lyme disease is easily treatable with the antibiotics tetracycline or penicillin. However, if not treated properly, Lyme disease may progress to a chronic stage, causing a crippling numbness of the limbs and severe exhaustion. Treatment at this stage generally requires intravenous antibiotics. The drug ceftriaxone, a highly active β-lactam antibiotic, is used to treat chronic Lyme disease because it is one of the few antibiotics that can cross the blood-brain barrier and can thus attack spirochetes residing in the central nervous system. If no treatment is obtained, cells of *B. burgdorferi* infecting the central nervous system may lie dormant for long periods before eliciting a variety of additional chronic symptoms, including visual dis-

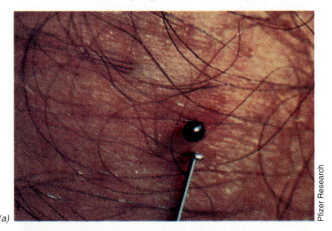

(a)

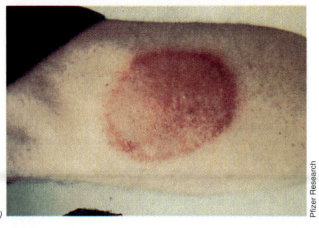

(b)

FIGURE 15.41 (a) Deer tick obtaining a blood meal from a human. (b) Characteristic circular rash associated with Lyme disease. The rash typically starts at the site of the bite and grows in a circular fashion over a period of several days. This rash is about 5 cm in diameter.

turbances, facial paralysis, and seizures. Pregnant women should be especially aware of Lyme disease and avoid tick exposure because there is some indication that Lyme disease can cause miscarriages and stillbirths.

Because the disease is so new—the causative agent was first identified in 1982—little is known about the pathogenesis of Lyme disease. No toxins or other virulence factors have yet been identified. In many respects the latent symptoms of Lyme disease resemble those of syphilis, caused by a different spirochete, *Treponema pallidum* (see Section 15.6). Indeed, some of the neurological symptoms of Lyme disease resemble those of chronic syphilis. However, unlike syphilis, Lyme disease has not been reported to be spread by sexual intercourse or other types of human contact. Small numbers of *B. burgdorferi* cells are shed in the urine of infected individuals and there is some indication that Lyme disease can spread through domestic animal populations, particularly cattle, by infected urine.

Detection and treatment of Lyme disease

An ELISA test has been developed for detection of antibodies to *B. burgdorferi*. Antibodies appear about four to six weeks after infection and can be detected by an indirect ELISA. Unfortunately, the immune response to Lyme disease is frequently not strong (as is also true of syphilis) and thus antibodies against the pathogen are difficult to detect in many cases of the disease. New diagnostic tools are thus being developed, including a nucleic acid probe assay that employs the polymerase chain reaction (see Sections 8.9 and 13.11) to increase the number of *B. burgdorferi*-specific DNA sequences in clinical specimens to a level sufficient to be detected by the probe. Other assays involve immunological methods to detect *B. burgdorferi* antigens in urine. Early detection and treatment of Lyme disease generally affords complete recovery without further or more progressive symptoms. In cases where a typical Lyme rash is observed surrounding a tick bite (Figure 15.41*b*), antibiotic therapy is usually begun without waiting for an antibody titer to develop. If no rash is apparent but Lyme disease is still suspected, the Lyme ELISA test is given after an appropriate interval.

Prevention of Lyme disease is straightforward and simply requires proper dressing to prevent tick attachment. In tick-infested areas like woods, tall grass, and brush, it is advisable to wear protective clothing including shoes, long pants, and a long-sleeved shirt with a snug collar and cuffs. Tucking the pants into tight-fitting socks worn with boots forms an effective barrier to tick attachment. After spending time in a tick-infested environment, individuals should check themselves carefully for ticks, gently removing any attached ticks (including the head) discovered. Insect repellants containing diethyl-*M*-toluamide (DEET) strongly discourage tick attachment, if applied to both skin and clothing. Care to avoid tick-infested areas or to prevent attachment of ticks is the best way to avoid contracting Lyme disease. Finally, several experimental vaccines are promising. In one case, vaccination of the white-footed mouse with *Borrelia* cell wall antigens results in protection of the mouse and also transfers bactericidal immunity, presumably by means of antibodies in the blood meal, to the tick. If this early study is confirmed, this vaccine will virtually eliminate the pathogen.

> **One of the most recently discovered infectious diseases is Lyme disease, caused by a spirochete and transmitted from wild animals to humans by ticks. Both prevention and treatment of Lyme disease are straightforward; the main problem is proper diagnosis.**

15.11   Insect-Transmitted Diseases: Malaria

Malaria is a disease caused by a protozoan, a member of the Sporozoa group. We discuss sporozoans as a group in Section 21.4. The malaria parasite is one of the most important human pathogens and has played an extremely significant role in the development and spread of human culture. Indeed, as we will see, malaria has even affected the human evolutionary process. The magnitude of the malaria problem can be appreciated when it is considered that over 100 *million* cases of malaria are estimated worldwide, and that at least 1 million people die from the disease each year. The historical importance of malaria can be appreciated by considering that almost half of all people who have died have died from malaria!

Ecology and pathogenesis of malaria

Four species of sporozoans infect humans, of which the most widespread is *Plasmodium vivax*. This parasite carries out part of its life cycle in humans, and part in the mosquito; the mosquito is the vector by which the parasite spreads from person to person. Only female mosquitoes of the genus *Anopheles* are involved, and since these primarily inhabit warmer parts of the world, malaria occurs predominantly in the tropics and subtropics. Malaria did not exist in the northern regions of North America prior to settlement by Europeans, but was a major problem in the South, where appropriate breeding grounds for the mosquito existed. The disease has always been associated with swampy low-lying areas, and the name *malaria* is derived from the Italian words for "bad air."

The life cycle of the malaria parasite is complex (Figure 15.42). One stage of the cycle occurs in humans, with the liberation into the bloodstream of a stage infective for mosquitoes. If a mosquito bites an infected person, the life cycle can be completed in the mosquito, with the formation of a stage infective for humans.

As illustrated in Figure 15.42, the human host is infected by plasmodial **sporozoites**, small elongated cells produced in the mosquito, which localize in the salivary gland of the insect. The sporozoites replicate in the liver, where they become transformed into a stage referred to as a *schizont*, which subsequently enlarges and segments into a number of small cells called **merozoites**; these cells are liberated from the liver into the blood stream. Some of the merozoites infect red blood cells (erythrocytes). The cycle in erythrocytes proceeds as in the liver and usually repeats at regular intervals, 48 hours in the case of *P. vivax*. It

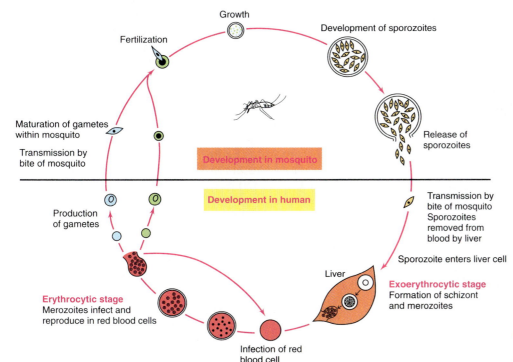

Growth

Fertilization

Development of sporozoites

Release of sporozoites

Maturation of gametes within mosquito

Transmission by bite of mosquito

Development in mosquito

Development in human

Transmission by bite of mosquito
Sporozoites removed from blood by liver

Sporozoite enters liver cell

Production of gametes

Liver

Exoerythrocytic stage
Formation of schizont and merozoites

Erythrocytic stage
Merozoites infect and reproduce in red blood cells

Infection of red blood cell

FIGURE 15.42
Life cycle of the malaria parasite, *Plasmodium vivax*.

is during this period that the characteristic symptoms of malaria occur, a fever of up to 40°C (104°F) followed by chills. The chills occur when a new brood of cells is liberated from erythrocytes. Vomiting and severe headache may accompany the fever-chill cycles, but asymptomatic periods generally alternate with periods in which the characteristic symptoms are present. Because of the loss of red blood cells, malaria generally causes anemia and some enlargement of the spleen as well.

Not all protozoal cells liberated from red blood cells are able to infect other erythrocytes; those which cannot, called *gametocytes*, are infective only for the mosquito. If these gametocytes happen to be ingested when another insect of the proper species of *Anopheles* bites the infected person, they mature within the mosquito into *gametes*. Two gametes fuse, and a zygote is formed; the zygote migrates by amoeboid motility to beneath the outer wall of the insect's intestine, where it enlarges and forms a number of sporozoites. These are liberated, some of them reaching the salivary gland of the mosquito, from where they can be inoculated into another person; and the cycle begins again.

Conclusive evidence for the diagnosis of malaria in humans is obtained by examining blood smears for the presence of infected erythrocytes (Figure 15.43). *Chloroquine* is the drug of choice for treating parasites *within* red blood cells, but this quinine derivative does not kill stages of the malarial parasite which reside *outside* erythrocytes. The related drug *primaquine* effectively eliminates the latter and thus treatment of malarial patients with chloroquine and primaquine together can affect a complete cure. However, because recurrences of malaria many years after a primary infection are not uncommon, it is likely that small numbers of sporozoites can reside in the liver protected from the effects of quinine drugs and release merozoites months or years later to reinitiate the disease.

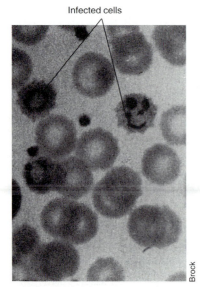

Infected cells

FIGURE 15.43 *Plasmodium vivax,* the causative agent of malaria, growing inside human red blood cells.

Plasmodium vivax has been very difficult to grow in culture, and because of its specificity for humans, experimental infection has also been difficult. For these reasons, most of the experimental work on malarial parasites has been done with species of *Plasmodium* that infect birds or rats, and it is with these forms that most of the studies on the development of new drugs have been carried out. However, scientists have been able to grow *P. falciparum* in laboratory culture and this organism has therefore served as a model system for experimental malarial research.

Eradication of malaria

Although quinine derivatives are effective in treating human cases of malaria, because of the obligatory alternation of hosts, *control* of malaria can be best effected by elimination of the *Anopheles* mosquito. Two approaches to mosquito control are possible: elimination of the habitat by drainage of swamps and similar breeding areas, and elimination of the mosquito by insecticides. Both approaches have been extensively used. The marked drop in incidence of malaria as a result of mosquito eradication programs in the United States is shown in Figure 15.44. During the 1930s, about 33,000 miles of ditches were dug in 16 southern states, removing 544,000 acres of mosquito breeding area. Millions of gallons of oil were also spread on swamps, with the purpose of cutting off the oxygen supply to mosquito larvae. With the discovery of the insecticide dichlorodiphenyltrichloroethane (DDT, see Figure 17.48), chemical control of both larvae and adult mosquitoes was possible. During World War II, the Public Health Service organized an Office of Mosquito Control in War Areas, and because many U.S. military bases were in the southern states, this organization carried out an extensive eradication program in the United States as well as overseas. In 1946, Congress established a five-year malaria eradication program, which involved drug treatment of humans and spraying of mosquito areas with DDT to a point where malaria transmission could not occur. As a further indication of the success of this program, in 1935 there were about 4000 deaths from malaria in the United States, and by 1952, the number of deaths had been reduced to 25.

In other parts of the world, eradication has been much more difficult, but the same control measures are used. Despite environmental problems (see Section 17.20), DDT is still an effective agent for mosquito control and has been most responsible for the control of malaria in large areas of Africa and South America.

> Malaria is one of the most widespread infectious diseases of humans but is confined primarily to warmer parts of the world, where its mosquito vector occurs. Malaria is a major cause of death in developing countries. Control is primarily through destruction of mosquitoes and mosquito habitat, although the disease can also be treated, at least in part, by use of quinine and related drugs.

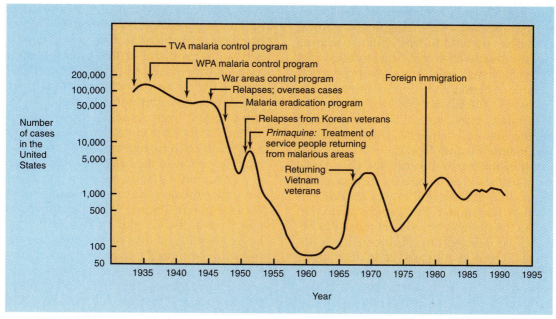

FIGURE 15.44 Dramatic decrease in incidence of malaria in the United States as a result of mosquito eradication programs. Note that the ordinate is logarithmic, so that the rise during the Vietnam War era is really quite minor.

Malaria and biochemical evolution of humans

One of the most interesting discoveries about malaria concerns the mechanism by which humans in regions of the world where the disease is endemic acquire resistance to *Plasmodium* infections. Malaria has undoubtedly been endemic in Africa for thousands of years. In West Africans, resistance to malaria caused by *P. falciparum* is associated with the presence in their red blood cells of hemoglobin S, which differs from normal hemoglobin A only in a single amino acid in each of the two identical halves of the molecule. In hemoglobin S, the neutral amino acid valine is substituted for the acidic amino acid glutamic acid. Red blood cells containing hemoglobin S have reduced affinity for oxygen, and the malaria parasite, having a highly aerobic metabolism, cannot grow as well in these red blood cells as it can in normal ones. An additional consequence is that individuals with hemoglobin S are less able to survive at high altitudes, where oxygen pressures are lower, but in tropical lowland Africa this disadvantage is not manifested. This altered hemoglobin S is responsible for the "sickle-cell" anemia trait. In West Africans, resistance to another malarial parasite, *P. vivax*, is associated with the presence of another abnormal hemoglobin, hemoglobin E. In certain Mediterranean regions where malaria is endemic, resistance to *P. falciparum* is associated with a deficiency in the red blood cells of the enzyme glucose-6-phosphate dehydrogenase (GPD). The faulty GPD leads to higher levels of oxidants which damage parasite membranes.

Another case in which the malaria parasite influences biochemical evolution involves the major histocompatibility complex (MHC) and the immune system (see Section 12.5). As discussed previously, the MHC class I and class II proteins are the molecules that present antigens to T cells for initiation of an immune response. In malaria-prone equatorial West Africa, individuals are very likely to have one particular MHC class I gene and one particular set of class II genes, selected from a total of perhaps 100 different genes and gene sets. As a result, these particular selected MHC genes are more common in the West African population than in any other human population group. Individuals who express these genes have as much resistance to severe fatal malaria infections as those with the hemoglobin S trait. These particular MHC proteins are exceptionally good antigen-presenting molecules for certain malarial antigens and confer protective resistance to *Plasmodium* sp. infection due to the powerful immune response they help initiate.

As in the hemoglobin variant cases, the parasite acts as a strong selection agent for individual genes important for host survival: individuals with these MHC genes have a measurable survival advantage and are thus more likely to reproduce and pass the resistance genes to their progeny. The malaria parasite has thus been a factor in the biochemical evolution of human beings. Other microbial parasites have also probably promoted evolutionary changes in their hosts, but in no case do we have such clear evidence as in the case of malaria.

> Malaria has had a demonstrable selective effect on human genes controlling traits as functionally disparate as globins (oxygen transport) and MHC proteins (antigen presentation). This disease is a good demonstration of the importance of pathogen-driven selection in human evolution.

15.12   Insect-Transmitted Diseases: Plague

Pandemic occurrences of **plague** in the past have been responsible for more human deaths than any other infectious disease, other than malaria. Plague is caused by a Gram-negative, facultatively aerobic rod called *Yersinia pestis* (see Section 19.20). Plague is a natural disease of domestic and wild rodents, but rats appear to be the primary disease reservoir. Most infected rats die soon after symptoms begin, but a low proportion develop a chronic infection and can serve as a source of virulent *Y. pestis*. The majority of cases of human plague in the United States occur in the southwestern states, where the disease is endemic among wild rodents (sylvatic plague) (Figure 15.45).

Plague is transmitted by the rat flea (*Xenopsylla cheopsis*), which ingests *Y. pestis* cells by sucking blood from an infected animal. Cells multiply in the flea's intestine and can be transmitted to a healthy animal in the next bite. As the disease spreads, rat mortality becomes so great that infected fleas seek new hosts, including humans. Once in humans, cells of *Y. pestis* usually travel to the lymph nodes, where they cause the formation of swollen areas referred to as *buboes*, and for this reason the disease is frequently referred to as **bubonic plague** (Figure 15.46*a*). The buboes become filled with *Y. pestis*, but the distinct capsule on cells of *Y. pestis* prevent them from being phagocytized. Secondary buboes form in peripheral lymph nodes and cells eventually enter the bloodstream, causing a generalized septicemia. Multiple hemorrhages produce dark splotches on the skin (Figure 15.46*b*). If not treated prior to the bacteremic stage, the symptoms of plague, including extreme lymph node pain, prostration, shock, and delirium, usually cause death within three to five days.

The pathogenesis of plague is not clearly understood, but it is known that cells of *Y. pestis* produce a number of antigenically distinct molecules, including toxins, that undoubtedly contribute to the disease process. The V and W antigens of *Y. pestis* cell walls are protein/lipoprotein complexes which serve to prevent phagocytosis. Other envelope proteins are also present. An exotoxin called *murine toxin*, because of its extreme toxicity for mice, is produced by all virulent strains of *Y. pestis*. Murine toxin serves as a respiratory inhibitor by blocking mitochondrial electron transport reactions at the point of coenzyme Q (see Section 4.10). Although it is not clear that murine toxin is involved in the pathogenesis of human plague (because murine toxin is highly toxic for certain animal species but not for others), the symptoms it produces in mice, systemic shock, liver damage, and respiratory distress, are similar to those in human cases of plague. *Y. pestis* also produces a highly immunogenic *endotoxin* which may play a role in the disease process as well.

Pneumonic plague occurs when cells of *Y. pestis* are either inhaled directly or reach the lungs during bubonic plague (Figure 15.46*c*). Symptoms are usually absent until the last day or two of the disease when large amounts of bloody sputum are emitted. Untreated cases rarely survive more than two days. Pneumonic plague, as one might expect, is a highly contagious disease, and can spread rapidly via the respiratory route if infected individuals are not immediately quarantined. **Septicemic plague** involves the rapid spread of *Y. pestis* throughout the body without the formation of buboes, and usually causes death before a diagnosis can be made.

Plague can be successfully treated if swiftly diagnosed. Although *Y. pestis* is naturally resistant to penicillin, most strains are sensitive to streptomycin, chloramphenicol, or the tetracyclines. If treatment is begun soon enough, mortality from bubonic plague can be reduced to as few as 1–5 percent of those infected. Pneumonic and septicemic plague can also be treated, but these forms progress so rapidly that antibiotic therapy in the latter stages of the disease is usually too late. Although potentially a devastating disease, an average of fewer than 20 cases of human plague are reported each year in the United States, and worldwide, there are fewer than 1500 confirmed cases and fewer than 300 deaths per year. This is undoubtedly due to improved sanitary practices and the overall control of rat populations.

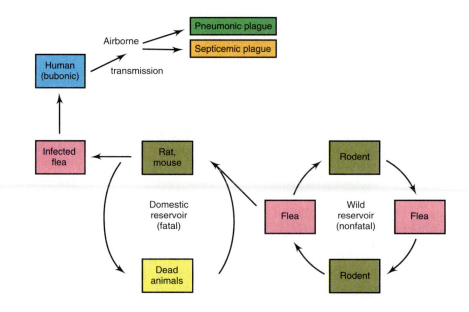

FIGURE 15.45
The ecology of plague. Plague in wild animals is generally a mild infection. Plague in rats and humans is frequently fatal.

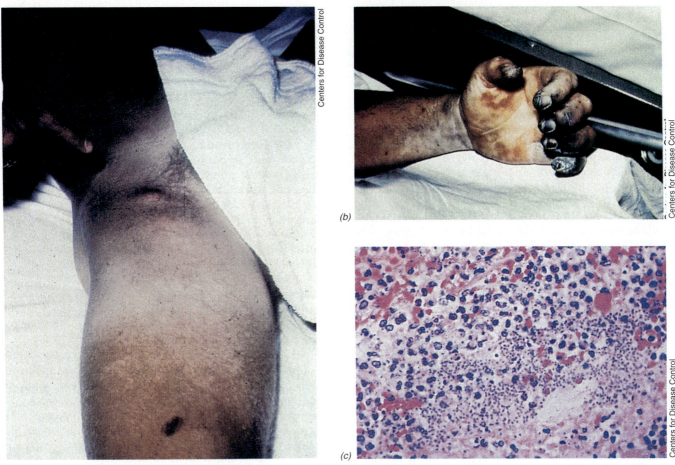

(a)

(b)

(c)

Centers for Disease Control

FIGURE 15.46 Plague in humans. (a) Bubo formed in the groin. (b) Gangrene and sloughing of skin in hand. (c) *Yersinia pestis*, the causative agent of plague. Cells are seen as very small blue cells from the lung tissue of a pneumonic plague victim.

15.13 Foodborne Diseases

There are two categories of foodborne disease, *food poisoning*, caused by toxins produced by microorganisms before eating, and *food infection*, caused by growth of the microorganisms in the human body after the contaminated food has been eaten. Table 15.4 gives a breakdown of foodborne illness in the United States. As seen, the majority of the cases are bacterial.

In addition to passive transfer of pathogens in food, active *growth* of a pathogen may also occur in foods (for example, due to improper storage), leading to marked increases in the microbial load. We have discussed some aspects of heat sterilization, pasteurization, and food microbiology in Sections 9.13 and 9.18. A summary of the major food-poisoning and food infection bacteria is given in Table 15.5. We begin our discussion here with the common food poisonings, *staphylococcal* and *perfringens*, and proceed to a discussion of the most serious of all poisonings, *botulism*. We will then turn our attention to common food infections including salmonellosis, *Campylobacter* infection, hepatitis, and traveler's diarrhea.

Staphylococcal food poisoning

The most common food poisoning is caused by the Gram-positive coccus, *Staphylococcus aureus*. This organism produces several *enterotoxins* (see Section 11.9)

that are released into the surrounding medium or food; if food containing the toxin is ingested, severe reactions are observed within 1–6 hours, including nausea with vomiting and diarrhea. Six types of *S. aureus* enterotoxin have been identified, A, B, C_1, C_2, D, and E. Enterotoxin A is most frequently associated

Table 15.4	Foodborne disease outbreaks due to microorganisms[a]	
Responsible microorganism	**Outbreaks, percent[b]**	**Cases, percent[c]**
Bacteria		
Salmonella spp.	39	31
Staphylococcus aureus	25	38
Clostridium perfringens	13.5	14
Clostridium botulinum	10.5	0.3
Shigella spp.	5	4.5
Campylobacter jejuni	2.5	7
Vibrio parahaemolyticus	2	0.2
Yersinia enterocolitica	0.5	4
Virus		
Hepatitis A	2	1

[a]*Compiled from data provided by the Centers for Disease Control.*
[b]*Of 600 outbreaks reported in recent years.*
[c]*Of 8000 cases in a single year.*

Table 15.5 The major food-poisoning and food infection bacteria	
Organism	**Products involved**
Salmonella spp.	Poultry, other meats; milk and cream, eggs
Staphylococcus aureus	Meat dishes, desserts
Clostridium perfringens	Cooked and reheated meats and meat products
Vibrio parahaemolyticus	Sea foods
Bacillus cereus	Rice and other starchy foods
Clostridium botulinum	Home-canned vegetables (especially beans and corn), smoked fish
Campylobacter jejuni	Poultry, milk
Yersinia enterocolitica	Pork, milk

with outbreaks of staphylococcal food poisoning. The mechanism of action of enterotoxin A is that of a superantigen (see Sections 12.14 and 15.2) and involves systemic stimulation of large numbers of T cells. *S. aureus* enterotoxin A is a small single peptide of 30,000 molecular weight that is encoded by a chromosomal gene. Cloning and sequencing of this gene, the *entA* gene, and of several other *S. aureus* enterotoxin genes, show that this family of toxins is genetically related. Although the *entA* gene is chromosomally located, there are indications that the B and C type *S. aureus* enterotoxins are plasmid or transposon encoded, or alternatively, are encoded by a lysogenic bacteriophage. We discussed the importance of accessory genetic elements such as plasmids and bacteriophages as vectors for toxin production in other pathogens in Sections 11.8 and 11.9.

The kinds of foods most commonly involved in *S. aureus* food poisoning are custard- and cream-filled baked goods, poultry, meat and meat products, gravies, egg and meat salads, puddings, and creamy salad dressings. If such foods are kept refrigerated after preparation, they remain relatively safe, as *Staphylococcus* is unable to grow at low temperatures. In many cases, however, foods of this type are kept warm for a period of hours after preparation, such as in warm kitchens or outdoors at summer picnics. Under these conditions, *Staphylococcus*, which might have entered the food from a food handler during preparation, grows and produces enterotoxin. Many of the foods involved in staphylococcal food poisoning are not cooked again before eating, but even if they are, this toxin is relatively stable to heat and may remain active. Staphylococcal food poisoning can be prevented by careful sanitation methods so that the food does not become inoculated, by storage of the food at low temperatures to prevent staphylococcal growth, and by the discarding of foods stored for any period of hours at warm temperatures.

Perfringens food poisoning

Clostridium perfringens is a major cause of food poisoning in the United States. *C. perfringens* produces an enterotoxin (see Section 11.9) which elicits diarrhea and intestinal cramps, but not vomiting, in those infected. The symptoms last for about 24 hours and fatalities

are rare. The disease results from the ingestion of a large dose ($>10^8$ cells) of *C. perfringens* and is most frequently associated with the consumption of tainted meat or meat products. The bacteria form spores in the intestine and during the sporulation process the enterotoxin is produced and released. The toxin alters the permeability of the intestinal epithelium which leads to the symptoms described above.

The onset of symptoms of perfringens food poisoning begins some 8–22 hours after consumption of the contaminated food. *C. perfringens* is quite common in a variety of cooked and uncooked foods, especially meat, poultry, and fish, and in soil and sewage. The organism is present naturally in low numbers in the human gut, but apparently is not a successful enough competitor to reach large numbers. Large doses of *C. perfringens* are most likely to be obtained from meat dishes where the meat has been cooked in bulk lots (heat penetration in these situations is often slow and insufficient) and then left at 20–40°C for short periods. Spores of *C. perfringens* germinate and grow quickly in the meat and can easily produce the required dosage of the enterotoxin in a brief period. Diagnosis of perfringens food poisoning is made by isolation of *C. perfringens* from the gut, or more reliably, by a direct ELISA to detect *C. perfringens* enterotoxin in feces.

Botulism

Botulism is the most severe type of food poisoning; it is usually fatal, and occurs following the consumption of food containing the exotoxin produced by the anaerobic bacterium *Clostridium botulinum*. This bacterium normally lives in soil or water, but its spores may contaminate raw foods before harvest or slaughter. If the foods are properly processed so that the *C. botulinum* spores are killed, no problem arises; but if viable spores are present, they may initiate growth and even a small amount of the resultant neurotoxin can render the food poisonous.

We discussed the nature and action of botulinum toxin in Section 11.8 (see also Figure 11.17). At least seven distinct types of botulinum toxin are known, most of which are toxic to humans. The toxins are destroyed by heat (80°C for 10 minutes) so that a properly cooked food should be harmless, even if it did originally contain toxin. Most cases of botulism occur as a result of eating foods that are not cooked after processing. Canned vegetables and beans are often used without cooking in making cold salads. Similarly, smoked fish and meat and most of the vacuum-packed sliced meats are often eaten directly, without heating. If these products contain the botulinum toxin, then ingestion of even a small amount will result in this severe and highly dangerous type of food poisoning. In Japan cases of botulism are often linked to the consumption of *sushi*, a raw fish preparation.

Infant botulism occurs occasionally when spores of *C. botulinum* are ingested. If the infant's normal flora is not well developed or if the infant is undergoing antibiotic therapy, the spores may germinate and *C. botulinum* cells grow and release toxin. Honey as a vehicle of infant botulism has been suspected in a number of cases. Most cases of infant botulism occur between the first week of life and two months of age; infant botulism is rare in infants older than six months.

Salmonellosis

Although sometimes called a food poisoning, gastrointestinal disease due to foodborne *Salmonella* is more aptly called *Salmonella* **food infection**, because symptoms arise only after the pathogen grows in the intestine (hence symptoms can be experienced as late as several *days* after eating a contaminated food). The symptoms of salmonellosis include the sudden onset of headache, chills, vomiting, and diarrhea, followed by a fever that lasts a few days. Diagnosis is from symptoms and by culture of the organism from feces. Virtually all species of *Salmonella* are pathogenic for humans: one, *S. typhi*, causes the serious human disease typhoid fever, while a small number of other species cause foodborne gastroenteritis. *Salmonella typhimurium* is the most common cause of salmonellosis in humans.

The ultimate sources of the foodborne salmonellas are humans and warm-blooded animals. The organism reaches food by contamination from food handlers; or in the case of foods such as eggs or meat, the animal which produced the food may be the source of contamination. The foods most commonly implicated are meats and meat products (such as meat pies, sausage, and cured meats), poultry, eggs, and milk and milk products. If the food is properly cooked, the organism will be killed and no problem will arise, but many of these products are eaten uncooked or partially cooked. *Salmonella* strains causing foodborne gastroenteritis are often traced to products made with *uncooked* eggs such as custards, cream cakes, meringues, pies, and eggnog. Previously cooked foods that have been warmed and held without refrigeration or canned foods held for a while after opening often support the growth of *Salmonella* if they have become contaminated by an infected food handler (see Typhoid Mary box in Chapter 14). *Salmonella* infection is more common in summer than in winter, probably because warm environmental conditions are more favorable for growth of microorganisms in foods. The incidence of salmonellosis has been steadily and significantly increasing over the last several decades, as shown in Figure 15.47.

Campylobacter

Campylobacter is a Gram-negative, curved rod which grows at reduced oxygen tensions, that is, as a microaerophile. Two major species are recognized, *C. jejuni* and *C. fetus*, and together these species are thought to account for the majority of cases of bacterial diarrhea in children. *C. fetus* is also of economic importance because it is a major cause of sterility and spontaneous abortion in cattle and sheep. The symptoms of *Campylobacter* infection include a high fever (usually greater than 104°F or 40°C), nausea, abdomi-

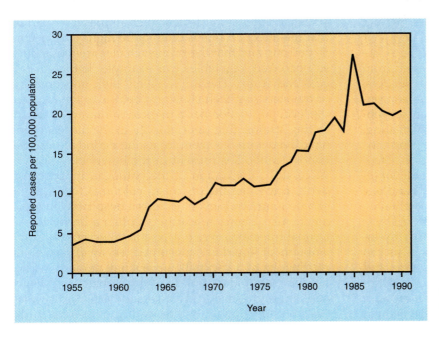

FIGURE 15.47
The incidence of salmonellosis in the United States. Data are from Centers for Disease Control, Atlanta, GA.

nal cramps, and a watery, frequently bloody, stool. Diagnosis requires isolation of the organism from stool samples and identification by growth-dependent tests or immunological assays. Because of the frequency with which *C. jejuni* infections are observed in infants, a variety of selective media and highly specific immunological methods have been developed (see Section 13.2) for positive identification of this organism.

Campylobacter is transmitted to humans via contaminated food, most frequently in poultry, pork, raw clams and other shellfish, or by a water route in surface waters not subjected to chlorination. Poultry is a major reservoir of *C. jejuni* and virtually all chicken and turkey carcasses contain this organism. Beef, on the other hand, is rarely a vehicle for *Campylobacter*. Proper washing of uncooked poultry (and any utensils coming in contact with uncooked poultry) and thorough cooking of the meat eliminates the problem of *Campylobacter* infection. *Campylobacter* species also infect domestic animals such as dogs, causing a milder form of diarrhea than that observed in humans. Infant cases of *Campylobacter* infection are frequently traced to contact with infected domestic animals, especially dogs.

Hepatitis

Infectious hepatitis (hepatitis A) is a viral-mediated inflammation of the liver caused by a picornavirus (positive single-strand RNA virus, see Section 6.15). The virus is transmitted primarily through fecal contamination of water, food, or milk. **Hepatitis A** infection can be subclinical in mild cases, or can lead to severe liver damage in chronic infections. The type A virus spreads from the intestine via the bloodstream to the liver and usually results in jaundice, a yellowing of the skin and eyes, and a browning of the urine due to stimulation of bile pigment production by infected liver cells. An immune response is initiated against the hepatitis A virus and this eventually brings the condition under control. However, in severe cases, permanent loss of a portion of liver function can occur.

The most significant food vehicles for type A hepatitis are shellfish (oysters and clams) harvested from waters polluted with human feces. As filter feeders, shellfish living in such environments tend to concentrate the hepatitis virus. Only *raw* shellfish are a problem because hepatitis A virus is destroyed by heating. Therefore the best means of controlling hepatitis transmission include sound sanitary practices, especially sewage treatment, the prevention of fecal contamination of food products by infected food handlers, and avoiding the consumption of uncooked shellfish.

Serum hepatitis (hepatitis B) is caused by a *DNA-containing* hepatitis virus transmitted primarily by infected blood or blood products. The type B virus can also be spread by maternal transmission in utero, and from one infected individual to another by sexual intercourse. Serum hepatitis frequently results in more severe liver damage than infectious hepatitis, leading to death in up to 10 percent of those infected; hepatitis

A rarely causes death. Long term infection with type B virus is also associated with increasing risk of liver cancer. Those at high risk for serum hepatitis include individuals that require frequent blood transfusions, dialysis patients, and intravenous drug abusers. As for infectious hepatitis, no specific therapy for serum hepatitis is currently available. However, due to the seriousness of the infection, a vaccine against serum hepatitis is now available. Use of the serum hepatitis vaccine is recommended for those in the high risk groups already mentioned, as well as for health personnel who come in frequent contact with blood or blood products. A variety of immunological tests, especially ELISAs (see Section 12.13), have been devised for the diagnosis of infectious and serum hepatitis by either antibody detection or direct viral detection methods.

A third type of hepatitis virus, unrelated to type A or type B and referred to as type C hepatitis virus, is also known. Although not yet reliably grown in tissue culture, the type C virus is widespread in the human population and is now the most common hepatitis virus encountered in cases of hepatitis mediated by blood transfusion. Like type B, type C hepatitis virus is primarily spread by contaminated blood and possibly also by sexual transmission. Although less likely to cause serious liver damage than type A or type B hepatitis viruses, the type C virus can cause chronic ailments such as cirrhosis.

Traveler's diarrhea

Traveler's diarrhea is an extremely common enteric infection in North Americans and Europeans traveling to developing countries. The primary causal agent is enteropathogenic *Escherichia coli* (EEC), although *Salmonella* and *Shigella* are sometimes implicated. The *E. coli* is an enterotoxin-producing strain, and as noted in Section 11.9, the gene for enterotoxin is often plasmid-encoded.

Several studies have been done on groups of U.S. citizens traveling in Mexico. Such studies have shown that the infection rate is often quite high, greater than 50 percent, and that the prime vehicles are foods, such as uncooked vegetables (for example, lettuce in salads) and water. The impressively high infection rate in travelers has been shown to be due to the fact that the local population has a marked immunity to the infecting strains, due undoubtedly to the fact that they have lived with the agent for a long period of time. Secretory antibodies present in the bowel may prevent successful colonization of the pathogen in local residents, but when the organism colonizes the intestine of a nonimmune person, it finds a hospitable environment. Also, stomach acidity, so often a barrier to intestinal infection, may not be able to act if only small amounts of liquid are consumed (as, for instance, the melting ice of a cocktail), since small amounts of liquid induce rapid emptying of the stomach and hence pass through so quickly that stomach acidity may have no effect on an enteric pathogen present in the liquid.

Assessing microbial content of foods

All fresh foods will have some viable microorganisms present. The purpose of assay methods is to detect evidence of abnormal microbial growth in foods or to detect the presence of specific organisms of public health concern, such as *Salmonella*, *Staphylococcus*, or *Clostridium botulinum*. We discussed in Section 13.11 the use of nucleic acid probes for the detection of specific foodborne pathogens, and these methods are finding increasing use. For cultural studies of nonliquid food products, preliminary treatment is usually required to suspend microorganisms embedded or entrapped within the food in a liquid medium. The most suitable method for treatment is high-speed blending. Examination of the food should be done as soon after sampling as possible, and if examination cannot begin within one hour of sampling, the food should be refrigerated. A frozen food should be thawed in its original container in a refrigerator and examined as soon as possible after thawing is complete. For *Salmonella* several selective media are available (see Section 13.2), and tests for its presence are most commonly done on animal food products, such as raw meat, poultry, eggs, and powdered milk, since *Salmonella* from the animal may contaminate the food. For staphylococcal counts, a medium high in salt (either sodium chloride or lithium chloride at a final concentration of 7.5 percent) is used. Of the organisms present in foods, staphylococci are the only common ones tolerant of such levels of salt. Since *Staphylococcus aureus* is responsible for one of the most common types of food poisoning, staphylococcal counts are of considerable importance. See Section 9.18 for a discussion of food preservation methods.

> Some of the most common microbial diseases of humans are foodborne. Some foodborne diseases are food poisonings in which the causal agent produces an exotoxin or enterotoxin during growth in food. This toxin is eaten with the food and causes the food poisoning symptoms. Examples are botulism and staphylococcal food poisoning. Other foodborne diseases are food infections, in which the food is the agent by which the pathogen is transmitted to the host. Examples include viral hepatitis and infections caused by the bacteria *Salmonella* and *Campylobacter*.

15.14   Waterborne Diseases

Organisms pathogenic to humans that are transmitted by water include bacteria, viruses, and protozoa (Table 15.6). Organisms transmitted by water usually grow in the intestinal tract and leave the body in the feces. Fecal pollution of water supplies may then occur, and if the water is not properly treated, the pathogens enter a new host when the water is consumed. Because water is consumed in large quantities, it may be infectious even if it contains only a small number of pathogenic organisms.

Pathogenic organisms transmitted by water

Probably the most important pathogenic *bacteria* transmitted by the water route are *Salmonella typhi*, the organism causing typhoid fever, and *Vibrio cholerae*, the organism causing cholera. Although the causal agent of typhoid fever may also be transmitted by contaminated food (see Section 15.13) and by direct contact from infected people, the most common and serious means of transmission is the water route. Typhoid

Table 15.6   Waterborne disease outbreaks due to microorganisms[a]

Disease	Causal agent	Outbreaks[b] (%)	Cases[c] (%)
Bacteria			
Typhoid fever	*Salmonella typhi*	10	0.5
Shigellosis	*Shigella* species	9	9
Salmonellosis	*Salmonella paratyphi*, etc.	3	12
Gastroenteritis	*Escherichia coli*	0.3	2.5
	Campylobacter species	0.3	2.5
Viruses			
Infectious hepatitis	Hepatitis A virus	11	1.6
Poliomyelitis	Poliovirus	0.2	0.01
Diarrhea	Norwalk virus	1.5	2
Protozoa			
Dysentery	*Entamoeba histolytica*	0.1	0.05
Giardiasis	*Giardia lamblia*	7	13
Unknown etiology			
Gastroenteritis		57	58

[a]Compiled from data provided by the Centers for Disease Control.
[b]Of 650 outbreaks in recent decades.
[c]Of 150,000 cases over the same period.

fever has been virtually eliminated in many parts of the world, primarily as a result of the development of effective water treatment methods. However, a breakdown in water purification methods, contamination of water during floods, earthquakes, and other disasters, or cross contamination of water pipes from leaking sewer lines, occasionally results in epidemics of typhoid fever. Water that has been contaminated can be rendered safe for drinking by boiling for 5 to 10 minutes or by adding chlorine as discussed later.

The causal agent of cholera is usually transmitted by the water route. However, during the recent cholera epidemic in the Americas, consumption of raw shellfish and raw vegetables was also implicated as a means of cholera spread. Presumably, the vegetables were washed in contaminated water and the shellfish beds were contaminated by raw sewage. At one time, cholera was common in Europe and North America, but the disease has virtually been eliminated from these areas by effective water purification. The disease is still common in Asia and in certain parts of Central and South America. Travelers to these areas are advised to be vaccinated for cholera. Both *V. cholerae* and *S. typhi* are eliminated from sewage during proper sewage treatment and hence do not enter water courses receiving treated sewage effluent. More frequent than typhoid, but less serious a disease, is salmonellosis caused by species of *Salmonella* other than *S. typhi* (see Section 15.13). As seen in Table 15.6, the largest number of cases of waterborne bacterial disease in the United States have been due to salmonellosis.

Enteric bacteria are effectively eliminated from water during the water purification process (see below), so that they should never be present in properly treated drinking water. Most outbreaks of waterborne disease in the United States are due to breakdowns in treatment systems, or as a result of post-contamination in the pipelines. This latter problem can be controlled by maintaining a detectable level of free chlorine in the pipelines.

Viruses transmitted by the water route include poliovirus and other viruses of the enterovirus group, as well as the virus causing infectious hepatitis (Table 15.6). Poliovirus has several modes of transmittal, and transmission by water may be of serious concern in some areas.

Because viruses are acellular, they are more stable in the environment and are not as easily killed as bacteria. However, both poliovirus and infectious hepatitis virus are eliminated from water by proper treatment practices, and the maintenance of 0.6 ppm free chlorine in a water supply will generally ensure its safety.

Cholera

We discussed the action of cholera enterotoxin in Section 11.9. The disease cholera is caused by *Vibrio cholerae*, a Gram-negative, curved rod transmitted almost exclusively via contaminated water. Cholera enterotoxin catalyzes a life-threatening diarrhea which

can result in dehydration and death unless the patient is given fluid and electrolyte therapy. The disease has swept the world in seven major pandemics, the most recent in 1961. Today the disease is virtually absent from developed countries, although it is still common in areas where sewage treatment is either not practiced or poorly performed.

Two major biotypes of *V. cholerae* have been recognized, the *classic* and the *El Tor* types. Each biotype has two major serotypes. The classic strain, such as the *V. cholerae* strain first isolated by Robert Koch in 1883 was more prevalent in cholera outbreaks before 1960, while the El Tor strain has been more frequently observed since that time. Following ingestion of a substantial inoculum, the *V. cholerae* cells take up residence in the *small* intestine. Studies with human volunteers have shown that the acidity of the stomach is responsible for the large inoculum needed to initiate cholera. Although the ingestion of 10^8–10^9 cholera vibrios is generally required to cause cholera, human volunteers given bicarbonate to neutralize gastric acidity developed the disease when only 10^4 cells were administered. Far lower cell numbers are required to initiate infection if administered with food. Cholera vibrios attach firmly to small intestinal epithelium and grow and release enterotoxin. The enterotoxin causes large loss of fluids, 20 liters per day not being uncommon in a fulminant case of cholera. If untreated, the mortality rate can be as high as 60 percent. Intravenous or oral liquid replacement therapy is the major means of treatment (see Section 11.9 and Figure 11.19). Streptomycin or tetracycline may shorten the course of cholera but antibiotics are of little benefit without simultaneous fluid replacement.

Control of cholera depends primarily on satisfactory sanitation measures, particularly in the treatment of sewage and the purification of drinking water. *V. cholerae* organisms can adhere to normal flora in fresh water and can survive for long periods of time (Figure 15.48). In the last ten years sporadic outbreaks of

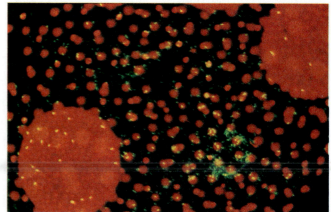

FIGURE 15.48 *Vibrio cholerae* attached to the surface of *Volvox*, a freshwater alga. The isolate was from a cholera-endemic area in Bangladesh. The green is *V. cholerae* cells stained with a monoclonal antibody to bacterial cell surface proteins. The red color is due to fluorescence of chlorophyll *a* in algal cells.

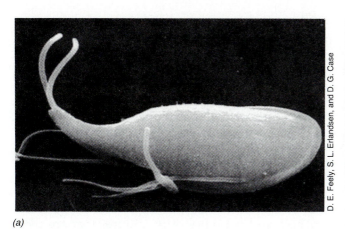

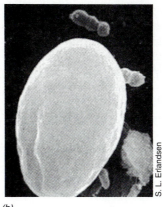

FIGURE 15.49
Scanning electron micrographs of the parasite *Giardia*. (a) Motile trophozoite. (b) Cyst.

cholera (less than 300 total cases) have been reported in the United States, and some evidence exists that raw shellfish may be an alternate vehicle, as mentioned previously. Cholera is heavily endemic in India, Pakistan, Bangladesh, and the Americas with occasional epidemics flourishing from time to time. Although most of these cases are due to water contaminated with human feces, cholera is also spread within households in these regions by fecally contaminated food.

Giardiasis

Giardiasis is an acute form of gastroenteritis caused by the protozoan parasite, *Giardia lamblia* (Figure 15.49). *G. lamblia* is a flagellated protozoan that is transmitted to humans primarily by contaminated water, although foodborne and even sexual transmission of giardiasis has been documented (see Section 21.4 for general discussion of flagellates). The protozoal cells, called trophozoites (Figure 15.49a) produce a resting stage called a cyst (Figure 15.49b), and this is the primary form transmitted by water. Cysts germinate in the gastrointestinal tract and bring about the symptoms of giardiasis: an explosive, foul-smelling, watery diarrhea, and intestinal cramps, flatulence, nausea, and malaise. The foul-smelling nature of the diarrhea and the absence of blood or mucus in the stool are diagnostically helpful in distinguishing giardiasis from diarrhea of bacterial or viral origin. The drugs quinacrine and metronidazole are useful in treating the disease.

Between 1965 and 1981, 53 waterborne outbreaks of giardiasis affecting over 20,000 people were reported in the United States. Most outbreaks occurred in undeveloped or mountainous regions where surface water sources have been used for drinking purposes. *Giardia* cysts are fairly resistant to chlorine and many outbreaks have been associated with water systems using only chlorination as a means of water purification. Water subjected to proper sedimentation and filtration (see below) in addition to being chlorinated is generally free of *Giardia* cysts. Many cases of giardiasis have been associated with drinking untreated water in wilderness areas. Although water may look "crystal clear," it may contain fecal material from animals harboring *Giardia*. Studies of wild animals have indicated that beavers and muskrats are

major carriers of *Giardia* and may transmit cells or cysts to water supplies. As a safety precaution, all water consumed from natural rivers and streams, for example during a camping or hiking trip, should either be filtered or boiled. Boiling is the preferred method of rendering water microbially safe.

Amebiasis

A number of different amoebas inhabit the tissues of humans and other vertebrates, usually in the oral cavity or intestinal tract. Some of these are pathogenic. We discuss the general properties of amoeboid protozoans in Section 21.4. *Entamoeba histolytica*, the causative agent of amebiasis, is a common pathogenic protozoan transmitted to humans primarily by contaminated water and occasionally by the foodborne route (Figure 15.50). *E. histolytica* is an *anaerobic* amoeba, the trophozoites lacking mitochondria. Like *Giardia*, the trophozoites of *Entamoeba* produce cysts. Cyst germination occurs in the intestine and cells grow both on and in intestinal mucosal cells. Continued growth leads to ulceration of intestinal mucosa, causing diarrhea and severe intestinal cramps. Diarrhea is replaced by a condition referred to as *dysentery*, characterized by the passage of intestinal exudates, blood, and mucus. If not treated, trophozoites of *E. histolytica* (Figure 15.50) can migrate to the liver, lung, and brain. Growth in these tissues can cause abscesses and other tissue damage.

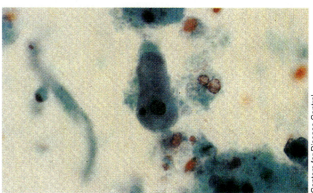

FIGURE 15.50 Trophozoites of *Entamoeba histolytica*, the causative agent of amebiasis. The small red structures are red blood cells.

Amebiasis can be treated with the drugs metronidazole and chloroquine, but experience has shown that amebicidal drugs are not always effective. Spontaneous cures do occur, implying that the host immune system plays some role in ending the infection. However, protective immunity is not afforded by primary infection because reinfection is not uncommon. Amebiasis is rather easily diagnosed by examining stool samples for the morphologically distinct cysts of *E. histolytica*. The disease occurs at very low incidence in those regions which practice sanitary sewage treatment. Ineffective sewage treatment or use of untreated surface waters for drinking purposes are the usual scenarios for cases of amebiasis.

The coliform test and water purity

As we have just discussed, a number of important diseases are waterborne. Even water which looks clear and pure may be sufficiently contaminated with pathogenic microorganisms to be a health hazard. Some means are necessary to ensure that drinking water is safe. One of the main tasks of water microbiology is the development of laboratory methods which can be used to detect the microbiological contaminants that may be present in drinking water. It is not usually practical to examine water directly for the various pathogenic organisms that may be present. As stated earlier, a wide variety of organisms may be present, including bacteria, viruses, and protozoa. To check each drinking water supply for each of these agents would be a difficult and time-consuming job. In practice, *indicator organisms* are used instead. These are organisms associated with the intestinal tract, whose presence in water indicate that the water has received contamination of an intestinal origin. The most widely used indicator is the **coliform group** of organisms. This group is defined in water bacteriology as all the aerobic and facultatively aerobic, Gram-negative, non-spore-forming, rod-shaped Bacteria that ferment lactose with gas formation within 48 hours at 35°C. This is an operational rather than a taxonomic definition, and the coliform group includes a variety of organisms, mostly of intestinal origin. In practice, the coliform organisms are almost always members of the enteric bacterial group (see Section 19.20). The coliform group includes the organism *Escherichia coli* a common intestinal organism, plus the organism *Klebsiella pneumoniae*, a less common intestinal organism. The definition also currently includes organisms of the species *Enterobacter aerogenes*, not generally associated with the intestine.

The coliform group of organisms are suitable as indicators because they are common inhabitants of the intestinal tract, both of humans and warm-blooded animals, and are generally present in the intestinal tract in large numbers. When excreted into the water environment, the coliform organisms eventually die, but they do not die at any faster rate than the pathogenic bacteria *Salmonella* and *Shigella*, and both the coliforms and the pathogens behave similarly during water purification processes. Thus it is likely that if coliforms are found in a water sample, the water has received fecal contamination and may be unsafe for drinking purposes. There are very few organisms in nature that meet the definition of the coliform group that are not associated with the intestinal tract. It should be emphasized that the coliform group includes organisms derived not only from humans but from other warm-blooded animals. Since many of the pathogens (for example, *Salmonella*, *Leptospira*) found in warm-blooded animals will also infect humans, an indicator of both human and animal pollution is desirable.

There are two types of procedures that are used for the coliform test. These are the **most-probable-number** (MPN) procedure and the **membrane filter** (MF) procedure. The MPN procedure employs liquid culture medium in test tubes, the samples of drinking water being added to the tubes of media. In the more common MF procedure, the sample of drinking water is passed through a sterile membrane filter, which removes the bacteria (see Figure 9.29), and the filter is then placed on a culture medium for incubation. When using the membrane filter method with drinking water, at least 100 ml of water should be filtered, although in clean water systems, even larger volumes could be filtered. After filtration of a known volume of water, the filter is placed on the surface of a plate of a culture medium highly selective for coliform organisms (see Section 13.2). The coliform colonies (Figure 15.51) are counted and from this value the number of coliforms in the original water sample can be determined. In well-regulated water supply systems, coliform tests will always be negative. If the coliform tests are not uniformly negative, a breakdown in the system has occurred (such as in chlorination) or in the distribution network (pipelines).

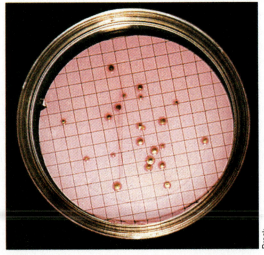

FIGURE 15.51 Coliform colonies growing on a membrane filter. A drinking water sample has been passed through the filter and the filter placed on a culture medium which is both selective and differential for lactose-fermenting bacteria (coliforms). The dark color of the colonies is characteristic of coliforms. A count of the number of colonies gives a measure of the coliform count of the original water sample.

FIGURE 15.52 Water purification plant. Aerial view of water treatment plant in Louisville, Kentucky. The arrows indicate direction of flow of water through the plant.

Water purification

It is a rare instance when available water is of such a clarity and purity that no treatment is necessary before use. Water treatment is carried out both to make the water safe microbiologically and to improve its utility for domestic and industrial purposes. Treatments are performed to remove pathogenic and potentially pathogenic microorganisms and also to decrease turbidity, eliminate taste and odor, reduce or eliminate nuisance chemicals such as iron or manganese, and soften the water to make it more useful for the laundry.

The kind of treatment that water is given before use must depend on the quality of water supply. A typical treatment installation for a large city using a source water of relatively poor quality is shown in Figure 15.52. Water is first pumped to **sedimentation basins**, where sand, gravel, and other large particles settle out. A sedimentation basin should be used only if the water supply is highly turbid, since it has the disadvantages that algal growth may occur in the basin, adding odors and flavors, and that pollution of the water by surface runoff may occur. Bacteria may grow in the bottom mud and add further problems.

Most water supplies are subjected to **coagulation**. Chemicals containing aluminum and iron are added, which under proper control of pH form a flocculent, insoluble precipitate that traps organisms, absorbs organic matter and sediment, and carries them out of the water. After the chemicals are added in a mixing basin, the water containing the coagulated material is transferred to a settling basin where it remains for about 6 hours, during which time the coagulum separates out. Around 80 percent of the turbid material, color, and bacteria are removed by this treatment.

After coagulation, the clarified water is usually filtered to remove the remaining suspended particles and microorganisms. Filters can be of the slow or rapid sand type. **Slow sand filters** are suitable for small installations such as resorts or rural places. The water is simply allowed to pass through a layer of sand 2–4 ft deep. Eventually the top of the sand filter will become clogged and the top layer must be removed and replaced with fresh sand. **Rapid sand filters** are used in large installations. The rate of water flow is kept high by maintaining a controlled height of water over the filter. When the filter becomes clogged, it is clarified by backwashing, which involves pumping water up through the filter from the bottom. From 98 to 99.5 percent of the total bacteria in raw water can be removed by proper settling and filtration.

Chlorination is the most common method of ensuring microbiological safety in a water supply. In sufficient doses it causes the death of most microorganisms within 30 minutes. In addition, chlorine reacts with organic compounds, oxidizing and effectively neutralizing them. Therefore, since most taste- and odor-producing compounds are organic in nature, chlorine treatment also improves water taste and smell. It also oxidizes soluble iron and manganese compounds, forming precipitates which can be removed. Chlorine can be added to water either from a concentrated solution of sodium or calcium hypochlorite or as a gas from pressure tanks. The latter method is used most commonly in large water treatment plants, as it is most amenable to automatic control.

When chlorine reacts with organic materials, it is used up. Therefore, if a water supply is high in organic materials, sufficient chlorine must be added so that there is a residual amount left to react with the microorganisms after all reactions with organic materials have occurred. The water plant operator must perform chlorine analyses on the treated water to determine the residual level of chlorine. A chlorine residual of about 0.2–0.6 µg/ml is an average level suitable for most water supplies.

After final treatment, the water is usually pumped to storage tanks, from which it flows by gravity to the consumer. Covered storage tanks are essential to ensure that algal growth does not take place (no light is available) and that dirt, insects, or birds do not enter.

Drinking water standards

Drinking water standards in the United States are specified under the Safe Drinking Water Act, which provides a framework for the development by the Environmental Protection Agency (EPA) of drinking water standards. Current standards prescribe that when the MF technique is used, 100-ml samples must be filtered, and the number of coliform bacteria shall not exceed any of the following: (1) 1 per 100 ml as the arithmetic mean of all samples examined per month; (2) 4 per 100 ml in more than one sample when less than 20 are examined per month; or (3) 4 per 100 ml in more than 5 percent of the samples when 20 or more are examined per month. Water utilities report their results to the EPA and if they are not meeting the prescribed standards they must notify the public and take steps to correct the problem. Many smaller communities and even large cities sometimes fail to meet the standards.

Public health significance of drinking water purification

Today the incidence of waterborne disease in developed countries is so low that it is difficult to appreciate the significance of treatment practices and drinking water standards. Most intestinal infection today is not due to transmission by the water route, but via food (see Section 15.13). It was not always so. Early in the twentieth century, effective water treatment practices did not exist, and there were no bacteriological methods for evaluating the health significance of polluted drinking water (see Snow on Cholera box in Chapter 14). The first coliform counting procedures were introduced about 1905. Until then, water purification, if practiced at all, was primarily for aesthetic purposes, to remove turbidity. Actually, turbidity removal by filtration provides a significant decrease in the microbial load of water, so filtration did play a part in providing safer drinking water. But filtration alone was of only partial value since many organisms passed through the filters. It was the discovery of the efficacy of chlorine as a water disinfectant, in about 1910, that had major impact. Chlorine is so effective and so inexpensive that its use spread widely; it is almost certain that the practice of chlorination was of major significance in reducing the incidence of waterborne disease. However, the effectiveness of chlorination would not have been realized, and the necessary doses could not have been determined, if standard methods for assessing the coliform content of drinking water had not been developed. Thus engineering and microbiology moved forward together.

The significance of filtration and chlorination for ensuring the safety of drinking water cannot be overemphasized. Figure 15.53 illustrates the dramatic drop in incidence of typhoid fever in a major American city after these two purification procedures were introduced. Similar results were obtained in other major cities. The dramatic improvement in the health of the American people in the early decades of the twentieth century was due to a large extent to the establishment of satisfactory water purification procedures.

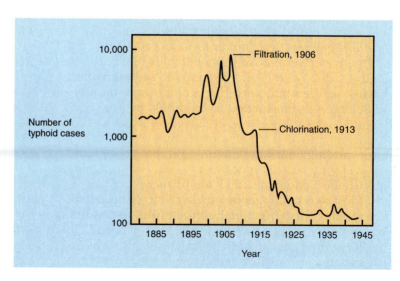

FIGURE 15.53
The dramatic effect of water purification on incidence of waterborne disease. The graph shows the incidence of typhoid fever in Philadelphia during the early part of the twentieth century. Note the marked reduction in incidence of the disease after the introduction of filtration and chlorination.

Pathogens that are associated with waterborne disease are the primary causes of intestinal infections. They are released from the host in feces and are transmitted in water that has not been properly treated. Public health measures for treatment of sewage and purification of drinking water are used to reduce or almost eliminate most waterborne diseases in developed countries, although these diseases are still prevalent in developing countries. Waterborne pathogens include bacteria such as *Salmonella* and *Escherichia*, viruses such as hepatitis and poliovirus, and protozoa such as *Entamoeba*. Bacterial indicators used to assess the quality of drinking water play a major role in the control of waterborne diseases.

15.15 Pathogenic Fungi

The pathogens discussed in this chapter are bacteria, viruses, and protozoa. However, some human diseases are caused by fungi. Discussion of these diseases has remained until last because the fungi are normally found in nature as free-living saprophytes and are primarily opportunistic pathogens. Thus, they are found in some form in nearly every ecological environment, usually growing on nonliving organic materials. The fungi include the eukaryotic organisms we commonly refer to as *yeasts*, which normally grow as single cells (Figure 15.54a), and *molds*, which grow in branching chains called *hyphae* (Figure 15.54b). The taxonomy and biological diversity of these organisms are discussed in Section 21.2. Fortunately, most fungi are harmless to humans. Only about 50 species cause human disease, and the overall incidence of serious fungal infections is rather low, although certain superficial fungal infections are quite common.

Fungi cause disease through three major mechanisms. First, some fungi cause immune responses that can result in allergic (hypersensitivity) reactions following exposure to specific fungal antigens (see Section 12.14). For example, the mold *Thermoactinomyces vulgaris* grows in stored hay. Farm workers often become sensitized to the mold and develop asthma when reexposed to moldy hay. This allergy, known as *Farmer's lung disease,* is one of a group of occupational respiratory allergies caused by fungi. Workers in sugar cane, brewing, and wood products industries can be similarly affected by fungal hypersensitivities.

A second fungal disease-producing mechanism involves the production and action of *mycotoxins*, a large and diverse group of fungal exotoxins. The best known examples of mycotoxins are produced by *Aspergillus flavus*, an organism that commonly grows on improperly stored food such as grain. The toxins produced by *A. flavus* are known as *aflatoxins* (Figure 15.55). Aflatoxins are highly toxic and induce tumors in some animals, especially in birds that feed on contaminated grain. Their direct role in human disease is not well defined.

The third fungal disease-producing mechanism is through infection. The growth of a fungus on or in the body is called a *mycosis* (plural *mycoses*). Mycoses can range in severity from relatively innocuous, superficial diseases to serious, life-threatening diseases.

Mycoses
Mycoses are subdivided into three categories. The first of these categories is the *superficial mycoses*. These diseases involve colonization of the skin, hair, or nails and infect only the surface layers (Figure 15.56a). Table 15.7 lists some of the common superficial fungi. In general, these diseases are relatively benign and self-limiting. Some, such as *Trichophyton* infections of the feet (athlete's foot), are quite common. Spread is by personal contact with an infected person or by con-

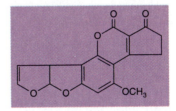

FIGURE 15.55
Structure of Aflatoxin B1. This toxin is one of a group of similar compounds produced by *Aspergillus flavus*.

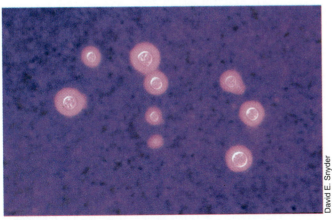

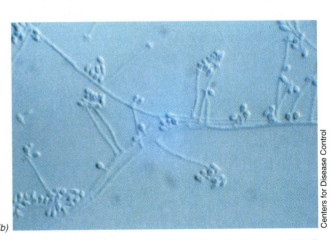

(a) David E. Snyder (b) Centers for Disease Control

FIGURE 15.54 Typical forms of pathogenic fungi. (a) Yeast form of encapsulated *Cryptococcus neoformans*, stained with India ink. The cell itself ranges from 4–20 μm. (b) *Sporothrix schenckii* showing the branching, or hyphae, characteristic of the mold form of fungi. The round basidiospores are about 2 μm in diameter.

tact with contaminated surfaces such as bathtubs or shower stalls or other contaminated shared articles such as towels or bed linens. Treatment for severe cases is with topical application of miconazole nitrate or griseofulvin. Griseofulvin is administered orally. After entering the bloodstream, it passes to the skin where it can inhibit fungal growth.

The *subcutaneous mycoses* involve deeper layers of skin (Figure 15.56*b*) and a different group of organisms (Table 15.7). One disease in this category is *sporotrichosis*, an occupational hazard of agricultural workers, miners, and other workers who come into contact with the soil. The causative organism is found as a ubiquitous saprophyte on wood and in soil. The lesions, usually initiated by fungal infection of a small wound or abrasion, resemble the ulcers and chancres seen in ulcerating diseases such as syphilis (compare Figures 15.56*b* and 15.26). The causative organism, *Sporothrix schenckii*, can readily be isolated from the lesion and cultured *in vitro*. Treatment is with oral potassium iodide or oral ketoconazole.

The *systemic mycoses* involve fungal growth in internal organs of the body. These are subclassified as *primary* or *secondary* infections. A primary infection is one resulting directly from the fungal pathogen in otherwise normal, healthy individuals. A secondary infection involves infection with the pathogen only in hosts with a predisposing condition such as antibiotic therapy or immunosuppression. In the United States, the most widespread primary fungal infections are *histoplasmosis*, caused by *Histoplasma capsulatum*, and *coccidioidomycosis* ("San Joaquin Valley fever"), caused by *Coccidioides immitis*. Both of these organisms normally live in soil. These are both respiratory diseases in which the host becomes infected by breathing in airborne spores, which germinate and grow in the lungs. Histoplasmosis is primarily a disease of rural areas in the midwestern United States, especially in the Ohio and Mississippi river valleys. Most cases are mild and can be mistaken for more common respiratory infections. San Joaquin Valley fever is generally restricted to the desert regions of the southwestern

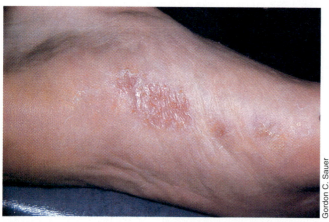

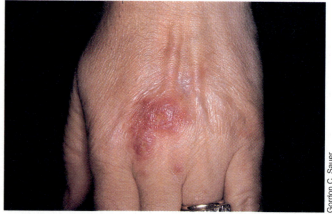

FIGURE 15.56   Fungal infections. (a) Superficial mycosis of the foot (athlete's foot) due to infection with *Trycophyton rubrum*. (b) Sporotrichosis, a subcutaneous infection due to infection by *Sporothrix schenckii*.

Table 15.7	Some pathogenic fungi and the diseases they cause	
Disease	**Causal organism**	**Main disease foci**
Superficial mycoses (dermatomycoses)		
Ringworm	*Microsporum*	Scalp of children
Favus	*Trichophyton*	Scalp
Athlete's foot	*Epidermophyton, Trichophyton*	Between toes, skin
Jock itch	*Trichophyton, Epidermophyton*	Genital region
Subcutaneous mycoses		
Sporotrichosis	*Sporothrix schenckii*	Arms, hands
Chromoblastomycosis	Several genera	Legs, feet
Systemic mycoses		
Cryptococcosis*	*Cryptococcus neoformans*	Lungs, meninges
Coccidioidomycosis*	*Coccidioides immitis*	Lungs
Histoplasmosis*	*Histoplasma capsulatum*	Lungs
Blastomycosis	*Blastomyces dermatitidis*	Lungs, skin
Candidiasis*	*Candida albicans*	Oral cavity, intestinal tract

Considered opportunistic pathogens and have been implicated in the pathogenesis of AIDS.

United States. The fungus lives in desert soils, and the spores are disseminated on dry, windblown particles that are inhaled. In some areas in the southwestern United States, as many as 80 percent of the inhabitants may be infected, although most individuals suffer no apparent ill effects.

A number of fungal infections, including histoplasmosis and coccidioidomycosis, are especially serious and common in individuals whose immune systems have been impaired, either by AIDS (see Section 15.7) or by immunosuppressive drugs. These are secondary fungal diseases, since normal individuals either do not get the disease or generally have a less severe form. Examples of other fungal organisms involved as secondary pathogens are given in Table 15.7. These fungi are known as *opportunistic* pathogens because of their particular ability to cause serious infections only in individuals with impaired defense mechanisms (in particular AIDS patients, see Section 15.7, the Definition of AIDS box, and Figure 15.30).

Effective chemotherapy against systemic fungal infections is very difficult. Most antibiotics that inhibit fungi also harm other eukaryotic organisms, including the human host. One of the most effective antibiotics, amphotericin B, is widely used to treat systemic fungal infections of humans, but serious side effects such as kidney toxicity may occur.

> Although fungal diseases are much less common than bacterial or viral diseases, they are often difficult to control because of a lack of suitable fungus-specific antibiotics. This emerges from the fact that fungi, like human cells, are eukaryotic. The incidence of fungal diseases seen as secondary infections is on the rise. Serious systemic infections due to the normally nonpathogenic *Candida albicans* and other fungi have been routinely seen in AIDS patients. The three classes of fungal infections (mycoses) include the superficial, the subcutaneous, and the systemic mycoses.

Study Questions

1. Give an explanation why many respiratory bacterial pathogens are Gram-positive.

2. Most sneezes generate an *aerosol* of mucus droplets containing microorganisms. Why are infections due to sneeze transmission more likely to be *upper* rather than *lower* respiratory tract infections?

3. Explain how *rheumatic fever* can be considered a type of *autoimmune* response.

4. The *DTP* and *MMR vaccines* are routinely administered to children at several times from infancy through early childhood (see Table 12.9). What is the composition of each of these so-called trivalent vaccines and to what diseases do they offer protection?

5. What are the typical symptoms of a staphylococcal skin infection and what properties of *Staphylococcus aureus* are responsible for these symptoms?

6. Describe the course of events leading to the disease toxic shock syndrome. How can this disease be prevented?

7. Why does the disease *tuberculosis* often lead to a permanent reduction in lung capacity while most other respiratory diseases cause only temporary respiratory problems?

8. From what has been described about the reservoir of *Legionella pneumophila*, explain why this disease shows an atypical disease cycle as compared with other respiratory diseases.

9. Why are colds and influenza such common respiratory diseases? Discuss at least five reasons why this might be so.

10. Discuss the molecular biology and immunological basis of *antigenic shift* and the reasons why this phenomenon effectively thwarts control measures for influenza. Why do cycles of influenza occur about every 3–5 years?

11. Compare antigenic *shift* to antigenic *drift*. Which mechanism is more important for the influenza virus? Which causes the greatest antigenic change? Which is most important for vaccine developers to be aware of?

12. Compare and contrast the diseases *measles, mumps,* and *rubella*. Include in your discussion a description of the pathogen, major symptoms encountered, and any potentially serious consequences of these infections. Why is it essential that females be vaccinated against rubella at an early age?

13. Discuss at least three reasons why the incidence of gonorrhea rose dramatically in the mid 1960s while that of syphilis did not.

14. Despite the ease with which gonorrhea can be diagnosed and cured, its incidence remains high. Why?

15. For the sexually transmitted diseases listed below, describe the methods for treating the infection. In each case, is the treatment an effective cure? Why or why not? Why is trichomoniasis not treatable with antibiotics such as penicillin and tetracycline?
    a. Gonorrhea
    b. Syphilis
    c. Chlamydial infections
    d. Genital herpes
    e. Trichomoniasis

16. Describe how HIV effectively shuts down most aspects of both humoral immunity and cell-mediated immunity. Are there any aspects of the immune system that remain functional in cases of AIDS?

17. What is the basis for treatment of AIDS? Why does it work? Is this treatment a cure? Why or why not?

18. Describe the sequence of events you would take if your child received an unprovoked bite from a stray dog.

19. Describe the various vaccine possibilities for AIDS. How would each vaccine actually confer immunity to the disease?

20. Discuss at least three common properties of the disease agents and/or the disease process for the following diseases: Rocky Mountain spotted fever, typhus fever, Q fever, Lyme disease.

21. Discuss the following as they relate to Lyme disease: causative agent, mode of transmission, symptoms, therapy, prevention, diagnosis.

22. The classic symptoms of malaria, fever followed by chills, are related to activities of the pathogen *Plasmodium vivax*. Describe the growth stages of this pathogen in the human host and relate these to the classic symptoms. Why might a person of Western European descent be more susceptible to malaria than a person of West African descent?

23. Contrast *food infection* and *food poisoning*. Give an example of each and indicate how each of these could be effectively controlled.

24. Write a scenario for a typical food poisoning outbreak due to *Staphylococcus*. (That is, what kind of conditions might lead to the development of the toxic food and how might the outbreak arise?)

25. Compare and contrast *perfringens* and *botulinum* food poisoning. Discuss similarities and differences between the organisms involved, the toxins produced, and the symptoms observed.

26. Compare and contrast the toxins responsible for cholera, botulism, and staphylococcal food poisoning. What is the mechanism of action of each toxin? Is *infection* with the causative organism necessary for the initiation of disease symptoms?

27. Why is the incidence of *Salmonella* infection higher in the human population in the summer than in the winter?

28. Why was the development of the coliform test necessary in order to perfect effective methods for drinking water purification?

29. Describe a typical water purification plant and indicate the role of the various stages in the process.

30. Describe the steps used to carry out a coliform test on a sample of drinking water. What are the acceptable limits for coliforms in drinking quality water?

31. What similarities does Lyme disease have with the diseases Rocky Mountain spotted fever and typhus? With syphilis?

32. Describe the modes of pathogenesis by fungi. With regard to infections, why are fungi often seen as secondary opportunistic pathogens in immunologically compromised individuals?

33. What are the three categories of fungal infections? Why are fungal infections particularly difficult to treat?

Supplementary Readings

Baron, S. 1992. *Medical Microbiology*. Addison-Wesley, Menlo Park, CA. An advanced level medical microbiology text; very complete and up to date.

Balows, A., W. J. Hausler, K. L. Hermann, H. D. Isenberg, and **H. J. Shadomy** (eds.) 1991. *Manual of Clinical Microbiology*, 5th edition. American Society for Microbiology, Washington, D.C. A comprehensive collection of information concerning isolation, culture, and diagnosis of pathogenic microorganisms. An excellent reference source.

Brock, T. D. (ed.) 1989. *Microorganisms: From Smallpox to Lyme Disease*. W. H. Freeman, New York. A collection of short articles on infectious diseases taken from the pages of Scientific American.

Finegold, S. M., and **W. L. George** (eds.) 1989. *Anaerobic Infections in Humans*. Academic Press, New York. Complete coverage of anaerobic pathogens.

Gallo, R. C., and **F. Wong-Staal.** 1990. *Retrovirus Biology and Human Disease*. Marcel Dekker, New York. An overview of retroviruses that infect humans and other animals, including a detailed treatment of the molecular biology of HIV and the pathogenesis of AIDS.

Holmes, K. K., P. Mardh, P. F. Sparling, and **P. J. Wiener** (eds.) 1990. *Sexually Transmitted Diseases*, 2nd Edition, McGraw-Hill, New York. A comprehensive, authoritative treatment of sexually transmitted diseases.

Joklik, W. K., H. P. Willett, D. B. Amos, and **C. M. Wilfert.** 1992. *Zinsser Microbiology*, 19th edition. Appleton-Century-Crofts, New York. A standard medical school text-book of microbiology. Quite detailed and of considerable reference value.

Massachusetts Medical Society. *Morbidity and Mortality Weekly Report*, Waltham, MA. The weekly report of the Centers for Disease Control. Regular issues and supplements are invaluable sources of information concerning infectious diseases in the United States.

McFeters, G. A. (ed.) 1990. *Drinking Water Microbiology*. Springer-Verlag, New York. A consideration of the problems and processes of potable water.

Murray, P. H., W. L. Drew, G. S. Kobayashi, and **J. H. Thompson**. 1990. *Medical Microbiology*. The C. V. Mosby Co., St. Louis, MO. An elementary textbook of medical microbiology.

Sauer, G. C. 1991. *Manual of Skin Diseases*, 6th edition. J. B. Lippincott Company, Philadelphia. An excellent source of information for fungal and bacterial diseases of the skin.

Schaechter, M., G. Medoff, and **B. Eisenstein.** 1993. *Mechanisms of Microbial Disease*, 2nd edition. Williams and Wilkins Co., Baltimore, MD. A medical school textbook that takes a case history approach to medical microbiology. Very readable.

Stine, G. J. 1993. *Acquired Immune Deficiency Syndrome*. Prentice Hall, Englewood Cliffs, NJ. An excellent monograph describing the current status of the AIDS epidemic.

World Health Organization. 1992. *Global Health Situation and Projections*. WHO, Geneva. A biennial survey of infectious diseases in the entire world. An invaluable source for global information.

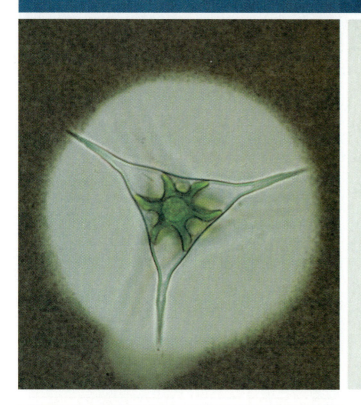

Metabolic Diversity among the Microorganisms

Up until now, we have discussed microorganisms that can be said to have a "conventional" metabolism. These are organisms that use organic compounds as energy sources, primarily via aerobic (respiratory) metabolism. In Chapter 4, we discussed briefly alternate modes of energy metabolism and indicated that there were interesting types of metabolism that we would return to later in this book. We discuss in this chapter some of these alternate microbial "life styles."

The organisms we are about to discuss are some of the most widespread and significant organisms on earth. They play major roles in the functioning of the whole biosphere and are of great importance in agriculture and other aspects of human affairs, as we will discuss in Chapter 17.

16.1 Energy Metabolism

Microorganisms show impressive diversity of metabolic processes. The discussion in this chapter will build on earlier discussions of microbial metabolism from Chapter 4. Figure 16.1 reviews the basic types of energy metabolism and lists some of the terms that have been used. As seen, energy metabolism can be divided into two broad categories, that in which the energy source is *chemical*, and that in which *light energy* is involved. Many microorganisms resemble animals and humans, using *organic* chemicals as energy sources. We call these organisms **chemoorganotrophs**. An interesting collection of microorganisms, exclusively prokaryotic, use *inorganic* chemicals as energy sources. These organisms are called **chemolithotrophs**.

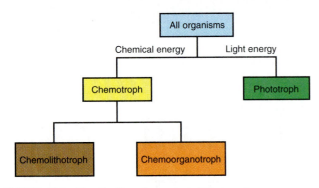

FIGURE 16.1 Classification of organisms in terms of energy.

Table 16.1 Reduction potentials of some redox pairs of biochemical and microbiological importance*

Redox pair	E_0' (V)
SO_4^{2-}/HSO_3^-	−0.52
$2H^+/H_2$	−0.41
$S_2O_3^{2-}/HS^- + HSO_3^-$	−0.40
Ferredoxin ox/red	−0.38
$NAD^+/NADH$ (or $NADP^+/NADPH$)	−0.32
Cytochrome c_3 ox/red	−0.29
CO_2/acetate	−0.29
S^0/HS^-	−0.27
CO_2/CH_4	−0.24
SO_4^{2-}/HS^-	−0.22
Pyruvate/lactate	−0.19
HSO_3^-/HS^-	−0.11
Fumarate/succinate	+0.033
Cytochrome b ox/red	+0.035
Ubiquinone ox/red	+0.115
Cytochrome c_1 ox/red	+0.23
NO_2^-/NO	+0.36
Cytochrome a_3 ox/red	+0.385
NO_3^-/NO_2^-	+0.43
Fe^{3+}/Fe^{2+}	+0.77
O_2/H_2O	+0.82
N_2O/N_2	+1.36

*See Appendix 1 for details.

A major class of living organisms, called **phototrophs**, use *light* as an energy source. Phototrophs include, of course, green plants, but also many microorganisms, both prokaryotic and eukaryotic. We will see that there are two distinct types of photosynthesis, that typified by the green plants but also found in both eukaryotic and certain prokaryotic microorganisms, and that found exclusively in special groups of bacteria, the purple and green bacteria.

Organisms that use inorganic chemicals or light as energy sources are frequently able to grow in the complete absence of organic materials, using *carbon dioxide* as their sole source of carbon. The term **autotroph** (meaning literally, *self-feeding*) is applied to organisms able to obtain all the carbon they need from inorganic sources. Note that autotrophy does not refer to the *energy* source used, but to the *carbon* source. Autotrophs are of great importance in the functioning of the biosphere because they are able to bring about the synthesis of organic matter from inorganic (nonliving) sources. Because humans and other animals require *organic* carbon, and are thus nutritionally called **heterotrophs**, the life of the biosphere itself is dependent upon the activities of autotrophic organisms.

It might be useful at this point to review briefly the concept of oxidation–reduction, especially the terms *oxidation–reduction, electron acceptor, electron donor,* and *electron transport.* We use these terms in this chapter and will assume that the student is familiar with them from Section 4.4. A summary of reduction potentials of some redox pairs important in the present chapter is given in Table 16.1 (full details can be found in Appendix 1).

In addition to alternate styles of energy metabolism, we also discuss in this chapter a number of specialized aspects of catabolism that we passed over lightly earlier. For instance, our discussion of respiratory metabolism centered primarily on O_2 as electron acceptor, yet there are a number of other electron acceptors that can participate in the electron transport process in particular groups of microorganisms. The process in which an electron acceptor other than molecular oxygen is used is called **anaerobic respiration** and will be one of the major features of metabolic diversity discussed here.

Moreover, our discussion of organic energy metabolism dealt primarily with the utilization of glucose and related sugars, but there are many other organic compounds that serve as energy sources for chemoorganotrophs. The utilization of these "unusual" organic compounds is not only of biochemical interest, but of applied interest as well, as we will see when we discuss petroleum microbiology, sewage treatment, and biodegradation in the next chapter.

Finally, although the major thrust of this chapter is carbon metabolism, our discussion would not be complete without a consideration of one of the important metabolic reactions of microorganisms that involves *nitrogen*. This is the utilization of nitrogen gas, N_2, as a source of nitrogen for biosynthesis, a process called **nitrogen fixation**. This important process contributes to the recycling of nitrogen into living matter and is important not only agriculturally but in the total function of the biosphere.

> Metabolic diversity is expressed most commonly in terms of energy metabolism. Depending on the electron donors and electron acceptors used, an organism can be categorized as a phototroph, chemolithotroph, or chemoorganotroph. Organisms capable of obtaining all of their carbon for cell synthesis from CO_2 are called autotrophs; organisms requiring organic carbon as a carbon source are called heterotrophs.

16.2 Photosynthesis

One of the most important biological processes on earth is photosynthesis, the conversion of light energy into chemical energy. Most phototrophic organisms

Miniglossary for Chapter 16

ANAEROBIC RESPIRATION   respiration in which some substance such as SO_4^{2-} or NO_3^- serves as terminal electron acceptor instead of O_2

ANOXYGENIC PHOTOSYNTHESIS   photosynthesis in which O_2 is not produced

BACTERIOCHLOROPHYLL   the chlorophyll pigment of anoxygenic phototrophs

CALVIN CYCLE   the biochemical route of CO_2 fixation in many autotrophic organisms

CAROTENOIDS   hydrophobic accessory pigments present along with chlorophyll in photosynthetic membranes

CHEMOLITHOTROPH   microorganisms capable of oxidizing inorganic compounds as energy sources

CHLOROPHYLL   the light-sensitive Mg-containing porphyrin of photosynthetic organisms which initiates the process of photophosphorylation

DENITRIFICATION   anaerobic respiration in which NO_3^- is reduced to gaseous nitrogen compounds, primarily N_2

DISPROPORTIONATION   splitting of a compound into two new compounds, one more oxidized and one more reduced, than the original compound

FERMENTATION   anaerobic catabolism of an organic compound in which the compound serves as both an electron donor and an electron acceptor and in which ATP is produced by substrate-level phosphorylation

HYDROGENASE   an enzyme, widely distributed in anaerobic microorganisms, capable of taking up or evolving H_2

METHANOGENESIS   the biological production of methane (CH_4)

MIXOTROPHIC   a nutritional state in which an inorganic compound serves as energy source (electron donor) while organic compounds serve as carbon source

MONOOXYGENASE   an enzyme which catalyzes the incorporation of one atom of O_2 into a substrate while the other atom is reduced to H_2O

NITRIFICATION   the microbial conversion of NH_3 to NO_3^-

NITROGENASE   an enzyme capable of reducing N_2 to NH_3 in the process of nitrogen fixation

NITROGEN FIXATION   the biological reduction of N_2 to NH_3 by nitrogenase

OXYGENIC PHOTOSYNTHESIS   photosynthesis carried out by cyanobacteria and green plants in which O_2 is evolved

PHOTOPHOSPHORYLATION   the production of ATP in photosynthesis

PHOTOTROPH   an organism capable of using light as an energy source

REACTION CENTER   a photosynthetic complex containing chlorophyll (or bacteriochlorophyll) and several other components within which occurs the initial electron transfer reactions of photophosphorylation

RECALCITRANT   resistant to microbial attack

REVERSED ELECTRON TRANSPORT   the energy-dependent movement of electrons against the thermodynamic gradient to form a strong reductant from a weaker electron donor

SYNTROPHY   a process whereby two or more microorganisms cooperate to degrade a substance neither can degrade alone

XENOBIOTIC   a man-made compound previously not present on earth

---

are autotrophs, capable of growing on CO_2 as sole carbon source. Energy from light is thus used in the *reduction* of CO_2 to organic compounds. The ability to photosynthesize is dependent on the presence of special light-sensitive pigments, the *chlorophylls*, found in plants, algae, and some bacteria. Light reaches phototrophic organisms in distinct units called *quanta*. Absorption of light quanta by chlorophyll pigments begins the process of photosynthetic energy conversion.

The growth of a phototrophic autotroph can be characterized by two distinct sets of reactions: the **light reactions**, in which light energy is converted into chemical energy, and the **dark reactions**, in which this chemical energy is used to reduce CO_2 to organic compounds. For autotrophic growth, energy is supplied in the form of ATP, while electrons for the reduction of CO_2 come from NADPH. The latter is produced by the reduction of $NADP^+$ by electrons originating from various electron donors to be discussed below.

The light reactions bring about the conversion of *light* energy into *chemical* energy in the form of ATP. Purple and green bacteria use light primarily to form ATP; they produce NADPH from reducing materials present in their environment, such as H_2S or organic compounds. Green plants, algae, and cyanobacteria, however, do not generally use H_2S or organic compounds to obtain reducing power. Instead, they obtain electrons for $NADP^+$ reduction by splitting water molecules, producing O_2 as a byproduct (Figure 16.2). The

reduction of $NADP^+$ to NADPH by these organisms is therefore a *light-mediated* event. Because molecular oxygen, O_2, is produced, the process of photosynthesis in these organisms is called **oxygenic** photosynthesis. In contrast, the purple and green bacteria do not produce oxygen; their process is called **anoxygenic** photosynthesis.

> The process of photosynthesis can be considered in terms of two separate sets of reactions, the light reactions, in which light energy is converted into chemical energy, and the dark reactions, in which this chemical energy is used for CO_2 fixation. Two groups of phototrophs are known, oxygenic and anoxygenic.

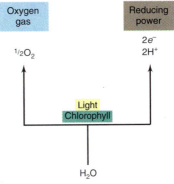

FIGURE 16.2 Production of O_2 and reducing power from H_2O by oxygenic phototrophs.

16.3   Role of Chlorophyll and Bacteriochlorophyll in Photosynthesis

Photosynthesis occurs only in organisms that possess some type of **chlorophyll**. Chlorophyll is a porphyrin, as are the cytochromes, but chlorophyll contains a *magnesium* atom instead of an iron atom at the center of the porphyrin ring. Chlorophyll also contains specific substituents bonded to the porphyrin ring as well as a long hydrophobic alcohol molecule. Because of this alcohol side chain, chlorophyll associates with lipid and hydrophobic proteins of photosynthetic membranes.

The structure of chlorophyll *a*, the principal chlorophyll of higher plants, most algae and of the cyanobacteria, is shown in Figure 16.3. Chlorophyll *a* is green in color because it *absorbs* red and blue light preferentially and *transmits* green light. We discussed the electromagnetic spectrum in Section 7.3 (see Figure 7.6). The spectral properties of any pigment can best be expressed by its *absorption spectrum*, which indicates the

degree to which the pigment absorbs light of different wavelengths. The absorption spectrum of cells containing chlorophyll *a* shows strong absorption of red light (maximum absorption at a wavelength of 680 nm) and blue light (maximum at 430 nm) (Figure 16.4a).

There are a number of chemically different chlorophylls that are distinguished by their different absorption spectra. Chlorophyll *b*, for instance, absorbs maximally at 660 nm rather than at 680 nm. Many plants have more than one chlorophyll, but the most common are chlorophylls *a* and *b*. Among prokaryotes, the cyanobacteria have chlorophyll *a*, but the purple and green bacteria have chlorophylls of slightly different structure, called **bacteriochlorophyll**. Bacteriochlorophyll *a* (Figure 16.3) from purple phototrophic bacteria absorbs maximally around 800 and 850–880 nm (Figure 16.4b); other bacteriochlorophylls absorb maximally at 720–790 nm, and one type of bacteriochlorophyll absorbs at 1020 nm (see Figure 19.1).

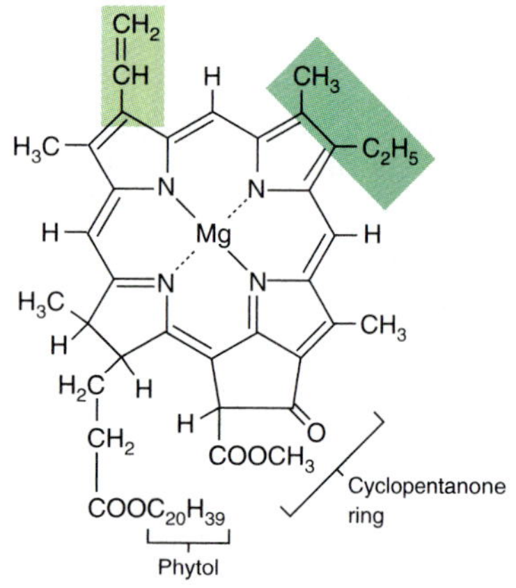

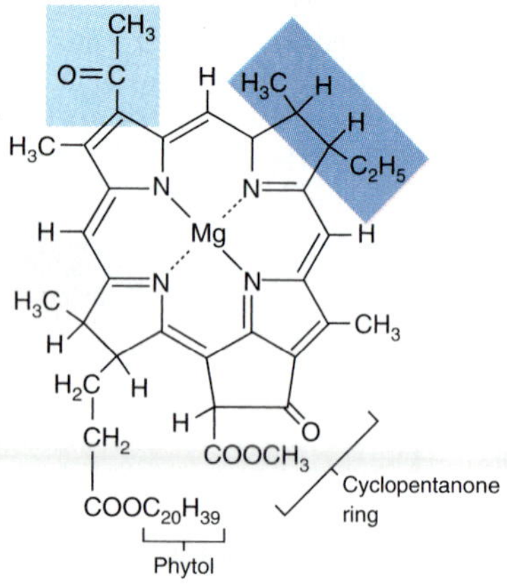

FIGURE 16.3 Structures of chlorophyll *a* and bacteriochlorophyll *a*. The two molecules are almost identical except for those portions shown in different colors.

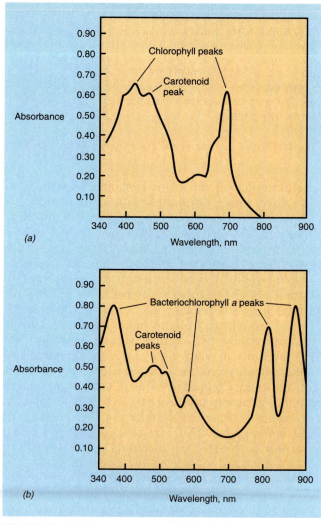

FIGURE 16.4 (a) Absorption spectrum of cells of the green alga *Chlamydomonas*. The peaks at 680 and 440 nm are due to chlorophyll *a*; the peak at 480 is due to carotenoids. (b) Absorption spectrum of cells of the phototrophic purple bacterium *Rhodopseudomonas palustris*. Peaks at 870, 800, 590, and 360 nm are due to bacteriochlorophyll *a* (870 = light harvesting I; 800 = light harvesting II); peaks at 525 and 475 are due to carotenoids.

Why do organisms have several kinds of chlorophylls absorbing light at different wavelengths? One reason appears to be to make it possible to use more of the energy of the electromagnetic spectrum. Only light energy that is *absorbed* will be used biologically, so that by having more than one chlorophyll, more of the incident light energy becomes available to the organism. By having different pigments, two unrelated organisms can coexist in a habitat, each using wavelengths of light that the other is not using. Thus, pigment diversity has ecological significance.

Photosynthetic membranes

Just where are the chlorophyll pigments located inside the cell? These pigments, and all the other components of the light-gathering apparatus, are associated with special membrane systems, the **photosynthetic membranes**. The location of the photosynthetic membranes within the cell differs between prokaryotic and eukaryotic microorganisms. In eukaryotes, photosynthesis is associated with special intracellular organelles, the **chloroplasts** (see Section 3.14 and Figures 3.71, 3.72, and 16.5*a*). The chlorophyll pigments are attached to sheetlike (lamellar) membrane structures of the chloroplast (Figure 16.5*b*). These photosynthetic membrane systems are called **thylakoids**; stacks of thylakoids are called *grana* (Figure 16.5*b*). The thylakoids are so arranged that the chloroplast is divided into two regions, the matrix space, which surrounds the thylakoids, and the inner space within the thylakoid array. This arrangement makes possible the development of a light-driven pH gradient and proton motive force which can be used to synthesize ATP, as will be described in the next section.

Within the thylakoid membrane, the chlorophyll molecules are associated in complexes consisting of about 200 to 300 molecules. Only relatively few of these chlorophyll molecules participate *directly* in the conversion of light energy to ATP. These special chlorophyll molecules are referred to as **reaction center chlorophyll** and receive energy by transfer from

the more numerous **light-harvesting** (or **antenna**) chlorophyll molecules (Figure 16.6). The chlorophyll molecules are bound to proteins which precisely control their orientation in the membrane so that energy absorbed by one chlorophyll molecule can be efficiently transferred to another.

In prokaryotes, chloroplasts are not present and the photosynthetic pigments are integrated into extensive internal membrane systems that arise from invagination of the cytoplasmic membrane; a single reaction center contains 25–30 light-harvesting bacteriochlorophyll molecules.

Antenna chlorophyll molecules make possible a dramatic increase in the rate at which photosynthesis

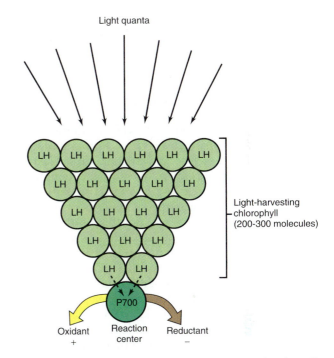

FIGURE 16.6 The photosynthetic unit and its associated reaction center in oxygenic phototrophs. Light energy, absorbed by light-harvesting chlorophyll molecules, travels to the reaction center where the actual ejection of an electron occurs, generating a charge separation.

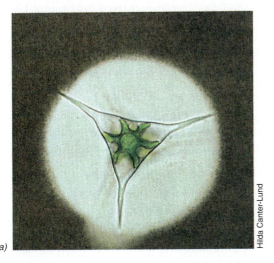

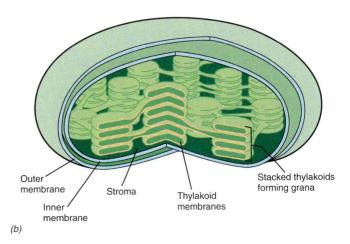

FIGURE 16.5 The chloroplast. (a) Photomicrograph of an algal cell showing chloroplasts. (b) Details of chloroplast structure, showing how the convolutions of the thylakoid membranes define an inner space called the stroma, and form membrane stacks called grana.

can be carried out. At the low light intensities often prevailing in nature, reaction centers can only be excited about once per second, which would not be sufficient to carry out a significant photosynthetic process. The additional antenna chlorophyll molecules permit collection of light energy at a much more rapid rate. Since reaction-center chlorophyll absorbs light energy only over a very narrow range of the spectrum, antenna pigments also perform the additional function of spreading out the spectral range available for use.

> **The central light-harvesting pigment of photosynthesis is chlorophyll (or bacteriochlorophyll in anoxygenic photosynthesis). Chlorophylls are located in special membrane systems called photosynthetic membranes, where the light reactions of photosynthesis are carried out. Most chlorophyll molecules serve only to harvest light energy and transfer it on to special molecules called reaction-center chlorophylls.**

16.4   Anoxygenic Photosynthesis

The process of light-mediated ATP synthesis in all phototrophic organisms involves electron transport through a sequence of electron carriers. These electron carriers are arranged in the photosynthetic membrane in series from those with negative to those with more positive potentials. Conceptually, the process of photosynthetic electron flow resembles that of respiratory electron flow. As a matter of fact, in phototrophic bacteria capable of aerobic (dark) growth, many of the same electron transport components are present in the membranes of cells grown in either the light (anaerobic) or dark (aerobic). We now consider the structure of the photosynthetic apparatus in anoxygenic phototrophs and the details of photosynthetic electron flow in purple bacteria, where much is known concerning the molecular events of photosynthesis.

Structure of the bacterial photosynthetic apparatus

The photosynthetic apparatus of purple phototrophic bacteria is contained in intracytoplasmic membrane systems of various morphologies. Membrane vesicles (chromatophores) (Figure 16.7) are a commonly observed membrane type. The photosynthetic apparatus consist of four membrane-bound pigment-protein complexes plus an ATPase complex which allows ATP synthesis at the expense of a proton gradient (see Section 4.11). Three of the four complexes specific to photosynthesis are the reaction center, light-harvesting I, and light-harvesting II components. The fourth complex of the photosynthetic apparatus, the cytochrome bc_1 complex, is common to both respiratory and photosynthetic electron flow. We discussed the structure and function of the cytochrome bc_1 complex in Section 4.11. In purple bacteria, the light-harvesting complexes (or *antenna complexes*, as they are often called) contain two distinct spectral forms of bacteriochlorophyll *a* called B870 (light-harvesting I) and B800–850 (light-harvesting II) (see Figure 16.4*b*). The

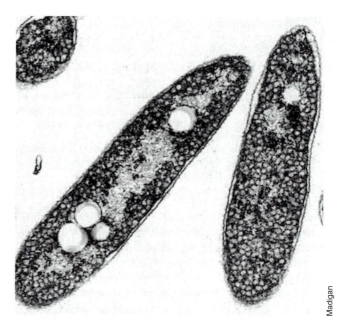

FIGURE 16.7 Chromatophores. Section through a cell of the phototrophic purple bacterium *Rhodobacter capsulatus*, containing an abundance of vesicular photosynthetic membranes called chromatophores. The circular membranes arise by invagination of the cytoplasmic membrane. The clear areas are regions in the cell in which the reserve polymer, poly-β-hydroxybutyrate was stored.

numbers refer to the wavelengths of radiation that are most strongly absorbed by these forms of bacteriochlorophyll in the cell. As previously described for oxygenic phototrophs (see Figure 16.6), the function of the antenna complexes is to absorb radiation and funnel the energy to the reaction center.

The purple bacterial photosynthetic reaction center has been crystallized and its structure determined to atomic resolution by X-ray diffraction (Figure 16.8*a*). Reaction centers of purple bacteria contain three polypeptides, designated the L, M, and H subunits. These proteins are firmly embedded in the photosynthetic membrane and traverse the membrane several times (Figure 16.8*b*). The L, M, and H polypeptides bind the reaction center photochemical complex, which consists of two molecules of bacteriochlorophyll *a* called the *special pair*, two additional bacteriochlorophyll *a* molecules whose function is unknown, two molecules of *bacteriopheophytin* (bacteriochlorophyll *a* minus its magnesium atom), two molecules of quinone, and two molecules of a carotenoid pigment (accessory photosynthetic pigment) (see Section 16.6) (Figure 16.8). All components of the reaction center are integrated in such a way that they can interact in very fast electron transfer reactions which, as we will see, ultimately results in ATP production.

Photosynthetic electron flow

It should be recalled that the photosynthetic reaction center is surrounded by light-harvesting antenna bacteriochlorophyll *a* molecules which function to funnel light energy to the reaction center (see above and Figure 16.8). Light energy is transferred from the antenna to the reaction center in packets called *excitons*, mobile electronic singlet states which migrate through the an-

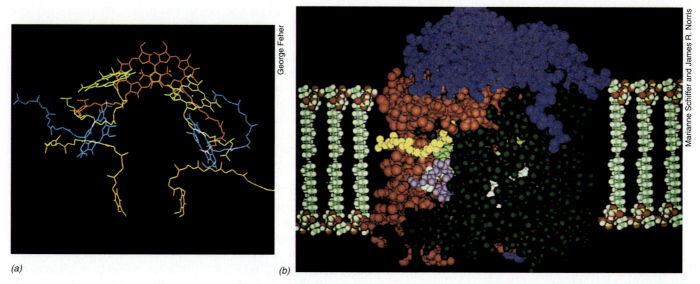

FIGURE 16.8 Structure of the reaction center of purple phototrophic bacteria. (a) Arrangement of components in the reaction center. The "special pair" of bacteriochlorophyll molecules are shown in red and molecules of quinone are in dark yellow. (b) Molecular model of the protein structure of the reaction center. The pigments discussed in (a) are bound to membranes by three reaction center proteins called protein H (blue), protein M (red), and protein L (green). The reaction center pigment–protein complex is integrated into the lipid bilayer.

tenna to the reaction center at high efficiency. Photosynthesis begins when exciton energy strikes the special pair bacteriochlorophyll *a* molecules (Figure 16.8*a*). The absorption of energy excites the special pair, converting it into a *strong reductant*, sufficiently strong to reduce an acceptor molecule of very low potential. This represents work done on the system by light energy.

Before excitation, the bacterial reaction center has an E_0' of about +0.5 volts; after excitation it has a potential of about −0.7 volts. The excited electron within the special pair proceeds to reduce a molecule of bacteriopheophytin within the reaction center (Figures 16.8*b* and 16.9). This transition occurs incredibly fast, taking about 3 trillionths of a second (3×10^{-12} seconds) to occur. Once reduced, bacteriopheophytin *a* reduces a

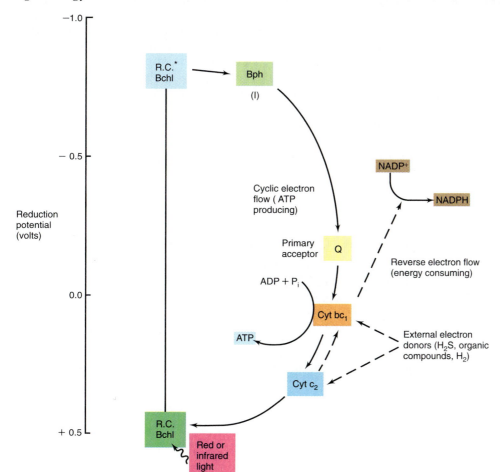

FIGURE 16.9

General scheme of electron flow in anoxygenic (bacterial) photosynthesis. Only a single light reaction occurs. RC, reaction center; Bchl, bacteriochlorophyll; Bph, bacteriopheophytin (intermediary carrier, I); Q, quinone; Cyt, cytochrome.

quinone molecule, which is part of the reaction center but is actually nearer the *outer* surface of the photosynthetic membrane. This transition is also very fast, taking less than 1 billionth of a second. The quinone is referred to as the *primary electron acceptor* (Figures 16.9 and 16.10). Relative to what has happened in the reaction center, further electron transport reactions occur rather slowly, on the order of microseconds to milliseconds. From the quinone, electrons are transported in the membrane through a series of iron-sulfur proteins and cytochromes (Figures 16.9 and 16.10), eventually returning to the reaction center. Key electron transport proteins include cytochrome bc_1 and cytochrome c_2 (Figure 16.9). Cytochrome c_2 is a periplasmic cytochrome and serves as an electron shuttle between the membrane-bound bc_1 complex and the reaction center (Figures 16.9 and 16.10).

Photophosphorylation

Synthesis of ATP during photosynthetic electron flow occurs as a result of the formation of a proton gradient generated by proton extrusion during electron transport and the activity of ATPases in coupling the dissipation of the proton gradient to ATP formation (see Section 4.11 for a discussion of proton-motive force). The reaction series is completed when cytochrome c_2 returns the electron to the special pair bacteriochlorophylls, returning these molecules to their original ground state potential (E_0' +0.5 volts). The reaction center is then capable of absorbing new energy and repeating the process. This method of making ATP is called **cyclic photophosphorylation**, since electrons are repeatedly moved around a closed circle; in cyclic photophosphorylation *there is no net input or consumption of electrons* as in respiration.

It is extremely important that the light-driven redox reaction just described takes place *across* the photosynthetic membrane. The spatial relationships of the electron transport components in the bacterial photosynthetic membrane are illustrated in Figure 16.10. Note that as in respiratory electron flow (see

Section 4.11), the cytochrome bc_1 complex interacts with the quinone pool during photosynthetic electron flow to allow functioning of the Q cycle (see Figure 4.18), which is a major means of generating the proton gradient across the membrane used to drive ATP synthesis (Figure 16.10). Thus, the major function of light in photosynthesis in purple bacteria is to generate a strong reductant that can transfer electrons through the reaction center to quinones; the remaining reactions leading to ATP synthesis have a strong parallel in respiratory electron flow.

Genetics of bacterial photosynthesis

Purple phototrophic bacteria are Gram-negative prokaryotes (see Section 19.1), and certain species are readily amenable to genetic manipulation. Species of the genus *Rhodobacter*, especially *R. capsulatus* and *R. sphaeroides*, have been the main subjects of genetic research in bacterial photosynthesis. In *R. capsulatus*, all genes involved in photosynthesis are clustered into several operons that span a 45–50 kb region of the chromosome that has been called the **photosynthetic gene cluster** (Figure 16.11). Genes in the photosynthetic gene cluster encode proteins involved in bacteriochlorophyll biosynthesis (*bch* genes) and in carotenoid biosynthesis (*crt* genes) and polypeptides that bind pigment molecules in the reaction center and light-harvesting complexes (*puf* and *puh* genes) (Figure 16.11).

As can be imagined, efficient synthesis of bacteriochlorophyll, carotenoids, and specific pigment-binding proteins in phototrophic bacteria must be a highly coordinated process. When new photosynthetic complexes are synthesized, the correct proportions of each of the components of the complex must be available within the cell for final assembly. Biochemical and genetic analyses of photosynthesis in *R. capsulatus* have shown that coordinate expression of photosynthetic components indeed does occur because the operons are arranged to form **superoperons**. Instead of terminating at the end of one operon, transcripts of pigment biosyn-

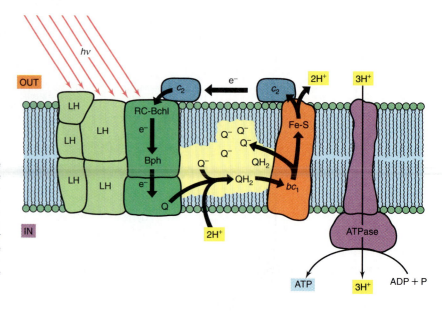

FIGURE 16.10
Electron flow leading to proton translocation in the membrane of a purple phototrophic bacterium. The light-generated proton gradient is used in the synthesis of ATP by the ATP synthase (ATPase). Abbreviations: LH, light-harvesting bacteriochlorophyll complexes; RC, reaction center; Bchl, bacteriochlorophyll; Bph, bacteriopheophytin; Q, quinone; FeS, iron–sulfur protein; bc_1, cytochrome bc_1 complex; c_2, cytochrome c_2. For description of the functioning of the cytochrome bc_1 complex and the Q cycle, see Figure 4.18.

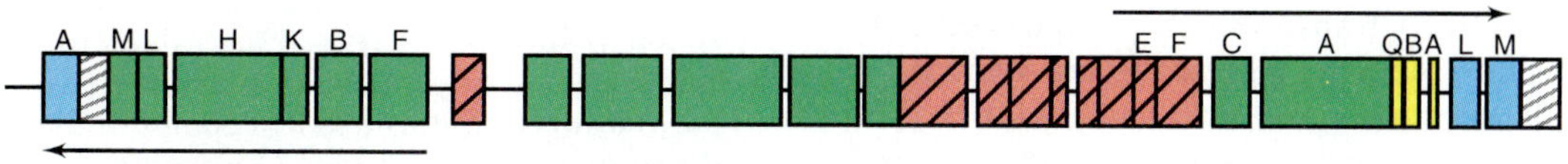

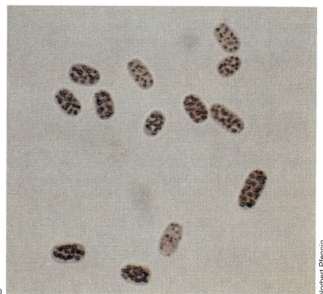

FIGURE 16.11 Map of the photosynthetic gene cluster of *Rhodobacter capsulatus*. Genes are arranged in superoperons where transcripts of pigment biosynthesis operons extend through to include transcription of polypeptides of the photosynthetic complexes. The *bch* genes, which code for bacteriochlorophyll synthesis proteins, are shown in green, while *crt* genes, which code for proteins that synthesize carotenoids, are shown in red. Genes encoding reaction center polypeptides (*puh* and *puf* genes) are shown in blue, and genes encoding light-harvesting I polypeptides (B870 complex) (*puf* genes) are shown in yellow. Genes shown by diagonal lines are of unknown function. Not all genes have been given letter designations. Arrows indicate direction of transcription.

thesis operons (*bch* and *crt*) (Figure 16.11) extend through the promoters and structural genes encoding the polypeptides of the photosynthetic complexes, yielding large transcripts coding for many proteins. Photosynthesis superoperons thus allow for transcription of many functionally related genes whose products interact and form the photosynthetic complexes that eventually integrate into the membrane. The master regulatory signal governing transcription of the photosynthetic gene cluster is O_2. Molecular oxygen represses pigment synthesis such that photosynthesis in anoxygenic phototrophs only occurs under *anaerobic* conditions. Oxygen presumably interacts with a regulatory protein that prevents transcription of the photosynthetic gene cluster until conditions become anoxic.

Anoxygenic photosynthesis shares a number of important features with the oxygenic process of green plants, and genetic analysis of purple bacteria has thus contributed an enormous amount to our understanding of basic photosynthetic processes, especially as regards light-mediated ATP production. The remarkable nutritional versatility of purple bacteria (see Section 19.1) has made it possible to isolate mutants unable to photosynthesize (such mutants can still grow by respiration or fermentation in darkness). Nonphotosynthetic mutants of *R. capsulatus* and *R. sphaeroides* in particular have been used to identify and genetically manipulate the key genes involved in photosynthesis (Figure 16.11).

Autotrophy and anoxygenic photosynthesis

As we have noted, purple and green bacteria do *not* produce O_2 during photosynthesis. Their photosynthesis is thus said to be *anoxygenic*. The reactions described above have led to the conversion of light energy into high-energy phosphate bonds of ATP. However, if an anoxygenic phototroph is going to grow with CO_2 as its sole or major carbon source, formation of ATP is not enough. Reducing power (NADPH) must also be made so that CO_2 can be reduced to the level of cell material. The source of electrons for anoxygenic phototrophs is some reduced substance from the environment; unlike oxygenic phototrophs, the source of reducing power in anoxygenic phototrophs is *not* water. Examples of electron donors used in anoxygenic photosynthesis are reduced sulfur compounds (for example, H_2S, elemental sulfur, thiosulfate), H_2, or organic compounds (for example, succinate, malate, butyrate). A few purple phototrophic bacteria can even use ferrous iron (Fe^{2+}) as photosynthetic electron donor, producing Fe^{3+} as the oxidized product. During autotrophic growth, anoxygenic phototrophs oxidize the external electron donor and reduce $NADP^+$ to NADPH. For instance, during growth on H_2S, elemental sulfur is produced: $6CO_2 + 12H_2S \rightarrow C_6H_{12}O_6 + 6H_2O + 12S^0$. The sulfur formed is deposited either outside the cells in phototrophic green bacteria (Figure 16.12*a*) or inside the cells in some of the purple bacteria (Figure 16.12*b*).

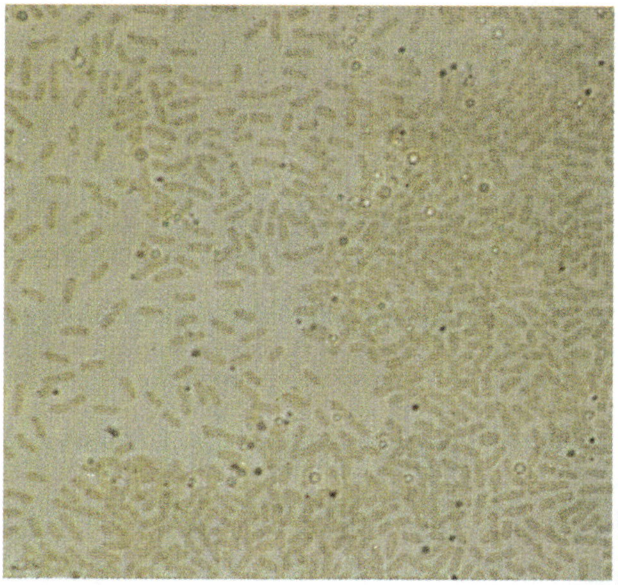

FIGURE 16.12 Photomicrographs of phototrophic bacteria taken by bright field microscopy. (a) Green bacterium: *Chlorobium limicola*. The refractile bodies are sulfur granules deposited *outside* the cell. (b) Purple bacterium: *Chromatium okenii*. Notice the sulfur granules deposited *inside* the cell.

(a)

(b)

Norbert Pfennig

Norbert Pfennig

How are electrons transferred from reduced substances to NADP+? At least two ways are known for using reduced substances to produce NADPH. The first involves direct transfer from a highly reduced compound to NADP+. The best example of an electron donor capable of direct transfer to NADP+ is H_2. The E_0' of H_2 is sufficiently low (-0.42 volts) to reduce NADP+ (-0.32 volts) directly, provided the organism has the enzyme *hydrogenase* to incorporate H_2 into the cell. However, many anoxygenic phototrophs can grow autotrophically using electron donors such as sulfide or thiosulfate, whose reduction potential is *more positive* than that of the NADP+/NADPH couple. Under these conditions, electrons usually enter the photosynthetic electron transport chain at the level of cytochromes and are forced backward (against the electropotential gradient) by the membrane potential, eventually reducing NADP+ to NADPH. This energy-consuming process, called **reversed electron transport**, is necessitated by the need for a low-potential reductant for CO_2 fixation, coupled with the reality of having to use electron donors of higher reduction potential (Figure 16.9). It is because electrons travel in a thermodynamically *unfavorable* direction during reverse electron flow that energy, in this case the membrane potential, is required to form NADPH from NADP+. Reversed electron transport is also a key feature of the synthesis of reducing power in chemolithotrophic prokaryotes (see Section 16.13).

A brief overview of the characteristics of the purple and green bacteria is given in Table 16.2. Many of the purple and green bacteria are strict anaerobes, but one subgroup each of purple and green bacteria is able to grow aerobically in the dark as chemoorganotrophs. Purple bacteria are nutritionally diverse and use a variety of organic compounds as carbon sources as well as remaining capable of autotrophic growth when necessary. The term *photoheterotroph* is frequently used to describe the nutritional state of the organism when phototrophic growth on organic compounds is occurring. Similarly, the term *photoautotroph* is used to describe growth phototrophically on CO_2 as sole carbon source.

Table 16.2   Characteristics of the purple and green bacteria

Purple bacteria	Green bacteria
Bacteriochlorophyll *a* or *b*	Bacteriochlorophyll *c, d,* or *e* (plus small amounts of bacteriochlorophyll *a*)
Purple- or red-colored carotenoids	Yellow or brown-colored carotenoids
Photosynthetic membrane system contains photosynthetic complexes	Majority of bacteriochlorophyll in chlorosomes; only bacteriochlorophyll *a* in cytoplasmic membrane

> Anoxygenic photosynthesis is carried out by the purple and green phototrophic bacteria, which use organic compounds or reduced sulfur compounds as photosynthetic electron donors for CO_2 fixation. A complex series of electron transport reactions occur in the photosynthetic reaction center which results in the formation of a proton motive force and the synthesis of ATP. Energy transfer from light-harvesting complexes to the reaction center ensures that sufficient energy is obtained from the available light. The reducing power for CO_2 fixation comes either from chemical reductants present in the environment or from the process of reversed electron transport, in which the energy of ATP is used to reduce NADP+ to NADPH using as reductant substances with E_0's higher than that of NADPH.

16.5   Oxygenic Photosynthesis

Electron flow in oxygenic phototrophs involves two distinct, but interconnected, photochemical reactions. Oxygenic phototrophs use light to generate both ATP *and* NADPH, the electrons for the latter arising from the splitting of *water* into oxygen and electrons (see Figure 16.2). The two systems of light reactions are called *photosystem I* and *photosystem II*, each photosystem having a spectrally distinct form of reaction center chlorophyll *a*. Photosystem I chlorophyll, called P700, absorbs light best at long wavelengths (far red light), whereas photosystem II chlorophyll, called P680, absorbs best at shorter wavelengths (near red light). Like anoxygenic photosynthesis, oxygenic photochemical reactions occur in membranes. In eukaryotic cells, these membranes are found in the *chloroplast*, while in cyanobacteria, photosynthetic membranes are arranged in stacks within the cytoplasm. In both groups of phototrophs, the two forms of chlorophyll *a* are attached to specific proteins in the membrane and interact as shown in Figure 16.13.

The path of electron flow in oxygenic phototrophs roughly resembles the letter Z turned on its side, and scientists studying oxygenic photosynthesis have come to refer to the electron flow of oxygenic phototrophs as the "Z" scheme. We should first note that the reduction potential of the P680 chlorophyll *a* molecule in photosystem II is very high, slightly higher than that of the O_2/H_2O couple (see Table 16.1). This is because the first step in oxygenic electron flow is the *splitting* of water into oxidizing and reducing equivalents (Figure 16.13), a thermodynamically unfavorable reaction. An electron from water is donated to the P680 molecule following the absorption of a quantum of light near 680 nm. Light energy converts P680 into a moderately strong reductant, capable of reducing an intermediary molecule of E_0' about -0.2 volts. The nature of this molecule is unknown, but it may be a pheophytin *a* molecule (chlorophyll *a* without the magnesium atom). From here the electron travels through several membrane carriers including quinones, cytochromes, and a copper-containing protein called **plastocyanin**; the latter donates electrons to photosystem I. The electron is accepted by the reaction

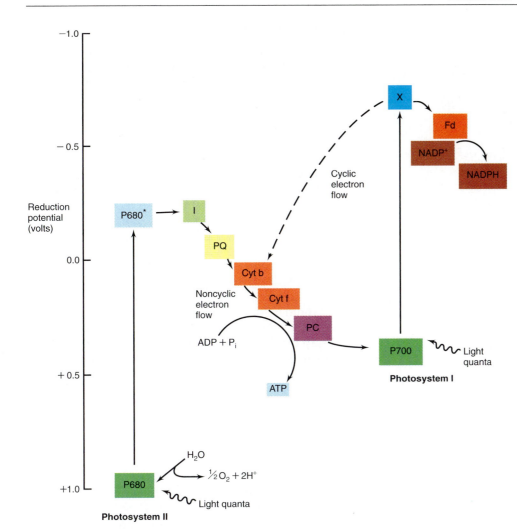

FIGURE 16.13
Electron flow in oxygenic (green plant) photosynthesis, the "Z" scheme. Two photosystems (PS) are involved, PSI and PSII. I, unidentified intermediary acceptor of PSII; PQ, plastoquinone; Cyt, cytochrome; PC, plastocyanin; X, unidentified acceptor of electrons in PSI; Fd, ferredoxin. P680 and P700 are the reaction center chlorophylls of PSII and PSI, respectively.

center chlorophyll of photosystem I, P700, which has previously absorbed light quanta and donated electrons to the primary acceptor of photosystem I; this acceptor has a very negative potential, E_0' of about -0.71 volts. As in photosystem II, the primary acceptor of electrons from photosystem I has not been positively identified, but is thought to be a free radical form of chlorophyll a. At any rate, the acceptor in photosystem I, once reduced, is at a reduction potential sufficiently negative to reduce the iron-sulfur protein **ferredoxin**, which then reduces $NADP^+$ to NADPH (Figure 16.13).

ATP synthesis in oxygenic photosynthesis

Besides the net synthesis of reducing power (i.e., NADPH), other important events occur while electrons flow in the membrane from one photosystem to another. During transfer of an electron from the acceptor in photosystem II to the reaction center chlorophyll molecule in photosystem I, electron transport occurs in a thermodynamically favorable (negative to positive) direction. This generates a membrane potential (a proton gradient) from which ATP can be produced. This type of ATP generation has been called *noncyclic photophosphorylation* because the electron travels a direct route from water to $NADP^+$. (Contrast cyclic photophosphorylation shown in Figure 16.9 with noncyclic photophosphorylation shown in Figure 16.13.) When sufficient reducing power is present, ATP can

also be produced in oxygenic phototrophs by *cyclic photophosphorylation* involving only photosystem I (Figure 16.13). This occurs when the primary acceptor of photosystem I, instead of reducing ferredoxin (and hence $NADP^+$), returns the electron to the P700 molecule via membrane-bound cytochromes b and f. This flow creates a membrane potential and synthesis of additional ATP (see dashed line in Figure 16.13).

Anoxygenic photosynthesis in oxygenic phototrophs

Photosystems I and II normally function together in the oxygenic process. However, under certain conditions many algae and some cyanobacteria are able to carry out cyclic photophosphorylation using only photosystem I, obtaining reducing power from sources other than water, in effect photosynthesizing *anoxygenically* as do purple and green bacteria. This alteration requires the presence of anaerobic conditions as well as a reducing substance, such as H_2 or H_2S. Under these conditions the electrons for CO_2 reduction come not from water but from the reducing substance. In algae, H_2 is generally the reductant, and following a period of adaptation to anaerobic conditions, the enzyme hydrogenase is made and is used to assimilate H_2 which reduces $NADP^+$ to NADPH directly.

A number of cyanobacteria can use H_2S as an electron donor for anoxygenic photosynthesis. When H_2S

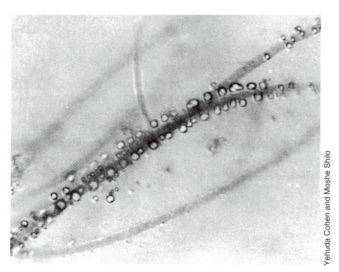

Yehuda Cohen and Moshe Shilo

FIGURE 16.14 Cells of the filamentous cyanobacterium *Oscillatoria limnetica* grown anaerobically on sulfide as photosynthetic electron donor. Note the globules of elemental sulfur, the oxidation product of sulfide, formed *outside* the cells.

is used, it is oxidized to elemental sulfur (S^0), and sulfur granules are deposited outside the cells similar to those produced by green sulfur bacteria (see Figure 16.12a). The filamentous cyanobacterium *Oscillatoria limnetica* has been found in sulfide-rich saline ponds where it carries out anoxygenic photosynthesis along with photosynthetic green and purple bacteria and produces sulfur as an oxidation product of sulfide (Figure 16.14). In cultures of *O. limnetica*, electron flow from photosystem II is strongly inhibited by H_2S, thus necessitating anoxygenic photosynthesis if the organism is to survive in its sulfide-rich environment.

From an evolutionary point of view, the existence of cyclic photophosphorylation indicates a close relationship between green-plant and bacterial photosynthesis. Although organisms such as *Oscillatoria limnetica* that carry out oxygenic photosynthesis have acquired photosystem II, and hence the ability to split H_2O, they still retain the ability under certain conditions to use photosystem I alone.

> In oxygenic photosynthesis, water is used as the source of electrons and oxygen is produced. Electron transport in oxygenic photosynthesis follows the Z scheme in which two separate light reactions are involved, photosystems I and II. Photosystem I resembles the system in anoxygenic photosynthesis. Photosystem II is responsible for splitting water to yield O_2 + H.

16.6   Additional Aspects of Photosynthetic Light Reactions

Although chlorophyll or bacteriochlorophyll is obligatory for photosynthesis, phototrophic organisms have other pigments involved in the capture of light energy. These include the **carotenoids** and the **phycobilins**. These function as accessory pigments and primarily play a photoprotective role (carotenoids) or serve as light-harvesting pigments (phycobilins). We consider each of these groups of pigments now.

Carotenoids

The most widespread accessory pigments are the **carotenoids**, which are almost always found in phototrophic organisms. Carotenoids are water insoluble pigments firmly embedded in the membrane; the structure of a typical carotenoid is shown in Figure 16.15. Carotenoids have long hydrocarbon chains with alternating C–C and C=C bonds, an arrangement called a *conjugated* double bond system. As a rule, carotenoids are yellow, red, brown, or green in color (see Figure 19.2b) and absorb light in the blue region of the spectrum (see Figure 16.4). Carotenoids are closely associated with chlorophyll in the photosynthetic membrane but do not act directly in photophosphorylation reactions. They can, however, transfer energy to the reaction center, and this transferred energy may be used in photophosphorylation in the same way as light energy captured directly by chlorophyll.

Another function of the carotenoids is as photoprotective agents. Bright light can often be harmful to cells, in that it causes various photooxidation reactions that can actually lead to the destruction of chlorophyll and of the photosynthetic apparatus itself. Carotenoids absorb much of this harmful light and thus provide a shield for the light-sensitive chlorophyll. Since phototrophic organisms must by their very nature live in the light, the photoprotective role of carotenoids is of obvious advantage.

Phycobilins and phycobilisomes

Cyanobacteria and red algae contain **phycobiliproteins**, which are the main light-harvesting pigments of these organisms. Phycobiliproteins are red or blue in color. The red pigment, called *phycoerythrin*, absorbs light most strongly at wavelengths around 550 nm, whereas the blue pigment, *phycocyanin*, absorbs most strongly at 620 nm. A third pigment, called *allophycocyanin* absorbs about 650 nm. Phycobiliproteins contain open-chain tetrapyrroles called *phycobilins* (Figure 16.16a), which are coupled to proteins. Phycobiliproteins occur as high molecular weight aggregates called **phycobilisomes,** attached to the photosynthetic membranes (Figure 16.16b). Phycobilisomes are constructed in such a way that the allophycocyanin molecules make physical contact with the photosynthetic membrane and are surrounded by molecules of phycocyanin and phycoerythrin. The latter pigments absorb shorter (higher energy) wavelengths of light and transfer the energy to allophycocyanin, which is closely linked to the reaction center chlorophyll and transfers energy to this site. The phycobilisome thus yields very efficient energy transfer from the biliprotein complex to chlorophyll *a*, which allows for

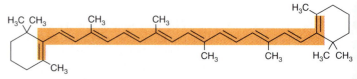

FIGURE 16.15 Structure of β-carotene, a typical carotenoid. The conjugated double-bond system is highlighted.

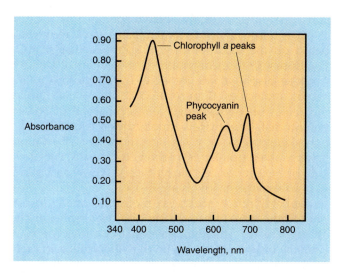

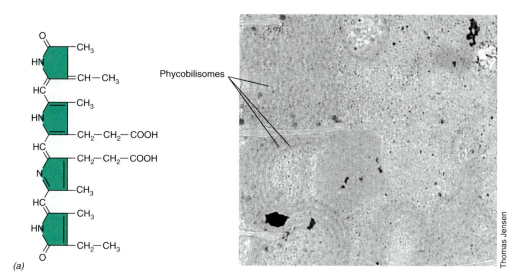

Phycobilisomes

Thomas Jensen

FIGURE 16.16
Phycobilins and phycobilisomes. (a) A typical phycobilin. This compound is an open chain tetrapyrrole, derived biosynthetically from a closed porphyrin ring by loss of one carbon atom as carbon monoxide. The structure shown is the prosthetic group of phycocyanin, a proteinaceous pigment found in cyanobacteria and red algae. (b) Electron micrograph of a thin section of the cyanobacterium *Microcoleus vaginatis*. Note the darkly staining ball-like phycobilisomes attached to the lamellar membranes.

FIGURE 16.17   The absorption spectrum of a cyanobacterium which has a phycobiliprotein (phycocyanin) as an accessory pigment. Note how the presence of phycocyanin broadens the wavelengths of usable light energy (between 600 and 700 nm). Compare with Figure 16.4.

growth of cyanobacteria at fairly low light intensities. Indeed, phycobilisome content *increases* in cells of cyanobacteria as light intensity *decreases*, such that phycobilisome-rich cells are those grown at the *lowest* light intensities.

The light-gathering function of accessory pigments like carotenoids and phycobilins is of obvious advantage to the organism. Light from the sun is distributed over the whole visible range, yet chlorophylls absorb well in only a part of this spectrum. By having accessory pigments, the organism is able to capture more of the available light (Figure 16.17).

> **Accessory pigments such as carotenoids and phycobilins can absorb light and transfer the energy to reaction center chlorophyll, thus broadening the wavelengths of light usable in photosynthesis. Carotenoids also play an important photoprotective role in preventing photooxidative damage to the phototrophic cell.**

Phototaxis

Many phototrophic microorganisms move toward light, a process called **phototaxis**. The advantage of phototaxis is that it allows the organism to orient itself for most efficient photosynthesis; indeed, if a light spectrum is spread across a microscope slide on which there are motile phototrophic bacteria, the bacteria accumulate at those wavelengths at which their pigments absorb (Figure 16.18).

We discussed in Section 3.8 the details of chemotaxis, the movement of motile bacteria toward and away from chemicals. The mechanism of phototaxis is somewhat different. The bacteria exhibit a kind of response called a *shock movement*. When a bacterial cell moves out of the light into the dark it stops suddenly (as if shocked) and reverses direction. This shock movement is the only response the bacterial cell makes to changing light conditions, but it is sufficient to keep the cell within light. How is that? Imagine a phototrophic bacterium moving in a spot of light. As it

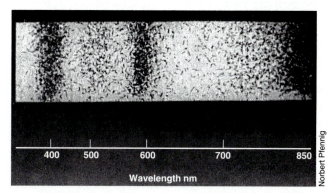

Norbert Pfennig

FIGURE 16.18   Phototactic accumulation of the phototrophic bacterium *Thiospirillum jenense* at light wavelengths at which its pigments absorb. A light spectrum was displayed on a microscope slide containing a dense suspension of the bacteria; after a period of time, the bacteria had accumulated selectively and the photomicrograph was taken. The wavelengths at which accumulations occur are those at which bacteriochlorophyll a absorbs (compare with Figure 16.4b).

swims around through the liquid, it may by accident wander out of the light field. This event signals the cell to stop, reverse its flagellar motor (and hence its direction), and thus reenter the light. The accumulation of bacteria at different wavelengths of light, as illustrated in Figure 16.18, results because the pigments of the bacteria absorb light at certain wavelengths better than at others, so that the bacteria "perceive" these regions of the spectrum as being brighter.

It is thought that transient changes in ATP levels or the state of energization of the membrane (extent of the proton motive force) serve in some way as molecular signals that affect flagellar rotation, perhaps in a manner similar to that of the chemoreceptor-driven scheme of chemotaxis (see Section 3.8). Under the microscope it is possible to detect phototrophic bacteria in enrichment cultures by quickly blocking the light source with a hand or piece of paper and then allowing light to return to the cell suspension. Motile phototrophic organisms will stop when the field is darkened and proceed in the opposite direction.

> **Even though phototrophic bacteria are unable to sense the presence of light, they are able to exhibit phototaxis because of the phenomenon of shock movement.**

16.7   Autotrophic CO_2 Fixation

In the first part of this chapter, we discussed the *light* reactions of phototrophic organisms. We now discuss the so-called *dark* reactions, the reactions by which autotrophs, both phototrophic and chemolithotrophic, convert CO_2 into organic matter.

Although all organisms require small amounts of their carbon for cellular biosynthesis to come from CO_2, autotrophs can obtain *all* of their carbon from CO_2. The overall process is called *CO_2 fixation*, and all of the reactions of CO_2 fixation can occur in complete darkness, using ATP and reducing power (NADPH) generated either during the light reactions of photosynthesis or during oxidation of inorganic compounds (see Section 16.8). Most autotrophs so far examined have a special pathway for CO_2 reduction, the **Calvin cycle**, also referred to as the *reductive pentose cycle*. This cycle is summarized later, but first let us consider some of the key individual reactions.

Key enzymes of the Calvin cycle

The first step in CO_2 reduction is the reaction catalyzed by the enzyme *ribulose bisphosphate carboxylase (RubisCO)*, which involves a reaction between CO_2 and ribulose bisphosphate (Figure 16.19*a*) leading to the formation of two molecules of *3-phosphoglyceric acid (PGA)*, one of which contains the carbon atom from CO_2. PGA constitutes the first identifiable intermediate in the CO_2 reductive process. The carbon atom in PGA is still at the same oxidation level as it was in CO_2; the next two steps involve reduction of PGA to the oxidation level of carbohydrate (Figure 16.19*b*). In these steps, *both* ATP and NADPH are required: the former

is involved in the phosphorylation reaction that activates the carboxyl group, the latter in the reduction itself. The carbon atom from CO_2 is now at the reduction level of carbohydrate (CH_2O), but only one of the carbon atoms of glyceraldehyde phosphate has been derived from CO_2, the other two having arisen from the ribulose bisphosphate. Most of the remaining reactions in the Calvin cycle involve rearrangements to yield new ribulose bisphosphate molecules with some of the carbon going to new cell synthesis.

Stoichiometry of the Calvin cycle

The overall scheme of the Calvin cycle is shown in Figure 16.20. The series of reactions leading to the synthesis of ribulose bisphosphate involve a number of sugar rearrangements. Through the action of enzymes which act on pentose phosphate compounds and enzymes of the glycolytic pathways, glyceraldehyde-3-phosphate is converted into ribulose-5-phosphate and subsequently into ribulose bisphosphate.

It is best to consider reactions of the Calvin cycle based on the incorporation of *six* molecules of CO_2. For RubisCO to incorporate *six* molecules of CO_2, *six* ribulose bisphosphate molecules are required as acceptor molecules (Figure 16.20). This yields *twelve* molecules of 3-phosphoglyceric acid (total of 36 carbon atoms). These twelve molecules serve as the carbon skeletons to form six *new* molecules of ribulose bisphosphate (total of 30 carbon atoms), and *one* molecule of hexose for cell biosynthesis. A complex series of rearrangements involving C-3, C-5, C-6, and C-7 intermediates finally yields six molecules of ribulose-5-phosphate, from which the six ribulose-bisphosphates are generated. The final step in the regeneration of ribulose bisphosphate is the phosphorylation of ribulose-5-phosphate with ATP by the enzyme *phosphoribulokinase* (Figure 16.19*c*). This enzyme is another catalyst unique to the Calvin cycle.

To summarize the Calvin cycle, six molecules of CO_2 and six molecules of ribulose bisphosphate (C-5) react to form twelve C-3 compounds. Two of the C-3 units go to biosynthesis as a 6-carbon sugar while the remaining ten are recycled to form six more C-5 acceptors. Since we begin and end the reaction sequence with the same number of acceptor molecules, the cyclic nature of the Calvin cycle is thus apparent.

Energy and reducing power needs of the Calvin cycle

Let us consider now the *overall* stoichiometry for conversion of six molecules of CO_2 into one molecule of fructose-6-phosphate (Figure 16.20). Twelve molecules each of ATP and NADPH are required for the reduction of twelve molecules of PGA to glyceraldehyde phosphate, and six ATP molecules are required for conversion of ribulose phosphate to ribulose bisphosphate. Thus twelve NADPH and eighteen ATP are required to manufacture *one* hexose molecule from CO_2. Hexose molecules can be converted to *storage polymers* such as glycogen, starch, or poly-β-hydroxyalkanoates (see Section 3.9) during periods when ATP and

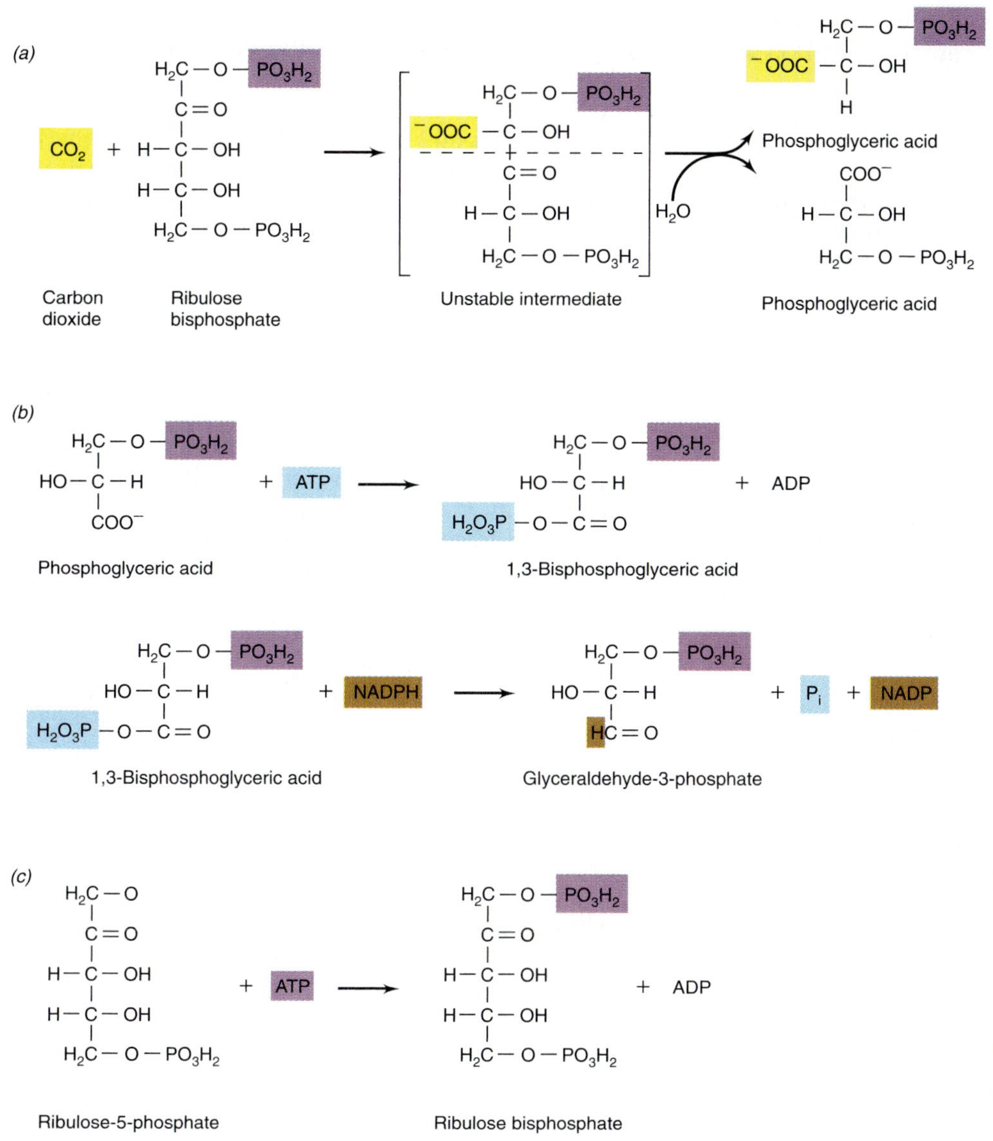

(a)

Carbon
dioxide
Ribulose
bisphosphate

Unstable intermediate

Phosphoglyceric acid

Phosphoglyceric acid

(b)

Phosphoglyceric acid

1,3-Bisphosphoglyceric acid

1,3-Bisphosphoglyceric acid

Glyceraldehyde-3-phosphate

(c)

Ribulose-5-phosphate

Ribulose bisphosphate

FIGURE 16.19
Key enzyme reactions of the Calvin cycle. (a) Reaction of the enzyme ribulose bisphosphate carboxylase. (b) Steps in the conversion of 3-phosphoglyceric acid (PGA) to glyceraldehyde-3-phosphate. Note that both ATP and NADPH are required. (c) Conversion of ribulose 5-phosphate to ribulose bisphosphate by the enzyme phosphoribulokinase.

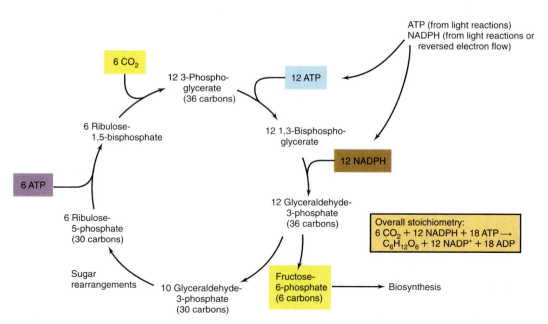

FIGURE 16.20   The Calvin cycle. For each *six* molecules of CO₂ incorporated, *one* fructose 6-phosphate is produced.

NADPH are abundant and then can be used in other periods, such as in darkness, as sources of carbon and energy.

Alternatives to the Calvin cycle

The key enzymes of the Calvin cycle, *ribulose bisphosphate carboxylase* and *phosphoribulokinase*, are unique to autotrophs which fix CO_2 via the Calvin cycle. These enzymes have been found in virtually all phototrophic organisms examined—plants, algae, and bacteria. They are also found in many chemolithotrophic Bacteria, such as the sulfur, iron, and nitrifying bacteria (see below). The enzyme ribulose bisphosphate carboxylase is sometimes stored inside the cell in large aggregates called *carboxysomes* (see Section 19.5)

However, several groups of autotrophs including the phototrophic green bacteria, the methanogenic Archaea, and the acetogenic bacteria, do not use the Calvin cycle for CO_2 fixation. In these groups, Calvin cycle enzymes are not present. We discuss the pathways by which these groups of prokaryotes fix CO_2 in Chapters 19 and 20.

> The fixation of CO_2 by most phototrophic and other autotrophic organisms involves a special biochemical pathway called the Calvin cycle, in which the enzyme ribulose bisphosphate carboxylase (RubisCO) plays a key role. The Calvin cycle is a reductive energy-demanding process in which reducing equivalents from NADPH and energy from ATP are used to reduce CO_2 to the oxidation state of cell material.

16.8 Chemolithotrophy: Energy from the Oxidation of Inorganic Electron Donors

Organisms that can obtain their energy from the oxidation of inorganic compounds are called **chemolithotrophs**. Most chemolithotrophic bacteria are also able to obtain all their carbon from CO_2 and they are therefore also autotrophs. As we have noted, two components are needed when an organism grows on CO_2 as sole carbon source: energy in the form of ATP and reducing power. In chemolithotrophs, ATP generation is in principle similar to that in chemoorganotrophs except that the electron donor is *inorganic* rather than organic. Thus ATP synthesis is coupled to the *oxidation* of the electron donor. Reducing power in chemolithotrophs is obtained either directly from the inorganic compound, if it has a sufficiently low reduction potential, or by *reverse electron transport reactions*, as discussed for phototrophic bacteria in Section 16.4

Energetics of chemolithotrophy

A review of redox potentials illustrated in Table 16.1 will indicate that a number of inorganic compounds can provide sufficient energy for ATP synthesis when O_2 is used as electron acceptor. Recall from Chapter 4 that the further apart two half reactions are, the more energy can be released. For instance, the difference in

reduction potential between the H^+/H_2 couple and the $\frac{1}{2} O_2/H_2O$ couple is -1.23 volts, which is equivalent to a free energy yield of -237 kJ/mole (see Appendix 1 for calculations). On the other hand, the potential difference between the H^+/H_2 couple and the NO_3^-/NO_2^- couple is less, -0.84 volts, equivalent to a free energy yield of -163 kJ/mole. This is still quite sufficient for the production of ATP (the high-energy phosphate bond of ATP has a free energy of around -31.8 kJ/mole). However, a similar calculation will show that there is insufficient energy available from the oxidation of H_2S using CO_2 as electron acceptor.

From such energy calculations, it is possible to predict the kinds of chemolithotrophs that might be found in nature. Since organisms obey the laws of thermodynamics, only reactions which are thermodynamically favorable will be potential energy-yielding reactions. However, just because a reaction is theoretically possible does not mean that it actually occurs. There may be ecological reasons why a particular reaction does not take place. Table 16.3 summarizes energy yields for some reactions which are known to be carried out by chemolithotrophic microorganisms. We discuss some of these processes briefly in the rest of this chapter, and the organisms involved are discussed in more detail in Chapter 19. We also discuss ecological aspects of chemolithotrophy in Chapter 17.

> A number of specialized prokaryotes, called chemolithotrophs, are able to use inorganic chemicals as their sole sources of energy. Most chemolithotrophs are able to grow autotrophically, fixing CO_2 by the Calvin cycle.

16.9 Hydrogen-Oxidizing Bacteria

Hydrogen (H_2) is a common product of microbial metabolism and a number of chemolithotrophs are able to use H_2 as an energy source. A wide variety of H_2-oxidizing bacteria are known, differing in the electron *acceptor* they use. Those H_2-oxidizing bacteria which use electron acceptors other than oxygen carry out a type of anaerobic respiration and are discussed later in this

Table 16.3 Energy yields from the oxidation of various inorganic electron donors*	
Reaction	**$\Delta G^{0'}$ (kJ/ reaction)**
$S^0 + 1\frac{1}{2}O_2 + H_2O \rightarrow SO_4^{2-} + 2H^+$	-587.1
$NH_4^+ + 1\frac{1}{2}O_2 \rightarrow NO_2^- + 2H^+ + H_2O$	-274.7
$H_2 + \frac{1}{2}O_2 \rightarrow H_2O$	-237.17
$HS^- + H^+ + \frac{1}{2}O_2 \rightarrow S^0 + H_2O$	-209.4
$NO_2^- + \frac{1}{2}O_2 \rightarrow NO_3^-$	-75.8
$Fe^{2+} + H^+ + \frac{1}{4}O_2 \rightarrow Fe^{3+} + \frac{1}{2}H_2O$	-31

*Data calculated from values in Appendix 1. Values for Fe^{2+} are for pH 2, others for pH 7.

chapter (Section 16.14). We discuss here the *aerobic* H₂-oxidizing species, which are the types most commonly referred to when the term *hydrogen bacteria* is used.

The hydrogen bacteria are facultative chemolithotrophs, capable of growing either as conventional chemoorganotrophs on organic compounds, or chemolithotrophically by H₂ oxidation. Taxonomically and phylogenetically, the hydrogen bacteria belong to various groups of Bacteria (see Section 19.6). When growing autotrophically, hydrogen bacteria show the following stoichiometry:

$$6H_2 + 2O_2 + CO_2 \rightarrow (CH_2O) + 5H_2O$$

where the formulation (CH_2O) signifies cell material at the oxidation state of carbohydrate.

We discussed the free energy change occurring in the above listed reaction earlier in this chapter. The enzyme **hydrogenase** catalyzes the initial oxidation of H₂, the electrons being transferred to either NAD⁺ or a quinone acceptor. Electrons introduced into the electron transport chain at the level of NADH or reduced quinone undergo electron transport reactions leading to the formation of a proton-motive force, as discussed earlier, and ATP formation occurs via the membrane-bound ATPase.

Many bacteria incapable of autotrophy can still oxidize H₂ as a sole energy source. For instance, *Escherichia coli* can use H₂ as an energy source but cannot grow on CO₂ as sole carbon source. To obtain growth on H₂ therefore, *E. coli* must be given an organic compound as carbon source. Organisms which use an inorganic compound as energy source and an organic compound as carbon source are sometimes called *mixotrophs*. Many sulfate-reducing bacteria, the homoacetogenic bacteria and most methanogens also use H₂ as electron donor, as discussed later in this chapter.

When growing autotrophically, hydrogen bacteria fix CO₂ by the Calvin cycle. If organic compounds are present, synthesis of the Calvin cycle enzymes are repressed and enzymes involved in the utilization of the organic compound induced. Regulation of chemolithotrophic metabolism in hydrogen bacteria is very precise and specific, with aspects of both carbon metabolism and energy metabolism playing important roles in gene expression.

The hydrogen bacteria use hydrogen gas (H₂) as energy source and fix CO₂ into cell carbon. Hydrogen bacteria are facultative chemolithotrophs, also capable of using certain organic compounds when they are present.

16.10   Sulfur Bacteria

Many reduced sulfur compounds can be used as electron donors by a variety of colorless sulfur bacteria (called "colorless" to distinguish them from the bacteriochlorophyll-containing [pigmented] green and purple sulfur bacteria discussed earlier in this chapter). The most common sulfur compounds used as energy sources are hydrogen sulfide (H₂S), elemental sulfur (S⁰), and thiosulfate ($S_2O_3^{2-}$). The final product of sulfur oxidation in most cases is sulfate (SO_4^{2-}), and the total number of electrons involved between H₂S (oxidation state, −2) and sulfate (oxidation state, +6) is 8 (see Table 16.6 later for a summary of sulfur oxidation states). Less energy is available when one of the intermediate sulfur oxidation states is used:

$$H_2S + 2O_2 \rightarrow SO_4^{2-} + 2H^+$$
$$\text{sulfide } -798.2 \text{ kJ/reaction}$$
$$S^0 + H_2O + 1\,1/2O_2 \rightarrow SO_4^{2-} + 2H^+$$
$$\text{sulfur } -587.1 \text{ kJ/reaction}$$
$$S_2O_3^{2-} + H_2O + 2O_2 \rightarrow 2SO_4^{2-} + 2H^+$$
$$\text{thiosulfate } -822.6 \text{ kJ/reaction}$$
$$(411 \text{ kJ/S atom oxidized})$$

The oxidation of the most reduced sulfur compound, H₂S, occurs in stages, and the first oxidation step results in the formation of elemental sulfur, S⁰. Some H₂S-oxidizing bacteria deposit the elemental sulfur formed inside the cell (Figure 16.21a). The sulfur deposited as a result of the initial oxidation is an energy reserve, and when the supply of H₂S has been depleted, additional energy can be obtained from the oxidation of sulfur to sulphate.

When elemental sulfur is provided externally as an electron donor, the organism must grow attached to the sulfur particle because of the extreme insolubility of elemental sulfur (Figure 16.21b). By adhering to the particle, the organism can efficiently obtain the atoms of sulfur needed, which slowly go into solution. As these atoms are oxidized, more dissolve; gradually the sulfur particle is consumed.

Note that in the sulfur-oxidation reactions shown above, one of the products is H⁺. Production of pro-

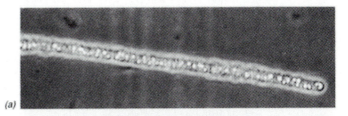

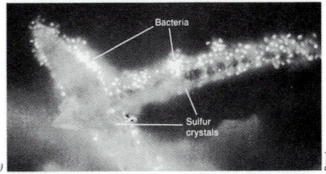

FIGURE 16.21 Sulfur bacteria. (a) Deposition of internal sulfur granules by *Beggiatoa*. (b) Attachment of the sulfur-oxidizing archaeon *Sulfolobus acidocaldarius* to a crystal of elemental sulfur. Visualized by fluorescence microscopy after staining the cells with the dye acridine orange. The sulfur crystal does not fluoresce.

tons results in a lowering of the pH, and one result of the oxidation of reduced sulfur species is the acidification of the medium. The acid formed by the sulfur bacteria is *sulfuric acid*, H_2SO_4, and sulfur bacteria are often able to bring about a marked reduction in pH of the medium. Some sulfur bacteria have been found to lower the pH to values less than 1!

Energetics and carbon metabolism in sulfur chemolithotrophs

The electron transport system of a typical sulfur-oxidizing bacterium is illustrated in Figure 16.22*a*. As shown, electrons from reduced sulfur compounds enter the chain at various points (depending upon their reduction potentials), and are transported to molecular oxygen, the whole process generating a *membrane potential* which leads to ATP synthesis by oxidative phosphorylation. Electrons for CO_2 reduction come from reverse electron transport reactions (see Section 16.4) eventually yielding NADPH (Figure 16.22*a*). Note in Figure 16.22 how the relatively high reduction potentials of compounds like $S_2O_3^{2-}$ or S^0 result in a rather abbreviated electron transport chain for energy purposes and a rather long route to the reduction of NADPH; the latter is, of course, an energy-dependent series of reactions. These biochemical limitations are in part responsible for the relatively poor growth yields of most sulfur chemolithotrophs.

When growing autotrophically, the sulfur bacteria use the Calvin cycle to fix CO_2. However, some sulfur bacteria are also able to grow using organic compounds. A few sulfur bacteria can grow anaerobically on reduced sulfur compounds, utilizing nitrate (NO_3^-) as electron acceptor. Under these latter conditions, a type of anaerobic respiration is carried out (see Section 16.14). Several sulfur-oxidizing bacteria appear to grow only mixotrophically, using H_2S as energy source and an organic compound as carbon source. In this category is *Beggiatoa*, a widespread organism in marine and freshwater sediments (see box).

> The sulfur bacteria are chemolithotrophs that can use reduced sulfur compounds such as hydrogen sulfide and elemental sulfur as electron donors. Most sulfur bacteria fix CO_2 via the Calvin cycle with reducing power coming from reverse electron flow.

16.11   Iron-Oxidizing Bacteria

The aerobic oxidation of iron from the ferrous (Fe^{2+}) to the ferric (Fe^{3+}) state is an energy-yielding reaction for a few bacteria. Only a small amount of energy is available from this oxidation (see Table 16.3), and for this reason the iron bacteria must oxidize large amounts of iron in order to grow. Ferric iron forms very insoluble ferric hydroxide [$Fe(OH)_3$] precipitates in water (Figure 16.23*a*). Many iron-oxidizing bacteria also oxidize sulfur and are thus obligate acidophiles.

This is in part because at neutral pH, ferrous iron rapidly oxidizes nonbiologically to the ferric state and is thus only stable under *anaerobic* conditions. At acid pH, however, ferrous iron is stable to chemical oxidation. Since spontaneous oxidation of ferrous iron in oxic environments at neutral pH is so rapid, significant amounts of ferrous iron do not accumulate under these conditions. This explains why most of the iron-oxidizing bacteria are obligately acidophilic.

The best known iron-oxidizing bacterium is *Thiobacillus ferrooxidans*, which is able to grow autotrophically using either ferrous iron (Figure 16.23*b*) or reduced sulfur compounds as electron donors. The electron transport system was illustrated in Figure 16.22*b*. This organism is very common in acid-polluted environments such as coal-mining dumps (Figure 16.23*a*), and we discuss its role in acid-mine pollution and mineral oxidation in Sections 17.16 and 17.17. Another iron-oxidizing prokaryote is the archaean *Sulfolobus*, which lives in hot, acid springs at temperatures up to the boiling point of water (see Figure 16.21*b*).

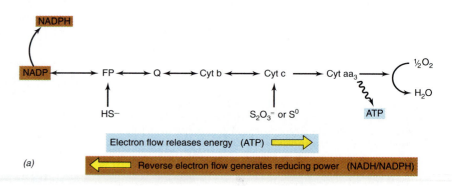

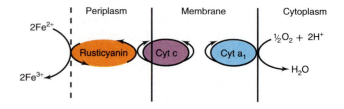

FIGURE 16.22 Electron flow in *Thiobacillus* grown on (a) sulfur, thiosulfate, sulfide, or (b) ferrous iron. In (a) note the direction of electron flow in ATP synthesis versus reverse electron flow.

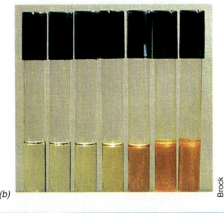

(b)

(a)

(c)

FIGURE 16.23 Iron-oxidizing bacteria. (a) Acid mine drainage, showing the confluence of a normal river and a creek draining a coal-mining area. The acidic creek is very high in ferrous iron. At low pH values, ferrous iron does not oxidize spontaneously in air, but *Thiobacillus ferrooxidans* carries out the oxidation. Insoluble ferric hydroxide and complex ferric salts precipitate, forming the precipitate called "yellow boy" by coal miners. (b) Cultures of *Thiobacillus ferrooxidans*. Shown is a dilution series, no growth in the tube on the left and increasing amounts of growth from left to right. Growth is evident from the production of Fe^{3+} which readily forms $Fe(OH)_3$ leading to the yellow orange color. (c) Extensive development of insoluble ferric hydroxide in a small pool draining a bog in Iceland. Iron deposits such as this are widespread in cooler parts of the world and are modern counterparts of the extensive bog-iron deposits of earlier geological eras. These ancient deposits are now the sources of much commercially mined iron ore. In the water-saturated bog soil, facultatively aerobic bacteria reduce ferric iron to the more soluble ferrous state. The ferrous iron leaches into the drainage area surrounding the bog, where oxidation occurs, either spontaneously or through the agency of iron-oxidizing bacteria (*Gallionella* and *Leptothrix*), and the insoluble ferric hydroxide deposit is formed.

Energy from ferrous iron

The bioenergetics of iron oxidation by *T. ferrooxidans* is of biochemical interest because of the high reduction potential of the Fe^{3+}/Fe^{2+} couple (+0.77 volts). The respiratory chain of *T. ferrooxidans* contains cytochromes of the c and a_1 types and a copper-containing protein called *rusticyanin*. Since the reduction potential of the Fe^{3+}/Fe^{2+} couple is so high, the route of electron transport to oxygen ($\frac{1}{2}\ O_2/H_2O$, $E_0' = +0.82$ volts) will obviously be very short. Electrons from the oxidation of Fe^{2+} cannot reduce NAD^+, FAD, or many of the other components of the electron transport chain. How, then, do acidophilic iron-oxidizing bacteria make ATP?

Biochemical studies of electron flow in *Thiobacillus ferrooxidans* suggest that acidophilic iron bacteria take advantage of the preexisting proton gradients of their environment for energy-generating purposes. Remember that the cytoplasm of any organism, including that of acidophiles, must remain near neutrality; in *T. ferrooxidans* the internal pH is about 6. However, the pH of the environment of *T. ferrooxidans* is much lower, near pH 2. This pH difference across the cytoplasmic membrane represents a natural proton motive force that can play a role in ATP synthesis. However, to retain a neutral pH environment, protons entering the cell through the proton translocating ATPase (driving the phosphorylation of ADP in the process) must be consumed. It is here that the oxidation of Fe^{2+} plays an important role.

The oxidation of Fe^{2+} to Fe^{3+} ($2Fe^{2+} + \frac{1}{2}\ O_2 + 2H^+ \rightarrow 2Fe^{3+} + H_2O$) is a *proton-consuming* reaction. Experimental evidence suggests that the reaction $\frac{1}{2}\ O_2 + 2H^+ \rightarrow H_2O$ occurs on the *inner* face of the cytoplasmic membrane while the reaction $Fe^{2+} \rightarrow Fe^{3+}$ occurs near the *outer* face of the membrane. Electrons from Fe^{2+} are accepted in the periplasmic space by rusticyanin (Figure 16.22b); as befits its periplasmic location, rusticyanin is acid stable and functions optimally at pH 2.

Winogradsky's Legacy

The discovery of autotrophy in chemolithotrophic bacteria was of major significance in the advance of our understanding of cell physiology, since it showed that CO_2 could be converted into organic carbon without the intervention of chlorophyll. Previously, it had been thought that only green plants converted CO_2 into organic form. The idea of chemolithotrophic autotrophy was first developed by the great Russian microbiologist Sergei Winogradsky. Winogradsky studied sulfur bacteria because certain of the colorless sulfur bacteria (*Beggiatoa, Thiothrix*) are very large and hence are easy to study even in the absence of pure cultures. Springs with waters rich in H_2S are fairly common around the world, and Winogradsky studied several such springs in the Bernese Oberland district of Switzerland. In the outflow channels of sulfur springs, vast populations of *Beggiatoa* and *Thiothrix* develop, and suitable material for microscopic and physiological studies could be obtained by merely lifting up the white filamentous masses. Pure cultures were not needed for many studies. As Winogradsky noted, "This study of demipurity would be poor in an ordinary culture but is sufficient for a culture under the microscope, since it is possible to observe the development from day to day, almost from hour to hour, and see easily the presence of contaminants." Winogradsky first showed that the colorless sulfur bacteria were present only in water containing H_2S. As the water flowed away from the source, the H_2S gradually dissipated, and sulfur bacteria were no longer present. This suggested to him that their development was dependent on the presence of H_2S. Winogradsky then showed that by starving *Beggiatoa* filaments for a while, they lost their sulfur granules; he found, however, that the granules were rapidly restored if a small amount of H_2S was added. He thus concluded that H_2S was being oxidized to elemental sulfur. But what happened to the sulfur granules when the filaments were starved of H_2S? Winogradsky showed by some clever microchemical tests that when the sulfur granules disappeared, sulfate appeared in the medium. Thus he formulated the idea that *Beggiatoa* (and by inference other colorless sulfur bacteria) oxidize H_2S to elemental sulfur and subsequently to sulfate. Because they seemed to require H_2S for development in the springs, he postulated that this oxidation was the principal source of energy for these organisms.

Studies on *Beggiatoa* provided the first evidence that an organism could oxidize an *inorganic* substance as a possible energy source, and this was the origin of the concept of chemolithotrophy. However, the sulfur bacteria proved difficult to work with, primarily because there are a number of spontaneous chemical changes in sulfur compounds which can also occur and confuse the study. Winogradsky thus turned to a study of the nitrifying bacteria, and it was with this group that he clearly was able to show that autotrophic fixation of CO_2 was coupled to the oxidation of an inorganic compound in the complete absence of light and chlorophyll. The process of nitrification had been known before Winogradsky's work from studies on the fate of sewage when added to soil. Two French soil scientists, T. Schloesing and A. Mutz, had shown that the process was due to living organisms. Winogradsky proceeded to isolate some of the bacteria, using completely mineral media in which CO_2 was the sole carbon source and ammonia was the sole electron donor. Because ammonia is chemically stable, it was easy to show that the oxidation of ammonia to nitrite, and subsequently to nitrate, is a strictly bacterial process. As no organic materials were present in the medium, it was also possible to show that organic matter (the bacterial cell material) was formed only from CO_2. If the ammonia was left out of the medium, no growth occurred. Careful chemical analyses showed that the amount of organic matter formed by the bacteria was proportional to the amount of ammonia or nitrite which they oxidized. Winogradsky concluded, "This [process] is contradictory to that fundamental doctrine of physiology which states that a complete synthesis of organic matter cannot take place in nature except through chlorophyll-containing plants by the action of light." His basic conclusion has been confirmed by a large number of subsequent studies. At least in one way, however, autotrophy in most chemolithotrophs and phototrophs is similar in that, in both processes, the pathway of CO_2 fixation follows the same biochemical steps (the Calvin cycle) involving the enzyme ribulose bisphosphate carboxylase.

Electrons are donated from rusticyanin to an unusually high-potential membrane-bound cytochrome *c* which subsequently transfers electrons to cytochrome a_1, the terminal oxidase (Figure 16.22*b*). Cytochrome a_1 donates electrons to $\frac{1}{2} O_2$ with the two protons required to form water coming from the cytoplasm. An influx of protons via the ATPase replenishes the proton supply, and as long as Fe^{2+} remains available, the natural proton motive force across the *T. ferrooxidans* membrane can continue to drive ATP synthesis. Thus, although energy conservation in *T. ferrooxidans* results from a classical chemiosmotic ATPase reaction which couples the entry of protons to the synthesis of ATP, the proton gradient in *T. ferrooxidans* is not established as a result of electron transport, but instead is a simple consequence of the habitat of the organism!

As we discussed earlier, autotrophic CO_2 fixation requires the presence of both ATP and reducing power. Reducing power in the form of NADH or NADPH is formed by the process of *reversed electron*

FIGURE 16.24 Phase contrast photomicrograph of empty iron-encrusted sheaths of *Sphaerotilus* collected from seepage at the edge of a small swamp.

transport discussed earlier in this chapter, using energy from ATP to reverse electron flow and form NADH or NADPH. Because of the high potential of the electron donor this process takes a lot of iron oxidation, and it is therefore understandable that the cell yields of iron-oxidizing bacteria are quite low. In an environment where these organisms are living, their presence is signalled not by the formation of much cell material, but by the presence of massive amounts of ferric iron precipitation (Figure 16.23*a* and *b*). We discuss the important ecological roles of the iron-oxidizing bacteria in Section 17.16, and their taxonomy is discussed in Section 19.5.

Iron oxidation at neutral pH

In addition to the iron-oxidizing bacteria just discussed, there are some other bacteria which have been traditionally considered to be iron bacteria but which grow at neutral pH. These bacteria live at near *neutral* pH and are commonly found in environments where ferrous iron is moving from anoxic to oxic conditions (see Figure 16.23*c* for a typical site for these bacteria). As we have discussed, at neutral pH, ferrous iron is not stable in the presence of oxygen, and is rapidly oxidized to the ferric (insoluble) state. Because of this, the only neutral pH environments where ferrous iron is present are *interfaces* between anoxic and oxic conditions. *Gallionella ferruginea, Sphaerotilus natans,* and *Leptothrix ochracea* live at such interfaces, and are generally seen mixed in with the characteristic deposits which they form (Figure 16.24). Only in *Gallionella* has true autotrophy been found. Cultures of *G. ferruginea* grow on CO_2 as sole carbon source via the Calvin cycle. Autotrophy in other neutral pH iron bacteria is still unproven. We discuss the taxonomy of these interesting organisms in Section 19.10.

In addition to ferrous iron, the chemically related metal *manganese* is also oxidized by a few bacteria at neutral pH. The most common manganese-oxidizing bacterium is *Leptothrix discophorus*, which is discussed in Section 19.14. We discuss the process of manganese oxidation itself in Section 17.16.

> The iron bacteria are a specialized group of chemo-lithotrophs that are able to use reduced iron (Fe^{2+}) as sole energy source. Most iron bacteria grow only at acid pH and are often associated with acid pollution from mineral and coal mining.

16.12   Ammonium and Nitrite-Oxidizing Bacteria

The most common *inorganic nitrogen compounds* used as electron donors are ammonia (NH_3) and nitrite (NO_2^-), which are oxidized aerobically by the **nitrifying bacteria.** The nitrifying bacteria are widely distributed in soil and water. One group of organisms (*Nitrosomonas* is one genus) oxidizes ammonia to nitrite, and another group (*Nitrobacter*) oxidizes nitrite to nitrate; the complete oxidation of ammonia to nitrate, an eight electron transfer (see Table 16.5 in Section 16.15), is thus carried out by members of these two groups of organisms acting in sequence. Nitrifying bacteria are widespread in soil, and their significance in soil fertility and in the nitrogen cycle is discussed in Section 17.14; their morphology, taxonomy, and physiology, is discussed in Section 19.4. We deal here with energetic principles.

Energy and autotrophy in nitrifying bacteria

The electrons from nitrogen compounds enter an electron transport chain and electron flow establishes a membrane potential and proton motive force which is linked to ATP synthesis. However, because of the reduction potential of their electron donors, nitrifying bacteria are faced with bioenergetic problems similar to those of the sulfur chemolithotrophs. The E_0' of the NH_2OH/NH_3 couple, the first step in the oxidation of NH_3, is about 0 volts. The E_0' of the NO_3^-/NO_2^- couple is very high, about +0.43 volts. These relatively high reduction potentials mean that nitrifying bacteria must input electrons to their electron transport chains at rather late steps in the overall process. This effectively limits the amount of ATP that can be produced from each pair of electrons introduced.

Several key enzymes are involved in oxidizing reduced nitrogen compounds. In ammonia-oxidizing bacteria, NH_3 is oxidized by *ammonia monooxygenase* (see Section 16.23) that produces NH_2OH and H_2O (Figure 16.25). *Hydroxylamine oxidoreductase* then oxidizes NH_2OH to NO_2^-, removing *four* electrons in the process. Ammonia monooxygenase is an integral membrane protein, while hydroxylamine oxidoreductase is periplasmic (Figure 16.25). In the reaction carried out by ammonia monooxygenase:

$$NH_3 + O_2 + 2H^+ + 2e^- \rightarrow NH_2OH + H_2O$$

there is a need for two exogenously supplied electrons to reduce one atom of dioxygen to water. These electrons originate from the oxidation of hydroxylamine and are supplied to ammonia monooxygenase from hydroxylamine oxidoreductase via cytochrome *c* (Figure

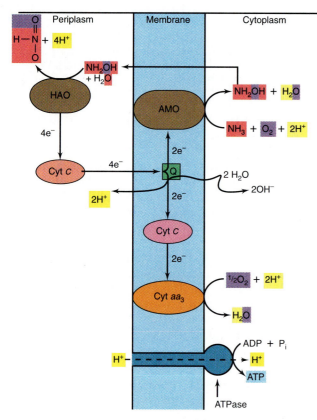

FIGURE 16.25 Oxidation of ammonia and electron flow in ammonia-oxidizing bacteria. The reactants and the products of this reaction series are highlighted. The cytochrome c (cyt c) in the periplasm is a different form of cyt c than that in the membrane. AMO, ammonia monooxygenase; HAO, hydroxylamine oxidoreductase.

16.25). Thus, for every four electrons generated from the oxidation of NH_3 to NO_2^-, only two actually reach the terminal oxidase (cytochromes aa_3, Figure 16.25).

Nitrite-oxidizing bacteria employ the enzyme *nitrite oxidase* to oxidize nitrite to nitrate, with electrons traveling a very short electron transport chain (because of the high potential of the NO_3^-/NO_2^- couple)

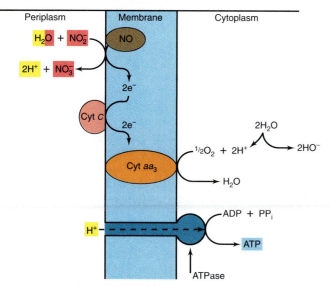

FIGURE 16.26 Oxidation of nitrite to nitrate by nitrifying bacteria. The reactants and products of this reaction series are highlighted. NO, nitrite oxidase.

to the terminal oxidase (Figure 16.26). Cytochromes of the *a* and *c* types are present in the electron transport chain of nitrite-oxidizers and generation of a membrane potential (which ultimately drives ATP synthesis) probably occurs during the oxidation of NO_2^- directly (Figure 16.26). And like the situation with iron oxidation (see Section 16.11), only small amounts of energy are available. Thus growth yields of nitrifying bacteria are relatively low.

Growth yields of nitrifying bacteria are also affected by the energy demands of CO_2 fixation. Like sulfur- and iron-oxidizing chemolithotrophs, nitrifiers employ the Calvin cycle for CO_2 fixation, and the ATP requirements of this process (18 ATP are required for every glucose molecule produced), places additional burdens on an already inefficient energy-generating system. The energetic constraints are particularly severe for nitrite oxidizers, and it is perhaps for this reason that most nitrite oxidizers can also grow chemoorganotrophically on glucose and certain other organic substrates (see Section 19.4).

> Two inorganic nitrogen compounds, ammonia (NH_3) and nitrite (NO_2^-) can be used as energy sources by specialized bacteria called the nitrifying bacteria. The ammonia-oxidizing bacteria produce nitrite which is then oxydized by the nitrite-oxidizing bacteria to nitrate (NO_3^-).

16.13   Summary of Autotrophy and ATP Production among Chemolithotrophs

As we have noted, when growing autotrophically chemolithotrophs must also produce reducing power in addition to ATP if they are to grow with CO_2 as sole carbon source. Although NAD^+ can be reduced to NADH directly by H_2, all of the other inorganic electron donors have reduction potentials *higher* than that of NADH (Table 16.1). If the reduction potential of an electron donor is higher than that of NADH, there is no way in which its oxidation can be directly coupled to the reduction of NAD^+ to NADH. Thus, NAD^+ must be reduced by energy-driven reversed electron transport as already discussed in Sections 16.4 and 16.11.

ATP production by chemolithotrophs

As noted in the discussions of the individual groups of chemolithotrophs, these bacteria all contain electron transport chains consisting of cytochromes, quinones, and many of the major components described in aerobic chemoorganotrophic organisms (see Sections 4.10 and 4.11). However, with the exception of H_2, no inorganic reductant has a sufficiently low E_0' to reduce NAD^+ directly, so electrons must feed into the electron transport chain at a point where the reduction potential of the inorganic electron donor is more *negative* than that of the acceptor molecule. In most cases the electron acceptor turns out to be a cytochrome of some type or an enzyme that feeds electrons into a cytochrome (see Figures 16.22, 16.25, and 16.26). As we noted, the situation is an extreme one for the iron-oxidizing and the nitrite-oxidizing bacteria because the reduction potentials of their electron donors is already very high. This directly affects growth yields.

Table 16.3 summarized energy yields from the aerobic oxidation of various inorganic electron donors. It can be seen that the energetics of hydrogen and sulfur oxidation are far more favorable than those of the nitrifying or iron-oxidizing bacteria, since ATP yield is directly proportional to the amount of energy released in a given oxidation and the amount of energy released is a direct function of the *difference* in reduction potential between the electron donor and electron acceptor couples, respectively. Because of the added burden of autotrophy, most chemolithotrophs synthesize only small amounts of cell material while oxidizing huge amounts of substrate; chemoorganotrophs, by contrast, do not have as severe a problem (Figure 16.27). From an ecological standpoint, however, it should be remembered that chemolithotrophs are utilizing electron donors not catabolized by chemoorganotrophic organisms. Hence, chemolithotrophs survive in nature without severe nutritional competition from the bulk of the microbial world.

Comparison of energy and carbon metabolism

A summary of energy and carbon metabolism and the terminology used to describe different nutritional groups of organisms is indicated in Table 16.4. The ability of certain chemolithotrophs to assimilate organic compounds as carbon sources while using inorganic oxidations for energy (Table 16.4) makes possible the growth of these organisms as chemolithotrophic heterotrophs. This type of growth is referred to as **mixotrophic**. Some chemolithotrophs grow best under mixotrophic conditions because they are free of the energy-demanding process of autotrophic CO_2 fixation. At least one sulfur bacterium,

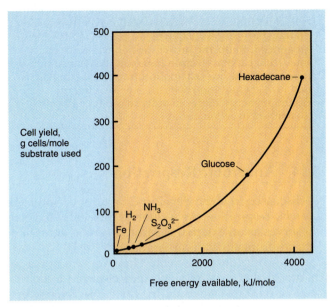

FIGURE 16.27 Comparison of cell yield (various microorganisms) on energy sources with different free energies of reaction. In all cases, the electron acceptor is O_2.

Beggiatoa, seems to require mixotrophic conditions to use reduced sulfur compounds as electron donors. Most strains of *Beggiatoa* are unable to grow on a completely inorganic medium with reduced sulfur compounds and CO_2, but will use reduced sulfur compounds as electron donors when acetate is present. *Beggiatoa* will also grow on acetate as sole source of carbon and energy in the absence of reduced sulfur compounds. Mixotrophy allows an organism the best of both chemolithotrophic and chemoorganotrophic worlds. However, growth studies in mixed culture have shown that most mixotrophs compete poorly with strict chemolithotrophs or strict chemoorganotrophs when the mixotrophic organism is forced to grow under one or the other contrasting nutritional regime. Thus, most mixotrophs are successful in nature only when *both* inorganic and organic compounds are present simultaneously.

Table 16.4   Summary of energy and carbon metabolism and nutritional groups of organisms

	Energy source		
Carbon source	**Light**	**Inorganic chemicals**	**Organic chemicals**
Inorganic (CO_2, HCO_3^-, CO_3^{2-})	Phototrophic eukaryotes; cyanobacteria (photoautotrophs, H_2O is electron donor) Purple and green bacteria (photoautotrophs, H_2S is electron donor)	Chemolithoautotrophs: hydrogen, sulfur, iron, and nitrifying bacteria	—
Organic	Purple and green bacteria using organic carbon sources (photoheterotrophs)	Chemolithoheterotrophs (mixotrophs)	Chemoorganotrophs (most prokaryotes and all nonphototrophic eukaryotes)

> The chemolithotrophic bacteria produce ATP via electron transport processes in which a proton motive force is generated. Reducing power for CO_2 fixation usually comes from reverse electron transport reactions. Since the energy yield from chemolithotrophic metabolism is often low, chemolithotrophs must oxidize a large amount of energy source in order to produce a significant amount of cell material.

16.14 Anaerobic Respiration

We discussed in some detail in Chapter 4 the process of *aerobic* respiration. As we noted, molecular oxygen (O_2) serves as an external electron acceptor, accepting electrons from electron carriers such as NADH by way of an electron transport chain. However, we noted in Chapter 4 that a variety of other electron acceptors could be used instead of O_2, in which case the process was called *anaerobic* respiration. We now discuss some of the anaerobic respiration processes that have been recognized.

In most cases, the energy source(s) used by organisms carrying out anaerobic respiration are *organic* compounds, but several chemolithotrophic organisms can also carry out this process. In addition, although most of the electron acceptors to be discussed are *inorganic*, several organic compounds can also serve as electron acceptors. Most organisms carrying out anaerobic respiration are prokaryotes and the chemical transformations which they carry out during their energy generation process are frequently of great importance ecologically or industrially, as we will discuss in Chapter 17.

The bacteria carrying out anaerobic respiration generally possess electron transport systems containing cytochromes, quinones, iron–sulfur proteins and other typical electron transport proteins. Their respiratory systems are thus analogous to those of conventional aerobes. In some cases, such as with the denitrifying bacteria, the anaerobic respiration process competes in the same organism with an aerobic one. In such cases, if O_2 is present, aerobic respiration is usually favored, and only when O_2 is depleted from the environment would the alternate electron acceptor be reduced. Other organisms carrying out anaerobic respiration are *obligate* anaerobes, and are unable to use O_2.

The energy released from the oxidation of an electron donor using O_2 as electron acceptor is higher than if the same compound is oxidized with an alternate electron acceptor. These energy differences are apparent if the reduction potentials listed in Table 16.1 are examined. As discussed at the beginning of this chapter, and in detail in Appendix 1, theoretical energy yields can be calculated using differences in reduction potential between the electron donor and the electron acceptor. Because the O_2/H_2O couple is the most oxidizing, more energy is available when O_2 is used than when another electron acceptor is used. As noted in Table 16.1, other electron acceptors which are near O_2

are Fe^{3+}, NO_3^-, and NO_2^-. Farther up the scale are S^0, CO_2 and SO_4^{2-}. A summary of the most common types of anaerobic respiration is given in Figure 16.28.

Assimilative and dissimilative metabolism

Inorganic compounds such as NO_3^-, SO_4^{2-}, and CO_2 are reduced by many organisms as sources of nitrogen, sulfur, and carbon, respectively. The end products of such reductions are amino groups (—NH_2), sulfhydryl groups (—SH), and organic carbon compounds (at the oxidation level of carbohydrate or lower), respectively. We discussed briefly the *nutrition* of microorganisms in Section 4.21 and noted that all organisms need sources of N, S, and C for growth. When an inorganic compound such as NO_3^-, SO_4^{2-}, and CO_2 is reduced for use as a nutrient source, it is said to be *assimilated*, and the reduction process is called *assimilative* metabolism. We emphasize here

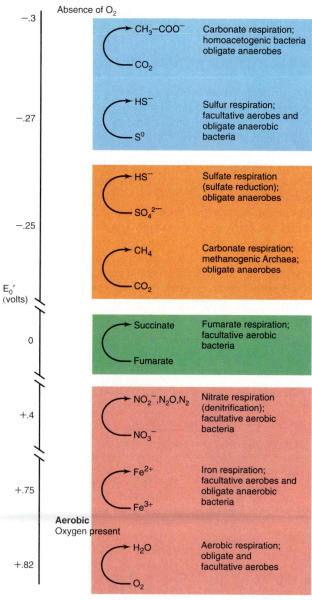

FIGURE 16.28 Various kinds of anaerobic respiration processes. The compounds are arranged in order from most reducing (top) to most oxidizing.

that assimilative metabolism of NO_3^-, SO_4^{2-}, and CO_2 is quite different from the use of these compounds as electron acceptors for *energy* metabolism. To distinguish these two kinds of reduction processes, the use of these compounds as electron acceptors in energy metabolism is called *dissimilative* metabolism.

Assimilative and dissimilative metabolism differ markedly. In assimilative metabolism, only enough of the compound (NO_3^-, SO_4^{2-}, or CO_2) is reduced to satisfy the needs of the nutrient for growth. The reduced atoms are eventually converted to cell material in the form of macromolecules. In dissimilative metabolism, a comparatively large amount of the electron acceptor is reduced, and the reduced product is *excreted* into the environment. Many organisms carry out assimilative metabolism of compounds such as NO_3^-, SO_4^{2-}, and CO_2 (for example, many Bacteria, Archaea, fungi, algae, and higher plants), whereas only a restricted variety of organisms carry out dissimilative metabolism (almost all are prokaryotes). We now discuss the main types of dissimilative metabolism.

> Although oxygen (O_2) is the most widely used electron acceptor in energy metabolism, a number of different kinds of bacteria are able to reduce other compounds and use them as electron acceptors. This process of anaerobic respiration is less energy efficient but makes it possible for these bacteria to respire in environments where oxygen is absent.

16.15   Nitrate Reduction and the Denitrification Process

Inorganic nitrogen compounds are some of the most common electron acceptors in anaerobic respiration. A summary of the various inorganic nitrogen species with their oxidation states is given in Table 16.5. The most widespread inorganic nitrogen species in nature are ammonia and nitrate, both of which are formed in the atmosphere by inorganic chemical processes, and nitrogen gas, N_2, also an atmospheric gas, which is the most stable form of nitrogen in nature. We discuss *nitrogen fixation*, the utilization of N_2 as a nitrogen source, later in this chapter.

Table 16.5 Oxidation states of key nitrogen compounds	
Compound	**Oxidation state**
Organic N (R—NH$_2$)	−3
Ammonia (NH$_3$)	−3
Nitrogen gas (N$_2$)	0
Nitrous oxide (N$_2$O)	+1 (average per N)
Nitrogen oxide (NO)	+2
Nitrite (NO$_2^-$)	+3
Nitrogen dioxide (NO$_2$)	+4
Nitrate (NO$_3^-$)	+5

One of the most common alternative electron acceptors is nitrate, NO_3^-, which is converted into more reduced forms of nitrogen, N_2O, NO, N_2. Because these products of nitrate reduction are all gaseous, they can easily be lost from the environment, and because of this the process is called **denitrification**.

Assimilative nitrate reduction, in which nitrate is reduced to the oxidation level of ammonia for use as a nitrogen source for growth, and *dissimilative nitrate reduction*, in which nitrate is used as an alternative electron acceptor in energy generation, are contrasted in Figure 16.29. Under most conditions, the end product of dissimilative nitrate reduction is N_2 or N_2O. The process is the main means by which gaseous N_2 is formed biologically, and since N_2 is much less readily available to organisms than nitrate as a source of nitrogen, for agricultural purposes at least, denitrification is a detrimental process. For sewage treatment, however, denitrification is beneficial because it converts NO_3^- to N_2, effectively decreasing the amount of available nitrogen in the sewage treatment effluent that could stimulate algal growth.

Biochemistry of dissimilative nitrate reduction

The enzyme involved in the first step of nitrate reduction, *nitrate reductase*, is a molybdenum-containing enzyme. In general, assimilative nitrate reductases are soluble proteins which are ammonia repressed, whereas dissimilative nitrate reductases are membrane-bound proteins which are repressed by O_2 and synthesized under anoxic conditions. Thus, the process of denitrification is strictly an *anaerobic* process, whereas assimilative nitrate reduction can occur quite well under fully aerobic conditions. Assimilative nitrate reduction occurs in all plants and most fungi, as well as in many prokaryotes, whereas dissimilative nitrate reduction is restricted to prokaryotes, although a wide diversity of such organisms can carry out this process.

In all cases, the first product of nitrate reduction is nitrite, NO_2^-, and another enzyme, *nitrite reductase*, is responsible for the next step. In the dissimilative process, two routes are possible, one to ammonia and the other to N_2. The route to ammonia is carried out by a fairly large number of bacteria, but is of less practical significance. There are also some bacteria which do not reduce nitrate but do reduce nitrite to ammonia. This may be a detoxification mechanism, since nitrite can be toxic under acidic conditions (nitrous acid is an effective mutagen, see Table 7.2). The pathway to nitrogen gas proceeds via two intermediate gaseous forms of nitrogen, nitric oxide (NO) and nitrous oxide (N_2O). Several organisms are known which produce only N_2O during the denitrification process, while other organisms produce N_2 as the gaseous product. Because of their global significance, the formation of gaseous nitrogen compounds by denitrifying bacteria has been under considerable study (see Section 17.14).

The biochemistry of nitrate reduction has been studied in most detail in *Escherichia coli*, an organism which is able to carry out only the first step of the

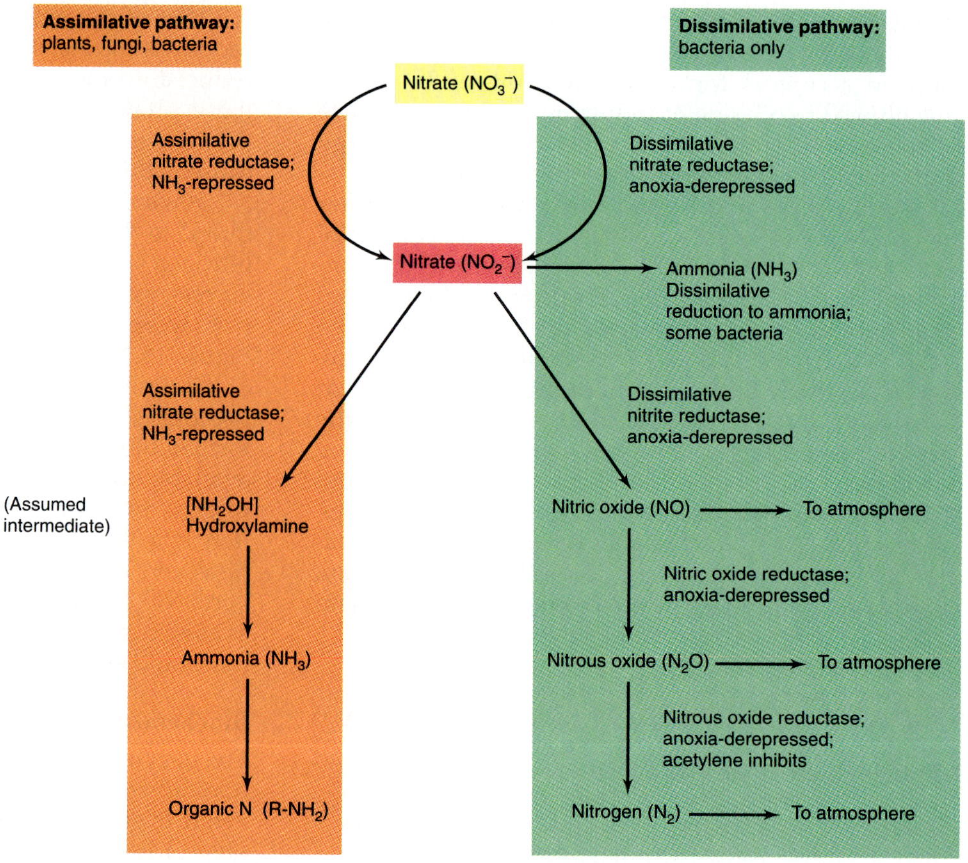

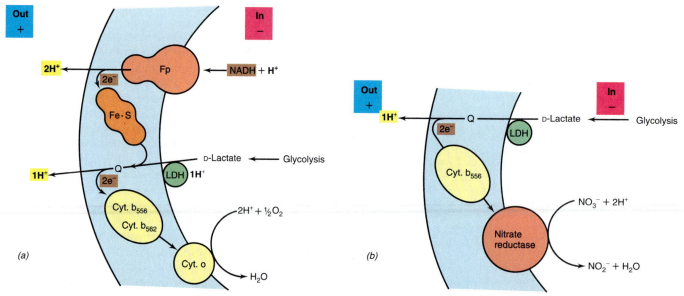

Assimilative pathway: plants, fungi, bacteria
Dissimilative pathway: bacteria only
Nitrate (NO3−)
Assimilative nitrate reductase; NH3-repressed
Dissimilative nitrate reductase; anoxia-derepressed
Nitrate (NO2−)
Ammonia (NH3) Dissimilative reduction to ammonia; some bacteria
Assimilative nitrate reductase; NH3-repressed
Dissimilative nitrite reductase; anoxia-derepressed
(Assumed intermediate)
[NH2OH] Hydroxylamine
Nitric oxide (NO)
To atmosphere
Nitric oxide reductase; anoxia-derepressed
Ammonia (NH3)
Nitrous oxide (N2O)
To atmosphere
Nitrous oxide reductase; anoxia-derepressed; acetylene inhibits
Organic N (R-NH2)
Nitrogen (N2)
To atmosphere

FIGURE 16.29 Comparison of assimilative and dissimilative processes for the reduction of nitrate.

process, the reduction of NO_3^- to NO_2^-. The enzyme nitrate reductase is membrane-bound and accepts electrons from a cytochrome *b* when O_2 is *absent* from the environment. A comparison of the electron transport chains in aerobic metabolism and nitrate respiration in *E. coli* is given in Figure 16.30. As seen, because of the reduction potential of the NO_3^-/NO_2^- couple, only one proton translocating process occurs instead of the two which occur in aerobic respiration. This is consistent with the lower reduction potential of the NO_3^-/NO_2^- couple as compared to the O_2/H_2O couple (see Table 16.1).

Isolation of denitrifying bacteria

Enrichment culture of denitrifying bacteria is straightforward. A defined culture medium is used in which potassium nitrate is added as an electron acceptor and anaerobic conditions maintained. An energy source

Out +
In −
2H+
Fp
NADH + H+
2e−
Fe·S
D-Lactate
Glycolysis
1H+
Q
2e−
LDH
1H+
Cyt. b556
Cyt. b562
2H+ + ½O2
(a)
Cyt. o
H2O
Out +
In −
1H+
Q
D-Lactate
Glycolysis
2e−
LDH
Cyt. b556
NO3− + 2H+
Nitrate reductase
(b)
NO2− + H2O

FIGURE 16.30 Electron transport processes in *Escherichia coli* when (a) O_2 or (b) NO_3^- are used as electron acceptors. Fp, flavoprotein; Q, coenzyme Q; LDH, lactate dehydrogenase.

(electron donor) must be added which is not readily fermentable (see Section 16.19) so that fermentative organisms will not be selected. Suitable energy sources include ethanol, acetate, succinate, or benzoate. Because most denitrifying organisms are facultative aerobes, great care to maintain anaerobic conditions is not necessary. A simple glass-stoppered incubation vessel is sufficient. The residual oxygen present in the culture medium will be quickly used up and then the population will switch to anaerobic respiration. Once good growth is obtained, the culture should become quite turbid. If nitrogen gas is produced, bubbles may appear under the glass stopper. The most common bacteria enriched in this way are *Pseudomonas* species, such as *Pseudomonas fluorescens*.

> Nitrate is a commonly used electron acceptor in anaerobic respiration. Its utilization involves participation of the enzyme nitrate reductase, a molybdenum-containing enzyme which is capable of reducing nitrate to nitrite. The bacteria which use nitrate in anaerobic respiration generally produce nitrogen gas (N_2), a process called denitrification.

16.16   Sulfate Reduction

Several inorganic sulfur compounds are important electron acceptors in anaerobic respiration. A summary of the oxidation states of the key sulfur compounds is given in Table 16.6. Sulfate, the most oxidized form of sulfur, is one of the major anions in seawater and is used by the *sulfate-reducing bacteria*, a group which is widely distributed in nature. The end product of sulfate reduction is H_2S, an important natural product which participates in many biogeochemical processes (see Section 17.15). Again, it is important to distinguish between assimilative and dissimilative sulfate reduction. Many organisms, including higher plants, algae, fungi, and most prokaryotes use sulfate as a sulfur source for biosynthesis. But the ability to utilize sulfate as an *electron acceptor* for energy-generating processes is restricted to a very special group of obligately anaerobic bacteria, the sulfate-reducing bacteria.

As the reduction potential in Table 16.1 shows, sulfate is a much less favorable electron acceptor than either O_2 or NO_3^-. However, sufficient energy to make ATP is available when an electron donor that yields NADH or FADH is used. Because of the less favorable energetics, growth yields are much lower for an organism growing on SO_4^{2-} than one growing on O_2 or NO_3^-. A list of some of the electron donors used by sulfate-reducing bacteria is given in Table 16.6. The first three compounds listed, H_2, lactate, and pyruvate, are used by a wide variety of sulfate-reducing bacteria; the others have more restricted use. However, a large variety of morphological and physiological types of sulfate-reducing bacteria are known; their characteristics and taxonomy are discussed in Section 19.8.

Biochemistry and energetics of sulfate reduction

The reduction of SO_4^{2-} to hydrogen sulfide, an 8-electron reduction, proceeds biochemically through a number of intermediate stages. The sulfate ion is fairly stable and cannot be used without first being activated. Sulfate is activated by means of ATP. The enzyme *ATP sulfurylase* catalyzes the attachment of the sulfate ion to a phosphate of ATP, leading to the formation of **adenosine phosphosulfate, APS**, as shown in Figure 16.31. In dissimilative sulfate reduction, the sulfate ion of APS is reduced directly to sulfite (SO_3^{2-}), but in assimilative reduction, another P is added to APS to form **phosphoadenosine phosphosulfate (PAPS)** (Figure 16.31b), and only then is the sulfate ion reduced. In both cases, the first product of sulfate reduction is *sulfite*, SO_3^{2-}. Once SO_3^{2-} is formed, the subsequent reductions proceed readily. Many organisms are able to reduce SO_3^{2-}, for use either as an electron acceptor or in the detoxification of sulfite (which is fairly toxic). In *assimilative* sulfate reduction, the H_2S formed is immediately converted into organic sulfur in the form of amino acids, etc., but in dissimilative sulfate reduction, the H_2S is excreted.

The sulfate-reducing bacteria carry out a cytochrome-based electron transport process, the electrons from the energy source being transferred to the sulfate ion in APS. The cytochrome of the sulfate-reducing bacteria is a unique low potential type called *cytochrome* c_3. This cytochrome is not found in organisms using other electron acceptors. Other electron carriers in the electron transport chain of the sulfate-reducing bacteria include ferredoxin and flavodoxin. Type II sulfate-reducers (species capable of degrading acetate and other fatty acids, see Section 19.8) also contain a cytochrome of the *b* type which is presumably

Table 16.6	Sulfur compounds and electron donors for sulfate reduction
Compound	**Oxidation state**
Oxidation states of key sulfur compounds	
Organic S (R—SH)	−2
Sulfide (H_2S)	−2
Elemental sulfur (S^0)	0
Thiosulfate ($S_2O_3^{2-}$)	+2 (average per S)
Tetrathionate ($S_4O_6^{2-}$)	+2.5 (average per S)
Sulfur dioxide (SO_2)	+4
Sulfite (SO_3^{2-})	+4
Sulfur trioxide (SO_3)	+6
Sulfate (SO_4^{2-})	+6
Some electron donors used for sulfate reduction	
H_2	Propionate
Lactate	Acetate
Pyruvate	Butyrate
Ethanol	Fatty acids
Fumarate	Benzoate
Malate	Indole
Choline	

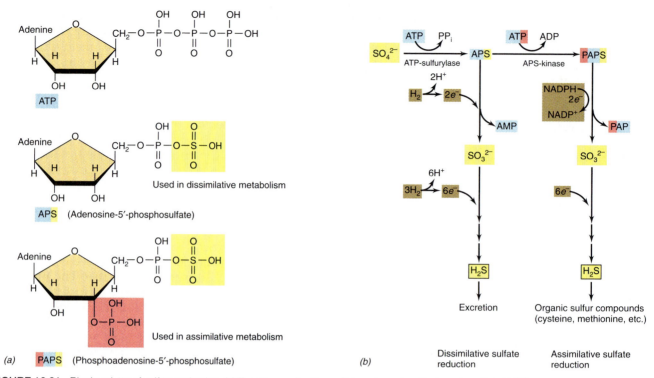

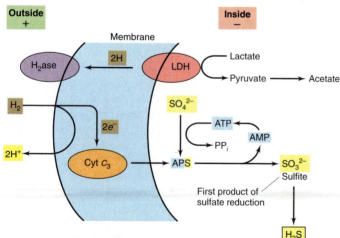

FIGURE 16.31 Biochemistry of sulfate reduction. (a) Two forms of *active sulfate*, adenosine-5'-phosphosulfate (APS) and phosphoadenosine-5'-phosphosulfate (PAPS). (b) Schemes of assimilative and dissimilative sulfate reduction.

involved in the electron transport chain of these species; cytochrome *b* is absent from species of sulfate-reducing bacteria which do not degrade fatty acids. The electron transport system is shown in Figure 16.32.

In electron transport in sulfate-reducing bacteria, hydrogen, H_2, either directly from the environment or generated from certain organic electron donors, transfers electrons to the enzyme *hydrogenase*, which is situated in the periplasm in close association with cytochrome c_3. Due to the spatial arrangement of the electron transport components in the membrane, when the H atoms of H_2 are oxidized, the protons (H^+) remain *outside* the membrane, whereas the electrons are transferred *across* the membrane. In this way, a

FIGURE 16.32 Electron transport and generation of a proton motive force in a sulfate-reducing bacterium. In addition to external hydrogen (H_2) arising from the metabolism of fermentative anaerobes, this compound can also originate from the catabolism of organic compounds, such as lactate.

proton-motive force is set up which can be used for the synthesis of ATP. In the cytoplasm, the electrons are used in the reduction of APS.

When sulfate-reducing bacteria grow on H_2/SO_4^{2-}, they are growing chemolithotrophically as H_2 bacteria (see Section 16.9). Some species can even grow autotrophically under these conditions, using CO_2 as sole source of carbon. However, most sulfate-reducing bacteria are chemoorganotrophs and use various organic compounds as electron donors (some of which were illustrated in Table 16.6).

Growth of sulfate-reducing bacteria on acetate

Many sulfate-reducing bacteria are known that are capable of growth on *acetate* as sole energy source; most such organisms are of marine origin (see Section 19.8). These organisms oxidize acetate completely to CO_2 and reduce sulfate to sulfide:

$$acetate^- + SO_4^{2-} + 3H^+ \rightarrow 2CO_2 + H_2S + 2H_2O$$
$$\Delta G^{0\prime} = -57.5 \text{ kJ/reaction}$$

Although the tricarboxylic acid cycle (Section 4.12) is the major means by which acetate is oxidized in living organisms, it does not function as such in acetate-oxidizing sulfate reducers. Instead, two biochemical mechanisms for acetate oxidation by sulfate-reducing bacteria are known, a modified tricarboxylic acid (TCA) cycle and the acetyl-CoA pathway.

Organisms like *Desulfobacter* use the modified TCA cycle for acetate oxidation. This series of reactions uses many but not all of the enzymes of the citric acid cycle. *Desulfobacter* contains an enzyme which first activates acetate by reaction with succinyl-CoA to yield succinate and acetyl-CoA; the latter then enters

the TCA cycle and is oxidized to CO_2 (Figure 16.33). As previously mentioned, because the SO_4^{2-}/SO_3^{2-} couple has such a low reduction potential (see Table 16.1), energy is actually *required* to drive this first step in sulfate reduction (see Figure 16.31). However, this energy requirement creates a special problem for acetate-oxidizing sulfate-reducers, because the ATP yield from the oxidation of acetate via the TCA cycle is about that required for activation of sulfate to form APS (and hence sulfite, see Figure 16.31). However, a novel enzyme present in *Desulfobacter* allows for net ATP synthesis (and thus growth) at the expense of acetate. *Desulfobacter* uses *citrate lyase*, an enzyme which can couple to ATP synthesis by *substrate-level phosphorylation* from the conversion of acetyl-CoA (via acetyl-P) to acetate during the production of citrate (Figure 16.33). This additional ATP makes growth on acetate possible.

Most acetate-oxidizing sulfate-reducing bacteria do not use the modified TCA cycle but instead employ the acetyl-CoA pathway for acetate oxidation. This pathway serves to oxidize acetate to CO_2 via a quite different series of reactions from that of the TCA cycle, employing the key enzyme *carbon monoxide dehydrogenase*. However, because the acetyl-CoA pathway was first discovered in *homoacetogenic* bacteria (discussed in Section 19.9) and its presence only later discovered in sulfate-reducing bacteria, we reserve discussion of this pathway for later (see Sections 19.8, 19.9, and 20.5).

Sulfate and thiosulfate disproportionation

Certain sulfate-reducing bacteria are capable of a unique form of energy metabolism using reduced sulfur compounds, called *disproportionation*. The term disproportionation refers to the splitting of a compound into two new compounds, one of which is *more oxi-dized* and one of which is *more reduced* than the original substrate. In the present discussion, we describe the disproportionation of thiosulfate ($S_2O_3^{2-}$) and sulfite (SO_3^{2-}).

Desulfovibrio sulfodismutans can disproportionate sulfur compounds as follows:

$$S_2O_3^{2-} + H_2O \rightarrow SO_4^{2-} + H_2S$$
$$\Delta G^{0'} = -21.9 \text{ kJ/reaction}$$

Note that one sulfur atom of $S_2O_3^{2-}$ becomes more oxidized (forming SO_4^{2-}) and the other more reduced (forming H_2S). Another disproportionation involves sulfite:

$$4SO_3^{2-} + 2H^+ \rightarrow 3SO_4^{2-} + H_2S$$
$$\Delta G^{0'} = -235.6 \text{ kJ/reaction}$$

In these reactions electrons from either $S_2O_3^{2-}$ or SO_3^{2-} enter the electron transport chain and eventually reduce other molecules of $S_2O_3^{2-}$ or SO_3^{2-}, respectively, to H_2S. Sulfur disproportionation can also be viewed as a type of chemolithotrophy, because the energy source (electron donor) is inorganic. Presumably, disproportionation serves as a means of energy metabolism when organic compounds are in short supply but when reduced sulfur compounds are still available. Some strains of sulfate-reducing bacteria can even grow autotrophically with energy coming from sulfur disproportionation, but most strains require organic compounds as carbon sources.

Although the enzymology of sulfur disproportionation has not been worked out, it is clear that electron transport events, presumably linked to proton gradient formation, are central to the bioenergetics of these energy-yielding reactions. In addition, however, in the case of SO_3^{2-} oxidation, substrate-level phosphorylation may serve to increase the ATP yield as SO_3^{2-} is oxidized to SO_4^{2-} via reversal of the APS reductase system (see Figure 16.31).

> The sulfate-reducing bacteria are a widely distributed but rather specialized group of organisms which are able to reduce sulfate to hydrogen sulfide. Because the reduction of sulfate requires first its activation by a reaction with ATP, the sulfate-reduction process is not very efficient energetically. Electron donors for sulfate reduction include hydrogen gas and a number of organic compounds, of which acetate and lactate are two of the most important.

16.17   Carbon Dioxide as an Electron Acceptor

Carbon dioxide, CO_2, is common in nature and is, of course, a major product of the energy metabolism of chemoorganotrophs. Several prokaryotic groups are able to use CO_2 as an electron acceptor in anaerobic respiration. The most important CO_2-reducing prokaryotes are the **methanogens**, a major group of Archaea. Some of these organisms utilize H_2 as the

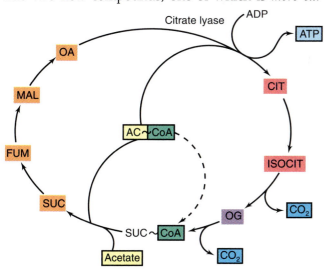

FIGURE 16.33 Mechanism of acetate oxidation to CO_2 in sulfate-reducing bacteria that use the citric acid cycle for acetate oxidation. Note how coenzyme-A is recycled in this reaction series through interaction of succinyl-CoA and acetate, and note the use of citrate lyase to form citrate. See text for details. Abbreviations: OA, oxalacetate; MAL, malate; FUM, fumarate; SUC, succinate; SUC~CoA, succinyl~Coenzyme A; OG, α-ketoglutarate; ISOCIT, isocitrate; CIT, citrate; AC~CoA, acetyl~coenzyme A.

electron donor (energy source); the overall reaction is:

$$4H_2 + H^+ + HCO_3^- \rightarrow CH_4 + 3H_2O$$
$$\Delta G^{0'} = -135.6 \text{ kJ/reaction}$$

When growing on H_2/CO_2, the methanogens are chemolithotrophic autotrophs. However, the process by which CO_2 is fixed is not the Calvin cycle of conventional autotrophs but instead the acetyl-CoA pathway (see Section 20.5). We discuss the ecology of methanogenesis in Section 17.12 and the methanogens themselves in Section 20.5.

Another group of CO_2-reducing bacteria are the **homoacetogens**, which produce *acetate* rather than CH_4 from CO_2 and H_2. The overall reaction of homoacetogenesis is

$$4H_2 + H^+ + 2HCO_3^- \rightarrow CH_3-COO^- + 4H_2O$$
$$\Delta G^{0'} = -104.6 \text{ kJ/reaction}$$

The homoacetogenic bacteria are not a defined taxonomic group, since they include such widely differing organisms as the Gram-positive spore-forming bacterium *Clostridium aceticum* and the Gram-negative nonspore-forming *Acetobacterium woodii*. Homoacetogenic bacteria are discussed in detail in Section 19.9.

When comparing the two equations above, it is seen that somewhat more energy is released when methane is formed from CO_2 reduction than when acetate is formed. When growing strictly on $H_2 + CO_2$, both methanogens and homoacetogens are growing autotrophically (they are also chemolithotrophs). Both groups are also obligate anaerobes. Note, however, that methanogens are not restricted to $H_2 + CO_2$; many methanogens grow and form methane from methanol, formic acid, and acetate, as discussed in Sections 17.12 and 20.5. In addition, as discussed in Section 19.9, most homoacetogenic bacteria can also grow chemoorganotrophically by the fermentation of sugar.

The bioenergetics of ATP synthesis in both methanogens and homoacetogens is an active area of research. In methanogenesis from $H_2 + CO_2$, good evidence exists for proton gradient formation as the means by which ATP is synthesized. In homoacetogens, both substrate-level phosphorylation and oxidative phosphorylation are thought to occur. We discuss details of the energetics of homoacetogenesis in Section 19.9 and of methanogenesis in Section 20.5.

A specialized group of prokaryotes called the methanogens are able to use carbon dioxide (CO_2) as an electron acceptor in a reaction in which hydrogen (H_2) serves as energy source. The methanogens produce methane (CH_4) but another group, the homoacetogens, convert H_2 plus CO_2 into acetic acid.

16.18   Other Electron Acceptors for Anaerobic Respiration

In addition to nitrate, sulfate, and carbon dioxide, a variety of other compounds, both organic and inor-

ganic, are used by one or another group of bacteria in anaerobic respiration. We discuss the most important in this section.

Ferric iron reduction

Ferric (Fe^{3+}) iron can be reduced to the ferrous (Fe^{2+}) state by a variety of microorganisms, and because Fe^{3+} is abundant in many microbial habitats, its reduction can be a major form of anaerobic respiration. Ferric iron can serve as an electron acceptor for energy metabolism in certain fungi and in a wide variety of both chemoorganotrophic and chemolithotrophic bacteria. The reduction potential of the Fe^{3+}/Fe^{2+} couple is very high ($E_0' = 0.77$ volts), and because of this, Fe^{3+} reduction can be coupled to the oxidation of a wide variety of both organic and inorganic electron donors. Various organic compounds including aromatic compounds can be oxidized anaerobically by ferric iron reducers with electrons presumably traveling through electron transport chains that terminate in a ferric iron reductase system. Such electron flow establishes a proton gradient that can be used to generate ATP. Most research on the energetics of ferric iron reduction has been done in the Gram-negative bacterium *Shewenella putrefaciens*, in which Fe^{3+}-dependent anaerobic growth occurs with various organic electron donors.

Ferric iron is one of the most common metals present in soils and rocks, and its reduction leads to the production of ferrous iron, a more soluble form of iron. Bacterial iron reduction can thus lead to solubilization of iron, an important geochemical process. As illustrated in Figure 16.23c, one type of iron deposit, called *bog iron*, is formed as a result of the activities of ferric iron-reducing microorganisms.

Manganese reduction

The metal manganese has a number of oxidation states, of which Mn^{4+} and Mn^{2+} are the most stable and biologically relevant. Anaerobic reduction of Mn^{4+} to Mn^{2+} is carried out by a variety of microorganisms, mostly chemoorganotrophs. However, in most cases it has been difficult to show that Mn^{4+} reduction is energetically beneficial because fortuitous reduction of Mn^{4+} also occurs in many organisms. However, in *S. putrefaciens* and a few other bacteria, anaerobic growth on acetate and several other nonfermentable carbon sources occurs with Mn^{4+} as electron acceptor. This indicates that Mn^{4+}-supported anaerobic respiration is possible, presumably at the expense of a proton motive force formed during electron transport from an electron donor to Mn^{4+} as electron acceptor.

Organic electron acceptors

Several organic compounds can participate in anaerobic respirations. Note that many organic compounds also serve as electron acceptors in conventional chemoorganotrophic metabolism. However, in these cases, for example during fermentation, the organic electron acceptor is produced *internally* as a part of the metabolic process and then reduced internally. In the present case, we are discussing organic compounds

that serve as electron acceptors when added *externally.* More importantly, in fermentation there is no electron transport process with accompanying oxidative phosphorylation as there is in anaerobic respiration.

Several organic compounds that have been recognized as common electron acceptors for anaerobic respiration are listed in Table 16.7. Of those listed, the compound which has been most extensively studied is **fumarate**, which is reduced to **succinate.** An examination of the *tricarboxylic acid cycle* in Figure 4.20 will indicate that fumarate and succinate are important intermediates. Fumarate's role as an electron acceptor for anaerobic respiration derives from the fact that the fumarate–succinate couple has an acceptable reduction potential (near 0 volts, see Table 16.1), which allows coupling of fumarate reduction with NADH oxidation. The energy yield is sufficient for the synthesis of 1 ATP. Bacteria able to use fumarate as an electron acceptor include *Wolinella succinogenes* (which can grow on H_2 as sole energy source using fumarate as electron acceptor), *Desulfovibrio gigas* (a sulfate-reducing bacterium which can also grow under nonsulfate-reducing conditions), some clostridia, *Escherichia coli,* and *Proteus rettgeri.* Another bacterium, *Streptococcus faecalis,* can use fumarate as an electron acceptor but does not couple this to oxidative phosphorylation. In the latter case, fumarate merely serves in the reoxidation of NADH that had been formed during glycolysis.

The compound **trimethylamine oxide** listed in Table 16.7 is an interesting electron acceptor. Trimethylamine oxide (TMAO) is an important osmotic solute in marine fish, where it serves in these animals as a means of excreting excess nitrogen, but a variety of bacteria are able to reduce TMAO to trimethylamine (TMA). TMA has a strong odor and flavor, and some of the spoiled odor that frequently occurs in marine fish is due to TMA produced by bacterial action. A variety of facultatively aerobic bacteria are able to utilize TMAO as an alternate electron acceptor. In addition, several phototrophic purple bacteria (see Section 19.1) are able to use TMAO as an electron acceptor for dark anaerobic metabolism. A compound analogous to TMAO is **dimethyl sulfoxide** (DMSO), which is reduced by a variety of bacteria to dimethyl sulfide (DMS). DMSO is a common natural product and is found in both marine and freshwater environments.

DMS has a strong, pungent odor, and bacterial reduction of DMSO to DMS is signaled by the presence of the characteristic odor of DMS. A variety of bacteria, including *Campylobacter, Escherichia,* and many purple bacteria, are able to use DMSO as an electron acceptor in energy generation (see Section 17.15 for further discussion of DMSO metabolism).

The reduction potential of the TMAO/TMA and DMSO/DMS couples are similar, about +0.15 volts, which means that any electron-transport chain that ends with TMAO or DMSO reduction must be rather brief. In most instances of TMAO and DMSO reduction, cytochromes of the *b*-type (with reduction potentials near 0 volts) have been identified as terminal oxidases.

16.19   Fermentations

Because oxygen is not highly soluble, many environments easily become anoxic. In such environments, decomposition of organic materials occurs anaerobically. If adequate supplies of the electron acceptors previously discussed are not available in such anaerobic environments, much of the carbon will be catabolized by fermentation. We discussed the overall process of fermentation in Section 4.8 and showed that it was an internally balanced oxidation–reduction process in which carbon from the same external organic compound was partially oxidized and partially reduced (Figure 16.34).

There are two problems an organism faces if it is to catabolize organic compounds in energy-yielding metabolism: (1) conserving some of the energy released as ATP and (2) disposing of electrons removed from the electron donor. In fermentation, ATP synthesis generally occurs by way of *substrate-level phosphorylation,* a mechanism by which high energy phosphate bonds from organic intermediates of the fermentation are transferred to ADP (see Section 4.8). The second problem, that of redox balance, is solved by production and excretion by the organism of *fermentation products* generated from the original substrate (Figure 16.34). We now consider these basic principles of fermentation in more detail and highlight the enormous diversity of microbial fermentations known.

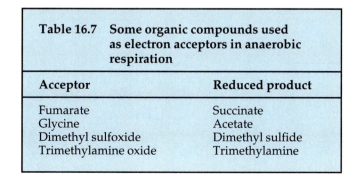

Table 16.7	Some organic compounds used as electron acceptors in anaerobic respiration
Acceptor	**Reduced product**
Fumarate	Succinate
Glycine	Acetate
Dimethyl sulfoxide	Dimethyl sulfide
Trimethylamine oxide	Trimethylamine

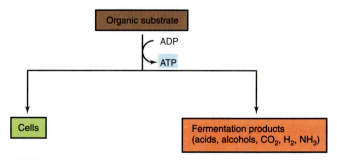

FIGURE 16.34 *Overall process of fermentation.*

High energy compounds and substrate-level phosphorylation

Energy can be obtained by substrate-level phosphorylation in many different ways. However, central to the mechanism of ATP synthesis is the production of one or another *high energy compound*. These are generally organic compounds containing a phosphate group or a coenzyme-A molecule, the hydrolysis of which is highly exergonic. A list of the major high energy intermediates is given in Table 16.8. This list is not complete, but it includes most recognized high energy intermediates known to be formed during biochemical processes. Since most of the compounds listed in Table 16.8 can couple directly to ATP synthesis (-31.8 kJ/mole), if an organism can form one or another of these compounds during fermentative metabolism, it can make ATP. Substrate-level phosphorylation is a more direct way of making ATP than via a proton motive force, but requires that the energy source couple directly to a high energy intermediate.

Pathways for the anaerobic breakdown of various fermentable substances to high-energy intermediates are summarized in Figure 16.35. It should be noted that this figure is organized by the high energy compounds listed in Table 16.8 and that either one of these compounds, or a related derivative, are generated in each case and lead to ATP synthesis. Thus, Figure 16.35 and Table 16.8 should be examined together.

Table 16.8	Energy-rich compounds involved in substrate-level phosphorylation*	
Name	**Free energy of hydrolysis, $-\Delta G^{0\prime}$ (kJ/mole)**
Acetyl-CoA	35.7
Propionyl-CoA	35.6
Butyryl-CoA	35.6
Succinyl-CoA	35.1
Acetylphosphate	44.8
Butyrylphosphate	44.8
1,3-Bisphosphoglycerate	51.9
Carbamyl phosphate	39.3
Phosphoenolpyruvate	51.6
Adenosine-phosphosulfate	88
N^{10}-formyltetrahydrofolate	23.4

Data from Thauer, R.K., K. Jungermann, and K. Decker. 1977. Bacteriol. Rev. 41:100–180.

Energy yields of fermentative organisms

How much ATP can be produced by a fermentative organism? As we have seen, glucose fermenters produce 2–3 ATPs per mole of glucose fermented in glycolysis (see Figure 4.11). This is about the maximum amount of ATP produced by fermentation; many other substrates provide less energy. The potential en-

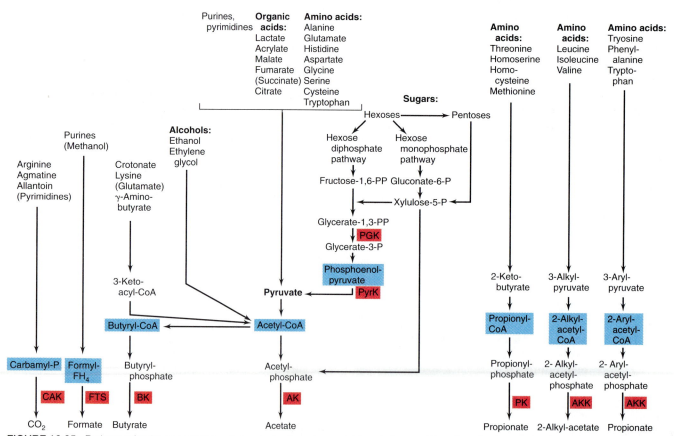

FIGURE 16.35 Pathways for the anaerobic breakdown of various fermentable substances. The sites of substrate-level phosphorylation are shown by the abbreviations: CAK, carbamyl phosphate kinase; FTS, formyltetrahydrofolate synthetase; AK, acetate kinase; PK, propionate kinase; BK, butyrate kinase; AKK, alkyl (aryl) acetate kinase; PGK, phosphoglycerate kinase; PyrK, pyruvate kinase. High energy CoA derivatives and other key high energy compounds and enzymes are highlighted. Refer back to Table 16.8.

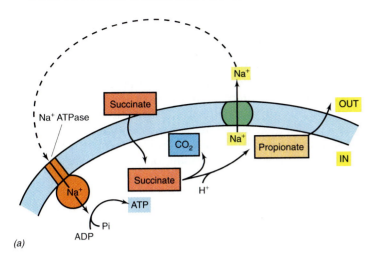

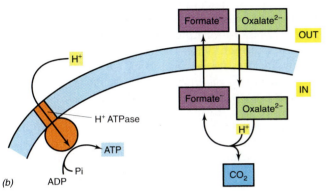

FIGURE 16.36 The unique fermentations of succinate and oxalate. (a) Succinate fermentation by *Propionigenium modestum*. A sodium-translocating ATPase produces ATP; sodium export is linked to succinate decarboxylation. (b) Oxalate fermentation by *Oxalobacter formigenes*. Oxalate import and formate export by a formate/oxalate antiporter consume protons. ATP synthesis is linked to a normal proton-driven ATPase.

ergy released from a particular fermentation can be calculated from the balanced reaction and from the free energy values that are given in Appendix 1. For instance, the fermentation of glucose to ethanol and CO_2 has a theoretical energy yield of -235 kJ/mole, enough to produce about 7 ATPs. However, only 2 ATPs are actually produced, which implies that the organism operates at considerably less than 100 percent efficiency (see Section 4.13 for a discussion of energy efficiency).

Fermentations without substrate-level phosphorylation

With certain substrates, there is insufficient energy released to couple to the synthesis of ATP directly by substrate-level phosphorylation, yet these compounds support fermentative growth of an organism. In these cases, catabolism of the substrate is linked to ion pumps that establish a proton or sodium gradient across the membrane. Examples of this include the fermentations of *Propionigenium modestum* and *Oxalobacter formigenes*; both of these organisms couple the fermentation of dicarboxylic acids to membrane-bound energy-linked ion pumps. *P. modestum* carries out the following reaction:

$$\text{succinate}^{2-} + H_2O \rightarrow \text{propionate}^- + HCO_3^-$$
$$\Delta G^{0\prime} = -20.5 \text{ kJ/reaction}$$

This reaction yields insufficient free energy to couple to ATP synthesis directly by substrate-level phosphorylation, but nevertheless, it serves as the sole energy-yielding reaction for growth of the organism. This is possible because the decarboxylation of succinate by *P. modestum* is coupled to the export of Na^+ across the cytoplasmic membrane (Figure 16.36a). A Na^+-translocating ATPase in the membrane of *P. modestum* employs this Na^+ gradient to drive ATP synthesis (Figure 16.36a). *O. formigenes* carries out the fermentation of oxalate:

$$\text{oxalate}^{2-} + H_2O \rightarrow \text{formate}^- + HCO_3^-$$
$$\Delta G^{0\prime} = -26.7 \text{ kJ/reaction}$$

At neutral pH, oxalate exists in the ionized form as oxalate^{2-} and its decarboxylation to formate$^-$ consumes one proton. The subsequent export of formate from the cell then builds a proton gradient which can be coupled to ATP synthesis by a proton-translocating ATPase in the membrane (Figure 16.36b).

The interesting and unique aspect of the metabolism of both *P. modestum* and *O. formigenes* is the fact that ATP synthesis occurs without substrate-level phosphorylation *or* electron-transport occurring; ATP is formed as a result of a Na^+/H^+ pump linked to export of organic acids. The lesson to be learned from these fermentations is clear: any chemical reaction that yields less than the 31.8 kJ required to make 1 ATP or that appears unable to couple to a substrate-level phosphorylation, cannot be automatically ruled out as a potential growth-supporting reaction for a bacterium. If the reaction can be coupled to an ion gradient, ATP production (and subsequently growth) remains a possibility.

> Although most organic compounds can be utilized when a suitable electron acceptor is present, in the absence of an external electron acceptor, organic compounds can only be utilized by the process of fermentation. To be fermentable, an organic compound must be neither too oxidized nor too reduced. In most fermentations, ATP is produced by substrate-level phosphorylation but certain organisms are able to create an ion gradient and produce ATP in this way when growing fermentatively.

Oxidation–reduction balance

In any fermentation reaction, there must be a *balance* between oxidation and reduction. The total number of electrons in the products on the right side of the equation must balance the number in the substrates on the left side of the equation. When fermentations are studied experimentally in the laboratory, it is conventional to calculate a *fermentation balance* to make certain that no products are missed. The fermentation balance can also be calculated theoretically from the oxidation states of the substrates and products (see Appendix 1 for the procedure for calculating oxidation states).

In a number of fermentations, electron balance is maintained by the production of molecular hydrogen, H_2. In H_2 production, protons (H^+) derived from water serve as electron acceptor. Production of H_2 is generally associated with the presence in the organism of an iron-sulfur protein called *ferredoxin*, an electron carrier of low redox potential. The transfer of electrons from ferredoxin to H^+ is catalyzed by the enzyme **hydrogenase**, as illustrated in Figure 16.37. We have already discussed the enzyme hydrogenase earlier in this chapter in reference to the *utilization* of hydrogen by sulfate-reducing bacteria. In the present case, hydrogenase is involved in the *production* of hydrogen.

The energetics of hydrogen production are actually somewhat unfavorable, so that most fermentative organisms only produce a relatively small amount of hydrogen along with other fermentation products. Hydrogen production thus primarily serves to maintain redox balance. If hydrogen production is prevented, for instance, then the oxidation-reduction balance of the other fermentation products will be directed more toward *reduced* products. Thus, many fermentative organisms that make H_2 produce both ethanol and acetate. Since ethanol is more reduced than acetate, its formation is favored when hydrogen production is inhibited.

Numerous anaerobic bacteria produce *acetate* as one of the products of fermentation. The production of acetate is energetically advantageous since it allows the organism to produce additional ATP by substrate-level phosphorylation. The key intermediate generated in acetate production is acetyl-CoA (see Table 16.8), a high-energy intermediate. Acetyl-CoA can be converted to acetyl phosphate (also listed in Table 16.8), and the high-energy phosphate group of acetyl phosphate subsequently transferred to ADP by acetate kinase, yielding ATP. One of the main substrates that is converted to acetyl-CoA is pyruvate, a major product of glycolysis. However, the conversion of pyruvate to acetyl-CoA is an oxidation reaction, since pyruvate is more reduced than acetyl-CoA. The excess electrons generated must either be used to make a more re-

duced end product or be used in the production of H_2 as discussed above.

Diversity of fermentations

Fermentations can be classified either in terms of the substrate fermented or in terms of the fermentation products formed. Many of the specific fermentation reactions of bacteria will be discussed when the individual groups are discussed in Chapters 19 and 20. Here we present an overview of common anaerobic fermentations.

Table 16.9 summarizes some of the main types of fermentations, as classified on the basis of *products formed*. Note some of the broad categories, such as alcohol, lactic acid, propionic acid, mixed acid, butyric acid, and homoacetic acid. A number of fermentations are classified on the basis of the *substrate fermented* rather than the fermentation product. For instance, many of the spore-forming anaerobic bacteria (genus *Clostridium*) ferment *amino acids* with the production of acetate, lactate, ammonia, and H_2. Other *Clostridium* species, such as *C. acidi-urici* and *C. purinolyticum*, ferment *purines* such as xanthine or adenine with the formation of acetate, formate, CO_2, and ammonia. Still other anaerobes ferment *aromatic* compounds. As an example, the bacterium *Pelobacter acidigallici* ferments the aromatic compound *phloroglucinol* (1,3,5-benzenetriol, $C_6H_6O_3$) via the following overall pathway:

$$\text{Phloroglucinol } (C_6H_6O_3) + 3H_2O \rightarrow 3 \text{ acetate}^- + 3H^+$$
$$\Delta G^{0\prime} = -142.5 \text{ kJ/reaction}$$

Many unusual fermentations are known that appear to be carried out by only a very restricted group of anaerobes and, in some cases, only by a single known bacterium. Some examples are listed in Table 16.10. Many of these bacteria can be considered metabolic specialists, having evolved biochemical capabilities to catabolize a substrate or substrates not catabolized by other bacteria. However, as for the substances listed in Table 16.9, successful fermentation of these more unusual substrates requires that the organism be able to produce a high energy intermediate, usually a coenzyme A derivative of the type listed in Table 16.8, during the fermentation in order to recover the energy released as ATP.

Methane as the final product of anaerobic decomposition processes

Many of the products of fermentative metabolism listed in Tables 16.9 and 16.10 are themselves energy sources for other fermentative organisms. It is reasonable that organisms might exist which are able to use the fermentation products of other organisms, since all these anaerobes would likely be found together in environments where fermentation was taking place. For example, succinate, lactate, and ethanol, produced from the fermentation of sugars, can themselves be fermented by other organisms. Fermentation of these "fermentation products" leads ultimately to the formation of acetate, H_2, and CO_2, substrates for the

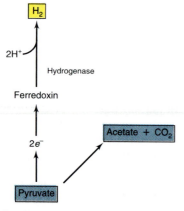

FIGURE 16.37 Production of molecular hydrogen from pyruvate.

Table 16.9 Examples of common bacterial fermentations and some of the organisms carrying them out

Type	Overall reaction	Organisms
Alcoholic fermentation	Hexoses → Ethanol + CO_2	Yeast *Zymomonas*
Homolactic fermentation	Hexose → Lactic acid	*Streptococcus* Some *Lactobacillus*
Heterolactic fermentation	Hexose → Lactic acid Ethanol CO_2	*Leuconostoc* Some *Lactobacillus*
Propionic acid	Lactate → Propionate Acetate CO_2	*Propionibacterium* *Clostridium propionicum*

Mixed acid	Hexoses → Ethanol
	2,3-Butanediol
	Succinate
	Lactate
	Acetate
	Formate
	$H_2 + CO_2$
		Enteric bacteria
		  Escherichia
		  Salmonella
		  Shigella
		  Klebsiella
		  Enterobacter

Butyric acid	Hexoses → Butyrate Acetate $H_2 + CO_2$	*Clostridium butyricum*
Butanol	Hexoses → Butanol Acetate Acetone Ethanol $H_2 + CO_2$	*C. butyricum*
Caproate	Ethanol + Acetate + CO_2 → Caproate + Butyrate + H_2	*C. kluyveri*
Homoacetic	$4H_2 + 2CO_2$ → Acetate	*C. aceticum* *Acetobacterium*
Methanogenic	Acetate → CH_4 + CO_2	*Methanothrix* *Methanosarcina*

methanogenic Archaea. However, two fermentation products listed in Tables 16.9 and 16.10 cannot be further fermented: CO_2 and CH_4, the most oxidized and the most reduced forms of carbon. Thus, the terminal products of anaerobic decomposition are CH_4 and CO_2. It is to these two carbon compounds, one the most reduced, the other the most oxidized, to which all anaerobic decomposition processes ultimately lead.

Table 16.10 Some unusual bacterial fermentations

Type	Overall reaction	Organisms
Acetylene	$2C_2H_2 + 3H_2O$ → ethanol + acetate + H^+	*Pelobacter acetylenicus*
Glycerol	4 Glycerol + $2HCO_3^-$ → 7 acetate + $5H^+$ + $4H_2O$	*Acetobacterium* spp.
Resorcinol (an aromatic compound)	$2C_6H_4(OH)_2 + 6H_2O$ → 4 acetate + butyrate + $5H^+$	*Clostridium* spp.
Cinnamate (an aromatic compound	$2C_9H_7O_2 + 2H_2O$ → $C_9H_9O_2$ + benzoate + acetate	*Acetivibrio multivorans*
Phloroglucinol (an aromatic compound)	$C_6H_6O_3 + 3H_2O$ → 3 acetate + $3H^+$	*Pelobacter massiliensis* *Pelobacter acidigallici*

Putrescine	$10C_4H_{14}N_2 + 26H_2O$ → 6 acetate + 7 butyrate
	  + $10NH_4^+$ + $16H_2$ + $13H^+$
		Unclassified Gram-positive
		  nonsporing anaerobes

Citrate	Citrate + $2H_2O$ → formate + 2 acetate + HCO_3^- + H^+	*Bacteroides* sp.
Glyoxylate	4 Glyoxylate + $3H^+$ + $3H_2O$ → $6CO_2$ + $5H_2$ + glycolate	Unclassified Gram-negative bacterium
Succinate	Succinate + H_2O → propionate + HCO_3^-	*Propionigenium modestum*
Oxalate	Oxalate + H^+ → formate + CO_2	*Oxalobacter formigenes*
Malonate	Malonate + H_2O → acetate + HCO_3^-	*Malonomonas rubra* *Sporomusa malonica*

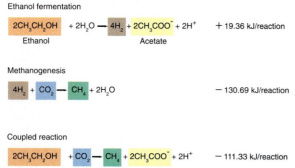

FIGURE 16.38 Fermentation of ethanol to methane and acetate by the coupling of an energetically unfavorable reaction (ethanol fermentation) with an energetically favorable reaction (methanogenesis). The overall energy yield is thus greater than −100 kJ/reaction. This is an example of *syntrophy* based on *interspecies hydrogen transfer*.

Coupled fermentation reactions

Even if a single fermentation reaction does not provide favorable energetics, this reaction may still be carried out by an organism if the *product* of the reaction, especially H_2, can be used by another organism in an energetically favorable reaction. We are dealing here with the phenomenon of **syntrophy** (mutual feeding), where two organisms do something together that neither can do separately. One of the best studied syntrophic systems in fermentative metabolism is the fermentation of ethanol to acetate and methane by two organisms, an ethanol fermenter and a methanogen. This coupled scheme is illustrated in Figure 16.38. As seen, the ethanol fermenter produces hydrogen and acetate, but this reaction has an unfavorable (positive), standard free energy balance. However, the H_2 produced by the ethanol fermenter is consumed by the methanogen in an energetically favorable reaction. When the energies of these two reactions are summed, the overall reaction is favorable energetically. The reaction summarized here is an example of a phenomenon called **interspecies hydrogen transfer**, which is discussed in more detail in Section 17.12. Note that the methanogen is unable by itself to utilize ethanol so that the ethanol fermenter and the methanogen both benefit from this relationship.

> A wide variety of fermentations are known, and fermentation products themselves are frequently subject to additional fermentation until only CH_4 and CO_2 remain. In some cases, an organic compound can only be fermented if its product can be used by another organism in a further fermentation process. Such coupled fermentation reactions usually involve H_2 and are called syntrophy (mutual feeding) and are important in the anaerobic degradation of organic matter.

Limits of anaerobic catabolism: recalcitrant compounds

It has been a watchword of general microbiology that the power of microorganisms to utilize organic compounds is impressive. But are there limits? Can *any* organic compound be fermented by a microorganism? Are there natural products that are stable to anaerobic metabolism?

Two groups of natural organic materials appear to be refractory to fermentative breakdown: lignin, and hydrocarbons. **Lignin** is a complex aromatic polymer of phenylpropane building blocks, held together by C—C and C—O—C (ether) linkages. Lignin is a major component of wood and is responsible for conferring rigidity on the cellulosic walls of woody plants. Lignin appears to be completely stable to anaerobic degradation and hence does not decompose in anoxic habitats. *Coal* is an organic material that is ultimately derived from woody plants. Coal would not exist today if lignin did not stabilize woody material to anaerobic decay.

The other major natural materials that are refractory to fermentation are the long-chain **aliphatic hydrocarbons**, such as hexadecane ($C_{16}H_{34}$) and octadecane ($C_{18}H_{38}$). These hydrocarbons, which are major constituents of petroleum, are either produced directly by plants and animals, or are derived chemically by reduction from long-chain fatty acids. Protected from decay by anoxic conditions, these hydrocarbons are stable in nature and become the basis of petroleum reserves that are a major foundation of modern society. However, although unable to serve as substrates for fermentation, saturated hydrocarbons can be slowly degraded anaerobically by certain sulfate-reducing bacteria.

Once lignin or hydrocarbons are brought into oxygen-containing environments, these compounds are readily broken down by microorganisms. Thus, it is only under anoxic conditions that these materials are stable in the biosphere. We discuss aerobic hydrocarbon oxidation in Section 16.24, the global carbon cycle in Section 17.11, and petroleum microbiology in Section 17.19.

In addition to the vast array of natural organic compounds, many organic compounds exist which have been produced artificially by chemical synthesis, for industrial or agricultural purposes. Although some of these synthetic compounds are similar to natural ones, others are quite different chemically from anything produced by living organisms. These unusual synthetic compounds are of special microbiological interest because, since these compounds have never existed in natural environments, the evolution of microorganisms capable of degrading these compounds may not have occurred. These unusual synthetic compounds are sometimes called **xenobiotics** (*xeno* is a combining form meaning *foreign*). However, in many cases careful microbiological studies have shown that xenobiotic compounds can be degraded, but only by a microbial consortium and not by pure cultures. For example, certain chlorinated hydrocarbons can only be dechlorinated anaerobically with the intermediates generated further metabolized only aerobically. A discussion of the metabolism of xenobiotics is given in Section 17.20.

16.20   Sugar Metabolism

We complete our discussion of chemoorganotrophic metabolism with consideration of a few special aspects of the catabolism of organic compounds, especially the use of polymeric substances that must first be hydrolyzed to monomeric units before energy-generating mechanisms can be employed. We begin with the microbial degradation of polysaccharides.

Hexose and polysaccharide utilization

Sugars with six carbon atoms, called **hexoses**, are the most important electron donors for many chemoorganotrophs and are also important structural components of microbial cell walls, capsules, slimes, and storage products. The most common hexose sources in nature are listed in Table 16.11 from which it can be seen that most are polysaccharides, although a few are disaccharides. Cellulose and starch are two of the most important natural polysaccharides.

Although both starch and cellulose are composed of glucose units, they are connected differently (Table 16.11), and this profoundly affects their properties. Cellulose is much more insoluble than starch and is usually less rapidly digested. Cellulose forms long fibrils, and organisms that digest cellulose are often found closely associated with them (Figure 16.39). Many fungi are able to digest cellulose and these are mainly responsible for decomposition of plant materials on the forest floor. Among bacteria, however, cellulose digestion is restricted to only a few groups, of which the gliding bacteria such as *Sporocytophaga* and *Cytophaga* (Figures 16.39 and 16.40), clostridia, and

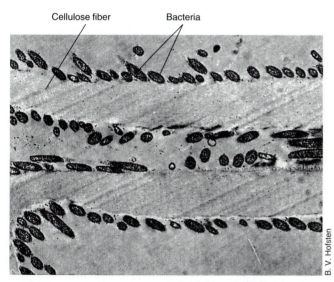

FIGURE 16.39 Transmission electron micrograph showing attachment of cellulose-digesting bacteria, *Sporocytophaga myxococcoides*, to cellulose fibers. Cells are about 0.5 μm in diameter.

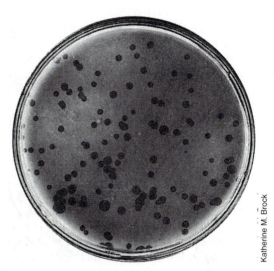

FIGURE 16.40 *Cytophaga hutchinsonii* colonies on a cellulose-agar plate. Clear areas are where cellulose has been digested.

Table 16.11   Sources of hexose sugars in nature*

Substance	Composition	Sources	Catabolic Enzymes
Cellulose	Glucose polymer (β-1,4-)	Plants (leaves, stems)	Cellulases (β,1-4 glucanases)
Starch	Glucose polymer (α-1,4-)	Plants (leaves, seeds)	Amylase
Glycogen	Glucose polymer (α-1,4- and α-1,6-)	Animals (muscle)	Amylase, phosphorylase
Laminarin	Glucose polymer (β-1,3-)	Marine algae (Phaeophyta)	β-1,3-Glucanase (laminarinase)
Paramylon	Glucose polymer (β-1,3-)	Algae (Euglenophyta and Xanthophyta)	β-1,3-Glucanase
Agar	Galactose and galacturonic acid polymer	Marine algae (Rhodophyta)	Agarase
Chitin	N-acetyl glucosamine polymer (β-1,4-)	Fungi (cell walls) Insects (exoskeletons)	Chitinase
Pectin	Galacturonic acid polymer (from galactose)	Plants (leaves, seeds)	Pectinase (polygalacturonase)
Sucrose	Glucose-fructose disaccharide	Plants (fruits, vegetables)	Invertase
Lactose	Glucose-galactose disaccharide	Milk	β-Galactosidase
Dextran	Glucose polymer	Capsules or slime layers of bacteria	Dextranase

Each of these is subject to degradation by microorganisms.

actinomycetes are the most common. Anaerobic digestion of cellulose is carried out by a few *Clostridium* species, which are common in lake sediments, animal intestinal tracts, and systems for anaerobic sewage digestion. Cellulose digestion is also a major process in the rumen of ruminant animals where *Bacteroides* and *Ruminococcus* species actively degrade cellulose (see Section 17.13).

Starch is digestible by many fungi and bacteria; this is illustrated for a laboratory culture in Figure 16.41. Starch-digesting enzymes, called *amylases*, are of considerable practical utility in many industrial situations where starch must be digested, such as the textile, laundry, paper, and food industries, and fungi and bacteria are the commercial sources of these enzymes (see Section 10.11).

All of the polysaccharides occurring extracellularly and utilized as substrates are broken down to monomeric units by hydrolysis. In contrast, the polysaccharides formed within cells as storage products are broken down not by hydrolysis but by **phosphorolysis**. This process, involving the addition of *inorganic* phosphate, results in the formation of hexose phosphate rather than the free hexose and may be summarized as follows for the degradation of starch, an α-1,4 polymer of glucose:

$$(C_6H_{12}O_6)_n + P_i \rightarrow (C_6H_{12}O_6)_{n-1} + \text{glucose-1-phosphate}$$

Because glucose-1-phosphate can be easily converted to glucose-6-phosphate, a key intermediate in glycolysis (see Figure 4.11), and no ATP is required to form it, phosphorolysis represents a net energy savings to the cell.

Many microorganisms can use *disaccharides* for growth (Table 16.11). *Lactose* utilization by microorganisms is of considerable economic importance because milk-souring organisms produce lactic acid from lactose. *Sucrose*, the common disaccharide of higher plants, is usually first hydrolyzed to its compo-

FIGURE 16.42 Slimy colony formed by the dextran-producing bacterium, *Leuconostoc mesenteroides*, growing on a sucrose-containing medium. When the same organism is grown on glucose, the colonies are small and not slimy.

nent monosaccharides (glucose and fructose) by the enzyme *invertase*, and the monomers are then metabolized by normal pathways. *Cellobiose*, β-1,4-diglucose and a major product of cellulose digestion, is also readily degraded by a variety of bacteria that cannot degrade the cellulose polymer itself.

The microbial polysaccharide *dextran* is synthesized by some bacteria using the enzyme *dextransucrase* and sucrose as starting material:

$$n \text{ sucrose} \rightarrow (\text{glucose})_n + n \text{ fructose}$$
$$\textit{dextran}$$

Dextran is formed in this way by the bacterium *Leuconostoc mesenteroides* and a few others, and the polymer formed accumulates around the cells as a massive slime or capsule (Figure 16.42). Since sucrose is required for dextran formation, no dextran is formed when the bacterium is cultured on a medium with glucose or fructose. In nature, when cells containing dextran or other polysaccharide capsules die, these materials once again become available for attack by fermentative or other chemoorganotrophic microorganisms.

16.21 Organic Acid Metabolism

A variety of organic acids can be utilized by microorganisms as carbon sources and electron donors. The acids of the tricarboxylic acid cycle such as *citrate, malate, fumarate,* and *succinate* are common natural products formed by plants and are also fermentation products of microorganisms. Because the tricarboxylic acid cycle has major *biosynthetic* (see Section 4.18) as well as *energetic* (see Section 4.13) functions, the complete cycle or major portions of it are nearly universal in microorganisms. Thus, it is not surprising that many microorganisms are able to utilize these acids as electron donors and carbon sources. Aerobic utilization of four-, five-, and six-carbon acids can be accomplished by means of enzymes of the tricarboxylic acid cycle, with ATP formation by oxidative phosphorylation.

Anaerobic utilization of organic acids usually involves conversion to pyruvate and ATP formation via the formation of acetate (see Section 16.19).

Glyoxylate cycle

Utilization of two- or three-carbon acids as carbon sources cannot occur by means of the tricarboxylic acid cycle alone. This cycle can continue to operate

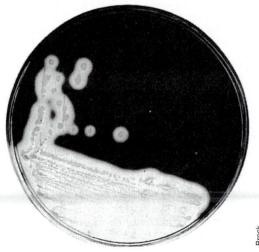

FIGURE 16.41 Demonstration of hydrolysis of starch by colonies of *Bacillus subtilis*. After incubation, the plate was flooded with Lugol's iodine solution. Where starch hydrolysis occurred, the characteristic purple color of the starch-iodine complex is absent. Hydrolysis of starch occurs at some distance from the bacterial colonies because of the production of extracellular amylase, which diffuses into the surrounding medium.

only if the acceptor molecule, the four-carbon acid *ox-alacetate*, is regenerated at each turn of the cycle; any removal of carbon compounds for biosynthetic reactions would prevent completion of the cycle. When acetate is utilized, the oxalacetate needed to continue the cycle is produced through the **glyoxylate cycle** (Figure 16.43), so-called because glyoxylate is a key intermediate. This cycle is composed of most of the TCA cycle reactions plus two additional enzymes: *isocitrate lyase*, which splits isocitrate to succinate and glyoxalate, and *malate synthase*, which converts glyoxylate and acetyl-CoA to malate.

Biosynthesis through the glyoxylate cycle occurs as follows. The splitting of isocitrate into succinate and glyoxylate allows the succinate molecule to be drawn off for biosynthesis, since glyoxylate combines with acetyl-CoA to yield malate. Malate can be converted to oxalacetate to maintain the cyclic nature of the TCA cycle despite the fact that a C-4 intermediate (succinate) has been drawn off. The succinate molecule can be used directly in the production of porphyrins, be oxidized to oxalacetate and serve as a carbon skeleton for C-4 amino acids, or be converted (via oxalacetate and phosphoenolpyruvate) into glucose.

Pyruvate and C-3 utilization

Three-carbon compounds such as pyruvate or compounds converted to pyruvate (for example, lactate or carbohydrates) also cannot be utilized as energy sources through the tricarboxylic acid cycle alone. Since some of the TCA cycle intermediates are used for biosynthesis, the oxalacetate needed to keep the cycle going is synthesized from pyruvate or phosphoenolpyruvate by the addition of a carbon atom from CO_2. In some organisms this step is catalyzed by the enzyme *pyruvate carboxylase*:

$$\text{Pyruvate} + \text{ATP} + CO_2 \rightarrow \text{oxalacetate} + \text{ADP} + P_i$$

whereas in others it is catalyzed by *phosphoenolpyruvate carboxylase*:

$$\text{Phosphoenolpyruvate} + CO_2 \rightarrow \text{oxalacetate} + P_i$$

These reactions replace oxalacetate that is lost when intermediates of the tricarboxylic acid cycle are removed for use in biosynthesis, and the cycle can continue to function.

16.22 Lipids as Microbial Nutrients

Fat and phospholipid hydrolysis

Fats are esters of glycerol and fatty acids (see Section 2.6). Microorganisms utilize fats only after hydrolysis of the ester bond, and extracellular enzymes called **lipases** are responsible for the reaction (Figure 16.44).

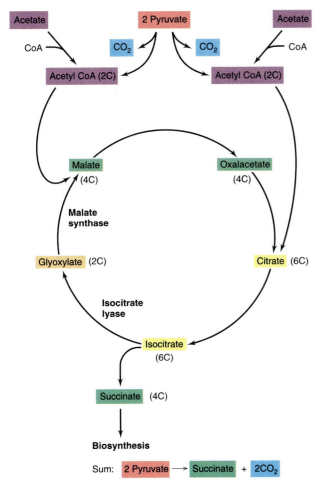

FIGURE 16.43 The glyoxylate cycle, leading to the synthesis of oxalacetate from acetate. Two unique reactions, isocitrate lyase and malate synthase, operate with a majority of the TCA cycle reactions. In addition to growth on pyruvate, the glyoxylate cycle can also operate during growth on acetate.

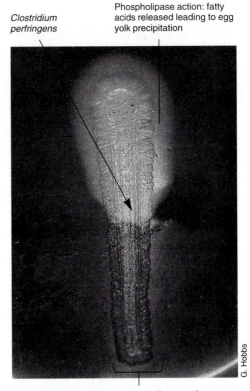

FIGURE 16.44 Action of phospholipase around streak of *Clostridium perfringens* growing on an agar medium containing egg yolk. On half of the plate, an inhibitor of phospholipase was added, preventing action of the enzyme.

The end result is formation of glycerol and free fatty acids (Figures 16.44 and 16.45). Lipases are not highly specific, and attack fats containing fatty acids of various chain lengths. Phospholipids are hydrolyzed by specific enzymes called *phospholipases*, given different letter designations depending on which ester bond they cleave (Figure 16.45). Phospholipases A and B cleave fatty acid esters and thus resemble the lipases described above, but phospholipases C and D cleave phosphate ester linkages and hence are quite different types of enzymes. The result of lipase action is the release of free fatty acids and glycerol, and all of these substances can be attacked both anaerobically as well as aerobically by various chemoorganotrophic microorganisms.

Fatty acid oxidation

Fatty acids are oxidized by a process called *beta oxidation*, in which two carbons of the fatty acid are split off at a time (Figure 16.46). In eukaryotes the enzymes are in the mitochondria, whereas in prokaryotes, they are cytoplasmic. The fatty acid is first activated with coenzyme A; oxidation results in the release of *acetyl-CoA* and the formation of a fatty acid shorter by two carbons (Figure 16.46). The process of beta oxidation is then repeated and another acetyl-CoA molecule is released. Two separate dehydrogenation reactions occur. In the first, electrons are transferred to flavin-adenine dinucleotide (FAD), whereas in the second they are transferred to NAD^+. Most fatty acids have an even number of carbon atoms, and complete oxidation yields only acetyl-CoA. The acetyl-CoA formed is then oxidized by way of the TCA cycle or is converted into hexose and other cell constituents via the glyoxylate cycle. Fatty acids are good electron donors. For example, the anaerobic oxidation of the 16-carbon fatty acid palmitic acid results in the net synthesis of 129 ATP molecules from electron-transport phosphorylation of electrons generated during the formation of acetyl-

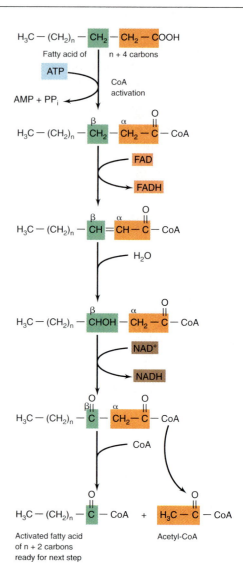

FIGURE 16.46 Mechanism of beta oxidation of a fatty acid, which leads to successive formation of two-carbon fragments of acetyl-CoA.

CoA from beta oxidations and from the oxidation of the acetyl-CoA units themselves through the TCA cycle.

> Fats are metabolized by hydrolysis by lipases or phospholipases to free fatty acids. The latter are oxidized by beta oxidation to acetyl-CoA units which are subsequently oxidized to CO_2 by the TCA cycle.

16.23   Molecular Oxygen (O_2) as a Reactant in Biochemical Processes

We have discussed in some detail the role of O_2 as an *electron acceptor* in energy-generating reactions. Although this is by far the most important role of O_2 in cellular metabolism, O_2 also plays an interesting and important role as a direct reactant in certain types of anabolic and catabolic processes.

Oxygenases are enzymes that catalyze the incorporation of oxygen from O_2 into organic compounds. There are two kinds of oxygenases: *dioxygenases*, that

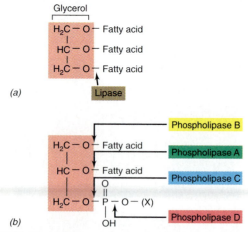

FIGURE 16.45 (a) Action of lipase on a fat. (b) Phospholipase action on phospholipid. The sites of action of the four distinct phospholipases A, B, C, and D are shown. X refers to a number of small organic molecules which may be at this position in different phospholipids. Compare this diagram to the more complete figure of a phospholipid in Figure 2.7.

catalyze the incorporation of *both* atoms of O_2 into the molecule; and *monooxygenases*, that catalyze the transfer of *only one* of the two O_2 atoms to an organic compound as a hydroxyl (OH) group, with the second atom of O_2 being reduced to water, H_2O. Because monooxygenases catalyze the formation of hydroxyl groups (OH) in organic compounds, they are sometimes called *hydroxylases*. In most monooxygenases, the electron donor is NADH or NADPH, although the direct coupling to O_2 is through a flavin which becomes reduced by the NADH or NADPH donor. In the case of ammonia monooxygenase discussed previously (see Section 16.12), the electron donor is cytochrome *c*.

There are several types of reactions in living organisms that require O_2 as a reactant. One of the best examples is the involvement of O_2 in sterol biosynthesis. The formation of the fused sterol ring system (see Figure 3.16) requires the participation of molecular oxygen. Such a reaction can obviously not take place under anaerobic conditions, so that organisms which grow anaerobically must either dispense with this reaction or obtain the required substance (sterol) preformed from their environment. The requirement of O_2 as a reactant in biosynthesis is of considerable evolutionary significance, as molecular O_2 was originally absent from the atmosphere of the earth when life evolved and only became available after the evolution of cyanobacteria, the first phototrophic organisms to produce O_2 (see Chapter 18). The role of O_2 in hydrocarbon utilization is discussed below.

> In addition to its role as an electron acceptor, oxygen (O_2) is also a chemical reactant in certain biochemical processes. Enzymes called oxygenases activate oxygen and convert it to a form in which the oxygen atom can be incorporated directly into a biochemical compound.

16.24   Hydrocarbon Transformations

Hydrocarbons are organic carbon compounds containing only carbon and hydrogen and are highly insoluble in water. Low-molecular-weight hydrocarbons are gases, whereas those of higher molecular weight are liquids or solids at room temperature. Some hydrocarbons are aliphatic compounds, a class of carbon compounds in which the carbon atoms are joined in open chains. There is a tremendous variation among aliphatic hydrocarbons in chain length, degree of branching, and number of double bonds. Another important group of hydrocarbons contains the aromatic ring and can be viewed as derivatives of benzene.

Aliphatic hydrocarbons

Only relatively few kinds of microorganisms (for example, *Nocardia*, *Pseudomonas*, *Mycobacterium*, and certain yeasts and molds) can utilize hydrocarbons for growth. For the most part, utilization of saturated aliphatic hydrocarbons is an *aerobic* process: in the absence of O_2, saturated hydrocarbons are virtually unaffected by microorganisms (a novel sulfate-reducing bacterium is an exception; see Section 16.19).

The initial oxidation step of saturated aliphatic hydrocarbons involves molecular oxygen (O_2) as a reactant, and one of the atoms of the oxygen molecule is incorporated into the oxidized hydrocarbon. This reaction is carried out by a monooxygenase (see Section 16.23), and a typical reaction sequence is that shown in Figure 16.47. The end product of the reaction sequence is acetyl-CoA. However, the initial oxidation is not at the terminal carbon in all cases. Oxidation may sometimes occur at the second carbon and then quite different subsequent reactions occur. *Unsaturated* aliphatic hydrocarbons containing a terminal double bond are not refractory to anaerobic decomposition and can be oxidized by certain sulfate-reducing and other anaerobic bacteria.

Aromatic hydrocarbons

Many aromatic hydrocarbons can be used as electron donors aerobically by microorganisms, of which bacteria of the genus *Pseudomonas* have been the best studied. It has been demonstrated that the metabolism of these compounds, some of which are quite complex, frequently has as its initial stage the formation of either of two molecules, *protocatechuate* or *catechol*, as shown in Figure 16.48*a*.

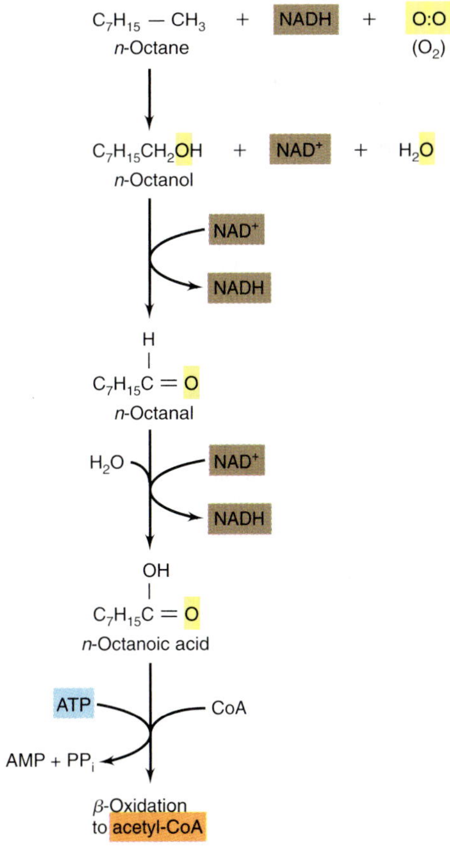

FIGURE 16.47 Steps in oxidation of an aliphatic hydrocarbon, the first step of which is catalyzed by a monooxygenase.

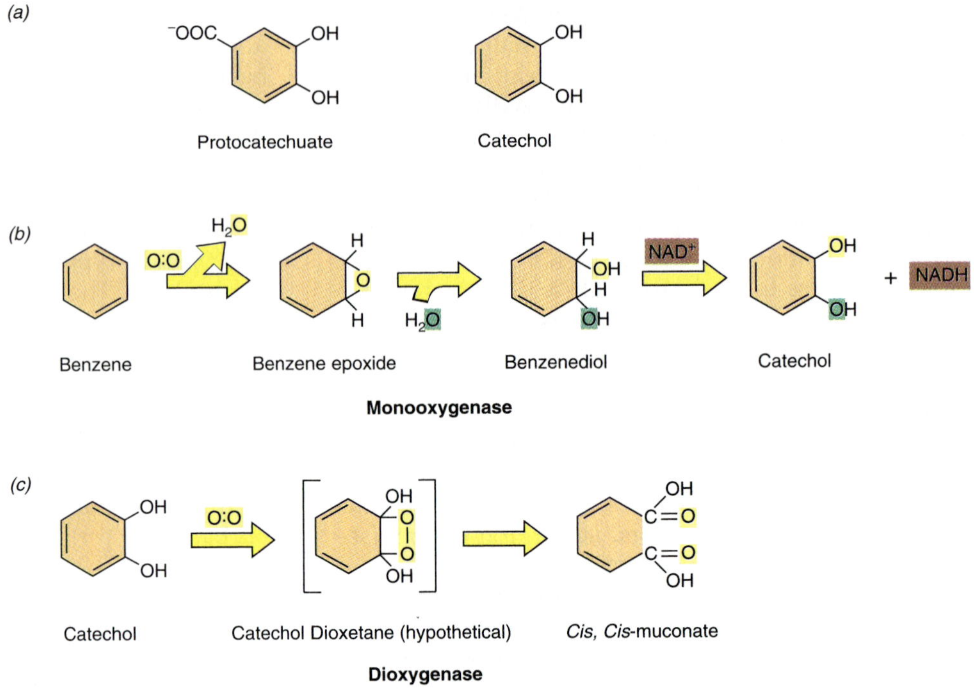

(a)

Protocatechuate   Catechol

(b)

Benzene   Benzene epoxide   Benzenediol   Catechol   + NADH

Monooxygenase

(c)

Catechol   Catechol Dioxetane (hypothetical)   *Cis, Cis*-muconate

Dioxygenase

FIGURE 16.48
Roles of oxygenases in catabolism of aromatic compounds. (a) Protocatechuate and catechol, two common oxidation products of the benzene ring. (b) Hydroxylation of benzene to catechol by a monooxygenase in which NADH is an electron donor. (c) Cleavage of catechol to *cis,cis*-muconate by a dioxygenase. Reactant oxygen atoms are shown in color in both reactions to demonstrate the different mechanisms.

These single ring compounds are referred to as *starting substrates*, since oxidative catabolism proceeds only after the complex aromatic molecules have been converted to these more simple forms. Protocatechuate and catechol may then be further degraded to compounds which can enter the TCA cycle: succinate, acetyl-CoA, pyruvate. Several steps in the catabolism of aromatic hydrocarbons usually require oxygenases. Figures 16.48b and c show two different oxygenase-catalyzed reactions, one using a monooxygenase and the other a dioxygenase.

Aromatic compounds can also be degraded anaerobically and quite readily if they already contain an atom of oxygen. Mixed bacterial cultures capable of degrading aromatic compounds have been shown to degrade benzoate and other substituted phenolic compounds yielding CH_4 and CO_2 as final products. Substituted phenolic compounds are also degraded by certain denitrifying, phototrophic, ferric iron-reducing, and sulfate-reducing bacteria. The anaerobic catabolism of aromatic compounds proceeds by *reductive* rather than oxidative ring cleavage (Figure 16.49). This is reasonable, since anaerobically oxygen cannot be involved in the oxidation of aromatic compounds. Anaerobic catabolism involves *ring reduction* followed by *ring cleavage* to yield a straight chain fatty acid or dicarboxylic acid. These intermediates can generally be converted to acetyl-CoA and used for both biosyn-

thetic and energy yielding purposes. Benzoate and benzoate derivatives are common natural products and are readily degraded anaerobically.

Some evidence for the anaerobic degradation of benzene and toluene, aromatic compounds lacking an oxygen atom, has also been obtained. Catabolism of benzene occurs by anaerobic microbial consortia leading to methanogenesis, but toluene oxidation to CO_2 can occur in pure culture. In the latter instance, growth on toluene is supported by anaerobic respiration coupled to the reduction of ferric iron or nitrate. The biochemistry of anaerobic toluene oxidation is unknown.

> Hydrocarbons are common pollutants in nature, associated with the production and utilization of petroleum resources. Many microorganisms are effective degraders of aliphatic and aromatic hydrocarbons. Rapid breakdown of a hydrocarbon requires the participation of an oxygenase enzyme and hence only occurs aerobically, although a few examples of anaerobic hydrocarbon degradation are known.

16.25 Nitrogen Metabolism

We have discussed nitrogen nutrition in general terms in Sections 4.18 and 4.21. Here we discuss some of the special biochemistry and genetics of nitrogen utiliza-

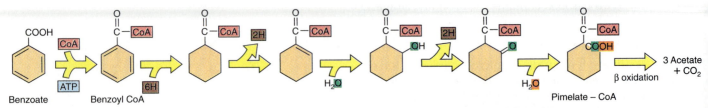

FIGURE 16.49 Anaerobic degradation of benzoate by reductive-ring cleavage. Note that all intermediates of the pathway are bound to coenzyme A.

tion, with emphasis on the process of **nitrogen fixation**. Nitrogen can be obtained by microorganisms either from inorganic or organic forms. The most common inorganic nitrogen sources are nitrate and ammonia, but other inorganic sources used by certain microorganisms include cyanide (CN^-), cyanate (OCN^-), thiocyanate (SCN^-), cyanamide (NCN^{2-}), nitrite (NO_2^-), and hydroxylamine (NH_2OH). As discussed in the next section, free nitrogen gas (N_2) is also used by a variety of bacteria.

We discussed the general biochemistry of the utilization of organic nitrogen in Section 4.18. There we discussed enzymes such as *transaminases* and *amino-acid dehydrogenases* which catalyze the reductive assimilation of ammonia from the environment. Another very important means of incorporating ammonia is via the enzyme **glutamine synthetase**. As shown in Figure 16.50, this enzyme catalyzes the addition of ammonia to glutamate to yield *glutamine*. Then in a second reaction, the enzyme **glutamate synthase** catalyzes the transfer of the amide group of glutamine to α-ketoglutarate, forming two molecules of *glutamate*. When these two reactions are summed, the net reaction is the conversion of one molecule of α-ketoglutarate and one molecule of ammonia to glutamate, with consumption of one high-energy phosphate bond from ATP. In addition to its role in ammonia assimilation, glutamine also functions as an amino donor for a variety of reactions, for example, in the synthesis of the purine ring (see Section 4.19).

A rather interesting metabolic control is exerted on the activity of the enzyme glutamine synthetase. Glutamine synthetase exists in two forms, an active form and an inactive form; the latter form occurs when adenyl residues from ATP are attached to the molecule, a process called *adenylylation*. Adenylylation of glutamine synthetase is promoted by high concentrations of ammonia, leading to inactivation of the enzyme. However, since glutamate dehydrogenase functions when ammonia concentrations are high, ammonia assimilation is shifted from the glutamine pathway to the glutamate pathway. When ammonia concentrations fall, glutamine synthetase becomes deadenylylated and becomes active again. Control via adenylylation thus involves a *covalent*

modification of the enzyme rather than merely a change in its conformation as described in Section 4.20. The presumed function of adenylylation is to prevent the wasteful consumption of ATP by glutamine synthetase when ammonia levels are sufficiently high for the operation of amino acid dehydrogenases; the latter do not require ATP for their function. We discussed the utilization of nitrate and the whole process of assimilative nitrate reduction earlier in this chapter (see Section 16.15).

16.26   Nitrogen Fixation

The utilization of nitrogen gas (N_2) as a source of nitrogen is called *nitrogen fixation* and is a property of only certain prokaryotes. An abbreviated list of nitrogen-fixing organisms is given in Table 16.12, from which it can be seen that a variety of prokaryotes, both anaerobic and aerobic, fix nitrogen. In addition, there are some bacteria, called *symbiotic*, that fix nitrogen only when present in nodules or on roots of specific host plants. As far as is currently known, no eukaryotic organisms fix nitrogen. Symbiotic nitrogen fixation will be discussed in Section 17.24.

Nitrogenase

In the fixation process, N_2 is *reduced* to ammonium and the ammonium converted into organic form. The reduction process is catalyzed by the enzyme complex **nitrogenase,** which consists of two separate proteins called *dinitrogenase* and *dinitrogenase reductase*. Both components contain iron, and dinitrogenase contains molybdenum as well. The molybdenum and iron in dinitrogenase are contributed by a cofactor known as *FeMoCo*, and the actual reduction of N_2 involves participation of this iron–molybdenum center. Some nitrogen-fixing bacteria can synthesize more than one nitrogenase under certain growth conditions, and these so-called *alternative nitrogenases* do not contain molybdenum but instead contain either vanadium (and iron) or iron only. Alternative nitrogenases are not synthesized when sufficient molybdenum is present, as the molybdenum nitrogenase is generally the main nitrogenase in the cell. Alternative nitrogenases presumably serve as a back-up mechanism to ensure

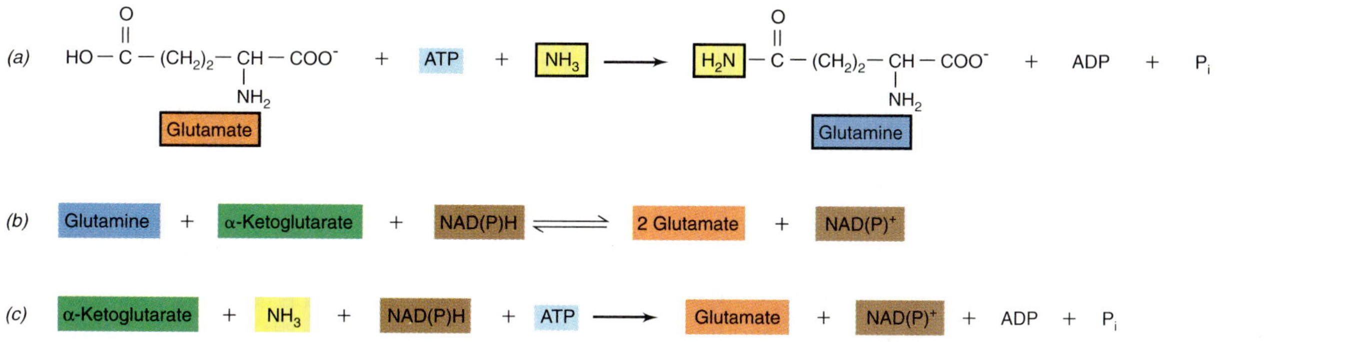

FIGURE 16.50 Role of glutamine synthetase in the assimilation of ammonia. (a) Glutamine synthetase reaction. (b) Glutamate synthase reaction. (c) Sum of the two reactions. Reaction (c) can also be carried out by the enzyme glutamate dehydrogenase in the absence of ATP but only at high ammonia concentrations. Glutamine synthetase and glutamate synthase operate at low ammonia concentrations.

Table 16.12 Some nitrogen-fixing organisms

Free-living aerobes		
Chemo-organotrophs	Phototrophs	Chemo-lithotrophs
Bacteria: *Azotobacter* spp. *Klebsiella* *Beijerinckia* *Bacillus polymyxa* *Mycobacterium flavum* *Azospirillum lipoferum* *Citrobacter freundii* Methylotrophs (various, but not all)	Cyanobacteria (various, but not all)	*Alcaligenes* *Thiobacillus* (some species)

Free-living anaerobes		
Chemo-organotrophs	Phototrophs	Chemo-lithotrophs
Bacteria: *Clostridium* spp. *Desulfovibrio* *Desulfoto-* *maculum*	Bacteria: *Chromatium* *Chlorobium* *Rhodospirillum* *Rhodopseudomonas* *Rhodomicrobium* *Rhodopila* *Rhodobacter* *Heliobacterium* *Heliobacillus*	Archaea: *Methanosarcina* *Methanococcus*

Symbiotic	
Leguminous plants	Nonleguminous plants
Soybeans, peas, clover, locust, etc., in association with a bacterium of the genus *Rhizobium* or *Bradyrhizobium*	*Alnus, Myrica, Ceanothus,* *Comptonia, Casuarina;* in association with actinomycetes of the genus *Frankia*

that N_2 fixation can still occur when molybdenum is limiting in the habitat (see Section 19.16).

Owing to the stability of the N≡N triple bond, N_2 is extremely inert and its activation is a very energy-demanding process. Six electrons must be transferred to reduce N_2 to $2NH_3$, and several intermediate steps might be visualized; but since no intermediates have ever been isolated, it is now assumed that the three successive reduction steps occur with the intermediates firmly bound to nitrogenase (Figure 16.51a). Nitrogen fixation is highly reductive in nature and the process is inhibited by oxygen since dinitrogenase reductase is rapidly and irreversibly inactivated by O_2 (even when isolated from *aerobic* nitrogen-fixers). In aerobic bacteria, N_2 fixation occurs in the presence of O_2 in whole cells, but not in purified enzyme preparations, and nitrogenase in such organisms is pro-

tected from O_2 inactivation either by removal of O_2 by respiration, the production of O_2-retarding slime layers, or by compartmentalizing nitrogenase in a special type of cell (the heterocyst, see Section 19.2).

Electron flow in nitrogen fixation

The sequence of electron transfer in nitrogenase is: electron donor → dinitrogenase reductase → dinitrogenase → N_2 → $2NH_3$. The electrons for nitrogen reduction are transferred to dinitrogenase reductase from ferredoxin or flavodoxin, low-potential iron–sulfur proteins. In *Clostridium pasteurianum*, ferredoxin is the electron donor and is reduced by phosphoroclastic splitting of pyruvate to acetyl-CoA + CO_2. In addition to reduced ferredoxin, ATP is required for N_2 fixation. The ATP requirement for nitrogen fixation is very high, about 4 to 5 ATPs hydrolyzed for each $2e^-$ transferred. ATP is required to lower the reduction potential of nitrogenase sufficiently so that N_2 may be reduced. The reduction potential of dinitrogenase reductase is -0.30 volts, and this is lowered to -0.40 volts when the electrons are transferred to the enzyme and the enzyme hydrolyzes ATP (Figure 16.51). This complex then combines with dinitrogenase and the latter becomes reduced. Reduced dinitrogenase then reduces N_2 to NH_3, with the actual reduction occurring at the FeMoCo center. Although only *six* electrons are necessary to reduce N_2 to $2NH_3$, careful measurements have shown that *eight* electrons are actually consumed in the process, *two* electrons being lost as hydrogen (H_2), for each mole of N_2 reduced (Figure 16.51a). The reason for this apparent wastage is not known, but evidence is strong that H_2 evolution is an intimate part of the reaction mechanism of nitrogenase.

Genetics and regulation of nitrogen fixation

The genes for dinitrogenase and dinitrogenase reductase in *Klebsiella pneumoniae*, a well-studied N_2 fixer, are part of a complex regulon (a large network of operons) called the *nif* regulon (Figure 16.51b and see box). In addition to nitrogenase structural genes, the genes for FeMoCo, genes controlling the electron transport proteins, and a number of regulatory genes are also present in the *nif* regulon. Dinitrogenase is a complex protein made up of two subunits, α (product of the *nifD* gene) and β (product of the *nifK* gene), each of which is present in two copies. Dinitrogenase reductase is a protein dimer consisting of two identical subunits, the product of *nifH*. FeMoCo is synthesized through the participation of five separate genes, *nifN, E,* and *B,* as well as *Q,* which controls a product involved in molybdenum uptake. *NifA* encodes a positive regulatory protein which serves to activate transcription of other *nif* genes.

Nitrogenase is subject to strict regulatory controls. Nitrogen fixation is inhibited by O_2 and by fixed nitrogen, including NH_3, NO_3^-, and certain amino acids. A major part of this regulation is at the level of transcription. The various transcriptional units of the *nif* regulon are shown in Figure 16.51b. While transcription of the *nif* structural genes is *activated* by the NifA protein

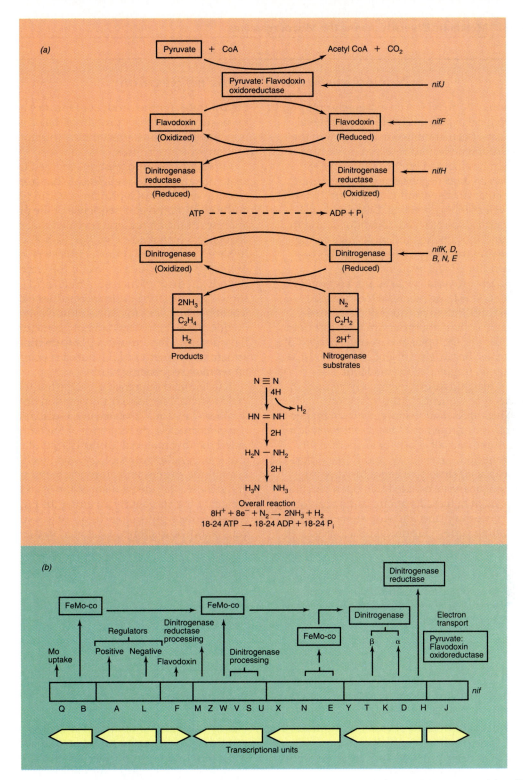

FIGURE 16.51 The nitrogenase system. (a) Steps in nitrogen fixation: reduction of N_2 to $2NH_3$. (b) The genetic structure of the *nif* regulon (operon) in *Klebsiella pneumoniae*, the best studied nitrogen-fixing organism. The functions of some of the genes are uncertain. Several other genes suspected of being *nif* genes are known but not shown in this figure. The mRNA transcripts (transcriptional units) are shown below the genes; arrows indicate the direction of transcription.

(positive regulation) it is *repressed* by the NifL protein (negative regulation), although the details of how these proteins regulate transcription are just becoming understood.

The ammonia produced by nitrogenase does not repress enzyme synthesis, of course, because as soon as it is made it is incorporated into organic form and used in biosynthesis. But when ammonia is in excess (as in environments high in ammonia), nitrogenase synthesis is quickly repressed. This prevents the wastage of ATP by not making a product already present in ample amounts. In certain nitrogen-fixing bacteria, especially phototrophic bacteria, nitrogenase *activity* is also regulated by ammonia, a phenomenon called the ammonia "switch-off" effect. In this case, excess ammonia catalyzes a covalent modification of dinitrogenase reductase, which results in a loss of enzyme activity. When ammonia again becomes limit-

Nucleotide Sequence of the *nif* Regulon

The reduction of N_2 to NH_3 is an extremely complex series of biochemical events, and at the genetic level over 30 genes have been identified as involved in the process. Much of the molecular research in the area of nitrogen fixation has focused on the organism *Klebsiella pneumoniae*, a Gram-negative enteric bacterium that fixes molecular nitrogen and is often found in soil and water as well as in the intestine. Like *E. coli*, the genetics of *K. pneumoniae* are well understood, and to better understand the genetics and molecular biology of nitrogen fixation, the entire *nif* regulon of this organism has been cloned and sequenced.*

The *nif* regulon of *K. pneumoniae* spans a region of 24,206 base pairs. This large piece of DNA was sequenced by first subcloning portions of a larger restriction fragment containing the entire *nif* region into 10 different plasmids; each plasmid insert consisted of just a few *nif* genes. Because sufficient overlap existed between the cloned fragments, the nucleotide sequence of all 10 plasmid inserts eventually yielded the complete nucleotide sequence of the *nif* region.

Interesting new information emerged from inspection of the nucleotide sequence of the *K. pneumoniae nif* regulon. First, three new *nif* genes (or suspected *nif* genes) were identified. Genes *Z*, *W*, and *T* (see Figure 16.51) are probably *nif* structural genes because nucleotide sequence analysis revealed that open-reading frames were present at these locations in the regulon along with the necessary start and stop signals typical of protein encoding genes. Curiously, previous genetic experiments yielded no evidence for genes *Z*, *W*, and *T*, possibly because mutations in them do not yield a discernible mutant phenotype that could serve as a clue to their existence. In addition to new genes, sequencing the *nif* regulon showed that some *nif* genes overlap, usually only by a few nucleotides, and that

binding sites for the product of the *nifA* gene, the master positive regulatory protein of the *nif* complex, exist in many *nif* operons, presumably to activate each operon for transcription.

Other types of information emerged from *nif* DNA sequence analysis as well. Recall that from the nucleotide sequence of a protein-encoding gene, one can deduce the amino acid sequence of the protein. Such analyses of the *nif* regulon revealed that an amino acid sequence which reads: Cys-X_2-Cys-X_2-Cys-X_3-Cys-Pro (where X can be any amino acid) exists in the product of the *nifJ* gene, which codes for the enzyme pyruvate-flavodoxin oxidoreductase. Amino acid sequences showing alternating cysteine residues like this are typical of binding sites for iron-sulfur clusters in ferredoxins. Similar clusters had not been known to exist in flavodoxins. Thus, important information about the structural properties of flavodoxins emerged from the *nif* regulon sequencing study.

Nucleic acid sequencing is clearly a tool that can yield essential molecular information not obtainable in other ways. In *K. pneumoniae*, elucidation of the complete nucleotide sequence of the *nif* regulon provided the raw material for the design of new experiments in genetics, physiology, and biochemistry. From these studies will emerge a better understanding of the molecular mechanism of nitrogen fixation. With biotechnologists eagerly awaiting the opportunity to transfer major physiological processes such as nitrogen fixation into plants, a detailed understanding of how *nif* works is an obvious prerequisite. The complete nucleotide sequence of the *nif* regulon was an important first step.

*Key reference: Arnold, W., Rump, A., Klipp, W., Preifer, U. B., and Puhler, A. 1988. *J. Molec. Biol.* 203:715–738.

ing, this modified protein is converted back to the active form and N_2 fixation resumes. Ammonia "switch-off" is thus a rapid and reversible method of controlling ATP consumption by nitrogenase.

Nitrogenase has been purified from a large number of nitrogen-fixing organisms and in all cases has been shown to be a two protein complex. It is of considerable evolutionary interest that within molybdenum nitrogenases, dinitrogenase from one organism usually will function with dinitrogenase reductase from another organism. This can be interpreted to mean that the structures of the nitrogenase components have not changed markedly during evolution, suggesting that the molecular requirements for N_2 reduction are fairly specific. Indeed, studies of the "HDK" gene cluster have confirmed this, in that molecular probes containing cloned *nifHDK* genes from one nitrogen-fixing bacterium will hybridize to DNA from virtually all N_2-fixers, but not to DNA from non-N_2-fixing bacteria.

Assaying nitrogenase

Nitrogenase is not specific for N_2 but will also reduce cyanide (CN^-), acetylene ($HC{\equiv}CH$), and several other triply-bonded compounds. The reduction of acetylene by nitrogenase is only a two-electron process and *ethylene* ($H_2C{=}CH_2$) is produced. The reduction of acetylene probably serves no useful purpose to the cell but provides the experimenter with a simple way of measuring the activity of nitrogen-fixing systems, since it is fairly easy to measure the reduction of acetylene to ethylene by gas chromatography. This technique is now widely used to detect nitrogen fixation in unknown systems. Previously, it was not easy to prove that an organism fixed N_2; indeed, many claims for nitrogen fixation in microorganisms have been erroneous. The growth of an organism in a medium to which no nitrogen compounds have been added does not mean that the organism is fixing nitrogen from the air since traces of nitrogen compounds often occur as contaminants in the ingredients of culture media or drift into the media in gaseous form or as dust particles.

A definitive procedure to prove N_2 fixation is to use an isotope of nitrogen, ^{15}N, as a tracer. (^{15}N is not a radioisotope but a stable isotope. It is detected with a mass spectrometer.) The gas phase of a culture is enriched with ^{15}N, and after incubation, the cells and medium are digested, the ammonia produced being distilled off and assayed for its ^{15}N content. If there has been a significant production of ^{15}N-labeled NH_3, it is proof of nitrogen fixation. However, the acetylene-reduction method is a more sensitive way of measuring nitrogen fixation. The sample, which may be soil, water, a culture, or a cell extract is incubated with acetylene, and the reaction mixture is later analyzed by gas chromatography for production of ethylene. This method is far simpler and faster than other methods.

> One of the most important biochemical processes in nature is nitrogen fixation, in which nitrogen gas (dinitrogen, N_2) from the atmosphere is reduced to ammonia and organic nitrogen compounds. Nitrogen fixation involves a complex enzyme system called nitrogenase, which consists of dinitrogenase and dinitrogenase reductase, metal-containing enzymes found only in certain prokaryotic organisms. Most nitrogenases contain molybdenum or vanadium and iron as metal cofactors, and the process of nitrogen fixation is highly energy-demanding. The enzyme system involved in nitrogen fixation is complex, involving numerous proteins, and is encoded by a large genetic region called the *nif* regulon. Although the natural substrate of the nitrogen fixation system is N_2, certain artificial substrates which are structurally similar to N_2, such as acetylene and cyanide, can also serve as substrates. Acetylene reduction is used as a sensitive and convenient assay for nitrogenase.

Study Questions

1. In what nutritional class would you place an organism that used glucose as sole carbon and energy source? Elemental sulfur as energy source? How would we refer to the latter organism if it grew with CO_2 as sole carbon source? What is the energy source for *phototrophic* organisms?

2. What is the role of light in the photosynthetic process of green and purple bacteria? Of cyanobacteria? Compare and contrast the photosynthesis process in these two groups of prokaryotes.

3. Compare and contrast the absorption spectrum of chlorophyll *a* and bacteriochlorophyll *a*. What wavelengths are preferentially absorbed by each pigment and how do the absorption properties of these molecules compare with the regions of the spectrum visible to our eye? Why are most plants green in color?

4. What are the functions of light harvesting and reaction center chlorophyll? Why would a mutant incapable of making light-harvesting chlorophylls (such mutants can be readily isolated) probably not be a successful competitor in nature?

5. How does light result in ATP production in an *anoxygenic phototroph*? In what ways are photosynthetic and respiratory electron flow similar? In what ways do they differ?

6. Where are the photosynthetic pigments located in a purple bacterium? A cyanobacterium? A green alga? Considering the function of chlorophyll pigments why couldn't they be located elsewhere in the cell, for example in the cytoplasm or in the cell wall?

7. How is reducing power made for autotrophic growth in a purple bacterium? In a cyanobacterium?

8. The growth rate of the phototrophic purple bacterium *Rhodobacter* is about twice as fast when the organism is grown phototrophically in a medium containing malate as carbon source than when the organism is grown with CO_2 as carbon source (with H_2 as electron donor). Discuss the reasons why this occurs and list the nutritional class we would place *Rhodobacter* in when growing under each of the two different conditions.

9. How does the reduction potential of chlorophyll *a* in photosystem I and photosystem II differ? Why must the reduction potential of photosystem II chlorophyll be so high?

10. Discuss the nature of the evidence obtained from studies of the photosynthetic process of certain cyanobacteria that supports the hypothesis that these organisms evolved from anoxygenic phototrophs.

11. What is the major function of carotenoids and phycobilins in phototrophic microorganisms?

12. Although physiologically distinct, chemolithotrophs and chemoorganotrophs share a number of features in common with respect to the production of ATP. Discuss these common features along with reasons for why the growth yield (grams of cells/mole of substrate) of a chemoorganotroph respiring glucose is so much higher than for a chemolithotroph respiring sulfur.

13. Based on a consideration of energetics, discuss why phototrophic bacteria are able to thrive in nature even though their photosynthetic pigments absorb in regions of the spectrum such as the infrared, which contains relatively little energy.

14. What two enzymes are unique to organisms which carry out the Calvin cycle? What reactions do these enzymes carry out? What would be the consequence if a mutant arose that lacked either of these enzymes?

15. For conversion of 6 molecules of CO_2 into one fructose molecule, 18 molecules of ATP are required. Where in the Calvin cycle reactions are these ATPs consumed?

16. Compare and contrast the utilization of H_2S by a purple phototrophic bacterium and by a colorless sulfur bacterium. What role does H_2S play in the metabolism of each organism?

17. Discuss why the growth yield (grams of cells/mole of substrate) of *Thiobacillus ferrooxidans* is considerably *greater* when the organism is growing aerobically on elemental sulfur than on ferrous iron as energy source (assume the organism is growing autotrophically in both cases).

18. Growth yields of *Nitrosomonas* are generally much higher than for *Nitrobacter*. Why might this be so?

19. What is a mixotroph? Why will an organism capable of mixotrophy frequently grow better mixotrophically than chemolithotrophically or even chemoorganotrophically?

20. In *Escherichia coli* synthesis of the enzyme *nitrate reductase* is strongly repressed by oxygen. Based on bioenergetic arguments, why do you think this repression phenomenon might have evolved?

21. Discuss at least three major differences between assimilative and dissimilative nitrate reduction.

22. Why is the following statement, if interpreted literally, incorrect: "Anaerobic respiration is simply a process where an alternative electron acceptor is substituted for O_2 in the respiratory process."

23. Define the term *substrate-level phosphorylation*. How does it differ from oxidative phosphorylation? Assuming an organism is facultative, what basic nutritional conditions will dictate whether the organism obtains energy from substrate-level rather than oxidative phosphorylation?

24. Although many different compounds are theoretically fermentable, in order to support a fermentative process most organic compounds must be eventually converted into one of a relatively small group of molecules. What are these molecules and why must they be produced?

25. To a culture of *Escherichia coli* growing fermentatively you add 1 g/liter of $NaNO_3$. Would you expect the growth *yield* of the culture to increase or decrease and why?

26. Why have hydrocarbons accumulated in large reservoirs on earth despite the fact that they are readily degradable microbiologically under certain conditions?

27. How are xenobiotic compounds defined? Give an example of a compound you think would qualify as a xenobiotic.

28. Compare and contrast the conversion of cellulose and intracellular starch to glucose units. What enzymes are involved and which process is the more energy efficient?

29. Although dextran is a glucose polymer, glucose cannot be used to make dextran. Explain. How is dextran synthesis important in oral hygiene (see Section 11.3)?

30. How do monooxygenases differ from dioxygenases in the reactions they catalyze?

31. *Pseudomonas fluorescens* can grow on benzoate aerobically while the phototrophic bacterium *Rhodopseudomonas palustris* can grow on benzoate anaerobically. Compare and contrast the metabolism of benzoate by these two species focusing on the following considerations: requirement for oxygenases, initial reactions leading to the opening of the ring, and product(s) formed that can feed into central metabolic pathways.

32. How do ammonia levels in a cell control the activities of the enzyme glutamine synthetase? How does such control benefit the cell? Although ammonia controls the *activity* of glutamine synthetase, this is not a typical "feedback" inhibition phenomenon. Explain.

33. Write out the reaction catalyzed by the enzyme *nitrogenase*. How many electrons are required in this reaction? How many are actually used? Explain.

34. What is the function of ATP in the nitrogen fixation process? How does ATP affect dinitrogenase reductase?

35. What metals are found in nitrogenase? What is the evidence that nitrogenase is a highly conserved protein in an evolutionary sense?

36. How can fermentations occur in the absence of substrate-level phosphorylation?

Supplementary Readings

Balows, A., H. G. Trüper, M. Dworkin, W. Harder, and **K. -H. Schleifer.** (eds.) 1992. *The Prokaryotes: A Handbook on the Biology of Bacteria, Ecophysiology, Isolation, Identification, and Applications*, 2nd edition. Springer-Verlag, New York. A complete reference source on all groups of bacteria. Excellent chapters on phototrophs, chemolithotrophs, methylotrophs, nitrogen fixers, and various fermentative anaerobes. In many cases the chapters on the different groups of organisms are prefaced by an overview chapter on the particular metabolic process characteristic of the group.

Dawes, I. W., and **I. W. Sutherland.** 1992. *Microbial Physiology*, 2nd edition. Blackwell Scientific Publications, Oxford, U.K. Brief but clear treatment of many of the physiological processes emphasized in this chapter.

Gottschalk, G. 1986. *Bacterial Metabolism*, 2nd edition. Springer-Verlag, New York. This standard textbook of bacterial physiology covers many of the physiological processes discussed in this chapter. Includes especially good discussion of fermentations.

Hopwood, D. A., and **K. E. Chater.** (eds.) 1989. *Genetics of Bacterial Diversity*. Academic Press, New York. A short treatment of the genetics behind such metabolic processes as photosynthesis, nitrogen fixation, aromatic catabolism, and plant-bacterial interactions.

Krulwich, T. A. (ed.) 1990. *Bacterial Energetics*. Academic Press, New York. Detailed coverage of bioenergetic principles in a wide variety of metabolic types of prokaryotes.

Nicholls, D. G., and **S. J. Ferguson.** (eds.) 1992. *Bioenergetics 2*. Academic Press, Orlando, FL. An advanced treatment of bioenergetic principles focusing on the generation of a proton motive force by various means and recovery of membrane energy as ATP.

Odom, J. M., and **R. Singleton, Jr.** (eds.) 1992. *The Sulfate-Reducing Bacteria*. Springer-Verlag, New York. A monograph devoted to all aspects of sulfate-reducing bacteria.

Schlegel, H. G., and **B. Bowien** (eds.) 1989. *Autotrophic Bacteria*. Science-Tech Publishers, Madison, WI. A series of chapters written by experts on all phases of the microbiology, physiology, and biochemistry of autotrophic bacteria.

Sebald, M. (ed.) 1992. *Genetics and Molecular Biology of Anaerobic Bacteria*. Springer-Verlag, New York. Genetics and other molecular information on a wide variety of anaerobic bacteria carrying out reactions that are described in this chapter.

Silver, S., Chakrabarty, A. M., Iglewski, B., and **S. Kaplan.** (eds.) 1990. *Pseudomonas: Biotransformations, Pathogenesis, and Evolving Biotechnology*. American Society for Microbiology, Washington, D.C. A compendium of papers on a very metabolically diverse group of bacteria.

Zehnder, A. J. B. (ed.) 1988. *Biology of Anaerobic Bacteria*. John Wiley and Sons, New York. A collection of review chapters covering all major groups of anaerobes and anaerobic processes.

17

Microbial Ecology

Up to this point we have considered microorganisms as laboratory entities or in their relationships to humans in disease. In this chapter we consider microorganisms in soil, water, and other environments and consider how microorganisms act to chemically change their environments. The term *environment* refers to everything surrounding a living organism: the chemical, physical, and biological factors and forces that act on a living organism. Microorganisms play a far more important ecological role in natural environments than their small size would suggest. From an ecological perspective, microorganisms are part of organismal communities called *ecosystems*. Each organism in an ecosystem interacts with its surroundings, in some cases greatly modifying the characteristics of the ecosystem in the process. This is particularly true of microorganisms where significant chemical changes can occur because of their activities.

Energy enters ecosystems mainly in the form of sunlight and is used by phototrophic organisms to synthesize new organic matter. The latter contains not only carbon, but also nitrogen, sulfur, phosphorus, iron, and a host of other elements. When an organism dies, its chemical constituents are released and once again become available to the ecosystem to support the growth of new organisms. Microorganisms play important synthetic and degradative roles. Some microorganisms are autotrophic, capable of generating new organic matter from CO_2, while most microorganisms are heterotrophic and participate in the biodegradation and recycling of organic and inorganic matter. For the key elements of living systems a *biogeo-*

chemical cycle can be defined in which the element undergoes changes in oxidation state as it moves through an ecosystem. Microorganisms are intimately involved in biogeochemical reactions, and for many elements are the *only* biological agents capable of regenerating forms of the elements usable by other organisms, particularly plants.

Microbial ecologists are primarily interested in two aspects of the microbiology of an ecosystem: (1) the microorganisms present and (2) the nature and extent of their metabolic activities. Knowledge of the microorganisms present in an ecosystem yields valuable information about the potential of that ecosystem for biogeochemical cycling. However, of equal importance is knowledge of the *activities* of those microorganisms. Just because organisms are present does not necessarily mean they are active; only *active* organisms have an ecological impact. The measurable occurrence of a specific biochemical transformation thus serves as good evidence that a specific group of bacteria is both present and metabolically active.

Why is microbial ecology important? Not only is it important in its own right as a basic science, but in future years an understanding of microbial ecology will likely be critical to solving a number of serious environmental problems. As the world's human population increases and natural resources diminish, environmental problems grow. However, humans may be able to harness the vast metabolic potential of microorganisms discussed throughout this book to solve some of these environmental problems. Thus, one of the goals of microbial ecology is to develop an understanding of the types and activities of microorganisms in nature and how they interact with one another and with higher organisms, and to apply this knowledge to dealing with environmental problems that may have microbial solutions.

In this chapter we discuss modern methods of assessing and tracking microbial communities in their natural habitats and consider how the activities of microorganisms in nature play important roles in biogeochemical cycling. We also consider a number of specialized microbial habitats where much is understood about the role of microorganisms in the ecosystem. We begin our treatment of microbial ecology with the special properties of microbial habitats since it is the *habitat* where the microorganism actually lives and carries out its important functions.

> Ecology is the study of organisms in their natural environments. Microorganisms are important parts of ecosystems and play major roles in energy transformations and biogeochemical processes. Methods for assessing the activities of microorganisms in ecosystems are available and are used to determine the ecological impact of microorganisms in particular habitats. Microbial ecology can contribute to the solution of environmental problems involving microorganisms.

17.1   Microorganisms in Nature

The natural habitats of microorganisms are exceedingly diverse. Any habitat suitable for the growth of higher organisms will also support growth of microorganisms. But in addition, there are many habitats where, because of some physical or chemical extreme, higher organisms are absent yet microorganisms exist and occasionally even flourish. Microorganisms inhabit the surfaces of higher organisms and in some cases actually live *within* plants and animals. Microorganisms frequently reach large numbers in such habitats and may benefit the plant or animal in a nutritionally significant way. On the other hand, as we saw in Chapters 11–15, some microorganisms are pathogenic and bring harm to the host. We focus now on the microbial habitat from the standpoint of the microorganism and emphasize the heterogenous and rapidly changing nature of typical microbial habitats.

The microorganism and its environment

Before considering growth of microorganisms in natural environments, it will be useful to recall some of the principles of the laboratory culture of microorganisms discussed in Chapters 4 and 9. Environmental factors affect growth of microorganisms in nature much as they do growth of organisms in culture. Thus temperature, pH, water availability, and oxygen all play a role in governing the composition and number of microorganisms in a particular habitat. In general, however, it is the concentration and nature of available *nutrients* in a given environment that will ultimately dictate the types and numbers of organisms present. Microorganisms in nature actively compete for nutrients and the most successful species in a given habitat will be those capable of best utilizing the available nutrients under the environmental conditions that prevail. However, because nutrient levels in nature are generally much *lower* than those in laboratory culture, bacterial growth rates and cell yields in most natural environments fall well below that of laboratory cultures.

Microenvironments

Because microorganisms are so small, their habitats will also be small. Microbial ecologists use the term *microenvironment* to describe where a microorganism actually lives. What may appear to a human to be a rather uniform environment, a crumb of soil for example, may actually consist of numerous chemically and physically distinct microenvironments, differing from one another in many important ways. This can be visualized by considering the distribution of an important microbial nutrient like oxygen in a soil particle. Using microelectrodes (a technique to be described in Section 17.5) it is possible to measure oxygen concentrations throughout small soil particles. As shown in data from an actual experiment in Figure 17.1, soil particles are not homogeneous in terms of their oxy-

Miniglossary for Chapter 17

ACID MINE DRAINAGE   acidic water containing H_2SO_4 derived from the microbial oxidation of iron sulfide minerals

ANOXIC   an oxygen-free environment, usually also highly reducing (low E_0')

BACTEROID   morphologically misshapen *Rhizobium* cells inside a leguminous plant root nodule; can fix N_2

BAROPHILIC   an organism that grows best when placed under a pressure greater than one atmosphere

BAROTOLERANT   an organism that can grow under elevated pressures but that grows best at atmospheric pressure

BIOFILM   colonies of microbial cells encased in slime and attached to a surface

BIOGEOCHEMISTRY   study of biologically mediated chemical transformations

BLACK SMOKER   an extremely hot (250–350°C) deep sea hot spring emitting both hot water and various minerals

CO-METABOLISM   metabolism of a compound in the presence of a second organic compound which is used as the primary energy source

ECOSYSTEM   a community of organisms and their natural environment

ENRICHMENT CULTURE   a means of obtaining cultures of microorganisms from a natural environment by using highly selective culture methods

HYDROTHERMAL VENT   deep sea warm or hot spring

INFECTION THREAD   in the formation of root nodules, a cellulosic tube through which *Rhizobium* cells can travel to reach and infect root cells

INTERSPECIES HYDROGEN TRANSFER   the production and subsequent consumption of H_2 by different groups of microorganisms which interact closely during anaerobic catabolism

LEACHING   solubilization and removable of metals from an ore by microbial attack

LICHEN   a fungus and an alga (or cyanobacterium) living in symbiotic association

MICROBIAL PLASTICS   biodegradable polymeric materials obtained from microorganisms that have properties similar to synthetic plastics

MICROENVIRONMENT   the immediate environmental surroundings of a microbial cell or group of cells

MYCORRHIZA   a symbiotic association between a fungus and the roots of a plant

OXIC   an oxygen-containing environment frequently possessing a high E_0'

PRIMARY PRODUCER   an organism that uses light to synthesize new organic material from CO_2

PYRITE   a common iron-containing ore, FeS_2

REDUCTIVE DECHLORINATION   removal of Cl as Cl^- from an organic compound by reducing the carbon atom from C-Cl to C-H

RHIZOSPHERE   the region immediately adjacent to plant roots

ROOT NODULE   a tumor-like growth on plant roots that contains symbiotic nitrogen-fixing bacteria

RUMEN   the first vessel in the multichambered stomach of ruminant animals in which cellulose digestion occurs

SECONDARY TREATMENT   in sewage treatment, either the aerobic or anaerobic decomposition of sewage following the removal of nondegradable objects by primary treatment

TI PLASMID   a conjugative plasmid present in the bacterium *Agrobacterium tumefaciens* which can transfer genes into plants

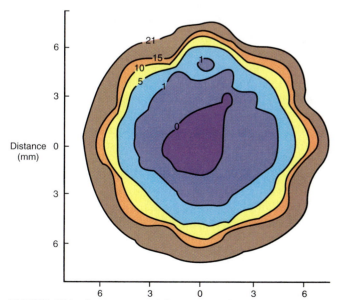

FIGURE 17.1   Contour map of O_2 concentrations in a soil particle. The axes show the dimensions of the particle. The numbers on the contours are O_2 concentrations (in percent; air is 21 percent O_2). In terms of oxygen relationships for microorganisms, each zone can be considered a different microenvironment.

gen content. The outer zones of a small soil particle may be fully oxic while the center remains completely anoxic (Figure 17.1). This suggests that various physiological types of microorganisms could coexist in such a soil particle. Anaerobic organisms could be active near the center of the particle shown in Figure 17.1, microaerophiles could be active further out, and obligately aerobic organisms could metabolize in the outer 2–3 mm of the particle; facultatively aerobic bacteria could be distributed throughout the particle.

Physicochemical conditions in the microenvironment can change rapidly. Because the oxygen concentrations shown in Figure 17.1 represent only "instantaneous" measurements, oxygen measurements taken following a period of microbial respiration or after an increase in soil water content, could show a drastically different distribution of oxygen across the microenvironment. It can thus be said that microenvironments are *heterogeneous* and that conditions in a given microenvironment can change rapidly. This probably accounts for the finding that distinctly different physiological groups of microorganisms (aerobes–anaerobes, phototrophs–chemoorganotrophs, etc.) can often be isolated from the same small sample of soil, mud, or water. In summary, because of microenvironments, high microbial diversity is possible in a relatively small physical area.

Surfaces and biofilms

Surfaces are often of considerable importance as microbial habitats because nutrients can adsorb to them; in the microenvironment of a surface, nutrient levels may be much higher than they are in the bulk solution. This phenomenon can greatly affect the rate of microbial metabolism. On surfaces microbial numbers and activity will usually be much greater than in free water due to adsorption effects. Microscope slides can serve as experimental surfaces upon which organisms can attach and grow. If a slide is immersed in a microbial habitat, left for a period of time, and then retrieved and examined by microscopy, the importance of the surface to microbial development is apparent (Figure 17.2). Microcolonies readily develop on such surfaces much as they do on natural surfaces in nature. Microscopic examination of immersed microscope slides can actually be used as a technique to measure growth rates of attached organisms in nature.

A surface may also itself be a nutrient, such as a particle of organic matter, where attached microorganisms catabolize organic or inorganic nutrients directly from the surface of the particle. Dead plant material, for example, is rapidly colonized by microorganisms in soil, and simple staining techniques can detect microbial populations attached to the solid surface (Figure 17.3).

Studies of microbial colonization of surfaces have shown that most microorganisms grow on surfaces enclosed in **biofilms**. These are encased microcolonies of bacterial cells attached to a surface by way of adhesive polysaccharides excreted by the cells. Biofilms trap nutrients for growth of the enclosed microbial population and help prevent detachment of cells on surfaces in flowing systems. Biofilms have significant implications in human medicine and commerce. In the human body, bacterial cells within a biofilm are made

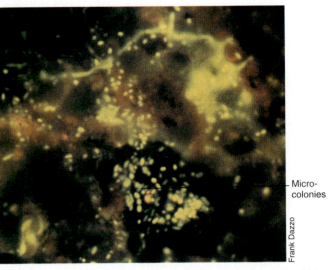

FIGURE 17.3 Fluorescence photomicrograph of a natural microbial community colonizing plant roots in soil. Note microcolony development. The preparation has been stained with acridine orange.

unavailable for attack by the immune system. This fact thus complicates the use of introduced artificial surfaces such as medical implants which can serve as sites of development of biofilms containing pathogenic microorganisms. Biofilms are also important in oral hygiene; dental plaque, a typical biofilm, contains acid-producing bacteria responsible for dental caries (see Section 11.3). In industrial situations, biofilms can slow the flow of water or oil through pipelines, accelerate the corrosion of the pipes themselves, and initiate degradation of submerged objects such as offshore oil rigs, boats, and shore line installations.

Nutrient levels and growth rates

Nutrients often enter an ecosystem in intermittent fashion. A large pulse of nutrients—for example an input of leaf litter or the carcass of a dead fish or animal—may be followed by a period of severe nutrient deprivation. Microorganisms in nature are thus often faced with a "feast or famine" type of existence, and many microorganisms have evolved biochemical systems for production of storage polymers that serve as reserve material. Reserve polymers store excess nutrients present under favorable growth conditions for use during periods of nutrient deprivation. Examples of reserve materials are poly-β-hydroxybutyrate and other alkanoates, polysaccharides, polyphosphate, and so on (see Section 3.9 for a discussion of reserve materials).

Extended periods of exponential growth in nature are rare. Growth more often occurs in spurts, linked closely to the availability of nutrients. Because physiochemical conditions in nature are rarely all optimal at the same time, growth rates of microorganisms in nature are generally well below the maximum growth rates recorded in the laboratory. For instance, the generation time of *Escherichia coli* in the intestinal tract is about 12 hours (two doublings per day) whereas in

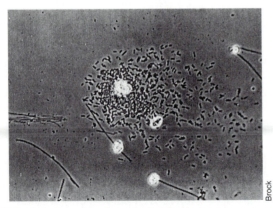

FIGURE 17.2 Bacterial microcolonies developing on a microscope slide immersed in a small river. The bright particles are mineral matter. The short, rod-shaped cells are about 3 μm long.

pure culture it grows much faster, with a minimum generation time of 20 minutes under the best of conditions. Estimates of the growth rate of certain soil bacteria have shown that they grow in nature at less than 1 percent of the maximal growth rate measured in the laboratory. On average, these slow growth rates are a reflection of the fact that: (1) nutrients are frequently in low supply, (2) the distribution of nutrients throughout the microbial habitat is not uniform, and (3) except for rare instances, microorganisms do not grow in pure culture in natural environments and thus must deal with the competitive effects of other microorganisms, a situation not encountered in growth in pure culture.

Microbial competition and cooperation

Competition among microorganisms for available nutrients may be intense. In the simplest case, the outcome of a competitive interaction will depend on rates of nutrient uptake, inherent metabolic rates, and, ultimately, growth rates. However, in some cases of microbial competition, one organism may inhibit growth or metabolism of other organisms. This may occur by excretion of a specific inhibitor such as an antibiotic, or may occur due to the physiological activities of an organism producing a toxic product, like acid from the fermentation of sugars.

Instead of competing for the same nutrient, some microorganisms work together to carry out a particular transformation that neither organism can carry out alone. These types of microbial interactions, called *syntrophy*, are crucial to the competitive success of certain anaerobic bacteria, as will be described in Section 17.12. Syntrophic relations generally require that the two or more organisms involved in the process share the same microenvironment, because the product of the metabolism of one organism must be easily accessible to the second.

> Microorganisms are very small and their natural environments are likewise small. Surfaces are often ideal microbial habitats. Microorganisms in nature usually live in a feast or famine existence; extended periods of exponential growth are rare. Competition among microorganisms for available nutrients is also intense so that only the best adapted organisms survive.

17.2   Methods in Microbial Ecology

As previously mentioned, microbial ecology focuses on two major issues: (1) the isolation, identification, and quantification of microorganisms in various habitats; and (2) the activities of microorganisms in these habitats. Although both types of studies are important, studies designed to measure microbial activities *in situ* have the advantage of allowing the microbial ecologist to make perturbations in a sample of the environment and measure what effects such chemical or physical changes have on microbial activity. However,

identification and isolation studies have a role in microbial ecology as well and go hand-in-hand with activity measurements to generate a more complete picture of the microbial ecology of a given habitat.

We begin here with a consideration of methods for enrichment and isolation of microorganisms, and then consider non-culture methods of identification and enumeration based on fluorescent antibodies and nucleic acid probes. We go from there to a consideration of microbial activity and describe the key methods that have been developed for measuring microbial activities directly in natural environments.

17.3   Enrichment and Isolation Methods

Rarely does a natural environment contain only a single type of microorganism. In most cases, a variety of organisms are present and it is particularly challenging for the microbiologist to devise methods and procedures which will permit the isolation and culture of organisms of interest. The most common approach to this goal is the **enrichment culture technique**. In this method, a medium and set of incubation conditions are used that are *selective* for the desired organism and that are counterselective for the undesired organisms. An overview of some successful enrichment culture procedures is given in Table 17.1.

Successful enrichment requires that an appropriate *inoculum* source be used. Thus, we begin an enrichment culture protocol by going to the appropriate habitat (see Table 17.1) and sampling to obtain an enrichment inoculum. Enrichment cultures are frequently established by placing the inoculum directly into a highly selective medium; many common prokaryotes can be isolated this way. For example, by pasteurizing the inoculum to kill vegetative cells, endospore-forming organisms can be isolated on various enrichment media. The nitrogen-fixing bacterium *Azotobacter* can be easily isolated aerobically in a mannitol mineral salts medium lacking fixed forms of nitrogen (see box), and so on. Literally hundreds of different enrichment approaches have been tried in the past and some of the more dependable ones are listed in Table 17.1. Both the culture medium *and* the general incubation conditions are listed because in devising enrichment culture protocols, one must optimize both parameters in order to have the best chance at obtaining the organism of interest.

The Winogradsky Column

For isolation of purple and green phototrophic bacteria and other anaerobes, the **Winogradsky column** has traditionally been used. Named for the famous Russian microbiologist, Sergei Winogradsky (see Winogradsky box, Chapter 16), the column was devised by him in the 1880s to study soil microorganisms. The column is a miniature anaerobic ecosystem that can be an excellent long term source of all types of prokaryotes involved in nutrient cycling.

A Winogradsky column can be prepared by filling

Table 17.1   Enrichment culture methods for prokaryotes

Light-phototrophic bacteria: main C source, CO_2

Aerobic incubation	**Organisms enriched**	**Inoculum**
N_2 as nitrogen source	Cyanobacteria	Pond or lake water, sulfide-rich muds, stagnant water, raw sewage, moist decomposing leaf litter, moist soil exposed to light, pasteurized soil (heliobacteria)
Anaerobic incubation		
H_2 or organic acids; N_2 as sole nitrogen source	Nonsulfur purple bacteria, heliobacteria	
H_2S as electron donor	Purple sulfur bacteria	
H_2S as electron donor	Green sulfur bacteria	

Dark-chemolithotrophic bacteria: main C source, CO_2 (medium must lack organic C)

Aerobic incubation **Electron donor**	**Electron acceptor**	**Organisms enriched**	**Inoculum**
NH_4^+	O_2	Nitrosofying bacteria (*Nitrosomonas*)	Soil, mud, sewage effluent
NO_2^-	O_2	Nitrifying bacteria (*Nitrobacter*)	
H_2	O_2	Hydrogen bacteria (various genera)	
$H_2S, S^0, S_2O_3^{2-}$	O_2	*Thiobacillus* spp.	
Fe^{2+}, low pH	O_2	*Thiobacillus ferrooxidans*	
Anaerobic incubation			
$S^0, S_2O_3^{2-}$	NO_3^-	*Thiobacillus denitrificans*	
H_2	NO_3^- + yeast extract	*Paracoccus denitrificans*	

Dark-chemoorganotrophic bacteria and methanogens: main C source, organic compounds

Aerobic incubation: respiration **Electron donor and nitrogen source**	**Electron acceptor**	**Organisms enriched**	**Inoculum**
Lactate + NH_4^+	O_2	*Pseudomonas fluorescens*	Soil, mud, lake sediments; decaying vegetation; pasteurize inoculum (80°C) for all *Bacillus* enrichments
Benzoate + NH_4^+	O_2	*Pseudomonas fluorescens*	
Starch + NH_4^+	O_2	*Bacillus polymyxa*, other *Bacillus* spp.	
Ethanol (4%) + 1% yeast extract pH 6.0	O_2	*Acetobacter, Gluconobacter*	
Urea (5%) + 1% yeast extract	O_2	*Sporosarcina ureae*	
Hydrocarbons (e.g., mineral oil) + NH_4^+	O_2	*Mycobacterium, Nocardia*	
Cellulose + NH_4^+	O_2	*Cytophaga, Sporocytophaga*	
Mannitol or benzoate, N_2 as N source	O_2	*Azotobacter*	
Anaerobic incubation: anaerobic respiration **Main ingredients**	**Electron acceptor**	**Organisms enriched**	**Inoculum**
Organic acids	KNO_3 (0.2%)	*Pseudomonas* (denitrifying species)	Soil, mud; lake sediments
Yeast extract	KNO_3 (1%)	*Bacillus* (denitrifying species)	

Table 17.1   *continued*

Organic acids	Na_2SO_4	*Desulfovibrio,* *Desulfotomaculum*	
Acetate, propionate, butyrate	Na_2SO_4	Type II sulfate reducers	As above, or sewage
Acetate, ethanol	S^0	*Desulfuromonas*	digestor sludge, rumen
H_2	Na_2CO_3	Methanogens (chemolithotrophic species only)	contents
CH_3OH	Na_2CO_3	*Methanosarcina barkeri*	
CH_3NH_2	KNO_3	*Hyphomicrobium*	

Anaerobic incubation: fermentation

Electron donor and nitrogen source	Electron acceptor	Organisms enriched	Inoculum
Glutamate or histidine	No exogenous electron acceptors added	*Clostridium tetanomorphum*	Mud, lake sediments, rotting plant material, dairy products (lactic and propionic acid bac- teria), rumen or intes- tinal contents; sewage sludge
Starch + NH_4^+	None	*Clostridium* spp.	
Starch, N_2 as N source	None	*Clostridium pasteurianum*	
Lactate + yeast extract	None	*Veillonella* spp.	
Glucose or lactose + NH_4^+	None	*Escherichia, Enterobacter,* other fermentative organisms	
Glucose + yeast extract (pH 5)	None	Lactic acid bacteria (*Lactobacillus*)	
Lactate + yeast extract	None	Propionic acid bacteria	
Succinate + NaCl	None	*Propionigenium*	

a large glass cylinder about one-third full with organic-rich, preferably sulfide-containing, mud (Figure 17.4). Carbon substrates are first mixed with the mud, with the exact composition of the final product being left somewhat to the imagination. Organic additions that have been successfully used in the past include hay, shredded newsprint, sawdust, shredded leaves or roots, ground meat, hard boiled eggs, and even dead animals! The mud is also supplemented with $CaCO_3$ and $CaSO_4$ to serve as a buffer and as a source of sulfate, respectively. The mud is packed tightly in the container, care being taken to avoid entrapping air. A small layer of unadulterated mud is then packed on top (Figure 17.4). The mud is then covered with lake, pond or ditch water and the top of the cylinder covered with aluminum foil. The cylinder is placed in a north window so as to receive adequate (but not excessive) sunlight and left to develop for a period of weeks (those in the Southern Hemisphere should use a south window).

In a typical Winogradsky column, algae and cyanobacteria appear quickly in the upper portions of the water column and by producing O_2 help to keep this zone oxic. Fermentative decomposition processes in the mud quickly lead to the production of organic acids, alcohols, and H_2, suitable substrates for sulfate-reducing bacteria. As a result of the production of sulfide, purple and green patches appear on the outer

layers of the mud exposed to light, the purple patches, consisting of purple sulfur bacteria, frequently developing in the upper layers and the green patches, consisting of green sulfur bacteria, in the lower layers (this occurs because of differences in sulfide tolerance between green and purple bacteria, see Section 19.1). At the mud–water interface the water is frequently quite turbid and may be colored due to the growth of purple sulfur and purple nonsulfur bacteria (Figure 17.4). Sampling of the column for phototrophic bacteria is performed by inserting a long thin pipette into the column and removing some colored mud or water. These can be used to inoculate enrichment media.

Winogradsky columns have been used to enrich for a variety of prokaryotes, both aerobes and anaerobes. The great advantage of a column, besides the ready availability of inocula for enrichment cultures, is that the column can be spiked with a particular compound whose degradation one wishes to study, and then allowed to select for an organism or organisms that can degrade it. In addition, since the Winogradsky column more nearly resembles the natural environment than does culture media, a variety of each physiological type of microorganism is more frequently observed in Winogradsky columns than in enrichments where a liquid culture medium is inoculated directly with a natural sample. In the latter approach, rapidly growing species frequently rise to

Rise of General Microbiology

The field of general (primarily nonmedical) microbiology owes a great debt to the Delft School of Microbiology, which was initiated by Martinus Beijerinck and continued by A. J. Kluyver and C. B. van Niel. Beijerinck, one of the world's greatest microbiologists (among other things, he was the first to characterize viruses), first devised the enrichment culture technique and used this technique in the isolation and characterization of a wide variety of bacteria. Subsequently, Kluyver and van Niel used the enrichment culture technique to isolate phototrophic and chemolithotrophic bacteria and to show the fascinating physiological diversity among the bacteria. Another important figure from the Dutch school was L. G. M. Baas-Becking, who carried out the first calculations of the energetics of phototrophic and chemolithotrophic bacteria and emphasized the importance of these organisms in geochemical processes. van Niel subsequently came to the United States, and was responsible for the training of a number of general bacteriologists who have carried on the Delft tradition, including the late R.Y. Stanier and M. Doudoroff, and Robert E. Hungate, now retired from the University of California at Davis. In addition, a number of scientists visited van Niel's laboratory at Pacific Grove, California, in the 1950s and 1960s and learned his methods and enrichment approaches. In recent years, the "Delft tradition" has been carried on in Germany by Norbert Pfennig, Bernhard Schink and Fritz Widdel, and by many others in Europe, the countries of the former Soviet Union, and the United States.

An excellent example of the application of the enrichment culture technique can be found in the isolation of *Azotobacter*, a nitrogen-fixing bacterium discovered by Beijerinck in 1901 (see Section 19.16). Beijerinck was interested in knowing whether any aerobic bacteria capable of fixing nitrogen existed in the soil. The only other nitrogen-fixing organism known at that time, *Clostridium pasteurianum*, was an anaerobe discovered several years earlier by Sergei Winogradsky. To look for an aerobic nitrogen fixer, Beijerinck added a small amount of soil to an Erlenmeyer flask containing a thin layer of mineral salts medium devoid of ammonia, nitrate, or any other form of fixed nitrogen, and which contained mannitol as carbon source. Within three days a thin film developed on the surface of the liquid and the liquid became quite turbid. Beijerinck observed large, rod-shaped cells which appeared quite distinct from the spore-containing rods of *C. pasteurianum*. Beijerinck streaked agar plates containing phosphate and mannitol with the turbid liquid, incubated aerobically, and within 48 hours obtained large slimy colonies typical of *Azotobacter*. Pure cultures were obtained by picking and restreaking well-isolated colonies a number of times. Beijerinck assumed his new organism was using N_2 from air as its source of cell nitrogen and later proved this by showing total nitrogen increases in pure cultures of *Azotobacter* grown in the absence of fixed nitrogen.

The selective pressure of the enrichment approach used here should be evident. By omitting from his medium any source of combined nitrogen and incubating aerobically, Beijerinck placed severe constraints on the microbial population. Any organism that developed had to be able to both fix its own nitrogen and tolerate the presence of molecular oxygen. Beijerinck noted that if he placed too much liquid in his flasks, or employed more readily fermentable organic substrates, such as glucose or sucrose, in place of mannitol, his enrichment would frequently turn anaerobic and favor the growth of *Clostridium* rather than *Azotobacter*. The addition of ammonia or nitrate to the original enrichment never resulted in the isolation of *Azotobacter*, but only to a variety of non-nitrogen-fixing bacteria instead. Hence Beijerinck showed that *Azotobacter* has a strong selective advantage over other soil bacteria under a specific set of nutritional conditions, and in the process defined the key physiological properties of *Azotobacter*: aerobiosis and nitrogen fixation. Beijerinck also demonstrated that the composition of the growth medium as well as the incubation conditions employed were of paramount importance in the proper development of an enrichment culture.

the forefront leaving the slower growing species behind.

From enrichments to pure cultures

Usually the objective of an enrichment culture study is to obtain a pure culture. In examining a culture microscopically, it is almost impossible to detect a stray contaminant because the sensitivity of the light microscope is low. Much more sensitive purity checks involve inoculation of a putative pure culture into a medium which favors growth of contaminants but not that of the culture of interest; growth in such media quickly indicates that the culture is not pure. On the other hand, absence of growth does not necessarily mean that contaminants are absent, since the culture medium used or incubation conditions chosen may not have been optimal for their growth. Thus, critical determination that a culture is pure is not easy.

Pure cultures can be obtained in many ways, but the most frequently employed means are the streak plate, the agar shake, and liquid dilution methods. For organisms that grow well on agar plates, the **streak plate** is the method of choice (see Section 4.22). By repeated picking and restreaking of a well-isolated colony, one usually obtains a pure culture which can then be transferred to a liquid medium. With proper incubation it is possible to purify both aerobes and anaerobes on agar plates via the streak plate method.

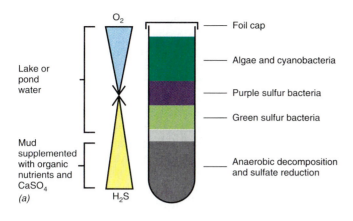

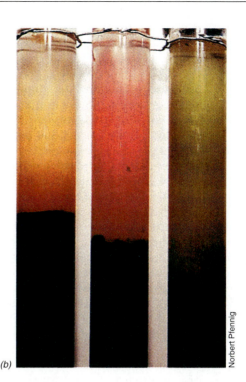

O₂

Lake or pond water

Mud supplemented with organic nutrients and CaSO₄

H₂S

(a)

Foil cap

Algae and cyanobacteria

Purple sulfur bacteria

Green sulfur bacteria

Anaerobic decomposition and sulfate reduction

(b)

Norbert Pfennig

FIGURE 17.4 The Winogradsky column. (a) Schematic view of a typical column. The column is placed so as to receive subdued sunlight. Chemoorganotrophic bacteria grow throughout the column, aerobes and microaerophiles in the upper regions, anaerobes in the zones containing H₂S. Anaerobic decomposition leading to sulfate reduction creates the gradient of H₂S. Green and purple sulfur bacteria stratify according to their tolerance for H₂S. (b) Photo of Winogradsky columns where blooms of three different phototrophic bacteria have occurred. Left to right: *Thiospirillum jenense, Chromatium okenii,* and *Chlorobium limicola.*

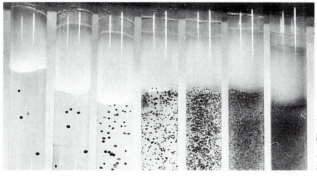

Norbert Pfennig

FIGURE 17.5 Agar shake technique for isolation of anaerobic bacteria in pure culture. A dilution series was established from right to left, eventually yielding well-isolated colonies. The tubes are sealed with a sterile mixture of parafin and mineral oil to maintain anaerobiosis.

The **agar shake tube method** (Figure 17.5) involves the dilution of a mixed culture in tubes of molten agar, resulting in colonies embedded *in* the agar rather than on the surface of a plate. The shake tube method has been found useful for purifying particular types of microorganisms (for example, phototrophic sulfur bacteria and sulfate-reducing bacteria). Purification can be obtained by successively diluting a cell suspension in tubes of liquid medium

> The enrichment culture technique is a simple means of selecting from natural samples microorganisms that are already present and are capable of carrying out specific reactions. The Winogradsky column is a type of enrichment culture that is suitable for obtaining phototrophic bacteria and other anaerobes. Once a vigorous enrichment culture has been obtained, a pure culture can usually be obtained by use of conventional bacteriological procedures.

until a dilution is reached in which well-isolated colonies are obtained (Figure 17.5). By repeating this procedure using the highest dilution showing growth as inoculum for a new set of dilutions, it is possible in most cases to eventually obtain pure cultures.

17.4 Identification and Quantification: Nucleic Acid Probes, Fluorescent Antibodies, and Viable Counts

Methods that have been devised for counting microorganisms in pure culture can easily be adapted for use on natural samples. For example, dilution and plate counts (see Section 9.3) can be used to enumerate bacteria from water or soil samples. However, depending on the culture medium and incubation conditions chosen, only a fraction of the total microbial community can ever be measured by viable counting methods. In certain instances, highly selective media may be available for counting a specific subset of the total community and such counts are usually more reliable estimates than total counts.

In some cases the microorganisms of interest in an ecosystem may not be culturable. If this is the case, methods are needed to at least detect that they are present. Several fairly specific methods have been developed for identification and enumeration, including methods based on the specificity of antibodies and nucleic acid sequences. However, these methods only yield information on the *kinds* of organisms present; they do not indicate whether the organisms are *active* in their natural habitat. Nevertheless, if activity measurements are not available, or only very little information is available about the type of organisms present, identification methods give some useful information about the microbiology of the habitat.

Staining and fluorescent antibodies

Microorganisms can occasionally be identified and enumerated by direct microscopic examination of the habitat. However, special procedures are needed to make microorganisms visible in opaque habitats. In soil, for instance, organisms can be stained with any of a variety of different fluorescent dyes specific for living cells (see Figure 17.3). Acridine orange is such a dye; this dye stains DNA and RNA. Direct staining techniques using acridine orange are widely used for estimating total microbial numbers in soil and water.

Fluorescent antibody staining is a method for identifying a single species (or even a single strain of a given species) in soil samples; in certain applications the technique can be used as a method of quantification as well. We discussed the theory of fluorescent antibodies in Section 12.12. The great specificity of antibodies prepared against cell surface constituents of a particular organism can be exploited to identify that organism in a complex habitat such as soil (Figure 17.6). Fluorescent antibodies are therefore most useful for tracking a single microbial species in soil or other habitats. For example, fluorescent antibodies were used successfully to identify the causative agent of Legionnaires disease, *Legionella pneumophila* (see Section 15.2 and Figure 13.8a), in a variety of natural samples.

Nucleic acid probes

A very powerful approach to the identification and quantification of microorganisms in nature is the use of nucleic acid probes. We discussed the use of nucleic acid probes in our description of molecular identification tools in Section 13.11 and extend that discussion here to the use of probes in an ecological context. Some probes that have been used employ 16S rRNA sequences as tools for differentiating microorganisms in natural environments. We discuss in Chapter 18 how 16S ribosomal RNA sequencing can be used to identify the primary domain (Bacteria, Archaea, Eukarya) to which a given microorganism belongs. And, because the number of available 16S rRNA sequences has increased greatly since the technique was first introduced, highly specific probes capable of differentiating organisms *within* a phylogenetic domain can now be constructed. By using a collection of such probes to test natural samples it is possible to identify different species in a single sample (see Figure 17.7).

Two types of labeling systems have been used in phylogenetic probes, *radioisotopes* and *fluorescent dyes*. The radioisotopic method employs ^{35}S or ^{32}P-labeled probes added to formaldehyde-fixed cells immobilized on glass fiber filters. Before treatment with a probe, the dried cells are treated in such a way that they are made permeable, to allow the probe to penetrate and find a ribosome to bind to. Labeling of individual cells is then visualized by autoradiography: the exposure of a photosensitive emulsion by the radioactivity in the ribosomes to which the probe bound. Following development, visible silver grains accumulate around the cells that have incorporated the probe (Figure 17.7a).

An alternative detection system for probe-based studies employs fluorescent dyes. The dyes are chemically attached to the probe and the cells visualized by fluorescent microscopy (Figure 17.7b). Because they are complementary to 16S rRNA, only relatively short (15–40 nucleotides) oligonucleotide probes are required to differentiate organisms at the domain level and only slightly longer probes are needed for distinguishing related organisms *within* a domain. Using fluorescent microscopy and probes labeled with various dyes that fluoresce at different wavelengths, it is possible to treat a sample with several different probes at once and then identify and quantify different cell types in the sample by use of the appropriate fluorescence filter (Figure 17.7b and c).

Study of the composition of microbial communities in which the organisms present have not yet been cultured can be done by use of ribosomal RNA probes and subsequent sequencing of cloned genes. The general approach used is outlined in Figure 17.8. Cells obtained from a natural environment are treated to extract bulk DNA, which is then fragmented by restriction enzymes and cloned into cells of *Escherichia coli* to form a "library" of extracted DNA (see Section 8.1 for a discussion of gene libraries). Genes encoding ribosomal RNA from the microbial community are amplified by the polymerase chain reaction (PCR) method using probes complementary to highly conserved regions of ribosomal RNA as PCR templates (see Section 8.9 for a discussion of the polymerase chain reaction). Following this, the amplified DNA in unique clones is sequenced and phylogenetic identification of community members made based on sequence comparisons (Figure 17.8) (see Chapter 18 for detailed discussion of ribosomal RNA sequencing and microbial phylogeny). From such phylogenetic data, it may be possible to predict certain physiological properties of the uncultured

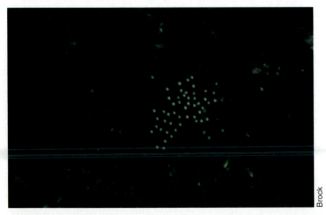

FIGURE 17.6 Visualization of bacterial microcolonies (bacterial cells appear as greenish-yellow dots) on the surface of soil particles by use of the fluorescent antibody technique. Cells are about 1 μm in diameter.

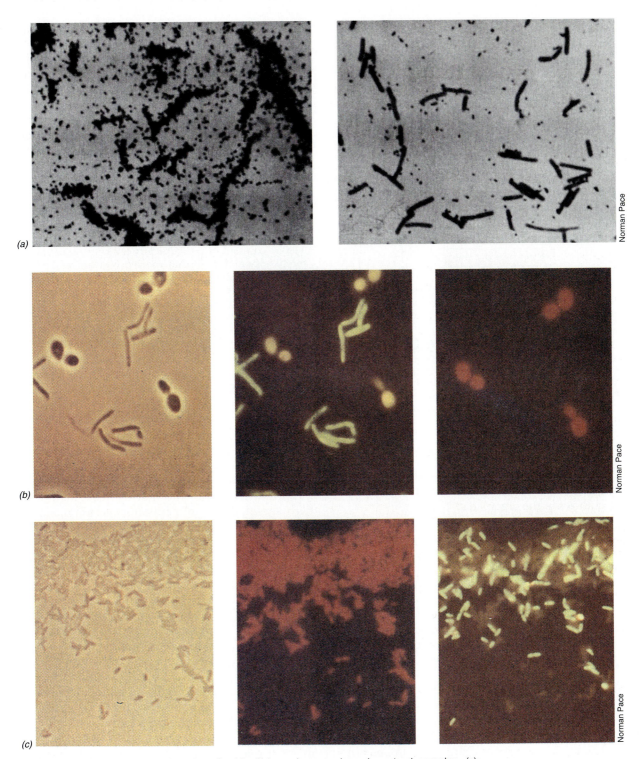

FIGURE 17.7 Nucleic acid probes for identifying microorganisms in natural samples. (a) Microautoradiographs of single cells of *Bacillus megaterium* hybridized with ribosomal RNA oligonucleotide probes complementary to specific sequences in the 16S rRNA of Bacteria (left) and to the 18S rRNA of Eukarya (right). Note the extensive number of silver grains around cells that have hybridized to the Bacterial probe. (b) Fluorescently labeled ribosomal RNA probes. Left, phase-contrast photomicrograph of *B. megaterium* and the yeast *Saccharomyces cerevisiae* (no probes present). Center, same field, cells stained with universal rRNA probe. Right, same field, cells stained with eukaryal probe. (c) Differentiation of closely related Gram-negative Bacteria. Left, phase micrograph of mixture of *Proteus vulgaris* and a related bacterium isolated from wasps. Center, same field stained with the bacterial probe. Right, same field stained with a probe specific for the bacterium from wasps. Cells of *B. megaterium* are about 1.5 μm in diameter, *S. cerevisiae* about 6 μm in diameter, and *P. vulgaris* about 0.8 μm in diameter.

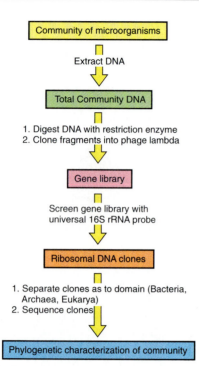

Community of microorganisms

Extract DNA

Total Community DNA

1. Digest DNA with restriction enzyme
2. Clone fragments into phage lambda

Gene library

Screen gene library with
universal 16S rRNA probe

Ribosomal DNA clones

1. Separate clones as to domain (Bacteria, Archaea, Eukarya)
2. Sequence clones

Phylogenetic characterization of community

FIGURE 17.8 Steps in phylogenetic analysis of a natural microbial community. Once ribosomal DNA clones have been obtained and separated as to domain, additional screening steps are usually required to identify unique clones and discard any duplicate clones. Sequencing of 16S rDNA yields the sequence of the corresponding rRNA that is used in phylogenetic analysis (see Chapter 18).

members of the microbial community, which in turn may help guide enrichment experiments for eventually obtaining cultures of the organisms.

One can also study the composition of a microbial habitat using *in vitro* DNA nucleic acid hybridization methods if the right probes are available. DNA is extracted directly from soil, water, or any other microbial habitat, denatured, and tested for hybridization to highly specific phylogenetic probes or probes to particular genes characteristic of one or more organisms. Cultivation of the organism(s) from the habitat is not required. Such hybridization techniques can be used to track a particular organism or group of organisms through an ecosystem or to screen a habitat for a given organism using one or more key genes as a probe. For example, genes encoding proteins involved in the biodegradation of a particular compound can be used as probes to detect the presence in a habitat of organisms containing these genes. Moreover, probes can be used to assess the competitiveness of an organism in a microbial ecosystem when perturbations of the habitat are experimentally induced. The latter technique has been used to study the efficacy of using introduced species rather than natural members of the microbial community (which may lack the genes tracked by the probe) for bioremediation of toxic pollutants.

The future of nucleic acid probes in microbial ecology appears bright, especially for use where culture methods may be unsatisfactory or where the organisms

of interest are present in numbers below the sensitivity of activity measurements. However, the major limitation in all of the methods described thus far is that the information obtained is limited to the detection, enumeration, or phylogenetic positioning of the organisms. In some types of ecological studies, this may be the main goal. However, in most microbial ecology studies, knowledge of microbial *activity* is needed as well to conclude whether the microorganisms present are of real ecological significance. We now discuss some methods for measuring microbial activities in nature.

> **Procedures for counting microbial populations in natural environments are available, and individual species can be counted microscopically if a fluorescent antibody is available for that species. Nucleic acid probes made radioactive or fluorescent can provide specific and highly sensitive means for identifying and tracking individual microorganisms in natural habitats. However, these procedures only measure the presence of an organism; they do not show that it is active.**

17.5 Measurements of Microbial Activity in Nature

Several methods are available for measuring microbial activity. We limit our discussion here to three that have found widespread use in microbial ecology: radioisotopes, microelectrodes and stable isotopes. An important feature of the first of these methods is its high *sensitivity*. If microorganisms are metabolically active in an environment but are present in relatively small numbers, activity assays must be sufficiently sensitive to measure whether a reaction has occurred. Sensitive detection systems also allow for short term incubation times, if this is necessary. These are desirable when, for example, a sample of soil, water, microbial mat, or other natural material is removed to do an experiment. Short assay periods guard against the tendency for samples removed from their surroundings to change in nonspecific ways with time. This helps to ensure that the measurements obtained are representative of the sample as it existed in nature.

Radioisotopes

In many cases, direct chemical or physical measurement of microbial activities in nature is possible. Instrumentation has improved greatly in recent years, and the sensitivity of many scientific instruments is such that only very small amounts of a given compound need be present for accurate measurements. Thus, chemical measurements are often satisfactory and in some cases even desirable in ecological studies. However, **radioisotopes** are very useful for measurement of specific microbial processes at very high sensitivity and to obtain information on the turnover rates of chemical species in nature. Radioisotopes provide extremely sensitive and highly specific means of

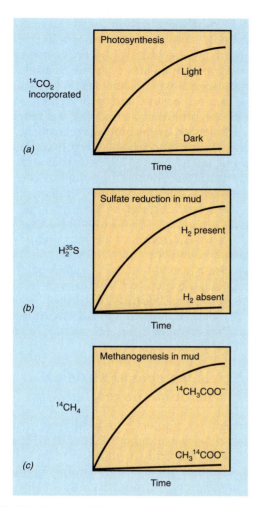

FIGURE 17.9 Use of radioisotopes to measure microbial activity in nature. (a) Photosynthesis measured in natural seawater with $^{14}CO_2$. (b) Sulfate reduction in mud measured with $^{35}SO_4^{2-}$. (c) Methanogenesis measured in mud with acetate labeled in either the methyl ($^{14}CH_3COO^-$) or the carboxyl ($CH_3{}^{14}COO^-$) carbon.

measuring chemical processes (Figure 17.9). For instance, if photoautotrophy is to be measured, the light-dependent uptake of $^{14}CO_2$ into microbial cells can be measured; if sulfate reduction is of interest, the rate of conversion of $^{35}SO_4^{2-}$ to $H_2{}^{35}S$ can be assessed (Figure 17.9). Methanogenesis in natural environments can be studied by measuring the conversion of $^{14}CO_2$ to $^{14}CH_4$ in the presence of a suitable reductant such as H_2, or by the conversion of methyl compounds such as ^{14}C-labeled methanol or acetate (Figure 17.9c) to $^{14}CH_4$. Chemoorganotrophic activity can be measured by following incorporation of ^{14}C-organic compounds. For example, the uptake of ^{14}C-glucose or ^{14}C-amino acids by chemoorganotrophs in lake water can be studied by filtering the bacterial cells from suspension following an incubation period with the isotope. Alternatively, in soil one might measure the extent of $^{14}CO_2$ production from the added radioisotope.

Isotope methods are widely used to evaluate the activity of microorganisms in nature. However, because there is always the possibility that some transformation of a labeled compound might be due to a strictly chemical, rather than microbial process, it is essential when using isotopes to employ proper controls. The key control necessary, of course, is the *killed cell control*. It is absolutely essential to show that the transformation being measured in nature is prevented by microbicidal agents or heat treatments that are known to block microbial action or kill the organisms. Formalin at a final concentration of 4 percent is frequently used as a chemical sterilant in microbial ecology studies.

It is possible to combine the use of radioisotopes with study of the activities of *single* microbial cells using *autoradiography*. We discussed autoradiography in connection with the use of nucleic acid probes in microbial ecology in Section 17.4 (see Figure 17.7a). Here we consider other uses of this method. Autoradiography can be used to measure the uptake of any substrate available in radioactive form, including nucleotides, amino acids, and many other organic compounds. If autoradiography is used in conjunction with an identification method, such as immunofluorescence or phylogenetic staining (see Figure 17.7), then information about both the activity and the taxonomy/phylogeny of a microorganism can be obtained simultaneously.

> The activity of microorganisms in natural samples often is assessed in a very sensitive manner by measuring the uptake of a radioactively labeled substrate. To measure the activity of single cells, autoradiographic procedures with radioactive substrates can be used.

Microelectrodes

We discussed earlier the fact that the environments of microorganisms are very small, that is, they are *microenvironments*. Microbial ecologists have used small glass electrodes, referred to as **microelectrodes**, to study the activity of microorganisms in their microenvironments (we discussed the use of O_2 microelectrodes in Section 17.1, see Figure 17.1). Although limited to measurements of those chemical species that can be detected with an electrode, several types of electrodes have found use in field studies. Currently, the most popular microelectrodes are those that measure pH, oxygen, and sulfide.

As the name implies, microelectrodes are very small, the tips of the electrode ranging in diameter from 2 to 100 μm; O_2 microelectrodes 2–3 μm in diameter can be routinely made (Figure 17.10). The electrodes are carefully inserted into the habitat using a micromanipulator, a device that allows for precise movement of the electrode through distances of a millimeter or less. Microelectrodes have been used extensively in the study of chemical transformations and photosynthesis in *microbial mats*. The latter are layered microbial communities usually containing cyanobacteria in the uppermost layer, anoxygenic phototrophic

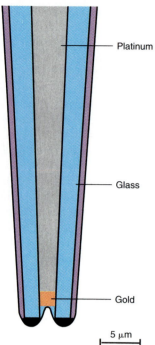

Platinum

Glass

Gold

5 µm

FIGURE 17.10 Schematic drawing of an oxygen microelectrode. Note the scale of the electrode.

(a)

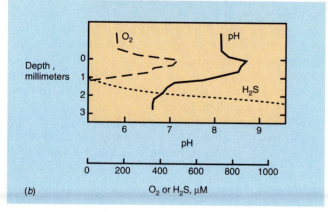

(b)

FIGURE 17.11 Microbial mats and the use of microelectrodes to study them. (a) Photograph of a core taken through the kind of hot spring microbial mat used in the experiment shown in part (b). Upper layer (dark green) contains cyanobacteria beneath which are several layers of anoxygenic phototrophic bacteria (orange and yellow). The whole thickness of the mat is about 2 cm. (b) Oxygen, sulfide, and pH microprofiles in a hot spring microbial mat. Note the millimeter scale on the ordinate.

bacteria in subsequent layers (until the mat becomes light-limited), and chemoorganotrophic, especially sulfate-reducing, bacteria in the lower layers (Figure 17.11). Microbial mats are found in a variety of environments, especially in hot springs (Figure 17.11a) and shallow marine basins and are dynamic systems where photosynthesis occurring in the upper layers of the mat is balanced by decomposition from below.

Using O_2 microelectrodes, oxygen concentrations in microbial mats (Figure 17.11b) or soil particles (Figure 17.1) can be sensitively measured over extremely fine intervals. Using a micromanipulator, electrodes can be immersed through a microbial habitat and readings taken every 0.05–0.1 mm (50–100 µm). Using a bank of microelectrodes, each sensitive to a different chemical, simultaneous measurements of a number of microbial transformations can be made at one time (Figure 17.11b). Oxygen and sulfide measurements are often taken together because gradients of both chemical species form in many microbial environments due to photosynthesis and sulfate reduction, respectively (Figure 17.11b).

17.6   Stable Isotopes and Their Use in Microbial Biogeochemistry

We learned in Chapter 2 that different isotopes of most elements exist and that while certain isotopes are unstable and break down (due to radioactive decay, see above), others are stable and simply contain a different number of neutrons than the major isotopic form. In microbial ecology geochemists have teamed up with microbiologists to study biogeochemical transformations using these *stable isotopes*.

The two chemical elements which have proven most useful for stable isotope studies are carbon and sulfur. Carbon exists in nature primarily as ^{12}C. However, a small amount of carbon exists as ^{13}C. Likewise, sulfur exists primarily as ^{32}S, although some sulfur exists as ^{34}S. All of these are stable isotopes. The reason stable isotope measurements are useful in microbial ecology is that most biochemical reactions involving carbon or sulfur favor the *lighter* isotope. That is, the heavier isotope is discriminated against when the elements are acted upon biochemically. Thus, when CO_2 of known isotopic composition is fed to a phototrophic organism, the cellular carbon becomes enriched in ^{12}C, while the CO_2 remaining in the medium becomes enriched in ^{13}C. Likewise, sulfide produced from the bacterial reduction of sulfate is much "lighter" than sulfide of strictly geochemical origin.

The typical isotopic composition of different carbon samples relative to a limestone standard is shown in Figure 17.12. It is easy to see that plant material and petroleum (which is derived from plant material), have similar isotopic compositions; carbon from both sources is isotopically lighter than the standard because it was fixed by a pathway that discriminated against $^{13}CO_2$. Methane of biological origin can be extremely light, indicating that CO_2-reducing meth-

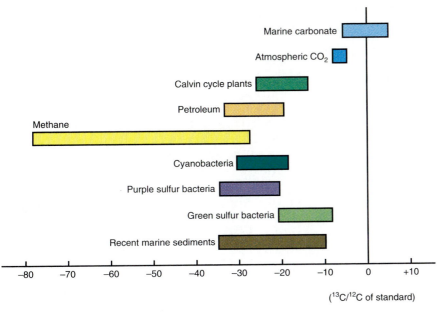

FIGURE 17.12
Carbon isotopic compositions of various substances. The values were calculated using the formula

$$\frac{(^{13}C/^{12}C \text{ sample}) - (^{13}C/^{12}C \text{ standard})}{(^{13}C/^{12}C \text{ standard})} \times 1000$$

The standard is a belemnite sample from the PeeDee rock formation.

anogens dramatically discriminate between isotopic forms of CO_2 (see Sections 17.12 and 20.5 for discussion of the biochemistry of methanogenesis). Marine carbonates, on the other hand, are clearly of geological origin, because they are isotopically much "heavier" than biogenic carbon (Figure 17.12). This large carbon reservoir has obviously not yet passed through phototrophic organisms. Because of the vast differences in the proportion of ^{12}C and ^{13}C in carbon of biological and geological origins, the $^{13}C/^{12}C$ isotopic ratio of various geological strata can be used to detect the onset of living (in this case, autotrophic) processes.

The key role of sulfate-reducing bacteria in the formation of sulfur deposits is known from studies of the fractionation of $^{34}S/^{32}S$ (Figure 17.13). As compared to a meteoritic sulfur standard, sedimentary sulfide is highly enriched in ^{32}S. Nonbiogenic sulfide (as, for instance, in igneous rocks from volcanic deposits) does not show this bias toward the lighter isotope (Figure 17.13). Also, biological oxidation of sulfide to sulfur or sulfate, either aerobically or anaerobically, shows a preference for the lighter isotope, although the fractionation is not nearly as great as that occurring during sulfate reduction (Table 17.2).

Stable isotope analyses have several uses in microbial ecology. Sulfur isotopes have been used to distinguish between biogenic and abiogenic ores (iron sulfides) and elemental sulfur deposits. Knowledge of the fractionations observed in sulfate reduction and subsequent sulfide oxidation by phototrophic or chemolithotrophic bacteria allows monitoring of important links in the sulfur cycle. Carbon isotopic analyses have been used to distinguish biogenic from abiogenic sediments, and oxygen analyses (using $^{18}O/^{16}O$ measurements of rocks of various ages) have been used to trace the earth's transition from an anoxic to an oxic environment (the earth's molecular oxygen originated from oxygenic photosynthesis by cyanobacteria, see Section 18.1).

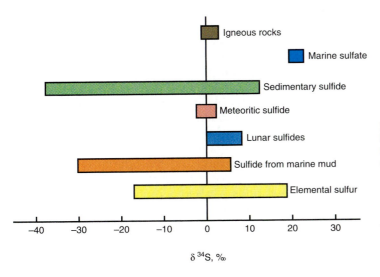

FIGURE 17.13
Summary of the isotope geochemistry of sulfur, indicating the range of values for ^{34}S and ^{32}S in various sulfur-containing substances. The values were calculated using the formula

$$\frac{(^{34}S/^{32}S \text{ sample}) - (^{34}S/^{32}S \text{ standard})}{(^{34}S/^{32}S \text{ standard})} \times 1000$$

The standard is an iron sulfide mineral from the Canyon Diabolo meteorite. Note that sulfide and sulfur of biogenic origin tend to be depleted in ^{34}S (enriched in ^{32}S).

Table 17.2 Selective utilization of ^{32}S over ^{34}S (isotope fractionation) in certain parts of the microbiological sulfur cycle

Process	Starting substance	End product	Isotope fractionation*
Dissimilatory sulfate reduction (*Desulfovibrio*)	SO_4^{2-}	HS⁻	−46.0
Dissimilatory sulfite reduction (*Desulfovibrio*)	SO_3^{2-}	HS⁻	−14.3
Assimilatory sulfate reduction (*E. coli*)	SO_4^{2-}	Organic S	−2.8
Putrefaction and desulfurylation (*Proteus*)	Organic S	HS⁻	−5.1
Chemolithotrophic sulfide oxidation (*Thiobacillus*)	HS⁻	S⁰	−2.5
	HS⁻	SO_4^{2-}	−18.0
Phototrophic sulfide oxidation (*Chromatium*)	HS⁻	S⁰	−10.0
	HS⁻	SO_4^{2-}	0

*Isotope fractionation is expressed as

$$\delta^{34}S(\text{‰}) = \frac{^{34}S/^{32}S\ sample - {}^{34}S/^{32}S\ standard}{^{34}S/^{32}S\ standard} \times 1000$$

Stable isotopic analyses have also been used as evidence for the lack of living processes on the moon. For example, the data of Figure 17.13 show that the sulfide isotopic composition of lunar rocks closely approximates that of meteoritic sulfide and not that typical of biogenic sulfide. Stable isotope analyses can also be used to track microbial and higher organism food chains. Because the isotopic composition of a given animal should approximate the isotopic composition of its major food source, it has been possible to trace the flow of carbon through an ecosystem in ways previously not possible.

17.7 Aquatic Habitats

Typical aquatic environments are the oceans, estuaries, salt marshes, lakes, ponds, rivers, and springs. Aquatic environments differ considerably in chemical and physical properties, and it is not surprising that their microbial species compositions also differ. The predominant phototrophic organisms in most aquatic environments are microorganisms; in oxic areas cyanobacteria and algae prevail, and in anoxic areas anoxygenic phototrophic bacteria are preponderant. Algae floating or suspended freely in the water are called **phytoplankton**; those attached to the bottom or sides are called **benthic algae**. Because these phototrophic organisms utilize energy from light in the initial production of organic matter, they are called **primary producers**. In the final analysis, the biological activity of an aquatic ecosystem is dependent on the rate of primary production by the phototrophic organisms.

The activities of primary producers are in turn affected by the physical environment (for example, temperature, pH, and light) and by the kinds and concentrations of nutrients available. Open oceans are very low in primary productivity, whereas inshore ocean areas are high, with some lakes and springs being highest of all. The open ocean is relatively infertile because the inorganic nutrients needed for phytoplankton growth are present only in low concentrations. The more fertile inshore ocean areas, on the other hand, receive extensive nutrient enrichment from rivers and other polluted water inputs (Figure 17.14). The amount of economically important crops such as fish or shellfish is determined ultimately by the rate of primary production; lakes and inshore ocean areas are high in primary production (Figure 17.14) and thus are the richest sources of fish and shellfish.

Oxygen relationships in lakes and rivers

We discussed oxygen requirements and anaerobiosis in Section 9.12, the production of oxygen via photosynthesis in Section 16.5, and oxygen in microenvironments in Section 17.1. Although oxygen is one of the most plentiful gases in the atmosphere (~21% of air), it has limited solubility in water, and in a large water mass its exchange with the atmosphere is slow. Significant photosynthetic production of oxygen occurs only in the *surface layers* of a lake or ocean, where light is available. Organic matter that is not consumed in these surface layers sinks to the depths and is decomposed by facultative microorganisms, using oxygen dissolved in the water. In lakes, once the oxygen is consumed, the deep layers become anoxic; here strictly aerobic organisms such as higher plants and animals cannot grow, and the bottom layers have a species composition restricted to anaerobic bacteria and a few kinds of microaerophilic animals. In addition, there is a conversion from a respiratory to a fermentative metabolism, with important consequences for the carbon cycle (see Sections 16.14–16.19 for a consideration of this transition).

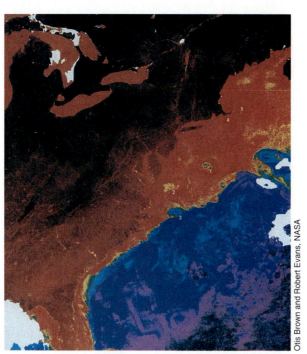

FIGURE 17.14 Distribution of chlorophyll in the western North Atlantic Ocean as recorded by satellite. The east coast of the United States from mid-Florida to northern Maine is shown. Near the center of the photo is Chesapeake Bay; the Great Lakes are in the upper left. Areas rich in phytoplankton are shown in red (>1 mg chlorophyll/m³); blue and purple areas have lower chlorophyll concentrations (<0.0/1 mg/m³). Note the high primary productivity of coastal areas.

Whether or not a body of water becomes depleted of oxygen depends on several factors. If organic matter is sparse, as it is in unproductive lakes or in the open ocean, there may be insufficient substrate available for chemoorganotrophs to consume all the oxy-

gen. Also important is how rapidly the water from the depths exchanges with surface water. Where strong currents or turbulence occurs, the water mass may be well mixed, and consequently oxygen may be transferred to the deeper layers. In many bodies of water in temperate climates, however, the water mass becomes *stratified* during the summer, with the warmer and less dense surface layers (referred to by limnologists as the *epilimnion*), separated from the colder and denser bottom layers (the *hypolimnion*, Figure 17.15). After stratification sets in, usually in early summer, the bottom layers become anoxic (Figure 17.15). In the late fall and early winter, the surface waters become colder and heavier than the bottom layers, and the water "turns over," leading to a reaeration of the bottom. Most lakes in temperate climates thus show an annual cycle in which the bottom layers of water pass from oxic to anoxic and back to oxic.

Rivers

The oxygen relations in a river are of particular interest, especially in regions where rivers receive much organic matter in the form of sewage and industrial pollution. Even though a river may be well mixed because of rapid water flow and turbulence, large amounts of added organic matter can lead to a marked oxygen deficit. This is illustrated in Figure 17.16. As the water moves away from a sewage outfall, organic matter is gradually consumed, and the oxygen content returns to normal. Oxygen depletion in a body of water is undesirable because higher animals require O_2 and die under even very temporary anoxic conditions. Further, conversion to anoxia results in the production by anaerobic bacteria of odoriferous compounds (for example, amines, H_2S, mer-

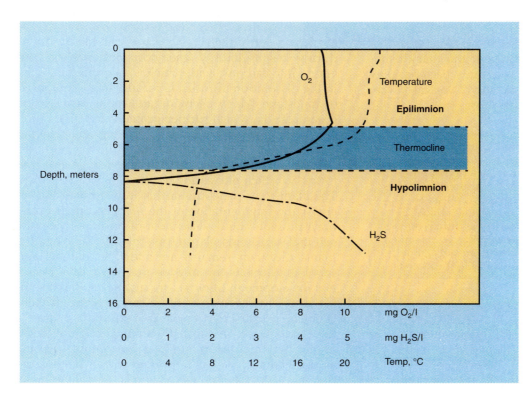

FIGURE 17.15
Development of anoxic conditions in the depths of a temperate climate lake as a result of summer stratification. The colder bottom waters are more dense and contain H_2S from bacterial sulfate reduction. The zone of rapid temperature change is referred to as the thermocline.

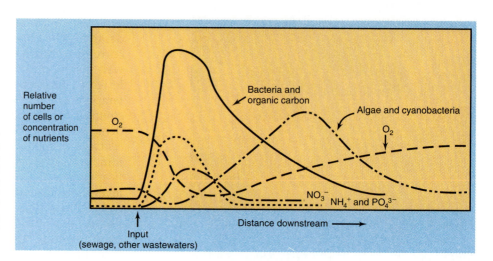

FIGURE 17.16
Effect of input of sewage or other organic-rich wastewaters into a river. Increase in heterotrophic bacterial numbers and decrease in O_2 levels occur immediately. If NH_4^+ is present in the input, for example from sewage, NH_4^+ is oxidized to NO_3^- by nitrifying bacteria. The rise in numbers of algae and cyanobacteria is primarily a response to inorganic nutrients, especially PO_4^{3-}.

captans, fatty acids), some of which are also toxic to higher organisms.

Biochemical oxygen demand

Sanitary engineers term the oxygen-consuming property of a body of water its **biochemical oxygen demand** (BOD). The BOD is determined by taking a sample of water, aerating it well, placing it in a sealed bottle, incubating for a standard period of time (usually 5 days at 20°C), and determining the residual oxygen in the water at the end of incubation. Although it is a crude method, a BOD determination gives some measure of the amount of organic material in the water that could be oxidized by microorganisms. As a river recovers from contamination with an organic pollutant, the drop in BOD is accompanied by a corresponding increase in dissolved oxygen (Figure 17.16). Government regulatory agencies specify permissible BOD levels for effluents to be released in rivers or lakes.

We thus see that the oxygen and carbon cycles in a water body are greatly intertwined, and that chemoorganotrophic microorganisms, mainly bacteria, play important roles in determining the biological nature and productivity of the body of water.

> In aquatic ecosystems, phototrophic microorganisms such as algae and cyanobacteria are usually the main primary producers. Most of the organic matter produced by the primary producers is consumed by bacteria, and bacterial respiratory activities can lead to depletion of oxygen in the environment. Polluted environments often receive excessive additions of organic matter and oxygen may be completely consumed, leading to harmful changes. The oxygen and carbon cycles of aquatic environments are thus highly related.

17.8   Terrestrial Environments

In the consideration of terrestrial environments, our attention inevitably turns to *soil* and *plants*, since it is within the soil and on or near plants that many of the key processes occur that influence the functioning of the ecosystem. The process of soil development involves complex interactions between the parent material (rock, sand, glacial drift, and so on), topography, climate, and living organisms. Soils can be divided into two broad groups—**mineral soils** and **organic soils**—depending on whether they derive initially from the weathering of rock and other inorganic material or from sedimentation in bogs and marshes, respectively. Our discussion will concentrate on mineral soils, the predominant soil in most areas.

Soil formation

Soils form as a result of combined physical, chemical, and biological processes. An examination of almost any exposed rock will reveal the presence of algae, lichens, or mosses. These organisms are able to remain dormant on the dry rock and then grow when moisture is present. They are phototrophic and produce organic matter, which supports the growth of chemoorganotrophic bacteria and fungi. The numbers of chemoorganotrophs increase directly with the degree of plant cover of the rocks. Carbon dioxide produced during respiration by chemoorganotrophs is converted into carbonic acid, which is an important agent in the dissolution of rocks, especially those composed of limestone. Many chemoorganotrophs also excrete organic acids, which further promote the dissolution of rock into smaller particles. Freezing and thawing and other physical processes lead to the development of cracks in the rocks. In these crevices, a raw soil forms, in which pioneering higher plants can develop. The plant roots penetrate farther into crevices and increase the fragmentation of the rock, and their excretions promote the development of a rhizosphere (the soil that surrounds plant roots) microflora. When the plants die, their remains are added to the soil and serve as nutrients for an even more extensive microbial development. Minerals are further rendered soluble, and as water percolates it carries some of these chemical substances deeper. As weathering proceeds, the soil increases in depth, thus permitting the development of larger plants and trees. Soil animals become established and play an important role in keeping the upper layers of the soil mixed and aerated. Eventually the movement of materials downward results in the

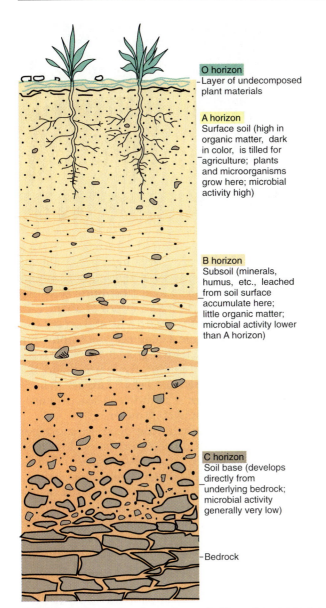

FIGURE 17.17 Profile of a mature soil. The soil horizons are soil zones as defined by soil scientists.

O horizon
Layer of undecomposed plant materials

A horizon
Surface soil (high in organic matter, dark in color, is tilled for agriculture; plants and microorganisms grow here; microbial activity high)

B horizon
Subsoil (minerals, humus, etc., leached from soil surface accumulate here; little organic matter; microbial activity lower than A horizon)

C horizon
Soil base (develops directly from underlying bedrock; microbial activity generally very low)

Bedrock

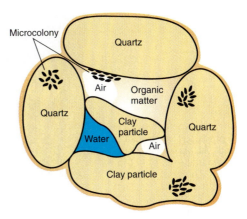

FIGURE 17.18 A soil aggregate composed of mineral and organic components, showing the localization of soil microorganisms. Very few microorganisms are found free in the soil solution; most of them occur as microcolonies attached to the soil particles.

Soil as a microbial habitat

The most extensive microbial growth takes place on the *surfaces* of soil particles (see Figures 17.3, 17.6, and 17.18). As pointed out in Section 17.1, even a small soil aggregate can have many differing microenvironments (compare Figures 17.1 and 17.18), and thus several different types of microorganisms may be present. To examine soil particles directly for microorganisms, fluorescence microscopes are often used, the organisms in the soil being stained with a dye that fluoresces (see Figure 17.3). To observe a *specific* microorganism in a soil particle, **fluorescent-antibody staining** (see Figure 17.6) can be used. Microorgan-

formation of layers, and a typical soil profile becomes outlined (Figure 17.17). The rate of development of a typical soil profile depends on climatic and other factors, but it is usually very slow, taking hundreds of years.

isms can also be visualized on such opaque surfaces as soil by means of the **scanning electron microscope** (Figure 17.19). The scanning electron microscope gives excellent information on the morphology of soil bacteria and can also be used to enumerate cells on soil particle surfaces.

One of the major factors affecting microbial activity in soil is the availability of *water*. Water is a highly variable component of soil, its presence depending on soil composition, rainfall, drainage, and plant cover. Water is held in the soil in two ways, by adsorption onto surfaces, or as free water existing in thin sheets or films between soil particles. The water present in soils has a variety of material dissolved in it; the whole mixture being referred to as the *soil solution*. In well-drained soils, air penetrates readily and oxygen concentrations can be high. In waterlogged soils, however, the only oxygen present is that dissolved in the water, and this is soon consumed by microorganisms. Such soils quickly become anoxic, showing profound changes in their biological properties. We discussed oxygen relationships in soil in Section 17.1 and Figure 17.1.

The *nutrient status* of a soil is the other major factor affecting microbial activity. The greatest microbial activity is in the organic-rich surface layers, especially in the region adjacent to plant roots (the rhizosphere). The numbers and activity of soil microorganisms depends to a great extent on the balance of nutrients present. In some soils carbon is not the limiting nutrient but instead the availability of *inorganic* nutrients such as phosphorus and nitrogen limit microbial productivity.

Deep subsurface microbiology

Recent interest in the chemistry of ground water and the potential leaching of pollutants and their transfer in ground water aquifers has led to consideration of the role microorganisms play in the *deep subsurface* terrestrial environment. The deep soil subsurface, which can extend for several hundred meters below the soil surface, is not a biological wasteland. A variety of mi-

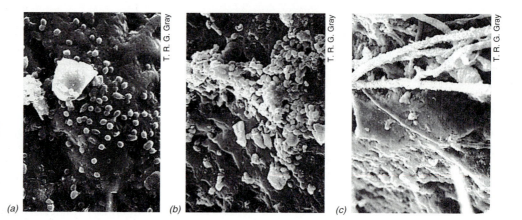

FIGURE 17.19
Visualization of microorganisms on the surface of soil particles by use of the scanning electron microscope. (a) Rod-shaped bacteria. (b) Actinomycete spores. The cells in (a) and the spores in (b) are about 1–2 μm wide. (c) Fungus hyphae. The fungal hyphae is about 4 μm wide.

(a)   *(b)*   *(c)*

croorganisms, primarily bacteria, are present in most deep underground soils. In samples collected aseptically from bore holes drilled down to 300 meters, a diverse array of bacteria have been found including anaerobes such as sulfate-reducing bacteria and various aerobes and facultative aerobes. Microorganisms in the deep subsurface presumably have access to key nutrients because ground water flows through their habitats, but activity measurements suggest that metabolic rates of these bacteria are rather low in their natural habitats. Compared to microorganisms in the upper layers of soil, the biogeochemical significance of deep subsurface microorganisms may be minimal. However, there is evidence that the metabolic activities of these buried microorganisms may over very long periods be responsible for some mineralization of organic compounds and release of products into groundwater. The potential for *in situ* bioremediation (see Section 17.19) of toxic substances leached from soil into groundwater (for example, benzenes and agricultural chemicals) by deep subsurface microorganisms is of particular current interest.

> The soil is a complex habitat with numerous microenvironments and niches. Microorganisms are present in the soil primarily attached to soil particles. The most important factor influencing microbial activity in surface soil is the availability of water whereas in deep soil (the subsurface environment) nutrient availability plays a major role.

17.9 Deep Sea Microbiology

The oceans cover over three-fourths of the earth's surface and marine microbiology is consequently an important area of study. However, because open ocean waters are relatively poor in nutrients (see Section 17.7), microbial activity in these areas is not extensive. Nevertheless, marine microorganisms are of interest for several reasons, including their ability to grow at low-nutrient concentrations and cold temperatures and, in microorganisms inhabiting the deep sea, to withstand enormous hydrostatic pressures. We focus in this section on microbial life in the deep sea.

What is the deep sea? Visible light penetrates no further than about 300 meters in open ocean waters; this upper region is referred to as the **photic zone**. Beneath the photic zone, down to a depth of about 1000 meters, considerable biological activity still occurs due to the action of animals and chemoorganotrophic microorganisms. Water at depths greater than 1000 meters is, by comparison, biologically inactive and has come to be known as the "deep sea." Greater than 75 percent of all ocean water is in the deep sea, primarily at depths between 1000 and 6000 meters.

Organisms which inhabit the deep sea are faced with three major environmental extremes: low temperature, high pressure, and low nutrient levels. Below depths of about 100 meters ocean water stays a constant 2–3°C. We discussed the responses of microorganisms to changes in temperature in Section 9.9. As would be expected, bacteria isolated from depths below 100 meters are *psychrophilic*. Some are *extreme* psychrophiles, growing only in a narrow range near the *in situ* temperature (see below). Deep sea microorganisms must also be able to withstand the enormous hydrostatic pressures associated with great depths. Pressure increases by *1 atmosphere* for every *10 meters* depth. Thus, an organism growing at a depth of 5000 meters must be able to withstand pressures of 500 atmospheres. We focus on this interesting aspect of life in the deep sea here.

Barotolerant and barophilic bacteria

Do deep sea bacteria simply tolerate high pressure (that is, are they *barotolerant*) or, are they highly adapted to life at high hydrostatic pressures (*barophilic*)? To answer this question, it is necessary to sample, maintain, and culture organisms under the pressures they would experience in their natural habitat. Studies of bacterial growth in special pressurized culture vessels (Figure 17.20) show that pure cultures of different chemoorganotrophic bacteria derived from deep sea samples (down to about 4000 meters) and maintained at high pressure during all isolation and culture procedures are indeed **barotolerant**; higher metabolic rates are observed at 1 atmosphere than at 400 atmospheres, although growth rates at the

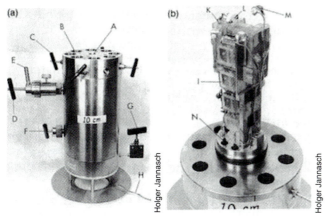

FIGURE 17.20 Photograph of a pressurized sampling and culture device developed by microbiologists at Woods Hole Oceanographic Institution. (a) Isolation chamber assembled: A, viewing window; B, bolts; C, valves for sample transfer; D and E, handles to control streaking loop; F, handle for moving agar plates; G, pressurization valve; H, wires for sterilizing loop. (b) Chamber disassembled: I, storage bin for agar plates; K, storage bin for liquid vials; L, streaking loop; M, lamps for illumination; N, seals for wires.

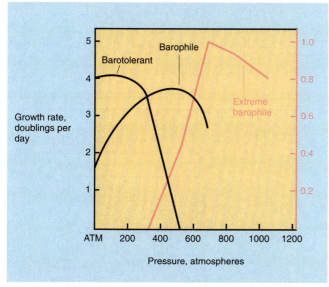

FIGURE 17.21 Growth of barotolerant, barophilic, and extremely barophilic bacteria. The extreme barophile was isolated from the Mariana Trench (10,500 meters). Note the much slower growth rate (at any pressure) of the extreme barophile (right ordinate) as compared to the barotolerant and barophilic bacteria (left ordinate). Note also the inability of the extreme barophile to grow at low pressures.

meters) have yielded cultures of **extreme (obligate) barophiles**. One strain studied in detail grew fastest at a pressure of 700–800 atmospheres, and grew nearly as well at 1035 atmospheres, the pressure it was experiencing in its natural habitat (see Figure 17.21). The unique aspect of this extremely barophilic isolate was that it not only *tolerated* pressure, it actually *required* pressure for growth; the isolate would not grow at pressures of less than about 400 atmospheres. Interestingly, however, this extreme barophile was not killed by decompression because it could tolerate moderate periods of decompression; however, viability was lost if the culture was left for several hours in a decompressed state.

Barotolerant and barophilic bacteria are also coldloving, that is, psychrophilic. This property appears to be more prevalent among extremely barophilic isolates. The extreme barophile described in Figure 17.21 was found to be sensitive to temperature; the optimal growth temperature was determined to be the environmental temperature of 2°C, and temperatures above 10°C significantly reduced viability.

Physiology of barophiles

Pressure is known to affect cellular physiology and biochemistry. It has been established that increased pressure decreases the binding capacity of enzymes for their substrates. Thus, the enzymes of extreme barophiles must be folded in such a way as to minimize these pressure-related effects. Other potential pressure-sensitive targets include protein synthesis and membrane phenomena such as transport. An organism grown under high pressure has an increase in the proportion of unsaturated fatty acids in its cytoplasmic membrane. This change is presumably of adaptive significance since it makes the membrane less likely to gel at high pressures. The rather slow growth rates of extreme barophiles (see Figure 17.21) are probably due to a combination of pressure effects on cellular biochemistry and to the fact that these organisms grow only at low temperatures, where reactions rates are decreased considerably to begin with.

Molecular genetic tools have given new insight into the physiology of barophilism. In barophiles capable of growth up to 500–600 atmospheres, it has been shown that growth at high pressure is accompanied by changes in the protein composition of the cell wall outer membrane. In one barophile studied in detail, a specific outer membrane protein called the OmpH protein (Omp stands for *outer membrane protein*) is synthesized in cells grown at high pressures but not in cells grown at 1 atmosphere pressure. The OmpH protein is a type of *porin*. Porins are structural proteins that form channels for the diffusion of organic molecules through the outer membrane and into the periplasm (see Section 3.5). Presumably, the porin present in cells of the barophile grown at low pressures cannot function properly at high pressure and thus a new type of porin molecule must be synthesized.

two pressures are about the same (Figure 17.21). However, barotolerant isolates do not grow at pressures above 500 atmospheres. On the other hand, cultures derived from samples taken at greater depths, 5000–6000 meters, are **barophilic**, growing optimally at pressures of about 400 atmospheres (Figure 17.21). Note that although barophiles grow best under pressure, they retain the ability to grow at 1 atmosphere (Figure 17.21).

Samples from even deeper oceanic areas (10,000

The *ompH* gene (which codes for the OmpH protein) from this barophile has been cloned into *Escherichia coli*. Studies of gene expression of the *ompH* gene in this experimental system have shown that pressure affects its transcription; messenger RNA coding for *ompH* is expressed in cells of *E. coli* grown at 200–300 atmospheres but not in cells grown at 1 atmosphere (*E. coli* can grow slowly at high pressure). Further, sequencing of the *ompH* gene has shown that the OmpH protein is related but clearly distinct from the porin produced at 1 atmosphere.

Studies of the *ompH* system thus show that pressure can affect gene expression in barophilic bacteria. How this happens is unclear, but could involve the activities of pressure-sensitive repressor proteins or pressure-dependent gene activators. However, it appears that relatively few proteins are controlled by pressure in barophiles because many proteins seem to be the same in cells grown at both high and low pressure; cell wall and related structural proteins seem to be the major variable components. Thus pressure acts selectively to turn on or off the transcription of specific genes coding for proteins needed for growth at high pressure. Control of most genes in barophilic bacteria is presumably affected by environmental and nutritional factors in the same manner as it is in non-barophilic bacteria (see Sections 5.10–5.13).

17.10 Hydrothermal Vents

The general conception of the deep sea as a remote, low temperature, high pressure environment capable of supporting only the slow growth of barotolerant and barophilic bacteria is generally correct, but there are some amazing exceptions. A number of dense, thriving *invertebrate* communities supported by the activities of microorganisms exist clustered about thermal springs in deep waters throughout the world. Geophysical measurements have identified thermal springs in several locations on both the Atlantic and Pacific Ocean floors. Geologically, these springs are associated with *ocean floor spreading centers*, regions where hot basalt and magma very near the sea floor cause the floor to slowly drift apart. Seawater seeping into these cracked regions mixes with hot minerals and is emitted from the springs (see Figure 17.22); because of their unique properties these springs have come to be known as **hydrothermal vents**.

Two major types of vents have been found. *Warm vents* emit hydrothermal fluid at temperatures of 6–23°C (into seawater at 2°C). *Hot vents*, usually referred to as "black smokers" because the mineral-rich

> The deep sea is a vast and little known habitat where hydrostatic pressure and low nutrient concentrations and pressure restrict microbial life. Certain specialized bacteria called barophiles are specifically adapted to the high pressures of the deep oceans.

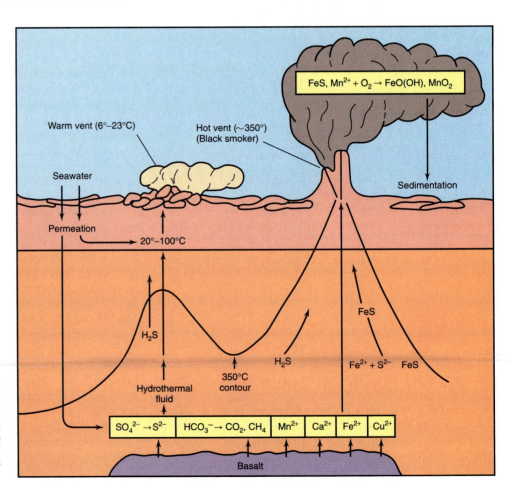

FIGURE 17.22
Schematic diagram showing the morphology and major chemical species occurring at warm vents and black smokers. At warm vents, the hot hydrothermal fluid is cooled by cold (2–3° C) seawater permeating the sediments. In black smokers, hot hydrothermal fluid reaches the sea floor directly.

hot water forms a dark cloud of precipitated material on mixing with seawater, emit hydrothermal fluid at 270–380°C (Figure 17.22). The flow rates of these two vent types are also characteristic; warm vents emit fluid at 0.5–2 cm/sec while hot vents have higher flow rates, about 1–2 meters/sec.

Using small pressurized submarines, it is possible to study the organisms associated with the thermal vents. Thriving invertebrate communities have been found, with tube worms over two meters in length, and large numbers of giant clams and mussels (Figure 17.23). Considering that other locations in the deep sea are so biologically unproductive, how do dense animal communities exist in the absence of phototrophic primary producers? What is the nature of the energy source(s)? Chemical analyses of hydrothermal fluid show large amounts of reduced inorganic materials, including H_2S, Mn^{2+}, H_2, and CO. Some vents contain little H_2S, but contain high levels of NH_4^+. Organic matter is not present in the fluid emitted from any of the hydrothermal vents thus far examined. From studies of vent chemistry and associated microbial processes, it is clear that the animals are dependent upon the activities of chemolithotrophic bacteria (see Sections 16.8–16.13), which grow at the expense of inorganic energy sources emitted from the vents. Carbon dioxide, abundant in seawater as CO_3^{2-} and HCO_3^-, is fixed into organic carbon by the chemolithotrophs, and the latter form the base of an extremely short food chain for hydrothermal vent animals. We now discuss these communities in detail.

Microorganisms in hydrothermal vents

Large numbers of sulfur-oxidizing chemolithotrophs such as *Thiobacillus*, *Thiomicrospira*, *Thiothrix*, and *Beggiatoa* (see Section 16.10) are present in and around the vents. Samples collected from near the vents have yielded cultures of these organisms, and *in situ* experiments have shown fixation of CO_2 and oxidation of H_2S and $S_2O_3^{2-}$ by natural populations of these bacteria. Other vents yield nitrifying bacteria, hydrogen-oxidizing bacteria, iron and manganese-oxidizing bacteria, and methylotrophic bacteria, the latter presumably growing on the methane and carbon monoxide (CO) emitted from the vents (see Chapter 16 and Section 19.7 for discussion of some of these physiological groups). Table 17.3 summarizes the electron donors and electron acceptors for chemolithotrophs suspected of playing a role in hydrothermal vent ecology. However, there is no direct evidence that the animals of the vents directly *eat* these chemolithotrophic bacteria. Instead, a different role of chemolithotrophs in animal nutrition occurs.

Nutrition of animals living near hydrothermal vents

Perhaps the most exciting discovery is that certain chemolithotrophs live directly in association with animals of the thermal vents. The 2-meter-long tube worms (see Figure 17.23) lack a mouth, gut, or anus, but contain a modified gastrointestinal tract consisting

FIGURE 17.23 Invertebrates from habitats near deep sea thermal vents. (a) Tube worms (family *Pogonophora*), showing the sheath (white) and plume (red) of the worm bodies. (b) Close-up photograph showing worm plume. (c) Mussel bed in vicinity of warm vent. Note yellow deposition of elemental sulfur in and around mussels.

primarily of spongy tissue called the **trophosome**. Making up about 50 percent of the weight of the worm, trophosome tissue is loaded with sulfur granules, and microscopy of sectioned trophosome tissue shows large numbers of prokaryotic cells (Figure 17.24), an average of 3.7×10^9 cells per gram of trophosome tissue. The large spherical cells observed in the trophosome are structurally similar to the ma-

Table 17.3   Chemolithotrophic prokaryotes of potential significance to hydrothermal vent primary productivity*

Chemolithotroph	Electron donor	Electron acceptor
Sulfur-oxidizing bacteria	HS^-, S^0, $S_2O_3^{2-}$	O_2, NO_3^-
Nitrifying bacteria	NH_4^+, NO_2^-	O_2
Sulfate-reducing bacteria	H_2	S^0, SO_4^{2-}
Methanogenic Archaea	H_2	CO_2
Hydrogen-oxidizing bacteria	H_2	O_2, NO_3^-
Iron and manganese-oxidizing bacteria	Fe^{2+}, Mn^{2+}	O_2
Methylotrophic bacteria	CH_4, CO	O_2

See Sections 16.8–16.13 for a discussion of chemolithotrophs.

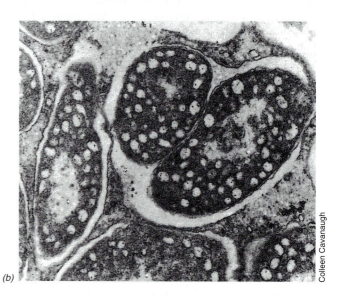

(a)   Colleen Cavanaugh   (b)   Colleen Cavanaugh

FIGURE 17.24 Chemolithotrophic sulfur-oxidizing bacteria associated with the trophosome tissue of tube worms from hydrothermal vents. (a) Scanning electron microscopy of trophosome tissue showing spherical chemolithotrophic sulfur-oxidizing bacteria. Cells are 3–5 μm in diameter. (b) Transmission electron micrograph of bacteria in trophosome tissue. The cells are frequently enclosed in pairs by an outer membrane of unknown origin.

rine sulfur-oxidizing bacterium *Thiovulum*. Trophosome tissue (containing symbiotic sulfur bacteria) also shows activity of the enzyme *rhodanese*, an enzyme capable of disproportionating $S_2O_3^{2-}$ to S^0 and SO_3^{2-}. Also present are enzymes of the *Calvin cycle*, the pathway by which most autotrophic organisms fix CO_2 into cellular material (see Section 16.7).

The chemolithotrophic bacteria supply the worm with its nourishment, the animal living off the excretory products and dead cells of its chemolithotrophic symbionts. The bright red plume (Figure 17.23*b*) is rich in blood vessels and serves as a trap for O_2 and H_2S (see below) for transport to chemolithotrophs in the trophosome. Similar conclusions can be reached concerning the nutrition of the giant clams and mussels (see Figure 17.23*c*) present around the vents, because sulfur-oxidizing bacterial communities are found in the gill tissues of these animals as well. Use of nucleic acid sequencing techniques (see Nucleic Acids box, Chapter 5) have shown that each vent ani-

mal harbors only one major species of bacterial symbiont and that the species of symbiont varies among the different animal types.

Further study of the tube worms have shown that these animals contain unusual soluble hemoglobins that bind H_2S as well as O_2 and transport both substrates to the trophosome where they are presumably released to the bacterial symbiont; trapping and transporting sulfide is necessary to prevent the H_2S from poisoning the animal. Furthermore, stable isotope analyses (see Section 17.6) of the elemental sulfur found within the bacterial symbionts have shown the $^{34}S/^{32}S$ isotope composition to be the same as the sulfide emitted from the vent. This ratio is distinctly different from that of seawater sulfate and serves as proof that geothermal sulfide is entering the worm.

The link between animal nutrition and other physiological groups of chemolithotrophs (for example, H_2-oxidizers and nitrifying bacteria) has been suggested but not yet proven. However, methylotrophic symbionts play a nutritional role in hydrothermal vent animals, since methane-oxidizing bacteria (see Section 19.7) have been found which live in symbiotic association with giant clams near natural gas seeps at relatively shallow depths in the Gulf of Mexico. Although not truly autotrophs (CH_4 is an organic compound), these symbionts support growth of the animal, in this case by the oxidation of CH_4 as an energy source.

Other chemolithotrophs, iron and manganese oxidizers, for example, are probably not animal symbionts, but instead exist as free-living bacteria growing at the expense of reduced substances emitted from the vents. Nevertheless, these chemolithotrophs probably contribute to overall primary productivity in the vent ecosystem. For example, it is possible that free-living chemolithotrophs serve as a food source for fish frequently observed swimming near the vents.

Black smokers

The great depths of the deep sea create huge hydrostatic pressures that affect the physical properties of water. At a depth of 2600 meters, water does not boil until it reaches a temperature of about 450°C. At certain vent sites superheated (but not boiling) hydrothermal fluid is emitted at temperatures of 270–380°C (see Figure 17.22) and could theoretically be a habitat for hyperthermophilic bacteria (see Section 9.9). The hydrothermal fluid emitted from black smokers contains abundant metal sulfides, especially iron sulfides, and cools quickly as it emerges into cold seawater. The precipitated metal sulfides form a tower referred to as a "chimney" about the source (Figure 17.25). Although it is very doubtful that prokaryotes

actually live in the hot (>250°C) hydrothermal fluid (see Section 20.8), good evidence exists for the presence of thermophilic/hyperthermophilic bacteria of various types in the seawater/hydrothermal fluid *gradient* that forms as the hot water blends with cold ocean water.

In efforts to define the upper temperature limits for life, attempts have been made to detect bacterial growth in the outflows of black smokers at various temperatures. By fitting a vent with a titanium "cap" (Figure 17.26) from which glass microscope slides and other surfaces for microbial colonization can be suspended in the emerging hot water, evidence for colonization and growth of prokaryotes has been obtained at temperatures above 125°C to about 140°C. Similar surfaces exposed to outflows at 200°C or higher showed no microbial attachment or growth. Although hyperthermophiles from water at 125°C or higher

> Hydrothermal vents are specialized environments in the deep sea where thermal (volcanic) activity leads to the presence of fluids containing large amounts of inorganic energy sources that can be used by chemolithotrophic bacteria. Numerous sulfur-oxidizing and methane-oxidizing bacteria are found associated with the deep sea hydrothermal vents. Some of these organisms are free-living whereas others live in symbiotic association with invertebrate animals. The chemolithotrophic bacteria fix CO_2 autotrophically into organic carbon, some of which is then used by the deep sea animals. The deep sea hydrothermal vents are thus habitats where the primary producers are chemolithotrophic rather than phototrophic.

(a)

(b)

FIGURE 17.25 Black smokers emitting sulfide- and mineral-rich water at temperatures of 350° C. (a) The chimney is quite large, about 1 meter in length. (b) The chimney is much smaller. Note the scientific equipment near the smoker in (b), giving a feeling for the relatively small size of the chimney.

Robert D. Ballard

Dudley Foster, Woods Hole Oceanographic Institution

Norman Pace

FIGURE 17.26 Methodology for studying possible extremely hyperthermophilic prokaryotes in black smokers. The titanium cap placed over the vent serves as a support for microscope slides submerged to regions of the vent of known temperature. With the vent cap system, some evidence for organisms growing at temperatures above 125° C has been obtained. The vent cap shown covers a 160° C hydrothermal vent located at a depth of 2000 meters in the Guayamas Basin (Gulf of California).

have not yet been cultured, the formation of microcolonies on artificial surfaces at these temperatures is similar to those observed at lower temperatures (see Figure 17.2), and good evidence that microbial growth is occurring here. These results suggest that the upper temperature limit for microbial life is probably under 150°C. We discuss some of the reasons for this upper temperature limit in Section 20.8.

17.11   Carbon Cycle

On a global basis, carbon is cycled through all of the earth's major carbon reservoirs: the atmosphere, the land, the oceans and other aquatic environments, and sediments and rocks (Figure 17.27). The largest carbon reservoir is present in the sediments and rocks of the earth's crust, but the turnover time is so long that flux out of this compartment is relatively insignificant on a human scale. From the viewpoint of living organisms,

a large amount of organic carbon is found in land plants. This represents the carbon of forests and grasslands and constitutes the major site of photosynthetic CO_2 fixation. However, more carbon is present in dead organic material, called *humus,* than in living organisms. **Humus** is a complex mixture of organic materials. It is derived partly from the protoplasmic constituents of soil microorganisms which have resisted decomposition, and partly from resistant plant material. Some humic substances are fairly stable, with a global turnover time of about 40 years, although certain other humic components decompose much more rapidly than this. The most chemically complex humic materials have residence times of hundreds of years.

The most rapid means of global transfer of carbon is via the CO_2 of the atmosphere. Atmospheric CO_2 constitutes a fairly significant reservoir, but its residence time is quite short. Carbon dioxide is removed from the atmosphere primarily by photosynthesis of land plants and is returned to the atmosphere by respiration of animals and chemoorganotrophic microorganisms. An analysis of the various processes suggests that the single most important contribution of CO_2 to the atmosphere is via microbial decomposition of dead organic material, including humus.

Importance of photosynthesis in the carbon cycle

The only major way in which new organic carbon is synthesized on earth is via photosynthesis. Phototrophic organisms are therefore at the basis of the carbon cycle (Figure 17.27). We discussed in Sections 16.4 and 16.5 the processes of anoxygenic and oxygenic photosynthesis. The bulk of the photosynthesis on earth is carried out by oxygenic phototrophs, so that the carbon cycle and oxygen cycle on earth are intimately intertwined.

Phototrophic organisms are found in nature almost exclusively in habitats where light is available. Thus, the deep sea and other permanently dark habitats are generally devoid of phototrophs. Phototrophic

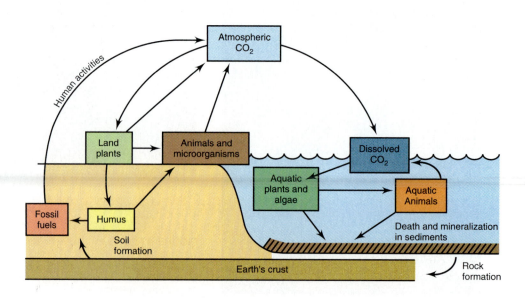

FIGURE 17.27
The carbon cycle.

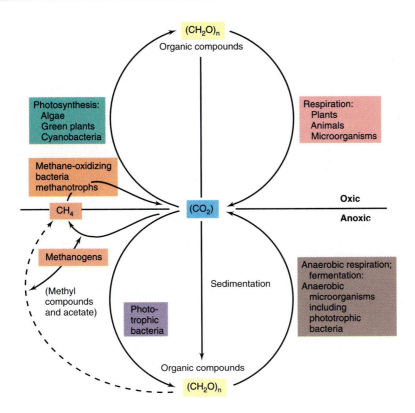

FIGURE 17.28
Redox cycle for carbon. Photosynthesis in oxic habitats is mainly oxygenic, whereas in anoxic environments it is mainly anoxygenic.

organisms can be divided into two major groups: higher plants and microorganisms. Higher plants are the dominant phototrophic organisms of terrestrial environments, whereas phototrophic microorganisms are the most abundant photosynthesizers of aquatic environments.

The redox cycle for carbon is shown in Figure 17.28. We begin with photosynthesis. The overall equation for oxygenic photosynthesis is

$$CO_2 + H_2O \xrightarrow{\text{light}} (CH_2O) + O_2$$

where (CH_2O) represents organic matter at the oxidation state of cell material such as polysaccharides (the main form in which photosynthesized organic matter is stored in the cell). Phototrophic organisms also carry out respiration, both in the light and the dark. The overall equation for respiration is the reverse of the above equation:

$$(CH_2O) + O_2 \xrightarrow{\text{light or dark}} CO_2 + H_2O$$

where (CH_2O) again represents storage polysaccharides. If an organism is to grow (that is, increase in cell number or mass) phototrophically, then the rate of photosynthesis must exceed the rate of respiration. If that occurs, then some of the carbon fixed from CO_2 into polysaccharide can become the starting material for biosynthesis. The whole carbon cycle is built on a net positive balance of the rate of photosynthesis over the rate of respiration.

Decomposition

Photosynthetically fixed carbon is eventually degraded by various organisms and two major oxidation states of carbon result: methane (CH_4) and carbon dioxide (CO_2) (Figure 17.28). These two gaseous products are formed from the activities of methanogens (CH_4), or from various chemoorganotrophs via fermentation, anaerobic respiration, or aerobic respiration (CO_2). When methane is transported to oxic environments, it is oxidized to CO_2 by methanotrophic bacteria (see Section 19.7). Hence all carbon eventually returns to CO_2 from which autotrophic metabolism once again begins the cycle.

The balance between the oxidative and reductive portions of the carbon cycle is critical; the products of metabolism of some organisms are the substrates for others. Any significant changes in levels of gaseous forms of carbon may have serious global consequences (as we are already experiencing from the increasing CO_2 levels in the atmosphere caused by deforestation and the burning of fossil fuels). We continue our treat-

On a global basis, the oxygen and carbon cycles are highly interrelated, primarily through the activities of oxygenic phototrophic microorganisms. The principal means by which the global transfer of carbon occurs is via CO_2, produced principally by the microbial decomposition of organic matter. The major way in which new organic carbon is synthesized on earth is via photosynthesis.

ment of the carbon cycle in the next two sections with detailed discussions of methanogenesis and carbon cycling in the rumen of ruminant animals.

17.12   Methane and Methanogenesis

Although methane (CH_4) is a relatively minor component of the global carbon cycle, it is of great importance in many localized situations. In addition, it is of considerable microbiological interest because it is a product of anaerobic microbial metabolism. Methane production is carried out by a highly specialized group of Archaea, the *methanogens,* which are obligate anaerobes. We discussed methane formation in Section 16.17 and will discuss the organisms themselves in Section 20.5. Most methanogens use CO_2 as their terminal electron acceptor in anaerobic respiration, reducing it to methane; the electron donor used in this process is generally hydrogen, H_2. The overall reaction of **methanogenesis** in this pathway is as follows:

$$4H_2 + CO_2 \rightarrow CH_4 + 2H_2O$$
$$\Delta G^{0\prime} = -130.7 \text{ kJ/reaction}$$

which shows that CO_2 reduction to methane is an eight-electron process (four H_2 molecules). A few other substrates can be converted to methane, including methanol, CH_3OH; formate, $HCOO^-$; methyl mer-

captan, CH_3SH; acetate, CH_3COO^-; and methylamines (see Section 20.5). It is of interest that despite the widespread production of methane, very few carbon compounds serve as direct precursors of methanogenesis. Thus, methanogenesis is dependent on the production of these few carbon compounds by other organisms from complex organic matter.

Anaerobic decomposition, interspecies hydrogen transfer, and syntrophy

We mentioned this concept briefly in Section 16.19 during our discussion of fermentative metabolism. Here we present the significance for the whole problem of the anaerobic carbon cycle. Large-molecular-weight substances such as polysaccharides, proteins, and fats, are converted to CH_4 by the cooperative interaction of several physiological groups of prokaryotes. In many anoxic environments the immediate precursors of CH_4 are H_2 and CO_2, these substrates being generated by the activities of fermentative anaerobes. For the conversion of a typical polysaccharide such as cellulose to methane as many as five major physiological groups of prokaryotes may be involved in the overall process (Figure 17.29 and Table 17.4). *Cellulolytic bacteria* cleave the high-molecular-weight cellulose molecule into cellobiose (glucose–glucose) and into free glucose. Glucose is then fermented by *fermen-*

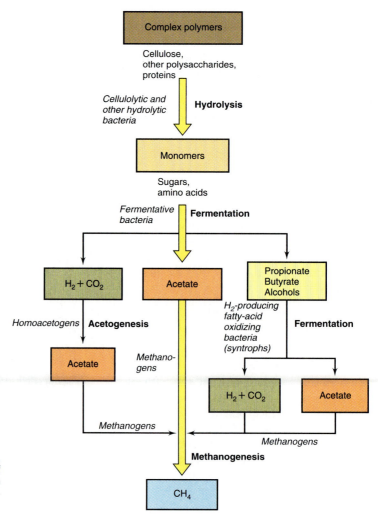

FIGURE 17.29
Overall process of anaerobic decomposition, showing the manner in which various groups of fermentative anaerobes act together in the conversion of complex organic materials ultimately to methane (CH_4) and CO_2.

tative anaerobes to a variety of fermentation products, acetate, propionate, butyrate, alcohols, H_2, and CO_2 being the major ones observed. Any H_2 produced in primary fermentative processes is immediately consumed by methanogens, homoacetogens, or sulfate-reducing bacteria. In addition, acetate can be converted to methane by certain methanogens.

Key organisms in the conversion of complex organic materials to methane are the *H_2-producing fatty acid-oxidizing bacteria*. These organisms use fatty acids or alcohols as energy sources but grow poorly or not at all on these substrates in pure culture. However, in association with a *H_2-consuming* organism (such as a methanogen or a sulfate-reducing bacterium), the H_2-producing bacteria grow luxuriantly. As explained below, H_2 consumption by a second organism is critically important to the growth of H_2-producing fatty acid-oxidizing bacteria. Examples of H_2 producers include *Syntrophomonas* and *Syntrophobacter*. The genus names of these organisms reflect their dependence on *syntrophic* relationships with other bacteria (in this case H_2-consuming bacteria) for growth (see Figure 16.38). **Syntrophy** means, literally, "eating together." *Syntrophomonas* oxidizes C-4 to C-7 fatty acids yielding acetate, CO_2 (if the fatty acid contained an odd number of carbon atoms), and H_2 (from the reduction of protons, see Table 17.4). *Syntrophobacter* specializes in propionate oxidation and generates acetate, CO_2, and H_2 (see Table 17.4). When written with all reactants at standard conditions (solutes, 1 *M*; gases, 1 atm), these fatty acid conversions yield free energy changes that are *positive* (see Table 17.4). That is, the $\Delta G^{0'}$ associated with these conversions are such that the reactions do *not* occur with the release of free energy.

Without energy release, how can these reactions support growth of the fatty acid-oxidizing bacteria? If growth of the H_2-producers occurs when H_2 is removed, the anaerobic oxidation of fatty acids to H_2 must somehow be coupled to ATP production. This implies that the presence or absence of H_2 affects the energetics of the reaction. A brief review of the principles of free energy given in Appendix 1 indicates that the *actual* concentration of reactants and products in a given reaction can drastically change the available free energy of a reaction. In a natural situation, reactants and products of fatty acid oxidation, acetate and H_2, are consumed by methanogenic Archaea (or sulfate-reducing bacteria); measurements of H_2 in actively methanogenic ecosystems are generally below 10^{-3} atmospheres.

As shown in Table 17.4, if the concentration of H_2 is kept very low by constant removal, the free energy change associated with the oxidation of fatty acids to acetate plus H_2 by H_2-producing fatty acid-oxidizers becomes *negative* in sign, indicating that energy is released. Although the mechanism of ATP production by syntrophic fatty acid oxidizers is not clear, it probably involves the conversion of acetyl CoA (generated by β-oxidation of the fatty acid) to acetate. Moreover, electron transport-mediated ATP production may be possible under certain circumstances because *Syntrophomonas* can be grown in pure culture on certain unsaturated fatty acids. Crotonate ($H_3C-CH=CH-COO^-$), for example, supports growth of *Syntrophomonas*, with some of the crotonate being oxidized to acetate and some reduced to butyrate. It is possible that crotonate reduction by *Syntrophomonas* is coupled to a proton gradient and ATP formation as in other types of anaerobic respirations employing organic compounds, such as fumarate reduction to succinate (see Section 16.18).

The term **interspecies hydrogen transfer** describes the interdependent series of reactions involved in the anaerobic conversion of complex polymers to methane (see also Section 16.19). With the exception of the cellulolytic microflora, all of the remaining microbial groups are in some way dependent upon each other, with interspecies transfer of hydrogen generally

Reaction type	Reaction	Free energy change (kJ/reaction)	
		ΔG^{0b}	ΔG^c
Fermentation of glucose to acetate, H_2 and CO_2	Glucose + $4H_2O \rightarrow 2CH_3COO^- + 2HCO_3^- + 4H^+ + 4H_2$	-207	-319
Fermentation of glucose to butyrate, CO_2, and H_2	Glucose + $2H_2O \rightarrow C_4H_7O_2^- + 2HCO_3^- + 2H_2 + 3H^+$	-135	-284
Fermentation of butyrate to acetate and H_2	Butyrate + $2H_2O \rightarrow 2CH_3COO^- + H^+ + H_2$	$+48.2$	-17.6
Fermentation of propionate to acetate, CO_2 and H_2	Propionate + $3H_2O \rightarrow CH_3COO^- + HCO_3^- + H^+ + H_2$	$+76.2$	-5.5
Methanogenesis from $H_2 + CO_2$	$4H_2 + HCO_3^- + H^+ \rightarrow CH_4 + 3H_2O$	-136	-3.2
Methanogenesis from acetate	Acetate + $H_2O \rightarrow CH_4 + HCO_3^-$	-31	-24.7
Acetogenesis from $H_2 + CO_2$	$4H_2 + 2HCO_3^- + H^+ \rightarrow CH_3COO^- + 2H_2O$	-105	-7.1

Table 17.4 Major reactions occurring in the anaerobic conversion of organic compounds to methane[a]

[a]*Data taken from Zinder, S. 1984. American Society for Microbiology News 50:294–298.*
[b]*Standard conditions: solutes, 1 molar; gases, 1 atmosphere.*
[c]*Concentrations of reactants in typical anaerobic ecosystem: fatty acids, 1 mM; HCO_3^-, 20 mM; glucose, 10 μM; CH_4, 0.6 atm; H_2, 10^{-4} atm.*

terminating at the methane "sink." In most anoxic ecosystems, the rate-limiting step in methanogenesis from organic compounds is not the terminal step of methane formation, but the steps involved in the production of acetate and H_2. Growth rates of syntrophic fatty acid-oxidizers are generally very slow. As soon as H_2 is formed during their fermentations, it is quickly consumed by a methanogen, a homoacetogen, or a sulfate-reducer. The only situations in which H_2 accumulates in nature are when methanogenesis and sulfate reduction are inhibited in some way.

Methanogenic habitats

Despite the obligate anaerobiosis and specialized metabolism of methanogens, they are quite widespread on earth. Although high levels of methanogenesis only occur in anoxic environments, such as swamps and marshes, or in the rumen (see Section 17.13), the process also occurs in habitats that normally might be considered oxic, such as forest and grassland soils. In such habitats methanogenesis occurs in anoxic microenvironments, for example, in the midst of soil crumbs (see Figure 17.1). An overview of the rates of methanogenesis in different kinds of habitats is given in Table 17.5. It should be noted that biogenic production of methane by the methanogenic Archaea exceeds the production rate from gas wells and other abiogenic sources. Eructation by ruminants (see Section 17.13) is the largest single source of biogenic methane. In fact, it is likely that the nearly 2 percent rise in atmospheric methane levels observed over the past few

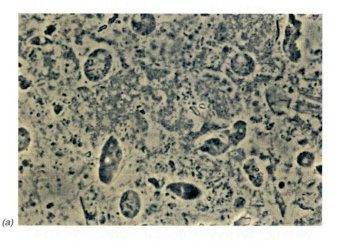

(a)

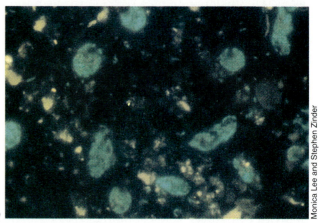

(b)

Monica Lee and Stephen Zinder

FIGURE 17.30 Microorganisms from the hindgut of the termite *Zootermopsis angusticolis*. A single microscope field was photographed by two different methods. (a) Phase-contrast. (b) Epifluorescence, showing color typical of methanogens due to the high content of the fluorescent coenzyme F_{420}. The methanogens are inside cells of the protozoan *Tricercomitis* sp. Plant particles fluoresce yellow. The average diameter of the protozoa is 15–20 μm.

years is due to an increase in the number of domestic ruminants on a worldwide basis. Methanogens are also found in the mammalian intestinal tract, the guts of wood-eating insects, and in most other anoxic habitats such as marshes, swamps, sediments, rice paddies, and landfills.

Methanogens have also been found living as endosymbionts of certain protozoa. Several types of protozoa, including free-living aquatic amoebae and flagellates found in the insect gut, have been shown to harbor methanogens. In termites, for example, methanogens are present within cells of several small trichomonal protozoa which inhabit the termite hindgut (Figure 17.30). Methanogenic symbionts of protozoa resemble rod-shaped species of the genus *Methanobacterium* or *Methanobrevibacter* (see Section 20.5) but their exact relationship to other methanogens is unclear. In the termite hindgut, endosymbiotic methanogens are thought to benefit their protozoan hosts by consuming H_2 generated from glucose fermentation by cellulolytic protozoans. Although the total amount of methane produced in the termite hindgut is relatively

Table 17.5 Sources of atmospheric methane[a]

Source	Mean annual production (Tg CH_4/year)[b]	
Biogenic		
Ruminants	85	
Paddy fields	45	
Swamps	35	
Termites	3.5	
Ocean/Lakes	4	
Other	7	
Abiogenic		
Biomass burning	75	
Pipeline losses	24	
Coal mining	30	
Automobile	1	
Volcanoes	0.5	
Totals:		% of total
Biogenic	180	58
Abiogenic	130	42

[a]Data adapted from Seiler, W. 1984. Contribution of Biological Processes to the Global Budget of CH_4 in the Atmosphere. Pages 468–477 in M. J. Klug and C. A. Reddy (eds.). Current Perspectives in Microbial Ecology, American Society for Microbiology, Washington, D.C.
[b]One Tg = 10^{12} g.

small, the very high numbers of endosymbiotic methanogens found in certain free-living amoeboid protozoans observed in methane-rich aquatic sediments and in ciliated protozoans common in municipal landfills, suggest that endosymbiotic methanogens could be a major source of methane in these habitats.

Methanogenesis in the oceans

Methanogenesis is more extensive in freshwater and terrestrial environments than in the oceans. The reason for this appears to be that marine waters and sediments contain rather high levels of sulfate, and sulfate-reducing bacteria (see Sections 16.16 and 19.8), which are abundant in marine sediments, effectively compete with the methanogenic population for available acetate and H_2:

$$4H_2 + SO_4^{2-} \rightarrow H_2S + 2H_2O + 2OH^-$$
$$\Delta G^{0\prime} = -155.13 \text{ kJ/reaction}$$

$$CH_3COO^- + SO_4^{2-} \rightarrow 2HCO_3^- + HS^-$$
$$\Delta G^{0\prime} = -47.6 \text{ kJ/reaction}$$

The biochemical basis for the success of sulfate-reducing bacteria in scavenging H_2 appears to lie in the increased *affinity* of sulfate-reducing bacteria have for H_2 as compared to methanogens. When H_2 levels get below 5 to 10 μM, as they often do in sulfate-rich environments, methanogens are no longer able to grow since their H_2 uptake systems do not function at such low H_2 concentrations. Sulfate reducers, on the other hand, can grow at these low pressures of H_2, effectively preventing H_2-mediated methanogenesis. Sulfate reduction can also be a significant process in fresh water, but because the sulfate concentration of fresh water is so low, sulfate is rapidly depleted at the surface of anoxic sediments; thus, throughout the bulk of the sediment, methanogenesis is the major process consuming H_2. Acetate utilization is also more efficient in sulfate-reducing bacteria. The affinity for acetate of some sulfate-reducers is over ten times that of methanogens.

Because of consumption of H_2 and acetate by sulfate-reducers, the major precursors of methane in marine environments are methylated substrates, such as methylamines and methanol, which are poorly used by sulfate-reducers. Trimethylamine, a major excretory product of marine animals, is readily converted

to CH_4 by the methanogens *Methanosarcina* and *Methanococcus*. Thus, it is of interest that methanogenesis in marine sediments is not supported by H_2 and acetate, the major methanogenic substrates in other methanogenic ecosystems.

17.13   Rumen Microbial Ecosystem

Ruminants are herbivorous mammals that possess a special organ, the **rumen**, within which the digestion of cellulose and other plant polysaccharides occurs through the activity of special microbial populations. Some of the most important domestic animals, the cow, sheep, and goat, are ruminants. Since the human food economy depends to a great extent on these animals, rumen microbiology is of considerable economic significance.

Rumen fermentation

The bulk of the organic matter in terrestrial plants is present in insoluble polysaccharides, of which *cellulose* is the most important. Mammals, and indeed almost all animals, lack the enzymes necessary to digest cellulose, but all mammals that subsist primarily on grasses and leafy plants can metabolize cellulose by making use of microorganisms as digestive agents. Unique features of the rumen as a site of cellulose digestion are its relatively large size (100 to 150 liters in a cow, 6 liters in a sheep) and its position in the alimentary tract as the organ where ingested food goes first. The high constant temperature (39°C), constant pH (6.5), and anoxic nature of the rumen are also important factors in overall rumen function. The rumen operates in a more or less continuous fashion, and in some ways can be considered analogous to a microbial chemostat (see Section 9.6).

The relationship of the rumen to other parts of the ruminant digestive system is shown in Figure 17.31. Food enters the rumen mixed with saliva containing bicarbonate and is churned in a rotary motion during which the microbial fermentation occurs. This peristaltic action grinds the cellulose into a fine suspension which assists in microbial attachment. The food mass then passes gradually into the reticulum, where it is formed into small clumps called cuds, which are regurgitated into the mouth where they are chewed again. The now finely divided solids, well mixed with saliva, are swallowed again, but this time the material passes to the omasum, finally ending in the abomasum, an organ more like a true (acidic) stomach. Here chemical digestive processes begin which continue in the small and large intestine.

Food remains in the rumen about 9 to 12 hours. During this period cellulolytic bacteria and cellulolytic protozoa hydrolyze cellulose to the disaccharide cellobiose and to free glucose units. The released glucose then undergoes a bacterial fermentation with the production of **volatile fatty acids** (VFAs), primarily *acetic*, *propionic*, and *butyric*, and the gases *carbon dioxide* and *methane* (Figure 17.32). The fatty acids pass through

---

Under anoxic conditions, organic matter is degraded principally to methane (CH_4) and CO_2. Methanogens are a unique group of organisms which use only a limited variety of substrates to make methane. They are thus dependent on the activities of other organisms which produce from the anaerobic decomposition process the substrates needed by the methanogens. Especially important are syntrophic fatty acid-oxidizing bacteria that produce H_2, which is immediately consumed by methanogens. In high sulfate environments such as the oceans, sulfate-reducing bacteria outcompete methanogens for the electron donors acetate and H_2.

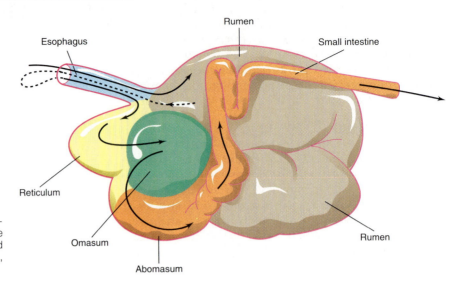

FIGURE 17.31
Schematic diagram of the rumen and gastro-intestinal system of a cow. Food travels from the esophagus to the rumen and is then regurgitated and travels to the reticulum, omasum, abomasum, and intestines, in that order. See text for details.

the rumen wall into the bloodstream and are oxidized by the animal as its main source of energy. In addition to their digestive functions, rumen microorganisms synthesize amino acids and vitamins that are the main source of these essential nutrients for the animal. The rumen contents consist of enormous numbers of microbial cells (10^{10}–10^{11} bacteria per milliliter of rumen fluid) plus partially digested plant materials; these proceed through the gastrointestinal tract of the ani-

mal where they undergo further digestive processes similar to those of nonruminants. Many microbial cells formed in the rumen are digested in the gastrointestinal tract and serve as a major source of proteins and vitamins for the animal. Since many of the microorganisms of the rumen are able to grow on urea as a sole nitrogen source, it is often supplied in cattle feed in order to promote microbial protein synthesis. The bulk of this protein will end up in the animal it-

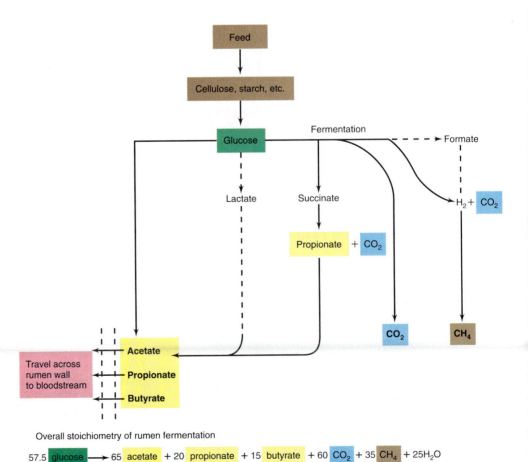

FIGURE 17.32
Biochemical reactions in the rumen. The major substrate, glucose, and end products are highlighted; dashed lines indicate minor pathways. Steady state rumen levels of volatile fatty acids (VFAs) are acetate, 60 mM; propionate, 20 mM; butyrate, 10 mM.

Overall stoichiometry of rumen fermentation

57.5 glucose $\longrightarrow$ 65 acetate + 20 propionate + 15 butyrate + 60 CO_2 + 35 CH_4 + 25H_2O

self. A ruminant is thus nutritionally superior to the nonruminant when subsisting on foods that are deficient in protein, such as grasses.

Rumen microorganisms

The biochemical reactions occurring in the rumen are complex and involve the combined activities of a variety of microorganisms. Since the reduction potential of the rumen is -0.4 volts (the O_2 concentration at this highly reducing potential is 10^{-22} M), anaerobic bacteria naturally dominate. Furthermore, since the conversion of cellulose to CO_2 and CH_4 involves a multistep microbial food chain, a variety of anaerobes can be expected (Table 17.6).

Several different rumen bacteria hydrolyze polymers such as cellulose to sugars and ferment the sugars to volatile fatty acids. *Bacteroides succinogenes* and *Ruminococcus albus* are the two most abundant cellulolytic rumen anaerobes. Although both organisms produce cellulases, *Bacteroides*, a Gram-negative bacterium, employs a periplasmic cellulase (see Section 3.5 for a discussion of the periplasm) to break down cellulose; thus the organism must remain attached to the cellulose fibril while digesting it. *Ruminococcus*, on the other hand, produces an extremely large (2×10^6 molecular weight) cellulase that is excreted into the rumen contents where it degrades cellulose outside the bacterial cell proper. However, the end result is the same in both cases; free glucose is made available for fermentative anaerobes. If a ruminant is gradually switched from cellulose to a diet high in starch (grain, for instance), then starch-digesting bacteria such as *Bacteroides amylophilus* and *Succinomonas amylolytica* develop; on a low starch diet these organisms are in a minority. If an animal is fed on legume hay, which is high in pectin, then the pectin-digesting bacterium *Lachnospira multiparus* is a common member of the rumen flora.

Some of the fermentation products of the saccharolytic rumen microflora are used as energy sources by other rumen bacteria. Thus *succinate* is converted to *propionate* and CO_2 (Figure 17.32), and *lactate* is fermented to *acetic* and other acids by *Selenomonas* and *Peptostreptococcus* (Table 17.6). A number of rumen bacteria produce ethanol as a fermentation product when grown in pure culture, yet ethanol rarely accumulates in the rumen because it can be fermented to acetate + H_2. Hydrogen produced in the rumen by fermentative processes never accumulates because it is quickly used to reduce CO_2 to CH_4 by the methanogen *Methanobrevibacter ruminantium*. Another source of H_2 and CO_2 for methanogens is *formate* (Figure 17.32). Large amounts of CH_4 and CO_2 accumulate in the rumen, the average gas composition being about 65 percent CO_2 and 35 percent CH_4. These gases leave the ruminant during eructation (belching). Acetate is not converted to CH_4 in the rumen because the retention time is too short for development of acetoclastic methanogens, which typically grow very slowly (see Section 20.5).

In addition to prokaryotes, the rumen has a characteristic protozoal fauna (about 10^6/ml), composed almost exclusively of ciliates (see Section 21.4 for a discussion of ciliated protozoa). Many of these protozoans are obligate anaerobes, a property which is rare among eukaryotes. Although protozoans are not essential for the rumen fermentation, they definitely contribute to the overall process. They are able to hy-

Table 17.6   Characteristics of some rumen bacteria

Organism	Gram stain	Shape	Motility	Fermentation products
Cellulose decomposers				
Bacteroides succinogenes	Neg.	Rod	−	Succinate, acetate, formate
Butyrivibrio fibrisolvens	Neg.	Curved rod	+	Acetate, formate, lactate, butyrate, H_2, CO_2
Ruminococcus albus	Pos.	Coccus	−	Acetate, formate, H_2, CO_2
Clostridium lochheadii	Pos.	Rod (spores)	+	Acetate, formate, butyrate, H_2, CO_2
Starch decomposers				
Bacteroides amylophilus	Neg.	Rod	−	Formate, acetate, succinate
Bacteroides ruminicola	Neg.	Rod	−	Formate, acetate, succinate
Selenomonas ruminantium	Neg.	Curved rod	+	Acetate, propionate, lactate
Succinomonas amylolytica	Neg.	Oval	+	Acetate, propionate, succinate
Streptococcus bovis	Pos.	Coccus	−	Lactate
Lactate decomposers				
Selenomonas lactilytica	Neg.	Curved rod	+	Acetate, succinate
Peptostreptococcus elsdenii	Pos.	Coccus	−	Acetate, propionate, butyrate, valerate, H_2, CO_2
Pectin decomposer				
Lachnospira multiparus	Pos.	Curved rod	+	Acetate, formate, lactate, H_2, CO_2
Methanogens				
Methanobrevibacter ruminantium	Pos.	Rod	−	CH_4 (from H_2 + CO_2 or formate)
Methanomicrobium mobile	Neg.	Rod	+	CH_4 (from H_2 + CO_2 or formate)

drolyze cellulose and starch and ferment glucose with the production of the same organic acids formed by the bacteria. Rumen protozoa ingest rumen bacteria and are thought to play a role in controlling rumen bacterial densities.

Anaerobic fungi also inhabitat the rumen and are known to play a role in ruminal digestive processes. Rumen fungi are generally species that alternate between a flagellated and a thallus form, and studies with pure cultures show that they can ferment cellulose to VFAs. Rumen fungi play an important role in the degradation of other plant polysaccharides as well, including a partial degradation of lignin (the strengthening agent in the cell walls of woody plants), hemicellulose, and pectins.

Dynamics of the rumen ecosystem

A major feature of the rumen ecosystem is its *constancy*. Studies of various ruminant species in different parts of the world show that most animals contain the same major rumen bacterial species, with the proportions of each species varying somewhat with diet. In addition, the nature and proportions of the volatile fatty acids produced and the levels of rumen CO_2 and CH_4 are relatively constant among different ruminant species.

Occasionally, changes in the microbial composition of the rumen cause illness or even death of the animal. For example, if a cow is changed abruptly from forage to a completely grain diet, an explosive growth of *Streptococcus bovis* is observed in the rumen; the normal level of *S. bovis*, about 10^7 cells/ml, quickly expands to over 10^{10} cells/ml. This occurs because *S. bovis* grows rapidly on starch and grain contains high levels of starch, while grasses contain mainly cellulose. Being a lactic acid bacterium, *S. bovis* produces large amounts of lactate from the fermentation of starch and this acidifies the rumen (a condition called *acidosis*) killing off the normal rumen flora. Severe acidosis can cause death of the animal. To avoid acidosis, animals are switched from forage rations to grain *gradually* over a period of a few days. A slow introduction of starch selects for volatile fatty acid-producing starch degraders instead of *S. bovis*, and thus normal rumen biochemical processes are not disrupted.

Other grazing animals

The familiar ruminants are cows and sheep. However, goats, camels, buffalo, deer, reindeer, caribou, and elk are also ruminants. There is even some evidence that baleen whales have a rumen-like fermentation. Baleen whales contain a multichambered stomach consisting of a forestomach similar to that of the rumen. Samples of forestomach material from gray and bowhead whales show abundant volatile fatty acid production in proportions typical of the volatile fatty acids observed in the rumen of cattle or sheep. The diet of baleen whales is primarily chitinous invertebrates, small fish and kelp. It is thought that *N*-acetylglucosamine, the major monomeric unit of chitin, is the primary energy source for the forestomach microbial fermentation of baleen whales.

At least one *bird* has been shown to have a foregut fermentation that resembles that of the rumen. The *hoatzin*, a tropical bird, is one of the only obligate folivorous (leaf-eating) birds known. Probably because of its restricted diet—cellulose is its sole carbon source—the hoatzin has evolved a rumen-like forestomach to allow for cellulose digestion. In the hoatzin, a structure called the *crop* functions as the major digestive organ. The pH of the crop is near neutrality and the organ contains high bacterial numbers and a volatile fatty acid content similar to that found in the rumen. After digestion in the crop, food travels through the esophagus and the proventriculus (an acidic organ analogous to a true stomach) to the small intestine.

Horses and rabbits are also herbivorous mammals, but they are not ruminants. Instead, these animals have only one stomach but use an organ called the **cecum,** a small digestive organ located posterior to the large intestine (just before the anus), as their cellulolytic fermentation vessel. The cecum contains a cellulolytic microflora and digestion of cellulose occurs here. The precise microflora of the cecum is not well understood, but the species involved are not thought to be the same as those in the rumen. Nutritionally, ruminants have an advantage over horses and rabbits in that the cellulolytic microflora of the ruminant eventually passes through a true (acidic) stomach and can serve as a protein source for the animal. In horses and rabbits, the cellulolytic microflora is passed out of the animal in the feces. To recover some of this lost protein, rabbits frequently practice coprophagy (eating of the feces).

> Ruminants are animals that have a special digestive organ, the rumen, which serves as a huge microbial culture vessel. The rumen is a unique ecosystem in which anaerobic microorganisms play the principle role in the digestion of insoluble feed materials such as cellulose and starch. The fermentative bacteria, protozoa, and fungi of the rumen produce volatile fatty acids which then serve as the principle energy source of the ruminant animal. In addition to their role in the digestive process, the rumen microorganisms synthesize vitamins and amino acids used by the ruminant.

17.14   Biogeochemical Cycles: Nitrogen

The element nitrogen, N, a key constituent of protoplasm, exists in a number of oxidation states (see Table 16.5). We discussed two major processes of microbial nitrogen transformation in Chapter 16: denitrification in Section 16.15 and nitrification in Section 16.12. These and several other nitrogen transformations are summarized in the redox cycle shown in Figure 17.33.

Several of the key redox reactions of nitrogen are carried out in nature almost exclusively by microorganisms, so that microbial involvement in the nitrogen

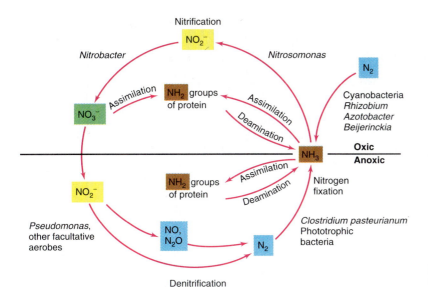

FIGURE 17.33
Redox cycle for nitrogen.

cycle is of great importance. Thermodynamically, nitrogen gas, N_2, is the most stable form of nitrogen, and it is to this form that nitrogen will revert under equilibrium conditions. This explains the fact that a major reservoir for nitrogen on earth is the atmosphere. This is in contrast to carbon, in which the atmosphere is a relatively minor reservoir (CO_2, CH_4). The high energy necessary to break the N≡N bond of molecular nitrogen (Section 16.26) means that the reduction of N_2 is an energy-demanding process. Only a relatively small number of organisms are able to utilize N_2, in the process called **nitrogen fixation**; thus the recycling of nitrogen on earth involves to a great extent the more easily available forms, ammonia and nitrate. However, because N_2 constitutes by far the greatest reservoir of nitrogen available to living organisms, the ability to utilize N_2 is of great ecological importance. In many environments, productivity is limited by the short supply of combined nitrogen compounds, putting a premium on biological nitrogen fixation.

The global nitrogen cycle is given in Figure 17.34. Transfer of nitrogen in and out of the atmosphere is to a great extent as N_2, with a smaller amount of transfer as nitrous oxides, N_2O and NO, and as gaseous ammonia, NH_3. Transfer between terrestrial and aquatic compartments is primarily as organic nitrogen, ammonium ion, and nitrate ion.

Nitrogen fixation

We discussed the biochemistry and microbiology of nitrogen fixation ($N_2 + 6H^+ + 6e^- \rightarrow 2NH_3$) in Section 16.26, and we discuss symbiotic nitrogen fixation by legumes in Section 17.24. Nitrogen fixation can also occur chemically in the atmosphere, to a small extent, via lightning discharges, and a certain amount of nitrogen fixation occurs in the industrial production of nitrogen fertilizers (labeled as industrial fixation in Figure 17.34). Some nitrogen fixation also occurs during artificial combustion processes, since air contains 78 percent N_2 by weight, and burning in air inevitably involves high temperature combustion of some N_2 (to nitrogen

oxides and ultimately to nitrate). However, as can be calculated from the fluxes given in Figure 17.34, about 85 percent of nitrogen fixation on earth is of *biological* origin. As can also be calculated from Figure 17.34, about 60 percent of biological nitrogen fixation occurs on land, and the other 40 percent in the oceans.

Denitrification

We discussed the role of nitrate as an alternative electron acceptor in Section 16.15. Assimilatory nitrate reduction, in which nitrate is reduced to the oxidation level of ammonia for use as a nitrogen source for growth, and dissimilatory nitrate reduction, in which nitrate is used as an alternative electron acceptor in energy generation, were contrasted in Figure 16.29. Under most conditions, the end product of dissimilatory nitrate reduction is N_2 or N_2O, and the conversion of nitrate to gaseous nitrogen compounds is called **denitrification**. This process is the main means by which gaseous N_2 is formed biologically, and since N_2 is much less readily available to organisms than nitrate as a source of nitrogen, denitrification is a detrimental process.

Ammonia fluxes and nitrification

Ammonia is produced during the decomposition of organic nitrogen compounds (**ammonification**) and exists at neutral pH as ammonium ion (NH_4^+). Under anoxic conditions, ammonia is stable, and it is in this form that nitrogen predominates in anoxic sediments. In soils, much of the ammonia released by aerobic decomposition is rapidly recycled and converted into amino acids in plants. Because ammonia is volatile, some loss can occur from soils (especially highly alkaline soils) by vaporization, and major losses of ammonia to the atmosphere occur in areas of dense animal populations (for example, cattle feedlots). On a global basis, ammonia constitutes only about 15 percent of the nitrogen released to the atmosphere, the majority of the rest being in the form of N_2 or N_2O (from denitrification).

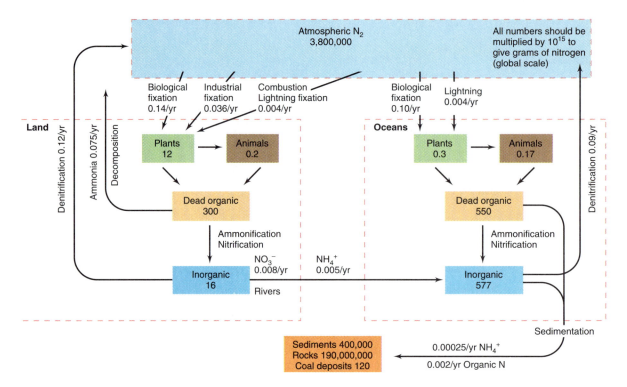

FIGURE 17.34   The global nitrogen cycle. Minor compartments and fluxes are not given.

In oxic environments ammonia can be oxidized to nitrogen oxides and nitrate, but ammonia is a rather stable compound and strong oxidizing agents or catalysts are usually needed for the chemical reaction. However, a specialized group of bacteria, the *nitrifying bacteria*, are biological catalysts, oxidizing ammonia to nitrate in a process called **nitrification** (see Section 16.12).

Nitrification is strictly an *aerobic* process and occurs readily in well-drained soils at neutral pH; it is inhibited by anoxic conditions or in highly acidic soils. If materials high in protein, such as manure or sewage, are added to soils, the rate of nitrification is increased. Although nitrate is readily assimilated by plants, it is very water soluble and is rapidly leached from soils receiving high rainfall. Consequently, nitrification is not beneficial in agricultural practice. Ammonia, on the other hand, is cationic and consequently is strongly adsorbed to negatively charged clay minerals.

Anhydrous ammonia is used extensively as a nitrogen fertilizer and chemicals are commonly added to the fertilizer to inhibit the nitrification process. One

of the most common inhibitors of nitrification is a substituted pyridine compound called *nitrapyrin* (2-chloro-6-trichloromethyl pyridine). Nitrapyrin specifically inhibits the first step in nitrification, the oxidation of NH_3 to NO_2^- (see Section 16.12), thus effectively inhibiting both steps in the nitrification process. The addition of nitrification inhibitors has served to greatly increase the efficiency of fertilization and helps to prevent pollution of waterways from nitrate leached from fertilized soils.

17.15   Biogeochemical Cycles: Sulfur

Sulfur transformations are even more complex than those of nitrogen due to the variety of oxidation states of sulfur and the fact that some transformations occur at significant rates *chemically* as well as biologically. We discussed the processes of sulfate reduction and chemolithotrophic sulfur oxidation in Sections 16.10 and 16.16. The redox cycle for sulfur and the involvement of microorganisms in sulfur transformations are given in Figure 17.35. Although a number of oxidation states are possible, only three form significant amounts of sulfur in nature, -2 (sulfhydryl, R–SH and sulfide, HS^-), 0 (elemental sulfur, S^0), and $+6$ (sulfate, SO_4^{2-}). The bulk of the sulfur of the earth is found in sediments and rocks, in the form of sulfate minerals (primarily gypsum, $CaSO_4$) and sulfide minerals (primarily pyrite, FeS_2), although the oceans constitute the most significant reservoir of sulfur for the biosphere (in the form of inorganic sulfate). The global transport cycle for sulfur is given in Figure 17.36, and some of the components of this cycle are discussed below.

> The principal nitrogen component on earth is N_2, a highly stable gas that can only be used as a nitrogen source by a restricted group of microorganisms, the nitrogen-fixing bacteria. Biological nitrogen fixation is an important process because it provides fixed nitrogen for plants and animals. Ammonia produced by nitrogen fixation or by ammonification from organic nitrogen compounds can be assimilated into organic matter or it can be oxidized to nitrate by the nitrifying bacteria. Losses of nitrogen from the biosphere occur as a result of denitrification, in which nitrate is converted back into N_2.

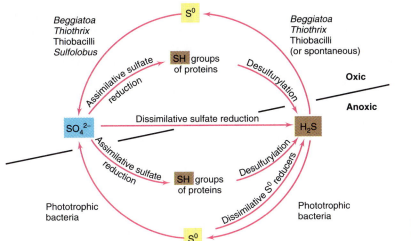

FIGURE 17.35
Redox cycle for sulfur.

Hydrogen sulfide and sulfate reduction

A major volatile sulfur gas is *hydrogen sulfide*. As we saw in Section 16.16, this substance is formed primarily by the bacterial reduction of sulfate:

$$SO_4^{2-} + 8e^- + 8H^+ \rightleftharpoons H_2S + 2H_2O + 2OH^-$$

The form in which sulfide is present in an environment depends on pH due to the following equilibria:

$$H_2S \rightleftharpoons HS^- \rightleftharpoons S^{2-}$$

Low pH   Neutral pH   High pH

At high pH, the dominant form is sulfide, S^{2-}. At neutral pH, HS^- predominates, and below pH 6, H_2S, a gaseous product, is the major species. HS^- and S^{2-} are very water soluble, but H_2S is not and readily volatilizes. Even at neutral pH, some volatilization of H_2S from HS^- can occur, because there is an equilibrium between HS^- and H_2S, and as volatilization occurs, the reaction is pulled toward H_2S.

A wide variety of organisms can use sulfate as a sulfur source and carry out *assimilative* sulfate reduction, converting the HS^- formed to organic sulfur, R—SH (see Figure 16.31b). HS^- is ultimately reformed from the decomposition of this organic sulfur by putrefaction and desulfurylation (Figure 17.35), and this is a significant source of HS^- in fresh water. In the marine environment, because of the vast amount of sulfate present, dissimilatory sulfate reduction is the main source of HS^-. Dissimilatory sulfate reduction, in which sulfate serves as an electron acceptor, is carried out by a variety of bacteria, collectively referred to as the *sulfate-reducing bacteria* (see Sections 16.16 and 19.8), which are all obligate anaerobes. It should be emphasized that the sulfate anion is very stable chemically, and its reduction does not occur spontaneously in nature under normal environmental conditions.

As discussed in Section 17.12, there is a competition in nature between methanogenic and sulfate-reducing bacteria for available electron donors, especially H_2 and acetate, and as long as sulfate is present

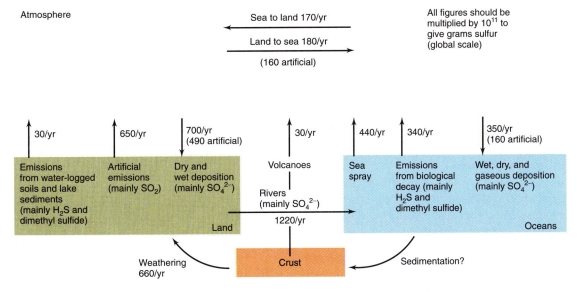

FIGURE 17.36   The global sulfur cycle. Artificial emmissions are derived from human activities.

the sulfate-reducing bacteria are favored. Because of the necessity of organic electron donors (or molecular H_2, which is itself derived from the fermentation of organic compounds), sulfate reduction only occurs extensively where significant amounts of organic matter are present. In many marine sediments, the rate of sulfate reduction is carbon limited, and the rate can be greatly increased by addition of organic matter. This is of considerable importance for marine pollution, since disposal of sewage, sewage sludge and garbage in the sea can lead to marked increases in organic matter in the sediments. Since HS^- is a toxic substance to many organisms, formation of HS^- by sulfate reduction is potentially detrimental. Sulfide is toxic because it combines with the iron of cytochromes and other essential iron-containing compounds in the cell. One common detoxification mechanism for sulfide in the environment is combination with iron, leading to the formation of the insoluble FeS. The black color of many sediments where sulfate reduction is taking place is due to the accumulation of FeS.

Sulfide and elemental sulfur oxidation

Aerobically, sulfide (HS^-) rapidly oxidizes spontaneously at neutral pH (see Section 16.10). Sulfur-oxidizing bacteria are also able to catalyze the oxidation of sulfide, but because of the rapid spontaneous reaction, bacterial oxidation of sulfide only occurs in areas in which H_2S rising from anoxic areas meets O_2 descending from oxic areas. If light is available, anaerobic oxidation of HS^- can also occur, catalyzed by the phototrophic sulfur bacteria (see Sections 16.4 and 19.1), but this only occurs in restricted areas, usually in lakes, where sufficient light can penetrate to anoxic zones.

Elemental sulfur, S^0, is chemically stable in most environments in the presence of oxygen, but is readily oxidized by sulfur-oxidizing bacteria. Although a number of sulfur-oxidizing bacteria are known, members of the genus *Thiobacillus* (see Section 16.10) are most commonly involved in elemental sulfur oxidation. Elemental sulfur is very insoluble, and the bacteria that oxidize it attach firmly to the sulfur crystals (see Figure 16.21). Oxidation of elemental sulfur results in the formation of sulfate and hydrogen ions, and sulfur oxidation characteristically results in a *lowering* of the pH. Elemental sulfur is sometimes added to alkaline soils to effect a lowering of the pH, reliance being placed on the ubiquitous thiobacilli to carry out the acidification process.

Organic sulfur compounds

In addition to the *inorganic* forms of sulfur whose biogeochemistry was just discussed, a vast array of *organic* sulfur compounds are also synthesized by living organisms, and these enter into biogeochemical sulfur cycling as well. The most abundant organic sulfur compound in nature is dimethyl sulfide ($H_3C-S-CH_3$). It is produced primarily in marine environments as a degradation product of dimethylsulfonium propionate, a major osmoregulatory solute in marine algae.

Dimethylsulfonium propionate can be used as a carbon and energy source by microorganisms and is catabolized to dimethyl sulfide and acrylate; the latter compound, a derivative of the fatty acid propionate, is used to support growth.

Microbial production of dimethyl sulfide in nature is dramatic, some 45 million tons being produced annually. Dimethyl sulfide released to the atmosphere undergoes photochemical oxidation to methane sulfonic acid ($CH_3SO_3^-$), SO_2, and SO_4^{2-}, but dimethyl sulfide produced in anoxic habitats can be used microbiologically as a substrate for methanogenesis (yielding CH_4 and H_2S), as an electron donor for photosynthetic CO_2 fixation in phototrophic purple bacteria (yielding dimethyl sulfoxide, or DMSO), and as an electron donor in energy metabolism in certain chemoorganotrophs and chemolithotrophs (also yielding DMSO). Anaerobically, DMSO can serve as an electron acceptor for anaerobic respiration (see Section 16.18), once again yielding dimethyl sulfide. Many other organic sulfur compounds impact on the global sulfur cycle, including methanethiol (CH_3SH), dimethyl disulfide ($H_3C-S-S-CH_3$), and carbon disulfide (CS_2), but on a global basis, dimethyl sulfide production and consumption is quantitatively the most significant.

> **Bacteria play major roles in both the oxidative and reductive sides of the sulfur cycle. Sulfur- and sulfide-oxidizing bacteria, which are often chemolithotrophs, produce sulfate, often in the form of sulfuric acid, and are hence responsible for the acidification of the environment. Sulfate-reducing bacteria use sulfate as electron acceptor in anaerobic respiration and produce hydrogen sulfide. Because hydrogen sulfide is toxic and also reacts with various metals, sulfate reduction is an important biogeochemical process. Dimethyl sulfide is the major organic sulfur compound of ecological significance in nature.**

17.16   Biogeochemical Cycles: Iron

Iron is one of the most abundant elements in the earth's crust, but is a relatively minor component in aquatic systems because of its relative insolubility in water. Iron exists in two oxidation states, ferrous (+II) and ferric (+III). We discussed iron oxidation and reduction in Sections 16.11 and 16.18, respectively. The form in which iron is found in nature is greatly influenced by pH and oxygen. Because of the high electrode potential of the Fe^{3+}/Fe^{2+} couple, +0.76 volts, the only electron acceptor able to oxidize ferrous iron is oxygen, O_2. At neutral pH, ferrous iron oxidizes spontaneously in air to ferric iron, which forms highly insoluble precipitates of ferric hydroxide and ferric oxides. Thus, at neutral pH, the only way that iron is maintained in solution is by chelation with organic materials.

Bacterial iron reduction and oxidation

The bacterial reduction of ferric iron to the ferrous state is a major means by which iron is solubilized in nature. As we noted in Section 16.18, a number of organisms can use ferric iron as an electron acceptor. In addition to the bacterially catalyzed reduction, if hydrogen sulfide is present, as it is in many anoxic environments (Section 17.15), ferric iron is also reduced chemically to FeS (ferrous sulfide). Thus, there are complex interactions in many environments between the iron and sulfur cycles.

Ferric iron reduction is very common in waterlogged soils, bogs, and anoxic lake sediments. Movement of iron-rich groundwater from anoxic bogs or waterlogged soils can result in the transport of considerable amounts of ferrous iron. Once this iron-laden water reaches oxic regions, the ferrous iron is quickly oxidized spontaneously and ferric compounds precipitate, leading to the formation of a brown deposit (see Figure 16.23c). The overall reaction of ferrous iron oxidation is as follows:

$$Fe^{2+} + \frac{1}{4}O_2 + H^+ \rightarrow Fe^{3+} + \frac{1}{2}H_2O$$
$$Fe^{3+} + 3OH^- \rightarrow Fe(OH)_3 \text{ precipitates}$$

Sum:

$$Fe^{2+} + \frac{1}{4}O_2 + 2OH^- + \frac{1}{2}H_2O \rightarrow Fe(OH)_3$$

Note that although the initial oxidation of ferrous iron consumes hydrogen ions and thus leads to a rise in pH, the hydrolysis of Fe^{3+} and formation of $Fe(OH)_3$ consumes hydroxyl ions and leads to an acidification of the medium. This is one way in which iron oxidation leads to the formation of acidic conditions in the environment.

Although ferric iron forms very insoluble hydroxides, some ferric iron can be kept in solution in natural waters by forming complexes with organic materials. If an organism is present that can oxidize the organic compound, then the iron present will precipitate. This is probably a major mechanism of iron precipitation in many neutral pH environments. In addition, at neutral pH, organisms such as *Gallionella* (see Figure 19.42) and *Leptothrix* (see Figure 19.64) contribute to the oxidation of ferrous iron, presumably in energy-yielding reactions.

Ferrous iron oxidation at acid pH

At low pH, chemical oxidation of ferrous iron to the ferric state is slow (see Section 16.11). However, the acidophilic chemolithotroph *Thiobacillus ferrooxidans* is able to catalyze the oxidation (Figure 17.37); *T. ferrooxidans* oxidizes ferrous iron as its primary energy-generating process. Because very little energy is generated in the oxidation of ferrous to ferric iron (see Section 16.11), these bacteria must oxidize large amounts of iron in order to grow, and consequently even a small number of cells can be responsible for precipitating a large amount of iron. This iron-oxidizing bacterium, which is a strict acidophile, is very common in acid mine drainages and in acid springs, and is probably responsible for most of the ferric iron precipitated at acid pH values.

Thiobacillus ferrooxidans lives in environments in which sulfuric acid is the dominant acid and large

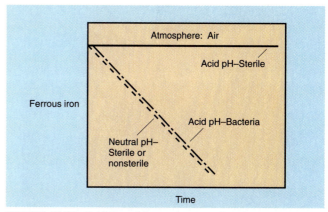

FIGURE 17.37 Oxidation of ferrous iron as a function of pH and the presence of the bacterium *Thiobacillus ferrooxidans*.

amounts of sulfate are present. Under these conditions, ferric iron does not precipitate as the hydroxide, but as a complex sulfate mineral called *jarosite*, (usually found as $HFe_3(SO_4)_2(OH)_6$). Jarosite is a yellowish or brownish precipitate and is responsible for one of the manifestations of acid mine drainage, an unsightly yellow stain called "yellow boy" by U. S. miners (see Figure 16.23a).

Pyrite oxidation

One of the most common forms of iron and sulfur in nature is *pyrite*, which has the overall formula FeS_2. Pyrite is formed from the reaction of sulfur with ferrous sulfide (FeS) to form a highly insoluble crystalline structure, and pyrite is very common in bituminous coals and in many ore bodies (Figure 17.38). The bacterial oxidation of pyrite is of great significance in the development of acidic conditions in mines and mine drainages. Additionally, oxidation of pyrite by bacteria is of considerable importance in the process called microbial leaching of ores (see below). The oxidation of pyrite is a combination of chemically and bacterially catalyzed reactions. Two electron acceptors for this process can function: molecular oxygen (O_2) and ferric ions (Fe^{3+}). However, ferric ions are only present when the solution is acidic, at pH values below about 2.5. At pH values above 2.5, ferric ion reacts with water to form the insoluble ferric hydroxide. When pyrite is first exposed, as in a mining operation, a slow chemical reaction with molecular oxygen occurs as shown in the following reaction:

$$FeS_2 + 3\frac{1}{2}O_2 + H_2O \rightarrow Fe^{2+} + 2SO_4^{2-} + 2H^+$$

This reaction, called the *initiator reaction*, leads to the development of acidic conditions under which the ferrous iron formed is relatively stable in the presence of oxygen. However, *T. ferrooxidans* catalyzes the oxidation of ferrous to ferric ions. The ferric ions formed under these acidic conditions, being soluble, can readily react spontaneously with more pyrite to oxidize the pyrite to ferrous ions plus sulfate ions:

$$FeS_2 + 14Fe^{3+} + 8H_2O \rightarrow 15Fe^{2+} + 2SO_4^{2-} + 16H^+$$

FIGURE 17.38 Pyrite-rich microbial habitats in bituminous coal and copper mining environments. (a) Bituminous coal mining operation in a strip mine. The shovel is removing the soil (overburden) to reach the coal seam. (b) The coal seam. Removal of the bituminous coal exposes the environment to air. The pyrite associated with the coal is colonized with iron-oxidizing bacteria. (c) Copper ore deposit rich in pyrite. Mining of the copper ore results in exposure of the formation to air.

The ferrous ions formed are again oxidized to ferric ions by the bacteria, and these ferric ions again react with more pyrite. Thus there is a progressive, rapidly increasing rate at which pyrite is oxidized, called the *propagation cycle*, as illustrated in Figure 17.39. Under natural conditions some of the ferrous iron generated by the bacteria leaches away, being carried by groundwater into surrounding streams. However, because oxygen is present in the aerated drainage, bacterial oxidation of the ferrous iron takes place in these outflows and an insoluble ferric precipitate is formed.

Acid mine drainage

Bacterial oxidation of sulfide minerals is the major factor in the formation of **acid mine drainage**, a common environmental problem in coal mining regions (Figure 17.40; see also Figure 16.23*a*). Acid mine drainage occurs because of the attack by *Thiobacillus ferrooxidans* on pyrite, following the steps outlined above.

Not all coal seams contain iron sulfide; thus acid mine drainage does not occur in all coal mining regions. Where acid mine drainage does occur, however, it is often a very serious problem. Mixing of acidic mine waters with natural waters in rivers and

FIGURE 17.40 Acid mine drainage from a bituminous coal region. Note the yellowish-red color due to precipitated iron oxides. See also Figure 16.23*a*.

lakes causes a serious degradation in the quality of the natural water, since both the acid and the dissolved metals are toxic to aquatic life (see Figures 16.23*a* and 17.40). In addition, such polluted waters are unsuitable for human consumption and industrial use.

We have outlined in Figure 17.39 the steps in the bacterial oxidation of pyrite, and we noted that certain of the steps occurred chemically, but that the rate-limiting step, the oxidation of ferrous to ferric iron, occurred at acid pH only in the presence of the bacterium. The breakdown of pyrite leads ultimately to

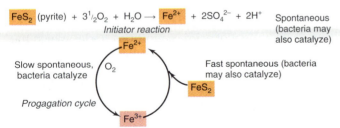

$$FeS_2 \text{ (pyrite)} + 3\tfrac{1}{2}O_2 + H_2O \longrightarrow Fe^{2+} + 2SO_4^{2-} + 2H^+$$
Initiator reaction

Spontaneous (bacteria may also catalyze)

Slow spontaneous, bacteria catalyze

Fast spontaneous (bacteria may also catalyze)

Progagation cycle

FIGURE 17.39 Role of iron-oxidizing bacteria in the oxidation of the mineral pyrite.

the formation of sulfuric acid and ferrous iron, and pH values can be as low as pH 2. The acid formed attacks other minerals in the rock associated with the coal and pyrite, causing breakdown of the whole rock fabric. A major rock-forming element, aluminum, is only soluble at low pH, and often several grams of Al^{3+} are present per liter.

The requirement for O_2 in the oxidation of ferrous to ferric iron helps to explain how acid mine drainage develops. As long as the coal is unmined, oxidation of pyrite cannot occur, since neither air nor the bacteria can reach it. When the coal seam is exposed, it quickly becomes contaminated with *T. ferrooxidans*, and O_2 is introduced, making oxidation of pyrite possible. The acid formed can then leach into the surrounding streams (Figure 17.40).

> Iron exists in two oxidation states, ferrous and ferric, and bacterial transformation of these cations is of great geological and ecological importance. Bacterial ferric iron reduction occurs in anoxic environments and results in the mobilization of iron from swamps, bogs, and other iron-rich aquatic habitats. Bacterial oxidation of ferrous iron occurs significantly only at low pH and is very common in coal-mining regions, where it results in a type of pollution called acid mine drainage.

17.17   Microbial Leaching

We consider here a situation in which acid production and metal solubility by acidophilic bacteria play a beneficial role in mining. Sulfide forms highly insoluble minerals with many metals, and many ores used as sources of these metals are sulfides. If the concentration of metal in the ore is low, it may not be economically feasible to concentrate the mineral by conventional chemical means. Under these conditions, **microbial leaching** is frequently practiced. Microbial leaching is especially useful for *copper* ores, because copper sulfate, formed during the oxidation of the

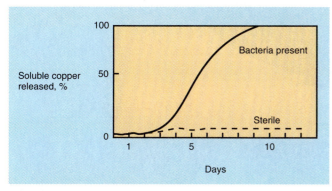

FIGURE 17.41   Effect of the bacterium *Thiobacillus ferrooxidans* on the leaching of copper from the mineral covellite. The leaching was done in a laboratory column and the acid leach solution contained inorganic nutrients necessary for the development of the bacterium. The leaching activity was monitored by assaying for soluble copper in the leach solution at the bottom of the column. The leach solution was continuously recirculated, maintaining an essentially closed system.

copper sulfide ores, is very water soluble. We have noted that sulfide itself, HS^-, oxidizes spontaneously in air. Most metal sulfides will also oxidize spontaneously, but the rate is very much slower than that of free sulfide. Bacteria such as *Thiobacillus ferrooxidans* are able to catalyze a much faster rate of oxidation of the sulfide minerals, thus aiding in solubilization of the metal. The relative rate of oxidation of a copper mineral in the presence and absence of bacteria is illustrated in Figure 17.41. The susceptibility to oxidation also varies among minerals, and those minerals which are most readily oxidized are most amenable to microbial leaching. Thus, iron and copper sulfide ores such as pyrhotite (FeS) and covellite (CuS) are readily leached, whereas lead and molybdenum ores are much less so.

Leaching process

In the general microbial leaching process, low-grade ore is dumped in a large pile (the leach dump), and a dilute sulfuric acid solution (pH around 2) is percolated down through the pile (Figure 17.42*a*). The liquid coming out of the bottom of the pile (Figure 17.42*b*), rich in the mineral, is collected and transported to a precipitation plant (Figure 17.42*c*) where the metal is reprecipitated and purified. The liquid is then pumped back to the top of the pile and the cycle is repeated. As needed, more acid is added to maintain the low pH.

There are several mechanisms by which the bacteria can catalyze oxidation of the sulfide minerals. To illustrate these, examples will be used of the oxidation of two copper minerals, chalcocite, Cu_2S, in which copper has a valence of $+1$, and covellite, CuS, in which copper has a valence of $+2$. As illustrated in Figure 17.43, *T. ferrooxidans* is able to oxidize Cu^+ in chalcocite (Cu_2S) to Cu^{2+}, thus removing some of the copper in the soluble form, Cu^{2+}, and forming the mineral covellite. Note that in this reaction, there is no change in the valence of sulfide, the bacteria utilizing the reaction Cu^+ to Cu^{2+} as a source of energy. This is analogous to the oxidation by the same bacterium of Fe^{2+} to Fe^{3+}. Covellite can then be oxidized releasing sulfate and soluble Cu^{2+} (Figure 17.43).

A second mechanism, and probably the most important in most mining operations, involves an indirect oxidation of the copper ore with *ferric* ions that were formed by the bacterial oxidation of ferrous ions (Figure 17.43). In almost any ore, pyrite is present, and the oxidation of this pyrite (see Section 17.16) leads to the formation of ferric iron. Ferric iron is a very good oxidant for sulfide minerals, and reaction of CuS with ferric iron results in the solubilization of the copper and the formation of ferrous iron. In the presence of O_2, at the acid pH values involved, *T. ferrooxidans* reoxidizes the ferrous iron back to the ferric form, so that it can oxidize more copper sulfide. Thus the process is kept going indirectly by the action of the bacterium on iron.

Another source of iron in leaching operations is at

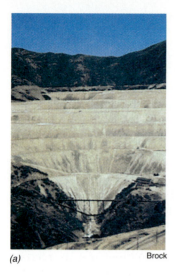
(a)   Brock

(b)   Brock

(c)   Brock

(d)   Brock

FIGURE 17.42 The leaching of low-grade copper ores using bacteria. (a) A typical leaching dump. The low-grade ore has been dumped in a large pile. Pipes distribute the acidic leach water over the surface of the pile. The acidic water slowly percolates through the pile and exits at the bottom. (b) Effluent from a copper leaching dump. The acidic water is very rich in dissolved copper. (c) Recovery of dissolved copper by passage of the copper-rich water over metallic iron in a long flume. (d) A small pile of recovered copper metal removed from the flume, ready for further purification.

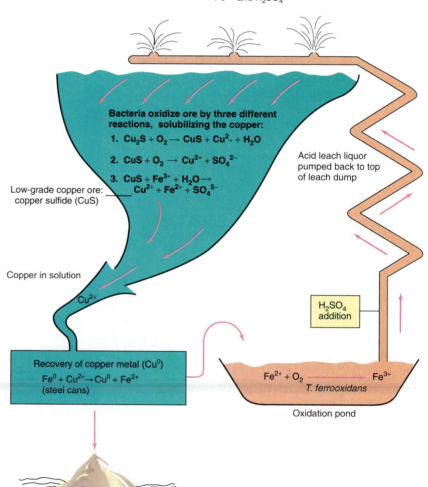

Sprinkling of acid leach liquor on copper ore:
Fe^{3+} and H_2SO_4

Bacteria oxidize ore by three different reactions, solubilizing the copper:

1. $Cu_2S + O_2 \longrightarrow CuS + Cu^{2+} + H_2O$

2. $CuS + O_2 \longrightarrow Cu^{2+} + SO_4^{2-}$

3. $CuS + Fe^{3+} + H_2O \longrightarrow Cu^{2+} + Fe^{2+} + SO_4^{2-}$

Low-grade copper ore: copper sulfide (CuS)

Acid leach liquor pumped back to top of leach dump

Copper in solution

Cu^{2+}

H_2SO_4 addition

Recovery of copper metal (Cu^0)
$Fe^0 + Cu^{2+} \rightarrow Cu^0 + Fe^{2+}$
(steel cans)

$Fe^{2+} + O_2 \longrightarrow Fe^{3+}$
T. ferrooxidans

Oxidation pond

Copper metal (Cu^0)

FIGURE 17.43
Arrangement of a leaching pile and reactions involved in the microbial leaching of copper sulfide minerals to yield Cu^0 (copper metal).

the precipitation plant used in the recovery of the soluble copper from the leaching solution (Figure 17.42*c* and *d*). Scrap iron, Fe^0, is used to recover copper from the leach liquid by the reaction shown in the lower part of Figure 17.43, and this results in the formation of considerable Fe^{2+}. In most leaching operations, the Fe^{2+}-rich liquid remaining after the copper is removed is conducted to an oxidation pond, where *T. ferrooxidans* proliferates and forms Fe^{3+}. Acid is added at the pond to keep the pH low, thus keeping the Fe^{3+} in solution, and this ferric-rich liquid is then pumped to the top of the pile and the Fe^{3+} is available to oxidize more sulfide mineral.

Because of the huge dimensions of copper leach dumps, penetration of oxygen from air is poor, and the interior of these piles usually becomes anoxic. Although most of the reactions written in Figure 17.43 require molecular O_2, because *T. ferrooxidans* can use Fe^{3+} as an electron *acceptor* in the absence of O_2, the oxidation reactions can also proceed anaerobically; the large amounts of Fe^{3+} added to the leach solution from scrap oxidized iron drive the process forward under anoxic conditions.

> Oxidation of copper ores by bacteria can lead to the solubilization of copper, a process called microbial leaching. Leaching is important in the recovery of copper and uranium from low-grade ores. Bacterial oxidation of iron in the iron sulfide mineral pyrite is also an important part of the microbial leaching process because the ferric iron produced is itself an oxidant of ores.

17.18 Biogeochemical Cycles: Trace Metals and Mercury

Trace elements are those elements that are present in low concentrations in rocks, waters, and the atmosphere. Some trace elements (for example, cobalt, copper, zinc, nickel, molybdenum) are nutrients (see Section 4.21), but a number of trace elements in high concentrations are actually toxic to organisms. Of these toxic elements, several are sufficiently volatile so that they exhibit significant atmospheric transport, and hence are of some environmental concern. These include mercury, lead, arsenic, cadmium, and selenium. Many of these trace elements undergo redox reactions catalyzed by microorganisms, and several are also converted into organic form via microbial action. Because of environmental concern and significant microbial involvement, we focus our discussion on the biogeochemistry of the element mercury.

Global cycling of mercury and methylmercury

Although mercury is present in extremely low concentrations in most natural environments, averaging about 1 ng/liter, it is a widely used industrial product, and is the active component of many pesticides which have been introduced into the environment. Because of its unusual ability to be concentrated in living tissues and its high toxicity, mercury is of considerable environmental importance. The mining of mercury ores and the burning of fossil fuels release about 40,000 *tons* of mercury into the environment each year; an even greater amount is released by geochemical processes.

The major form of mercury in the atmosphere is elemental mercury (Hg^0), which is volatile and gets oxidized to mercuric ion (Hg^{2+}) photochemically; most of the mercury entering aquatic environments is thus Hg^{2+} (Figure 17.44). Mercuric ion readily adsorbs to particulate matter and can be acted upon from there by microorganisms. The major microbial reaction observed is the *methylation* of mercury, yielding methylmercury, CH_3Hg^+ (Figure 17.44). Methylmercury is soluble and can be concentrated in the aquatic food chain, primarily in fish, or further methylated by microorganisms to yield the volatile compound called dimethylmercury, CH_3-Hg-CH_3. Metabolically, methylation of mercury occurs by donation of methyl groups from CH_3-B_{12}.

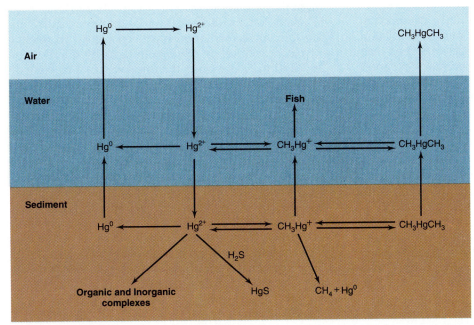

FIGURE 17.44
Biogeochemical cycling of mercury. The major reservoirs of mercury are in water and in sediments where it can be concentrated in animal tissues or precipitated out as HgS.

Both methylmercury and dimethylmercury bind to proteins and tend to accumulate in animal tissues, especially muscle. Methylmercury is about 100 times more toxic than Hg^0 or Hg^{2+} and can be concentrated over 1 million times in fish, where it acts as a potent neurotoxin, eventually causing death. Methylmercury is thus a potent environmental toxin, and its accumulation seems to be a particular problem in freshwater lakes where enhanced levels of methylmercury have been observed in fish caught for human consumption.

Several other mercury transformations occur on a global scale, including reactions involving sulfate-reducing bacteria ($H_2S + Hg^{2+} \rightarrow HgS$) and methanogens ($CH_3Hg^+ \rightarrow CH_4 + Hg^0$) (see Figure 17.44). The solubility of HgS is very low, so that in anoxic sulfate-reducing sediments, most mercury is found as HgS. But upon aeration, oxidation of HgS can occur, primarily by thiobacilli, leading to the formation of Hg^{2+} and eventually methylmercury.

At sufficiently high concentrations, Hg^{2+} and CH_3Hg^+ can be toxic not only to higher organisms but also to microorganisms. Several methods of detoxifying toxic mercury species exist. An NADPH-linked enzyme called *mercuric reductase* transfers two electrons to Hg^{2+}, reducing it to Hg^0. The Hg^0 produced in this reaction is volatile but is essentially nontoxic to humans and microorganisms, compared with Hg^{2+}. Bacterial conversion of Hg^{2+} to Hg^0 then allows more CH_3Hg^+ to be converted to Hg^{2+}. Some organisms that produce a mercuric reductase, such as mercury-resistant *Pseudomonas* species, also produce a periplasmic Hg^{2+}-trapping protein that binds free Hg^{2+} and transports it across the membrane where it is reduced by mercuric reductase. Such a trapping mechanism prevents the incorporation of Hg^{2+} into other cellular proteins that could denature them. The ability of mercuric compounds to act as antiseptics and disinfectants (see Section 9.16) is due to their protein-denaturing effects; Hg^{2+} combines with SH groups of cysteines in proteins to destroy the higher order structure of the protein. Mercury resistance involving a Hg^{2+}-trapping protein and mercuric reductase is a plasmid-borne resistance due to resistance plasmid *R100* (see Section 7.9), a plasmid widespread among mercury-resistant pseudomonads. As many as six genes called *mer* genes, present in this plasmid, are necessary to confer mercury resistance on the cell. Besides mercuric reductase and the Hg^{2+}-trapping protein, *mer* genes encoding a regulatory protein, Hg^{2+}-transport protein, and other factors are necessary for resistance to Hg^{2+}.

Resistance to other heavy metals

A variety of plasmids isolated from both Gram-positive and Gram-negative Bacteria have been found to code for resistance to the effects of heavy metals. Certain antibiotic resistance plasmids also have genes for resistance to mercury and arsenic. Other plasmids code only for heavy-metal resistances. A large plasmid isolated from *Staphylococcus aureus* has been found to code for resistance to mercury, cadmium, arsenate, and arsenite. The mechanism of resistance to any specific metal varies. For example, arsenate and cadmium resistances are due to the action of enzymes that immediately pump out any arsenate or cadmium ions incorporated, thus preventing the metals from denaturing proteins.

Studies of nickel and cobalt resistant bacteria have shown that in most cases the resistance genes are plasmid-borne; resistance to both metals on a single plasmid is typical. Enrichment culture studies have shown that nickel-resistant bacteria are uncommon in soils and other environments in which this metal is absent in significant amounts. Bacteria highly resistant to nickel or other metals are most common in waste waters of the metal processing industry or in mining operations where heavy metals are leached out along with iron or copper ores.

> Trace metals such as mercury and arsenic are often mobilized by bacteria capable of oxidizing them. A major toxic form of mercury is methylmercury. The ability of bacteria to resist the toxicity of heavy metals is often due to the presence of specific plasmids that code for enzymes capable of detoxifying the metals.

17.19   Petroleum and Natural Gas (Methane) Biodegradation

Petroleum

Microbial decomposition of petroleum and petroleum products is of considerable economic and environmental importance. Since petroleum is a rich source of organic matter and the hydrocarbons within it are readily attacked aerobically by a variety of microorganisms, it is not surprising that when petroleum is brought into contact with air and moisture, it is subject to microbial attack. Under some circumstances such as in bulk storage tanks, microbial growth is not desirable. However, in other situations, such as in oil spills, microbial utilization of oil is desirable and may even be promoted by the addition of inorganic nutrients. The term bioremediation has been coined to refer to the cleanup of oil or other pollutants by microorganisms and in recent years the importance of bioremediation in oil spills has been amply demonstrated in several major crude oil spills in the marine environment.

Hydrocarbon-oxidizing bacteria and fungi are the main agents responsible for decomposition of oil and oil products. A wide variety of bacteria, several molds and yeasts, and certain cyanobacteria and green algae have been shown to be able to oxidize hydrocarbons. Bacteria and yeasts, however, appear to be the prevalent hydrocarbon degraders in aquatic ecosystems. Small scale oil pollution of aquatic and terrestrial ecosystems from human as well as natural activities is very common, and hence it is not surprising that a diverse microbial community exists capable of using hydrocarbons as an electron donor.

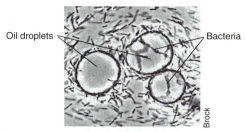

FIGURE 17.45 Hydrocarbon-oxidizing bacteria in association with oil droplets. The bacteria are concentrated in large numbers at the oil-water interface but are not within the droplet.

Hydrocarbon-oxidizing microorganisms develop rapidly on oil films and slicks. However, as we saw in Section 16.24, aliphatic hydrocarbons are not fermentable. Thus, significant aliphatic hydrocarbon oxidation occurs only in the presence of O_2; if the oil gets carried into anoxic sediments, it will decompose very slowly and may remain in place for many years (natural oil deposits in anoxic environments are millions of years old). Even in oxic environments, hydrocarbon-oxidizing microorganisms can act only if other environmental conditions, such as temperature, pH, and inorganic nutrients, are adequate. Because oil is insoluble in water and is less dense, it floats to the surface and forms slicks. Hydrocarbon-oxidizing bacteria are able to attach to insoluble oil droplets, and can often be seen there in large numbers (Figure 17.45). The action of these bacteria eventually leads to decomposition of the oil and dispersal of the slick. A wide variety of microorganisms are capable of petroleum biodegradation, including pseudomonads, various corynebacteria and mycobacteria, and even some yeasts.

Microorganisms participate in oil spill cleanups by oxidizing the oil to CO_2. In large spills, volatile hydrocarbon fractions evaporate quickly, leaving longer chain aliphatic and aromatic components for cleanup crews or microorganisms to tackle. In oil spills where careful bioremediational studies have been performed, it has been shown that hydrocarbon-oxidizing bacteria increase in number 10^3–10^6 times shortly after the oil spill. Experiments using radioisotopic hydrocarbons as tracers or O_2-uptake as a measure of heterotrophic activity have shown that under ideal conditions up to 80 percent of the nonvolatile components are oxidized by bacteria within 6 months to a year of the spill. Certain fractions, such as branched chained and polycyclic hydrocarbons, however, remain in the environment much longer. Spilled oil that travels to the sediments is only slowly degraded and can have a significant long term impact on fisheries and related activities which depend on unpolluted waters for productive yields.

Large oil spills are not common, but when they do occur, such as the 11 million gallon spill from the supertanker Exxon Valdez that ran aground near Prince William Sound, Alaska, in March 1989, the input of oil to the environment can be ecologically devastating (Figure 17.46a) and cleanup costs staggering (estimated at $1.5 billion).

Addition of inorganic nutrients such as phosphorous and nitrogen to oil spill areas can increase bioremediation rates significantly. In the Exxon Valdez spill, for example, accelerated bioremediation of oil washed up on beaches was observed following spraying of the beaches with a mixture of inorganic nutrients (Figure 17.46b). For most large scale oil spills,

(a)

(b)

FIGURE 17.46
Environmental consequences of large oil spills in the marine environment and the effect of bioremediation. (a) A contaminated beach along the coast of Alaska containing oil from the Exxon Valdez spill of 1989. (b) The center rectangular plot was treated with inorganic nutrients to stimulate bioremediation of spilled oil by microorganisms, whereas areas to the left and right were untreated.

FIGURE 17.47 Bulk fuel storage tanks, where massive microbial growth may occur at oil-water interfaces.

however, a combination of both human and microbial efforts are needed to effect a satisfactory cleanup in a reasonable amount of time.

Interfaces where oil and water meet often occur on a massive scale. It is virtually impossible to keep moisture from bulk storage tanks; it accumulates as a layer of water beneath the petroleum. Gasoline storage tanks (Figure 17.47) are thus potential habitats for hydrocarbon-oxidizing microorganisms, which can accumulate and grow at the oil-water interface. Gasoline generally has additional chemicals (added to aid combustion and inhibit corrosion in engines) which inhibit the growth of microorganisms, but not all fuels have such additives.

Methane oxidation

Methane is the simplest hydrocarbon. We discussed the ecology of methane formation in Section 17.12. Methane is an important microbial substrate for a specific group of bacteria, the **methanotrophic bacteria**. We discuss the taxonomy and diversity of methanotrophic bacteria in Section 19.7; here we note some of their important ecological activities. Methane is to a great extent stable in anoxic environments, and the methane formed through the action of methanogenic bacteria often accumulates in deep subsurface deposits called *natural gas*. It is only when methane is brought back to the surface, where it is exposed to oxygen, O_2, that it is subject to attack by methanotrophs. Methanotrophs are responsible for the recycling of carbon from CH_4 back into CO_2 (see Figure 17.28). Methane is also of interest because it can be used as an energy source for the large-scale growth of microorganisms for use as possible food sources (single-cell protein, see Section 10.16).

> Hydrocarbons associated with petroleum are generally stable in anoxic environments but are subject to bacterial attack in oxic habitats. Although bacteria can thus play an important role in the clean-up of oil spills, it is often essential that nutrients such as inorganic nitrogen and phosphorus be added so that the hydrocarbon-oxidizing bacteria will be able to flourish.

17.20 Biodegradation of Xenobiotics

We defined the term *xenobiotic* in Section 16.19, noting that xenobiotics were chemically synthesized compounds that have never existed naturally. Thus, organisms capable of using them may not exist in nature. Some of the most widely distributed xenobiotics are the **pesticides** which are common components of toxic wastes. Over 1000 pesticides have been marketed for chemical pest control purposes. These include primarily *herbicides, insecticides,* and *fungicides.* Figure 17.48 shows the structures of some commonly used pesticides.

Pesticides are of a wide variety of chemical types, such as chlorophenoxyalkyl carboxylic acids, substituted ureas, nitrophenols, triazines, phenylcarbamates, organochlorines and organophosphates, and others (Figure 17.48). Some of these substances are suitable as carbon sources and electron donors for certain soil microorganisms, whereas others are not. If a substance can be attacked by microorganisms, it will eventually disappear from the soil. Such degradation in the soil is usually desirable, since toxic accumulations of the compound are avoided. However, even closely related compounds may differ remarkably in their degradability, as is shown for the relative persis-

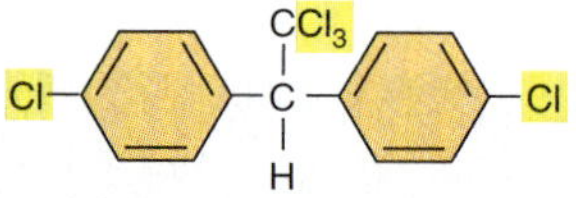

DDT; dichlorodiphenyltrichloroethane
(an organochlorine)

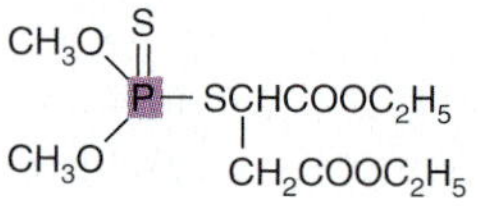

Malathion; mercaptosuccinic acid diethyl ester
(an organophosphate)

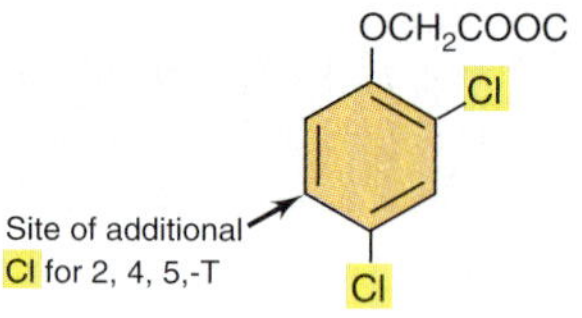

2,4-D; 2,4-dichlorophenoxy acetic acid
(a chlorophenoxy acetic acid derivative)

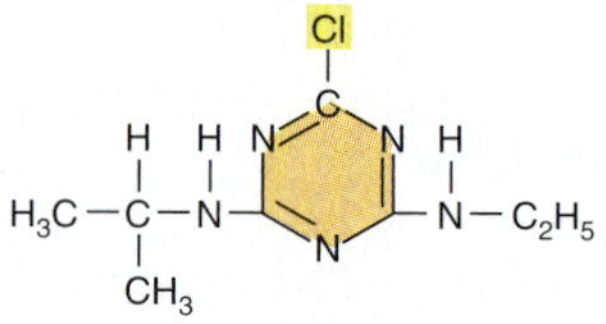

Atrazine; 2-chloro-4-ethylamino-6-isopropylamino-triazine
(a triazine derivative)

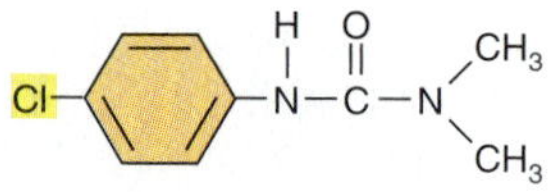

Monuron; 3-(4-chlorophenyl)-1,1-dimethylurea
(a substituted urea)

FIGURE 17.48 Some xenobiotic compounds. Although none of these compounds exist naturally, various microorganisms exist (or have been developed experimentally) which will break them down.

Table 17.7	Persistence of herbicides and insecticides in soils	
Substance		**Time for 75–100% disappearance**
Chlorinated insecticides		
DDT[1,1,1-trichloro-2,2-bis-(*p*-chlorophenyl)ethane]		4 years
Aldrin		3 years
Chlordane		5 years
Heptachlor		2 years
Lindane (hexachloro-cyclohexane)		3 years
Organophosphate insecticides		
Diazinon		12 weeks
Malathion		1 week
Parathion		1 week
Herbicides		
2,4-D(2,4-dichloro-phenoxy-acetic acid)		4 weeks
2,4,5-T(2,4,5-trichlorophenoxy-acetic acid)		20 weeks
Dalapin		8 weeks
Atrazine		40 weeks
Simazine		48 weeks
Propazine		1.5 years

tence rates of a number of herbicides in Table 17.7. However, these figures are only approximate since a variety of environmental factors, such as temperature, pH, aeration, and organic matter content of the soil, influence decomposition. Some of the chlorinated insecticides are so indestructible that they have persisted for over 10 years. Disappearance of a pesticide from an ecosystem does not necessarily mean that it was degraded by microorganisms, since pesticide loss can also occur by volatilization, leaching, or spontaneous chemical breakdown.

The organisms that are able to metabolize pesticides and herbicides are fairly diverse, including genera of both bacteria and fungi. Some pesticides serve well as carbon and energy sources and are oxidized completely to CO_2. However, other compounds are much more recalcitrant, and are attacked only slightly or not at all, although they may often be degraded either partially or totally provided some other organic material is present as primary energy source, a phenomenon called **co-metabolism**. However, when the breakdown is only partial, the microbial degradation product of a pesticide may sometimes be even more toxic than the original compound.

Reductive dechlorination

It is now believed that significant chlorinated pesticide degradation occurs in anoxic environments. In these cases, anaerobic biodegradation is linked to *reductive dechlorination* of the molecule, the dechlorinated derivative being much less toxic than the original chlorinated molecule. A sulfate-reducing bacterium implicated in this process, *Desulfomonile*, reduces 3-

chlorobenzoate (used as a model compound for studies of chlorinated pesticide degradation) to benzoate and Cl^-:

$$C_7H_4O_2Cl^- + 2H \rightarrow C_7H_5O_2^- + HCl$$
3-chlorobenzoate   benzoate

Electrons for this reduction can come from acetate, formate, or H_2, and the reductive reaction is linked to establishment of a proton gradient across the cytoplasmic membrane which *Desulfomonile* uses to drive ATP synthesis. Thus, the reductive dechlorination of 3-chlorobenzoate is a type of anaerobic respiration (see Section 16.14). Reductive dechlorination of a variety of chlorinated aromatic compounds in nature is suspected, and direct evidence for reductive dechlorinations has been obtained in the case of dichloroethylene, trichloroethylene, and tetrachloroethylene (perchloroethylene). These chloroaromatic compounds, some of which (particularly trichloroethylene) are suspected of being carcinogenic, are widely used as industrial solvents and degreasing agents and are among the most frequently detected groundwater contaminants in the United States. Aerobic dechlorination of chlorinated organic compounds also occurs, probably by different biochemical mechanisms than for anaerobic biodegradation, but anaerobic dechlorination is of particular environmental interest because of the rapidity in which anoxic conditions can develop in polluted microbial habitats in nature.

Biodegradation of synthetic polymers and the landfill crisis

A major area of environmental concern is the disposal of solid wastes. In 1992 the plastics industry alone produced nearly 70 *billion* pounds of plastics, 40 percent of which was discarded in landfills. Solid wastes, which beside plastics include packaging materials, polyurethane, polystyrenes, and other polymeric substances, are rapidly filling up available landfill sites. The chemical structures of the building blocks of some of these synthetic polymers are shown in Figure 17.49a.

It is now recognized that many synthetic polymers are highly recalcitrant to microbial degradation and remain essentially unaltered for decades in landfills and other refuse dumps. This problem has fueled the search for *biodegradable* alternatives to the common synthetic polymers now in use. Some improvements have been made in this area, including photodegradable, starch-linked, and "microbial" plastics (see below). Photobiodegradable plastics consist of material whose polymeric structure is altered by exposure to ultraviolet radiation (from sunlight), generating modified polymers amenable to microbial attack. Other biodegradable plastic bags have been made which incorporate starch to hold together short fragments of a biodegradable polymer. This design accelerates biodegradation because starch-digesting bacteria in soil attack the starch, releasing polymer fragments which are degraded by other microorganisms.

Extensive use of biodegradable synthetic poly-

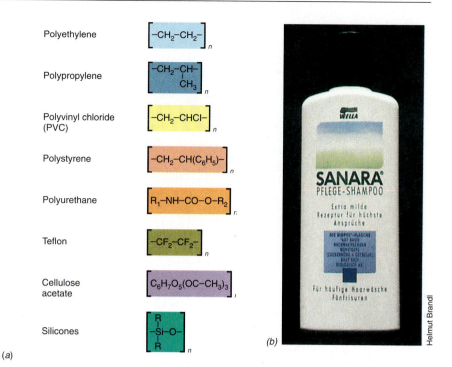

Polyethylene	$-CH_2-CH_2-]_n$
Polypropylene	$-CH_2-CH- \atop CH_3]_n$
Polyvinyl chloride (PVC)	$-CH_2-CHCl-]_n$
Polystyrene	$-CH_2-CH(C_6H_5)-]_n$
Polyurethane	$R_1-NH-CO-O-R_2]_n$
Teflon	$-CF_2-CF_2-]_n$
Cellulose acetate	$C_6H_7O_5(OC-CH_3)_3]_n$
Silicones	$R \atop -Si-O- \atop R]_n$

(a)

(b)

Helmut Brandl

FIGURE 17.49
Synthetic and microbial polymers. (a) Monomeric structure of a number of common synthetic polymers. (b) A brand of shampoo marketed in Europe and packaged in a bottle made of "microbial plastic." The bottle consists of a copolymer of poly-β-hydroxybutyrate and poly-β-hydroxyvalerate. Because this material is a natural product, the bottle readily degrades both aerobically and anaerobically.

mers could help alleviate the current landfill crisis in either of two ways. If biodegradable synthetic materials were separated from other solid wastes, the materials could be microbially pretreated much as wastewater is treated in the sewage treatment industry (see Section 17.21). This would result in a significant reduction in the total amount of waste sent to the landfill. Alternatively, biodegradable synthetic polymers could be degraded directly in the landfill site, although degradation using this approach would undoubtedly take longer than pretreatment methods. However, direct disposal of biodegradable polymers would still allow for more efficient use of available landfill sites because the total amount of solid waste disposed at a given site could be increased if it was known that in a reasonable time frame mineralization of synthetic polymer fractions was to occur.

Bacterial plastics

Besides *degrading* synthetic polymers, microorganisms have been recruited to *synthesize* biodegradable plastics. *Microbial plastics,* as this area has come to be called, represents an exciting new area of research where naturally synthesized bacterial polymers such as the lipid storage material poly-β-hydroxybutyrate (PHB) (see Section 3.9) are being used as raw materials for plastic-based packaging materials (Figure 17.49b and see Biodegradable Polymers box). Research in PHB production has shown that the precise chemical nature (and hence chemical properties) of the polymer produced by a bacterium can be controlled by varying the substrates used to grow the organism. For example, in certain bacteria acetate and butyrate lead to poly-β-hydroxybutyrate (C-4) production, whereas caproate (a C-6 fatty acid) leads to a polymer containing C-6 units and valerate (a C-5 fatty acid) yields a

polymer containing C-5 units. *Copolymers,* containing alternating repeats of various monomeric units, can also be synthesized. Microbial plastics are particularly appealing because they have been shown to undergo rapid biodegradation under both aerobic and anaerobic conditions.

Biodegradation and microbial evolution

The existence of organisms able to metabolize xenobiotics is of considerable evolutionary interest, since these compounds are completely new to the earth in the past 50 years or so. Observations on the rapidity with which organisms metabolizing new compounds arise can give us some idea of the rates of microbial evolution in general. In Section 7.9 we discussed the evolution in past decades of plasmids conferring resistance to antibiotics. The evolution of pesticide-degrading bacteria seems to be a similar case. For example, enrichment cultures (see Section 17.3) yield bacteria capable of degrading 2,4,5-T and other recalcitrant pesticides. These organisms appear to be common *Pseudomonas* species which are now capable of growing on these pesticides as sole sources of carbon and energy (Figure 17.50). In a study of 2,4,5-T and 2,4-D biodegradation it has been shown that portions of plasmids that code for 2,4-D biodegradation are recruited to form new plasmids conferring the ability to degrade 2,4,5-T. Because this can happen relatively quickly, it can be concluded that if biodegradation of a particular xenobiotic compound is possible, evolutionary events will move rather rapidly to establish microorganisms with new genetic properties to allow for breakdown of the compound. This of course assumes that a sufficient amount of the compound is present in the environment to maintain a selective advantage for biodegradation potential in the new population.

Biodegradable Polymers

Although humans in developed countries have grown accustomed to life in a "plastic society," it is now clear that municipal landfills cannot continue to be filled up with synthetic (xenobiotic) plastics. Because no significant biodegradation of many of these materials has occurred in landfills over relatively long periods of time, naturally occurring polymeric materials have been sought as substitutes for xenobiotic plastics. Because any substance of biological origin can be degraded by one or another microorganism, the widespread use of biologically produced plastics could solve major solid waste problems.

For several reasons, poly-β-hydroxyalkanoates (PHAs) appear to be the most ideal candidates for synthetic plastic substituents (see Figure 17.49b). PHAs are microbially produced storage polymers that have many of the general properties of synthetic plastics and which can be synthesized by cells in various chemical forms. Depending on the length of the side chain in the monomeric units of the PHA polymer (a property which can be adjusted by modifying the composition of the growth medium or by genetically modifying the producing bacterium), PHAs of varying melting points, crystallinity, flexibility, and tensile strength can be obtained. Various bacteria have been tested for use as catalysts for commercial production of PHAs, but as previously discussed in Chapter 10, an industrial fermentation is only commercially feasible if the organism can be made to overproduce the product and can be grown on a large scale using inexpensive carbon sources such as glucose or ethanol as chemical feedstocks. Thus far, only a relatively few bacterial species fit these criteria.

At a manufacturing plant in Billingham, England, the British chemical giant Imperial Chemical Industries (ICI) is producing PHAs and marketing the material as a packaging polymer under the trade name *Biopol*. ICI uses the bacterium *Alcaligenes eutrophus* to produce PHAs using glucose as feedstock. With *A. eutrophus*, ICI obtains yields of a poly-β-hydroxybutyrate/poly-β-hydroxyvalerate (PHB/PHV) copolymer of greater than 80 percent of cell dry weight. The polymer is extracted from the cells and milled to a powder or pellet form for use in the production of injection-molded articles such as the shampoo bottle shown in Figure 17.49b. Because PHA production by ICI is on a relatively small scale right now, about 500 tons a year, the current cost of a PHB/PHV bottle is several times that of a synthetic plastic bottle. However, plant expansion to a capacity 20 times this size could put PHAs on a cost competitive basis with petroleum-based plastics, especially in markets for environmentally conscious consumers. Moreover, because the metabolic diversity of bacteria is so substantial (see Chapter 16), novel polymers probably await discovery, and thus far several variations on the basic PHB or PHV pattern have already been discovered. Commercial production of these polymers could boost the future market share of bacterial plastics, especially in the speciality plastics markets where materials of specific physical properties are often needed.

It is of considerable scientific interest that the field of polymer science, which up until recently has been dominated by chemical and physical scientists, might conceivably be strengthened by microbiologists who offer microbial solutions to two of the most serious problems in the synthetic polymer industry today: (1) the ecological problem surrounding the recalcitrance of many synthetic disposable plastics, and (2) the production of novel polymers possessing unusual physical properties not obtainable by strictly chemical syntheses.

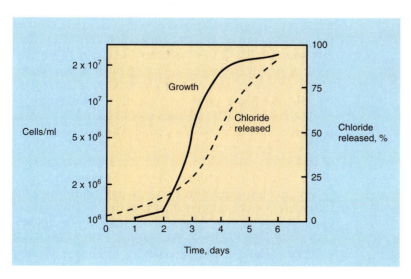

FIGURE 17.50

Growth of *Pseudomonas cepacia* on the herbicide 2,4,5-T as sole source of carbon and energy. The strain was enriched from nature using a chemostat to keep the concentration of herbicide low. Growth here is on 1.5 g/liter of 2,4,5-T. The release of chloride from the molecule is indicative of degradation.

Many chemically synthesized compounds such as insecticides, herbicides, and plastics (all called xenobiotics) are completely foreign to the environments in which they are used and can persist because microorganisms capable of degrading these xenobiotics may not naturally occur. Although appropriate microorganisms may eventually evolve, it is desirable to use materials that are biodegradable and thus readily degraded by indigenous microorganisms.

17.21   Sewage and Wastewater Microbiology

Wastewater treatments are processes in which microorganisms play crucial roles, and illustrate well some of the principles of biogeochemistry discussed above. Wastewaters are materials derived from domestic sewage or industrial processes, which for reasons of public health and for recreational, economic, and aesthetic considerations cannot be disposed of merely by discarding them untreated into convenient lakes or streams. Rather, the undesirable and toxic materials in the water must first be either removed or rendered harmless. Inorganic materials such as clay, silt, and other debris are removed by mechanical and chemical methods, and microorganisms participate only casually or not at all. If the material to be removed is *organic* in nature, however, treatment usually involves the activities of microorganisms, which oxidize and convert the organic matter to CO_2. Wastewater treatment usually also results in the elimination

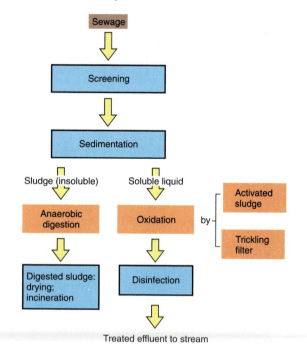

| Biological processes | Nonbiological processes |

FIGURE 17.51   An overview of sewage treatment processes.

of pathogenic microorganisms, thus preventing these organisms from getting into rivers or other supply sources.

Levels of sewage treatment

Sewage treatment is generally a multistep process employing both *physical* and biological treatment steps (Figure 17.51). **Primary treatment** of sewage consists only of physical separations. Sewage entering the treatment plant is passed through a series of grates and screens that remove large objects and then the effluent is left to settle for a number of hours to allow suspended solids to sediment.

Because of the high nutrient loads that remain in sewage effluent following primary treatment, those municipalities that treat sewage no further than the primary stage usually suffer from extremely polluted water when the sewage is dumped into adjacent waterways. This is why the majority of sewage processing plants employ **secondary treatment** processes to reduce the biochemical oxygen demand (BOD) (see Section 17.7) of the sewage to acceptable levels before releasing it to natural waterways. Secondary treatment is intimately tied to microbiological processes as described in the following sections.

Tertiary treatment is the most complete method of treating sewage, but has not been widely adopted because it is so expensive. Tertiary treatment is a physicochemical process employing precipitation, filtration, and chlorination to sharply reduce the levels of inorganic nutrients, especially phosphate and nitrate, from the final effluent. Wastewater receiving proper tertiary treatment is unable to support extensive microbial growth, and in many cases, it is of such high quality that it can be pumped directly into the municipal water supply! We focus our discussion here on secondary treatment processes, for it is these that have the most microbial components and are the most widely practiced in sewage treatment facilities today.

Anaerobic secondary treatment processes

Anaerobic sewage treatment involves a complex series of digestive and fermentative reactions carried out by a host of different bacterial species. The net result is the conversion of organic materials into CO_2 and methane gas (CH_4) (see Sections 16.19 and 17.12), the latter of which can be removed and burned as a source of energy. Since both end products, CO_2 and CH_4, are volatile, the liquid effluent is greatly decreased in organic substances. The efficiency of a treatment process is expressed in terms of the percent decrease of the initial biochemical oxygen demand (BOD); the efficiency of a well-operated plant can be 90 percent or greater, depending on the nature of the organic waste.

Anaerobic decomposition is usually employed for the treatment of materials that have much *insoluble* organic matter, such as fiber and cellulose, or for concentrated industrial wastes. The degradation process depends on interspecies hydrogen transfer reactions (see Section 17.12) and can be summarized as follows: (1)

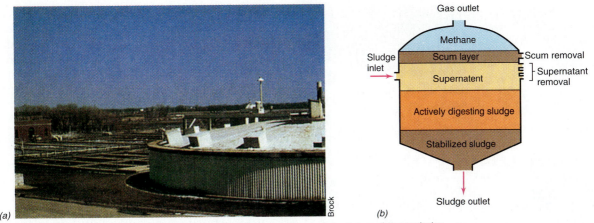

FIGURE 17.52 (a) Anaerobic sludge digestor. Only the top of the tank is shown; the remainder is underground. (b) Inner workings of an anaerobic sludge digestor.

initial digestion of the macromolecular materials by extracellular polysaccharidases, proteases, and lipases to soluble materials; (2) fermentation of the soluble materials to fatty acids, H_2 and CO_2, by acid-producing fermentative organisms; (3) fermentation of the fatty acids to acetate, CO_2, and H_2; and (4) conversion of H_2 plus CO_2 and acetate to CH_4 by methanogens. Unlike in the rumen, where H_2 is the main precursor for methanogenesis, both acetate-utilizing and H_2-utilizing methanogens are active in anoxic sludge digestors.

Anaerobic decomposition processes operate semi-continuously in large enclosed tanks called **sludge digestors** or **bioreactors**, into which the untreated material is introduced and from which the treated material is removed at intervals (Figure 17.52). The retention time in the tank is on the order of two weeks to a month. The solid residue consisting of indigestible material and bacterial cells is allowed to settle and is removed periodically and dried for subsequent burning or burial.

Aerobic secondary treatment process

There are several kinds of aerobic decomposition processes used in sewage treatment, but the trickling filter and activated sludge methods are the most common. A **trickling filter** is basically a bed of crushed rocks, about 2 meters thick, on top of which the liquid containing organic matter is sprayed. The liquid slowly trickles through the bed, the organic matter adsorbs to the rocks, and microbial growth takes place. The complete mineralization of organic matter to carbon dioxide, ammonia, nitrate, sulfate, and phosphate occurs.

The most common aerobic treatment system is the **activated sludge** process. Here, the wastewater to be treated is mixed and aerated in a large tank (Figure 17.53). Slime-forming bacteria (primarily a species called *Zoogloea ramigera*) grow and form flocs (so-called **zoogloeas**, Figure 17.54), and these flocs form the substratum to which protozoa and other animals attach. Occasionally, filamentous bacteria and fungi also are present. The basic process of oxidation is simi-

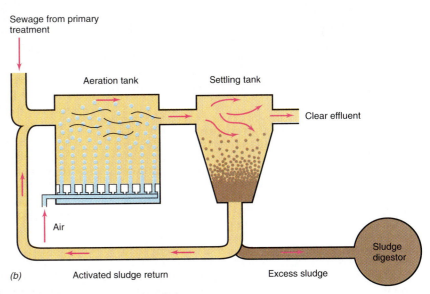

FIGURE 17.53 Activated sludge process. (a) Aeration tank of an activated sludge installation. (b) Inner workings of an activated sludge installation.

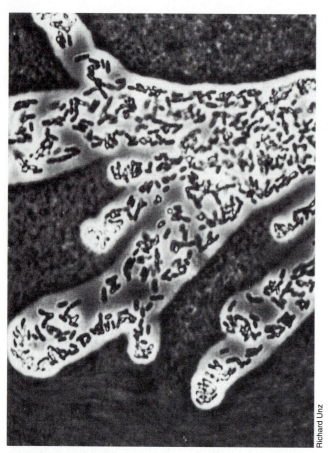

Richard Unz

FIGURE 17.54 Photomicrograph of a floc formed by *Zoogloea ramigera*, the characteristic organism of the activated-sludge process. Note the large number of small, rod-shaped bacteria surrounded by a polysaccharide slime and the characteristic fingerlike projections of the floc. Negative stain using India ink.

lar to that in a trickling filter. The effluent containing the flocs is pumped into a holding tank or clarifier, where the flocs settle. Some of the floc material is then returned to the aerator to serve as inoculum, while the rest is sent to the sludge digestor. The residence time in an activated sludge tank is generally 5–10 hours, too short for complete oxidation of organic matter. The main process occurring during this short time is *adsorption* of soluble organic matter to the floc, and incorporation of some of the soluble materials into mi-

Domestic and industrial wastes are rich in organic materials and must be treated in some way before they are disposed in the environment. Sewage treatment processes are essentially large-scale microbial culture systems in which the organic materials of the sewage are broken down (mineralized) to CO_2, CH_4, and inorganic nutrients. Two kinds of sewage treatment processes are used: anoxic, in which organic materials are converted principally to methane and carbon dioxide, and oxic, in which organic materials are converted into microbial cells and carbon dioxide. As a result of the sewage treatment process, the effluent is stabilized and its content of toxic materials is reduced; it can thus be disposed into the environment with less difficulty.

crobial cell material. The BOD of the liquid is thus considerably reduced by this process (75–90 percent), but the overall BOD (liquid plus solids) is only slightly reduced, because most of the adsorbed organic matter still resides in the floc. The main process of BOD reduction thus occurs in the sludge digestor to which the floc is transferred.

17.22   Plant–Microorganism Interactions

As microbial habitats, plants are clearly vastly different from animals (see Chapter 11). Compared with warm-blooded animals, plants vary greatly in temperature, both diurnally and throughout the year, and compared with the complex circulatory system of animals, the internal communication system of the plant is only poorly developed, so that transfer of microorganisms within the plant is relatively inefficient. The aboveground parts of the plant, especially the leaves and stems, are subjected to frequent drying, and for this reason many plants have developed waxy coatings that retain moisture and serve to keep out microorganisms. The roots, on the other hand, exist in an environment in which moisture is less variable and nutrient concentrations are higher. For this reason, the roots of plants are a main area of microbial action.

The **rhizosphere** is the region immediately outside the root; it is a zone where microbial activity is usually high. The bacterial count is almost always higher in the rhizosphere than it is in regions of the soil devoid of roots, often many times higher. This is because roots excrete significant amounts of sugars, amino acids, hormones, and vitamins, which promote such an extensive growth of bacteria and fungi that these organisms often form microcolonies on the root surface. There is some evidence that rhizosphere microorganisms benefit the plant by promoting the absorption of nutrients, but this is probably only of minor importance. The numbers and kinds of microorganisms on or near root surfaces is highly variable depending on the plant type and soil nutrient and water status.

The **phyllosphere** is the surface of the plant leaf, and under conditions of high humidity, as in wet forests in tropical and temperate zones, the microbial flora of leaves may be quite high. In tropical rain forests, the phyllosphere bacteria often produce gums or slimes, which presumably help bacterial cells stick to leaves under the onslaught of heavy tropical rains. Many of the bacteria on leaves fix nitrogen (see Sections 16.26 and 17.24), and nitrogen fixation presumably aids these organisms in growing with the predominantly carbohydrate nutrients provided by leaves. Some of the nitrogen fixed by these bacteria

Plants provide numerous habitats for the growth of microorganisms, including the roots (rhizosphere) and leaves (phyllosphere). Plant roots excrete various organic substances useable as energy sources by bacteria.

may be absorbed by the plant, although there is little evidence that this occurs on a large scale, except for the stem-nodulating rhizobia discussed in Section 17.24.

Lichens

Lichens are leafy or encrusting growths that are widespread in nature and are often found growing on bare rocks, tree trunks, house roofs, and surfaces of bare soils (Figure 17.55). The lichen plant consists of two organisms, a fungus and an alga. However, little specificity resides in the relationship, as a given fungus can establish the lichen symbiosis with several different algae and vice versa. The alga is phototrophic and is able to produce organic matter which is then used for nutrition of the fungus. Since the fungus is unable to carry out photosynthesis, its ability to live in nature is dependent on the activity of its algal partner. Lichens are usually found in environments where other organisms do not grow, and it is almost certain that their success in colonizing such extreme environments is due to the mutual interrelationships between the alga and fungus partners.

Lichens consist of a tight association of many fungal cells, within which the algal cells are embedded (Figure 17.56). The shape of the lichen is determined primarily by the fungal partner, and a wide variety of fungi are able to form lichen associations. The diversity of algal types is much smaller, and many different kinds of lichens may have the same algal component.

Some lichens contain cyanobacteria instead of algae as the phototrophic component. The algae or cyanobacteria are usually present in defined layers or clumps within the lichen structure.

The fungus clearly benefits from associating with the alga, but how does the alga benefit? The fungus provides a firm anchor within which the alga can grow protected from erosion by rain or wind. In addition, the fungus facilitates the uptake of water and absorbs from the rock or other substrate upon which the lichen is living the inorganic nutrients essential for the growth of the alga. *Lichen acids*, complex organic compounds excreted by the fungus, promote the dissolution and chelation of nutrients. Another role of the fungus is to protect the alga from drying; most of the habitats in which lichens live are dry (rock, bare soil, roof tops, see Figure 17.55), and fungi are in general much better able to tolerate dry conditions than are algae.

Most lichens grow extremely slowly—a 2-cm lichen observed on the surface of a rock may be actually several years old. Measurements of lichen growth vary from 1 mm or less per year to over 3 cm per year, depending on the organisms composing the symbiosis, the amount of rainfall and sunlight received, and general weather conditions. Although lichens live in nature under rather harsh conditions, they are extremely sensitive to air pollution and quickly disappear in areas experiencing heavy air pollution. One

(a)

(b)

FIGURE 17.55 Lichens. (a) A lichen growing on a branch of a dead tree. (b) Lichens coating the surface of a large rock. (c) Several different lichens growing on the surface of a rock.

(c)

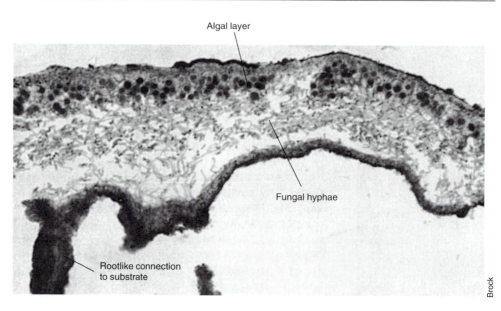

Algal layer

Fungal hyphae

Rootlike connection to substrate

Brock

FIGURE 17.56
Photomicrograph of a cross section through a lichen.

reason for this sensitivity is that they absorb and concentrate materials from rainwater and air and have no means for excreting them, so that lethal concentrations of compounds such as SO_2 are easily reached.

> Lichens are a unique type of association between a fungus and a phototrophic microorganism, either an alga or a cyanobacterium. The phototrophic organism provides the organic matter and the fungus provides the structure and serves as an organ for the absorption of nutrients. Lichens are capable of living in dry environments where other kinds of plants cannot grow.

Mycorrhiza

Mycorrhiza literally means "root fungus" and refers to the symbiotic association that exists between plant roots and fungi. Probably the roots of the majority of terrestrial plants are mycorrhizal. There are two general classes of mycorrhiza: **ectomycorrhiza,** in which fungal cells form an extensive sheath around the outside of the root with only little penetration into the root tissue itself, and **ericoid mycorrhiza,** in which the fungal mycelium is embedded in the root tissue. The present discussion will deal only with ectomycorrhizae.

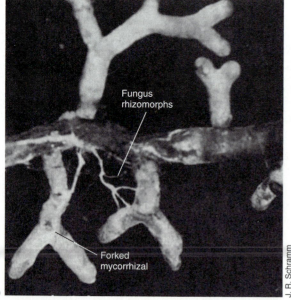

Fungus rhizomorphs

Forked mycorrhizal

J. R. Schramm

(a)

D.J. Read

FIGURE 17.57 Mycorrhiza. (a) Typical mycorrhizal root of the pine, *Pinus rigida,* with rhizomorphs of the fungus *Thelophora terrestris.* (b) Cross section of a seedling of *Pinus contorta* (lodgepole pine), showing extensive development of mycorrhiza. The seedling extends about 4 cm above the soil surface.

(b)

Ectomycorrhizae are found mainly in forest trees, especially conifers, beeches, and oaks, and are most highly developed in temperate forests. In a forest, almost every root of every tree will be mycorrhizal. The root system of a mycorrhizal tree is composed of both long and short roots. The short roots, which are characteristically dichotomously branched (Figure 17.57), show the typical fungal sheath, whereas long roots are usually uninfected. Most mycorrhizal fungi do not attack cellulose and leaf litter, but instead use simple carbohydrates for growth and usually have one or more vitamin requirements; they obtain their nutrients from root secretions. The mycorrhizal fungi are never found in nature except in association with roots and hence can be considered obligate symbionts. These fungi produce plant growth substances that induce morphological alterations in the roots, causing characteristically short dichotomously branched mycorrhizal roots to be formed. Despite the close relationship between fungus and root, there is little species specificity involved; a single species of pine can form mycorrhizae with over 40 species of fungi.

The beneficial effect on the plant of the mycorrhizal fungus is best observed in poor soils, where trees that are mycorrhizal will thrive, but nonmycorrhizal ones will not (Figure 17.58). If trees are planted in prairie soils, which ordinarily lack a suitable fungal inoculum, trees that were artificially inoculated at the time of planting grow much more rapidly than uninoculated trees (Figure 17.58). It is well established that the mycorrhizal plant is able to absorb nutrients from its environment more efficiently than does a nonmycorrhizal one. This improved nutrient absorption is probably due to the greater surface area provided by the fungal mycelium (Figure 17.57).

> The mycorrhizal relationship is one in which a fungus lives in association with the roots of a tree. The plant provides the organic materials needed by the fungus and the fungus serves as a nutrient-absorptive organ for the plant.

Plant disease resistance

Plants are subject to a variety of microbial diseases and have evolved several defenses to counter the harmful effects of microorganisms. At least three major microbial defense mechanisms have been identified in plants: nonspecific and specific antimicrobial agents, and the hypersensitive response. Many plants routinely produce toxic chemicals that either prevent microbial growth or actually kill invading microorganisms; examples include the phenols produced by onions and many cereal and root plants, hydrogen cyanide produced by flax, and the allyl sulfides produced by onions.

Phytoalexins are an important class of antimicrobial compounds produced by a variety of plants. Phytoalexins are a chemically heterogenous group of low-molecular-weight aromatic compounds, a number of which are derived from isoflavanoids (Figure 17.59). Unlike toxic compounds which are produced by the plant nonspecifically, phytoalexins are absent in healthy plants and are only produced in response to microbial invasion. Phytoalexin synthesis is induced by a variety of microbial components, including fungal β-glucans, glycoproteins, and acidic cell wall components. Phytoalexin synthesis is also elicited by pectin components released from the plant's own cell wall. The pectin elicitors are all low-molecular-weight polysaccharides, frequently of unusual chemical structure. Presumably these pectin fragments serve as a chemical signal to the plant that the plant cell wall has been damaged by the action of microbial pectinases.

Although poorly understood, plants also protect themselves from microbial attack by rapidly killing their own cells in the vicinity of an infection site. This mechanism has been referred to as the **plant hypersensitive response**, and has been observed in a num-

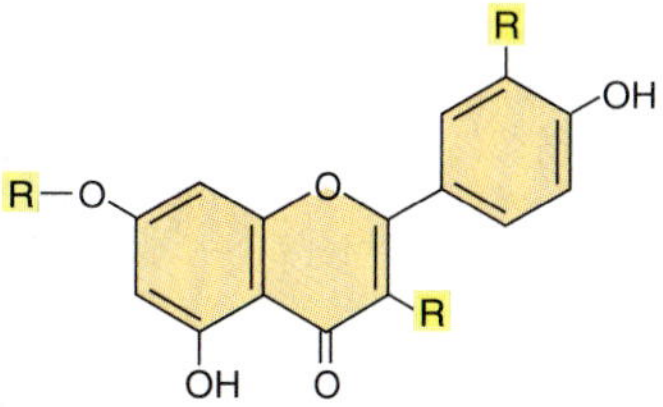

FIGURE 17.58 Six-month-old seedlings of Monterey Pine (*Pinus radiate*) growing in prairie soil: left, nonmycorrhizal; right, mycorrhizal.

S. A. Wilde

FIGURE 17.59 Parent structure of the isoflavanoid-derived phytoalexins, plant antimicrobial agents. The R groups represent substituents that can vary from one compound to another.

ber of plant species. The hypersensitive response is characterized by water loss from the tissue surrounding the infection site leading to a collapse and flattening of the cells. The invading microorganisms are then trapped within the collapsed area, thus protecting the surrounding healthy tissues.

Thus, although plants lack the sophisticated microbial defense mechanisms of warm-blooded animals (see Chapters 11 and 12), they do have a variety of mechanisms for countering microbial invasion.

Bacterial insecticides

Several *Bacillus* species, most notably *B. popilliae* and *B. thuringiensis*, function as biological control agents for plants by serving as insect larvicides. *B. popilliae* causes a fatal disease called *milky disease* in Japanese beetle larvae and larvae of closely related beetles of the family Scarabaeidae. *B. thuringiensis* causes a fatal disease of larvae of many different groups of insects, although individual strains are specific as to host affected. Strains exist that are specific for lepidopterans, such as the silkworm, the cabbage worm, the tent caterpillar, and the gypsy moth. Some strains kill dipterans such as mosquitoes and black flies. Others kill coleopterans such as Colorado potato beetles. Strains of *B. thuringiensis* have also been discovered that are toxic to Japanese beetles.

The disease caused by *B. popilliae* is a septicemia, whereas the disease caused by *B. thuringiensis* is essentially an intoxication. Both of these insect pathogens form a crystalline protein during sporulation called the *parasporal body*, which is deposited within the sporangium but outside the spore proper (Figure 17.60). In the case of *B. thuringiensis*, the crystal (parasporal body) protein is a protoxin that is ultimately responsible for the fatal insect disease. The function of the crystal protein in *B. popilliae* is uncertain. Purified crystals do not cause the disease. Also, a closely related *Bacillus* species, *B. lentimorbus*, causes essentially the same disease as *B. popilliae*, but it does not produce parasporal crystals.

The crystal protoxin protein of *B. thuringiensis* is converted to a toxin by proteolytic cleavage in the larval gut. The toxin binds to intestinal epithelial cells and induces pore formation that causes leakage of the host cells followed by lysis. Death is probably due to starvation. Thus, the disease is essentially an intoxication. Genes encoding crystal proteins from several *B.*

thuringiensis strains have been isolated. Separate domains of the toxin molecule have been identified that are responsible for toxicity and binding to host cells. Genetic engineering is being used to modify host specificity of the crystal proteins. The genes for the *B. thuringiensis* crystal protein (known commercially as "Bt–toxin") have been introduced into plants to render the plants "naturally" resistant to insects. This strategy has been shown to be effective in controlled situations and a variety of Bt–toxins are being developed by genetic engineering (see Section 8.14).

Spore (and parasporal crystal) preparations derived from insect pathogens are widely used as biological insecticides. Interest in such biological control is currently high because of declining use of chemical insecticides for human food crops.

> A number of insects which cause serious diseases of crop plants can be controlled by the use of bacteria that are pathogenic to the insects. The use of one organism (in this case, a bacterium) to control the growth of another organism (the insect) is an example of a process called biological control. The crystal protein of *Bacillus thuringiensis* is a widely used biological control and its insecticidal mechanism is the best understood.

17.23 *Agrobacterium* and Plant Interactions: Crown Gall and Hairy Root

The genus *Agrobacterium* comprises organisms that cause the formation of tumorous growths on a wide variety of plants. The two species most widely studied are *A. tumefaciens*, which causes *crown gall*, and *A. rhizogenes*, which causes *hairy root*. Although plants often

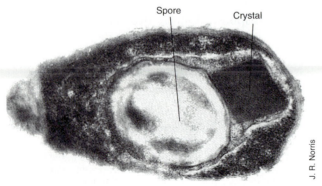

FIGURE 17.60 Formation of the toxic parasporal crystal in the insect pathogen *Bacillus thuringiensis*. Electron micrograph of a thin section.

FIGURE 17.61 Photograph of tumor on a tobacco plant caused by crown gall bacteria of the genus *Agrobacterium*.

form benign accumulations of tissue, called a **callus**, when wounded, the growth induced by *A. tumefaciens* (Figure 17.61) is different in that the callus shows uncontrolled growth. It thus resembles tumor growth in animals, and considerable research on crown gall has been carried out with the idea that it may provide a model for how malignant growths occur in humans. *A. rhizogenes* causes a malignant mass of roots to emanate from the infected wound site, creating the condition known as *hairy root*. Interestingly, once induced, these tumors continue to grow in the absence of *Agrobacterium* cells. Thus, once *Agrobacterium* has brought about the induction of the tumorous condition, its presence is no longer necessary.

An overview of the events in crown gall formation is given in Figure 17.62. It is now well established that a large plasmid called the *Ti (tumor induction) plasmid* (Figure 17.63) must be present in the *Agrobacterium* cells if they are to induce tumor formation. In *A. rhizogenes*, a similar plasmid called the *Ri plasmid* is necessary for induction of hairy root. However, the best studied system is that of the *A. tumefaciens*/crown gall disease so we focus on that here. Following infection, a part of the Ti plasmid, called the *transfer DNA (T-DNA)*, is integrated into the plant's genome. T-DNA carries the genes for tumor formation and also for the production of a number of modified amino acids called **opines**. *Octopine* [N²-(1,3-dicarboxyethyl)-L-arginine] and *nopaline* [N²-(1,3-dicarboxypropyl)-L-arginine] are the two most common opines. Opines are produced by plant cells transformed by T-DNA and serve as a source of carbon and nitrogen for *Agrobacterium* cells (Figure 17.62). The Ti plasmid is also used in genetic engineering (see Section 8.14).

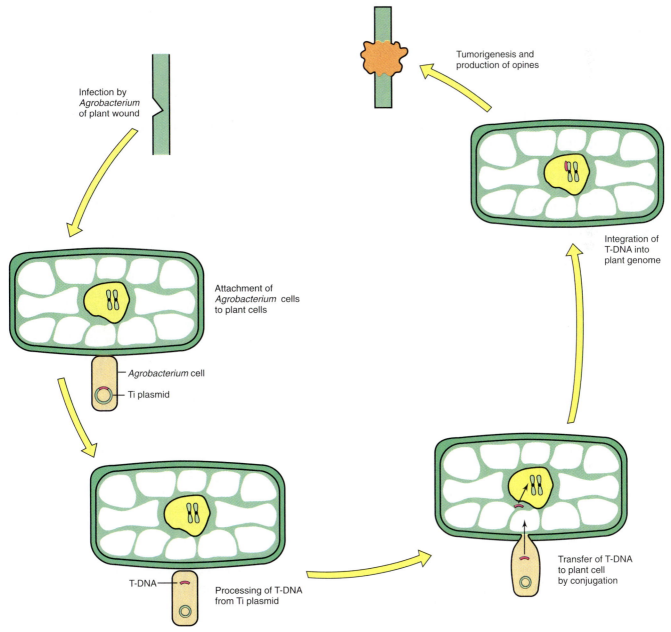

Infection by
Agrobacterium
of plant wound

Tumorigenesis and
production of opines

Attachment of
Agrobacterium cells
to plant cells

Integration of
T-DNA into
plant genome

Agrobacterium cell

Ti plasmid

Transfer of T-DNA
to plant cell
by conjugation

T-DNA

Processing of T-DNA
from Ti plasmid

FIGURE 17.62 Overview of events in crown gall disease following infection of a susceptible plant by *Agrobacterium tumefaciens*.

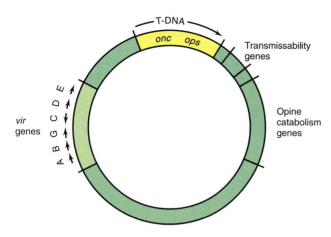

FIGURE 17.63 Structure of the Ti plasmid of *Agrobacterium tumefaciens*. T-DNA is the region actually transferred to the plant. *vir*, virulence genes; *onc*, oncogenes; *ops*, opine synthesis genes. Arrows indicate the direction of transcription of each gene. The entire Ti plasmid is about 200 kb of DNA and the T-DNA about 20 kb.

Recognition of *Agrobacterium* by the plant

To initiate the tumorous state, cells of *Agrobacterium* must first attach to a wound site on the plant. The recognition of *Agrobacterium* by plant tissue involves complementary receptor molecules on the surfaces of the bacterial and plant cells. It is thought that the plant receptor molecule is a type of *pectin* (a complex polysaccharide) and that the bacterial receptor is a type of polysaccharide containing β-glucans, imbedded in the cell wall lipopolysaccharide.

Studies with nontumorigenic mutants of *A. tumefaciens* have shown clearly that most functions necessary for attachment of the bacterium to plant are borne on the bacterial chromosome. Presumably β-glucans modify the bacterial cell wall in a way that facilitates binding of the bacterial cell to the plant cell. In addition to β-glucan, attachment of *A. tumefaciens* to certain plant types, notably to carrot or tobacco tissue, is mediated by bacterial production of *cellulose* microfibrils. Following recognition and initial binding (mediated by LPS and β-glucan), the rapid synthesis of cellulose microfibrils anchors the inoculum to the wound site and literally entraps the bacterial cells, forming large bacterial aggregates on the plant cell surface. Although helpful in mediating colonization and rapid growth of agrobacteria at the site of infection, studies with mutants of *A. tumefaciens* incapable of synthesizing cellulose indicates that cellulose production is not necessary for induction of the tumorous state.

Plasmid transfer and tumorigenesis

The structure of the Ti plasmid is given in Figure 17.63. Note that although a number of genes are needed for infectivity, only a smaller portion is needed for tumorigenesis. Before the onset of disease symptoms in a plant, Ti plasmid DNA must be transferred from the bacterium to the plant. Only a small portion of the Ti plasmid, a region called the T-DNA (Figures 17.62 and 17.63), is actually transferred to the

plant. The T-DNA contains oncogenes that direct events leading to tumorigenesis. The *vir* genes which reside on the Ti plasmid code for proteins that are essential for T-DNA transfer (Figure 17.63). *Vir* gene expression is induced by plant signal molecules that are synthesized by wounded plant tissues. Some inducers that have been identified include the phenolic compounds acetosyringone, para-hydroxybenzoic acid, and vanillin.

The *vir* genes are the key to T-DNA transfer. The *virA* gene codes for a protein kinase that phosphorylates the product of the *virG* gene (Figure 17.64). The latter becomes activated by the phosphorylation event and functions to activate other *vir* genes. The product of the *virD* gene has endonuclease activity and nicks DNA in the Ti plasmid in a region adjacent to the T-DNA (Figures 17.63 and 17.64). The product of the *virE* gene is a single-stranded DNA binding protein that binds the *single strand* of T-DNA generated from endonuclease activity and transports this small fragment of DNA into the plant cell. The *virB* gene product is located in the bacterial membrane and serves to mediate transfer of the single strand of DNA between bacterium and plant.

T-DNA transfer occurs in a process that resembles bacterial conjugation, the *virB* protein facilitating the transfer of T-DNA across the membrane through a porelike structure (Figures 17.62 and 17.64). The latter becomes inserted into the nuclear genome of the plant (apparently plant cell organellar genomes are not infected) where integration can occur at any number of sites where specific inverted or direct tandem repeats are present. The oncogenes of the Ti plasmid (Figure 17.63) code for enzymes involved in plant hormone production and for at least one key enzyme of opine biosynthesis. Expression of these oncogenes leads to tumor formation.

The Ri plasmid involved in hairy root disease also contains oncogenes. In this case, the oncogenes confer increased auxin responsiveness to the plant cells, which leads to overproduction of root tissue, resulting in the symptoms of the disease. The Ri plasmid also codes for several opine biosynthetic enzymes.

Despite differences in tumor morphology, the similar nature of the molecular events involved in crown gall and hairy root disease suggest a close molecular relationship among the two diseases and among the two infecting plasmids. From the standpoint of microbiology both diseases involve a unique type of plant–microorganism interaction in which bacterial DNA serves to transform plant cells. When this was recognized, the usefulness of this natural plant transformation system as a vector for the introduction of genetically engineered DNA into plants became immediately obvious.

Genetic engineering with the Ti plasmid

The Ti and Ri plasmid systems have revolutionized plant genetics and have done much to open the field of plant biotechnology. We discussed in Section 8.14

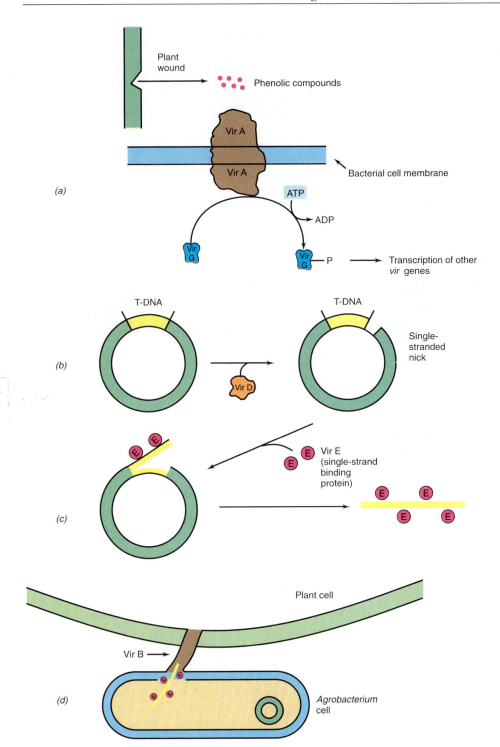

FIGURE 17.64
Mechanism of transfer of T-DNA to the plant cell by *Agrobacterium tumefaciens*. (a) Levels of VirA protein increase dramatically upon stimulation with plant phenolic inducer molecules. VirA activates VirG by phosphorylation and VirG activates transcription of other Vir proteins. (b) VirD is an endonuclease. (c) VirE is a single-stranded binding protein. (d) VirB acts as a conjugation bridge between *Agrobacterium* and the plant cell.

the use of *Agrobacterium* to genetically transform plants and discussed some of the applications of this technology to genetic engineering. Here we briefly consider a novel use of the *A. tumefaciens* Ti plasmid system in plant molecular biology. Besides using *Agrobacterium* to transfer genes for herbicide and drought resistance into plants or for attempts to improve the nutritional composition of plants (see Section 8.14), agrobacteria have served as vectors for introducing reporter genes in plants as easily visible markers of gene expression. For example, *A. tumefa-*

ciens has been used to introduce the firefly luciferase gene into tobacco plants, yielding plants that are now bioluminescent (Figure 17.65). Used as a reporter and linked to other key genes of interest, luciferase has served as a visible marker of gene expression, as a genetic marker in classical plant breeding, and as a general tool for studying plant molecular biology. Experience has shown that plants are fairly difficult to transform by other means. Thus, the Ti and Ri plasmids have served to open up plant biology to the power of molecular genetics.

FIGURE 17.65 Leaf from a tobacco plant, *Nicotiana tabacum*, that had been genetically engineered by insertion of the luciferase gene from the firefly, *Photinus pyralis*. Note that the leaf is now luminescent due to expression of the luciferase gene.

The crown gall bacterium *Agrobacterium* enters into a unique molecular relationship with higher plants. A plasmid in the bacterium (the Ti plasmid) is able to transfer part of itself to the genome of the plant, in this way bringing about the production of the crown gall disease. The crown gall plasmid has also found extensive use in the genetic engineering of crop plants.

17.24 Root Nodule Bacteria and Symbiosis with Legumes

One of the most interesting and important plant bacterial interactions is that between leguminous plants and bacteria of the genera *Rhizobium* and *Bradyrhizobium*. Legumes are a large group that includes such economically important plants as soybeans, clover, alfalfa, beans, and peas, and are defined as plants which bear seeds in pods. *Rhizobium* and *Bradyrhizobium* are Gram-negative motile rods. Infection of the roots of a leguminous plant with the appropriate species of *Rhizobium* or *Bradyrhizobium* leads to the formation of **root nodules** (Figure 17.66), which are able to convert gaseous nitrogen into combined nitrogen, a process called *nitrogen fixation* (see Section 16.26). Nitrogen fix-

FIGURE 17.66 Soybean root nodules. The nodules develop by infection with *Bradyrhizobium japonicum*.

ation by the legume-*Rhizobium* symbiosis is of considerable agricultural importance, as it leads to very significant increases in combined nitrogen in the soil. Since nitrogen deficiencies often occur in unfertilized bare soils, nodulated legumes are at a selective advantage under such conditions and can grow well in areas where other plants cannot (Figure 17.67).

Under normal conditions, neither legume nor *Rhizobium* alone is able to fix nitrogen; yet the interaction between the two leads to the development of nitrogen-fixing ability. In pure culture, the *Rhizobium* is able to fix N_2 alone when grown under strictly controlled *microaerophilic* conditions. Apparently *Rhizobium* needs some O_2 to generate energy for N_2 fixation, yet its nitrogenase (like those of other nitrogen-fixing organisms, see Section 16.26) is inactivated by O_2. In the nodule, precise O_2 levels are controlled by the O_2-binding protein **leghemoglobin**. This is a red, iron-containing protein which is always found in healthy, N_2-fixing nodules (Figure 17.68). Neither plant nor *Rhizobium* alone synthesizes leghemoglobin, but formation is induced through the interaction of these two organisms. The heme and globin portions of leghemoglobin are synthesized by the bacterium and plant, respectively. Leghemoglobin serves as an "oxygen buffer" cycling between the oxidized (Fe^{3+}) and reduced (Fe^{2+}) forms, to keep free O_2 levels within the nodule at a low but constant level. The ratio of leghemoglobin-bound O_2 to free O_2 in the root nodule is on the order of 10,000:1.

About 90 percent of all leguminous plant species are capable of becoming nodulated. However, there is a marked specificity between species of legume and strains of *Rhizobium*. A single *Rhizobium* strain is generally able to infect certain species of legumes and not

Ben B. Bohlool

FIGURE 17.67 A field of unnodulated (left) and nodulated soybean plants growing in nitrogen-poor soil.

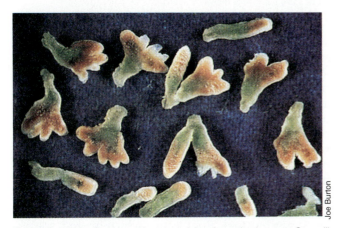

Joe Burton

FIGURE 17.68 Sections of root nodules from the legume *Coronilla varia*, showing the reddish pigment leghemoglobin.

Table 17.8	Major cross-inoculation groups of leguminous plants	
Plant	**Nodulated by**	**Growth rate***
Pea	*Rhizobium leguminosarum*	Fast
Bean	*Rhizobium phaseoli*	Fast
Clover	*Rhizobium trifolii*	Fast
Alfalfa	*Rhizobium meliloti*	Fast
Soybean	*Bradyrhizobium japonicum*	Slow

*In yeast extract-mannitol agar. Fast species generally have a generation time of less than 4 hours yielding gummy raised colonies 2–4 mm in diameter within 3–5 days. Slow species generally have a generation time of greater than 8 hours and yield granular colonies not exceeding 1 mm in diameter within 5–7 days.

others. A group of *Rhizobium* strains able to infect a group of related legumes is called a *cross-inoculation group*. The major rhizobial cross-inoculation groups are listed in Table 17.8. Even if a *Rhizobium* strain is able to infect a certain legume, it is not always able to bring about the production of nitrogen-fixing nodules. If the strain is *ineffective,* the nodules formed will be small, greenish-white, and incapable of fixing nitrogen; if the strain is *effective*, on the other hand, the nodule will be large, reddish (Figure 17.68), and nitrogen-fixing. Effectiveness is determined by genes in the bacterium that can be lost by mutation or gained by genetic transformation (see "Genetics of nodule formation: *nod* genes," on page 686).

Stages in nodule formation

The stages in the infection and development of root nodules are now fairly well understood (Figure 17.69). They include

1. **Recognition** of the correct partner on the part of both plant and bacterium and **attachment** of the bacterium to root hairs.

2. **Invasion** of the root hair by the bacterial formation of an infection thread.

3. **Travel** to the main root via the infection thread.

4. Formation of deformed bacterial cells, **bacteroids**, within the plant cells and the development of the nitrogen-fixing state.

5. Continued plant and bacterial division and formation of the mature **root nodule**.

We now discuss some of these stages in nodule formation in more detail.

The roots of leguminous plants secrete a variety of organic materials which stimulate the growth of a rhizosphere microflora. This stimulation is not restricted to the rhizobia but occurs with a variety of rhizosphere bacteria. If there are rhizobia in the soil, they grow in the rhizosphere and build up to high population densities. Attachment in the legume–*Rhizobium* symbiosis depends upon proper recognition of surface macromolecules of the root hair surface which interact with cell surface polysaccharides of the *Rhizobium* cell. This process has been studied in most detail in the white clover–*R. trifolii* symbiosis. The cells of *R. trifolii* have

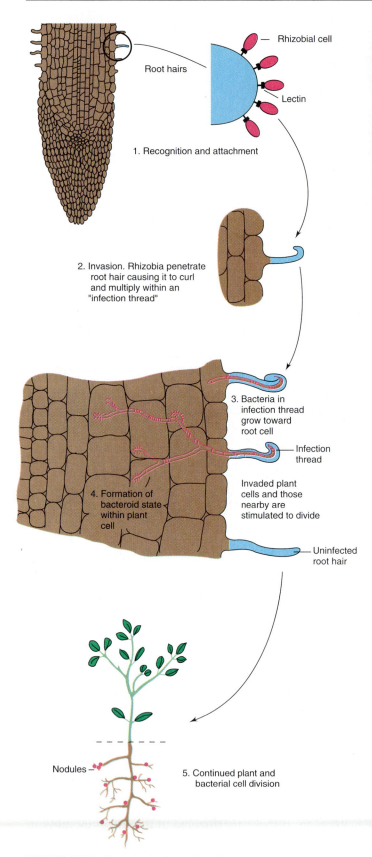

Root hairs

1. Recognition and attachment

Rhizobial cell

Lectin

2. Invasion. Rhizobia penetrate root hair causing it to curl and multiply within an "infection thread"

3. Bacteria in infection thread grow toward root cell

Infection thread

4. Formation of bacteroid state within plant cell

Invaded plant cells and those nearby are stimulated to divide

Uninfected root hair

Nodules

5. Continued plant and bacterial cell division

FIGURE 17.69 Steps in the formation of a root nodule in a legume infected by *Rhizobium*.

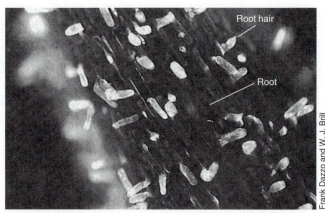

Frank Dazzo and W. J. Brill

Root hair

Root

FIGURE 17.70 Specific attachment to clover root hairs of the capsular lipopolysaccharide from *Rhizobium trifolii*. The LPS extracted from the bacteria was rendered fluorescent by conjugation with fluorescein isothiocyanate, and after treating roots with the preparation, they were viewed under a fluorescence microscope. The root hair tips are especially heavily stained.

an extensive capsule and it has been shown that there is a specific binding of the bacterial cells to the root hairs of white clover via capsular polysaccharide. *R. trifolii* does not bind strongly to root hairs of other legumes, and other species of *Rhizobium*, not infective for white clover, do not bind strongly to white clover root hairs. A substance can be extracted and purified from white clover roots that specifically agglutinates *R. trifolii* cells, and this substance, a **lectin** called *trifolin* is the agent that binds *R. trifolii* capsular polysaccharide and attached cells to the root hair (Figure 17.70). Lectins are plant proteins with a high affinity for specific sugar residues. Root lectins which interact specifically with their respective *Rhizobium* species have also been reported for soybean, sweet clover, and pea.

Binding in the *R. trifolii*–clover symbiosis involves a polysaccharide portion of the *R. trifolii* outer LPS layer, which contains the sugar 2-deoxyglucose. If 2-deoxyglucose is added to roots, the binding of *R. trifolii* is specifically blocked. The interaction of the polysaccharide capsular material from *R. trifolii* with the root hair surface has been studied by converting it into a fluorescent derivative and then observing fluorescence of root hairs when treated with this material. As seen in Figure 17.70, the root hairs bind this material more strongly than do the cells beneath the root hairs, and the binding seems to be especially strong near the root hair tips. In addition to 2-deoxyglucose, a cyclic polysaccharide containing β-1,2 linked glucose units is also necessary for effective nodulation to occur.

It is known that the initial penetration of *Rhizobium* cells into the root hair is via the root hair tip. Following binding, the root hair curls and bacteria enter the root hair and induce formation by the plant of a cellulosic **infection thread**, which spreads down the root hair (Figure 17.71). Root cells adjacent to the root hairs subsequently become infected by rhizobia. If these cells are normal plant diploid cells they are usually destroyed by the infection, undergoing necrosis and degeneration; if they are *tetraploid* cells, however,

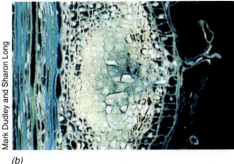

Ben B. Bohlool

(a)

Mark Dudley and Sharon Long

(b)

FIGURE 17.71 (a) An infection thread formed by *Rhizobium trifolii* on white clover (*Trifolium repens*). A number of bacteria can still be seen attached to the root hair. The infection thread consists of a cellulose tube through which the bacteria move to the cortical cells of the root. Cells of *R. trifolii* are about 2 μm long. (b) Emerging alfalfa root nodule, resulting from *Rhizobium meliloti* infection. The infection thread can be seen as a blue staining cell in the upper right of the photograph.

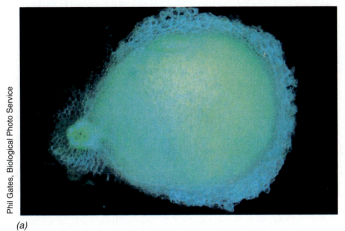

Phil Gates, Biological Photo Service

(a)

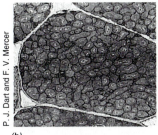

P. J. Dart and F. V. Mercer

(b)

FIGURE 17.72
(a) Cross section through a legume root nodule, as seen by fluorescence microscopy. The darkly stained region contains plant cells filled with bacteria. (b) Electron micrograph of a thin section through a single bacteria-filled cell of a subterranean clover nodule. Cells of *R. trifolii* are about 2 μm long.

they can become the forerunner of a nodule. There are always a small number of tetraploid cells of spontaneous origin in the root, and if one of these cells becomes infected with *Rhizobium* cells, it is stimulated to divide. Progressive divisions of such infected cells leads to the production of the tumorlike nodule (Figures 17.66 and 17.72*a*). In culture, rhizobia produce substances called *cytokinins* which cause tetraploid cells to divide, and it is likely that production of cytokinins also occurs in the infected cells.

The bacteria multiply rapidly within the tetraploid cells and are transformed into swollen, misshapen, and branched forms called **bacteroids**. Bacteroids become surrounded singly or in small groups by portions of the plant cell membrane, called the *peribacteroid membrane*. Only after the formation of bacteroids does nitrogen fixation begin. (Effective nitrogen-fixing nodules can be detected by acetylene reduction, see Section 16.25.) Eventually the nodule deteriorates, releasing bacteria into the soil. The bacteroid forms are incapable of division, but there are always a small number of dormant rod-shaped cells present. These now proliferate, using some of the products of the deteriorating nodule as nutrients, and the bacteria can initiate the infection in other roots or maintain a free-living existence in the soil.

Biochemistry of nitrogen fixation in nodules

As discussed in Section 16.26, nitrogen fixation involves the activity of the enzyme **nitrogenase**, a large two-component protein containing iron and molybdenum. Nitrogenase in root nodules has characteristics similar to the enzyme from free-living N_2-fixing bacteria, including O_2 sensitivity and ability to reduce acetylene as well as N_2. Nitrogenase is localized within the bacteroids themselves and is not released into the plant cytosol.

Bacteroids are totally dependent on the plant for supplying them with energy sources for N_2 fixation. The major organic compounds transported across the peribacteroid membrane and into the bacteroid proper are citric acid cycle intermediates, in particular the C-4 acids *succinate, malate,* and *fumarate* (Figure 17.73). These serve as electron donors for ATP production and, following conversion to pyruvate, as the ultimate source of electrons for the reduction of N_2.

The first stable product of N_2 fixation is *ammonia*, and several lines of evidence suggest that assimilation of ammonia into organic nitrogen compounds in the root nodule is primarily carried out by the plant. Although bacteroids can assimilate some ammonia into organic form, the levels of ammonia-assimilatory enzymes in bacteroids are quite low. By contrast, the am-

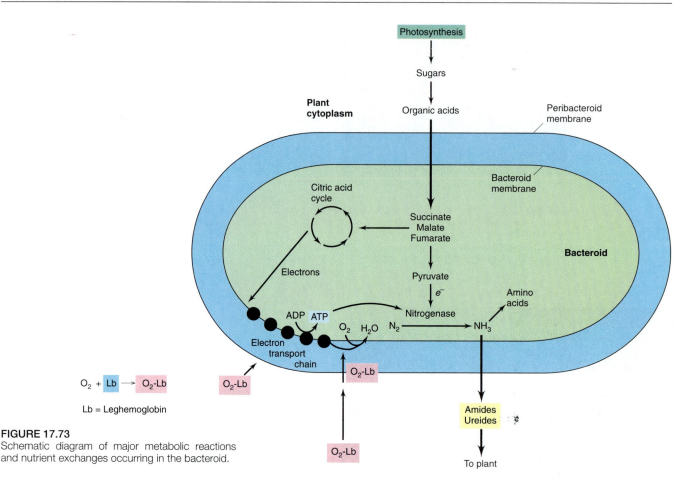

O₂ + Lb → O₂-Lb

Lb = Leghemoglobin

FIGURE 17.73
Schematic diagram of major metabolic reactions and nutrient exchanges occurring in the bacteroid.

monia assimilating enzyme *glutamine synthetase* (see Figure 16.50) is present in high levels in the plant cell cytoplasm. Hence, ammonia, transported from the bacteroid to the plant cell can be assimilated by the plant as the amino acid *glutamine*. Besides glutamine, other nitrogenous compounds, in particular other amino acid amides, such as asparagine and 4-methylene glutamine, and the ureides *allantoin* and *allantoic acid*, are synthesized by the plant and subsequently transported to plant tissues (see Figure 17.73).

Genetics of nodule formation: *nod* genes

Genes directing specific steps in nodulation of a legume by a strain of *Rhizobium* have been called *nod genes*. Many *nod* genes from different *Rhizobium* species are highly conserved and are borne on large plasmids called *Sym plasmids*. In addition to *nod* genes which direct specific nodulation events, Sym plasmids contain *specificity genes*, which restrict a strain of *Rhizobium* to a particular host plant. Indeed, cross-inoculation group specificity can be transferred across species of *Rhizobium* by simply transferring its Sym plasmid. For example, when the Sym plasmid of *R. leguminosarum* (whose host is the pea) is transferred to *R. trifolii* (whose host is clover), cells of the latter species will effectively nodulate pea. Interestingly, Sym plasmids can also be transferred between bacterial genera. If a Sym plasmid is transferred from *Rhizobium* to the closely related bacterium *Agrobacterium*, the latter organism forms root nodules. Although pure cultures of

Agrobacterium will fix N₂, it is not clear whether *Agrobacterium*-induced nodules will fix N₂.

In the Sym plasmid of *R. leguminosarum*, *nod* genes are located between two clusters of genes for nitrogen fixation, the *nif* genes (in this species and in certain other *Rhizobium* species, *nif* genes are plasmid borne). The arrangement of *nod* genes in the *R. leguminosarum* Sym plasmid is shown in Figure 17.74. Ten *nod* genes have been identified in this species. The entire *nod* region has been sequenced and the function of many *nod* proteins is known. The genes *nodABC* direct root hair curling by production of specific polysaccharide signal molecules called *NOD factors*. The *nodD* gene encodes a regulatory protein that controls expression of other *nod* genes. In the presence of specific nodulation signals (see below), the *nodD* gene product controls transcription of other *nod* genes. In most rhizobial species the NodD protein is constitutive and regulates transcription of other *nod* functions by a simple induction type of control (see below and Section 5.10). The NodF protein is involved in the synthesis of a lipid or polysaccharide needed for effective nodulation and is thought to be involved in the transport of substances needed to begin nodulation. The *nodE* and *nodL* genes function in determining host range (cross-inoculation groups, see Table 17.8), *nodM* encodes a protein involved in purine biosynthesis (whose link to nodulation is not clear), and *nodI* and *nodJ* encode membrane proteins whose function is unknown.

Cloning and sequencing of *nod* genes in *R. legumi-*

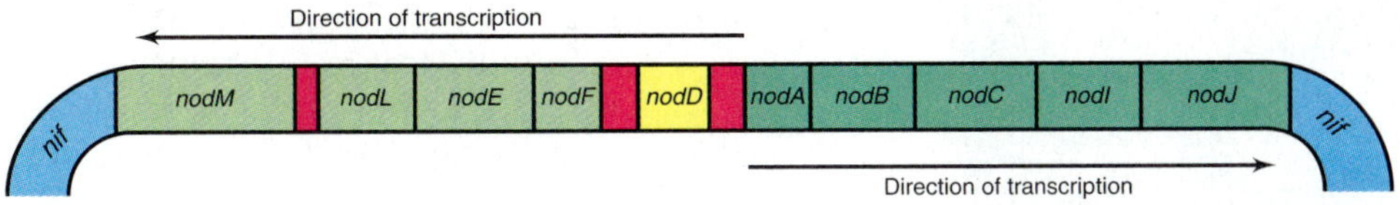

Direction of transcription

| nif | nodM | nodL | nodE | nodF | nodD | nodA | nodB | nodC | nodI | nodJ | nif |

Direction of transcription

FIGURE 17.74 Organization of the *nod* gene cluster on the Sym plasmid of *Rhizobium leguminosarum*. The product of the *nodD* gene controls transcription of other *nod* genes. The *nod* boxes are highlighted and the arrows indicate the direction of transcription of *nod* genes. See text for further details.

nosarum suggests that they are arranged in at least three operons (Figure 17.74). As a regulatory protein, the NodD protein binds to a highly conserved 35 base-pair DNA sequence which preceeds each operon. These regions, referred to as *nod boxes,* are presumably operator genes to which the NodD protein can bind. Following interaction with inducer molecules, the conformation of the NodD protein changes, causing it to fall off the nod box DNA. This event presumably exposes a promoter site for RNA polymerase binding which initiates transcription of *nod* structural genes. Several inducer molecules have been identified. In most cases these are plant *flavonoids,* complex organic molecules that are widespread plant products (Figure 17.75a and b). Flavonoids have many functions in plants, including growth regulation and attraction of pollinating animals. However, leguminous plants are unusual because unlike other plants, their *roots* secrete large amounts of flavonoids, presumably to trigger *nod* gene expression in nearby rhizobial cells in the soil. Interestingly, some flavonoids that are structurally very closely related to *nodD* inducers (Figure 17.75c) strongly *inhibit* induction of *nod* genes in certain rhizobial species, suggesting that part of the specificity observed between plant and bacterium in the *Rhizobium*–legume symbiosis could lie in the chemical nature of the flavonoids excreted by a particular plant.

Genetic cooperativity in the *Rhizobium*–legume symbiosis

Other genetically directed functions crucial to the *Rhizobium*–legume symbiosis have also been identified. As previously noted, the important O_2-binding protein in the root nodule, *leghemoglobin,* is genetically encoded in part by both the plant and the bacterium. The globin (protein) portion of leghemoglobin is encoded by plant DNA while heme synthesis is coded for by bacterial genes. Nitrogen fixation itself is clearly a process genetically directed by the bacterium; **nif** genes reside either on Sym plasmids or, in some *Rhizobium* species, on the chromosome itself. By contrast, lectin and flavonoid synthesis are clearly genetic properties of the plant.

The gene for the enzyme hydrogenase, the *hup* gene, is bacterial and is frequently encoded on the Sym plasmid. The role of hydrogenase in rhizobia is to incorporate H_2 evolved by the activity of nitrogenase (see Section 16.26). Screening of wild-type strains of various species of *Rhizobium* have shown that only

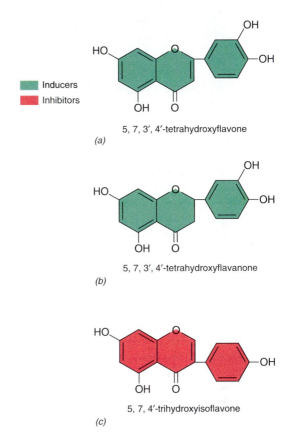

Inducers
Inhibitors

5, 7, 3′, 4′-tetrahydroxyflavone
(a)

5, 7, 3′, 4′-tetrahydroxyflavanone
(b)

5, 7, 4′-trihydroxyisoflavone
(c)

FIGURE 17.75 Structure of flavonoid molecules that act as (a) and (b) inducers of *nod* gene expression, and (c) inhibitors of *nod* gene expression in *Rhizobium leguminosarum*. Note the similarities in the structures of all three molecules. The common name of the structure shown in (a) is *luteolin* and it is a flavone derivative. The structure in (b) is called *eriodictyol* and is chemically a flavanone. The structure in (c) is called *genistein* and it is an isoflavone derivative. Note also the similarity in structure between these molecules and the antibacterial molecules shown in Figure 17.59.

about 20 percent of all *R. leguminosarum* and *B. japonicum* strains contain *hup* genes and produce hydrogenase, and that this important enzyme is totally absent from field strains of *R. meliloti* and *R. trifolii*. Yield studies have shown that strains of rhizobia containing hydrogenase increase total plant nitrogen levels significantly over strains that lack hydrogenase activity because some of the otherwise wasted H_2 (from the activity of nitrogenase) is recycled. This finding has stimulated efforts to genetically engineer *hup* genes into all strains of *Rhizobium* used for field inoculation.

Stem-nodulating rhizobia

Although most leguminous plants form nitrogen-fixing nodules on their *roots,* a few legume species bear

FIGURE 17.76 Stem nodules caused by stem-nodulating *Rhizobium*. The photograph shows the stem of the tropical legume *Sesbania rostrata*. On the left side of the stem are uninoculated sites, on the right identical sites inoculated with stem-nodulating rhizobia.

nodules on their *stems*. Stem-nodulated leguminous plants are quite wide-spread in tropical regions where soils are often nitrogen deficient due to leaching and intense biological activity. Although thought to be members of the genus *Rhizobium*, stem-nodulating bacteria do not form root nodules on any normally root-nodulated legumes. Stem nodules usually form in the submerged portion of the stems or on portions of the stem just above the water level (Figure 17.76). From studies carried out thus far, the general sequence of events in formation of stem nodules strongly resembles that of root nodules. An infection thread is formed and bacteroid formation predates the N_2-fixing state.

One curious finding is that some stem-nodulating rhizobia produce bacteriochlorophyll *a* and thus may have the potential to carry out anoxygenic photosynthesis. Bacteriochlorophyll-containing rhizobia are ap-

parently widespread in nature, particularly in association with tropical legumes, but it is not yet clear whether the bacteriochlorophyll-containing cells are actually capable of photophosphorylation (light-mediated ATP synthesis) (see Section 16.4). If so, this would significantly alleviate the energy drain on the plant imposed by symbiotic N_2 fixation because light could drive at least part of the energy-demanding process of N_2 fixation in the bacterium.

Nonlegume nitrogen-fixing symbioses

In addition to the legume–*Rhizobium* relationship, nitrogen-fixing symbioses occur in a variety of nonleguminous plants, involving microorganisms other than rhizobia. Nitrogen-fixing cyanobacteria form symbioses with a variety of plants. The water fern *Azolla* contains a species of heterocystous N_2-fixing cyanobacteria called *Anabaena azollae* within small pores of its fronds (Figure 17.77). *Azolla* has been used for centuries to enrich rice paddies with fixed nitrogen. Before planting rice, the farmer allows the surface of the rice paddy to become densely covered with the fern. As the rice plants grow they eventually crowd out the *Azolla–Anabaena* mixture, leading to death of the fern and release of fixed nitrogen which is assimilated by the rice plants. By repeating this process each growing season, the farmer can obtain high yields of rice without the need for nitrogenous fertilizers.

The alder tree (genus *Alnus*) has nitrogen-fixing root nodules (Figure 17.78) which harbor a filamentous, streptomycete-like organism called *Frankia*. *Frankia* grows very slowly in the laboratory but has been shown to fix nitrogen in culture. Although when assayed in cell extracts the nitrogenase of *Frankia* is sensitive to molecular oxygen, like intact cells of *Azotobacter*, intact cells of *Frankia* fix N_2 at full oxygen tensions. This is because *Frankia* protects its nitrogenase by localizing it in terminal swellings on the cells called

(a)

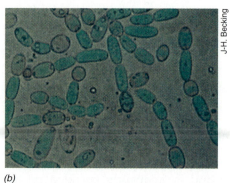

(b)

FIGURE 17.77 *Azolla–Anabaena* symbiosis. (a) Intact association showing a single plant of *Azolla pinnata*. The diameter of the plant is approximately 1 cm. (b) Cyanobacterial symbiont *Anabaena azollae* as observed in crushed leaves of *A. pinnata*. Single cells of *Anabaena azollae* are about 5 μm wide. Note the oval-shaped *heterocysts* (lighter color), the site of nitrogen fixation in the cyanobacterium.

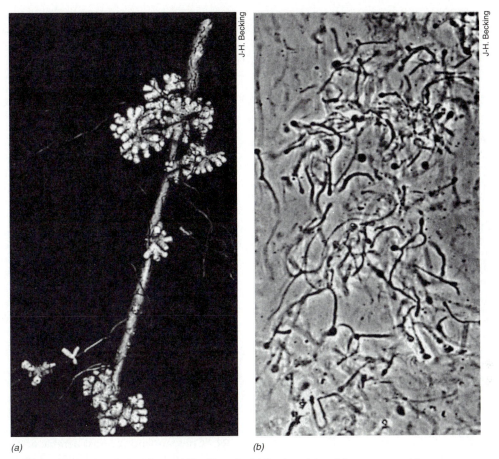

J-H. Becking

J-H. Becking

(a)   *(b)*

FIGURE 17.78 *Frankia* nodules and *Frankia* cells. (a) Root nodules of the common alder *Alnus glutinosa*. (b) *Frankia* culture purified from nodules of *Comptonia peregrina*. Note vesicles (spherical structures) on the tips of hyphal filaments.

vesicles (Figure 17.78*b*). The vesicles contain thick walls that act as a barrier to O_2 diffusion, thus maintaining the O_2 tension within vesicles at levels compatible with nitrogenase activity.

Alder is a characteristic pioneer tree, able to colonize bare soils in nutrient-poor sites, and this is likely due to its ability to enter into a symbiotic nitrogen-fixing relationship with *Frankia*. A number of other small woody plants are nodulated by *Frankia*. This root nodule symbiosis has been reported in at least eight families of plants, many of which show no evolutionary relationships to one another. This suggests that the nodulation process in the *Frankia* symbiosis may be more of a generalized phenomenon than the highly specific process known in the *Rhizobium*–legume symbiosis.

Azospirillum lipoferum is a N_2-fixing bacterium that lives in a rather casual association with roots of tropical grasses. It is found in the rhizosphere, where it grows on products excreted from the roots. It also has the ability to grow around the roots of cultivated grasses, such as corn (*Zea mays*), and inoculation of corn with *A. lipoferum* may lead to small increases (about 10 percent) in growth yield of the plant. Recent evidence suggests that *Azospirillum* can actually penetrate roots and grow *inter*cellularly between root cells. In such an environment it is likely that considerable

fixed nitrogen and organic carbon flow from bacterium to plant and plant to bacterium, respectively. It should be recalled that the *Rhizobium*–legume symbiosis involves *intra*cellular (rather than *inter*cellular) growth of the bacterium. It is likely that many casual relationships of the *Azospirillum* type exist between bacteria and plants, although only in the case of the *Rhizobium* or *Frankia* nodule has direct evidence for a symbiotic relationship been found.

One of the most widespread and important plant–microbial relationships is that between legumes and certain nitrogen-fixing bacteria. The bacteria induce the formation of root nodules within which the nitrogen-fixing process occurs. Under normal conditions, neither the plant nor the bacterium alone can fix nitrogen; only when the two are associated can this process occur. Details of the molecular biology of the bacterium–legume symbiosis are well known and the process is under strict genetic control. The plant provides the organic energy source needed by the root nodule bacteria, and the bacteria provide fixed nitrogen for the growth of the plant. In this way, plants with root nodules can often grow in nitrogen-poor environments where other plants are unable to compete. The root nodule bacteria play an important agricultural role because many important crop plants are legumes (alfalfa, clover, soybeans, peas, etc.).

Study Questions

1. Compare and contrast a lake ecosystem with a hydrothermal vent ecosystem. How does energy enter each ecosystem? What is the basis of primary production in each ecosystem? What nutritional classes of organisms exist in each ecosystem and how do they feed themselves?

2. What is the basis of the *enrichment culture technique*? Why is an enrichment medium usually only suitable for the enrichment of a certain group or groups of bacteria?

3. What is the principle of the *Winogradsky column* and what types of organisms does it serve to enrich? How might a Winogradsky column be used to study the breakdown of a xenobiotic compound?

4. Compare and contrast the use of fluorescent dyes and fluorescent antibodies for use in enumerating bacteria in natural environments. What advantages and limitations do the use of each have?

5. Can nucleic acid probes in microbial ecology be as sensitive as culturing methods? If so, how? What advantages do nucleic acid probe methods have over culture methods? What disadvantages?

6. What are the major advantages of *radioisotopic methods* for the study of microbial ecology? What type of controls (discuss at least two) would you include in a radioisotopic experiment to show $^{14}CO_2$ incorporation by phototrophic bacteria or to show $^{35}SO_4^{2-}$ reduction by sulfate-reducing bacteria?

7. What information is obtained from knowledge of the *biochemical oxygen demand* (BOD) of a water source? Which should have a higher BOD, drinking water or raw sewage? Why?

8. How is the *deep sea* defined and why is an understanding of its biology of importance?

9. What is the difference between *barotolerant* and *barophilic* bacteria? Between these two groups and *extreme barophiles*? What properties do barotolerant, barophilic, and extremely barophilic bacteria have in common?

10. ^{14}C-labeled cellulose is added to a vial containing a small amount of sewage sludge and sealed under anaerobic conditions. A few hours later $^{14}CH_4$ appears in the vial. Discuss what has happened to yield such a result.

11. How can organisms such as *Syntrophobacter* and *Syntrophomonas* grow when their metabolism is based on thermodynamically unfavorable reactions? How are these organisms grown in laboratory culture?

12. Why is sulfate reduction the main form of anaerobic respiration in marine environments whereas methanogenesis dominates in fresh waters? Does any methanogenesis occur in the marine environment? If so, how?

13. What is the *rumen* and how do the digestive processes operate in the ruminant digestive tract? What are the major benefits and the disadvantages of a rumen system? How does a cecal animal compare with a ruminant?

14. Compare and contrast the microbiological steps involved in the conversion of cellulose to methane in the rumen as compared to cellulose conversion in a sewage digestor or in lake sediments. What organisms are involved in each ecosystem and why?

15. Why can urea or ammonia serve as a nitrogen source for ruminants but not for humans?

16. Compare and contrast the processes of *nitrification* and *denitrification* in terms of the organisms involved, the environmental conditions which favor each process, and the changes in nutrient availability which accompany each process.

17. What organisms are involved in cycling sulfur compounds anaerobically? If sulfur chemolithotrophs had never evolved, would there be a problem in the microbial cycling of sulfur compounds? What *organic* sulfur compounds are of interest in nature?

18. Why are all iron-oxidizing chemolithotrophs obligate aerobes and why are most iron oxidizers acidophilic?

19. Explain how spontaneous chemical reactions can acidify a coal seam and how both chemical reactions and *Thiobacillus ferrooxidans* continues the production of acid thereafter.

20. How is *Thiobacillus ferrooxidans* useful in the mining of copper ores? What crucial step in the *indirect* oxidation of copper ores is carried out by *Thiobacillus ferrooxidans*? How is copper recovered from copper solutions produced by leaching?

21. What physical and chemical conditions are necessary for the rapid microbial degradation of oil in aquatic environments? Design an experiment that would allow you to test what conditions optimized the oil oxidation process.

22. What are *xenobiotic compounds* and why might microorganisms have difficulty catabolizing them?

23. Why is sewage that has received only primary treatment both a serious health hazard and a source of environmental pollution?

24. Compare and contrast aerobic and anaerobic means of *secondary* sewage treatment. What types of compounds are best degraded by each method? What are the major products of the treatment process in each case?

25. What methods are used by plants to resist attack by pathogens? How do these differ from the substances produced by animals for the same purpose?

26. Compare and contrast the production of a plant tumor by *Agrobacterium tumefaciens* and a root nodule by a *Rhizobium* species. In what ways are these structures similar? In what ways are they different? Of what importance are plasmids to the development of both structures?

27. What genetic information resides on the *Ti plasmid*? Which gene(s) could be deleted from the Ti plasmid without affecting tumorigenesis?

28. Describe the steps involved in tumorigenesis by *Agrobacterium tumefaciens*.

29. What ecological advantages do leguminous plants have over nonlegumes?

30. What substances of plant origin are found in root nodules of legumes that are required for the rhizobial symbiont to fix nitrogen?

31. Describe the steps in the development of root nodules on a leguminous plant. What is the nature of the recognition between plant and bacterium? How does this compare with recognition in the *Agrobacterium*–plant system?

32. How does nitrogen fixed by *Rhizobium* become part of the plant's proteins?

33. Describe the *nod* operons of a typical *Rhizobium* species such as *R. leguminosarum*. Explain how *nod* genes are regulated.

34. Explain why both obligately anaerobic and obligately aerobic bacteria could be isolated from the same soil sample.

35. You wish to identify in soil samples any organism capable of autotrophic growth using the Calvin cycle. Keeping in mind that such autotrophs contain a unique enzyme, ribulose-bisphosphate carboxylase, design a procedure that could be used to identify such organisms in nature.

36. Suppose you discovered a new animal that consumed only grass in its diet. You suspect it to be a ruminant and have available a specimen for anatomical inspection. If this animal were a ruminant, describe the position and basic components of the digestive tract you expect to find, and any key microorganisms and substances you might look for.

37. Why is bioremediation a more attractive method of

cleaning up oil spills or other toxic wastes than more conventional methods of attacking these problems?

38. Compare and contrast lichens and mycorrhizae in terms of the symbiotic relationship that each exhibits.

39. What are microbial plastics and why are they of interest?

40. Assuming that all necessary molecular tools were available, design an experiment that would allow you to determine whether a genetically engineered bacterium you introduced into the environment was becoming successfully established.

41. Describe how an organism such as *Desulfomonile* is of benefit to solving environmental pollution problems.

Supplementary Readings

Ahmadjian, V. 1993. *The Lichen Symbiosis.* John Wiley & Sons, Inc., New York. The most complete reference to experimental studies on lichens.

Andrews, J. H. 1991. *Comparative Ecology of Microorganisms and Macroorganisms.* Springer-Verlag, New York. A comparative view of basic ecological principles as they pertain to organisms of dramatically different size.

Atlas, R. M., and **R. Bartha.** 1993. *Microbial Ecology—Fundamentals and Applications.* The Benjamin/Cummings Publishing Co., Redwood City, CA. An elementary textbook of microbial ecology.

Ford, T. E. (ed.) 1993. *Aquatic Microbiology: An Ecological Approach.* Blackwell Scientific Publishing, Cambridge, MA. A series of chapters on various aspects of microorganisms and their activities in aquatic ecosystems.

Fry, J. C., G. M. Gadd, R. A. Herbert, C. W. Jones, and **I. A. Watson-Craik** (eds.) 1992. *Microbial Control of Pollution.* Cambridge University Press, New York. Examples of the uses of microorganisms in various biodegradation processes.

Hobson, P. N. (ed.) 1988. *The Rumen Microbial Ecosystem.* Elsevier Applied Science, London. A detailed treatment of the organisms and microbial processes of the rumen.

Lederberg, J. (ed.) 1992. *Encyclopedia of Microbiology.* Academic Press, San Diego. Several good chapters on activities of microorganisms in nature.

Paul, E. A., and **F. E. Clark.** 1989. *Soil Microbiology and Biochemistry.* Academic Press, New York. An advanced undergraduate textbook with an emphasis on chemical and physical processes.

Poole, R. K., and **G. M. Cadd.** 1989. *Metal-Microbe Interactions.* IRL Press, Oxford, England. A short book discussing microbial interactions with various metals and mechanisms of toxic metal resistance.

Rogers, J. E., and **W. B. Whitman** (eds.) 1991. *Microbial Production and Consumption of Greenhouse Gases.* American Society for Microbiology, Washington, D.C. A consideration of production and sinks for methane, nitrogen oxides, and halogenated gases in natural ecosystems.

Saltzman, E. S., and **W. J. Cooper** (eds.) 1989. *Biogenic Sulfur in the Environment.* American Chemical Society, Washington, D.C. An ACS series publication focusing on biological aspects of the sulfur cycle.

18

Molecular Systematics and Microbial Evolution

A recurrent theme throughout this book is the enormous diversity of microorganisms. How did such diversity arise? It should be clear by now that the broad morphological, physiological, and ecological characteristics of the various groups of microorganisms are ultimately controlled by the *genetic* constitution of the organisms interacting with the *environments* of which they are a part. This microbial diversity arose as a result of evolution: mutation and genetic recombination operating on microorganisms capable of growth in various habitats. As new microbial habitats arise, they are subject to microbial colonization, either by preexisting organisms or by new organisms that have arisen by genetic means. Those organisms best able to grow and compete will become established in these new habitats. Thus, the environment is constantly selecting for the best adapted organisms. The great physiochemical diversity of habitats on earth allows for the great microbial diversity we see today.

In this chapter we consider the properties of the early earth upon which life arose and the subsequent evolution and diversification of microorganisms due to their continual growth, genetic change, and dispersal. We also discuss in detail the molecular tools that have been used to determine phylogenetic relationships among present day living organisms.

18.1 Evolution of the Earth and Earliest Life Forms

Origin of the earth

The earth is about 4.6 billion years old as determined by radiodating measurements. Our solar system is thought to have been formed about 4.6 billion years ago when a large, very hot star exploded, generating a new star (our sun), and the other components of our galaxy. Although no rocks dating to this period have yet been discovered on earth, rocks dating back to nearly 4 *billion* years ago have been found in several locations on earth. The oldest rocks discovered thus far are those of the Isua formation in Greenland, which date to about 3.8 billion years ago. The Isua rocks are of three types: sedimentary, volcanic, and carbonate. The sedimentary composition of the Isua rocks is of particular evolutionary interest, because from our understanding of how modern sedimentary rocks are formed, the presence of sedimentary rocks 3.8 billion years old strongly suggests that *liquid water* was present at that time. The presence of liquid water in turn implies that conditions on earth at that time were likely to be compatible with life as we know it. Other rocks of ancient origin include the Warawoona series, Towers Formation, and Early Archaean Apex Basalt in western Australia and the Swaziland series in southern Africa; both rock formations are about 3.5 billion years old.

Evidence for microbial life on the early earth

Fossil evidence of microbial life exists in rocks 3.5 billion years old and younger. Most of the microfossils in the oldest rocks resemble simple rod-shaped bacteria. But in certain rocks of this age, stromatolitic microfossils containing morphologically diverse prokaryotes are present in abundance. *Stromatolites* are fossilized microbial mats consisting of layers of filamentous prokaryotes containing trapped sediment (Figure 18.1a and b). We discussed some characteristics of microbial mats in Section 17.5 (see Figure 17.11a). By comparison with modern stromatolites growing in shallow marine basins (Figure 18.1c,d and e) and in hot springs in various locations in the world (Figures 17.11a and 18.1f), it is assumed that ancient stromatolites consisted of filamentous phototrophic bacteria. Although modern stromatolites are frequently composed of filamentous *cyanobacteria*, this would not have been the case in the oldest stromatolites. Because the earth was still anoxic at this time, stromatolites dating from 3 billion years or older were probably made exclusively by *anoxygenic* phototrophs (purple and green bacteria) rather than O_2-evolving cyanobacteria.

(a)

(b)

(c)

(d)

(e)

(f)

FIGURE 18.1 Ancient and modern stromatolites. (a) The oldest known stromatolite, found in a rock around 3.5 billion years old, from the Warrawoona Group in Western Australia. Shown is a vertical section through a laminated, hemispheroidal structure, which has been preserved in the rock. Scale, 10 cm. (b) Stromatolites of conical shape from 1.6 billion year old dolomite rock of the McArthur Basin of the Northern Territory of Australia. (c) Modern stromatolites in a warm marine bay, Shark Bay, Western Australia. (d) Another view of large modern stromatolites from Shark Bay. Note the resemblance to the ancient stromatolites shown in (b). (e) Underwater photograph of modern stromatolites growing in Shark Bay. The diver indicates the scale. Shown are large columns formed by a complex community of diatoms, cyanobacteria, and green algae, to which are attached various macroscopic algae. (f) Modern stromatolites composed of thermophilic cyanobacteria growing in a thermal pool in Yellowstone National Park.

In rocks younger than about 2 billion years old, the morphological diversity of fossil microorganisms is considerably greater than that of 3.5 billion year old rocks. Figure 18.2 shows some photomicrographs of thin sections of rocks containing structures remarkably similar to modern filamentous bacteria. Structures such as those shown in Figure 18.2*a* are surprisingly similar to filamentous green phototrophic bacteria and cyanobacteria and provide strong evidence that prokaryotes as a group had evolved an impressive morphological diversity long before the advent of modern eukaryotic cells.

The primitive earth

The atmosphere of the early earth was devoid of significant amounts of O_2 and hence constituted a *reducing* environment. Besides H_2O, a variety of gases were present, the most abundant being CH_4, CO_2, N_2, and NH_3. In addition, trace amounts of CO and H_2 existed, as well as considerable amounts of sulfide, much of it as FeS. It is also likely that a considerable amount of hydrogen cyanide, HCN, was produced on the early earth when NH_3 and CH_4 reacted chemically to yield HCN. Geochemical estimates of the temperature of the early earth also suggest that it was a much hotter

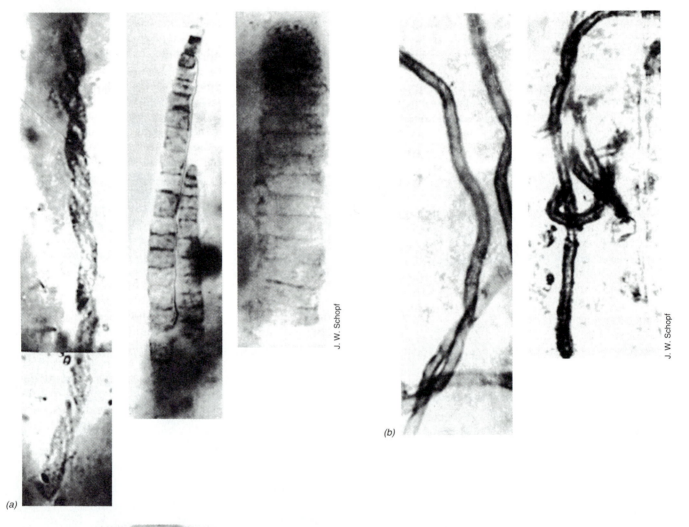

(a)

J. W. Schopf

(b)

J. W. Schopf

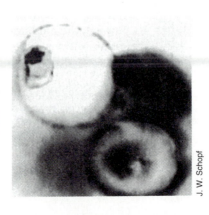

(c)

J. W. Schopf

FIGURE 18.2
Fossil prokaryotes. The five photographs in (a) (magnification, 2000×) and (b) (magnification, 920×) show fossil prokaryotic microorganisms found in the Bitter Springs formation, a rock formation in central Australia about 1 billion years old. These forms bear a striking resemblance to modern cyanobacteria, anoxygenic phototrophs, or filamentous sulfur chemolithotrophs. The two photographs in (c) (magnification, 2000×) show microfossils possibly of a eukaryotic alga. The cellular structure is remarkably similar to that of certain modern green algae, such as *Chlorella* sp. These are from the same rock formation as are the prokaryotic organisms.

Miniglossary for Chapter 18

ARCHAEA   a group of phlyogenetically related prokaryotes distinct from Bacteria

BACTERIA   a group of phylogenetically related prokaryotes distinct from Archaea

DOMAIN   in a taxonomic sense, the highest level of biological classification (see alternative usage in Chapter 12)

ENDOSYMBIOSIS   a theory which states that the mitochondrion and chloroplast were originally free-living Bacteria that established stable residence in primitive eukaryotic cells, eventually yielding the modern eukaryotic cell

EUKARYA   all eukaryotic cells: algae, protozoa, fungi, slime molds, plant and animal cells

EVOLUTIONARY DISTANCE   in phylogenetic trees, the sum of the physical distance on a tree separating organisms; this distance is inversely proportional to evolutionary relatedness

G + C BASE RATIO   in DNA (or RNA) from any organism, the percent of the total nucleic acid which consists of guanine and cytosine bases

GENUS   a collection of different species, each sharing one or more major properties

PHYLOGENY   the evolutionary history of organisms

PROTEOBACTERIA   a term used to describe a large group of phylogenetically related Gram-negative Bacteria

RNA LIFE   a life form lacking DNA and protein which may have existed on early earth and in which RNA served both a genetic coding and a catalytic function

SIGNATURE SEQUENCE   short oligonucleotides of defined sequence in 16S or 18S rRNA characteristic of specific organisms or a group of related organisms

16S rRNA   a large polynucleotide (~1500 bases) which functions as a part of the small subunit of the ribosome of prokaryotes and from whose sequence evolutionary information can be obtained; eukaryotic counterpart, 18S rRNA

SPECIES   in microbiology, a collection of strains which all share the same major properties but which differ in one or more significant properties from other collections of strains

STROMATOLITES   laminated microbial mats, typically built from layers of filamentous and other microorganisms which can become fossilized

TAXONOMY   the science of identification classification and nomenclature of organisms

---

planet than it is today. For the first half billion years of its existence it is likely that the surface of the earth was greater than 100°C; thus free water probably did not exist on the early earth, but only accumulated later as the earth cooled. It is thus likely that life originated on an earth much hotter than it is today, and it therefore follows that the earliest life forms were hyperthermophilic or at least very thermophilic (this has implications for microbial evolution as discussed in Sections 18.7 and 20.8).

Origin of life

It is now well established that the synthesis of biologically important molecules can occur if reducing atmospheres containing the aforementioned gases are subjected to intense energy sources. Of the energy sources available on the primitive earth, the most important was probably ultraviolet (UV) radiation from the sun, but lightning discharges, radioactivity, and thermal energy from volcanic activity were also available. If gaseous mixtures resembling those thought to be present on the primitive earth are irradiated with UV or subjected to electric discharges in the laboratory, a wide variety of biochemically important molecules can be made, such as sugars, amino acids, purines, pyrimidines, various nucleotides, and fatty acids. It has also been shown that under prebiological conditions some of these biochemical building blocks could have polymerized, leading to the formation of polypeptides, polynucleotides, and the like. We can therefore imagine that on the primitive earth a rich mixture of organic compounds eventually accumulated, but in the absence of living organisms these compounds would have been stable and should have persisted for countless years. Thus, with time, there should have been an extensive accumulation of organic materials, setting the stage for biological evolution.

When considering the origin of living organisms, any speculation must be consistent with the physical setting of the early earth (high temperature and anoxic conditions) as well as the chemical realities of abiological macromolecular synthesis. In the latter connection, a major difficulty with most hypotheses of prebiotic synthesis surrounds how macromolecules could have originated spontaneously in an aqueous environment. From a chemical standpoint, nucleic acids and proteins are polymerized via *dehydration* reactions, thus it is difficult to reconcile how macromolecules could have arisen in a completely aqueous setting such as an organic-rich "prebiotic soup." To avoid this problem, it has been hypothesized that relatively anhydrous *exposed surfaces* such as clays, pyrite, or basaltic glasses were the supports for prebiotic polymerization reactions. Such surfaces would have provided a stable, relatively dry environment for the synthesis and accumulation of macromolecules into organic films from which a primitive cellular structure could have emerged. Pyrite in particular is favored for this process because of the crucial role it may have played in early energy-generating systems (see next section).

> The earth is about 4.5 billion years old, and the first evidence of microbial life is found in rocks about 3.5 billion years old. For most of the time that the earth has existed, life on earth has consisted solely of microorganisms. The atmosphere of the primitive earth was reducing and the first biochemically related compounds that formed spontaneously in this anoxic environment led to the formation of the first very primitive cells and set the stage for biological evolution. Surfaces, rather than an aqueous milieu, were likely the sites upon which the first cells arose.

18.2 Primitive Organisms

What was the first living organism like? One can envisage several paths leading to the first living organism. However, even very primitive organisms must have possessed the following: (1) **metabolism**, that is, the ability to accumulate, convert, and transform nutrients and energy; and (2) a **hereditary mechanism**, the ability to replicate and transfer its properties to its offspring. Both of these features as we currently understand them require the development of a *cellular* structure. Such structures probably arose through the spontaneous aggregation of lipid and protein molecules to form membranous structures, within which were trapped polynucleotides, polypeptides, and other substances. This step may have occurred countless times on the early earth to no avail, but just once the proper set of constituents could have become associated, and a primitive organism arose. This primordial organism would certainly have been structurally simple (that is, resembling *prokaryotic* cells) and would have depended on very simple energy-generating and self-replicating mechanisms.

Although no precise time period can be given for when primitive, self-replicating entities first arose, the fact that prokaryotic microfossils are found in rocks 3.5 billion years old (see Figure 18.1), implies that the earliest life forms probably date between 3.5 and 4 billion years ago. Remarkably, this means that after the early earth cooled to the point at which liquid water was present (thought to be about 4.0–4.2 billion years ago), life arose relatively rapidly, perhaps in as little as 200–400 million years.

Metabolism in primitive organisms

Because of the reducing conditions prevailing on the early earth, primitive organisms must have been able to carry out some form of *anaerobic* metabolism. However, from our consideration of metabolic diversity, this constraint rules out very little; early organisms could have used chemoorganotrophic, chemolithotrophic, or phototrophic metabolism because all of these energy-generating processes can occur anaerobically (see Chapter 16).

One potential energy source that has been seriously considered for early organisms is the reaction of ferrous iron, such as ferrous carbonate or ferrous sulfide, with H_2S to form pyrite and H_2:

$$FeCO_3 + 2H_2S \rightarrow FeS_2 + H_2 + H_2O + CO_2$$
$$\Delta G^{0\prime} = -61.7 \text{ kJ/reaction}$$
$$FeS + H_2S \rightarrow FeS_2 + H_2$$
$$\Delta G^{0\prime} = -41.9 \text{ kJ/reaction}$$

Ferrous iron and H_2S are thought to have been abundant on the early earth (see Section 18.1), and thus primitive cells would have had a readily available energy source for initiating energy-yielding metabolism (catabolism). Both of the above reactions are exergonic and yield sufficient free energy to couple to the formation of ATP or other energy-trapping molecules that

may have been used by early cells. Interestingly, both of the reactions shown also yield H_2, and it has been further proposed that H_2 produced from these reactions could have served as an electron donor for reduction of elemental sulfur (S^0) to H_2S (Figure 18.3). Energy coupling in this chemolithotrophic reaction could have occurred by separating H_2 into electrons and protons across a simple cytoplasmic membrane, thereby establishing a proton gradient to drive a primitive ATPase (Figure 18.3). Interestingly, most hyperthermophilic Archaea, which are probably the closest extant relatives of earth's earliest organisms (see Section 18.5), are able to reduce S^0 with H_2 and form H_2S; most can also produce pyrite when supplied with Fe^{2+} (see Section 20.8). Perhaps these reactions are modern remnants of primitive metabolic schemes employed on the early earth by the first living organisms.

In addition to these chemolithotrophic energy-yielding reactions, the earliest cells could also have obtained energy by chemoorganotrophic mechanisms, most likely simple fermentations. Photosynthesis is also a possibility but seems less likely than chemolithotrophic or chemoorganotrophic reactions because of the biochemical requirements and complexity of photosynthetic reactions. The attractive feature of the pyrite-based energy scheme discussed above is that it would probably have required very few enzymes other than a primitive hydrogenase and an ATPase to trap energy released from the reaction (Figure 18.3).

Regardless of the energy sources for life on early earth, the organisms present were undoubtedly biochemically very simple, in the sense of possessing very few enzymes. Nutritional requirements would likely have been complex. No cytochrome system would have been present, and no flagella or other special morphological properties would have been necessary. To make survival in osmotically varying environments possible, a cell wall probably evolved fairly early. The earliest organisms would probably have re-

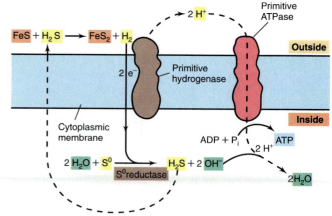

FIGURE 18.3 Hypothetical energy-generating scheme for primitive cells. Formation of pyrite leads to H_2 production and S^0 reduction, which fuels a primitive ATPase. Note how H_2S plays only a catalytic role; the net substrates would be FeS and S^0. Also note how few different proteins would be required. The $\Delta G^{0\prime}$ of the reaction FeS + H_2S → $FeS_2 + H_2 = -41.9$ kJ.

quired a variety of growth factors and other complex organic nutrients. However, as time went on, mutation and selection would have yielded new organisms with greater biosynthetic capacities, better adapted to the changing chemical environment.

Further metabolic evolution and photosynthesis

Another highlight in metabolic evolution would have been the evolution of the first porphyrins since this, along with refinements in membrane structure over time, could have led to the construction of cytochromes and an electron-transport chain able to carry on electron-transport phosphorylation. This could have opened up many new possibilities for anaerobic respiration (see Chapter 16).

The evolution of porphyrins would also have allowed development of light-sensitive magnesium tetrapyroles, the bacteriochlorophylls, and thus photosynthesis. Following this major event, a great explosion of life could have occurred because of the availability of the enormous amount of energy from the radiation of the sun. The first phototrophic organisms were no doubt anoxygenic, probably using light only for ATP synthesis and using reduced compounds from its environment—such as H_2S—as sources of reducing power. Such an organism may have resembled one of the modern day purple or green sulfur bacteria.

A monumental step in microbial evolution occurred with development of the second light reaction of photosynthesis, making it possible to use the plentiful supply of H_2O as an electron donor. Since both ATP and reduced pyridine nucleotides (NADH, NADPH) could now be made photosynthetically, light energy could be used more efficiently. Such organisms would have considerable competitive advantage over other phototrophic organisms. The first of this type was prokaryotic, similar to cyanobacteria. The evolution of oxygenic photosynthesis also had enormous consequences for the environment of the earth since, as O_2 gradually accumulated, the atmosphere changed from a reducing to an oxidizing one (Figure 18.4). With O_2 now available as an electron acceptor, aerobic organisms evolved; these were able to obtain much more energy from the oxidation of organic compounds than could anaerobes (see Chapter 4). More energy was made available, and higher population densities could develop, increasing the chances for the appearance of new types of organisms. There is good evidence from the fossil record that, at about the time that the earth's atmosphere became highly oxidizing, there was an enormous burst in the rate of evolution, leading to the appearance of eukaryotic microorganisms with organelles and from them to metazoans (multi-celled organisms) and eventually to higher animals and plants (see Section 18.8).

Another major consequence of the appearance of O_2 was the formation of ozone (O_3), a substance that provides a barrier preventing the intense ultraviolet radiation of the sun from reaching the earth. When O_2 is subject to short-wavelength ultraviolet radiation, it is converted to O_3, which strongly absorbs wavelengths up to 300 nm. Until an ozone shield devel-

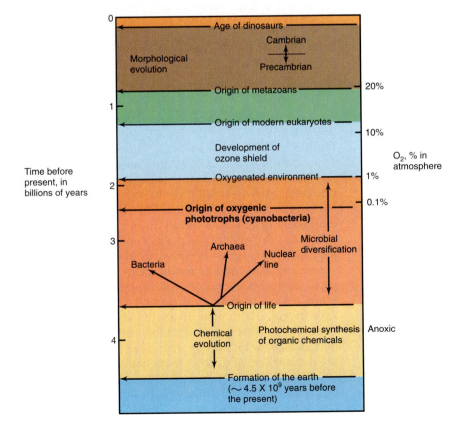

FIGURE 18.4

Major landmarks in biological evolution. The positions of the stages on the time scale are approximate. Note how the oxygenation of the atmosphere was a gradual process, occurring over a period of about a billion years.

oped, evolution could only have continued in environments protected from direct radiation from the sun, such as under rocks or in the oceans. After the photosynthetic production of O_2 and development of an ozone shield, organisms could have ranged generally over the surface of the earth, permitting evolution of a greater diversity of living organisms. A summary of the steps that could have occurred in biological evolution is shown in Figure 18.4.

> The first organism must have had two major properties: metabolism, the ability to convert energy and nutrients, and heredity, the ability to divide and transfer its properties to offspring. Primitive metabolism was anaerobic and was likely either chemoorganotrophic or chemolithotrophic or both. Pyrite (FeS_2) formation might have been an early metabolic strategy. Eventually, porphyrin-based photosynthesis arose. At first, photosynthesis was anoxygenic, but eventually oxygenic photosynthesis led to the development of an oxygenated atmosphere where higher organisms could evolve.

The era of RNA life and evolution of the genome

Following the discovery that certain types of RNA are catalytic (see discussion of ribozymes in Section 5.6), it has been suggested that the earliest life forms may have lacked DNA and proteins and depended on RNA as the sole macromolecule in the cell. That is, RNA would have served in both an informational and an enzymatic context, preceding protein as chief biochemical catalyst in primitive organisms. This then would have been the age of "RNA life." Such life forms would have been very simple by present-day standards and could be considered intermediate structures between nonliving systems and life as we know it today.

How would RNA life forms have accomplished all the reactions we associate with living systems? Without proteins, biochemistry in "RNA-organisms" would by necessity have been very simple compared with that in modern cells. RNA has poor catalytic specificity, and in modern cells those few reactions catalyzed by RNA are biochemically simple compared with typical reactions catalyzed by protein enzymes (see Section 5.6). Thus, before life diversified significantly, a few types of RNA may have been able to catalyze the relatively few reactions required. However, as living organisms became biochemically more complex, an evolutionary push to *proteins* as cellular biocatalysts would probably have occurred. It is likely that proteins appeared gradually in cells, perhaps at first complexed with RNA, and as evolution selected for more rapid and precise catalysts, RNA was eventually replaced by protein as major cellular enzymes.

The establishment of DNA as the genome of the cell may have resulted from the need to store genetic information in a more stable form. By storing all the genetic information in one place in the cell and processing only what was needed under a specific set of growth conditions, cells would have saved energy, which would have increased their competitive fitness. However, an additional reason for the origin of DNA as the genetic material may have been the highly error-prone nature of enzymes that copy RNA. For unknown reasons, enzymes that copy RNA are inherently less precise than DNA polymerases. Maintaining RNA as the genetic material and relying on inherently error-prone copying systems for its replication would thus have been incompatible with increasing cellular complexity and the genetic demands this would entail. Evolution would therefore have favored transfer of genetic storage to a form that had inherently high replicative fidelity, like DNA, and this would have eventually eliminated the less precise RNA life forms.

Somewhere in the early stages of microbial evolution, the three part system—DNA, RNA, and protein—became fixed in cellular life as the best solution to biological information flow. That this system was an evolutionary success can be attested to by the fact that all extant cells contain all three kinds of informational macromolecules. Thus, although modern life forms employ DNA and proteins as essential parts of cellular function, early life forms may have accomplished all of this with only RNA, employing a much simpler system of information storage and a rather nonspecific and inefficient means of biochemical catalysis.

> Primitive organisms may have used RNA in both a genetic and an enzymatic capacity, but since catalytic RNA is not very efficient, evolution of protein catalysts would have led to a marked improvement in cellular efficiency. DNA may have arisen because it provides a more stable form of genetic information than RNA and one that can be copied more accurately. That the DNA, RNA, protein system was an evolutionary success is shown by the fact that modern cells contain all three kinds of informational macromolecules.

18.3 Origin of Modern Eukaryotes

From comparative nucleic acid sequencing studies we now know that the three main lines of descent, *Bacteria*, *Archaea*, and *Eukarya* (nuclear), were established relatively early in cellular evolution (see Section 18.4). However, the eukaryotic cell we know today undoubtedly differs structurally from primitive eukaryotic cells. The nuclear line of descent (primitive eukaryotic cells) originally consisted of structurally quite simple cells, resembling modern day prokaryotic cells in lacking mitochondria, chloroplasts, and a membrane-bound nucleus (see Section 18.8). Modern eukaryotic cells, with their distinctive cellular organelles, were the result of endosymbiotic events (see below) that took place billions of years after the divergence of the nuclear line of descent from the universal ancestor (see Figure 18.4 and Section 18.8).

It seems likely that the eukaryotic nucleus and mitotic apparatus arose as a necessity for ensuring the replication and orderly partitioning of DNA once the genome size had increased to the point where replication as one molecule was no longer feasible. The widespread occurrence of plasmids in bacteria suggests that even in prokaryotes there are evolutionary advantages for segregation of genetic information into more than one DNA molecule. It is thus possible to imagine how separate chromosomes might have arisen in a primitive eukaryotic cell from plasmid-like structures and become segregated within the cell into a membrane-enclosed nucleus. Probably spindle fibers and the mitotic apparatus would also have had to evolve at the same time. There is no obvious reason why this primitive eukaryote would have needed other typical eukaryotic organelles, and these could have arisen later.

Endosymbiosis

Strong evidence now exists that the modern eukaryotic cell evolved in steps through the incorporation into cells from the nuclear line of descent of chemoorganotrophic and phototrophic symbionts. This theory, referred to as the **endosymbiotic theory** of eukaryotic evolution has, through the years, gathered increasing experimental support (see Sections 3.14 and

7.15). The theory postulates that an aerobic bacterium established residency within the cytoplasm of a primitive eukaryote and supplied the larger cell with energy in exchange for a stable, protected environment and a ready supply of nutrients (Figure 18.5). This aerobic bacterium would represent the forerunner of the present eukaryotic mitochondrion. In similar fashion, the endosymbiotic uptake of an oxygenic phototrophic prokaryote would have made the primitive eukaryote photosynthetic and no longer dependent on organic compounds for energy production. The phototrophic endosymbiont would then be considered the forerunner of the chloroplast (Figure 18.5). Following the acquisition of prokaryotic endosymbionts, eukaryotic cells underwent an explosion in biological diversity. The period from about 1.5 billion years ago to the present saw the rise of the metazoans and diversification of this highly successful group culminating in the structurally complex higher plants and animals (see Section 18.8). Studies of the nucleic acids and ribosomes of eukaryotic cells and their organelles (see Sections 5.1 and 18.5) have built an impressive case for the theory of endosymbiosis.

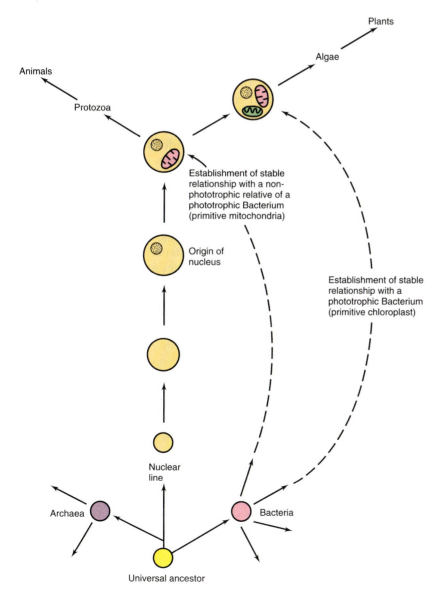

FIGURE 18.5
Origin of modern eukaryotes by endosymbiotic events. Note how eukaryotic cell organelles originated from Bacteria rather than Archaea.

in microbial ecology (see Section 17.4) and in clinical diagnostics (see Section 13.11).

> Ribosomal RNAs are ancient molecules found in all cellular organisms and serve as excellent evolutionary chronometers. Comparisons of sequences of ribosomal RNA can be used to determine the relationships between organisms. Phylogenetic trees based on ribosomal RNA have now been prepared for all of the major prokaryotic and eukaryotic groups.

18.5   Microbial Phylogeny as Revealed by Ribosomal RNA Sequencing

Molecular sequencing has revealed a previously unsuspected microbial phylogeny, a phylogeny quite different from previous ones based primarily on phenotypic relationships. Biologists have often grouped the living world into five kingdoms, only one group of which is prokaryotic, based on structural similarities between organisms. Molecular phylogeny, on the other hand, has revealed that the five kingdoms do not represent five major evolutionary lines. Instead, cellular life on earth has evolved along *three* major lineages, two of which are exclusively microbial and composed only of prokaryotic cells. The third line constitutes the eukaryotic lineage (Figure 18.10). The two prokaryotic lines are the *Bacteria* (formerly *eubacteria*) and the *Archaea* (formerly *archaebacteria*). The eukaryotic line is called the *Eukarya* (Figure 18.10). These terms define the three **domains** of life, the domain supplanting the kingdom as the highest of biological taxons. Thus plants, animals, fungi, and protists are all kingdoms within the domain Eukarya.

The universal tree of life

The universal phylogenetic tree (Figure 18.10) shows the relative evolutionary positions of major groups of living organisms. Interestingly, the tree shows that the Eukarya are not of recent origin but instead are as ancient as either of the prokaryotic lineages. Thus, before endosymbiotic events led to the modern eukaryotic cell (that is, with membrane-bound organelles) (see Section 18.3), organelle-less Eukarya inhabited earth and these organisms shared common ancestry with the other two evolutionary lines (Figure 18.10).

The *root* of the universal tree (Figure 18.10) represents a point in evolutionary time when all extant life on earth shared a common ancestor. The root of the tree has been determined by rRNA sequencing and by related macromolecular sequencing methods, but it clearly indicates that initial evolution from the universal ancestor was at first in *two* directions, the Bacteria

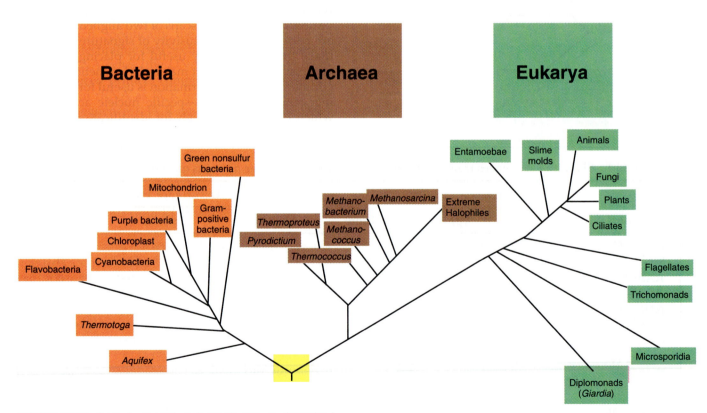

FIGURE 18.10 Rooted universal phylogenetic tree as determined from comparative ribosomal RNA sequencing. The data support the separation of three domains, two of which (Bacteria and Archaea) contain only prokaryotic representatives. The location highlighted in yellow is the root of the tree, which represents the position of the universal ancestor of all cells.

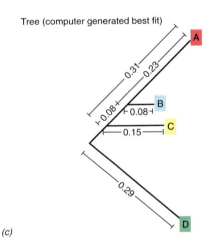

Organism   Sequence

A CGUAGACCUGAC

B CCUAGAGCUGGC

C CCAAGACGUGGC

D GCUAGAUGUGCC

(a)

Evolutionary distance   Corrected evolutionary distance

E_D A ⟶ B   0.25   0.30

E_D A ⟶ C   0.33   0.44

E_D A ⟶ D   0.42   0.61

E_D B ⟶ C   0.25   0.30

E_D B ⟶ D   0.33   0.44

E_D C ⟶ D   0.33   0.44

(b)

Tree (computer generated best fit)

A
0.31
0.23
0.08
B
0.08
C
0.15
0.29
D

(c)

FIGURE 18.9

Preparing a phylogenetic distance-matrix tree from 16S ribosomal RNA sequences. For illustrative purposes, only short sequences are shown, but it should be considered that the sequences shown in (a) are representative of the entire 16S rRNA. The evolutionary distance (E_D) shown in (b) is calculated as the percent of *nonhomologous* sequence between the RNAs of any total two organisms. The corrected E_D is a statistical correction necessary to account for either back mutations to the original genotype or additional forward mutations at the same site that could have occurred. The tree (c) is ultimately generated by computer analysis of the data to give the best fit. The length of the branches separating any two organisms is proportional to the evolutionary distance between them.

amount of evolutionary change needed to diverge two lineages from a common ancestor occurred in their evolutionary divergence. Parsimony analyses thus generate phylogenetic trees based on evolutionary minimization. The appearance and interpretation of trees derived from parisomony analyses, however, is similar to that of distance matrix trees.

Signature sequences

Computer analysis of ribosomal RNA sequences has revealed **signature sequences**, short oligonucleotides unique to a certain group or groups of organisms. Oligonucleotide signatures defining each of the three primary domains (Table 18.1) have been identified. Other signatures defining the major taxa within each domain have also been detected (see Sections 18.6 and 18.7). The "signatures" are generally found in defined regions of the 16S rRNA molecule, but are only revealed when computer scans of two aligned sequences are performed. The exclusivity of signature sequences make them particularly useful for placing unknown organisms in the correct major phylogenetic group. In addition, signature information is useful for constructing genus and species-specific nucleic acid probes, which are used extensively for identification purposes

Table 18.1 Signature sequences from 16S or 18S rRNA defining the three domains of living organisms

Oligonucleotide sequence[a]	Approximate position[b]	Occurrence among[c]		
		Archaea	Bacteria	Eukarya
CACYYG	315	0	>95	0
CYAAYUNYG	510	0	>95	0
AAACUCAAA	910	3	100	0
AAACUUAAAG	910	100	0	100
NUUAAUUCG	960	0	>95	0
YUYAAUUG	960	100	< 1	100
CAACCYYCR	1110	0	>95	0
UUCCCG	1380	0	>95	0
UCCCUG	1380	>95	0	100
CUCCUUG	1390	>95	0	0
UACACACCG	1400	0	>99	100
CACACACCG	1400	100	0	0

[a] Y = any pyrimidine; R = any purine; N = any purine or pyrimidine.
[b] Refer to Figure 18.7 for numbering scheme of 16S rRNA.
[c] Occurrence refers to percentage of organisms examined in any domain.

in microbial ecology (see Section 17.4) and in clinical diagnostics (see Section 13.11).

> Ribosomal RNAs are ancient molecules found in all cellular organisms and serve as excellent evolutionary chronometers. Comparisons of sequences of ribosomal RNA can be used to determine the relationships between organisms. Phylogenetic trees based on ribosomal RNA have now been prepared for all of the major prokaryotic and eukaryotic groups.

18.5 Microbial Phylogeny as Revealed by Ribosomal RNA Sequencing

Molecular sequencing has revealed a previously unsuspected microbial phylogeny, a phylogeny quite different from previous ones based primarily on phenotypic relationships. Biologists have often grouped the living world into five kingdoms, only one group of which is prokaryotic, based on structural similarities between organisms. Molecular phylogeny, on the other hand, has revealed that the five kingdoms do not represent five major evolutionary lines. Instead, cellular life on earth has evolved along *three* major lineages, two of which are exclusively microbial and composed only of prokaryotic cells. The third line con-

stitutes the eukaryotic lineage (Figure 18.10). The two prokaryotic lines are the *Bacteria* (formerly *eubacteria*) and the *Archaea* (formerly *archaebacteria*). The eukaryotic line is called the *Eukarya* (Figure 18.10). These terms define the three **domains** of life, the domain supplanting the kingdom as the highest of biological taxons. Thus plants, animals, fungi, and protists are all kingdoms within the domain Eukarya.

The universal tree of life

The universal phylogenetic tree (Figure 18.10) shows the relative evolutionary positions of major groups of living organisms. Interestingly, the tree shows that the Eukarya are not of recent origin but instead are as ancient as either of the prokaryotic lineages. Thus, before endosymbiotic events led to the modern eukaryotic cell (that is, with membrane-bound organelles) (see Section 18.3), organelle-less Eukarya inhabited earth and these organisms shared common ancestry with the other two evolutionary lines (Figure 18.10).

The *root* of the universal tree (Figure 18.10) represents a point in evolutionary time when all extant life on earth shared a common ancestor. The root of the tree has been determined by rRNA sequencing and by related macromolecular sequencing methods, but it clearly indicates that initial evolution from the universal ancestor was at first in *two* directions, the Bacteria

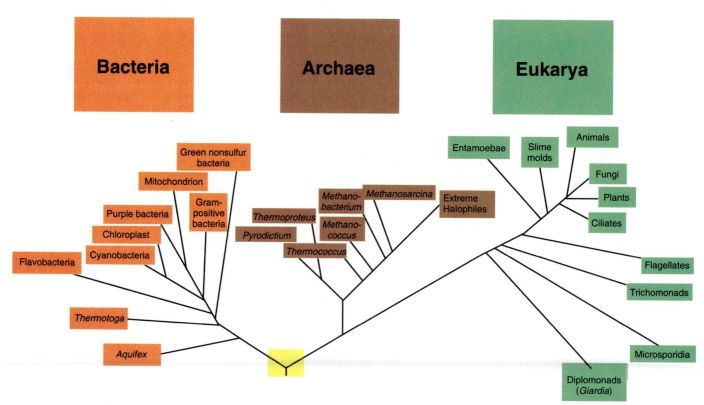

FIGURE 18.10 Rooted universal phylogenetic tree as determined from comparative ribosomal RNA sequencing. The data support the separation of three domains, two of which (Bacteria and Archaea) contain only prokaryotic representatives. The location highlighted in yellow is the root of the tree, which represents the position of the universal ancestor of all cells.

Sequence methodology

Ribosomal RNAs are now relatively easy to sequence because they can be directly sequenced from crude cell extracts using reverse transcriptase and the dideoxy sequencing method (see Nucleic Acids box, Chapter 5). A typical rRNA sequencing experiment involves phenol extraction of a relatively small (0.3–0.5 g) amount of cells to remove protein and other non-nucleic acid material and release the RNA. Total RNA is then precipitated with alcohol and salt, and then a small DNA oligonucleotide primer of about 15–20 nucleotides in length *complementary* in base sequence to some highly conserved region of the 16S rRNA molecule, is added to the mixture (Figure 18.8). The enzyme reverse transcriptase (see box, Chapter 5, and Section 6.22) is then added along with a ^{32}P-labeled deoxyadenosine triphosphate and the other unlabeled deoxyribonucleotides, and then the mixture is divided into four identical portions. To each portion, a small amount of a different 2′,3′-*dideoxynucleotide* is added. Reverse transcriptase will read the 16S rRNA template and begin making a DNA copy terminated at various spots by the incorporation of *dideoxynucleotides*. The fragments are then separated by electrophoresis (see Nucleic Acids box, Chapter 5). From knowledge of the cDNA sequence obtained by this dideoxy sequencing, the sequence of the original 16S rRNA can be deduced.

Although direct sequencing of ribosomal RNA is still used, newer methods are beginning to supplant this approach. Specifically, the polymerase chain reaction (PCR) technique (see Section 8.9) is being used to amplify rRNA *genes* (the DNA itself) using synthetically produced primers complementary to conserved sequences in rRNA as PCR templates (see section on signature sequences). PCR amplification of the DNA coding for rRNA requires less cell material than direct rRNA sequencing and is more rapid and convenient for large scale studies than the direct method. The amplified DNA is then sequenced directly using the dideoxy sequencing method.

Phylogenetic trees from RNA sequences

A major method of generating phylogenetic trees from rRNA sequences is the so-called *distance-matrix method* (Figure 18.9). Two rRNA sequences are aligned and an **evolutionary distance** (E_D) is calculated by recording the number of positions in the sequence at which the two *differ*. A statistical correction factor is then used to account for the possibility that multiple changes might have occurred which would lead back to the same sequence. A matrix of evolutionary distances generated from sequence comparisons is then run through a computer program designed to produce phylogenetic trees from E_D measurements. In effect, this amounts to examining all possible branching arrangements for the set of distances compared, arranging the branch lengths for each branching arrangement to optimally fit the data, and then selecting the best arrangement as the correct phylogenetic tree. The E_D separating any two organisms is directly proportional to the *total*

length of the branches separating them. The principles of phylogenetic tree formation are shown in Figure 18.9 with a sample tree constructed from sequence data from four hypothetical organisms.

Phylogenetic analyses can also be done by using parsimony analysis. **Parsimony** is a method of analyzing sequence data that generates evolutionary trees based on the assumption that only the *minimal*

1. Centrifuge cells (300-500 mg wet weight) and break open cells in the presence of DNAase. All DNA destroyed.

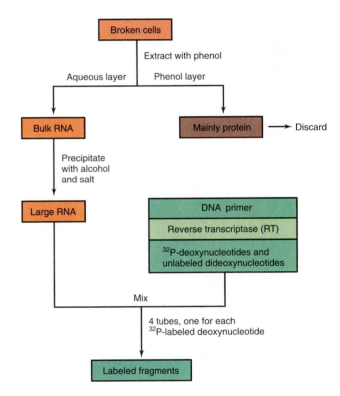

2. Preparation of labeled cDNA fragments for sequence determination

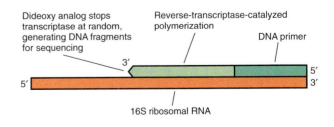

3. Determine sequences by the Sanger method (see Box, Chapter 5) and deduce the ribosomal RNA sequence from the cDNA sequence.

FIGURE 18.8 Steps in sequencing a 16S rRNA using the retroviral enzyme reverse transcriptase and the dideoxy sequencing method.

FIGURE 18.6 Electron micrograph of 70S ribosomes from the bacterium *Escherichia coli*.

Recall the structure of the ribosome (Figure 18.6). There are three ribosomal RNA molecules, which in prokaryotes have sizes of 5S, 16S, and 23S. The large bacterial rRNAs, 16S (Figure 18.7) and 23S rRNA (approximately 1500 and 3000 nucleotides, respectively) contain several regions of highly conserved sequence useful for obtaining proper sequence alignments, yet contain sufficient sequence variability in other regions of the molecule to serve as excellent phylogenetic chronometers (Figure 18.7).

The 5S rRNA has also been used for phylogenetic measurements, but its small size (~120 nucleotides) limits the information obtainable from this molecule. Because 16S RNA is more experimentally manageable than 23S RNA, it has been used extensively to develop the phylogeny of prokaryotes and to some extent, the eukaryotes (using the 18S rRNA counterpart of prokaryotic 16S rRNA). The accumulating data base of rRNA sequences now comprises over 2000 different organisms. Use of 16S rRNA as a phylogenetic tool was pioneered in the early 1970s by Carl Woese at the University of Illinois, and the method is now widely used.

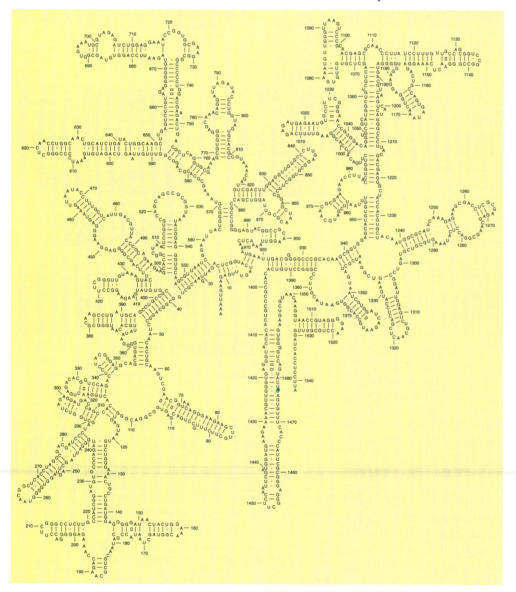

FIGURE 18.7
Primary and secondary structure of 16S ribosomal RNA (rRNA). This is the 16S rRNA from *Escherichia coli* (Bacteria). 16S rRNA from Archaea has general similarities in secondary structure (folding) but numerous differences in primary structure (sequence). The counterpart to 16S rRNA in eukaryotes is 18S rRNA present in cytoplasmic ribosomes.

genome size had increased to the point where replication as one molecule was no longer feasible. The widespread occurrence of plasmids in bacteria suggests that even in prokaryotes there are evolutionary advantages for segregation of genetic information into more than one DNA molecule. It is thus possible to imagine how separate chromosomes might have arisen in a primitive eukaryotic cell from plasmid-like structures and become segregated within the cell into a membrane-enclosed nucleus. Probably spindle fibers and the mitotic apparatus would also have had to evolve at the same time. There is no obvious reason why this primitive eukaryote would have needed other typical eukaryotic organelles, and these could have arisen later.

Endosymbiosis

Strong evidence now exists that the modern eukaryotic cell evolved in steps through the incorporation into cells from the nuclear line of descent of chemoorganotrophic and phototrophic symbionts. This theory, referred to as the **endosymbiotic theory** of eukaryotic evolution has, through the years, gathered increasing experimental support (see Sections 3.14 and

7.15). The theory postulates that an aerobic bacterium established residency within the cytoplasm of a primitive eukaryote and supplied the larger cell with energy in exchange for a stable, protected environment and a ready supply of nutrients (Figure 18.5). This aerobic bacterium would represent the forerunner of the present eukaryotic mitochondrion. In similar fashion, the endosymbiotic uptake of an oxygenic phototrophic prokaryote would have made the primitive eukaryote photosynthetic and no longer dependent on organic compounds for energy production. The phototrophic endosymbiont would then be considered the forerunner of the chloroplast (Figure 18.5). Following the acquisition of prokaryotic endosymbionts, eukaryotic cells underwent an explosion in biological diversity. The period from about 1.5 billion years ago to the present saw the rise of the metazoans and diversification of this highly successful group culminating in the structurally complex higher plants and animals (see Section 18.8). Studies of the nucleic acids and ribosomes of eukaryotic cells and their organelles (see Sections 5.1 and 18.5) have built an impressive case for the theory of endosymbiosis.

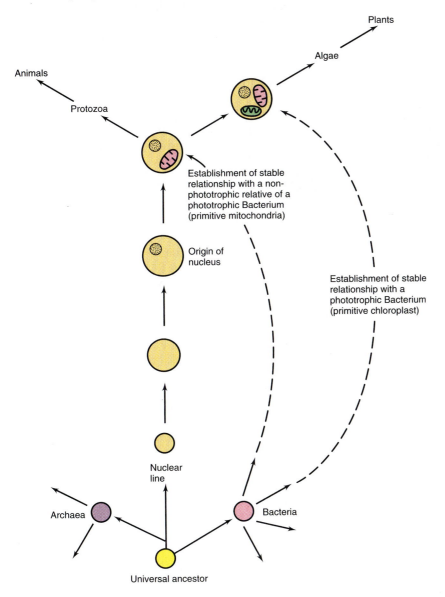

FIGURE 18.5
Origin of modern eukaryotes by endosymbiotic events. Note how eukaryotic cell organelles originated from Bacteria rather than Archaea.

However, to evolve to the modern eukaryotic stage, primitive eukaryotic cells had to abandon several advantageous features of prokaryotic cells such as a haploid genome (and the concomitant ability to adapt rapidly to new environments), and structural simplicity. Also, the greater complexity of the eukaryotic cell meant that it would face difficulty in adapting to life in extreme environments, such as thermal areas, where the prokaryote is today preeminent. For these reasons the evolution of the eukaryote did not toll the death bell for the prokaryotes. All three cell lines continued to evolve, carving out their ecological niches and serving as the forerunners of the various species we know today.

A point that deserves emphasis here is that *none* of the organisms living today are *primitive*. All extant life forms are *modern* organisms, well adapted to, and successful in, their ecological niches. Certain of these organisms may indeed be phenotypically similar to primitive organisms and may represent stems of the evolutionary tree that have changed little for millions if not billions of years (see Section 20.9); in this respect they are evolutionarily *related* to primitive organisms, but they are not themselves primitive.

Biological evolution and geological time scales

In terms of geological time, the period from the origin of the first metazoans to the present represents only one-sixth of the total time that life has existed on earth. Put another way, five-sixths of the earth's history was restricted to *microbial* life, the bulk of this period to *prokaryotes only* (see Figure 18.4). However, because metazoans have left a considerable and highly diverse fossil record, our understanding of biological evolution from the time of the metazoans is much greater than our knowledge of evolutionary relationships among prokaryotes. Fortunately, this has changed with the advent of molecular methods for discerning evolutionary relationships. The objective of the remainder of this chapter is to present in some detail the modern approaches to an understanding of microbial evolution which are based on molecular biology and genetics. The emphasis will be on the molecular evolution of prokaryotes, because prokaryotes are the organisms with the greatest molecular diversity and for which the greatest amount of molecular biological information is known.

> The eukaryotic nucleus and mitotic apparatus probably arose as a necessity for ensuring the orderly partitioning of DNA in large-genome organisms. Mitochondria and chloroplasts, the principle organelles of eukaryotes, probably arose from prokaryotes that became symbiotic within eukaryotic cells. Mitochondria probably arose from aerobic bacteria and chloroplasts arose from oxygenic phototrophs. Once these endosymbionts had become well established, they became increasingly specialized so that today the independent genetic competence of these organelles is quite restricted.

18.4 Evolutionary Chronometers

It is now clear that certain cellular macromolecules can serve as evolutionary chronometers—actual measures of evolutionary change. From studies of the sequence of monomers in informational macromolecules, it has been shown that the *evolutionary distance* between two species can be measured by differences in the nucleotide or amino acid sequence of *homologous* macromolecules found in the two species. This is so because the number of sequence differences in a molecule is proportional to the number of stable mutational changes fixed in the DNA coding for that molecule in species that diverged from a common ancestor. As different mutations become fixed in different populations, biological evolution results.

Choosing the right chronometer

In order to determine true evolutionary relationships between species, it is essential that the correct molecules be chosen for sequencing studies. This is important for several reasons. First, the molecule should be *universally distributed* across the group chosen for study. Second, it must be *functionally homologous* in each organism; phylogenetic comparisons must start with molecules of *identical* function. One cannot compare the amino acid sequences of enzymes that carry out different reactions, or the nucleotide sequences of nucleic acids of different function, because functionally unrelated molecules would not be expected to show sequence similarities. Third, it is crucial in sequence comparisons to be able to properly *align* the two molecules in order to identify regions of sequence homology and sequence heterogeneity. Finally, the sequence of the molecule chosen should change at a rate commensurate with the evolutionary distance measured. And in fact the broader the phylogenetic distance being measured, the *slower* must be the rate at which the sequence changes. A molecule which has undergone too many sequence changes would be essentially useless as a tool for determining evolutionary relationships because regions of common sequence would eventually be lost.

Because of the antiquity of the protein-synthesizing process, ribosomal RNAs turn out to be excellent molecules for discerning evolutionary relationships among living organisms. Ribosomal RNAs (rRNAs) are ancient molecules, functionally constant, universally distributed, and moderately well conserved across broad phylogenetic distances. Also, because the *number* of different possible sequences of large molecules such as ribosomal RNAs is so large, similarity in two sequences always indicates *some* phylogenetic relationship. However, it is the *degree* of similarity in ribosomal RNA sequences between two organisms that indicates their relative evolutionary relatedness. From comparative sequence analyses, molecular genealogies can be constructed leading to phylogenetic trees that show the true evolutionary position of organisms relative to one another (see Figure 18.10).

versus the Archaea/Eukarya line. Later, the Archaea and Eukarya diverged from one another to yield two major lineages. Archaea and Eukarya are therefore phylogenetically more related to each other than either group is to the Bacteria. Thus, from an evolutionary standpoint, a group of prokaryotes (the Archaea) share more in common with a group of eukaryotes (the Eukarya) than they do with a second group of prokaryotes (the Bacteria).

Notice also how the universal phylogenetic tree has the Archaea branching off the tree at a point closest to the root (Figure 18.10). This suggests that Archaea remain, even today, as the *most primitive* (least evolved) of organisms in the three domains, while the Eukarya, now the farthest away from the universal ancestor, are the *least primitive,* that is, the most evolved (Figure 18.10). This suggests that the eukaryotic lineage evolved some important characteristic or set of characteristics that allowed for rapid, albeit late, evolutionary diversity. Perhaps this evolutionary event was something as simple as a larger cell size; this would have facilitated the larger amounts of DNA needed for the development of metazoan organisms, and produced a cell sufficiently large to accept prokaryotic endosymbionts. Placement of the Archaea nearest the universal ancestor is supported by the fact that many Archaea inhabit extreme environments, such as high temperature, low pH, high salinity, and so on (see Chapter 20); these environmental extremes, especially that of high temperature, reflect the environmental conditions under which life originated (see Section 18.1). Thus, members of the Archaea may well be evolutionary relics of the earth's earliest life forms.

Organelles

The overall phylogenetic picture presented in Figure 18.10 also tells us about other evolutionary events. For example, it is clear that mitochondria and chloroplasts arose from endosymbiotic Bacteria that established stable relationships, perhaps several different times, with cells from the nuclear line of descent (see also Figure 18.5). Mitochondria appear to have arisen from a small group of organisms that includes the modern prokaryotes *Agrobacterium, Rhizobium*, and the rickettsias. The latter organisms belong to a larger group of Bacteria called the purple bacteria (organisms such as *Rhodopseudomonas* and *Rhodobacter*). (This whole group is sometimes called the *Proteobacteria*.) It is interesting to note that the mitochondrion, itself an intracellular symbiont, is specifically related to organisms (*Agrobacterium, Rhizobium*, and the rickettsias) that are capable of an intracellular existence (see Sections 17.23 and 17.24). The other eukaryotic endosymbiont, the chloroplast, originated from Bacteria as well. The universal evolutionary tree in Figure 18.10 shows that the chloroplast and cyanobacteria *shared* a common ancestor. Thus molecular sequencing supports the endosymbiotic theory and has independently shown that the evolutionary roots of the mitochondrion and the chloroplast are within the Bacteria.

> The major conclusion from phylogenies derived from ribosomal RNA sequences is that life on earth is based on three main lines of cellular descent, all derived from a universal ancestor. Two of the major groups are prokaryotic, the Bacteria and the Archaea, and the third, the Eukarya, contains all the eukaryotes. The universal phylogenetic tree also shows that the Archaea are actually more closely related to Eukarya than they are to the Bacteria and that, of all known organisms, Archaea are the least evolved from the universal ancestor.

18.6   Bacteria (Eubacteria)

Phylogenetic analyses of Bacteria have shown that at least 12 distinct groups exist. Each group can be considered a *phylum*, akin to usage of this term in zoology. Some groups previously established by phenotypic criteria such as physiology or morphology, for example, green nonsulfur bacteria or spirochetes, form distinct phylogenetic groups as well. However, most groups are a collage of physiological and morphological types of bacteria, emphasizing the fact that phenotypic properties are often poor phylogenetic markers.

We summarize here the Bacteria and proceed to describe them in detail in Chapter 19. The detailed phylogeny of Bacteria is summarized in Figure 18.11.

1. Purple phototrophic bacteria and their relatives (the Proteobacteria)

This group is the largest and most physiologically diverse of all Bacteria. The group comprises five subdivisions, referred to as the *alpha, beta, gamma, delta,* and *epsilon* purple bacterial groups. Phototrophic purple bacteria form the heart of three of the subdivisions (thus the name purple bacteria for the group as a whole), but chemoorganotrophic and chemolithotrophic representatives are distributed across the groups as well. The *delta* and *epsilon* subdivisions contain only nonphototrophic bacteria. The major representatives of each group are shown in Table 18.2.

Many organisms grouped into the purple bacteria are phototrophic (Table 18.2). It has been proposed that the common ancestor of the purple bacterial group was itself phototrophic and that the physiological diversity characteristic of the purple bacterial group arose as the result of the exchange of photosynthetic capacity for other physiological processes (such as sulfate reduction and chemolithotrophy) better fitted to new and expanding ecological niches (Figure 18.12). Such a hypothesis nicely accounts for the specific phylogenetic relatedness of organisms like *Rhodopseudomonas* (a budding *phototrophic* bacterium containing polar stacks of lamellar membranes) and *Nitrobacter* (a budding *chemolithotrophic* bacterium containing polar stacks of lamellar membranes), or the specific phylogenetic relatedness of phototrophic and certain chemoorganotrophic spirilla.

Nonsulfur purple bacteria (see Section 19.1) are found only in the *alpha* and *beta* subdivisions of the purple bacterial group, while purple sulfur bacteria

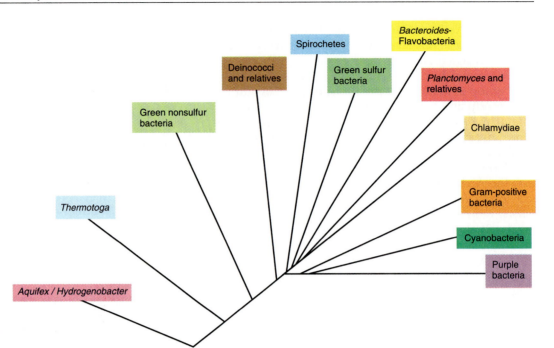

FIGURE 18.11
Detailed phylogenetic tree of Bacteria based upon 16S ribosomal RNA sequence comparisons. The relative positions of branches on this detailed tree differ slightly (for statistical reasons) from those shown in Figure 18.10, but the branch lengths on the tree remain proportional to the corrected evolutionary distances calculated between any two groups. The purple bacteria are also called the Proteobacteria.

are found only in the *gamma* subdivision (Table 18.2). Certain chemolithotrophic sulfide-oxidizing bacteria which store sulfur intracellularly, for example members of the genus *Beggiatoa* (see Sections 16.10 and 19.5), are also found within the *gamma* purple bacteria group. Sulfur chemolithotrophs carry out the oxidation of sulfide to sulfur and sulfate just as purple sulfur bacteria do, but do so as a means of obtaining *energy* rather than simply as a source of reducing power

for CO_2 reduction (see Section 16.10). If the storage of sulfur intracellularly is a unique biochemical property, then one could imagine how the process of phototrophic sulfur storage could have been recruited as a mechanism to store sulfur for use as an *energy* source by chemolithotrophs. Indeed, experiments have shown that in darkness a variety of phototrophic sulfur bacteria are capable of chemolithotrophic growth at the expense of reduced sulfur compounds (see Section 19.1). Hence, the phylogenetic relationships between phototrophic and chemolithotrophic sulfide oxidizers is not surprising.

The *delta* subdivision of purple bacteria is an extremely broad one that consists of physiologically diverse organisms that appear to have no recognizable phenotypic thread among them. The major subgroups include (a) the sulfate- and sulfur-reducing bacteria, (b) the fruiting myxobacteria, and (c) the prokaryotic predator, *Bdellovibrio*. The sulfate/sulfur reducers are strictly anaerobic chemoorganotrophs that oxidize a variety of organic electron donors and reduce sulfate or sulfur to H_2S (see Sections 16.16 and 19.8). The fruiting myxobacteria are obligate aerobes that move by gliding motility and which aggregate under certain nutritional conditions to form complex arrangements of cells called fruiting bodies (see Section 19.13). *Bdellovibrio* is a small, spiral-shaped Gram-negative bacterium which acts as a predator of other Gram-negative Bacteria, attaching and eventually moving into the periplasmic space of its prey and destroying it (see Section 19.11). The *epsilon* subdivision contains *Thiovulum* (a chemolithotroph), and the chemoorganotrophs *Wolinella*, *Campylobacter*, and *Helicobacter*. The latter two organisms are human pathogens (see Chapter 15).

Finally, all of the enteric bacteric, most *Pseudomonas* species, the free-living and symbiotic N_2-fixers, and most chemolithotrophs are also phylo-

Table 18.2	Major genera that can be grouped in the purple Bacteria
Group	**Genera**
Alpha group	*Rhodospirillum*, Rhodopseudomonas*, Rhodobacter*, Rhodomicrobium*, Rhodopila*, Rhizobium, Nitrobacter, Agrobacterium, Aquaspirillum, Hyphomicrobium, Acetobacter, Beijerinckia, Paracoccus, Pseudomonas* (some species)
Beta group	*Rhodocyclus*, Rhodoferax*, Rhodovivax*, Spirillum, Nitrosomonas, Sphaerotilus, Thiobacillus, Alcaligenes, Pseudomonas, Bordetella, Neisseria*
Gamma group	*Chromatium*, Thiospirillum*,* and other purple sulfur bacteria*, *Beggiatoa, Leucothrix, Escherichia* and other enteric bacteria, *Legionella, Azotobacter,* fluorescent *Pseudomonas* species, *Vibrio*
Delta group	*Myxococcus, Bdellovibrio, Desulfovibrio* and other sulfate-reducing bacteria, *Desulfuromonas*
Epsilon group	*Thiovulum, Wolinella, Campylobacter, Helicobacter*

*Phototrophic representatives are shown by an asterisk after the name.

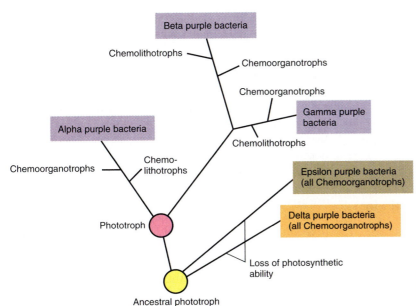

FIGURE 18.12
Evolutionary divergence of the purple Bacteria (Proteobacteria) to give rise to nonphototrophic relatives.

genetically related to the purple bacteria. Hence the purple bacterial lineage gave rise to a wide diversity of ecologically and physiologically significant bacteria. The 16S rRNA oligonucleotide sequence AAAUUCG is diagnostic of *alpha* purple bacteria, CYUUACACAUG (where Y is any pyrimidine), is characteristic of the *beta* group, and ACUAAAACUCAAAG is present in the 16S RNA of most *delta* purple bacteria; no specific signature has been identified for the *gamma* or *epsilon* subdivisions.

2. Green sulfur bacteria

The phototrophic green sulfur bacterium *Chlorobium* is related to other Bacteria (including purple phototrophic bacteria) at only the deepest phylogenetic levels (Figure 18.11). This lack of close relationship is not too surprising in light of the unique pigments, light-harvesting apparatus (chlorosomes), and autotrophic mechanisms used by green bacteria (see Section 19.1). However, the finding that photosynthesis once again has deep evolutionary roots further supports the hypothesis that photosynthesis was an early invention that spread among several groups of Bacteria before the major lines of Bacterial descent within the lineage were defined. Included in the green bacterial group besides *Chlorobium* is the genus *Chloroherpeton,* a gliding green bacterium; all species of *Chlorobium* are immotile. The sequence AUACAAUG provides a signature for green bacteria.

3. Green nonsulfur bacteria: *Chloroflexus*

The physiology, pigments, and fine structure of *Chloroflexus* (see Section 19.1) suggest a close relationship to green sulfur bacteria (*Chlorobium*). Phylogenetically, however, no such relationship exists. The *Chloroflexus* group, which includes two nonphototrophic genera, *Herpetosiphon,* and *Thermomicrobium,* represents a distinct and ancient line of Bacteria (see Figure 18.11).

Chloroflexus is clearly an evolutionary puzzle because of its lack of phylogenetic relationship to green sulfur bacteria with which it shares a number of phenotypic properties (see Section 19.1). Because most of the members of the *Chloroflexus* group are thermophiles, the thermophilic property may be of more significance in defining the phylogeny of *Chloroflexus* and its relatives than is their relationship to green bacteria. The sequence CCUAAUG provides a signature for green nonsulfur Bacteria.

4. Cyanobacteria

The cyanobacteria were discussed previously in connection with the origin of the chloroplast (see Section 18.5). Although the taxonomy of cyanobacteria is presently weighted toward morphological criteria (see Section 19.2), only a few morphologically defined cyanobacterial groups have firm phylogenetic standing. Both the branching and the heterocystous cyanobacteria appear to form rather tight phylogenetic groups; unicellular cyanobacteria by contrast are phylogenetically diverse. The branch of the tree leading to the cyanobacteria is also the branch within which the precursor of the green plant chloroplast arose (see Figure 18.10). The nucleotide sequence AAUUUUYCG is a signature sequence for cyanobacteria.

5. *Planctomyces–Pirella*

This group contains budding organisms which lack peptidoglycan, contain a proteinaceous cell wall, are obligately aerobic, and require very dilute culture media for growth. Many species are found in nature attached to surfaces by stalks or holdfasts. The *Planctomyces–Pirella* line forms a major branch of Bacteria (see Figure 18.11), albeit a branch quite distinct from the other groups (see also Section 19.10). The oligonucleotide CUUAAUUCG provides a signature for members of the *Planctomyces–Pirella* group.

6. Spirochetes

The main spirochete cluster contains the genera *Spirochaeta*, *Borrelia*, and *Treponema*. These organisms may be free-living or host-associated and many are well-known pathogens. The genus *Leptospira* forms a distinct branch within the spirochete group. Spirochetes have traditionally been grouped on the basis of their unique morphology and mode of motility (see Section 19.12), and these traits turn out to be valid phylogenetic indicators as well. The signature sequence AAUCUUG is highly diagnostic for one subgroup of spirochetes; the sequence UCACACYAY-CYG (where Y is any pyrimidine) is diagnostic for most, but not all, spirochetes.

7. *Bacteroides–Flavobacterium*

This group consists of a number of genera forming a major phylogenetic line of Gram-negative Bacteria. Members of the genus *Bacteroides* form one subline within the group, while *Cytophaga* and *Flavobacterium* define a second. A mixture of physiological types shows up here, obligate anaerobes (*Bacteroides*) along with obligate aerobes (*Sporocytophaga*). However, the two subgroups are themselves relatively homogeneous. *Bacteroides* and *Fusobacterium* are the major genera in one subgroup; both are obligately anaerobic, primarily fermentative rods. The *Bacteroides–Flavobacterium* group also shows a close affinity with the green sulfur bacteria (Figure 18.11), indicating that the two groups shared a specific common ancestor. The *Flavobacterium* subgroup includes *Cytophaga*, *Flexibacter*, *Sporocytophaga*, and *Saprospira*, all rod-shaped or filamentous organisms, primarily respiratory in physiology, but which move by gliding motility. The sequence UUACAAUG is the signature for members of the *Bacteroides* group; the sequence CCCCCACACUG is the signature for the gliding group.

8. *Chlamydia*

This group consists thus far only of species of the obligate intracellular parasite, *Chlamydia*, agent of a variety of sexually transmitted infections and a form of acquired blindness called *trachoma* (see Section 15.6). On the basis of signature sequence comparisons, the chlamydial group shows a distant relationship to members of the *Planctomyces* group. It is interesting to note that members of both of these groups have cell walls that lack peptidoglycan. The biology of the chlamydia is discussed in Section 19.23.

9. *Deinococcus–Thermus*

This group really consists of only two well-characterized genera, the Gram-positive, highly radiation-resistant *Deinococcus* (see Section 19.24) and the Gram-negative chemoorganotrophic thermophile *Thermus*. Both *Deinococcus* and *Thermus* have an atypical cell wall that contains ornithine in place of diaminopimelic acid in its peptidoglycan (*Chloroflexus* also contains ornithine in its peptidoglycan). Sequence analyses indicate that the group is tightly clustered and probably of

relatively recent origin. The oligonucleotide CU-UAAG is a signature for the *Deinococcus–Thermus* group.

10. Gram-positive Bacteria

The Gram-positive Bacteria, both rods and cocci, form a coherent phylogenetic cluster. Included here are the endospore formers, lactic acid bacteria, most anaerobic and aerobic cocci, the coryneform bacteria, the actinomycetes, and even the majority of mycoplasmas, which phylogenetically are simply clostridia that lack cell walls. Surprisingly, certain phototrophic bacteria are related to these Gram-positive Bacteria. Heliobacteria, a group consisting of *Heliobacterium* and *Heliobacillus* (see Section 19.1), are relatives of the clostridia. This finding is significant because we have seen that phototrophic representatives are present in other groups (green sulfur, green nonsulfur, and purple bacteria and cyanobacteria) as well. This once again points out the probable importance of photosynthesis as an early energy-generating process.

Two main classes of Gram-positive Bacteria are recognized on the basis of rRNA sequencing, those containing relatively low GC base ratios (clostridia, bacilli, lactic acid bacteria, see Sections 19.24 through 19.26), and those with high GC ratios (actinomycete group, see Section 19.28). The oligonucleotide signature CUAAAACUCAAAG is highly diagnostic for the high GC group. No low GC group oligonucleotide signature has been found. The major genera placed in the Gram-positive group are shown in Table 18.3.

11. *Thermotoga–Thermosipho*

This group of Bacteria consists of only two genera, *Thermotoga* and *Thermosipho*. Isolated thus far only from geothermally heated marine sediments, *Thermo-*

Table 18.3	Major genera of phylogenetically related Gram-positive Bacteria
Group	**Genera**
Low GC group	*Clostridium* and relatives: *Clostridium, Bacillus, Desulfotomaculum, Thermoactinomyces, Sporosarcina, Acetobacterium, Streptococcus, Enterococcus, Peptococcus, Lactobacillus, Lactococcus, Staphylococcus, Ruminococcus, Planococcus, Mycoplasma, Acholeplasma, Spiroplasma*
High GC group	Actinomycetes: *Actinomyces, Bifidobacterium, Propionibacterium, Streptomyces, Nocardia, Actinoplanes, Arthrobacter, Corynebacterium, Mycobacterium, Micromonospora, Frankia, Cellulomonas, Brevibacterium*
Phototrophic subdivision	*Heliobacterium, Heliobacillus*

toga and *Thermosipho* are hyperthermophilic Bacteria. *Thermotoga* is a strictly anaerobic fermentative organism, capable of growth at temperatures from 55 to 90°C, with an optimum of about 80°C. *Thermotoga* produces unique lipids containing extremely long-chain fatty acids but its cell wall contains peptidoglycan, the signature molecule of Bacteria. *Thermosipho* differs from *Thermotoga* primarily in morphology and its lower GC content. Although clearly Bacteria, the *Thermotoga* group is phylogenetically positioned far from other Bacteria (Figure 18.11), and in the universal tree (Figure 18.10), it seems to occupy an intermediate phylogenetic position between the remaining Bacteria and the Archaea. Because of their extremely thermophilic nature and proximity to the root of the universal phylogenetic tree (Figure 18.10), the discovery of *Thermotoga* and *Thermosipho* lends support to the hypothesis that early organisms were thermophilic (see Sections 20.6 and 20.8).

12. *Aquifex–Hydrogenobacter*

The final group of Bacteria to be discussed also contains extremely thermophilic organisms: *Aquifex pyrophilus* and *Hydrogenobacter thermophilus*. *Aquifex* is an obligately chemolithotrophic hyperthermophile that oxidizes H_2 or reduced sulfur compounds *aerobically* under microaerophilic conditions. *Hydrogenobacter* oxidizes only H_2. *Aquifex* has a growth temperature optimum of about 85°C and grows up to 95°C, while *Hydrogenobacter* is not quite as thermophilic. Phylogenetically, *Aquifex* and *Hydrogenobacter* form the deepest lineage on the phylogenetic tree of Bacteria (Figure 18.10). The discovery and characterization of these hyperthermophilic chemolithotrophic H_2 oxidizers further supports the argument that the earliest Bacteria were hyperthermophilic, consistent with the likely physical conditions of the early earth (see Section 18.2).

> Ribosomal RNA sequence analysis suggests that at least 12 distinct groups of Bacteria exist. Some of these groups are phylogenetically diverse whereas others are highly related. One of the largest groups of Bacteria is the purple bacteria, which contains most of the Gram-negative Bacteria. Another large and important grouping contains most of the Gram-positive Bacteria.

18.7   Archaea (Archaebacteria)

Phylogenetically, Archaea can be divided into four subgroups, one consisting of some methanogens and the extremely halophilic Archaea, one consisting of methanogens only, one containing the hyperthermophiles, and one containing only the genus *Thermoplasma* (Figure 18.13).

The extremely halophilic Archaea, such as *Halobacterium* and its relatives, group near the methanogens, more specifically near the genus *Methanobacterium*. *Thermoplasma acidophilum*, an acidophilic, thermophilic mycoplasma (see Section 20.7) is also found near the methanogen branch of the Archaea. The sequence AYUAAG is a signature for the methanogens while the sequence AAUUAG provides a signature for extreme halophiles. The sequences AAAACUG and ACCCCA are signatures for *Thermoplasma*.

The other subgroup of Archaea is the hyperthermophilic (sometimes called sulfur-dependent) Archaea (Figure 18.13). Phenotypically, this subgroup shows a variety of striking properties. First, they mostly have growth temperature optima above 80°C, sometimes even above 100°C; the hyperthermophile *Pyrodictium* has a temperature optimum of 105°C (see Section 20.6). Second, most of these Archaea require *elemental sulfur* for optimal growth. This is because they are for the most part anaerobes that use sulfur as an electron *acceptor*, oxidizing various organic compounds or H_2 as

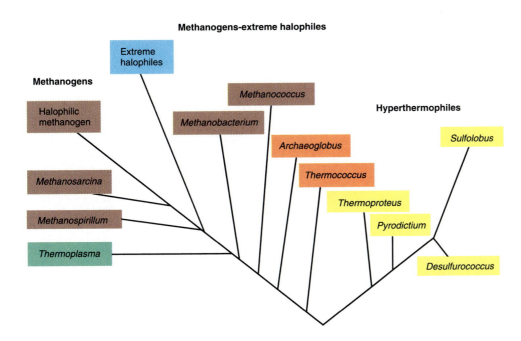

FIGURE 18.13
Detailed phylogenetic tree of the Archaea based upon 16S ribosomal RNA sequence comparisons. See note in Figure 18.11 for comparisons of detailed tree with the universal tree.

energy sources. One hyperthermophile, *Sulfolobus*, uses elemental sulfur not as an electron acceptor but as an electron *donor* with oxygen or ferric iron serving as an electron acceptor (see Sections 16.10, 16.18, and 20.6). *Archaeoglobus*, a hyperthermophilic *sulfate*-reducing archaean, represents a phenotypic bridge between the sulfur-reducing hyperthermophiles and the methanogens; *Archaeoglobus* both reduces sulfate and makes methane (see Section 20.6).

Sulfur plays an intimate role in the biochemistry of hyperthermophilic Archaea. This fact, coupled with the thermophilic character of all members of this group, supports the contention that these Archaea represent an ancient metabolic phenotype (see Section 18.2 and Figure 18.3). Since hyperthermophiles appear to have evolved *more slowly* than other Archaea and are closer to the root of the universal phylogenetic tree (see Figures 18.10 and 18.13), their phenotype may resemble more closely the phenotype of ancient organisms than does that of other organisms (see Section 20.9). Such ideas are consistent with what we know of early earth geochemistry, and the probable environmental conditions within which life arose (see Section 18.1). The sequences UAACACCAG and CACCACAAG are oligonucleotide signatures for hyperthermophilic Archaea.

> Archaea can be divided into four relatively broad groups: the methanogens, the halophiles, the hyperthermophiles, and the lone genus *Thermoplasma*. All of these organisms live in relatively extreme environments.

18.8 Eukarya (Eukaryotes)

Molecular sequencing methods have been applied to eukaryotic organisms and have greatly improved the construction of evolutionary trees for eukaryotes, especially microbial eukaryotes. A eukaryotic phylogenetic tree based on ribosomal RNA sequence comparisons is shown in Figure 18.14. Because eukaryotes contain *18S* rRNA instead of *16S* rRNA in their small ribosomal subunits, the tree shown in Figure 18.14 is constructed from sequence comparisons of this slightly larger, but functionally equivalent, RNA species.

Details of the eukaryotic tree: Early branches

The eukaryotic tree suggests that evolution along the nuclear line of descent was not a continuous process, but instead occurred in major epochs. By comparing Figures 18.10 and 18.14, it is clear that *microsporidia* and *diplomonads* (for example, *Giardia*) (see Section 15.14) are modern relatives of the earliest major eukaryotic cell lines, and that following the development of these organisms, several groups of eukaryotes radiated off the nuclear line of descent, culminating in the major eukaryotic lineages of plants and animals (Figure 18.14).

Microsporidia and diplomonads are obligate parasites which live in association with representatives of virtually all groups of eukaryotes, from microorganisms to humans. Although they contain a membrane-bound nucleus, these organisms lack mitochondria; in this connection they resemble the type of cell that might have first accepted stable endosymbionts (see Figure 18.5). Besides lacking mitochondria, microsporidia and diplomonads differ from modern eukaryotes in many other ways. These organisms lack 5.8S rRNA, the small RNA species found in the large subunit of the ribosome of all other eukaryotic organisms examined. In addition, the small ribosomal subunit of these primitive eukaryotes contains an RNA of about 16S (that is, of bacterial dimensions) rather than the 18S rRNA of other eukaryotes. It thus appears that microsporidia and diplomonads represent an evolutionary relic of the nuclear line of descent, the latter of which is now dominated by organisms containing mitochondria, chloroplasts, and other internal structures (see Figure 18.5).

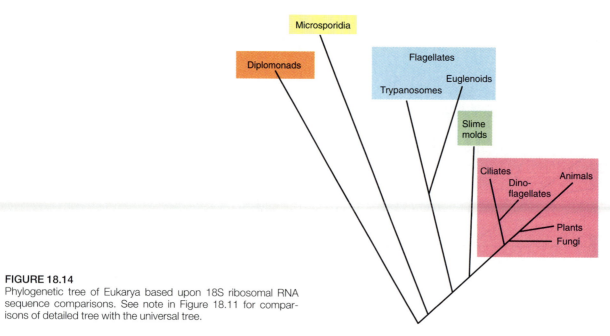

FIGURE 18.14
Phylogenetic tree of Eukarya based upon 18S ribosomal RNA sequence comparisons. See note in Figure 18.11 for comparisons of detailed tree with the universal tree.

Evolution of plants and animals

The radial branches of the eukaryotic tree (Figure 18.14) represent later evolving organisms such as the plants and animals. The rapid radiation observed in this region of the nuclear lineage was likely triggered by some major geochemical or geophysical event. When the fossil record is compared with the eukaryotic phylogenetic tree derived from molecular sequencing, this rapid evolutionary radiation can be dated to about 1.5 billion years ago (see Figure 18.4). Geochemical evidence suggests that this is about the period in the earth's history that significant *oxygen* levels were accumulating. Perhaps the rapid evolutionary radiation of eukaryotes (Figure 18.14) was catalyzed by the onset of oxic conditions on the planet. The accumulation of significant O_2 levels would have allowed the development of an ozone shield (see Figure 18.4) which in turn would have greatly expanded the number of habitats available for colonization (see Section 18.1). The explosive development of eukaryotes could then simply have been a response to the availability of new oxic habitats for evolutionary development.

Interestingly, a closer examination of the evolutionary tree for animals (Figure 18.15) shows that the evolutionary distance between phenotypically quite different animals is actually very small. For example, based on rRNA sequences, the evolutionary distance between green sulfur bacteria and enteric bacteria (purple bacterial group), two groups of prokaryotes that are morphologically very similar, is far greater (Figure 18.11) than the evolutionary distance between earthworms and humans (Figure 18.15), organisms which show obvious morphological differences. These findings, which at first seem surprising, can be better understood when it is recalled that molecular sequencing measures evolutionary relationships between *cells*, while classical evolutionary approaches look at relationships from the perspective of the *whole organism*.

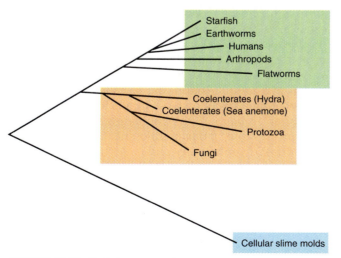

FIGURE 18.15 Evolutionary tree for animals based on sequences of 18S ribosomal RNAs. Three major evolutionary radiations are apparent.

> The evolution of eukaryotes was not a continuous process but occurred in major epochs. Two specialized groups, the microsporidia and the diplomonads, are modern relatives of the earliest major eukaryotic lines. Plants, animals, and fungi constitute another major eukaryotic lineage. The rapid radiation of this latter major line was probably triggered by a major geochemical event in the earth's history, most likely the accumulation in the atmosphere of significant levels of molecular oxygen. Among macroorganisms, the true evolutionary distance between morphologically distinct organisms is often very small.

18.9 Characteristics of the Primary Domains

Although the primary domains—Bacteria, Archaea, and Eukarya—were originally defined on the basis of comparative ribosomal RNA sequencing, subsequent studies have shown that each domain has associated with it a number of *phenotypic* properties. Some of these characteristics are unique to one domain, while others are found in only two out of the three domains. Although no phenotypic property *shared* between domains is of value for placing an organism in a particular domain, certain traits that are *unique* to one or another domain can be used for placing unknown organisms into the correct primary domain. We present here an overview of major phenotypic traits of phylogenetic value.

Cell walls

Virtually all Bacteria have cell walls containing peptidoglycan (see Section 3.5). The only known exceptions are members of the *Planctomyces–Pirella* group (see Section 18.6), whose cell walls are made of protein, and the *Chlamydia* and *Mycoplasma* groups, which lack cell walls altogether. Eukaryotes never have peptidoglycan; their cell walls are composed of either cellulose or chitin (or are absent in protozoa and animal cells, see Chapter 21).

Likewise, none of the Archaea have peptidoglycan in their cell walls. Instead, various cell wall types are known (see Section 20.2). Most extreme halophiles, as well as the hyperthermophiles and a few methanogens, contain a glycoprotein wall. Certain extreme halophiles produce a sulfated heteropolysaccharide wall. A few methanogens make an extremely acidic (negatively charged) polysaccharide cell wall, but the majority of methanogens have a cell wall consisting only of protein. One group of methanogens produces a modified peptidoglycan. This material, referred to as *pseudopeptidoglycan*, resembles the peptidoglycan of Bacteria but differs in the chemical nature of the cross-linking amino acids and the fact that it lacks muramic acid, the key amino sugar in peptidoglycan (see Figure 20.4 for the structure of pseudopeptidoglycan). Thus, cell wall structure has phylogenetic significance: peptidoglycan can be considered a signature molecule for Bacteria and glycoprotein for most Archaea.

Lipids

The chemical nature of membrane lipids is perhaps the most useful of all nongenetic criteria for differentiating Archaea from Bacteria. Bacteria and eukaryotes synthesize membrane lipids with a backbone consisting of fatty acids hooked in *ester* linkage to a molecule of glycerol (see Figure 2.7). Although the nature of the fatty acid can be highly variable, the key point is that the chemical linkage to glycerol is an **ester link.** By contrast, archaeal lipids consist of **ether-linked** molecules (see Figure 20.1). In ester-linked lipids, the fatty acids are straight chain (linear) molecules, whereas in ether-linked lipids, branched chained hydrocarbons are present. In Archaea, long chain branched hydrocarbons, either of the phytanyl or biphytanyl type, are bonded by ether linkage to glycerol molecules (see Figure 20.1).

In addition to the differences in chemical *linkage* between the hydrocarbon and alcohol portions of archaeal and bacterial/eukaryotic lipids, the chirality of the glycerol moiety differs in the lipids of representatives of the three domains. The central carbon atom of the glycerol molecule is stereoisomerically of the R form in Bacteria/eukaryotes and of the L form in Archaea (see Section 2.9 for more discussion about stereoisomerism and life).

RNA polymerase

Transcription is carried out by DNA-dependent RNA polymerases in all organisms; DNA is the template, RNA is the product (see Section 5.6). Cells of Bacteria contain a single type of RNA polymerase of rather simple quaternary structure. This is the classic RNA polymerase, containing *four* polypeptides, α, β, β', and one of a variety of different σ factors, combined in a ratio of 2:1:1:1, respectively, in the active polymerase (Figure 18.16; see also Section 5.6).

Archael RNA polymerases are of several types and are structurally more complex than those of Bacteria. The RNA polymerases of methanogens and halophiles contain *eight* polypeptides, five large ones and three smaller ones (Figure 18.16). Hyperthermophilic Archaea contain an even more complex RNA polymerase, consisting of at least *10* distinct polypeptides (Figure 18.16). The major RNA polymerase of eukaryotes (there are several, only one of which makes messenger RNA) contains 10–12 polypeptides, and the relative sizes of the peptides coincide most closely with those from the hyperthermophilic Archaea (Figure 18.16). Two other RNA polymerases are known in eukaryotes, and each specializes in transcribing certain regions of the genome, one for rRNA and one for tRNA. In terms of phylogenetic signatures, the $\alpha_2\beta\beta'\sigma$ polymerase of Bacteria is highly diagnostic while the remaining polymerases are too complex to be phylogenetically definitive. It should also be noted that the antibiotic rifamycin, which acts by specifically interfering with the beta subunit of RNA polymerase, inhibits only Bacteria because Archaea and Eukarya lack this form of RNA polymerase.

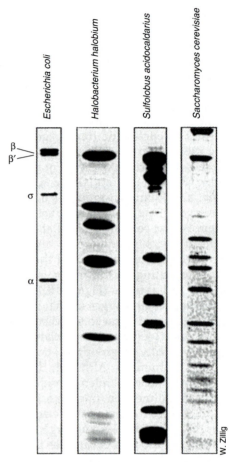

FIGURE 18.16 RNA polymerases from representatives of the three domains, *Escherichia coli* (Bacteria), *Halobacterium halobium* (Archaea), *Sulfolobus acidocaldarius* (Archaea), and *Saccharomyces cerevisiae* (Eukarya). The purified RNA polymerase proteins have been denatured and separated by electrophoresis on a polyacrylamide gel. Largest subunits on the top, smallest subunits are on the bottom. Only Bacteria contain the simple (four polypeptide) RNA polymerase.

Features of protein synthesis

Because of differences in ribosomal RNA sequences and several protein synthesis factors, it is not surprising that certain aspects of the protein synthesizing machinery differ in representatives of the three domains. Although ribosomes of Archaea and Bacteria are the same size (70S, as compared with the 80S ribosomes in the cytoplasm of eukaryotes), several steps in archael protein synthesis more strongly resemble those of eukaryotes than Bacteria. Recall that translation always begins at a unique codon, the so-called *start codon*. In Bacteria this start codon (AUG) calls for the incorporation of an initiator tRNA containing a modified methionine residue, *formyl*methionine (see Section 5.8). By contrast, in eukaryotes and in Archaea, the initiator tRNA carries an *unmodified* methionine.

The exotoxin produced by *Corynebacterium diphtheriae* is a potent inhibitor of eukaryotic protein synthesis because it ADP-ribosylates (adds ADP to) an elongation factor required to translocate the ribosome along the mRNA; the modified elongation factor is inactive (see Section 11.8). Diphtheria toxin also inhibits protein synthesis in Archaea, but does not affect this process in Bacteria.

Most antibiotics that specifically affect Bacterial protein synthesis do not affect archaeal (or eukaryotic) protein synthesis. The sensitivity of representatives of the three domains to various protein synthesis inhibitors is shown in Table 18.4 (see also Section 5.8), where various antibiotics are grouped according to their modes of action in blocking protein synthesis in various domains of organisms. Knowledge of the mode of action of specific antibiotics and the organisms whose growth they inhibit (Table 18.4), has yielded valuable information about the evolution of the translational apparatus.

Functional similarities in the translational machinery of Archaea and Eukarya have been exposed by measurements of *in vitro* protein synthesis by *hybrid ribosomes* made up of ribosomal components from organisms of different domains. Specifically, hybrid ribosomes composed of the large subunit (50S) of ribosomes from *Sulfolobus* (Archaea) and the small subunit (40S) of ribosomes from yeast (Eukarya) have been shown to be functional *in vitro* and to actually produce polypeptides. However, if large ribosomal subunits from *Escherichia coli* (Bacteria) pair with small ribosomal subunits from yeast, no protein synthesizing activity occurs. These results suggest that ribosomal proteins from cells of Archaea and Eukarya are more similar to each other than they are to ribosomal proteins from Bacteria. Such results tend to reinforce conclusions about the relationships of various domains to one another (Figure 18.10).

Other features defining the domains

A number of other more subtle differences can be listed which assist in delineating organisms at the domain level. For example, the number of modified nucleotides present in eukaryotic 18S rRNA (the func-

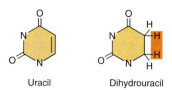

FIGURE 18.17 Comparison of the structures of uracil and dihydrouracil. Dihydrouracil is found in the RNAs of Bacteria and Eukarya, but not in the RNAs of Archaea (with one exception).

tional equivalent of prokaryotic 16S rRNA) is considerable. Depending on the eukaryote, 25–35 or more nucleotides are chemically modified. A similar number of modified nucleotides are found in the 16S rRNA of hyperthermophilic Archaea. By contrast, Bacteria and the methanogen/halophile/*Thermoplasma* archaeal line, contain only 4–8 modified residues in their 16S rRNA. The modified base dihydrouracil (Figure 18.17) is present in the tRNAs of all Bacteria and eukaryotes but is absent from Archaea, except for one group of methanogens.

Table 18.5 summarizes some major points that define the three domains. In most cases the differences are in fundamental biological molecules of major structural and functional importance. This suggests that the evolution of the three cellular lines from the universal ancestor of all cells was accompanied by significant biochemical changes in each lineage as the result of genetic changes to produce organisms optimally adapted to their environment.

> Although the three domains of living organisms were originally defined by ribosomal RNA sequencing, subsequent studies have shown that they differ in many other ways. Particularly, the Bacteria and Archaea differ extensively in cell wall and lipid chemistry, and in features of transcription and protein synthesis.

Table 18.4 Sensitivity of representatives of the three domains to various protein synthesis inhibitors*

| Antibiotics | Mode of action | Archaea | | Bacteria | Eukarya |
		Methanobacterium	*Sulfolobus*	*Escherichia coli*	*Saccharomyces cerevisiae*
Fusidic acid, Sparsomycin	Inhibit elongation steps	+	−	+	+
Anisomycin, Narciclasine	Inhibit peptidyl transfer	+	−	−	+
Cycloheximide	Blocks initiation	−	−	−	+
Erythromycin, Streptomycin, Chloramphenicol	Increase error frequencies and other effects	−	−	+	−
Virginiamycin, Pulvomycin	Inhibit elongation steps	+	−	+	−
Neomycin, Puromycin	Cause premature termination	+	+	+	+
Rifamycin	Inhibit RNA polymerase	−	−	+	−

*A "+" indicates that protein synthesis (and growth) is inhibited. Data adapted from Bock, A., J. H. Hummel, and G. Schmid. 1985. Evolution of translation, p. 73–90. In: K.-H. Schleifer and E. Stackebrandt (eds.), Evolution of Prokaryotes, Academic Press, London.

Table 18.5   Summary of major differentiating features among Bacteria, Archaea, and Eukarya

Characteristic	Bacteria	Archaea	Eukarya
Membrane-bound nucleus	Absent	Absent	Present
Cell wall	Muramic acid present	Muramic acid absent	Muramic acid absent
Membrane lipids	Ester linked	Ether linked	Ester linked
Ribosomes	70S	70S	80S
Initiator tRNA	Formylmethionine	Methionine	Methionine
Introns in tRNA genes	Yes	Yes	Yes
Operons	Yes	Yes	No
Capping and poly A tailing of mRNA	No	No	Yes
Plasmids	Yes	Yes	Rare
Ribosome sensitivity to diphtheria toxin	No	Yes	Yes
RNA polymerases (see Figure 18.16)	One (4 subunits)	Several (8–12 subunits each)	Three (12–14 subunits each)
Sensitivity to chloramphenicol, streptomycin, and kanamycin	Yes	No	No
Methanogenesis	No	Yes	No
Reduction of S^0 to H_2S	Yes	Yes	No
Nitrogen fixation	Yes	Yes	No
Chorophyll-based photosynthesis	Yes	No	Yes

18.10   Taxonomy, Nomenclature, and *Bergey's Manual*

Taxonomy is the science of classification and consists of two major subdisciplines, *identification* and *nomenclature*. It is important to distinguish between bacterial *taxonomy* and the main topic of this chapter up to this point, bacterial *phylogeny,* for the terms really mean different things. Bacterial taxonomy has traditionally relied on *phenotypic* analyses as the basis of classification. By contrast, because bacteria are so small and contain relatively few structural clues to their evolutionary roots, phylogenetic relationships between prokaryotes have only emerged from the *genotypic* analyses discussed in the previous sections. Phenotypic analyses have traditionally played an important role in bacterial identification and classification, especially in practical situations where identification may be an end in itself, for example, in clinical diagnostic microbiology (see Chapter 13). However, even in an area such as diagnostic microbiology where traditional microbiological methods have ruled for decades, genotypic methods (for example, nucleic acid probes) are taking over because of the sensitivity, rapidity, and specificity of these identification methods (see Chapter 13).

Classification and the species concept

Because microbiologists need to identify and classify bacteria for a variety of practical reasons, bacterial taxonomy is still an important discipline. In microbiology, the basic taxonomic unit is the **species**. A species can be operationally defined as a collection of similar strains that differ sufficiently from other groups of strains to warrant recognition as a basic taxonomic unit. A species is usually defined from the characterization of

several strains or clones. The use of the word *clone* in this sense can be taken to mean a population of *genetically identical* cells derived from a single cell. The species concept is important because it gives the collected strains formal taxonomic identity. Groups of species are collected into **genera** (singular, **genus**). By analogy to the species, a genus can be defined as a collection of different species, each sharing some major property or properties that define the genus, but differing from one another at the species level by the presence or absence of other (usually less significant) characteristics. Groups of genera are collected into **families,** which is generally the highest taxonomic unit used routinely in taxonomic studies of prokaryotes.

Nomenclature and formal taxonomic standing

Following the **binomial system** of nomenclature, all bacteria are given genus and species names. The genus name of an organism is usually abbreviated to a single (capital) letter; the species name is never abbreviated. Thus, *Escherichia coli* is usually written *E. coli*. The genus and species names are either Latin or Greek derivatives of some descriptive property appropriate for the species, and are set in print in *italics*. For example, several species of the genus *Bacillus* have been described, including *Bacillus (B.) subtilis, B. cereus, B. stearothermophilus,* and *B. acidocaldarius*. The species names mean "slender," "waxen," "heat-loving," and "acid-thermal," respectively, and in each case refer to key morphological or physiological/ecological traits characteristic of each organism. The rules for bacterial nomenclature are fixed in a publication called *The International Code of Nomenclature of Bacteria*, which spells out rules for naming newly isolated organisms as new genera or new species. The code governs nomenclatural policy for all prokaryotes, both Bacteria and Archaea.

When a new organism is isolated and thought to be unique, a decision must be made as to whether it is sufficiently different from other species to be described as a new species, or perhaps even sufficiently different from all described genera to warrant description as a new genus (in which a new species is automatically created). In order to achieve formal taxonomic standing as a new genus or species, a description of the isolate and the proposed name is published and a pure culture of the organism is deposited in an approved culture collection, usually the American Type Culture Collection (ATCC) or the Deutsche Sammlung für Mikroorganismen (DSM, German Collection for Microorganisms). (See Section 10.1 for a discussion of microbial culture collections.) The deposited strain serves as the *type* strain of the new species or genus/species, and remains as the standard by which other strains thought to be the same can be compared. Although not required for proposing a new species, it is desirable to publish the 16S rRNA sequence of the new organism along with a phenotypic description. A difference of 1.5–2.0 percent or greater in 16S rRNA sequence between a newly proposed species and its closest relative, is generally considered the minimum for describing a new isolate as a new species.

Culture collections preserve the deposited culture, generally by freezing or freeze drying. This practice is very different from the botanical or zoological approach. These disciplines employ preserved (dead) specimens (either dried herbarium material or chemically fixed animal specimens) as the basis for comparison with proposed new species. Microbiologists rely on a *living type strain*, and this approach allows for more detailed and reproducible comparisons, especially at the molecular level.

Conventional bacterial taxonomy

There are many ways in which to group prokaryotes. In conventional bacterial taxonomy, a variety of characteristics of different strains or species are measured, and these traits are then used to group the organisms. Characteristics of taxonomic value that are widely used include: morphology, Gram reaction, nutritional classification (phototroph, chemoorganotroph, chemolithotroph), cell-wall chemistry, presence of cell inclusions and storage products, capsule chemistry, pigments, nutritional requirements, ability to use various carbon, nitrogen, and sulfur sources, fermentation products, gaseous needs, temperature and pH requirements (and tolerances), antibiotic sensitivity, pathogenicity, symbiotic relationships, immunological characteristics, and habitat.

The *Gram stain* reaction is an especially useful differentiating characteristic. In traditional taxonomic arrangements, bacteria have been grouped together as Gram-positive rods, Gram-negative cocci, Gram-negative rods, and so on. Interestingly, the Gram reaction turns out to be a property of fundamental importance for classifying bacteria phylogenetically as well as taxonomically (see Section 18.6), and it is usually one of

the first tests performed in classifying a new bacterial isolate. Cell morphology and special cell structures such as spores are widely used for taxonomic purposes. The bacterial endospore, for example, is a structure of such unique characteristics and practical importance that all endospore-forming bacteria are classified together (both taxonomically and phylogenetically). However, as pointed out earlier in this chapter, cell morphology is frequently not a good predictor of phylogenetic relationships, so should be used with caution in deriving taxonomic or phylogenetic conclusions.

Molecular taxonomy: G+C ratios

Nucleic acid analyses have become part of conventional taxonomy, but the techniques involved differ from the nucleic acid *sequencing* methods discussed in Section 18.4 and 18.5. Determination of the **guanine plus cytosine base composition** of the DNA of an organism is required in order to name it as a new species:

DNA base ratios, expressed as mole percent GC:

$$\frac{G+C}{A+T+G+C} \times 100\%$$

vary over a wide range (Figure 18.18). GC ratios as low as 20 percent and as high as 78 percent are known among prokaryotes, a somewhat broader range than for eukaryotes (Figure 18.18).

Base compositions of DNA have been determined (for methods see Nucleic Acids box, Chapter 5) for a wide variety of organisms and several correlations can be observed. (1) Organisms with highly similar phenotypes often but not always possess similar DNA base ratios. (2) If two organisms thought to be closely related by phenotypic criteria are found to have widely different base ratios, closer examination usually indicates that the organisms are not so closely related as was supposed. (3) Two organisms can have identical base ratios and yet be quite unrelated (both taxonomically and phylogenetically), since a variety of *sequences* is possible with DNA of a given base composition.

Determination of the base ratio of an organism's DNA is required for formal description as either a new genus or new species. In reality, however, the GC

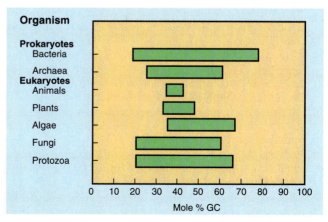

FIGURE 18.18 *Ranges of DNA base composition of various organisms.*

ratio is an exclusionary determinant rather than an inclusive one, because two organisms with identical GC *ratios* need not have similar DNA *sequences*. On the other hand, organisms which show significantly different GC base ratios seldom have any DNA sequences in common. Thus, GC ratios are useful for determining *unrelatedness*, but cannot be used to support the claim that two organisms are *related*.

Bergey's Manual and The Prokaryotes

Bergey's Manual of Systematic Bacteriology is a compendium of classical and molecular information on all recognized species of prokaryotes and contains a number of dichotomous keys that are useful for identification purposes. Although still a taxonomic (rather than phylogenetic) treatise, the latest edition of *Bergey's Manual* has incorporated some molecular sequencing information in the descriptions of various bacterial groups. *Bergey's Manual* consists of four volumes. Volume I covers Gram-negative Bacteria of medical or industrial significance; Volume II, Gram-positive Bacteria of medical or industrial significance; Volume III, the remaining Gram-negative Bacteria, the Archaea and cyanobacteria; and Volume IV, the Gram-positive, filamentous spore-forming Bacteria (Actinomycetes). The genera described in *Bergey's Manual* are given in Appendix 3 of this book.

Although *Bergey's Manual* recognizes the contribution of molecular phylogeny to bacterial evolution, *Bergey's Manual* retains many classical taxonomic groupings because it is not yet clear how molecular sequencing studies will affect the nomenclatural aspects of taxonomy. A more extensive coverage of prokaryotic taxonomy with a stronger emphasis on phylogeny is the four-volume treatise, *The Prokaryotes*, 2nd edition. This work of over 4000 pages is the most complete reference source on prokaryotes available today.

> Although phylogeny permits the grouping of organisms based on evolutionary lines of descent, taxonomy groups organisms for convenience of laboratory study and hence focuses on phenotypic differences. Several properties of an organism including the G+C base ratio of its DNA are determined in order to classify the organism in a taxonomic sense. There is a wide variation in DNA base composition between organisms, and two organisms that differ significantly in DNA base composition are usually found to be taxonomically unrelated, although they may still show significant phylogenetic relationship. The species concept has been applied to prokaryotes, although names are used primarily for identification purposes. *Bergey's Manual* is mainly a taxonomic rather than a phylogenetic compilation of all known prokaryotes.

Study Questions

1. From the following terms, select a combination that could best describe the earliest types of organisms on earth and discuss why or why not each term would be appropriate to describe early life forms: phototroph, hyperthermophile, eukaryote, chemolithotroph, acidophile, chemoorganotroph, psychrophile, autotroph, halophile, mesophile, aerobe, facultative, prokaryote, obligate intracellular parasite, free-living, anaerobe, microaerophile.

2. Why was the evolution of cyanobacteria of such importance to the further evolution of life on earth?

3. What properties of RNA could have made possible an era of "RNA life"? If RNA life forms ever existed, why is there no trace of them today?

4. Describe the evolutionary events leading to the rise of modern eukaryotes. What molecular evidence supports this hypothesis?

5. Why are macromolecules like nucleic acids or proteins excellent phylogenetic markers whereas polysaccharides and lipids are not? (You may want to review the material in Chapter 2 before answering this question.)

6. Why are ribosomal RNAs better molecules for phylogenetic studies than proteins like ferredoxins, cytochromes, or specific enzymes?

7. What are signature sequences and of what phylogenetic value are they? How are signature sequences discerned?

8. Describe the methods involved in obtaining 16S rRNA sequences. How has the polymerase chain reaction benefitted molecular phylogeny?

9. What major evolutionary finding emerged from the study of ribosomal RNA sequences? How did this modify the classic view of evolution? How has this discovery served to change our thinking on the origin of eukaryotic organisms?

10. What major phylogenetic group led to the evolution of familiar Gram-negative Bacteria such as the enteric bacteria and the pseudomonads? Describe how such groups could have been derived from this major group of Bacteria.

11. What major lesson has microbiology learned from RNA sequencing concerning the use of *phenotypic* criteria in establishing evolutionary relationships?

12. Describe the two basic groups of Gram-positive Bacteria and discuss what phenotypic and genotypic properties each group displays.

13. What major physiological and biochemical properties do Archaea share with Eukarya? With Bacteria?

14. What major phenotypic properties are used to group organisms in classical bacterial taxonomy? Which, if any, of these properties have phylogenetic predictive value?

15. Why aren't GC base ratios useful for making phylogenetic determinations? In what situations are GC base ratios of use in taxonomic studies?

16. What is the percentage of A plus T in DNA of 40% G plus C content?

Supplementary Readings

Balows, A., H. G. Trüper, M. Dworkin, W. Harder, and **K.-H. Schleifer** (eds.) 1992. *The Prokaryotes*, 2nd edition. Springer-Verlag, New York. The most up-to-date treatment of the biology of the prokaryotes which blends both taxonomic and phylogenetic information. Highly recommended for beginning a detailed study of any prokaryotic group.

Goodfellow, M., and **A.G. O'Donnell** (eds.). 1993. *Handbook of New Bacterial Systematics*. Academic Press, Orlando, FL. A comprehensive treatment of modern bacterial systematics including traditional as well as modern molecular methods.

Hillis, D. M., and **C. Moritz** (eds.) 1990. *Molecular Systematics.* Sinauer Associates, Sunderland, MA. A monograph on evolutionary methods giving detailed protocols for molecular analyses.

Holt, J. G. (editor-in-chief). *Bergey's Manual of Systematic Bacteriology.* Vol I, 1984; Vol II, 1986; Vol III and Vol IV, 1989. Williams and Wilkins, Baltimore. The major taxonomic reference of the prokaryotes.

Li, W-H., and **D. Graw.** 1991. *Fundamentals of Molecular Evolution.* Sinauer Associates, Sunderland, MA. A textbook of molecular sequencing explaining many of the theoretical concepts behind going from sequences to evolutionary principles.

Schleifer, K.-H., and **E. Stackebrandt** (eds.) 1985. *Evolution of Prokaryotes.* Academic Press, Inc., Orlando, FL. The proceedings of a symposium on various aspects of molecular evolution.

Schopf, J. W., and **C. Klein** (eds.) 1992. *The Proterozoic Biosphere—A Multidisciplinary Study.* Cambridge University Press, New York. A detailed treatment of the paleobiology of the early earth including geology, biogeochemistry, origin of life, microbial mats, and related topics.

Warren, L., and **H. Koprowski** (eds.) 1991. *New Perspectives in Evolution.* Wiley-Liss, New York. Chapters written by experts on both cellular and macroorganismal evolution.

Woese, C. R. 1987. Bacterial evolution. *Microbiological Reviews.* 51:221–271. An excellent summary of what ribosomal RNA sequencing has taught us about the evolution of both prokaryotes and eukaryotes. Highly recommended as a starting point for a detailed study of molecular evolution.

Woese, C. R., and **R. S. Wolfe** (eds.) 1985. *Archaebacteria. Volume 8 of The Bacteria—A treatise on structure and function.* Academic Press, Inc., Orlando, FL. A series of chapters, each written by an expert, on the biology of Archaea. Highly recommended.

19 The Bacteria

I n the preceding chapter, we stressed the evolutionary relationships among microorganisms. In this and the next chapter, we describe the major groups of prokaryotes based primarily on physiological, ecological, and morphological criteria, as is currently stressed in *Bergey's Manual of Systematic Bacteriology,* a recognized sourcebook of prokaryotic taxonomy. The present chapter deals with Bacteria, while Chapter 20 focuses on Archaea. We have divided each chapter into several sections, each dealing with a specific group of prokaryotes. Additional information on each of these groups can be found in *Bergey's Manual* and in the four-volume treatise entitled *The Prokaryotes,* 2nd edition. A complete list of genera of prokaryotes listed in *Bergey's Manual* is given in Appendix 3.

19.1 Purple and Green (Phototrophic) Bacteria

Bacteria able to use light as an energy source comprise a large and heterogeneous group of organisms, grouped together primarily because they possess one or more pigments called *chlorophylls* and are able to carry out light-mediated generation of ATP, a process called **photophosphorylation**. Two major groups are recognized, the **purple and green bacteria** as one group, and the **cyanobacteria** as the other group. The basic distinction between the purple and green bacteria and the cyanobacteria is based on the photopigments and overall photosynthetic process. Cyanobacteria are *oxygenic* phototrophs, employing chlorophyll *a* and two photosystems in their photosynthetic process. Purple and green bacteria are *anoxygenic* pho-

Miniglossary for Chapter 19

ACID–ALCOHOL FAST (ACID FAST)   a property of *Mycobacterium* species in which cells stained with the dye basic fuchsin resist decolorization with acidic alcohol

CARBOXYSOMES   polyhedral cellular inclusions of crystalline ribulose bisphosphate carboxylase, the key enzyme of the Calvin cycle

CHEMOLITHOTROPHS   organisms able to oxidize inorganic compounds as energy sources

CHLOROSOMES   cigar-shaped structures bounded by a nonunit membrane and containing the light harvesting bacteriochlorophyll (*c*, *d*, or *e*) in green bacteria and *Chloroflexus*

CYANOBACTERIA   prokaryotic oxygenic phototrophs which contain chlorophyll *a* and phycobilins but not chlorophyll *b*

ENTERIC BACTERIA   a large group of Gram-negative rod-shaped Bacteria characterized by a facultatively aerobic metabolism

GREEN BACTERIA   anoxygenic phototrophs containing chlorosomes and bacteriochlorophyll *c*, c_s, *d*, or *e* as light-harvesting chlorophyll

HELIOBACTERIA   anoxygenic phototrophs containing bacteriochlorophyll *g*

HETEROCYST   a differentiated cyanobacterial cell which carries out nitrogen fixation but not oxygenic photosynthesis

HETEROFERMENTATIVE   in reference to lactic acid bacteria, capable of making more than one fermentation product

HOMOACETOGEN   an obligately anaerobic bacterium which produces only acetate from fermentation of sugars or from CO_2 reduction with H_2

HOMOFERMENTATIVE   in reference to lactic acid bacteria, producing only lactic acid as a fermentation product

METHANOTROPH   an organism capable of oxidizing methane (CH_4)

METHYLOTROPH   an organism capable of oxidizing organic compounds which do not contain carbon–carbon bonds; if able to oxidize CH_4, also a methanotroph

NITRIFYING BACTERIA   chemolithotrophs capable of carrying out the transformations: $NH_3 \rightarrow NO_2^-$ or $NO_2^- \rightarrow NO_3^-$

NONSULFUR PURPLE BACTERIA   a group of phototrophic prokaryotes containing bacteriochlorophylls *a* or *b* which grow best as photoheterotrophs and have a relatively low tolerance for H_2S

PROCHLOROPHYTE   a prokaryotic oxygenic phototroph that contains chlorophylls *a* and *b* but which lack phycobilins

PROSTHECAE   an extrusion of cytoplasm often forming a distinct appendage, bounded by the cell wall

PSEUDOMONAD   member of the genus *Pseudomonas*, a large group of Gram-negative, obligately respiratory (never fermentative) Bacteria

PURPLE SULFUR BACTERIA   a group of phototrophic prokaryotes containing bacteriochlorophylls *a* or *b* and characterized by the ability to oxidize H_2S and store elemental sulfur inside the cells

SPIROCHETE   a slender, tightly-coiled Gram-negative prokaryote characterized by possession of axial filaments used for motility

STICKLAND REACTION   fermentation of an amino acid pair in which one amino acid serves as an electron donor while a second serves as an electron acceptor

SULFATE-REDUCING BACTERIA   a large group of anaerobic Bacteria which respire anaerobically with SO_4^{2-} as electron acceptor, producing H_2S

---

totrophs, employing bacteriochlorophyll (of several different types), but only *one* photosystem in their photosynthetic process.

The biochemistry of photosynthesis was discussed in some detail in Chapter 16. Green plant photosynthesis, which is also exhibited by the cyanobacteria, involves two light reactions, and H_2O serves as the electron donor. As a result of the photolysis of water, O_2 is produced. By contrast, during photosynthesis by the purple and green bacteria water is not photolysed, and O_2 is not produced. Because the purple and green bacteria are unable to photolyse water, they must obtain their reducing power for CO_2 fixation from a reduced substance in their environment. This can be either an organic compound, a reduced sulfur compound, or H_2.

Purple and green bacteria can only grow phototrophically under *anaerobic* conditions because pigment synthesis in these organisms is repressed by O_2. Cyanobacteria, on the other hand, develop readily under aerobic conditions with no reduced compounds present. Many purple and green bacteria can grow autotrophically with CO_2 as carbon source and a reduced sulfur compound or H_2 as reductant under anaerobic conditions in the light. Under microaerobic condi-

tions, these same reductants serve as electron donors to support chemolithotrophic growth of many purple bacteria. Phototrophic growth is also possible, however, with light serving as energy source and an organic compound as a carbon source (photoheterotrophy); under such conditions, CO_2 is usually only a minor source of cell carbon. Some purple and green bacteria can also grow chemoorganotrophically in the dark using an organic compound as electron donor, but under these conditions a suitable electron acceptor is necessary.

Bacteriochlorophylls

We compared a typical bacteriochlorophyll with algal chlorophyll in Figure 16.3. A number of bacteriochlorophylls exist, differing in substituents on various parts of the porphyrin ring, and these are outlined in Figure 19.1. The various modifications lead to changes in the absorption spectra of the bacteriochlorophylls, so that an organism containing a certain bacteriochlorophyll is best able to utilize light of particular wavelengths. The ecological significance of ability to utilize different wavelengths of light was discussed in Section 16.5; it is likely that this selective absorption provides an evolutionary pressure for the

Pigment	R₁	R₂	R₃	R₄	R₅	R₆	R₇	Infrared absorption maxima (nm)	
								In vivo	Extract (methanol)
Bacterio-chlorophyll *a*	$-\overset{\displaystyle O}{\underset{\displaystyle \|}{C}}-CH_3$	$-CH_3$ [b]	$-CH_2-CH_3$	$-CH_3$	$-\overset{\displaystyle O}{\underset{\displaystyle \|}{C}}-O-CH_3$	P/Gg [a]	$-H$	805 830–890	771
Bacterio-chlorophyll *b*	$-\overset{\displaystyle O}{\underset{\displaystyle \|}{C}}-CH_3$	$-CH_3$ [c]	$=\overset{\displaystyle H}{\underset{\displaystyle \|}{C}}-CH_3$	$-CH_3$	$-\overset{\displaystyle O}{\underset{\displaystyle \|}{C}}-O-CH_3$	P	$-H$	835–850 1020–1040	794
Bacterio-chlorophyll *c*	$-\overset{\displaystyle H}{\underset{\displaystyle OH}{C}}-CH_3$	$-CH_3$	$-C_2H_5$ $-C_3H_7$ [d] $-C_4H_9$	$-C_2H_5$ $-CH_3$	$-H$	F	$-CH_3$	745–755	660–669
Bacterio-chlorophyll *cₛ*	$-\overset{\displaystyle H}{\underset{\displaystyle OH}{C}}-CH_3$	$-CH_3$	$-C_2H_5$	$-CH_3$	$-H$	S	$-CH_3$	740	667
Bacterio-chlorophyll *d*	$-\overset{\displaystyle H}{\underset{\displaystyle OH}{C}}-CH_3$	$-CH_3$	$-C_2H_5$ $-C_3H_7$ $-C_4H_9$	$-C_2H_5$ $-CH_3$	$-H$	F	$-H$	705–740	654
Bacterio-chlorophyll *e*	$-\overset{\displaystyle H}{\underset{\displaystyle OH}{C}}-CH_3$	$-\overset{\displaystyle O}{\underset{\displaystyle \|}{C}}-H$	$-C_2H_5$ $-C_3H_7$ $-C_4H_9$	$-C_2H_5$ $-CH_3$	$-H$	F	$-CH_3$	719–726	646
Bacterio-chlorophyll *g*	$-\overset{\displaystyle H}{\underset{\displaystyle \|}{C}}=CH_2$	$-CH_3$	$-C_2H_5$	$-CH_3$	$-\overset{\displaystyle O}{\underset{\displaystyle \|}{C}}-O-CH_3$	F	$-H$	670, 788	765

[a] *P = phytyl ester ($C_{20}H_{39}O-$); F = farnesyl ester ($C_{15}H_{25}O-$); Gg = geranylgeraniol ester ($C_{10}H_{17}O-$); S = stearyl alcohol ($C_{18}H_{37}O-$).*
[b] *No double bond between C-3 and C-4; additional H-atoms are in position C-3 and C-4.*
[c] *No double bond between C-3 and C-4; an additional H-atom is in position C-3.*
[d] *Bacteriochlorophylls c, d, and e consist of isomeric mixtures with the different substituents on R₃ as shown. From Gloe, A., N. Pfennig, H. Brockmann, and W. Trowitsch. 1975. Archives of Microbiology 102:103–109, and Gloe, A., and N. Risch. 1978. Archives of Microbiology 118:153–156.*

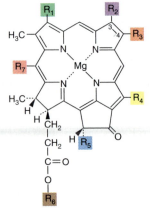

FIGURE 19.1 Structure of all known bacteriochlorophylls. The different substituents present in the positions R₁ to R₇ are given in the accompanying table.

development of organisms with various chlorophylls. The long wavelength absorption maxima of the *bacteriochlorophylls* are the most characteristic, and these are given in Figure 19.1, as measured in the living cell (*in vivo*) and in solvent extract. Although the *in vivo* absorption maximum is of most significance ecologically, from the viewpoint of characterizing the various purple and green bacteria taxonomically, the absorption spectrum in solvent extract is most convenient, because its measurement is easier. Thus, to characterize the bacteriochlorophyll of a new isolate of a purple or green bacterium, a cell extract in methanol or other organic solvent is made and the absorption spectrum determined. From the long wavelength maximum listed in Figure 19.1, the likely identification of the bacteriochlorophyll can be made.

Classification

The purple and green bacteria are a diverse group morphologically and phylogenetically (see Section 18.6). Morphologically, cocci, rods, vibrios, spirals, budding, and gliding types are known. It thus seems likely that the ability to grow phototrophically has developed in a wide variety of bacterial types, and that the only unifying thread among the whole group is the ability to carry out photophosphorylation. Anoxygenic phototrophic bacteria have been classified into three major groups: purple bacteria, green bacteria, and heliobacteria. The major properties of each group are as follows:

Group	Bacterio-chlorophylls	Photosynthetic membrane systems
Purple bacteria	Bchl *a*, Bchl *b*	Lamellae, tubes, or vesicles continuous with cytoplasmic membrane (see Figure 19.3*a* and *b*)
Green bacteria	Bchl *c*, Bchl *d*, or Bchl *e*, plus small amounts of Bchl *a*	Chlorosomes, attached to but not continuous with cytoplasmic membrane (see Figure 19.3*c*)
Helio-bacteria	Bchl *g*	Cytoplasmic membrane only (see Figure 19.3*d*)

Anoxygenic phototrophic bacteria also produce **carotenoid pigments** (Figure 19.2), and the carotenoids of the purple bacteria generally differ from those of the green bacteria and heliobacteria. Carotenoid pigments are responsible for the purple color of the purple bacteria, and mutants lacking carotenoids are blue-green in color, reflecting the actual color of bchl *a* (see Figure 19.2*b*). In fact, purple phototrophic bacteria are frequently not purple, but brown, pink, brown-red, or purple-violet, depending on their carotenoid pigments (see Figure 19.2*b*). Also, many of the "green" bacteria are actually brown in color, due to their complement of carotenoids (Figure 19.2). Thus color is not a good criterion for use in identifying isolates as either green bacteria or purple bacteria.

Photosynthetic membrane systems

A major difference between the green and purple bacteria is in the nature of the photosynthetic membrane system. In the purple bacteria, the photosynthetic pigments are part of an elaborate internal membrane system, connected to and produced from the cytoplasmic membrane; the membrane often occupies much of the cell interior. In some cases, the membrane system is an array of flat sheets called **lamellae** (Figure 19.3*a*), whereas in others it consists of round tubes referred to as **vesicles** (Figure 19.3*b*). The membrane content of the cell varies with pigment content, which is itself affected by light intensity and presence of O_2. When cells are grown aerobically, synthesis of bacteri-ochlorophyll is repressed, and the organisms may be virtually devoid of photopigments as well as the internal membrane systems. Consequently, phototrophic growth is only possible under *anaerobic* conditions, where bacteriochlorophyll synthesis can occur.

Superimposed upon this O_2 effect is an effect of light intensity. Even under anaerobic conditions, when synthesis of the photosynthetic apparatus is not repressed, the level of the photopigments and internal membranes is affected by light intensity. At *high* light intensity, the synthesis of the photosynthetic apparatus is inhibited, whereas when cells are grown at *low* light intensity the bacteriochlorophyll content is high, and the cells are packed with membranes. (This increase in cell pigment content at low light intensities allows the organism to better utilize the available light.) Usually carotenoid pigment synthesis is coordinately regulated with the bacteriochlorophyll content (see Figure 16.11).

In the green bacteria, the photosynthetic apparatus is structurally quite different, consisting of a series of cylindrically shaped structures called **chlorosomes** underlying and attached to the cytoplasmic membrane (Figure 19.3*c*). These vesicles are enclosed within a thin membrane that does not have the usual bilayered appearance (it is referred to as a *nonunit* membrane). Bacteriochlorophylls *c*, c_s, *d*, or *e* (depending on the species) are present inside the chlorosomes, while most of the Bchl *a* and components of the photosynthetic electron transport chain are located in the cytoplasmic membrane.

In heliobacteria, a group of strictly anaerobic phototrophs that are green in color but not phylogenetically related to any other green bacteria, neither internal membranes, typical of purple bacteria, nor chlorosomes, typical of green bacteria, are observed. Bacteriochlorophyll in heliobacteria is associated with the cytoplasmic membrane (Figure 19.3*d*). High resolution electron microscopy of thin sections of heliobacteria have shown the membrane to be studded with particles that may be the site of bacteriochlorophyll:protein complexes in these organisms.

Aerobic anoxygenic phototrophs

An interesting group of bacteria from marine environments produce bacteriochlorophyll only under *aerobic* conditions. These aerobic phototrophs resemble nonsulfur purple bacteria, but they will not grow or produce bacteriochlorophyll anaerobically. Significant pigment levels are only observed when cells are grown aerobically in either the light or the dark. The function of bacteriochlorophyll in the aerobic phototrophs is presumably to carry out photophosphorylation and it is of interest that O_2 regulates pigment synthesis in these organisms in a manner opposite from that of typical purple bacteria. The majority of aerobic phototrophic bacteria have been placed in the genus *Erythrobacter*, but a few isolates have been classified within the genus *Pseudomonas*.

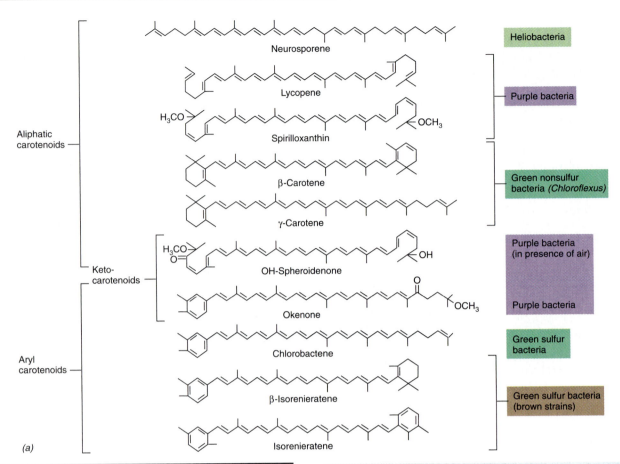

(a)

(c)

Norbert Pfennig

Group	Names	Color of organisms
1	Lycopene, rhodopin, spirilloxanthin, neurosporene	Orange-brown brownish-red, pink, purple-red, green
2	Spheroidene, hydroxyspheroidene, spheroidenone, hydroxyspheroidenone, spirilloxanthin	Red (aerobic) Brownish red to purple (anaerobic)
3	Okenone, methoxylated keto carotenoids (*Rhodopila globiformis* only)	Purple-red
4	Lycopenal, lycopenol, rhodopin, rhodopinal, rhodopinol	Purple-violet
5	Chlorobactene, hydroxycholoro-bactene, β-isorenieratene, isorenieratene	Green (chlorobactene) Brown (isorenieratene)
6	β-carotene, γ-carotene	Orange-green

(b)

Based on Schmidt, K. 1978. Pages 729-750, and S. Liaaen-Jensen, 1978. Pages 233-247 in R. K. Clayton and W. R. Sistrom (eds.), The Photosynthetic Bacteria, Plenum Press, New York, N.Y.

FIGURE 19.2 (a) Structures of major carotenoids of phototrophic bacteria. (b) This table gives a few representative structures of carotenoids of the various groups. A number of variants on the above structures also occur and are listed in the accompanying table. (c) Photographs of mass cultures of phototrophic bacteria showing the color of strains with various carotenoid pigments. The blue culture is a carotenoid-less mutant derivative of *Rhodospirillum rubrum* showing how bacteriochlorophyll *a* is actually *blue* in color. The bottle on the far right (*R. sphaeroides* strain G) lacks one of the carotenoids of the wild type and thus is more green in color.

Enrichment culture

The purple and green sulfur bacteria are generally found in *anaerobic* zones of aquatic habitats, often where H_2S accumulates. These orgnisms can be enriched by duplicating the habitat in the laboratory (Table 19.1). A basal mineral salts medium is used, to which bicarbonate is added as a source of CO_2. Since many phototrophic bacteria require vitamin B_{12}, this vitamin is usually added. The optimum pH range is 6 to 8, and the optimum temperature for mesophilic species is 30°C. Incubation is anaerobic, and a small amount of sodium sulfide (0.05–0.1 percent $Na_2S\cdot9H_2O$) is added as photosynthetic electron donor. Selection of appropriate light conditions is im-

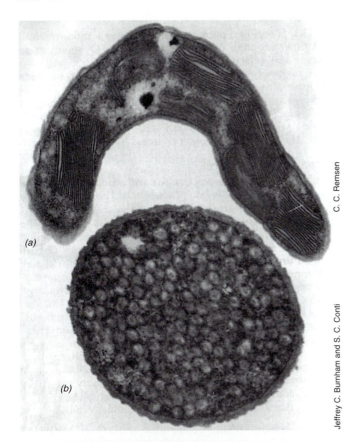

C. C. Remsen

(a)

(b)

Jeffrey C. Burnham and S. C. Conti

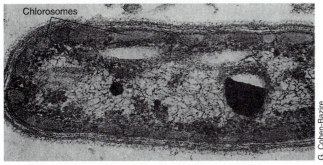

Chlorosomes

(c)

G. Cohen-Bazire

(d)

F. Rudy Turner and Howard Gest

FIGURE 19.3 Membrane systems of phototrophic bacteria as revealed by the electron microscope. (a) Purple phototrophic bacterium, *Ectothiorhodospira mobilis*, showing the photosynthetic membranes in flat sheets. (b) *Chromatium* sp., strain D, another purple photosynthetic bacterium, showing the phototrophic membranes as individual vesicles. (c) Green phototrophic bacterium, *Pelodictyon* sp., showing location of chlorosomes. (d) *Heliobacterium chlorum*, showing the lack of internal membranes.

portant. Light intensities should not be too high, since these bacteria usually live in deep areas of lakes or other anaerobic habitats where light is low. Intensities between 200 and 1000 lux are adequate (about the intensity 1–2 feet away from a 40W incandescent light bulb). The quality of light is also important. Purple and green bacteria use radiation of the near infrared, 700–1000 nm, which can be most easily obtained in the laboratory by use of conventional tungsten light bulbs. Fluorescent bulbs are usually deficient in infrared radiation and are not as satisfactory, especially for purple bacteria. Purple bacteria which contain bacteriochlorophyll *b* can be enriched by using infrared filters which pass wavelengths greater than 900 nm, since bacteriochlorophyll *b* has an absorption maximum at 1020 nm. After inoculation, culture tubes or bottles are incubated for several weeks and examined periodically for signs of visible growth. Enrichment cultures should appear pigmented, and microscopic examination of positive cultures should reveal organisms resembling purple and green bacteria. Pure cultures can be obtained from positive enrichments by shake tube methods (see Section 17.3 and Figure 17.5), care being taken, of course, to keep cultures anaerobic throughout. However, the purple and green bacteria are not as sensitive to oxygen as some other anaerobes, so that extreme precautions to maintain anaerobic conditions are usually not necessary. A widely used medium for the culture of purple and green sulfur bacteria is given in Table 19.1.

For *nonsulfur* purple bacteria, the sulfide concentration of the medium should be reduced to a very low level, 0.005–0.01 percent $Na_2S \cdot 9H_2O$ (or sulfide can be eliminated altogether), and an *organic* compound added to provide a carbon source and/or electron donor. Because many nonsulfur purple bacteria have multiple growth-factor requirements, usually one or more B vitamins, addition of 0.01 percent yeast extract as a source of growth factors is recommended. The organic substrate used should be nonfermentable (to avoid competition in the enrichment from fermentative organisms); acetate, ethanol, benzoate, isopropanol, butyrate or dicarboxylic acids such as succinate are ideal.

Heliobacteria can be isolated by taking advantage of a major property apparently shared by all representatives. All known heliobacteria grow well at temperatures up to 42°C and most species produce endospores. This temperature is restrictive for most other phototrophic bacteria. Hence, to enrich heliobacteria, anaerobic enrichments containing dilute yeast extract and lacking sulfide are inoculated with soil (heliobacteria are very resistant to drying and thus dry soil will often yield strains of these organisms) pasteurized to eliminate purple bacteria, and incubated in the light at 40–42°C.

It should be noted that many purple and green bacteria are incapable of assimilatory sulfate reduction, so that they must be given a source of reduced sulfur. The sulfide added to the medium not only uses up the remaining O_2 and serves as an electron donor, but provides this source of reduced sulfur. If an electron donor such as H_2 or an organic compound is to be used, and sulfide must for some reason be avoided, then it is necessary to add another source of reduced sulfur, such as methionine or cysteine (if yeast extract

Table 19.1 Pfennig's medium for the culture of phototrophic sulfur bacteria*

Solution	Medium
Solution 1	0.83 g $CaCl_2 \cdot 2H_2O$ in 2.5 liters H_2O. For marine organisms, add NaCl, 130 g.
Solution 2	H_2O, 67 ml; KH_2PO_4, 1 g; NH_4Cl, 1 g; $MgCl_2 \cdot 2H_2O$, 1 g; KCl, 1 g; 3 ml vitamin B_{12} solution (2 mg/100 ml H_2O); 30 ml trace element solution (1 liter H_2O; ethylene diamine tetraacetic acid, 500 mg; $FeSO_4 \cdot 7H_2O$, 200 mg; $ZnSO_4 \cdot 7H_2O$, 10 mg; $MnCl_2 \cdot 4H_2O$, 3 mg; H_3BO_3, 30 mg; $CoCl_2 \cdot 6H_2O$, 20 mg; $CuCl_2 \cdot 2H_2O$, 1 mg; $NiCl_2 \cdot 6H_2O$, 2 mg; $Na_2MoO_4 \cdot 2H_2O$, 3 mg, pH 3).
Solution 3	Na_2CO_3, 3 g; H_2O, 900 ml. Autoclave in a container suitable for gassing aseptically with CO_2.
Solution 4	$Na_2S \cdot 9H_2O$, 3 g; H_2O, 200 ml. Autoclave in flask containing a Teflon-covered magnetic stirring rod.

*Dispense the bulk of solution 1 in 67-ml aliquots in 30 screw-capped bottles of 100 ml capacity, and autoclave with screw caps loosely on. Autoclave the other solutions in bulk. After autoclaving, cool all solutions rapidly by placing the bottles in a cold water bath (to prevent lengthy exposure to air). Solution 3 is then gassed with CO_2 gas until it is saturated (about 30 minutes; pH drops to 6.2). Add this to cooled solution 2 and aseptically place 33 ml of this mixture in each bottle containing solution 1. Solution 4 is partially neutralized by adding dropwise (while stirring on a magnetic stirrer) 1.5 ml of sterile 2 M H_2SO_4. Add 5-ml portions of solution 4 to each bottle, fill the bottles completely with remaining solution 1, and tightly close. The final pH should be between 6.7 and 7.2. Store overnight to consume residual oxygen before using. (For some organisms, the sulfide concentration should be reduced. Use 2.5 ml instead of 5 ml of solution 4 per bottle.) It may be necessary to "feed" the cultures with sulfide (solution 4) from time to time, as the sulfide is used up during growth. To do this, remove 2.5 or 5 ml of liquid from a bottle and refill with an equal amount of sterile, neutralized solution 4.
Originally described by Pfennig, N. 1965. Zentrabl. Bakteriol. Parasitenkd. Infektionskr. Hyg. Abt. 1, Supplementh. 1; 179. For an English version, see van Niel, C. B. 1971. Methods Enzymol. 23:3–28.

is added, it often serves as a source of reduced sulfur as well as a growth-factor source).

Nonsulfur purple bacteria

These bacteria have been called the *nonsulfur* purple bacteria because it was originally thought that they were unable to use sulfide as an electron donor for the reduction of CO_2 to cell material. However, sulfide can be used by most species provided the concentration is maintained at a low level. It appears that levels of sulfide utilized well by green or purple *sulfur* bacteria are toxic to most *nonsulfur* purple bacteria. Some nonsulfur purple bacteria can also grow anaerobically in the dark using fermentative metabolism and most can grow aerobically in darkness by respiration. Under the latter conditions, the electron donor can be an organic compound, or in some species even an inorganic compound such as H_2. Nonsulfur purple bacteria are thus among the most energetically versatile of all prokaryotes. However, because of their great photoheterotrophic abilities, these organisms are generally enriched and cultured under phototrophic conditions with an organic compound as major carbon source. Most members of this group also require vitamins, so that yeast extract or some other source of vitamins is usually provided.

The morphological diversity of this group is typical of that of other purple and green bacteria (Table 19.2 and Figure 19.4), and it is clearly a heterogeneous group as it contains both polarly and peritrichously flagellated genera, the latter growing by budding.

Some nonsulfur purple bacteria have the ability to utilize methanol as sole carbon source for phototrophic growth. When growing anaerobically with methanol, some CO_2 fixation is necessary, and the following stoichiometry is observed:

$$2CH_3OH + CO_2 \rightarrow 3(CH_2O) + H_2O$$
$$\textit{cell material}$$

Table 19.2 Genera and characteristics of nonsulfur purple bacteria

Characteristics	Genus	16S rRNA group*	DNA (mole % GC)
Spirilla, polarly flagellated	*Rhodospirillum*	Alpha purple	62–66
Rods, polarly flagellated; divide by budding	*Rhodopseudomonas*	Alpha purple	62–72
Rods, divide by binary fission	*Rhodobacter*	Alpha purple	65–71
Ovals, peritrichously flagellated; growth by budding and hypha formation	*Rhodomicrobium*	Alpha purple	61–63
Large spheres, pH optimum 5	*Rhodopila*	Alpha purple	66
Ring-shaped or spirilla	*Rhodocyclus*	Beta purple	64–72
Curved rods	*Rhodovivax*	Beta purple	70–72
Curved rods	*Rhodoferax*	Beta purple	59–60

*See Section 18.6

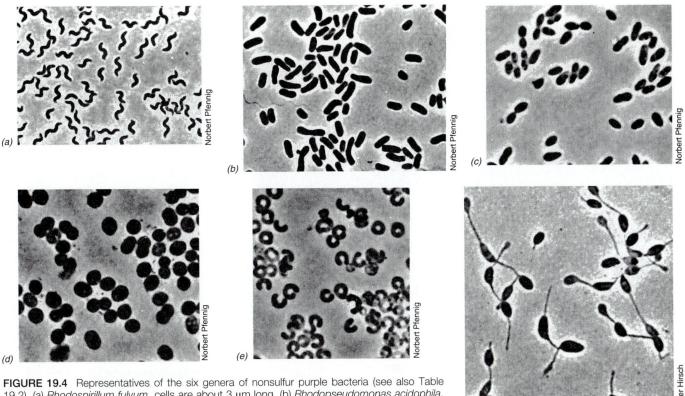

FIGURE 19.4   Representatives of the six genera of nonsulfur purple bacteria (see also Table 19.2). (a) *Rhodospirillum fulvum*, cells are about 3 μm long. (b) *Rhodopseudomonas acidophila*, cells are about 4 μm long. (c) *Rhodobacter sphaeroides*, cells are about 1.5 μm wide. (d) *Rhodopila globiformis*, cells are about 1.6 μm wide. (e) *Rhodocyclus purpureus*, cells are about 0.7 μm in diameter. (f) *Rhodomicrobium vannielii*, cells are about 1.2 μm wide.

CO_2 is required because methanol is at a more reduced oxidation state than cell material, and the CO_2 serves as an electron sink (electron acceptor).

Enrichments for nonsulfur purple bacteria can be made highly selective by omitting fixed nitrogen sources such as ammonia or nitrate from the medium, and substituting an ample supply of gaseous nitrogen, N_2. Most nonsulfur purple bacteria are active N_2 fixers and grow well in a medium in which N_2 is the sole nitrogen source.

Purple sulfur bacteria

Purple bacteria that deposit sulfur and oxidize it to sulfate are morphologically diverse (Table 19.3). The cell is usually larger than that of green bacteria, and in sulfide-rich environments may be packed with sulfur granules (Figure 19.5a and b; see also Figure 19.13), although in the smaller-celled genera the sulfur granules may not be so obvious (Figure 19.5c). Purple sulfur bacteria are commonly found in anaerobic zones of lakes as well as in sulfur springs; because of their conspicuous purple color they are easily visible as large blooms or masses (see Figures 19.13, 19.14, and 19.16). The genus *Ectothiorhodospira* is of interest because it deposits sulfur externally and is also halophilic, growing at sodium chloride concentrations approaching saturation, often at very high pH. It is found in saline lakes, soda lakes, salterns, and other bodies of water high in salt (see Section 20.4).

Purple sulfur bacteria have a limited ability to uti-lize organic compounds as carbon sources for phototrophic growth. Acetate and pyruvate are utilized by all species, and some species will use a few other organic compounds. A few purple sulfur bacteria will grow chemolithotrophically in darkness with thiosulfate as electron donor, and *Thiocapsa* will grow chemoorganotrophically on acetate.

Green bacteria

Green bacteria are morphologically quite diverse, including nonmotile rods, spirals, and spheres (green sulfur bacteria), and motile filamentous gliding forms (green nonsulfur or *Chloroflexus* group) (see Figure 19.6 and Table 19.4). Green bacteria are also very phylogenetically diverse, as the green sulfur bacteria and *Chloroflexus* represent two distinct branches of Bacteria (see Section 18.6). Several other green bacteria have complex appendages called **prosthecae**, and in this connection resemble the budding and/or appendaged bacteria (Section 19.10).

The green sulfur bacteria that live planktonically in lakes generally possess gas vesicles, whereas the species that live in the outflow of sulfur and hot springs, or in other benthic habitats, are not gas vesiculate. Members of one genus, *Pelodictyon*, consist of rods that undergo branching, and since the rods remain attached, a three-dimensional network is formed (Figure 19.6b).

Green sulfur bacteria are strictly anaerobic and obligately phototrophic, being unable to carry out res-

Table 19.3   Genera and characteristics of purple sulfur bacteria*

Characteristics	Genus	DNA (mole % GC)
Sulfur deposited externally:		
Spirilla, polar flagella	*Ectothiorhodospira*	62–70
Sulfur deposited internally:		
Do not contain gas vesicles		
Ovals or rods, polar flagella	*Chromatium*	48–70
Spheres, diplococci, tetrads, nonmotile	*Thiocapsa*	63–70
Spheres or ovals, polar flagella	*Thiocystis*	62–68
Large spirilla, polar flagella	*Thiospirillum*	45
Small spirilla	*Thiorhodovibrio*	61
Contain gas vesicles		
Irregular spheres, ovals, nonmotile	*Amoebobacter*	65
Rods	*Lamprobacter*	64
Spheres, ovals, polar flagella	*Lamprocystis*	64
Rods; nonmotile; forming irregular network	*Thiodictyon*	65–66
Spheres; nonmotile; forming flat sheets of tetrads; some species not available in pure culture	*Thiopedia*	62–64

*From a phylogenetic standpoint, all are members of the gamma subdivision, of the purple Bacteria (Proteobacteria) (see Section 18.6).

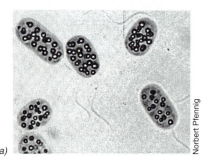
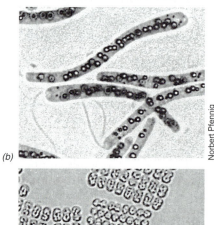
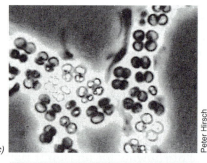
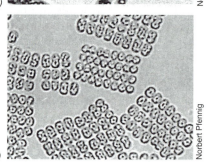

FIGURE 19.5 Bright field photomicrographs of purple sulfur bacteria. (a) *Chromatium okenii*, cells are about 5 µm wide. Note the globules of elemental sulfur inside the cells. (b) *Thiospirillum jenense*, a very large, polarly flagellated spiral, cells are about 30 µm long. Note the sulfur globules. (c) *Thiocapsa*, cells are about 2 µm wide. (d) *Thiopedia rosea*, cells are about 1.5 µm wide. (e) Scanning electron micrograph of a sheet of 16 cells of *Thiopedia rosea* showing the major division planes.

piratory metabolism in the dark. Most green sulfur bacteria can assimilate simple organic substances for phototrophic growth, provided that a reduced sulfur compound is present as a sulfur source (since they are incapable of assimilatory sulfate reduction). Organic compounds used by these species include acetate, propionate, pyruvate, and lactate. *Chloroflexus* is much more versatile than green sulfur bacteria, being able to grow chemoorganotrophically in the dark under aerobic conditions, as well as phototrophically on a wide variety of sugars, amino acids, and organic acids, or photoautotrophically with H_2S or H_2 and CO_2. Although capable of oxidizing sulfide, *Chloroflexus* grows best as a photoheterotroph, and thus, in anal-

ogy to the nutritional situation in purple nonsulfur bacteria, it has been given the designation green *nonsulfur* bacterium.

Green sulfur bacteria share some photochemical characteristics with heliobacteria (see below for discussion of the latter group). The chain of electron carriers between the primary electron acceptor and the bacteriochlorophyll of green bacteria and heliobacteria is quite similar. Of particular interest is the fact that the primary electron acceptor in green bacteria and heliobacteria is poised at a reduction potential of about −0.5 volts. This is much more reducing than the primary electron acceptor of purple bacteria, which has a reduction potential of about −0.15 volts (the primary

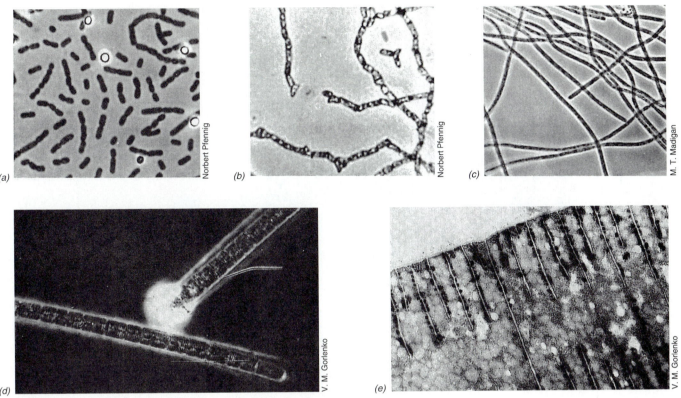

FIGURE 19.6 Green phototrophic bacteria. (a) *Chlorobium limicola*, cells are about 0.8 μm wide. Note the sulfur granules deposited *extra*cellularly. (b) *Pelodictyon clathratiforme*, a bacterium forming a three-dimensional network, cells are about 0.8 μm wide. (c) *Chloroflexus aurantiacus*, a filamentous gliding bacterium, cells are about 1 μm wide. (d) *Oscillochloris*, a large filamentous, gliding green bacterium, cells are about 5 μm wide. Phase contrast. The brightly contrasting material is the holdfast. (e) Electron micrograph of *Oscillochloris*. The chlorosomes in this preparation are darkly stained.

Table 19.4   Genera and characteristics of green phototrophic bacteria

Characteristics	Genus	16S rRNA group*	DNA (mole % GC)
No gas vesicles:			
Straight or curved rods, nonmotile	*Chlorobium*	Green sulfur	49–58
Spheres and ovals, nonmotile, forming prosthecae (appendages)	*Prosthecochloris*	Green sulfur	50–56
Filamentous, gliding	*Chloroflexus*	Green nonsulfur	53–55
	Heliothrix	Green nonsulfur	—
Filamentous, gliding, large diameter (2–5 μm)	*Oscillochloris*	Green nonsulfur	59
Contain gas vesicles:			
Branching nonmotile rods, in loose irregular network	*Pelodictyon*	Green sulfur	48–58
Spheres with prosthecae	*Ancalochloris*	Green sulfur	—
Rods, gliding	*Chloroherpeton*	Green sulfur	45–48
Filamentous, gliding, large diameter (2–2.5 μm)	*Chloronema*	—	—

See Section 18.6. Organisms shown with a dash are not available in pure culture.

electron acceptor in *Chloroflexus* closely resembles that of purple bacteria). The significance of the low potential primary acceptor in green sulfur bacteria and heliobacteria lies in the fact that such an acceptor, once reduced, is sufficiently negative to reduce NAD⁺ directly; this alleviates the need for reverse electron flow as a means of reducing NAD⁺ (Figure 19.7). Reverse electron flow (see Section 16.3) is presumably the mechanism by which purple bacteria and *Chloroflexus* produce NADH (Figure 19.7). Thus, although they employ completely different bacteriochlorophylls in their photosynthetic reactions (see Table 19.1), green sulfur bacteria and heliobacteria are remarkably similar in terms of primary photochemical events.

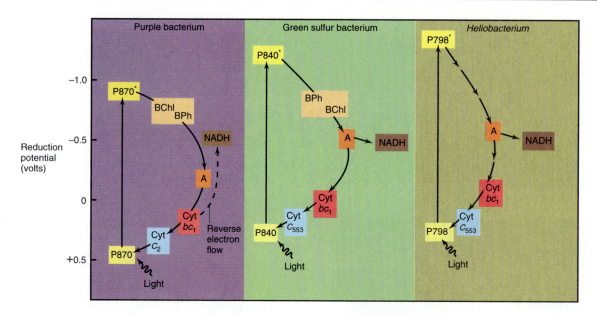

FIGURE 19.7 A comparison of electron flow in green sulfur bacteria, heliobacteria, and purple bacteria. Note how reverse electron flow in purple bacteria is necessitated by the fact that the primary acceptor (A) is more positive in potential than the NAD⁺/NADH couple. BChl, bacteriochlorophyll; BPh, bacteriopheophytin. P870 and P840 are reaction centers of purple and green bacteria, respectively, and consist of BChl *a*. The reaction center of heliobacteria (P798) contains BChl *g*. The reaction center of *Chloroflexus* is similar to that of purple bacteria.

Heliobacteria

The heliobacteria are an entirely separate group of anoxygenic phototrophic bacteria which contain a structurally distinct form of bacteriochlorophyll, bacteriochlorophyll *g* (see Figure 19.1). Bacteriochlorophyll *a* is absent. These bacteria are both physiologically and phylogenetically unique from the other phototrophic bacteria. The group consists of two genera, *Heliobacterium* and *Heliobacillus*. *Heliobacterium* (Figure 19.3*d*) contains gliding rod and motile spirilla species, while *Heliobacillus* (Figure 19.8) is an actively motile rod. Nutritionally, heliobacteria are remarkably restricted. The only carbon sources utilized by these organisms are pyruvate and lactate (*Heliobacillus* will also use acetate and butyrate) and none of the heliobacteria appear capable of autotrophic growth. Although physiologically similar to nonsulfur purple bacteria, heliobacteria are strictly anaerobic phototrophs and are unable to grow by respiratory means as is typical of nonsulfur purple bacteria.

Two species of *Heliobacterium* produce true endospores. These structures contain dipicolinic acid and elevated Ca^{2+} levels, typical of the endospores of *Bacillus* or *Clostridium* (see Section 3.11). Heliobacteria seem most abundant in tropical soils, especially in rice soils where extremes of heat and alternating flooding and drying might favor sporulating phototrophs. Interestingly, phylogenetic studies of heliobacteria show them to be closely related to clostridia (see Section 18.6). Thus, endospore-formation by heliobacteria is explainable in terms of their evolutionary roots.

Figure 19.9 shows the structure of bacteriochlorophyll *g* and that of chlorophyll *a*. Unlike all other bac-

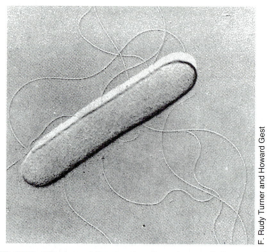

FIGURE 19.8 Electron micrograph of a representative species of heliobacteria, *Heliobacillus mobilis*. Cells are about 1 μm wide. Notice the peritrichous flagellation.

teriochlorophylls, but like chlorophyll *a*, bacteriochlorophyll *g* contains a vinyl ($H_2C{=}CH_2$) group on ring I of the tetrapyrrole molecule (Figures 19.1 and 19.9). The only structural differences between bacteriochlorophyll *g* and chlorophyll *a* therefore lie in ring II of the tetrapyrrole. Like other bacteriochlorophylls, ring II of bacteriochlorophyll *g* is *reduced*, unlike that of chlorophyll *a*. However, if cells of *Heliobacterium* (which are originally a brown-green in color) are exposed to air and light, a bond in ring II of bacteriochlorophyll *g* becomes oxidized, and cultures of the organism slowly turn an emerald green in color, due

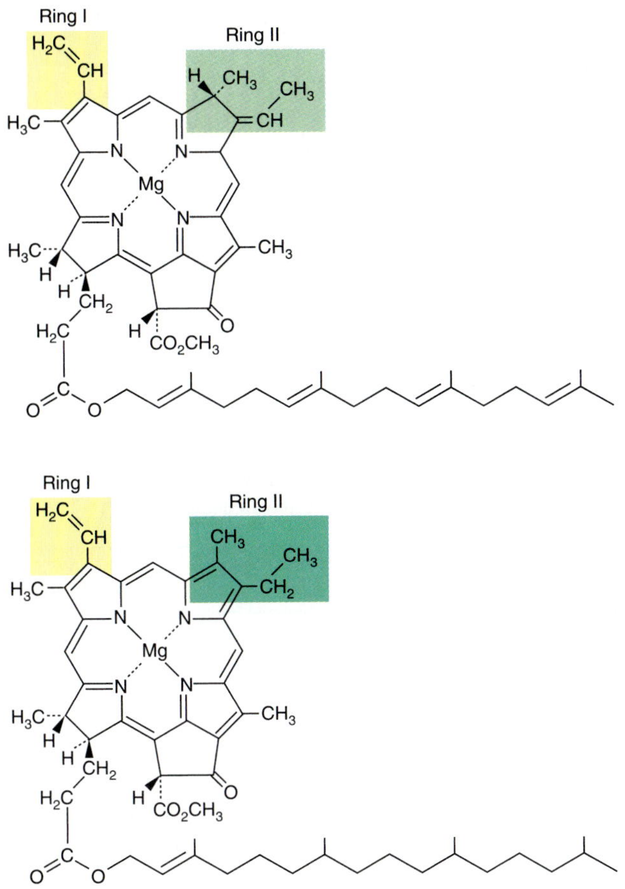

FIGURE 19.9 Comparison of the structures of bacteriochlorophyll *g* (top) and chlorophyll *a* (bottom). Note the differences on ring II of the tetrapyrrole but the similarities on ring I and elsewhere. No other bacteriochlorophyll contains a vinyl group ($H_2C=CH-$) on ring I.

to the conversion of *bacteriochlorophyll g* to *chlorophyll a* (Figure 19.9). This transition is apparently irreversible and leads to loss of cell viability.

Detailed studies of heliobacterial photosynthetic reaction centers have shown that a modified form of chlorophyll *a* called hydroxychlorophyll *a* is present in normal heliobacterial membranes and that it plays a role in photosynthetic electron flow. Thus, although clearly anoxygenic phototrophs, heliobacteria contain a slightly modified form of the major pigment of oxygenic phototrophs. The phylogenetic standing of heliobacteria may be important in the pigment complement of heliobacteria because the Gram-positive roots of these organisms are also shared by the cyanobacteria, probably the first oxygenic phototrophs to have evolved (see Section 18.2).

Physiology of phototrophic growth

In anoxygenic phototrophs light is used primarily in the generation of ATP, and is not generally involved in the generation of reducing power as it is in organisms exhibiting oxygenic photosynthesis. Possible exceptions to this rule exist with green sulfur bacteria

and heliobacteria, as discussed above (see Figure 19.7). However, whether or not a source of reducing power is actually needed depends upon the carbon source supplied. If CO_2 is the sole carbon source, then reducing power is needed, and this can come from a reduced sulfur compound or H_2. Reducing power for CO_2 fixation may also come from an organic compound, but if an organic compound is supplied (photoheterotrophy) the situation is more complex. This is because the organic compound will serve as a carbon source itself, so that CO_2 need not necessarily be reduced. Whether or not CO_2 is reduced when an organic compound is added will depend at least in part on the oxidation state of the organic compound. Compounds such as acetate, glucose, and pyruvate are at about the oxidation level of cell material and can thus be assimilated directly as carbon sources with no requirement for either oxidation or reduction. Fatty acids longer than acetate (for example, propionate, butyrate, caprylate) are more reduced than cell material, and some means of disposing of excess electrons is necessary, such as the reduction of CO_2. Thus the amount of CO_2 fixed by a purple or green bacterium growing with an organic compound will depend upon the oxidation state of the compound, and whether or not inorganic electron donors such as sulfide are also present. Many purple and green bacteria can grow phototrophically using H_2 as sole electron donor, with CO_2 as carbon source. These organisms have a *hydrogenase* for activating H_2 for CO_2 reduction. Most species also fix N_2 and contain a typical nitrogenase system for this purpose (see Section 16.26).

Autotrophy in green bacteria

Although the Calvin cycle (see Section 16.6) is used for CO_2 fixation in phototrophic purple bacteria, it is not the mechanism by which CO_2 is fixed in green sulfur bacteria or in *Chloroflexus*. In *Chlorobium*, CO_2 fixation occurs by a reversal of steps in the tricarboxylic acid cycle, a pathway referred to as the **reverse TCA cycle** (Figure 19.10a). *Chlorobium* contains two ferredoxin-linked enzymes, which catalyze a reductive fixation of CO_2 into intermediates of the tricarboxylic acid cycle (Figure 19.10a). The two ferredoxin-linked reactions involve the carboxylation of succinyl-CoA to α-ketoglutarate and the carboxylation of acetyl-CoA to pyruvate. Most of the other reactions in the reverse citric acid cycle in *Chlorobium* are enzymes of the cycle working in reverse of the normal oxidative direction of the cycle. One exception is *citrate lyase*, an ATP-dependent enzyme which cleaves citrate into acetyl-CoA and oxalacetate in green sulfur bacteria. In the oxidative direction, citrate is produced from these same components by the enzyme *citrate synthase*.

Chloroflexus will grow autotrophically with either H_2 or H_2S as electron donor. However, neither the Calvin cycle nor reverse TCA cycle operate in this organism. Instead, two molecules of CO_2 are converted into glyoxylate by a unique autotrophic pathway, the

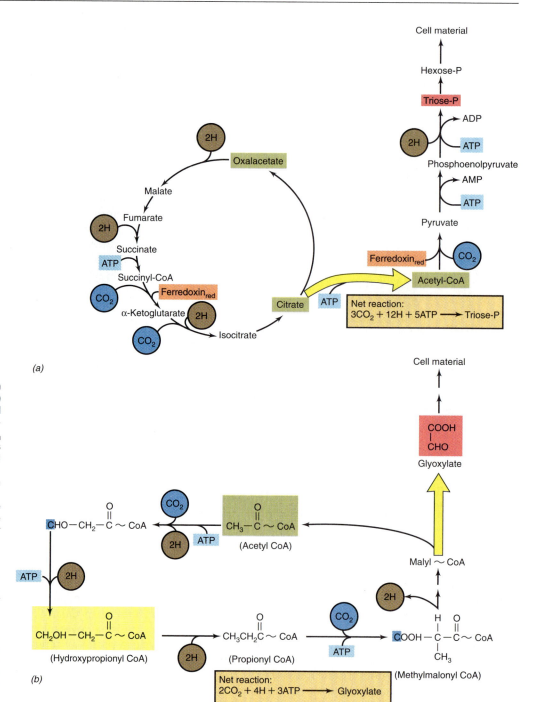

FIGURE 19.10
Unique autotrophic pathways in phototrophic green bacteria. (a) The reverse tricarboxylic acid cycle is the mechanism of CO_2 fixation in *Chlorobium*. Ferredoxin$_{red}$ indicates carboxylation reactions requiring reduced ferredoxin (2H each). Starting from oxalacetate, each turn of the cycle results in three molecules of CO_2 being incorporated and pyruvate as the product. The cleavage of citrate regenerates the C-4 acceptor oxalacetate, and produces acetyl-CoA for biosynthesis. The conversion of pyruvate to phosphoenolpyruvate consumes two ~P equivalents. (b) The hydroxypropionate pathway is the means of autotrophy in *Chloroflexus*. Acetyl-CoA is carboxylated twice to yield methylmalonyl-CoA. This intermediate is rearranged to yield acetyl CoA and glyoxylate. The latter is converted to cell material probably through a serine or glycine intermediate.

hydroxypropionate pathway (Figure 19.10b). This pathway leads to the synthesis of hydroxypropionate as a key intermediate. It is of evolutionary interest that *Chloroflexus*, a phylogenetically ancient member of the Bacteria (see Section 18.6), has a mechanism for autotrophy unknown in any other organism. This suggests that the hydroxypropionate pathway may have been one of the earliest attempts at autotrophy by evolving Bacteria.

Sulfur metabolism in the purple and green bacteria

Most of the purple and green bacteria are able to oxidize reduced sulfur compounds under anaerobic conditions, with the formation of sulfate. The most common reduced sulfur compounds used are *sulfide* and *thiosulfate*. Elemental sulfur is frequently formed during the oxidation of sulfide or thiosulfate, and is either deposited inside or outside the cells. Elemental sulfur deposited inside the cells (see Figure 19.5) is readily available as a further source of electrons. The overall pathway of oxidation of reduced sulfur compounds in purple and green bacteria is shown in Figure 19.11. As seen, either sulfide or thiosulfate is oxidized first to sulfite (SO_3^{2-}) and the enzymes adenosylphosphosulfate reductase (APS reductase) and ADP-sulfurylase catalyze the oxidation of sulfite to sulfate. Note that a substrate-level phosphorylation occurs at this step, providing for the synthesis of a high-energy phosphate bond (in ADP).

It seems likely that elemental sulfur is not an

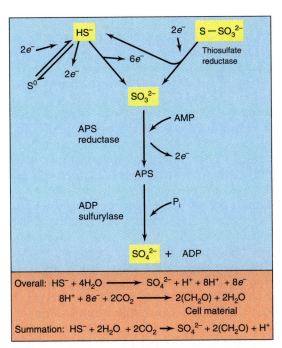

FIGURE 19.11 Pathways of oxidation of reduced sulfur compounds in phototrophic bacteria. Note that a substrate-level phosphorylation occurs via the enzyme ADP sulfurylase.

obligatory intermediate between sulfide and sulfate, but merely a side product. Elemental sulfur is a *storage* product, formed when sulfide concentrations in the environment are high. Indeed, under limiting sulfide levels, sulfide is oxidized directly to sulfate without formation of elemental sulfur. The formation of elemental sulfur as a storage product has been most clearly shown in *Chromatium*, where it is deposited *inside* the cells (Figure 19.5*a*; see also Figure 16.8*b*). Intracellular elemental sulfur in *Chromatium* can serve as an electron donor for phototrophic growth when sulfide is absent. In green bacteria, sulfur is formed *outside* the cells but often remains attached to the outer surface of the cell and can still be used as an electron donor.

Symbiotic associations and mixed culture interactions

A two-membered system in which each organism does something for the benefit of the other has been called a **consortium**. An association of two organisms consisting of a large, colorless central bacterium, which is polarly flagellated, surrounded by 12–24 smaller, ovoid- to rod-shaped green sulfur bacteria arranged in rows has been called *Chlorochromatium aggregatum* (Figure 19.12*a*). Electron micrographs of thin sections of this association clearly show the chlorosomes of the green bacterial partner and the intimate relationship which exists between the two components of *C. aggregatum* (Figure 19.12*b*). Cell division in the *C. aggregatum* symbiosis is synchronous, suggesting that the two cell types have some means of communicating with each other.

The genus and species name *Chlorochromatium aggregatum* are invalid in formal taxonomy because they refer not to a single organism but to an association of two or more organisms, yet the association is seen

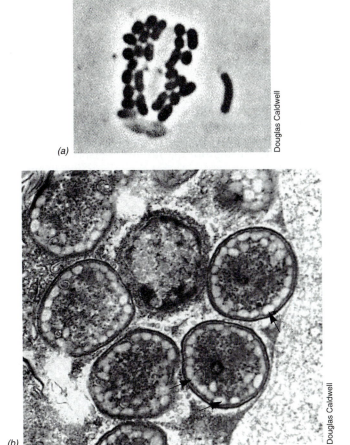

FIGURE 19.12 (a) Phase contrast micrograph and (b) transmission electron micrograph of the green bacterial consortium, *Chlorochromatium aggregatum*. In (a) the nonphotosynthetic central organism is much lighter in color than the pigmented phototrophic bacteria. Note the chlorosomes (arrows) in (b). The entire consortium is about 3 by 6 μm.

quite commonly in lakes and in muds and is probably a rather specific one. The green organism in this association has been cultured and resembles *Chlorobium limicola*, but the large colorless central organism has not been cultured. A similar association in which the colored organism is a brown-pigmented *Chlorobium* has been called *Pelochromatium roseum*. The symbiosis presumably functions to allow the normally nonmotile green bacterium to move up and down in the water column in response to gradients of light and sulfide.

A clue to the possible role of the colorless central organism comes from the common observation that the sulfide needed by purple and green bacteria can be derived from sulfate- or sulfur-reducing bacteria associated with them. Sulfate- and sulfur-reducing bacteria use organic compounds such as ethanol, lactate, formate, or fatty acids as electron donors and reduce sulfate or sulfur to sulfide. The sulfide produced can then be used by associated phototrophic bacteria, which in the light oxidize the sulfide back to sulfate. If some of the organic matter produced by the phototrophic green bacterium is used by the sulfate reducer, a self-feeding system can develop, driven by light energy. This interrelationship may also control motility of the consortium.

Ecology

Masses of purple and green bacteria are frequently found in the depths of certain kinds of lakes where stable conditions for growth occur. We discussed the development of thermal stratification in lakes in Chapter 17 (see Figure 17.15). After stratification occurs, stable anoxic conditions may continue in the deep waters throughout the summer season. If there is a sufficient supply of H_2S and the lake water is sufficiently clear so that light penetrates to the anoxic zone,

a massive layer of purple or green bacteria can develop (Figure 19.13). This layer, hidden from view of the observer on the surface, can be studied by sampling water at various depths (Figure 19.14). The bacteria often form a distinct layer just at the depth where H_2S is first present (Figure 19.15). The most favorable lakes for development of these bacteria are those called **meromictic**, which are permanently stratified due to the presence of denser (usually saline) water in the bottom, or **holomictic** lakes that stratify seasonally because of thermal differences. In meromictic lakes, the bloom of purple or green bacteria may be present throughout the year. Often the photosynthetic activity of these bacteria is sufficiently great so that it is an important source of organic matter to the lake ecosystem. In some cases, more photosynthesis may occur in the bacterial layer than in the surface algal layer. In general, blooms of purple and green bacteria are more common in small lakes and ponds than in large bodies of water, mainly because the stratification of large bodies of water is more affected by wind.

One of the easiest places to observe purple and green bacteria in nature is in sulfur springs, where massive blooms are often present a few inches below the surface of the water (Figure 19.16). Along the seacoast, blooms also occur in warm shallow pools of seawater not connected to the open ocean, where the activities of sulfate-reducing bacteria lead to production of large amounts of H_2S.

Hot springs are habitats for a variety of purple and green bacteria. *Chloroflexus* forms thick mats (frequently in association with cyanobacteria) in hot springs at temperatures from 40 to 70°C (see Figure 17.11a for photograph of a microbial mat). *Chlorobium* mats are occasionally found in high sulfide acidic hot springs, generally at temperatures below 50°C. One species of the purple sulfur bacterium *Chromatium*, *Chromatium tepidum*, grows in high sulfide springs at temperatures up to 60°C. A filamentous, phototrophic gliding bacterium that resembles *Chloroflexus* but that lacks chlorosomes and bacteriochlorophyll c_s grows in certain hot springs in association with *Chloroflexus*. This organism, named *Heliothrix*, is unusual among phototrophic bacteria because it requires extremely high light intensities for growth. Although *Heliothrix* contains only bacteriochlorophyll *a* and thus should technically be considered a purple bacterium, the morphological similarity between *Heliothrix* and *Chlo-*

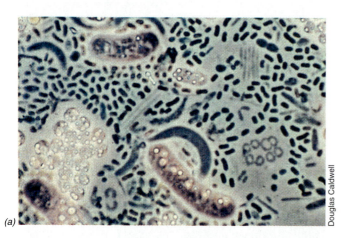

(a)

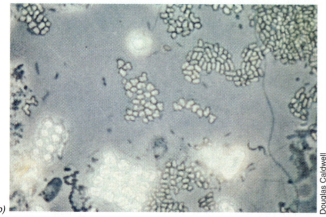

(b)

FIGURE 19.13 (a) Phase contrast photomicrographs of layers containing purple sulfur bacteria and (b) green sulfur bacteria from the water column of a small, stratified lake in Michigan. The purple bacteria include large-celled *Chromatium* species and *Thiocystis*. The green bacteria are predominantly cells of *Ancalochloris* filled with gas vesicles.

FIGURE 19.14
Water samples from various depths of a small meromictic alpine lake in Switzerland. During the summer months, a layer of phototrophic purple sulfur bacteria forms at a depth of 13–13.5 meters.

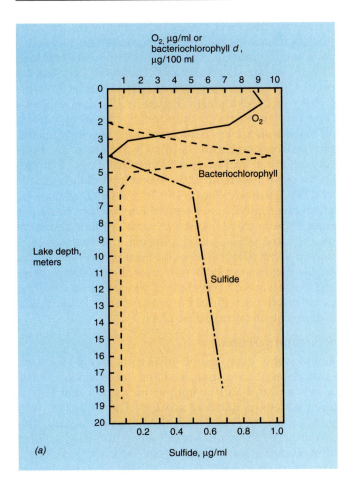

(a)

FIGURE 19.16 Massive accumulation of purple sulfur bacteria, *Thiopedia roseopersicinia*, in a spring in Madison, Wisconsin. The bacteria grow near the bottom of the spring pool and float to the top (by virtue of their gas vesicles) when disturbed. The green color is from cells of the eukaryotic alga *Spirogyra*.

FIGURE 19.17 Cross section through a bacterial mat in the Sippewisset salt marsh near Woods Hole, Massachusetts. The layers are as follows: top, cyanobacteria; pink, phototrophic purple sulfur bacteria attached to sand grains; black, sulfate-reducing bacteria; peach, bacteriochlorophyll *b*-containing cells of *Thiocapsa pfennigii*.

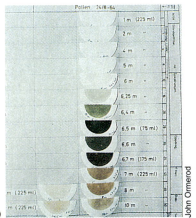

(b)

FIGURE 19.15 Vertical distribution of *Chlorobium* in stratified lakes. (a) Lake Mary, Wisconsin. The phototrophic bacterium forms a layer just at the top of the anaerobic zone. The bacterial population size is quantified by measuring bacteriochlorophyll *d* concentration. (b) Stratified Norwegian lake. Series of membrane filters through which were passed water samples taken at varying depths. At 6.4–6.6 meters a heavy cyanobacterial bloom is present, and at 6.7–8 meters the green sulfur bacterium *Pelodictyon luteolum* reaches its highest population density.

roflexus, the ability to glide, and the fact that the two organisms have nearly identical carotenoids have been used to provisionally classify *Heliothrix* in with the other gliding green bacteria. Nucleic acid sequencing of ribosomal RNA (see Chapter 18) from *Heliothrix* and *Chloroflexus* supports this classification.

Mats of phototrophic bacteria also form in nonthermal environments. Cyanobacterial-phototrophic bacterial mats are common in salt marsh and marine intertidal regions (Figure 19.17). The top layer of the mat usually contains filamentous cyanobacteria, and phototrophic purple (generally purple sulfur) bacteria lie underneath (Figure 19.17). Some marine mats also contain *Chloroflexus*. The lower layers of these mat communities are typically black, due to active sulfate reduction which generates the sulfide necessary for development of the purple bacteria.

Purple and green bacteria are one of the most diverse groups of bacteria known and are of interest for a wide variety of reasons. They have provided extremely useful systems for studying fundamental aspects of photosynthesis (see Chapter 16), and are of great evolutionary significance (see Chapter 18). Their ecological roles may also be important, and their extreme diversity challenges the bacterial taxonomist.

19.2 Cyanobacteria

The **cyanobacteria** comprise a large and heterogeneous group of phototrophic Bacteria. Cyanobacteria differ in fundamental ways from purple and green bacteria, most notably in the fact they are *oxygenic* phototrophs. Cyanobacteria represent one of the major phylogenetic lines of Bacteria and show a distant relationship to Gram-positive Bacteria (see Section 18.6).

Structure

The morphological diversity of the cyanobacteria is considerable (Figure 19.18). Both unicellular and filamentous forms are known, and considerable variation within these morphological types occurs. *Bergey's Manual* has divided the cyanobacteria into five morphological groups: unicellular dividing by binary fission (see Figure 19.18*a*); unicellular dividing by multiple fission (colonial, see Figure 19.18*b*); filamentous containing differentiated cells called heterocysts which function in nitrogen fixation (see Figures 19.18*d* and 19.20); filamentous nonheterocystous forms (see Figure 19.18*c*); and branching filamentous types (see Figure 19.18*e*). Cyanobacterial cells range in size from those of typical bacteria (0.5–1 μm in diameter) to cells as large as 60 μm in diameter (in the species *Oscillatoria princeps*).

The cyanobacteria differ in fatty acid composition from all other prokaryotes. Other bacteria contain almost exclusively saturated and monounsaturated fatty acids (one double bond), but the cyanobacteria frequently contain unsaturated fatty acids with two or more double bonds.

The fine structure of the cell wall of some cyanobacteria is similar to that of Gram-negative Bacteria and peptidoglycan can be detected in the walls. Many cyanobacteria produce extensive mucilaginous envelopes, or sheaths, that bind groups of cells or filaments together (see, for example, Figure 19.18*a*). The photosynthetic lamellar membrane system is often complex and multilayered (see Figure 16.16*b*), although in some of the simpler cyanobacteria the lamellae are regularly arranged in concentric circles around the periphery of the cytoplasm (Figure 19.19). Cyanobacteria have only one form of chlorophyll, chlorophyll *a*, and all of them also have characteristic biliprotein pigments, **phycobilins** (see Figure 16.16), which function as accessory pigments in photosynthesis. One class of phycobilins, *phycocyanins*, are blue, absorbing light maximally at 625–630 nm (see Figure 16.17), and together with the green chlorophyll *a*, are responsible for the blue-green color of the bacteria. However, some cyanobacteria produce *phycoerythrin*, a red phycobilin absorbing light maximally at 570–580 nm, and bacteria possessing this pigment are red or brown in color. Even more confusing, the eukaryotic red algae (Rhodophyta) are red because of phycoerythrin, but some species have phycocyanin instead and are blue-green.

Structural variation

Among the cytoplasmic structures seen in many cyanobacteria are **gas vesicles** (see Section 3.10), which are especially common in species that live in open waters (planktonic species). Their function is to provide the organism with flotation (see Figure 3.53) so that it may remain where there is most light. Some cyanobacteria form **heterocysts**, which are rounded, seemingly more or less empty cells, usually distributed individually along a filament or at one end of a filament (Figure 19.20). Heterocysts arise from differentiation of vegetative cells and are the sole sites of *nitrogen fixation* in heterocystous cyanobacteria. In *Anabaena*, a well studied heterocystous cyanobacterium, complex gene rearrangements occur within the heterocyst to yield a contiguous cluster of *nif* genes that can be expressed as a unit (see Section 16.26).

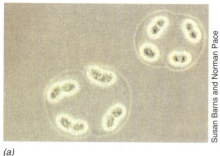
(a)

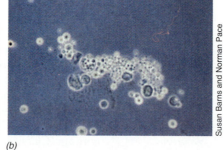
(b)

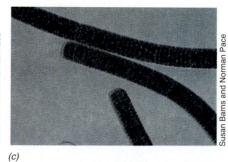
(c)

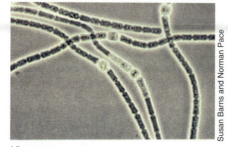
(d)

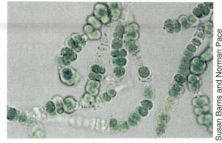
(e)

FIGURE 19.18 Morphological diversity among the cyanobacteria: the five major morphological types of cyanobacteria. (a) Unicellular, *Gloeothece*, phase contrast; a single cell measures 5–6 μm in diameter; (b) colonial, *Dermocarpa*, phase contrast; (c) filamentous, *Oscillatoria*, bright field; a single cell measures about 15 μm wide; (d) filamentous heterocystous, *Anabaena*, phase contrast; a single cell measures about 5 μm wide; (e) filamentous branching, *Fischerella*, bright field.

Susan Barns and Norman Pace

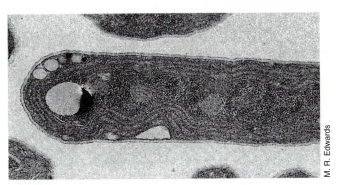

FIGURE 19.19 Electron micrograph of a thin section of the cyanobacterium *Synechococcus lividus*.

Heterocyst

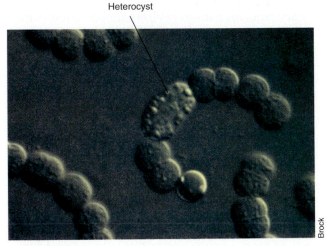

FIGURE 19.20 Heterocysts in the cyanobacterium *Anabaena* sp. Heterocysts are the sole site of nitrogen fixation in heterocystous cyanobacteria.

Heterocysts have intercellular connections with adjacent vegetative cells, and there is mutual exchange of materials between these cells, with products of photosynthesis moving from vegetative cells to heterocysts and products of nitrogen fixation moving from heterocysts to vegetative cells. Heterocysts are low in phycobilin pigments and *lack* photosystem II, the oxygen-evolving photosystem. They are also surrounded by a thickened cell wall containing large amounts of glycolipid, which serves to slow the diffusion of O_2 into the cell. Because of the oxygen lability of the enzyme nitrogenase (see Section 16.26), it seems likely that the heterocyst, by maintaining an anoxic environment, stabilizes the nitrogen-fixing system in organisms that are not only aerobic but also oxygen-producing. Indeed, some nonheterocystous filamentous cyanobacteria produce nitrogenase and fix nitrogen in normal vegetative cells if they are grown anaerobically. However, a few unicellular cyanobacteria of the sheath-forming *Gloeothece* type (Figure 19.18a) do not produce heterocysts but nevertheless fix nitrogen aerobically. Some marine *Oscillatoria* also fix nitrogen without heterocysts and seem to produce a series of cells in the center of the filament that lack photosystem II activity; nitrogen fixation apparently occurs in this O_2-nonproducing region.

A structure called **cyanophycin** can be seen in electron micrographs of many cyanobacteria. This structure is a simple polymer of aspartic acid, with each aspartate residue containing an arginine molecule:

$$
\begin{array}{ccccc}
\text{asp} & \text{asp} & \text{asp} & \text{asp} & \text{asp} \\
| & | & | & | & | \\
\text{arg} & \text{arg} & \text{arg} & \text{arg} & \text{arg}
\end{array}
$$

and can constitute up to 10 percent of the cell mass. This co-polymer serves as a nitrogen storage product in many cyanobacteria, and when nitrogen in the environment becomes deficient, this polymer is broken down and used. The phycobilin pigments can also constitute a major portion of the cell mass, up to 10 percent, and likewise serve as a nitrogen storage material, being broken down under nitrogen starvation. Because of this, nitrogen-starved cyanobacteria often appear green instead of blue-green in color. Cyanophycin can apparently also serve as an energy reserve in cyanobacteria. Arginine, derived from cyanophycin, can be hydrolyzed to yield ornithine, with the production of ATP through the action of the enzyme *arginine dihydrolase* with carbamyl phosphate (see Section 16.19) occurring as an intermediate:

$$
\begin{array}{l}
\text{Arginine} + \text{ADP} + \text{P}_i + \text{H}_2\text{O} \rightarrow \\
\text{Ornithine} + 2\text{NH}_3 + \text{CO}_2 + \text{ATP}
\end{array}
$$

Arginine dihydrolase is present in many cyanobacteria and may function there as a source of ATP for maintenance purposes during dark periods.

Many, but by no means all, cyanobacteria exhibit gliding motility; flagella have never been found. Gliding occurs only when the cell or filament is in contact with a solid surface or with another cell or filament. In some cyanobacteria gliding is not a simple translational movement but is accompanied by rotations, reversals, and flexings of filaments. Most gliding forms exhibit directed movement in response to light (phototaxis); it is usually positive although negative movement from bright light may also occur. Chemotaxis (see Section 3.8) may occur as well. One species of marine cyanobacteria is capable of swimming motility, but the mechanism of motility, which does not involve flagella, is unknown.

Among the filamentous cyanobacteria, fragmentation of the filaments often occurs by formation of **hormogonia** (Figure 19.21a and b), which break away from the filaments and glide off. In some species resting spores or **akinetes** (Figure 19.21c) are formed, which protect the organism during periods of darkness, drying, or freezing. These are cells with thickened outer walls; they germinate through the breakdown of the outer wall and outgrowth of a new vegetative filament. However, even the vegetative cells of many cyanobacteria are relatively resistant to drying or low temperatures.

Physiology

The nutrition of cyanobacteria is simple. Vitamins are not required, and nitrate or ammonia is used as nitro-

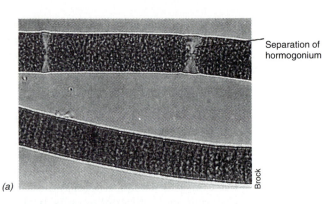

Separation of hormogonium

Brock

(a)

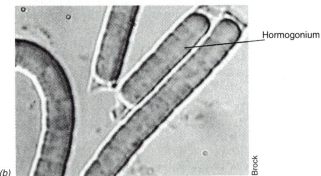

Hormogonium

Brock

(b)

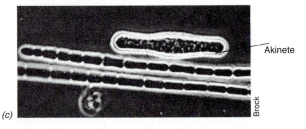

Akinete

Brock

(c)

FIGURE 19.21 Structural differentiation in filamentous cyanobacteria. (a) Initial stage of hormogonium formation in *Oscillatoria*. Notice the empty spaces where the hormogonium is separating from the filament. (b) Hormogonium of a smaller *Oscillatoria* species. Notice that the cells at both ends are rounded. Nomarski interference contrast microscopy. (c) Akinete (resting spore) of *Anabaena* sp. by phase contrast.

needed to cause death do not occur extensively, although subclinical manifestations of cyanobacterial water blooms may be a common, but unobserved, occurrence. Many cyanobacteria are also responsible for the production of earthy odors and flavors in fresh waters, and if such waters are used as drinking water sources, esthetic problems may arise. The compound produced is **geosmin** (*trans*-1,10-dimethyl-*trans*-9-decalol). This substance is also produced by many actinomycetes (see the discussion in Section 19.32) and is responsible for the distinctive "earthy" odor of soil.

Ecology and evolution

Cyanobacteria are widely distributed in nature in terrestrial, freshwater, and marine habitats. In general they are more tolerant to environmental extremes than are algae and are often the dominant or sole photosynthetic organisms in hot springs (Table 9.2), saline lakes, and other extreme environments. Many members are found on the surfaces of rocks or soil and occasionally even within rocks themselves. In desert soils subject to intense sunlight, cyanobacteria often form extensive crusts over the surface, remaining dormant during most of the year and growing during the brief winter and spring rains. In shallow marine bays, where relatively warm seawater temperatures exist, cyanobacterial mats of considerable thickness may form. Freshwater lakes, especially those that are fairly rich in nutrients, may develop blooms of cyanobacteria (see Figure 3.53). A few cyanobacteria are symbionts of liverworts, ferns, and cycads; a number are found as the photosynthetic component of lichens. In the case of the water fern *Azolla* (see Section 17.24), it has been shown that the cyanobacterial endophyte (a species of *Anabaena*) fixes nitrogen that becomes available to the plant.

Base compositions of DNA of a variety of cyanobacteria have been determined. Those of the unicellular forms vary from 35 to 71 percent GC, a range so wide as to suggest that this group contains many members with little relationship to each other. On the other hand, the values for the heterocyst formers vary much less, from 39 to 47 percent GC. Phylogenetically, cyanobacteria group along morphological lines in most cases (see Section 18.6). Filamentous heterocystous and nonheterocystous species form distinct groups, as do the branching forms. However, unicellular cyanobacteria are phylogenetically highly diverse, as representatives show evolutionary relationships to several different groups and show little phylogenetic relationship to one another.

The evolutionary significance of cyanobacteria was discussed in Section 18.2, and it is nearly certain that these organisms were the first oxygen-evolving phototrophic organisms and were responsible for the initial conversion of the atmosphere of the earth from anoxic to oxic. Microfossil evidence of cyano-bacterial-like organisms over 3 billion years old is good (see Section 18.1), and there is evidence that cyanobacteria occupied vast areas of the earth in those ancient times.

gen source. Nitrogen-fixing species are also common. Most species tested are obligate phototrophs, being unable to grow in the dark on organic compounds. However, some cyanobacteria can assimilate simple organic compounds such as glucose and acetate if light is present. Apparently they are unable to make ATP by oxidation of organic compounds, but if ATP is provided by means of photophosphorylation, organic compounds can be utilized as carbon sources. Some species, mainly filamentous forms, can grow in the dark on glucose or other sugars, using the organic material as both carbon and energy source. As we discussed in Section 16.5, a number of cyanobacteria can carry out anoxygenic photosynthesis using only photosystem I when sulfide is present in the environment.

Several metabolic products of cyanobacteria are of considerable practical importance. Many cyanobacteria produce potent neurotoxins, and during water blooms when massive accumulations of cyanobacteria may develop, animals ingesting such water may succumb rapidly. Fortunately, the massive accumulations

Although quantitatively much less significant today, the cyanobacteria still exist in considerable numbers and can be the dominant microbial member of certain communities, such as microbial mats (see Section 17.5).

19.3   Prochlorophytes

Prochlorophytes are prokaryotic phototrophs that contain chlorophyll *a* and *b* but that do *not* contain phycobilins. Prochlorophytes therefore resemble both cyanobacteria (because they are prokaryotic and produce chlorophyll *a*) and the plant chloroplast (because they contain chlorophyll *b* instead of phycobilins). *Prochloron*, the first prochlorophyte discovered, generated obvious excitement and stimulated attempts to trace the evolutionary origin of the chloroplast using molecular approaches to phylogeny (see Section 18.4). Phenotypically, *Prochloron* is the best candidate yet discovered for the type of organism that led, following endosymbiotic events (see Section 18.3), to the green plant chloroplast.

Unfortunately, progress in the study of *Prochloron* has been rather slow because the organism has not yet been grown in laboratory culture. *Prochloron* is found in nature as a symbiont of marine invertebrates (didemnid ascidians) and all studies of the organism to date have relied on material collected from natural samples. Cells of *Prochloron* expressed from the cavities of didemnid tissue are roughly spherical in morphology (Figure 19.22), 8–10 μm in diameter. Electron micrographs of thin sections (Figure 19.22) show that *Prochloron* has an extensive thylakoid membrane system similar to that observed in the chloroplast (see Figure 3.72). Evidence that *Prochloron* is phylogenetically a member of the Bacteria is the presence of muramic acid in the cell walls, indicating that peptidoglycan is present.

The ratio of chlorophyll *a*:chlorophyll *b* in cells of *Prochloron* is about 4–7:1, somewhat higher than the value of 1–2:1 typical of green algae. The carotenoids of *Prochloron* are similar to those of cyanobacteria, predominantly β-carotene and zeaxanthin. The GC base ratio of different samples of *Prochloron* isolated from different ascidians vary from 31 to 41 percent, indicating a fair bit of genetic heterogeneity. Different species of *Prochloron* probably exist, but confirmation of this must await laboratory culture and study of pure strains.

Another prochlorophyte has been discovered that can be grown in pure culture. This prochlorophyte is filamentous (Figure 19.23), and grows as one of the dominant phototrophic bacteria in several shallow Dutch lakes. This filamentous prochlorophyte has been given the genus name *Prochlorothrix*. Like *Prochloron*, *Prochlorothrix* is prokaryotic and contains chlorophylls *a* and *b* and lacks phycobilins. The ratio of chlorophyll *a* to chlorophyll *b* is somewhat higher in *Prochlorothrix* (about 8:1 to 9:1) than in *Prochloron* and the thylakoid membranes are much less developed than in *Prochloron* (compare Figures 19.22 and 19.23*b*).

A novel type of prochlorophyte has been found in the deep euphotic zone of the open oceans. These phototrophs are extremely small cocci, measuring less than 1 μm in diameter, and because of this have been referred to as *picoplankton*. Like other prochlorophytes, picoplankton are prokaryotic and contain chlorophyll *b*. However, these phototrophs lack true chlorophyll *a*, and produce instead a modified form of chlorophyll *a* called *divinyl-chlorophyll a*. Picoplankton also contain α- (instead of β-) carotene, a pigment previously unknown in prokaryotes. Interestingly, the chlorophyll *a/b* ratio of picoplankton is near 1, similar to that of marine green algae. Although not yet obtained in laboratory culture, picoplankton prochlorophytes suggest that considerable diversity may be present in the prochlorophyte group. Also, because their numbers in the oceans are relatively large (10^4–10^5 cells/ml), picoplankton prochlorophytes may have considerable ecological significance as primary producers in open ocean waters.

Recalling our discussion of endosymbiosis (see Section 18.3), the evolutionary significance of prochlorophytes should be apparent. Until the discovery of prochlorophytes it was always assumed that the chloroplast originated from endosymbiotic association of *cyanobacteria* with a primitive eukaryotic cell. However, this hypothesis has never been scientifically satisfying for one major reason: how did the plant chloroplast evolve the pigment complement it has today if it originated from a *cyanobacterial* endosymbiont that contained *phycobilins* instead of chlorophyll *b*? The hypothesis that *prochlorophytes* instead of cyanobacteria were the ancestors of the green plant chloroplast eliminates this major point of contention. However, comparative sequencing of 16S rRNA (see Section 18.6) does not show *Prochloron* or *Prochlorothrix* to be the immediate ancestors of the green plant chloroplast. Instead, prochlorophytes, cyanobacteria and the plant chloroplast all *share* a common ancestor. Thus, although prochlorophytes have arisen from the same evolutionary roots as the chloroplast and are phenotypically highly related to chloroplasts, it appears that the closest free-living ancestor of chloroplasts has thus far defied isolation and culture.

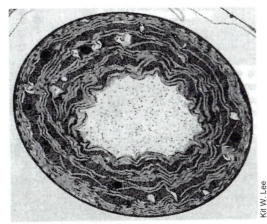

Kit W. Lee

FIGURE 19.22 Electron micrograph of the prochlorophyte, *Prochloron*. Note the extensive intracytoplasmic membranes (thylakoids). Cells are about 10 μm in diameter.

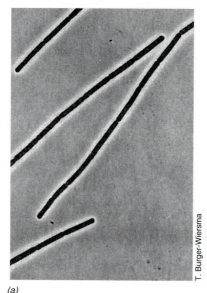
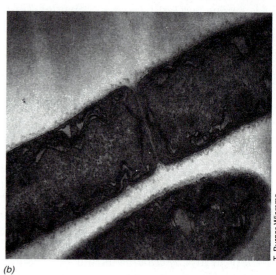

FIGURE 19.23
Phase and electron micrographs of the filamentous prochlorophyte, *Prochlorothrix*. (a) Phase contrast. (b) Electron micrograph of thin section showing arrangement of membranes. The diameter of cells is approximately 1–1.5 μm.

(a)

(b)

T. Burger-Wiersma

19.4 Chemolithotrophs: The Nitrifying Bacteria

We discussed in Chapter 16 the conceptual basis of chemolithotrophy. Various chemolithotrophic bacteria are known, but they are all physiologically united by their ability to utilize *inorganic* electron donors as energy sources. Most chemolithotrophs are also capable of autotrophic growth and in this way share a major physiological trait with phototrophic bacteria and cyanobacteria. The best studied chemolithotrophs are those capable of oxidizing reduced sulfur and nitrogen compounds, and the hydrogen-oxidizing bacteria, and we focus here and in the next two sections on these groups.

Bacteria able to grow chemolithotrophically at the expense of reduced inorganic nitrogen compounds are called **nitrifying bacteria**. Several genera are recognized on the basis of morphology and the particular steps in the oxidation sequences that they carry out (Table 19.5). No chemolithotroph is known that will carry out the complete oxidation of ammonia to ni-

Table 19.5 Characteristics of the nitrifying Bacteria*

Characteristics	Genus	DNA (mole % GC)	Habitats
Oxidize ammonia:			
Gram-negative short to long rods, motile (polar flagella) or nonmotile; peripheral membrane systems	*Nitrosomonas*	45–53	Soil, sewage, freshwater, marine
Large cocci, motile; vesicular or peripheral membranes	*Nitrosococcus*	49–50	Freshwater, marine
Spirals, motile (peritrichous flagella); no obvious membrane system	*Nitrosospira*	54	Soil
Pleomorphic, lobular, compartmented cells; motile (peritrichous flagella)	*Nitrosolobus*	54	Soil
Slender curved rods	*Nitrosovibrio*	54	Soil
Oxidize nitrite:			
Short rods, reproduce by budding, occasionally motile (single subterminal flagellum); membrane system arranged as a polar cap	*Nitrobacter*	59–62	Soil, freshwater, marine
Long, slender rods, nonmotile; no obvious membrane system	*Nitrospina*	58	Marine
Large cocci, motile (one or two subterminal flagella); membrane system randomly arranged in tubes	*Nitrococcus*	61	Marine
Helical to vibrioid shaped cells; nonmotile; no internal membranes	*Nitrospira*	50	Marine

*Phylogenetically, all nitrifying bacteria thus far examined are either α or β purple Bacteria (see Section 18.6).

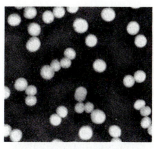
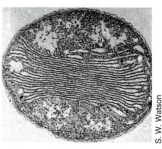

FIGURE 19.24 Phase contrast photomicrograph and electron micrograph of the nitrosofying bacterium *Nitrosococcus oceanus*. A single cell is about 2 μm in diameter.

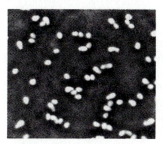
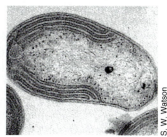

FIGURE 19.25 Phase contrast photomicrograph and electron micrograph of the nitrifying bacterium *Nitrobacter winogradskyi*. A cell is about 0.7 μm in diameter.

Nitrosofying bacteria

1. $NH_3 + O_2 + NADH + H^+ \longrightarrow NH_2OH + H_2O + NAD^+$

2. $NH_2OH + O_2 \longrightarrow NO_2^- + H_2O + H^+$

Sum: $NH_3 + 2O_2 + NADH \longrightarrow NO_2^- + 2H_2O + NAD^+$

$\Delta G^{0'} = -275$ kJ/reaction

Nitrifying bacteria

$NO_2^- + 1/2O_2 \longrightarrow NO_3^-$

$\Delta G^{0'} = -76$ kJ/reaction

FIGURE 19.26 Reactions involved in the oxidation of inorganic nitrogen compounds by chemolithotrophic nitrifying bacteria.

trate; thus **nitrification** of ammonia in nature results from the sequential action of two separate groups of organisms, the **ammonia-oxidizing bacteria**, the **nitrosofyers** (Figure 19.24), and the **nitrite-oxidizing bacteria**, the true nitrifying (nitrate-producing) bacteria (Figure 19.25). Some chemoorganotrophic bacteria and fungi will oxidize ammonia completely to nitrate, but the rate of the process is much less than that accomplished by the chemolithotrophic nitrifying bacteria, and may not be significant ecologically. Historically, the nitrifying bacteria were the first organisms to be shown to grow chemolithotrophically; Winogradsky showed that they were able to produce organic matter and cell mass when provided with CO_2 as sole carbon source.

Many of the nitrifying bacteria have remarkably complex internal membrane systems (see Figures 19.24 and 19.25), although not all genera have such membranes.

Biochemistry

Molecular oxygen is required for ammonia oxidation, the initial step involving a *monooxygenase*, which uses NADH as electron donor. The first product of ammonia oxidation is *hydroxylamine*, NH_2OH, and no energy is generated in this step (energy is actually used up via the oxidation of NADH) (Figure 19.26). Hydroxylamine is then oxidized to nitrite and ATP formation occurs at this step via electron-transport phosphorylation through a cytochrome system (Figure 19.26; see also Figure 16.25).

There are some interesting similarities between the ammonia-oxidizing bacteria and the methane-oxidizing bacteria. As noted in Section 19.7, methane-oxidizing bacteria generally oxidize ammonia to nitrite, and ammonia inhibits methane oxidation. In a similar fashion, certain ammonia oxidizers are capable of oxidizing methane and can incorporate significant amounts of carbon originating from methane into cell material. However, methane will *not* serve as sole carbon and electron donor for growth of ammonia oxidizers. Nitrite oxidizers do not oxidize methane. It appears that the two substrates methane and ammonia have some structural similarities, so that an enzyme which recognizes one can also combine with, and be inhibited by, the other. Another similarity between ammonia oxidizers and methane oxidizers is that both groups generally possess extensive internal membrane systems (compare Figure 19.24 with Figure 19.31).

Only a single step is involved in the oxidation of nitrite to nitrate by the nitrite-oxidizing bacteria (Figure 19.26; see also Figure 16.25). This reaction is carried out by a *nitrite oxidase* system, the electrons being transported to O_2 via cytochromes, with ATP being generated by electron-transport phosphorylation. The energy available from the oxidation of nitrite to nitrate is only 76 kJ/mole, which is sufficient for the formation of two ATPs. However, careful measurements of molar growth yields suggest that only *one* ATP is produced per each NO_2^- oxidized to NO_3^-. As we discussed in Section 16.8, the generation of reducing power (NADPH) for the reduction of CO_2 to organic compounds comes from ATP-driven *reversed electron transport* reactions, since the NO_3^-/NO_2^- reduction potential is too high to reduce $NADP^+$ directly.

Ecology

Although rarely present in large numbers, the nitrifying bacteria are widespread in soil and water. They can be expected to be present in highest numbers in habitats where considerable amounts of ammonia are present, such as sites where extensive protein decomposition occurs (ammonification). Nitrifying bacteria develop especially well in lakes and streams which receive inputs of untreated (or even treated) sewage, because sewage effluents are generally high in ammonia. Because O_2 is required for ammonia oxidation, ammonia tends to accumulate in anaerobic habitats, and in stratified lakes nitrifying bacteria may develop especially well at the thermocline, where both ammonia and O_2 are present. Nitrification results in acidification of the habitat, due to the buildup of nitric acid (the situation is analogous to the buildup of sulfuric acid through the activities of sulfur-oxidizing bacteria) (Section 17.15). Since nitrous acid can form at acid pH values as a result of ammonia oxidation, the accumulation of this toxic (and mutagenic) (see Table 7.2) agent in acidic environments can result in inhibition of further nitrification. In general, nitrification is much more extensive in neutral and alkaline than in acidic habitats.

Enrichment and Culture

Enrichment cultures of nitrifying bacteria are readily obtained by using selective media containing ammonia or nitrite as electron donor, and bicarbonate as sole carbon source. Because of the inefficiency of growth of these organisms, visible turbidity may not develop even after extensive nitrification has occurred, so that the best means of monitoring growth is to assay for production of nitrite (with ammonia as electron donor) or disappearance of nitrite (with nitrite as electron donor). After one or two weeks incubation, chemical assays will reveal whether a successful enrichment has been obtained, and attempts can be made to get pure cultures by streaking on agar plates. Because many common chemoorganotrophs present in the enrichment will grow rapidly on the traces of organic matter present in the medium, purification of nitrifying bacteria must be done by repeated picking and streaking, followed by testing to be certain that chemoorganotrophic contaminants have not been taken. In addition, many nitrifying bacteria, especially the ammonia oxidizers, appear to be inhibited by the traces of organic material present in most agar preparations. Culture of these organisms on solid media sometimes requires the use of extensively washed, high purity agar, or the completely *inorganic* solidifying agent, **silica gel**. Because most nitrifying bacteria do not grow on organic compounds, the simplest way of checking for contamination is to inoculate a presumed nitrifying isolate into media containing organic matter. If growth occurs, it can be assumed that the isolate was not a nitrifier.

Most of the nitrifying bacteria appear to be obligate chemolithotrophs. *Nitrobacter* is an exception, however, and is able to grow, although slowly, on acetate or pyruvate as sole carbon and energy source. None of these bacteria require growth factors. Although the group is somewhat heterogeneous morphologically, it seems to be more homogeneous than the sulfur-oxidizing or phototrophic bacteria, as shown by the fairly narrow range of DNA base composition (Table 19.5) and the similar biochemical properties. Phylogenetically, nitrifying bacteria are members of the purple bacteria (see Section 18.6).

19.5 Chemolithotrophs: Sulfur- and Iron-Oxidizing Bacteria

The ability to grow chemolithotrophically on reduced sulfur compounds is a property of a diverse group of microorganisms. However, only six genera, *Thiobacil-*

lus, Thiosphaera, Thiomicrospira, Thermothrix, Beggiatoa, and *Sulfolobus* have been consistently cultured so that our discussion here is restricted to these genera. With the exception of *Sulfolobus,* which is an archaeon and is discussed in Section 20.6, the remaining sulfur chemolithotrophs are Bacteria and phylogenetically fall into the purple bacteria group (see Sections 18.5 and 18.6). Two broad ecological classes of sulfur-oxidizing bacteria can be discerned, those living at neutral pH and those living at acid pH. Many of the forms living at acid pH also have the ability to grow chemolithotrophically using ferrous iron as electron donor. We discussed the biogeochemistry of these acidophilic sulfur- and iron-oxidizing bacteria in Sections 16.10 and 16.11.

Thiobacillus

The genus *Thiobacillus* contains those Gram-negative, polarly flagellated rods that are able to derive their energy from the oxidation of elemental sulfur, sulfides, and thiosulfate (Table 19.6). Some pseudomonads have been confused with members of the genus *Thiobacillus*. Morphologically, *Thiobacillus* (Figure 19.27) is similar to *Pseudomonas,* but the two differ in that *Thiobacillus* can grow chemolithotrophically using reduced sulfur compounds. A few thiobacilli can also grow chemoorganotrophically with organic electron donors, and under such conditions resemble pseudomonads. Further, a few pseudomonads (both marine and freshwater) can oxidize thiosulfate to tetrathionate:

$$2S\!-\!SO_3^{2-} \rightarrow {}^-O_3S\!-\!S\!-\!S\!-\!SO_3^- + 2e^-$$

and although they cannot grow solely from the energy obtained in this reaction, they do obtain some slight growth advantage from this process when growing on sugars or other organic compounds (mixotrophy, see Section 16.13).

The biochemical steps in the oxidation of various sulfur compounds are summarized in Figure 19.28. Oxidation of sulfide and sulfur involves first the reaction of these substances with sulfhydryl groups of the cell, such as glutathione with formation of a sulfide-sulfhydryl complex (Figure 19.28). The sulfide ion is then oxidized to sulfite (SO_3^{2-}) by the enzyme *sulfide oxidase.* There are two ways in which sulfite can be oxidized to produce high-energy phosphate bonds. In one, sulfite is oxidized to sulfate by a cytochrome-linked sulfite oxidase, with the formation of ATP via electron-transport phosphorylation. This pathway is universally present in thiobacilli. In the second, sulfite reacts with AMP; two electrons are removed, and adenosine phosphosulfate (APS) is formed. The electrons removed in either case are transferred to O_2 via the cytochrome system, leading to the formation of high-energy phosphate bonds through electron-transport phosphorylation. In addition, a substrate-level phosphorylation occurs, APS reacting with P_i and being converted to ADP and sulfate. With the enzyme adenylate kinase, two ADP can be converted to one ATP and one AMP. Thus oxidation of two sulfite ions

Table 19.6   Physiological characteristics of sulfur-oxidizing chemolithotrophic prokaryotes[a]

	Inorganic electron donor	Range of pH for growth	DNA (mole % GC)
***Thiobacillus* species growing poorly in organic media:**			
1. *T. thioparus*	H_2S, sulfides, S^0. $S_2O_3^{2-}$	6–8	61–66
2. *T. denitrificans*[b]	H_2S, S^0, $S_2O_3^{2-}$	6–8	63–68
3. *T. neapolitanus*	S^0, $S_2O_3^{2-}$	6–8	52–56
4. *T. thiooxidans*	S^0	2–4	51–53
5. *T. ferrooxidans*	S^0, sulfides, Fe^{2+}	2–4	55–65
***Thiobacillus* species growing well in organic media:**			
1. *T. novellus*	$S_2O_3^{2-}$	6–8	66–68
2. *T. intermedius*	$S_2O_3^{2-}$	3–7	64
Filamentous sulfur chemolithotrophs			
Beggiatoa	H_2S, $S_2O_3^{2-}$	6–8	37–43
Thiothrix	H_2S	6–8	—
Other genera			
Thiomicrospira[c]	$S_2O_3^{2-}$, H_2S	6–8	36–44
Thiosphaera	H_2S, $S_2O_3^{2-}$, H_2	6–8	66
Thermothrix[a]	H_2S, $S_2O_3^{2-}$, SO_3^-	6.5–7.5	—
Sulfolobus[d]	H_2S, S^0	1–5	37
Acidianus[d]	S^0	1–5	31

[a]*Phylogenetically, all sulfur-oxidizing Bacteria are α or β purple Bacteria (see Section 18.6).*
[b]*Facultative aerobes; use NO_3^- as electron acceptor anaerobically.*
[c]*One of its species is capable of using NO_3^- anaerobically.*
[d]*Thermophilic or hyperthermophilic Archaea, see Section 20.6.*

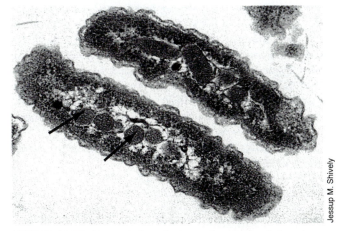

FIGURE 19.27 Transmission electron micrograph of the chemolithotrophic sulfur oxidizer *Thiobacillus neapolitanus*. A single cell is about 0.4 μm in diameter. Note the polyhedral bodies (carboxysomes) distributed throughout the cell (arrows). See text for details.

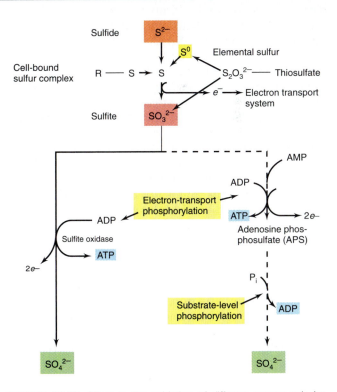

FIGURE 19.28 Steps in the oxidation of different compounds by thiobacilli. The sulfite oxidase pathway is thought to account for the majority of sulfite oxidized.

via this system produces three ATP, two via electron-transport phosphorylation and one via substrate-level phosphorylation. The significance of the APS pathway in thiobacilli in general is unclear, however, since it has only been found in a few *Thiobacillus* species.

Thiosulfate ($S_2O_3^{2-}$), is split into sulfite and sulfur (Figure 19.28). The sulfite is oxidized to sulfate with production of ATP, and the other sulfur atom is converted into insoluble elemental sulfur. Thus, when they oxidize thiosulfate, the thiobacilli produce elemental sulfur, but when they oxidize sulfides they do not. The elemental sulfur produced can itself be oxidized later when the thiosulfate supply is exhausted. If thiosulfate is low, elemental sulfur does not accumulate, probably being oxidized as soon as it is formed.

Enrichment cultures of thiobacilli are easy to prepare. Sulfur or thiosulfate is added to a basal salts medium with NH_4^+ as a nitrogen source and bicarbonate as a carbon source, and the medium is then inoculated with a sample of soil or mud. After aerobic incubation at room temperature for a few days, the liquid should appear turbid owing to the growth of thiobacilli. If thiosulfate has been used, droplets of amorphous sulfur will also be present. If elemental sulfur is used, many of the bacteria may be attached to the insoluble sulfur crystals, and the edges of such crystals should be examined for the presence of bacteria. The bacteria also attach to the crystals of metal sulfides such as PbS, HgS, or CuS if these are used as energy sources. From thiosulfate enrichment cultures, pure cultures can be obtained by streaking onto a solidified medium of the same composition. However, pure cultures are fairly difficult to obtain, since growth is usually slow and chemoorganotrophic contaminants grow on small amounts of organic matter released by the thiobacilli. Another problem in obtaining pure cultures of thiobacilli that grow best at neutral pH, such as *T. thioparus*, is that sulfur oxidation leads to sulfuric acid production and a drop in pH, resulting in the death of the culture. Thus highly buffered media (2–10 g/liter of a mixture of K_2HPO_4 and KH_2PO_4) and frequent transfers of the culture are necessary. On the other hand, isolation of the acidophilic *T. thiooxidans* is relatively easy, since this organism is resistant to the acid it produces and most chemoorganotrophic contaminants cannot grow at low pH values (2–5) where *T. thiooxidans* thrives.

Many isolates of acidophilic thiobacilli can also oxidize ferrous iron. The geochemical aspects of iron oxidation were discussed in Sections 17.16 and 17.17. At acid pH, ferrous iron is not readily oxidized spontaneously, and the acid-tolerant thiobacilli are the main agents in nature for the oxidation of ferrous iron in acidic environments. Those isolates which oxidize both iron and sulfur compounds are currently classified as the species *T. ferrooxidans*, the species *T. thiooxidans* being restricted to those acid-tolerant thiobacilli which cannot oxidize iron.

Thiosphaera is a coccoid-shaped relative of *Thiobacillus*, which is capable of growing chemolithotrophically on H_2 as well as reduced sulfur compounds, and can also grow chemoorganotrophically.

Thiomicrospira and *Thermothrix*

Members of the genera *Thiomicrospira* and *Thermothrix* are thiosulfate-oxidizing bacteria which grow at neutral pH. *Thiomicrospira*, as its name implies, is a tiny, spiral-shaped bacterium. In fact, this organism is so small that it can be selectively enriched by inoculating a thiosulfate or hydrogen sulfide-containing medium with the filtrate remaining from filtering a mud slurry

through a membrane filter with a pore size of just 0.22 μm. Such a filter would retain organisms the size of *Thiobacillus*. The genus *Thiomicrospira* consists of two species; both are obligate chemolithotrophs and one species can grow anaerobically with nitrate as electron acceptor. *Thermothrix* is a filamentous bacterium which inhabits hot sulfur springs having a neutral or slightly acidic pH. Cultures of *Thermothrix* grow aerobically or anaerobically (with nitrate as electron acceptor) at temperatures between 55 and 85°C (optimum at 70°C). *Thermothrix* is capable of chemoorganotrophic growth and utilizes a variety of organic compounds, including glucose, acetate, and amino acids as electron donors.

Carbon metabolism and carboxysomes

Several carbon nutritional classes of chemolithotrophs are known, from pure autotrophs to pure heterotrophs. The sulfur chemolithotrophs discussed thus far, such as *Thiobacillus*, are generally capable of *autotrophic* growth with CO_2 as sole carbon source. However, certain sulfur chemolithotrophs are incapable of autotrophic growth. Organisms such as most species of *Beggiatoa*, certain pseudomonads, and one species of *Thiobacillus*, *T. perometabolis*, fall into this category. Most of the remaining chemolithotrophs are true autotrophs and presumably grow autotrophically in nature.

Studies of CO_2 fixation in *Thiobacillus* have shown that reactions of the Calvin cycle (see Section 16.7) are responsible for CO_2 fixation in this chemolithotroph. Thin sections of cells of various thiobacilli have revealed polyhedral cell inclusions scattered throughout the cell (see Figure 19.27). The inclusions are about 100 nm in diameter (Figure 19.29) and are surrounded by a thin, nonunit membrane. The polyhedral bodies have been referred to as *carboxysomes* and have been shown to consist of molecules of the enzyme *ribulose-1,5-bisphosphate carboxylase* (the key enzyme of the Calvin cycle) in crystalline form.

Although the reason for carboxysome formation is not clear, they may serve to increase the amount of ribulose bisphosphate carboxylase in the cell. Since the enzyme is in an insoluble form, the carboxysome is an effective mechanism for concentrating enzyme in the cell without affecting osmolarity. Carboxysomes have been observed in nitrifying bacteria, various sulfur chemolithotrophs, and in the cyanobacteria, but not in

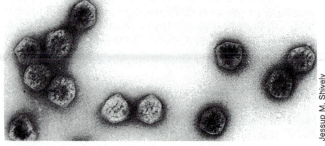

FIGURE 19.29 Polyhedral bodies (carboxysomes) purified from the chemolithotrophic sulfur oxidizer *Thiobacillus neapolitanus*. The bodies are about 100 nm in diameter.

Jessup M. Shively

chemolithotrophic bacteria that require organic compounds as a carbon source or in phototrophic purple bacteria, many of which can grow chemoorganotrophically. This suggests that carboxysomes are unique to organisms that have specialized in the autotrophic way of life.

19.6   Chemolithotrophs: Hydrogen-Oxidizing Bacteria

A wide variety of bacteria are capable of growing with H_2 as sole electron donor and O_2 as electron acceptor using the "knallgas" reaction, the reduction of O_2 with H_2:

$$2H_2 + O_2 \rightarrow 2H_2O$$

Many, but not all, of these organisms can also grow autotrophically (using reactions of the Calvin cycle to incorporate CO_2) and are grouped together here as the chemolithotrophic *hydrogen-oxidizing bacteria*. Both Gram-positive and Gram-negative hydrogen Bacteria are known, with the best studied representatives clas-

sified in the genus *Pseudomonas, Paracoccus,* and *Alcaligenes* (Table 19.7). All hydrogen-oxidizing bacteria contain one or more *hydrogenase* enzymes that function to bind H_2 and use it to either produce ATP or serve as reducing power for autotrophic growth.

Almost all hydrogen bacteria are *facultative chemolithotrophs,* meaning that they can also grow chemoorganotrophically with organic compounds as energy sources. This is a major distinction between hydrogen chemolithotrophs and many sulfur chemolithotrophs or nitrifying bacteria; most representatives of the latter two groups are *obligate chemolithotrophs*—growth does not occur in the absence of the inorganic energy source. By contrast, hydrogen chemolithotrophs can switch between chemolithotrophic and chemoorganotrophic modes of metabolism and presumably do so in nature as nutritional conditions warrant.

Energetics of hydrogen oxidation

We briefly described the energetics of hydrogen-oxidizing bacteria in Section 16.9. The majority of hydrogen-oxidizing bacteria are obligate aerobes; a few can

Table 19.7   Differential characteristics of species of hydrogen-oxidizing bacteria*

Genus/species	Denitri-fication	Growth on fructose	Motility	DNA (mole% GC)	Other characteristics
Gram-negative					
Acidovorax facilus	−	+	+	64	Membrane-bound hydrogenase
Alcaligenes eutrophus	+	+	+	66	Membrane-bound and cytoplasmic hydrogenases
Alcaligenes ruhlandii	−	−	+	−	Membrane-bound and cytoplasmic hydrogenases
Aquaspirillum autotrophicum	−	−	+	61	Only membrane-bound hydrogenase present
Pseudomonas carboxydovorans	−	−	+	60	Only membrane-bound hydrogenase present; also oxidizes CO
Hydrogenophaga flava	−	+	+	67	Colonies are bright yellow
Seliberia carboxydohydrogena	−	?	+	58	Also oxidizes CO
Paracoccus denitrificans	+	+	−	66	Only membrane-bound hydrogenase present. Strong denitrifyer
Aquifex pyrophilus	+	−	+	65	Hyperthermophile, grows microaerophilically or anaerobically (with NO_3^-) obligate chemo-lithotroph, also uses S^0 or $S_2O_3^{2-}$ as electron donor
Hydrogenobacter	−	−	−	37–46	As for *Aquifex,* but obligate aerobe (microaerophile)
Gram-positive					
Bacillus schlegelii	−	−	+	66	Produces endospores
Arthrobacter sp.	−	+	−	70	Only membrane-bound hydrogenase present
Mycobacterium gordonae	−	?	−	−	Acid fast, colonies yellow to orange

*Phylogenetically, most H_2-oxidizing bacteria are α or β purple bacteria, except for the Gram-positive organisms; Aquifex and Hydrogenobacter represent one of the major lineages of Bacteria (see Section 18.6).

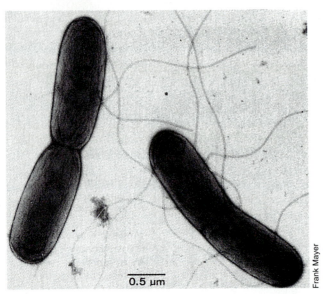

Frank Mayer

FIGURE 19.30 Transmission electron micrograph of negatively stained cells of the hydrogen-oxidizing chemolithotroph *Alcaligenes eutrophus.*

grow anaerobically with nitrate as electron acceptor. H_2 is taken up by hydrogenase and electrons shunted through an electron transport chain leading to the establishment of a membrane potential and a proton gradient; ATP is then produced by chemiosmotic mechanisms (see Section 4.11 for discussion of chemiosmosis and related topics). In *Alcaligenes eutrophus* (Figure 19.30), one of the best studied hydrogen bacteria, two distinct hydrogenases are present, one membrane bound and the other cytoplasmic (soluble). The membrane-bound enzyme is involved in energetics. Following binding of H_2 to the enzyme, hydrogen atoms are transferred from the hydrogenase to a quinone and from there through a series of cytochromes to O_2. Cytochromes of the *c* and *a/a₃* types are present in *A. eutrophus.* In other organisms, such as *Paracoccus denitrificans*, a similar sequence is observed, but *b*-type cytochromes are also present.

The second hydrogenase of *A. eutrophus* is involved in the production of *reducing power* for autotrophic growth. This soluble hydrogenase takes up H_2 and reduces NAD⁺ to NADH directly. NADH can be converted into NADPH (by enzymes called *transhydrogenases*) for use as a reductant in the Calvin cycle, but many hydrogen chemolithotrophs use NADH itself, instead of NADPH, as the reductant in Calvin cycle reactions. Because the reduction potential of H_2 is so low (−0.42 volts) it can be used to reduce NAD⁺ directly in hydrogen bacteria like *A. eutrophus* that contain two hydrogenases. Thus in hydrogen bacteria that contain a soluble hydrogenase, ATP-dependent reverse electron flow reactions (see Section 16.8) are *not* required. Because of this, growth rates and cell yields of organisms like *A. eutrophus* are typically much higher than in chemolithotrophic bacteria that must use ATP to make NAD(P)H. Many hydrogen bacteria contain only a membrane-bound hydrogenase (see Table 19.7), however, and this enzyme does

not couple to the reduction of NAD⁺. In these organisms reverse electron flow operates as a means of generating NADH.

Physiology and isolation of hydrogen bacteria

Most hydrogen bacteria prefer microaerophilic conditions when growing chemolithotrophically on H_2 because hydrogenases are oxygen-sensitive enzymes. Typically, oxygen levels of about 10 percent (half the level in air) support best growth. *Nickel* must be present in the medium for growth of hydrogen bacteria, because virtually all hydrogenases have Ni^{2+} in the active enzyme. A few hydrogen bacteria also fix molecular N_2, and when growing on N_2, the organisms are quite oxygen sensitive because the enzyme nitrogenase needed for the reduction of molecular nitrogen (see Section 16.26) is an oxygen-sensitive enzyme. Unlike nitrifying bacteria or methanotrophs, hydrogen-oxidizing bacteria generally lack internal membranes; apparently the cytoplasmic membrane of hydrogen bacteria can integrate sufficient levels of hydrogenase to maintain high rates of H_2 oxidation.

All species of H_2 bacteria will incorporate organic compounds if they are added to the medium. In some hydrogen bacteria organic compounds such as glucose or acetate strongly repress the synthesis of both Calvin cycle enzymes and hydrogenase. Some hydrogen bacteria will also grow on carbon monoxide, CO, as energy source, with electrons from the oxidation of CO to CO_2 entering the electron transport chain to drive ATP synthesis. CO-oxidizing bacteria, which are known as *carboxydotrophic* bacteria, grow autotrophically using Calvin cycle reactions to fix the CO_2 generated from the oxidation of CO. CO is oxidized to CO_2 by the enzyme *carbon monoxide dehydrogenase*, which is a molybdenum-containing enzyme. The molybdenum in CO dehydrogenase is bound to a small cofactor consisting of a multi-ringed structure called a *pterin*, similar to the situation in the enzyme nitrate reductase (see Section 16.15).

Hydrogen-oxidizing bacteria are fairly easy to isolate. A small amount of mineral salts medium containing trace metals (especially Ni^{2+} and Fe^{2+}) is inoculated with soil or water and incubated in a large sealed flask containing a head space of 5 percent O_2, 10 percent CO_2, and 85 percent H_2. When the liquid becomes turbid, plates of the same medium are streaked and incubated in a glass jar containing the same gas mixture. Pure cultures are easily obtained by picking and restreaking colonies several times (one must exercise care in mixing gas phases, however—mixtures of O_2 and H_2 are potentially explosive).

Ecology of hydrogen bacteria and carboxydotrophic bacteria

The ecological importance of aerobic hydrogen-oxidizing bacteria is difficult to assess. This is because in natural environments there is great competition for H_2 among various anaerobic bacteria. Methanogens, homoacetogens, sulfate-reducing bacteria, phototrophic bacteria, and certain denitrifying bacteria all compete for H_2 produced from fermentation. And, since hydrogen levels in actively methanogenic or sulfide-rich ecosystems are frequently very low, it is doubtful whether large amounts of H_2 ever reach oxic habitats from sources in anoxic environments. It is perhaps for this reason that most aerobic hydrogen bacteria also grow chemoorganotrophically; a "backup" system is required if the primary energy source is not always available.

CO consumption by carboxydotrophic bacteria is a very significant ecological process. Although much CO is generated from various sources, measurements of CO levels in air have shown that they have not risen significantly over many years. Microbial CO consumption is probably the reason why. Under anoxic conditions CO is oxidized by certain phototrophic bacteria, methanogens and sulfate-reducers. However, the most significant releases of CO (primarily from automobile exhaust, incomplete combustion of fossil fuels, and the catabolism of lignin) occur in oxic environments. Thus carboxydotrophic bacteria in the upper layers of soil probably represent the most significant sink for CO in nature. Carboxydotrophic bacteria include both Gram-negative and Gram-positive types. Some of the best studied carboxydobacteria include *Pseudomonas carboxydovorans*, *Bacillus schlegelii*, and *Alcaligenes carboxydus*. At least one carboxydobacterium can grow on CO anaerobically with nitrate as electron acceptor, but this does not seem to be a widespread property of the group. Like the hydrogen bacteria, all isolates of carboxydotrophic bacteria, with the exception of a thermophilic *Streptomyces* species, grow chemoorganotrophically on organic substrates as well as on CO.

19.7   Methanotrophs and Methylotrophs

Methane, CH_4, is found extensively in nature. It is produced in anoxic environments by methanogenic bacteria (see Section 17.12) and is a major gas of anaerobic muds, marshes, anoxic zones of lakes, the rumen, and the mammalian intestinal tract. Methane is the major constituent of natural gas and is also present in many coal formations. It is a relatively stable molecule; but a variety of bacteria, the **methanotrophs,** oxidize it readily, utilizing methane and a few other one-carbon compounds as electron donors for energy generation and as sole sources of carbon. These bacteria are all aerobes, and are widespread in nature in soil and water. They are also of a diversity of morphological types, seemingly related only in their ability to oxidize methane. There has been considerable interest in methane-utilizing bacteria in recent years because of the possibility of using this simple and widely available energy source to produce bacterial protein as a food or feed supplement (single-cell protein).

In addition to methane, a number of other one-carbon compounds are known to be utilized by microorganisms. A list of these compounds is given in Table 19.8. From a biochemical viewpoint, these compounds share a key characteristic: they contain no carbon-carbon bonds. Thus all carbon-carbon bonds of the cell must be synthesized de novo. Organisms that can grow using only one-carbon compounds are generally called **methylotrophs.** Many, but not all methylotrophs are also methanotrophs. From the viewpoint of carbon assimilation, methanotrophs and methylotrophs have something in common with autotrophs (Chapter 16), which also use a carbon compound lacking a carbon-carbon bond, CO_2. The two groups differ, however, in that the methylotrophs require a carbon compound *more reduced* than CO_2.

It is important to distinguish between methylotrophs and methanotrophs. A wide variety of bacteria are known which will grow on methanol, methylamine or formate, but not methane, and these bacteria are members of various genera of chemoorganotrophs: *Hyphomicrobium*, *Pseudomonas*, *Bacillus*, and *Vibrio*. In contrast, methanotrophs are unique in that they can grow not only on some of the more oxidized

Table 19.8 Substrates used by methylotrophic bacteria*	
Substrates used for growth	**Substrates oxidized, but not used for growth (co-metabolism)**
Methane, CH_4 Methanol, CH_3OH Methylamine, CH_3NH_2 Dimethylamine, $(CH_3)_2NH$ Trimethylamine, $(CH_3)_3N$ Tetramethylammonium, $(CH_3)_4N^+$ Trimethylamine N-oxide, $(CH_3)_3NO$ Trimethylsulfonium, $(CH_3)_3S^+$ Formate, $HCOO^-$ Formamide, $HCONH_2$ Carbon monoxide, CO Dimethyl ether, $(CH_3)_2O$ Dimethyl carbonate, $CH_3OCOOCH_3$	Ammonium, NH_4^+ Ethylene, $H_2C{=}CH_2$ Chloromethane, CH_3Cl Bromomethane, CH_3Br Higher hydrocarbons (ethane, propane)

A single isolate does not use all of the above, but at least one methylotrophic bacterium has been reported to oxidize each of the listed compounds.

one-carbon compounds, but also on methane. The methane-oxidizing bacteria possess a specific enzyme system, *methane monooxygenase* for the introduction of an oxygen atom into the methane molecule, leading to the formation of methanol. It should be noted that although the methanotrophs can also oxidize more oxidized one-carbon compounds such as methanol and formate, initial isolation from nature requires the use of methane as a sole electron donor, since if one of these other one-carbon compounds is used in initial enrichment, a nonmethanotrophic methylotroph will almost certainly be isolated. All methanotrophs appear to be obligate C-1 utilizers, unable to utilize compounds with carbon-carbon bonds. By contrast, many nonmethanotrophic methylotrophs are able to utilize organic acids, ethanol, and sugars.

Methane-oxidizing bacteria are also unique among prokaryotes in possessing relatively large amounts of **sterols.** As we noted in Chapter 3, sterols are found in eukaryotes as a functional part of the membrane system, but seem to be absent from most prokaryotes. In methanotrophs, sterols may be an essential part of the complex internal membrane system (see below) that is involved in methane oxidation.

Classification

An overview of the classification of methanotrophs is given in Table 19.9. These bacteria were initially distinguished on the basis of morphology and formation of resting stages, but it was then found that they could be divided into two major groups depending on their internal cell structure and carbon assimilation pathway. Type I organisms assimilate one-carbon compounds via a unique pathway, the **ribulose mono-phosphate cycle**, whereas Type II organisms assimilate C-1 intermediates via the **serine pathway**. The requirement for O_2 as a reactant in the initial oxidation of methane explains why all methanotrophs are obligate aerobes, whereas some organisms using methanol as electron donor can grow anaerobically (with nitrate or sulfate as electron acceptor).

Both groups of methanotrophs contain extensive internal membrane systems, which appear to be related to their methane-oxidizing ability. Type I bacteria are characterized by internal membranes arranged as bundles of disk-shaped vesicles distributed throughout the organism (Figure 19.31*b*) whereas Type II bacteria possess paired membranes running along the periphery of the cell (Figure 19.31*a*). Type I methanotrophs are also characterized by a lack of a complete tricarboxylic acid cycle (the enzyme *α-ketoglutarate dehydrogenase* is absent), whereas Type II organisms possess a complete cycle. Absence of a complete tricarboxylic acid cycle greatly diminishes the ability of an organism to grow chemoorganotrophically. If these reactions cannot be run as a cycle, NADH cannot be generated from reactions of the cycle, thus preventing growth at the expense of organic compounds metabolized through the tricarboxylic acid cycle.

Biochemistry of methane oxidation

Two aspects of the biochemistry of methanotrophs are of interest, the manner of oxidation of methane to CO_2 (and how this is coupled to ATP synthesis), and the manner by which one-carbon compounds are assimilated into cell material. The overall pathway of methane oxidation involves stepwise, two-electron oxidations:

Table 19.9   Some characteristics of methanotrophic bacteria

Organism	Morphology	16S r RNA group[a]	Resting stage	Internal membranes[b]	Citric Acid Cycle[c]	Carbon assimilation pathway[d]	N₂ Fixation	DNA (mole % GC)
Methylomonas	Rod	Gamma purple bacteria	Cystlike body	I	Incomplete	Ribulose monophosphate	No	50–54
Methylobacter	Rod	Gamma purple bacteria	Cystlike body	I	Incomplete	Ribulose monophosphate	No	50–54
Methylococcus	Coccus	Gamma purple bacteria	Cystlike body	II	Incomplete	Ribulose monophosphate	No	62
Methylosinus	Rod or vibrioid	Alpha purple bacteria	Exospore	II	Complete	Serine	Yes	63–66
Methylocystis	Rod	Alpha purple bacteria	Exospore	II	Complete	Serine	Yes	62

[a]*See Section 18.6*
[b]*Internal membranes: Type I, bundles of disk-shaped vesicles distributed throughout the organism; Type II, paired membranes running along the periphery of the cell. See Figure 19.31.*
[c]*Organisms with an incomplete citric acid cycle lack the enzyme α-ketoglutarate dehydrogenase.*
[d]*See Figures 19.32 and 19.33.*

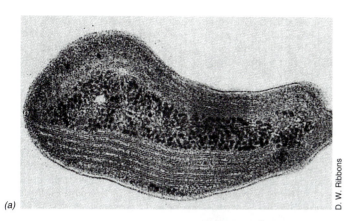

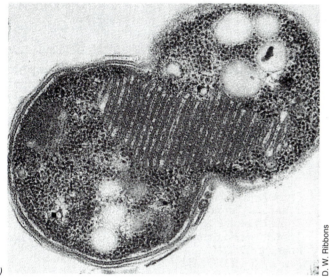

FIGURE 19.31   Electron micrographs of methane-oxidizing bacteria. (a) *Methylomonas methanica*, illustrating a type II membrane system. Cells are about 0.6 μm in diameter. (b) *Methylococcus capsulatus*, illustrating type I membrane system. Cells are about 1 μm in diameter.

$$CH_4 \xrightarrow[-126 \text{ kJ}]{} CH_3OH \xrightarrow[-193 \text{ kJ}]{} HCHO \xrightarrow[-214 \text{ kJ}]{}$$

Methane   Methanol   Formaldehyde

$$HCOOH \xrightarrow[-239 \text{ kJ}]{} HCO_3^-$$

Formate   Bicarbonate

The initial step in the oxidation of methane involves an enzyme called **methane monooxygenase**. As we discussed in Section 16.23, oxygenase enzymes catalyze the incorporation of oxygen from O_2 into carbon compounds and seem to be widely involved in the metabolism of hydrocarbons. Oxygenases require a source of reducing power, usually NADH, but in *Methylosinus*, where the process has been most thoroughly studied, electrons come not from NADH but from cytochrome *c*. This cytochrome *c* is involved in recycling of electrons from methanol dehydrogenase (and possibly from formaldehyde dehydrogenase). No ATP synthesis occurs during the first step, the oxidation of methane to methanol, and this is consistent with the fact that growth yields of methane-oxidizing bacteria are the same whether methane or methanol is used as substrate. Thus, although considerable energy

is potentially available in the oxidation of CH_4 to CH_3OH (about 126 kJ/mole), this energy is not available to organisms, apparently because no biochemical mechanism is available for conserving the energy of methane oxidation.

Biochemistry of one-carbon assimilation

As noted above, all Type I methanotrophs possess the ribulose monophosphate pathway for carbon assimilation, whereas Type II methanotrophs have the serine pathway.

The serine pathway is outlined in Figure 19.32. It is not only present in Type II methanotrophs, but is also the pathway for C-1 incorporation in facultative methylotrophs (*Hyphomicrobium* and *Pseudomonas*, for example). In this pathway, a two-carbon unit, acetyl-CoA, is synthesized from one molecule of formaldehyde and one molecule of CO_2. The pathway requires the introduction of reducing power and energy in the form of *two* molecules each of NADH and ATP for each acetyl-CoA synthesized.

The ribulose monophosphate pathway, present in Type I methanotrophs, is outlined in Figure 19.33. It is more efficient than the serine pathway in that *all* of the carbon atoms for cell material are derived from formaldehyde, and since formaldehyde is at the same oxidation level as cell material no reducing power is needed. The ribulose monophosphate pathway requires the introduction of energy in the form of *one* molecule of ATP for each molecule of 3-phosphoglyceric acid synthesized (Figure 19.33).

Consistent with the lower energy requirements of the ribulose monophosphate pathway, the cell yield of Type I organisms from a given amount of methane is *higher* than the cell yield of Type II organisms. It seems clear that if a methane-oxidizing bacterium were to be used as a source of single-cell protein as previously discussed, that an organism with a ribulose monophosphate pathway would be preferable, since cell yields on methane are higher. However, since no energy is gained from the methane to methanol step, it is not certain that there is any advantage of producing single-cell protein from methane as opposed to methanol. Methanol, not being a gas, is much more convenient to use industrially than methane, so that methanol may be the preferred energy source for single-cell protein.

Ecology and isolation

Methanotrophs are widespread in aquatic and terrestrial environments, being found wherever stable sources of methane are present. Methane produced in the anoxic regions of lakes rises through the water column, and methanotrophs are often concentrated in a narrow band at the thermocline, where methane from the anaerobic zone meets oxygen from the aerobic zone. There is some evidence that although methane oxidizers are obligate aerobes, they are sensitive to O_2 at normal concentrations, and prefer microaerophilic habitats for development. One reason for this O_2 sensitivity may be that many aquatic methanotrophs simul-

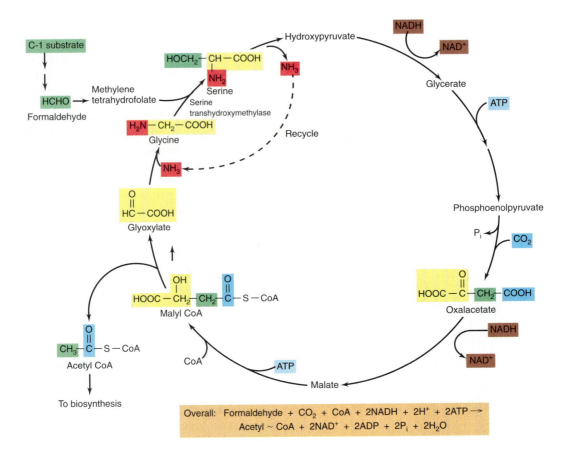

FIGURE 19.32 The serine pathway for the assimilation of C-1 units into cell material by Type II methylotrophic bacteria. The product of the pathway, acetyl-CoA, is used as the starting point for making new cell material. The key enzyme of the pathway is serine transhydroxymethylase.

taneously fix N_2, and nitrogenase is O_2 sensitive, so that optimal development occurs where O_2 concentrations are reduced. Methane-oxidizing bacteria play a small but probably important role in the carbon cycle, converting methane derived from anaerobic decomposition back into cell material (and CO_2).

The initial enrichment of methanotrophs is relatively easy, and all that is needed is a mineral salts medium over which an atmosphere of 80 percent methane and 20 percent air is maintained. Once good growth is obtained, purification is carried out by streaking on mineral salts agar plates, which are incubated in a jar with the methane-air mixture. Colonies appearing on the plates will be of two types, common chemoorganotrophs growing on traces of organic matter in the medium, which appear in one to two days, and methane oxidizers, which appear after about a week. The colonies of many methane oxidizers are pink in color. Colonies of methane oxidizers should be picked when small, and purification by continued picking and restreaking is essential. It should be emphasized that cultures of methane oxidizers are often contaminated with chemoorganotrophs, or with methanol oxidizers, so that careful attention should be paid to the purity of an isolate before any detailed studies are undertaken.

Methanotrophs are able to oxidize ammonia, although they apparently cannot grow chemolithotrophically using ammonia as sole electron donor. Methane monooxygenase also functions as an ammonia oxygenase, and a competitive interaction between the two substrates exists. For this reason, ammonia is generally toxic to methanotrophs, and the preferred nitrogen source is nitrate. It has been speculated that perhaps the methane-oxidizing bacteria could have arisen from the nitrifying bacteria via genetic changes causing the conversion of ammonia oxidase to methane oxidase. The fact that both groups of bacteria have elaborate internal membrane systems (see Section 19.4) supports such a theory.

Methanotrophic symbionts of animals

A symbiotic association between methanotrophic bacteria and marine mussels is known to occur. Mussels live in the vicinity of hydrocarbon seeps where methane is released in substantial amounts. Intact mussels as well as isolated mussel gill tissue consume methane at high rates in the presence of O_2. In the gill tissue of the mussel, coccoid-shaped bacteria are present in high numbers (Figure 19.34). The bacterial symbionts contain stacks of intracytoplasmic membranes (Figure 19.34b) typical of Type I methanotrophs. The

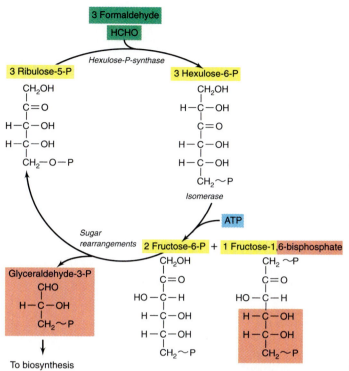

Overall: 3 Formaldehyde + ATP ⟶ glyceraldehyde-3-P + ADP

FIGURE 19.33 The ribulose monophosphate pathway for assimilation of one-carbon compounds, as found in Type I methylotrophic bacteria. The complete name of the hexulose sugar is D-erythro-L-glycero-3-hexulose-6-phosphate. Three formaldehydes are needed to carry the cycle to completion, with the net result being one molecule of glyceraldehyde-3-P. Regeneration of ribulose-5-phosphate and formation of phosphoglyceraldehyde occur via a series of pentose phosphate reactions.

symbionts are found in vacuoles within animal cells near the gill surface, which probably ensures an effective gaseous exchange with seawater. The oxidation of methane by mussel gill tissue is strictly O_2 dependent and totally inhibited by acetylene, a known poison of biological methane oxidation. This is consistent with the hypothesis that methane oxidation is carried out by the methanotrophic symbiont.

Stable isotope studies of mussel tissue suggest that methane serves as the major food source for the animal. Using $^{13}C/^{12}C$ analyses (see Section 17.6), it was shown that both gill tissues and deep tissues of the mussel had $^{13}C/^{12}C$ isotopic compositions nearly identical to that of the methane being consumed from its environment. Presumably methane assimilated by the methanotrophs is distributed throughout the animal by the excretion of carbon compounds by the methanotrophs. The methanotrophic symbiosis is therefore conceptually similar to the symbiosis established between sulfide-oxidizing chemolithotrophs and hydrothermal vent tube worms and giant clams discussed in Section 17.10. Animal–bacterial symbioses, such as the methanotrophic mussel symbiosis and the sulfide-oxidizing vent animal symbioses, thus show that prokaryotic cells can occasionally constitute the basis of a one-step food chain.

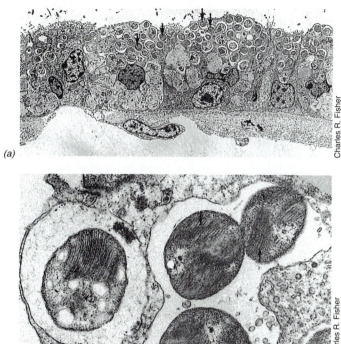

FIGURE 19.34 Methanotrophic symbionts of marine mussels. (a) Electron micrograph of a thin section at low magnification of gill tissue of a marine mussel which lives near hydrocarbon seeps in the Gulf of Mexico. Note the symbiotic methanotrophs (arrows) in the tissues. (b) High magnification view of gill tissue showing Type I methanotrophs. Note membrane bundles (arrows). The methanotrophs are about 1 μm in diameter.

19.8   Sulfate- and Sulfur-Reducing Bacteria

Sulfate is used as a terminal electron acceptor under anaerobic conditions by a heterogeneous assemblage of bacteria which utilize organic acids, fatty acids, alcohols and H_2 as electron donors. Although morphologically diverse, the **sulfate-reducing bacteria** can be considered a physiologically unified group, in the same manner as the phototrophic or methanotrophic bacteria. Thirteen genera of dissimilatory sulfate-reducing bacteria are currently recognized and can be placed in two broad physiological subgroups as outlined in Table 19.10. The genera in group I, *Desulfovibrio* (Figure 19.35a), *Desulfomonas*, *Desulfotomaculum*, and *Desulfobulbus* (Figure 19.35c), utilize lactate, pyruvate, ethanol, or certain fatty acids as carbon and energy sources, reducing sulfate to hydrogen sulfide. The genera in group II, *Desulfobacter* (Figure 19.35d), *Desulfococcus*, *Desulfosarcina* (Figure 19.35e), and *Desulfonema* (Figure 19.35b), specialize in the oxidation of fatty acids, particularly *acetate*, reducing sulfate to sulfide. The sulfate-reducing bacteria are all obligate anaerobes, and strict anaerobic techniques must be used in their cultivation, although they are not as fastidious with respect to oxygen as the methanogenic bacteria (see Section 20.5).

Sulfate-reducing bacteria are widespread in

Table 19.10 Characteristics of sulfate- and sulfur-reducing bacteria*

Genus	Characteristics	DNA (mole% GC)
Group I sulfate reducers: Nonacetate oxidizers		
Desulfovibrio	Polarly flagellated, curved rods, no spores; Gram negative; contain desulfoviridin; twelve species, one thermophilic	46–61
Desulfomicrobium	Motile rods, no spores; Gram negative; desulfoviridin absent; two species	52–57
Desulfobotulus	*Vibrios*; Gram negative; motile; desulfoviridin absent; one species	53
Desulfotomaculum	Straight or curved rods; motile by peritrichous or polar flagellation; Gram negative; desulfoviridin absent; produce endospores; four species, one thermophilic; one species capable of utilizing acetate as energy source	37–46
Archaeoglobus	Archaeon; hyperthermophile, temperature optimum, 83°C; contains some unique coenzymes of methanogenic bacteria, makes small amount of methane during growth; H_2, formate, glucose, lactate, and pyruvate serve as electron donors, SO_4^{2-}, $S_2O_3^{2-}$, or SO_3^{2-} as electron acceptor; one species (see Section 20.6)	46
Desulfobulbus	Ovoid or lemon-shaped cells; no spores; Gram negative; desulfoviridin absent; if motile, by single polar flagellum; utilizes only propionate; three species as electron donor with acetate + CO_2 as products; one species	59–60
Thermodesulfobacterium	Small, Gram-negative rods; desulfoviridin present; thermophilic, optimum growth at 70°C	34
Group II sulfate reducers: Acetate oxidizers		
Desulfobacter	Rods; no spores, Gram negative; desulfoviridin absent; if motile, by single polar flagellum; utilizes only acetate as electron donor and oxidizes it to CO_2 via the citric acid cycle; four species	45–46
Desulfobacterium	Rods, some with gas vesicles, marine; capable of autotrophic growth via the acetyl-CoA pathway, eight species	41–59
Desulfococcus	Spherical cells; nonmotile; Gram negative; desulfoviridin present, no spores; utilizes C_1–C_{14} fatty acids as electron donor with complete oxidation to CO_2; capable of autotrophic growth via the acetyl-CoA pathway; two species	57
Desulfonema	Large filamentous gliding bacteria; Gram positive; no spores; desulfoviridin present or absent; utilizes C_2–C_{12} fatty acids as electron donor with complete oxidation to CO_2; capable of autotrophic growth via the acetyl-CoA pathway (H_2 as electron donor); two species	35–42
Desulfosarcina	Cells in packets (sarcina arrangement); Gram negative; no spores; desulfoviridin absent; utilizes C_2–C_{14} fatty acids as electron donor with complete oxidation to CO_2; capable of autotrophic growth via the acetyl-CoA pathway (H_2 as electron donor); one species	51
Desulfoarculus	*Vibrios*; Gram negative; motile; desulfoviridin absent; utilizes only C_1–C_{18} fatty acids as electron donor	66
Dissimilatory sulfur reducers:		
Desulfuromonas	Straight rods, single lateral flagellum; no spores; Gram negative; does not reduce sulfate; acetate, succinate, ethanol or propanol used as electron donor; obligate anaerobe; four species	50–63
Desulfurella	Motile short rods; Gram negative; requires acetate; thermophilic	31
Campylobacter	Curved, vibrio-shaped rods; polar flagella; Gram negative; no spores; unable to reduce sulfate but can reduce sulfur, sulfite, thiosulfate, nitrate, or fumarate anaerobically with acetate or a variety of other carbon/electron donor sources; facultative aerobe	40–42
Extremely thermophilic Archaea (see Section 20.6)		

*Phylogenetically, most sulfate and sulfur-reducing bacteria are delta purple Bacteria (see Section 18.6).

aquatic and terrestrial environments that become anoxic due to microbial decomposition processes. The best known genus is *Desulfovibrio* (Figure 19.35a), which is common in aquatic habitats or waterlogged soils containing abundant organic material and sufficient levels of sulfate. *Desulfotomaculum* consists of endospore-forming rods primarily found in soil, and one species is thermophilic. Growth and reduction of sulfate by *Desulfotomaculum* in certain canned foods leads to a type of spoilage called *sulfide stinker*. The remaining genera of sulfate reducers are indigenous to freshwater or marine anoxic environments; *Desulfomonas* can also be isolated from the intestine.

The sulfate- and sulfur-reducing bacteria represent a rather broad phylogenetic group. By 16S rRNA sequence analysis the endospore-forming sulfate reducer *Desulfotomaculum* groups, as expected, with the *Clostridium* subdivision of the Gram-positive Bacteria (see Sections 18.6 and 19.26). The remaining sulfate reducers align most closely with other Gram-negative

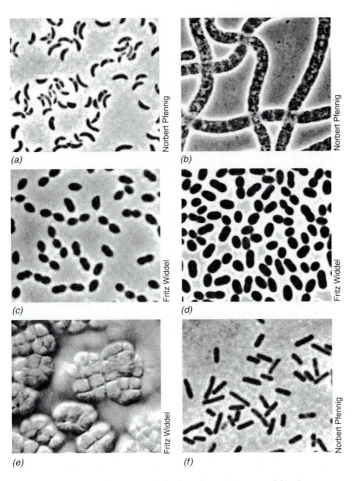

(a) — Norbert Pfennig

(b) — Norbert Pfennig

(c) — Fritz Widdel

(d) — Fritz Widdel

(e) — Fritz Widdel

(f) — Norbert Pfennig

FIGURE 19.35 Photomicrographs by phase contrast of (a–e) representative sulfate-reducing and (f) sulfur-reducing bacteria. (a) *Desulfovibrio desulfuricans*, cell diameter about 0.7 μm. (b) *Desulfonema limicola*, cell diameter 3 μm. (c) *Desulfobulbus propionicus*, cell diameter about 1.2 μm. (d) *Desulfobacter postgatei*, cell diameter about 1.5 μm. (e) *Desulfosarcina variabilis* (interference contrast microscopy), cell diameter about 1.25 μm. (f) *Desulfuromonas acetoxidans*, cell di-

Bacteria in the "delta" subdivision of the purple bacterial group. The sulfate reducers *Desulfovibrio, Desulfobacter,* etc. are phylogenetically related to the bacterial predator *Bdellovibrio* (see Section 19.11) and the gliding myxobacteria (see Section 19.13). By contrast, species of the sulfur-reducing organism *Desulfuromonas* (see below) are very closely related to one another, but are not closely related (in an evolutionary sense) to sulfate-reducing bacteria.

Sulfur reduction

A variety of bacteria are known that can reduce elemental sulfur to sulfide but are unable to reduce sulfate to sulfide. These organisms are referred to as **sulfur-reducing bacteria**. Members of the genus *Desulfuromonas* (Figure 19.35*f*) can grow anaerobically by coupling the *oxidation* of substrates such as acetate or ethanol to the *reduction* of elemental sulfur to hydrogen sulfide. However, the ability to reduce elemental sulfur, as well as other sulfur compounds such as thiosulfate, sulfite, or dimethyl sulfoxide (DMSO), is a widespread property of a variety of chemoorgan-

otrophic, generally facultatively aerobic bacteria (for example, *Proteus, Campylobacter, Pseudomonas,* and *Salmonella*). *Desulfuromonas* differs from the latter in that it is an obligate anaerobe and utilizes only sulfur as an electron acceptor (see Table 19.10). In addition, certain sulfate-reducing bacteria are capable of substituting sulfur for sulfate as an electron acceptor for anaerobic growth.

Studies of sulfur-rich thermal environments have led to the discovery of a host of extremely thermophilic dissimilatory sulfur-reducing Archaea (see Section 20.6). Many of these organisms have growth temperature optima near 90°C or higher, and these extremely thermophilic anaerobes represent a distinct branch of dissimilatory sulfur reducers, sharing little with organisms like *Desulfuromonas* except the ability to use sulfur as an electron acceptor. One extremely thermophilic archaeon, *Archaeoglobus*, has been shown to reduce sulfate but is incapable of reducing elemental sulfur (see Section 20.6).

Physiology

The range of electron donors used by sulfate-reducing bacteria is broad. Lactate and pyruvate are almost universally used, and many group I species will utilize malate, formate, and certain primary alcohols (for example, methanol, ethanol, propanol, and butanol). Some strains of *Desulfotomaculum* will utilize glucose but this is rather rare among sulfate reducers in general. Group I sulfate reducers oxidize their energy source to the level of acetate and excrete this fatty acid as an end product. Group II organisms differ from those in group I by their ability to oxidize fatty acids, lactate, succinate, and even benzoate in some cases, all the way to CO_2. *Desulfosarcina, Desulfonema, Desulfococcus, Desulfobacterium, Desulfotomaculum,* and certain species of *Desulfovibrio* (see Figure 19.35), are unique in their ability to grow chemolithotrophically with H_2 as electron donor, sulfate as electron acceptor, and CO_2 as sole carbon source.

In addition to using sulfate as an electron acceptor, many sulfate-reducing bacteria grow using *nitrate* (NO_3^-) as an electron acceptor, reducing NO_3^- to ammonia, NH_3, or can use certain organic compounds for energy generation by fermentative pathways in the complete absence of sulfate or other terminal electron acceptors. The most common fermentable organic compound is *pyruvate*, which is converted via the phosphoroclastic reaction to acetate, CO_2, and H_2. With lactate or ethanol, insufficient energy is available from fermentation, so that sulfate is required. The value of sulfate as an electron acceptor can be readily demonstrated by comparing growth yields on pyruvate with and without sulfate. In the presence of sulfate, the growth yield is higher, because of the larger amount of energy available when pyruvate utilization is coupled with sulfate reduction.

Certain sulfate-reducing bacteria can fix N_2, but this property is not uniformly distributed across the group. Species of *Desulfovibrio* and *Desulfobacter* are most likely to fix N_2.

Biochemistry

The enzymology of sulfate reduction was discussed in Section 16.16. Recall that the first step in dissimilatory sulfate reduction is the formation of *adenosine phosphosulfate* (APS) from ATP and sulfate; the enzyme APS reductase then catalyzes the reduction of the sulfate moiety to sulfite (see Figure 16.31). Sulfite is then reduced to sulfide in a series of steps of which little is known biochemically. A sulfite reductase has been identified, but it is not clear whether this enzyme serves to reduce all or only some of the potential intermediates between sulfite and sulfide.

Sulfate-reducing bacteria contain a number of electron transfer proteins including cytochromes of the c_3- and *b*-types (see Section 16.16), ferredoxins, flavodoxins, and hydrogenases, although the actual function of many of the components (other than hydrogenase) remains obscure. Because many sulfate-reducers can use H_2 as an energy source, however, it is clear that the reduction of sulfate is linked to ATP synthesis via electron transport and chemiosmotic mechanisms (see Section 16.16 and Figure 16.32). The utilization of organic compounds as electron donors presumably involves the transfer of electrons through the same electron transport chain, generating a membrane potential.

Growth yield studies with sulfate-reducing bacteria have been used to determine the stoichiometry between ATP production and sulfate reduction. Studies with *Desulfovibrio* suggest that one (net) ATP is produced per sulfate reduced to sulfide and that three (net) ATPs are produced per sulfite reduced to sulfide. These values are in line with what is known concerning the conversion of sulfate to sulfite. During this conversion, two high energy phosphate bond equivalents are consumed since ATP and sulfate are the initial reactants and AMP and sulfite are the final products (see Section 16.16).

Certain sulfate reducers will grow completely autotrophically with CO_2 as sole carbon source, H_2 as electron donor, and sulfate as electron acceptor. As discussed in Section 19.9 in connection with the homoacetogenic bacteria, autotrophic sulfate reducers use the *acetyl-CoA pathway* for fixation of CO_2 into cell material (see Table 19.10). In addition, most group II sulfate reducers (those capable of oxidizing acetate to CO_2) *reverse* the steps of the acetyl-CoA pathway to yield CO_2 from acetate, and do not employ the more common citric acid cycle for this purpose.

Isolation and ecology

The enrichment of *Desulfovibrio* is relatively easy on a lactate-sulfate medium to which ferrous iron is added. A reducing agent such as thioglycolate or ascorbate is also added. The sulfide formed from sulfate reduction combines with the ferrous iron to form black insoluble ferrous sulfide. This blackening not only serves as an indicator of the presence of sulfate reduction, but the iron ties up and detoxifies the sulfide, making possible growth to higher cell yields. The conventional procedure is to set up liquid enrichments, and after some

growth has occurred as evidenced by blackening of the medium, purification is accomplished by streaking onto a tube coated on the inside surface with a thin layer of agar (called roll tubes) or on Petri plates in an anaerobic glove box. Although the sulfate reducers are obligate anaerobes, they are not as rapidly inactivated by oxygen as methanogens (see Section 20.5), so that streaking on plates can be done in air, provided a reducing agent is present in the medium and plates are quickly incubated in an anaerobic environment.

Alternatively, agar "shake tubes" can be used for purification purposes. In the *shake tube method* a small amount of liquid from the original enrichment is added to a tube of molten agar growth medium, mixed thoroughly, and sequentially diluted through a series of molten agar tubes (see Section 17.3 and Figure 17.5). Upon solidification, individual cells distributed throughout the agar form colonies which can be removed aseptically, and the whole process repeated until pure cultures are obtained. Colonies of sulfate-reducing bacteria are recognized by the black deposit of ferrous sulfide, and purified by further streaking.

19.9   Homoacetogenic Bacteria

Homoacetogenic bacteria are obligate anaerobes that utilize CO_2 as a terminal electron acceptor, producing acetate as the sole product of anaerobic respiration; we discussed the overall formation of acetate by these organisms in Section 16.17. Electrons for the reduction of CO_2 to acetate can come from H_2, a variety of C-1 compounds, sugars, organic acids, alcohols, amino acids and certain nitrogenous bases. However the major unifying thread among homoacetogens is the pathway of CO_2 reduction. It appears that most, if not all, homoacetogens convert CO_2 to acetate by the **acetyl-CoA pathway** (see below), and in many acetogens autotrophic growth via the acetyl-CoA pathway also occurs. A list of the major organisms that produce acetate or oxidize acetate via the acetyl-CoA pathway are listed in Table 19.11. Organisms such as *Acetobacterium woodii* and *Clostridium aceticum* can grow both chemoorganotrophically or chemolithotrophically by carrying out a homoacetic acid fermentation of sugars or through the reduction of CO_2 to acetate with H_2 as electron donor, respectively. The stoichiometries observed are as follows:

1. $C_6H_{12}O_6 \rightarrow 3CH_3COOH$
2. $2CO_2 + 4H_2 \rightarrow CH_3COOH + 2H_2O$

Homoacetogens ferment glucose via the glycolytic pathway converting glucose into two molecules of pyruvate and two molecules of NADH (the equivalent of 4H). From this point, two molecules of acetate are produced as follows:

3. 2 pyruvate $\rightarrow$ 2 acetate + $2CO_2$ + 4H

The third acetate of the homoacetate fermentation comes from the reduction of the two molecules of CO_2 generated in reaction three above, using the four elec-

<table>
<tr><td>

Table 19.11 Organisms employing the acetyl-CoA pathway of CO_2 fixation

I. Acetate synthesis the result of energy metabolism
  Acetoanaerobium noterae
  Acetobacterium woodii
  Acetobacterium wieringae
  Acetogenium kivui
  Clostridium aceticum
  Clostridium thermoaceticum
  Clostridium formicoaceticum
  Desulfotomaculum orientis
  Sporomusa paucivorans
II. Acetate synthesis in autotrophic metabolism
  Autotrophic homoacetogenic bacteria
  Autotrophic methanogens
    (see Section 20.5)
  Autotrophic sulfate-reducing bacteria
    (see Section 19.8)
III. Acetate oxidation in energy metabolism
  Reaction: Acetate + $2H_2O \rightarrow 2CO_2$ + 8H
  Group II sulfate reducers (other than
    Desulfobacter)
  Reaction: Acetate $\rightarrow CO_2$ + CH_4
  Acetoclastic methanogens (*Methanosarcina,*
    Methanothrix)

</td></tr>
</table>

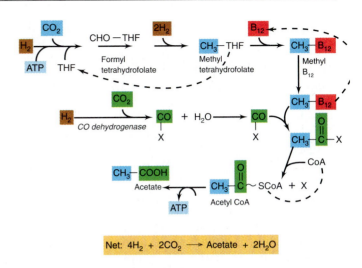

Net: $4H_2 + 2CO_2 \longrightarrow$ Acetate + $2H_2O$

FIGURE 19.36 Reactions of the acetyl-CoA pathway, the mechanism of autotrophy in homoacetogenic, sulfate-reducing, and methanogenic bacteria. THF stands for tetrahydrofolate and B_{12} is vitamin B_{12} in an enzyme-bound intermediate. X also represents an enzyme-bound form of the compound. Note how the pathway is direct rather than cyclical.

trons generated from glycolysis *plus* the four electrons produced during the oxidation of two pyruvates to two acetates (reaction 3). Starting from pyruvate, then, the overall production of acetate can be written as

$$2 \text{ pyruvate} + 4H \rightarrow 3 \text{ acetate}$$

All homoacetogenic bacteria that produce and excrete acetate in *energy metabolism* are Gram-positive and many are classified in the genus *Clostridium*. A few other Gram-positive and many different Gram-negative Bacteria utilize the acetyl-CoA pathway for *autotrophic* purposes, reducing CO_2 to acetate, which then serves as a source of cell carbon. The acetyl-CoA pathway functions in autotrophic growth for those sulfate-reducers capable of $H_2 + CO_2 + SO_4^{2-}$-mediated autotrophy (see Section 19.8), and is also used by the methanogenic bacteria, most of which can grow autotrophically on $H_2 + CO_2$ (see Section 20.5 and Table 19.11). By contrast, certain bacteria employ the reactions of the acetyl-CoA pathway primarily in the *reverse* direction, as a means of *oxidizing* acetate. These include acetoclastic methanogens (see Section 20.5) and most type II (acetate-oxidizing) sulfate-reducers (see Section 19.8).

Reactions of the acetyl-CoA pathway

We will consider the acetyl-CoA pathway here from the standpoint of organisms using H_2 as electron donor and CO_2 as electron acceptor for autotrophic growth. Unlike other autotrophic pathways, such as the Calvin cycle (see Section 16.7) and the reverse citric acid cycle (see Section 19.1), the acetyl-CoA pathway of CO_2 fixation is *not* a cycle (Figure 19.36). In-

stead it involves the direct reduction of CO_2 to acetyl-CoA; one molecule of CO_2 is reduced to the methyl group and the other molecule of CO_2 is reduced to the carbonyl group of acetate. The acetyl-CoA pathway has thus far only been found in certain obligate anaerobes, and it appears to require H_2 as an electron donor. A key enzyme of the acetyl-CoA pathway is *carbon monoxide (CO) dehydrogenase*. CO dehydrogenase catalyzes the following reaction:

$$CO_2 + H_2 \rightleftharpoons CO + H_2O$$

The importance of CO dehydrogenase in the acetyl-CoA pathway is that it acts to reduce CO_2 to the level of CO, the CO ending up in the *carbonyl* position of acetate. The reactions of the acetyl-CoA pathway are shown in Figure 19.36. The methyl group of acetate originates from the reduction of CO_2 to a methyl group by a series of enzymatic reactions involving the coenzyme *tetrahydrofolate*. The latter is a common C-1 carrier in all organisms, but plays a special role in homoacetogenic organisms. CO_2 is first reduced to formate by the enzyme formate dehydrogenase, and then the formate is converted to formyl tetrahydrofolate (Figure 19.36). The addition of two more pairs of electrons yields *methyl tetrahydrofolate*. The methyl group is then transferred from methyl tetrahydrofolate to an enzyme containing vitamin B_{12} as cofactor, and it is the latter compound that serves as the methyl donor in the final assembly of the acetyl-CoA molecule (Figure 19.36). The methyl corrinoid combines with CO to form an acetyl group bound to an enzyme (coenzyme A is then added to yield the final product, acetyl-CoA) (Figure 19.36).

An interesting homoacetogen capable of methylotrophic growth has also been described. This organism, *Sporomusa paucivorans,* converts methanol into acetate, presumably by oxidizing methanol to CO_2 + 6H, and then using the acetyl-CoA pathway to reduce CO_2

to acetate. However, because methyl groups are directly available from CH_3OH, *Sporomusa* probably puts methyl groups in at the terminal part of the tetrahydrofolate branch of the acetyl-CoA pathway (Figure 19.36) and joins the methyl group to CO, produced from the reduction of CO_2 by the enzyme CO dehydrogenase.

Because homoacetogens can grow at the expense of reactions of the acetyl-CoA pathway, this reaction sequence must be an overall energy-releasing one. One potential site of ATP synthesis in the reduction of CO_2 to acetate is the terminal step, the conversion of acetyl-CoA to acetate and ATP (via acetyl-P, see Section 16.19). However, additional energy-linked steps may be involved, as evidence is good that a Na^+ gradient is established across the cytoplasmic membrane during growth of homoacetogens. This gradient could drive ATP synthesis as discussed previously for the succinate fermenter, *Propionigenium* (see Section 16.19).

19.10 Budding and Appendaged (Prosthecate) Bacteria

This large and rather heterogeneous group contains bacteria which form various kinds of cytoplasmic extrusions: *stalks, hyphae,* or *appendages* (Table 19.12). Extrusions of these kinds, which are smaller in diameter than the mature cell and which contain cytoplasm and

are bounded by the cell wall, are called **prosthecae** (singular, **prostheca**) (Figure 19.37). Of considerable interest in this group of bacteria is that cell division often occurs as a result of unequal cell growth. In contrast to cell division in the typical bacterium, which occurs by *binary fission* and results in the formation of two equivalent cells (Figure 19.38), cell division in the stalked and budding bacteria involves the formation of a new daughter cell with the mother cell retaining its identity after the cell division process is completed (Figure 19.38). The genera from Table 19.12 which show this unequal cell-division process are indicated in Figure 19.38.

The critical difference between these bacteria and conventional bacteria is not the formation of buds or stalks, but the formation of a new cell wall from a *single point* (polar growth) rather than throughout the whole cell (intercalary growth). Several genera not normally considered to be budding bacteria show polar growth without differentiation of cell size (Figure 19.38). An important consequence of **polar growth** is that internal structures, such as membrane complexes, are not involved in the cell-division process, thus permitting the formation of more complex internal structures than in cells undergoing intercalary growth. Several additional consequences of polar growth include: *aging* of the mother cell occurs; cells are mortal (rather than immortal as in intercalary growth); cell division may be asymmetrical; the

Table 19.12   Characteristics of stalked, appendaged (prosthecate), and budding bacteria*

Characteristics	Genus	DNA (mole % GC)
Stalked bacteria:		
Stalk an extension of the cytoplasm and involved in cell division	*Caulobacter*	62–67
Stalked fusiform-shaped cells	*Prosthecobacter*	54–60
Stalked, but stalk an excretory product, not containing cytoplasm:		
Stalk depositing iron, cell vibrioid	*Gallionella*	55
Laterally excreted gelatinous stalk, not depositing iron	*Nevskia*	60
Pear-shaped or globular cells with long stalk	*Planctomyces*	50
Appendaged (prosthecate) bacteria:		
Single or double prosthecae	*Asticcacaulis*	55–61
Multiple prosthecae		
Short prosthecae, multiply by fission, some with gas vesicles	*Prosthecomicrobium*	64–70
Flat, star-shaped cells, some with gas vesicles	*Stella*	69–74
Long prosthecae, multiply by budding, some with gas vesicles	*Ancalomicrobium*	70–71
Phototrophic	*Prosthechochloris*	50–56
With gas vesicles	*Ancalochloris*	–
Budding bacteria:		
Phototrophic, produce hyphae	*Rhodomicrobium*	61–64
Phototrophic, budding without hyphae	*Rhodopseudomonas*	61–64
Chemoorganotrophic, pear-shaped, stalks lacking	*Pirella*	57
Chemoorganotrophic, rod-shaped cells	*Blastobacter*	59–66
Chemoorganotrophic, buds on tips of slender hyphae:		
Single hyphae from parent cell	*Hyphomicrobium*	59–65
Multiple hyphae from parent cell	*Pedomicrobium*	62–67

*Phylogenetically, Planctomyces and Pirella *form a distinct line of Bacteria; most other budding/stalked organisms are members of the gamma or alpha subdivisions of the purple Bacteria (Proteobacteria) (see Section 18.6).*

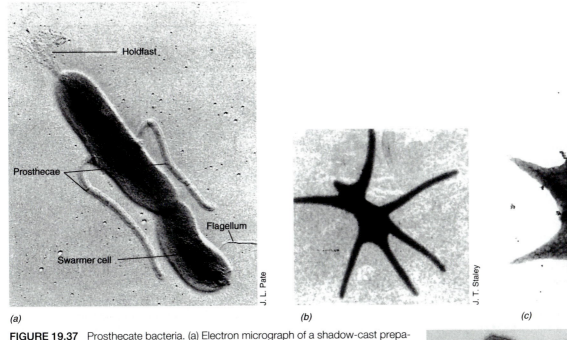

(a)   (b)   (c)

FIGURE 19.37   Prosthecate bacteria. (a) Electron micrograph of a shadow-cast preparation of *Asticcacaulis biprosthecum*, illustrating the location and arrangement of the prosthecae. Cells are about 0.6 μm wide. Note also the holdfast material and the swarmer cell in process of differentiation. (b) Electron micrograph of a negative-stained preparation of a cell of the prosthecate bacterium, *Ancalomicrobium adetum*. The appendages are cellular (prosthecae) because they are bounded by the cell wall and contain cytoplasm and are about 0.2 μm in diameter. (c) Electron micrograph of a whole cell of a prosthecate phototrophic green bacterium *Ancalochloris perfilievii*, cells are about 0.7 μm in diameter. The structures seen within the cell are gas vesicles. (d) Electron micrograph of the star-shaped bacterium *Stella*. Cells are about 0.8 μm in diameter.

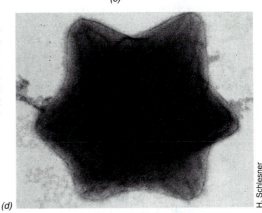

(d)

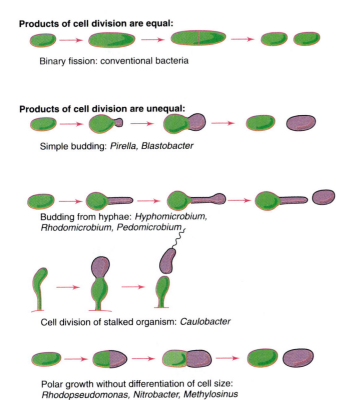

Products of cell division are equal:

Binary fission: conventional bacteria

Products of cell division are unequal:

Simple budding: *Pirella, Blastobacter*

Budding from hyphae: *Hyphomicrobium, Rhodomicrobium, Pedomicrobium*

Cell division of stalked organism: *Caulobacter*

Polar growth without differentiation of cell size: *Rhodopseudomonas, Nitrobacter, Methylosinus*

FIGURE 19.38   Contrast between cell division in conventional bacteria and in budding and stalked bacteria.

daughter cell at division is immature and must form internal or budding structures before it can divide; organisms showing polar growth have a potential for morphogenetic evolution not possible in cells with intercalary growth. Thus some of the most complex morphogenetic processes in the prokaryotes are found in the budding and stalked bacteria.

Although the majority of budding and/or appendaged bacteria phylogenetically group with the purple Bacteria group, specifically the alpha purple Bacteria (see Section 18.6), the genera *Planctomyces* and *Pirella* do not. These two organisms are related to each other, but are phylogenetically unrelated to any other group of Bacteria. *Planctomyces* and *Pirella* are also unusual because their cell walls consist primarily of protein. The walls of *Planctomyces* and *Pirella* contain large amounts of cysteine (crosslinked as cystine) and proline, but lack peptidoglycan. As would be expected of organisms lacking peptidoglycan, *Planctomyces* and *Pirella* are totally resistant to the antibiotics penicillin, cephalosporin, and cycloserine, all drugs which affect peptidoglycan synthesis.

Most budding and/or appendaged bacteria are aquatic; in nature many live attached to surfaces, their stalks or appendages serving as attachment sites. Many of the prosthecate forms are free-floating, and it is thought that their appendages serve as absorptive organs, making possible more efficient growth in the nutritionally dilute aquatic environment. Calculations of surface:volume ratios (see Section 3.2) of appendaged bacteria suggest that a major function of bacterial appendages may be to *increase* the surface:volume ratio; such structures would serve to increase the cell's surface area and thus be of competitive advantage for survival in dilute environments. Many of the free-floating forms have gas vesicles, presumably an adaptation to the planktonic existence.

Stalked bacteria

The stalked bacteria comprise a group of Gram-negative, polarly flagellated rods that possess a **stalk,** a structure by which they attach to solid substrates. Most members of this group are classified in the genus *Caulobacter*. Stalked bacteria are frequently seen in aquatic environments attached to particulate matter, plant materials, or other microorganisms; generally they are found attached to microscope slides that have been immersed in lake or pond water for a few days. When many *Caulobacter* cells are present in the suspension, groups of stalked cells are seen attached, exhibiting the formation of *rosettes* (Figure 19.39a). Electron microscopic studies reveal that the stalk is not an excretion product but is an outgrowth of the cell since it contains cytoplasm surrounded by cell wall and cytoplasmic membrane (Figure 19.39b and c). The *holdfast* by which the stalk attaches the cell to a solid substrate is at the tip of the stalk, and, once attached, the cell usually remains permanently fixed. Since the stalk is cytoplasmic, it is also a prostheca. A stalk which is *not* a prostheca can be seen in an electron micrograph of the budding bacterium *Planctomyces* (Figure 19.40). In this organism the stalk contains no cytoplasm or cell wall and is probably proteinaceous in nature. The stalk functions to anchor the cell to surfaces via a holdfast located at the tip of the stalk.

The *Caulobacter* cell division cycle (Figure 19.41) is of special interest because it involves a process of *unequal binary fission*. Cell division occurs by elongation of the cell followed by fission, a single flagellum forming at the pole opposite the stalk. The flagellated cell so formed, called a *swarmer*, separates from the nonflagellated mother cell, swims around, and settles down on a new surface, forming a new stalk at the flagellated pole; the flagellum then disappears (Figure 19.41). Stalk formation is a necessary precursor to cell division and is coordinated with DNA synthesis (Figure 19.41). The time span for the division of the stalked cell is thus shorter than the time span for division of the swarmer cell, owing to the requirement that a swarmer (flagellated) cell must synthesize a stalk before it divides (Figure 19.41). The cell division cycle in *Caulobacter* is thus more complex than simple binary fission since the stalked and swarmer cells have polar differentiation, and the cells themselves are structurally different.

Caulobacters are chemoorganotrophic aerobes; they usually have one or several vitamin requirements, but are able to grow on a variety of organic carbon compounds as sole sources of carbon and energy. Various amino acids serve as nitrogen sources. The enrichment culture for caulobacters makes use of the fact that they occur quite commonly in the organic film that develops at the surface of an undisturbed liquid. If pond, lake, or seawater is mixed with a small amount of organic material such as 0.01 percent peptone, and incubated at 20–25°C for two to three days, a surface film consisting of bacteria, fungi, and protozoa develops, and in this microbial film caulobacters are common. A sample of the surface film is then streaked on an agar medium containing 0.05 percent peptone, and after three to four days the plates are examined under a dissecting microscope for the presence of microcolonies which are typical of *Caulobacter*. These colonies are then picked and streaked on fresh medium containing a higher concentration of organic matter (for example, 0.5 percent peptone + 0.1 percent

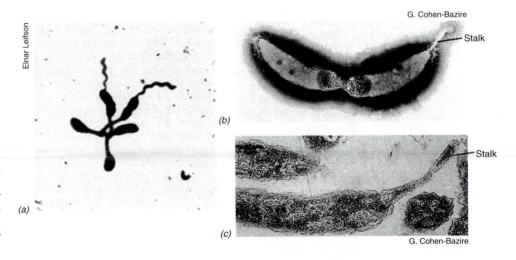

FIGURE 19.39
(a) A *Caulobacter* rosette. A single cell is about 0.5 µm wide. The five cells are attached by their stalks (prosthecae). Two of the cells have divided and the daughter cells have formed flagella. (b, c) Electron micrographs of *Caulobacter* cells. (b) Negatively stained preparation of a cell in division. (c) A thin section. Notice that cytoplasmic constituents are present in the stalk region.

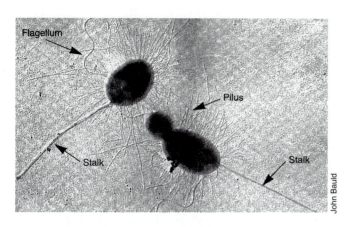

FIGURE 19.40 An electron micrograph of a metal-shadowed preparation of *Planctomyces maris*. A single cell is about 1–1.5 μm long. Note the fibrillar nature of the stalk. Pili are also abundant. Note also the flagella (curly appendages) on each cell and the bud that is developing from the nonstalked pole of one cell.

yeast extract), and the resulting colonies are examined microscopically for stalked bacteria. Water low in organic matter is a good source of caulobacters; tap or distilled water that has been left undisturbed usually shows good *Caulobacter* development in the surface film. The ability to grow in dilute media is a common property of organisms that attach to solid substrates; in the case of stalked organisms such as *Caulobacter* the stalk itself may also function as an absorptive organ, making it possible for the organism to acquire larger amounts of the restricted supply of organic nutrients than can organisms lacking these appendages.

A stalked organism sometimes classified with the caulobacters is *Gallionella*, which forms a twisted stalk containing ferric hydroxide (Figure 19.42). However, the stalk of *Gallionella* is not an integral part of the cell but is *excreted* from the cell surface. It contains an organic matrix on which the ferric hydroxide accumulates. *Gallionella* is frequently found in the waters draining bogs, iron springs, and other habitats where ferrous iron is present, usually in association with sheathed bacteria such as *Sphaerotilus*. In very acidic

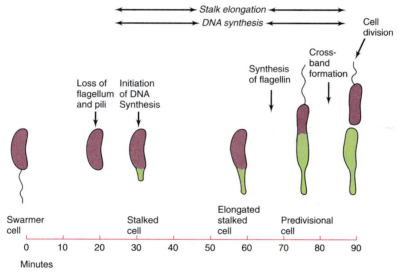

Caulobacter cell cycle beginning with swarmer cell

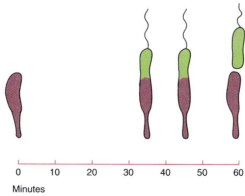

Caulobacter cell cycle beginning with stalked cell

FIGURE 19.41
Stages in the *Caulobacter* cell cycle.

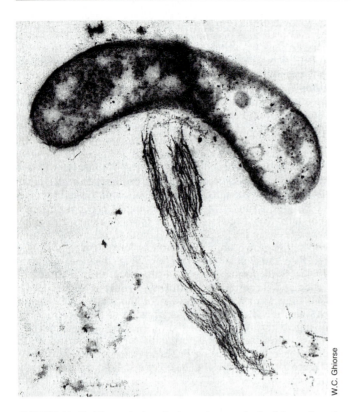

FIGURE 19.42 Transmission electron micrograph of a thin section of the iron bacterium *Gallionella ferruginea*. Cells are about 0.6 μm wide. Note the twisted stalk of ferric hydroxide emanating from the center of the cell.

waters containing iron, *Gallionella* is not present, and acid-tolerant thiobacilli replace it.

Budding bacteria

The two best studied budding bacteria are *Hyphomicrobium*, which is chemoorganotrophic, and *Rhodomicrobium*, which is phototrophic; both organisms release buds from the ends of long thin hyphae. The process of reproduction in a budding bacterium is illustrated in Figure 19.43. The mother cell, which is often attached by its base to a solid substrate, forms a thin outgrowth that lengthens to become a hypha, and at the end of the hypha a bud forms. This bud enlarges, forms a flagellum, breaks loose from the mother cell, and swims away. Later, the daughter cell loses its flagellum and after a period of maturation forms a hypha and buds. Further buds can also form at the hyphal tip of the mother cell. Many variations on this cycle are possible. In some cases the daughter cell does not break away from the mother cell but forms a hypha from its other pole. Complex arrays of cells connected by hyphae are frequently seen (Figure 19.44). In some cases a bud begins to form directly from the mother cell without the intervening formation of hypha, whereas in other cases a single cell forms hyphae from each end (Figure 19.44). The hypha is a direct cellular extension of the mother cell (Figure 19.45), containing cell wall, cytoplasmic membrane, ribosomes, and occasionally DNA.

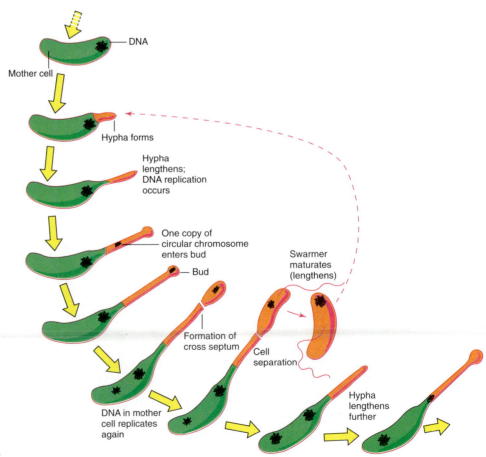

FIGURE 19.43
Stages in the *Hyphomicrobium* cell cycle.

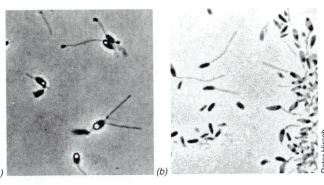

(a)   (b)

Peter Hirsch

FIGURE 19.44 Photomicrographs of cells of *Hyphomicrobium*. Cells are about 0.7 μm wide. By phase contrast, showing typical fields. Notice the long hyphae and the occasional budding cell.

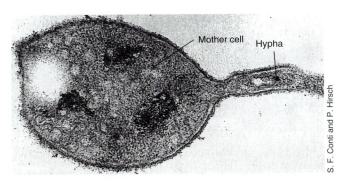

Mother cell   Hypha

S. F. Conti and P. Hirsch

FIGURE 19.45 Electron micrograph of a thin section of a single *Hyphomicrobium* cell. The hypha is about 0.2 μm wide.

Chromosomal replication events during the budding cycle are of interest (Figure 19.43). The DNA located in the mother cell replicates and then once the bud has formed, a copy of the circular chromosome is moved down the length of the hypha and into the bud. A cross-septum then forms, separating the still developing bud from the hypha and mother cell.

Hyphomicrobium is a methylotrophic bacterium. Preferred carbon sources are *one-carbon* compounds such as methanol, methylamine, formaldehyde and formate. Growth on acetate, ethanol, or higher aliphatic compounds is usually slow, and growth is poor on sugars or most amino acids. Urea, amides, ammonia, nitrite, and nitrate can be utilized as nitrogen sources; no vitamins are required. *Hyphomicrobium* is widespread in freshwater, marine, and terrestrial habitats. Initial enrichment cultures can be prepared using a mineral-salts medium lacking organic carbon and nitrogen, to which a sample of natural material is added. After several weeks incubation, the surface film that develops is streaked out on agar medium containing methylamine or methanol as a sole carbon source. Colonies are then checked microscopically for the characteristic *Hyphomicrobium* cellular morphology. A fairly specific enrichment procedure for *Hyphomicrobium* uses methanol as electron donor with nitrate as electron acceptor under anaerobic conditions. Virtually the only denitrifying organism using methanol is *Hyphomicrobium*, so that this

procedure selects this organism out of a wide variety of environments.

19.11   Spirilla

The **spirilla** are Gram-negative, motile, spiral-shaped bacteria with a wide variety of physiological attributes. Some of the key taxonomic criteria used are cell shape, size, kind of polar flagellation (single or multiple), relation to oxygen (obligately aerobic, microaerophilic, facultative), relationship to plants (as symbionts or plant pathogens) or animals (as pathogens), fermentative ability, and certain other physiological characteristics (nitrogen-fixing ability, ability to utilize methane as electron donor, halophilic nature, thermophilic nature, luminescence). Phylogenetically, spirilla constitute a line of purple Bacteria. The genera to be covered in the present discussion are given in Table 19.13.

Spirillum, Aquaspirillum, Oceanospirillum, and Azospirillum

These are helically curved rods, which are motile by means of polar flagella (usually tufts at both poles, see Figure 19.46). The number of turns in the helix may vary from less than one complete turn (in which case the organism looks like a vibrio; see Section 19.19) to many turns. Spirilla with many turns can superficially resemble spirochetes (see Section 19.12), but differ in that they do not have an outer sheath and axial filaments, but instead contain typical bacterial flagella. Some spirilla are very large bacteria, and were seen by early microscopists. It is likely that van Leeuwenhoek first described *Spirillum* in the 1670s, and the genus was first created by the protozoologist Ehrenberg in 1832. The organism seen by these workers is now called *Spirillum volutans* and is a rather large bacterium (Figure 19.46*a* and *b*). A phototrophic organism resembling *S. volutans* is *Thiospirillum* (see Figure 19.5*b*). *S. volutans* is microaerophilic, requiring O_2, but is inhibited by O_2 at normal levels. The simplest way of achieving microaerophilic growth of *S. volutans* is to use a medium such as nutrient broth in a closed vessel with a large head (gas) space, with an atmosphere of N_2, and to then inject sufficient O_2 from a hypodermic syringe to occupy 1–5 percent of the head space volume. Because *S. volutans* is dependent on O_2 for growth, high cell yields cannot be expected in this way, because the O_2 is quickly used up. For cultivation for physiological studies, a continuous stream of 1 percent O_2 in N_2 can be passed through the culture vessel. Another characteristic of *S. volutans* is the formation of prominent granules (volutin granules) consisting of polyphosphate (Figure 19.46*b* and Section 3.9).

Azospirillum lipoferum is a nitrogen-fixing organism, which was originally described and named *Sprillum lipoferum* by Beijerinck in 1922. It has become of considerable interest in recent years because this bacterium has been found to enter into a loose symbiotic

Table 19.13 **Characteristics of the genera of spiral-shaped bacteria**[a]

Genus	16S rRNA group[b]	Characteristics	DNA (mole % GC)
Spirillum	Beta	Cell diameter 1.7 µm; microaerophilic; fresh water	36–38
Aquaspirillum	Alpha or beta	Cell diameter 0.2–1.5 µm; aerobic; fresh water	49–66
Oceanospirillum	Gamma	Cell diameter 0.3–1.2 µm; aerobic; marine (require 3% NaCl)	42–51
Azospirillum	Alpha	Cell diameter 1 µm; microaerophilic; soil and rhizosphere; fixes N_2	68–70
Herbaspirillum	—	Cell diameter 0.6–0.7 µm; microaerophilic; soil and rhizosphere; fixes N_2	66–67
Campylobacter	Epsilon	Cell diameter 0.2–0.8 µm; microaerophilic to anaerobic; pathogenic or commensal in humans and animals; single polar flagellum	30–38
Helicobacter	Epsilon	Cell diameter 0.5–1 µm; tuft of polar flagella; associated with pyloric ulcers in humans	—
Bdellovibrio	Delta	Cell diameter 0.25–0.4 µm; aerobic; predatory on other bacteria; single polar sheathed flagellum	33–52
Spirosoma	*Bacteroides/ Flavobacterium* group	Cell diameter 0.5 µm; curved rods forming rings; nonmotile; aerobic; sometimes gas vesicuolate	66–69

[a]*All are Gram-negative and respiratory but never fermentative.*
[b]*All indicated genera are members of the purple Bacteria (Proteobacteria) (see Section 18.6).*

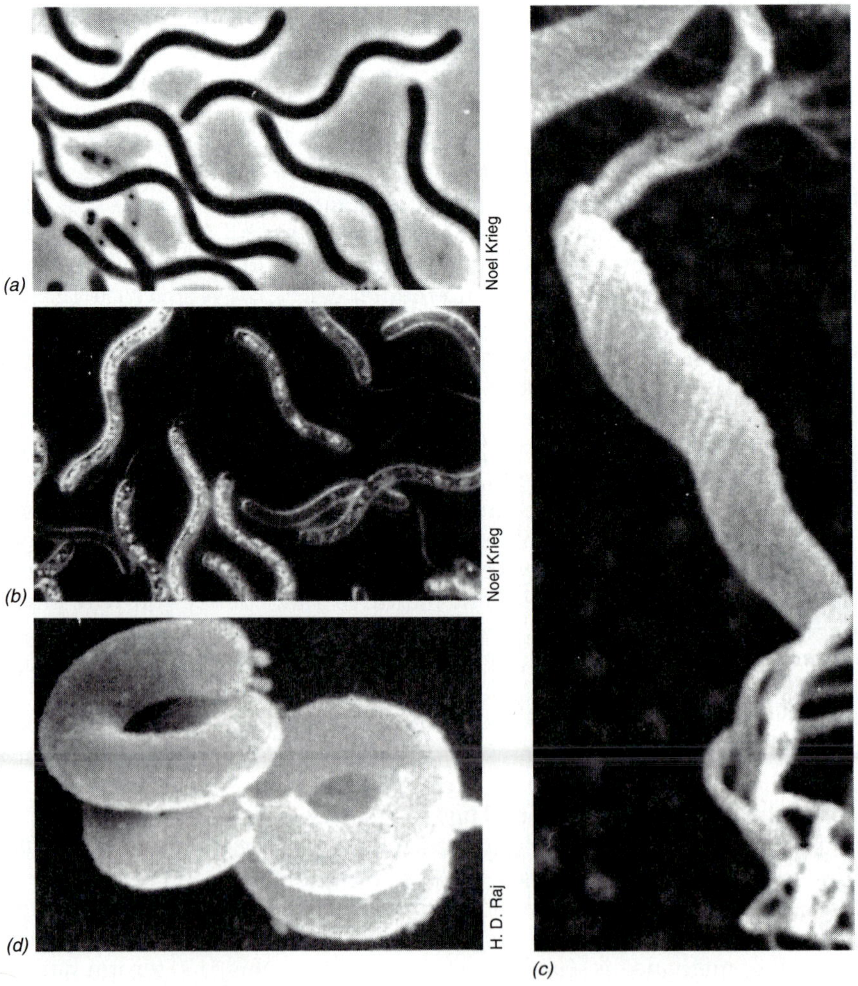

FIGURE 19.46
(a) Photomicrograph by phase contrast of *Spirillum volutans*, the largest spirillum. Cells are about 1.6 by 20–50 µm. (b) *S. volutans*, by dark-field microscopy, showing flagellar bundles and volutin (polyphosphate) granules. (c) Scanning electron micrograph of an intestinal spirillum. Note the polar flagellar tufts and the spiral structure of the cell surface. (d) Scanning electron micrograph of cells of *Spirosoma linguale*. Cells are about 0.5 µm in diameter.

relationship with tropical grasses and grain crops (see Section 17.24).

The genus *Spirillum* includes only a single species, *S. volutans*, characterized by its microaerophilic character, large size, and formation of volutin granules. The small-diameter spirilla (which are not microaerophilic) have been separated into two genera, *Aquaspirillum* and *Oceanospirillum*, the former for freshwater forms and the latter for those living in seawater and requiring NaCl for growth (Table 19.13). At least 17 species of *Aquaspirillum* have been described and 9 species of *Oceanospirillum*, the various species being separated on physiological grounds. The student isolating chemoorganotrophic bacteria from freshwater and marine environments will almost certainly obtain isolates of these genera, as they are among the most common organisms appearing in chemoorganotrophic enrichments, and they are easily purified by streaking on agar. These organisms undoubtedly play an important role in the recycling of organic matter in aquatic environments.

Highly motile microaerophilic spirilla have been isolated from freshwater habitats. These organisms demonstrate a dramatic directed movement in a magnetic field referred to as **magnetotaxis**. In an artificial magnetic field magnetotactic spirilla quickly orient their long axis along the north-south magnetic moment of the field. Within the cells, chains of 5–40 magnetic particles consisting of Fe_3O_4 called **magnetosomes** (see Section 3.9) are present (Figure 19.47), and

these serve as internal magnets that orient the cells along a specific magnetic field. Magnetotactic bacteria can have one of two magnetic polarities depending upon the orientation of magnetosomes within the cell. Cells in the northern hemisphere have the north-seeking pole of their magnetosomes forward with respect to their flagella and thus move in a northward direction. Cells in the southern hemisphere have the opposite polarity and move southward. Although the ecological role of bacterial magnets is unclear, it has been suggested that the ability to orient in a magnetic field would be of selective advantage in directing these microaerophilic organisms downwards toward microaerobic zones near the sediments.

Bdellovibrio

These small vibroid organisms have the unusual property of preying on other bacteria, using as nutrients the cytoplasmic constituents of their hosts. These bacterial predators are small, highly motile cells, which stick to the surfaces of their prey cells. Because of the latter property, they have been given the name *Bdellovibrio* (bdello- is a combining form meaning "leech"). Other predatory bacteria have been isolated and given such names as *Vampirococcus*. However, it appears that *Bdellovibrio* has a unique mode of attack and development intraperiplasmatically. After attachment of a *Bdellovibrio* cell to its prey, the predator penetrates through the prey wall and replicates in the space between the prey wall and membrane (the periplasmic space), eventually forming a spherical structure called a **bdelloplast**. The stages of attachment and penetration are shown in electron micrographs in Figure 19.48 and diagramatically in Figure 19.49. A wide variety of Gram-negative Bacteria can be attacked by a single *Bdellovibrio*; Gram-positive cells are not attacked.

As originally isolated, *Bdellovibrio* cells grow only upon the living prey, but it is possible to isolate mutants that are prey independent and are able to grow on complex organic media such as yeast extract-peptone. These strains, like the wild-type strain, are unable to utilize sugars as electron donors, but are proteolytic and can oxidize the amino acids liberated by protein digestion. Prey-dependent revertants can be reisolated from the prey-independent mutants by introducing a host strain.

Bdellovibrio is an obligate aerobe, obtaining its energy from the oxidation of amino acids and acetate (via the citric acid cycle). It apparently is unable to utilize sugars as electron donors. Studies have been carried out to determine the growth efficiency of *Bdellovibrio* as indicated by the Y_{ATP} value. As we noted in Section 9.5, most organisms exhibit a Y_{ATP} value of around 10. *Bdellovibrio*, on the other hand, exhibits a Y_{ATP} value of 18 to 26, considerably higher than that of free-living bacteria. This higher efficiency of conversion of substrates into ATP may be due to the more efficient growth process of *Bdellovibrio* cells in the periplasmic environment of its host, and to the impor-

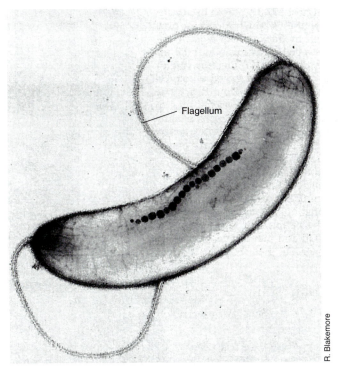

FIGURE 19.47 Negatively stained electron micrograph of a magnetotactic spirillum, *Aquaspirillum magnetotacticum*. A cell measures 0.3 by 2 µm. This bacterium contains particles of Fe_3O_4 (magnetite) called magnetosomes arranged in a chain; the particles serve to align the cell along geomagnetic lines. The organism was isolated from a water treatment plant in Durham, New Hampshire.

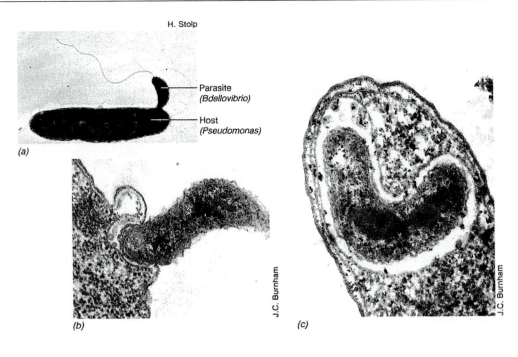

H. Stolp

Parasite
(Bdellovibrio)

Host
(Pseudomonas)

(a)

(b)

(c)

J.C. Burnham

J.C. Burnham

FIGURE 19.48
Stages of attachment and penetration of a prey cell by *Bdellovibrio*. A *Bdellovibrio* cell measure about 0.3 μm in diameter. (a) Electron micrograph of a shadowed whole-cell preparation showing *B. bacteriovorus* attacking *Pseudomonas*. (b, c) Electron micrographs of thin sections of *Bdellovibrio* attacking *Escherichia coli*: (b) early penetration; (c) complete penetration. The *Bdellovibrio* cell is enclosed in a membranous infolding of the prey cell (the bdelloplast), and replicates in the periplasmic space between wall and membrane.

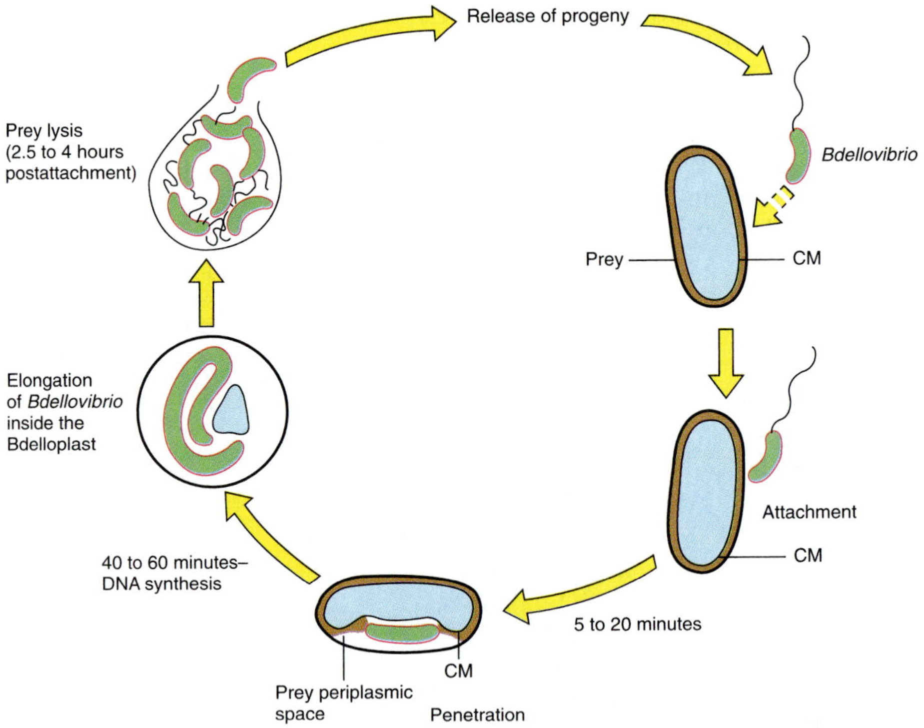

Release of progeny

Prey lysis
(2.5 to 4 hours
postattachment)

Bdellovibrio

Prey

CM

Elongation
of *Bdellovibrio*
inside the
Bdelloplast

Attachment

CM

40 to 60 minutes—
DNA synthesis

5 to 20 minutes

CM

Prey periplasmic
space

Penetration

FIGURE 19.49
Developmental cycle of the bacterial predator, *Bdellovibrio bacteriovorus*. Following primary contact between a highly motile *Bdellovibrio* cell and a Gram-negative Bacterium, attachment and penetration into the prey periplasmic space occurs. Once inside, *Bdellovibrio* elongates and within 4 hours progeny cells are released. The number of progeny cells released varies with the size of the prey bacterium. For example, 5 to 6 bdellovibrios are released from each infected *E. coli* cell, 20 to 30 for *Spirillum serpens*. CM = prey cytoplasmic membrane.

tant fact that *Bdellovibrio* assimilates nucleoside phosphates, fatty acids, and even whole proteins in some cases directly from its host without first breaking them down. It is clear that the predatory mode of existence has involved the development in *Bdellovibrio* of interesting and unusual biochemical processes.

Phylogenetically, bdellovibrios fall into the delta group of purple Bacteria (see Section 18.6). Taxonomically, three species of *Bdellovibrio* are recognized as shown in Table 19.14. Nucleic acid studies in bdellovibrios have shown them to be a rather heterogeneous group. Both GC base ratio and DNA:DNA hybridization analyses suggest that bdellovibrios are genetically diverse (Table 19.14), even though they

share a unique life style. In addition to being predators themselves, bdellovibrios, as for most bacteria, are subject to attack by bacteriophages. Nicknamed *bdellophages*, most phages that plaque on lawns of *Bdellovibrio* cells are strictly lytic single-stranded DNA phages.

Members of the genus *Bdellovibrio* are widespread in soil and water, including the marine environment. Their detection and isolation require methods reminiscent of those used in the study of bacterial viruses. Prey bacteria are spread on the surface of an agar plate to form a lawn, and the surface is inoculated with a small amount of soil suspension that has been filtered through a membrane filter; the latter retains most bac-

Table 19.14 Genetic relationships among species of *Bdellovibrio**					
		Genome size (kilobase pairs)	Genomic DNA hybridization to		
Species	DNA GC (mole %)		*B. bacteriovorus*	*B. stolpii*	*B. starrii*
Bdellovibrio:					
B. bacteriovorus	50	2000	100	28	26
B. stolpii	42	2270	28	100	16
B. starrii	44	2575	23	16	100
***Escherichia* (for comparison)**					
E. coli	51	4700	—	—	—

*Data from Torrella, F., R. Guerrero, and R. J. Seidler. 1978. Can. J. Microbiol. 24:1387–1394, and Ruby, E. G. 1992. The genus Bdellovibrio. In: A. Balows, et al. The Prokaryotes, 2nd edition. Springer-Verlag, New York.

teria but allows the small *Bdellovibrio* cells to pass. Upon incubation of the agar plate, plaques analogous to those produced by bacteriophages are formed at locations where *Bdellovibrio* cells are growing. However, unlike phage plaques, which continue to enlarge only as long as the bacterial host is growing, *Bdellovibrio* plaques continue to enlarge even after the prey has stopped growing, resulting in large plaques on the agar surface. Pure cultures of *Bdellovibrio* can then be isolated from these plaques. *Bdellovibrio* cultures have been obtained from a wide variety of soils and are thus common members of the soil population. As yet, the ecological role of bdellovibrios is not known, but it seems likely that they play some role in regulating the population densities of their prey.

Spirosoma

Members of the genus *Spirosoma* are ring-shaped, nonmotile, chemoorganotrophic bacteria (Figure 19.46*d*). They resemble very tightly curved vibrios and are widely distributed in aquatic environments. A phototrophic counterpart to *Spirosoma* exists in the genus *Rhodocyclus* (see Section 19.1 and Figure 19.4*e*).

19.12   Spirochetes

Spirochetes are bacteria with a unique morphology and mechanism of motility. They are widespread in aquatic environments and in the bodies of animals. Some of them cause diseases of animals and humans, of which the most important is *syphilis*, caused by *Treponema pallidum*. The spirochete cell is typically slen-

der, flexuous, helical (coiled) in shape, and often rather long (Figure 19.50). The "protoplasmic cylinder," consisting of the regions enclosed by the cytoplasmic membrane and the cell wall, constitutes the major portion of the spirochetal cell. Fibrils, referred to as **axial fibrils** or **axial filaments**, are attached to the cell poles and wrapped around the coiled protoplasmic cylinder (Figure 19.51). Both the axial fibrils and the protoplasmic cylinder are surrounded by a three-layered membrane called the *outer sheath* or *outer cell envelope* (Figure 19.51). The outer sheath and axial fibrils are usually not visible by light microscopy, but are observable in negatively stained preparations or thin sections examined by electron microscopy (see, for example, Figure 19.52).

From two to more than 100 axial fibrils are present per cell, depending on the type of spirochete. The ultrastructure and the chemical composition of axial fibrils are similar to those of bacterial flagella (see Section 3.7). As in typical bacterial flagella (Figure 3.41), basal hooks and paired disks are present at the insertion end. The shaft of each fibril is composed of a core surrounded by an "axial fibril sheath," so that the spirochete axial fibrils are in a sense analogous to sheathed flagella.

The axial fibrils play a significant role in spirochete motility. Each fibril is anchored at one end and extends for approximately two-thirds of the length of the cell. It is thought that the axial fibrils rotate rigidly, as do bacterial flagella (see Section 3.7). Since the protoplasmic cylinder is also rigid, whereas the outer sheath is flexible, if both axial fibrils rotate in the same

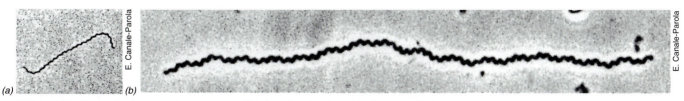

(a)   (b)   E. Canale-Parola

FIGURE 19.50   Two spirochetes at the same magnification, showing the wide size range in the group. (a) *Spirochaeta stenostrepta*, by phase-contrast microscopy. A single cell is 0.25 μm in diameter. (b) *Spirochaeta plicatilis*. A single cell is 0.75 μm in diameter and can be up to 250 μm (0.25 mm) in length.

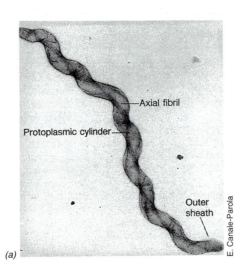

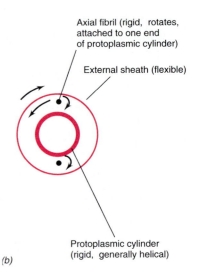

FIGURE 19.51
(a) Electron micrograph of a negatively stained preparation of *Spirochaeta zuelzerae*, showing the position of the axial fibril. A single cell is about 0.3 μm in diameter. (b) Cross section of a spirochete cell, showing the arrangement of the protoplasmic cylinder, axial fibrils, and external sheath, and the manner by which the rotation of the rigid axial fibril can generate rotation of the protoplasmic cylinder and (in opposite direction) rotation of the external sheath. If the sheath is free, the cell will rotate about its longitudinal axis and move along it. If the sheath is in contact with a solid surface, the cell will creep forward. See text for details.

(a)

(b)

Axial fibril (rigid, rotates, attached to one end of protoplasmic cylinder)

External sheath (flexible)

Protoplasmic cylinder (rigid, generally helical)

E. Canale-Parola

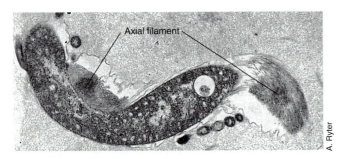

Axial filament

A. Ryter

FIGURE 19.52 Electron micrograph of a thin section of *Cristispira*, a very large spirochete. A cell measures about 2 μm in diameter. Notice the numerous fibrils in the axial filament.

direction, the protoplasmic cylinder will rotate in the *opposite* direction, as illustrated in Figure 19.51*b*. If the sheath is not in contact with a surface, it also rotates (see arrow in Figure 19.51*b*). This simple mechanism is all that is needed to generate the wide variety of motions exhibited by spirochetes. If the protoplasmic cylinder is helical (as in most spirochetes), then forward motion will be generated when the sheath is moving through a liquid or semisolid medium by the circumferential slip of the helix through the medium. If the sheath is in contact along its length with a solid surface, the protoplasmic cylinder may not be able to rotate, so that the roll of the sheath will cause the cell to slide in a direction nearly parallel to the axis of the helix, generating a "creeping" motility. In free liquid, many narrow-diameter spirochetes show flexing or lashing motions due to torque exerted at the ends of the protoplasmic cylinder by the twisting axial fibrils. In this model of motility, the external sheath plays a central role, as do the opposing axial fibrils. Similarly, the rigidity of the protoplasmic cylinder is essential in this mechanism. It thus appears that despite superficial differences, spirochetes have fundamentally the same motility mechanism as other bacteria, namely the rotation of rigid flagellar fibrils attached in the cell membrane via a basal hook.

Spirochetes are classified into five genera primarily on the basis of habitat, pathogenicity, and morphological and physiological characteristics. Table 19.15 lists the major genera and their characteristics. Molecular sequencing studies of 16S rRNA from various spirochetes show that spirochetes form a cluster, albeit a loose one, phylogenetically. In addition, it has been found that most spirochetes are naturally resistant to the antibiotic rifampicin, an indication that their RNA polymerases may differ from those of other bacteria (see Section 5.6).

Spirochaeta and *Cristispira*

The genus *Spirochaeta* includes free-living, anaerobic, and facultatively aerobic spirochetes. These organisms are common in aquatic environments, such as the water and mud of rivers, ponds, lakes, and oceans. One species of the genus *Spirochaeta* is *S. plicatilis* (Figure 19.50*b*), a fairly large spirochete found in freshwater and marine H_2S-containing habitats and is probably anaerobic. The axial fibrils of *S. plicatilis* are arranged in a bundle that winds around the coiled protoplasmic cylinder. From 18 to 20 axial fibrils are inserted at each pole of this spirochete. Another species, *S. stenostrepta*, has been cultured, and is shown in Figure 19.50*a*. It is an obligate anaerobe commonly found in H_2S-rich, black muds. It ferments sugars via the glycolytic pathway to ethanol, acetate, lactate, CO_2, and H_2. The species *S. aurantia* is an orange-pigmented facultative aerobe, fermenting sugars via the glycolytic pathway under anaerobic conditions, and oxidizing sugars aerobically mainly to CO_2 and acetate.

The genus *Cristispira* (Figure 19.52) contains organisms with a unique distribution, being found in nature primarily in the *crystalline style* of certain molluscs, such as clams and oysters. The crystalline style is a flexible, semisolid rod seated in a sac and rotated against a hard surface of the digestive tract, thereby mixing with and grinding the small particles of food. Being large spirochetes, the cristispiras can readily be seen microscopically within the style as they rapidly

Table 19.15   Genera of spirochetes and their characteristics*

Genus	Dimensions (µm)	Number of species recognized	General characteristics	Number of axial fibrils	DNA (mole % GC)	Habitat	Diseases
Cristispira	30–150 × 0.5–3.0	1	3–10 complete coils; bundle of axial fibrils visible by phase-contrast microscopy	>100	–	Digestive tract of molluscs; has not been cultured	None known
Spirochaeta	5–250 × 0.2–0.75	7	Anaerobic or facultatively aerobic; tightly or loosely coiled	2–40	50–65	Aquatic, free-living	None known
Treponema	5–15 × 0.1–0.4	13	Microaerophilic or anaerobic, coil amplitude up to 0.5 µm	2–15	25–53	Commensal or parasitic in humans, other animals	Syphilis, yaws, swine dysentery
Borrelia	8–30 × 0.2–0.5	15	Anaerobic; 5–7 coils of approx. 1 µm amplitude	Unknown	46	Humans and other mammals, arthropods	Relapsing fever, Lyme disease
Leptospira	6–20 × 0.1	2	Aerobic; tightly coiled, with bent or hooked ends; requires long chain fatty acids	2	33–43	Free-living or parasitic in humans, other mammals	Leptospirosis
Leptonema	6–20 × 0.1	1	Aerobic; does not require long chain fatty acids	—	54	Free-living	None known

Phylogenetically, spirochetes form a distinct line of Bacteria (see Section 18.6).

rotate forward and backward in corkscrew fashion. *Cristispira* may occur in both freshwater and marine molluscs, but not all species of molluscs possess them. Unfortunately, *Cristispira* has not been cultured, so that the physiological reason for its restriction to this unique habitat is not known. There is no evidence that *Cristispira* is harmful to its host; in fact, the organism may be more common in healthy than in diseased molluscs.

Treponema

Anaerobic, host-associated spirochetes that are commensals or parasites of humans and animals are placed in the genus *Treponema*. *Treponema pallidum*, the causal agent of syphilis (see Section 15.6), is the best known species of *Treponema*. It differs in morphology from other spirochetes; the cell is not helical, but has a flat wave form. Furthermore, electron microscopy does not show the presence of an outer sheath surrounding both the axial fibrils and the protoplasmic cylinder. Apparently the axial fibrils of *T. pallidum* lie on the outside of the organism. The *T. pallidum* cell is remarkably thin, measuring approximately 0.2 µm in diameter. Living cells are clearly visible in the dark-field microscope or after staining with fluorescent antibody; dark-field microscopy has long been used to examine exudates from suspected syphilitic lesions (see Figure 15.25). In nature *T. pallidum* is restricted to humans, although artificial infections have been established in rabbits and monkeys. Although never grown in laboratory culture, it has now been established from

animal studies that virulent *T. pallidum* cells (purified from infected rabbits) contain a cytochrome system, and are in fact microaerophiles. This property further separates *T. pallidum* from the other species of *Treponema*, which, as noted, are obligate anaerobes. Meaningful taxonomic studies on this species will have to await cultivation away from its host.

Treponema pallidum is quite sensitive to increased temperature, being rapidly killed by exposure to 41.5–42.0°C. The heat sensitivity of *T. pallidum* is also reflected in the fact that the organism becomes most easily established in cooler sites of the body, such as the male genital organs, although once established in other areas of the body it will multiply there. The organism is rapidly killed by drying, and this at least partially explains why transmission of *T. pallidum* between persons is only by direct contact, usually sexual intercourse.

Other species of the genus *Treponema* are common commensal organisms in the oral cavity of humans and can generally be seen in material scraped from between the teeth and from the narrow space between the gums and the teeth. Various oral spirochetes can be cultivated anaerobically in complex media containing serum. Three species, *T. denticola*, *T. macrodentium*, and *T. oralis*, have been described, differing in morphology and physiological characteristics. *Treponema denticola* ferments amino acids such as cysteine and serine, forming acetate as the major fermentation acid as well as CO_2, NH_3, and H_2S. This spirochete can also ferment glucose, but in media containing both glucose

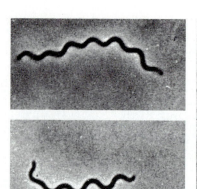
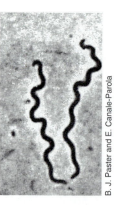

B. J. Paster and E. Canale-Parola

FIGURE 19.53 Phase contrast photomicrographs of *Treponema saccharophilum*, a large pectinolytic spirochete from the bovine rumen. A cell measures about 0.4 μm in diameter. Left, regularly coiled cells. Right, irregularly coiled cells.

and amino acids, the amino acids are used preferentially. The true relationship between *T. pallidum* and the remaining members of the genus *Treponema* may be a distant one, since the GC base ratio of *T. pallidum* is about 53 percent while other species of this genus cluster between 38 and 40 percent or 25 and 26 percent.

Spirochetes are also found in the rumen. *Treponema saccharophilum* (Figure 19.53) is a large, *pectinolytic* spirochete found in the bovine rumen. *T. saccharophilum* is an obligately anaerobic bacterium that ferments pectin, starch, inulin, and other plant polysaccharides. This and other spirochetes may play an important role in the conversion of plant polysaccharides to volatile fatty acids, usable as energy sources by the ruminant (see Section 17.13).

Leptospira and *Leptonema*

The genera *Leptospira* and *Leptonema* contain strictly *aerobic* spirochetes that use long-chain fatty acids (for example, oleic acid) as electron donor and carbon sources. With few exceptions, these are the only substrates utilized by leptospiras for growth. The leptospira cell is thin, finely coiled, and usually bent at each end into a semicircular hook. At present, several species are recognized in this group, some free-living and some parasitic. Two major species are *Leptospira interrogans* (parasitic) and *L. biflexa* (free living). Strains of *L. interrogans* are parasitic for humans and animals. Different strains are distinguished serologically by agglutination tests, and a large number of serotypes have been recognized. Rodents are the natural hosts of most leptospiras, although dogs and pigs are also important carriers of certain strains. In humans the most common leptospiral syndrome is *Weil's disease*; in this disorder the organism usually localizes in the kidney. Leptospiras ordinarily enter the body through the mucous membranes or through breaks in the skin. After a transient multiplication in various parts of the body the organism localizes in the kidney and liver, causing nephritis and jaundice. The organism passes out of the body in the urine and infection of another individual is most commonly by contact with

infected urine. Therapy with penicillin, streptomycin, or the tetracyclines is possible but may require extended courses to eliminate the organism from the kidney; this is probably because of the slow growth and protected location of the leptospiras. Domestic animals are vaccinated against **leptospirosis** with a killed virulent strain; dogs are usually immunized routinely with a combined distemper-leptospira-hepatitis vaccine. In humans, prevention is effected primarily by elimination of the disease from animals. The serotype that infects dogs, *L. interrogans* serotype *canicola*, does not ordinarily infect humans, but the strain attacking rodents, *L. interrogans* serotype *icterohaemorrhagiae*, does; hence elimination of rats from human habitation is of considerable aid in preventing the organism from reaching human beings.

Borrelia

The majority of species in the genus *Borrelia* are animal or human pathogens. *B. recurrentis* is the causative agent of **relapsing fever** in humans and is transmitted via an insect vector, usually by the human body louse. Relapsing fever is characterized by a high fever and generalized muscular pain which lasts for three to seven days followed by a recovery period of seven to nine days. Left untreated, the fever returns in two to three more cycles (hence the name relapsing fever) and causes death in up to 40 percent of those infected. Fortunately, the organism is quite sensitive to tetracycline, and if the disease is correctly diagnosed, treatment is straightforward. Other borrelia are of veterinary importance causing diseases in cattle, sheep, horses, and birds. In most of these diseases the organism is transmitted by ticks. *B. burgdorferi* is the causative agent of the tick-borne disease called *Lyme disease*, which infects humans and other animals. Lyme disease was discussed in Section 15.10. *B. burgdorferi* is also of interest because it is as yet the only known prokaryote with a *linear* (as opposed to a *circular*) chromosome (see Section 5.5).

19.13 Gliding Bacteria

A variety of bacteria exhibit gliding motility. These organisms have no flagella, but are able to move when in contact with surfaces. All **gliding bacteria** are Gram-negative. One group of gliding bacteria, the fruiting myxobacteria, possesses the interesting property of forming multicellular structures of complex morphology called **fruiting bodies**. Table 19.16 gives a brief outline of some of the genera of gliding bacteria.

The mechanism of gliding by gliding bacteria appears complex, and it is likely that more than one mechanism is responsible. Gliding is most apparent when cells are on a solid surface. Some gliding bacteria rotate along their long axis while moving while others seem to keep only one side of the cell in contact with the surface while moving. Gliding motility is generally much slower than flagellar motility, but absolute rates of gliding are somewhat dependent on cell

Table 19.16   Characteristics of some genera of chemoorganotrophic gliding bacteria

Characteristics	Genus	16S rRNA group*	DNA (mole % GC)
Rods and nonseptate filaments: Unicellular, rod shaped, many digest cellulose, chitin, or agar	*Cytophaga* (no microcysts)	*Bacteroides/Flavobacterium*	30–40
	Sporocytophaga (microcysts formed)	*Bacteroides/Flavobacterium*	36
Inhabit human oral cavity; facultative aerobes	*Capnocytophaga*	*Bacteroides/Flavobacterium*	33-41
Helical or spiral shaped	*Saprospira*	*Bacteroides/Flavobacterium*	35–48
Filamentous	*Microscilla, Flexibacter*		37–47
Can lyse a variety of microorganisms	*Lysobacter*	Gamma purple Bacteria	65–70
Septate filaments: Filamentous, chemoorganotrophic or chemolithotrophic, producing S⁰ granules from H_2S	*Beggiatoa*	Gamma purple Bacteria	37–51
Filamentous, chemoorganotrophic; life cycle involving gonidia and rosette formation	*Leucothrix*	Gamma purple Bacteria	46–51
Filamentous, chemolithotrophic sulfur oxidizer; life cycle like *Leucothrix*	*Thiothrix*	Gamma purple Bacteria	52
Cells in chains or short filaments; or chemoganotrophic; occurs in oral cavity or digestive tract of man and other animals	*Simonsiella, Alysiella*	Beta purple Bacteria	41–55
Filamentous	*Vitreoscilla*	Beta purple Bacteria	44–45
Rods, forming fruiting bodies: Unicellular, rod shaped; life cycle involving aggregation, fruiting-body formation, and myxospore formation	Fruiting myxobacteria; *Archangium, Chondromyces, Myxococcus, Polyangium, etc.* (see Table 19.17)	Delta purple Bacteria	67–71

*See Section 18.6

length; filamentous gliding bacteria usually move much faster than unicellular gliding bacteria. In *Cytophaga*, there is evidence that small rotating particles, presumably made of protein, lie between the cytoplasmic membrane and the Gram-negative outer membrane. Acting like miniature ball bearings, these particles rotate, presumably at the expense of a membrane potential or perhaps ATP directly, and this rotary motion slides the cell along the solid surface. By contrast, in *Myxococcus*, gliding seems to occur as a result of the secretion of a chemical surfactant which affects surface tension forces between the cell and the solid surface, and is apparently sufficient to slide the cell along the surface.

Cytophaga and related genera

Organisms of the genus *Cytophaga* are long slender rods, often with pointed ends, which move by gliding. Many digest cellulose, agar, or chitin. They are widespread in the soil and water, often being present in great abundance. The cellulose decomposers can be easily isolated by placing small crumbs of soil on pieces of cellulose filter paper laid on the surface of mineral agar. The bacteria attach to and digest the cellulose fibers, forming transparent spreading colonies that are usually yellow or orange in color. Microscopic examination reveals the bacteria aligned upon the surface of the cellulose fibrils. The cytophagas do not produce soluble, extracellular, cellulose-digesting enzymes (cellulases); the enzymes probably remain attached to the cell envelope, accounting for the fact that the cells must adhere to cellulose fibrils in order to digest them.

Organisms of the genus *Sporocytophaga* are similar to *Cytophaga* in morphology and physiology, but form resting spherical structures called *microcysts*, similar to those produced by some fruiting myxobacteria (see below), although they are produced without formation of fruiting bodies. Despite its ability to form microcysts, *Sporocytophaga* is not related to the fruiting myxobacteria; its DNA base composition is similar to that of *Cytophaga*, but far removed from those of the fruiting myxobacteria (Table 19.16). In pure culture, *Cytophaga* can be cultured on agar containing embedded cellulose fibers, the presence of the organism being indicated by the clearing that occurs as the cellulose is digested (Figure 16.40). *Lysobacter*, a gliding rod-shaped bacterium, has the ability to lyse both bacteria and fungi through the action of an array of proteases, chitinases, and other lytic enzymes excreted into the medium.

Beggiatoa

Organisms of this genus are morphologically similar to filamentous cyanobacteria. The filaments of *Beggiatoa* are usually quite large in diameter and long, consisting of many short cells attached end-to-end (Figure 19.54). In addition to moving by gliding, they can flex and twist so that many filaments may become intertwined to form a complex tuft. *Beggiatoa* is found in nature primarily in habitats rich in H_2S, such as sulfur springs, decaying seaweed beds, mud layers of lakes, and waters polluted with sewage, and in these habitats the filaments of *Beggiatoa* are usually filled with sulfur granules (Figures 19.55 and 19.56). *Beggiatoa* are also common inhabitants of hydrothermal vents (see Section 17.10). It was with *Beggiatoa* that Winogradsky first demonstrated that a living organism could oxidize H_2S to S^0 and then to SO_4^{2-}, leading him to formulate the concept of chemolithotrophy. However, most pure cultures of *Beggiatoa* so far isolated grow best *chemoorganotrophically* on compounds such as acetate, succinate, and glucose, and when H_2S is provided as an electron donor, they still require organic substances for growth. On the other hand, some marine strains of *Beggiatoa* have been isolated and shown to be true chemolithotrophic autotrophs, but this does not seem to be common.

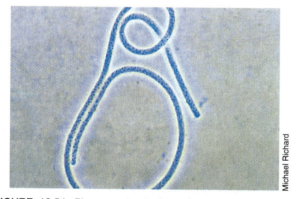

FIGURE 19.54 Phase contrast photomicrograph of a *Beggiatoa* species. A cell is about 2 μm in diameter.

FIGURE 19.55 Filamentous sulfur-oxidizing bacteria in a small stream. The filamentous cells twist together to form thick streamers and the white color is due to the abundant elemental sulfur content of the cells.

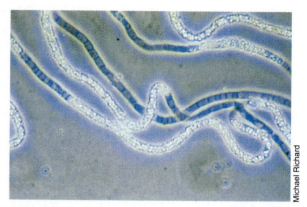

FIGURE 19.56 Phase contrast photomicrograph of the filamentous sulfur-oxidizing bacterium isolated from a sewage treatment plant. Note the abundant elemental sulfur granules in some of the cells.

An interesting habitat of *Beggiatoa* is the rhizosphere of plants (rice, cattails, and other swamp plants) living in flooded, and hence anoxic, soils. Such plants pump oxygen down into their roots, so that a sharply defined boundary develops at the root surface between O_2 on the root and H_2S in the soil. *Beggiatoa* (and probably other sulfur bacteria) develops at this boundary, and it has been suggested that *Beggiatoa* plays a beneficial role for the plant by oxidizing and thus detoxifying hydrogen sulfide. The growth of *Beggiatoa* is greatly stimulated by the addition to culture media of the enzyme *catalase* (which converts hydrogen peroxide into water and oxygen), and since plant roots contain catalase, it has been suggested that the plant promotes the growth of *Beggiatoa* in its rhizosphere via catalase production, thus leading to the development of a loose mutualistic relationship between the plant and the bacterium.

Beggiatoa and related filamentous bacteria (Figure 19.56) can cause major settling problems in sewage treatment facilities and in industrial waste lagoons such as from canning, paper pulping, brewing, and milling. These problems are generally referred to as *bulking* and occur when filamentous bacteria overgrow the normal flora of the waste system, producing a loose detrital floc instead of the normal and more easily settling tight floc. If bulking occurs, the wastewater remains improperly treated because the effluent discharged is still high in BOD; in sewage treatment, for example, bulking occurs when *Beggiatoa* or other filamentous bacteria replace *Zoogloea* in the activated sludge process (see Section 17.21).

Leucothrix and Thiothrix

These two genera are related in cell structure and life cycle. *Thiothrix*, a chemolithotroph that oxidizes H_2S, grows only as a mixotroph (requiring both H_2S and an organic compound). *Leucothrix* is the chemoorganotrophic counterpart of *Thiothrix*, and since its members have been more amenable to cultivation, the details of its life cycle and physiology are fairly well established. *Leucothrix* is a filamentous organism that has been found in nature only in marine environments, where it grows most commonly as an epiphyte

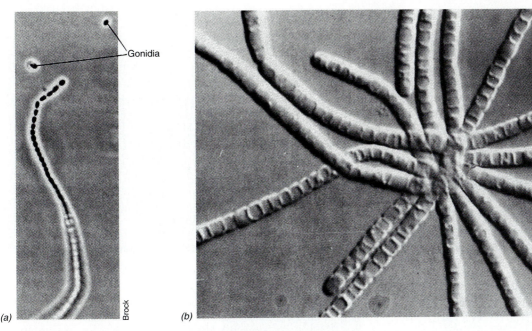

FIGURE 19.57 *Leucothrix mucor.* A cell is about 2 μm in diameter. (a) Filaments showing multicellular nature and release of gonidia. (b) Rosette composed of several multicellular filaments. Nomarski interference contrast.

on marine algae. *Leucothrix* filaments are 2–5 μm in diameter and may reach lengths of 0.1–0.5 cm. The filaments have clearly visible cross walls, and cell division is not restricted to either end but occurs throughout the length of the filament. The free filaments never glide (thus distinguishing them from *Beggiatoa*), although they occasionally wave back and forth in a jerky fashion. Under environmental conditions unfavorable to rapid growth, individual cells of the filaments become round and form ovoid structures called *gonidia*, which are released individually, often from the tips of the filaments (Figure 19.57*a*). The gonidia are able to glide in a jerky manner when they come into contact with a solid surface. They settle down on solid surfaces, synthesize a holdfast, and through growth and successive cell divisions form new filaments. Presumably in nature the gonidia are elements of dispersal, enabling the organism to spread to other areas. If there are high concentrations of gonidia, individual cells may aggregate, probably because of mutual attraction; they then synthesize a holdfast that causes their ends to adhere in a rosette, and new filaments grow out (Figure 19.57*b*). Rosette formation is found in both *Leucothrix* and *Thiothrix* and is an important means of distinguishing these organisms from many other filamentous bacteria.

Fruiting myxobacteria

The fruiting myxobacteria exhibit the most complex behavioral patterns and life cycles of all known prokaryotic organisms. In keeping with this complexity, the chromosome size of some myxobacteria can be rather large. *Myxococcus xanthus*, for example, has a single circular chromosome of 9500 kilobase pairs, *twice* as large as that of *Escherichia coli*. Indeed, this is two-thirds the size of the entire yeast genome which is

contained on 17 chromosomes (see Section 7.14). The vegetative cells of the fruiting myxobacteria are simple, nonflagellated, Gram-negative rods that glide across surfaces and obtain their nutrients primarily by causing the lysis of other bacteria. Under appropriate conditions, a swarm of vegetative cells aggregate and construct "fruiting bodies," within which some of the cells become converted into resting structures called **myxospores**. (A myxospore is a resting cell contained in a fruiting body.) It is the ability to form complex fruiting bodies that distinguishes the myxobacteria from all other prokaryotes. Since the vegetative cells of fruiting myxobacteria look like those of nonfruiting gliding bacteria, it is only through observation of the fruiting bodies that these organisms can be identified (see Table 19.17).

The fruiting bodies of the myxobacteria vary from simple globular masses of myxospores in loose slime to complex forms with a fruiting-body wall and a stalk (Figure 19.58). The fruiting bodies are often strikingly colored (Table 19.17). Occasionally they can be seen with a hand lens or dissecting microscope on pieces of decaying wood or plant material. Fruiting bodies of myxobacteria often develop on dung pellets (for example, those of the sheep or rabbit) after they have been incubated for a few days in a moist chamber. Although the vegetative cells are common in soils, the fruiting bodies themselves are less common. An effective means of isolating fruiting myxobacteria is to prepare Petri plates of water agar (1.5 percent agar in distilled water with no added nutrients) on which is spread a heavy suspension of any of several bacteria that the myxobacteria can lyse and use as a source of nutrients (for example, *Micrococcus luteus* or *E. coli*). In the center of the plate a small amount of soil, decaying bark, or other natural material is placed. Myxobacteria

Typical fruiting bodies of selected myxobacteria

Myxococcus fulvus, about 125 μm high

Myxococcus stipitatus, about 170 μm high

Mellitanigium erectum, about 50 μm high

Chon-dromyces crocatus, about 560 μm high

Stigmatella aurantiaca, about 150 μm high

Hans Reichenbach

Table 19.17 Classification of the fruiting myxobacteria*		
Characteristics	**Genus**	**DNA (mole % GC)**
Vegetative cells tapered		
Spherical or oval myxospores, fruiting bodies usually soft and slimy without well-defined sporangia or stalks	*Myxococcus*	68–71
Tough, cartilaginous ridged fruiting bodies	*Corallococcus*	—
Rod-shaped myxospores:		
Myxospores not contained in sporangia, fruiting bodies without stalks	*Archangium*	67–68
Myxospores embedded in slime envelope:		
Fruiting bodies without stalks	*Cystobacter*	68
Stalked fruiting bodies, single sporangia	*Melittangium*	—
Stalked fruiting bodies, multiple sporangia	*Stigmatella*	68–69
Vegetative cells not tapered (blunt rounded ends); myxospores resemble vegetative cells; sporangia always produced:		
Fruiting bodies without stalks, myxospores rod shaped	*Polyangium*	69
Fruiting bodies without stalks; myxospores oval; highly cellulolytic	*Sorangium*	—
Fruiting bodies without stalks, myxospores coccoid	*Nannocystis*	70–72
Large solitary yellow fruiting bodies with netlike surface	*Haploangium*	—
Stalked fruiting bodies	*Chondromyces*	69–70

Phylogenetically, those species examined fall into the delta subdivision of the purple Bacteria (Proteobacteria).

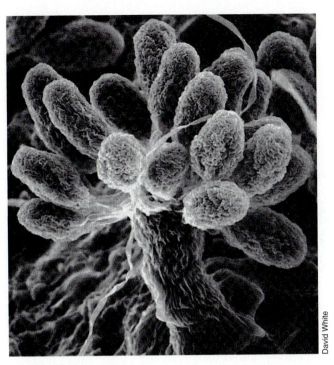

FIGURE 19.58 Scanning electron micrograph of a fruiting body of the gliding myxobacterium *Stigmatella aurantiaca* growing on a piece of wood. Note the individual cells visible in each fruiting structure.

in the inoculum lyse the bacterial cells and use their liberated products as nutrients; as they grow, they swarm out across the plate from the inoculum site. After several days to a week, the plates are examined under a dissecting microscope for myxobacterial swarms or fruiting bodies, and pure cultures are obtained by transfer of cells from the fruiting bodies or from the edge of the swarm to organic media.

The life cycle of a typical fruiting myxobacterium is shown in Figure 19.59. The vegetative cells are typical Gram-negative rods and do not reveal in their fine structure any clue as to their gliding motility or to their ability to aggregate and form fruiting structures. The vegetative cells of many strains grow poorly or not at all when first dispersed in liquid medium but can often be adapted to growth in liquid by making several passages through shaken liquid growth medium. A vegetative cell usually excretes slime, and as it moves across a solid surface it leaves a slime trail behind (Figure 19.60*a*). This trail is preferentially used by other cells in the swarm so that often a characteristic radiating pattern is soon created, with cells migrating along slime trails (Figure 19.60*b*). The fruiting body ultimately formed (Figure 19.60*c*) is a complex structure formed by the differentiation of cells in the

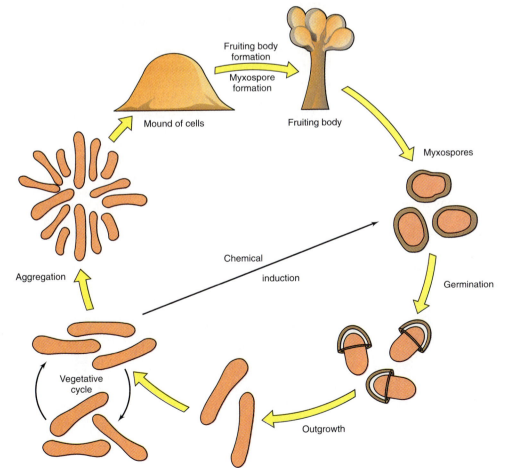

FIGURE 19.59

Life cycle of *Myxococcus xanthus*. Aggregation serves to assemble vegetative cells for fruiting body formation. Vegetative cells undergo morphogenesis to resting cells called myxospores. The latter germinate under favorable nutritional and physical conditions to yield vegetative cells. Vegetative cells can be converted directly to myxospores without fruiting body formation by certain chemical inducers, notably high concentrations of glycerol. See photo of *Myxococcus* fruiting bodies in Table 19.17.

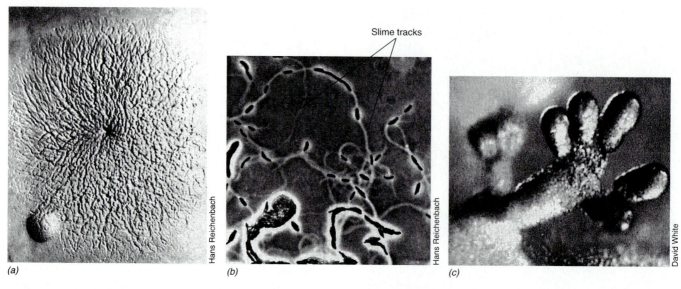

Slime tracks

Hans Reichenbach

Hans Reichenbach

David White

(a)   *(b)*   *(c)*

FIGURE 19.60 (a) Photomicrograph of a swarming colony (9-mm diameter) of *Myxococcus xanthus* on agar. (b) Single cells of *M. fulvus* from an actively gliding culture, showing the characteristic slime tracks on the agar. (c) Scanning electron micrograph of a fruiting body of *Stigmatella aurantiaca*.

stalk region and in the myxospore-bearing head. A wide variety of Gram-positive and Gram-negative Bacteria, as well as fungi, yeasts, and algae, can be used as food sources. A few fruiting myxobacteria can also use cellulose. Many myxobacteria can be grown in the laboratory on media containing peptone or casein hydrolysate, which provides organic nutrients in the form of amino acids or small peptides. The organisms are typical aerobes with a complete tricarboxylic acid cycle and cytochrome system.

Fruiting-body formation does not occur so long as adequate nutrients for vegetative growth are present, but upon exhaustion of amino acids, the vegetative swarms begin to fruit. Cells aggregate, possibly through a chemotactic response, with the cells migrating toward each other and forming mounds or heaps (Figure 19.61*a*). A single fruiting body may have 10⁹ or more cells. As the cell mounds become higher, the differentiation of the fruiting body into stalk and head begins (Figure 19.61*b* and *c*). Figure 19.61*d* clearly illustrates the differentiation of the fruiting body into stalk

and head. The stalk is composed of slime, within which a few cells may be trapped. The majority of the cells accumulate in the fruiting-body head and undergo differentiation into *myxospores* (Figure 19.62 and Table 19.17). And, in some genera, the myxospores are enclosed in larger walled structures called **cysts**. Compared to the vegetative cell, the myxospore is more resistant to drying, sonic vibration, UV radiation, and heat, but the degree of heat resistance is much less than that of the bacterial endospore. It seems likely that the main function of encysted myxospores is to enable the organism to survive desiccation during dispersal or during drying of the habitat. The myxospore eventually germinates by a localized rupture of the capsule, with the growth and emergence of a typical vegetative rod.

Myxobacteria are usually colored by carotenoid pigments (see Table 19.17), and the main pigments are carotenoid glycosides. Pigment formation is promoted by light, and at least one function of the pigment is photoprotection. Since in nature, the myxobacteria usually form fruiting bodies in the light, the presence

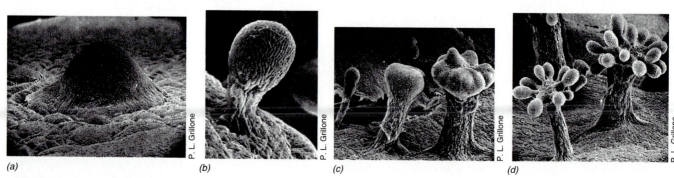

P. L. Grillone

P. L. Grillone

P. L. Grillone

P. L. Grillone

(a)   *(b)*   *(c)*   *(d)*

FIGURE 19.61 Scanning electron micrographs of fruiting body formation in *Chondromyces crocatus*. (a) Early stage, showing aggregation and mound formation. (b) Initial stage of stalk formation. Slime formation in the head has not yet begun so that the cells of which the head is composed are still visible. (c) Three stages in head formation. Note that the diameter of the stalk also increases. (d) Mature fruiting bodies. The entire fruiting structure is about 700 μm in height.

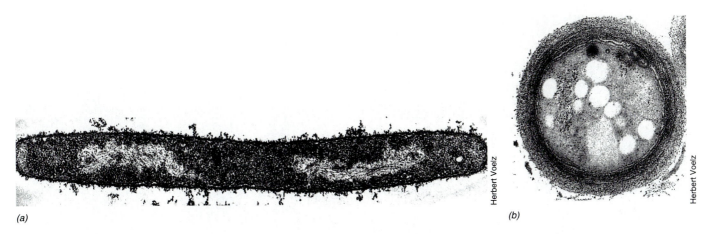

Herbert Voelz

(a)   *(b)*   Herbert Voelz

FIGURE 19.62   (a) Electron micrograph of a thin section of a vegetative cell of *M. xanthus*. A cell measure about 0.75 µm wide. (b) Myxospore of *M. xanthus*, showing the multilayered outer wall. Myxospores measure about 2 µm in diameter.

of these photoprotective pigments is understandable. In the genus *Stigmatella*, light greatly stimulates fruiting body formation, and it is thought that light catalyzes production of a pheromone that initiates the aggregation step. The fruiting myxobacteria are classified primarily on morphological grounds using characteristics of the vegetative cells, the myxospores, and fruiting body structure (Table 19.17). Phylogenetically, gliding myxobacteria belong to the purple Bacteria group (see Section 18.6).

19.14   Sheathed Bacteria

Sheathed bacteria are filamentous organisms with a unique life cycle involving formation of flagellated swarmer cells within a long tube or sheath. Under certain (generally unfavorable) conditions, the swarmer cells move out and become dispersed to new environments, leaving behind the empty sheath. Under favorable conditions, vegetative growth occurs within the filament, leading to the formation of long, cell-packed sheaths. Sheathed bacteria are common in freshwater habitats that are rich in organic matter, such as polluted streams, trickling filters, and activated sludge plants, being found primarily in flowing waters. In habitats where reduced iron or manganese compounds are present, the sheaths may become coated with ferric hydroxide or manganese oxide (see, for example, Figure 16.24). Iron precipitation is probably due to chemical reactions, but some sheathed bacteria have the ability to oxidize manganous ions to manganese oxide. Three genera are currently recognized: *Sphaerotilus*, in which manganese oxidation does not occur, *Leptothrix*, whose members do oxidize Mn^{2+}, and *Crenothrix*, which contains a multilayered sheath and whose unicellular forms are nonmotile. A single species of *Sphaerotilus* is recognized, *S. natans*, but several species of *Leptothrix* have been discerned, distinguished primarily on size, flagellation of swarmers, and some other morphological characteristics. Most of our discussion will concern *S. natans*, the organism which has been most extensively studied.

The *Sphaerotilus* filament is composed of a chain of rod-shaped cells with rounded ends enclosed in a closely fitting sheath. This thin and transparent sheath is difficult to see when it is filled with cells, but when the filament is partially empty, the sheath can easily be seen by phase-contrast microscopy (Figure 19.63a) or by staining. The cells within the sheath divide by binary fission (Figure 19.63b), and the new cells pushed out at the end synthesize new sheath material. Thus the sheath is always formed at the *tips* of the filaments. Individual cells are 1–2 µm wide by 3–8 µm long and stain Gram-negatively. Eventually cells are liberated from the sheaths, probably when the nutrient supply is low. These free cells are actively motile, the flagella being arranged lophotrichously (in a bundle at one pole) (Figure 19.63c). Probably the flagella are synthesized before the cells leave the sheath and, if so, may even aid in their liberation. It is thought that the swarmer cells then migrate, settle down, and begin to grow, each swarmer being the forerunner of a new filament. The sheath, which is devoid of muramic acid or other components of the peptidoglycan cell wall, is a protein-polysaccharide-lipid complex, possibly analogous to the capsules formed by many Gram-negative Bacteria but differing in that it forms a linear structure.

Sphaerotilus cultures are nutritionally versatile, able to use a wide variety of simple organic compounds as carbon and energy sources, with inorganic nitrogen sources. Many strains require vitamin B_{12}, a substance frequently needed by aquatic microorganisms. Befitting its habitat in flowing waters, *Sphaerotilus* is an obligate aerobe.

As we noted, *Sphaerotilus* is widespread in nature in aquatic environments receiving rich organic matter. *Sphaerotilus* blooms often occur in the fall of the year in streams and brooks when leaf litter causes a temporary increase in the organic content of the water. Its filaments are the main component of a microbial complex that sanitary engineers call "sewage fungus," which is the fungus-like filamentous slime found on the rocks in streams receiving sewage pollution. In activated sludge plants (see Section 17.21), *Sphaerotilus*

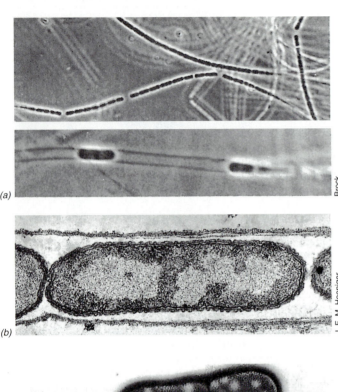

(a)

(b)

Brock

J. F. M. Hoeniger

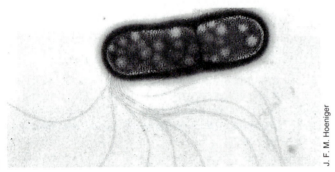

(c)

J. F. M. Hoeniger

FIGURE 19.63 *Sphaerotilus natans*. A single cell is about 2 μm wide. (a) Phase contrast photomicrographs of material collected from a polluted stream. Active growth stage (above) and swarmer cells leaving the sheath. (b) Electron micrograph of a thin section through a filament. (c) Electron micrograph of a negatively stained swarmer cell. Notice the polar flagellar tuft.

growth, like that of *Beggiatoa* (see Section 19.13), is often responsible for the detrimental condition called "bulking." The tangled masses of *Sphaerotilus* filaments so increase the bulk of the sludge that it does not settle properly, thus presenting difficulties in sludge clarification.

However, the ability of *Sphaerotilus* and *Leptothrix* to cause precipitation of iron oxides on their sheaths is well established. Such iron-encrusted sheaths are frequently seen in iron-rich waters (Figure 19.64). The process whereby iron deposition occurs is as follows: in iron-rich waters, ferrous iron is often held in solution as a chelate with organic materials such as humic and tannic acids. The sheathed bacteria can take up these soluble chelates, oxidize the organic compound, liberating the ferrous ions, which then oxidize spontaneously and become precipitated, generally in the region of the sheath (see Figure 19.64).

In the case of Mn^{2+}, specific oxidation by *Leptothrix*, but not by *Sphaerotilus*, is known to occur. (Oxidation of Mn^{2+} is not unique to *Leptothrix*, as a number of other bacteria, as well as fungi and yeasts, are able

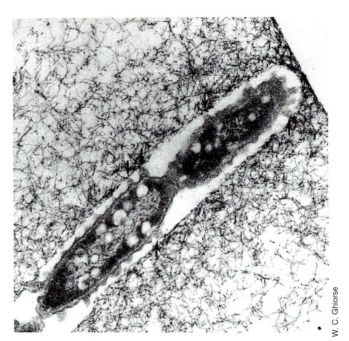

W. C. Ghiorse

FIGURE 19.64 Transmission electron micrograph of a thin section of *Leptothrix* sp. in a sample from a ferromanganese film in a swamp in Ithaca, New York. A single cell measures about 0.9 μm in diameter. Note the protuberances of the cell envelope which contact the sheath (arrows).

to carry out this process; see Section 16.18.) At pH values below 8, Mn^{2+} does not oxidize spontaneously, so that a significant biological oxidation is possible. However, even with Mn^{2+}, there is no evidence that *Leptothrix* obtains *energy* from the process, either chemolithotrophically or mixotrophically. Although *Leptothrix* produces an enzyme which catalyzes Mn^{2+} oxidation, it is unclear as yet what benefit the organism derives from this oxidation process.

19.15   Pseudomonads

All of the genera in this group are straight or slightly curved rods with *polar* flagella (Figure 19.65). Pseudomonads are a major group of chemoorganotrophic aerobic Gram-negative rods that never show a fermentative metabolism. (Fermentative organisms with polar flagella are generally classified in the genera *Aeromonas* or *Vibrio*, as indicated in Section 19.19). The important genera are *Pseudomonas* and *Commamonas*, discussed in some detail here. Other genera include *Xanthomonas*, primarily a plant pathogen responsible for a number of necrotic plant lesions and which is characterized by its yellow-colored pigments; *Zoogloea*, characterized by its formation of an extracellular fibrillar polymer, which causes the cells to aggregate into distinctive flocs (this organism is a dominant component of activated sludge, see Section 17.21), and *Gluconobacter*, characterized by its incomplete oxidation of sugars or alcohols to acids, such as the oxidation of glucose to gluconic acid or ethanol to acetic acid (this organism is discussed briefly with the other acetic acid bacteria in Section 19.17). Phylogenetically, the various genera of pseudomonads scatter within the purple Bacteria (see Table 19.19 and Section 18.6). Presumably the

(a)

(b)

James Shapiro

Arthur Kelman

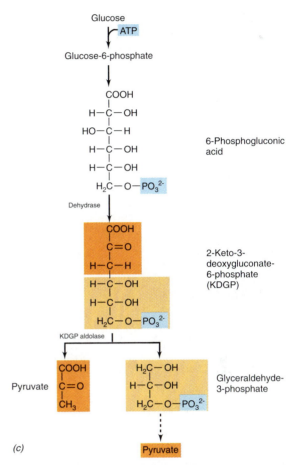

(c)

FIGURE 19.65 Typical pseudomonad colony and cell morphology and a biochemical pathway common to pseudomonads. (a) Photo of colonies of *Pseudomonas cepacia* on an agar plate. (b) Shadow cast preparation of a *Pseudomonas* species. The cell measures about 1 μm in diameter. (c) The Entner–Doudoroff pathway, the major means of glucose catabolism in pseudomonads.

pseudomonads are derived from ancestral phototrophic bacteria that dispensed with the property of photosynthesis in evolving to colonize habitats in which the ability to carry out anoxygenic photosynthesis was not a significant advantage (for example, in soil and on the surfaces of plants and animals).

Genus *Pseudomonas*

The distinguishing characteristics of the *Pseudomonas/Commamonas* group are given in Table 19.18. Also given in this table are the minimal characteristics needed to identify an organism as a pseudomonad. Key identifying characteristics are the absence of gas formation from glucose, and the positive oxidase test, both of which help to distinguish pseudomonads from enteric bacteria (Section 19.20). Also, the genus *Rhodopseudomonas* (see Section 19.1) is distinguished from *Pseudomonas* by the presence of photosynthetic pigments when the former is grown anaerobically in the light. (Since *Rhodopseudomonas* will grow aerobically in the dark without producing photosynthetic pigments, it could be mistaken for a pseudomonad.)

The species of the genus *Pseudomonas* are defined on the basis of various physiological characteristics, as outlined in Tables 19.19 and 19.20. Some species (for

---

Table 19.18   Genus *Pseudomonas/Commamonas*

General characteristics:
Straight or curved rods but not vibrioid; size 0.5–1.0 μm by 1.5–4.0 μm; no spores; Gram-negative; polar flagella: single or multiple; no sheaths, appendages or buds; respiratory metabolism, never fermentative, although may produce small amounts of acid from glucose aerobically; use low-molecular-weight organic compounds, not polymers; some are chemolithotrophic, using H₂ or CO as sole electron donor; some can use nitrate as electron acceptor anaerobically; some can use arginine as energy source anaerobically.

Minimal characteristics for identification:
Gram-negative, straight or slightly curved; no spores; motile (always); polar flagella (flagellar stain); oxidative-fermentative medium with glucose: open, acid produced; closed, acid not produced; gas not produced from glucose (distinguishes them easily from enteric bacteria and *Aeromonas*); oxidase, almost always positive (enterics are oxidase negative); catalase always positive; photosynthetic pigments absent (distinguishes them from nonsulfur purple bacteria); indole negative; methyl red negative; Voges-Proskauer negative

Table 19.19   Characteristics of subgroups and species of the genus *Pseudomonas/Commamonas*

Group	16s rRNA group*	Characteristics	DNA (mole % GC)
Fluorescent subgroup	Gamma	Most produce water-soluble, yellow-green fluorescent pigments; do not form poly-β-hydroxybutyrate; single DNA homology group	
P. aeruginosa		Pyocyanin production, growth at up to 43°C, single polar flagellum, denitrification	67
P. fluorescens		Does not produce pyocyanin or grow at 43°C; tuft of polar flagella	59–61
P. putida		Similar to *P. fluorescens* but does not liquefy gelatin and does grow on benzylamine	60–63
P. syringae		Lacks arginine dihydrolase, oxidase negative, pathogenic to plants	58–60
Acidovorans subgroup	Beta	Nonpigmented, forms poly-β-hydroxybutyrate, tuft of polar flagella, does not use carbohydrates; single DNA homology group	
Commamonas acidovorans		Uses muconic acid as sole carbon source and electron donor	67
Commamonas testosteroni		Uses testosterone as sole carbon source	62
Pseudomallei-cepacia subgroup	Beta	No fluorescent pigments, tuft of polar flagella, forms poly-β-hydroxybutyrate; single DNA homology group	62
P. cepacia		Extreme nutritional versatility; some strains pathogenic to plants	67
P. pseudomallei		Causes melioidosis in animals; nutritionally versatile	69
P. mallei		Causes glanders in animals; nonmotile; nutritionally restricted	69
Diminuta-vesicularis subgroup	Alpha	Single flagellum of very short wavelength, require vitamins (pantothenate, biotin, B_{12})	
P. diminuta		Nonpigmented, does not use sugars	66–67
P. vesicularis		Carotenoid pigment, uses sugars	66
Miscellaneous species	—		
P. solanacearum		Plant pathogen	66–68
P. saccharophila		Grows chemolithotrophically with H_2, digests starch	69
P. maltophilia		Requires methionine, does not use NO_3^- as N source, oxidase negative	67

*All pseudomonads are members of the purple Bacteria (Proteobacteria) (see Section 18.6).

example, *P. aeruginosa*) are quite homogeneous, so that all isolates fit into a very narrow range of distribution of characteristics, whereas other species are much more heterogeneous. The taxonomy of the genus *Pseudomonas* has been greatly clarified by DNA hybridization studies, and the subgroups given in Table 19.19 are supported by 16S rRNA sequencing.

Representatives of the genus *Pseudomonas* have in most cases very simple nutritional requirements and grow chemoorganotrophically at neutral pH and at temperatures in the mesophilic range. One of the striking properties of the pseudomonads is the wide variety of organic compounds used as carbon sources and as electron donors for energy generation. Some species utilize over *100* different compounds, and only a few species utilize fewer than 20. As an example of this versatility, a single strain of *P. aeruginosa* can make use of many different sugars, fatty acids, dicarboxylic acids, tricarboxylic acids, alcohols, polyalco-

hols, glycols, aromatic compounds, amino acids, and amines, plus miscellaneous organic compounds not fitting into any of the above categories. On the other hand, pseudomonads are generally unable to break down polymers into their component monomers. Nutritionally versatile pseudomonads typically contain numerous inducible operons because the catabolism of unusual organic substrates often requires the activity of several different enzymes. The pseudomonads are ecologically important organisms in soil and water and are probably responsible for the aerobic degradation of many soluble compounds derived from the breakdown of plant and animal materials.

A few pseudomonads are pathogenic (Table 19.20). Among the fluorescent pseudomonads, the species *P. aeruginosa* is frequently associated with infections of the urinary and respiratory tracts in humans. *Pseudomonas aeruginosa* infections are also common in patients receiving treatment for severe burns or other traumatic skin

Table 19.20 Pathogenic pseudomonads	
Species	**Relationship to disease**
Animal pathogens	
P. aeruginosa	Opportunistic pathogen, especially in hospitals; in patients with metabolic, hematologic, and malignant diseases; hospital-acquired infections from catheterizations, tracheostomies, lumbar punctures, and intravenous infusions; in patients given prolonged treatment with immunosuppressive agents, corticosteroids, antibiotics and radiation; may contaminate surgical wounds, abscesses, burns, ear infections, lungs of patients treated with antibiotics; primarily a soil organism
P. fluorescens	Rarely pathogenic, as does not grow well at 37°C; may grow in and contaminate blood and blood products under refrigeration
P. maltophilia	An ubiquitous, free-living organism that is a common nosocomial pathogen
P. cepacia	Onion bulb-rot; has also been isolated from humans and from environmental sources of medical importance
P. pseudomallei	Causes melioidosis, a disease endemic in animals and humans in southeast Asia
P. mallei	Causes glanders, a disease of horses that is occasionally transmitted to humans
P. stutzeri	Often isolated from humans and environmental sources; may live saprophytically in the body
Plant pathogens	
P. solanacearum	Causes wilts of many cultivated plants (e.g., potato, tomato, tobacco, peanut)
P. syringae	Attacks foliage, causing chlorosis and necrotic lesions on leaves; rarely found free in soil
P. marginalis	Causes soft rot of various plants; active pectinolytic species
Xanthomonas	Causes necrotic lesions on foliage, stems, fruits; also causes wilts and tissue rots; rarely found free in soil

damage. *Pseudomonas aeruginosa* is not an obligate parasite, however, since it can be readily isolated from soil, and as a denitrifier (see Section 17.14) it plays an important role in the nitrogen cycle in nature. As a pathogen it appears to be primarily an opportunist, initiating infections in individuals whose resistance is low. In addition to urinary infections, it can also cause systemic infections, usually in individuals who have experienced extensive skin damage. The organism is naturally resistant to many of the widely used antibiotics, so that chemotherapy is often difficult. Resistance is frequently due to a *resistance transfer plasmid* (*R plasmid*; see Section 7.9), which is a plasmid carrying genes coding for detoxification of various antibiotics. *P. aeruginosa* is commonly found in the hospital environment and can easily infect patients receiving treatment for other illnesses (see Section 14.7 for a discussion of hospital-acquired infections). Polymyxin, an antibiotic not ordinarily used in human therapy because of its toxicity, is effective against *P. aeruginosa* and can be used with caution.

Many pseudomonads, as well as a variety of other Gram-negative Bacteria, metabolize glucose via the Entner-Doudoroff pathway (Figure 19.65c). Two key enzymes of the Entner-Doudoroff pathway are *6-phosphogluconate dehydrase* and *ketodeoxyglucosephosphate aldolase* (Figure 19.65c). A survey for the presence of these enzymes in a wide variety of bacteria has shown that they are absent from all Gram-positive Bacteria (except a few *Nocardia* isolates) and are generally present in bacteria of the genera *Pseudomonas*, *Rhizobium*, and *Agrobacterium*, as well as in some isolates of several other genera of Gram-negative Bacteria.

Phytopathogens

Certain species of *Pseudomonas* and the genus *Xanthomonas* are well-known plant pathogens (phytopathogens, see Table 19.20). In many cases these organisms are so highly adapted to the plant environment that they can rarely be isolated from other habitats, including soil. Phytopathogens frequently inhabit nonhost plants (where disease symptoms are not apparent) and from here become transmitted to host plants and initiate infection. Disease symptoms vary considerably depending on the particular phytopathogen and host plant, and are generally due to the release by the bacterium of plant toxins, lytic enzymes, plant growth factors, and other substances that destroy or distort plant tissue. In many cases the disease symptoms are highly diagnostic of the type of phytopathogen, and are actually used in the taxonomy of phytopathogenic pseudomonads. Thus, *P. syringae* is frequently isolated from leaves showing chlorotic lesions, whereas *P. marginalis* is a typical "soft-rot" pathogen, infecting stems and shoots, but rarely leaves.

19.16   Free-Living Aerobic Nitrogen-Fixing Bacteria

A variety of organisms which primarily inhabit the soil are capable of fixing N_2 aerobically (Table 19.21). The genus *Azotobacter* comprises large, Gram-negative, obligately aerobic rods capable of fixing N_2 nonsymbiotically (Figure 19.66). The first member of this genus was

Table 19.21   Genera of free-living aerobic nitrogen-fixing bacteria*

Genus	Characteristics	DNA (mole % GC)
Azotobacter	Large rod; produces cysts; primarily found in neutral to alkaline soils	63–67
Azomonas	Large rod; no cysts; primarily aquatic	52–59
Azospirillum	Microaerophilic rod; associates with plants	69–71
Beijerinckia	Pear-shaped rods with large lipid bodies at each end; produces extensive slime; inhabits acidic soils	54–59
Derxia	Rods; form coarse, wrinkled colonies	69–73

*All species examined are members of the purple Bacteria, primarily the alpha and gamma subdivisions.

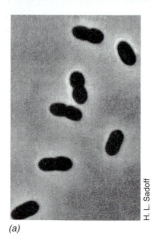
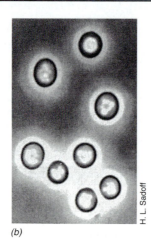

(a)   *(b)*

FIGURE 19.66 *Azotobacter vinelandii*: (a) vegetative cells and (b) cysts by phase contrast microscopy. A cell measures about 2 μm in diameter and a cyst about 3 μm.

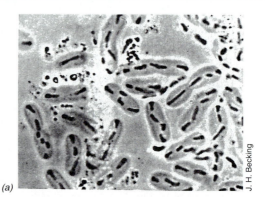

(a)

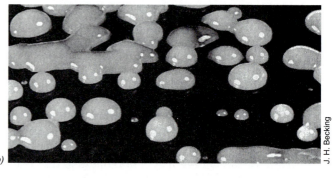

(b)

FIGURE 19.67 Examples of slime production by free-living N_2-fixing bacteria. (a) Cells of *Derxia gummosa* encased in slime. Cells are about 1–1.2 μm wide. (b) Colonies of *Beijerinckia* species growing on a carbohydrate-containing medium. Note the raised glistening appearance of the colonies due to abundant capsular slime.

discovered by the Dutch microbiologist M. W. Beijerinck early in the twentieth century, using an enrichment culture technique with a medium devoid of a combined nitrogen source (see Rise of General Microbiology box, Chapter 17). Although capable of growth on N_2, *Azotobacter* grows more rapidly on NH_3; indeed, adding NH_3 actually *represses* nitrogen fixation (see Section 16.26). Much work has been done in seeking to evaluate the role of *Azotobacter* in nitrogen fixation in nature, especially in comparison to the anaerobic organism *Clostridium pasteurianum* and the symbiotic organisms of the genus *Rhizobium*. *Azotobacter* is also of interest because it has the highest respiratory rate (measured as the rate of O_2 uptake) of any living organism. In addition to its ecological and physiological importance, *Azotobacter* is of interest because of its ability to form an unusual resting structure called a *cyst* (Figure 19.66b). Phylogenetically, *Azotobacter* belongs to the gamma group of purple Bacteria (see Section 18.6).

Azotobacter cells are rather large for prokaryotes, many isolates being almost the size of yeasts, with diameters of 2–4 μm or more. Pleomorphism is common, and a variety of cell shapes and sizes have been described. Some strains are motile by peritrichous flagella. On carbohydrate-containing media, extensive capsules or slime layers are produced by free-living

N_2-fixing bacteria (Figure 19.67). *Azotobacter* is able to grow on a wide variety of carbohydrates, alcohols, and organic acids. The metabolism of carbon compounds is strictly oxidative, and acids or other fermentation products are rarely produced. All members fix nitrogen, but growth also occurs on simple forms of combined nitrogen: ammonia, urea, and nitrate.

Despite the fact that *Azotobacter* is an obligate aerobe, its nitrogenase is as O_2 sensitive as all other nitrogenases (Section 16.26). It is thought that the high respiratory rate of *Azotobacter* (mentioned above) has something to do with protection of nitrogenase from O_2. The intracellular O_2 concentration is kept low enough by metabolism so that inactivation of nitrogenase does not occur.

The remaining genera of free-living N_2 fixers in-

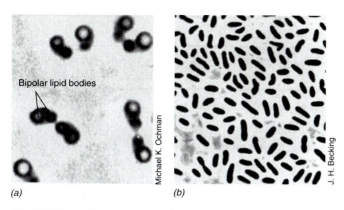

(a)   (b)

Michael K. Ochman

J. H. Becking

FIGURE 19.68 Phase contrast photomicrographs of two genera of acid-tolerant, free-living N$_2$-fixing bacteria. (a) *Beijerinckia indica*. The cells are roughly pear-shaped, about 0.8 μm in diameter, and contain a large globule of poly-β-hydroxybutyrate at each end. (b) *Derxia gummosa*.

clude *Azomonas*, a genus of large rod-shaped bacteria that resemble *Azotobacter* except that they do not produce cysts and are primarily aquatic, and *Beijerinckia* and *Derxia* (Figure 19.68), two genera that grow well in acidic soils.

Like bacterial endospores, *Azotobacter* cysts (Figure 19.66**b**) show negligible endogenous respiration and are resistant to desiccation, mechanical disintegration, and ultraviolet and ionizing radiation. In contrast to endospores, however, they are *not* especially heat resistant, and they are not completely dormant since they rapidly oxidize exogenous energy sources. The carbon source of the medium greatly influences the extent of cyst formation, butanol being especially favorable; compounds related to butanol, such as β-hydroxybutyrate, also promote cyst formation.

The species *Azotobacter chroococcum* was the first N$_2$-fixing bacterium shown capable of growth on N$_2$ in the complete *absence* of molybdenum. When cells of *A. chroococcum* are placed in a mineral medium lacking ammonia and molybdenum but containing the metal vanadium, a second form of nitrogenase containing *vanadium* in place of *molybdenum* is produced and functions to support growth of the organism on N$_2$. Like the molybdenum enzyme (see Section 16.26), vanadium nitrogenase consists of *two* proteins, one of which contains iron, the second iron and vanadium, and reduces N$_2$ to NH$_3$, H$^+$ to H$_2$, and C$_2$H$_2$ to C$_2$H$_4$. However, the rates of substrate reduction are considerably *lower* by the vanadium nitrogenase than by the molybdenum enzyme, indicating that the vanadium enzyme is a less effective catalyst. Interestingly, the vanadium nitrogenase also reduces C$_2$H$_2$ to ethane, C$_2$H$_6$. Although only about 3 percent of the acetylene reduced goes to ethane, the ability of the vanadium nitrogenase to catalyze the complete reduction of C$_2$H$_2$ to C$_2$H$_6$ can serve as a test for detecting N$_2$-fixing bacteria capable of synthesizing this alternative nitrogenase. Molybdenum nitrogenases produce ethylene, C$_2$H$_4$, but not ethane, C$_2$H$_6$. *Azotobacter* can also produce a nitrogenase containing only *iron*. This nitrogenase is produced if both molybdenum and vanadium

are absent, but it is a very poor N$_2$-reducing catalyst. The iron-only nitrogenase also produces some C$_2$H$_6$.

Vanadium nitrogenase is not the major nitrogenase of *A. chroococcum*, because addition of molybdenum to the culture medium represses synthesis of both the vanadium and iron-only nitrogenases, and instead the conventional nitrogenase is formed. Vanadium nitrogenases are not universal among N$_2$-fixing bacteria. However, it is known that vanadium stimulates N$_2$ fixation in a number of organisms including various species of *Azotobacter*, some cyanobacteria and phototrophic bacteria, and *Clostridium pasteurianum*. Hence, vanadium- and iron-based nitrogenases may be a widespread "backup" system for fixing N$_2$ in molybdenum-deficient habitats.

19.17   Acetic Acid Bacteria

As originally defined, the *acetic acid bacteria* comprised a group of Gram-negative, aerobic, motile rods that carried out an *incomplete* oxidation of alcohols, leading to the accumulation of organic acids as end products. With *ethanol* as a substrate, *acetic acid* is produced; hence the derivation of the common name for these bacteria. Another property is the relatively high tolerance to acidic conditions, most strains being able to grow at pH values lower than five. This acid tolerance should of course be essential for an organism producing large amounts of acid. The acetic acid bacteria are a heterogeneous assemblage, comprising both peritrichously and polarly flagellated organisms. The *polarly* flagellated organisms are related to the pseudomonads, differing mainly in their acid tolerance and their inability to carry out a complete oxidation of alcohols. These organisms are now classified in the genus *Gluconobacter* (Table 19.22). All acetic acid bacteria group phylogenetically with the purple Bacteria.

The genus *Acetobacter* comprises the *peritrichously* flagellated organisms. In addition to flagellation, *Acetobacter* differs from *Gluconobacter* in being able to further oxidize the acetic acid it forms to CO$_2$. This difference in ability to oxidize acetic acid is related to the presence of a complete citric acid cycle. *Gluconobacter*, which *lacks* a complete citric acid cycle, is unable to oxidize acetic acid, whereas *Acetobacter*, which has all enzymes of the cycle, can oxidize it. Because of these differences in oxidative potential, gluconobacters are sometimes called *underoxidizers*; the acetobacters, *overoxidizers*.

The acetic acid bacteria are frequently found in association with alcoholic juices, and probably evolved from pseudomonads inhabiting sugar-rich flowers and fruits where a yeast-mediated alcoholic fermentation is common. Acetic acid bacteria can often be isolated from alcoholic fruit juices such as cider or wine. Colonies of acetic acid bacteria can be recognized on CaCO$_3$-agar plates containing ethanol, the acetic acid produced causing a dissolution and clearing of the insoluble CaCO$_3$ (Figure 19.69). Cultures of acetic acid bacteria are used in commercial production of vinegar (see Section 10.12).

Table 19.22   Differentiation of *Acetobacter*, *Gluconobacter*, and *Pseudomonas*

Character	Acetobacter	Gluconobacter	Pseudomonas
Flagellation	Peritrichous	Polar	Polar
Growth at pH 4.5	+	+	−
Oxidation of ethanol to acetic acid at pH 4.5	+	+	−
Complete tricarboxylic acid cycle	Present	Absent	Present
DNA (mole % GC)	53–65	56–64	58–70
Number of species	7	3	27
16S rRNA group*	Alpha purple Bacteria	Alpha purple Bacteria	Alpha, beta, or gamma purple Bacteria

*See Section 18.6.

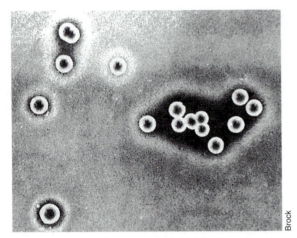

FIGURE 19.69 Photograph of colonies of *Acetobacter aceti* on calcium carbonate agar containing ethanol as energy source. Note the clearing around the colonies due to the dissolution of calcium carbonate by the acetic acid produced by the bacteria.

In addition to ethanol, these organisms carry out an incomplete oxidation of such organic compounds as higher alcohols and sugars. For instance, glucose is oxidized only to gluconic acid, galactose to galactonic acid, arabinose to arabonic acid, and so on. This property of underoxidation is exploited in the manufacture of ascorbic acid (vitamin C). Ascorbic acid can be formed from sorbose, but sorbose is difficult to synthesize chemically. It is, however, conveniently obtainable microbiologically from acetic acid bacteria, which oxidize sorbitol (a readily available sugar alcohol) only to sorbose, a process called *bioconversion*, see Section 10.10. The use of acetic acid bacteria makes the manufacture of ascorbic acid economically feasible.

Another interesting property of some acetic acid bacteria is their ability to synthesize *cellulose*. The cellulose formed does not differ significantly from that of plant cellulose, but instead of being a part of the cell wall, the bacterial cellulose is formed as a matrix outside the wall and the bacteria become embedded in the tangled mass of cellulose microfibrils. When these species of acetic acid bacteria grow in an unshaken vessel, they form a surface pellicle of cellulose, in which the bacteria develop. Since these bacteria are obligate aerobes, the ability to form such a pellicle may be a means by which the organisms are assured of remaining at the surface of the liquid, where oxygen is readily available.

19.18   *Zymomonas* and *Chromobacterium*

The genera *Zymomonas* and *Chromobacterium* are grouped together as facultatively aerobic Gram-negative rods of uncertain taxonomic affiliation. Phylogenetically, both organisms are members of the beta purple Bacteria group (see Section 18.6). The best known *Chromobacterium* species is *C. violaceum*, a bright purple pigmented organism (Figure 19.70) found in soil and water and occasionally in pyogenic infections of man and other animals. *C. violaceum* and a few other chromobacteria produce the purple pigment *violacein* (Figure 19.71), a water insoluble pigment that has antibiotic properties and is only produced in media containing the amino acid tryptophan. *Chromobacterium* is a facultative aerobe, growing fermentatively on sugars and aerobically on a variety of carbon sources.

The genus *Zymomonas* consists of large Gram-negative rods (2–6 μm long by 1–1.4 μm wide), which carry out a vigorous fermentation of sugars to ethanol. Although not all strains are motile, if motility occurs it is by lophotrichous flagella. *Zymomonas* is a common organism involved in alcoholic fermentation of various plant saps, and in many tropical areas of South and Central America, Africa, and Asia, it occupies a position similar to that of *Saccharomyces cerevisiae* (yeast) in North America and Europe. *Zymomonas* has been shown to be involved in the alcoholic fermentation of agave in Mexico, and palm sap in many tropical areas. It also carries out an alcoholic fermentation of sugar cane juice and honey. Although *Zymomonas* is rarely the sole organism involved in these alcoholic fermentations, it is often the dominant organism, and is probably responsible for the production of most of the ethanol (the desired product) in these beverages. *Zymomonas* is also responsible for spoilage of fruit juices such as apple cider and perry. It also may be a constituent of the bacterial flora of spoiled beer, and

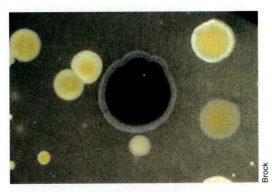

FIGURE 19.70 A large colony of *Chromobacterium violaceum* growing among other colonies on an agar plate.

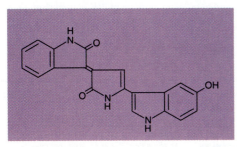

FIGURE 19.71 Structure of the pigment violacein, produced by various species of the genus *Chromobacterium*.

may be responsible for the production in beer of an unpleasant odor of rotten apples (due to traces of acetaldehyde and H_2S).

Zymomonas is distinguished from *Pseudomonas* by its fermentative metabolism, its microaerophilic to anaerobic nature, oxidase negativity, and by other molecular taxonomic characteristics. It resembles most closely the acetic-acid bacteria, specifically *Gluconobacter* because of its polar flagellation, and it is often found in nature associated with the acetic-acid bacteria. This is of interest because *Zymomonas* ferments glucose to ethanol whereas the acetic acid bacteria oxidize ethanol to acetic acid. Thus, the acetic-acid bacteria may depend upon the activity of *Zymomonas* for the production of their growth substrate, ethanol. Like the acetic-acid bacteria, *Zymomonas* is quite tolerant of low pH. Unlike yeast, which ferments glucose to ethanol via the Embden-Meyerhof (glycolytic) pathway, *Zymomonas* employs the Entner-Doudoroff pathway. This pathway is active in many pseudomonads as a means of catabolizing glucose (see Figure 19.65).

Zymomonas is of interest to the ethanol industry because it shows higher rates of glucose uptake and ethanol production and gives a higher yield of ethanol than many yeasts. *Zymomonas* is also rather tolerant of high ethanol concentrations (up to 10 percent), but is not quite as tolerant as some of the best yeast strains which can grow up to 12–15 percent ethanol. However, the fact that *Zymomonas* is a *Gram-negative* bacterium, and can thus be readily manipulated genetically, makes it an attractive candidate for use by ethanol production industries.

19.19 *Vibrio* and Related Genera

The *Vibrio* group contains Gram-negative, facultatively aerobic rods and curved rods which possess a fermentative metabolism. Most of the members of the *Vibrio* group are polarly flagellated, although some members are peritrichously flagellated. One key difference between the *Vibrio* group and enteric bacteria is that members of the former are oxidase *positive* (see Table 13.3), whereas members of the latter are oxidase *negative*. Although *Pseudomonas* is also polarly flagellated and oxidase positive, it is *not* fermentative, and hence can be separated from the vibrios by simple sugar fermentation tests.

Most vibrios and related bacteria are aquatic, found either in freshwater or marine habitats, although one important organism, *Vibrio cholerae*, is pathogenic for humans. The group contains four genera, *Vibrio*, *Aeromonas*, *Photobacterium*, and *Plesiomonas*, and the major characteristics of each are given in Table 19.23.

Vibrio cholerae is the specific cause of the disease *cholera* in humans (see Section 15.14); the organism does not normally infect other hosts. Cholera is one of the most common infectious human diseases in undeveloped countries and one that has had a long history. The organism is transmitted almost exclusively via water, and studies on its distribution in the nineteenth century played a major role in demonstrating the importance of water purification in urban areas (see Snow on Cholera box, Chapter 14). We discussed the pathogenesis of *V. cholerae* in Section 11.9. *Vibrio cholerae* is capable of good growth at a pH of over 9, and this characteristic is frequently employed in the selective isolation and identification of this organism.

Vibrio parahemolyticus is a marine organism. It is a major cause of gastroenteritis in Japan (where raw fish is widely consumed) and has also been implicated in outbreaks of gastroenteritis in other parts of the world, including the United States. The organism can be frequently isolated from seawater or from shellfish and crustaceans, and its primary habitat is probably marine animals, with human infection being a secondary development (see Section 15.14).

Luminescent bacteria

A number of Gram-negative, polarly flagellated rods possess the interesting property of emitting light (luminescence). Most of these bacteria have been classified in the genus *Photobacterium* (Table 19.23, see Figure 19.72*b*), but a few *Vibrio* isolates are also luminescent (Figure 19.72*a*). Most **luminescent bacteria** are marine forms, usually found associated with fish. Some fish possess a special organ in which luminescent bacteria grow (Figure 19.72*c–f*). Other luminescent marine bacteria live saprophytically on dead fish. A good way of isolating luminescent bacteria is to incubate a dead marine fish for one or two days at 10–20°C; the luminescent bacterial colonies that usually appear on the surface of the fish can be easily seen

Table 19.23   Distinguishing characteristics of genera of the family Vibrionaceae and related genera*

Characteristic	Vibrio	Aeromonas	Photobacterium	Plesiomonas
Morphology	Straight or curved rods, single sheathed polar flagellum	Straight rods, single polar flagellum	Straight rods, 1–3 polar flagella	Straight rods with round ends, 2–5 polar flagella
Sodium required for growth	+	–	+	–
Gas production	–	+	+	–
Sensitivity to vibriostat (2,4-diamino-6,7-diisopropyl pteridine)	+	–	+	+
Fermentation of mannitol	+	+	–	–
Luminescence	+ or –	–	+	–
DNA (mole % GC)	38–51	57–63	40–44	51
Pigment	None	None or brown	None	None
Cause disease in humans	+	+	–	+

All are Gram-negative rods, straight or curved, facultative aerobes with a fermentative metabolism (O/F medium), but oxidase positive. They are predominantly aquatic organisms and are phylogenetically members of the gamma purple Bacteria.

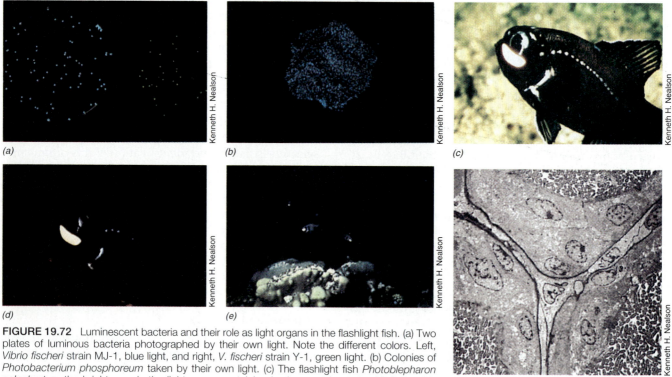

(a) (b) (c)

(d) (e) (f)

Kenneth H. Nealson

FIGURE 19.72 Luminescent bacteria and their role as light organs in the flashlight fish. (a) Two plates of luminous bacteria photographed by their own light. Note the different colors. Left, *Vibrio fischeri* strain MJ-1, blue light, and right, *V. fischeri* strain Y-1, green light. (b) Colonies of *Photobacterium phosphoreum* taken by their own light. (c) The flashlight fish *Photoblepharon palpabratus*; the bright area is the light organ containing luminescent bacteria. (d) Same fish photographed by its own light. (e) Underwater photograph taken at night of *P. palpabratus* in coral reefs in the Gulf of Eilat. (f) Electron micrograph of a thin section through the light-emitting organ of *P. palpabratus*, showing the dense array of luminescent bacteria.

and isolated (Figure 19.72a and b). (To see luminescence readily, one should observe the material in a completely dark room after the eyes have become adapted to the dark.)

Although *Photobacterium* isolates are facultative aerobes, they are luminescent only when O_2 is present. Several components are needed for bacterial luminescence: the enzyme **luciferase**, and a long-chain ali-

phatic aldehyde (for example, *dodecanal*); flavin mononucleotide (FMN) and O_2 are also involved. The primary electron donor is NADH, and the electrons pass through FMN to the luciferase. The reaction can be expressed as

$$FMNH_2 + O_2 + RCHO \xrightarrow{luciferase} FMN +$$

$$RCOOH + H_2O + \text{light}$$

The light-generating system constitutes a bypass route for shunting electrons from $FMNH_2$ to O_2, without involving other electron carriers such as quinones and cytochromes.

The enzyme luciferase shows a unique kind of regulatory synthesis called **autoinduction**. The luminous bacteria produce a specific substance, the *autoinducer*, which accumulates in the culture medium during growth, and when the amount of this substance has reached a critical level, induction of the enzyme occurs. The autoinducer in *V. fischeri* has been identified as *N*-β-ketocaproylhomoserine lactone. Thus, cultures of luminous bacteria at low cell density are not luminous, but become luminous when growth reaches a sufficiently high density so that the autoinducer can accumulate and function. Because of the autoinduction phenomenon, it is obvious that free-living luminescent bacteria in seawater will not be luminous because the autoinducer could not accumulate, and luminescence only develops when conditions are favorable for the development of high population densities. Although it is not clear why luminescence is density-dependent in free-living bacteria, in symbiotic strains of luminescent bacteria (see Figure 19.72), the rationale for density-dependent luminescence is clear: luminescence only develops when sufficiently high population densities are reached in the light organ of the fish to allow a visible flash of light.

Much new information about bioluminescence has emerged from studies of the genetics of this process. Several *lux* operons have been identified in luminescent *Vibrio* species and the key structural genes have been cloned and sequenced. The *luxA* and *luxB* genes code for the a and b subunits, respectively, of bacterial luciferase. The *luxC*, *luxD*, and *luxE* genes code for polypeptides that function in the bioluminescence reaction and in the generation and activation of fatty acids for the luminescence system. Light-emitting strains of *Escherichia coli* have been constructed by inserting *lux* genes cloned from a *Vibrio* species into cells of *E. coli* and then placing their expression under the control of specific and easily manipulable *E. coli* promoters. This has served to accelerate the pace of research on the regulatory aspects of bioluminescence because many *E. coli* genetic tools can be employed to probe aspects of *lux* gene control. In addition, cloned *lux* genes have stimulated biotechnological exploitation of bioluminescence in clinical diagnostics and other biomedical fields, although the cloned *lux* genes need not be from Bacteria (see Figure 8.19).

19.20   Facultatively Aerobic Gram-Negative Rods

We focus here on the **enteric bacteria** which comprise a relatively homogeneous phylogenetic group within the gamma purple Bacteria (see Section 18.6), and are characterized phenotypically as follows: Gram-negative, nonsporulating rods, nonmotile or motile by *peritrichous* flagella (Figure 19.73), facultative aerobes, oxidase *negative* with relatively simple nutritional re-

(a)

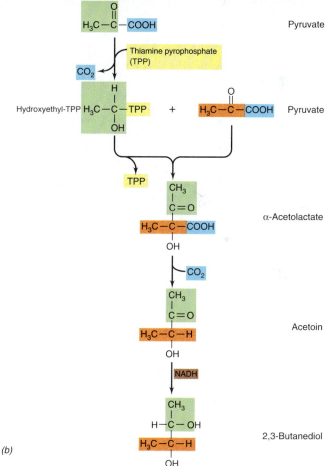

(b)

FIGURE 19.73 (a) Electron micrograph of a shadow cast preparation of cells of the butanediol-producing enteric bacterium, *Erwinia carotovora*. Cells are about 0.8 μm wide. Note the peritrichously arranged flagella. (b) Biochemical pathway for formation of butanediol from two molecules of pyruvate by butanediol fermenters.

quirements, fermenting sugars to a variety of end products. The phenotypic characteristics used to separate the enteric bacteria from other bacteria of similar morphology and physiology are given in Table 19.24.

Among the enteric bacteria are many strains pathogenic to humans, animals, or plants as well as other strains of industrial importance. Probably more is known about *Escherichia coli* than about any other bacterial species (see box, Chapter 7).

Because of the medical importance of the enteric bacteria, an extremely large number of isolates have been studied and characterized, and a fair number of distinct genera have been defined. Despite the fact that there is marked genetic relatedness between many of the enteric bacteria, as shown by DNA homologies and genetic recombination, separate genera are maintained, largely for practical reasons. Since these organisms are frequently cultured from diseased states, some means of identifying them is necessary.

One of the key taxonomic characteristics separating the various genera of enteric bacteria is the type and proportion of *fermentation products* produced by anaerobic fermentation of glucose. Two broad patterns are recognized, the **mixed-acid fermentation** and the **2,3-butanediol fermentation** (Figure 19.74). In mixed-acid fermentation, *three* acids are formed in significant amounts—acetic, lactic, and succinic—and ethanol, CO_2, and H_2 are also formed, but *not* butanediol. In butanediol fermentation, *smaller* amounts of acids are formed, and butanediol, ethanol, CO_2, and H_2 are the main products. As a result of a mixed-acid fermentation *equal* amounts of CO_2 and H_2 are produced, whereas with a butanediol fermentation *more* CO_2 than H_2 is produced. This is because mixed-acid fermenters produce CO_2 only from formic acid by means of the enzyme system *formic hydrogenlyase*:

$$HCOOH \rightarrow H_2 + CO_2$$

and this reaction results in equal amounts of CO_2 and H_2. The butanediol fermenters also produce CO_2 and H_2 from formic acid, but they produce additional CO_2 during the reactions that lead to butanediol (Figure 19.73b).

A variety of diagnostic tests and differential

Table 19.24   Defining characteristics of the enteric bacteria

General characteristics:

Gram-negative straight rods; motile by peritrichous flagella, or nonmotile; nonsporulating; facultative aerobes, producing acid from glucose; sodium neither required nor stimulatory; catalase positive; oxidase negative; usually reduce nitrate to nitrite (not to N_2); 16S rRNA of gamma section of purple Bacteria (Proteobacteria)

Key tests to distinguish enteric bacteria from other bacteria of similar morphology:

Oxidase test, enterics always negative—separates enterics from oxidase-positive bacteria of genera *Pseudomonas, Aeromonas, Vibrio, Alcaligenes, Achromobacter, Flavobacterium, Cardiobacterium*, which may have similar morphology; nitrate reduced only to nitrite, (assay for nitrite after growth)—distinguishes enteric bacteria from bacteria that reduce nitrate to N_2 (gas formation detected), such as *Pseudomonas* and many other oxidase-positive bacteria; ability to ferment glucose—distinguishes enterics from obligately aerobic bacteria

---

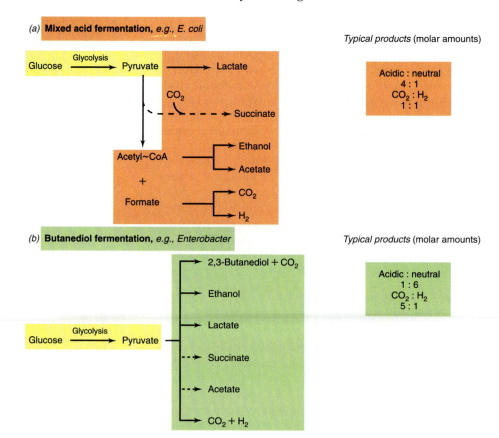

FIGURE 19.74
Distinction between mixed acid and butanediol fermentation in enteric bacteria. The bold arrows indicate reactions leading to major products. Dashed arrows are minor products.

Table 19.25   Key differential media and tests for classifying enteric bacteria*

Media and Tests	Composition and Procedure
Triple sugar iron (TSI) agar	Butt reactions: acid (yellow), gas
	Slant reactions: acid (yellow), alkaline (red)
	H$_2$S reaction: blackening of butt (in the absence of acid butt reaction)
Urea medium	Urea agar with phenol red indicator; look for alkaline reaction due to urease action on urea, liberating ammonia
Citrate medium	Growth on citrate as sole energy source
Indole test	Peptone broth with high tryptophan content; assay for indole
Voges-Proskauer (VP) test	Glucose-containing broth; assay for acetoin
Methyl red (MR) test	Buffered glucose-peptone broth; measure pH drop with methyl red indicator
KCN medium	Broth with 75 μg/ml potassium cyanide; look for growth
Phenylalanine agar	Contains 0.1 percent phenylalanine; after growth, look for phenylpyruvic acid production (indicative of phenylalanine deaminase) by adding ferric chloride solution and looking for green color formation
β-Galactosidase	Medium containing the lactose analog, ortho-nitrophenylgalactoside (ONPG). Positive strains cleave this to yield yellow product
Utilization of carbon sources	Mannitol, tartrate, mucate, acetate, dulcitol, sorbitol, adonitol, inositol

Several of these tests are depicted in color in Figure 13.5.

media are used to separate the various genera in the two broad groups of enteric bacteria, and these are listed in Table 19.25. On the basis of these and other tests, the genera can be defined, as outlined in Tables 19.26 and 19.27.

Positive identification of enteric bacteria often presents considerable difficulty because of the large number of strains that have been characterized. Almost any small set of diagnostic characteristics will fail to provide clear-cut distinctions between genera if a large number of strains are tested, because exceptional strains will be isolated. In advanced work in clinical laboratories, identification is now based on computer analysis of a large number of diagnostic tests, carried out using miniaturized rapid diagnostic media kits and immunological and nucleic acid probes (see Figure 13.1), with consideration being given for variable reactions of exceptional strains. Thus the separation of genera given in Tables 19.26 and 19.27 must be considered approximate, for it will always be possible to isolate a strain that does not possess one or another characteristic normally considered positive for the genus as a whole. With these limitations in mind, an even more simplified separation of the key genera is found in Figure 19.75. This key will permit a quick decision on the likely *genus* in which to place a new isolate.

Escherichia

Members of the genus *Escherichia* are almost universal inhabitants of the intestinal tract of humans and warm-blooded animals, although they are by no means the dominant organisms in these habitats. *Escherichia* may play a nutritional role in the intestinal tract by synthesizing vitamins, particularly vitamin K. As a facultative aerobe, this organism probably also helps consume oxygen, thus rendering the large intestine anaerobic. Wild-type *Escherichia* strains rarely

show any growth-factor requirements and are able to grow on a wide variety of carbon and energy sources such as sugars, amino acids, organic acids, and so on. Some strains of *Escherichia* are pathogenic. The latter strains of *Escherichia* have been implicated in diarrhea in infants, occasionally occurring in epidemic proportions in children's nurseries or obstetric wards, and *Escherichia* may also cause urinary tract infections in older persons or in those whose resistance has been lowered by surgical treatment or by exposure to ionizing radiation. Enteropathogenic strains of *E. coli* are becoming more frequently implicated in dysentery-like infections and generalized fevers (see Sections 11.9 and 15.13). As noted, these strains form *K antigen*, permitting attachment and colonization of the *small* intestine, and *enterotoxin*, responsible for the symptoms of diarrhea.

Shigella

The shigellas are genetically very closely related to *Escherichia*; in fact they are so similar that they are able to undergo genetic recombination with each other and are susceptible to some of the same bacteriophages. Tests for DNA homology show strains of *Shigella* having 70 to nearly 100 percent homology with *E. coli*. In contrast to *Escherichia*, however, *Shigella* is commonly pathogenic to humans, causing a rather severe gastroenteritis usually called *bacillary dysentery*. *S. dysenteriae* is transmitted by food and waterborne routes and is capable of invading intestinal epithelial cells. Once established, it produces both an endotoxin and a neurotoxin that exhibits enterotoxic effects.

Salmonella

Salmonella and *Escherichia* are quite closely related; the two genera have about 45 to 50 percent of their DNA sequences in common. However, in contrast to *Es-*

Table 19.26   Key diagnostic reactions[a] used to separate the various genera of enteric bacteria

Genus	H₂S(TSI)	Urease	VP	Indole	Motility	Gas from glucose	β-Galac-tosidase
Escherichia	−	−	−	+	+ or −	+	+
Enterobacter	−	−	+	−	+	+	+
Shigella	−	−	−	+ or −	−	−	+ or −
Edwardsiella	+	−	−	+	+	+	−
Salmonella	+	−	−	−	+	+	+ or −
Klebsiella	−	+	+ or −	−	−	+	+
Arizona	+	−	−	−	+	+	+
Citrobacter	+ or −	−	−	−	+	+	+
Proteus	+ or −	+	−	+ or −	+	+ or −	−
Providencia	−	−	−	+	+	−	−
Yersinia	−	+	−	−	+[b]	−	+
Hafnia	−	−	+	−	+	+	+ or −

Genus	KCN	Citrate	Mucate	Phenyl-Methyl Red	Tartrate	Alanine deaminase	DNA (mole % GC)
Escherichia	−	−	+	+	+	−	48–52
Enterobacter	+	+	+	−	−	−	52–60
Shigella	−	−	−	+	−	−	50
Edwardsiella	−	−	−	+ or −	−	−	53–59
Salmonella	−	+ or −	+ or −	+	+ or −	−	50–53
Klebsiella	+	+	+	−	+ or −	−	53–58
Arizona	−	+	−	+	−	−	50
Citrobacter	+ or −	+	+	+	+	−	50–52
Proteus	+	+ or −	−	+	+	+	38–41
Providencia	+	+	−	+	+	+	39–42
Yersinia	−	−	−	+	−	−	46–50
Hafnia	+	+	−	+	−	−	48–49

[a]See Tables 19.24 and 19.25 for the procedures for these diagnostic reactions. Tartrate: utilization as carbon source; mucate: fermentation.
[b]Motile when grown at room temperature; nonmotile at 37°C.

Table 19.27   Key diagnostic reactions used to separate the various genera of 2,3-butanediol producers

Genus	Ornithine decarboxy-lase	Gelatin hydrol-ysis	Temperature optimum (°C)	Pigmenta-tion	Motility	Lactose	DNase	Sorbitol	DNA (mole % GC)
Klebsiella	−	−	37–40	None	−	+	−	+	53–58
Entero-bacter	+	slow	37–40	Yellow (or none)	+	+	−	+	52–60
Serratia	+	+	37–40	Red (or none)	+	−	+	−	52–60
Erwinia	−	+ or −	27–30	Yellow (or none)	+	+ or −	−	+	50–58
Hafnia	+	−	35	None	+	−	−	−	48–49

cherichia, members of the genus *Salmonella* are usually pathogenic, either to man or to other warm-blooded animals. In humans the most common diseases caused by salmonellas are *typhoid fever* and *gastroenteritis* (see Sections 15.13 and 15.14). The salmonellas are characterized immunologically on the basis of three cell-surface antigens, the O, or cell-wall (somatic) antigen; the H, or flagellar, antigen; and the Vi (outer polysaccharide layer) antigen, found primarily in strains of *Sal-monella* causing typhoid fever. The O antigens are complex lipopolysaccharides that are part of the endotoxin structure of these organisms (see Section 11.10). We discussed the chemical structure of these lipopolysaccharides in Section 3.5 (see Figure 3.30). The genus *Salmonella* contains over 1000 distinct types having different antigenic specificities in their O antigens. Additional antigenic subdivisions are based on the antigenic specificities of the flagellar (H) antigens.

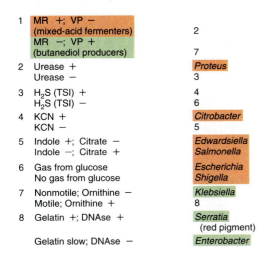

1	MR +; VP — (mixed-acid fermenters)	2
	MR —; VP + (butanediol producers)	7
2	Urease +	*Proteus*
	Urease —	3
3	H$_2$S (TSI) +	4
	H$_2$S (TSI) —	6
4	KCN +	*Citrobacter*
	KCN —	5
5	Indole +; Citrate —	*Edwardsiella* *Salmonella*
	Indole —; Citrate +	
6	Gas from glucose	*Escherichia* *Shigella*
	No gas from glucose	
7	Nonmotile; Ornithine —	*Klebsiella*
	Motile; Ornithine +	8
8	Gelatin +; DNAse +	*Serratia* (red pigment)
	Gelatin slow; DNAse —	*Enterobacter*

FIGURE 19.75 A simplified key to the main genera of enteric bacteria. Only the most common genera are given. See text for precautions in use of this key. Diagnostic tests for use with this figure are given in Table 19.25. Other characteristics of the genera are given in Tables 19.26 and 19.27. Color coding is as in Figure 19.74.

There is little or no correlation between the antigenic type of a *Salmonella* and the disease symptoms elicited, but antigenic typing permits tracing a single strain involved in an epidemic.

Proteus

The genus *Proteus* is characterized by rapid motility and by production of the enzyme *urease*. By DNA homology it shows only a distant relationship to *E. coli*. *Proteus* is a frequent cause of urinary-tract infections in people and rarely may cause enteritis. Because of its active urea-splitting ability, *Proteus* has been implicated in several kidney infections (see Section 11.6). The species of *Proteus* probably do not form a homogeneous group, as is indicated by the fact that DNA base compositions vary over a fairly wide range (39–50 percent GC). Because of the rapid motility of *Proteus* cells, colonies growing on agar plates often exhibit a characteristic **swarming** phenomenon. Cells at the edge of the growing colony are more rapidly motile than are those in the center of the colony; the former move a short distance away from the colony in a mass, then undergo a reduction in motility, settle down, and divide, forming a new crop of motile cells that again swarm. As a result, the mature colony appears as a series of concentric rings, with higher concentrations of cells alternating with lower concentrations.

Although all enteric bacteria can use nitrate as alternate electron acceptor anaerobically, *Proteus* has the additional ability to use several *sulfur* compounds as electron acceptors for anaerobic growth: thiosulfate, tetrathionate, and dimethylsulfoxide.

Yersinia

The genus *Yersinia* has been created to accommodate former members of the genus *Pasteurella* that are obviously members of the enteric group. The genus *Yersinia* consists of three species, *Y. pestis*, the causal agent of the ancient and dreaded disease bubonic

plague (see Section 15.12); *Y. pseudotuberculosis*, the causal agent of a tuberculosis-like disease of the lymph nodes in animals (and rarely in humans); and *Y. enterocolitica*, the causal agent of an intestinal infection (and also occasionally systemic infections) in humans and animals. Although the latter two species are rarely involved in fatal infections, *Yersinia pestis* was responsible for the so-called Black Death that ravaged Europe during the fourteenth century killing over *one-fourth* of the European population. Other pandemics of bubonic plague occurred in subsequent centuries causing great suffering and countless thousands of deaths. The differentiation of *Yersinia* from other mixed-acid fermenters is given in Table 19.26. In recent years, the prevalence of *Y. enterocolitica* as a waterborne and foodborne pathogen has been realized (see Section 15.13), and now that methods are available for its rapid isolation and recognition, it is being detected with increasing frequency.

Butanediol fermenters

The butanediol fermenters are genetically more closely related to each other than to the mixed-acid fermenters, a finding that is in agreement with the observed physiological differences. Their DNA base composition is higher, 53–58 percent GC, and genetic recombination does not occur between the two groups. However, it is possible to transfer plasmids from a mixed-acid fermenter to a butanediol fermenter; the plasmid replicates in the latter but does not become integrated into the chromosome. A current classification of this group is outlined in Table 19.27.

One species of *Klebsiella*, *K. pneumoniae*, occasionally causes pneumonia in humans, but klebsiellas are most commonly found in soil and water. Most *Klebsiella* strains fix N$_2$ when growing anaerobically, a property not found among other enteric bacteria. The genus *Serratia* is also a butanediol producer, and forms a series of red pyrrole-containing pigments called **prodigiosins**. This pigment, a linear tripyrrole, is of interest because it contains the pyrrole ring also found in the pigments involved in energy transfer: porphyrins, chlorophylls, and phycobilins. There is no evidence that prodigiosin plays any role in energy transfer, however, and its exact function is unknown. Species of *Serratia* can be isolated from water and soil as well as from the gut of various insects and vertebrates and occasionally from the intestines of humans.

19.21   *Neisseria* and Other Gram-Negative Cocci

This group comprises a diverse collection of organisms, related by Gram stain, morphology, lack of motility, nonfermentative aerobic metabolism, and similar DNA base composition. The four genera *Neisseria, Moraxella, Kingella,* and *Acinetobacter* are distinguished as outlined in Table 19.28. A major distinction is made on the basis of oxidase reaction. The oxidase-positive organisms (such as most isolates of *Neisseria*

Table 19.28   Characteristics of the genera of Gram-negative cocci*

Characteristics	Genus	DNA (mole % GC)
I. Oxidase-positive, penicillin sensitive:		
Cocci; complex nutrition, utilize carbohydrates, obligate aerobes	*Neisseria*	49–55
Rods or cocci; generally no growth factor requirements, generally do not utilize carbohydrates; do not contain flagella but some species exhibit "twitching" motility	*Moraxella*	40–47
	Kingella	47–55
II. Oxidase-negative, penicillin resistant: some strains can utilize a restricted range of sugars, and some exhibit "twitching" motility	*Acinetobacter*	38–47

Phylogenetically, those species examined are members of the beta purple Bacteria (see Section 18.6).

gonorrhoeae, see Figure 13.3*b*) are unusually sensitive to penicillin, being inhibited by 1 µg/ml, differing in this way from most other Gram-negative Bacteria, including *Acinetobacter.*

In the genus *Neisseria,* the cells are always cocci (see Figure 15.24*a*), whereas the cells of the other genera are rod shaped, becoming coccoid in the stationary phase of growth. This has led to designation of these organisms as **coccobacilli.** Organisms of the genera *Neisseria, Kingella,* and *Moraxella* are commonly isolated from animals, and some of them are pathogenic, whereas organisms of the genus *Acinetobacter* are common soil and water organisms, although they are occasionally found as parasites of certain animals and have been implicated in nosocomial infections. Some strains of *Moraxella* and *Acinetobacter* possess the interesting property of **twitching motility,** exhibited as brief translocative movements or "jumps," covering distances of about 1–5 µm. We discussed the clinical microbiology of *Neisseria gonorrhoeae* in Section 13.1 and the disease gonorrhea in Section 15.6.

19.22   Rickettsias

The rickettsias are Gram-negative, coccoid or rod-shaped cells in the size range of 0.3–0.7 µm wide by 1–2 µm long. They are, with one exception, *obligate intracellular parasites* (Figure 19.76*a*) and have not yet been cultivated in the absence of host cells. Rickettsias are the causative agents of such diseases as typhus fever, Rocky Mountain spotted fever, and Q fever (see Section 15.9). Electron micrographs of thin sections of rickettsias show profiles with a normal bacterial morphology (Figure 19.76*b*); both cell wall and cell membrane are visible. The cell wall contains muramic acid and diaminopimelic acid. Both RNA and DNA are present, and the DNA is in the normal double-stranded form, with a GC content varying from 29 to 33 percent in various species of the genus *Rickettsia* and 43 percent for *Coxiella burnetii* (the causal agent of Q fever). The rickettsias divide by normal binary fission, with doubling times of about 8 hours. The penetration of a host cell by a rickettsial cell is an active process, requiring both host and parasite to be alive

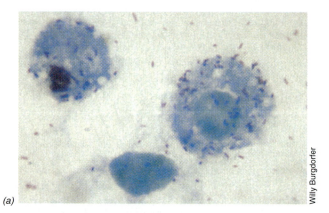

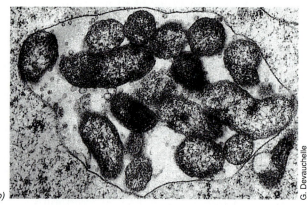

FIGURE 19.76 Rickettsias growing within host cells. (a) *Rickettsia rickettsii* in tunica vaginalis cells of the vole, *Microtus pennsylvanicus.* Cells are about 0.3 µm in diameter. (b) Electron micrograph of cells of *Rickettsiella popilliae* within a blood cell of its host, the beetle *Melolontha melolontha.* Notice that the bacteria are growing within a vacuole within the host cell.

and metabolically active. Once inside the phagocytic cell, the bacteria multiply primarily in the cytoplasm and continue replicating until the host cell is loaded with parasites (see Figures 11.28 and 19.76), at which time the host cell bursts and liberates the bacteria into the surrounding fluid.

Metabolism and pathogenesis

Much attention has been directed to the metabolic activities and biochemical pathways of rickettsias in an

Table 19.29   Major properties of rickettsias and comparison of rickettsias with chlamydias and viruses

Property	Rickettsias	Chlamydias	Viruses
Structural			
Nucleic acid	RNA and DNA	RNA and DNA	Either RNA or DNA, never both
Ribosomes	Present	Present	Absent
Cell wall	Muramic acid, DAP	Muramic acid not present	No wall
Structural integrity during multiplication	Maintained	Maintained	Lost
Metabolic capacities			
Macromolecular synthesis	Carried out	Carried out	Only with use of host machinery
ATP-generating system	Present	Absent	Absent
Capable of oxidizing glutamate	Yes	No	No
Sensitivity to antibacterial antibiotics	Sensitive	Sensitive (except for penicillin)	Resistant

attempt to explain why they are obligate intracellular parasites. Many rickettsias possess a highly distinctive energy metabolism, being able to oxidize only glutamate or glutamine and being unable to oxidize glucose, glucose-6-phosphate, or organic acids. However, *Coxiella burnetii* is able to utilize both glucose and pyruvate as electron donors. Rickettsias possess a respiratory chain complete with cytochromes and are able to carry out electron-transport phosphorylation, using NADH as electron donor. They are also able to synthesize at least some of the small molecules needed for macromolecular synthesis and growth, while they obtain the rest of their nutrients from the host cell. There is some suggestion that the host also provides some key coenzymes, such as NAD^+ and coenzyme A. A summary of the biochemical properties of rickettsias is given in Table 19.29.

Rickettsias do not survive long outside their hosts and this may explain why they must be transmitted from animal to animal by arthropod vectors. When the arthropod obtains a blood meal from an infected vertebrate, rickettsias present in the blood are inoculated directly into the arthropod, where they penetrate to the epithelial cells of the gastrointestinal tract, multiply, and appear later in the feces. When the arthropod feeds upon an uninfected individual, it then transmits the rickettsias either directly with its mouthparts or by contaminating the bite with its feces. However, the causal agent of Q fever, *C. burnetii* (see Section 15.9), can also be transmitted to the respiratory system by aerosols. *Coxiella burnetii* is the most resistant of the rickettsias to physical damage, probably because it produces an endospore-like form, and this explains its ability to survive in air. *Rochalimaea* is an atypical rickettsia because it can be grown in culture and is thus not an obligate intracellular parasite. In addition, when growing in tissue culture, cells of *Rochalimaea* grow on the *outside surface* of the eukaryotic host cells

rather than within the cytoplasm or the nucleus. *R. quintana* is the causative agent of *trench fever*, a disease that decimated troops in World War I. A summary of the major properties of species of *Rickettsia*, *Rochalimaea*, and *Coxiella* are given in Table 19.30.

Phylogeny of rickettsias

Despite their intracellular existence, the phylogenetic relationships of the rickettsias have been determined. By 16S rRNA sequencing, the rickettsias clearly group within the purple Bacteria (see Section 18.6), more specifically with the plant pathogen *Agrobacterium*. Although at first this may appear puzzling, it should be remembered that *Agrobacterium*, like the rickettsias, has evolved close associations with eukaryotic cells, being responsible for the plant disease crown gall (see Section 17.23). The 16S rRNA sequences of rickettsias and *Agrobacterium tumefaciens* are about 95 percent homologous; those of *Escherichia coli* or *Bacillus subtilis* and the rickettsias are less than 80 percent homologous. The evolutionary relationships between rickettsias and intracellular symbionts such as *A. tumefaciens* therefore suggests that rickettsias evolved from plant-associated bacteria. Perhaps the rickettsias were originally *plant* pathogens that evolved to be associated with animals following transfer from plants to animals by insect vehicles.

19.23   Chlamydias

Organisms of the genus *Chlamydia* probably represent a further stage in degenerate evolution from that discussed above for the rickettsias, since the chlamydias are obligate parasites in which there has been an even greater loss of metabolic function. Three species of *Chlamydia* are recognized (Table 19.31): *C. psittaci*, the causative agent of the disease *psittacosis*; *C. trachomatis*, the causative agent of trachoma and a variety of other

Table 19.30   Characteristics of representative *Rickettsia*, *Rochalimaea*, and *Coxiella* species*

Species	Rickettsial group	Alternate host	Cellular location	DNA (mole % GC)	DNA percent hybridization to *R. rickettsii* DNA
Rickettsia					
R. rickettsii	Spotted Fever	Tick	Cytoplasm and nucleus	32–33	100
R. prowazekii	Typhus	Louse	Cytoplasm	29–30	53
R. typhi	Typhus	Flea	Cytoplasm	29–30	36
Rochalimaea					
R. quintana	Trench fever	Louse	Epicellular	39	30
R. vinsonii	—	Vole	Epicellular	39	30
Coxiella					
C. burnetii	Q Fever	Tick	Vacuoles	43	—

Phylogenetically, those species examined fall within the alpha purple Bacteria (see Section 18.6).

Table 19.31   Differential characteristics of species of the genus *Chlamydia*

Characteristic	*C. trachomatis*	*C. psittaci*	*C. pneumoniae*
Hosts	Humans	Birds, lower mammals, occasionally humans	Humans
Usual site of infection	Mucous membrane	Multiple sites	Respiratory mucosa
Human-to-human transmission	Common	Rare	Probable
Mole % GC	42–45	39–43	40
Percent homology to *C. trachomatis* DNA by DNA: DNA hybridization	100	10	10
DNA, kilobase pairs/genome (*E. coli* = 4700)	1000	550	~1000

Phylogenetically, Chlamydia form a distinct line of Bacteria (see Section 18.6).

Table 19.32   Chlamydial diseases

Human diseases caused by *C. trachomatis*
 Trachoma
 Inclusion conjunctivitis
 Otitis media
 Infant pneumonia
 Nongonococcal urethritis (males)
 Urethral inflammation (females)
 Lymphogranuloma venereum
 Cervicitis

Human diseases caused by *C. pneumoniae*
 Respiratory syndromes probably transmitted from person-to-person

Human diseases caused by *C. psittaci*
 Psittacosis

Diseases caused by *C. psittaci* in nonhuman hosts
 Avian chlamydiosis (Parrots, parakeets, pigeons, turkeys, geese, other birds)
 Seminal vesiculitis (Sheep, cattle)
 Pneumonia (Kittens, lambs, calves, piglets, foals)
 Conjunctivitis (Lambs, calves, piglets, cats)
 Synovial tissue arthritis (Lambs, calves, foals, piglets)

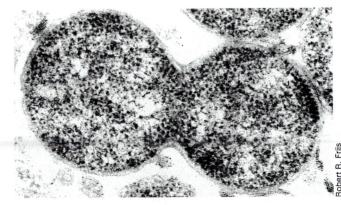

Robert R. Friis

FIGURE 19.77 Electron micrograph of a thin section of a dividing cell (reticulate body) of *Chlamydia psittaci*, a member of the psittacosis group, within a mouse tissue culture cell. A single chlamydial cell is about 1 μm in diameter.

Elementary body

Size: ~0.3μm
Rigid cell wall
Infectious
RNA : DNA = 1: 1
Nongrowing form

Reticulate body

Size: ~1μm
Fragile cell wall, cells pleomorphic
Non-infectioius
RNA : DNA = 3: 1
Growing form

1. Elementary body attacks host cell

2. Phagocytosis of elementary body

3. Conversion to reticulate body

4. Multiplication of reticulate bodies

5. Conversion to elementary bodies

6. Release of elementary bodies

(a)

FIGURE 19.78 The infectious cycle of chlamydia. (a) Schematic diagram of the cycle; the whole cycle takes about 48 hours. (b) Human chlamydial infection. An infected fallopian tube is bursting, releasing mature elementary bodies.

human diseases; and *C. pneumoniae*, the cause of a variety of respiratory syndromes (Table 19.32). **Psittacosis** is an epidemic disease of birds that is occasionally transmitted to humans and causes pneumonia-like symptoms. **Trachoma** is a debilitating disease of the eye characterized by vascularization and scarring of the cornea. Trachoma is the leading cause of blindness in humans. Other strains of *C. trachomatis* infect the genitourinary tract, and chlamydial infections are one of the leading sexually transmitted diseases today (see Section 15.6). A comparison of the properties of *C. psittaci*, *C. trachomatis*, and *C. pneumoniae* is shown in Table 19.31.

Besides being disease entities, the chlamydias are intriguing because of the biological and evolutionary problems they pose. Biochemical studies show that the chlamydias have Gram-negative cell walls (although peptidoglycan is absent) and they have both DNA and RNA. Electron microscopy of thin sections of infected cells shows forms that clearly are undergoing binary fission (Figure 19.77). The biosynthetic capacities of the chlamydias are even more restricted than that of the rickettsias. This raises the interesting question of the limits to which evolutionary loss of function can be pushed while independence of macromolecular function is still retained. Phylogenetically, chlamydia form a major branch of Bacteria, totally unrelated to rickettsias or other Gram-negative Bacteria (see Section 18.6).

Life cycle of *Chlamydia*

The life cycle of a typical member of the genus *Chlamydia* is shown in Figure 19.78. Two cellular types are seen in a typical life cycle: a small, dense cell, called an **elementary body**, which is relatively resistant to drying and is the means of *dispersal* of the agent, and a larger, less dense cell, called a **reticulate body**, which divides by binary fission and is the *vegetative* form. El-

Elementary bodies

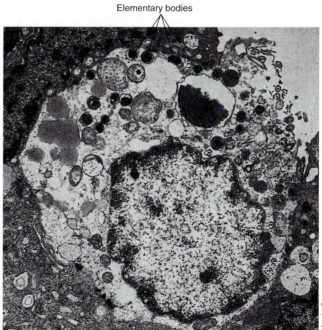

(b)

Morris Cooper

ementary bodies are nonmultiplying cells specialized for transmission, whereas reticulate bodies are noninfectious forms that specialize in intracellular multiplication. Unlike the rickettsias just discussed above, the chlamydias are not transmitted by arthropods but are primarily *airborne* invaders of the respiratory system–hence the significance of resistance to drying of elementary bodies. When a virus infects a cell, it loses its structural integrity and liberates nucleic acid. When an elementary body enters a cell, however, although it changes form, it remains a structural unit and enlarges and begins to undergo binary fission. A reticulate body is seen in Figure 19.77. After a number of divisions, the vegetative cells are converted into elementary bodies that are released when the host cell disintegrates and can then infect other cells. Generation times of two to three hours have been reported, which are considerably faster than those found for the rickettsias.

Molecular and metabolic properties

As we noted, chlamydia cells have a chemical composition similar to that of other Bacteria. Both RNA and DNA are present. The DNA content of a chlamydia corresponds to about twice that of vaccinia virus and one-eighth to one-fifth that of *Escherichia coli*; genome sizes range between 500 and 1000 kilobase pairs. At least some of the RNA is in the form of ribosomes, and, like those of other prokaryotes, the ribosomes are 70S particles composed of one 50S and one 30S unit.

The metabolic properties of chlamydias purified from infected cells have been studied by methods similar to those used with the rickettsias. The biosynthetic capacities of the chlamydias are much more limited than are those of the rickettsias (see Table 19.29). Although macromolecular syntheses occur in the chlamydias, no energy-generating system is present and the cells are thus "energy parasites" on their host cells. From the limited biosynthetic and catabolic capacities of chlamydias it is easy to see why they are obligate parasites. The chlamydias probably have the simplest biochemical abilities of all cellular organisms.

19.24   Gram-Positive Bacteria: Cocci

With the exception of *Deinococcus*, all Gram-positive Bacteria form a phylogenetically coherent group. However, two major subgroups of Gram-positive Bacteria emerge from ribosomal RNA sequencing studies: the "Clostridium" subgroup, consisting of the endosporeformers, the lactic acid bacteria, and most Gram-positive cocci; these organisms generally have a fairly *low* mole percent GC in their DNA, and the "Actinomycetes" subgroup, which essentially consists of genera of actinomycetes, most of which have a rather *high* GC content, and the genus *Propionibacterium*. We begin here with the Gram-positive cocci, common organisms to the beginning student in microbiology.

The Gram-positive cocci (Figure 19.79) include bacteria with widely differing physiological characteristics. The major genera of Gram-positive cocci and

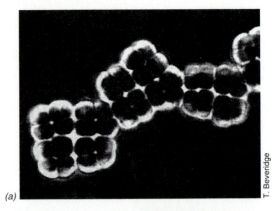
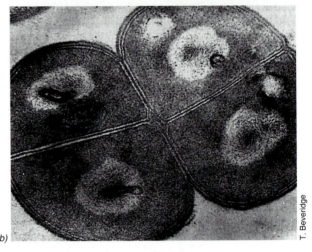

FIGURE 19.79   (a) Phase contrast photomicrograph of cells of a typical Gram-positive coccus *Sarcina* sp. A single cell is about 2 μm in diameter. (b) Electron micrograph of a thin section.

some differentiating characteristics are given in Table 19.33. The genus *Streptococcus* is not considered here but instead is discussed in Section 19.25 along with the other lactic acid bacteria.

Staphylococcus and *Micrococcus*

Staphylococcus and *Micrococcus* (Figure 19.80) are both aerobic organisms with a typical respiratory metabolism. They are catalase positive, and this test permits their distinction from *Streptococcus* and some other genera of Gram-positive cocci.

As we discussed in Section 9.11, the Gram-positive cocci are relatively resistant to reduced water potential, and tolerate drying and high salt fairly well. Their ability to grow in media with high salt provides a simple means for isolation. If an inoculum is spread on an agar plate with a fairly rich medium containing about 7.5 percent NaCl, and the plate incubated aerobically, Gram-positive cocci will often form the predominant colonies. Often, these organisms are pigmented, and this provides an additional aid in selecting Gram-positive cocci.

The two genera *Micrococcus* and *Staphylococcus* can easily be separated based on the oxidation/fermentation (O/F) (Table 13.3) test. *Micrococcus* is an obligate aerobe, and produces acid from glucose only aerobically, whereas *Staphylococcus* is a facultative aerobe,

Table 19.33   Distinguishing features of Gram-positive cocci*

Genus	Motility	Arrangement of cells	Growth by fermentation	DNA (mole % GC)	16S rRNA group*	Other characteristics
Micrococcus	−	Clusters, tetrads	−	66–73	High GC	Strict aerobes
Staphylococcus	−	Clusters, pairs	+	30–39	Low GC	Only genus to contain teichoic acid in cell wall
Stomatococcus	−	Clusters, pairs	+	56–60	—	Only genus containing a capsule
Planococcus	+	Pairs, tetrads	−	39–52	Low GC	Motile, primarily marine
Sarcina	−	Cuboidal packets of 8 or more cells	+	28–31	Low GC	Extremely acid tolerant; cellulose in cell wall
Deinococcus	−	Pairs, tetrads	−	62–70	Deinococcus/Thermus	Phylogenetically unique
Ruminococcus	+	Pairs, chains	+	39–46	Low GC	Obligate anaerobe; inhabits rumen, cecum, and large intestine of many animals
Peptococcus	−	Clusters, pairs	+	50–51	Low GC	Obligate anaerobe; ferments peptone but not sugars
Peptostreptococcus	−	Clumps, short chains	+	28–37	Low GC	Obligate anaerobe; ferments peptone; common member of human normal flora, skin, intestine, vagina; also isolated from vaginal and purulent discharges

*With the exception of Deinococcus, all organisms listed belong to the "High GC" or "Low GC" Gram-positive Bacteria (see Section 18.6).

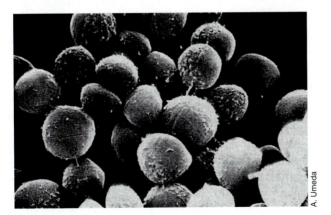

FIGURE 19.80 Scanning electron micrograph of typical *Staphylococcus*, showing the irregular arrangement of the cell clusters. Individual cells are about 0.8 μm in diameter.

A. Umeda

and produces acid from glucose both aerobically and anaerobically. Their DNA base compositions are also widely different: *Micrococcus*, species have very high GC ratios whereas *Staphylococcus* species are rather low (Table 19.33).

Staphylococci are common parasites of humans and animals, and occasionally cause serious infections. In humans, two major forms are recognized, *S. epidermidis*, a nonpigmented, nonpathogenic form that is usually found on the skin or mucous membranes, and *S. aureus*, a yellow pigmented form that is most commonly associated with pathological conditions, including boils, pimples, pneumonia, osteomyelitis, meningitis, and arthritis. We listed the exotoxins of *S. aureus* in Table 11.3. One of the significant exotoxins is **coagulase**, an enzymelike factor that causes fibrin to coagulate and form a clot. Strains of *S. aureus* are generally coagulase *positive*, whereas *S. epidermidis* is coagulase *negative*. We discussed the possible role of the yellow carotenoid pigment of *S. aureus* in resistance to phagocytosis in Section 11.13, and staphylococcal food poisoning was discussed in Section 15.13.

Sarcina

The genus *Sarcina* contains two species of bacteria that divide in three perpendicular planes to yield packets of eight cells or more (Figure 19.79a). *Sarcina* are obligate anaerobes and are extremely acid tolerant, being able to ferment sugars and grow down to pH 2. Cells of one species, *S. ventriculi*, contain a thick fibrous layer of cellulose surrounding the cell wall (Figure 19.79b). The cellulose layers of adjacent cells become attached and this functions as a cementing material to hold together packets of *S. ventriculi* cells. *Sarcina* can be isolated from soil, mud, feces, and stomach contents. Because of its extreme acid tolerance, *S. ventriculi* is one of the few bacteria that can actually grow in the stomach of humans and other monogastric animals. Rapid growth of *S. ventriculi* is observed in the stomach of humans suffering from certain gastrointestinal pathological conditions (such as pyloric ulcerations) which retard the flow of food to the intestine.

Deinococcus

The genus *Deinococcus* is an unusual genus of Gram-positive cocci. Besides their unique phylogenetic stature (see Section 18.6), species of *Deinococcus* differ from other Gram-positive cocci in a number of interesting chemical and physiological properties. The cell walls of deinococci are structurally complex and consist of several layers, including an *outer membrane* layer (Figure 19.81) normally only present in Gram-

negative Bacteria (see Section 3.5). However, the outer membrane of *Deinococcus* is chemically unique and does not contain heptoses and lipid A typical of that of Gram-negative Bacteria.

Most deinococci are bright red or pink in color because of the variety of carotenoids found in these organisms, and many strains are highly resistant to ultraviolet radiation and to desiccation. Resistance to radiation can be used to advantage for isolating deinococci. These remarkable organisms can be isolated from soil, ground meat, dust, and filtered air following exposure of the sample to intense ultraviolet (or even gamma) radiation and plating on a rich medium containing tryptone and yeast extract. Since many strains of *Deinococcus radiodurans* are even more resistant to radiation than bacterial endospores, treatment of a sample with strong doses (1–2 megarads) of radiation effectively sterilizes the sample of organisms other than *D. radiodurans*, making isolation of deinococci relatively straightforward.

In addition to impressive radiation resistance, *D. radiodurans* is resistant to the mutagenic effects of many highly mutagenic chemicals. Studies of the mutability of *D. radiodurans* have shown it to be highly efficient in repairing damaged DNA. Several different DNA repair enzymes exist in *D. radiodurans* to repair breaks in single- or double-stranded DNA and to excise and repair thymine dimers formed by the action of ultraviolet light. These efficient repair mechanisms have actually hindered studies of the molecular genetics of *D. radiodurans*. Because it is difficult to induce stable mutations in the DNA of *D. radiodurans*, it has been nearly impossible to obtain *mutants* of this organism for genetic study! The only chemical mutagens that seem to work on *D. radiodurans* are agents like nitrosoguanidine, which tend to induce *deletions* in DNA; deletions are apparently not repaired as efficiently as point mutations in this organism. Consistent with the fact that *D. radiodurans* is an extremely radiation-resistant bacterium, strains of this organism have been isolated from near atomic reactors and other potentially lethal radiation sources.

19.25 Lactic Acid Bacteria

The lactic acid bacteria are characterized as Gram-positive, usually nonmotile, nonsporulating Bacteria that produce lactic acid as a major or sole product of fermentative metabolism. Members of this group lack porphyrins and cytochromes, do not carry out electron-transport phosphorylation, and hence obtain energy only by *substrate-level phosphorylation*. All of the lactic acid bacteria grow anaerobically. Unlike many anaerobes, however, most lactics are not sensitive to O_2 and can grow in its presence as well as in its absence; thus they are **aerotolerant anaerobes**. Some strains are able to take up O_2 through the mediation of flavoprotein oxidase systems, producing H_2O_2, although most strains lack catalase and most dispose of H_2O_2 via alternative enzymes referred to as *peroxidases*

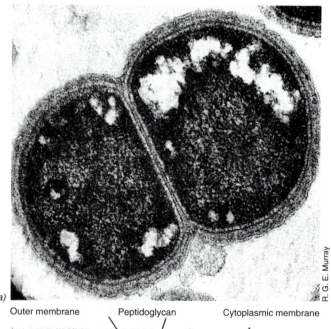

(a)

R. G. E. Murray

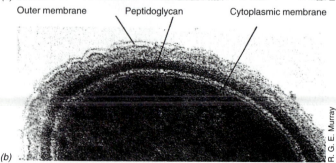

Outer membrane   Peptidoglycan   Cytoplasmic membrane

(b)

R. G. E. Murray

FIGURE 19.81 The radiation-resistant coccus, *Deinococcus radiodurans*. An individual cell is about 2.5 μm in diameter. (a) Transmission electron micrograph of *D. radiodurans*. Note the outer membrane layer. (b) High magnification micrograph of wall layer.

(see Section 9.12). No ATP is formed in the flavoprotein oxidase reaction, but the oxidase system can be used for reoxidation of NADH generated during fermentation. Most lactic acid bacteria can obtain energy only from the metabolism of sugars and related fermentable compounds, and hence are usually restricted to habitats in which sugars are present. They usually have only limited biosynthetic ability, and their complex nutritional requirements include needs for amino acids, vitamins, purines, and pyrimidines.

Traditionally, the lactic acid bacteria were considered a single group containing both cocci and rods. Some workers prefer to separate the group; however, we continue to keep both rods and cocci together in the present text, recognizing the obvious physiological and phylogenetic similarities of the group as a whole.

Homo- and heterofermentation

One important difference between subgroups of the lactic acid bacteria lies in the nature of the products formed during the fermentation of sugars. One group, called **homofermentative**, produces virtually a single fermentation product, *lactic acid*, whereas the other group, called **heterofermentative**, produces other products, mainly *ethanol* and *CO_2* as well as lactate. Abbreviated pathways for the fermentation of glucose by a homo- and a heterofermentative organism are shown in Figure 19.82. The differences observed in the

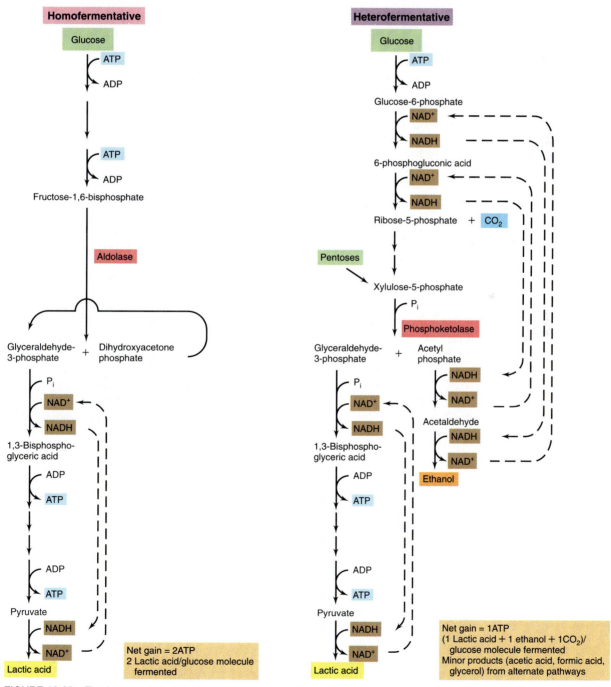

FIGURE 19.82   The fermentation of glucose in homofermentative and heterofermentative lactic acid bacteria. Note that no ATP is made in reactions leading to ethanol formation.

fermentation products are determined by the presence or absence of the enzyme **aldolase**, one of the key enzymes in *glycolysis* (see Figure 4.11). The heterofermenters, lacking aldolase, cannot break down fructose bisphosphate to triose phosphate. Instead, they oxidize glucose-6-phosphate to 6-phosphogluconate and then decarboxylate this to pentose phosphate, which is broken down to triose phosphate and acetylphosphate by means of the enzyme **phosphoketolase**.

Triose phosphate is converted ultimately to lactic acid with the production of 1 mole of ATP, while the acetylphosphate accepts electrons from the NADH generated during the production of pentose phosphate and is thereby converted to ethanol *without* yielding ATP. Because of this, heterofermenters produce only *1 mole* of ATP from glucose instead of the 2 moles produced by homofermenters. This difference in ATP yield from glucose is reflected in the fact that homofermenters produce twice as much cell mass as heterofermenters from the same amount of glucose. Because the heterofermenters decarboxylate 6-phosphogluconate, they produce CO_2 as a fermentation product, whereas the homofermenters produce little or no CO_2; therefore one simple way of detecting a heterofermenter is to observe for production of CO_2 in laboratory cultures. At the enzyme level, heterofermenters are characterized by the lack of aldolase and the presence of phosphoketolase. Many strains of heterofermenters can use O_2 as an electron acceptor with a reduced flavoprotein serving as electron donor. In this reaction, half of the NADH generated from the oxidation of glucose to ribose is transferred to a flavin and on to O_2. Acetylphosphate can then be converted to acetate instead of being reduced to ethanol, and an additional ATP is synthesized.

The various genera of lactic acid bacteria have been defined on the basis of cell morphology, DNA base composition, and type of fermentative metabolism, as is shown in Table 19.34. Members of the genera *Streptococcus*, *Enterococcus*, *Lactococcus*, *Leuconostoc*, and *Pediococcus* have fairly similar DNA base ratio compositions; in addition, there is very little variation from strain to strain. The genus *Lactobacillus*, on the other hand, has members with widely diverse DNA compositions and hence does not constitute a homogeneous group.

Streptococcus and other cocci

The genus *Streptococcus* (Figure 19.83) contains a wide variety of species with quite distinct habitats, whose activities are of considerable practical importance to humans. Some members are pathogenic to people and animals (see Section 15.2). As producers of lactic acid, certain streptococci play important roles in the production of buttermilk, silage, and other fermented products (see Section 9.18). To distinguish generally nonpathogenic streptococci from human pathogenic species, the genus *Streptococcus* has been split into three genera. The genus *Lactococcus* contains those streptococci of dairy significance while the genus *Enterococcus* has been created to group streptococci that are primarily of fecal origin.

Organisms remaining in the genus *Streptococcus* have been further divided into two groups of related species on the basis of characteristics enumerated in Table 19.35. *Hemolysis* on blood agar is of considerable importance in the subdivision of the genus. Colonies of those strains producing streptolysin O or S are surrounded by a large zone of complete red blood cell hemolysis, a condition called β **hemolysis**. On the other

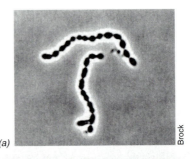
(a) Brock

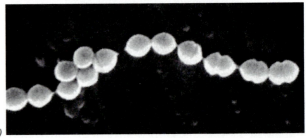
(b) Bryan Larsen

FIGURE 19.83 Phase contrast and scanning electron micrographs of *Streptococcus* species. (a) *Streptococcus lactis*. (b) *Streptococcus* sp. Cells in both cases are 0.5–1 µm in diameter.

Table 19.34   Differentiation of the principal genera of lactic acid bacteria*

Genus	Cell form and arrangement	Fermentation	DNA (mole % GC)
Streptococcus	Cocci in chains	Homofermentative	34–46
Leuconostoc	Cocci in chains	Heterofermentative	38–41
Pediococcus	Cocci in tetrads	Homofermentative	34–42
Lactobacillus	(1) Rods, usually in chains	Homofermentative	32–53
	(2) Rods, usually in chains	Heterofermentative	34–53
Enterococcus	Cocci in chains	Homofermentative	38–40
Lactococcus	Cocci in chains	Homofermentative	38–41

**Phylogenetically, all organisms are members of the "low GC Gram-positive subdivision" of the Gram-positive Bacteria (see Section 18.6).*

Table 19.35   Differential characteristics of streptococci, lactococci, and enterococci

Group	Antigenic (Lancefield) groups	Representative species	Type of hemolysis on blood agar	Good growth at 10°C	Good growth at 45°C	Survive 60°C for 30 min	Growth in Milk with 0.1% methylene blue	Growth in Broth with 40% bile	Habitat
Streptococci									
Pyogenes subgroup	A,B,C,F,G	Streptococcus pyogenes	Lysis (β)	−	−	−	−	−	Respiratory tract, systemic
Viridans subgroup	—	Streptococcus mutans	Greening (α)	−	+	−	−	−	Mouth, intestine
Enterococci	D	Enterococcus faecalis	Lysis (β), greening (α), or none	+	+	+	+	+	Intestine, vagina, plants
Lactococci	N	Lactococcus lactis	None	+	−	+	+	+	Plants, dairy products

hand, many streptococci and the lactococci and enterococci do not produce hemolysins and cause the formation of a greenish or brownish zone around colonies on blood agar, which is due not to true hemolysis but to discoloration and loss of potassium from the red cells. This type of reaction has classically been referred to as α **hemolysis**. Streptococci and related cocci are also divided into *immunological* groups based on the presence of specific carbohydrate antigens. These antigenic groups (or **Lancefield groups** as they are commonly known, named for Rebecca Lancefield, a pioneer in *Streptococcus* taxonomy), are designated by letters; A through O are currently recognized. Those β-hemolytic streptococci found in human beings usually contain the group A antigen, which is a cell-wall polymer containing *N*-acetylglucosamine and rhamnose. Enterococci contain the group D antigen, a glycerol teichoic acid containing glucose side chains. Group B streptococci are usually found in association with animals and are a cause of mastitis in cows, but have also been implicated in human infections. Lactococci are of antigen group N and are not pathogenic.

Placed in the genus *Leuconostoc* are cocci that are morphologically similar to streptococci but are *heterofermentative*. Strains of *Leuconostoc* also produce the flavoring ingredients diacetyl and acetoin by breakdown of citrate and have been used as starter cultures in dairy fermentations, but their place has now been taken by *S. diacetilactis*. Some strains of *Leuconostoc* produce large amounts of dextran polysaccharides (α-1,6-glucan) when cultured on sucrose (see Figure 16.42). Dextrans produced by *Leuconostoc* have found some medical use as plasma extenders in blood transfusions. Other strains of *Leuconostoc* produce fructose polymers called *levans*.

Lactobacillus

Lactobacilli are typically rod-shaped, varying from long and slender to short, bent rods (Figure 19.84). Most species are homofermentative, but some are heterofermentative. The genus has been divided into three major subgroups (Table 19.36).

Lactobacilli are often found in dairy products, and some strains are used in the preparation of fermented products. For instance, *L. delbrueckii* is used in the preparation of yogurt, *L. acidophilus* (Figure 19.84) in the production of acidophilus milk, and other species are involved in the production of sauerkraut, silage, and pickles (see Section 9.18). The lactobacilli are usually more resistant to acidic conditions than are the other lactic acid bacteria, being able to grow well at pH values around 4–5. Because of this, they can be selectively isolated from natural materials by use of carbohydrate-containing media of acid pH, such as tomato juice-peptone agar. The acid resistance of the lactobacilli enables them to continue growing during natural lactic fermentations when the pH value has dropped too low for other lactic acid bacteria to grow, and the lactobacilli are therefore responsible for the final stages of most lactic acid fermentations. The lactobacilli are rarely if ever pathogenic.

19.26   Endospore-Forming Gram-Positive Rods and Cocci

The structure and heat resistance of the bacterial endospore along with the process of spore formation itself was discussed in Section 3.11. Several genera of endospore-forming bacteria have been recognized, distinguished on the basis of morphology, relationship to O_2, and energy metabolism (Table 19.37). The two genera most frequently studied are *Bacillus*, the species of which are aerobic or facultatively aerobic, and *Clostridium*, which contains the strictly anaerobic species. The genera *Bacillus* and *Clostridium* consist of Gram-positive or Gram-variable rods, which are usually motile, possessing peritrichous flagella. A major property of taxonomic value for distinguishing between species of *Bacillus* and *Clostridium* is the shape

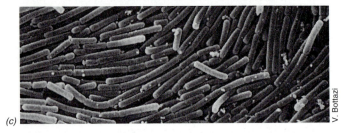

FIGURE 19.84 Phase contrast and electron micrographs of *Lactobacillus* species. (a) *L. acidophilus*. Cells are about 0.75 μm wide. (b) *L. brevis*, transmission electron micrograph. Cells measure about 0.8 by 2 μm. (c) *L. delbrueckii*, scanning electron micrograph. Cells are about 0.7 μm in diameter.

and position of endospores. These properties vary considerably among different species of endospore formers (Figure 19.85), and are useful starting points for beginning a taxonomic study of the group. Most endospore-forming bacteria produce one spore per cell. Sporulation is thus a means of *survival*, not reproduction. *Anaerobacter polyendosporus* is unusual in this connection because up to 5 endospores can be produced per cell. Members of the genus *Bacillus* produce the enzymes catalase and superoxide dismutase. Clostridia do not produce catalase and produce only low levels of superoxide dismutase, and it is thought that one reason they are obligately anaerobic is that they have no way of getting rid of the toxic H_2O_2 and O_2^- produced from molecular oxygen (see Section 9.12).

Molecular taxonomic studies of the genus *Bacillus* have shown that it is a heterogeneous group and can hardly be considered an assemblage of closely related organisms. The DNA base composition of *Bacillus* species vary from about 30 to nearly 70 percent GC, suggesting considerable genetic heterogeneity. Even though they are not closely related genetically, all the endospore-forming bacteria are *ecologically* related since they are found in nature primarily in soil. Even those species that are pathogenic to humans or animals are primarily saprophytic soil organisms, and infect hosts only incidentally. Spore formation should be advantageous for a soil microorganism because the soil is a highly variable environment. Although at some times nutrient supply is in excess, at other times it is deficient. Soil temperatures can be quite high in summer, especially at the surface. Thus a heat-resistant dormant structure should offer considerable survival value in nature. On the other hand, the ability to germinate and grow quickly when nutrients become available is also of value as it enables the organism to capitalize on a transitory food supply. The longevity of bacterial endospores is noteworthy. Records of spores surviving in the dormant state for over 50 years

Characteristics	Species	DNA (mole % GC)
Table 19.36 Characteristics of subgroups in the genus *Lactobacillus*		
Homofermentative: Lactic acid the major product (>85% from glucose) No gas from glucose; aldolase present		
(1) Grow at 45°C but not at 15°C, long rods; glycerol teichoic acid	*L. delbrueckii*	49–51
	L. acidophilus	34–37
(2) Grow at 15°C, variable growth at 45°C; short rods and coryneforms; ribitol and glycerol teichoic acids; can produce more oxidized fermentation products if O_2 is present	*L. casei, L. plantarum*	45–46
	L. curvatus	42–44
Heterofermentative: Produce about 50% lactic acid from glucose; produce CO_2 and ethanol; aldolase absent; phosphoketolase present; long and short rods; glycerol teichoic acid	*L. fermentum*	52–54
	L. brevis, L. buchneri	44–47
	L. kefir	41–42

Table 19.37   Genera of endospore-forming bacteria*

Characteristics	Genus	DNA (mole % GC)
Rods:		
Aerobic or facultative, catalase produced	*Bacillus*	32–69
Microaerophilic, no catalase; homofermentative lactic acid producer	*Sporolactobacillus*	46–47
Anaerobic:		
Sulfate-reducing	*Desulfotomaculum*	38–50
Does not reduce sulfate, fermentative	*Clostridium*	21–54
Gram-negative; can grow as homoacetogen on $H_2 + CO_2$	*Sporomusa*	41–49
Halophile, isolated from the Dead Sea	*Sporohalobacter*	31
Produces up to five spores per cell; fixes N_2	*Anaerobacter*	29
Cocci (usually arranged in tetrads or packets), aerobic	*Sporosarcina*	40–41

Phylogenetically, all organisms are members of the "low GC" subdivision of the Gram-positive Bacteria (see Section 18.6).

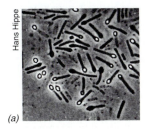 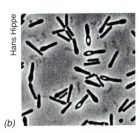 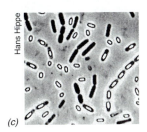

(a)   (b)   (c)

FIGURE 19.85
Phase contrast photomicrographs of various *Clostridium* species showing the different location of the endospore. (a) *C. cadaveris*, terminal spores. Cells are about 0.9 μm wide. (b) *C. sporogenes*, subterminal spores. Cells are about 1 μm wide. (c) *C. bifermentans*, central spores. Cells are about 1.2 μm wide.

are well established, and isolations of thermophilic spore-forming bacilli from marine sediment cores nearly 6000 years old suggest that spores can exist for literally *thousands* of years under the appropriate conditions!

Bacillus

Members of the genus *Bacillus* are easy to isolate from soil or air and are among the most common organisms to appear when soil samples are streaked on agar plates containing various nutrient media. Sporeformers can be selectively isolated from soil, food, or other material by exposing the sample to 80°C for 10 minutes, a treatment that effectively destroys vegetative cells while many of the spores present remain viable. When such pasteurized samples are streaked on plates and incubated aerobically, the colonies that develop are almost exclusively of the genus *Bacillus*. Bacilli usually grow well on synthetic media containing sugars, organic acids, alcohols, and so on, as sole carbon sources and ammonium as the sole nitrogen source; a few isolates have vitamin requirements. Many bacilli produce extracellular *hydrolytic enzymes* that break down polysaccharides, nucleic acids, and lipids, permitting the organisms to use these products as carbon sources and electron donors. Many bacilli produce antibiotics, of which bacitracin, polymyxin, tyrocidin, gramicidin, and circulin are examples. In most cases, antibiotic production seems to be related to the sporulation process, the antibiotic being released when the culture enters the stationary phase of growth and after it is committed to sporulation. An outline of the subdivision of the genus *Bacillus* is given in Table 19.38.

Certain bacilli, most notably *B. thuringiensis*, are insect pathogens due to a toxic crystalline protein produced during sporulation (Figure 19.86). The protein interferes with insect development and is used as a biological insecticide (see Sections 8.14 and 17.22).

Clostridium

The clostridia lack a cytochrome system and a mechanism for electron-transport phosphorylation, and hence they obtain ATP *only* by substrate-level phosphorylation. A wide variety of anaerobic energy-yielding mechanisms are known in the clostridia (fermentative diversity is discussed in Section 16.19); indeed, the separation of the genus into subgroups is based primarily on these properties and on the nature of the electron donors used (Table 19.39).

A number of clostridia ferment sugars, producing as a major end product *butyric acid*. Some of these also produce *acetone* and *butanol*, and at one time the acetone-butanol fermentation by clostridia was of great industrial importance as it was the main commercial source of these products. Today, however, the chemical synthesis of acetone and butanol from petroleum products has mostly replaced the microbiological process. Some clostridia of the acetone-butanol type fix N_2; the most vigorous N_2 fixer is *C. pasteurianum*, which probably is responsible for most anaerobic nitrogen fixation in the soil. One group of clostridia ferments cellulose with the formation of acids and alcohols, and these are the main organisms decomposing cellulose anaerobically in soil. There is considerable industrial interest in the production of ethanol (an automotive fuel additive) by the clostridial fermentation

Table 19.38   Characteristics of representative species of the genus *Bacillus*

Characteristics	Species	Spore position	DNA (mole % GC)
I. Spores oval or cylindrical, facultative aerobes, casein and starch hydrolyzed; sporangia not swollen, spore wall thin			
Thermophiles and acidophiles	*B. coagulans*	Central/terminal	47
	B. acidocaldarius	Terminal	60
Mesophiles	*B. licheniformis*	Central	46
	B. cereus	Central	35
	B. anthracis	Central	33
	B. megaterium	Central	37
	B. subtilis	Central	43
Insect pathogen	*B. thuringiensis*	Central	34
Sporangia distinctly swollen, spore wall thick			
Thermophile	*B. stearothermophilus*	Terminal	52
Mesophiles	*B. polymyxa*	Terminal	44
	B. macerans	Terminal	52
	B. circulans	Central/terminal	35
Insect pathogens	*B. larvae*	Central/terminal	—
	B. popilliae	Central	41
II. Spores spherical, obligate aerobes, casein and starch not hydrolyzed			
Sporangia swollen	*B. sphaericus*	Terminal	37
Sporangia not swollen	*B. pasteurii*	Terminal	38

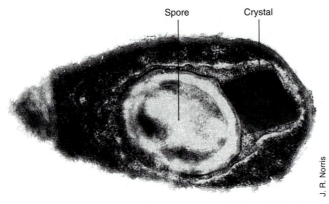

Spore   Crystal

J. R. Norris

FIGURE 19.86 Formation of the toxic parasporal crystal in the insect pathogen *Bacillus thuringiensis*. Electron micrograph of a thin section.

of *cellulose*, and genetic studies are underway to increase the yield of ethanol and reduce the formation of acidic fermentation products, the goal being to use waste cellulose as a motor fuel.

The biochemical steps in the formation of butyric acid and butanol from sugars are well understood (Figure 19.87). Glucose is converted to pyruvate via the Embden-Meyerhof pathway, and pyruvate is split to acetyl-CoA, CO_2, and reducing equivalents (reduced ferredoxin) by a phosphoroclastic reaction (see Section 16.19). Acetyl-CoA is then reduced to fermentation products using the NADH derived from glycolytic reactions. The proportions of the various products are influenced by the duration and the conditions of the fermentation. During the early stages, butyric and acetic acids are the predominant products, but as the pH of the medium drops, synthesis of acids ceases and the neutral products acetone and butanol begin to accumulate. If the medium is kept alkaline with $CaCO_3$, very little of the neutral products are formed and the fermentation products consist of about three parts butyric and one part acetic acid. Studies of the enzymology of solvent production by clostridia indicate that acid production in some way depresses synthesis of enzymes involved in acetone and butanol synthesis.

Another group of clostridia obtain their energy by fermenting *amino acids*. Some strains do not ferment single amino acids, but only amino acid *pairs*. In this situation one functions as the electron *donor* and is *oxidized*, while the other acts as the electron *acceptor* and is *reduced*. This type of coupled decomposition is known as the **Stickland reaction**. For instance, *C. sporogenes* will catabolize a mixture of glycine and alanine, as outlined in Figure 19.88.

Various amino acids that can function as either oxidants or reductants in Stickland reactions are listed in Table 19.40. The products of Stickland oxidation are always NH_3, CO_2, and a carboxylic acid with one *less* carbon atom than the amino acid which is oxidized.

Some amino acids can be fermented singly, rather than in a Stickland-type reaction. These are alanine, cysteine, glutamate, glycine, histidine, serine, and threonine. The products are generally acetate, butyrate, CO_2 and H_2. It is usually found that each group of clostridia is specific in the kinds of substances it can ferment; usually either sugars or amino acids are utilized, although some strains ferment both. Many of the products of amino acid fermentation by clostridia are foul-smelling substances, and the odor that results

Table 19.39   Characteristics of some groups of the genus *Clostridium*

Key characteristics	Other characteristics	Species	DNA (mole % GC)
I. Ferment carbohydrates:			
Ferment cellulose	Fermentation products: acetate, lactate, succinate, ethanol, CO_2, H_2	*C. cellobioparum* *C. thermocellum*	28 38–39
Ferment sugars, starch, and pectin	Fermentation products: acetone, butanol, ethanol, isopropanol, butyrate, acetate, propionate, succinate, CO_2, H_2; some fix N_2	*C. butyricum* *C. acetobutylicum* *C. pasteurianum* *C. perfringens* *C. thermosulfurogenes*	27–28 28–29 26–28 24–27 33
Ferment sugars primarily to acetic acid	Total synthesis of acetate from CO_2; cytochromes present in some species	*C. aceticum* *C. thermoaceticum* *C. formicoaceticum*	33 54 34
Ferments only pentoses or methylpentoses	Ring-shaped cells form left-handed, helical chains; fermentation products: acetate, propionate, *n*-propanol, CO_2, H_2	*C. methylpentosum*	46
II. Ferment proteins or amino acids	Fermentation products: acetate, other fatty acids, NH_3, CO_2, sometimes H_2; some also ferment sugars to butyrate and acetate; may produce exotoxins	*C. sporogenes* *C. tetani* *C. botulinum* *C. tetanomorphum*	26 25–26 26–28 25–28
	Ferments three carbon compounds to propionate, acetate, and CO_2	*C. propionicum*	35
III. Ferments carbohydrates or amino acids	Fermentation products from glucose: acetate, formate, small amounts of isobutyrate and isovalerate	*C. bifermentans*	27
IV. Purine fermenters	Ferments uric acid and other purines, forming acetate, CO_2, NH_3	*C. acidurici*	27–30
V. Ethanol fermentation to fatty acids	Produces butyrate, caproate, and H_2; requires acetate as electron acceptor; does not use sugars, amino acids, or purines	*C. kluyveri*	30

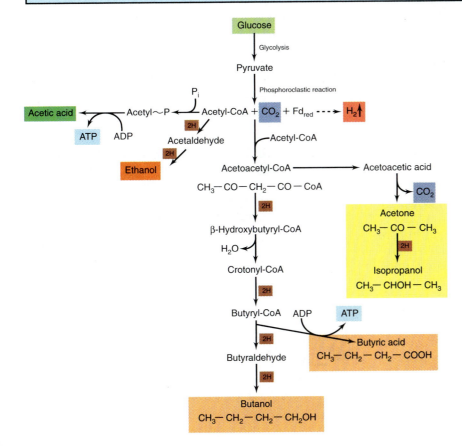

FIGURE 19.87

Pathway of formation of fermentation products from the butyric acid group of clostridia. The designation "2H" represents two electrons from one molecule of NADH.

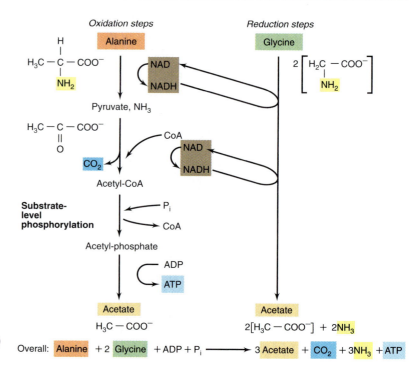

FIGURE 19.88
Coupled oxidation–reduction reaction (Stickland reaction) between alanine and glycine in *Clostridium sporogenes*.

Table 19.40	**Amino acids participating in coupled fermentations (Stickland reaction)**

Amino acids oxidized:
Alanine
Leucine
Isoleucine
Valine
Histidine

Amino acids reduced:
Glycine
Proline
Hydroxyproline
Tryptophan
Arginine

These pathogenic clostridia seem in no way unusual metabolically, but are distinct in that they produce specific toxins or, in those causing gas gangrene, a group of toxins (see Section 11.8 and Table 11.3). Many gas-gangrene clostridia also cause diseases in domestic animals, and botulism occurs in sheep and ducks, and a variety of other animals. An unsolved ecological problem is what role these toxins play in the natural habitat of the organism.

Sporosarcina

The genus *Sporosarcina* is unique among endospore-formers because cells are *cocci* instead of rods. *Sporosarcina* consists of spherical to oval cells which divide in two or three perpendicular planes to form tetrads or packets of eight or more cells (see Figure 19.89). The organism is motile and strictly aerobic.

from putrefaction is a result mainly of clostridial action. In addition to butyric acid, other odoriferous compounds produced are isobutyric acid, isovaleric acid, caproic acid, hydrogen sulfide, methylmercaptan (from sulfur amino acids), cadaverine (from lysine), putrescine (from ornithine), and ammonia.

The main habitat of clostridia is the soil, where they live primarily in anoxic "pockets," made anaerobic primarily by facultative organisms acting upon various organic compounds present. In addition, a number of clostridia have adapted to the anoxic environment of the mammalian intestinal tract. Also, as was discussed in Section 11.8, several clostridia that live primarily in soil are capable of causing disease in humans under specialized conditions. Botulism is caused by *C. botulinum*, tetanus by *C. tetani*, and gas gangrene by *C. perfringens* and a number of other clostridia, both sugar and amino acid fermenters.

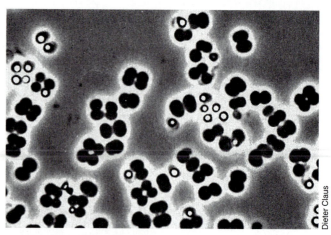

FIGURE 19.89 Phase contrast photomicrograph of *Sporosarcina ureae*. A single cell is about 2 μm wide. Note bright refractile endospores.

Two species of *Sporosarcina* are known, *S. ureae* and *S. halophila*. The latter species is of marine origin and differs from *S. ureae* primarily in its requirement for sodium ions for growth. Analyses of cell walls have also shown that *S. ureae* and *S. halophila* differ in the nature of the diamino acid found in the crosslinking polypeptide of peptidoglycan; *S. ureae* contains lysine and *S. halophila* contains ornithine. The endospores of *Sporosarcina* are highly refractile, centrally located (see Figure 19.89), and contain dipicolinic acid, typical of endospores from other genera.

Sporosarcina ureae can easily be enriched from soil by plating dilutions of a pasteurized soil sample on nutrient agar supplemented with 8 percent urea. Most soil bacteria are strongly inhibited by as little as 2 percent urea. However, *S. ureae* actively decomposes urea to CO_2 and NH_3, and in so doing can dramatically raise the pH of unbuffered media (*S. ureae* is remarkably alkaline tolerant and will grow in media up to pH 10–11). *S. ureae* is common in soils, and studies of its distribution suggest that numbers of *S. ureae* are greatest in soils that receive inputs of urine (a source of urea), such as soils in which animals periodically urinate. Since many soil organisms are quite urea sensitive, these results suggest that *S. ureae* is ecologically important as a major urea degrader in nature.

19.27   Mycoplasmas

Although seeming at first to be out of place in a discussion of Gram-positive Bacteria because their lack of cell walls yields no reaction in the Gram stain, phylogenetically mycoplasmas belong with the Gram-positive Bacteria because of their close evolutionary ties to the clostridia (see Section 18.6). The mycoplasmas are organisms without cell walls that do not revert to walled organisms. They are probably the smallest organisms capable of autonomous growth and are of special evolutionary interest because of their extremely simple cell structure.

Properties of mycoplasmas

The lack of cell walls in the mycoplasmas observed by electron microscopy has been reinforced by chemical analysis, the latter showing that the key wall components, muramic acid and diaminopimelic acid, are missing. In Chapter 3 we discussed protoplasts and showed how these structures can be formed when cell-wall-digesting enzymes act on cells that are in an osmotically protected medium, and that when the osmotic stabilizer is removed, protoplasts take up water, swell, and burst. Mycoplasmas resemble protoplasts in their lack of a cell wall, but they are more resistant to osmotic lysis and are able to survive conditions under which protoplasts lyse. This ability to resist osmotic lysis is at least partially determined by the nature of the mycoplasma cytoplasmic membrane, which is more stable than that of other prokaryotes. In one group of mycoplasmas, the membrane contains

sterols that seem to be responsible for stability, whereas in other mycoplasmas, carotenoids or other compounds may be involved instead. Those mycoplasmas possessing sterols in their membrane do not synthesize them but require them preformed in the culture medium. This sterol requirement is a current basis for separating the mycoplasmas into two groups (Table 19.41). *Thermoplasma* is phylogenetically unrelated to all other mycoplasmas because it is an archaeon. *Thermoplasma* is thermophilic and acidophilic, growing optimally at 55°C and pH 2; the properties of *Thermoplasma* are considered in more detail in Section 20.7.

Certain mycoplasmas contain compounds called **lipoglycans** (see Table 19.41). Lipoglycans are long-chain heteropolysaccharides covalently linked to membrane lipids and embedded in the cytoplasmic membrane of many mycoplasmas. Lipoglycans resemble the lipopolysaccharides of Gram-negative Bacteria (see Section 3.5) except that they lack the lipid A backbone and the phosphate typical of bacterial LPS. Lipoglycans presumably function to help stabilize the membrane, and have also been identified as facilitating attachment of mycoplasmas to cell surface receptors of animal cells. Like LPS, lipoglycans stimulate antibody production when injected into experimental animals.

Growth of mycoplasmas

Mycoplasma cells are usually fairly small, and they are highly pleomorphic, a consequence of their lack of rigidity. A single culture may exhibit small coccoid elements, larger, swollen forms, and filamentous forms of variable lengths, often highly branched (Figure 19.90). It is from the production of filamentous, funguslike forms that the name *Mycoplasma* (*myco* means "fungus") derives. A common growth form is seen in cultures that divide by budding; division occurs with the cells remaining either directly attached or connected by thin hyphae (Figure 19.91).

The small coccoid elements (0.2–0.3 µm in size) are the smallest mycoplasma units capable of independent growth. Because of flexibility due to lack of a cell wall, mycoplasma cells pass through filters with pore sizes smaller than the true diameter of the cells, and this has led to erroneous estimates of the minimum cell size capable of growth. Cellular elements of diameters close to 0.1 µm exist in mycoplasma cultures, but these are not capable of growth. Even so, the minimum reproductive unit of 0.2–0.3 µm probably represents the smallest *free-living* cell. Additionally, the genome size of mycoplasmas is also smaller than that of most prokaryotes, between 600 and 1100 kilobase pairs of DNA, which is comparable to that of the obligately parasitic chlamydia and rickettsia (see Sections 19.22 and 19.23), and about one-fifth to one-fourth that of *Escherichia coli*.

The mode of growth of mycoplasmas differs in liquid and agar cultures. On agar, there is a tendency for the organisms to grow so that they become embed-

Table 19.41   Major characteristics of mycoplasmas

Genus	Number of recognized species	Properties	DNA (mole % GC)	Genome size (kilobase pairs)	Presence of lipoglycans
I. Require sterols					
Mycoplasma	87	Many pathogenic; require sterols; facultative aerobes	23–41	750	+
Anaeroplasma	4	May or may not require sterols; obligate anaerobes; degrade starch, producing acetic, lactic, and formic acids plus ethanol and CO_2; inhibited by thallium acetate	29	1500	+
Spiroplasma	11	Spiral to corkscrew-shaped cells; associated with various phyto-pathogenic (plant disease) conditions	25–31	1500	–
Ureaplasma	5	Coccoid cell; occasional clusters and short chains; growth optimal at pH 6; strong urease reaction; associated with certain urinary tract infections in humans; inhibited by thallium acetate	27–30	750	–
II. Do not require sterols					
Acholeplasma	11	Facultative aerobes	27–36	1500	+
Asteroleplasma	1	Obligate anaerobe; isolated from the bovine rumen	40	1500	+
Thermoplasma	1	Thermophilic, acidophilic archaeon found in heated coal refuse sites; no sterol requirement; DNA contains histones	46	1500	+

Phylogenetically, all mycoplasmas (except for Thermoplasma*) are members of the "low GC" subdivision of the Gram-positive Bacteria (see Section 18.6).*

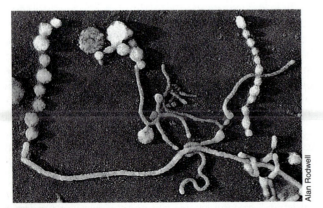

FIGURE 19.90 Electron micrograph of a metal-shadowed preparation of *Mycoplasma mycoides*. Note the coccoid and hyphalike elements. The average diameter of cells in chains is about 0.5 μm.

ded in the medium, and the fibrous nature of the agar gel seems to affect the division process, perhaps by promoting separation of units from the growing mass. Colonies of mycoplasmas on agar exhibit a characteristic "fried-egg" appearance because of the formation of a dense central core, which penetrates downward into the agar, surrounded by a circular spreading area that is lighter in color (Figure 19.92). Growth of mycoplasmas is not inhibited by penicillin, cycloserine, or other antibiotics that inhibit cell wall synthesis, but the organisms are as sensitive as other bacteria to antibiotics that act on targets other than the cell wall. Use is made of the natural penicillin resistance of mycoplasmas in preparing selective media for their isolation from natural materials. The culture media used for the growth of most mycoplasmas have usually been quite complex. Growth is poor or absent even in complex

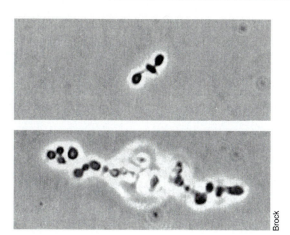

FIGURE 19.91 Photomicrograph by phase contrast microscopy of a mycoplasma culture, showing irregular cell arrangement.

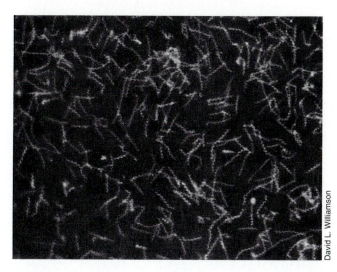

FIGURE 19.93 Dark field micrograph of the "sex ratio" spiroplasma removed from the hemolymph of the fly, *Drosophila pseudoobscura*. Female flies infected with the sex ratio spiroplasma bear only female progeny. Individual spiroplasma cells are about 0.15 μm in diameter.

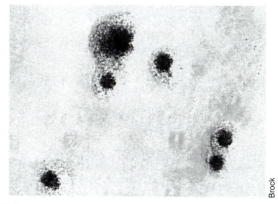

FIGURE 19.92 Typical "fried egg" appearance of mycoplasma colonies on agar. The colonies are around 0.5 mm in diameter.

yeast extract-peptone-beef heart infusion media unless fresh serum or ascitic fluid is added. The main constituents provided by these two adjuncts are unsaturated fatty acids and sterols. Some mycoplasmas can be cultivated on relatively simple media, however, and synthetic media have been developed for some strains. Most mycoplasmas use carbohydrates as energy sources, and require a range of vitamins, amino acids, purines, and pyrimidines as growth factors. The energy metabolism of mycoplasmas is not unique. Some species are oxidative, possessing a cytochrome system and making ATP by electron-transport phosphorylation. Other species resemble the lactic acid bacteria in being strictly fermentative, producing energy by substrate-level phosphorylation and yielding lactic acid as the final product of sugar fermentation. Members of the genus *Anaeroplasma* are obligate anaerobes which ferment glucose or starch to a variety of acidic products.

Spiroplasma

The genus *Spiroplasma* consists of pleomorphic cells, spherical or slightly ovoid, which are often helical or spiral in shape (Figure 19.93). Although they lack a cell wall and flagella, they are motile by means of a rotary (screw) motion or a slow undulation. Intracellular

fibrils that are thought to play a role in motility have been demonstrated. The organism has been isolated from ticks, the hemolymph and gut of insects, vascular plant fluids and insects that feed on fluids, and from the surfaces of flowers and other plant parts. *S. citri* has been isolated from the leaves of citrus plants, where it causes a disease called *citrus stubborn disease* and from corn plants suffering from *corn stunt disease*. A number of other mycoplasma-like bodies have been detected in diseased plants by electron microscopy, which indicates that a large group of plant-associated mycoplasmas may exist. Four species of *Spiroplasma* are recognized that cause a variety of animal diseases such as *honeybee spiroplasmosis, suckling mouse cataract disease*, and *lethargy disease* of the beetle *Melolontha*.

19.28   High GC Gram-Positive Bacteria: "Actinomycetes"

An extremely large variety of bacteria falls under this heading, as evidenced by the entire volume (Volume IV) of *Bergey's Manual*, which is devoted to the filamentous actinomycetes, and major portions of Volume II devoted to the rod-shaped relatives of this group. Phylogenetically, the actinomycetes form a subdivision of Gram-positive Bacteria, distinct from the endosporeformers and Gram-positive cocci and asporogenous rods. Despite great morphological variability, the actinomycetes form a tight phylogenetic unit and most representatives have GC ratios in the 60s and 70s mole percent. There are considerable difficulties in drawing clear-cut distinctions between various genera of actinomycetes. All of these organisms show a few common features: they are Gram-positive, rod-shaped to filamentous, and generally nonmotile in the vegetative phase (although motile stages are known). A continuum exists from simple rod-shaped organisms to rod-shaped organisms which occasionally grow in a filamentous manner, to strictly filamen-

Table 19.42 Actinomycetes and related genera (all Gram-positive)*

Major groups	DNA (mole % GC)
Coryneform group of bacteria: Rods, often club-shaped, morphologically variable; not acid-fast or filamentous; snapping cell division	
Corynebacterium: irregularly staining segments, sometimes granules; club-shaped swellings frequent; animal and plant pathogens, also soil saprophytes	51–65
Arthrobacter: coccus-rod morphogenesis; soil organisms	59–70
Cellulomonas: coryneform morphology; cellulose digested; facultative aerobe	71–73
Kurthia: rods with rounded ends occurring in chains; coccoid later	36–38
Brevibacterium: coccus-rod morphogenesis; cheese, skin	60–67
Propionic acid bacteria: anaerobic to aerotolerant; rods or filaments, branching	
Propionibacterium: nonmotile; anaerobic to aerotolerant; produce propionic acid and acetic acid; dairy products (Swiss cheese); skin, may be pathogenic	53–68
Eubacterium: obligate anaerobes; produce mixture of organic acids, including butyric, acetic, formic, and lactic; intestine, infections of soft tissue, soil; may be pathogenic; probably the predominant member of the intestinal flora	–
Obligate Anaerobes	
Bifidobacterium: smooth microcolony, no filaments; coryneform cells common; found in intestinal tract of breast-fed infants	55–67
Acetobacterium: homoacetogen; sediments and sewage	39–43
Butyrivibrio: curved rods; rumen	36–42
Thermoanaerobacter: rods; thermophilic found in hot springs	37–39
Actinomycetes: filamentous, often branching; highly diverse	
Group I. Actinomycetes: not acid-alcohol-fast; facultatively aerobic; mycelium not formed; branching filaments may be produced; rod, coccoid, or coryneform cells	
Actinomyces: anaerobic to facultatively aerobic; filamentous microcolony, but filaments transitory and fragment into coryneform cells; may be pathogenic for humans or animals; teeth	57–69
Other genera: *Arachnia, Bacterionema, Rothia, Agromyces*	
Group II. Mycobacteria: acid-alcohol-fast, filaments transitory	
Mycobacterium: pathogens, saprophytes; obligate aerobes; lipid content of cells and cell walls high; waxes, mycolic acids; simple nutrition; growth slow; tuberculosis, leprosy, granulomas, avian tuberculosis; also soil organisms; hydrocarbon oxidizers	62–70
Group III. Nitrogen-fixing actinomycetes: nitrogen-fixing symbionts of plants; true mycelium produced	
Frankia: forms nodules of two types on various plant roots; probably microaerophilic; grows slowly; fixes N$_2$	–
Group IV. Actinoplanes: true mycelium produced; spores formed, borne inside sporangia	
Actinoplanes, Streptosporangium	69–71
Group V. Dermatophilus: mycelial filaments divide transversely, and in at least two longitudinal planes, to form masses of motile, coccoid elements; aerial mycelium absent; occasionally responsible for epidermal infections	
Dermatophilus, Geodermatophilus	–
Group VI. Nocardias: mycelial filaments commonly fragment to form coccoid or elongate elements; aerial spores occasionally produced; sometimes acid-alcohol-fast	
Nocardia: common soil organisms; obligate aerobes; many hydrocarbon utilizers	61–72
Rhodococcus: soil saprophytes, also common in gut of various insects	59–69
Group VII. Streptomycetes: mycelium remains intact, abundant aerial mycelium and long spore chains	69–75
Streptomyces: Nearly 500 recognized species, many produce antibiotics	
Other genera (differentiated morphologically): *Streptoverticillium, Sporichthya, Microcellobosporia, Kitasatoa, Chainia*	67–73
Group VIII. Micromonosporas: mycelium remains intact; spores formed singly, in pairs, or short chains; several thermophilic; saprophytes found in soil, rotting plant debris; one species produces endospores	
Micromonospora, Thermoactinomyces, Thermomonospora	54–79

Phylogenetically, all species (except for Acetobacterium, Butyrivibrio, *and* Thermoanaerobacter) *fall into the "high GC" subdivision of the Gram-positive Bacteria (see Section 18.6).*

tous forms. Table 19.42 provides an overview of this group. In the following sections, we discuss some of the more interesting and important genera.

19.29   Coryneform Bacteria

The coryneform bacteria are Gram-positive, aerobic, nonmotile, rod-shaped organisms that have the characteristic of forming irregular-shaped, club-shaped, or V-shaped cell arrangements during normal growth. V-shaped cell groups arise as a result of a snapping movement that occurs just after cell division (called post-fission snapping movement or, simply, *snapping division*) (Figure 19.94). **Snapping division** has been shown to occur in one species because the cell wall consists of two layers; only the inner layer participates in cross-wall formation, so that after the cross-wall is formed, the two daughter cells remain attached by the outer layer of the cell wall. Localized rupture of this outer layer on one side results in a bending of the two cells away from the ruptured side (Figure 19.95), and thus development of V-shaped forms.

The main genera of coryneform bacteria are *Corynebacterium* and *Arthrobacter*. The genus *Corynebacterium* consists of an extremely diverse group of bacteria, including animal and plant pathogens as well as saprophytes. The genus *Arthrobacter*, consisting primarily of soil organisms, is distinguished from *Corynebacterium* on the basis of a cycle of development in *Arthrobacter* involving conversion from rod to sphere and back to rod again (Figure 19.96). However, some corynebacteria are pleomorphic and form coccoid elements during growth, so that the distinction between the two genera on the basis of life cycle is not absolute. The *Corynebacterium* cell frequently has a swollen end, so that it has a club-shaped appearance (hence the name of the genus: *koryne* is the Greek word for "club"), whereas *Arthrobacter* is less commonly club shaped.

Organisms of the genus *Arthrobacter* are among the most common of all soil bacteria. They are remarkably resistant to desiccation and starvation, despite the fact that they do not form spores or other resting cells. Arthrobacters are a heterogeneous group that have considerable nutritional versatility, and strains

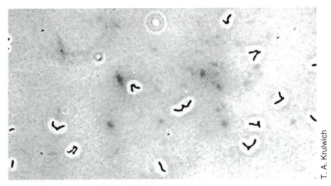

FIGURE 19.94 Photomicrograph of characteristic V-shaped cell groups in *Arthrobacter crystallopoietes*, resulting from snapping division. Cells are about 0.9 μm in diameter.

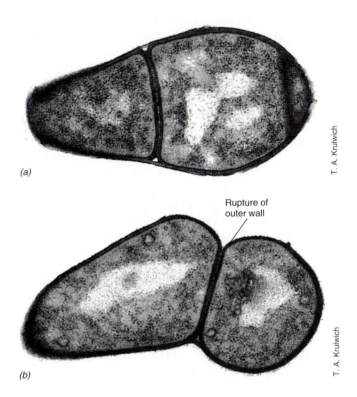

FIGURE 19.95 Electron micrograph of cell division in *Arthrobacter crystallopoietes*, illustrating how snapping division and V-shaped cell groups arise. (a) Before rupture of the outer cell-wall layer. (b) After rupture of the outer layer on one side. Cells are 0.9–1 μm in diameter.

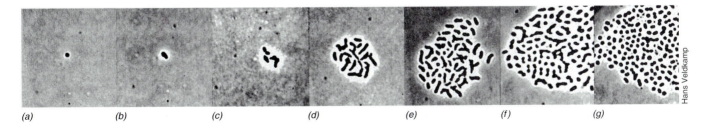

FIGURE 19.96 Stages in the life cycle of *Arthrobacter globiformis* as observed in slide culture; (a) single coccoid element; (b–e) conversion to rod and growth of microcolony consisting predominantly of rods; (f–g) conversion of rods to coccoid forms. Cells are about 0.9 μm in diameter.

have been isolated that decompose herbicides, caffeine, nicotine, phenol, and other unusual organic compounds.

19.30   Propionic Acid Bacteria

The propionic acid bacteria (genus *Propionibacterium*) were first discovered as inhabitants of Swiss (Emmentaler) cheese, where their fermentative production of CO_2 produces the characteristic holes; the presence of propionic acid is at least partly responsible for the unique flavor of the cheese. Although this acid is produced by some other bacteria, its production in large amounts by the propionic acid bacteria is a distinguishing characteristic of the genus. The bacteria in this group are Gram-positive, pleomorphic, nonsporulating rods, nonmotile and anaerobic. They ferment lactic acid, carbohydrates, and polyhydroxy alcohols, producing propionic acid, succinic acid, acetic acid, and CO_2. Their nutritional requirements are complex, and they usually grow rather slowly. In some taxonomic schemes, the facultatively aerobic coryneforms are also classified as propionic acid bacteria.

The enzymatic reactions leading from glucose to propionic acid are of interest (Figure 19.97). The initial catabolism of glucose to pyruvate follows the Embden-Meyerhof pathway as in the lactic acid bacteria, but the NADH formed is reoxidized as one part of a cycle in which *propionic acid* is formed. Pyruvate accepts a carboxyl group from methylmalonyl-CoA by a transcarboxylase reaction, leading to the formation of oxalacetate and propionyl-CoA. The latter substance reacts with succinate in a step catalyzed by a CoA transferase, producing succinyl-CoA and propionate. The succinyl-CoA is then isomerized to methylmalonyl-CoA, and the cycle is complete (Figure 19.97). Reoxidation of NADH occurs in the steps between oxalacetate and succinate, and the oxidation-reduction balance is restored.

Most propionic acid bacteria also ferment lactate with the production of propionate, acetate, and CO_2. The anaerobic fermentation of lactic acid to propionate is of interest because lactic acid itself is an end product of fermentation for many bacteria. The propionic acid bacteria are thus able to obtain energy anaerobically from a substance that other bacteria are producing by carrying out a secondary fermentation.

It is the fermentation of lactate to propionate that is important in Swiss cheese manufacture. The starter culture consists of a mixture of homofermentative streptococci and lactobacilli, plus propionic acid bacteria. The initial fermentation of lactose to lactic acid during formation of the curd is carried out by the homofermentative organisms. After the curd (protein and fat) has been drained, the propionic acid bacteria develop rapidly and usually reach numbers of 10^8 per gram by the time the cheese is two months old. Swiss cheese "eyes" (holes) are formed by the accumulation of CO_2, the gas diffusing through the curd and gathering at weak points.

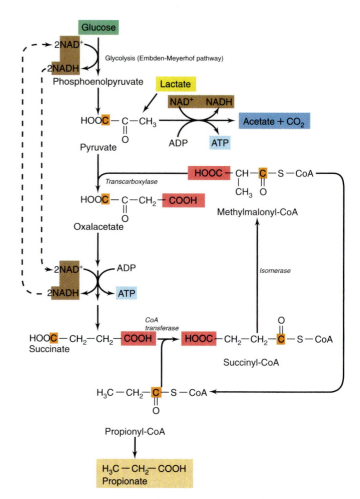

FIGURE 19.97 The formation of propionic acid by *Propionibacterium*. Either lactate, produced by the fermentative activities of other bacteria, or glucose can serve as starting substrate in the propionate fermentation. ATP synthesis is associated with electron transport reactions occurring during the formation of succinate and by substrate level phosphorylation in the production of acetate.

Propionigenium

A second bacterium, quite unrelated to *Propionibacterium*, also produces propionic acid fermentatively. *Propionigenium* is a Gram-negative, strictly anaerobic bacterium that ferments *succinate* to propionate and CO_2:

$$\text{Succinate} + H_2O \rightarrow \text{Propionate} + HCO_3^-$$
$$\Delta G^{0'} = -20.5 \text{ kJ/reaction}$$

We discussed the bioenergetics of succinate fermentation by *Propionigenium* in Section 16.19, pointing out that this fermentation is unusual because of the role Na^+ plays in establishing an ion gradient that ultimately drives ATP synthesis. *Propionigenium* also grows on fumarate, malate, aspartate, oxaloacetate, and pyruvate as sole energy sources, but will not ferment sugars or carry out anaerobic respiration linked to nitrate, sulfate, or other potential electron acceptors. In nature, *Propionigenium* presumably specializes in succinate fermentation, receiving a steady supply of substrate from other fermentations or from organisms that produce succinate during anaerobic respiration with fumarate as electron acceptor.

19.31 *Mycobacterium*

The genus *Mycobacterium* consists of rod-shaped organisms, which at some stage of their growth cycle possess the distinctive staining property called **acid–alcohol fastness**. This property is due to the presence on the surface of the mycobacterial cell of unique lipid components called **mycolic acids** and is found only in the genus *Mycobacterium*. First discovered by Robert Koch during his pioneering investigations on tuberculosis, this unique staining property permitted the identification of the organism in tuberculous lesions; it has subsequently proved to be of great taxonomic use in defining the genus *Mycobacterium*.

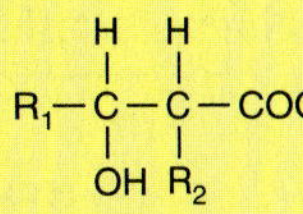

Acid–alcohol fastness (*Ziehl–Neelsen stain*)

A mixture of the dye basic fuchsin and phenol is used in the acid–alcohol staining procedure, the stain being driven into the cells by slow heating of the microscope slide to the steaming point for two to three minutes. The role of the phenol is to enhance penetration of the fuchsin into the lipids. After washing in distilled water, the preparation is decolorized with acid alcohol (3 percent HCl in 95 percent ethanol); this procedure removes the fuchsin dye from other organisms, but it is retained by the mycobacteria. After another wash in water, a final counterstain of methylene blue is used. Acid–alcohol-fast organisms on the final preparation appear *red* whereas the background and nonacid-alcohol-fast organisms appear *blue*.

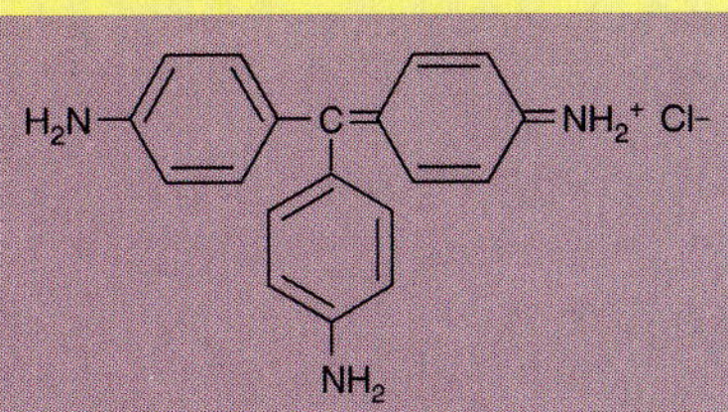

As noted, the key component necessary for acid-alcohol fastness is a unique lipid fraction of mycobacterial cells called *mycolic acid*. Mycolic acid is actually a group of complex branched-chain hydroxy lipids with the overall structure shown in Figure 19.98a. The carboxylic acid group of the mycolic acid must be free (unesterified), and it reacts on a one-to-one basis with the fuchsin dye (Figure 19.98b). The mycolic acid is complexed to the peptidoglycan of the mycobacterial wall, and this complex somehow prevents approach of the acid–alcohol solvent during the decolorization step. It was also first demonstrated by Koch that disruption of cellular integrity destroys the acid-alcohol-fast property; thus cellular integrity is a necessary prerequisite of this property.

Mycobacteria are not readily stained by the Gram method because of the high surface lipid content, but if the lipoidal portion of the cell is removed with alkaline ethanol (1 percent KOH in absolute ethanol), the intact cell remaining is non-acid–alcohol fast but instead is Gram-positive. However, if the lipoidal portion is not removed, then the cells are resistant to decolorization by the Gram procedure even when stained with crystal violet alone (in the absence of io-

FIGURE 19.98 Structure of (a) mycolic acid and (b) basic fuchsin, the dye used in the acid–alcohol fast stain. The fuchsin dye probably combines with the mycolic acid via ionic bonds between COO^- and NH_2^+.

dine), whereas iodine is essential for the conventional Gram-staining procedure (see Figure 3.5). *Mycobacterium* can thus be considered a true Gram-positive Bacterium and phylogenetic studies have confirmed this. *Mycobacterium* groups with other "high GC" actinomycetes by 16S ribosomal RNA sequencing (see Section 18.6).

Characteristics of mycobacteria

Mycobacteria are generally rather pleomorphic, and may undergo branching or filamentous growth. However, in contrast to the actinomycetes, filaments of the mycobacteria become fragmented into rods or coccoid elements upon slight disturbance; a true mycelium is not formed. In general, mycobacteria can be separated into two major groups, *slow growers* and *fast growers* (Table 19.43). *M. tuberculosis* is a typical slow grower, and visible colonies are produced from dilute inoculum only after days to weeks of incubation. (The reason Koch was successful in first isolating *M. tuberculosis* was that he waited long enough after inoculating media.) When growing on solid media, mycobacteria generally form tight, compact, often wrinkled colonies, the organisms piling up in a mass rather than spreading out over the surface of the agar (Figure 19.99a). This formation is probably due to the high lipid content and hydrophobic nature of the cell surface. The characteristic slow growth of most mycobacteria is probably also due, at least in part, to the *hydrophobic* character of the cell surface, which renders the cells strongly impermeable to nutrients; species having less lipid grow considerably more rapidly.

For the most part, mycobacteria have relatively simple nutritional requirements. Growth often occurs in simple mineral salts medium with ammonium as nitrogen source and glycerol or acetate as sole carbon source and electron donor. Growth of *M. tuberculosis* is stimulated by lipids and fatty acids, and egg yolk (a

Table 19.43   Some characteristics of representative mycobacteria

Species	Growth in 5% NaCl	Nitrate reduction	Growth at 45°C	Human pathogen	Pigmentation
Slow-growing species					
M. tuberculosis	−	+	−	+	None
M. avium	−	−	−	+	Old colonies pigmented (see Figure 19.99c)
M. bovis	−	−	+	+	None
M. kansasii	−	+	−	+	Photochromogenic
Fast-growing species					
M. smegmatis	+	+	+	−	None
M. phlei	+	+	+	−	Pigmented
M. chelonae	+	−	−	+	None
M. parafortuitum	+	+	−	−	Photochromogenic

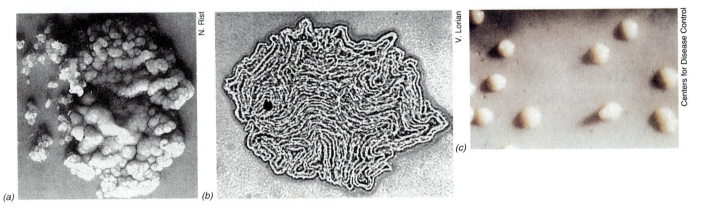

(a)   N. Rist   *(b)*   V. Lorian   *(c)*   Centers for Disease Control

FIGURE 19.99 Characteristic colony morphology of mycobacteria. (a) *Mycobacterium tuberculosis*, showing the compact, wrinkled appearance of the colony. The colony is about 7 mm in diameter. (b) A colony of virulent *M. tuberculosis* at an early stage, showing the characteristic cordlike growth. Individual cells are about 0.5 μm in diameter. (c) Colonies of *M. avium* from a strain of this organism isolated as an opportunistic pathogen from an AIDS patient.

good source of lipids) is often added to culture media to achieve more luxuriant growth. A glycerol-whole egg medium (Lowenstein-Jensen medium) is often used in primary isolation of *M. tuberculosis* from pathological materials. Perhaps because of the high lipid content of its cell walls, *M. tuberculosis* is able to resist such chemical agents as alkali or phenol for considerable periods of time, and this property is used in the selective isolation of the organism from sputum and other materials that are grossly contaminated. The sputum is first treated with 1 *N* NaOH for 30 minutes, then neutralized and streaked onto isolation medium.

A characteristic of many mycobacteria is their ability to form yellow carotenoid pigments (Figure 19.99c). Based on pigmentation, the mycobacteria can be classified into three groups: nonpigmented (including *M. tuberculosis*, *M. bovis*); forming pigment only when cultured in the light, a property called **photochromogenesis** (including *M. kansasii*, *M. marinum*); and forming pigment even when cultured in the dark, a property called **scotochromogenesis** (including *M. gordonae*, *M. paraffinicum*). The property of photochromogenesis is of some interest and has been extensively studied. This property is not unique to mycobacteria, as it also occurs in a number of fungi.

Photoinduction of carotenoid formation involves short-wavelength (blue) light, and only occurs in the presence of O_2. The evidence indicates that the critical event in photoinduction is a light-catalyzed oxidation event, and it appears that one of the early enzymes in carotenoid biosynthesis is photoinduced. As with other carotenoid-containing bacteria, it has been suggested that carotenoids protect mycobacteria against oxidative damage involving singlet oxygen (see Section 9.12).

The cell walls of mycobacteria contain a peptidoglycan that is covalently bound to an arabinose-galactose-mycolic acid polymer, and it is this lipid-polysaccharide-peptidoglycan complex which confers the hydrophobic character to the mycobacterial cell surface. In addition to this lipid component, mycobacteria form a wide variety of other lipids, providing lipid chemists with a fascinating amount of material.

Virulence of *M. tuberculosis* cultures has been correlated with the formation of long cordlike structures (Figure 19.99) on agar or in liquid medium, due to side-to-side aggregation and intertwining of long chains of bacteria. Growth in cords reflects the presence on the cell surface of a characteristic lipid, the **cord factor**, which is a glycolipid (Figure 19.100). The

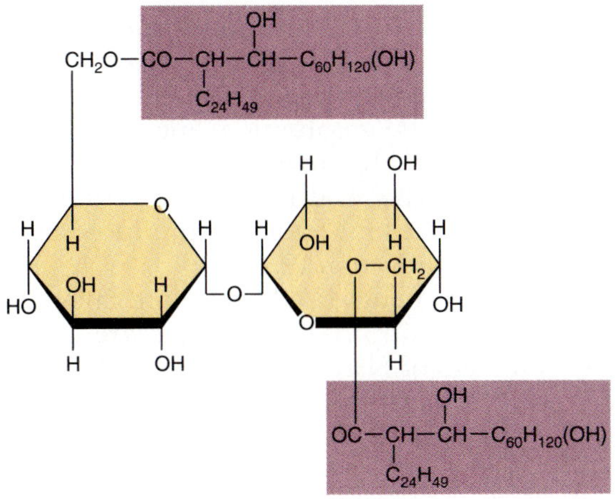

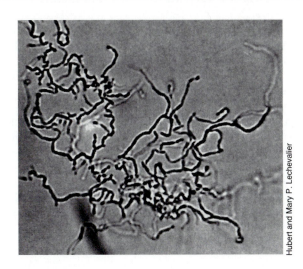

FIGURE 19.100 Structure of cord factor, a mycobacterial glycolipid: 6,6'-dimycolyltrehalose. The two identical long-chain dialcohol groups are shown in color.

FIGURE 19.101 A young colony of an actinomycete of the genus, *Nocardia*, showing typical filamentous cellular structure (mycelium). Each filament is about 0.8–1 µm in diameter.

pathogenesis of the disease tuberculosis was discussed in detail in Section 15.3.

19.32 Filamentous Actinomycetes

The actinomycetes are a large group of filamentous Bacteria, usually Gram-positive, which form branching filaments. As a result of successful growth and branching, a ramifying network of filaments is formed, called a *mycelium* (Figure 19.101). Although it is of bacterial dimensions, the mycelium is in some ways analogous to the mycelium formed by the filamentous fungi. Most actinomycetes form spores; the manner of spore formation varies and is used in separating subgroups, as outlined in Table 19.42. The genus *Mycobacterium*, members of which often show a tendency to form branches, is also placed in the actinomycetes by some taxonomists. The DNA base compositions of all members of the actinomycetes fall within a relatively narrow range of 63 to 78 percent GC. Organisms at the upper end of this range have the highest GC percentage of any bacteria known. Phylogenetically, the filamentous actinomycetes form a coherent group; thus the mycelial spore-forming habit is of both phylogenetic as well as taxonomic importance. In the present discussion we concentrate on the genus *Streptomyces*.

Streptomyces

Streptomyces is a genus represented by a large number of species and varieties. Over 500 species of *Streptomyces* are recognized by *Bergey's Manual*, although GC base ratios cluster tightly between 69 and 73 mole percent. *Streptomyces* filaments are usually 0.5–1.0 µm in diameter and of indefinite length, and often lack cross walls in the vegetative phase. Growth occurs at the tips of the filaments and is often accompanied by branching so that the vegetative phase consists of a complex, tightly woven matrix, resulting in a compact, convoluted colony. As the colony ages, characteristic aerial filaments called *sporophores* are formed, which project above the surface of the colony and give rise to spores (Figure 19.102). *Streptomyces* spores, usually called **conidia**, are not related in any way to the endospores of *Bacillus* and *Clostridium* since the streptomycete spores are produced simply by the formation of cross walls in the multinucleate sporophores followed by separation of the individual cells directly into spores (Figure 19.103). The surface of the conidial wall often has convoluted projections, the nature of which is characteristic of each species. Differences in shape and arrangement of aerial filaments and spore-bearing structures of various species are among the fundamental features used in separating the *Strepto-*

Sporophores

Chain of conidia

FIGURE 19.102
Photomicrographs of several spore-bearing structures of actinomycetes. (a) *Streptomyces*, a monoverticillate type. (b) *Streptomyces*, a spiral type. Filaments are about 0.8 µm wide in both cases.

myces groups (Figure 19.104). The conidia and sporophores are often pigmented and contribute a characteristic color to the mature colony; in addition, pigments sometimes are produced by the substrate mycelium and contribute to the final color of the colony (Figure 19.105*a*). The dusty appearance of the mature colony, its compact nature, and its color make detection of *Streptomyces* colonies on agar plates relatively easy (Figure 19.105*b*).

Ecology and isolation of *Streptomyces*

Although a few streptomycetes can be found in aquatic habitats, they are primarily *soil* organisms. In fact, the characteristic earthy odor of soil is caused by the production of a series of streptomycete metabolites called **geosmins**. These substances are sesquiterpenoid compounds, unsaturated ring compounds of carbon, oxygen, and hydrogen. The geosmin first discovered has the chemical name *trans*-1,10-dimethyl-*trans*-9-decalol. Geosmins are also produced by some cyanobacteria (see Section 19.2).

Alkaline and neutral soils are more favorable for the development of *Streptomyces* than are acid soils. Higher numbers of *Streptomyces* are usually found in well-drained soils (such as sandy loams, or soils covering limestone), and there is some evidence to sug-

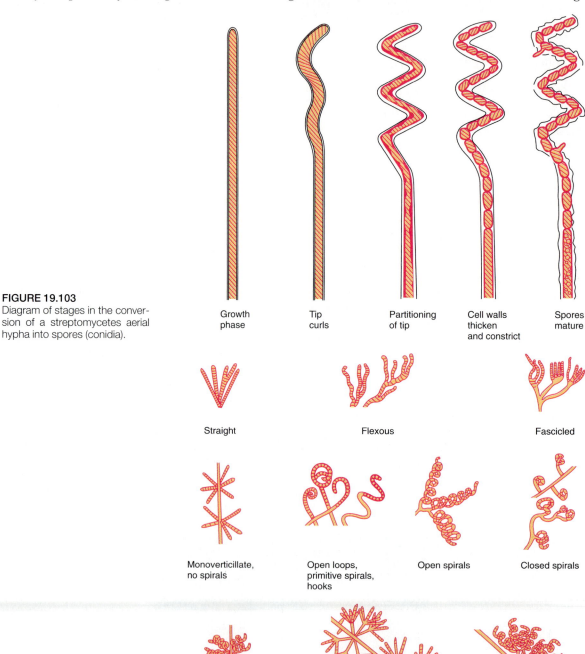

FIGURE 19.103
Diagram of stages in the conversion of a streptomycetes aerial hypha into spores (conidia).

Growth phase

Tip curls

Partitioning of tip

Cell walls thicken and constrict

Spores mature

Straight

Flexous

Fascicled

Monoverticillate, no spirals

Open loops, primitive spirals, hooks

Open spirals

Closed spirals

FIGURE 19.104
Various types of spore-bearing structures in the streptomycetes.

Monoverticillate, with spirals

Biverticillate, no spirals

Biverticillate, with spirals

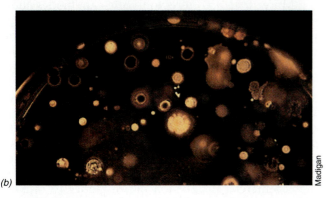

Alma Dietz

Madigan

FIGURE 19.105 *Streptomyces.* (a) Typical appearance of a strepto-mycete growing on agar slants. Varying degrees of pigmentation are shown on different culture media. The coloration results from both the production of soluble pigments that diffuse into the agar and from the production of pigmented spores. (b) Colonies of *Streptomyces* and other soil bacteria derived from spreading a soil dilution on a starch agar plate. The *Streptomyces* colonies are of various colors (several black *Streptomyces* colonies are in the foreground) but can easily be identified by their opaque, rough, and nonspreading morphology.

gest that *Streptomyces* require a lower water potential for growth than many other soil bacteria. Isolation of large numbers of *Streptomyces* from soil is relatively easy: a suspension of soil in sterile water is diluted and spread on selective agar medium, and the plates are incubated at 25°C. Media often selective for *Streptomyces* contain the usual assortment of inorganic salts to which starch, asparagine, or calcium malate is added as a carbon source and undigested casein or

potassium nitrate as nitrogen source. After incubation for five to seven days, the plates are examined for the presence of the characteristic *Streptomyces* colonies (Figure 19.105b) and spores of interesting colonies can be streaked and pure cultures isolated.

Nutritionally, the streptomycetes are quite versa-tile. Growth-factor requirements are rare, and a wide variety of carbon sources, such as sugars, alcohols, or-ganic acids, amino acids, and some aromatic com-pounds, can be utilized. Most isolates produce extra-cellular hydrolytic enzymes that permit utilization of polysaccharides (starch, cellulose, hemicellulose), pro-teins, and fats, and some strains can use hydrocar-bons, lignin, tannin, or even rubber. *Streptomyces* can often be obtained by spreading a soil dilution on an agar medium containing polymers such as casein and starch (Figure 19.105b). A single isolate may be able to break down over 50 distinct carbon sources. Strepto-mycetes are strict aerobes, whose growth in liquid cul-ture is usually markedly stimulated by forced aera-tion. Sporulation usually does not take place in liquid culture but only when the organism is growing on the surface of agar or another solid substrate; it can occur, however, when organisms form a pellicle on the sur-face of an unshaken liquid culture.

Antibiotics of *Streptomyces*

Perhaps the most striking property of the strepto-mycetes is the extent to which they produce **antibi-otics** (Table 19.44). Evidence for antibiotic production is often seen on the agar plates used in the initial isola-tion of *Streptomyces*: adjacent colonies of other bacteria show zones of inhibition (Figure 19.106). In some studies close to 50 percent of all *Streptomyces* isolated have proved to be antibiotic producers. Because of the great economic and medical importance of many streptomycete antibiotics, an enormous amount of work has been done on these producers. Over 500 dis-tinct antibiotic substances have been shown to be pro-

Table 19.44   Some common antibiotics synthesized by species of *Streptomyces*

Chemical class	Common name	Produced by	Active against*
Aminoglycosides	Streptomycin	*S. griseus*	Most Gram-negative Bacteria
	Spectinomycin	*Streptomyces spp.*	*M. tuberculosis*, penicillinase-producing *N. gonorrhoeae*
	Neomycin	*S. fradiae*	Broad spectrum, usually used in topical applications due to toxicity
Tetracyclines	Tetracycline	*S. aureofaciens*	Broad spectrum, Gram-positive and Gram-negative Bacteria, rickettsias and chlamydias, *Mycoplasma*
	Chlortetracycline	*S. aureofaciens*	As for tetracycline
Macrolides	Erythromycin	*S. erythreus*	Most Gram-positive Bacteria, frequently used in place of penicillin, *Legionella*
	Clindamycin	*S. lincolnensis*	Effective against obligate anaerobes, especially *Bacteroides fragilis*
Polyenes	Nystatin	*S. noursei*	Fungi, especially *Candida* infections
	Amphocetin B	*S. nodosus*	Fungi
None	Chloramphenicol	*S. venezuelae*	Broad spectrum; drug of choice for typhoid fever

Most antibiotics are effective against several different Bacteria. The entries in this column refer to the most frequent clinical application of a given antibiotic.

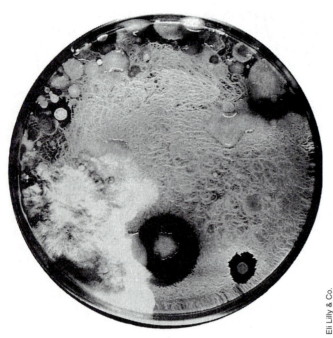

FIGURE 19.106 Antibiotic action of soil microorganisms on a crowded plate. The smaller colonies surrounded by inhibition zones are streptomycetes; the larger, spreading colonies are *Bacillus* species.

Eli Lilly & Co.

The genetics of antibiotic production is now an active area of research and was briefly discussed in Sections 10.7 and 10.8. Interestingly, several *Streptomyces* sp. have been found to contain large *linear* plasmids of over 500 kilobases in length and some of these contain genes for antibiotic biosynthesis. Linear plasmids have occasionally been found in other bacteria but are apparently widespread among *Streptomyces* species.

More than 50 streptomycete antibiotics have found practical application in human and veterinary medicine, agriculture, and industry. Some of the more common antibiotics of *Streptomyces* origin are listed in Table 19.44. They are grouped into classes based on the chemical structure of the parent molecule. The search for new streptomycete antibiotics continues, since many infectious diseases are still not adequately controlled by existing antibiotics. Also, the development of antibiotic-resistant strains requires the continual discovery of new agents. We discussed the antibiotic industry in general and the role of *Streptomyces* in the commercial production of antibiotics in Sections 10.7 and 10.8. Ironically, despite the extensive work on antibiotic-producing streptomycetes, and the fact that the antibiotic industry is a multi-billion dollar enterprise, the ecology of *Streptomyces* remains poorly understood. The ecological rationale for why antibiotics are produced is not clear. However, one hypothesis for why *Streptomyces* species produce antibiotics is that antibiotic production, which is linked to sporulation (a process itself triggered by nutrient depletion), serves to inhibit the growth of other organisms competing with differentiating *Streptomyces* cells for limiting nutrients. Whatever the ecological reasoning for antibiotic production by *Streptomyces* sp. and other antibiotic producers, humans have reaped the benefits of this process and it has revolutionized clinical medicine.

duced by streptomycetes, and a large number of these have been studied chemically. Some organisms produce more than one antibiotic, and often the several kinds produced by one organism are not even chemically related. The same antibiotic may be formed by different species found in widely scattered parts of the world. A change in nutrition of the organism may result in a change in the nature of the antibiotic produced. The organisms are usually resistant to their own antibiotics, but they may be sensitive to antibiotics produced by other streptomycetes.

Supplementary Readings

Balows, A., H. G. Trüper, M. Dworkin, W. Harder, and **K.-H. Schleifer** (eds.) 1992. *The Prokaryotes,* 2nd edition. Published in four volumes. Springer-Verlag, New York. The most complete reference on the characteristics of bacteria. Also includes formulation of various media and isolation procedures for virtually every prokaryotic group known. Well illustrated.

Holt, J. G. (editor-in-chief). *Bergey's Manual of Systematic Bacteriology.* Williams and Wilkins, Baltimore, MD. Volume I, 1984. Gram-negative Bacteria of medical or industrial importance. Volume II, 1986. Gram-positive Bacteria of medical or industrial importance. Volume III, 1989. Other Gram-negative Bacteria, cyanobacteria, Archaea. Volume IV, 1989. Other Gram-positive Bacteria. This manual is a recognized authority on bacterial taxonomy. A good place to begin a literature survey on a specific bacterial group.

Lederberg, J. (ed.) 1992. *Encyclopedia of Microbiology.* Academic Press, San Diego, CA. A multivolume collection of articles on microorganisms, their habitats, metabolism, activities in nature, and human significance.

Schlegel, H. G., and **B. Bowien** (eds.) 1989. *Autotrophic Bacteria.* Science Tech Publishers, Madison, WI, and Springer-Verlag, Heidelberg. An excellent treatment of all autotrophic bacteria. Also has a nice historical survey.

Zehnder, A. J. B. (ed.) 1988. *Ecology of Anaerobic Microorganisms.* John Wiley and Sons, New York. A collection of chapters detailing the microbiology of the major groups of anaerobes: anoxygenic phototrophs, sulfate-reducers, methanogens, homoacetogens, denitrifyers, etc.

Archaea

$\mathbf{W}$e now consider members of the domain *Archaea*, previously known as the *archaebacteria*. We emphasized in Chapter 18 the profound evolutionary differences between Bacteria and Archaea. Despite the fact that they are prokaryotes, by molecular sequencing criteria Archaea are no more closely related to Bacteria than they are to Eukarya. Hence, we should expect some unusual phenotypes among the Archaea and indeed that is what we will see. However, before we begin our treatment of the classification of Archaea, we discuss some key structural properties of these organisms, focusing on cell walls and membranes, and then briefly consider the intermediary metabolism and genetics of Archaea, comparing and contrasting these systems with those of Bacteria and Eukarya.

20.1 Archaeal Membranes

We mentioned in Chapter 18 how membranes of Archaea lack fatty acids and instead have hydrocarbon moieties bonded to glycerol by *ether* (instead of ester) linkages (see Figure 20.1 for the chemical differences between ester and ether linkages). Glycerol *diethers* and diglycerol *tetraethers* (Figure 20.1) are the major classes of lipids present in Archaea. *Polar* forms of archaeal lipids include phospholipids, sulfolipids, and glycolipids (see Section 2.6 for a consideration of polar lipids). In addition, Archaea contain substantial amounts of *nonpolar* lipids, most of which are derivatives of the isoprenoid compound squalene (Figure 20.2). Methanogens and hyperthermophilic Archaea contain a variety of isoprenoid compounds ranging from C_{15} up to C_{30}. Literally dozens of other isoprenoid derivatives, including the carotenoids α- and β-carotene (Figure 20.2) and geranylgeraniol, have been

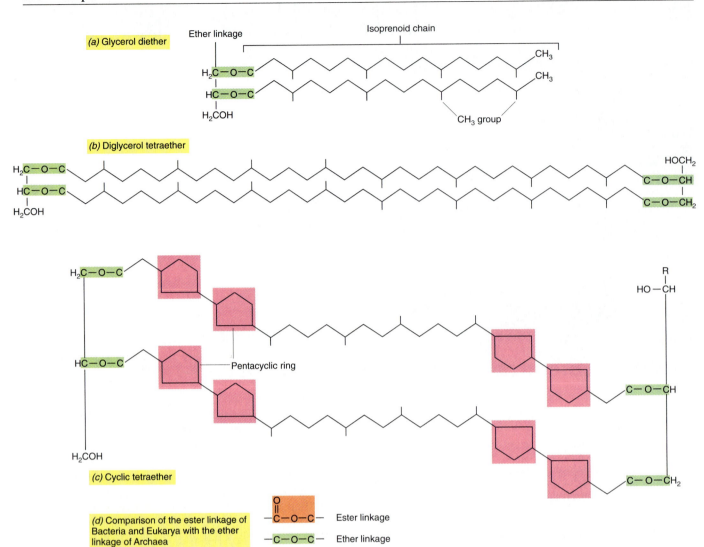

FIGURE 20.1 Major lipids of Archaea. (a) Glycerol diethers. (b) Diglycerol tetraethers. (c) Tetrapentacyclic diglycerol tetraethers. Note that in all cases, the hydrocarbon is attached to the glycerol by *ether* linkages. Hydrocarbon in (a) phytanyl; (b) and (c) dibiphytanyl. Note also how the introduction of pentacyclic rings in (c) shortens the length of the hydrocarbon chain. (d) Comparison of the ester linkage of Bacteria and Eukarya with the ether linkage of Archaea.

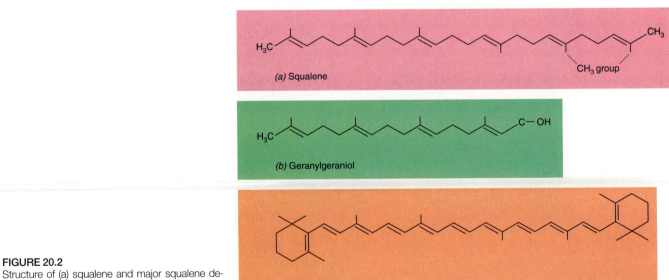

FIGURE 20.2
Structure of (a) squalene and major squalene derivatives, (b) geranylgeraniol, and (c) β-carotene, found in the lipids of extremely halophilic Archaea.

Miniglossary for Chapter 20

ACETOCLASTIS   splitting of acetate into CH_4 plus CO_2 by certain methanogens

ACETYL-COA PATHWAY   a pathway of autotrophic CO_2 fixation widespread in obligate anaerobes including methanogens, homoacetogens, and sulfate-reducing bacteria

BACTERIORHODOPSIN   a membrane protein containing retinal, a pigment produced by certain extreme halophiles and capable of light-mediated ATP synthesis

COENZYME M   a coenzyme involved in the terminal step of methanogenesis

CORRINOID   a porphyrin-like compound containing cobalt and involved in methyl transfers in methanogenesis

EXTREME HALOPHILE   an organism whose growth is dependent on large amounts (generally >10%) of NaCl

FACTOR$_{420}$ (F$_{420}$)   an electron donor involved in certain steps in the reduction of CO_2 to CH_4

HYPERTHERMOPHILE   a prokaryote with a growth temperature optimum of 80°C or greater

METHANOGEN   a methane-producing prokaryote

PHYTANYL   a branched-chain hydrocarbon containing 20 carbon atoms and commonly found in the lipids of Archaea

SULFATARA   a hot, sulfur-rich, generally acidic environment commonly inhabited by hyperthermophilic Archaea

---

identified in various Archaea.

By comparison with cells of Bacteria and Eukarya, membranes of Archaea are indeed chemically unique. However, archaeal membranes are *structurally* arranged in a fashion similar to membranes of Bacteria and Eukarya. That is, they contain *polar* inner and outer surfaces (glycerol molecules) and a *nonpolar* interior; as previously described (see Section 3.3), such structures tend to spontaneously form lipid bilayers. Glycerol diether lipids (Figure 20.1a) form a true bilayer typical of membranes made of fatty acids (Figure 20.3a). However, some Archaea lack diether lipids and instead contain mainly glycerol tetraether lipids (Figure 20.1b). How would membranes be constructed from these molecules? Tetraethers are arranged in Archaea to form lipid *monolayer* instead of lipid *bilayer* membranes (Figure 20.3). Monolayer membranes are the structural equivalent of lipid bilayers, the major difference being that in monolayer membranes the hydrophobic moieties pointing inward toward one another are *covalently linked*. Lipid monolayer membranes occur in certain methanogens and are widespread among hyperthermophilic Archaea.

Since the length of the hydrocarbon moiety of di- and tetraethers is fixed at 20 carbons or 40 carbons, respectively, it would appear at first glance that Archaea are limited in their ability to adjust the width of their membranes to varying physiological conditions (this would be analogous to changes in the length and degree of saturation of fatty acids in lipid bilayers of Bacteria grown at different temperatures, see Section 9.9). However, the length of archaeal tetraethers (and hence the thickness of the monolayer membrane formed from them) can be changed by the introduction of 5-membered (pentacyclic) rings into the hydrocarbon chain (see Figure 20.1). The synthesis or removal of pentacyclic rings does not change the *number* of carbon atoms in the lipid monolayer, only the *length* of the chain. This allows for some flexibility in Archaeal membrane composition and presumably affects chemical properties of the membrane.

The production of ether-linked lipids is a hallmark of Archaea—no archaeon has been shown to lack them. These structures are so distinctive that they

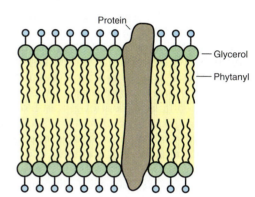

(a) Lipid bilayer

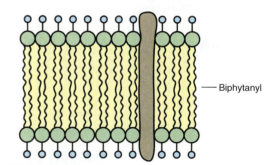

(b) Lipid monolayer

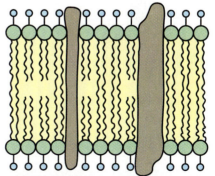

(c) Lipid monolayer/bilayer mixtures

FIGURE 20.3   (a) Glycerol diether lipid bilayer membrane structure. (b) Diglycerol tetraether lipid monolayer membrane. (c) Mixed membrane type.

have been used as biomarkers for detecting fossilized Archaea in micropaleontological studies of rocks, sediment cores, and other ancient materials. The chemical differences between archaeal and bacterial lipids once again emphasizes the major evolutionary distance between the two domains Archaea and Bacteria (see Chapter 18), despite the fact that structurally both groups consist of prokaryotic cells.

20.2 Archaeal Cell Walls

We have just emphasized the profound differences between the cell *membranes* of Archaea and Bacteria. What about cell walls? We know that Bacteria have either Gram-positive or Gram-negative cell walls and that cell walls of Bacteria contain peptidoglycan (see Section 3.5). By contrast, although cells of various Archaea may stain Gram-positive or Gram-negative, no archaeon has been shown to contain muramic acid and D-amino acids, the "signature molecules" of peptidoglycan, in their cell walls.

Among methanogenic and the extremely halophilic Archaea, where cell wall studies have been most extensive, several wall types have been described. *Methanobacterium*, *Methanobrevibacter*, and *Methanothermus* species contain a peptidoglycan-like material referred to as *pseudopeptidoglycan*. Pseudopeptidoglycan is composed of alternating repeats of two amino sugars, *N*-acetlyglucosamine and *N*-acetyltalosaminuronic acid (Figure 20.4). The latter compound is unique to Archaea while the former makes up one-half of the backbone of normal peptidoglycan.

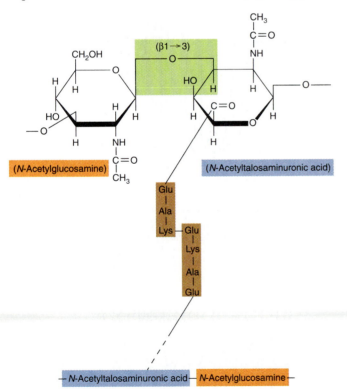

FIGURE 20.4 Structure of pseudopeptidoglycan, the cell wall polymer of *Methanobacterium* species. Note the resemblance to the structure of peptidoglycan shown in Figure 3.28.

The acetyltalosaminuronic acid residues of pseudopeptidoglycan are cross-linked by amino acids (as is peptidoglycan) but the amino acids are all of the L-form instead of the D-amino acids found in true peptidoglycan. The glycosidic linkage between amino sugars in pseudopeptidoglycan is a $(1 \rightarrow 3)$ bond instead of the $(1 \rightarrow 4)$ bond formed between acetylglucosamine and muramic acid residues of peptidoglycan; since these $1 \rightarrow 4$ glycosidic bonds are the ones broken by the enzyme *lysozyme* (see Section 3.5), methanogens (and all other Archaea) are resistant to the action of lysozyme.

Methanosarcina species contain a thick cell wall that stains Gram-positive and consists exclusively of *polysaccharide*. The major sugars present are galactosamine, glucuronic acid, and glucose; a considerable amount of acetate is also present. Phosphate and sulfate are absent. Although the polysaccharide wall of *Methanosarcina* is complex, the major structural elements suggest a relationship to chondroitin, a major component of the connective tissue of animal cells. Unlike chondroitin, however, the polysaccharide wall of *Methanosarcina* is not sulfated. The cell-wall of *Halococcus*, an extremely halophilic archaeon, is a sulfated polysaccharide which otherwise resembles that of *Methanosarcina*. The chemical composition of *Halococcus* cell walls is complex. Glucose, mannose, and galactose are present both as neutral sugars and in amino form. Uronic acids and acetate are abundant and most sugar residues are sulfated. A proposed structure for the *Halococcus* cell wall is shown in Figure 20.5.

Certain methanogens, the extremely halophilic *Halobacterium*, and the hyperthermophilic Archaea have cell walls made of *glycoprotein*. The carbohydrates include hexoses such as glucose, glucosamine, galactose, and mannose, and pentoses such as ribose and arabinose. In *Halobacterium* cell walls there is a great excess of acidic (negatively-charged) amino acids, especially aspartate, and these serve to balance the abundance of positive charges by the high concentration of sodium in the organism's environment (see Section 20.4). Cells of most *Halobacterium* species require 20–25 percent NaCl to maintain an intact cell wall. As the salt concentration is reduced to about 15 percent, *Halobacterium* cells (which are normally rod-shaped) begin to take on a coccoid shape, presumably due to cell wall components that are unstable at this low salt concentration. At even lower NaCl concentrations, the cells actually lyse.

A few methanogens lack carbohydrate in their cell walls and instead have walls consisting only of protein. For example, *Methanococcus* and *Methanomicrobium* cell walls consist of several distinct proteins, whereas *Methanospirillum* contains a single protein subunit wall which forms a sheath around the organism. The cell walls of hyperthermophilic Archaea consist of protein or glycoprotein. The cell wall of *Sulfolobus*, for example, is formed from a regular array of glycoprotein subunits (see Figure 20.20a). This complex is extremely tough because purified *Sulfolobus*

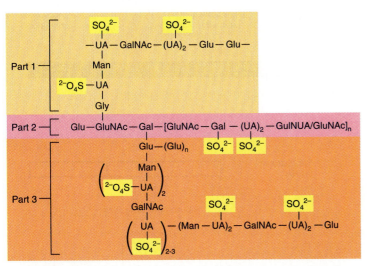

FIGURE 20.5 Cell wall structure of *Halococcus*, an extremely halophilic archaeon. The wall basically consists of a repeating three-part pattern. Note the abundance of sulfate groups. Abbreviations: UA, uronic acid; Glu, glucose; Gal, galactose; GluNAc, *N*-acetylglucosamine; GalNAc, *N*-acetylgalactosamine; Gly, glycine; GulNUA, *N*-acetylgulosaminuronic acid; Man, mannose.

cell walls remain intact in boiling detergent solutions! The cell walls of *Pyrodictium*, currently the most thermophilic of all known prokaryotes with a growth temperature maximum of 113°C (see Section 20.6), is also made of glycoprotein. The wall of *Desulfurococcus*, a sulfur-reducing hyperthermophile, consists only of protein. Pure polysaccharide cell walls have not been detected in any hyperthermophiles.

20.3 Metabolism and Genetics of Archaea

As we consider the various groups of Archaea we will occasionally encounter highly specialized metabolic pathways. However, we focus here on what is known concerning general catabolic and anabolic pathways in Archaea and consider gene structure and expression in these evolutionarily distinct prokaryotes.

Energy metabolism and central metabolic pathways in the Archaea

Several Archaea are chemoorganotrophic and thus use organic compounds as energy sources for growth. Catabolism of glucose in extremely halophilic and hyperthermophilic Archaea proceeds via slight modifications of the Entner–Duodoroff (E–D) pathway (see Section 19.15 for a description of this pathway in Bacteria). All enzymatic steps leading up to formation of the key intermediate of the E–D pathway, 2-keto-3-deoxy-6-phosphogluconate, are not the same in all Archaea. However, 2-keto-3-deoxy-6-phosphogluconate, or a nonphosphorylated derivative, is eventually produced during glucose catabolism, and the remaining steps of the E–D pathway in Archaea occur as for Bacteria which utilize this reaction sequence. By contrast, gluconeogenesis, the production of glucose from noncarbohydrate precursors, proceeds in Archaea by reversal of steps in the Embden–Meyerhof pathway (glycolysis, see Section 4.8).

Oxidative catabolic sequences in Archaea proceed through the tricarboxylic acid cycle (see Section 4.12) or some slight variation of this reaction series. Little is known concerning biosynthesis of amino acids and other macromolecular precursors in Archaea, but presumably key monomers are produced from the central biosynthetic intermediates discussed previously for Bacteria (see Section 4.16 and Figure 4.24). Well-developed electron transport chains including cytochromes of the *a*, *b*, and *c* types exist in extreme halophiles and *a*-type cytochromes are present in the membranes of certain hyperthermophiles. With these and other electron carriers in place, chemoorganotrophic metabolism in these Archaea probably proceeds by introduction of electrons from organic electron donors into the electron transport chain leading to the reduction of O_2, S^0, or some other electron acceptor, with concurrent establishment of a proton motive force which drives ATP synthesis by membrane-bound ATPases (see Section 4.11 for a description of ATPase function). Chemolithotrophy is also well established in the Archaea, with H_2 being a common electron donor (see Sections 20.5 and 20.6).

Autotrophy is widespread among the Archaea and occurs by several different means. In methanogens and presumably in most chemolithotrophic hyperthermophiles, CO_2 is incorporated via the acetyl-CoA pathway or some modification thereof (see Section 19.9 for a description of the acetyl-CoA pathway). In the hyperthermophiles *Sulfolobus* and *Thermoproteus*, however, CO_2 fixation occurs via the reverse tricarboxylic acid cycle, a reaction series previously described as the autotrophic pathway in the green sulfur Bacteria (see Section 19.1). Because extreme halophiles are all strict chemoorganotrophs, autotrophic pathways would not be expected. Nevertheless, cultures of various species of extreme halophiles have been shown to fix significant amounts of CO_2, and the biochemical route of fixation is apparently the Calvin cycle, the most widespread autotrophic pathway in Bacteria and eukaryotes (see Section 16.13).

We thus see that many of the major catabolic and anabolic sequences in the Archaea are familiar ones from our study of these processes in various Bacteria. Although the enzymatic details of major pathways in Archaea sometimes differ from those observed in Bacteria, with the exception of the reactions leading to methanogenesis (see Section 20.5), it appears that most metabolic sequences in Archaea have counterparts in various Bacteria.

Genetics of the Archaea

Like those of most other prokaryotes, the chromosomes of Archaea that have so far been investigated consist of a single covalently closed circular DNA molecule (Figure 20.6a). Restriction maps have been constructed for several of these, and the chromosomes range in size from 1900 kilobase pairs (both *Methanococcus voltae* and *Thermococcus celer*) to 2900 kilobase pairs (*Haloferax volcanii*), considerably smaller than the

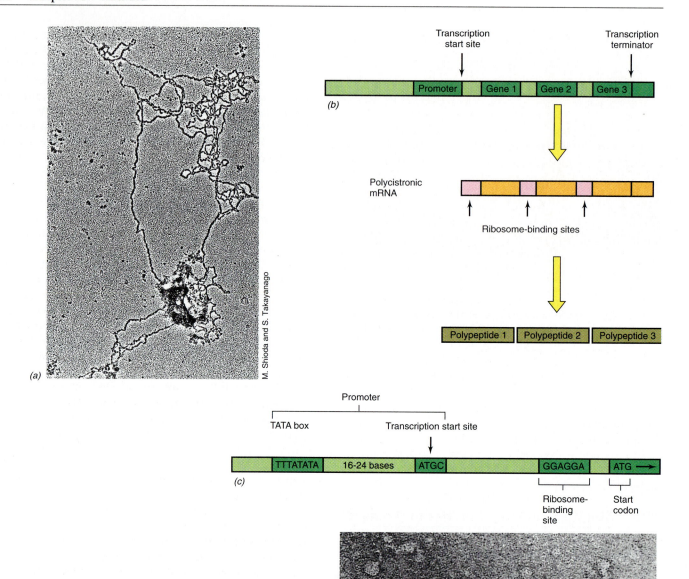

(a) M. Shioda and S. Takayanago

FIGURE 20.6
The chromosome and molecular biological processes in Archaea. (a) Circular chromosome isolated from a cell of the hyperthermophile *Sulfolobus*. (b) Genes are arranged in operons on the covalently closed and circular archaeal chromosome. (c) Polycistronic mRNA molecules are produced containing Shine–Dalgarno ribosome binding sequences just upstream of the coding sequences. In the figure, the ribosome-binding sites are shown outside of the coding region, but the ribosome-binding site for the second gene is often part of the coding region of the first one, and so forth. (d) Bacteriophage ψM1, a virus that infects the methanogen *Methanobacterium thermoautotrophicum*. Although lytic, phage ψM1 also acts as a transducing phage and can be used for genetic exchange experiments in methanogens. The phage is about 250 nm long.

(d) Daniel Studer and Thomas Leisinger

4700 kilobase pair genome of *Escherichia coli*, but not unusually small compared to some other Bacteria. Plasmids of various sizes (including some as large as 700 kilobase pairs) have been identified, and various types of transposable elements are known. Some of the genomes of halophilic Archaea are particularly rich in insertion sequences. At least some of the Archaea also contain restriction modification systems, a common feature in other prokaryotes (see Section 5.3).

Some species of Archaea can be transformed by plasmids and by genomic DNA, and cloning vectors have been developed. By the isolation and analysis of mutants and by cloning and sequencing genes, considerable advances have been made in the understanding of gene expression in these organisms. It is clear that the Archaea contain operons and that many genes are transcribed into polycistronic messages, as is the case for Bacteria but not for eukaryotes (see Section 5.8). A "typical" operon from an archaeon is shown in Figure 20.6b. Except for differences in sequence, this operon has the same basic structure as in Bacteria.

Cells of Archaea have a single RNA polymerase, as is the case for the Bacteria, but their subunit structure resembles that of the RNA polymerase II of eukaryotes and is much different from bacterial RNA polymerases (see Section 18.9). The promoter sequences from a wide variety of Archaea have been compared and most contain a conserved A-T rich sequence of 6 to 8 base pairs about 18 to 27 base pairs upstream of the transcription start site (Figure 20.6c). This sequence is similar to that of the Pribnow box found in promoters of Bacteria, but is more similar in sequence and is located at about the same position as an important element in the promoters recognized by eukaryotic RNA polymerase II (see Section 5.6). In fact, the archaeal sequence is now called a *TATA box*, the same name as originally given to the sequence from eukaryotic promoters. Less is known about transcription termination signals in Archaea. In some genes, it seems clear that inverted repeats (which lead to stem loops in the RNA) followed by A-T rich sequences are involved, sequences very similar to those found in many transcription terminators in Bacteria (see Section 5.6). However, such sequences are not found in other archaeal genes. One other possible type of transcription terminator contains no inverted repeats but rather repeated stretches containing runs of T's.

As for Bacteria, Shine–Dalgarno (ribosome-binding) nucleotide sequences are present in transcripts of cells of Archaea. They are located a few nucleotides upstream from the initiation codon and are complementary to the 3' end of the 16S rRNA, a molecule that plays an intimate role in translation. Recall how in Bacteria the complementarity between 16S rRNA and mRNA ensures a precise fit of the message on the ribosome such that the correct reading frame is aligned to begin translation (see Section 5.8). A similar relationship between ribosomal RNA and mRNA is obviously necessary to ensure faithful translation in Archaea as well.

The genetics of the Archaea is currently an area of intense research activity, not only because of the evolutionary importance of these organisms but also because of the great interest in the unusual phenotypic properties of these organisms (growth over 100°C, production of natural gas, growth in saturated salt, and so on). Genetic exchange systems have been reported for some Archaea (Figure 20.6d), and exploitation of these systems is improving our understanding of genetic phenomena in these organisms. At this time, genetic manipulations are most easily accomplished with the extreme halophiles because they are easier to grow than other Archaea and can be readily transformed and because cloning vectors with selectable markers are available. Cloning and sequencing of archaeal genes into bacterial genetic systems (usually *Escherichia coli*) has greatly accelerated the pace of

research in this area and will probably remain a major tool for studying archaeal genetics until facile genetic systems are more widespread among Archaea.

With this introduction to the basic biology of Archaea, we now turn our attention to the organisms themselves, retaining in our discussion of the major groups of Archaea the physiological and ecological approach we used to describe groups of Bacteria in the previous chapter.

20.4   Extremely Halophilic Archaea

We begin our discussion with a consideration of the extreme halophiles. **Extremely halophilic Archaea** are a diverse group of prokaryotes that inhabit highly saline environments such as solar salt evaporation ponds and natural salt lakes, or artificial saline habitats such as the surfaces of heavily salted foods like certain fish or meats. Such habitats are often called *hypersaline*. The term *extreme* halophile is used to indicate that these organisms are not only halophilic, but that their requirement for salt is *very high*, in some cases near that of saturation. A general definition of an extreme halophile is that the organism requires at least 1.5 M (about 9 percent) NaCl for growth but most species require 2–4 M NaCl (12–23 percent) for optimal growth. Virtually all extreme halophiles can grow at 5.5 M NaCl (32 percent, the limits of saturation for this salt), although some species grow only very slowly at this salinity. Until recently only two genera of halophilic Archaea were formally recognized, the rod-shaped *Halobacterium* and the coccus *Halococcus*. However, enrichment culture studies of highly saline habitats have yielded several new extreme halophiles; most of these organisms group phylogenetically within the Archaea (see Section 18.7). We begin with a consideration of the habitats of these unusual prokaryotes, and proceed to discuss the taxonomy and physiology of the group. Finally, we consider a unique mechanism of light-driven ATP synthesis which does not involve chlorophyll or bacteriochlorophyll.

Saline environments

Salty habitats are common throughout the world, but *extremely* saline habitats are rather rare. Most extremely saline environments are in hot, dry areas of the world, and such climatic conditions encourage evaporation and further concentration of the salts. Salt lakes can vary considerably in ionic composition. The predominant ions in a saline lake depend to a major extent on the surrounding topography, geology, and general climatic conditions. The Great Salt Lake in Utah (USA) (Figure 20.7a), for example, is essentially concentrated seawater because the relative proportions of the various ions are those of seawater, although the overall concentration of ions is much higher. Sodium is the predominant cation in Great Salt Lake while chloride is the predominant anion; significant levels of sulfate are also present at a slightly alka-

FIGURE 20.7 Hypersaline habitats. (a) Great Salt Lake, Utah, a hypersaline lake in which the ratio of ions is similar to seawater but in which absolute concentrations of ions are about ten times that of seawater. (b) Aerial view near San Francisco Bay, California, of a series of seawater evaporating ponds where solar salt is prepared. The red/purple color is predominantly due to bacterioruberins and bacteriorhodopsin in cells of halobacteria.

line pH (Table 20.1). By contrast, another very saline basin, the Dead Sea, is relatively low in sodium but contains high levels of magnesium because of the abundance of magnesium minerals in the surrounding rocks (Table 20.1). The water chemistry of soda lakes resembles that of saline lakes such as Great Salt Lake, but because high levels of *carbonate* minerals are present in the surrounding rocks, the pH of soda lakes is quite high; pH values of 10–12 are not uncommon in these environments (Table 20.1).

Despite what may seem like rather harsh conditions, salt lakes can be highly productive ecosystems. Archaea are not the only microorganisms found. The eukaryotic alga *Dunaliella* is the major, if not sole oxy-

genic phototroph in most salt lakes. In highly alkaline soda lakes where *Dunaliella* is absent, anoxygenic phototrophic purple bacteria of the genus *Ectothiorhodospira* (see Section 19.1) predominate. Because sulfate is generally abundant in salt lakes (Table 20.1), significant sulfate reduction can occur, leading to the formation of the sulfide needed for development of purple bacteria. At the high salt concentrations found in saline lakes, no higher plants can develop, although in addition to microorganisms a few types of animals are capable of living (brine shrimp, brine flies). Organic matter originating from primary production by oxygenic or anoxygenic phototrophs then sets the stage for development of the extremely halophilic Archaea,

Table 20.1 Ionic composition of some highly saline environments

Ion	Concentration (g/liter)			
	Great Salt Lake	**Dead Sea**	**Typical soda lake**	**Seawater (for comparison)**
Na^+	105	40.1	142	10.6
K^+	6.7	7.7	2.3	0.38
Mg^{2+}	11	44	<0.1	1.27
Ca^{2+}	0.3	17.2	<0.1	0.40
Cl^-	181	225	155	18.9
Br^-	0.2	5.3	—	0.065
SO_4^{2-}	27	0.5	23	2.65
HCO_3^-/CO_3^{2-}	0.7	0.2	67	0.14
pH	7.7	6.1	11	8.1

all of which are chemoorganotrophic. In addition, a few extremely halophilic chemoorganotrophic Bacteria, such as *Haloanaerobium* and *Halobacteroides,* are present in such environments.

Marine salterns are also habitats for extremely halophilic prokaryotes. Marine salterns are small basins filled with seawater which are left to evaporate, yielding NaCl and other salts of commercial value (Figure 20.7b). As salterns approach the minimum salinity limits for extreme halophiles, the waters turn a reddish purple in color indicative of the massive growth (called a *bloom*) of halophilic Archaea (the red coloration apparent in Figure 20.7b comes from carotenoids and other pigments of the halobacteria discussed below). Extreme halophiles have also been found in high salt foods such as certain sausages, marine fish, and salt pork. Other than causing some esthetic problems, growth of extreme halophiles in foods is of little consequence, since no extreme halophile has been shown to cause foodborne illness.

Taxonomy of extremely halophilic Archaea

Table 20.2 lists the currently recognized species of extremely halophilic Archaea. 16S ribosomal RNA sequencing and other studies have defined six genera of extreme halophiles: *Halobacterium, Halococcus, Haloferax, Haloarcula, Natronobacterium,* and *Natronococcus.* The extremely halophilic Archaea are frequently referred to collectively as "halobacteria," probably because the genus *Halobacterium* was the first in this group to be described and is still the best studied representative of the group. As Archaea, halobacteria lack peptidoglycan in their cell walls, contain ether-linked lipids, and archaeal RNA polymerases. They also are insensitive to most antibiotics, and possess the other general attributes of representatives of this domain (see Section 18.9). *Natronobacterium* and *Natronococcus* differ from other extreme halophiles in being extremely *alkaliphilic* as well as halophilic. As befits their soda lake habitat (see Table 20.1), growth of natronobacteria is optimal at very low Mg^{2+} concentrations and high pH (9–11). Natronobacteria also contain unusual diether lipids not found in other extreme halophiles and cluster nicely as a phylogenetic group.

All halophilic Archaea stain Gram *negatively,* reproduce by binary fission, and do not form resting stages or spores. Most halobacteria are nonmotile, but a few strains are weakly motile by lophotrichous flagella. The genomic organization of *Halobacterium* and *Halococcus* is highly unusual in that large plasmids containing up to 25–30 percent of the total cellular DNA are frequently present and the GC base ratio of these plasmids (57–60 percent GC) is significantly different from that of chromosomal DNA (66–68 percent GC). Plasmids from extreme halophiles are among the largest naturally occurring plasmids known. In addition to the large amount of nonchromosomal DNA, the *Halobacterium* genome also contains considerable amounts of highly repetitive DNA, the function of which is unknown.

Table 20.2   Taxonomy of some extremely halophilic Archaea

Genus	Morphology	DNA (mole % GC)	Habitat and Comments
Halobacterium	Rods		
H. salinarium		66–71	Isolated from salted fish, hides, hypersaline lakes; related organisms, probably all the same species are *H. halobium* and *H. cutirubrum*
H. sodomense		68	Dead Sea; requires high Mg^{2+}
H. saccharovorum		71	Salterns; uses sugars
Haloferax	Flattened disc		
H. volcanii	or cup-shaped	63–66	Dead Sea; requires high Mg^{2+}
H. mediterranei		60-62	Salterns; uses starch
H. gibbonsii		62	Marine salterns in Spain
H. denitrificans		64	Baja California saltern; capable of dissimilative nitrate reduction
Haloarcula	Irregular disks,		
H. vallismortis	triangles,	65	Death Valley, CA
H. hispanica	rectangles	63	Marine salterns in Spain
Halococcus	Cocci		
H. morrhuae		60–66	Salted fish
H. saccharolyticus		59	Saltern; uses sugars
Natronobacterium	Rods		
N. gregoryi		65	All from highly saline soda lakes; optimum pH for growth, 9.5
N. magadii		63	
N. pharaonis		64	
Natronococcus	Cocci		
N. occultus		64	Highly saline soda lake

Physiology of extremely halophilic Archaea

All extremely halophilic Archaea are chemoorganotrophs and most species are obligate *aerobes*. Most halobacteria use amino acids or organic acids as energy sources and require a number of growth factors (mainly vitamins) for optimal growth. A few *Halobacterium* species will oxidize carbohydrates but this ability is relatively rare. Electron transport chains containing cytochromes *a, b,* and *c* are present in *Halobacterium* and energy is conserved during aerobic growth via a proton motive force arising from membrane-mediated chemiosmotic events. Some strains of *Halobacterium* have been shown to grow anaerobically. Anaerobic growth at the expense of sugar fermentation and by anaerobic respiration (see Section 16.14) linked to the reduction of nitrate, fumarate, elemental sulfur, or thiosulfate has been demonstrated in certain strains. The finding that elemental sulfur serves as an electron acceptor for *Halobacterium* is especially interesting in light of the fact that both methanogens (see Section 20.5) and hyperthermophilic Archaea (see Section 20.6) can also employ S^0 as an electron acceptor; sulfur reduction is thus a universal trait among Archaea and may represent one of the earliest types of anaerobic respiration to develop on earth (see Sections 18.2 and 20.9 and Figure 18.3).

All extremely halophilic Archaea require large amounts of sodium for growth. In the case of *Halobacterium*, where detailed salinity studies have been performed, the requirement for Na^+ cannot be satisfied by replacement with another ion, even with the chemically related ion K^+. We learned in Section 9.11 that certain microorganisms can withstand the osmotic forces that accompany life in a high solute environment by accumulating organic compounds *intracellularly*; the latter compounds are referred to as **compatible solutes**. These compounds counteract the tendency of the cell to become dehydrated under conditions of high osmotic strength by placing the cell in positive water balance with its surroundings (see Section 9.11). Ironically, however, although *Halobacterium* only thrives in an osmotically stressful environment, it produces no *organic* compatible solute. Instead, cells of *Halobacterium* pump large amounts of K^+ from the environment into the cell such that the concentration of K^+ *inside* the cell is considerably greater than the concentration of Na^+ *outside* the cell; thus *Halobacterium* employs an *inorganic ion* as its compatible solute and remains in positive water balance (Table 20.3).

The cell wall of *Halobacterium* is stabilized by sodium ions; in low-sodium environments the cell wall breaks down. In electron micrographs of thin sections of *Halobacterium* (Figure 20.8), the organism appears similar to Gram-negative Bacteria in many respects, but the cell wall is quite different (compare Figure 20.8 with Figure 3.26). Na^+ binds to the outer surface of the *Halobacterium* wall and is absolutely essential for maintaining cellular integrity; when insufficient Na^+ is present the cell wall breaks apart and the cell lyses. Peptidoglycan is absent in the cell wall of

Table 20.3	Levels of specific ions in the environment and in cells of *Halobacterium salinarium**	
Ion	Concentration in medium (molar)	Concentration in cells (molar)
Na^+	3.3	0.8
K^+	0.05	5.3
Mg^{2+}	0.13	0.12
Cl^-	3.3	3.3

*Data from Matheson, A. T., G. D. Sprott, I. J. McDonald, and H. Tessie. 1976. Can. J. Microbiol. 22:780–786.

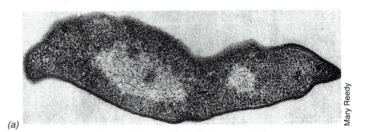
(a)

(b)

FIGURE 20.8 Electron micrograph of thin sections of the extreme halophile *Halobacterium salinarium*. A cell is about 0.8 μm in diameter. (a) Longitudinal section. (b) High magnification electron micrograph showing the regular structure of the cell wall.

Halobacterium (as it is in the walls of all Archaea, see Section 20.2), and instead the cell wall is composed of *glycoprotein*. This protein has an exceptionally high content of the *acidic* (negatively charged) amino acids aspartate and glutamate. The negative charges contributed by the carboxyl groups of these amino acids in the cell wall glycoprotein are shielded by Na^+; when Na^+ is diluted away, the negatively charged parts of the proteins actively repel each other, leading to cell lysis.

Cytoplasmic proteins of *Halobacterium* are also highly acidic, but studies of several halobacterial enzymes have shown that K^+, not Na^+, is required for activity. This, of course, is not surprising when it is recalled that K^+ is the predominate internal cation in cells of *Halobacterium* (Table 20.3). Besides a high acidic amino acid composition, halobacterial cytoplasmic proteins typically contain very low levels of *hydrophobic* amino acids. This phenomenon probably represents an evolutionary adaptation to the highly ionic cytoplasm of *Halobacterium*; in such an environment highly polar proteins would tend to remain in solution, whereas nonpolar molecules would tend to cluster and perhaps lose activity. The ribosomes of *Halobacterium* also require high K$^+$ levels for stability (ribosomes of nonhalophiles have no K$^+$ requirement). It thus appears that the extremely halophilic Archaea are highly adapted, both internally and externally, to life in a highly ionic environment. Cellular components exposed to the external environment require high Na$^+$ for stability while internal components require high K$^+$. In no other group of prokaryotes do we find this unique requirement for specific cations in such high amounts.

Bacteriorhodopsin and light-mediated ATP synthesis

Certain species of extreme halophiles have the additional interesting property of showing a *light-mediated* synthesis of ATP that does not involve chlorophyll pigments. As discussed in Section 16.4, one of the key aspects of photosynthesis is *photophosphorylation*, the synthesis of ATP as a consequence of the light-dependent generation of a proton gradient across the photosynthetic membrane. We have seen the highly pigmented nature of extremely halophilic Archaea in Figure 20.7. Pigmentation is due to red- and orange-colored carotenoids, primarily C$_{50}$ carotenoids called *bacterioruberins*, and also to inducible pigments involved in energy metabolism. Although lacking chlorophylls or bacteriochlorophylls, under conditions of low aeration *Halobacterium salinarium* and certain other extreme halophiles synthesize and insert a protein called **bacteriorhodopsin** into their membranes. Bacteriorhodopsin was named because of its functional similarity to the visual pigment of the eye, *rhodopsin*. Conjugated to bacteriorhodopsin is a molecule of *retinal*, a carotenoid-like molecule that can absorb light and catalyze the transfer of protons across the cytoplasmic membrane (Figure 20.9). Due to its

retinal content, bacteriorhodopsin is purple in color, and cells of *Halobacterium* switched from growth under conditions of high aeration to oxygen-limiting conditions gradually change from an orange or red color to more of a reddish purple in color because of the insertion of bacteriorhodopsin into the cytoplasmic membrane. Isolated purple membranes of *Halobacterium* contain about 25 percent lipid and 75 percent protein, and the purple membrane component is inserted at random on the surface of the cytoplasmic membrane; apparently bacteriorhodopsin is wedged into preexisting cytoplasmic membranes during the switch to low O$_2$.

Bacteriorhodopsin absorbs light strongly in the green region of the spectrum at about 570 nm. The retinal chromophore of bacteriorhodopsin, which normally exists in an all *trans* configuration, is excited and temporarily converted to the *cis* form following the absorption of light (Figure 20.9). This transformation results in the transfer of protons to the *outside* surface of the membrane. The retinal molecule then relaxes and returns to its more stable all *trans* isomer in the dark following the uptake of a proton from the cytoplasm, thus completing the cycle (Figure 20.9). As protons accumulate on the outer surface of the membrane, the proton motive force (see Section 4.11) increases until the membrane is sufficiently "charged" to drive ATP synthesis through action of the membrane-bound ATPase (Figure 20.9).

Light-mediated ATP production in *H. salinarium* has been shown to support slow growth of this organism anaerobically under nutritional conditions in which other energy-generating reactions do not occur, and light has been shown to maintain the viability of cultures of *Halobacterium* incubated anaerobically in the absence of organic energy sources. The light-stimulated proton pump of *H. salinarium* also functions to

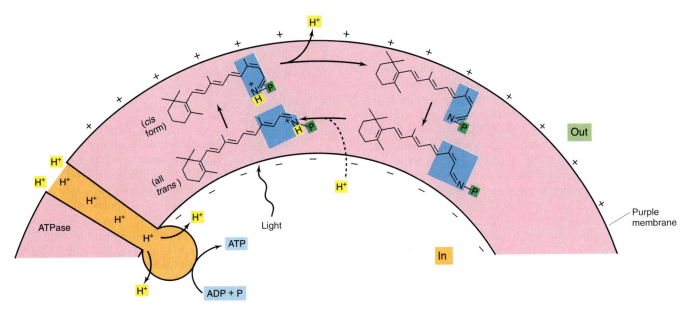

FIGURE 20.9   Model of the light-mediated bacteriorhodopsin proton pump in the purple membrane of *Halobacterium*. The P stands for the protein to which the chromophore retinal is attached. Out and In designate opposite sides of the cytoplasmic membrane.

pump Na^+ out of the cell by action of a Na^+/H^+ antiport system (see Section 3.4), and to drive the uptake of a variety of nutrients, including the K^+ needed for osmotic balance. The uptake of amino acids by *H. salinarium* has been shown to be indirectly driven by light, because the transport of amino acids occurs with Na^+ uptake by an amino acid/Na^+ symporter (see Section 3.4). Continued uptake depends on the removal of Na^+ via the (light-driven) Na^+/H^+ antiporter. A separate light-driven pump called *halorhodopsin* is employed by halobacteria to pump Cl^- into the cell to act as a counter ion to K^+.

Evolution of extreme halophiles

Although they share virtually nothing with them from a *phenotypic* standpoint, the extreme halophiles are phylogenetically quite closely related to methanogens (see Figure 18.13). This at first seems odd because most extreme halophiles are obligate aerobes whereas methanogenic bacteria are all obligate anaerobes (see below). However, more detailed studies of halophilic Archaea have shown them to be remarkably versatile, including many species capable of anaerobic growth by fermentation or anaerobic respiration. Indeed, even extremely halophilic and alkaliphilic methanogens have now been isolated, so that the connection between the two groups may not be as unusual as first thought. We will return to the theme of the evolution of Archaea in Section 20.9, after we have considered the remaining groups of Archaea.

20.5   Methane-Producing Archaea: Methanogens

We described the overall process of methanogenesis in Section 16.17 and the ecology of methanogenesis in Section 17.12, and have noted that the biological production of methane is carried out by a unique group of prokaryotes, the **methanogens**. Methane formation occurs only under strictly *anaerobic* conditions. Thus, methanogenesis is restricted to habitats which are anoxic.

Ten substrates have been shown to be converted to methane by one or another methanogen. Carbon dioxide, CO_2, is a nearly universal substrate for methanogens, the needed electrons usually being derived from H_2. When growing on $H_2 + CO_2$, the methanogens are *autotrophic*, with CO_2 serving as both carbon source and electron acceptor. In addition to CO_2, however, a variety of other compounds can be converted to methane by certain methanogenic species; these substrates are listed in Table 20.4. As described in Section 16.17, methane formation from $H_2 + CO_2$ can be viewed as a type of *anaerobic respiration* in which CO_2 serves as the electron acceptor. However, biochemical studies of methanogenesis indicate that a conventional electron-transport system involving cytochromes and quinones is absent from methanogens grown on $H_2 + CO_2$. Thus, electron transport to CO_2 involves different electron carriers than that of other anaerobic respiratory processes such as nitrate and sulfate reduction. Although

---

Table 20.4   Substrates converted to methane by various methanogenic Archaea

CO_2-type substrates
  Carbon dioxide CO_2 (with electrons derived from H_2 or certain alcohols)
  Formate $HCOO^-$
  Carbon monoxide CO
Methyl substrates
  Methanol CH_3OH
  Methylamine $CH_3NH_3^+$
  Dimethylamine $(CH_3)_2NH_2^+$
  Trimethylamine $(CH_3)_3NH^+$
  Methylmercaptan CH_3SH
  Dimethylsulfide $(CH_3)_2S$
Acetoclastic substrate
  Acetate CH_3COO^-

---

our understanding of methanogenesis is far from complete, the electron carriers involved in the reduction of CO_2 to methane are now well understood. This process requires a host of specific coenzymes unique to methanogens which serve as cofactors for enzymes which sequentially reduce C-1 intermediates starting with CO_2 and yielding CH_4.

Three *classes* of methanogenic substrates are known (see Table 20.4) and as expected, all are used with the release of free energy suitable for ATP synthesis. The first involves the use of *CO_2-type substrates*:

$$CO_2 + 4H_2 \rightarrow CH_4 + 2H_2O$$
$$\Delta G^{0\prime}\ -131\ kJ/reaction$$
$$4HCOOH \rightarrow CH_4 + 3CO_2 + 2H_2O$$
$$\Delta G^{0\prime}\ -304\ kJ/reaction$$
$$4CO + 2H_2O \rightarrow CH_4 + 3CO_2$$
$$\Delta G^{0\prime}\ -210\ kJ/reaction$$

The second class of reaction involves reduction of the *methyl group* of methyl-containing compounds to methane. In the case of methanol or methylamine, the overall reaction to methane has the following stoichiometry:

$$4CH_3OH \rightarrow 3CH_4 + CO_2 + 2H_2O$$
$$\Delta G^{0\prime}\ -319\ kJ/reaction$$
$$4CH_3NH_3Cl + 2H_2O \rightarrow 3CH_4 + CO_2 + 4NH_4Cl$$
$$\Delta G^{0\prime}\ -230\ kJ/reaction$$

In these reactions, some molecules of the substrate serve as an electron *donor* and are oxidized to CO_2, whereas other molecules are reduced and serve as electron *acceptor*. During growth on methyl compounds the reducing power for methanogenesis can also come from H_2. In fact, one species of methanogen, *Methanosphaera stadtmanae*, can grow on methanol *only* in the presence of H_2. In this case the stoichiometry would be:

$$CH_3OH + H_2 \rightarrow CH_4 + H_2O$$
$$\Delta G^{0\prime}\ -113\ kJ/reaction$$

A few methanogens will grow on alcohols other than methanol. *Methanospirillum* and *Methanogenium*, for example, will grow with 2-propanol as electron donor, four molecules of 2-propanol being converted to four molecules of acetone along with reduction of CO_2 to CH_4. Ethanol, 1-propanol, and 1-butanol are also oxidized by certain strains, the products being either acetate, propionate, or butyrate, and CH_4 (from the reduction of CO_2). The final methanogenic reaction is **acetoclastis**, the cleavage of acetate to CH_4 plus CO_2:

$$CH_3COO^- + H_2O \rightarrow CH_4 + HCO_3^-$$
$$\Delta G^{0\prime} -31 \text{ kJ/reaction}$$

Only two genera of methanogens, *Methanosarcina* and *Methanothrix*, have representatives able to carry out the acetoclastic reaction. The conversion of acetate to methane appears to be a very significant ecological process, especially in sewage digestors and in freshwater anoxic environments where competition for acetate between sulfate-reducing bacteria and methanogenic bacteria is not extensive (see Section 17.12).

Diversity and physiology of methanogenic Archaea

A variety of morphological types of methanogenic bacteria have been isolated and studies of their physiology and molecular properties have served to classify methanogens into seven major groups containing a total of 17 genera (Table 20.5). Short and long rods, cocci in various cell arrangements, plate-shaped cells, and filamentous methanogens are all known (Figures 20.10 through 20.12). Both Gram-positive and Gram-negative methanogens are known (see Table 20.5), thus the Gram stain is of little use in classifying these organisms. The current taxonomy of methanogens relies primarily on molecular methods, in particular on 16S rRNA sequence comparisons (see Chapter 18 for a discussion of molecular phylogeny), and, to a lesser degree, on immunological methods. The creation of seven major groups (Table 20.5) has emerged from extensive 16S ribosomal RNA sequence analyses.

Table 20.5   Characteristics of methanogenic Archaea

Genus	Morphology	Gram reaction	Number of species	Substrates for methanogenesis	DNA (mole % GC)
GROUP I					
Methanobacterium	Long rods	+ or −	8	$H_2 + CO_2$, formate	29–61
Methanobrevibacter	Short rods	+	3	$H_2 + CO_2$, formate	27–31
Methanosphaera	Cocci	+	1	Methanol + H_2 (both needed)	26
GROUP II					
Methanothermus	Rods	+	2	$H_2 + CO_2$; also reduce S^0	33
GROUP III					
Methanococcus	Irregular cocci	−	5	$H_2 + CO_2$, formate	29–34
GROUP IV					
Methanomicrobium	Short rods	−	2	$H_2 + CO_2$, formate	45–49
Methanogenium	Irregular cocci	−	3	$H_2 + CO_2$, formate	51–61
Methanospirillum	Spirilla	−	1	$H_2 + CO_2$, formate	46–50
Methanoplanus	Plate-shaped cells—occurring as thin plates with sharp edges	−	2	$H_2 + CO_2$, formate	38–47
GROUP V					
Methanosarcina	Large irregular cocci in packets	+	5	$H_2 + CO_2$, methanol, methylamines, acetate	41–43
Methanolobus	Irregular cocci in aggregates	−	3	Methanol, methylamines	38–42
Methanoculleus	Irregular cocci	−	4	$H_2 + CO_2$, alcohols, formate	54–62
Methanococcoides	Irregular cocci	−	1	Methanol, methylamines	42
Methanohalophilus	Irregular cocci	−	3	Methanol, methylamines, methyl sulfides; halophile	41
Methanothrix	Long rods to filaments	−	3	Acetate	52–61
GROUP VI					
Methanopyrus	Rods in chains	+	1	$H_2 + CO_2$; extreme thermophile, growth at 110°C	60
GROUP VII					
Methanocorpusculum	Irregular cocci	−	3	$H_2 + CO_2$, formate, alcohols	48–52

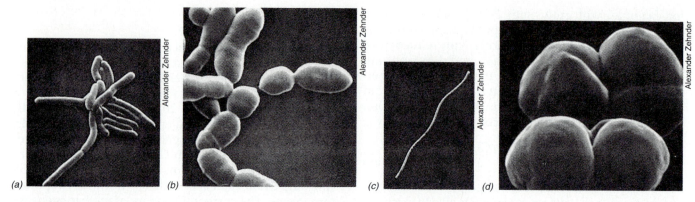

FIGURE 20.10 Scanning electron micrographs of cells of methanogenic Archaea, showing the considerable morphological diversity. (a) *Methanobrevibacter ruminantium*. A cell is about 0.7 μm in diameter. (b) *Methanobacterium* strain AZ. A cell is about 1 μm in diameter. (c) *Methanospirillum hungatii*. A cell is about 0.4 μm in diameter. (d) *Methanosarcina barkeri*. A cell is about 1.7 μm wide.

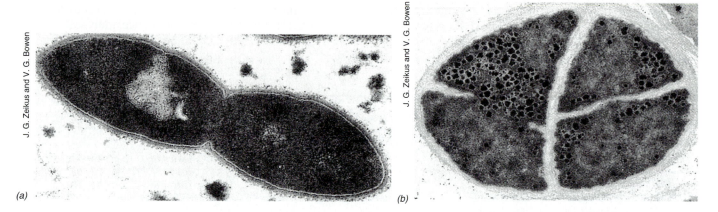

FIGURE 20.11 Transmission electron micrographs of thin sections of methanogenic Archaea. (a) *Methanobrevibacter ruminantium*. A cell is 0.7 μm in diameter. (b) *Methanosarcina barkeri*, showing the thick cell wall and the manner of cell segmentation and cross-wall formation. A cell is 1.7 μm in diameter.

When growing autotrophically, CO_2 serves as carbon source for methanogens and those species capable of metabolizing organic compounds to methane use intermediates generated during the catabolic process as biosynthetic precursors (see section on Autotrophy in methanogens). Growth of virtually all methanogens is stimulated by acetate and the growth of some species is also stimulated by certain amino acids. For culture, many methanogens require complex additions such as yeast extract or casein digests, and some rumen methanogens require a mixture of branched chain fatty acids (see Section 17.13). All methanogens use NH_4^+ as a nitrogen source and a few species are known to fix molecular nitrogen (N_2 fixation, see Section 16.26). The trace metal *nickel* is required by all methanogens; it is a component of an important methanogenic coenzyme, *Factor$_{430}$*, and is also present in the enzymes *hydrogenase* and *carbon monoxide dehydrogenase* (discussion follows). Iron and cobalt are also important trace metals for methanogens, and along with nickel, have been the only trace metals shown to be absolutely required for growth of these organisms.

Unique methanogenic coenzymes

A number of coenzymes have been found in methanogens that are unique to this group of bacteria. These coenzymes play important roles in the biochemistry of methanogenesis. Because of their importance in the energy yielding pathway, many of these coenzymes are present at far *higher* levels in methanogens than many of the common coenzymes such as NAD^+ or FMN are in other prokaryotes. We detail here the structure and properties of each of these coenzymes as a prelude to our discussion of the biochemistry of methanogenesis.

Coenzymes that serve as C_1 carriers

Methanofuran is a low-molecular-weight coenzyme that interacts in the *first* step of methanogenesis from CO_2 (see below). Methanofuran consists of a molecule of phenol, two glutamic acid molecules, an unusual long chain dicarboxylic fatty acid, and a furan ring (Figure 20.13*c*). CO_2 is reduced to the formyl level and bound by the amino side chain of the furan in the initial step of methanogenesis and is subsequently transferred to a second coenzyme in later steps of the pathway (see below).

Methanopterin is a methanogenic coenzyme containing a substituted pterin ring (Figure 20.13*d*). Methanopterin exhibits a bright blue fluorescence following absorbance at 342 nm. Methanopterin resembles the vitamin *folic acid* (see Figure 9.33*c*) and serves as a C_1 car-

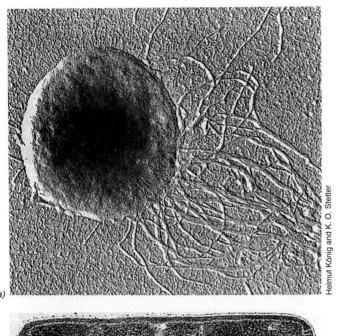

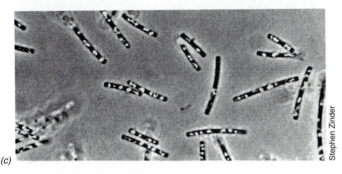

(a)   Helmut König and K. O. Stetter

(b)   Helmut König and K. O. Stetter

(c)   Stephen Zinder

FIGURE 20.12 Hyperthermophilic and thermophilic methanogens. (a) *Methanococcus jannaschii* (temperature optimum, 85°C), shadowed preparation electron micrograph. A cell is about 1 μm in diameter. (b) *Methanothermus fervidus* (temperature optimum, 83°C), thin sectioned electron micrograph. A cell is about 0.4 μm in diameter. (c) *Methanothrix* sp. (temperature optimum, 60°C), phase contrast micrograph. A cell is about 1 μm in diameter. The refractile bodies inside the cells are gas vesicles.

rier during the reduction of CO_2 to CH_4. The nitrogen atoms highlighted in Figure 20.13*d* are the atoms to which the C_1 intermediate binds. Methanopterin carries the C_1 unit during the majority of reductive steps in the methanogenic pathway, from the formyl (–CHO) level to the methyl (–CH_3) level (see below). *In vivo*, the reduced form of methanopterin, *tetrahydromethanopterin*, is the active form of the coenzyme.

Coenzyme M is involved in the *final step* in methane formation. Coenzyme M (Figure 20.13*e*), a very simple structure, has the chemical name 2-mercaptoethanesulfonic acid. The coenzyme is the carrier of the *methyl* group which is reduced to methane by the methyl reductase-F_{430} enzyme complex in the final step of methanogenesis:

$$CH_3—S—CoM + 2H \rightarrow HS—CoM + CH_4$$

Despite the structural simplicity of coenzyme M, it is highly specific in the methyl reductase reaction. A number of closely related analogs of coenzyme M have been found to be inactive in methanogenesis *in vitro*. The rumen methanogen, *Methanobrevibacter ruminantium,* has the interesting property of requiring coenzyme M as a growth factor. In this way coenzyme M can be considered to be a *vitamin*. Some methanogenic bacteria excrete coenzyme M, and this is apparently the source of coenzyme M for *M. ruminantium* in the rumen (see Section 17.13). Coenzyme M is so active as a vitamin for *M. ruminantium* that the organism shows a growth response at concentrations as low as $5 nM$ ($5 \times 10^{-9} M$).

A potent inhibitory analog of coenzyme M is *bromoethanesulfonic acid*, $Br—CH_2—CH_2—SO_3H$. This compound causes 50 percent inhibition of methyl reductase activity at a concentration of 10^{-6} M, and it also inhibits growth of methanogenic bacteria. Because coenzyme M is restricted to methanogenic bacteria, the bromo analog can be used experimentally to specifically inhibit methanogenesis in natural environments in ecological studies of the anaerobic degradation of organic matter to methane.

Although not a C_1 carrier, **coenzyme F_{430}** is a yellow, soluble, nickel-containing tetrapyrrole (Figure 20.13*b*) and, like CoM, plays an intimate role in the *terminal* step of methanogenesis as part of the methyl reductase system. Coenzyme F_{430} absorbs light strongly at 430 nm, but, unlike F_{420}, does not fluoresce. The nickel requirement for growth of methanogens reflects the abundance of F_{430} in the cells because most of the nickel in cells of methanogens is associated with the F_{430}–methyl reductase system.

Coenzymes involved in redox reactions

Coenzyme F_{420} and the coenzyme 7-mercaptoheptanoylthreonine phosphate (abbreviated **HS-HTP**) serve as electron donors in methanogenesis. Coenzyme F_{420} is a flavin derivative, structurally resembling the common flavin coenzyme, FMN (see Figure 4.13). The structure of F_{420} (Figure 20.13*a*) resembles that of FMN, but F_{420} lacks one of the nitrogen atoms of FMN in its middle ring and also lacks the methyl groups found on the benzene ring typical of true flavins (see Figure 4.13). F_{420} is a *two* electron carrier of rather low reduction potential ($E_0{}' = -0.37$ volts). Coenzyme F_{420} interacts with a number of different enzymes in methanogens including hydrogenase and NADP$^+$ reductase. F_{420} also plays a role in methanogenesis as the electron donor in at least one of the steps of CO_2 reduction (see below). The oxidized form of F_{420} absorbs light at 420 nm and fluoresces blue-green (Figure 20.14); upon reduction the coenzyme becomes colorless. The fluorescence of F_{420} is a useful tool for preliminary identification of an organism as a methanogen (Figure 20.14). Although important in methanogenesis, F_{420} may have other biological functions because cofactors similar to F_{420} have been detected in sulfate-reducing Archaea and in low levels in various Bacteria, including *Streptomyces* and some cyanobacteria.

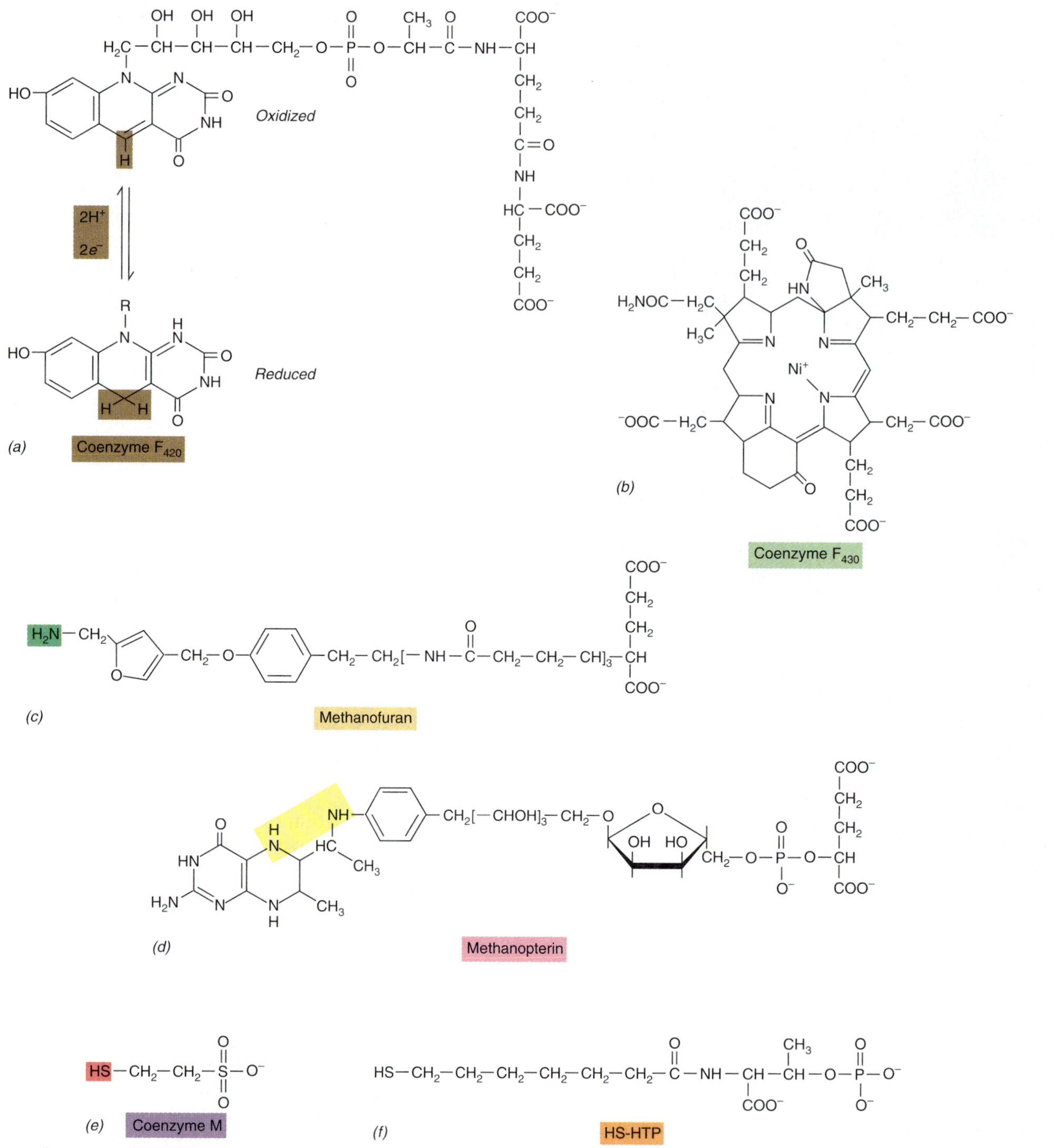

FIGURE 20.13 Coenzymes unique to methanogenic Archaea. The atoms shaded in color are the sites of oxidation-reduction reactions (F_{420}) or the position to which the C_1 moiety is attached during the reduction of CO_2 to CH_4 (methanofuran, methanopterin, and coenzyme M). The colors used to highlight a particular coenzyme itself (HS-HTP is orange, for example) are used throughout in Figures 20.15 through 20.17, and can be used to follow the reactions in each figure.

HS-HTP (7-mercaptoheptanoyl threonine phosphate) is the final unique coenzyme of the methanogens to be discussed. Like coenzyme M, this cofactor is involved in the *terminal* step of methanogenesis catalyzed by the methyl reductase system. As shown in Figure 20.13f, the structure of HS-HTP is rather simple. It is a phosphorylated derivative of the amino acid threonine containing a fatty acid side chain with a terminal SH group. HS-HTP resembles the vitamin *pantothenic acid* (part of acetyl-CoA, see Figure 4.21), and serves as an electron *donor* to the methyl reductase

system. HS-HTP reduction of coenzyme M (discussed later) is at the core of energy conservation in methanogenesis.

Biochemistry of CO_2 reduction to CH_4

Now that we have discussed the major coenzymes involved in methanogenesis from CO_2 plus H_2, we consider how these molecules interact with their specific enzymes in the conversion of CO_2 to CH_4. The reduction of CO_2 to CH_4 is generally H_2-dependent but formate, carbon monoxide, and even elemental iron (Fe^0)

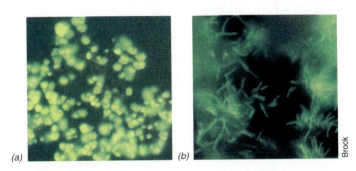

(a)   (b)

FIGURE 20.14 (a) Autofluorescence of the methanogen *Methanosarcina barkeri* due to the presence of the unique electron carrier F_{420}. A single cell is about 1.7 μm in diameter. The organisms were visualized with blue light in a fluorescence microscope. (b) F_{420} fluorescence in the methanogen *Methanobacterium formicicum*. A single cell is about 0.6 μm in diameter.

can serve as electron donors for methanogenesis. In the latter case, Fe^0 is oxidized to Fe^{2+}, with the electrons released combining with protons to form H_2 that serves as the immediate reductant in methanogenesis. In a few methanogens even certain simple *organic* compounds such as alcohols can supply the electrons for CO_2 reduction. For example, 2-propanol can be oxidized to acetone yielding electrons for methanogenesis in some species. But in general, the production of CH_4 from CO_2 is driven by molecular hydrogen (H_2).

The steps in CO_2 reduction, shown in Figure 20.15, are summarized as follows:

1. CO_2 is activated by methanofuran and subsequently reduced to the *formyl* level.
2. The formyl group is transferred from methanofuran to tetrahydromethanopterin (MP in Figure

20.15) and subsequently dehydrated and reduced in two separate steps to the *methylene* and *methyl* levels.

3. The methyl group is transferred from methanopterin to coenzyme M.
4. Methyl-coenzyme M is reduced to *methane* by the methyl reductase system in which F_{430} and HS-HTP are involved. The electron donor for this reaction is HS-HTP and the product of the reaction, beside CH_4, is a disulfide of CoM and HTP (CoM-S-S-HTP). Free CoM and HS-HTP are regenerated by reduction with H_2.

In the steps of methanogenesis shown in Figure 20.15, reduced F_{420} serves as electron donor in the reduction of the methenyl to the methylene group but the nature of the electron donor in the initial step of CO_2 reduction and in the step between methylene and methyl are unknown.

Autotrophy in methanogens

Carbon dioxide is converted into organic form in methanogenic Archaea through the reactions of the **acetyl-CoA pathway** used also by homoacetogens and sulfate-reducing bacteria. (See Section 19.9 for detailed discussion of this pathway.) However, unlike the other anaerobes which use this pathway, methanogens growing on H_2 + CO_2 integrate their biosynthetic and bioenergetic pathways because common intermediates are shared. This is possible because both the acetyl-CoA pathway and the methanogenic pathway lead to the production of CH_3 groups. As shown in Figure 20.16, autotrophically grown methanogens lack that part of the acetyl-CoA pathway leading to the

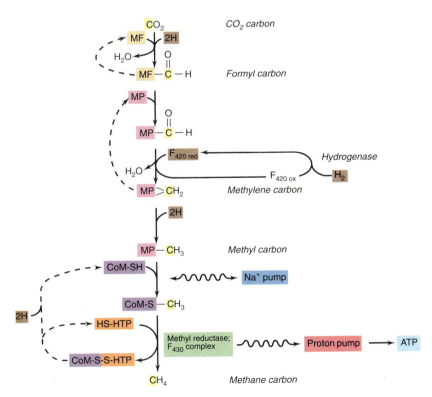

FIGURE 20.15
Pathway of methanogenesis from CO_2. Abbreviations: MF, methanofuran; MP, tetrahydromethanopterin; CoM, coenzyme M; F_{420}, coenzyme F_{420}; F_{430}, coenzyme F_{430}; HS-HTP, 7-mercaptoheptanoylthreonine phosphate. The carbon atom reduced is shown in yellow and the source of electrons are highlighted in brown. See Figure 20.13 for the structures of the coenzymes and the text for discussion of the reversible Na^+ pump.

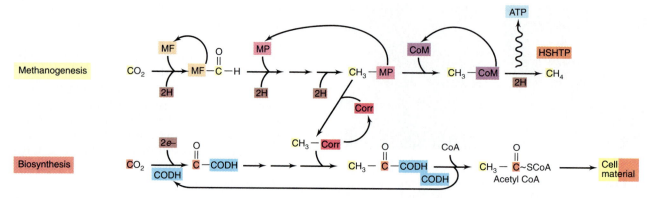

FIGURE 20.16 How autotrophic methanogens combine aspects of biosynthesis and bioenergetics. Abbreviations and color coding are as in Figure 20.13. CODH = carbon monoxide dehydrogenase; Corr = C_1-carrying corrinoid protein. Note how half of the acetyl-CoA molecule produced comes from reactions leading to methanogenesis.

production of methyl groups via tetrahydrofolate intermediates, and instead obtain methyl groups for the production of acetate for biosynthesis from the methanogenic pathway (compare Figure 20.16 with Figure 19.36). Specifically, methyl tetrahydromethanopterin donates methyl groups to a corrinoid-containing enzyme (see below) to yield CH_3-corrinoid (Figure 20.16). The CH_3 group is then transferred to carbon monoxide dehydrogenase (which has previously reduced CO_2 to the level of CO), to eventually yield acetyl-CoA (Figure 20.16). Since, relative to methanogenesis, only a small amount of CO_2 is incorporated into cell material, this small drain on the methanogenic pathway is of no consequence, and the merging of the two pathways probably effects an energy savings to the organism, since synthesis of additional enzymes of the acetyl-CoA pathway is not necessary in order to make the CH_3 group of acetate for biosynthesis.

Methanogenesis from methyl compounds and acetate

Reactions of the acetyl-CoA pathway discussed above are intimately involved in the production of methane from methyl compounds and from acetate. Methyl compounds such as methanol are catabolized by donating methyl groups to a corrinoid protein to form CH_3-corrinoid (Figure 20.17a). Corrinoids are the parent structures of such compounds as vitamin B_{12}, which contain a porphyrin-like corrin ring with a central cobalt atom. The CH_3-corrinoid complex donates the methyl group to CoM to give CH_3-CoM from which methane is obtained by reduction with electrons derived from oxidation of other molecules of methanol to CO_2 (Figure 20.17a). Carbon for biosynthesis originates by formation of CO from the oxidation of a methyl group from methanol and the combination of CO and a methyl group by carbon monoxide dehydrogenase to yield acetate (Figure 20.17a).

Growth of acetoclastic (acetate-degrading) methanogens is also tied to reactions of the acetyl-CoA pathway. In acetoclastic methanogens acetate is used directly for biosynthesis. For energy purposes, acetate is also the energy source. Acetate is activated to acetyl-

CoA which can interact with carbon monoxide dehydrogenase, following which the methyl group of acetate is transferred to the corrinoid enzyme of the acetyl-CoA pathway to yield CH_3-corrinoid (Figure 20.17b). From here the methyl group is transferred to tetrahydromethanopterin and then to coenzyme M to yield CH_3-CoM. The latter is then reduced to CH_4 using electrons generated from the oxidation of CO to CO_2 by CO dehydrogenase (Figure 20.17b). Energy conservation occurs in the acetoclastic reaction as a result of proton gradient formation during the methyl reductase step, just as it does for $H_2 + CO_2$ or methylotrophically grown methanogens (see below).

Energetics of methanogenesis

Under standard conditions, the free energy change of the reduction of CO_2 to CH_4 with H_2 is -131 kJ/mole. However, concentrations of H_2 in methanogenic habitats are usually quite low, no higher than 10 μM and because of the influence of concentration of reactants on free energy change (see Appendix 1), the free energy for the reaction forming CH_4 from $H_2 + CO_2$ by methanogens in their natural habitat is probably much lower, about -30 kJ. Thus probably only one ATP will be formed during CO_2 reduction to CH_4, and this agrees with molar growth yield data obtained for methanogens growing on $H_2 + CO_2$.

The *terminal step* of methanogenesis, in which the methyl reductase enzyme complex reduces CH_3-CoM to CH_4 (Figure 20.15), is the energy conservation step. The interaction of HS-HTP with CH_3-CoM in this terminal step forms CH_4 and CoM-S-S-HTP (see Figure 20.15). The latter is then reduced with electrons derived from reduced F_{420} (or possibly from H_2 itself), to yield CoM-SH and HS-HTP (Figure 20.18). This reduction, carried out by the enzyme *heterodisulfide reductase*, is exergonic and is associated with the extrusion of protons across the membrane, creating a proton motive force (Figure 20.18). Dissipation of the proton gradient by a membrane-integrated proton-translocating ATPase (see Section 4.11 for a discussion of ATPases) drives ATP synthesis during methanogenesis in the same way this process occurs in other forms of respiratory metabolism.

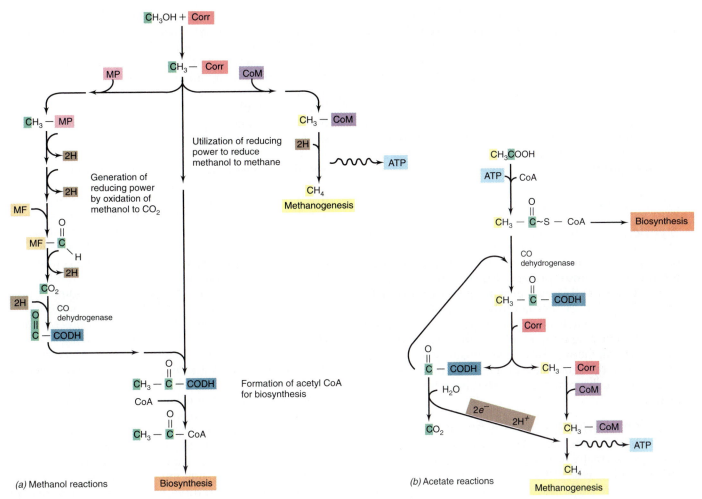

FIGURE 20.17   Utilization of reactions of the acetyl-CoA pathway during growth on methanol (a) or acetate (b) by methanogenic Archaea. Abbreviations and color coding are as in Figures 20.13, 20.15, and 20.16.

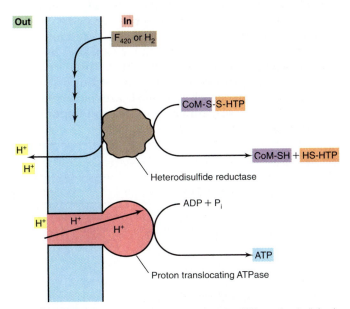

FIGURE 20.18   Energetics of methanogenesis. ATP synthesis linked to a proton gradient established during the terminal step of methanogenesis. In some methanogens, the electron transport steps shown by arrows from F_{420} to the heterodisulfide reductase are mediated by b- and c-type cytochromes. Refer to Figure 20.13 for further details.

Growth of methanogens on methyl compounds is also linked to the heterodisulfide reductase proton pump, but an additional factor is involved. In the absence of H_2, methanogenesis at the expense of methylated compounds requires that some of the substrate be oxidized to generate the electrons needed for methyl reduction to methane. This occurs at the expense of a membrane-bound *sodium pump*, which establishes a sodium gradient across the cytoplasmic membrane to drive the oxidation of methyl groups. The sodium pump is linked to the interconversion of CH_3–tetrahydromethanopterin and methyl-CoM during methanogenesis; production of CH_3-CoM establishes the sodium gradient, while production of CH_3–tetrahydromethanopterin from CH_3-CoM dissipates the gradient (see Figure 20.15).

The oxidation of methyl groups by methylotrophic methanogens is driven by energy released by dissipation of the Na^+ gradient that drives the first step in methyl group oxidation, an endergonic reaction. Further oxidative steps in the conversion of methyl groups to CO_2 to yield electrons for methyl reduction to methane proceed by reversal of the enzymatic steps leading to CH_4 formation from CO_2 (Figure 20.15) and show no energy requirement. Energy from the Na^+ gradient is also required to drive the carboxylation of methanofuran in methanogenesis (see Figure 20.15). This reaction is sufficiently endergonic that an energetic push (the Na^+ gradient) is needed to drive the reaction. Thus, in methanogens we see two types of ion pumps in play: a typical proton pump used to drive ATP synthesis and a reversible Na^+ pump that can function to drive methyl group oxidation and methanofuran carboxylation.

20.6   Hyperthermophilic Archaea

A phylogenetically distinct branch of Archaea consists of organisms unified by their extremely thermophilic nature. This branch, referred to as the **hyperthermophiles**, contains representatives which are the most thermophilic of all known prokaryotes (see Section 9.9 for a discussion of thermophily). Several hyperthermophiles are capable of growth at temperatures above the normal boiling point of water and all have temperature optima above 80°C! We begin with a brief overview of the group and proceed to a more detailed discussion of the major genera that have been characterized.

Overview of hyperthermophilic Archaea

Most hyperthermophilic Archaea have been isolated from geothermally heated soils or waters containing elemental sulfur and sulfides, and most species metabolize sulfur in some way. Elemental sulfur is formed from geothermal H_2S either by the spontaneous oxidation of H_2S with O_2 or by reaction of H_2S with SO_2 (the latter is a common component of volcanic gases). In terrestrial environments, sulfur-rich springs, mud pots, and soils may have temperatures up to 100°C and are generally mildly to extremely acidic due to production of sulfuric acid, H_2SO_4, from the biological oxidation of H_2S and S^0 (see Section 17.15). Such hot, sulfur-rich environments are called *sulfataras*, and solfataric fields are found throughout the world (Figure 20.19a). Extensive solfataras are found in Italy, Iceland, New Zealand, and Yellowstone National Park in Wyoming. Depending on the surrounding geology, solfataric en-

vironments may be either slightly alkaline to mildly acidic, pH 5–8, or extremely acidic, with pH values below 1 not uncommon. Hyperthermophiles have been obtained from both types of environments but the majority of these organisms inhabitat neutral or mildly acidic habitats. In addition to these natural habitats, hyperthermophilic Archaea also thrive within artificial thermal habitats, in particular the boiling outflows of geothermal power plants.

With only a few exceptions, hyperthermophiles are *obligate anaerobes*. Their energy-yielding metabolism is either chemoorganotrophic or chemolithotrophic. The requirement for sulfur, nearly universal across the group, is based on the need for an electron *acceptor* to carry out anaerobic respiration or as an electron *donor* for chemolithotrophic metabolism. Elemental sulfur (S^0) is reduced to H_2S using electrons derived from the oxidation of organic compounds or from H_2. The energy-yielding reactions of some hyperthermophiles are shown in Table 20.6. Chemoorganotrophs, such as those of the genera *Thermococcus* and *Thermoproteus*, oxidize a variety of organic compounds, in particular small peptides, glucose, and starch anaerobically in the presence of S^0 as electron acceptor. *Sulfolobus*, on the other hand, will utilize a variety of organic compounds (as well as S^0) as energy sources with O_2 as electron acceptor.

Many hyperthermophilic Archaea can grow *chemolithotrophically* with H_2 as energy source (Table 20.6). *Pyrodictium*, for example, grows strictly anaerobically in a mineral salts medium supplemented with H_2 and S^0 at temperatures up to 113°C. Cultures are usually grown in sealed glass or stainless steel vessels to which H_2 is added under pressure. Growth of *Pyrodictium* is stimulated by the addition of organic compounds, but the latter cannot replace H_2. *Acidianus* and *Sulfolobus* can grow aerobically with H_2 as electron donor. Other chemolithotrophic growth modes among hyperthermophiles include the oxidation of elemental sulfur and ferrous iron (aerobically) by various *Sulfolobus* species (see Table 20.6). Thus, a variety of respiratory processes can be carried out by hyperthermophilic Archaea, but in most cases elemental sulfur plays a key role, either as an electron donor or an electron acceptor.

Table 20.6   Energy-yielding reactions of hyperthermophilic Archaea

Nutritional class	Energy-yielding reaction	Example
Chemoorganotrophic	Organic compound $+ S^0 \rightarrow H_2S + CO_2$	*Thermoproteus; Thermococcus; Desulfurococcus; Thermofilum; Pyrococcus*
	Organic compound $+ SO_4^{2-} \rightarrow H_2S + CO_2$	*Archaeoglobus*
	Organic compound $+ O_2 \rightarrow H_2O + CO_2$	*Sulfolobus*
	Organic compound $\rightarrow CO_2 +$ fatty acids	*Staphylothermus*
	Organic compound $\rightarrow CO_2 + H_2$	*Pyrococcus*
Chemolithotrophic	$H_2 + S^0 \rightarrow H_2S$	*Acidianus; Pyrodictium; Thermoproteus*
	$2H_2 + O_2 \rightarrow 2H_2O$	*Acidianus; Sulfolobus*
	$2S + 3O_2 + 2H_2O \rightarrow 2H_2SO_4$	*Sulfolobus; Acidianus*
	$4FeS_2 + 15O_2 + 2H_2O \rightarrow 2Fe_2(SO_4)_3 + 2H_2SO_4$	*Sulfolobus*
	$4H_2 + SO_4^{2-} + 2H^+ \rightarrow 4H_2O + H_2S$	*Archaeoglobus*
	$4H_2 + CO_2 \rightarrow CH_4 + 2H_2O$	*Methanopyrus*

(a)

(b)

(c)

(d)

FIGURE 20.19

Habitats of hyperthermophilic Archaea. (a) A typical solfatara in Yellowstone National Park. Steam rich in hydrogen sulfide rises to the surface of the earth. Because of the heat and acidity, higher forms of life do not develop. (b) Sulfur-rich hot spring, a habitat where *Sulfolobus* grows in profusion. (c) A typical boiling spring of neutral pH in Yellowstone Park; Imperial Geyser. (d) Iron-rich geothermal spring, another *Sulfolobus* habitat.

Hyperthermophiles from volcanic habitats

As mentioned previously, volcanic habitats can have temperatures as high as 100°C and are thus suitable habitats for hyperthermophilic Archaea. The first such organism discovered, *Sulfolobus*, grows in sulfur-rich hot acid springs (Figure 20.19*b*) at temperatures up to 90°C and at pH values of 1–5. *Sulfolobus* (Figure 20.20*a*) is an obligate aerobe capable of oxidizing H_2S or S^0 to H_2SO_4 and fixing CO_2 as carbon source. *Sulfolobus* can also grow chemoorganotrophically. Cells of *Sulfolobus* are generally spherical but form distinct lobes (Figure 20.20*a*). Cells adhere tightly to sulfur crystals where they can be visualized microscopically by use of fluorescent dyes (see Figure 16.21*b*). Besides an active *aerobic* metabolism, *Sulfolobus* is also known to reduce Fe^{3+} to Fe^{2+} (but not grow) *anaerobically*. The ability of *Sulfolobus* to oxidize Fe^{2+} to Fe^{3+} aerobically

(Figure 20.19*c*), however, has been used quite successfully in the high temperature leaching of iron and copper ores (see Section 17.17). As mentioned in Section 20.3, autotrophy in *Sulfolobus* occurs via the reverse tricarboxylic acid cycle.

A facultative aerobe resembling *Sulfolobus* is also present in acidic solfataric springs. This organism, named *Acidianus* (Figure 20.20*b*), differs from *Sulfolobus* primarily by virtue of its ability to grow *anaerobically*. Remarkably, *Acidianus* is able to use S^0 in both its aerobic and anaerobic metabolism. Under *aerobic* conditions the organism uses S^0 as an electron *donor*, oxidizing S^0 to H_2SO_4. Anaerobically, *Acidianus* uses S^0 as an electron *acceptor* (with H_2 as electron *donor*) forming H_2S as the reduced product. Thus, the metabolic fate of S^0 in cultures of *Acidianus* depends upon the presence of O_2 and/or an electron donor.

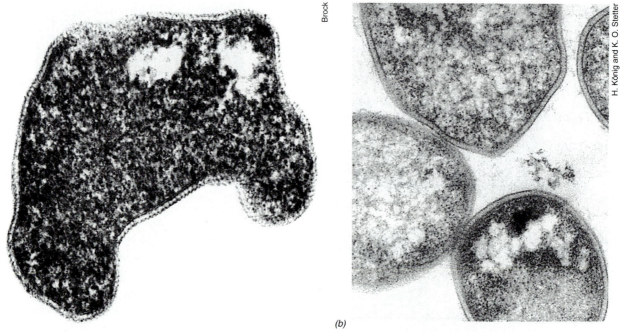

(a) Brock

(b) H. König and K. O. Stetter

FIGURE 20.20 Acidophilic hyperthermophilic Archaea. (a) *Sulfolobus acidocaldarius*. Electron micrograph of a thin section. (b) *Acidianus infernus*. Electron micrograph of a thin section. Cells of both organisms vary from 0.8 to 2 μm in diameter.

Like *Sulfolobus, Acidianus* is roughly spherical in shape (Figure 20.20*b*). It grows at temperatures from about 65°C up to a maximum of 95°C, with an optimum of about 90°C. Another property shared by *Sulfolobus* and *Acidianus* is an unusually low GC base ratio. The DNA of *Sulfolobus* is about 38 percent GC, while that of *Acidianus* is even lower, about 31 percent; many other hyperthermophiles have DNA of low GC content as well (see Table 20.7). These low GC base ratios are intriguing when one considers the hyperther-

mophilic nature of these organisms; how do they prevent their DNA from melting? In the test tube, DNA of 30–40 percent GC content would melt almost instantly at 90°C. Obviously hyperthermophiles have evolved protective mechanisms to prevent DNA melting *in vivo* (see Section 20.8).

The genera *Thermoproteus* and *Thermofilum* consist of *rod-shaped* cells which inhabit neutral or slightly acidic hot springs. Cells of *Thermoproteus* are stiff rods about 0.5 μm in diameter and are highly variable in

Table 20.7   Properties of hyperthermophilic Archaea

Genus	Morphology	Number of species	DNA (mole % GC)	Temperature (°C) Minimum	Temperature (°C) Optimum	Temperature (°C) Maximum	Optimum pH
Terrestrial volcanic isolates							
Sulfolobus	Lobed sphere	4	37	55	75–85	87	2–3
Acidianus	Sphere	2	31	65	85–90	95	2
Thermoproteus	Rod	2	56	60	88	96	6
Thermofilum	Rod	2	57	60	88	96	5.5
Desulfurococcus	Sphere	2	51	60	85	93	6
Desulfurolobus	Lobed sphere	1	32	65	80	87	2.5
Pyrobaculum	Rod	2	46	74	100	102	6
Methanothermus	Rod	2	33	60	83–88	97	6–7
Submarine volcanic isolates							
Pyrodictium	Disc-shape with attached filaments	3	62	82	105	113	6
Pyrococcus	Sphere	2	38	70	100	106	6–8
Thermodiscus	Disc-shaped	1	49	75	90	98	5.5
Staphylothermus	Spheres which aggregate into clumps	1	35	65	92	98	6–7
Thermococcus	Sphere	3	38–57	70	88	98	6–7
Methanopyrus	Rods	1	60	85	100	110	6.5
Arehaeoglobus	Cocci	2	46	64	83	95	7

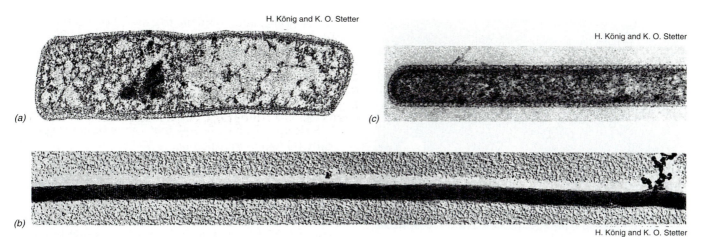

H. König and K. O. Stetter

H. König and K. O. Stetter

H. König and K. O. Stetter

FIGURE 20.21   Rod-shaped hyperthermophilic Archaea from terrestrial volcanic habitats. (a) *Thermoproteus neutrophilus*. Electron micrograph of a thin section. A cell is about 0.5 μm in diameter. (b) *Thermofilum librum*. A cell is only about 0.25 μm in diameter. Electron micrograph of shadowed cells. (c) *T. librum*. Electron micrograph of a thin section.

length, ranging from short cells of 1–2 μm up to filaments of 70–80 μm in length (Figure 20.21*a*). Filaments of *Thermofilum* are thinner, some 0.17–0.35 μm in width with filament length ranging up to 100 μm (Figure 20.21*b*). Both *Thermoproteus* and *Thermofilum* are strict anaerobes which carry out a S⁰-based anaerobic respiration. Unlike most hyperthermophiles, the oxygen sensitivity of *Thermoproteus* and *Thermofilum* is extreme, comparable to that of the methanogens (see Section 20.5); thus, elaborate precautions must be taken in their culture. Most *Thermoproteus* isolates can grow chemolithotrophically on H₂ or chemoorganotrophically on complex carbon substrates such as yeast extract, small peptides, starch, glucose, ethanol, malate, fumarate, or formate. *Thermofilum* can be grown in either mixed culture with *Thermoproteus* or in pure culture only by the addition of a highly polar lipid fraction isolated from cells of *Thermoproteus*; in nature the two organisms probably coexist in close association. Both *Thermoproteus* and *Thermofilum* have similar GC base ratios (56–58 percent GC), but appear to be phylogenetically distinct by DNA:RNA hybridization analyses.

Desulfurococcus (Figure 20.22) is a spherical, obligately anaerobic, S⁰-respiring organism. *Desulfurococcus* grows best at neutral pH and 80–90°C. The major features which differentiate *Desulfurococcus* from other hyperthermophilic Archaea are *genotypic* rather than phenotypic. DNA from *Desulfurococcus* contains 51 percent GC and hybridizes at only low levels with nucleic acid from *Thermoproteus*.

Extreme thermophiles from submarine volcanic areas

We now turn our attention to *submarine* volcanic habitats where a phylogenetically distinct set of hyperthermophilic Archaea exist. Although these underwater microbial habitats are generally quite shallow, the pressure of even a few meters of water can raise the boiling point of water sufficiently to select for organisms capable of growth above 100°C.

Geothermally heated sea floors exist in various locations around the world, and from a series of such habitats in the Mediterranean several genera of hyperthermophilic Archaea have been isolated. In submarine solfatara fields located in 2–10 meters of water off the coast of Vulcano, Italy, the sea floor consists of sandy sediments with cracks and holes from which geothermally heated (but not yet boiling) water is emitted at temperatures up to 103°C. From such waters the following genera of hyperthermophiles have been isolated: *Pyrodictium*, *Thermococcus*, *Thermodiscus*, *Pyrococcus*, and *Staphylothermus*.

Pyrodictium* and *Thermodiscus

Pyrodictium is perhaps the most fascinating of all submarine volcanic hyperthermophiles because its growth temperature optimum, 105°C, is the highest of any known organism. Cells of *Pyrodictium* are irregularly disc- and dish-shaped (Figure 20.23) and grow in culture as a mold-like layer upon sulfur crystals suspended in the medium. The cell mass consists of a huge network of fibers to which individual cells are attached (Figure 20.23). The fibers are hollow and con-

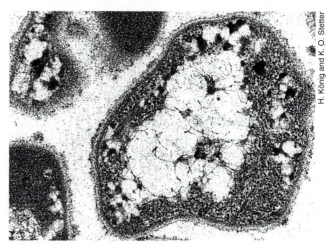

H. König and K. O. Stetter

FIGURE 20.22   *Desulfurococcus*. Electron micrograph of a thin section of *D. sacchavorans*. A cell is about 0.7 μm in diameter.

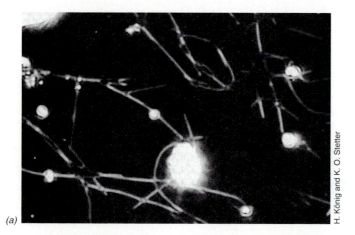

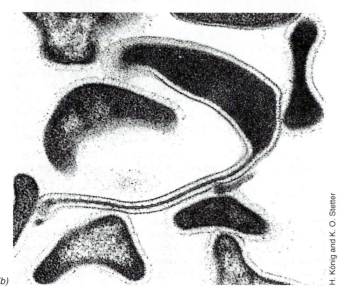

H. König and K. O. Stetter

FIGURE 20.23 *Pyrodictium occultum*, the most thermophilic of all known prokaryotes. (a) Dark field micrograph. (b) Electron micrograph of a thin section. A cell is highly variable in diameter, from 0.3 to 2.5 μm.

sist of proteinaceous subunits arranged in a fashion similar to that of the flagellin protein of the bacterial flagellum (see Section 3.7). The filaments of *Pyrodictium* do not function in motility but apparently serve to *attach* the cells to a solid substratum. *Pyrodictium* is a strict anaerobe that grows chemolithotrophically at neutral pH on H₂ with S⁰ as electron acceptor. Growth occurs between 82° and 113°C (the highest recorded temperature for growth of any hyperthermophile), and is stimulated by the addition of organic compounds. The cell envelope of *Pyrodictium* consists of glycoprotein or protein and lacks peptidoglycan (as do all archaeal cell walls, see Section 20.2). Phylogenetically, *Pyrodictium* shows affinities to *Sulfolobus*. However, due to the very large number of posttranscriptionally modified bases in the ribosomal RNAs of *Pyrodictium*, the exact phylogenetic position of this organism is unclear. DNA from *Pyrodictium* shows a substantially higher base ratio (62 percent GC) than that of any hyperthermophilic archaean from terrestrial habitats.

Thermodiscus resembles *Pyrodictium* morphologically but is *chemoorganotrophic* rather than chemo-

lithotrophic, and is unable to grow above 100°C. *Thermodiscus* is an obligate chemoorganotroph, with growth occurring on complex organic mixtures with S⁰ as electron acceptor at temperatures between 75° and 98°C (optimum around 90°C). The DNA of *Thermodiscus* contains a significantly lower GC base ratio (49 percent) than that of *Pyrodictium*, but analyses of 16S rRNA sequences show that *Thermodiscus* is specifically related to *Pyrodictium*.

Thermococcus, *Pyrococcus*, and *Staphylothermus*

Thermococcus is a spherical hyperthermophilic archaeon indigenous to submarine thermal waters in various locations throughout the world. The spherical cells contain a tuft of polar flagella and are thus highly motile (Figure 20.24*a*). *Thermococcus* is an obligately anaerobic chemoorganotroph that grows on proteins and other complex organic mixtures (including some sugars) with S⁰ as electron acceptor. It is not as thermotolerant as *Pyrodictium*, the optimum growth temperature being only 88°C, with a temperature range of 70–95°C. *Thermococcus* is an especially interesting organism from a *phylogenetic* standpoint. As shown in Figure 18.13, 16S rRNA sequence comparisons show *Thermococcus* to be a member of a very *slowly evolving* archaeon lineage. This means that *Thermococcus* has evolved the least (of known organisms) from the universal ancestor of all living organisms, and this implies that it should more closely resemble ancient phenotypes than other extant organisms. Indeed, the fact that *Thermococcus* is a hyperthermophilic sulfur-respiring organism fits nicely with hypotheses of what the chemical and physical conditions on earth were like at the time life originated (see Sections 18.1 and 20.8).

An organism morphologically quite similar to *Thermococcusis* is *Pyrococcus* (Figure 20.24*b*). *Pyrococcus* (the Latin derivation literally means "fireball") differs from *Thermococcus* primarily by its significantly different GC ratio (see Table 20.7) and its higher temperature requirements; *Pyrococcus* grows between 70° and 106°C with an optimum of 100°C. However, metabolically *Thermococcus* and *Pyrococcus* are quite similar:

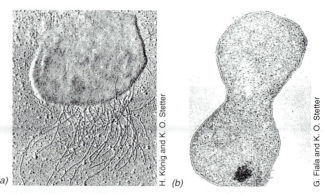

FIGURE 20.24 Spherical-shaped hyperthermophilic Archaea from submarine volcanic areas. (a) *Thermococcus celer*. Electron micrograph of shadowed cells. (b) Dividing cell of *Pyrococcus furiosus*. Electron micrograph of thin section. Cells of both organisms are about 0.8 μm in diameter.

FIGURE 20.25 *Staphylothermus marinus.* Electron micrograph of shadowed cells. A single cell is about 1 μm in diameter.

proteins, starch, or maltose are oxidized as energy sources and S^0 is reduced to H_2S. Comparisons of 16S rRNA sequences (see Figure 18.13) show *Thermococcus* and *Pyrococcus* to be specifically related, despite their significantly different GC base ratios and optimal growth temperatures (see Table 20.7).

The genus *Staphylothermus* consists of spherical cells about 1 μm in diameter which form aggregates of up to 100 cells, resembling those of *Staphylococcus* (Figure 20.25). *Staphylothermus* is a strictly anaerobic hyperthermophile growing optimally at 92°C and capable of growth between 65° and 98°C. Although S^0 is required for growth, oxidation of complex organic compounds is not tightly coupled to S^0 reduction in *Staphylothermus* as it is in most other hyperthermophiles; fatty acids such as acetate and isovalerate are produced during growth by this organism, suggesting that a fermentative metabolism is occurring. The GC base ratio of *Staphylothermus* is quite low, about 35 percent, and 16S rRNA sequence studies have not shown this organism to be closely related to any other hyperthermophile.

Isolates of *Staphylothermus* have been obtained from both a shallow marine hydrothermal vent and from samples taken from a "black smoker" at a depth of 2500 meters (see Section 17.9 for a discussion of black smokers). Since the temperature of the black smoker yielding the *Staphylothermus* isolate was some 330°C, it is likely that the organism originated from thermally heated seawater and not from the superheated spring water itself (see Section 20.8). Nevertheless, *Staphylothermus* may be widely distributed near hot hydrothermal vents and may be an ecologically important microbial consumer in these environments.

Table 20.7 summarizes the major properties of both submarine and terrestrial hyperthermophilic Archaea.

Archaeoglobus

In the preceding discussion we have emphasized that hyperthermophilic Archaea share a major metabolic theme: the use of elemental sulfur (S^0) as an electron acceptor for anaerobic growth. Interestingly, however, all hyperthermophilic Archaea discussed thus far are unable to use *sulfate* as an electron acceptor. *Archaeoglobus,* however, is a true sulfate-reducing hyperthermophile. This organism, isolated from hot sediments from marine hydrothermal vents, couples the oxidation of H_2, lactate, pyruvate, glucose, or complex organic mixtures to the reduction of sulfate to sulfide (sulfite or thiosulfate are also reduced as electron acceptors). Cells of *Archaeoglobus* are irregular motile spheres and cultures of this organism grow between 64° and 92°C with an optimum of 83°C. *Archaeoglobus* clearly belongs to the Archaea as adjudged from ribosomal RNA analyses, its lack of peptidoglycan, and its pattern of antibiotic resistance (see Section 18.9). However, one of the most remarkable aspects of this organism (besides its ability to reduce sulfate) is that it contains certain coenzymes previously found only in *methanogenic* Archaea. Specifically, coenzyme F_{420} and methanopterin, coenzymes intimately involved in biochemical reactions leading to the production of methane (see Section 20.5), are also present in cells of *Archaeoglobus*. In fact, *Archaeoglobus* actually produces small amounts of methane during growth as a sulfate-reducing organism!

Archaeoglobus thus shares some metabolic features with methanogens, despite its major phenotypic property of reducing sulfate. In addition, a specific relationship between *Archaeoglobus* and methanogens has been detected by molecular sequencing methods. In fact, analyses of 16S rRNA suggest that *Archaeoglobus* may be a "missing link" among Archaea. The phylogenetic position of *Archaeoglobus* on the archaeal tree (see Figure 18.13) lies approximately between the hyperthermophilic Archaea and the methanogens. Thus, sulfate reduction might have been a transitional type of metabolism important in the metabolic diversification of Archaea from sulfur-respiring phenotypes to methanogens and extreme halophiles.

Methanopyrus

A methanogenic prokaryote capable of growth over 100°C is also known. *Methanopyrus*, a Gram-positive rod-shaped methanogen has been isolated from sediments near submarine hot vents (see Section 17.10 for a discussion of thermal vents). *Methanopyrus* produces methane only from $H_2 + CO_2$ and grows at temperatures between 85° and 110°C. The organism grows rapidly (generation time less than one hour) at its temperature optimum of 100°C. Unlike most methanogens, however, cells of *Methanopyrus* have large amounts of the glycolytic derivative, cyclic 2,3-diphosphoglycerate, dissolved in the cytoplasm. This compound, present at over 1M concentration in *Methanopyrus* and other hyperthermophilic methanogens such as *Methanothermus,* is thought to act as a thermostabilizing agent to prevent denaturation of enzymes and DNA inside the cell.

The discovery of *Methanopyrus* may explain the origin of hydrocarbon-like materials in hot oceanic sediments previously thought to be too hot to support biogenic methanogenesis. In addition, at the depth at which *Methanopyrus* was found, approximately 2000 meters, water remains liquid at temperatures up to 350°C, suggesting that other hyperthermophilic methanogens may exist capable of growth at 110°C or at perhaps even higher temperatures.

20.7   *Thermoplasma*: A Cell-Wall-less Archaean

Thermoplasma is a cell-wall-less prokaryote that in this respect resembles the mycoplasmas. However, phylogenetically, *Thermoplasma* is a member of the Archaea. *Thermoplasma* (Figure 20.26) is an acidophilic, aerobic chemoorganotroph and is also thermophilic. *Thermoplasma* grows optimally at 55°C and pH 2 in complex media. With one exception, all strains of *Thermoplasma* have been obtained from self-heating coal refuse piles (Figure 20.27). Coal refuse contains coal fragments, pyrite, and other organic materials extracted from

coal, and when dumped into piles in coal mining operations, tends to self heat by spontaneous combustion (Figure 20.27). This sets the stage for growth of *Thermoplasma* which apparently metabolizes organic compounds leached from the hot coal refuse. Because coal refuse piles are transitory environments, other potential habitats for *Thermoplasma* have been sought. Although acid hot springs would appear to be ideal habitats for this organism, only one isolate of *Thermoplasma* has ever been obtained from such an environment, suggesting that coal refuse piles, or perhaps coal itself, is the primary habitat of *Thermoplasma*.

To survive the osmotic stresses of life without a cell wall and to withstand the dual environmental extremes of low pH and high temperature, *Thermoplasma* has evolved a cell membrane of chemically unique structure. The membrane contains lipopolysaccharide (referred to as *lipoglycan* in mycoplasmas, see Section 19.27) consisting of a *tetraether* lipid with mannose and glucose units (Figure 20.28). This molecule constitutes a major fraction of the total lipid composition of *Thermoplasma*. The membrane also contains glycoproteins but not sterols. Together these and other molecules render the *Thermoplasma* membrane stable to hot acid conditions.

The genome of *Thermoplasma* is of interest. *Thermoplasma* contains an extremely small genome, perhaps the smallest of all free-living prokaryotes. The DNA of *Thermoplasma* (about 1100 kilobase pairs) has a GC content of 46 percent and is surrounded by a highly basic DNA-binding protein that organizes the DNA into globular particles that resemble the nucleosomes of eukaryotic cells (see Section 3.13 for a discussion of the arrangement of DNA in eukaryotes). This protein strongly resembles the basic histone proteins of eukaryotic cells, and comparisons of amino acid sequences between the *Thermoplasma* protein and eukaryotic nuclear histones show significant sequence homology. Phylogenetically, *Thermoplasma* fits nearer the methanogen/extreme halophile branch than to the hyperthermophiles (see Figure 18.13).

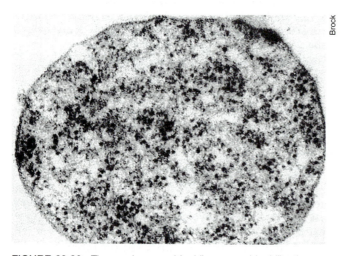

FIGURE 20.26 *Thermoplasma acidophilum*, an acidophilic, thermophilic mycoplasma-like archaean. Electron micrograph of thin section. The diameter of cells is highly variable from 0.2 to 5 µm. The cell shown is about 1 µm in diameter.

FIGURE 20.27
A typical self-heating coal refuse pile, habitat of *Thermoplasma*. (a) Spontaneous heat production can ignite nearby vegetation. (b) Photo of a large hot refuse pile.

(a)   *(b)*

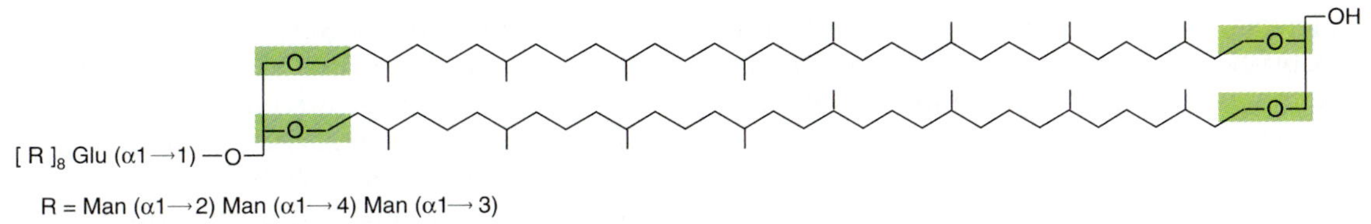

FIGURE 20.28 Structure of the tetraether lipoglycan of *Thermoplasma acidophilum*. Glu = glucose; Man = mannose. Note the ether linkages (shown in color) and compare with Figure 20.1.

20.8   Limits of Microbial Existence: Temperature

In the case of the hyperthermophilic Archaea, we see growth at temperatures far higher than those supporting growth of any other prokaryotes. What is the nature of this extreme heat tolerance and what are the temperature limits beyond which life is impossible? As we discussed in Section 9.9, the *macromolecules* of thermophiles and hyperthermophiles are stable to heat, and in the case of proteins, this stability resides in the unique folding of the molecules; the latter, of course, ultimately depends upon the *sequence* of amino acids in the protein. But is this all there is to it? No. Life at extremely high temperatures apparently requires a number of molecular adaptations, some of which are just beginning to become understood. Protein stability in hyperthermophiles may be assisted by the accumulation of solutes, such as the cyclic 2,3-diphosphoglycerate found in large amounts in cells of *Methanopyrus* and other thermophilic methanogens (see Section 20.7). In addition, special proteins may be necessary for growth at the highest temperatures. For example, in *Pyrodictium*, cells grown at 110°C produce 80 percent of their total protein as a single protein that has two enzymatic activities. Besides possessing ATPase activity, this protein functions as a molecular chaperonin, stabilizing other cellular proteins by refolding them as they begin to denature near the upper temperature limits for growth (the function of chaperonin proteins was discussed in Section 5.13). At 100°C (near the optimum for growth of *Pyrodictium*), very little of this chaperonin protein is made, suggesting that only at very extreme temperatures close to the upper limit do the otherwise thermally stable proteins of this organism begin to denature.

Opening and closing of the DNA helix in hyperthermophiles is also an interesting question. How does DNA in hyperthermophiles keep from denaturing? How is the helix opened up to allow exposure of the strands for transcription? This problem has been studied most extensively in *Methanothermus*, a hyperthermophilic methanogen. A DNA-binding protein has been found in *Methanothermus* cells that is closely related to the histone proteins of eukaryotic cells. This protein binds tightly to DNA and probably functions to open up the helix for transcription. It is not known how DNA is prevented from melting, especially in very low G + C hyperthermophiles (such as *Pyrococcus*, whose DNA of 38 percent G + C should quickly melt at its optimum growth temperature of 100°C). It

may involve a combination of high cytoplasmic solute concentrations (the melting temperature of DNA increases as solute concentration increases) and the activity of specific DNA-binding proteins that somehow prevent DNA from melting, perhaps by folding the DNA into a conformation consistent with thermal stability.

Besides the stability of *macromolecules*, however, a second consideration is important in governing the upper temperature limits for growth of hyperthermophilic bacteria: the thermal lability of *monomers*. At temperatures above 100°C, a number of biologically significant monomers show some degree of heat lability. And, no matter how stable the macromolecules are, life is impossible in an environment in which the basic building blocks themselves are unstable. The thermal stability of biomolecules therefore may dictate the upper temperature for life, and at temperatures as low as 100°C, some important biomolecules are destroyed. For example, molecules such as ATP and NAD$^+$ hydrolyze quite rapidly at high temperature; the half-life of both of these molecules is less than 30 minutes at 100°C. However, temperatures up to 113°C are clearly still compatible with life because at least one hyperthermophile (a *Pyrodictium* species) grows at this temperature.

The upper temperature limits for life

Since life depends on liquid water, microbial habitats in excess of 100°C are limited to environments where cells are under *pressure*, such as the sea floor. Indeed, all hyperthermophilic Archaea capable of growth at 100°C or higher seem restricted to these environments. However, another possible habitat is the deep subsurface of the earth where heat from the earth's interior coupled with immense pressures creates conditions suitable for growth of hyperthermophiles. Although such organisms have not been extensively sought out, various lines of evidence, primarily geochemical, suggest that hyperthermophiles may also exist kilometers below the earth's crust.

As discussed in Section 17.9, extremely hot (but not boiling) water is emitted from deep sea thermal vents (black smokers) at temperatures from 250° to 350°C. Will prokaryotes capable of growth at 250–350°C ever be isolated? Probably not. Life in such environments seems impossible. Although hyperthermophiles have been isolated from very near black smoker emissions, presumably existing in the gradient of hot water formed as the hydrothermal fluid mixes with 2°C seawater, at 250°C or higher, macromolecules and simple organic molecules such as amino acids and nucleotides are spontaneously hydrolyzed so quickly that life *as we know it* could not occur at such temperatures. Hence, although the upper temperature for life has not yet been defined, the discovery of hyperthermophilic Archaea has shown that this limit is not lower than 113°C, and is possibly even higher. Laboratory experiments on the heat stability of biomolecules suggest that living processes could be maintained at temperatures as high as 140–150°C, but that above this temperature organisms would probably not be able to overcome the heat lability of the important biomolecules of life. Indeed, some evidence for prokaryotes growing at 140°C was previously discussed (see Section 17.10), but chemical principles tell us that if such organisms exist, they must be living near the limits of microbial existence.

20.9   Archaea: Earliest Life Forms?

Now that we have considered the various groups of Archaea, what can we say about their role in microbial evolution? Although metabolically diverse organisms, a common theme running through the Archaea is *adaptation to environmental extremes.* High salt, high temperature, low pH, strictly anaerobic conditions— these environmental obstacles have all been overcome by one or another species of Archaea. Recalling our description of the geochemical conditions prevalent on earth at the time that life presumably arose (Section 18.2), do any particular groups of Archaea stand out as candidates for the earliest life forms? Yes, the hyperthermophiles. Recall that the surface of the early earth was likely much *hotter* than it is today, perhaps as high as 100°C or higher when living organisms first appeared. This implies that the earliest life forms would have had to be hyperthermophilic, perhaps not unlike present day hyperthermophilic Archaea.

Current phylogenetic evidence deduced from comparison of 16S rRNA sequences (see Chapter 18) suggest that Archaea have evolved *slower* than either Bacteria or the eukaryotes. This is especially true of hyperthermophilic Archaea (see Figure 18.13). It is not known why Archaea are the slowest evolving of the three domains, but it may be related to their inhabiting extreme environments. It is not hard to imagine that organisms living in thermal environments must maintain those genes which specify phenotypic characteristics critical to life at high temperatures; these genes cannot be significantly changed during evolution if the organism is to maintain itself in these environments. Thus, organisms like the hyperthermophilic Archaea are likely to have been among the earliest life forms. The phenotypic properties of this group, thermophilicity, and anaerobic chemoorganotrophic/chemolithotrophic metabolism, agree well with the phenotype of primitive organisms predicted from a consideration of early earth geochemical conditions (see Section 18.2). From sulfur-based respiration, it is possible that the methanogenic subline of Archaea arose, perhaps via transitionary forms like *Archaeoglobus.* Finally, extreme halophiles, thought to be the most derived (least ancient) of all Archaea, arose from methanogenic ancestors (see Figure 18.13).

Discovery of the Archaea has given us a new picture of evolution and has forced a reevaluation of even the most basic of evolutionary questions. But most excitingly, with the help of molecular sequencing (see Chapter 18), true evolutionary relationships can now be determined. Is a fourth domain of living organisms looming on the horizon? In the next few years more precise answers to this and a variety of other evolutionary questions should be forthcoming using molecular methods of phylogenetic analysis.

Supplementary Readings

Balows, A., H. G. Trüper, M. Dworkin, W. Harder, and **K. -H. Schleifer.** (eds.) 1992. *The Prokaryotes.* 2nd edition. Springer-Verlag, New York. This large compendium on the biology of bacteria has several chapters devoted to Archaea.

Danson, M. J., D. W. Hough, and **G. G. Lunt.** 1992. *Archaebacteria: Biochemistry and Biotechnology.* Biochemical Society Symposium, No. 58., Portland Press, Ltd., Colchester, U. K. A volume devoted to the biochemistry and potential biotechnological applications of Archaea. Balanced coverage of extreme halophiles, methanogens, and hyperthermophiles.

Kristjansson, J. K. (ed.) 1992. *Thermophilic Bacteria.* CRC Press, Inc., Boca Raton, FL. Focuses on thermophilic and hyperthermophilic Bacteria (rather than Archaea), but has good coverage of a number of areas relating to life at high temperatures.

Staley, J. T. (ed.) 1989 *Bergey's Manual of Systematic Bacteriology,* Vol. III. Williams and Wilkins, Baltimore, MD. This volume of *Bergey's Manual* includes the Archaea.

Woese, C. R., and **Wolfe, R. S.** (eds.) 1985. *Archaebacteria.* Volume VIII of *The Bacteria—a Treatise on Structure and Function.* Academic Press, New York. Although slightly dated, still a well-illustrated series of reviews on the basic biology of Archaea.

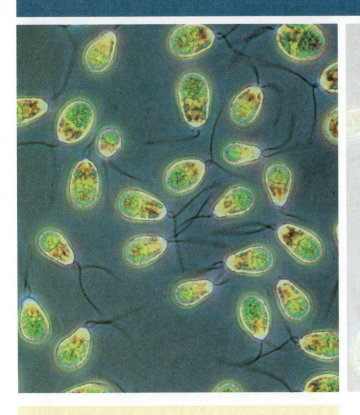

*E*ukarya: Eukaryotic Microorganisms

21

We complete our tour of the microbial world with a consideration of eukaryotic microorganisms. We have seen in various sections of this book how eukaryotes differ from Bacteria and Archaea in many fundamental ways, including cell size and internal structure (Chapter 3), genetic properties (Chapters 5 through 8), and evolutionary history (Chapter 18). We now consider eukaryotic microorganisms themselves and the major properties that are used to separate them into groups. We discuss here the algae, fungi (molds and yeasts), slime molds, and protozoans.

21.1 Algae

The term **algae** refers to a large and diverse assemblage of eukaryotic organisms that contain *chlorophyll* and carry out oxygenic photosynthesis (Figure 21.1). Algae should not be confused with *cyanobacteria*, which are also oxygenic phototrophs, but which are Bacteria and thus evolutionarily quite distinct from algae (see Section 19.2). Although most algae are of microscopic size and hence are clearly microorganisms, a number of forms are macroscopic, some seaweeds growing to over 100 feet in length.

Algae are either unicellular (Figure 21.1*a* and *c*) or colonial, the latter occurring as aggregates of cells (Figure 21.1*b*). When the cells are arranged end-to-end, the alga is said to be filamentous (Figure 21.1*d*). Among the filamentous forms, both unbranched filaments and more intricate branched filaments occur. Most algae contain chlorophyll and are thus green in color. However, a few kinds of common algae are not green but appear brown or red because in addition to chlorophyll, other pigments such as carotenoids are

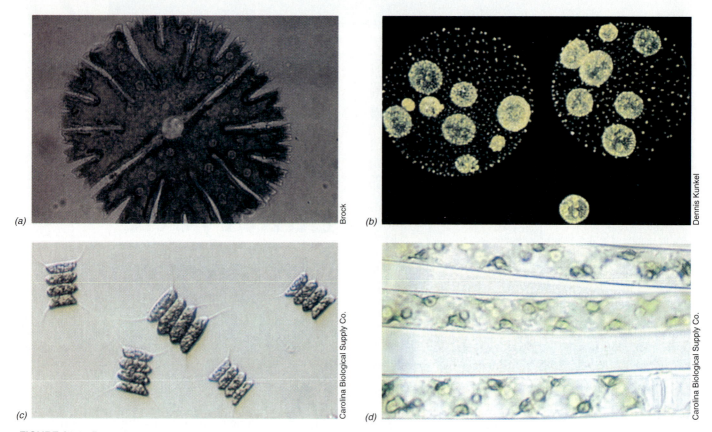

FIGURE 21.1 Representative green algae. (a) *Micrasterias*. A single cell. (b) *Volvox* colony, containing a large number of cells. (c) *Scenedesmus*, a packet of four cells. (d) *Spirogyra*, a filamentous alga. Note the green spiral-shaped chloroplasts.

present which mask the green color. Algal cells contain one or more **chloroplasts**, membranous structures that house the photosynthetic pigments. Chloroplasts can often be recognized microscopically within algal cells by their distinct green color (Figure 21.1). We discussed the general structure and properties of chloroplasts in Sections 3.14 and 3.15.

Characteristics used in classifying algae

Several characteristics are used to classify algae, including the nature of the chlorophyll(s) present, the carbon reserve polymers produced, the cell wall structure, and the type of motility. All algae contain chlorophyll *a*. Some, however, also contain other chlorophylls that differ in minor ways from chlorophyll *a*. The presence of these additional chlorophylls is characteristic of particular algal groups. Red algae are of special interest from the standpoint of pigment complement because their chloroplasts contain phycobiliproteins, a clear indication of their cyanobacterial roots (see Section 18.6). The distribution of chlorophylls and other photosynthetic pigments in algae is summarized in Table 21.1.

All algae carry out oxygen-evolving photosynthesis, using H_2O as an electron donor. In addition, some algae can use H_2 to carry out photosynthesis without yielding oxygen. Many algae are obligate phototrophs, and are thus unable to grow in darkness on organic carbon compounds. However, some algae can grow chemoorganotrophically and catabolize simple

sugars or organic acids in the dark. One of the organic compounds most widely used by algae is acetate, which can be used as a sole carbon and energy source by many flagellates and chlorophytes. In addition, some algae can assimilate simple organic compounds in the light (photoheterotrophy) but cannot grow on them as sole energy sources. One of the key characteristics used in the classification of algal groups is the nature of the **reserve polymer** synthesized as a result of photosynthesis. Algae of the division Chlorophyta produce starch (α-1,4-glucose) in a form very similar to that of higher plants. By contrast, algae of other groups produce a variety of reserve substances, some polymeric and some as free monomers, and the major ones are listed in Table 21.1.

Algae show considerable diversity in the structure and chemistry of their cell walls. In many cases the cell wall is composed of a network of cellulose fibrils, but it is usually modified by the addition of other polysaccharides such as pectin (highly hydrated polygalacturonic acid containing small amounts of the hexose rhamnose), xylans, mannans, alginic acids, or fucinic acid. In some algae, the wall is additionally strengthened by the deposition of calcium carbonate; these forms are often called "calcareous" or "coralline" (coral-like) algae. Sometimes chitin, a polymer of *N*-acetylglucosamine, is also present in the cell wall. In euglenids a cell wall is absent, which complicates the classification of these organisms: if the chloroplast of a euglenid is lost (as occurs on occasion), the organism is indistinguishable

Miniglossary for Chapter 21

ALGAE   phototrophic eukaryotic microorganisms

AMOEBOID MOVEMENT   a type of motility in which cytoplasmic streaming moves the organism forward

CHITIN   a polymer of *N*-acetylglucosamine commonly found in the cell walls of algae and fungi

CHLOROPLAST   the photosynthetic organelle of eukaryotic phototrophs

CILIATES   a group of protozoa characterized by rapid motility driven by numerous short appendages called cilia

CONIDIA   asexual spores of fungi

EUKARYA   all eukaryotic organisms

FLAGELLATES   a group of protozoa characterized by motility driven by the whiplike action of one or more long, thin appendages called flagella

FUNGI   nonphotosynthetic eukaryotic microorganisms that contain rigid cell walls

MITOCHONDRION   the respiratory organelle of eukaryotic organisms

MOLDS   filamentous fungi

MUSHROOMS   filamentous fungi that produce large, often edible structures called fruiting bodies

PHAGOCYTOSIS   a mechanism for ingesting particulate food in which a portion of the cell membrane surrounds the particle and brings it into the cell

PROTOZOA   unicellular eukaryotic microorganisms that lack cell walls

SLIME MOLDS   nonphototrophic eukaryotic microorganisms lacking cell walls which aggregate to form fruiting structures (cellular slime molds) or masses of protoplasm (acellular slime molds)

SPOROZOA   nonmotile parasitic protozoa

YEASTS   unicellular fungi

from a flagellated protozoan (see Section 21.4). In diatoms (Figure 21.2*b*) the cell wall is composed of *silica*, to which protein and polysaccharide are added. Even after the diatom dies and the organic materials have disappeared, the external structure remains, showing that the siliceous component is indeed responsible for the rigidity of the cell. Because of the extreme resistance to decay of these diatom frustules, they remain intact for long periods of time and constitute some of the best algal fossils ever found. From this excellent fossil record, it is known that diatoms first appeared on earth about 200 million years ago. Diatomaceous earth, an industrial filtering agent, is composed of fossilized diatom cells. This material is mined in large quantities from ancient sea beds, emphasizing the large numbers of these organisms found in prehistoric seas.

Table 21.1   Properties of major groups of algae

Algal group	Common name	Morphology	Pigments	Typical representative	Carbon reserve materials	Cell wall	Major habitats
Chlorophyta	Green algae	Unicellular to leafy	Chlorophylls *a,b*	*Chlamydomonas*	Starch (α-1, 4-glucan), sucrose	Cellulose	Freshwater, moist soils, a few marine
Euglenophyta	Euglenids	Unicellular	Chlorophylls *a,b*	*Euglena*	Paramylon (β-1,2-glucan)	No wall present	Freshwater
Chrysophyta	Golden-brown algae, diatoms	Unicellular	Chlorophylls *a,c,e*	*Navicula*	Lipids	Many have two overlapping components made of silica	Freshwater, marine, soil
Phaeophyta	Brown algae	Filamentous to leafy, occasionally massive and plantlike	Chlorophylls *a,c*, xanthophylls	*Laminaria*	Laminarin (β-1, 3-glucan), mannitol	Cellulose	Marine
Pyrrophyta	Dino-flagellates	Unicellular flagellated	Chlorophylls *a,c*	*Gonyaulax*	Starch (α-1, 4-glucan)	Cellulose	Freshwater, marine
Rhodophyta	Red algae	Unicellular, filamentous to leafy	Chlorophylls *a,d*, phycocyanin, phycoerythrin	*Polysiphonia*	Floridean starch (α-1, 4, and α-1,6, glucan), fluoridoside (glycerolgalactoside)	Cellulose	Marine

Algal cell walls are freely permeable to low molecular weight constituents such as water, ions, gases, and other nutrients. Their cell walls are essentially impermeable, however, to larger molecules or to macromolecules. Algal cell walls contain pores 3–5 nm wide which are sufficiently small to pass molecules of a molecular weight of about 15,000 or less. Thus, although animal cells can eat particulate matter, a process called *phagocytosis* (see Section 21.4), phagocytic activities are impossible in algae; particles large enough to be phagocytized never reach the cytoplasmic membrane because they are unable to penetrate the cell wall.

A number of algae are motile, usually because of flagella; cilia (see Sections 3.7 and 21.4) do not occur in algae. Simple flagellate forms, such as *Euglena* (see Figure 21.2*a*), have a single polar flagellum, while flagellated representatives of the Chlorophyta have either two or four polar flagella. Dinoflagellates (Figure 21.2*c*) have two flagella of different lengths and with different points of insertion into the cell. The transverse flagellum is attached laterally, whereas the longitudinal flagellum originates from the lateral groove of the cell and extends lengthwise. In many cases, algae are nonmotile in the vegetative state and form motile gametes only during sexual reproduction. Gliding motility, widespread among filamentous cyano-

bacteria (see Section 19.2), is present in only one group of eukaryotic algae, the diatoms.

Algae abound in nature in aquatic habitats, both freshwater and marine. Algae are also found in moist soils and artificial aquatic habitats like fish tanks and swimming pools, and in temporary pools of water formed from rainwater runoff. Evolutionary studies suggest that land plants are derived from green algae, although extant free-living close relatives of the type of green algae that became the forerunner of the green plants have not yet been discovered. A similar situation holds for the prokaryotic progenitor of the chloroplast (see Section 19.3).

21.2 Fungi

In contrast to the algae, the fungi lack chlorophyll. Fungi can be differentiated from prokaryotes by the fact that fungal cells are usually much larger and contain a nucleus, vacuoles, and mitochondria, typical of eukaryotic cells. Although the fungi are a large and diverse group of eukaryotic microorganisms, three groups of fungi have major practical importance: the **molds, yeasts**, and **mushrooms**.

The habitats of fungi are quite diverse. Some are aquatic, living primarily in fresh water, and a few ma-

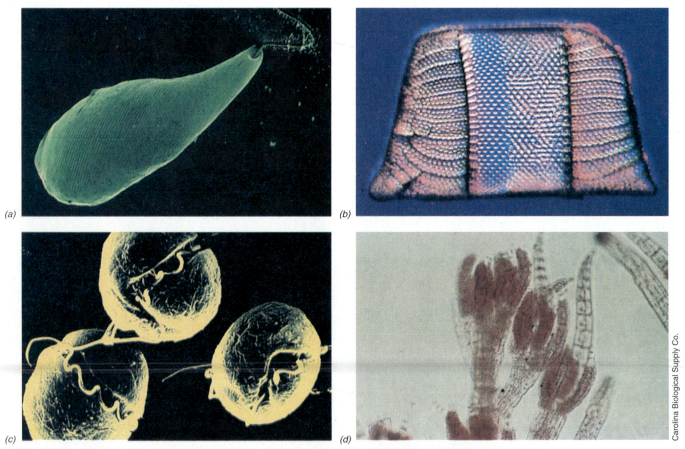

(a)

(b)

(c)

(d)

FIGURE 21.2 Scanning electron micrographs of algae other than Chlorophyta. (a) *Euglena*, a member of the Euglenophyta. (b) *Isthmix*, a diatom (Chrysophyta). (c) *Gonyaulax*, a dinoflagellate (Pyrrophyta). (d) *Polysiphonia*. This marine red alga grows attached to the surfaces of various marine plants.

Carolina Biological Supply Co.

rine fungi are also known. Most fungi, however, have terrestrial habitats, in soil or on dead plant matter, and these types often play crucial roles in the mineralization of organic carbon in nature. A large number of fungi are parasites of terrestrial plants. Indeed, fungi cause the majority of economically significant diseases of crop plants (see Table 21.2). A few fungi are parasitic on animals, including humans, although in general fungi are less significant as animal pathogens than are bacteria and viruses (see Section 15.15 for a discussion of pathogenic fungi).

Like algae, fungi also contain rigid cell walls. Fungal cell walls resemble plant cell walls architecturally, but not chemically. Although cellulose is present in the walls of certain fungi, many fungi have noncellulosic walls. **Chitin** is a common constituent of fungal cell walls. It is laid down in microfibrillar bundles like cellulose; other glucans such as mannans, galactosans, and chitosans replace chitin in some fungal cell walls. Fungal cell walls are generally 80–90 percent polysaccharide, with proteins, lipids, polyphosphates and inorganic ions making up the wall-cementing matrix. An understanding of fungal cell wall chemistry is important because of the extensive biotechnological uses of fungi (see Chapter 10) and because the chemical nature of the fungal cell wall has been useful in classifying fungi for research and industrial purposes.

All fungi are chemoorganotrophs. Lacking chlorophyll, they of course cannot photosynthesize, and the group also lacks chemolithotrophic forms. When compared to bacteria, the fungi in general have fairly simple nutritional requirements, and their metabolic and biosynthetic processes are not particularly diverse or unusual. It is in their morphological properties and in their sexual life cycles that the fungi exhibit considerable diversity; hence, it is on the basis of these characteristics that the fungi are classified. We present an overview of fungal classification in Table 21.2.

Molds

The molds are *filamentous* fungi. They are widespread in nature and are commonly seen on stale bread, cheese, or fruit. Each filament grows mainly at the tip, by extension of the terminal cell (Figure 21.3). A single filament is called a *hypha* (plural *hyphae*). Hyphae usually grow together across a surface and form compact tufts, collectively called a *mycelium*, which can be seen easily without a microscope. The mycelium arises because the individual hyphae form branches as they grow, and these branches interwine, resulting in a compact mat. In most cases, the vegetative cell of a fungal hypha contains more than one nucleus—often hundreds of nuclei are present. Thus, a typical hypha is a nucleated tube containing cytoplasm (referred to as *coenocytic*). Usually there is extensive cytoplasmic movement within a hypha, generally in a direction toward the hyphal tip, and the older portions of the hypha usually become vacuolated and virtually de-

	Table 21.2	Classification and major properties of fungi					
Group	Common name	Hyphae	Typical representatives	Type of sexual spore	Habitats	Common diseases
Ascomycetes	Sac fungi	Septate	*Neurospora, Saccharomyces, Morchella* (morels)	Ascospore	Soil, decaying plant material	Dutch elm, chestnut blight, ergot, rots
Basidiomycetes	Club fungi, mushrooms	Septate	*Amanita* (poisonous mushroom), *Agaricus* (edible mushroom)	Basidiospore	Soil, decaying plant material	Black stem, wheat rust, corn smut
Zygomycetes	Bread molds	Coenocytic	*Mucor, Rhizopus* (common bread mold)	Zygospore	Soil, decaying plant material	Food spoilage; rarely involved in parasitic disease
Oomycetes	Water molds	Coenocytic	*Allomyces*	Oospore	Aquatic	Potato blight, certain fish diseases
Deuteromycetes	Fungi imperfecti	Septate	*Penicillium, Aspergillus*	None	Soil, decaying plant material, surfaces of animal bodies	Plant wilt, fungal infections of animals such as ringworm, athlete's foot, and other dermatomycoses

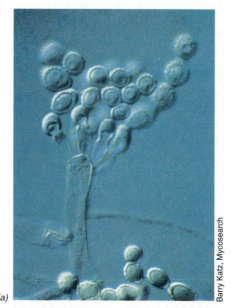

(a)

Barry Katz, Mycosearch

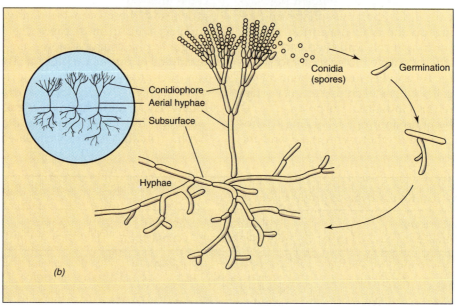

Conidiophore
Aerial hyphae
Subsurface

Hyphae

Conidia
(spores)

Germination

(b)

FIGURE 21.3 Mold structure and growth. (a) Photomicrograph of a typical mold. Conidia are seen as the spherical structures at the ends of aerial hyphae. (b) Diagram of a mold life cycle.

void of cytoplasm. Even if a hypha has cross walls, cytoplasmic movement is often not prevented, as there is usually a pore in the center of the septum, through which nuclei and cytoplasmic particles can move.

From the fungal mycelium, other hyphal branches may reach up into the air above the surface, and on these aerial branches spores called *conidia* are formed (Figure 21.3a). Conidia are *asexual* spores, often highly pigmented (see Figure 7.1a) and resistant to drying, and function in the dispersal of the fungus to new habitats. When conidia form, the white color of the mycelium changes, taking on the color of the conidia, which may be black, blue-green, red, yellow, or brown. The presence of these spores gives the mycelial mat a rather dusty appearance. Conidia are considered asexual spores because no sexual reproduction is involved in their formation. Because these spores are so numerous and spread so easily through the air, molds are common

laboratory contaminants. Airborne mold spores are also often responsible for allergies (see Section 12.14).

Some molds also produce *sexual* spores, formed as a result of sexual reproduction (Table 21.2). The latter occur from either the fusion of unicellular gametes or of specialized hyphae called *gametangia*. Alternatively, sexual spores can originate from the fusion of two haploid cells to yield a diploid cell which then undergoes meiosis and mitosis to yield individual spores. Depending upon the group to which a particular fungus belongs (see Table 21.2), different types of sexual spores are produced. Spores formed within an enclosed sac *(ascus)* are called *ascospores* and those produced on the ends of a club-shaped structure *(basidium)* are *basidiospores* (Table 21.2). Sexual spores of fungi are usually resistant to drying, heating, freezing, and some chemical agents. However, fungal sexual spores are not as resistant to heat as bacterial endospores. Either an asexual or sexual spore of a fungus can germinate and develop into a new hypha and mycelium.

A major ecological activity of many fungi, especially members of the Basidiomycetes (see Table 21.2), is the decomposition of wood, paper, cloth, and other products derived from natural sources. Basidiomycetes that attack these products are able to utilize cellulose or lignin from the product as carbon and energy sources. Lignin is a complex polymer in which the building blocks are phenolic compounds. It is an important constituent of woody plants, and in association with cellulose it confers rigidity on them. The decomposition of lignin in nature occurs almost exclusively through the action of certain Basidiomycetes called *wood-rotting fungi*. Two types of wood rots are known: *brown rot*, in which the cellulose is attacked preferentially and the lignin left unchanged, and *white rot*, in which both cellulose and lignin are decomposed. The white rot fungi are of considerable ecological interest since they play such an important role in decomposing woody material in forests.

Yeasts

The yeasts are *unicellular* fungi and most of them are classified with the Ascomycetes. Yeast cells are usually spherical, oval, or cylindrical, and cell division generally takes place by budding (Figure 21.4). In the budding process, a new cell forms as a small outgrowth of the old cell; the bud gradually enlarges and then separates (Figures 21.4 and 21.5). Yeasts usually do not form filaments or a mycelium, and the population of yeast cells remains a collection of single cells.

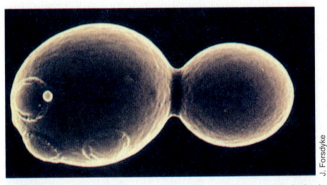

FIGURE 21.4   Scanning electron micrograph of the common baker's and brewer's yeast, *Saccharomyces cerevisiae*. Note the budding division and previous bud scars.

However, some yeasts can form a filamentous phase which in some cases is essential for expression of certain characteristics. For example, in the yeast *Candida albicans*, a pathogenic yeast causing vaginal infections, thrush, lung infections, and systemic tissue damage in AIDS patients (see Figure 15.30e), the filamentous phase is essential for pathogencity. Even the baker's yeast *Saccharomyces cerevisiae* has been shown to be capable of forming a filamentous phase.

Yeast cells are much larger than bacterial cells, and can be distinguished from bacteria by their size and by the obvious presence of internal cell structures. Some yeasts exhibit sexual reproduction by a process called *mating*, in which two yeast cells fuse. Within the fused cell, called a *zygote*, ascospores are eventually formed. We discussed the sexual cycle of a typical yeast, *Saccharomyces*, including the important property of *mating types*, in Section 7.14.

Yeasts usually flourish in habitats where sugars are present, such as fruits, flowers, and the bark of trees. A number of yeast species live symbiotically with animals, especially insects, and a few species are pathogenic for animals and humans (see Section 15.15). The most important commercial yeasts are the baker's and brewer's yeasts, which are members of the genus *Saccharomyces*. The original habitats of these yeasts were undoubtedly fruits and fruit juices, but the commercial yeasts of today are probably quite different from wild strains since they have been greatly improved through the years by careful selection and genetic manipulation by industrial microbiologists. Baker's and brewer's yeasts are probably the best-known scientifically of all fungi because they are easily manipulable eukaryotic cells, and they thus serve as excellent models for the study of many important problems in eukaryotic biology (see Section 7.14).

Mushrooms

Mushrooms are filamentous fungi that typically form large structures called *fruiting bodies*, the edible part of the mushroom (Figure 21.6). Many mushrooms live as mycorrhiza (see Section 17.22), whereas others live on dead organic matter in the soil or on the trunks of trees. Sexual spores called *basidiospores* are produced by mushrooms (Figure 21.6b) and are dispersed through the air and initiate mycelial growth on favorable substrates. The resulting haploid mycelium may grow extensively, but it will not form a fruiting body,

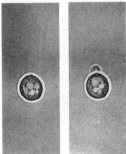
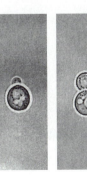
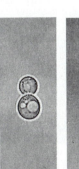

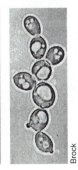

FIGURE 21.5
Growth by budding division in *Saccharomyces cerevisiae*.

(b)

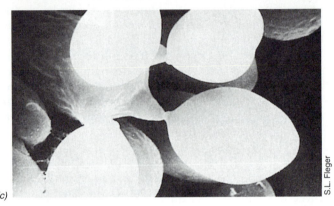

(c)

FIGURE 21.6
Mushrooms. (a) *Amanita*, a highly poisonous mushroom. (b) Gills on the underside of the mushroom contain the spore bearing basidia. (c) Scanning electron micrograph of basidiospores, released from mushroom basidia.

the latter occurring only after the fusion of haploid mycelia. Once fusion has occurred, each cell contains two nuclei and is called *dicaryotic*. Dicaryotic hyphae grow together and form small button-like structures which constitute the mushroom primordia. Buttons may remain underground for long periods until favorable conditions, usually heavy rains, stimulate enlargement and development of mature fruiting bodies. This expansion can occur rapidly, within a few hours or days, and is primarily due to the uptake of water. If many fruiting bodies mature simultaneously, a so-called *flush* is produced (see Figure 10.28) to yield a large number of mushrooms in a given locality which may have been devoid of these fruiting bodies the previous day.

Because many mushrooms are edible, and in some cases sought as delicacies, the large scale production of mushrooms is an important industrial process. We therefore considered other aspects of mushroom biology, particularly their growth as food products, in Section 10.16.

21.3   Slime Molds

Slime molds are nonphototrophic eukaryotic microorganisms that have some similarity to both fungi and protozoa. The slime molds can be divided into two groups, the *cellular slime molds*, whose vegetative forms are composed of single amoebalike cells, and the *acellular slime molds*, whose vegetative forms are naked masses of protoplasm of indefinite size and shape called *plasmodia*. Slime molds live primarily on decaying plant matter, such as leaf litter, logs, and soil (Figure 21.7). Their food consists mainly of other mi-

(a)

(b)

FIGURE 21.7   Slime molds. Plasmodia of acellular slime molds (a) growing on a decaying log and (b) growing on an agar surface.

croorganisms, especially bacteria, which they ingest by phagocytosis. One of the easiest ways of detecting and isolating slime molds is to bring small pieces of rotting logs into the laboratory and place them in moist chambers; the amoebas or plasmodia will proliferate (Figure 21.7), migrate over the surface of the wood, and eventually form fruiting bodies, which can then be observed under a dissecting microscope.

Cellular slime molds: *Dictyostelium*

Dictyostelium discoideum, a cellular slime mold, undergoes a remarkable life cycle in which vegetative cells aggregate, migrate as a cell mass, and eventually produce fruiting bodies in which cells differentiate and form spores (Figures 21.8 and 21.9). As cells of *Dictyostelium* become starved, they aggregate and form a *pseudoplasmodium,* a structure in which the cells lose their individuality, but do not fuse (Figures 21.8 and 21.9). This aggregation is triggered by the production of two compounds, *cyclic adenosine monophosphate* (cAMP) and a specific glycoprotein, both of which act as chemotactic agents. Those cells which are the first to produce these compounds serve as centers for the attraction of other vegetative cells, leading to aggregating masses of cells which come together and form a slimy migrating mass referred to as a *slug* (Figure 21.8*d*).

Fruiting-body formation begins when the slug ceases to migrate and becomes vertically oriented (Figures 21.8 and 21.9). The fruiting body then be-comes differentiated into a stalk and a head; cells in the forward end of the slug become stalk cells and those in the posterior end become spores. Cells that form stalk cells begin to secrete cellulose, which provides the rigidity of the stalk. Cells from the rear of the slug swarm up the stalk to the tip and form the head. Most of these posterior cells differentiate into spores. Upon maturation of the head the spores are released and dispersed. Each spore germinates and becomes a vegetative amoeba.

The cycle of fruiting-body and spore formation in *Dictyostelium* is an *asexual* process. However *sexual* spores called *macrocysts* are also produced by this organism and by other cellular slime molds. Macrocysts are structures of multicellular origin that develop from simple aggregates of amoebas, which, when mature, are enclosed in a thick cellulose wall. As a result of the conjugation of two amoebas, a single large amoeba develops near the center of the cell mass and becomes actively phagocytic. This phagocytic cell continues to enlarge and engulf amoebas until eventually all surrounding cells have been engulfed. At this stage a cellulose wall develops around the now greatly enlarged amoeba and the mature macrocyst is formed. During a period of dormancy, the diploid nucleus undergoes meiosis, haploid nuclei are formed, and by progressive nuclear divisions and cytoplasmic cleavage, a new generation of vegetative amoebas is formed. The macrocyst then germinates and releases amoebas that reinitiate vegetative growth.

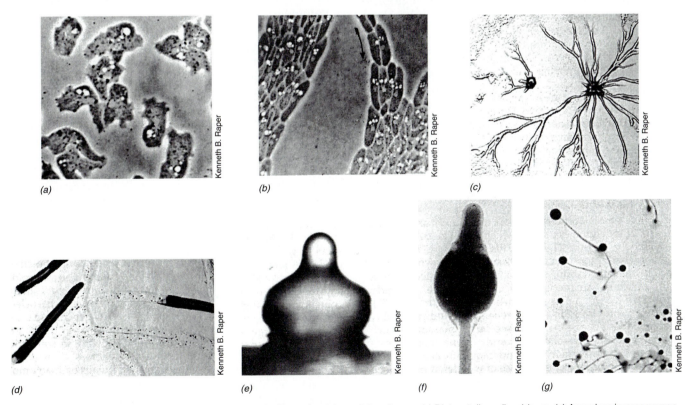

FIGURE 21.8 Photomicrographs of various stages in the life cycle of the cellular slime mold *Dictyostelium discoideum.* (a) Amoebas in preaggregation stage. Note irregular shape and lack of orientation. (b) Aggregating amoebas. Notice the regular shape and orientation. The cells are moving in streams in one direction. (c) Low-power view of aggregating amoebas. (d) Migrating pseudoplasmodia (slugs) moving on an agar surface and leaving trails of slime in their wake. (e, f) Early stage of fruiting body. (g) Mature fruiting bodies. See Figure 21.9 for sizes of these structures.

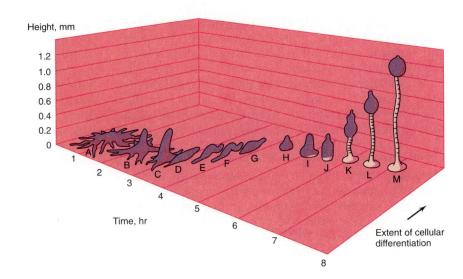

FIGURE 21.9
Stages in fruiting body formation in the cellular slime mold *D. discoideum*: (A–C) aggregation of amoebas; (D–G) migration of the slug formed from aggregated amoeba; (H–L) culmination and formation of the fruiting body; (M) mature fruiting body.

Acellular slime molds

In the vegetative phase, acellular slime molds exist as a mass of protoplasm of indefinite extent, which might be compared to a giant amoeba (Figure 21.7). This structure is actively motile by *amoeboid motion,* the plasmodium flowing over the surface of the substratum, engulfing food particles as it moves. Amoeboid movement is the result of cytoplasmic streaming. Cytoplasm flows forward because the tip of the plasmodium is less contracted and viscous and thus cytoplasm takes the path of least resistance. Cytoplasmic streaming is facilitated by filaments of a protein called *actin,* which exists in a thin layer just beneath the cytoplasmic membrane of eukaryotic cells. In acellular slime molds, cytoplasmic streaming occurs in definite strands, each surrounded by the thin cytoplasmic membrane (Figure 21.7*b*). The strands of protoplasm do not retain their individuality indefinitely, but fuse occasionally into larger masses which then separate again into smaller strands. Slime mold plasmodia are often brilliantly colored and frequently can be seen spreading across the surface of a piece of wood that has been kept moist (Figure 21.7).

The acellular slime mold plasmodium is diploid. From this, two kinds of differentiated structures are produced, *sporangia* and *sclerotia.* Sporangia arise as part of a sexual cycle; their production is initiated by aggregation of the protoplasmic mass into a compact unit and culminates in secretion by the plasmodium of a stalk, usually devoid of protoplasm, and in the formation of a complex head within which spores develop. Soon after the formation of the head, nuclei in this structure undergo meiosis, and these haploid nuclei then become forerunners of the spores. The protoplasmic mass in the head collects around the haploid nuclei, and a thick cell wall is formed around each. The spores thus formed are discharged when the noncellular part of the fruiting body disintegrates. These resting spores are relatively resistant to drying and other unfavorable conditions and are able to remain dormant for long periods of time. Under favorable conditions, germination occurs. Each spore forms one

to four swarm cells that eventually produce more swarm cells by binary fission. Eventually swarm cells conjugate and are converted into a diploid amoeboid cell. This cell then grows and undergoes successive nuclear divisions, producing the new diploid plasmodium.

Under starvation conditions, some acellular slime molds form another kind of resting structure, the sclerotium, an irregular, hard, dry mass of protoplasm, resistant to drying and desiccation. No sexual process is involved in this conversion, but the resistant sclerotium, having enabled the organism to overwinter, can form a new plasmodium when conditions are favorable.

21.4   Protozoa

Protozoa are unicellular eukaryotic microorganisms that lack cell walls (Figure 21.10). They are generally colorless and motile. **Protozoa** are distinguished from prokaryotes by their usually greater size and eukaryotic nature, from algae by their lack of chlorophyll, from yeasts and other fungi by their motility and absence of a cell wall, and from the slime molds by their lack of fruiting body formation. Protozoa usually obtain food by ingesting other organisms or organic particles. Protozoa are found in a variety of freshwater and marine habitats; a large number are parasitic in other animals, including humans, and some are found growing in soil or in aerial habitats, such as on the surfaces of trees.

Protozoans feed by ingesting particulate or macromolecular materials. The uptake of macromolecules in solution occurs by a process called pinocytosis. Fluid droplets are sucked into a channel formed by the invagination of the cell membrane, and when portions of this channel are pinched off, the fluid becomes enclosed within a membrane-bound vacuole. Most protozoa are also able to ingest particulate material by phagocytosis, a process of surrounding a food particle with a portion of their flexible cell membrane to engulf the particle and bring it into the cell. Some protozoa

(a)

(b)

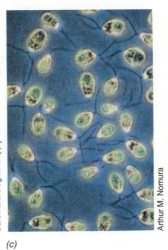
(c)

Carolina Biological Supply Co.

Arthur M. Nomura

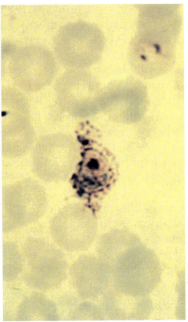
(d)

Arthur M. Siegelman

FIGURE 21.10   Typical protozoa. (a) *Amoeba*. (b) A typical ciliate, *Paramecium*.
(c) A flagellate, *Dunaliella* (this flagellate contains chloroplasts). (d) *Plasmodium vivax*, a sporozoan,
growing in red blood cells.

can literally swallow particulate matter (such as bacterial cells) by operation of a special structure called a gullet (see below, Figure 21.14).

As is appropriate for organisms that "catch" their own food, most protozoa are motile. Indeed, their mechanisms of motility are key characteristics used to divide them into major taxonomic groups (Table 21.3). Protozoa that move by amoeboid motion are called Sarcodina; those using flagella, the Mastigophora; and those using cilia, the Ciliophora. The Sporozoa, a fourth group, are generally nonmotile and are all parasitic for higher animals.

Mastigophora: The Flagellates

Members of this protozoal group are motile by the action of flagella (Figure 21.10c and Figure 21.11). Although many flagellated protozoans are free-living organisms, a number are parasitic in, or pathogenic for, animals, including humans. The most important path-

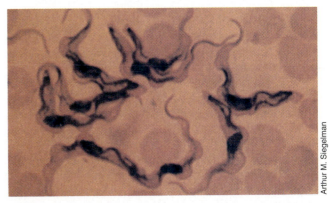

Arthur M. Siegelman

FIGURE   21.11   Photomicrograph   of   the   flagellated   protozoan, *Trypanosoma gambiense*, the causative agent of African sleeping sickness.

Table 21.3   Characteristics of the major groups of protozoa

Group	Common name	Typical representatives	Habitats	Common diseases
Mastigophora	Flagellates	*Trypanosoma, Giardia, Leishmania*	Freshwater; parasites of animals	African sleeping sickness, giardiasis, leishmaniasis
Sarcodina	Amoebas	*Amoeba, Entamoeba*	Freshwater and marine; animal parasites	Amoebic dysentery (Amebiasis)
Ciliophora	Ciliates	*Balantidium, Paramecium*	Freshwater and marine; animal parasites; rumen	Dysentery
Sporozoa	Sporozoans	*Plasmodium, Toxoplasma*	Primarily animal parasites; insects (vectors for parasitic diseases)	Malaria, toxoplasmosis

ogenic Mastigophora are the *trypanosomes*. These organisms cause a number of serious diseases in humans and vertebrate animals, including the feared disease *African sleeping sickness*. In *Trypanosoma*, the genus infecting humans, the protozoa are rather small, about 20 µm in length, and are thin, crescent-shaped organisms. They have a single flagellum that originates in a basal body and folds back laterally across the cell where it is enclosed by a flap of surface membrane (Figure 21.11). Both the flagellum and the membrane participate in propelling the organism, making effective movement possible even in blood, which is rather viscous. *Trypanosoma gambiense* is the species that causes the chronic and usually fatal African sleeping sickness. In humans, the parasite lives and grows primarily in the bloodstream, but in the later stages of the disease, invasion of the central nervous system occurs, causing an inflammation of the brain and spinal cord that is responsible for the characteristic neurological symptoms of the disease. The parasite is transmitted from host to host by the tsetse fly, *Glossina* sp., a bloodsucking fly found only in certain parts of Africa. The parasite proliferates in the intestinal tract of the fly and invades the insect's salivary glands and mouthparts from which it can be transferred to a new human host following a single fly bite.

Sarcodina: The Amoebas

Among the sarcodines are organisms such as *Amoeba* which are always naked in the vegetative phase (Figure 21.10a), and the foraminefera, amoebas which secrete a shell during vegetative growth (see Figure 21.12). A wide variety of naked amoebas are parasites of humans and other vertebrates, and their usual habitat is the oral cavity or the intestinal tract. They move in these habitats by *amoeboid movement* (Figure 21.13), a mechanism previously discussed for the acellular slime molds (Section 21.3). *Entamoeba histolytica* (see Figure 15.48) is a good example of a parasitic amoeba. In many cases infection causes no obvious symptoms, but in some individuals it produces ulceration of the intestinal tract, which results in a diarrheal condition called *amoebic dysentery* (amebiasis). The organism is transmitted from person to person in the cyst form by

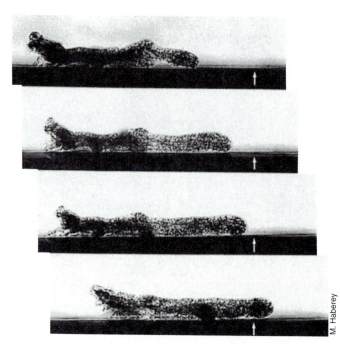

FIGURE 21.13 Side view of a moving amoeba, *Amoeba proteus*, taken from a film, the time interval between frames being 2 seconds. The arrows point to a fixed spot on the surface. A single cell is about 80 µm in diameter.

M. Haberey

fecal contamination of water and food. We discussed the etiology and pathogenesis of amoebic dysentery in Section 15.14.

Shelled sarcodines present a variety of interesting morphological forms. The best known of the shelled forms are the *foraminifera*. Foraminifera are exclusively marine organisms, living primarily in coastal waters. The shells, called *tests*, of different species show distinctive characteristics and are often of great beauty (Figure 21.12). Tests are usually made of calcium carbonate. The cell is usually not firmly attached to the test, and the amoeba cell may extend partway out of the shell during feeding. However, because of the weight of the test, the cell usually sinks to the bottom, and it is thought that the organisms feed on particulate deposits in the sediments, primarily bacteria and detritus. The shells of foraminifera are relatively resistant to decay and hence readily become fossilized (the White Cliffs of Dover, England, are composed to a great extent of foraminiferal shells.) Because of the excellent fossil record these organisms leave, we have a better idea of their distribution through geological time than of virtually any other protozoa.

Ciliophora: The Ciliates

Ciliates are defined as those protozoa that, in some stage of their life cycle, possess cilia (Figure 21.14a). They are also unique among protozoa in having *two kinds* of nuclei: the *micronucleus*, which is concerned only with inheritance and sexual reproduction; and the *macronucleus*, which is only involved in the production of mRNA for various aspects of cell growth and function. Probably the best known and most widely distributed of the ciliates are those of the genus *Paramecium*

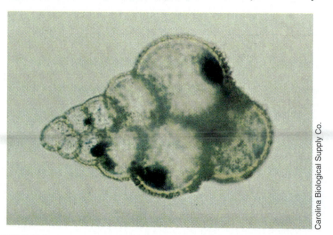

FIGURE 21.12 Shelled amoebae: foraminifera.

Carolina Biological Supply Co.

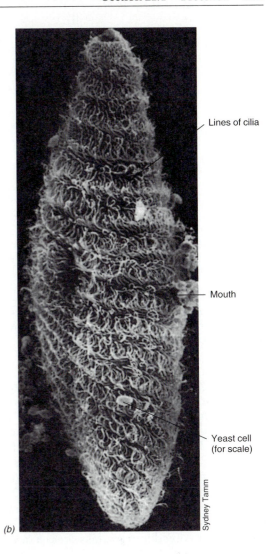

Lines of cilia

Mouth

Yeast cell (for scale)

Sydney Tamm

(b)

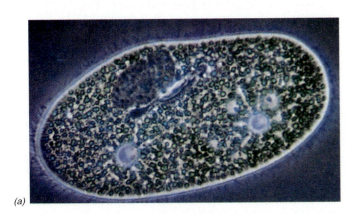

(a)

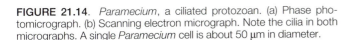

FIGURE 21.14. *Paramecium*, a ciliated protozoan. (a) Phase photomicrograph. (b) Scanning electron micrograph. Note the cilia in both micrographs. A single *Paramecium* cell is about 50 μm in diameter.

(Figure 21.14*a*) which will be used here as an example of the group. Most ciliates obtain their food by ingesting particulate materials through a distinct oral region or mouth which is connected to an underlying gullet (Figure 21.14*b*). Once inside, the food particles are carried down the gullet and into the cytoplasm where they are enclosed in a food vacuole, a structure into which digestive enzymes are secreted. In addition to cilia, which function in motility, many ciliates have *trichocysts*, which are long, thin filaments of a contractile nature, anchored beneath the surface of the outer cell layer. These enable the protozoa to attach to a surface and can aid in defense by indicating to the cell that it is being attacked by a predator. In the case of predatory ciliates, such as *Didinium*, trichocysts hold on to and paralyze the prey as a prelude to ingestion.

The two kinds of nuclei in the ciliates set this group aside from all other protozoa. The micronucleus is diploid, but the macronucleus is polyploid, containing from about 40 to over 500 times as much DNA as does the micronucleus. The micronucleus plays no direct role in vegetative growth and cell division since strains can be developed that lack micronuclei but which continue to divide normally. However, if the macronucleus is removed, the cell quickly dies.

Sexual reproduction in ciliates involves conjugation between two cells of opposite mating type. Conjugation involves exchange and fusion of micronuclei with the net result being the formation of exconjugants that are hybrid for certain genes of the two strains.

Many *Paramecium* species (as well as many other protozoa) contain endosymbiotic bacteria living in the cytoplasm or the macronucleus. In some cases, evidence exists that these endosymbionts play a nutritional role for the host, synthesizing vitamins or other growth factors that would otherwise have to be obtained from the environment. In the case of protozoans existing in the termite gut, we previously saw how endosymbiotic methanogens remove H_2 produced from fermentation, yielding CH_4 which is released to the atmosphere (see Section 17.12 and Figure 17.30).

Although a few ciliates are parasitic for animals, this mode of existence is less extensively developed in the ciliates than it is in other groups of protozoa. The species *Balantidium coli* (Figure 21.15) is primarily a parasite of domestic animals, but occasionally it infects the intestinal tract of humans, producing intestinal symptoms similar to those caused by *Entamoeba histolytica*. In addition, there is usually a characteristic fauna of *oblig-*

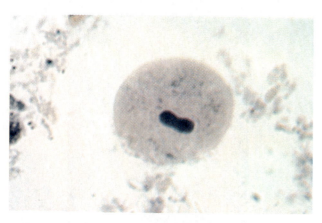

FIGURE 21.15. *Balantidium coli*, a ciliated protozoan that causes a dysentery-like disease in humans. The dark blue stained structure is the macronucleus.

ately anaerobic ciliates in the rumen, the forestomach of ruminant animals (see Section 17.13); these protozoa are thought to play a beneficial role in the digestive and fermentative processes which occur there.

Sporozoa

The Sporozoa comprise a large group of protozoa, all of which are obligate parasites. They are characterized by a *lack* of motile adult stages and by a nutritional mode of life in which food is generally not ingested but instead is absorbed in soluble form through the outer wall, such as occurs in prokaryotes and in fungi. Although the name *Sporozoa* implies the formation of spores, these organisms do not form true resting spores, like those of bacteria, algae, and fungi, but instead produce analogous structures called *sporozoites*, which are involved in transmission to a new host. Nu-

merous kinds of vertebrates and invertebrates serve as hosts for Sporozoa, and in some cases an alternation of hosts occurs, with some stages of the life cycle occurring in one host and some in another. The most important members of the sporozoa are the coccidia, usually parasites of birds, and the plasmodia (malaria parasites, Figure 21.10*d*), which infect birds and mammals, including humans. Because malaria is a major disease of humans, especially in developing countries, we devoted a considerable discussion to this disease and the properties of malarial parasites in Section 15.11.

21.5   Eukaryotes in the Microbial World

The above discussion provides only a brief overview of eukaryotic microbial diversity, but it is hoped that the material will serve as a springboard for those inclined to study a particular group in more depth. Because an enormous amount of information is available on the biology of the prokaryotes and their activities are of such immense importance, we have emphasized these organisms throughout this book. But this is not to say that eukaryotic microorganisms are less important. From molecular biology to industrial microbiology, and from medical microbiology to environmental microbiology, microorganisms of all types, including eukaryotes, have served as superb tools for basic and applied biological studies. We hope that with this treatment of eukaryotic microorganisms, which completes our discussion of the microbial world, that we have infected the reader with some of the delight and excitement in microorganisms that the microbiologist experiences, and have stimulated a desire to explore further the fascinating world of microbial life.

Supplementary Readings

Bold, H. C., and **M. J. Wynne.** 1985. *Introduction to the Algae,* 2nd edition. Prentice-Hall, Englewood Cliffs, NJ. A basic introduction to all groups of algae.

Farmer, J. N. 1980. *The Protozoa: Introduction to Protozoology.* C. V. Mosby, St. Louis.

Moore-Landecker, E. 1990. *Fundamentals of the Fungi,* 3rd edition. Prentice-Hall, Englewood Cliffs, NJ. A general textbook of fungi.

Rowan, K. S. 1989. *Photosynthetic Pigments of Algae.* Cambridge University Press, Cambridge, U.K. Algal pigments from a taxonomic viewpoint.

Energy Calculations in Microbial Bioenergetics

The information in Appendix 1 is given to help students calculate changes in free energy accompanying chemical reactions carried out by microorganisms. It begins with definitions of the terms required to make such calculations and proceeds to show how knowledge of redox state, atomic and charge balance, and other factors are necessary to calculate free energy problems successfully.

Definitions

1. ΔG^0 = standard free energy change of the reaction, at 1 atm pressure and 1 M concentrations; ΔG = free energy change under the conditions used; $\Delta G^{0\prime}$ = free energy change under standard conditions at pH 7.

2. Calculation of ΔG^0 for a chemical reaction from the free energy of formation, G_f^0, of products and reactants:

    $$\Delta G^0 = \Sigma\, \Delta G_f^0 \text{ (products)} - \Sigma\, \Delta G_f^0 \text{ (reactants)}$$

    That is, sum the ΔG_f^0 of products, sum the ΔG_f^0 of reactants, and subtract the latter from the former.

3. For energy-yielding reactions involving H^+, converting from standard conditions (pH 0) to biochemical conditions (pH 7):

    $$\Delta G^{0\prime} = \Delta G^0 + m\, \Delta G'_f (H^+)$$

    where m is the net number of protons in the reaction (m is negative when more protons are consumed than formed), and $\Delta G'_f (H^+)$ is the free energy of formation of a proton at pH 7 = -39.87 kJ at 25°C.

4. Effect of concentrations on ΔG: with soluble substrates, the concentration ratios of products formed to exogenous substrates used are generally equal to or greater than 10^{-2} at the beginning of growth, and equal to or less than 10^{-2} at the end of growth. From the relation between ΔG and equilibrium constant (see item 8 below), it can be calculated that ΔG for the free energy yield in practical situations differs from free energy yield under standard conditions by at most 11.7 kJ, a rather small amount, so that for a first approximation, standard free energy yields can be used in most situations. However, with H_2 as a product, H_2-consuming bacteria present may keep the concentration of H_2 so low that the free energy yield is significantly affected. Thus, in the fermentation of ethanol to acetate and H_2 ($C_2H_5OH + H_2O \rightarrow C_2H_3O_2^- + 2H_2 + H^+$), the $\Delta G^{0\prime}$ at 1 atm H_2 is +9.68 kJ, but at 10^{-4} atm H_2 it is -36.03 kJ. With H_2-consuming bacteria present, therefore, the ethanol fermentation becomes useful. (See also item 9 below.)

5. Reduction potentials: by convention, electrode equations are written in the direction, oxidant + $ne^- \rightarrow$ reductant (that is, as reductions), where n is the number of electrons transferred. The standard potential (E_0) of the hydrogen electrode, $2H^+ + 2e^- \rightarrow H_2$ is set by definition at 0.0 V at 1.0 atm pressure of H_2 gas and 1.0 M (molar) H^+, at 25°C. E'_0 is the standard reduction potential at pH 7. See also Table A1.2.

6. Relation of free energy to reduction potential:

    $$\Delta G^{0\prime} = -nF\, \Delta E'_0$$

    where n is the number of electrons transferred, F is the Faraday constant (96.48 kJ/V), and $\Delta E'_0$ is the E'_0 of the electron-accepting couple minus the E'_0 of the electron-donating couple.

7. Equilibrium constant, K. For the generalized reaction $aA + bB \rightleftharpoons cC + dD$

    $$K = \frac{[C]^c[D]^d}{[A]^a[B]^b}$$

    where A, B, C, and D represent reactants and products; a, b, c, and d represent number of molecules of each; and brackets indicate concentrations. This is true only when the chemical system is in equilibrium.

8. Relation of equilibrium constant, K, to free energy change. At constant temperature and pressure,

$$\Delta G = \Delta G^0 + RT \ln K$$

where R is a constant (8.29 joules·mole^{-1}K^{-1}) and T is the absolute temperature (^{0}K).

9. Two substances can be coupled in a redox reaction even if the standard potentials are unfavorable, provided that the concentrations are appropriate.

Assume that normally the reduced form of A would donate electrons to the oxidized form of B. However, if the concentration of the reduced form of A were low and the concentration of the reduced form of B were high, it would be possible for the reduced form of B to donate electrons to the oxidized form of A. Thus the redox couple would function in the reverse direction from that predicted from standard potentials. A practical example of this is the utilization of H^+ as an electron acceptor to produce H_2. Normally, H_2 production in fermentative bacteria is not extensive because H^+ is a poor electron acceptor; the midpoint potential of the $2H^+/H_2$ pair is -0.42 V. However, if the concentration of H_2 is kept low by continually removing it (a process done by methanogenic prokaryotes, which use $H_2 + CO_2$ to produce methane, CH_4, or by many other anaerobes capable of consuming H_2 anaerobically), the potential will be more positive and then H^+ will serve as a suitable electron acceptor.

Oxidation State or Number

1. The oxidation state of an element in an elementary substance (i.e., H_2, O_2) is zero.

2. The oxidation state of the ion of an element is equal to its charge (i.e., $Na^+ = +1$, $Fe^{3+} = +3$, $O^{2-} = -2$).

3. The sum of oxidation numbers of all atoms in a neutral molecule is zero. Thus, H_2O is neutral because it has two H at $+1$ each and one O at -2.

4. In an ion, the sum of oxidation numbers of all atoms is equal to the charge on that ion. Thus, in the OH^- ion, $O(-2) + H(+1) = -1$.

5. In compounds, the oxidation state of O is virtually always -2, and that of H is $+1$.

6. In simple carbon compounds, the oxidation state of C can be calculated by adding up the H and O atoms present and using the oxidation states of these elements as given in item 5, since in a neutral compound the sum of all oxidation numbers must be zero. Thus the oxidation state of carbon in methane, CH_4, is -4 (4H at $+1$ each $= +4$); in carbon dioxide, CO_2, the oxidation state of carbon is $+4$ (2O at -2 each $= -4$).

7. In organic compounds with more than one C atom, it may not be possible to assign a specific oxidation number to each C atom, but it is still useful to calculate the oxidation state of the compound as a whole. The same conventions are used. Thus the oxidation state of carbon in glucose, $C_6H_{12}O_6$, is zero (12H at $+1 = 12$; 6O at $-2 = -12$) and the oxidation state of carbon in ethanol, C_2H_6O, is -4 (6H at $+1 = +6$; one O at -2).

8. In all oxidation-reduction reactions there is a balance between the oxidized and reduced products. To calculate an oxidation-reduction balance, the number of molecules of each product is multiplied by its oxidation state. For instance, in calculating the oxidation-reduction balance for the alcoholic fermentation, there are two molecules of ethanol at $-4 = -8$ and two molecules of CO_2 at $+4 = +8$, so that the net balance is zero. When constructing model reactions, it is useful to calculate redox balances to be certain that the reaction is possible.

Calculating Free Energy Yields for Hypothetical Reactions

Energy yields can be calculated either from free energy of formations of the reactants and products or from differences in reduction potentials of electron-donating and electron-accepting partial reactions.

Calculations from free energy

Free energies of formation are given in Table A1.1. The procedure to use for calculating energy yields of reactions follows.

1. *Balancing reactions*   In all cases, it is essential to ascertain that the coupled oxidation-reaction is balanced. Balancing involves three things: (*a*) the total number of each kind of atom must be identical on both sides of the equation; (*b*) there must be an ionic balance, so that when positive and negative ions are added up on the right side of the equation, the total ionic charge (whether positive, negative, or neutral) exactly balances the ionic charge on the left side of the equation; and (*c*) there must be an oxidation-reduction balance, so that all of the electrons removed from one substance must be transferred to another substance. In general, when constructing balanced reactions, one proceeds in the reverse of the three steps listed above. Usually, if steps (*c*) and (*b*) have been properly handled, step (*a*) becomes correct automatically.

2. *Examples*   (*a*) What is the balanced reaction for the oxidation of H_2S to SO_4^{2-} with O_2? First, decide how many electrons are involved in the oxidation of H_2S to SO_4^{2-}. This can be most easily calculated from the oxidation states of the compounds, using the rules given above. Since H has an oxidation state of $+1$, the oxidation state of S in H_2S is -2. Since O has an oxidation state of -2, the oxidation state of S in SO_4^{2-} is $+6$ (since it is an ion, using the rules given in items 4 and 5 above). Thus, the oxidation of H_2S to SO_4^{2-} involves an eight-electron transfer (from -2 to $+6$). Since each O atom can accept two electrons (the oxidation state of O in O_2

Table A1.1   Free energies of formation (G^0_f) for some substances (kJ/mole)*

Carbon compound	Metal	Nonmetal	Nitrogen compound
CO, −137.34	Cu⁺, +50.28	H_2, 0	N_2, 0
CO_2, −394.4	Cu²⁺, +64.94	H⁺, 0 at pH 0	NO, +86.57
CH_4, −50.75	CuS, −49.02	−5.69 per pH unit	NO_2, +51.95
H_2CO_3, −623.16	Fe²⁺, −78.87	O_2, 0	NO_2^-, −37.2
HCO_3^-, −586.85	Fe³⁺, −4.6	OH⁻, −157.3 at pH 14	NO_3^-, −111.34
CO_3^{2-}, −527.90	FeS_2, −150.84	−198.76 at pH 7	NH_4^+, −79.37
Acetate, −369.41	$FeSO_4$, −829.62	−237.57 at pH 0	N_2O, +104.18
Alanine, −371.54	PbS, −92.59	H_2O, −237.17	
Aspartate, −700.4	Mn²⁺, −227.93	H_2O_2, −134.1	
Benzoate, −245.6	Mn³⁺, −82.12	PO_4^{3-}, −1026.55	
Butyrate, −352.63	MnO_4^{2-}, −506.57	Se⁰, 0	
Caproate, −335.96	MnO_2, −456.71	H_2Se, −77.09	
Citrate, −1168.34	$MnSO_4$, −955.32	SeO_4^{2-}, −439.95	
Crotonate, −277.4	HgS, −49.02	S⁰, 0	
Ethanol, −181.75	MoS_2, −225.42	SO_3^{2-}, −486.6	
Formaldehyde, −130.54	ZnS, −198.60	SO_4^{2-}, −744.6	
Formate, −351.04		$S_2O_3^{2-}$, −513.4	
Fructose, −915.38		H_2S, −27.87	
Fumarate, −604.21		HS⁻, +12.05	
Gluconate, −1128.3		S²⁻, +85.8	
Glucose, −917.22			
Glutamate, −699.6			
Glycerate, −658.1			
Glycerol, −488.52			
Glycine, −314.96			
Glycolate, −530.95			
Guanine, +46.99			
Lactate, −517.81			
Lactose, −1515.24			
Malate, −845.08			
Mannitol, −942.61			
Methanol, −175.39			
Oxalate, −674.04			
n-Propanol, −175.81			
Propionate, −361.08			
Pyruvate, −474.63			
Ribose, −757.3			
Succinate, −690.23			
Urea, −203.76			
Valerate, −344.34			

*Values for free energy of formation of various compounds can be found in Dean, J. A. 1973. Lange's Handbook of Chemistry, 11th ed, McGraw-Hill, New York; Garrels, R. M., and C. L. Christ. 1965. Solutions, Minerals, and Equilibria. Harper & Row, New York; Burton, K. 1957. Appendix in Krebs, H. A., and H. L. Komberg, Energy Transformations in Living Matter, Ergebnisse der Physiologie. Springer-Verlag, Berlin; and Thauer, R. K., K. Jungerman, and K. Decker, 1977. Bacteriol. Rev. Vol. 41, pp 100–180.

is zero, but in H_2O is −2), this means that two molecules of molecular oxygen, O_2, will be required to provide sufficient electron-accepting capacity. Thus, at this point, we know that the reaction requires one H_2S and $2O_2$ on the left side of the equation, and one SO_4^{2-} on the right side. To achieve an ionic balance, we must have two positive charges on the right side of the equation to balance the two negative charges of SO_4^{2-}. Thus, two H⁺ must be added to the right side of the equation, making the overall reaction

$$H_2S + 2O_2 \rightarrow SO_4^{2-} + 2H^+$$

By inspection, it can be seen that this equation is also balanced in terms of the total number of atoms of each kind on each side of the equation.

(b) What is the balanced reaction for the oxidation of H_2S to SO_4^{2-} with Fe³⁺ as electron acceptor? We have just ascertained that the oxidation of H_2S to SO_4^{2-} is an eight-electron transfer. Since the reduction of Fe³⁺ to Fe²⁺ is only a one-electron transfer, 8Fe³⁺ will be required. At this point, the reaction looks like this:

$$H_2S + 8Fe^{3+} \rightarrow 8Fe^{2+} + SO_4^{2-} \qquad \text{(not balanced)}$$

We note that the ionic balance is incorrect. We have 24 positive charges on the left and 14 positive charges on the right (16+ from Fe, 2− from sulfate). To equalize the charges, we add 10H⁺ on the right. Now our equation looks like this:

$$H_2S + 8Fe^{3+} \rightarrow 8Fe^{2+} + 10H^+ + SO_4^{2-} \qquad \text{(not balanced)}$$

To provide the necessary hydrogen for the H^+ and oxygen for the sulfate, we add $4H_2O$ to the left and find that the equation is now balanced:

$$H_2S + 4H_2O + 8Fe^{3+} \rightarrow 8Fe^{2+} + 10H^+ + SO_4^{2-}$$

(balanced)

In general, in microbiological reactions, ionic balance can be achieved by adding H^+ or OH^- to the left or right side of the equation, and since all reactions take place in an aqueous medium, H_2O molecules can be added where needed. Whether H^+ or OH^- is added will generally depend upon whether the reaction is taking place in acid or alkaline conditions.

3. *Calculation of energy yield for balanced equations from free energies of formation*   Once an equation has been balanced, the free energy yield can be calculated by inserting the values for the free energy of formation of each reactant and product from Table A1.1, and using the formula in item 2 of the first section of this Appendix.

For instance, for the equation

$$H_2S + 2O_2 \rightarrow SO_4^{2-} + 2H^+$$

G'_f values $\rightarrow (-27.87) + (0)$ $(-744.6) + 2\ (-39.83)$

(assuming pH 7)

$\Delta G^{0\prime} = -796.39$ kJ/mole

The values on the right are summed and subtracted from the values on the left, taking care to ensure that the signs are correct. From the data in Table A1.1, a wide variety of free energy yields for reactions of microbiological interest can be calculated.

Calculation of free energy yield from reduction potential

Reduction potentials of some important redox pairs are given in Table A1.2. The amount of energy that can be released when two half-reactions are coupled can be calculated from the differences in reduction potentials of the two reactions and from the number of electrons transferred. The further apart the two half-reactions are, and the greater the number of electrons, the more energy released. The conversion of potential difference to free energy is given by the formula $\Delta G^0 = -nF\ \Delta E'_0$, where n is the number of electrons, F is the Faraday constant (96.48 kJ/V) and $\Delta E'_0$ is the difference in potentials. Thus the $2H^+/H_2$ redox pair has a potential of -0.41 and the $\frac{1}{2}O_2/H_2O$ pair has a potential of $+0.82$, so that the potential difference is 1.23, which (since two electrons are involved) is equivalent to a free energy yield (ΔG^0) of -237.34 kJ/mole. On the other hand, the potential difference between the $2H^+/H_2$ and the NO_3^-/NO_2^- reactions is less, 0.84 V, which is equivalent to a free energy yield of -162.08 kJ/mole.

Because many biochemical reactions are two-electron transfers, it is often useful to give energy yields for two-electron reactions, even if more electrons are involved. Thus the SO_4^{2-}/H_2 redox pair involves eight

electrons, and complete reduction of SO_4^{2-} with H_2 would require $4H_2$ (equivalent to eight electrons). From the reduction potential difference between $2H^+/H_2$ and SO_4^{2-}/H_2S (0.19 V), a free-energy yield of -146.64 kJ is calculated, or -36.66 kJ per two electrons. By convention, reduction potentials are given for conditions in which equal concentrations of oxidized and reduced forms are present. In actual practice, the concentrations of these two forms may be quite different, As discussed earlier in this appendix (item 9, first section), it is possible to couple half-reactions even if the potential difference is unfavorable, providing the concentrations of the reacting species are appropriate.

Table A1.2 Reduction potentials of some redox pairs of biochemical and microbiological importance	
Redox pair	E'_0**(V)**
SO_4^{2-}/HSO_3^-	-0.52
$CO_2/$formate$^-$	-0.43
H^+/H_2	-0.41
$S_2O_3^{2-}/HS^- + HSO_3^-$	-0.40
Ferredoxin ox/red	-0.39
Flavodoxin ox/red*	-0.37
$NAD^+/NADH$	-0.32
Cytochrome c_3 ox/red	-0.29
$CO_2/$acetate$^-$	-0.29
S^0/HS^-	-0.27
CO_2/CH_4	-0.24
FAD/FADH	-0.22
SO_4^{2-}/HS^-	-0.217
$HSO_3^-/S_3O_6^{2-}$	-0.20
Acetaldehyde/ethanol	-0.197
Pyruvate$^-$/lactate$^-$	-0.19
FMN/FMNH	-0.19
Dihydroxyacetone phosphate/ glycerolphosphate	-0.19
$HSO_3^-/S_3O_6^{2-}$	-0.17
Flavodoxin ox/red*	-0.12
HSO_3^-/HS^-	-0.116
Menaquinone ox/red	-0.075
APS/AMP + HSO_3^-	-0.060
Rubredoxin ox/red	-0.057
Acrylyl-CoA/propionyl-CoA	-0.015
Glycine/acetate$^-$ + NH_4^+	-0.010
$S_4O_6^{2-}/S_2O_3^{2-}$	$+0.024$
Fumarate$^-$/succinate$^-$	$+0.033$
Cytochrome b ox/red	$+0.035$
Ubiquinone ox/red	$+0.113$
$S_3O_6^{2-}/S_2O_3^{2-} + HSO_3^-$	$+0.225$
Cytochrome c_1 ox/red	$+0.23$
NO_2^-/NO	$+0.36$
Cytochrome a_3 ox/red	$+0.385$
NO_3^-/NO_2^-	$+0.43$
Fe^{3+}/Fe^{2+}	$+0.77$
O_2/H_2O	$+0.82$
NO/N_2O	$+1.18$
N_2O/N_2	$+1.36$

Data from Thauer, R. K., K. Jungermann, and K. Decker, 1977. Bacteriol. Rev. 41:100–180.
*Separate potentials are given for each electron transfer in this potentially two-electron transfer.

Mathematics of Microbial Growth and Chemostat Operation

Exponential Growth

Although many analyses of microbial growth can be done graphically or algebraically, as discussed in Chapter 9, for some purposes it is convenient to use a differential equation that expresses the quantitative relationships of growth:

$$\frac{dX}{dt} = \mu X \tag{1}$$

where X may be cell number or some specific cellular component such as protein and μ is the *instantaneous growth-rate constant*. If this equation is integrated, we obtain a form which reflects the activities of a typical batch culture population:

$$\ln X = \ln X_0 + \mu(t) \tag{2}$$

where ln refers to the natural logarithm (logarithm base e), X_0 is cell number at time 0, X is cell number at time t, and t is elapsed time during which growth is measured. This equation fits experimental data from the exponential phase of bacterial cultures very well, such as those in Figure 9.2. Taking the antilogarithm of each side gives

$$X = X_0 e^{\mu t} \tag{3}$$

This equation is useful because it allows prediction of population density at a future time by knowing the present value and μ, the growth-rate constant. As discussed in Section 9.2, an important constant parameter for an exponentially growing population is the doubling (or generation) time. Doubling of the population has occurred when $X/X_0 = 2$. Rearranging and substituting this value into equation 3 gives

$$2 = e^{\mu(t_{gen})} \tag{4}$$

Taking the natural logarithm of each side and rearranging gives

$$\mu = \frac{\ln 2}{t_{gen}} = \frac{0.693}{t_{gen}} \tag{5}$$

The generation time t_{gen} may be used to define another growth parameter, k, as follows:

$$k = \frac{1}{t_{gen}} \tag{6}$$

where k is the growth-rate constant for a batch culture. Combining Equations 5 and 6 shows that the two growth-rate constants, μ and k, are related:

$$\mu = 0.693k$$

It is important to understand that μ and k are both reflections of the same growth process of an exponentially increasing population. The difference between them may be seen in their derivation above: μ is the *instantaneous rate constant* and k is an *average* value for the population over a finite period of time. (See Table A2.1 for calculation of k values from experimental results.) This distinction is more than a mathematical point. As was emphasized in Chapter 9, microbial growth studies must deal with population phenomena, not the activities of individual cells. The constant k reflects this averaging assumption. However, the constant μ, being instantaneous, is a closer approximation of the rate at which individual activities are occurring. Further, the instantaneous constant μ allows us to consider bacterial growth dynamics in a theoretical framework separate from the traditional batch culture.

Mathematical Relationships of Chemostats

An especially important application of the instantaneous growth-rate constant μ is, of course, the chemostat, a culture device (Section 9.6) in which population size and growth rate may be maintained at constant values of the experimenter's choosing, over a wide range of values. As we saw in Section 9.5, the rate of bacterial growth is a function of nutrient concentration. This function represents a saturation process which may be described by the equation

$$\mu = \mu_{max} \frac{S}{K_s + S} \tag{8}$$

where μ_{max} is the growth rate at nutrient saturation, S is the concentration of nutrient,* and K_s is a saturation constant which is numerically equal to the nutrient concentration at which $\mu = \frac{1}{2}\,\mu_{max}$. Equation 8 is formally equivalent to the Michaelis–Menten equation used in enzyme kinetic analyses. Estimates of the unknown parameters μ_{max} and K_s can be made by plotting $1/\mu$ on the y axis versus $1/s$ on the x axis. A straight line will be obtained which intercepts the y axis, and the value at the intercept is $1/\mu_{max}$. The slope of the line is K_s/μ_{max}, and since μ_{max} is known, K_s can then be calculated. Values for K_s as a rule are very low. A few examples are 2.1×10^{-5} M for glucose (*Escherichia coli*), 1.34×10^{-6} M for oxygen (*Candida utilis*), 3.5×10^{-8} M for phosphate (*Spirillum* spp.), and 4.7×10^{-13} M for thiamine (*Cryptococcus albidus*).

The key to the operation of the chemostat is that, in a steady-state population, the nutrient concentration is very low, usually very near 0. Therefore, each drop of fresh medium entering the chemostat supplies nutrient which is consumed almost instantaneously by the population and the experimenter has direct control over the growth rate: if nutrient is added more rapidly, μ increases; if it is added more slowly, μ decreases. In a chemostat the rate at which medium enters is equal to the rate at which spent medium and cells exit through the overflow. This rate is called the *dilution rate*, D, and is defined:

$$D = \frac{F}{V} \qquad (9)$$

where F is the flow rate (in units of volume per time, usually ml/hour) and V is chemostat volume (in ml). Therefore, D has units of time^{-1} (usually hour^{-1}) as does μ. In fact, at steady state, $\mu = D$. That μ must equal D may be seen by consideration of Equations 8 and 9 and the discussion above. When S is low, μ is a direct function of S. But since nutrient is instantly consumed upon addition, the value of S at any moment is controlled by the rate of medium addition, that is, the dilution rate D. A further crucial consequence of the relation $\mu = D$ is that a chemostat is a *self-regulating* system within broad limits of D. Consider a steady-state chemostat functioning with $\mu = D$. If μ were to increase momentarily due to some variation in the culture, S would necessarily decline, as the increased growth caused increased nutrient consumption. The lower value of S would in turn cause μ to decrease back to its stable level. Conversely, if μ were to decrease momentarily, then S would increase as "extra" nutrient went unconsumed. This rise in S would lead to higher values of μ until steady state is again reached. This biological feedback system functions well in practice, as it does in theory. Chemostat populations may be maintained at constant growth rates (μ) for long periods of time: months or even years.

There is another important aspect to chemostat studies and that is *population density*. The relation between cell growth and nutrient consumption may be defined:

$$Y = \frac{dX}{dS} \qquad (10)$$

where Y is the yield constant for the particular organism on the particular medium and dX/dS is the amount of cell increase per unit of substrate consumed. The yield constant is in units of

$$Y = \frac{\text{weight of bacteria formed}}{\text{weight of substrate consumed}}$$

and is a measure of the efficiency with which cells convert nutrient to more cell material. The composition of the medium in the reservoir is central to the successful steady-state operation of the chemostat. All components in this medium are present in excess, except one, the growth-limiting substrate. The symbol for the concentration of that substrate in the reservoir is S_R, whereas that for the same substrate in the culture vessel is S. Taking into account Equations 1, 8, and 10, the following considerations can be made.

When a chemostat is inoculated with a small number of bacteria, and the dilution rate does not exceed a certain critical value (D_c), the number of organisms will start to increase. The increase is given by

---

Table A2.1. Calculation of k values

Bacterial population densities are expressed in scientific notation with powers of 10, so Equation 3 may be converted to terms of logarithm base 10 and k substituted for the instantaneous constant μ:

$$k = \frac{\log_{10}X_t - \log_{10}X_0}{0.301t}$$

Example 1:
$X_0 = 1000\ (= 10^3)\ \log_{10}$ of $1000 = 3$
$X_t = 100{,}000\ (= 10^5)\ \log_{10}$ of $100{,}000 = 5$
$t = 4$ hours
$$k = \frac{5 - 3}{(0.301)4} = \frac{2}{1.204}$$
$k = 1.66$ doublings per hour
$t_{gen} = 0.60$ hour (36 minutes) for population to double

Example 2:
$X_0 = 1000\ (= 10^3)\ \log_{10}$ of $1000 = 3$
$X_t = 100{,}000{,}000\ (= 10^8)\ \log_{10}$ of $10^8 = 8$
$t = 120$ hours
$$k = \frac{8 - 3}{(0.301)120} = \frac{5}{36.12}$$
$k = 0.138$ doubling per hour
$t_{gen} = 7.2$ hours (430 minutes) for population to double

---

*It is usual to speak of a *limiting nutrient*, that is, some component of the growth medium (often the electron donor) which, if increased in concentration, causes an increase in growth rate. One must be careful with this terminology, however, for, as Section 9.5 showed, there is usually observed a saturation level for each nutrient. At saturation values, the particular nutrient under consideration no longer limits growth and some other component or factor becomes limiting.

$$\frac{dX}{dt} = \text{growth} - \text{output} \quad \text{or} \quad \frac{dX}{dt} = \mu X - DX \quad (11)$$

since initially $\mu > D$, dX/dt is positive. However, as the population density increases, the concentration of the growth-limiting substrate decreases, causing a decrease of μ (Equation 8). As discussed above, if D is kept constant, a steady state will be reached in which $\mu = D$ and $dX/dt = 0$.

Steady states are possible only when the dilution rate does not exceed a critical value D_c. This value depends on the concentration of the growth-limiting substrate in the reservoir (S_R):

$$D_c = \mu_{max}\left(\frac{S_R}{K_s + S_R}\right) \quad (12)$$

Thus, when $S_R \gg K_s$ (which is usually the case), steady states can be obtained at growth rates close to μ_{max}. The change in S is given by

$$\frac{dS}{dt} = \text{input} - \text{output} - \text{consumption}$$

Since consumption is growth divided by yield constant (see Equation 10), we have

$$\frac{dS}{dt} = S_R D - sD - \frac{\mu X}{Y} \quad (13)$$

When a steady state is reached, $dS/dt = 0$.

From Equations 11 and 13, the steady-state values of organism concentration ($\bar{X}$) and growth-limiting substrate ($\bar{S}$) can be calculated:

$$\bar{X} = Y(S_R - S) \quad (14)$$

$$\bar{S} = K_s\left(\frac{D}{\mu_{max} - D}\right) \quad (15)$$

These equations have the constants, S_R, Y, K_s, and μ_{max}. Once their values are known, the steady-state values of $\bar{X}$ and $\bar{S}$ can be predicted for any dilution rate. As can be seen from Equation 15, the steady-state substrate concentration ($\bar{S}$) depends on the dilution rate (D) applied, and is independent of the substrate concentration in the reservoir (S_R). Thus D determines $\bar{S}$, and $\bar{S}$ determines $\bar{X}$. A specific example of the use of the equations to predict chemostat behavior can be obtained by study of Figure 9.10. In that figure, the constants are $\mu_{max} = 1.0$ hour^{-1}; $Y = 0.5$; $K_s = 0.2$ g/liter; $S_R = 10$ g/liter. With these constants, the curves in Figure 9.10 can be generated.

Experimental Uses of the Chemostat

One of the major theoretical advantages of a chemostat is that this device allows the experimenter to control growth rate (μ) and population density ($\bar{X}$) independently of each other. Over rather wide ranges (see above and Section 9.6) any desired value of μ can be obtained by alteration of D. In practice this means changing the flow rate, since volume is usually fixed (Equation 9). Similarly, the population density $\bar{X}$ may be determined by varying nutrient concentration in

the reservoir (Equation 14). This independent control of these two crucial growth parameters is not possible with conventional batch cultures.

A practical advantage to the chemostat is that a population may be maintained in a desired growth condition (μ and $\bar{X}$) for long periods of time. Therefore, experiments can be planned in detail and performed whenever most convenient. Comparable studies with batch cultures are essentially impossible, since specific growth conditions are constantly changing.

Mathematics of the Stationary Phase

The differential equation (Equation 1) expressing the exponential growth phase can be modified to include the trend toward the stationary phase, which occurs at high population density. A second term is added to the equation that expresses the maximum population attainable for that organism under the environmental conditions specified, here called X_m. The equation then assumes a form often called the *logistics equation*:

$$\frac{dX}{dt} = \mu X - \frac{\mu}{X_m}X^2 \quad (16)$$

The second term in this equation essentially expresses self-crowding effects, such as nutrient depletion or inhibitor buildup. In integrated form, this equation will graph as a typical microbial growth curve, showing exponential and stationary phases. The length of the exponential phase will be determined by the size of X_m. When X is small, the second term will have little effect, whereas as X approaches X_m, the growth rate (dX/dt) will approach zero. The logistics equation is widely used by ecologists studying higher organisms to express the growth rate and carrying capacity of the habitat for animals and plants. The equation is equally applicable to microbial situations.

Mathematics of Disinfection

An equation can also be developed expressing the death rate of unicellular organisms upon treatment with a disinfectant or other lethal agent. If X_d represents the number of dead cells at time t, then $X_0 - X_d$ will represent the survivors. The death rate is proportional to the survivors:

$$\frac{dX}{dt} = k(X_0 - X_d) \quad (17)$$

Upon integration, this equation gives a typical first-order relationship:

$$kt = \ln\frac{X_0}{X_0 - X_d} \quad (18)$$

and when the logarithm of $X_0 - X_d$ (that is, the number of survivors) is plotted against time, a straight line will be obtained. This equation is useful when determining the length of time necessary to allow a disinfectant to completely sterilize something (for example, the contact time for chlorine gas in a water purification plant).

Bergey's Classification of Bacteria

Bergey's Manual of Systematic Bacteriology is a recognized authority on bacterial taxonomy.* The *Manual* is divided into four volumes. Each volume contains several sections, and each section contains a number of related genera. In brief, the contents of each volume is as follows:

Volume I **1984. Gram-negative Bacteria of medical and commercial importance:** spirochetes, spiral and curved Bacteria, Gram-negative aerobic and facultatively aerobic rods, Gram-negative obligate anaerobes, Gram-negative aerobic and anaerobic cocci, sulfate and sulfur-reducing Bacteria, rickettsias and chlamydias, mycoplasmas.

Volume II **1986. Gram-positive Bacteria of medical and commercial importance:** Gram-positive cocci, Gram-positive endospore-forming and nonsporing rods, mycobacteria, nonfilamentous actinomycetes.

Volume III **1989. Remaining Gram-negative Bacteria and the Archaea:** phototrophic, gliding, sheathed, budding and appendaged Bacteria, cyanobacteria, chemolithotrophic Bacteria; methanogens, extreme halophiles, hyperthermophiles, *Thermoplasma* and other Archaea.

Volume IV 1989. Filamentous actinomycetes and related bacteria.

The detailed list of genera in *Bergey's Manual of Systematic Bacteriology* follows. Names given in quotation marks refer to organisms in which the nomenclature is not yet clear.

*Besides *Bergey's Manual,* a four volume treatise called *The Prokaryotes,* 2nd edition (edited by A. Balows, H.G. Trüper, M. Dworkin, W. Harder, and K.-H. Schleifer, 1992) is available and should be consulted for more recent developments in bacterial classification.

Volume I

SECTION 1
The Spirochetes
Order I: Spirochaetales
    Family I: Spirochaetaceae
        Genus I: *Spirochaeta*
        Genus II: *Cristispira*
        Genus III: *Treponema*
        Genus IV: *Borrelia*
    Family II: Leptospiraceae
        Genus I: *Leptospira*
Other Organisms
    Hindgut Spirochetes of Termites
        and *Cryptocercus punctulatus*

SECTION 2
Aerobic/Microaerophilic, Motile, Helical/Vibrioid Gram-Negative Bacteria
        Genus: *Aquaspirillum*
        Genus: *Spirillum*
        Genus: *Azospirillum*
        Genus: *Oceanospirillum*
        Genus: *Campylobacter*
        Genus: *Bdellovibrio*
        Genus: *Vampirovibrio*

SECTION 3
Nonmotile (or Rarely Motile), Gram-Negative Curved Bacteria
    Family I: Spirosomaceae
        Genus I: *Spirosoma*
        Genus II: *Runella*
        Genus III: *Flectobacillus*
    Other Genera
        Genus: *Microcyclus*
        Genus: *Meniscus*
        Genus: *Brachyarcus*
        Genus: *Pelosigma*

SECTION 4
Gram-Negative Aerobic Rods and Cocci
    Family I: Pseudomonadaceae
        Genus I: *Pseudomonas*
        Genus II: *Xanthomonas*
        Genus III: *Frateuria*
        Genus IV: *Zoogloea*
    Family II: Azotobacteraceae
        Genus I: *Azotobacter*
        Genus II: *Azomonas*
    Family III: Rhizobiaceae
        Genus I: *Rhizobium*
        Genus II: *Bradyrhizobium*
        Genus III: *Agrobacterium*
        Genus IV: *Phyllobacterium*

Family IV: Methylococcaceae
  Genus I: *Methylococcus*
  Genus II: *Methylomonas*
Family V: Halobacteriaceae
  Genus I: *Halobacterium*
  Genus II: *Halococcus*
Family VI: Acetobacteraceae
  Genus I: *Acetobacter*
  Genus II: *Gluconobacter*
Family VII: Legionellaceae
  Genus I: *Legionella*
Family VIII: Neisseriaceae
  Genus I: *Neisseria*
  Genus II: *Moraxella*
  Genus III: *Acinetobacter*
  Genus IV: *Kingella*
Other Genera
  Genus: *Beijerinckia*
  Genus: *Derxia*
  Genus: *Xanthobacter*
  Genus: *Thermus*
  Genus: *Thermomicrobium*
  Genus: *Halomonas*
  Genus: *Alteromonas*
  Genus: *Flavobacterium*
  Genus: *Alcaligenes*
  Genus: *Serpens*
  Genus: *Janthinobacterium*
  Genus: *Brucella*
  Genus: *Bordetella*
  Genus: *Francisella*
  Genus: *Paracoccus*
  Genus: *Lampropedia*

SECTION 5
Facultatively Anaerobic
Gram-Negative Rods
Family I: Enterobacteriaceae
  Genus I: *Escherichia*
  Genus II: *Shigella*
  Genus III: *Salmonella*
  Genus IV: *Citrobacter*
  Genus V: *Klebsiella*
  Genus VI: *Enterobacter*
  Genus VII: *Erwinia*
  Genus VIII: *Serratia*
  Genus IX: *Hafnia*
  Genus X: *Edwardsiella*
  Genus XI: *Proteus*
  Genus XII: *Providencia*
  Genus XIII: *Morganella*
  Genus XIV: *Yersina*
Other Genera of the Family
Enterobacteriaceae
  Genus: *Obesumbacterium*
  Genus: *Xenorhabdus*
  Genus: *Kluyvera*
  Genus: *Rahnella*
  Genus: *Cedecea*
  Genus: *Tatumella*
Family II: Vibrionaceae
  Genus I: *Vibrio*
  Genus II: *Photobacterium*
  Genus III: *Aeromonas*
  Genus IV: *Plesiomonas*
Family III: Pasteurellaceae
  Genus I: *Pasteurella*
  Genus II: *Haemophilus*
  Genus III: *Actinobacillus*
Other Genera
  Genus: *Zymomonas*
  Genus: *Chromobacterium*
  Genus: *Cardiobacterium*
  Genus: *Calymmatobacterium*
  Genus: *Gardnerella*
  Genus: *Eikenella*
  Genus: *Streptobacillus*

SECTION 6
Anaerobic Gram-Negative Straight,
Curved and Helical Rods
Family I: Bacteroidaceae
  Genus I: *Bacteroides*
  Genus II: *Fusobacterium*
  Genus III: *Leptotrichia*
  Genus IV: *Butyrivibrio*
  Genus V: *Succinimonas*
  Genus VI: *Succinivibrio*
  Genus VII: *Anaerobiospirillum*
  Genus VIII: *Wolinella*
  Genus IX: *Selenomonas*
  Genus X: *Anaerovibrio*
  Genus XI: *Pectinatus*
  Genus XII: *Acetivibrio*
  Genus XIII: *Lachnospira*

SECTION 7
Dissimilatory Sulfate- or
Sulfur-Reducing Bacteria
  Genus: *Desulfuromonas*
  Genus: *Desulfovibrio*
  Genus: *Desulfomonas*
  Genus: *Desulfococcus*
  Genus: *Desulfobacter*
  Genus: *Desulfobulbus*
  Genus: *Desulfosarcina*

SECTION 8
Anaerobic Gram-Negative Cocci
Family I: Veillonellaceae
  Genus I: *Veillonella*
  Genus II: *Acidaminococcus*
  Genus III: *Megasphaera*

SECTION 9
The Rickettsias and Chlamydias
Order I: Rickettsiales
Family I: Rickettsiaceae
  Tribe I: *Rickettsieae*
    Genus I: *Rickettsia*
    Genus II: *Rochalimaea*
    Genus III: *Coxiella*
  Tribe II: *Ehrlichieae*
    Genus IV: *Ehrlichia*
    Genus V: *Cowdria*
    Genus VI: *Neorickettsia*
  Tribe III: *Wolbachieae*
    Genus VII: *Wolbachia*
    Genus VIII: *Rickettsiella*
Family II: Bartonellaceae
  Genus I: *Bartonella*
  Genus II: *Grahamella*
Family III: Anaplasmataceae
  Genus I: *Anaplasma*
  Genus II: *Aegyptianella*
  Genus III: *Haemobartonella*
  Genus IV: *Eperythrozoon*
Order II: Chlamydiales
Family I: Chlamydiaceae
  Genus I: *Chlamydia*

SECTION 10
The Mycoplasmas
Order I: Mycoplasmatales
Family I: Mycoplasmataceae
  Genus I: *Mycoplasma*
  Genus II: *Ureaplasma*
Family II: Acholeplasmataceae
  Genus I: *Acholeplasma*
Family III: Spiroplasmataceae
  Genus I: *Spiroplasma*
Other Genera
  Genus: *Anaeroplasma*
  Genus: *Thermoplasma*
  Mycoplasma-like Organisms of Plants
  and Invertebrates

SECTION 11
Endosymbionts
A: Endosymbionts of Protozoa
  Endosymbionts of ciliates
  Endosymbionts of flagellates
  Endosymbionts of amoebas
  Taxa of endosymbionts:
    Genus I: *Holospora*
    Genus II: *Caedibacter*
    Genus III: *Pseudocaedibacter*
    Genus IV: *Lyticum*
    Genus V: *Tectibacter*
B: Endosymbionts of Insects
  Blood-sucking insects
  Plant sap-sucking insects
  Cellulose and stored grain feeders
  Insects feeding on complex diets
  Taxon of endosymbionts:
    Genus: *Blattabacterium*
C: Endosymbionts of Fungi and
Invertebrates other than Arthropods
  Fungi
  Sponges
  Coelenterates
  Helminthes
  Annelids
  Marine worms and mollusks

Volume II

SECTION 12
Gram-Positive Cocci
Family I: Micrococcaceae
  Genus I: *Micrococcus*
  Genus II: *Stomatococcus*
  Genus III: *Planococcus*
  Genus IV: *Staphylococcus*
Family II: Deinococcaceae
  Genus I: *Deinococcus*
Other Genera
  Genus: *Streptococcus*
    Pyogenic Hemolytic Streptococci
    Oral Streptococci
    Enterococci
    Lactic Acid Streptococci
    Anaerobic Streptococci
    Other Streptococci
  Genus: *Leuconostoc*
  Genus: *Pediococcus*
  Genus: *Aerococcus*
  Genus: *Gemella*
  Genus: *Peptococcus*
  Genus: *Peptostreptococcus*
  Genus: *Ruminococcus*
  Genus: *Coprococcus*
  Genus: *Sarcina*

SECTION 13
Endospore-forming Gram-Positive
Rods and Cocci
  Genus: *Bacillus*
  Genus: *Sporolactobacillus*
  Genus: *Clostridium*
  Genus: *Desulfotomaculum*
  Genus: *Sporosarcina*
  Genus: *Oscillospira*

SECTION 14
Regular, Nonsporing, Gram-
Positive Rods
  Genus: *Lactobacillus*
  Genus: *Listeria*
  Genus: *Erysipelothrix*
  Genus: *Brochothrix*
  Genus: *Renibacterium*
  Genus: *Kurthia*
  Genus: *Caryophanon*

SECTION 15
Irregular, Nonsporing, Gram-Positive Rods
Genus: *Corynebacterium*
Plant Pathogenic Species of
Corynebacterium
Genus: *Gardnerella*
Genus: *Arcanobacterium*
Genus: *Arthrobacter*
Genus: *Brevibacterium*
Genus: *Curtobacterium*
Genus: *Caseobacter*
Genus: *Microbacterium*
Genus: *Aureobacterium*
Genus: *Cellulomonas*
Genus: *Agromyces*
Genus: *Arachnia*
Genus: *Rothia*
Genus: *Propionibacterium*
Genus: *Eubacterium*
Genus: *Acetobacterium*
Genus: *Lachnospira*
Genus: *Butyrivibrio*
Genus: *Thermoanaerobacter*
Genus: *Actinomyces*
Genus: *Bifidobacterium*

SECTION 16
The Mycobacteria
Family: Mycobacteriaceae
Genus: *Mycobacterium*

SECTION 17
Nocardioforms
Genus: *Nocardia*
Genus: *Rhodococcus*
Genus: *Nocardioides*
Genus: *Pseudonocardia*
Genus: *Oerskovia*
Genus: *Saccharopolyspora*
Genus: *Micropolyspora*
Genus: *Promicromonospora*
Genus: *Intrasporangium*

Volume III

SECTION 18
Anoxygenic Phototrophic Bacteria
Purple bacteria
Family I: Chromatiaceae
Genus I: *Chromatium*
Genus II: *Thiocystis*
Genus III: *Thiospirillum*
Genus IV: *Thiocapsa*
Genus V: *Lamprobacter*
Genus VI: *Lamprocystis*
Genus VII: *Thiodictyon*
Genus VIII: *Amoebobacter*
Genus IX: *Thiopedia*
Family II: Ectothiorhodospiraceae
Genus: *Ectothiorhodospira*
Purple Nonsulfur Bacteria
Genus: *Rhodospirillum*
Genus: *Rhodopila*
Genus: *Rhodobacter*
Genus: *Rhodopseudomonas*
Genus: *Rhodomicrobium*
Genus: *Rhodocyclus*
Green Bacteria
Green Sulfur Bacteria
Genus: *Chlorobium*
Genus: *Prosthecochloris*
Genus: *Pelodictyon*
Genus: *Ancalochloris*
Genus: *Chloroherpeton*
Symbiotic Consortia
Multicellular, Filamentous,
Green Bacteria
Genus: *Chloroflexus*

Genus: *Heliothrix*
Genus: *"Oscillochloris"*
Genus: *Chloronema*
Genera Incertae Sedis
Genus: *Heliobacterium*
Genus: *Erythrobacter*

SECTION 19
Oxygenic Photosynthetic Bacteria
Group I: Cyanobacteria
Subsection I: Order Chroococcales
Genus I: *Chamaesiphon*
Genus II: *Gloeobacter*
Genus III: *Gloeothece*
Subsection II: Order Pleurocapsales
Genus I: *Dermocarpa*
Genus II: *Xenococcus*
Genus III: *Dermocarpella*
Genus IV: *Myxosarcina*
Genus V: *Chroococcidiopsis*
Subsection III: Order Oscillatoriales
Genus I: *Spirulina*
Genus II: *Arthrospira*
Genus III: *Oscillatoria*
Genus IV: *Lyngbya*
Genus V: *Pseudanabaena*
Genus VI: *Starria*
Genus VII: *Crinalium*
Genus VIII: *Microcoleus*
Subsection IV: Order Nostocales
Family I: Nostocaceae
Genus I: *Anabaena*
Genus II: *Aphanizomenon*
Genus III: *Nodularia*
Genus IV: *Cylindrospermum*
Genus V: *Nostoc*
Family II: Scytonemataceae
Genus I: *Scytonema*
Family III: Rivulariaceae
Genus I: *Calothrix*
Subsection V: Order Stigonematales
Genus I: *Chlorogloeopsis*
Genus II: *Fischerella*
Genus III: *Stigonema*
Genus IV: *Geitleria*
Group II: Order Prochlorales
Family I: Prochloraceae
Genus: *Prochloron*
Other taxa:
Genus: *"Prochlorothrix"*

SECTION 20
Aerobic Chemolithotrophic
Bacteria and Associated Organisms
A: Nitrifying Bacteria
Family: Nitrobacteraceae
Genus I: *Nitrobacter*
Genus II: *Nitrospina*
Genus III: *Nitrococcus*
Genus IV: *Nitrospira*
Genus V: *Nitrosomonas*
Genus VI: *Nitrosococcus*
Genus VII: *Nitrosospira*
Genus VIII: *Nitrosolobus*
Genus IX: *Nitrosovibrio*
B: Colorless Sulfur Bacteria
Genus: *Thiobacterium*
Genus: *Macromonas*
Genus: *Thiospira*
Genus: *Thiovulum*
Genus: *Thiobacillus*
Genus: *Thiomicrospira*
Genus: *Thiosphaera*
Genus: *Acidiphilium*
Genus: *Thermothrix*
C: Obligate Hydrogen Oxidizers
Genus: *Hydrogenobacter*

D: Iron and Manganese Oxidizing and/or
Depositing Bacteria
Family: Siderocapsaceae
Genus I: *Siderocapsa*
Genus II: *Naumanniella*
Genus III: *Siderococcus*
Genus IV: *Ochrobium*
E. Magnetotactic Bacteria
Genus: *Aquaspirillum (A.magnetotacticum)*
Genus: *Bilophococcus*

SECTION 21
Budding and/or Appendaged Bacteria
A. Prosthecate Bacteria
1. Budding Bacteria
Genus: *Hyphomicrobium*
Genus: *Hyphomonas*
Genus: *Pedomicrobium*
Genus: *Ancalomicrobium*
Genus: *Prosthecomicrobium*
Genus: *Stella*
Genus: *Labrys*
2. Nonbudding Bacteria
Genus: *Caulobacter*
Genus: *Asticcacaulis*
Genus: *Prosthecobacter*
B. Nonprosthecate Bacteria
1. Budding Bacteria
Genus: *Planctomyces*
Genus: *"Isosphaera"*
Genus: *Blastobacter*
Genus: *Angulomicrobium*
Genus: *Gemmiger*
Genus: *Ensifer*
2. Nonbudding Bacteria
Genus: *Gallionella*
Genus: *Nevskia*
C. Morphologically Unusual Budding
Bacteria (involved in iron and manganese
deposition)
Genus: *Seliberia*
Genus: *Metallogenium*
Genus: *Caulococcus*
Genus: *Kuznezovia*
Genus: *Thiodendron*
D. Others
Spinate bacteria

SECTION 22
Sheathed Bacteria
Genus: *Sphaerotilus*
Genus: *Leptothrix*
Genus: *Haliscominobacter*
Genus: *"Lieskeella"*
Genus: *"Phragmidiothrix"*
Genus: *Crenothrix*
Genus: *"Clonothrix"*

SECTION 23
Nonphotosynthetic, Nonfruiting, Gliding
Bacteria
Order I: Cytophagales
Family: Cytophagaceae
Genus I: *Cytophaga*
Genus II: *Sporocytophaga*
Genus III: *Capnocytophaga*
Genus IV: *Flexithrix*
Other genera
Genus: *Flexibacter*
Genus: *Chitinophaga*
Genus: *Microscilla*
Order II: Lysobacterales
Family: Lysobacteriaceae
Genus: *Lysobacter*
Order III: Beggiatoales
Family: Beggiatoaceae
Genus I: *Beggiatoa*

Genus II: *Thioploca*
Genus III: *"Thiospirillopsis"*
Genus IV: *Thiothrix*
Others
Family I: Simonsiellaceae
Genus I: *Simonsiella*
Genus II: *Alysiella*
Family II: "Pelonemataceae"
Genus I: *"Pelonema"*
Genus II: *"Peloploca"*
Genus III: *"Achroonema"*
Genus IV: *"Desmanthus"*
Other genera
Genus: *Herpetosiphon*
Genus: *Toxothrix*
Genus: *Leucothrix*
Genus: *Vitreoscilla*
Genus: *Desulfonema*
Genus: *Agitococcus*
Genus: *Achromatium*

SECTION 24
Gliding, Fruiting Bacteria
Order: Myxococcales
Family I: Myxococcaceae
Genus: *Myxococcus*
Family II: Archangiaceae
Genus: *Archangium*
Family III: Cystobacteraceae
Genus I: *Cystobacter*
Genus II: *Melittangium*
Genus III: *Stigmatella*
Family IV: Polyangiaceae
Genus I: *Polyangium*
Genus II: *Nannocystis*
Genus III: *Chondromyces*

SECTION 25
Archaeobacteria*
Group I: Methanogenic
Archaeobacteria
Order I: Methanobacteriales
Family I: Methanobacteriaceae
Genus I: *Methanobacterium*
Genus II: *Methanobrevibacter*
Family II: Methanothermaceae
Genus: *Methanothermus*
Order II: Methanococcales
Family: Methanococcaceae
Genus: *Methanococcus*
Order III: Methanomicrobiales
Family I: Methanomicrobiaceae
Genus I: *Methanomicrobium*
Genus II: *Methanogenium*
Genus III: *Methanospirillum*

*An older term for the Archaea.

Family II: Methanosarcinaceae
Genus I: *Methanosarcina*
Genus II: *Methanococcoides*
Genus III: *Methanolobus*
Genus IV: *Methanothrix*
Other taxa
Family: Methanoplanaceae
Genus: *Methanoplanus*
Others
Genus: *Methanosphaera*
Group II: Archaeobacterial
Sulfate-Reducers
Order: Archaeoglobales
Family: "Archaeoglobaceae"
Genus: *Archaeoglobus*
Group III: Extremely Halophilic
Archaeobacteria
Order: Halobacteriales
Family: Halobacteriaceae
Genus I: *Halococcus*
Genus II: *Halobacterium*
Genus III: *Haloferax*
Genus IV: *Haloarcula*
Genus V: *Natronococcus*
Genus VI: *Natronobacterium*
Group IV: Cell Wall-less
Archaeobacteria
Genus: *Thermoplasma*
Group V: Extremely Thermophilic
S⁰-Metabolizers
Order I: Thermoproteales
Family I: Thermoproteaceae
Genus I: *Thermoproteus*
Genus II: *Thermophilum*
Family II: Desulfurococcaceae
Genus I: *Desulfurococcus*
Genus II: *Thermococcus*
Genus III: *Thermodiscus*
Genus IV: *Pyrodictium*
Order II: Thermococcales
Family: Thermococcaceae
Genus I. *Thermococcus*
Genus II. *Pyrococcus*
Order III: Sulfolobales
Family: Sulfolobaceae
Genus: *Sulfolobus*
Genus: *Acidianus*

Volume IV

SECTION 26
Nocardioform Actinomycetes
Genus: *Nocardia*
Genus: *Rhodococcus*
Genus: *Nocardioides*
Genus: *Pseudonocardia*
Genus: *Oerskovia*

Genus: *Saccharopolyspora*
Genus: *Faenia (Micropolyspora)*
Genus: *Promicromonospora*
Genus: *Intrasporangium*
Genus: *Actinopolyspora*
Genus: *Saccharomonospora*
Genus: *Amycolatopsis*
Genus: *Amycolata*

SECTION 27
Actinomycetes with Multi-Locular
Sporangia
Genus: *Geodermatophilus*
Genus: *Dermatophilus*
Genus: *Frankia*

SECTION 28
Actinoplanetes
Genus: *Actinoplanes*
Genus: *Ampullariella*
Genus: *Pilimelia*
Genus: *Dactylosporangium*
Genus: *Micromonospora*

SECTION 29
Streptomycetes and Related Genera
Genus: *Streptomyces*
Genus: *Streptoverticillium*
Genus: *Kineosporia*
Genus: *Sporichthya*

SECTION 30
Maduromycetes
Genus: *Actinomadura*
Genus: *Microbispora*
Genus: *Microtetraspora*
Genus: *Planobispora*
Genus: *Planomonospora*
Genus: *Spirillospora*
Genus: *Streptosporangium*

SECTION 31
Thermomonospora and Related Genera
Genus: *Thermomonospora*
Genus: *Actinosynnema*
Genus: *Nocardiopsis*
Genus: *Streptoalloteichus*

SECTION 32
Thermoactinomycetes
Genus: *Thermoactinomyces*

SECTION 33
Other Genera
Genus: *Glycomyces*
Genus: *Kibdelosporangium*
Genus: *Kitasatosporia*
Genus: *Saccharothrix*

Glossary

Only the major terms and concepts are included. If a term is not here, consult the index.

Abscess   A localized infection characterized by production of pus.

Acetoclastis   Splitting of acetate into CH_4 plus CO_2 by certain methanogens.

Acetyl-CoA Pathway   A pathway of autotrophic CO_2 fixation widespread in obligate anaerobes including methanogens, homoacetogens, and sulfate-reducing bacteria.

Acetylene reduction assay   Method of measuring activity of nitrogenase by substituting acetylene for the natural substrate of the enzyme, N_2. Acetylene is reduced to ethylene or ethane.

Acid mine drainage   Acidic water containing H_2SO_4 derived from the microbial oxidation of iron sulfide minerals.

Acid-alcohol fastness   A staining property of *Mycobacterium* species where cells stained with hot carbolfuschin will not decolorize with acid-alcohol.

Acidophile   Organism that grows best at acidic pH values.

Activation energy   Energy needed to make substrate molecules more reactive; enzymes function by lowering activation energy.

Activator protein   A regulatory protein that binds to specific sites on DNA and stimulates transcription; involved in positive control.

Active immunity   An immune state achieved by self-production of antibodies. Compare with *Passive immunity*.

Active site   The region of an enzyme that is directly involved in binding substrate(s).

Active transport   The energy-dependent process of transporting substances into or out of the cell in which the transported substances are chemically unchanged.

Acute   In reference to infections, short-term, usually characterized by dramatic onset and rapid recovery.

Adherence   A property of bacteria that allows them to stick to host surfaces.

Aerobe   An organism that grows in the presence of O_2; may be facultative or obligate.

Aerosol   Suspension of particles in airborne water droplets.

Aerotolerant   Of an anaerobe, not being inhibited by O_2.

Agglutination   Reaction between antibody and particle-bound antigen resulting in clumping of the particles.

Algae   Phototrophic eukaryotic microorganisms.

Alkaliphile   An organism that grows best at high pH.

Allergy   A harmful immune reaction, usually caused by a foreign antigen in food, pollen, or chemicals; immediate-type or delayed-type hypersensitivity.

Allosteric   Characteristic of some proteins, especially enzymes, in which a compound combines with a site on the protein other than the active site. The result is a change in conformation at the active site; useful in regulating activity.

Amoeboid movement   A type of motility in which cytoplasmic streaming moves the organism forward.

Anabolism   The biochemical processes involved in the synthesis of cell constituents from simpler molecules, usually requiring energy.

Anaerobe   An organism that grows in the absence of O_2. Some may even be killed by O_2.

Anaerobic respiration   Use of an electron acceptor other than O_2 in an electron-transport-based oxidation. Most common anaerobic electron acceptors are nitrate, sulfate, and carbonate.

Anaphylatoxins   The C3a and C5a fractions of complement which act to mimic some of the reactions of anaphylaxis.

Anaphylaxis (anaphylactic shock)   A violent allergic reaction caused by an antigen-antibody reaction.

Anoxic   Absence of oxygen. Usually used in reference to a microbial habitat.

Anoxygenic photosynthesis   Use of light energy to synthesize ATP by cyclic photophosphorylation without O_2 production in green and purple bacteria.

Antibiotic   A chemical agent produced by one organism that is harmful to other organisms.

Antibody   A protein present in serum or other body fluid that combines specifically with antigen. An immunoglobulin.

Anticodon   A sequence of three bases in transfer RNA that base pairs with a codon in messenger RNA.

Antigen   A substance, usually macromolecular, that induces a specific immune response.

Antigen-presenting cell (APC)   Cells that process and present antigen to T lymphocytes.

Antigenic determinants   The portion of an antigen that interacts with an immunoglobulin or T cell receptor.

Antigenic drift   In influenza virus, minor changes in viral proteins (antigens) due to gene mutation.

Antigenic shift   In influenza virus, major changes in viral proteins (antigens) due to gene reassortment.

Antimicrobial   Harmful to microorganisms by either killing or inhibiting growth.

Antiparallel   In reference to double-stranded DNA, the orientation of the two strands: one strand runs $5' \rightarrow 3'$, the other $3' \rightarrow 5'$.

Antiseptic   An agent that kills or inhibits microbial growth, but is not harmful to human tissue.

Antiserum   A serum containing antibodies.

Antitoxin   An antibody that specifically interacts with and neutralizes a toxin.

Archaea   An evolutionarily distinct group (domain) of prokaryotes consisting of the methanogens, most extreme halophiles and hyperthermophiles, and *Thermoplasma*.

Archaebacteria   An old term for the Archaea.

Aseptic technique   Manipulation of sterile instruments or culture media in such a way as to maintain sterility.

Atomic weight   The total mass of protons and neutrons in an atom.

ATP   Adenosine triphosphate, the principal energy carrier of the cell.

Attenuation   Selection of nonvirulent strains of a pathogen still capable of immunizing. Also: a process that plays a role in the regulation of enzymes involved in amino acid biosynthesis.

Autoantibody   Antibodies that react to self antigens.

Autoimmunity   Immune reactions of a host against its own self constituents.

Autolysis   Spontaneous lysis.

Autoradiography   Detection of radioactivity in a sample, for example a cell or gel, by placing it in contact with a photographic film.

Autotroph   Organism able to utilize CO_2 as sole source of carbon.

Auxotroph   A mutant that has a growth factor requirement. Contrast with a *Prototroph*.

Bacteremia   The transient appearance of bacteria in the blood.

Bacteria   All prokaryotes that are not members of the domain Archaea.

Bactericidal   Capable of killing bacteria.

Bacteriocin Agents produced by certain bacteria that inhibit or kill closely related species.

Bacteriophage A virus that infects a prokaryote.

Bacteriorhodopsin A protein containing retinal found in the membranes of certain extremely halophilic Archaea and which is involved in light-mediated ATP synthesis.

Bacteriostatic Capable of inhibiting bacterial growth without killing.

Bacteroid A swollen, deformed *Rhizobium* cell, found in the root nodule; capable of nitrogen fixation.

Barophile An organism able to live optimally at high hydrostatic pressure.

Barotolerant An organism able to tolerate high hydrostatic pressure, although growing better at normal pressures.

Base composition In reference to nucleic acids, the proportion of the total bases consisting of guanine plus cytosine or thymine plus adenine base pairs. Usually expressed as a guanine + cytosine (G + C) value, e.g. 60% G + C.

Bioconversion In industrial microbiology, use of microorganisms to convert an added chemical to a chemically modified form.

Biofilm Microbial colonies encased in an adhesive, usually polysaccharide material, and attached to a surface.

Biogeochemistry Study of microbially mediated chemical transformations of geochemical interest, e.g. nitrogen or sulfur cycling.

Bioremediation Use of microorganisms to remove or detoxify toxic or unwanted chemicals from an environment.

Biosynthesis The production of needed cellular constituents from other (usually simpler) molecules.

Biotechnology The use of living organisms to carry out defined chemical processes for industrial application.

Black smoker Thermal vent emitting very hot (270–380°C) water and minerals.

B lymphocyte A cell of the immune system that differentiates into an immunoglobulin-producing cell.

Brewing The manufacture of alcoholic beverages such as beer from the fermentation of malted grains.

Calvin cycle The biochemical route of CO_2 fixation in many autotrophic organisms.

Capsid The protein coat of a virus.

Capsomere An individual protein subunit of the virus capsid.

Capsule A compact layer of polysaccharide exterior to the cell wall in some bacteria. See also *Glycocalyx* and *Slime layer*.

Carboxysomes Polyhedral cellular inclusions of crystalline ribulose bisphosphate carboxylase (RubisCO), the key enzyme of the Calvin Cycle.

Carcinogen A substance that causes the initiation of tumor formation. Frequently a mutagen.

Carrier An individual that continually releases infective organisms but does not show symptoms of disease.

Catabolism The biochemical processes involved in the breakdown of organic compounds, usually leading to the production of energy.

Catabolite repression Repression of a variety of unrelated enzymes when cells are grown in a medium containing glucose.

Catalysis Increase in rate of a chemical reaction.

Catalyst A substance that promotes a chemical reaction without itself being changed in the end.

CD4 cells T helper cells which are targets for HIV infection.

Cell The fundamental unit of living matter.

Cell-mediated immunity An immune response generated by the activities of non-antibody-producing cells such as T cells. Compare with *Humoral immunity*.

Chaperonins Proteins that aid in the correct folding of other proteins and the assembly of multisubunit structures.

Chemiosmosis The use of ion gradients across membranes, especially proton gradients, to generate ATP. See *Proton motive force*.

Chemolithotroph Organism obtaining its energy from the oxidation of inorganic compounds.

Chemoorganotroph Organism obtaining its energy from the oxidation of organic compounds.

Chemostat A continuous culture device controlled by the concentration of limiting nutrient and dilution rate.

Chemotaxis Movement toward or away from a chemical.

Chemotherapy Treatment of infectious disease with chemicals or antibiotics.

Chlorination A highly effective disinfectant procedure for drinking water using chlorine gas or other chlorine-containing compounds as disinfectant.

Chlorophyll and bacteriochlorophyll Pigments of phototrophic organisms consisting of light-sensitive magnesium tetrapyrroles.

Chloroplast The chlorophyll-containing organelle of photosynthetic eukaryotes.

Chlorosomes Cigar-shaped structures bounded by a nonunit membrane and containing the light harvesting bacteriochlorophyll (c, c_s, d, or e) in green sulfur bacteria and in *Chloroflexus*.

Chromogenic Producing color; a chromogenic colony is a pigmented colony.

Chromosome A genetic element carrying genes essential to cellular metabolism. Prokaryotes typically have a single chromosome, consisting of a circular DNA molecule. Eukaryotic cells contain several chromosomes, each containing a linear DNA molecule complexed with specific proteins.

Chronic Long-term infection.

Cilium Short, filamentous structure that beats with many others to make a cell move.

Class I MHC proteins Antigen-presenting molecules found on all nucleated vertebrate cells.

Class II MHC proteins Antigen-presenting molecules found primarily on macrophages and B lymphocytes.

Clonal selection A theory that each B or T lymphocyte, when stimulated by antigen, divides to form a clone of itself.

Clone A population of cells all descended from a single cell; a number of copies of a DNA fragment obtained by allowing an inserted DNA fragment to be replicated by a phage or plasmid.

Cloning vector A DNA molecule that is able to bring about the replication of foreign DNA fragments.

Coccoid Sphere-shaped.

Coccus A spherical bacterium.

Codon A sequence of three purine and/or pyrimidine bases in messenger RNA that encodes a specific amino acid.

Coenzyme A low-molecular-weight chemical which participates in an enzymatic reaction by accepting and donating electrons or functional groups. Examples: NAD^+, FAD.

Coliforms Gram-negative, nonsporing, facultative rods that ferment lactose with gas formation within 48 hours at 35°C.

Colonization Multiplication of a microorganism after it has attached to host tissues or other surfaces.

Colony A macroscopically visible population of cells growing on solid medium, arising from a single cell.

Cometabolism The metabolic transformation of a substance while a second substance serves as primary energy or carbon source.

Commodity chemical Chemicals such as ethanol that have low monetary value and are thus sold primarily in bulk.

Common-source epidemic An epidemic resulting from infection of a large number of people from a single contaminated source.

Compatible solutes Organic compounds which serve as cytoplasmic solutes to balance water relations for cells growing in environments of high salt or sugar.

Competence   Ability to take up DNA and become genetically transformed.

Complement   A complex of proteins in the blood serum that interacts sequentially with specific antibody in certain kinds of antigen–antibody reactions.

Complement fixation   The consumption of complement by an antibody–antigen reaction.

Complementary   Nucleic acid sequences that can base pair with each other.

Complex media   Culture media whose precise chemical composition is unknown. Also called *undefined media*.

Concatamer   A DNA molecule consisting of two or more separate molecules linked end-to-end to form a long linear structure.

Congenital syphilis   Syphilis contracted by an infant from its mother during birth.

Conjugation   In prokaryotes, transfer of genetic information from a donor cell to a recipient cell by cell-to-cell contact.

Consensus sequence   A nucleic acid sequence in which the base present in a given position is that base most commonly found when many experimentally determined sequences are compared.

Consortium   A two- (or more) membered bacterial culture (or natural assemblage) in which each organism benefits from the other.

Contagious   Of a disease, transmissible.

Cortex   The region inside the spore coat of an endospore, around the core.

Covalent   A nonionic chemical bond formed by a sharing of electrons between two atoms.

Crista   Inner membrane in a mitochondrion, site of respiration.

Culture   A particular strain or kind of organism growing in a laboratory medium.

Culture media   An aqueous solution of various nutrients suitable for the growth of microorganisms.

Cutaneous   Relating to the skin.

Cyanobacteria   Prokaryotic oxygenic phototrophs containing chlorophyll *a* and phycobilins.

Cyst   A resting stage formed by some bacteria and protozoa in which the whole cell is surrounded by a protective layer; not the same as spore.

Cytochrome   Iron-containing porphyrin rings complexed with proteins, which act as electron carriers in the electron-transport system.

Cytokine   A soluble immune response modulator produced by cells other than lymphocytes, usually phagocytic cells.

Cytoplasm   Cellular contents inside the cytoplasmic membrane, excluding the nucleus.

Cytoplasmic membrane   The permeability barrier of the cell, separating the cytoplasm from the environment.

Defined media   Culture media whose exact chemical composition is known. Compare with *Complex media*.

Degeneracy   In relation to the genetic code, the fact that more than one codon can code for the same amino acid.

Deletion   A removal of a portion of a gene.

Denaturation   Irreversible destruction of a macromolecule, as for example the destruction of a protein by heat.

Denitrification   Conversion of nitrate into nitrogen gases under anaerobic conditions, resulting in loss of nitrogen from ecosystems.

Dental plaque   Bacterial cells encased in a matrix of extracellular polymers and salivary products, found on the teeth.

Deoxyribonucleic acid (DNA)   A polymer of nucleotides connected via a phosphate-deoxyribose sugar backbone; the genetic material of the cell.

Desiccation   Drying.

Dideoxynucleotide   A nucleotide lacking the 3'-hydroxyl group on the deoxyribose sugar. Used in the Sanger method of DNA sequencing.

Differentiation   The modification of a cell in terms of structure and or function occurring during the course of development.

Diploid   In eukaryotes, an organism or cell with two chromosome complements, one derived from each haploid gamete.

Disinfectant   An agent that kills microorganisms, but may be harmful to human tissue.

Disproportionation   The splitting of a chemical compound into two new compounds, one more oxidized and one more reduced than the original compound.

DNA fingerprinting   Use of the techniques of genetic engineering to determine the origin of DNA in a sample of tissue.

DNA library   A collection of cloned DNA fragments which in total contain genes from the entire genome of an organism; also called a *gene library*.

Domain   The highest level of biological classification, superseding kingdoms. The three domains of biological organisms are the Bacteria, the Archaea, and the Eukarya. Also used to describe a region of a protein having a distinct function.

Doubling time   The time needed for a population to double. See also *Generation time*.

Downstream position   Refers to nucleic acid sequences on the 3'-side of a given site on the DNA or RNA molecule. Compare with *Upstream position*.

Ecology   Study of the interrelationships between organisms and their environments.

Ecosystem   A community of organisms and their natural environment.

Electron acceptor   A substance that accepts electrons during an oxidation-reduction reaction. An electron acceptor is an oxidant.

Electron donor   A compound that donates electrons in an oxidation-reduction reaction. An electron donor is a reductant.

Electron-transport phosphorylation   Synthesis of ATP involving a membrane-associated electron transport chain and the creation of a proton-motive force. Also called oxidative phosphorylation. See also *Chemiosmosis*.

Electrophoresis   Separation of charged molecules in an electrical field.

Electroporation   The use of an electrical pulse to enable cells to take up DNA.

ELISA   Enzyme-linked immunosorbent assay. An immunoassay that uses specific antibodies to detect antigens or antibodies in body fluids. The antibody-containing complexes are visualized through enzyme coupled to the antibody. Addition of substrate to the enzyme-antibody-antigen complex results in a colored product.

Endemic   A disease that is constantly present in low numbers in a population. Compare with *Epidemic*.

Endergonic   A chemical reaction requiring input of energy to proceed.

Endocytosis   A process in which a particle such as a virus is taken intact into an animal cell. Phagocytosis and pinocytosis are two kinds of endocytosis.

Endoplasmic reticulum   An extensive array of internal membranes in eukaryotes.

Endospore   A differentiated cell formed within the cells of certain gram-positive bacteria and extremely resistant to heat as well as to other harmful agents.

Endosymbiosis   The hypothesis that mitochondria and chloroplasts are the descendants of ancient prokaryotic organisms from the domain Bacteria.

Endotoxin   A toxin not released from the cell; bound to the cell surface or intracellular. Compare with *Exotoxin*.

Enrichment culture   Use of selective culture media and incubation conditions to isolate microorganisms directly from nature.

Enteric   Intestinal.

Enterotoxin   A toxin affecting the intestine.

Enzyme   A protein functioning as a catalyst in living organisms, which promotes specific reactions or groups of reactions.

Epidemic   A disease occurring in an unusually high number of individuals in a population at the same time. Compare with *Endemic*.

Epidemiology   The study of the incidence and prevalence of disease in populations.

Epitope   Antigenic determinant.

Eubacteria   An old term for the Bacteria.

Eukarya   The phylogenetic domain containing all eukaryotic organisms.

Eukaryote   A cell or organism having a unit membrane-bound (true) nucleus and usually other organnelles.

Evolutionary distance   In phylogenetic trees, the sum of the physical distance on a tree separating organisms; this distance is inversely proportional to evolutionary relatedness.

Exergonic reaction   A chemical reaction that proceeds with the liberation of energy.

Exon   The coding sequences in a split gene. Contrast with *Introns*, the intervening noncoding regions.

Exotoxin   A toxin released extracellularly. Compare with *Endotoxin*.

Exponential growth   Growth of a microorganism where the cell number doubles within a fixed time period.

Exponential phase   A period during the growth cycle of a population in which growth increases at an exponential rate.

Expression   The ability of a gene to function within a cell in such a way that the gene product is formed.

Expression vector   A cloning vector that contains the necessary regulatory sequences to allow transcription and translation of a cloned gene or genes.

Extreme halophile   An organism whose growth is dependent on large amounts (generally > 10%) of NaCl.

Facultative   A qualifying adjective indicating that an organism is able to grow either in the presence or absence of an environmental factor (e.g., "facultative aerobe").

Feedback inhibition   Inhibition by an end product of the biosynthetic pathway involved in its synthesis.

Fermentation   Catabolic reactions producing ATP in which organic compounds serve as both primary electron donor and ultimate electron acceptor.

Fermentation (industrial)   A large-scale microbial process.

Fermenter   An organism which carries out the process of fermentation.

Fermentor   A large growth vessel used to culture microorganisms on a large scale frequently for the production of some commercially valuable product.

Ferredoxin   An electron carrier of low reduction potential; small protein containing iron-sulfer clusters.

Fever   A rise of body temperature above the normal.

Filamentous   In the form of very long rods, many times longer than wide.

Fimbria (plural fimbriae)   Short filamentous structure on a bacterial cell; although flagella-like in structure, generally present in many copies and not involved in motility. Plays a role in adherence to surfaces and in the formation of pellicles. See also *Pilus*.

Flagellum (plural flagella)   An organ of motility.

Flavoprotein   A protein containing a derivative of riboflavin, which acts as electron carrier in the electron-transport system.

Fluorescent   Having the ability to emit light of a certain wavelength when activated by light of another wavelength.

Fluorescent antibody   Immunoglobulin molecule which has been coupled with a fluorescent molecule so that it exhibits fluorescence.

Fomites   Inanimate objects which, when contaminated with a viable pathogen, can transfer the pathogen to a host.

Food infection   Microbial infection resulting from ingestion of contaminated food.

Food poisoning   Disease resulting from ingestion of food contaminated with a toxin produced by a microorganism.

Frameshift   Since the genetic code is read three bases at a time, if reading begins at either the second or third base of a codon, a faulty product usually results. This is called a frameshift (the reading frame refers to the pattern of reading).

Free energy   Energy available to do useful work.

Fruiting body   A macroscopic reproductive structure produced by some fungi (e.g., mushrooms) and some Bacteria (e.g., myxobacteria). Fruiting bodies are distinct in size, shape, and coloration for each species.

Fungi   Nonphototrophic eukaryotic microorganisms that contain rigid cell walls.

Fusion protein   The result of translation of two or more genes joined such that they retain their correct reading frames but make a single protein.

G+C base ratio   In DNA (or RNA) from any organism, the percent of the total nucleic acid that consists of guanine plus cytosine bases.

Gametes   In eukaryotes, haploid cells analogous to sperm and egg, which result from meiosis.

Gas vesicle   A gas-filled structure in certain prokaryotes that confers ability to float.

Gel   An inert polymer, usually made of agarose or polyacrylamide, used for separating macromolecules such as nucleic acids or proteins by electrophoresis.

Gene   A unit of heredity; a segment of DNA specifying a particular protein or polypeptide chain, a tRNA or a rRNA.

Gene cloning   The isolation of a desired gene from one organism and its incorporation into a suitable vector for the production of large amounts of the gene.

Gene disruption   Use of both *in vitro* and *in vivo* recombination to substitute an easily selected mutant gene for a wild-type gene.

Gene library   A collection of cloned DNA fragments that contains all the genetic information of a particular organism.

Gene therapy   Replacement or augmentation of a dysfunctional gene for medical purposes.

Generation time   Time needed for a population to double. See also *Doubling time*.

Genetic engineering   The use of *in vitro* techniques in the isolation, manipulation, recombination, and expression of DNA.

Genetic map   The physical arrangement and order of genes on the chromosome.

Genetics   Heredity and variation of organisms.

Genome   The complete set of genes present in an organism.

Genotype   The precise genetic constitution of an organism. Compare with *Phenotype*.

Genus   A group of related species.

Germicide   A substance that inhibits or kills microorganisms.

Glycocalyx   General term for polysaccharide components outside the bacterial cell wall. See also *Capsule* and *Slime layer*.

Glycolysis   Reactions of the Embden-Meyerhof pathway in which glucose is oxidized to pyruvate.

Glycosidic bond   A type of covalent bond that links sugar units together in a polysaccharide.

Gonococcus   *Neisseria gonorrhoeae*, the Gram-negative diplococcus that causes the disease gonorrhea.

Gram-negative   A prokaryotic cell whose cell wall contains relatively little peptidoglycan but contains an outer membrane composed of lipopolysaccharide (LPS), lipoprotein, and other complex macromolecules.

Gram-positive   A prokaryotic cell whose cell wall consists chiefly of peptidoglycan and lacks the outer membrane of Gram-negative cells.

Growth   In microbiology, an increase in cell number.

Growth rate   The rate at which growth occurs, usually expressed as the generation time.

Habitat   The location where an organism resides.

Halophile   An organism requiring salt (NaCl) for growth.

Halotolerant   An organism capable of growing in the presence of NaCl but not requiring it.

Haploid   In eukaryotes, an organism or cell containing one chromosome complement and the same number of chromosomes as the gametes.

Hapten   A substance not inducing antibody formation but able to combine with a specific antibody.

Helix   A spiral structure in a macromolecule that contains a repeating pattern.

Hemagglutination   Agglutination of red blood cells.

Hemolysins   Bacterial toxins capable of lysing red blood cells.

Hemolysis   Lysis of red blood cells.

Herd immunity   Resistance of a group to a pathogen due to immunity of a large proportion of the group to that pathogen.

Heterocyst   A differentiated cyanobacterial cell which carries out nitrogen fixation.

Heteroduplex   A double-stranded DNA in which one strand is from one source and the other strand from another, usually related, source.

Heterofermentation   Fermentation of glucose or other sugar to a mixture of products.

Heterotroph   Chemoorganotroph.

Homoacetogens   Bacteria that produce acetate as sole product of sugar fermentation or from $H_2 + CO_2$.

Homofermentation   Fermentation of glucose or other sugar leading to virtually a single product, lactic acid.

Homologous antigen   An antigen reacting with the antibody it had induced.

Host   An organism capable of supporting the growth of a virus or other parasite.

Humoral immunity   An immune response involving antibodies.

Hybridization   The natural formation or artificial construction of a duplex nucleic acid molecule by complementary base pairing between two nucleic acid strands derived from different sources.

Hybridoma   The fusion of an immortal cell with a single B lymphocyte to produce an immortal lymphocyte which produces a monoclonal antibody.

Hydrogen bond   A weak chemical bond between a hydrogen atom and a second, more electronegative element, usually an oxygen or nitrogen atom.

Hydrolysis   Breakdown of a polymer into smaller units, usually monomers, by addition of water; digestion.

Hydrophobic interactions   Attractive forces between molecules due to the close positioning of nonhydrophilic portions of the two molecules.

Hydrothermal vents   Warm or hot water emitting springs associated with crustal spreading centers on the sea floor.

Hypersensitivity   An immune reaction, usually harmful to the animal, caused either by antigen-antibody reactions or cellular immune processes (see *Allergy*).

Hyperthermophile   A prokaryote having a growth temperature optimum of 80°C or higher.

Icosahedron   A geometrical shape occurring in many virus particles, with 20 triangular faces and 12 corners.

Immobilized enzyme   An enzyme attached to a solid support over which substrate is passed and is converted into product.

Immune   Able to resist infectious disease.

Immunity   The ability of an organism to resist infection.

Immunization   Induction of specific immunity by injecting antigens, antibodies, or immune cells.

Immunoblot (Western blot)   Detection of proteins immobilized on a filter by complementary reaction with specific antibody. Compare with *Southern* and *Northern blot*.

Immunodeficiency   Having a dysfunctional or completely nonfunctional immune system.

Immunogen   An antigen that will induce the production of an immune response.

Immunoglobulin   Antibody.

Immunologic memory   The ability to rapidly produce large quantities of specific immune cells after subsequent exposure to a previously encountered antigen.

In vitro   In glass, away from the living organism.

In vivo   In the body, in a living organism.

Incidence   In reference to disease transmission, the number of cases of the disease in a specific subset of the population.

Induced enzyme   An enzyme subject to induction.

Induction   The process by which an enzyme is synthesized in response to the presence of an external substance, the inducer.

Infection   Growth of an organism within the body.

Infection thread   In the formation of root nodules, a cellulosic tube through which *Rhizobium* cells can travel to reach and infect root cells.

Inflammation   Characteristic reaction to foreign particles and noxious stimuli, resulting in redness, swelling, heat, and pain.

Inhibition   Prevention of growth or function.

Inoculum   Material used to initiate a microbial culture.

Insertion   A genetic phenomenon in which a piece of DNA is inserted into the middle of a gene.

Insertion sequence (IS elements)   The simplest type of transposable element. Has only genes involved in transposition.

Integration   The process by which a DNA molecule becomes incorporated into another genome.

Interferon   A protein produced by cells as a result of virus infection which interferes with virus replication.

Interspecies hydrogen transfer   The process in which organic matter is degraded anaerobically by the interaction of several groups of microorganisms in which H_2 production and H_2 consumption are closely coupled.

Introns   The intervening noncoding sequences in a split gene. Contrasted with *Exons*, the coding sequences.

Invasiveness   Degree to which an organism is able to spread through the body from a focus of infection.

Ionophore   A compound which can cause the leakage of ions across membranes.

Isotopes   Different forms of the same element containing the same number of protons and electrons, but differing in the number of neutrons.

Joule   A unit of work or energy equal to 10^7 ergs; 1000 joules equals 1 kilojoule (kJ).

Kilobase (kb)   A 1000 base fragment of nucleic acid. A *kilobase pair* is a fragment containing 1000 base pairs.

Lag phase   The period after inoculation of a population before growth begins.

Latent virus   A virus present in a cell, yet not causing any detectable effect.

Leaching   Removal of valuable metals from ores by microbial action.

Leukocidin   A substance able to destroy phagocytes.

Leukocyte   A white blood cell, usually a phagocyte.

Lichen   A fungus and an alga (or a cyanobacterium) living in symbiotic association.

Lipid   Water-insoluble organic molecules important in structure of the cytoplasmic membrane and (in some organisms) the cell wall. See also *Phospholipid*.

Lipopolysaccharide (LPS)   Complex lipid structure containing unusual sugars and fatty acids found in many Gram-negative Bacteria, and constituting the chemical structure of the outer layer.

Lophotrichous   Having a tuft of polar flagella.

Lower respiratory tract   Trachea, bronchi, and lungs.

Luminescence   Production of light.

Lymph   A clear yellowish fluid found in the lymphatic vessels that carries various white (but not red) blood cells.

Lymphocyte   A white blood cell involved in antibody formation or cellular immune responses.

Lymphokines   Substances secreted from T lymphocytes which stimulate the activity of other cells.

Lysin   An antibody that induces lysis.

Lysis   Rupture of a cell, resulting in loss of cell contents.

Lysogen   A prokaryote containing a prophage. See also *Temperate virus*.

Lysosome   A cell organelle containing digestive enzymes.

Macromolecule   A large molecule formed from the connection of a number of small molecules. A polymer. Informational macromolecules play a role in transfer of genetic information.

Macrophage   Large noncirculating phagocytic cells involved in both phagocytosis and the antibody production process.

Magnetosomes   Small particles of Fe_3O_4 present in cells that exhibit magnetotaxis.

Magnetotaxis   Directed movement of bacterial cells by a magnetic field.

Major histocompatability complex (MHC)   A cluster of genes coding for cell surface proteins important in antigen presentation to T cells.

Malignant   In reference to a tumor, an infiltrating metastasizing growth no longer under normal growth control.

Mast cell   Tissue cells adjoining blood vessels throughout the body that contain granules with inflammatory mediators.

Medium (plural media)   In microbiology, the nutrient solution(s) used to grow microorganisms.

Meiosis   In eukaryotes, reduction division, the process by which the change from diploid to haploid occurs.

Membrane   Any thin sheet or layer. See especially *Cytoplasmic membrane*.

Memory cell   A differentiated B lymphocyte capable of rapid conversion to an antibody-producing plasma cell upon subsequent stimulation with antigen.

Mesophile   Organism living in the temperature range around that of warm-blooded animals.

Messenger RNA (mRNA)   An RNA molecule transcribed from DNA which contains the information to direct the synthesis of a particular protein.

Metabolism   All biochemical reactions in a cell, both anabolic and catabolic.

Methanogen   A methane-producing prokaryote; member of the Archaea.

Methanogenesis   The biological production of methane (CH_4).

Methanotroph   An organism capable of oxidizing methane.

Methylotroph   An organism capable of oxidizing organic compounds which do not contain carbon-carbon bonds; if able to oxidize CH_4, also a methanotroph.

Microaerophilic   Requiring O_2 but at a level lower than atmospheric.

Microenvironment   The immediate physical and chemical surroundings of a microorganism.

Micrometer   One-millionth of a meter, or 10^{-6} m (abbreviated μm), the unit used for measuring microorganisms.

Microtubules   Tubes that are the structural entity for eukaryotic flagella, have a role in maintaining cell shape, and function as mitotic spindle fibers.

Minus (negative)-strand nucleic acid   An RNA or DNA strand which has the opposite sense of (would be complementary to) the mRNA of a virus.

Mitochondrion   Eukaryotic organelle responsible for processes of respiration and electron-transport phosphorylation.

Mitosis   A highly ordered process by which the nucleus divides in eukaryotes.

Mixotroph   An organism able to assimilate organic compounds as carbon sources while using inorganic compounds as electron donors.

Molds   Filamentous fungi.

Molecule   The result of two or more atoms combining by chemical bonding.

Monoclonal antibody   An antibody produced from a single clone of cells. This antibody has uniform structure and specificity.

Monocyte   Circulating white blood cells which contain many lysosomes and can differentiate into macrophages.

Monomer   A building block of a polymer.

Monotrichous   The state of having a single polar flagellum.

Morbidity   Incidence of disease in a population, including both fatal and nonfatal cases.

Mortality   Incidence of death in a population.

Motility   The property of movement of a cell under its own power.

Mushrooms   Filamentous fungi that produce large, often edible structures, called fruiting bodies.

Mutagen   An agent that induces mutation, such as radiation or certain chemicals.

Mutant   A strain differing from its parent because of mutation.

Mutation   An inheritable change in the base sequence of the DNA of an organism.

Mycorrhiza   A symbiotic association between a fungus and the roots of a plant.

Mycoses   Infections caused by fungi.

Myeloma   A malignant tumor of a plasma cell (antibody-producing cell).

Natural killer (NK) cell   A specialized lymphocyte that recognizes and destroys foreign cells or infected host cells in a nonspecific manner.

Nitrification   The conversion of ammonia to nitrate.

Nitrogen fixation   Reduction of nitrogen gas to ammonia by the enzyme nitrogenase.

Nodule   A tumorlike structure produced by the roots of symbiotic nitrogen-fixing plants. Contains the nitrogen-fixing microbial component of the symbiosis.

Nonpolar   Possessing hydrophobic (water repelling) characteristics and not easily dissolved in water.

Nonsense mutation   A mutation that changes a sense codon into one which does not code for an amino acid.

Northern blot   Hybridization of a single strand of nucleic acid (DNA or RNA) to RNA fragments immobilized on a filter. Compare with *Southern* and *Western blot*.

Nosocomial infection   Hospital-acquired infection.

Nucleic acid   A polymer of nucleotides. See *Deoxyribonucleic acid* and *Ribonucleic acid*.

Nucleic acid probe   A strand of nucleic acid which can be labeled and used to hybridize to a complementary molecule from a mixture of other nucleic acids. In clinical microbiology, short oligonucleotides of unique sequences used as hybridization probes for identifying pathogens.

Nucleoid   The aggregated mass of DNA that makes up the chromosome of prokaryotic cells.

Nucleoside   A nucleotide minus phosphate.

Nucleotide   A monomeric unit of nucleic acid, consisting of a sugar, a phosphate, and a nitrogenous base.

Nucleus   A membrane-enclosed structure containing the genetic material (DNA) organized in chromosomes.

Nutrient   A substance taken by a cell from its environment and used in catabolic or anabolic reactions.

Obligate   A qualifying adjective referring to an environmental factor always required for growth (e.g., "obligate anaerobe").

Oligonucleotide   A short nucleic acid molecule, either obtained from an organism or synthesized chemically.

Oligotrophic   Describing a body of water in which nutrients are in low supply.

Oncogene   A gene whose expression causes formation of a tumor.

Open reading frame (ORF)   The entire length of a DNA molecule that starts with a start codon and ends with a stop codon.

Operator   A specific region of the DNA at the initial end of a gene, where the repressor protein binds and blocks mRNA synthesis.

Operon   A cluster of genes whose expression is controlled by a single operator. Typical of prokaryotic cells.

Opsonization   Promotion of phagocytosis by a specific antibody in combination with complement.

Organelle   A membrane-enclosed body specialized for carrying out certain functions; found in eukaryotic cells.

Osmosis   Diffusion of water through a membrane from a region of low solute concentration to one of higher concentration.

Outbreak   The occurrence of a large number of cases of a disease in a short period of time.

Oxic   Containing oxygen; aerobic. Usually used in reference to a microbial habitat.

Oxidation   A process by which a compound gives up electrons, acting as an electron donor, and becomes oxidized.

Oxidation–reduction (redox) reaction   A coupled pair of reactions, in which one compound becomes oxidized, while another becomes reduced and takes up the electrons released in the oxidation reaction.

Oxidative (electron transport) phosphorylation   The non-phototrophic production of ATP at the expense of a proton motive force formed by electron transport.

Oxygenic photosynthesis   Use of light energy to synthesize ATP and NADPH by noncyclic photophosphorylation with the production of O_2 from water.

Palindrome   A nucleotide sequence on a DNA molecule in which the same sequence is found on each strand, but in the opposite direction.

Pandemic   A worldwide epidemic.

Parasite   An organism able to live on and cause damage to another organism.

Passive immunity   Immunity resulting from transfer of antibodies or immune cells from an immune to a nonimmune individual.

Pasteurization   A process using mild heat to reduce the microbial level in heat-sensitive materials.

Pathogen   An organism able to inflict damage on a host it infects.

Pathogenicity   The ability of a parasite to inflict damage on the host.

Peptide bond   A type of covalent bond joining amino acids in a polypeptide.

Peptidoglycan   The rigid layer of the cell walls of Bacteria, a thin sheet composed of N-acetylglucosamine, N-acetylmuramic acid, and a few amino acids. Also called *Murein*.

Periplasmic space   The area between the cytoplasmic membrane and the cell wall in Gram-negative Bacteria, containing certain enzymes involved in nutrition.

Peritrichous flagellation   Having flagella attached to many places on the cell surface.

Person to person epidemic   An epidemic resulting from person to person contact, characterized by a gradual rise and fall in number of cases.

Phage   See Bacteriophage.

Phagemid   A cloning vector that can replicate either as a plasmid or as a bacteriophage.

Phagocyte   A body cell able to ingest and digest foreign particles.

Phagocytosis   Ingestion of particulate material such as bacteria by protozoa and phagocytic cells of higher organisms.

Phenotype   The observable properties of an organism. Compare with *Genotype*.

Phosphodiester bond   A type of covalent bond linking nucleotides together in a polynucleotide.

Phospholipid   Lipids containing a substituted phosphate group and two fatty acid chains on a glycerol backbone.

Photoautotroph   An organism able to use light as its sole source of energy and CO_2 as sole carbon source.

Photoheterotroph   An organism using light as a source of energy and organic materials as carbon source.

Photophosphorylation   Synthesis of high-energy phosphate bonds as ATP, using light energy.

Photosynthesis   The use of light energy to drive the incorporation of CO_2 into cell material. See also *anoxygenic* photosynthesis and *oxygenic* photosynthesis.

Phototaxis   Movement toward light.

Phototroph   An organism that obtains energy from light.

Phylogeny   The ordering of species into higher taxa and the construction of evolutionary trees based on evolutionary (natural) relationships.

Phytanyl   A branched-chain hydrocarbon containing 20 carbon atoms, commonly found in the lipids of Archaea.

Pilus   A fimbria-like structure that is present on fertile cells, both Hfr and F⁺, and is involved in DNA transfer during conjugation. Sometimes called *Sex pilus*. See also *Fimbria*.

Pinocytosis   In eukaryotes, phagocytosis of soluble molecules.

Plaque   A localized area of lysis or cell inhibition caused by virus infection on a lawn of cells.

Plasma   The noncellular portion of blood.

Plasma cell   A large differentiated and short-lived B lymphocyte specializing in abundant (but short term) antibody production.

Plasmid   An extrachromosomal genetic element not essential for growth and which has no extracellular form.

Platelet   A noncellular disc-shaped structure containing protoplasm found in large numbers in blood and functioning in the blood clotting process.

Plus-strand nucleic acid   An RNA or DNA strand which has the same sense as the mRNA of a virus.

Polar   Possessing hydrophilic characteristics and generally water soluble.

Polar flagellation   Condition of having flagella attached at one end or both ends of the cell.

Poly-β-hydroxybutyrate (PHB)   A common storage material of prokaryotic cells consisting of a polymer of β-hydroxybutyrate (PHB) or other β-alkanoic acids (PHA).

Polyclonal antiserum   A mixture of antibodies to a variety of antigens or to a variety of determinants on a single antigen.

Polymer   A large molecule formed by polymerization of monomeric units.

Polymerase chain reaction (PCR)   A method for amplifying DNA *in vitro*, involving the use of oligonucleotide primers complementary to nucleotide sequences in a target gene and the copying of the target sequences by the action of DNA polymerase.

Polymorphonuclear leukocyte (PMN) Small, actively motile white blood cells containing many lysosomes and specializing in phagocytosis.

Polynucleotide A polymer of nucleotides bonded to one another by phosphodiester bonds.

Polypeptide Several amino acids linked together by peptide bonds.

Polysaccharide A long chain of monosaccharides (sugars) linked by glycosidic bonds.

Porins Protein channels in the lipopolysaccharide layer of Gram-negative Bacteria.

Porters Membrane proteins that function to transport substances into and out of the cell.

Precipitation A reaction between antibody and soluble antigen resulting in a visible mass of antibody–antigen complexes.

Prevalence The *proportion* of individuals in a population having a disease.

Pribnow box The consensus sequence TATAAT located approximately 10 base pairs upstream from the transcriptional start site. A binding site for RNA polymerase.

Primary antibody response Antibodies made upon first exposure to antigen; mostly of the class IgM.

Primary metabolite A metabolite excreted during the growth phase.

Primary producer An organism that uses light to synthesize new organic material from CO_2.

Primary structure In an informational macromolecule, such as a polypeptide or a nucleic acid, the *precise sequence* of monomeric units.

Primary transcript An unprocessed RNA molecule which is the direct product of transcription.

Primer A molecule (usually a polynucleotide) to which DNA polymerase can attach the first nucleotide during DNA replication.

Prion An infectious agent whose extracellular form may contain no nucleic acid.

Probe See *Nucleic Acid Probe.*

Prochlorophyte A prokaryotic oxygenic phototroph that contains chlorophylls *a* and *b* but which lack phycobilins.

Prokaryote A cell or organism lacking a unit membrane-bound (true) nucleus and other organnelles, usually having its DNA in a single circular molecule.

Promoter The site on DNA where the RNA polymerase binds and begins transcription.

Prophage The state of the genome of a temperate virus when it is replicating in synchrony with that of the host, typically integrated into the host genome.

Prophylactic Treatment, usually immunologic or chemotherapeutic, designed to protect an individual from a future attack by a pathogen.

Prostheca A cytoplasmic extrusion from a cell such as a bud, hypha, or stalk.

Prosthetic group The tightly bound, nonprotein portion of an enzyme; not the same as *Coenzyme.*

Protein A polymeric molecule consisting of one or more polypeptides.

Proton motive force An energized state of a membrane created by expulsion of protons through action of an electron transport chain. See also *Chemiosmosis.*

Protoplasm The complete cellular contents, cytoplasmic membrane, cytoplasm, and nucleus; usually considered to be the living portion of the cell, thus excluding those layers peripheral to the cytoplasmic membrane.

Protoplast A cell from which the wall has been removed.

Prototroph The parent from which an auxotrophic mutant has been derived. Contrast with *Auxotroph.*

Protozoa Unicellular eukaryotic microorganisms that lack cell walls.

Provirus See *Prophage.*

Psychrophile An organism able to grow at low temperatures and showing a growth temperature optimum of < 15°C.

Psychrotolerant An organism able to grow at low temperature but having a growth temperature optimum of >15°C.

Public health The health of the population as a whole.

Pure culture A culture containing a single kind of microorganism.

Pyogenic Pus-forming; causing abscesses.

Pyrite A common iron ore, FeS_2.

Pyrogenic Fever-inducing.

Quarantine The limitation on the freedom of movement of an individual, to prevent spread of a disease to other members of a population.

Quaternary structure In proteins, the number and arrangement of individual polypeptides in the final protein molecule.

Radioimmunoassay An immunological assay employing radioactive antibody or antigen for the detection of certain substances in body fluids.

Radioisotope An isotope of an element that undergoes spontaneous decay with the release of radioactive particles.

Reaction center A photosynthetic complex containing chlorophyll (or bacteriochlorophyll) and other components, within which occurs the initial electron transfer reactions of photophosphorylation.

Reading-frame shift See *Frameshift.*

Recalcitrant Resistant to microbial attack.

Recombinant DNA A DNA molecule containing DNA originating from two or more sources.

Recombination Process by which genetic elements in two separate genomes are brought together in one unit.

Redox See *Oxidation–reduction reaction.*

Reduction A process by which a compound accepts electrons to become reduced.

Reduction potential (E_0') The inherent tendency of a compound to act as an electron donor or an electron acceptor. Measured in volts.

Reductive dechlorination Removal of Cl as Cl^- from an organic compound by reducing the carbon atom from C-Cl to C-H.

Regulation Processes that control the rates of synthesis of proteins. Induction and repression are examples of regulation.

Regulon A set of operons that are all controlled by the same regulatory protein (repressor or activator).

Replacement vector A cloning vector, such as a bacteriophage, in which some of the DNA of the vector can be replaced with foreign DNA.

Replication Conversion of one double-stranded DNA molecule into two identical double-stranded DNA molecules.

Repression The process by which the synthesis of an enzyme is inhibited by the presence of an external substance, the repressor.

Repressor protein A regulatory protein that binds to specific sites on DNA and blocks transcription.

Reservoir In epidemiology, the organism or environment that normally harbors a pathogen.

Respiration Catabolic reactions producing ATP in which either organic or inorganic compounds are primary electron donors and organic or inorganic compounds are ultimate electron acceptors.

Restriction endonucleases (restriction enzymes) Enzymes that recognize and cleave specific DNA sequences, generating either blunt or single-stranded (sticky) ends.

Retrovirus Virus containing single-stranded RNA as its genetic material and which produces a complementary DNA by action of the enzyme reverse transcriptase.

Reverse electron transport The energy-dependent movement of electrons against the thermodynamic gradient to form a strong reductant from a weaker electron donor.

Reverse transcription The process of copying information found in RNA into DNA.

Rheumatic fever An inflammatory autoimmune disease triggered by an immune response to infection by *Streptococcus pyogenes*.

Rhizosphere The region immediately adjacent to plant roots.

Ribonucleic acid (RNA) A polymer of nucleotides connected via a phosphate-ribose backbone, involved in protein synthesis.

Ribosomal RNA (rRNA) Types of RNA found in the ribosome; some participate actively in the process of protein synthesis.

Ribosome A cytoplasmic particle composed of ribosomal RNA and protein, which is part of the protein-synthesizing machinery of the cell.

Ribozyme An RNA molecule that can catalyze a chemical reaction.

Rickettsias Obligate intracellular parasites which cause a variety of diseases, including typhus and Rocky Mountain spotted fever.

RNA life A hypothetical life form lacking DNA and protein which may have existed on early earth and in which RNA served both a genetic coding and a catalytic function.

RNA processing The conversion of a precursor RNA into its mature form.

Root nodule A tumor-like growth on plant roots that contains symbiotic nitrogen-fixing bacteria.

Rumen The forestomach of ruminant animals in which cellulose digestion occurs.

Scale-up Conversion of an industrial process from a small laboratory setup to a large commercial fermentation.

Scarlet fever Characteristic reddish skin rash resulting from an exotoxin produced by cells of *Streptococcus pyogenes*.

Secondary antibody response Antibody made upon second (subsequent) exposure to antigen; mostly of the class IgG.

Secondary metabolite A product excreted by a microorganism near the end of the growth phase or during stationary phase.

Secondary structure The initial pattern of folding of a polypeptide or a polynucleotide, usually the result of hydrogen bonding.

Secondary treatment In sewage treatment, either the aerobic or anaerobic decomposition of sewage following the removal of nondegradable objects by primary treatment.

Secretion vector A DNA vector in which the protein product is both expressed and secreted (excreted) from the cell.

Selection Placing organisms under conditions where the growth of those with a particular genotype will be favored.

Semiconservative replication DNA synthesis yielding new double helices, each consisting of one parental and one progeny strand.

Septicemia Infection of the bloodstream by microorganisms.

Serology The study of antigen–antibody reactions *in vitro*.

Serum Fluid portion of blood remaining after the blood cells and materials responsible for clotting are removed.

Sexually transmitted disease (STD) A disease whose usual means of transmission is by sexual contact.

Shine–Dalgarno sequence A short stretch of nucleotides on a prokaryotic mRNA molecule upstream of the translational start site, that serves to bind to ribosomal RNA and thereby bring the ribosome to the initiation codon on the mRNA.

Shuttle vector A DNA vector which can replicate in two different organisms, used for moving DNA between unrelated organisms.

Sickle-cell anemia A genetic trait which confers resistance to malaria but which causes a reduction in numbers of red blood cells.

Signal sequence A short stretch of amino acids found at the beginning of proteins that are typically excreted from the cell. The signal sequence is usually rich in hydrophobic amino acids which helps transport the entire polypeptide through the membrane.

Signature sequence Short oligonucleotides of unique sequence found in 16S ribosomal RNA of a particular group of prokaryotes.

Single-cell protein Protein derived from microbial cells for use as food or a food supplement.

Site-directed mutagenesis The insertion of a different nucleotide at a specific site in a DNA molecule using synthetic DNA methodology.

16S rRNA A large polynucleotide (~1500 bases) which functions as a part of the small subunit of the ribosome of prokaryotes and from whose sequence evolutionary information can be obtained; eukaryotic counterpart, 18S rRNA.

Slime layer A diffuse layer of polysaccharide exterior to the cell wall in some prokaryotes. See also *Capsule* and *Glycocalyx*.

Slime molds Nonphototrophic eukaryotic microorganisms lacking cell walls, which aggregate to form fruiting structures (cellular slime molds) or simply masses of protoplasm (acellular slime molds).

Southern blot Hybridization of a single strand of nucleic acid (DNA or RNA) to DNA fragments immobilized on a filter. Compare with *Northern* and *Western blot*.

Species A collection of closely related strains sufficiently different from all other strains to be recognized as a distinct unit.

Specificity The ability of the immune response to interact with individual antigens.

Spheroplast A spherical, osmotically sensitive cell derived from a bacterium by loss of some but not all of the rigid wall layer. If all the rigid wall layer has been completely lost, the structure is called a *Protoplast*.

Spontaneous generation The hypothesis that living organisms can originate from nonliving matter.

Spore A general term for resistant resting structures formed by many prokaryotes and fungi.

Sporozoa Nonmotile parasitic protozoa.

Stalk An elongate structure, either cellular or excreted, which anchors a cell to a surface.

Stationary phase The period during the growth cycle of a population in which growth ceases.

Stereoisomers Mirror image forms of two molecules having the same molecular and structural formulas.

Sterile Free of living organisms and viruses.

Sterilization Treatment resulting in death of all living organisms and viruses in a material.

Strain A population of cells all descended from a single cell; a clone.

Stromatolites Laminated microbial mats, typically built from layers of filamentous and other microorganisms which can become fossilized.

Substrate The molecule undergoing reaction with an enzyme.

Substrate-level phosphorylation Synthesis of high-energy phosphate bonds through reaction of inorganic phosphate with an activated (usually) organic substrate.

Sulfatara A hot, sulfur-rich, generally acidic environment, commonly inhabited by hyperthermophilic Archaea.

Supercoil Highly twisted form of circular DNA.

Superoxide anion (O_2^-) A harmful derivative of O_2 capable of oxidative destruction of cell components.

Suppressor   A mutation that restores wild-type phenotype without altering the original mutation, usually arising by mutation in another gene.

Symbiosis   A relationship between two organisms.

Syntrophy   A nutritional situation in which two or more organisms combine their metabolic capabilities to catabolize a substance not capable of being catabolized by either one alone.

Systemic   Not localized in the body; an infection disseminated widely through the body is said to be systemic.

Taxis   Movement toward or away from a stimulus.

Taxonomy   The study of scientific classification and nomenclature.

T cell receptor   The antigen-specific receptor on the surface of T lymphocytes.

T-DNA   The segment of the *Agrobacterium* Ti plasmid which is transferred to plant cells.

T lymphocyte   A type of immune cell responsible for many cell-mediated immune responses plus stimulation of differentiation of antibody-producing (B) lymphocytes during the humoral immune response.

Temperate virus   A virus which upon infection of a host does not necessarily cause lysis but whose genome may replicate in synchrony with that of the host. See *Lysogen*.

Tertiary structure   The final folded structure of a polypeptide that has previously attained secondary structure.

Thermocline   Zone of water in a stratified lake in which temperature and oxygen concentration drop precipitously with depth.

Thermophile   An organism living at high temperature, usually with a growth temperature optimum between 45 and 80°C.

Ti plasmid   A conjugative plasmid present in the bacterium *Agrobacterium tumefaciens* which can transfer genes into plants.

Titer   Measure of antibody quantity.

Tolerance   In reference to immunology, the acquisition of nonresponsiveness to a molecule normally recognized by the immune system.

Toxic shock syndrome   Acute shock resulting from host response to an exotoxin produced by *Staphylococcus aureus*.

Toxigenicity   The degree to which an organism is able to elicit toxic symptoms.

Toxin   A microbial substance able to induce host damage.

Toxoid   A toxin modified so that it is no longer toxic but is still able to induce antibody formation.

Transcription   Synthesis of an RNA molecule complementary to one of the two strands of a double-stranded DNA molecule.

Transduction   Transfer of host genetic information via a virus particle.

Transfection   The transformation of a prokaryotic cell by DNA or RNA from a virus. Used also to describe the process of genetic transformation in eukaryotic cells.

Transfer RNA (tRNA)   A type of RNA that carries amino acids to the ribosome during translation; contains anticodon.

Transformation   Transfer of genetic information via free DNA. Also, a process, sometimes initiated by infection with certain viruses, whereby a normal animal cell becomes a cancer cell.

Transgenic   Used to describe genetically modified plants or animals containing foreign genes inserted by recombinant DNA means.

Translation   The synthesis of protein using the genetic information in a messenger RNA as a template.

Transpeptidation   The formation of peptide bonds between the short peptides present in the cell wall polymer, peptidoglycan.

Transposable element   A genetic element that has the ability to move (transpose) from one site on a chromosome to another.

Transposon   A type of transposable element which, in addition to genes involved in transposition, carries other genes; often genes conferring selectable phenotypes such as antibiotic resistance.

Transposon mutagenesis   Insertion of a transposon into a gene; this inactivates the host gene leading to a mutant phenotype and also confers the phenotype associated with the transposon gene.

Tricarboxylic acid cycle (citric acid cycle, Krebs cycle)   A series of steps by which pyruvate is oxidized completely to CO_2, also forming NADH, which allows ATP production.

Tuberculin test   A test for previous infection with *Mycobacterium tuberculosis*.

Upper respiratory tract   The nasopharynx, oral cavity, and throat.

Upstream position   Refers to nucleic acid sequences on the 5'-side of a given site on a DNA or RNA molecule. Compare with *Downstream position*.

Vaccination   Artificial introduction of a killed or attenuated pathogen to promote protective immunity. See *Vaccine*.

Vaccine   Material used to induce specific protective immunity to a pathogen.

Vacuole   A small space in a cell containing fluid and surrounded by a membrane. In contrast to a vesicle, a vacuole is not rigid.

Vector   An agent, usually an insect or other animal, able to carry pathogens from one host to another. Also: A genetic element able to incorporate DNA and cause it to be replicated in another cell.

Vehicle   Nonliving source of pathogens which infect large numbers of individuals; common vehicles are food and water.

Viable   Alive; able to reproduce.

Viable count   Measurement of the concentration of live cells in a microbial population.

Virion   A virus particle; the virus nucleic acid surrounded by protein coat and in some cases other material.

Viroid   A small RNA molecule with viruslike properties.

Virulence   Degree of pathogenicity of a parasite.

Virus   A genetic element containing either DNA or RNA that is able to alternate between intracellular and extracellular states, the latter being the infectious state.

Water activity (a_w)   An expression of the relative availability of water in a substance. Pure water has an a_w of 1.000.

Western blot   See *Immunoblot*.

Wild type   A strain of microorganism isolated from nature. The usual or native form of a gene or organism.

Wobble   In reference to reading the genetic code, the concept that nonstandard base pairing is allowed between the anticodon and the third position of the codon.

Xenobiotic   A completely synthetic chemical compound not naturally occurring on earth.

Xerophile   An organism adapted to growth at very low water potentials.

Yeasts   Unicellular fungi.

Zoonoses   Diseases primarily of animals which are occasionally transmitted to humans.

Zygote   In eukaryotes, the single diploid cell resulting from the union of two haploid gametes.

CHAPTER 1—**1.1** Thomas D. Brock. **1.2a** Thomas D. Brock; **1.2b** Herbert Voelz. **1.5** Richard Feldmann. **1.6a** Richard Feldmann. **1.9b** Howard Gest; **1.9d** Lonnie Russell. **1.10a** Councilman Morgan. **1.13a** Thomas D. Brock; **1.13b** D.E. Caldwell. **1.14** NASA. **1.17** Thomas D. Brock. **1.19** Thomas D. Brock.

CHAPTER 3—**3.2a** Thomas D. Brock; **3.2b** Norbert Pfennig; **3.3** Thomas Brock. **3.5b** Leon J. LeBeau. **3.7a** Stanley C. Holt; **3.7b** M.T. Madigan and R. Turner; **3.8** Norbert Pfennig. **3.9a** Bryan Larsen; **3.9b** A. Umeda, et al. **3.9c** A. Umeda and K. Amako. **3.10** S.F. Conti and T.D. Brock. **3.14a** G. Wanner. **3.26c** J.L. Pate; **3.26d** T.D. Brock and S.F. Conti. **3.26e,f** A. Umeda and K. Amako. **3.32** Terry Beveridge. **3.34b** A. Umeda and K. Amako. **3.37** D.E. Snyder. **3.38** Arthur Kelman. **3.39** E. Leifson. **3.40** R. Jarosch. **3.48** J.P. Duguid and J.F. Wilkinson. **3.49** C. Brinton. **3.50a** Elliot Juni; **3.50b** Frank Dazzo and Richard Heinzen. **3.51b** F.R. Turner and Michael T. Madigan. **3.52** R. Blakemore, N. Blakemore, W. O'Brien. **3.53** Thomas D. Brock. **3.54** A.E. Konopka and J.T. Staley. **3.55a,b** A.E. Walsby. **3.56** H. Hippe. **3.57** H.S. Pankratz, T.C. Beaman, and Philipp Gerhardt. **3.60** Judith Hoeniger and C.L. Headley. **3.63** E. Kellenberger and J.A. Hobot. **3.64** E. Kellenberger. **3.65** B. Arnold-Schultz-Gahmen. **3.67a** E. Guth, T. Hashimoto, and S.F. Conti; **3.67b** D.W. Fawcett. **3.68** Carolina Biological Supply Co. **3.70b,c** D.W. Fawcett. **3.71** Thomas D. Brock. **3.72** T. Slankis and S. Gibbs. **3.73** Jian-Ming Li and Nancy Martin.

CHAPTER 4—**4.4** Richard Feldmann. **4.14d** Richard Feldmann. **4.33** Thomas D. Brock. **4.37c** Thomas D. Brock.

CHAPTER 5—**5.5** Stephen Edmondson. **5.16b** Steve Jordan, Karl Pabo, and John Reidhaar-Olson. **5.27b** Sarah French. **5.34** Dino Moras. Science 252:1682–1689 (1991). **5.44** Thomas A. Steitz and Steve Schultz. Science 253:1001–1007 (1991). **Box, page 138, 139** Pharmacia, Inc.; **gel, page 138** Jack Parker; **page 141** Schleicher and Schnell, Inc.; **gel and blot** Carl Marrs and Martha Howe; **page 142** Jack Parker.

CHAPTER 6—**6.2a** J.T. Finch. **6.4b** Thomas D. Brock. **6.5a** W.F. Noyes; **6.5b** P.W. Choppin and W. Stoeckenius; **6.5c** M. Wurtz. **6.6b** Thomas D. Brock. **6.7** Paul Kaplan, Thomas D. Brock. **6.8** Paul Kaplan. **6.15** R.C. Valentine. **6.26** D. Kaiser. **6.32** Frank Grundy and Martha Howe. **6.34b** G. Chardonnet and S. Dales. **6.37a** Arthur J. Olson. **6.38** Erskine Caldwell. **6.40a** P.W. Choppin and W. Stoeckenius. **6.41** Heather Mayor. **6.42** Alexander Erb and Jerome Vinograd. **6.45a** R.W. Horne. **6.46** D. Dales and F. Fenner.

CHAPTER 7—**7.1a** Peter Borgia; **7.1b** Thomas D. Brock. **7.1c** James Shapiro **7.2b** Thomas D. Brock. **7.8** Thomas D. Brock. **7.13** Hunington Potter and David Dressler. **7.22** C. Brinton. **7.36** S.F. Conti and T.D. Brock. **7.38b** David Clayton.

CHAPTER 8—**8.6b,c** Jack Parker. **8.19** Keith V. Wood.

CHAPTER 9—**9.14a** Katherine M. Brock; **9.14b** Thomas D. Brock. **9.15** Thomas D. Brock. **9.16** Thomas D. Brock. **9.20** Coy Laboratories, Inc. **9.23** Thomas D. Brock. **9.28** Carlos Pedros-Alio and T.D. Brock. **9.31** Thomas D. Brock. **9.37** Thomas D. Brock.

CHAPTER 10—**10.4a** Queue Systems, Inc.; **10.4b,c** Novo Industri. **10.5** Novo Industri. **10.6** Elmer L. Gaden, Jr. **10.8b** Pfizer Research. **10.24** The Christian Brothers Winery. **10.28a,b** American Mushroom Institute; **10.28c,d** Bob Harris.

CHAPTER 11—**11.5** T. Lie. **11.6** C. Lai, M.A. Listgarten, and B. Rosan. **11.7** I.L. Schechmeister and J. Bozzolla. **11.9** Dwayne Savage. **11.10** Dwayne Savage and R.V.H. Blumershine. **11.12b** John Durham/Science Photo Library/Photo Researchers Inc. **11.13a** E.T. Nelson, J.D. Clements, and R.A. Finkelstein; **11.13b** J.W. Costerton. **11.14** James A. Roberts. **11.15a** Thomas D. Brock; **11.15b** Leon J. LeBeau. **11.18** H.W. Smith and S. Halls. **11.20** A.O. Tzianabos and R.D. Millham. **11.25** Thomas D. Brock. **11.26** J.G. Hirsch.

CHAPTER 12—**12.6a** Richard J. Feldmann; **12.6b** Craig Smith. **12.11** Based on art from Don Wiley. **12.23** E. Munn. **12.26** Richard S. Lewis. **12.27** R.J. Poljak. **12.28** C. Weibull, W.D. Bickel, W.T. Hashius, K.C. Milner, and E. Ribi; **12.28d** Norman L. Morris, American Red Cross. **12.29** Wellcome Research Laboratories. **12.31** Geoffrey Codd. **12.35** Ken Greer/Visuals Unlimited. **Box, page 460** John Martinko. **Box, page 469** Perkin-Elmer Cetus. **Box, page 464** Keystone Diagnostics, Inc.

CHAPTER 13—**13.3a** Theodor Rosebury; **13.3b** Leon J. LeBeau. **13.4** Thomas D. Brock. **13.5** Leon J. LeBeau. **13.6d** Miles/Science VU/Visuals Unlimited; **13.6e** Leon J. LeBeau. **13.8a** William B. Cherry; **13.8b,c** Dharam Ablashi and Robert C. Gallo. **13.9** Grace Bradley. **13.11b** Victor Tsang. **13.12** Lucy S. Tompkins. **13.15** S.E. Miller.

CHAPTER 14—NONE.

CHAPTER 15—**15.1** Thomas D. Brock. **15.3** Franklin H. Top, Communicable and Infectious Diseases, 6th ed. C.V. Mosby Co., St. Louis, MO. 1968. **15.4** Franklin H. Top, Communicable and Infectious Diseases, 6th ed. C.V. Mosby Co., St. Louis, MO. 1968. **15.5** Isaac Schechmeister. **15.7a** Biophoto Assoc./Photo Researchers Inc.; **15.7b** Franklin H. Top, Communicable and Infectious Diseases, 6th ed. C.V. Mosby Co., St. Louis, MO. 1968. **15.8** Carey S. Callaway. **15.10** Aaron Friedman. **15.12** C.H. Binford **in** Top, F.H. and P.F. Wehrle, (ed.), Communicable and Infectious Diseases, 7th ed., C.V. Mosby Co. St., Louis, MO. 1972. **15.14a** Heather Davies and D. Tyrell; **15.14b** Heather Davies and D. Tyrell. **15.16** Irene T. Schulze. **15.18** Centers for Disease Control. **15.20** Centers for Disease Control. **15.22** Centers for Disease Control. **15.24a** Morris D. Cooper. **15.25a** Theodor Rosebury; **15.25b** Centers for Disease Control. **15.26a** Centers for Disease Control. **15.26b** S. Olansky and L.W. Shaffer **in** Top, F.H. and P.F. Wehrle (ed.), Communicable and Infectious Diseases, 7th ed., C.V. Mosby Co, St. Louis, MO. 1972; **15.27** Morris D. Cooper. **15.28a** Gordon A. Tuffli; **15.28b** Centers for Disease Control; **15.28c** David Phillips/Visuals Unlimited. **15.30a-f,h** Centers for Disease Control; **15.30g** Jonathan W.M. Gold. **15.31** Centers for Disease Control. **15.32** Centers for Disease Control. **15.37a,b** Willy Burgdorfer; **15.37c** University of Virginia Dermatology Department. **15.38** Dano Corwin. **15.39** Pfizer Research. **15.41** Pfizer Research. **15.43** Thomas D. Brock. **15.46** Centers for Disease Control. **15.48** Mark L. Tamplin. **15.49a** D.E. Feely, S.L. Erlandsen, and D.G. Case; **15.49b** S.L. Erlandsen. **15.50** Centers for Disease Control. **15.51** Thomas D. Brock. **15.52** Louisville Water Company.

CHAPTER 16—**16.5a** Hilda Canter-Lund. **16.7** F. Rudy Turner and Michael T. Madigan. **16.8a** George Feher. **16.8b** Marianne Schiffer and James A. Norris. **16.12** Norbert Pfennig. **16.14** Yehuda Cohen and Moshe Shilo. **16.16b** Thomas Jensen. **16.18** Norbert Pfennig. **16.21** Thomas D. Brock. **16.23a** Bill Strode; **16.23b,c** Thomas D. Brock. **16.24** Thomas D. Brock. **16.39** B.V. Hofsten. **16.40** Katherine M. Brock. **16.41** Thomas D. Brock. **16.42** Thomas D. Brock.

CHAPTER 17—**17.2** Thomas D. Brock. **17.3** Frank Dazzo. **17.4b** Norbert Pfennig. **17.5** Norbert Pfennig. **17.6** Thomas D. Brock. **17.7** Norman Pace. **17.11a** David M. Ward. **17.14** Otis Brown and Robert Evans, NASA. **17.19** T.R.G. Gray. **17.20** Jannasch, H.W., C.O. Wirsen and C.D. Taylor. 1982. Science 216:1315–1317. **17.23a** Dudley Foster, Woods Hole Oceanographic Institution; **17.23b** James Agviar, Woods Hole Oceanographic Institution; **17.23c** Carl Wirsen, Woods Hole Oceanographic Institution. **17.24a** Colleen Cavanaugh. 1981. Science. 209:340–342; **17.24b** Colleen Cavanaugh. 1983. Nature. 302:58–61. **17.25a** Robert D. Ballard; **17.25b** Dudley Foster, Woods Hole Oceanographic Institution. **17.26** Norman Pace. **17.30** Monica Lee and Stephen Zinder. **17.38** Thomas D. Brock. **17.40** Thomas D. Brock. **17.42** Thomas D. Brock. **17.45** Thomas D. Brock. **17.46** US Environmental Protection Agency. **17.47** Standard Oil Company. **17.49b** Helmut Brandl. **17.52a** Thomas D. Brock. **17.53a** Thomas D. Brock. **17.54** Richard Unz. **17.55a** Thomas D. Brock; **17.55b** Michael T. Madigan. **17.55c** D.R. Kuhn. **17.56** Thomas D. Brock. **17.57a** J.R. Schramm. **17.57b** D.J. Read. **17.58** S.A. Wilde. **17.60** J.R. Norris. **17.61** Jo Handelsman. **17.65** Wood, Keith V., D.W. Ow, M. DeLuca, J.R. deWet. D.R. Heliniski, and S.H. Howell. 1986. Science. 234:856–859. **17.66** Joe Burton. **17.64** Ben B. Bohlool, NifTAL Project, University of Hawaii. **17.68** Joe Burton. **17.70** F.B. Dazzo and W.J. Brill. 1977. Applied and Environmental Microbiology. 33:132–136. **17.71a** Ben B. Bohlool; **17.71b** Mark Dudley and Sharon Long. **17.72a** Phil Gates, Biological Photo Service; **17.72b** P.J. Dart and F.V. Mercer. **17.76** B. Dreyfus. **17.77** J-H. Becking. **17.78** J-H. Becking.

CHAPTER 18—**18.1a-d** Malcolm Walter; **18.1f** Thomas D. Brock. **18.2** J.W. Schopf. 1968. Journal of Paleontology Pt. 1 42:651–688. **18.6** James A. Lake. **18.7** Carl R. Woese. **18.16** W. Zillig.

CHAPTER 19—**19.2b** Norbert Pfennig. **19.3a** Remsen, C.C., S.W. Watson, J.B. Waterbury, and H.G. Trüper. 1968. Journal of Bacteriology. 95:2374–2392; **19.3b** Jeffery C. Burnham and S.F. Conti; **19.3c** G. Cohen-Bazire; **19.3d** F. Rudy Turner and Howard Gest. **19.4a-e** Norbert Pfennig; **19.4f** Peter Hirsch. **19.5a,b,d** Norbert Pfennig; **19.5c** Peter Hirsch; **19.5e** R. Scherrer. **19.6a,b** Norbert Pfennig; **19.6c** Madigan, M.T. and T.D. Brock. 1977. Journal of General Microbiology. 102:279–285; **19.6d,e** Gorlenko, V.M. and T.A. Pivovarova. 1977. Bulletin of Academy of Sciences of U.S.S.R. Biological Series, No. 3, 396–409. **19.8** F. Rudy Turner and Howard Gest. **19.12** Douglas Caldwell. 1975. Canadian Journal of Microbiology. 21:362–376. **19.13** Douglas Caldwell. **19.14** Kurt Hanselmann. **19.15** John Ormerod and the Norwegian Institute for Water Research. **19.16** Thomas D. Brock. **19.17** Kurt Hanselmann. **19.18** Susan Barns and Norman Pace. **19.19** Edwards, M.R., D.S. Berns, W.D. Ghiorse, and S.C. Holt. 1968. Journal of Phycology. 4:283–298. **19.20** Thomas D. Brock. **19.21** Thomas D. Brock. **19.22** Kit W. Lee. **19.23** T. Burger-Wiersma. **19.24** S.W. Watson. 1971. International Journal of Systemic Bacteriology. 21:254–270. **19.25** S.W. Watson. 1971. International Journal of Systemic Bacteriology. 21:254–270. **19.27** Jessup M. Shively. **19.29** Jessup M. Shively. **19.30** Frank Mayer. **19.31a** Smith, U. and D.W. Ribbons. 1970. Archiv Mikrobiologie. 74:116–122; **19.31b** Smith, U., D.W. Ribbons, and D.S. Smith. 1970. Tissue and Cell. 2:513–520. **19.34** Charles R. Fisher. **19.35a,f** Norbert Pfennig; **19.35c-e** Fritz Widdel; **19.35b** N. Pfennig, F. Widdel, and H.G. Trüper. 1981. The dissimilatory sulfate-reducing bacteria. Pages 926–940. **In** Starr, M.P., H. Stolp, H.G. Trüper, A. Balows, and H.G. Schlegel, (eds.). The Prokaryotes—A Handbook of Habitats, Isolation, and Identification of Bacteria. Springer-Verlag, Berlin. **19.37a** Pate, J.L. and E.J. Ordal. 1965. Journal of Cell Biology. 27:133–150; **19.37b** J.T. Staley. 1968. Journal of Bacteriology. 95:1921–1942; **19.37c** V.M. Gorlenko. **19.39a** Einar Leifson; **19.39b,c** Cohen-Bazire, G., R. Kunisawa, and J.S. Poindexter. 1966. Journal of General Microbiology. 42:301–308. **19.40** Bauld, J. and J.T. Staley. 1976. Journal of General Microbiology. 97:45–55. **19.42** W.C. Ghiorse. 1984. Annual Review of Microbiology. 38:515. **19.44** Peter Hirsch. **19.45** Conti, S.F. and P. Hirsch. 1965. Journal of Bacteriology. 89:503–512. **19.46a,b** Noel Krieg; **19.46c** Stanley Erlandsen; **19.46d** H.D. Raj. **19.47** Blakemore R.P. and R.B. Frankel. 1981. Scientific American 245:58–65. **19.48a** Stolp, H. and H. Petzold. 1962. Phytopathologische Zeitschrift. 45:373; **19.48b,c** Burnham, J.C., T. Hashi moto, and S.F. Conti. 1968. Journal of Bacteriology. 96:1366–1381. **19.50a** Canale-Parola, E., S.C. Holt, and Z. Udris. 1967. Archiv Mikrobiologie. 59:41–48; **19.50b** Blakemore, R.P. and E. Canale-Parola. 1973. Archiv Mikrobiologie. 89:273–289. **19.51a** Joseph R. and E. Canale-Parola. 1972. Archiv Mikrobiologie. 81:146–168. **19.52** Ryter, A. and J. Pillot. 1965. Annale de l'Institut Pasteur 109:552–562. **19.53** Paster, B.J. and E. Canale-Parola. 1985. Applied Environmental Microbiology. 50:212–219. **19.54** Michael Richard. **19.55** Thomas D. Brock. **19.56** Michael Richard. **19.57** Thomas D. Brock. **19.58** David White. **19.60a,b** Hans Reichenbach; **19.60c** Stephens, K., G.D. Hegeman, and D. White. 1982. Journal of Bacteriology. 149:739–747. **19.61** Grillone, P.L. and J. Pangborn. 1975. Journal of Bacteriology. 124:1558–1565. **19.62** Herbert Voelz. **19.63a** Thomas D. Brock; **19.63b,c** Hoeniger, J.F.M., H.D. Tauschel, and J.L. Stokes. 1973. Canadian Journal of Microbiology. 19:309–313 and plates I–VII. By permission of the National Research Council of Canada. **19.64** W.C. Ghiorse. 1984. Annual Review of Microbiology. 38:515. **19.65a** James Shapiro; **19.65b** Arthur Kelman. **19.66** L.P. Lin and H.L. Sadoff. **19.67a,b** J.H. Becking. 1981. The family Azotobacteraceae. **In** Starr, M.P., H. Stolp, H.G. Trüper, A. Balows, and H.G. Schlegel (eds.). 1981. The Prokaryotes—A Handbook on Habitats, Isolation, and Identification of Bacteria. Springer-Verlag, Berlin. **19.68a** Michael K. Ochman; **19.68b** J.H. Becking. 1981. The family Azotobacteriaceae. **In** Starr, M.P., H. Stolp, H.G. Trüper, A. Balows, and H.G. Schlegel

(eds.). 1981. The Prokaryotes—A Handbook on Habitats, Isolation, and Identification of Bacteria. Springer-Verlag, Berlin. **19.69** Thomas D. Brock. **19.70** Thomas D. Brock. **19.72** Kenneth H. Nealson. **19.37a** Arthur Kelman. **19.76a** Willy Burgdorfer; **19.76b** Devauchelle, G., G. Meynadier, and C. Vago. 1972. Journal of Ultrastructure Research. 38:134–148. **19.77** Robert R. Friis. **19.78b** Morris Cooper. **19.79** T. Beveridge. 1980. Canadian Journal of Microbiology. 26:235–242. **19.80** A. Umeda, et al. 1980. Journal of Bacteriology. 141:838–844. **19.81** D. Moyles and R.G.E. Murray. **19.83a** Thomas D. Brock; **19.83b** Bryan Larsen, University of Iowa. **19.84a,b** Otto Kandler; **19.84c** Vittorio Bottazzi. **19.85** Hans Hippe. **19.86** J.R. Norris. **19.89** Dieter Claus. **19.90** Alan Rodwell. **19.91** Thomas D. Brock. **19.92** Thomas D. Brock. **19.93** David L. Williamson. **19.94** Krulwich, T.A. and J.L. Pate. 1971. Journal of Bacteriology. 105:408–412. **19.95** Krulwich, T.A. and J.L. Pate. 1971. Journal of Bacteriology. 105:408–412. **19.96** Veldkamp, Hans, G. van den Berg, and L.P.T.M. Zevenhuizen. 1963. Antonie van Leeuwenhoek 29:35–51. **19.99a** N. Rist, Pasteur Institute, Paris; **19.99b** V. Lorian. 1968. Annual Review of Respiratory Diseases. 97:1133–1135; **19.99c** Centers for Disease Control, Atlanta, GA. **19.101** Hubert and Mary P. Lechevalier. **19.102a** Peter Hirsch; **19.102b** Hubert and Mary P. Lechevalier. **19.105a** Alma Dietz; **19.105b** Michael T. Madigan. **19.106** Eli Lilly & Co. **Table 19.17** Hans Reichenbach.

CHAPTER 20—**20.6a** M. Shioda and S. Takayanogo. **20.6d** Daniel Studer and Thomas Leisinger. **20.7a** Thomas D. Brock; **20.7b** NASA. **20.8** Mary Reedy. **20.10** Alexander Zehnder. **20.11** Zeikus, J.G. and V.G. Bowen. 1975. Canadian Journal of Microbiology. 21:121–129. Reproduced by permission of the National Research Council of Canada. **20.12a,b** Helmut König and K.O. Stetter; **20.12c** Stephen Zinder. **20.14** Thomas D. Brock. **20.19** Thomas D. Brock. **20.20a** Brock, T.D., K.M. Brock, R.T. Belly, and R.L. Weiss. 1972. Archiv Mikrobiologie. 84:54–68. **20.20b** H. König and K.O. Stetter. **20.21** H. König and K.O. Stetter. **20.22** H. König and K.O. Stetter. **20.23a** H. König and K.O. Stetter. **20.23b** Stetter, K.O., H. König, and E. Stackebrandt. 1983. Systematic and Applied Microbiology. 4:535–551. **20.24a** H. König and K.O. Stetter; **20.24b** Fiala, G. and K.O. Stetter. 1986. Archives of Microbiology. 145:56–61. **20.25** H. König and K.O. Stetter. **20.26** Thomas D. Brock. **20.27** Thomas D. Brock.

CHAPTER 21—**21.1a** Thomas D. Brock; **21.1b** Dennis Kunkel. **21.1c,d** Carolina Biological Supply. **21.2a,c** David M. Phillips/Visuals Unlimited; **21.2b** CBS/Visuals Unlimited; **21.2d** Carolina Biological Supply. **21.3** Barry Katz, Mycosearch. **21.4** J. Forsdyke/Science Photo Library/Visuals Unlimited. **21.5** Thomas D. Brock. **21.6a** W. Ormerod/Visuals Unlimited; **21.6b** US Department of Agriculture; **21.6c** Stanley L. Flegler/Visuals Unlimited. **21.7a** Kerry T. Givens/Tom Stack and Associates; **21.7b** Carolina Biological Supply. **21.8** Kenneth B. Raper. **21.10a,b** Carolina Biological Supply; **21.10c** Arthur M. Nomura; **21.10d** Arthur M. Siegelman/Visuals Unlimited. **21.11** Arthur M. Siegelman/Visuals Unlimited. **21.12** Carolina Biological Supply. **21.13** M. Haberey. **21.14b** Sydney Tamm. **21.15** American Society for Clinical Pathologists.

Line art by Hans & Cassady, Inc.